Some Physical Constants

Gravitational constant	G	6.673×10^{-11} N·m/kg^2
Speed of light	c	2.998×10^8 m/s
Fundamental charge	e	1.602×10^{-19} C
Avogadro's number	N_A	6.022×10^{23} particles/mol
Gas constant	R	8.315 J/mol·K 1.987 cal/mol·K
Boltzmann's constant	$k = R/N_A$	1.381×10^{-23} J/K 8.617×10^{-5} eV/K
Unified mass unit	$u = (1/N_A)$	1.661×10^{-24} g
Coulomb's constant	$k = \dfrac{1}{4\pi\varepsilon_0}$	8.988×10^9 N·m^2/C^2
Permittivity of free space	ε_0	8.854×10^{-12} C^2/N·m^2
Permeability of free space	μ_0	$4\pi \times 10^{-7}$ N/A^2 1.257×10^{-6} N/A^2
Planck's constant	h	6.626×10^{-34} J·s 4.136×10^{-15} eV·s
	$h = h/2\pi$	1.055×10^{-34} J·s 6.582×10^{-16} eV·s
Mass of electron	m_e	9.109×10^{-31} kg 511.0 keV/c^2
Mass of proton	m_p	1.673×10^{-27} kg 938.3 MeV/c^2
Mass of neutron	m_n	1.675×10^{-27} kg 939.6 MeV/c^2
Bohr magneton	$m_B = e\hbar/2m_e$	9.274×10^{-24} J/T 5.788×10^{-5} eV/T
Nuclear magneton	$m_n = e\hbar/2m_p$	5.051×10^{-27} J/T 3.152×10^{-8} eV/T
Magnetic flux quantum	$\phi_0 = h/2e$	2.068×10^{-15} T·m^2
Quantized Hall resistance	$R_K = h/e^2$	2.581×10^4 Ω
Rydberg constant	R_H	1.097×10^7 m^{-1}
Josephson frequency–voltage quotient	$2e/h$	4.836×10^{14} Hz/V
Compton wavelength	$\lambda_C = h/m_e C$	2.426×10^{-12} m

Physical Data Often Used

Mass of Earth, M_E	5.98×10^{24} kg
Radius of Earth, R_E, mean	6.37×10^6 m
Mass of Moon, M_M	7.35×10^{22} kg
Radius of Moon, R_M	1.74×10^6 m
Mass of Sun, M_S	1.99×10^{30} kg
Radius of Sun, R_S	6.96×10^8 m
Earth–Sun distance, $D_{E\text{-}S}$, mean	1.496×10^{11} m
Earth–Moon distance, $D_{E\text{-}M}$, mean	3.844×10^8 m
One light year (the distance light travels in one year)	9.47×10^{15} m
Acceleration of gravity on Earth, g (standard value)	9.81 m/s²
At sea level, at equator	9.78 m/s²
At sea level, at poles	9.83 m/s²
Acceleration of gravity on Moon, g_M	1.62 m/s²
Orbital speed:	
Earth	2.98×10^4 m/s
Moon	1.02×10^3 m/s
Period:	
Earth (sidereal year)	365.26 days
Moon (sidereal month)	27.3 days
Escape velocity:	
Earth, $v_{E\text{-esc}}$	1.12×10^4 m/s
Moon, $v_{M\text{-esc}}$	2.4×10^3 m/s
Solar constant	1.35 kW/m²
Time it takes light to travel the distance:	
from Earth to Sun	498 s = 8.3 min
from Earth to Moon	1.28 s
around Earth	0.13 s
equal to one light year	1 year = 3.16×10^7 s
Milky Way Galaxy data:	
Diameter	100 000 to 120 000 light years ($\approx 10^{21}$ m)
Thickness (average)	1 000 light years ($\approx 10^{19}$ m)
Number of stars	100 to 400 billion
Estimated mass	$5.8 \times 10^{11}\ M_S$
Useful ratios	$\dfrac{M_S}{M_E} = 3.33 \times 10^5;\ \dfrac{M_E}{M_M} = 81.3$
	$\dfrac{R_S}{R_E} = 109;\ \dfrac{R_E}{R_M} = 3.66$
	$\dfrac{D_{E\text{-}S}}{D_{E\text{-}M}} = 389;\ \dfrac{D_{E\text{-}M}}{R_E} = 60.4;\ \dfrac{D_{E\text{-}S}}{R_E} = 2.35 \times 10^4;$
	$\dfrac{g_E}{g_M} = 6.1;\ \dfrac{v_{E\text{-esc}}}{v_{M\text{-esc}}} = 4.67$

ENHANCED
WebAssign

Increased Engagement.
Improved Outcomes.
Superior Service.

Exclusively from Nelson Education and Cengage Learning, **Enhanced WebAssign®** combines the exceptional content that you know and love with the most powerful and flexible online homework solution, **WebAssign**. **Enhanced WebAssign** engages students with immediate feedback, rich tutorial content, and interactive eBooks, helping students to develop a deeper conceptual understanding of subject matter. Instructors can build online assignments by selecting from thousands of text-specific problems or supplementing with problems from any Nelson Education or Cengage Learning textbook in our collection.

With **Enhanced WebAssign**, you can

- Help students stay on task with the class by requiring regularly scheduled assignments using problems from the textbook.

- Provide students with access to a personal study plan to help them identify areas of weakness and offer remediation.

- Focus on teaching your course and not on grading assignments. Use **Enhanced WebAssign** item analysis to easily identify the problems that students are struggling with.

- Make the textbook a destination in the course by customizing your Cengage YouBook to include sharable notes and highlights, links to media resources, and more.

- Design your course to meet the unique needs of traditional, lab-based, or distance learning environments.

- Easily share or collaborate on assignments with other faculty or create a master course to help ensure a consistent student experience across multiple sections.

- Minimize the risk of cheating by offering algorithmic versions of problems to each student or fix assignment values to encourage group collaboration.

Learn more at www.cengage.com/ewa

NELSON EDUCATION

PHYSICS
FOR SCIENTISTS AND ENGINEERS

AN INTERACTIVE APPROACH

PHYSICS
FOR SCIENTISTS AND ENGINEERS

AN INTERACTIVE APPROACH

ROBERT HAWKES
Mount Allison University

JAVED IQBAL
University of British Columbia

FIRAS MANSOUR
University of Waterloo

MARINA MILNER-BOLOTIN
University of British Columbia

PETER WILLIAMS
Acadia University

NELSON / EDUCATION

NELSON / EDUCATION

Physics for Scientists and Engineers: An Interactive Approach

by Robert Hawkes, Javed Iqbal, Firas Mansour, Marina Milner-Bolotin, and Peter Williams

Vice President, Editorial Higher Education:
Anne Williams

Publisher:
Paul Fam

Marketing Manager:
Leanne Newell

Technical Reviewer:
Santo D'Agostino

Senior Developmental Editor:
Mark Grzeskowiak

Photo Researcher:
Kristiina Paul

Permissions Coordinator:
Kristiina Paul

Content Production Manager:
Christine Gilbert

Production Service:
Integra Software Services Pvt. Ltd.

Copy Editor:
Betty Robinson

Proofreader:
Donna Collingwood

Indexer:
Ursula Acton

Senior Manufacturing Coordinator:
Ferial Suleman

Design Director:
Ken Phipps

Managing Designer:
Franca Amore

Interior Design:
Peter Papayanakis

Cover Design:
Trinh Truong

Cover Image:
© 2012 – Jim Kramer

Compositor:
Integra Software Services Pvt. Ltd.

Printer:
RR Donnelley

Library and Archives Canada Cataloguing in Publication

Physics for scientists and engineers: an interactive approach / Robert Hawkes ... [et al.]. — 1st ed.

Includes index.
ISBN-13: 978-0-17-666950-8

1. Physics—Textbooks.
I. Hawkes, Robert Lewis, 1951-

QC23.2.P39 2014   530
C2012-908600-2

PKG ISBN-13: 978-0-17-666950-8
PKG ISBN-10: 0-17-666950-7

Brief Table of Contents

Contents

Preface

THE NEED FOR THIS RESOURCE AND ITS APPROACH

Physics for Scientists and Engineers: An Interactive Approach provides a more relevant approach to the subject to match the Canadian curriculum and will better reflect this fundamental, multidisciplinary, inquisitive, and inspirational science as it applies to Canadian students and instructors. The text also capitalizes on the growing interest in and appreciation of (PER) Physics Education Research. Too many students are either intimidated by physics or not interested in it, and our goal is to simply change that. Physics should be seen as the creative process that it is, and the reader should be helped to feel the thrill of discovery. *Physics for Scientists and Engineers: An Interactive Approach* reflects Canada's revitalized physics curriculum, providing a fresh approach to teaching and more exciting and relevant pedagogy for students. The story is supported by the following key changes that are occurring in the discipline and the market:

- The text is written with Physics Education Research (PER) findings in mind, and in a way to encourage and support instructional strategies based on PER. The author team brings considerable PER expertise to the project, including direct experience with a variety of PER-supported instructional strategies, such as classroom response systems (clickers), peer instruction, studio physics, online simulations, interactive lecture demonstrations, online tutorial systems, collaborative learning, project-based approaches, student reflection, and PSI (personalized system of instruction) approaches.

- While the text encourages PER-based approaches, **it does not support only a single instructional strategy**. Instructors who use traditional lecture + laboratory approaches, those who use audience-response systems in peer instruction, those who favour interactive lecture demonstrations, or indeed use other approaches, will find the text well suited for their needs.

- One idea that spans a number of the PER-supported instructional strategies is **the value of student collaboration**. It is clear that learning is deeper when students develop ideas in collaboration with peers and work together both in brainstorming approaches and in developing solutions. While no book can ensure such approaches, the text has been written in a manner to encourage collaborative learning. For example, the open-ended problems are ideally suited to a group approach. The conceptual problems in each chapter are well suited for use in audience-response peer-instruction approaches, or in studio-style classrooms. In some places we have moved derivations from chapters to problems to encourage student discovery of key relationships. The simulations and experiment suggestions will encourage interactive, engaged learning. If student collaboration is valuable in a classroom, think how powerful collaboration more broadly can be.

- As the Canadian Association of Physicists (CAP), Division of Physics Education (DPE) and others have pointed out, the work of physicists—and the use of physics by other scientists, engineers, and professionals from related fields—is increasingly interdisciplinary. We feel this text is one of the richest in terms of **promoting the interdisciplinary nature of physics** beyond simply having problem applications from various fields. Chapter content is presented with a rich interdisciplinary feel and stresses the need to use ideas from other sciences and related professions. The diverse backgrounds of the author team help create this rich interdisciplinary environment.

- One key differentiator between expert and novice views of physics is that a professional physicist views all of physics as a unified, small set of concepts, principles, and approaches that can be applied to a very diverse set of problems, while the novice sees physics as a huge number of loosely related ideas. Almost all introductory physics texts are much too long and do not adequately illuminate the key principles. This text is **concise in wording and emphasizes the unifying principles and techniques**.

- Closely related to this is the importance of **mastering a number of key concepts**, as opposed to learning incompletely a much broader body of knowledge. As mentioned earlier, the text uses more concise writing in each chapter; in addition, to the degree possible, we have made chapters self-contained so that each instructor can select which content is covered in a course. A carefully selected set of problems, both conceptual and quantitative, helps to reinforce mastery of key concepts.

- Modern computational tools play key roles in the lives of physicists and have been shown to be effective in promoting the learning of physics concepts. Where applicable, we have encouraged explorations through the PhET and similar animation and simulation products. These **allow students to develop their own unifying ideas by manipulating variables in the simulated environments**.

One role of any book is to inspire students to appreciate the beauty of the subject and go on to contribute and become leaders in the field. In order for this to be achieved, students must see the relevance of the subject.

The strong interdisciplinary focus throughout the book will help achieve this goal. But it is also important that **students can see themselves as future physicists**. This is the first broad-market introductory physics text written by a Canadian author team, and we have used Canadian and international examples (discoveries, applications, and current scientists) throughout, including an emphasis on contributions from young Canadians.

Current students place high value on **learning that will help them contribute to society**. For example, service learning is more popular than ever before, and a high number of students set goals of medical or social development careers. Also, there is strong public interest in such fundamental areas as particle physics, quantum mechanics, relativity, string theory, and cosmology.

This book is richer than competitive texts (in the science and engineering physics market) with respect to these topics. We have also employed many examples from the medical physics world, from alternative energy, and from sustainable environments. Thus, the text is richer than most in coverage of areas such as relativity, particle physics, quantum physics, and cosmology. While all physics texts strive to provide **"real-world problems"**, we feel that we have achieved this to a higher degree. Our open problems are modelled on how the world really is: a key part of applying physics is deciding what is relevant, and to make reasonable approximations as needed. Closed-form problems, which in most textbooks are the only type used, portray an artificial situation in which what is relevant—and only that—is given to the student. Our data-rich problems and encouragement of graphing, statistical, and numerical solution software help reinforce realistic situations.

We feel that traditional physics texts have not employed the power of images to the degree possible. Too-early adoption of highly abstract summaries may interfere with learning of concepts through visual images. Physics texts tend to have idealized drawings and a set of images drawn mainly from royalty-free sources. As appropriate, **students are encouraged to explore deeply the messages in rich and compelling visual images, including video**.

Ultimately students must take ownership of their learning; that is essentially the goal of all education. The strong link of objectives, sections, and checkpoint questions will provide an efficient environment for students to achieve this. We view our role in terms of maximizing interest and engagement and eliminating obstacles.

Robert Hawkes
Javed Iqbal
Firas Mansour
Marina Milner-Bolotin
Peter Williams
March 2013

About the Authors

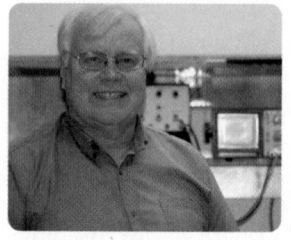

Bob Hawkes

ROBERT HAWKES Dr. Robert Hawkes is a professor of physics at Mount Allison University. In addition to having extensive experience in teaching introductory physics, he has taught upper-level courses in mechanics, relativity, electricity and magnetism, electronics, and astrophysics, as well as education courses in science methods and technology in education. His research program is in solar system astrophysics using advanced electro-optical devices to study atmospheric meteor ablation, as well as complementary lab-based techniques such as laser ablation. Dr. Hawkes received his B.Sc. (1972) and a B.Ed. (1978) at Mount Allison University, and his M.Sc. (1974) and Ph.D. (1979) in physics from the University of Western Ontario. He has won a number of teaching awards, including a 3M STLHE National Teaching Fellow, the Canadian Association of Physicists Medal for Excellence in Undergraduate Teaching, and the Science Atlantic University Teaching Award. He has been an early adopter of a number of interactive teaching techniques, in particular collaborative learning in both introductory and advanced courses. He was the co-editor of the *Physics in Canada* special issue on physics education.

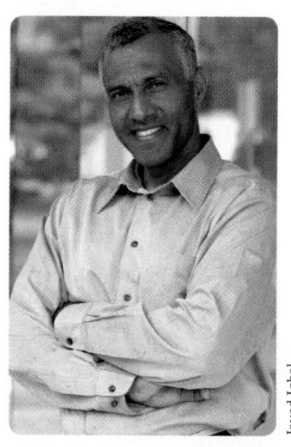

Javed Iqbal

JAVED IQBAL Dr. Javed Iqbal is the director of the Science Co-op Program and an adjunct professor of physics at the University of British Columbia. At UBC he has taught first-year physics for 20 years and has been instrumental in promoting the use of "clickers" at UBC and other Canadian universities. In 2004, he was awarded the Faculty of Science Excellence in Teaching Award. In 2012, he was awarded the Killam Teaching Prize. His research areas include theoretical nuclear physics, computational modelling of light scattering from nanostructures, and computational physics. Dr. Iqbal received his Doctoral Degree in Theoretical Nuclear Physics from Indiana University.

Photo by Hany M. D. Ibrahim

FIRAS MANSOUR As a lecturer in the Department of Physics and Astronomy at the University of Waterloo since 2000, Firas Mansour has gained respect and praise from his students for his exceptional teaching style. He currently teaches first-year physics classes to engineers, life science, and physical science students and has taught upper-year elective physics courses in the past. He is highly regarded for his quality of teaching, his enthusiasm in teaching, and his understanding of students' needs. His dedication to teaching is exemplary, as is his interest in outreach activities in taking scientific knowledge beyond the university boundary. He is a 2012 Distinguished Teaching Award recipient at the University of Waterloo.

Marina Milner-Bolotin

MARINA MILNER-BOLOTIN Dr. Marina Milner-Bolotin is a science educator within the Department of Curriculum and Pedagogy at the University of British Columbia. She specializes in science (physics) teaching and studies ways of using technology to promote student interest in science. For the last 20 years she has been teaching science and mathematics in Israel, the United States (Texas and New Jersey), and Canada. She has taught physics and mathematics to a wide range of students, from elementary gifted students to university undergraduates in science programs and future teachers. She has also led a number of professional development activities for science in-service and pre-service teachers and university faculty—from LoggerPro training workshops, to clicker and tablet training, to physics content presentations at conferences and professional development days. She has been engaged in science education research since 1994. Dr. Milner-Bolotin earned her M.Sc. in theoretical physics at Kharkiv National University, Ukraine, in 1991 and completed her M.A. and Ph.D. in science education at the University of Texas at Austin in 2001. At UT Austin, she investigated how project-based instruction in science courses for future elementary teachers affected their interest in science and their ability to do and teach science. Before joining UBC she was an assistant professor of physics at Ryerson University in Toronto. Dr. Milner-Bolotin is a member of the executive board of the American Association of Physics Teachers (AAPT) and the president and AAPT section representative for the BC Association of Physics Teachers. In 2010, Dr. Milner-Bolotin was awarded the CAP Medal for Excellence in Teaching Undergraduate Physics, the first woman to receive this award.

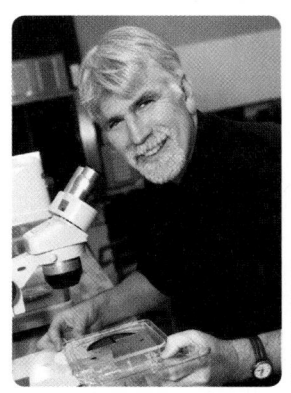
Deborah Nicholson

PETER WILLIAMS Dr. Peter Williams is dean of the Faculty of Pure and Applied Science and professor of physics at Acadia University and won the 2006 Canadian Association of Physicists (CAP) Medal for Excellence in Teaching. Dr. Williams has developed and taught a great diversity of courses, has shown innovation in the classroom, and has published a number of articles in teaching journals. Dr. Williams received his B.Sc (1985), M.Sc (1988), and Ph.D (1991), from McMaster University. Dr. Williams devotes significant efforts to improving secondary and postsecondary physics education in Atlantic Canada. He is one of those rare individuals who can effectively combine the best of technology-enhanced educational techniques while maintaining a strong personal approach to teaching. He has played a critical role in the development of studio physics modes of instruction at Acadia University and has developed several innovative courses, including most recently a Physics of Music course at Acadia. Furthermore, Dr. Williams has provided regional, national, and international leadership in new modes of physics teaching through his writing and presentations and set an exemplary model for applying research methodology to evaluation of the effectiveness of different modes of physics instruction.

ENRICHED MEDIA

WebAssign Nelson Education's Enhanced WebAssign (EWA) allows you to easily assign, collect, grade, and record homework assignments via the Internet. This proven and reliable system uses pedagogy and content found in Cengage Learning and Nelson Education's best-selling physics textbooks, then enhances them to help your students visualize the problem-solving process and reinforce mathematical concepts more effectively. EWA encourages active learning and time on task, and respects diverse ways of learning.

Used by more than 2.2 million students, Enhanced WebAssign is the leading homework system for math and science. Created by physicists for physics instructors, Enhanced WebAssign is easy to use and works with all major operating systems and browsers. Enhanced WebAssign adds interactive features to go far beyond simply duplicating text problems online. Students can watch solution videos, see problems solved step-by-step, and receive feedback as they complete their homework.

Nelson YouBook

The innovative YouBook available through Enhanced WebAssign allows full instructor customization and robust class-communication tools, including animations, interactive flashcards, and an online glossary to motivate student review, and additional materials only available online.

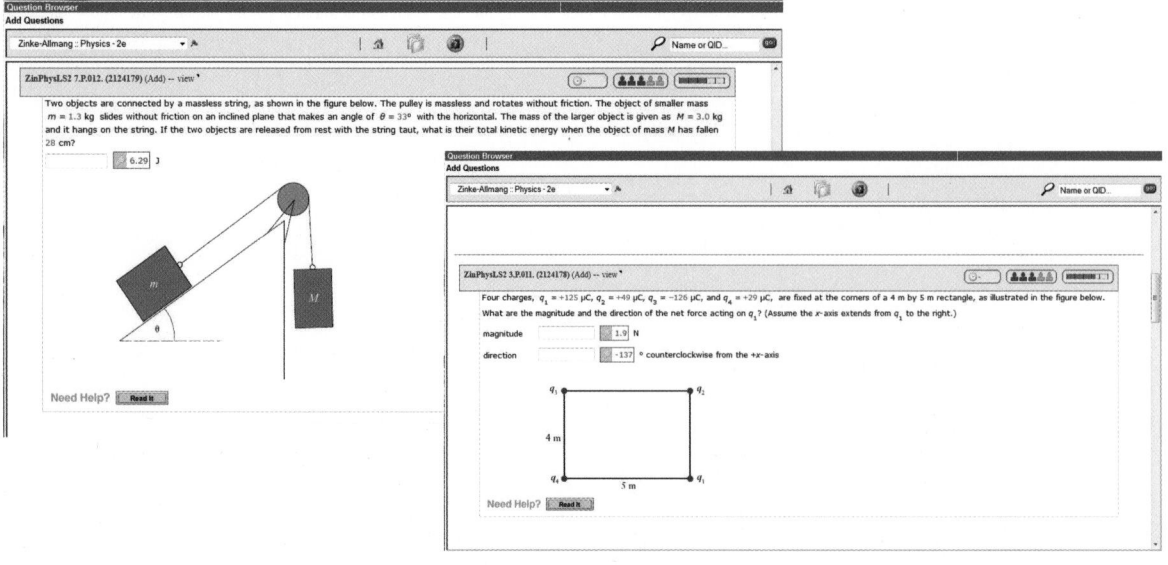

THE BIG PICTURE

Each chapter of *Physics for Scientists and Engineers: An Interactive Approach* is carefully organized according to learning objectives so you can stay focused on the most important concepts. Easy-to-use learning tools point out the topics covered in each chapter, show why they are important, and help you learn the material.

Look Ahead Section These narratives at the beginning of each chapter introduce topics through an interesting and engaging real-life example that pertains to the chapter's subject matter. Each Look Ahead section is illustrated with an engaging photo.

Learning Objectives Learning objectives are brief numbered and directive goals or outcomes that the student should take away from the chapter. Each learning objective corresponds to a major heading within the chapter.

Icons

Major sections are directly tied to **learning objectives**; section elements are visually indicated by individual **icons**.

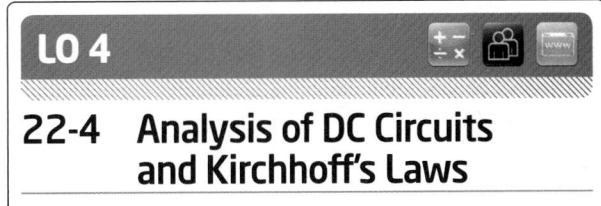

22-4 Analysis of DC Circuits and Kirchhoff's Laws

A **direct current (DC)** circuit is a circuit in which electric current flows in one direction. A simple DC electric circuit consists of circuit elements, such as batteries, resistors, switches, capacitors, and diodes, connected by conducting wires that allow the flow of electric current. We will use the concepts of electric current I, resistance R, power P, and voltage V to describe the operation of these circuits. First, we consider the source of electric potential energy.

Making Connections Making Connections boxes are provided in a narrative format and contain concise examples from international contexts, the history of physics, daily life, and other sciences.

MAKING CONNECTIONS

How Do Touch-Sensitive Lamps Work?

Touch-sensitive lamps commonly have a power-level switch controlled by an RC circuit. Your body has some capacitance. When you touch the metal body of a touch-sensitive lamp, you add capacitance to the control circuit and usually some static charge from your body, as well. These changes in the capacitance trigger the control circuit such that it changes the voltage supplied to the lamp's bulb. Often, touch-sensitive switches cycle through three power levels: off-low-medium-high, then back to off.

Peer to Peer Written by students for students, Peer to Peer boxes provide useful tips for navigating difficult concepts.

Online Activity Online Activity boxes provide interactive activities, such as computer simulations, that help with concept development. Many of these are matched to the physics education research-validated PhETs.

PEER TO PEER

I always thought that in a parallel circuit, the current splits equally between the circuit's parallel branches. Then I realized that this is only true when the branches have equal resistance. The current will split in a way such that the more resistance the branch has, the less current will flow through it. For example, if 6 A of current enters the junction and splits between two branches of 2 Ω and 4 Ω, the branch that has twice the resistance will receive half the current. Thus, the current will split into 4 A and 2 A, respectively. Keeping this in mind always helps me determine whether my circuit solution is reasonable.

ONLINE ACTIVITY

Simple and Not So Simple DC Circuits

The e-resource that accompanies every new copy of this textbook contains an Online Activity using the PhET simulation "Circuit Construction Kit." Work through the simulation and accompanying questions to gain an understanding of DC circuits.

Key Equations It is important for students to differentiate fundamental relationships from equations that are used in steps of derivations and examples. Key equations are clearly indicated.

KEY EQUATION

$$I = I_e \cdot e = (n_e v_d A)e = n_e v_d A (1.6 \times 10^{-19} \, C) \quad (22\text{-}4)$$

Examples Each example is numbered and corresponds to each major concept introduced in the section. Examples include a statement of the problem, the solution, and "Making sense of the result." Within the example, the authors have modelled desired traits, such as care with units and consideration of appropriate significant figures. "Making sense of the result" is one of the most important features, in which authors model the idea of always considering what has been calculated to determine whether it is reasonable.

EXAMPLE 22-2

Applying Ohm's Law to a Current-Carrying Wire

A potential difference of 100 mV is applied across a 5.0 m-long metal wire that has a diameter of 0.50 mm. The resulting electric current is 66.7 mA.

(a) Find the resistance of the wire.
(b) Find the resistivity of the wire, and suggest what metal it could be made of.

SOLUTION

(a) We assume that the metal wire obeys Ohm's law. Therefore, we can find its resistance, R, using Equation (22-15):

$$R = \frac{V}{I} = \frac{100 \text{ mV}}{66.7 \text{ mA}} = \frac{100 \times 10^{-3} \text{ V}}{66.7 \times 10^{-3} \text{ A}} = 1.50 \ \Omega$$

(b) Rearranging Equation (22-14), we can find the resistivity of the wire:

$$\rho = \frac{RA}{l} = \frac{(1.5 \ \Omega) \cdot \pi (0.25 \times 10^{-3} \text{ m})^2}{5.0 \text{ m}} = 5.9 \times 10^{-8} \ \Omega \cdot \text{m}$$

Comparing this value to the resistivities listed in Table 22-2, we find that the wire could be made of zinc.

Making sense of the result:

Resistance and resistivity are different physical quantities. The resistance provides a description of how a particular piece of wire resists the flow of electric current. The resistivity is the resistance of a hypothetical wire that is 1 m long and has a cross-sectional area of 1 m².

Checkpoints Each learning objective has a Checkpoint box to test students' understanding of the material they have just read. Checkpoint boxes include questions in different formats, followed immediately by the answer placed upside-down at the end of the box. While different formats are used, in all cases these checkpoints have a single clear answer so that they can be self-administered and students know whether they have mastered the concept before moving on to dependent material. The close linking of sections, learning objectives, and checkpoints is a major feature of the text.

☑ CHECKPOINT

C-22-6 The Relative Brightness of Two Light Bulbs in Series and Parallel Electric Circuits

A 25 W light bulb and a 40 W light bulb are first connected in series. Then they are connected in parallel. Which statement describes their relative brightness?

(a) The 25 W light bulb is brighter in a series circuit and dimmer in a parallel circuit.
(b) The 25 W light bulb is brighter in a parallel circuit and dimmer in a series circuit.
(c) The 25 W light bulb is always brighter.
(d) The 40 W light bulb is always brighter.

C-22-6 (b) The light bulbs are designed to be used in a parallel circuit; therefore, the higher nominal power (power rating), the less the resistance of the light bulb (higher-rated light bulbs have thicker filaments).

Putting It All Together The Putting It All Together section at the end of each chapter provides students with an opportunity to review the key concepts discussed in the chapter. Care has been taken to make these concise and yet at the same time cover all core ideas.

Questions, Problems by Section, and Comprehensive Problems are provided at the end of each chapter to test the students' understanding of the material. For the problems, star ratings are used (*, **, or ***), with more stars indicating more challenging problems.

QUESTIONS

1. Which of the following is *not* a characteristic of physics?
   (a) Physics seeks to explain phenomena in the simplest way possible and in a way that has the widest application.
   (b) While some phenomena may be described qualitatively, in general, physics seeks quantitative descriptions.
   (c) Physics is primarily involved with developing a large number of equations so that there is one equation relevant for any given situation.
   (d) Physics requires creativity.

2. Assume that measurements of a time for some process are normally distributed, with no systematic errors. About two-thirds of the time the value is between 3.4 and 4.8 s. What are the mean and standard deviation of the time measurement?

3. Assume that measurements of a length are normally distributed, with no systematic errors. About 19 times out of 20, the value is between 2.2 and 5.8 m. What are the mean and standard deviation of the length measurement?

4. How many significant figures are in the number 230 100?
   (a) 3
   (b) 4
   (c) 5
   (d) 6

5. Complete the calculation $\dfrac{12.0 \times 5.0}{4.000}$. According to the rules for significant digits, how should the answer be written?
   (a) 15
   (b) 15.0
   (c) 15.00
   (d) 15.000

6. Which of the following is *not* a base SI unit?
   (a) kg
   (b) s
   (c) N
   (d) cd

7. Which of the following is *not* a base SI unit?
   (a) kg
   (b) m
   (c) J
   (d) K

8. Work (or energy) is equal to force times x. Use Tables 1-2 and 1-3 to identify the units of x.

9. Consider the graph of Figure 1-4. Taking into account the various points made in this chapter, critically evaluate the graphical presentation, indicating all aspects that could be improved.

10. (a) Use a reference source to look up how the length of the metre is currently defined.
    (b) From this definition, what must be the speed of light (to the number of significant figures inherent in the definition)?
    (c) Provide a brief overview of some of the previous ways that have been used to define the length of the metre.

11. Use a reliable reference to determine how the second is currently defined.

12. (a) Use a reliable reference to determine the method currently used to define the kilogram.
    (b) Provide a brief overview of some other ways that have been used to define the kilogram.

13. Describe the current method used to define the ampere. Include a reference to the source of your information.

14. Use a reliable reference source to identify the method currently used to define the candela, as well as the method that was historically used.

15. How many cubic centimetres are in one cubic metre?
    (a) 100
    (b) 1 000
    (c) 10 000
    (d) 1 000 000

16. Two data sets, A and B, have the same mean value, but data set B has a larger standard deviation by a factor of 2. There are 16 data points in data set A. How many data points are needed in set B for the SDOM to be identical for the two sets?

17. If we multiply a force (N) by a velocity (m/s), which of the following will be the units of the resulting quantity?
    (a) kg/s
    (b) J
    (c) Pa
    (d) W

PROBLEMS BY SECTION

For problems, star ratings will be used, (*, **, or ***), with more stars meaning more challenging problems.

Section 1-3 Mean, Standard Deviation, and SDOM

18. * A set of 16 measurements has a mean of 4.525 and a standard deviation of 1.2. How should the mean be written with an uncertainty given by the SDOM?

19. * The SDOM is 1.2, the mean is 4.8, and the standard deviation is 2.2. Approximately how many data points must be in the set?

20. * According to a journal article, experiments suggest that 19 times out of 20, the measurement obtained is within ± 0.5 of 10.5. What must be the standard deviation and mean values, assuming that the data is normally distributed?

Section 1-4 Significant Figures

21. * How many significant figures are in each of the following?
    (a) 324.2
    (b) 0.032
    (c) 9.004
    (d) 12 000
    (e) 85.0
    (f) 6000.

22. ** Suppose that the uncertainty in the length 17.386 22 m is ± 0.04 m. Round the length to the correct number of significant figures.

Section 1-5 Scientific Notation

23. * Write each of the following in scientific notation.
    (a) 2452
    (b) 0.592
    (c) 12 000
    (d) 0.000 045

24. * Perform the following calculation, and express the final result in scientific notation with the correct number of significant figures.

$$\frac{(3.20 \times 10^5)(1.5 \times 10^4)}{6.400 \times 10^3}$$

ANCILLARIES

An extensive array of supplemental materials is available to accompany the first edition of *Physics for Scientists and Engineers: An Interactive Approach*. The supplements are designed to make teaching and learning more effective. For more information on any of these resources, please contact your local Nelson Education sales representative or call Nelson Education Ltd. Customer Support at 1-800-268-2222.

Instructor Ancillaries

The **Nelson Education Teaching Advantage (NETA)** program delivers research-based instructor resources that promote student engagement and higher-order thinking to enable the success of Canadian students and educators.

Instructors today face many challenges. Resources are limited, time is scarce, and a new kind of student has emerged: one who is juggling school with work, has gaps in his or her basic knowledge, and is immersed in technology in a way that has led to a completely new style of learning. In response, Nelson Education has gathered a group of dedicated instructors to advise us on the creation of richer and more flexible ancillaries that respond to the needs of today's teaching environments.

The members of our editorial advisory board have experience across a variety of disciplines and are recognized for their commitment to teaching. They include:

Norman Althouse, Haskayne School of Business, University of Calgary

Brenda Chant-Smith, Department of Psychology, Trent University

Scott Follows, Manning School of Business Administration, Acadia University

Jon Houseman, Department of Biology, University of Ottawa

Glen Loppnow, Department of Chemistry, University of Alberta

Tanya Noel, Department of Biology, York University

Gary Poole, Senior Scholar, Centre for Health Education Scholarship, and Associate Director, School of Population and Public Health, University of British Columbia

Dan Pratt, Department of Educational Studies, University of British Columbia

Mercedes Rowinsky-Geurts, Department of Languages and Literatures, Wilfrid Laurier University

David DiBattista, Department of Psychology, Brock University

Roger Fisher, PhD

In consultation with the editorial advisory board, Nelson Education has completely rethought the structure, approaches, and formats of our key textbook ancillaries.

We've also increased our investment in editorial support for our ancillary authors. The result is the Nelson Education Teaching Advantage and its key components: *NETA Engagement*, *NETA Assessment*, and *NETA Presentation*. Each component includes one or more ancillaries prepared according to our best practices, and a document explaining the theory behind the practices.

NETA Engagement presents materials that help instructors deliver engaging content and activities to their classes. Instead of instructor's manuals that regurgitate chapter outlines and key terms from the text, NETA Enriched Instructor's Manuals (EIMs) provide genuine assistance to teachers. The EIMs answer questions like *What should students learn?, Why should students care?*, and *What are some common student misconceptions and stumbling blocks?* EIMs not only identify the topics that cause students the most difficulty, but also describe techniques and resources to help students master these concepts. Dr. Roger Fisher's *Instructor's Guide to Classroom Engagement (IGCE)* accompanies every Enriched Instructor's Manual. (Information about the NETA Enriched Instructor's Manual prepared for *Physics for Scientists and Engineers: An Interactive Approach* is included in the description of the IRDVD below.)

NETA Assessment relates to testing materials: not just Nelson's Test Banks and Computerized Test Banks, but also in-text self-tests, Study Guides and web quizzes, and homework programs like CNOW. Under *NETA Assessment*, Nelson's authors create multiple-choice questions that reflect research-based best practices for constructing effective questions and testing not just recall but also higher-order thinking. Our guidelines were developed by David DiBattista, a 3M National Teaching Fellow whose recent research as a professor of psychology at Brock University has focused on multiple-choice testing. All Test Bank authors receive training at workshops conducted by Prof. DiBattista, as do the copyeditors assigned to each Test Bank. A copy of *Multiple Choice Tests: Getting Beyond Remembering*, Prof. DiBattista's guide to writing effective tests, is included with every Nelson Test Bank/Computerized Test Bank package. (Information about the NETA Test Bank prepared for *Physics for Scientists and Engineers: An Interactive Approach* is included in the description of the IRDVD below.)

NETA Presentation has been developed to help instructors make the best use of PowerPoint® in their classrooms. With a clean and uncluttered design developed by Maureen Stone of StoneSoup Consulting, NETA Presentation features slides with improved readability, more multi-media and graphic materials, activities to use in class, and tips for instructors on the Notes page. A copy of *NETA Guidelines for Classroom Presentations* by Maureen Stone is included with each set of PowerPoint slides. (Information about the NETA PowerPoint® prepared for *Physics for Scientists and Engineers: An Interactive Approach* is included in the description of the IRDVD below.)

IRDVD Key instructor ancillaries are provided on the *Instructor's Resource DVD* (ISBN 978-0-17-666168-7), giving instructors the ultimate tool for customizing lectures and presentations. (Downloadable web versions will also be available at **www.physics1e.nelson.com**.) The IRDVD includes

- **NETA Engagement:** The Enriched Instructor's Manual is organized according to the textbook chapters and addresses eight key educational concerns, such as typical stumbling blocks students face and how to address them. Other features include tips on teaching using cases, as well as suggestions on how to present material and use technology and other resources effectively, integrating the other supplements available to both students and instructors. This manual doesn't simply reinvent what's currently in the text; it helps the instructor make the material relevant and engaging to students. The Enriched Instructor's Manual was written by Dan McQueen.

- **NETA Assessment**: The Test Bank includes multiple-choice questions written according to NETA guidelines for effective construction and development of higher-order questions. Also included are true/false and completion-type questions. Test Bank files are provided in Word format for easy editing and in PDF format for convenient printing whatever your system. The Computerized Test Bank by ExamView® includes all the questions from the Test Bank. The easy-to-use ExamView software is compatible with Microsoft Windows and Mac. Create tests by selecting questions from the question bank, modifying these questions as desired, and adding new questions you write yourself. You can administer quizzes online and export tests to WebCT, Blackboard, and other formats. The NETA Test Bank was written by Naeem Syed Ahmed of Laurentian University.

- **NETA Presentation:** Microsoft® PowerPoint® lecture slides are available for every chapter and include key figures, tables, and photographs from *Physics for Scientists and Engineers: An Interactive Approach*. NETA principles of clear design and engaging content have been incorporated throughout. The NETA Presentation was created by Jason Harlow of the University of Toronto.

- **Image Library:** This resource consists of digital copies of figures, short tables, and photographs used in the book. Instructors may use these jpegs to create their own PowerPoint presentations.

- **Day One:** Day One—Prof InClass is a PowerPoint presentation that you can customize to orient your students to the class and their text at the beginning of the course.

- **TurningPoint®:** Another valuable resource for instructors is **TurningPoint® classroom response software** customized for *Physics for Scientists and Engineers: An Interactive Approach*. Now you can author, deliver, show, access, and grade, all in PowerPoint … with no toggling back and forth between screens! JoinIn on Turning Point is the only classroom response software tool that gives you true PowerPoint integration. With JoinIn, you are no longer tied to your computer. You can walk about your classroom as you lecture, showing slides and collecting and displaying responses with ease. There is simply no easier or more effective way to turn your lecture hall into a personal, fully interactive experience for your students. If you can use PowerPoint, you can use JoinIn on TurningPoint! (Contact your Nelson Education publishing representative for details.) These contain poll slides and pre- and post-test slides for each chapter in the text.

- **Instructor's Solutions Manual:** This manual has been independently checked for accuracy. It contains complete solutions to all end-of-chapter problems in the book. The Instructor's Solutions Manual was written by Naeem Syed Ahmed of Laurentian University.

Nelson YouBook The innovative YouBook available through Enhanced WebAssign allows full instructor customization and robust class-communication tools, including videos, animations, interactive flashcards, online glossary, stimulating games to motivate student review, and additional materials only available online.

Student Ancillaries

A groundbreaking homework management system, Enhanced WebAssign (EWA) combines our exceptional physics content with the most flexible online homework solution. Powerful, effective, yet simple to use, EWA provides easy problem selection along with automatic grading and interactive tutorial assistance for students that reinforce the book's pedagogical approach.

EWA engages students with immediate feedback, rich tutorial content, and interactive e-books, helping them develop a deeper conceptual understanding of the subject matter. EWA offers interactive features that go far beyond simply duplicating text problems online. Students can watch solution videos, see problems solved step-by-step, and receive feedback as they complete their homework. Online assignments can be built by selecting from thousands of text-specific problems or supplemented with problems from specific Cengage Learning and Nelson Education textbooks. EWA promotes active learning and time on task, and respects diverse ways of learning.

The more you study, the better the results. Make the most of your study time by accessing everything you need to succeed in one place. Read your textbook, take notes, watch videos, and take practice quizzes—online with Premium Website.

Student Solutions Manual (ISBN 978-0-17-667213-3) The Student Solutions Manual includes solutions to all odd-numbered questions.

Acknowledgements

The production of a "ground up" new physics textbook is a major endeavour and would not have been possible without the expert work and assistance of many people.

The Nelson Education team has been amazing—this book would never have been completed without their expertise, attention to detail, flexibility, and above all emphasis on producing a high quality and innovative text. Particular credit goes to Paul Fam, Publisher—Higher Education who has so enthusiastically guided and supported the project from the earliest days. During the key stages of the project the Senior Developmental Editor was Mark Grzeskowiak. His extensive experience and professionalism were critical in moving us from rough drafts to finished manuscript. Because of the length of the project, in the early stages we worked with two different editors; first Tracy Yan and then Alwynn Pinard. Their leadership helped with critical decisions on the format and composition of the book. In particular we note that it was Alwynn who first suggested that each section of the text should be tied to exactly one learning objective—a key feature of the text.

A text written by a team of physicists who had not previously worked together posed a challenge in making the final book have a common voice and a consistent approach. The success we have achieved in that regard is due in large part to our substantive editor, David Peebles, and our copy editor, Betty Robinson. Words cannot adequately express the debt we owe to these two incredible professionals. David's background in physics, along with his editing experience, helped him to play a major role in suggesting alternative and clearer ways to explain concepts. He helped us to write more concisely and more clearly. Betty demonstrated a care and expertise that improved every single page of the book. As authors we were time and time again amazed at the care to detail she demonstrated. The fact that she is also such an efficient and personable individual made her contribution even more special.

The production stage was more demanding and complex than the authors realized at the outset. We thank the many people who helped us through this process—often under tight deadlines—especially Christine Gilbert, Content Production Manager, who had primary responsibility for overall production issues. Kristiina Paul, our photo researcher, worked hard to get permissions for our first choices for images, and, when they were not available, to find suitable alternatives. As a first edition the task of creating the various figures was challenging. We thank in particular Indhu Gunasekaran, and her team at Integra, and Deborah Crowle of the CrowleArt group. The publisher and the author team would also like to convey their thanks to Santo D'Agostino for his technical edit of the text. Santo's edit ensured consistency in key areas, such as the use of significant digits, accuracy in the figures, tables, and photos, and making sure all of the steps were accounted for in the examples presented.

The text is of course only part of the learning package. We acknowledge the leadership of Roberta Osborne, Senior Development Editor, in that component of the work, and the valuable work of the various ancillary authors. The authors would also like to thank Naeem Syed Ahmed, Mihai Gherase, and Santo D'Agostino for their work on the solutions manual to accompany this text, and the help they provided in reviewing the many end-of-chapter problems contained herein.

Thanks go to those who reviewed the text. Collectively these professionals offered ideas, and occasional corrections, that helped make the book more accurate and improved clarity. The diversity of their views of physics and how it should be taught—while occasionally resulting in that not all suggestions could be incorporated in this printing—provided us with a broader view than that of the five authors alone. We give thanks to the following individuals:

Ben Newling, University of New Brunswick
Syed Ahmed Naeem, Laurentian University
Tetyana Antimirova, Ryerson University
Mark Laidlaw, University of Victoria
Calvin Kalman, Concordia University
Erik Rosolowsky, University of British Columbia—Okanagan
Salam Tawfiq, University of Toronto
Stan Greenspoon, Capilano University
Mahi Singh, University of Western Ontario
Roman Koniuk, Concordia University
Neil Alberding, Simon Fraser University
Ruxandra M. Serbanescu, University of Toronto

The Student Advisory Board (SAB) members helped us remain grounded in what sort of text students wanted and would use, and most of them also contributed to the Peer to Peer boxes ensuring that the material in this text is presented from a truly "student" perspective. The members of the Mount Allison-related SAB were Kevin Alexander, Jeremy Bourque, Emma Driedger, Darin Eddy, James Ehrman, Ashley Faloon, Monica Firminger, Theo Heffler, Kyle Hill, Catherine Hughes, Luyao Li, Ellen Milley, R. Bruce Muir, Andrew Nelles, Ryan Robski, Paras Satja, Ari Silburt, Asmita Sodhi, Sarah Thomas, Evelyn Wainewright, David Watson, and Rory Woods. The members of the Waterloo-based SAB were Andrew Boyko, Xin Jin, Mathew Michael, Ali Ali-Mahmoud, Mukund Vermuri, Yun Zhe Li, James Gerinea, Scott Duncan, David Lu, Calvin Lu, Kurt McCarrell, Aaamir Jawaid, and Lawrence Lu.

We thank Sean Chamberland, Executive Marketing Manager, and Leanne Newell, Marketing Manager at

Nelson Education for their skilled promotion of the book, along with the team of publisher's sales representatives across the country.

To be honest, this book has taken more of our time and energy than any of the authors ever anticipated. All of the authors combined the writing of this text with other career demands, and as a result many weekends and evenings were devoted to this text. We thank our family members for their understanding and encouragement. We thank our colleagues for their support in various ways during the course of this project. The authors would also like to acknowledge Rohan Jayasundera and Simarjeet Saini for enlightening discussions. The authors would like to thank Olga Myhaylovska for her valuable input and advice. The authors would also like to thank Reema Deol and Renee Chu for conducting background research on some of the material used in the text.

The authors would like to thank professor David F. Measday (UBC) for providing valuable feedback for Chapter 33, Introduction to Nuclear Physics.

Daily in our classrooms we learn from our students, and are rejuvenated by their enthusiasm, creativity, and energy. Our view of the teaching and learning of physics owes a great deal to them, probably more than they realize. Similarly, our interactions with colleagues, both at our own institution and beyond, have shown us new and more effective ways to approach difficult concepts, and helped in our own education as physicists. We would like to acknowledge the valuable work done by organizations such as the Canadian Association of Physicists Division of Physics Education (CAP DPE), and the American Association of Physics Teachers (AAPT).

Chapter 1
Introduction to Physics

Why should you study physics? To start with, physics helps us answer amazing questions. For example, physics has provided a remarkably detailed picture of what the universe is like and how it has developed over time. The Hubble Space Telescope produced the image in Figure 1-1. The image shows an area of the sky equivalent to what you would cover if you held a 1 mm square at arm's length. Yet this image shows about 10 thousand galaxies, and each galaxy typically contains 100 billion stars. In the hundred years since the first evidence of the existence of galaxies, observations and theoretical calculations by numerous physicists have provided strong evidence that the evolution of the universe started in a "big bang" about 13.7 billion years ago, when the universe was infinitesimally small, almost infinitely dense, and incredibly hot.

Like cosmology, aspects of research in particle physics, quantum mechanics, and relativity can be fascinating because they challenge our commonsense ideas. However, discoveries in these fields have also led to numerous extremely useful applications. For example, atomic physics underlies medical imaging technologies from simple X-rays to the latest MRIs and CAT scans. In fact, physics concepts are the basis for almost all technologies, including energy production and telecommunications. As you explore this book, each chapter will bring a new answer to the question, "Why should physics matter to me?"

LOOK AHEAD

R. Williams (STScI), the Hubble Deep Field Team and NASA

Figure 1-1 A tiny part of the night sky captured with incredible detail by the Hubble Space Telescope

Learning Objectives

When you have completed this chapter you should be able to:

1 Explain the distinguishing characteristics of physics.

2 Differentiate between precision and accuracy, and list the types of possible errors.

3 Calculate the mean, standard deviation, and standard deviation of the mean (SDOM) for data sets, and correctly use $\pm$ notation.

4 Relate uncertainties to significant figures, and state the number of significant figures in expressions.

5 Use scientific notation, and convert quantities to and from scientific notation.

6 Correctly apply significant figure rules to calculated quantities.

7 Use and interpret error bars to express uncertainties in values.

8 State the SI units for physical quantities, and write the units and their abbreviations correctly.

9 Apply dimensional analysis to help determine whether a relationship is correct.

10 Perform unit conversions.

11 Make reasonable order-of-magnitude estimates in solving open problems.

12 Comment on what a professional physicist does and how one becomes a physicist.

 For additional Making Connections, Examples, and Checkpoints, as well as activities and experiments to help increase your understanding of the chapter's concepts, please go to the text's online resources at www.physics1e.nelson.com.

LO 1

1-1 What Is Physics?

We will start this chapter by considering some defining characteristics of physics.

The domain of physics is the physical universe Physics does not seek to answer questions of religion, literature, or social organizations.

Physics is a quantitative discipline Although there are a few topics in physics where our understanding is mainly qualitative, overall, measurement and calculation play critical roles in developing and testing physics ideas.

Physics is strongly model-driven Physicists develop hypotheses and models, use them to make predictions, and then test those predictions with experiments and observations. Through the study of physics you can learn analytical skills that can be applied beyond the sciences and engineering. For example, a number of physicists find employment developing economic and investment models for financial institutions.

Physicists need to be creative Physicists design experiments, find applications for physical principles, and develop new models and theories. Some philosophers of science have asked whether the electron was invented or discovered. Such questioning stresses that, while there is a part of physics that is independent of the observer, the specific models we develop to help understand nature critically depend on the creativity and imagination of physicists. It is not surprising that many physicists also work in other creative areas, such as music and art.

Physics interfaces strongly with other sciences The study of physics is increasingly interdisciplinary, with many physicists working in areas that span physics and other disciplines, for example, medical physics, biophysics, chemical physics, materials science, geophysics, or physical environmental science.

Physics is a highly collaborative discipline Most physicists routinely work with colleagues from other countries, often using international research facilities. Pick up a physics research journal, and you will see that the majority of papers are written by collaborations of scientists from different institutions and countries.

Physics is both deeply theoretical and highly applied Some physicists work exclusively in fundamental areas that have no immediate application, whereas others concentrate on solving applied problems. Sometimes work initially deemed to have little practical use turns out to

MAKING CONNECTIONS

Creative Physicist and Musician

Canadian Diane Nalini de Kerckhove is both a brilliant physicist and a highly acclaimed jazz musician. Growing up, she studied ballet, modern dance, and jazz and considered a career in one of those fields. After an undergraduate degree in physics at McGill University, Montréal, Quebec, she studied at Oxford University, United Kingdom, as a Rhodes Scholar, where she did her graduate work using a powerful analytical instrument called PIXE (proton-induced X-ray emission). At the University of Guelph, her research projects included strain measurements in semiconductors, ion optics, and trace element analysis of human hair samples.

She has performed as a jazz singer on national radio and has released several CDs, including *Kiss Me Like That*, which is based on music created for the International Year of Astronomy (2009), and *Sweet Fire*, which contains Shakespeare sonnets put to music.

Guelph Mercury–Nicki Corrigall/The Canadian Press

Figure 1-2 Physicist and musician Diane Nalini de Kerckhove, a Rhodes Scholar, a researcher in high-resolution nuclear microprobe technology, and an acclaimed jazz musician

have important applications. For example, in 1915, Albert Einstein published a new theory of gravitation called general relativity. When general relativity was developed, it had no practical application. Today, however, the Global Positioning System (GPS) would be hopelessly inaccurate without corrections for the gravitational effects predicted by general relativity theory.

Physics seeks explanations with the greatest simplicity and widest realm of application Physics is not a collection of a huge number of laws. Rather, physics seeks to explain the physical universe and all that it contains, using a limited number of relationships. For example, we

The CCD: Applied Physics and a Nobel Prize

The 2009 Nobel Prize in Physics was awarded to three scientists: the late Canadian Willard S. Boyle and American George E. Smith for the invention of the charge coupled device (CCD), and Charles Kuen Kao from China for work leading to fibre-optic communication. Born in Amherst, Nova Scotia, Boyle studied at McGill University before working at Bell Laboratories in New Jersey, where he and Smith made the first CCD.

The CCD is a semiconductor device with rows of tiny cells that accumulate an electric charge proportional to the light intensity at each cell. The cells are arranged such that a sequence of voltage pulses applied to the cells causes the charge in each cell to transfer to the next cell in the row. The charge leaving the last cell produces a signal that corresponds to the light that was focused on the cells. This signal is amplified to make an electronic record of the image that was stored as charge on the CCD.

CCD imaging has many advantages over film, including substantially greater sensitivity, which is of great importance to astronomers. CCDs are the heart of the Hubble Space Telescope—indeed, space telescopes would not be feasible without them. CCDs in camcorders and digital cameras have revolutionized photography, making photographic film almost obsolete. CCDs have also enabled significant advances in medical and scientific instrumentation.

Reprinted with permission of Alcatel-Lucent USA Inc

Figure 1-3 Willard Boyle (left) and George Smith in 1970, shortly after their invention of the charge coupled device (CCD)

only need to invoke four types of interactions to explain all "forces" in physics: gravitation, electromagnetism, weak nuclear, and strong nuclear. In your study of physics it is critical to keep in mind this goal of applying core theoretical ideas in a wide variety of situations. When you have finished this book, perhaps you may be able to list the key ideas of physics on a single page.

Physics uses equations extensively to express ideas You should view physics equations as a shorthand notation for the theories and relationships they represent. Applying physics is not simply selecting from a large pool of established equations. You will need to develop proficiency in manipulating equations and deriving relationships from basic principles.

LO 2

1-2 Measurement and Uncertainties

Why do we measure? Often we make measurements to test hypotheses. For example, if a technician proposes that a new material expands less in response to temperature changes than some other material, we can use physics measurements to test this claim.

Sometimes, we make measurements to learn about relationships between quantities and to help develop models and theories. For example, in the 1920s, astronomers Edwin Hubble and Milton Humason made measurements of the light from 46 galaxies and found that the light from the more distant galaxies has a greater redshift. (A *redshift* is a shift to longer wavelengths caused by motion away from the observer.) These measurements showed that the relationship between velocity of recession and distance was approximately linear, as shown in Figure 1-4. This relationship, in turn, led to the startling conclusion that the universe is expanding.

CHECKPOINT

C-1-1 What Is Physics?

Which of the following activities is *not* consistent with our description of physics.

(a) testing models for subatomic particles with experiments

(b) collecting observations of interacting galaxies

(c) developing a plan for an experiment to test a new theory of gravity

(d) creating a new mathematics function

C-1-1 (d) Physics should in some way be linked to the physical universe, not simply the world of mathematical definitions (although later physicists might use that mathematical function).

Edwin Hubble, "A Relation between Distance and Radial Velocity among Extra-Galactic Nebulae." PNAS March 15, 1929 vol. 15 no. 3, pp. 168–173

Figure 1-4 The plot of expansion velocity (in km/s) versus distance (in parsecs) published by Hubble and Humason in 1925. A parsec is the distance that light travels in 3.26 years, so the galaxies in the middle of the graph are about 3 million light years away.

Sometimes, we make measurements not to develop a theory, but rather to test its predictions. At other times, we make measurements to obtain precise values for quantities such as the speed of light and the charge on an electron.

Now, suppose that measurements of the expansion of two materials showed that for a given increase in temperature, the length of one material increased by 0.0025% and the length of the other material increased by 0.0030%. Do these measurements mean that the first material has a lower expansion coefficient? The answer depends not only on the measurements, but on the uncertainties in those measurements.

There are three different classes of potential errors in measurements of physical quantities: true errors, systematic errors, and inherent uncertainties.

True errors are mistakes made in the measurement. These errors can in general be identified by conducting more measurements, ideally by experimenters using different techniques. Examples of true error are an experimenter reading a scale on a measurement incorrectly, or transcribing a reading into a notebook incorrectly, or connecting electrical equipment incorrectly so that the measurement made is not the one intended.

Systematic errors result from some bias in the way that the quantity is being measured. For example, if you are measuring an electrical current with a meter that is not accurately calibrated, all the measured values might be somewhat higher (or lower) than the true values.

Inherent uncertainties (also called **random errors**) result from limitations in the precision of the measurements. No matter which technique is used, a measurement cannot be made with absolute precision, so there is always some uncertainty. In Chapter 31, you will learn how the uncertainty principle determines the minimum uncertainty inherent in all measurements.

We need to distinguish between the terms "precision" and "accuracy." **Accuracy** is how close the measurement of some quantity is to the true value. **Precision** is a measure of how close to one-another repeated measurements are. For example, if you measured the length of a box and got 100 values all between 1.259 and 1.260 m, then the precision would be quite good. However, if the actual length of the box is 1.204 m, then the accuracy is rather poor (and you need to look for the cause of the systematic error).

MAKING CONNECTIONS

Becoming a Physicist: Conventional and Unconventional Paths

Edwin Hubble and Milton Humason (Figure 1-5) came to careers in physics by radically different paths. Hubble took an undergraduate degree in astronomy and mathematics and then studied law as one of the first Rhodes Scholars. He practised law in Kentucky but soon abandoned law for a doctorate in astronomy at the University of Chicago. He graduated in 1917. After serving in World War I, Hubble got a job at the Mount Wilson Observatory in California.

Humason dropped out of school at the age of 14. He was a mule driver and hauled materials and equipment up Mount Wilson when the observatory was being built. Humason joined the observatory as a janitor and then volunteered to work as a night assistant. The quality of his work led to an appointment of staff astronomer despite his lack of academic credentials. Among Humason's discoveries is the giant comet C/1961 R1, which is named after him.

Today, the vast majority of physicists follow a path of undergraduate degree followed by graduate work, but there are still a few rare exceptions of individuals who come to physics by unconventional routes, as Humason did.

© Bettmann/CORBIS

Figure 1-5 A group of distinguished physicists at Mount Wilson Observatory in 1931. Milton Humason and Edwin Hubble are at the left. Just to the left of Einstein is Albert Michelson, whose work is discussed in Chapters 28 and 29.

MECHANICS

The best way to state the uncertainty in a measured quantity is to provide an estimate of the uncertainty. That is, we could write a length measurement as 1.15 ± 0.08 m, which means that the true value is likely between 1.07 and 1.23 m. Often the ± value is selected so that the true value is within the stated range about two-thirds of the time. In some fields, the criterion is 95% certainty, meaning, in this case, that 95 times out of 100 the true value will be between 1.07 and 1.23 m. We will consider these criteria in more detail in the next section, along with the question of how to decide whether two different mean values are significantly different.

 CHECKPOINT

C-1-2 Comparing Means

The mean of a set of length measurements for object A is 1.25 ± 0.05 m, and the mean of similar measurements for the length of object B is 1.32 ± 0.20 m. Can we conclude that B is longer than A? Explain your reasoning.

C-1-2 No. When the range of uncertainties is taken into account, A could be between 1.20 and 1.30 m long, whereas B is between 1.12 and 1.52 m long. Therefore, we cannot conclude that B is longer than A.

1-3 Mean, Standard Deviation, and SDOM

The first step in determining the uncertainty inherent in a set of measurements is to calculate the **mean** (or arithmetic average). For a set of N measurements with values x_i, the mean, μ, is defined as follows:

KEY EQUATION
$$\mu = \bar{x} = \frac{1}{N} \sum_{i=1..N} x_i \qquad (1\text{-}1)$$

Consider the two sets of measurements shown in Table 1-1 and plotted in Figure 1-6. They both have the same mean value (5.0), but they have very different distributions: set A has values with a much smaller spread about the mean value.

Table 1-1 Two Sets of Data with the Same Mean

Set A	Set B
5.5	3.0
5.0	8.5
4.7	7.0
4.8	1.5
4.4	5.5
5.6	4.5
Mean = 5.0	Mean = 5.0

Figure 1-6 The two distributions have the same mean value (5.0), but the data shown with green bars has much more scatter around this mean value.

We can use the standard deviation, σ, to specify how near the values cluster around the mean value. The **standard deviation** is defined as follows:

KEY EQUATION
$$\sigma = \sqrt{\sum_{i=1..N} \frac{(x_i - \bar{x})^2}{N-1}} \qquad (1\text{-}2)$$

The smaller the standard deviation, the more closely the measured values cluster around their mean. For example, data set A in Table 1-1 has a standard deviation of 0.47, and data set B has a standard deviation of 2.57.

For many types of measurements, intrinsic uncertainties (random errors) follow a **normal distribution** (also called a Gaussian distribution). This distribution is shown in Figure 1-7. Approximately 68% of the measurements from a normally distributed set lie within ±1 standard deviation of the mean value, about 95% of the values lie within ±2 standard deviations of the mean, and 99.7% lie within ±3 standard deviations of the mean.

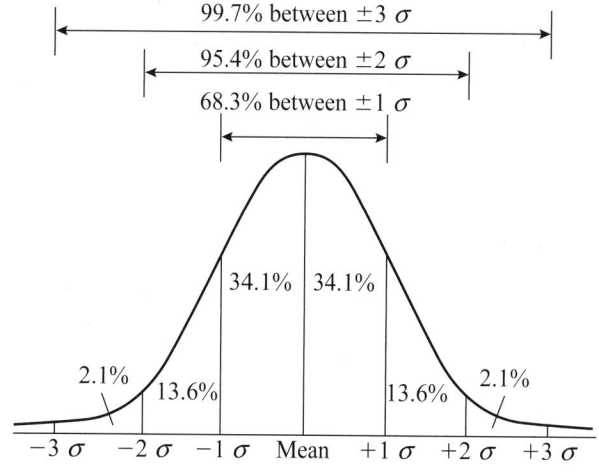

Figure 1-7 An example of a normal distribution (also called a Gaussian distribution)

CHECKPOINT

C-1-3 Standard Deviation

The results of a public opinion poll are reported as being accurate to within 4%, 19 times out of 20. Expressed as a percentage, what is the implied standard deviation?

Therefore, one standard deviation is 2% uncertainty.

C-1-3 *Nineteen times out of 20 is about 95%, so it must correspond to two standard deviations.*

The more measurements that we take, the closer the mean of the sampled values in our set of measurements will come to the true mean of the distribution. Now, we consider how to determine whether differences between two mean values are significant. You might think we could use the standard deviations of the two sets of data, but these standard deviations are actually quite misleading for comparing different sets of data. If the mean is based on a larger number of observations, its value is more certain, even if the standard deviation of the set is the same as the standard deviation of a smaller set.

Therefore, we use a quantity called the **standard deviation of the mean (SDOM)**, which is defined in terms of the number of observations (N) used for the mean

and the standard deviation (σ). The SDOM is also called the *standard error of the mean*.

KEY EQUATION
$$\text{SDOM} = \frac{\sigma}{\sqrt{N}} \qquad (1\text{-}3)$$

For example, in the data in Table 1-1, set B has a SDOM of $\frac{2.57}{\sqrt{6}}$, or about 1.0.

It is easy to make mistakes when interpreting the standard deviation and the SDOM. Remember that the standard deviation is a measure of the uncertainty of a single value, and the SDOM is a measure of the uncertainty of a mean. For example, about 68% of the time, a single measurement will be within ± 1 standard deviation of the mean value, and about 68% of the time, the mean of a set of measurements will be within ± 1 SDOM of the true value. Therefore, the SDOM is used when you have taken the average value from multiple samples of two different populations, and you want to determine whether the sets are statistically different. Most literature in physics uses the notation with the form mean $\pm$ SDOM when expressing a quantity, but you will sometimes see a notation with the SDOM in parentheses after the mean value.

EXAMPLE 1-1

Significant Difference?

In an experiment, the decay time of a subatomic particle (the time the particle takes to break down into other particles) was measured repeatedly, producing the following values: 0.45, 0.37, 0.67, 0.84, 0.38, 0.55, and 0.48 μs. A different experiment measured decay times of 0.68, 0.48, 0.88, 0.87, and 0.46 μs. Can we conclude that the decay times in the two experiments are statistically different? Note: In this chapter we will use a criterion of ± 1 standard deviation, unless otherwise stated.

SOLUTION

First, we calculate the mean of each set of measurements using Equation (1-1). For set A (plotted in blue in Figure 1-8), we have a mean decay time of 0.53 μs, and for set B we have a mean decay time of 0.67 μs.

Next, we calculate the standard deviation of each set, using Equation (1-2), and we obtain the values 0.17 μs and 0.20 μs.

Now, we calculate the two SDOM values. Set A has $N = 7$, and set B has $N = 5$. Using Equation (1-3), we obtain SDOM values of 0.06 μs for set A and 0.09 μs for set B.

The mean of set A can be written as 0.53 ± 0.06 μs, and the mean of set B can be written as 0.67 ± 0.09 μs. So, we are reasonably confident that the true value of the mean of A lies between 0.47 and 0.59 μs, and also reasonably confident that the true value of the mean of B lies between

0.58 and 0.76 μs. Since these two ranges overlap slightly, we cannot conclude that the difference between the sets of data is statistically significant.

Figure 1-8 Comparison of the two sets of data for particle decay times. The blue bars represent the values from set A, and the green bars represent the values from set B. The shaded rectangles show the areas defined by the mean $\pm$SDOM for each distribution.

Making sense of the result:

Note that, although we cannot conclude that the differences between the means are significant, we have not proven that the means are the same. If we took more measurements, we might find that the mean decay times are significantly different.

MECHANICS

LO 4

1-4 Significant Figures

When uncertainty in values is not stated explicitly using a $\pm$ range or standard deviation, we assume that the uncertainty is indicated by the **significant figures** (sometimes called *significant digits*) in the data. For example, two measurements of a time interval are stated as 1.6 and 1.5984 s; the latter measurement implies a much greater precision. The first measurement indicates that the true value is somewhere between 1.55 and 1.65 s; the second measurement implies a much narrower range, between 1.598 35 and 1.599 45 s. We say that there are more significant digits in the second measurement than in the first. It is important not to state a measurement with more significant digits than justified by the true uncertainty in the value.

Unfortunately, conventions for determining the number of significant figures vary somewhat. In this book, we use the following rules:

1. All nonzero digits are always significant. For example, 345.6 has 4 significant digits.

2. Zeros between nonzero digits are always significant. For example, 1204 has 4 significant digits.

3. Zeros at the end of a number are significant only if they are to the right of the decimal point *or* followed by a decimal point. For example, both 450 and 45 000 have 2 significant digits, 450. has 3 significant digits, and both 450.0 and 1.560 have 4 significant digits.

4. Zeros at the beginning of a number are not significant. This rule makes the number of significant digits consistent when unit prefixes are changed. For example, the measurements 0.067 m and 6.7 cm express the same length, and both have 2 significant digits. Example 1-2 demonstrates the application of these rules.

EXAMPLE 1-2

Significant Figures

How many significant figures are in each of the following?
(a) 5.46 m
(b) 0.024 575 m
(c) 320 m
(d) 24.260 m
(e) 204 m

SOLUTION

(a) The measurement 5.46 m has 3 significant figures according to rule 1.

(b) The measurement 0.024 575 m has 5 significant figures (neither 0 is significant according to rule 4).

(c) The measurement 320 m has 2 significant figures according to rule 3.

(d) The measurement 24.260 has 5 significant figures according to rules 1 and 3.

(e) The measurement 204 m has 3 significant figures according to rule 2.

✓ CHECKPOINT

C-1-4 Significant Digits

How many significant digits are in each of the following numbers?
(a) 340.03
(b) 245 000
(c) 12.00

C-1-4 (a) 5 significant digits according to rule 1. (b) 3 significant digits according to rule 3. (c) 4 significant digits according to rule 3.

LO 5

1-5 Scientific Notation

Scientific notation eliminates any confusion over whether zeros at the end of a number are significant digits. Scientific notation also makes it easier to work with the very large and the very small numbers that we frequently encounter in physics.

In scientific notation we write the number 46 580 as 4.658×10^4. To convert a number to scientific notation, we determine the power of 10 factor that will let us express the number in a form with just one nonzero digit to the left of the decimal point. Example 1-3 demonstrates conversions to and from scientific notation.

EXAMPLE 1-3

Scientific Notation

(a) The approximate distance from Earth to the Sun is 150 000 000 000 m. Express this distance in scientific notation.
(b) How would 0.023 be written in scientific notation?
(c) Write the number 2.45×10^{-3} in regular notation.

SOLUTION

(a) We move the decimal point 11 places to the left to get 1.5×10^{11} m.
(b) We move the decimal point 2 places to the right to get 2.3×10^{-2}.
(c) We move the decimal point 3 places to the left to eliminate the multiplier, giving 0.002 45.

CHECKPOINT

C-1-5 Scientific Notation and Significant Digits

How many significant digits are in the number 3.50×10^{-2}?

C-1-5 There are 3 significant digits in the number 3.50×10^{-2}.

LO 6

1-6 Significant Figures and Mathematical Operations

There are two basic rules for dealing with significant figures in calculations.

Multiplication and Division When multiplying and dividing quantities, the number of significant figures in the final answer should be the same as the least precise quantity used in the calculation, as demonstrated in Example 1-4. Note that integers and mathematical constants are assumed to have infinite significant figures. For example, in the expression for the surface area of a sphere, $4\pi r^2$, the factors 4 and π do not affect the number of significant figures in the calculated area.

 EXAMPLE 1-4

Significant Figures in Multiplication and Division

Express the answer to the following calculation with the correct number of significant figures:

$$\frac{32.55 \times 0.021}{465}$$

SOLUTION

When we enter these numbers into a calculator, the answer will likely be displayed as 0.001 47 or 0.001 470. However, the number 32.55 has 4 significant figures, the number 0.021 has 2 significant figures, and the number 465 has 3 significant figures. Therefore, the final answer should have only 2 significant figures, and the calculated number should be rounded to 0.0015.

Addition and Subtraction When adding and subtracting quantities, we keep the same number of significant figures after (or in front of) the decimal point as in the least precise of the quantities used in the calculation. For example, the number 2.45 is assumed to have an uncertainty of the order of 0.01. If we subtract the number 2.30 from 2.45, we obtain 0.15—the uncertainty is still of the order of 0.01. This uncertainty stemming

from the first measurement would not be reduced if the second number had a higher precision, such as 2.3043. Example 1-5 demonstrates the application of this rule.

 EXAMPLE 1-5

Significant Figures in Addition and Subtraction

You measure the total length of two objects together as 2.45 m. You know that one of the objects has a length of 2.248 m. What is the length of the other object?

SOLUTION

A calculator will give an answer of 0.202 m. However, the measurement 2.45 m has 2 significant figures to the right of the decimal point. Therefore, the final answer should be written as 0.20 m.

Making sense of the result:

The measured lengths have 3 and 4 significant digits, but the calculated length has only 2. When we subtract two quantities that are approximately equal, the result will have fewer significant digits than the original quantities.

 CHECKPOINT

C-1-6 Computations and Significant Digits

How should the answer be written for the following calculation?

$$\frac{37.45 - 37.020}{124.67}$$

(a) 0.003 449
(b) 0.003 45
(c) 0.0034
(d) 0.003

C-1-6 (c) According to the subtraction rule, the operation in the numerator gives us 2 significant figures. Therefore, the final answer should have the lesser of 2 and 5 significant figures.

Note that the significant figure techniques described above give only an approximation of the uncertainty in measured and calculated quantities. More advanced courses in physics will show you more precise methods for dealing with the uncertainty in calculated quantities.

LO 7

1-7 Error Bars

When graphing data, we can indicate the uncertainty in the data with **error bars**. In Figure 1-9, the error bars are symmetric about the data points, and the uncertainty is the same for all the data values. However, uncertainties are often not symmetric, and they can vary with the data values. In some instances, there are significant uncertainties in the x-values as well, in which case the graph would have both vertical and horizontal error bars.

Figure 1-9 The uncertainty in y-values of measured quantities is indicated with error bars. When a data point is obtained by finding the mean of values, it is important to realize that we use the SDOM in determining the length of the error bar. Sometimes, however, the error bar is determined from an estimate of the uncertainty rather than a calculation.

 CHECKPOINT

C-1-7 Error Bars

A data point on a plot has the x-value determined by taking the mean of the numbers 2.4, 2.7, 3.0, 3.3, and 3.6. The y-value is obtained by taking the mean of the values 3.0, 4.0, 5.0, 6.0, and 7.0. Describe how the error bars should be drawn for this case.

C-1-7 First, consider the x-values, which have a mean of 3.0 and a standard deviation of 0.47. The error bar is drawn using the SDOM, however, which will be 0.47 divided by the square root of N or 2.2 in this case. Therefore, the error bar will extend a horizontal distance of 0.2 (in both directions). Similar arguments lead to a mean of 5.0 in the y-direction and a SDOM of 0.7, so the error bars in the y-direction extend 0.7 (in both directions) from the point.

Error bars are important enough to be included in most experimental results published in physics. These bars give a clear indication of uncertainty in the individual values and how much confidence we can have in any apparent trend in the data.

 ONLINE ACTIVITY

Curve Fitting and Error Bars

The e-resource that accompanies every new copy of this textbook contains an Online Activity using the PhET simulation "Curve Fitting." Work through the simulation and accompanying questions to gain an understanding of error bars, curve fitting, and uncertainty.

LO 8

1-8 SI Units

In this text, we use almost exclusively the International System of Units, or **SI**. The abbreviation comes from the French title, *le Système internationale d'unités*. This system is used worldwide now; in fact, only three countries have not adopted SI for commerce as well as for science: Burma (also known as Myanmar), Liberia, and the United States.

SI is an integrated metric system, designed so that it can be used readily by all the sciences. SI has seven base units, chosen such that all quantities can be expressed in some combination of the base units. SI also has a set of derived units, defined in terms of the base units, to provide convenient measures of particular quantities. Table 1-2 lists SI base units and the quantities they measure.

Table 1-2 The Base SI Units

Unit Name	Unit Symbol	Measures
metre	m	length
second	s	time
kilogram	kg	mass
kelvin	K	temperature
mole	mol	amount of substance
ampere	A	electric current
candela	cd	luminous intensity

The *metre*, *second*, and *kilogram* are units that you likely know well from everyday use. For historical reasons, the kilogram—not the gram—is the SI base unit of mass. The kilogram is the only base unit that has a prefix.

A *kelvin* is equal to a Celsius degree, but the zero point of the Kelvin temperature scale, denoted 0 K, is absolute zero, the lowest temperature possible (approximately $-273°C$).

A *mole* is the amount of a substance that contains the same number of elementary entities (e.g., atoms or molecules) as exactly 12 g of the isotope carbon-12, about 6.022×10^{23}. In other words, 1 mol contains Avogadro's number of atoms or molecules.

The *ampere* is a measure of electric current, the flow of electric charge past a given point. When you study electromagnetism later in this book, you will see why the designated base unit for electrical quantities is based on current rather than charge.

The *candela* is the unit of luminous intensity, a measure of the light output of a source. As you might guess from the name, the candela is based on the light output of a standard candle with a specified composition and rate of combustion. In fact, a typical candle has brightness in the order of 1 cd. The candela is a precisely defined unit that replaces older units.

As well as the base units, there are a number of fundamental derived SI units. We list the most important ones used in this book in Table 1-3.

There are a number of frequently used units not shown in Table 1-3. Some of these are simply combinations of the units listed. For example, speed and velocity have units of m/s, and acceleration (the rate of change of velocity) has units of m/s^2.

There are a number of commonly used non-SI units, including the minute (min), the solar day (d), the year (a or y), and the light year (ly), which is the distance light travels in one year.

Table 1-3 Derived SI Units Commonly Used in First-Year Physics Courses

Unit Name	Unit Symbol	Quantity	Base Unit Equivalent
newton	N	force	$m \cdot kg \cdot s^{-2}$
joule	J	energy, work	$m^2 \cdot kg \cdot s^{-2}$
watt	W	power	$m^2 \cdot kg \cdot s^{-3}$
hertz	Hz	frequency	s^{-1}
radian	rad	angle	$m \cdot m^{-1}$ (dimensionless)
steradian	sr	solid angle	$m^2 \cdot m^{-2}$ (dimensionless)
pascal	Pa	pressure	$m^{-1} \cdot kg \cdot s^{-2}$
coulomb	C	electric charge	$s \cdot A$
volt	V	electric potential difference	$m^2 \cdot kg \cdot s^{-3} \cdot A^{-1}$
ohm	Ω	electric resistance	$m^2 \cdot kg \cdot s^{-3} \cdot A^{-2}$
farad	F	electric capacitance	$m^{-2} \cdot kg^{-1} \cdot s^4 \cdot A^2$
henry	H	electromagnetic inductance	$m^2 \cdot kg \cdot s^{-2} \cdot A^{-2}$
tesla	T	magnetic field strength	$kg \cdot s^{-2} \cdot A^{-1}$
weber	Wb	magnetic flux	$m^2 \cdot kg \cdot s^{-2} \cdot A^{-1}$
becquerel	Bq	radioactivity	s^{-1}
lumen	lm	luminous flux	cd
lux	lx	luminous flux per unit area	$m^{-2} \cdot cd$

SI Prefixes

SI has a set of prefixes that indicate different powers of 10; for example, a kilometre is 1000 m, and a millimetre is 0.001 m. Table 1-4 lists the names and abbreviations for the most commonly used SI prefixes. Note that the prefix symbols greater than "kilo" are capitalized.

Table 1-4 The Most Common SI Prefixes

Name	Symbol	Multiplication
zetta	Z	10^{21}
exa	E	10^{18}
peta	P	10^{15}
tera	T	10^{12}
giga	G	10^9
mega	M	10^6
kilo	k	10^3
hecto	h	10^2
deca	da	10^1
deci	d	10^{-1}
centi	c	10^{-2}
milli	m	10^{-3}
micro	μ	10^{-6}
nano	n	10^{-9}
pico	p	10^{-12}
femto	f	10^{-15}
atto	a	10^{-18}
zepto	z	10^{-21}

SI specifies rules for writing measurements in SI units. These rules combine aspects of traditional styles used in various countries into a single, unambiguous convention that is recognized worldwide.

- There is always a space between the number and the unit symbol (e.g., 10 m, not 10m).
- Abbreviations for units are capitalized when they are named after a person but not otherwise (e.g., N for newton but kg for kilogram).
- Units are not capitalized, even when named after a person, when written out in full (e.g., newton not Newton). This rule prevents confusion over whether one is referring to the person or the unit.
- SI unit symbols are not considered abbreviations, so they do not end with a period (e.g., W and cd, not W. and cd.).
- Unit symbols stand for both single units and multiples of the unit, so we never add an "s" to make a symbol plural (e.g., 10 m means ten metres, but 10 ms means ten milliseconds).
- The unit symbols are written in normal (roman or upright) type to distinguish them from variables, which are written in italic (e.g., "m" always represents the unit metre, while m usually represents the variable mass).

- When writing numbers with many digits, sets of 3 digits on either side of the decimal point are separated by spaces instead of commas. In a number with 4 digits, the space is optional (e.g., 52 469 450, 7.630 941 23, and 4123 or 4 123).

✓ CHECKPOINT

C-1-8 Writing SI

Identify all the errors in this statement: "The force applied was 23Ns. and the displacement ...."

C-1-8 One should not use an "s" with the unit symbol for the newton, N. There should be a space between the number and the unit. There should not be a period after the unit symbol. So the statement should be, "The force applied was 23 N and the displacement ...."

When we use SI units consistently and carry the units through in calculations, the results will have appropriate SI units. The website of the National Institute of Standards and Technology (a U.S. body) is a helpful reference for learning more about the definitions, derivations, and uses of SI units (http://physics.nist.gov/cuu/Units/).

LO 9

1-9 Dimensional Analysis

We can consider the units in a problem to help determine whether a proposed relationship is possible; this is called **dimensional analysis**. Note that dimensional analysis determines only whether the units in a relationship are equivalent; further data is needed to prove that the relationship is actually valid. We demonstrate this technique in Example 1-6.

EXAMPLE 1-6

Dimensional Analysis for the Period of a Pendulum

Simple observations show that the period, T, of a simple pendulum depends on the length of the pendulum and on the acceleration due to gravity, g (about 9.8 m/s^2 on Earth's surface). Use dimensional analysis to determine a possible equation for period.

SOLUTION

The period must have units of time, s. The SI unit for length is the metre, and acceleration due to gravity has units of m/s^2. We can see that the only way to get units of

time by combining the units of length and acceleration is as follows:

$$T = \sqrt{\frac{\ell}{g}} \leftrightarrow \sqrt{\frac{m}{m/s^2}} = \sqrt{s^2} = s$$

In this case, the real expression has a numerical factor (see below).

Making sense of the result:

Observations confirm that a longer pendulum will oscillate more slowly and hence have a longer period, and a faster acceleration due to gravity would make the period shorter. In fact, the actual relationship for the period does have the form we determined above, but the relationship also contains a dimensionless numerical factor:

$$T = 2\pi\sqrt{\frac{\ell}{g}}$$

✓ CHECKPOINT

C-1-9 Dimensional Analysis

Later in the text, you will use kinematic relationships between displacement (in m), velocity, acceleration, and time. Use dimensional analysis to determine whether the following relationship is correct (x is in m, v_0 is in m/s, and a is in m/s^2). If the relationship is not correct, suggest how it could be altered.

$$x (=?) \, x_0 + v_0 t^2 + at$$

C-1-9 This relationship is not possible. The term $v_0 t^2$ has units of m · s, and the term at has units of m/s. We can rewrite the relationship in the following form, which does meet the dimensional analysis requirements:

$$x (=?) x_0 + v_0 t + at^2$$

As you will see in later chapters, the real expression (for constant acceleration situations) is $x = x + v_0 t + \frac{1}{2} at^2$. Remember that dimensional analysis cannot give you numerical factors, such as the $\frac{1}{2}$.

LO 10

1-10 Unit Conversion

You can easily do some **unit conversions** in your head; for example, 25 mm is the same as 2.5 cm, and 12 km is the same as 12 000 m. However, for conversions that involve more than a simple power of 10, we recommend that you use the formal unit conversion method outlined in Example 1-7. This method is based on multiplying by factors in which the quantities in the numerator and denominator are equivalent. Since these factors equal 1, the multiplication does not change value of the quantity. We can cancel units to confirm that the conversion is correct.

MECHANICS

EXAMPLE 1-7

Unit Conversion

Express the speed 72 km/h in SI base units.

SOLUTION

Speed is a displacement (unit of length) divided by a time, so the SI base units are metres per second.

We do the conversion by writing factors so that our units cancel into the desired final units of m/s, with each factor having quantities in the numerator and the denominator that are equivalent.

$$\frac{72 \text{ km}}{1 \text{ h}} \times \frac{1000 \text{ m}}{1 \text{ km}} \times \frac{1 \text{ h}}{60 \text{ min}} \times \frac{1 \text{ min}}{60 \text{ s}} = 20 \text{ m/s}$$

Making sense of the result:

If we travelled at 1 m/s, we would go 3600 m, or 3.6 km, in an hour. Thus, it makes sense that the speed in m/s will be numerically slightly less than one-third of 72.

PEER TO PEER

Be careful when converting units that are squared. A useful technique is to write the squared unit as a product using the conversion factor twice. For example, to convert 70 mm² to cm², write

$$70 \text{ mm}^2 = 70 \text{ mm mm} \times \frac{1 \text{ cm}}{10 \text{ mm}} \times \frac{1 \text{ cm}}{10 \text{ mm}} = 0.70 \text{ cm}^2$$

 CHECKPOINT

C-1-10 Converting Units

The rate at which a process deposits mass on a surface is given as 25 g/cm². What would this rate be in units of kg/m²?

(a) 2.5
(b) 250
(c) 2500
(d) 25 000

C-1-10 (b) We multiply by two factors: 1 kg/1000 g and (100 cm/1 m)².

MAKING CONNECTIONS

Costly Mistakes in Unit Conversion

In 1999, the Mars Climate Orbiter (Figure 1-10) provided a spectacular and very expensive demonstration of the critical importance of unit conversions. This spacecraft was intended to orbit Mars and collect data on the planet's climate. However, on September 23, 1999, the orbiter angled far too low into the Martian atmosphere and disintegrated. The reason for the trajectory error was a misunderstanding between NASA staff and contractors on the units used in thrust calculations.

On July 3, 1983, an Air Canada jetliner ran out of fuel while on a flight from Montréal to Edmonton. The fuel gauge system on the aircraft was malfunctioning, so the ground crew and pilots had to calculate the fuel for the flight manually. These calculations involved both the volume and mass of the fuel. The aircraft was a new model calibrated with metric units, but the paperwork for refuelling was based on gallons and pounds, the units used by most aircraft at the time. Although the pilots double-checked the calculations, one of the conversion factors was inverted. Fortunately, the captain was an experienced glider pilot and managed to land with only minor damage on a drag strip at a former military base at Gimli, Manitoba. That aircraft has gone down in aviation history as the Gimli Glider.

Figure 1-10 The ill-fated Mars Climate Orbiter (artist's rendition)

1-11 Fermi Problems

Often in physics (and in engineering and other sciences), we encounter situations in which we do not have enough information to calculate precise values, but we still may be able to make useful approximations. Such estimates often involve reasonable guesses for the values of unknown quantities, probabilities, and dimensional analysis if a relevant physical law is not known. The famous physicist Enrico Fermi (1901–54) skillfully used such techniques, and problems involving such estimates are now called **Fermi problems** in his honour.

When doing Fermi problems, scientific notation is very useful because it separates the magnitude of the number (in exponent) from the precise value. Often, we are just seeking an order-of-magnitude estimate and can ignore everything other than the powers of 10 in coming up with an estimate.

Physicists regularly work in situations in which they cannot necessarily establish even a best estimate for a number, but they can determine a maximum possible value (or minimum possible value).

MAKING CONNECTIONS

Enrico Fermi

Enrico Fermi (Figure 1-11) was an Italian-born physicist who worked in the United States for the last 16 years of his life. He made important contributions to quantum mechanics, nuclear physics, statistical mechanics, and particle physics. He was an experimentalist and a theorist, and he was awarded the Nobel Prize in Physics in 1938. Fermi was a master of "back of the envelope" calculations. In one often-quoted example, he estimated the energy released by the Trinity bomb test in 1945 from a measurement of how much the blast deflected bits of paper at a position well away from the test site. His estimate was accurate to within one order of magnitude.

Department of Energy, Office of Public Affairs

Figure 1-11 Enrico Fermi

EXAMPLE 1-8

The Wheels on the Bus Go Round and Round

How many revolutions do the wheels on a bus make during a trip from Toronto to Montréal?

SOLUTION

We first estimate (or look up) the driving distance from Toronto to Montréal. This distance is about 540 km.

Next, we need to measure, estimate, or look up the radius, r, of a typical bus wheel. Tire sizes for buses vary somewhat, but a reasonable value is 50.0 cm.

With each revolution, the bus tire will move a distance of $2\pi r$ along the road. Using N as the number of revolutions and D as the total distance travelled, then

$$N(2\pi r) = D$$

We rearrange this equation to solve for N. Before we substitute the values for our estimates, we convert the distance and radius estimates to SI units (m) using the techniques of the earlier section. Then we solve for N:

$$N = \frac{D}{2\pi r} = \frac{540\,000 \text{ m}}{2\pi \times 0.50 \text{ m}} = 1.7 \times 10^5$$

Therefore, the wheels on the bus turn about 170 000 times during the trip. Note that our input data is probably only valid to 2 significant figures, so we round the final answer to that precision.

Making sense of the result:

For each kilometre travelled, the wheel makes about 300 revolutions, which seems reasonable. If the wheels are rated 60 000 km between wheel-bearing servicings, this would imply that the wheels turn more than 18 million times between servicings.

CHECKPOINT

C-1-11 Standing Side by Side

Suppose the entire population of the world stood next to each other, forming a square. Calculate the length of the side of the square, in kilometres.

(a) 50 km

(b) 250 km

(c) 500 km

(d) 5000 km

C-1-11 (a) Earth's population is about 7 billion people. If we assume that the area occupied by each person averages about 1 m × 1/3 m, or 0.3 m², then the total area required is about 2.3 billion square metres. This is about 2300 square kilometres. One side of the square is the square root of this, which is about 48 km. Therefore, (a) is the best choice.

We usually represent a **probability** using a numerical value between 0 and 1, with 0 representing no chance of something happening and 1 indicating that it is certain to happen. For example, if you flip a coin, the probability of heads is 0.5, and so is the probability of tails. Similarly, if you roll a six-sided unbiased die, the probability of any particular number, for example, a 4, will be 1/6, or about 0.17.

If we want to determine the overall probability of some event that depends on multiple independent events, then we simply multiply the probabilities of each of the events. For example, the probability of rolling two 6s with a pair of unbiased dice is (1/6)(1/6), or about 0.028. Note that this technique is valid only if the individual probabilities are independent.

 EXAMPLE 1-9

Habitable Planets

Estimate the number of habitable planets in the universe.

SOLUTION

Start with an estimate of the number of stars in our galaxy, the Milky Way. Informed estimates range from about 100 to 400 billion. We will use 150 billion for our calculation.

Now, consider the image of galaxies shown at the beginning of this chapter. Extrapolating from this sample region, we can estimate that the total number of galaxies in all directions is about 100 billion. It is possible that the image does not show some of the fainter, more distant galaxies, so this estimate may be low.

Therefore, the total number of stars in the visible galaxies is roughly 150 billion times 100 billion, or about 1.5×10^{22}.

Next, we need to estimate what fraction of stars will have planetary systems. Astrophysicists believe that the same process that produces stars from clouds of dust and gas also creates planets. However, some of these stars will be in binary (two-star) or multiple-star systems that may not have planets. With considerable uncertainty, we estimate that one-tenth of the stars have planetary systems, so there could be about 1.5×10^{21} planetary systems.

Now, we must consider the probable number of habitable planets per star. We know that our solar system has eight major planets. However, only Mars and Earth have conditions that could support life forms that depend on water. We might consider other forms of life, but many experts regard them as unlikely. In any case, we are looking only for an order-of-magnitude estimate. So, we have two habitable planets out of eight. Many stars will have energy outputs much different from that of the Sun, which could substantially reduce the number of planets with habitable temperatures. So, we estimate that there is an average of perhaps 0.2 habitable planets per planetary system.

Then the number of habitable planets in the universe is in the order of

$$0.2 \times 1.5 \times 10^{21} = 3.0 \times 10^{20}$$

Making sense of the result:

This is an almost unimaginably large number. For example, if a human could count one planet per second, and if the entire world population of about 7 billion people counts simultaneously (for 10 h/day, 6 days/week), about

$$365 \times (6/7) \times 10 \times 60 \times 60 \times 7 \times 10^9 = 7.9 \times 10^{16} \text{ planets}$$

could be counted per year. In other words, it would take the entire population about 3800 years to count this many planets at one per second, 6 days a week, 10 h/day.

LO 12

1-12 Becoming a Physicist

We hope that as you study physics in this text you will consider becoming a physicist, or using physics in some other science or engineering career. The Canadian Association of Physicists (CAP) has a useful website (www.cap.ca) that includes a section on careers in physics. The corresponding organization in the United States is the American Physical Society (APS) (www.aps.org). An international organization with particular strengths in Europe is the Institute of Physics (IOP) (www.iop.org).

It is important to realize that many physicists work as part of interdisciplinary teams and apply physics in areas such as medical physics, astrophysics, biophysics, geophysics, materials science, and many areas of engineering.

MECHANICS

MAKING CONNECTIONS

Professional Physicist Certification

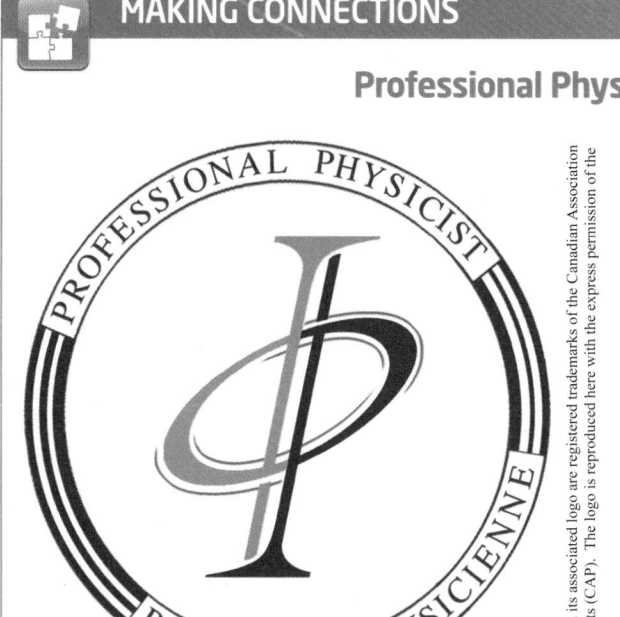

P.Phys. and its associated logo are registered trademarks of the Canadian Association of Physicists (CAP). The logo is reproduced here with the express permission of the CAP

The Canadian Association of Physicists has a professional certification program called Professional Physicist (PPhys). To qualify for a PPhys certification, you must complete minimum education and experience requirements, agree to a code of ethical conduct, and pass a professional examination that tests your ability to interpret physics in applied situations, as well as your sensitivity to ethical aspects of the profession. See the CAP website (www.cap.ca) for additional information.

Figure 1-12

The typical route to becoming a research physicist is to complete an undergraduate degree (usually an honours degree) in physics or a closely related area, such as applied mathematics, geophysics, or astrophysics, followed by graduate degrees (masters and doctorate). Many physicists then complete several years of postdoctoral work prior to taking a permanent position in research, development, teaching, or administration. Yes, that is a long road. However, graduate students in physics are almost always fully supported by graduate scholarships and research and teaching assistantships, so following your undergraduate degree you will probably be financially independent.

While most research positions require a doctorate, a B.Sc. is sufficient for many careers, including technical positions in research or development labs, application of physics in professional programs such as patent law, teaching at the high school level, marketing of physics-related materials, and science journalism. Others use skills learned in physics courses such as quantitative analysis, problem solving, model formation, and experimental design in a broad array of fields, including administration.

Before finishing this chapter, we want to leave you with some suggestions for doing well in your study of physics. You only really learn physics by mentally interacting with the ideas. Don't skip suggested interactive activities; they are key to your learning. While the chapter summaries are valuable, you should read the entire chapter at least once. Remember that your goal in physics is not to remember a large number of relationships or algorithms for solving problems of a certain type. Rather, it is to get a sound grasp of a surprisingly small number of key concepts. Those, along with expertise in the appropriate mathematical techniques, will allow you to tackle novel problems. Physics can, for all of us, feel frustrating at times, but the more problems you try on your own, the more your expertise will grow. It is important that you try the whole range of questions and problems, including conceptual questions, problems by section, the broader comprehensive problems, and problems that involve data analysis and the open problems that will require you to make reasonable assumptions and decide what is relevant to the situation.

FUNDAMENTAL CONCEPTS AND RELATIONSHIPS

Physics deals with physical phenomena on all scales from subatomic particles to the universe as a whole.

Accuracy is how close a measurement is to its true value, and precision is a measure of how close to one-another repeated measurements are.

For data that is normally distributed, about two-thirds of the data lie within ± 1 standard deviation of the mean, and about 95% of the data lie within ± 2 standard deviations of the mean.

SI is a unified system of units recognized worldwide and applicable for all fields of science and engineering.

Unit conversions are performed by multiplying factors with equivalent expressions in the numerator and the denominator.

Nature of Physics

Physics has a number of distinguishing characteristics, including its quantitative nature, the strong role of models, both theoretical and applied aspects, the need for creativity and a collaborative work environment, and, perhaps most important, the goal of explaining physical phenomena in the simplest way possible.

Precision and Accuracy

Precision specifies how narrowly the values range around a central value. Accuracy refers to how close that central value is to the "correct" value. There are three types of errors: true errors (mistakes), random errors (spread about the central value), and systematic errors (e.g., if the technique overestimated all values).

Mean and Standard Deviation

The mean, μ, and standard deviation, σ, of N measurements are defined by the following:

$$\mu = \bar{x} = \frac{1}{N} \sum_{i=1..N} x_i \qquad \sigma = \sqrt{\sum_{i=1..N} \frac{(x_i - \bar{x})^2}{N-1}}$$

Standard Deviation of the Mean

When determining whether the mean values from two different data sets are statistically different, we use the standard deviation of the mean (SDOM), also called the standard error of the mean:

$$\text{SDOM} = \frac{\sigma}{\sqrt{N}}$$

Significant Figures

Significant figures indicate the approximate precision of a value. When multiplying and dividing, you should have the same number of significant digits in the result as in the least significant entry number. When adding and subtracting, keep the same number of significant figures after the decimal point as in the least precise of the quantities used in the calculation.

SI Units

SI has seven base units: metre, kilogram, second, ampere, candela, kelvin, and mole. Many other SI units are derived from these base units. If you use only SI units when solving problems, the answer will always be another SI unit.

Unit Conversion

To convert from one unit to another, you multiply by a series of factors, each with a numerator and a denominator that are equivalent. The factors are arranged so that the original units cancel out, leaving the desired final units.

Dimensional Analysis

Dimensional analysis can be used to determine possible relationships among physical quantities. Dimensional analysis can rule out an incorrect equation, but it is not sufficient to prove whether a proposed equation is correct.

Fermi Problems

We can combine realistic estimates, dimensional analysis, and other techniques to obtain order-of-magnitude estimates for values that cannot be calculated precisely.

Combining Probabilities

Independent probabilities can be multiplied to obtain the probability of a combined event.

APPLICATIONS

Applications: determination of the significance of the differences between mean values for any two sets of measurements (e.g., is the average temperature different from an airport location and one near the centre of the city), unit conversion using the multiplication by equal terms ratio method (e.g., converting speeds from km/h to m/s, or areas from square centimetres to square metres), using dimensional analysis to check or propose functional relationships, quantitative approximation techniques to provide estimates for situations in real life (e.g., if I turn down the thermostat in a building by 2°C, how much energy will I conserve over a year, and what does that mean for greenhouse gas emissions).

Key Terms: accuracy, dimensional analysis, error bars, Fermi problems, mean, normal distribution, precision, probability, random error, SI, significant figures, standard deviation of the mean (SDOM), standard deviation, unit conversions

PUTTING IT ALL TOGETHER

QUESTIONS

1. Which of the following is *not* a characteristic of physics?
   - (a) Physics seeks to explain phenomena in the simplest way possible and in a way that has the widest application.
   - (b) While some phenomena may be described qualitatively, in general, physics seeks quantitative descriptions.
   - (c) Physics is primarily involved with developing a large number of equations so that there is one equation relevant for any given situation.
   - (d) Physics requires creativity.

2. Assume that measurements of a time for some process are normally distributed, with no systematic errors. About two-thirds of the time the value is between 3.4 and 4.8 s. What are the mean and standard deviation of the time measurement?

3. Assume that measurements of a length are normally distributed, with no systematic errors. About 19 times out of 20, the value is between 2.2 and 5.8 m. What are the mean and standard deviation of the length measurement?

4. How many significant figures are in the number 230 100?
   - (a) 3
   - (b) 4
   - (c) 5
   - (d) 6

5. Complete the calculation $\dfrac{12.0 \times 5.0}{4.000}$. According to the rules for significant digits, how should the answer be written?
   - (a) 15
   - (b) 15.0
   - (c) 15.00
   - (d) 15.000

6. Which of the following is *not* a base SI unit?
   - (a) kg
   - (b) s
   - (c) N
   - (d) cd

7. Which of the following is *not* a base SI unit?
   - (a) kg
   - (b) m
   - (c) J
   - (d) K

8. Work (or energy) is equal to force times x. Use Tables 1-2 and 1-3 to identify the units of x.

9. Consider the graph of Figure 1-4. Taking into account the various points made in this chapter, critically evaluate the graphical presentation, indicating all aspects that could be improved.

10. (a) Use a reference source to look up how the length of the metre is currently defined.
    - (b) From this definition, what must be the speed of light (to the number of significant figures inherent in the definition)?
    - (c) Provide a brief overview of some of the previous ways that have been used to define the length of the metre.

11. Use a reliable reference to determine how the second is currently defined.

12. (a) Use a reliable reference to determine the method currently used to define the kilogram.
    - (b) Provide a brief overview of some other ways that have been used to define the kilogram.

13. Describe the current method used to define the ampere. Include a reference to the source of your information.

14. Use a reliable reference source to identify the method currently used to define the candela, as well as the method that was historically used.

15. How many cubic centimetres are in one cubic metre?
    - (a) 100
    - (b) 1 000
    - (c) 10 000
    - (d) 1 000 000

16. Two data sets, A and B, have the same mean value, but data set B has a larger standard deviation by a factor of 2. There are 16 data points in data set A. How many data points are needed in set B for the SDOM to be identical for the two sets?

17. If we multiply a force (N) by a velocity (m/s), which of the following will be the units of the resulting quantity?
    - (a) kg/s
    - (b) J
    - (c) Pa
    - (d) W

PROBLEMS BY SECTION

For problems, star ratings will be used, (✱, ✱✱, or ✱✱✱), with more stars meaning more challenging problems.

Section 1-3   Mean, Standard Deviation, and SDOM

18. ✱   A set of 16 measurements has a mean of 4.525 and a standard deviation of 1.2. How should the mean be written with an uncertainty given by the SDOM?

19. ✱   The SDOM is 1.2, the mean is 4.8, and the standard deviation is 2.2. Approximately how many data points must be in the set?

20. ✱   According to a journal article, experiments suggest that 19 times out of 20, the measurement obtained is within ±0.5 of 10.5. What must be the standard deviation and mean values, assuming that the data is normally distributed?

Section 1-4   Significant Figures

21. ✱   How many significant figures are in each of the following?
    - (a) 324.2
    - (b) 0.032
    - (c) 9.004
    - (d) 12 000
    - (e) 85.0
    - (f) 6000.

22. ✱✱   Suppose that the uncertainty in the length 17.386 22 m is ±0.04 m. Round the length to the correct number of significant figures.

Section 1-5   Scientific Notation

23. ✱   Write each of the following in scientific notation.
    - (a) 2452
    - (b) 0.592
    - (c) 12 000
    - (d) 0.000 045

24. ✱   Perform the following calculation, and express the final result in scientific notation with the correct number of significant figures.

$$\frac{(3.20 \times 10^5)(1.5 \times 10^4)}{6.400 \times 10^3}$$

Section 1-6 Significant Figures and Mathematical Operations

25. ✷ Perform the following calculation. Express the answer with the correct number of significant figures.

$$\frac{12\ 400 \times 0.025\ 63}{12.34}$$

26. ✷ Perform the following calculations. Express the answer with the correct number of significant figures.
(a) 12.456 m − 11.9 m
(b) 6.542 m + 0.0092 m

Section 1-7 Error Bars

27. (a) ✷✷ Plot the following data points, with error bars, on a graph. Assume that there is no uncertainty in the x-value and that each y-value has an uncertainty of 1 m: (0,0), (1,2), (2,5), (3,7).
(b) Is there a line of best fit that passes through the data points, within the uncertainty of each data point? If so, plot it approximately on your graph.
(c) Add a new data point: (4,6). What must be the error bar for that point in order for the line of best fit to include that point as well?

Section 1-8 SI Units

28. ✷ What are the units of RC, where R is electrical resistance and C is electrical capacitance? (Use Tables 1-2 and 1-3.)

29. ✷✷ Acceleration due to gravity, g, has dimensions of acceleration (m/s^2). Assuming that the gravitational potential energy depends only on mass m, g, and height h above Earth h in some way, what must be the form of the equation? Use units to verify this.

30. ✷✷ In thermodynamics, we have the equation $PV = nRT$, where P is pressure, V is volume, n is the number of moles, T is the temperature (in kelvins), and R is a constant. Use the data in Tables 1-2 and 1-3 to determine the units for the constant R.

Section 1-9 Dimensional Analysis

31. ✷✷ Use dimensional analysis to determine a relationship between electric potential difference (voltage), current, and electrical resistance. Consider the units given in Tables 1-2 and 1-3.

32. ✷✷ The aerodynamic drag force, F, for fast relative motion (e.g., a car at highway speeds) has the form $F = -cv^2$, where v is speed (m/s). What must be the units (in terms of SI base units) of the variable c?

33. ✷✷ Aerodynamic drag for slow relative motion can be expressed in the following form: $F = -bv$, where F is a force in newtons and v is a speed (m/s). Use dimensional analysis to determine the units (in terms of SI base units) of the variable b.

Section 1-10 Unit Conversion

34. ✷ Scientists believe that continental drift in the Atlantic happens at the rate of about 1 to 2 cm/year. Convert this rate into nm/s.

35. ✷✷ A cell of baker's yeast is typically 4.0 μm in diameter. (Assume that the cells are spherical, although in reality they have a variety of shapes.) If you had 25 cm^3 of yeast (i.e., one-tenth of a measuring cup), about how many cells would it contain?

Section 1-11 Fermi Problems

36. ✷✷ Volunteers are searching a wooded area for a piece of forensic evidence. It is estimated that searchers should cross on a grid with a separation no larger than 4.0 m between searchers and that the searchers average a walking speed of 0.25 m/s. How long would it take a search party of 10 people to search an area of 1 km^2?

37. ✷✷ One of the famous Fermi problems is a calculation of how many piano tuners there were in Chicago. (If you are interested in the problem, search the Internet. There are many solutions available.) Using similar methods, estimate the number of goalies there are in Canada at any one time (from all types of hockey teams).

38. ✷✷✷ Using reasonable estimates, determine how many molecules there are in one litre of milk.

COMPREHENSIVE PROBLEMS

39. ✷✷ The HIV virus is approximately spherical, and it has a diameter of about 130 nm. If (reasonably) the mass density is about the same as water, 1000. kg/m^3, what is the mass of one HIV virus? Express your answer in pg.

40. ✷✷ An Intel Core 2 Duo processor has the equivalent of 291 million transistors. The dimensions of the integrated circuit chip are approximately 4.0 mm × 4.0 mm. What are the dimensions of one transistor on the integrated circuit?

41. ✷✷ Based on the data in Table 1-5, can we conclude that the teams of the years 1990–99 were significantly better than the teams of the years 2000–09? Clearly explain your reasoning.

Table 1-5 Win–Loss Statistics for the Toronto Blue Jays Baseball Team

Year	Win–Loss Percentage	Year	Win–Loss Percentage
1990	0.531	2000	0.512
1991	0.562	2001	0.494
1992	0.593	2002	0.481
1993	0.586	2003	0.531
1994	0.478	2004	0.416
1995	0.389	2005	0.494
1996	0.457	2006	0.537
1997	0.469	2007	0.512
1998	0.543	2008	0.531
1999	0.519	2009	0.463

42. ✷✷✷ The expansion of the universe mentioned at the beginning of this chapter is often represented by the equation $v = H_0 d$, where v is the speed of recession of the galaxy, d is the distance to the galaxy, and H_0 is the Hubble constant. In this equation, astronomers usually express v in km/s and d in parsecs.

(a) What are the units of H_0 in terms of these units?

(b) The parsec is equal to 3.26 ly (light years). A light year is the distance travelled by light in one year. Write the number of metres in one parsec in scientific notation.

(c) Use unit conversion to write the units of H_0 in SI base units.

43. ✷✷ Scientists estimate that the flux of meteoritic material on the entire Earth is about 20 000 tonnes per year. One tonne is 1000 kg.

(a) Convert this flux to mass per square metre of Earth's surface.

(b) Micrometeorites are meteorites that are small enough to float to Earth without being ablated significantly. They come in a variety of sizes, but for this problem we assume that a typical micrometeorite has a diameter of 25 μm. If 25% of the total meteoritic annual mass was in the form of micrometeorites of this size, how many micrometeorites (on average) would be deposited on each square metre of Earth annually?

44. ✷✷✷ Look up the values and units for the following three fundamental physical constants: G (the universal gravitational constant), c (the speed of light), and $\hbar$ (Planck's constant divided by 2π).

(a) Use dimensional analysis to determine an expression using only G, c, and h that has the units of length. Hint: The expression will have a square root, and one of the

constants will be to a power, such as squared or cubed. The other two will be just the constant.

(b) The expression that you obtained is called the Planck length. Calculate the numerical value of the Planck length.

(c) Use Internet or other resources to find objects that might have dimensions of the order of the Planck length.

45. (a) ✷✷ Find a combination of the same three physical constants listed in question 44 (G, h, and c) that has units of time. This is called the Planck time. Some physicists consider the Planck time to be the time after the Big Bang, which we can first study using physics.

(b) Calculate a numerical value for the Planck time.

46. ✷✷ A typical oil molecule has an effective diameter of 0.25 nm. If 25 L (one litre is the same as 1000 cm³) of oil is spilled on a lake, and if the oil is spread so that it is only a single molecule thick, what will be the area of the oil slick? Assume that the oil molecules are just touching. Hint: Determine the cross-sectional area of an oil molecule.

 See the text online resources at www.physics1e.nelson.com for Open Problems and Data-Rich Problems related to this chapter.

Learning Objectives

When you have completed this chapter you should be able to:

1 Determine whether a physical quantity is a vector or a scalar.

2 Describe vector quantities in terms of components and in Cartesian and polar notation.

3 Add and subtract vector quantities, and find the product of a vector and a scalar.

4 Define a dot (scalar) product between two vectors, and use dot products to determine angles between vectors and projections of a vector on an arbitrary axis.

5 Define and calculate a cross (vector) product between two vectors using either Cartesian notation or the algebraic definition and the right-hand rule.

Chapter 2
Scalars and Vectors

Philosophy is written in this grand book, the universe which stands continually open to our gaze. But the book cannot be understood unless one first learns to comprehend the language and read the letters in which it is composed. It is written in the language of mathematics, and its characters are triangles, circles and other geometric figures without which it is humanly impossible to understand a single word of it; without these, one wanders about in a dark labyrinth.

Galileo Galilei in *Assayer*

In this chapter, we introduce the concepts of scalars and vectors and develop the mathematical language that will be used throughout the book to describe these concepts.

Have you ever wondered why two different words—speed and velocity—are used in English to describe how fast you are moving? The word "speed" has Old English roots, and the word "velocity" has Latin roots. Speed refers to how fast you are moving. If you are asked about your driving speed, your answer might be 55 km/h. However, if you are asked about your velocity, the 55 km/h value does not suffice: velocity requires additional information about the direction of motion, for example, 55 km/h north on Main St. For many physical quantities, not knowing the direction makes the information incomplete. For example, if you have been flying from Ottawa for 6 h with an average speed of 500 km/h, your final destination could be anywhere within the circle shown in Figure 2-1. To be able to pinpoint the exact destination, you need to know the direction of your flight. If your plane headed directly northwest your destination could be Yellowknife, the capital of Northwest Territories.

Figure 2-1 If you were to fly from Ottawa for 6 h with an average speed of 500 km/h, what would your final destination be?

 For additional Making Connections, Examples, and Checkpoints, as well as activities and experiments to help increase your understanding of the chapter's concepts, please go to the text's online resources at www.physics1e.nelson.com.

LO 1

2-1 Definitions of Scalars and Vectors

The physical quantities we use throughout this book can be divided into two categories: scalars and vectors.

Scalar quantities, or **scalars,** require only a number (either positive or negative) and a unit for their description. For example, temperature, mass, length, area, volume, time, distance, speed, work, and energy are all scalars. Scalar symbols are denoted using *italics.* For example, the quantities listed above are usually represented by the symbols *T, m, L, A, V, t, d, v, W,* and *E.*

Vector quantities, or **vectors,** require a positive number, called a **vector magnitude,** a unit, and a direction for their description. For example, displacement, velocity, acceleration, force, momentum, angular momentum, impulse, and magnetic field are all vectors. To describe vectors we need a **frame of reference** (a convention about what is "at rest"), and a coordinate system. For example, a two-dimensional (2-D) coordinate system can consist of an *x*-axis and a *y*-axis, as shown in Figure 2-2. A three-dimensional (3-D) coordinate system has an additional axis, the *z*-axis.

In this textbook, we use primarily rectangular (orthogonal) coordinate systems, where the coordinate axes are perpendicular to each other. In Figure 2-2, vector $\vec{v_1}$ has a magnitude of 30 m/s and a direction of 25° above the horizontal ($\theta_1 = 25°$); the second velocity, $\vec{v_2}$, has a magnitude of 15 m/s and a direction of $\theta_2 = -10° = 350°$. By convention, a positive angle is measured in the counterclockwise sense from the positive *x*-axis, and a negative angle is measured in the clockwise sense from the positive *x*-axis.

By convention, the symbols for vector quantities are either italic letters with arrows over them or bold letters. Throughout this textbook, vector quantities will be denoted using the arrow notation, for example, $\vec{v}$ for velocity, $\vec{p}$ for momentum, $\vec{F}$ for force, and $\vec{a}$ for acceleration. The magnitude of a vector is always a positive scalar quantity and can be denoted using the absolute value sign or by using the same letter as the vector without the arrow or bolding. For example, the magnitude of the velocity vector $\vec{v}$ (speed) can be denoted as either $|\vec{v}|$ or *v.*

Vectors are represented graphically using arrows. The length of the arrow represents the vector's magnitude; the vector's direction coincides with the arrow's direction. Note that a vector is not attached to any particular location in space: moving a vector parallel to itself does not affect it. The end-point of a vector is called the tail of the **vector,** and the arrow-end of a vector is called the **head of the vector** (Figure 2-2). In order to describe a vector, we have to know its magnitude and direction. A 2-D vector can be described in terms of a magnitude and an angle measured counterclockwise from the positive *x*-axis. Such a description is referred to as a **polar notation** or **polar coordinates.** For example, the polar notation for vector $\vec{v_1}$ in Figure 2-2 can be written as 30 m/s with a direction of 25° or as 30 m/s [25°].

✓ CHECKPOINT

C-2-1 Describing Vectors

Which of the following is the most accurate description of a vector quantity?
(a) vector magnitude, direction, and the location of its tail relative to the origin of the coordinate system
(b) vector magnitude, including its units, and direction
(c) vector magnitude, including its units, and location of its head relative to the origin
(d) vector magnitude, including its units, direction, and location of its head relative to the origin
(e) none of the above

Vectors can also be described in terms of their **components,** which are scalar quantities. A 2-D vector has two components, v_x and v_y, as shown in Figure 2-3. These components are **projections** of the vector onto the *x*- and *y*-axes:

$$\begin{cases} v_x = v\cos\theta \\ v_y = v\sin\theta \end{cases} \Rightarrow \begin{cases} v = \sqrt{v_x^2 + v_y^2} \\ \tan\theta = \dfrac{v_y}{v_x} \end{cases} \quad (2\text{-}1)$$

where $|\vec{v}| \equiv v$

Using the Pythagorean theorem, we can see that the magnitude of a vector is equal to the square root of the

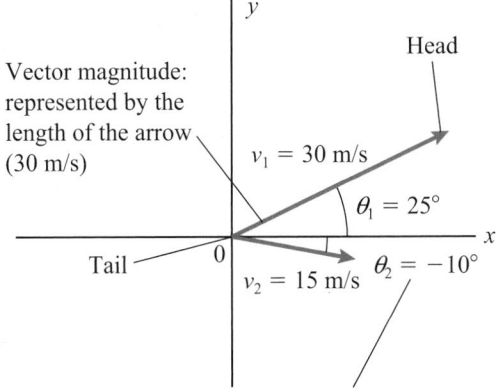

Vector magnitude: represented by the length of the arrow (30 m/s)

$v_1 = 30$ m/s

$\theta_1 = 25°$

Tail — 0

$v_2 = 15$ m/s $\theta_2 = -10°$

Head

y

x

Vector direction: represented by the angle θ measured from the positive *x*-axis

Figure 2-2 Graphical representation of two 2-D velocity vectors

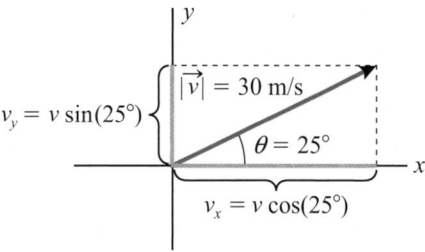

$|\vec{v}| = 30$ m/s

$v_y = v \sin(25°)$

$\theta = 25°$

$v_x = v \cos(25°)$

Figure 2-3   Vector components of velocity vector $\vec{v_1}$

sum of the squares of its components. By convention, all angles are measured in a counterclockwise direction from the positive x-axis. However, it is sometimes convenient to use the complementary angle θ^*, which measures the angle clockwise from the vertical y-axis, such as $\theta^* = 90° - \theta$. Hence,

$$\begin{cases} v_x = v \sin\theta^* \\ v_y = v \cos\theta^* \end{cases} \Rightarrow \begin{cases} v = \sqrt{v_x^2 + v_y^2} \\ \tan\theta^* = \dfrac{v_x}{v_y} \end{cases} \quad (2\text{-}2)$$

where $|\vec{v}| \equiv v$

Representing a vector by using its components is often called a **scalar notation**. The vector component in the x-direction is positive when the vector is fully or partially oriented in the positive x-direction. Similarly,

the x-component is negative when the vector is fully or partially oriented in the negative x-direction. The same principle applies to any other axis. Therefore, the notation $v_x = 20$ m/s means that a projection of the object's velocity vector onto the x-axis is $+20$ m/s. However, an endless number of vectors have this x-component, for example, 20 m/s [0°] and 40 m/s [60°]. We need to give both components to fully define a 2-D vector. It takes three components to represent a 3-D vector. In general, the description of a vector using components is called **vector resolution into scalar components**. Example 2-1 demonstrates how to resolve a vector into scalar components.

EXAMPLE 2-1

Vector Resolution

Force $\vec{F_1}$ has a magnitude of 20.0 N and is directed at a 45.0° angle to the positive x-axis. What will happen to the components of the force when

(a) the magnitude of the force remains the same, but the angle decreases to 30.0°;
(b) the magnitude of the force doubles, but the direction of the force remains the same?

SOLUTION

Let us express the components of force $\vec{F_1}$ in scalar notation (2-1). The magnitudes of forces will be denoted as F_1 and F_2 respectively:

$$\begin{cases} F_{1x} = F_1 \cos\theta \\ F_{1y} = F_1 \sin\theta \end{cases} \Rightarrow \begin{cases} F_{1x} = (20.0 \text{ N}) \cos(45.0°) = 14.14 \text{ N} \\ F_{1y} = (20.0 \text{ N}) \sin(45.0°) = 14.14 \text{ N} \end{cases}$$

(a) Let $\vec{F_2}$ denote the force when the angle decreases to 30.0°:

$$\begin{cases} F_{2x} = F_2 \cos\theta \\ F_{2y} = F_2 \sin\theta \end{cases} \Rightarrow \begin{cases} F_{2x} = (20.0 \text{ N}) \cos(30.0°) = 17.32 \text{ N} \\ F_{2y} = (20.0 \text{ N}) \sin(30.0°) = 10.0 \text{ N} \end{cases}$$

We can check that force $\vec{F_2}$ has the same magnitude as the original force:

$$F_2 = \sqrt{F_{2x}^2 + F_{2y}^2} = \sqrt{(17.32 \text{ N})^2 + (10.0 \text{ N})^2}$$

$$F_2 = 20.0 \text{ N}$$

(b) Let $\vec{F_3}$ denote the force when the magnitude of the force doubles while the direction remains the same. Then,

$$\begin{cases} F_{3x} = F_3 \cos\theta_2 \\ F_{3y} = F_3 \sin\theta_2 \end{cases} \Rightarrow \begin{cases} F_{3x} = (40.0 \text{ N}) \cos(45.0°) = 28.28 \text{ N} \\ F_{3y} = (40.0 \text{ N}) \sin(45.0°) = 28.28 \text{ N} \end{cases}$$

We can verify that these components describe a vector with twice the magnitude of $\vec{F_1}$.

$$F_3 = \sqrt{F_{3x}^2 + F_{3y}^2} = 40.0 \text{ N}$$

Making sense of the result:

Changing the angle such that the vector is closer to the x-axis increases the x-component and decreases the y-component. Doubling the magnitude doubles both components. You can confirm these relationships in a vector diagram.

2-2 Vector Addition: Geometric and Algebraic Approaches

From everyday experience, we intuitively know that the sum of two forces is also a force. The sum of two equal forces pulling in opposite directions is zero, and the sum of two equal forces pulling in the same direction is a force with double the magnitude. Therefore, it is not surprising that the sum of two or more vectors depends on both their magnitudes and directions.

The Geometric Addition of Vectors

There are two commonly used rules for the **geometric method for vector addition**: the **parallelogram rule** and the **triangle construction rule**. Figure 2-4 shows how these rules can be used to find the resultant vector when

two students pull on the arms of a friend. Both rules use the fact that a vector can be translated as long as its magnitude and direction are not changed.

Parallelogram Rule Move the vectors so that they are joined at their tails. Then, build a parallelogram with the two vectors forming adjacent sides, as shown in Figure 2-4(b). The vector sum of $\vec{F_1}$ and $\vec{F_2}$ is represented by the diagonal of this parallelogram, $\vec{F_R}$ (the subscript R stands for resultant). The tail of $\vec{F_R}$ coincides with the tails of $\vec{F_1}$ and $\vec{F_2}$, and the head of $\vec{F_R}$ is at the opposite vertex of the parallelogram.

Triangle Construction Rule Move the vectors so that the head of the first vector is joined with the tail of the second vector. Then, connect the tail of the first vector with the head of the second vector, as shown in Figure 2-4(c). The vector sum ($\vec{F_R}$) of forces $\vec{F_1}$ and $\vec{F_2}$ is represented by the third side of the triangle, where the tail of $\vec{F_R}$ coincides with the tail of $\vec{F_1}$, and its head coincides with the head of $\vec{F_2}$. The same method can be applied to add a chain of three or more vectors.

As an exercise, prove that the parallelogram and triangle construction rules produce the same result.

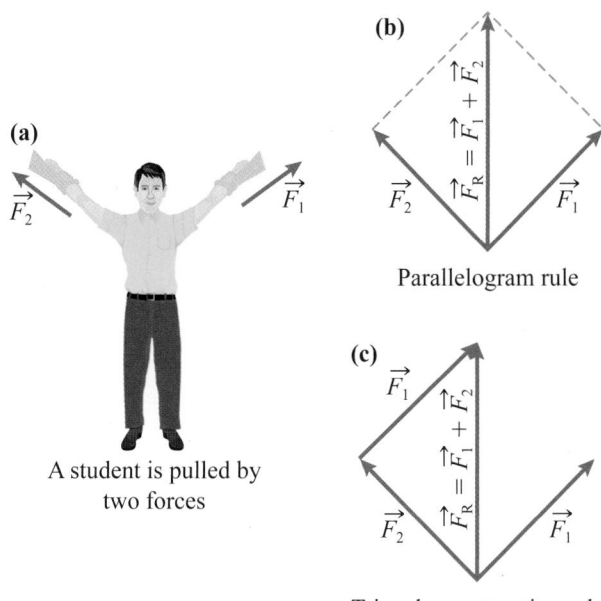

A student is pulled by two forces

Parallelogram rule

Triangle construction rule

Figure 2-4 Vector addition. Two forces are exerted on a person. The resultant of these two forces is found using (a) the parallelogram rule and (b) the triangle construction rule.

✓ CHECKPOINT

C-2-2 Geometrical Addition of Vectors

Which diagram in Figure 2-5 (on page 24) correctly represents the vector sum of three vectors $\vec{F_1}$, $\vec{F_2}$, and $\vec{F_3}$? Explain.

(a) diagram (a) only
(b) diagram (b) only
(c) diagrams (c) and (d)
(d) diagram (c) only
(e) diagram (d) only
(f) diagrams (b) and (c)
(g) diagrams (a) and (b)

C-2-2 (c) Only diagrams (c) and (d) properly apply the triangle construction rule and the parallelogram rule to vector addition.

👥 PEER TO PEER

I find that adding vectors geometrically helps me get a sense of how the result of vector addition or subtraction should look. Even though drawing by hand might not be exact, it is often helpful for understanding.

(a)

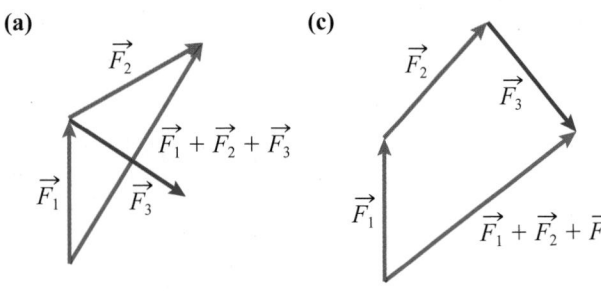

(c)

(b)

(d)

Figure 2-5   C-2-2 Geometrical Addition of Vectors

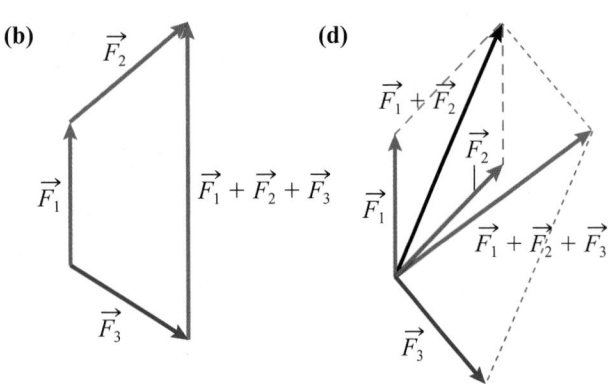

The Algebraic Addition of Vectors

The geometric approach to vector addition is useful for visualizing the relationships between vectors, but it is difficult to draw and measure the diagrams with high precision. Example 2-2 illustrates an algebraic approach to vector addition.

The method for finding the resultant vector for the two perpendicular 2-D vectors in Example 2-2 can be generalized to apply to any number of vectors with arbitrary orientations:

KEY EQUATION

$$\begin{cases} F_{R,x} = F_{1x} + F_{2x} + \cdots + F_{N,x} \\ F_{R,y} = F_{1y} + F_{2y} + \cdots + F_{N,y} \\ F_R = \sqrt{F_{R,x}^2 + F_{R,y}^2} \\ \tan \theta = \dfrac{F_{R,y}}{F_{R,x}} \end{cases} \qquad (2\text{-}3)$$

EXAMPLE 2-2

Finding the Resultant of Perpendicular Forces

Two 40.0 N forces, $\vec{F}_1$ and $\vec{F}_2$, directed at 90.0° to each other act on a hook, as shown in Figure 2-6.

(a)   Use the geometric approach to vector addition to find the resultant force acting on the hook.
(b)   Find the components of each force along the x- and y-axes.
(c)   Find the components of the net force (resultant vector).
(d)   Compare your results from parts (b) and (c). What conclusions can you make?

$\vec{F}_1 = 40.0$ N

$\vec{F}_2 = 40.0$ N

Figure 2-6   Two 40.0 N forces are acting on a hook

SOLUTION

(a)   To add the two vectors geometrically, we can use the parallelogram rule, as shown in Figure 2-7. Since the original forces have equal magnitudes, the resultant force, $\vec{F}_R$, is represented by a diagonal of a square. Therefore,

$$\vec{F}_R = \sqrt{2}(40.0 \text{ N})$$

and $\theta = -45.0°$

$$\vec{F}_R = 56.6 \text{ N}$$

(Make sure you can justify these results.)

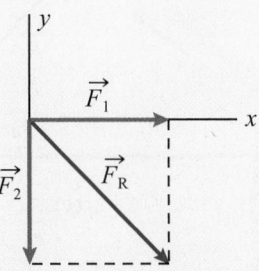

Figure 2-7   Vector addition for Example 2-2

(b)   To find the components of forces $\vec{F}_1$ and $\vec{F}_2$, we first choose a coordinate system, then we use Equation (2-1):

$$\begin{cases} F_{1x} = 40.0 \text{ N} \\ F_{1y} = 0 \end{cases} \text{ and } \begin{cases} F_{2x} = 0 \\ F_{2y} = -40.0 \text{ N} \end{cases}$$

(c)   To find the components of the resultant vector $\vec{F}_R$, we once again use Equation (2-1):

$$\begin{cases} F_{R,x} = 40.0 \text{ N} \\ F_{R,y} = -40.0 \text{ N} \end{cases} \text{ and } F_R = \sqrt{(40.0 \text{ N})^2 + (40.0 \text{ N})^2}$$

$$= \sqrt{2}\,(40.0 \text{ N}) = 56.6 \text{ N}$$

(d)   Comparing the results of (b) and (c), we see that the x-components of the two individual vectors sum to 40.0 N and their y-components sum to −40.0 N. These sums match the components of the resultant vector.

Making sense of the result:

The components of the resultant vector are equal to the sums of the corresponding components of the individual vectors.

These relationships give us an **algebraic method for vector addition** (Figure 2-8):

1. Resolve the vectors into components.

2. Add components pointing along the same axis to obtain the corresponding components of the resultant vector.

3. Use the components of the resultant vector to determine its magnitude and direction.

By resolving vectors into components, we can use scalar calculations to perform operations with vectors.

The algebraic method can also be extended to 3-D vectors:

$$\begin{cases} F_{R,x} = F_{1x} + F_{2x} + \cdots + F_{N,x} \\ F_{R,y} = F_{1y} + F_{2y} + \cdots + F_{N,y} \\ F_{R,z} = F_{1z} + F_{2z} + \cdots + F_{N,z} \end{cases}$$

$$F_R = \sqrt{F_{R,x}^2 + F_{R,y}^2 + F_{R,z}^2} \qquad (2\text{-}4)$$

Example 2-3 demonstrates algebraic addition of two arbitrary vectors. It also shows how a **free body diagram (FBD)** can help you visualize the forces acting on an object.

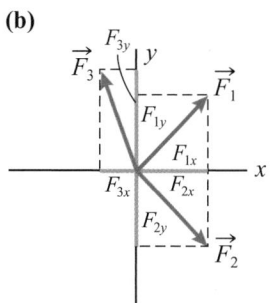

Figure 2-8 (a) Visual representation of the addition of three vectors . (b) Vector resolution of the three vectors (c) Vector addition using vector components

EXAMPLE 2-3

Finding the Resultant Force Acting on a Student Standing on a Ramp

A 51.0 kg student steps onto a slippery 10.0° ramp. Two forces are acting on her: the force of gravity exerted by Earth (her weight), which is $F_g = 500$ N directed down, and the normal force exerted by the ramp, which is $N = 493$ N, directed perpendicular to the ramp, as shown in Figure 2-9.

(a) Find the components of the student's weight along x- and y-axes directed along the incline and perpendicular to the incline, respectively.

(b) Find the components of the normal force along these axes.

(c) Find the resultant force acting on the student.

Figure 2-9 (a) A lifelike problem representation (b) The FBD shows the forces acting on the student. (c) Geometrical representation of the net force acting on the student

(continued)

SOLUTION

When dealing with two or more forces, it is often helpful to draw a detailed FBD, such as the one in Figure 2-9(b).

(a) To find the components of each vector, we have to find the angle that the vector forms with the coordinate axes. The angle between the force of gravity $\vec{F_g}$ and the y-axis equals the angle of the ramp. (Make sure you can see why these two angles are equal.) The force of gravity vector, marked in green, forms the hypotenuse of the right-angled triangle in Figure 2-9(b). In this triangle, the side adjacent to the 10.0° angle represents the y-component of $\vec{F_g}$, and the opposite side represents the x-component. Therefore, the components of the force of gravity are as follows, by Equation (2-1):

$$\begin{cases} F_{g,x} = F_g \sin(10.0°) = (500 \text{ N})(0.1736) = 86.8 \text{ N} \\ F_{g,y} = -F_g \cos(10.0°) = -(500 \text{ N})(0.9848) = -493 \text{ N} \end{cases}$$

The negative sign indicates that the y-component of the force of gravity is directed in the negative y-direction. In this case, it is convenient to measure the angle from the negative y-axis; we compare the direction of the vector with the direction of the positive x- and y-axes to determine the signs on the x- and y-components.

(b) The normal force is directed in the positive y-direction. Consequently, the angle between the normal force and the x-axis is 90.0°, and

$$\begin{cases} N_x = N \cos(90.0°) = (493 \text{ N})(0) = 0 \\ N_y = N \sin(90.0°) = 493 \text{ N} \end{cases}$$

(c) From Figure 2-9(c) we can see that if the vectors are not drawn exactly to scale, or if their directions are slightly inaccurate, we will not be able to get a precise value of the resultant force from measuring it on the diagram. Therefore, we use the algebraic approach. Adding the x- and y-components of the gravitational and normal forces, we get

$$\begin{cases} F_{R,x} = 86.8 \text{ N} + 0 = 86.8 \text{ N} \\ F_{R,y} = -493 \text{ N} + 493 \text{ N} = 0 \end{cases}$$

The resultant force has a nonzero x-component and a zero y-component. Therefore, it points along the x-axis in the positive x-direction—down the ramp. Since the y-component is zero, the magnitude of the resultant force is equal to the x-component, 86.8 N.

Making sense of the result:

The resultant force in this problem is not zero. As we will discuss later in the chapters on Newton's laws, an unbalanced net force (the resultant of all forces acting on the object) means that the object's velocity is changing. Therefore, we should expect the student to slide down the ramp with increasing speed. In this example, we ignored the force of friction between the student and the ramp (since it was a slippery ramp). If present, a friction force would have been directed up the ramp. Choosing an x-axis along the ramp simplified the calculations because the resultant force is directed down the ramp. It is often convenient to have the resultant force directed along one of the axes.

Now let us consider a few consequences of the vector addition rules:

Multiplying a Vector by a Scalar Adding a vector to itself preserves its direction while doubling the magnitude (Figure 2-10(a)). In general, adding N identical vectors means preserving the original direction and multiplying the original magnitude of the vector by N, as shown in Figure 2-10(b). We can generalize this process to a multiplication of a vector by any scalar, not just integers. The product of a scalar and a vector is a vector with a magnitude equal to the product of the magnitude of the original vector and the absolute value of the scalar. The direction of the resulting vector coincides with the original direction of the vector if the scalar is positive, and is opposite to the original vector direction if the scalar is negative:

$$|a\vec{F}| = |a||\vec{F}|$$

$$\begin{cases} \text{if } a > 0, a\vec{F} \text{ has the same direction as } \vec{F} \\ \text{if } a < 0, a\vec{F} \text{ has the opposite direction to } \vec{F} \\ \text{if } a = 0, a\vec{F} \text{ has a magnitude of zero and no direction} \end{cases}$$

(2-5)

(a)

$$\vec{F} + \vec{F} = 2\vec{F}$$

(b)

$$\underbrace{\vec{F} + \vec{F} + ... \vec{F} + \vec{F}}_{n \text{ times}} = n\vec{F}$$

Figure 2-10 Multiplication of a vector by a scalar (a) Adding two identical vectors (b) Adding n identical vectors

Opposite Vectors Vectors $\vec{F_1}$ and $\vec{F_2}$ are **opposite** if they have equal magnitudes and opposite directions:

$$\vec{F_1} = -\vec{F_2} \qquad (2-6)$$

Subtracting Vectors We can perform vector subtraction by adding an opposite vector, as shown in Figure 2-11:

$$\vec{F_2} - \vec{F_1} = \vec{F_2} + (-\vec{F_1}) \qquad (2-7)$$

We can also find the difference between two vectors, $\vec{F_2} - \vec{F_1}$, by moving the two vectors parallel to themselves so that their tails coincide, and then drawing a vector from the head of $\vec{F_2}$ to the head of $\vec{F_1}$. If you know the magnitudes of the two vectors and the angle between them, you can use the law of cosines to find the magnitude of the difference vector.

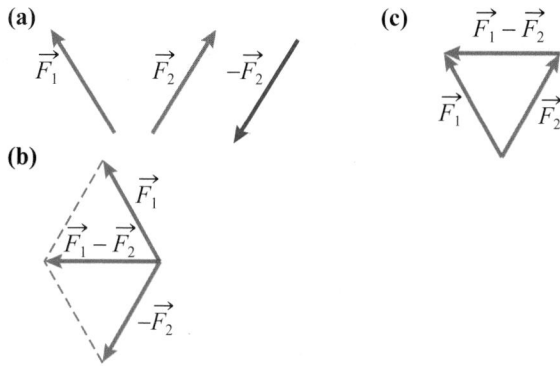

(a)
$\vec{F_1}$ $\vec{F_2}$ $-\vec{F_2}$

(b)
$\vec{F_1}$
$\vec{F_1} - \vec{F_2}$
$-\vec{F_2}$

(c)
$\vec{F_1} - \vec{F_2}$
$\vec{F_1}$ $\vec{F_2}$

Figure 2-11   Vector subtraction. To find $\vec{F_1} - \vec{F_2}$, first construct a vector opposite to $\vec{F_2}$ and then add $\vec{F_1}$ and $(-\vec{F_2})$.

 CHECKPOINT

C-2-3 Operations with Vectors

Which, if any, of the following statements is true? Explain.
(a) Opposite vectors always have opposite magnitudes and are directed along the same line.
(b) The sum of two vectors is also a vector directed in the same direction as one of the vectors being added.
(c) The difference between two vectors is a vector that is always perpendicular to the sum of the two vectors.
(d) The magnitude of the sum of two vectors always equals the sum of the magnitudes of the vectors being added.
(e) The magnitude of the product of a vector and a scalar is always larger than the magnitude of the original vector.

C-2-3 All the statements are false.

LO 3

2-3   Cartesian Vector Notation

Cartesian vector notation is another method for handling vector components. This notation is based on the concept of **unit vectors**, specifically, unit vectors in the x-, y-, and z-directions. A unit vector has a dimensionless magnitude of 1. Unit vectors in the positive x-, y-, and z-directions are denoted as $\hat{i}$, $\hat{j}$, and $\hat{k}$, respectively (Figure 2-12). Some books use $\hat{x}$, $\hat{y}$, and $\hat{z}$ as an alternative notation for unit vectors. A unit vector in the direction of an arbitrary vector $\vec{F}$ is denoted as $\hat{u}_F$. A "hat" above an italic letter indicates a unit vector.

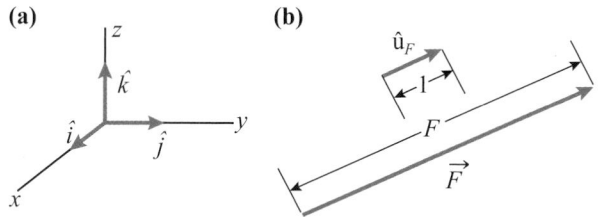

Figure 2-12   (a) Cartesian unit vectors: $\hat{i}$, $\hat{j}$, and $\hat{k}$ (b) A unit vector in the direction of the force vector $\vec{F}$

 PEER TO PEER

Leaving the "hat" off a unit vector can be confusing. Sometimes I forget that a unit vector is a vector and not a scalar. Although the magnitude of a unit vector is always one, the direction might vary, and omitting the hat makes it easy to forget that it's a vector.

In Cartesian notation, any 3-D vector can be represented as follows:

$$\vec{F} = F_x \hat{i} + F_y \hat{j} + F_z \hat{k} \qquad (2\text{-}8)$$

where F_x, F_y, and F_z are the usual scalar vector components: projections of the vector $\vec{F}$ onto the x-, y-, and z-axes, respectively.

CHECKPOINT

C-2-4 Cartesian Vector Notation

Which of the following statements are true? Explain.
(a) $3\hat{i} + 4\hat{j} = 7\hat{k}$
(b) $(3\hat{i} + 4\hat{j} + 2\hat{k}) + (2\hat{i} - 2\hat{j} - 2\hat{k}) = 5\hat{i} + 2\hat{j} + 4\hat{k}$
(c) $(3\hat{i} + 4\hat{j}) + (2\hat{i} - 2\hat{k}) = 3\hat{i} + 4\hat{j}$
(d) $(3\hat{i} + 4\hat{j} + 2\hat{k}) + (2\hat{i} - 2\hat{j} - 2\hat{k}) = 5\hat{i} + 2\hat{j}$
(e) $3\hat{i} - 3\hat{j} + 3\hat{k} = 3\hat{k}$

C-2-4 (d) $(3\hat{i} + 4\hat{j} + 2\hat{k}) + (2\hat{i} - 2\hat{j} - 2\hat{k}) = (3\hat{i} + 2\hat{i}) + (4\hat{j} - 2\hat{j}) + (2\hat{k} - 2\hat{k}) = (3 + 2)\hat{i} + (4 - 2)\hat{j} + (2 - 2)\hat{k} = 5\hat{i} + 2\hat{j}$

Cartesian vector notation allows us to simplify vector addition significantly:

KEY EQUATION

$$\vec{F}_R = \vec{F_1} + \vec{F_2} + \vec{F_3} + \cdots + \vec{F_N}$$
$$\vec{F}_R = F_{1x}\hat{i} + F_{1y}\hat{j} + F_{1z}\hat{k} + F_{2x}\hat{i} + F_{2y}\hat{j} + F_{2z}\hat{k}$$
$$+ F_{3x}\hat{i} + F_{3y}\hat{j} + F_{3z}\hat{k} + \cdots + F_{N,x}\hat{i} + F_{N,y}\hat{j} + F_{N,z}\hat{k}$$
$$\vec{F}_R = [F_{1x} + F_{2x} + F_{3x} + \cdots + F_{N,x}]\hat{i}$$
$$+ [F_{1y} + F_{2y} + F_{3y} + \cdots + F_{N,y}]\hat{j}$$
$$+ [F_{1z} + F_{2z} + F_{3z} + \cdots + F_{N,z}]\hat{k}$$
$$\vec{F}_R = \left(\sum_{l=1}^{N} F_{l,x}\right)\hat{i} + \left(\sum_{l=1}^{N} F_{l,y}\right)\hat{j} + \left(\sum_{l=1}^{N} F_{l,z}\right)\hat{k} \quad (2\text{-}9)$$

Equation (2-9) states that the sum of all the x-components of the original vectors equals the x-component of the resultant vector, the sum of all the y-components of the original vectors equals the y-component of the resultant vector, and the sum of all the z-components of the original vectors equals the z-component of the resultant vector. Example 2-4 illustrates the use of Cartesian notation.

EXAMPLE 2-4

Finding a Unit Vector in the Direction of a Given Force

A force vector is described as $\vec{F} = (3\hat{i} + 4\hat{j} - 5\hat{k})$ N. Express the unit vector in the $\vec{F}$ direction using Cartesian notation $(\hat{i}, \hat{j}, \hat{k})$.

SOLUTION

A unit vector must have a magnitude of 1. Therefore, if we divide the force vector $\vec{F}$ by its magnitude (which is a scalar), the resultant vector will have a magnitude of 1 and the same direction as the force $\vec{F}$. To find the magnitude of the force vector, we use the Pythagorean theorem:

$$F = \sqrt{F_x^2 + F_y^2 + F_z^2} = \sqrt{(3 \text{ N})^2 + (4 \text{ N})^2 + (-5 \text{ N})^2}$$

$$= \sqrt{50} \text{ N} = 5\sqrt{2} \text{ N}$$

The unit vector in the direction of $\vec{F}$ is

$$\hat{u}_F = \frac{\vec{F}}{F} = \frac{(3\hat{i} + 4\hat{j} - 5\hat{k}) \text{ N}}{(5\sqrt{2}) \text{ N}} = \frac{3}{5\sqrt{2}}\hat{i} + \frac{4}{5\sqrt{2}}\hat{j} - \frac{1}{\sqrt{2}}\hat{k}$$

Making sense of the result:

Since the units divide out, $\hat{u}_F$ is dimensionless, as required for a unit vector and its magnitudes equals 1.

The method used in Example 2-4 can be applied to any vector:

KEY EQUATION

$$\hat{u}_F = \frac{\vec{F}}{F} = \frac{F_x}{F}\hat{i} + \frac{F_y}{F}\hat{j} + \frac{F_z}{F}\hat{k}, \quad \text{which is equivalent to}$$

$$\vec{F} = F\hat{u}_F \tag{2-10}$$

where $\hat{u}_F$ is a dimensionless unit vector, and F is the magnitude of vector $\vec{F}$.

 CHECKPOINT

C-2-5 Identifying Unit Vectors

Which of the following vectors is a unit vector? Explain.

(a) $\frac{3}{2\sqrt{2}}\hat{i} + \frac{3}{2\sqrt{2}}\hat{j} - \frac{2}{2\sqrt{2}}\hat{k}$

(b) $3\hat{i} + 3\hat{j} - 5\hat{k}$

(c) $\frac{3}{\sqrt{14}}\hat{i} + \frac{3}{\sqrt{14}}\hat{j} - \frac{2}{\sqrt{14}}\hat{k}$

(d) $\frac{3}{\sqrt{22}}\hat{i} + \frac{3}{\sqrt{22}}\hat{j} - \frac{2}{\sqrt{22}}\hat{k}$

(e) $\frac{4}{\sqrt{12}}\hat{i} - \frac{4}{\sqrt{12}}\hat{j} - \frac{4}{\sqrt{12}}\hat{k}$

C-2-5 (d) The unit vector must have a magnitude of 1.

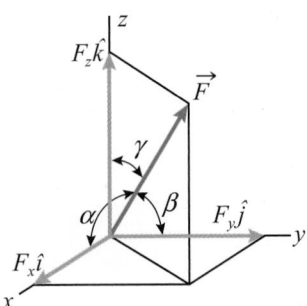 **CHECKPOINT**

C-2-6 Adding Vectors Using Cartesian Notation

Which of the following vectors represents the sum of the two vectors $\vec{F}_1 = 3\hat{i} + 3\hat{j} - 5\hat{k}$ and $\vec{F}_2 = -3\hat{i} + 6\hat{j} - 5\hat{k}$? Explain.

(a) $\vec{F}_R = 9\hat{j} - 10\hat{k}$

(b) $\vec{F}_R = 6\hat{i} + 9\hat{j} - 10\hat{k}$

(c) $\vec{F}_R = 6\hat{i} - 3\hat{j}$

(d) $\vec{F}_R = -10\hat{k}$

(e) $\vec{F}_R = -3\hat{i} - 10\hat{k}$

C-2-6 (a) This follows from the vector addition rule, Equation (2-9).

We can specify the direction of a 3-D vector by giving the angles the vector makes with each of the coordinate axes. These angles are called **coordinate direction angles** and are denoted α, β, and γ, as shown in Figure 2-13. In Example 2-5, we will see that these three angles are not independent of each other.

Figure 2-13 The angles describing the orientation of vector $\vec{F}$ relative to coordinate axes x, y, and z

Position Vector A special vector widely used in physics is the **position vector**, $\vec{r}$ (Figure 2-15). A position vector describes the location of a point in space relative to some fixed point (often, the origin). The magnitude of the position vector corresponds to the distance between the origin and point A. The position vector of point A (A_x, A_y, A_z) can be expressed as

$$\vec{r}_A = (A_x - 0)\hat{i} + (A_y - 0)\hat{j} + (A_z - 0)\hat{k}$$
$$= A_x\hat{i} + A_y\hat{j} + A_z\hat{k} \tag{2-15}$$

When the position of an object is described relative to an arbitrary point (not necessarily the origin), for example, from point B (B_x, B_y, B_z), the position vector is called a **displacement vector** because it refers to how much an object was displaced while moving from point A to point B.

MECHANICS

EXAMPLE 2-5

Relationships Between the Coordinate Direction Angles

Prove that the coordinate direction angles for any 3-D vector $\vec{F}$ satisfy the equation

$$\cos^2 \alpha + \cos^2 \beta + \cos^2 \gamma = 1 \quad (2\text{-}11)$$

SOLUTION

From Figure 2-14 we see that

$$\cos \alpha = \frac{F_x}{F}; \quad \cos \beta = \frac{F_y}{F}; \quad \cos \gamma = \frac{F_z}{F} \quad (2\text{-}12)$$

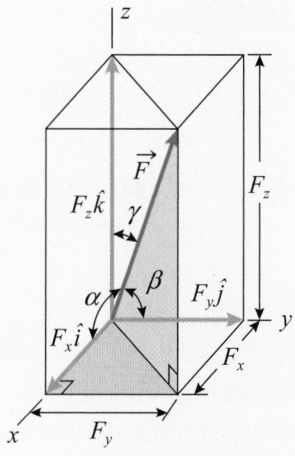

Figure 2-14 Vector $\vec{F}$ expressed in Cartesian vector notation

Substituting expressions (2-12) into the equation for a unit vector, Equation (2-10), gives

$$\hat{u}_F = \frac{\vec{F}}{F} = \frac{F_x}{F}\hat{i} + \frac{F_y}{F}\hat{j} + \frac{F_z}{F}\hat{k}$$

$$= (\cos \alpha)\hat{i} + (\cos \beta)\hat{j} + (\cos \gamma)\hat{k} \quad (2\text{-}13)$$

The magnitude of the unit vector is 1. Therefore,

$$(\cos \alpha)^2 + (\cos \beta)^2 + (\cos \gamma)^2 = 1$$

Making sense of the result:

This result means that the three directional angles are not independent. In fact, given two of the angles, we can find the absolute value of the cosine of the third angle, which limits the value of the unknown angle to four possible values. This result also applies to 2-D vectors. In the case of 2-D vectors, $\gamma = 90°$, $\cos \gamma = 0$; therefore, $\cos^2 \alpha + \cos^2 \beta = 1$. However, for 2-D vectors, the angles α and β are complementary angles; therefore, $\alpha = 90° - \beta$. Consequently,

$$\cos^2 \alpha + \cos^2 \beta = \cos^2 \alpha + \cos^2(90° - \alpha) = \cos^2 \alpha + \sin^2 \alpha = 1$$

$$(2\text{-}14)$$

This is a well-known result. Thus, expression (2-11) applies to both three- and two-dimensional vectors.

$$\vec{r}_{A \to B} = (B_x - A_x)\hat{i} + (B_y - A_y)\hat{j} + (B_z - A_z)\hat{k} \quad (2\text{-}16)$$

The magnitude of the displacement vector $\vec{r}_{A \to B}$ corresponds to the distance between point A and point B.

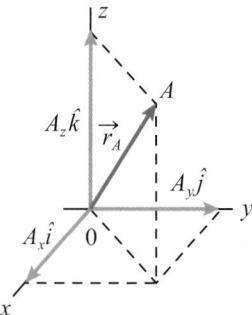

Figure 2-15 Position vector $\vec{r}_A$ represents the location of point A relative to the origin. Its magnitude corresponds to the distance between the origin and point A.

✓ CHECKPOINT

C-2-7 Cartesian Vector Notation and Coordinate Direction Angles

Which, if any, of the following statements do you agree with? Explain.

(a) The sum of the coordinate direction angles is always equal to 1.
(b) A unit vector must always be directed along one of the three coordinate axes.
(c) A position vector and a force vector must always be directed along the same line.
(d) When adding two force vectors, the resultant vector cannot be perpendicular to either one of them.
(e) The magnitude of the unit vector is always 1, and its dimension is the newton (N).
(f) all of the above
(g) none of the above

C-2-7 (g)

LO 4

2-4 The Dot Product of Two Vectors

Many physics and engineering applications involve the projection of a vector onto a given axis. Although this projection can be done geometrically, it is often more practical to use an algebraic approach. The **dot**, or **scalar**, **product** is a scalar quantity that represents the product of the magnitudes of a vector and the component of another vector projected onto it.

The scalar product of vectors $\vec{A}$ and $\vec{B}$ is defined as the product of their magnitudes and the cosine of the angle between the vectors:

KEY EQUATION $\qquad \vec{A} \cdot \vec{B} = AB \cos\theta$, where $0° \leq \theta \leq 180°$

(2-17)

and θ is the angle between the vectors, as shown in Figure 2-16.

The product $\vec{A} \cdot \vec{B}$ is commonly read as "A dot B."

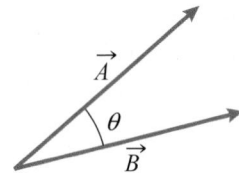

Figure 2-16 The dot product of vectors $\vec{A}$ and $\vec{B}$ is the product of their magnitudes and the cosine of the angle between them.

From the definition above, we can see that a dot product has the following properties:

- A dot product has the same units as the product of the vectors being multiplied.

- A dot product of two perpendicular vectors equals zero because $\cos(90°) = 0$.

- A dot product of two vectors of given magnitudes is greatest when the vectors are parallel to each other because $\cos(0°) = 1$. The dot product is smallest when the vectors are antiparallel to each other because $\cos(180°) = -1$.

- The dot product of an arbitrary vector $\vec{A}$ and $\hat{u}_B$, a unit vector in the direction of vector $\vec{B}$, represents a projection of vector $\vec{A}$ onto the direction of vector $\vec{B}$.

- A dot product is commutative, that is,

$$\vec{A} \cdot \vec{B} = \vec{B} \cdot \vec{A} = AB \cos\theta \qquad (2\text{-}18)$$

- Multiplication of a dot product by a scalar can be expressed as follows:

$$a(\vec{A} \cdot \vec{B}) = (a\vec{B}) \cdot \vec{A} = \vec{A} \cdot (a\vec{B}) = (\vec{A} \cdot \vec{B})a = aAB \cos\theta$$

(2-19)

- The dot product is distributive:

$$\vec{A} \cdot (\vec{B} + \vec{C}) = \vec{A} \cdot \vec{B} + \vec{A} \cdot \vec{C}$$
$$= AB \cos(\theta_{A-B}) + AC \cos(\theta_{A-C}), \qquad (2\text{-}20)$$

where θ_{A-B} and θ_{A-C} are the angles between vectors $\vec{A}$ and $\vec{B}$, and $\vec{A}$ and $\vec{C}$, respectively.

A dot product has two important uses: it can be used to find a vector component or a projection of a vector onto a given direction, and it can be used to find the angle between any two given vectors.

The Dot Product and Unit Vectors

The dot products of the various combinations of unit vectors are as follows:

KEY EQUATION

$$\hat{i} \cdot \hat{i} = 1 \cdot 1 \cos(0°) = 1; \ \hat{i} \cdot \hat{j} = 1 \cdot 1 \cos(90°) = 0; \ \hat{i} \cdot \hat{k} = 1 \cdot 1 \cos(90°) = 0$$
$$\hat{j} \cdot \hat{j} = 1 \cdot 1 \cos(0°) = 1; \ \hat{j} \cdot \hat{i} = 1 \cdot 1 \cos(90°) = 0; \ \hat{j} \cdot \hat{k} = 1 \cdot 1 \cos(90°) = 0$$
$$\hat{k} \cdot \hat{k} = 1 \cdot 1 \cos(0°) = 1; \ \hat{k} \cdot \hat{i} = 1 \cdot 1 \cos(90°) = 0; \ \hat{k} \cdot \hat{j} = 1 \cdot 1 \cos(90°) = 0$$

(2-21)

These dot products can be written in a more compact form, as a dot (scalar) multiplication table:

KEY EQUATION

$\cdot$	$\hat{i}$	$\hat{j}$	$\hat{k}$
$\hat{i}$	1	0	0
$\hat{j}$	0	1	0
$\hat{k}$	0	0	1

(2-22)

The dot product of two unit vectors is at the intersection of the row for the first unit vector and the column for the second unit vector. For example, the dot product of $\hat{i}$ and $\hat{i}$ is 1. The dot products $\hat{i}$ of $\hat{i}$ and $\hat{j}$, and $\hat{i}$ and $\hat{k}$ are 0. Since the dot product is commutative, the table is symmetric about its main diagonal.

Now we are ready to define the dot product for two arbitrary vectors using Cartesian notation:

$$\vec{A} \cdot \vec{B} = (A_x\hat{i} + A_y\hat{j} + A_z\hat{k}) \cdot (B_x\hat{i} + B_y\hat{j} + B_z\hat{k})$$
$$= A_xB_x(\hat{i} \cdot \hat{i}) + A_xB_y(\hat{i} \cdot \hat{j}) + A_xB_z(\hat{i} \cdot \hat{k})$$
$$+ A_yB_x(\hat{j} \cdot \hat{i}) + A_yB_y(\hat{j} \cdot \hat{j}) + A_yB_z(\hat{j} \cdot \hat{k})$$
$$+ A_zB_x(\hat{k} \cdot \hat{i}) + A_zB_y(\hat{k} \cdot \hat{j}) + A_zB_z(\hat{k} \cdot \hat{k})$$
$$= A_xB_x + A_yB_y + A_zB_z \qquad (2\text{-}23)$$

PEER TO PEER

I have to remind myself that the dot product of two vectors is just a number (a scalar); therefore, it does *not* have components. However, the dot product has a unit, and the unit is the product of the units of the two vectors. I find it helpful to call the dot product "the scalar product."

Therefore, the dot product of two vectors can be defined in terms of their components:

KEY EQUATION $\quad \vec{A} \cdot \vec{B} = A_x B_x + A_y B_y + A_z B_z \qquad$ (2-24)

Since the magnitude of a unit vector is 1, the dot product of vector $\vec{A}$ and a unit vector gives the component of $\vec{A}$ along an axis in the direction of the unit vector:

KEY EQUATION $\qquad \vec{A} \cdot \hat{u}_l = A_l \cdot 1 = A_l \qquad$ (2-25)

Where $\hat{u}_l$ is a unit vector in an arbitrary direction l, and A_l is the projection of vector $\vec{A}$ onto the direction l.

✓ CHECKPOINT

C-2-8 The Dot Product of Two Vectors

Which of the following expressions correctly represents the dot product of vectors $\vec{A}$ and $\vec{B}$, where $\vec{A} = 3\hat{i} + 4\hat{j} - 5\hat{k}$ and $\vec{B} = -2\hat{i} + \hat{j} - 2\hat{k}$? Explain.

(a) -1
(b) 8
(c) 20
(d) $-6\hat{i} + 4\hat{j} + 10\hat{k}$
(e) other

C-2-8 (b) Using Equation (2-23) we find
$\vec{A} \cdot \vec{B} = (3\hat{i} + 4\hat{j} - 5\hat{k})(-2\hat{i} + \hat{j} - 2\hat{k}) = -6(\hat{i} \cdot \hat{i}) + 4(\hat{j} \cdot \hat{j}) + 10(\hat{k} \cdot \hat{k})$
$= -6 + 4 + 10 = 8$

EXAMPLE 2-6

The Dot Product and Simple Coordinate Transformations

Two forces, $\vec{F}_1$ and $\vec{F}_2$, are described as $\vec{F}_1 = (3.00\hat{i} - 2.00\hat{j})\,\text{N}$ and $\vec{F}_2 = (-1.00\hat{i} + 3.00\hat{j})\,\text{N}$.

(a) Find the angle between the forces.
(b) Verify that the x- and y-components of the forces, $F_{1x}, F_{1y}, F_{2x},$ and F_{2y}, are their projections onto the x- and y-axes.
(c) Find the components of forces $\vec{F}_1$ and $\vec{F}_2$ in the orthogonal $x'y'$-coordinate system that is rotated by 30° counterclockwise about the origin, as shown in Figure 2-17(a).
(d) Calculate the magnitudes of the forces using their components in both coordinate systems, and compare the results.
(e) Suppose that we move the xy-coordinate system in a plane such that the new coordinate axes remain parallel to the original axes, but the origin O moves to a new location (Figure 2-17(b)). Find the components and the magnitudes of forces $\vec{F}_1$ and $\vec{F}_2$ in the $x''y''$-coordinate system.

SOLUTION

(a) To find the angle between the forces, we use Equation (2-17):

$$\cos\theta = \frac{\vec{F}_1 \cdot \vec{F}_2}{F_1 F_2} = \frac{F_{1x}F_{2x} + F_{1y}F_{2y}}{\left(\sqrt{F_{1x}^2 + F_{1y}^2}\right)\left(\sqrt{F_{2x}^2 + F_{2y}^2}\right)}$$

$$= \frac{(-3.00 - 6.00)\,\text{N}^2}{\sqrt{13\,\text{N}^2}\sqrt{10\,\text{N}^2}} = \frac{-9.00\,\text{N}^2}{\sqrt{130\,\text{N}^2}} = -0.789$$

$$\theta = \cos^{-1}\left(\frac{-9.00}{\sqrt{130}}\right) = \cos^{-1}(-0.789) = 142°$$

(b) We can use dot products with the $\hat{i}$ and $\hat{j}$ unit vectors to find the projections of the forces onto the x- and y-axes:

$$\begin{cases} F_{1x} = \vec{F}_1 \cdot \hat{i} = [(3.00\hat{i} - 2.00\hat{j}) \cdot \hat{i}]\,\text{N} = 3.00\,\text{N} \\ F_{1y} = \vec{F}_1 \cdot \hat{j} = [(3.00\hat{i} - 2.00\hat{j}) \cdot \hat{j}]\,\text{N} = -2.00\,\text{N} \end{cases}$$

Dot products for $\vec{F}_2$ indicate that its x- and y-projections are -1.00 N and 3.00 N, respectively.

$$\begin{cases} F_{2x} = \vec{F}_2 \cdot \hat{i} = [(-\hat{i} + 3.00\hat{j}) \cdot \hat{i}]\,\text{N} = -1.00\,\text{N} \\ F_{2y} = \vec{F}_2 \cdot \hat{j} = [(-\hat{i} + 3.00\hat{j}) \cdot \hat{j}]\,\text{N} = 3.00\,\text{N} \end{cases}$$

(c) To find the projections of forces $\vec{F}_1$ and $\vec{F}_2$ onto the orthogonal x'- and y'-axes, we define unit vectors $\hat{u}_{x'}$ and $\hat{u}_{y'}$ directed along the x'- and y'-axes, as shown in Figure 2-18(a):

$$\hat{u}_{x'} = \cos(30°)\hat{i} + \sin(30°)\hat{j} = \frac{\sqrt{3}}{2}\hat{i} + \frac{1}{2}\hat{j}$$

$$\hat{u}_{y'} = -\sin(30°)\hat{i} + \cos(30°)\hat{j} = -\frac{1}{2}\hat{i} + \frac{\sqrt{3}}{2}\hat{j}$$

Then, we use Equation (2-25) to find the projections of the forces:

$$F_{1x'} = \vec{F}_1 \cdot \hat{u}_{x'} = \left(3.00\hat{i} - 2.00\hat{j}\right)\left(\frac{\sqrt{3}}{2}\hat{i} + \frac{1}{2}\hat{j}\right)\,\text{N}$$

$$= \left(\frac{3\sqrt{3}}{2} - 1\right)\text{N} = 1.60\,\text{N}$$

(continued)

(a)

(b)

Figure 2-17 (a) The xy-coordinate system undergoes rotation in a plane (by 30°) around its origin transforming into an $x'y'$-coordinate system. (b) The xy-coordinate system undergoes translation in a plane such that the new coordinate axes (x'', y'') remain parallel to the original coordinate axes (x, y).

$$F_{1y'} = \vec{F_1} \cdot \hat{u}_{y'} = (3.00\hat{i} - 2.00\hat{j})\left(-\frac{1}{2}\hat{i} + \frac{\sqrt{3}}{2}\hat{j}\right)N = \left(-\frac{3}{2} - \sqrt{3}\right)N$$
$$= -3.23\,N$$

$$F_{2x'} = \vec{F_2} \cdot \hat{u}_{x'} = (-\hat{i} + 3.00\hat{j})\left(\frac{\sqrt{3}}{2}\hat{i} + \frac{1}{2}\hat{j}\right)N = \left(-\frac{\sqrt{3}}{2} + \frac{3}{2}\right)N$$
$$= 0.634\,N$$

$$F_{2y'} = \vec{F} \cdot \hat{u}_{y'} = (-\hat{i} + 3.00\hat{j})\left(-\frac{1}{2}\hat{i} + \frac{\sqrt{3}}{2}\hat{j}\right)N = \left(\frac{1}{2} + \frac{3\sqrt{3}}{2}\right)N$$
$$= 3.10\,N$$

(d)

$$\left.\begin{array}{l} F_1 = \left(\sqrt{F_{1x}^2 + F_{1y}^2}\right) = (\sqrt{13})\,N = 3.61\,N \\[12pt] F_2 = \left(\sqrt{F_{2x}^2 + F_{2y}^2}\right) = (\sqrt{10})\,N = 3.16\,N \end{array}\right\} \quad \begin{array}{l}\text{the } xy\text{-}\\ \text{coordinate}\\ \text{system}\end{array}$$

$$\left.\begin{array}{l} F_1 = \left(\sqrt{F_{1x'}^2 + F_{1y'}^2}\right) = \left(\sqrt{(1.60)^2 + (3.23)^2}\right)N = 3.61\,N \\[12pt] F_2 = \left(\sqrt{F_{2x'}^2 + F_{2y'}^2}\right) = \left(\sqrt{(0.634)^2 + (3.16)^2}\right)N = 3.16\,N \end{array}\right\} \quad \begin{array}{l}\text{the } x'y'\text{-}\\ \text{coordinate}\\ \text{system}\end{array}$$

(e) When a coordinate system is moved parallel to itself—that is, the coordinate system undergoes pure translation—such that its origin is moved to a different location but the directions of its axes remain unchanged, the components of a vector in the new coordinate system $(x''y'')$ will be the same as its components in the old coordinate system (xy). The reason for this is that the magnitude of a vector, as well as the angle between the vectors and the coordinate axes, will not change under this coordinate system transformation (Figure 2-18(b)). Therefore, applying Equation (2-1) we find

$$\begin{cases} F_{1x} = F_{1x''} = 3.00\,N \\ F_{1y} = F_{1y''} = -2.00\,N \end{cases} \text{ and } \begin{cases} F_{2x} = F_{2x''} = -1.00\,N \\ F_{2y} = F_{2y''} = 3.00\,N \end{cases}$$

Making sense of the result:

Coordinate systems can undergo transformations, such as coordinate axes translation or rotation. When a coordinate system undergoes pure translation, the coordinates of the vectors remain unchanged. When coordinate axes are rotated about the origin (the coordinate system undergoes a rotational transformation), the components of the vectors change, but the magnitudes of the vectors do not.

✓ **CHECKPOINT**

C-2-9 Identifying Orthogonal Vectors

Which of the following pairs of vectors are orthogonal to one another? Explain.

(a) $\vec{A} = 3\hat{i} + 4\hat{j} - \hat{k}$ and $\vec{B} = -2\hat{i} + \hat{j} - 2\hat{k}$
(b) $\vec{A} = 2\hat{i} - 3\hat{j} - 5\hat{k}$ and $\vec{B} = -2\hat{i} + \hat{j} - 2\hat{k}$
(c) $\vec{A} = -2\hat{i} - 3\hat{j} - 5\hat{k}$ and $\vec{B} = 4\hat{i} + 6\hat{j} - 2\hat{k}$
(d) $\vec{A} = 2\hat{i} - 2\hat{j} - 2\hat{k}$ and $\vec{B} = -2\hat{i} + 2\hat{j} + 2\hat{k}$
(e) $\vec{A} = 2\hat{i} - 3\hat{j} - 5\hat{k}$ and $\vec{B} = -5\hat{i} + 2\hat{j} - 3\hat{k}$

C-2-9 (a) The dot product of two orthogonal vectors equals zero.

LO 5 ±× 👥

2-5 The Cross Product of Vectors

A number of physics and engineering applications involve vectors that are perpendicular to two given vectors. As you will see in later chapters, the angular momentum vector is perpendicular to both the position vector and the velocity vector of an object, and the vector for the magnetic force acting on a charged particle is perpendicular to both the magnetic field vector and the velocity vector of the particle. The **cross**, or **vector**, **product** of two vectors defines a vector that is perpendicular to the two given vectors.

The magnitude of the cross product, $\vec{C}$, of vectors $\vec{A}$ and $\vec{B}$ is defined as the product of the magnitudes of $\vec{A}$ and $\vec{B}$ and the sine of the angle θ between them

(Figure 2-18(a)). The direction of the cross product is determined using the **right-hand rule**: when you curl the fingers of your right hand from vector $\vec{A}$ to vector $\vec{B}$, your thumb will point in the same direction as the cross product, $\vec{C}$ (Figure 2-18(b)).

 KEY EQUATION
$$\vec{A} \times \vec{B} = \vec{C}$$
$$C = AB\sin\theta, \text{ where } 0° \leq \theta \leq 180° \qquad (2\text{-}26)$$

(a)

(b)

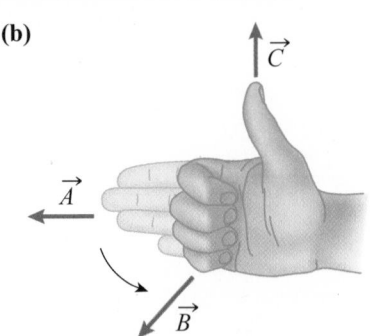

Figure 2-18 (a) Visual representation of the cross product of two vectors, $\vec{A}$ and $\vec{B}$, the area of the parallelogram represents the magnitude of the cross product (b) The right-hand rule for determining the direction of the cross product

PEER TO PEER

I have to remind myself that the cross product of two vectors is also a vector (it has magnitude, direction, and a unit). Its unit is the product of the units of two original vectors. I find it helpful to call a cross product "a vector product" so I remember it.

A cross product has the following properties:

- The cross product is a vector.

- The unit of a cross product is the product of the units of the vectors being multiplied.

- The magnitude of the cross product corresponds to the area of a parallelogram that has vectors $\vec{A}$ and $\vec{B}$ as its sides.

- The cross product of two parallel or two antiparallel vectors equals zero because $\sin(0°) = \sin(180°) = 0$. The cross product of two vectors of a given magnitude is greatest when the vectors are perpendicular to each other because $\sin(90°) = 1$.

- The cross product is anticommutative:

$$\vec{A} \times \vec{B} = -\vec{B} \times \vec{A} \qquad (2\text{-}27)$$

You can verify this property by applying the right-hand rule to the vector products in the equation above.

- The cross product can be multiplied by a scalar. From the definition of the vector product, we can see that

$$a(\vec{A} \times \vec{B}) = (a\vec{A}) \times \vec{B} = \vec{A} \times (a\vec{B}) = (\vec{A} \times \vec{B})a$$
$$= aAB\sin(\theta) \qquad (2\text{-}28)$$

- The cross product operation is distributive:

$$\vec{A} \times (\vec{B} + \vec{C}) = (\vec{A} \times \vec{B}) + (\vec{A} \times \vec{C}) = \vec{A} \times \vec{B} + \vec{A} \times \vec{C} \qquad (2\text{-}29)$$

The Cross Product and Unit Vectors

Consider the vector products of the unit vectors $\hat{i}$, $\hat{j}$, and $\hat{k}$. Since the unit vectors are orthogonal and have a magnitude of 1,

$$\hat{i} \times \hat{i} = 0; \quad \hat{i} \times \hat{j} = \hat{k}; \quad \hat{i} \times \hat{k} = -\hat{j}$$
$$\hat{j} \times \hat{j} = 0; \quad \hat{j} \times \hat{i} = -\hat{k}; \quad \hat{j} \times \hat{k} = \hat{i}$$
$$\hat{k} \times \hat{k} = 0; \quad \hat{k} \times \hat{i} = \hat{j}; \quad \hat{k} \times \hat{j} = -\hat{i} \qquad (2\text{-}30)$$

As with the dot products in Section 2-4, we can write cross products as a cross product multiplication table:

KEY EQUATION

$\times$	$\hat{i}$	$\hat{j}$	$\hat{k}$
$\hat{i}$	0	$\hat{k}$	$-\hat{j}$
$\hat{j}$	$-\hat{k}$	0	$\hat{i}$
$\hat{k}$	$\hat{j}$	$-\hat{i}$	0

$$(2\text{-}31)$$

The cross product of two unit vectors is at the intersection of the row for the first vector and the column for

the second vector. Since the cross product is anticommutative, the order in which vectors are multiplied changes the result. Consequently, unlike a dot product multiplication table (see Equation 2-2), this table is *not* symmetrical about its main diagonal.

You can use a **cross-multiplication circle** to remember the signs of the cross products of the unit vectors. Arrange the unit vectors in a counterclockwise order around a circle, as shown in Figure 2-19. To find a cross product of two unit vectors, locate the first vector and then move along the circle to the second vector of the product. If you move counterclockwise (in the direction of the arrows), the result is the third vector along the circle. If you have to move clockwise, the result is the vector opposite the third vector. For example, to find the cross product $\hat{i} \times \hat{j}$, you move counterclockwise along the circle, so the result is $\hat{k}$. However, to find $\hat{j} \times \hat{i}$, you move clockwise, so the result is $-\hat{k}$.

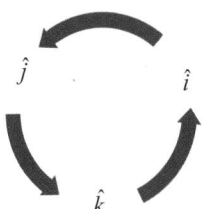

Figure 2-19 The unit cross-multiplication circle

 CHECKPOINT

C-2-10 Practising Cross Multiplication

Which of the following represents the result of $(\hat{i} \times \hat{k}) \times \hat{i}$? Explain.

(a) $\hat{i}$
(b) $-\hat{i}$
(c) $\hat{j}$
(d) $-\hat{j}$
(e) $\hat{k}$
(f) $-\hat{k}$
(g) 1
(h) 0

C-2-10 (e) $(\hat{i} \times \hat{k}) \times \hat{i} = (-\hat{j}) \times \hat{i} = -(\hat{j} \times \hat{i}) = -(-\hat{k}) = \hat{k}$

We can define a cross product for two arbitrary vectors in terms of their components and unit vectors:

$$\vec{A} \times \vec{B} = (A_x\hat{i} + A_y\hat{j} + A_z\hat{k}) \times (B_x\hat{i} + B_y\hat{j} + B_z\hat{k})$$
$$= A_xB_x(\hat{i} \times \hat{i}) + A_xB_y(\hat{i} \times \hat{j}) + A_xB_z(\hat{i} \times \hat{k})$$
$$+ A_yB_x(\hat{j} \times \hat{i}) + A_yB_y(\hat{j} \times \hat{j}) + A_yB_z(\hat{j} \times \hat{k})$$
$$+ A_zB_x(\hat{k} \times \hat{i}) + A_zB_y(\hat{k} \times \hat{j}) + A_zB_z(\hat{k} \times \hat{k})$$
$$= A_xB_y\hat{k} + A_xB_z(-\hat{j}) + A_yB_x(-\hat{k}) + A_yB_z(\hat{i})$$
$$+ A_zB_x(\hat{j}) + A_zB_y(-\hat{i})$$
$$= (A_yB_z - A_zB_y)\hat{i} + (A_zB_x - A_xB_z)\hat{j}$$
$$+ (A_xB_y - A_yB_x)\hat{k} \qquad (2\text{-}32)$$

Therefore, using Cartesian notation the cross product of two vectors can be expressed as follows:

KEY EQUATION
$$\vec{A} \times \vec{B} = (A_y B_z - A_z B_y)\hat{i} + (A_z B_x - A_x B_z)\hat{j} + (A_x B_y - A_y B_x)\hat{k} \qquad (2\text{-}33)$$

The **determinant form of a vector product** is a more elegant way of writing Equation (2-33):

KEY EQUATION
$$\vec{A} \times \vec{B} = \begin{vmatrix} \hat{i} & \hat{j} & \hat{k} \\ A_x & A_y & A_z \\ B_x & B_y & B_z \end{vmatrix} \qquad (2\text{-}34)$$

The first row of the matrix lists the unit vectors, the second row lists the components of the first vector to be multiplied, and the bottom row lists the components of the

second vector. Expand the determinant in Equation (2-34) about its first row and you will reproduce Equation (2-33).

 CHECKPOINT

C-2-11 Vector Cross Products

Which of the following expressions does *not* make sense? Explain.

(a) $(\vec{A} \times \vec{B}) \times \vec{C}$

(b) $(\vec{A} \times \vec{B}) \cdot \vec{C}$

(c) $\vec{A} \times (\vec{B} \cdot \vec{C})$

(d) $\vec{A} \cdot (\vec{B} \times \vec{C}) \cdot \vec{D}$

(e) $\vec{A} \cdot (\vec{B} + \vec{C})$

C-2-11 (c) The vector in (c) is cross multiplied by a scalar, which is not possible. The expression in (d) also makes no sense, because it is the dot product of three vectors, which is not defined.

 EXAMPLE 2-7

Finding a Unit Vector Perpendicular to Two Given Vectors

Two forces are described as $\vec{F}_1 = (\hat{i} + 2\hat{j} - 3\hat{k})$ N and $\vec{F}_2 = (-2\hat{i} + 3\hat{j})$ N. Find an expression for a unit vector directed perpendicular to both of them.

SOLUTION

By definition, the cross product of any two vectors is a vector that is perpendicular to both vectors. Therefore, the unit vector we are looking for is directed along the line of the cross product of the two given vectors. Let us first find the cross product:

$$\vec{C} = \vec{F}_1 \times \vec{F}_2 = [(\hat{i} + 2\hat{j} - 3\hat{k})\,\text{N}] \times [(-2\hat{i} + 3\hat{j})\,\text{N}]$$

$$= [3\hat{k} + 4\hat{k} + 6\hat{j} + 9\hat{i}]\,\text{N}^2 = [9\hat{i} + 6\hat{j} + 7\hat{k}]\,\text{N}^2$$

Now, we find a unit vector along the direction of vector $\vec{C}$:

$$\hat{u}_C = \frac{\vec{C}}{C} = \frac{[9\hat{i} + 6\hat{j} + 7\hat{k}]\,\text{N}^2}{[\sqrt{81 + 36 + 49}]\,\text{N}^2} = \frac{1}{\sqrt{166}}(9\hat{i} + 6\hat{j} + 7\hat{k})$$

Making sense of the result:

We can check geometrically that our answer is correct. We can also use scalar products to confirm that the unit vector is perpendicular to the force vectors:

$$\hat{u}_C \cdot \vec{F}_1 = \frac{9\hat{i} + 6\hat{j} + 7\hat{k}}{\sqrt{166}} \cdot (\hat{i} + 2\hat{j} - 3\hat{k})\,\text{N} = \frac{9 + 12 - 21}{\sqrt{166}}\,\text{N} = 0$$

$$\hat{u}_C \cdot \vec{F}_2 = \frac{9\hat{i} + 6\hat{j} + 7\hat{k}}{\sqrt{166}} \cdot (-2\hat{i} + 3\hat{j})\,\text{N} = \frac{-18 + 18}{\sqrt{166}}\,\text{N} = 0$$

Since the dot products are zero, the unit vector is perpendicular to both force vectors. Note that the vector opposite to $\hat{u}$ is also perpendicular to both force vectors.

FUNDAMENTAL CONCEPTS AND RELATIONSHIPS

Physical quantities used in this book can be categorized as scalars and vectors. Scalars require only a number (either positive or negative) and a unit for their description. Vectors require a number, a unit, and a direction. Vector operations include addition and subtraction, multiplication of a vector by a scalar, and dot (scalar) and cross (vector) products of two or more vectors.

Vector Components

Vector components are scalar quantities that represent vector projections onto the coordinate axes. In the case of 2-D vectors:

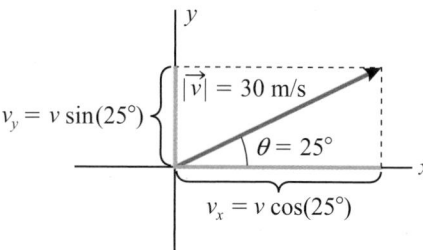

$$\begin{cases} v_x = v\cos\theta \\ v_y = v\sin\theta \end{cases} \Rightarrow \begin{cases} v = \sqrt{v_x^2 + v_y^2} \\ \tan\theta = \dfrac{v_y}{v_x} \end{cases}$$

$v_y = v\sin(25°)$ $\quad |\vec{v}| = 30 \text{ m/s}$

$\theta = 25°$

$v_x = v\cos(25°)$

Cartesian Notation

$\vec{F} = F_x\hat{i} + F_y\hat{j} + F_z\hat{k}$, where $\hat{i}$, $\hat{j}$, and $\hat{k}$ are unit vectors

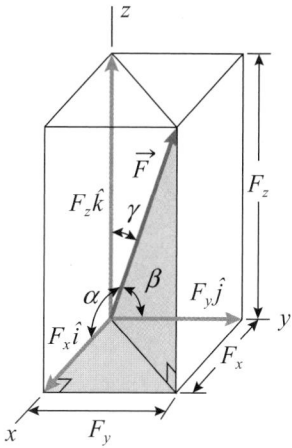

Vector Magnitude and Simple Coordinate System Transformations

When a coordinate system undergoes either pure translation or rotation, the magnitude of a vector in the transformed coordinate system remains unchanged. The vector components in the translated coordinate system do not change

either. However, the vector components in a rotated coordinate system will be different from the vector components in the original coordinate system.

Vector Addition and Subtraction

Vector Addition: Vectors can be added using either a geometric or an algebraic approach.

Geometric Approach: triangle construction or parallelogram rule

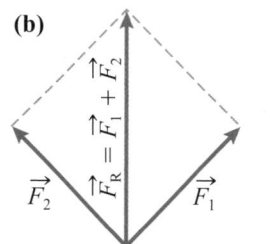

Parallelogram rule   Triangle construction rule

Vector Subtraction: $\vec{F}_1 - \vec{F}_2 = \vec{F}_1 + (-\vec{F}_2)$

Algebraic Approach:

$$\begin{cases} F_{R,x} = F_{1x} + F_{2x} + \cdots + F_{N,x} \\ F_{R,y} = F_{1y} + F_{2y} + \cdots + F_{N,y} \\ F_R = \sqrt{F_{R,x}^2 + F_{R,y}^2} \\ \tan\theta = \dfrac{F_{R,y}}{F_{R,x}} \end{cases}$$

Vector Multiplication by a Scalar

$|a\vec{F}| = |a||\vec{F}|$

$\begin{cases} \text{if } a > 0, \ a\vec{F} \text{ has the same direction as } \vec{F} \\ \text{if } a < 0, \ a\vec{F} \text{ has the opposite direction to } \vec{F} \\ \text{if } a = 0, \ a\vec{F} \text{ has a magnitude of zero and no direction} \end{cases}$

Products of Vectors

Dot Product: $\vec{A} \cdot \vec{B} = AB\cos\theta$, where $0° \le \theta \le 180°$ and

$$\vec{A} \cdot \vec{B} = A_xB_x + A_yB_y + A_zB_z$$

Cross Product: $\vec{A} \times \vec{B} = \vec{C}$

$C = AB\sin\theta$, and $\vec{C}$ is perpendicular to both $\vec{A}$ and $\vec{B}$, where $0° \le \theta \le 180°$

$$\vec{A} \times \vec{B} = (A_yB_z - A_zB_y)\hat{i} + (A_zB_x - A_xB_z)\hat{j} + (A_xB_y - A_yB_x)\hat{k}$$

(continued)

Applications: Vectors are used to describe physical quantities that require magnitude, unit, and direction for their description, such as velocity, force, acceleration, and momentum. The dot product is used to determine the projections of vectors onto various directions. The cross product of vectors can be used to define or calculate physical quantities, such as angular momentum, torque, and magnetic force.

Key Terms: algebraic method for vector addition, Cartesian vector notation, coordinate direction angles, cross-multiplication circle, cross (vector) product, determinant form of a vector product, displacement vector, dot (scalar) product, free body diagram, geometric method for vector addition, opposite vectors, parallelogram rule, polar coordinates, polar notation, position vector, reference frame, right-hand rule, scalar notation, scalars, triangle construction rule, unit vectors, vector addition, vector components, vector head, vector magnitude and direction, vector projection, vector resolution into scalar components, vector subtraction, vector tail, vectors

QUESTIONS

1. A given force $\vec{F}$ is located in the horizontal plane. It has a magnitude of 30 N and is directed at a 45.0° angle clockwise from the positive x-axis. Describe $\vec{F}$ using
   (a) polar notation;
   (b) its x- and y-components;
   (c) Cartesian notation.

2. A given force $\vec{F}$ is located in the yz-plane. It has a magnitude of 15 N and is directed at a 30.0° angle clockwise from the positive y-axis. Describe $\vec{F}$ using
   (a) polar notation;
   (b) its y- and z-components;
   (c) Cartesian notation.

3. A force acting on a 5.00 kg object is located in the xy-plane. It has a magnitude of 8.00 N and is directed parallel to the line $y = -3x + 4$. Express this force in Cartesian notation. How many results do you have? Explain why.

4. An object is located at the point (3.00 m, –4.00 m, 0 m). A position vector connects this object to the origin. Find the magnitude of the position vector and the angle it makes with the positive x-axis.

5. A displacement vector connects two points, A (–2.00 m, 2.00 m, 0 m) and B (3.00 m, 0 m, –2.00 m). Find the magnitude of the displacement vector.

6. An object travelling in space passes five different points: A (2.00 m, 3.00 m, –1.00 m), B (2.00 m, 6.00 m, –1.00 m), C (2.00 m, 6.00 m, –8.00 m), D (–4.00 m, 6.00 m, –8.00 m), and E (–11.00 m, 6.00 m, –1.00 m). The object returns to its initial location, point A.
   (a) Rank the displacement vectors $(\vec{r}_{AB}, \vec{r}_{BC}, \vec{r}_{CD}, \vec{r}_{DE}, \vec{r}_{EA})$ from the vector that has the largest magnitude to the vector that has the smallest magnitude.
   Largest 1____ 2____ 3____ 4____ 5____ Smallest
   (b) Are the displacements the same in all cases?
   (c) Is the displacement zero in all cases?
   (d) If the displacement is the same for two or more cases, clearly indicate it on the ranking scheme. Explain your reasoning.

7. Three vectors are described as $\vec{A}(1, -3, 4)$, $\vec{B}(5, -2, 1)$, and $\vec{C}(-2, 3, 6)$. Find the following vectors:
   (a) $\vec{D} = \vec{A} + \vec{B} + \vec{C}$
   (b) $\vec{E} = -\vec{A} - \vec{B} - \vec{C}$
   (c) $\vec{F} = 2\vec{A} - 4\vec{B} + 3\vec{C}$
   (d) $\vec{G} = 2(\vec{A} - 2\vec{B}) + \vec{C}$

8. Two horizontal forces of magnitude 10 N each are acting on a horizontal force board ring (Figure 2-20). A third force acting on the ring perfectly balances these two forces. What *must be* true about this force?
   (a) The third force must be 10 N, and it must be at an angle of 120 degrees to each one of the forces.
   (b) The magnitude of the third force should be more than 10 N.
   (c) The magnitude of the third force must be less than 10 N.
   (d) The magnitude of the third force must be between –20 N and 0 N.
   (e) The third force must be directed at a 45° angle to the two given forces.

Figure 2-20 Question 8. A bird's eye view of a force board

9. Which of the following statement(s) do you agree with? Explain.
   (a) Vector components must always be smaller than the vector magnitudes.
   (b) In order to find vector magnitude, one must add the vector components.
   (c) Vector magnitude does not depend on the choice of the coordinate system, even though vector components do.
   (d) Vector components do not depend on the choice of the coordinate system, but vector magnitude does.
   (e) If two vectors have equal magnitudes, they must either be opposite vectors or have the same direction.

10. Which of the following vector pairs are orthogonal (perpendicular to each other)? Explain.
    (a) (2, 3, –6) and (–2, 3, –6)
    (b) (2, 3, –1) and (–2, –1, –7)
    (c) (–2, –3, –1) and (–2, –1, 7)

(d) (1, 0, –6) and (–2, 3, 0)

(e) (2, 3, 0) and (0, –3, 6)

11. Two students have an argument. Student 1 claims that any two vectors located in orthogonal planes must also be orthogonal, and student 2 disagrees. Who do you agree with and why? (Hint: It is sufficient to have one counterexample to disprove a statement.)

12. Two objects are moving at constant velocities: $\vec{v}_A = 2\hat{i} - 5\hat{j}$ and $\vec{v}_B = -2\hat{i} + 4\hat{j}$.
(a) Which object is moving faster?
(b) Find the angle between objects' trajectories.

13. Two vectors are described as $\vec{A}$ (2, 2, –2) and $\vec{B}$ (–1, 3, –2).
(a) Find the projections of these vectors on the x-, y-, and z-coordinate axes.
(b) Determine the scalar product of these two vectors.
(c) Find the vector products of these two vectors.

14. Find the coordinate direction angles α, β, and γ for the position vector $\vec{r} = (2\hat{i} - 5\hat{j} - 3\hat{k})$ m. Verify that expression (2-11) holds true.

15. You overhear your classmate uttering the following statement: "Since the magnitude of a unit vector equals 1, the components of a unit vector must also equal 1." Do you agree or disagree with this statement? Explain why.

16. Three forces are acting on an object: $\vec{F}_1 = (3.00\hat{i} + 4.00\hat{j} - 5.00\hat{k})$ N; $\vec{F}_2 = (\hat{i} - 4.00\hat{j} + 2.00\hat{k})$ N; $\vec{F}_3 = (-2.00\hat{i} + 3.00\hat{j} - 3.00\hat{k})$ N. Find the expression for the net force acting on it, and then calculate the magnitude of the net force.

17. Complete the following vector addition (the forces are all measured in N), and find the magnitude of each force, as well as the magnitude of the net force:

+	$\hat{i}$	$\hat{j}$	$\hat{k}$
$\vec{F}_1$	1.00	–3.00	1.00
$\vec{F}_2$	–2.00	?	3.00
$\vec{F}_3$	?	2.00	0.00
$\vec{F}_R$	0.00	–5.00	?

18. Prove that the area of a parallelogram that has vectors $\vec{A}$ and $\vec{B}$ for two of its sides can be expressed as $|\vec{A} \times \vec{B}|$.

19. Your classmate claims that, in order to add two vectors using Cartesian notation, you need to add corresponding vector components; therefore, the same can be done using polar notation for vectors. Do you agree with your classmate? Explain why or why not.

20. List the advantages and disadvantages of using polar versus Cartesian vector notation. Illustrate your result with relevant examples.

21. The head-to-tail rule for adding vectors is as follows: "Start with the first vector in the sum, and then arrange the rest of the vectors so that every vector's tail touches the previous vector's head. The sum can be represented as a vector connecting the tail of the first vector in the sum to the head of the last one." Prove that when you add more than two vectors you can still use the head-to-tail rule.

22. Find the relationship between the coordinate direction angles of any two opposite vectors.

23. Explain why the scalar product is commutative and the vector product is not.

24. If the scalar product of two nonzero vectors equals zero, then the vectors are orthogonal. What does it mean when we say that the vector product of two vectors equals zero? Explain how your result corresponds to the right-hand rule for finding the vector product of two vectors.

25. Force $\vec{F}$ has a magnitude of 5.00 N; its coordinate direction angles are $\alpha = 30.0°$, $\beta = -40.0°$, and $\gamma = -60.0°$. Describe $\vec{F}$ using Cartesian notation.

PROBLEMS BY SECTION

For problems, star ratings will be used, (✶, ✶✶, or ✶✶✶), with more stars meaning more challenging problems.

Section 2-2   Vector Addition: Geometric and Algebraic Approaches

26. ✶   All of the force vectors shown in Figure 2-21 lie in the xy-, yz-, or xz-planes.
(a) Estimate the x-, y-, and z-components of each force as accurately as you can from the figure.
(b) Find the angles each force forms with the x- and y-axes.
(c) Express each force in terms of Cartesian unit vectors.
(d) Determine the magnitude of each force.
(e) Write the expressions for the forces $\vec{F}_A$, $\vec{F}_B$, $\vec{F}_C$, and $\vec{F}_D$ in Cartesian notation using an x'y'z' coordinate system that is rotated 40° counterclockwise about the x-axis relative to the original xyz-coordinate system.
(f) Calculate the magnitude of each force in the x'y'z'-coordinate system.
(g) Compare your results to the previous parts. How do you know if your results make sense? What do they tell you?

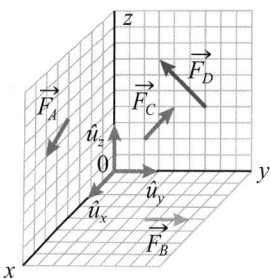

All the vectors are located in the x-y, x-z and y-z planes respectively.

Figure 2-21   Problem 26

27. ✶   Find the sum of the forces in Figure 2-21. Express the net force in Cartesian notation. Check your result using the PhET computer simulation "Vector Addition" (http://phet.colorado.edu/simulations/sims.php?sim=Vector_Addition). Explain why it is not practical to add these forces by hand using the geometric method.

28. ✶   Four 2-D displacement vectors describing consecutive displacements of an object moving on a flat surface are shown in Figure 2-22.
(a) Use the geometric approach to vector addition to find the magnitude and direction of the total displacement of the object.

(b) Estimate the components of the four displacement vectors as accurately as you can. Then use the algebraic approach to find the total displacement. Compare your results to your results in part (a).

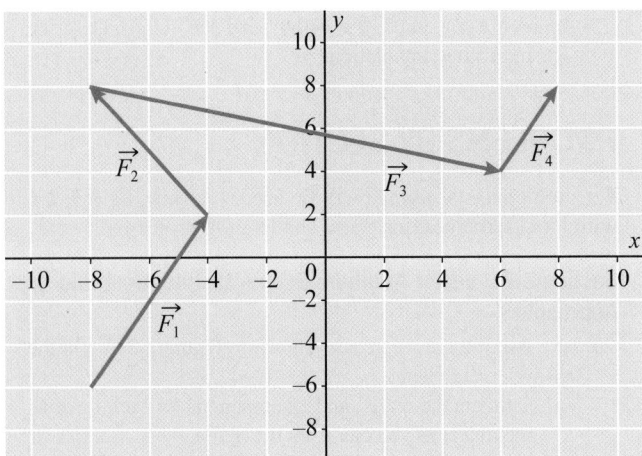

Figure 2-22 Problem 28

29. ✷✷ A cat travels 10.0 m in the direction of 20.0° [W of N]. Then the cat travels 7.0 m [W] and runs 20.0 m in the direction of 30.0° [W of S]. Find the magnitude of the total displacement of the cat using a geometric method, then check your result using the algebraic approach.

30. ✷✷ A child is sliding down a steep 60.0° water slide. The magnitude of the gravitational force exerted on the child by Earth equals 300 N, and the normal force exerted on the child by the slide is 150 N.
    (a) Draw an FBD representing the problem.
    (b) Find the components of the gravitational force along the x- and y-axes directed along the slide (x) and perpendicular to it (y).
    (c) Find the components of the normal force along the x- and y-axes in (b).
    (d) Find the net force acting on the child. What does your result imply?

31. ✷✷ The magnitudes of the vectors shown in Figure 2-23 are $A = 20.0$ units, $B = 15.0$ units, and $C = 25.0$ units, and the corresponding angles are $\theta_1 = -15.0°$, $\theta_2 = 35.0°$, and $\theta_3 = 125°$.
    (a) Determine the components of each vector.
    (b) Find the sum of the three vectors, and express the sum in both Cartesian and polar notation (magnitude and direction).
    (c) Find the magnitude and direction of $\vec{D} = 2\vec{A} - 3\vec{B} + \vec{C}$.

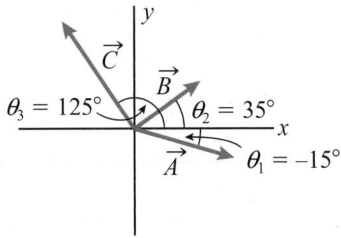

Figure 2-23 Problem 31

Section 2-3 Cartesian Vector Notation

32. ✷✷ Find the magnitude and the coordinate direction angles for the following vectors:
    (a) $\vec{F} = (3.00\hat{i} - 5.00\hat{j} + 6.00\hat{k})$ N
    (b) $\vec{F} = (-3.00\hat{i} + 5.00\hat{j} - 6.00\hat{k})$ N
    (c) Compare your results from (a) and (b). What does your comparison mean?
    (d) Find the sum of the squares of the cosines of the corresponding coordinate direction angles found above. How do you know if your result makes sense?

33. ✷✷ Find the Cartesian expression of a unit vector directed
    (a) parallel to the force $\vec{F} = (3.00\hat{i} - 5.00\hat{j} + 6.00\hat{k})$ N;
    (b) perpendicular to the force $\vec{F} = (3.00\hat{i} - 5.00\hat{j} + 6.00\hat{k})$ N;
    (c) parallel to the plane defined by $3x + 2y - 4z = 0$;
    (d) perpendicular to the plane defined by $3x + 2y - 4z = 0$.

34. ✷✷ The average velocity of an object as it is moving from position A to position B is defined as $\vec{v}_{avg} = \dfrac{\vec{r}_B - \vec{r}_A}{\Delta t}$, where Δt is the time it takes for the object to move from A to B. The object's positions are described as $\vec{r}_A = 3.00\hat{i} + 2.00\hat{j}$ and $\vec{r}_B = -3.00\hat{i} + 2.00\hat{j}$, and it took 5.00 s to move from A to B.
    (a) Find the object's average velocity.
    (b) What other information do you need to find the object's average speed?

35. ✷✷ Solve the following vector equations:
    (a) $F_x\hat{i} + 3\hat{j} + \sqrt{2}\hat{i} - F_y\hat{j} + F_z\hat{k} - 5\hat{k} = 0$
    (b) $3_x\hat{i} - 5\hat{j} + F\hat{i} - 2F_y\hat{j} + F_z\hat{k} - 3\hat{k} = 0$

Section 2-4 The Dot Product of Two Vectors

36. ✷ Calculate the dot product of the following two vectors: $\vec{A}$ (3, 4, –5) and $\vec{B}$ (2, –2, 4).

37. ✷✷ Find a unit vector located in the xy-plane and perpendicular to the given vector $\vec{A} = 3\hat{i} - 2\hat{j} + 5\hat{k}$. How many results do you have? Explain why.

38. ✷✷ Solve the following vector equation:

$$(3\hat{i} - 5\hat{j} + F_z\hat{k}) \cdot (4\hat{i} + 4\hat{j} + F_z\hat{k}) = 0$$

39. ✷✷ Two vectors, $\vec{A}$ and $\vec{B}$, have magnitudes 10 units and 5 units, respectively, and are located in the xy-plane. The angle between them is 20°. Find the dot product of these two vectors.

40. ✷✷ Calculate the following dot products: $\vec{A} \cdot \vec{B}$; $\vec{B} \cdot \vec{C}$; $\vec{A} \cdot \vec{C}$; $\vec{B} \cdot \vec{A}$; $\vec{C} \cdot \vec{B}$ and $\vec{C} \cdot \vec{A}$ for vectors $\vec{A}$, $\vec{B}$, and $\vec{C}$ in Figure 2-24.

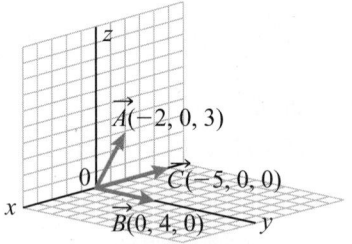

Figure 2-24 Problem 40

41. ✷✷ Find the projection of the 2-D force vector $\vec{F}$(3.00, 4.00) N onto the line $y = -2x + 5$.

42. ✹✹ Find the angle between the two vectors $\vec{A}$ and $\vec{B}$, where $\vec{A} = -2\hat{i} + 3\hat{j} - 5\hat{k}$ and $\vec{B} = -2\hat{i} + 2\hat{j} - 5\hat{k}$.

Section 2-5 The Cross Product of Vectors

43. ✹ Calculate the vector cross product of the following two vectors: $\vec{A}(3, 4, -5)$ and $\vec{B}(2, -2, 4)$.

44. ✹✹ Three vectors, $\vec{A}(4, 3, 1)$, $\vec{B}(-2, 4, -2)$, and $\vec{C}(-1, -3, 5)$, form the sides of a parallelogram prism.
    (a) Find the volume of the prism.
    (b) Find the surface area of the prism.

45. ✹✹ Two vectors, $\vec{A}$ and $\vec{B}$, have magnitudes of 10 units and 5 units, respectively, and are located in the xz-plane. The angle between them is 20°. Find the magnitude and direction of the vector product of these two vectors.

46. ✹✹ For the vectors $\vec{A}$, $\vec{B}$, and $\vec{C}$ in Figure 2-23, calculate the following cross products:

$$\vec{A} \times \vec{B}; \vec{B} \times \vec{C}; \vec{A} \times \vec{C}; \vec{B} \times \vec{A}; \vec{C} \times \vec{B} \text{ and } \vec{C} \times \vec{A}$$

47. ✹✹ Solve the following vector equations:
    (a) $2\hat{k} \times \left(\dfrac{1}{4} F_x\hat{i} + 7\hat{j} \right) + F_y\hat{i} - \sqrt{2}\hat{j} = 0$
    (b) $4\hat{j} \times (F_x\hat{i} + 5\hat{k}) + F_z\hat{k} + \sqrt{3}F_x\hat{i} = 0$

48. ✹✹✹ Determine the angle between the diagonal of a cube and each one of its edges.

COMPREHENSIVE PROBLEMS

49. ✹✹ An orthogonal coordinate system, xy, undergoes the following transformation: it moves parallel to itself such that the new location of its origin is O′(4, 5). It is then rotated 60° counterclockwise. The new coordinate system is $x'y'$.
    (a) Derive an expression for the coordinates of an arbitrary point under this transformation. Express how the new coordinates depend on the old coordinates.
    (b) Derive an expression that describes how vector components change under this transformation. Express how the new vector components depend on the old vector components.

50. ✹✹ For the three vectors defined as $\vec{A} = (a_x, a_y, a_z)$, $\vec{B} = (b_x, b_y, b_z)$, $\vec{C} = (c_x, c_y, c_z)$, prove that

$$\vec{A} \cdot (\vec{B} \times \vec{C}) = \begin{vmatrix} a_x & a_y & a_z \\ b_x & b_y & b_z \\ c_x & c_y & c_z \end{vmatrix}$$

51. ✹✹ Prove that for any two non-zero vectors $\vec{A}$ and $\vec{B}$, both $\vec{A}$ and $\vec{B}$ are perpendicular to their cross product $\vec{A} \times \vec{B}$.

52. ✹✹ Prove the following relationships:
$$\vec{A} \times \vec{B} \times \vec{C} = \vec{C} \times \vec{A} \times \vec{B} = \vec{B} \times \vec{C} \times \vec{A};$$
$$\vec{A} \times \vec{B} \times \vec{C} = -(\vec{A} \times \vec{C} \times \vec{B}) = -(\vec{C} \times \vec{B} \times \vec{A}) = -(\vec{B} \times \vec{A} \times \vec{C})$$

53. ✹✹✹ Prove that the algebraic expression for the cross product of two vectors using the determinant form (2-34) is equivalent to the cross product calculation using the cross product definition (2-26). (Hint: Find the magnitudes of each vector, and then find the angle between them using the dot product.)

54. ✹✹ Prove that the algebraic expression for the cross product of two vectors using the determinant form (2-34) is equivalent to the cross product calculation using Cartesian notation (2.32) and the expressions for the cross product for the unit vectors (2-30).

55. ✹✹ Find the cross product of two vectors $\vec{A} = 3\hat{i} + 2\hat{j} - 5\hat{k}$ and $\vec{B} = 3\hat{i} + 2\hat{j} - 5\hat{k}$ using the determinant form, Equation (2-34).

56. ✹✹ Two straight lines, a and b, are oriented in the xy-plane and described as a: $y = m_1x + b_1$ and b: $y = m_2x + b_2$, where neither m_1 nor m_2 is 0. Prove that, in order to be perpendicular to each other, the following must be true: $m_1m_2 = -1$. (Hint: Prove that any two arbitrary vectors directed along these lines are perpendicular to each other.)

57. ✹✹✹ A triangle is built on the diameter of a circle such that all three of its vertices lie on the circumference of the circle. Prove that the triangle is a right triangle. (Hint: Express the sides of the triangle as vectors, and use a scalar product to prove that these vectors are orthogonal.)

58. ✹✹✹ Show that $|\vec{A} \times \vec{B}| = |\vec{A}|^2|\vec{B}|^2 - (\vec{A} \cdot \vec{B})^2$.

59. ✹✹✹ The law of cosines states that for any triangle, the following holds true: $C^2 = A^2 + B^2 - 2AB\cos\theta$, where θ is the angle between sides A and B of the triangle. Prove the law of cosines using the triangle construction method of vector addition and the fact that a square of the magnitude of a vector is equal to the dot product of the vector with itself.

60. ✹✹✹ Prove the formula for the differentiation of the scalar product: $d(\vec{A} \cdot \vec{B}) = (d\vec{A}) \cdot \vec{B} + \vec{A} \cdot (d\vec{B})$. Use your proof to find an expression for $d(\vec{v} \cdot \vec{v})$, where $\vec{v}$ is the velocity of a particle.

61. ✹✹✹ The velocity of a particle moving in the xy-plane can be described as follows: $\vec{v} = v_{0x}\hat{i} + (v_{0y} - gt)\hat{j}$, where g is a positive constant called the acceleration due to gravity, t is time, and v_{0x} and v_{0y} are the components of the velocity on the x- and y-axes when $t = 0$. For Earth, $g \approx 9.81$ m/s².
    (a) Find the expression for the speed of the particle as a function of time.
    (b) Draw a diagram that illustrates how the velocity of the particle changes with time.
    (c) Determine the time when the speed of the particle is the smallest.
    (d) This mathematical model is appropriate to describe a common, everyday phenomenon. Suggest what that phenomenon might be.

62. ✹✹✹ The work of a constant force W_F on an object can be described as a scalar product of the force and the object's displacement, $W_F = \vec{F} \cdot \Delta\vec{r}$, and is measured in joules when the force is measured in newtons and the displacement in metres. A constant force $\vec{F} = (3.00\hat{i} + 4.00\hat{j} - 5.00\hat{k})$ N acts on an object moving along the line $\vec{r} = \vec{a} + t\vec{b}$, with $\vec{a} = 2.00\hat{i} - 3.00\hat{j} - 4.00\hat{k}$ and $\vec{b} = 3.00\hat{i} - 2.00\hat{j} + 4.00\hat{k}$ (t is a parameter represented by a real number: $t \in R$). Find the work done on the object by the force $\vec{F}$ while the object moved from point A (1.00 m, 1.00 m, 0.00 m) to point B (3.00 m, –3.00 m, 6.00 m).

63. ✹✹✹ Newton's second law states that the acceleration of a moving object is proportional to the net force acting

on it and inversely proportional to the object's mass: $\vec{a} = \dfrac{\sum \vec{F}}{m} = \dfrac{\vec{F}_{net}}{m}$. There are three forces acting on a particle:

$\vec{F}_1 = (3.00\hat{i} + 2.00\hat{j} - 5.00\hat{k})$ N; $\vec{F}_2 = (-\hat{i} + 4.00\hat{j} - 2.00\hat{k})$ N; $\vec{F}_3 = (-2.00\hat{i} + 5.00\hat{j} - 7.00\hat{k})$ N.

The particle's mass is 0.100 kg.
(a) Find the magnitude of the particle's acceleration.
(b) Find the coordinate direction angles of its acceleration vector.
(c) Write the expression for the particle's acceleration in Cartesian notation.
(d) Determine the projections of the particle's acceleration onto each of the coordinate planes xy, xz, and yz.

64. ✶✶✶ Three vectors are described as $\vec{A}(2, 3, -5)$, $\vec{B}(-1, 4, -2)$, and $\vec{C}(1, 2, 1)$. Find the following products:
(a) $\vec{D} = \vec{A} \times \vec{B}$
(b) $\vec{E} = \vec{A} \times \vec{B} + \vec{C}$
(c) $\vec{F} = 2\vec{A} \times \vec{B} + \vec{C}$
(d) $G = (\vec{A} \times \vec{B}) \cdot \vec{C}$

65. ✶✶✶ A trigonometric identity involving the cosine function can be proven using the dot product operation. Imagine a unit circle, which is a circle whose radius is 1 unit. Unit vectors $\hat{u}$ and $\hat{v}$ form angles α and β, respectively, with the horizontal axis. Use the unit circle to prove that $\cos(\beta - \alpha) = \cos\alpha\cos\beta + \sin\alpha\sin\beta$.

66. ✶✶✶ Three vectors are described as $\vec{A}(2, -3, -6)$, $\vec{B}(1, -4, 2)$, and $\vec{C}(1, -2, 3)$. Find all the vectors that are perpendicular to vectors $\vec{A}$ and $\vec{B}$ and have the same magnitude as vector $\vec{C}$.

67. ✶✶✶ The carbon tetrachloride molecule, CCl_4 (Figure 2-25(a)), has the shape of a tetrahedron. Chlorine atoms are located at the vertices of the tetrahedron, and the carbon atom is located in the centre. The bond angle (θ) is an angle of three connected atoms: Cl–C–Cl. A tetrahedron is formed by connecting alternating vertices of a cube, as shown in Figure 2-25(b). Find the bond angle (θ) of the carbon tetrachloride molecule.

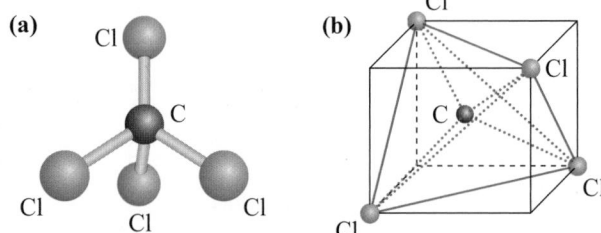

Figure 2-25 Problem 67 (a) A carbon tetrachloride molecule (b) Atom locations in the carbon tetrachloride molecule

 See the text online resources at www.physics1e.nelson.com for Open Problems and Data-Rich Problems related to this chapter.

Chapter 3
Motion in One Dimension

Learning Objectives

When you have completed this chapter you should be able to:

1 Define distance and displacement.

2 Define velocity and average velocity, and distinguish between velocity and speed.

3 Define acceleration and average acceleration, and calculate acceleration from velocity and position information.

4 Develop and use kinematics equations for constant acceleration.

5 Define and calculate relative motion quantities in one dimension.

LOOK AHEAD

High accelerations can cause serious injuries. During emergency landings, a plane and its passengers can experience high accelerations as the plane comes to a stop over a short distance. In the United States, passenger planes designed between 1952 and 1988 had to have 9 g seats, which means that the seats had to be able to withstand accelerations 9 times the acceleration due to gravity. However, analysis of aircraft crashes equipped with these seats indicated that often the majority of the fatalities and serious injuries in a crash resulted from the seats separating from the tracks on which they were mounted. In 1988, the U.S. Federal Aviation Administration (FAA) introduced regulation requiring commercial aircraft designed after 1988 to be equipped with seats that could withstand accelerations of 16 g. This measure appears to have saved numerous lives during emergency landings and runway overshoots. For example, in 2009, American Airlines flight 331 overshot a runway in Jamaica without any fatalities, even though the Boeing 737 cracked its fuselage (Figure 3-1).

Similarly, there were no fatalities and only 12 serious injuries when an Air France Airbus 340 with 309 people onboard ran off a runway into a ravine at Pearson Airport, Toronto, Ontario, while landing in bad weather in 2005. All passenger aircraft in the United States built after October 27, 2009, are required to have the 16 g seats, and older aircraft are to be retrofitted with 16 g seats by 2016.

ANDREW P. SMITH/Reuters/Landov

Figure 3-1   American Airlines flight 331 crashed on landing at Kingston's Norman Manley International Airport. The Boeing 737 cracked its fuselage, stopping just before it hit the water. There were no fatalities among the 154 passengers.

 For additional Making Connections, Examples, and Checkpoints, as well as activities and experiments to help increase your understanding of the chapter's concepts, please go to the text's online resources at www.physics1e.nelson.com.

Kinematics is the study of motion. In kinematics, we analyze how an object's position, velocity, and acceleration relate to one another, and how they change with time. We restrict the discussion in this chapter to motion in one dimension, and extend it to two and three dimensions in the next chapter. We will derive the general relationships for kinematics and the equations that describe motion under constant acceleration.

LO 1

3-1 Distance and Displacement

The **displacement** of an object is the change in its **position**. As shown in Figure 3-2, the displacement is the difference between the final position vector and the initial position vector:

$$\Delta \vec{x} = \vec{x}_2 - \vec{x}_1 \qquad (3\text{-}1)$$

where $\vec{x}_1$ and $\vec{x}_2$ are, respectively, the initial and final positions of the object as measured from a reference point.

Figure 3-2 Displacement is the change in the position vector.

The SI unit for both distance and displacement is the metre (m). The displacement of an object does not depend on the details of the path taken; it depends only on the initial and final positions. The distance travelled by an object, however, *does* depend on the details of the journey. Distance is a scalar and does not depend on direction.

The distance travelled by an object can be defined as the integral of the lengths of infinitesimal segments of the path:

$$d = \int_{x_1}^{x_2} |d\vec{x}| \qquad (3\text{-}2)$$

Figure 3-3 illustrates the difference between displacement and distance. A person walks from point A to

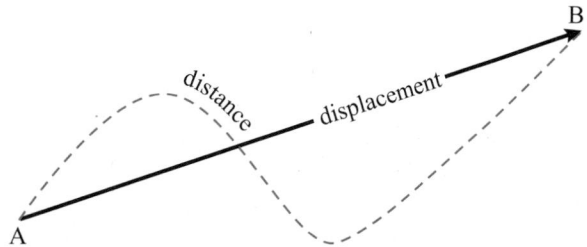

Figure 3-3 Distance (dashed line) and displacement (arrow from A to B)

point B along the path represented by the dashed line—the displacement is the shortest distance between the initial and final positions; however, the distance travelled is the total length of the path. Examples 3-1 and 3-2 highlight the difference between distance and displacement.

EXAMPLE 3-1

Distance and Displacement

A water taxi starts from a dock on the west bank of a channel, delivers passengers to a dock on the east bank, and then returns to its starting point on the west bank. The two docks are 230 m apart.

(a) Determine the displacement of the water taxi.
(b) Calculate the distance travelled by the taxi.

SOLUTION

(a) At the end of the round trip, the taxi ends up where it started, so its displacement is zero.
(b) The taxi covers 230 m each way, for a total distance travelled of 460 m.

Making sense of the result:

Displacement is not the same as distance travelled.

EXAMPLE 3-2

Calculating Distance and Displacement

A tarantula is on a wall 20 cm below a nail. It moves down to a point 60 cm below the nail to catch a fly, and then takes the fly straight up to a position 30 cm above the nail, as shown in Figure 3-4.

(a) Find the displacement of the tarantula.
(b) Find the distance covered by the tarantula.

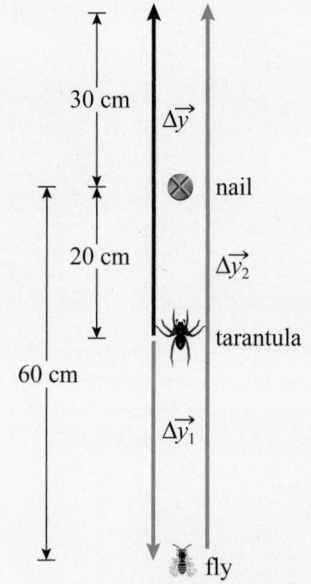

Figure 3-4 Displacement of the tarantula

SOLUTION

(a) We will use the nail as our origin. We define up to be the positive y direction. Therefore, y_1, the initial position of the tarantula as measured from the nail, is negative, and y_2, the final position of the tarantula, is positive. Therefore, the displacement is given by

$$\Delta \vec{y} = \vec{y}_2 - \vec{y}_1 = 30\hat{j}\ \text{cm} - (-20)\hat{j}\ \text{cm} = +50\hat{j}\ \text{cm}$$

(b) On the way down, the tarantula moves from an initial position of 20 cm below the nail to 60 cm below the nail. The distance covered on the way down equals the magnitude of the displacement:

$$\left| \Delta \vec{y}_{\text{down}} \right| = 40\ \text{cm}$$

On the way up, the tarantula moves from an initial position of 60 cm below the nail to a position of 30 cm above the nail, for a total upward distance of

$$\left| \Delta \vec{y}_{\text{up}} \right| = 90\ \text{cm}$$

The total distance covered by the tarantula is the sum of the two distances, 130 cm.

Making sense of the result:

For displacement all we care about are initial and final positions, the distance travelled takes into account the details of the path taken.

Position and displacement are defined as vector quantities: they have a magnitude and a direction. For one-dimensional motion, only two directions are possible. Since these two directions are opposite, we can define one of the directions as positive, and then indicate the direction of vector quantities by simply putting the appropriate positive or negative sign in front of the magnitude.

 CHECKPOINT

C-3-1 Distance and Displacement

When a Formula One race car drives a full lap around a circular track 210 m in radius,

(a) the displacement of the car is the circumference of the track;

(b) the distance travelled by the car is the circumference of the track;

(c) the distance travelled by the car is zero;

(d) the displacement and distance travelled are equal.

C-3-1 (b) displacement is the difference between final and initial positions, and is therefore zero.

 MAKING CONNECTIONS

Range of Distances in Nature and Everyday Life

Table 3-1 Range of Distances in Nature and Everyday Life

Distance	Description
1.6162×10^{-35} m	Planck length, the shortest distance that can theoretically be measured
0.8768 fm = 8.768×10^{-16} m	Radius of a nucleon
12.4 pm = 12.4×10^{-12} m	Upper limit for the wavelength of gamma rays
52.9 pm = 52.9×10^{-12} m	Radius of a hydrogen atom
~0.1 nm	Covalent bond length
~0.2 nm	Hydrogen bond length
0.1–10 nm or 0.1×10^{-9}–10×10^{-9} m	Wavelength of X-rays
580–590 nm	Wavelength of yellow light
30.48 cm	1 ft.
1.609 344 km	1 mile
1.852 km	1 nautical mile
79.8 m	Wingspan of Airbus 380-800
1737 km	Mean radius of Moon
6371 km	Mean radius of Earth
~35.4 solar radii	Radius of R136a1, the most massive known star
1800–2100 solar radii	Radius of VY Canis Majoris, possibly the largest known star,

(continued)

Distance	Description
384 400 km	Mean distance between Earth and the Moon
6.955×10^8 m	Radius of the Sun
149 597 871 km	1 astronomical unit, the mean Earth–Sun distance
4.545 billion km	Mean Neptune–Sun distance
$9.460\,528\,4 \times 10^{15}$ m	1 light year, the distance that light travels in one year
100 000 light years	Approximate diameter of the Milky Way galaxy
93 billion light years	Approximate diameter of the observable universe

LO 2

3-2 Velocity and Speed

The terms "velocity" and "speed" are often used inter-changeably in everyday life. However, in physics, **speed** is a scalar quantity and **velocity** is a vector. Speed tells us only how fast an object is moving; velocity also tells us the direction of the motion.

Average Velocity

For an object that moves from position $\vec{x_1}$ to position $\vec{x_2}$ during an amount of time Δt, the **average velocity** of the object is given by its displacement divided by the time it takes to complete the displacement:

$$\vec{v}_{avg} = \frac{\Delta \vec{x}}{\Delta t} = \frac{\vec{x_2} - \vec{x_1}}{\Delta t} \qquad (3\text{-}3)$$

The average velocity does not depend on the details of the journey between two points. A car driver might decide to drive along a highway at a given velocity for some time, stop for a coffee, and then continue to a destination, quite possibly at a different velocity, depending on traffic and road conditions. The total displacement divided by the total elapsed time, including the time for the coffee break, gives us the average velocity for the trip. For someone driving along a straight, flat highway, the average velocity between two points is represented by the slope of the line connecting the corresponding two points on the position-time plot; that is, the rise, Δx, divided by the run, Δt, as shown in Figure 3-5.

The average speed gives us more detail regarding how fast the car was going throughout the trip. **Average speed** is defined as the distance travelled divided by the elapsed time.

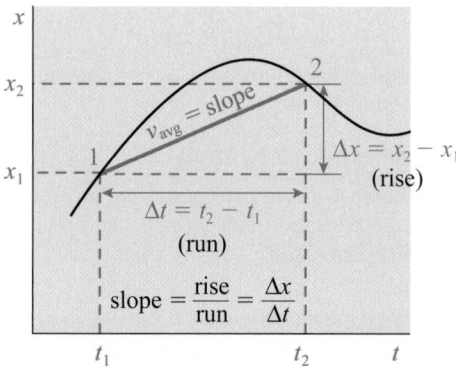

Figure 3-5 On a position versus time plot, the average velocity between two points is given by the slope (rise over run) of the line connecting the two points.

EXAMPLE 3-3

Average Velocity

A plane takes off from Prague in the Czech Republic and lands in Brussels, Belgium, 710 km northwest 1.20 h later. What is the average velocity of the plane?

SOLUTION

The average velocity is given by Equation (3-3):

$$\vec{v}_{avg} = \frac{\Delta \vec{x}}{\Delta t} = \frac{710\hat{i} \text{ km}}{1.20 \text{ h}} = 592\hat{i} \text{ km/h}$$

$$= 592\hat{i} \, \frac{\text{km}}{\text{h}} \left(\frac{1000 \text{ m/km}}{3600 \text{ s/h}} \right) = 164\hat{i} \text{ m/s}$$

where we have chosen the positive x-axis to be along the direction of motion of the plane.

Making sense of the result:

The magnitude of the average velocity is approximately 600 km/h, which is a reasonable speed for a passenger jet.

EXAMPLE 3-4

Speed and Velocity

A plane flies straight from Stockholm, Sweden, to arrive over Moscow, Russia, 1240 km southeast, 1.70 h later. The pilot is then told to return to Stockholm. The return trip to Stockholm takes 2.10 h, but the plane is then diverted to Oslo, Norway, 425 km to the northwest of Stockholm. The plane finally lands 0.900 h later. The three cities lie along a straight line, as shown in Figure 3-6.

(a) Find the average velocity of the plane.
(b) Find the average speed of the plane for the entire trip.

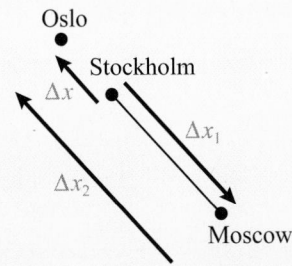

Figure 3-6 Flight path of the plane in Example 3-4

SOLUTION

(a) The average velocity depends on the net displacement and the total elapsed time:

$$\vec{v}_{avg} = \frac{\Delta \vec{x}}{\Delta t} = \frac{425\hat{i} \text{ km}}{1.70 \text{ h} + 2.10 \text{ h} + 0.900 \text{ h}} = 90.4\hat{i} \text{ km/h}$$

(b) The average speed is the total distance travelled divided by the total time:

$$v_{avg} = \frac{d}{\Delta t} = \frac{1240 \text{ km} + 1240 \text{ km} + 425 \text{ km}}{1.70 \text{ h} + 2.10 \text{ h} + 0.900 \text{ h}}$$

$$= \frac{2905 \text{ km}}{4.70 \text{ h}} = 618 \text{ km/h}$$

where we have chosen the positive x-axis to point from Moscow to Oslo.

Making sense of the result:

Unlike the average speed, which tells us how fast the plane was moving in the air, the average velocity depends only on the starting and ending positions.

✓ CHECKPOINT

C-3-2 Average Speed

You make a bungee jump off a 300 m-high bridge. After 20 s, you have fallen 270 m and bounced 230 m back up. What is your average speed?

(a) 2 m/s
(b) 3.5 m/s
(c) 25 m/s
(d) none of the above

C-3-2 (c) The distance covered is 270 m + 230 m = 500 m. Covering 500 m in 20 s gives an average speed of 25 m/s.

Instantaneous Velocity

If we want to know details of how an object's position is changing at a given location along its trajectory, we consider the **instantaneous velocity**, the velocity of the object at a given moment. To calculate the instantaneous velocity, we find the limit of the average velocity as the change in time goes to zero. In other words, the

instantaneous velocity is the derivative of the position with respect to time:

$$\vec{v} = \lim_{\Delta t \to 0} \vec{v}_{avg} = \lim_{\Delta t \to 0} \frac{\Delta \vec{x}}{\Delta t} = \frac{d\vec{x}}{dt} \qquad (3\text{-}4)$$

Similarly, **instantaneous speed** is the magnitude of the instantaneous velocity. On a plot of position versus time, the instantaneous speed corresponds to the slope of the tangent to the plot at that point (Figure 3-7).

Note, however, that the average speed is not necessarily equal to the magnitude of the average velocity as demonstrated by Example 3-4 above.

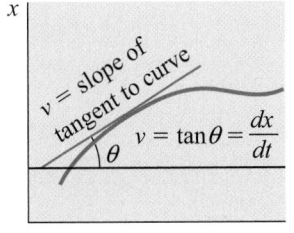

Figure 3-7 Instantaneous velocity is given by the slope of the tangent on the position versus time plot.

MAKING CONNECTIONS

Range of Speeds in Nature and Everyday Life

Table 3-2 Range of Speeds in Nature and Everyday Life

Speed	Description
~1 m in 10^{10} y	Theoretical rate of flow of glass at room temperature
~1 m in 800 y	Theoretical rate of flow of glass at 414°C
~1 cm/y	Speed of continental drift
0.2 cm/s	Speed of a small earth worm
1 cm/s	Speed of a molecule in a Bose–Einstein condensate (occurs at very low temperatures)
1.852 km/h	1 knot, the customary unit for the speed of ships
~10 m/s	Sprint speed in a 100 m race
41 km/h	Highest speed ever achieved by a sailing ship
52 km/h	Top running speed of a bear
82.78 km/h	Top speed of K222, an Anchar (Papa) class Soviet submarine
112 km/h	Top speed of cheetahs and sailfish
325 km/h	Top swooping speed of a peregrine falcon
360 km/h	Top speed of Formula One cars at Monza, Italy
431.07 km/h	Top speed of a Bugatti Veyron race car
574.8 km/h	Speed record for the V150 version of Alstom's TGV high-speed passenger train
581 km/h	Speed record for the MLX01 magnetic levitation train (Japan, 2003)
945 km/h	Maximum operating speed of an Airbus 380 at cruising altitude
330 m/s	Maximum rate of climb of a Mig-29 fighter plane
340 m/s	Speed of sound at sea level at 15°C
1040 km/h	Top speed of the Yokosuka MXY7 Ohka attack plane
1352 m/s	Root mean square speed of a helium atom at room temperature
465 m/s or 1674 km/h	Equatorial rotation speed of Earth
2172 km/h	Top speed of the Concorde supersonic jet at cruising altitude
Mach 2.8	Speed of the Brahmos antiship missile
3630 km/h	Maximum speed of a Mig 25 fighter plane
1 km/s or 3600 km/h	Speed of the Moon around Earth
2 km/s	Equatorial rotation speed of the Sun
Mach 7	Speed of the Brahmos II missile
7275 km/h	Top speed of the X-15, the fastest piloted rocket plane
12 144 km/h	Maximum speed of the X-43, an unpiloted experimental hypersonic jet
28 000 km/h	Speed of the space shuttle on reentry
16.26 km/s or 58 536 km/h	Speed of NASA's *New Horizons* spacecraft

Speed	Description
29.77 km/s	Speed of Earth around the Sun
220 km/s	Speed of the Sun around the galaxy centre
~600 km/s	Speed of the Milky Way galaxy (relative to cosmic microwave background)
2100 km/s	Equatorial rotation speed of the Crab Pulsar
3×10^8 m/s	Speed of light

 EXAMPLE 3-5

Velocity and Speed Plots

Figure 3-8(a) shows a position versus time graph for a car. Use the graph to plot the velocity vs. the time t, and the speed versus time for the time intervals shown.

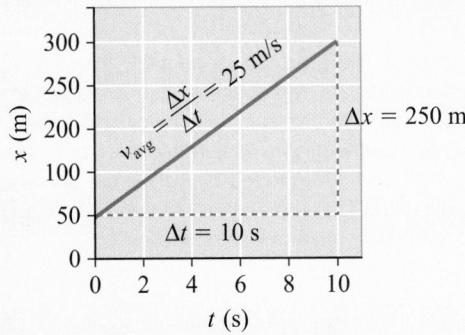

Figure 3-8(a) Position versus time for a car

SOLUTION

The slope of the position graph is constant, indicating that the velocity and the speed are constant. Hence, the instantaneous velocity is equal to the average velocity for the 10 s interval shown on the graph. Since the displacement is 250 m in the positive direction during this time,

$$|\vec{v}_{\text{avg}}| = \frac{\Delta x}{\Delta t} = \frac{250 \text{ m}}{10 \text{ s}} = 25 \text{ m/s}$$

A plot of the instantaneous velocity and speed is simply a horizontal line with a y-intercept of 25, as shown in Figure 3-8(b).

Figure 3-8(b) Velocity versus time graph for the car

Making sense of the result:

The x vs t plot is linear, which means the speed is constant.

 MAKING CONNECTIONS

Fastest Planes in World War II

Toward the end of World War II, the major powers on both sides began to produce jet- and rocket-powered warplanes. The German Messerschmitt Me262 was the first operational jet fighter (Figure 3-9). It had a maximum speed of 900 km/h. The Allies' first operational jet was the British Gloster Meteor, which had a top speed of 965 km/h. The fastest of these new aircraft was the Yokosuka MXY7 Ohka (Figure 3-10), used by Thunder God Corps of the Japanese air force. The Ohka was essentially a rocket-powered warhead designed for suicide attacks on Allied warships. When diving with its three rocket engines at full power, the Ohka could reach speeds as high as 1040 km/h. However, the speed of the MXY7 in level flight was 649 km/h, and it had a range of only 36 km.

© Airpix/Alamy

Figure 3-9 The Me 262 in the collection of the National Air and Space Museum, Steven F. Udvar-Hazy Center, Washington, D.C.

© Maurice Savage/Alamy

Figure 3-10 The Yokosuka MXY7 Ohka on display in the Air & Space Gallery, Museum of Science & Industry, Manchester, U.K.

EXAMPLE 3-6

Average and Instantaneous Velocity

An object moves in a straight line such that its position (in metres) is given by $x = 0.250\,t^3$, where t is measured in seconds.

(a) Plot the position versus time for the interval $t = 0$ s to $t = 2.00$ s.
(b) Find the average velocity during the interval $t = 1.00$ s to $t = 2$ s, and include the corresponding line on your graph.
(c) Find the average velocity during the interval 1.25 s to 1.75 s, and include the corresponding line on your graph.
(d) Find the instantaneous velocity at $t = 1.50$ s, and include the corresponding line on your graph.

SOLUTION

(a) To create the plot, we can pick convenient values along the t-axis, calculate the corresponding x-coordinates, and sketch a smooth curve from the resulting points, as shown in Figure 3-11.
(b) We can apply Equation (3-3) to find the average velocity between $t = 1$ s and $t = 2$ s:

$$v_{avg} = \frac{\Delta x}{\Delta t} = \frac{x_2 - x_1}{\Delta t} = \frac{0.25(2^3)\ \text{m} - 0.25(1^3)\ \text{m}}{2\ \text{s} - 1\ \text{s}} = +1.75\ \text{m/s}$$

This velocity is illustrated by an orange line on the graph. The average velocity between $t = 1$ s and $t = 2$ s

is actually the slope of the line connecting the points at $t = 1$ s and $t = 2$ s on the position curve.

(c) The average velocity between $t = 1.25$ s and 1.75 s is

$$v_{avg} = \frac{\Delta x}{\Delta t} = \frac{x_2 - x_1}{\Delta t} = \frac{0.25(1.75^3)\ \text{m} - 0.25(1.25^3)\ \text{m}}{1.75\ \text{s} - 1.25\ \text{s}}$$

$$= +1.70\ \text{m/s}$$

This velocity corresponds to the slope of the line connecting the points at $t = 1.25$ s and $t = 1.75$ s also shown in orange.

(d) To get the instantaneous velocity at $t = 1.50$ s, we use Equation (3-4):

$$v = \frac{dx}{dt} = \frac{d(0.25t^3)}{dt} = 3(0.25)t^2 = 0.75(1.5)^2 = +1.69\ \text{m/s}$$

This velocity corresponds to the slope of the tangent to the x versus t graph at $t = 1.5$ s shown in green.

Making sense of the result:

We see that as the time, Δt, separating the two points gets smaller, the line representing the average velocity approaches the instantaneous velocity given by the slope of the tangent to the graph, until in the limit as Δt goes to zero, the two points separated by Δt merge, and we get the tangent to the line at the point of interest, who's slope is given by the derivative of x as a function of t.

Figure 3-11 The closer that two points are on a position versus time plot, the closer the average velocity gets to the instantaneous velocity.

C-3-3 Average Velocity

The average velocity of an object

(a) must be greater than its instantaneous velocity;

(b) must be less than its instantaneous velocity;

(c) cannot be equal to its instantaneous velocity;

(d) none of the above.

C-3-3 (d) the average velocity can be greater or equal or less than the instantaneous velocity.

MAKING CONNECTIONS

Cosmic Cannonball

The Chandra X-ray Observatory detected a neutron star, RX J0822-4300, which is moving away from the centre of Pupis A, a supernova remnant about 7000 ly away (Figure 3-12). Believed to be propelled by the strength of the lopsided supernova explosion that created it, this neutron star is moving at a speed of about 4.8 million km/h, putting it among the fastest-moving stars ever observed. At this speed, its trajectory will take it out of the Milky Way galaxy in a few million years. Astronomers were able to estimate its speed by measuring its position over a period of 5 years.

Chandra: NASA/CXC/Middlebury College/F. Winkler; ROSAT: NASA/GSFC/S.Snowden et al.; Optical: NOAO/CTIO/Middlebury College/F. Winkler et al.

Figure 3-12 Supernova remnant RX J0822-4300

LO 3

3-3 Acceleration

The rate of change of the velocity with time is called **acceleration**. The **average acceleration** of an object is given by the change in its velocity divided by the time over which it changes:

$$\vec{a}_{\text{avg}} = \frac{\Delta \vec{v}}{\Delta t} = \frac{\vec{v}_2 - \vec{v}_1}{\Delta t} \qquad (3\text{-}5)$$

The average acceleration of an object between two points depends on only the final and initial velocities and the elapsed time. The details of the object's motion between the two points do not affect the average acceleration.

Taking the limit of the average acceleration as Δt goes to zero effectively gives us the **instantaneous acceleration** at a specific point. Thus, the instantaneous acceleration is the derivative of the velocity as a function of time:

$$\vec{a} = \lim_{\Delta t \to 0} \frac{\Delta \vec{v}}{\Delta t} = \frac{d\vec{v}}{dt} \qquad (3\text{-}6)$$

Since the instantaneous velocity is equal to the derivative of the position with respect to time, the instantaneous acceleration is equal to the *second* derivative of position with respect to time:

$$\vec{a} = \frac{d^2 \vec{x}}{dt^2} \qquad (3\text{-}7)$$

C-3-4 Acceleration

If the acceleration vector of a moving object points in the same direction as the motion but is decreasing in magnitude, the speed of the object

(a) is increasing;

(b) is decreasing;

(c) is constant;

(d) could be either increasing or decreasing, depending on the magnitude of the velocity.

C-3-4 (a) there is non zero forward acceleration, the object's speed will increase.

 ONLINE ACTIVITY

The e-resource that accompanies every new copy of this textbook contains an Online Activity using the PhET simulation "Moving Man." Work through the simulation and accompanying questions to gain an understanding of the relationships between the quantities that describe motion.

MECHANICS

MAKING CONNECTIONS

Acceleration and Safety Standards

Until the end of World War II, the prevalent belief among military aircraft engineers was that subjecting the human body to accelerations in excess of 18 g would be fatal. Therefore, airplane cockpits and seats did not have to be built to withstand accelerations greater than 18 g. However, research led by Colonel John Stapp (Figure 3-13(a)) of the U.S. Army Air Force showed that pilots could survive much greater accelerations if the seats and restraining harnesses could withstand the forces involved. The most notable tests used a rocket sled called the "Gee Whiz" (Figure 3-13(b)) to subject test dummies and human volunteers to accelerations comparable to those during a crash landing.

Stapp often rode the Gee Whiz himself and sustained broken bones and burst blood vessels in some of the tests. In 1954, he survived 46.2 g, likely still the world record for such tests. As a result of Stapp's research, new standards were adopted to require fighter jet seats to be able to withstand accelerations of 32 g. Stapp also developed improved harnesses for pilots and paratroopers, and advocated strongly for the installation of seat belts in cars. His work led to the 1966 law that made the provision of seat belts mandatory for cars sold in the United States.

(a)

(b)

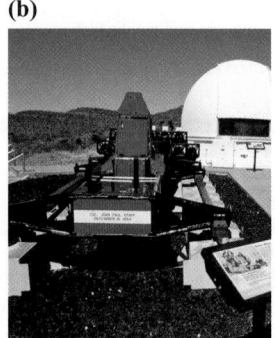

Figure 3-13 (a) Col. John P. Stapp secures Capt. Edward Magid into "the vertical accelerator." (b) The Gee Whiz, at the International Space Hall of Fame Museum, New Mexico

MAKING CONNECTIONS

Noteworthy Accelerations

Table 3-3 Noteworthy Accelerations

0.033 73 m/s^2	Radial acceleration at Earth's equator due to Earth's rotation
9.81 m/s^2	Typical acceleration due to gravity at Earth's surface at sea level (g)
1.7 g	Typical transverse acceleration of a Formula One car on a curve
~4 g	Acceleration experienced by astronauts during takeoff and reentry for both the space shuttle and the Soyuz
5.2 g	Typical acceleration of a luge sled at Whistler, B.C., during the Vancouver Olympics
5.3 g	Typical acceleration of a Top Fuel drag racer
6 g	Lateral acceleration, Formula One, Suzuka circuit; Maximum deceleration, Formula One
8–9 g	Tolerable sustained acceleration during inside loop for pilots
12 g	Threshold acceleration for sustained inside loop (vertical), with G-suit
16 g	Commercial aircraft seat standard as of 1988
18 g	Injury threshold for upward acceleration for a duration of 100 ms
20–25 g	Spinal injury threshold for upward acceleration for a duration of 100 ms
20 g	Injury threshold for lateral acceleration
32 g	Tolerance standard for U.S. military aircraft seats
~214 g	Highest acceleration survived during a crash
18 000 g	Approximate maximum acceleration of a golf ball in competition (Jason Zuback, 2009)
2 000 000 g	Radial acceleration in an ultracentrifuge
3.3 × 10^{10} g	Radial acceleration at the equator of the fastest known spinning neutron star[1]
~7 × 10^{11} g	Gravitational acceleration on the surface of a neutron star

[1]At 716 Hz (716 rotations per second), PSR J1748-2446ad is the fastest *confirmed* pulsar.

EXAMPLE 3-7

Using Derivatives to Calculate Acceleration

The position in metres of a particle on the x-axis is given by $x = -0.400t^4 + 4.00t^2 - 16.0t$, where time is measured in seconds. Find

(a) the velocity of the particle at $t = 2.00$ s;
(b) the acceleration of the particle at $t = 1.00$ s;
(c) the average acceleration of the particle between $t = 0.900$ s and $t = 1.10$ s;
(d) the maximum velocity the particle will reach.

SOLUTION

(a) To find the velocity from the position, we use Equation (3-4):

$$v = \frac{dx}{dt} = -4(0.4)t^3 + 2(4)t - 16 = -1.6t^3 + 8t - 16$$

At $t = 2$ s,

$$v = \frac{dx}{dt} = -1.6(2^3) + 8(2) - 16 = -12.8 \text{ m/s}$$

(b) To find the acceleration, we take the derivative of the velocity with respect to time:

$$a = \frac{dv}{dt} = \frac{d^2x}{dt^2} = -12(0.4)t^2 + 2(4) = -4.8t^2 + 8$$

At $t = 1$ s,

$$a = -4.8(1)^2 + 8 = +3.20 \text{ m/s}^2$$

(c) For the average acceleration, we use Equation (3-6):

$$a_{\text{avg}} = \frac{\Delta v}{\Delta t} = \frac{v_2 - v_1}{\Delta t}$$

Here, v_2 is the velocity at 1.1 s, and v_1 is the velocity at 0.9 s, so

$$v_2 = -1.6(1.1)^3 + 8(1.1) - 16 = -9.3296 \text{ m/s}$$

$$v_1 = -1.6(0.9)^3 + 8(0.9) - 16 = -9.9664 \text{ m/s}$$

$$a_{\text{avg}} = \frac{v_2 - v_1}{\Delta t} = \frac{-9.3296 - (-9.9664)}{0.2} = +3.18 \text{ m/s}^2$$

(d) The maximum (or minimum) speed occurs when the derivative of the velocity is zero, which occurs when the acceleration is zero:

$$a = -4.8t^2 + 8 = 0$$

$$\Rightarrow t = \sqrt{\frac{8}{4.8}} = 1.291 \text{ s}$$

At this time, the velocity is

$$v = -4(0.4)(1.291)^3 + 2(4)(1.291) - 16 = -9.115 \text{ m/s}$$

The maximum velocity is therefore -9.12 m/s.

Making sense of the result:

In the interim calculations we keep more significant digits than is required for the question. For parts b and c, since the time interval is so small, the average acceleration between $t = 0.9$ s and $t = 1.1$ s is very close to the instantaneous acceleration at $t = 1$ s. Note that the maximum velocity does not correspond to the maximum speed.

CHECKPOINT

C-3-5 Average Acceleration

An object is thrown straight down with a speed of 10 m/s. The object then bounces back up from the ground, reaching its initial location 4 s later with the same speed. The average acceleration of the object is

(a) 9.8 m/s²;
(b) zero;
(c) 5 m/s²;
(d) None of the above.

C-3-5 (c) the change in velocity is 20 m/s, the time elapsed is 4 s.

LO 4

3-4 Kinematics Equations

General Framework for Kinematics Equations

In the previous section, we obtained velocity and acceleration from position information for a given object. Conversely, if we know how the acceleration of an object varies with time, we can determine how the velocity and the position of the object change with time. If the acceleration is given by a continuous smooth function $a(t)$, then we can rearrange Equation (3-6) and integrate to find the velocity of the object:

$$d\vec{v} = \vec{a}\, dt \tag{3-8}$$

$$\Delta \vec{v} = \vec{v}(t) - \vec{v}_0 = \int_0^t \vec{a}\, dt \tag{3-9}$$

or

$$\vec{v}(t) = \vec{v}_0 + \int_0^t \vec{a}\, dt \tag{3-10}$$

where $\vec{v}_0$ is the velocity of the object at time $t = 0$, that is, the initial velocity of the object.

According to Equation (3-10), $\vec{v}(t)$, the velocity of the object as a function of time, is given by the vector sum of the initial velocity of the object and the change in the velocity due to the acceleration term.

Similarly, rearranging Equation (3-4) and integrating gives $\vec{x}(t)$, the position of the object as a function of time:

$$d\vec{x} = \vec{v}\, dt \tag{3-11}$$

$$\Delta x = \vec{x}(t) - \vec{x_0} = \int_0^t \vec{v}\, dt \qquad (3\text{-}12)$$

or

$$\vec{x}(t) = \vec{x_0} + \int_0^t \vec{v}\, dt \qquad (3\text{-}13)$$

where $\vec{x_0}$ is the initial position of the object.

EXAMPLE 3-8

Applying Kinematics Equations

A Formula One car is moving at a speed of 98.0 m/s along a straight track when the driver brings the car to a stop, gradually applying the brakes such that the acceleration increases linearly with time. The car takes 3.40 s to stop.

(a) Express the car's acceleration as a function of time.
(b) Calculate the speed of the car 3 s after the driver starts braking.
(c) Find the total distance covered by the car while slowing.
(d) Find the highest acceleration of the car.

SOLUTION

(a) Since the acceleration changes linearly with time and has an initial value of zero, we can write the acceleration as

$$a(t) = ct \qquad (1)$$

where c is the *rate of change* of the acceleration (often called *jerk* or *jolt*).
Substituting this acceleration function into Equation (3-10), we have

$$v(t) = v_0 + \int_0^t a(t)dt = v_0 + \frac{1}{2}ct^2 - \frac{1}{2}c(0)^2 = v_0 + \frac{1}{2}ct^2 \quad (2)$$

Choosing the direction of motion as positive, and seeing that the speed is zero at the end of 3.4 s,

$$0 = v_0 + \frac{1}{2}ct^2 \Rightarrow c = \frac{-2v_0}{t^2} = \frac{-2(98 \text{ m/s})}{(3.4 \text{ s})^2} = -16.96 \text{ m/s}^3 \quad (3)$$

Combining Equations (1) and (3):

$$a(t) = (-17.0 \text{ m/s}^3)t$$

to 3 significant figures.

(b) For the speed of the car 3 s into the acceleration, we use Equation (3):

$$v(t) = v_0 + \frac{1}{2}ct^2 = 98 \text{ m/s} + \frac{1}{2}(-16.96 \text{ m/s}^3)(3 \text{ s})^2 = +21.7 \text{ m/s}$$

(c) For the distance covered by the car, we use Equation (3-12):

$$\Delta x = x(t) - x_0 = \int_0^t v\, dt = \int_0^t \left(v_0 + \frac{1}{2}ct^2\right) dt$$

$$= (v_0 t - v_0(0)) + \left(\frac{1}{6}ct^3 - \frac{1}{6}c(0)\right) = v_0 t + \frac{1}{6}ct^3$$

taking the direction of motion to be positive,

$$\Delta x = v_0 t + \frac{1}{6}ct^3 = 98(3.4) + \frac{1}{6}(-16.96)(3.4)^3 = 222.1 \text{ m}$$

(d) From Equation (1), the highest acceleration is

$$a = -16.96 \text{ m/s}^3 \times 3.4 \text{ s}$$

$$= -57.7 \text{ m/s}^2$$

$$a = -5.88 \text{ g}$$

Making sense of the result:

The constant c has the units of m/s³, and the acceleration as a function of time is the product of that constant with time, leaving us with m/s² as the units for the acceleration, as expected. The initial speed of the car is about 353 km/h, so the fairly long stopping distance seems reasonable. The maximum acceleration is equal to 5.88 g, not unheard of in Formula One racing.

The body's tolerance to acceleration depends on the direction of the acceleration as well as the rate of change of the acceleration, the convention illustrated in Figure 3-14 is applied. This is discussed in more detail in the text online resources.

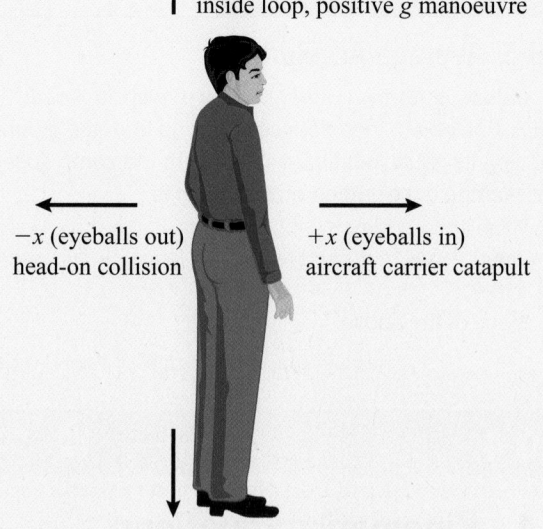

Figure 3-14 Conventions used for discussing the tolerance of the human body to acceleration

☑ CHECKPOINT

C-3-6 Acceleration-Time Graph

Figure 3-15 shows an acceleration versus time graph for an object. Positive is taken to be right. At $t = 3$ s, the object is

(a) moving to the right;
(b) moving to the left;
(c) not moving at all.
(d) Any of the above could be correct.

C-3-6 (d) Answer could be any of the above.

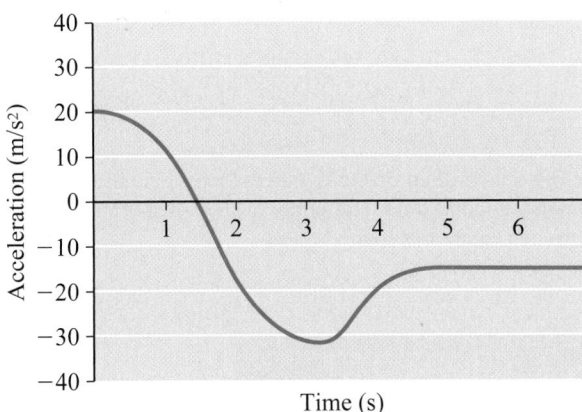

Figure 3-15 Acceleration-Time Graph

(continued)

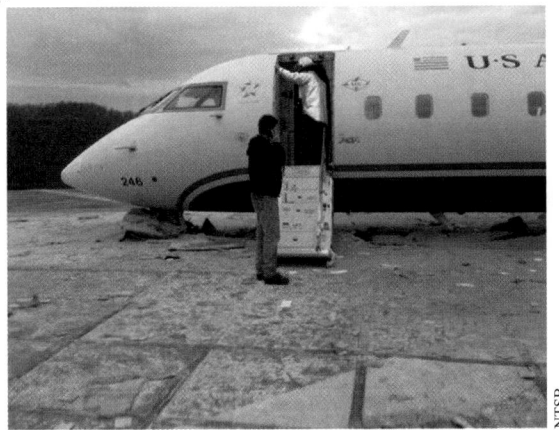

Figure 3-16 A PSA Bombardier CRJ200 jet stopped in the EMAS installation at Charleston, South Carolina, on January 19, 2010.

☑ CHECKPOINT

C-3-7 Velocity-Time Graph

Figure 3-17 shows a velocity versus time graph for a speedboat. For the time when the velocity curve is below the x-axis,

(a) the average speed is zero;
(b) the average acceleration is zero;
(c) the instantaneous acceleration is always negative.
(d) None of the above.

C-3-7 (b) change in velocity for interval is zero

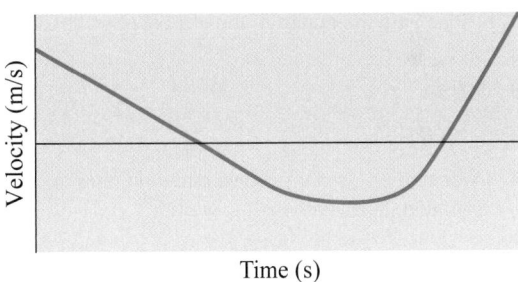

Figure 3-17 Velocity versus time graph for a speedboat

MAKING CONNECTIONS

EMAS Runway Extensions

Modern jet aircraft require longer runways than propeller aircraft. However, at some older airports, runway expansion is not feasible. Some of the older runways are long enough for large jets under normal conditions, but do not meet safety standards, which must allow for conditions such as high winds, storms, and brake failure. One solution used at some airports with serious space constraints is a runway extension made with an engineered material arresting system (EMAS), such as "foamcrete." These materials are designed to crumble under the weight of an aircraft to slow the plane down enough to prevent it from running off the end of the extension. In a number of incidents, EMAS materials have stopped aircraft from running off the end of a runway while causing few injuries to the passengers and limited damage to the aircraft. Figure 3-16 shows a Bombardier CRJ 200 that ran off a runway into an EMAS installation when the takeoff was aborted. None of the 34 occupants were injured.

(continued)

Kinematics Equations for Constant Acceleration

Although most situations in life do not involve constant acceleration, the acceleration we experience due to gravity is very nearly constant as long as we stay close to Earth's surface. When you let a ball drop, it falls with almost constant acceleration. Air drag can slow the acceleration, but this effect is negligible at slow speeds. The acceleration due to gravity at Earth's surface is about 9.81 m/s^2. The force of gravity is discussed in detail in Chapter 11.

When acceleration is constant, Equation (3-10) becomes

$$\vec{v}(t) = \vec{v_0} + \int_{t_0}^{t} \vec{a}\, dt = \vec{a}(t - t_0) + \vec{v_0} \qquad (3\text{-}14)$$

If we start at $t = 0$, Equation (3-14) can be written as follows.

KEY EQUATION First time-dependent kinematics equation:

$$\vec{v}(t) = \vec{a}t + \vec{v}_0 \qquad (3\text{-}15)$$

Combining Equations (3-13) and (3-15) gives us the position as a function of time:

KEY EQUATION
$$\vec{x}(t) = \vec{x}_0 + \int_{t_0}^{t} \vec{v}\,dt = \vec{x}_0 + \int_{t_0}^{t} (\vec{a}t + \vec{v}_0)\,dt$$
$$(3\text{-}16)$$

or

KEY EQUATION Second time-dependent kinematics equation:

$$\vec{x}(t) = \frac{1}{2}\vec{a}t^2 + \vec{v}_0 t + \vec{x}_0 \qquad (3\text{-}17)$$

which can also be written as

$$\vec{x}(t) - \vec{x}_0 = \frac{1}{2}\vec{a}t^2 + \vec{v}_0 t \qquad (3\text{-}18)$$

or

$$\Delta\vec{x} = \frac{1}{2}\vec{a}t^2 + \vec{v}_0 t \qquad (3\text{-}19)$$

 EXAMPLE 3-9

Gravitational Acceleration

A child throws a marble straight up in the air with a speed of 21.0 m/s. Find the speed of the marble after

(a) 1.00 s;
(b) 3.00 s.

SOLUTION

(a) Choosing up as our positive direction, and applying Equation (3-15) with $t = 1$ s, we have

$$\vec{v}(t) = \vec{a}t + \vec{v}_0 = 21\hat{j}\ \text{m/s} + (-9.81\hat{j}\ \text{m/s}^2)(1\ \text{s}) = +11.2\hat{j}\ \text{m/s}$$

(b) Similarly, after 3 s,

$$\vec{v}(t) = 21\hat{j}\ \text{m/s} + (-9.81\hat{j}\ \text{m/s}^2)(3\ \text{s}) = -8.43\hat{j}\ \text{m/s}$$

Making sense of the result:

Every second the velocity changes by 9.81 m/s. the difference in speed between $t = 1$s and $t = 3$s should be 19.62 m/s, which it is, within the accuracy of the data.

 EXAMPLE 3-10

Speed of a Falling Object

You drop a ball from a point 16 m above the ground. Find the speed of the ball just before it hits the ground.

SOLUTION

We know the acceleration, the initial speed (zero), and the displacement but not the elapsed time. The time it takes for an object to fall to the ground from rest depends on the height from which it falls. Choosing up as the positive direction and applying Equation (3-19), we have

$$\Delta\vec{y} = \frac{1}{2}\vec{a}t^2 + \vec{v}_0 t$$

$$-16.0\hat{j}\ \text{m} = \frac{1}{2}(-9.81\hat{j}\ \text{m/s}^2)t^2$$

$$t = \sqrt{\frac{2 \times 16.0\ \text{m}}{9.81\ \text{m/s}^2}} = 1.81\ \text{s}$$

Now that we know the elapsed time, we can apply Equation (3-15):

$$\vec{v}(t) = \vec{v}_0 + \vec{a}t = 0 - (9.81\hat{i}\ \text{m/s}^2)(1.81\ \text{s}) = -17.8\hat{i}\ \text{m/s}$$

Making sense of the result:

The speed of an object in free fall increases by 9.81 m/s every second. After just under 2 s, we expect the speed to be slightly less than 19.6 m/s.

✓ CHECKPOINT

C-3-8 Relative Speed

You throw two balls from the top of a building at the same time: a red one with a speed of 7 m/s straight up, and a grey one with a speed of 14 m/s straight down. After 2 s,
(a) the relative speed of the two balls has increased;
(b) the relative speed of the two balls has decreased;
(c) the relative speed of the two balls has not changed;
(d) the grey ball is accelerating faster than the red ball.

C-3-8 (c) the two objects have the same acceleration and hence zero relative acceleration.

In Example 3-10, we found the time it took the ball to fall first, then we applied Equation (3-15). Since the question did not explicitly ask about time, one wonders if there is a way to find how fast the object is falling without having to calculate the elapsed time. We do know that the greater the distance an object falls, the higher its speed at impact will be.

In problem 71 you will be asked to substitute Equation (3-15) into Equation (3-19) in order to obtain the third kinematics equation. The third kinematics equation is time-independent and relates displacement to velocity and acceleration:

KEY EQUATION Third kinematics equation:

$$v^2 - v_0^2 = 2\vec{a} \cdot \Delta\vec{x} \qquad (3\text{-}20)$$

where v is the final speed of the object.

Note that the term on the right side of Equation (3-20) is a scalar product. Unlike the two time-dependent kinematics, Equations (3-15) and (3-19), the third kinematics equation is not a vector equation. In problem 119, you will be asked to prove that Equation (3-20) holds for nonconstant acceleration as well.

The last three examples each applied one of the kinematics equations. Next, we look at more involved examples.

EXAMPLE 3-11

Launching Lunch

A construction worker standing on a beam asks his friend on the ground 9.00 m below to throw his lunch bag up to him. The friend wants to throw the bag straight up such that it has a speed of 0.500 m/s when it reaches the worker on the beam. At what speed should the friend throw the lunch bag?

SOLUTION

Here, we know the distance covered, the acceleration, and the final speed. We need to find the initial speed. We can apply Equation (3-20) here. Note that the acceleration and the initial velocity have opposite directions.

$$v^2 - v_0^2 = 2\vec{a} \cdot \Delta\vec{y} = |\vec{a}||\Delta\vec{y}| \sin(180°)$$

$$(0.50 \text{ m/s})^2 - v_0^2 = 2(9.81 \text{ m/s}^2)(9 \text{ m})(-1)$$

$$v_0 = \sqrt{(0.5 \text{ m/s})^2 + 2(9.81 \text{ m/s}^2)(9 \text{ m})} = 13.3 \text{ m/s}$$

where we have used Equation 2-18 for the scalar product of $\vec{a}$ and $\Delta\vec{y}$.

Making sense of the result:

The minus sign in the brackets indicates that the acceleration is opposite to the displacement.

EXAMPLE 3-12

Falling Firecrackers

A boy reaches over the railing around the top of a building and throws a firecracker straight up in the air, with a speed of 17.0 m/s.

(a) Find the time it takes for the firecracker to reach the ground 24.0 m below.
(b) Find the speed of the firecracker just before it touches the ground.
(c) How would the answer to (b) change if the boy threw the firecracker straight down at the same speed of 17.0 m/s?

SOLUTION

(a) We know the displacement, the acceleration, and the initial velocity of the firecracker, and we need to find the elapsed time. Therefore, Equation (3-19) is ideal for this part.

$$\Delta\vec{y} = \frac{1}{2}\vec{a}t^2 + \vec{v}_0 t$$

Choosing up as the positive direction, we can write

$$-24\hat{j} \text{ m} = \frac{1}{2}(-9.81 \text{ m/s}^2)t^2 + (17 \text{ m/s})t$$

We now have a quadratic equation for t. We rearrange this equation in the form $ax^2 + bx + c = 0$, and then use the formula for the roots of a quadratic equation:

$$4.905t^2 - 17t - 24 = 0$$

$$t = \frac{-b \pm \sqrt{b^2 - 4ac}}{2a} \quad \text{where } a = 4.905, b = 17, \text{ and } c = 24$$

the roots are:

$$t = 4.54 \text{ s} \quad \text{or} \quad t = -1.08 \text{ s}$$

We discard the negative root, so the firecracker takes 4.54 s to reach the ground.

(b) We have the initial speed, the displacement, and the acceleration. So we can use either Equation (3-15) or Equation (3-20) to find the speed of the firecracker before it hits the ground. Applying Equation (3-20) and noting that the acceleration and the displacement are both directed downward,

$$v^2 - v_0^2 = 2\vec{a} \cdot \Delta\vec{y}$$

$$v^2 = v_0^2 + 2\vec{a} \cdot \Delta\vec{y} = (17 \text{ m/s})^2 + 2(9.81 \text{ m/s}^2)(24 \text{ m})(1)$$

$$v = \sqrt{17.0^2 + 2(9.81)(24.0)(1)} = 27.6 \text{ m/s}$$

where we have used Equation 2-18 for the scalar product of $\vec{a}$ and $\Delta\vec{y}$.

(c) The expression for v^2 above depends on the square of the initial speed, and not on the initial velocity. Provided the initial speed is the same, the final speed will not be affected by whether the firecracker is thrown straight up or straight down.

Making sense of the result:

If you throw the firecracker upward, it will reach a maximum height and then come back down toward you. When it reaches your position again, it will have the same speed downward as its initial speed upward (by conservation of energy). It then has the same motion as if it had been thrown downward at the same speed.

CHAPTER 3 | MOTION IN ONE DIMENSION

EXAMPLE 3-13

Rocket Sled

Figure 3-18 shows an acceleration versus time graph for a rocket sled equipped with thrusters on both ends. Take right to be the positive x-direction. The magnitude of the acceleration is shown in the figure and the sled is slowing down at $t = 0$.

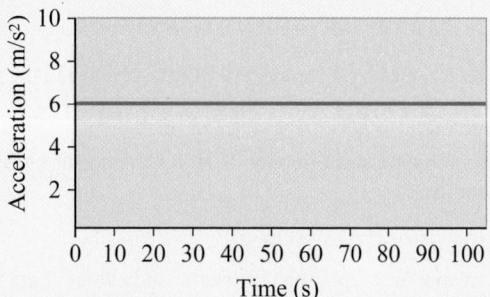

Figure 3-18 Acceleration versus time graph for a rocket sled

(a) Plot the velocity versus time graph until the sled comes to a stop for the first time, knowing that at $t = 0$ the velocity was 250 m/s to the right.
(b) For what value of t will $v = 0$?
(c) Plot the position versus time graph for the rocket sled as measured from the observaion tower, knowing that at $t = 0$ the sled was 700 m to the left of the tower. Your plot should show the points where $x = 0$ and $v = 0$.

SOLUTION

(a) The graph shows that the acceleration is constant at 6 m/s². Since the acceleration is constant, we can use Equation (3-15) to find an expression for the velocity as a function of time:

$$\vec{v}(t) = \vec{v}_0 + \vec{a}t = +250\hat{i} \text{ m/s} + (-6\hat{i} \text{ m/s}^2)t \quad (1)$$

The velocity versus time plot is linear, as shown in Figure 3-19.

Figure 3-19 Velocity versus time plot

(b) The change in velocity is the area under the acceleration curve. For v to be zero, that area must be 250 m/s. Using Equation (3-16), we then have

$$at = 250 \Rightarrow t = \frac{250 \text{ m/s}}{6 \text{ m/s}^2}$$

$$t = 41.7 \text{ s} \quad (2)$$

we report this as 41.7 s but keep more digits for the next part.

(c) The dependence of the position on time is given by Equation (3-17):

$$\vec{x}(t) = \frac{1}{2}\vec{a}t^2 + \vec{v}_0 t + \vec{x}_0 \quad (3)$$

or

$$\vec{x}(t) = \frac{1}{2}(-6\hat{i} \text{ m/s}^2)t^2 + (250\hat{i} \text{ m/s})t - 700\hat{i} \text{ m} \quad (4)$$

The graph of this equation is a parabola. At $t = 0$, $x = -700$ m. Let us pick a few more points to plot the position. The position x will be zero when the expression in Equation (4) is 0. The roots of this quadratic are given by

$$t = \frac{-b \pm \sqrt{b^2 - 4ac}}{2a} = \frac{-250 \pm \sqrt{(+250)^2 - 4(3)(700)}}{-6} \quad (5)$$

So, $t = 2.90$ s or 80.4 s when $x = 0$.

Since the velocity is the derivative of the position, a maximum or minimum for x will occur at the point of zero velocity. From part (b), the velocity is zero when $t = 41.7$ s. At this time,

$$\vec{x}(t) = ((-3 \text{ m/s}^2)(41.67 \text{ s})^2 + (250 \text{ m/s})(41.67 \text{ s}) - 700 \text{ m})\hat{i}$$

$$= 4510\hat{i} \text{ m} \quad (6)$$

A plot of $x(t)$ is shown in Figure 3-20.

Figure 3-20 Position versus time for the rocket sled

Making sense of the result:

The rocket sled will go past the control tower at $t = 2.90$ s. The speed of the sled will continue to drop until it stops momentarily at $t = 41.7$ s. When the sled stops, it will be 4510 m to the right of the tower. The sled will then reverse direction, reaching the tower again at $t = 80.4$ s.

LO 5

3-5 Relative Motion in One Dimension

When you are driving, trees and other objects beside the road appear to be moving past you. Although these objects are not moving relative to the ground, they do have motion relative to you. Similarly, oncoming traffic seems to be moving toward you at a much faster speed. In this section, we introduce methods for describing such **relative motion**. First, a note about notation in relation to relative motion: we use double subscripts to indicate relative velocities. For example, $\vec{v}_{1g}$ is the velocity of object 1 with respect to the ground, and $\vec{v}_{12}$ is the velocity of object 1 with respect to object 2.

Relative Velocity

In general, the velocity of object 1 with respect to object 2 is given by

$$\vec{v}_{12} = \vec{v}_{1g} - \vec{v}_{2g} \qquad (3\text{-}21)$$

Here, we have used the ground as our frame of reference, but we could choose any reference frame instead.

EXAMPLE 3-14

Relative Velocities

You are driving east on the Trans-Siberia highway at 100 km/h. Take east to be the positive x-direction.

(a) Express the velocity of a tree on the side of the highway with respect to you, using double-subscript notation.

(b) Find the speed of a tree on the side of the highway with respect to you.

(c) Find your velocity with respect to the tree.

(d) Find the velocity, with respect to you, of a bus driving west at 115 km/h. Find the velocity of your car with respect to the bus.

SOLUTION

We will use the subscripts c, t, and b for your car, the tree, and the bus, respectively. We choose east as the positive x-direction.

(a) For the velocity of a tree relative to your car, we use the relative velocity Equation (3-21):

$$\vec{v}_{tc} = \vec{v}_{tg} - \vec{v}_{cg} = 0 - 100\hat{i} \text{ km/h} = -100\hat{i} \text{ km/h}$$

(b) The speed of the tree relative to you equals the magnitude of the relative velocity found in part (a): 100 km/h.

(c) The velocity of your car with respect to the tree is

$$\vec{v}_{ct} = \vec{v}_{cg} - \vec{v}_{tg} = 100\hat{i} \text{ km/h} - 0 = 100\hat{i} \text{ km/h} \qquad (3)$$

(d) The velocity, with respect to you, of the bus moving in the opposite direction is

$$\vec{v}_{bc} = \vec{v}_{bg} - \vec{v}_{cg} = -115\hat{i} \text{ km/h} - (+100\hat{i} \text{ km/h})$$
$$= -215\hat{i} \text{ km/h}$$

Similarly, your velocity with respect to the bus is

$$\vec{v}_{cb} = \vec{v}_{cg} - \vec{v}_{bg} = 100\hat{i} \text{ km/h} - (-115\hat{i} \text{ km/h}) = 215\hat{i} \text{ km/h}$$

Making sense of the result:

Vehicles coming in the direction opposite to yours often seem to be moving twice as fast.

✓ CHECKPOINT

C-3-9 Relative Velocity

You are in the free-fall stage of a skydive when a skydiver below you opens her parachute. A few seconds later, she passes you at a relative velocity of 50 km/h upward. She has a velocity with respect to the air of 80 km/h downward. What is your velocity with respect to the air?

(a) 30 km/h down;

(b) 30 km/h up;

(c) 130 km/h down;

(d) 130 km/h up.

C-3-9 (c) your velocity is 50 km/h down with respect to her, her velocity is 80 km/h relative to the air.

Relative Acceleration and Position

The equation for the relative velocity of objects is easily adapted to apply to other vector quantities that describe motion. We simply differentiate Equation (3-21) to find relative acceleration (see problem 96):

$$\vec{a}_{12} = \vec{a}_{1g} - \vec{a}_{2g} \qquad (3\text{-}22)$$

Similarly, integrating Equation (3-21) gives relative position (see problem 96):

$$\vec{x}_{12} = \vec{x}_{1g} - \vec{x}_{2g} \qquad (3\text{-}23)$$

For displacement, we have

$$\Delta \vec{x}_{12} = \Delta \vec{x}_{1g} - \Delta \vec{x}_{2g} \qquad (3\text{-}24)$$

CHECKPOINT

C-3-10 Acceleration and Relative Velocity

You throw a ball straight down with a speed of 20 m/s, wait 1.0 s, and throw another ball down with a speed of 20 m/s. Use down as the positive direction. The velocity of the second ball relative to the first ball will

(a) always be −9.8 m/s;
(b) always be +9.8 m/s;
(c) change with time.
(d) None of the above.

C-3-10 (a) after 1s the speed of first ball has increased by 9.8 m/s. Then both speeds increase at same rate.

 EXAMPLE 3-15

Relative Position

At the annual auto show in Frankfurt, Germany, an Aston Martin is 7.0 m to left of a post and a Maserati is 12 m to the left of the same post. Find the position of the Aston Martin relative to the Maserati.

SOLUTION

We use Equation (3-23) and choose right as the positive x-direction. We will use the post as our origin; therefore, left of the post is the negative direction. Both x_A and x_M are negative because they are both left of the origin (Figure 3-21).

$$\vec{x}_{AM} = \vec{x}_A - \vec{x}_M$$

Here we drop the subscript g for simplicity as both positions are taken with respect to ground by default.

$$\vec{x}_{AM} = -7.0\hat{i}\ m - (-12\hat{i}\ m) = 5.0\hat{i}\ m$$

Figure 3-21 The position of one car relative to the other

Making sense of the result:

The Aston Martin is to the right of the Maserati, hence its position relative to the Maserati should be positive.

Derivation of the General Kinematics Equations for Relative Motion

We now derive the general kinematics equations for one object with respect to another when both objects have constant accelerations. Here again we drop the subscript g, as all quantities are taken relative to ground.

Relative Position The position of object 1 with respect to object 2 is given by Equation (3-23):

$$\vec{x}_{12} = \vec{x}_1 - \vec{x}_2$$

The position of each object as a function of time is given by Equation (3-17):

$$\vec{x}(t) = \frac{1}{2}\vec{a}t^2 + \vec{v}_0 t + \vec{x}_0$$

Combining these two equations, we get the following:

$$\vec{x}_{12}(t) = \frac{1}{2}\vec{a}_1 t^2 + \vec{v}_{01} t + \vec{x}_{01} - \left(\frac{1}{2}\vec{a}_2 t^2 + \vec{v}_{02} t + \vec{x}_{02}\right)$$

$$\vec{x}_{12}(t) = \left(\frac{1}{2}\vec{a}_1 - \frac{1}{2}\vec{a}_2\right)t^2 + (\vec{v}_{01} - \vec{v}_{02})t + (\vec{x}_{01} - \vec{x}_{02})$$

$$\vec{x}_{12}(t) = \frac{1}{2}\vec{a}_{12}t^2 + \vec{v}_{012}t + \vec{x}_{012} \qquad (3\text{-}25)$$

Relative Velocity We begin with Equation (3-21):

$$\vec{v}_{12} = \vec{v}_1 - \vec{v}_2$$

 EXAMPLE 3-16

Relative Acceleration

You are driving a pickup truck when the driver of a motorcycle in front of you suddenly brakes to slow down for a bump on the road. Being a smaller vehicle with good brakes, the motorcycle can slow down at an acceleration of 19 m/s². Your heavy pickup truck slows at a rate of 12 m/s².

(a) What is the acceleration of the motorcycle with respect to the pickup truck?
(b) What is the acceleration of the pickup truck with respect to the motorcycle?

SOLUTION

For this example, we use Equation (3-22), and choose the direction of motion as the positive x-direction. Therefore, both accelerations are negative. We use the subscript p for the pickup truck and m for the motorcycle.

(a) The acceleration of the motorcycle with respect to the pickup truck is

$$\vec{a}_{mp} = \vec{a}_{mg} - \vec{a}_{pg} = -19\hat{i}\ m/s^2 - (-12\hat{i}\ m/s^2) = -7.0\hat{i}\ m/s^2$$

(b) The acceleration of the pickup truck with respect to the motorcycle is

$$\vec{a}_{pm} = \vec{a}_{pg} - \vec{a}_{mg} = -12\hat{i}\ m/s^2 - (-19\hat{i}\ m/s^2) = +7.0\hat{i}\ m/s^2$$

Making sense of the result:

Since the motorcycle is slowing down faster than the pickup truck, from the point of view of the pickup truck the motorcycle is accelerating toward it. The acceleration of the motorcycle relative to the truck will point towards the truck which is the negative direction as per our choice of coordinate system.

Then using Equation (3-15) for the velocity of an object moving with constant acceleration, we have

$$\vec{v}_{12}(t) = \vec{a}_1 t + \vec{v}_{01} - (\vec{a}_2 t + \vec{v}_{02})$$

$$\vec{v}_{12}(t) = (\vec{a}_1 - \vec{a}_2)t + \vec{v}_{01} - \vec{v}_{02}$$

$$\vec{v}_{12}(t) = \vec{a}_{12} t + \vec{v}_{012} \qquad (3\text{-}26)$$

EXAMPLE 3-17

Relative Motion with Constant Acceleration

At the same time, you throw a red ball straight up with a speed of 21.0 m/s and a blue ball straight down with a speed of 17.0 m/s. Find the acceleration, speed, and position of the blue ball with respect to the red ball 3.00 s later. Assume that air resistance is negligible.

SOLUTION

Once thrown, both balls have the same constant acceleration of 9.81 m/s².

We calculate the relative speed using Equation (3-26). We use the subscript r for the red ball and the subscript b for the blue ball:

$$\vec{v}_{br}(t) = \vec{a}_{br} t + \vec{v}_{0br} \qquad (1)$$

Using Equations (3-21) and (3-22), we can rewrite (1) as

$$\vec{v}_{br} = (\vec{a}_b t - \vec{a}_r t) + (\vec{v}_{0b} - \vec{v}_{0r}) = \vec{v}_{0b} - \vec{v}_{0r} \qquad (2)$$

The terms in the first parentheses cancel because the two balls have the same constant acceleration. Consequently, the relative velocity of the two balls is constant because it depends on only their initial velocities. Taking down to be the positive y-direction, we can write

We leave it as an exercise in problem 72 to apply relative motion notation to the time-independent kinematics Equation (3-20) to show that

$$v_{12}^2 - v_{012}^2 = 2\vec{a}_{12} \cdot \Delta \vec{x}_{12} \qquad (3\text{-}27)$$

$$v_{br} = v_{0b} - (-v_{0r}) = v_{0b} + v_{0r} = 17 \text{ m/s} + 21 \text{ m/s} = 38.0 \text{ m/s} \quad (3)$$

where the variables in Equation (3) denote magnitdues of the quantities in Equation (2)

For relative position we use Equation (3-24):

$$\vec{y}_{12}(t) = \frac{1}{2}\vec{a}_{12} t^2 + \vec{v}_{012} t + \vec{y}_{012} \qquad (4)$$

We use Equations (3-21) and (3-23) to rewrite (4) as

$$\vec{y}_{br} = \left(\frac{1}{2}\vec{a}_b - \frac{1}{2}\vec{a}_r\right)t^2 + (\vec{v}_{0b} - \vec{v}_{0r})t + (\vec{y}_{0b} - \vec{y}_{0r})$$

$$= (\vec{v}_{0b} - \vec{v}_{0r})t \qquad (5)$$

Again, the acceleration terms cancel out. The last two terms also cancel because the initial position of the two balls is the same. Since we have chosen down to be positive,

$$y_{br} = (v_{0b} - (-v_{0r}))t = (17 \text{ m/s} + 21 \text{ m/s})3 \text{ s} = 114 \text{ m} \qquad (6)$$

where the variables in Equation (6) denote magnitudes of the vectors in Equation (5)

Making sense of the result:

Since gravity contributes to the change in velocity and position of both objects identically, the acceleration due to gravity makes no contribution to the relative quantities.

EXAMPLE 3-18

Applying the Kinematics Equations to Relative Motion

A Mercedes Formula One car is moving at a speed of 330 km/h, 50.0 m directly behind a Renault moving at a speed of 220 km/h. Preparing for a sharp turn ahead, the Renault driver slows down with an acceleration 2.62 times the acceleration due to gravity.

(a) Find the minimum acceleration that the Mercedes must have to avoid running into the Renault. Assume that the Mercedes brakes at the same time as the Renault, and assume constant acceleration for both cars.

(b) How long does it take the Mercedes to reach the Renault if the Mercedes does brake with the minimum acceleration needed to avoid a collision? How far did each car move during this time?

SOLUTION

Since both cars are accelerating, it could be quite involved to calculate the displacement and velocity of each car, and then compare them. However, the solution is much easier if we think in terms of the relative motion of the two cars.

(a) To avoid just colliding with the Renault, the Mercedes must reach the same speed as the Renault a moment before catching up to it. In other words, the relative speed of the two cars must be zero by the time the Mercedes closes the distance between them.

We choose the direction of motion as the positive x-direction, which makes the accelerations for both cars negative. Since the Mercedes is initially moving faster than the Renault, the Mercedes will have to slow more rapidly (i.e., have greater acceleration in the negative direction) than the Renault in order to match

(continued)

CHAPTER 3 | **MOTION IN ONE DIMENSION**

its speed. Therefore, the relative acceleration of the Mercedes, with respect to the Renault, $\vec{a}_{MR} = \vec{a}_M - \vec{a}_R$, also points in the negative direction.

The initial speed of the Mercedes with respect to the Renault is

$$\vec{v}_{0MR} = \vec{v}_{0M} - \vec{v}_{0R} = 330\hat{i} \text{ km/h} - 220\hat{i} \text{ km/h}$$

$$= 110\hat{i} \text{ km/h} = 110\hat{i} \text{ km/h} \times \frac{1000 \text{ m}}{3600 \text{ s}} = 30.56\hat{i} \text{ m/s}$$

From the point of view of the Renault, the Mercedes is getting closer. Hence the direction of the displacement of the Mercedes relative to the Renault, $\Delta \vec{x}_{MR}$, as per our choice of coorindate system, is positive.

Applying the third kinematics equation for relative motion, Equation (3-27), we have

$$v_{MR}^2 - v_{0MR}^2 = 2\vec{a}_{MR} \cdot \Delta \vec{x}_{MR}$$

where the subscript MR refers to the Mercedes with respect to the Renault.

The two vectors in this equation have opposite directions, so using Equation 2-18

$$v_{MR}^2 - v_{0MR}^2 = 2a_{MR}\Delta x_{MR}(\cos(180°)) = 2a_{MR}\Delta x_{MR}(-1)$$

$$\Rightarrow 0 - (30.56 \text{ m/s})^2 = 2a_{MR}(50 \text{ m})(-1)$$

$$a_{MR} = 9.336 \text{ m/s}^2 = 0.952g$$

$$a_M - a_R = 0.952g \Rightarrow a_M = a_R + 0.952g$$

$$= 2.62g + 0.952g = 3.57g$$

(b) To find the time it takes for the relative velocity to be zero, we use Equation (3-26):

$$\vec{v}_{MR}(t) = \vec{a}_{MR}t + \vec{v}_{0MR}$$

$$0 = (\vec{a}_M - \vec{a}_R)t + \vec{v}_{0M} - \vec{v}_{0R}$$

With our choice of positive direction, both accelerations are negative and both initial velocities are positive. In the approach we take below, a_M and a_R are introduced as variables that denote the magnitudes of the accelerations. This approach is more adequate for the level of complexity of the problem. Therefore,

$$(-a_M - (-a_R))t\hat{i} + (v_{0M} - v_{0R})\hat{i} = 0$$

Rearranging and using the values calculated in part (a), we have

$$t = \frac{v_{0R} - v_{0M}}{a_R - a_M} = \frac{-30.56 \text{ m/s}}{-9.336 \text{ m/s}^2} = 3.273 \text{ s}$$

During that time, the Mercedes would have moved by

$$\Delta \vec{x} = \frac{1}{2}\vec{a}t^2 + \vec{v}_0 t$$

$$= \frac{1}{2}(-35.039\hat{i} \text{ m/s}^2)(3.273 \text{ s})^2 + (91.667\hat{i} \text{ m/s})(3.273 \text{ s})$$

$$= 112.348\hat{i} \text{ m} = 112\hat{i} \text{ m}$$

to 3 significant figures. For the Renault,

$$\Delta \vec{x} = \frac{1}{2}\vec{a}t^2 + \vec{v}_0 t$$

$$= \frac{1}{2}(-25.702\hat{i} \text{ m/s}^2)(3.273 \text{ s})^2 + (61.111\hat{i} \text{ m/s})(3.273 \text{ s})$$

$$= 62.349\hat{i} \text{ m} = 62.4\hat{i} \text{ m/s}$$

to 3 significant figures.

Making sense of the result:

The distance covered by the Mercedes should be 50 m longer than the distance covered by the Renault. The result does agree to within the accuracy of the given data.

Although the relative motion Equations (3-21) to (3-24) have been developed using one-dimensional examples, these vector equations are readily applied to two- and three-dimensional situations, as we will see in Section 4-4.

✓ CHECKPOINT

C-3-11 Relative Acceleration

You are slowing your car, which at a given instant is moving at 34 m/s with an acceleration of 23 m/s². Behind you is a truck moving in the same direction as you at a speed of 43 m/s and slowing down with an acceleration of 16 m/s². At that instant, the truck's acceleration relative to you is

(a) parallel to the truck's relative displacement with respect to you;
(b) antiparallel to the truck's relative position with respect to you at that moment;
(c) parallel to the truck's relative velocity with respect to you;
(d) antiparallel to the truck's relative velocity with respect to you.

C-3-11 (c) truck has larger speed than you, and a smaller acceleration than you in opposite direction to motion. Look at equations, draw diagrams if in doubt.

PUTTING IT ALL TOGETHER

Kinematics is the study of motion. In kinematics, we study the relationships between an object's position, displacement, velocity, and acceleration and their dependence on time. We also examine relative motion.

Position, Velocity, Acceleration, and Time

The displacement of an object is the change in its position vector:

$$\Delta \vec{x} = \vec{x}_2 - \vec{x}_1 \tag{3-1}$$

Distance is the path length covered by an object:

$$d = \int_{x_1}^{x_2} |d\vec{x}| \tag{3-2}$$

The average velocity is the displacement divided by the time elapsed:

$$\vec{v}_{\text{avg}} = \frac{\Delta \vec{x}}{\Delta t} = \frac{\vec{x}_2 - \vec{x}_1}{\Delta t} \tag{3-3}$$

The average speed is the distance covered divided by the time elapsed The instantaneous velocity is the derivative of the position with respect to time:

$$\vec{v} = \lim_{\Delta t \to 0} \frac{\Delta \vec{x}}{\Delta t} = \frac{d\vec{x}}{dt} \tag{3-4}$$

The instantaneous speed is the magnitude of the instantaneous velocity.

The average acceleration is the change in the velocity divided by the time elapsed:

$$\vec{a}_{\text{avg}} = \frac{\Delta \vec{v}}{\Delta t} = \frac{\vec{v}_2 - \vec{v}_1}{\Delta t} \tag{3-5}$$

The instantaneous acceleration is the derivative of velocity with respect to time:

$$\vec{a} = \frac{d\vec{v}}{dt} = \frac{d^2\vec{x}}{dt^2} \tag{3-6}, \tag{3-7}$$

Kinematics Equations

The velocity of an object can be obtained by integrating the acceleration with respect to time:

$$\vec{v}(t) = \vec{v}_0 + \int_{t_0}^{t} \vec{a}\, dt \tag{3-9}$$

The position (and displacement) of an object can be obtained by integrating velocity with respect to time:

$$\vec{x}(t) = \vec{x}_0 + \int_{t_0}^{t} \vec{v}\, dt \tag{3-13}$$

For constant acceleration, the velocity of an object as a function of time is given by

$$\vec{v}(t) = \vec{a}t + \vec{v}_0 \tag{3-15}$$

The position of an object moving with constant acceleration is given by

$$\vec{x}(t) = \frac{1}{2}\vec{a}t^2 + \vec{v}_0 t + \vec{x}_0 \tag{3-17}$$

The time-independent kinematics equation is given by

$$v^2 - v_0^2 = 2\vec{a} \cdot \Delta \vec{x} \tag{3-20}$$

Relative Motion

The relative velocity of object 1 with respect to object 2 is

$$\vec{v}_{12} = \vec{v}_{1g} - \vec{v}_{2g} \tag{3-21}$$

The relative acceleration is given by

$$\vec{a}_{12} = \vec{a}_{1g} - \vec{a}_{2g} \tag{3-22}$$

The relative displacement is given by

$$\Delta \vec{x}_{12} = \Delta \vec{x}_{1g} - \Delta \vec{x}_{2g} \tag{3-24}$$

The relative position is expressed as

$$\vec{x}_{12} = \vec{x}_1 - \vec{x}_2$$

APPLICATIONS

Applications: airline safety, motor vehicle safety, travel times, runway construction, relative motion

Key Terms: acceleration, average acceleration, average speed, average velocity, displacement, instantaneous acceleration, instantaneous speed, instantaneous velocity, kinematics, position, relative motion, speed, velocity

QUESTIONS

1. Figure 3-22 shows a plot of the position versus time for two objects.
   (a) Which object has the higher average velocity for the interval shown?
   (b) Which object has the higher speed at $t = 30$ s? Which has the higher speed at $t = 90$ s?
   (c) Do the objects ever have the same speed? If so, at what time do you estimate this will happen?

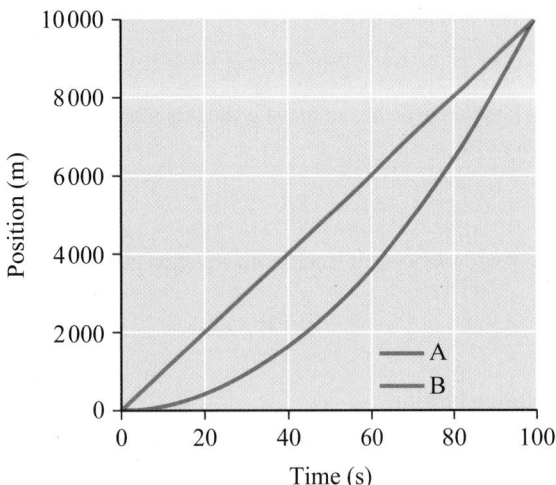

Figure 3-22 Question 1

2. Figure 3-23 shows a plot of the position versus time for two objects, A and B.
   (a) Which object has the higher average speed for the interval shown?
   (b) Estimate the average speeds of object A and object B for the interval shown.
   (c) At the points indicated by the arrows, do the objects have the same speed? The same acceleration?
   (d) Which object has the higher speed at the position indicated by arrow 1? Is the speed of object A ever higher than the speed of object B?
   (e) Which object has the higher speed at the position indicated by arrow 2?

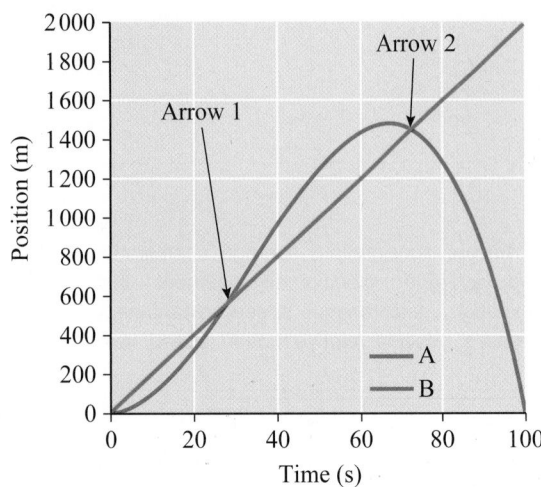

Figure 3-23 Question 2

3. A plot of velocity versus time for two objects is shown in Figure 3-24.
   (a) Starting at the same point, are the two objects ever at the same position again?
   (b) Do the two objects have the same speed? If so, when does that occur?
   (c) Which object has the higher acceleration for the interval shown? Estimate the magnitude of the maximum acceleration for each object.

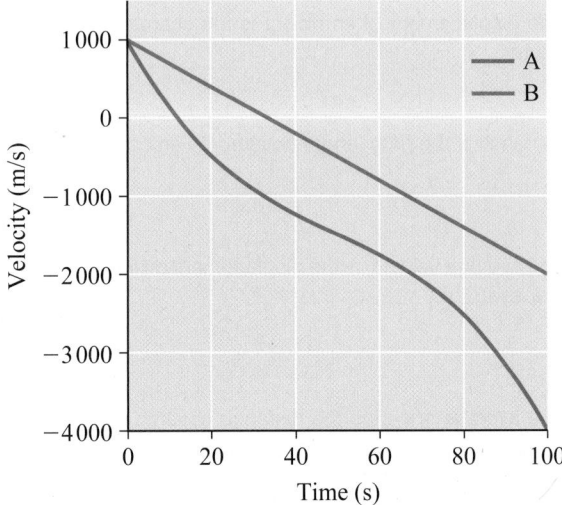

Figure 3-24 Question 3

4. Figure 3-25 shows a plot of the speed versus time for two boats.
   (a) For the interval shown, which boat covers more distance?
   (b) Do the boats ever have the same acceleration?

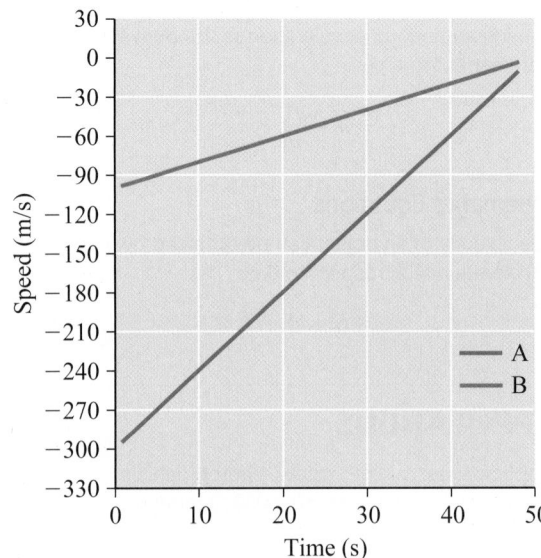

Figure 3-25 Question 4

5. Figure 3-26 shows a velocity versus time plot for two objects; they are at the same point at $t = 0$.
   (a) For the interval shown, which object has the higher average acceleration?

(b) Which object covers more distance?

(c) Which object has the higher acceleration at the point of intersection for the two curves?

(d) Which object is ahead of the other at the point of intersection of the two curves?

(e) Do the objects ever have the same displacement and/or acceleration? If so, where does that happen?

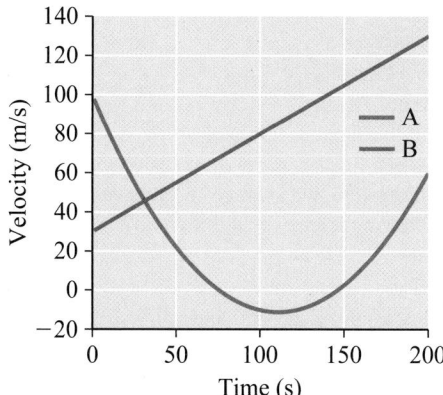

Figure 3-26 Question 5

6. Figure 3-27 show a plot of an object's position versus time.
   (a) Where does this object have its highest speed?
   (b) What is the object's lowest speed?
   (c) Infer where the object's acceleration is zero.
   (d) Is the object ever at rest?
   (e) Does the object cover more distance between $t = 0$ and $t = 140$ s than it does between $t = 140$ s and $t = 200$ s? Explain your reasoning.

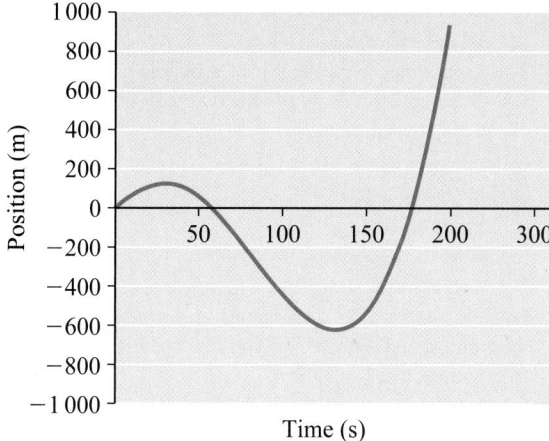

Figure 3-27 Question 6

7. The velocity versus time graph for an object is shown in Figure 3-28.
   (a) Is the acceleration of this object ever zero?
   (b) Where is this object's magnitude of acceleration the highest? The lowest?

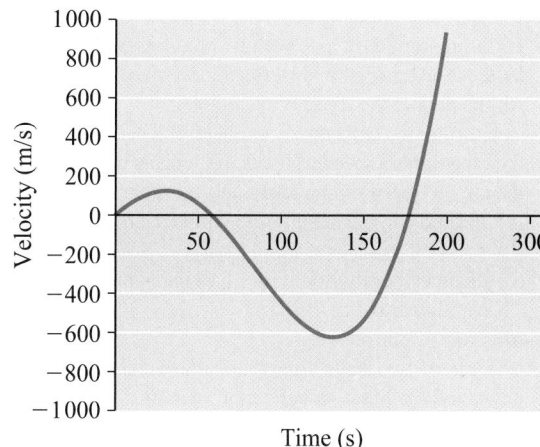

Figure 3-28 Question 7

8. Two balls are thrown simultaneously, one up and one down, both with an initial speed of 27 m/s. They pass each other 3 s after they are thrown. What is the speed of one ball with respect to the other ball as they pass each other?
   (a) 27 m/s
   (b) 54 m/s
   (c) It depends on the distance the balls travel.
   (d) None of the above.

9. You drop a tennis ball from a hot-air balloon that is going upward at a speed of 4 m/s. Relative to ground, the ball
   (a) falls straight down;
   (b) goes up a short distance, stops, and then goes down;
   (c) hangs in the air for a short time, then goes down.
   (d) None of the above.

10. You are standing in an open construction elevator at a site. The elevator is going up at a constant speed v. You throw a stone up at speed $2v$ with respect to the elevator, while your friend on the ground throws a stone up at speed $2v$ relative to ground. Who will get their stone back first?
    (a) you
    (b) your friend
    (c) You both get them back at the same time.
    (d) The answer cannot be determined without a value for v.

11. You throw two stones at a speed of 24 m/s: a small one straight up and a larger one at an angle of 42° above the horizontal. The stones are launched from and land at the same height. Which one lands with the greater speed? Ignore air resistance.
    (a) the small one
    (b) the large one
    (c) The speeds will be the same.
    (d) The answer cannot be determined from the given data.

12. You are piloting a plane at a speed of 980 km/h relative to the ground. Another plane has a speed of zero with respect to you. Relative to the ground, the other plane is
    (a) stationary;
    (b) moving at 980 km/h in the opposite direction;
    (c) moving at 980 km/h in some arbitrary direction.
    (d) None of the above.

13. When a Formula One race car completes two laps around a track,
    (a) the distance covered by the car is zero;
    (b) the car's displacement is zero;
    (c) the car's displacement is twice the circumference of the track.
    (d) None of the above.

14. Evaluate this statement: When you throw two balls from a 10 m-high building, one with twice the speed of the other, by the time they get to the ground, the speed of the one will still be double the speed of the other.

15. You throw two balls down, one with a speed of 10 m/s and the other with a speed of 6 m/s. As long as both balls are in the air, the speed of one relative to the other
    (a) depends on the height from which they fall;
    (b) stays the same as it was initially;
    (c) changes because the faster ball gains more speed than the slower ball.
    (d) None of the above.

16. You throw two balls from the same height, one straight down and the other straight up at the same speed *v*. Which of the following are correct?
    (a) As long as both balls are in the air, the ball you throw straight down will have twice the speed of the ball you throw straight up.
    (b) As long as both balls are in the air, the relative speed of the two balls is 2*v*.
    (c) The balls will hit the floor with the same speed.
    (d) Both (b) and (c) are correct.

17. If a rocket sled has an acceleration to the right three times the acceleration due to gravity,
    (a) the rocket sled must be moving;
    (b) the rocket sled must be moving to the left;
    (c) the rocket sled is stationary.
    (d) All the above are possible.

18. An object moves with an acceleration that increases in magnitude. If the object is not stationary, its speed
    (a) will increase;
    (b) will decrease;
    (c) can be zero.
    (d) All the above are possible.

19. Can an object's average acceleration be equal to its instantaneous acceleration? Explain.

20. Can an object's average speed be equal to its instantaneous velocity? Explain.

21. Can an object's instantaneous velocity be equal to its average velocity? Explain.

22. When an object moves to the right and its acceleration vector points to the right, then
    (a) the object's speed will decrease;
    (b) the object's speed will increase.
    (c) Both (a) and (b) are possible.
    (d) None of the above.

23. An object moves to the right with increasing speed. Define left to be the positive direction. The acceleration of the object
    (a) is negative;
    (b) is positive;
    (c) could be either negative or positive.
    (d) None of the above.

24. If an object's velocity increases, then
    (a) its speed must increase;
    (b) its speed must decrease.
    (c) Both (a) and (b) are possible.
    (d) None of the above.

25. An object moves to the right, but its acceleration vector is directed to the left. Define left to be the positive direction. Which of the following is true?
    (a) The object's velocity increases.
    (b) The object's velocity decreases.

(c) The object's speed increases.
(d) More one than of the above is possible.

26. Can your speed be zero when your velocity is nonzero? Explain.

27. Can your speed be zero when your acceleration is nonzero? Explain.

28. Can your acceleration be zero when your speed is nonzero? Explain.

29. When an object's instantaneous velocity is zero, its acceleration must be
    (a) positive;
    (b) negative;
    (c) zero.
    (d) All the above are possible.

30. A person's displacement over a given interval of time is zero. Therefore, the average speed must be zero. Which of the following is true about this claim?
    (a) true
    (b) false
    (c) It depends on the start and finish speeds.
    (d) It depends on the distance that the person moved during the interval.

31. You start from rest at point A, then go to point B, where you stop. Which of the following is true?
    (a) Your average speed is zero.
    (b) Your average velocity is zero.
    (c) Your average displacement is zero.
    (d) None of the above are true.

32. If your net displacement is zero, then
    (a) the distance you travel must be zero;
    (b) your average speed must be zero;
    (c) your average velocity is zero.
    (d) All the above are possible.

33. If your average speed over a given interval is zero, then
    (a) you did not move;
    (b) you returned to the same place;
    (c) your displacement can be nonzero.
    (d) All the above are possible.

34. You drop a ball from the top of a building. Then 3 s later you drop another ball. While the two balls are in the air, the distance between them
    (a) increases as the square of *t*;
    (b) increases linearly with *t*;
    (c) stays constant.
    (d) None of the above.

35. You throw a ball straight down with a speed of 20 m/s. It bounces off the floor and comes up to the same height at the same speed with which you launched it. Which of the following is true?
    (a) The average acceleration is 9.8 m/s.
    (b) The average speed is 20 m/s.
    (c) The average velocity is zero.
    (d) None of the above are true.

36. Two friends start from points that are 30 m apart and change positions. It takes them 6 s to do so. Which of the following is true?
    (a) Their relative displacement is zero.
    (b) Their average relative velocity is zero.
    (c) Their average relative velocity is 10 m/s.
    (d) Their relative displacement is 30 m.

PROBLEMS BY SECTION

For problems, star ratings will be used, (✱, ✱✱, or ✱✱✱), with more stars meaning more challenging problems.

Section 3-1 Distance and Displacement

37. ✱ You walk from your house 1100 m down the street, then you reverse direction and walk 1300 m up the street.
(a) What is the distance you covered?
(b) What is your total displacement? Choose down the street as the positive direction.

38. ✱ Starting from a trail 3 m above water level, you climb straight up a cliff to a point 11 m above the water. Then you dive into the water and climb back to the trail. Find your displacement as well as the distance you covered.

39. ✱ A grasshopper jumps straight up from a leaf that is 80 cm above the ground, reaches a maximum height of 75 cm above the leaf, and lands on the ground. Find the distance covered by the grasshopper as well as its displacement.

Section 3-2 Velocity and Speed

40. ✱ A barge takes a load of processed fish from a fishing boat to the harbour 3.2 km away, and then doubles back to the fishing boat along the same direction. In the meantime, the fishing boat has moved 2.0 km from the dock. The whole journey takes 20 min. Find
(a) the displacement of the barge;
(b) the average velocity of the barge;
(c) the average velocity of the fishing boat;
(d) the average speed of the barge;
(e) the average speed of the fishing boat.

41. ✱ On a distant planet, a ball is dropped from rest from a height of 21 m. It hits the ground and bounces back up until it is 1 m below its original position. The trip down and back up takes 10 s. Find the ball's average speed, displacement, and average velocity.

42. ✱✱ The sailfish has a maximum speed of 112 km/h. A sailfish moves in a straight line at a maximum speed for 3.00 s, slows down, and then comes to a stop 2.00 s later, having covered a total distance of 125 m.
(a) Find the acceleration during the last 2 s, assuming constant acceleration.
(b) Find the average acceleration during the entire 5 s.
(c) Find the average velocity during the entire 5 s.
(d) The sailfish now takes 8.00 s to return to its starting point, accelerating to its maximum speed as it does so. Find its average speed and average velocity for the round trip.

43. ✱✱ A stone is released at the top of a 1.2 m-tall barrel of liquid. The stone reaches a speed of 1.7 m/s just before it hits the bottom of the barrel 0.90 s later. Assuming constant acceleration, find the stone's average speed and average velocity.

44. ✱✱ Find the magnitude of the average velocity of Earth with respect to the Sun over one month. Also find the average speed and average acceleration. Assume that Earth's orbit is circular.

45. ✱ The maximum speed of an Airbus 380 at cruising altitude is 945 km/h. How long would this plane take to go from Paris to New York at that speed?

46. ✱ The fastest known rotation of a neutron star is an equatorial speed that is 0.24 of the speed of light. How long is a "day" on this star? Assume a radius of 16 km.

47. ✱ The distance between the Sun and Earth is one astronomical unit, 149 597 871 km. Calculate the speed of Earth

around the Sun, assuming that Earth has a nearly circular orbit and that the Sun is at the centre of Earth's orbit.

48. ✱ The speed of the Moon around Earth is approximately 1 km/s. Calculate the distance between the Moon and Earth.

49. ✱ Lightning flashes in your neighbourhood, and 3.00 s later you hear thunder. How far away was the lightning flash? Assume the speed of sound to be 345 m/s.

50. ✱ The speed of the space shuttle in low Earth orbit is about 28 000 km/h. When the altitude of the shuttle was 300 km, how long did it take to orbit Earth once?

51. ✱ A rocket car covers a distance of 10 km in 1.0 min, turns around, which takes 5.0 min, and goes back halfway to its starting position in 1.5 min.
(a) Find the displacement of the rocket car.
(b) Find the distance covered by the rocket car.
(c) Find (i) the average velocity and (ii) the average speed of the rocket car.

52. ✱ The maximum speed of the Concorde at cruising altitude was 2172 km/h. How long would it take the Concorde to travel from London, England, to Montréal, Québec, at that speed?

53. ✱✱ Figure 3-29 shows a velocity versus time plot for a motorcycle.
(a) Find the displacement of the motorcycle between $t = 0$ s and $t = 3$ s, between $t = 3$ s and $t = 7$ s, and between $t = 7$ s and $t = 9$ s.
(b) What is the acceleration of the motorcycle between $t = 2$ s and $t = 3$ s?

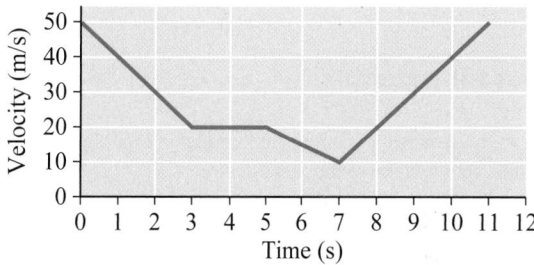

Figure 3-29 Question 53

54. ✱✱ Figure 3-30 shows a position versus time plot for a speedboat.

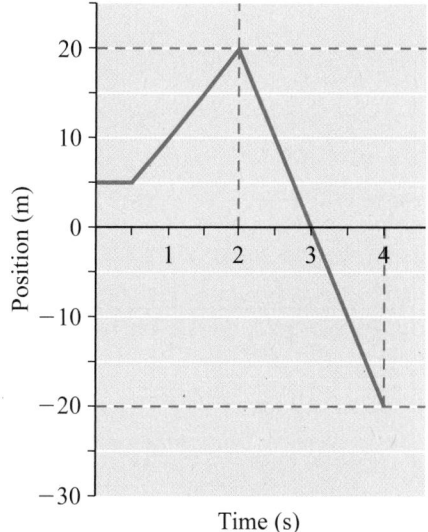

Figure 3-30 Problem 54

(a) Find the average speed of the speedboat over the entire interval shown.

(b) Find the average velocity of the speedboat over the entire interval.

(c) Find the velocity of the speedboat between $t = 1$ s and $t = 2$ s.

55. ✳ (a) How long does light take to reach us from the Sun?

(b) Roughly how long does it take light to reach us from the star Sirius?

(c) How long would it take light to cross the diameter of the Milky Way galaxy?

Section 3-3 Acceleration

56. ✳ The space shuttle had a takeoff acceleration approximately 3.5 times the acceleration due to gravity. What was the speed of the space shuttle 100 s after takeoff when it maintained that acceleration?

57. ✳✳ Figure 3-31 shows a position versus time plot for a particle. Find the speed and velocity of the particle.

Figure 3-31 Problem 57

58. ✳✳ A sunspot is on the Sun's equator. Find the velocity of this spot relative to the Sun's centre.

59. ✳✳ Starting with a speed of 4 m/s, a stone falls toward a container of molasses, reaching a speed of 21 m/s by the time it hits the surface of the molasses. Over 2 s, the molasses slows the stone down to a speed of 4 m/s, its terminal velocity. Find the average speed of the stone during this fall.

60. ✳✳ Figure 3-32 shows a velocity versus time plot for a car.

(a) Find the average acceleration of the car over the first 4 s.

(b) Find the acceleration at $t = 11$ s.

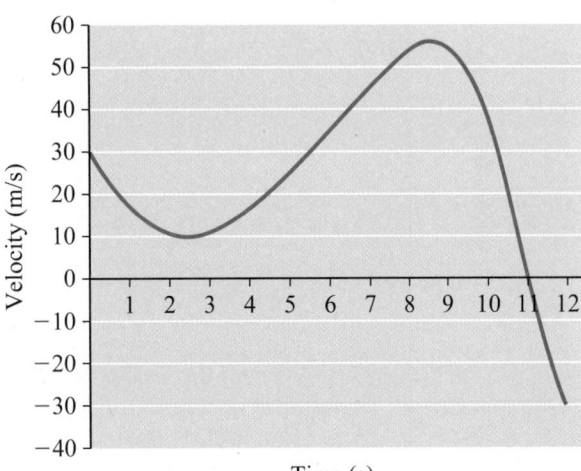

Figure 3-32 Problem 60

(c) Find the average acceleration of the car over the entire interval shown.

(d) Estimate the distance covered by the car between $t = 0$ s and $t = 12$ s.

61. ✳✳✳ Figure 3-33 shows an acceleration versus time plot for a particle undergoing a form of cyclic motion called simple harmonic motion. The particle is at $x = 0$ when $t = 0$.

(a) Find the average speed of the particle between $t = 2$ s and $t = 3$ s.

(b) Find the speed of the particle at $t = 3$ s.

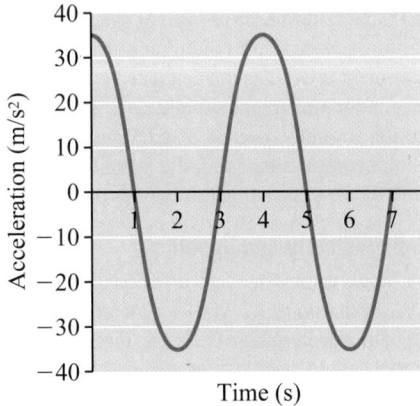

Figure 3-33 Problem 61

62. ✳ Using SI base units, a particle's position as a function of time is given by $x(t) = 7t^3 - 4t^2 + 6$. At what time will the particle's acceleration be zero? Find the speed of the particle at that time.

Section 3-4 Kinematics Equations

For this section some data is given as obtained from NASA. Convert all units to SI units and report your answers in SI units. mph = miles per hour ft/s = feet/s.

63. ✳✳ During the second-stage burn of an *Apollo* launch, the rocket's speed increased from 7882.9 ft./s to 21 377.0 ft./s. This burn lasted 384.22 s. Assuming constant acceleration and a straight path, find

(a) the average acceleration

(b) the distance travelled during this second-stage burn.

64. ✳✳ Use Equations (3-15) and (3-19) to derive Equation (3-20).

65. ✳✳ Show that Equation (3-27) is valid.

66. ✳✳✳ You throw two balls down from a height of 90 m, 3.0 s apart. You throw the first ball at a speed of 8.0 m/s. At what speed must you throw the second ball if it is to reach the ground

(a) with the same speed as the first ball;

(b) at the same time as the first ball?

67. ✳✳ You throw a stone straight down at a speed of 8.0 m/s from a height of 70 m. What is the average acceleration of the stone while it is in the air?

68. ✳ Typically, a golf club is in contact with a golf ball for about 0.5 ms. Jason Zuback is reported to have hit a golf ball with a speed of 357 km/h. Assuming constant acceleration, estimate the acceleration of Zuback's golf ball and the distance his club travelled while in contact with the ball.

69. ✱ A Formula One car has an acceleration 1.45 times the acceleration due to gravity. Find the time the car takes to go from 0 to 100 km/h, assuming constant acceleration.

70. ✱✱ You throw a marble at a speed of 10 m/s straight down from the top of a 30 m-tall building, and follow it with a second marble 0.50 s later. At what speed must you throw the second marble so that the two marbles arrive at the same time?

71. ✱✱ Your throw a ball straight up in the air with an initial speed of 32 m/s, and catch it at the moment it has the same speed with which you launched it. Find the ball's average acceleration, average speed, and average velocity.

72. ✱✱ A peregrine falcon flies in a straight line at 78 km/h for 3.0 s and spots a mouse on the ground 310 m below. The falcon dives at an angle of 30° to the vertical, reaching a maximum speed of 325 km/h in 4.0 s. The falcon maintains that speed as it makes a circular arc of radius 73 m to fly horizontally for another 2.0 s at 1.0 m above the ground. Find the average speed and the magnitude of the average velocity of the falcon during this flight.

73. ✱✱ The Bugatti Veyron can go from 0 to 100 km/h in 2.46 s (Figure 3-34). It takes another 7.34 s to reach the speed 240 km/h. Assume constant acceleration.
   (a) Calculate the acceleration of the car over the first 2.46 s.
   (b) How much distance does the Veyron have to cover to reach a speed of 240 km/h?
   (c) The speed of the car is limited to 415 km/h to protect the tires. When the car maintains the same acceleration that it had at 240 km/h, how long would it take the car to reach 415 km/h?

© culture-images GmbH/Alamy

Figure 3-34 Problem 73

74. ✱✱ The greatest acceleration experienced by Col. Stapp on the Gee Whiz rocket sled was 46.2 g as the sled braked from a speed of 632 mph. Over what distance would the sled come to a stop if the acceleration was constant? How long would the sled take to come to a stop?

75. ✱✱ You throw a ball up straight in the air, and your friend, who is 14 m above you, catches it when its speed is 3 m/s on the way back down.
   (a) How long does it take for the ball to pass your friend on the way up?
   (b) What is the average speed of the ball between the time you let it go and the time your friend catches it?
   (c) What is the average speed of the ball from launch to maximum height, and what is that maximum height?

76. ✱✱ Engineered material arresting systems (EMASs) are sometimes used to stop airplanes when they overshoot a runway. Consider a worst-case scenario in which an Airbus 380 moving at its full landing speed of 250 km/h hits the EMAS. What is the minimum stopping distance that will keep the acceleration below the 16 g rating for the seats on the plane?

77. ✱✱ Crumple zones in cars increase the time and the distance over which a car stops during a collision. The length of one crumple zone is 1.6 m. Find the maximum speed that a car could have when hitting a wall head-on without subjecting the passenger to an acceleration of more than 46.2 g. How long would that collision last?

78. ✱✱ Kenny Brack survived a 214 g car crash during an Indy car race in Fort Worth, Texas, in 2003. His speed is reported to have been 354 km/h when he crashed into a wall. If his final speed was zero, how long did the crash last and over what distance did he stop? Assume constant acceleration during the crash.

79. ✱ Some ejection seats have an acceleration of 14 g upward. How fast would the pilot be going relative to the plane at the end of a 1.3 s ejection burst?

80. ✱✱ The Peugeot EX1 electric sports car set a few speed records in 2010. Starting from rest, it covered the first 201.2 m (1/8 mile) in 8.89 s.
   (a) Assuming constant acceleration, find the time the EX1 takes to go from 0 to 100 km/h.
   (b) The car took a total of 16.81 s to travel 500 m starting from rest. Find the average acceleration for the period after it covered the first 1/8 mile.

81. ✱✱ A rocket is launched such that for the first 40.0 s its acceleration is upward at 5.00 g. For the following 25.0 s, its acceleration is upward at 3.50 g. The final stage lasts 90.0 s with an acceleration of 3.20 g.
   (a) What is the speed of the rocket at the end of the second stage ($t = 65.0$ s)?
   (b) What is the total distance covered by the rocket in the third stage?
   (c) What is the total distance covered by the rocket during the three stages?

82. ✱✱ A lunar lander 200 m above the surface of the Moon has a downward speed of 50 m/s. Find the constant acceleration required for the lander to arrive at the surface with a speed of 1.1 m/s.
   (a) What must the acceleration of the lander be so that it arrives at the surface with a speed of 1.1 m/s?

83. ✱ A rocket is launched vertically such that its acceleration upward is 4.50 g. If it can maintain that acceleration, how much time would it take to reach the orbit of the International Space Station 300 km above Earth's surface?

84. ✱✱✱ A stone is tossed straight down from the top of a building. During the third second of its flight, the stone covered a distance of 45.0 m. How much distance did the stone cover during the first 2.00 s? What was the stone's initial speed?

85. ✱✱ You throw a ball straight up in the air and catch it 7.00 s later. Find the average speed and average acceleration of the ball, the total distance covered by the ball, and its displacement.

86. ✱✱ You are standing on top of a building and are challenged by a friend to throw a ball such that it covers the 1.7 m distance between the top and the bottom of his window in 0.3 s. The top of the window is 13 m below the point where you release the all. Is this throw possible? If so, what is the maximum amount of time during which the ball can cover the distance?

Section 3-5 Relative Motion in One Dimension

87. ✳✳✳ A child standing on a balcony drops a marble from a height of 16 m above the ground at the same time that another child throws a tennis ball up from a height of 1 m above the ground.
  (a) What must be the initial speed of the tennis ball for the two to meet at the highest point in the ball's trajectory?
  (b) What was the speed of the ball relative to the marble just before the marble hit the ground?

88. ✳✳ Derive the relative velocity, acceleration, displacement, and position for two objects moving with constant accelerations.

89. ✳ A Mini Cooper moving at 135 km/h passes a Mustang moving in the same direction at 90 km/h. Calculate the velocity of the Mustang relative to the Mini, and the speed of the Mustang relative to the Mini.

90. ✳✳ Two young men on a balcony are trying to attract the attention of a young lady on the ground 11 m below. One young man drops a flower. The other young man wants his flower to land first. After 0.7 s, how fast should he throw it down?

91. ✳✳ A ball is dropped from the top of a 56.0 m-tall building. After 2.00 s, another ball is thrown upward from the ground. When the two balls pass the same point, they have the same speed. How fast was the upward-bound ball thrown?

COMPREHENSIVE PROBLEMS

92. ✳✳ A stone is dropped from a high cliff such that its speed just before impact on the ground below is 190 m/s. How much distance did the stone cover during the interval $5.0 \text{ s} \leq t \leq 15.0 \text{ s}$? How far did the stone travel in its last second of flight?

93. ✳✳✳ You are riding in an elevator with an open roof. You are moving down at a speed of 4.00 m/s when you throw an apple up into the air at a speed of 6.00 m/s relative to the elevator.
  (a) How long does it take for you to catch the apple?
  (b) What is the maximum height the apple reaches above the elevator floor? How does this height change if the elevator is moving up?
  (c) A friend standing on the ground throws an apple up at a speed of 6.00 m/s. How long does it take for that apple to return to his hand?

94. ✳✳ (a) A Formula One race car can go from 0 to 100 km/h in 1.70 s, from 0 to 200 km/h in 3.80 s, and from 0 to 300 km/h in 8.60 s. Assuming constant acceleration during the intervals 0 s to 1.70 s, 1.70 s to 3.80 s, and 3.80 to 8.60 s, find the distance that the car covers in the first 1.70 s and between 3.80 s and 8.60 s.
  ✳✳ (b) A Formula One race car starts from rest, reaches a speed of 100 km/h in 1.70 s, reaches a speed of 325 km/h over the next 7.30 s, then slows with an acceleration of 4.00 g for the next 100 m. Find the average speed, average acceleration, and average velocity of the car for the entire interval.

95. ✳✳ How much distance does a Formula One race car initially travelling at 360 km/h cover while it brakes at 6.00 g for 1.20 s?

96. ✳✳ You are sitting well back from the windows in the library at the Bamfield Marine station on Vancouver Island. A bald eagle hovers overhead with a fish in its grip. The fish wriggles loose and falls straight down. You happen to be looking out the window and notice that the fish is in sight for 0.14 s. The windows are 1.2 m high. How high was the eagle when it dropped the fish? What is the speed of the fish as it passes the window sill?

97. ✳✳ A particle's position is given by $x(t) = x_0 \cos(\omega t + \phi)$, where x_0 and ϕ are constants. At $t = 0$, the particle is at $x = 0$.
  (a) Find the possible values of ϕ.
  (b) Given that $x_0 = 20$ cm, plot the position of the particle over the interval $t = 0$ s to 10 s with (i) $\omega = \pi$ and $\phi = \pi$, (ii) $\omega = \pi/2$ and $\phi = 0$, and (iii) $\omega = \pi/2$ and $\phi = \pi/2$.
  (c) Write an expression for the particle's velocity as a function of time.
  (d) Write an expression for the particle's acceleration as a function of time.
  (e) Deduce a relationship between the acceleration and the position of a particle executing this motion (called simple harmonic motion).

98. ✳✳ Find the maximum speed and the magnitude of the maximum acceleration of the particle in problem 97. Use mathematical equations to show where these maxima occur.

99. ✳✳ An ocean liner approaches a pier at 20 km/h. When the ocean liner is 10 km away, the captain sends a speedboat to the pier to pick up the harbour pilot, who will help dock the ocean liner. The speedboat takes 12 min to get to the pier.
  (a) How long would the speedboat take to return to the ocean liner if it maintains the same speed?
  (b) Find the displacement of the speedboat during the round trip and the distance covered by the speedboat during the round trip.

100. ✳✳ A speedboat accelerates from rest at the start line of a race to reach a speed of 30 km/h over a distance of 70 m, then continues with the same constant acceleration for another 11 s before slowing down to a speed of 25 km/h over a distance of 100 m. Another speedboat begins to accelerate from rest at the start line and catches up to the first boat within 30 s.
  (a) Find the acceleration of the second speedboat.
  (b) Find the time it takes the first speedboat to cover the first 70 m.
  (c) Find the speed of the second speedboat when it catches up to the first speedboat.

101. ✳✳ You are in a hot-air balloon that is accelerating upward at 4.00 m/s². The balloon is at a height of 47.0 m when its speed upward is 9.00 m/s. You drop an apple from the side. Find the maximum height the apple will reach above the ground, and the speed of the apple when it hits the ground.

102. ✳✳✳ A child has a tennis ball and a baseball. He lets go of the tennis ball from the top of a high building, waits, and then throws the baseball straight down with a speed such that the two balls have a relative speed of 10 m/s when they meet 40 m below. How long did the child wait before throwing the baseball?

103. ✳✳✳ You throw a ball up in the air with a speed such that it just reaches your friend on a balcony 30.0 m above. At the same time, your friend throws another ball down to you such that it reaches you with twice the speed with which you threw your ball. What is the relative velocity of the two balls when they pass each other?

104. ✱✱ Two children are playing a game in which one of them throws a baseball down from the top of his 6.0 m-high tree house and the other is supposed to hit the baseball with a tennis ball. If the child in the tree house throws the baseball with a downward speed of 1.2 m/s, what is the speed the child on the ground has to throw his baseball with from a height of 1.1 m so the baseballs meet halfway?

105. ✱✱✱ Two children are playing a game in which one of them throws a volleyball up in the air and the other, standing very close to her friend, throws a tennis ball up to hit the volleyball as it reaches its highest point. The first child throws the volleyball at a speed of 12 m/s, and the second child throws the tennis ball at a speed of 25 m. How long after the volleyball is thrown should the second child wait to throw the tennis ball?

106. ✱✱✱ Show that Equation (3-20) is valid for nonconstant acceleration.

107. ✱✱✱ Two friends live beside a 25.0 m-high cliff, one at the top of the cliff and the other directly below, at the bottom. While they are making dinner, the friend at the top of the cliff throws a loaf of bread down with a speed of 7.0 m/s at the same time that the other friend throws a lemon up with a speed of 30.0 m/s. Find the relative speed of the bread and the lemon when they pass one another.

108. ✱✱ Madrid, Spain, and Beijing, China, both lie on the 40th parallel. Beijing is at 116°23′ E, and Madrid is at 3°40′ W. Find the average speed of a plane that takes 7.0 h to travel from Beijing to Madrid.

109. ✱✱✱ Figure 3-35 shows data for the velocity of a meteor fragment entering the atmosphere as a function of altitude.
  (a) Estimate the rate of change of the velocity at altitudes of 120 km and 92 km.
  (b) Assuming that gravity and friction with the atmosphere are the principal forces acting on the meteor fragment, what does the graph reveal about the relative strength of the two forces?
  (c) What does the shape of the velocity graph suggest about the upper atmosphere?

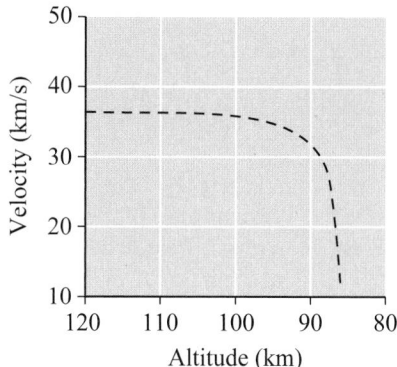

Figure 3-35 Problem 109

110. ✱✱✱ Figure 3-36(a) shows a graph of altitude versus time for the *Soyuz* launch vehicle.

  (a) Estimate the radial speed (i.e., the magnitude of the vertical component of velocity) of the vehicle at (i) 100 s, (ii) 200 s, (iii) 300 s, and (iv) 400 s.
  (b) Find the mean radial velocity between (i) 0 s and 100 s and (ii) between 300 s and 400 s.

Figure 3-36 (a) Problem 110

Figure 3-36 (b) Problem 110

111. ✱✱✱ Figure 3-37 shows a velocity versus time plot for the *Soyuz* launch vehicle.
  (a) Estimate the peak acceleration at $t = 120$ s, $t = 280$ s, and $t = 540$ s.
  (b) What is the average acceleration of the vehicle for the entire flight time shown?
  (c) Find the average acceleration for each of the three stages ending at 120 s, 280 s, and 540 s.

Arianespace. 2006. Soyuz CSG User's Manual, Issue 1.

Figure 3-37   Problem 111

112. ✱✱✱   Use Figure 3-37 to estimate the distance covered by the *Soyuz* vehicle in the first 120 s. How much distance did the vehicle cover by the end of 280 s? Compare these results to the graph of Figure 3-36(a), and explain the difference.

113. ✱✱✱   Use the velocity versus time graph for the *Soyuz* launch vehicle (Figure 3-37) to construct an acceleration versus time graph. Use two scales to show your results in m/s² and in multiples of g. Compare your graph to Figure 3-38 below.

114. ✱✱✱   Figure 3-38 shows an acceleration versus time plot for the *Soyuz* launch vehicle. Using the graph, calculate the speed of the vehicle at each of the three peak accelerations shown, and construct a velocity versus time graph for the vehicle. Compare your results with Figure 3-37.

Arianespace. 2006. Soyuz CSG User's Manual, Issue 1.

Figure 3-38   Problem 114

See the text online resources at
www.physics1e.nelson.com for Open Problems and Data-Rich Problems related to this chapter.

Chapter 4
Motion in Two and Three Dimensions

The equatorial speed of a star, also called rotational speed, can vary significantly from one star to another. The Sun's equatorial speed is only a few kilometres per second. The equatorial speed for some stars can exceed 200 km/s. Some neutron stars, which are vastly smaller than stars like our Sun, have rotational speeds as high as 38 000 km/s. With an equatorial speed of approximately 600 km/s, the fastest spinning full-sized star observed to date is VFTS 102. VFTS 102 was discovered by astronomers using the Very Large Telescope at the European Southern Observatory in Paranal, Chile (Figure 4-1).

VFTS 102 is 25 times more massive and 100 times brighter than the Sun. The star is located in the Tarantula nebula within the Large Magellanic Cloud, about 160 000 ly from Earth. The high radial acceleration resulting from the high equatorial speed has VFTS 102 close to tearing itself apart. The rotational and translational speeds of this star suggest that it was ejected from a binary star system when its companion exploded in a supernova. If VFTS 102 does tear itself apart, it is expected to produce one of the brightest stellar explosions in the universe.

Learning Objectives

When you have completed this chapter you should be able to:

1 Express displacement, velocity, and acceleration in two and three dimensions.

2 Describe and analyze the motion of a projectile.

3 Derive the radial acceleration vector for an object moving in a circle.

4 Apply relative motion principles to two and higher dimensions.

MECHANICS

ESO/M.-R. Cioni/VISTA Magellanic Cloud survey. Acknowledgment: Cambridge Astronomical Survey Unit

Figure 4-1 VFTS 102, the fastest spinning star found to date, is at the centre of the Tarantula nebula in the Large Magellanic Cloud.

 For additional Making Connections, Examples, and Checkpoints, as well as activities and experiments to help increase your understanding of the chapter's concepts, please go to the text's online resources at www.physics1e.nelson.com.

4-1 Displacement, Velocity, and Acceleration in Two and Three Dimensions

We can use the vector algebra tools from Chapter 2 to extend our analysis of motion to two and three dimensions. As described in Section 2-2, two- and three-dimensional vectors can be written in terms of components. For example, in two dimensions the position vector $\vec{r}$ can be expressed as

$$\vec{r} = \vec{r}_x + \vec{r}_y = x\hat{i} + y\hat{j} \qquad (4\text{-}1a)$$

In three dimensions, the position vector is given by

$$\vec{r} = x\hat{i} + y\hat{j} + z\hat{k} \qquad (4\text{-}1b)$$

Equation 4-1a and 4-1b give the position vector as the sum of its scalar components, multiplied by the corresponding unit vectors. This will, in general, be the approach we take to representing vectors in this text. However, occasionally it proves to be more elegant to represent a given vector as the sum of its vector components. In three dimensions the position vector given in equation 4-1b can be written as

$$\vec{r} = \vec{x} + \vec{y} + \vec{z} \qquad (4\text{-}1c)$$

Similarly, the displacement vector between two points can be written as

$$\Delta\vec{r} = x\hat{i} + y\hat{j} \qquad (4\text{-}2a)$$

and in three dimensions

$$\Delta\vec{r} = x\hat{i} + y\hat{j} + z\hat{k} \qquad (4\text{-}2b)$$

The average and instantaneous velocities of an object are then given by

$$\vec{v}_{avg} = \frac{\Delta\vec{r}}{\Delta t} = v_{avg,x}\hat{i} + v_{avg,y}\hat{j} \qquad (4\text{-}3a)$$

in two dimensions and

$$\vec{v}_{avg} = \frac{\Delta\vec{r}}{\Delta t} = v_{avg,x}\hat{i} + v_{avg,y}\hat{j} + v_{avg,z}\hat{k} \qquad (4\text{-}3b)$$

in three dimensions; and

$$\vec{v} = \lim_{\Delta t \to 0} \frac{\Delta\vec{r}}{\Delta t} = \frac{d\vec{r}}{dt} = v_x\hat{i} + v_y\hat{j} \qquad (4\text{-}4a)$$

in two dimensions and

$$\vec{v} = \lim_{\Delta t \to 0} \frac{\Delta\vec{r}}{\Delta t} = \frac{d\vec{r}}{dt} = v_x\hat{i} + v_y\hat{j} + v_z\hat{k} \qquad (4\text{-}4b)$$

in three dimensions.
Alternatively, one can write

$$\vec{v} = \vec{v}_x + \vec{v}_y + \vec{v}_z \qquad (4\text{-}4c)$$

The average and instantaneous accelerations are written as

$$\vec{a}_{avg} = \frac{\Delta\vec{v}}{\Delta t} = a_{avg,x}\hat{i} + a_{avg,y}\hat{j}$$

$$\vec{a} = \lim_{\Delta t \to 0} \frac{\Delta\vec{v}}{\Delta t} = \frac{d\vec{v}}{dt} = \vec{a}_x + \vec{a}_y$$

in two dimensions, and

$$\vec{a}_{avg} = \frac{\Delta\vec{v}}{\Delta t} = a_{avg,x}\hat{i} + a_{avg,y}\hat{j} + a_{avg,z}\hat{k} \qquad (4\text{-}5)$$

and

$$\vec{a} = \lim_{\Delta t \to 0} \frac{\Delta\vec{v}}{\Delta t} = \frac{d\vec{v}}{dt} = a_x\hat{i} + a_y\hat{j} + a_z\hat{k} \qquad (4\text{-}6)$$

in three dimensions.
Alternatively, one can write

$$\vec{a} = \vec{a}_x + \vec{a}_y + \vec{a}_z \qquad (4\text{-}4c)$$

Although we will use the scalar form for the components of a vector in general, there will be occasions where it will prove more elegant, convenient, and advantageous to treat the components of a vector as vectors. These occurrences will be clearly pointed out.

EXAMPLE 4-1

Average Speed and Average Velocity

An airplane flying at a constant speed takes 1.20 h to go from Amsterdam, the Netherlands, to Berne, Switzerland, 630 km to the southwest. The airplane is then diverted to Paris, France, 432 km northwest. It takes 0.900 h to get there, travelling at the same constant speed.

(a) Find the average speed of the airplane during the trip.
(b) Find the average velocity of the airplane, given that Amsterdam and Paris are 440 km apart.
(c) In a diagram, show the difference between the displacement of the airplane and the distance covered by the airplane throughout the journey.

SOLUTION

(a) The average speed is the total *distance* covered by the airplane divided by the total time elapsed. Since distance is a scalar quantity, the total distance is the sum of the distances in each part of the journey:

$$d = d_1 + d_2 = 630 \text{ km} + 432 \text{ km} = 1062 \text{ km}$$

The total time taken is 1.2 h + 0.9 h = 2.1 h. Therefore, the average speed is

$$v_{avg} = \frac{1062 \text{ km}}{2.1 \text{ h}} = 506 \text{ km/h}$$

(b) Since the airplane started over Amsterdam and ended over Paris 440 km away, the displacement is 440 km pointing directly from Amsterdam to Paris. The average velocity is given by the displacement divided by the elapsed time:

$$\vec{v}_{avg} = \frac{\Delta x \hat{i}}{\Delta t} = \frac{440 \hat{i}\ km}{2.1\ h} = 210 \hat{i}\ km/h$$

where the direction from Amsterdam to Paris is chosen as the positive x-axis

(c) The distance covered by the plane is shown in Figure 4-2 with the dashed line that follows the airplane's path. The displacement is the vector connecting the start and finish points.

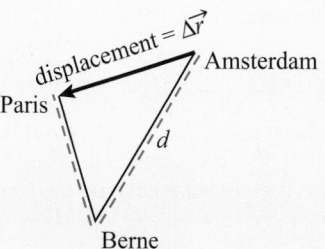

Figure 4-2 The difference between distance covered (dashed line) and the displacement of the plane

Making sense of the result:

Unlike the average speed, which tells us how fast the airplane is moving, the average velocity depends only on the direct distance between the start and finish points and the time elapsed on the way there.

 CHECKPOINT

C-4-1 Distance and Displacement

When you walk halfway around a circle with a radius of 1.2 m,

(a) your displacement is 1.2π m;

(b) you travel a distance of 2.4 m;

(c) your displacement is 2.4 m directed straight across the circle;

(d) your displacement and the distance travelled are equal.

C-4-1 (c) Displacement depends on initial and final positions only.

 EXAMPLE 4-2

Components of Velocity

At a practice, football players run 100 m at top speed in one direction, make a 90° turn, and run another 120 m, again at top speed (Figure 4-3). One player runs the first leg in 12.0 s and the second leg in 18.0 s.

(a) Ignoring the time taken to change directions, find the player's average velocity. Express your answer in polar and Cartesian components.

(b) Calculate the average velocities during the first and the second legs of the run.

(c) Calculate the player's average speed for the entire run.

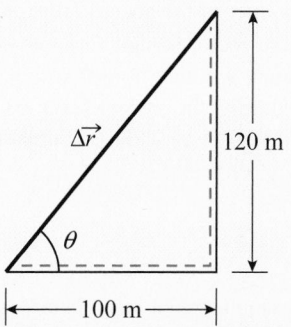

Figure 4-3 The path distance (dashed line) and the displacement (vector arrow) for the football player in Example 4-1

SOLUTION

(a) Let us pick the positive x-axis to lie along the first 100 m and the positive y-axis to lie along the second direction.

The total displacement is

$$\Delta \vec{r} = \Delta x \hat{i} + \Delta y \hat{j} \qquad (1)$$

Since the angle between the two legs of the trip is 90°,

$$\Delta r = \sqrt{\Delta x^2 + \Delta y^2} = \sqrt{100^2 + 120^2} = 156.2\ m \qquad (2)$$

to 3 significant figures, this is 156 m, but for the calculations below we keep the value from Equation 2.

The total time elapsed is $12\ s + 18\ s = 30\ s$, so the magnitude of the average velocity is

$$v = \frac{\Delta r}{\Delta t} = \frac{156.2}{30} = 5.21\ m/s$$

The Cartesian components of the average velocity are

$$v_{avg,x} = v_{avg} \cos\theta = 5.21 \cos(50.19°) = 3.34\ m/s$$

$$v_{avg,y} = v_{avg} \sin\theta = 5.21 \sin(50.19°) = 4.00\ m/s$$

(b) The velocity during the first leg of the run is the displacement along the x-axis divided by the time elapsed:

$$\vec{v}_1 = \frac{\Delta x \hat{i}}{\Delta t} = \frac{100 \hat{i}\ m}{12.0\ s} = 8.33\ m/s\ in\ the\ +x\text{-direction}$$

(continued)

CHAPTER 4 | **MOTION IN TWO AND THREE DIMENSIONS**

During the second leg of the run,

$$\vec{v_2} = \frac{\Delta y \hat{j}}{\Delta t} = \frac{120 \hat{j} \text{ m}}{18.0 \text{ s}} = 6.67 \text{ m/s in the } +y\text{-direction}$$

(c) The average speed for the entire run is the total distance (not the displacement) divided by the time elapsed. The distance covered is $100 \text{ m} + 120 \text{ m} = 220 \text{ m}$, and the total time is 30 s. Therefore,

$$v_{\text{avg}} = \frac{d}{\Delta t} = \frac{220 \text{ m}}{30 \text{ s}} = 7.33 \text{ m/s}$$

Making sense of the result:

Since displacement is not the same as distance, we get different answers for parts (a) and (b).

✓ CHECKPOINT

C-4-2 Components of Velocity

Why are the components of the average velocity for the entire run in Example 4-2 not equal to the average velocities along each direction for each leg of the run?

C-4-2 The average velocities along each direction are not the components of the overall average velocity vector.

PEER TO PEER

The average speed is the total distance travelled divided by the time elapsed; however, the average velocity is the displacement vector divided by the time elapsed.

LO 2

4-2 Projectile Motion

Figure 4-4 shows the trajectory of a projectile under the influence of gravity only. Other factors, such as the effect of air resistance, are ignored. The launch point is used as the origin of the coordinate system. The projectile was launched with an initial speed v_0, making an angle θ above the horizontal. The initial velocity can be resolved into a horizontal component equal to $v_0 \cos\theta$ and a vertical component equal to $v_0 \sin\theta$.

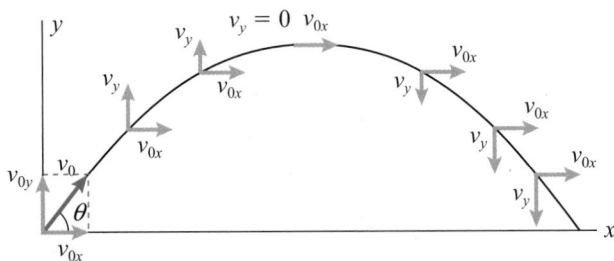

Figure 4-4 Typical projectile trajectory

Since there is no acceleration in the horizontal direction, the horizontal component of the velocity is constant throughout the motion, as indicated by the length of the horizontal arrows representing this component in Figure 4-4. In the vertical direction, the force of gravity is

acting on the projectile, so the vertical component of the velocity decreases until it reaches zero momentarily at the top of the trajectory, and then increases in the downward direction. The overall speed of the projectile therefore reaches a minimum at the top of the trajectory.

The horizontal distance covered by the projectile depends on the speed in the horizontal direction and the time of flight:

$$\Delta x = v_x t \tag{4-7}$$

In the y-direction, we can apply the results we obtained for motion in one dimension under constant acceleration. Using Equation (3-19), the vertical displacement can be written as

$$\Delta \vec{y} = \frac{1}{2}\vec{a}t^2 + \vec{v}_{0y}t \tag{4-8}$$

Similarly, Equation (3-15) gives a vertical velocity of

$$\vec{v}_y(t) = \vec{a}t + \vec{v}_{0y} \tag{4-9}$$

and the time-independent kinematics Equation (3-20) gives

$$v_y^2 - v_{0y}^2 = 2\vec{a} \cdot \Delta\vec{y} \tag{4-10}$$

Note that we are copying the approach we have taken in Chapter 3 in dealing with velocity, displacement, and acceleration as vectors. The dot product form used on the right-hand side of Equation 4-10 is an elegant way of representing the product of the acceleration and the displacement. We keep the vector format here for relevance to later chapters. However, we also note that this can equally be done by representing the components of the quantities along the x- and y-axes as scalars. This latter approach is used frequently in this text.

EXAMPLE 4-3

Football Trajectory

A football is kicked from a level field at a velocity of 19.0 m/s at an angle of 41.0° above the horizontal.

(a) Find the horizontal distance covered by the football.

(b) Find the maximum height reached by the football.
(c) Find the time the football takes to reach its maximum height.
(d) Find the minimum speed of the football during its flight.

SOLUTION

The trajectory of the soccer ball is shown in Figure 4-5. We choose up and right as the positive directions.

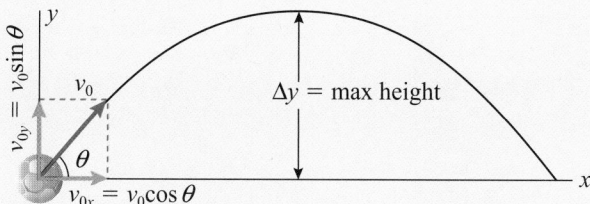

Figure 4-5 Example 4-3

(a) The velocity in the horizontal direction is constant. Therefore, the horizontal distance covered by the football is given by $\Delta x = v_x t$. The horizontal component of the velocity, v_x, is

$$v_x = v_0 \cos\theta = (19 \text{ m/s})(\cos(41°))$$

Now we need to find the time, t, that the football is in the air. At the end of this time, the ball hits the ground, so its height, y, is zero. From Equation (4-8) we have

$$\Delta \vec{y} = \frac{1}{2}\vec{a}t^2 + \vec{v}_0 t$$

$$0 = \frac{1}{2}(-9.81\hat{j}) \text{ m/s}^2 (t^2) + (19 \sin(41°)\hat{j}) \text{ m/s}(t)$$

This quadratic equation gives us two solutions: $t = 0$ and $t = 2.5413$ s. Clearly, the later time is the landing time for the football. Using this time in the expression for v_x gives

$$x = v_0 \cos(41°)t = (19 \cos(41°) \text{ m/s})(2.5413 \text{ s}) = 36.4 \text{ m}$$

(b) To find the maximum height reached by the ball, we apply Equation (4-10) to the moment the ball reaches the top of its trajectory:

$$v_y^2 - v_{0y}^2 = 2\vec{a} \cdot \Delta \vec{y}$$

At the maximum height the vertical speed is zero. We know that the initial speed in the y-direction is $v_{0,y} = 19 \sin(41°)$. Since the acceleration vector points downward as the ball moves upward, the vector dot product $\vec{a} \cdot \Delta \vec{y}$ is negative. Therefore, using Equation 2-18 for the scalar product of $\vec{a}$ and $\Delta \vec{y}$

$$0 - (19 \sin(41°) \text{ m/s})^2 = 2(9.81 \text{ m/s}^2)(\Delta y)(-1)$$

$$\Delta y = 7.92 \text{ m}$$

(c) To find the time the football takes to reach the maximum height, we use Equation (4-9):

$$\vec{v}_y(t) = \vec{v}_{0y} + \vec{a}t$$

$$0 = (19 \sin(41°)\hat{j}) \text{ m/s} + (-9.81)\hat{j} \text{ m/s}^2 (t)$$

$$\Rightarrow t = \frac{19 \sin(41°) \text{ m/s}}{9.81 \text{ m/s}^2} = 1.27 \text{ s}$$

(d) The minimum speed occurs at the top of the trajectory, where the vertical component of the velocity is zero. We only have the horizontal velocity $v_x = 19 \cos(41°) = 14.339$ m/s, which is 14.3 m/s to three significant figures.

Making sense of the result:

The height reached by the football is reasonable, and the time it takes for the football to reach the maximum height is half its flight time, which makes sense due to symmetry.

✓ CHECKPOINT

C-4-3 Minimum Speed of a Projectile

What is the minimum speed of a projectile launched with a velocity of 200 m/s at a 60° angle above the horizontal?

(a) zero
(b) 100 m/s
(c) $100\sqrt{3}$ m/s
(d) none of the above

C-4-3 (b) min speed is the speed in the x direction $= 200 \cos(60°) = 100$ m/s.

ONLINE ACTIVITY

Projectile Motion

The e-resource that accompanies every new copy of this textbook contains an Online Activity using the PhET simulation "Projectile Motion." Work through the simulation and accompanying questions to gain an understanding of projectile motion.

MAKING CONNECTIONS

Giant Artillery

One of the world's largest artillery pieces was "Dora," a railway-mounted siege gun built for the German army during World War II (Figure 4-6). This cannon had a calibre (bore diameter) of 800 mm, and could fire a 4800 kg shell up to 48 km. Because of air resistance, the firing angle had to be slightly greater than 45° to get the maximum range.

Figure 4-6 The Dora rail cannon

EXAMPLE 4-4

Projectile Motion

You are standing on a roof 13.0 m above the ground, and you throw a tennis ball to your friend, who is on the ground 20 m from the building. Your hand is 2.00 m above the roof when you release the ball. Your friend reaches up such that her hand is 2.10 m above the ground. The ball leaves your hand with a velocity of 16.0 m/s at an angle of 15.0° below the horizontal.

(a) By how much vertical distance does the ball miss your friend's hand?

(b) What is the speed of the ball as it passes your friend's hand?

SOLUTION

(a) Since we are given the horizontal distance and the horizontal speed, we can calculate the time the ball takes to cover the distance. We can then find the height of the ball at this time. We choose up and right as the positive directions.

From Equation (4-7) and Figure 4-7,

$$\Delta x = v_x t = (v \cos\theta) t$$

$$t = \frac{\Delta x}{v \cos\theta} = \frac{20 \text{ m}}{16 \cos(15°) \text{ m/s}} = 1.2941 \text{ s}$$

The vertical displacement is given by Equation (4-8):

$$\Delta \vec{y} = \frac{1}{2} \vec{a} t^2 + \vec{v}_{0,y} t$$

$$\Delta y = \frac{1}{2}(-9.81 \hat{j} \text{ m/s}^2)(1.2941^2) - (16 \sin(15°)) \hat{j} \text{ m/s} (1.2941 \text{ s})$$

$$= -13.57 \hat{j} \text{ m}$$

The ball has therefore dropped a total of 13.57 m from its launch height, which was 15 m above the ground.

Figure 4-7 Trajectory of the ball in Example 4-4

Thus, the ball is at a height of 1.43 m when it gets to your friend. Since her hand is 2.1 m above the ground, the ball misses by 67.0 cm.

(b) As the ball passes your friend, its velocity has a constant horizontal component of $v_0 \cos(15°)$ and a vertical component given by

$$\vec{v}(t) = \vec{a} t + \vec{v}_0$$

$$v_y(t) = -9.81 \text{ m/s}^2 (1.2941 \text{ s}) - 16 \sin(15°) \text{ m/s} = -16.836 \text{ m/s}$$

$$v = \sqrt{v_x^2 + v_y^2} = \sqrt{(16 \cos(15°) \text{ m/s})^2 + (16.836 \text{ m/s})^2}$$

$$= 22.9 \text{ m/s}$$

Making sense of the result:

For the distances given, the amount by which your friend misses the ball is reasonable.

4-3 Circular Motion

Uniform Circular Motion

An object moving with **uniform circular motion** moves along a circular path of fixed radius at a constant speed. Since the direction of motion of the object is constantly changing, its velocity is also constantly changing. Therefore, the object must be continuously accelerating.

Figure 4-8 shows an object moving around a circle at a constant speed. The velocity at point A is $\vec{v}_A$, and the velocity at point B is $\vec{v}_B$. The velocities have exactly the same magnitude, v. The position vector at point A is $\vec{r}_A$, and the position vector at point B is $\vec{r}_B$. Both position vectors have magnitude r, as measured from the centre of the circle. At every point on the circle, the velocity of the object is tangential to the circle and perpendicular to the position vector. The average acceleration is

$$\vec{a}_{\text{avg}} = \frac{\Delta \vec{v}}{\Delta t}$$

Therefore, the direction of average acceleration is parallel to $\Delta \vec{v}$.

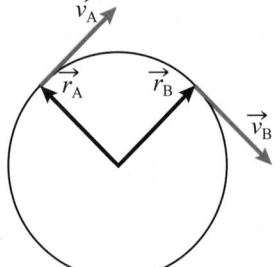

Figure 4-8 An object in uniform circular motion

Now, we redraw the position and velocity vectors using the principle that a vector can be translated as long as its magnitude and direction do not change. First, we construct a triangle formed by the two position vectors and the

Uranium Enrichment and Centrifugation

Highly enriched uranium fuels are used in nuclear reactors. Gas centrifuges are often used to enrich the uranium by separating uranium's various isotopes. In a gas centrifuge, vaporized material is spun at very high speeds to separate components with different densities. As early as the end of the 19th century, chemists used simple hand-cranked centrifuges to help precipitate fine particles from liquids. Ultracentrifuge rotors spin at very high speeds, producing accelerations up to 2 million times the acceleration due to gravity.

Theodor Svedberg invented the first ultracentrifuge in 1925, and he received the Nobel Prize for his research on colloids and proteins with his invention. The Russians, using the expertise of Austrian scientist Gernot Zippe, were the first to mass-produce enriched uranium using centrifuge technology (Figure 4-9). The same centrifuge principles used to produce fuel for nuclear reactors can be used to produce the more highly enriched weapons-grade uranium.

Fastfission/Wikipedia

www.sutton-images.com

Figure 4-9 (a) A schematic of the Zippe centrifuge (b) A cascade of gas centrifuges used to successively increase the concentration of the fissionable isotope of uranium

change in position Δr, as shown in Figure 4-10. We then construct a triangle formed by the two velocity vectors and the change in velocity. Since the velocity vector is perpendicular to the position vector at either location, we can conclude that the change in velocity and, consequently, the average acceleration vector, are perpendicular to Δr.

Next, we take the limit of the average acceleration as Δt goes to zero to get the instantaneous acceleration:

$$\vec{a} = \lim_{\Delta t \to 0} \frac{\Delta \vec{v}}{\Delta t} = \frac{d\vec{v}}{dt}$$

As the time interval Δt shrinks, so does the displacement between A and B. In the limit, the two points merge. As the position triangle collapses, so does the velocity triangle. However, the acceleration which is parallel $\Delta \vec{v}$ remains perpendicular to $\Delta \vec{r}$, which is a chord of the circle. The acceleration vector, being perpendicular to that chord, points along the radial to the centre of the circle. In the limit as Δt goes to zero, the instantaneous acceleration

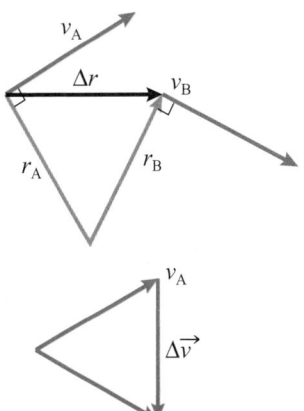

Figure 4-10 The change in velocity and, therefore, the acceleration vector, are perpendicular to the change in the position vector.

points along the radius. Therefore, an object moving in a circle must continuously accelerate toward the centre of the circle. Acceleration toward the centre of the circle is called **radial acceleration**.

To find the magnitude of the acceleration, consider the following ratios of the two similar triangles formed by the position and velocity vectors:

$$\frac{\Delta v}{v} = \frac{\Delta r}{r}$$

$$\Delta v = \frac{v \Delta r}{r}$$

Dividing both sides by Δt, the change in time, gives

$$\frac{\Delta v}{\Delta t} = \frac{v \Delta r}{r \Delta t}$$

As Δt goes to zero,

$$\frac{dv}{dt} = \frac{v dr}{r dt}$$

KEY EQUATION $$a = \frac{v^2}{r}\qquad (4\text{-}11)$$

We have thus established both the direction and the magnitude for the radial acceleration of an object moving around a circle at a constant speed.

 CHECKPOINT

C-4-4 Uniform Circular Motion

For an object moving in uniform circular motion,
(a) the velocity of the object is constant;
(b) the speed of the object is constant.
(c) Both (a) and (b) are true.
(d) None of the above are true.

C-4-4 (b) the speed must be constant for uniform circular motion, but the velocity is not constant.

ONLINE ACTIVITY

The e-resource that accompanies every new copy of this textbook contains an Online Activity using the PhET simulation "Circular Motion." Work through the simulation and accompanying questions to gain an understanding of circular motion.

Nonuniform Circular Motion

An object moving in a circle does not necessarily maintain a constant speed. For example, if you let a marble roll down the inside of a hemispherical bowl, the marble will speed up while travelling down the circular surface. When the speed of an object changes while travelling along a circle, then the velocity of the object changes

Acceleration on Formula One Tracks

On Formula One (F1) racetracks, the acceleration during turns can be harder on the cars and drivers than accelerations when starting and when braking on a straight section of the track. When accelerating from rest, the driver is pushed back into the seat, and during braking the driver is pushed forward into a harness. However, during turns, the accelerating force acts sideways on the driver. F1 cars can experience accelerations as high as 6 *g* on the more challenging turns, such as the Blanchimont turn in Circuit de Spa-Francorchamps in Belgium and turn R130 of the Suzuka circuit in Japan. The duration of the acceleration is also critical. Many F1 drivers say that turn 8 at the Istanbul Grand Prix (Figure 4-11) is the most difficult to negotiate despite the lower acceleration required. For this approximately 200° turn, drivers typically have to sustain speeds of approximately 280 km/h, resulting in accelerations of 4.0 *g* to 5.5 *g* lasting 7 s, the longest duration for a hard turn on the F1 circuit.

www.Sutton-images.com

Figure 4-11   Michael Schumacher losing control on turn 8 in Istanbul during qualifying in the 2010 Istanbul Grand Prix.

along the direction of motion, which is tangential to the circle. Such an object is said to have **tangential acceleration**. Therefore, the **total acceleration** of this object is the *vector sum* of its tangential and radial accelerations (Figure 4-12).

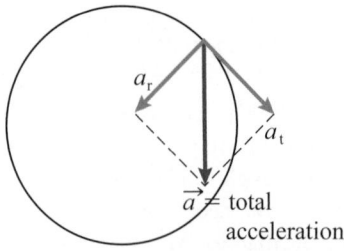

Figure 4-12   The total acceleration for an object moving in a circle at a changing speed is the vector sum of the radial and tangential acceleration vectors.

MAKING CONNECTIONS

Sustained versus Sudden Acceleration

The magnitude of the acceleration that the human body is subjected to is not the only factor in determining the effects of acceleration on the body: the direction and duration of the acceleration are also important. The body reacts differently and has different tolerance thresholds to abrupt as opposed to sustained acceleration. The resulting symptoms can be quite different. The body can cope with rather high accelerations applied over a short time. For example, the body can tolerate peak vertical accelerations up to $25\ g$ in the $+z$-direction over a period of $100\ \mathrm{ms}$ without incurring spinal injury.

However, much smaller accelerations can produce adverse effects if applied over a longer time. For example, pilots can sustain the short period of high acceleration in an ejection (Figure 4-13) seat without injury (ejection seat accelerations can reach $13\ g$). Accelerations as small as $5\ g$, however, can result in blackouts if sustained for a few seconds in manoeuvres such as an inside loop. The effects of sustained acceleration depend, in part, on the rate of change of the acceleration, called *onset*. For example, if the acceleration during an inside loop manoeuvre were to increase from zero to about $5\ g$ over a period of $5\ \mathrm{s}$, it would produce loss of consciousness. It would take longer for symptoms to occur if the onset were lower.

Figure 4-13 A pilot experiences $+z$-acceleration when deploying an ejection seat.

EXAMPLE 4-5

Nonuniform Circular Motion

Starting from rest, an experimental mini-drone (remote-controlled aircraft) flies around a circle of radius $2.00\ \mathrm{m}$ three times in $3.00\ \mathrm{s}$. The mini-drone has constant tangential acceleration. Find the magnitude of the mini-drone's acceleration at the end of $0.50\ \mathrm{s}$.

SOLUTION

Since the mini-drone is accelerating along the circumference of the circle, its acceleration has both tangential and radial components. We know the initial speed, the total distance travelled in the three circuits, and the time taken. We pick the direction of motion as positive.

To find the tangential acceleration, we apply Equation (3-19):

$$d = \frac{1}{2}a_t t^2 + v_0 t$$

$$3(2\pi)2\ \mathrm{m} = \frac{1}{2}a_t(3\ \mathrm{s})^2 + 0$$

$$a_t = \frac{8\pi}{3} = 8.378\ \mathrm{m/s^2}$$

From Equation (4-11), we know that the radial acceleration depends on speed. Since the speed is increasing continuously, the magnitude of the radial acceleration of the mini-drone is also increasing. The speed at the end of $0.50\ \mathrm{s}$ is given by Equation (3-15):

$$v = a_t t + v_0 = 8.378\ \mathrm{m/s^2}(0.50\ \mathrm{s}) + 0 = 4.189\ \mathrm{m/s}$$

Substituting this speed into Equation (4-11) gives

$$a_r = \frac{v^2}{r} = \frac{(4.189\ \mathrm{m/s})^2}{2\ \mathrm{m}} = 8.773\ \mathrm{m/s^2}$$

The total acceleration in this particular case is the vector sum of the radial and the tangential accelerations, which are perpendicular to each other:

$$\vec{a} = \vec{a_r} + \vec{a_t}$$

$$a = \sqrt{a_r^2 + a_t^2} = \sqrt{(8.773\ \mathrm{m/s^2})^2 + (8.378\ \mathrm{m/s^2})^2} = 12.1\ \mathrm{m/s^2}$$

C-4-5 Total Acceleration

An object under the influence of gravity and frictional forces is moving at a constant speed along the hemispherical bowl shown in Figure 4-14. The direction of the total acceleration is

(a) toward the centre of the circle;
(b) straight down;
(c) in the direction of the vector sum of the gravitational acceleration and the radial acceleration;
(d) impossible to determine without knowing the directions of the frictional forces.

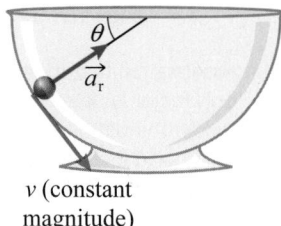

v (constant magnitude)

Figure 4-14 The object is moving in a bowl at a constant speed.

C-4-5 (a) since the object has constant speed the only acceleration it has is radial.

MAKING CONNECTIONS

G-LOC, Grey-outs, and Red-outs

G-LOC, induced loss of consciousness caused by acceleration, was first reported toward the end of World War I. Pilots who underwent accelerations of approximately 4.5 g often experienced G-LOC for about 20 s. During a grey-out, the pilot is conscious but has partial loss of vision. When the body is accelerated upward, blood is forced away from the brain, resulting in reduced consciousness. Upward acceleration results from positive G-manoeuvres, such as the inside loop shown in Figure 4-15(a). Before the onset of G-LOC there are other symptoms, such as blurry vision, tunnel vision, and grey-outs. An outside loop (Figure 4-15(b)) produces –z-acceleration and can cause a red-out, which is a reddening in the visual field as excess blood is forced into the brain. Human tolerance to –z-acceleration is less than human tolerance to +z-acceleration. For high-speed aerobatic manoeuvres, pilots wear a G-suit, which is a suit that is designed to restrict blood flow from the head to the body. With a G-suit, the threshold for vertical acceleration during an inside loop is about 12 g, and 8 g accelerations can be sustained over relatively long periods.

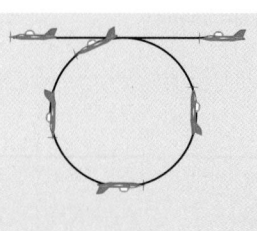

Figure 4-15 Acceleration during (a) an inside loop and (b) an outside loop

LO 4

4-4 Relative Motion in Two and Three Dimensions

A ferry crosses the English Channel from Calais, France, to Hastings, England, 110 km west-southwest. However, there is a strong flood tide flowing northeast. If the captain simply sets a course straight for Hastings, the ferry will be too far north when it reaches the English coast. To make the crossing in the least amount of time, the captain must adjust the course to allow for the tidal current. In Section 3-5, we introduced the concept of relative motion needed to deal with such problems. Since we developed the equations in vector form, we can readily adapt them to relative motion in two and three dimensions.

Equation (3-23) was written for the relative position along the x-axis, but we can also apply it to the components of the relative position of the two objects in three dimensions:

$$\vec{x}_{12} = \vec{x}_{1g} - \vec{x}_{2g} \qquad (4\text{-}12a)$$

$$\vec{y}_{12} = \vec{y}_{1g} - \vec{y}_{2g} \qquad (4\text{-}12b)$$

$$\vec{z}_{12} = \vec{z}_{1g} - \vec{z}_{2g} \qquad (4\text{-}12c)$$

Adding the three equations above gives

$$\vec{r}_{12} = \vec{r}_{1g} - \vec{r}_{2g} \qquad (4\text{-}13a)$$

$$\vec{r}_{12} = x_{12}\hat{i} + y_{12}\hat{j} + z_{12}\hat{k} \qquad (4\text{-}13b)$$

Similarly, the formulas for relative velocity and relative acceleration for two dimensions are

$$\vec{v}_{12} = \vec{v}_{1g} - \vec{v}_{2g} \qquad (4\text{-}14)$$

$$\vec{a}_{12} = \vec{a}_{1g} - \vec{a}_{2g} \qquad (4\text{-}15)$$

and are easily extended to motion in three dimensions.

EXAMPLE 4-6

Flight Plan

A plane is scheduled to fly from Casablanca, Morocco, to Lisbon, Portugal, 581 km [13° W of N]. A Poniente wind is blowing from Gibraltar at 34 km/h [15° N of W]. When the plane's airspeed is 560 km/h, and the pilot flies on a direct bearing to Lisbon, how long does the flight take?

SOLUTION

If the pilot were to point the plane along the straight line from Casablanca to Lisbon, the wind would push the plane off course to the west. To compensate, the pilot needs to point the plane somewhat to the east of the direct heading to Lisbon.

Applying Equation (4-14) to find the velocity of the plane (p) with respect to the wind (w) gives

$$\vec{v}_{pw} = \vec{v}_{pg} - \vec{v}_{wg}$$

$$\vec{v}_{pg} = \vec{v}_{pw} - \vec{v}_{wg}$$

Thus, the velocity of the plane with respect to the ground equals the sum of the velocity of the plane in the air and the velocity of the air with respect to the ground (i.e., the wind velocity). Let us construct a diagram of these velocities (Figure 4-16).

Although we do not know yet in which direction the plane should be oriented, we do know that the sum of the velocities of the plane and the wind point along the line connecting the two cities. Since the wind speed is small compared to the airspeed of the plane, the airspeed and the speed of the plane with respect to ground will not be very different, as we can see from our scale diagram.

We label the angles between the velocities α, β, and γ as shown. In Figure 3-16, v_{pg} makes an angle of 13° with the vertical, and v_{wg} makes an angle of 15° with the horizontal. Then,

$$\alpha = 90° - (13° + 15°) = 62°$$

Applying the sine law to the velocity triangle, we have

$$\frac{\sin \alpha}{v_{pw}} = \frac{\sin \beta}{v_{wg}}$$

$$\sin \beta = \frac{v_{wg}(\sin \alpha)}{v_{pw}} = \frac{34 \text{ km/hr }(\sin(62°))}{560 \text{ km/hr}} = 0.0536$$

$$\beta = 3.07°$$

Therefore,

$$\gamma = 180 - (\alpha + \beta) = 180° - (3.07° + 62°) = 114.93°$$

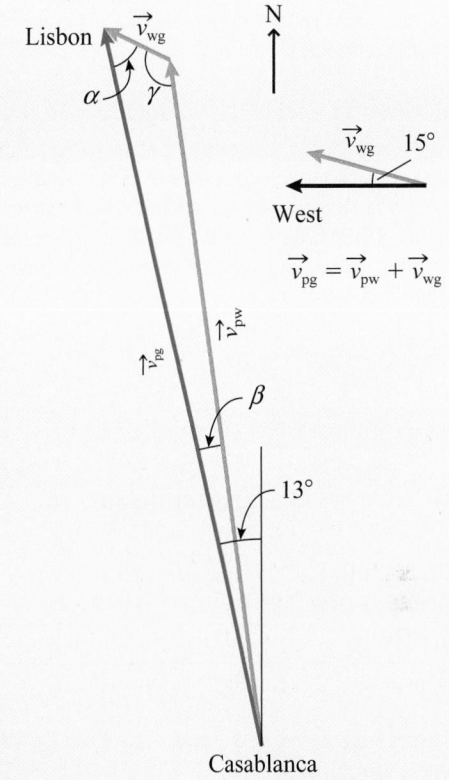

Figure 4-16 Velocity diagram for Example 4-6

To three significant digits the value is 115°.

Using the sine law once again, we have

$$\frac{\sin \alpha}{v_{pw}} = \frac{\sin \gamma}{v_{pg}} \Rightarrow v_{pg} = \frac{v_{pw} \sin \gamma}{\sin \alpha} = \frac{(560 \text{ km/hr}) \sin(114.93°)}{\sin(62°)}$$

$$= 575.16 \text{ km/hr}$$

The time it takes the plane to arrive at its destination is

$$t = \frac{581 \text{ km}}{575.14 \text{ km/h}} = 1.01 \text{ h}$$

Making sense of the result:

The plane's speed is very large compared to speed of the wind, so the small heading correction of 3.07° is reasonable.

EXAMPLE 4-7

Military Mud Ball

A French patrol on a cliff 50.0 m above the St. Lawrence River during the 1759 Battle for Québec spots a British scout boat leaving the seemingly undefended beach below.

Not wanting to give away their position by firing guns, the French soldiers prepare to launch a large mud ball from an improvised catapult that operates almost silently. The catapult launches mud balls at a velocity of 35.0 m/s at an

(continued)

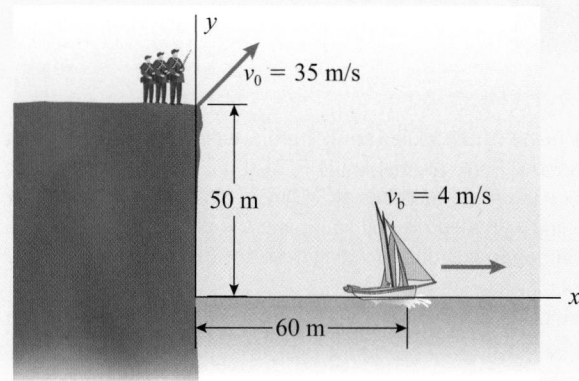

angle of 41.0° above the horizontal. The British scout boat starts 60.0 m from the point directly below the French patrol, and has a speed of 4.00 m/s. The French take 5.00 s to aim the catapult. How much longer should they wait before releasing the catapult?

SOLUTION

Figure 4-17 shows the trajectory needed to hit the British boat. We know the launch velocity and the change in height, but we need to determine how long the mud ball will be in the air. This transit time is determined by the y-component of the motion. Choosing up as the positive direction, we apply Equation (4-8):

$$\Delta \vec{y} = \frac{1}{2}\vec{a}t^2 + \vec{v}_{0y}t$$

$$-50\hat{j}\ \text{m} = \frac{1}{2}(-9.81\hat{j}\ \text{m/s}^2)t^2 + (35\sin(41°))\hat{j}\ \text{m/s}\,(t)$$

$$0 = \frac{1}{2}(9.81\ \text{m/s}^2)t^2 - (35\sin(41°))\ \text{m/s}\,(t) - 50$$

$$t = \frac{-b \pm \sqrt{b^2 - 4ac}}{2a}$$

$$= \frac{35\sin(41°)\ \text{m/s} \pm \sqrt{(35\sin(41°)\ \text{m/s})^2 + 4(4.905\ \text{m/s}^2)(50\ \text{m})}}{2(4.905\ \text{m/s}^2)}$$

The two roots are $t = 6.2995$ s and $t = -1.6182$ s. We take the positive root.

The distance that the mud ball travels in the x-direction is given by Equation (4-7):

$$x = v_xt = v\cos\theta t = 35\cos(41°)\ \text{m/s}\,(6.2995\ \text{s}) = 166.40\ \text{m}$$

Initially, the British soldiers are 60 m from the cliff. By the time the catapult is ready to fire, the additional distance covered by the British soldiers is

Figure 4-17 Schematic for Example 4-7

$$x = v_xt = 4\ \text{m/s}\,(5\ \text{s}) = 20.0\ \text{m}$$

Thus, they are a distance of 60 m + 20 m = 80.0 m from the cliff by the time the French are ready to fire. In the time the mud ball takes to reach the river, the British will travel an additional distance of

$$x = 6.2995\ \text{s}\,(4\ \text{m/s}) = 25.198\ \text{m}$$

Now the British boat will be a total horizontal distance of 80 m + 25.198 m = 105.198 m from the cliff. If the French fire when ready, the mud ball will overshoot the boat by 166.40 m − 105.198 m = 61.203 m. Before releasing the catapult, the French should wait for

$$t = \frac{61.203\ \text{m}}{4\ \text{m/s}} = 15.3\ \text{s}$$

Making sense of the result:

If the catapult fires only at the given angle then the French Patrol has to wait. The final results are reported to three significant figures. However, in the intermediate values needed to obtain the final answer, more significant figures are kept.

EXAMPLE 4-8

Relative Speed

Two children launch marbles from spring-loaded toy cannons placed side by side. The slower of the two cannons fires a marble at a velocity of 17.0 m/s at an angle of 50.0° above the horizontal; the other cannon fires a marble at a velocity of 21.0 m/s at an angle of 45.0° above the horizontal. Find the relative speed of the two marbles 1.70 s into the flight.

SOLUTION

We can find the relative speed from the components of the relative velocity. The relative velocity is given by Equation (4-14):

$$\vec{v}_{12} = \vec{v}_{1g} - \vec{v}_{2g}$$

In the x-direction,

$$\vec{v}_{12,x} = \vec{v}_{1g,x} - \vec{v}_{2g,x}$$

Taking the x-direction of motion of the faster marble to be positive,

$$\vec{v}_{12,x} = 21\cos(45°)\hat{i}\ \text{m/s} - 17\cos(50°)\hat{i}\ \text{m/s} = 3.922\hat{i}\ \text{m/s}$$

In the y-direction, the vertical relative velocity between the two marbles remains constant because both marbles undergo the same acceleration due to gravity. Thus, we can find the vertical relative velocity from the vertical components of the initial velocities:

$$\vec{v}_{12,y} = \vec{v}_{1g,y,0} - \vec{v}_{2g,y,0}$$

Choosing up to be positive,

$$v_{12y} = 21\sin(45°)\ \text{m/s} - 17\sin(50°)\ \text{m/s} = 1.826\ \text{m/s}$$

Therefore, the relative speed is given by

$$v_{12} = \sqrt{(v_{12,x})^2 + (v_{12,y})^2} = \sqrt{(3.922\ \text{m/s})^2 + (1.826\ \text{m/s})^2}$$

$$= 4.33\ \text{m/s}^2$$

Making sense of the result:

The relative speed between the two marbles will be constant, since the acceleration contribution is the same for both.

EXAMPLE 4-9

Torpedo and Boat

In a naval tactics competition, a contestant is launching a torpedo at another contestant's boat. The boat and torpedo launch from positions 70.0 m apart on a straight shore that runs east to west, with the torpedo launched east of the boat. The boat starts from rest with a constant acceleration of 0.60 m/s^2 in a direction $21.0°$ west of north. At the same time, the torpedo is launched at an angle of $31.0°$ to the shore with an initial speed of 3.00 m/s. What constant acceleration must the torpedo have to hit the boat?

SOLUTION

We will break up the motion into x- and y-components and choose an x-axis parallel to the shore, as shown in Figure 4-18. The distance between the torpedo (designated t) and the boat (designated b) is given by Equation (4-13):

$$\vec{r}_{tb} = \vec{r}_t - \vec{r}_b \tag{1}$$

Figure 4-18   Paths of the torpedo and boat in Example 4-9

The x-component is given by Equation (4-12a):

$$\vec{x}_{tb} = \vec{x}_t - \vec{x}_b \tag{2}$$

From Equation (3-25) we can express this component as

$$\vec{x}_{tb}(t) = \frac{1}{2}\vec{a}_{tb,x}t^2 + \vec{v}_{tb,0,x}t + \vec{x}_{tb,0} \tag{3}$$

We choose west as the positive direction. The boat is 70 m west of the torpedo at the time of launch. Therefore, $x_{0tb} = -70$ m. The x-component of the relative velocity is

$$\vec{v}_{tb,0,x} = \vec{v}_{t,0,x} - \vec{v}_{b,0,x} \tag{4}$$

The initial speed of the boat is zero. Since the torpedo is headed in a westerly direction, the x-component of its velocity is positive and is equal to $v_{t,0,x} = v_{t,0}\cos(31°)$, as we can see from Figure 4-18.

$$\vec{v}_{tb,0,x} = (v_{t,0,x} - v_{b,0,x})\hat{i} = v_{t,0}\cos(31°)\hat{i} - v_{b,0}\sin(21°)\hat{i}$$

$$= 3\cos(31°)\hat{i} \text{ m/s} - 0 = 2.572\hat{i} \text{ m/s} \tag{5}$$

The acceleration in the x-direction is given by

$$\vec{a}_{tb,x} = \vec{a}_{t,x} - \vec{a}_{b,x} \tag{6}$$

Since both accelerations point west, their x-components will be positive and

$$\vec{a}_{tb,x} = a_t\cos(31°)\hat{i} - 0.6\sin(21°)\hat{i} \tag{7}$$

Substituting (5) and (7) into (3), we have

$$x_{tb} = \frac{1}{2}(a_t\cos(31°) - 0.215 \text{ m/s}^2)t^2 + 2.572t - 70 = 0 \tag{8}$$

In the y-direction, we can apply Equation (4-12b) and Equation (3-25):

$$\vec{y}_{tb} = \vec{y}_t - \vec{y}_b \tag{9}$$

$$\vec{y}_{tb}(t) = \frac{1}{2}\vec{a}_{tb,y}t^2 + \vec{v}_{tb,0,y}t + \vec{y}_{tb,0} \tag{10}$$

Similarly, using Figure 4-18, we can write

$$y_{tb} = \frac{1}{2}(a_t\sin(31°) - a_b\cos(21°))t^2$$

$$+ (v_{t,0}\sin(31°) - v_{b,0}\cos(21°))t + 0 = 0 \tag{11}$$

Equations (8) and (11) have two unknowns: the acceleration of the torpedo and the time it takes to reach the boat. We can find the acceleration in terms of the time from Equation (11) and substitute into Equation (8). Equation (11) is a quadratic equation, but the constant term (c) in the quadratic formula) is zero, so there is only one root:

$$a_t = -2\frac{(v_{t,0}\sin(31°) - v_{b,0}\cos(21°))}{t\sin(31°)} + a_b\frac{\cos(21°)}{\sin(31°)}$$

$$= -\frac{2(3 \text{ m/s}\sin(31°))}{t\sin(31°)} + \frac{0.6 \text{ m/s}\cos(21°)}{\sin(31°)}$$

$$= -\frac{6 \text{ m/s}}{t} + 1.088 \text{ m/s}^2 \tag{12}$$

Substituting into Equation (8), we have

$$x_{tb} = \frac{1}{2}\left[\left(-\frac{6 \text{ m/s}}{t} + 1.088 \text{ m/s}^2\right)\cos(31°) - 0.215 \text{ m/s}^2\right]t^2$$

$$+ 2.572 \text{ m/s } t - 70 \text{ m} = 0$$

$$x_{tb} = -2.572 \text{ m/s } t + 0.466 \text{ m/s}^2t^2 - 0.1075 \text{ m/s}^2 t^2$$

$$+ 2.572 \text{ m/s } t - 70 \text{ m} = 0$$

$$t = \sqrt{\frac{70 \text{ m}}{0.3585 \text{ m/s}^2}} = 13.97 \text{ s}$$

Substituting this value for t into Equation (12), we get

$$a_t = -\frac{6 \text{ m/s}}{13.97 \text{ s}} + 1.088 \text{ m/s}^2 = 0.658 \text{ m/s}^2$$

Making sense of the result:

Both x and y components of the motion need to be accounted for as the torpedo approaches the boat. While the torpedo has the advantage of an initial speed on the boat, it has an extra 70 m to travel, which explains why it has to have a slightly higher acceleration than the boat.

FUNDAMENTAL CONCEPTS AND RELATIONSHIPS

Position, Velocity, Acceleration, and Time in Three Dimensions

Position:

$$\vec{r} = x\hat{i} + y\hat{j} + z\hat{k} \qquad (4\text{-}1)$$

Displacement:

$$\Delta\vec{r} = \Delta x\hat{i} + \Delta y\hat{j} + \Delta z\hat{k} \qquad (4\text{-}2)$$

Distance covered = path length
Average velocity:

$$\vec{v}_{avg} = \frac{\Delta\vec{r}}{\Delta t} = v_{avg,x}\hat{i} + v_{avg,y}\hat{j} + v_{avg,z}\hat{k} \qquad (4\text{-}3)$$

Instantaneous velocity:

$$\vec{v} = \lim_{\Delta t \to 0}\frac{\Delta\vec{r}}{\Delta t} = \frac{d\vec{r}}{dt} = v_x\hat{i} + v_y\hat{j} + v_z\hat{k} \qquad (4\text{-}4)$$

Average acceleration:

$$\vec{a}_{avg} = \frac{\Delta\vec{v}}{\Delta t} = a_{avg,x}\hat{i} + a_{avg,y}\hat{j} + a_{avg,z}\hat{k} \qquad (4\text{-}5)$$

Instantaneous acceleration:

$$\vec{a} = \lim_{\Delta t \to 0}\frac{\Delta\vec{v}}{\Delta t} = \frac{d\vec{v}}{dt} = a_x\hat{i} + a_y\hat{j} + a_z\hat{k} \qquad (4\text{-}6)$$

Projectile Motion

Horizontal distance for a projectile:

$$\Delta x = v_x t \qquad (4\text{-}7)$$

Vertical displacement:

$$\Delta\vec{y} = \frac{1}{2}\vec{a}t^2 + \vec{v}_{0y}t \qquad (4\text{-}8)$$

Vertical velocity:

$$\vec{v}_y(t) = \vec{a}t + \vec{v}_{0y} \qquad (4\text{-}9)$$

Time-independent kinematics equation for motion in the vertical direction:

$$v_y^2 - v_{0y}^2 = 2\vec{a}\cdot\Delta\vec{y} \qquad (4\text{-}10)$$

Circular Motion

Radial acceleration of an object moving in a circle of radius r at speed v:

$$a = \frac{v^2}{r} \qquad (4\text{-}11)$$

Total acceleration of an object moving in a circle:

$$\vec{a} = \vec{a}_r + \vec{a}_t$$

Relative Motion

The relative position of object 1 as seen by object 2:

$$\vec{r}_{12} = \vec{r}_{1g} - \vec{r}_{2g} \qquad (4\text{-}13)$$

Relative velocity:

$$\vec{v}_{12} = \vec{v}_{1g} - \vec{v}_{2g} \qquad (4\text{-}14)$$

Relative acceleration:

$$\vec{a}_{12} = \vec{a}_{1g} - \vec{a}_{2g} \qquad (4\text{-}15)$$

Applications: projectile motion, centrifuges, race car acceleration, effects of acceleration on the human body, military tactics

Key Terms: nonuniform circular motion, radial acceleration, tangential acceleration, total acceleration, uniform circular motion

QUESTIONS

1. A projectile is launched horizontally from a cliff at a very high speed with a horizontal acceleration that slows the projectile down but decreases with time. Which trajectory in Figure 4-19 best describes the motion of the projectile?

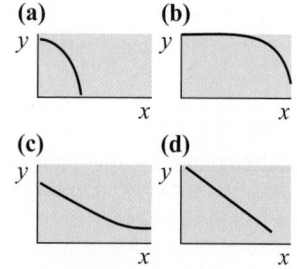

Figure 4-19   Question 1

2. A missile is launched horizontally with a low speed. The missile's engine accelerates it in the horizontal direction only. The acceleration provided by the engine increases with time. Which path in Figure 4-20 best describes its trajectory?

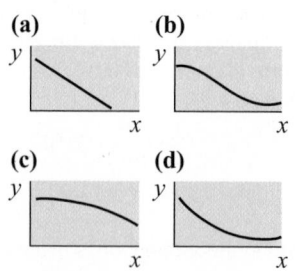

Figure 4-20   Question 2

3. You release a marble at the top edge of a bowl. The marble rolls down the inside of the bowl and continues up the opposite side. What is the direction of the marble's acceleration when the marble comes to a momentary stop at its highest point on the side of the bowl?
   (a) down
   (b) normal to the bowl
   (c) tangent to the bowl
   (d) The acceleration is zero, so it has no direction.

4. A rocket is launched horizontally off a cliff. The rocket engine provides a constant horizontal acceleration with a magnitude greater than the magnitude of the acceleration due to gravity. Which path in Figure 4-21 best corresponds to the trajectory of the rocket?

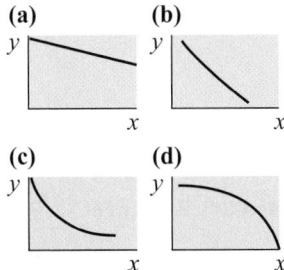

Figure 4-21 Question 4

5. A missile is launched from a plane travelling horizontally. The missile leaves the plane in a horizontal direction, with an acceleration that increases linearly with time. Which path in Figure 4-22 best describes its trajectory?

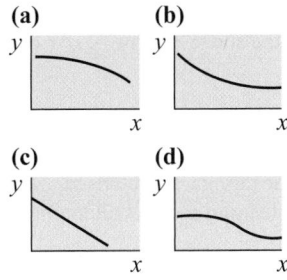

Figure 4-22 Question 5

6. You intend to cross a river that flows due south. Your speed with respect to the water is twice the speed of the river current. In which direction relative to the current should you swim to cross in the shortest time?
   (a) at 30° to the river
   (b) at 60° to the river
   (c) at 45° to the river
   (d) perpendicular to the river

7. You have a battery-powered model boat that has a fixed rudder so that it always moves straight ahead. You launch the boat in a pond, aiming straight at a point 60 m away on the opposite side of the pond. The boat reaches the shore 12 m north of the target point because of a current in the pond. On the next run, you aim the boat 12 m south of your target point. The boat will
   (a) arrive at the target;
   (b) arrive north of the target;
   (c) arrive south of the target.
   (d) Cannot answer without more information.

8. Two objects, A and B, travelling along the same line have speeds v_A and v_B, respectively. The speed of object A with respect to object B is
   (a) $v_B - v_A$;
   (b) $v_A - v_B$;
   (c) $v_A + v_B$.
   (d) Could be any of the above.

9. Two objects move along the same circular trajectory with the same speed. They start at the same time, one on the x-axis and the other on the y-axis. When the objects move in the same direction around the circle,
   (a) their relative position vector will always point in the same direction;
   (b) their relative acceleration vector will not change;
   (c) the magnitude of their relative acceleration will always equal $\dfrac{v^2}{r}$.
   (d) None of the above.

10. An object falls straight down while another object moves in a vertical circular trajectory at a speed that is not necessarily constant. The acceleration vector of the free-falling object can be equal to
    (a) the radial acceleration vector of the object in the circle;
    (b) the linear acceleration vector of the object in the circle;
    (c) the total acceleration vector of the object in the circle.
    (d) Any of the above.

11. You slide a marble along your kitchen counter toward a friend who stands 1 m from the end of the counter. Your friend is holding a marble in her hand at the height of the counter. She lets go of her marble just as your marble leaves the countertop. The speed of your marble is such that it will travel 1 m horizontally before your friend's marble hits the floor, as shown in Figure 4-23. Which of the following statements is true?
    (a) The two marbles will never collide.
    (b) Your marble will pass over your friend's marble.
    (c) Your marble will pass below your friend's marble.
    (d) The marbles will collide before they hit the floor.

Figure 4-23 Question 11

12. A projectile is fired at an angle of 30° with an initial speed of 340 m/s. It lands at the same elevation from which it was fired. While in the air,
    (a) the projectile's average acceleration is zero;
    (b) the projectile's average acceleration is 9.81 m/s^2;
    (c) the projectile's average velocity in the y-direction is zero.
    (d) Both (b) and (c) are true.

13. An object completes 2.25 revolutions around a circle of radius $4/\pi$ m. Which of the following statements is true?
    (a) The distance covered by the object is 4 m.
    (b) The displacement of the object is $A = \dfrac{2\sqrt{2}}{\pi}$.
    (c) Both (a) and (b) are correct.
    (d) None of the above are true.

14. An object travelling at a constant speed takes 3 s to complete three full cycles around a circle of radius 1 m. Which of the following statements is true?
    (a) The object's average velocity is zero.
    (b) The object's average speed is 2π m/s.
    (c) The object's average acceleration is $4\pi^2$ m/s^2.
    (d) Both (a) and (b) are true.

15. A sharpshooter has a rifle that fires bullets at a speed of 1100 m/s. A trainer 100 m away throws an apple 30 m straight up in the air. The sharpshooter tracks the apple and fires directly at it when it is at its maximum height. Which of the following statements is true?
    (a) The bullet will hit the apple at the apple's maximum height.
    (b) The bullet will pass under the apple on its way down.
    (c) The bullet will pass over the apple on its way down.
    (d) The bullet will hit the apple on its way down.

16. Two archers aim for the same coconut on a tree 14 m above the ground. One archer stands 50 m away and shoots an arrow at 60 m/s. The other archer stands 70 m away and shoots an arrow at 87 m/s. They both aim directly at the coconut and release their arrows at the moment the coconut begins to fall from the tree. Which of the following statements is true?
    (a) The two arrows will hit the coconut at the same time.
    (b) Neither arrow will hit the coconut.
    (c) The faster arrow will hit the coconut but not the slower one.
    (d) Both arrows will hit the coconut but at different times.

17. Two cannons, side by side on a cliff, fire projectiles at the same speed, one horizontal and one at an angle of 30° above the horizontal. Which of the following statements is true?
    (a) The vertical distance between the two projectiles is $v\sin(30°)t$.
    (b) The vertical distance between the two projectiles is $v\sin(30°)t - 4.9t^2$.
    (c) The horizontal distance between the two projectiles is constant.
    (d) Both (a) and (c) are correct.

18. NASA's 20 g centrifuge is made to spin at 17 g. Which of the following statements is true?
    (a) The acceleration is constant.
    (b) The magnitude of the acceleration is constant.
    (c) The velocity is constant.
    (d) None of the above statements are true.

19. If you throw a tennis ball straight up (relative to yourself) while running at a constant speed, the ball will land
    (a) in your hand;
    (b) slightly ahead of you;
    (c) slightly behind you.
    (d) I cannot tell.

20. While a projectile is in the air, can its velocity and acceleration vectors ever be parallel or perpendicular? Explain.

21. One of Robin Hood's men shoots an arrow horizontally at a target, with a speed of 75 m/s. At the same time, the shooter drops an apple he is holding while shooting the arrow. Ignoring air resistance, which object will hit the ground first? Explain.

22. A projectile is launched at speed v, making an angle θ with the horizontal. What is the projectile's acceleration
    (a) right after the launch;
    (b) at the top of its trajectory?

23. A bead sits on top of an overturned hemispherical bowl and begins to slide down, as shown in Figure 4-24. What are the directions of the tangential acceleration and radial acceleration of the bead when the bead's height changes by one-quarter of the radius of the bowl?

Figure 4-24 Question 23

24. An astronaut gets bored while on an extravehicular activity outside a space station. She happens to have a piece of putty, which she throws at the station, with a speed of 20 m/s. The putty covers a distance of 50 m and sticks to the wall of the station. The recoil of the station is negligible. Which of the following is true?
    (a) The average speed of the putty is 10 m/s.
    (b) The average acceleration of the putty is 2 m/s^2.
    (c) The average acceleration of the putty is zero.
    (d) None of the above are true.

PROBLEMS BY SECTION

For problems, star ratings will be used, (✱, ✱✱, or ✱✱✱), with more stars meaning more challenging problems.

Section 4-1 Displacement, Velocity, and Acceleration in Two and Three Dimensions

25. ✱ A fireman hears the station alarm, runs 3.0 m straight to the fire station pole in 1.7 s, and descends 4.0 m down the pole in 2.0 s.
    (a) Find the magnitude of the fireman's overall average velocity.
    (b) Find the fireman's overall average speed.
    (c) Find the fireman's speed during the run.
    (d) Find the fireman's speed during the descent down the pole.
    (e) Can the quantities from (c) and (d) be considered components of the fireman's overall average velocity? Explain why or why not.

26. ✱✱ A cheetah chases a gazelle at a speed of 112 km/h for 3.00 s, then turns 42° from its original direction, runs at 78.0 m/s for 2.00 s, and finally tackles the gazelle, coming to a stop in 11.0 m.
    (a) Find the cheetah's average speed.
    (b) Find the cheetah's average velocity.
    (c) Find the cheetah's average acceleration.
    (d) How long does it take the cheetah to stop in the last 11.0 m? Assume constant acceleration.

27. ✱✱ A particle has the coordinates $x = r\cos(8t)$ and $y = r\sin(8t)$.
(a) Find the x- and y-components of the velocity of this particle at $t = 0.20$ s.
(b) Find the velocity at $t = 0.20$ s.
(c) Describe the trajectory of the particle. Express the position, velocity, and acceleration of the particle as a function of time using Cartesian notation.

28. ✱✱ A particle has coordinates given by $x = 11\cos(4t)$, $y = 11\sin(4t)$, and $z = 0.7t$.
(a) Describe the trajectory of the particle.
(b) Find the speed of the particle at $t = 4.0$ s.
(c) Express the particle's position, velocity, and acceleration as a function of time, using Cartesian notation.
(d) When will the acceleration vector make an angle of $30°$ with the x-axis? Is that time affected by the z-motion? Explain.

29. ✱✱ A particle has coordinates given by $x = 8t$ and $y = \dfrac{1}{16t}$. Describe the trajectory of the particle, and find the particle's speed and acceleration at $t = 3$ s.

30. ✱✱ A particle has coordinates given by $x = 20\cos(5t)$, $y = 5\sin(0.5t)$, and $z = 7t$.
(a) Express the particle's position, velocity, and acceleration as a function of time using Cartesian notation.
(b) What is the speed of the particle at $t = 0$?
(c) Find the velocity of the particle at (i) $t = \pi$ s, (ii) $t = 2\pi$ s, and (iii) $t = 3\pi$ s.
(d) When will the velocity make an angle of $45°$ with the positive x-axis? Does that happen only once? Explain, and describe the trajectory of the particle.

31. ✱✱ Starting from rest, a race car moves straight up an incline with a $37.0°$ angle at a constant acceleration for 5.00 s such that its speed reaches 23.0 m/s. The road becomes horizontal, and the car slows down with constant acceleration as it follows a circular curve that changes its direction by $90°$. The radius of the curve is 121 m, and the car is moving at a speed of 7.00 m/s at the end of the curve.
(a) Find the total displacement of the race car.
(b) Find the average velocity of the race car from the start until the end of the circular curve.
(c) Find the average acceleration of the race car during this trip.

Section 4-2 Projectile Motion

32. ✱ A projectile is fired from ground level and lands 43.0 km away 50.0 s later. Find the projectile's launch speed, minimum speed, angle of launch, average speed, average acceleration while in the air, and average velocity.

33. ✱ A child flings a marble such that it leaves the horizontal surface of a table that is 70 cm high. The marble lands with a speed of 7.0 m/s. How fast did the child fling the marble?

34. ✱ A particle is launched at an initial speed v making an angle θ above the horizontal. Prove that the trajectory of the particle is a parabola.

35. ✱✱ A stunt car driver drives off a 7.0 m-high cliff into a lake with a horizontal speed of 12 m/s. The car needs to clear an 8.0 m-long ledge that is 5.0 m below the edge of the cliff, as shown in Figure 4-25. The car just misses the ledge on the way down.
(a) With what speed does the car hit the water?

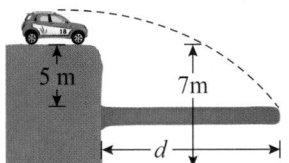

Figure 4-25 Problem 35

Section 4-3 Circular Motion

36. ✱✱ If Earth had no atmosphere, what would be the speed of a satellite in a circular orbit that just clears the 8848 m summit of Mount Everest?

37. ✱✱ The distance between the cities of Libreville, Gabon, and Kampala, Uganda, is 2570 km. Both cities are almost at the equator.
(a) What is the time zone difference between the two cities?
(b) You board a plane that takes off from Kampala toward Libreville. How fast does the plane have to travel so you do not have to adjust your clock when you arrive?

38. ✱✱ The distance between São Tomé off the coast of Africa and the city of Quito, Ecuador, is approximately 9500 km. Both cities are almost on the equator.
(a) How long does it take Earth to rotate to cover the distance between the two cities?
(b) You board a plane in São Tomé and fly west to Quito at a speed of 970 km/h, landing 12 h later. By how many hours do you need to adjust your watch? What is your radial acceleration when the plane's altitude is 10 000 m?

39. ✱ Pilots can sustain accelerations as high as 13 g during an inside loop manoeuvre. When a plane is moving at a speed of 830 km/h, what is the minimum radius for an inside loop turn the plane can fly without its acceleration exceeding 13 g?

40. ✱ Calculate the radial acceleration of a point on the Sun's equator.

41. ✱✱ A bead on top of an overturned hemispherical bowl of radius 32 cm starts sliding down the side of the bowl, as shown in Figure 4-24. By the time the bead has moved by $30°$, its speed is 0.92 m/s. Find the bead's tangential acceleration, radial acceleration, and total acceleration.

42. ✱ At the Istanbul Grand Prix, drivers sustain lateral accelerations (sideways) about five times the acceleration due to gravity while driving around curves at speeds of 280 km/h. Assuming a circular curve, find the radius of that section of the track.

43. ✱ An ultracentrifuge produces accelerations that are 2 million times the acceleration due to gravity. Find the speed of rotation for the ultracentrifuge, which has a radius of 4.5 cm.

44. ✱ The P2 centrifuge, used to enrich uranium, has a diameter of approximately 15 cm, a length of 1.0 m, and a peripheral speed of approximately 500 m/s.
(a) What is the acceleration of a point on the circumference of a P2 centrifuge?
(b) What is the acceleration of a point midway between the axis and the circumference of a P2 centrifuge?
(c) How many turns does the P2 centrifuge complete in 1 s?

CHAPTER 4 | **MOTION IN TWO AND THREE DIMENSIONS**

45. ✱✱ An object moves in a vertical circle of radius 4.0 m. The speed of the object is 12 m/s when the object is 30° above the lowest point along the circle, and 8.0 m/s when the object is at the top of the circle. The tangential acceleration is uniform. What is the total acceleration of the object when its speed is 8.0 m/s?

46. ✱✱ A particle moves along a circular path of radius r at a constant speed v. The particle is at the angular position ϕ above the positive x-axis, as shown in Figure 4-26. Without using calculus, write expressions for
    (a) the projection of the particle's position onto the x-axis as a function of time;
    (b) the velocity of the projection as a function of time;
    (c) the acceleration of the projection as a function of time.

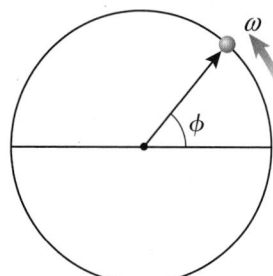

Figure 4-26 Problem 46

47. ✱✱ A steel ball that is accelerated on a circular track of radius 72 cm reaches a speed of 4.0 m/s over three circuits, starting from rest. How long after the ball starts to move does its radial acceleration equal its tangential acceleration? How many turns does the ball take for that to happen? Assume constant tangential acceleration.

48. ✱✱✱ An object moves on a circular track of radius 29 m such that the object's radial acceleration doubles every 2.0 s. The object completes its first circuit in 4.0 s. The object starts from rest, at which point its radial acceleration is zero.
    (a) Write the equation for the object's speed as a function of time.
    (b) What is the total acceleration of the object 6.0 s into the motion?
    (c) What is the distance covered as a function of time?

Section 4-4   Relative Motion in Two and Three Dimensions

49. ✱✱ A parade car moving at 10 km/h shoots an acrobat out of a cannon that is mounted at 90° to Earth's surface. The cannon fires the acrobat with a speed of 14 m/s. Where will the acrobat land relative to the parade car? Find the magnitude of the displacement of the acrobat relative to the ground. Ignore air resistance.

50. ✱✱ A child has a cannon mounted on a battery-powered car. While the car is moving at a speed of 2.00 m/s, the cannon fires a marble with a speed of 7.00 m/s with respect to the car, at an angle of 27.0° above the horizontal. The car's speed does not change noticeably.
    (a) What angle does the marble make with the horizontal when it lands on the ground?
    (b) How far will the marble land from the position from which it was launched?
    (c) How far is the marble from the car when it lands?

51. ✱✱ A swimmer swims in a river against the current for 5 min, and then swims with the current for another 5 min, and ends up 500 m away from where he started. Find the speed of the current in the river if the swimmer's speed relative to the water is 2.6 m/s.

52. ✱✱✱ Two cannons at the same location launch balls at a speed of 450 m/s at the same time. One cannon is aimed east and fires its ball at an angle of 31° above the horizontal, and the other is aimed north and fires its ball at an angle of 21° above the horizontal.
    (a) Find the relative speed of the two cannonballs 2.0 s and 5.0 s after firing.
    (b) Find the distance between the two cannonballs 3.0 s and 23.0 s after firing.

53. ✱✱ A director wants to produce a scene in which an archer shoots a watermelon that is dropped from a 30 m-high bridge. To compensate for gravity, the director asks the archer to aim 5.0 m above the watermelon, which is released at the same time that the archer releases his arrow. The initial speed of the arrow is 160 m/s, and the distance to the watermelon is 90 m. How close will the arrow get to the watermelon?

54. ✱✱ A child has mounted a spring-loaded cannon on a toy car such that the cannon fires backward from the car at an angle of 42° above the horizontal. While the car is moving forward at 1.1 m/s, the cannon fires a marble with speed v with respect to the car. The marble lands at an angle of 89° with the level ground. What is the speed of the marble with respect to the toy car?

55. ✱✱ A parade car has a cannon that shoots teddy bears into the crowd. As the car moves forward at 2.7 m/s, the cannon fires horizontally at an angle of 37° from the direction of motion of the car. The speed of the teddy bears with respect to the cannon is 6.0 m/s, the cannon is 3.0 m above the ground, and the recoil of the cannon is negligible. Find the speed of the teddy bear with respect to the ground.

56. ✱✱✱ During a clay pigeon shooting match, a clay pigeon is fired with an initial speed of 82 km/h at an angle of 39° above the horizontal. Then 1.7 s later, the shooter fires a shot that has a speed of 404 m/s. When the shooter is standing straight behind the clay pigeon launcher, at what angle should she aim?

57. ✱✱✱ You notice an eagle soaring at an altitude of 25 m, eyeing a rabbit. You want to scare the eagle away, so 1.0 s after the eagle flies directly overhead at 34 km/h, you throw a stone at an angle of 38° above the horizontal. You want the stone to pass 50 cm in front of the eagle. With what speed should you throw the stone? Assume your hand to be 2.0 m above the ground when you throw the stone.

58. ✱✱✱ Your opponent is operating a model helicopter in a simulated battle. You have a gun that fires paintballs at a speed of 40 m/s. The helicopter is hovering 12 m above the ground and is 17 m from you. Just before you fire, the helicopter dives at twice the acceleration due to gravity. You take 0.2 s to correct your aim. What angle must your gun make with the horizontal to hit the helicopter on its way down?

59. ✱✱✱ A bird flies 25 m above a hunter at a speed of 39 km/h. The hunter's gun fires a shot with a speed of 378 m/s. The hunter waits 4.0 s to judge the speed of the bird before firing. At what angle above the horizontal should the hunter aim?

COMPREHENSIVE PROBLEMS

60. ✱ A spinning neutron star with a radius of 16 km completes 712 rev/s.
   (a) Find the average speed of a point on the star's equator over one-third of a revolution.
   (b) Find the average acceleration of a point on the star's circumference over three-quarters of a revolution.
   (c) Find the distance covered by a point on the star's equator in 1 s.
   (d) Find the displacement of the point in part (c).

61. ✱✱ A monkey is 14 m above the ground in a tree and drops a papaya. At the same time, two other monkeys, one 11 m from the base of the tree and the other 16 m from the base of the tree, throw small stones at the papaya such that all three objects collide 0.9 s before the papaya hits the ground. Find the speed of each stone as it hits the papaya. Assume both stones were thrown from ground level.

62. ✱✱✱ Starting from rest, a particle moves under constant acceleration in a horizontal line such that its speed 3.00 s later is 1900 m/s. The particle is deflected by 37.0° in the horizontal plane without changing its speed. In this new direction, the particle accelerates such that it reduces its speed by half in 1.00 s. The particle is then deflected straight up, and immediately accelerates uniformly such that its velocity after 3.00 s is 9000 m/s straight up. For the whole 7.00 s interval, find the particle's total displacement, average velocity, average speed, and average acceleration. Express these quantities using magnitude and angles measured with respect to the positive coordinate axes.

63. ✱✱ A missile is launched with an initial speed of 410 m/s, making an angle of 34° above the horizontal. Radar tracking shows that the altitude of the missile is given by $60t^2 - 7t^3 + 4t + 120$, and the horizontal distance from the point of launch is given by $1400 + 3t$.
   (a) Determine the maximum height reached by the missile.
   (b) Find the speed and horizontal displacement of the missile just before it hits the ground 120 m below the point of launch.
   (c) Where does the maximum acceleration of the missile occur?

64. ✱✱ A sounding rocket has an engine that allows it to maintain a constant horizontal speed of 190 m/s while providing it with a constant vertical acceleration of 32 m/s² until it reaches a height of 60 km. The engine then shuts off, and the rocket continues under the influence of just the gravitational force. Assume that air resistance and the change in gravitational acceleration with altitude are negligible.
   (a) What is the maximum height reached by the rocket, and how long does it take to reach that height?
   (b) What angle does its velocity make with the horizontal just before the landing?

65. ✱✱ A projectile is fired at speed 140 m/s such that the projection of its velocity onto the xy-plane makes an angle of 40° with the positive x-axis. The initial velocity vector makes an angle of 29° with the positive z-axis. Express the x-, y-, and z-coordinates of the projectile as a function of time, taking the launch point as the origin (0, 0, 0).

66. ✱✱ A fireworks rocket is fired horizontally from a height of 17 m above the ground. The rocket has a horizontal acceleration that is twice as large as the acceleration due to gravity.
   (a) What trajectory will the rocket follow?
   (b) What will its speed be just before it hits the ground?
   (c) Find the horizontal distance between the launch and landing points.

67. ✱ A child fires a tomato from an improvised catapult from a tree house 4.00 m above the ground. The tomato leaves the catapult at a speed of 35.0 m/s, making an angle of 37° with the horizontal.
   (a) Find the maximum height reached by the tomato.
   (b) How far from the tree house will the tomato land?
   (c) How fast is the tomato moving just before it lands?

68. ✱✱ A plane is approaching a remote community in northern Manitoba to drop a crate of supplies when clear turbulence buffets the plane, pushing it suddenly upward. The crate tears loose at an altitude of 160 m, leaving the plane with a velocity of 290 km/h directed at an angle of 23° above the horizontal. What is the speed of the crate just before it lands?

69. ✱✱ The location of Toronto's Lester B. Pearson International Airport is 43°40′ N latitude and 79°36′ W longitude, and the Beijing Capital International Airport is situated at 40°44′ N and 116°36′ E. A plane takes you from Toronto to Beijing in 11.0 h.
   (a) How fast is the plane?
   (b) By how much do you have to adjust your watch?

70. ✱ How fast would a plane have to fly at an altitude of 13 000 m above the equator so that passengers would not have to adjust their watches? Would the plane be heading east or west?

71. ✱✱ You are onboard a plane flying west at an altitude of 600 m above the 32nd parallel. You fly over a given point at 12:00 p.m. local time, and arrive at an airport seven time zones away such that you only have to adjust your watch by 1 h. Estimate the airspeed of the plane.

72. ✱ Express y as a function of x for a projectile that is launched with speed v_0 at an angle θ and lands at the same level from which it was launched. Derive expressions for the angle for which the maximum distance is covered, the distance at which the maximum height occurs for that angle, and the maximum height.

73. ✱✱ A particle's position is given as $(2 + (0.5) \sin(0.4t))\hat{i} + (3 + (0.5) \sin(0.4t))\hat{j} + 2t^2\hat{k}$.
   (a) What is the position of the particle at $t = 0$?
   (b) Find the speed of the particle at $t = 0.20$ s.
   (c) Find the average velocity of the particle between $t = 0.20$ s and $t = 0.4$ s.
   (d) Find the average acceleration of the particle between $t = 0.2$ s and $t = 0.4$ s.
   (e) Find the displacement of the particle between $t = 0$ s and $t = 1.1$ s.
   (f) What is the trajectory of the particle?

74. ✱✱ Two disks of radius 32 cm are placed side by side, and the two closest points on the disks are painted red. The disks are then made to spin about fixed axes in opposite directions at 10 rpm.
   (a) Write an expression for the relative acceleration of one red point with respect to the other.
   (b) Write expressions for the relative velocity and relative speed of the two points.
   (c) Given that the two points are in contact at the start, write an expression that describes their relative position.

75. ✱✱ A ring lies in a horizontal plane and spins about its symmetry axis at a rate of 3 rev/s. Find the position, acceleration, and velocity of a point A on the ring with respect to a point B that is in the same plane and is a distance $3r$ from A, given that the distance between A and B is $2r$ at $t = 0$.

76. ✱✱ A Boeing 747 is flying 790 km/h at an altitude of 20 km in a direction 11° [N of W]. A smaller jet at the same altitude is 12 km away 17° [N of E] from the Boeing. The velocity of the smaller jet is 430 km/h 23° [W of S]. Find the distance of closest approach between the planes and the time they take to reach that distance.

77. ✱✱ NASA reports that the position of a space shuttle during the first few minutes of its flight can be represented by the following equations:

altitude: $h(t) = 658.8 - 0.154t^3 + 18.3t^2 - 345t$

down-range distance: $R(T) = 1426.5e^{0.029t}$

Find the position, speed, and acceleration of the shuttle 1 min after takeoff.

78. ✱✱ A test pilot is placed in a high-g training pod that follows a horizontal path of radius 20 m. The pod accelerates from rest to achieve an acceleration six times the acceleration due to gravity in the radial direction over five turns. The pod has constant tangential acceleration during these five turns. The pod then continues at a constant speed for 2.0 s, and then undergoes constant deceleration, coming to a stop in the next 6.0 s.
   (a) How long does the pod take to finish the first five turns?
   (b) What is the total acceleration of the pod 3.0 s after it begins to decelerate?
   (c) What is the total number of turns the pod makes?
   (d) Find the magnitude of the average acceleration between 3.0 s and 6.0 s.
   (e) Find the magnitude of the average velocity between 4.0 s and 6.0 s.
   (f) Find the magnitude of the displacement of the pod between $t = 2.0$ s and $t = 5.0$ s.

79. ✱✱ You are in the passenger seat of a car driving on a country road at 43 km/h. You throw an acorn out the window at an angle of 26° above the horizontal with a speed of 4.0 m/s relative to the car. The projection of the acorn's velocity onto the horizontal makes an angle of 34° with the velocity of the car.
   (a) How far ahead of the car is the acorn 1 s later?
   (b) What is the speed of the acorn in flight with respect to the ground?
   (c) What angle with respect to the horizontal does the acorn's velocity make just before the acorn lands? What angle does the projection of the acorn's velocity make with the road?

80. ✱✱ Two monkeys standing 20 m apart are both trying to use stones to knock down a coconut hanging 7.0 m up in a tree that is directly between them. One monkey is 7.0 m from the tree, and the other is 13 m from the tree. They aim straight for the coconut, not accounting for the acceleration due to gravity. The close monkey throws a stone at speed v, and the far monkey throws a stone at $2v$. The stone thrown by the close monkey strikes the ground 3.0 s after it leaves her hand. By what vertical distance do the two stones miss each other?

81. ✱✱✱ In a robot demolition derby, your friend's robot, Sir Burnalot, is stuck and is being charged by an opponent's more powerful robot, Sir Crunchalot. You have a cannon that fires a heavy ball at a speed of 7.0 m/s. Crunchalot is moving at 1.1 m/s on course for a head-on collision with Burnalot 11 m in front of it. Your cannon is 4.0 m behind Burnalot, which is 50 cm tall. At what angle must you fire the ball to knock over the opposing robot before it gets to Burnalot?

82. ✱✱ A particle's position along the x-axis is given by $x = 20\cos(0.4t)$, and its position along the y-axis is given by $y = 20\sin(0.4t)$.
   (a) Describe the trajectory of the particle.
   (b) Find the time it takes the particle to finish a complete cycle. (This time is the period of the motion.)
   (c) Find the tangential speed of the particle.
   (d) Find the magnitude of the acceleration of the particle at a given time t.

83. ✱✱ Given that the particle in problem 82 is at $x = 20$ at $t = 0$, find
   (a) the velocity of the particle at $t = 12$ s;
   (b) the average velocity and the average acceleration of the particle between $t = 12$ s and $t = 20$ s.

84. ✱✱✱ You are playing an innovative fortress defence game. You build a 4.0 m-high fortress out of wood and cover the front wall with a thin layer of playdough. Your opponents build their fortress 21 m from yours and mount a home-built cannon at its edge, 7.0 m above the ground. Their cannon fires baseballs with a speed of 15.0 m/s. They fire a baseball at your fortress at an angle of 19° above the horizontal. The opponents score points with every mark their baseball leaves on your playdough. You have a defensive cannon that shoots golf balls at a speed of 52.0 m/s. If your cannon fires from ground level 0.80 s after their cannon fires, at what angle must you fire your golf ball to hit the incoming baseball?

85. ✱✱ A pilot points her plane in the direction of an airfield that is 300 km east and 210 km north of her current location. Since the winds are light and varying, the pilot does not attempt to correct her bearing for wind speed. Near the end of her flight, she finds that her course will take her 34 km north of the airfield. The plane's airspeed is 270 km/h, and the flight would take 98 min by the most direct route. Find the bearing that would have taken her straight to the airfield.

86. ✱✱ In a marine-rescue exercise, a small boat moves at velocity 13 km/h [37° N of E]. The rescue boat has a speed of 30 km/h and is positioned 60 m [42° S of E] from the small boat.
   (a) In what direction must the rescue boat head to intercept the small boat?

87. ✱✱✱ A pheasant flying at 28 m/s at a height of 20 m passes directly over a hunter. The hunter's shotgun fires pellets at a speed of 332 m/s. At what angle should the hunter aim to hit the pheasant if he fires just as it passes overhead? Assume that the effects of wind and air resistance are negligible.

88. ✱✱✱ Figure 4-27 is a speed versus time graph for a meteoroid fragment entering the atmosphere. Focus on the region where the speed drops from 45 to 35 km/s. Use the graph to construct approximate distance versus time and acceleration versus time graphs.

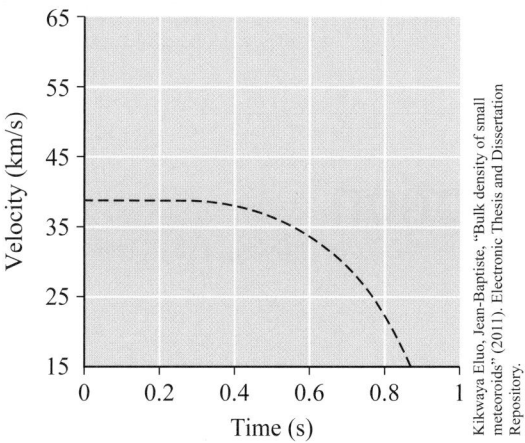

Figure 4-27   Problem 88

89. ✷✷ Figure 4-28 shows an altitude versus time plot for the *Arianne 5*'s ascent to space.
(a) Estimate the average radial velocity (i.e., vertical component of velocity) of the rocket between launch time and the separation of the second rocket stage (point H2 on the graph).
(b) Estimate the radial speed at $t = 200$ s, $t = 400$ s, $t = 1500$ s, and $t = 2000$ s.
(c) Describe the radial acceleration of the Arianne 5 after point H3 on the graph (separation of stage 3 of the rocket).
(d) Why does the slope of the graph in Figure 4-28 not correspond to the total speed of the rocket?

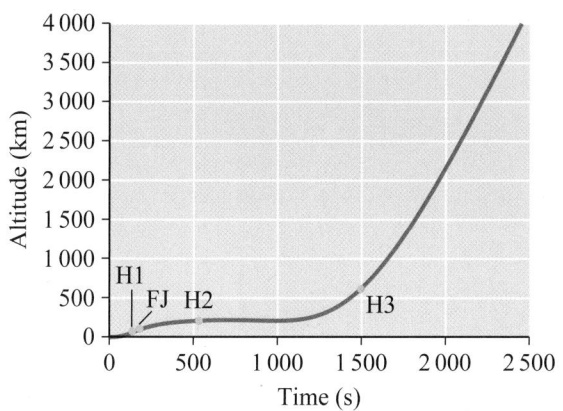

Figure 4-28   Problem 89

90. ✷✷ Figure 4-29(a) shows the acceleration curve for a Ferrari Scuderia (Figure 4-29(b)).
(a) What is the maximum acceleration on the curve?
(b) Calculate the acceleration at $t = 7$ s.

(c) What is the average acceleration between $t = 0$ and $t = 14$ s?
(d) How far does the car go before reaching maximum speed?
(e) How far does the car go between $t = 14$ and $t = 16$ s?
(f) What is the total distance covered by the car during the whole interval shown?

Figure 4-29(a)   Problem 90

Figure 4-29(b)   Problem 90

91. ✷ You want a golf ball on its way to the hole to clear a 12 m-tall tree 40 m away. You estimate that the golf ball will leave the ground at an angle of 47°. What initial speed must the golf ball have to just clear the tree? Is there only one answer to this problem? Explain.

See the text online resources at
www.physics1e.nelson.com for Open Problems
and Data-Rich Problems related to this chapter.

CHAPTER 4 | **MOTION IN TWO AND THREE DIMENSIONS**

Learning Objectives

When you have completed this chapter you should be able to:

1 Represent forces as vectors, and resolve and add forces.

2 Explain Newton's laws and the relationship between force, mass, and motion.

3 Draw free body diagrams to resolve forces into components, and write and solve equations of motion.

4 Explain weight, *g* forces, and weightlessness.

5 Incorporate normal forces into free body diagrams.

6 Differentiate between tension and compression, and explain the action of tension in a string.

7 Distinguish between static and kinetic friction, and define the maximum force of static friction.

8 Explain Hooke's law and the forces exerted by an ideal spring.

9 Identify the forces contributing to the circular motion of an object.

10 Explain fictitious forces and the difference between inertial and noninertial reference frames.

e For additional Making Connections, Examples, and Checkpoints, as well as activities and experiments to help increase your understanding of the chapter's concepts, please go to the text's online resources at www.physics1e.nelson.com.

Chapter 5
Forces and Motion

During the reentry of a space capsule into Earth's atmosphere (Figure 5-1), the two main forces acting on the capsule are gravity and drag resulting from collisions with air molecules. Reentry capsules are shaped to provide some lift to slow their descent if they enter the atmosphere at the correct angle. If the trajectory is too steep, the drag force, or the force of air resistance, can overheat the hull of the capsule and decelerate the capsule so rapidly that the crew may be harmed. On April 19, 2008, a malfunction caused the *Soyuz TMA-11* capsule to enter the atmosphere at an abnormally steep angle. Although the three astronauts onboard survived, they experienced forces as high as 11 times the force of gravity on Earth's surface.

© Sergei Remezov/Reuters/Corbis

Figure 5-1 Image of Soyuz TMA 11 after re-entry

LO 1

5-1 Force and Net Force

We first discuss the basic properties of forces and then their relation to mass and motion in the context of Newton's laws. Later in the chapter we outline and apply strategies for solving problems involving the more common types of mechanical forces.

Forces Are Vectors

Forces are vectors; therefore, they have both magnitude and direction. When you push against a wall, you are exerting a force. When you pull a door handle, you are also exerting a force. How hard you push or pull determines the magnitude of the force you exert. The direction in which you push or pull determines the direction of the force vector. The application of force on an object results in the motion of the object, the deformation of the object, or both. In this chapter, we will deal primarily with objects that do not deform.

Fundamental and Nonfundamental Forces

In this chapter, for the most part we will focus on forces that require physical contact between objects, except for the force of gravity. Such contact forces include friction, tension, elastic forces, normal forces, and applied forces. By "applied force" we specifically refer to the force applied by you, me, or a donkey, for example, on a crate, a box, or some other object. These forces are typically called **nonfundamental forces**. In contrast, the

fundamental forces in nature – gravitational attraction, the electromagnetic force, and the two nuclear forces – do not require contact between objects and can act at a distance. Although the exact nature of the four fundamental forces is not fully known, they are real, identifiable, and measurable. What we sometimes refer to as nonfundamental forces are, in fact, macroscopic manifestations of the fundamental forces. The fundamental forces underlie the nonfundamental forces.

Forces must have real physical identifiable causes. They must be one of the fundamental or nonfundamental forces, a few of which are mentioned above.

Net Force

The net force acting on an object is the sum of all the forces acting on that object. Forces are vectors, so by "sum" we mean the "vector sum."

> ✓ **CHECKPOINT**
>
> ### C-5-1 Net Force of "Squeeze"
>
> You squeeze a baseball between your hands with a force of 210 N from each hand. Assuming no deformation, the net force acting on the ball is
> (a) 420 N;
> (b) zero;
> (c) 210 N.
> (d) None of the above.

C-5-1 (b) the two forces add up to zero.

 ## EXAMPLE 5-1

Adding Forces

Calculate the net force on the object in Figure 5-2(a). Give the magnitude of the resultant force and the angle it makes with the direction of $\vec{F}_1$.

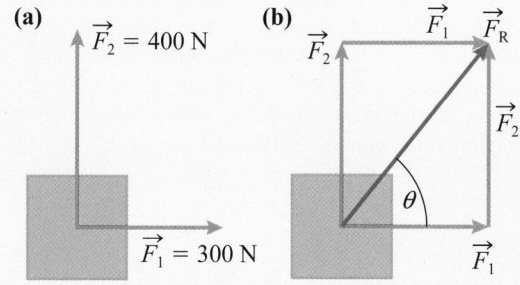

Figure 5-2   Example 5-1

SOLUTION

$$\vec{F}_R = \vec{F}_1 + \vec{F}_2$$

Since the forces act at 90° to each other, we can apply Pythagoras' theorem:

$$F_R^2 = F_1^2 + F_2^2 = (300 \text{ N})^2 + (400 \text{ N})^2 = 2500 \text{ N}^2$$

$$F_R = 500 \text{ N}$$

giving us a magnitude of 500 N for F_R.

From Figure 5-2(b), we can see that the angle that the resultant makes with $\vec{F}_1$ is given by

$$\tan\theta = \frac{400 \text{ N}}{300 \text{ N}} \Rightarrow \theta = 53.1°$$

Making sense of the result:

The forces make a 3;4;5 triangle, therefore the value of 500 N is expected.

MECHANICS

EXAMPLE 5-2

Calculating Net Force

Calculate the net force acting on the object in Figure 5-3.

$$\overrightarrow{F_2} = 700 \text{ N} \qquad \overrightarrow{F_1} = 1100 \text{ N}$$

Figure 5-3   Example 5-2

SOLUTION

$$\overrightarrow{F}_R = \overrightarrow{F}_1 + \overrightarrow{F}_2 \qquad (1)$$

we will pick the direction of $\overrightarrow{F}_1$ to be positive. Therefore,

$$F_R = F_1 - F_2 = 1100 \text{ N} - 700 \text{ N} = 400 \text{ N} \qquad (2)$$

where the quantities in Equation 2 are variables that denote the magnitudes of the forces in Equation (1).

Making sense of the result:

The vector sum results in the magnitudes of the forces being subtracted.

LO 2

5-2   Newton's Laws

Newton's First Law

When a hockey player's stick strikes a puck, the puck moves in a straight line on the ice, and its speed does not change significantly as it travels across the rink if the ice is smooth. In fact, if there were no friction between the puck and the ice, the velocity of the puck would not change at all. If you want to change the speed or the direction of the moving puck, you have to exert a force on it. In addition, if you were to set the puck gently on the horizontal ice surface, the puck would remain where you placed it. The puck will remain at rest unless a force is exerted on it.

These examples demonstrate **Newton's first law:** If no net force is exerted on an object, the object's velocity will not change.

✓ CHECKPOINT

C-5-2 Dropped Ball

A ball is dropped from rest and bounces off the floor. Which of the following statements applies to the ball on its way up?

(a) The ball's acceleration is upward because it is moving upward.

(b) There must be a force causing the ball to continue to move upward.

(c) The ball is only under the influence of the force of gravity as it moves upward.

(d) None of the above apply.

C-5-2 (c) the only force acting on the ball, in the absence of air resistance, is the force of gravity.

Newton's Second Law

If a hockey puck is at rest on the ice and you push it with a horizontal force, the puck will move in the direction of the force, and the speed of the puck will continue to increase as long as you apply the force. The response of the puck to the force you apply to it is given by Newton's second law, which is often presented in equation form:

KEY EQUATION
$$\overrightarrow{F}_{\text{net}} = m\overrightarrow{a} \qquad (5\text{-}1)$$

EXAMPLE 5-3

Which Way Does the Puck Move?

The puck in Figure 5-4 sits on frictionless ice and is acted upon by the forces as shown. What can we say about the motion of the puck: is the puck moving to the left, to the right, or is it stationary?

SOLUTION

The direction of the net force on an object does not necessarily dictate the direction in which the object is moving. We do not know which way the puck is moving. For example, when you throw a ball up in the air, it is under the influence of only the force of gravity as it moves up. The net force on the ball is downward as the ball moves up. The force of gravity will eventually cause the ball to slow down and move downward, but at any given moment the ball could be moving up (right after you throw it), down, or it could be in such a position that its speed is zero. The same holds for the puck: it could be moving to the right, to the left, or, at the instant shown, not be moving at all (at the point where it changes direction of motion from left to right).

$$\overrightarrow{F}_1 = 60 \text{ N} \qquad\qquad \overrightarrow{F}_2 = 110 \text{ N}$$

Figure 5-4   Example 5-3

Equation (5-1) states that the acceleration of an object is directly proportional to the net force exerted on it. If you double the force on the puck in the above example, the acceleration will double. If you double the mass while keeping the force constant, the acceleration will drop by a factor of 2. It is much easier to push a hockey puck across the ice than push a more massive object such as a hockey player. Ultimately, how massive an object is depends on how much material it contains.

EXAMPLE 5-4

Force and Motion

An object moving to the right is under the effect of a force that points to the right but whose magnitude decreases linearly with time. What can we say about the speed of the object: is it decreasing or increasing as a result of this force?

SOLUTION

Despite the fact that the magnitude of the force is decreasing with time, at any given point the force is non-zero and pointing to the right. According to Newton's second law in Equation (5-1), the acceleration is to the right; therefore, the speed is increasing. Of course, this applies as long as the force does not reach a value of zero, at which point the acceleration will be zero and the puck's speed will be constant.

Making sense of the result:

The fact that the magnitude of the force decreases with time means that the acceleration decreases with time, but as long as the acceleration is not zero, the puck will continue to accelerate and its speed will continue to increase.

Inertia refers to an object's resistance to being accelerated. The quantitative measure of inertia is mass. A net force of 1 N will cause a 1 kg mass to accelerate at 1 m/s²:

$$1 \text{ N} = 1 \text{ kg m/s}^2$$

Point Mass A point mass is an object that has mass but whose dimensions are infinitesimally small. The concept of the point mass does not take into account the dimensions and shape of the body. For all intents and purposes, all the objects in this chapter will be treated as point masses. This will serve us well as long as we do not consider any rotation of these objects that could be caused by the forces acting on them. In the diagrams we draw, however, the objects will be shown as having some dimensions and not just as point masses for clarity and sometimes ease of perception.

MAKING CONNECTIONS

The International Standard for Mass

Currently, SI defines the base unit for mass—the kilogram—as the mass of the International Prototype Kilogram (IPK), a cylinder machined out of a platinum iridium alloy, first produced in 1879 (Figure 5-5). Although precise copies were given to other nations for reference, the original IPK is kept in Sèvre, France. The IPK is currently the only SI unit based on an actual artifact.

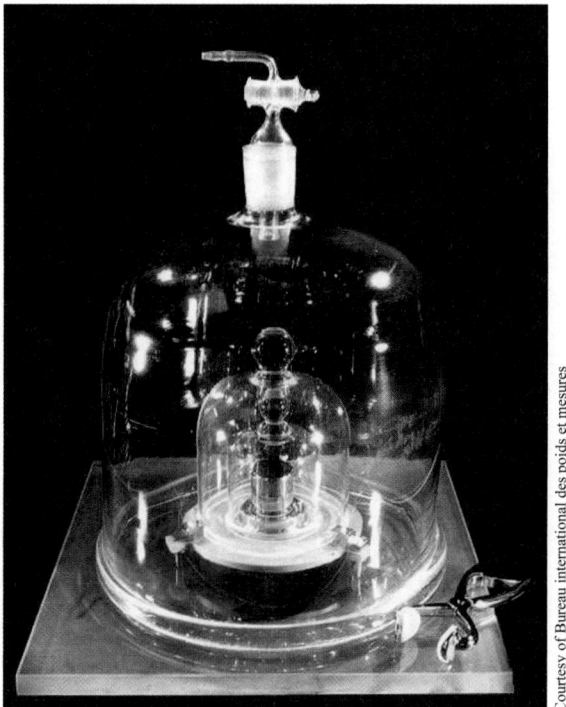

Courtesy of Bureau international des poids et mesures

Figure 5-5 The IPK, held at the Bureau International des Poids et Mesures

✓ CHECKPOINT

C-5-3 Net Force

A cat of mass 2.4 kg enters an elevator, which moves upward with an acceleration of 4 m/s². The net force on the cat is

(a) 33.1 N;

(b) 23.5 N;

(c) 13.9 N;

(d) 9.6 N.

C-5-3 (d) By Newton's second law, to get the net force on an object one multiplies its mass with its acceleration.

EXAMPLE 5-5

Force and Acceleration

A Formula One race car is at rest at the start line. When the race starts, the car attains a speed of 135 km/h in 4.50 s. Find the net force on the driver, who has a mass of 67.0 kg. Assume constant acceleration.

SOLUTION

Newton's second law relates the net force on the driver and his acceleration:

$$F_{net} = ma$$

The driver's acceleration is the same as the car's acceleration, and from Equation 3-15

$$v(t) = at + v_0$$

The initial speed is zero, and the final speed is

$$v(t) = 135 \text{ km/h} = 135\left(\frac{1000 \text{ m}}{3600 \text{ s}}\right) = 37.5 \text{ m/s}$$

Hence,

$$a = \frac{v(t) - v_0}{t} = \frac{37.5 \text{ m/s} - 0}{4.5 \text{ s}} = 8.33 \text{ m/s}^2$$

Therefore,

$$F_{net} = 67.0 \text{ kg } (8.33 \text{ m/s}^2)$$

$$F_{net} = 558 \text{ N}$$

Making sense of the result:

All we need to know is an object's mass and its acceleration for the net force acting on it.

EXAMPLE 5-6

Braking Distance

A 120 tonne metro train is brought to a stop from a speed of 23.0 m/s using a constant braking force of 400 kN. Over what distance does the train come to a stop? Assume that the only force stopping the train is the force from the brakes.

SOLUTION

Since the only force acting on the train is from the brakes, we choose the positive x-axis to be the direction of motion of the train.

$$F_{net} = ma \Rightarrow a = \frac{F_{net}}{m} = \frac{400\,000 \text{ N}}{120\,000 \text{ kg}} = 3.33 \text{ m/s}^2 \quad (1)$$

We can use the time-independent kinematic equation for constant acceleration to find the distance that the train travels while braking:

$$v_f^2 - v_i^2 = 2\vec{a} \cdot \Delta\vec{x} \quad (2)$$

The final speed is zero. Since the train is slowing down, the acceleration and the displacement have opposite directions. Therefore, using Equation 2-18, $\vec{a} \cdot \Delta\vec{x} = |\vec{a}||\Delta\vec{x}| \cos(180°)$ and Equation (2) becomes

$$0 - (23 \text{ m/s})^2 = -2(3.33 \text{ m/s}^2)\Delta x \Rightarrow \Delta x = 79.43 \text{ m} = 79.4 \text{ m}$$

where Δx denotes the magnitude of the displacement

Making sense of the result:

400 KN is a large force, but the train is also very massive.

EXAMPLE 5-7

Force in Circular Motion

You are sitting in the rear seat of a car, which is going around a track with a radius of 120 m at a speed of 26.0 m/s. The track is banked at an angle of 21°. The coefficient of static friction between the car and the road is 0.34. The car is on the verge of sliding up the bank. A friend sitting next to you pushes you against the door with a force of 290 N. Another friend, sitting in the seat in front of you, pulls you forward, squeezing you against the back of her seat with a force of 360 N. Your mass is 68 kg. Calculate the net force acting on you.

SOLUTION

From the statement of the problem, you are evidently not moving relative to either of your friends or the car; therefore, your acceleration is the same as the car's acceleration.

According to Newton's second law, the net force on you is simply your mass times your acceleration, and since the car is moving in a circle, that acceleration, as we saw in Chapter 4, is equal to v^2/r.

$$F_{net} = ma = \frac{mv^2}{r} = \frac{(68 \text{ kg}) (26.0 \text{ m/s})^2}{120 \text{ m}} = 383 \text{ N}$$

Making sense of the result:

Here, Newton's second law allows us to calculate the net force without considering the individual forces. There is extra information given in the question, but that information is not needed. The point here is that all you need is the mass and the acceleration to determine the net force.

Newton's Laws Before Newton?

Before Isaac Newton (1642–1727), Western civilization's perception of motion was dominated by Aristotle's view dating back to the 4th century BCE. According to Aristotle, in the absence of an external force, or "motive power," an object would come to rest. Consequently, an object moving at a constant velocity would have a net force acting on it. Once that force is removed, the object will begin to slow down until it comes to rest.

According to Newton's second law, when a net force acts on an object, the object's velocity will not be a constant; the object will accelerate. Newton's first law states that an object will move at a constant velocity only if no net force acts on it.

Statements summarizing the principle contained within Newton's first law date as far back as the 5th century BCE in China and the 11th century CE in parts of the Islamic world. Newton first published his laws of motion in 1687 as part of his three-volume work *Philosophiæ Naturalis Principia Mathematica*, or *Principia*. *Principia* also includes his theory of universal gravitation and a mathematical derivation of Kepler's laws of planetary motion. Newton's laws of motion laid the foundation for classical mechanics and dominated physicists' view of the universe for the next three centuries. Go to the text's online resources for an interesting story on the connections between coffee, Newton's Laws and a dutch spyglass.

Newton's Third Law

Now consider what you will feel if you stand firmly on a nonslippery surface beside an ice rink and push with your hand on a friend who is standing on the ice. You will feel a force from your friend on your hand, and your friend will feel the force of your push, which causes your friend to slide away from you. This interaction demonstrates

Newton's third law: Whenever you exert a force on an object, that object will exert a force on you that is equal in magnitude and opposite in direction to the force you exert on it, whether or not motion is involved. In vector form, one can say $\vec{F}_{12} = -\vec{F}_{21}$.

 EXAMPLE 5-8

Who Is Pushing Harder?

A polar bear cub and his mother are standing still on a frictionless icy patch in the Gulf of Boothia, Nunavut. The 50.0 kg cub pushes the 400 kg mother bear with a constant force. What happens to the cub?

SOLUTION

According to Newton's third law, the force exerted by the cub on the mother must be equal in magnitude and opposite in direction to the force exerted by the mother on the cub. The acceleration of each is given by Newton's second law

$$a = \frac{F_{net}}{m}$$

Since they both experience forces of equal magnitude but have different masses, their accelerations will not be the same. In fact, when the cub pushes his mother on the ice, he will end up moving backward at a much higher speed than she will end up moving in the direction he pushed her.

Making sense of the result:

In everyday life, when a lighter person pushes a heavier person, the lighter person does not expect to move backward. Enough friction is often present to prevent the lighter person from sliding backwards.

 EXAMPLE 5-9

The Force of a Serve

A volleyball player serves a 249 g volleyball at a speed of 90.0 km/h. Assuming that the player's hand contacts the ball for 42.0 ms, find the average force exerted by the ball on the player's hand during that time.

SOLUTION

We have no knowledge of the mass of the player's hand or its change in speed during the collision. However, Newton's third law tells us that the magnitude of the force on the player's hand equals the magnitude of the force on the volleyball. The average acceleration of the volleyball is given by

$$a_{avg} = \frac{\Delta v}{\Delta t} = \frac{25 \text{ m/s}}{0.042 \text{ s}} = 595.2381 \text{ m/s}^2$$

$$F_{avg} = ma_{avg} = (0.249 \text{ kg})(595.2381 \text{ m/s}^2) = 148 \text{ N}$$

Making sense of the result:

The force on the player's hand is roughly sixty times the weight of the volleyball, which is reasonable because serving the volleyball takes a lot more effort than holding it stationary. Also note how the final answer is kept to 3 significant figures, whereas the interim value of a, needed to calculate F, is kept to more than just 3. Ideally, for an interim calculation you keep as many digits as possible.

✓ CHECKPOINT

C-5-4 When Are the Forces Equal?

When you are standing in an elevator that is accelerating upward,

(a) the force that the elevator exerts on you is less than the force that you exert on the elevator;

(b) the force that the elevator exerts on you is more than the force that you exert on the elevator;

(c) the force that the elevator exerts on you is equal to the force that you exert on the elevator;

(d) the forces will be equal only when the acceleration is constant.

C-5-4 (c) as given by Newton's third law.

LO 3

5-3 Free Body Diagrams, Acceleration, and Equations of Motion

Free Body Diagrams

Once the forces acting on an object have been identified, the next step is to draw a **free body diagram (FBD)**, which is a diagram that shows only the object and the forces acting on it.

  EXAMPLE 5-10

Drawing FBDs

In an FBD, show all the forces acting on a crate that is sliding down a frictionless incline.

SOLUTION

The forces are shown in the FBD in Figure 5-6. We have the force of gravity, acting vertically downward, and the normal force from the surface of the incline acting normal (perpendicular) to the surface. We will discuss these forces in detail below.

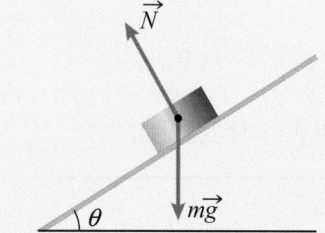

Figure 5-6 The FBD of a crate on a frictionless incline

The Direction of Acceleration and the Choice of Coordinate System

Often, we deal with forces that act in different directions, like those in Figure 5-6. To solve problems, we need to resolve these forces into components. The Cartesian system of coordinates is generally the most convenient for this purpose. Any set of perpendicular axes can be used to resolve vectors into Cartesian components. The choice of axes as such does not affect the physics of the problem. However, we can simplify the calculations considerably by having *one of the Cartesian axes point in the same direction as the acceleration of the object.*

For example, when an object slides down an incline, the acceleration is directed down the incline; therefore, we choose an *x*-axis along the incline and a *y*-axis normal to the incline, as shown in Figure 5-7(a). (Example 5-10 demonstrates the advantage of this choice of axes.) Figure 5-7(b) shows a car going around a frictionless banked curve at just the right speed to avoid sliding up or down the incline. Here, the acceleration is a horizontal radial acceleration, so it is better to have a horizontal axis instead of an axis aligned with the incline.

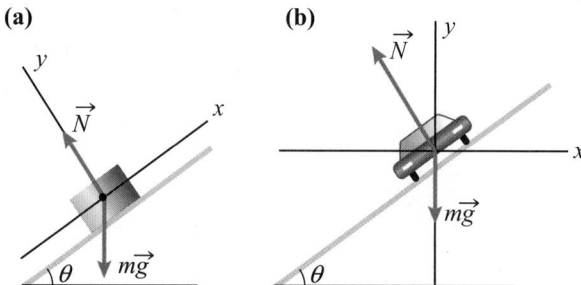

Figure 5-7 Comparing choice of axes

Equations of Motion

When the forces are resolved and broken into components, we can write equations of motion by simply applying Newton's second law along the directions of each axis in the coordinate system. The equations of motion can then be solved individually or simultaneously to extract the desired information, as will be demonstrated in examples throughout this chapter.

A Problem-Solving Strategy for Applying Newton's Laws

We can combine the graphical and mathematical techniques described above into a powerful strategy for analyzing and solving problems involving Newton's laws:

Step 1. Identify all the forces acting on the object.

Step 2. Draw a free body diagram.

Step 3. Establish the direction of acceleration.

Step 4. Set up a coordinate system with one axis oriented along the direction of the acceleration.[1]

Step 5. Resolve the forces into components along the axes of the coordinate system.

Step 6. Write the equations of motion along each axis.

Step 7. Solve the equations of motion.

✓ CHECKPOINT

C-5-5 Choosing Coordinate Axes

Figure 5-8 shows a FBD for a polar bear cub sliding down an igloo. The polar bear is not on the verge of leaving the igloo's surface. Which of the following statements is true?

(a) The axes should be vertical and horizontal.

(b) The normal should point in the opposite direction to the one shown.

(c) The acceleration is along the direction of the force of gravity.

(d) none of the above.

C-5-5 (b) the axis should be as shown, the normal points as it should, and the acceleration points to centre of circle, so answer is d).

Figure 5-8 A FBD of a polar bear sliding down an igloo. The axes shown would be used typically to resolve the forces into components.

LO 4

5-4 The Force of Gravity

Among Newton's greatest contributions was the formulation of the law of universal gravitational attraction, which is detailed in Chapter 11. For this chapter, we will consider only the gravitational attraction between Earth and objects at

[1]Sometimes, it is convenient to decompose the acceleration of an object into two components, for example, in the case of an object moving freely in a vertical circle. In this case, one axis can point along the direction of one of the components of the acceleration, and the other axis will then point along the direction of the second component.

or near its surface. The attractive force that Earth exerts on an object is called the weight of the object and is given by

$$\vec{W} = \vec{F} = m\vec{g}$$

where m is the mass of the object and g is the gravitational acceleration (a measure of the strength of Earth's gravitational field). The value of g varies somewhat with altitude and with latitude because the force of gravity depends on the distance from Earth's centre. For calculations in this chapter, we will assume that g has a constant value of 9.81 m/s^2 near Earth's surface. This approximation is reasonably accurate, especially for objects near sea level.

By Newton's third law, the gravitational force that Earth exerts on an object is equal in magnitude and opposite in direction to the gravitational force that the object exerts on Earth.

✓ CHECKPOINT

C-5-6 The Force of Gravity

A 45 kg mountain goat is drinking water from Shunda Lake in Nordegg, Alberta. The acceleration due to gravity at that elevation (1378 m) is approximately 9.8024 m/s^2. What is the force with which the goat pulls Earth?

(a) negligible

(b) $M_E(9.8024 \text{ m/s}^2)$, where M_E is the mass of Earth

(c) $(45 \text{ kg}) (9.8024 \text{ N})$

(d) There is no way that a goat can pull Earth.

C-5-6 (c)

🧩 MAKING CONNECTIONS

The g Force

The **g force** can be defined as the sum of nongravitational forces acting on an object divided by the object's mass. Often, the nongravitational force of concern is the normal force. For example, an astronaut in a space shuttle before launch experiences a g force of 1 (from the normal force exerted by the astronaut's seat). If the shuttle takes off with an acceleration that is equal to three times the acceleration due to gravity, or $3\,g$, the g force experienced by the astronaut in the shuttle is $4\,g$.

LO 5

5-5 The Normal Force

When you are standing indoors, two forces act on you: the force of gravity, directed downward (toward the centre of Earth), and a **normal force**, directed upward, exerted by the floor under your feet. Since you are not accelerating, the vector sum of the two forces is zero – the floor must be pushing upward on your body with a force equal to your weight.

When two objects are in physical contact, often they exert a normal force on each other. Newton's third law states that this pair of forces is equal in magnitude and opposite in direction. Therefore, the normal force that the objects exert on each other are equal in magnitude and opposite in

direction. Further, the normal force acts in a direction that is perpendicular to the surface of contact between the two objects. The normal force is sometimes called the contact force, or the force of contact. We use the term "normal force" to avoid confusion with friction, which also results from contact between objects. However, frictional forces are directed *parallel* to the surface of contact.

CHECKPOINT

C-5-7 Newton's Third Law

You are standing on the floor in your study space. Your mass is m. Which of the following statements is true?

(a) Earth pulls you, and you pull Earth with a force of the same magnitude, mg, by Newton's third law.

(b) You push the floor, and the floor pushes you with a force of the same magnitude by Newton's third law.

(c) Since the sum of the forces on you is zero, $N = mg$.

(d) All the above statements are true.

C-5-7 (d) all the statements are true.

EXAMPLE 5-11

Apparent Weight in an Elevator

A person with a mass of 61.0 kg stands on a scale in an elevator that is accelerating upward at 3.00 m/s². Find the reading of the scale, in newtons.

SOLUTION

The scale reads the force that the person exerts on it. This force is the normal force exerted by the person on the scale and has the same magnitude, N, as the force exerted by the scale on the person. The FBD in Figure 5-9 shows the forces acting on the person.

Figure 5-9   The FBD of a person in an elevator

According to Newton's second law

$$\sum \vec{F} = m\vec{a} \tag{1}$$

Summing forces in the y direction

$$\sum F_y = N_y + mg_y = ma_y \tag{2}$$

Taking up as the positive direction,

$$N - mg = ma \tag{3}$$

Therefore, the reading of the scale is

$$N = mg + ma = m(g + a) = 61.0(3.00 \text{ m/s}^2 + 9.81 \text{ m/s}^2)$$

$$= 781.41 \text{ N} = 781 \text{ N}$$

Please note in the approach above, we have written the sum of the forces in the y direction in scalar form. This was done in Equation 2. Another important aspect of the approach above that we will use frequently is in step 3. In step 3, the symbols are variables that denote magnitudes, or absolute values, of the quantities in step (2). Alternatively, the solution could be done using full vector representation, using unit vectors, in which case, Newton's second law can be expressed as

$$\vec{N} + m\vec{g} = m\vec{a}$$

which can then be written as

$$N\hat{j} - mg\hat{j} = ma\hat{j}$$

Making sense of the result:

The scale supports the person's weight and transfers the force from the elevator to accelerate the person upward. Thus, it makes sense that the force exerted by the scale is greater than mg. You have most likely experienced the increase in the force on your feet as an elevator accelerates upward, making you feel heavier.

How would the scale reading in Example 5-11 change if the elevator were accelerating downward? We can show that the magnitude of the normal force is given by

$$N = mg - ma$$

If the acceleration of the elevator equals g (i.e., the elevator is free-falling), the normal force on the passenger inside is zero. This normal force is often called apparent weight, as it is the value shown by the scale. When there is no normal force acting on us, we experience a feeling of weightlessness. It is important to remember that our weight in the elevator is not affected by the elevator's acceleration, but the force we are pushing down on a scale (apparent weight) is. (On certain amusement park rides, this feeling of apparent weightlessness is often accompanied by loud screams.)

 MAKING CONNECTIONS

Osteoporosis and Microgravity

Osteoporosis, or "porous bones," is a condition that afflicts a significant fraction of the aging population, especially women. Osteoporosis weakens the bones, increasing the risk of fractures and a host of related complications. Among the various causes of osteoporosis are physical inactivity and reduced mobility. Exposure to the microgravity of space also results in a significant loss of bone mass. NASA reports an average bone mass loss of about 1% per month for astronauts during space missions. There are various remedies for earthbound osteoporosis, depending on the cause. Subjecting the body to higher forces generally increases the loading on the bones so as to promote bone mass increase. In order to mitigate the effects of space bound bone mass loss, astronauts are required to exercise for two hours per day using various devices; however the current measures are not sufficient. One promising proposed remedy is to have the astronauts endure vibrations for 20 to 30 minutes per day, thus subjecting their bodies to higher accelerations, and subsequently more force for the purpose of stimulating bone growth in the body (Figure 5-10).

Figure 5-10 Bioengineering Professor Dr. Clinton Rubin, of the State University of New York, sitting opposite a turkey on a vibration plate used for experiments on bone loss.

 EXAMPLE 5-12

Horizontal Normal Forces

Block 1 in Figure 5-11(a) has a mass of 2.00 kg, and block 2 has a mass of 8.00 kg. The two blocks are sitting on a horizontal frictionless surface. You push block 1 to the right with a force of 100 N.

(a) Find the normal force between the two blocks.
(b) Verify that the normal forces exerted by the two blocks have equal magnitudes, as predicted by Newton's third law.
(c) You now push block 2 to the left with a 100 N force. Find the normal force between the blocks.

(a)

Figure 5-11(a) Example 5-12

SOLUTION

(a) Let us begin by drawing an FBD of the two blocks combined (Figure 5-11(b)).

In the horizontal direction, the only force acting on the system of two blocks is the 100 N force directed right. This will be the direction of the positive x-axis. Hence the acceleration of the system is

(b)

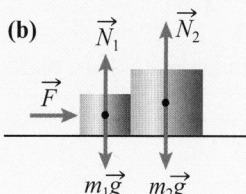

Figure 5-11(b) The FBD of block 1 and block 2 together

$$\sum F_x = F_x = m_T a_x$$

$$a = \frac{F}{m_T}$$

According to the choice of the coordinate axis we made above,

$$a = \frac{F}{m_T} = \frac{100 \text{ N}}{10 \text{ kg}} = 10.0 \text{ m/s}^2$$

We now draw the FBD of the larger block, block 2, shown in Figure 5-11(c). The only force acting on block 2 in the horizontal direction is the normal force from the 2.0 kg block. Denoting the magnitude of that normal force as F_{12}, we get

$$\sum F_x = F_{12,x} = m_2 a_x = (8 \text{ kg})(10 \text{ m/s}^2) = 80.0 \text{ N}$$

(continued)

(c)

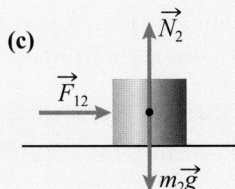

Figure 5-11(c)   The FBD of block 2

(b)   Now we draw the FBD of block 1 (Figure 5-11(d)).

(d)

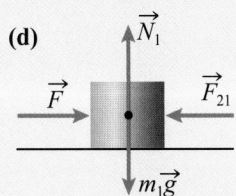

Figure 5-11(d)   The FBD of block 1

$$\sum F_x = F_x + F_{21,x} = m_1 a_x$$

Choosing right as the positive direction,

$$F - F_{21} = m_1 a$$

$$F_{21} = F - m_1 a = 100 - 20 = 80.0 \text{ N}$$

We note here that F, F_{21}, and a are variables that denote the magnitudes of the forces and the acceleration.

Thus in magnitude,

$$F_{21} = F_{12}$$

Recall however that these forces are an action-reaction pair, and though they are equal in magnitude they are opposite in direction.

(c)   Now we are pushing m_2, which in turn pushes m_1. Since the total mass is unchanged and the magnitude of the applied force is the same, the acceleration of the two-mass system is still 10.0 m/s^2, to the left. The normal force between the two masses is the only force acting on the smaller mass (Figure 5-12). We can show that by summing the forces on m_1 in the x-direction,

$$\sum F_x = F_{21,x} = m_1 a_x = (2 \text{ kg})(10 \text{ m/s}^2) = 20.0 \text{ N}$$

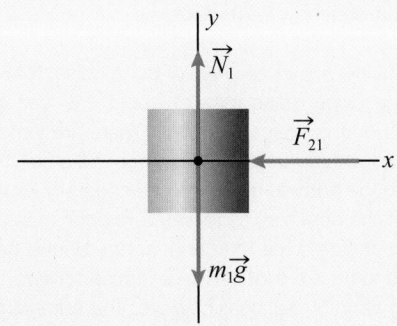

Figure 5-12   The FBD of block 1, now being pushed to the left

Making sense of the result:

The ratio of the net forces on the blocks equals the ratios of their masses: 1 : 4.

Inclined Surfaces

What dictates how fast an object slides down an incline? Intuitively, we know that the steeper the incline, the higher the acceleration. In Example 5-13, we quantify the motion of an object on an inclined surface. This section deals with frictionless surfaces, but later in the chapter we will consider motion with friction.

 EXAMPLE 5-13

Motion on a Frictionless Incline

A mass is placed on a frictionless surface inclined at an angle θ above the horizontal (Figure 5-13). Determine the acceleration of the mass when it is allowed to slide freely, and calculate the magnitude of the normal force acting on the mass.

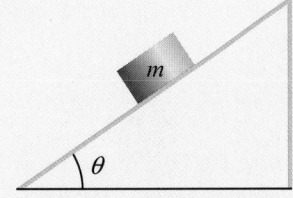

Figure 5-13   Example 5-13

SOLUTION

Here, we apply the analytical strategy outlined in Section 5-3.

Step 1.   Identify all the forces acting on the object.

The forces acting on the mass are the force of gravity, which pulls the mass downward, and the normal force, which acts perpendicular to the incline.

Step 2:   Draw an FBD.

The FBD is shown in Figure 5-14. Notice that we have shown the surface of the incline in the FBD. This will make it easier to resolve the forces into components and determine the angles and the components; in general,

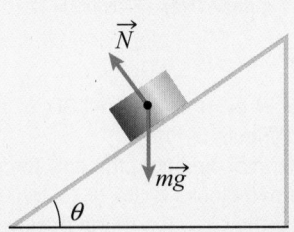

Figure 5-14   The FBD of the mass on an incline in Example 5-13

showing the surface also aids with the aesthetics of the diagram.

Steps 3 and 4: Establish the direction of the acceleration, and set up a coordinate system with one axis oriented along the direction of this acceleration.

The mass will accelerate down the incline. Therefore, one of the axes should be oriented in that direction. We refer to that axis as the x-axis (to the motion). We will refer to the second axis as the y-axis. Alternatively, these axes can be referred to as the parallel and perpendicular axes, respectively. We choose the centre of the object as the origin for the coordinate system.

Step 5: Resolve the forces into components along the chosen axes.

The normal force, $\vec{N}$, from the incline on the block points along the perpendicular axis (or y-axis), so the y-component is equal to the magnitude of the normal force, and the x-component is zero.

To resolve the force of gravity on the block into components, we draw two lines from the tip of the mg vector, one parallel to the incline (along the x-axis) and one perpendicular to the incline (along the y-axis). These lines intersect the axes at the tips of the components of mg. We designate these components mg_x and mg_y, respectively (Figure 5-15). Vector mg is perpendicular to the horizontal, and the perpendicular component of mg is perpendicular to the incline. The angle between mg and mg_y, then, is the same as the angle between the horizontal and the incline, θ.

Step 6: Write the equations of motion along each axis.

The sum for the forces along the direction normal to the incline (i.e., along the y-axis) is

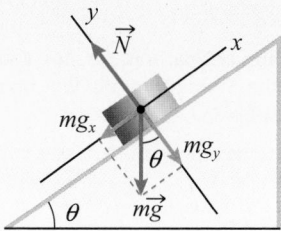

Figure 5-15   Components of the forces along the chosen axes

$$\sum F_y = ma_y \qquad (1)$$

or

$$N_y + mg_y = ma_y \qquad (2)$$

To analyze the motion along the incline, we sum the forces in the direction of the incline (parallel to the x-axis):

$$\sum F_x = ma_x$$

Therefore,

$$mg_x = ma_x \qquad (3)$$

Step 7: Solve the equations of motion.

The components of $m\vec{g}$ parallel to the incline and, therefore, to the motion of the mass on the incline depend on θ. In the right-angled triangle formed by the force of gravity and its components, mg_x is opposite the angle θ, mg_y is adjacent to the angle θ, and mg is the hypotenuse. Therefore, in terms of magnitudes,

$$|mg_x| = mg \sin\theta \qquad (4)$$

and

$$|mg_y| = mg \cos\theta \qquad (5)$$

Therefore, from Equation (2), keeping in mind that the N points along the positive y-direction, and mg_y points along the negative y-direction we have

$$N - mg \cos\theta = 0 \qquad (6)$$

or

$$N = mg \cos\theta$$

And from Equations (2) and (3) we get

$$a = g \sin\theta \qquad (7)$$

where we use the variable a to denote the magnitude of the acceleration.

Making sense of the result:

When the mass is placed on a horizontal surface, the acceleration is zero. An incline angle of 90° means that the mass is effectively placed against a vertical wall, in which case it will simply free-fall with an acceleration equal to g, as given by Step 6. According to step 5, the normal will be zero for a vertical incline.

✓ CHECKPOINT

C-5-8 The Normal Force

For the box in Figure 5-16, the magnitude of the normal force

(a) equals mg;

(b) is greater than mg;

(c) is less than mg;

(d) may be less than, equal to, or greater than mg, depending on the value of θ.

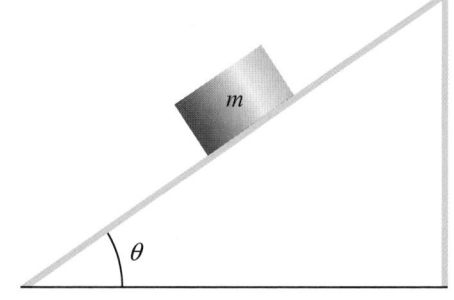

Figure 5-16   C-5-8

C-5-8 (c) box sits freely, $N = mg \cos\theta$.

EXAMPLE 5-14

Horizontal Force Acting on an Object on an Inclined Plane

Figure 5-17 shows a worker, wearing special boots for added traction, pushing a crate of mass m up an incline with a slope of angle θ using a horizontal force F. Find the normal force from the incline on the crate, and determine the acceleration of the crate. Assume that the friction between the crate and the incline is negligible.

Figure 5-17 Example 5-14

SOLUTION

Here, again, we invoke our strategy for solving problems involving Newton's laws.

Step 1. Identify the forces acting on the object.

There are three forces acting on the crate: the force of gravity, the normal force exerted by the incline on the crate, and the force applied by the worker.

Step 2. Draw an FBD for the crate.

The FBD is shown in Figure 5-18.

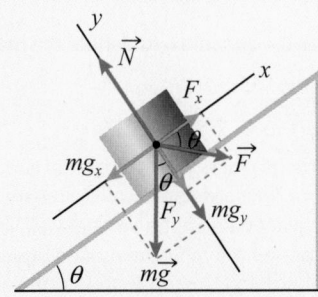

Figure 5-18 The FBD of the crate

Step 3. Determine the direction of the acceleration.

The acceleration of the crate is along the incline.

Step 4. Choose your system of axes.

We choose the x-axis to be parallel to the incline and the y-axis to be perpendicular to the incline.

Step 5. Resolve the forces into components along the chosen axes.

Here, again, the normal force points along the y-axis and has an x-component of zero. As in Example 5-13, the parallel component of the gravitational force is $mg\sin\theta$, and the perpendicular component is $mg\cos\theta$. In Figure 5-18, in the triangle formed by F we can see the hypotenuse F and the components F_x and F_y, where F_x is the adjacent side and F_y is the side opposite the angle θ. Therefore, in terms of magnitudes,

$$|F_x| = F\cos\theta \quad \text{and} \quad |F_y| = F\sin\theta$$

Step 6. Write the equations of motion along each axis. Along the x-axis we have

$$\sum F_x = ma_x \Rightarrow F_x + mg_x = ma_x \tag{1}$$

If we choose up the incline as the positive direction, then

$$F\cos\theta - mg\sin\theta = ma \tag{2}$$

where F, g and a in Equation (2) are used as variables to denote magnitudes of the quantities in Equation (1).

Along the y-axis we have

$$\sum F_y = ma_y \Rightarrow F_y + mg_y + N_y = 0 \tag{3}$$

Taking the direction of the normal force to be positive,

$$-F\sin\theta - mg\cos\theta + N = 0 \tag{4}$$

Step 7. Solve the equations of motion.

We can solve Equation (4) for the normal force:

$$N = F\sin\theta + mg\cos\theta \tag{5}$$

Note that here the normal force is not equal to $mg\cos\theta$. To determine the normal force, we had to sum the y-components of the forces.

We can find the acceleration of the crate from Equation (2):

$$a = \frac{F\cos\theta - mg\sin\theta}{m} \tag{6}$$

Making sense of the result:

Since the worker is pushing the crate to a certain degree against the face of the incline, the y-component of the applied force contributes to the normal force. Consequently, the magnitude of the normal force is greater than $mg\cos\theta$. The expression for the acceleration is the net force divided by the mass, as we would expect.

PEER TO PEER

A misconception I always had was that the normal force on an object sitting on an inclined surface was always equal to $mg\sin\theta$. The value of the normal force can vary, depending on the details of the problem, and should be obtained by solving the equations of motion.

LO 6

5-6 Tension

When you hold a string stationary with a weight attached to it, the string is under **tension** because the weight is pulling the lower end of the string downward while you are pulling the top end upward to prevent it from moving down.

The string is pulling the weight up and pulling your hand down at the same time. The force of tension in a string (or rope) pulls the weight toward the string. Generally, as you pull objects with a string, the string exerts a force of equal magnitude on you and on the object, in accordance with Newton's third law.

Similarly, if you grip both ends of a pen and try to pull the ends apart, the pen is under tension. Your hands are being pulled toward one another by the tension in the pen. Alternatively, you can say that the force of tension prevents your hands from moving apart. If you press on both ends of the pen, the force of **compression** in the pen prevents your hands from coming toward each other, and under compression. However, you cannot have compression in a rope: if you try to push the two ends of a rope together, the rope will buckle. In other words, for almost all practical applications you cannot push on a flexible rope.

The force of tension at either end of an accelerating string will not be the same; neither will the force of compression on the above pen be the same, if the string or the pen is accelerating. The net force, whose magnitude is the difference between the magnitudes of the forces at either end, is, by Newton's second law, equal to the mass of the string or pen multiplied by its acceleration. Therefore, for the tension or compression forces at either end to be equal, we require that either the string (or pen) have zero acceleration or zero mass. In this chapter, we will deal almost exclusively with massless strings.

MAKING CONNECTIONS

Tensile Strength

The ultimate tensile strength (UTS) indicates the maximum tension force per unit area that a material can withstand without structural failure. Structural steel has a UTS of about 400 MN/m^2. Although concrete is widely used in elements under compression, its UTS is relatively weak, with a typical value of approximately 3 MN/m^2. Silk has 2.5 times the tensile strength of structural steel, and some multiwalled carbon nanotubes have been made with a UTS exceeding 60 GN/m^2. For a more detailed discussion of this topic, please see chapter 10.

Certain armed-forces safety regulations require that, when a vehicle is being winched, all nonessential personnel stay at least a cable length away. This rule reduces the risk of injuries should the winching cable snap, because the amount of tension that develops in a towing line can be quite significant.

CHECKPOINT

C-5-9 Pulling on a Rope

A 40 kg seal corners you ($m = 80$ kg) on a frozen shelf in the Davis Strait. He grabs a rope with his mouth and brings it to you, insisting with repeated head nods that the way out for you is to play a game where you pull the rope he is holding with his teeth. You pull the rope and notice that the seal accelerates with respect to the frictionless ice at 3 m/s^2. Which of the following statements is true? Assume you experience no friction.

(a) The net force on you is twice the force on the seal.
(b) Your acceleration is the same as the seal's acceleration.
(c) The net force on you is 240 N.
(d) The net force on you 480 N.
(e) None of the above are true.

C-5-9 (e) force on you equals that on seal, none of the answers adds up to that.

EXAMPLE 5-15

Towing Tension

Three masses ($m_1 = 20.0$ kg, $m_2 = 50.0$ kg, and $m_3 = 30.0$ kg) resting on a horizontal frictionless surface are connected by ropes as shown in Figure 5-19. A force of 200.0 N pulls m_3 to the right. Find the tension in the rope connecting m_1 and m_2. The mass of the ropes is negligible.

Figure 5-19 Example 5-15

SOLUTION

From the FBD for m_1 in Figure 5-20, we see that the only force acting on m_1 in the x-direction is the tension in the rope connecting it to m_2. Using T_1 to denote the magnitude of this force, we have

$$\sum F_x = T_{1x} = m_1 a_x \Rightarrow T_1 = m_1 a$$

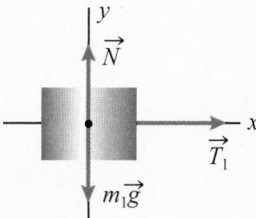

Figure 5-20 The FBD of m_1

Since m_1 is known, we can solve for T_1 if we have a value for a. The acceleration of all three masses must be the same, as we can see from the FBD of the whole system in Figure 5-21.

(continued)

Figure 5-21 The FBD of the three masses in Example 5-15

We take the positive x-axis to point along the line of action of the force F. Since the only force acting on the three-mass system in the x-direction is the 200 N applied force,

$$\sum F_x = F_x = m_T a_x \Rightarrow F = m_T a$$

and

$$a = \frac{F}{m_T} = \frac{200 \text{ N}}{100 \text{ kg}} = 2.00 \text{ m/s}^2$$

Therefore,

$$T_1 = m_1 a = 20 \text{ kg } (2 \text{ m/s}^2) = 40.0 \text{ N}$$

we note here that the variables F and a denote magnitudes of the force and the acceleration.

Making sense of the result:

We are able to treat the three masses like one mass because they are connected by taut ropes (which we assume do not stretch). Although the applied force acts directly on m_3 only, in effect, the force is pulling all three masses.

 CHECKPOINT

C-5-10 How Much Tension?

What is the tension in the string in Figure 5-22?
(a) 196 N
(b) 98 N
(c) zero
(d) We need to know whether or not the system is moving.
(e) None of the above.

Figure 5-22 C-5-10

C-5-10 (b) for either block, $mg = T = 98$ N, tension is same in rope everywhere.

 EXAMPLE 5-16

Simple Rope and Pulley System

An 8.00 kg mass (m_1) is resting on a horizontal frictionless surface. A rope that runs over a massless pulley connects m_1 to a 6.00 kg mass (m_2), as shown in Figure 5-23. The two-mass system is initially at rest. Find the acceleration of m_1 and the tension in the rope.

Figure 5-23 Example 5-16

SOLUTION

Here we have two unknowns, the tension force and the acceleration, so we will need two equations to solve this problem. We begin by drawing a FBD for each object (Figure 5-24).

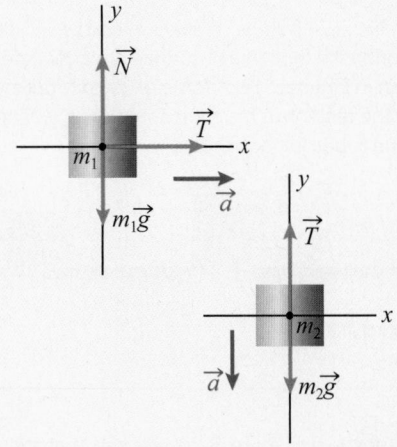

Figure 5-24 The FBD of the two masses in Example 5-16

Since the only force acting on m_1 in the x-direction is the tension in the rope,

$$\sum F_x = T_x = m_1 a_x \Rightarrow T = m_1 a \qquad (1)$$

From the FBD of m_2, we get

$$\sum F_y = T_y + m_2 g_y = m_2 a_y$$

Choosing up as our positive direction gives

$$T - m_2 g = -m_2 a \qquad (2)$$

Note that we have introduced a variable, a, to denote the magnitude of the acceleration, the same is true for T and g.

Substituting Equation (1) into Equation (2), we get

$$m_1 a - m_2 g = -m_2 a \Rightarrow a = \frac{m_2 g}{m_1 + m_2}$$

$$= \frac{(6 \text{ kg}) (9.81 \text{ m/s}^2)}{(6 + 8) \text{ kg}} = 4.20 \text{ m/s}^2$$

To find the tension in the rope, we substitute this value for a into either Equation (1) or Equation (2). From Equation (1) we get

$$T = m_1 a = 8 \text{ kg} (4.2 \text{ m/s}^2) = 33.6 \text{ N}$$

Making sense of the result:

The acceleration of the system is less than g because the weight of m_2 is pulling both m_1 and m_2. Note that the tension is not equal to $m_2 g$, as it would have been were m_2 not accelerating. The tension in the horizontal and the vertical segments of the rope is the same (but would not be if the pulley has mass, as we will see in Chapter 8). We also note than when $m_1 = 0$, the acceleration of m_2 is equal to g.

EXAMPLE 5-17

Train Acceleration

A bored but observant child on a moving train ties a ball to one end of a string and ties the other end to a light fixture on the ceiling of the train. When the train slows down, the child notices that the string is no longer hanging straight down but instead makes an angle of $19°$ with the vertical. The mass of the ball is 175 g.
(a) Will the string tilt forward or backward? Find the acceleration of the train.
(b) What is the tension in the string?

SOLUTION

(a) The string tilts forward. As the train slows down, the only forces acting on the ball are the tension in the string and the force of gravity. The acceleration is opposite to direction of motion of the ball (and the train). So, we choose a horizontal x-axis and a vertical y-axis. The FBD in Figure 5-25 shows the forces resolved into components along these axes.

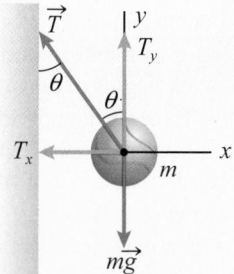

Figure 5-25 The FBD of the suspended ball in Example 5-17

Summing the forces on the ball in the y-direction, we get

$$\sum F_y = T_y + m g_y = 0$$

Choosing up as our positive direction gives

$$T \cos\theta - mg = 0, \quad \text{or} \quad T \cos\theta = mg \qquad (1)$$

Here, we have used the fact that the vertical component of the force of tension is the side adjacent to θ in the triangle formed by T, T_x, and T_y in Figure 5-25. Since the acceleration of the ball is in the horizontal direction,

$$\sum F_x = T_x = ma_x \Rightarrow T \sin\theta = ma \qquad (2)$$

We have two unknowns—the tension and the acceleration—and we now have two equations involving these unknowns. Dividing Equation (2) by Equation (1), we get

$$\tan\theta = \frac{a}{g} \Rightarrow a = g \tan\theta = (9.81 \text{ m/s}^2) \tan(19°) = 3.38 \text{ m/s}^2$$

(b) For the tension in the rope we can use either Equation (1) or Equation (2). Substituting for the known values in Equation (1) gives

$$T = \frac{mg}{\cos\theta} = \frac{0.175 \text{ kg} \times 9.81 \text{ m/s}^2}{\cos(19°)} = 1.82 \text{ N}$$

Making sense of the result:

The tension is the only force acting to slow the ball down. The tension is not much greater than mg because the acceleration is relatively small, as evidenced by the small angle the string makes with the vertical.

MECHANICS

✓ CHECKPOINT

C-5-11 Which Mass Is Faster?

Which of the following statements is true about the accelerations of the two masses in Figure 5-26?

(a) They both have the same acceleration.

(b) The acceleration of m_1 is twice the acceleration of m_2.

(c) The acceleration of m_2 is greater than the acceleration of m_1.

(d) None of the above.

C-5-11 (b) If the vertical mass moves by 2d, the horizontal mass will move by d.

Figure 5-26 C-5-11

LO 7

5-7 Friction

If you push gently on the side of a heavy textbook lying on a table, the book does not move even though you are exerting a horizontal force on it. According to Newton's second law, the net force on the book must be zero because the acceleration is zero. Therefore, another force must be acting on the book to exactly balance the horizontal force you are applying. If you decrease the force you are applying, increase it slightly, or change which side of the book you push on, the book still does not move. The force that is preventing the book from moving is changing to remain exactly opposite to the force you apply. The force that prevents the book from moving is **friction**. Friction that acts between objects that are not moving relative to each other is called **static friction**.

If you now gradually increase the horizontal force you apply to the book, you will find that the static friction reaches a maximum just before the book starts to move relative to the table. The maximum force of static friction between two objects in contact with each other always occurs when the objects are on the verge of moving relative to each other.

If you were to place a second, identical book on top of the first book and gradually increase the horizontal

force you apply to the bottom book, you will notice that the force you have to exert on the bottom book to bring it to the verge of moving has increased. In fact, if you pull the book with a spring scale, you will find that the force required to bring the book to the verge of moving is doubled when you place the second book on top of it. The maximum force of static friction between two surfaces is directly proportional to the normal force between the two surfaces:

KEY EQUATION
$$f_s^{max} = \mu_s N \qquad (5\text{-}2)$$

where μ_s is the coefficient of static friction.

The force of static friction therefore takes on values between zero and the maximum force of static friction:

KEY EQUATION
$$0 \le f_s \le f_s^{max} \qquad (5\text{-}3)$$

✓ CHECKPOINT

C-5-12 The Force of Static Friction

What is the minimum value for the force of static friction?

C-5-12 In general, zero.

✓ CHECKPOINT

C-5-13 Force of Static Friction

You place a book on the horizontal desk in front of you. The book's weight is 20 N, and the coefficient of static friction between the desk and the book is 0.25. You push the book with a horizontal force of 2 N. What is the force of static friction on the book?

(a) zero

(b) 2 N

(c) 5 N

(d) none of the above

C-5-13 (b) book is not moving so the force of friction is equal to the force you apply.

When you apply a force to the book discussed above that is slightly more than the maximum force of static friction, the book will begin to move and continue to accelerate. As soon as the book begins to move, you will notice a sudden drop in the force that opposes you. To keep the book moving at a constant speed, you will have to decrease the applied force. The force of friction decreases somewhat when an object starts moving.

Friction that acts on a moving object is called **kinetic friction**. Kinetic friction is directly proportional to the normal force between an object and the surface on which it moves:

 KEY EQUATION
$$f_k = \mu_k N \tag{5-4}$$

where μ_k is the coefficient of kinetic friction.

For a given object and surface, the coefficient of kinetic friction is usually less than the coefficient of static friction.

 ## ONLINE ACTIVITY

Friction

The e-resource that accompanies every new copy of this textbook contains an Online Activity using the PhET simulation "Motion." Work through the simulation and accompanying questions to gain an understanding of the force of friction.

EXAMPLE 5-18

Coefficient of Kinetic Friction

Your friend is sitting on a sled on a level snow-covered surface. The sled has steel runners. You give the sled one push, which causes it to move away from you at a speed of 2.10 m/s. The sled stops 7.00 s later. Find the coefficient of kinetic friction between the sled and the snow-covered surface.

SOLUTION

Since we know the initial speed and elapsed time, we can calculate the acceleration, which will give us information about the forces acting on the sled.

The FBD of the sled in Figure 5-27 shows that the only horizontal force acting on the sled after you stop pushing it is the force of kinetic friction, f_k. Therefore, using Equation 5-4,

$$\sum F_x = m_1 a_x \Rightarrow f_k = ma \Rightarrow \mu_k N = ma \tag{1}$$

For the normal force, we sum the forces in the vertical direction. Choosing up as our positive direction, we get

$$\sum F_y = ma_y \Rightarrow N - mg = 0$$
$$\Rightarrow N = mg \tag{2}$$

Substituting Equation (2) into Equation (1), we get

$$\mu_k mg = ma \Rightarrow a = g\mu_k \tag{3}$$

Figure 5-27 The FBD for the sled in Example 5-18

Choosing the direction of the sled displacement as positive, we have

$$\vec{v}(t) = \vec{a}t + \vec{v_0}$$
$$0 = -a(7\,s) + 2.1\,\frac{m}{s} \Rightarrow a = \frac{2.1\ m/s}{7\ s} = 0.300\ m/s^2 \tag{4}$$

note that a in the line above is a variable that denotes the magnitude of the acceleration.

Substituting Equation (4) into Equation (3) gives

$$\mu_k = \frac{a}{g} = \frac{0.3\ m/s^2}{9.81\ m/s^2} = 0.0306$$

Making sense of the result:

A sled with steel runners slides easily on snow, so a low value for μ_k seems reasonable. We also note that μ_k has no units as expected.

EXAMPLE 5-19

Static or Kinetic?

You are pushing a 26.0 kg crate on a horizontal floor, where the coefficient of static friction is 0.170 and the coefficient of kinetic friction is 0.130. You apply a 140 N force at an angle of 27.0° below horizontal.

(a) Will the crate move?
(b) If the crate does move, find its acceleration.

SOLUTION

First, we draw the FBD of the crate; see Figure 5-28.

(a) The only force that opposes the motion of the crate is the force of static friction. If the horizontal component of the applied force is greater than the maximum force of static friction, then the crate will move. The maximum force of static friction depends on the normal force between the surface and the crate, which

(continued)

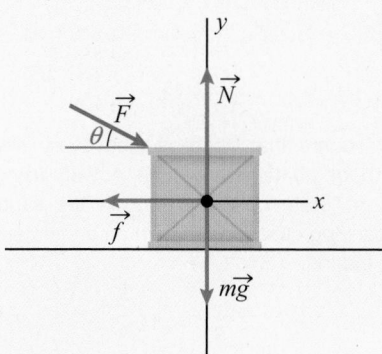

Figure 5-28 The FBD of the crate in Example 5-19

depends, in part, on the vertical component of the applied force.

The forces acting on the crate are the normal force, the force of gravity, the applied force, and the force of friction ($\vec{f}$), as shown in the FBD in Figure 5-28. The applied force, $\vec{F}$, is the only force that requires calculations to determine its x- and y-components.

Summing the forces in the x-direction gives

$$\sum F_x = m_1 a_x \Rightarrow F_x + f_x = ma_x \tag{1}$$

Choosing right as our positive direction,

$$F\cos\theta - f = ma \tag{2}$$

Next, we sum forces in the y-direction:

$$\sum F_y = F_y + mg_y + N_y = 0 \tag{3}$$

Choosing up as our positive direction, we get

$$-F\sin\theta - mg + N = 0 \tag{4}$$

$$N = mg + F\sin\theta \tag{5}$$

$$N = F\sin\theta + mg \tag{6}$$

where the variables in Equations (2) and (4) denote magnitudes of the quantities in Equations (1) and (3) respectively.

The maximum force of static friction is, by Equation 5-2,

$$f_s^{\max} = \mu_s N = \mu_s(F\sin\theta + mg)$$

$$= 0.17\,\text{N}\,[140\sin(27°) + (26\,\text{kg})(9.81\,\text{m/s}^2)]$$

$$= 54.2\,\text{N} \tag{7}$$

The x-component of the applied force is

$$F_x = F\cos(27°) = (140\,\text{N})\cos(27°)$$

$$F_x = 125\,\text{N}$$

Since the x-component of the applied force is greater than the maximum force of static friction, the crate will move.

(b) We use the force of kinetic friction in the calculation of the acceleration. Summing the forces in the x-direction and using Equation (2) we can write

$$F\cos\theta - \mu_k N = ma \tag{8}$$

$$F\cos\theta - \mu_k(F\sin\theta + mg) = ma$$

Therefore,

$$a = \frac{F\cos\theta - \mu_k(F\sin\theta + mg)}{m}$$

$$= \frac{140\cos(27°) - 0.13[140\,\text{N}\sin(27°) + (26\,\text{kg})(9.81\,\text{m/s}^2)]}{26\,\text{kg}}$$

$$= 3.20\,\text{m/s}^2$$

Making sense of the result:

Since the vertical component of the applied force contributes to the normal force, the magnitude of the normal force is greater than mg. This, in turn, affects the magnitude of the force of friction. The units for acceleration make sense. In limit of no friction, the acceleration reduces to the force applied in the direction of motion divided by the mass.

EXAMPLE 5-20

Which Way Will the System Move?

A 4.20 kg mass, m_1, hangs from a rope that runs over a massless pulley and is connected to a 6.70 kg mass, m_2, resting on the surface at an incline of 31.0°, as shown in Figure 5-29(a). The coefficient of static friction between the incline and m_2 is 0.09, and the coefficient of kinetic friction

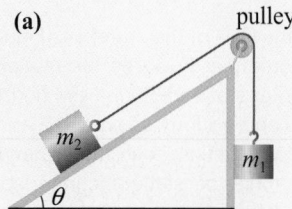

Figure 5-29(a) Example 5-20

is 0.07. Initially, the system is at rest. Find the acceleration of the masses and the tension in the string.

SOLUTION

First, we determine which way the masses will move. On one side we have the weight of m_1 at 41.2 N [(4.2 kg)(9.81 m/s²)]; on the other side we have the component of the weight of m_2 along the incline at 33.9 N [(6.7 kg)(9.81 m/s²)(sin(31°))]. The weight of m_1 is opposed by the weight of m_2 and the force of friction. The force of friction, as a quick check, is given by $N\mu_k$. Mass m_2 is on the incline, so we can say that the normal force between m_2 and the incline is $N = m_2 g\cos\theta$. Therefore, $f_k = m_2 g\cos\theta\mu_k = 5.07$ N. The sum of the component of the weight of m_2 along the incline and the force for friction is less than the weight of m_1, which means that m_1 will move down and m_2 will move up the incline.

In problems involving friction, it is best to do the above check because the direction of motion will affect the form of the equations of motion. This is not necessary in the absence of friction.

For the FBD for m_1 we have the force of gravity and the force of tension acting in the y-direction (Figure 5-29(b)).

(b)

Figure 5-29(b) The FBD for m_1

$$\sum F_y = T_y + m_1 g_y = ma_y \tag{1}$$

Choosing up as our positive direction, we get

$$T - m_1 g = -m_1 a \tag{2}$$

The acceleration of m_1 shows up as negative because up is our choice for a positive direction and the mass is moving down. Note that we have introduced 'a' as a variable to denote the magnitude of the acceleration.

The forces acting on m_2 are the force of friction, the normal force from the incline, the tension in the string, and the force of gravity. Figure 5-29(c) shows these forces resolved into components along the coordinate axes, one of which, the parallel, or the x-axis, points along the direction of the incline, the direction of the acceleration.

Summing the forces in the perpendicular, or y-direction, we get

$$\sum F_y = N_y + m_2 g_y = 0 \tag{3}$$

Choosing the direction of the normal force to be positive, we get

$$N - m_2 g \cos\theta = 0, \text{ hence } N = m_2 g \cos\theta \tag{4}$$

Summing the forces in the parallel, or x-, direction yields

$$\sum F_x = T_x + f_{kx} + m_2 g_x = ma_x \tag{5}$$

Choosing up the incline as our positive direction,

(c)

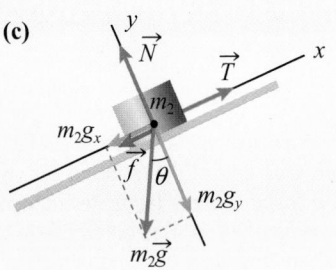

Figure 5-29(c) The FBD for m_2

$$T - f_k - m_2 g \sin\theta = m_2 a \tag{6}$$

where the variables in Equations (4) and (6) denote magnitude of the quantities in Equations (3) and (5) respectively.

We multiply Equation (2) by –1 and add it to Equation (6) to get

$$m_1 g - f_k - m_2 g \sin\theta = m_1 a + m_2 a \tag{7}$$

Therefore,

$$a = \frac{m_1 g - m_2 g \mu_k \cos\theta - m_2 g \sin\theta}{m_1 + m_2}$$

$$= \frac{(4.2)(9.81) - (6.7)(9.81)(0.07)\cos(31°) - (6.7)(9.81)\sin(31°)}{4.2 + 6.7} \tag{8}$$

Substituting the values, we get $a = 0.313 \text{ m/s}^2$, and

$$T = m_1 g - m_1 a = m_1(g - a) = 4.2 \text{ kg } (9.81 - 0.313) \text{ m/s}^2$$
$$= 39.9 \ N \tag{9}$$

Making sense of the result:

Since the acceleration is rather small, we would expect the value of the tension in the rope to be close to $m_1 g$. In the limiting case, where the system is not moving (or moving at a constant speed), the tension would be exactly equal to $m_1 g$.

It is worthwhile to determine whether the system will move when let go initially. For this, we have to look at the maximum force of static friction and see whether it is large enough to prevent the system from moving. In this case it is not. You can verify that yourself by using the value of μ_s provided.

ONLINE ACTIVITY

Ramp: Forces and Motion

The e-resource that accompanies every new copy of this textbook contains an Online Activity using the PhET simulation "Motion." Work through the simulation and accompanying questions to gain an understanding of the forces acting on an object and the object's resulting motion.

MAKING CONNECTIONS

The Power of Friction

While mooring a ship, sailors can exert a relatively small tension force on their end of the mooring line, while the other end sustains a significantly larger tension due to the pull by the ship being moored. This difference in tension results from the tremendous force of friction that can be generated by wrapping the rope a few times around a bollard. The force of friction between the bollard and the rope increases exponentially with the angle subtended by the rope as it hugs the bollard. (See Example 5-21 below.) For a coefficient of friction of 0.5 between the bollard and the rope, if the sailor wraps the rope only one time around the bollard, the force of tension on the other end will be over 20 times that with which the sailor is pulling. If, however, the sailor manages to wrap the rope 4 times around the bollard, the force on the other end can be over 12 000 times larger than the force with which the sailor pulls. Bollards are located on the docks or on the ship's deck.

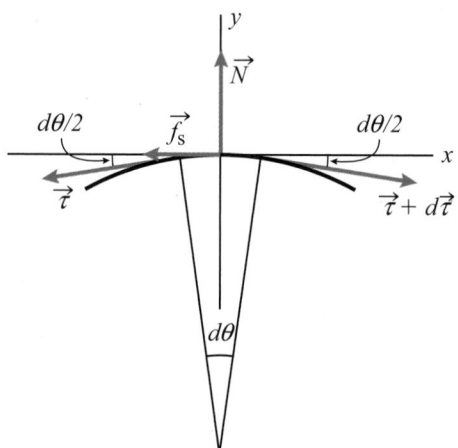

Figure 5-30   Forces acting on rope. This is used in Example 5-21 below.

EXAMPLE 5-21

Mooring a Ship: Rope and Bollard

The problem of calculating the effect of the force of friction between the rope and the bollard is complicated by the fact that the force of friction depends on the tension in the rope in the first place. For a given number of turns, the force of friction increases as the tension increases. A sailor moors a ship by wrapping one of its mooring lines around a bollard. Develop a relationship between the tension forces on each side of the bollard.

SOLUTION

Let us consider a small segment of the bollard subtended by the rope. On one side, the tension is $\tau_1 = \tau$, and it increases by an amount of $d\tau$ over an angle $d\theta$, giving us a value of $\tau_2 = \tau + d\tau$ on the other side (See Figure 5-30). The other force is the normal force exerted by the bollard on the rope. It can be shown that the angle between either end of the rope segment and the tangent to the bollard at the midpoint of the segment is $\dfrac{d\theta}{2}$.

$$N - \tau \sin \frac{d\theta}{2} - (\tau + d\tau) \sin \frac{d\theta}{2} = 0 \qquad (1)$$

For a very small angle $d\theta$, $\sin d\theta \approx d\theta$. Therefore, Equation (1) becomes

$$N - \tau \frac{d\theta}{2} - (\tau + d\tau) \frac{d\theta}{2} = 0 \qquad (2)$$

The product of the two infinitesimal quantities $d\theta$ and $d\tau$ becomes negligible compared to the other quantities in the problem. In the limit, we can ignore it. Therefore,

$$N \sim 2\tau \frac{d\theta}{2} = \tau d\theta \qquad (3)$$

The maximum force of static friction occurs when the rope is on the verge of slipping on the bollard. For this limiting case we have

$$f_s^{\max} = \mu_s N = \mu_s \tau d\theta \qquad (4)$$

Next, we sum the forces in the x-direction. Choosing right as positive we have

$$(\tau + d\tau) \cos \frac{d\theta}{2} - f_s^{\max} - \tau \cos \frac{d\theta}{2} = 0 \qquad (5)$$

where, for an infinitesimally small length of the rope, we can say that the force of friction acts in the x-direction.

We will use the small-angle approximation here again, for which $\cos d\theta \approx 1$. Using Equation (4), we get

$$(\tau + d\tau) - \mu_s \tau d\theta - \tau = 0 \qquad (6)$$

or

$$d\tau = \mu_s \tau d\theta \Rightarrow \frac{d\tau}{\tau} = \mu_s d\theta \qquad (7)$$

Integrating we get

$$\int_{\tau_1}^{\tau_2} \frac{d\tau}{\tau} = \mu_s \int_{\theta_1}^{\theta_2} d\theta$$

or

$$\ln \frac{\tau_2}{\tau_1} = \mu_s \Delta\theta \qquad (5\text{-}5a)$$

or

KEY EQUATION

$$\frac{\tau_2}{\tau_1} = e^{\mu_s \Delta\theta} \qquad (5\text{-}5b)$$

Making sense of the result:

The force of friction increases tremendously with each turn.

LO 8

5-8 The Force of a Spring

The force required to extend or compress an "ideal spring" is directly proportional to the displacement of the free end of the spring. This relationship is called **Hooke's law** and is often expressed as

$$F_s = -kx \qquad (5\text{-}6)$$

where k is the spring constant.

The greater the value of k, the more force required to stretch the spring by a given amount and, therefore, the stiffer the spring will be. Hooke's law is a useful approximation for many elastic materials.

✓ CHECKPOINT

C-5-14 Mass on a Vertical Spring

A spring with a spring constant k hangs vertically from a ceiling. You attach a 2 kg mass to the free end of the spring, pull the mass down, and then release it. When will the acceleration of the mass be zero?

(a) Never, because the mass is always under the effect of gravity.

(b) When the mass stops momentarily at the maximum extension of the spring.

(c) When the mass returns to the top and stops momentarily.

(d) When the extension of the spring is mg/k.

C-5-14 (d) This is the point where the net force is zero.

 ## EXAMPLE 5-22

Hooke's Law

A spring hangs freely from the ceiling of an elevator. While the elevator is accelerating upward at 2.90 m/s², you attach a 300 g mass to the spring and find that the spring stretches by 7.00 cm. Calculate the spring constant.

SOLUTION

The forces acting on the mass are the force of the spring and the force of gravity, as shown in the FBD in Figure 5-31. Summing the forces in the vertical direction we get

Figure 5-31 The FBD of the mass in Example 5-22

$$\sum F_y = F_{sy} + mg_y = ma \qquad (1)$$

Choosing up as the positive direction, we have

$$F_s - mg = ma, \quad \text{or} \quad F_s = mg + ma \qquad (2)$$

where the variables in Equation 2 denote magnitudes of the force of the spring, the acceleration due to gravity and the acceleration of the mass.

Applying Hooke's law, we get

$$kx = mg + ma$$

$$k = \frac{mg + ma}{x} = \frac{0.3 \text{ kg } (9.81 \text{ m/s}^2 + 2.9 \text{ m/s}^2)}{0.070 \text{ N/m}} = 54.5 \text{ N/m}$$

Making sense of the result:

The force of the spring counters the force of gravity acting on the mass and transmits the acceleration of the elevator to the mass. Recall that Hooke's law, $F = -kx$, means that the spring force points in the opposite direction to that of the stretch or compression. The actual sign of the force (kx) in the solution above depends on our choice of positive and negative directions.

 ## MAKING CONNECTIONS

Inertial Navigation

An inertial navigation system uses mechanical forces to calculate the speed and position of a moving object, such as a ship or an airplane. Force sensors disclose data on the acceleration of an object. The acceleration information is then integrated to calculate the object's speed and location relative to its starting point. Since inertial navigation systems do not require signals from navigation beacons or satellites, such systems are immune to signal reception problems and electronic jamming. A rather simple motion (or acceleration) sensor can, in principle, be built using springs and masses.

5-9 Circular Motion and Centripetal Forces

In Chapter 3, we saw that an object moving with speed v along a circular trajectory of radius r has a radial acceleration with a magnitude of

$$a_r = \frac{v^2}{r} \qquad (5\text{-}7)$$

According to Newton's second law, the object must then be experiencing a net force that points toward the centre of the circle, as well. This net force is the vector sum of all the forces acting on the object in the radial direction. This net force is often called a centre-seeking force, or a **centripetal force**.

PEER TO PEER

I always thought that there was a separate physical entity called the "centripetal force" that existed in addition to the real and actual forces acting on an object moving in a circular trajectory. However, the centripetal force is really the net force, or the sum of all the actual physical forces, acting along the radial direction. Without these real physical forces, such as the force of tension, the normal force, and other real forces such as the ones discussed in this chapter, there would be no centripetal force.

EXAMPLE 5-23

Traffic Circle Speed Limit

A car is going around a traffic circle (roundabout) that has a radius of 42.0 m. The traffic circle is not banked, and the coefficient of static friction between the tires and the road is 0.270.
(a) Find the maximum speed that the car can travel on the traffic circle without sliding.
(b) Comment on the centripetal force in this problem.

SOLUTION

The FBD in Figure 5-32 shows the forces as viewed from in front of the car. The force of friction, as shown, points to the centre of the circle.

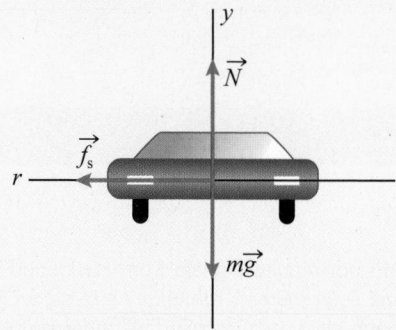

Figure 5-32 The FBD of the car in Example 5-23

(a) Since the direction of the acceleration is toward the middle of the traffic circle, we choose a horizontal x-axis or r-axis. We choose the direction pointing to the centre of the circle as the positive r-axis. Summing the forces in the horizontal (or radial) direction, we get

$$\sum F_r = ma_r \qquad (1)$$

The only force acting on the car is the force of friction, so static friction provides all the centripetal force. Since the car is on the verge of slipping, the force of static friction is at its maximum value, using Equation 5-2,

$$f_s^{max} = N\mu_s = \frac{mv^2}{r} \qquad (2)$$

To solve for v, we need to calculate the value for N. Summing the forces in the y-direction

$$\sum F_y = N_y + mg_y = ma_y = 0 \qquad (3)$$

Choosing up as our positive direction, we get

$$N - mg = 0 \Rightarrow N = mg \qquad (4)$$

Combining Equations (2) and (3), we get

$$mg\mu_s = \frac{mv^2}{r} \Rightarrow \qquad (5)$$

$$v^2 = rg\mu_s = (42 \text{ m})(9.81 \text{ m/s}^2) \times 0.27 = 111.24 \text{ m}^2/\text{s}^2$$

$$v = 10.5 \text{ m/s}$$

(b) The centripetal force in this case is the force of friction because it is the only physical force pointing toward the centre of the circle.

Making sense of the result:

When the coefficient of friction is zero, the speed of the car, for it not to slip, must be zero! The car would not be able to turn, no matter how low its speed is.

 ## EXAMPLE 5-24

Staying in the Loop

You swing a 310 g metal ball at the end of a 1.20 m-long string in a vertical circle. The mass has a speed of 17.0 m/s at the lowest point in the circle.
(a) Find the tension in the string.
(b) Comment on the centripetal force in this problem.

SOLUTION

(a) The FBD of the ball in Figure 5-33 shows that only the tension in the string and the force of gravity act on the mass.

Summing the forces in the vertical, or radial, direction gives

$$\sum F_y = T_y + mg_y = ma_y \quad (1)$$

Taking the direction toward the centre of the circle to be positive, we get

$$T - mg = ma \Rightarrow T = mg + \frac{mv^2}{r}$$

$$= 0.310 \text{ kg}\left(9.81 \text{ m/s}^2 + \frac{17^2}{1.2 \text{ m}} \text{ m}^2/\text{s}^2\right) = 77.7 \text{ N} \quad (2)$$

where the variables T, g, and a, in Equation (2) denote magnitudes of the tension, the acceleration due to gravity and the acceleration of the ball repsectively.

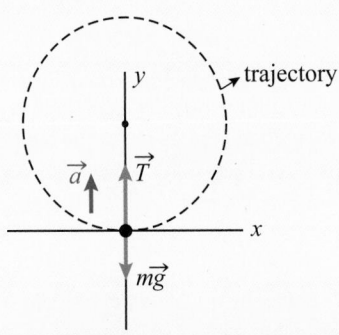

Figure 5-33 The FBD for the mass in Example 5-24

(b) Note that labelling the vertical axis as the r-axis would have also been appropriate. In vector form, the centripetal force is the vector sum of the force of gravity and the tension force. The magnitude of the centripetal force is given in Equation (2).

Making sense of the result:

The tension in the string is much greater than the weight of the ball as it supports the weight of the ball and provides its acceleration. When the speed of the ball is zero, the tension will be equal to mg, as expected.

 ## EXAMPLE 5-25

Banked Racetrack

A race car is on a slippery circular track of radius 190 m. The racetrack is banked at an angle of 27° with respect to the horizontal.
(a) Find the speed at which the race car will not slide up or down the track, assuming that friction with the track surface is negligible.
(b) Identify the centripetal force in this problem.

SOLUTION

(a) Step 1: Identify all the forces acting on the object.

The forces acting on the race car are the force of gravity and the normal force from the incline.

Step 2: Draw an FBD.

Figure 5-34 shows the forces as viewed from in front of the car.

Steps 3 and 4: Establish the direction of the acceleration, and set up a coordinate system with one axis pointing along the direction of the acceleration.

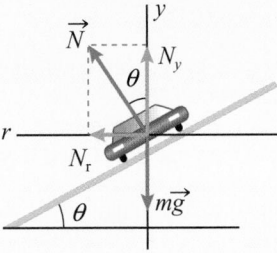

Figure 5-34 The FBD for the car on a banked curve in Example 5-25

The car will move in a horizontal circle. The acceleration is therefore directed horizontally toward the centre of the circle. So we choose a horizontal axis and a vertical y-axis. Either x-axis or r-axis is an appropriate label for the horizontal axis. However, since centripetal or radial acceleration is involved, we will refer to this as the r-axis, for radial.

Step 5. Resolve the forces into components along the chosen axes.

(continued)

The force of gravity points along the vertical axis. We use simple trigonometry to resolve the normal force from the incline as a radial component (N_r) and a vertical component (N_y), as shown in Figure 5-34.

Step 6. Write equations for the motion along each axis. The sum of the forces along the vertical direction is

$$\sum F_y = N_y + mg_y = 0 \qquad (1)$$

Taking up to be positive, from Figure 5-34 we have

$$N \cos\theta - mg = 0 \qquad (2)$$

where the variables in Equation (2) are used to denote magnitudes of the quantities in Equation (1).

Next, we sum the forces along the radial direction:

$$\sum F_r = N_r = ma_r \Rightarrow$$

$$N \sin\theta = ma_r = m\frac{v^2}{r} \qquad (3)$$

Step 7. Solve the equations of motion.
From Equation (2),

$$N \cos\theta = mg \qquad (4)$$

We have three unknowns: the mass of the car, m, the speed of the car, v, and the normal force, N. If we divide Equation (3) by Equation (4), we get

$$\tan\theta = \frac{v^2}{rg} \Rightarrow v = \sqrt{rg \tan\theta} = \sqrt{(190 \text{ m})(9.81 \text{ m/s}^2)\tan(27°)}$$

$$v = 30.8 \text{ m/s}$$

(b) The centripetal force is the vector sum of all the forces acting along the radial direction pointing toward the centre. In this case, it is the horizontal component of the normal force, or $N\sin\theta$.

Making sense of the result:

Notice the difference between this example and Example 5-13 where the direction of the acceleration was along the surface of the incline for a mass sliding down the incline. Therefore in that example, one of the axes pointed along the incline. We also note here that the normal force from the track on the car is given by Equation (4) as $N = \dfrac{mg}{\cos\theta}$. For a banking angle of zero, the car, without friction, will not be able to move along the track without sliding.

EXAMPLE 5-26

Friction on a Banked Curve

A car is travelling on a curve of radius r banked at an angle θ. The coefficient of static friction between the car and the road is μ_s.

(a) Derive an expression for the maximum speed that the car can travel without slipping.

(b) Derive an expression for the normal force, and comment on the expression you get for the normal force in this example compared to the expressions you obtained for the normal force in Examples 5-13, 5-14, and 5-25.

(c) Identify the centripetal force acting on the car.

SOLUTION

(a) Steps 1 and 2: Identify all the forces acting on the object, and draw an FBD.

The forces acting on the car are the force of gravity, the normal force from the incline, and the force of friction, as shown in Figure 5-35.

Steps 3 and 4: Establish the direction of the acceleration, and set up a coordinate system with one axis parallel to the acceleration.

Figure 5-35 The FBD for the car as it goes around a banked curve for Example 5-26. Notice the direction of the force of friction

✓ CHECKPOINT

C-5-16 Friction Direction

Why is the force of friction acting downward on the incline?

As in Example 5-25, the car moves in a horizontal circle, and the acceleration points toward the centre of the circular path. Therefore, we again choose a horizontal (radial) r-axis and a vertical y-axis.

Step 5: Resolve the forces into components along the axes of the coordinate system.

In Figure 5-36, we can see that the components of the normal force and the force of friction can be found using the sine and cosine of the incline angle θ, see step 6 for details.

Step 6. Write the equations of motion along each axis.

The sum for the forces along the vertical direction gives us

$$\sum F_y = N_y + mg_y + f_y = ma_y$$

Taking up as positive

$$N \cos\theta - mg - f\sin\theta = 0 \qquad (1)$$

If the car is travelling at the maximum speed possible without slipping, then it is on the verge of slipping, and

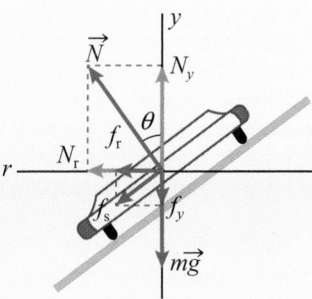

Figure 5-36 The FBD for the car in Example 5-26 with the components of the forces decomposed along chosen axes

the force of static friction is at its maximum value of $N\mu_s$. Therefore, using Equation 5-2, we have

$$N\cos\theta - mg - N\mu_s\sin\theta = 0 \tag{2}$$

Next, we sum the forces along the radial direction:

$$\sum F_r = N_r + f_r = ma_r$$

Taking the radial direction toward the centre of the circle to be positive, we have

$$N\sin\theta + N\mu_s\cos\theta = \frac{mv^2}{r} \tag{3}$$

Alternatively, one can also say that we have taken the positive r-axis to point to the centre of the circle.

Step 7: Solve the equations of motion.
Rearranging Equation (2) gives

$$N\cos\theta - N\mu_s\sin\theta = mg \tag{4}$$

We divide Equation (3) by Equation (4) to eliminate N:

$$\frac{\sin\theta + \mu_s\cos\theta}{\cos\theta - \mu_s\sin\theta} = \frac{v^2}{rg}$$

$$v^2 = \frac{rg(\sin\theta + \mu_s\cos\theta)}{\cos\theta - \mu_s\sin\theta} \quad \text{or} \quad v = \sqrt{\frac{rg(\sin\theta + \mu_s\cos\theta)}{\cos\theta - \mu_s\sin\theta}}$$

note that the variables N, g and f in Equations (1), (2) and (3) denote magnitudes.

(b) As for the normal force as given by Equation (4), we can write

$$N = \frac{mg}{\cos\theta - \mu_s\sin\theta} \tag{5}$$

Each of the examples involving inclined planes (Examples 5-13, 5-14, 5-20, 5-25, and 5-26) gives us a different value for the normal force, depending on the situation involved. It is a common error to assume that the normal force is always equal to mg or $mg\cos\theta$. Now you can see that most often it is not. See the examples for details.

(c) The centripetal force is the sum of all the forces acting along the radial direction. In vector form, the sum is given by $\sum \vec{F}_r = \vec{N}_r + \vec{f}_r$. The magnitude of the sum is given by Equation (3) as $N\sin\theta + N\mu_s\cos\theta$.

Making sense of the result:

When the coefficient of friction is zero, the value of the normal force reduces to $mg/\cos\theta$ and v^2 reduces to $rg\tan\theta$ as we have seen the previous example, (example 5-25).

PEER TO PEER

I was under the impression that the normal force for a car going around a banked curve always equals $mg/\cos\theta$. This expression does not give the normal force in all cases, although it does apply for a car going around a banked curve where there is no friction. The correct way to deal with the normal force is to treat it as an unknown and obtain its value if needed by solving the relevant equations of motion.

LO 10

5-10   Frames of Reference and Fictitious Forces

A Thought Experiment

A passenger is standing still in a stationary bus. The bus then accelerates before the passenger can sit down. At that moment, the passenger is not holding on to any of the handles or the support poles on the bus. You happen

---

MAKING CONNECTIONS

Maximum *g* Forces

A *g* force is a force expressed as a multiple of the force of gravity at Earth's surface. Aerobatics pilots can experience considerable centripetal forces, especially when flying at jet speeds. For an inside loop, the plane flies in a vertical circular path loop with the canopy (and the pilot's head) facing toward the centre of the loop (Figure 5-37). The *g* forces experienced during an inside loop are called positive *g* forces. Positive *g*

forces drive blood away from the brain. During a high positive *g* manoeuvre, pilots can experience grey-outs or black-outs as a result of reduced oxygen in the brain. With special training and a G-suit, which helps maintain blood flow to the brain, a pilot can withstand a positive *g* force of $9g$.

During an outside loop, the top of the plane faces away from the centre of the vertical loop. Depending on

(continued)

Photography Perspectives – Jeff Smith/Shutterstock.com

Figure 5-37 A plane in a positive *g* loop

the acceleration, an outside loop can cause negative *g* forces, which force more blood into the brain and are often more dangerous than positive *g* forces. Generally, the maximum tolerable negative *g* force is about −3*g*. For a high *g* manoeuvre, a pilot will position the plane to experience a positive rather than a negative *g* force, for example, rolling the plane to turn an outside loop into an inside loop.

In contrast, astronauts experience *g* forces mostly during liftoff, at the end of a launch, and during reentry, due to linear rather than rotational accelerations. The blood is "driven" either to the front or to the back of the body. As the body accelerates, the blood's inertia will cause it to lag behind the body parts, owing to blood's fluid nature, which would cause the blood to be pushed toward the front or the back relative to the accelerating body. The *g* forces experienced in this way are called horizontal *g* forces and can be as high as 12*g* or 17*g* for short periods of time, depending on the direction of the force causing the acceleration. In 1954, Colonel John Stapp is reported to have sustained a *g* force of 46.2*g* in a rocket sled as it was slowing down. Higher values have been reported during accidents as well. Refer to Chapters 3 and 4 for a more detailed and in-depth discussion.

to be standing outside the bus while a friend of yours is already sitting comfortably inside the bus. Let us say for the purpose of demonstration that the floor of the bus is so slippery that there is effectively no friction between the bus floor and the shoes of the standing passenger.

The bus accelerates, and from your perspective, as it moves, the passenger simply stays still with respect to you. Since the passenger is not holding on to anything, and since there is no friction between his shoes and the floor of the bus, the net force on the passenger in the direction of motion of the bus is zero. Therefore, that person will not move.

Your frame of reference is called an **inertial frame of reference**. For the passenger to accelerate, he would have to be acted upon either by friction from the floor or by the force from the pole.

In an inertial frame of reference, an object will not accelerate unless there is a real, physical, identifiable force acting on it.

How do things look from the point of view of your friend who is sitting comfortably in her seat on the bus? Suppose that the acceleration is not too great and that she does not feel much of a force from the seat as the bus accelerates. What she will notice, however, is the standing passenger accelerating backward with respect to her. It would then only be natural for her to assume that he is

under the influence of a force pulling him backward. The magnitude of the force will be, naturally, to her, the person's mass times his acceleration. The force she perceives to be there is called a **fictitious force** and only exists with respect to her, in her frame of reference. It is a fictitious force because there is no identifiable agency, or cause, for that force. In her frame of reference the passenger is accelerating without any real, physical identifiable forces acting on him, forces such as the fundamental force (e.g., gravity) or the nonfundamental forces (e.g., tension and the normal force). A frame of reference in which objects accelerate in the absence of real, identifiable physical forces, and in which one must account for fictitious forces to explain physical phenomena is called a **noninertial frame of reference**.

When the bus accelerated, the passenger was standing still. This makes sense from your perspective because there are no forces to move him forward. However, as soon as your friend stretched her arm back to grab the passenger, the passenger started to move forward. From your perspective, this also makes sense. Suddenly, there is a real net force acting on the passenger in the direction of motion of the bus; this is the force your friend is exerting on the passenger and is the only reason the person moves forward from your perspective is that there is a very real force pulling him, and that is the force exerted by your friend's arm on him.

FUNDAMENTAL CONCEPTS AND RELATIONSHIPS

Forces are vectors. Forces are caused by physical interactions between objects, such as gravitational attraction between you and Earth. There are fundamental and nonfundamental forces in nature. With the exception of the force of gravity, all the forces reviewed in this chapter are nonfundamental forces that require physical contact. A net force acting on an object causes the object to accelerate, deform, or both. Cartesian axes are often the most convenient type for analyzing forces and motion. Problems become simpler to solve when one of the axes is chosen to lie along the direction of the acceleration.

Forces

1. Forces are vectors that have magnitude and direction.

2. Forces have real, physical causes.

3. The four fundamental forces underlie all the nonfundamental forces.

4. Identifying and quantifying the forces acting on an object is necessary to predict the object's trajectory.

Newton's Second Law

Mass is a measure of an object's inertia. Inertia determines the ability of an object to resist being accelerated. The mass of an object depends on the amount of material contained within that object.

The net force acting on an object results in the acceleration of the object:

$$\vec{F}_{net} = m\vec{a} \qquad (5\text{-}1)$$

Acceleration and Free Body Diagrams

A free body diagram shows all the forces acting on a single object.

Force vectors can be resolved into components.

When solving problems, the object's acceleration dictates the choice of coordinate system.

Newton's second law can be used to generate equations of motion which can be solved and used to predict trajectories.

APPLICATIONS

Applications: *g*-forces and medical conditions, car racing, driving safety, effects of friction, tension safety, inertial navigation

Key Terms: centripetal force, compression, fictitious force, force, free body diagram, friction, fundamental force, *g* force, Hooke's law, inertia, inertial frame of reference, kinetic friction, mass, Newton's first law, Newton's second law, Newton's third law, nonfundamental force, noninertial frame of reference, normal force, static friction, tension

QUESTIONS

Note that more than one answer can be true

1. A polar bear cub pushes his mother on an icy, frictionless surface in Lancaster Sound, Nunavut. The cub and his mother are originally at rest, and the cub's weight is 1/6 of the mother's weight. She moves away from the cub at 1.1 m/s, and the cub then grabs her arm just before she gets out of his range, and hangs on. What is the final speed of the cub and his mother?
   (a) zero
   (b) 1.1/7 m/s
   (c) The mother stops, and the cub moves at 6.6 m/s.
   (d) The mother slows down slightly while the cub's speed increases.

2. Figure 5-38 shows the FBD of an object. What is the direction of motion of the object?
   (a) The object is moving to the left.
   (b) The object is moving to the right.
   (c) The object is not moving.
   (d) There is not enough information to answer the question.

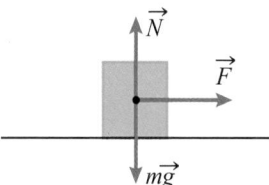

Figure 5-38   Question 2

3. An object moves to the right while subject to a force whose magnitude increases with time but points to the left. Which of the following statements is true?
   (a) The particle will reverse direction as soon as the force is applied.
   (b) The particle will slow down and eventually reverse direction.
   (c) The particle will not slow down because the magnitude of the force decreases with time.
   (d) The particle will not reverse direction.

4. A puck is moving freely on frictionless ice along the positive x-axis when a force is applied to the puck. The direction of the force is maintained along the positive x-axis, but its magnitude drops exponentially with time. Which of the following statements is true?
   (a) The puck's speed increases with time.
   (b) The puck's speed decreases with time.
   (c) The puck's acceleration increases with time.
   (d) None of the above are true.

5. A particle moving to the right is acted upon by a constant net force pointing to the left. The particle moves along the x-axis, and the force is applied at $t = 0$, when the particle's position is to the right of the origin (positive). We can say that the magnitude of the particle's position with respect to time
   (a) changes like a parabola that opens upward;
   (b) changes like a parabola that opens downward;
   (c) decreases linearly;
   (d) increases linearly.

6. You ($m = 76$ kg) are riding in a car moving at a constant speed of 12 m/s on a curved road of radius 10 m. What is the net force acting on you?
   (a) The net force cannot be determined without knowing the force of friction between the road and the car.
   (b) The net force cannot be determined without knowing the force of friction between you and the car.
   (c) The net force cannot be determined without knowing the normal force from the seat cushion on you.
   (d) None of the above.

7. You pull your friend with a horizontal rope such that the force you apply to the rope is 160 N. Your friend stands his ground without pulling back. Neither of you move. Which of the following is true?
   (a) The net force on your friend is 160 N.
   (b) The force of friction on your friend is 160 N.
   (c) The tension in the rope is 320 N.
   (d) None of the above are true.

8. In Figure 5-39, the combined mass of the monkey, bananas, and platform is m. In order to keep the platform still, the monkey has to pull the rope with which force?
   (a) mg
   (b) $\dfrac{mg}{2}$
   (c) $\dfrac{mg}{4}$
   (d) $2mg$

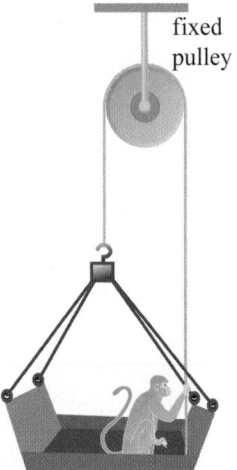

Figure 5-39 Question 8

9. What can be said about the initial accelerations of the two masses in Figure 5-40?

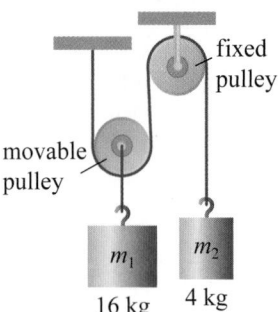

Figure 5-40 Question 9

10. What can be said about the initial accelerations of the three masses in Figure 5-41?

Figure 5-41 Question 10

11. You are on a frictionless skating rink. Your mass is 90 kg. You give a rope to a friend whose mass is 60 kg. You pull the rope, and your friend hangs on as she accelerates toward you at 3 m/s² without pulling on the rope. Which of the following statements is true?
    (a) You do not move.
    (b) The force acting on you is 270 N.
    (c) Your acceleration is 3 m/s².
    (d) The force acting on you is 180 N.

12. Two friends ($m_1 = 52$ kg and $m_2 = 87$ kg) pull one another using a rope. They each apply a force of 230 N to the horizontal rope, but neither move. Which of the following statements is true?
    (a) The tension in the rope is 230 N.
    (b) The tension in the rope is 460 N.
    (c) The force of friction on each friend is 460 N.
    (d) The heavier friend experiences a larger force of friction as they both pull.

13. Two friends are standing on an ice rink, one of them pulling the other with a rope with a force of 145 N. The friend being pulled $m_1 = 50$ kg) moves toward the friend doing the pulling $m_2 = 75$ kg) at a constant speed of 6 m/s. Which of the following statements is true?
    (a) The speed of the friend doing the pulling must be 4 m/s.
    (b) The force of kinetic friction on the friend being pulled is 145 N.
    (c) The net force on the friend being pulled is 145 N.
    (d) The force of static or kinetic friction on the friend doing the pulling is 145 N.

14. Two children are riding in the back of a truck. The truck comes to an emergency stop and both children slide with respect to the truck. Assuming the same coefficient of friction between each child and the back of the truck, which child will slide first—the lighter or the heavier child ?

15. Figure 5-42 shows the FBD of an object. The object must
    (a) be moving up;
    (b) be accelerating up
    (c) momentarily stationary
    (d) be accelerating down

Figure 5-42 Question 15

16. Two boxes, one twice the mass of the other, slide down the same incline, starting from rest at the same height. The coefficient of friction between each of the boxes and the incline is the same. Compare the times the boxes take to go down the incline.

17. You are standing in an elevator that is accelerating upward. The force that you exert on the elevator
    (a) is larger than the force that the elevator exerts on you;
    (b) is smaller than the force that the elevator exerts on you;
    (c) is equal to the force that the elevator exerts on you.
    (d) none of the above.

18. You are sitting in a 1000 kg truck, which is slowing down at 10 m/s². Your mass is 50 kg. Which of the following is true?
    (a) The force that you exert on the truck is 500 N.
    (b) The net force on the truck is 9500 N.
    (c) The force that the truck exerts on you is 500 N.
    (d) The vertical force that the truck exerts on you is 490 N.

19. A crate is sitting on a horizontal surface where the coefficient of static friction is 0.6. A worker pushes horizontally on a crate, which has a mass of 45 kg. The crate does not move, and there are no other forces acting on it. Which of the following statements is true?
    (a) The force of friction on the crate is $mg\mu_s$.
    (b) The force of friction on the crate is greater than the force that the worker exerts on the crate.
    (c) The force exerted on the worker by the crate is equal in magnitude to the force exerted by the crate on the surface.

20. Two friends are pushing a book in opposite directions, each pushing with a force of 320 N. One friend says the net force on the book is 640 N. Is this correct? Explain.

21. Two friends push a sled on snow in opposite directions. One pushes with a force of 190 N, and the other pushes with a force of 130 N. The sled does not move. What is the net force on the sled?
    (a) 320 N
    (b) 60 N
    (c) zero
    (d) The net force cannot be determined.

22. You are sitting in a car that is rounding a 40 m-radius curve banked at 21°, with a speed of 12 m/s. Which of the following statements is true?
    (a) The horizontal force exerted on you by the car is larger than the horizontal force exerted on the car by you.
    (b) The net force on the car is $\dfrac{Mv^2}{r} - \dfrac{mv}{r}$.
    (c) The net force on you is $\dfrac{mv^2}{r}$.
    (d) The net force on the car is $\dfrac{mv^2}{r}$.

23. A lion trainer is in a train car looking after his 120 kg lion named Pouncer. Pouncer is not in the mood to play and is sitting still. The trainer prods Pouncer gently with a force of 200 N in a direction toward the front of the train, but it is not enough to get Pouncer to move. The train is slowing down at an acceleration of 5 m/s². The net force on Pouncer is
    (a) 720 N;
    (b) 600 N;
    (c) 480 N;
    (d) 120 N;

24. A rope connects M_1 hanging vertically to M_2 resting on top of a horizontal frictionless surface. The system is originally at rest.
    (a) Is the tension in the rope larger than, smaller than, or equal to M_1g?
    (b) If the surface that M_2 rests on provides enough friction to prevent M_2 from moving, what is the tension in the rope?

25. You are riding in a car moving at a speed of 20.0 m/s around a curve of radius 40.0 m. The mass of the car is 1300 kg, and your mass is 60.0 kg. Find the magnitude of
    (a) the force that you exert on the car
    (b) the net force on the car
    (c) the net force on you
    (d) the horizontal force that the car exerts on you
    (e) the force that the car exerts on the road

26. Two buckets, each of mass m, are suspended from a rope that goes around the fixed massless frictionless pulley in Figure 5-43. The system is at rest. What is the tension in the rope?
    (a) mg
    (b) $2mg$
    (c) zero
    (d) mg^2

Figure 5-43 Question 26

27. Two buckets of equal mass are suspended vertically from a rope that goes around the fixed massless frictionless pulley as shown in Figure 5-43. The pulley rotates clockwise so

that the masses are moving at a constant speed. Which of the following statements is true?

(a) The tension in the rope is equal to mg.

(b) The tension in the rope is equal to $2mg$.

(c) The tension in the rope is zero.

(d) The tension cannot be determined without knowing the masses.

28. The bucket (mass m) hanging from the right side of the massless frictionless pulley in Figure 5-43 is heavier than the bucket hanging from the rope on the left side of the pulley. The tension in the left side of the rope is

(a) mg;

(b) greater than mg;

(c) less than mg;

(d) zero.

29. A centipede of mass m is climbing a thread that is hanging vertically from the edge of a table. When the centipede begins its climb, its acceleration is 2 m/s² upward. The tension in the rope is

(a) mg;

(b) ma;

(c) $mg + ma$;

(d) $mg - ma$.

30. A monkey uses a fixed rope to climb down from a coconut tree. The monkey races down with an acceleration of $1.5g$ downward to avoid a coconut dropped by another monkey. Taking up as the positive direction, the tension in the fixed rope is

(a) $0.5mg$;

(b) $1.5mg$;

(c) $2.5mg$;

(d) $-0.5mg$.

31. A child swings a string with three beads attached to it in a vertical circle. The bead closest to his hand (distance l away) has a mass of m, the bead after that (distance $2l$ away) has a mass of $2m$, and the farthest bead from his hand (distance $3l$) has a mass of $3m$. The tension in the rope is the highest in the section

(a) closest to the child's hand;

(b) between m and $2m$;

(c) between $2m$ and $3m$.

32. Mass m_1 hangs from the massless movable pulley shown in Figure 5-44. One end of the rope supporting the pulley is attached to the ceiling; the other goes over the fixed pulley and is attached to the mass on the frictionless surface. When the system is let go, the acceleration of m_1 is

(a) half the acceleration of m_2;

(b) double the acceleration of m_2;

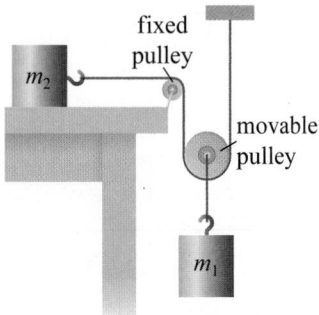

Figure 5-44   Question 32

(c) the same as the acceleration of m_2.

(d) The relationship between the two accelerations depends on the masses.

33. Mass m_1 hangs vertically from one end of a rope as shown in Figure 5-45. The rope goes over the fixed pulley and around the movable pulley that is attached to m_2. The other end of the rope is fixed to the protrusion that is attached to the horizontal surface. As the system moves, the acceleration of m_1 is

(a) twice the acceleration of m_2;

(b) half the acceleration of m_2;

(c) the same as the acceleration of m_2.

(d) More information is needed to answer the question.

Figure 5-45   Question 33

34. Four blocks rest on top of each other such that $m_1 = 2m_2 = 4m_3 = 8m_4$. The coefficients of friction between the blocks are $\mu_{s3} = 2\mu_{s2} = 4\mu_{s1}$. The bottom block is the heaviest of the four blocks. It is pushed with a horizontal force strong enough that at least one pair of the blocks slips. Which block will be the first to slide with respect to the block beneath it?

(a) m_1

(b) m_2

(c) m_3

(d) They will all slide simultaneously.

35. A bug is standing stationary on a spinning turntable. Some oil is poured on the turntable to render the surface frictionless. As soon as the surface becomes frictionless, the bug will

(a) continue to move in a circle, but the radius of the circle will get larger and larger;

(b) move in a straight line tangent to the circular trajectory it was on;

(c) move radially outward with respect to the turntable;

(d) move radially outward toward the edge of the turntable;

36. A child is swinging a string with a ball attached to its end in a circle that is nearly horizontal. Ignoring gravity, if the string is cut, in which direction will the ball go?

PROBLEMS BY SECTION

For problems, star ratings will be used, (✶, ✶✶, or ✶✶✶), with more stars meaning more challenging problems.

Section 5-1   Force and Net Force

37. ✶   (a) Resolve each force in Figure 5-46 into Cartesian components.

(b) Calculate the net force in Figure 5-46.

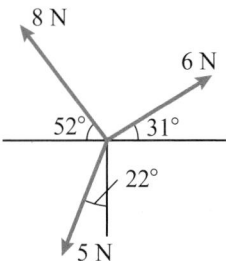

Figure 5-46 Problem 37

38. ✷ Find the magnitude and direction of the forces listed below. Report the direction as an angle measured counterclockwise with respect to the *x*-axis.
(a) $6\hat{i} + 2\hat{j}$ N
(b) $-21\hat{i} + 16\hat{j}$ N
(c) $7\hat{i} - 4\hat{j}$ N

39. ✷ Add the following forces. Express the resultant forces in Cartesian form with unit vectors.
(a) $-4\hat{i} - 2\hat{j} + 0.5\hat{k}$ N, $3\hat{j} - 4\hat{k}$ N, $3\hat{i} + 2\hat{j}$ N
(b) 14.0 N [57°], $(3.0\hat{i} - 4.0\hat{j})$ N, 8.0 N [236°]
(c) $(4\hat{i}, 2\hat{j}, -5\hat{k})$, $(-9\hat{i}, 14\hat{j}, 11\hat{k})$
(d) $r\theta\phi$ notation, (21 N, 32°, 27°), (17 N, 160°, 295°)

40. ✷ Add the following forces, which are all in newtons. Express your answer in both polar and Cartesian notation.
(a) $(4\hat{i}$ N, $2\hat{j}$ N$)$, $(6\hat{i}$ N, $-4\hat{j}$ N$)$, $(10\hat{i}$ N, $-4\hat{j}$ N$)$
(b) (16 N, 7° N of E), (5 N, 289°), (24 N, −25°)
(c) $(5\hat{i}$ N, $-2\hat{j}$ N$)$, (2 N, 230°)

Section 5-2 Newton's Laws

41. ✷ A constant force of 130 N is applied to a block of mass 9.00 kg, which is sitting on a frictionless horizontal surface. Find the speed of the mass after it has moved 14.0 m, starting from rest.

42. ✷ The last part of a roller coaster's track is horizontal. A roller coaster is brought to rest using a constant braking force of 4300 N. It takes 9.00 s to bring the roller coaster to a stop from a speed of 23.0 m/s. Find the mass of the roller coaster.

43. ✷ Assume that a particular crash test expert can withstand 36 times the acceleration due to gravity. Find the shortest distance over which her car can be brought to a stop from a speed of 70 km/h without the acceleration exceeding 36*g*.

44. ✷ A car of mass 2100 kg is brought to an emergency stop from a speed of 43 m/s over a distance of 21 m. What is the average net force experienced by the car during this stop?

45. ✷ You are sitting in the passenger seat of a car driving on the highway at a speed of 110 km/h when the driver ($m = 79$ kg) exits on a circular ramp of radius 63.0 m. The mass of the car is 1300 kg, and your mass is 71 kg. Find
(a) the horizontal force that the road exerts on the car;
(b) the horizontal force that you exert on the car;
(c) the net horizontal force on the car.

46. ✷ You are sitting in the back seat of a car moving on the highway, when the driver hits the brakes. The net force on you is 320 N. Find the net force that you exert on the car.

Section 5-3 Free Body Diagrams, Acceleration, and Equations of Motion

47. ✷✷ For each of the following scenarios, determine the direction of the acceleration. Assign your system of axes, resolve the forces into components, and write the equations of motions for each scenario.
(a) A child slides down a frictionless incline with a slope of angle θ.
(b) A child slides down an incline with a slope of angle θ.
(c) A conical pendulum.
(d) A man pushes his daughter of mass *m* up the frictionless surface of a slide at an angle θ using a horizontal force *F*. Draw the FBD of the daughter. Assume she is moving up the incline as a result of being pushed by her father.
(e) A worker pulls a crate of mass *m* on a horizontal surface, where the coefficient of friction is μ_k. She uses a force of magnitude *F* directed at an angle θ above the horizontal. Assume the crate is moving as a result of being pushed.
(f) A mass moving at a constant speed at the end of a string rotated in a vertical circle when the mass is
  (i) at the top;
  (ii) at the bottom;
(g) A car banking a frictionless curve with radius *r* and banking angle θ.
(h) A car banking a curve with radius *r*, banking angle θ, and coefficient of static friction μ_s.

Section 5-4 The Force of Gravity

48. (a) ✷ If your weight is 670 N, what is your mass?
(b) How much would you weigh on the Moon, where the acceleration due to the Moon's gravity is approximately 1/6 the acceleration due to gravity at Earth's surface?
(c) If you can jump 60 cm straight up on Earth, how high can you jump on the Moon?

49. ✷✷ You ($m_1 = 81$ kg) are standing in an elevator ($m_2 = 2300$ kg) that is moving downward with a speed of 9.0 m/s. The elevator comes to a stop with a constant acceleration over a distance of 4.0 m. Find
(a) the net force on you;
(b) the net force on the elevator;
(c) the force that you exert on the elevator;
(d) the gravitational force on the elevator.

Section 5-5 The Normal Force

50. ✷✷ Four blocks of masses 20 kg, 30 kg, 40 kg, and 50 kg are stacked on top of each other in an elevator in order of decreasing mass, with the lightest mass on top. The elevator moves down with an acceleration of 3.2 m/s². Find the contact force between the 30 kg and the 40 kg blocks.

51. ✷✷ A woman is standing in an elevator carrying a 2.7 kg cat. The elevator begins to accelerate upward at 3.2 ms². Find the force that the cat exerts on the woman.

52. ✷✷ You are pushing four blocks of masses 4.00 kg, 7.00 kg, 11.00 kg, and 17.00 kg positioned next to each other on a frictionless surface as shown in Figure 5-47 (Page 124). Find the force of contact between the 7.00 kg and 11.00 kg blocks when
(a) you push on the lightest mass with a 190 N force to the left;
(b) you push on the heaviest block with a 190 N force to the right.

Figure 5-47 Problem 52

53. ✱✱ Starting from rest, a car skids 15 m down an ice-covered road, where the friction is negligible. The speed of the car at the end of 15 m is 3.2 m/s. Find the slope angle of the road.

54. ✱✱ Children are enjoying a constant slope water slide at an amusement park. They run toward the top of the 6 m-long slide and lie flat on the slide surface. A child who arrives at the top of the slide with a speed of 1.4 m/s reaches a speed of 3.6 m/s at the bottom of the slide. Find the slope angle of the slide, assuming that its friction is negligible.

55. ✱ A polar bear cub slides 31 m toward his mother down a snow-covered 37° slope starting from rest. At the end of the 31 m, the cub's speed is 4.7 m/s. Find the coefficient of friction between the cub and the snow-covered surface.

56. ✱✱ You have just had the apples in your orchard harvested into a large number of boxes with masses of 130 kg each. You wish to lift these boxes 12.0 m up to the fourth floor of your packaging facility. Since you have only one person to help, you construct a slippery surface that is inclined 30° above the horizontal, and then fix a pulley at the top corner as shown in Figure 5-48. You hang from the rope at the top of the incline and let the force of gravity pull the crates up the ramp. The friction between the crates and the inclined surface is negligible.
    (a) What must your minimum mass be for this system to work?
    (b) If you do not wish to hit the ground with a speed greater than 2.0 m/s, what must your maximum mass be?
    (c) How long will it take to lift one crate to the fourth floor if you have the mass calculated in part (b)?

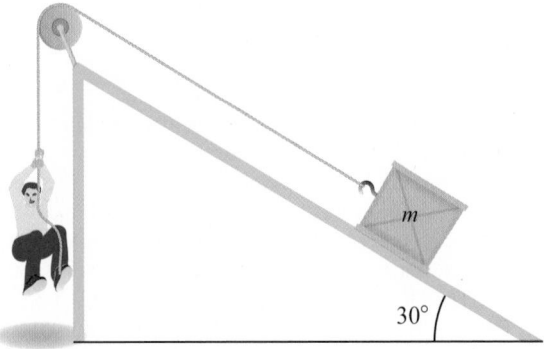

Figure 5-48 Problem 56

Section 5-6 Tension

57. ✱✱ Mass M_1, sitting on a horizontal frictionless surface, is connected by a rope that runs over a massless pulley to mass M_2, which is hanging vertically. The system is originally at rest and is then let go.

(a) Find the tension in the rope when $M_2 > M_1$.
(b) How would your answer to part (a) change if the masses were interchanged?
(c) Obtain numerical values for parts (a) and (b) using $M_1 = 2.0$ kg and $M_2 = 12.0$ kg.

58. ✱✱ Four food packages ($m_1 = 7.00$ kg, $m_2 = 5.00$ kg, $m_3 = 8.00$ kg, and $m_4 = 3.00$ kg) are connected to each other using ropes as shown in Figure 5-49. The four masses rest on a frozen lake surface in Labrador. A curious husky pulls on a rope that is attached to m_4 and pulls the four masses to the right.
    (a) The tension in the rope between m_1 and m_2 is 40.0 N. Calculate the tension in the rope connecting m_2 and m_3.
    (b) What is the magnitude of the force with which the husky is pulling on m_4?

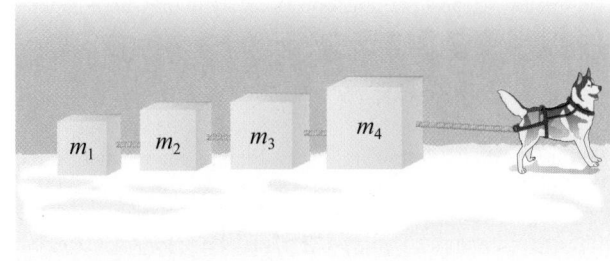

Figure 5-49 Problem 58

59. ✱✱ You are holding a special pendulum with two masses, m_1 and m_2, instead of one, connected by a rope as shown in Figure 5-50. You lower the pendulum such that the tension in the rope connecting the two masses is half the weight of the bottom mass.
    (a) Find the acceleration with which you lower the pendulum.
    (b) For what tension in the rope between you and m_1 would the tension in the rope between the two masses be zero?

Figure 5-50 Problem 59

Section 5-7 Friction

60. ✱ A hockey puck is shot across an ice rink at a speed of 3 m/s and comes to a stop over a distance of 11 m. Find the coefficient of friction between the puck and the ice.

Section 5-8 Force of a Spring

61. ✳ In order to stretch a spring by 10 cm, you have to apply a force of 120 N. How much would the spring stretch if a 10 kg mass were to hang from it vertically?

62. ✳ The left side of a mass resting freely on a frictionless surface is attached to an unstretched spring. The mass is pulled 20 cm to the right and then released. Write an expression for the acceleration of the mass when it has moved 15 cm back toward its original position.

63. ✳ A spring is suspended from an elevator, and a 12.0 kg mass is then attached to the spring. When the elevator moves down at an acceleration of 3.00 m/s², the net elongation of the spring is 30.0 cm. Find the spring constant.

Section 5-9 Circular Motion and Centripetal Forces

64. ✳ You swing a marble attached to the end of a string in a horizontal circle as shown in Figure 5-51. The angle that the string makes with the vertical is 37°.
    (a) Find the speed of the marble when the string is 44.0 cm long.
    (b) Write an expression for the tension in the string.

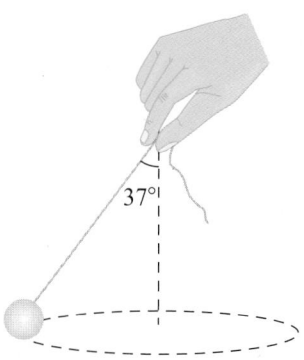

37°

Figure 5-51   Problem 64

65. ✳ A child of mass 20 kg stands on a merry-go-round 3.0 m from the centre. You push the merry-go-round until the child is about to slip on the merry-go-round surface. The coefficient of static friction between the child and the surface is 0.45.
    (a) Find the child's speed at the moment she is about to slip.
    (b) Write an expression for the tension in the string.

66. ✳ You are testing a remote-controlled toy truck in an empty parking lot. The coefficient of static friction between the surface and the truck is 0.34. Find the radius of the smallest circle the truck can move in without slipping at its maximum speed of 33 km/h.

67. ✳✳ You are driving on a circular ramp of radius 72.0 m on an icy January day. The ice-covered ramp is banked at a 23.0° angle, and you anticipate that the ramp is not going to provide any significant friction to help keep your car on the road.
    (a) What must your speed be so that your car does not slide off the road?
    (b) What are the direction and magnitude of the net force acting on you at that point? Your mass is 72.0 kg.

68. ✳✳ Your niece finds her father's watch. The light watch chain has a length of 32 cm, and the mass of the watch is 210 g. Your niece swings the watch in a vertical circle, maintaining the speed of the watch at 2.8 m/s. Find the tension in the chain when it makes an angle of 43° with respect to the vertical.

Section 5-10 Frames of Reference and Fictitious Forces

69. ✳✳ You are sitting comfortably in a train. A passenger standing beside you is loading her luggage into the overhead compartment when the train starts moving forward with an acceleration of 5.0 m/s². Find the coefficient of static friction that is necessary to prevent the person from sliding with respect to the train. Solve the problem by considering the scenario from the train's frame of reference, and then again using the train station's frame of reference.

70. ✳✳ Derive the equations of motion for a conical pendulum using the pendulum mass's frame of reference and then again using a stationary frame of reference.

71. ✳✳ A 30 kg bag of mortar mix is sitting in the back of a truck rounding a traffic circle of radius 23 m at a speed of 67 km/h. Using the truck's frame of reference, find the force of contact between the bag and the side of the truck. Assume that the floor at the back of the truck is frictionless.

72. ✳✳ An astronaut sits in a spacecraft orbiting Earth. Explain the feeling of weightlessness that the astronaut experiences. Is it true that an astronaut living on the International Space Station (ISS) is too far from Earth to experience the force of gravity? What is the altitude of the ISS's orbit? Research the value of the force of gravity at that altitude.

COMPREHENSIVE PROBLEMS

73. ✳✳ A curious child finds a rope hanging vertically from the ceiling of a large storage hangar. The child grabs the rope and starts running in a circle. The length of the rope is 17.0 m. When the child runs in a circle of radius 6.0 m, the child is about to lose contact with the floor. How fast is the child running at that time?

74. ✳✳✳ A massless spring with a spring constant $k = 490$ N/m is suspended from a hook on the ceiling of a cargo truck moving in a straight line on the highway. A child reaches over from her seat and suspends a 1.2 kg toy on the free end of the spring. The child notices that the spring is now 4.0 cm longer than when it was hanging freely. Find the acceleration for the truck.

75. ✳✳ A cruise ship moving at a constant speed of 30 km/h is equipped with a gymnasium. A gymnast is climbing a rope suspended from the ceiling of the gym. The captain of the ship puts the ship into a circular turn of radius 150 m to get closer to a pod of whales. Find the angle that the rope makes with the vertical.

76. ✳✳ A child sits in the back of a car with a yo-yo freely suspended from his fingers. The car is leaving the highway at a speed of 78 km/h on a ramp of radius 210 m banked at a 23° angle.
    (a) Find the angle that the yo-yo makes with the vertical.
    (b) Find the angle that the yo-yo makes with the normal to the roof of the car.

77. ✳✳✳ The truck in Figure 5-52 (page 126) is going around a circular track of radius 72 m, banked at a 60° angle. A spider rests on the inside wall of the truck as shown. The coefficient of static friction between the truck wall and the spider is 0.91. Find the maximum speed that the truck can have before the spider begins to slip down the wall.

spider

Figure 5-52 Problem 77

78. ✷✷ A toy car is made to run the bowl-shaped track shown in Figure 5-53. As the car's speed increases, the car moves farther and farther up the track walls. In the end, the car is able to round the vertical walls of the track, such that its height from the bottom of the track does not change. The radius of the bowl is 1.9 m. Find minimum speed that the car needs to not slip when rounding the vertical walls. The coefficient of static friction between the car and the wall is 0.39.

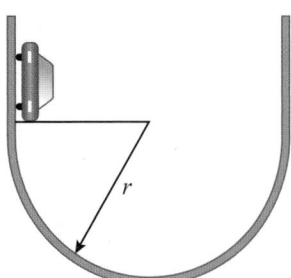

Figure 5-53 Problem 78

79. ✷✷ Two masses (m_1 = 12.0 kg and m_2 = 14.0 kg) are hanging from a rope that passes over a massless pulley, as shown in Figure 5-54. The two masses are originally held in place and then let go.
(a) Find the speed of each mass after one has moved 4.00 m.
(b) What is the tension in the rope?

Figure 5-54 Problem 79

80. ✷✷ A roller coaster car is upside down at the top of a vertical circular loop with a radius of 21 m. The speed of the car is 24 m/s, and the coefficient of kinetic friction between the car and the track is 0.16. Find the radial acceleration of the car.

81. ✷✷ A marble slides down a hemispherical bowl starting from rest as shown in Figure 5-55 starting from rest. When the marble is halfway to the bottom, its speed is 2.3 m/s. Write an expression for the magnitude of the total acceleration of the marble.

Figure 5-55 Problem 81

82. ✷✷✷ A block of mass m is held in place on a ramp of mass M with a slope of angle θ, which in turn is held in place as shown in Figure 5-56. The ramp is free to move on the surface beneath it. There is no friction anywhere. The block and the ramp are then released. Derive an expression for the acceleration of the ramp.

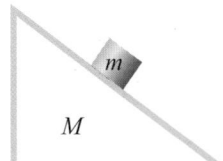

Figure 5-56 Problem 82

83. ✷✷✷ A block of mass m is held in place on a ramp of mass M with a slope of angle θ, which in turn is held in place as shown in Figure 5-57. The ramp is free to move on the surface beneath it. There is no friction anywhere. The ramp is then pushed with a horizontal force so that the block does not slide on the ramp. Derive an expression for the magnitude of this force.

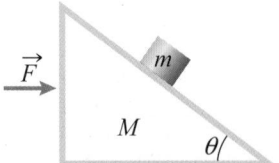

Figure 5-57 Problem 83

84. ✷✷✷ The coefficient of static friction between the inclined face of a wedge (mass M) and the mass m sitting on its inclined edge is μ_s. The coefficient of kinetic friction is μ_k. Derive an expression for the maximum horizontal force that you can apply to the vertical face of the wedge so that the mass m does not move with respect to the wedge.

85. ✷✷✷ The coefficient of kinetic friction between the inclined face of a wedge of mass M and a block of mass m sitting on its inclined face is μ_k (Figure 5-58). The vertical face of the wedge is pushed with a horizontal force such that m slides up the inclined surface with an acceleration a with respect to the inclined surface. Derive an expression for the magnitude of the force.

MECHANICS

MECHANICS

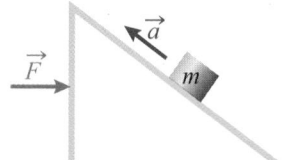

Figure 5-58   Problem 85

86. ★★★   Block A of mass m_A = 25 kg rests on a friction-less surface as shown in Figure 5-59. Block B is suspended from a rope that runs over the massless pulley and is then attached to block A. Block C of mass m_C = 6 kg rests on top of block A. The coefficient of static friction between blocks A and C is 0.23. The system is initially held in place. Find the maximum value for the mass of block B so that when the system is let go, block C does not slide on block A.

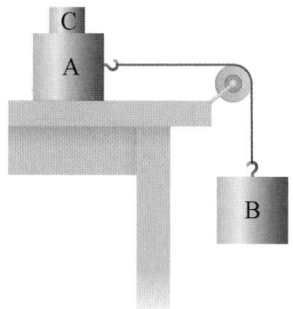

Figure 5-59   Problem 86

87. ★★★   Mass m rests on mass M as shown in Figure 5-60. Mass m is connected with a rope that runs over a massless pulley to mass $3m$. There is no friction anywhere, and the system is originally held in place. When the system is let go, mass M is pushed with a horizontal force such that $3m$ does not move vertically. Derive an expression for the magnitude of the force.

Figure 5-60   Problem 87

88. ★★★   Block B (m_B = 1.00 kg) is held in contact with block A (m_A = 3.00 kg) as shown in Figure 5-61. When block B is let go, block A is pushed to the right with force of magnitude 160 N. The coefficient of static friction between the two blocks is 0.50.
(a) Find the force of friction on block B.
(b) Find the normal force exerted by block A on block B.
(c) Find the vertical force exerted by block B on block A.

Figure 5-61   Problem 88

89. ★★★   Block B (m_B = 4.00 kg) sits on a horizontal frictionless surface. Block A (m_A = 2.00 kg) sits on top of block B and is attached to a rope that runs over the massless pulley as shown in Figure 5-62. Block C (m_C = 1.00 kg) hangs from the rope. There is no friction anywhere.
(a) With what horizontal force $\vec{F}$ must you push block B so that block C rises at an acceleration of 3.00 m/s²?
(b) For part (a), find the net force on blocks B, A, and C. Why is the net force on block B different from the force used to push block B?

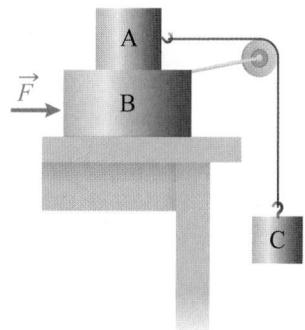

Figure 5-62   Problem 89

90. ★★   You swing a 1.2 m-long string with a metal ball attached to its end in a vertical circle, such that the speed of the ball does not change but the rope is always taut. The tension in the string when the ball is at the bottom of the circle is 60 N more than the tension when the ball is at the top.
(a) Find the mass of the ball.
(b) How does the difference in tension change when the speed is doubled?

91. ★★★   The driver of a race car of mass M wishes to test a wide, banked turn at angle θ on a circular race track of radius R. First, the surface of the banked curve is covered with ice. The driver places his car at the top of the bank and observes it as it slides down the bank. The engine at this stage is off, and the car is originally at rest.
(a) Derive an expression for the magnitude of the normal force on the car as it slides downhill.
(b) With the bank still covered in ice, the driver next drives around the track at the exact speed needed to keep the car from sliding up or down the bank. Derive an expression for the magnitude of the normal force acting on the car.
(c) For the last test, the driver asks for the ice to be melted, and he drives his car around the track at the maximum speed that will not cause the car to slide. The coefficient of static friction is μ_s. Derive an expression for the magnitude of the normal force on the car from the track.

92. ★★★ Four blocks have masses as follows: $M_A = 11.0$ kg, $M_B = 7.0$ kg, $M_C = 5.0$ kg, and $M_D = 3.0$ kg. The coefficient of static friction between blocks A and B is 0.12, the coefficient of static friction between blocks B and C is 0.16, and the coefficient of static friction between blocks C and D is 0.18. There is no friction between block A and the surface on which it sits. A horizontal force, F, is applied to block A. Find the maximum value of F for which none of the blocks slide with respect to one another.

93. ★★ An astronaut-training device is made of a ramp connected to heavy rope. The other end of the rope is fixed. The ramp is made to spin in a horizontal circle. The astronaut sits on the ramp facing the centre of the circle. The coefficient of static friction between the astronaut and the ramp is 0.25. Derive an expression for the maximum tension in the rope as the ramp spins so that the astronaut does not slip on the ramp. There is no friction between the ramp and the surface beneath it.

94. ★★ You want to push your 53 kg friend up the 33° ramp of a slide in a nearby park. You do so by pushing your friend with a 510 N horizontal force. The coefficient of kinetic friction between your friend and the slide is 0.17. Find the acceleration of your friend up the slide. What is the normal force from the ramp on your friend?

95. ★★★ The face of block M in Figure 5-63 is shaped like a semicircular bowl of radius R. A mass m is placed at the top-left corner of the bowl and then let go. Find the acceleration of block M when m is a distance of $0.2R$ from the bottom of the bowl. There is no friction between M and m, or between M and the surface on which it sits.

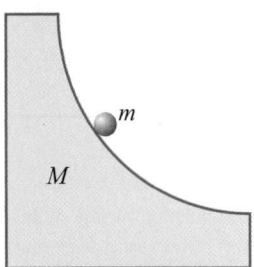

Figure 5-63   Problem 95

96. ★★★ The face of block M is shaped like a semicircular bowl of radius R. There is no friction between M and m, or between M and the surface on which it sits. Derive an expression for the force you need to apply to the vertical face so that a mass that placed on the semicircular face a distance of $0.5R$ above the bottom of the bowl does not slide with respect to M.

97. ★★★ You carve out a piece of ice in the shape of a circular dome of mass M and radius R. You then place the carved piece on an ice surface, and place an ice cube on your dome, at a height h above the ice surface. Derive an expression for the force you need to apply to this dome so that the cube does not slide on it.

98. ★ Mass m_1 hangs from a massless movable pulley as shown in Figure 5-64. The rope rounding the movable pulley passes over the fixed pulley and is attached to mass m_1. There is no friction between m_2 and the surface supporting it. The system is originally held at rest and then let go. Derive expressions for the acceleration of m_1 and the tension in rope.

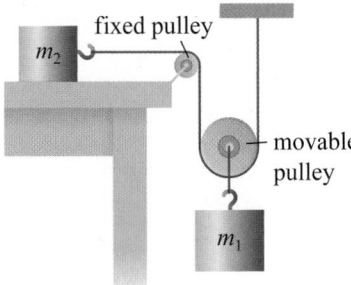

Figure 5-64   Problem 98

99. ★★ Figure 5-65 shows two friends testing a stunt ride they invented. One friend ($m_1 = 56.0$ kg) sits on a seat hanging from a cliff above water, and the other friend ($m_2 = 78.0$ kg) stands on a frictionless surface at the top of the cliff while holding a moveable pulley. The upper end of the rope that passes over the fixed pulley and the movable pulley is fastened to a shock-absorbing block and the lower end to the suspended seat as shown. They start with the seat at rest 19.0 m above the water. Find the acceleration of each friend, and the speed at which the seat hits the water.

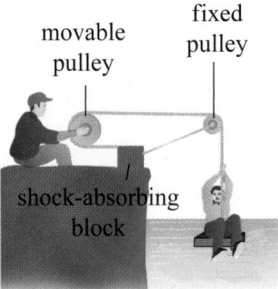

Figure 5-65   Problem 99

100. ★ A 1300 kg car takes a 44 m-radius roundabout at a speed of 7.0 m/s. The road is not banked. The coefficient of static friction between the car and the road is 0.41. Find
(a) the force of friction on the car;
(b) the net force on the car;
(c) the maximum force of static friction on the car.

101. ★★ You are sitting on the back of a truck in the Prairies when the driver decides to take you on a joy ride. Your mass is 69 kg, and the coefficient of static friction between you and the truck is 0.34. The truck follows a circular path of radius 54 m at a speed of 11 m/s. Find
(a) the force of friction on you;
(b) the force that you exert on the truck;
(c) the net force on you;

102. ★★ Blocks A (20.0 kg), B (10.0 kg), C (40.0 kg), and D (30.0 kg) are in contact beside each other on a horizontal frictionless surface as shown in Figure 5-66. Block A is being pushed to the left with force F such that the force of contact between the C and D blocks is 60.0 N. Find the force of contact between blocks B and C. Also find the force F and the acceleration of each block.

Figure 5-66 Problem 102

103. ✷✷ You place a book on the seat next to you, and you then drive your car at a speed of 110 km/h on the highway.

You suddenly have to slow down. The book is just beginning to move on the seat. You stop over a distance of 120 m. What is the coefficient of friction between the seat and the book?

See the text online resources at www.physics1e.nelson.com for Open Problems and Data-Rich Problems related to this chapter.

Chapter 6
Energy

Learning Objectives

When you have completed this chapter you should be able to:

1. Explain the work–kinetic energy theorem, and define the work done by a force.

2. Calculate the work done by a constant force in one or more dimensions.

3. Calculate the work done by a variable force.

4. Calculate the work done by a spring.

5. Differentiate between conservative and nonconservative forces, and relate the change in potential energy to the work done by conservative forces.

6. Explain the principle of conservation of mechanical energy, and compare the work–energy approach to solving problems to the approach used when solving problems using Newton's laws.

7. Relate potential energy to reference points.

8. Define equipotential surfaces, and explain their relationship to conservative forces.

9. Define power and calculate power output.

e For additional Making Connections, Examples, and Checkpoints, as well as activities and experiments to help increase your understanding of the chapter's concepts, please go to the text's online resources at www.physics1e.nelson.com.

Most of the energy we use on Earth originates from the Sun. The rivers that run hydroelectric generators are fed by precipitation, which results from water being evaporated by the Sun's energy. When we burn fossil fuels, we use chemical energy from compounds originally created by photosynthesis millions of years ago, another process powered by the Sun.

An unintended consequence of the massive use of fossil fuels and some large-scale industrial processes is a significant increase in the quantities of greenhouse gases, such as carbon dioxide and methane, in the atmosphere. As a result, a greater proportion of the energy radiated by the Sun is absorbed by Earth.

The consensus among scientists is that human activity is having a measurable and accelerating effect on the climate. For example, satellite images from 2007 show that the extent of Arctic ice in the summer had shrunk by over one-third from the average size measured from 1979 to 2000.

Figure 6-1 is a satellite image taken by NASA in 2003. The red line indicates a possible route for the long-sought-after Northwest Passage through the Canadian Arctic. Due to pack ice, this route was considered navigable only by icebreakers and other specially equipped vessels. However, in August 2007, enough pack ice melted to make a route feasible for ordinary cargo vessels. It is conceivable that the increased absorption of energy from the Sun could open the Northwest Passage year-round before the end of the century. While a boon for shipping, this climate change could have disastrous consequences elsewhere.

In this chapter, we will discuss the transformation of several types of energy and solve problems related to energy.

Figure 6-1 A possible route for the Northwest Passage through the Canadian Arctic

MAKING CONNECTIONS

Going Carbon-Neutral

On November 7, 2007, the residents of the town of Eden Mills, Ontario, launched an initiative to make their town the first carbon-neutral community in North America. The project includes measuring greenhouse emissions and taking active steps to reduce these emissions at both the town and the household levels. The residents of Eden Mills hope that this project will pave the way for other communities to see that going carbon-neutral can be a reality. It is also a testimony to future generations that they have done their share, or a satisfactory portion of it, in this regard.

Figure 6-2 shows the pond at Eden Mills. Waterpowered mills such as the one located at Eden Mills pond rely on a

Figure 6-2 The pond at Eden Mills, Ontario

direct conversion of the gravitational potential energy of the water–Earth system to kinetic energy to power processes such as sawing and finishing lumber and grinding grain into flour.

LO 1

6-1 Work and Kinetic Energy for Constant Forces

Work converts energy from one form to another. For example, as water stored behind a hydroelectric dam falls to activate generator turbines, the work done by the force of gravity transforms the gravitational potential energy into kinetic energy.

To the best of our knowledge energy is conserved. We have yet to see this principle violated. The conservation of energy is one of the fundamental laws of modern science. Energy takes on different forms and is manifested in various ways, but it is always conserved.

Consider an object being pushed by a constant net force. The object's acceleration is constant by Newton's second law. The time-independent kinematics equation for motion in one dimension with constant acceleration relates the speed, acceleration, and displacement of an object:

$$v^2 - v_0^2 = 2\overrightarrow{a} \cdot \Delta \overrightarrow{x} \qquad (3\text{-}18)$$

Dividing each side of the equation by two, and multiplying by the mass of the object gives

$$\frac{1}{2}mv^2 - \frac{1}{2}mv_0^2 = m\overrightarrow{a} \cdot \Delta \overrightarrow{x} \qquad (6\text{-}1)$$

By Newton's second law, $m\overrightarrow{a}$ is the net force acting on the object, $\overrightarrow{F}_{net}$. Therefore,

$$\frac{1}{2}mv^2 - \frac{1}{2}mv_0^2 = \overrightarrow{F}_{net} \cdot \Delta \overrightarrow{x} \qquad (6\text{-}2)$$

Since $\overrightarrow{F}_{net}$ is the sum of all the forces acting on an object, the term on the right-hand side of Equation (6-2)

is called the **total work**, W_T, done on the accelerating the object by the net force acting on it. It can also be thought of as the sum-total of the work done by all the forces acting on the object.

KEY EQUATION
$$W_T = \overrightarrow{F}_{net} \cdot \Delta \overrightarrow{x} \qquad (6\text{-}3)$$

We define the **kinetic energy** of an object of mass m moving at speed v as

$$K = \frac{1}{2}mv^2 \qquad (6\text{-}4)$$

Therefore, Equation (6-2) can be rewritten as follows:

KEY EQUATION   The total work–kinetic energy theorem:

$$W_T = \Delta K = K_f - K_i \qquad (6\text{-}5)$$

Equation (6-5) is a statement of the **work–kinetic energy theorem**, which simply states that the *total* work done on an object (i.e., the work done by *all* the forces acting on the object) is equal to the change in its kinetic energy. Since work is the scalar product of the force vector and the displacement vector, work is a scalar and can be positive or negative.

For the work done by a single constant force, we can write

KEY EQUATION
$$W_F = \overrightarrow{F} \cdot \Delta \overrightarrow{x} \qquad (6\text{-}6)$$

We note here that for Equation (6-6), we have chosen the x-axis as our default direction. However, the same expression applies to the y-direction, the z-direction, and any other direction.

Expanding the scalar product gives

$$W_F = |\overrightarrow{F}||\Delta \overrightarrow{x}|\cos\theta \qquad (6\text{-}7)$$

The first two terms on the right-hand side of Equation (6-7) are magnitudes and are therefore always positive. The term that determines the sign for the work done is therefore $\cos\theta$, where θ is the angle between the force vector and the displacement vector in the tail-to-tail representation. For example, when the force points in the same direction as the displacement the angle between the force and the displacement is $\theta = 0$, since $\cos\theta = 1$, the work is positive. When the force points in the direction opposite to the displacement, the angle between the force and the displacement is $180°$, hence $\cos\theta = -1$ and the work done by the force is negative.

We can also see that work can be positive or negative in light of Equation (6-5), the total work–kinetic energy theorem, as illustrated in Example 6-1.

 CHECKPOINT

C-6-1 Work and Total Work

What is the difference between the quantity W_T in Equation (6-5) and the quantity W_F in Equation (6-6)?

C-6-1 W_T is the sum of the work done by all the forces on an object, W_F is the work done by an individual force.

SI Energy Units

The SI unit for work and energy is the joule (J), which is equal to a newton metre:

$$1\ J = 1\ N\cdot m = 1\ kg\ m^2/s^2$$

 EXAMPLE 6-1

Work and Kinetic Energy

You push a crate of mass 22.0 kg on a horizontal frictionless surface over a distance of 8.00 m using a constant horizontal force of 100 N.

(a) Find the work done you as you push the crate.
(b) Find the speed of the crate at the end of the 8 m push.

SOLUTION

(a) Since the force you use is constant, we can use Equation (6-6):

$$W_F = \vec{F}\cdot\Delta\vec{x} = F\Delta x \cos\theta$$

The crate will naturally move in the direction of your push; therefore, the angle between the force and the displacement vectors is zero, and the work for the force you apply can be expressed as

$$W_F = F\Delta x\,(1) = 100\ N\,(8\ m) = 800\ J$$

(b) To calculate the speed of the crate, we use Equation (6-5). The total work done is the work done by you because there are no other forces doing work on the crate. Here, we also note that the normal force and the force of gravity do no work. This is covered in more detail in the next few sections.

$$W_T = W_F = \Delta K = K_f - K_f$$

$$\Rightarrow W_F = F\Delta x = \frac{1}{2}mv^2 - 0 \Rightarrow v^2 = \frac{2W_F}{m}$$

Since the crate starts from rest, its initial kinetic energy is zero and

$$v^2 = \frac{2W_F}{m} \Rightarrow v = \sqrt{\frac{2W_F}{m}} = \sqrt{\frac{2(800\ J)}{22\ kg}} = 8.53\ m/s$$

Making sense of the result:

The work done by you is positive, since the crate moves in the same direction in which you push.

 EXAMPLE 6-2

Work Against Gravity

You lift your 2.20 kg laptop a total vertical distance of 70.0 cm above the table on which it rests. How much work do you do?

SOLUTION

We do not know if the force you applied was constant, so we cannot assume that the force is equal to mg. Therefore, we cannot use Equation (6-6). However, we can still use Equation (6-5):

$$W_T = \Delta K = 0 \qquad (1)$$

We are assuming that the laptop starts at rest and ends at rest, so the change in its kinetic energy is zero.

The total work in Equation (1) is the work done by all the forces involved; this is the work done by you or the force you apply and the work done by the force of gravity:

$$W_T = W_F + W_g = 0 \qquad (2)$$

Figure 6-3 shows the laptop being lifted upward with the forces acting on it.

Figure 6-3 Example 6-2

For the work done by the force of gravity, we use Equation (6-6):

$$W_g = m\vec{g} \cdot \Delta\vec{y} = (mg)(\Delta y)\cos\theta \qquad (3)$$

The laptop is lifted upward and the force of gravity points downward, so the angle between the two vectors is $180°$ (Figure 6-3). Therefore,

$$W_g = m\vec{g} \cdot \Delta\vec{y} = (mg)(\Delta y)(-1) = -(mg)(\Delta y) \qquad (4)$$

From Equation (2):

$$W_F = -W_g = mg\Delta y = (2.2 \text{ kg})(9.81 \text{ m/s}^2)(0.7 \text{ m}) = 15.1 \text{ J}$$

Making sense of the result:

The work done by gravity is negative. This is the case whenever the force points in the direction opposite to the displacement. We will discuss this in more detail in the next example.

EXAMPLE 6-3

Negative Work

In a "strong man contest," a car with a mass of 1200 kg rolls toward a contestant at a speed of 0.80 m/s. The contestant pushes against the car with a constant force (Figure 6-4) and brings it to a stop over a distance of 0.5 m.

(a) Find the work that the contestant does on the car.
(b) Find the force exerted by the contestant.

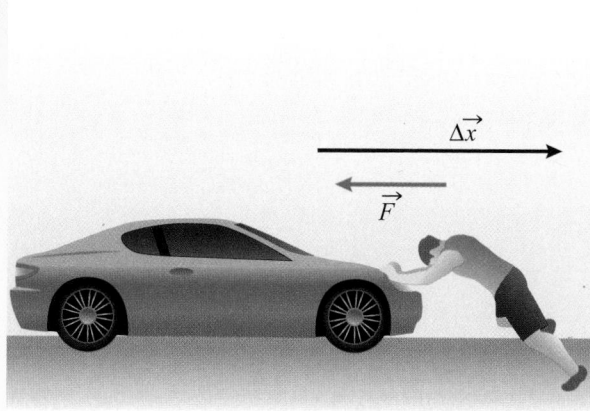

Figure 6-4 In a "strong man contest," a contestant pushes against a car.

SOLUTION

(a) The contestant is the only agent doing work. The normal force and the force of gravity act normal to the direction of motion, and the work done by each of these forces is zero. The total work done on the car is the work done by the contestant. From Equation (6-5),

$$W_c = W_T = \Delta K = K_f - K_i = 0 - \frac{1}{2}mv^2$$

$$= -\frac{1}{2}(1200 \text{ kg})(0.8^2) \text{ m}^2/\text{s}^2 = -384 \text{ J} \qquad (1)$$

(b) From Equation (6-7), the work done by the contestant is given by

$$W_c = |\vec{F}||\Delta\vec{x}|\cos\theta \qquad (2)$$

Since the contestant pushes in the direction opposite to that of the displacement of the car, the angle between the force and the displacement is $180°$. Thus,

$$W_c = |\vec{F}||\Delta\vec{x}|\cos\theta = F\Delta x(-1) \qquad (3)$$

Combining Equations (1) and (3), we have

$$F\Delta x(-1) = -384 \text{ J} \Rightarrow F(5 \text{ m})(-1) = -384 \text{ J}$$

$$F = \frac{-384 \text{ J}}{-0.5 \text{ m}} = 768 \text{ N}$$

Making sense of the result:

The final kinetic energy of the car ($K_f = 0$) is less than the kinetic energy that the car had to begin with. Therefore, the difference, $\Delta K = K_f - K_i$, is negative, and the work done on the car by the contestant is also negative by Equation 6-5.

We can also look at the sign of the work done in light of the scalar product, given by Equation (6-7). To slow the car down, the contestant will naturally have to push in a direction that is opposite to the direction of motion of the car; hence $\theta = 180°$. Since $\cos(180°) = -1$, the work done by the contestant is negative. A force of 768 N is what one would use to lift a weight of ~78 Kg.

Other Energy Units

The joule is the internationally recognized SI unit for energy, but there are several pre-SI units for energy that are still in common use.

One such traditional unit is the calorie (cal), defined as the energy required to raise the temperature of 1 g of water by 1°C. This unit is also called the gram calorie and the small calorie. However, the energy required for a 1°C increase in the temperature of a given quantity of water varies slightly with the starting temperature. Various definitions of the calorie give it a value ranging from 4.182 J to 4.204 J. (SI was developed to eliminate this sort of confusion over unit definitions.) For precise measurements it is necessary to specify the type of calorie. One of the more commonly used calorie units is the thermochemical calorie, denoted cal_{th}:

$$1 \ cal_{th} = 4.184 \ J$$

A related unit is the Calorie, also called the large calorie, the kilogram calorie, and the nutritionist calorie. (However, the nutritional sciences calorie is defined as 4.182 J.) The Calorie is denoted Cal or kcal; this unit is defined as the amount of energy required to raise the temperature of 1 kg of water by 1°C.

Table 6.1 Appreciating the Joule: Energy Orders of Magnitude

Energy (J)	Object
6.068×10^{-21}	Average kinetic energy of a molecule of an ideal gas at room temperature
1.6×10^{-19}	1 eV (electron volt)[1]; the kinetic energy of an electron that has been accelerated through a potential difference of 1 V
2.614×10^{-19} to 5.28×10^{-19}	Energy range of a photon of visible light
8.187×10^{-14}	0.511 MeV (mega electron volt); the rest mass energy of an electron[2]
3.205×10^{-6}	20 TeV (terra electron volt); the proposed beam energy in the Superconducting Supercollider[3]
1	Approximate kinetic energy gained by a 1 kg weight falling a distance of 10 cm
4.184	Thermochemical calorie; the energy required to raise the temperature of 1 g of water by 1°C
7×10^2	Approximate kinetic energy of an average person after 1 m of free fall
1.0551×10^3	1 BTU (British thermal unit); the amount of energy needed to raise the temperature of 1 lb of water by 1°F[4]
3.6×10^3	1 W·h (watt hour)[5]; see Section 6-9
4.184×10^3	1 $kcal_{th}$; the energy required to raise the temperature of 1 kg of water by 1°C
3.858×10^5	Kinetic energy of a 1000 kg vehicle moving at 100 km/h
5–50 kJ/mol	Range of hydrogen bond energies
~23 kJ/mol	Hydrogen bond energy in water
413 kJ/mol	Hydrogen–carbon covalent bond energy
4.1868×10^{10}	1 toe; the energy released by burning 1 T of crude oil (1 tonne of oil equivalent, or 1 toe)
3.6×10^6	1 kW·h (kilowatt hour)
4.184×10^{12}	1 kilotonne of TNT, the energy released in the detonation of 1000 tonnes of TNT
5.1×10^{20}	12 267.38 Mtoe; the world's primary energy supply for 2008
4×10^{23}	Energy of the meteor impact that formed the Chicxulub crater

[1] The electron volt is commonly used as an energy unit in quantum physics. See Chapter 31 for details.

[2] The rest mass energy of a particle is the energy equivalent of the mass of the particle, in effect, the energy required to create the particle, using Einstein's famous equation, $E = mc^2$. See Chapter 29.

[3] The Superconducting Supercollider was to be the largest particle collider built during the 20th century. The project was scheduled to start operating in 2019 but was cancelled by the U.S. Congress in 1993.

[4] The value presented here is the international table BTU.

[5] Unit commonly used in power production and consumption.

Walking It Off

An average fast food burger without cheese has 590 Cal (kcal) and provides 52% of the daily limit of fat for an average recommended daily caloric intake of 2000 Cal. A medium-sized cola drink (473 mL) has 210 Cal, all from sugar, and a medium order of fries has 453 Cal and provides 33% of the daily fat limit.

On average, a 70 kg person burns 390 Cal/h during a very brisk 7 km/h walk. At this rate, it would take close to 3 h (21 km) to burn off the calories in the fast food described above.

EXAMPLE 6-4

Work Against Friction

A fisherman pushes a crate full of fish ($m = 123$ kg) from the edge of a fishing boat to the middle of the boat with a constant force of 560 N over a total distance of 7.00 m. Over this distance, the speed of the crate changes from zero to 4 m/s. Find the coefficient of kinetic friction between the crate and the boat surface.

SOLUTION

Figure 6-5 shows the FBD for the crate. We can use the work–kinetic energy theorem, Equation (6-5):

$$W_T = \Delta K = K_f - K_i \qquad (1)$$

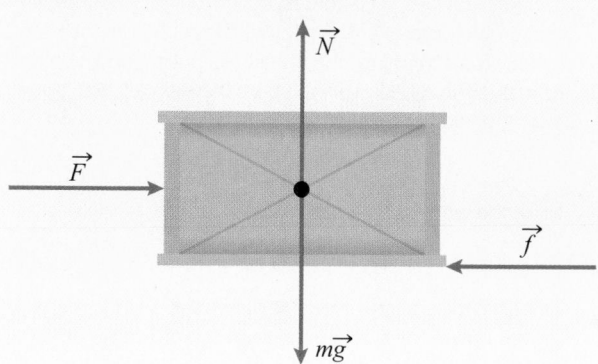

Figure 6-5 Example 6-4

Neither the force of gravity nor the normal force do any work because they both act normal to the direction of motion. The total work done is the work done by the force of the fisherman and the work done by the force of friction. The initial kinetic energy of the crate is zero, then

$$W_F + W_f = K_f - 0 \qquad (2)$$

The work done by each force is given by Equation (6-6). The force of the fisherman acts in the same direction as the force of the crate's motion, and the angle between the force and the displacement vectors is zero:

$$W_F = \vec{F} \cdot \Delta \vec{x} = F\Delta x \cos\theta = F\Delta x \qquad (3)$$

The force of friction acts opposite to the direction of motion, so the angle between the force and the displacement vectors is 180°:

$$W_f = \vec{f_k} \cdot \Delta \vec{x} = f_k \Delta x \cos(180°) = -f_k \Delta x \qquad (4)$$

Putting Equations (3) and (4) into Equation (2) yields

$$F\Delta x - f_k \Delta x = K_f = \frac{1}{2}mv^2 \qquad (5)$$

Therefore,

$$f_k \Delta x = F\Delta x - \frac{1}{2}mv^2 \qquad (6)$$

The force of friction between the boat's surface and the crate is given by Equation (5-4) as $f_k = \mu_k N$. From Figure 6-5, the normal force on the crate is equal to its weight. Therefore, $f_k = \mu_k mg$, and Equation (6) becomes

$$\mu_k mg\Delta x = F\Delta x - \frac{1}{2}mv^2$$

or

$$\mu_k = \frac{F\Delta x - \frac{1}{2}mv^2}{mg\Delta x} = \frac{(560 \text{ N})(7 \text{ m}) - \frac{1}{2}(123 \text{ kg})(4 \text{ m/s})^2}{(123 \text{ kg})(9.81 \text{ m/s}^2)(7 \text{ m})} = 0.348$$

Making sense of the result:

The coefficient of kinetic friction has no units, as expected. In addition, a value between 0 and 1 is expected.

PEER TO PEER

Something I often ran into trouble with is the concept of kinetic energy. I have to remind myself that kinetic energy is always a positive scalar and it doesn't depend on the direction of motion. If an object of mass m moves along the positive x-axis with speed v, its kinetic energy is $\frac{1}{2}mv^2$. If that object then moves along the negative x-axis with the same speed, its kinetic energy is the same, $\frac{1}{2}mv^2$. The difference in kinetic energy between the two cases is zero.

✓ CHECKPOINT

C-6-2 Work and Kinetic Energy

You are standing on a flat, horizontal icy stretch in the Arctic, when a baby seal comes sliding slowly toward you. You are wearing shoes with cleats that prevent you from sliding. You place your hand gently against the seal to slow it down and then push it back toward its mother at the same speed it was coming toward you, 0.5 m/s. The seal's mass is 8 kg. Assume that you use a constant force and that you are in contact with the seal for a total of 1.1 s. What is the total work done by you on the seal during the entire process?

(a) 1 J

(b) 2 J

(c) -2 J

(d) -1 J

(e) none of the above

C-6-2 (e) The incoming speed is the same as the outgoing speed, the change in kinetic energy is zero, hence the total work done is zero.

EXAMPLE 6-5

Work and Gravity

You lift a 420 g glass of iced coffee from your desk to a shelf 35.0 cm above the desk. It takes you 0.700 s to lift the glass, and the maximum speed that the glass reaches on its way up is 0.900 m/s. What is the total work done on the glass of coffee during its journey upward?

SOLUTION

Since the glass starts from rest and ends up at rest, the change in its kinetic energy is zero. Therefore, the total work done on the glass is zero according to Equation (6-5).

Making sense of the result:

The work you do is positive, and the work done by the force of gravity is negative. These two quantities comprise the total work and up to zero.

EXAMPLE 6-6

Eager Worker

You are training with two friends at the gym. One friend throws a 9.00 kg medicine ball at you at a speed of 2.00 m/s. You stop the ball over a distance of 30.0 cm, and throw it to your other friend in a direction that makes an angle of $120°$ with the incident direction. The ball leaves your hands with the same speed it had when you intercepted it. Determine the total work done by you on the ball.

SOLUTION

You slide the ball with the same speed that you receive it, so the overall change in ball's kinetic energy is zero.

Therefore, the total work done on the ball is zero. Since you are the only agency doing work, the work done by you is the total work done on the ball, which is zero.

Making sense of the result:

Although you exert a force on the ball in two different directions, the *total* work done by you is zero. When you slow the ball, you are doing negative work. When you slide the ball, you are doing positive work. The sum total of the work done is zero.

EXAMPLE 6-7

What Is Doing Work?

A 17.0 kg monkey climbs up a rope up to a coconut tree, starting from ground level. The monkey is originally at rest. When the monkey is 6.00 m above the ground, his speed is 2.30 m/s. What force moves the monkey upward? How much work is done by that force?

SOLUTION

The monkey climbs by holding the rope and pulling down on it. The force of friction from the rope pushes the monkey upward. The forces acting on the monkey are the force of gravity and the force of friction from the rope; by Newton's third law, the monkey is pulling down on the rope and the rope is pulling up on the monkey with equal and opposite force. Using Equation 6-5

$$W_T = W_f + W_g = \Delta K \qquad (1)$$

Now, the work done by the force of gravity is

$$W_g = m\vec{g} \cdot \Delta\vec{y} \qquad (2)$$

Since the force of gravity points downward and the monkey moves upward, the angle between the gravitational force and the displacement is $180°$ and

$$W_g = |m\vec{g}||\Delta\vec{y}|\cos(180°) = |m\vec{g}||\vec{h}|\cos(180°) = -mgh \quad (3)$$

We can rewrite Equation (1) as

$$W_f - mgh = \Delta K = \frac{1}{2}mv^2 - \frac{1}{2}mv_0^2 \qquad (4)$$

Since the monkey starts from rest, $v_0 = 0$ and

$$W_f = mgh + \frac{1}{2}mv^2 = (17 \text{ kg})(9.81 \text{ m/s}^2)(6 \text{ m})$$

$$+ \frac{1}{2}(17 \text{ kg})(2.3 \text{ m/s}) = 1040 \text{ J} \qquad (5)$$

Making sense of the result:

The work done by the force of friction has to account for the work needed to lift the monkey against the force of gravity (mgh) as well as for the increase in the monkey's kinetic energy.

C-6-3 Who Is Doing the Work?

For Example 6-7, which of the following statements is true?

(a) The work is done by the force of static friction.

(b) The work is done by the force of kinetic friction on the monkey because the monkey is moving.

(c) The work is done by the monkey on the monkey to lift the monkey up.

(d) None of the above are true.

C-6-3 (b) Of course, we are assuming that the monkey's hands do not slip on the rope.

 MAKING CONNECTIONS

Impact Mass Extinction Events

During a mass extinction event, a significant number of species become extinct. A large meteor impact is perhaps the most dramatic type of such an event. A meteor impact releases enormous amounts of energy, which can cause massive forest fires and gigantic tsunamis. The smoke and dust released into the atmosphere prevent the process of photosynthesis, thus catastrophically disrupting food chains through a domino effect. There is strong evidence that the dinosaurs became extinct as a direct result of a 10 to 20 km-wide meteor striking the Yucatán peninsula about 65 million years ago (Figure 6-6). This impact is estimated to have released an amount of energy equivalent to 100 billion megatonnes of TNT. The impact threw trillions of tonnes of dust into the atmosphere, which may have caused global climate changes in addition to the food chain disruptions mentioned above. To put this in perspective, the atomic bomb that destroyed the city of Hiroshima in 1945 had a yield of only 13 to 18 kilotonnes.

Figure 6-6 An artist's rendition of what the 180 km-wide Chicxulub impact crater on the Yucatán Peninsula would have looked like shortly after it was created.

6-2 Work Done by Constant Forces in Two and Three Dimensions

✓ CHECKPOINT

C-6-4 Work Done by the force of Gravity

A satellite of mass m completes half a turn around Earth's axis in low Earth orbit. Use R as the radius of Earth, h as the altitude of the satellite, and g as the value of the acceleration due to gravity at that altitude. How much work is done by the force of gravity on the satellite as it completes the half turn?

(a) $mg(R + h)$
(b) $2mg(R + h)$
(c) zero
(d) none of the above

C-6-4 (c) The change in the kinetic energy is zero hence the total work, which in this case the is the work done by the force of gravity is zero.

Thus far, we have considered examples where the force and the displacement are directed along the same direction, so our calculations have involved only one dimension. However, forces are frequently not parallel to the direction of motion, as in the example shown in Figure 6-7(a). The work done by a constant force that is not parallel to the displacement can be defined as

$$W_F = \vec{F} \cdot \Delta \vec{s} \qquad (6\text{-}8)$$

where $\Delta \vec{s}$ denotes displacement in more than one dimension. Work, as given in equation 6-8, is the scalar product between the two vectors $\vec{F}$ and $\Delta \vec{s}$.

Note that this equation applies only when the force is constant and when the angle between the force and the displacement remains constant.

(a)

Figure 6-7(a) A crate being pulled on the floor. The crate moves along a horizontal surface, and the force acts in a direction above the horizontal, making an angle θ with the horizontal.

We will highlight three ways to calculate the work done in situations involving two or more dimensions. We emphasize here that the expression for the work done by

a constant force is always given by Equation (6-8) above. The three methods that we develop below are simply three ways of looking at the scalar product of Equation (6-8).

Approach 1. For an object moving in a straight line, expanding the scalar product in Equation (6-8) gives

$$W_F = |\vec{F}||\Delta \vec{s}|\cos\theta \qquad (6\text{-}9)$$

We can then calculate the work done by using values for the magnitude of the force, the magnitude of the displacement, and the cosine of the angle between the force and displacement vectors, as illustrated in Figure 6-7(b).

(b)

Figure 6-7(b) Calculating the work done by force $\vec{F}$

Approach 2. In the second approach, we rewrite Equation (6-9) as

$$W_F = \{|\vec{F}|\cos\theta\}|\Delta \vec{s}| = F_{\parallel}\Delta s \qquad (6\text{-}10a)$$

We can derive this equation more formally by writing the force vector as the sum of two components: one parallel to the displacement and the other perpendicular to the displacement. Since the scalar product of perpendicular vectors is zero, the expression for the work done by force $\vec{F}$ becomes

$$W_F = (\vec{F_{\parallel}} + \vec{F_{\perp}}) \cdot \Delta \vec{s} = \vec{F_{\parallel}} \cdot \Delta \vec{s} = \{|\vec{F}|\cos\theta\}|\Delta \vec{s}| \quad (6\text{-}10b)$$

Thus, only the component of $\vec{F}$ parallel to the displacement does work. The work will be positive when $\vec{F_{\parallel}}$ and $\Delta \vec{s}$ point in the same direction and negative when they point in opposite directions (Figure 6-7(c)).

(c)

Figure 6-7(c) Only the component of $\vec{F}$ parallel to the displacement does work.

(d)

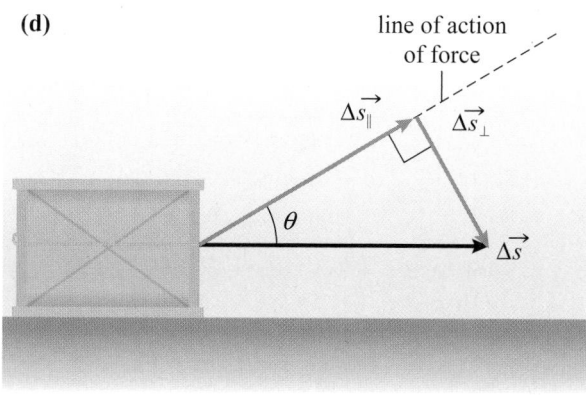

line of action of force

Figure 6-7(d) Only the component of *F* parallel to the displacement does work.

Approach 3. In the third approach (Figure 6-7(d)), we rewrite Equation (6-9) as

$$W_F = |\vec{F}| \{ |\Delta \vec{s}| \cos \theta \} = F \Delta s_{\parallel} \qquad \text{(6-11a)}$$

As with the force vector, we can express the displacement vector as the sum of a component parallel to the force and a component perpendicular to the force:

$$W_F = \vec{F} \cdot (\Delta \vec{s}_{\parallel} + \Delta \vec{s}_{\perp}) = \vec{F} \cdot \Delta \vec{s}_{\parallel} \qquad \text{(6-11b)}$$

 EXAMPLE 6-8

Work Done by the Force of Gravity

Starting from rest, a block slides down a frictionless ramp of slope angle 32.0°, as shown in Figure 6-8. Use the three representations introduced above to find the speed of the block at the bottom of the ramp.

Figure 6-8 A block slides down a frictionless incline.

SOLUTION

The normal force does no work because the angle between the force and the displacement is 90°. Since the force of gravity is the only force doing work,

$$W_g = W_T = \Delta K$$

Once we find the work done on the block, we can use its kinetic energy to find its speed.

APPROACH 1.

Using the scalar product of the force and displacement vectors, from Equation 6-9:

✅ CHECKPOINT

C-6-5 A Sliding Bear

A polar bear cub runs toward an icy hill and slides freely up the icy slope. Which of the following statements is accurate?

(a) The force of gravity does positive work on the cub.
(b) The force of gravity does negative work on the cub.
(c) Work is a scalar, not a vector, and cannot be positive or negative.
(d) There is no work done because there is no friction.

C-6-5 (b) the vertical displacement is up while the force of gravity points down.

Thus, only the component of the displacement that is parallel to the force contributes to the work done. Here, again, when $\vec{F}$ and $\Delta \vec{s}_{\parallel}$ point in the same direction, the work done by the force is positive, and when they point in opposite directions, the work is negative.

Equations (6-9), (6-10), and (6-11) offer three representations of the same principle. Depending on the problem being solved, using one of these representations might be more convenient than another.

Example 6-8 demonstrates that the work done by the force of gravity does not depend on the path taken; only the vertical component of the displacement matters.

$$W_g = m\vec{g} \cdot \Delta \vec{s} = |m\vec{g}| |\Delta \vec{s}| \cos \beta$$

where β is the angle between the displacement vector and the force vector.

From Figure 6-9(a), $\beta = 90° - \theta$, $\cos \beta = \sin \theta$, and $\sin \theta = h/\Delta s$. Therefore,

$$W_g = |m\vec{g}| |\Delta \vec{s}| \sin \theta = mgh$$

(a)

β

$\Delta \vec{s}$

$m\vec{g}$

$\theta = 32°$

Figure 6-9(a) Using the scalar product to find the work done

APPROACH 2.

Using the parallel component of the force, from Equation 6-10b:

$$W_g = m\vec{g}_{\parallel} \cdot \Delta \vec{s}$$

From Figure 6-9(b), the parallel component of mg is $mg \sin \theta$. Therefore,

$$W_g = mg \sin \theta (\Delta s) = mgh$$

(continued)

(b)

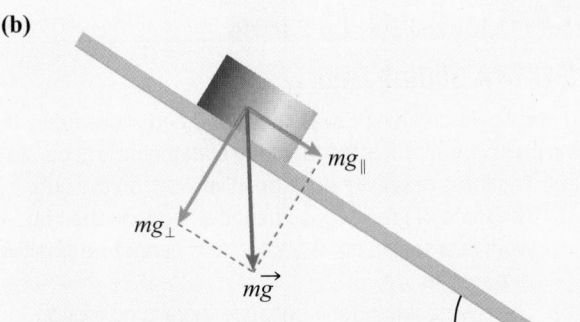

Figure 6-9(b) Only the component of mg along the incline contributes to the work done by gravity.

APPROACH 3.

Using the parallel component of the displacement, from Equation 6-11b:

$$W_g = m\vec{g} \cdot \Delta \vec{s}_\parallel$$

Since $m\vec{g}$ is vertical, the parallel component of Δs is also vertical. From Figure 6-9(c), this component is $\Delta s (\sin\theta)$. Therefore,

$$W_g = mg\Delta s (\sin\theta) = mgh$$

All three approaches give the same expression for W_g, as expected. We now solve for the speed at the bottom of the ramp:

(c)

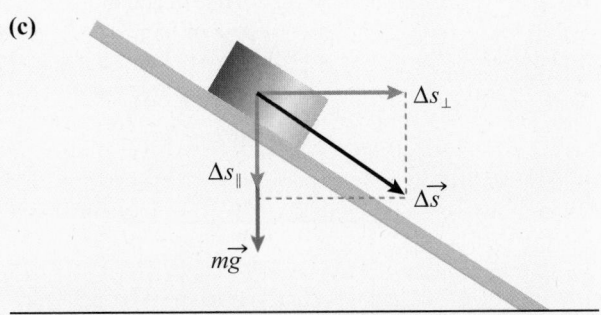

Figure 6-9(c) Only the vertical component of the displacement contributes to the work done by the force of gravity.

$$W_g = \Delta K$$

The work done by the force of gravity by all three approaches is mgh:

$$mgh = \frac{1}{2}mv^2 \Rightarrow v^2 = 2gh$$

Making sense of the result:

This problem can also be solved using Newton's second law, but the work–kinetic energy approach solves it more easily. The result obtained is the same as the result obtained using kinematics equations for an object that falls straight down a distance h.

ONLINE ACTIVITY

Work and Energy

The e-resource that accompanies every new copy of this textbook contains an Online Activity using the PhET simulation "The Ramp." Work through the simulation and accompanying questions to gain an understanding of the concepts of work and energy.

MAKING CONNECTIONS

Turbines on Kites

Christopher Furlong/Staff/Getty Images News/Getty Images

wind flow

Based on Magenn Air Rotor System (M.A.R.S.)

Figure 6-10 (a) An offshore wind farm. Wind turbines convert kinetic energy to electric energy. (b) An artist's rendition of a high-altitude wind turbine.

(continued)

Wind energy reportedly has the potential to provide the world's total need for energy many times over. Although renewable, the amount of energy generated from terrestrial and offshore wind turbines (Figure 6-10(a)) remains relatively modest due to the limited wind speed at low altitudes. Research is being conducted into developing technologies to harvest wind energy at high altitudes (Figure 6-10(b)). Due to the tremendous wind speeds at higher altitudes, the energy available is much higher than what is available closer to Earth's surface. The work done by the wind on the turbine goes into the turbine's kinetic energy, which generates electricity. Figure 6-10(b) shows an artist's rendition of a proposed airborne high-altitude wind turbine. While the European Union seems to be leading the way in terms of harvesting energy from wind, Canada and Russia have the largest potential in this regard. Germany is on schedule to have 80% of its electrical power needs met by renewable energy by 2050, 44% of which is anticipated to come from wind energy.

EXAMPLE 6-9

Pulling a Crate

The worker in Figure 6-11(a) pulls on a rope with a force of 610 N to drag a 90 kg crate along a horizontal surface. The rope makes a 41° angle with the horizontal. The coefficient of kinetic friction between the surface and the crate is 0.24. What is the speed of the crate after having been pulled a total distance of 4.3 m?

(a)

Figure 6-11(a)   A worker pulls a crate on a rough surface.

SOLUTION

This is another example where we can use Newton's second law and the kinematics equations to solve the problem. However, since the work–kinetic energy theorem combines both, it is ideal to use it to solve this problem.

The total work done on the crate comes from the work done by the applied force and the work done by friction:

$$W_F + W_f = \Delta K \qquad (1)$$

There is no term in Equation (1) for either the work done by the force of gravity or the work done by the normal force because these forces act normal to the direction of motion. Therefore, by Equations (6-8) and (6-9), the work done by these forces is zero.

Using Equation (6-9), examining Figure 6-11(b), and considering that the two forces act over the distance Δs, we have

$$F\Delta s \cos\theta - f_k d = \frac{1}{2} mv^2 \qquad (2)$$

The work done by F is positive because the horizontal component of F points along the same direction as the displacement. The force of friction points in a direction that is opposite to the displacement's direction. Therefore, the work done by friction is negative.

(b)

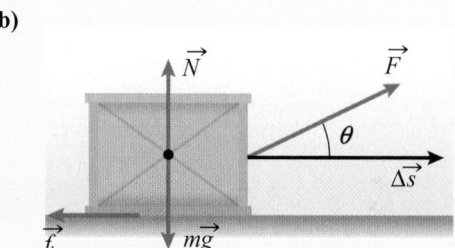

Figure 6-11(b)   Forces acting on the crate being pulled

To calculate v from Equation (2), we need to determine the force of friction. We can use $f_k = N\mu_k$ because we can find N, the normal force on the crate, by summing forces in the vertical direction:

$$N + F\sin\theta - mg = 0$$

Therefore,

$$N = mg - F\sin\theta$$

Substituting into Equation (2) gives

$$F d \cos\theta - (mg - F\sin\theta)\mu_k d = \frac{1}{2} mv^2 \qquad (3)$$

$$v^2 = \frac{2d}{m}\{F\cos\theta - (mg - F\sin\theta)\mu_k\}$$

$$= \frac{2(4.3\text{ m})}{90\text{ kg}}\{(610\text{ N})(\cos(41°))$$

$$- [(90\text{ kg})(9.81\text{ kg m/s}^2) - (610\text{ N})(\sin(41°))](0.24)\}$$

and

$$v = 5.74\text{ m/s}$$

Making sense of the result:

Here, we had to keep in mind that the magnitude of normal force was not simply equal to mg; rather, it had to be obtained from the summation of the forces in the vertical direction. Without the force of friction, we would have $Fd\cos\theta = \frac{1}{2}mv^2$ and the speed would have been 6.63 m/s. Friction dissipates some of the kinetic energy, resulting in a lower speed.

6-3   Work Done by Variable Forces

So far, we have dealt with work done by constant forces. However, most of the forces we encounter in daily life are not constant. Standing up, sitting down, starting and stopping a car, skiing, and canoeing all involve forces that vary. The fundamental forces in nature—gravity, the electrostatic force, and the two nuclear forces—all vary with distance. How is the calculation of work done by a constant force different from the calculation of work done by a variable force?

Graphical Representation of Work

Consider an object being pushed over a distance Δs by a constant force that points in the direction of the motion. Since this force does not change, the force versus displacement curve is simply a horizontal line, as shown in Figure 6-12.

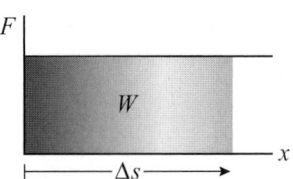

Figure 6-12   Force versus distance graph for a constant force

From Equation (6-12a), the work done by a constant force parallel to the displacement is $F\Delta s$. In Figure 6-12, this quantity is represented by the area of the rectangle under the F versus Δs graph.

Figure 6-13 shows a force versus distance graph for a force with a continuously changing magnitude. As in Figure 6-12, the area under the curve represents the work done by the force. If the force does not have the same direction as the displacement, we can graph just the component of the force parallel to the displacement. Then the area under the graph will correspond to the work done by the force. Alternatively, we can graph each component of the force against its corresponding component of the displacement and sum all the areas to calculate the work done.

Figure 6-13   The work done by the force over the displacement is the area under the curve.

We can break up the area under the curve into thin rectangles, or trapezoids, and then sum their areas (Figure 6-14):

$$W \approx \sum \vec{F} \cdot \Delta \vec{x} \qquad (6\text{-}12)$$

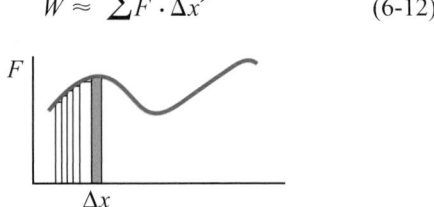

Figure 6-14

The narrower the width of the rectangles, the more accurate the result will be. As Δx approaches zero, that is, for an infinitesimally small value of Δx (denoted dx), the area calculated becomes infinitely accurate. The sum $\sum \vec{F} \cdot \Delta \vec{x}$ then becomes the integral of the force over the displacement. The area obtained using integration is exactly equal to the work done by the variable force over the displacement from x_1 to x_2 in Figure 6-15:

$$W_{\text{F}} = \int_{x_1}^{x_2} \vec{F} \cdot d\vec{x} \qquad (6\text{-}13\text{a})$$

Figure 6-15   The integral gives the exact area under the curve.

For more than one dimension, the expression becomes

$$W_{\text{F}} = \int_{a}^{b} \vec{F} \cdot d\vec{s} \qquad (6\text{-}13\text{b})$$

where $d\vec{s}$ is the infinitesimal displacement along the trajectory of the object acted upon by force $\vec{F}$ while moving from point a to point b.

The Work–Kinetic Energy Theorem for Variable Forces

We introduced the total work–kinetic energy theorem with Equations (6-2) and (6-5). Although Equation (6-5) is universal, Equation (6-2) holds only for constant forces. We now present a proof that the work–kinetic energy theorem does hold for variable forces.

In three dimensions, the definition of the total work done on an object by variable forces can be expressed as

$$W_{\text{T}} = \int_{r_1}^{r_2} \vec{F}_{\text{net}} \cdot d\vec{s} = \int_{r_1}^{r_2} F_s \, ds \qquad (6\text{-}14)$$

Where F_s is the component of the force parallel to the displacement. Since $F_s = ma_s$,

$$W_{\text{T}} = \int_{r_1}^{r_2} ma_s \, ds$$

Where r_1 and r_2 are the initial and final positions of the object.

Using the chain rule for integration yields

$$a = \frac{dv}{dt} = \frac{dv}{ds}\frac{ds}{dt} = v\frac{dv}{ds}$$

Therefore,

$$W_T = \int_{r_1}^{r_2} m\left(v\frac{dv}{ds}\right)ds$$

and $\quad W_T = \int_{v_1}^{v_2} mv\,dv = \frac{mv^2}{2}\bigg|_{v_1}^{v_2} = \frac{1}{2}mv_2^2 - \frac{1}{2}mv_1^2$

Thus, the total work–kinetic energy theorem applies for both constant and variable forces.

LO 4

6-4   Work Done by a Spring

MAKING CONNECTIONS

Buildings on Springs

Buildings located near seismically active geological formations are often constructed with special features to minimize the damage from an earthquake. One such measure is base isolation, which refers to isolating the base of the building from the moving surface. For example, some buildings sit on rollers or heavy rubber support columns, which allow the buildings to remain relatively still while the ground beneath them moves laterally. Massive springs can be used to protect a structure from vertical motion. The main complex of NORAD's facility at Cheyenne Mountain sits on top of 1000 giant springs that allow the building to move by 2.5 cm in any direction, in case of a massive earthquake (or a nuclear attack) (Figure 6-16).

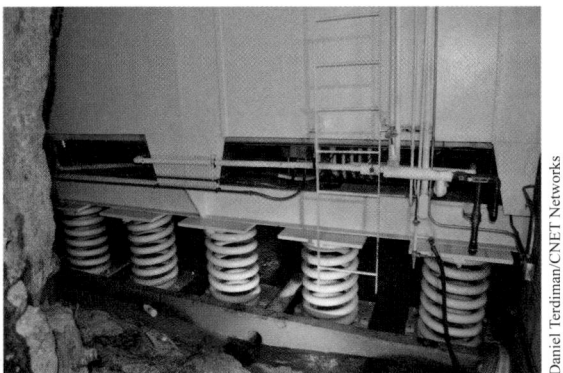

Figure 6-16   Springs supporting a building inside the Cheyenne Mountain complex

A simple example of work done by a variable force is the work done by the force of a spring. As described in Chapter 5, the force exerted by a spring is given by Hooke's law:

$$\vec{F}_s = -k\vec{x} \qquad (6\text{-}15)$$

where $\vec{x}$ is the elongation of the spring from its unstretched length, and k is the spring constant.

Hooke's law is a good approximation for springs within the elastic limit. (Springs stretched beyond this limit become nonlinear and may permanently deform.) The larger the value of k, the stiffer the spring, that is, the more force it takes to stretch the spring by a given amount.

Consider the spring resting on the horizontal surface in Figure 6-17. The left end of the spring is fixed. If we move the right end by a distance x to the right of the unstretched (equilibrium) position, the spring will exert a force on us to the left. If we double the distance by which the spring is stretched, the force that the spring exerts on us doubles. The negative sign indicates that if you move one end of a spring away from the unstretched position, the spring will exert a force on you to pull you back. Thus, the force of a spring is a **restoring force**.

Figure 6-17   The restoring nature of the spring force: when stretched to the right, the spring pulls to the left

✓ CHECKPOINT

C-6-6 Work Done by a Spring

When is the work done by a spring positive?
(a) as the spring is being compressed from equilibrium
(b) as the spring is being stretched from equilibrium
(c) as the spring returns to equilibrium
(d) as the spring is being moved away from equilibrium

C-6-6 (c) this is the only instance when the force of the spring points in the direction of motion, and not opposite to it.

Work Done by a Spring

Let us move the free end of a spring such as the one in Figure 6-17, whose other end is fixed, from position x_1 to position x_2, both relative to the position where the spring is unstretched. Since the force of the spring is not a constant, we use calculus to calculate the work done by a spring:

$$W_s = \int_{x_1}^{x_2} \vec{F}_s \cdot d\vec{x} \qquad (6\text{-}16)$$

The force that the spring exerts is given by

$$\vec{F}_s = -k\vec{x}$$

Substituting for $\vec{F}_s$ in the expression for the work done gives

$$W_s = \int_{x_1}^{x_2} -k\vec{x} \cdot d\vec{x}$$

Since $\vec{x}$ and $d\vec{x}$ have the same direction,

$$W_s = -k \int_{x_1}^{x_2} x \, dx = \left(-\frac{kx_2^2}{2}\right) - \left(-\frac{kx_1^2}{2}\right)$$

and

$$W_s = \frac{1}{2} kx_1^2 - \frac{1}{2} kx_2^2 \qquad (6\text{-}17)$$

While deriving Equation (6-17), we did not need to specify whether the spring was being compressed or stretched. Nor did we consider whether the spring might have already been stretched at position x_1 and then either stretched farther to position x_2, or perhaps moved through equilibrium to become compressed at position x_2. The initial and final positions, as measured from the unstretched position, determine the work done by the spring, regardless of *how* the spring moved between these positions.

CHECKPOINT

C-6-7 $(\Delta x)^2$ or $\Delta(x^2)$?

A spring of spring constant $k = 690$ N/m is held at a position such that it is compressed by 10.0 cm from its unstretched position. The spring is then compressed by another 90.0 cm. The work done by the spring is

(a) 221 J;
(b) −221 J;
(c) 276 J;
(d) −276 J;
(e) none of the above.

C-6-7 (d) the first two answers use $(0.8)^2$, and are wrong, the work should be negative.

EXAMPLE 6-10

Collision with a Spring

A 17.0 kg block is sliding on a frictionless horizontal surface at a speed of 3.80 m/s when it strikes a spring that is already compressed by 21.0 cm. Given that the spring constant is 710 N/m, find the distance by which the spring is further compressed during the collision.

SOLUTION

The force of the spring and, therefore, the acceleration of the mass are not constant. The kinematics equations we developed in Chapter 3 are not applicable here because they apply only for constant acceleration. However, we easily apply a work–energy approach because the work done by the spring depends only on the initial and final positions.

The only force doing work on the block is the force of the spring; therefore, $W_T = W_s = \Delta K$. Since the final speed of the block is zero at maximum spring compression,

$$W_s = \frac{1}{2} kx_1^2 - \frac{1}{2} kx_2^2 = 0 - \frac{1}{2} mv^2 \qquad (1)$$

Then, $$\frac{1}{2} kx_2^2 = \frac{1}{2} mv^2 + \frac{1}{2} kx_1^2 \qquad (2)$$

and $$x_2^2 = \frac{1}{k}(mv^2 + kx_1^2)$$

$$= \frac{1}{710}(17 \text{ kg } (3.80 \text{ m/s})^2 + 710 \text{ N/m } (0.21 \text{ m})^2)$$

$$x_2 = 0.624 \text{ m}$$

The distance by which the spring compresses after the block strikes it is the difference between the final and initial positions:

$$\Delta x = x_2 - x_1 = 0.624 \text{ m} - 0.210 \text{ m} = 0.414 \text{ m}$$

Making sense of the result:

If the spring had not been compressed initially, the result would have been

$$\frac{1}{2} kx_2^2 = \frac{1}{2} mv^2 \quad \text{and} \quad \Delta x = 0.588 \text{ m}$$

The more compressed a spring is, the harder it becomes to compress it further.

EXAMPLE 6-11

Spring Trampoline

While testing springs in your backyard, you jump from a height of 4.00 m down onto a platform sitting on top of a spring, as shown in Figure 6-18. The spring constant is 501 N/m. Find the maximum compression of the spring. Your mass is 72.0 kg, and assume that the mass of the platform is negligible.

SOLUTION

The work–energy approach is once again ideal to use to solve this problem involving springs. Two forces do work on you: the force of gravity and the force of the spring. You start from rest and are momentarily at rest again at maximum spring compression. Therefore, the change in kinetic energy of the platform is zero, and

$m = 72$ kg

$v_i = 0$

4 m

x_{max} $k = 360$ N/m

Figure 6-18   Example 6-11

$$W_T = W_g + W_s = \Delta K = 0$$

Since the spring is originally unstretched, $x_1 = 0$. Substituting into Equation (6-17) gives

$$W_s = \frac{1}{2} kx_1^2 - \frac{1}{2} kx_2^2 = -\frac{1}{2} kx_2^2$$

The force of gravity does work on you as long as you are moving downward. Thus,

$$W_g = mgh + mgx_2$$

The work done by the force of gravity is positive as given by Equations (6-8) and (6-9) because the force of gravity points along the same direction as your displacement before and during spring compression. Therefore,

$$mgx_2 + mgh - \frac{1}{2} kx_2^2 = 0$$

$$\Rightarrow 72 \text{ kg } (9.81 \text{ m/s}^2)x_2 + 72 \text{ kg } (9.81 \text{ m/s}^2)4 - 250.5x_2^2 = 0$$

Solving this quadratic equation gives a maximum compression of $x_{max} = x_2 = 5.05$ m.

Making sense of the result:

The key here is to keep in mind that the force of gravity does work as the spring is being compressed. If we neglect the mgx_2 term (the work done by the force of gravity while the spring is being compressed), we get a maximum compression of 3.36 m much less than the actual value.

LO 5

6-5 Conservative Forces and Potential Energy

Suppose you lift a book a distance d straight up from a table and then let the book drop. The book will fall back to its original position, hitting the table with a certain speed. Now suppose you slide the book the same distance d horizontally along the table and then release it. The book does not return to its original position. What is the difference between the two cases?

In the first case, you had to do work against the force of gravity to lift the book. In the second case, you did work against the force of friction. The work you did against friction was largely converted into thermal energy, which, for practical purposes, you cannot recover. When you let the book drop after lifting it, not only does it return to its original position, but it gains some kinetic energy. We can show that you can recover all the work that you put in when you work against the force of gravity.

Conservative and Nonconservative Forces

Forces in the physical universe can be either **conservative** or **nonconservative**. For example, the force of friction is a dissipative, or a nonconservative, force because the work you do against friction is dissipated, primarily as thermal energy (we use this fact to warm our hands by rubbing them together). The dissipated energy is "lost" and cannot

be recovered as kinetic energy. Common examples of nonconservative forces are applied forces, which are forces that we would apply, for instance, to push a crate on an incline, compress a spring, or lift a weight.

The work you do against conservative forces can be recovered in the form of kinetic energy. When you stretch a spring and release it, it "springs" back to its equilibrium position. In the case of charged particles, the closer you bring two opposite charges together, the more kinetic energy they will have when they move away from each other after being released. The more work you do in bringing them together, the more energy is stored in the system that can be recovered in the form of kinetic energy. This stored energy is called **potential energy**.

Potential Energy

When you lift a book a distance h as shown in Figure 6-19, beginning and ending with the book at rest, the change in kinetic energy is zero, and

$$W_T = \Delta K = 0$$

The work, W_F, that you do on the book as you lift it upward is positive because the force that you exert has the same direction as the displacement. However, the *total* work done on the book also includes the work done by the force of gravity, W_g, which is negative:

$$W_T = W_F + W_g = \Delta K = 0$$

Therefore, $W_F + W_g = 0$

and $W_F = -W_g$

$v_f = 0$

$\vec{F}$

h

$v_i = 0$ m

$m\vec{g}$

Figure 6-19 The work done by $\vec{F}$ is positive. The work done by gravity is negative

We can use Equation (6-9) to find the work done by the constant force of gravity:

$$W_g = \vec{F} \cdot \vec{h} = m\vec{g} \cdot \vec{h} \qquad (6\text{-}18)$$

Expanding the scalar product, we get

$$W_g = \vec{F} \cdot \vec{h} = m|\vec{g}||\vec{h}|\cos\theta \qquad (6\text{-}19)$$

Since the angle between the vectors $\vec{g}$ and $\vec{h}$ is 180°,

$$W_g = mgh\cos(180°) = -mgh$$

and

$$W_F = -W_g = mgh$$

Gravitational Potential Energy

When you lift an object of mass m a height h, the gravitational potential energy of the object–Earth system increases by mgh. The conventional symbol for potential energy is U. Thus, in this situation,

$$\Delta U = mgh$$

Note that the quantity mgh is also the negative of the work done by the force of gravity. In fact, the change in gravitational potential energy is defined in terms of the work done by the force of gravity:

KEY EQUATION
$$\Delta U_g = -W_g \qquad (6\text{-}20a)$$

A common misconception is that the change in potential energy is given by the work done by you as you lift the book. The only reason the work done by you in this case is equal to the change in potential energy is that you started and ended with the book at rest. If this is not the case, the work done by you would not be equal to the change in potential energy. We look at this in detail shortly.

If we move an object up, the work done by the force of gravity is negative, and the change in potential energy is positive. The gravitational potential energy increases as the height increases. Conversely, if we move the object down, the change in potential energy is negative.

![MAKING CONNECTIONS]

Green Cars: Hybrids, Biofuels, and Electric Cars

According to the International Energy Agency, the global transportation sector used approximately 61.4% of the oil consumed worldwide and produced around 20% of global carbon dioxide (CO_2) emissions in 2008. Alternatives to fossil fuel–powered cars are being developed worldwide. Hybrid vehicles driven in the city do reduce fuel consumption and reduce CO_2 emissions; however, the bulk of the energy to drive the hybrid vehicle still comes from fossil fuels.

Biofuel, in the form of bioethanol and biodiesel, is another alternative. In Brazil, ethanol use already exceeds gasoline use as a car fuel. Theoretically, biofuels put no more CO_2 back into the atmosphere than what the crops use to produce the biofuels; unlike fossil fuels, biofuels are renewable. The use of biofuel, however, is questionable because of competition with food crops and limitations in the capacity to meet global demand.

Electric cars remain a viable alternative, depending on circumstances. In the United States and China, the bulk of electricity is produced by burning coal. One issue here is that burning coal releases more CO_2 than burning gasoline. Considering the efficiency limits inherent in the energy-conversion stages required to put an electric car in motion (from thermal energy from burning the coal, to electrical to electrochemical energy in the car's battery, to the mechanical energy of the moving car), do you think it would be advisable to operate an electric car in a region in which electricity is produced by burning fossil fuels?

Canada is the largest producer of hydroelectricity in North America and the third largest producer in the world after China and Brazil. In 2008, close to 70% of Canada's electric power came from hydroelectricity. It would be advantageous to operate an electric car in a region of the country that is powered by a hydroelectric generating plant. Figure 6-20 shows the Tesla Roadster 2.5, a premium, fully electric car with impressive performance.

Olga Besnard/Shutterstock.com

Figure 6-20 This Tesla Roadster 2.5 boasts a 288 horsepower output; it accelerates from 0 to 100 km/h in 3.7 s and has a range per charge of approximately 400 km.

EXAMPLE 6-12

Work Done by a Variable Force

You push a block of mass m up a quarter-circle track as shown in Figure 6-21(a) using a variable force. You start from rest such that the mass arrives at the top of the track with a speed of zero. Calculate the work done by the force of gravity and the change in the block's potential energy.

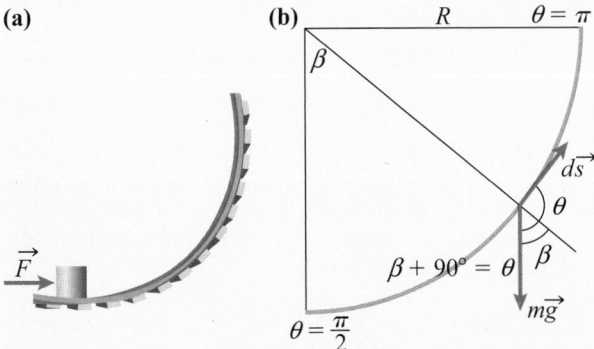

Figure 6-21 (a) You push a block up a semicircular track. (b) The FBD of the block moving up the track

SOLUTION

From Equation (6-20), the change in potential energy of the block is the negative of the work done by the force of gravity. The force of gravity is a constant and points downward. The displacement is directly along the track. Therefore, the angle θ between the force of gravity vector and the displacement changes continuously (Figure 6-21(b)). Since the

angle is not constant, we have to use integration to calculate the work done by the force of gravity:

$$W_g = \int_{s_1}^{s_2} \vec{F} \cdot d\vec{s} = \int_{s_1}^{s_2} F ds \cos\theta \qquad (1)$$

When the block is at the bottom of the track, θ is $90°$; at the top of the track, θ is $0°$. We will use β for the angular position of the block with respect to the vertical, as measured from the centre of the circular track. From Figure 6-21(b), we can see that

$$\beta = \theta - 90° \qquad (2)$$

The infinitesimal arc length ds is given by

$$ds = R d\beta = R d\theta \qquad (3)$$

Substituting for ds in the integral gives

$$W_g = \int_{\pi/2}^{\pi} mgR d\theta \cos\theta = mgR \int_{\pi/2}^{\pi} \cos\theta \, d\theta$$

$$W_g = mgR\,[\sin(\pi)] - \{mgR\,[\sin(\pi/2)]\} = -mgR$$

The change in potential energy is the negative of the work done by the force of gravity. Therefore using Equation 6-20a,

$$\Delta U = mgR$$

Making sense of the result:

The change in the height of the mass is equal to R, the radius of the track. So, this example illustrates the fact that the change in gravitational potential energy, ΔU, is path-independent and depends only on the endpoints of the path.

The key relationship given in Equation (6-20a) applies for any conservative force:

The change in potential energy is defined as the negative of the work done by the conservative force:

KEY EQUATION
$$\Delta U_c = -W_c \qquad (6\text{-}20b)$$

The Change in Potential Energy Is Path-Independent

In Section 6-3, we described the work done by the force of gravity as path-independent because it is dependent only on the change in vertical position. Equation (6-20) illustrates that the change in gravitational potential energy is path-independent and depends only on the height of the initial and final positions. In fact, the change in potential energy associated with all conservative forces, and not just the force of gravity, depends on the initial and final states and not on the path taken between these states. The chapters on thermodynamics discuss the concepts of "states" and "state functions" in more detail.

Potential Energy of a Spring

The work done by the force of a spring was derived in Section 6-4-1:

$$W_s = \frac{1}{2}kx_1^2 - \frac{1}{2}kx_2^2 \qquad (6\text{-}17)$$

where x is the distance from the equilibrium position.

Using the definition of the change in potential energy from Equation (6-20b), we conclude that the change in the potential energy of a spring is given by

$$\Delta U_s = -W_s = \frac{1}{2}kx_2^2 - \frac{1}{2}kx_1^2 \qquad (6\text{-}20c)$$

The change in the potential energy of the spring depends only on the initial and final positions, and is not affected by the path taken to get from one state to another. We can define the potential energy stored in a spring as

$$U_s = \frac{1}{2}kx^2 \qquad (6\text{-}21)$$

Then, Equation (6-20c) becomes

$$\Delta U_s = U_{s_2} - U_{s_1}$$

CHECKPOINT

C-6-8 Zero of Potential Energy

Compare the expression for the elastic potential energy of a spring in Equation (6-21) and the expression for the gravitational potential energy in Equation (6-20a). Why is there a uniquely identifiable zero for the elastic potential energy of a spring but not for the force of gravity?

C-6-8 The potential energy for a spring is zero when it is not stretched. The zero for gravitational potential energy near Earth can be set arbitrarily.

LO 6

6-6 Nonconservative Forces and Mechanical Energy

When you carry a box up a ladder, the change in the potential energy of the box might not equal the work done by you. If the final speed of the box does not equal the initial speed, you do work to change the box's kinetic energy in addition to the work you do against the force of gravity.

Now, let us apply the work–kinetic energy theorem to a situation in which you lift a stationary basketball straight up from the floor and prepare to throw it. The initial speed of the basketball is zero. You accelerate the basketball such that its speed upward when it reaches chest height is nonzero. Then,

$$W_T = \Delta K = W_F + W_g$$

where W_F is the work done by you and W_g is the work done by gravity.

We have already established that the force of gravity is a conservative force. Let us consider why the applied force is nonconservative. If you do work against a spring, by compressing it, you can get the work you do back when the spring goes back to equilibrium. If you do work by lifting a book, you can get the work you did back in the form of kinetic energy when you let the book go. Both the spring force and the force of gravity are conservative forces. If however, you were to bend your friend's arm, or push a book on a rough surface, the work you do is not recoverable in the form of mechanical energy. Hence

the force of friction and the applied force (force exerted by your friend, or by you) are nonconservative. The force applied by you is a dissipative, or nonconservative, force. In general, since all forces can be classified as conservative or nonconservative, we can express the statement of the total work–kinetic energy theorem as follows:

$$W_T = W_{nc} + W_c = \Delta K \qquad (6\text{-}22)$$

where the subscripts nc, and c, stand for nonconservative and conservative respectively. From Equation 6-20b, the work done by a conservative force is the negative of the change in potential energy.

Substituting $W_c = -\Delta U_c$ in Equation 6-22, we get

$$W_{nc} - \Delta U_c = \Delta K \qquad (6\text{-}23)$$

or

$$W_{nc} = \Delta K + \Delta U_c \qquad (6\text{-}24)$$

The sum of an object's potential and kinetic energy is its **total mechanical energy**:

KEY EQUATION
$$E_m = U + K \qquad (6\text{-}25)$$

Equation (6-24) can be rewritten as follows:

KEY EQUATION
$$W_{nc} = \Delta E_m \qquad (6\text{-}26)$$

Therefore, the action of a nonconservative force changes the total mechanical energy of an object. When we do positive work on an object, its mechanical energy increases. Examples are pushing a car (increasing its kinetic energy) and lifting a book (increasing its potential energy). When we do negative work on an object, its mechanical energy decreases. Examples are slowing down an object, such as when applying the brakes in a moving car, and taking a book down from a high shelf.

CHECKPOINT

C-6-9 Lifting Weights

How much work do you have to do to lift a 1.00 kg dumbbell from the floor by 1.00 m so that it has a speed of 3.00 m/s at that height?

C-6-9 $W_F = \frac{1}{2}mv^2 + mgh = \frac{1}{2}(1)(9)^2 + 1(9.81)(1) = 14.3$ J

EXAMPLE 6-13

Stretching a Spring

A block of mass 7.20 kg lies on a horizontal frictionless surface. It is attached to a horizontal unstretched spring with a spring constant $k = 190$ N/m. The other end of the spring is fixed. You apply a constant force $\vec{F}$ to the mass to pull it away from the spring. When the mass has moved by 40 cm, its speed is 3.10 m/s. Find the force $\vec{F}$. Assume that the force acts in the direction of motion of the mass.

SOLUTION

Since the force you apply is a nonconservative force, we use Equation (6-24):

$$W_{nc} = \Delta K + \Delta U_c \qquad (1)$$

The change in the kinetic energy of the block is

$$\Delta K = K_f - K_i = \frac{1}{2}mv^2 - 0 \qquad (2)$$

The change in the spring's potential energy is given by Equation (6-20c):

$$\Delta U_s = \frac{1}{2}kx_2^2 - \frac{1}{2}kx_1^2 = \frac{1}{2}kx^2 - 0 \qquad (3)$$

The work done by the force you apply is

$$W_{nc} = W_F = \vec{F} \cdot \vec{x} = Fx \cos(0°) = Fx \qquad (4)$$

where x is the distance over which the force acts. Putting together the above equations, we have

$$Fx = \frac{1}{2}mv^2 + \frac{1}{2}kx^2 \qquad (5)$$

$$F = \frac{\frac{1}{2}mv^2 + \frac{1}{2}kx^2}{x}$$

$$= \frac{\frac{1}{2}(7.2 \text{ kg})(3.1 \text{ m/s})^2 + \frac{1}{2}(190 \text{ Nm}^{-1})(0.4 \text{ m})^2}{0.4 \text{ m}} = 124 \text{ N}$$

Making sense of the result:

The plus sign in Equation (5) indicates that the work done by the applied force is responsible for both stretching the spring and increasing the kinetic energy of the block, as expected.

Conservation of Mechanical Energy

Equation (6-26) has a profound consequence: if there are no nonconservative forces (or the work of the nonconservative forces sums to zero), then the change in the object's mechanical energy will be zero:

$$\Delta E_m = 0 \quad \text{if} \quad W_{nc} = 0 \qquad (6\text{-}27)$$

Since the mechanical energy does not change in the absence of the work done by nonconservative forces, we can say the mechanical energy of an object is conserved:

$$E_m = constant \qquad (6\text{-}28)$$

Consequently, the mechanical energy of an object at any point on its trajectory is the same:

$$E_{m1} = E_{m2} = E_{m3} = \cdots \qquad (6\text{-}29)$$

Since the mechanical energy of an object is the sum of its potential and kinetic energies, the **law of conservation of mechanical energy** can also be stated as follows:

In the absence of nonconservative forces,

$$0 = \Delta K + \Delta U_c \qquad (6\text{-}30)$$

Therefore,

$$\Delta K = -\Delta U_c \qquad (6\text{-}31)$$

Equation (6-31) indicates that for an object under the influence of conservative forces only, whenever potential energy is lost, it is converted into kinetic energy, and vice versa.

✓ CHECKPOINT

C-6-10 What Is Conserved?

Which of the following statements is true?
(a) Potential energy is not conserved.
(b) Kinetic energy is conserved.
(c) Mechanical energy is always conserved.
(d) None of the above are true.

C-6-10 (d) Mechanical energy is only conserved when the work of non conservative forces is zero.

EXAMPLE 6-14

Block on an Incline

A 1.20 kg block slides down a frictionless incline with a slope angle of 42°, starting from a height of 2.30 m above the bottom of the incline, as shown in Figure 6-22. The incline meets a frictionless horizontal surface, at the end of which is an unstretched spring ($k = 460$ N/m) used to stop the block. Find the maximum compression of the spring.

SOLUTION

Since there are no nonconservative forces, we can apply the principle of conservation of mechanical energy. We begin with Equation (6-31):

$$\Delta K = -\Delta U_c$$

The block starts at rest, and when the spring is at maximum compression, the block is also momentarily at rest.

Figure 6-22 Example 6-14

Therefore, the change in kinetic energy, ΔK, is zero. Therefore, the change in potential energy, ΔU, is also zero.

The change in potential energy has two components: the change in the elastic potential energy of the spring and

(continued)

the change in the gravitational potential energy of what is referred to as the block–Earth system.

$$\Delta U_s + \Delta U_g = 0$$

When the block reaches the bottom of the incline, the change in gravitational potential energy is $\Delta U_g = -mgh$. The change in the spring's potential energy is given by

$$\Delta U_s = \frac{1}{2}kx_2^2 - \frac{1}{2}kx_1^2 = \frac{1}{2}kx_2^2$$

Since the spring is originally unstretched,

$$\frac{1}{2}kx^2 - mgh = 0$$

and

$$x^2 = \frac{2mgh}{k}$$

Therefore,

$$x = \sqrt{\frac{2(1.2 \text{ kg})(9.81 \text{ m/s}^2)(2.3 \text{ m})}{460 \text{ N/m}}} = 0.343 \text{ m}$$

Making sense of the result:

The gravitational potential energy is converted into kinetic energy, which is converted into spring elastic potential energy. However, there was no net change in kinetic energy, so we were able to bypass the kinetic energy step altogether.

ONLINE ACTIVITY

The e-resource that accompanies every new copy of this textbook contains an Online Activity using the PhET simulation "Energy Skate Park." Work through the simulation and accompanying questions to gain an understanding of the concepts of the conservation of energy.

EXAMPLE 6-15

Marble on a Track

The marble in Figure 6-23 starts from rest and slides down a frictionless incline from a height h above the ground, and then goes up the circular track of radius r as shown. Derive an expression for the minimum value of the height, h, so that the marble stays in contact with the circular track at the highest point on the track.

Figure 6-23   Example 6-15

SOLUTION

We are demanding that the marble barely be in contact with the track at the top. The force of contact between the marble and the top of the track is equal to zero. So, we begin with an FBD of the marble at the top of the track (Figure 6-24). Applying Newton's second law to the marble, we have, in the vertical direction,

$$\vec{N} + m\vec{g} = m\vec{a} \tag{1}$$

Choosing down as our positive direction, we get

$$N + mg = \frac{mv^2}{r} \tag{2}$$

$$N + mg = \frac{mv^2}{r}$$

Figure 6-24   The FBD of the marble

Since N is zero,

$$v^2 = rg \tag{3}$$

For the marble to have this speed at the top of the track, some of the gravitational potential energy was converted to kinetic energy. Since there are no dissipative forces, we can apply the conservation of mechanical energy:

$$\Delta K = -\Delta U_c \tag{6-32}$$

The change in height is $h - 2r$. Since the marble ends up at a lower height, the gravitational potential energy decreases, and the change in potential energy is negative:

$$\Delta U_c = -mg(h - 2r) \tag{4}$$

Since the marble starts from rest, the change in its kinetic energy is

$$\Delta K = \frac{1}{2}mv^2 - 0$$

where v is the speed of the marble at the top of the track. Therefore,

$$\frac{1}{2}mv^2 = mg(h - 2r) \tag{5}$$

Using the result from Equation (3), we get

$$\frac{1}{2}mrg = mg(h - 2r) \tag{6}$$

and, therefore,

$$h = 2.5\,r \tag{7}$$

Making sense of the result:

The speed at the top of the circular track cannot be zero because the marble needs enough speed to make it across the top without falling. Therefore, the initial height of the marble should be higher than the height of the loop ($2r$).

EXAMPLE 6-16

Sliding on a Spherical Surface

A polar bear cub sits on top of an igloo. The surface of the igloo has a spherical curvature and is smooth enough to have negligible friction. The cub begins to slide down the igloo. Determine where the cub will leave the surface of the igloo.

SOLUTION

As the bear cub slides down, the Earth-cub's potential energy is converted to kinetic energy and the cub gains speed. Let us draw the FBD of the cub at the point where the cub leaves the surface of the igloo. We will measure the angular position of the bear from the vertical. The forces acting on the bear are the force of gravity and the normal force. Since the cub's motion is circular, the principal acceleration is radial. Thus, we choose our axes so that one axis points along the radial direction and the other axis is tangent to the igloo. We then resolve the forces into components along these axes, and write the equation for motion along each axis.

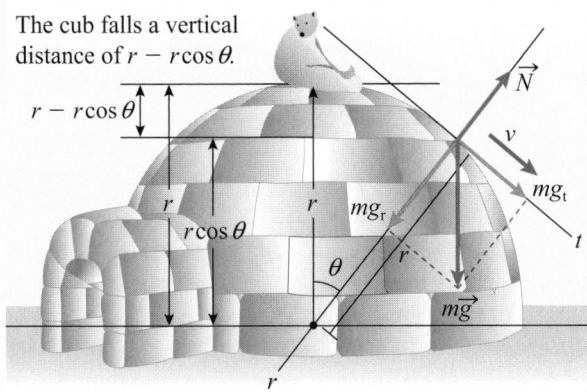

The cub falls a vertical distance of $r - r\cos\theta$.

Figure 6-25   Example 6-16

In the radial direction we have

$$N_r + mg_r = ma_r \tag{1}$$

Choosing the inward direction as positive gives

$$mg\cos\theta - N = \frac{mv^2}{r} \tag{2}$$

where N used in Equation (2) denotes the magnitude of the normal force.

When the cub is on the verge of leaving the surface, the normal force is zero and

$$mg\cos\theta = \frac{mv^2}{r} \tag{3}$$

We have two unknowns—the speed, v, and the angle, θ, at which the cub leaves the surface. Since we need one more relevant equation, we now consider using the mechanical energy of the cub.

As the cub slides down the surface, some potential energy is converted to kinetic energy. From Equation (6-32),

$$\Delta K = -\Delta U_g \tag{4}$$

As we can see from Figure 6-25, the change in height is $r - r\cos\theta$. Since the bear moves to a lower height, the change in potential energy is negative:

$$\Delta U_g = -mg(r - r\cos\theta) \tag{5}$$

The cub starts from rest, so the change in kinetic energy is

$$\Delta K = \frac{1}{2}mv^2 - 0 \tag{6}$$

Substituting Equations (5) and (6) into Equation (4), we have

$$\frac{1}{2}mv^2 = mg(r - r\cos\theta) \tag{7}$$

Using the result from Equation (3), we get

$$\frac{1}{2}mgr\cos\theta = mg(r - r\cos\theta) \tag{8}$$

$$\cos\theta = \frac{2}{3} \tag{9}$$

$$\theta = 48.19°$$

Thus, the cub will leave the surface at an angle of about 48.2° from the vertical.

Making sense of the result:

This example again demonstrates the effectiveness of combining mechanical energy considerations with Newton's laws. The normal force is zero when the cub is about to leave.

EXAMPLE 6-17

Pendulum Tension

The small marble in Figure 6-26 is attached to a massless string, which is free to rotate about the pivot at its other end. The marble is initially held such that the string is horizontal and then let go. Find the angle that the string makes with the vertical when the tension in the string is equal to the weight of the marble.

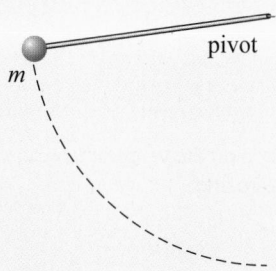

Figure 6-26   Example 6-17

SOLUTION

When the string is horizontal and the mass is at rest, the tension is zero because the gravitational force is directed straight downward, and thus has no horizontal component. As the mass moves down, its speed increases. At the bottom of the mass's path, the string both supports the mass and provides the centripetal acceleration, making the tension greater than mg. Therefore, at some angle θ between $0°$ and $90°$ from the vertical, the tension will be equal to mg.

As shown in the FBD in Figure 6-27, the forces acting on the mass are tension and the force of gravity. Since the pendulum swings in a circular arc, the principal acceleration

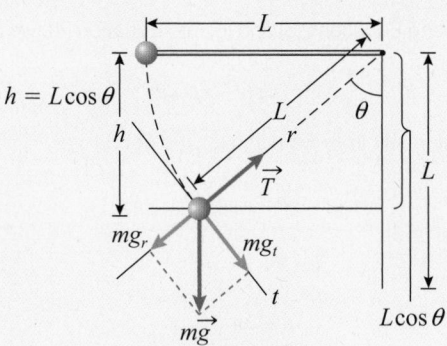

Figure 6-27   The FBD of the marble

is along the radial direction. Therefore, we choose one axis along the radial direction and the other axis perpendicular to it and tangent to the path of the mass.
If we sum the forces along the radial direction r-axis, we get

$$T_r + mg_r = ma_r \tag{1}$$

Taking the radial direction (direction of the tension) as the positive r-axis, we have

$$T - mg\cos\theta = \frac{mv^2}{r} \tag{2}$$

When the tension equals mg,

$$mg - mg\cos\theta = \frac{mv^2}{L} \tag{3}$$

We see that the mass cancels out, thus, we have two unknowns: the speed of the mass and the angle of the string.

The mass gains speed as it falls because gravitational potential energy is being converted into kinetic energy. Since there are no dissipative forces, mechanical energy is conserved and, using Equation 6-31

$$\Delta K = -\Delta U_g \tag{6-32}$$

As the pendulum swings down, ΔK is positive and ΔU_g is negative. From Figure 6-27 we can see that $h = L\cos\theta$, where h is the vertical distance from the initial position of the mass and L is the length of the string. Therefore,

$$\Delta K = \frac{1}{2}mv^2 = -(-mgL\cos\theta) \tag{5}$$

and

$$mv^2 = 2mgL\cos\theta$$

Substituting for mv^2 in Equation (2) we have

$$mg - mg\cos\theta = 2mg\cos\theta$$

$$\cos\theta = \frac{1}{3}$$

$$\theta = 70.53°$$

Making sense of the result:

As a quick check setting θ to zero in Equation (5) means that the marble has fallen a total vertical distance of L. The speed of the marble would be $v = \sqrt{2gL}$, as one would expect for an object falling straight down a total distance of L.

✓ CHECKPOINT

C-6-11 $T = mg$?

Why is the tension not equal to mg when the string is vertical?

C-6-11 The tension needs to support the weight and provide radial acceleration.

EXAMPLE 6-18

Using the Work-Energy Approach to Find Acceleration

A block of mass M rests on top of a table with a horizontal frictionless surface. The block is attached to a hanging mass m with a rope that passes over the massless, frictionless pulley as shown in Figure 6-28.

(a) Find the speed of mass M after m has fallen a total distance h, when the system starts from rest.
(b) Use your result from part (a) to find the acceleration of the system.
(c) Repeat parts (a) and (b) when the coefficient of friction between M and the surface of the table is μ_k.
(d) Compare this approach to the approach used in Chapter 5, Example 5-19.

Figure 6-28 Example 6-18

SOLUTION

(a) Since there is no friction, we can use the work–kinetic energy theorem or one of the expressions based on the principle of conservation of mechanical energy. We will use the latter here. Since the total mechanical energy is conserved, the gravitational potential energy is converted as m drops into kinetic energy of the system.

$$\Delta K = -\Delta U_c \qquad (6\text{-}32)$$

ΔK is the change in the kinetic energy of *both* masses:

$$\Delta K = \Delta K_1 + \Delta K_2 = \frac{1}{2} M v_f^2 - \frac{1}{2} M v_i^2 + \frac{1}{2} m v_f^2 - \frac{1}{2} m v_i^2 \qquad (1)$$

Since the system starts at rest, $v_i = 0$:

$$\Delta K = \frac{1}{2} M v_f^2 + \frac{1}{2} m v_f^2 \qquad (2)$$

The only change in potential energy comes from m descending a distance h (Figure 6-28). So,

$$\Delta U = -mgh \qquad (3)$$

and

$$\frac{1}{2} M v^2 + \frac{1}{2} m v^2 = mgh \qquad (4)$$

We note here that both masses move at the same speed.

Solving Equation (4) for v, we obtain

$$v = \sqrt{\frac{2mgh}{M + m}}$$

(b) To determine the acceleration, we can take the derivative of Equation (4) with respect to time. Using the chain rule gives

$$\frac{d}{dt} v^2 = \left(\frac{d}{dv} v^2 \right) \left(\frac{dv}{dt} \right) = 2va$$

Since the derivative of the height with respect to time is $dh/dt = v$,

$$Mva + mva = mgv \qquad (5)$$

$$Ma + ma = mg \qquad (6)$$

$$a = \frac{mg}{M + m} \qquad (7)$$

(c) Since friction is a nonconservative force, we use Equation (6-26):

$$W_{nc} = \Delta E_m = \Delta U + \Delta K \qquad (8)$$

$$W_{nc} = -mgh + \frac{1}{2} M v^2 + \frac{1}{2} m v^2 \qquad (9)$$

As M moves to the right, the force of friction from the surface on M is directed to the left and therefore, by Equation 6-7, does negative work:

$$W_{nc} = \vec{f_k} \cdot \Delta \vec{x} = -f \Delta x = -N \mu_k h = -M g \mu_k h \qquad (10)$$

Therefore,

$$-M g \mu_k h = -mgh + \frac{1}{2} M v^2 + \frac{1}{2} m v^2 \qquad (11)$$

Solving for v, we obtain

$$v = \sqrt{\frac{2mgh - 2Mg\mu_k h}{M + m}} \qquad (12)$$

To get the acceleration, we take the derivative of Equation 11 with respect to time, as in part (b). The resulting expression for the acceleration is

$$a = \frac{mg - Mg\mu_k}{M + m}$$

(d) If we used the approach that we used in Chapter 5, Example 5-19, we would have to deal with a system of two equations for the unknowns, a and T.

Making sense of the result:

The system is propelled to move by the weight of the hanging mass, and is slowed down by the force of friction on the horizontal mass. It makes sense for the numerator in the expression for the acceleration to be the net force, or the difference between the two forces and for the denominator to be the total mass.

LO 7

6-7 Potential Energy and Reference Points

As we saw in Section 6-5, the elastic potential energy of the spring can be expressed as

$$U_s = \frac{1}{2} kx^2$$

where x is the distance by which the spring is stretched or compressed as measured from its unstretched length.

The elastic potential energy stored in a spring is position-dependent. The spring's potential energy is zero when $x = 0$. *This point is unique and cannot be arbitrarily assigned.* It is the point where the spring is unstretched. The work done by a spring and, therefore, the elastic potential energy stored in a spring depend on both the starting and ending positions.

 EXAMPLE 6-19

Spring Stiffness

You are given a spring of spring constant $k = 1000$ N/m and asked to stretch it 10 cm from the unstretched length. You then repeat the exercise with the same spring, but this time the spring is already stretched by 30 cm before you stretch an extra 10 cm. Calculate the change in the spring's potential energy in each case. Which case involves a larger change in the spring's potential energy?

SOLUTION

The change in the potential energy of the spring is given by Equation (6-20c):

$$\Delta U_s = -W_s = \frac{1}{2}(kx_2^2 - kx_1^2) \qquad (6\text{-}20c)$$

We begin with the initially unstretched spring, $x_1 = 0$, and $x_2 = 10$ cm:

$$\Delta U_s = \frac{1}{2}(1000\text{ N/m}(0.1\text{ m})^2 - 0) = 5\text{ J}$$

When the spring is already stretched by 30 cm, this will be our x_1, and when we stretch it another 10 cm, x_2 is (30 cm + 10 cm) = 40 cm. Then $\Delta U_s = \frac{1}{2}(1000\text{ N/m}(0.4\text{ m})^2 - 1000\text{ N/m }(0.3\text{ m})^2) = 35$ J.

Making sense of the result:

The spring that is already stretched will require more work to stretch it farther even though the amount by which the distance is stretched is the same in both cases.

Clearly, in the previous example stretching the spring by the same amount did not result in the same amount of change in potential energy. The change in potential energy depends, of course, on the amount by which the spring is stretched. However, it also depends on the starting point.

Now, suppose you carry out an exercise similar to the one in Example 6-19, but this time, we lift an object against the force of gravity. Say you lift a 1 kg book a distance of 50 cm. First, you start from the surface of a table in front of you. Then you start from the roof of a 100 m-tall building. Which exercise involves a greater change in the gravitational potential energy of the book–Earth system?

The answer, of course, is that the change in the potential energy is the same in both cases.

For the book being lifted against the force of gravity, all that matters is the change in height; the starting point is immaterial. However, the change in potential energy of a spring that is stretched by a given amount depends on the starting position. The point of zero potential energy in the case of the spring is uniquely defined: this is the point where the spring is not stretched. In the case of the force of gravity, we can choose any point to be the point of zero potential energy.

LO 8

6-8 Equipotential Surfaces and Field Lines

We have defined the change in potential energy as the negative of the work of the conservative force:

$$W_c = -\Delta U_c \qquad (6\text{-}20b)$$

For a constant conservative force, the difference in potential energy between two points $\overrightarrow{\Delta r}$ apart is

$$\overrightarrow{F_c} \cdot \overrightarrow{\Delta r} = -\Delta U_c$$

If the conservative force is not constant, we can use the more general form of this relationship:

$$\int_{r_1}^{r_2} \overrightarrow{F_c} \cdot \overrightarrow{\Delta r} = -\Delta U_c$$

We can then express the conservative force as a vector derivative, in the elegant language of calculus:

KEY EQUATION $\qquad \overrightarrow{F_c} = -\nabla U = -grad(U) \qquad (6\text{-}33a)$

The vector differential operator ∇ is called "del." In this context, del denotes the *gradient* of the scalar potential energy. In Cartesian form, the expression in Equation (6-33) is written as follows:

KEY EQUATION $\qquad \overrightarrow{F_c} = -\frac{\partial U}{\partial x}\hat{i} - \frac{\partial U}{\partial y}\hat{j} - \frac{\partial U}{\partial z}\hat{k} \qquad (6\text{-}33b)$

The *partial derivative* operator ∂ indicates that we are taking the derivative of a multivariate function (a function that depends on more than one variable) with respect to one of the variables while keeping the other variables constant.

In one dimension, say, along the x-axis, Equation (6-33b) reduces to

$$\vec{F}_c = -\frac{dU}{dx}\,\hat{i} \qquad (6\text{-}34)$$

EXAMPLE 6-20

Force of a Spring

Use the potential energy function for a spring to derive an expression for the force exerted by a spring.

SOLUTION

In Section 6-5, we found that the potential energy stored in a spring is $U = \frac{1}{2}kx^2$. Since the expression depends on only one variable, we can use Equation (6-34):

$$\vec{F}_c = -\frac{dU}{dx}\,\hat{i} = -\frac{d}{dx}\left(\frac{1}{2}kx^2\right)\hat{i} = -kx\,\hat{i}$$

Making sense of the result:

The derived expression for the force is the same as the equation for Hooke's law.

EXAMPLE 6-21

Using a Gradient to Determine Force

A particle is in a region in which its potential energy is given by $U = (3x^2 - 2xy^2 + zx)$ J. Find an expression for the force acting on the particle, and determine the magnitude of the force at point A $(0.5\text{ m}, -1\text{ m}, 0.5\text{ m})$.

SOLUTION

Here, we use Equation (6-33b):

$$\vec{F}_c = -\frac{\partial U}{\partial x}\,\hat{i} - \frac{\partial U}{\partial y}\,\hat{j} - \frac{\partial U}{\partial z}\,\hat{k}$$

which gives

$$\vec{F}_c = ((-6x + 2y^2 - z)\hat{i} + 4xy\hat{j} - x\hat{k})\text{J/m}$$

Substituting the coordinates of point A gives

$$\vec{F}_c = ((-1.5)\hat{i} - 2\hat{j} - 0.5\hat{k})\text{ J/m}$$

The magnitude of the force at point A is

$$|\vec{F}_c| = \left(\sqrt{(-1.5)^2 + (-2)^2 + (-0.5)^2}\right)\text{J/m} = \sqrt{6.5}\text{ N} = 2.5\text{ N}$$

Equipotential Surfaces and Field Lines

An **equipotential surface** is a surface along which the potential energy does not change. For example, the surface of Earth is an approximate gravitational equipotential surface. If you walk from point A to point B without any

change in elevation, the potential energy of the interaction between you and Earth does not change. In the absence of dissipative forces, we do not need to do work to move along an equipotential surface.

One of the consequences of Equation (6-33) is that a conservative force must be perpendicular to the equipotential surface (see Example 6-22).

Around Earth, the gravitational equipotential surfaces are (almost) perfectly spherical shells with Earth at their centre. As we know from everyday experience, the force of gravity points radially toward Earth's centre, as shown in Figure 6-29. Thus, the force of gravity is perpendicular to these equipotential surfaces.

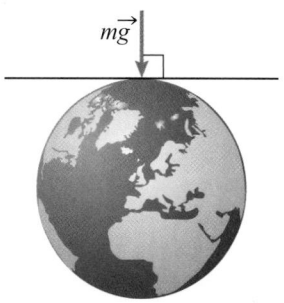

Figure 6-29 The force of gravity points radially toward Earth's centre.

Further, the gravitational acceleration, or field strength, $\vec{g}$, is parallel to the force of gravity because $\vec{F} = m\vec{g}$. We thus conclude that the gravitational field is also perpendicular to the gravitational equipotential surface. This fundamental relationship does not apply to gravity only; it applies to any equipotential surface and its corresponding conservative force: *conservative force vectors are perpendicular to equipotential surfaces.*

EXAMPLE 6-22

Force and Equipotential Surfaces

Prove that the conservative force lines are perpendicular to a straight equipotential line. Assume that the conservative force producing the equipotential line is constant.

SOLUTION

Consider the equipotential surface in the field of a conservative force, as shown in Figure 6-30. We will assume

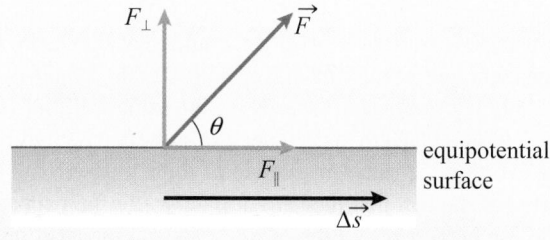

Figure 6-30 Conservative force pointing in an arbitrary direction relative to equipotential surface.

(continued)

that the conservative force (F) points in some arbitrary direction relative to the equipotential surface.

The force vector can be resolved into one component parallel to the equipotential line and another component perpendicular to the equipotential line. The force makes an angle θ with its projection onto the equipotential surface as shown. The work, W_c, done by the conservative force is an object moves along the equipotential line from point A to point B is given by Equations 6-8 and 6-10a

$$W_c = \vec{F} \cdot \Delta \vec{s} = F_{\parallel} \Delta s$$

where $\Delta \vec{s}$ is the distance between points A and B. The change in potential energy between points A and B is given by Equation 6-20a

$$\Delta U = -W_c = -F_{\parallel} \Delta s$$

However, since points A and B lie on the same equipotential surface, the potential energy at A must be equal to the potential energy at B, and the potential energy difference between the two points must be zero. Then $F_{\parallel}$ must be zero. Since the conservative force cannot have a component in a direction parallel to the equipotential surface, the force must be perpendicular to the equipotential surface.

Making sense of the result:

This example also shows that the field of a conservative force is perpendicular to an equipotential surface generated by that force or field for the geometry we chose.

CHECKPOINT

C-6-12 The Conservative Force

Is the following statement true or false? A conservative force points in the direction of decreasing potential energy.

C-6-12 True

LO 9

6-9 Power

Consider two otherwise identical dump trucks loaded with equal amounts of oil sand in Northern Alberta. Both trucks are moving at constant speeds up the same hill, but one truck moves faster than the other. We know that both trucks are exerting the same magnitude of force on the road, by virtue of the fact that they are moving at constant speeds. The force exerted by each of the trucks on the road must be equal to $mg\sin\theta$ to result in zero acceleration. Further, we know that the two trucks, by the time they deliver their loads at the top of the hill, will have done the same amount

of work, namely, mgh, where h is the height of the hill. However, since one of the trucks is moving faster than the other, the *rate* at which they do work is different.

Power is the rate at which work is done. Average power is defined as

$$P_{avg} = \frac{W}{\Delta t} \qquad (6\text{-}35)$$

The limit of the average power as Δt approaches zero is the instantaneous power:

KEY EQUATION
$$P = \frac{dW}{dt} \qquad (6\text{-}36)$$

where dW is the infinitesimal amount of work done during the infinitesimal amount of time dt.

The power delivered by a variable force acting in one dimension between points x_1 and x_2 can be written as

$$P = \frac{d}{dt}\left(\int_{x_1}^{x_2} \vec{F} \cdot d\vec{x} \right) \qquad (6\text{-}37)$$

If the force is constant, the expression for dW reduces to $\vec{F} \cdot d\vec{x}$, and the expression for instantaneous power becomes

$$P = \vec{F} \cdot \frac{d\vec{x}}{dt}$$

For forces acting in more than one dimension,

$$P = \vec{F} \cdot \frac{d\vec{s}}{dt}$$

or

$$P = \vec{F} \cdot \vec{v} \qquad (6\text{-}38)$$

The SI unit for power is the watt (W), which is defined as

$$1\ \text{W} = 1\ \text{J/s}$$

Another commonly used unit to measure power is horsepower (hp). Unfortunately, the horsepower (like the calorie) has a number of different definitions. One commonly used standard for horsepower is

$$1\ \text{hp} = 746\ \text{W}$$

EXAMPLE 6-23

Water Power

At 14 GW, the Itaipu Dam and generating plant has the second largest installed power generation capacity in the world, after the Three Gorges Dam in China. Itaipu also holds the world record for the most power generated in one year (2008: 94.86 TW). Located on the Upper Paraná River at the border between Paraguay and Brazil, the dam is 196 m high and has an average annual river flow of 12 000 m³/s. Calculate the theoretical maximum possible power generation for the water flowing through the dam.

SOLUTION

As shown in Example 6-8, the work done by the force of gravity as an amount of water of mass m falls through a height h is

$$W_g = mgh$$

The average power from Equation (6-35) is given by

$$P_{avg} = \frac{W_g}{\Delta t}$$

The average amount of water flowing every second through the dam is 12 000 m³. Using 1000 kg/m³ for the density of water, we have 12×10^6 kg of water flowing every second. The work done by the force of gravity is then

$$W_g = mgh = 12 \times 10^6 \text{ kg } (9.81 \text{ m/s}^2)(196 \text{ m}) = 2.3073 \times 10^{10} \text{ J}$$

This amount of work is done by the force of gravity in one second. Therefore,

$$P_{avg} = \frac{W_g}{\Delta t} = \frac{2.3049 \times 10^{10} \text{ J}}{1 \text{ s}} = 23.0 \text{ GW}$$

Making sense of the result:

This is not much larger than the installed capacity. Hydro plants often run short of the maximum theoretical power generation for various reasons. One of these reasons is the "installed capacity," which generally refers to the infrastructure in place to harness the available energy. Another reason is the issue of the inherent efficiency of a given device or system in converting energy from one form to another. This efficiency is rarely 100%.

The efficiency of a given system or device is measured as the ratio of the power output to the power input:

$$e\,(\%) = \frac{P_{out}}{P_{in}} \times 100\%$$

MAKING CONNECTIONS

Potential Energy and Hydroelectricity

As water accelerates in a waterfall, potential energy is converted to kinetic energy through the work done by the force of gravity. This kinetic energy can be harnessed as a clean and renewable source of energy for running electric generators. The water in the Robert Bourassa Reservoir powers the LG-2 (Figure 6-31) and LG-2-A generating stations on the La Grande River in Québec, with a combined output capacity of 7.326 GW. This is the largest installed capacity in North America, and the fifth largest in the world. To date, the Three Gorges Dam in China is the most powerful hydroelectric generating station in the world, with an installed capacity of 18.2 GW, expandable to over 22 GW. However, this power comes at the expense of extensive flooding, loss of cropland, and environmental damage.

Photo Hydro-Québec

Figure 6-31 The main hall at the LG-2 generating station (Robert Bourassa) on La-Grande River in Québec

✓ CHECKPOINT

C-6-13 The Work Done by an Escalator

An escalator takes people up a few floors at the airport. A steward stands still on the escalator, and a pilot with the same mass as the steward walks up the escalator. The pilot reaches the top in less time than the steward. Which of the following statements is true?

(a) The escalator does more work to move the steward upward than it does to move the pilot.

(b) The escalator does the same amount of work to move both the pilot and the steward upward.

(c) The escalator does less work to move the steward than it does to move the pilot.

(d) None of the above are necessarily true.

C-6-13 (c) the pilot does work against the force of gravity to go up.

MECHANICS

Forces are classified as conservative and nonconservative. The work you do against conservative forces, such as the force of gravity and the force of a spring, is stored in the form of potential energy. The work you do against nonconservative forces, such as friction, is dissipated and cannot be recovered in the form of kinetic energy. In the absence of nonconservative forces, the mechanical energy of an object is conserved. The work of a nonconservative force on an object results in a change in its total mechanical energy. The work–mechanical energy approach can be a very powerful tool for solving problems and is often more convenient than the approach using Newton's second law.

The Work-Kinetic Energy Theorem

The total work done on an object is equal to the change in its kinetic energy:

$$W_T = \Delta K \qquad (6\text{-}5)$$

Work done by a force:
in one dimension:

$$|\vec{F}||\Delta\vec{x}|\cos\theta \qquad (6\text{-}5), (6\text{-}7)$$

$$W_F = \int_{x_1}^{x_2} \vec{F} \cdot d\vec{x} \qquad (6\text{-}13a)$$

Total work done by a constant force:

$$W_T = \vec{F}_{net} \cdot \Delta\vec{x} \qquad (6\text{-}3)$$

Total work done by a varying force:

$$W_T = \int_{x_1}^{x_2} \vec{F}_{net} \cdot d\vec{x}$$

In more than one dimension

$$W_F = \int_{a}^{b} \vec{F} \cdot d\vec{s} \qquad (6\text{-}13b)$$

Work done by a spring:

$$W_s = \frac{1}{2}kx_1^2 - \frac{1}{2}kx_2^2 \qquad (6\text{-}17)$$

Power is the rate at which work is delivered:

$$P = \frac{dW}{dt} \qquad (6\text{-}36)$$

Conservative Forces, Potential Energy, and Mechanical Energy

The work you do against a conservative force is stored as potential energy. The change in potential energy is defined as the negative of the work done by the conservative force:

$$\Delta U_c = -W_c \qquad (6\text{-}20b)$$

The force can be obtained from the potential energy by using

$$\vec{F}_c = -\frac{dU}{dx}\hat{i} \qquad (6\text{-}34)$$

The potential energy is the same everywhere on an equipotential surface.
The force lines are perpendicular to the equipotential surface.
The change in gravitational potential energy is given by

$$\Delta U_g = mg\Delta h$$

The change in the elastic potential energy of a spring is given by

$$\Delta U_s = \frac{1}{2}kx_2^2 - \frac{1}{2}kx_1^2 \qquad (6\text{-}20c)$$

The Conservation of Mechanical Energy

The work done by nonconservative forces on an object results in a change in the total mechanical energy of the object:

$$W_{nc} = \Delta E_m$$

In the absence of nonconservative forces, mechanical energy is conserved.

$$\text{If } W_{nc} = 0, \quad \text{then} \quad \Delta E_m = 0$$

The conservation of mechanical energy can be expressed as

$$E_m = \text{constant} \qquad (6\text{-}28)$$

or

$$\Delta K = -\Delta U_c \qquad (6\text{-}31)$$

Applications: conservation of energy, alternate fuel sources, efficient energy use, renewable energy, the global energy crisis, power generation, energy conversion, energy storage, earthquake engineering, shock absorbers, auto safety, engines, space flight, metabolism, electrostatic interactions, gravitation, friction losses, air resistance, nuclear interactions, thermodynamics, quantum mechanics.

Key Terms: conservation of mechanical energy, conservative force, equipotential surface, kinetic energy, mechanical energy, nonconservative force, potential energy, power, restoring force, total work, total work–kinetic energy theorem, work

QUESTIONS

1. Does the work–kinetic energy theorem hold for forces that are not constant? Explain.
2. When you drop a stone from the top of a building, does the instantaneous power delivered by gravity increase, decrease, or stay the same as the stone falls?
3. True (T) or false (F): A spring is already stretched 10 cm when you decide to stretch it another 10 cm. The spring constant k is 1000 N/m. The work you do is positive.
4. The total work done as you stretch a massless spring is
   (a) the change in the potential energy of the spring;
   (b) the negative of the work done by the spring;
   (c) parts (a) and (b);
   (d) none of the above.
5. You stretch a spring that is already stretched. Is the work done by the spring positive or negative?
6. A spring is compressed from equilibrium. Is the work done by the spring positive or negative?
7. Can the work done by a force ever be negative? Explain.
8. Can the total work done on an object ever be negative? Explain.
9. When you do positive work on a moving car, does the car's speed increase or decrease? Explain.
10. You ride on a Ferris wheel. Your mass is 73 kg, and the radius of the Ferris wheel is 45 m. What is the total work done on you during your journey from the lowest to the highest points? What is the work done by gravity?
11. What is the total work done by Earth on a 22 000 kg satellite as it moves around Earth by 180°, at a constant altitude? What is the work done by gravity during that time?
12. You spin the bob (mass m) at the end of a pendulum of length L at a constant speed v in a vertical circle. What is the work done by the tension in the string as the bob is brought from the lowest to the highest position?
    (a) $-2mgL$
    (b) $2mgL$
    (c) zero
    (d) The answer could be either (a) or (b), depending on the choice of positive direction.
13. A hockey player intercepts a 170 g puck moving toward him on ice at 23 m/s and sends it at the same speed toward the opponent's goalie. Find the work done by the hockey player.
14. Your friend is holding on to a rope that is suspended from a tree branch, such that his feet are 60 cm off the ground. You push him with a horizontal force of variable magnitude such that he moves at a constant speed as his feet end up being lifted by an extra 40 cm off the ground. Your friend's mass is 69 kg. Find the work done by your variable force.
15. You drop a 2.0 kg weight from a height of 43 cm above a scale. The weight comes to rest after falling 48 cm. Find the total work done on the weight during its downward journey.
16. You compress a spring from equilibrium by an amount Δx. We can say that the change in potential energy of the spring is equal to
    (a) $\frac{1}{2}k(\Delta x)^2$;
    (b) $-\frac{1}{2}k(\Delta x)^2$;
    (c) part (a) or (b), depending on our choice of a positive direction.
    (d) We need to know either the initial or final length of the spring to answer this question.

17. You stretch a spring from an already stretched position. Which of the following statements is true?
    (a) The work you do is negative.
    (b) The work done by the spring is positive.
    (c) The change in the spring's potential energy is negative.
    (d) None of the above are true.
18. You intercept a hockey puck shot at you at a speed of 15 m/s by kicking it such that it deflects at the same speed at an angle of 65° to the incoming direction. Assume the average force that you exert on the puck is 90 N and that it acts over a total distance of 12 cm in one direction. What is the total work done on the puck by you? Ignore gravity.
    (a) 10.8 J
    (b) −10.8 J
    (c) 5.04 N
    (d) −5.04 N
    (e) none of the above
19. A spring cannon fires a projectile at an angle of 47° above the horizontal. The cannon is fired at $t = 0$, and the projectile reaches its maximum height a time Δt later. Which of the following statements is true?
    (a) During the time Δt, the work done by the force of gravity is larger than the work done by the spring.
    (b) During the time Δt, the work done by the spring is larger than the work done by gravity.
    (c) During the time Δt, the work done by the spring is equal to and opposite in sign to the work done by the force of gravity.
    (d) None of the above are true.
20. You lift your backpack of mass m a vertical distance h. What is the definition of the change in potential energy?
    (a) the work done by you
    (b) the negative of the work done by you
    (c) the negative of the work done by the force of gravity
    (d) the work done by the force of gravity
21. The work done by a spring as it is being compressed from its unstretched position is
    (a) positive;
    (b) negative;
    (c) either positive or negative, depending on where you start;
    (d) depends on how fast you are moving;
    (e) none of the above because work is a scalar that cannot be referred to as positive or negative.
22. A spring is already stretched 20 cm from its equilibrium position (Figure 6-32). You pull it so that it is now stretched 40 cm. While the spring is being stretched,
    (a) you do negative work;
    (b) the spring does positive work;
    (c) the sign of the work done by the spring on you depends on whether we choose right or left to be positive;
    (d) the work done by the spring is always positive because of the term $(dx)^2$;
    (e) none of the above.

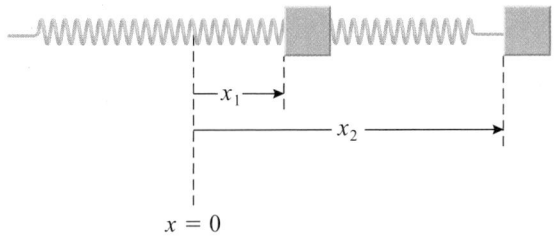

Figure 6-32 Question 22

23. A child moves down a slide at a park at a constant speed. Find a relation between the slope angle of the slide and the coefficient of kinetic friction between the child and the slide's surface. What can be said about the relationship between the work done by gravity and the work done by friction?

24. You slide down a snowy hill on a toboggan from a height of 7 m. The hill has a constant slope such that your speed remains constant at 5 m/s throughout the entire journey. Air resistance is negligible. Which of the following statements is true?
    (a) The magnitude of the work done by gravity is larger than the magnitude of the work done by friction.
    (b) The work done by friction is equal to the change in the gravitational potential energy.
    (c) The work done by friction cannot be determined.
    (d) Friction is not significant because you actually move.
    (e) None of the above is true.

25. A compressed spring is used to launch a marble onto a surface that starts out horizontal and then slopes downward at a constant angle. When the marble moves down the incline at a constant speed,
    (a) the total work done on the marble is zero;
    (b) the work done by the spring is greater than the work done by gravity;
    (c) the work done by friction is equal to the work done by gravity;
    (d) the work done by the spring is greater than the work done by friction.
    (e) This motion is impossible.

26. You move a spring such that the work done by the spring is negative. This means that you
    (a) stretched the spring from equilibrium;
    (b) compressed the spring from equilibrium;
    (c) stretched an already stretched spring;
    (d) any of the above;
    (e) none of the above.

27. You move a spring such that the work done by the spring is negative. This means that
    (a) you allow the spring to move toward equilibrium;
    (b) the potential energy of the spring decreases;
    (c) you do negative work.
    (d) There is not enough information to answer this question.
    (e) None of the above.

28. A spring is compressed by 10 cm from its equilibrium position. When you move the spring so that it is now stretched by 10 cm,
    (a) the work that you do is positive;
    (b) the total work done on the spring is positive;
    (c) the work that you do is negative;
    (d) the work done by the spring is negative;
    (e) none of the above.

29. You are sitting at your desk. You decide to set your desktop as the point of zero gravitational potential energy, then you lift a book from a distance h below the desk to a distance h above the desk. Which of the following statements is true?
    (a) The change in potential energy is zero.
    (b) The final potential energy is $2mgh$.
    (c) The final potential energy is mgh.
    (d) The change in potential energy is $2mgh$.

30. A spring is already stretched 10 cm when you decide to stretch it another 10 cm. The spring constant is 200 N/m. As you stretch the spring,
    (a) the work done by the spring is −1 J;
    (b) the work done by the spring is −2 J;
    (c) the work done by the spring is −3 J;
    (d) the work done by the spring is −4 J.
    (e) None of the above.

31. You drop a 2 kg mass from a height of 7 m above a spring whose spring constant is 100 N/m at its equilibrium position. During the object's downward journey,
    (a) the magnitude of the work done by gravity is less than $14g$;
    (b) the magnitude of the work done by the spring is less than $14g$;
    (c) the total work done on the mass is zero.
    (d) I really don't think this makes sense.
    (e) None of the above.

32. Which of the statements below is true when you catch a baseball that comes at you horizontally? More than one statement might apply.
    (a) The work that you do is positive.
    (b) The work that you do is negative.
    (c) Work is not a vector and cannot be referred to as positive or negative.
    (d) Work is a scalar product expressed in terms of magnitudes, so it is never negative.
    (e) Work is expressed in terms of kinetic energy, which must be positive.

33. A spring is used to launch a marble up a frictionless incline such that the marble reaches the top of the incline with a speed of zero. Which of the following statements is true?
    (a) The total work done on the marble is positive.
    (b) The total work done on the marble is negative.
    (c) The work done on the marble by the spring is equal in magnitude and opposite in sign to the change in the gravitational potential energy.
    (d) The change in the gravitational potential energy is smaller than the change in elastic potential energy.
    (e) None of the above are true.

PROBLEMS BY SECTION

For problems, star ratings will be used, (∗, ∗∗, or ∗∗∗), with more stars meaning more challenging problems.

Section 6-1   Work and Kinetic Energy for Constant Forces

34. ∗   You throw a stone with a mass of 215 g straight up in the air. It reaches a maximum height of 12 m, and then you catch it at the same level from which you threw it.
    (a) Find the work done by the force of gravity on the stone on its way up.
    (b) Find the work done by the force of gravity on the stone on its way down.
    (c) Find the total work done by the force of gravity over the entire trip.

35. ∗   You lift your 400 g cup of coffee from the table a total distance of 45 cm and hold it there.

(a) Find the work done by the force of gravity.

(b) What is the work done by you on the cup?

(c) What is the total work done on the cup?

36. ✳ You push a 10.0 kg wagon with a horizontal force of 145 N over a total distance of 7.00 m.

(a) How much work is done by you?

(b) Now you move so that you are opposing the motion of the wagon. You stop the wagon by applying a force that stops it in 2.00 m. How much work do you do as you slow the wagon?

37. ✳ While ice fishing, your friend slides a 5.0 kg sausage bucket over the ice surface to you at a speed v. You stop the bucket over a distance of 90 cm by applying a force of 20 N on the bucket. How much work did you do?

38. ✳ A tennis player strikes an incoming tennis ball of mass m and speed v such that the ball leaves the racket at speed v as well. How much work was done by the tennis player to reverse the direction of the ball? Explain. Did the tennis player do any negative work?

39. ✳ You throw an 8.0 kg medicine ball straight up in the air and then catch it on the way down. Just before it hits your hands, it is moving at a speed of 3.0 m/s. You stop it over a distance of 50 cm.

(a) How much work did you do to stop it?

(b) How much work was done by the force of gravity while you stopped the ball?

40. ✳ A worker tosses a 1.5 kg lunch box to a friend who is on a platform 12 m above. The lunch box is in contact with the worker's hands for a total distance of 35 cm. How much work did the worker do to accelerate the lunch box?

41. ✳✳ An athlete picks up a 20.0 kg sandbag from the ground and throws it straight up in the air. It leaves her hands 1.50 m above the ground and reaches a height of 7.00 m.

(a) How much work did the athlete do?

(b) How much work was done by the force of gravity while the bag was in contact with the athlete?

42. ✳ A block of mass 26 kg rests on a frictionless inclined surface with a slope of angle 37°. A horizontal force of 510 N is used to push the block a distance of 4.6 m along the inclined surface, starting from rest, as shown in Figure 6-33. Find the work done by the force $\vec{F}$.

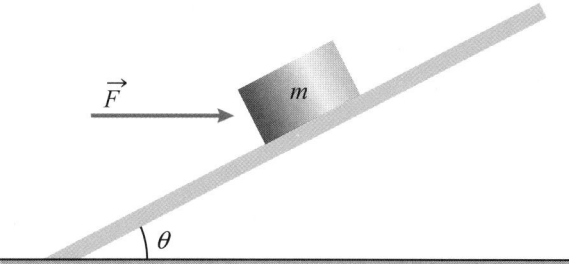

Figure 6-33 Problem 42

43. ✳✳ Two masses, $M = 16.8$ kg and $m = 4.8$ kg, are attached to each other with a rope that winds around a massless hanging pulley as shown in Figure 6-34. Mass m is attached to one end of an unstretched spring whose other end is attached to the ground. The system is initially at rest.

(a) Derive an expression for the speed of mass m after M has fallen 45 cm.

(b) Find the maximum speed of m.

(c) Find the speed of m when $k = 96$ N/m.

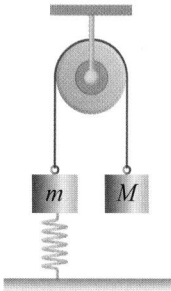

Figure 6-34 Problem 43

44. ✳ A child starts from rest and slides down a snow-covered hill with a slope of angle 47° from a height of 1.7 m above the bottom of the hill. The speed of the child at the bottom of the hill is 3.2 m/s. Find the coefficient of friction between the hill and the child.

45. ✳ An object of mass m is pushed along a frictionless horizontal surface by a force F, over a distance d, such that the speed is constant.

(a) Find an expression for the coefficient of kinetic friction between the object and the surface.

(b) Find the work done by F.

(c) Friction is removed, and the force F is now applied along an incline to cause the mass to move up the incline at a constant speed. Does the force F do more or less work over the same distance?

(d) How does your answer to part (c) change if the mass were to accelerate up the incline? (Hint: We make the incline angle smaller.)

(e) How does your answer to part (c) change if friction were brought back such that the object still moved up the incline?

Section 6-2 Work Done by Constant Forces in Two or Three Dimensions

46. ✳ A worker leans on a crate and pushes it on a horizontal surface with a force of 230 N in a direction that makes an angle of 67° below the horizontal over a distance of 12 m. How much work is done by the worker?

47. ✳✳ A 22 kg sled comes loose during a sled race and is stopped by an observer over a horizontal distance of 1.3 m. The observer is pushing down on the sled with a force that makes an angle of 42° with the horizontal. The sled's speed is 4.0 m/s. How much work does the observer do? What is the magnitude of the force used by the observer?

48. ✳ A child pulls a 3.0 kg wooden bus on a horizontal surface over a distance of 3.0 m using a rope that makes a 47° angle above the horizontal starting from rest. The speed of the bus at the end of the pull is 5.0 m/s. Determine the work done by the child and the force exerted by the child. How much work is done by the force of gravity during the pull?

49. ✳✳ A child uses a rope to pull her brother on a toboggan on a horizontal, snow-covered surface. The combined mass of her brother and the toboggan is 34 kg. The rope makes an angle of 41° with the surface. The horizontal projection of the rope makes an angle of 27° with the direction of motion. The child is pulling the toboggan with a force of 60 N, and the toboggan moves at a constant speed. Find the coefficient of friction between the toboggan and the snow.

50. ✳✳ A person applies a force $F = 4\hat{i} - 2\hat{j} + 7\hat{k}$ N to move a crate on a track in the xz-plane that angles 32° above

the positive *x*-axis. Find the work done by the person over the first 3.0 m of the motion.

51. ✵✵ A particle moves on a track between points A $(3, 2, -1)$ and B $(5, -4, 3)$ as a result of being pushed by a force of magnitude 170 N that rises 32° above the *xy*-plane such that its projection makes an angle of 195° with the positive *x*-axis.
    (a) The mass of the particle is 1.1 kg, and the particle starts from rest. Find the work done by the force and find the particle's speed at the end of its trajectory.
    (b) Repeat part (a), but include gravity and use the coefficient of kinetic friction between the particle and the track, $\mu_k = 0.15$.

52. ✵ You pull a crate of mass 7.00 kg a distance of 3.00 m up a frictionless incline with a slope of angle 32.0° by applying a force of magnitude 300 N along the incline.
    (a) Find the amount of work you do.
    (b) You now push the same mass on a frictionless horizontal surface for the same distance as in part (a), using a force of the same magnitude as in (a) but now pointing horizontally in the same direction of motion. Determine the work done by you.

Section 6-3 Work Done by Variable Forces

53. ✵✵ A marble of mass *m* sits at the bottom of a bowl of radius *R*. The surface of the bowl is frictionless, and the bowl is not allowed to move. You push the marble in the horizontal direction using a force of variable magnitude until the marble is a height *h* above the bottom of the bowl, such that $h < R$. You bring the marble to a stop. Find the work done by the variable force.

Section 6-4 Work Done by a Spring

54. ✵✵ Two identical springs, each of spring constant 720 N/m, are set up in parallel horizontally such that they are compressed by 32.0 cm. The two springs are used to eject a 500 g mass onto a frictionless surface. The mass then moves 1.90 m up a frictionless incline with a slope of angle 32.0°, where it encounters a combination of springs identical to the springs used to eject it, lying along the incline. How high above the horizontal surface is the mass when it is finally stopped?

55. ✵✵ A 10.0 kg block slides 3.00 m down a frictionless surface inclined 30° above the horizontal, before being stopped by a spring of spring constant $k = 340$ N/m secured to the inclined surface. Find the maximum compression of the spring.

56. ✵✵✵ A perforated bead of mass 37 g moves on a frictionless horizontal circular track of radius $R = 0.65$ m. The track runs through the bead (Figure 6-35). When the bead is at rest in the horizontal position, it is tied to a spring ($k = 230$ N/m), the other end of which is fixed at a pivot at the centre of the track as shown. The unstretched length of the spring is $R/3$. The bead is given a gentle nudge downward.
    (a) Find the speed of the bead.
    (b) As the bead moves, the track is tilted so that it is now sitting vertically. Find the difference between the bead's maximum and minimum speeds.
    (c) Find the work done by the spring in bringing the bead from the lowest to the highest points on the now vertical track.

(d) Find the work done by gravity as the bead goes from the lowest to the highest points along the track.

Figure 6-35 Problem 56

Section 6-5 Conservative Forces and Potential Energy

57. ✵✵ You push a crate of mass 27 kg 15 m up a ramp that is sloped 19° above the horizontal. The coefficient of friction between the ramp and the crate is 0.15, and the speed of the crate, which starts from rest, is 1.7 m/s at the end of the 15 m. To accomplish the task, you use a variable force whose magnitude and direction both change.
    (a) Calculate the change in the crate's gravitational potential energy (crate-Earth system).
    (b) Calculate the work done by the force.

Section 6-6 Nonconservative Forces and Mechanical Energy

58. ✵✵ A mass ($m = 1.20$ kg), originally at rest, sits on a frictionless surface. It is attached to one end of an unstretched spring ($k = 790$ N/m), the other end of which is fixed to a wall (Figure 6-36). The mass is then pushed with a constant force to stretch the spring. As a result, the system comes to a momentary stop after the mass moves 14.0 cm. Find the
    (a) work done by the constant force;
    (b) change in potential energy of the spring;
    (c) change in mechanical energy of the system;
    (d) total work done by nonconservative forces in the problem;
    (e) speed of the mass 4.00 cm into the motion;
    (f) new equilibrium position of the system;
    (g) maximum speed of the mass and where that occurs.

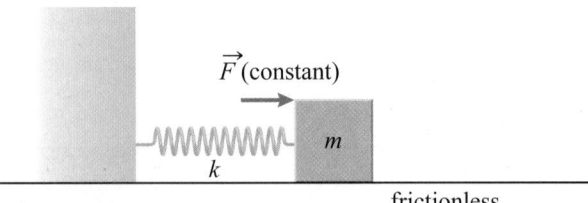

$\vec{F}$ (constant)

m

k

frictionless

Figure 6-36 Problem 58

59. ✵✵ A mass ($m = 900$ g) rests on a 63.0° incline, where it is attached to one end of a spring, also resting on the incline as shown in Figure 6-37. The other end of the spring is fixed. The mass is held in place so that the spring is unstretched. The mass is then released. The coefficient of kinetic friction between the surface and the mass is 0.0800, and the spring constant is 110 N/m. Find the maximum compression of the spring.

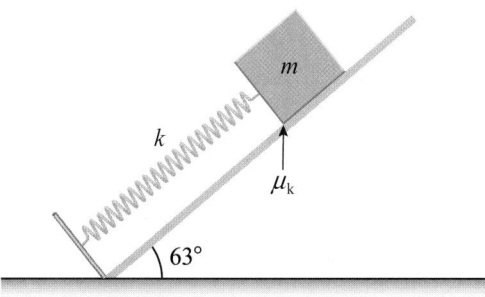

Figure 6-37 Problem 59

Section 6-7   Potential Energy and Reference Points

60. ✱✱✱ Newton's law of gravitation states that the force of attraction between two bodies is proportional to the product of their masses and is inversely proportional to the square of the distance between them. The force is given by $\dfrac{Gm_1m_2}{r^2}$, where G is a constant. (See Chapter 11.)

(a) Find an expression for the gravitational potential energy between two objects.

(b) Is the expression positive or negative?

(c) What would your escape velocity have to be if you weighed 800 N on Earth's surface? The escape velocity is the velocity with which you have to be launched so you do not come back to Earth. Ignore losses due to air resistance.

(d) Find the speed of a 1000 kg satellite orbiting Earth at an altitude of 1000 km.

(e) Find the total mechanical energy for the satellite in part (d).

(f) What total mechanical energy does the satellite need to just break free from Earth's gravitational field?

(g) The binding energy of a satellite is the additional energy needed for it to escape Earth's gravity. Find the binding energy for the satellite in part (d).

Section 6-8   Equipotential Surfaces and Field Lines

61. ✱✱ The force of gravitational attraction on an object beneath Earth's surface is proportional to r, the distance from the centre. Beyond Earth's surface, the force of attraction becomes inversely proportional to r^2. Plot the gravitational potential energy for an object as it moves from the centre of Earth to a point that is three times the radius of Earth as measured from Earth's centre.

62. ✱✱ A particle enters a region where the potential energy function is given by $\alpha x^{-2} + \beta xy - \gamma z^2x$. Find an expression for the force in Cartesian unit vector notation.

63. ✱✱✱ The electrostatic potential energy between two charges is given by KQq/r. Charge q is brought to within a distance R of charge Q and then released from rest. Charge Q is fixed in space.

(a) Ignoring gravity, derive an expression for the speed of charge q after it has moved a distance of Δr from Q.

(b) Using $m_q = 51$ g, $q = 16\ \mu C$, $Q = 47\ \mu C$, $K = 8.99 \times 10^9\ Nm/C^2$, and $R = 1.1$ cm, find a value for the speed in part (a).

64. ✱✱ The potential energy of an interaction between two objects is given by $-A(r^3 + r)$, where A has a positive value.

(a) Is the force between these two objects repulsive or attractive? How can you tell?

(b) Plot a potential energy versus r graph for $A = 120$ N/m, over the range $r = 0$ to 10 m.

(c) Derive an expression for the force between the two objects as a function of r.

(d) One of the objects is held at rest, while the other is placed a distance of 20.0 m from the fixed object, and then released from rest. The mass of the moving object is 11.0 kg. Find its speed after it has moved 3.00 m.

Section 6-9   Power

65. ✱ The Tupolev Tu-144 supersonic passenger jet had four Kolesov RD-36-51 afterburning turbojets that each provide a maximum thrust of 200 kN. The maximum flying speed of the Tu-144 was Mach 2.2 at an altitude of 120 000 ft. Find the force of air resistance on the airplane at that speed.

66. ✱✱ A tractor of mass 5400 kg pulls a 1400 kg crate up a hill with a slope of angle 12.0° above the horizontal at a constant speed of 11.0 m/s.

(a) What acceleration would the tractor have on a level road at that speed? Ignore any frictional losses.

(b) Repeat part (a), but now consider that the coefficient of friction between the crate and the ground is 0.070. Assume that all frictional losses come from the friction between the crate being pulled and the ground, and not from the tractor.

67. ✱ A sports car has a maximum speed of 320 km/h and a maximum operating engine power output of 490 hp. Find the force of air resistance on the car at that speed. Ignore all other sources of friction and resistance to the motion.

68. ✱✱ You kick a 200 g rock from the edge of a cliff while hiking along the edge of the Grand Canyon.

(a) Ignoring air resistance, find the average power delivered by gravity during the first 4 s of the fall.

(b) Find the instantaneous power delivered by gravity at the end of the fifth second of the fall.

COMPREHENSIVE PROBLEMS

69. ✱✱✱ A bead slides down a frictionless vertical helical track as shown in Figure 6-38. The helical track has 1 turn/cm of length of the helical axis, and the loops have a radius of 17.0 cm each. There are 7 turns in the helix.

Figure 6-38 Problem 69

(a) Estimate the slope angle experienced by the bead along the helix.

(b) Find the speed of the bead at the bottom of the helix.

70. ✷✷ You swing a massless string of length L with a mass m attached to its end in a vertical circle, such that the mass has more than enough speed to keep the string taut all the time. You then hold the end of the string steady while the other end with the mass attached to it continues to swing freely. Derive an expression for

(a) the difference between the maximum and the minimum tensions in the string as the mass moves in its circular path.

(b) the work done by the tension during a full revolution.

71. ✷✷ A marble slides on the frictionless ramp starting at rest from height h shown in Figure 6-39. The marble then moves to the circular track of radius r. Write an expression for the minimum value of the height h so that the normal force on the marble is three times its weight when it is at the same height as the centre of the circular track.

Figure 6-39   Problem 71

72. ✷✷✷ You happen upon what looks like an abandoned hemispherical igloo of radius R. You climb the igloo and sit at the top (Figure 6-40). A few moments later you feel a broom very gently prodding you off the roof. The push you receive from the disgruntled igloo dweller is just enough to cause you to begin to move.

(a) Find the angle that your position vector as measured from the centre of the igloo makes with the vertical at the point where you slide off the surface of the igloo.

(b) at what angle is the normal force equal to 1/3 of you weight.

Figure 6-40   Problem 72

73. ✷✷✷ You happen upon a perfectly shaped ceramic semi-spherical bowl at home, with a smooth, frictionless surface. You place a small bead at the top of the inside edge of the bowl and let it slide on the inside of the bowl under the effect of gravity (Figure 6-41). Locate the point along the bowl where the normal force from the bowl on the bead equals the bead's weight. What is force of contact between the bead and the bowl when the bead is passing across the bottom of the bowl?

Figure 6-41   Problem 73

74. ✷✷ The simple pendulum in Figure 6-42 is made to spin in a vertical circle such that it has sufficient kinetic energy to just stay taut at the top of its trajectory. When the string is in the vertical position shown, it is intercepted by a peg at a distance of $2L/3$ from the centre of the vertical circle, and the mass continues along the new trajectory as shown.

(a) Find the speed of the mass(m) at the lowest point of the new trajectory.

(b) Find the speed of the mass at the top of the new trajectory.

(c) Find the tension in the string when it makes an angle of 40° with the vertical before hitting the peg.

(d) Find the tension in the string at the top of the new trajectory.

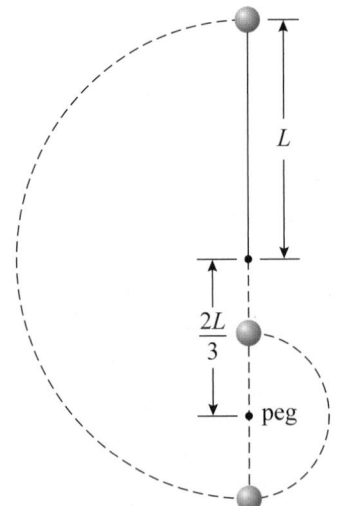

Figure 6-42   Problem 74

75. ✷✷ A 200 g cube slides down a ramp starting from rest as shown in Figure 6-43. The ramp has a 46° slope. After falling a distance of 92 cm, the cube strikes a spring of spring constant $k = 20$ N/m. Find the maximum compression of the spring when the coefficient of kinetic friction between the cube and the ramp is 0.17.

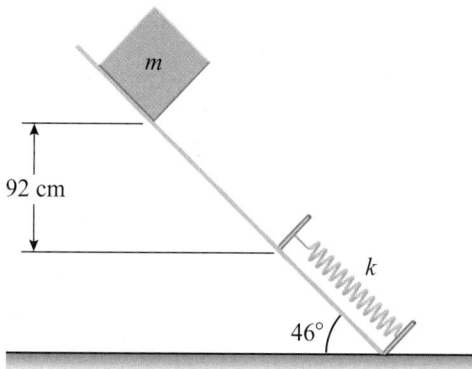

Figure 6-43 Problem 75

76. **✷✷** You are sitting in a roller coaster car that begins its descent, from rest, down the main track from a height of 62.0 m. The roller coaster then climbs up the inside of a vertical circle of radius 20.0 m to a point two-thirds of the circle's radius above the bottom of the circle.
 (a) Find the normal force exerted on you by the seat if your mass were 57.0 kg. The normal direction is taken to be perpendicular to the track.
 (b) What is the total force exerted on you by the seat?

77. **✷✷** The string in Figure 6-44 is wound around the nail as shown. The end of the string is attached to a mass m that is released from the horizontal position, starting from rest. Derive an expression for
 (a) the tension in the string when it makes an angle of 23° with the vertical.
 (b) the total work done to bring the string to the position in part (a).
 (c) the work done by the tension.

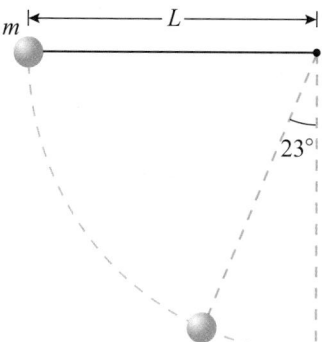

Figure 6-44 Problem 77

78. **✷✷✷** A block of mass M sits on top of a horizontal frictionless bench. One end of the block is connected to an originally unstretched spring of spring constant k. The other end is connected to a vertically hanging mass m by a string that runs over a massless pulley. Find an expression for
 (a) the maximum distance that m will fall, starting from rest.
 (b) the speed of mass m after it has fallen a distance h

79. **✷✷** An unstretched spring of spring constant k hangs vertically from the ceiling. You now attach a mass m to it and let it extend slowly until the mass stops.
 (a) By how much does the spring stretch?
 (b) Next, you take the mass back up to the original unstretched position and let it go. How far will the mass fall? What is the maximum speed of the mass?

80. **✷✷** A spring of spring constant $k = 710$ N/m is suspended vertically from a ceiling. You attach a 32 kg mass to the spring and let it drop. Find the speed of the mass when it is three-quarters of the way down to the position where it momentarily stops.

81. **✷✷✷** A block of mass $m = 3.2$ kg rests on top of another block of mass $M = 16.4$ kg, which rests on top of a horizontal frictionless surface. The coefficient of static friction between the two blocks is 0.23, and the coefficient of kinetic friction is 0.17. Block M is pushed with a horizontal force of 37 N. Find the energy lost to friction during the time it takes M to move 4.0 m, starting from rest.

82. **✷✷** You throw a 240 g orange straight up in the air from a height of 90 cm above the ground starting from rest. You let go of the orange when it is 140 cm above the ground and it reaches a maximum height of 14 m.
 (a) How much work is done by you?
 (b) How much work is done by the force of gravity while your hand is in contact with the orange?
 (c) Find the power delivered by the force of gravity during the last second before the orange reaches it maximum height.

83. **✷✷✷** A block of mass $M = 37$ kg rests on top of a horizontal bench, where the coefficient of friction between the block and the surface is $\mu_k = 0.20$. A string attached to M runs over a massless, frictionless pulley and is attached to the vertically hanging mass ($m = 11$ kg). The system is originally held at rest and then released.
 (a) Find the speed of M after m has fallen a distance of 60 cm.
 (b) Using energy considerations, derive an expression for the acceleration of the system as a function of time.

84. **✷✷** A block of mass $M = 17$ kg rests on top of a frictionless horizontal bench. A string attached to M runs over a massless frictionless pulley and is attached to a vertically hanging mass ($m = 4.0$ g). The system is originally held at rest and then released. Find the speed of M after m has fallen a distance of 30 cm. Using energy considerations, find the acceleration of the system.

85. **✷✷✷** The force between two particles is measured to be $F = -Ar^{-3}$.
 (a) Develop an expression for the change in the potential energy of interaction between the two particles as the distance between them changes from r_1 to r_2.
 (b) You are doing work to change the distance between the two particles. Develop an expression for the work done by you when the distance changes at a constant speed.
 (c) What is the expression for the work done by the force between the two charges?
 (d) Is the force between the particles attractive or repulsive?
 (e) One of the particles is fixed in space, and the other is let go from a distance of 110 cm from the fixed particles. The mass of the moving particle is 12 g. Find its speed when it has moved 70 cm from the position indicated above. The experiment is conducted on a horizontal frictionless surface.

86. ✶✶✶ The worker in Figure 6-45 has to pull a crate over a concrete surface over a distance of 9.0 m, as shown.
  (a) Find the angle that will maximize the effectiveness of the worker in pulling the crate. (Hint: Minimize the time for the task.)
  (b) The crate's mass is 340 kg, and the coefficient of kinetic friction between the crate and the floor is 0.20. For a given angle θ, what is the minimum force needed to cause the crate to start moving when the coefficient of static friction between the crate and the surface is 0.50?
  (c) Once the block starts moving, the worker applies just enough force to keep it moving at a constant speed. Find the work done by the worker.
  (d) Find the work done by friction.
  (e) Find the work done by gravity.
  (f) Find the total work done on the crate.

Figure 6-45 Problem 86

87. ✶✶ A particle enters a field where the potential energy is given by $U = \dfrac{-7}{r^2 + 6}$ J. Derive an expression for the force experienced by this particle in this region as a function of r. Find the speed of the particle 3 s into the field.

88. ✶✶✶ A 67 kg skydiver free-falls for the first 3 min, reaching 99% of her terminal speed of 210 km/h in 15 s.
  (a) Find the work done by gravity during the last 2 min of the fall.
  (b) Find the work done by the air resistance during the last 2 min of the fall.
  (c) Find the power delivered by gravity during that time.
  (d) Find the power delivered by air resistance during that time.
  (e) Find the average force of air resistance on the skydiver during the first 15 s of the fall.

89. ✶✶✶ A particle with a mass of 310 g, moving in one dimension enters a progressively damped region where the force it experiences is given by $F = -kx^3$. Find how far the particle enters the region when it enters with a speed of 120 m/s. The force constant is 320 N/m³.

90. ✶✶✶ A test particle's position as a function of time is given by $x = -\alpha t^3 + \beta t^2 + \gamma t^{-4}$ as a result of being under the influence of a variable force, F.

  (a) Find the position of the particle at $t = 0$.
  (b) Find the position of the particle at $t = 3.5$ s for $\alpha = 0.70$ m/s³, $\beta = 1.1$ m/s², and $\gamma = 1.5$ ms⁴.
  (c) Find the value of the force at $t = 0$.
  (d) Find the value of F at $t = 3.5$ s.
  (e) Evidently, the acceleration is not constant. Can we still use the work–kinetic energy theorem for this particle? Explain.

91. ✶✶✶ Construct a proof to show that the force lines or the field lines, as given by Equation (6-34), follow the path of steepest descent between equipotential surfaces.

92. ✶✶✶ Repeat problem 71 with friction when the coefficient of kinetic friction between the marble and the ramp is 0.15 and the mass of the marble is 120 g.

93. ✶✶✶ Repeat problem 72 when the coefficient of kinetic friction between you and the igloo is 0.2, and your mass is 72 kg.

94. ✶✶✶ The air resistance on a free-falling stone is proportional to the square of the speed of the stone. The stone falls from rest and reaches 99% of its terminal velocity in the first 15 s of the fall. Find the coefficient of proportionality between the resistive force and the square of the speed. The terminal speed is 230 km/h. Find the work done by air resistance during the first 2 s of the fall.

95. ✶ The Avro Arrow jet fighter had a maximum takeoff weight of 62 430 pounds and an operating ceiling of 58 500 feet. Calculate an estimate of the change in gravitational potential energy when a fully loaded Avro Arrow climbs to its operating ceiling. List any assumptions you make for your estimate.
  (a) Find the combined power output of its two Pratt & Whitney turbojet engines, the thrust of which was 23 450 lbs each when cruising at maximum speed.
  (b) Using the fact that the force of air resistance on an object is proportional to the square of its speed, find the maximum acceleration of the plane when at a cruising speed of 701 mph.
  (c) Assuming a constant climb rate, find the work done by the engines to get the plane to its operating ceiling, moving at its maximum speed.

96. ✶✶✶ A typical banana nut muffin or a sesame seed bagel with cream cheese has around 400 Cal. Determine the height of a mountain you would have to climb to do that much work against gravity. Use a mass of 72 kg. How many kilometres of walking does that equal if you are climbing a 20° hill? This initial estimate does not take into account the fact that you actually burn calories walking on a perfectly horizontal surface. A 72 kg person walking briskly at 6.0 km/h will burn approximately 285 Cal on level ground in 1.0 h. How many kilometres do you have to walk if you factor that fact in?

 See the text online resources at **www.physics1e.nelson.com** for Open Problems and Data-Rich Problems related to this chapter.

Chapter 7
Linear Momentum, Collisions, and Systems of Particles

In cars, the combination of crumple zones and strengthened passenger compartments is a direct application of Newton's laws that has proved to be a major contribution to passenger safety (Figure 7-1). The crumple zone is a part of a car that is controllably weakened in such a way that it deforms during a collision, which maximizes the dissipation of the kinetic energy from the collision. The crumpling also increases the time over which the car loses speed and, hence, decreases the acceleration and the force experienced by the passengers during an impact. First introduced by Mercedes Benz in the 1950s, crumple zones are generally installed at both the front and rear of modern cars. Cars with crumple zones generally have a hardened passenger cabin, which is designed to maintain its integrity during a collision and protect passengers against intrusions from the crumpling exterior.

Older cars have rigid bodies that do not deform much during a collision. Such a design makes these cars more likely to rebound from each other during a collision. There is minimum damage to the car bodies but there are high rebound accelerations. High rebound accelerations cause more serious passenger injuries than the reduced collision accelerations experienced in cars with crumple zones.

Learning Objectives

When you have completed this chapter you should be able to:

1 Define linear momentum, and calculate and compare momenta of various real-life objects.

2 Express Newton's laws in terms of rates of change of linear momentum.

3 Define and calculate impulse, and establish the conditions for the conservation of linear momentum.

4 Find the centre of mass of a system of particles.

5 Distinguish between internal and external forces on a given system, and explain how they affect the motion of the system's centre of mass.

6 Distinguish between elastic and inelastic collisions in terms of energy and momentum conservation.

7 Apply the laws of conservation of momentum and energy to elastic collisions.

8 Derive the ideal rocket equation, and explain the basics of jet and rocket propulsion.

Figure 7-1 A car's crumple zone

For additional Making Connections, Examples, and Checkpoints, as well as activities and experiments to help increase your understanding of the chapter's concepts, please go to the text's online resources at www.physics1e.nelson.com.

7-1 Linear Momentum

In Chapter 6, we stated that in the absence of nonconservative forces, the mechanical energy of an object is conserved.

In this chapter, we introduce another conserved quantity: **linear momentum**, $\vec{p}$. The linear momentum of an object is defined as the product of its mass and its velocity:

$$\vec{p} = m\vec{v} \qquad (7\text{-}1)$$

Momentum is a vector quantity and has units of kilogram metres per second.

Table 7-1 lists momenta for different objects. The table gives an indication of the wide range of possible values for momentum.

Table 7-1 Appreciating Momentum

Momentum (kg · m/s)	Object
~1.66×10^{-29}	An atom in a Bose–Einstein condensate (a very low-energy state of matter)
3.67×10^{-24}	A free neutron at 290 K
2.123×10^{-23}	An atom of argon at room temperature
2.733×10^{-23}	An electron moving at 1/10 the speed of light
5.01×10^{-20}	A proton moving at 1/10 the speed of light
0.014	A 7.0 g marble moving at 2.0 m/s
3.0	A 7.5 g bullet moving at 400 m/s
2.8×10^4	A 1.0 tonne car moving at 100 km/h
14.5×10^6	A fully loaded Avro Arrow flying at 2100 km/h
1.6×10^9	Super Carrier (100 000 tonnes at full speed, 54 knots)
1.776×10^{29}	Earth as it orbits the Sun

EXAMPLE 7-1

Rocket Momentum

Find the horizontal and the vertical components of the momentum of a space shuttle at the point following its takeoff where the mass of the shuttle (including fuel) is 1.90×10^6 kg and its velocity is 4200 km/h at an angle of $11.0°$ from the vertical.

SOLUTION

The magnitude of the space shuttle's momentum at that point is given by

$$p = mv = 1.90 \times 10^6 \text{ kg} \times 4200 \left(\frac{1000 \text{ m}}{3600 \text{ s}}\right)$$

$$= 2.22 \times 10^9 \text{ kg m/s}$$

The expression for the momentum vector is given by

$$\vec{p} = \vec{p}_x + \vec{p}_y$$

Using Figure 7-2 we can express the momentum as

$$\vec{p} = p\sin(11°)\hat{i} + p\cos(11°)\hat{j}$$

$$= (4.23 \times 10^8\,\hat{i} + 2.18 \times 10^9\hat{j}) \text{ kg m/s}$$

Making sense of the result:

The vertical component is significantly greater than the horizontal component, as we would expect, because the angle from the vertical is relatively small.

Figure 7-2 The horizontal and vertical components of the linear momentum of a space shuttle

---

✓ **CHECKPOINT**

C-7-1 Speed versus Momentum

A car enters a roundabout at a constant speed and exits in the direction opposite to the one in which it entered the roundabout. The change in momentum of the car is

(a) zero;
(b) twice the initial momentum;
(c) twice the final momentum.
(d) None of the above.

C-7-1 (c) $p_f - p_i = p_f - (-p_f) = 2p_f$.

EXAMPLE 7-2

Change of Momentum

A marble with a mass of 10.0 g is sliding at a constant speed of $v = 2.00$ m/s along a circular, horizontal frictionless track of radius $r = 0.90$ m, as shown in Figure 7-3(a). Find the change in the marble's momentum as it goes from point A to point B. Express the change in vector form, and find the magnitude of the momentum change. Illustrate your solution with a diagram.

SOLUTION

The change in momentum is the difference between the final and the initial momenta. When the marble is in position A, all of its momentum is in the y-direction. Therefore,

$$\vec{p}_A = 0\hat{i} + (mv)\hat{j} = 0\hat{i} + (0.010 \text{ kg})(2 \text{ m/s})\hat{j}$$

$$= (0.020 \text{ kg m/s})\hat{j} \qquad (1)$$

At position B, the magnitude of the momentum has not changed, but it is now pointing in the x-direction:

$$\vec{p}_B = (mv)\hat{i} + 0\hat{j} = (0.020 \text{ kg m/s})\hat{i}$$

The change in momentum as a vector is equal to

$$\Delta\vec{p} = \vec{p}_B - \vec{p}_A = 0.020\hat{i} - 0.020\hat{j}$$

The change in momentum can be expressed as

$$\Delta\vec{p} = \Delta\vec{p}_x + \Delta\vec{p}_y = 0.020\hat{i} - 0.020\hat{j}$$

The magnitude of the momentum vector is

$$\Delta p = \sqrt{(\Delta p_x)^2 + (\Delta p_y)^2} = \sqrt{0.020^2 + 0.020^2} = 0.028 \text{ kg m/s}$$

The diagram for the Solution is shown in Figure 7-3(b).

 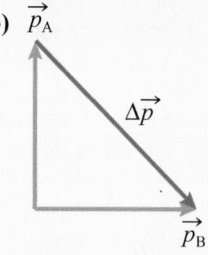

(a) **(b)**

Figure 7-3 (a) The linear momentum has the same magnitude but different directions at A and B. (b) The change in momentum is shown as a difference of vectors.

Making sense of the result:

The change in momentum must point down and to the right to give us the final momentum which points straight to the right.

☑ **CHECKPOINT**

C-7-2 Change in Momentum

A high-g training capsule of mass m moves counterclockwise around a circular track at a constant speed v. What is the change in the capsule's momentum by the time it has completed one-quarter of a turn if it starts in the positive y-direction?

(a) $mv\hat{i} + mv\hat{j}$
(b) $-mv\hat{i} - mv\hat{j}$
(c) $mv\hat{i} - mv\hat{j}$
(d) $-mv\hat{i} + mv\hat{j}$

C-7-2 (b) horizontal momentum goes from zero to $-mv$, vertical momentum goes from $+mv$ to zero.

Momentum and Kinetic Energy

The kinetic energy of a particle is given by

$$K = \frac{1}{2}mv^2 \qquad (6\text{-}4)$$

From the definition of momentum (Equation (7-1)), we get

$$\vec{v} = \frac{\vec{p}}{m}$$

Then,

$$K = \frac{p^2}{2m} \quad \text{or}$$

$$p = \sqrt{2mK} \qquad (7\text{-}2)$$

Thus, an object that has momentum also carries kinetic energy.

☑ **CHECKPOINT**

C-7-3 Kinetic Energy and Momentum

When the momentum of a particle doubles, and its mass remains constant, its kinetic energy
(a) doubles;
(b) quadruples;
(c) increases by $\sqrt{2}$.
(d) Both (a) and (b) are possible.

C-7-3 (b).

CHAPTER 7 | LINEAR MOMENTUM, COLLISIONS, AND SYSTEMS OF PARTICLES

MECHANICS

MAKING CONNECTIONS

Momentum Without Mass

Light has a dual nature: it is both a particle and a wave. Its wave nature can be demonstrated through interference and diffraction measurements, the most famous of which is Young's double-slit experiment, carried out in 1803 (Chapter 28). The particle nature of light was conclusively demonstrated by Einstein in his paper on the photoelectric effect (Chapter 30) in 1905. Although a photon (a particle of light) has zero mass, it still has momentum. A photon's momentum is defined in terms of its ability to carry energy. The momentum of a photon is inversely proportional to the light's wavelength and is defined in terms of its ability to carry energy.

LO 2

7-2 Rate of Change of Linear Momentum and Newton's Laws

MAKING CONNECTIONS

Fatal Momentum

The *Titanic* received six warnings of "imminent ice ahead" before its fateful collision with an iceberg. However, the captain maintained full speed, in accordance with the shipping industry's standard operating procedure of maintaining full cruising speed until ice is actually sighted. Up until this time, this procedure had proven efficient and safe. This procedure was reasonable for smaller and more manoeuvrable ships; however, *Titanic* was a 52 000 tonne leviathan. Its large momentum could not be changed quickly, and the corresponding kinetic energy made the collision with the iceberg more damaging than it would have been for a smaller ship travelling at the same speed. By the time the ice was sighted, it was too late to change course enough to save the ship. Ironically, it is believed that had the *Titanic* maintained its course and hit the iceberg head on, the ship might not have sunk, because the bow structure would have distributed the force of the collision lengthwise along the ship's hull. However, the angled impact along the side of the ship broke a line of hull plates loose at their riveted joints, dooming the ship.

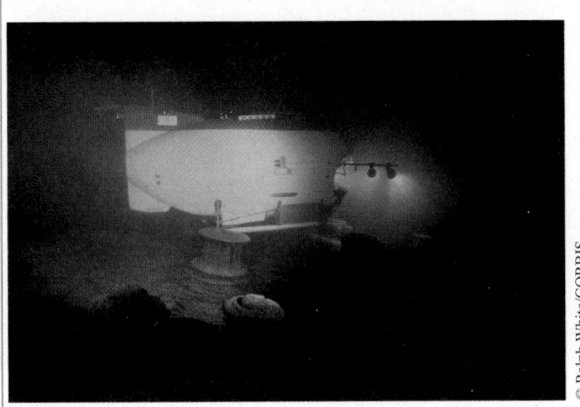

Figure 7-4 The remains of the *Titanic* on the floor of the Atlantic Ocean

Taking the derivative of Equation (7-1) with respect to time gives

$$\frac{d\vec{p}}{dt} = \frac{d}{dt}(m\vec{v}) \tag{7-3a}$$

This equation looks similar to Newton's second law, but its form is different from the one we saw in Chapter 5. For a constant mass, however, Equation (7-3) does reduce to the familiar form of Newton's second law. When the mass is constant, it can be taken out of the differentiation and

$$\frac{d\vec{p}}{dt} = m\frac{d}{dt}\vec{v} = m\vec{a} \tag{7-3b}$$

Equation (7-3a) is the form of Newton's second law that we must use when mass varies. An example in which a mass can vary is the squid: a squid ejects water as a means of propulsion, so the total mass of the squid decreases as it releases the water. Similarly, jets and rockets lose fuel mass as they propel themselves.

We therefore use a form of Newton's law that can address the acceleration of an object that has changing mass:

$$\vec{F}_{net} = \frac{d\vec{p}}{dt} \tag{7-4a}$$

Here, the net force acting on an object is defined as the rate of change of the momentum of the object. The greater the change in momentum, the greater the applied force. Equation (7-4a) also lets us define the average net force acting on an object in terms of the total change in its momentum:

$$\vec{F}_{net}^{\,avg} = \frac{\Delta\vec{p}}{\Delta t} \tag{7-4b}$$

Equation (7-4a) gives us the instantaneous force acting on an object, and Equation (7-4b) gives us the average net force.

MAKING CONNECTIONS

Cushioning the Blow: Airbags and Seat Belts

A critical factor for passengers in a car collision is the time over which their momentum changes, which determines the net force experienced as the car slows down. The dashboard is often padded with a firm rubber foam, which deforms when a passenger hits it, reducing the rate of change of the passenger's momentum. Seat belts are also used to restrain the passengers and prevent them from colliding with the harder components on the car's interior. An essential element of a seat belt is the elasticity, or deformability, which helps increase the time over which the momentum of the passenger changes during a collision. Emergency tension-limiting devices are also used to give the seat belt a controlled amount of "slack" during the collision to help reduce injuries. Sensors in the car seat measure the weight of the passenger and adjust the rate at which the built-in air bag inflates and then deflates to optimize the time over which the passenger's momentum changes, again reducing the net force experienced by the passenger.

EXAMPLE 7-3

Forces in a Car Crash

A 1700 kg car moving at a speed of 90.0 km/h runs into a tree, which stops the car in 100 ms.

(a) Find the average force of impact on a 75.0 kg passenger.
(b) Calculate the acceleration experienced as a result of this force, express in multiples of g.
(c) At what rate should an airbag deflate to keep the force experienced by the passenger below $17g$ (commonly regarded as fatal)?

SOLUTION

We choose the x-axis to be along the direction of motion.
(a) The average force of impact:

$$\vec{F}_{net}^{avg} = \frac{\Delta \vec{p}}{\Delta t} = 0 - \frac{(75 \text{ kg}) (90) \left(\frac{1000}{3600}\right) \hat{i} \text{ m/s}}{0.1 \text{ s}} = -18750\hat{i} \text{ N}$$

(b) The average acceleration that the car undergoes is given by Newton's second law, in magnitude:

$$a_{avg} = \frac{F_{avg}}{m} = \frac{18750}{75} = 250 \text{ m/s}^2 = \frac{250}{9.81} g = 25.5 \, g$$

(c) Working backward, we need a maximum g value of 17:

$$a_{avg} = 17 \times 9.81 \text{ m/s}^2 = 166.77 \text{ m/s}^2$$

The average force experienced by the passenger is then

$$F_{avg} = ma_{avg} = 75 \text{ kg} (166.77 \text{ m/s}) = 12508 \text{ N}$$

Therefore, the time the airbag takes to deflate must be at least

$$\Delta t = \frac{\Delta p}{F_{net}^{avg}} = \frac{(75 \text{ kg}) (90 \text{ km/h}) \left(\frac{1000 \text{ m}}{3600 \text{ s}}\right)}{12508 \text{ N}} = 0.150 \text{ s} = 150 \text{ ms}$$

Making sense of the result:

In an actual collision with a modern car, deformation of the car body would make the duration of the collision longer than 100 ms, thus reducing the force of the impact. Airbags often deploy within 60 ms, and then deflate within approximately 120 ms following full inflation. The average force during a collision is not the maximum force (as we will see below); the time over which the actual maximum force is in effect is very small. Therefore, the average force during the collision is taken to be representative of the force during a collision.

CHECKPOINT

C-7-4 Bouncing Ball

When a ball hits a wall horizontally and bounces back with the same speed,
(a) the average net force on the ball is zero;
(b) the momentum of the ball does not change;
(c) the force on the ball is parallel to the ball's incident momentum.
(d) None of the above.

CHECKPOINT

C-7-5 The Force on a Satellite

A satellite of mass m orbits Earth at speed v in an orbital period of T. What is the magnitude of the average force on the satellite during half an orbit?
(a) zero
(b) mv/T
(c) $2mv/T$
(d) $4mv/T$

C-7-4 (d) momentum does change, force is not zero and is parallel to final momentum.

C-7-5 (c) change in momentum is 2 mv.

MECHANICS

LO 3

7-3 Impulse and the Conservation of Linear Momentum

In comic books and movies, Superman, Spiderman, and the Flash sweep people off their feet and pick them up while in mid-fall to save them from peril. What is often missed is the peril that lies in the details of the rescue: It is not uncommon to see a victim falling from the top of a skyscraper, only to have the superhero swoop up from below to reverse the direction of motion almost instantaneously. In real life, changing the speed or the momentum at this rate might well be fatal. Recent movies show the action more realistically, with the superhero gently bringing a falling person or vehicle to a stop before reversing the direction of motion.

In this section, we examine in more detail the behaviour of the force on an object during an impact or collision. We begin by rewriting Equation (7-3) as

$$d\vec{p} = \vec{F}_{net}\, dt \qquad (7\text{-}5)$$

According to Equation (7-5), the infinitesimal change in the momentum of an object equals the product of the net force acting on that object and the infinitesimal amount of time over which the force acts on the object. Integrating Equation (7-5) gives the change in momentum, which is called the **impulse, I**:

$$\vec{I} = \Delta\vec{p} = \int_{t_1}^{t_2} \vec{F}_{net}\, dt \qquad (7\text{-}6a)$$

We can also say that

$$\vec{I} = \Delta\vec{p} = \vec{F}_{net}^{\,avg}\, \Delta t \qquad (7\text{-}6b)$$

where $\vec{F}_{net}^{\,avg}$ is the average net force acting on the object during the time Δt.

When the net force on an object is zero, the impulse is also zero. The fact that an object's linear momentum

✓ CHECKPOINT

C-7-6 Direction of Impulse

A 17 g marble moving at a constant speed on a horizontal surface strikes a peg and deflects $37°$ below its original direction of motion without losing any of its kinetic energy. If the marble was originally moving to the right, the impulse that it receives

(a) has a magnitude of zero;
(b) is nonzero and points downward to the right;
(c) is nonzero and points straight downward;
(d) is nonzero and points downward and the left.

C-7-6 (d) this can be seen by drawing the momentum diagram.

EXAMPLE 7-4

Impulse of a Falling Object

Find the impulse experienced by a 200 g stone when it falls freely for 10.0 s near Earth's surface.

SOLUTION

The force acting on the stone has a constant magnitude, $F = mg$. Taking down as the positive direction and applying Equation (7-6b), we get

$$\Delta\vec{p} = \vec{F}_{net}\, \Delta t \Rightarrow p_f - p_i = mg\, \Delta t$$

Since the stone falls from rest, the initial momentum is zero and

$$\Delta p = mg\, \Delta t = (200 \times 10^{-3}\ \text{kg})\,(9.81\ \text{m/s}^2)\,(10\ \text{s})$$

$$\Delta p = 19.6\ \text{kg m/s}$$

Making sense of the result:

The force of gravity changes the momentum of the stone.

remains constant in the absence of a net force acting on the object is in keeping with Newton's first law.

The Force of Impact

What takes place when a tennis ball bounces off a wall? For simplicity, we will consider a ball travelling horizontally when it hits a vertical wall. We also assume that the collision is elastic, (neither the wall nor the tennis ball is irrecoverably deformed). Before the impact, the wall does not exert a force on the ball. As the ball strikes the wall, the force of the wall on the ball increases sharply to a peak value. During this first phase of the impact, the ball deforms somewhat. We assume that the wall is hard enough to not deform at all. The second phase of the impact starts after the force reaches its peak. The ball begins to reverse direction as it recovers from the deformation sustained during the collision. If the collision is completely elastic, the recovery of the ball will be complete. This profile for the behaviour of the magnitude of the force with time is typical of elastic collisions.

The area under the force versus time curve corresponds to the integral of the force over the time during the collision. According to Equation (7-6a), the integral is the total change in the momentum of the ball.

Linear Approximation for the Force of Impact

Since the variation of force with time during a collision is complex, integration of real-time data like those shown in Figure 7-5(a) is not easy, so the graph in Figure 7-5(a) is of limited practical use. However, we can often use

(a)

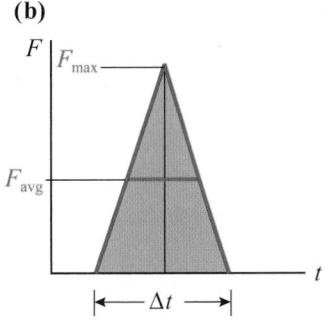

(b)

Figure 7-5 (a) Actual force versus time graph for a collision (b) Linear model for force versus time

the relatively simple linear model to calculate reasonable approximations.

Criteria The two parameters for our linear model are the duration of the impact and the total change in momentum. Our simplified model should produce a reasonable estimate of the average force of the impact. This model is based on a triangular graph, as shown in Figure 7-5(b). Using the properties of similar triangles, we can show that in this model, the maximum force is twice as great as the average force. Example 7-5 will give us some sense of the accuracy of this approximation.

EXAMPLE 7-5

Linear Model of Impact Force

A golf ball with a mass of 46.0 g strikes a wall horizontally at a speed of 130 km/h, and bounces straight back from the wall with the same speed. The collision takes 0.900 ms.

(a) Find the impulse delivered to the ball by the wall.
(b) Find the average force experienced by the ball during the collision.
(c) Assuming that the force changes linearly with time, plot a force versus time graph.
(d) Estimate the maximum value of the force of impact.

SOLUTION

The situation is depicted in Figure 7-6(a). The initial momentum of the golf ball points to the right; after the collision, the golf ball moves to the left.

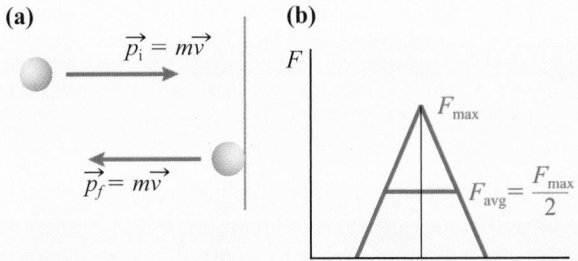

Figure 7-6 (a) The golf ball before and after the collision (b) The linear model for the behaviour of force versus time

(a) For convenience, we take left as our positive direction and convert the speed to metres per second. Then,

$$\vec{p}_f = m\vec{v} = -\vec{p}_i, \text{ then} \tag{1}$$

$$\Delta p = p_f - (-p_i) = 2mv$$

$$= 2(46 \times 10^{-3}) \text{ kg } (130) \left(\frac{1000 \text{ m}}{3600 \text{ s}} \right) = 3.32 \text{ kg m/s} \tag{2}$$

where the variables used for momentum in Equation (2) denote magnitudes

(b) From Equation (7-6b), we have

$$\Delta p = F_{net}^{avg} \Delta t$$

$$\Rightarrow F_{net}^{avg} = \frac{\Delta p}{\Delta t} = \frac{3.32 \text{ kg m/s}}{0.9 \times 10^{-3} \text{ s}} = 3.69 \text{ kN} \tag{3}$$

(c) Figure 7-6(b) shows the force as a function of time, taken to be linear.
(d) The average force is half the maximum value for the linear approximation in Figure 7-6(b); therefore, the maximum value of the force of impact is 7.38 kN.

Making sense of the result:

The force calculated is tremendous—equivalent to the weight of a three-quarter tonne truck. However, this force exists for only a very short time.

Figure 7-7 High-speed footage of a golf ball colliding with a wall. The frame-by-frame images (taken at small intervals) reveal details that human vision cannot otherwise detect.

The degree to which an object can deform during a perfectly elastic collision can be quite spectacular. The sequence of images shown in Figure 7-7 shows a golf ball during various stages of impact with a metal plate.

✓ CHECKPOINT

C-7-7 Collision Duration

Figure 7-8 shows the normal force versus time for a 90.0 g sponge ball as it hits the ground at a speed of 7.00 m/s. Find the speed with which the ball leaves the ground. The peak force is 1150 N, and the duration of the collision is 2.00 ms.

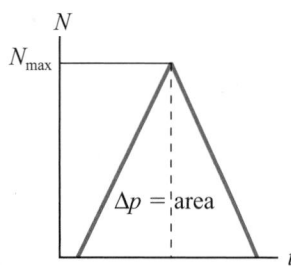

Figure 7-8 Simplified model showing the linear behaviour of force versus time.

C-7-7 5.77 m/s (force of gravity acts during collision)

LO 4

7-4 Systems of Particles and Centre of Mass

From Newton's third law, we know that when a golf ball strikes a wall, the force exerted by the wall on the golf ball is equal in magnitude and opposite in direction to the force exerted by the golf ball on the wall. If the wall were free-standing on a frictionless surface, then the wall would accelerate away from the ball during the collision. When the initial velocity of the ball is normal to the wall, the change in momentum and speed for the wall are as follows:

$$\Delta \vec{p} = m\Delta \vec{v}$$

$$\Delta \vec{v} = \frac{\vec{N}\Delta t}{m} \tag{7-7}$$

Since the mass of the wall is much greater than the mass of the golf ball, the recoil speed of the wall would be much less than the recoil speed of the ball. However, an actual wall is most likely firmly attached to the ground, so the change in momentum will result in an immeasurably small acceleration or deformation of the wall, its foundations, and the ground around it.

✓ CHECKPOINT

C-7-8 Earth-Moving Marble

When you drop a marble from the top of a tall building, and the marble lands on sand without bouncing,
(a) after the marble lands, Earth will gain some speed;
(b) whatever speed Earth gains as the marble falls, it will lose when the marble hits the ground;
(c) Earth does not gain speed as the marble speeds up;
(d) Earth's position does not change at all.

C-7-8 (b) conservation of momentum

All the objects considered so far have been involved in a collision with other objects so massive that the effect on the larger object is negligible. For example, when a 20 kg dog moves on Earth's surface, the force that causes him to move also acts on Earth's surface, pushing Earth's surface back. We can see the dog move, but the corresponding motion of Earth is far too small for us to detect. What about the motion of objects compared with objects much less massive than Earth? For example, when the same dog walks along the length of a 30 kg canoe, however, the canoe's motion is visible despite the friction from the water around it. Also, when two cars of comparable mass collide, the changes in momentum and velocity for the cars will often be obvious.

Systems of Particles: From Point Masses to Rigid Bodies

We now examine the principles of mechanics as they apply to systems of particles, or objects. So far, we have treated all objects as point masses, or particles. A **point mass** is an object with a nonzero mass but with very small dimensions, or, ideally, zero size. Although this approach was not realistic, it provided a simple platform for the application of physical principles. Now that we have mastered the basics of Newton's mechanics and the work–mechanical energy approach, we are in a good position to move on to a more realistic description of physical objects. Here, we will consider a system of particles, or discrete distributions of mass. Systems of particles are in effect a step in the transition from dealing with point masses to rigid objects. In the next chapter, we will deal with continuous distributions of mass in various rigid objects.

Centre of Mass The first tool in our study of systems of particles is the concept of the **centre of mass**, the point corresponding to the mean position of the mass in a body or a system. For many calculations, the mass of a system or an object can be treated as a point mass, located at the centre of mass. Interactions between the particles or objects in a system do not change the total momentum of the system.

In one dimension, the position of the centre of mass of a system of n particles is given by

$$M_T \vec{x}_{cm} = m_1 \vec{x}_1 + m_2 \vec{x}_2 + \cdots + m_n \vec{x}_n \quad (7\text{-}8a)$$

where m_i is the mass of each particle, $\vec{x}_i$ is the position vector of each particle, and M_T is the total mass of all the particles in the system: $M_T = m_1 + m_2 + \cdots + m_n$

Therefore, the location of the centre of mass of the system is

$$\vec{x}_{cm} = \frac{m_1 \vec{x}_1 + m_2 \vec{x}_2 + \cdots + m_n \vec{x}_n}{M_T} \quad (7\text{-}8b)$$

Similarly, for systems in more than one dimension,

$$M_T \vec{r}_{cm} = m_1 \vec{r}_1 + m_2 \vec{r}_2 + \cdots + m_n \vec{r}_n \quad (7\text{-}9a)$$

and

$$\vec{r}_{cm} = \left(\frac{\sum_{i=1}^{n} m_i \vec{r}_i}{\sum_{i=1}^{n} m_i} \right) \quad (7\text{-}9b)$$

In three dimensions, the $\vec{r}$ vector has independent components in the x-, y-, and z-directions. Thus, we can have equations with the y- and z-components of $\vec{r}_{cm}$ with the same form as Equation (7-8).

The expression for the centre of mass of a *continuous* distribution of mass is

$$\vec{r}_{cm} = \frac{\int_V \vec{r}\, dm}{\int_V dm} \quad (7\text{-}9c)$$

where the integration limit V indicates that the integral is taken over the volume of an object or a continuous distribution of particles.

Equations (7-8) and (7-9) are all vector equations, so it is important to define a frame of reference and choose an appropriate coordinate system when dealing with centre of mass, as the next two examples demonstrate.

EXAMPLE 7-6

Centre of Mass of Two Particles

Find the centre of mass of the system of two particles of masses $m_1 = 5m$ and $m_2 = m$ located a distance L apart as shown in Figure 7-9(a).

(a)

Figure 7-9(a) Two masses a distance L apart

SOLUTION

Our first step is to choose a coordinate system. For convenience, we choose the line connecting the two masses as the x-axis. We use the heavier of the two masses, m_1, as our origin (see Figure 7-9(b)) and choose right as the positive direction.

Using Equation 7-8(b) we have

$$\vec{x}_{cm} = \frac{m_1 \vec{x}_1 + m_2 \vec{x}_2}{m_1 + m_2} = \frac{5m(0)\hat{i} + m(L)\hat{i}}{5m + m} = \frac{L\hat{i}}{6}$$

We will now solve this same problem using the lighter of the two masses, m_2, as our origin (Figure 7-9(c)), but still taking right as the positive direction. Then,

(continued)

(b)

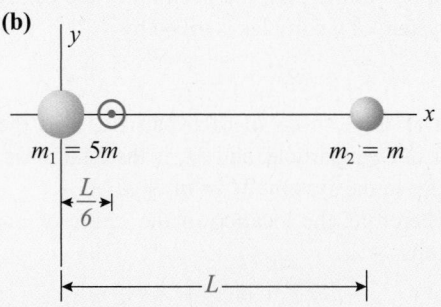

Figure 7-9(b) Mass m_1 is the origin of the coordinate system. The centre of mass is $L/6$ to the right.

$$x_{cm} = \frac{5m(-L)\hat{i} + m(0)\hat{i}}{5m + m} = -\frac{5L\hat{i}}{6}$$

Here, the negative value for x_{cm} tells us that the centre of mass is to the left of the origin.

(c)

Figure 7-9(c) Mass m_1 is the origin of the coordinate system.

Making sense of the result:

As we would expect, the centre of mass is between the two objects and closer to the more massive object. Whether we place our origin at the heavier or the lighter mass, the centre of mass will be a distance of $L/6$ from the heavier mass and $5L/6$ from the lighter mass.

 MAKING CONNECTIONS

What Is Orbiting What?

Earth does not really orbit the Sun, nor does the Moon orbit Earth. In a system consisting of just Earth and the Sun, both objects orbit the centre of mass of the system. The gravitational force that holds the Earth–Sun system together is internal to the system. In the absence of external forces, the centre of mass remains stationary. The centre of mass is much closer to the centre of the Sun than to the centre of Earth.

 EXAMPLE 7-7

Three Equally Spaced Masses

Find the centre of mass of three particles of masses m_1, m_2, and m_3 located respectively at the left, right and top vertices of an equilateral triangle as shown in Figure 7-10(a). The length of each side of the triangle is L. All three particles have the same mass, m.

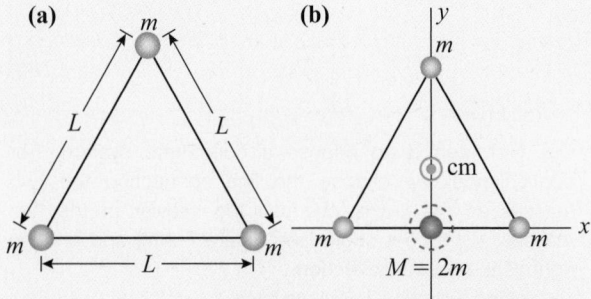

Figure 7-10 (a) Three equal masses at the corners of an equilateral triangle (b) The centre of mass of the system of three equal masses

SOLUTION

First, we find the centre of mass of the two masses at the base of the triangle. From the symmetry of the triangle, we can see this centre of mass is located in the middle of the base. To find the centre of mass of the whole system, we can replace masses m_1 and m_2 with a combined mass $M = 2m$ at the middle of the base of the triangle. We pick this location as the origin of the coordinate system (Figure 7-10(b)).

Now, we have only two objects to deal with: mass $M = 2m$, located at the origin, and mass m_3, located at the top of the triangle. From Figure 7-9(b) we can see that the distance between m_3 and the origin is

$$y = L \cos(30°) = L\frac{\sqrt{3}}{2}$$

Applying Equation (7-1) in the y-direction, we get

$$y_{cm} = \frac{2m(0) + mL\dfrac{\sqrt{3}}{2}}{3m} = \frac{\sqrt{3}}{6}L$$

Thus, the coordinates of the centre of mass of the system of three masses are $\left(0, \dfrac{\sqrt{3}}{6}L\right)$.

Making sense of the result:

It makes sense for the centre of mass to be closer to the base than the apex.

In Example 7-7, our first step was to use symmetry to locate the centre of mass of the two masses at the base, m_1, and m_2. Our second step was to choose that location as our origin of coordinates for the problem. Alternatively, our first step could have been to pick the location of our origin, perhaps to coincide with one of the masses, and then calculate the location of the centre of mass of the system.

LO 5

7-5 Newton's Laws Applied to Systems of Particles

We begin this discussion by using momentum to define the velocity, $\vec{v}_{cm}$, of the centre of mass of a system of particles. Taking the derivative of Equation (7-8a) with respect to time gives

$$M_T \vec{v}_{cm} = m_1 \vec{v}_1 + m_2 \vec{v}_2 + \cdots + m_n \vec{v}_n \qquad (7\text{-}10a)$$

or

$$\vec{p}_{cm} = \vec{p}_1 + \vec{p}_2 + \cdots + \vec{p}_n \qquad (7\text{-}10b)$$

From Equation (7-10a), we see that the velocity of the centre of mass of a system is really the weighted average of the velocities of the particles making up the system. Taking the derivative of Equation (7-10b) with respect to time, we get

$$\frac{d\vec{p}_{cm}}{dt} = \frac{d\vec{p}_1}{dt} + \frac{d\vec{p}_2}{dt} + \cdots + \frac{d\vec{p}_n}{dt} \qquad (7\text{-}11a)$$

For a system with constant mass, this is equivalent to taking the derivative of Equation (7-10a):

$$M_T \frac{d^2 \vec{r}_{cm}}{dt^2} = m_1 \frac{d^2 \vec{r}_1}{dt^2} + m_2 \frac{d^2 \vec{r}_2}{dt^2} + \cdots + m_n \frac{d^2 \vec{r}_n}{dt^2} \quad (7\text{-}11b)$$

In both Equations (7-11a) and (7-11b), the right-hand side is the sum of all the individual net forces acting on the individual particles of the system. This sum is the net force acting on the system:

$$\frac{d\vec{p}_{cm}}{dt} = \vec{F}_{net} \qquad (7\text{-}12a)$$

The significance of Equation (7-12a) is that, in the absence of a net force on a system of particles, the acceleration of the centre of mass of these particles is zero.

Internal Forces and Systems of Particles

For a system of particles, the internal forces acting between particles do not contribute to the net force acting on the system as presented in Equations (7-12a). **Internal forces** between any pair of particles are action–reaction pairs. Action–reaction forces are, by definition, equal in magnitude and opposite in direction, so when

we add all the forces acting on the individual particles in a system, the total is zero. Therefore, internal forces do not affect the acceleration of the centre of mass of a system, and Equation (7-12a) can be written unambiguously as

$$\frac{d\vec{p}_{cm}}{dt} = \vec{F}_{ext} \qquad (7\text{-}12b)$$

The implication of Equation (7-12b) is that where the **external force** is zero, the momentum of the system of particles does not change:

$$if \ \vec{F}_{ext} = 0, \ then \ \frac{d\vec{p}_{cm}}{dt} = 0 \Rightarrow \vec{p}_i = \vec{p}_f \qquad (7\text{-}12c)$$

Later in the chapter, we will apply this concept to collisions.

System Identification

It is important to clearly define the system we are working with, and to identify which of the forces acting on the system are external and which are internal.

During a collision between two cars, the force of the collision changes the momentum of each car. Therefore, the momentum of each car is not conserved during the collision, although the *total* momentum of the two cars is conserved. The forces exerted by the cars on each other are internal forces, which have no effect on the total momentum of the system that is the two cars, here the two cars are the system.

For the linear momentum of a system to be conserved, there can be no unbalanced external forces acting on it. For problems involving conservation of momentum, we will therefore choose systems such that the forces involved are internal to the systems whenever possible. The total linear momentum of such a system is conserved.

For example, when a dog walks in a canoe, the forces between the dog and the canoe are internal to the dog–canoe system and do not affect the system's total momentum. Similarly, when an astronaut takes a walk in the International Space Station, both the astronaut and the space station move relative to the centre of mass of the astronaut–station system, but the momentum of the system does not change as a result of the force they exert on one another.

However, an external force acting on a system will change its total momentum, as indicated by Equation (7-12a). For example, the total momentum changes in the case when two cars collide on a surface where friction is not zero, and the momentum of the astronaut–space station system changes continuously as the space station (and the astronaut) is held in an orbit around Earth by the force of gravity.

 CHECKPOINT

C-7-10 Jumping off a Sled

You are standing on a dogsled, which sits on an icy surface just outside of Oslo, Norway, in early December. You jump off the back of the sled to tend to your dogs. The sled, not tethered to the dogs, moves forward. Ignoring the friction between the sled and the ice, the centre of mass of the system consisting of you and the sled

(a) remains still;

(b) moves forward;

(c) moves backward;

(d) may move either forward or backward depending on whether you or the sled is more massive.

 EXAMPLE 7-8

Stranded Astronaut

An astronaut's manoeuvring unit fails while he is inspecting the nose of a space shuttle in orbit. Another astronaut uses the orbiter's (shuttle's) robotic arm to move the stranded astronaut from the nose section to the airlock. Derive an expression for how far the orbiter moves as a result of this action.

SOLUTION

The force that the robotic arm exerts on the astronaut (and, therefore, the force that the astronaut exerts on the arm and the orbiter) is internal to the astronaut–orbiter system. According to Equation (7-9), the acceleration of the centre of mass of the astronaut–orbiter system is zero, because the external, or net, force acting on the system is zero. Since the centre of mass of the two objects is originally stationary (in their orbiting frame of reference), the centre of mass of the orbiter-astronaut system will remain stationary.

Let us draw a diagram of the orbiter and the astronaut before and after the astronaut moves (Figure 7-11).

We can use Equation (7-8b) to locate the centre of mass of the astronaut and the orbiter before and after the robotic arm moves the astronaut:

$$M_T \vec{x}_{cm} = m_a \vec{x}_a + m_o \vec{x}_o \tag{1}$$

$$M_T \vec{x}_{cm}' = m_a \vec{x}_a' + m_o \vec{x}_o' \tag{2}$$

Subtract Equation (1) from Equation (2):

$$M_T \vec{x}_{cm}' - M_T \vec{x}_{cm} = m_a \vec{x}_a' - m_a \vec{x}_a + m_o \vec{x}_o' - m_o \vec{x}_o$$

$$M_T(\vec{x}_{cm}' - \vec{x}_{cm}) = m_a(\vec{x}_a' - \vec{x}_a) + m_o(\vec{x}_o' - \vec{x}_o)$$

$$M_T \Delta \vec{x}_{cm} = m_a \Delta \vec{x}_a + m_o \Delta \vec{x}_o \tag{3}$$

We have already established that the centre of mass of the system does not move; therefore $\Delta \vec{x}_{cm} = 0$. We know that

initial orbiter centre of mass position

final orbiter centre of mass position

(a)

initial astronaut position

orbiter centre of mass

final astronaut position

(b)

orbiter centre of mass

reference line
(position of combined centre of mass)
this does not change as astronaut and orbiter move

Figure 7-11 The orbiter and the astronaut (a) before and (b) after the astronaut moves.

the orbiter will move by an amount $\Delta \vec{x}_o$ when the astronaut moves by an amount $\Delta \vec{x}_a$. We do not know what these displacements are yet, but they are related by Equation (3).

In principle, we have two unknowns, $\Delta \vec{x}_a$ and $\Delta \vec{x}_o$, and one equation. The relative displacement of the astronaut with respect to the orbiter is given by Equation (3-24):

$$\Delta \vec{x}_{ao} = \Delta \vec{x}_a - \Delta \vec{x}_o \tag{4}$$

The relative displacement of the astronaut with respect to the orbiter is equal to the distance from the tail to the airlock, L. We pick right, as viewed in our diagram, as the positive x-direction. The orbiter moves to the right, so its displacement is positive. The astronaut moves to the left, so his displacement is negative. Since the displacement of the astronaut with respect to the orbiter is to the left, $\Delta \vec{x}_{ao}$ is negative. Therefore,

$$\Delta \vec{x}_{ao} = -L\hat{i}$$

Equations (3) and (4) can be written as

$$M_T \Delta x_{cm} = 0 = -m_a \Delta x_a \hat{i} + m_o \Delta x_o \hat{i} \qquad (3a)$$

$$\Delta x_{ao} \hat{i} = -\Delta x_a \hat{i} - \Delta x_o \hat{i} \qquad (4a)$$

Where the Δx's in Equations (3a) and (4a) indicate magnitudes of the displacements.

From Equation (3a) we have

$$\Delta x_a = \frac{m_o}{m_a} \Delta x_o$$

Substituting into Equation (4a) gives

$$-\Delta x_{ao} = \frac{m_o}{m_a} \Delta x_o + \Delta x_o$$

$$L = \frac{m_o}{m_a} \Delta x_o + \Delta x_o$$

$$\Delta x_o = \frac{L}{m_o/m_a + 1}, \quad \text{and} \quad \Delta x_a = \frac{L}{m_a/m_o + 1}$$

Making sense of the result:

Since the mass of the orbiter is much greater than the mass of the astronaut, $m_0 \gg m_a$, and $\Delta x_o \ll \Delta x_a \approx L$.

 CHECKPOINT

C-7-11 Sun Orbits Earth

If we assume that the external forces acting on the Sun–Earth system are negligible, we can say that, as Earth orbits the Sun,

(a) the Sun remains stationary with respect to the centre of mass of the Sun–Earth system;

(b) the centre of mass of the Sun–Earth system moves back and forth;

(c) the Sun orbits the centre of mass of the Earth–Sun system;

(d) the motion of Earth does not affect the Sun's motion.

C-7-11 (c) both Earth and Sun orbit the centre of mass.

 PEER TO PEER

In Example 7-8, it is important to distinguish between the displacement of the astronaut, $\Delta \vec{x}_a$, and the displacement of the astronaut with respect to the orbiter, $\Delta \vec{x}_{ao} = \Delta \vec{x}_a - \Delta \vec{x}_o$. The former is what an external observer in the centre of mass frame of reference would see—the actual distance covered by the astronaut relative to that frame of reference. The latter is the distance covered by the astronaut relative to the orbiter. Since the orbiter also moves within the centre of mass frame of reference, the value for $\Delta \vec{x}_{ao}$ depends on both the motion of the orbiter and the motion of the astronaut. For example, if the astronaut moves by 3.0 m and the orbiter moves 0.1 m in the opposite direction, then the astronaut would have moved by 3.1 m with respect to the orbiter.

 ### EXAMPLE 7-9

Applying Conservation of Momentum

A bear cub ($m = 21.0$ kg) is sitting on a wagon ($M = 31.0$ kg), which is resting on a horizontal surface. A larger bear pushes the wagon with the bear cub on it until the wagon is moving at a speed of $v = 2.00$ m/s. The cub then jumps backward off the wagon with a speed of 3.00 m/s with respect to the wagon. Ignoring the effects of friction, find

(a) the speed of the cub with respect to the ground;

(b) the speed of the wagon with respect to the ground.

SOLUTION

The problem involves conservation of momentum, so we will consider the system consisting of the bear cub and

the wagon, just before and just after the cub jumps off, as shown in Figure 7-12.

Figure 7-12 The wagon and the bear cub (a) just before the cub jumps and (b) just after the cub jumps.

(continued)

(a) When the bear cub jumps off the wagon, the force that the cub exerts on the wagon and the force that the wagon exerts on the cub are internal to the cub–wagon system. Since we are ignoring the effects of friction and there are no other external forces acting on the system at this time, the linear momentum of the cub–wagon system is conserved as the bear cub jumps off the wagon:

$$\vec{p}_i = \vec{p}_f \tag{1}$$

Initially, the bear cub and the wagon are both moving with the same velocity $\vec{v}$, so

$$(m + M)\vec{v} = m\vec{v}_b + M\vec{v}_w \tag{2}$$

where $\vec{v}_b$ and $\vec{v}_w$ are, respectively, the velocities of the bear cub and the wagon after the cub jumps off. Both quantities are unknown, and so far we have one equation only, we need another.

The relative velocity of the bear cub with respect to the wagon after the bear cub jumps is (Equation (3-21)):

$$\vec{v}_{bw} = \vec{v}_b - \vec{v}_w \tag{3}$$

After the bear cub jumps off the wagon, the velocity of the bear cub with respect to the wagon is the difference between the velocity of the bear cub and the velocity of the wagon (both with respect to the ground). We have used the result for relative velocity from Equation (3-21).

Let us pick right as the positive x-direction. The wagon is originally moving to the right, so v is positive. The bear cub jumps to the left, which is expected to increase the speed of the wagon as it moves to the right. Therefore, v_w should be positive. When the bear cub is in the air, he is moving to the left with respect to the wagon, so v_{bw} will be negative. We will have to guess whether or not the bear cub is moving to the right or to the left with respect to the ground. We emphasize here that *the assumption we make will not make a difference.* If our choice is incorrect, we will end up with a negative value for the speed of the bear cub, but the magnitude will be correct. Let us assume that the bear cub is moving to the left with respect to the ground.

Applying our choice of positive direction and our assumption for the direction of v_b to Equation (2), we get

$$(m + M)v\hat{i} = -mv_b\hat{i} + Mv_w\hat{i} \tag{4}$$

And Equation (3) becomes

$$-v_{bw}\hat{i} = -v_b\hat{i} - v_w\hat{i} \text{ or}$$

$$v_w = -v_b + v_{bw} \tag{5}$$

Again, the variables used in Equations (4) and (5) denote magnitudes of the velocity vectors in Equations (2) and (3) respectively. Substituting Equation (5) into Equation (4):

$$(m + M)v = -mv_b + M(-v_b + v_{bw}) = -mv_b - Mv_b + Mv_{bw}$$

The speed of the cub is given by

$$v_b = \frac{(m + M)v - Mv_{bw}}{-(m + M)}$$

$$= \frac{(21 \text{ kg} + 31 \text{ kg})(2 \text{ m/s}) - 31 \text{ kg} (3 \text{ m/s})}{-(21 \text{ kg} + 31 \text{ kg})} = -0.212 \text{ m/s}$$

Since the answer is negative, our choice for the direction of the velocity of the bear cub with respect to ground was incorrect. The bear cub is then actually moving to the right relative to the ground. We will keep than in mind and use v_b as a negative value next.

Had we assumed that the bear cub was moving to the right with respect to the ground, we would have ended up with the same number, but with a negative sign telling us we picked the incorrect direction.

(b) To get the speed of the wagon, we use Equation (5):

$$v_w = -v_b + v_{bw}$$

$$v_w = -(-0.212 \text{ m/s}) + 3 \text{ m/s} = 3.21 \text{ m/s}$$

Note that we can also use unit vector notation to solve this problem. The signs in the vector equations would then indicate the direction relative to the direction chosen for the x-axis unit vector.

Making sense of the result:

As a quick check, the change in momentum for the bear should be equal to and opposite to the change in momentum of the wagon. In this case they both work out to ~37.5 kg m/s in magnitude. It is not always easy to predict the direction of motion of the bear cub relative to ground. We also note that a correct prediction is not necessary.

✓ CHECKPOINT

C-7-12 Leaping Bear

In Example 7-9,

(a) the system had more kinetic energy after the bear cub jumped;

(b) the system had more kinetic energy before the bear cub jumped;

(c) the system's kinetic energy does not change when the frictional forces are negligible;

(d) the bear cub exerts more force on the wagon than the wagon exerts on the bear cub.

C-7-12 (a) Bear does positive work.

LO 6

7-6 Inelastic Collisions

An **inelastic collision** is a collision during which the colliding objects lose kinetic energy. The energy is lost in the form of thermal energy, sound, and irreversible deformation of the objects. In a completely inelastic collision, the two colliding objects stick together. Although kinetic energy is lost during the collision, the momentum of the system is conserved. The less elastic a collision is, the greater the proportion of the kinetic energy lost.

C-7-13 Adding Momentum

Two cars of equal mass (1100 kg) and speed (36 km/h) collide head-on in a completely inelastic collision. What is the total momentum of the system of two cars after the collision?

(a) 11 000 kg·m/s

(b) 22 000 kg·m/s

(c) zero

(d) none of the above

C-7-13 (c) the vector sum of the momenta is zero.

EXAMPLE 7-10

Vehicles in a Collision

In a collision-testing facility, a car with mass $m_1 = 1150$ kg moving at speed of 10.0 m/s collides head-on with a truck of mass $m_2 = 2800$ kg moving at speed of 7.00 m/s. The collision leaves the two vehicles joined in a heap. Find

(a) the speed of the vehicles after the collision;

(b) the kinetic energy lost in the collision.

SOLUTION

(a) The forces that exist between the vehicles during the collision are internal to the two-vehicle system. Therefore, we can say that the total vector momentum of the system is conserved during the collision:

$$\vec{p}_i = \vec{p}_f \quad \text{or} \quad (1)$$

$$m_1\vec{v}_1 + m_2\vec{v}_2 = (m_1 + m_2)\vec{v}_f \quad (2)$$

The two vehicles were initially moving separately, and after the collision they moved together as one object of mass.

$$M = m_1 + m_2$$

Assume that the car is moving to the right, and choose right to be the positive x-direction. (Note that the choice of direction does not affect the physics of the problem.) Then, Equation (2) can be written as

$$m_1 v_1 \hat{i} - m_2 v_2 \hat{i} = (m_1 + m_2) v_f \hat{i} \quad (3)$$

where the velocity symbols without the vector signs are variables that indicate the magnitudes of the velocity vectors. We will assume that the two objects move to the right after the collision; so we write v_f as a positive quantity in Equation (3):

$$v_f = \frac{m_1 v_1 - m_2 v_2}{m_1 + m_1}$$

$$= \frac{1150 \text{ kg} (10 \text{ m/s}) - 2800 \text{ kg} (7 \text{ m/s})}{(1150 + 2800) \text{ kg}}$$

$$= -2.05 \text{ m/s} \quad (4)$$

(b) For the energy lost in the collision, we will compare the initial and final kinetic energies of the system. The sum of the kinetic energies of the two vehicles before the collision is the initial kinetic energy of the system. The final kinetic energy of the system is the kinetic energy of the heap they form after the collision.

The energy lost is then

$$\frac{1}{2} m_1 v_1^2 + \frac{1}{2} m_2 v_2^2 - \frac{1}{2} (m_1 + m_2) v_f^2$$

$$= \frac{1}{2} 1150 \text{ kg} (10 \text{ m/s})^2 + \frac{1}{2} 2800 \text{ kg} (7 \text{ m/s})^2$$

$$- \frac{1}{2} (1150 \text{ kg} + 2800 \text{ kg})(2.05 \text{ m/s})^2$$

$$= 118\ 000 \text{ J}$$

Making sense of the results:

The minus sign for the speed of the two vehicles together in Equation (4) indicates that our assumption that the heap is moving to the right is incorrect; the heap is moving to the left. The energy is lost to thermal energy, deformation, and sound.

MAKING CONNECTIONS

Cosmic Collisions

Stars readily form in the arms of a spiral galaxy because these regions have relatively high densities and strong gravitational forces (Chapter 11). Spiral galaxies are subclassified by their structure. As shown in Figure 7-13, the Whirlpool galaxy (catalogued as NGC 5194 and as M51) has only two prominent spiral arms, which makes it a Grand Design spiral galaxy, in contrast to spiral galaxies that have many arms and often a less-defined spiral structure.

(continued)

The clear spiral structure of the Whirlpool galaxy was likely enhanced by an interaction with the smaller nearby galaxy, NGC 5195. The two galaxies may have undergone two collisions, one a few hundred million years ago and the other about 100 million years ago, as NGC 5195 crossed the disk of the Whirlpool galaxy. The interaction between the two galaxies created regions of even higher gravitational attraction, promoting spectacular star formation near the lagging edge of the arms of the spirals of the Whirlpool galaxy. The Whirlpool galaxy is about 31 million light years away, in the constellation Canis Venatici (the Hunting Dogs). Its orientation with respect to Earth and its relative closeness have given astronomers a great opportunity to study the birth of stars within spiral galaxies.

Figure 7-13 One of the sharper images of NGC 5194 and 5195 taken in January 2005 with the Advanced Camera for Surveys (ACS) aboard the Hubble Space Telescope. The pink colour coding represents intense star-forming regions.

EXAMPLE 7-11

Inelastic Collision

Children are playing with a small ball of steel of mass m_b suspended on a string such that the ball hangs just above a slippery plastic mat. The children slide a magnet of mass m_M along the mat to strike the steel ball. When the magnet strikes the steel ball at speed v_M, the magnet sticks to the ball, causing it to swing in an arc from its rest position. Assume that all friction in the system is negligible.

(a) Derive an expression for the height, h, that the ball and the magnet reach after the collision.
(b) Calculate the height using the following values: $m_M = 92.0$ g, $m_b = 190$ g, and $v_M = 7.00$ m/s.

SOLUTION

(a) We would be tempted to say that the kinetic energy of the incoming magnet is completely converted to the potential energy of interaction between the rising ball and magnet and Earth. However, since the magnet sticks to the steel ball, the collision is by definition completely inelastic. Kinetic energy is lost during an inelastic collision. Momentum, however, is conserved.

As Far as the collision is concerned, the system of concern to us is the magnet and the steel ball. The forces that occur as a result of the collision are internal to the system, so the total linear momentum of the system is conserved during the collision.

During the collision, the magnet and the steel ball effectively become one object of mass $M = m_M + m_b$ moving with velocity $\vec{v}$. Therefore,

$$\vec{p}_i = \vec{p}_f \tag{1}$$

$$m_M \vec{v}_M + m_b \vec{v}_b = (m_M + m_b)\vec{v} \tag{2}$$

choosing the direction of motion as the positive x-axis,

$$m_M v_M \hat{i} = (m_M + m_b)v\hat{i} \tag{3}$$

After the collision, the only force doing work on the ball-magnet system is the force of gravity. Since no nonconservative forces are doing work on the system, the total mechanical energy is conserved:

$$\Delta E_m = \Delta U + \Delta K = 0 \tag{4}$$

The kinetic energy of the system is converted into gravitational potential energy as the ball with the magnet attached swings upward:

$$\Delta U = -\Delta K \tag{5}$$

$$(m_M + m_b)gh = -(K_f - K_i) = -\left(0 - \frac{1}{2}(m_M + m_b)\right)v^2$$

$$= \frac{1}{2}(m_M + m_b)v^2 \tag{6}$$

$$h = \frac{v^2}{2g} \tag{7}$$

(b) Substituting the given values into Equation (3) gives

$$v = \frac{m_M v_M}{(m_M + m_b)} = \frac{(0.092 \text{ kg})(7 \text{ m/s})}{(0.092 + 0.190) \text{ kg}} = 2.284 \text{ m/s} \tag{8}$$

and

$$h = \frac{v^2}{2g} = \frac{(2.284 \text{ m/s})^2}{2(9.81 \text{ m/s}^2)} = 0.266 \text{ m} \tag{9}$$

Making sense of the result:

The mechanical energy is not conserved during the collision. Had energy been conserved, then

$$h = \frac{1}{2}m_M(v_M)^2/(m_M + m_m)g = 0.815 \text{ m}$$

EXAMPLE 7-12

Shock-Absorber Test

In one particular collision test, a 1450 kg car is driven at 48.0 km/h toward a stationary car of mass 1780 kg. A spring with spring constant 88.0 kN/m is used to simulate the effect of shock-absorbing features that could be added to the vehicles, and such a spring is attached to the stationary car. Assuming that friction with the ground can be ignored, find

(a) the maximum compression in the spring;
(b) the maximum force on either car during the collision;
(c) the maximum acceleration experienced by the lighter car during the collision.

SOLUTION

(a) As the incoming car strikes the spring attached to the stationary car, the spring begins to compress, thus exerting a force on each car. As a result, the incoming car slows down, and the stationary car accelerates until both cars reach the same speed, v. At the instant the two cars have the same speed the distance between the two cars does not change, and the spring reaches its maximum compression.

The system is comprised of the two cars and the spring. Since the forces during the collision are internal to this system, the momentum is conserved, and

$$\vec{p_i} = \vec{p_f}$$

where $\vec{p_i}$ is the momentum of the system before the collision, and $\vec{p_f}$ is the momentum at the instant when the spring is at maximum compression. A speed of 48 km/hr reads 13.3 m/s.

Then

$$m_1\vec{v_1} = (m_1 + m_2)\vec{v}$$

$$v = \frac{m_1 v_1}{(m_1 + m_2)} = \frac{(1450 \text{ kg})(13.3 \text{ m/s})}{1450 \text{ kg} + 1780 \text{ kg}} = 5.99 \text{ m/s}$$

Since the spring cushions the impact and the spring force is conservative, the total mechanical energy of the system is conserved. Some kinetic energy changes into elastic potential energy in the spring, but this energy is not lost from the system. Thus, when the cars are moving at the same speed, we can say

$$\frac{1}{2}m_1 v_1^2 = \frac{1}{2}kx^2 + \frac{1}{2}(m_1 + m_2)v^2$$

$$x = \sqrt{\frac{m_1 v_1^2 - (m_1 + m_2)v^2}{k}}$$

$$= \sqrt{\frac{(1450 \text{ kg})(13.3 \text{ m/s})^2 - (1450 \text{ kg} + 1780 \text{ kg})(5.99 \text{ m/s})^2}{88\,000 \text{ kg/s}^2}}$$

$$= 1.27 \text{ m}$$

(b) The maximum force exerted by one car on the other equals the force of the spring at maximum compression. The magnitude of the force can be obtained using Hooke's law:

$$F = kx = (88\,000 \text{ N/m})(1.27 \text{ m}) = 112 \text{ kN}$$

(c) The maximum acceleration experienced by the lighter car is

$$a = \frac{kx}{m} = \frac{(88\,000 \text{ N/m})(1.27 \text{ m})}{1450 \text{ kg}} = 77.1 \text{ m/s}^2$$

Making sense of the result:

The maximum acceleration experienced by a passenger would be about 8 g, which is high but not lethal in a head-on collision. Note that at the maximum compression both cars are moving, the spring will eventually release its potential energy and the two cars will go at different speeds.

✓ CHECKPOINT

C-7-14 Conserved Momentum

You throw a bouncy ball directly at a vertical wall (at 90° to the wall), and the ball bounces back at the same speed with which it struck the wall. Which of the following statements is the most correct?

(a) The momentum of the ball is conserved as it bounces back.
(b) As the ball moves toward the wall, the wall's position with respect to Earth's mantle remains constant.
(c) Earth's crust recoils slightly on the mantle as a result of the impact of the ball with the wall, which is rigidly attached to the crust
(d) The surface of Earth is too massive to be affected.

C-7-14 (c) the system's momentum is conserved, and none of the other statements are true.

LO 7

7-7 Elastic Collisions

During an **elastic collision**, both the total linear momentum and the total kinetic energy of the system are conserved. We demonstrate a strategy for analyzing elastic collisions by considering an elastic collision in one dimension.

Suppose a marble of mass m moving to the right with speed v collides elastically with a marble of mass M moving to the left at speed u (Figure 7-14(a)). Depending on the masses and the initial speeds, after the collision, the two marbles will bounce away from one another (Figure 7-14(b)); both move to the left, or both move to the right. Since linear momentum and velocity are vectors, we need to establish a coordinate system. We take right as positive.

Figure 7-14 (a) The two marbles head toward each other. (b) The two marbles bounce away from each other.

Conservation of Momentum

The total linear momentum of the system of two marbles is the vector sum of their individual momenta:

$$\vec{p_i} = m\vec{v} + M\vec{u}$$

After the collision, the expression of the final momentum of the system is

$$\vec{p_f} = m\vec{v'} + M\vec{u'}$$

where the primed variables indicate the final values.

The forces involved during the collision between the two marbles are internal to the two-marble system. Since there are no external forces involved, the total linear momentum of the system is conserved during the collision, and

$$\vec{p_i} = \vec{p_f} \quad or$$

$$m\vec{v} + M\vec{u} = m\vec{v'} + M\vec{u'} \qquad (7\text{-}13)$$

We could try to predict which of the three scenarios will occur after the collision, but the magnitude of the speeds we calculate after the collision will be the same no matter which assumption we make. Only the sign of the quantity will change depending on whether or not we choose the "correct" scenario. A negative value for a quantity indicates that the direction of the corresponding vector is opposite to what we had assumed.

We choose right to be the direction of the positive x-axis. The quantities in Equation (7-13) that point to the right will be positive, and the quantities that point to the left will be negative. Therefore,

$$mv\hat{i} - Mu\hat{i} = -mv'\hat{i} + Mu'\hat{i} \qquad (7\text{-}14)$$

The velocity symbols in Equation 7-14 denote the magnitudes of the velocity vectors.

In Equation (7-14), we see that we have two unknowns: the speeds of the two masses. Notice here that we say speeds and not velocities. The quantities that appear in Equation (7-13) are velocities. In principle, these speeds should be positive. However, if they turn out to be negative, it is because we made the "incorrect" assumption about the post-collision directions of motion. The signs depend on our assumptions, but the magnitudes of these quantities do not change.

To determine v' and u' from the masses and the initial velocities, we need a second equation relating v' and u'.

Conservation of Energy Since the collision is elastic, kinetic energy is conserved. Therefore, the kinetic energy of the system before the collision is equal to the kinetic energy after the collision:

$$\frac{1}{2}mv^2 + \frac{1}{2}Mu^2 = \frac{1}{2}mv'^2 + \frac{1}{2}Mu'^2 \qquad (7\text{-}15)$$

This equation involves the squares of the final speeds. Although we can solve for the final speeds, it would be simpler to deal with a system of two linear equations. Gathering the terms in Equation (7-15) by mass, we have

$$m(v^2 - v'^2) = M(u'^2 - u^2) \qquad (7\text{-}16)$$

Factoring the differences of squares, we get

$$m(v - v')(v + v') = M(u' - u)(u' + u) \qquad (7\text{-}17)$$

Now we regroup the terms in Equation (7-14) in the same way:

$$m(v + v') = M(u' + u) \qquad (7\text{-}18)$$

If we divide Equation (7-17) by Equation (7-18), we have our second linear equation:

$$v - v' = u' - u \qquad (7\text{-}19)$$

Between Equations (7-18) and (7-19) we now have two equations with our two unknowns. We multiply Equation (7-19) by m and add it to Equation (7-18). Expanding all the brackets, we get

$$mv + mv' + mv - mv' = Mu' + Mu + mu' - mu \qquad (7\text{-}20)$$

$$u' = \frac{2mv + (m - M)u}{M + m} \qquad (7\text{-}21)$$

Similarly, solving for v' gives

$$v' = \frac{2Mu + (M - m)v}{M + m} \qquad (7\text{-}22)$$

The above steps are useful for analyzing an elastic collision. Equations (7-16), (7-17), and (7-18) are universal and should look the same for every solution involving elastic collisions. The steps relating to momentum conservation, however, depend on the assumptions we make about the directions of the unknown quantities and our choice of a positive direction. Ultimately, the final results in Equations (7-21) and (7-22) also depend on our choices; these two equations are not universal formulas.

✓ CHECKPOINT

C-7-15 Something's Got to Give

In a collision between a tennis ball and a tennis racquet, is it possible for the tennis player to hold the racquet hard enough so that the racquet does not recoil at all relative to the player?

(a) No, the racket must have a recoil speed depending on its mass.

(b) Yes, it is possible for the racquet to stay in place. There is no need for recoil if the player holds the racquet with enough force in which case the player must recoil.

(c) If the player does not recoil then the ground she stands on must recoil.

(d) b or c can be correct.

C-7-15 (d) if momentum is to be conserved then something has to give.

ONLINE ACTIVITY

Collisions

The e-resource that accompanies every new copy of this textbook contains an Online Activity using the PhET simulation "Collision Lab". Work through the simulation and accompanying questions to gain an understanding of collisions.

✓ CHECKPOINT

C-7-16 What Went Wrong?

A student carries out the solution for an elastic collision between two billiard balls as outlined above, and makes assumptions about the unknown speeds of the two balls after the collision. The student obtains negative values for the speeds of both balls after the collision. Which of the following statements is correct?

(a) The student made a mistake when picking the negative direction.

(b) The student made an incorrect assumption about the direction of motion of one ball, but not both balls after the collision.

(c) The student made an incorrect assumption about the direction of motion of both balls after the collision.

(d) None of the above statements are correct.

C-7-16 (c) according to the approach we highlight above, a negative value for an unknown speed means the wrong direction was assumed for the velocity.

MAKING CONNECTIONS

Laser Cooling of Atoms

A direct application using the momentum of a photon is the laser cooling of atoms. The higher the temperature of a gas, the more kinetic energy its atoms will have and, hence, the higher their speed. Similarly, gas molecules move slower at lower temperatures. Since photons have momentum, a laser can be used to cool a gas by slowing down its atoms. The laser is fired at atoms in the gas. Incoming photons at the correct wavelength are absorbed by atomic electrons that are moving toward the laser. The momentum of the absorbed photon transfers to the atom, thus slowing the atom. The target atom cools further every time it absorbs a photon. The electron that absorbs the photon moves to a higher state of energy. If the photon is subsequently re-emitted by the excited atom, the momentum of the photon has a random direction. Energy is conserved in this process, and the collision is elastic. Carefully calibrated lasers (such as those in Figure 7-15) can cool atoms to a few tenths of a microkelvin and slow them down to the point where they are effectively "trapped". Researchers began cooling atoms with lasers in 1978 and reached temperatures below 40 K. They achieved temperatures a million times colder just 10 years later. This technology eventually led to better atomic clocks and the observation of an ultracold state of matter. The 1997 Nobel Prize in Physics was awarded to Steven Chu, Claude Cohen-Tannoudji, and William D. Phillips in 1997 for their work in developing technologies to use laser light to cool and trap atoms and the 2001 Nobel prize in physics was awarded to Eric A. Cornell, Wolfgang Ketterie and Carl E. Wieman for the achievement of Bose Einstein condensation.

The lowest temperaturs ever recorded to date is 1/2 of a nano Kelvin reached by an MIT research team lead by Ketterie, Leanhardt and Pritchard.

Figure 7-15 A cloud of cold sodium atoms (bright spot at centre) floats in a trap at the National Institute of Standards and Technology.

7-8 Variable Mass and Rocket Propulsion

Rocket propulsion provides an example of a system with variable mass. Here, the *system* consists of the rocket, its fuel, and the exhaust gas. In the rocket engine, the burning fuel turns into hot exhaust gases, primarily carbon dioxide and water vapour. When this exhaust is directed out the back of the rocket, the rocket accelerates forward. The more fuel that the rocket burns, the more mass it loses. The forces generated by the combustion of the fuel are internal to the system, so the linear momentum of the system is conserved as the fuel burns.

Consider a rocket that fires its engine in deep space. Assume that the engine burns fuel at a constant rate R, and that the exit (or effective) speed, v_{er}, of the exhaust gas relative to the rocket is also constant. Since the rocket is in deep space, we do not have to consider external forces, such as air drag, which would be significant in a launch from Earth. We also assume that gravitational forces are negligible.

At a given instant, the total mass of the rocket including the remaining fuel is M as shown in Figure 7-16(a). Figure 7-16(b) shows the rocket at a time dt later. The speed of the rocket has increased to $v + dv$, the mass of the exhaust released during the interval dt is dm, and the velocity of the exhaust gas is $\vec{v_e}$. The mass of the rocket plus fuel has decreased by dm to $M - dm$. The choice for the direction of $\vec{v_e}$ in Figure 7-16 (b) will be explained shortly. We note here that we assume the mass loss rate is a constant.

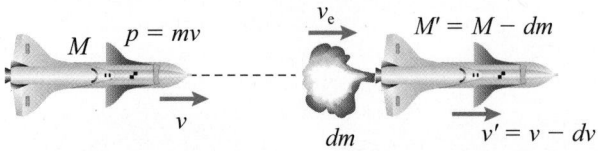

Figure 7-16 (a) A rocket moving at time t (b) The rocket moving at time $t + dt$

Since there are no external forces acting on the rocket–fuel–exhaust system and all the forces generated by burning the fuel are internal to the system, the total momentum of the system is conserved. Therefore, the momentum at time $t + dt$ at time t is

$$\vec{p}_{(t)} = \vec{p}_{(t+dt)} \qquad (7\text{-}23)$$

At time t, the momentum of the rocket and unburned fuel is Mv. After the interval dt, the mass of the rocket and the remaining fuel is $M' = M - dm$, and their velocity is $\vec{v'} = \vec{v} + d\vec{v}$. During the same interval, the mass of exhaust gas increases by dm. Using these expressions to find the momenta in Equation (7-23) gives

$$M\vec{v} = (M - dm)(\vec{v} + d\vec{v}) + dm(\vec{v_e}) \qquad (7\text{-}24)$$

Assuming that we know the initial mass and speed of the rocket system, the unknowns are the new rocket velocity, $\vec{v'} = \vec{v} + d\vec{v}$, and the velocity of the exhaust $\vec{v_e}$. The change in mass can be determined from the given rate of fuel consumption. However, to solve for the two unknown quantities we need another equation.

The equation can be found by considering the relative velocities of the components of the system. We know the exit speed of the exhaust relative to the rocket, $\vec{v}_{er}$. Applying Equation (3-21), we can readily see that

$$\vec{v}_{er} = \vec{v_e} - \vec{v'} = \vec{v_e} - (\vec{v} + d\vec{v}) \qquad (7\text{-}25)$$

Now, we choose the direction of the acceleration of the rocket (right) as the positive direction. The velocity of the exhaust with respect to the rocket, $\vec{v}_{er}$, is undoubtedly negative because it points to the left. However, the direction of the velocity of the exhaust, with respect to a stationary observer, depends on how fast the rocket is moving to the right and how fast the exhaust moves to the left with respect to the rocket.

Rockets in deep space reach speeds of up to 30 times the speed of sound near sea level. When the speed of the rocket is low, a stationary observer would expect to see the rocket move to the right and the exhaust gas move to the left. However, when the speed of the rocket is greater than the exit speed of the exhaust, both the rocket and the exhaust move to the right relative to a stationary observer. We assume that the exhaust is indeed moving to the right with respect to a stationary observer. We can consider the direction of motion of the rocket to align with the positive x-axis. Therefore, we can rewrite Equations (7-24) and (7-25) as

$$Mv\hat{i} = (M - dm)(v + dv)\hat{i} - dm(v_e)\hat{i} \qquad (7\text{-}26)$$

$$-v_{er}\hat{i} = -v_e\hat{i} - v'\hat{i} \quad \text{or} \quad -v_{er} = -v_e - (v + dv) \qquad (7\text{-}27)$$

$$v_e = v_{er} - (v + dv) \qquad (7\text{-}28)$$

Where the velocity symbols in the above three equations denote magnitudes.

Substituting the value of v_e from Equation (7-28) into (7-26) gives

$$Mv = (M - dm)(v + dv) - dm(v_{er} - v - dv) \qquad (7\text{-}29)$$

$$Mv = Mv + Mdv - vdm - dmdv - dmv_{er} + vdm + dmdv \qquad (7\text{-}30)$$

$$Mdv = dmv_{er} \qquad (7\text{-}31)$$

The decrease in the mass of the rocket and the remaining fuel, M, is the same as the increase in the total mass of exhaust, m. Thus

$$dM = -dm$$

Substituting the above into Equation (7-31), we get

$$Mdv = -dMv_{er} \qquad (7\text{-}32)$$

Equation (7-32) gives a relation between the incremental increase of the mass and the rate of change of the speed. Rearranging this equation gives

$$dv = -\frac{dM}{M} v_{er} \qquad (7\text{-}33)$$

To find the change in speed of the rocket after a given amount of fuel has been burned, we integrate Equation (7-33) with the initial and final mass as our integral limits:

$$\Delta v = \int_{v_i}^{v_f} dv = -v_{er} \int_{M_i}^{M_f} \frac{dM}{M} \qquad (7\text{-}34a)$$

Since

$$\Delta v = v_f - v_i$$

$$v_f - v_i = -v_{er} \ln\left(\frac{M_f}{M_i}\right) = v_{er} \ln\left(\frac{M_i}{M_f}\right) \qquad (7\text{-}34b)$$

Equation (7-34b) is called the **ideal rocket equation**. The change in speed is proportional to the logarithm of the ratio of the masses. Since the initial mass of the rocket with fuel is greater than the final mass, the final speed will be faster than the initial speed. Among other simplifications, the ideal rocket equation does not include the effects of gravity or air resistance.

The speed of the rocket depends on the mass of the remaining fuel, which is a function of time, since

$$\frac{dM}{dt} = R \qquad (7\text{-}35)$$

Thus, we can also drive an equation for the time dependence of the speed of rocket.

To find the instantaneous acceleration of the rocket, we take the derivative with respect to time of both sides of Equation (7-31):

$$M\frac{dv}{dt} = \frac{dm}{dt} v_{er} \qquad (7\text{-}36a)$$

$$Ma = Rv_{er} \qquad (7\text{-}36b)$$

The term Rv_{er} is called the thrust of the rocket engine. For a given mass of the rocket, the instantaneous acceleration is given by Equation (7-36b).

✅ **CHECKPOINT**

C-7-17 What Drives a Rocket?

When a rocket burns fuel for propulsion,

(a) the ejected exhaust is external to the rocket–fuel system, and the total momentum of the rocket and fuel decreases;
(b) the driving force is external to the system;
(c) the rocket is propelled by the speed of the exhaust, not the reduction in mass;
(d) The rocket accelerates as a result of the reduction in mass as the exhaust is ejected.

C-7-17 (d) according to 7-36 a, the reduction in mass as the exhaust is ejected provides the force that accelerates the rocket.

MAKING CONNECTIONS

Rockets and the Space Age

In the spring of 1241, the Mongolian empire defeated the kingdom of Hungary at the battle of Mohi. It is reported that Mongols used gun powder–propelled rockets during that battle, as well as during the rest of their 13th-century invasion of Europe. The Mongols adopted rocket technology from the Chinese, who had been reportedly using rockets for various purposes as early as the 14th century BCE. Chinese rocket experts were captured/hired by the Mongols to supply the rocket power for military conquest.

Before and during World War II, major advancements in rocket technology took place, mostly due to military demands. The advancements in rocket technology continued after World War II during the full-fledged rocket and space conquest race between the East and the West. Some landmark events that resulted include sending a human into space and landing on the Moon. Rocket technology made space exploration a reality and space colonization a possibility. Not free of tragedy, the story of the space age boasts many triumphs, not the least of which is the transformation of what started as an international quest for space supremacy into an endeavour that brings out the best of human achievement through collaboration. Collaborative efforts among the leading nations in the world continue to reach new heights as is evidenced by the International Space Station and other feats.

EXAMPLE 7-13

Applying the Ideal Rocket Equation

A space probe has a total mass of 1300 kg, 520 kg of which is fuel. The probe is near the edge of the solar system. The probe fires its rocket thrusters to make a final course correction, burning the fuel at a constant rate until it is gone.

The burn lasts 100 s and produces a constant thrust of 11.0 kN. Estimate

(a) the initial acceleration of the probe;
(b) the speed of the probe 80.0 s into the burn;
(c) the maximum speed of the probe;
(d) the maximum acceleration of the probe.

(continued)

SOLUTION

If we assume that all external forces acting on the probe are negligible, we can apply the ideal rocket equation. To do so, we need to find the rate of fuel consumption, R, and the effective exhaust speed, v_{er}. Since the rate of fuel consumption is constant and the rocket burns 520 kg of fuel in 100 s,

$$R = \frac{dm}{dt} = \frac{520 \text{ kg}}{100 \text{ s}} = 5.20 \text{ kg/s} \qquad (1)$$

We can determine the effective exhaust speed from the thrust, T:

$$T = Rv_{er} \Rightarrow 11\,000 \text{ N} = (5.20 \text{ kg/s})(v_{er})$$

$$v_{er} = \frac{11\,000}{5.20} = 2.115 \text{ km/s}$$

(a) Assume the probe is so far away from any celestial bodies such that the only significant force acting on the probe is the thrust from the rocket. Therefore,

$$F = Rv_{er} = ma$$

Since the initial mass of the rocket and fuel is 1300 kg, the initial acceleration is

$$a = \frac{F}{m} = \frac{Rv_{er}}{m} = \frac{11\,000 \text{ N}}{1300 \text{ kg}} = 8.46 \text{ m/s}^2$$

(b) The mass of the probe and the remaining fuel after 80.0 s is

$$M_f = M_i - Rt = 1300 \text{ kg} - 5.20 \text{ kgs}^{-1}(80.0) = 884 \text{ kg} \qquad (2)$$

Substituting this mass gives

$$v_f - v_i = v_{er} \ln\left(\frac{M_i}{M_f}\right) = 2115 \ln\left(\frac{1300}{884}\right) = 816 \text{ m/s}$$

(c) The maximum speed of the rocket occurs when the rocket has burned all its fuel, at which point the mass of the probe is 1300 kg − 520 kg = 780 kg. Then,

$$v_f - v_i = v_{er} \ln\left(\frac{M_i}{M_f}\right) = 2115 \ln\left(\frac{1300}{780}\right) = 1080 \text{ m/s}$$

(d) The probe's maximum acceleration occurs just as the rocket runs out of fuel. At this instant, the mass of the probe is the least it can be:

$$a = \frac{F}{m} = \frac{Rv_{er}}{m} = \frac{11\,000 \text{ N}}{780 \text{ kg}} = 14.1 \text{ m/s}^2$$

Making sense of the result:

The acceleration is about ~1.5 g, which is reasonable for rockets sent into space. As the mass drops the acceleration will increase, since the thrust is assumed to be constant.

MAKING CONNECTIONS

Flight of the Komet

Unlike its contemporary jet-powered counterparts, of which there were a few during World War II, the Messerschmitt ME 163 (the Komet, seen in Figure 7-17) was a full-fledged rocket plane, powered solely by rocket fuel. While not the first rocket plane to be built, it was the first to become operational during World War II. The Komet's rocket engines gave it a few minutes of powered flight after which it went into a glide. The performance was extraordinary to the point that plans to have a few of these as point defence interceptors against bombers in various locations were put in place, but never implemented. There were, however, some serious drawbacks, mostly due to the high speed of the plane relative to its bomber targets, which allowed the pilot a small window of opportunity to fire on Allied bombers. While almost impossible to hit during its fuel burst, the Komet reportedly made an easy target for fighters during its landing and gliding manoeuvres. The speeds reached by the Komet would only be surpassed in 1947.

Figure 7-17   An ME 163 rocket plane at an airfield in the 1940s

© Bettmann/CORBIS

Momentum

Momentum is a vector:

$$\vec{p} = m\vec{v} \tag{7-1}$$

The net force on an object results in a change in the object's momentum:

$$\vec{F}_{net} = \frac{d\vec{p}}{dt} \tag{7-4a}$$

or

$$\vec{F}_{net}^{\,avg} = \frac{\Delta\vec{p}}{\Delta t} \tag{7-4b}$$

Impulse is equal to the change in an object's momentum:

$$\vec{I} = \Delta\vec{p} = \int_{t_i}^{t_2} \vec{F}_{net}\, dt \tag{7-5b}$$

or

$$\vec{I} = \Delta\vec{p} = \vec{F}_{net}^{\,avg}\Delta t \tag{7-6b}$$

When the net force on an object is zero, the object's momentum does not change.

$$\text{if } \vec{F}_{net} = 0 \quad \text{then} \quad \Delta\vec{p} = 0 \Rightarrow \vec{p}_i = \vec{p}_f \tag{7-6c}$$

Systems of Particles

The centre of mass in one dimension is given by

$$M_T\vec{x}_{cm} = m_1\vec{x}_1 + m_2\vec{x}_2 + \cdots + m_n\vec{x}_n \tag{7-8a}$$

or

$$\vec{x}_{cm} = \frac{m_1\vec{x}_1 + m_2\vec{x}_2 + \cdots + m_n\vec{x}_n}{M_T} \tag{7-8b}$$

The centre of mass in more than one dimension is given by

$$\vec{r}_{cm} = \left(\frac{\sum_{i=1}^{n} m_i\vec{r}_i}{\sum_{i=1}^{n} m_i}\right) \tag{7-9b}$$

Continuous Distribution

$$\vec{r}_{cm} = \frac{\int_v \vec{r}\, dm}{\int_v dm} \tag{7-9c}$$

External Force and Centre of Mass

$$\frac{d\vec{p}_{cm}}{dt} = \vec{F}_{net} \tag{7-12a}$$

or

$$\frac{d\vec{p}_{cm}}{dt} = \vec{F}_{ext} \tag{7-12b}$$

Internal forces do not affect the momentum of a system of particles:

$$\text{if } \vec{F}_{ext} = 0, \quad \text{then} \quad \frac{d\vec{p}_{cm}}{dt} = 0 \Rightarrow \vec{p}_i = \vec{p}_f \tag{7-12c}$$

Rocket Propulsion

The ideal rocket equation is

$$v_f - v_i = -v_{er} \ln\left(\frac{M_f}{M_i}\right) = v_{er} \ln\left(\frac{M_i}{M_f}\right) \tag{7-34b}$$

The mass flow rate is

$$\frac{dM}{dt} = R \tag{7-35}$$

Rocket engine thrust is given by

$$Ma = Rv_{er} \tag{7-36b}$$

COLLISIONS

Inelastic Collisions

The momentum of a system comprised of colliding particles is conserved:

$$\vec{p}_i = \vec{p}_f$$

Kinetic energy is not conserved in an inelastic collision.

Elastic Collisions

The momentum of a system comprised of colliding particles is conserved:

$$\vec{p}_i = \vec{p}_f$$

Kinetic energy is conserved in an elastic collision:

$$K_i = K_f$$

Applications: rocket propulsion, collisions

Key Terms: centre of mass, conservation of momentum, elastic collision, external forces, ideal rocket equation, inelastic collision, impulse, internal forces, linear momentum, point mass, rocket propulsion

QUESTIONS

For all questions and problems involving springs, assume that the masses of the springs are negligible unless otherwise stated.

1. To double the momentum of a particle, you could
   (a) double the particle's mass;
   (b) double the particle's speed;
   (c) quadruple the particle's kinetic energy.
   (d) All of the above.

2. A bullet is stopped by a wall. Which of the following statements is true?
   (a) The change in the bullet's momentum is opposite in direction to the incoming momentum.
   (b) The change in the bullet's momentum is equal to the incoming momentum.
   (c) Both (a) and (b) are correct.
   (d) None of the above.

3. A car makes a three-quarter a turn around a traffic circle at a constant speed. The car is originally moving in the positive *x*-direction. Which of the following indicates the car's change in momentum?
   (a) $mv\hat{i} + mv\hat{j}$
   (b) $mv\hat{i} - mv\hat{j}$
   (c) $-mv\hat{i} + mv\hat{j}$
   (d) $-mv\hat{i} - mv\hat{j}$

4. A hockey puck of mass *m* moves at speed *v* on ice. A hockey player hits the puck along a direction perpendicular to its velocity. The puck's speed stays the same, but the puck is now moving at 60° with respect to its incident direction. What is the magnitude of the impulse?
   (a) zero
   (b) $mv \sin(30°)$
   (c) mv
   (d) $mv \cos(30°)$

5. If the puck in question 4 were moving along the positive *x*-axis before the shot, and the shot was directed along the positive *y*-direction, then the final momentum can be written as
   (a) $\frac{mv}{2}\hat{i} - \frac{\sqrt{3}mv}{2}\hat{j}$
   (b) $\frac{\sqrt{3}mv}{2}\hat{i} + \frac{1mv}{2}\hat{j}$
   (c) $\frac{\sqrt{3}mv}{2}\hat{i} - \frac{1mv}{2}\hat{j}$
   (d) None of the above.

6. For the puck in question 5, the impulse is
   (a) $\frac{mv}{2}\hat{i} + \frac{\sqrt{3}mv}{2}\hat{j}$
   (b) $-\frac{1mv}{2}\hat{i} + \frac{1mv}{2}\hat{j}$
   (c) $-\frac{1mv}{2}\hat{i} + \frac{\sqrt{3}mv}{2}\hat{j}$
   (d) $\frac{1mv}{2}\hat{i} - \frac{\sqrt{3}mv}{2}$

7. A bouncy ball of mass $m = 100$ g and speed $v = 10$ m/s is incident on a wall at an angle of 30°. It then bounces off the wall at the same speed, still making an angle of 30° with the wall after the collision. The collision takes 1 ms. The average force experienced by the ball during the collision is

   (a) 1000 N into the wall;
   (b) 1000 N out of the wall;
   (c) zero because the speed does not change;
   (d) $\sqrt{3}(1000)$ N parallel to the wall.

8. A bouncy ball of mass *m* and speed *v* impacts on a wall at an angle of 60°. The component of its velocity perpendicular to the wall does not change, and the angle it makes with the wall after the collision is 30°. Is this possible? Explain.

9. A slightly deflated ball hits a wall at an angle of 60° and leaves at an angle of 30°. The ratio of the outgoing momentum to the incoming momentum is
   (a) $\frac{\sqrt{3}}{2}$
   (b) $\frac{1}{\sqrt{3}}$
   (c) $\frac{2\sqrt{13}}{13}$
   (d) None of the above.

10. A sponge ball strikes a wall and leaves with a speed that is $\frac{\sqrt{13}}{4}$ of its incoming speed. If the incident ball makes an angle of 30° with the wall, then the component of its momentum perpendicular to the wall changes by factor of
    (a) $\frac{\sqrt{3}}{2}$
    (b) $\frac{\sqrt{13}}{4}$
    (c) $\frac{1}{2}d$
    (d) None of the above.

11. You balance a firecracker on the tip of a needle and then light it. What happens to the firecracker's centre of mass when it explodes?

12. When you drop a tennis ball onto Earth's surface, the collision is completely elastic and lasts a few milliseconds. Will the ball bounce back to the same height? Explain.

13. Assuming Earth does not move at all during the fall of a bouncy ball and the subsequent collision with Earth's surface, would you expect the bouncy ball to go back to the same height it fell from if the collision is completely elastic? Explain.

14. How can you explain that a bouncy ball, in a completely elastic collision with Earth's surface, bounces back to exactly the same height it fell from despite the fact that the force of gravity acts on the ball during the collision and thus the ball has a lesser rebound speed than the speed it impacted with?

15. You drop a 30 g bouncy ball from a height of 11 m onto a flat, horizontal surface. The ball is in contact with the surface for 20 ms. Find the rebound speed of the ball. Why is this not exactly the same as the impact speed?

16. Suppose Earth were the only planet in the solar system. As Earth orbits the Sun,
    (a) the Sun would remain stationary;
    (b) the Sun would always move in the same direction as Earth;
    (c) the Earth–Sun centre of mass would move back and forth;
    (d) both Earth and the Sun would orbit the stationary Earth–Sun centre of mass.

17. If one of the stars in a binary system explodes without directly damaging the second star, immediately after the explosion
   (a) the centre of mass of the two stars will move closer to the star that remained intact;
   (b) the centre of mass of the two stars will be at the centre of the intact star;
   (c) the centre of mass will shift, depending on the explosion;
   (d) the orbit of the intact star will not be affected by the explosion if the intact star does not does not absorb any material from the other star.

18. If one of the stars in a binary star system explodes, and the intact star absorbs a good portion of the material from the first star, then
   (a) the centre of mass of the exploding star is not affected by the explosion;
   (b) the centre of mass of the two-star system is not affected by the explosion;
   (c) the centre of mass of the two-star system will move closer to the intact star as the material from the exploding star gets absorbed by intact star.

19. A spring is held compressed between two masses, one having 10 times the mass of the other. The spring is then released. When the spring is at maximum extension,
   (a) the smaller mass is moving toward the larger mass with 9 times the speed of the larger mass;
   (b) the larger mass is moving away from the smaller mass;
   (c) the two masses are stationary;
   (d) only the larger mass is stationary.

20. When you throw a ball straight down so that it bounces perfectly elastically from the surface of Earth, the duration of the impact will dictate how high the ball goes. True or false? Explain.

21. A flamingo lands vertically on the back of a floating hippopotamus, then takes a few steps along the hippopotamus's back toward its head. Which of the following statements is true?
   (a) The hippopotamus will move in a direction opposite to the flamingo's motion.
   (b) The hippopotamus will not move.
   (c) The centre of mass of the hippopotamus–flamingo system will move in the same direction that the hippopotamus moves.
   (d) The centre of mass of the hippopotamus–flamingo system will move in the same direction as the flamingo's motion.

22. A fish is stuck in a closed bottle that is half-filled with water. The bottle floats on its side in a still lake. The fish swims to the left. Which of the following is true?
   (a) The bottle will move to the left.
   (b) The bottle will move to the right.
   (c) The bottle will not move at all.
   (d) The centre of mass of the fish, water, and bottle will move in the same direction as the fish.

23. A bar is held vertically on ice. You let the bar drop such that friction prevents its rough bottom edge from moving. When the bar is horizontal, however, its bottom edge is no longer in contact with the ice ground, and there is little friction between the smooth side of the bar and the ice. After the bar falls down completely,
   (a) it moves on the ice;
   (b) it is stationary;

   (c) its centre of mass has not moved at all.
   (d) None of the above.

24. A child fits a spring on the front of a stationary block of wood and places it on a frictionless surface. He then throws another block at it such that the spring cushions the blow. After a few moments, the two blocks are moving at the same speed. Which of the following statements is true?
   (a) The collision is elastic; therefore, the kinetic energy of the system does not change.
   (b) Since a spring is involved, we cannot say whether the momentum of the two masses is conserved during the collision.
   (c) The kinetic energy of the system changes during the collision.
   (d) The total mechanical energy of the system is not conserved.

25. A child has a spring-loaded cannon that launches small marbles horizontally. The cannon is mounted on little skis and is placed on an icy surface in the backyard during the winter. When the marble is in the air, the cannon moves as a result of firing. Which of the following statements is true?
   (a) The sum of the kinetic energy of the cannon and the flying marble is not equal to the energy that was stored in the spring, so the collision is inelastic.
   (b) The sum of the kinetic energies of the cannon and the marble is equal to the energy stored in the spring.
   (c) The spring will exert a greater force on the marble than it does on the cannon.
   (d) None of the above are true.

26. You are standing on a sled in the Arctic and decide to jump forward off the sled such that your horizontal speed with respect to the sled is 4 m/s. Which of the following statements is correct? (Hint: Distinguish between speed and velocity.)
   (a) Your speed relative to the sled is the sum of both speeds.
   (b) Your velocity vector with respect to the sled is defined as the sum of both velocity vectors.
   (c) Your velocity relative to the ground is the same as your velocity relative to the sled.
   (d) Your velocity relative to the ground is the sum of the velocity of the sled and your velocity with respect to the sled.

27. A toy car is moving forward. The car has a mounted cannon that fires horizontal projectiles backward. Which of the following statements is true?
   (a) The projectiles must be moving backward relative to the ground.
   (b) The projectiles must move forward with respect to the ground.
   (c) Could be (a) or (b), depending on the specifics.
   (d) None of the above are true.

28. You go for a sprint across a soccer field, heading east. Assuming that you are the only mobile object on Earth's surface, Earth's rotation would be
   (a) slower than usual;
   (b) faster than usual;
   (c) unaffected;
   (d) faster or slower, depending on the frame of reference.

29. A cannon fires a probe into the Moon such that the probe leaves Earth's surface at 90° to the surface. Which of the following statements is true?
   (a) Earth will not be affected at all.
   (b) Earth will recoil slightly.

CHAPTER 7 | LINEAR MOMENTUM, COLLISIONS, AND SYSTEMS OF PARTICLES

(c) Since the force of gravity is involved, momentum is not conserved.

(d) None of the above are true.

30. A child jumps onboard a stationary wagon. The losses due to friction between the wagon and the floor and in the wagon's wheels are negligible. Neither the child's body nor the wagon sustains an irrecoverable deformation, and the child and the wagon move as one. Which of the following statements is true?

(a) The collision is elastic because there is no irrecoverable deformation.

(b) The two objects stick to one another, so the collision must be inelastic.

(c) The kinetic energy of the child is reduced by an amount that is smaller than the kinetic energy gained by the wagon.

(d) The child's momentum is reduced by an amount that is larger than the amount by which the wagon's momentum increases.

31. For question 30, is the following statement true (T) or false (F): The only mechanism for the kinetic energy to be lost is friction between the child and the wagon. (Ignore losses to sound.)

32. A man standing on a wagon that is originally at rest begins to run. If there are no losses in the wagon motion to friction,

(a) the momentum of the man relative to the ground is the same in magnitude as the momentum of the wagon with respect to the ground;

(b) the momentum of the man with respect to the wagon is the same in magnitude as the momentum of the wagon with respect to the ground;

(c) the speed of the man with respect to the ground is greater than the speed of the man with respect to the wagon;

(d) it depends on the friction between the man and the wagon.

33. A well-behaved cat goes on a canoeing trip with her owners. She is left in the canoe as they set up camp. She sees a fish jump in front of the canoe and leaps forward to try to catch it. The canoe is originally at rest, and we take the direction of the canoe with respect to ground to be negative. Which of the following statements is true?

(a) The speed of the cat with respect to the ground is positive.

(b) The speed of the cat with respect to the ground is negative.

(c) Whether (a) or (b) is correct depends on how fast the canoe moves.

(d) Whether (a) or (b) is correct depends on the ratio of the cat's mass to the canoe's mass.

34. A sled is moving on horizontal frictionless ice when a dog onboard leaps onto the ice, leaving the sled horizontally backward. The direction of motion of the sled is taken to be positive. Whether the velocity of the dog with respect to the ground is positive or negative depends on

(a) the masses of the sled and the dog;

(b) the original speed of the sled and the relative speed of the dog with respect to the sled;

(c) both (a) and (b).

(d) None of the above are true.

35. A rocket is fired vertically upward from Earth's surface. If you were to choose a system in which momentum is conserved during the process, the most correct system to pick would be

(a) rocket + fuel/exhaust;

(b) rocket + fuel/exhaust + atmosphere;

(c) rocket + fuel/exhaust + atmosphere + Earth;

(d) rocket + fuel/exhaust + atmosphere + Milky Way galaxy.

36. A heavy cannon fires a shell horizontally. Which of the following statements is true?

(a) It is possible for the cannon not to recoil.

(b) If the cannon does not recoil at all, then the ground it is sitting on must move a little.

(c) If the ground does not move, then Earth's crust must move on the mantel.

(d) Both (b) and (c) are correct.

37. A canoeist notices her canoe drifting toward a waterfall. She has a battery-powered water pump fitted with a hose that she can use to direct the water from the pump. The pump draws water vertically up from the river. Which of the following statements is true?

(a) If the pump shoots out enough water at a high enough speed, the canoeist can angle the hose so that she can move away from the waterfall.

(b) The pump cannot propel the canoe because the momentum of the system must be conserved.

(c) Conservation of momentum does not apply here because the pump has a power source.

(d) None of the above are true.

PROBLEMS BY SECTION

For problems, star ratings will be used, (∗, ∗∗, or ∗∗∗), with more stars meaning more challenging problems.

Section 7-1   Linear Momentum

38. ∗ Calculate the linear momentum of a space shuttle of mass 2040 tonnes moving at a speed of 27 000 km/h.

39. ∗ Find the momentum of a 60 g stone that starts from rest and falls to the ground from a height of 20 m.

Section 7-2   Rate of Change of Linear Momentum and Newton's Laws

40. ∗∗ The engine of a fire rescue boat gets damaged during a rescue operation. The captain decides to use the fire-fighting water-pumping system to push the boat back to the harbour. The pump is mounted on the deck and draws the water vertically out of the ocean. The crew points the nozzle of the hose toward the rear of the boat. The mass of the boat is 180 000 kg, and the pump sprays water at a rate of 330 kg/min with a muzzle speed of 20 m/s. Ignoring water resistance, find the acceleration of the rescue boat.

41. ∗∗ An athlete of mass 74 kg is running inside a train car of mass 2100 kg, at a speed of 9.0 m/s with respect to the ground. While the athlete is running, the car is stationary. The athlete then comes to a stop over 0.20 s. Find the average force exerted on the train car by the athlete during this time.

(a) Find the speed of the train car when the athlete comes to a full stop on the car.

(b) The athlete comes to a stop over 0.2 s. During this time, the force of kinetic friction is the only force between her feet and the train car. Find the average force exerted on the train car by the athlete.

Section 7-3   Impulse and the Conservation of Linear Momentum

42. ✲   A soccer player kicks a 425 g soccer ball straight on with an average force of 78.0 N. The speed of the soccer ball right after the collision is 41.0 m/s. Find the duration of the collision. Ignore the effects of gravity.

43. ✲   Tennis balls, whose individual masses are 58 g, are fired at a wall by a tennis-ball launcher such that they impact the wall horizontally in perfectly elastic collisions at a speed of 117 km/h.
    (a) The launcher fires the balls at a rate of 55 balls/min. Find the average force exerted on the wall as a result.
    (b) The average collision time is 25 ms. Find the average force per collision.

44. ✲✲   A somewhat deflated volleyball ($m = 220$ g) falls on the floor from a height of 2.0 m, starting from rest. The collision between the ball and the floor takes 0.30 s and is perfectly elastic.
    (a) Find the speed of the ball on the way up after the impact.
    (b) Draw the profile for the net force acting on the ball as a function of time.
    (c) Find the net force acting on the ball as a function of time, using the linear behaviour model.
    (d) Find the maximum values for (b) and (c).

45. ✲✲   Starting from rest, a bouncy ball falls from a height of 10 m onto a flat, horizontal surface and rebounds with 99% of its incident speed in a perfectly elastic collision. Find the amount of time that the ball was in contact with the ground.

46. ✲   Figure 7-18 shows the net force versus time profile for an object colliding with a horizontal surface. The object's mass is 300 g, and its incident speed is 13 m/s. Find the recoil speed.

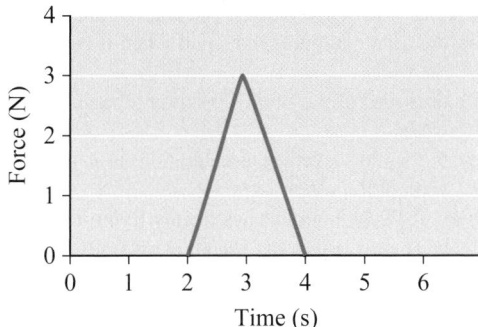

Figure 7-18   Problem 46

Section 7-4   Systems of Particles and Centre of Mass

47. ✲   Sixteen identical point masses are placed at equal intervals such that they outline the circumference of a circle of radius R. By how much does the centre of mass shift when one of those masses is removed?

48. ✲   Find the centre of mass of the particles shown in Figure 7-19.

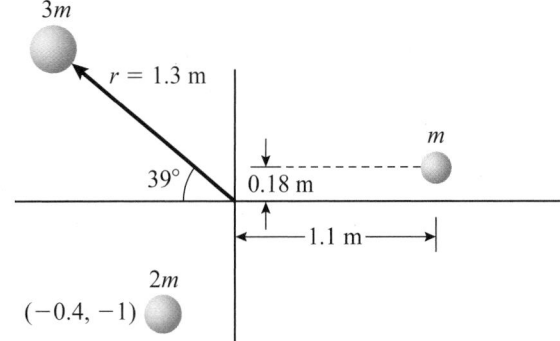

Figure 7-19   Problem 48

49. ✲   Masses m, $2m$, $3m$, and $4m$ are located at the corners of a square of side L as shown in Figure 7-20. Locate the centre of mass of this system.

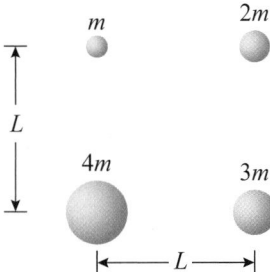

Figure 7-20   Problem 49

50. ✲   A pendulum is fashioned out of a thin bar of length 0.55 m and mass 1.9 kg, welded to a sphere of radius 0.11 m and mass 0.86 kg. Find the centre of mass of the composite object.

Section 7-5   Newton's Laws Applied to Systems of Particles

51. ✲✲   A 4.0 kg dog runs at constant speed from one end of a 21 kg canoe to the other, a distance of 6.7 m, in 3.1 s. Assuming negligible water resistance, how far does the canoe move?

52. ✲✲   An astronaut finds herself stranded on a 43 kg platform designed for short hops around a space station, when the platform loses its propulsion system. The platform is moving away from the space station at a speed of 2 m/s and is oriented such that the astronaut's head points toward the station as she stands straight. She then decides to kick her feet down in the hopes of acquiring enough momentum to return to the station. If the mass of the astronaut with her spacesuit is 105 kg, what must the astronaut's minimum speed relative to the platform be if she is to succeed?

53. ✲✲   A child starts to throw a water balloon at a friend, but the balloon bursts while still in the child's hand. At that instant, the balloon is 90 cm above the ground and moving horizontally at a speed of 2.3 m/s. Where is the centre of mass of the water when all the water has hit the ground? (The balloon is very light.)

54. ✱✱ A 71.0 kg astronaut inside a 2.00 t spacecraft at rest kicks off from one end at a speed of 2.50 m/s, and stops when he reaches the other end, travelling a distance of 9.40 m. How far does the spacecraft move?

Section 7-6   Inelastic Collisions

55. ✱ Two huskies collide on frictionless ice. Anyu, with a mass of 27 kg, is heading due east at a speed of 6.5 m/s. Anut, with a mass of 21 kg, is heading due north at a speed of 4.0 m/s. Find the speed of the two dogs after they collide. Assume the collision to be completely inelastic.

56. ✱ Two cars in a completely inelastic head-on collision are at rest immediately after the collision. The mass of one car is 1.7 times the mass of other car. Find the ratio of their initial speeds.

57. ✱✱ An 11 g bullet is fired with speed v into a 68 kg ballistic pendulum. The wires holding the pendulum deviate from the vertical by 27°. The vertical distance from the point of impact to the suspension points for the wires is 92 cm. Find the speed with which the bullet hits the pendulum.

58. ✱✱ A tennis ball hits a wall at an angle of 48° with respect to the wall and loses 25% of its energy in the collision. Find the angle the ball makes with the wall on the way out.

59. ✱✱✱ Derive an expression for the fractional loss of energy for an object that impacts a wall at an angle θ and leaves it an angle θ'.

60. ✱ Two hockey players race for the puck. Player 1 has a mass of 69 kg and a velocity 5.6 m/s at an angle of 32° with the rink board. Player 2 has a mass of 78 kg and a velocity of 7.1 m/s at an angle of 16° with the same board. The players are locked in a completely inelastic collision. Find the energy lost in the collision.

61. ✱✱ A medicine ball of mass m and speed v is sliding on frictionless ice when it strikes a stationary medicine ball of mass $5m/4$. After the collision, the kinetic energy of the first ball is three-quarters of its initial energy, and this ball moves at an angle of 21° with the incident direction. The second medicine ball moves in a direction that makes an angle of 43° with the final momentum of the first ball. Express the speed of the second ball in terms of v.

62. ✱✱ Two balls collide head-on; one ball (m_1) has twice the speed of the other ball (m_1). The collision is completely inelastic. Derive an expression for the speed of the two balls after the collision.

Section 7-7   Elastic Collisions

63. ✱✱ A 3.0 kg mass is attached to a spring of spring constant $k = 185$ N/m resting on a horizontal frictionless surface. The other end of the spring is fixed to a wall. A 1.6 kg mass is thrown toward the 3.0 kg mass and collides with it in a perfectly elastic collision. The maximum compression of the spring is 80 cm. Find the speed of the incoming mass.

64. ✱✱ A billiard ball of mass m is shot at speed v to strike two adjacent stationary billiard balls with the same mass as the incident ball. The two stationary balls are in contact, and the incoming ball strikes them along the line connecting their centre of mass. The collisions are perfectly elastic. Derive an expression for the final speed of each ball.

65. ✱✱✱ Two balls, one of which is stationary, collide elastically. The speed of one ball after the collision is half the speed of the other. Find the ratio of the masses.

66. ✱✱✱ A moving ball (speed v) collides with another at rest. After the collision, the first ball is at rest. Find the ratio of their masses if the collision is perfectly elastic. What is the speed of the second ball?

67. ✱✱✱ Two balls collide elastically head-on; one ball has mass m and is moving at speed v, the other ball is moving at $v/10$. After the collision, the lighter mass moves at speed $20v$. Find the ratio of the masses and express the final speed of the heavier ball in terms of v.

68. ✱✱ A billiard ball of mass m and speed v strikes another ball of mass M and speed u as shown in Figure 7-21. The collision is perfectly elastic. Derive expressions for the speed of each ball after the collision.

Figure 7-21   Problem 68

Section 7-8   Variable Mass and Rocket Propulsion

69. ✱✱ A rocket is launched vertically. The fuel constitutes half the mass of the rocket. After 21.0 s of flight, half the fuel is burned and the speed of the rocket is 190 m/s. The rocket burns fuel at a rate of 72.0 kg/min. Find the thrust of the rocket and the relative speed of the exhaust with respect to the rocket. Ignore air resistance.

70. ✱✱ A rocket's engine burns fuel at a rate of 65 kg/s. The rocket is launched in deep space. Twenty seconds later, only half the mass of the fuel is left, and the rocket has reached a speed of 9300 km/h.
(a) Find the original mass of the rocket.
(b) Find the average acceleration of the rocket over the first 10 s.
(c) Find the rocket's thrust.

71. ✱✱ A giant squid of mass 230 kg propels itself horizontally by swallowing 56 kg of water and then expelling it in 6.0 s. The squid is initially at rest with zero buoyancy. Assuming that the water resistance is negligible and that the squid expels the water at a constant rate through a circular opening 7.0 cm in diameter, calculate estimates for
(a) the acceleration of the squid when it begins expelling the water
(b) the speed of the squid 4.0 s after it begins expelling the water
(c) the squid's average acceleration as it expels the last sixth of the water

72. ✱✱ A 72 kg man exhales deeply trying to blow a wasp away. The air leaves his mouth at rate of 0.70 L/min at a speed of 9.0 m/s. What force must his neck muscles exert on his head to prevent it from moving?

COMPREHENSIVE PROBLEMS

73. ✱✱ A planet orbits its star in an elliptical orbit, with a major-to-minor axis ratio of 1.4. The mass of the planet is 1/100 the mass of the star. The major axis is 7 billion kilometres. Predict the length of the long axes for the orbit of the star. Draw a scale diagram of the two orbits. Assume that there are no other objects in the vicinity.

74. ✳✳✳ A 2.0 m-long bar is held vertical on a frictionless surface. The top is tipped very gently, and the bar begins to swing. When the effects of the gentle push on the bar are ignored, what is the speed of the bar when it is fully horizontal?

75. ✳ A bar of length 1.16 m is tilted 43.0° from the vertical. Find the horizontal displacement of the bar's centre of mass.

76. ✳✳ A child of mass 34 kg is standing on a parade wagon (empty mass $M = 120$ kg), which is loaded with 600 gift baskets, each with a mass of 500 g. She throws the baskets off the side one at a time, such that they leave in a horizontal direction, making an angle of 57° with the length of the wagon. There are no frictional losses between the wagon and the road, and the wagon can move only forward and backward. The wagon is originally at rest. By the time the 70th basket is thrown, the speed of the wagon is 2.1 m/s. Find the magnitude of the relative *velocity* of the baskets with respect to the wagon. Assume this velocity is constant.

77. ✳✳ A 5 g bullet is fired at a speed of 265 m/s into a 9.2 kg watermelon sitting on a thin vertical rod. The bullet shatters the watermelon and leaves with a speed of 235 m/s. Find the speed of the centre of mass of the watermelon immediately after the collision. What is the acceleration of the centre of mass of water melon after the collision?

78. ✳✳✳ A child mounts a toy spring cannon on a sled, and takes this toy out to play on the icy surface of a nearby lake. The mass of the cannon plus sled is *M*. The projectile that the cannon shoots has a mass *m* and leaves with a muzzle speed *v*. The cannon faces forward at an angle θ with respect to the horizontal. The sled-mounted cannon is moving at speed v_c when the projectile is launched.
(a) Derive an expression for the speed of the cannon immediately after it launches the projectile.
(b) Find the speed of the cannon, given that $M = 915$ g, $m = 32.0$ g, $v = 17.0$ m/s, $v_c = 5.00$ m/s and $\theta = 41°$.

79. ✳✳✳ Starting from rest, a 2.4 t elephant in a 7.4 t boxcar runs at a speed of 3.0 m/s with respect to the boxcar and then throws herself against a spring with $k = 17.5$ kN/m mounted on the other end of the boxcar, thus compressing the spring. The boxcar is mounted on smooth rails, so losses due to friction as the car moves are minimal. Initially the spring is neither stretched nor compressed and the boxcar is at rest, and the spring is unstretched.
(a) Find the maximum compression of the spring.
(b) What is the speed of the boxcar when the spring is at maximum compression?

80. ✳✳✳ An engineer working with an expedition in the Arctic needs to move a heavy rectangular package across the frozen ground. Initially, the package is standing on its edge. He puts two pegs in the ground at one edge of the package and lets the package drop by tilting it gently until it begins to fall. The package has a mass of 120 kg, and travels a total distance of 25 m on the ground, where the coefficient of friction is 0.05. How tall is the package?

81. ✳ A child of mass 19 kg lands on a 34 kg wagon. The wagon then moves with a speed of 2.11 m/s. Find the horizontal speed with which the child lands.

82. ✳✳ Two balls, one twice the mass of the other, are thrown straight up in the air at a speed of 20 m/s, 0.70 s apart. The heavier ball is thrown second.

(a) Where is the centre of mass of the two-ball system 0.20 s after the second ball is thrown?
(b) What is the acceleration of the centre of mass at the following times: (i) before the second ball is thrown, (ii) when the two balls are moving up, (iii) when the lighter ball reaches its maximum height, and (iv) 0.10 s after the lighter ball lands?

83. ✳✳ You are sitting on one end of a canoe ($m = 19$ kg) in calm water. You then throw your 15 kg backpack to your friend at the other end of the canoe, a distance of 3.0 m. Your friend's mass is 65 kg, and your mass is 72 kg. How much would the canoe move if the water resistance were negligible?

84. ✳✳✳ A research team in Antarctica is launching a 32 kg weather probe into the air above the Ross Ice Shelf. They use a special cannon of mass 1600 kg, which launches the probe with a muzzle speed of 710 m/s. The cannon points at an angle of 57° above the horizontal. The team forgets to secure the cannon, which slides backward on the ice after the probe is fired. Find the angle that the probe makes with respect to the ground when it lands. Ignore air resistance, and assume that the probe lands at the same altitude from which it was launched.

85. ✳✳ A low Earth orbit observation station, made of 11 modules each of mass 13 000 kg, moves at a speed of 4.0 km/s. It is equipped with an emergency collision avoidance system whereby an explosive charge can be detonated to release one of the modules. The collision alarm sounds, and the commander detonates a charge, releasing one of the modules at a speed of 1400 km/h with respect to the station, directed opposite to the original direction of travel of the station. Find the resulting change in the speed of the station.

86. ✳✳ During the Iditarod race in Alaska, your sled gets detached from your snow dogs. The sled moves at 4.1 m/s, headed for thin ice. You jump backward off the sled with a speed of 2.7 m/s with respect to the sled. What is your horizontal speed with respect to the ground while you are in the air?

87. ✳ An astronaut holds a 20 m-long bar with one 40 kg package on each end. If she rotates the bar by 180°, by how much does she move?

88. ✳✳✳ A man stands on a relatively long sled resting on snow. The sled, of mass *M*, is tethered with a bungee cord, which can be treated like a spring of spring constant *k*. The man then moves on the sled toward the bungee cord end, at speed *v* with respect to the sled. The maximum extension of the cord is *L*. Derive an expression for the coefficient of friction between the sled and the snow.

89. ✳✳✳ A child launches a fireworks rocket at an angle θ from the vertical. When the rocket is at its highest point, its two parts separate by detonating a charge such that the lighter part falls straight down. The speed of the rocket before the separation is *v*, and the lighter part is one-third the mass of the heavier part. Derive an expression for the speed of the heavier part after the separation.

90. ✳✳ A hockey puck on a frictionless ice strikes the rink board at an angle of 65° with respect to the normal to the board. The collision is partly elastic, and the puck loses 17% of its kinetic energy. Find the angle that the puck makes with the normal to the board after the collision.

91. ✷✷✷ A performance-testing device is made of a long cart of length L and mass M, connected to a wall by a spring of spring constant k. Athletes stand on the end of the cart farthest from the spring and run at maximum speed toward the end where the spring is attached. The maximum extension of the spring is X. Derive an expression for the maximum speed of an athlete of mass m.

92. ✷✷ A polar bear cub is sliding toward his sister on frictionless ice in the Arctic. The sister is stationary before the collision. After the collision, the brother is stationary. Find the ratio of their masses.

93. ✷✷ An archer shoots an arrow of mass 110 g through a vertical plank of cork that is firmly fixed in place on top of a stationary cart. The plank is parallel to the length of the cart, and the cart is free to move on wheels with frictionless bearings. The incident arrow is horizontal when it strikes the vertical plank and makes an angle of 40° with the plane of the plank. When the arrow emerges, it makes an angle of 43° with respect to the now moving plank. The combined mass of the plank and the cart is 3.78 kg, and the incident speed of the arrow is 90 m/s.
(a) Find the energy lost during the collision.
(b) Find the speed of the cart immediately after the collision.

94. ✷✷ A pellet gun is used to shoot 0.58 g pellets at a speed of 280 m/s into a piece of wood of mass M, resting on a horizontal frictionless surface. The surface then curves upward into the shape of a hemispherical bowl of radius $R = 70$ cm. The block then moves up the bowl-shaped surface until its angular position measured from the vertical is 37°. The pellet gets completely lodged in the piece of wood. Find M.

95. ✷✷ You are standing on top of Mount Everest with a solid ball made out of neutrons (density 4×10^{17} kg/m³). The ball is 20 cm in radius. You then let the neutron ball roll down to the foothills, 7000 m below.
(a) How far would Earth have moved along the line connecting you to the centre of Earth? Take Earth's mass to be 6×10^{24} kg.
(b) Assume there were no frictional losses, and determine the change in Earth's speed.
(c) If the collision is completely elastic, find the rebound speed of Earth from the point of view of an observer stationary with respect to the centre of mass of Earth–neutron ball system.

96. ✷✷✷ A solid disk of radius R and mass M has a hollow part of radius r, centred at distance x from the centre of the otherwise solid disk, as shown in Figure 7-22. Find the centre of mass of the disk.

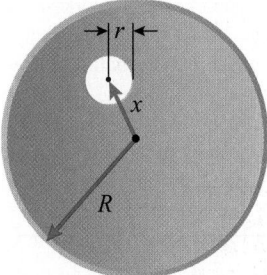

Figure 7-22   Problem 96

97. ✷✷✷ A bar has a linearly increasing density, such that the density at one end is five times the density at the other end. Find the location of the centre of mass as measured from the lighter end.

98. ✷✷✷ A figure skater of mass 3/2 m is holding on to the middle of a bar that connects two other figure skaters of masses m and M. The figure skater in the middle then spins the bar at a constant angular speed. Derive an expression for the radius of the circular trajectory that the centre figure skater will undergo.

99. ✷✷ A spring ($k = 430$ N/m) is compressed 9.0 cm between two masses, $m = 910$ g and $M = 1.4$ kg. While the system moves along a frictionless surface at a constant speed v, the spring is released. What is the speed of the lighter mass when the spring is stretched 7.0 cm?

100. ✷✷✷ A child has a toy made of a spring (spring constant k) connected to a base of mass M to a bobbing monkey head of mass m. When the spring is fully cocked, it is compressed 9.0 cm from equilibrium. The child sits her toy on its base such that the spring is vertical and the bobbing monkey head is attached to the other end. The child then presses a release knob that releases the compressed spring. Find the maximum height reached by the base after the spring is released using the following values: $M = 87$ g, $m = 29$ g, and $k = 92$ N/m.

101. ✷✷ A fire breaks out on the fourth floor of a building. Firefighters bring in a platform of mass 300 kg propped up on four springs, each with a spring constant of 410 N/m. A 67 kg man jumps from a height of 7.0 m above the platform and lands on the platform in such a way that he stays with the platform. Find the maximum compression of the springs.

102. ✷✷✷ You drop a small ball with a mass of 32 g from the roof of a building 40 m above the ground. One second later, you throw a volleyball with a mass of 226 g down with an initial speed such that it hits the small ball 5.0 m above the ground. All the collisions are elastic.
(a) Find the speed of the small ball immediately after the collision.
(b) Find the speed of the small ball after it collides with the ground.

103. ✷✷✷ Mass m_1 moves with speed v_1 toward stationary mass m_2. One end of a spring of spring constant k and unstretched length L is attached to the front of mass m_1. Mass m_2 sticks to the other end of the spring when they make contact.
(a) Derive an expression for the maximum distance between the two masses after the collision.
(b) Calculate this distance using the values $m_1 = 100$ g, $m_2 = 300$ g, $v_1 = 6.00$ m/s, $L = 28.5$ cm and $k = 50.0$ N/m.

104. ✷✷✷ Where is the centre of mass of the Earth-Moon system relative to the surface of Earth?

105. ✷✷✷ A block of mass m is fitted with a horizontal spring (spring constant k) at on end. The mass is made to move at a speed v, when it strikes a another block of mass M that was moving toward it at speed u. The collision is cushioned by the spring. Derive expressions for
(a) the maximum compression of the spring;

(b) the speed of m when the spring is at its maximum compression;

(c) the final speed of each mass;

(d) the speed of each mass at maximum spring compression.

106. ✸✸✸ Estimate how far the crew of a submerged submarine could move the vessel by all gathering at one end of it.

107. ✸✸ A child places a spring between two of his toy cars. He holds the two cars together on a frictionless surface in such a way that they compress the spring. He causes the two cars with the spring in-between to move at speed v, keeping the spring compressed between the cars; then he lets the two cars go. Find the final speed of the car in the front. The spring constant is k, and the mass of each car is m.

108. ✸✸ A 7.45 g bullet moving at a speed of 1100 m/s strikes a 27.8 kg block of wood and emerges from the other end to become lodged in 2 kg plate. The plate rests on a horizontal frictionless surface and is attached to a horizontal spring ($k = 1340$ N/m) that is compressed 11 cm as a result of the collision. Find the speed of the block immediately after the bullet emerges from it.

109. ✸✸ Your cat, Boots ($m = 3.15$ kg), is standing in the middle of a cart with frictionless wheels. The cart is at rest on a frictionless surface. Boots then jumps to one end of the cart such that the cart is moving at 0.30 m/s while the cat is in the air. The cat is in the air for 0.8 s.

(a) Find the mass of the cart.

(b) What is the speed of the cart after the cat lands?

110. ✸✸✸ Two blocks with masses m_1 and m_2 are thrown toward each other on a horizontal frictionless surface with respective speeds v_1 and v_2, respectively. The collision between them is cushioned by a spring of spring constant k that is attached to one of the masses.

(a) Derive an expression for the maximum compression of the spring during the collision.

(b) Calculate the maximum compression in part (a) for $m_1 = 3.20$ kg, $m_2 = 5.24$ kg, $v_1 = 2.09$ m/s, $v_2 = 3.76$ m/s, and $k = 615$ N/m.

111. ✸✸✸ A 32 kg child runs at a speed of 3.00 m/s and jumps onto a wagon of mass $m = 167$ kg sitting on frictionless wheels. The child slides for a distance d on the wagon before coming to a stop relative to the wagon. If the coefficient of kinetic friction between the child and the wagon is 0.7, find a) the final speed of the cart, b) the distance the child slides on the wagon (relative to the wagon).

See the text online resources at www.physics1e.nelson.com for Open Problems and Data-Rich Problems related to this chapter.

Learning Objectives

When you have completed this chapter you should be able to:

1 Define angular quantities, and relate angular variables to linear variables.

2 Solve kinematics equations for rotation.

3 Define torque, and compare different representations of torque.

4 Explain the relationship between torque, moment of inertia, and angular motion.

5 Calculate moments of inertia for rigid objects, and solve equations of motion for rotation about a fixed axis using Newton's laws.

6 Use the work–mechanical energy approach to solve rotational dynamics problems, and compare that approach to the force–torque approach for solving rotational dynamics problems.

7 Define angular momentum, conservation of angular momentum, and change of angular momentum in terms of Newton's second law.

8 Solve problems using conservation of angular momentum.

For additional Making Connections, Examples, and Checkpoints, as well as activities and experiments to help increase your understanding of the chapter's concepts, please go to the text's online resources at www.physics1e.nelson.com.

Chapter 8
Rotational Dynamics

The Crab pulsar, a pulsating neutron star within the Crab Nebula, is the result of a type II supernova, that was recorded by both Chinese and Arab astronomers in 1054 CE. The light from the supernova explosion was visible to the naked eye for close to 22 months in the night sky before fading from view. The bulk of the exploding star's substance was ejected during the supernova creating the crab nebula, the rest of the stellar mass was crushed together to form the neutron star. The original star was spinning when it exploded, and the mass that became the neutron star continues to spin.

The radius of this neutron star is only 10 km, a tiny fraction (likely less than 1/100 000) of the radius of the progenitor star. As a result, the Crab pulsar spins very fast: the rotational period is only 33.1 ms. The conservation of angular momentum, the principle behind the incredibly high rotational speed of the pulsar (also often observed during figure skating competitions) is covered in this chapter. A spinning neutron star emits electromagnetic radiation along the axis of its magnetic field. If the rotation axis and the magnetic axis do not coincide, the radiation beam is visible to a distant observer once during each revolution, much like the beam of light in a lighthouse. The pulsed nature of the beam and the radio frequency component of the radiation result in the name pulsating radio star, or pulsar.

Figure 8-1 The Crab Nebula (left) and the shock waves that bombard it as a result of the radiation from the pulsar within (right)

8-1 Angular Variables

Consider a particle moving along the circumference of a circle of radius R. When the particle moves from point A to point B (Figure 8-2), the length of the arc travelled by the particle is

KEY EQUATION
$$\Delta s = R\Delta\theta \tag{8-1}$$

where $\Delta\theta$ is the **angular displacement**.

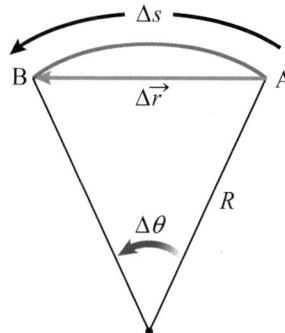

Figure 8-2 (a) Linear displacement: the displacement vector $\overrightarrow{\Delta r}$ is a straight line connecting A and B. (b) Tangential displacement: Δs is the full arc length covered by the particle.

In Equation (8-1), R is measured in metres, and θ is measured in radians, which are dimensionless. Hence, Δs has dimensions of length. However, R is a simple length (a line segment), and Δs is an arc length (a segment of a circle).

The differential, or infinitesimal, displacement along the circumference can be written in terms of the infinitesimal angular displacement:

$$ds = Rd\theta \tag{8-2}$$

It follows that the instantaneous rate of change with respect to time of the position of the particle along the circumference is related to the instantaneous rate of change of the angular position:

$$\frac{ds}{dt} = R\frac{d\theta}{dt} \tag{8-3a}$$

The derivative $\dfrac{ds}{dt}$ is the instantaneous linear speed, v, of the particle along the circumference, or the particle's instantaneous tangential speed (Figure 8-3).

The derivative $\dfrac{d\theta}{dt}$ is the instantaneous rate of change of the angular position with respect to time, or, more

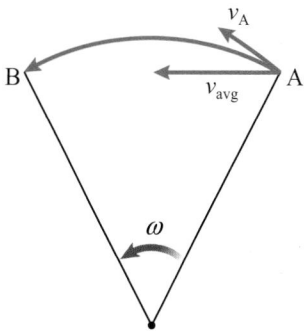

Figure 8-3 The instantaneous linear velocity is tangent to the circle, and the average velocity points directly along the straight line from A to B.

simply, the **angular speed**. The conventional symbol for angular speed is ω:

$$\omega = \frac{d\theta}{dt} \tag{8-3b}$$

The SI units for ω are radians per second. Like linear speed, angular speed can be a function of time. Equation (8-3a), therefore, can be rewritten as follows:

KEY EQUATION
$$v(t) = \omega(t)R \tag{8-3c}$$

Taking the derivative of Equation (8-3a) with respect to time, we have

$$\frac{d^2s}{dt^2} = R\frac{d^2\theta}{dt^2} \tag{8-4a}$$

The derivative on the left-hand side of Equation (8-4a) is the tangential acceleration of the object, not to be confused with the radial acceleration (see Figure 8-4). The derivative on the right-hand side is the **angular acceleration**. The standard symbol for angular acceleration is α:

$$\alpha = \frac{d^2\theta}{dt^2} = \frac{d\omega}{dt} \tag{8-4b}$$

Since a and α are not necessarily constant, they are shown as functions of time, t. Thus, Equation (8-4a) can be rewritten as follows:

KEY EQUATION
$$a_t(t) = \alpha(t)R \tag{8-4c}$$

PEER TO PEER

An object moving with a constant angular speed has an angular acceleration of zero; therefore, the tangential acceleration is also zero. The radial, or centripetal, acceleration is not zero, however, and is given by $a_r = \dfrac{v^2}{r} = \omega^2 r$.

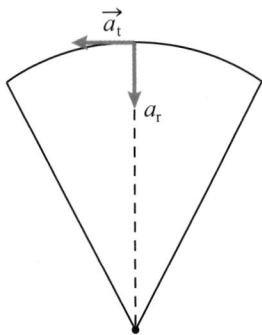

Figure 8-4 An object travelling at constant speed along the circumference has radial, or centripetal, acceleration but no tangential acceleration. If the speed also changes then the object will also have a tangential acceleration

Ladybug Revolution

The e-resource that accompanies every new copy of this textbook contains an Online Activity using the PhET simulation "Ladybug Revolution." Work through the simulation and accompanying questions to gain an understanding of the relation between angular speed and the acceleration vector for an object moving in a circle.

 ## EXAMPLE 8-1

Rotation of Earth

(a) Determine the values for the linear speed, angular velocity, tangential acceleration, and radial acceleration of a point at the equator. Assume that Earth is a sphere of radius 6340 km.

(b) Calculate how far the rotation of Earth moves a person sitting on an equatorial beach during a 2.00 h sunbath.

SOLUTION

(a) Earth completes a full revolution (2π rad) in about 24 h, so the angular speed is

$$\omega = \frac{2\pi \, \text{rad}}{(24 \, \text{h})(3600 \, \text{s/h})} = 7.2722 \times 10^{-5} \, \text{rad/s} = 7.27 \times 10^{-5} \, \text{rad/s}$$

to 3 significant figures

The linear speed for a point on the equator is given by Equation (8-3):

$$v = \omega R = (7.2722 \times 10^{-5} \, \text{rad/s})(6340 \, \text{km})$$

$$= 461.1 \, \text{m/s} = 461 \, \text{m/s}$$

to 3 significant figures

Earth's angular velocity is nearly constant; the tangential acceleration as given by Equation (8-4c) is approximately zero.

The radial, or centripetal, acceleration is given by Equation (4-11):

$$a_r = \frac{v^2}{r} = \omega^2 r = (7.2722 \times 10^{-5} \, \text{rad/s})^2 (6340{,}000 \, \text{m})$$

$$= 3.35 \times 10^{-2} \, \text{m/s}^2$$

(b) We can use the linear speed calculated in part (a) to find the distance travelled by a point on the equator in 2 h:

$$s = vt = (461.1 \, \text{m/s})(2)(3600 \, \text{s}) = 3.32 \times 10^6 \, \text{m}$$

Making sense of the result:

The distance covered by a point on Earth's equator during one day is Earth's circumference given by $2\pi r = 2\pi(6380 \, \text{km})$ which is approximately 40,080 km, this is covered in 24 hours. if we divide that by 12 we get 3340 km, the distance covered in 2 hours.

 ## CHECKPOINT

C-8-1 Geostationary Orbits

Satellites in a geostationary orbit are directly above Earth's equator and appear motionless in the sky. Such orbits are particularly useful for communication and television satellites. Which of the following statements are true for a satellite in a geostationary orbit?

(a) The linear velocity of the satellite, with respect to Earth's centre, is the same as the linear velocity of a stationary object on Earth's surface.

(b) The angular velocity of the satellite, with respect to Earth's centre, is the same as the angular velocity of a stationary object on Earth's surface at equator.

(c) The tangential acceleration of the satellite is nonzero.

(d) The radial acceleration of the satellite is zero.

C-8-1 (b) all points along line connecting satellite to Earth centre move at same angular speed.

LO 2

8-2 Kinematics Equations for Rotation

Following a treatment similar to the one we used in Section 3-4 to derive the kinematics equations for constant acceleration, we can derive kinematics equations for rotation with a constant angular acceleration.

We can define angular acceleration as

$$\frac{d^2\theta(t)}{dt^2} = \frac{d\omega(t)}{dt} = \alpha \qquad (8\text{-}5)$$

Integrating Equation (8-5) gives an expression for angular velocity as a function of time:

$$\omega(t) = \alpha t + \omega_0 \qquad (8\text{-}6)$$

Equation (8-6) shows that the angular speed at any time t is determined by the initial value at $t = 0$ plus the change incurred as a result of the acceleration.

When we integrate Equation (8-6), we can show that the angular position as a function of time is given by

$$\theta(t) = \frac{1}{2}\alpha t^2 + \omega_0 t + \theta_0 \qquad (8\text{-}7)$$

Combining Equations (8-6) and (8-7) gives us the rotational version of the time-independent kinematics equation, identical in form to Equation (3-20):

$$\omega^2 = \omega_0^2 + 2\alpha\Delta\theta \qquad (8\text{-}8)$$

It is not surprising that Equations (8-6) and (8-7) are almost identical in form to Equations (3-15) and (3-17)

because they are derived in the same fashion and hold for the same condition of constant acceleration. Note that the product of α and $\Delta\theta$ is positive when the two quantities are parallel (i.e., the object speeds up) and negative when the quantities are antiparallel (i.e., the object slows down).

✓ **CHECKPOINT**

C-8-2 Cranking Your Bike

On a bicycle, a large sprocket is driven by a pedal on crank arms and is connected to the small sprocket on the rear wheel by a chain. Assume that the radius of the large sprocket is R, the radius of the small sprocket is $R/2$, the radius of the wheel is $3R$, and the length of each pedal crank arm is $2R$. Which of the following statements is true?

(a) The rear wheel has the same angular acceleration as the crank arms.

(b) The small sprocket has the same angular speed as the large sprocket.

(c) The radial acceleration of a point on the circumference of the small sprocket is half the acceleration of a point on the circumference of the large sprocket.

(d) The linear speed of a point on the circumference of the small sprocket is the same as the linear speed of a point on the circumference of the large sprocket.

C-8-2 (d) the speed of the chain is the same as that of the tangential speed of each sprocket.

EXAMPLE 8-2

Spinning Flywheel

A flywheel of radius 0.720 m is rotating clockwise at 3.60 rad/s.

(a) Find the constant acceleration needed to slow the wheel to an angular speed of 2.10 rad/s over 1.80 s.

(b) Find the number of revolutions the wheel will go through as it slows down.

SOLUTION

(a) Equation (8-6) relates angular speed, acceleration, and time:

$$\omega = \alpha t + \omega_0$$

We choose the direction of rotation (clockwise) as positive, so ω and ω_0 will be positive. Since the wheel is slowing down, α is directed counterclockwise and is negative according to our sign convention:

$$\alpha = \frac{\omega - \omega_0}{t} = \frac{2.1 - 3.6}{1.8} = \frac{-1.5}{1.8} = -0.833 \text{ rad/s}^2$$

(b) Now we can use either Equation (8-7) or Equation (8-8) to determine the angular displacement of the wheel as it slows down. Here, we use Equation (8-8):

$$\omega^2 = \omega_0^2 + 2\alpha\Delta\theta$$

$$\Delta\theta = \frac{\omega^2 - \omega_0^2}{2\alpha} = \frac{2.1^2 - 3.6^2}{2(-1.5/1.8)} = 5.13 \text{ rad} = 0.816 \text{ rev}$$

Making sense of the result:

The units for θ are $(\text{rad}^2/\text{s}^2)/(\text{rad/s}^2) = \text{rad}$. We divide by 2π to get the number of revolutions.

8-3 Torque

When you tighten a bolt using a wrench, as in Figure 8-5, three factors determine the effectiveness of the force you apply:

- the strength of the applied force
- the angle θ between the applied force and the wrench
- the distance $|\vec{r}|$ between the bolt and the point where you apply the force

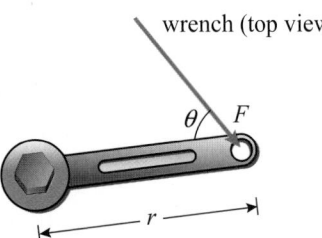

wrench (top view)

Figure 8-5 Torque depends on three parameters.

These three factors determining the **torque** of a force are elegantly summarized using the vector cross product introduced in Chapter 2. The torque of a force about a point, or pivot, is defined as the cross product of the position vector of the point of application of the force (as measured from the point) and the applied force:

KEY EQUATION
$$\vec{\tau} = \vec{r} \times \vec{F} \qquad (8\text{-}9)$$

The magnitude of the cross product in Equation (8-9) is given by

$$|\vec{\tau}| = |\vec{r}||\vec{F}| \sin\theta \qquad (8\text{-}10)$$

where θ is the angle measured counterclockwise from $\vec{r}$ to $\vec{F}$.

ONLINE ACTIVITY

Torque

The e-resource that accompanies every new copy of this textbook contains an Online Activity using the PhET simulation "Torque." Work through the simulation and accompanying questions to gain an understanding of angular momentum.

Supersonic Blades

The propellers on turboprop planes turn at high rotational speeds. The linear speed of the tip of the propeller blades depends on the rotational speed and the distance from the tip to the propeller centre. The outer part of the blades can reach supersonic speeds. However, aircraft designers usually keep blade speeds below the speed of sound to avoid supersonic shockwaves and the increased drag they cause. The Tupolev Tu 114 (Figure 8-6) holds the record for the fastest turboprop passenger plane with a maximum recorded airspeed of 877.21 km/h, and remains one of the fastest passenger planes in use. It has special propellers designed to sustain supersonic blade–tip speeds and cope with the resulting shockwave drag.

Richard Seaman

Figure 8-6 The Tupolev Tu-114

Direction of a Torque Vector

The direction of a torque vector can be determined using the right-hand rule (Figure 8-7 a, b). When the palm of your hand is curled in the direction of rotation, with the thumb aligned with the rotation axis, the thumb points in the direction of the torque vector (Figure 8-7). The direction of the torque vector can also be determined using the Cartesian cross product rules outlined in Chapter 2. It should be mentioned that the angular velocity can be regarded as a vector, whose direction is also determined using the right-hand rule as shown in Figure 8-7(c).

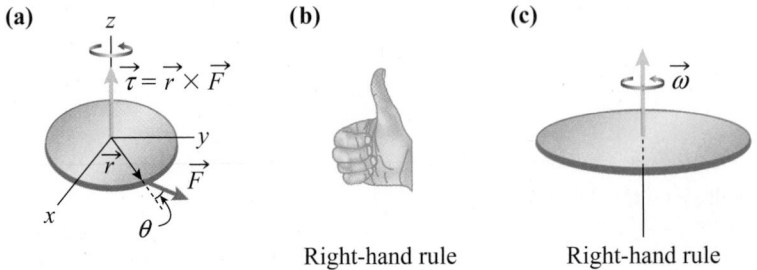

(a) **(b)** **(c)**

Right-hand rule   Right-hand rule

Figure 8-7 Demonstrating the right-hand rule for cross products

EXAMPLE 8-3

Torque Without Contact

A forest ranger is using a mallet to drive a peg into the ground to set up her tent in a level area. She drives the peg straight into the ground with an average force of 420 N. Another forest ranger is standing 2.00 m away. Can we say that the force she applies to the peg has a torque about the foot of the other ranger? If so, find the magnitude of the torque.

SOLUTION

The magnitude of the torque is given by $\tau = rF\sin\theta$. Since the peg is being driven straight into the ground, we can say that $\theta = \pi/2$. Therefore, $\sin\theta = 1$ and

$$\tau = rF = 2.0 \text{ m} \times 420 \text{ N} = 840 \text{ N} \cdot \text{m}$$

Making sense of the result:

For the force to exert a torque about a pivot or an axis, it is not necessary that there be a physical *moment arm* between the force and the pivot. (The component of the distance to the pivot that is perpendicular to the force is called the **moment arm** of the force.) So, despite the fact that there is no physical object connecting the mallet to the second ranger, the force used by the first ranger still exerts a torque about the second ranger. Here we are looking at the mathematical definition of torque.

EXAMPLE 8-4

Calculating Torque

Calculate the torque of the force shown in Figure 8-8 about the pivot at point O, given that $r = 1.70$ m, $F = 16.3$ N, and $\theta = 37.0°$.

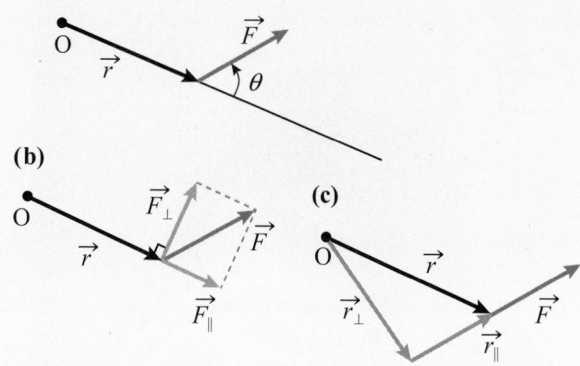

(a)

(b)

(c)

Figure 8-8 Calculating torque (a) as a vector cross product, (b) as a product between *r* and the component of *F* perpendicular to *r*, and (c) as a product between *F* and the component of *r* perpendicular to *F*

SOLUTION

We will carry out this calculation using three equivalent methods. These methods are essentially different perspectives of the way we view the definition of torque as given in Equations (8-9) and (8-10).

METHOD A: USING A CROSS PRODUCT

From Equation (8-10), the magnitude of the torque is

$$\vec{\tau} = \vec{r} \times \vec{F} = |\vec{\tau}| = |\vec{r}||\vec{F}|\sin\theta \qquad (1)$$

Here, we use the definition of torque and simply substitute the given values of $|r|$, F and θ into Equation (1) (see Figure 8-8(a)):

$$|\vec{\tau}| = |\vec{r}||\vec{F}|\sin\theta = (1.7)(16.3)\sin(37°) = 16.7 \text{ N} \cdot \text{m}$$

METHOD B: USING THE PERPENDICULAR COMPONENT OF THE FORCE

We can consider the product in Equation (8-10) as a product of $|r|$ and $|F|\sin\theta$ (Figure 8-8(b)). The term $F\sin\theta$ gives the component of F that is perpendicular to the position vector of the force as measured from point O.

$$|\vec{\tau}| = |\vec{r}|\{|\vec{F}|\sin\theta\} = rF_{\perp} \qquad (8\text{-}11)$$

Alternatively, recognizing that the force can be expressed as the sum of these two components, the expression for the cross product can then be rewritten as

$$\vec{\tau} = \vec{r} \times (\vec{F}_{//} + \vec{F}_{\perp}) = \vec{r} \times \vec{F}_{//} + \vec{r} \times \vec{F}_{\perp} \qquad (8\text{-}12)$$

The first term in the expression is identically zero because the angle between $F_{//}$ and r is 0, and only the second term survives:

$$|\vec{\tau}| = |\vec{r} \times \vec{F}| = \sin(90°) = rF_{\perp} \qquad (8\text{-}13)$$

Both Equations (8-11) and (8-13) tell us that the force component perpendicular to the displacement vector contributes to the cross product for the torque, and the parallel component of F does not. Note that we are loosely using r and F in Equations (8-11) and (8-13) to represent the magnitudes of the position vector and the force vector.

METHOD C: USING THE PERPENDICULAR COMPONENT OF THE DISPLACEMENT

A third way of looking at the cross product is to consider the magnitude of the torque as the product of $r\sin\theta$ and F.

(continued)

As shown in Figure 8-8(c), $r\sin\theta$ gives the component of r that is perpendicular to the force F, and the component of r that is parallel to the force does not contribute to the torque:

$$|\vec{\tau}| = |\vec{r}||\vec{F}|\sin\theta = |\vec{r}|\sin\theta|\vec{F}| = r_\perp F \qquad (8\text{-}14)$$

Another way to derive this relationship is to rewrite the expression for the torque as

$$\vec{\tau} = (\vec{r}_{//} + \vec{r}_\perp) \times \vec{F} = \vec{r}_{//} \times \vec{F} + \vec{r}_\perp \times \vec{F} \qquad (8\text{-}15)$$

Here, again, the first term vanishes, so

$$|\vec{\tau}| = |\vec{r} \times \vec{F}| = |\vec{r}||\vec{F}|\sin(90°) = r_\perp F \qquad (8\text{-}16)$$

You can verify that Equations (8-14) and (8-16) yield the same value of $16.7\ \text{N}\cdot\text{m}$ for the required torque.

Making sense of the result:

Note that in this example we have treated the components of a vector as vectors themselves to facilitate a more elegant treatment of the vector product. Only perpendicular components matter.

 CHECKPOINT

C-8-3 The Greatest Torque

The forces in Figure 8-9 have the same magnitude and direction. Which force exerts the greatest torque on P?

(a) $\vec{F}_1$

(b) $\vec{F}_2$

(c) $\vec{F}_3$

(d) They are all the same.

Figure 8-9   C-8-3

C-8-3 (d) They all have the same perpendicular distance to P.

MAKING CONNECTIONS

Gears and Transmission Systems

A simple transmission consists of two gears coupled together (Figure 8-10). Typically, one gear (the drive gear) is connected to a power source, such as the engine of a car, and couples to the second gear (the driven gear). If the two gears are not the same size, they will rotate at different speeds. An automotive variable transmission has a mechanism that couples the driven gear to different-sized gears driven by the engine. When the smallest engine gear is coupled to the wheel gear, a relatively high engine rpm results in a relatively low wheel rotation speed. This gear ratio is useful when the car encounters resistance, such as when starting from rest or travelling uphill. Putting the car in low gear effectively assigns a smaller moment arm to the external resisting force, thus reducing the resistive torque. The engine can then turn the wheels with less torque. Once the car is moving, it is more efficient to couple to a larger engine gear. Coupling gears by a chain as shown is only one type of transmission mechanism, in cars this is seen for example in the timing chain (or belt). Other common mechanisms involve direct coupling between gears of different size.

Figure 8-10   A simple transmission with two gears coupled by a chain

8-4 Moment of Inertia of a Point Mass

Now, we examine how torque relates to angular motion. We begin by considering the rotating point mass in Figure 8-11.

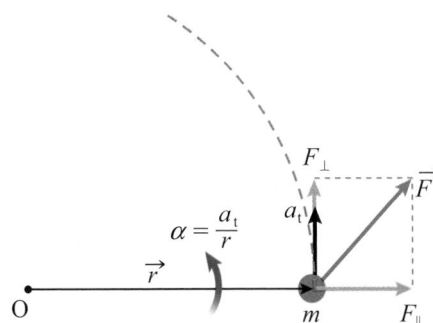

$$\alpha = \frac{a_t}{r}$$

Figure 8-11 A force is applied to a point mass at the end of a rigid massless rod. The radial acceleration is not shown.

The mass in Figure 8-11 is rigidly attached by a massless rod to the pivot at point O. The point mass and the rod are originally at rest and are free to rotate in the plane of the page about the pivot, or, more precisely, about an axis perpendicular to the page running through the pivot. We now apply a force F to the mass as shown. The mass will move upward initially, and, because of the rigid rod, it will also move in a circular trajectory about the axis through O. We will keep the magnitude of the force and its direction relative to the massless rod constant. Note that we do not focus on radial acceleration in this chapter. Since our focus is rotational motion we only consider the tangential direction as tangential acceleration relates directly to angular acceleration as per Equation (8-4).

According to Newton's second law, $F_\perp = ma_t$. Note that the parallel component of the force does not contribute to the motion of the mass. The acceleration a in this case is the tangential acceleration of the point mass m. From Equation (8-4), the angular acceleration of m is given by $\alpha = a_t/r$, where r is the length of the rod. The force F exerts a torque about the pivot such that $|\tau| = |r \times F| = |r| |F_\perp|$, which reduces to $|r||F_\perp| = |r||ma_t|$. Since $a_t = \alpha r$,

KEY EQUATION
$$\vec{\tau}_{net} = mr^2 \vec{\alpha} \qquad (8\text{-}17a)$$

Moment of Inertia

Equation (8-17a) is the rotational equivalent of Newton's second law for a point mass, $F_{net} = ma$. The torque takes the place of the force, and the angular acceleration replaces the linear acceleration. The mass, m, is replaced by the quantity mr^2.

An object's mass is a measure of its inertia, or resistance to acceleration in a straight line. Newton's second

law tells us that the net force required for a given acceleration of an object, such as the box of Figure 8-12(a) depends solely on the mass of the object. However, if we hold the object at a constant distance from an axis and rotate it about the axis with a given angular acceleration as in Figure 8-12, Equation (8-17) tells us that the torque needed depends not only on the mass of the object, but also on the square of the distance from the rotation axis.

The product mr^2 is called the rotational inertia, or the moment of inertia, of a point mass. The **moment of inertia** is a measure of an object's resistance to rotational acceleration, similar to the way mass is a measure of an object's resistance to linear acceleration. The symbol for moment of inertia is I. For a point mass, Equation (8-17) can be written as follows:

KEY EQUATION
$$\vec{\tau}_{net} = I\vec{\alpha} \qquad (8\text{-}17b)$$

(a)

(b)

Figure 8-12 (a) In linear motion, resistance to acceleration depends only on the mass of an object. (b) In rotational motion, resistance to acceleration also depends on distance: the farther the object is from the rotation axis, the greater the resistance to rotational acceleration.

CHECKPOINT

C-8-4 Who's Faster?

Two identical point masses are each connected to a pivot by massless rods. The rods are free to rotate about the pivots and are held in a horizontal position. One of the rods is twice as long as the other. At the instant when the rods are released,

(a) the linear acceleration of both masses is the same;
(b) the mass connected to the longer rod will have a smaller linear acceleration;
(c) the bars will have the same radial acceleration;
(d) the longer bar will have the greater angular acceleration.

C-8-4 (a) since the rods are massless, the masses will accelerate down at $a = g$.

LO 5

8-5 Moments of Inertia of Rigid Bodies

Thus far, we have dealt with objects as single point masses. In Chapter 7, we touched upon the idea of a system of particles. We now make the transition from treating point masses and systems of particles to treating rigid bodies. The treatment of a rigid body takes into account the geometry of the body and the mass distribution within it.

Systems of Point Masses

Consider the system of three particles shown in Figure 8-13. The particles are attached to a massless rod that pivots around a perpendicular axis through point O.

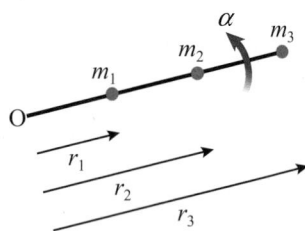

Figure 8-13 Three point masses connected to a rigid massless rod

Let us now apply a torque to the rod to rotate it with an angular acceleration α. The net torque on the system of three masses is the sum of the torques on the individual masses using $\vec{\tau}_{net} = mr^2\vec{\alpha}$ from Equation 8-17a for each of the masses:

$$\vec{\tau}_{net} = \vec{\tau}_1 + \vec{\tau}_2 + \vec{\tau}_3 = m_1 r_1^2 \vec{\alpha} + m_2 r_2^2 \vec{\alpha} + m_3 r_3^2 \vec{\alpha} \quad (8\text{-}18a)$$

we note that the angular accelerations of the 3 masses are the same, $\alpha_1 = \alpha_2 = \alpha_3 = \alpha$. Therefore, Equation 8-18a can be written as

$$\vec{\tau} = m_1 r_1^2 \vec{\alpha} + m_2 r_2^2 \vec{\alpha} + m_3 r_3^2 \vec{\alpha}$$

$$= (m_1 r_1^2 + m_2 r_2^2 + m_3 r_3^2)\vec{\alpha} \quad (8\text{-}18b)$$

This equation can now be written in the following compact form:

$$\vec{\tau}_{net} = I_{tot} \vec{\alpha} \quad (8\text{-}19)$$

where $I_{tot} = m_1 r_1^2 + m_2 r_2^2 + m_3 r_3^2$ is the total moment of inertia of the system of rigidly connected point masses.

In general, the moment of inertia of a two-dimensional system of particles (or point masses) about an axis perpendicular to the plane of the particles is equal to the sum of the individual moments of inertia of each of the particles about that axis.

 KEY EQUATION

$$I = \sum_{i=1}^{N} m_i r_i^2 \quad (8\text{-}20)$$

where N is the number of particles.

The summation notation in Equation (8-20) holds as long as we are dealing with a system of discrete (noncontinuous) point masses. We can use this property to derive moments of inertia of some simple objects. For a continuous distribution of mass, the sum is replaced by an integral of infinitesimally small moments of inertia taken over the volume of the object. Then the expression for the total moment of inertia becomes

KEY EQUATION

$$I = \int r^2 dm \quad (8\text{-}21)$$

where r is the distance between each infinitesimal amount of mass dm and the axis of rotation.

 CHECKPOINT

C-8-5 Inertia

Consider a thin ring and a solid disk both of mass M and radius R. Both objects are homogeneous (mass is uniformly distributed), and both rotate freely at the same angular speed about fixed frictionless axes running through their respective centres of mass. Which object is more difficult to stop?

C-8-5 The ring has a higher moment of inertia and is therefore more difficult to stop.

ONLINE ACTIVITY

The e-resource that accompanies every new copy of this textbook contains an Online Activity using the PhET simulation "Torque." Work through the simulation and accompanying questions to gain an understanding of the relation between torque and moment of inertia T.

Rigid Bodies

A **rigid body** is a body that does not deform when a force is applied to it. In our study of rigid bodies, we begin by considering a uniform ring with mass M, radius R, and negligible width. We can consider the ring to be composed of identical point masses distributed uniformly along the rim. Let us assign a mass m_i to each of these point masses (Figure 8-14). The moment of inertia of the entire ring about a given rotation axis is simply the sum of all the moments of inertia of the individual point masses about that axis. For the axis that passes through the centre of mass of the ring and is perpendicular to the plane of the ring, the moment of inertia, I_{cm}, can found using Equation (8-20):

$$I_{cm} = \sum m_i r_i^2 = m_1 r_1^2 + m_2 r_2^2 + m_3 r_3^2 + \cdots + m_n r_n^2$$

Since all the point masses are located on the circumference at a distance R from the centre of the ring,

$$I_{cm} = m_1 R^2 + m_2 R^2 + m_3 R^2 + \cdots + m_n R^2$$
$$= (m_1 + m + m_3 + \cdots + m_n) R^2$$

The sum of all the point masses making up the ring adds to the total mass of the ring:

$$I_{cm} = MR^2 \qquad (8\text{-}22)$$

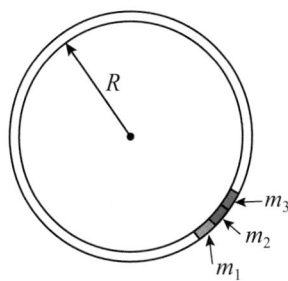

Figure 8-14 Mass distribution in a uniform ring

The distribution of mass in the ring is continuous, so the moment of inertia of the ring is more appropriately calculated using integration. As shown in Figure 8-15, the moment of inertia about the centre of the ring for an infinitesimally small differential element of mass, dm, is

$$dI_{cm} = R^2 dm \qquad (8\text{-}23)$$

Since the ring is uniform, it has uniform linear mass density, λ (mass per unit length). So, $dm = \lambda dl$, where dl is the infinitesimally small length of the mass dm on the ring, and

$$dI = R^2 dm = R^2 (\lambda dl) \qquad (8\text{-}24a)$$

Now, dl can be written as $R d\theta$, where $d\theta$ is the infinitesimally small angle subtending the infinitesimal ring segment of length dl. Substituting in Equation (8-24a) gives

$$dI = R^2 dm$$
$$= R^2 (\lambda dl)$$
$$= R^2 \lambda R d\theta = \lambda R^3 d\theta \qquad (8\text{-}24b)$$

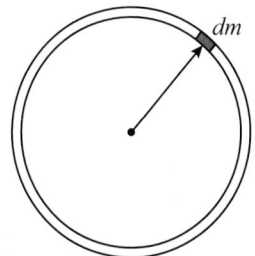

Figure 8-15 Differential elements for the calculation of the moment of inertia of a ring about its centre of mass

To obtain the moment of inertia of the entire ring, we integrate dI over the circumference of the ring:

$$I_{cm} = \int R^2 dm = \int_0^{2\pi} R^3 \lambda d\theta = R^3 \lambda \int_0^{2\pi} d\theta = R^3 \lambda 2\pi \quad (8\text{-}25)$$

Since the total mass of the ring is $M = \lambda(2\pi R)$,

$$I_{cm} = MR^2$$

This expression is the same as Equation (8-22).

We now consider the moment of inertia of a thin uniform disk shown in contrast to that of a ring in Figure 8-16 about an axis perpendicular to its plane, through its centre. The mass per unit area of this disk is a constant, σ, with units of kilogram per square metre.

Figure 8-16 Ring and disk with the same radius

The differential element we begin with is a ring of radius r and infinitesimal width dr (Figure 8-17). The area of this ring is $dA = 2\pi r dr$, and the mass of the ring is $dm = \sigma dA = \sigma 2\pi r dr$. As shown in Equation (8-21), the moment of inertia of an infinitesimally thin ring is mr^2. Therefore, for the ring of width dr,

$$dI = r^2 dm = r^2 (2\pi r \sigma dr) = 2\pi r^3 \sigma dr \qquad (8\text{-}26)$$

We integrate this differential element over the range of radii of the rings to obtain the moment of inertia of the disk about its centre of mass:

$$I_{cm} = \int r^2 dm = \int_0^R 2\pi r^3 \sigma dr = 2\pi\sigma \int_0^R r^3 dr = 2\pi\sigma \frac{R^4}{4}$$
$$(8\text{-}27a)$$

Since the total mass of the disk is $\pi R^2 \sigma$, the expression of the total moment of inertia simplifies to

$$I_{cm} = \frac{MR^2}{2} \qquad (8\text{-}27b)$$

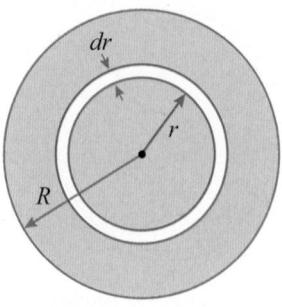

Figure 8-17　The differential element for the calculation of the moment of inertia of a uniform disk

If we think of a disk as being made up of point masses, the point masses located at the edge of the disk will each have a moment of inertia of $m_i R^2$, the same as the point masses making up a ring. However, the rest of the point masses making up the disk are closer to the centre of the disk and will have correspondingly smaller moments of inertia. For example, the point mass located at the centre of the disc has zero moment of inertia. Thus, the moment of inertia of a disk about its centre of mass is less than the moment of inertia of a ring of the same mass and radius.

MAKING CONNECTIONS

If Earth Did Not Rotate

Our survival on Earth's surface depends, in part, on the effect that Earth's rotation has on its molten core. The outer core is a fluid mass of iron, nickel, and sulfur. A combination of Earth's rotation, convection currents, and buoyancy and compositional currents causes hotter fluid close to the inner core to rise in helical columns aligned more or less with Earth's axis (Figure 8-18). The resulting electrical currents produce the bulk of Earth's magnetic field, which shields Earth's surface from the potentially devastating effects of the stream of charged particles ejected from the Sun (the solar wind). The interaction of the charged particles with Earth's magnetic field also causes the aurora borealis and the aurora australis.

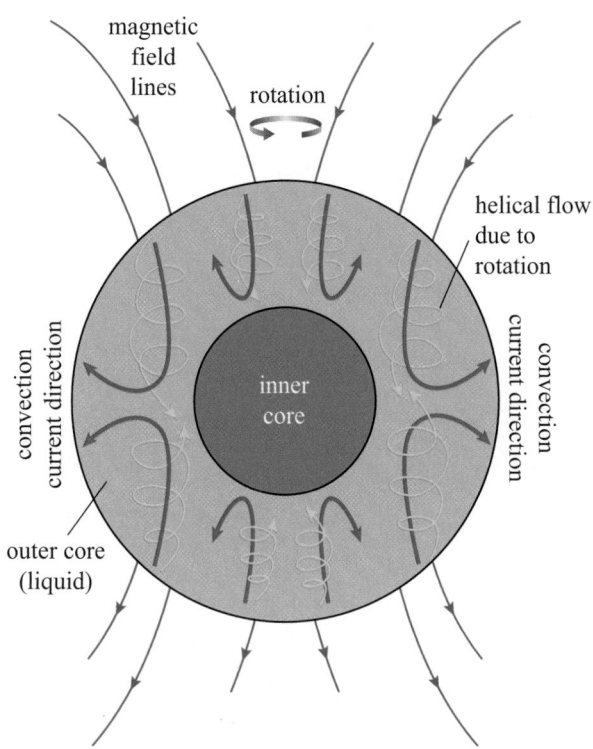

Figure 8-18　The aurora borealis. Earth's rotation ultimately produces currents whose magnetic fields shield Earth from the solar wind.

EXAMPLE 8-5

Clever Monkey

The solid drum in Figure 8-19(a) is free to rotate about an axle through its centre. A 23.0 kg monkey decides to ride the rope on the drum down to the ground 12.0 m below. The rope is at rest when the monkey grabs onto it. The drum is a uniform cylinder of radius 0.180 m and mass 190 kg. Find the acceleration of the monkey and the tension in the rope. Assume that the mass of the rope is negligible.

SOLUTION

Let us begin by drawing the FBD for the monkey (Figure 8-19(b)). The forces acting on the monkey are its

Figure 8-19   (a) Example 8-5. (b) The FBD of the pulley. (c) The FBD of the monkey

weight (mg) and the tension in the rope. Therefore, summing the forces in the y-direction, we have

$$\sum F_y = T_y + mg_y = ma_y$$

We choose up as the positive direction, so

$$T - mg = -ma \qquad (1)$$

where we introduce a as a variable to indicate the magnitude of the acceleration. The same holds for T and g in Equation (1) above.

We have two unknowns so far: T and a. Note that $T = mg$ only when the acceleration of the mass is zero.

The only force acting on the drum is the tension in the rope (Figure 8-19(c)). The tension is applied at the circumference of the drum, so the resulting torque rotates the drum. Next, we sum torques about centre of drum. Using Equations (8-10) and (8-19), and noting that the only force exerting a torque on the drum is the tension in the rope, we have

$$\tau = rT \sin\theta = I\alpha$$

The solid drum has the same basic shape as a solid disk, so its moment of inertia is $\dfrac{Mr^2}{2}$. The angle between the tension

and its moment arm to the centre is $90°$. So, the expression for the torque reduces to

$$\tau = rT = I\alpha = \frac{Mr^2}{2}\alpha \qquad (2)$$

Here, T and α are unknown, so we have a total of three unknowns in two equations. We therefore have to find another relationship between the variables. From Figure 8-19(a) we can see that the linear acceleration of the monkey and the rotational acceleration of the drum are related by Equation (8-4c):

$$a = \alpha r \qquad (3)$$

You can now solve the system of three equations with three unknowns by whatever method you prefer. We will simply substitute the value of T obtained from Equation (1), and the value of a from Equation (3) into Equation (2):

$$r(mg - ma) = \frac{Ia}{r}$$

Solving for a gives

$$a = \frac{mg}{m + I/r^2} = \frac{mg}{m + \frac{1}{2}M}$$

$$= \frac{23 \text{ kg} (9.81 \text{ m/s}^2)}{23 \text{ kg} + \frac{1}{2}(190 \text{ kg})} = 1.91 \text{ m/s}^2$$

for the magnitude of the tension, substituting this value for the acceleration into Equation (1), we get

$$T = mg - ma = m(g - a) = 23 \text{ kg}(9.81 - 1.91) \text{ m/s}^2 = 182 \text{ N}$$

Making sense of the result:

The acceleration has the units of m/s². The acceleration is about one-quarter of g, which seems reasonable because the monkey is not in free fall. The moment of inertia of the drum slows the acceleration of the monkey.

 CHECKPOINT

C-8-6 $F = ma$

Suppose we replace the drum in Example 8-5 with a hoop. What is the expression for a now?

C-8-6 The expression for a is now $a = \dfrac{Mg}{m + M_{ring}}$ because all the mass in the ring would have the same linear acceleration.

Next, we will determine moments of inertia about axes that do not run through the centre of mass. Table 8-1 lists the moments of inertia about the axes of some common geometrical objects.

The Parallel-Axis Theorem

Hold the middle of a horizontal metre stick between your thumb and index finger, and rotate the metre stick back and forth horizontally about an axis perpendicular to its

length. Keep the angle by which you rotate the metre stick relatively small. Gauge the amount of resistance the metre stick presents as you accelerate it. Next, hold the metre stick horizontal with the one end between your thumb and index finger. Swing the metre stick back and forth through approximately the same angular displacement as when you rotated it horizontally. You will notice that you encounter more resistance and need to use more force when you hold the stick at its end. This difference indicates that the moment of inertia of the metre stick is greater about an axis running through its end than about an axis through its centre of mass.

To understand how the choice of axis affects the moment of inertia, we begin by calculating the moment of inertia of a uniform bar about an axis running through its centre of mass perpendicular to its length. The moment of inertia of the differential element of mass dm (Figure 8-20) about the centre-of-mass axis running perpendicular to the length of the bar is

Table 8-1 Moments of Inertia About the Axes of Some Common Objects

Thin rod, mass M and length L; centre-of-mass axis perpendicular to bar

$$I_{cm}^{bar} = \frac{1}{12} ML^2$$

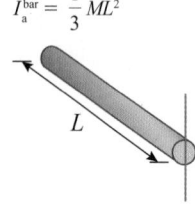

Thin rod, mass M and length L; axis running through edge perpendicular to bar

$$I_a^{bar} = \frac{1}{3} ML^2$$

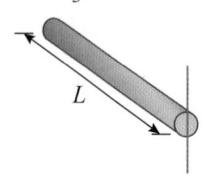

Rectangular block, about centre-of-mass axis perpendicular to plane of rectangle

$$I = \frac{1}{12} M(a^2 + b^2)$$

Rectangular block, mass M, side lengths a and b; about axis along one of the sides of length b.

$$I = \frac{1}{12} Ma^2$$

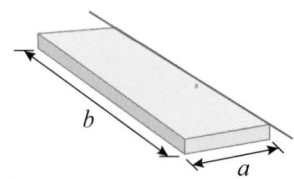

Cylinder, mass M and radius R; axis of symmetry

$$I_{cm}^{cyl} = \frac{1}{2} MR_2^2 - \frac{1}{2} MR_1^2$$

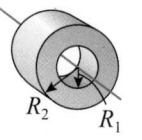

Solid cylinder, mass M, radius r, and length L;

$$I_{cm}^{cyl} = \frac{1}{2} MR^2$$

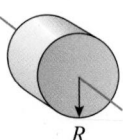

Ring, mass M and radius R,

$$I_{cm}^{ring} = MR^2$$

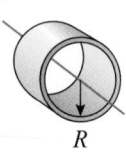

Sphere, mass M and radius R

$$I_{cm}^{sphere} = \frac{2}{5} MR^2$$

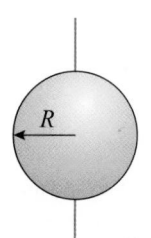

Thin spherical shell, mass M and radius R

$$I_{cm}^{ss} = \frac{2}{3} MR^2$$

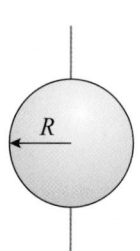

$r^2 dm$. Using λ as the linear density of the rod, we can express dm as λdr. The moment of inertia of the differential element is then $dI = r^2 \lambda dr$. We obtain the moment of inertia of the bar by integrating over the length of the bar:

$$I_{cm}^{bar} = \int_{length} r^2 dm = \int_{-L/2}^{L/2} r^2 \lambda\, dr = \frac{\lambda L^3}{12} \qquad (8\text{-}28)$$

The total mass of the rod is given by $M = \lambda L$, so

$$I_{cm}^{bar} = \frac{1}{12} ML^2 \qquad (8\text{-}29)$$

We next consider the moment of inertia of the bar about an axis, a, perpendicular to the bar and running through one end of it. The axis is parallel to the centre-of-mass axis. The moment of inertia of the differential element of mass dm about the axis a is given by $r'^2 dm$, where r' is the distance from dm to axis a.

The distance r' can be written as $r' = r + d$, where d is the distance between the centre-of-mass axis and the axis a. Then, the moment of inertia of the differential element of mass dm is $(r + d)^2 dm$. We integrate this differential over the length of the bar to obtain the moment of inertia of the bar about axis a:

$$I_a^{bar} = \int_{length} r'^2 dm = \int_{bar} (r + d)^2 dm = \int_{bar} (r^2 + d^2 + 2rd)dm \quad (8\text{-}30a)$$

$$I_a^{bar} = \int_{bar} r^2 dm + \int_{bar} d^2 dm + \int_{bar} 2(d)(r)dm \qquad (8\text{-}30b)$$

The first part of the expression is the moment of inertia of the bar about the centre-of-mass axis, as in Equation (8-29). The integral of dm over the length of the bar gives the total mass of the bar, M; therefore, the second term in Equation (8-30b) equals Md^2. The third term can be written as $2d \int_{bar} r dm$. The integral $\int_{bar} r dm$ is Mr_{cm}, as we saw in Chapter 7 Equation (7-9c). Since we have chosen our coordinate system so that the centre of mass of the bar is at the origin, the third term in Equation (8-30b) reduces to zero. So, the expression of the moment of inertia of the bar about axis a can be written as

$$I_a^{bar} = I_{cm}^{bar} + Md^2 \qquad (8\text{-}30c)$$

Figure 8-20 The moment of inertia of a differential element of a bar about an axis (a) through the centre of mass and (b) through the end of the bar and parallel to the axis through the centre of mass

Equation (8-30c) is an expression of the **parallel-axis theorem**. This theorem applies for any object and for any parallel axis at any arbitrary distance. The theorem can be stated as follows:

KEY EQUATION $$I_a^{object} = I_{cm}^{object} + Md^2 \qquad (8\text{-}31a)$$

Equation (8-31a) states that the moment of inertia of an object about an arbitrary axis is equal to the moment of inertia of that object about a parallel axis running through its centre of mass, plus the product of the total mass of the object and the square of the distance between the centre-of-mass axis and the arbitrary axis.

For the above bar, $d = L/2$, and

$$I_a^{bar} = \frac{1}{12}ML^2 + M\left(\frac{L}{2}\right)^2 = \frac{1}{12}ML^2 + M\frac{L^2}{4} = \frac{1}{3}ML^2$$

$$(8\text{-}31b)$$

Note that this moment of inertia is four times greater than the moment of inertia of the bar about the centre-of-mass axis.

✓ CHECKPOINT

C-8-7 Reference Axis

Given that the moment of inertia of an object about an arbitrary axis (not the centre-of-mass axis) is I_a, can we say that the moment of inertia about another axis, axis b, parallel to axis a, is $I_b = I_a + md^2$, where d is the distance between the two parallel axes?

C-8-7 No. $I_b = I_{cm} + md^2$, where r is the distance between the centre-of-mass axis and axis b.

LO 6

8-6 Rotational Kinetic Energy and Work

We next develop an expression for **rotational kinetic energy**, which is the kinetic energy of a rotating object. We start with a point mass connected to a massless rod, rotating freely about the pivot at point O (Figure 8-21). The linear speed of the point mass at the position shown

is v. This mass is under the influence of an external force directed *perpendicular* to the massless rod. (The mass also experiences a centripetal force exerted by the rod, but this radial force does no work on the mass.)

The kinetic energy of the mass is given by $KE = \frac{1}{2}mv^2$.

Since the particle is restricted to move in a circle, it has a rotational speed of $v = \omega r$. Substituting for v in the expression for kinetic energy gives an expression for the rotational kinetic energy of the point mass:

$$K_{rot} = \frac{1}{2}m\omega^2 r^2 = \frac{1}{2}I_o\omega^2 \qquad (8\text{-}32)$$

where I_o is the moment of inertia of the point mass about the pivot at O.

We can follow the same reasoning as we did in Section 8-5 to show that Equation (8-32) applies for the kinetic energy of any rigid rotating object.

✓ CHECKPOINT

C-8-8 Wheel and Disk

When a wheel and a disk with the same radius spin such that they have the same kinetic energy,
(a) they have the same mass;
(b) they have the same angular speed;
(c) the mass of the disk must be twice as large as the mass of the ring;
(d) all the above are possible.

C-8-8 (d) all the above are possible, we are demanding the same kinetic energy which depends on a few variables.

The work done by the force $\vec{F}$ in Figure 8-22 can be expressed in terms of angular variables. Since the magnitude of the force is constant in this example, the work done is

$$W_F = \vec{F} \cdot \Delta\vec{s} = |\vec{F}||\Delta\vec{s}|\cos\theta \qquad (8\text{-}33)$$

where θ is the angle between the force and the displacement.

For simplicity, we consider a force that remains perpendicular to the massless rod (Figure 8-22). Then, θ in

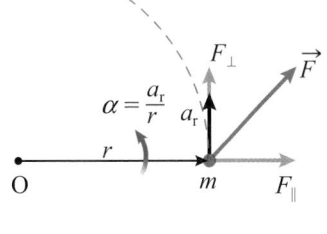

Figure 8-21 The point mass has a linear speed of v and an angular speed of $\omega = v/r$

Equation 8-33 is zero and $\cos \theta = 1$. Since $\Delta s = r\Delta \theta$, the work done by force F over distance Δs is

$$W_F = F\Delta s = Fr\Delta \theta \qquad (8\text{-}33')$$

The magnitude of the torque is rF in this case, so the work done by F can be expressed as

$$W_F = F\Delta s = Fr\Delta \theta = \tau \Delta \theta \qquad (8\text{-}34a)$$

$$W_\tau = \tau \Delta \theta \qquad (8\text{-}34b)$$

In the case of a nonconstant force F, the work done by the varying torque is

$$W_F = \int_{\Delta s} Fds = \int_{\theta_1}^{\theta_2} \tau d\theta \qquad (8\text{-}34c)$$

If the force were not perpendicular to the rod, only the perpendicular component of the force would contribute to the torque, and the parallel component would not contribute to the work done. We can consider the force F or the torque τ as doing work on the rotating point mass.

The universal relation between the total work done on an object and the change in its kinetic energy is

$$W_T = \Delta K \qquad (8\text{-}35)$$

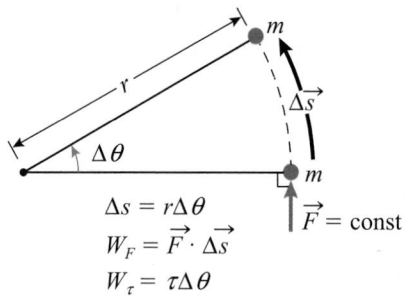

$$\Delta s = r\Delta \theta$$
$$W_F = \vec{F} \cdot \vec{\Delta s}$$
$$W_\tau = \tau \Delta \theta$$

Figure 8-22 Work done on a rotating point mass

Applying this relationship to rotational motion with constant torque gives the following:

KEY EQUATION   for constant torque:

$$W_T = \tau \Delta \theta = \frac{1}{2} I\omega_f^2 - \frac{1}{2} I\omega_i^2 \qquad (8\text{-}36a)$$

and

for varying torque: $W_T = \int \tau d\theta = \frac{1}{2} I\omega_f^2 - \frac{1}{2} I\omega_i^2$

$$(8\text{-}36b)$$

EXAMPLE 8-6

Energy and Angular Acceleration

Two masses ($m_1 = 31.0$ kg and $m_2 = 23.0$ kg) are connected with a massless rope across a pulley as shown in Figure 8-23(a). The pulley is free to rotate about an axis through its centre. There is no friction between mass m_1 and the incline. The angle of the incline is $\theta = 32.0°$. The pulley has a radius of 0.300 m and a moment of inertia of 5.00 kg·m². Determine the acceleration of the blocks.

SOLUTION

METHOD A: NEWTON'S SECOND LAW APPLIED TO ROTATION

A quick check shows that m_2g is greater than $m_1g\sin\theta$. Therefore, the pulley rotates clockwise.

Since $\sum F_y = ma_y$ for the forces acting on m_2,

$$T_{2y} + m_2g_y = m_2a_y$$

Choosing up as our positive direction, we get

$$T_2 - m_2g = -m_2a \qquad (1)$$

where we use a as a variable to denote the magnitude of the acceleration. The same holds for T_2 and g. Equation (1) has two unknowns, T_2 and a. For the forces acting on m_1, $\sum F_x = m_1a_x$, so

$$T_{1x} + m_1g_x = m_1a_x$$

Choosing up the incline as our positive direction, we get

$$T_1 - m_1g\sin\theta = m_1a \qquad (2)$$

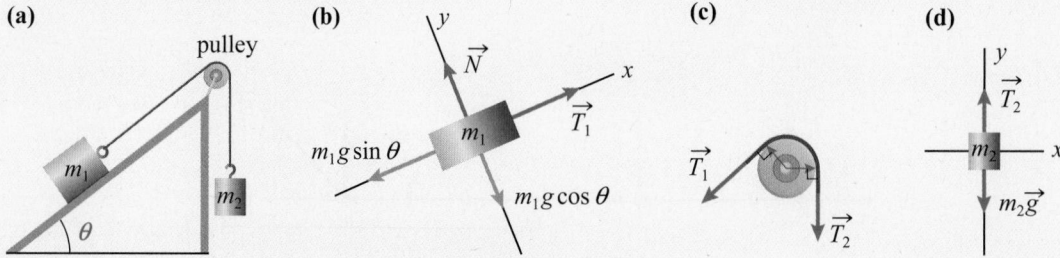

Figure 8-23   (a) Energy and angular acceleration. (b) to (d) The FBDs for Example 8-6

Note that we considered the tension in the rope to be different on each side of the pulley.

In Equations (1) and (2) we have three unknowns. We therefore need at least one more equation. For the pulley, we use the rotational equivalent of Newton's second law. Summing the torques about the centre of mass of the pulley, which we designate as o, we get

$$\sum \vec{\tau}_o = I_o \vec{\alpha}$$

$$\vec{\tau}_{o1} + \vec{\tau}_{o2} = I_o \vec{\alpha}$$

Choosing counterclockwise as our positive direction, we get

$$-\tau_{o1} + \tau_{o2} = I_o \alpha$$

where the variables used in the line above for torque and angular acceleration denote magnitudes. Since the rope applies force tangentially to the circumference of the pulley, the angle between the tension forces and the distance to the centre of rotation (the pivot point) is $\pi/2$ for both forces. Therefore,

$$-rT_1 + rT_2 = I_o \alpha \qquad (3)$$

We now have a total of four unknowns: T_1, T_2, α, and a. From the diagram, we can see another relationship: the acceleration of the rope is the same as the tangential acceleration along the circumference of the pulley:

$$a = \alpha r \qquad (4)$$

Here, again, you have a choice of several methods for solving the system of equations. We will substitute the expressions for T_1 from Equation (1), T_2 from Equation (2), and α from Equation (4) into Equation (3).

$$r(m_2 g - m_2 a) - r(m_1 g \sin\theta + m_1 a) = I_o(a/r)$$

Solving for a gives

$$a = \frac{m_2 g - m_1 g \sin\theta}{m_1 + m_2 + I_o/r^2}$$

$$= \frac{23\,\text{kg}\,(9.81)\,\text{m/s}^2 - 31\,\text{kg}\,(9.81\,\text{m/s}^2)(\sin(32°))}{23\,\text{kg} + 31\,\text{kg} + (5\,\text{kgm}^2/0.3^2\,\text{m}^2)} = 0.583\ \text{m/s}^2$$

Let us examine in detail why the two tension forces need to be different. According to Equation (3), the two tensions can be equal in one of two cases: Either the angular acceleration of the pulley is zero (i.e., the system is in equilibrium) or the moment of inertia of the pulley is zero (i.e., the mass of the pulley is zero). You may recall that in all the problems involving pulleys in previous chapters we assumed that the pulley was frictionless and massless. This assumption guarantees that $T_1 = T_2$. Note that friction at the axle of the pulley would add an extra torque to the equation, increasing the difference between the tension forces on either side of the pulley.

METHOD B: THE WORK-MECHANICAL ENERGY APPROACH

When m_2 descends a distance h, the gravitational potential energy of interaction between m_2 and Earth potential energy

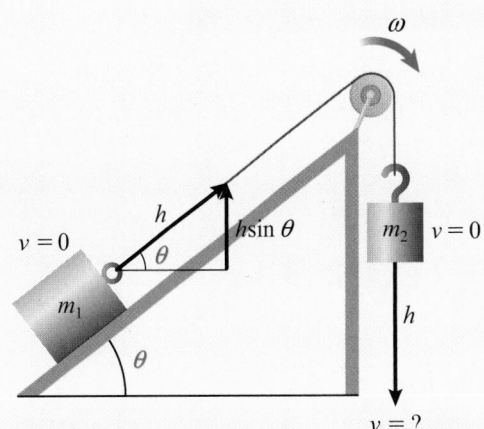

Figure 8-23(e) As m_2 descends and m_1 goes up the incline, some potential energy of the Earth-m_2 system is converted into potential energy for the Earth-m_2 system and into kinetic energy of the blocks and pulley.

decreases by m_2gh. This loss in potential energy goes into increasing the potential energy of the Earth-m_1 system by $m_1 g \sin\theta h$ and increasing the kinetic energies of the pulley and the two masses as shown in Figure 8-23(e). If the two masses are initially at rest,

$$m_2 gh = m_1 g \sin\theta h + \frac{1}{2}m_1 v^2 + \frac{1}{2}m_2 v^2 + \frac{1}{2}I\omega^2 \qquad (5)$$

When we apply the chain rule to take the derivative of Equation (5) with respect to time, noting that h is a function of time. we get

$$m_2 gv = m_1 g \sin\theta v + m_1 va + m_2 va + I\omega\alpha$$

Here, we have used the fact that $\dfrac{dv^2}{dt} = \dfrac{dv^2}{dv}\dfrac{dv}{dt} = 2v\dfrac{dv}{dt} = 2va$ as given by the chain rule. Substituting a/r for α and v/r for ω yields

$$m_2 gv = m_1 g \sin\theta v + m_1 va + m_2 va + Iva/r^2$$

Dividing this equation by v and solving for a, we have

$$a = \frac{m_2 g - m_1 g \sin\theta}{m_1 + m_2 + I/r^2}$$

This is the same expression as the one derived using Method A.

Making sense of the result:

As a quick check, we can verify the units for the acceleration, a. Every term in the numerator has units of $\text{kg}\cdot\text{m/s}^2$, and the units of I/r^2 are $\text{kg}\cdot\text{m}^2/\text{m}^2 = \text{kg}$, so the units for the denominator are kg. Therefore, the units for the acceleration a are m/s^2, as expected. We were able to use the principle of conservation of energy here because there are no nonconservative forces involved.

For a pulley with negligible mass, I is zero. If we were to substitute this zero value into the expression for a above, we would obtain the same expression that we had for the same problem in Chapter 5, where the pulley was assumed to be massless and frictionless.

CHECKPOINT

C-8-9 Same Rope = Same Tension?

If the pulley in Example 8-6 is accelerating, can the forces of tension on each side be equal to each other?

C-8-9 No. According to Equation (3), $-rT_1 + rT_2 = I_0 \alpha$, if the tensions are equal, then either the pulley has a zero moment of inertia or is not accelerating.

EXAMPLE 8-7

Falling Bar

The vertical bar shown in Figure 8-24 pivots about its lower end. Starting from rest, the bar swings downward. Find the speed of the free end when the bar has rotated through angle θ.

SOLUTION

Assuming that friction in the system is negligible, all the gravitational potential energy lost as the bar falls the bar will be converted into kinetic energy. From Chapter 6, we have

$$\Delta K = -\Delta U \qquad (1)$$

Since the bar starts from rest, the change in kinetic energy is

$$\Delta K = K_f - K_i = \frac{1}{2}I\omega^2 - 0 = \frac{1}{2}I\omega^2 \qquad (2)$$

To calculate the change in potential energy, we need to find the distance through which the centre of the mass of the bar falls, Δh_{cm}. We can see from Figure 8-24 that this distance is

$$\Delta h_{cm} = \frac{L}{2} - \frac{L}{2}\cos\theta \qquad (3)$$

Therefore,

$$\frac{1}{2}I\omega^2 = Mg\left(\frac{L}{2} - \frac{L}{2}\cos\theta\right) = Mg\frac{L}{2}(1 - \cos\theta) \qquad (4)$$

The moment of inertia of the bar about the pivot at its end is given by Equation (8-31b):

$$I_{cm} = \frac{1}{3}ML^2$$

Substituting into Equation (4) gives

$$\frac{1}{2}\left(\frac{1}{3}ML^2\right)\omega^2 = Mg\frac{L}{2}(1 - \cos\theta)$$

Figure 8-24 The potential energy from the fall of the bar is converted into kinetic energy.

$$L^2\omega^2 = 3gL(1 - \cos\theta)$$

Since the linear speed of the free end of the bar is $v = \omega L$,

$$v^2 = 3gL(1 - \cos\theta)$$

Making sense of the result:

For $\theta = 90°$, the top of the bar would have fallen by a full bar length, giving us $v^2 = 3gL$. If, instead of a swinging bar we had a swinging pendulum with a massless rod and a point mass at its end, such that the mass falls by a height L, the result, as we saw in Chapter 6, would be $v^2 = 2gL$. Since the bar has a moment of inertia that is less than the moment of inertia of a point mass of the same mass, we expect the tip of the bar to be easier to accelerate.

MAKING CONNECTIONS

Mechanical Hybrids and Continuously Variable Transmissions

In 1900, Ferdinand Porsche developed the first electric hybrid vehicle, the Lohner-Porsche Semper Vivus (Figure 8-25). Most automotive brakes use friction to slow a car, converting most of the kinetic energy into thermal energy. Some hybrid cars use magnetic braking, where kinetic energy is converted first into electrical energy and then

into electrochemical energy, which is stored in the car's batteries. However, as in any energy conversion process, there is significant energy loss in the conversion processes of this system. A flywheel with a relatively high moment of inertia can, in principle, be used to store the car's mechanical energy. Such systems generally require a continuously variable transmission for the precise control of energy between the car and the flywheel. Although no fully mechanical hybrids are in production at the time of writing, continuously variable transmissions are already available on a few cars. Fittingly perhaps, the Porsche 911 GT3 R is the first hybrid vehicle to use mechanical flywheel energy storage instead of the conventional electrochemical battery storage. However, this car still uses magnetic braking to transfer energy to the flywheel which acts as a generator when called upon.

Figure 8-25   (a) The Lohner-Porsche Semper Vivus, the first electrical hybrid, on display. (b) The Porsche 911 GT3 R, the first semi mechanical hybrid, on a racetrack.

LO 7, LO 8

8-7   Angular Momentum

The Angular Momentum of a Point Mass

Let us again consider a point mass attached with a massless rod to a pivot at point O and rotating with speed v as shown in Figure 8-26. In the absence of an external force, the mass will continue to move at a constant speed. The tension in the rod acts perpendicular to the direction of motion of the mass and does not change its speed.

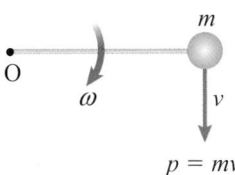

Figure 8-26   A point mass rotating about a pivot

As shown in Figure 8-27, the **angular momentum** of the point mass about an axis through point O, perpendicular to the page, is defined as

$$\vec{L_o} = \vec{r} \times \vec{p} \qquad (8\text{-}37)$$

where $\vec{p}$ is the linear momentum of the point mass, and $\vec{r}$ is the position vector of the object with respect to the pivot or axis.

Any particle that has linear momentum also has angular momentum about any specified axis. The angular momentum is zero when $\vec{r}$ and $\vec{p}$ are collinear. Note that a particle (or an object) can have angular momentum with respect to an axis even when the particle does not rotate about that axis.

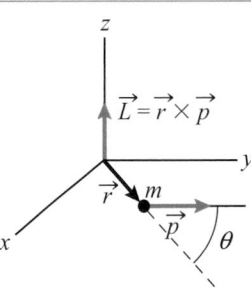

Figure 8-27   The angular momentum for a point mass

✓ CHECKPOINT

C-8-10 Angular Momentum on a Straight Line?

Two cars of the same mass M are moving in straight line in opposite directions on a highway at the same speed v. When the cars pass each other, the distance between them is d. What is the magnitude of the angular momentum of each about the other car as they pass each other?

(a) Mvd
(b) $2Mvd$
(c)   cannot be defined as both are moving
(d)   none of the above

C-8-10 (a) $L = rp \sin \theta$

✓ CHECKPOINT

C-8-11 Higher Angular Momentum

The mass in Figure 8-28 is travelling at a constant speed v. Compare the angular momentum about point A for the mass at the three locations shown at the three different times, t_1, t_2 and t_3.

C-8-11 The angular momentum is the same because only the perpendicular component of the position vector (as measured from the pivot) matters.

Figure 8-28 A point mass moving in a straight line at a constant speed

For the point mass in Figure 8-27, r and p are perpendicular, so

$$|\vec{L}_o| = |\vec{r} \times \vec{p}| = rp \sin \theta = rp = rmv$$
$$= rm\omega r = mr^2 \omega = I_o \omega \qquad (8\text{-}38a)$$

Stated in vector form:

$$\vec{L}_o = I_o \vec{\omega} \qquad (8\text{-}38b)$$

The subscript O on the symbol for angular momentum indicates that the angular momentum is taken about the point or pivot O (as with torques and moments of inertia).

Since I is a scalar, the angular momentum vector has the same direction as the angular velocity vector. The direction of the angular velocity vector is determined using the right-hand rule as illustrated earlier in Section 8-3.

The Angular Momentum of a Rotating Rigid Body

We can demonstrate that Equation (8-38) applies for any rigid body rotating about a point or an axis. Figure 8-29 shows a disk spinning about its axis of symmetry at angular velocity $\vec{\omega}$. The right-hand rule gives the direction of the angular momentum vector, which, according to Equation 8-38b, is parallel to the angular velocity vector.

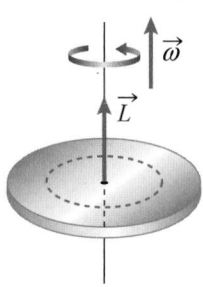

Figure 8-29 The angular momentum of a rigid object

The Rate of Change of Angular Momentum

In Chapter 7, we derived the relationship between the net force on a point mass and the change in its linear momentum:

$$\vec{F}_{net} = \frac{d\vec{p}}{dt} \quad \text{and} \quad \vec{F}_{net}^{avg} = \frac{\Delta \vec{p}}{\Delta t} \qquad (8\text{-}39)$$

EXAMPLE 8-8

The Angular Momentum of a Spherical Shell

A child spins the brass globe on her parent's desk at 3.00 rev/s. The globe is a thin spherical shell with a mass of 1.10 kg and a radius of 23.0 cm. Find the angular momentum of the spinning globe.

SOLUTION

The angular momentum of a rigid body about an axis of rotation is given by Equation (8-38):

$$L = I\omega$$

From Table 8-1, the moment of inertia of a spherical shell about an axis through its centre of mass is $2/3\ MR^2$. There are 2π radians in one revolution, so the angular speed is

$$\omega = (2\pi \text{ rad/rev})(3 \text{ rev/s}) = 6.00\ \pi \text{ rad/s}$$

Substituting into the expression for angular momentum, we get

$$L = I\omega = \frac{2}{3}(1.1 \text{ kg})(0.23 \text{ m})^2(6\pi \text{ rad/s}) = 0.731 \text{ kg} \cdot \text{m}^2/\text{s}$$

Making sense of the result:

The units for L are as expected.

When we take the cross product of r with both sides of Equation (8-39), we get

$$\vec{\tau}_{net} = \vec{r} \times \vec{F}_{net} = \vec{r} \times \frac{d\vec{p}}{dt} = \frac{d}{dt}(\vec{r} \times \vec{p}) = \frac{d\vec{L}}{dt} \qquad (8\text{-}40a)$$

and

$$\vec{\tau}_{avg}^{net} = \frac{\Delta \vec{L}_{net}}{\Delta t} \qquad (8\text{-}40b)$$

Thus, a net torque acting on an object changes the angular momentum of the object. Rearranging Equation (8-40) and integrating gives

$$d\vec{L}_o = \vec{\tau}_{net}\ dt$$
$$\Delta \vec{L} = \int_{interval} \vec{\tau}_{net}\ dt = \vec{\tau}_{net}\ \Delta t \qquad (8\text{-}41)$$

The change in angular momentum, $\Delta \vec{L}$, is called the **angular impulse**. The symbol J is often used for this quantity.

Conservation of Angular Momentum

Equation (8-40) states that the change in the angular momentum of an object depends on the net torque experienced by that object. In the absence of a net torque on an object, the angular momentum of the object does not change and, therefore, is conserved. This principle can be extended to any system of objects.

This principle is demonstrated by the increase in the rate at which a figure skater spins when she reduces her moment of inertia by bringing her arms in close to her torso (the product $I_o\omega$ remains constant).

EXAMPLE 8-9

Shuttle Payload

A space shuttle orbiter is carrying a 21 000 kg satellite, which has a cylindrical shape with a radius of 1.90 m (Figure 8-30). The satellite needs to be spinning about its axis before it is released into space, so just before the satellite is released, a motor in the cargo bay spins the satellite until the satellite has an angular speed of 45.0 rad/s. The moment of inertia of the orbiter about the axis of rotation of the satellite is 4.20×10^6 kg·m².

(a) Find the angular speed of the orbiter just before the spinning satellite is released.
(b) What effect does releasing the satellite have on the angular speed of the orbiter?
(c) Does the rate at which the satellite is accelerated have an effect on the final angular speed of the orbiter?

SOLUTION

(a) We apply the concept of conservation of angular momentum to the system consisting of the orbiter and the satellite. The force (or torque) exerted by the orbiter on the satellite is internal to the satellite–orbiter system. When the motor in the obiter exerts a force (or torque) on the satellite, the satellite exerts an equal but opposite force (or torque) on the orbiter, in accordance with Newton's third law. Since the action–reaction pair is internal to the system, the total angular momentum is conserved. Therefore, $L_i = L_f$ for the system.

No part of the system was spinning before the motor was turned on, so the angular momentum of the system about the axis of rotation of the satellite is zero, and

$$\vec{L}_i = \vec{L}_f = 0$$

The final angular momentum is the sum of the angular momentum of the satellite and the angular momentum of the orbiter:

$$\vec{L}_f = \vec{L}_{sat} + \vec{L}_{orb} = 0$$

Therefore, the angular momenta of the satellite and the orbiter have equal magnitudes but opposite directions in magnitude we can say:

$$L_{orb} = I_{orb}\omega_{orb} = L_{sat} = I\omega_{sat} = \frac{1}{2}M_{sat}R_{sat}^2\omega_{sat}$$

Figure 8-30 Example 8-9

Solving for ω_{orb} and then substituting the given values, we get

$$\omega_{orb} = \frac{\frac{1}{2}M_{sat}R_{sat}^2\omega_{sat}}{I_{obr}}$$

$$\omega_{orb} = \frac{\frac{1}{2}(21\ 000\ \text{kg})(1.9\ \text{m})^2(45\ \text{rad/s})}{4.2 \times 10^6\ \text{kg}\cdot\text{m}^2}$$

$$= 0.406\ \text{rad/s} = 23.3°/\text{s}$$

(b) Releasing the satellite has no effect on the angular speed of the orbiter because the satellite was already spinning when it was released. The effect of the satellite's rotation on the orbiter was solved in part (a).
(c) The rate at which the satellite is accelerated has no effect on the final angular speed of the orbiter. All that matters are the initial and final states.

Making sense of the result:

The heavier the orbiter the slower it will spin, according to the answer from (a) an infinitely heavy orbiter will not spin at all and a relatively light payload will not significantly affect the orbiter's rotation.

MAKING CONNECTIONS

Tides and Precise Time

Due to tidal friction, the Earth's rotation loses about 2.3 ms every century compared to atomic clocks. In 2010, NIST researchers introduced a quantum logic clock (Figure 8-31), which uses the vibrations of laser-cooled aluminum and magnesium ions. The new quantum clock is expected to be accurate to within 1 s over 37 billion years. Pulsars can also be used to keep time. Even though pulsar periods are about 100 times less accurate than the quantum clock, pulsars are useful for long-term time measurements because they will likely last much longer that the service life of a clock on Earth. Sudden changes in pulsar periods may indicate gravitational waves originating from black holes.

(continued)

Figure 8-31 The ion trap is a key component of the NIST (National Institute of Standards and Technology) aluminum ion clock.

J. Koelemeij/NIST

ONLINE ACTIVITY

Angular Momentum

The e-resource that accompanies every new copy of this textbook contains an Online Activity using the PhET simulation "Torque." Work through the simulation and accompanying questions to gain an understanding of the conservation of angular momentum.

 CHECKPOINT

C-8-12 Conservation of Angular Momentum

A child steps onto a stationary merry-go-round and starts walking along its edge. Which of the following statements is true? Assume that the friction in the mechanism of the merry-go-round is negligible.

(a) The total angular momentum of the merry-go-round–child system increases as a result of the child walking.

(b) The child has no angular momentum because she walks in a tangential direction around the merry-go-round.

(c) The kinetic energy of the merry-go-round–child system does not change as a result of the child walking.

(d) The angular momentum of the merry-go-round alone does not change.

(e) The child's motion on the merry-go-round has no effect on the total angular momentum.

☑ **CHECKPOINT**

C-8-13 Rotational Stabilization

You are standing on a platform that is free to rotate while holding a bicycle wheel that is spinning about its axle, which is horizontal. You then turn the axle vertical (Figure 8-32).

(a) Why do you need to exert a torque to tilt the axle?

(b) When the wheel slows down, what happens to the torque you have to exert to tilt the spinning axle?

(c) Why does the platform you are standing on begin to spin as you tilt the wheel axle?

(a)

The student stands on a stationary platform holding a bicycle wheel that spins counterclockwise; the bicycle wheel's angular momentum points horizontally.

(b)

She flips the moving wheel, reversing its angular momentum. The total angular momentum is conserved, so the wheel and the student must rotate the other way.

Figure 8-32 (a) Initially, the total angular momentum points horizontally. (b) When you turn the axis of rotation of the wheel to the vertical, the direction of the total angular momentum changes

C-8-13 a) Any change to the angular momentum requires a torque. b) the torque you apply will be less if you wish to tilt the axle at the same rate. c) Angular momentum is conserved; because the vertical component of angular momentum was zero initially, it must be zero in the final situation.

C-8-12 (e) The force or torque exerted by the child on the merry-go-round is internal to the merry-go-round–child system. The total angular momentum is conserved.

EXAMPLE 8-10

Magnetic Boots

Two astronauts are installing equipment on an orbiting platform that is rotating freely at a rate of 1.10 rad/s about its centre of mass, as shown in Figure 8-33. Astronaut 1 is standing at the centre of the platform, and astronaut 2 is at the edge. The platform has a radius of 11.0 m. The moment of inertia of the platform without the astronauts is $I_p = 2400 \text{ kg} \cdot \text{m}^2$. Astronaut 1 walks to the edge and hands a container of mass $m_c = 24.0 \text{ kg}$ to astronaut 2 and then returns to the centre. Both astronauts are wearing magnetic boots that keep them on the surface of the platform. Compared to the dimensions of the platform, you can consider the astronauts to be thin cylinders or point masses of mass $m_a = 81.0 \text{ kg}$ each. The container can also be considered a point mass.

(a) Find the resulting angular speed of the platform after astronaut 1 has returned to the centre.
(b) Find the total work done by the astronaut(s) in the process.
(c) If astronaut 1 keeps walking back and forth between the edge of the platform and its centre, will he eventually do enough work on the platform to lose all its kinetic energy and come to a stop? If so, how many times does astronaut 1 have to walk back and forth?

Figure 8-33 Example 8-10

SOLUTION

Any forces that astronaut 1 exerts on the platform to walk to the edge, deliver the container, and then return, along with any force the astronauts exert on the container during transport and delivery, are all internal to the astronauts–platform–container system. There is no net external torque on the system, so the angular momentum of the system does not change. Therefore,

$$\vec{L}_i = \vec{L}_f \quad \text{and} \quad I_i \vec{\omega}_i = I_f \vec{\omega}_f$$

The initial moment of inertia of the system is the total of the moments of inertia of the platform, astronaut 1 and the container at the centre, and astronaut 2 at the edge. Since we can treat the astronauts and the container as point masses, their moments of inertia about the axis are simply mr^2, and r is zero for the astronaut and the container when at the centre. Therefore,

$$I_i = m_a(0)^2 + m_c(0)^2 + I_p + m_a r^2$$
$$= 0 + 0 + 2400 + 81(11)^2 = 12\,201 \text{ kg} \cdot \text{m}^2$$

The final moment of inertia is equal to the initial moment of inertia plus the moment of inertia of the container when at the edge of the platform:

$$I_f = I_p + m_a r^2 + m_c r^2 = 2400 + 81(11)^2 + 24(11)^2$$
$$= 15\,105 \text{ kg} \cdot \text{m}^2$$

So, our conservation of angular momentum equation becomes

$$(12\,201 \text{ kg} \cdot \text{m}^2)(1.1 \text{ rad/s}) = (15\,105 \text{ kg} \cdot \text{m}^2)\omega_f$$

and $\omega_f = 0.889 \text{ rad/s}$

(b) The work done by astronaut 1 is the only work done on the system. Therefore,

$$W_T = \Delta K_E = \frac{1}{2}I_f \omega_f^2 - \frac{1}{2}I_i \omega_i^2$$

$$= \frac{1}{2}(15\,105 \text{ kg} \cdot \text{m}^2)(0.889 \text{ rad/s})^2$$

$$- \frac{1}{2}(12\,201 \text{ kg} \cdot \text{m}^2)(1.1 \text{ rad/s})^2$$

$$= -1420 \text{ J}$$

(c) Since there are no mechanisms for losing energy, such as an external force of friction, the energy of the whole system is conserved. The work done by astronaut 1 as he walks back and forth on the platform alternates between being positive and negative, such that in each return trip the total work done is zero.

Making sense of the result:

Since moving the container to the edge of the platform results in an increase in the total moment of inertia of the isolated system, we would expect the angular speed of the system to decrease to keep the total angular momentum constant.

☑ **CHECKPOINT**

C-8-14 Negative Work?

In Example 8-10, How would the result have been different if the astronaut on the inside chose to slide the container on the surface of the platform to his comrade instead of delivering it himself? How would the result have changed if the astronaut inside threw the package to the other astronaut who caught it in the air?

C-8-14 The final result would be the same in each case, because the initial and final states are the same.

☑ **CHECKPOINT**

C-8-15 Constant Force?

If the astronauts in Example 8-10 were to walk from the centre of the platform at a constant speed, would the force they exert on the platform surface be constant?

C-8-15 No, the centripetal force they experience increases as they move away from the centre.

FUNDAMENTAL CONCEPTS AND RELATIONSHIPS

Table 8-2 summarizes the main concepts and quantities of rotational dynamics.

Table 8-2   Comparison of Angular and Linear Quantities and Concepts

Linear quantity	Symbol/Expression	Angular quantity	Symbol/Expression
Linear position	x	Angular position	θ
Linear velocity	v	Angular velocity	ω
Linear acceleration	a	Angular acceleration	α
Kinematics equations for constant a	$x(t) = \frac{1}{2}at^2 + v_0 t + x_0$ $v(t) = at + v_0$ $v^2 = v_0^2 + 2a \cdot \Delta x$	Kinematics equations for constant α	$\theta(t) = \frac{1}{2}\alpha t^2 + \omega_0 t + \theta_0$ $\omega(t) = \alpha t + \omega_0$ $\omega^2 = \omega_0^2 + 2\alpha \cdot \Delta \theta$
Mass	M	Moment of inertia	I
Force, average force, and net force	$\vec{F}, \vec{F}_{avg}, \vec{F}_{net}$	Torque, average torque, and net torque	$\vec{\tau}, \vec{\tau}_{avg}, \vec{\tau}_{net}$
Work of a force	constant force $W = \vec{F} \cdot \Delta \vec{s}$ variable force: $W = \int \vec{F} \cdot d\vec{s}$	Work of a torque	constant torque $W = \tau \Delta \theta$ variable torque: $W = \int \tau d\theta$
Work–kinetic energy theorem	$W_{Total} = \Delta K$	Work–kinetic energy theorem	$W_{Total} = \Delta K$
Linear momentum	$\vec{p} = m\vec{v}$	Angular momentum	$\vec{L} = \vec{r} \times \vec{p}, \vec{L} = I\vec{\omega}$
Linear impulse	$\vec{I} = \Delta \vec{p}$	Angular impulse	$\vec{J} = \Delta \vec{L}$
Conservation of linear momentum	if $\vec{F}_{ext} = 0, \vec{p}_i = \vec{p}_f, \Delta \vec{p} = 0$ (isolated system)	Conservation of angular momentum	If $\vec{\tau}_{net} = 0, \vec{L}_i = \vec{L}_f, \Delta \vec{L} = 0$ (isolated system)
Net force: change in linear momentum	average $\vec{\tau}_{ext} = \dfrac{\Delta \vec{L}}{\Delta t}$ instantaneous $\vec{F}_{ext} = \dfrac{d\vec{p}}{dt}$	Net torque: change in angular momentum	average $\vec{F}_{ext} = \dfrac{\Delta \vec{p}}{\Delta t}$ instantaneous $\vec{\tau}_{ext} = \dfrac{d\vec{L}}{dt}$

Key Applications: navigation, GPS, flywheels, mechanical hybrids, variable transmissions, engines and propellers, tidal friction, Earth's rotation, microgravity, biophysical systems, nuclear rotation, atomic angular momentum, spin angular momentum, atomic and molecular spectra, laser cooling of atoms, nuclear magnetic resonance and MRI, neutron stars, rotating black holes, black Saturns, gravity waves

Key Terms: angular acceleration, angular displacement, angular impulse, angular momentum, angular speed, moment arm, moment of inertia, parallel-axis theorem, rigid body, rotational kinetic energy, torque

QUESTIONS

1. Two spheres of equal mass and radius are spinning at the same angular speed about axes through their centres. One of the spheres is solid, and the other is a spherical shell. Which sphere has the higher kinetic energy?
2. Can one talk about an object's moment of inertia without specifying an axis? Explain.
3. A bicycle wheel mounted on a vertical axis is being accelerated from its initial angular speed at a constant acceleration. Compare the tangential acceleration of a point on the circumference to a point midway between the circumference and the centre. Also compare the speed and the radial acceleration of the two points.
4. Two globes of identical mass and radius are spinning about axes through their centres of symmetry at the same speed. One of the globes is a solid sphere, and the other has a hollow space at its centre. Which globe is easier to stop? Explain.

5. Two disks of different radii are connected by a belt running along their circumference, and driven by a motor that spins a larger disk at a constant angular speed. Can we say that the disks have the same angular speed? Is the radial acceleration of points on the circumference of the two disks the same? What do the two disks have as a common property?

6. Consider the heavy pulley setup in Figure 8-34. Can we say that the tension in the belt is the same everywhere? Which of the two masses would you expect to have the higher acceleration?

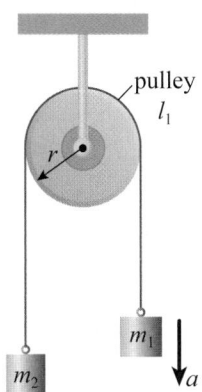

Figure 8-34   Question 6

7. You are standing on a freely spinning disk while holding on to a medicine ball, with your arms extended over the edge of the disk. You then throw the medicine ball such that its speed has only a vertical component when it lands on the ground. Did your and the disk's angular speed increase or decrease after throwing the ball?

8. Two thin rods of the same length are pivoted about a free hinge at their bottom edge while being held vertically. One rod is heavier than the other. The two rods are let go simultaneously, and they begin to fall. Which rod will be rotating faster at the bottom of its trajectory?

9. Two thin rods of the same mass are pivoted about a free hinge at their bottom edge while being held vertically. One rod is longer than the other. The two rods are let go simultaneously, and they begin to fall. Which rod will be rotating faster at the bottom of its trajectory?

10. Two pendulums are held in a horizontal position and then let go. Both pendulums are bars made of the same material, but one is thinner and has a sphere at the end. The masses and overall lengths of the pendulums are the same. Which pendulum will have a higher angular speed when it reaches a vertical position?

11. For the kinematics Equation (8-7), do you have to use radians, or can you use degrees or other units of measuring angles? Explain.

12. The propeller blades of single-engine plane spin at a constant speed. Compare the angular speed in rad/s of a point at the tip of one of the blades to the angular speed of a point at the middle of the blade. Does either of the two points accelerate? If so, compare their accelerations.

13. Your bike is upside down on the floor, and you are fixing the chain. The gear setting is such that the chain winds around the small gear at the rear wheel and around the large gear attached to the pedals (Figure 8-35).

(a) When you spin the pedals at a constant speed, will the wheel necessarily be turning at the same rotational speed as the pedals?

(b) You mark a point with a white marker on the rear tire, and you watch that point as you spin the wheel at a constant speed. Can you say that the white mark accelerates while the wheel spins at a constant speed?

Figure 8-35   Question 13

14. How could you tell a raw egg from a boiled egg without cracking them open?

15. A figure skater is executing a high-speed spin routine with her arms tucked close to her body (Figure 8-36). When she spreads her arms out, is she doing positive or negative work? When she brings her arms in closer to her body again, is she doing positive or negative work? Assuming zero friction, how would her initial angular speed compare to her angular speed after extending her arms out and then bringing then back in?

Figure 8-36   Question 15

16. Would the expression for the moment of inertia of a thin rod about an axis through its centre hold if the diameter of the rod were no longer negligible?

17. Use the moment of inertia for a ring to derive the moment of inertia for a thin cylindrical shell.

18. You are sitting on a freely spinning pedestal holding a long bar upright close to your body. You then tilt the bar until it becomes horizontal.
    (a) What happens to your angular speed?
    (b) Is the work you do as you rotate the bar positive or negative?

19. An insect ($m = 11$ g) is standing on the rim of a horizontal ring of mass $M = 0.90$ kg and radius $R = 9.0$ cm. The ring is free to spin about its centre-of-mass axis. The insect then moves around the ring and reaches a maximum speed of 15 cm/s with respect to the ring. The insect completes two full cycles around the ring and comes to a stop with respect to the ring. Find the final speed of the insect.

20. Why is the moment of inertia of a solid sphere only an approximate model for calculating the moment of inertia of Earth?

21. Using the vector model for angular momentum, explain why torque is required to tilt the axis of a spinning bicycle wheel.

22. Two cars are driving toward each other at the same speed on the highway. The lanes they are travelling in are each a distance d away from the median that divides the highway. Does the angular momentum of the two-car system about a point midway between them change as the cars pass each other?

23. A rope over a pulley with a moment of inertia I has a heavy load m_1 attached to one end and a counterweight m_2 tied to the other end, as in Figure 8-37. Is the tension ever the same in the rope on both sides of the pulley?

24. In the system shown in Figure 8-37, m_2 is greater than m_1, and the pulley has a moment of inertia I. Mass m_1 is given an initial speed downward, and the system slows due to gravity. When the system momentarily stops, which part of the rope (on either side of the pulley) has the higher tension? Does the pulley have an acceleration at that moment?

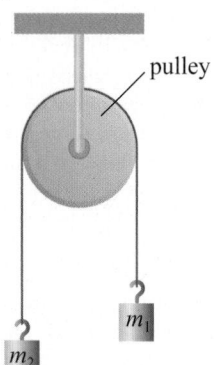

Figure 8-37   Question 24

25. Can a wheel have an angular acceleration if it is not moving? Explain.

26. Can a rotating object have an angular acceleration in one direction while rotating in the other? Explain.

27. Three people are pushing a stalled car toward a gas station. Can they create a torque about a tree a few hundred metres away? Explain.

28. A particle is flying in a straight line at a constant speed. Does the particle have angular momentum? Can we make conclusions regarding the direction of its angular momentum, if it exists? Explain.

29. Can we define the torque of a force without referring to a pivot point or rotation axis? Can we speak of force without referring to a frame of reference coordinate system? Explain.

PROBLEMS BY SECTION

For problems, star ratings will be used, (✱, ✱✱, or ✱✱✱), with more stars meaning more challenging problems.

Section 8-1   Angular Variables

30. ✱ What is the angular speed, in rad/s, of (a) a point on the equator and (b) a point in Antarctica close to the South Pole?

31. ✱ Taking Earth to be a perfect sphere, find the linear speed of a point located on the 32nd parallel, as a result of Earth's rotation.

32. A car is undergoing an emissions test. The car is stationary on rollers while the gas pedal is depressed until the speedometer reads 60 km/h. The wheels are 54 cm in diameter; find their angular speed.

33. ✱ Blades of a plane propeller need to be designed such that the top speed of the blade tip is just under the speed of sound. When the maximum engine speed is 7000 rev/min, determine what length the blade should be.

34. ✱ A Ferris wheel has a radius of 37.0 m. When a point on its circumference has moved 21.0 m, how many degrees did the Ferris wheel rotate?

35. ✱ Find the angular speed of the hour hand of an analog clock.

36. ✱ Astronauts often have to train in high-g environments before they embark on space missions. One way to simulate a high-g environment is to place astronauts in rotating capsules, such as the one in Figure 8-38, used by NASA. What angular speed is required when the radius of the track is 72 m and the acceleration target is $7.0g$?

NASA/Photo Tony Landis

Figure 8-38   Problem 36

Section 8-2   Kinematics Equations for Rotation

37. ✱ You have your bicycle upside down while you adjust the chain. The chain is wound around the rear wheel's small gear, which has a radius of 4.0 cm. At the pedals, the chain is wound around the large gear, which has a radius of 11.0 cm. The radius of the rear wheel is 35.0 cm. You spin the pedals initially in such a way that they finish two revolutions in 1.0 s, starting from rest. Assume constant acceleration.
    (a) Find the angular acceleration of the rear wheel in rad/s².

(b) Find the total acceleration of a point at the edge of the small gear at the end of 1.5 s.
(c) What is the tangential acceleration for that point during the first second?
(d) Calculate the linear acceleration of the chain.

38. ✱ A space-training module consists of a capsule that rotates inside a horizontal circular track of radius 110 m. The rotational speed increases at a constant rate from rest such that the capsule finishes its first cycle in 9.00 s.
(a) How long does it take from the start of the motion for the radial acceleration to be equal to six times the acceleration due to gravity?
(b) How long does it take for the radial acceleration to be equal in magnitude to the linear acceleration?

39. ✱✱ The propeller blades of an airplane are 2.1 m long. The plane is getting ready for takeoff, and the propeller starts turning from rest at a constant angular acceleration. The propeller blades go through two revolutions between the fifth and the seventh second of the rotation. Find the angular speed at the end of 8.0 s.

40. ✱✱ A wheel is spinning about a fixed axis at an angular frequency of 29 rad/s. A braking mechanism reduces the speed by half over the first five revolutions. Assume that the braking force is constant.
(a) Find the angular speed 1.6 s into the braking phase.
(b) Calculate the angular displacement during that time.
(c) Find the angular speed at the end of 4.0 s.

41. ✱✱ A spinning globe is slowed at a constant acceleration of 1.3 rad/s² until it stops. One of the points on the equator moves 23° in the first 0.70 s of the slowing phase.
(a) Find the total angular displacement of the globe during the acceleration phase.
(b) Find the initial angular speed of the globe.

42. ✱ A bicycle wheel initially spinning at 7.00 rad/s is slowed at a constant acceleration such that it reaches a speed of 2.50 rad/s over an angular displacement of 11.2 rad. Find the angular acceleration.

43. ✱ A horizontally mounted bicycle wheel 42 cm in radius is accelerated from rest at 3.5 rad/s² for 7.0 s, after which time it rotates at a constant speed.
(a) Find the radial acceleration of a point on its circumference 3.0 s into the motion.
(b) Find the radial acceleration of the same point 9.0 s into the motion.

Section 8-3   Torque

44. ✱ A carpenter is driving a nail straight into the northern wall of house, with an average force of 1200 N. Find the magnitude of the torque of this force about a point on the wall 3.0 m to the right of the nail.

45. ✱ Find the cross product of $\vec{r}(2\hat{i}, 4\hat{j}, 0)$ m and $\vec{F}(4, 133°, 90°)$ N.

46. ✱ A force $\vec{F} = (12.0\hat{i} + 4.0\hat{j} - 16.0\hat{k})$ N is applied at the point A (−3.0 m, 5.0 m, 2.0 m). Find the torque of $\vec{F}$ about point B (−3.0 m, 2.0 m, −6.0 m).

47. ✱ Calculate the torque about the origin of force $\vec{F} = (11.0\hat{i} + 6.0\hat{j} - 14.0\hat{k})$ N applied at point A (2.0 m, −5.0 m, 7.0 m).

48. ✱ Determine the force F needed to prevent a beam ($m = 1100$ kg) from swinging down under the effect of its own weight (Figure 8-39).

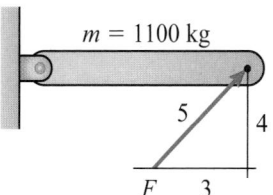

Figure 8-39 Problem 48

49. ✱ Calculate the magnitude of the torque about the origin of force $\vec{F}$ (670 N, 37°) applied in the xy-plane at a point on the x-axis 2.00 m to the right of the origin.

50. ✱ Find the cross product $\vec{r} \times \vec{F}$ where $\vec{r} = $ (17.0 m, 33.0°, 90°) m and $\vec{F} = $ (430 N, 81.0°, 90°) N. The vectors are given in spherical coordinate notation. Report the result in both spherical and Cartesian notation.

51. ✱ Calculate the vector product of $\vec{r} = (2\hat{i} - 3\hat{j} - 5\hat{k})$ m and $\vec{F} = (-3\hat{i} - 5\hat{j} + \hat{k})$ N. Give the answer in both Cartesian and magnitude angle notation.

Section 8-4   Moment of Inertia of a Point Mass

52. ✱ Three masses, each $m = 12$ kg, are located at the corners of an equilateral triangle. Calculate the moment of inertia of the system about an axis running through one of the point masses perpendicular to the plane of the triangle. The length of one side of the triangle is 0.94 m.

53. ✱ Four 400 g masses are located at the corners of a square with sides 1.1 m long. Find the moment of inertia about one of the diagonals.

Section 8-5   Moment of Inertia of Rigid Bodies

54. ✱✱✱ Using integration, calculate the moment of inertia of a solid disk of mass M and radius R about an axis through its centre of mass when the axis lies in the plane of the disk.

55. ✱ A ceiling fan is made from a cylindrical plate with a mass of 900 g and a radius of 11 cm. Three rectangular blades, 1.1 m in length and 17 cm in width, are attached to the circumference of the cylindrical plate. The total mass of the fan is 1.32 kg. Find the moment of inertia of the fan about an axis through its centre.

56. ✱✱✱ A solid, uniform disk of radius R and mass M has a smaller disk of radius $R/8$ removed from it (Figure 8-40). Find the moment of inertia of the resulting partially hollow disk about an axis running perpendicular to the plane of the disk through (a) the centre of the solid disk and (b) the centre of the void section, in terms of the moment of inertia of the large solid disk. The distance between the centre of the large disk and the centre of the hollow part is $R/4$.

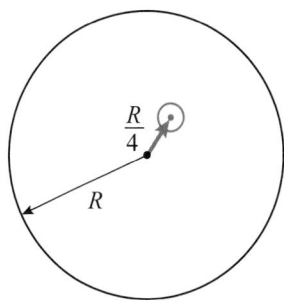

Figure 8-40 Problem 56

57. ✶✶✶ Use integration to calculate the moment of inertia of a sphere about an axis through its centre of mass.

Section 8-6   Rotational Kinetic Energy and Work

58. ✶✶ A cylindrical bar with a sphere firmly attached to one end is originally at rest in a horizontal position. The bar is pivoted about a free hinge at one end (Figure 8-41). The bar is 4 m long and has a mass of 231 kg. The sphere has a radius of 60 cm and a mass of 53 kg. A variable torque is applied to lift the bar by rotating it 37° about the pivot at a constant angular speed. Calculate the work done by the variable torque.

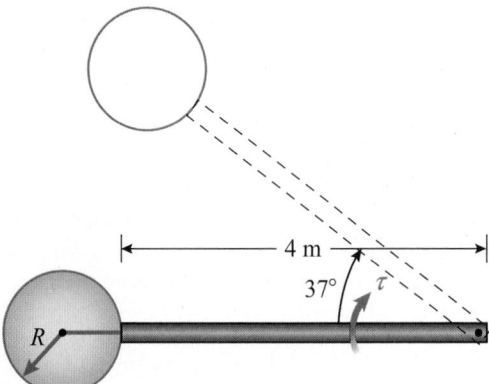

Figure 8-41   Problem 58

59. ✶✶ Use energy considerations to derive an expression for the angular acceleration of the pulley in Figure 8-42. Assume that the mass of the pulley and the friction between it and its axle are negligible. Also assume that the friction between the masses and the surfaces is negligible.

Figure 8-42   Problem 59

60. ✶✶ You are trying to get a better feel for the effect of geometry and mass distribution on the moment of inertia. You have a solid disk and a thin ring, each of radius $r = 1.3$ m and mass $m = 73$ kg. You mount both on fixed, horizontal frictionless axes about which they can spin freely. Then you spin them both.
(a) How much work do you need to do to get each object to spin at 3.00 rad/s?
(b) Let us assume that you have been causing them to spin by using a constant force applied tangentially to their circumferences. If the above speed is to be reached within 0.700 s, what is the magnitude of the force you need to apply to each object?

(c) You next attempt to stop each object by pressing your finger on the side of each object, right at its outer edge. The coefficient of kinetic friction between your finger and the surface of each object is 0.30. Find the minimum force you have to apply to stop each object within 1.0 min?

Section 8-7   Angular Momentum

61. ✶ A car is travelling in the middle lane of a highway at 133 km/h. A police officer has a speed trap set up 210 m down the road. The police officer is standing on the side of the road, 11.0 m from the centre of the middle lane. The car weighs 1250 N. Calculate the angular momentum of the car about the officer. What would the angular momentum of the car about the officer be at the moment the car passes the officer?

62. ✶✶ A pellet gun is fired at a solid disk that is free to rotate about a fixed horizontal axis as shown in Figure 8-43. The disk is 21 cm in diameter and has a mass of 370 g. The pellet ($m = 30$ g) gets lodged in the disk and causes it to spin at rate of 11 rad/s. Find the speed of the pellet just before it hits the wheel.

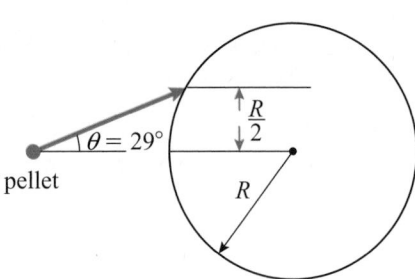

Figure 8-43   Problem 62

63. ✶✶✶ A paint ball gun fires a ball of putty at the pendulum shown in Figure 8-44. The putty has a mass of 53 g and strikes the pendulum at a speed of 14 m/s. The pendulum is made of a thin bar that is 51 cm in length and has a mass of 310 g. The sphere fixed to the end of the pendulum is 17 cm in radius and has a mass of 190 g. The pendulum is originally at rest in the vertical position, and pivots about a free hinge at its top. The putty sticks to the pendulum at the point of impact shown. Find the maximum angle that the pendulum makes with the vertical after the collision. Consider the putty to be a point mass.

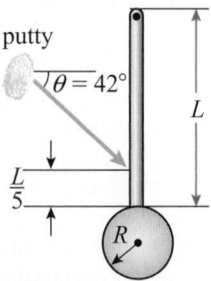

Figure 8-44   Problem 63

COMPREHENSIVE PROBLEMS

64. ✱✱✱ For a particle moving with uniform circular motion, show that $\vec{a}_r = \vec{\omega} \times \vec{v}$. Demonstrate this relationship for a bead moving in a circular path of radius 0.42 m and speed of 3.7 m/s.

65. ✱✱✱ Consider a bicycle wheel that consists of a thin ring of mass 200 g and radius 45 cm along with has 12 thin spokes of mass 21 g each. The wheel is made to spin freely about its axis at an angular speed of 0.30 rad/s. An insect with a mass of 31 g sits at the centre of the ring. The insect then walks radially outward (with respect to the ring) and stops at the rim of the wheel. The insect does not slip.
    (a) Find the final angular speed of the wheel when the insect reaches the rim.
    (b) How much work did the insect do to reach the edge? The coefficient of friction between the insect and the wheel is 0.23.

66. ✱✱✱ A solid disk of radius $R = 89$ cm is spinning about a vertical axis through its centre of mass. The disk is driven by a motor that maintains its angular speed at 4.2 rad/s. The moment of inertia of the disk about the axis is 0.30 kg/m².
    (a) How much work does a 12 g cockroach have to do to walk from the centre of the disk to its edge?
    (b) Is the work in part (a) positive or negative?
    (c) Determine the minimum coefficient of friction needed for the cockroach to make trip in part (a). Consider the cockroach to be a point mass.

67. ✱✱ Calculate the cross product V of the following two vectors: $\vec{V}_1 = 2\hat{i} + 4\hat{j} + 5\hat{k}$ and $\vec{V}_2 = 7\hat{i} + 4\hat{j} - 2\hat{k}$. Verify that $\vec{V} = \vec{V}_1 \times \vec{V}_2$ is perpendicular to the plane containing $\vec{V}_1$ and $\vec{V}_2$. Show that $\vec{V}' = \vec{V}_2 \times \vec{V}_1 = -\vec{V}$.

68. ✱✱ Show that the magnitude of the cross product between two vectors is equal to twice the area of the triangle they define. Assume that the two vectors have the same origin.

69. ✱✱ A worker is felling a 21.0 m-high tree by cutting it at a point 1.1 m above the ground, as shown in Figure 8-45. The mass of the tree is 4200 kg. Approximate the tree as a cylinder of radius 0.27 m.
    (a) Find the speed of the tree's centre of mass when the tree trunk is parallel to the ground on its way down. Assume that the tree rotates about the apex of the cut.
    (b) How much kinetic energy does the tree have at that point? How fast would a 700 kg car have to be moving to have that kinetic energy?

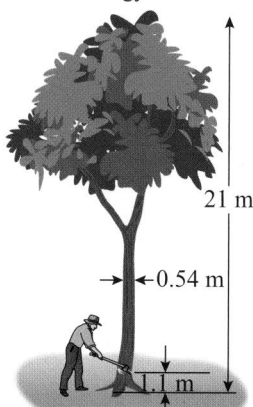

Figure 8-45 Problem 69

70. ✱✱ Find the moment of inertia of a ring about an axis tangent to the ring.

71. ✱ In reduced-gravity environments, astronauts suffer from loss of bone mass and muscle mass. One solution is to rotating sections in space stations and deep space vehicles to simulate Earth's gravity. If a cylindrical segment has a radius of 104 m to the inside surface of its outer wall, what rotational speed is required to create a 1.1*g* environment at this surface?

72. ✱✱ A movie stunt requires a tiny car with a mass of 380 kg moving at a high speed to collide with a powerful electromagnet. The electromagnet is fixed to the end of a vertical steel bar, which is free to rotate about a fixed axis through its top edge. The steel bar is 7.0 m in length and has a mass of 21.3 t. The magnet is cylindrical in shape and has a mass of 210 kg. When the car collides with the magnet, it completely sticks to it. Treat the car and the electromagnet as point masses. Find approximate values for
    (a) the maximum height that the car reaches when its speed before the collision is 69 km/h;
    (b) the maximum angle the bar makes with the vertical.

73. ✱✱ A disk is rotating at an angular speed of 63 rad/s. It is then slowed according to $\alpha = -12t - 2.5t^3$. How long will it take for the disk to reach half its speed? What is the angular displacement during that time?

74. ✱✱ The angular position of a point on a gear in a machine is given by $3t^3 + 5t - 32$. Determine the angular speed and angular acceleration of this point at $t = 5.0$ s.

75. ✱✱✱ Two disks are spinning freely about axes that run through their respective centres (Figure 8-46). The larger disk ($R_1 = 1.42$ m) has a moment of inertia of 1120 kg·m² and an angular speed of 5.0 rad/s. The smaller disk ($R_2 = 0.60$ m) has a moment of inertia of 910 kg·m² and an angular speed of 8.0 rad/s. The smaller disk is rotating in a direction that is opposite to the larger disk. The edges of the two disks are brought into contact with each other while keeping their axes parallel. They initially slip against each other until the friction between the two disks eventually stops the slipping. How much energy is lost to friction?

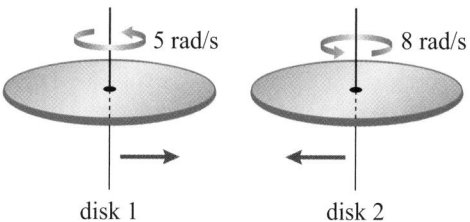

Figure 8-46 Problem 75

76. ✱✱✱ In Figure 8-47, two pulleys are mounted on fixed axles that have negligible friction. The small pulley has a moment of inertia of 9.0 kg·m²; it is made up of two cylinders welded together, one of radius 7.0 cm and the other of radius 15.0 cm. The large pulley has a radius of 41.0 cm and a moment of inertia of 84 kg·m²; the pulleys are coupled using a light belt. A 7.00 kg mass hangs from the smaller pulley by a rope that is wound around the smaller cylinder. The system is initially at rest. . . The mass is let go and starts to fall.

(a) Find the acceleration of the mass.
(b) Find the tension in the rope.
(c) Is the tension in the belt is the same everywhere? Explain.

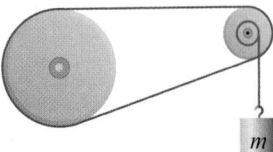

Figure 8-47   Problem 76

77. ✶✶ Calculate the moment of inertia of the object shown in Figure 8-48 about the axis *a* through one of the five thin bars of mass *m* and length *L* each, as shown.
(a) Derive the expression for the moment of inertia of each of the bars parallel to the *a* axis about that axis.
(b) Can you use the parallel-axis theorem to obtain that expression?
(c) Why is that expression identical in form to the expression for a point mass?

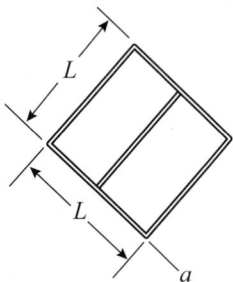

Figure 8-48   Problem 77

78. ✶✶ Using calculus, find the moment of inertia of a solid cylinder about its axis of symmetry. Repeat the same calculation by considering the cylinder to be a stack of disks.
79. ✶✶✶ Calculate the moment of inertia of a thin rectangular plate of length *L* and width *W*, about an axis parallel to *W*, passing through the centre of the plate. How will this change if the plate thickness is *T* and no longer negligible?
80. ✶✶ Two performers, of masses 62 kg and 84 kg, are standing on two fixed platforms holding on to the ends of a rope wound around the pulley shown in Figure 8-49. This pulley has a moment of inertia of 23 kg·m² and a radius of 37 cm. The platforms are then simultaneously removed, and the pulley starts to rotate.
(a) Find the angular speed of the pulley 3.0 s after the platforms are removed.
(b) After 3.0 s, the lighter performer begins to climb up the rope. What must this performer's acceleration be so that the pulley stops momentarily 7.0 s later?
(c) Find the tension in the rope for parts (a) and (b). Assume no slippage between the rope and pulley.

performer 1   performer 2

Figure 8-49   Problem 80

81. ✶✶ Two gears mounted on fixed axes are connected by a belt as shown in Figure 8-50. The large gear has a radius of 52 cm. The smaller gear has a radius of 17 cm. The smaller gear is driven by a motor that causes it to spin at 7.0 rad/s.
(a) Find the linear speed of a point on the circumference of the small gear.
(b) Find the linear speed of a point on the circumference of the larger gear.
(c) Find the linear speed of the belt.
(d) Find the angular speed of the larger gear.
(e) Find the ratio of the acceleration of a point on the circumference of the large gear to the acceleration of a point on the circumference of the small gear.

Figure 8-50   Problem 81

82. ✶✶ The pulley in Figure 8-51 has a moment of inertia of 6.4 kg·m² about the axis of rotation and a radius of 0.26 m. The masses are as follows: $m_1 = 31$ kg and $m_2 = 13$ kg. Assume that the tabletop and the pulley axle are frictionless.
(a) Determine the angular speed of the pulley when the system is allowed to move from rest under the effect of gravity 3 s into the motion.
(b) Find the tension force(s) in the rope.
(c) Is the tension force the same on each side of the pulley?

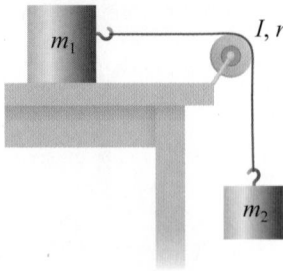

Figure 8-51   Problem 82

83. ✸✸✸ A solid, uniform disk of mass 12 kg and radius 0.91 m is pivoted about an axis through its edge parallel to its axis of symmetry. Initially, the disk is held such that its centre of mass is directly above the axis. The highest point on the disk is marked with a dot. The disk is then let go, and it swings from rest under the effect of gravity.
   (a) Find the speed of the centre of mass when the centre of mass is (i) at the same horizontal level as the axis and (ii) directly below the axis.
   (b) Find the speed of the dot when the centre of mass is 40° below the horizontal.

84. ✸✸✸ A child (m = 34 kg) is standing on a merry-go-round midway between the centre and the edge. Both the child and the merry-go-round are initially at rest. The child starts running around the merry-go-round without changing her distance from the centre until she reaches a speed of 3.0 m/s with respect to the merry-go-round surface underneath her feet. The moment of inertia of the merry-go-round is 1600 kg·m², and its radius is 4.0 m. Assume that friction between the merry-go-round and its mounting is negligible.
   (a) Find the final angular speed of the merry-go-round when the child is running.
   (b) What will the speed of the merry-go-round be when the child stops running?
   (c) How much work did the child do to get the merry-go-round to rotate at the angular speed in part (a)?
   (d) Calculate the total work done by the child.

85. ✸✸ The object shown in Figure 8-52 is made from two solid disks: m_1 = 23 kg and m_2 = 39 kg, and r_1 = 17 cm and r_2 = 33 cm. The disks are connected by a thin rod of mass M = 11 kg and length L = 67 cm. The object is held in a horizontal position and then let go. The object is free to rotate about a pivot through the centre of mass of the larger disk. Find the speed of the centre of mass of the smaller disk when it has swung 90° from its original position.

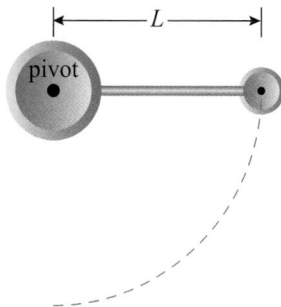

Figure 8-52 Problem 85

86. ✸✸✸ For problem 85 plot the net torque about the pivot as a function of (a) the angle θ that the rod makes with the horizontal and (b) time. How long does it take the bar to reach the vertical position?

87. ✸✸ You are trying to pry two pieces of wood apart by jamming a 37 cm-long crowbar between them. You drive one end of the crowbar in 0.2 cm in between the pieces of wood, which are held together by six nails. The frictional force from each piece of wood onto a nail is 110 N. With the crowbar perpendicular to the edge of the pieces of wood, what is the force needed if you push at the other end of the crowbar? A clamp on the lower piece of wood keeps it in place.

88. ✸✸ In Figure 8-53, the block of mass m_1 = 16 kg sits on the surface of a bench such that the coefficient of static friction between the block and the surface is 0.20. Find the maximum weight that can hang down from the outer cylinder (R = 34 cm) of the pulley without causing m_1 to move. The radius, r, of the inner cylinder is 12 cm.

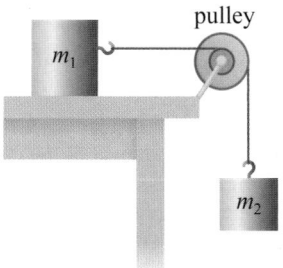

Figure 8-53 Problem 88

89. ✸ For the system shown in Figure 8-54, the coefficient of kinetic friction between the mass and the incline is μ_k, and the pulley's moment of inertia is I and radius is r. Derive an expression for the speed of m_1 after m_2 has fallen a distance d from rest along the incline. Assume m_2 is sufficiently larger than m_1 to cause the system to move from rest.

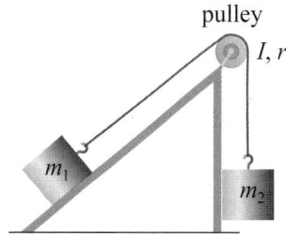

Figure 8-54 Problem 89

90. ✸✸✸ Two mice jump onto the edge of a record turntable, causing it to rotate at the angular speed ω = 2.7 rad/s. After a few seconds, the two mice jump off the turntable along a direction that is tangential to the turntable and opposite to its direction of motion. They jump one after the other at a speed of 2.7 m/s with respect to the turntable. Consider the turntable to be a disk of mass 510 g and radius 18 cm. The mice are 37 g each. Find the final speed of the turntable after the second mouse has jumped off. Consider the mice to be point masses.

91. ✸✸✸ A solid disk with a mass of 32 kg and a radius of 27 cm is spinning counterclockwise at 17 rad/s in the horizontal plane about its axis of symmetry. Another disk with half the mass and half the radius is spinning at 34 rad/s in the opposite direction when it drops onto the first disk, as shown in Figure 8-55.
   (a) Find the final angular speed of the two disks when they eventually spin together.
   (b) How much energy is lost to friction?

Figure 8-55 Problem 91

92. ★★★ The beam in Figure 8-56 pivots about one end, swings down from its horizontal position, and strikes a small box, as shown. The collision is perfectly elastic. The beam has a mass of $M = 160$ kg and a length of 1.3 m. Treat the box as a point mass ($m = 32$ kg). The friction at the pivot is negligible.
   (a) Find the speed of the box after the collision.
   (b) Find the maximum angle the bar makes with the vertical after the collision.

Figure 8-56 Problem 92

93. ★★★ A child of mass $m = 28$ kg is standing at the edge of a playground merry-go-round (moment of inertia $I_{cm} = 2700$ kg·m² and radius $R = 3.0$ m), which is initially turning at 6.0 rev/min. The child walks opposite to the direction of motion of the merry-go-round with a speed of 2.0 m/s with respect to it. He walks along its edge and completes two full revolutions, stopping at his starting point.
   (a) Find the speed of the merry-go-round after the child stops.
   (b) The child then starts walking along the edge of the merry-go-round until the merry-go-round is spinning with an angular speed of 0.80 rad/s. He then jumps off the merry-go-round by leaping outward in a radial direction with respect to the merry-go-round. Find the angular speed of the merry-go-round just after the child leaps off. The coefficient of static friction between the child and the merry-go-round is 0.50.

94. ★★★ An archaeological research team is attempting to simulate a catapult that might have been used in Europe 12 centuries ago. The catapult arm—a bar of length 3.00 m and mass 170 kg—is pulled from its vertical position against specially mounted springs fixed to the wall as shown in Figure 8-57. At the vertical position, the springs

are unstretched. The catapult arm is then pulled back until it is in the horizontal position. A stone of mass $m = 92.0$ kg is loaded into a light basket at the end of the arm, and the arm is released. At 51.0° above the horizontal, the catapult arm is stopped by a mechanism.
   (a) Consider the stone to be a point mass. What should the effective spring constant be to launch the stone at a speed of 190 m/s?
   (b) Consider the stone to be a sphere of radius 25.0 cm, and recalculate the required spring constant. Is it reasonable to approximate the stone as a point mass? What is the percent difference between the two results?

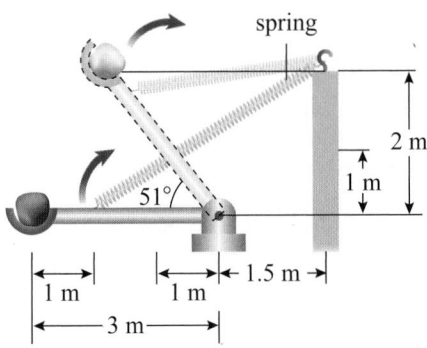

Figure 8-57 Problem 94

95. ★★★ A flywheel is turning at an angular speed of 35 rad/s. It is then coupled with a load that causes it to slow down with a time-dependent acceleration given by $\alpha = -4e^{-2t}$. How many revolutions does the wheel go through before it stops?

96. ★★★ A cylindrical observation station of mass 5200 kg and radius 11 m is spinning freely in space about its central axis. The platform is attended by one astronaut ($m = 72$ kg). She is initially at the centre of the cylinder, but then starts slowly walking toward the edge to look out from one of the windows at the Sun rising behind Earth. The platform is rotating freely at 10 rev/min when the astronaut is at the centre. The cylinder's moment of inertia about the axis is 15 300 kg·m². Find the angular speed of the platform when the astronaut reaches the edge. How much work did the astronaut have to do to reach the edge? What was the net change in the rotational kinetic energy?

97. ★★ In Figure 8-58, a rope is wound around a horizontal disk ($M_d = 2.9$ kg and $R_d = 20$ cm), passed across a pulley ($I_{cm} = 48$ kg·cm² and $R_p = 8.0$ cm), and connected to a hanging 4.0 kg mass. Initially the mass is held in place, and the system is at rest. The mass is then released and descends in response to the force of gravity. The rope does not slip on the pulley or the horizontal disk. Both the pulley and disk rotate on bearings that are effectively frictionless.
   (a) Find the angular speed of the pulley after the mass has descended 1.2 m.
   (b) Using energy considerations, calculate the acceleration of the mass without writing the equations of motion.

disk

pulley

m

Figure 8-58   Problem 97

98. ✷✷   You are sitting on a pedestal that is free to rotate about its axis of symmetry; your arms are stretched out fully in front of you. The combined moment of inertia of you and the pedestal is 21 kg·m². You are then handed a wheel of mass $m = 15$ kg and radius $R = 27$ cm, spinning at 21 rad/s in a vertical plane. You tilt the wheel axis from the horizontal to the vertical at an angular speed of 0.15 rad/s.
  (a) Find the torque you have to apply as you tilt the wheel.
  (b) The wheel axis eventually becomes fully vertical. Find the angular speed of the pedestal at that point. When the axis of the wheel is vertical, the distance between the axis of the wheel and the axis of the pedestal is 1.1 m.

99. ✷✷✷   Two astronauts are standing on a spinning, disk-shaped platform in space. The platform has a radius $R = 17$ m and a mass $M = 15\ 000$ kg, and is spinning at an angular speed of 1.3 rad/s. One astronaut ($m_1 = 76$ kg) is standing at the edge of the platform, and the other astronaut ($m_2 = 82$ kg) is standing at the centre. The astronaut at the edge walks toward the centre.
  (a) Does the astronaut do positive or negative work while walking toward the centre?
  (b) As the astronaut walks in, would you expect the angular speed of the platform to increase or decrease?
  (c) Calculate the spinning speed of the platform when the astronaut at the edge reaches the centre.
  (d) Next, the two astronauts walk in opposite directions to the edge of the platform. Do you expect the kinetic energy of the system (platform + two astronauts) to increase or decrease?
  (e) Find the work done by the two astronauts to reach the edge.

100. ✷✷✷   By how much does Earth's angular momentum change as a result of tidal friction over one year? What is the average torque involved?

e   See the text online resources at
www.physics1e.nelson.com for Open Problems-Ended and Data-Rich Problems related to this chapter.

Chapter 9
Rolling Motion

Bear Claw

During the winter in cold-climate countries, roads are often covered with snow, ice, or slush. Ongoing research into skid control, antilock braking systems, and tires aims to improve traction in extreme weather conditions. Winter tires often have more "aggressive" treads with channels that are designed to allow the snow, slush, mud, and water to leave the tires more effectively. However, even winter tires may not have an adequate grip on snowy or icy roads in mountainous terrains. In some parts of the world it is not uncommon to see car tires with studs that bite into the ice. The term "Bear Claw" is sometimes used to describe sharp edged, claw-like projections on tread blocks which enhance winter tire performance. In addition, tires with chains for added traction are required by law on mountain roads in certain areas, such as the Rockies of British Columbia (Figure 9-1).

Wayne R Bilenduke/Stone/Getty Images

Blaz Kure/Shutterstock   vesilvio/Shutterstock   Valentin Mosichev/Shutterstock   iStockphoto/Thinkstock

Figure 9-1 Examples of enhancements used to improve winter tire performance

9-1 Rolling and Slipping

In Chapter 8, we discussed the rotation of an object about a fixed axis. However, many applications involve rotation about a moving axis, such as a ball rolling down hill. Before we begin our discussion however, we need to define a few terms.

Rolling is rotation about an axis that is translating. For example, when you use a baker's rolling pin to flatten dough, you push the axle forward, causing the centre of mass of the rolling pin to translate forward. At the same time, the pin rotates about an axis through its centre of mass.

Spinning occurs when the rotational speed of the surface of an object is too high in comparison to the translational speed of the axis about which the object rotates. For example, when a truck is trying to emerge from a muddy hole, you often see the wheels spinning at relatively high speeds while the truck moves slowly on the muddy surface or does not move at all. Similarly, when a race car driver makes a fast start, the high torque of the powerful engine causes the wheels to spin, even on dry asphalt, and the friction of the asphalt on the spinning tire can burn the surface of the rubber tire (Figure 9-2).

© fastcars/Alamy   © James Warren/Alamy

Figure 9-2 (a) A Porsche Carrera 2 starting rapidly with tire burnout. (b) A Formula 1 car wheel spin

Skidding occurs when the translational speed of a rolling object is too high compared to the rotational speed of its surface. When you are driving a car, it is possible for the wheels to lock when you depress the brake pedal, in which case the wheels stop rotating completely while the car is still moving. The car will also skid without the wheels locking if the tires rotate at a speed slower than is necessary to keep up with the translational speed of the car (Figure 9-3).

Slipping, or the presence of kinetic friction between the tire and the road, refers to both spinning and skidding. A car moving on asphalt such that its wheels are skidding or spinning often leaves skid marks on the road. When a wheel slips, the part of the tire wheel that is touching the asphalt rubs against the asphalt. The asphalt

© David Madison/Corbis

Figure 9-3 Race car skidding into a pit box

actually strips part of the black rubber from the tire in the same way that paper strips away some of a rubber eraser when you erase pencil marks on a page. Slipping often indicates trouble unless, of course, it is done deliberately, such as during a mountain rally or other type of car race (Figure 9-4).

© Olivier Prevosto/TempSport/Corbis

Figure 9-4 French driver Marcel Tarres puts his Citroen Xsara into a controlled skid during the final of the famous car race on ice, *Trophée Andros*, at the Stade de France.

If an object neither spins nor skids as it rolls, its motion is called **rolling without slipping** (also called **ideal rolling** or **perfect rolling**). The relative speed between that part of the wheel in contact with the road and the pavement surface is zero. Figure 9-5(b) shows the motion of a white dot painted on the edge of a tire. Notice that the dot momentarily stops with respect to the ground just as the two come in contact.

In Figure 9-5(a), the car leaves clear tire marks because the part of the tire that touches the sand or dirt is actually stationary with respect to the ground as the car moves. Otherwise, the tire marks would be smeared.

© John Nakata/Corbis

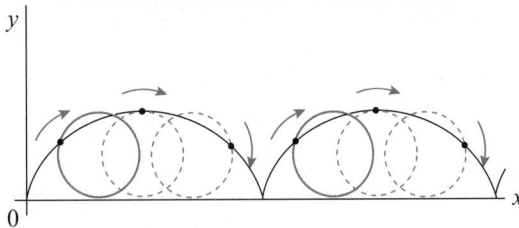

Figure 9-5 (a) Rolling without slipping produces a clear tire track in the Mojave Desert. (b) The trajectory of a point on the edge of a tire

✓ CHECKPOINT

C-9-1 Moving or Not?

A car is travelling at a speed of 98 km/h. Its tires have a radius of 31 cm. The speed of a point on the tire that is momentarily touching the ground is

(a) 98 km/h
(b) −98 km/h
(c) 98 km/h + ωr
(d) zero

C-9-1 (d) The speed of the point on the tire in contact with the road is zero for ideal rolling.

✓ CHECKPOINT

C-9-2 Why Do Cars Move?

What force moves a car forward when it accelerates from rest without spinning the tires?

(a) static friction
(b) kinetic friction
(c) the torque delivered by the engine
(d) another force

C-9-2 (a) No spinning means static friction.

LO 2

9-2 Relationships Between Rotation and Translation for a Rolling Object

Consider a wheel of radius r rotating in a clockwise direction about a fixed axis through its centre of symmetry (Figure 9-6). Since the axis is fixed, the velocity of the centre of mass, v_{cm}, is zero. Choosing right as the positive direction, the velocity of a point at the top of the wheel, given by Equation 8-2, is $v_{top} = +\omega r$. The velocity of a point at the bottom of the wheel is then $v_{bottom} = -\omega r$. Since the velocity of the centre of mass is zero, the velocity of a point at the top is ωr greater than the velocity of the centre of mass, and the velocity of a point at the bottom is ωr less than the velocity of the centre of mass:

$$v_{top} = v_{cm} + \omega r$$

$$v_{bottom} = v_{cm} - \omega r$$

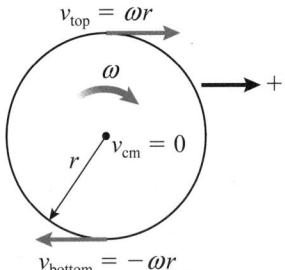

Figure 9-6 Rotation about a fixed axis

Rearranging and combining the above equations give the following relationships:

$$v_{cm} = v_{bottom} + \omega r \qquad (9\text{-}1)$$

$$v_{cm} = v_{top} - \omega r \qquad (9\text{-}2)$$

$$v_{top} = v_{bottom} + 2\omega r \qquad (9\text{-}3)$$

The same reasoning applies for the relation between the acceleration of the centre of mass and the tangential component of the acceleration at the top and the bottom of any rotating object. Taking the derivatives of Equations (9-1), (9-2), and (9-3) with respect to time gives the following:

$$a_{cm} = a_{bottom} + \alpha r \qquad (9\text{-}4)$$

$$a_{cm} = a_{top} - \alpha r \qquad (9\text{-}5)$$

$$a_{top} = a_{bottom} + 2\alpha r \qquad (9\text{-}6)$$

These relationships apply for any rotating object, regardless of whether the rotation axis is stationary or moving.

EXAMPLE 9-1

Stuck Truck

While trying to free a dump truck stuck in the mud, you notice that all four wheels are spinning at approximately 5.00 rev/s, and that the truck is moving forward at 5.00 m/s. The diameter of the wheels is about 1.50 m.

(a) Calculate the velocity with respect to the ground, of the point on the tire that contacts the ground.
(b) Calculate the velocity of the centre of mass of the tire with respect to the ground.
(c) Determine the velocity of a point at the top of a wheel.

SOLUTION

a) From equation (9-1)

$$v_{bottom} = v_{cm} - \omega r$$
$$= 5 \text{ m/s} - 5(2\pi)(0.75)$$
$$= -18.6 \text{ m/s}$$

b) Speed of center of mass of tire is the same as that of the truck = 5 m/s

c) For the top of the wheel we can use Equation 9-2

$$v_{top} = v_{cm} + \omega r$$
$$= 28.6 \text{ m/s}$$

Making sense of the result:

The difference between the velocity at the top and bottom is $2\omega r = 15\pi$.

 CHECKPOINT

C-9-3 Quick Calculation

If you drive a car at the constant speed 10 m/s such that the wheels do not slip, then

(a) the speed of the point at the bottom of the wheel (while in contact with the ground) is −10 m/s;
(b) the velocity of a point on the wheel at the same height as the centre of mass of the wheel is 10 m/s and points either up or down;
(c) the speed of a point on the wheel at the same height as the centre of mass of the wheel is 14.14 m/s;
(d) the acceleration of the point at the top of the wheel is zero.

C-9-3 (c) vertical speed is 10 m/s, horizontal speed is 10 m/s, using Pythagoras we get (c).

LO 3

9-3 Rolling Motion: Two Perspectives

In this section, we will discuss two approaches to analyzing the motion and dynamics of an object rolling without slipping. In the first approach, we consider rolling as a

rotation about a moving centre of mass axis. In the second approach, we use the fact that the point on the rolling body in contact with the surface is momentarily at rest, which allows us to consider rolling as a continuous series of successive rotations of the body about that point.

Rolling as a Rotation About the Moving Centre of Mass

For an object that rolls on a horizontal surface without slipping (Figure 9-7), the velocity of the part of the object in contact with the surface is zero: $v_{bottom} = 0$. Substituting into Equation (9-1) gives

$$v_{cm} = v_{bottom} + \omega r$$
$$v_{cm} = 0 + \omega r \qquad (9-6)$$

Hence for an object rolling without slipping,

KEY EQUATION $$v_{cm} = \omega r \qquad (9-7)$$

$$v_{top} = v_{bottom} + 2\omega r = 2\omega r$$

$$v_{cm} = v_{bottom} + \omega r = \omega r$$

$$v_{bottom} = 0$$

Figure 9-7 Rolling without slipping

Similarly, substituting into Equation (9-3) gives the speed at the top of the object, or more generally, the speed of the point opposite to the point of contact with the surface on which the object is rolling:

$$v_{top} = v_{bottom} + 2\omega r$$
$$= 0 + 2\omega r$$
$$v_{top} = 2\omega r \qquad (9-8)$$

To determine the acceleration of the centre of mass, we use Equation (9-4) to get

KEY EQUATION $$a_{cm} = \alpha r \qquad (9-9)$$

At the top of the object, the tangential component of the acceleration is

$$a_{top} = 2\alpha r \qquad (9-10a)$$

For the point on the object that is touching the ground, the tangential acceleration is

$$a_{bottom} = 0 \qquad (9-10b)$$

In this approach, rolling is regarded as a translation of the centre of mass, combined with rotation of the object

CHECKPOINT

C-9-4

Which of the following statements is *not* true for a wheel rolling without slipping on a flat horizontal surface?

(a) The total acceleration of the point touching the ground is zero.

(b) The radial acceleration of the point touching the ground is zero.

(c) The vertical velocity of the point touching the ground is zero.

(d) The total speed of the point touching the ground is zero.

C-9-4 (a) a is the only false statement.

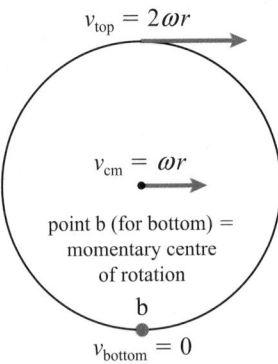

Figure 9-8 Rolling as successive rotations about the contact point b

about an axis through the centre of mass. The equations of motion will have to address both the translational and the rotational components of the motion.

For the translational dynamics of the centre of mass, we can apply Newton's second law:

KEY EQUATION
$$\sum \vec{F} = M\vec{a}_{cm} \qquad (9\text{-}11)$$

For the rotational dynamics, we apply the equivalent of Newton's second law for rotation about the centre of mass (Equation (8-17)):

KEY EQUATION
$$\sum \vec{\tau}_{cm} = I_{cm}\vec{\alpha} \qquad (9\text{-}12)$$

As you will see in the examples, Equations (9-11) and (9-12) often have to be used in conjunction with Equation (9-9).

Rolling as a Rotation about the Point of Contact Between the Object and the Surface

As an object rolls, successive points on its circumference touch the surface on which the object is rolling. If the object does not slip, each of the successive points will be stationary with respect to the surface at the moment that the point touches the surface. Consequently, rolling can be considered as a continuous series of rotations about a contact point, which we label point b (for bottom), as shown in Figure 9-8. This contact point is called a **momentary pivot**.

Since point b is at rest, we can sum the torques about this pivot point and apply the equivalent of Newton's second law for rotation. Although less intuitive than the approach using Equations (9-11) and (9-12), this approach

is an effective method of analysis; it is also often the more convenient approach.

Applying the rotational equivalent of Newton's second law by summing torques about point b gives the following:

KEY EQUATION
$$\sum \vec{\tau}_b = I_b\vec{\alpha} \qquad (9\text{-}13)$$

The object's rotational speed is ω, so the centre of mass located a distance r from the momentary pivot has a speed of $v_{cm} = \omega r$. The top of the object is located a distance $2r$ from the momentary pivot and has the speed $v_{top} = 2\omega r$. These speeds are identical to the speeds obtained in Equations (9-7) and (9-8). The expressions for the tangential acceleration of the centre of mass and the top of the object given in Figure 9-8 and in Equations (9-9) and (9-10) can be obtained using the same reasoning.

Notice how we have labelled the torques and the moment of inertia in Equation (9-13). Since we are summing the torques about point b, we must use I_b, the object's moment of inertia about point b, rather than I_{cm}, the moment of inertia about the object's centre of mass.

LO 4

9-4 Newton's Second Law and Rolling

Next, we study the forces involved in rolling. Let us first analyze the motion of an object rolling freely without slipping down an inclined plane.

 EXAMPLE 9-2

Rolling Down an Incline

Consider a solid uniform disk that starts from rest and rolls without slipping down a ramp sloping at an angle θ as

shown in Figure 9-9(a). Derive expressions for the acceleration of the centre of mass of the disk and for the force of friction between the surface of the ramp and the disk.

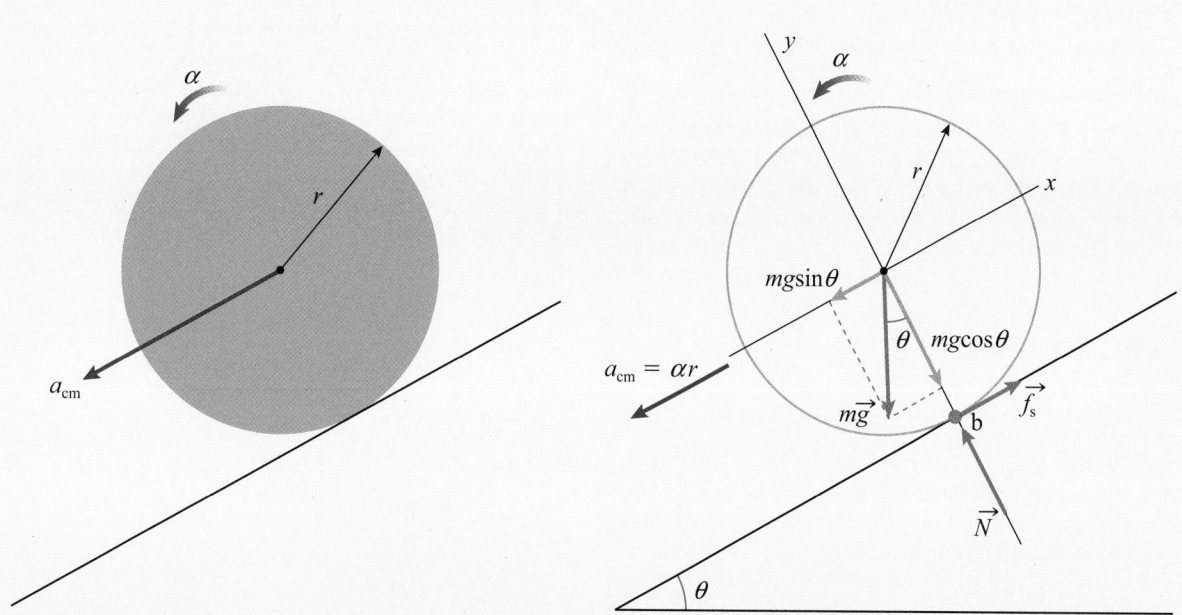

Figure 9-9 (a) A solid uniform disk rolling without slipping down an incline. (b) Free body diagram of the disk in (a)

SOLUTION

Let us begin by drawing the free body diagram of the disk. The forces acting on it are the force of gravity, the normal force from the ramp, and the friction from the surface of the ramp as shown in Figure 9-9(b).

We can analyze this situation using either of the approaches described above. Here, we consider the rolling as a translation of the centre of mass combined with rotation of the object about the centre of mass.

In the free body diagram in Figure 9-9(b), we choose our axes to reflect the direction of the acceleration, which is down the ramp. Applying Newton's second law to translation along the ramp gives

$$\sum F_x = ma_{cm,\,x}$$

$$f_{sx} + mg_x = ma_{cm,\,x}$$

Choosing up the incline as the positive direction, we get

$$f_s - mg\sin\theta = -ma_{cm} \qquad (1)$$

Like f_s and g in Equation (1), a_{cm} is a variable that denotes the magnitude of the acceleration of the centre of mass.

Next, we apply the rotational equivalent of Newton's second law by summing the torques about the centre of mass, noting that only the force of friction exerts a torque about the centre of mass:

$$\sum \vec{\tau}_{cm} = I_{cm}\vec{\alpha}$$

$$rf_s = I_{cm}\alpha \qquad (2)$$

So far, we have two equations and three unknowns. The third equation is the relationship between angular and linear acceleration for an object rolling without slipping (Equation 9-9):

$$a_{cm} = \alpha r \qquad (3)$$

As with most rotation problems, the simplest way to solve a system of three equations is substitution: we substitute expressions for f_s and a into Equation (2), and then solve for a_{cm}, as we did many times in Chapter 8. The key here is to substitute into the torque equation, using the expressions obtained from the other equations.

From Equation (1), $\qquad f_s = mg\sin\theta - ma_{cm}$

From Equation (3), $\qquad \alpha = \dfrac{a_{cm}}{r}$

Substituting these expressions into Equation (2) gives

$$r(mg\sin\theta - ma_{cm}) = I_{cm}\frac{a_{cm}}{r}$$

$$a_{cm} = \frac{mg\sin\theta}{m + \dfrac{I_{cm}}{r^2}} \qquad (4)$$

The moment of inertia of a disk about its centre-of-mass axis is $I_{cm} = \dfrac{mr^2}{2}$. Substituting into the expression for a_{cm} gives

$$a_{cm} = \frac{2}{3}g\sin\theta \qquad (5)$$

To calculate the force of friction, we can substitute this expression for a_{cm} in either Equation (1) or Equation (2) to obtain

$$f_s = \frac{1}{3}mg\sin\theta$$

The same results can be derived by considering rotation about the point of contact between the inclined surface and the disk. This is done in full detail online in Online Example 9-1. It would be worthwhile for you to solve the problem using this approach before you see the online example

(continued)

Making sense of the result:

The expression for the acceleration of the centre of mass in Equation (4) has the form of $a = \dfrac{F_{net}}{m}$ and, hence, has the units N/kg, as it should. The driving force is the component of the force of gravity along the ramp. The disk's ability to resist being accelerated comes from its mass, which indicates its ability to resist being accelerated linearly, and from its moment of inertia, which is a measure of its ability to resist being accelerated rotationally. The force of friction was not the maximum force of static friction because we are not told that the object is on the verge of slipping. When the object slides without rotation, we recover $a = g\sin\theta$ which we obtained in Chapter 5, Example 5-13 for a point mass sliding down an incline. This would happen here if the force of friction was zero. If the force of friction is zero, there will be no torque to cause the object to rotate and hence the object will not roll and will simply slide.

EXAMPLE 9-3

The Yo-yo

A yo-yo consists of a disk with a string wound around it as shown in Figure 9-10(a). Derive an expression for the acceleration of the yo-yo as the string unwinds.

SOLUTION

In this example, we consider rolling as a rotation about the point where the string is unwinding from the disk.

Let us begin by drawing the free body diagram of the yo-yo as shown in Figure 9-10(b). The forces acting in the

Figure 9-10 (a) A yo-yo. (b) The free body diagram of the forces acting on the yo-yo in part (a)

vertical direction are the weight of the yo-yo and the tension in the string. Since this yo-yo does not slip on the string, the point at which the string leaves the yo-yo is momentarily at rest and can be treated as a temporary, or momentary, pivot, where $\sum \vec{\tau}_b = I_b \vec{\alpha}$. We label this point b but note that this pivot is not at the bottom of the yo-yo.

The only force that exerts a torque about b is the weight of the yo-yo, which acts at the centre of mass, a distance r from point b. Here again we emphasize that in this approach both the torque and the moment of inertia are taken about point b and *not* about the centre of mass. Noting that mg acts at $90°$ to r, we have

$$rmg = I_b \alpha$$

and $\alpha = \dfrac{rmg}{I_b}$

Since $a_{cm} = \alpha r$, we have

$$a_{cm} = \alpha r = \dfrac{r^2 mg}{I_b}$$

Using the parallel-axis theorem gives

$$I_b = I_{cm} + md^2 = I_{cm} + mr^2$$

and $a_{cm} = \dfrac{r^2 mg}{I_b} = \dfrac{r^2 mg}{I_{cm} + mr^2} = \dfrac{mg}{\dfrac{I_{cm}}{r^2} + m}$

$$= \dfrac{mg}{\dfrac{m}{2} + m} = \dfrac{2g}{3}$$

where a_{cm} is a variable denoting the magnitude of the acceleration of the centre of mass. We have used the fact the moment of inertia of a disk about its centre of mass is $mr^2/2$.

Note that by summing torques about b, we are able to calculate the angular acceleration in a single step instead of solving a system of three equations as we did in Example 9-2.

You are encouraged to solve this problem by considering rolling as a rotation about the moving centre-of-mass axis. This is done fully online in Online Example 9-2. With this approach, you can also show that the tension in the rope is $T = \dfrac{mg}{3}$.

Making sense of the result:

Again, the expression for the acceleration has the correct units of m/s². The terms in the denominator reflect the object's resistance to acceleration: mass (m), for resistance to translational acceleration, and the moment of inertia $\left(\dfrac{I}{r^2}\right)$, for resistance to rotational acceleration. The equation for a_{cm} indicates that acceleration will decrease if either the mass or the moment of inertia of the yo-yo increases, which is reasonable.

CHECKPOINT

C-9-5 Who's Faster?

Compare three yo-yos: one made with a sphere, one made with a cylinder, and one made with a ring, all with the same mass and the same radius. Which of the following is true?

(a) The ring yo-yo will have the smallest downward acceleration.

(b) The sphere yo-yo will have the smallest angular acceleration.

(c) The force for tension will be the highest in the case of the sphere yo-yo.

(d) The tension on the string will be the same for all three yo-yos.

C-9-5 (a) the ring will have the highest moment of inertia.

MAKING CONNECTIONS

Rolling, Friction, and Antilock Braking Systems

When applying the brakes in a car, unless the wheels slip the external force that slows the car is static friction. The direction of the force of static friction exerted by the road on the wheels is opposite to the direction of motion of the car. However, when you accelerate, the force of static friction on the drive wheels is in the direction of motion of the car. If the car is moving at a constant speed without slipping, the friction exerted on the tires by the road is zero.

When a car goes into a skid while trying to stop on an icy surface (Figure 9-11), the braking distance is greatly increased. Antilock braking systems (ABSs) are designed to maximize the force of friction between the road and the tires by keeping the tires on the verge of skidding. Ideally, the ABS will control the car's speed to keep the force of static friction close to its maximum possible value and avoid skidding, which would cause the weaker force of kinetic friction to act between the road and the tires. As described in Section 5-7, the force of static friction can take on values between zero and a maximum value determined by the normal force and the coefficient of friction:

$$0 \leq f_s \leq \mu_s N \tag{9-14a}$$

When the tires are on the verge of skidding but not actually skidding, the force of friction between the tires and road reaches the maximum value:

$$f_{s,\max} = \mu_s N \tag{9-14b}$$

The ABS pumps the brakes automatically, repeatedly reducing the braking force just before the car goes into a skid, and then increasing the braking again. A feedback mechanism regulates the braking to maximize the force of static friction acting on the tires.

Figure 9-11 Extreme driving conditions: a Volvo on an ice-covered road

When slipping does occur, the force of friction is given by

$$f_k = \mu_k N \tag{9-14c}$$

For rubber on ice, the coefficient of kinetic friction can be considerably less than the coefficient for static friction, which makes having ABS quite desirable. Gabriel Voisin is credited with the introduction of the first ABS, used on airplanes in 1929. Figure 9-12 shows a sketch of the Avions Voisin AJ-T.

Figure 9-12 Sketch of the Voisin AJ-T, by Avions Voisin, a Dutch automobile manufacturing firm named in honour of Gabrel Voisin, reputed to be the inventor of the ABS system

MECHANICS

✓ CHECKPOINT

C-9-6 Rolling and Friction

Which of the following statements is not correct?

(a) Rolling cannot occur without friction.

(b) Friction is often necessary to cause objects to start rolling initially, but once they do, friction is not always necessary to maintain rolling.

(c) If you throw a marble onto a surface with a horizontal speed v, and give it a forward rotational speed of $\frac{v}{r}$, then the marble will roll with or without friction.

(d) Under ideal conditions, the force of friction on a rigid object, rolling freely on a solid, smooth (not frictionless) horizontal surface, will be zero.

C-9-6 (a) You can cause an object to roll, without friction; see answer (c).

LO 5

9-5 Mechanical Energy and Rolling

How does the kinetic energy of an object rolling without slipping on a smooth horizontal surface differ from the kinetic energy of the same object sliding without rotation on a frictionless surface or that of the same object spinning about a fixed axis? We will first analyze this question using the rotation/translation approach first, and then we will apply the momentary-pivot approach to similar systems.

Kinetic Energy of a Rolling Object

The object in Figure 9-13 is rolling without slipping. Since the speed of the centre of mass is v_{cm}, the kinetic energy of translation is

KEY EQUATION
$$K_{trans} = \frac{1}{2} m v_{cm}^2 \qquad (9\text{-}15)$$

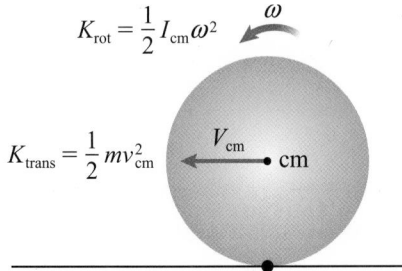

Figure 9-13 Rotation/translation analysis of the kinetic energy of a rolling object

The object is also rotating about the moving centre-of-mass axis at the angular frequency ω. The kinetic energy for this rotation is given by

KEY EQUATION
$$K_{rot} = \frac{1}{2} I_{cm} \omega^2 \qquad (9\text{-}16)$$

The total kinetic energy is therefore the sum of the translational and the rotational kinetic energies:

KEY EQUATION
$$K = \frac{1}{2} m v_{cm}^2 + \frac{1}{2} I_{cm} \omega^2 \qquad (9\text{-}17)$$

EXAMPLE 9-4

Kinetic Energy of a Rolling Wheel

A spoked wheel of mass m and radius r rolls without slipping. Most of the mass of the wheel is distributed along its rim, so the wheel can be treated as a hoop or a ring.

(a) Derive an expression for the total kinetic energy of the wheel when the speed of the centre of mass is v_{cm}.

(b) Starting from rest, the wheel rolls without slipping down a ramp from a height h. Derive an expression for the speed of the centre of mass of the wheel at the bottom of the ramp.

SOLUTION

(a) Since the wheel rolls without slipping, Equation 9-7 gives us $v_{cm} = \omega r$, or $\omega = \frac{v_{cm}}{r}$. We can rewrite the expression for the total kinetic energy (Equation (9-17)) as

$$K = \frac{1}{2} m v_{cm}^2 + \frac{1}{2} I_{cm} \omega^2 \qquad (1)$$

The moment of inertia of a ring or a hoop about its centre of mass is $I_{cm} = mr^2$. Substituting this into Equation (1) and using Equation 9-7 we get

$$K = \frac{1}{2} m v_{cm}^2 + \frac{1}{2} \frac{mr^2}{r^2} v_{cm}^2 = \frac{1}{2} m v_{cm}^2 + \frac{1}{2} m v_{cm}^2 = m v_{cm}^2 \qquad (2)$$

(b) The conservation of mechanical energy requires that the gravitational potential energy lost as the wheel rolls down the ramp equal the increase in kinetic energy, from Equation 6-31:

$$\Delta K = -\Delta U \qquad (3)$$

Since gravitational potential energy is lost as the wheel rolls down the ramp,

$$\Delta U = -mgh \qquad (4)$$

Substituting Equations (2) and (4) into Equation (3), we get

$$m v_{cm}^2 = mgh$$

Therefore,

$$v_{\text{cm}} = \sqrt{gh}$$

Making sense of the result:

For a ring that rolls without slipping, the rotational kinetic energy is equal to the translational kinetic energy. However, this relationship stems from the expression for the moment of inertia of the ring. We saw in chapter 6, example 6-8, that the expression for an object sliding down a frictionless ramp from height h, the speed was more by a factor of $\sqrt{2}$. This is because part of the potential energy lost as the wheel rolls down the incline goes into the rotational kinetic energy of the object, in addition to the translational kinetic energy. When the object slides down without rolling we recover the result from Example 6-8.

Kinetic Energy Using the Momentary-Pivot Approach

We now apply the momentary-pivot approach to analyze the kinetic energy of rolling objects. As before, the point on the rolling object that contacts the ground or surface is momentarily stationary with respect to the ground, and can be treated as a pivot (Figure 9-14). Using this method, the kinetic energy of a rolling sphere can written as a single term:

KEY EQUATION
$$K = \frac{1}{2} I_{\text{b}} \omega^2 \qquad (9\text{-}18)$$

All the rolling motion of an object is accounted for by a rotation about point b. With this approach, there is no need for a separate term for the translational motion.

Notice that Equations (9-17) and (9-18) give two different expressions for the *same* quantity: the total kinetic energy of a rolling object. Therefore, these two expressions must be equal. (See problem 86.)

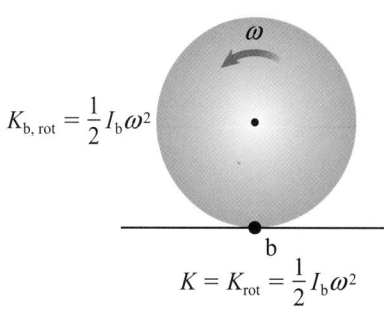

$$K_{\text{b, rot}} = \frac{1}{2} I_{\text{b}} \omega^2$$

$$K = K_{\text{rot}} = \frac{1}{2} I_{\text{b}} \omega^2$$

Figure 9-14 Momentary-pivot analysis of the kinetic energy of a rolling object

✓ CHECKPOINT

C-9-7 Ring or Disk?

When a ring and a disk with the same mass and radius start from rest and roll down a ramp without slipping,
(a) the ring will have the greater angular acceleration;
(b) the ring will have the greater linear acceleration;
(c) the two objects will have the same linear acceleration because they have the same mass;
(d) none of the above.

C-9-7 (d) The ring has the greater moment of inertia, so its angular acceleration is less, which also means that its linear acceleration is less.

EXAMPLE 9-5

The Speed of a Sphere Rolling Downhill

Starting from rest, a solid uniform sphere with mass m rolls without slipping down a ramp with height h (Figure 9-15). Derive an expression for the speed of the sphere at the bottom of the ramp.

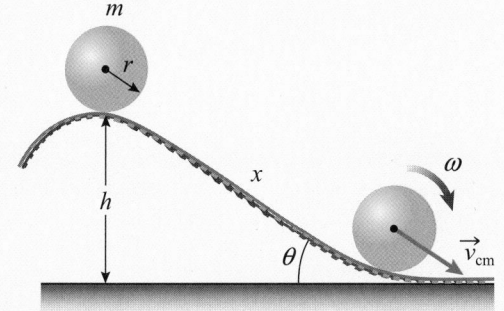

Figure 9-15 Gravitational potential energy is converted to kinetic energy as the sphere rolls down the ramp

SOLUTION

Here, we use the momentary-pivot approach. From Equation (9-18) we have

$$K = \frac{1}{2} I_{\text{b}} \omega^2 \qquad (1)$$

Conservation of mechanical energy gives

$$mgh = \frac{1}{2} I_{\text{b}} \omega^2 \qquad (2)$$

From the parallel-axis theorem,

$$I_{\text{b}} = I_{\text{cm}} + md^2 = I_{\text{cm}} + mr^2 \qquad (3)$$

$$mgh = \frac{1}{2}(I_{\text{cm}} + mr^2)\omega^2 = \frac{1}{2}I_{\text{cm}}\omega^2 + \frac{1}{2}mr^2\omega^2 \qquad (4)$$

Substituting $v_{\text{cm}} = \omega r$ gives

$$mgh = \frac{1}{2}I_{\text{cm}}\omega^2 + \frac{1}{2}mv_{\text{cm}}^2 \qquad (5)$$

(continued)

MECHANICS

NEL

CHAPTER 9 | ROLLING MOTION   239

For a sphere, $I_{cm} = \frac{2}{5}mr^2$. Substituting this expression into Equation (5) yields

$$mgh = \left(\frac{1}{2}\right)\frac{2}{5}mv_{cm}^2 + \frac{1}{2}mv_{cm}^2 = \frac{7}{10}mv_{cm}^2$$

and

$$v_{cm} = \sqrt{\frac{10}{7}gh}$$

To see this same example done by considering rolling as a rotation about the moving centre of mass, go to Online Example 9-4.

Making sense of the result:

While Equation (1) does not explicitly contain an expression for the translational kinetic energy of the sphere, this energy does appear in Equation (5) as a result of applying the parallel-axis theorem in Equation (3). Note: Equation (5) applies to any object of mass m and moment of inertia I starting from rest and rolling downhill without slipping. We will apply this result later in the chapter. Again, we see when the object does not rotate, we recover the scenario in example 6-8.

EXAMPLE 9-6

The Acceleration of a Sphere Rolling Down a Ramp

(a) Using energy considerations, derive an expression for the linear acceleration of an object of mass m, moment of inertia I, and radius r as it rolls without slipping down a ramp with a slope of angle θ.

(b) Apply your results to a solid uniform sphere and to a solid uniform disk.

(c) Compare your results for the disk to the results in Example 9-2, which was done using Newton's second law.

SOLUTION

(a) In Equation (5) in Example 9-5, we applied the conservation of energy and the parallel-axis theorem to show that the increase in kinetic energy for an object rolling down a ramp without slipping is

$$mgh = \frac{1}{2}I_{cm}\omega^2 + \frac{1}{2}mv_{cm}^2 \qquad (1)$$

If the object rolls a distance x down the ramp, the change in height is $h = x\sin\theta$. So,

$$\Delta K = -\Delta U \qquad (2)$$

and

$$mgh = mgx\sin\theta = \frac{1}{2}mv^2 + \frac{1}{2}I_{cm}\omega^2 \qquad (3)$$

As we saw in Chapters 6 and 8, taking the derivative of the energy conservation equation with respect to time gives an equation with acceleration as one of the terms. Because we are solving for linear acceleration, it is convenient to replace the angular quantities before taking the derivative. (If we were solving for angular acceleration, it would make more sense to replace the linear quantities.) Since the object is rolling without slipping, the speed of the centre of mass is

$$v_{cm} = \omega r \qquad (4)$$

Substituting this expression into Equation (3) gives

$$mgh = mgx\sin\theta = \frac{1}{2}mv_{cm}^2 + \frac{1}{2}\frac{I_{cm}}{r^2}v_{cm}^2 \qquad (5)$$

Taking the derivative of both sides with respect to time yields

$$mgv_{cm}\sin\theta = mv_{cm}a_{cm} + \frac{I_{cm}}{r^2}v_{cm}a_{cm} \qquad (6)$$

We have used the fact that the derivative with respect to time of the distance travelled by the object's centre of mass along the ramp is the velocity of the centre of mass, v_{cm}. We have also used the chain rule to obtain the derivative of the square of the velocity with respect to time:

$$\frac{dv^2}{dt} = \frac{dv^2}{dv}\frac{dv}{dt} = 2va \qquad (7)$$

Cancelling v on both sides of Equation (6) leaves us with only one unknown, a_{cm}. Here, the momentary-pivot approach has rather elegantly given us a single equation with the desired variable as the only unknown. Solving for a_{cm}, we get

$$a_{cm} = \frac{mg\sin\theta}{m + \dfrac{I_{cm}}{r^2}} \qquad (8)$$

(b) For a uniform solid sphere, $I_{cm} = \frac{2}{5}mr^2$. Substituting this expression into Equation (8), we get

$$a_{cm} = \frac{5}{7}g\sin\theta$$

We can consider a solid uniform disk to be a solid cylinder. For a uniform solid cylinder, $I_{cm} = \frac{1}{2}mr^2$, and the acceleration becomes

$$a_{cm} = \frac{2}{3}g\sin\theta$$

(c) The expression for the acceleration of the uniform solid disk is identical to the expression obtained in Example 9-2 for a cylinder rolling downhill.

Making sense of the result:

From Equation 8, we see that if the object was a point mass and had no moment of inertia, a reduces to $g\sin\theta$, as we saw in chapter 5, example 5-13.

C-9-8 Sphere, Cylinder, and Wheel

Starting from rest, a sphere, a solid cylinder, and a wheel with the same mass and radius roll, without slipping, down the same hill from the same height. At the bottom of the hill,

(a) the three objects have the same linear speed;

(b) the ring has the highest angular speed;

(c) the three objects have the same translational kinetic energy;

(d) the three objects have the same rotational kinetic energy;

(e) the three objects have the same total kinetic energy.

C-9-8 (e): As given by conservation of energy.

EXAMPLE 9-7

Acceleration in a Pulley System

Figure 9-16 shows a rope that is wound around a solid cylinder of mass m_c and radius R. The free end of the rope passes over a pulley of radius r and moment of inertia I, and is attached to a hanging mass M. The system is initially at rest. Mass M is then allowed to fall, pulling the rope and causing the pulley to rotate and the cylinder to roll as the rope unwinds from the cylinder. The cylinder rolls without slipping. Derive an expression for the acceleration of the hanging mass.

Figure 9-16 Example 9-7

SOLUTION

We begin by drawing free body diagrams for the three objects involved (Figure 9-17).

The forces acting on the hanging mass are its weight and the tension in the rope, T_1. Applying Newton's second law to the hanging mass, in the vertical direction we have

$$\sum F_y = Ma_y$$

$$T_{1y} + Mg_y = Ma_y$$

Choosing upward as the positive direction gives

$$T_1 - Mg = -Ma \qquad (1)$$

where a is a variable that denotes the magnitude of the acceleration. The same is true for T_1 and g in Equation (1).

Next, we sum the torques about the pulley's centre of mass. The torques acting on the pulley are τ_1 exerted by T_1 and τ_2 exerted by T_2:

$$\sum \vec{\tau}_{cm} = I_{cm}\vec{\alpha}$$

$$\vec{\tau}_1 + \vec{\tau}_2 = I\vec{\alpha}$$

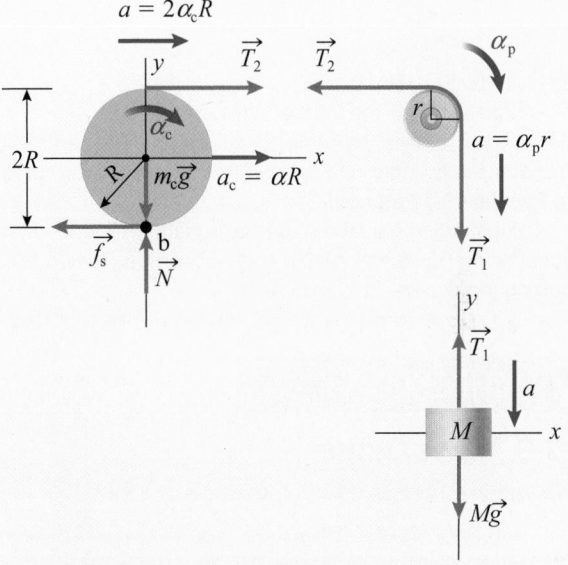

Figure 9-17 Free body diagrams of the (a) cylinder, (b) pulley, and (c) hanging mass

Choosing clockwise as positive,

$$\tau_1 - \tau_2 = rT_1 - rT_2 = I\alpha \qquad (2)$$

the variables in Equation (2) denote magnitudes of the vectors in Equation (1).

For the rolling cylinder, the simplest approach is to consider the torques about the point of contact between the cylinder and the surface, point b:

$$\sum \tau_b = I_b \alpha_c$$

Since the only force that exerts a torque about b is the tension T_2,

$$2RT_2 = I_b \alpha_c \qquad (3)$$

We can use the parallel-axis theorem to determine the moment of inertia of a cylinder about point b:

$$I_b = I_{cm} + md^2$$

The moment of inertia of a cylinder about its symmetry axis is $I_{cm} = \frac{1}{2}mR^2$, and the distance between the centre of mass of the cylinder and the point b is R. Hence,

$$I_b = \frac{m_c R^2}{2} + m_c R^2 = \frac{3}{2}m_c R^2$$

(continued)

Substituting for I_b in Equation (3) gives

$$2RT_2 = \frac{3}{2} m_c R^2 \alpha_c$$

$$2T_2 = \frac{3}{2} m_c R \alpha_c \qquad (3a)$$

So far, we have three equations and five unknowns: the tension forces T_1 and T_2; the angular acceleration of the pulley, α; the angular acceleration of the cylinder α_c; and the linear acceleration of the hanging mass, a. However, we can also write equations relating the linear acceleration of the mass to the angular acceleration of the pulley and that of the cylinder:

For the pulley, $\qquad a = \alpha r \qquad (4)$

For the cylinder, $\qquad a = \alpha_c 2R \qquad (5)$

as given by Equation 9-10a.

Note that the top of the cylinder, the rope, and the hanging mass have the same acceleration a. The centre of mass of the cylinder has acceleration a_c, which according to Equation 9-10 equals $a/2$.

Substituting Equation (5) into Equation (3) and then substituting Equations (1), (3), and (4) into Equation (2) and solving for a gives

$$a = \frac{Mg}{M + \frac{3}{8} m_c + \frac{I}{r^2}}$$

For the force of friction on the cylinder, we can sum the forces acting on the cylinder in the x-direction. We leave it to you to find the tension forces and the force of friction acting on the cylinder.

You can also solve this problem by considering rolling as a rotation about the moving centre of mass, and by using the work–mechanical energy approach as in Online Example 9-5 in the text online resources.

Making sense of the result:

The numerator in the expression for a has units of force, and the denominator has that of mass, as expected. When the pulley is light, and the cylinder does not roll, we recover the expression we obtained for the similar problem in Chapter 5, Example 5-16.

 CHECKPOINT

C-9-9 Top vs Centre

In Equation (5) in Example 9-7, why is $a = \alpha_c 2R$?

C-9-9 For an object of radius R rolling without slipping with angular acceleration α, the acceleration of the point at the top of an object (or of the point diametrically opposite to the point of contact between the object and the surface) is $\alpha 2R$. (Refer to Equations (9-3) and (9-10a).)

EXAMPLE 9-8

Ball on a Vertical Circular Track

The metal ball in Figure 9-18 rolls down the ramp as shown, then up the vertical circular track. Find the minimum height h for which the ball will go all the way around the circular part of the track. Assume perfect rolling throughout. The radius of the track is R, and the radius of the ball is r.

Figure 9-18 Example 9-8

SOLUTION

Let us begin by drawing the free body diagram of the ball at the top of the vertical circular track (Figure 9-19).

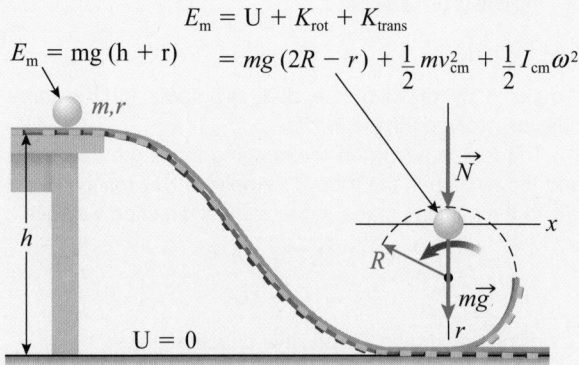

$$E_m = U + K_{rot} + K_{trans}$$
$$E_m = mg(h + r) = mg(2R - r) + \frac{1}{2} mv_{cm}^2 + \frac{1}{2} I_{cm}\omega^2$$

Figure 9-19 Free body diagram of the metal ball at the top of the circular track

We apply Newton's second law. In the vertical (r) direction we have

$$\sum F_r = ma_r$$
$$N_r + mg_r = ma_r$$

When travelling around the circular track, the centre of mass of the ball has a trajectory with radius $R - r$. Using the expression for centripetal force for circular motion and choosing down as the positive r-direction, we get

$$N + mg = m\frac{v^2}{R - r}$$

The problem requires that the ball just barely touch the top of the circular track. So, at this point, the track exerts no force on the ball (i.e., the normal force is zero). Hence,

$$mg = m\frac{v_{cm}^2}{R - r}$$

$$g(R - r) = v_{cm}^2 \qquad (1)$$

Conservation of energy requires an increase in the ball's kinetic energy to match the decrease in the Earth-ball's gravitational potential energy. The ball's centre of mass starts at a height of $h + r$ and ends up at the top of the track at a height of $2R - r$. The net change in the ball's height is $2R - 2r - h$; hence, the net change in the ball's potential energy is $-mg(h - 2(R - r))$. Therefore, using Equation 6-31, $\Delta K = -\Delta U$

$$mg(h - 2(R - r)) = \frac{1}{2}mv_{cm}^2 + \frac{1}{2}I_{cm}\omega^2$$

$$= \frac{1}{2}mv_{cm}^2 + \frac{1}{2}\left(\frac{2}{5}\right)mr^2\omega^2 \qquad (2)$$

Since the ball is rolling without slipping, $v_{cm} = \omega r$. Substituting for ωr in Equation (2) gives

$$mg(h - 2(R - r)) = \frac{7}{10}mv_{cm}^2 \qquad (3)$$

Using Equation (1) to substitute for v_{cm}^2 in Equation (3) gives

$$mg(h - 2(R - r)) = \frac{7}{10}mg(R - r)$$

$$h = 2.7(R - r)$$

When $r \ll R$, the expression becomes $h = 2.7R$.

Making sense of the result:

The result we obtained for the related example in Chapter 6, Example 6-15, showed that the minimum height required for an object *sliding* on a frictionless track was $h = 2.5r$. Thus, we can see that it takes more energy for the object to roll (rotational in addition to translational kinetic energy) along the track than to slide without rotating it. For the object to reach the speed necessary, it must start at a larger height when it is rolling as opposed to just sliding. When the object does not roll or when it is a point mass ($r = 0$), we get the same result as in Example 6-15.

  EXAMPLE 9-9

Marble on an Inverted Bowl

A bowl of radius R is sitting upside down on a horizontal surface (Figure 9-20). A child holds a marble of radius r at the top of the outside surface of the bowl. The child lets go of the marble, and it rolls without slipping. Derive an expression for the location at which the marble will fly off the surface of the bowl. Comment on the difference between this example and Example 6-16, Sliding on a Spherical Surface, in Chapter 6.

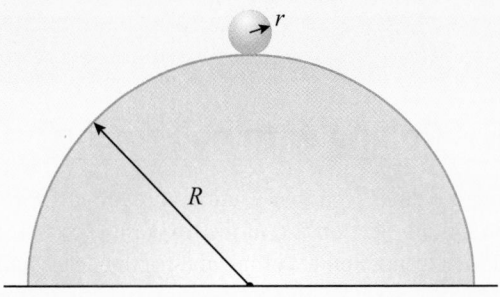

Figure 9-20   Example 9-9

SOLUTION

As the Earth-marble system loses gravitational potential energy, it will gain speed and kinetic energy, which might cause it to lose contact with the surface. We begin by drawing the free body diagram of the marble at the location where it is about to leave the surface of the bowl (Figure 9-21).

Since the marble moves on a semispherical bowl, the acceleration of interest is the radial acceleration. Hence, we need to determine the radial and tangential components of the forces. We apply Newton's second law. In the tangential direction we have

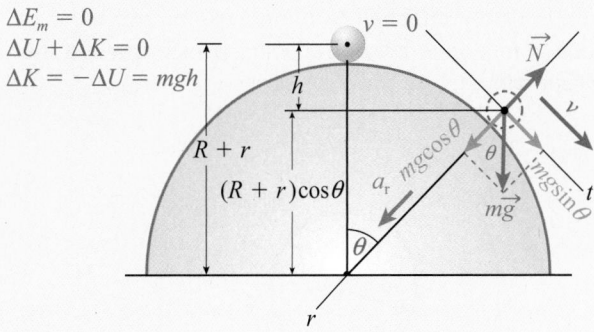

Figure 9-21   Free body diagram of the marble about to leave the surface of the bowl

$$\sum F_t = ma_t$$

$$mg_t = ma_t$$

$$mg\sin\theta = ma_t$$

In the radial direction we have

$$\sum F_r = ma_r$$

$$N_r + mg_r = ma_r$$

Choosing the radial direction toward the centre of the bowl as the positive direction, we get

$$-N + mg\cos\theta = m\frac{v^2}{R + r} \qquad (1)$$

Again, here N is a variable that indicates the magnitude of the normal force.

When the marble is on the verge of leaving the surface, the normal force is zero; hence,

(continued)

$$mg\cos\theta = m\frac{v_{cm}^2}{R+r} \quad \text{or} \quad g\cos\theta = \frac{v_{cm}^2}{R+r} \qquad (2)$$

So far, we have two unknowns and one equation. Since the potential energy lost by the marble is transformed into kinetic energy, using Equation 6-31,

$$mgh = \frac{1}{2}mv_{cm}^2 + \frac{1}{2}I_{cm}\omega^2$$

$$mgh = \frac{1}{2}mv_{cm}^2 + \frac{1}{2}\left(\frac{2}{5}\right)mr^2\omega^2 \qquad (2)$$

Since the marble is rolling without slipping,

$$v_{cm} = \omega r$$

From Figure 9-21, we can see that the height from which the marble drops is

$$h = (R+r)(1 - \cos\theta)$$

Substituting for ωr and h in Equation (2), we have

$$mg(R+r)(1 - \cos\theta) = \frac{7}{10}mv_{cm}^2 \qquad (3)$$

Using Equation (1) to substitute for v_{cm}^2 in Equation (3) gives

$$mg(R+r)(1 - \cos\theta) = \frac{7}{10}mg\cos\theta(R+r)$$

$$\cos\theta = \frac{10}{17}$$

$$\theta = 54.0°$$

In Chapter 6, example 6-16, we obtained a value for θ of 48.2° for an object sliding down a frictionless spherical surface.

Making sense of the result:

If no rotational motion were involved, all the lost potential energy would all be converted to translational kinetic energy, resulting in the object achieving a higher speed as it falls, and hence leaving the surface earlier as we have seen in example 6-16.

✔ **CHECKPOINT**

C-9-10 Left or Right?

In Figure 9-22, a rope is wound around an axle. The axle is rigidly attached to a large drum, and there is enough friction to prevent slipping. Which way will the object roll?

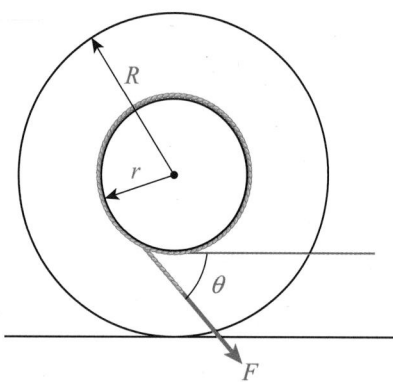

Figure 9-22   Which way will the object roll?

C-9-10 Clockwise. Hint: Sum torques about point of contact between object and surface.

✔ **CHECKPOINT**

C-9-11 Which Way Should You Push?

The drum shown in Figure 9-23 is machined from a single block such that a narrower cylindrical part protrudes from the centre of one face. A rope is wound around the protruding part. Which of the forces shown will result in the object rolling to the right?

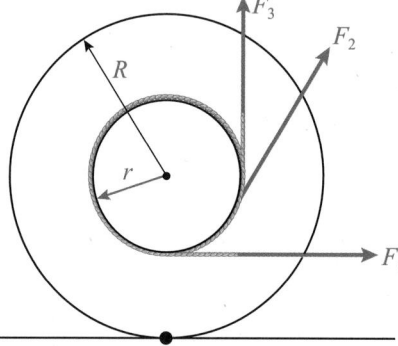

Figure 9-23   Which of the forces would cause the object to move to the right

C-9-11 only F_1. Hint: Sum torques about point of contact between object and surface.

LO 6

9-6 Rolling with no Friction?

So far, we have seen that friction is required for rolling. Go to Online Section 9-6 in the text online resources for a detailed discussion of rolling and friction, especially the concept of perfect rolling, in situations where the force of friction is zero.

LO 7

9-7 Rolling Friction

Go to Online Section 9-7 in the text online resources for a discussion of rolling friction and the effect of surface and object deformation on rolling. This discussion offers a more realistic model of the forces encountered when a rigid object rolls on a smooth surface.

ROLLING

Rolling is a combination of translational and rotational motion. For rolling without slipping,

$$v_{cm} = \omega R$$

$$a_{cm} = \alpha R$$

The part of the object in touch with the surface is momentarily stationary with respect to the surface.

$$v_{bottom} = 0$$

Rolling can also be treated as a series of successive rotations about the point of contact between the object and the surface.

The Dynamics of Rolling

Translation/Rotation Approach

$$\sum \vec{F} = m\vec{a}_{cm} \qquad (9\text{-}11)$$

$$\sum \vec{\tau}_{cm} = \vec{I}_{cm}\alpha \qquad (9\text{-}12)$$

Momentary-Pivot Approach (b refers to point of contact between object and surface)

$$\sum \vec{\tau}_{b} = I_{b}\vec{\alpha} \qquad (9\text{-}13)$$

The force of static friction on an object rolling without slipping is usually an unknown:

$$0 \le f_{s} \le \mu_{s}N \qquad (9\text{-}14a)$$

When an object is on the verge of slipping, the force of friction becomes

$$f_{s,max} = \mu_{s}N \qquad (9\text{-}14b)$$

If an object slips as it rolls, the force of friction becomes the force of kinetic friction:

$$f_{k} = \mu_{k}N \qquad (9\text{-}14c)$$

WORK AND ENERGY

A. Translation/Rotation Approach

$$K_{translation} = \frac{1}{2}mv_{cm}^{2} \qquad (9\text{-}15)$$

$$K_{rotation} = \frac{1}{2}I_{cm}\omega^{2} \qquad (9\text{-}16)$$

Total kinetic energy:

$$K = \frac{1}{2}mv_{cm}^{2} + \frac{1}{2}I_{cm}\omega^{2} \qquad (9\text{-}17)$$

B. Momentary-Pivot Approach

$$K = \frac{1}{2}I_{b}\omega^{2} \qquad (9\text{-}18)$$

APPLICATIONS

Applications: ABS, traction control, tire design, drive trains, rolling element bearings

Key Terms: ideal rolling, momentary pivot, perfect rolling, rolling without slipping, rolling, skidding, slipping, spinning

QUESTIONS

1. A solid uniform ring and a solid uniform disk of the same radius and mass roll without slipping at the same speed on a horizontal surface, which turns into a ramp sloping upward from the horizontal. Which object will go higher up the incline?

2. A uniform solid sphere and a uniform hollow cylinder roll without slipping at the same speed on a horizontal surface. The surface then becomes frictionless and curves upward into a semicylindrical track of radius r. Will one of these objects go higher up the track? If so, which one? Assume that both objects are in contact with the track at all times and that neither of the objects has enough kinetic energy to make it to a height r above the bottom of the track.

3. A uniform solid cylinder and a uniform solid sphere of the same mass and radius roll without slipping downhill from the same height.

   (a) Which object will have the higher rotational kinetic energy at the bottom of the hill?
   (b) Which object will have the higher angular speed at the bottom of the hill?

4. Consider a uniform solid sphere rolling without slipping down a ramp of angle θ. How will the force of friction on this object change if you double its mass without changing its radius?

5. What is the main function of an antilock braking system?

6. In the absence of ABS, why is a driver advised to "pump" the brakes when stopping on a slippery surface?

7. A car coasts on the highway at a speed of 130 km/h. Is the friction between the tires and the road kinetic or static?

8. The ball in Figure 9-24 rolls on a flat horizontal surface without slipping. Establish the direction of the total acceleration of points A, B, and C on the wheel at the instant shown.

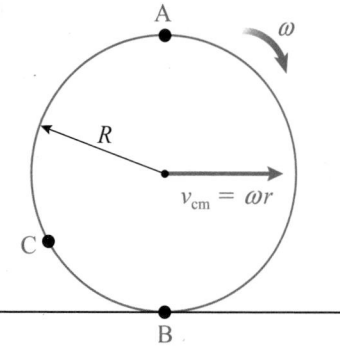

Figure 9-24 Question 8 and Problem 25

9. A solid uniform ring, disk, and sphere with equal masses and equal radii roll without slipping down a smooth ramp from the same height. Which of the three objects will reach the bottom of the ramp first?

10. A ring rolls down a ramp without slipping. At any point, calculate the ratio of the rotational energy to the translational kinetic energy. Repeat the exercise for a uniform solid disk and for a uniform solid sphere.

11. An object rolls down a hill such that $\frac{4}{6}$ of its kinetic energy is rotational. Determine an expression for the object's moment of inertia in terms of its mass and radius.

12. Can an object's rotational and translational kinetic energy ever be equal? Explain.

13. Two uniform solid disks, disk 1 and disk 2, of the same radius and thickness, roll down a ramp from the same height. The mass density of disk 2 is twice the density of disk 1. Will one disk reach the bottom of the hill before the other? If so, which one?

14. Two uniform solid spheres, sphere 1 and sphere 2, are at rest at the top of a hill. The spheres have the same mass but different radii: the radius of sphere 1 is greater than the radius of sphere 2. They roll without slipping downhill from the same height. Will one sphere reach the bottom of the hill before the other? If so, which one?

15. Two cylindrical shells, shell A and shell B, are at rest at the top of a hill. The mass of shell A is 16 times the mass of shell B, and the radius of shell A is one-half the radius of shell B. Both shells roll downhill from the same height. Will one shell reach the bottom of the hill before the other? If so, which one?

16. Two spheres, sphere A and sphere B, are at rest at the same height at the top of a hill. The mass of sphere A is 5 times the mass of sphere B, and the radius of sphere A is $\frac{1}{\sqrt{2}}$ the radius of sphere B. Both spheres roll without slipping downhill. Will one sphere reach the bottom of the hill before the other? If so, which one?

17. Is it possible for a spherical shell and a solid uniform cylinder, originally at rest and at the same height on top of a hill, to roll downhill without slipping and reach the bottom of the hill together? If so, find the ratio of their radii.

18. A cylinder and a sphere of the same mass roll freely down the same slope without slipping. Which of the two objects has a greater force of friction acting on it?

19. A sphere and a ring roll without slipping down a hill. The force of friction acting on the ring is 10 times the force of friction acting on the sphere. Find the mass ratio of the two objects.

20. A cylinder and another object roll without slipping down a slope. The mass of the second object is twice the mass of the cylinder. The force of friction acting on the cylinder is twice the force of friction acting on the other object. Find an expression for the moment of inertia of the second object in terms of its mass and radius.

21. Consider a typical wheel rolling freely on a flat horizontal surface. What could cause the wheel to slow down?

22. A marble rolls on a horizontal, nondeformable, smooth glass surface. Do you expect the marble to slow down because of friction? Explain why or why not.

23. A ball rolls down a hill without slipping. Assume ideal conditions, that is, the object and surface are nondeformable, and the surface is smooth. In which direction does the force of friction act on the ball?

PROBLEMS BY SECTION

For problems, star ratings will be used, (✷, ✷✷, or ✷✷✷), with more stars meaning more challenging problems.

Section 9-1 Rolling and Slipping

24. ✷ A rear-wheel-drive car accelerates from rest, and the wheels turn (roll) without slipping. Determine the direction of the force of friction on the car's front and rear wheels.

Section 9-2 Relationships Between Rotation and Translation for a Rolling Object

25. ✷ The wheel in Figure 9-24 rolls on a horizontal surface without slipping at a constant speed of 15 m/s. The wheel is 74 cm in diameter.
    (a) Determine the speed and acceleration of point A at the top of the wheel.
    (b) Find the magnitude of the total speed and acceleration of the point one-quarter of a period later and one-half of a period later.

26. ✷✷ A tundra buggy, which is a bus fitted with oversized wheels, is stuck in Churchill, Manitoba, on slippery ice. The wheel radius is 0.87 m. The speedometer goes from 0 to 27 km/h while the buggy moves a total distance of 5.0 m in 7.0 s. Find the tangential acceleration and the total acceleration of a point at the bottom of the wheel at the end of 3.0 s.

27. ✷ While driving on the frozen surface of a lake, you notice that your truck is moving at 10 km/h, but the speedometer says 56 km/h. The wheel radius is 46 cm. (Hint: The speedometer reading is obtained from the rotational speed of the tires.)
    (a) Find the speed of a point at the top of a tire.
    (b) Is there a point on the tire where the velocity is zero? If so, where is it located?

28. ✷✷ A cylinder of radius 1.2 m rolls at a constant speed. You apply a brake mechanism that slows the cylinder. At a certain instant, the speed of the cylinder's centre of mass is 11 m/s and the speed of the top of the cylinder is 14 m/s. Calculate the angular speed of the cylinder at that instant.

29. ✷✷ A wheel undergoes acceleration from rest, such that it spins while rolling. The acceleration of the centre of mass is measured to be 2.30 m/s², and the angular acceleration is measured to be 7.10 rad/s². The wheel's radius is 80.0 cm.

(a) Find the magnitude of the tangential acceleration of a point at the top of the wheel.

(b) Find the magnitude and direction of the total acceleration of a point at the bottom of the wheel at 1.70 s into the motion.

Section 9-3 Rolling Motion: Two Perspectives

For the problems in this section, approaching rolling as a rotation about point b, which is the point at which the rolling object is in contact with the surface on which it sits.

30. ✳ Show that it is valid to consider rolling as a rotation about point b.

31. ✳ If we consider rolling to be a rotation about point b, that leaves us with an apparent paradox: If the object were moving at a constant speed, then the centre of mass would be accelerating with respect to b. How do you resolve this apparent paradox?

32. ✳✳ Find the acceleration of the point on a rolling object in contact with the surface on which it rolls.

33. ✳✳ Calculate the maximum angle that the force F can make with the horizontal to cause the object in Figure 9-25 to roll to the right. Assume there is enough friction to ensure rolling without slipping.

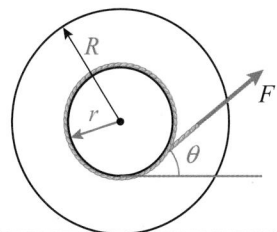

Figure 9-25 Problem 33

34. ✳✳ The object in Figure 9-26 has a moment of inertia of 17 kg·m². The object is pulled by a force of 160 N in the direction shown. There is enough friction to prevent slipping. Calculate the angular acceleration of the object, given that $R = 0.91$ m and $r = 0.32$ m.

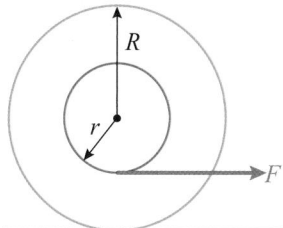

Figure 9-26 Problem 34

35. ✳✳ A form of yo-yo consists of a thin cylindrical shell with a string wound around it. Find the acceleration of this yo-yo when a child lets it go while holding the end of the string.

36. ✳✳ A wheel of radius 72 cm rolls without slipping down a 32° slope starting from rest (Figure 9-27). Point A is shown at $t = 0$. Find the speed and acceleration of point A at 3.0 s into the motion.

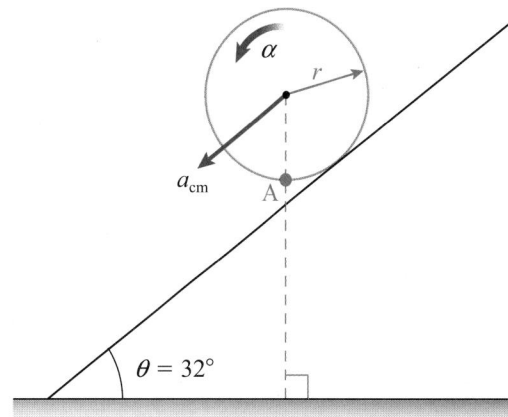

Figure 9-27 Problem 36

37. ✳ Indicate the direction in which the object in Figure 9-28 will roll for each of the given forces.

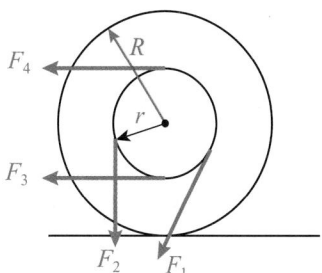

Figure 9-28 Problem 37

38. ✳✳ A car is moving at a constant speed of 29 m/s on a flat horizontal road. The car wheels, 41 cm in radius, roll without slipping. You paint a dot on the edge of one of the tires and monitor the dot's motion. Find the magnitude and direction of the acceleration of the dot when

(a) its speed is zero;

(b) its horizontal speed is zero;

(c) its vertical speed is zero;

(d) the point is at the top of the wheel.

Section 9-4 Newton's Second Law and Rolling

39. ✳✳ A solid uniform spherical shell of mass 1.2 kg rolls from rest without slipping down a hill with a 32° slope. Find the force of friction acting on the spherical shell. What would happen to the force of friction if you tripled the radius of the shell but kept the mass the same?

40. ✳✳ Consider a yo-yo as a uniform solid disk of mass 210 g and radius 6.0 cm. You release the yo-yo while holding the end of the string wound around it.

(a) Find the tension in the string.

(b) Find the acceleration of the centre of mass of the yo-yo.

(c) Find the angular acceleration.

41. ✳✳ Consider a cylindrical shell of radius 21 cm and mass 1.3 kg that rolls without slipping down a 37° hill. Find the force of friction between the surface and the cylindrical shell by taking torques about

(a) the centre of mass;

(b) the stationary point on the shell (the point in contact with the ground).

42. ✱✱✱ A rope is wound around a small, solid uniform ball, which you release from rest and allow to fall vertically while holding the rope (like a yo-yo). The length of the rope is such that the ball comes loose when it hits the ground. The ball is seen to spin for 11 m before coming to a perfect roll. The coefficient of static friction between the ball and the ground is 0.15.
    (a) Determine the height from which the ball was released.
    (b) Find the angular displacement of the ball before it comes to a perfect roll.
    (c) Calculate the displacement of the centre of mass before the ball comes to a perfect roll.
    (d) Does $a_{cm} = \alpha r$ before the ball starts rolling without slipping? Explain.
    (e) How much energy is lost to thermal energy before the ball comes to a perfect roll?
    (f) How long does the ball spin before coming to a perfect roll?
    (g) What is the final speed of the ball?

43. ✱✱✱ The ancient Egyptians may have used the method shown in Figure 9-29 to transfer large stones. The stone in Figure 9-29 of mass M is resting on two logs such that the weight is equally distributed between the logs. The stone is then pushed forward with a constant horizontal force F. Find an expression for the acceleration of the stone. The logs can be considered as ideal cylinders of mass m and radius R. Assume there is enough friction everywhere to prevent slipping.

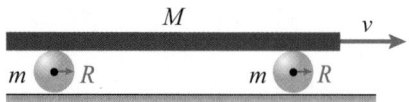

Figure 9-29 Problem 43

44. ✱✱✱ The coefficient of friction between each brake pad and brake disk on the wheels of a sports car is 0.25. The brake pad presses against the disk at a point that is 17 cm from the centre of the disk. The car has a mass of 1630 kg, and the weight is equally distributed on the four wheels. The car is moving at a constant speed on the road where the coefficient of static friction between the tires and the road is 0.56.
    (a) What is the maximum normal force that the brakes can exert on the brake disk so that the wheels do not skid? The wheels have a moment of inertia of 34 kg·m² and a radius of 29 cm.
    (b) Repeat part (a) with the car driving at constant speed up a 23° slope.

45. ✱✱ A uniform solid cylinder sits on a 4 m high hill with a 39° slope. The coefficient of kinetic friction is 0.12, and the coefficient of static friction is 0.17. The mass of the cylinder is 340 g, and its radius is 3 cm.
    (a) Will the cylinder roll without slipping? Explain.
    (b) Find the angular acceleration of the cylinder.
    (c) The cylinder travels from the top to the bottom of the hill. How much energy is lost to friction?

46. ✱✱✱ On an icy road, the coefficient of static friction between the tires of a bicycle and the road is 0.050. The wheels can be considered to be disks of radius 32 cm and mass 3.0 kg. The combined weight of the bicycle and the cyclist is 870 N and is distributed equally between the two wheels. What is the maximum torque that can be delivered to the rear wheel without causing it to spin?

47. ✱✱ You place a sphere on an inclined plane. The coefficient of static friction between the sphere and the plane is 0.30. Determine the maximum angle that the inclined plane can make with the horizontal so that the sphere does not skid. The sphere starts from rest.

48. ✱✱ A car rolls freely down a hill without slipping.
    (a) In which direction must the force of friction act on the tires?
    (b) The driver of the car puts the car in gear and depresses the gas pedal. What must the acceleration of the car be if the force of friction on the wheels is zero? Consider the car wheels to be disks.

49. ✱✱✱ A 52 kg cyclist rides his 7.0 kg bicycle down a 17° hill. The normal force from the road on the bicycle is equally distributed between the two wheels. Each wheel is 27 cm in radius and has a mass of 900 g. Two brake pads act at the rim of each of the wheels. The wheels can be considered as ideal rings. What is the normal force required between the brake pads and the wheels so that the force of friction between the tires and the road is zero?

50. ✱✱✱ A car driver waiting at a traffic light depresses the gas pedal as soon as the light turns green. Find the maximum torque that can be delivered to the wheels of the car without causing them to spin. The diameter of the tires is 65 cm. Assume that the weight of the 2200 kg car is evenly distributed on the four wheels and that the coefficient of static friction between the asphalt and the tires is 0.70.

Section 9-5   Mechanical Energy and Rolling

51. ✱ A uniform solid cylinder of mass 1.2 kg rolls without slipping down a hill from a height of 3.4 m.
    (a) Calculate the cylinder's rotational and translational kinetic energies at the bottom of the hill.
    (b) Repeat the problem for a ring of the same mass.

52. ✱ A sphere and a cylinder, both of radius 17 cm and mass 34 kg, roll without slipping down a 4.0 m high. Both objects start from rest.
    (a) Calculate the total kinetic energy of each object at the bottom of the hill.
    (b) Calculate the angular speed of each object at the bottom of the hill.
    (c) Calculate the linear speed of the centre of mass of each object at the bottom of the hill.
    (d) Determine the ratio of the rotational energy to the translational kinetic energy for each object. Explain your results.

53. ✱✱ A solid, uniform bowling ball, 12.7 cm in diameter, starts from rest and rolls down a 7.0 m high ramp with a 37° slope. There is enough friction between the surface of the ramp and the bowling ball to prevent the ball from skidding. At the bottom of the ramp the ball goes up a frictionless ramp with the same slope as the first ramp.
    (a) Determine the maximum height that the ball can reach on the frictionless ramp.
    (b) Is mechanical energy conserved? Explain.
    (c) What is the angular speed of the bowling ball when it reaches its maximum height on the frictionless ramp?

54. ✱✱ A 7.0 kg steel ball of radius 4.0 cm rolls without slipping from rest down a hill of height 3.0 m. The steel ball then rolls on a horizontal surface for a certain distance before striking a spring whose spring constant k is 72 N/m. The spring is originally unstretched. The steel ball

compresses the spring until it comes to a momentary stop. Find the maximum compression of the spring.

55. ✷✷ A solid uniform ball of mass m and radius r rolls down a semispherical bowl, starting from a height h above the bottom of the bowl (Figure 9-30). The surface of the left half of the bowl provides enough friction to prevent the ball from slipping on the way down. The surface of the right half of the bowl is frictionless.
   (a) Derive an expression for the maximum height the ball will reach on the frictionless side. Is energy not conserved? Explain.
   (b) Find the translational kinetic energy of the ball at its highest point on the frictionless side.
   (c) What is the rotational kinetic energy at that point?
   (d) Explain how would your answer to (a) will change if we replace the ball with an ideal ring.

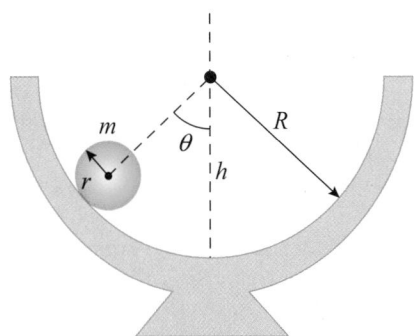

Figure 9-30 Problem 55

56. ✷✷ A sphere of mass m starts from rest and rolls without slipping down a ramp from a height h. At the bottom of the ramp, the sphere continues to move on a flat horizontal frictionless surface, and then collides with a horizontal spring whose spring constant is k. The spring is originally unstretched and is fitted with a light, frictionless plate at its end, which the sphere collides with (Figure 9-31).
   (a) Derive an expression for the maximum compression of the spring.
   (b) Derive an expression for the angular speed of the sphere when it strikes the spring.
   (c) What would the result be in part (a) if there were no friction anywhere?

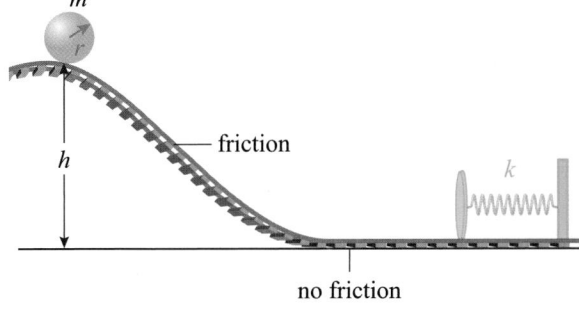

Figure 9-31 Problem 56

57. ✷✷ Use the work–mechanical energy approach to solve problem 44.

COMPREHENSIVE PROBLEMS

58. ✷✷ A uniform solid sphere of mass m and radius R rolls down an ideally flat ramp with a slope of angle θ with respect to the horizontal, as shown in Figure 9-32. Use energy considerations to derive an expression for the acceleration of the sphere. Do you have to account for friction? Explain. Compare the energy approach to the Newtonian equation of motion approach.

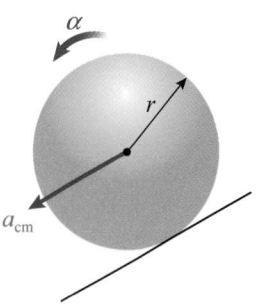

Figure 9-32 Problem 58

59. ✷✷✷ In Figure 9-33, a massless string is wound around a sphere of mass M_s and radius R_s. The string passes over a pulley with moment of inertia I_1 and radius r_1; the string is attached to the mass m on top of the table as shown. The mass m, in turn, is attached to another massless string, which passes over a second pulley with moment of inertia I_2 and radius R_2, and is attached to the centre of a solid uniform cylinder of mass M_c and radius R_c. The system starts from rest, and then the cylinder accelerates downward. Find an expression for the acceleration of the sphere's centre of mass by applying (a) Newton's second law and (b) work–mechanical energy considerations. Assume perfect rolling without slipping anywhere.

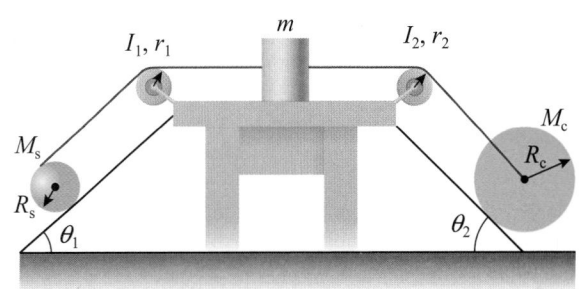

Figure 9-33 Problem 59

60. ★★ The marble of mass m and radius R in Figure 9-34 starts from rest at height h and rolls down the track without slipping. The marble then rolls up the semicircular part of the track, which has a radius r equal to $h/4$. The coefficient of static friction between the marble and the track is μ_s. At the position shown, the normal force from the track on the marble is equal to one-half the marble's weight.
  (a) Find an expression for the moment of inertia of the marble.
  (b) Determine the force of friction on the marble.

Figure 9-34 Problem 60

61. ★★ The object in Figure 9-35 is machined out of a single piece of metal such that its moment of inertia is $\dfrac{mr^2}{3}$. You pull on the rope horizontally as shown.
  (a) In which direction will the object roll, assuming there is enough friction between the object and the surface to prevent slipping?
  (b) For what value of r will you be able to pull the rope and cause the cylinder to roll such that the force of friction is zero?

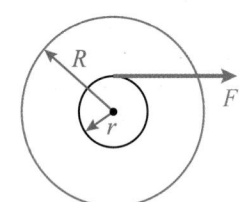

Figure 9-35 Problem 61

62. ★★ A pilot applies the wheel brakes upon landing so that his plane skids to a stop on a runway from its landing speed of 130 km/h. The 1.9 m diameter wheels were rolling without slipping before the brakes were applied. Assume constant acceleration of the plane's centre of mass and constant angular acceleration of its wheels. The plane comes to a stop over a distance of 270 m.
  (a) Find the speed of a point at the bottom of the wheels 5.4 s into the landing.
  (b) Find the speed of a point at the top of a wheel 110 m after the pilot started braking.

  (c) Find the total acceleration of a point on the front edge of a wheel at the same height as the centre of the wheel after 3.0 s of braking.

63. ★★★ A uniform solid ball of mass $m = 1.6$ kg and radius $r = 0.11$ m rolls down a 4.3 m hill without slipping. At the bottom of the hill, the surface is horizontal and the ball continues to roll until it strikes a stationary block of mass $M = 9.0$ kg (Figure 9-36). The collision is perfectly elastic; therefore, no energy is lost during the collision. How fast is the ball moving after the collision?

Figure 9-36 Problem 63

64. ★★★ You kick a soccer ball of mass 0.90 kg and radius 14 cm, as shown in Figure 9-37. The kick has a duration of 91 ms and an average force of 161 N. Your foot strikes the ball horizontally 14 cm above the floor. The coefficient of kinetic friction between the ball and the ground is 0.090.
  (a) Find the linear speed of the ball immediately after the collision.
  (b) What is the angular speed of the ball immediately after the collision?
  (c) Determine the linear speed of the ball when it starts rolling.
  (d) Find the total amount of energy lost to friction.

$\mu_k = 0.09$

Figure 9-37 Problem 64

65. ★★ A ball, which can be considered a uniform spherical shell, of mass m and radius r is placed on the surface of a hemispherical bowl of radius R and rolls down the bowl without slipping (Figure 9-30).

(a) The ball starts from rest at an angle of 72° with the vertical (top of the bowl). What is the normal force on the ball when it is at the bottom of the bowl?

(b) Repeat part (a) with $r \ll R$.

(c) Repeat part (a) with $m = 0.67$ kg, $r = 0.070$ m, and $R = 9.0$ m. Does the approximation $r \ll R$ hold?

66. ✷✷✷ Starting from rest, a ball of mass 0.63 kg and radius 0.17 m slides from the top to the bottom of a circular track of radius 1.1 m (Figure 9-38). The surface of the track is frictionless. The ball then moves up a ramp that is at an incline of 33° from the horizontal. The coefficient of friction between the ball and the ramp is 0.12.

(a) Find the acceleration of the ball as it skids up the ramp.

(b) What is the angular speed of the ball when it starts rolling perfectly?

(c) What is the acceleration of the ball as it rolls perfectly?

(d) Use energy considerations to find the maximum height the ball will reach above the bottom of the ramp.

Figure 9-38 Problem 66

67. ✷✷✷ You throw a solid uniform sphere of radius 11 cm and mass 2.1 kg onto a smooth horizontal surface with a horizontal speed of 9.0 m/s. The sphere skids initially and then starts rolling without slipping. The coefficient of static friction between the sphere and the surface is 0.18 and the coefficient of kinetic friction is 0.14.

(a) Find the acceleration of the sphere when it starts rolling without slipping on the horizontal surface.

(b) For how long does the sphere skid?

(c) How much energy is lost to friction?

68. ✷✷ The ideal ring shown in Figure 9-39 sits on a horizontal surface. There is enough friction between the surface and the ring to prevent slipping. What should the radius of the ring be so that the force of friction on the ring is zero as it is being pulled?

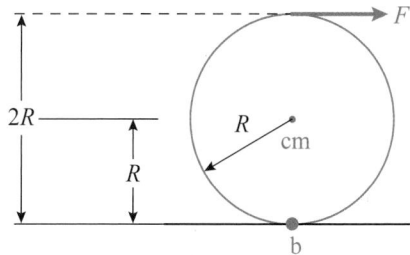

Figure 9-39 Problem 68

69. ✷✷ Redo Example 9-7, this time summing the torques about the point on the cylinder touching the ground.

70. ✷✷ A 2.0 kg uniform solid bowling ball of radius 0.32 m rolls without slipping down a 43° slope with a smooth surface. At the bottom of the hill, the surface is horizontal.

(a) Find the acceleration of the ball on its way down the hill and when it is rolling on the horizontal surface.

(b) How would the acceleration of the ball on the horizontal surface change if the horizontal surface is covered with glossy ice such that it is almost frictionless?

71. ✷✷✷ Is it possible for an object to roll without slipping on completely frictionless ice? Explain.

72. ✷✷ A solid uniform cylinder is rolling without slipping at a constant speed of 10 m/s on a horizontal surface. The cylinder then rolls 7.0 m down a frictionless 17° slope. Determine the ratio of the rotational kinetic energy to the translational kinetic energy of the cylinder at the bottom of the slope.

73. ✷✷ A marble starts from rest and rolls without slipping down a smooth incline from height h. At the bottom of the incline, the surface is horizontal and frictionless. Will the marble eventually stop rolling on the frictionless surface? If so, what force would be responsible for changing the marble's angular speed?

74. ✷✷✷ You spin a bowling ball of mass m and radius r in your hands until it reaches an angular frequency ω. You then place it gently on a surface where the coefficient of friction between the surface and the ball is μ_k. The ball initially spins on the surface as it picks up linear speed because of kinetic friction (Figure 9-40).

(a) How much time passes before the ball starts rolling without spinning? How many revolutions does the ball go through during that time?

(b) Calculate the distance x that the centre of mass of the ball travels during the time you calculated in part (a).

(c) Does $\Delta x_{cm} = r(\Delta \theta)$ during that time?

(d) Find the final linear speed of the centre of mass of the ball when it starts to roll without slipping.

(e) What is the force of friction on the ball as it rolls on the horizontal surface without slipping?

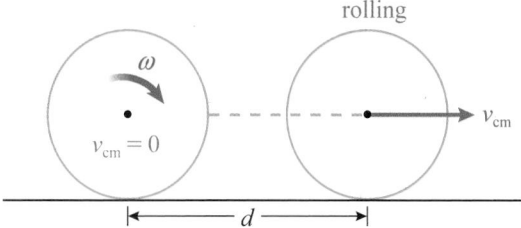

Figure 9-40 Problem 74

75. ✷ Equations (9-17) and (9-18) give two different expressions for the total kinetic energy of a rolling object. Show that these two expressions are always equal.

76. ✷✷✷ Develop parametric equations to describe the motion of a point at the edge of a rolling object.

77. ✷✷ A plank of mass M rests on two solid uniform cylinders as shown in Figure 9-33. The coefficient of static friction between the plank and the cylinders is 0.35, and the coefficient of static friction between the cylinders the horizontal surface they roll on is 0.55. Assume that the weight of the plank is equally distributed on the two cylinders.

(a) Find the maximum force F you can push the plank with so that no slipping occurs anywhere.

(b) What is the acceleration of the plank in part (a)?

78. ✱✱✱ A ball of diameter 24 cm is released from rest on a slope that is inclined 42° (Figure 9-41). The surface of the slope is frictionless for the first 7.0 m, and then becomes uniformly rough with some friction.

(a) Will the ball eventually start rolling without slipping if the rough surface has a coefficient of kinetic friction of 0.13? Explain.

(b) What is the minimum coefficient of kinetic friction required for the ball to roll without slipping 14 m after it meets the rough surface?

(c) How many turns will the ball in part (b) undergo before rolling without slipping?

(d) How long will it take the ball in part (b) to start rolling without slipping?

(e) Find the angular speed of the ball in part (b) when it starts rolling without slipping.

Figure 9-41 Problem 78

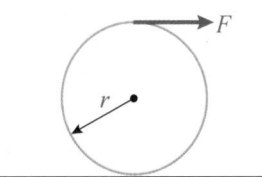

79. ✱✱✱ Suppose that the object in Figure 9-42 has a moment of inertia greater than mr^2. Show that the force of friction will actually push the object forward as you pull on it. Assume that there is enough friction between the surface and the object to prevent slipping.

Figure 9-42 Problem 79

80. ✱✱✱ A ball of mass 940 g and radius 5.0 cm rolls down a semicylindrical surface of radius 0.70 m as shown in Figure 9-43. The coefficient of static friction between the surface and the ball is 0.15.

(a) Determine the location at which the ball begins to skid or slip on the way down.

(b) What is the normal force exerted by the surface on the ball at that point?

(c) Find the position where the ball leaves the surface.

(d) Had the surface been completely frictionless, would the ball have left it at the same position as in part (a)? Explain.

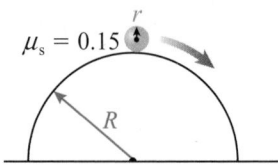

Figure 9-43 Problem 80

81. ✱✱ A rolling marble of radius r enters a hemispherical bowl of radius R through a hole as shown in Figure 9-44. The marble is moving with translational speed v and angular speed $\frac{v}{r}$. Assume that friction between the marble and the bowl is negligible.

(a) What speed v will cause the marble to maintain a height h above the bottom of the bowl?

(b) What would happen if the marble entered the bowl at a point slightly above or below the desired height at the same speed as in part (a)?

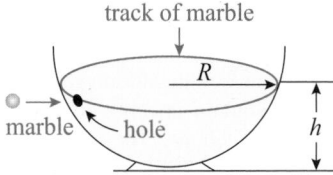

Figure 9-44 Problem 81

82. ✱✱ A hemispherical shell of radius R and mass m is made to spin about a fixed vertical axis as shown in Figure 9-45. As the spinning speed increases very slowly from zero, a small metal ball of radius r at the bottom of the shell starts to move upward. The shell's spinning speed is maintained at a constant value when the ball reaches a height h above the bottom of the bowl.

(a) Find the spinning speed of the shell in terms of the other variables in the problem.

(b) Find the work done on the metal ball as the shell accelerates. Assume that the ball rolls without slipping.

(c) Why does the speed need to be increased slowly? Had the speed been changed quickly, what would you expect the ball to do, assuming it never goes above the rim?

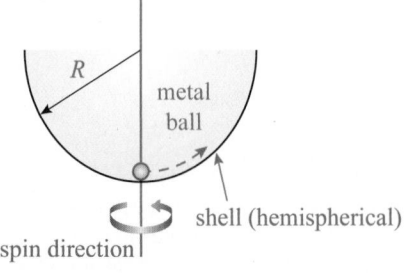

Figure 9-45 Problem 82

83. ✳✳ (a) Use and compare the work–energy and Newton's second law approaches for finding the angular acceleration of the sphere of radius R and mass m as it rolls up the incline in Figure 9-46 without slipping. The pulley has a moment of inertia 0.40 kg m², and radius 11 cm, $m = 1.1$ kg, $R = 23$ cm, and $M = 5.6$ kg. Assume that the hanging mass M accelerates downward.

(b) Find the force of friction between the sphere and the incline.

$\theta = 32°$

Figure 9-46   Problem 83

84. ✳✳✳ A solid uniform cylinder of mass 0.65 kg and radius 11 cm is held on a horizontal surface where it compresses a spring ($k = 1210$ N/m) by a distance of 41 cm from the equilibrium position. The cylinder is then let go, and it gains kinetic energy from the spring as the spring returns to equilibrium. Initially the surface is frictionless and then it becomes rough suddenly with a coefficient of friction between the surface and the cylinder of 0.12.

(a) What is the speed of the cylinder when it hits the rough part of the surface?

(b) How far does the cylinder go before it stops skidding?

(c) What is the angular displacement of the cylinder before it starts rolling without slipping?

(d) How much energy is lost to friction?

(e) Find the final angular speed of the cylinder.

See the text online resources at
**www.physics1e.nelson.com for Open Problems
and Data-Rich Problems related to this chapter.**

When you have completed this chapter you should be able to:

1 Define equilibrium, and distinguish between static equilibrium and dynamic equilibrium.

2 Distinguish between centre of mass and centre of gravity.

3 Apply the conditions for static equilibrium.

4 Identify and calculate support forces acting on the parts of a structure.

5 Explain stress, strain, Young's modulus, elastic deformation, and plastic deformation.

6 Describe the behaviour of ductile and brittle materials.

Chapter 10
Equilibrium and Elasticity

Soft Strength

It is desirable to use structural materials that can sag or stretch substantially before fracturing under extreme loading, especially for buildings that may be subject to earthquakes or hurricane-force winds. Generally, the more rigid steel alloys are also more brittle. Most construction-grade steel is designed to sag a significant amount before fracturing. During a severe earthquake, a building supported by this type of steel collapses more slowly than a building with a harder, more rigid steel, thus giving the occupants more time to evacuate.

Researchers at Stanford University have taken earthquake protection a step further by designing special building braces that can be incorporated into new buildings or added to existing buildings to make them more earthquake-resistant. The blue braces in Figure 10-1 are intended to be flexible enough to rock without damage. Strong steel cables hold the brace frames together and help restore them to their original configuration when they rock. As the frame of braces rocks, the bulk of the energy absorbed by the structure from the earthquake goes into deforming arrays of leaf-like steel "fuses" (shown in yellow) at the base of the frame. As the blue central column moves up and down or sideways, it causes the fuses to bend and yield with the motion, thus absorbing kinetic energy from the building. The deformed fuses can be replaced after an earthquake to restore the protection. Go to the text's online resources for a link to a video showing how these structural braces at work.

For additional Making Connections, Examples, and Checkpoints, as well as activities and experiments to help increase your understanding of the chapter's concepts, please go to the text's online resources at www.physics1e.nelson.com.

Courtesy of Brad Wilkins

vertical post tensioning

flexible braces

fuses at the base deform to absorb energy

Figure 10-1 During an earthquake, the structure absorbs energy from the earthquake, and the energy deforms the fuses at the base of the frame.

LO 1

10-1 Static and Dynamic Equilibrium

For an object to be in equilibrium two conditions must be met:

1. The vector sum of the external forces acting on the object must be zero:

$$\sum \vec{F} = 0 \qquad (10\text{-}1)$$

2. The vector sum of all the external torques acting on the object about its centre of mass or any other given point must be zero:

$$\sum \vec{\tau} = 0 \qquad (10\text{-}2)$$

These two conditions can be restated in terms of momenta:

1. The linear momentum of the object is constant:

$$\frac{d\vec{p}}{dt} = 0 \qquad (10\text{-}3)$$

2. The angular momentum of the object about a given point is constant:

$$\frac{d\vec{L}}{dt} = 0 \qquad (10\text{-}4)$$

✓ CHECKPOINT

C-10-1 Equilibrium

A marble slides back and forth inside a frictionless bowl, as shown in Figure 10-2. At what point(s) is the marble in equilibrium?

(a) At points A and B
(b) At points A, B, and C
(c) Only point C because the acceleration there is zero
(d) The marble is never in equilibrium because it is always under the effect of a nonzero external force.

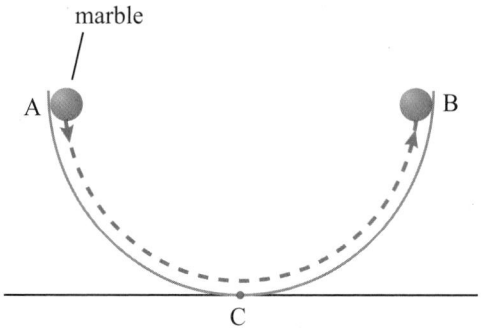

Figure 10-2 C-10-1

When an object is at rest and the above conditions for equilibrium are met, the object is in a state of **static equilibrium**. When the object is moving and the conditions for equilibrium are met, the object is in a state of **dynamic equilibrium**.

LO 2

10-2 Centre of Gravity

The force of gravity is one of the external forces that act on an object. For most applications, we can treat the force of gravity (the object's weight) as acting at a single point on the object. That point is called the **centre of gravity**. The location of the centre of gravity is important when analyzing the equilibrium conditions for a given object, as well as when determining the response of a rigid object to external forces. When the acceleration due to gravity is uniform over the dimensions of the object, the centre of gravity coincides with the centre of mass. The proof is left as an exercise (see problem 47). For most practical applications in this textbook, the centre of gravity coincides with the centre of mass. However, Example 10-1 deals with a situation in which the two centres do *not* coincide.

✓ CHECKPOINT

C-10-2 Tippy Refrigerator

You are pushing a refrigerator across a room at a constant speed (Figure 10-3). There is friction between the floor and the refrigerator legs. Is the refrigerator more likely to tip if its centre of gravity is higher?

Figure 10-3 C-10-2

EXAMPLE 10-1

Determining the Centre of Gravity

The force of gravity near a small asteroid changes appreciably over short distances. A point mass $M_2 = m$ is a distance L directly above an identical point mass $M_1 = m$ on the surface of an asteroid (Figure 10-4). The acceleration due to gravity at the elevation of M_2 is 0.80 times the acceleration due to gravity at the elevation of M_1. Find the centre of gravity of the two masses.

Figure 10-4 Where is the centre of gravity?

SOLUTION

Let Y_g, Y_1, and Y_2 represent the positions of the centre of gravity, M_1, and M_2, respectively. Since M_2 is directly above

M_1 and the line connecting the two masses is parallel to the direction of the force of gravity,

$$Y_g = \frac{Y_1 M_1 g_1 + Y_2 M_2 g_2}{M_1 g_1 + M_2 g_2}$$

Choosing the position of M_1 as our origin and up as the positive direction,

$$Y_g = \frac{(0)M_1 g_1 + L M_2 g_2}{M_1 g_1 + M_2 g_2} = \frac{L M_2 g_2}{M_1 g_1 + M_2 g_2}$$

Since $g_2 = 0.80g_1$ and $M_2 = M_1 = m$,

$$Y_g = \frac{Lm(0.80)g_1}{mg_1 + m(0.80)g_1} = \frac{L(0.80)mg_1}{1.8mg_1} = 0.444\ L$$

Making sense of the result:

Since the gravitational attraction is stronger at the lower mass, the centre of gravity is closer to the lower mass than to the upper mass, and is below the centre of mass.

For masses that are not positioned such that one is directly above the other, we can find the centre of gravity by summing torques about a given point (see problem 44).

MAKING CONNECTIONS

Balance and Centre of Mass

At first glance, you might think that the fork and spoon in Figure 10-5 should immediately drop off the side of the wineglass. However, the system consisting of the fork, spoon, and toothpick wedged together is actually in equilibrium. If you consider this system carefully, you will realize that its centre of mass is just above the rim of the glass, so the system can remain balanced there indefinitely.

Figure 10-5 How is it possible that a fork and spoon can be suspended on a toothpick totally on the outside of the edge of a glass?

Stable and Unstable Equilibrium

Consider the worker holding the beam in Figure 10-6(a). If the worker lets the beam swing down until it is hanging vertically from the pivot (Figure 10-6(b)), the worker has moved the beam to a state of **stable equilibrium**. The centre of mass and the centre of gravity of the beam are now below the pivot. A sideways nudge on the beam will move the beam slightly from its stable equilibrium

position, but the force of gravity will pull the beam back to this stable position. However, if the worker tilts the beam up so that its centre of mass is directly above the pivot (Figure 10-6(c)), the beam is in a state of **unstable equilibrium**. A sideways nudge will cause the beam to swing down. If there were no frictional losses, the beam would continue rotating about the pivot indefinitely.

(a)

(b)   **(c)**

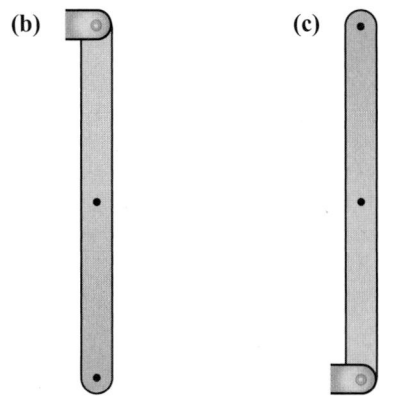

Figure 10-6 (a) A worker holding a beam (b) Stable equilibrium. (c) Unstable equilibrium

 EXAMPLE 10-2

A Horizontal Beam

The beam in Figure 10-7 has a mass of 120 kg. Assume that the hinge at the left end of the beam is frictionless.

(a) How much force must the worker exert on the beam for it to be in equilibrium?

(b) How much force does the beam then exert on the hinge?

Figure 10-7 Example 10-2

SOLUTION

(a) We start by drawing the FBD of the beam. The FBD in Figure 10-8 shows three forces acting on the beam: the force of gravity at the centre of mass of the beam, the

10-3 Applying the Conditions for Equilibrium

We can use the conditions of equilibrium to determine the forces acting on a mechanical system. We could apply either condition of equilibrium first; however, as we will see throughout the chapter, we can almost always simplify the solution to a given by carefully choosing which condition of equilibrium to consider first. We illustrate this technique in the following example.

A strategy: Note that in the approach we take below, we will be treating the horizontal and vertical reaction forces at a given hinge or support as separate, independent variables rather than as components of a reaction force. This approach is simpler and will prove more efficient, as is used widely in engineering applications. We note that the total reaction force can be obtained from its component using simple vector algebra. We also note that although in the approach we take below the components of forces along a given direction will be treated as scalars, it is also valid to treat them as vectors, as is done elsewhere. We will often sum the forces along a certain direction, choose a positive direction and then use variables to denote the magnitudes of the forces. This approach will prove valuable when we have coupled equations and multiple unknowns.

Figure 10-8 The FBD of the hinged horizontal beam

applied vertical force exerted by the worker at the right end, and the reaction force exerted by the hinge at the left end. We will use A_h and A_v to represent the horizontal and vertical components of the reaction force.

When the beam is in equilibrium, the sum of the forces in the x-direction must be zero, so $A_h = 0$. We then have two unknowns: the vertical reaction force, A_v, exerted by the hinge and the force, $\vec{F}$, exerted by the worker. The two conditions for equilibrium give us two equations, so we can solve for the unknown support forces.

Summing the forces in the y-direction, we have

$$\sum F_y = F_y + Mg_y + A_{v,y} = 0 \qquad (1)$$

(continued)

Choosing up as our positive direction, we have

$$F - Mg + A_v = 0 \qquad (2)$$

where the quantities in Equation (2) are variables that denote the magnitudes of the forces in Equation (1).

We have two unknowns in this equation. We have deliberately chosen to begin with the sum of the forces in the y direction to illustrate that although the equation is valid and needed that this is not the most efficient choice for a first step. We would like, whenever possible to begin with an equation that would hand us one of the unknowns. If the worker were to let go of the beam, it would rotate about the pivot at A. The force exerted by the worker prevents the beam from swinging. Summing the torques about point A gives

$$\sum \vec{\tau} = \vec{\tau}_{Mg} + \vec{\tau}_F = 0 \qquad (3)$$

The support forces at the hinge at A exert no torque about A; the moment arm (the perpendicular distance from the pivot to the force) of both supports, A_h and A_v is zero and hence the torque they exert about the point at which they act is zero. Taking clockwise to be positive, we have

$$\tau_{Mg} - \tau_F = 0 \qquad (4)$$

where the quantities in Equation (4) are variable that denote the magnitudes of the torques in Equation (3).

From Equation (8-13), we know that the torque due to a force about a pivot equals the product of the force and the perpendicular distance between the force and the pivot. Therefore,

$$Mg\left(\frac{L}{2}\right) - F(L) = 0$$

$$F = \frac{Mg}{2} = \frac{(120 \text{ kg})(9.81 \text{ m/s}^2)}{2} = 589 \text{ N} \qquad (5)$$

and so considering the torques first is more efficient than starting with the sum of the forces as we did above.

(b) Substituting the value of F from Equation (5) into Equation (2) gives

$$\frac{Mg}{2} - Mg + A_v = 0$$

$$A_v = \frac{Mg}{2} = 589 \text{ N}$$

Alternative Solution

The second condition for equilibrium (Equation (10-2)) states that the sum of the torques about *any* point must be zero. Summing the torques about a pivot or hinge is often the simplest approach; however, we can sum the torques about any other convenient point.

Let us consider the torques about the end of the beam that the worker is holding. The forces exerting torques about this point are the weight of the beam and the reaction force A_v. The worker exerts no torque at this end of the beam because the moment arm is zero. Therefore,

$$\sum \vec{\tau} = \vec{\tau}_{Mg} + \vec{\tau}_{A_v} = 0$$

Taking counterclockwise to be positive, we have

$$\tau_{Mg} - \tau_{A_v} = 0$$

Using the definition for torque (Equation (8-13)), we have

$$Mg\left(\frac{L}{2}\right) - A_v(L) = 0$$

and $\qquad A_v = \frac{Mg}{2} = 589 \text{ N}$

Making sense of the result:

Because of the symmetry of the problem, one would expect the weight of the beam to be equally distributed between the hinge and the worker.

PEER TO PEER

To solve a given problem, you need to have at least as many equations as there are unknowns. For equilibrium problems, try to begin with the condition of equilibrium that will give you one of the unknowns immediately.

EXAMPLE 10-3

A Sloping Beam

The beam in Figure 10-9 has a mass of 2200 kg. It is supported at one end by a frictionless hinge, and the other end leans against a wall. Find the force that the wall exerts on the beam, and the reaction forces at the hinge. Assume there is no friction between the wall and the beam.

SOLUTION

We begin by drawing the FBD for the beam, which is shown in Figure 10-10. Since the wall is frictionless, the reaction force from the wall has no vertical component. Since this force is normal to the wall, we label it N. A_h and A_v are the horizontal and vertical components of the support force at the hinge, and the weight of the beam, Mg, acts at the middle of the beam.

no friction

L

$42.0°$

Figure 10-9 Example 10-3

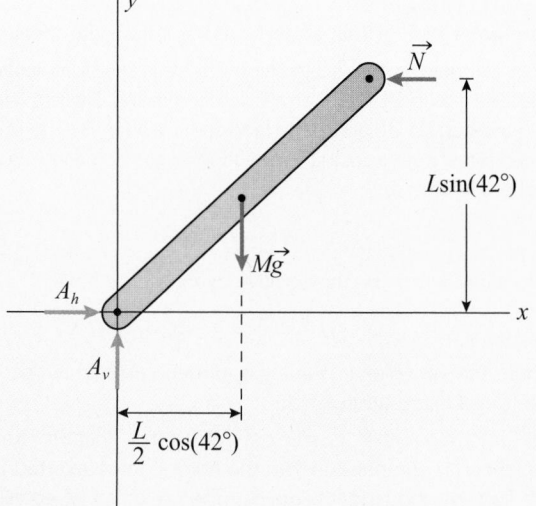

y

$\vec{N}$

$L\sin(42°)$

$\vec{Mg}$

A_h

x

A_v

$\dfrac{L}{2}\cos(42°)$

Figure 10-10 The FBD of the forces on the beam

We begin by summing the torques about the hinge at the base of the beam. The forces that exert torques about this point are the weight of the beam and the normal force from the wall, the support forces at the hinge exert no torque about the hinge as their moment arms are zero:

$$\sum \vec{\tau}_A = \vec{\tau}_{Mg} + \vec{\tau}_N = 0 \tag{1}$$

Taking clockwise to be positive, we have

$$\tau_{Mg} - \tau_N = 0 \tag{2}$$

where the variables in Equation (2) denote magnitudes of the torques in Equation (1).

From Figure 10-10, we see that the perpendicular distance between the gravitational force and the pivot point of the hinge is $\dfrac{L}{2}\cos(42°)$, and the perpendicular distance between the normal force and the pivot is $L\sin(42°)$. We can express the sum of the torques at the pivot in terms of these moment arms:

$$Mg\left(\frac{L}{2}\cos(42°)\right) - N(L\sin(42°)) = 0 \tag{3}$$

Solving for N gives

$$N = \frac{Mg}{2\tan(42°)} = 12.0 \text{ kN} \tag{4}$$

Summing the forces acting on the beam in the y-direction,

$$\sum F_y = 0 \tag{5}$$

$$A_{v,y} + Mg_y = 0 \tag{6}$$

Choosing up as our positive direction, we have

$$A_v - Mg = 0 \tag{7}$$

$$A_v = Mg = 21.6 \text{ kN} \tag{8}$$

Taking the sum of the forces in the horizontal direction gives

$$\sum F_x = 0 \tag{9}$$

$$A_{h,x} + N_x = 0 \tag{10}$$

Choosing right as the positive direction and substituting from Equation (4),

$$A_h - N = 0 \Rightarrow A_h = N = \frac{Mg}{2\tan(42°)} = 12.0 \text{ kN} \tag{11}$$

we note here that the quantities in Equations (7) and (11) are variables that denote the magnitudes of the forces in Equations (6) and (10) respectively.

Making sense of the result:

The smaller the angle at the base of the beam, the smaller the moment arm of the normal force from the wall, and the greater the normal force has to be to counterbalance the torque from the force of gravity.

PEER TO PEER

In order to figure out the direction of action of a torque about a pivot, it helps to find the moment arm for the force generating the torque. The moment arm is the perpendicular distance from the pivot to the force. Once you have the moment arm is becomes easy to see whether the torque acts in a clockwise or a counterclockwise fashion.

EXAMPLE 10-4

Ladder Safety

The worker of mass M in Figure 10-11 wants to climb three-quarters of the way up a 4.00 m–long ladder of mass m. The ladder makes an angle of $65.0°$ with the floor. Find the minimum coefficient of friction between the ladder and the floor needed to prevent the ladder from sliding on the floor. Assume that the wall is frictionless.

(continued)

CHAPTER 10 | EQUILIBRIUM AND ELASTICITY

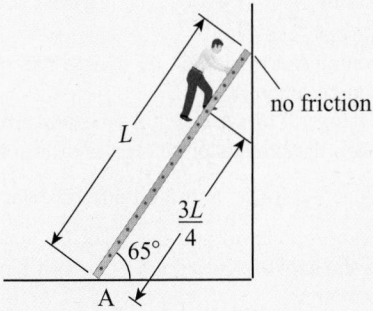

Figure 10-11 Example 10-4

SOLUTION

If the ladder does not slide, the friction between it and the floor is static friction. Equation (5-2) relates the maximum force of static friction to the normal force and the coefficient of static friction:

$$f_s = N_g \mu_s \tag{1}$$

We start by drawing the FBD of the ladder. Five forces are acting on the ladder: the weight of the ladder (mg), the weight of the worker (Mg), the normal support force (N_g) exerted by the floor, the force of static friction (f), and a horizontal support force (N_w) exerted by the wall (Figure 10-12). Since the wall is frictionless, it cannot exert a vertical force on the ladder.

Figure 10-12 The FBD of the ladder

Now consider the torques about A, the point of contact between the ladder and the floor. The force of friction and the normal force from the floor do not exert a torque about point A because their moment arms are zero. Since the ladder is in equilibrium if it does not slide, the sum of the torques of the other three forces must be zero:

$$\sum \vec{\tau}_A = \vec{\tau}_{mg} + \vec{\tau}_{Mg} + \vec{\tau}_{N_w} = 0 \tag{2}$$

Taking clockwise to be positive,

$$\tau_{mg} + \tau_{Mg} - \tau_{N_w} = 0 \tag{3}$$

where the quantities in Equation (3) denote magnitudes of the torques in Equation (2)

The torque exerted by the worker's weight can be written as the product of Mg and the perpendicular distance between the force and the foot of the ladder. From Figure 10-12

we see that the distance is $\dfrac{3L}{4}\cos(65°)$. Similarly, the moment arm of the weight of the ladder is $\dfrac{L}{2}\cos(65°)$, and the moment arm of the normal force from the wall is $L\sin(65°)$. Substituting these values into Equation (3) gives

$$mg\left(\frac{L}{2}\cos(65°)\right) + Mg\left(\frac{3L}{4}\cos(65°)\right) - N_w(L\sin(65°)) = 0$$

and

$$N_w = \frac{mg\left(\dfrac{L}{2}\cos(65°)\right) + Mg\left(\dfrac{3L}{4}\cos(65°)\right)}{L\sin(65°)}$$

$$= \frac{\dfrac{mg}{2} + \dfrac{3}{4}Mg}{2\tan(65°)} \tag{4}$$

The factor of $\dfrac{3}{4}$ in this equation comes from the location of the worker on the ladder. Note that this factor indicates that the value of the normal force exerted by the wall depends on how high up the ladder the worker is. The only other force acting in the x-direction is the force of static friction. Therefore,

$$\sum F_x = N_{wx} + f_x = 0 \tag{5}$$

Choosing right to be the positive direction, we have

$$-N_w + f = 0 \quad \Rightarrow \quad f = N_w \tag{6}$$

where the variables in Equation (6) denote magnitudes of the forces in Equation (5)

To determine the coefficient of static friction from Equation (1), we need to find the normal force exerted by the floor on the ladder. Applying the condition of equilibrium for forces in the y-direction, we get

$$\sum F_y = Mg_y + mg_y + N_{gy} = 0 \tag{7}$$

Taking up to be positive,

$$-Mg - mg + N_g = 0 \quad \text{and} \quad N_g = Mg + mg \tag{8}$$

Now that we have what we need, putting the value of f from Equation (6) and that of N_g from Equation (8) into Equation (1), we have

$$\mu_s = \frac{f}{N_g} = \frac{\dfrac{mg}{2} + \dfrac{3}{4}Mg}{2(Mg + mg)\tan(65°)} = \frac{2m + 3M}{8(M + m)\tan(65°)}$$

Note here that the quantities in Equations (6) and (8) are variables that denote the magnitudes of the forces in Equations (5) and (7) respectively.

Making sense of the result:

The larger the angle that the ladder makes with the vertical, the less friction there needs to be between the floor and the ladder. When the ladder is vertical, it does not exert a horizontal force on the ground, and no friction is needed. However, this position would be an unstable equilibrium because any outward force will cause the ladder to fall away from the wall. For safety, a ladder is normally positioned such that the distance from the foot of the ladder to a vertical wall is one-quarter of the height of the point of contact of the ladder with the wall.

MECHANICS

PEER TO PEER

EXAMPLE 10-5

A Beam Supported by a Cable

A 500 kg crate sits 3.00 m from a hinge on a 4.00 m–long horizontal beam, as shown in Figure 10-13. The beam is supported at one end with a cable that is attached to a wall; the cable makes an angle of 42.0° with the horizontal. The beam has a mass of 270 kg. Determine the tension in the cable and the reaction force at the hinge.

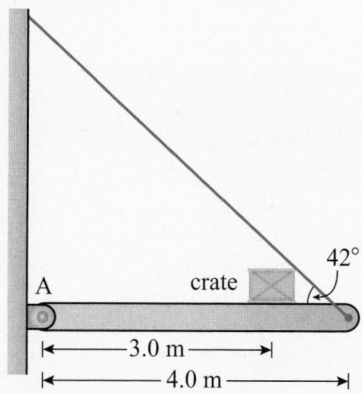

Figure 10-13 Example 10-5

SOLUTION

The weight of the beam, the weight of the crate, and the tension in the cable, as well as the horizontal and vertical components of the reaction force at the hinge are shown in the FBD of Figure 10-14.

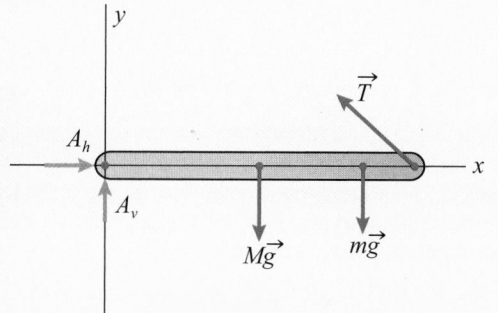

Figure 10-14 The FBD of the beam

We apply the condition for equilibrium for the torques about the hinge at A. The forces that exert torques about A

are the weight of the beam (Mg), the weight of the crate (mg), and the tension in the cable (T). A_h and A_v do not exert a torque about A because the moment arm is zero. Therefore,

$$\sum \vec{\tau}_A = \vec{\tau}_{Mg} + \vec{\tau}_{mg} + \vec{\tau}_T = 0$$

Taking clockwise to be positive,

$$\tau_{Mg} + \tau_{mg} - \tau_T = 0$$

the variables in the line above denote magnitude of the torques. From Figure 10-14, using Equation 8-10 for the torque exerted by the tension and Equation 8-16 for the torque exerted by Mg and mg, and taking the weight of the beam to be at its centre, we get

$$Mg(2 \text{ m}) + mg(3 \text{ m}) - T(4 \text{ m}) \sin(42°) = 0$$

Solving for T gives

$$T = \frac{(500 \text{ kg})(9.81 \text{ m/s}^2)(3 \text{ m}) + (270 \text{ kg})(9.81 \text{ m/s}^2)(2 \text{ m})}{(4 \text{ m}) \sin(42°)}$$

$$= 7477 \text{ N} = 7480 \text{ N}$$

Next, we sum the forces in the y-direction. Here, we have the weight of the beam, the weight of the crate, the vertical component of the tension, and the vertical component (A_v) of the reaction force at the hinge:

$$\sum F_y = Mg_y + mg_y + T_y + A_{v,y} = 0$$

From Figure 10-14, the y-component of the tension force is $T\sin(42°)$.

Choosing up as our positive direction, we have

$$-Mg - mg + T\sin(42°) + A_v = 0$$

and

$$A_v = Mg + mg - T\sin(42°)$$

$$= (500 \text{ kg})(9.81 \text{ m/s}^2) + (270 \text{ kg})(9.81 \text{ m/s}^2) - 7477 \sin(42°) \text{ N}$$

$$= 2550.6 \text{ N} = 2550 \text{ N}$$

to 3 significant figures.

Now, the only unknown is the horizontal component, A_h, of the reaction force at the hinge. We take the sum of the forces acting on the beam in the x-direction, A_h, and the horizontal component of the tension, T:

$$\sum F_x = T_x + A_{h,x} = 0$$

(continued)

From Figure 10-14, the x-component of the tension is $T\cos(42°)$.

Choosing right as our positive direction,

$$-T\cos(42°) + A_h = 0 \quad \text{or} \quad A_h = T\cos(42°)$$

Therefore,

$$A_h = T\cos(42°) = (7477 \text{ N})\cos(42°) = 5556.5 \text{ N} = 5560 \text{ N}$$

Making sense of the result:

The cable supports part of the weight of the beam, but since the cable is on an angle, part of the force it exerts is horizontal, which results in the horizontal component of the reaction force at the hinge. A quick check would be to sum torques about the other end of the beam, which gives $A_v = 2550$ N as well.

✓ CHECKPOINT

C-10-3 Torque Signs

A torque is positive or negative depending on
(a) whether the force causing the torque points up or down, and whether you choose up or down as positive;
(b) whether the force causing the torque points left or right, and whether you choose left or right as positive;
(c) whether the "twisting" action of the torque is clockwise or counterclockwise, and whether you choose clockwise or counterclockwise as positive.
(d) All of the above are true.

C-10-3 (c) Only c is true.

10-4 Working with Unknown Forces

For most of the problems in this chapter, the directions of unknown forces are fairly obvious. However, for more complex situations, it can be difficult to predict the direction of a given unknown force. For such problems, we can simply assume a direction. If the magnitude of the force then comes out to be positive, our assumption was correct. However, if we get a negative value for the magnitude of the force, then the direction of the force is opposite to what we had assumed.

 EXAMPLE 10-6

Working with a Force with an Unknown Direction

Assume that the weight of the V-shaped structural element shown in Figure 10-15 is negligible compared to the other forces acting on it. Find the reaction forces acting on this element and the force it exerts on the roller. Assume that the roller is frictionless and that $F_1 = 600.0$ N, $F_2 = 400.0$ N, and $F_3 = 500.0$ N.

Figure 10-15 Example 10-6

SOLUTION

We begin with the FBD of the structural element (Figure 10-16). At the hinged end, we have the reaction force with horizontal and vertical components, A_h and A_v. At the roller end, we have only a normal force, N, since there is no friction. We do not know the directions of A_h or A_v, we will assume that A_v points down and that A_h points to the left.

Figure 10-16 The FBD of the structural element

Let us begin by taking the sum of the torques about the hinge. The forces that exert torques about A are the forces F_1, F_2, F_3 and the normal force at the roller end, N. The reaction forces at A exert no torques about A because their moment arm is zero. Thus,

$$\sum \vec{\tau}_A = \vec{\tau}_{F_1} + \vec{\tau}_{F_2} + \vec{\tau}_{F_3} + \vec{\tau}_N = 0 \quad (1)$$

Taking clockwise to be positive,

$$\tau_{F_1} + \tau_{F_2} + \tau_{F_3} - \tau_N = 0 \quad (2)$$

where the variables in Equation (2) denote the magnitudes of the torques in Equation (1)

then

$$F_1(1.00\,\text{m}) + F_2(1.50\,\text{m}) + F_3(2.00\,\text{m})\sin(37°) - N(3.00\,\text{m}) = 0$$

Note that for the torques exerted by F_1, F_2 and N, we have used the moment arm concept (perpendicular distance from pivot to force) as given in Equation 8-13 and 8-16. For torque exerted by F_3 we have used the definition of torque as given by Equation 8-10.

Solving for N gives

$$N = \frac{(600\,\text{N})(1.00\,\text{m}) + (400\,\text{N})(1.50\,\text{N}) + (500\,\text{N})(2.00\,\text{m})\sin(37°)}{3.00\,\text{m}}$$

$$= 600.6\,\text{N} \tag{3}$$

Next, we sum the forces in the y-direction:

$$\sum F_y = A_{v,y} + F_{2y} + F_{3y} + N_y = 0 \tag{4}$$

Choosing up as our positive direction,

$$-A_v - F_2 - F_3 \sin(37°) + N = 0$$

in the line above the variables denote the magnitudes of the forces in Equation (4), then,

$$A_v = -F_2 - F_3 \sin(37°) + N$$

$$= -400\,\text{N} - (500\,\text{N})\sin(37°) + 600.6\,\text{N} = -100.3\,\text{N}$$

The negative value for A_v indicates that our guess for the direction of this component is not correct, and that the force points up, opposite to what we had guessed.

To determine A_h, we apply the condition for equilibrium for forces in the x-directions:

$$\sum F_x = A_{h,x} + F_{1x} + F_{3x} = 0 \tag{5}$$

Choosing right as our positive direction,

$$-A_h + F_1 + F_3 \cos(37°) = 0 \tag{6}$$

$$A_h = F_1 + F_3 \cos(37°) = 600\,\text{N} + (500\,\text{N})\cos(37°) = 999.3\,\text{N} \tag{7}$$

we note here that the variables in Equations (6) and (7) denote the magnitudes of the quantities in Equation (5).

Making sense of the result:

The normal force has a longer moment arm than all the other forces exerting a torque about the hinge. This explains its relatively low value considering that it is the only force exerting a torque in a direction opposite to those exerted by the other forces.

EXAMPLE 10-7

Analyzing a Multipart Structure

The shorter beam in Figure 10-17 has a mass $m_1 = 1100$ kg and is supported by a free hinge at point A. The longer beam has a mass $m_2 = 2100$ kg and is supported by a free hinge at point B. The two beams are fixed to each other by a free hinge at C. Find the reaction forces at the hinges.

Figure 10-17 Example 10-7

SOLUTION

Let us first look at the FBD of the two-beam structure as a whole, which is shown in Figure 10-18. The forces acting on this system are the reaction forces at points A and B and the weights of the two beams. We do not consider the force acting on the hinge at C because this force is internal to the system. An FBD shows only the external forces acting on an object or a system.

We could try to infer the direction of the components of the reaction forces, but we will simply take a guess. If

Figure 10-18 The FBDs of both beams

our choice for the direction of these components is correct, then the values that we obtain for the forces will be positive; if the value that we obtain for a component is negative, then we know that our choice for that component is incorrect. An incorrect choice of direction for a component will not affect the magnitude of the force.

If we take the sum of the forces in the x-direction at this stage, we will end up with two unknowns, A_h and B_h, in one equation. Similarly, if we take the sum of the forces in the y-direction, both A_v and B_v will be in the same equation. Although these are valid equations, starting with the sum of the torques will prove to be a simpler and more efficient approach.

(continued)

The forces that exert torques about B in figure 10-18 are the weights of the beams, m_1g and m_2g, as well as A_h, the horizontal component of the reaction force at A. The components of the reaction force at B, B_h and B_v exert no torque about B because the moment arm of these components about B is zero. The vertical component of the reaction force at A exerts no torque about point B because the line of action of this component passes through the hinge at B; therefore, the moment arm for A_v about B is also zero. So, the sum of the torques about point B is

$$\sum \vec{\tau}_B = \vec{\tau}_{m_1g} + \vec{\tau}_{m_2g} + \vec{\tau}_{A_h} = 0$$

Taking clockwise to be positive, the torques of m_1g and m_2g are positive, and the torque of A_h is negative. Thus,

$$\tau_{m_1g} + \tau_{m_2g} - \tau_{A_h} = 0$$

where the variables in the line above denote the magnitudes of the torques.

From Figure 10-18, the perpendicular distances from the pivot at B to m_1g and m_2g are 1.5 m and 4.0 m, respectively. Therefore,

$$m_1g(1.5\text{m}) + m_2g(4.0 \text{ m}) - A_h(2.5 \text{ m}) = 0$$

$$A_h = \frac{m_1g(1.5 \text{ m}) + m_2g(4 \text{ m})}{2.5 \text{ m}}$$

$$= \frac{1100 \text{ kg } (9.81 \text{ m/s}^2)(1.5 \text{ m}) + 2100 \text{ kg } (9.81 \text{ m/s}^2)(4 \text{ m})}{2.5 \text{ m}}$$

$$= 39\,436 \text{ N} = 39.4 \text{ kN} \qquad (1)$$

We can now immediately determine another unknown by summing the forces in the x-direction. The only forces acting in the x-direction are the horizontal components of the reaction forces at A and B, A_h and B_h. Therefore,

$$\sum F_x = A_{h,x} + B_{h,x} = 0 \qquad (2)$$

Choosing right as our positive direction,

$$B_h = -A_h = -39.4 \text{ kN} \qquad (3)$$

where B_h and A_h are variables denoting the magnitudes of the forces in (2).

The negative value for B_h means that we have chosen the incorrect direction for the force. We simply treat the variable B_h as a negative quantity every time it comes up in the calculations.

No other unknown can be found by looking at the entire two-beam structure as one body, so we now apply the conditions for equilibrium to the components of the structure. We still need to find the vertical reactions at each hinge at A and B, as well as the forces acting on the hinge at C.

We now draw the FBDs for each beam (Figure 10-19). The forces acting on the long beam are its weight, the horizontal and vertical components of reaction force at B, B_h and B_v, and the force acting on the hinge at C. Since we do not know the direction of the force at C, we make the assumptions shown in Figure 10-19.

Examine closely how C_h and C_v, the components of the force at C, act one way on the top beam and the opposite way on the bottom beam. According to our assumption, C_v acts down on the top beam, which means that the bottom beam pulls the top beam down. Consequently, the

Figure 10-19 (a) The FBD of the top beam (b) The FBD of the bottom beam

top beam must pull the bottom beam up with an equal but opposite force in accordance with Newton's third law.

Note that to illustrate what happens when you make the incorrect assumption about the direction of an unknown force, we have made an absurd assumption that the vertical component of the reaction force at C acts downward on the top beam. In fact, C_v must be directed upward so that it counters the torque due to the weight of the top beam and hence hold the beam up. However, choosing the incorrect direction will not affect the end result of the analysis. Nonetheless, the directions we choose have to be consistent with each other. The forces with magnitude C_v in Figure 10-19a and b are an action–reaction pair.

Summing the torques about B in the FBD for the top beam gives us one of the unknowns. Since the forces that exert torques about B are the weight of the beam (m_1g) and the vertical component of the reaction force at point C,

$$\vec{\tau}_{m_1g} + \vec{\tau}_{C_v} = 0 \qquad (4)$$

If we take clockwise to be positive, both torques

$$\tau_{m_1g} + \tau_{C_v} = 0 \qquad (5)$$

$$m_2g(4 \text{ m}) + C_v(3 \text{ m}) = 0 \qquad (6)$$

which gives us

$$C_v = \frac{-m_2g(4 \text{ m})}{3 \text{ m}} = \frac{-2100 \text{ kg}(9.81 \text{ m/s}^2)(4 \text{ m})}{3 \text{ m}}$$

$$= -27\,468 \text{ N} = -27.5 \text{ kN} \qquad (7)$$

where the variables in Equation (5) denote magnitudes of the torques in Equation (4)

The negative value for the variable C_v means that our guess for the direction of C_v is incorrect. We can redraw the FBD to show the correct direction of C_v or we can keep the FBD as is and use the negative value for C_v in all subsequent calculations involving that variable. We will always choose to do the latter.

To find the vertical reaction at the hinge at B, we apply the equilibrium conditions for the vertical forces acting on the top beam. Here, we have the weight of the beam, the vertical reaction at C, and the vertical reaction at B, therefore

$$\sum F_y = m_2 g_y + C_{v,y} + B_{v,y} = 0$$

Choosing up as our positive direction, we have

$$-m_2 g - C_v + B_v = 0$$

$$B_v = m_2 g + C_v$$

where the variables B_v and C_v denote magnitudes.

Keeping in mind that the value for C_v is negative, we have

$$B_v = Mg + C_v = 2100\ \text{kg}(9.81\ \text{m/s}^2) + (-27\ 468\ \text{N})$$

$$-6867\ \text{N} = -6870\ \text{N} \qquad (8)$$

The negative value for the variable B_v means that our assumption regarding the direction for B_v in Figure 10-18 was not correct.

Next, we find the horizontal component of the force acting on the hinge at C, C_h, by summing the forces acting on the top beam in the x-direction. Since the only forces here are the reaction forces at B and C,

$$\sum F_x = C_{h,x} + B_{h,x} = 0$$

Choosing right as our positive direction,

$$C_h + B_h = 0 \Rightarrow C_h = -B_h = -(-39.4\ \text{kN}) = 39.4\ \text{kN}$$

where the variables C_h and B_h in the line above denote magnitudes of their respective forces.

Note that we have used the value of B_h as -39.4 kN as obtained in Equation 3 above. The positive value indicates that we picked the correct direction for C_h in the FBD in Figure 10-19. Now we can go back to the FBD for the entire system. The only unknown we have left is the vertical component of the reaction force at A. We take the sum of the vertical forces on the entire structure: the weight of the top beam ($m_2 g$), the weight of the bottom beam ($m_1 g$), and the vertical components of the reaction forces at A and B, A_v and B_v. Thus,

$$\sum F_y = m_1 g_y + m_2 g_y + A_{v,y} + B_{v,y} = 0$$

Choosing up as our positive direction,

$$-m_1 g - m_2 g + A_v + B_v = 0 \qquad (9)$$

$$A_v = m_1 g + m_2 g - B_v$$

$$= 1100\ \text{kg}\ (9.81\ \text{m/s}^2) + 2100\ \text{N}\ (9.81\ \text{m/s}^2) - (-6860)\ \text{N}$$

$$= 38\ 259\ \text{N} = 38.2\ \text{kN}$$

where the variables in Equation (9) denote magnitudes of the forces.

Making sense of the result:

One expects the vertical component at A to be larger than the vertical component at B because the bottom hinge is supporting the weight of the bottom beam, and some of the weight of the top beam is transferred to the bottom beam through the hinge at C.

 CHECKPOINT

C-10-4 Internal Forces

Why do we not include the reaction force at C when summing the forces on the entire system in Example 10-7?

C-10-4 The reaction force at C is internal to the system.

 PEER TO PEER

Sometimes, it helps to look at the FBD for the whole structure before focusing on individual components of the structure.

LO 5

10-5 Deformation and Elasticity

So far, we have treated structural components such as beams and supports as rigid objects that do not deform. However, deformation often occurs when external forces act on objects. If sufficient force is applied, steel and concrete structural elements will deform, break, or buckle.

The three types of forces of primary interest to us are tension, compression, and shear.

When you pull each end of a bar away from the centre of the bar, the bar is under **tension.** This is shown in Figure 10-20(a). When you push each end of a bar toward the centre, you are exerting a **compression force** as in Figure 10-20(b). When you clamp one end of a bar and push on the side of the free section, you are exerting a **shear force** as shown in Figure 10-20(c). The forces exerted by scissors and tin snips when cutting paper and sheet metal are examples of shear forces.

Stress

Consider the cylinder of length L and cross section A shown in Figure 10-21(a). Two forces equal in magnitude (F) and opposite in direction pull the ends of the cylinder away from the centre. This tension on the cylinder causes it to elongate. Similarly, if the forces push the ends of the cylinder toward the centre, the compression causes the cylinder to shorten as in Figure 10-21(b).

The effect of the force exerted on a beam depends on the cross section of the beam. We are interested in the force per cross section of the beam, called **stress**, σ:

$$\sigma = \frac{F}{A} \qquad (10\text{-}5)$$

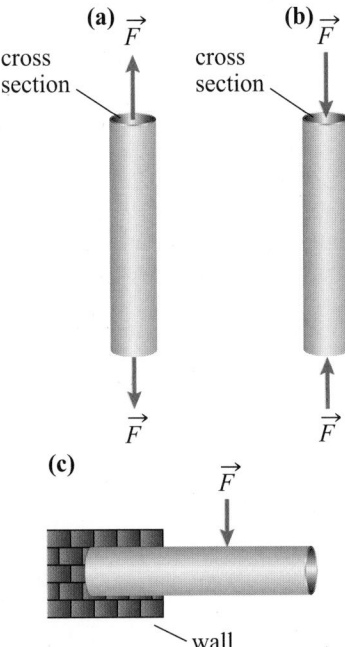

Figure 10-20 (a) Beam is under tension. (b) Beam is under compression. (c) Shear force

Figure 10-21 (a) Tension causes the cylinder to elongate. (b) Compression causes the cylinder to shorten.

Strain, Elastic Deformation, and the Proportional Limit

The deformation of an object in response to stress is characterized by a quantity called **strain**, which is the fractional change in the length of the object. A symbol commonly used for strain is the Greek letter epsilon:

$$\epsilon = \frac{\Delta L}{L} \qquad (10\text{-}6)$$

An **elastic deformation** is a deformation that is completely reversible: when the stress is removed, the object returns to its original length. It is common for

objects to deform elastically in response to applied stress as long as the stress is not too large. For many materials of interest, such as steel alloys, the elastic response region has two parts. The first part is characterized by a linear relation between stress and strain, and the material follows Hooke's law:

$$\sigma = Y\epsilon \qquad (10\text{-}7)$$

or

$$\frac{F}{A} = Y\frac{\Delta L}{L} \qquad (10\text{-}8)$$

The parameter Y is called **Young's modulus**, which is a measure of a material's elasticity. For many materials, especially steel alloys, the value of Young's modulus is the same for compressive and tensile stress. However, the response of a composite material, such as concrete, to tension can differ drastically from its response to compression.

Although tension and compression forces act in a direction parallel to the length of the object in figure 10-20, (perpendicular to the cross section), a shear force acts perpendicular to the length of the object, (parallel to the cross section) as shown in Figure 10-20(c). The deformation of the object along the direction of the shear force is given by

$$\frac{F}{A} = G\frac{\Delta L}{L}$$

The constant of proportionality between the shear stress and the strain is called the **shear modulus**, G.

MAKING CONNECTIONS

Classic Arches

Stone, like concrete, is a lot weaker in tension than it is in compression. Therefore, stone buildings are designed to have the structural elements under compression rather than tension. This is true of the stones in a classic arch. This design has been used since the times of the Etruscans and the ancient Greeks.

The maximum stress for which the response of the material is linear is called the **proportional limit**. This point is labelled P in the typical stress versus strain curve of Figure 10-22. Past the proportional limit the slope of the stress versus strain curve changes: The response of the material is no longer linear, and Hooke's law does not apply. However, the deformation remains elastic past the proportional point until the stress reaches a value called the **yield point**, **yield strength**, or the **elastic limit**, labelled E in Figure 10-22. Beyond this limit, the deformation is not completely reversible.

Figure 10-22 Typical stress versus strain curve for a ductile material

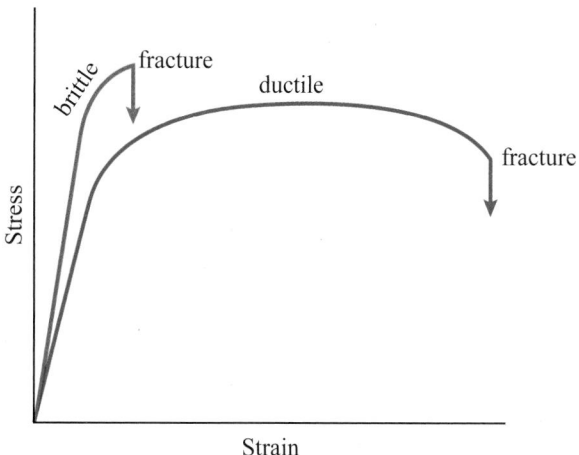

Figure 10-23 Ductile versus brittle behaviour

✓ **CHECKPOINT**

C-10-5 The Yield Point

The yield point of a given material is

(a) its ultimate tensile stress;
(b) the maximum stress that the material will take such that its strain is directly proportional to the applied stress;
(c) slightly lower than the proportional point;
(d) the maximum stress for which the deformation of the material is completely, or almost completely, reversible.

C-10-5 (d) d is the definition of the yield point.

If the material is strained beyond the yield point, the material does not return to its original length when the stress is removed. Past the yield point, the material undergoes **plastic deformation**, which is not completely reversible. The material has a **permanent set**, meaning that a permanent deformation has occurred. The greatest stress value on the graph of stress versus tensile strain is called the **ultimate tensile strength**. Increasing the strain beyond the ultimate tensile strength point fractures the object.

LO 6

10-6 Ductile and Brittle Materials

If a significant amount of plastic deformation occurs before the material fractures, the material is called a **ductile** material. If fracturing occurs shortly beyond the yield point, the material is called **brittle** (Figure 10-23). For example, the harder steel alloys tend to be more brittle, and the softer alloys tend to be more ductile. The behaviour of a material in compression may differ from its behaviour under tension. Steel alloys are made in various classes: usually the harder (or stronger) the steel, the more brittle it is.

MAKING CONNECTIONS

The *Titanic*

There is evidence that the steel used in the hull of the *Titanic*, while ductile at room temperature, might have become brittle at lower temperatures. This loss of ductility may have substantially increased the size of the fracture in the hull when the ship hit the iceberg in the icy Atlantic waters.

 EXAMPLE 10-8

Animal versus Mineral

The yield point of both A36 steel (a grade of structural steel) and spider silk is 250 MPa, and the yield point for human tendon is 28.0 MPa. Compare the maximum mass than can be carried in tension by a steel rod or spider silk, without permanent deformation, with the mass that can be supported by a human tendon of the same size. The rod (and spider silk) is 2.00 m long and has a radius of 2.00 cm.

SOLUTION

According to Equation (10-5), $F = \sigma A = Mg$.
For steel and silk,

$$M = \frac{\sigma A}{g} = \frac{(250 \text{ MPa})(\pi)(0.020 \text{ m})^2}{9.81 \text{ m/s}^2} = 3.20 \times 10^4 \text{ kg}$$

For tendon,

$$M = \frac{\sigma A}{g} = \frac{(28 \text{ MPa})(\pi)(0.020 \text{ m})^2}{9.81 \text{ m/s}^2} = 3.59 \times 10^3 \text{ kg}$$

This is an order of magnitude less than the mass that can be carried by an equivalent piece of silk or steel.

Making sense of the result:

The mean diameter of an Achilles tendon is approximately 5 mm, which makes it capable of carrying approximately 60 kg.

CHAPTER 10 | EQUILIBRIUM AND ELASTICITY 267

MAKING CONNECTIONS

Is Silk Stronger than Steel?

The ultimate tensile strength (UTS) for many kinds of steel is of the order of 500 MPa, while some spider silk threads have a UTS upward of 1.5 GPa. Thus, a strand of spider silk can support, in tension, three times the weight that can be supported by a strand of steel of the same size.

 CHECKPOINT

C-10-6 Ductile Metals

A ductile metal
(a) is in general harder than a brittle metal;
(b) is not elastic;
(c) displays more plastic behaviour before fracture than a brittle material.
(d) None of the above.

C-10-6 (c) a ductile metal undergoes a more significant amount of plastic deformation before it fractures in comparison to a brittle metal.

Failure Modes in Compression and Tension

Whether under compression or tension, concrete and stone structures will fail by fracturing as shown in Figure 10-24(a). Under sufficient tension, a steel beam will eventually snap in two as shown in Figure 10-24(b). Steel under compression has several possible failure modes. Brittle steel will likely crack as shown in Figure 10-24(c). More ductile steel will bulge out as shown in Figure 10-24(d), and plastic deformation will take place, a process called **ductile failure**. Ductile failure is applied in fabricating processes such as extruding and stamping metals as shown in Figure 10-24(e). However, a steel beam is more likely to buckle under compression as shown in Figure 10-24(f) before reaching its compressive strength limit and undergoing ductile failure.

Beyond the ultimate tensile strength, a steel bar, for example, will experience what is called necking before

(a)

(b)

© sciencephotos/Alamy

(c) **(d)**

(e)

iStockphoto/Thinkstock

(f)

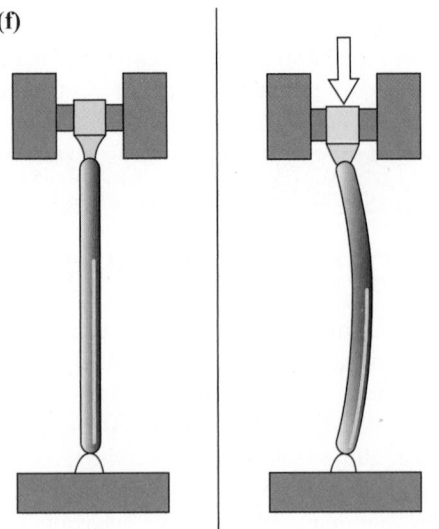

Figure 10-24 (a) Concrete failure by fracture (b) Failure of steel under tension (c) Failure of brittle steel by cracking (d) Ductile failure (e) Ductile failure used in stamping (f) Buckling of steel

Figure 10-25   A metal strip experiences necking before snapping

it eventually snaps (Figure 10-25). During **necking**, the cross section of the material shrinks as it elongates due to plastic deformation past the ultimate tensile strength.

Apparent stress, also called **engineering stress**, is defined as the force per unit area of the original cross section. However, the apparent stress is not the real stress that the material undergoes. **True stress** is defined as the force divided by the actual cross section as the material undergoes deformation. True stress continues to increase until fracture, and apparent stress may drop somewhat, as shown in Figure 10-26. Depending on the application,

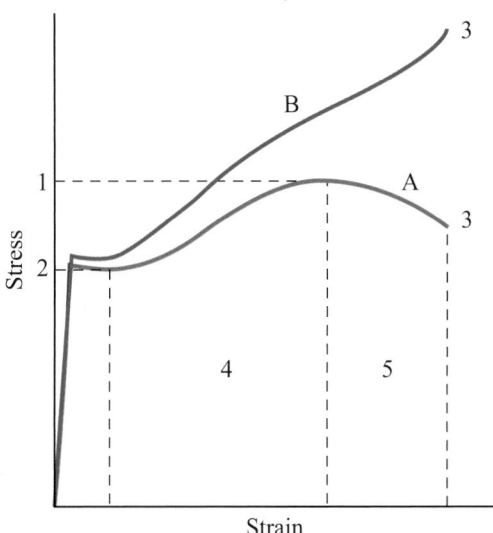

Figure 10-26   Engineering stress (red) versus true (blue) stress

MAKING CONNECTIONS

Steel and Concrete Structural Materials

The response of steel to tensile stress is similar to the response of steel under tension. The ultimate tensile strength of the American Society for Testing and Materials International (ASTM) A36 steel is 400 to 550 MPa. The tensile yield strength is 250 MPa, and the compressive yield point is 152 MPa, indicating that this alloy withstands tension better than it withstands compression. However, the ultimate tensile strength for concrete is only 2 to 5 MPa, and its compressive strength (the stress at which the concrete shatters) is 20 to 40 MPa. While steel can be used in elements under compression or tension almost equally effectively, concrete is almost exclusively used in elements under compression.

stress versus strain curves may show either the true stress or the apparent stress.

Maximum Tensile and Compressive Strength

For most metals, for example, steel and copper, the stress versus strain curve under compression is similar to the stress versus strain curve under tension. The maximum stress that a material can take under compression, called the **compressive strength** (not the ultimate compressive strength), cannot always be as precisely defined as the ultimate tensile strength. For materials that fail by shattering under compression, such as concrete, the compressive strength is defined as the strain at which the fracture occurs. However, for materials that do not shatter under compression, the definition of the compressive strength has to be defined in terms of the stress required to cause a somewhat arbitrarily chosen amount of permanent deformation. A deformation of 0.2% is used as the standard for many applications. Table 10-1 lists the tensile and compressive properties for a variety of materials.

Table 10-1   The Tensile and Compressive Properties for Various Materials

Material	Yield strength	Ultimate tensile strength	Young's modulus	Compressive strength
A36 steel	250 MPa	400–550 MPa	200 GPa	152 MPa
Stainless steel A316L	290 MPa	579 MPa	193 GPa	170 MPa
Aluminum	95 MPa	110 MPa	69 GPa	
Rubber		15 MPa		
Concrete		5 MPa	30 GPa (compression)	14–42 MPa
Wood		~88–150 MPa	6–15 GPa	~14–50 MPa (parallel to grain)
Tendon			0.6–1.8 GPa	
Hair	~200 MPa	380 MPa	~5 GPa	
Spider silk	225 MPa	Up to 1.9 GPa	1.5 GPa	
Sandstone		4–25 MPa	1–20 GPa	20–170 MPa
Granite		7–25 MPa	10–70 GPa	100–250 MPa
Femur bone (along axis)	~100 MPa	135 MPa	17.4 GPa	190–205 MPa

CHAPTER 10 | **EQUILIBRIUM AND ELASTICITY**

✓ CHECKPOINT

C-10-7 Compressive Strength

Which of the following statements are true?

(a) One cannot define a compressive strength for most steel alloys.

(b) The compressive strength for steel is the point at which steel fractures.

(c) The compressive strength for steel is the stress point for which the steel goes back to within an arbitrary percentage of original length, often 0.2%.

(d) Statement (b) is true for brittle steel, and statement (c) is true for ductile steel.

C-10-7 (c) The accepted arbitrary limit is set to 0.2% for most standards.

EXAMPLE 10-9

The Strength of Human Bone

Young's modulus for the human femur bone for longitudinal stress is approximately 17.0 GPa, and the compressive strength is approximately 190 MPa. The average length of a female femur is 48.0 cm, and the average diameter is 25.4 mm.

(a) Compare the maximum mass that can be carried by the femur to the maximum mass that can be carried by a column of concrete that has the same dimensions and a compressive strength of 42.0 MPa.

(b) Assuming a linear relation between stress and strain until the point of maximum compression, estimate the compressive strain at which the femur will fracture.

SOLUTION

(a) At a given stress, the force is given by Equation (10-5):

$$F = \sigma A$$

The force in this context is the weight carried by the material. So,

$$M = \frac{\sigma A}{g}$$

Since we are looking at the maximum mass, we use the maximum value of the stress, which is the compressive stress.

For the femur:

$$M = \frac{\sigma A}{g} = \frac{(190\text{ MPa})(\pi)(0.0127\text{ m})^2}{9.81\text{ m/s}^2} = 9.81 \times 10^3\text{ kg}$$

For concrete:

$$M = \frac{\sigma A}{g} = \frac{(42\text{ MPa})(\pi)(0.0127\text{ m})^2}{9.81\text{ m/s}^2} = 2.17 \times 10^3\text{ kg}$$

(b) Using Equation (10-7) or Equation (10-8),

$$\sigma = Y\epsilon \quad \text{or} \quad \frac{F}{A} = \frac{Y\Delta L}{L}$$

$$\epsilon = \frac{\sigma}{Y} = \frac{190\text{ MPa}}{17\text{ GPa}} = \frac{190 \times 10^6\text{ Pa}}{17 \times 10^9\text{ Pa}} = 1.12\%$$

Making sense of the result:

The bone will compress by 1.1% of the length at fracture. Since the tensile strain of bone at ultimate tensile strength is slightly over 1%, the value makes sense.

✓ CHECKPOINT

C-10-8 Compressive Strength

Which of the following statements is true?

(a) The compressive strength for concrete is the point at which the concrete returns to within 0.2% of its original length.

(b) Concrete shatters at its compression strength.

(c) One cannot define a compression strength for concrete.

(d) None of the above statements are true.

C-10-8 (b) The compressive strength for concrete and other composite materials is defined as the point at which the material shatters.

Steel Grades

Steel is an alloy composed primarily of iron. Depending on the desired properties, steel contains various amounts of other elements, such as carbon, manganese, chromium, nickel, copper, aluminum, boron, and molybdenum. For example, 18-8 stainless steel, often used for quality cookware, contains 18% chromium and 8% nickel by mass. All types of stainless steel contain at least 10.5% chromium by mass.

The higher the carbon content of a steel alloy, the greater the strength, hardness, rigidity, and brittleness of the alloy. The carbon content of steel alloys is generally between 0.2% and 2.1%. Alloys with a carbon content greater than 2.1% are called cast iron. Such alloys are easy to cast but rather brittle; a hard impact can crack or shatter cast iron. The alloy used for wrought iron contains minimal carbon and is quite malleable, making it easy to work but less strong.

A36 is an ASTM designation for an alloy widely used for construction and structural supports in the United States and Europe (Figure 10-27). The number 36 comes from the alloy's yield point of 36 000 psi, which is equivalent to 250 MPa. A36 was the preferred grade for

Figure 10-27 A36 steel beams

will elongate by a minimum of 23% of its original length before fracturing, and a 20 cm–long piece will stretch by 20% before fracture. The amount by which an element stretches before snapping is called **elongation at facture**. The stress at which a steel beam buckles depends on its length and cross section. The buckling stress for a cylindrical A36 beam with a length 10 times its radius is approximately 145 MPa, but an A36 steel beam with a length 100 times its radius buckles as little as 52 MPa under stress.

construction steel for a long time, but it has now given way to better-performing alloys. Canada's version of A36 steel is an alloy designated 44W. This alloy has slightly better properties, including a greater yield strength and more elongation before snapping.

Steel will eventually fracture once the ultimate tensile strength is reached. A 5 cm–long piece of A36 steel

✓ CHECKPOINT

C-10-9 A36 Steel

A36 steel has

(a) an ultimate tensile strength of 36 000 psi;

(b) a yield point of 36 000 psi;

(c) an ultimate tensile strength of 36 MPa;

(d) a yield point of 36 MPa.

C-10-9 (b) The yield point of A36 steel is 36,000 psi.

THE CONDITIONS OF EQUILIBRIUM

When an object is in equilibrium, the vector sum of all the external forces acting on the object must be zero:

$$\sum \vec{F} = 0 \qquad (10\text{-}1)$$

When an object is in equilibrium, the vector sum of all the external torques acting on the object about its centre of mass or any other given point must be zero:

$$\sum \vec{\tau} = 0 \qquad (10\text{-}2)$$

In terms of momenta:
The linear momentum of an object in equilibrium is constant:

$$\frac{d\vec{p}}{dt} = 0 \qquad (10\text{-}3)$$

The angular momentum of an object in equilibrium about a given point is constant:

$$\frac{d\vec{L}}{dt} = 0 \qquad (10\text{-}4)$$

Elasticity

Stress, σ, is the force per unit area:

$$\sigma = \frac{F}{A} \qquad (10\text{-}5)$$

Strain is the fractional change in the length of an object in response to stress:

$$\epsilon = \frac{\Delta L}{L} \qquad (10\text{-}6)$$

The linear part of an elastic deformation is characterized by a linear relation between stress and strain:

$$\sigma = Y\epsilon \qquad (10\text{-}7)$$

or

$$\frac{F}{A} = Y\frac{\Delta L}{L} \qquad (10\text{-}8)$$

Young's modulus is the same for compressive and tensile stress.

The response of certain materials to tension can differ drastically from the materials' response to compression.

The deformation of an object along the direction of a shear force is given by

$$\frac{F}{A} = G\frac{\Delta L}{L} \qquad (10\text{-}9)$$

Plastic Deformation and Failure Modes

The yield point is the stress beyond which deformation is irreversible. Plastic deformation occurs beyond the yield point.

Ductile materials endure significant plastic deformation before fracture. Brittle materials fracture shortly past the yield point.

Materials can fail by fracture, buckling or ductile failure.

APPLICATIONS

Applications: earthquake-resistant structures; construction materials; cast iron and stainless steel

Key Terms: apparent stress, brittle, centre of gravity, compression force, compressive strength, ductile, ductile failure, dynamic equilibrium, elastic deformation, elastic limit, elongation at facture, necking, permanent set, plastic deformation, proportional limit, shear force, shear modulus, stable equilibrium, static equilibrium, strain, stress, tension, true stress, ultimate tensile strength, unstable equilibrium, yield point, yield strength, Young's modulus

QUESTIONS

1. Will a ball thrown straight up be in equilibrium at any point while in the air? In particular, will it be in equilibrium when it reaches its maximum height?
2. Is a car moving on level ground at a constant speed in equilibrium?
3. When a soccer ball rolls down a hill at a constant speed,
   (a) the ball cannot be in equilibrium because it is going downhill;
   (b) the sum of the forces on the ball is never zero;
   (c) the ball is in equilibrium.
   (d) None of the above are true.

4. A particle's position as a function of time is given by $x = x_0\cos\omega t$. Which of the following statements is true?
   (a) The particle is in equilibrium when $x = x_0$.
   (b) The particle is in equilibrium when $x = 0$.
   (c) The particle is never in equilibrium.
   (d) None of the above are true.
5. When you are standing in an elevator moving straight up at a constant speed, you are
   (a) not in equilibrium because the elevator is moving;
   (b) not in equilibrium because you are moving;
   (c) in equilibrium because your velocity is constant;
   (d) in equilibrium because your acceleration is constant.

6. You step onto a platform resting on top of a spring with a spring constant of k. The spring is fixed to the ground and is then compressed until it comes to a momentary stop. Which of the following statements is true?
   (a) You are in equilibrium throughout.
   (b) You are only in equilibrium when you stop momentarily.
   (c) You are in equilibrium when you are halfway to the point where you stop momentarily.
   (d) You are never in equilibrium during this process.

7. With one hand you hold one end of a string fixed. With the other hand you hold a small mass, which is attached to the other end of the string. You are holding the string such that it is initially horizontal. Then you let the mass go. The mass
   (a) is in equilibrium;
   (b) is in equilibrium at the bottom of its trajectory;
   (c) is in equilibrium when it stops momentarily at either end of its trajectory.
   (d) None of the above are true.

8. A rocket is launched and accelerates vertically, lifted by the thrust of its engine. When the engine runs out of fuel, the rocket begins to slow down under the effect of the force of gravity. Which of the following statements is true?
   (a) After ignition, the rocket is in equilibrium at only one point in its journey upward.
   (b) The rocket is in equilibrium at two points after ignition.
   (c) The rocket is only in equilibrium when it reaches its maximum height and stops momentarily.
   (d) The rocket is in equilibrium as long as it is on the ground, and again at another point between liftoff and maximum height.

9. Is an object that is undergoing uniform circular motion in equilibrium?

10. A disk is spinning about its centre of mass axis at a constant angular speed. Which of the following statements is true?
    (a) The disk is in equilibrium.
    (b) Any given point on the disk is not in equilibrium.
    (c) Both (a) and (b) are true.
    (d) None of the above are true.

11. Can an object's centre of gravity lie outside the physical dimensions of the object? Explain.

12. Two point masses are tied to a massless rod, one at the centre of the rod, the other at its end. The rod is made to spin about the end with no mass on it at a fixed angular speed. The force of gravity where this is done is zero. Since there is an acceleration experienced by each mass, one can define a centre of gravity. To which mass will the centre of gravity be closer?

13. A ball rolls downhill at a constant speed. Aside from the centre of mass of the ball, is any other point on the ball in equilibrium?

14. Calculate the centre of gravity for a uniform bar spinning in a zero gravity environment about a pivot through one end.

15. Two Earth-sized planets orbiting the same star are r and $2r$ from the centre of the star. Find the centre of gravity of the planets when the three objects are collinear with both planets on the same side of the star.

16. Why do you have to lean forward when walking against the current in a river?

17. Does doubling the cross section of a steel rod increase its Young's modulus?

18. Is bone stronger than concrete?

19. Which is tougher, steel or human hair?

20. Which of the following statements is true?
    (a) Steel is mined in steel mines.
    (b) Steel is an alloy that can contain any amount of carbon.
    (c) The amount of carbon in steel alloys ranges between 0.2% and 2.1%; the higher the carbon content, the stronger but more brittle the steel.
    (d) High-carbon steel is more ductile than medium- and low-carbon steel.

21. Does concrete behave better in compression or in tension?

22. The compressive strength for a material that fails by shattering is the strain in which
    (a) you will get the original length back if you remove the load
    (b) the material fractures;
    (c) a permanent set of 0.2% occurs.
    (d) None of the above are true.

23. The yield point for a ductile material is
    (a) the highest stress for which the material will return to its original length after the stress has been removed;
    (b) the stress for which the material will return to within 0.2% of its original length after the stress has been removed.
    (c) Both (a) and (b) can be true.
    (d) Neither (a) nor (b) is true.

24. Compression ductile failure is
    (a) plastic deformation in which a material shrinks in the direction of an applied force and expands along the transverse direction;
    (b) deformation similar to the deformation seen when a train runs over a coin;
    (c) actually "ducktile" failure, because the material makes a distinct "quack" sound like that of a duck.
    (d) Both (a) and (b) are true.

25. Which of the following statements is true?
    (a) Steel goes into plastic behaviour before reaching ultimate tensile strength.
    (b) Steel goes into plastic behaviour after reaching ultimate tensile strength.
    (c) Steel will begin to fail after reaching ultimate tensile strength.
    (d) None of the above are true.

26. Under compression,
    (a) ductile steel fails by cracking;
    (b) ductile steel fails by buckling;
    (c) ductile steel fails by undergoing ductile failure.
    (d) Any of the above could be true.

27. A brittle steel can fail by
    (a) ductile failure;
    (b) fracturing;
    (c) buckling.
    (d) None of the above are true.

28. Which of the following statements is true?
    (a) Steel behaves in a similar way under both compression and tension.
    (b) The compressive yield point of steel is often slightly higher than its tensile yield point.
    (c) Steel will most likely buckle before reaching its compression strength.
    (d) Only (a) and (c) are true.

29. The proportional limit is
   (a) the highest point at which the stress versus strain curve is linear;
   (b) the point at which a material goes into plastic deformation;
   (c) outside the elastic limit.
   (d) None of the above are true.

PROBLEMS BY SECTION

For problems, star ratings will be used, (✱, ✱✱, or ✱✱✱), with more stars meaning more challenging problems.

Section 10-1   Static and Dynamic Equilibrium

30. ✱ Consider the simplified model of an arm shown in Figure 10-28. What must be the force T applied by the muscle when the arm is in static equilibrium? Treat the elbow as the pivot, and assume that all forces are perpendicular to the arm as shown. Use the following data: $L = 5.00$ cm, $c = 15.0$ cm, $d = 30.0$ cm, the weight of the forearm alone is 20.0 N, and the weight of the moving mass is 10.0 N.

Figure 10-28   Problem 30

31. ✱ A 5.0 kg crate hangs vertically from a rope wound around the pulley as shown in Figure 10-29. The other end of the rope is tied to a 12 kg mass sitting on the horizontal surface. Find the coefficient of static friction between the second mass and the surface it sits on, given that the mass is on the verge of slipping.

Figure 10-29   Problem 31

32. ✱ A 10 kg mass hangs from a rope as shown in Figure 10-30. The rope passes over a fixed pulley, then under a movable pulley, and is attached to the ceiling as shown. The movable pulley has a mass m attached to it and moves at a constant speed. Find the tension in the rope. Assume that the masses of the pulleys are negligible.

Figure 10-30   Problem 32

33. ✱ A 65 kg climber in a slippery crevice wedges an adjustable bar between the two icy, vertical walls (Figure 10-31). What must the compression force in the bar be to support the climber's weight? The coefficient of friction between the bar and the walls is 0.12.

Figure 10-31   Problem 33

34. ✱ While climbing a chute inside a cave, what force must you apply against the walls to support your weight? The coefficient of static friction between your boots and the wall is 0.56, and that between your gloves and the wall is 0.39. Assume that you exert equal force with your hands and feet.

35. ✱ The bridge in Figure 10-32 is supported at its end by forces F_1 and F_2. Solve for these forces, given that the mass of the bridge is 14.0 t, its length is 5.00 m long, and its mass is distributed uniformly.

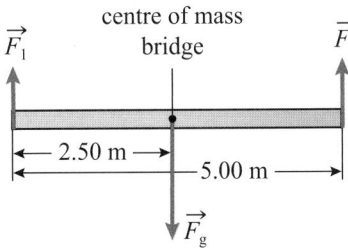

Figure 10-32 Problem 35

36. ✻ The 62 kg athlete in Figure 10-33 hangs from a 3.0 m–long horizontal bar. The athlete is one-fifth of the length from one end. Find the support forces at each end of the bar, assuming that the mass of the bar is negligible.

Figure 10-33 Problem 36

37. ✻✻ A child rests two branches against each other as shown in Figure 10-34. There is enough friction between the branches and the ground to prevent them from slipping. The longer of the two branches is 1.6 m in length and has a mass of 4.0 kg; the shorter branch has a mass of 2.7 kg. The shorter branch makes an angle of 67° with the horizontal ground and an angle of 105° with the longer branch. What must the coefficient of friction between the two branches be so that the structure does not collapse?

Figure 10-34 Problem 37

38. ✻ The 110 kg sign in Figure 10-35 is suspended by two cables as shown. Find the tension in each cable.

Figure 10-35 Problem 38

39. ✻ The 300 kg horizontal bar in Figure 10-36 is supported by two cables as shown. Find the tension in each cable, given that the bar is 7.0 m long.

Figure 10-36 Problem 39

40. ✻ A person's leg is suspended to lie horizontally as shown in Figure 10-37. All lengths shown are measured from the left edge of the leg, which is supported by the rest of the body. The mass of the leg alone is 6.50 kg, and its centre of mass is a distance of 0.450 m from the left end. A 2.00 kg mass is attached to the leg 0.750 m from the left end as shown. The leg is supported by the person's body through a vertical force F_p and by a tension T in a vertical cord attached at a distance of 0.650 m.
  (a) Find the torque due to the weight of the leg alone.
  (b) Find the torque due to the 2.0 kg mass alone.
  (c) Find the tension T.
  (d) Find the magnitude of the vertical force F_p exerted at the left end of the leg.
  (e) If the 2.0 kg mass were moved farther to the right, would the required tension T increase or decrease?

Figure 10-37 Problem 40

Section 10-2 Centre of Gravity

41. ✻✻ A 75 kg passenger on a cruise ship leans forward into a strong horizontal head wind. When his centre of gravity is 24 cm in front of his ankles, he does not have to exert any effort to keep his balance. His centre of gravity is 110 cm above the deck when he stands upright. What force does the wind exert on this passenger?

42. ✻ Find the distance between the centre of mass and the centre of gravity of a 30 m–long vertical bar on the surface of an asteroid where the gravitational gradient is such that the acceleration due to gravity at the top of the bar is 90% of that at the bottom of the bar.

43. ✻✻ The gravitational force near an anomalous phenomenon in space varies along the z-direction as $1400e^{-(r-r_0)/120}$, where $r_0 = 0$ is the reference point where the acceleration due to gravity is equal to the constant A. Find the distance between the centre of mass and the centre of gravity for a bar of length L that is aligned parallel to the z-axis and has its lower end at $r = r_0$.

44. **⁂** Show that the centre of mass coincides with the centre of gravity for an object if the acceleration due to gravity is constant over the dimensions of the object. (Hint: Add the torques due to the point masses that make up the object about a point outside the object.)

45. **⁂** Find the centre of gravity for two equal masses, one located at the surface of a small neutron star, where the gravitational field is 110.0 g, and the other located 100.0 m from the first mass along the surface of the star and 200.0 m above the surface, where the gravitational field is 107 g.

Section 10-3 Applying the Conditions for Equilibrium

46. **⁂** For the 20 kg sphere shown in Figure 10-38, find the normal forces from each surface. Assume that all the surfaces are frictionless.

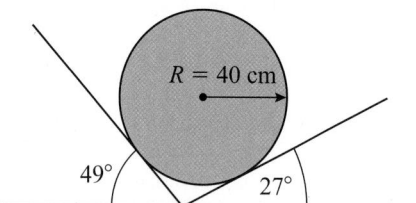

Figure 10-38 Problem 46

47. **⁂** Determine the tension in the rope in Figure 10-39. The beam is 4.0 m long, has a mass of 210 kg, and is supported by a free hinge at one end and a rope at the other end, making an angle of 46° with the horizontal.

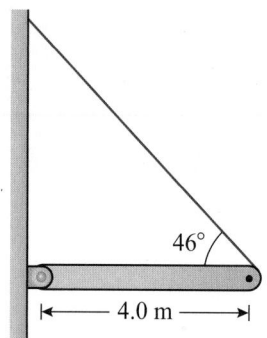

Figure 10-39 Problem 47

48. **⁂** Determine the tension in the horizontal rope supporting the beam in Figure 10-40. The 320 kg beam is 2.4 m long.

Figure 10-40 Problem 48

49. **⁎** If the system in Figure 10-41 is in equilibrium, how are the masses m_1 and m_2 related? Assume that the rope and pulleys have negligible mass.

Figure 10-41 Problem 49

50. **⁎** The 4.0 kg monkey in Figure 10-42 hangs from a rope that is wound around a fixed pulley and around a pulley attached to a banana crate as shown. The crate is on the verge of moving. Find the tension in the rope.

Figure 10-42 Problem 50

51. **⁂** The worker in Figure 10-43 is trying to push the 120 kg wheel over a curb 20 cm high. What force must he apply horizontally if he pushes on the wheel at the point shown in the figure? The wheel has a radius of 40 cm.

Figure 10-43 Problem 51

52. **⁂** A winch used to lift a car (mass 1350 kg) is supported by a cable as shown in Figure 10-44. The 450 kg mass of the winch arm is uniformly distributed. Find the tension in the supporting cable and the reaction forces at the hinge.

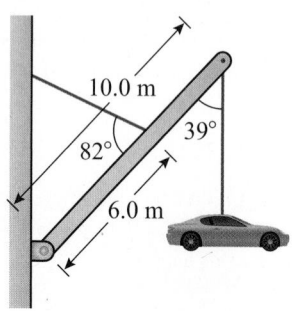

Figure 10-44 Problem 52

Section 10-4 Working with Unknown Forces

53. ✶✶✶ The mass of the upper beam in Figure 10-45 is 2100.0 kg and that of the lower beam is 700.0 kg. Find the horizontal and vertical reaction forces at the hinge at C and at the free hinges at A and B.

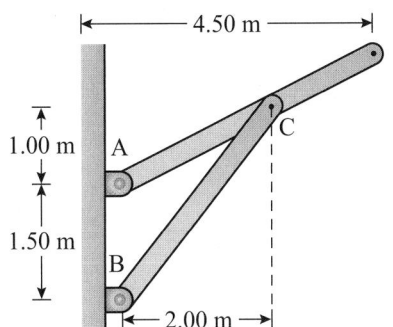

Figure 10-45 Problem 53

54. ✶✶✶ You are designing the crosspiece for the A-frame structure in Figure 10-46. Beams AB and AC are 4.00 m long and have a mass of 300.0 kg each. How much tension must the crosspiece EF withstand? Assume that the mass of the crosspiece and the friction at points A and B are negligible.

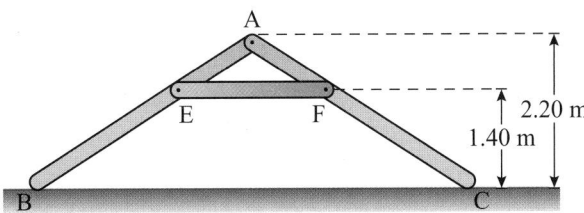

Figure 10-46 Problem 54

55. ✶✶✶ Each beam in Figure 10-47 has a mass of 1100 kg. The length of the shorter beam is 4.0 m. The assembly is in equilibrium. Find the force of friction on each beam, and the horizontal and vertical forces exerted by one beam on the other.

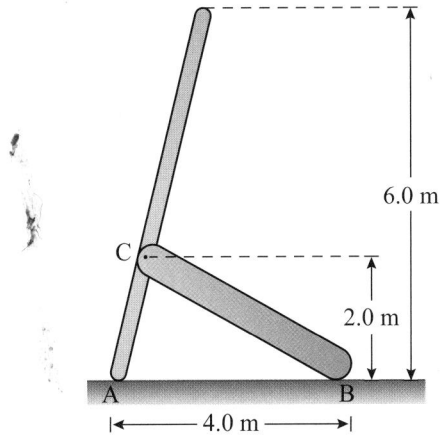

Figure 10-47 Problem 55

Section 10-5 Deformation and Elasticity

56. ✶ Below the proportional point, within the elastic limit, a steel rod can be considered like a spring. Find the effective spring constant of a steel rod of cross section A and length L, given that Young's modulus is Y. How much work would you have to do to stretch the rod by an amount x?

57. ✶ At a construction site, a 4.0 m–long beam supports a winch at its end (Figure 10-48). When the winch lifts a crate up from the ground, the beam deflects by 13 cm. The cross section of the beam is 25 cm², and its shear modulus is 190 MPa. Find the mass of the crate.

Figure 10-48 Problem 57

58. ✶ Vertebrate muscle typically produces approximately 25 N of tensile force per square centimetre of cross-sectional area. Calculate the stress.

59. ✶✶ The shear strength of limestone is approximately 25 MPa. Find the maximum load that can be supported at the end of a limestone beam that has a square cross section of 100 cm² and extends horizontally 2.3 m from its supporting structure?

60. ✶✶ A 22.0 tonne platform is supported by four vertical beams of concrete with a compressive strength of 42 MPa. What diameter of beams must be used?

61. ✶ A 3.0 m–long cylindrical alloy beam is 21 cm in radius. It is found to stretch by 1.5 cm under a tension of 2.00 kN. Find its Young's modulus.

62. ✶✶ Derive an expression for the elastic potential energy stored in a metal rod in terms of its length (L), cross-sectional area (A), Young's modulus (Y), and elongation (ΔL) within the elastic limit.

Section 10-6 Ductile and Brittle Materials

63. ✶✶ A36 structural steel has a yield point of 250 MPa and an ultimate tensile strength of approximately 450 MPa.
(a) What is the maximum mass that you can suspend from a 6.0 cm–diameter, 2.0 m–long A36 steel bar before the bar gets a permanent set?
(b) What is the maximum mass that you can suspend from the same bar such that it does not break?
(c) What diameter of spider silk thread would you need to suspend the same mass without the thread reaching its ultimate tensile strength of 1000 MPa?

64. ✶ Rubber has an ultimate tensile strength of 15 MPa. Find the maximum force that you can apply to a strip of rubber 10 cm long with a cross section of 5.0 mm². Assume that the cross section of the rubber strip does not change appreciably as you pull on it.

65. ✶✶ What proportion of the yield strength of bone is the stress on the femur of an 82 kg person standing on one leg?

66. ✶✶ If the concrete in a concrete-only building were replaced with enough bone to have the same strength, how much lighter would the building be? What would the change be if the concrete were replaced with steel instead? A typical femur bone has a length of 48 cm, an average diameter of 25 mm, and a mass of 260 g. The density of steel is approximately 8000 kg/m³, and the density of concrete is 1680 to 3000 kg/m³.

67. ✷✷ A shipment of limestone blocks has a density of 2700 kg/m³ and a compressive strength of 250 MPa. The blocks are cubical with a side length of 30.0 cm. How many of these blocks can you stack up before the bottom one cracks?

68. ✷✷ Some types of granite have a Young's modulus of 70 GPa and a compressive strength of 250 MPa. Assume linear behaviour until the compressive strength point.
    (a) Find the amount by which granite compresses before shattering.
    (b) How much weight can a 1 m–long, 20 cm–radius granite beam take in compression?

69. ✷✷✷ A sample of basalt has a compressive strength of 300 MPa and a Young's modulus of 73 GPa.
    (a) Assuming linear stress versus strain behaviour, estimate the amount of work it would take to break a column of basalt 2.0 m in length and 49 cm in diameter.
    (b) What is the minimum mass that you can drop from a height of 20 m to cause the column to shatter?

70. ✷ The ultimate tensile strength of concrete is 3.0 MPa, and the compressive strength is 42 MPa. Find the maximum weight that can be supported by a concrete pillar 50 cm in diameter and 4.0 m long under both tension and compression.

COMPREHENSIVE PROBLEMS

71. ✷✷ The 900 kg, 3.7 m–long beam in Figure 10-49 is pivoted at one end, and the other end rests on a sloped, frictionless incline. A 200 kg mass hangs from the beam 3.0 m from the hinge. Determine the normal force from the surface on the beam and the reaction force at the hinge.

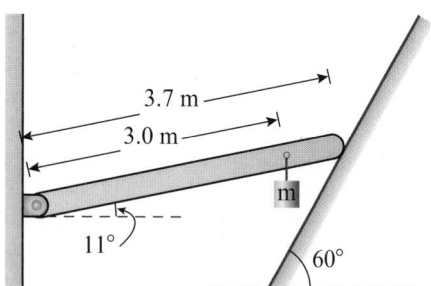

Figure 10-49   Problem 71

72. ✷✷ A disk rests on two other disks as shown in Figure 10-50. All three disks are the same size, and each has a mass of 600.0 kg. There is negligible friction between the disks, but enough friction between the lower disks and the floor to prevent the two disks from moving.
    (a) Find the normal force between the top disk and one of the lower disks.
    (b) Find the magnitudes of the force of friction between each lower disk and the floor, and the normal force from the floor on each disk.

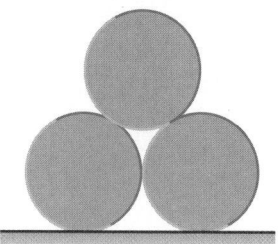

Figure 10-50   Problem 72

73. ✷✷✷ In Figure 10-51, a sphere of radius R is placed on a frictionless slope. A rope of length 6R is attached to the top of the sphere. The other end of the rope is attached to the slope above the sphere. What angle does the rope make with the slope when the sphere reaches a position of stable equilibrium?

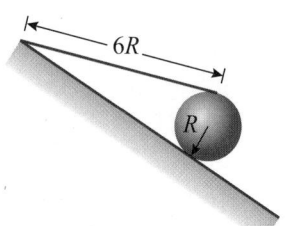

Figure 10-51   Problem 73

74. ✷✷ A 125 kg wheel with a radius of 2.5 m is being pulled up an incline by a 550 kg horse (Figure 10-52). There is a 45 cm–high protrusion in the incline, as shown. What is the tension in the rope when the wheel is just beginning to lift off the incline to go over the protrusion? What is the minimum possible coefficient of friction between the horse and the incline for the horse to be able to pull the wheel over the protrusion?

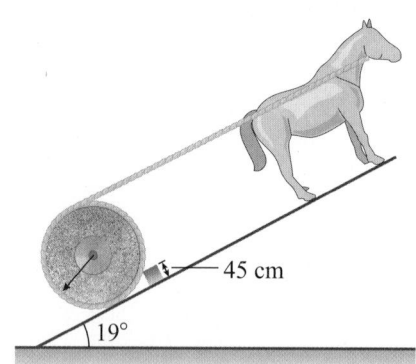

Figure 10-52   Problem 74

75. ✷ The 3.0 kg mass in Figure 10-53 rests on a horizontal surface. The mass is attached to one end of a rope, most of which is wound around a drum of radius R on the side of a pulley, as shown. Another rope is wound around the pulley, which has a radius of 2R. The coefficient of friction between the mass and the surface is 0.35. What is the largest mass than can hang from the vertical rope without causing the pulley to rotate?

Figure 10-53 Problem 75

76. ✷✷✷ The 80 kg painter in Figure 10-54 is using a large stepladder. The base of the ladder is 3 m wide, and the ladder is 7 m high. The painter is on a step that is 2.0 m above the floor. The horizontal bar of the ladder has a mass of 10 kg, and the entire ladder has a mass of 60 kg. Assume that the floor is frictionless.
   (a) Find the force of tension in the horizontal bar and the normal forces that the floor exerts on the ladder.
   (b) Find the reaction forces at the hinges A, B, and C.

Figure 10-54 Problem 76

77. ✷✷ The beam in Figure 10-55 is 3.4 m long and has a mass of 760 kg. A 220 kg crate is suspended from the beam seven-eighths of its length from the free hinge at the left end of the beam. A rope supports the beam as shown. Find the tension in the rope and the components of the reaction force at the hinge.

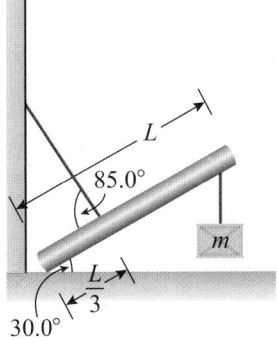

Figure 10-55 Problem 77

78. ✷✷ A stick of mass m and length L is supported at one end by a string, and the other end leans against a wall as shown in Figure 10-56. The string makes an angle θ with the wall, and the stick is horizontal. Derive expressions for the force of friction on the stick, and the minimum coefficient of friction that will stop the stick from sliding down the wall.

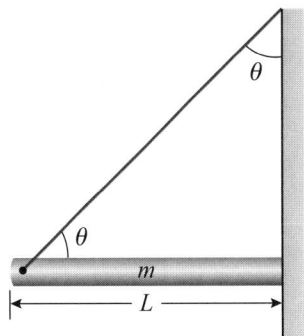

Figure 10-56 Problem 78

79. ✷✷✷ The sphere in Figure 10-57 rests on a friction-less floor. At the point of contact between the sphere and the wall, the coefficient of static friction is μ. Derive an expression for the maximum height d at which a horizontal force F on the sphere will not cause it to spin.

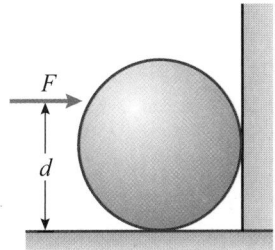

Figure 10-57 Problem 79

80. ✷✷ The drum-and-axle assembly in Figure 10-58 has mass M. The radius of the drum is R, and the radius of the axle is r. You pull on a rope wound around the axle with a force F parallel to the surface of the incline. The drum does not move, but it is about to slide. Derive an expression for the coefficient of friction between the incline and the drum.

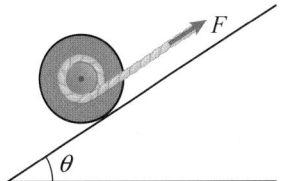

Figure 10-58 Problem 80

81. ✷✷✷ You pull the rope wound around the axle of the drum-and-axle assembly in Figure 10-59 with a force of 120 N directed horizontally as shown. The coefficient of friction between the drum and the inclined surface is 0.20. The mass of the drum is 4.0 kg. The axle radius is 20 cm, and the drum radius is 50 cm. The drum is on the verge of spinning in place. Find the coefficient of friction between the drum and the horizontal surface.

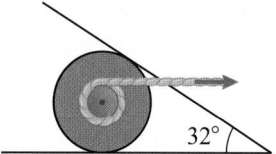

Figure 10-59   Problem 81

82. ★★★ The coefficient of friction between the pole and the incline in Figure 10-60 is 0.45. A guy wire runs from the top of the pole to the top of the incline. You push the 45 kg pole at its centre with a force parallel to the incline and directed down the incline. What is the maximum magnitude of the force that will not cause the pole to slide on the incline?

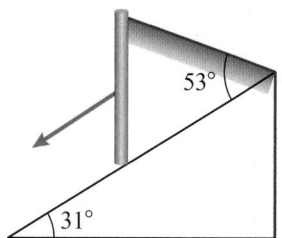

Figure 10-60   Problem 82

83. ★★ A 4.0 m–long bar is supported by a string as shown in Figure 10-61. What must be the coefficient of friction between the bar and the floor if the bar is on the verge of slipping?

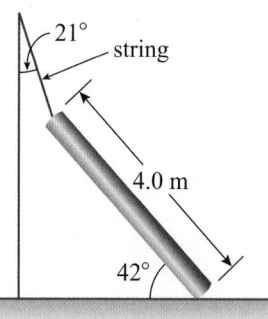

Figure 10-61   Problem 83

84. ★★★ The vertical bar of length L in Figure 10-62 is supported at the top by a string that makes an angle θ with the vertical. The bottom of the bar rests on a surface where the coefficient of friction is μ_s. You have another string that you can move up and down the bar and use it to pull horizontally on the bar.
   (a) Derive an expression for the maximum distance y above the floor at which you can position the string and still cause the bottom of the bar to slide on the floor.
   (b) You now bring the string to the middle of the bar. Derive an expression for the maximum horizontal force exerted on the string without the bar sliding on the floor.

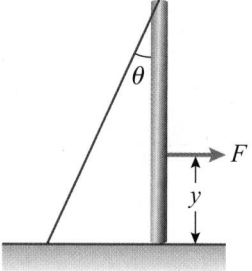

Figure 10-62   Problem 84

85. ★★★ You are pushing a 92 kg refrigerator on a horizontal floor (Figure 10-63). The coefficient of static friction between the refrigerator legs and the floor is 0.34, and the refrigerator's centre of gravity is 80 cm above the floor.
   (a) Find the horizontal force that causes the refrigerator to just move when applied 30 cm above the centre of gravity.
   (b) Will the refrigerator tip or slide first?
   (c) What is the maximum height at which you can apply the horizontal force without tipping the refrigerator?

Figure 10-63   Problem 85

86. ★★★ A forklift driver does not notice a truck right behind him and attempts to back up, pushing against the truck with a horizontal force F (Figure 10-64). The wheels of the truck are 1.9 m apart, and its mass is 4800 kg. The friction between the truck's wheels and the pavement prevents the truck from slipping. The point of contact between the truck and the forklift is 1.6 m above the pavement. The truck's centre of gravity is 0.9 m above the pavement. What is the minimum magnitude of the force F that will tip the truck?

Figure 10-64   Problem 86

87. ★★★ A child pulls on an empty 15 kg bookcase as shown in Figure 10-65. She pulls on a shelf 1.4 m above the floor and exerts a force at angle of 30° below the vertical. What is the maximum force the child can exert without causing the bookcase to tip? Assume that the mass of the bookcase is distributed symmetrically around its geometric centre and that there is enough friction to prevent the bookcase from sliding on the floor.

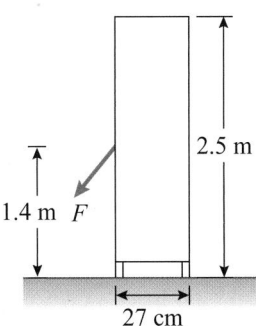

Figure 10-65   Problem 87

88. ★★★ The bar of length L in Figure 10-66 extends by a distance x over the edge of a table.
   (a) What is the maximum distance x for which the bar will not fall?
   (b) You put an identical bar on top of the first bar. What is the maximum distance that the right edge of the second bar can extend over the right edge of the first bar so that the system is stable?
   (c) Find the corresponding position for a third bar.

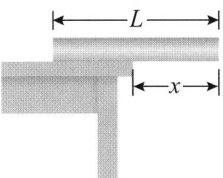

Figure 10-66   Problem 88

89. ★★ Figure 10-67 shows a typical stress versus strain curve for spider silk.
   (a) Find Young's modulus for spider silk by looking at the stress between 0 and 200 MPa.
   (b) Draw the force versus elongation curve for a thread diameter of 1.5 μm and an original length of 11 cm.
   (c) How much energy does it take to stretch the thread by 1% of its original length?
   (d) Identify the proportional limit on the curve.
   (e) Estimate the minimum speed that a 50 mg fly needs to break a strand of spider silk 20 cm in length. Is this speed practical?

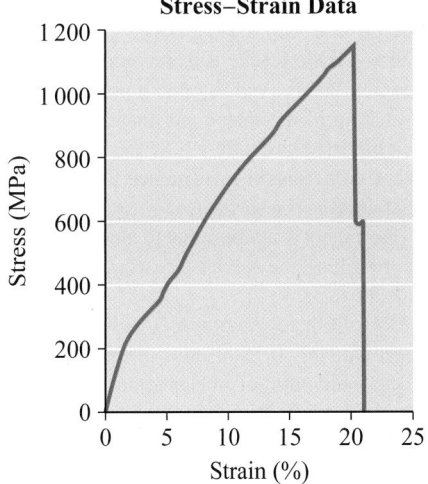

Figure 10-67   Problem 89

90. ★ For the curve in Figure 10-67, the silk snaps at a stress of approximately 1150 MPa. Find the maximum mass than can be supported by a 3.0 cm–diameter silk thread. What mass can an A36 steel beam of the same diameter support? (A36 has an ultimate tensile strength of approximately 440 MPa.)

91. ★★ Figure 10-68 shows the linear region of the stress versus strain curve for a spider silk thread 15 mm long and 1 μm in diameter. Find the spring constant for this material. (Within the elastic limit, the material behaves like a spring.)

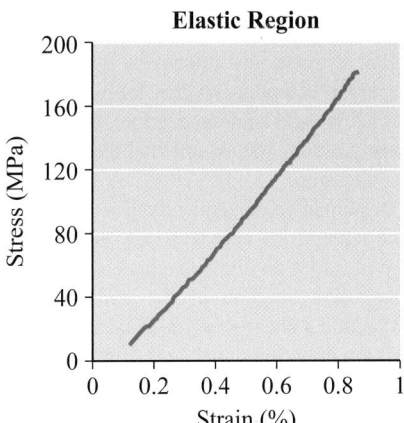

Figure 10-68   Problem 91

92. ★★★ Use Figure 10-67 to estimate the kinetic energy needed to break a 15 mm thread. (You may have to reconstruct the force versus elongation graph for the thread.)

93. ★★ Four vertical steel bars, each 2.0 m in length and 10 cm in diameter, support a 120 kg platform with a 4500 kg load of marble slabs on top of it. Find the stress and strain in each bar.

94. ★★ Kevlar is used in bulletproof vests. The yield point of Kevlar is approximately 3600 MPa, and its Young's modulus is approximately 100 GPa.

(a) What diameter of a 30 cm–long Kevlar thread would one need to stop a 7.5 g bullet moving at a speed of 420 m/s if the Kevlar is to stay within its elastic limit? Assume that the only stopping power comes from the stretching of the thread and that the bullet strikes the taut thread at its centre.

(b) What is the maximum distance that the bullet moves after striking the Kevlar?

(c) If one did not want the bullet to move more than 5.0 mm after striking the Kevlar thread, what length of thread must be used?

95. ✳✳ The ultimate shear strengths of many steel alloys are approximately 0.75 of their ultimate tensile strength. The shear yield point of steel is approximately 0.58 of the tensile yield point. If you wanted to construct a bullet-proof vest using a square steel plate 4.0 cm on a side, how thick should the plate be to remain within its shear elastic limit while stopping a 7.5 g bullet moving at a speed of 390 m/s?

96. ✳✳ A steel cable 10 cm in diameter secures a ship to a bollard. Wind and wave action combine to push the ship and stretch the 12 m–long cable by 1.0 cm. Find the elastic energy stored in the cable.

97. ✳✳✳ Human leg bone typically has a Young's modulus of 17 GPa, and a compressive strength of 190 MPa.

(a) From what height can a 70 kg athlete jump without fracturing a femur? Assume an impact time of 220 ms and that the femur has a length of 52 cm and a radius of 1.7 cm.

(b) How much can the femur compress before fracturing? Use the average force from the impact time, and assume that all the weight of the athlete is supported by the femur.

98. ✳✳ Some types of bone have a compressive yield strength of 190 MPa and a tensile yield strength of 130 MPa. Estimate the amount by which the femur of a 90 kg person compresses when the person jumps from a height of 2.0 m and lands on one foot. Use the data from the previous question, and assume that the person takes 40 ms to come to a stop.

99. ✳✳ A human femur typically has a longitudinal compressive strength of approximately 200 MPa, a strain at fracture of 0.019, an ultimate tensile strength of 135 MPa, a compressive strength in the transverse direction of 131 MPa, and a shear strength of 65 to 71 MPa. Estimate the maximum longitudinal and transverse forces that a femur can support. Assume a length of 53 cm and a radius of 1.5 cm.

100. ✳✳✳ Figure 10-69 shows the true compressive stress versus true strain curve for AISI 304L stainless steel.

(a) Find Young's modulus for this steel.

(b) What is the maximum elongation shown on the curve?

(c) Find the spring constant for the elastic part of this curve for a 1.7 m–long cylindrical beam that is 3.0 cm in radius.

(d) Plot the force versus elongation for this steel.

(e) Find the energy that it takes to compress this beam by 6% of its original length.

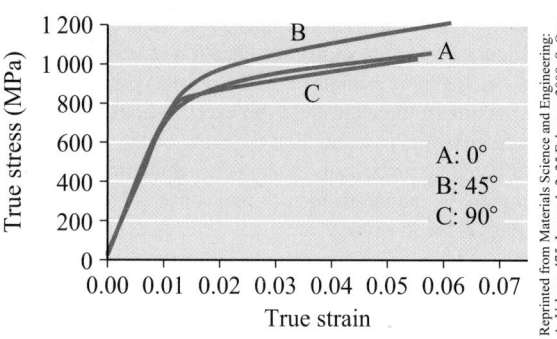

Figure 10-69 Problem 100

101. ✳✳ Figure 10-70(a) show the stress versus strain curve of stainless steel that has undergone severe plastic deformation in a process called equal-channel angular pressing. The steel is pushed down a vertical channel and undergoes severe deformation while passing through the junction with a horizontal channel (Figure 10-70(b)).

(a) Describe how this process changes the ultimate tensile strength of the steel.

(b) Why would this change be desirable for construction grade steel?

(a)

(b)

Figure 10-70 Problem 101

 See the text online resources at www.physics1e.nelson.com for Open Problems and Data-Rich Problems related to this chapter.

Reprinted from Materials Science and Engineering: A, Volume 475, Issues 1–2, 25 February 2008, S. Qua, C.X. Huanga, Y.L. Gaob, G. Yangb, S.D. Wu a, Q.S. Zanga, Z.F. Zhang, "Tensile and compressive properties of AISI 304L stainless steel subjected to equal channel angular pressing," Pages 207–216, Copyright 2008, with permission from Elsevier.

Reprinted from Materials Science and Engineering: A, Volume 475, Issues 1–2, 25 February 2008, S. Qua, C.X. Huanga, Y.L. Gaob, G. Yangb, S.D. Wu a, Q.S. Zanga, Z.F. Zhang, "Tensile and compressive properties of AISI 304L stainless steel subjected to equal channel angular pressing," Pages 207–216, Copyright 2008, with permission from Elsevier.

Chapter 11
Gravitation

Learning Objectives

When you have completed this chapter you should be able to:

1 Solve problems using Newton's law of universal gravitation.

2 Derive and apply an expression for acceleration due to gravity near a surface.

3 Explain what is meant by weightlessness, and describe several ways that weightlessness can be achieved.

4 Use expressions for gravitational potential energy in the general astrophysical case.

5 Calculate force components by taking the derivatives of potential energy functions.

6 Derive and apply an expression for escape from a planet or other object.

7 State Kepler's laws, and solve related problems.

8 Classify orbits as bound or unbound using energy considerations.

9 Calculate the centre of mass of a two-body system, and explain how masses of exoplanets are calculated.

There are more than a million asteroids with diameters larger than a kilometre (Figure 11-1). Most asteroids are located in the region between Mars and Jupiter; however, some have orbits that bring them close to Earth. These near-Earth asteroids pose a small, but not insignificant, hazard to life on Earth. It is believed that, 65 million years ago, an asteroid or a comet impact caused the demise of the dinosaurs—and about three-quarters of all other species. Asteroids also have potential benefits. Within your lifetime we may well begin to harness resources from asteroids.

Understanding the physics of gravitation is essential for assessing the risks from asteroids and for exploiting their resources. Gravitational theory enables you to answer questions such as, What is the gravitational acceleration at the surface of an asteroid? How fast would you need to throw an object for it to escape from the gravitational attraction of the asteroid? How long does an asteroid take to make one orbit around the Sun? When will an asteroid be closest to Earth?

As you will learn in this chapter, gravitational theory already has many practical applications here on Earth.

Courtesy USGS/NASA/JPL

Figure 11-1 Asteroid 951 Gaspra is about 18 km long. Would you be able to jump from the surface fast enough to escape from Gaspra's gravitational pull?

For additional Making Connections, Examples, and Checkpoints, as well as activities and experiments to help increase your understanding of the chapter's concepts, please go to the text's online resources at www.physics1e.nelson.com.

LO 1

11-1 Universal Gravitation

One of Isaac Newton's many contributions to physics is the **law of universal gravitation**: Every mass in the universe attracts every other mass with a force directed along a line joining the centres of the two masses, and with a magnitude given by the following equation:

KEY EQUATION
$$|\vec{F}_G| = \frac{GMm}{r^2} \qquad (11\text{-}1)$$

where M and m are the values of the masses (in kilograms), r is the distance between the centres of mass of the two objects (in metres), and $|\vec{F}_G|$ is the magnitude of the gravitational force (in newtons).

The **centre of mass** is the mean position of the mass in an object or a system. The centre of mass of a homogeneous sphere is at the centre of the sphere. (Newton proved this while developing his theory of universal gravitation.) The value G represents the gravitational constant. In SI units, G has the value 6.672×10^{-11} N m²/kg². You can find the gravitational force for objects of arbitrary shape by integrating over differential mass elements.

As shown in Figure 11-2, the direction of the force is toward the centre of the masses causing the gravitational force. The strength of the gravitational force depends directly on the two masses, M and m, and inversely on the square of the distance between the two masses, r.

Figure 11-2 The gravitational force that mass A exerts on mass B, $\vec{F}_{A\,on\,B}$, is equal in magnitude but opposite in direction to the force that mass B exerts on mass A, $\vec{F}_{B\,on\,A}$.

Every pair of objects in the universe exerts gravitational forces on each other, and the magnitudes of these forces can be determined using Equation (11-1). For example, you can use Equation (11-1) to calculate both the force that Earth exerts on your body (i.e., your weight on Earth) and the force that Earth exerts on the Moon. Since Newton's third law states that for every force exerted by one object on another, there is an equal and opposite reaction force, you can similarly calculate the force that you exert on Earth (or on any other object in the universe).

MAKING CONNECTIONS

Isaac Newton

Sir Isaac Newton (1643–1727) developed the key ideas for the law of universal gravitation during a two-year period, shortly after graduating with his first university degree in 1665. However, his work was not published until much later. Newton had planned to go to graduate school immediately, but the University of Cambridge was closed to help prevent the spread of the plague. So, Newton worked largely alone at his home (Figure 11-4). In addition to his work on universal gravitation, he made major contributions to the development of calculus, optics, and mechanics during this time.

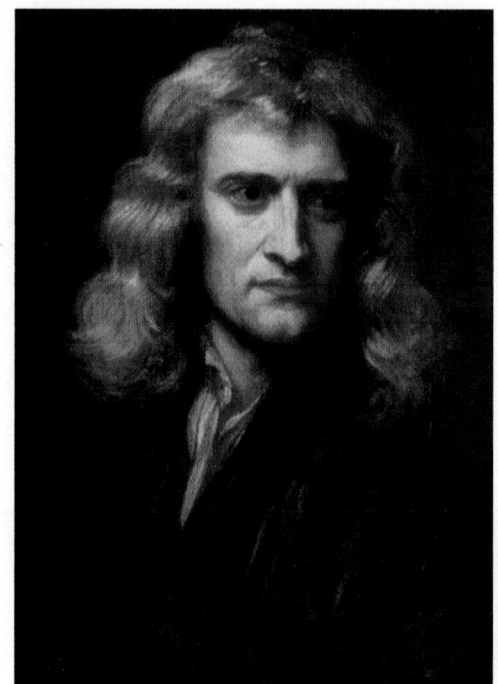

Godfrey Kneller, 1689

Figure 11-3 Sir Isaac Newton at age 46

Awe Inspiring Images/Shutterstock.com

Figure 11-4 Woolsthorpe Manor, where Sir Isaac Newton developed his most important contributions to physics

EXAMPLE 11-1

The Sun's Gravitational Force

(a) Find the gravitational force exerted by the Sun on a 50.0 kg person on Earth.
(b) Find the gravitational force exerted by Earth on the same person.

SOLUTION

(a) We can use the law of universal gravitation, substituting the person's mass (50.0 kg) for m, the mass of the Sun (1.989×10^{30} kg) for M, and the distance from Earth to the Sun (1.496×10^{11} m) for r. Since r is approximately 10^4 times larger than the diameter of Earth, the particular location of the person on Earth has negligible effect on the magnitude of the force in this calculation:

$$|\vec{F}_G| = \frac{GMm}{r^2}$$

$$= \frac{(6.672 \times 10^{-11} \text{ N·m}^2/\text{kg}^2)(1.989 \times 10^{30} \text{ kg})(50.0 \text{ kg})}{(1.496 \times 10^{11} \text{m})^2}$$

$$= 0.296 \text{ N}$$

(b) We use the same method as in part (a) and substitute Earth's mass and radius (6380 km) for M and r, respectively:

$$|\vec{F}_G| = \frac{GMm}{r^2}$$

$$= \frac{(6.672 \times 10^{-11} \text{ N·m}^2/\text{kg}^2)(5.97 \times 10^{24} \text{ kg})(50.0 \text{kg})}{(6.38 \times 10^6 \text{m})^2}$$

$$= 489 \text{ N}$$

Making sense of the result:

The gravitational force of the Sun on a person is much less than the gravitational force of Earth on the person (about 0.06%). However, for high-precision calculations the gravitational force exerted by the Sun on the person is not negligible compared to the gravitational force exerted by Earth.

 CHECKPOINT

C-11-1 Forces on a Satellite

A satellite is orbiting Earth. Which of the following statements is true?
(a) The magnitude of the gravitational force exerted by Earth on the satellite is larger than the gravitational force exerted by the satellite on Earth.
(b) The magnitude of the gravitational force exerted by the satellite on Earth is larger than the gravitational force exerted by Earth on the satellite.
(c) Both forces have exactly the same magnitude.
(d) The relationship between the magnitudes of the two forces depends on the distance between Earth and the satellite.

You encountered mass earlier when considering Newton's second law. In that context, we had **inertial mass**, which is defined as the amount of force required to accelerate the object divided by the acceleration. In this chapter, we have **gravitational mass**, which is a measure of the gravitational influence on other objects through the law of universal gravitation. While inertial mass and gravitational mass might be different, experiments suggest that they are equal.

In this chapter, we will refer to a number of celestial objects. A **planet** is a solid approximately spherical celestial object that orbits a star. A **satellite** is an object, either natural or artificial, that is in orbit around a planet. Smaller objects in orbits around stars are often called **asteroids**. A **meteoroid** is a smaller object in orbit about a star (typically from meter-sized to dust-sized).

The term "tidal force" refers to the difference in gravitational forces across any body, not just oceans. For example, if a spacecraft were near an extremely dense star, the difference between the gravitational forces on the side facing the star and the side facing away would produce a tidal force strong enough to tear the spacecraft apart. In Chapter 29, you will see how Einstein's general theory of relativity presents a radically different explanation of the nature of gravity.

MAKING CONNECTIONS

Tides

For understanding ocean tides, we can reasonably approximate Earth as a rigid body that responds to gravitational forces as if its mass were concentrated at its centre. Water on the side of Earth facing the Moon is somewhat closer to the Moon than Earth's centre of mass, and therefore experiences a corresponding greater gravitational acceleration. Conversely, water on the side of Earth away from the Moon experiences a lesser gravitational acceleration and therefore gets "left behind." As shown in Figure 11-5, high tides occur at opposite points on Earth. A given point has two high tides each day as Earth rotates.

For a more complete understanding of tides, we have to take into account ocean currents and the gravitational force exerted by the Sun. Tides are higher at times of new and full moon because the gravitational forces of the Sun and the Moon approximately align. The most extreme tides (called spring tides) occur when the Sun, the Moon, and Earth are in a line. The mean tidal range of the oceans is about 1 m, but the shape and orientation of coastlines affect the local differences between the water level at low and high tide. For example, in the Bay of Fundy a combination of resonance and funnel effects produces a tidal range of up to 16 m (Figure 11-6).

Courtesy Chignecto Creative Services

Figure 11-6   High and low tide at Alma, New Brunswick, in the Bay of Fundy

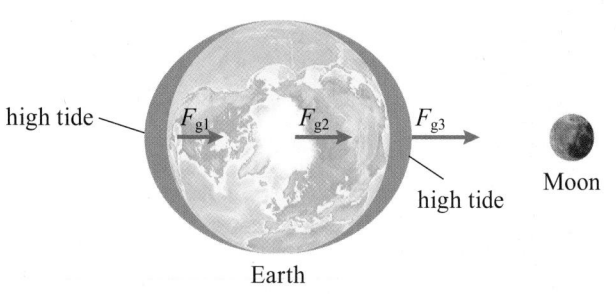

Figure 11-5   The pattern of two tides each day results from the relative magnitudes of gravitational acceleration due to the gravitational force of the Moon on water on the side of Earth nearer the Moon, the solid Earth, and water on the far side of Earth.

LO 2

11-2   Acceleration Due to Gravity

We designate the **acceleration due to gravity** near the surface of a planet (or a satellite of a planet) as the vector quantity $\vec{g}$. This vector is directed toward the centre of the planet, for example, toward the centre of Earth if we are dealing with acceleration due to gravity near Earth's surface. Without a vector sign, g represents the magnitude of the acceleration due to gravity. On Earth, g is approximately 9.81 m/s² at sea level. We can also write the acceleration due to gravity near Earth's surface as $-g\hat{k}$, where $+\hat{k}$ is a unit vector in the upward direction.

In a region where the acceleration due to gravity is $\vec{g}$, the gravitational force on a mass m is given by the following relationship:

$$\vec{F}_{G} = m\vec{g} \tag{11-2}$$

This force is called the **weight** of the mass. We now have two ways to calculate $\vec{F}_{G}$: When we know the local acceleration due to gravity, $\vec{g}$, we can use Equation (11-2), and when we know the mass and radius of the large object causing the gravitational force, we can use Equation (11-1). In Example 11-2, we demonstrate how these two methods are related.

When we measure the gravitational force on a small test mass, and then divide by that mass, we have the vector $\vec{g}$. This vector represents two equivalent

EXAMPLE 11-2

Finding the Acceleration Due to Gravity, g

Show that the observed value of approximately 9.81 m/s² for the magnitude of the gravitational acceleration at Earth's surface is consistent with Newton's law of universal gravitation.

SOLUTION

From Newton's law of universal gravitation, we have the following equation for the magnitude of the force due to gravity:

$$F_G = \frac{GMm}{r^2}$$

Equating this expression with $F_G = mg$ and then solving for g gives

$$g = \frac{GM}{r^2}$$

Next, we substitute values for Earth's radius and mass:

$$g = \frac{(6.672 \times 10^{-11}\,\text{N} \cdot \text{m}^2/\text{kg}^2)(5.974 \times 10^{24}\,\text{kg})}{(6.378 \times 10^6\,\text{m})^2} = 9.80\,\text{m/s}^2$$

In fact, this value for g applies only at the equator because Earth is slightly oblate—its polar radius is approximately 6357 km, and its equatorial radius is approximately 6378 km. As a result, the value of g at Earth's surface varies slightly with latitude.

Making sense of the result:

As expected, the value for g calculated using Newton's law of universal gravitation agrees with observed values.

---

quantities: the gravitational field strength (in newtons per kilogram) and the gravitational acceleration (in metres per second squared). A plot of this vector at different points around a mass shows the gravitational field created by the mass. In Chapter 20, we will apply the same concept to describe electrical fields around charged objects.

We can use the method in Example 11-2 to see how g varies at different altitudes above Earth's surface. In fact, the equation below applies for locations on or near any planet, satellite, or other body.

KEY EQUATION
$$g = \frac{GM}{r^2} \qquad (11\text{-}3)$$

It is important to realize that G is a fundamental constant of physics and, as far as we know, does not depend on where or when you measure it. Some physicists are currently working on theories in which G is not strictly

 CHECKPOINT

C-11-2 What Do We Call the Force of Gravity?

Which of the following terms describes the force of gravity on an object by a planet?
(a) the gravitational acceleration, g
(b) the universal gravitational constant, G
(c) the weight of the object, W
(d) the mass of the object, m

C-11-2 (c) Weight is a force, depending on both mass and the local acceleration due to gravity.

constant at different times, but the preponderance of evidence suggests that G is constant. On the other hand g—the magnitude of the acceleration due to gravity—is definitely not a universal constant. It is quite different on the Moon and Mars, for example, and even varies somewhat for different locations on Earth's surface.

EXAMPLE 11-3

Finding g on an Asteroid

(a) Find the value of g on the surface of an asteroid that has a density of 3200 kg/m³ and a radius of 5.00 km. Assume that the asteroid is homogeneous and spherical.
(b) How long would it take an object starting from rest to fall through a height of 1.0 m above the surface of the asteroid? Assume that g is constant during the fall.
(c) Show that the assumption in part (b) is reasonable.

SOLUTION

(a) First, find the mass of the asteroid:

$$M = \rho V = \rho \frac{4\pi r^3}{3} = 3200\,\text{kg/m}^3 \frac{4\pi(5.00 \times 10^3\,\text{m})^3}{3}$$
$$= 1.68 \times 10^{15}\,\text{kg}$$

Now, calculate the acceleration due to gravity at the asteroid's surface:

$$g = \frac{GM}{r^2} = \frac{(6.672 \times 10^{-11}\,\text{N} \cdot \text{m}^2/\text{kg}^2)(1.68 \times 10^{15}\,\text{kg})}{(5.00 \times 10^3\,\text{m})^2}$$
$$= 4.47 \times 10^{-3}\,\text{m/s}^2$$

(continued)

(b) From Section 3-4 we have

$$y - y_0 = v_0 t + \frac{1}{2} a t^2$$

We choose a convention with positive upward, so $y - y_0 = -1.0$ m. Since the object starts from rest, $v_0 = 0$. The acceleration is -4.48×10^{-3} m/s^2. The acceleration is negative because it is in the downward direction. Therefore,

$$-1.0 \text{ m} = 0 + \frac{1}{2}(-4.48 \times 10^{-3} \text{ m/s}^2)t^2$$

$$t^2 = \frac{2(-1.0 \text{ m})}{-4.48 \times 10^{-3} \text{ m/s}^2}$$

$$t = 21 \text{ s}$$

It would take 21 s for an object to fall 1.0 m on the asteroid.

(c) If we repeat part (a) with $r = 5.001 \times 10^3$ m, we obtain a value for g of 4.47×10^{-3} m/s^2. Therefore, the assumption that g is approximately constant is justified.

Making sense of the result:

As expected, we found that the gravitational acceleration is also much less than near the surface of Earth. Since 1.0 m is small compared to the radius of the asteroid, it is not surprising that the change in g would not affect calculations that have a precision of three significant figures.

MAKING CONNECTIONS

Mapping with *g*

The acceleration due to gravity, g, varies slightly from place to place due to different subsurface rock densities. Scientists use precise maps of the acceleration due to gravity to study geologic features. Figure 11-7 shows a gravity map of the region around the Chicxulub crater on the Yucatán peninsula in Mexico. This crater was created by the impact of an asteroid or a comet 65 million years ago, an event that is linked to the mass extinction of the dinosaurs and many other species living at that time. Dr. Alan Hildebrand of the University of Calgary discovered the crater, which is not visible on the surface.

Image courtesy of V. L. Sharpton

Figure 11-7 A gravity map of part of the Yucatán peninsula in Mexico. The curving white line shows the boundary between land and sea. The circular pattern in the measurements of *g* reveals the buried remnants of the Chicxulub crater.

LO 3

11-3 Orbits and Weightlessness

It is easy to see that an object in deep space, far from any other masses, is almost totally weightless. The gravitational forces acting on it are too weak to measure. Figure 11-8 shows the International Space Station (ISS). Inside the ISS objects appear to be weightless. Is the gravitational force actually zero inside the ISS? Example 11-4 answers this question.

Example 11-4 demonstrates that the force of gravity can provide the centripetal acceleration necessary to keep an object in a circular orbit. Figure 11-9 shows the

STS-114 Crew, NASA

Figure 11-8 The ISS is an orbiting laboratory built by a consortium of many countries, including Canada.

EXAMPLE 11-4

Gravity at the ISS

(a) Calculate the gravitational force exerted by Earth on a 1.00 kg object at the ISS. Assume that the ISS is in a circular orbit at an altitude of 353 km above Earth's surface.

(b) Calculate the value of g at the ISS.

(c) Calculate the centripetal acceleration required to keep an object in the circular orbit of the ISS, given that the orbital speed of the ISS is 27 740 km/h.

SOLUTION

(a) We can apply the law of universal gravitation with the mass of the object (1.00 kg) for m and the mass of Earth (5.974×10^{24} kg) for M. The distance r is the sum of the radius of Earth (we will use a mean value of 6371 km) and the altitude of the ISS:

$$r = 6371 \text{ km} + 353 \text{ km} = 6724 \text{ km} = 6.724 \times 10^6 \text{ m}$$

$$F_{\text{G}} = \frac{GMm}{r^2}$$

$$= \frac{(6.672 \times 10^{-11} \text{ N} \cdot \text{m}^2/\text{kg}^2)(5.974 \times 10^{24} \text{ kg})(1.00 \text{ kg})}{(6.724 \times 10^6 \text{ m})^2}$$

$$= 8.82 \text{ N}$$

(b) Using Equation (11-3),

$$g = \frac{GM}{r^2} = \frac{(6.672 \times 10^{-11} \text{ N} \cdot \text{m}^2/\text{kg}^2)(5.974 \times 10^{24} \text{ kg})}{(6.724 \times 10^6 \text{ m})^2}$$

$$= 8.816 \text{ m/s}^2$$

(c) We convert the speed to units of metres per second, and then apply the formula for centripetal acceleration:

$$v = 27\,740 \, \frac{\text{km}}{\text{h}} \times \frac{1000 \text{ m}}{1 \text{ km}} \times \frac{1 \text{ h}}{3600 \text{ s}} = 7.7056 \times 10^3 \text{ m/s}$$

$$a_{\text{c}} = \frac{v^2}{r} = \frac{(7.7056 \times 10^3 \text{ m/s})^2}{6.724 \times 10^6 \text{ m}} = 8.830 \text{ m/s}^2$$

Making sense of the result:

The gravitational force on the 1.00 kg object is approximately 10% less than the 9.81 N force that the object would experience on Earth's surface. The gravitational force at the ISS is certainly not zero. The centripetal acceleration is (within the precision of data) equal to the gravitational acceleration. Thus, the gravitational force provides just enough acceleration to keep the object in a circular orbit with the ISS. As a result, the object appears weightless in the ISS frame of reference.

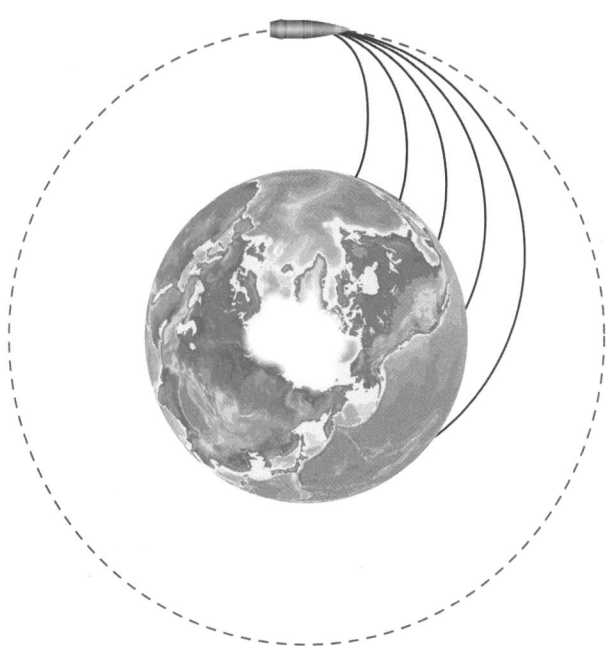

Figure 11-9 Projectiles with higher tangential speeds travel a greater distance before reaching Earth's surface. At just the right speed, the path of the projectile will curve such that the projectile remains at a constant distance from Earth.

trajectories of objects with various initial velocities. The greater the initial tangential velocity, the farther the object travels before it falls to Earth. However, when the object has just the right speed, it will fall toward Earth at the same rate that Earth's surface curves away, and the object will circle Earth at a constant altitude. So, an object in a circular orbit is falling continuously and appears weightless, similar to the way that you would feel weightless if you were in free fall. It is important to realize that it is an apparent weightlessness, and there is still a weight force.

✓ CHECKPOINT

C-11-3 Weightless in Orbit

Which of the following, if any, have a zero value for an object in a circular orbit?

(a) the gravitational force on the object

(b) the weight of the object

(c) the net force on the object

(d) the difference between the gravitational force on the object and its mass times its acceleration

C-11-3 (d)

MAKING CONNECTIONS

Achieving Weightlessness

Aircraft like the one in Figure 11-10 produce weightless conditions more readily and at a lower cost than in an orbiting spacecraft, but only for brief periods, typically 15 to 20 s. The aircraft accelerates steeply upward, as shown in Figure 11-11, then the engines are idled and the aircraft follows a parabolic path like a projectile. During the parabolic portion of the flight path, passengers in the aircraft experience weightlessness, although usually not quite complete weightlessness. These flights are called parabolic flights, or sometimes suborbital zero gravity flights. The European Space Agency (ESA), NASA, the Canadian Space Agency (CSA), and others use these flights to achieve weightless conditions for short periods.

Figure 11-10   The NASA parabolic flight aircraft during ascent

Figure 11-11   This flight path produces a brief period of weightlessness. The acceleration is expressed in units of g (equal to 9.81 m/s^2).

LO 4

11-4   Gravitational Potential Energy

Because it takes work to move objects in gravitational fields there must be **gravitational potential energy**, which is the work required to move from some zero point to a given configuration of masses. In Chapter 6, we derived the relationship $U = mgh$ for the gravitational potential energy (U) of an object with mass m at a height h above the surface of a much larger mass that causes a gravitational acceleration of g. This derivation assumed that g is constant over the distance h, and set $U = 0$ at the surface of the larger mass. It used the concept that the amount of potential energy increase between two points is equal to the work done in moving between the points. That work is found by integrating the area under the force versus displacement graph.

You can integrate the force of universal gravitation to obtain an expression for the gravitational potential energy difference. The gravitational force is infinite for the case

of zero distance, so we start the integration at an infinite distance (where the gravitational potential energy is defined as zero). That integration (which you complete in problem 60) produces the following more general definition of gravitational potential energy that does not assume that g is constant.

KEY EQUATION
$$U = -\frac{GMm}{r} \qquad (11\text{-}4)$$

As in earlier equations, M and m are the masses of the two objects, and r is the distance between their centres of mass. With this definition, the gravitational potential energy of the mass m has its greatest possible value—zero—when the masses are infinitely far apart and becomes increasing negative as r decreases. This choice for the zero point may seem strange, but reflection will show that it is a very reasonable definition for a universal zero point.

Although the two definitions of gravitational potential energy have different zero points, the two techniques should largely agree for calculations of changes in potential energy that occur in a region where g is approximately constant. Example 11-5 compares the results of one such calculation.

PEER TO PEER

At first, I found the idea of gravitational potential energy always being negative strange. It just seemed to me that an energy should be positive. Then I began thinking of it as the amount of energy needed to get away (kind of like being in debt by that amount). Really that is how it works. If at a certain position the gravitational potential energy is -15 J then you will need $+15$ J of kinetic energy to just barely escape. It is important to remember that potential energies in physics are always relative to some defined zero level.

✓ CHECKPOINT

C-11-4 Gravitational Potential Energy

(a) When you increase the distance between two masses, does the gravitational potential energy increase or decrease?

(b) Do you need to do work to increase the distance between two masses?

C-11-4 (a) When you increase the distance between two masses, the gravitational potential energy will *increase*. The increased distance results in a negative energy with a smaller magnitude, so the energy increases. Since it takes work to move the objects farther apart, the potential energy must increase. (b) Yes, you need to do work to increase the distance between two masses. The force required to separate the masses has the same direction as the increase in displacement, so the work is positive.

EXAMPLE 11-5

Calculating Changes in Gravitational Potential Energy

Calculate the work required to lift a 1.00 kg object from sea level to the top of Mount Everest, which has an elevation of 8848 m. Assume that Earth is spherical with a radius of 6371 km. Perform the calculation in the following two ways.

(a) Use the *mgh* expression for potential energy, and assume that *g* remains constant at its value at sea level, 9.81 m/s².

(b) Use Equation (11-4).

SOLUTION

(a) Since $U = 0$ at sea level,

$$\Delta U = mgh - 0 = 1.00 \text{ kg} \times 9.81 \text{ m/s}^2 \times 8848 \text{ m} = 86.8 \text{ MJ}$$

The gravitational potential energy of the object is greater at the top of the mountain. Therefore, it takes 86.8 MJ of work to lift the object to the top of Mount Everest.

(b) We find the difference between the gravitational potential energy at the top of the mountain and at sea level. The distance from the centre of Earth to the top of Mount Everest is

$$r_m = 6.371 \times 10^6 \text{ m} + 8848 \text{ m}$$

$$r_m = 6.3798 \times 10^6 \text{ m}$$

Using r_s to represent the distance from the centre of Earth to the surface at sea level, we can express the difference in gravitational potential energy as

$$\Delta U = -\frac{GMm}{r_m} - \left(-\frac{GMm}{r_s}\right) = GMm\left(\frac{1}{r_s} - \frac{1}{r_m}\right)$$

Substituting in the known values gives

$$\Delta U = (6.672 \times 10^{-11} \text{ N} \cdot \text{m}^2/\text{kg}^2)(5.974 \times 10^{24} \text{ kg} \times 1.00 \text{ kg})$$
$$\times \left(\frac{1}{6.3710 \times 10^6 \text{ m}} - \frac{1}{6.3798 \times 10^6 \text{ m}}\right)$$

$$= 86.8 \text{ MJ}$$

Therefore, the work required to lift the 1.00 kg mass to the mountaintop is 86.8 MJ.

Making sense of the result:

As expected, the two ways of calculating the work required give similar results. Since *g* decreases with height, the assumption that *g* remains equal to its value at sea level gives a result that slightly overestimates the work required.

EXAMPLE 11-6

Meteor Speeds

Estimate the speed that a meteoroid attains just before reaching Earth's surface and becoming a meteor (see Figure 11-12). Use the following simplifying assumptions in obtaining the estimate:

■ The meteoroid started from rest at such a great distance from Earth that the two can be considered to have been infinitely far apart.

■ The influence of all other objects (e.g., the Sun and the Moon) is negligible.

■ Earth's atmosphere does not significantly slow the meteoroid.

While the last assumption in particular may seem unreasonable, in fact all three are valid to the first order (see the Solution for more details).

(continued)

Courtesy R.L. Hawkes

Figure 11-12 A meteor is the bright trail or streak caused by a meteorite entering Earth's atmosphere at high speed.

SOLUTION

When the meteoroid is infinitely far away, its gravitational potential energy is 0.

At Earth's surface, the meteoroid's gravitational potential energy is $-\dfrac{GMm}{r}$, where r is the radius of Earth, M is the mass of Earth, and m is the unknown mass of the meteoroid. Since we are assuming that there is no loss to aerodynamic drag, the amount of gravitational potential energy converted to kinetic energy as the meteorite falls to Earth is $\dfrac{GMm}{r}$. Using v_s to represent the speed of the meteorite at Earth's surface, we have

$$\frac{1}{2}mv_s^2 = \frac{GMm}{r}$$

$$v_s = \sqrt{\frac{2GM}{r}} = \sqrt{\frac{2(6.672 \times 10^{-11}\,\text{N}\cdot\text{m}^2/\text{kg}^2)(5.974 \times 10^{24}\,\text{kg})}{6.3710 \times 10^6\,\text{m}}}$$

$$= 1.119 \times 10^4\ \text{m/s}$$

Making sense of the result:

A meteoroid that fell without air resistance to Earth's surface would accelerate due to the Earth's gravity to a speed of approximately 11.2 km/s. In fact, this approximation is within a few percent of the observed minimum speed of meteors in the upper atmosphere; most meteors have higher speeds. Astronomers who measure properties of meteors do need to provide a small correction for the aerodynamic drag in the high atmosphere.

LO 5

11-5 Force from Potential Energy

When a potential energy function is defined, we can determine force from potential energy using derivatives, as we will show in this section. Alternatively, we can determine a potential energy function using integration when we know the force everywhere. The concept that derivatives of potential energy functions represent the associated forces is applied in many branches of physics, for example, electromagnetism.

When a potential energy function, U, exists, we can obtain the force components through the following relationships:

$$F_x = -\frac{\partial U}{\partial x} \tag{11-5}$$

$$F_y = -\frac{\partial U}{\partial y} \tag{11-6}$$

$$F_z = -\frac{\partial U}{\partial z} \tag{11-7}$$

Note that these are partial derivatives, which means that the differentiation is only done with respect to an *explicit* appearance of the variable. This means that if, for example, U does not explicitly depend on y, then the partial derivative with respect to y will be zero and there will be no force in the y-direction. If you have only one-dimensional situations, you can think of the partial derivatives just like ordinary derivatives. We show in Example 11-7 how to obtain the gravitational force near Earth's surface from these relationships.

✓ CHECKPOINT

C-11-5 Force from Potential Energy

What are the direction and magnitude of the force when the potential energy function is $U = 3z^2$?

C-11-5 The partial derivatives in the x- and y-directions are 0, so there are no force components in those directions. The z-component is

$$F_z = -\frac{\partial U}{\partial z} = -6z$$

Therefore, when z is positive, the force is in the negative z-direction, and when z is negative, the force is in the positive z-direction.

We can extend the relationship between force components and derivatives to situations in which a potential energy function, $U(r)$, depends only on radial distance. In the following equation, $\hat{r}$ is a unit vector in the direction of increasing r:

$$\vec{F} = -\frac{\partial U(r)}{\partial r}\,\hat{r} \tag{11-8}$$

EXAMPLE 11-7

Gravitational Force Near Earth's Surface

Find the gravitational force near Earth's surface where the potential energy function is given by $U = mgy$.

SOLUTION

The y-component of the force is

$$F_y = -\frac{\partial U}{\partial y} = -mg$$

Since U does not depend on x or z, the partial derivatives yield a 0 result for these force components:

$$F_x = 0 \qquad F_z = 0$$

Making sense of the result:

As expected, we found that the direction of the gravitational force is downward (negative y-direction) and the magnitude of the force is simply the weight of the object at the surface, mg.

We demonstrate in Example 11-8 how the gravitational force can be obtained from the expression for gravitational potential energy.

EXAMPLE 11-8

Deriving Gravitational Force from Gravitational Potential Energy

Determine the gravitational force, given that the function for gravitational potential energy can be written as $U(r) = -\dfrac{GMm}{r}$.

SOLUTION

Substituting $U(r)$ into Equation (11-8), we get

$$\vec{F} = -\frac{\partial U(r)}{\partial r}\hat{r} = -\frac{\partial\left(-\dfrac{GMm}{r}\right)}{\partial r}\hat{r}$$

$$= GMm\frac{\partial\left(\dfrac{1}{r}\right)}{\partial r}\hat{r} = -\frac{GMm}{r^2}\hat{r}$$

Making sense of the result:

The derivative gives the magnitude of the force introduced at the beginning of the chapter for the force of universal gravitation. The force on mass m points in the $-\hat{r}$ direction, back toward the mass M because $\hat{r}$ points from M to m.

11-6 Escape Speed

In this section, we will consider the **escape speed**, which is the speed needed for an object to escape the gravitational pull of another object. What do we mean by "escape"? If you throw an object straight up, it will rise for a time, slowing as its kinetic energy decreases and its gravitational potential energy increases. The object will stop momentarily and then reverse direction. If we give the object a somewhat greater initial speed, it will go higher but will still eventually slow to a stop and then return. However, gravitational escape can occur when the object has sufficient initial speed that it does not ever stop completely, and instead travels infinitely far away.

For an object to escape, the total orbital energy must be positive: the object must have enough positive kinetic energy from its orbital motion to overcome its negative gravitational potential energy. Then the object will still have some kinetic energy when it reaches a point infinitely far away, where its gravitational potential energy is zero.

We call the orbit or trajectory "bound" when the total orbital energy is negative and "unbound" when the total is positive. To determine the minimum speed for escape, we set the total orbital energy equal to zero:

$$E_{\text{total}} = T + U = \frac{1}{2}mv_{\text{esc}}^2 - \frac{GMm}{r} = 0 \qquad (11\text{-}9)$$

Solving for the escape speed, we derive the relationship given in Equation (11-10) below, where r is the initial distance of the escaping object from the centre of the larger object, and M is the mass of the larger object. Note that the mass of the escaping object, m, cancels out and does not affect the escape speed.

KEY EQUATION

$$v_{\text{esc}} = \sqrt{\frac{2GM}{r}} \qquad (11\text{-}10)$$

✓ CHECKPOINT

C-11-6 Total Orbital Energy

Is the sum of the kinetic energy and the gravitational potential energy of Earth in its orbit about the Sun positive, zero, or negative?

C-11-6 The total energy is negative. We can show this starting with $F = ma$ and substituting the centripetal acceleration for a and the gravitational force for F, then compare the gravitational potential energy and the kinetic energy. We see in the next section that a circular orbit (or an elliptical orbit) has a negative total orbital energy.

EXAMPLE 11-9

Escape from Earth

Find the speed required to escape from Earth's surface. Neglect any work needed to overcome aerodynamic drag in the atmosphere.

SOLUTION

We use Equation (11-10), substituting the mass of Earth for M and the radius of Earth for r:

$$v_{esc} = \sqrt{\frac{2GM}{r}}$$

$$= \sqrt{\frac{2(6.672 \times 10^{-11}\,\text{N} \cdot \text{m}^2/\text{kg}^2)\,(5.974 \times 10^{24}\,\text{kg})}{6.3710 \times 10^6\,\text{m}}}$$

$$= 1.119 \times 10^4\,\text{m/s}$$

Making sense of the result:

The speed needed to escape from Earth's surface is about 11.2 km/s. Note that this escape speed matches the speed for the falling meteoroid in Example 11-8. This makes sense because the potential energy that an object loses while falling from an infinite distance is equal to the energy needed to take the object to an infinite distance from the surface. Since the aerodynamic drag on an object passing through Earth's atmosphere is not negligible, the speed required to escape from Earth's surface is actually somewhat greater than 11.2 km/s.

Looking at Equation (11-10), we can see that the escape speed for a planet varies with $\sqrt{\frac{M}{r}}$, the mass of the planet divided by its radius. The volume of a sphere increases with the cube of its radius, so a large planet will have a greater escape speed than a smaller planet with the same density. In Example 11-10, we calculate the escape speed of a mid-sized asteroid.

While the calculations in this section have dealt with escape from the surface of an object, the same techniques can be applied for cases of escape starting from an altitude above the surface of an object.

LO 7

11-7 Kepler's Laws

Historically, the orbits of celestial objects were first understood empirically and only later explained in terms of Newtonian mechanics. The German mathematician and astronomer Johannes Kepler (1572–1630) deduced

 EXAMPLE 11-10

Escape from an Asteroid

What speed is needed to escape from the surface of a spherical, homogeneous asteroid with a radius of 5.00 km and a density of 3200 kg/m³? The asteroid has no atmosphere.

SOLUTION

First, find the mass of the asteroid:

$$M = \rho V = \rho \frac{4\pi r^3}{3}$$

$$= 3200\,\text{kg/m}^3\,\frac{4\pi(5.00 \times 10^3\,\text{m})^3}{3} = 1.68 \times 10^{15}\,\text{kg}$$

Now, substitute values for M and r in Equation (11-6):

$$v_{esc} = \sqrt{\frac{2GM}{r}}$$

$$= \sqrt{\frac{2(6.672 \times 10^{-11}\,\text{N} \cdot \text{m}^2/\text{kg}^2)\,(1.68 \times 10^{15}\,\text{kg})}{5.00 \times 10^3\,\text{m}}}$$

$$= 6.69\,\text{m/s}$$

Making sense of the result:

As expected, the escape speed from the asteroid is very small compared to the escape speed from Earth.

three laws of motion that describe the orbital motion of objects with negative total orbital energy. Although Kepler devised his laws to describe the orbits of planets in the solar system, these laws apply to any object in a bound orbit.

 ONLINE ACTIVITY

Exploring Orbital Motion

The e-resource that accompanies every new copy of this textbook contains an Online Simulation Activity using the PhET simulation "My Solar System." Work through the simulation and accompanying questions to gain an understanding of planetary orbits.

Kepler's Laws

1. Each planet moves in an elliptical orbit with the Sun at one of the two foci of the ellipse (Figure 11-13).

2. A line drawn from the Sun to a planet will sweep out the area inside the elliptical orbit at a constant rate (Figure 11-14).

3. The square of the period of the orbit is proportional to the cube of the semi-major axis of the orbit. The semi-major axis (a) is one-half of the longest dimension of the elliptical orbit.

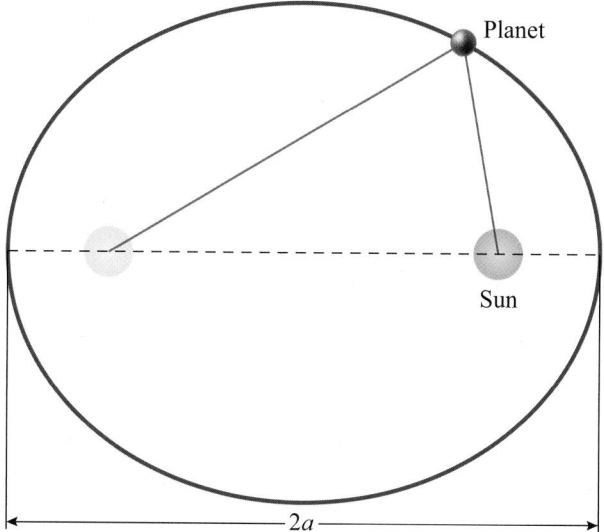

Figure 11-13 Kepler's first law: Planets move in elliptical orbits with the Sun at one focus.

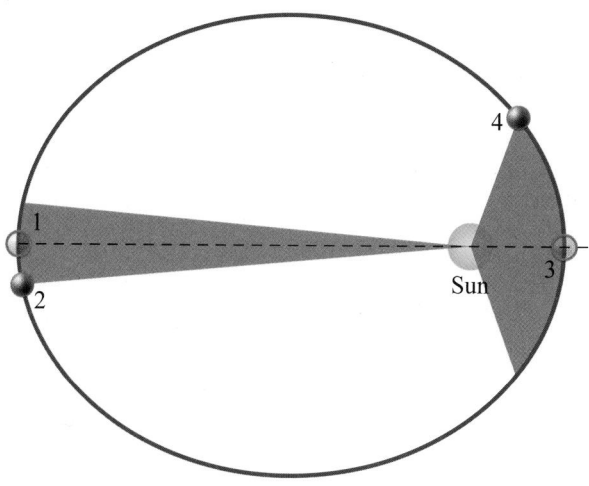

Figure 11-14 Kepler's second law: A line from the Sun to a planet will sweep equal areas in equal time periods. The areas shown in red are equal; therefore, the planet moves from point 1 to point 2 and from point 3 to point 4 in the same length of time.

The motion is confined to a single orbital plane because of the central nature (along the line joining the centres) of the gravitational force. This means that torque out of that plane is not possible. Kepler's second law is a consequence of the law of conservation of angular momentum. In order for angular momentum to remain constant, the planet must move faster when it is closer to the Sun. Conservation of energy also indicates that the kinetic energy must be greater when the planet is closer to the Sun and has less gravitational potential energy. In fact, the speed of an orbiting object a distance r from the Sun (or other relatively large mass) is given by the following relationship, where a is the semi-major axis of the ellipse:

$$v = \sqrt{GM\left(\frac{2}{r} - \frac{1}{a}\right)} \qquad (11\text{-}11)$$

If we take the case of a circular orbit where $r = a$, this can readily be shown to be equivalent to that obtained by using $F = ma$ with the centripetal acceleration for a.

Newton's law of universal gravitation gave astronomers an expression for the constant of proportionality in Kepler's third law. This constant of proportionality depends on the sum of the masses of the two objects (M and m), as shown in Equation (11-12) below.

KEY EQUATION $\qquad T^2 = \dfrac{4\pi^2}{G(M+m)}\,a^3 \qquad (11\text{-}12)$

When one of the masses is much less than the other, we can replace the sum of the masses with the mass of the larger object:

$$T^2 = \frac{4\pi^2}{GM}\,a^3 \quad \text{where } M \gg m \qquad (11\text{-}13)$$

The Sun is much more massive than other objects in the solar system (e.g., the mass of Earth is about one-millionth the mass of the Sun), so this approximation gives accurate results for orbits around the Sun. Example 11-11 shows the derivation of Equation (11-13) for a circular orbit.

EXAMPLE 11-11

Kepler's Third Law

Derive Kepler's third law for a planet moving in a circular orbit around the Sun. Assume that the mass of the planet is negligible compared to the mass of the Sun.

SOLUTION

The gravitational force provides the centripetal acceleration to maintain the circular orbit. Since both the force and the acceleration are in the same direction,

$$F_{\mathrm{G}} = \frac{GMm}{r^2} = \frac{mv^2}{r}$$

Simplifying gives

$$\frac{GM}{r} = v^2$$

The period is the time for one orbit, and the circumference of a circle is $2\pi r$, so

$$v = \frac{2\pi r}{T}$$

Substituting for v in the above simplified equation gives

$$\frac{GM}{r} = \frac{4\pi^2 r^2}{T^2}$$

(continued)

$$T^2 = \frac{4\pi^2 r^3}{GM}$$

For a circular orbit, the radius r is the semi-major axis a, and

$$T^2 = \frac{4\pi^2}{GM} a^3$$

Making sense of the result:

We have proved that the square of the period of a circular orbit is proportional to the cube of the semi-major axis, as stated in Kepler's third law.

Kepler's third law is a universal relationship and can be applied to any two masses in a bound orbit, including asteroids, comets, and meteoroids in elliptical orbits around the Sun. Kepler's laws also apply to the orbit of a satellite around a planet, and we can use measurements of the satellite's motion to determine the mass of the planet (see Example 11-12).

Any orbit can be completely specified with six parameters. The most important of these are the semi-major

EXAMPLE 11-12

Mass of Saturn

Titan, the largest natural satellite of Saturn, takes 15.945 Earth days for one complete orbit. The semi-major axis of Titan's orbit around Saturn is 1.22×10^6 km. Find the mass of Saturn. Assume that the mass of Titan is much less than the mass of Saturn.

SOLUTION

First, convert the period into SI units of seconds:

$$T = 15.945 \text{ days} \times \frac{24 \text{ h}}{1 \text{ day}} \times \frac{60 \text{ m}}{1 \text{ h}} \times \frac{60 \text{ s}}{1 \text{ m}}$$

$$= 1.3776 \times 10^6 \text{ s}$$

When one mass is considerably smaller than the other, Kepler's third law becomes

$$T^2 = \frac{4\pi^2}{GM} a^3$$

Rearranging to solve for M and substituting the given data, we have

$$M = \frac{4\pi^2}{GT^2} a^3 = \frac{4\pi^2 (1.22 \times 10^9 \text{ m})^3}{(6.672 \times 10^{-11} \text{ N} \cdot \text{m}^2/\text{kg}^2)(1.3776 \times 10^6 \text{ s})^2}$$

$$= 5.66 \times 10^{26} \text{ kg}$$

Making sense of the result:

This result agrees with published values for the mass of Saturn. In fact, the published values were obtained using Kepler's third law with measurements of various natural and artificial satellites.

✓ CHECKPOINT

C-11-7 Exoplanet Orbits

An exoplanet around some star is at radius R and has a circular orbit of period T. The same star has a second planet in a circular orbit of radius $4R$. What is the period of the planet?

(a) $\dfrac{T}{8}$

(b) $\dfrac{T}{4}$

(c) $4T$

(d) $8T$

$$\frac{T_2^2}{T^2} = \frac{(4R)^3}{R^3}$$

which yields $8T$ for the unknown period:

C-11-7 (d) From Kepler's third law, the ratio of the square of the period over the cube of the semi-major axis (equal to radius for a circular orbit) is constant, so we have the following relationship,

axis, the eccentricity, and the inclination, represented by the symbols a, e, and i, respectively. The **semi-major axis** specifies the size of the orbit, and the eccentricity specifies the shape. If $e = 0$, the orbit is circular; for $0 < e < 1$, the orbit is elliptical; if $e = 1$, the orbit is parabolic. The closer the eccentricity is to 1, the more elongated the shape of the orbit.

The inclination is the angle of the orbit relative to some reference plane. For motion around the Sun, the reference plane is the plane of Earth's orbit, which is called the ecliptic.

The **perihelion**, or closest point to the Sun, of an orbit and the **aphelion**, or most distant point from the Sun, are readily calculated from the following relationships:

Perihelion: $\quad r_{\text{per}} = a(1 - e) \quad$ (11-14)

Aphelion: $\quad r_{\text{aph}} = a(1 + e) \quad$ (11-15)

If $e = 0$, the orbit is circular, and the perihelion and aphelion are equal.

We frequently express solar system distances in terms of the astronomical unit (AU). One astronomical unit is defined as the mean distance between the Sun and Earth, which is approximately 1.50×10^{11} m.

11-8 Types of Orbits

The total orbital energy (kinetic plus gravitational potential) is used to classify the type of orbit. Gravitational orbits are classified as one of the types of conic sections: ellipse, parabola, or hyperbola, with the circle being a special case of the ellipse. As the name implies, a **conic section** is a planar geometric shape corresponding to one of the ways that a cone can be cut (Figure 11-15). Cutting parallel to the base of a cone produces a circle. Cutting at an angle less than the angle of a cone's edge produces an ellipse. Cutting parallel to a cone's edge produces a parabola, and cutting at an even greater angle produces a hyperbola. The shape of each conic section can be defined by a relatively simple equation.

Figure 11-15 A cone can be cut at different angles to produce four types of conic sections. These sections correspond to the possible types of orbits.

EXAMPLE 11-13

Interstellar Visitor

A spacecraft is observed to be at a distance of 2.00 AU from the Sun and is and moving at a speed of 38.0 km/s.

(a) What type of orbit (relative to the gravitational field of the Sun) is the spacecraft following?

(b) Could the spacecraft have come from interstellar space?

ONLINE ACTIVITY

Types of Orbits

The e-resource that accompanies every new copy of this textbook contains an Online Simulation Activity using the PhET simulation "Types of Orbits." Work through the simulation and accompanying questions to gain an understanding of different types of orbits.

The type of gravitational orbit is determined by the total orbital energy:

- When the total orbital energy is negative, the orbit is an ellipse.

- When the total orbital energy is zero, the orbit is a parabola.

- When the total orbital energy is positive, the orbit is a hyperbola.

A **hyperbolic orbit** is an unbound orbit, and an **elliptical orbit** is a bound orbit; a **parabolic orbit** is on the borderline between bound and unbound orbits. When an object is in a hyperbolic or parabolic orbit with respect to the Sun, the object will sweep past the Sun once and never return. Later chapters will describe how electromagnetic and nuclear forces can create similar types of motion for subatomic particles. In Example 11-13 we demonstrate how to determine the type of orbit.

✓ CHECKPOINT

C-11-8 Total Orbital Energy

Classify the orbit of each of the following:

(a) A meteoroid approaches Earth with a geocentric (Earth-relative) speed of 25 km/s.

(b) The altitude of a satellite orbiting Earth's varies from 1000 km to 3000 km.

C-11-8 (a) The meteoroid is approaching Earth with a speed higher than the escape velocity for Earth, so it is clear that the orbit must be hyperbolic (relative to the Earth). (b) The satellite orbit is elliptical from the description given.

SOLUTION

(a) To determine the type of orbit, we need to know if the total orbital energy of the spacecraft is negative, zero, or positive. Although the mass of the spacecraft is unknown, we can calculate the total orbital energy per unit mass:

$$\frac{E_{\text{total}}}{m} = \frac{T}{m} + \frac{U}{m} = \frac{1}{2}v^2 - \frac{GM}{r}$$

(continued)

where M is the mass of the Sun, $v = 3.80 \times 10^4$ m/s², and $r = 2.00 \times 1.496 \times 10^{11}$ m $= 2.99 \times 10^{11}$ m. Substituting into the above equation gives

$$\frac{E_{total}}{m} = \frac{1}{2}(3.8 \times 10^4 \text{ m/s})^2$$

$$- \frac{(6.672 \times 10^{-11} \text{ N} \cdot \text{m}^2/\text{kg}^2)(1.989 \times 10^{30} \text{ kg})}{2.992 \times 10^{11} \text{ m}}$$

$$= +2.78 \times 10^8 \text{ J/kg}$$

The total orbital energy is positive, so the object must be in a hyperbolic orbit.

(b) Since the orbit is hyperbolic, the spacecraft would have come from interstellar space unless its engines (or booster rocket) provided enough kinetic energy to move it into its current orbit.

Making sense of the result:

Earth moves at a speed of approximately 30 km/s in its near-circular orbit. Since the spacecraft is moving faster at twice the distance from the Sun, it is not unreasonable that the orbit is hyperbolic. Note that we did not need to know the direction that the spacecraft was moving to determine the type of orbit.

 MAKING CONNECTIONS

Space Debris and Meteoroids

Both meteoroids and space debris pose small, but not insignificant, threats to space operations. The impact of even a tiny object can be disastrous for a satellite. Space debris impacts other objects in orbit with a lower relative speed than meteoroids, but the probability of impact is greater for space debris. The term "space debris" refers to artificial material that has ended up in Earth orbit as a result of our use of space. Such debris varies from specks of paint to much larger pieces of spacecraft. In 2008, an astronaut on a space walk accidentally released a tool kit that ended up in Earth orbit. NASA and other space agencies track thousands of pieces of space debris larger than a centimetre across, but the majority of space debris is much smaller. Space debris eventually spirals down toward Earth. Even at orbital altitudes, there is a tenuous atmosphere that produces a tiny drag force that ultimately results in the debris falling from orbit. Relative to Earth, space debris has negative total orbital energy, and meteoroids have positive total orbital energy.

 MAKING CONNECTIONS

Finding Debris from Other Planetary Systems

Planets form along with their parent stars from a cloud of gas and dust called a nebula. The formation process likely also produces large numbers of small meteoroids, some of which escape the gravitational pull of the star (through collisions, near collisions, and radiation processes). We can detect escaped meteoroids that reach our solar system by finding meteoroids not gravitationally bound to the Sun. Canada is a world leader in the study of meteoroids, including the search for debris from other planetary systems. Researchers Peter Brown, Margaret Campbell-Brown, and Rob Weryk of the University of Western Ontario use the Canadian Meteor Orbit Radar (CMOR) and an automated super-sensitive digital camera system to compute meteor orbits and to isolate those on hyperbolic orbits. Figure 11-16 shows a map of meteor origins from a single day of observations with this facility. Each year, close to one million meteoroids are detected by this facility. A tiny fraction of these are not in orbits within our solar system, but are material ejected from another planetary system.

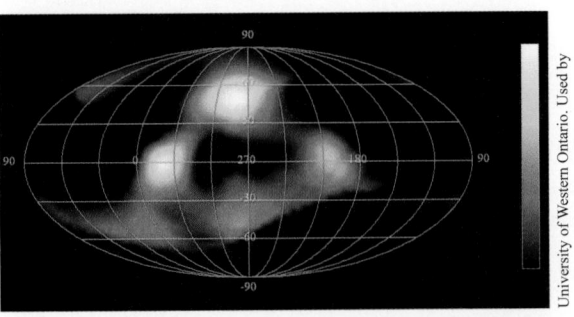

Figure 11-16 A plot of meteor origin directions from the CMOR meteor radar: the brighter colour indicates more meteors from that direction.

LO 9

11-9 Detection of Exoplanets

The distinguished Russian American astronomer Otto Struve (1897–1963) made numerous advances, one of which is his technique for detecting exoplanets; **exoplanets** are planets that orbit stars other than the Sun. Twenty-five years after his death, a Canadian research team used his technique to provide the first evidence that exoplanets do indeed exist. There are now more than 500 known exoplanets, and more are being discovered every year, many of them by NASA's Kepler space telescope.

While at least seven techniques have been used to detect exoplanets, the most successful so far has been Doppler spectroscopy (also called radial velocity spectroscopy). When a planet and a star are in orbit, we often consider the position of the star to be fixed because the star is so much more massive than the planet. However, it is the centre of mass of the system

ONLINE ACTIVITY

Star and Planet Orbital Motion

The e-resource that accompanies every new copy of this textbook contains an Online Activity using the PhET simulation "My Solar System." Work through the simulation and accompanying questions to gain an understanding of the orbital motion of stars and planets.

MAKING CONNECTIONS

The Discovery of Exoplanets

In 1988, Canadian scientists published a paper in the prestigious *Astrophysical Journal*, which presented clear evidence that seven of the stars they studied had objects in orbit with masses that could be planets. Led by Bruce Campbell, then of the University of Victoria and the Herzberg Institute of Astrophysics, the discovery team included Gordon Walker of the University of British Columbia and Stephenson Yang, who was then working at the Canada-France-Hawaii Telescope (CFHT) but is now at the University of Victoria. They found very small but regular shifts in the wavelengths of light from the stars. These shifts showed that the stars were in orbital motion, most likely with unseen planets.

In 1992, Dale Frail, a graduate of Acadia University and the University of Toronto working at the National Radio Astronomy Observatory (NRAO) in New Mexico, conclusively detected two planets in orbit around a pulsar (a rotating neutron star). Some astronomers consider 1992 to be the date of the first detection of an exoplanet because Frail made a more definitive claim of planet detection.

that is fixed, and both the star and the planet move in elliptical orbits about this centre. The star's orbit is very small compared to the planet's orbit. In Example 11-14, we examine the orbital motion of the Sun caused by its interaction with Jupiter.

We cannot directly see a planet orbiting a distant star, but we can infer that the planet is there from the slight

EXAMPLE 11-14

The Sun-Jupiter System

The mass of the Sun, M_S, is 1.99×10^{30} kg; the mass of Jupiter, M_J, is 1.90×10^{27} kg; and the mean distance between them, centre to centre, is 7.80×10^{11} m.

(a) Find the location of the centre of mass of the Sun–Jupiter system.
(b) Find the orbital velocity of the Sun about this centre of mass.

SOLUTION

We use x to represent the distance from the centre of the Sun to the centre of mass of the Sun–Jupiter system, and d to represent the distance from the centre of the Sun to the centre of Jupiter, as shown in Figure 11-17.

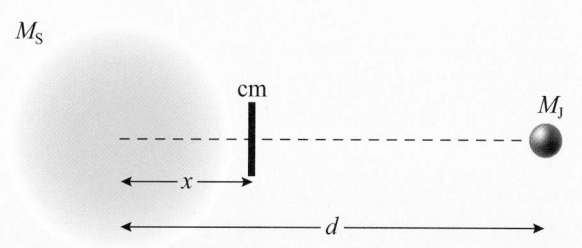

Figure 11-17 Example 11-14: The Sun–Jupiter system

(continued)

NEL

CHAPTER 11 | **GRAVITATION** 299

(a) For a system of just two masses, the product of one mass times its distance from the centre of mass must equal the corresponding product for the other mass. Therefore,

$$xM_S = (d - x)M_J$$

and

$$x = \frac{dM_J}{M_S + M_J} = \frac{(7.80 \times 10^{11}\ \text{m})(1.90 \times 10^{27}\ \text{kg})}{1.99 \times 10^{30}\ \text{kg} + 1.90 \times 10^{27}\ \text{kg}}$$

$$= 7.44 \times 10^8\ \text{m}$$

(b) The Sun makes a complete circuit of its orbit in one orbital period. The period of the Sun and Jupiter are the same because the two masses are in a joint orbit about a common centre. This period can be determined from Kepler's third law, or we can simply use the observed value, which is 11.86 y.

$$T = 11.86\ \text{yr} \times \frac{365.25 \times 24 \times 60 \times 60\ \text{s}}{1\ \text{yr}} = 3.74 \times 10^8\ \text{s}$$

Assuming that the Sun's orbit is circular with radius x, we have

$$v = \frac{2\pi x}{T} = \frac{2\pi(7.44 \times 10^8\ \text{m})}{3.74 \times 10^8\ \text{s}} = 12.5\ \text{m/s}$$

Making sense of the result:

(a) The radius of the Sun is 6.96×10^8 m, so the centre of mass is close to the surface of the Sun. This location is reasonable because the mass of the Sun is so much larger than the mass of Jupiter. (b) The orbital speed of the Sun is just 12.5 m/s. The speed of light is about 30 million times greater, so the shift caused by the presence of Jupiter is correspondingly small.

motion of the star–as the star orbits the centre of mass of the system, the star periodically moves toward us and away from us (Figure 11-18).

The light from a star moving away from us shifts slightly to a lower frequency (and a correspondingly longer wavelength); we call this effect redshift. (This is similar to a sound having a lower frequency when the source is moving away from you.) When the star moves toward us, its light shifts to shorter wavelengths, a blueshift (Figure 11-19). Note that the term "redshift" does not mean that the light becomes red; it means the light will have a longer wavelength (more toward the red end of the spectrum). We can use redshift and blueshift to determine star motions, and from that infer the presence of orbiting exoplanets. You can learn how to calculate the mass of an exoplanet with the online explanation and example in the accompanying eBook.

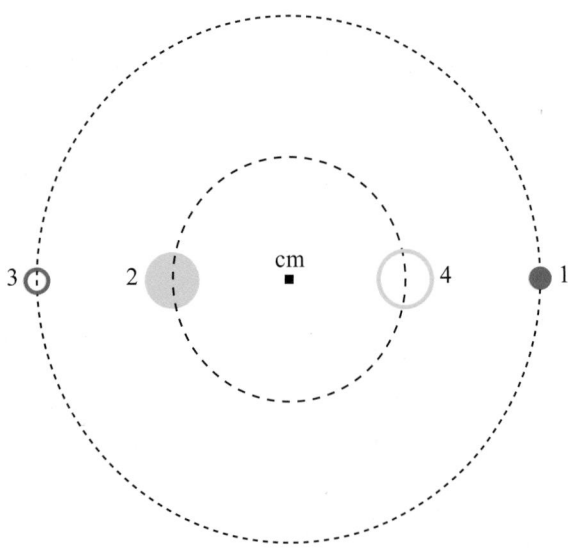

Figure 11-18 The motion of an exoplanet and a star about their centre of mass. When the planet (orange) is at point 1 on the right of the centre of mass, the star (yellow) will be on the left of the centre of mass at point 2. Similarly, when the planet is on the left at point 3, the star will be at point 4 on the right of the centre of mass. Note that the diagram is not to scale: the relative size of the orbit of the star will be much smaller than shown here.

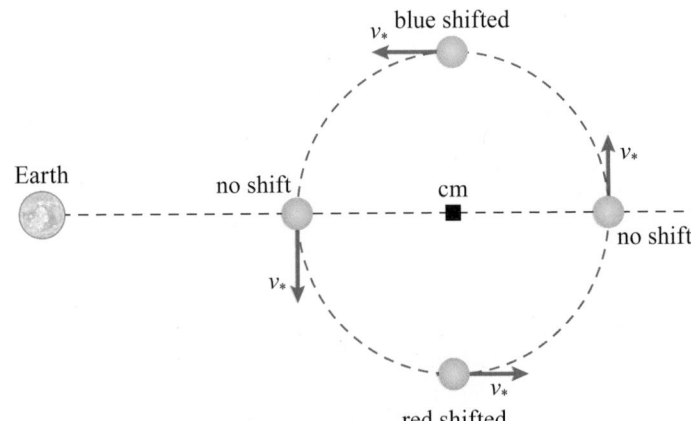

Figure 11-19 A star–planet system observed from Earth. The light from the star is blueshifted when the star moves toward us and redshifted when the star moves away from us.

According to the law of universal gravitation, every object in the universe attracts every other object with a force that depends directly on the two masses and inversely on the square of the distance between the centres of the objects. A corresponding universal form of gravitational potential energy can be derived from the law of universal gravitation, and has a negative value for all finite distances. The orbit of an object can be classified by the total of the object's kinetic energy and gravitational potential energy. Gravitational escape requires a positive total orbital energy.

Universal Gravitational Force

The gravitational force between any pair of masses in the universe is $|\vec{F}_G| = \dfrac{GMm}{r^2}$,

where M and m are the masses, r is the distance between their centres, and G is the universal gravitational constant. The force is directed toward the centre of the attracting mass.

The acceleration due to gravity near a point a distance r from the centre of a spherically symmetric mass can be readily shown from Newton's second law and the universal gravitational force to have the following magnitude: $g = \dfrac{GM}{r^2}$.

Force from Potential Energy

The derivatives of a potential energy function U give the components of the associated force, for example, $F_x = -\dfrac{\partial U}{\partial x}$, $F_y = -\dfrac{\partial U}{\partial y}$, $F_z = -\dfrac{\partial U}{\partial z}$, and $F_r = -\dfrac{\partial U}{\partial r}$.

Gravitational Potential Energy

The general expression for gravitational potential energy between spherically symmetric masses M and m, with centres separated by a distance r, is given by $U = -\dfrac{GMm}{r}$.

The gravitational potential energy is zero when the masses are infinitely far apart and negative everywhere else.

Total Orbital Energy

By calculating the sum of an object's kinetic energy and gravitational potential energy, we can determine the type of orbit. If the total is negative, the orbit is elliptical; if the total is zero, the orbit is parabolic; if the total is positive, the orbit is hyperbolic. A positive total also corresponds to gravitational escape.

Kepler's Laws

1. Planetary orbits are ellipses with the Sun at one focus.

2. A line joining a planet to the Sun sweeps out equal areas in equal times.

3. The square of the orbital period (T) is proportional to the cube of the semi-major axis (a) of the orbit. For a pair of objects with masses M_1, and M_2 orbiting around their centre of mass, this law is expressed as

$$T^2 = \frac{4\pi^2}{G(M_1 + M_2)} a^3 \qquad (11\text{-}12)$$

Orbital Speed

The speed of an object in an elliptical orbit with semi-major axis a about a mass M varies with distance r according to the following relationship:

$$v = \sqrt{GM\left(\frac{2}{r} - \frac{1}{a}\right)} \qquad (11\text{-}11)$$

Key Applications: gravitational field mapping in geophysics, satellite orbits, interplanetary travel, predicting tides, detecting exoplanets, predicting the paths of comets and asteroids, producing weightlessness

Key Terms: acceleration due to gravity, aphelion, asteroid, centre of mass, conic sections, elliptical orbit, escape speed, exoplanet, gravitational mass, gravitational potential energy, hyperbolic orbit, inertial mass, Kepler's laws, law of universal gravitation, meteoroid, parabolic orbit, perihelion, planet, satellite, semi-major axis, weight

QUESTIONS

1. When Earth exerts a force of 550 N on a person, what is the magnitude of the force that the person exerts on Earth?
   (a) exactly 0
   (b) 550 N
   (c) 550 N times the ratio of the mass of the person to Earth's mass
   (d) 550 N times the ratio of the mass of Earth to the mass of the person

2. Consider a planet that has the same mass as Earth, but eight times the average mass density. What would your weight on this planet be if your weight on Earth is W?
   (a) W
   (b) $2W$
   (c) $4W$
   (d) $8W$

3. Weigh scales are often marked in units of kilograms. Why is this marking incorrect?

4. What is your weight at the centre of Earth?

5. Consider two spherical homogeneous planets, both with exactly the same density. The radius of one planet is twice as large as the radius of the other planet. Which of the following statements is true?
    (a) The two planets have exactly the same gravitational acceleration at their surfaces.
    (b) The larger planet has a greater gravitational acceleration at its surface.
    (c) The smaller planet has a greater gravitational acceleration at its surface.
    (d) More information is needed to compare the gravitational accelerations.

6. On a planet that has the same mass as Earth but a smaller radius, the escape speed from the surface is
    (a) less than the escape speed on Earth;
    (b) more than the escape speed on Earth;
    (c) the same as on Earth;
    (d) impossible to compare without more information.

7. At the surface of the planet in the previous question, the gravitational acceleration is
    (a) < 9.81 m/s^2;
    (b) ≈ 9.81 m/s^2;
    (c) > 9.81 m/s^2;
    (d) impossible to compare without more information.

8. Will it require less energy to launch a satellite to orbit from east to west or from west to east? Explain.

9. Which of the following statements is true?
    (a) G is constant, but g varies with location.
    (b) Both G and g vary with location.
    (c) Both G and g are constant everywhere in the universe.
    (d) G varies with location, and g is constant.

10. Three objects fall to Earth from a location some distance away. Object A starts from rest; object B has an initial velocity directed toward Earth; and object C has the same initial speed but directed away from Earth. Which of the following statements correctly describes the speeds of the objects when they reach Earth's surface?
    (a) All three objects will have exactly the same speed.
    (b) Object A will have the greatest speed.
    (c) Objects B and C will have exactly the same speed, which will be greater than the speed of object A.
    (d) Object C will have the greatest speed, and object B will have the lowest speed.

11. Does the speed of a satellite increase or decrease when atmospheric drag causes it to lose altitude? Explain your reasoning.

12. (a) If you want to produce a straight-line graph by plotting a set of data for orbital periods and semi-major axes, what function of the periods would you plot on the y-axis? What function of the semi-major axes would you plot on the x-axis?
    (b) If you want the slope of the graph from part (a) to be 1, what units should be used for periods and semi-major axes?

13. A planet is in a circular orbit around the Sun. Its distance from the Sun is four times the average distance between Earth and the Sun. The period of this planet, in Earth years, is
    (a) 4;
    (b) 8;
    (c) 16;
    (d) 64.

14. When we are observing a planetary system (assume a single planet and star) at a certain instant in time, the star is observed to be in motion toward us. What is the direction of motion of the exoplanet at this time?
    (a) toward us
    (b) away from us
    (c) stationary relative to us
    (d) cannot be determined from the information given

15. Consider a star and an exoplanet that orbit in a plane perpendicular to the direction from the star to Earth. What shifts would we observe in the light from the star?
    (a) The light would be redshifted at all positions in the star's orbit.
    (b) The light would be blueshifted at all positions of the orbit.
    (c) There would be no observable shift from the orbital motion.
    (d) The light would be redshifted at one side of the orbit and blueshifted at the opposite side.

16. In 2005, a satellite was discovered in orbit around the distant dwarf planet Eris. What key information about Eris can we learn by observing this satellite?

17. Suppose that a powerful rocket engine is set up on an asteroid in an elliptical orbit around the Sun. When the asteroid is at its perihelion, the engine is fired for a moment to increase the speed of the asteroid. Which of the following statements correctly describes the subsequent motion of the asteroid, assuming that no other forces act to alter its orbit. Justify the answer you select.
    (a) The perihelion, aphelion, and semi-major axis of the orbit will change.
    (b) The perihelion and semi-major axis, but not the aphelion, will change.
    (c) The aphelion and semi-major axis, but not the perihelion, will change.
    (d) The perihelion, aphelion, and semi-major axis are unchanged.

18. In 2006, the International Astronomical Union redefined "planet." As a result, Pluto is no longer considered to be one of the major planets.
    (a) Use print and/or Internet resources to research and summarize the criteria for an object to be considered a planet.
    (b) One of these criteria is that the object pull itself into an approximately spherical shape. Using concepts from this chapter, explain why a molten body in space will assume a spherical shape.

PROBLEMS BY SECTION

For problems, star ratings will be used, (✷, ✷✷, or ✷✷✷), with more stars meaning more challenging problems.

Section 11-2 Acceleration Due to Gravity

19. ✷✷ What force does a person of mass 50 kg exert on the Andromeda Galaxy, M31? The Andromeda Galaxy is believed to have a total mass of approximately one trillion times the mass of the Sun and is about 2.3 million light years away. (A light year is the distance that lights travels in one year.)

20. ✷✷ Assume that Earth's orbit is circular with a radius of 1.50×10^{11} m. From first principles, calculate the approximate speed of Earth in its orbit.

21. ✷ Calculate the difference in force per kilogram for water on the side of Earth nearest the Moon compared to water on the side farthest from the Moon. Use 6378 km as Earth's equatorial radius, and assume that the centre of the Moon is 3.84×10^8 m from the centre of Earth.

22. ✷✷ Assume that you are 25 km from the centre of a neutron star that has a radius of 10.0 km and a mass of 4.4×10^{30} kg. Your feet are pointed at the neutron star, and your head is 1.5 m above your feet. What is the *difference* in gravitational force per kilogram on your head compared to your feet?

Section 11-3 Orbits and Weightlessness

23. ✷ Calculate the value of the acceleration due to gravity on the surface of Mars, given that the mass of Mars is 6.42×10^{23} kg. Assume that Mars is a homogeneous sphere with a radius of 3.40×10^6 m.

24. ✷ Find the radius of a spherical asteroid that has a density of 2900 kg/m³ and an acceleration due to gravity of 0.10 m/s² at the surface.

25. ✷✷✷ Assume that the normal value of g in a region is 9.800 m/s². However, one part of the region has an ore deposit directly below the surface. This deposit has a density of 7800 kg/m³, and the surrounding material has a mean density of 3000 kg/m³. Assume that the ore deposit is spherical with a radius of 500 m. Calculate the value of g above the ore deposit.

26. ✷✷ The acceleration due to gravity at the surface of a fictional planet is 3.11 m/s², and the mean density of the planet is 880 kg/m³. Find the radius and mass of the planet. Assume that it is spherical.

Section 11-4 Gravitational Potential Energy

27. ✷✷ A satellite is in a circular orbit 3000 km above the surface of a fictional planet. The satellite has a speed of 8000 m/s and is 8000 km from the centre of the planet.
(a) Find g at the altitude of the satellite.
(b) Find g at the surface of the planet.
(c) Find the mass of the planet.

28. ✷✷ If a parabolic flight aircraft starts the parabolic portion of a flight at an angle of 40° above the horizontal, and this portion of the flight lasts 20 s, what was the speed of the aircraft when it started the parabolic portion? Assume that the air resistance is negligible.

Section 11-5 Force from Potential Energy

29. ✷✷ Olympus Mons is an extinct volcano on Mars that is approximately 22 km high (almost three times the height of Mount Everest). Calculate the work required to raise a 25 kg object from the surface of Mars to the top of Olympus Mons. Assume that Mars is effectively spherical with a radius of 3400 km. The mass of Mars is 6.42×10^{23} kg.

30. ✷✷ An object is fired vertically from the surface of the Moon with a speed of 1000 m/s. What is the maximum height that it will reach?

Section 11-6 Escape Speed

31. ✷ Find the components of the force in a region where the potential energy function is $U = ky^2$.

32. ✷✷ If gravitational potential energy had the form $U(r) = -\dfrac{GMm}{\sqrt{r}}$, what would the gravitational force be?

Section 11-7 Kepler's Laws

33. ✷ Calculate the escape speed from the surface of the Moon.

34. ✷✷ The dwarf planet Ceres is approximately spherical with a radius of 488 km. The density of Ceres is 2078 kg/m³.
(a) What is the escape velocity from the surface of Ceres?
(b) If an object was launched vertically from the surface of Ceres with a speed that is exactly one-half the escape speed, what height would it reach above the surface of Ceres?

Section 11-8 Types of Orbits

35. ✷✷ It is believed that there is a black hole with a mass of approximately four million times the mass of the Sun near the centre of the Milky Way galaxy. Some stars are relatively close to this black hole. Assume that the semi-major axis of the orbit of one of these stars is 3.5 light days (i.e., the distance that light would travel in 3.5 days).
(a) Find the orbital period of the star.
(b) Find the orbital speed of the star, assuming that it is in a circular orbit.

36. ✷✷ An asteroid has a perihelion distance of 0.55 AU (astronomical unit) and an aphelion distance of 1.65 AU. Predict the orbital period of the asteroid.

37. ✷ Since the Moon has virtually no atmosphere, an object could orbit just above the Moon's surface without experiencing any aerodynamic drag. What speed would be needed to launch an object so that it would orbit just slightly above the Moon's surface? Treat the Moon as a smooth sphere for this question.

Section 11-9 Detection of Exoplanets

38. ✷✷✷ A sensitive Doppler radar instrument determines that a small object at a height of 1800 km above Earth's surface has a radial velocity component of 1.3 km/s directed away from Earth's centre. Night-vision images indicate that the object has a transverse velocity component of 6.7 km/s. Assuming that the transverse and radial components are perpendicular, determine whether the object is in a gravitationally bound orbit with Earth or in a hyperbolic orbit that will cause it to escape Earth's gravitational field. Assume no upper atmosphere drag, and neglect the influence of any other masses such as the Moon and the Sun.

39. ✷✷ (a) What is the centre of mass of the Sun-Earth system?
(b) Express your answer from the first part as a fraction of the radius of the Sun.

40. ✷✷✷ Spectroscopy of a star produces radial velocity plots suggesting a peak stellar speed of 34 m/s. The period of the cycle is 25 days. Other measurements give an estimate of 4.0×10^{30} kg for the mass of the star. Calculate the mass of the exoplanet that would account for the radial velocity of the star. Consult the eBook section on exoplanet mass determination for help in doing this problem.

COMPREHENSIVE PROBLEMS

41. ✳✳✳ Mars has two small satellites called Phobos and Deimos. Assume that these satellites are spherical with a mean density of 1880 kg/m³ and that the diameter of Phobos is 21 km.
(a) If you dropped an object from a height of 1.5 m above the surface of Phobos, how long would it take the object to fall to the surface?
(b) Phobos orbits Mars with a period of 0.319 days at a mean distance of 9400 km from the centre of Mars. Calculate the mass of Phobos.

42. ✳ For a museum exhibit, you want to make a set of scales that show the weight a person would have on different planets. If you start with scales that measure 0 to 1000 N, how should you alter the markings on the scales for each of the following celestial objects?
(a) Mars
(b) Venus
(c) Jupiter (if it had a surface you could stand on)
(d) Earth's Moon
(e) Pluto
(f) Mercury
Use reference works to find the information needed for each planet.

43. ✳✳✳ This problem involves calculations related to the Cavendish experiment. Refer to the online example for help with these calculations.
(a) In the original Cavendish experiment, two small spheres with a mass, m, of about 0.73 kg each, were attached to the end of a balance arm 1.8 m in length. What is the moment of inertia, I, of the two balls as they rotate about the centre of the balance? Ignore the mass of the balance arm.
(b) The period of a torsion balance is given by the following relationship:

$$T = 2\pi\sqrt{\frac{I}{\kappa}}$$

where T is the period, I is the moment of inertia, and κ is the torsion coefficient of the suspending wire. The observed period in the Cavendish experiment was about 420 s. Calculate the torsion coefficient for the apparatus.
(c) The large masses, M, each had a mass of 158 kg each. Calculate the gravitational force between a small sphere and the adjacent large sphere when the distance between their centres is 230 mm.
(d) The moment arm for each force is half the length of the balance arm, 0.90 m. Calculate the torque associated with one of the masses, using the force from part (c). The total torque will be double this value (since there is an equal torque at each end of the arm).
(e) The rotational analog to Hooke's law is given by $\tau = -\kappa\theta$, where τ is the torque, κ is the torsion coefficient, and θ is the deflection angle. Predict the deflection angle for the apparatus described above. (Remember that the total torque acting on the balance arm is double the value calculated in part (d).)

44. ✳✳ On Earth, a person has a weight of 520 N. What weight would the person have on Mars? The radius of Mars is 53% of the radius of Earth, and its mass is 11% of the mass of Earth.

45. ✳✳✳ Consider a planet that has a gravitational acceleration of 1.8 m/s² at its surface. Assume that the planet is spherical with a radius of 4300 km.
(a) Calculate the mass of the planet.
(b) Find the escape speed from the surface of this planet, assuming no air resistance.

46. ✳✳✳ Consider a geosynchronous satellite with a mass of 140 kg. Geosynchronous satellites orbit Earth with a period of approximately 24 h. The mass of Earth is 5.97×10^{24} kg. Assume that Earth is exactly spherical with a radius of 6378 km.
(a) At what height above Earth's surface will the geosynchronous satellite orbit?
(b) Calculate the angular velocity for the orbit.
(c) Calculate the angular momentum of the satellite in orbit.
(d) How much additional gravitational potential energy does the satellite have when it is in orbit, compared to when it is on Earth's surface?
(e) The orbital period of a geosynchronous satellite is not precisely 24 h. Explain why.

47. ✳✳ What is the fastest speed at which an object 1.0 AU from the Sun can move and still be gravitationally bound to the Sun? The object is positioned such that the gravitational forces exerted by Earth and other planets are negligible.

48. ✳✳ Europa is one of the large "moons" of Jupiter and is believed to have a major ocean under a thick frozen surface. Europa has a radius of 1570 km and a mass of 4.87×10^{22} kg. Assume that Europa is completely spherical.
(a) Find the escape speed for an object at the surface of Europa.
(b) Find the acceleration due to gravity, g, at the surface of Europa.
(c) What would the weight of a 64 kg person be on the surface of Europa?

49. ✳✳✳ The Sun subtends an angle of about 0.47° when photographed from Earth. (This angle varies slightly because Earth follows a slightly elliptical orbit.) The Sun is a distance of 1.50×10^{11} m from Earth. Calculate the mean density of the Sun, assuming that it is spherical. (Hint: The period of Earth's orbit is one year.)

50. ✳✳ Pluto has at least three satellites, the largest of which is Charon. Charon orbits Pluto once every 6.387 days, and the semi-major axis of the orbit is 19 570 km. Find the sum of the masses of Pluto and Charon.

51. ✳✳ Derive the relationship $M = \dfrac{rv^2}{G}$ for a small mass, m, in a circular orbit at a distance r from the centre of a much larger mass, M. The speed of mass m is v.

52. ✳✳✳ The solar system orbits the centre of our Milky Way galaxy with a speed of approximately 230 km/s at a distance from the centre of the galaxy of about 26 000 light years.
(a) Use the relationship from problem 51 to estimate the mass within the orbit of the solar system around the centre of our galaxy.
(b) Express the mass calculated in part (a) in units of solar mass, that is, as a multiple of the mass of the Sun.

(c) How long will it take our solar system to make one orbit around the galaxy? Express your answer in years.

53. ✷✷ Imagine that a space colony living in a large vessel is in a circular orbit at a distance of 55 km from the centre of a neutron star. The neutron star has a radius of 10.0 km and a mass of 4.4×10^{30} kg.
(a) Find the velocity of the vessel in its orbit.
(b) Find the period of the orbit.

54. ✷✷ A comet discovered in an orbit around the Sun has a semi-major axis of 36 AU (astronomical units).
(a) Use Kepler's third law to estimate the period for the orbit of this comet. Express your answer in years.
(b) If the same comet were in orbit around an exoplanet of mass 8.6×10^{29} kg with the same semi-major axis of 36 AU, what would be the period of the orbit? Convert the final answer to years.
(c) Would the orbit about the exoplanet be circular? Justify your answer.

55. ✷✷ Go to the PhET site (http://phet.colorado.edu/), and download the "My Solar System" simulation. Select Trojan asteroids from the pulldown menu. Start the simulation, and describe the subsequent motion. Describe as precisely as possible the location of the Trojans relative to the planet as viewed from the Sun. Read the online material on Lagrange points and write an explanation of how the positions of the Trojans might relate to Lagrange points.

56. ✷✷ Stars lose mass from gradual processes, such as the solar wind, and from the ejection of material in eruptions and explosions. Consider a star with several planets in circular orbits.
(a) Describe qualitatively what would happen to the orbits of the planets when the star loses a significant part of its mass (say 20%).
(b) How much of its original mass would a star have to lose to make the orbits of the planets hyperbolic?

57. ✷✷ Consult the e-book online material on exoplanet mass determination for background in doing this problem. Consider an exoplanet system with one star and two planets. Doppler spectroscopy shows that one planet causes motion of the star with a peak radial speed of 40 m/s and a period of 10 days, and the other planet causes motion with a peak radial speed of 20 m/s and a period of 50 days.
(a) Which planet is closer to the star?
(b) Which planet is more massive?

58. ✷✷✷ Consider an exoplanet system with a star of mass 2.5×10^{30} kg and a single planet of mass 1.5×10^{28} kg in a circular orbit around their centre of mass. The orbital period is 25 days.
(a) Calculate the semi-major axis of the planet's orbit.
(b) Where is the centre of mass of the system located?
(c) Find the approximate orbital speed of the planet.
(d) Find the approximate orbital speed of the star.
(e) What would be the range of observed wavelengths for a spectral line with a wavelength of 656.280 nm emitted by the star in this system?

59. ✷✷ Some meteorites come from the Moon. Calculate the minimum ejection speed needed to move a stone from the surface of the Moon to Earth, ignoring the orbital motion of the Moon and Earth. (Hint: First find the point on the line from the Moon to Earth where the gravitational forces balance.)

60. ✷✷✷ Using work–energy principles, derive the expression for gravitational potential energy at distance r from a symmetric spherical mass M.

61. ✷✷ Suppose that the potential energy between two masses, M and m, had the form $U(r) = -\dfrac{kMm}{r^3}$, where k is a constant. Use calculus to find an expression for the magnitude of the related force.

62. ✷✷✷ A visible star with an unseen companion star often emits X-rays. The visible star in one such pair has spectral characteristics that indicate it has a mass of about 7.6 solar masses (i.e., 7.6 times the mass of the Sun). After careful observations you have determined that the semi-major axis of the orbit of the visible star is 26 million km. The orbital period of that star is 8.5 Earth days.
(a) Use the general version of Kepler's third law to determine the mass of the unseen star.
(b) Theory suggests that a stellar remnant will not form a black hole with a mass less than 3.0 solar masses. Is the unseen object a black hole?

63. ✷✷ Assume that the Sun started as a thin, spherical shell of material with the same mass that it has now and with a radius of one light year.
(a) If we take three-quarters of the current radius of the Sun as the average distance of its mass from the centre of mass, how much gravitational potential energy would be released as the Sun contracts from the large shell to its current size?
(b) For what period of time could the released energy power the Sun at its current output of about 3.85×10^{26} W?

64. ✷✷ Assume that an object is perfectly spherical and has a uniform density equal to the density of water (1000 kg/m³).
(a) Find the radius and mass of the object that would make the escape speed at the surface equal to the speed of light.
(b) What does the answer to part (a) tell you about black holes?

65. ✷✷✷ Access the e-book online resources on exoplanet mass determination for background in doing this problem. Consider an exoplanet system that consists of a star of mass 1.8×10^{30} kg and a single planet of mass 4.5×10^{26} kg in a circular orbit around their centre of mass. The orbital period is 12 days. Find
(a) the maximum orbital speed of the star;
(b) the range of observed values for a spectral line that has an unshifted wavelength of 396.781 00 nm on Earth.

DATA RICH PROBLEM

66. ✷✷✷ A space probe to a fictitious planet accurately measures the radius of the (assumed perfectly spherical) planet as 5038 km. The space probe measures the g value at a number of heights above the surface of the planet as given in Table 11-1. From these data determine
(a) the mass of the planet;
(b) the mean density of the planet;
(c) the value of g at the surface of the planet.

Table 11-1   Problem 66

Height (m)	g (m/s²)
10 000	3.92583
15 000	3.91807
20 000	3.91033
25 000	3.90261
30 000	3.89491
35 000	3.88724
40 000	3.87959
45 000	3.87196
50 000	3.86435
55 000	3.85677
60 000	3.84921
65 000	3.84167
70 000	3.83415
75 000	3.82666

OPEN PROBLEM

67. ✱✱✱ At the beginning of the chapter, it is suggested that the event 65 million years ago that resulted in the extinction of the dinosaurs and perhaps 75% of all species was probably caused by a comet or an asteroid impact.

(a) Under a worst-case scenario, what would be the approximate maximum speed that a comet could impact Earth? Show calculations to support your answer, and assume that the comet is gravitationally bound to the solar system.

(b) Consult sources to suggest a reasonable radius and density for a comet, and use this information to calculate the comet's mass.

(c) Using your results from parts (a) and (b), find the kinetic energy of the comet just before impact. To put this in perspective, convert the result into the energy equivalent of a number of tonnes of TNT.

See the text online resources at www.physics1e.nelson.com for Open Problems and Data-Rich Problems related to this chapter.

Chapter 12
Fluids

The human cardiovascular system consists of a pump (the heart), a fluid (blood), and blood vessels of various cross-sectional areas and structure (Figure 12-1). Oxygen-rich blood leaves the heart at a pressure greater than atmospheric pressure. As blood flows through smaller arteries, frictional forces result in a loss of pressure. By the time the blood enters the capillaries, it is almost at atmospheric pressure. Clogged arteries restrict blood from flowing freely, and the heart has to pump harder to maintain circulation, resulting in higher blood pressure. Thus, high blood pressure is an indicator of clogged arteries. To understand how the pressure and flow of fluids are related, we will analyze some basic properties of static and moving fluids.

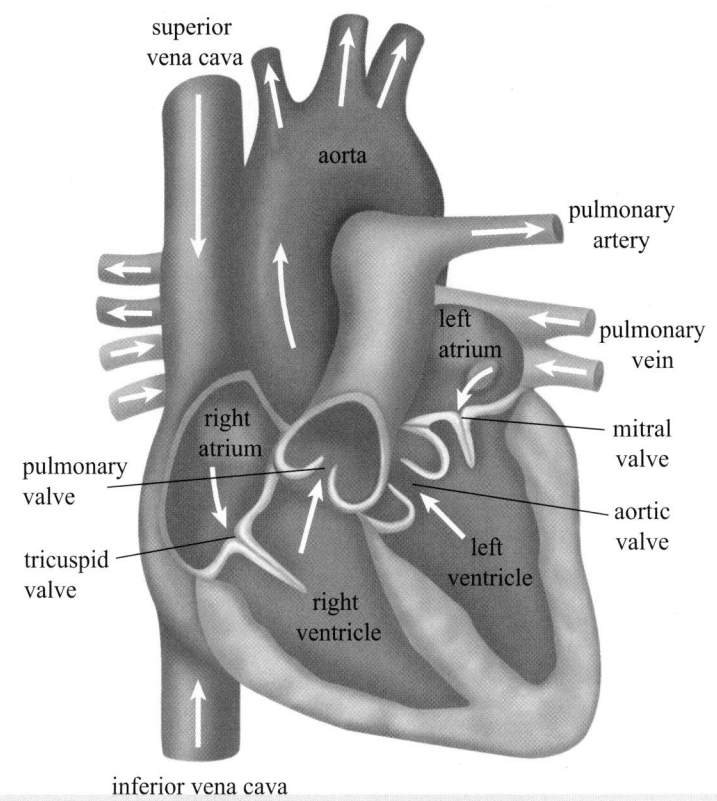

Figure 12-1 The human heart is a powerful muscle that controls the blood flow through the body.

MECHANICS

 For additional Making Connections, Examples, and Checkpoints, as well as activities and experiments to help increase your understanding of the chapter's concepts, please go to the text's online resources at www.physics1e.nelson.com.

LO 1

12-1   Phases of Matter

Most matter at everyday temperatures and pressures exists in three **phases** (or **states**), classified as gases, liquids, and solids (Figure 12-2). The physical properties of a given phase are determined by the average distance between its constituents (atoms and molecules) and the strength of the electromagnetic forces between them.

solid keeps its shape and does not deform easily. However, solid molecules are not completely stationary: they vibrate rapidly about their equilibrium positions.

There is a fourth phase of matter, called **plasma**. Plasmas form at very high temperatures and in strong electric fields. In this phase, collisions between molecules are so strong that the electrons can be knocked from the outer atomic orbits. Consequently, plasmas are a combination of ions and electrons. This state of matter exists in lightning, fluorescent lamps, neon signs, the solar wind, and stars.

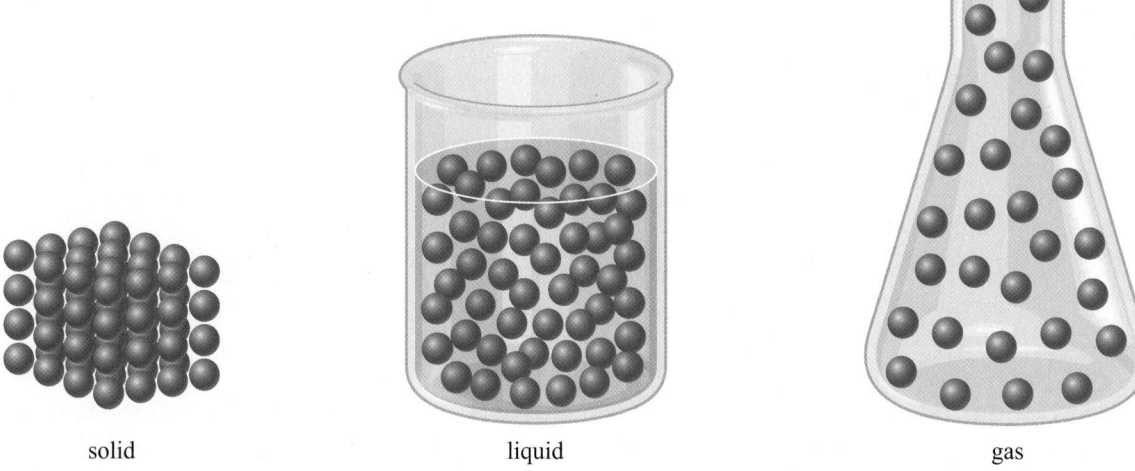

| solid | liquid | gas |

Figure 12-2   Solid, liquid, and gas states of matter

In gases, the average distance between the molecules is large compared to the size of the atoms; hence, the electromagnetic forces between the molecules are so weak that the molecules do not bind with each other and are able to move freely. Only when two molecules are very close to each other do they experience an appreciable force, and we say that "a collision has occurred between the two molecules." Such collisions obey Newton's laws of motion. At 20°C and sea level, the average distance between air molecules is 33×10^{-10} m, which is approximately 10 times larger than the diameter of a nitrogen molecule. Because of the large distances between gas molecules, gases are easily compressed.

In liquids, the molecules are very close to each other, and the electromagnetic forces between them are sufficiently strong to weakly bind the molecules but not strong enough to hold them at fixed positions. The molecules are still free to slide around each other without leaving the body of the liquid, enabling a liquid to flow under the influence of an external force. The average distance between water molecules is approximately 3×10^{-10} m, which is about the same as the diameter of a water molecule. Because the molecules in liquids are very close to each other, liquids are difficult to compress.

In solids, the electromagnetic forces hold the molecules at fixed positions relative to each other. Therefore, a

Gases and liquids are collectively called **fluids** (from the Latin word *fluidus*, meaning "to flow") because of their ability to flow. Fluids do not retain a specific shape, so we will define several new variables to describe their physical properties. Although the mathematical equations describing the behaviour and motion of fluids look different from what we have studied so far, the underlying physics is still based on Newton's laws and principles of conservation of energy and momentum.

LO 2

12-2   Density and Pressure

Mass Density

The mass density, ρ, of a substance is the mass per unit volume of the substance. For an object of mass M that occupies a volume V,

$$\rho = \frac{M}{V} \qquad (12\text{-}1)$$

Mass density is a scalar quantity. Note that Equation (12-1) gives the **average mass density** of an object. When we use the word "density" by itself, we are referring to the mass density. (The concept of density can be applied to other physical quantities, such as charge and current.) The SI units for density are kilograms per cubic metre (kg/m^3). Densities are also often expressed in grams per cubic centimetre (g/cm^3):

$$1\frac{kg}{m^3} = 1\frac{10^3\,g}{(10^2\,cm)^3} = 10^{-3}\frac{g}{cm^3} \quad or \quad 1\frac{g}{cm^3} = 1000\frac{kg}{m^3}$$

The density of a substance depends on the mass of its constituents (atoms) and the spacing between them. Table 12-1 lists the densities of various substances. The densities of gases are typically thousands of times less than the densities of liquids because the molecules are much farther apart in a gas than in a liquid. By definition, the density of a pure substance is independent of its volume. A steel nail has exactly the same density as a large block of the same material.

The density of an unconfined gas depends on its temperature. As the temperature of a gas increases, its volume increases and, therefore, its density decreases. When quoting the density of a gas, we will assume that the density is given at 0°C. The temperature dependence of density for liquids and solids is much less than that for gases.

Table 12-1 Mass densities of common substances

Substance		Mass density, ρ (kg/m³)
Solids		
Aluminum		2 700
Copper		8 890
Diamond		3 520
Gold		19 300
Ice		917
Lead		11 300
Silver		10 500
Wood (various types)	Balsa	160
	Maple	600–750
Liquids		
Alcohol		806
Blood (at 37°C)		1 060
Mercury		13 600
Water (at 4°C)		1 000
Gases		
Air		1.29
Helium		0.179
Hydrogen		0.0899
Nitrogen		1.25
Oxygen		1.43

EXAMPLE 12-1

The Density of the Sun

The mass of the Sun can be accurately determined from the length of the year on Earth, the distance between Earth and the Sun, and the value of the gravitational constant, G. The mass of the Sun is 1.99×10^{30} kg, and its radius is approximately 6.96×10^8 m. What is the average density of the Sun? Assume that the Sun is spherical.

SOLUTION

The Sun is assumed to be a sphere of radius 6.96×10^8 m. Therefore, the volume of the Sun is

$$V_{sun} = \frac{4\pi}{3}(6.96 \times 10^8\,m)^3 = 1.41 \times 10^{27}\,m^3$$

The mass density of the Sun, ρ_{sun}, is now calculated from Equation (12-1):

$$\rho_{sun} = \frac{M_{sun}}{V_{sun}} = \frac{1.99 \times 10^{30}\,kg}{1.41 \times 10^{27}\,m^3} \cong 1410\,kg/m^3$$

Making sense of the result:

The *average* density of the Sun is approximately 40% greater than the density of water. The inner core of the Sun has a density of approximately 1.5×10^5 kg/m³; the outer corona consists of hot gases with an average density of approximately 10^{-11} kg/m³. For intermediate layers, the density varies between these values. The value of 1410 kg/m³ therefore represents an average density for the entire Sun.

 CHECKPOINT

C-12-1 The Density of a Wooden Block

Three blocks of wood are cut from the same rectangular plank of wood, as shown in Figure 12-3. Which of the following statements about the densities of the blocks is correct?

(a) Block A has the highest density.
(b) Block B has the highest density.
(c) Block C has the highest density.
(d) All three blocks have the same density.

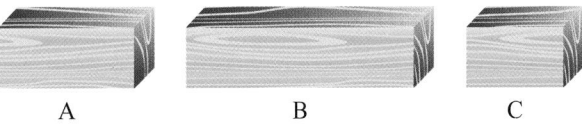

Figure 12-3 C-12-1

C-12-1 (d)

Specific gravity

The **specific gravity** (s.g.) of a material is the ratio of the mass of the material to the mass of an equal volume of water. Since mass = density × volume, the specific gravity of a material can also be defined as the ratio of its density to the density of pure water at the same temperature:

$$\text{s.g.} = \frac{\rho_{\text{material}}}{\rho_{\text{water}}} \qquad (12\text{-}2)$$

As you can see, specific gravity is a ratio of two quantities that have the same units, so it is a dimensionless quantity.

In Table 12-1, the density of blood is listed as 1060 kg/m³. Therefore, its specific gravity is

$$\text{s.g.} = \frac{\rho_{\text{material}}}{\rho_{\text{water}}} = \frac{1060 \text{ kg/m}^3}{1000 \text{ kg/m}^3} = 1.06$$

Pressure

When you push against a wall with your hand, the force exerted by your hand on the wall is distributed over the area of your hand that is touching the wall. **Pressure**, P, is defined as the magnitude of the component of the force perpendicular to a surface divided by the area of the surface on which the force acts.

Consider a force $\vec{F}$ acting on a surface of area A. Let us denote by F_n the component of $\vec{F}$ that is perpendicular to the surface. Then, the pressure exerted by the force on the surface is given by

$$P = \frac{F_n}{A} \qquad (12\text{-}3)$$

Since the component of a force and the area are both scalars, pressure is a scalar quantity. The SI unit of pressure is the **pascal** (symbol Pa), which is equal to one newton per square metre (N/m²):

$$1 \text{ Pa} = 1 \text{ N/m}^2$$

The name "pascal" honours the French mathematician, philosopher, and physicist Blaise Pascal (1623–62), who, among many other things in his short life, did some of the pioneering work on the properties of fluids.

For a given magnitude of force, pressure can be increased by applying the force over a smaller surface area. When you stand on one foot, the pressure that your foot exerts on the ground is twice the pressure that it exerts when you stand on both feet.

EXAMPLE 12-2

Calculating Pressure

A 0.500 kg book is resting on a table. Its cover has a surface area of 0.060 m². You push down on the book with your hand with a force of magnitude 50.0 N, with your arm inclined at an angle of 40.0° with respect to the normal to the book's surface (Figure 12-4). The area of your hand in contact with the book, A_{hand}, is 0.012 m².

(a) Find the pressure exerted by your hand on the book.
(b) What is the pressure exerted on the table by the book when your hand is pressing on the book?

Figure 12-4 The force exerted by the hand on the book is represented by an arrow that makes an angle of 40° with respect to the normal to the book.

SOLUTION

(a) The pressure exerted by your hand on the book is equal to the normal component of your hand's force on the book, $\vec{F}_{\text{hand-book}}$, divided by the area of the hand in contact with the book.

From Figure 12-4, the normal component of the force exerted by the hand on the book $F_{n,\text{hand-book}}$, is given by

$$F_{n,\text{hand-book}} = 50.0 \text{ N} \times \cos(40°) = 38.3 \text{ N}$$

Therefore, the pressure exerted by the hand on the book is

$$P_{\text{hand-book}} = \frac{F_{n,\text{hand-book}}}{A_{\text{hand}}} = \frac{38.3 \text{ N}}{0.012 \text{ m}^2} = 3.19 \text{ kN/m}^2 = 3.19 \text{ kPa}$$

(b) The net force exerted on the table by the book is the vector sum of the forces exerted by the hand on the book and the weight of the book:

$$\vec{F}_{\text{book-table}} = \vec{F}_{\text{hand-book}} + \vec{W}_{\text{book}}$$

The normal component of $\vec{F}_{\text{book-table}}$ is the sum of the normal components of the two forces on the right-hand side of the above equation:

$$F_{n,\text{book-table}} = F_{n,\text{hand-book}} + m_{\text{book}}g$$
$$= 50.0 \text{ N} \cos(40°) + 0.5 \text{ kg} \times 9.81 \text{ m/s}^2$$
$$F_{n,\text{book-table}} = 43.2 \text{ N}$$

The pressure exerted by the book on the table is equal to the normal component of the net force exerted by the book on the table, divided by the cross-sectional area of the book:

$$P_{\text{book-table}} = \frac{F_{n,\text{book-table}}}{A_{\text{book}}} = \frac{43.2 \text{ N}}{0.060 \text{ m}^2} = 720 \text{ Pa}$$

Making sense of the results:

The pressure in part (b) is considerably smaller than the pressure calculated in part (a), even though more force is being exerted than in part (a). This result is reasonable because the contact area of the book on the table is greater than the contact area between the hand and the book.

C-12-2 The Pressure on a Surface

Figure 12-5 shows four different situations in which a force of magnitude 45.0 N is exerted on a surface of area $0.10 \, \text{m}^2$. The direction of the force is represented by the direction of the arrow. Which of the following statements correctly describes the relationship between the pressures P_A, P_B, P_C, and P_D being exerted on the surface?

(a) $P_A < P_B < P_C < P_D$
(b) $P_B < P_A < P_C < P_D$
(c) $P_C < P_B < P_A < P_D$
(d) $P_A = P_B = P_C = P_D$

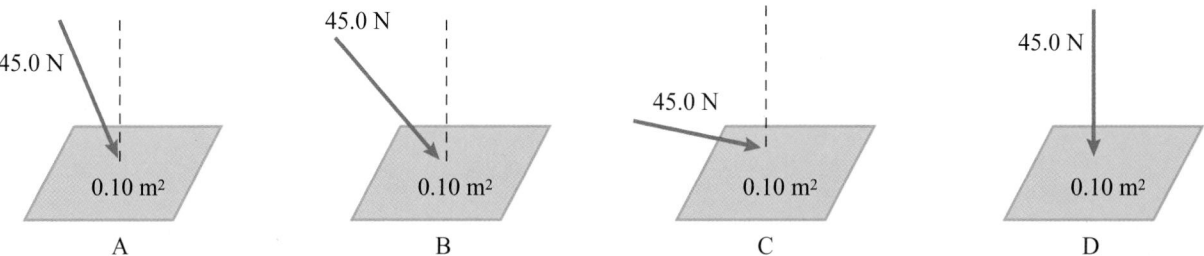

Figure 12-5 A force of magnitude 45.0 N is exerted at different angles on a surface.

C-12-2 (c)

LO 3

12-3 Pressure in Fluids

A fluid exerts pressure at every point within it and at every point where it contacts a surface, such as the wall of a container. This pressure is called **fluid pressure**. Fluid pressure results from the weight of the fluid and from the constant collisions of the molecules with each other and with the walls of the containing vessel. For liquids,

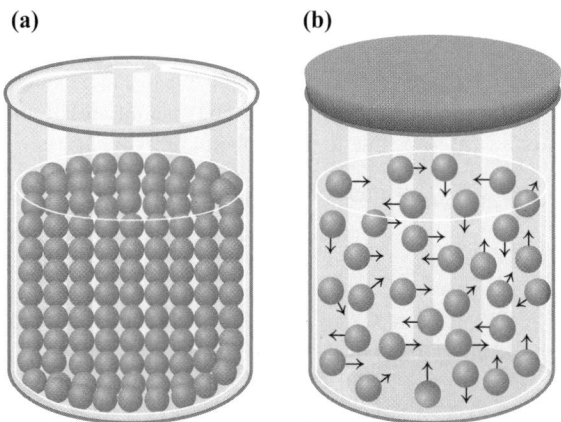

Figure 12-6 (a) In a liquid, the major source of pressure is the weight of the liquid. (b) In a gas, where densities are approximately 1000 times smaller than the densities of the liquids, pressure exists because of collisions between gas particles.

the major contribution to fluid pressure is the weight of the liquid. In contrast, the pressure exerted by a gas usually arises mainly from molecular collisions because the molecules of a gas are free to move and the densities of gases are much less than the densities of liquids (Figure 12-6). However, due to the enormous size of the atmosphere, the weight of air is the dominant contributor to atmospheric pressure.

Atmospheric Pressure

Earth's atmosphere consists mostly of nitrogen (78.1%), oxygen (20.9%), and argon (0.9%). Although there is no well-defined boundary between the atmosphere and space, it is generally agreed that Earth's atmosphere extends to about 120 km above sea level. The weight of the air molecules exerts a pressure at every point in the atmosphere and on Earth's surface. This pressure is called **atmospheric pressure** (atm). The atmospheric pressure at sea level at 20°C is experimentally determined to be $1.013 \times 10^5 \, \text{Pa}$ (≈ 101 kPa).

Objects on Earth are subject to atmospheric pressure. When you hold your hand flat, the air above the hand exerts a downward force on top of your hand. For a hand with an area of $130 \, \text{cm}^2$, the magnitude of this force is approximately $1.3 \times 10^3 \, \text{N}$, which is equivalent to holding a 130 kg object on your palm. Why are you able to hold your hand flat without feeling this force? It is because the air below the palm exerts an upward force on the other side of your hand, almost exactly balancing the downward force from the top (the difference is minuscule).

The density of air decreases with height above sea level, causing a decrease in atmospheric pressure with increasing height. Table 12-2 shows how atmospheric pressure varies with height above sea level.

Table 12-2 Variation of Atmospheric Pressure with Altitude

Height above sea level (m)	Atmospheric pressure* (kPa)
0	101.3
10 000	31.7
20 000	9.9
30 000	3.1
40 000	0.98
50 000	0.30
60 000	0.095

*These values were calculated assuming an exponential decrease in air density with increasing altitude.

EXAMPLE 12-3

Measuring the Mass of Earth's Atmosphere

Given that atmospheric pressure at Earth's surface is 1.01×10^5 Pa and the mean radius of Earth is about 6370×10^3 m, calculate the approximate mass of the air surrounding Earth. What assumptions do you need to make to calculate the mass of the air?

SOLUTION

The atmospheric pressure at Earth's surface results from the weight of the air. If we assume that Earth is spherical, we can calculate Earth's surface area from its radius. The total downward force exerted by the air, which is equal to the weight of the entire atmosphere, is the product of Earth's surface area and the atmospheric pressure at the sea level.

Assuming that the acceleration due to gravity from sea level to the top of the atmosphere remains constant, the magnitude of the downward force on Earth's surface due to the weight of the air is

$$F_{air,down} = M_{air}g \qquad (12\text{-}4)$$

This force is equal to the product of the atmospheric pressure at sea level and Earth's surface area:

$$F_{air,down} = P_0 \times 4\pi R_{earth}^2 \qquad (12\text{-}5)$$

From Equations (12-4) and (12-5), we get

$$M_{air} = \frac{P_0 \times 4\pi R_{earth}^2}{g} = \frac{1.01 \times 10^5 \frac{N}{m^2} \times 4\pi \times (6.370 \times 10^6 m)^2}{9.81 \text{ m/s}^2}$$

$$\cong 5.2 \times 10^{18} \text{ kg} \qquad (12\text{-}6)$$

Hydrostatic Pressure

Consider a container that is open to the atmosphere and filled with a fluid of density ρ_f (Figure 12-7). Imagine a cylindrical element of this fluid with cross-sectional area A, with its top end at the surface and the bottom end at a depth d below the surface. The following external forces are acting on this element in the vertical direction:

- The weight ($\vec{W}$) of the fluid element acting vertically downward. The magnitude of the weight is

  W = mass of the fluid element $\times g$

  = volume of the fluid $\times$ density of the fluid $\times g$

  = $(A\,d)\,\rho_f\,g$

- The downward force ($\vec{F}_{down}$) on the top surface, exerted by the atmospheric pressure. The magnitude of the downward force due to atmospheric pressure is

  F_{down} = atmospheric pressure

  $\times$ cross-sectional area of the cylinder

  = $P_0 A$

- The upward force ($\vec{F}_{up}$) that the rest of the fluid exerts on the bottom end of the fluid element.

  Let P be the pressure in the container at a depth d below its surface. Then the magnitude of the upward force is

  F_{up} = Pressure at the bottom of fluid element

  $\times$ cross-sectional area of the fluid element

  = PA

MECHANICS

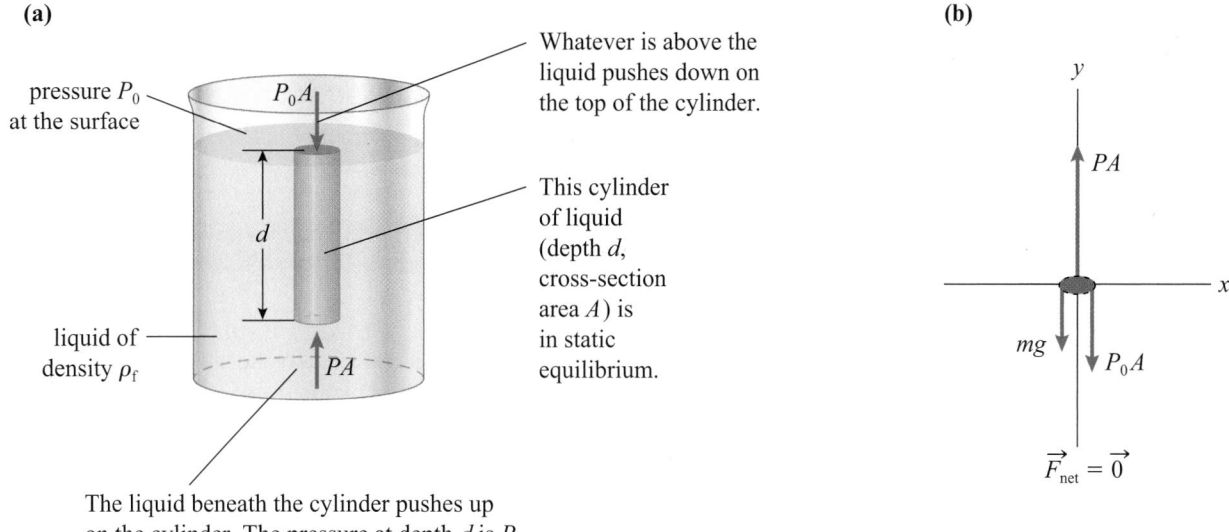

(a)

pressure P_0 at the surface

P_0A

Whatever is above the liquid pushes down on the top of the cylinder.

This cylinder of liquid (depth d, cross-section area A) is in static equilibrium.

d

liquid of density ρ_f

PA

The liquid beneath the cylinder pushes up on the cylinder. The pressure at depth d is P.

(b)

y

PA

mg

P_0A

x

$\vec{F}_{net} = \vec{0}$

MECHANICS

Figure 12-7 (a) Forces acting on a cylindrical element of a fluid in a container (b) The FBD of the column of liquid

There is no net horizontal flow of fluid because the *horizontal* force exerted on any point of the cylinder is exactly balanced by an equal horizontal force on the opposite side of the point. Taking the y-axis along the vertical direction with the positive y-axis upward, the net vertical force on the cylindrical volume is given by

$$\sum F_y = F_{up} - F_{down} - W$$
$$\sum F_y = PA - P_0A - \rho_f g\,(A\ d) \qquad (12\text{-}7)$$

Since the fluid element is stationary, the downward forces due to the weight of the fluid and the atmospheric pressure must be balanced by an upward force exerted by the rest of the fluid. Therefore, $\sum F_y = 0$, and we obtain the following relationship for the fluid pressure at a depth d below the surface of a fluid that is open to the atmosphere:

KEY EQUATION $\qquad P = P_0 + \rho_f g d \qquad (12\text{-}8)$

The second term on the right-hand side of Equation (12-8) arises due to the weight of the fluid. Notice two important consequences of this simple equation:

- The pressure within a fluid increases with depth.
- All points at a given depth within a connected body of fluid are at the same pressure.

The second point means that the pressure along a horizontal line through a connected body of a fluid is the same, no matter how the container is shaped. The pressure depends only on the depth below the surface. The pressure at a depth of 5 m inside a narrow tube filled with water is exactly the same as the pressure at a depth of 5 m in a large swimming pool, for example.

Equation (12-7) assumes that the fluid density does not change with depth. Therefore, it is not applicable for

the atmosphere, where the density of air changes with elevation.

Water Keeps Its Level

Water level in two arms of a U-shaped tube always rises to the same level (Figure 12-8). Why? Suppose that the water level is higher in the left arm. According to Equation (12-8), the pressure at the bottom of the left arm is higher than the pressure on the right side. The higher pressure at the left forces the water to flow from left to right until the level is the same in the two arms.

atmospheric pressure

B $\cdots\cdots\cdots\cdots\cdots$ A

Figure 12-8 The pressure at the bottom of the two arms of the tube is the same, so the height of the fluid in the two arms must be equal.

Suppose that the cross-sectional area of one arm of the tube is twice the cross-sectional area of the other (Figure 12-9). Is it possible to have an equilibrium situation where the fluid height in the wider arm is half the fluid height in the narrower arm? The volume and, therefore, the weight of the fluid in the two arms are then equal. Therefore, a force of equal magnitude is exerted at the bottom of the two arms of the tube. You might think that the fluid is in static equilibrium. However, with unequal levels of fluid in the two arms, the pressure at the bottom of the narrower arm is greater, causing the fluid to flow toward the wider arm until the fluid level in the two arms is the same.

Figure 12-9 Is it possible for a fluid at rest in a tube with two arms of unequal cross-sectional area to be at two different levels?

Equation (12-8) can be used to find the pressure difference between two points within a fluid. Let P_1 be the pressure at a point that is located at a depth d_1 within a fluid and P_2 be the pressure at a depth d_2. The pressure difference between the two points is

$$P_2 - P_1 = (P_0 + \rho_f g d_2) - (P_0 + \rho_f g d_1) = \rho_f g(d_2 - d_1)$$

$$(12\text{-}9)$$

In a room that is 5 m high, the increase in the air pressure at the floor level, compared to the pressure at the ceiling, is $(1.3 \text{ kg/m}^3) \times (9.81 \text{ m/s}^2) \times (5.0 \text{ m}) = 63.8$ Pa. This increase is a tiny fraction of the average atmospheric pressure of 1.01×10^5 Pa.

Gauge Pressure

Pressure is usually measured with respect to a reference pressure, usually taken to be atmospheric pressure. The absolute pressure (P), the reference pressure (P_0), and the measured or **gauge pressure** (P_G) are related as follows:

absolute pressure = measured pressure

+ reference pressure

$$P = P_G + P_0$$

A tire gauge measures the gauge pressure inside a tire with respect to the atmospheric pressure. So, a reading of zero on a tire gauge indicates that the pressure inside the tire is equal to the atmospheric pressure, indicating that the tire is flat. Similarly, blood pressure measured by a sphygmomanometer is the gauge pressure measured relative to atmospheric pressure.

CHECKPOINT

C-12-3 Pressure and Force

Four cylinders of different cross-sectional areas but the same height are completely filled with water, as shown in Figure 12-10.

(a) Order the cylinders, from highest to lowest, in terms of the pressure exerted by the water at the bottom of each cylinder.
Highest _____ _____ _____ _____ Lowest

(b) Order the cylinders, from highest to lowest, in terms of the force exerted by the water at the bottom of each cylinder.
Highest _____ _____ _____ _____ Lowest

Figure 12-10 Four cylinders of equal height and different cross-sectional areas

C-12-3 (a) A = B = C = D; (b) CADB

Measuring Pressure

A common instrument used to measure gauge pressure is a manometer. A manometer consists of a U-shaped glass tube of uniform cross-sectional area, partially filled with a liquid of known density (Figure 12-11). If the two arms of the manometer tube are at different pressures, then the fluid in the arm at the higher pressure is pressed downward, and the fluid level in the other arm rises.

Figure 12-11 A U-shaped manometer. The height difference in the column of fluid between the two arms is proportional to the pressure difference between the two sides.

Consider a manometer of uniform cross-sectional area partially filled with a liquid of density ρ_f (Figure 12-12). One arm (A) of the manometer is connected to a gas-filled cylinder that is maintained at a pressure P. The other arm (B) is open to the atmosphere and under atmospheric pressure P_0, which is assumed to be less than P. Let h be the difference in the heights of columns of fluids in the two arms. The pressure difference $P - P_0$ between the two arms of the manometer is proportional to h. To see this, draw a straight horizontal line that passes through a point (a) at the gas–fluid interface in arm A and a corresponding point (b) in arm B. Since the pressure in a static fluid is the same at all points at given level,

Figure 12-12 A U-shaped manometer connected to a pressurized device

$$\text{pressure at a } (P_a) = \text{pressure at b } (P_b) \qquad (12\text{-}10)$$

The pressure at the top of arm A is P; therefore,

$$P_a = P + \text{pressure because of the weight}$$
$$\text{of the gas in the left arm}$$

$$P_a = P + \rho_g g h_a$$

where h_a is the height of the column of gas in arm A, and ρ_g is the gas density. The pressure at point b is the sum of the atmospheric pressure and the weight of the column of liquid above the point b:

$$P_b = P_0 + \rho_f g h$$

Since $P_a = P_b$, we get

$$P + \rho_g g h_a = P_0 + \rho_f g h$$

Therefore,

$$P - P_0 = \rho_f g h - \rho_g g h_a \qquad (12\text{-}11)$$

The quantity on the left-hand side is the gauge pressure. Since liquid densities are generally approximately 1000 times greater than gas densities, the second term on the right-hand side of Equation (12-11) can be ignored. With this approximation,

$$P - P_0 = \rho_f g h \qquad (12\text{-}12)$$

This simple equation can be used to determine the pressure in a gas or a liquid pipeline by using a manometer.

EXAMPLE 12-4

Pipeline Pressure

The total pressure in a gas pipeline is to be maintained at three times atmospheric pressure. Figure 12-12 shows the cross-sectional area of the pipeline that is connected to a manometer filled with mercury. An inspector measures a difference of 1.25 m between the two columns of mercury. Is the gas pressure being maintained at the desired value? The density of mercury is 13 600 kg/m³.

SOLUTION

The open end (B) of the manometer is at atmospheric pressure. We do not know the height of the gas column in arm A, but we can ignore the pressure from the weight of the gas because the gas is much less dense than mercury.

Consider a horizontal line that passes through the gas–fluid interface in arm A at point a and through point b on arm B. The pressures at these two points are equal. If the pipeline is at the desired pressure then,

$$3P_0 = P_0 + \rho_f g h$$

Solving for h, we get

$$h = \frac{2P_0}{\rho_f g} = \frac{2 \times 1.01 \times 10^5 \text{ N/m}^2}{(13.6 \times 10^3 \text{ kg/m}^3)(9.81 \text{ m/s}^2)} = 1.51 \text{ m}$$

The measured height is only 1.25 m, so the gas pressure is below the desired value of $3P_0$.

ONLINE ACTIVITY

Variation of pressure with depth

The e-resource that accompanies every new copy of this textbook contains an Online Activity using the PhET simulation "Studying Variation of Fluid Pressure with Depth." Work through the simulation and accompanying questions to gain an understanding of the variation of pressure within the body of a fluid.

MAKING CONNECTIONS

Bourdon Pressure Gauges

Many mechanical pressure gauges use a Bourdon tube, a coiled metal tube that straightens when the pressure in it increases (patented in France by Eugene Bourdon in 1849). The end of the tube is connected to a gear mechanism that rotates a pointer on a calibrated dial (Figure 12-13). Alternatively, a Bourdon tube can be connected to a transducer to give an electronic readout. Bourdon gauges are used to measure pressure in bicycle tires, boilers, and tanks of compressed gases.

Figure 12-13 A Bourdon-type pressure meter

✓ CHECKPOINT

C-12-4 Pressure in a Closed Tube

In Figure 12-14, the tube is open at one end, closed at the other end, and filled with water. The pressure at the closed end is

(a) greater than the atmospheric pressure;
(b) equal to the atmospheric pressure;
(c) less than the atmospheric pressure but nonzero;
(d) zero.

Figure 12-14 C-12-4

C-12-4 (a)

LO 4

12-4 Pascal's Principle

Consider a container that is fitted with a movable piston and filled with a liquid (Figure 12-15).

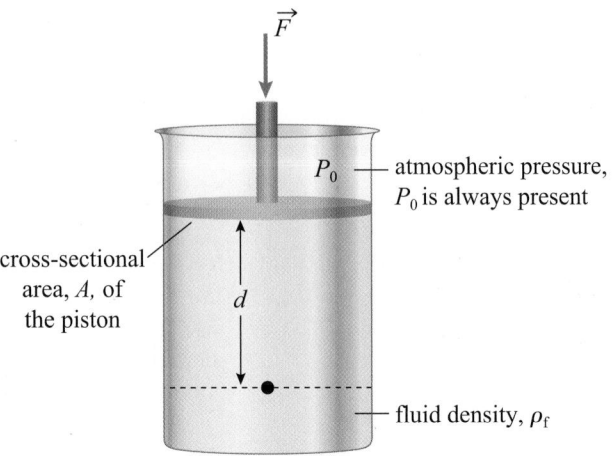

Figure 12-15 The total pressure at any point within the fluid is equal to the sum of the pressures due to the atmosphere, the weight of the fluid, and the external force.

When the piston is pushed downward, the layer of liquid in contact with it experiences an increase in pressure. The molecules of a liquid interact with each other and are free to move, so the increase in the pressure is quickly transferred to every part of the liquid and to the walls of the container. This simple fact was first realized and stated by Blaise Pascal, and is called **Pascal's principle**:

A change in pressure at any point of an enclosed fluid that is at rest is transmitted, without any loss, to every point of the fluid, including the walls of the container.

Notice that for Pascal's principle to hold true, the fluid must be confined. We encounter Pascal's principle in action all around us. When you squeeze the end of a toothpaste tube, the resulting increase in pressure is transmitted throughout the body of the tube, forcing the toothpaste through the opening (Figure 12-16).

Hydraulic Systems

A hydraulic system, based on Pascal's principle (Figure 12-17), is an essential component of many modern-day machines, for example, excavators, pumps, turbines, automobile braking systems (Figure 12-18), and hydraulic elevators. A simple hydraulic system consists of a sealed tube filled with a liquid and fitted with two pistons, usually called an input piston and an output piston. A force exerted on the input piston results in an increase in pressure within the fluid and at the output piston.

Consider two cylindrical containers of cross-sectional areas A_1 and A_2, connected by a tube as shown in Figure 12-19. The cylinders are fitted with movable pistons and are partially filled with a liquid of density ρ_f.

Figure 12-16 Pressure applied at one end of the tube is transmitted to all parts of the tube, forcing the toothpaste out at the open end.

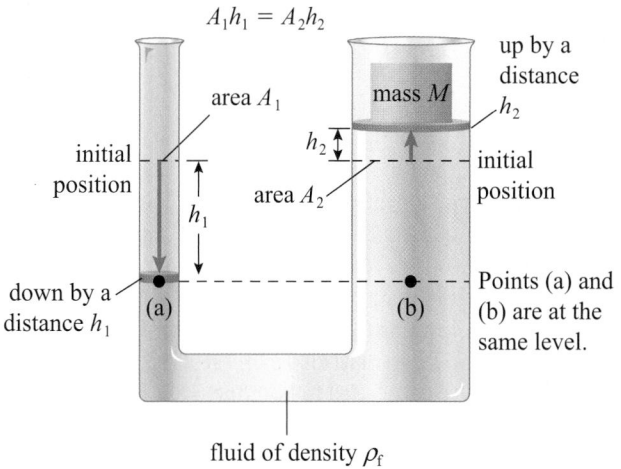

$$A_1 h_1 = A_2 h_2$$

Figure 12-19 Applying a downward force to the input piston raises the mass on the output arm of a hydraulic lift.

Figure 12-17 A simple hydraulic lift

Figure 12-18 Hydraulic brakes in a car. When the brake pedal is depressed, it pushes on the input piston in the hydraulic cylinder, causing the output pistons to move pads at each wheel.

This system is a **hydraulic lift** in its simplest form. To analyze this system, we will assume that the pistons are massless and that the frictional forces are negligible.

First, consider the case when the only force exerted on the two pistons is due to atmospheric pressure. The total pressure (atmospheric plus pressure due to the weight of the fluid) is the same for all points in the cylinder that are on the same horizontal line. Therefore, the height of the liquid in the two cylinders is the same, and the bottom surfaces of the two pistons are level with each other. Now, assume that the input piston is pushed down by applying

a force of magnitude F_1. The increase in the pressure, ΔP, on the input piston is

$$\Delta P = \frac{F_1}{A_1} \qquad (12\text{-}13)$$

By Pascal's principle, the pressure within the fluid increases by the same amount. This increase in the pressure results in an upward force exerted on the output piston by the fluid. The magnitude of the upward force, F_2, is

$$F_2 = \Delta P \times A_2$$

$$= \left(\frac{A_2}{A_1}\right) \times F_1 \qquad (12\text{-}14)$$

Note the amplification factor A_2/A_1 on the right-hand side of (12-14). By making the cross-sectional area of the output piston larger than the cross-sectional area of the input piston, a small amount of force can be amplified into a large force. This amplification is one of the most useful features of hydraulic systems.

Does the amplification of the applied force violate the law of conservation of energy? Consider the situation shown in Figure 12-19. An object (e.g., an automobile) of mass M is placed on the output piston. To hold the two pistons level, a force of magnitude $F_1 = \left(\frac{A_1}{A_2}\right) Mg$ must be exerted on the input piston. Suppose we need to elevate the object by a distance h_2. For this, the input piston needs to be pushed downward to force the fluid into the output end. Since the fluid is neither leaving nor entering the system, the decrease in the volume of the fluid in the input cylinder equals an increase in the volume in the output cylinder.

Let h_1 be the distance by which the input piston needs to be moved downward to raise the output piston by a distance h_2. Then,

$$A_1 \times h_1 = A_2 \times h_2$$

and

$$h_1 = \left(\frac{A_2}{A_1}\right)h_2 \qquad (12\text{-}15)$$

If $A_2 > A_1$, then $h_1 > h_2$. For example, if A_2 is 20 times larger than A_1 and the output piston is to be raised 10 cm, then the input piston must be pushed down 200 cm. Equation (12-15) is a consequence of the law of conservation of energy. The work done by the input piston on the liquid is equal to the work done by the liquid on the output piston.

As the elevation of the output piston increases, the force at the input end must also increase. Let F_1' be the magnitude of the force needed to hold the mass M at height h_2 with respect to the initial equilibrium position. In this configuration, the difference between the levels of the two pistons is $h_1 + h_2$. Consider points a and b on a horizontal line that passes through the fluid–piston interface in the input cylinder, as shown in Figure 12-19. The two points are at the same pressure; therefore,

$$P_0 + \frac{F_1'}{A_1} = P_0 + \frac{Mg}{A_2} + \rho_f g(h_1 + h_2) \qquad (12\text{-}16)$$

and

$$F_1' = \left(\frac{A_1}{A_2}\right)Mg + \rho_f g A_1 (h_1 + h_2) \qquad (12\text{-}17)$$

The first term on the right-hand side of Equation (12-17) is the force needed to keep the two pistons at the same level. The second term is the additional force needed to balance the weight of the volume $A_1(h_1 + h_2)$ of the fluid above the horizontal line that is level with the input arm. The force exerted at the input arm must balance the combined weight of the object and the lifted fluid.

✓ CHECKPOINT

C-12-5 Hydrostatic Pressure

A cylindrical U tube, with both arms open to the atmosphere, is partially filled with water and oil (Figure 12-20). The density of oil is less than the density of water. Consider the pressure within the two fluids at points A, B, C, D, E, F, G, and H. Rank the points, from highest to lowest, in terms of the total pressure at these points. Use an equality sign where needed.

Highest _____ _____ _____ _____ Lowest

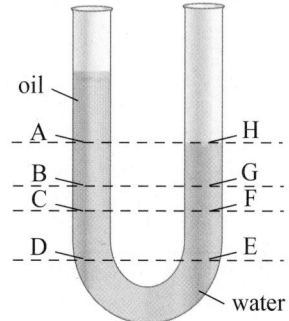

Figure 12-20 A U-shaped manometer

C-12-5 D = EC = FBGAH

 EXAMPLE 12-5

Hydraulic Car Jack

A hydraulic car jack has an output cylinder 30.0 cm in diameter and an input cylinder 5.00 cm in diameter, and is filled with oil of density 900 kg/m³. An automobile of mass 1100 kg needs to be elevated 40.0 cm.

(a) How far does the piston of the input cylinder need to move?
(b) What is the magnitude of the force on the input piston required to keep the car elevated?

SOLUTION

From Equation (12-15),

$$h_1 = \left(\frac{A_2}{A_1}\right)h_2$$

$$h_1 = \frac{\pi(0.15\text{ m})^2}{\pi(0.025\text{ m})^2} \times 0.4\text{ m} = 14.4\text{ m}$$

The magnitude of the force that needs to be applied on the input piston to keep the car elevated is given by Equation (12-17):

$$F_1' = \left(\frac{A_1}{A_2}\right)Mg + \rho_f g A_1 (h_1 + h_2)$$

$$F_1' = (1/36) \times (1100\text{ kg}) \times (9.81\text{ m/s}^2) + (900\text{ kg/m}^3)$$
$$\times (9.81\text{ m/s}^2) \times \pi(0.025\text{ m})^2 \times (14.8\text{ m})$$

$$= 299.7\text{ N} + 256.6\text{ N} = 556.3\text{ N}$$

The second term in Equation (12-17) is usually negligible if the fluid is a compressed gas.

LO 5

12-5 Buoyancy and Archimedes' Principle

If you hold a table tennis ball underwater and then release it, the ball accelerates upward and shoots above the surface. Clearly, the surrounding water must exert an upward force on the ball.

What causes this upward force? The answer lies in the facts that pressure exists at all points within a fluid and that this pressure increases with depth. Therefore, a fluid exerts forces that are normal to every point on the surface of an immersed object. The forces exerted by the water at various points on the surface of a fully immersed ball are shown in Figure 12-21. Since the lower half of the ball is at a greater depth, upward forces on the lower half of the ball are stronger than the downward forces on the upper half. The vector sum of all these forces is a single force that acts upward on the ball.

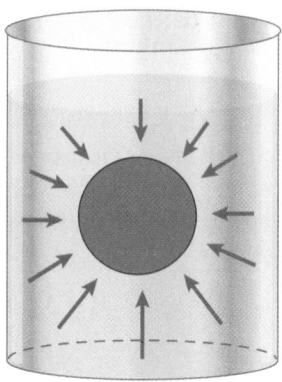

Figure 12-21 The forces due to fluid pressure on a sphere that is fully immersed in a liquid

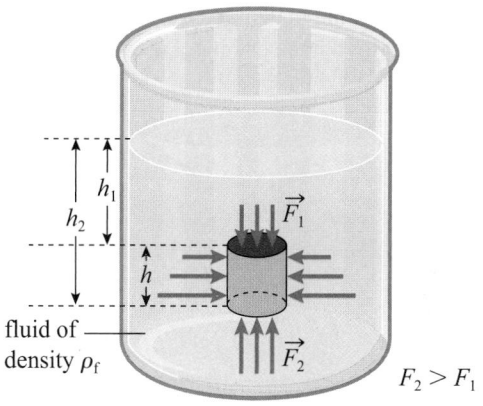

Figure 12-22 The forces exerted by the fluid on a fully submerged cylinder

We can conclude that the horizontal components of forces all cancel out because there is no horizontal motion when the ball is released.

The upward force exerted on an object that is fully or partially immersed in a fluid is called the **buoyant force**. We will denote the buoyant force by the symbol $\vec{F}_B$ and its magnitude by the symbol F_B. The direction and the magnitude of the buoyant force are stated by **Archimedes' principle**:

When an object is fully or partially immersed in a fluid, the fluid exerts an upward force (called the buoyant force) on the object. The magnitude of this upward force is equal to the weight of the fluid displaced by the object.

The genius of Archimedes was in realizing that the magnitude of the buoyant force is equal to the weight of the fluid displaced by the object. The more you immerse a table tennis ball into the water, the greater the volume of water displaced by the ball and the greater the upward force that the water exerts on it. Once the ball is completely submerged, it has displaced the maximum amount of water. If the ball is then moved anywhere within the body of the water, while remaining below the surface, it experiences exactly the same buoyant force (assuming that the density of the water does not change with depth and the ball does not deform).

Imagine a vertical cylinder of height h and cross-sectional area A that is fully immersed in an open container of fluid of density ρ_f (Figure 12-22). The top of the cylinder is at a depth h_1 below the fluid surface, and the bottom of the cylinder is at a depth $h_2 = h_1 + h$ below the surface.

The pressure at the top end of the cylinder, P_1, is

$$P_1 = P_0 + \rho_f g h_1 \qquad (12\text{-}18)$$

Therefore, the magnitude of the downward force, F_1, on the top of the cylinder is

$$F_1 = (P_0 + \rho_f g h_1)A \qquad (12\text{-}19)$$

The pressure at the bottom end of the cylinder, P_2, is

$$P_2 = P_0 + \rho_f g h_2 \qquad (12\text{-}20)$$

The magnitude of the upward force, F_2, on the bottom end of the cylinder is

$$F_2 = (P_0 + \rho_f g h_2)A \qquad (12\text{-}21)$$

The horizontal force at any given point on the side of the cylinder is cancelled by the corresponding force on the diametrically opposite point. Since $h_2 > h_1$, the magnitude of the upward force is greater than the magnitude of the downward force. The upward force exerted on the cylinder by the fluid, F_B, is given by

$$\begin{aligned}
F_B &= F_2 - F_1 = (P_0 + \rho_f g h_2)A - (P_0 + \rho_f g h_1)A \\
&= (\rho_f g h_2 - \rho_f g h_1)A \qquad (12\text{-}22) \\
&= \rho_f g (h_2 - h_1)A = \rho_f g V_o
\end{aligned}$$

The factor $V_o = (h_2 - h_1)A$ is the volume of the cylinder and is equal to the volume of the fluid displaced by the cylinder. The quantity $\rho_f V_o$ is the mass of the fluid displaced by the cylinder, and $\rho_f V_o g$ is the weight of the fluid displaced by the cylinder. Therefore, the magnitude of the buoyant force exerted by the fluid on the cylinder is equal to the weight of the fluid displaced by the cylinder.

KEY EQUATION
$$\begin{aligned}
F_B &= \text{weight of the displaced fluid} \\
&= (\rho_f V_o)g \qquad (12\text{-}23)
\end{aligned}$$

Although we have derived this result for a vertically immersed cylinder, the general result holds for any orientation or shape of an object. Consider a container filled with a fluid, and imagine that a three-dimensional segment of this fluid is suspended within it (Figure 12-23). The fluid is stationary, so this segment is in static equilibrium with the rest of the fluid. Therefore, the buoyant force exerted on this segment by the rest of the fluid must be equal to the weight of the fluid within the segment. Now, suppose that we replace this segment by a solid object of exactly the same shape. The fluid around the object would exert exactly the same buoyant force on the object. Therefore, the magnitude of the buoyant force on a fully or partially immersed object is equal to the weight of the fluid displaced by that object.

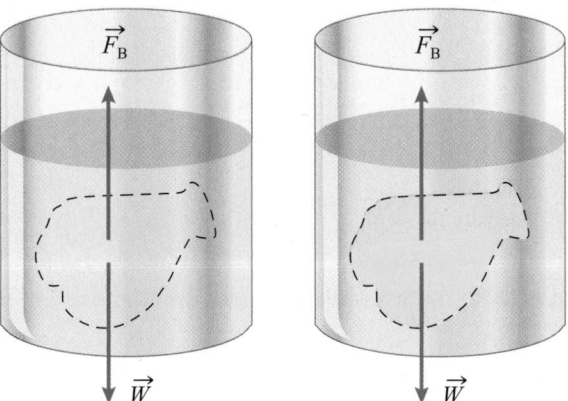

Figure 12-23 The buoyant force on an object immersed in a fluid is equal to the weight of the fluid displaced by the object.

✓ **CHECKPOINT**

C-12-6 The Buoyant Force on Spheres

Three identical spheres of different materials are held fully immersed in a container of water, as shown in Figure 12-24. Which of the following statements concerning the magnitude of buoyant force on each sphere is correct?

(a) $F_{B,cork} < F_{B,aluminum} < F_{B,lead}$
(b) $F_{B,cork} = F_{B,aluminum} = F_{B,lead}$
(c) $F_{B,cork} < F_{B,aluminum} = F_{B,lead}$
(d) $F_{B,cork} > F_{B,aluminum} > F_{B,lead}$

Figure 12-24 C-12-6

C-12-6 (b)

Flotation

The net vertical force on an object floating in a fluid is zero:

$$\vec{F}_{net} = \vec{F}_B + \vec{W} = 0 \quad \text{(for a floating object)} \quad (12\text{-}24)$$

Since the buoyant force and the weight act in opposite directions, the magnitude of the buoyant force on a floating object must be equal to the magnitude of the object's weight, W (Figure 12-25):

$$F_B = W \quad \text{(for a floating object)} \quad (12\text{-}25)$$

Note that Equation (12-24) is derived from Newton's second law of motion, and confirms that Archimedes' principle can be applied to floating objects.

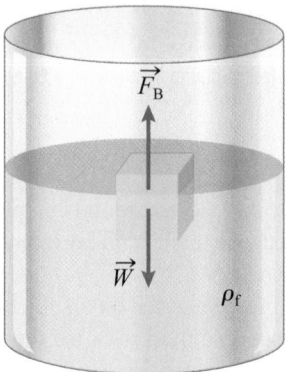

Figure 12-25 For a floating object, the magnitude of the upward buoyant force exerted on the object by the fluid is equal to the magnitude of the weight of the object.

Consider an object of volume V_o and density ρ_o that is floating in a fluid of density ρ_f. Assume that a volume V_{sub} of the object is submerged in the fluid. Then

$$W = \text{weight of the object} = (\rho_o V_o)\,g$$

$F_B = $ bouyant force on the object

$\quad$ = weight of the fluid displaced by the object

$\quad$ = (mass of fluid displaced by the object) g

$\quad = (\rho_f V_{sub})g$

Substituting into Equation (12-25), we get

$$\rho_o V_o g = \rho_f V_{sub} g$$

$$\text{and} \quad \frac{V_{sub}}{V_o} = \frac{\rho_o}{\rho_f} \quad (12\text{-}26)$$

Therefore, the fraction of the volume of a floating object below a fluid's surface is equal to the ratio of the density of the object to the density of the fluid.

The density of ocean water is approximately 1020 kg/m³, and the density of sea ice is 920 kg/m³. Therefore,

$$\frac{V_{sub}}{V_o} = \frac{920 \text{ kg/m}^3}{1020 \text{ kg/m}^3} = 0.902$$

approximately 90% of an iceberg is beneath the ocean's surface (Figure 12-26).

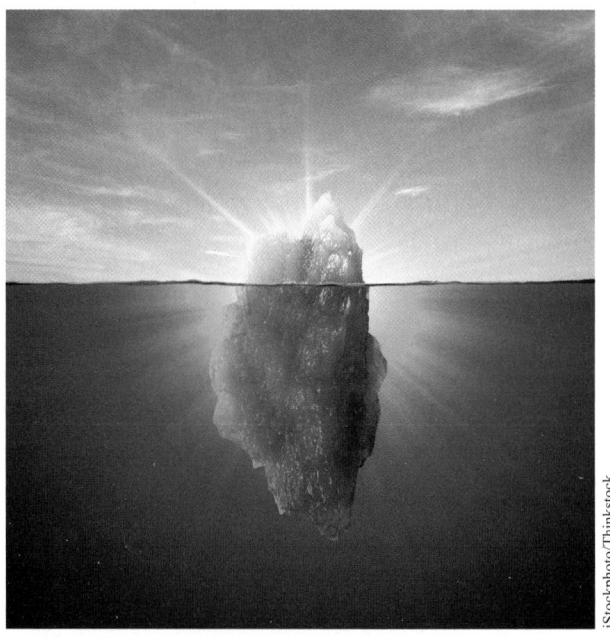

Figure 12-26 Approximately 90% of the volume of an iceberg is beneath the ocean's surface.

iStockphoto/Thinkstock

Apparent Weight in a Fluid

If the average density of an object is greater than the density of a fluid, the object will submerge in that fluid. The weight of the object is greater than the weight of the displaced fluid. The **apparent weight**, $W_{apparent}$, of the fully submerged object is defined as the difference between the weight of the object when not immersed and the buoyant force exerted on the object when it is immersed in the fluid. The magnitude of the apparent weight is given by

$$W_{apparent} = W - F_B \qquad (12\text{-}30)$$

A convenient method to determine the buoyant force on an object is to first weigh the object outside the fluid and then when it is completely immersed in a fluid. The difference between these two weights equals the magnitude of the buoyant force on the object. This method can also be used to determine the average density of an object.

EXAMPLE 12-6

How Large Is That Helium Balloon?

A balloon is carrying two people in a basket. The total mass of the people, the basket, and the material of the balloon, but excluding the mass of helium gas, is 280 kg. What volume of helium is needed to lift the balloon (Figure 12-27)? The density of the helium is 0.18 kg/m³, and the density of the air is 1.30 kg/m³.

Figure 12-27 A helium (or hot air) filled balloon floats when the buoyant force exerted on the balloon by the surrounding air is equal to the weight of the balloon plus the enclosed gas.

Karl R. Martin/Shutterstock.com

SOLUTION

When a balloon is filled with helium gas, the surrounding air exerts a buoyant force on the balloon and the gas inside it. As more helium is added, the volume of the balloon increases, so more air is displaced and the buoyant force on the balloon increases. When the buoyant force equals the total weight of the balloon (including the helium), the balloon will begin to float; adding any more helium will cause the balloon to accelerate upward.

The total weight of the balloon is

W = weight of balloon's contents
  + weight of the helium gas filling the balloon

$$W = 280\,g + m_{He}\,g = 280\,g + (V_{He}\,\rho_{He})\,g \qquad (12\text{-}27)$$

In Equation (12-27), V_{He} is the volume of the helium filled-balloon, and ρ_{He} is the density of helium gas. The buoyant force exerted by the air on the balloon is

F_B = weight of the air displaced by the balloon

$$= m_{air}\,g = (V_{He}\,\rho_{air})\,g \qquad (12\text{-}28)$$

At liftoff, the buoyant force equals the total weight of the balloon, so

$$280\,g + V_{He}\,\rho_{He}\,g = (V_{He}\,\rho_{air})\,g$$

and

$$V_{He} = \frac{280\ \text{kg}}{(\rho_{air} - \rho_{He})} = \frac{280\ \text{kg}}{(1.30 - 0.18)\ \text{kg/m}^3} = 250\ \text{m}^3$$

$$(12\text{-}29)$$

With 250 m³ of helium, the balloon is ready to lift off. To initiate an upward velocity, the amount of helium pumped into the balloon must be greater than the above calculated value.

EXAMPLE 12-7

Is the Ring Made of Pure Gold?

Your friend asks you to determine whether the ring that she recently bought is made of pure gold. (The density of gold is 19.3×10^3 kg/m³.) Using a thread of negligible mass and volume, you tie the ring to a very accurate spring balance and determine that in air it weighs 6.40×10^{-3} N. You then fully immerse the ring in a beaker full of water and find that its apparent weight is 6.00×10^{-3} N. Is the ring made of pure gold?

SOLUTION

To solve this problem, we apply the key points of this section as follows:

1. The difference between the weight of the ring in air and the weight of the ring in water is equal to the magnitude of the buoyant force on the ring.
2. The magnitude of the buoyant force is equal to the weight of the water displaced by the ring.
3. Knowing the density of the water and the value of g, the volume of water displaced by the ring can be calculated.
4. The volume of the ring is equal to the volume of the displaced water.

5. The weight of the ring is known; therefore, we can calculate the density of the ring.

Let us now do these calculations step by step:

1. $F_B = W_{ring,air} - W_{ring,water}$

   $= 6.40 \times 10^{-3}\,\text{N} - 6.00 \times 10^{-3}\,\text{N} = 0.40 \times 10^{-3}\,\text{N}$

2. $W_{displaced}$ = Weight of displaced water

   $= \rho_{water} V_{water}\, g = 0.40 \times 10^{-3}\,\text{N}$

   $V_{water} = \dfrac{0.40 \times 10^{-3}\,\text{N}}{\rho_{water}\, g} = \dfrac{0.40 \times 10^{-3}\,\text{N}}{1000\,\text{kg/m}^3 \times 9.81\,\text{m/s}^2}$

   $= 4.08 \times 10^{-8}\,\text{m}^3$

3. $V_{ring} = V_{water} = 4.08 \times 10^{-8}\,\text{m}^3$

4. $W_{ring,air} = m_{ring,air}\, g = (V_{ring}\, \rho_{ring})g$

5. $\rho_{ring} = \dfrac{W_{ring,air}}{V_{ring}\, g} = \dfrac{6.4 \times 10^{-3}\,\text{N}}{(4.08 \times 10^{-8}\,\text{m}^3) \times 9.81\,\text{m/s}^2}$

   $= 16.0 \times 10^3\,\text{kg/m}^3$

Since the density of the ring is considerably less than the density of pure gold (19.3×10^3 kg/m³), the ring is not made of pure gold.

✓ CHECKPOINT

C-12-7 Pressure on the Scale

A container, partially filled with water, is resting on a scale that measures its weight. If you immerse your hand in the water, without touching any part of the container and not spilling any water, what happens to the reading on the scale? Explain your reasoning.

(a) It stays the same.
(b) It increases.
(c) It decreases.
(d) Insufficient information is provided to answer this question.

C-12-7 (b)

LO 6

12-6 Fluids in Motion

There is an obvious difference in the energy of water in a large wave and stationary water in a cup. The water in a wave has significant kinetic energy. Water in a cup is at rest and has no net kinetic energy. Any physical description of a moving fluid must include its kinetic energy. We start our discussion of moving fluids by first introducing the concepts of kinetic energy per unit volume and potential energy per unit volume for a fluid.

Kinetic and Potential Energy Per Unit Volume

Consider a fluid of uniform density ρ moving with a velocity $\vec{v}$. The mass of a small volume ΔV of this fluid is $\Delta m = \rho \Delta V$, and its kinetic energy is

$$\Delta K = \frac{1}{2}(\Delta m)v^2 = \frac{1}{2}\rho v^2(\Delta V) \qquad (12\text{-}31)$$

The kinetic energy per unit volume of the fluid is

$$\frac{\Delta K}{\Delta V} = \frac{1}{2}\rho v^2 \qquad (12\text{-}32)$$

The SI unit for kinetic energy per unit volume is the pascal (Pa), the same as pressure:

$$\frac{\text{kg}}{\text{m}^3} \times \frac{\text{m}^2}{\text{s}^2} = \frac{\text{kg}}{\text{ms}^2} = \text{Pa}$$

We can similarly define potential energy per unit volume for a fluid. The gravitational potential energy of a small element of fluid that has volume ΔV and is located at a height h from the ground is

$$\Delta U = (\Delta m)\,gh = \rho gh(\Delta V) \qquad (12\text{-}33)$$

The potential energy per unit volume of the fluid is then

$$\frac{\Delta U}{\Delta V} = \rho gh \qquad (12\text{-}34)$$

The SI unit for potential energy per unit volume is also the pascal.

Ideal Fluids

A general description of moving fluids is complicated because fluids exhibit a wide variety of behaviour that depends on the type of fluid and its speed. The motion of water smoothly emerging from a tap is very different from the chaotic flow of water over Niagara Falls. Similarly, it is easier to suck water through a straw then it is to suck honey through the same straw. However, we can develop a fairly reasonable description of moving fluids by making some key assumptions about the physical properties of the fluids and how the fluids flow. These assumptions are as follows:

A fluid is incompressible

We assume that the density of a fluid remains constant and does not change with pressure. This assumption closely approximates the behaviour of liquids. For example, the density of water decreases by approximately 1% at a depth of 1000 m below the ocean's surface, where the absolute pressure is 100 times greater than the atmospheric pressure. The assumption that a gas is incompressible is only valid if the change in the gas's pressure is small.

A fluid is inviscid

Viscosity is the measure of a fluid's resistance to flow and to any deformation of its shape. For example, water flows much more readily than honey. The assumption that a fluid is inviscid implies that frictional forces within the fluid, as well as between the fluid and the walls of the containing vessel, are negligible. A frictionless flow is an idealization because all real fluids have nonzero friction. We will discuss viscous flow later in the chapter. A fluid that is incompressible and inviscid is called an **ideal fluid**.

The flow of the fluid is steady

A **steady flow** is smooth and uniform. In a steady flow, the velocity of the fluid at any *given* point within the region of flow has a fixed magnitude and direction and remains constant with time. The velocities at *different* points in the region of flow may differ from each other. Water slowly flowing from a tap (Figure 12-28) is an example of a steady flow. In this case, the velocity of water at a fixed distance below the tap is well-defined and does not change with time as long as the flow rate through the tap remains the same.

A flow that is not steady is called **turbulent**. In a turbulent flow, the velocity at a given point in the region of flow varies with time. The flow of gases and ash from the eruption of Mount St. Helens is an example of turbulent flow (Figure 12-29). The mathematical description of turbulent flows is quite complex.

The flow of the fluid is irrotational

In an irrotational flow, a fluid element does not move in a circular path. An example of irrotational flow is water moving in a straight pipe or falling straight downward under the influence of gravity.

Comstock/Thinkstock

Figure 12-28 Water gently flowing from a tap is an example of a steady flow.

Chaotic flow from the mouth of Mount St. Helens (Washington, U.S.A.). Compare this flow to the flow of water slowly emerging from a tap.

Photograph courtesy of the U.S. Geological Survey

Figure 12-29 The eruption of Mount St. Helens. The flow from the volcano is chaotic and turbulent.

Streamlines and Flow Tubes

When the flow is steady, fluid particles move along well-defined paths called streamlines. A **streamline** is a curve drawn in the body of a moving fluid such that the velocity of a particle at any point along the streamline is tangent to the streamline at that point. Figure 12-30 shows velocity vectors at various points along a streamline. The instantaneous velocity of a particle must be unique, and for steady flow the velocity at any point in the region of the flow does not change with time. *Therefore, in a steady flow, streamlines do not cross each other.* A steady flow is also called a **streamline**, or **laminar flow**.

Imagine a set of neighbouring streamlines that form a closed path in a given region of a moving fluid. The particles of the fluid that are either on the boundary of this closed path or are enclosed within it would follow their respective streamlines as they move along the fluid. These neighbouring streamlines form a tube, called a **flow tube** (Figure 12-30). The fluid within a flow tube must remain within the tube. There is no mixing of fluid from different flow tubes. The cross-sectional area of a given flow tube may change along its length, but the amount of the fluid that enters at one end of a flow tube must be equal to the amount that leaves through another end down the stream. You can envision a flow tube as an invisible pipe that keeps the fluid within it from mixing with the fluid outside the pipe.

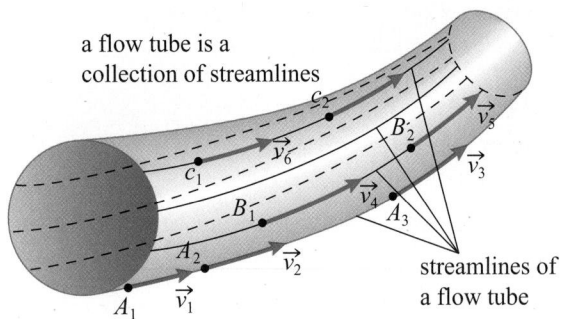

Figure 12-30 A flow tube. Solid lines represent streamlines. The magnitude and direction of the flow velocity at any point along the flow line is given by the length and direction of the arrow at that point. The direction of the arrow is tangential to the streamlines. The group of streamlines forms a stream tube (or flow tube).

There are several ways to visualize streamlines and flow tubes. Smoke in gases and dyes in liquids are often used by engineers and scientists to study flow patterns of fluids around objects (Figure 12-31). With computers, scientists can calculate the pattern of streamlines around objects. A study of such patterns provides information about the resistance (or drag) experienced by an object as it moves through a fluid. Smooth and continuous streamlines indicate a steady flow with minimum drag. Closely spaced streamlines indicate high speed.

Keep in mind that a description of an object moving through a stationary fluid is the same as that of a fluid that is moving past a stationary object. That is why streamlines created by moving air around a stationary automobile in a wind tunnel are exactly the same as the streamlines of an automobile moving through stationary air.

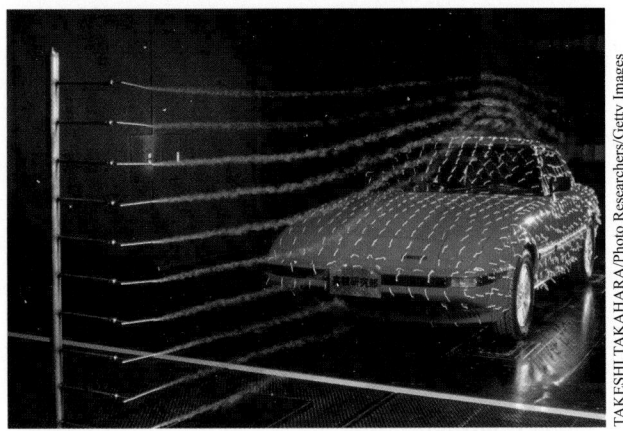

Figure 12-31 Streamlines of smoke passing over an automobile in a wind tunnel

Flow Rate

The **flow rate** (Q) is the volume of fluid that flows through a cross-sectional area per unit time:

$$Q = \frac{\text{volume of fluid that flows through a surface}}{\text{time taken}}$$

(12-35)

The SI units for flow rate are cubic metres per second (m^3/s). Other convenient units are litres per second (L/s) and cubic centimetres per second (cm^3/s):

$$1\ m^3/s = 10^6\ cm^3/s = 1000\ L/s$$

For a fluid moving with a speed v through a cross-sectional area A that is perpendicular to the direction of flow, the volume flow rate is the product of A and v:

$$Q = Av$$

(12-36)

LO 7

12-7 The Continuity Equation: Conservation of Fluid Mass

Consider water flowing through a pipe of variable cross-sectional area (Figure 12-32). If there are no other sources of water and no holes in the pipe, then the amount of water that enters the pipe at one end in some time Δt must be equal to the amount that leaves at the other end in the same time. In other words, what comes in must go out. The mathematical equation that describes this "law of conservation of fluid mass" is called the **equation of continuity**.

Consider a section of a pipe (or a flow tube) where the cross-sectional area is A_1, the speed of the fluid is v_1, and the density of the fluid is ρ_1. In time Δt, a length $\Delta x_1 = v_1\ \Delta t$ of the fluid passes through A_1. The volume, ΔV_1, of the

Figure 12-32 Fluid flowing through a horizontal pipe of variable cross-sectional area

fluid passing through A_1 is $\Delta V_1 = A_1(v_1 \Delta t)$. The mass of this volume of the fluid is

$$\Delta M_1 = \rho_1 \Delta V_1 = (\rho_1 A_1 v_1)\Delta t \qquad (12\text{-}37)$$

In the same time interval, a mass ΔM_2 of the fluid must pass through another section of the pipe where the cross-sectional area is A_2, the fluid speed is v_2, and the fluid density is ρ_2:

$$\Delta M_2 = \rho_2 \Delta V_2 = (\rho_2 A_2 v_2)\Delta t \qquad (12\text{-}38)$$

Since the mass of the fluid entering through A_1 must be equal to the mass leaving through A_2, we must have

$$\rho_1 A_1 v_1 = \rho_2 A_2 v_2 \qquad (12\text{-}39)$$

The quantity ρAv has the dimensions of kg/s and is called the **mass flow rate**. For incompressible fluids $(\rho_1 = \rho_2)$ the above equation reduces to a simpler relationship:

KEY EQUATION $A_1 v_1 = A_2 v_2$ (for incompressible fluids)
$$(12\text{-}40)$$

The quantity Av is the volume flow rate, Q. The continuity equation states that for incompressible fluids, the volume flow rate remains constant through all sections of a flow tube. An important consequence of the equation of continuity is that the fluid moves faster through a narrower section of a flow tube and slows when moving through a wider section (Figure 12-33).

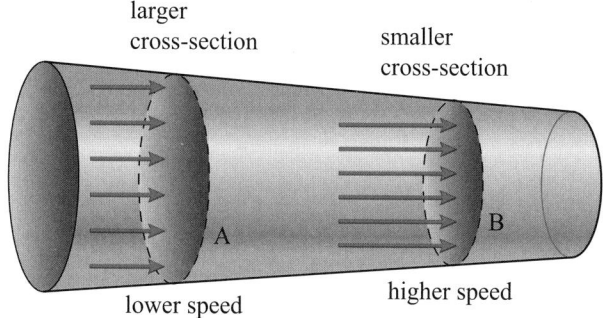

Figure 12-33 When a fluid moves through a horizontal pipe that gradually narrows, the fluid speed increases as the cross-sectional area of the pipe decreases.

 MAKING CONNECTIONS

Water Flowing Through a Tap

When you turn on a water tap, keeping the flow smooth and slow, you may notice that as the stream of water flows downward, it continuously narrows (Figure 12-28). Why does the stream narrow as it accelerates downward? The water flows out of the tap with an initial flow rate that is determined by the cross-sectional area of the mouth of the tap and the speed with which the water leaves the tap. Water accelerates due to gravity, so its speed increases. Since the flow rate must remain constant, the cross-sectional area of the stream must decrease with increasing speed of water.

The same phenomenon occurs when you gently tilt a spoonful of honey and watch the stream of honey as it accelerates downward.

✓ CHECKPOINT

C-12-8 Flow Rate Through Pipes

Water flows continuously in and out of an arrangement of pipes as shown in Figure 12-34. What is the flow rate from the pipe A? Is the water flowing into or out of this pipe?

Figure 12-34 Water flow through an arrangement of pipes

C-12-8 5 cm³/s; the water is flowing out of the pipe.

LO 8

12-8 Conservation of Energy for Moving Fluids

According to the equation of continuity, as a fluid passes through a narrower section of a pipe (or a flow tube), its speed increases. For an element of fluid to speed up, the fluid behind it must exert a force on it. Therefore, the pressure in the wider section of the pipe must be higher than the pressure in the narrower section. We thus reach an interesting and nonintuitive conclusion: *In the regions of flow where the fluid moves faster, the pressure within the fluid is lower than the pressure in the regions where the fluid speed is slower.* In this section, we will show that this relationship is a consequence of the law of conservation of energy as applied to moving fluids.

Consider an ideal fluid flowing through a flow tube of variable cross-sectional area and variable height (measured with respect to an arbitrary ground level). The fluid is moving from left to right (Figure 12-35). Let us focus on the body of the fluid that is contained between two sections (1 and 2) of the flow tube. We will calculate the work done on this "test volume" of fluid by the surrounding fluid (Figure 12-36). At section 1, the pressure in the fluid is P_1, the speed of the fluid is v_1, the cross-sectional area of the tube is A_1, and the height of the tube is y_1. Corresponding quantities at section 2 are

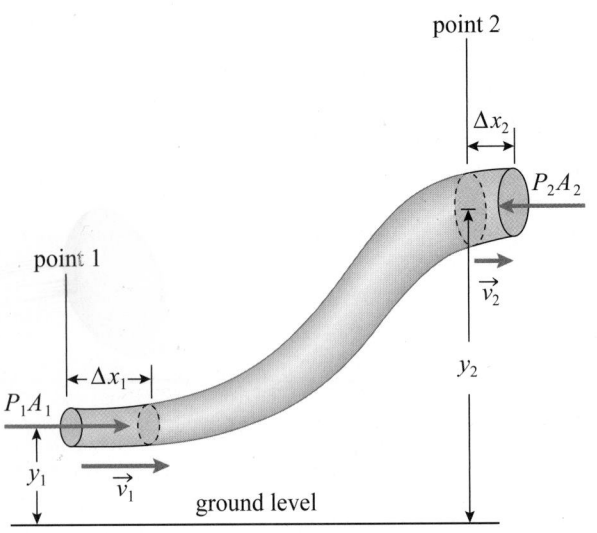

Figure 12-35 An ideal fluid flowing through a pipe of variable height and cross-sectional area

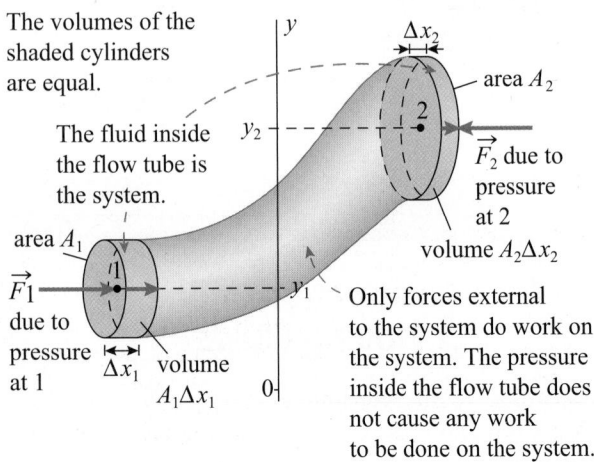

Figure 12-36 Fluid flow through a flow tube of variable cross-sectional area and height

denoted by P_2, v_2, A_2, and y_2. According to the work–energy theorem, in the absence of frictional forces, the net work done by external forces on an object is equal to the change in the object's total energy:

$$W_{net} = \Delta K + \Delta U \tag{12-41}$$

Here, W_{net} is the net work done on the object by external forces, ΔK is the change in the kinetic energy of the object, and ΔU is the change in the object's potential energy. Note that only external forces are considered when calculating the work done on the object because internal forces always come in action–reaction pairs and cancel out. Also, the total energy of the object is the sum of its kinetic and potential energies. Although potential energy could be of several forms, in dealing with fluids we will only consider gravitational potential energy.

In time Δt, a length $\Delta x_1 = v_1 \Delta t$ of the fluid enters through section 1. The volume of the entering fluid is

$$\Delta V = A_1 (v_1 \Delta t) \tag{12-42}$$

In the same time, the volume of fluid that leaves through section 2 is $A_2(v_2 \Delta t)$. By the equation of continuity,

$$\Delta V = A_1 v_1 \Delta t = A_2 v_2 \Delta t \tag{12-43}$$

The entering volume has speed v_1, and the speed of the leaving volume is v_2. Therefore, the change in the kinetic energy (ΔK) of the test volume is

$$\Delta K = (\text{kinetic energy of leaving fluid})$$
$$- (\text{kinetic energy of entering fluid})$$

$$\Delta K = \frac{1}{2}(\rho \Delta V)v_2^2 - \frac{1}{2}(\rho \Delta V)v_1^2 \tag{12-44}$$

The fluid enters at a height y_1 and leaves at a height y_2. Therefore, in the time Δt, the change in the potential energy (ΔU) of the test volume is

$$\Delta U = (\text{potential energy of leaving fluid})$$
$$- (\text{potential energy of entering fluid})$$

$$\Delta U = (\rho \Delta V) g y_2 - (\rho \Delta V) g y_1 \tag{12-45}$$

The fluid entering through section 1 exerts a force on the test volume, displacing it toward the right by a distance $v_1 \Delta t$ in time Δt. The magnitude of the force exerted by the entering fluid is $P_1 A_1$. Therefore, the **work done on the test volume** by the entering fluid, W_1, is

$$W_1 = P_1 A_1 (v_1 \Delta t) \tag{12-46}$$

At section 2, the magnitude of the force exerted by the external fluid on the enclosed test volume is $P_2 A_2$. As the test volume moves to the right, it does work on the fluid in front of it. The **work done by the leaving fluid**, W_2, is

$$W_2 = -P_2 A_2 (v_2 \Delta t) \tag{12-47}$$

The minus sign in W_2 implies that the force exerted on the enclosed test volume is in a direction opposite to that of fluid flow. The net work done on the test volume by the surrounding fluid in time Δt is therefore

$$W_{net} = W_1 + W_2 = P_1 A_1 (v_1 \Delta t) - P_2 A_2 (v_2 \Delta t) \quad (12\text{-}48)$$

Using Equation (12-43), we can write Equation (12-48) as

$$W_{net} = (P_1 - P_2)\Delta V \quad (12\text{-}49)$$

Inserting Equations (12-44), (12-45), and (12-49) into Equation (12-41), we obtain

$$(P_1 - P_2)\Delta V = \frac{1}{2}(\rho \Delta V)v_2^2 - \frac{1}{2}(\rho \Delta V)v_1^2$$
$$+ (\rho \Delta V)gy_2 - (\rho \Delta V)gy_1 \quad (12\text{-}50)$$

Dividing by ΔV gives

$$P_1 - P_2 = \frac{1}{2}\rho v_2^2 - \frac{1}{2}\rho v_1^2 + \rho g y_2 - \rho g y_1 \quad (12\text{-}51)$$

Rearranging so that all terms corresponding to a given section are on the same side,

KEY EQUATION $\quad P_1 + \frac{1}{2}\rho v_1^2 + \rho g y_1 = P_2 + \frac{1}{2}\rho v_2^2 + \rho g y_2$

$$(12\text{-}52)$$

Equation (12-52) is called **Bernoulli's equation**. Bernoulli's equation is a statement of the law of conservation of energy for ideal fluids in the absence of frictional forces. It states that during the steady flow of an ideal fluid, the sum of the fluid pressure, the kinetic energy per unit volume, and the potential energy per unit volume remains constant throughout the body of the fluid. Since points 1 and 2 can be located anywhere along the flow tube, we can also write Bernoulli's equation in the following form:

$$P + \frac{1}{2}\rho v^2 + \rho g y = \text{constant}$$

(for every point along a streamline in a flow tube)

$$(12\text{-}53)$$

Equations (12-40) and (12-52) are the two fundamental equations describing the flow of ideal fluids. In many situations a description of the flow requires using both of these equations. Let us examine two special cases.

Fluid flow through a horizontal pipe Consider an ideal fluid flowing through a horizontal pipe of variable cross-sectional area, as shown in Figure 12-37. Imagine a streamline that passes through the centre of the pipe with point 1 located in the wider section and point 2 in the narrower section of the pipe. Since the pipe is horizontal, $y_1 = y_2$, and the gravitational potential energy of

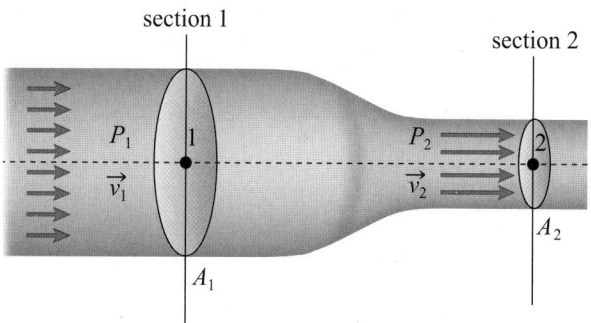

Since $v_1 < v_2$, P_1 must be greater than P_2. Bernoulli's equation predicts that the pressure is lower in the section of a tube in which the fluid moves faster.

Figure 12-37 Fluid flow along a horizontal pipe of variable cross-sectional area

the fluid does not change. Bernoulli's equation simplifies to the following:

$$P_1 + \frac{1}{2}\rho v_1^2 = P_2 + \frac{1}{2}\rho v_2^2 \quad (12\text{-}54)$$

Equation (12-54) implies that the sum of pressure and kinetic energy per unit volume remains constant in a horizontal flow. When passing through the narrow section of the pipe, the fluid speeds up (because of the continuity equation) and, hence, the pressure within the fluid must decrease. Conversely, when a fluid slows down, the pressure within the fluid must increase. You might think that as a fluid speeds up, the pressure within the fluid should also increase, but the opposite is true.

Applying the equation of continuity at points 1 and 2, we get $v_2 = (A_1/A_2)\,v_1$, and the pressure difference between the two sections of the pipe is

$$P_1 - P_2 = \frac{1}{2}\rho v_1^2 \left(\frac{A_1^2}{A_2^2} - 1\right) \quad (12\text{-}55)$$

Thus, $P_1 > P_2$ if $A_1 > A_2$. Bernoulli's equation explains many commonly observed phenomena and has many practical applications. Some of these are discussed below.

MAKING CONNECTIONS

Blowing Over a Paper Strip

Cut a small strip of paper (about 2 cm × 16 cm), and place it close to the lower lip of your mouth, holding it with your fingers. The strip will hang downward because of its weight. Now blow hard such that the air moves horizontal to the top surface of the strip. The strip will rise. As you blow, the air above the top surface moves faster, causing the pressure above the strip to drop below atmospheric pressure. Since pressure beneath the strip remains equal to atmospheric pressure, an upward force acts on the strip, lifting it up.

MAKING CONNECTIONS

The Venturi Meter

As a fluid passes through a constricted region, its speed increases (equation of continuity) and pressure within the fluid drops (Bernoulli's equation). A Venturi meter uses these principles to measure the speed of a fluid. One type of Venturi meter (Figure 12-38) consists of a horizontal pipe of area A_1 that is gradually constricted at the centre where the area is A_2. A fluid of density ρ_0, such as air, enters the horizontal tube at one end and leaves through the other. A U-tube of uniform area and partially filled with a liquid of density ρ is attached to the horizontal tube.

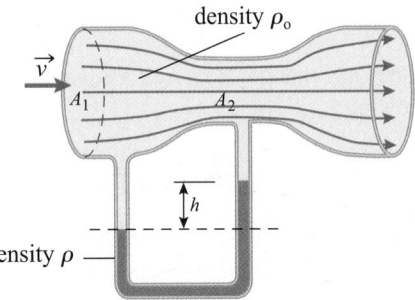

Figure 12-38 A Venturi meter

When air in the horizontal tube is stationary, the height of the liquid in the two arms of the U-tube is the same. Suppose the air flows through A_1 with speed v. Its speed will increase as it passes through the constricted region, resulting in a pressure drop at the constriction, and the liquid in that arm of the U-tube will rise. The following equation relates the air speed, v, and the difference between the levels of the liquid in the two arms, h:

$$v = \sqrt{2g\left(\frac{\rho - \rho_0}{\rho_0}\right)\left(\frac{A_2^2}{A_1^2 - A_2^2}\right)} \sqrt{h}$$

For a given Venturi meter, all quantities in the square root of the first factor are known, so the air speed is determined by measuring h.

CHECKPOINT

C-12-9 Blood Pressure in a Clogged Artery

Blood passes through a section of human artery that is partially blocked due to cholesterol buildup. As the blood enters from the unblocked (healthy) part of the artery to the blocked part, the blood pressure in the blocked artery

(a) decreases;
(b) remains the same;
(c) increases;
(d) may increase or decrease, depending on the pressure in the unblocked section.

C-12-9 (a)

Constant velocity flow For flow through a pipe of uniform cross section, the equation of continuity requires that the fluid speed remain constant. For this case, we obtain a simplified form of Bernoulli's equation:

$$P_1 + \rho g y_1 = P_2 + \rho g y_2 \qquad (12\text{-}56)$$

Here, the sum of pressure and the gravitational potential energy per unit volume remains constant along a streamline. As the height of the fluid increases, the pressure within the fluid decreases. Conversely, a decrease in the height results in an increase in fluid pressure. Water pressure is lower on higher floors of a building compared to the ground floor.

EXAMPLE 12-8

Water Pressure in a Home

Water flows with a constant speed through a pipe of constant cross-sectional area in the ground floor of a house. The pipe rises 5.00 m to reach a showerhead on the second floor (Figure 12-39). The gauge pressure at ground level is 140 kPa.

(a) What is the gauge pressure at the shower head?
(b) What is the maximum height that water from ground level could reach?

SOLUTION

(a) Since the water is moving, its behaviour is described (assuming a steady flow) by both the continuity equation and Bernoulli's equation. The area of the pipe does not change, so the flow speed remains constant.

We designate the segment of the pipe at ground level as 1 and the segment at the height of the showerhead as 2.

328   SECTION ONE | MECHANICS   NEL

Figure 12-39 Water flow to a showerhead

gauge pressure at ground level, $P_1 = 140.0 \times 10^3$ Pa

height of the section at ground level $= y_1$

height of the section at the showerhead level $= y_2$

gauge pressure at the showerhead level, P_2 is to be determined

For a pipe of constant cross section, Bernoulli's equation reduces to the form

$$P_1 + \rho g y_1 = P_2 + \rho g y_2$$
$$P_2 = P_1 + \rho g y_1 - \rho g y_2 \qquad (12\text{-}57)$$

Inserting the given values of the quantities on the right-hand side yields

$$P_2 = 140 \times 10^3 + (1000 \text{ kg/m}^3)(9.81 \text{ m/s}^2)(-5 \text{ m})$$

$$P_2 = 91.0 \times 10^3 \text{ Pa}$$

(b) At the maximum height, the gauge pressure is zero. Let us denote this height by y_{max}. Then, $P_2 = 0$ at $y_2 = y_{max}$. From Equation (12-57) we get

$$0 = P_1 - \rho g (y_{max} - y_1)$$

and $\quad y_{max} - y_1 = \dfrac{P_1}{\rho g} = \dfrac{140.0 \times 10^3 \text{ Pa}}{(1000 \text{ kg/m}^3)(9.81 \text{ m/s}^2)} = 14.3 \text{ m}$

Making sense of the result:

The potential energy per unit volume of the fluid increases as the fluid moves upward. There is a corresponding decrease in fluid's pressure and that is why P_2 is less than P_1.

EXAMPLE 12-9

Water Flow Through a Pressurized Tank

Water flows from a large enclosed tank through a horizontal pipe of variable cross-sectional area, as shown in Figure 12-40. The upper end of the tank is maintained at an absolute pressure of 1.20×10^5 N/m² by pumping compressed air into it. The horizontal outlet of the pipe has a cross-sectional area of 4.00×10^{-2} m² in the larger section and 2.00×10^{-2} m² in the smaller section. A vertical tube of cross-sectional area 4.00×10^{-2} m² is connected to the horizontal tube and open to the atmosphere.

Figure 12-40 Example 12-9

(a) What is the speed of the fluid leaving the outlet when the height of the water in the large tank is 2.00 m?

(b) What is the speed of the water in the horizontal pipe of cross-sectional area 4.00×10^{-2} m²?

(c) To what height does the water stand in the vertical tube when the water height in the large tank is 2.00 m?

SOLUTION

To determine the unknown quantities for parts (a), (b), and (c), we consider a streamline that starts at the top surface of the water in the large tank, passes through the horizontal pipe, and leaves through the outlet. We label three points on this streamline as follows: Point 1 is at the top of the water surface in the large tank because we know the height and the pressure at this point. Point 2 is at the centre of that portion of the vertical tube that is at the bottom of the vertical tube, because we need to determine the height of the column of water in the vertical tube. Point 3 is at the outlet where the water leaves the horizontal pipe, because we know that the pressure at the outlet must be equal to the atmospheric pressure.

The variables in Bernoulli's equation and the continuity equation are pressure, speed, height, and the cross-sectional area of the surface perpendicular to the fluid flow. The values of these quantities at the points of interest are:

point 1:

$$P_1 = 1.20 \times 10^5 \text{ Pa}, \quad v_1 = \text{unknown},$$
$$y_1 = 2.00 \text{ m}, \quad A_1 = \text{unknown}$$

point 2:

$$P_2 = \text{unknown}, \quad v_2 = \text{unknown},$$
$$y_2 = 0.00 \text{ m}, \quad A_2 = 4.00 \times 10^{-2} \text{ m}^2$$

(continued)

point 3:

$$P_3 = P_0, \quad v_3 = \text{unknown},$$

$$y_3 = 0.00 \text{ m}, \quad A_3 = 2.00 \times 10^{-2} \text{ m}^2$$

Here, P_0 is atmospheric pressure, and we have chosen to measure all heights with respect to the streamline that passes through points 2 and 3. There are five unknowns: v_1, A_1, P_2, v_2, and v_3. We can use Bernoulli's equation between points 1 and 2 (or 3) and between points 2 and 3, with two corresponding continuity equations, giving a total of four equations so there is not enough information to solve for five unknown quantities and we have to make an approximation.

Since the tank is large, the rate at which the level of the water in the tank decreases is negligible. Therefore, we can make the approximation that the speed at which the water is falling in the tank is negligible:

$$v_1 \approx 0 \quad \text{(large cross-sectional area approximation)}$$

Notice that once we assume that $v_1 \approx 0$, we cannot use the continuity equation between points 1 and 2 (or between points 1 and 3) to determine v_2 or A_2 as both v_1 and A_1 have dropped out of the problem.

Since all the variables except v_3 are known at point 3, we apply Bernoulli's equation between points 1 and 3:

$$P_1 + \frac{1}{2}\rho v_1^2 + \rho g y_1 = P_3 + \frac{1}{2}\rho v_3^2 + \rho g y_3 \qquad (12\text{-}58)$$

Inserting the given values and the approximation $v_1 \approx 0$, and solving for v_3 we get,

$$v_3 = \sqrt{\frac{2}{\rho}(P_1 - P_0 + \rho g y_1)} \qquad (12\text{-}59)$$

Note that the speed v_3 depends on both the pressure difference $(P_1 - P_0)$ and the height difference $(y_1 - 0)$. Inserting the given values, we get

$$v_3 = \sqrt{\frac{2}{1000 \text{ kg/m}^3}(1.20 \times 10^5 \text{ Pa} - 1.01 \times 10^5 \text{ Pa} + (1000 \text{ kg/m}^3) \times (9.81 \text{ m/s}^2) \times (2.00 \text{ m}))}$$

$$v_3 = 8.80 \text{ m/s} \qquad (12\text{-}60)$$

(b) Having calculated v_3, we can now calculate the speed and the pressure at point 2. To calculate v_2, we apply the continuity equation between points 2 and 3:

$$A_2 v_2 = A_3 v_3$$

$$v_2 = \frac{A_3}{A_2}v_3 = \frac{2.00 \times 10^{-2} \text{ m}^2}{4.00 \times 10^{-2} \text{ m}^2} \times 8.80 \text{ m/s} = 4.40 \text{ m/s}$$

(c) In order to find the height, h, of the column of water in the vertical tube, we need to calculate the pressure at a point that is directly underneath the vertical tube. That is why we chose point 2 to be under the vertical tube. The water in the vertical tube is stationary, so the pressure P_2 at the bottom of the tube and the height h are related by

$$P_2 = P_0 + \rho g h$$

Pressure P_2 can be calculated by applying Bernoulli's equation between points 2 and 3:

$$P_2 + \frac{1}{2}\rho v_2^2 + \rho g y_2 = P_3 + \frac{1}{2}\rho v_3^2 + \rho g y_3$$

$$P_2 = P_3 + \left(\frac{1}{2}\rho v_3^2 - \frac{1}{2}\rho v_2^2\right) + (\rho g y_3 - \rho g y_2)$$

Inserting $P_3 = P_0$, $y_2 = y_3 = 0$, $v_2 = 4.39$ m/s, $v_3 = 8.80$ m/s, we get

$$P_2 = 1.01 \times 10^5 \text{ Pa} + \frac{1}{2} \times (1000 \text{ kg/m}^3)$$

$$\times ((8.80 \text{ m/s})^2 - (4.40 \text{ m/s})^2)$$

$$P_2 = 1.30 \times 10^5 \text{ Pa}$$

The height, h, of the water in the vertical tube can now be calculated:

$$h = \frac{P_2 - P_0}{\rho g} = \frac{1.30 \times 10^5 \text{ Pa} - 1.01 \times 10^5 \text{ Pa}}{(1000 \text{ kg/m}^3) \times (9.81 \text{ m/s}^2)}$$

$$h = 2.95 \text{ m}$$

ONLINE ACTIVITY

Non-viscous flow through a horizontal pipe

The e-resource that accompanies every new copy of this textbook contains an Online Activity using the PhET simulation "Non-Viscous Flow Through a Horizontal Pipe of Variable Cross-Sectional Area." Work through the simulation and accompanying questions to gain an understanding of the flow of a non-viscous fluid through a horizontal pipe of variable cross-sectional area.

 CHECKPOINT

C-12-10 Pressure and Area

If the cross-sectional area of the vertical tube in Example 12-9 is doubled, the height of the water in the tube would

(a) increase by a factor of 4;
(b) increase by a factor of 2;
(c) remain the same;
(d) decrease by a factor of 2.

C-12-10 (c)

MECHANICS

LO 9

12-9 Conservation of Fluid Momentum

Consider a fluid moving through a straight horizontal tube that gradually narrows (Figure 12-37). As the cross-sectional area of the tube decreases, the speed of the fluid increases. According to Newton's second law of motion, any change in a fluid's velocity indicates that an external force is acting on the fluid. Similarly, there must be a net force on a fluid that passes through a bend in a tube because its direction changes. To study situations in which a fluid's momentum changes, we need to understand how Newton's second law of motion is applied to moving fluids.

Consider an element of fluid flowing between two sections of a tube of variable cross-sectional area (Figure 12-37). At section 1, the fluid pressure is P_1, the velocity is $\vec{v}_1$, and the cross-sectional area of the tube is A_1. In time Δt, the mass of the fluid that enters section 1 from the left is

$$\Delta M_1 = \rho(A_1\, v_1\, \Delta t)$$

The momentum of the entering fluid is

$$\Delta \vec{p}_1 = \Delta M_1\, \vec{v}_1 = (\rho A_1\, v_1 \Delta t)\, \vec{v}_1$$

In the same time, the mass of fluid leaving the tube at section 2 is

$$\Delta M_2 = \rho(A_2\, v_2\, \Delta t)$$

and the momentum of the leaving fluid is

$$\Delta \vec{p}_2 = \Delta M_2\, \vec{v}_2 = (\rho A_2\, v_2 \Delta t)\, \vec{v}_2$$

The rate of change of momentum of the fluid between the two sections of the tube is, therefore,

$$\frac{\Delta \vec{p}_2 - \Delta \vec{p}_1}{\Delta t} = (\rho A_2\, v_2)\vec{v}_2 - (\rho A_1\, v_1)\, \vec{v}_1 \qquad (12\text{-}61)$$

By the equation of continuity, $A_1 v_1 = A_2 v_2 \equiv Q$, and the rate of change of the fluid's momentum is

$$\frac{\Delta \vec{p}_2 - \Delta \vec{p}_1}{\Delta t} = \rho Q\, (\vec{v}_2 - \vec{v}_1) \qquad (12\text{-}62)$$

The rate of change of momentum of a physical object is equal to the sum of all external forces acting on that object. We can then write Equation (12-62) in the form of Newton's second law of motion:

$$\sum \vec{F} = \rho Q\, (\vec{v}_2 - \vec{v}_1) \qquad (12\text{-}63)$$

Here, $\sum \vec{F}$ denotes the vector sum of all external forces acting on the fluid element. Equation (12-63) is Newton's second law of motion as applied to moving fluids. This equation indicates that a fluid element accelerates or decelerates in response to external forces. By Newton's third law of motion, the fluid exerts an equal and opposite force on the surrounding fluid or the containing vessel.

EXAMPLE 12-10

Keeping a Fire Hose Stationary

Consider a fire hose that has an inner diameter of 4.40 cm. The diameter of the nozzle connected to the end of the hose tapers from 4.40 cm to 2.00 cm. Water is flowing out of the nozzle at a rate of 10 L/s, and the gauge pressure at the wide end of the nozzle is 6.80×10^5 Pa. What force must a firefighter exert to keep the nozzle stationary? (Ignore the weight of the nozzle.)

SOLUTION

To hold the nozzle steady, the firefighter has to exert a force to counter the force that the water exerts on the nozzle. Imagine a flow tube that consists of the fluid inside the nozzle. The forces acting on the element of water within the flow tube are as follows (See Figure 12-41):

- $\vec{F}_1$ is the force exerted from the left by the water entering the nozzle;
- $\vec{F}_2$ is the force exerted from the right by the atmospheric pressure P_0 on the water that is leaving the nozzle; and
- $\vec{F}_w$ is the force that the nozzle exerts on the water.

For this situation, Equation (12-63) gives

$$\vec{F}_1 + \vec{F}_2 + \vec{F}_w = \rho Q(\vec{v}_2 - \vec{v}_1) \qquad (1)$$

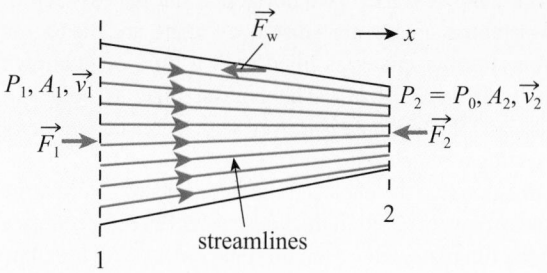

Figure 12-41 Example 12-10

Taking the flow direction as the positive x-axis,

$$\vec{F}_1 = P_1 A_1 \hat{x}$$
$$\vec{F}_2 = -P_2 A_2 \hat{x} = -P_0 A_2 \hat{x}$$
$$\vec{F}_w = F_w \hat{x}$$
$$\vec{v}_1 = v_1 \hat{x}, \quad \vec{v}_2 = v_2 \hat{x}$$

Substituting into Equation (1) gives

$$P_1 A_1 - P_0 A_2 + F_w = \rho Q(v_2 - v_1)$$

and the magnitude of the force exerted by the nozzle on the water is

$$F_w = -P_1 A_1 + P_0 A_2 + \rho Q(v_2 - v_1) \qquad (2)$$

(continued)

All the quantities on the right-hand side of Equation (2) are known:

$$P_0 = 1.01 \times 10^5 \text{ Pa}$$

$$P_1 = P_0 + 6.80 \times 10^5 \text{ Pa} = 7.81 \times 10^5 \text{ Pa}$$

$$A_1 = \frac{\pi D_1^2}{4} = \frac{\pi (4.40 \times 10^{-2} \text{m})^2}{4} = 1.52 \times 10^{-3} \text{ m}^2$$

$$A_2 = \frac{\pi D_2^2}{4} = \frac{\pi (2.00 \times 10^{-2} \text{m})^2}{4} = 3.14 \times 10^{-4} \text{ m}^2$$

$$Q = 10.0 \times 10^{-3} \text{ m}^3/\text{s}$$

$$\rho = 1000 \text{ kg/m}^3$$

$$v_1 = \frac{Q}{A_1} = \frac{10^{-2} \text{ m}^3/\text{s}}{1.52 \times 10^{-3} \text{ m}^2} = 6.58 \text{ m/s}$$

$$v_2 = \frac{Q}{A_2} = \frac{10^{-2} \text{ m}^3/\text{s}}{3.14 \times 10^{-4} \text{ m}^2} = 31.8 \text{ m/s}$$

Substituting these quantities into Equation (2) gives

$$F_\text{w} = -(7.81 \times 10^5 \text{ Pa}) \times (1.52 \times 10^{-3} \text{ m}^2)$$

$$+ (1.01 \times 10^5 \text{ Pa}) \times (3.14 \times 10^{-4} \text{ m}^2)$$

$$+ (1000 \text{ kg/m}^3) \times (10^{-2} \text{ m}^3/\text{s}) \times (31.8 \text{ m/s} - 6.58 \text{ m/s})$$

$$F_\text{w} = -1187 \text{ N} + 32 \text{ N} + 252 \text{ N} = -903 \text{ N}$$

Making sense of the result:

The sign of F_w is negative; therefore, the force exerted by the nozzle on the water is toward the negative x-axis, opposite to the direction of water flow, as it should be. How large is this force? It is equivalent to lifting a 92 kg weight. The firefighters need to be strong and fit.

LO 10

12-10   Viscous Flow

It takes less force to stir a spoon in a cupful of water than in a cupful of honey. We say that the honey is more *viscous* than water. Recall from Section 12-6 that viscosity is a measure of the resistance of a fluid to flow or change in its shape. A fluid with nonzero viscosity is called a **viscous fluid**. Frictional forces are always present (except for a few liquids at very low temperatures), so ignoring the viscosity of a moving fluid can give inaccurate results in many situations.

Viscosity can be measured in different ways. Consider the following scenario. Two horizontal flat plates, each of cross-sectional area A, are a distance d apart, and the region between the two plates is filled with a fluid of unknown viscosity (Figure 12-42). The bottom plate is held stationary, and the top plate is moved with a constant velocity by exerting a force on it.

Imagine that the enclosed fluid consists of a very large number of layers of small thickness. Fluid speed decreases from the top plate, where the fluid in contact with the plate moves with speed v, to zero for the fluid that is in contact with the stationary bottom plate. This variation in the fluid

speed is a consequence of the frictional forces between the fluid and the plates and between various layers of the fluid.

The speed of a given layer increases with its vertical distance from the bottom plate. The magnitude of the force F required to move the top plate with a speed v is

- proportional to the cross-sectional area A of the plates;
- inversely proportional to the distance between the plates; and
- proportional to the viscosity of the fluid between the plates.

Therefore,

$$F = \mu \left(\frac{A}{d} \right) v \qquad (12\text{-}64)$$

The constant of proportionality, μ, is called the viscosity (or the coefficient of viscosity) of the fluid. From Equation (12-64), we see that the units of viscosity are as follows:

$$\frac{\text{N} \cdot \text{m}}{\text{m}^2 \cdot (\text{m/s})} = \left(\frac{\text{N}}{\text{m}^2} \right) \text{s} = \text{Pa} \cdot \text{s} = \text{kg} \cdot \text{m}^{-1} \cdot \text{s}^{-1}$$

If a fluid with a viscosity of 1 Pa · s is placed between the two plates, and the top plate is pushed by applying a stress of 1 Pa, then in 1 s the plate will move a distance equal to the separation between the two plates.

A commonly used unit for viscosity is the poise (P), named after French physicist Jean Poiseuille:

$$1 \text{ P} = 1 \text{ g} \cdot \text{cm}^{-1} \cdot \text{s}^{-1} = 0.1 \text{ kg} \cdot \text{m}^{-1} \cdot \text{s}^{-1} = 0.1 \text{ Pa} \cdot \text{s}$$

Therefore,

$$10 \text{ P} = 1 \text{ Pa} \cdot \text{s}$$

Another commonly used measurement unit is the centipoise (cP):

$$1 \text{ cP} = 10^{-2} \text{ P} = 10^{-3} \text{ Pa} \cdot \text{s}$$

In general, viscosities of liquids are greater than viscosities of gases. In addition, the viscosity of liquids decreases with increasing temperature. In gases, viscosity arises from the diffusion of gas molecules within the gas.

Figure 12-42   Measuring the viscosity of a viscous fluid between two parallel plates

Table 12-3 Table of Viscosities

Material	Viscosity (Pa·s)	Material	Viscosity (Pa·s)
Air (0°C)	17.4×10^{-6}	Honey	2–10
Hydrogen (0°C)	8.4×10^{-6}	Corn syrup	1.38
Xenon (0°C)	2.12×10^{-5}	Olive oil	0.081
Water (20°C)	8.94×10^{-4}	Peanut butter	~250
Blood	$3 \text{ to } 4 \times 10^{-3}$	Ketchup	50–100

Therefore, the viscosity of a gas increases with temperature. Table 12-3 lists the viscosities of several fluids.

Poiseuille's Law for Viscous Flow

Consider streamlined flow of a viscous fluid flow through a pipe of length L and radius R. As discussed above, the layer in contact with the walls of the pipe is almost at rest. The layer adjacent to it moves with a slightly higher speed. The farther a layer is from the walls, the faster it moves, with the fluid at the centre of the pipe moving with the highest speed. The speed profile of a fluid is shown in Figure 12-43.

Because of viscosity, the kinetic energy and, hence, the speed of the fluid, decreases as the fluid moves through the pipe. Therefore, work must be done on the fluid by applying an external pressure to maintain a constant flow rate. Bernoulli's equation contains no information about frictional forces, so it is not applicable in this case. Poiseuille showed that the volume flow rate Q of a viscous fluid through a horizontal and cylindrical pipe is related to the pressure difference between the two ends of the pipe by the following equation:

Figure 12-43 The speed profile of a fluid moving through a horizontal tube. The length of the arrows is proportional to the speed of the fluid. The layers of the fluid that are close to the walls of the pipe move slowly. The farther a layer is from the walls, the faster its speed. Fluid moving along the central axis of the pipe has the highest speed.

$$P_1 - P_2 = 8\mu \frac{LQ}{\pi R^4} \qquad (12\text{-}65)$$

Here, $P_1 - P_2$ is the pressure difference between the two ends of the pipe. The quantity $\frac{8}{\pi}\left(\frac{\mu L}{R^4}\right)$ can be interpreted as the resistance offered to fluid flow by a pipe of length L and radius R. This quantity is called the total peripheral resistance (TPR). Equation (12-65) shows that in order to maintain a constant volume flow rate, a nonzero pressure difference needs to be maintained between the ends of a pipe. The required pressure difference is proportional to the length of the pipe and inversely proportional to its radius. The longer the pipe, the greater the energy loss; hence, a greater pressure difference is required to maintain a constant flow rate. For smaller radii, a large fraction of the fluid is close to the inner surface of the pipe, where fluid speed is small. Poiseuille's equation can be derived from Newton's laws of motion.

Note that the pressure difference required to maintain a constant flow rate is inversely proportional to the fourth power of the radius of the pipe. As radius increases, the required pressure difference decreases rapidly. Consequently, Bernoulli's equation can be used for flow through pipes of large cross-sectional areas with negligible error.

 ## EXAMPLE 12-11

Drawing Blood Through a Needle

A needle of inner diameter 0.50 mm is used to draw blood from a patient's forearm for a medical test. The average blood pressure in the patient's arm is 85 mm Hg, and the viscosity of the blood is 3.5×10^{-4} Pa·s. The length of the needle is 6.0 cm, and the blood needs to be drawn at a constant flow rate of 0.10 cm³/s. What pressure must be applied to the plunger?

SOLUTION

The diameter of the needle is small, so the viscosity of the blood cannot be ignored. We must use Poiseuille's equation to determine the pressure difference needed to maintain the required flow rate. Let P_{arm} be the pressure at the end of the needle that is inserted in the patient's arm and

P_{pull} the pressure that needs to be maintained to pull the plunger outward. We will assume that the pressure at the end of the needle that is closer to the plunger is P_{pull}.

Known quantities:

radius of the needle, $R = 0.50 \text{ mm}/2 = 0.25 \times 10^{-3}$ m

length of the needle, $L = 6.0 \text{ cm} = 6.0 \times 10^{-2}$ m

volume flow rate of the blood through the needle,
$Q = 0.1 \text{ cm}^3/\text{s} = 0.1 \times (10^{-2} \text{ m})^3/\text{s} = 10^{-7} \text{ m}^3/\text{s}$

blood viscosity, $\mu = 3.5 \times 10^{-4}$ Pa·s

pressure at the end of the needle that is inserted in the arm, $P_{\text{arm}} = 85$ mm Hg

(continued)

MECHANICS

Since 760 mm Hg = 1.0 atmospheric pressure
= 1.01×10^5 Pa,

85 mm Hg = 11 300 Pa

P_{pull} can be calculated from Equation (12-65), which for the present situation can be written as

$$P_{pull} = P_{arm} + 8\mu \frac{LQ}{\pi R^4}$$

P_{pull} = 11 300 Pa

$$+ \frac{8 \times (3.5 \times 10^{-4}\ Pa \cdot s) \times (6.0 \times 10^{-2}\ m) \times (10^{-7}\ m^3/s)}{\pi(0.25 \times 10^{-3}\ m)^4}$$

P_{pull} = 11 300 Pa + 1 370 Pa = 12 670 Pa

Making sense of the result:

As the radius of the needle is very small the viscosity of the blood cannot be ignored. Therefore, the pressure at the plunger side needs to higher than the blood pressure in the forearm to maintain a flow rate.

✓ **CHECKPOINT**

C-12-11 Viscous Flow Through a Pipe

Figure 12-44 shows viscous fluid in a horizontal pipe and two connected vertical tubes that are open to the atmosphere. Which of the following statements correctly describes the motion of the fluid in the pipe? Explain your reasoning.

(a) The fluid is not moving.
(b) The fluid is moving from left to right.
(c) The fluid is moving from right to left.
(d) The motion cannot be determined from the diagram.

Figure 12-44 C-12-11

C-12-11 (b)

🧩 **MAKING CONNECTIONS**

Blood Pressure and Blood Flow

The laws of fluid dynamics apply to the flow of blood through the human body. The human heart has two ventricles, which are chambers that contract to pump blood to the rest of the body (Figure 12-45). With each contraction (heartbeat), the pressure in the arteries—which carry blood from the heart—cycles between a minimum *diastolic* pressure (when the ventricles relax) and a maximum *systolic* pressure (when the ventricles contract). In a healthy adult, the systolic pressure is approximately 16 kPa (120 mm Hg), and the diastolic pressure is approximately 10 kPa (80 mm Hg).

The heart of a healthy adult at rest pumps blood at a rate of approximately 5 L/min, giving an average flow rate of approximately 80 cm³/s. During strenuous physical activity, the flow rate can increase by a factor of 5. Since the cross-sectional area of the major arteries does not increase significantly during exercise, the blood speed in the major arteries must increase by approximately the same factor.

When a person is standing, the weight of the blood has a substantial effect on the blood pressure throughout the body. Let h_{head} be the vertical distance between the head and the heart, and let h_{feet} be the distance between the feet and the heart. From Equation (12-8), the blood pressure at the head and the feet level is related to the pressure at the heart level as follows:

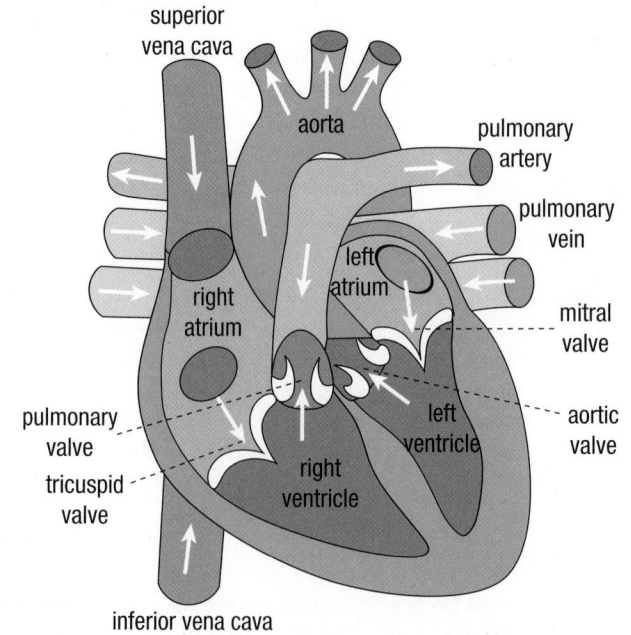

Figure 12-45 An interior view of the human heart showing major blood vessels

(*continued*)

$$P_{\text{head}} = P_{\text{heart}} - \rho_{\text{blood}} g h_{\text{head}}$$

$$P_{\text{feet}} = P_{\text{heart}} + \rho_{\text{blood}} g h_{\text{feet}}$$

Taking typical values of $P_{\text{heart}} = 13$ kPa, $\rho_{\text{blood}} = 1060$ kg/m³, $h_{\text{head}} = 50$ cm, and $h_{\text{feet}} = 130$ cm, we find that $P_{\text{head}} = 7.8$ kPa and $P_{\text{feet}} = 26.5$ kPa. This large difference in relative pressure explains why lying down increases blood flow to the brain and can help relieve swelling in the feet.

Because of viscosity, the blood pressure drops as the blood moves through the arteries. For example, taking the viscosity of blood to be 4×10^{-3} Pa·s and using a flow rate of 80 cm³/s, the pressure drop per unit length in an artery of radius 1.0 cm is

$$\frac{P_1 - P_2}{L} = \frac{8\mu}{\pi} \frac{Q}{R^4} = \frac{8 \times 4.0 \times 10^{-3} \text{ Pa·s}}{\pi}$$

$$\times \frac{80 \times 10^{-6} \text{ m}^3/\text{s}}{(10^{-2} \text{ m})^4} \approx 80 \text{ Pa/m}$$

By the time the blood reaches the small capillaries, it is at atmospheric pressure. Since changing R greatly affects the total peripheral resistance, a buildup of fats in and around the walls of arteries greatly reduces the blood flow, and the heart has to work harder to maintain the needed flow rate.

Fluids are substances that deform and flow under an applied stress. Liquids and gases are collectively called fluids. Physical laws describing fluid motion are derived using Newton's laws of motion along with the concepts of momentum and energy conservation. The concepts of density and pressure are useful for describing the properties of fluids.

Density

The density of a material is defined as its mass per unit volume. The average density is given by

$$\rho = \frac{M}{V}$$

The units of density are kg/m^3.

Pressure

The pressure exerted by a force on a surface element of area A is defined as the component of the force perpendicular to the surface (F_n) divided by the area of the surface:

$$P = \frac{F_n}{A}$$

The SI unit of pressure is the pascal ($1 \text{ Pa} = 1 \text{ N/m}^2$).

Pressure in Fluids

A fluid exerts pressure within itself and on the walls of the containing vessel. This pressure arises due to the weight of the fluid and due to collisions between the constituents of the fluids. In liquids, the primary source of pressure is the weight of the liquid. Gas pressure arises mainly due to collisions between gas molecules and between the gas and the walls of the containing vessel.

Atmospheric Pressure

Atmospheric pressure is due to the weight of the air above Earth's surface.

Gauge Pressure

Gauge pressure is the difference between the actual pressure and a reference pressure, usually taken to be atmospheric pressure.

Pressure Variation with Depth

Pressure within the body of a fluid increases with depth. If P_a denotes the pressure at a point a within a fluid of density ρ, and P_b is the pressure at a point b that is at a depth d below a, then

$$P_b = P_a + \rho g d$$

When point a is at the surface of the fluid where the pressure is equal to atmospheric pressure, P_0, then

$$P_b = P_0 + \rho g d$$

Pascal's Principle

Pascal's principle states that any change in the pressure of an enclosed fluid is transmitted undiminished to every part of the fluid and to the walls of the containing vessel.

Archimedes' Principle

Archimedes' principle states that, when an object is fully or partially immersed in a fluid, the fluid exerts a net upward force on the object (called the buoyant force); the magnitude of the buoyant force is equal to the weight of the fluid displaced by the object.

Flotation

An object will float in a fluid when the average density of the object is less than or equal to the density of the fluid.

Ideal Fluids

An ideal fluid is an incompressible fluid that has no viscosity.

Ideal Fluid Flow

A flow that is steady and nonturbulent is called an ideal flow.

Volume Flow Rate (Q)

The volume flow rate of a fluid through a surface that is normal to the direction of flow is equal to the volume of the fluid that passes through the surface per unit time:

$$Q = Av$$

Here, A is the cross-sectional area of the surface perpendicular to the direction of flow, and v is the speed of the fluid.

Streamlines

A streamline is the trajectory taken by an infinitesimal element of fluid in the region of flow. In a steady flow, streamlines never cross.

Continuity Equation

The amount of fluid that passes through a flow tube remains constant:

$$A_1 v_1 = A_2 v_2$$

Bernoulli's Equation

For the flow of an ideal fluid, at any point along a streamline through the fluid, the sum of pressure, kinetic energy per unit volume, and potential energy per unit volume remains constant:

$$P + \frac{1}{2}\rho v^2 + \rho g y = \text{constant}$$

Poiseuille's Law for Viscous Flow

When viscosity is taken into account, the volume flow rate Q of a fluid of viscosity μ from a pipe of length L and radius R is related to the pressure difference between the two ends of the pipe by the following relationship:

$$P_1 - P_2 = 8\mu \frac{LQ}{\pi R^4}$$

Applications: airplanes and hot-air balloons, spray bottles, hydrometers, hydraulic lifts and brakes, water distribution and irrigation systems, sphygmomanometers, Venturi flow meters, submarines

Key Terms: apparent weight, Archimedes' principle, atmospheric pressure, average mass density, Bernoulli's equation, buoyant force, equation of continuity, flow rate, flow tube, fluid pressure, fluids, gauge pressure, hydraulic lift, ideal fluid, mass flow rate, pascal, Pascal's principle, phase (state), plasma, pressure, specific gravity, steady flow, streamline (laminar flow), viscosity, viscous fluid, work done by the leaving fluid, work done on a test volume

QUESTIONS

1. Helium gas contracts when cooled. A rubber balloon is filled with helium gas at room temperature and closed tightly so that the gas cannot escape. When the balloon is cooled, the density of helium inside the balloon
   (a) decreases;
   (b) remains the same;
   (c) increases.

2. A cylindrical glass is filled with water. The gauge pressure at the bottom of the glass is P. A second cylindrical glass with twice the diameter is filled with water to the same height. The gauge pressure at the bottom of the second glass is
   (a) $P/2$
   (b) P
   (c) $2P$
   (d) $4P$

3. A cylindrical glass is filled with water. The magnitude of the net force at the bottom of the glass due to the weight of the water is F. A second cylindrical glass with twice the diameter is filled with water to the same height. The magnitude of the net force at the bottom of the second glass due to the weight of the water is
   (a) $F/2$
   (b) F
   (c) $2F$
   (d) $4F$

4. Is the following statement true or false? Explain your reasoning: The magnitude of the buoyant force on an object that is fully immersed in a fluid can never be greater than the magnitude of the weight of the object.

5. Figure 12-46 shows mass versus volume plots for three fluids, A, B, and C. Order the fluids from highest to lowest densities.

Figure 12-46 Plots of mass versus volume for three different materials

6. Four objects with identical volumes are submerged in a fluid, as shown in Figure 12-47. Which of the following statements is true about their densities?
   (a) $\rho_1 < \rho_2 < \rho_3 < \rho_4$
   (b) $\rho_1 = \rho_2 = \rho_3 = \rho_4$
   (c) $\rho_1 > \rho_2 > \rho_3 > \rho_4$
   (d) $\rho_1 < \rho_2 = \rho_3 < \rho_4$

Figure 12-47 Question 6

7. A plastic toy boat carrying a small iron cube is floating in a glass container. The cube is then gently dropped into the water. The level of the water in the container
   (a) goes down;
   (b) goes up;
   (c) stays the same.

   Explain your reasoning. Now perform this experiment to determine whether or not your prediction agrees with your observations.

8. A rubber ball with density 500 kg/m³ is floating in a container of water. Oil with a density less than the density of water but greater than the ball's density is slowly poured in the container until it completely covers the ball. When completely immersed in the oil, what fraction of the volume of the ball is submerged in the water? Explain your reasoning.
   (a) half (the same as before the oil was poured)
   (b) less than half
   (c) more than half

9. An ice cube is floating in a partially filled glass of water. After the ice cube melts, the level of the water in the glass
   (a) goes up;
   (b) goes down;
   (c) remains the same.

10. Three identical rectangular pieces of metal are held submerged under water in different orientations, as shown in Figure 12-48. The magnitudes of the buoyant forces on these pieces are $F_{B,1}$, $F_{B,2}$, and $F_{B,3}$. Which of the following statements about the magnitudes of the buoyant forces is correct?
    (a) $F_{B,1} = F_{B,2} = F_{B,3}$
    (b) $F_{B,1} < F_{B,2} = F_{B,3}$
    (c) $F_{B,1} = F_{B,2} < F_{B,3}$
    (d) $F_{B,1} < F_{B,2} < F_{B,3}$

Figure 12-48   Question 10

11. A container, partially filled with water, is resting on a scale that measures its initial weight. A solid iron ball is placed in the container, without spilling any water. What happens to the reading on the scale after the iron ball is placed in the container?
    (a) The reading does not change because the weight of the iron ball is exactly balanced by the upward buoyant force exerted by the water on the ball.
    (b) The reading increases by an amount exactly equal to the weight of the ball.
    (c) The reading increases by an amount that is less than the weight of the ball because the buoyant force cancels a fraction of the weight of the ball.
    (d) The reading increases by an amount that is greater than the weight of the ball because the water on top of the ball further pushes the ball down, thus increasing its weight.

(e) The reading decreases because the buoyant force on the ball is greater than its weight.

12. A gas-filled balloon of mass 2.25 kg (including the mass of the gas) and volume 5.00 m³ is released from a tall building. The density of air is 1.30 kg/m³. Which of the following statements is correct? Explain your reasoning.
    (a) The balloon will remain stationary.
    (b) The balloon will fall.
    (c) The balloon will rise.

13. Three containers of different shapes but equal base area are filled with water to the same height (Figure 12-49).

    Is the hydrostatic pressure at the base of each container the same? Explain your reasoning.

Figure 12-49   Question 13

14. A container filled with a liquid has a hole in its side through which the liquid is coming out. If the container (with the liquid in it) is dropped from a height and allowed to fall freely, what will happen to the liquid emerging from the container?
    (a) It will keep coming out following a similar trajectory as before.
    (b) It will start flowing upward.
    (c) It will stop flowing.

15. A piece of solid metal of mass M has a density four times the density of water. It is attached to a string and suspended in water, fully immersed. The tension in the string is
    (a) $\frac{3}{4} Mg$
    (b) Mg
    (c) $3Mg$
    (d) $4Mg$

16. Scuba divers control buoyancy by wearing a vest that can be inflated or deflated by adding or removing air. The density of water slightly increases with depth. A scuba diver is in a state of neutral buoyancy (neither sinking or rising) at a certain depth. To establish neutral buoyancy at a greater depth the diver should
    (a) inflate its vest;
    (b) deflate its vest;
    (c) do nothing as he/she is already in a state of neutral buoyancy

PROBLEMS BY SECTION

For problems, star ratings will be used, (✷, ✷✷, or ✷✷✷), with more stars meaning more challenging problems.

Section 12-1   Phases of Matter

17. ✷✷   Water has a density of 1.0 g/cm³, and 18 g of water contains 6.02×10^{23} molecules. From this information, estimate the average distance between water molecules.

18. ✷✷ Given that the molar mass of dry air is 29 g/mol and at STP the density of air is 1.30 kg/m³, estimate the average separation between air molecules.

Section 12-2 Density and Pressure

19. ✷ Your mass is 70 kg. The area of the parts of your shoes that are in contact with the ground is 0.010 m². While standing stationary on your two feet, what is the pressure that you exert on the ground?

20. ✷✷ Two rectangular blocks of masses M_1 and M_2 and cross-sectional areas A_1 and A_2, respectively, are lying on a smooth surface as shown in Figure 12-50. What is the pressure exerted by the blocks on the surface?

(a) $\dfrac{M_1 g}{A_1} + \dfrac{M_2 g}{A_2}$

(b) $\dfrac{M_1 g}{A_1} + \dfrac{M_2 g}{A_1}$

(c) $\dfrac{M_1 g}{A_2} + \dfrac{M_2 g}{A_2}$

(d) $\dfrac{M_1 g}{A_1} - \dfrac{M_2 g}{A_2}$

Figure 12-50 Problem 20

21. ✷ A neutron star results when a star in its final stages collapses due to gravitational pressure, forcing the electrons to combine with the protons in the nucleus and converting them into neutrons.
    (a) Assuming that a neutron star has a mass of 3.00×10^{30} kg and a radius of 1.20×10^3 m, determine the density of a neutron star.
    (b) How much would a 1.0 cm³ (the size of a sugar cube) of this material weigh at Earth's surface? Do you think you would be able to lift it?

22. ✷ Wind blowing at 25.0 m/s hits a glass window in a high-rise office building. The cross-sectional area of the window is 4.0 m². Calculate the magnitude of the force exerted on the window by the wind.

23. ✷✷ A spherical metal shell has an outer radius of 15.0 cm and an inner radius of 14.0 cm. The density of the metal is 6000 kg/m³. What is the average density of the shell? Will this shell float when immersed in water? If yes, what fraction of the volume of the shell will be beneath the water's surface?

Section 12-3 Pressure in Fluids

24. ✷ The air provided by a diver's scuba equipment is at the same pressure as the surrounding water. At absolute pressures greater than approximately 1.0×10^6 Pa, the nitrogen in the air becomes dangerously poisonous. At what depth in water does nitrogen become dangerous? The density of the ocean water is 1030 kg/m³.

25. ✷ A submarine is at a depth of 45.0 m under the ocean surface. The inside of the submarine is kept at atmospheric pressure. What is the net force being exerted on a circular window of the submarine that has a radius of 20 cm? The density of the ocean water is 1030 kg/m³.

26. ✷✷ A glass tube of cross-sectional area 10^{-4} m² is partially filled with water. An oil with a density of 800 kg/m³ is slowly poured into the tube so that it does not mix with the water and floats on top of it. The height of the oil above the water surface is 8 cm.
    (a) What is the change in pressure at a depth of 10 cm below the water surface?
    (b) Would the pressure at this depth increase or decrease if you were to add more oil to the tube? Why?

27. ✷ A cylindrical tube closed at one end and open at the other is filled with water, as shown in Figure 12-51. Determine the pressure exerted by the water on the closed end of the tube.

Figure 12-51 Problem 27

28. ✷✷ A container of fluid is accelerated vertically downward with acceleration a. Show that the pressure in the fluid at a depth h below the surface is given by

$$P = P_0 + \rho h(g - a)$$

where P_0 is the atmospheric pressure and ρ is the density. If the container is allowed to fall freely ($a = g$), why does the pressure not change with depth?

Section 12-4 Pascal's Principle

29. Two pistons of a hydraulic lift have radii of 2.67 cm and 20.0 cm. A mass of 2.00 t is placed on the larger piston. Calculate the minimum downward force needed to be exerted on the smaller piston to hold the larger piston at level with the smaller piston.

30. Consider the two hydraulic presses shown in Figure 12-52 (page 340). A force of 1000 N needs to be maintained at the output arm of each press. What is the relationship between the magnitudes of the forces that need to be exerted on the two input arms? Explain your result.
    (a) $F_1 = F_2$
    (b) $F_1 < F_2$
    (c) $F_1 > F_2$

Figure 12-52 Problem 30

Section 12-5 Buoyancy and Archimedes' Principle

31. ✷✷ A rubber ball of uniform density 0.7 g/cm³ is floating in a glass of water. Oil, with a density less than the density of the water, is slowly poured into the glass so that the oil floats on top of the water, and the ball is completely submerged under the oil. What fraction of the volume of the ball is now immersed in the water?
    (a) less than 0.7
    (b) 0.7
    (c) between 0.7 and 1.0
    (d) 1.0 (i.e., the ball is completely immersed in the water)

32. ✷✷ An object with specific gravity 0.85 is immersed in a liquid with specific gravity 1.2. What is the fraction of the volume of the object that submerges in the liquid?

33. ✷✷ A typical supertanker has a mass of 2.0×10^6 kg and carries oil of mass 4.0×10^6 kg. When empty, 9.0 m of the tanker is submerged in water. What is the minimum water depth needed for it to float when full of oil? Assume the sides of the supertanker are vertical and its bottom is flat.

34. ✷✷ An object weighs 500.0 N in air. A force of 30.0 N is needed to push the object down to completely submerge it under water. What is the average density of the object?

35. ✷✷ A 0.48 kg piece of wood floats in water but is found to sink in alcohol. When in the alcohol it has an apparent weight of 0.46 N. What is the density of the wood? The density of alcohol is 790 kg/m³.

36. ✷✷ A solid sphere of mass m floats in a container of water, half submerged, as shown in Figure 12-53. A second solid sphere of the same material but twice the mass is placed in a second container of water.
    (a) Sketch the final position of the second sphere.
    (b) What is the magnitude of the buoyant force on the first sphere?
    (c) What is the magnitude of the buoyant force on the second sphere?
    (d) Is the buoyant force acting on the second sphere larger, smaller, or equal to the buoyant force on the first sphere?
    (e) Is the volume of water displaced by the second sphere larger, smaller, or equal to the volume displaced by the first sphere?

37. A piece of cork with a mass of 10 g is held in place under water by a string tied to the bottom of the container. What is the tension in the string? The density of cork is 0.33 g/cm³.

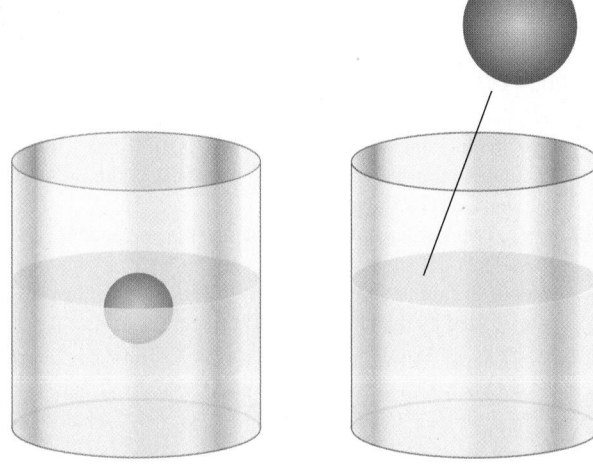

Figure 12-53 Problem 36

38. ✷✷✷ A hydrometer consists of a spherical bulb with a cylindrical stem (Figure 12-54). The cross-sectional area of the stem is 0.40 cm². The total volume of bulb and stem is 13.2 cm³. When immersed in water, the hydrometer floats with 8.0 cm of the stem above the water surface. In alcohol, 1.0 cm of the stem is above the water surface. Find the density of alcohol.

Figure 12-54 Problem 38

39. ✷✷✷ A glass beaker is placed in a large container of water. The beaker itself has a mass of 390.0 g and an interior volume of 500.0 cm³. A student starts filling the beaker with water and finds that when the beaker is less than half full, it floats (Figure 12-55). But when it is more than half full, it sinks to the bottom. What is the density of the glass making up the beaker?

Figure 12-55 Problem 39

40. ✷✷ You need to construct a hollow cubical box of aluminum that has an outside length of 20 cm and an average density of 1000 kg/m³. What must be the thickness of the aluminum sheet used to construct this box? The density of aluminum is 2700 kg/m³. Did you have to make any approximations to solve this problem?

41. ✷✷✷ A wooden boat has a density of 700 kg/m³ and a volume of 5.5 m³.
    (a) What fraction of volume of the boat is submerged in water when empty?
    (b) The boat needs to carry a number of survivors from a desert island. The average mass of each survivor is 65 kg. How many survivors can the boat carry without sinking?
    (c) If some of the survivors sit at the edge of the boat with their legs partially submerged in the water, would the boat be able to carry more people? Why?

42. ✷✷ An evacuated hollow spherical shell of aluminum floats almost completely submerged under water. The outer diameter of the shell is 100.0 cm, and the density of aluminum is 2.7 g/cm³. What is the inner diameter of the shell? Would your answer increase or decrease if you were to assume that the inside of the shell is filled with air?

43 ✷✷ A solid sphere of aluminum (density 2.7 g/cm³) is gently dropped into a deep ocean. (The density of ocean water is approximately 1.03 g/cm³.) Calculate the sphere's acceleration at the point where it is completely submerged in the ocean. As the sphere drops deeper into the ocean, does the acceleration of the sphere increase or decrease compared to its acceleration just beneath the surface? Explain your reasoning.

44. ✷✷✷ A rectangular block of wood weighs 4.0 N in air and has a density of 600 kg/m³. The block is immersed in water.
    (a) What fraction of the volume of the block is immersed in the water?
    (b) A piece of iron (density 6000 kg/m³) is placed on top of the block so that 85% of the block is now immersed in the water. What is the weight of the iron placed on the block?
    (c) The same piece of iron is now attached to the bottom of wood that is immersed in the water. What fraction of volume of the block is now immersed in the water?

45. ✷✷ Most balloons use hot air to fly. At 50°C the density of hot air is approximately 0.85 kg/m³. What would be the volume of the balloon if hot air is used in the Example 12-6? Assume the density of the ambient air to be 1.28 kg/m³.

Section 12-7 The Continuity Equation: Conservation of Fluid Mass

46. ✷ Water is travelling at a speed of 0.1 m/s in a portion of pipe that has a radius of 2.0 cm. The water enters another section of the pipe that has a radius of 1.0 cm. What is the speed of the water in this section of the pipe?

47. ✷✷ The cross-sectional area of the lake behind a dam is approximately 1.00×10^8 m². Water due to heavy rains enters the lake at a rate of 2.5×10^4 m³/s and leaves through the dam at a rate of 1.9×10^4 m³/s. By how much will the water level increase in the lake in a period of 6.0 h?

48. ✷✷ A large spherical balloon made of flexible rubber is filled with water at the rate of 0.1 L/s. What is the rate of increase of the radius of the balloon at an instant when its radius is 20.0 cm?

49. ✷✷ A water faucet has an inner area of 3.0 cm². The flow of water through the faucet is such that it fills a 500 mL container in 15 s.
    (a) What is the flow rate of the water as it comes out of the faucet?
    (b) What is the velocity with which the water emerges from the faucet?
    (c) What is the velocity of the water 20 cm below the faucet?
    (d) What is the area of the water stream at 20 cm below the faucet?

Section 12-8 Conservation of Energy for Moving Fluids

50. ✷✷ Water in a main pipe at street level is at a gauge pressure of 170 kPa and is moving with negligible speed. A pipe connected to the main pipe is used to deliver water to a kitchen located on the third floor of a building, at a 15.0 m height above street level (Figure 12-56). What is the maximum possible speed with which the water can emerge from an open kitchen faucet?

15.0 m

Figure 12-56 Problem 50

51. ✷ Water flows through a horizontal pipe that gradually narrows so that the final inside diameter of the pipe is one-half of the original diameter. In the wide section, the flow speed is 11.4 m/s, and the water pressure is five times the atmospheric pressure. Determine
    (a) the flow speed of the water in the narrower section;
    (b) the water pressure in the narrower section.

52. ✷ A pipe, which carries water with a speed of 5.0 cm/s, is smoothly connected to a pipe with a smaller cross-sectional area. The pressure in the small pipe is 7.0 kPa lower than the pressure in the large pipe. What is the velocity of the water in the small pipe?

53. ✷ Normal blood pressure in a human artery is 13.6 kPa; the blood flows with a speed of 0.12 m/s. Suppose that part of the artery is clogged such that only 20% of the artery's area is open.
    (a) What is the blood pressure in the clogged part of the artery?
    (b) What is the percentage change in the blood pressure in the clogged part?

54. ✷✷ A garden hose has an inside cross-sectional area of 4.00 cm², and the opening in the nozzle has a cross-sectional area of 0.50 cm². The water speed is 0.50 m/s in a segment of the hose that lies on the ground.
    (a) With what speed does the water leave the nozzle when it is held 1.00 m above the ground?
    (b) What is the water pressure in the hose on the ground?

55. ✱✱✱ A 12.0 m long outlet pipe with a diameter of 4.0 cm stands vertically in a water tank that has a diameter of 4.0 m and is open to the atmosphere. The section of the pipe above the water surface is 5.0 m long. Water is pumped from the tank by lowering the pressure at the top end of the pipe. The desired flow rate is 4.0 L/s.
   (a) By how much should the pressure at the top end of the pipe differ from atmospheric pressure?
   (b) Assuming that the pressure at the top of the outlet pipe remains constant, what is the lowest level from which the water can be pumped from the tank?

56. ✱✱ Water flows through a horizontal pipe of radius 40.0 cm that is located 20.0 m below the surface of a large lake next to a dam. The end of the pipe is connected to a nozzle of radius 10.0 cm, and the outlet end of the nozzle is open to the atmosphere. Assuming that the water behaves as an ideal fluid, find
   (a) the speed with which the water leaves the nozzle;
   (b) the water pressure inside the pipe.

57. ✱✱ In problem 56, if the nozzle is closed so that there is no water flow through it, would the pressure in the pipe be less than, greater than, or equal to the pressure calculated in part (b) of problem 53? Ignore viscosity.

58. ✱✱ Air is blown through a circular and horizontal Venturi tube, as shown in Figure 12-57. The diameters of the narrow and wider sections of the tube are given, and the height h of the mercury column is measured to be 1.0 mm. What are the air speeds in the wider and the narrower sections of the tube? The density of mercury is 13 600 kg/m³.

Figure 12-57   Problem 58

59. ✱✱ A tank of cross-sectional area A_2 is filled to a height y_2 with a fluid of density ρ. The tank has a small hole of cross-sectional area A_1 in its side at y_1 from the base. The pressure at the top of the fluid is kept at P_t. Determine an expression for the speed with which the fluid leaves the hole.

60. ✱✱ A very large tank, closed from above and not open to the atmosphere, is partially filled with water. The pressure at the top surface of the water is kept at three times atmospheric pressure by constantly injecting pressurized air from above. Water emerges from the tank through a circular nozzle that is located close to the base of the tank and is open to the atmosphere. At the instant when the water stands 3.0 m high above the nozzle, what is the water flow rate from the nozzle? The cross-sectional area of the nozzle is 2.0×10^{-4} m².

Section 12-9   Conservation of Fluid Momentum

61. ✱✱✱ A fluid of density ρ is placed in a cylindrical container which is rotated about its symmetry axis with a constant angular velocity ω.
   (a) Show that in the radial direction, the variation of pressure within the fluid is given by

$$\frac{dp}{dr} = \rho\,\omega^2 r$$

   where r is the radial distance from the axis of rotation.
   (b) Suppose the pressure along the axis of rotation is p_0. Show that the pressure at a radial distance R from the central axis is

$$p_R = p_0 + \frac{1}{2}\rho\omega^2 R^2$$

   (c) Show that the surface of the rotating fluid forms a paraboloid.

62. ✱✱✱ Water passes through a horizontal pipe of uniform cross-sectional area that has a 90° bend. The diameter of the pipe is 10.0 cm, and the volume flow rate is 50.0 L/s. Find the magnitude and direction of the net radial force exerted on the bend as the water passes through it.

Section 12-10   Viscous Flow

63. ✱ A cylindrical pipe of length 5.0 m and cross-sectional area 1.0×10^{-4} m² needs to deliver oil at a rate of 5.0×10^{-4} m³/s. What must be the pressure difference between the two ends of the pipe if the viscosity of the oil is 1.00×10^{-3} Pa·s?

64. ✱✱ An ostrich's neck is approximately 1.2 m long. Assume that the artery supplying blood to the ostrich's head has a radius of 2.0 mm and that the blood flows through it with an average speed of 2.0 cm/s. If the blood arrives at the ostrich's head (when upright) at one atmospheric pressure, with what pressure should the ostrich's heart pump the blood? Assume that the density of the ostrich's blood is 1060 kg/m³ and that the blood's viscosity is 4.0×10^{-3} Pa·s.

COMPREHENSIVE PROBLEMS

65. A pitot-static tube is a device that is used to measure the speed of gases. Its uses include measuring the speed of air flow in a wind tunnel and the speed of an airplane in air. In its simplest form, a pitot-static tube consists of a U-shaped tube that is open at both ends and is partially filled with a liquid, as shown in Figure 12-58. One end of the tube (A) is perpendicular to the incoming flow, and the other end (B) is horizontal to the direction of flow. The tube is closed in the middle (it contains liquid), so the air impinging on end A is brought to rest. The pressure at this end is then the total pressure (also called dynamic pressure) of the flowing air. Air is not flowing right at the edge of the end B, and the pressure at this end is equal to the atmospheric pressure (also called static pressure). Because of the pressure difference at the two ends, the liquid rises in the arm that is at the lower pressure. The height difference, h, between the two arms of the tube is related to the pressure difference between the two ends.
   (a) Consider the pitot-static tube in Figure 12-58, which is attached to a pipe through which air is flowing with speed v. The tube is partially filled with a liquid of

density ρ_f. Show that, if the difference between the heights of the fluid column in the two arms of the tube is h, then the air speed is related to h by the following relationship:

$$v = \sqrt{2gh\left(\frac{\rho_f}{\rho_{air}}\right)}$$

Thus, by measuring h and knowing the densities, flow speed can be calculated. When pitot-static tubes are used to measure the speed of an airplane, ρ_{air} is the density of the air at the flight altitude and needs to be measured or calculated.

(b) In modern pitot-static tubes, instead of using a liquid, an electronic pressure-measuring instrument is placed between the two ends of the tube. Show that the speed, v, of airflow is related to the measured pressure difference ΔP by the following relationship:

$$v = \sqrt{\frac{2(\Delta P)}{\rho_{air}}}$$

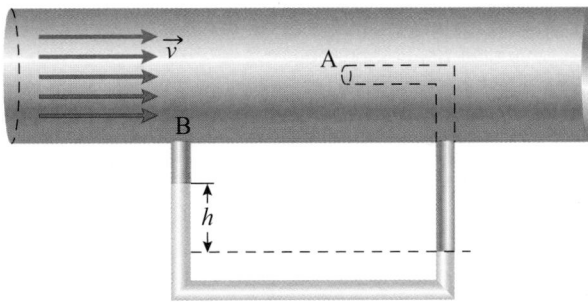

Figure 12-58 A simple pitot-static tube. The height difference of the column of liquid in the two arms of the tube is proportional to the speed of the air flowing through the tube.

66. The density of body fat in humans is approximately 900 kg/m³, and the density of lean muscle is 1100 kg/m³. Suppose a person weighs 790.0 N in air and 50.0 N when completely immersed in water (density 1000 kg/m³).
    (a) Find the person's volume.
    (b) What is the person's density?
    (c) Show that the fraction x_F of a person's body mass that is fat is given by

$$x_F = \frac{\rho_F}{\rho_P}\left(\frac{\rho_L - \rho_P}{\rho_L - \rho_F}\right)$$

where ρ_F is the density of body fat, ρ_L is the density of the lean muscle, and ρ_P is the average density of a person.

    (d) What approximations did you have to make to derive the above result?
    (e) What is x_F for the given data?

67. A rotating garden sprinkler consists of a pipe of diameter 2.0 cm that is open and slightly bent at both ends so that the water rushes out from the ends. The pipe is free to rotate around its centre (Figure 12-59). The flow rate from each opening of the pipe is 2.0 L/s. What is the net torque generated by the sprinkler?

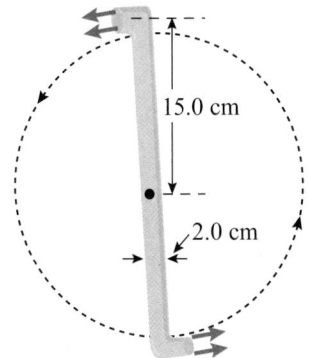

15.0 cm

2.0 cm

Figure 12-59 Problem 67

68. A wooden cube is floating at rest in a glass of water with a certain fraction of the cube's volume immersed. In this equilibrium position, the upward buoyant force exerted on the cube by the water is balanced by the weight of the cube's weight. The cube is now gently pressed farther into the water, held stationary, and then released. It bobs up and down before coming to rest in its original configuration.
    (a) Show that when the cube is gently pressed into the water and released, the net force exerted on the cube is proportional to the height of the cube immersed in the water.
    (b) Show, ignoring the damping effect of the water, that the resulting motion of the cube is simple harmonic motion.
    (c) Find the period of oscillation of the cube in terms of the height of the cube pressed down and the acceleration due to gravity.

69. A container of water has a block of wood floating in it (Figure 12-60). The block is half immersed in the water. The container is now placed in an elevator and accelerated upward with a constant acceleration. The fraction of the block that is immersed in the water
    (a) remains the same as before;
    (b) increases (i.e., more than half is immersed);
    (c) decreases (i.e., less than half is immersed).
    (d) There is not enough information to answer this question.

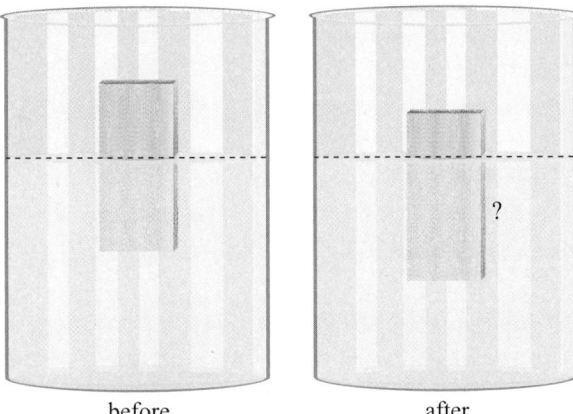

before   after

Figure 12-60 Problem 69

See the text online resources at
www.physics1e.nelson.com for Open Problems and Data-Rich Problems related to this chapter.

Learning Objectives

When you have completed this chapter you should be able to:

1. Interpret oscillations as periodic motion of a single particle.

2. Describe simple harmonic motion mathematically, and calculate displacement, velocity, and acceleration of a simple harmonic oscillator.

3. Describe uniform circular motion, and compare it to simple harmonic motion.

4. Explain what a restoring force is and why it causes a system to undergo simple harmonic motion.

5. Describe the relationship between the spring constant, k, and the natural frequency of a mass–spring oscillator.

6. Calculate the frequency and the period for simple systems that undergo simple harmonic motion.

7. Apply conservation of energy to calculate potential and kinetic energy in simple harmonic motion.

8. Describe the relationship between the length of a pendulum and its period.

9. Describe a physical pendulum, and calculate the period for a physical pendulum.

10. Interpret displacement versus time and velocity versus time graphs for simple harmonic motion.

11. Compare ideal simple harmonic motion to damped and driven oscillations.

12. Qualitatively explain resonance.

 For additional Making Connections, Examples, and Checkpoints, as well as activities and experiments to help increase your understanding of the chapter's concepts, please go to the text's online resources at www.physics1e.nelson.com.

Chapter 13
Oscillations

Located between New Brunswick and Nova Scotia, the Bay of Fundy is home to some of the highest tides in the world. The average tide height worldwide is about 1 m, and Bay of Fundy tides rise as high as 16 m.

The high tides at the Bay of Fundy result from the shape of the bay (Figure 13-1). The mouth of the bay is 100 km wide and between 120 m and 215 m deep. At its end near Hopewell Rocks, the bay is about 2.5 km wide and 14 m deep, at low tide. This shape funnels the incoming tidal water to create spectacular high tides at the end of the bay. The time it takes for the tidal water to travel the 290 km length of the bay is approximately equal to the time it takes for a new tide to come from the ocean. As a result, the previous tide reinforces the incoming tide in a phenomenon called tidal resonance.

The rise and fall of the water height at inlets like the Bay of Fundy follow a pattern called periodic motion. In this chapter, we will examine the nature of this type of motion.

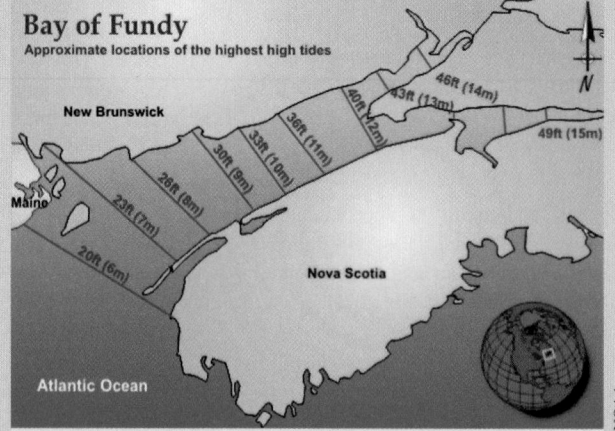

Figure 13-1 The width of the Bay of Fundy at various locations

LO 1

13-1 Periodic Motion

A motion that repeats after a finite amount of time is called **periodic motion**, or **harmonic motion**. The amount of time after which the motion repeats is called the **period** of the motion, and is denoted by the symbol T. Thus, if a motion is repeated after 20.0 s, then $T = 20.0$ s.

From very small to extremely large scales, examples of periodic motions are all around us. The motion of an electron around an atom, of atoms in molecules, of a mass attached to a spring, of Earth around the Sun, and of the Sun around the centre of Milk Way are all examples of periodic motion. Although the forces underlying these periodic motions may be different, the mathematical equations describing them are similar. The mathematical descriptions developed in this chapter apply to many types of periodic motion. In later chapters, we will use the descriptions to understand sound, electromagnetic waves, and atomic structure.

We will investigate periodic motion in one dimension. Although motion in one dimension can be discussed using vector notation, the underlying physics is more transparent by avoiding vector notation and distinguishing motion in the positive and negative directions by using plus and minus signs. We will use vector notation where necessary.

The statement that an object repeats its motion periodically means that both the displacement and the velocity of the object will be the same at the beginning and at the end of any interval of T seconds. Let $x(t)$ denote the position of the object, as measured from an equilibrium point, and $v(t)$ its velocity. Then, for any integer n,

$$x(t + nT) = x(t) \qquad (13\text{-}1)$$

$$v(t + nT) = v(t) \qquad (13\text{-}2)$$

One complete back-and-forth motion is called an **oscillation**. The number of oscillations completed in one second is called the **frequency**, f, of the periodic motion. The frequency is related to the period of the motion:

$$\text{frequency} = \frac{1}{\text{period}}$$

KEY EQUATION

$$f = \frac{1}{T} \qquad (13\text{-}3)$$

The SI unit for frequency is the hertz (Hz):

$$1 \text{ Hz} = 1 \text{ s}^{-1} = \text{one oscillation per second}$$

For periodic motion, a plot of the displacement $x(t)$ versus time t shows a curve whose shape repeats after every T seconds. Thus, the period and the frequency of a periodic motion can be determined if we are given a graph of its displacement versus time. Two examples of periodic

EXAMPLE 13-1

Periodic Motion of Earth

Find the period and frequency of the motion of Earth around the Sun.

SOLUTION

There are approximately 365.25 days in a year. Therefore,

$$T = 365.25 \text{ days} = 365.25 \text{ days} \times 24 \text{ h/day} \times 3600 \text{ s/h}$$

$$= 3.1557 \times 10^7 \text{ s}$$

$$f = 1/T = 1/(3.1557 \times 10^7 \text{ s}) = 3.1688 \times 10^{-8} \text{ Hz}$$

motion are shown in Figures 13-2 and 13-3: Figure 13-2 shows an electrocardiogram (ECG) as a function of time, and Figure 13-3 shows a graph of the displacement of a mass that is attached to a vertical spring and is oscillating about an equilibrium position (the other end of the spring is attached to a rigid support). The graph in Figure 13-3 ignores frictional forces.

Figure 13-2 An ECG. A plot of voltage as a function of time

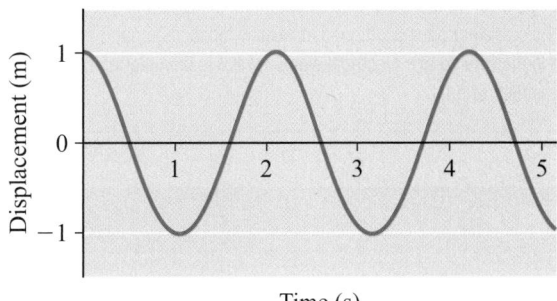

Figure 13-3 A plot of the displacement of a mass, attached to a spring, as a function of time.

Notice two things from these plots. First, both motions are periodic, so we can determine their periods from the graphs. Second, the ECG graph is more complicated than the mass–spring graph. It would be difficult to write an equation that describes the shape of the ECG graph as a function of time, whereas the displacement of the mass–spring system can be described by a cosine or a sine function. In this chapter, we consider a particular type of periodic motion that can be described by a sine or a cosine function, called simple harmonic motion.

LO 2

13-2 Simple Harmonic Motion

When the displacement of an oscillating object from an equilibrium position varies with time as a cosine (or a sine) function, the periodic motion is called **simple harmonic motion**. Therefore, for simple harmonic motion,

KEY EQUATION
$$x(t) = A \cos(\omega t + \phi) \qquad (13\text{-}4)$$

Note that we have chosen to describe the simple harmonic motion by a cosine function. A description using a sine function is equivalent since the two functions

are related by $\cos\left(\theta - \dfrac{\pi}{2}\right) = \sin(\theta)$. The cosine and the sine functions are periodic and have a period of 2π rad. An object that is undergoing simple harmonic motion is called a **simple harmonic oscillator**. The argument of a cosine function is a dimensionless quantity and is measured in radians. The quantities in Equation (13-4) are defined as follows.

A is called the **amplitude**. It is a positive number and describes the maximum displacement of the oscillator from its equilibrium position. As the object oscillates, its displacement from the equilibrium position continuously varies between the values $-A$ and $+A$ because the cosine function oscillates between the values -1 and $+1$. The SI unit for amplitude is the metre (m).

ω is called the **angular frequency**. Angular frequency is related to the frequency, f, of the simple harmonic motion as follows:

$$\omega = 2\pi f = \frac{2\pi}{T} \qquad (13\text{-}5)$$

Angular frequency is measured in radians per second (rad/s); since radians are dimensionless, we can simply write per second (s^{-1}).

ϕ is called the **phase constant**. The phase constant is an angle, measured in radians, and its value is determined by the position and the velocity of the motion at a given time, usually taken to be $t = 0$.

We will now explore the role of the amplitude, angular frequency, and the phase constant in describing simple harmonic motion. Figure 13-5 shows the displacements of two oscillators described by $x_1(t) = (1.0 \text{ m})\cos t$ and $x_2(t) = (2.0 \text{ m})\cos t$. For these motions, we have chosen $\omega = 1$ rad/s and $\phi = 0$ rad. Comparing with Equation (13-4), the first oscillator has amplitude $A = 1.0$ m, and the second oscillator has amplitude $A = 2.0$ m. The displacement of the first oscillates between the value $x = +1.0$ m and $x = -1.0$ m, and the displacement of the second oscillates between $x = +2.0$ m and $x = -2.0$ m. Both oscillators reach the maximum displacement, pass through

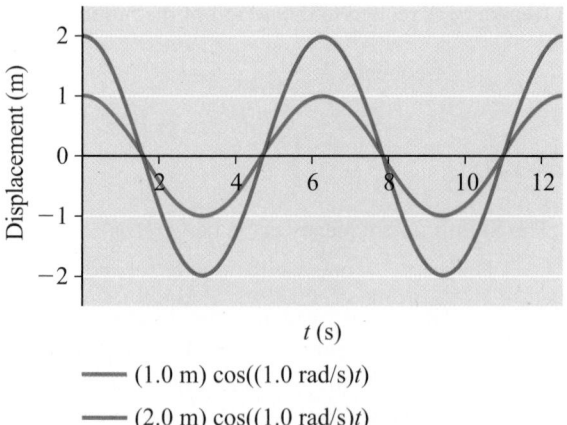

— $(1.0 \text{ m}) \cos((1.0 \text{ rad/s})t)$

— $(2.0 \text{ m}) \cos((1.0 \text{ rad/s})t)$

Figure 13-5 Plots of $(1.0 \text{ m}) \cos t$ and $(2.0 \text{ m}) \cos t$ as a function of time.

the equilibrium position, and then reach the maximum displacement at the opposite side of the equilibrium position at the same time.

For a given simple harmonic motion, the amplitude is a *constant* and does not change with time. We will see later that the amplitude of a simple harmonic oscillator is related to the energy of the oscillator: a particle with greater amplitude has more energy.

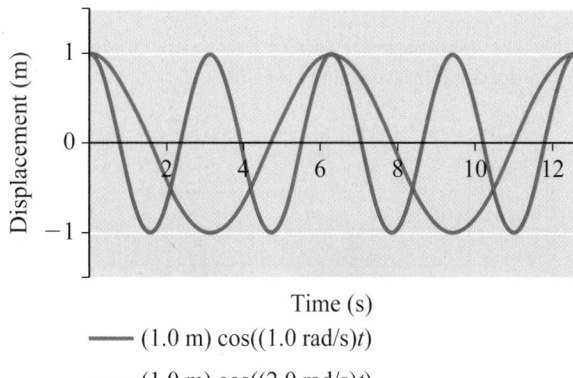

Figure 13-6 Plots of (1.0 m) cos*t* and (1.0 m) cos2*t* as a function of time

✓ CHECKPOINT

C-13-1 Amplitude of a Simple Harmonic Oscillator

Can the amplitude of a simple harmonic motion be negative?

C-13-1 No.

Now, we derive the relationship (13-5) between the angular frequency, ω, and the frequency, f. By the definition of the period,

$$x(t + T) = x(t)$$

Inserting Equation (13-4) in the above relationship, we get

$$A \cos(\omega(t + T) + \phi) = A \cos(\omega t + \phi) \qquad (13\text{-}6)$$

$$\cos(\omega t + \omega T + \phi) = \cos(\omega t + \phi) \qquad (13\text{-}7)$$

The cosine function repeats after 2π rad, so Equation (13-7) can only be satisfied when

$$\omega T = 2\pi \text{ rad} \qquad (13\text{-}8)$$

This condition implies that

$$\omega = \frac{2\pi}{T} = 2\pi f \text{ rad/s} \qquad (13\text{-}9)$$

The angular frequency is 2π times the frequency. By adjusting ω, we can describe fast and slow oscillations. Figure 13-6 shows graphs of two simple harmonic

motions, with $\omega = 1.0$ rad/s and $\omega = 2.0$ rad/s, $A = 1.0$ m, and $\phi = 0$ rad for both. For $\omega = 1.0$ rad/s, $T = 6.28$ s; for $\omega = 2.0$ rad/s, $T = 3.14$ s. The values of the periods can be read from the graphs.

✓ CHECKPOINT

C-13-2 Period of a Simple Harmonic Oscillator

What is the period of a simple harmonic motion described by $x(t) = (3.5 \text{ m})\cos(2\pi t)$? Is the period different at $t = 2.0$ s?

C-13-2 1 s. No, the period is not different at $t = 2.0$ s because the period does not change with time.

We now examine the meaning of the phase constant, ϕ. Setting $t = 0$ in Equation (13-4) we get

$$x(0) = A \cos(\phi)$$

Therefore, the phase constant is related to the position of the oscillatory motion at $t = 0$. Different values of ϕ describe different possible initial locations of the oscillator, as shown in the following example.

 EXAMPLE 13-2

Calculating the Phase Constant

A simple harmonic oscillator has $T = 2$ s and $A = 0.5$ m. Calculate the values of the phase constant, ϕ, for the following starting ($t = 0$) positions of the oscillatory motion:

(a) $x(0) = 0.5$ m;
(b) $x(0) = -0.5$ m;
(c) $x(0) = 0.3$ m;
(d) $x(0) = 0.0$ m.

SOLUTION

For the given values, $T = 2$ s, $\omega = \pi$ rad/s, and $A = 0.5$ m, the equation for simple harmonic motion is given by

$$x(t) = (0.5 \text{ m}) \cos(\pi t + \phi)$$

At $t = 0$,

$$x(0) = (0.5 \text{ m}) \cos \phi$$

For each starting position, the corresponding value of the phase constant can be determined by setting the above equation to the given value.

(a) $x(0) = 0.5$ m, then

$$0.5 \text{ m} = (0.5 \text{ m})\cos\phi$$

$$\cos\phi = 1 \Rightarrow \phi = \cos^{-1}(1) = 0 \text{ rad or } 2\pi \text{ rad}$$

The oscillatory motion of the particle is described by the equation

$$x(t) = (0.5 \text{ m}) \cos(\pi t)$$

The displacement versus time graph for the above equation is shown in Figure 13-7.

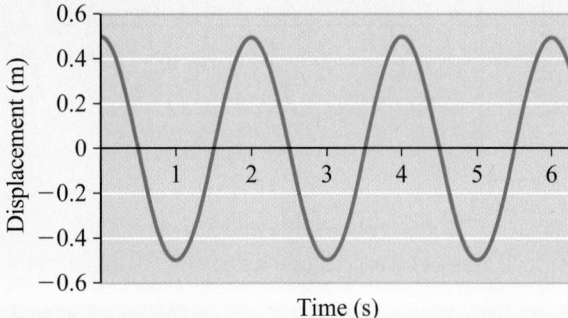

Figure 13-7 A plot of $(0.5 \text{ m}) \cos(\pi t)$ as a function of time

(b) $x(0) = -0.5$ m

$-0.5 \text{ m} = (0.5 \text{ m})\cos\phi$

$\cos\phi = -1 \Rightarrow \phi = \cos^{-1}(-1) = \pi \text{ rad}$

The oscillatory motion of the particle is described by the equation

$$x(t) = (0.5 \text{ m}) \cos(\pi t + \pi) \qquad (13\text{-}10)$$

A graph of this equation is shown in Figure 13-8. Note that the displacements described by (a) and (b) are equal and opposite for all times.

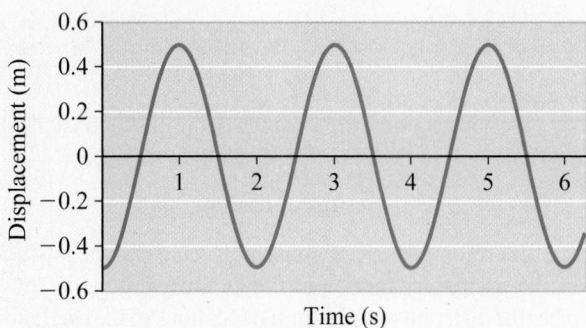

Figure 13-8 A plot of $(0.5 \text{ m}) \cos(\pi t + \pi)$ as a function of time

(c) $x(0) = +0.3$ m

$+0.3 \text{ m} = (0.5 \text{ m})\cos\phi$

$\cos\phi = \dfrac{0.3}{0.5} = 0.6$

$\phi = \cos^{-1}(0.6) = 0.93 \text{ rad or } (2\pi - 0.93) \text{ rad}$

Since $\cos\theta = \cos(2\pi - \theta)$ therefore, there are two possible values for the phase constant. To distinguish between the two values, additional information is needed, for example, the oscillator's velocity at $t = 0$. In general, a unique determination of the phase constant requires information about the oscillator's position and velocity at $t = 0$.

(d) $x(0) = 0.0$ m

$0.0 \text{ m} = (0.5 \text{ m})\cos\phi$

$\cos\phi = \dfrac{0.0}{0.5} = 0$

$\phi = \cos^{-1}(0) = \dfrac{\pi}{2} \text{ or } \left(2\pi - \dfrac{\pi}{2}\right) = \dfrac{3\pi}{2} \text{ rad}$

Again, there are two possible values for ϕ, and we need additional information to determine a unique value. Figure 13-9 shows a plot of $x(t)$ versus t for $\phi = \pi/2$ and $\phi = 3\pi/2$.

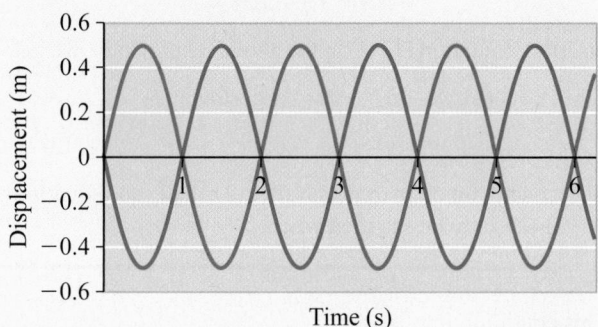

Figure 13-9 A plot of $x(t) = (0.5 \text{ m}) \cos(\pi t + \phi)$ for $\phi = \pi/2$ (blue) and $\phi = 3\pi/2$ (red) as a function of time

✓ CHECKPOINT

C-13-3 Displacement Versus Time Graph

What are the amplitude (A) and angular frequency (ω) of the displacement versus time graph shown in Figure 13-10?
(a) $A = 8$ m, $\omega = 4$ rad/s
(b) $A = 4$ m, $\omega = 2\pi/3$ rad/s
(c) $A = 0$ m, $\omega = 2\pi$ rad/s
(d) $A = -4$ m, $\omega = 2\pi/3$ rad/s

C-13-3 (b).

Figure 13-10 C-13-3

PEER TO PEER

The argument of the cosine function, $\omega t + \phi$, is called the *phase* of the simple harmonic motion. Do not confuse the *phase constant* ϕ, which does not change with time, and the *phase*, $\omega t + \phi$, which does change with time.

The Velocity of a Simple Harmonic Oscillator

The velocity $v(t)$ of a simple harmonic oscillator is obtained by taking the derivative of the displacement $x(t)$ with respect to time, t:

$$v(t) = \frac{dx(t)}{dt} = \frac{d}{dt}[A \cos(\omega t + \phi)]$$

$$= A \frac{d}{dt}[\cos(\omega t + \phi)] = A[-\omega \sin(\omega t + \phi)]$$

$$= -(\omega A) \sin(\omega t + \phi) \qquad (13\text{-}11)$$

KEY EQUATION $\qquad v(t) = -(\omega A) \sin(\omega t + \phi) \qquad (13\text{-}12)$

The sine function oscillates between -1 and $+1$, so the velocity of the oscillator varies between the values $+\omega A$ and $-\omega A$; its maximum value (v_{max}) is ωA.

$$v_{max} = \omega A \qquad (13\text{-}13)$$

In a displacement versus time graph, the direction of the velocity is given by the slope of the graph at that time.

The Phase Difference between an Oscillator's Velocity and Position The velocity of a simple harmonic oscillator varies with time as a sine function and the displacement varies as a cosine function. Therefore the position and velocity of the oscillator are $\pi/2$ rad (90°) out of phase. We can derive this result by using the trigonometric identity.

$$-\sin\theta = \cos\left(\theta + \frac{\pi}{2}\right)$$

Equation (13-12) for $v(t)$ can now be written as follows:

$$v(t) = \omega A \cos\left(\omega t + \phi + \frac{\pi}{2}\right) \qquad (13\text{-}14)$$

Both $x(t)$ and $v(t)$ are now written in terms of the cosine function, so the phase difference between $x(t)$ and $v(t)$ is

phase of $v(t)$ − phase of $x(t)$

$$= \left(\omega t + \phi + \frac{\pi}{2}\right) - (\omega t + \phi) = \frac{\pi}{2} \text{ rad} \qquad (13\text{-}15)$$

Graphs of $x(t)$ and $v(t)$ for an oscillator with $A = 1.0$ m, $\phi = 0$ rad, and $\omega = 2.0$ rad/s are shown in Figure 13-11.

Note that when an oscillator's displacement from equilibrium is maximum ($x = \pm A$), its velocity is zero, and when an oscillator passes through the equilibrium position ($x = 0$), its velocity is maximum ($v = \pm \omega A$). We will see later that this relationship between the position and the velocity is a result of conservation of the oscillator's energy.

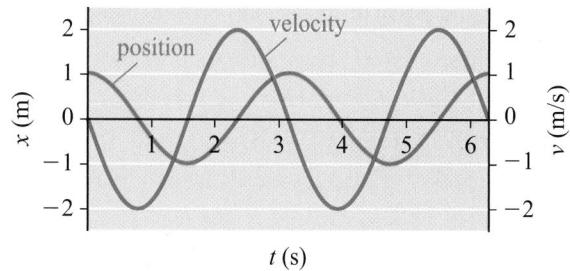

Figure 13-11 Position (blue) and velocity (red) plots for a harmonic oscillator with $A = 1.0$ m, $\phi = 0$ rad, and $\omega = 2.0$ rad/s.

The Acceleration of a Simple Harmonic Oscillator

The acceleration $a(t)$ of a simple harmonic oscillator is obtained by taking the derivative of the velocity, $v(t)$, with respect to time, t:

$$a(t) = \frac{dv(t)}{dt} = \frac{d}{dt}[-\omega A \sin(\omega t + \phi)]$$

$$= -\omega A \frac{d}{dt}[\sin(\omega t + \phi)] = -\omega A[\omega \cos(\omega t + \phi)]$$

$$= -(\omega^2 A) \cos(\omega t + \phi) \qquad (13\text{-}16)$$

KEY EQUATION $\qquad a(t) = -(\omega^2 A) \cos(\omega t + \phi) \qquad (13\text{-}17)$

The cosine function oscillates between -1 and $+1$, so the acceleration of the oscillator varies between the values $-\omega^2 A$ and $+\omega^2 A$; its maximum value (a_{max}) is $\omega^2 A$.

Using Equation (13-17), we can write the acceleration of a simple harmonic oscillator in terms of its displacement from equilibrium as

KEY EQUATION
$$a(t) = -\omega^2(A \cos(\omega t + \phi))$$

$$= -\omega^2 x(t) \qquad (13\text{-}18)$$

This relationship between the acceleration and the displacement is a fundamental property of all simple harmonic oscillators: *For a motion to be simple harmonic motion, the acceleration of the object must be proportional to its displacement from the equilibrium position and opposite in direction, at all times.*

The constant of proportionality between the acceleration and the displacement of a simple harmonic oscillator is equal to the square of its angular frequency.

The Phase Difference Between the Acceleration and Position of an Oscillator The minus sign in Equation (13-17) means that the acceleration and the displacement are π rad (180°) out of phase. Using the trigonometric identity

$$-\cos(\theta) = \cos(\theta + \pi)$$

we can write

$$a(t) = \omega^2 A \cos(\omega t + \phi + \pi) \qquad (13\text{-}19)$$

Therefore, the phase difference between $a(t)$ and $x(t)$ is

phase of $a(t)$ − phase of $x(t)$

$$= (\omega t + \phi + \pi) - (\omega t + \phi) = \pi \text{ rad} \qquad (13\text{-}20)$$

Graphs of $x(t)$ and $a(t)$ for an oscillator with $A = 1.0$ m, $\phi = 0$ rad, and $\omega = 2.0$ rad/s are shown in Figure 13-14.

The Restoring Force and Simple Harmonic Motion

As an object executes simple harmonic motion, its acceleration changes with time, so there must be a time-dependent force acting on the object. What is the nature of this force? Using equation (13-18) for the acceleration of a simple harmonic oscillator, a mass m that executes simple harmonic motion experiences a force,

$$F(t) = ma(t) = -m\omega^2 x(t) \qquad (13\text{-}21)$$

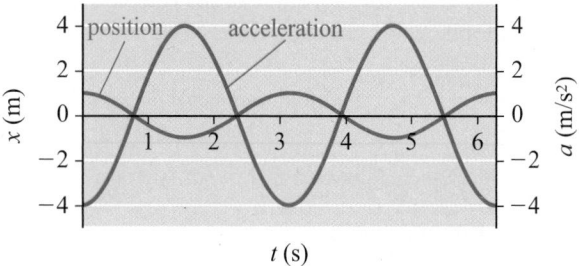

Figure 13-12 Position (blue) and acceleration (red) plots for a harmonic oscillator with $A = 1.0$ m, $\phi = 0$ rad, and $\omega = 2.0$ rad/s.

Since both the mass, m, and the angular frequency, ω, of the motion are fixed, the quantity $m\omega^2$ is constant for a given oscillator. Therefore,

$$F(t) \propto -x(t) \qquad (13\text{-}22)$$

Figure 13-13 shows the relationship between the direction of the force on the oscillator and its displacement from equilibrium. When the displacement is positive (x along the positive x-axis), the force is negative, pointing toward the origin. When the displacement is negative, the force is positive and again points toward the origin. The relationship described by Equation (13-22) is called **Hooke's law**. Elastic objects (e.g., springs) obey Hooke's law when they are deformed, as long as the deformation is small. A force that causes an object to return to its equilibrium position is called a **restoring force**. *Simple*

harmonic motion is motion by an object that is subjected to a restoring force.

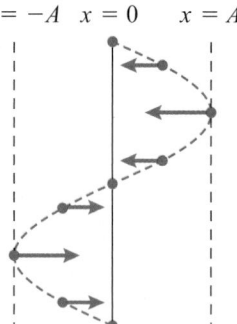

Figure 13-13 The direction of the restoring force for various displacements of a simple harmonic oscillator. The restoring force is always directed toward the equilibrium position.

To determine whether an object will undergo simple harmonic motion, we can use the following procedure:

1. Displace the object from the equilibrium position by a small amount, x.
2. Calculate the net force, F, on the object at the displaced position.
3. Determine whether F can be written as

$$F = -\text{constant} \times x \qquad (13\text{-}23)$$

If the displacement from the equilibrium position of the object is proportional to the net force exerted on the object but opposite in sign, the object will undergo simple harmonic motion. The angular frequency of oscillation is determined by comparing Equations (13-21) and (13-23):

$$\text{constant} = m\omega^2$$

$$\omega = \sqrt{\frac{\text{constant}}{m}} \qquad (13\text{-}24)$$

This procedure also applies for oscillations in two and three dimensions.

✓ CHECKPOINT

C-13-4 A Ball Bouncing on a Floor

Consider an ideal situation where a ball is bouncing elastically on a perfectly elastic floor without losing any energy. The motion of the ball is an example of
(a) periodic motion that is also simple harmonic motion;
(b) periodic motion that is not simple harmonic motion;
(c) simple harmonic motion that is not periodic motion;
(d) motion that is neither simple harmonic motion nor periodic motion;
Explain your reasoning.

C-13-4 (b)

13-3 Uniform Circular Motion and Simple Harmonic Motion

A point moving around a circle with uniform angular speed undergoes periodic motion. When this motion is projected onto the diameter of a circle, the point appears to be moving back and forth in periodic motion. As the point completes one revolution around the circle, its projection onto the diameter completes one oscillation. This analogy between the two motions provides a geometric meaning of the phase constant, ϕ, and the angular frequency, ω.

Consider a point P moving in a circle of radius R with an angular speed of ω rad/s. Let Q be the projection of P onto the x-axis. The time taken by P to complete one revolution (2π rad) is equal to $\frac{2\pi}{\omega}$ s, which is equal to the period of a simple harmonic oscillator of angular frequency ω rad/s. Therefore, the angular frequency of a simple harmonic oscillator is the same as the angular speed of a point moving around a circle.

Suppose at $t = 0$, P makes an angle ϕ with respect to the x-axis, as shown in Figure 13-14. A time t later, it undergoes an angular displacement of ωt and makes an angle of $(\omega t + \phi)$ with respect to the x-axis. The x-coordinate of P at time t is

$$x(t) = OQ = R\cos(\omega t + \phi) \tag{13-25}$$

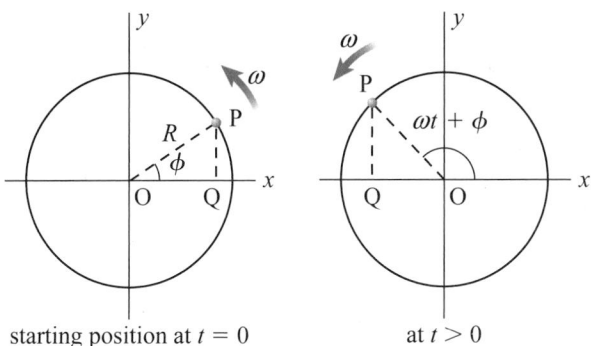

starting position at $t = 0$ at $t > 0$

Figure 13-14 A particle moving with an angular speed ω in a circle of radius R. At $t = 0$, the particle makes an angle ϕ with respect to the x-axis.

Equation (13-25) states that as P moves in a circle, its projection onto the x-axis undergoes simple harmonic motion. The phase constant, ϕ, for simple harmonic motion is therefore analogous to the starting angle of a circular motion.

Table 13-1 compares uniform circular motion and simple harmonic motion.

Table 13-1 Comparison Between Uniform Circular Motion and Simple Harmonic Motion

Uniform circular motion	Simple harmonic motion
Radius: R	Amplitude: A
Angular speed: ω	Angular frequency: ω
One complete revolution	One complete oscillation
Time taken for one complete revolution $= 2\pi/\omega$	Time period for one oscillation $= 2\pi/\omega$
Starting angle: ϕ	Phase constant: ϕ

✓ **CHECKPOINT**

C-13-5 Circular Motion and the Phase Constant

Starting positions ($t = 0$) for four particles moving around a circle of radius R are shown in Figure 13-15. Rank the starting positions in order of increasing phase constant of the corresponding simple harmonic motion.

Lowest 1. _____   2. _____   3. _____

4. _____ Highest

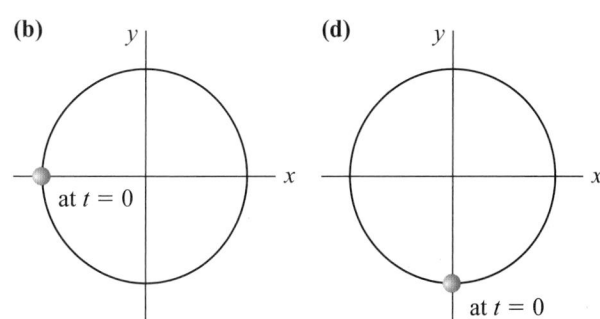

Figure 13-15 C-13-5. Starting positions for four particles moving around a circle of radius R

C-13-5 (a), (c), (b), (d)

LO 4

13-4 Mass-Spring Systems

A Horizontal Mass-Spring System

Figure 13-16 shows a mass, m, attached to one end of a spring of unstretched length L_0 and resting on a horizontal, frictionless surface. The other end of the spring is attached to a rigid support. The mass is restricted to move along the horizontal direction, so gravity does not play any role in its motion. We choose a coordinate system with the origin located at the equilibrium position of the mass and the positive x-axis toward the right. Suppose a force is exerted on the mass (by pulling on the mass) to displace it toward the positive x-axis by an amount x that does not exceed the elastic limit of the spring. At the displaced position, the change in the length of the spring is x, and the spring exerts an elastic restoring force on the mass, trying to return the mass to the equilibrium position. The elastic restoring force of a spring follows Hooke's law:

$$F_{spring,\,x} = -kx \qquad (13\text{-}26)$$

$x < 0 \quad x = 0 \quad x > 0$
spring–mass system (rest length)

spring–mass system (extension)

spring–mass system (compression)

Figure 13-16 A mass attached to a horizontal spring on a frictionless surface

The minus sign in Equation (13-26) indicates that the direction of the force exerted by the spring on the mass is opposite to the direction of the displacement of the mass. As long as the mass is held stationary by the external force, the net force on the mass is zero:

$$F_{net,\,x} = F_{spring,\,x} + F_{external,\,x} = 0$$

When the external force is removed, the only force acting on the mass is the elastic restoring force of the spring. Applying Newton's second law of motion,

$$F_{net,\,x} = F_{spring,\,x} = -kx = ma_x \qquad (13\text{-}27)$$

$$a_x = -\left(\frac{k}{m}\right)x \qquad (13\text{-}28)$$

The acceleration of the mass is proportional to the displacement of the mass from the equilibrium position and opposite in sign. The mass–spring system will undergo simple harmonic motion, with the displacement of the mass given by the standard equation:

$$x(t) = A\cos(\omega t + \phi) \qquad (13\text{-}4)$$

The square of the angular frequency of the motion is obtained by comparing Equations (13-18) and (13-28):

$$\omega^2 = \frac{k}{m}$$

Therefore, for a horizontal mass–spring system,

KEY EQUATION
$$\omega = \sqrt{\frac{k}{m}} \qquad (13\text{-}29)$$

KEY EQUATION $\quad T = 2\pi\sqrt{\dfrac{m}{k}} \qquad f = \dfrac{1}{2\pi}\sqrt{\dfrac{k}{m}} \qquad (13\text{-}30)$

Equation (13-30) tells us that the frequency of oscillation depends on the stiffness of the spring (k) and the mass of the oscillating object. A light mass attached to a stiff spring has a large frequency and, hence, a small period. Note that the frequency of oscillation is independent of the amplitude of motion.

EXAMPLE 13-3

A Horizontal Mass-Spring System

A 2.00 kg mass, resting on a frictionless table, is attached to a spring of spring constant 20.0 N/m. The spring is stretched 20.0 cm by pulling on the mass, which is held stationary and then released.

(a) Calculate the angular frequency, frequency, and period.
(b) What are the amplitude and the phase constant of oscillations?
(c) What are the maximum speed and maximum acceleration of the mass?
(d) What is the displacement of the mass from equilibrium as a function of time?

SOLUTION

The mass–spring system forms a simple harmonic oscillator, and when the mass is displaced from its equilibrium, it will undergo simple harmonic motion.

We choose a coordinate system with the origin at the equilibrium position of the mass and the positive x-axis toward the right. The initial conditions for this problem are as follows:

$$x(0) = 0.200 \text{ m} \quad \text{and} \quad v(0) = 0 \text{ m/s}$$

(a) The angular frequency, frequency, and period of the system are as follows:

$$\omega = \sqrt{\frac{20.0 \text{ N/m}}{2.00 \text{ kg}}} = 3.16 \text{ rad/s}$$

$$f = \frac{\omega}{2\pi} = \frac{3.16 \text{ rad/s}}{2\pi \text{ rad}} = 0.503 \text{ Hz}$$

$$T = \frac{1}{f} = 1.99 \text{ s}$$

(b) As the system oscillates, the position and velocity of the mass at any time t are given by

$$x(t) = A \cos(\omega t + \phi)$$

$$v(t) = -\omega A \sin(\omega t + \phi)$$

To determine A and ϕ, we use the given initial position and velocity of the mass.
Evaluating $x(t)$ and $v(t)$ at $t = 0$, we obtain

$$x(0) = A \cos\phi \tag{1}$$

$$v(0) = -\omega A \sin\phi \tag{2}$$

Squaring and adding Equations (1) and (2) and using the trigonometric identity $\sin^2\phi + \cos^2\phi = 1$, we can determine A:

$$A = \sqrt{x^2(0) + \frac{v^2(0)}{\omega^2}} \tag{3}$$

To determine ϕ, divide Equation (2) by Equation (1):

$$\tan\phi = \frac{\sin\phi}{\cos\phi} = -\frac{v(0)}{\omega x(0)} \tag{4}$$

$$\phi = \tan^{-1}\left(-\frac{v(0)}{\omega x(0)}\right) \tag{5}$$

Equations (3) and (5) tell us that A and ϕ can be calculated when we know the position and the velocity of an oscillator at $t = 0$.
Inserting the values for $x(0)$ and $v(0)$ into Equations (3) and (5), we get

$$A = \sqrt{(0.200 \text{ m})^2 + \frac{(0 \text{ m/s})^2}{(3.16 \text{ rad/s})^2}} = 0.200 \text{ m}$$

$$\phi = \tan^{-1}\left(-\frac{0 \text{ m/s}}{(3.16 \text{ rad/s}) \times (0.20 \text{ m})}\right) = \tan^{-1}(0) = 0 \text{ rad}$$

(c) The maximum speed of the mass is

$$v_{\text{max}} = \omega A = 3.16 \text{ rad/s} \times 0.200 \text{ m} = 0.632 \text{ m/s} \tag{6}$$

and its maximum acceleration is

$$a_{\text{max}} = \omega^2 A = (3.16 \text{ rad/s})^2 \times 0.200 \text{ m} = 2.00 \text{ m/s}^2$$

(d) The displacement of the mass from the equilibrium position at any time t is given by

$$x(t) = (0.200 \text{ m}) \cos(3.16 \, t) \tag{7}$$

Notice that a unique determination of the phase constant requires information about the oscillator's displacement and velocity at the start. A plot of the function $\tan^{-1}(x)$ is shown in Figure 13-17. For any x there is a single value between $-\pi/2$ and $\pi/2$.

Figure 13-17 A plot of the function $\tan^{-1}(x)$ versus x, where x is a real number.

✓ **CHECKPOINT**

C-13-6 A Horizontal Mass-Spring System

The amplitude of a simple harmonic oscillator is A. In one period,
(a) What is the total distance travelled by the oscillator?
(b) What is the net displacement of the oscillator from the starting position?

C-13-6 (a) $4A$; (b) 0

A Vertical Mass-Spring System

Consider a mass, m, suspended vertically from a spring of unstretched length L_0. The hanging mass experiences the elastic restoring forces of the spring and gravity. We take the positive y-axis vertically upward along the spring with $y = 0$ at the equilibrium position of the mass, as shown in Figure 13-18. At the equilibrium position, the spring stretches by the length ΔL such that the elastic restoring force on the mass acting upward balances the weight of the mass acting downward:

$$k\Delta L - mg = 0 \tag{13-31}$$

Figure 13-18 A mass attached to a vertically hanging spring (a) An unstretched spring (b) A mass, m, hanging free and motionless with the spring, with $y = 0$ taken as the equilibrium position of the mass. (c) The mass displaced from the equilibrium position by an external force.

Therefore, at the equilibrium position,

$$\Delta L = \frac{mg}{k} \qquad (13\text{-}32)$$

Suppose we displace the mass from the equilibrium position and release it. Would the mass–spring system undergo simple harmonic motion? Let the position of the displaced mass relative to the equilibrium position be y. Compared to the unstretched length L_0, the change in the length of the spring is $\Delta L - y$. Note that y is positive when the mass is above the equilibrium position, that is, when the spring has compressed. When y is negative, the spring has stretched. The net force on the displaced mass is the sum of the weight and elastic restoring force that could be acting either upward or downward:

$$F_{\text{net},y} = k(\Delta L - y) - mg = k\Delta L - ky - mg$$

Using Equation (13-31) to substitute for ΔL, we get

$$F_{\text{net},y} = -ky \qquad (13\text{-}33)$$

When displaced from the equilibrium position, the net force on the mass is proportional to its displacement and in a direction that is opposite to the displacement. The motion of the mass is therefore simple harmonic motion. Comparing Equation (13-33) to Equation (13-21), we see that

$$m\omega^2 = k$$

and that the angular frequency of oscillation is therefore the same as the angular frequency of a horizontal mass–spring oscillator:

$$\omega = \sqrt{\frac{k}{m}} \qquad (13\text{-}34)$$

$$T = 2\pi\sqrt{\frac{m}{k}} \qquad f = \frac{1}{2\pi}\sqrt{\frac{k}{m}} \qquad (13\text{-}35)$$

The displacement of the mass from its equilibrium is given by the standard equation for a simple harmonic oscillator:

$$y(t) = A\cos(\omega t + \phi) \qquad (13\text{-}36)$$

Notice that by choosing to measure the change in the length of the spring with respect to the equilibrium position, we have removed the effect of gravity from this problem. While solving a problem, always think whether a particular choice of coordinate system will make the problem easier to solve.

ONLINE ACTIVITY

The Spring Constant and the Period

The e-resource that accompanies every new copy of this textbook contains an Online Activity using the PhET simulation "Determining the spring constant and the period of a mass spring oscillator." Work through the simulation and accompanying questions to gain an understanding of the motion of mass spring oscillators.

MAKING CONNECTIONS

A Vertical Mass-Spring System

Equation (13-32) provides a convenient way of measuring the spring constant. Attach one end of a spring to a rigid support, and hang a known mass to its free end. Let the freely hanging mass come to a rest, and measure the change in the length (ΔL) of the spring in this equilibrium position. The spring constant is equal to the weight of the mass divided by ΔL.

EXAMPLE 13-4

A Vertical Mass-Spring System

A mass attached to a vertical spring is undergoing simple harmonic motion. A plot of the height of the mass measured with respect to the ground as a function of time, is shown in Figure 13-19.

(a) Find the height of the mass from the ground when the mass is at the equilibrium position.
(b) Find the amplitude and the period.
(c) Write an equation that describes the height of the mass with respect to the ground as a function of time.

SOLUTION

(a) This situation is slightly different in that the height, $h(t)$, of the mass is measured with respect to the ground

Figure 13-19   Position versus time plot of an oscillating vertical mass–spring system

and not with respect to the equilibrium position of the mass. From the graph, $h(t)$ varies between 6.5 m and 5.5 m. The equilibrium position of the mass is halfway between these two heights. Therefore, the height of the mass from the ground when the mass is in its equilibrium position is 6.0 m.

(b) From the plot, the period and the amplitude of the oscillations are

$$T = 1.5 \text{ s} \quad \text{and} \quad A = 0.50 \text{ m}$$

(c) To write a general equation for $h(t)$, we need to determine the phase constant, ϕ, from the graph. The equation for $h(t)$ can be written as follows:

$$h(t) = (0.50 \text{ m}) \cos\left(\frac{2\pi}{1.5} t + \phi\right) + 6.0 \text{ m} \qquad (1)$$

Note that we have added the equilibrium height of the mass to the standard equation for a simple harmonic oscillator. Equation (1) describes a **displaced harmonic oscillator**, where the origin of the coordinate system is not located at the equilibrium position of the oscillator.

To determine ϕ, we note from the graph that $h(0)$ is approximately 6.3 m. Inserting this value and $t = 0$ into the equation for $h(t)$ gives

$$h(0) = (0.50 \text{ m})\cos\phi + 6.0 \text{ m} = 6.3 \text{ m}$$

Therefore,

$$0.50 \cos\phi = 0.30$$

$$\cos\phi = 0.60$$

$$\phi = \cos^{-1}(0.60) \approx +0.93 \text{ rad}$$

$$\text{or } (2\pi - 0.93) \text{ rad}$$

The correct value of the phase constant can be determined from Figure 13-21. Choose a point, for example, at $t = 0.5$ s. From the plot, we note that $h(0.50) = 5.5$ m. Now, evaluate $h(0.50)$ from Equation (1) using both values of ϕ:

For $\phi = +0.93$ rad:

$$h(0.50) = (0.50 \text{ m}) \cos\left(\frac{2\pi}{1.5} \times 0.50 + 0.93\right)$$
$$+ 6.0 \text{ m} = 5.5 \text{ m}$$

For $\phi = (2\pi - 0.93)$ rad:

$$h(0.50) = (0.50 \text{ m}) \cos\left(\frac{2\pi}{1.5} \times 0.50 + 2\pi - 0.93\right)$$
$$+ 6.0 \text{ m} = 6.2 \text{ m}$$

Thus, $\phi = 0.93$ rad is the correct value for the phase constant because it reproduces the given value for $h(0.50)$. The general equation for $h(t)$ is

$$h(t) = (0.50 \text{ m}) \cos\left(\frac{2\pi}{1.5} t + 0.93\right) + 6.0 \text{ m}$$

LO 5

13-5 Energy Conservation in Simple Harmonic Motion

In the absence of frictional forces, the total energy (E) of an oscillating horizontal mass–spring system is the sum of the kinetic energy (K) of the mass and the elastic potential energy (U) of the spring (Figure 13-20).

Suppose that the mass is pulled from its equilibrium position to $x = A$ and held stationary. In doing so, work has been done to stretch the spring by a length A. This work is stored as the elastic potential energy of the spring. The total energy of the mass–spring system at $x = A$ is

$$E = K + U = 0 + \frac{1}{2} kA^2 \quad \text{(Total energy at } x = A\text{)} \quad (13\text{-}37)$$

where we have used the fact that the work done in stretching (or compressing) a spring by a length x is equal to $\frac{1}{2} kx^2$. When the mass is released, it accelerates toward the equilibrium position due to the elastic restoring force of the spring. Therefore, its velocity increases, and the length of the spring decreases. Let $x(t)$ be the position of the mass and $v(t)$ its velocity at some time t. At this instant, the kinetic energy of the mass is $\frac{1}{2} mv^2(t)$, and the elastic potential energy of the spring is $\frac{1}{2} kx^2(t)$. The total energy of the mass–spring system is

$$E = \frac{1}{2} mv^2(t) + \frac{1}{2} kx^2(t) \qquad (13\text{-}38)$$

(a)
$$v = 0, \ E = \frac{1}{2} kA^2$$
$x = -A \quad x = 0 \quad x = A$

(b)
$$v = v_{\text{max}}, \ E = \frac{1}{2} mv_{\text{max}}^2$$
$x = -A \quad x = 0 \quad x = A$

(c)
$$v = 0, \ E = \frac{1}{2} kA^2$$
$x = -A \quad x = 0 \quad x = A$

(d)
$$E = \frac{1}{2} mv^2 + \frac{1}{2} kx^2$$
$x = -A \quad x = 0 \quad x = A$

Figure 13-20 A mass–spring system oscillating in a straight line on a horizontal, frictionless surface (a) The spring is fully stretched, and the mass is at rest. (b) The spring is unstretched, and the mass is passing through the equilibrium position with maximum speed. (c) The spring is fully compressed, and the mass is at rest. (d) The spring is not fully stretched, and the mass has a nonzero speed.

Equating Equations (13-38) and (13-37), we get

$$\frac{1}{2}kA^2 = \frac{1}{2}mv^2(t) + \frac{1}{2}kx^2(t) \qquad (13\text{-}39)$$

At the equilibrium position, the spring is unstretched and the total energy of the system is stored as the kinetic energy of the mass. So the mass must move with maximum speed as it passes through the equilibrium position. Let v_m denote the maximum speed of the mass. Then, at equilibrium,

$$E = \frac{1}{2}mv_m^2 \quad (\text{Total energy at } x = 0) \qquad (13\text{-}40)$$

We can express v_m in terms of the maximum displacement A by using Equations (13-37) and (13-40):

$$E = \frac{1}{2}kA^2 = \frac{1}{2}mv_m^2$$

$$v_m = \sqrt{\frac{k}{m}}\,A = \omega A \qquad (13\text{-}41)$$

This expression for maximum speed is the same as Equation (13-13), which was derived previously. We see that the relationship between the amplitude and the maximum speed of a simple harmonic oscillator is a consequence of the conservation of energy.

Because of inertia, the mass overshoots the equilibrium position and continues to move toward the left, compressing the spring. The resulting restoring force slows the mass until it comes to rest at $x = -A$, where all the energy is again stored as elastic potential energy in the spring:

$$E = K + U = 0 + \frac{1}{2}k(-A)^2 = \frac{1}{2}kA^2$$

$$(\text{Total energy at } x = -A) \qquad (13\text{-}42)$$

This continuous transformation of energy between different modes (kinetic energy and elastic potential energy) is a fundamental property of all harmonic motions. In the absence of frictional forces, a harmonic oscillator, once set in motion, will oscillate forever. However, in reality, frictional forces are always present and oscillations eventually stop.

Figure 13-21 shows a plot of the potential, kinetic, and total energy of the mass–spring oscillator. For any

position between $-A$ and $+A$, the values of U and K can be read from the corresponding curves. Note that the sum of U and K is always equal to $\frac{1}{2}kA^2$. Using Equation (13-39), we can find the velocity of the mass for a given location x:

$$v = \pm\sqrt{\frac{k}{m}(A^2 - x^2)} \qquad (13\text{-}43)$$

Similarly, for a given value of velocity, the location of the mass is given by

$$x = \pm\sqrt{A^2 - \frac{m}{k}v^2} \qquad (13\text{-}44)$$

The variation of kinetic and potential energy with time can be determined by using

$$x(t) = A\,\cos(\omega t + \phi)$$

$$v(t) = -\omega A\,\sin(\omega t + \phi)$$

Then,

$$U = \frac{1}{2}kx^2(t) = \frac{1}{2}kA^2\cos^2(\omega t + \phi)$$

$$K = \frac{1}{2}mv^2(t) = \frac{1}{2}m\omega^2 A^2\sin^2(\omega t + \phi)$$

$$= \frac{1}{2}kA^2\sin^2(\omega t + \phi) \qquad (13\text{-}45)$$

Note that, individually, both U and K vary with time. However, the sum of the two is always equal to the total energy, E:

$$U + K = \frac{1}{2}kA^2\cos^2(\omega t + \phi) + \frac{1}{2}kA^2\sin^2(\omega t + \phi)$$

$$= \frac{1}{2}kA^2(\cos^2(\omega t + \phi) + \sin^2(\omega t + \phi))$$

$$= \frac{1}{2}kA^2 \qquad (13\text{-}46)$$

The total energy of a mass–spring oscillator is proportional to the square of the amplitude. Doubling the amplitude requires increasing the energy of the oscillator by a factor of 4.

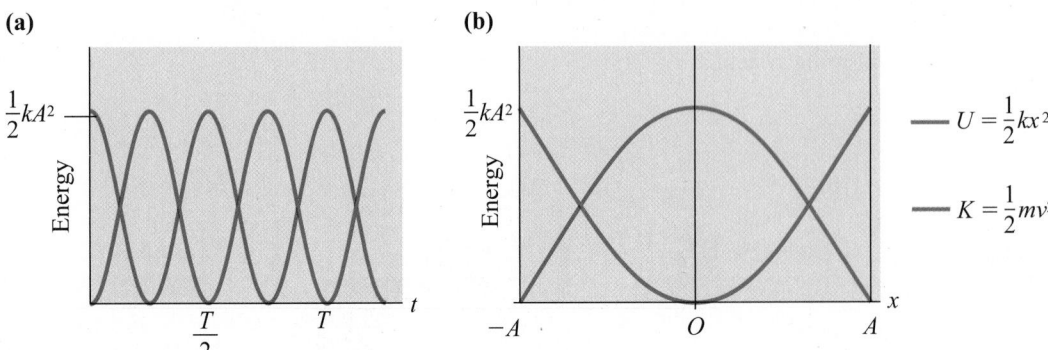

(a) **(b)**

$$U = \frac{1}{2}kx^2$$

$$K = \frac{1}{2}mv^2$$

Figure 13-21 (a) A plot of kinetic energy (K) and elastic potential energy (U) of a simple harmonic oscillator as a function of time. (b) A plot of K and U as a function of displacement

EXAMPLE 13-5

Energy of a Mass-Spring System

A mass–spring oscillator consists of a 1.00 kg mass and a spring with a spring constant of 200 N/m. The oscillator is set in motion by stretching the spring 0.300 m and releasing the mass from rest.

(a) What is the total energy of the mass–spring system?
(b) What is the maximum speed of the mass as it oscillates?
(c) What is the speed of the mass when it is 0.100 m from the equilibrium position?
(d) For which displacements of the mass is the kinetic energy half its total energy?

SOLUTION

We choose a coordinate system with the equilibrium position at the origin and the positive x-axis toward the right. The mass is released from rest at $t = 0$. Therefore,

$$x(0) = 0.300 \text{ m} \quad \text{and} \quad v(0) = 0.00 \text{ m/s}$$

(a) The total energy (E) is the sum of the kinetic energy (K) of the mass and the elastic potential energy (U) of the spring. Using the values of displacement and velocity at $t = 0$,

$$K = \frac{1}{2}mv(0)^2 = 0.00 \text{ J} \quad \text{and}$$

$$U = \frac{1}{2}kx(0)^2 = \frac{1}{2} \times (200 \text{ N/m}) \times (0.300 \text{ m})^2 = 9.00 \text{ J}$$

Therefore, $E = K + U = 9.00$ J.

(b) When the mass is moving with its maximum speed, the total energy of the oscillator is in the form of the kinetic energy of the mass. From Equation (13-41):

$$v_{\text{m}} = \sqrt{\frac{2E}{m}} = \sqrt{\frac{2 \times 9.00 \text{ J}}{1.00 \text{ kg}}} = 4.24 \text{ m/s}$$

(c) The total energy of the mass–spring oscillator is given by Equation (13-38). Rearranging this equation to solve for v and then evaluating for $x = 0.100$ m, we get

$$E = \frac{1}{2}mv^2 + \frac{1}{2}kx^2$$

$$v = \sqrt{\frac{2E}{m} - \frac{k}{m}x^2}$$

$$v = \sqrt{\frac{2 \times 9.00 \text{ J}}{1.00 \text{ kg}} - \frac{200 \text{ N/m}}{1.00 \text{ kg}}(0.100 \text{ m})^2}$$

$$= 4.00 \text{ m/s}$$

(d) When the kinetic energy of the mass is 4.50 J, the potential energy of the spring must also equal 4.50 J, since the sum of the two must equal 9.00 J. Therefore,

$$4.50 \text{ J} = \frac{1}{2}kx^2$$

$$x = \pm\sqrt{\frac{2 \times 4.50 \text{ J}}{200 \text{ N/m}}} = \pm 0.212 \text{ m}$$

Making sense of the results:

The speed calculated in part (c) when the mass is not at the equilibrium position is less than the maximum speed of 4.24 m/s, and the displacement calculated in part (d) is less than the amplitude of oscillation.

✓ **CHECKPOINT**

C-13-7 Energy of a Mass-Spring System

A mass attached to a spring is undergoing simple harmonic motion with an amplitude of 20 cm. When the mass is 10 cm from the equilibrium position, what fraction of its total energy is the kinetic energy of the mass?

(a) 1/4
(b) 1/2
(c) 3/4
(d) 3/8

C-13-7 (c)

 LO 6

13-6 The Simple Pendulum

A **simple pendulum** consists of a point mass attached to a string of negligible mass that does not stretch. The other end of the string is tied to a rigid and frictionless support. When displaced from the vertical equilibrium position and released, the pendulum swings back and forth in a plane under the influence of gravity, which acts as the restoring force. The motion of the pendulum is periodic. Is the motion simple harmonic?

Figure 13-22 shows a mass, m, hanging from a string of length L and making an angle θ with respect to the equilibrium position ($\theta = 0$). Angular displacement is taken to be positive to the right of the equilibrium position

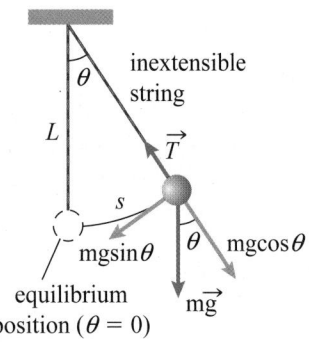

Figure 13-22 A mass attached to an inextensible string of length L. The equilibrium position is the vertical position at which $\theta = 0$.

and negative to the left. The displacement, s, along the arc of the circle is related to the angular displacement as:

$$s = L\theta \qquad (13\text{-}47)$$

The forces acting on the mass are $\vec{mg}$, its weight, acting vertically downward, and $\vec{T}$, the tension in the string. To analyze the pendulum's motion, we choose a coordinate system with one axis, called the radial axis, along the length of the string. The other axis, called the tangential axis, is tangent to the circular motion of the mass. Note that the radial and the tangential axes are perpendicular to each other and their directions change as the mass oscillates. This choice of coordinate system simplifies the analysis because the mass does not move along the radial axis.

The Radial Axis The weight, $\vec{mg}$, makes an angle θ with the radial axis. Therefore, the component of the weight along the radial axis is $mg\cos\theta$, and it points away from the suspension point. The tension, $\vec{T}$, is directed toward the suspension point. There is no motion along the radial axis, so the radial components of the weight and the tension are equal. Therefore,

$$T - mg\cos\theta = 0$$
$$T = mg\cos\theta \qquad (13\text{-}48)$$

Note that Equation (13-48) gives the correct result for $\theta = 0$ rad ($T = mg$) and $\theta = \pi/2$ rad ($T = 0$)

The Tangential Axis The tangential component of the weight is $mg\sin\theta$, and it provides the restoring force that tends to pull the mass toward the equilibrium position. Newton's second law of motion for the tangential component is given by

$$F_{\text{net},t} = -mg\sin\theta = ma_t \qquad (13\text{-}49)$$

where a_t denotes the tangential component of the acceleration. Since $\theta = s/L$,

$$a_t = -g\sin\theta = -g\sin\left(\frac{s}{L}\right) \qquad (13\text{-}50)$$

Note that the acceleration is not proportional to the displacement, s, but to $\sin(s/L)$. In general, pendulum motion is not simple harmonic motion. However, if the angle is small, we can use the small-angle approximation for the sine function:

$$\sin x = x \quad \text{(small-angle approximation)} \quad (13\text{-}51)$$

Figure 13-23 shows a plot of $\sin x$ versus x. We see that the small-angle approximation is reasonable for $x \cong 0.2$ rad (approximately $12°$).

Figure 13-23 A plot of $\sin x$ and x versus x (in rad)

Using the small-angle approximation,

$$\sin(s/L) = s/L \qquad (13\text{-}52)$$

Therefore,

$$a_t = -\left(\frac{g}{L}\right)s \quad \text{(small-angle approximation)} \quad (13\text{-}53)$$

For small angular displacements, the acceleration of the mass is proportional to the displacement and opposite in sign. Therefore, the motion of the pendulum is simple harmonic motion. The angular frequency is obtained by comparing Equation (13-53) with the standard equation for simple harmonic motion:

KEY EQUATION
$$\omega = \sqrt{\frac{g}{L}} \qquad (13\text{-}54)$$

The period and the frequency of oscillation of a pendulum are given by

$$T = \frac{2\pi}{\omega} = 2\pi\sqrt{\frac{L}{g}}$$

$$\text{and} \quad f = \frac{\omega}{2\pi} = \frac{1}{2\pi}\sqrt{\frac{g}{L}} \qquad (13\text{-}55)$$

Notice that the period of a simple pendulum depends on the length of the pendulum and the acceleration due to gravity. The value of g varies slightly across Earth's surface and with height from sea level, so the period of a pendulum of fixed length will vary slightly with the value of g at a given location. Time can be measured fairly accurately with a simple stopwatch, so we can use a pendulum to measure the variation in the value of g across Earth's surface.

CHECKPOINT

C-13-8 Tangential Acceleration

What is the tangential acceleration of the above mass when $\theta = 0$ rad and $\theta = \pi/2$ rad? Is your answer sensible?

C-13-8 0, −g

ONLINE ACTIVITY

Exploring a Simple Pendulum

The e-resource that accompanies every new copy of this textbook contains an Online Activity using the PhET simulation "Exploring a Simple Pendulum." Work through the simulation and accompanying questions to gain an understanding of the motion of a simple pendulum.

EXAMPLE 13-6

Period and Length of a Simple Pendulum

You need to design a simple pendulum that has a period of 1.00 s. The acceleration due to gravity in your lab is 9.81 m/s². What should the length of the string be?

SOLUTION

The period of a simple pendulum is given by

$$T = 2\pi\sqrt{\frac{L}{g}}$$

Solving for L, we get

$$L = \frac{gT^2}{4\pi^2}$$

$$L = \frac{9.81 \text{ m/s}^2 \times (1.00 \text{ s})^2}{4 \times (3.14)^2} = 2.48 \times 10^{-2} \text{m}$$

CHECKPOINT

C-13-9 Decreasing a Pendulum's Length

A pendulum of length L is undergoing simple harmonic motion. When the length of the string is halved, the period of the pendulum
(a) increases by a factor of $\sqrt{2}$;
(b) increases by a factor of 2;
(c) decreases by a factor of $\sqrt{2}$;
(d) decreases by a factor of 2.

C-13-9 (c)

LO 7

13-7   The Physical Pendulum

An extended object that, when displaced from its equilibrium position, oscillates about an axis is called a **physical pendulum**. This includes objects that oscillate about a pivot point, such as a sphere attached to the end of a rod or a string. Swings and baseball bats can also be considered as physical pendulums.

Consider a sphere of uniform density, radius R, and mass M attached to a rod of length L and negligible mass. The system is free to rotate about an axis that is perpendicular to the plane of the paper and passes through the other end of the rod at the pivot point P (Figure 13-24). Since the rod is massless, the centre of mass, C, of the rod–sphere system is located at the centre of the sphere. In the equilibrium position, C is directly below the pivot point P, and there is no torque on the sphere due to its weight.

The sphere is then displaced by an angle θ. In the displaced position (Figure 13-24(b)), the gravitational force exerts a torque $\vec{\tau}$ on the sphere about the pivot point, P.

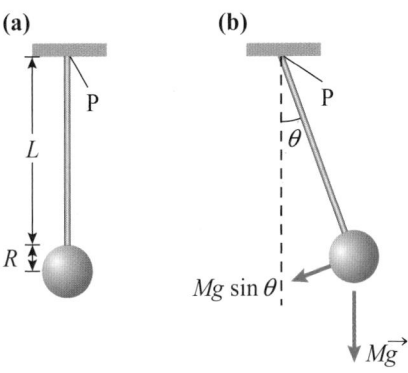

Figure 13-24   A sphere of mass M and radius R is attached to a massless rod of length L, which is hanging from a fixed, frictionless pivot point P. The rod is free to rotate about P. (a) The equilibrium position (b) The rod is displaced by an angle θ from the equilibrium position.

The direction of the torque is perpendicular to the plane of the paper, and the magnitude of the torque is given by

$$\tau = -(Mg)(L_{cm}) \sin\theta \qquad (13\text{-}56)$$

Here, $L_{cm} = L + R$ is the distance between the pivot point and the centre of mass of the rod–sphere pendulum. The minus sign indicates that the direction of the torque is opposite to the direction of the angular displacement, θ. Therefore, the torque tends to rotate the sphere toward the equilibrium position. According to Newton's second law of motion for rotational motion, the magnitude of the torque on the sphere is equal to the product of the angular acceleration, α, of the sphere and its moment of inertia, I_p, about the axis passing through the pivot point, P. Therefore,

$$I_p\alpha = -MgL_{cm}\sin\theta$$

$$\alpha = -\left(\frac{MgL_{cm}}{I_p}\right)\sin\theta \qquad (13\text{-}57)$$

By the parallel-axis theorem, the sphere's moment of inertia about the axis through P is

$$I_p = I_{cm} + M(L_{cm})^2$$

where $I_{cm} = \dfrac{2}{5}MR^2$ is the moment of inertia of the sphere about its centre of mass. When the angular displacement is small, we can use the small-angle approximation $\sin\theta = \theta$, in which case Equation (13-57) becomes

$$\alpha = -\left(\frac{MgL_{cm}}{I_p}\right)\theta \qquad (13\text{-}58)$$

This relationship between the angular acceleration, α, and the angular displacement, θ, is similar to the relationship between the acceleration, a, and displacement, x, of a simple harmonic oscillator. The angular acceleration is proportional to the angular displacement and opposite in sign. Therefore, for small angular displacement, the sphere will undergo simple harmonic motion about the pivot point, P. The angular frequency of oscillation is the square root of the constant of proportionality between α and θ:

KEY EQUATION
$$\omega = \sqrt{\frac{MgL_{cm}}{I_p}} \qquad (13\text{-}59)$$

and the period is

$$T = 2\pi\sqrt{\frac{I_p}{MgL_{cm}}} \qquad (13\text{-}60)$$

Inserting the expressions for I_p and L_{cm}, the period for a rod–sphere physical pendulum is

$$T = 2\pi\sqrt{\frac{\dfrac{2}{5}MR^2 + M(L+R)^2}{Mg(L+R)}}$$

Notice that for $R = 0$, the above expression reduces to $T = 2\pi\sqrt{\dfrac{L}{g}}$, the period for a simple pendulum. The angular displacement of the oscillating sphere is described by an equation analogous to that for linear oscillation:

$$\theta(t) = \theta_m\cos(\omega t + \phi) \qquad (13\text{-}61)$$

Here, θ_m is the maximum angular displacement of the sphere and is the analog of the amplitude A for linear oscillations.

The moment of inertia I_p in Equation (13-60) is calculated with respect to the pivot point, P, and a particular axis of rotation. If either or both of these change, I_p and hence ω and T will change. By comparing Equations (13-60) and (13-55), we notice that a physical pendulum and a simple pendulum have the same time period when

$$L = \frac{I_p}{ML_{cm}} \qquad (13\text{-}62)$$

EXAMPLE 13-7

A Cardboard Sheet as a Physical Pendulum

Consider a square sheet of cardboard with a mass of 20.0 g and a length of 28.0 cm. The sheet is free to rotate about a pivot point, P, that is located directly above the centre of mass of the sheet and 1.00 cm from the top edge. Calculate the period for small angular displacement about the pivot point P.

SOLUTION

A sheet of cardboard that freely rotates about a frictionless pivot point forms a physical pendulum. The period of the sheet is given by Equation (13-60). The sheet's moment of inertia about the point P is

$$I_p = I_{cm} + ML_{cm}^2$$

The moment of inertia of a square sheet of mass M and length L about its centre of mass is

$$I_{cm} = \frac{1}{12}ML^2$$

The distance L_{cm} is the distance from the centre of the mass of the sheet to the pivot point, so

$$L_{cm} = 14.0 \text{ cm} - 1.00 \text{ cm} = 13.0 \text{ cm}$$

Using the given values for the mass and the length of the sheet, we get

$$I_p = \frac{1}{12} \times (0.0200\,\text{kg}) \times (0.280\,\text{m})^2 + (0.0200\,\text{kg}) \times (0.130\,\text{m})^2$$

$$= 4.69 \times 10^{-4}\,\text{kg m}^2$$

Substituting into Equation (13-60) gives

$$T = 2\pi\sqrt{\frac{4.69 \times 10^{-4}\,\text{kg m}^2}{(0.0200\,\text{kg}) \times (9.81\,\text{m/s}^2) \times (0.130\,\text{m})}} = 0.852\text{ s}$$

Walking Motion and the Physical Pendulum

The natural walking motion of animals can be modelled as the free swinging motion of legs pivoting about the hip joints, as shown in Figure 13-25. A leg can be approximated as a physical pendulum of length L and moment of inertia I. During the walking motion, one foot remains in contact with the ground, and the other foot swings through from behind ($\theta = -\theta_0$) to the front of the planted foot ($\theta = +\theta_0$). This motion constitutes one-half the period of the pendulum. For most animals, the maximum deflection angle, θ_0, is typically 20°.

(a)

(b)

Figure 13-25 (a) Free walking motion of a person (b) A schematic representation of free walking motion, where a physical pendulum of length L oscillates between $\theta = -\theta_0$ and $\theta = +\theta_0$.

The period of a physical pendulum of length L, mass M, and moment of inertia I is

$$T = 2\pi \sqrt{\frac{I}{MgL_{cm}}}$$

where L_{cm} is the distance from the centre of mass of the pendulum from the pivot point, and I is calculated about an axis that is perpendicular to the plane of rotation and passes through the pivot point. Let us consider a simple model, where a leg is approximated as a rod, and the mass of the leg is uniformly distributed along the length of the rod. This model is reasonable for animals with legs that do not taper much, for example, humans, elephants, and ostriches. The moment of inertia for a rod of mass M and length L that is free to rotate about an axis passing through an end point of the rod is $\frac{1}{3}ML^2$. The centre of mass of the rod is at its centre, so $L_{cm} = L/2$. Therefore, the period of the rod is

$$T = 2\pi \sqrt{\frac{ML^2/3}{Mg(L/2)}} = \sqrt{\frac{2}{3}} \times 2\pi \sqrt{\frac{L}{g}} \qquad (13\text{-}63)$$

The amplitude of oscillation of the rod is given by

$$A = L\sin\theta_0 \qquad (13\text{-}64)$$

where θ_0 is the maximum angle of deflection. The maximum walking speed is, therefore,

$$v_{walk} = \omega A = \frac{2\pi}{T} \times L\sin\theta_0 = \sqrt{\frac{3}{2}} \sin\theta_0 \times \sqrt{gL} \qquad (13\text{-}65)$$

Taking $L = 1.0$ m as the average length of an adult human leg and $\theta_0 = 20°$, we obtain $v_{walk} = 1.3$ m/s $\cong 4.7$ km/h, which is a reasonable result for the maximum walking speed of an adult.

Several interesting points emerge from this simple model:

■ The walking speed is proportional to $\sqrt{g}$. On the Moon, $g = 1.6$ m/s². Using the numbers for leg length and displacement angle as above, we obtain the value $v_{walk} = 0.5$ m/s $\cong 1.8$ km/h, which is approximately 2.6 times slower than the walking speed on Earth. Have you ever watched a video of astronauts on the Moon and wondered why they were walking so slowly?

■ The walking speed depends on the moment of inertia of the leg. Fast animals (e.g., horses and lions) have legs that are thick at the top and slender at the bottom. A greater fraction of the leg's mass for these animals is located closer to the hip joint, thus decreasing the moment of inertia of the leg about the hip joint. This distribution decreases the period of the leg's oscillations, resulting in quicker steps and a faster walk.

■ As a leopard chases its prey, it hunches down and pounds hard on the ground with its feet (Figure 13-26). By hunching down, the leopard decreases its moment of inertia, and by pounding hard on the ground, the leopard increases its effective g as it accelerates upward. Both of these effects decrease the period of the legs, thus helping the leopard to run faster.

■ The walking speed is proportional to the square root of the length of the leg. Animals with longer legs have a faster natural walking speed. The walking speed of a child with legs half the length of an adult is reduced by a factor of $\sqrt{1/2}$, or approximately 30% compared to that of the adult.

Figure 13-26 A leaping leopard

C-13-10 Walking Speed

Two animals have similarly shaped legs of the same length. One animal's legs are twice as thick as the other animal's legs. Which animal has the faster walking speed?

(a) the animal with thicker legs;

(b) the animal with thinner legs;

(c) They both have the same walking speed;

(d) This question cannot be answered without knowing the leg mass for each animal.

C-13-10 (b)

 EXAMPLE 13-8

Determining an Equation for Simple Harmonic Motion from a Displacement Graph

Figure 13-27 shows a graph of displacement $x(t)$ versus time t for a particle. From this graph, determine an equation for $x(t)$.

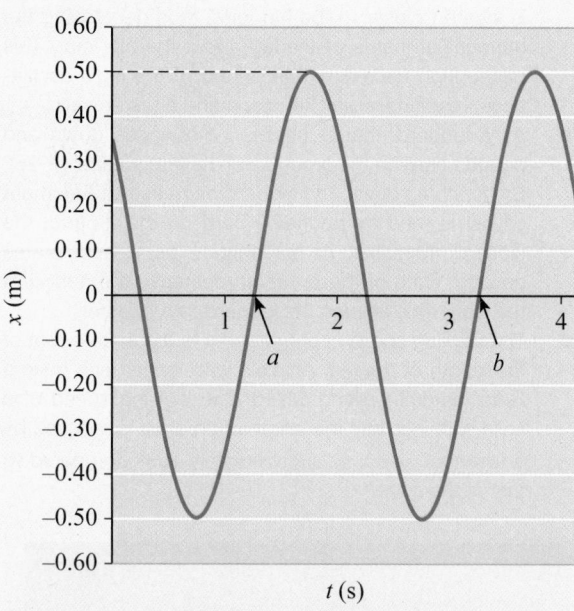

Figure 13-27   The displacement versus time graph for a particle

SOLUTION

We can see from Figure 13-27 that the graph is a sinusoidal function. Therefore, the motion is simple harmonic motion. We can also see that the maximum displacement of the particle is ±0.50 m. Therefore, the amplitude of the oscillation is 0.50 m.

We can determine the period by calculating the time taken to complete one oscillation. While any point on the curve can be taken as the starting point, using points where

 LO 8

13-8   Time Plots for Simple Harmonic Motion

The period, amplitude, and phase constant of a simple harmonic motion can also be determined from the graph of the displacement, the velocity, or the acceleration as a function of time. In this section, we show how this is done through two examples.

the displacement is zero is often convenient. Using t_a and t_b as the start and end times of an oscillation, the period is

$$T = t_b - t_a = 3.2\text{ s} - 1.2\text{ s} = 2.0\text{ s}$$

The angular frequency is

$$\omega = \frac{2\pi}{T} = \frac{2\pi\text{ rad}}{2.0\text{ s}} = 3.1\text{ rad/s}$$

To determine the phase constant, ϕ, we write the standard equation for $x(t)$ with $A = 0.5$ m and $\omega = 3.1$ rad/s:

$$x(t) = (0.50\text{ m})\cos(3.1t + \phi) \qquad (1)$$

From the graph, $x(0)$ is approximately 0.35 m. Substituting this value into Equation (1):

$$x(0) = 0.35\text{ m} = (0.50\text{ m})\cos\phi$$

$$\cos\phi = \frac{0.35\text{ m}}{0.50\text{ m}} = 0.70$$

$$\phi = \cos^{-1}(0.70) = 0.79\text{ rad}\quad\text{or}\quad(2\pi - 0.79)\text{ rad}$$

To determine the correct sign of ϕ, we need another piece of information, such as the velocity of the particle at $t = 0$ or its position at a time near $t = 0$. Let us choose $t = 0.1$ s. From the graph, $x(0.1) = 0.2$ m, approximately. Calculating $x(0.1)$ from Equation (1) for $\phi = 0.79$ rad and $\phi = (2\pi - 0.79)$ rad:

$$\phi = +0.79\text{ rad}$$

$$x(0.1) = (0.50\text{ m})\cos(3.1 \times 0.10 + 0.79) = 0.23\text{ m}$$

$$\phi = (2\pi - 0.79)\text{ rad}$$

$$x(0.1) = (0.50\text{ m})\cos(3.1 \times 0.10 + 2\pi - 0.79) = 0.44\text{ m}$$

Comparing these two values of $x(0.1)$ with the value from the graph, we see that +0.79 rad is the correct value for the phase constant. Thus, the equation for the displacement shown in Figure 13-29 is

$$x(t) = (0.50\text{ m})\cos(3.1t + 0.79) \qquad (2)$$

EXAMPLE 13-9

Determining an Equation for Simple Harmonic Motion from Position and Velocity Graphs

Position and velocity graphs of a simple harmonic oscillator are shown in Figure 13-28. Determine the equation of motion for the oscillator from these graphs.

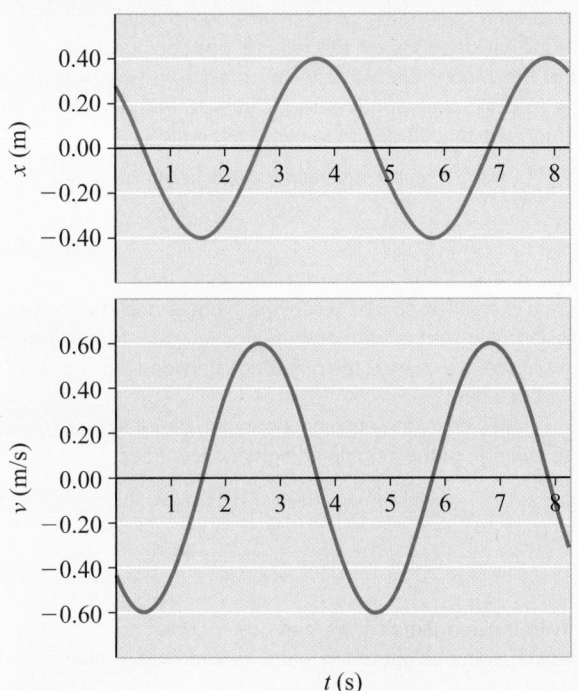

Figure 13-28 Position (blue) and velocity (red) plots for a harmonic oscillator

SOLUTION

From the position graph, the amplitude of the oscillations is $A = 0.40$ m, and the period is approximately $T = 4.2$ s, which gives the following value for angular frequency:

$$\omega = \frac{2\pi}{4.2 \text{ s}} = 1.5 \text{ rad/s}$$

The equations for $x(t)$ and $v(t)$ are as follows:

$$x(t) = (0.40 \text{ m}) \cos(1.5t + \phi)$$

$$v(t) = -(1.5 \times 0.40 \text{ m/s}) \sin(1.5t + \phi) \qquad (1)$$

We can use Equation (1) together with the graphs to determine ϕ. From the plots, approximately, $x(0) = 0.28$ m and $v(0) = -0.42$ m/s. Evaluating $x(t)$ and $v(t)$ for $t = 0$, we get

$$x(0) = (0.40 \text{ m}) \cos\phi = 0.28 \text{ m}$$

$$v(0) = -(0.60 \text{ m/s}) \sin\phi = -0.42 \text{ m/s}$$

Thus

$$\cos\phi = 0.70$$

$$\sin\phi = 0.70$$

and

$$\tan\phi = \frac{\sin\phi}{\cos\phi} = 1.0 \Rightarrow \phi = \tan^{-1}(1.0) = 0.79 \text{ rad}$$

Thus, the equation of motion for the oscillator is

$$x(t) = (0.40 \text{ m}) \cos(1.5t + 0.79)$$

✓ CHECKPOINT

C-13-11 Determining the Direction of Motion from a Displacement Graph

A graph of the displacement $x(t)$ versus time for a simple harmonic motion is given in Figure 13-29. Which of the following statements is true for the velocity of the particle at $t = 1$ s?

(a) The velocity is positive at $t = 1$ s.
(b) The velocity is zero at $t = 1$ s.
(c) The velocity is negative at $t = 1$ s.
(d) A displacement versus time graph has no information about the velocity of the oscillator.

C-13-11 (c)

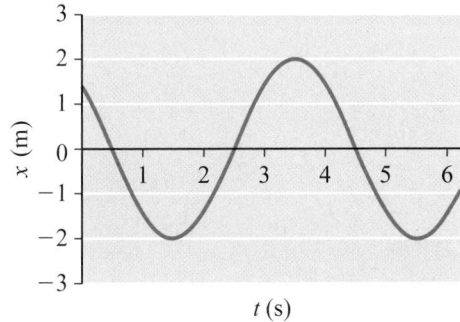

Figure 13-29 A displacement versus time plot for a simple harmonic motion

EXAMPLE 13-10

A Boat Stuck at Hopewell Rocks During Low Tide

At 09:00, you find that your sailboat is stuck in 0.3 m-deep water during low tide at Hopewell Rocks in the Bay of Fundy (Figure 13-30). You look at the tide table and verify that the next high tide is at 15:15, when the water height will be 13.1 m. Your boat needs a water depth of at least 3.0 m to float. Assuming that the tides cause the water height at Hopewell Rocks to change in harmonic motion, estimate the earliest time when your boat will be able to float. Use the tide data given in Table 13-2.

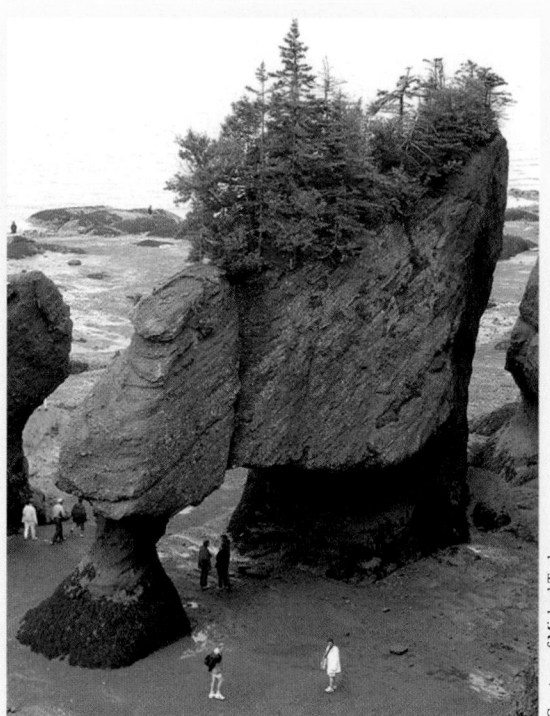

Figure 13-30 Hopewell Rocks, Bay of Fundy, Nova Scotia, at low tide

Courtesy of Michael Taylor

Table 13-2 Tides Table at Hopewell Rocks for April 9, 2008

April 9, 2008 (Wednesday)	
Time (ADT)	Height (m)
02:48	13.7
09:23	0.3
15:15	13.1
21:44	0.8

Fisheries and Oceans Canada, http://www.lau.chs-shc.gc.ca/english/Canada.shtml. Reproduced with the permission of the Canadian Hydrographic Service. Not to be used for navigation.

SOLUTION

Let us assume that the water height, h, at Hopewell Rocks varies in simple harmonic motion. We can then use the given data to determine the amplitude, the period, and the phase constant for this motion. From the equation for $h(t)$ we can determine the time when the height will be 3.0 m.

The period of the tides is the time between two successive high tides or two successive low tides. Taking the time between two successive high tides, we get the following:

$$T = 15:15 - 02:48 = 12 \text{ h} \quad \text{and} \quad 27 \text{ min} = 44\,820 \text{ s}$$

Let us determine the amplitude of the tides. Note that there is a variation in the height of successive high and low tides—the generation of tides is a complicated phenomenon that depends on the relative positioning of the Sun and the Moon. We will take the maximum tide height to be 13.1 m, corresponding to 15:15 ADT (Atlantic Daylight Time). The amplitude of the tides is, therefore,

$$A = \frac{\text{water height at high tide } - \text{ water height at low tide}}{2}$$

$$= \frac{13.1 \text{ m} - 0.30 \text{ m}}{2} = 6.4 \text{ m}$$

Since the difference between the highest and the lowest positions in simple harmonic motion is equal to twice the amplitude, we divided the difference between the high and low tides by 2.

In this case, the equilibrium height, denoted by $h(0)$, is the average of the heights at high and low tides.

$$h_0 = \frac{\text{height at high tide + height at low tide}}{2}$$

$$= \frac{13.1 \text{ m} + 0.30 \text{ m}}{2} = 6.7 \text{ m}$$

Given these quantities, we can now write an equation that describes the height of the water at Hopewell Rocks:

$$h(t) = 6.7 \text{ m} + (6.4 \text{ m}) \cos\left(\frac{2\pi}{T} t + \phi\right) \qquad (1)$$

The first term in Equation (1) is the equilibrium height, h_0. *The variation in the water height is measured with respect to h_0.* This term is absent when $x = 0$ is taken as the equilibrium point.

From the previous examples, we know that we can determine ϕ if we know $h(t)$ at a given time. For ease of calculating, we take the morning low tide time, 09:23, as the start time $t = 0$. At this time, the sea is 0.30 m high. Inserting this information in the above equation yields

$$h(0) = 6.7 \text{ m} + (6.4 \text{ m}) \cos\phi = 0.30 \text{ m}$$

Therefore,

$$\cos\phi = \frac{0.30 \text{ m} - 6.7 \text{ m}}{6.4 \text{ m}} = -1.0$$

$$\phi = \cos^{-1}(-1.0) = 3.1 \text{ rad}$$

Now we can write Equation (1) for the water height:

$$h(t) = 6.7 \text{ m} + (6.4 \text{ m}) \cos\left(\frac{2\pi}{T} t + 3.1\right) \qquad (2)$$

The last task is to determine the time when the water height is 3.0 m. Setting $h(t) = 3.0$ m in Equation (2) and using the trigonometric identity $\cos(\theta + \pi) = -\cos(\theta)$, we get

$$3.0 \text{ m} = 6.7 \text{ m} + (6.4 \text{ m}) \cos\left(\frac{2\pi}{T} t + 3.1\right)$$

$$= 6.7 \text{ m} - (6.4 \text{ m}) \cos\left(\frac{2\pi}{T} t\right)$$

Solving for t gives

$$\frac{2\pi}{T}t = \cos^{-1}\left(\frac{6.7\text{ m} - 3.0\text{ m}}{6.4\text{ m}}\right) = 0.95$$

$$t = \frac{T}{2\pi} \times 0.95 = \frac{44\,820\text{ s} \times 0.95}{2\pi} \approx 6.8 \times 10^3\text{ s} = 1\text{ h }53\text{ min}$$

Since $t = 0$ was at 09:23, your boat should be ready to float again at about 11:16. Happy sailing, and don't leave your tide table at home!

LO 9

13-9 Damped Oscillations

So far in our study of simple harmonic motion, we have ignored frictional forces. As a result, our model predicts that once a simple harmonic oscillator is set in motion, it will oscillate indefinitely. However, frictional forces are always present. These forces transform a fraction of an oscillator's kinetic energy into thermal energy, thus decreasing the amplitude of the oscillation. Eventually, all the energy of the oscillator is transformed into thermal energy, and it comes to rest.

Frictional forces can be quite complicated. For a vertical mass–spring oscillator in air, the frictional force between the mass and the surrounding air depends on the shape of the mass and the speed of oscillation. There is also friction at the contact points where the mass is attached to the spring and where the spring is connected to the rigid support. Furthermore, there are internal frictional forces between the molecules of the spring as it stretches and compresses. A spring that is continuously stretched and compressed gets hot. (Try stretching and compressing a metal spring.) Although an exact description of frictional forces is complicated, two features are common for frictional forces experienced by objects executing simple harmonic motion:

1. The magnitude of the frictional force is proportional to the speed of the object.

2. The direction of the frictional force is opposite to the direction of motion of the object.

Oscillations in the presence of frictional forces are called **damped oscillations**. In the study of fluids, such a frictional force is called a **drag force**, or **fluid resistance**. The simplest form of the drag force, $\vec{F}_D$, that an object experiences when moving with a velocity $\vec{v}$ in a medium can be written as

$$\vec{F}_D = -b\vec{v} \qquad (13\text{-}66)$$

The minus sign indicates that the direction of the drag force is opposite to the direction of the velocity of the object. The proportionality constant, b, is a measure of the strength of the drag force, and for oscillating objects it is called the **damping constant**, or **drag constant**. Its value is positive and depends on the properties of the medium in which the object is moving, as well as the shape of the object (Figure 13-31).

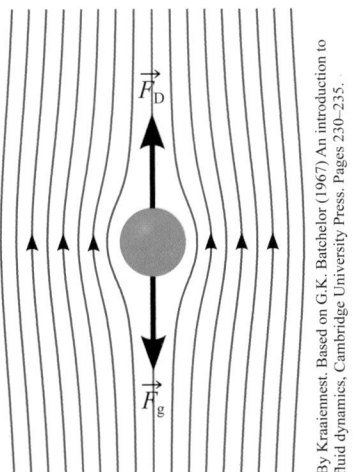

Figure 13-31 The drag force ($\vec{F}_D$) on a sphere falling through the air. Note that the direction of the drag force is opposite to the direction of the gravitational force, $\vec{F}_g$.

By Kraaiennest. Based on G.K. Batchelor (1967) An introduction to fluid dynamics, Cambridge University Press. Pages 230–235.

Since the dimensions on both sides of an equation must be same, we can use Equation (13-66) and dimensional analysis to find the dimension of b, denoted $[b]$:

$$\text{kg m/s}^2 = [b] \times \text{m/s}$$

$$[b] = \frac{\text{kg}}{\text{s}}$$

Equation (13-66) is only one of many forms for drag forces, and is valid only for special conditions, such as the air resistance on small, slowly moving objects. A fast-moving baseball experiences air resistance that is proportional to the square of the speed of the ball.

Let us analyze the motion of a mass–spring system, oscillating in the air, in the presence of the drag force of Equation (13-66). The net force, $\vec{F}_{net}$, on the oscillating mass is the sum of the elastic restoring force exerted by the spring and the drag force exerted by the air:

$$\vec{F}_{net} = \vec{F}_{spring} + \vec{F}_D \qquad (13\text{-}67)$$

Taking the x-axis along the direction of motion,

$$F_{net,x} = F_{spring,x} + F_{D,x} = -kx + (-bv_x) \qquad (13\text{-}68)$$

Using Newton's second law of motion, the acceleration of the mass is

$$ma_x = -kx - bv_x$$

$$a_x = -\frac{k}{m}x - \frac{b}{m}v_x \qquad (13\text{-}69)$$

Will the mass–spring system undergo simple harmonic motion in the presence of the damping force? Comparing Equation (13-69) with Equation (13-18), we see that in this case the acceleration has an additional term that depends on the velocity of the oscillator. So, in general, the motion is not simple harmonic motion. However, if the velocity-dependent term in Equation (13-69) is small compared to the displacement term, then the acceleration of the mass is "approximately" proportional to the displacement, and the motion will be nearly simple harmonic motion. Using calculus, it can be shown (see Online Section 13-1) that the damping force in Equation (13-69) modifies the motion of a simple harmonic oscillator as follows:

1. The amplitude of a damped oscillator decreases exponentially with time.

2. The oscillation frequency of a damped oscillator is lower than the oscillation frequency of an undamped oscillator.

A decrease with time in the amplitude of a damped oscillator is expected. The energy of a simple harmonic oscillator is proportional to the square of its amplitude. In the presence of a drag force, the oscillator loses its energy by doing work against the drag force. As its energy decreases, so does the amplitude of oscillation. The decrease in the oscillation frequency is due to the fact that as it loses energy, a damped oscillator slows down and therefore takes more time to complete a cycle.

As shown in Online Section 13-1, the displacement $x(t)$ of the damped oscillator described by Equation (13-69) is given by

KEY EQUATION $\quad x(t) = A e^{-bt/2m} \cos(\omega_D t + \phi)$ (13-70)

where

$$\omega_D = \sqrt{\frac{k}{m} - \frac{b^2}{4m^2}} = \sqrt{\omega_0^2 - \left(\frac{b}{2m}\right)^2} \quad (13\text{-}71)$$

and $\omega_0 = \sqrt{k/m}$

In Equation (13-71), ω_D is the oscillation frequency in the presence of the drag force, and ω_0 is the oscillation frequency when $b = 0$. The angular frequency ω_0 is called the **natural frequency** of the oscillation. The constant e is the base of the natural logarithm, and $e^{-bt/2m}$ denotes an exponential function that decreases with time. Note that $\omega_D < \omega_0$, so a damped oscillator oscillates at a lower frequency than a corresponding undamped oscillator. We can combine the amplitude A and the exponential term to define a time-dependent amplitude $A(t)$ for a damped oscillator and write Equation (13-70) in a form that is similar to the form for an undamped oscillator:

KEY EQUATION $\quad A(t) = A e^{-bt/2m}$ (13-72)

KEY EQUATION $\quad x(t) = A(t) \cos(\omega_D t + \phi)$ (13-73)

We now examine the motion of a damped oscillator over the range of possible values for the damping constant, b.

An Underdamped Oscillator $\left(\omega_0 > \dfrac{b}{2m}\right)$

In this case ω_D is positive, and the argument of the cosine function is a real number. The system oscillates with the frequency ω_D, and the oscillations are called **underdamped**. For example, when $k = 4.0$ N/m, $m = 0.20$ kg, and $b = 0.20$ kg/s,

$\omega_0 = 4.47$ rad/s, $b/2m = 0.50$ rad/s, and $\omega_D = 4.44$ rad/s

A plot of $x(t)$ for $A = 0.2$ m and $\phi = 0$ rad is shown in Figure 13-32.

For underdamped oscillations, the main effect of the drag force is to decrease the amplitude exponentially with time. The oscillation frequency, ω_D, is close to the natural frequency, ω_0. The dashed curve in Figure 13-32 is called the **envelope** of the damped oscillations. As the damping constant is increased, the amplitude decreases more rapidly with time, and the difference between ω_D and ω_0 increases. Figure 13-33 shows a plot of $x(t)$ with $b = 1.50$ kg/s, with other parameters kept the same as above. Increasing the value of b by a factor of 3 causes only a small decrease in ω_D (4.21 rad/s), but the effect on the amplitude is quite pronounced. The displacement of the oscillator is unnoticeable after two oscillations.

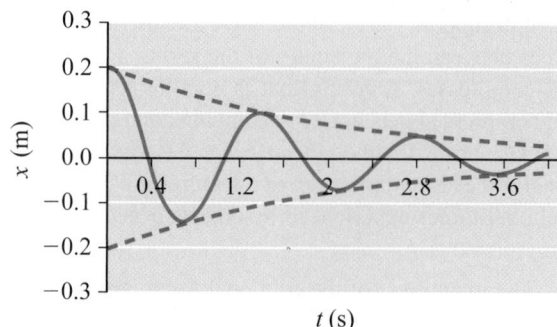

Figure 13-32 Displacement versus time graph for a damped harmonic oscillator, with $k = 4.0$ N/m, $m = 0.20$ kg, and $b = 0.20$ kg/s. The dashed lines are called the "envelope" of the oscillation and are shown to emphasize the damping.

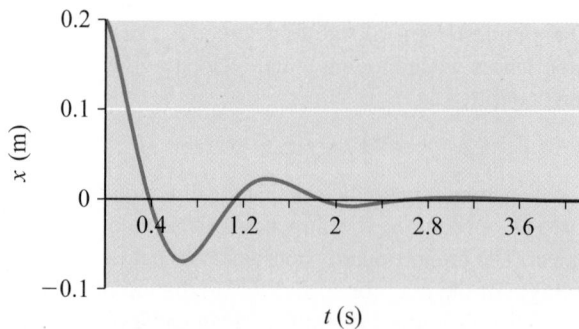

Figure 13-33 A displacement versus time graph for a damped harmonic oscillator, with $k = 4.0$ N/m, $m = 0.20$ kg, and $b = 0.60$ kg/s.

Most everyday oscillations that you observe—for example, a swinging pendulum and a child on a swing—are situations with low damping, where $\omega_0 \ll b/2m$. These systems lose a fraction of their energy with each oscillation, until finally all the energy has been lost. The greater the damping, the quicker the system comes to a stop.

✓ CHECKPOINT

C-13-12 Damped Mass-Spring Oscillator

Four identical mass–spring systems are subjected to damping forces of the form $F_x = -bv$, with a different value of b for each system. Each system is displaced by 0.10 m from equilibrium and released. The amplitude, $A(t)$, for each is shown in Figure 13-34. Rank in order, from smallest to largest, the damping constants for these systems.

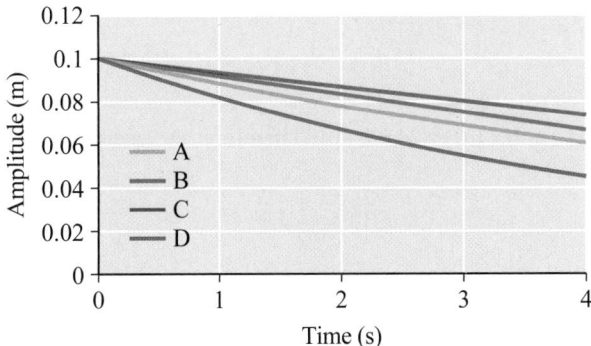

Figure 13-34 Amplitude as a function of time for four different values of damping constant b.

Oscillators from smallest to largest b:
Smallest ---------- ---------- ----------- ----------- Largest

C-13-12: B, C, A, D

A Critically Damped Oscillator $\left(\omega_0 = \dfrac{b}{2m}\right)$

As b increases, the frequency of the damped oscillator decreases, and the oscillator takes more and more time to complete an oscillation. The resulting motion is called **critically damped** when $b = 2\sqrt{mk}$ (i.e., $b/2m = \omega_0$). In this case, $\omega_D = 0$; therefore, the argument of the cosine function does not change with time, and there is no back-and-forth motion of the oscillator. Using calculus, it can be shown that for a critically damped oscillator the displacement from the equilibrium position as a function of time is given by

$$x(t) = (a_1 + a_2 t)e^{-\omega_0 t} \qquad (13\text{-}74)$$

Constants a_1 and a_2 are determined by the initial conditions, $x(0)$ and $v(0)$. When a critically damped oscillator is displaced, it returns to its equilibrium position without completing an oscillation. The time taken for the oscillator to return to the equilibrium position is related to the natural frequency of oscillations, ω_0. *A critically damped oscillator reaches its equilibrium position, without oscillating, faster than for any other value of the damping constant, b.* Figure 13-35 shows a displacement graph for a critically damped mass–spring system. For the graph, we chose $x(0) = 1.0$ m, $v(0) = 0.0$ m/s, $m = 0.20$ kg, and $k = 4.0$ N/m. With these values, $\omega_0 = \sqrt{20}$ rad/s, and critical damping occurs for $b = 2\sqrt{0.20 \times 4.0} = 1.8$ kg/s.

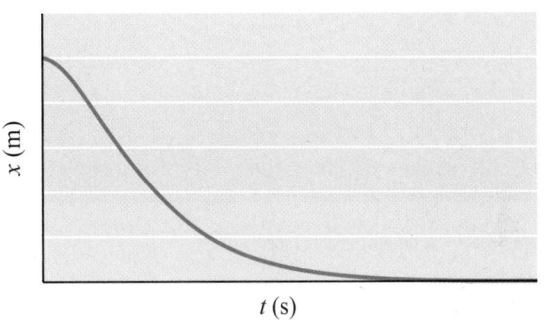

Figure 13-35 A displacement versus time graph for a critically damped oscillator with the initial conditions $x(0) = 1.0$ m, $v(0) = 0.0$ m/s, $\omega_0 = \sqrt{20}$ rad/s, and $b = 1.8$ kg/s.

Shock absorbers in cars are a good example of a critically damped system (Figure 13-36). When a car goes over a bump in the road, properly damped shock absorbers should settle the car back to its normal state in a minimal amount of time.

Figure 13-36 A typical spring-based shock absorber. The inside hydraulic piston contains oil that acts as a damping medium, which absorbs the vibrational energy of the spring and dissipates vibrations.

Another example of a critically damped oscillator is a **tuned-mass damper** (Figure 13-37), which is used in skyscrapers. Tuned-mass dampers are huge blocks of concrete or metal, suspended inside buildings to damp strong vibrations of the building caused by the wind and earthquakes. A detailed explanation is complicated, but basically, the damper absorbs the vibrational energy from the structure when the oscillation frequency of the damper matches the oscillation frequency of the structure.

Figure 13-37 A tuned-mass damper

Alvinku/Shutterstock

An Overdamped Oscillator $\left(\omega_0 > \dfrac{b}{2m}\right)$

When the damping constant increases beyond the critically damped value, the oscillator is called an **overdamped** oscillator. There is still no oscillatory motion, but as b increases beyond the critical value, the oscillator takes more and more time to return to its equilibrium position. Heavy doors in public places have hydraulic devices that are overdamped. This keeps the doors from slamming quickly.

Energy in a Damped Harmonic Oscillator

The energy of an oscillator is proportional to the square of its amplitude, so the energy of a damped harmonic oscillator also decays exponentially with time. Consider an

 ONLINE ACTIVITY

Determining the Damping Constant of a Damped Oscillator

The e-resource that accompanies every new copy of this textbook contains an Online Activity using the PhET simulation "Determining the Damping Constant of a Damped Oscillator." Work through the simulation and accompanying questions to gain an understanding of a damped harmonic oscillator.

underdamped mass–spring oscillator with an amplitude given by Equation (13-72). The energy of the oscillator at a given instant is

$$E(t) = \frac{1}{2}kA^2(t)$$

$$= \frac{1}{2}kA^2(e^{-bt/2m})^2$$

$$E(t) = \frac{1}{2}kA^2 e^{-bt/m} \qquad (13\text{-}75)$$

At $t = 0$, $A(0) = Ae^{-0} = A$. Therefore, the energy at time $t = 0$ is

$$E(0) = \frac{1}{2}kA^2 \qquad (13\text{-}76)$$

After one period, $T_D = 2\pi/\omega_D$, the energy of the oscillator is

$$E(T_D) = \frac{1}{2}kA^2 e^{-bT_D/m} = E(0)e^{-bT_D/m} \qquad (13\text{-}77)$$

The **fractional loss of energy** in one oscillation is therefore

$$\frac{E(0) - E(T_D)}{E(0)} \equiv \frac{\Delta E}{E(0)} = 1 - e^{-bT_D/m} \qquad (13\text{-}78)$$

 ## EXAMPLE 13-11

A Damped Harmonic Oscillator

A damped harmonic oscillator consisting of a block of mass 2.00 kg and a spring of spring constant 10.0 N/m has an amplitude of 25.0 cm at $t = 0$ s. The amplitude falls to 75.0% of its initial value after four oscillations. Assuming that the damping force is of the form $F_{D,x} = -bv_x$, calculate
(a) the value of the damping constant, b;
(b) the amount of energy lost during four oscillations.

SOLUTION

(a) We are given that the oscillations are damped. Since we know the spring constant and the amplitude at

$t = 0$, we can calculate the initial total energy of the oscillator:

$$E(0) = \frac{1}{2}kA^2 = \frac{1}{2} \times (10.0 \text{ N/m}) \times (0.250 \text{ m})^2 = 0.312 \text{ J}$$

We need to determine the period. Since we do not know the damping constant, ω_D is not known. Let us assume that ω_D is approximately equal to ω_0. This is a reasonable assumption because the oscillator loses only 25% of its initial energy in four oscillations so the damping must be weak. Therefore, the time period of the damped oscillator can be approximated by using the natural frequency, ω_0. Once we have determined

WAVES AND OSCILLATIONS

the damping constant, we will check the validity of this approximation. Calculating the period, we get

$$T = \frac{2\pi}{\omega_0} = 2\pi\sqrt{\frac{m}{k}} = 2\pi\sqrt{\frac{2.00 \text{ kg}}{10.0 \text{ N/m}}} = 2.81 \text{ s}$$

To determine the damping constant, we use the information that the amplitude falls to 75% of its initial value in four oscillations. Substituting into Equation (13-72) gives

$$A(4T) = Ae^{-4bt/2m} = \frac{3}{4}A$$

$$e^{-2bt/m} = \frac{3}{4}$$

Taking the natural logarithm (ln) of both sides of the above equation, we get

$$-\frac{2bT}{m} = \ln\left(\frac{3}{4}\right)$$

$$b = -\frac{m}{2T} \times \ln\left(\frac{3}{4}\right)$$

$$= -\frac{2.00 \text{ kg}}{2 \times 2.81 \text{ s}} \times \ln\left(\frac{3}{4}\right) = 0.102 \text{ kg/s}$$

Let us check whether our original assumption that ω_D is approximately equal to ω_0 was valid. Using Equation (13-71),

$$\omega_D = \sqrt{\frac{k}{m} - \frac{b^2}{4m^2}}$$

$$= \sqrt{\frac{10.0 \text{ N/m}}{2.00 \text{ kg}} - \frac{(0.102 \text{ kg/s})^2}{4 \times (2.00 \text{ kg})^2}} = 2.23 \text{ rad/s}$$

This value is very close to the undamped value, $\omega_0 = \sqrt{5}$ rad/s, so our approximation was excellent!

(b) Using Equation (13-77), the energy of the oscillator after four cycles is related to its energy at $t = 0$:

$$E(4T) = E(0)e^{-4bt/m}$$

Using values of b, T, and m:

$$E(4T) = (0.312 \text{ J})e^{-4 \times 0.100 \text{ kg/s} \times 2.81 \text{ s}/2.00 \text{ kg}} = 0.178 \text{ J}$$

This is the energy of the oscillator after four cycles. Therefore, the energy lost in four cycles is 0.312 J − 0.178 J = 0.134 J.

Note two limiting cases of Equation (13-78):

1. When $b = 0$, $\dfrac{\Delta E}{E(0)} = 1 - 1 = 0$, indicating that there is no loss of energy as the system oscillates.

2. At the critical damping limit, $\omega_D = 0$ and therefore $T_D \to \infty$. Since $e^{-\infty} \to 0$, we obtain the result $\Delta E = E(0)$, indicating that a critically damped oscillator loses all its energy in one oscillation.

PEER TO PEER

If you make an approximation to solve a problem, after solving the problem, evaluate your initial approximation to ensure that it is valid.

LO 10

13-10 Resonance and Driven Harmonic Oscillators

Consider a simple mechanical system, for example, a swing, with a natural frequency of oscillation, f_0. Left on its own, an oscillating swing loses energy due to frictional forces and comes to a rest after a while. The motion can be sustained if energy is transferred to the swing by an external mechanism, for example, by pushing the swing periodically. If the swing is pushed after each oscillation, then the time between successive pushes (t_{push}) is the same as the period of the swing ($t_{push} = T = 1/f_0$). In this case, the frequency of the externally applied force, called the **driving frequency**, is the same as the natural frequency of the swing.

The swing can also absorb energy at other driving frequencies. If the swing is pushed after every other oscillation, the period of successive pushes is twice the period of the swing, and the driving frequency is $f_0/2$. Other possible driving frequencies are $f_0/3$, $f_0/4$, $f_0/5$, ….

A mechanical system (such as a swing, a wine glass, a bridge, or a building) is said to be in **resonance** with an externally applied force when the frequency of the externally applied force ($f_{driving}$) matches the natural frequency (f_0) of the system.

$$f_{driving} = f_0 \text{ (resonance condition)} \quad (13\text{-}79)$$

When the driving and the resonance frequencies match, there is an efficient transfer of energy from the external force to the mechanical system.

There are numerous examples of mechanical resonances: resonating strings in a string instrument such as a violin, resonating membranes of a drum, resonance of the basilar membrane in the ear, and breaking a wine glass using sound waves. Resonance phenomena also play a key role in electromagnetism: when the input circuit of a radio receiver is tuned to match the frequency broadcast by a particular radio station, the amplitude of the received signal is greatly increased.

Consider the damped mass–spring system of Section 13-9 that is subjected to an external harmonic force. Such an oscillator is called a **driven oscillator**, and the resulting oscillations are called **forced oscillations**, or **driven oscillations**. The harmonic driving force has the form

$$F_{driven,x} = F_0 \cos(\omega_{driven} t) \qquad (13\text{-}80)$$

Here, F_0 is the magnitude of the applied force, and ω_{driven} is its angular frequency. The net force on the mass is the sum of the elastic restoring force exerted by the spring, the damping force exerted by the surface, and the external driving force:

$$\Sigma F_x = F_{spring,x} + F_{damping,x} + F_{driven,x}$$

$$\Sigma F_x = -kx + (-bv_x) + F_0 \cos(\omega_{driven} t) \qquad (13\text{-}81)$$

The acceleration of the mass is therefore given by

$$a_x = -\frac{k}{m}x - \frac{b}{m}v_x + \frac{F_0}{m}\cos(\omega_{driving} t) \qquad (13\text{-}82)$$

where $\omega_0 = \sqrt{k/m}$ is the natural frequency of the oscillator.

What is the equation for the displacement of the oscillator as a function of time? Using calculus, it can be shown that the displacement, $x(t)$, of the driven oscillator described by Equation (13-82) is given by

$$x(t) = A_{driven}\cos(\omega_{driven} t + \phi_{driven}) \qquad (13\text{-}83)$$

where

$$A_{driven} = \frac{F_0}{\sqrt{m^2(\omega_{driven}^2 - \omega_0^2)^2 + b^2\omega_{driven}^2}}$$

$$\phi_{driven} = \tan^{-1}\left(\frac{b\omega_{driven}}{m(\omega_{driven}^2 - \omega_0^2)}\right) \qquad (13\text{-}84)$$

A_{driven} and ϕ_{driven} are, respectively, the amplitude and the phase constant of the driven oscillator. Equation (13-83) is valid after transient effects have died down. (Transient effects are temporary and depend on how the initial motion was started.) Note several properties of driven oscillations:

1. An oscillator that is subjected to a sinusoidal driving force oscillates at the frequency of the driving force, *not* at the system's natural frequency. For example, an adult human ear resonates at approximately 3000 Hz. However, when an ear is subjected to a sound, for example, at a frequency of 500 Hz, the eardrum vibrates at 500 Hz rather than at its natural frequency.

2. The amplitude and the phase constant of the driven oscillations are determined not by the initial conditions, but by the parameters of the system: the natural frequency of the oscillator, the driving frequency of the external force, the magnitude of the driving force, the damping constant, and the mass of the oscillator. This situation is different from a simple harmonic oscillator, where the amplitude and the phase constant depend on the initial position and velocity of the oscillator.

3. As ω_{driven} approaches ω_0, the amplitude increases and reaches a maximum at $\omega_{driven} = \omega_0$. As ω_{driven} is increased past the value ω_0, the amplitude decreases again. If the damping constant, b, is small, the amplitude becomes quite large as ω_{driven} approaches ω_0. For $b = 0$, A_{driven} goes to infinity as $\omega_{driven} \to \omega_0$. In real life, damping is never zero, and the amplitude of driven oscillations can become large but not infinite. This dependence of A_{driven} on ω_{driven} forms the basis of resonance phenomena. For example, a wine glass can shatter when subjected to sound waves whose frequency is very close to the natural frequency of oscillation of the wine glass. Constructing earthquake-resistant buildings requires incorporating enough damping in the structures so that when a building is subjected to oscillations of frequency close to its natural oscillation frequency, the amplitude never becomes large. A plot of the amplitude as a function of b clearly shows the effect of resonance near the natural frequency of the oscillator. For the plot in Figure 13-38, we chose $m = 1.0$ kg, $\omega_0 = 1.0$ rad/s, and $F_0 = 1.0$ N and then plotted A_{driven} for $b = 0.01, 0.1, 0.5,$ and 0.8 kg/s. Near the resonance frequency, the amplitude is strongly dependent on the damping constant. Notice that the position of the peak amplitude shifts slightly away from the natural frequency as b increases. From Equation (13-84) it can be shown (by finding the maximum of A_{driven} as a function of ω_{driven}) that the amplitude is maximum at

$$\omega_{driven} = \sqrt{\omega_0^2 - \frac{b^2}{2m^2}} \qquad (13\text{-}85)$$

Figure 13-38 The amplitude for a forced harmonic oscillator for various values of damping constant, b, and $\omega_0 = 20$ rad/s.

Simple Harmonic Motion

A motion that repeats after a finite amount of time is called periodic motion. The time that it takes for the motion to repeat is called the period (T). One complete back-and-forth movement is called an oscillation. The number of oscillations per second completed by an object is the oscillation frequency (f) of the object and is related to the period:

$$f = \frac{1}{T} \qquad (13\text{-}3)$$

Periodic motion is called simple harmonic motion when, as a function of time, the displacement $x(t)$ of an object from the equilibrium position is given by a cosine (or a sine) function:

$$x(t) = A\,\cos(\omega t + \phi) \qquad (13\text{-}4)$$

The amplitude, A, and the phase constant, ϕ, are determined by the position and the velocity of the oscillator at $t = 0$. The angular velocity, ω, depends on the physical properties of the oscillating object and is related to the period:

$$\omega = 2\pi/T \qquad (13\text{-}5)$$

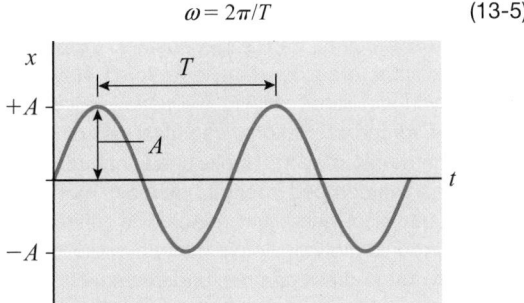

Figure 13-39 Simple harmonic motion; displacement as a function of time for simple harmonic motion.

The velocity and acceleration of a simple harmonic oscillator are determined by taking first and second derivatives, respectively, of Equation (13-4) as a function of time:

$$v(t) = -\omega A\,\sin(\omega t + \phi) \qquad (13\text{-}12)$$

$$a(t) = -\omega^2 A\,\cos(\omega t + \phi) \qquad (13\text{-}17)$$

An object undergoes simple harmonic motion when the net force acting on the object is proportional to its displacement from the equilibrium position and is opposite in sign of the displacement. Such a force is called a restoring force.

For horizontal and vertical mass–spring systems,

$$\omega = \sqrt{\frac{k}{m}}; \quad T = 2\pi\sqrt{\frac{m}{k}} \qquad (13\text{-}34),\ (13\text{-}35)$$

For a pendulum,

$$\omega = \sqrt{\frac{g}{L}}; \quad T = 2\pi\sqrt{\frac{L}{g}} \qquad (13\text{-}54),\ (13\text{-}55)$$

SHM and Uniform Circular Motion

The projection of a motion with a constant angular velocity along the circumference of a circle is mathematically equivalent to simple harmonic motion. The radius of the circle

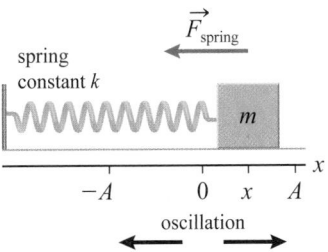

Figure 13-40 A horizontal mass–spring simple harmonic oscillator

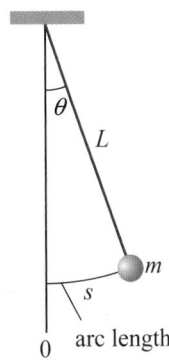

Figure 13-41 A simple pendulum

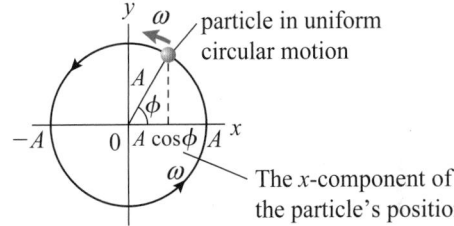

Figure 13-42 Comparing simple harmonic motion and circular motion

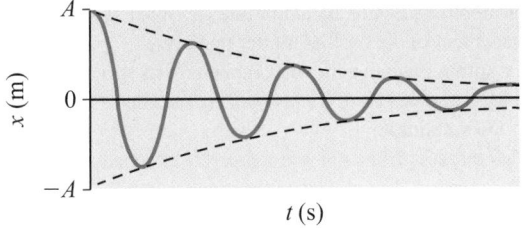

Figure 13-43 Displacement versus time plot for a damped harmonic oscillator

corresponds to the amplitude of the simple harmonic motion, and the angular velocity along the circle corresponds to the angular frequency of simple harmonic motion.

Damped Harmonic Oscillator

When frictional forces are present, the oscillations are damped and the oscillator is called a damped harmonic oscillator. For a velocity-dependent damping force ($\vec{F}_\mathrm{D} = -b\vec{v}$), the displacement of a damped harmonic oscillator is given by

$$x(t) = Ae^{-bt/2m}\cos(\omega_\mathrm{D} t + \phi) \qquad (13\text{-}70)$$

(continued)

where b is the damping constant, and the oscillation frequency ω_D is given by

$$\omega_D = \sqrt{\frac{k}{m} - \frac{b^2}{4m^2}} \qquad (13\text{-}71)$$

Energy

The total energy, E, of a simple harmonic oscillator is proportional to the square of its amplitude. In the absence of damping forces, the total energy of a harmonic oscillator remains constant and continuously transforms between kinetic energy, K, and potential energy, U:

$$E = K + U = \frac{1}{2}kA^2 \qquad (13\text{-}37)$$

Energy for a damped harmonic oscillator decreases exponentially with time and is given by

$$E(t) = \frac{1}{2}kA^2e^{-bt/m} \qquad (13\text{-}75)$$

where A is the amplitude of the damped oscillator at $t = 0$.

Resonance

A harmonic oscillator that is subjected to a sinusoidal external force oscillates with the frequency of the force. A resonance occurs when the frequency of the external force equals the natural frequency of the oscillator:

$$f_{driving} = f_0 \qquad (13\text{-}79)$$

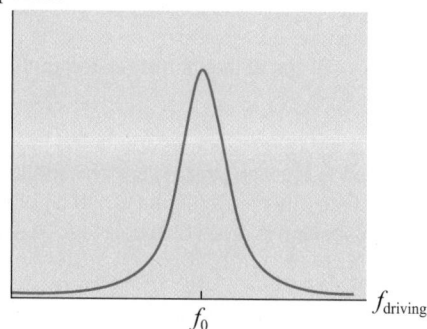

Figure 13-44 The amplitude of a driven oscillator as a function of driving frequency

---

Applications: pendulum clocks, shock absorbers, tune-mass dampers, seismometers

Key Terms: amplitude, angular frequency, critically damped, damped oscillations, damping constant, displaced harmonic oscillator, drag constant, drag force, driven oscillations, driven oscillator, driving frequency, envelope, fluid resistance, forced oscillations, fractional loss of energy, frequency, harmonic motion, Hooke's law, natural frequency, oscillation, overdamped, period, periodic motion, phase constant, physical pendulum, resonance, restoring force, simple harmonic motion, simple harmonic oscillator, simple pendulum, tuned-mass damper, underdamped

QUESTIONS

1. What is the total distance travelled in one cycle by a simple harmonic oscillator of amplitude A? What is the net displacement of the oscillator after one cycle?
2. If a spring is cut in half, what happens to its spring constant?
3. Why is the period of a mass–spring oscillator independent of the amplitude?
4. Two masses, m and $4m$, are separately attached to two identical springs. The period of the smaller mass is T. What is the period of the larger mass?
   (a) $T/2$ (b) T (c) $2T$ (d) $4T$
5. What would happen to the period of a mass–spring oscillator if the mass of the spring is not ignored?
6. As a pendulum oscillates, is there a point along its path where both its velocity and tangential acceleration are zero? Is there a point where both its displacement and tangential acceleration are zero?
7. What would happen to the period of a simple pendulum if the mass of the string is not ignored?
8. If the maximum angular displacement of a pendulum is increased, what happens to its speed as it passes through the equilibrium?
9. A pendulum is suspended from the ceiling of an elevator. When the elevator is at rest, the period of the pendulum is T. When the elevator accelerates upward, the period of the pendulum
   (a) decreases;   (b) remains T;   (c) increases.

10. You displace the mass of a mass–spring oscillator from equilibrium by 0.10 m, hold it stationary, and then release it. If you were to give the mass a kick as you released it, would the resulting amplitude be less than, equal to, or greater than 0.1 m? Why? Does it matter if the kick is toward the equilibrium or away from it? Explain.
11. A child is swinging on a playground swing while sitting. When the child stands up, what happens to the period of the swing?
12. A child is swinging on a playground swing while sitting. Another child sits on the same swing. What happens to the period of the swing? What happens to the speed of the swing as it passes through the equilibrium position?
13. The period of a pendulum is 2.00 s in a vacuum. Its period in air is
    (a) less than 2.00 s;
    (b) 2.00 s;
    (c) greater than 2.00 s.
14. In the following equations, x is displacement and t is time. Which equation describes simple harmonic motion?
    (a) $t(x) = (2.0)\cos(3.0x + \pi/4)$
    (b) $x(t) = 1.0/\sin(2t + \pi)$
    (c) $x(t) = 1.0\cos^2(4t)$
    (d) $x(t) = 3.0 + 5.0\cos(2t)$
15. Which of the following forces, acting on a particle of mass m, will *not* result in simple harmonic motion? The displacement of the particle from the equilibrium position is denoted by $x(t)$.

(a) $F(t) = -2x^2(t)$
(b) $F(t) = +2x^2(t)$
(c) $F(t) = +2x(t)$
(d) $F(t) = -2x(t)$

16. A mass attached to a spring is undergoing simple harmonic motion with an amplitude of 10.0 cm. When the mass is 5.0 cm from the equilibrium position, what fraction of the total energy is the *kinetic energy* of the mass?
(a) 0.25   (b) 0.50   (c) 0.75   (d) 1.00

17. When the amplitude of a simple harmonic oscillator is doubled, the total energy of the oscillator
(a) remains the same
(b) increases by a factor of 2;
(c) decreases by a factor of 2;
(d) increases by a factor of 4;
(e) decreases by a factor of 4.

18. The graph in Figure 13-45 shows the displacement versus time plot of an object that is undergoing simple harmonic motion. For the displacement at $t = 3$ s, which of the following statements is correct?
(a) the velocity of the particle is positive, and its acceleration is positive;
(b) the velocity of the particle is positive, and its acceleration is negative;
(c) the velocity of the particle is negative, and its acceleration is positive;
(d) the velocity of the particle is negative, and its acceleration is negative.

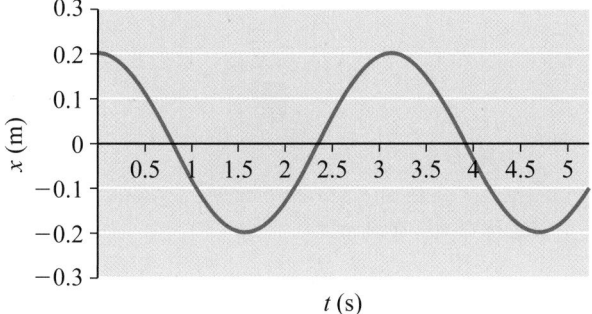

Figure 13-45   Question 18

PROBLEMS BY SECTION

For problems, star ratings will be used, (✱, ✱✱, or ✱✱✱), with more stars meaning more challenging problems.

Section 13-2   Simple Harmonic Motion

19. ✱ A mass of 2.0 kg is connected to a spring with a spring constant of 5.0 N/m. The mass is oscillating on a horizontal, frictionless surface. At time $t = 0$, the mass is 0.3 m from the equilibrium position and is moving with a velocity of 1.0 m/s. Assume that the phase constant is zero.
(a) What is the amplitude?
(b) What is the maximum speed of the mass?
(c) What is the maximum acceleration of the mass?
(d) Write an equation that describes the displacement of the mass from the equilibrium position as a function of time.

20. ✱ A mass–spring system has $k = 18.0$ N/m and $m = 0.71$ kg. The system oscillates with an amplitude of 0.54 m.
(a) Determine the angular frequency of the oscillations.
(b) What is the speed of the mass at $x = 0.034$ m?
(c) What is the displacement of the mass from the equilibrium position when its speed is 0.18 m/s?

21. ✱✱✱ A mass attached to a horizontal spring is pulled to the right and released at $t = 0$ s. The mass passes through $x = +10.0$ cm at $t = 0.3$ s with a velocity of 85.7 cm/s.
(a) What is the frequency of the mass–spring system?
(b) What is the amplitude?
(c) What is the phase constant?
(d) Write an equation that describes the position of the mass from the equilibrium position as a function of time.

22. ✱✱✱ Two identical mass–spring oscillators are $\pi/3$ rad out of phase and are undergoing simple harmonic motion with amplitude A about the same origin. One of the oscillators is at $x = +A$ at $t = 2.0$ s. When will the second oscillator next reach $x = +A$?

23. ✱✱ A mass–spring oscillator undergoing simple harmonic motion about $x = 0$ has a period of 2.0 s. At $t = 0$ s, the mass is at $x = 0$; at $t = 0.5$ s, the mass is at $x = -0.40$ m.
(a) Write an equation that describes the position of the mass as a function of time.
(b) What are the maximum velocity and acceleration of the mass?
(c) The mass is 0.10 kg. What is the total energy of the oscillator?

24. ✱✱✱ Two particles oscillate in simple harmonic motion with amplitude A, about the centre of a common straight line of length $2A$. Each particle has a period of 1.5 s, and their phase constants differ by $\pi/6$ rad.
(a) How far apart are the particles from each other (in terms of A) 0.50 s after the lagging particle leaves one end of the path?
(b) What are their velocities at that time?

25. ✱✱✱ Can we determine the amplitude, A, and the phase constant, ϕ, for a simple harmonic motion given the values of displacement at $t = 0$ and $t = T/2$, where T is the period? Explain your reasoning.

26. ✱✱ An object, oscillating with simple harmonic motion, has a period of 4.0 s and an amplitude of 0.20 m.
(a) How long does the object take to move from $x = 0.0$ m to $x = 0.07$ m?
(b) Would the object take the same time to move from $x = 0.07$ m to $x = 0.14$ m? Explain your answer without any calculation.

27. ✱✱ Equations for five simple harmonic oscillators are given below:

$$x_1(t) = 0.2 \cos(5t + \pi/4)$$
$$x_2(t) = 0.1 \cos(2t + \pi)$$
$$x_3(t) = 0.3 \cos(t/5)$$
$$x_4(t) = 0.2 \cos(7t/2 + \pi/2)$$
$$x_5(t) = 0.6 \cos(15t)$$

(a) Rank the oscillators, from high to low, in order of period:
High ---------- ---------- ---------- ---------- ---------- Low

(b) Rank the oscillators, from high to low, in order of maximum velocity:

High ---------- ---------- ---------- ---------- ---------- Low

(c) Rank the oscillators, from large to small, in order of the phase constant:

Large ---------- ---------- ---------- ---------- ---------- Small

(d) Rank the oscillators, from high to low, in order of maximum acceleration:

High ---------- ---------- ---------- ---------- ---------- Low

28. ✳✳✳ Two particles are executing simple harmonic motion with same amplitude and frequency about the same equilibrium position. They pass each other moving in opposite direction, each time their displacement is half their amplitude. What is the difference between their phase constants?

Section 13-3 Uniform Circular Motion and Simple Harmonic Motion

29. ✳ Consider a point located on the equator of Earth. Find the angular speed (in rad/s) and the period of this point.

30. ✳ A particle initially at rest on the x-axis, at $x = 3.0$ m, starts moving with a constant angular velocity of 2.0 rad/s in a counterclockwise direction in a circle of radius 3.0 m around the origin. The x-component of the particle's position at any time t is given by which of the following equations?

(a) $x(t) = 2.0 \sin(3.0t)$
(b) $x(t) = 3.0 \cos(2.0t)$
(c) $x(t) = 3.0 \sin(2.0t)$
(e) $x(t) = 3.0 + \cos(2.0t)$

Section 13-4 Mass–Spring Systems

31. ✳ The spring constants and masses for six mass–spring oscillators are given in Table 13-3. Rank the oscillators in the order of their periods, from small to large.

Small ---------- ---------- ---------- ---------- ----------Large

Table 13-3 Spring Constants and Masses for Six Mass–Spring Oscillators

Oscillator	A	B	C	D	E	F
k (N/m)	2.0	4.5	1.0	7.0	11.6	4.6
m (kg)	0.5	1.2	2.0	0.3	3.2	3.1

32. ✳ The displacement from equilibrium position of an oscillator is given by the equation $x(t) = (11.0 \text{ cm}) \cos(3.0\,t + 5.0)$

(a) What are the period and the phase constant of the oscillation?

(b) What is the phase of the oscillator at $t = 0$ s and $t = 2.0$ s?

(c) What are the position and the velocity of the oscillator at $t = 0$ s?

(d) At what time after $t = 0$ s is the oscillator at $x = 0$ cm?

(e) What is the total distance covered by the oscillator between $t = 1.3$ s and $t = 1.5$ s?

33. ✳✳ A simple harmonic oscillator consists of a 0.10 kg mass attached to a spring with a spring constant of 100 N/m. The mass is displaced 0.20 m from the equilibrium position, held motionless, and then released. Calculate the following quantities for this system:

(a) the angular frequency and the period;

(b) the maximum speed and the maximum acceleration;

(c) the total energy of the mass–spring system.

34. ✳✳ In problem 33, if the mass is released at time $t = 0$ s, at what time will the kinetic energy of the mass be equal to the potential energy of the spring?

35. ✳✳ A 2.0 kg mass rests on a smooth, horizontal surface and is attached to a spring, with the other end of the spring attached to a rigid support. An applied force of 10.0 N causes a displacement of 5.0 cm from the equilibrium.

(a) Find the spring constant of the spring.

(b) What is the frequency of the mass–spring system?

(c) When oscillating, the mass is at a maximum displacement of 5.0 cm at $t = 0.50$ s. Write an equation that describes the position of the mass from the equilibrium position as a function of time.

(d) What is the total distance travelled by the mass in the interval $t = 0.5$ s to $t = 2.0$ s?

36. ✳✳✳ Figure 13-46 shows a 0.20 kg mass resting on top of a 2.0 kg mass. The lower mass is attached to a spring with a spring constant of 100 N/m. The two masses are oscillating on a frictionless surface. The coefficient of static friction between the two masses is 0.60. What is the maximum amplitude for which the upper block does not slip?

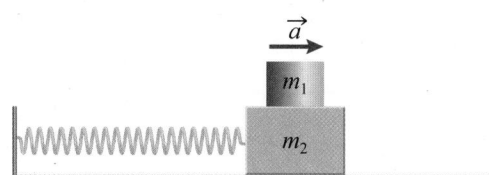

Figure 13-46 Problem 36

37. ✳✳✳ A 500 g block slides along a frictionless surface at a speed of 0.35 m/s. It runs into a horizontal massless spring that extends outward from a wall; the spring constant is 50 N/m. The block compresses the spring and is then pushed back in the opposite direction by the spring, eventually losing contact with the spring.

(a) How long does the block remain in contact with the spring?

(b) How would your answer to part (a) change if the block's initial speed were doubled?

38. ✳✳✳ A 0.20 kg block attached to a horizontal spring is oscillating with an amplitude of 5.0 cm and a frequency of 2.0 Hz. Just as it passes through the equilibrium point, moving to the right, a sharp blow directed to the left exerts a 10 N force for 1.0 ms. Calculate the new frequency and amplitude of oscillations.

39. ✳✳ An empty compact car has a mass of 1200 kg. Assume that the car has one spring on each wheel, the springs are identical, and the mass is equally distributed over the four springs.

(a) What is the spring constant of each spring if the car bounces up and down 2.0 times per second?

(b) What will be the car's oscillation frequency while carrying four passengers each of mass 70 kg?

40. ✷✷✷ Two masses, m_1 and m_2, are connected to two ends of a spring with a spring constant k. The masses are free to slide on a horizontal, frictionless surface. The spring is compressed and then released. Find an expression for the frequency of the system.

41. ✷✷✷ A block of mass m, resting on a frictionless surface, is connected to two springs, 1 and 2, with spring constants k_1 and k_2, respectively, as shown in Figure 13-47. The block is displaced from the equilibrium position and then released.

(a) Show that the resulting motion of the block is simple harmonic motion.

(b) Show that the frequency of oscillations of the block is given by

$$f = \sqrt{\frac{f_1^2 f_2^2}{f_1^2 + f_2^2}}$$

where f_1 is the frequency of the mass when only spring 1 is attached to the mass, and f_2 is the frequency of the mass when only spring 2 is attached to the mass.

(c) Let U_1 and U_2 denote the elastic potential energies stored in the two springs as the mass oscillates. Show that

$$\frac{U_1}{U_2} = \frac{k_2}{k_1}$$

(d) As the mass oscillates, is there a position of the mass where only one of the springs is stretched (or compressed) and the other spring is in its equilibrium position? Explain your reasoning.

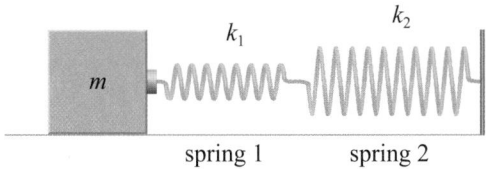

Figure 13-47 Problem 41

42. ✷✷✷ A block of mass m, resting on a frictionless surface, is connected to two springs, 1 and 2, with spring constants k_1 and k_2, respectively, as shown in Figure 13-48. The block is displaced from the equilibrium position and then released.

(a) Show that the resulting motion of the block is simple harmonic motion.

(b) Show that the frequency of the block is given by

$$f = \sqrt{f_1^2 + f_2^2}$$

where f_1 is the frequency of the mass when only spring 1 is attached to the mass, and f_2 is the frequency when only spring 2 is attached to the mass.

(c) Let U_1 and U_2 denote the elastic potential energies stored in the two springs as the mass oscillates. Show that

$$\frac{U_1}{U_2} = \frac{k_1}{k_2}$$

(d) As the mass oscillates, is there a position of the mass where only one of the springs is stretched (or compressed) and the other spring is in its equilibrium position? Explain your reasoning.

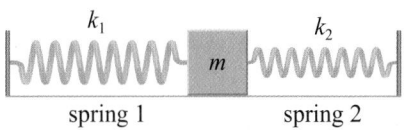

Figure 13-48 Problem 42

43. ✷✷ A spring hangs vertically from a fixed support. A mass is then attached to the lower end of a spring. When this system undergoes simple harmonic motion, it has a period of 0.50 s. By how much is the spring stretched from its initial length when the mass and spring are hanging motionless?

44. ✷✷✷ A mass, m, is attached to a spring with a spring constant k. The other end of the spring is connected to a fixed support. Initially, the mass is supported from underneath so that the spring remains in its unstretched configuration. The support is then removed, and the mass falls vertically downward under the influence of gravity.

(a) Show, using conservation of energy, that the maximum distance, ΔL, that the mass falls from its initial position is given by $\Delta L = \dfrac{2mg}{k}$.

(b) Once dropped, the mass will undergo simple harmonic motion about an equilibrium position. Show that the equilibrium position is at a distance $\dfrac{mg}{k}$ below the unstretched position of the spring, the amplitude of oscillations is $\dfrac{mg}{k}$, and period is $T = 2\pi\sqrt{\dfrac{\Delta L/2}{g}}$.

45. ✷✷ A 1.0 kg mass is hanging motionless from a spring that is attached to a fixed support. The mass is pulled down 5.0 cm, held motionless, and then released. The mass–spring system begins to oscillate with a period of 2.0 s.

(a) What is the spring constant of the spring?

(b) What is the total energy of the mass–spring system?

(c) What fraction of the system's total energy is the kinetic energy of the mass at $x = 3.0$ cm from the equilibrium position?

(d) Write an equation that describes the position of the mass from the equilibrium position as a function of time.

Section 13-5 Energy Conservation in Simple Harmonic Motion

46. ✷ A 2.0 kg object, attached to a spring, oscillates with simple harmonic motion. The spring constant is 200 N/m, and the maximum force acting on the object is measured to be 3.5 N.

(a) What is the amplitude of the oscillation?

(b) What is the total energy of the mass–spring system as it oscillates?

(c) What is the maximum speed of the mass?

47. ✱✱ An archer pulls her bow string back by 0.50 m before releasing the arrow. The bow–string system has an effective spring constant of 500 N/m.
(a) What is the elastic potential energy of the drawn bow?
(b) The arrow has a mass of 30.0 g. What is the speed of the arrow as it leaves the bow?

48. ✱✱ A particle undergoes simple harmonic motion with amplitude A. At what position (in terms of A) is the kinetic energy of the particle equal to its potential energy?

49. ✱✱ A mass–spring system is oscillating with an amplitude of 10.0 cm. What is the speed of the mass at a location where the kinetic energy and the potential energy of the mass are equal?

50. ✱✱ A student observes that when a mass of 2.0 kg is attached to one end of a vertical spring, the other end of which is connected to a fixed support, the spring stretches 10.0 cm. The student then pulls the mass farther down by 5.0 cm from the equilibrium position and lets it go. As expected, he observes that the mass–spring system undergoes simple harmonic motion.
(a) What is the amplitude and frequency of oscillations?
(b) What is the kinetic energy of the mass when it is 3.00 cm below the equilibrium position?
(c) Assuming that the upward direction is positive, write an equation for the displacement, y, from the equilibrium position, assuming that the mass was released at $t = 0$. Evaluate all parameters.
(d) If the student had connected two springs identical to the spring above, and then hung a mass of 2.0 kg on them, by how much would the combined springs stretch?

Section 13-6   The Simple Pendulum

51. ✱✱ The lengths and masses for six pendulums are listed in Table 13-4. Rank the pendulums in order of their periods, from small to large.
Small ---------- ---------- ---------- ---------- ---------- Large

Table 13-4 Lengths and Masses for Six Pendulums

Pendulum	A	B	C	D	E	F
L (m)	0.8	1.1	1.7	0.9	2.1	3.2
m (kg)	0.5	1.2	2.0	0.3	3.2	3.1

52. ✱✱ A simple pendulum of length 1.00 m is measured to have a period of 2.00 s.
(a) What is the value of g at the location of the pendulum?
(b) The pendulum is now taken to the top of Mount Everest. Does the period increase or decrease? Justify your answer.
(c) What would be the period of this pendulum on the Moon?
(d) What would be the period of this pendulum on the space shuttle at the International Space Station?

53. ✱✱ A pendulum with a length of precisely 1.000 m can be used to measure the acceleration due to gravity, g. Such a device is called a *gravimeter*.
(a) How long do 100 oscillations take at sea level, where $g = 9.81$ m/s^2?

(b) Suppose you take a gravimeter to the top of Mount Everest, where the value of g is 0.28% less than the value at sea level. How long do 100 oscillations take at the top of Mount Everest? Would you be able to measure the difference with a stopwatch that can measure time differences of 0.1 s?

54. ✱✱ A simple pendulum, consisting of a 0.30 kg mass attached to a string of length 1.0 m, is undergoing simple harmonic motion with a maximum angular displacement of 15°.
(a) What is the total energy of the pendulum?
(b) What is the speed of the mass as it passes through the equilibrium position?

55. ✱✱ (a) Consider a simple pendulum of length L. Show that a change in the length by an amount dL changes the pendulum's period by an amount dT where

$$dT = \left(\frac{T}{2L}\right)dL$$

(b) If the pendulum clock runs faster by 5 s/h, how should the length be adjusted?

Section 13-7   The Physical Pendulum

56. ✱✱✱ A thin circular ring of mass m and radius R hangs freely on a small peg at its rim. The ring is gently displaced from its equilibrium position and released. Derive an expression for the frequency of small oscillations of the ring.

57. ✱✱✱ A physical pendulum consists of a sphere of mass m and radius R that is attached to a massless string. The other end of the string is attached to a fixed support, and the distance between the centre of the bob and the fixed support is L. Show that the period of small oscillations for this pendulum is given by

$$T = T_0 \sqrt{1 + \frac{2R^2}{5L^2}}$$

where $T_0 = 2\pi\sqrt{L/g}$ is the period of a simple pendulum of length L.

58. ✱✱✱ A physical pendulum consists of a sphere of mass m_s and radius R that is attached to a uniform rod of mass m_r and length L. The other end of the rod is attached to a fixed support, and the rod is free to rotate in a plane about an axis that is perpendicular to the plane of rotation and passes through the fixed support.
(a) Show that the period of small oscillations for this physical pendulum is given by

$$T = 2\pi \sqrt{\frac{\frac{2}{5}m_s R^2 + m_r\frac{L^2}{3} + m_s(L + R)^2}{g\left(m_s(L + R) + m_r\frac{L}{2}\right)}}$$

(b) Calculate the period for the following parameters: $m_s = 1.00$ kg, $m_r = 0.20$ kg, $R = 5.0$ cm, and $L = 1.0$ m. Compare this period to the period of a simple pendulum with $L = 1.05$ m. What is the percentage difference between the two periods? Why is the physical pendulum faster than an equivalent simple pendulum?

Section 13-8   Time Plots for Simple Harmonic Motion

59. ✳   The position versus time graph for a simple harmonic oscillator is shown in Figure 13-49.
  (a) What are the amplitude and the period?
  (b) What is the phase constant?
  (c) What is the phase at $t = 1$ s and $t = 2$ s?
  (d) Write an equation for the oscillator that describes its position $x(t)$ as a function of time.
  (e) What is the velocity of the oscillator at $t = 1.0$ s?
  (f) What is the acceleration of the oscillator at $t = 2$ s?

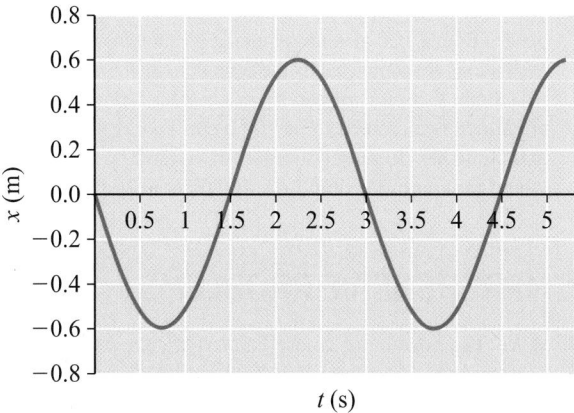

Figure 13-49   Problem 59

60. ✳✳   The velocity versus time graph for a simple harmonic oscillator is shown in Figure 13-50.
  (a) What are the maximum speed and the period?
  (b) What is the amplitude?
  (c) What is the phase constant?
  (d) Write an equation for the oscillator that describes its velocity $v(t)$ as a function of time.
  (e) What is the velocity of the oscillator at $t = 1.0$ s?
  (f) What is the position of the oscillator at $t = 3.0$ s?

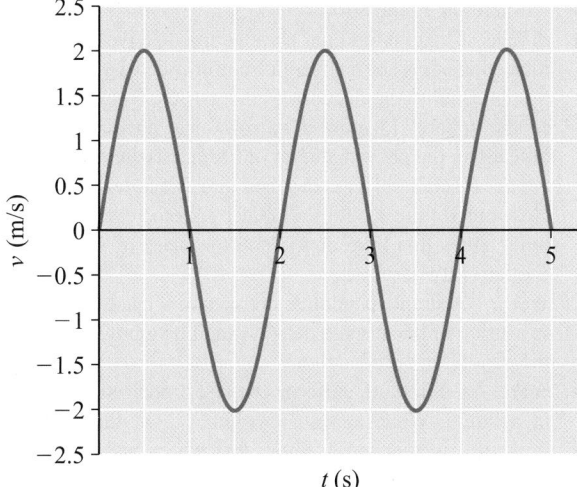

Figure 13-50   Problem 60

61. ✳✳   The position versus time graph for a simple harmonic oscillator is shown in Figure 13-51.

  (a) What are the amplitude and the period?
  (b) What is the equilibrium position?
  (c) What is the phase constant?
  (d) Write an equation for the oscillator that describes its position $x(t)$ as a function of time with respect to the point $x = 0$.
  (e) Write an equation for a simple harmonic oscillator that is $\pi/2$ rad out of phase with this oscillator.
  (f) What is the acceleration of the oscillator at $t = 1.0$ s?

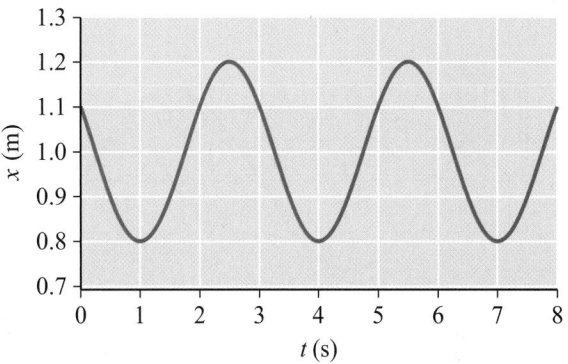

Figure 13-51   Problem 61

62. ✳✳   The position versus time plots of two oscillators are shown in Figure 13-52.
  (a) What is the phase difference between the two oscillators at $t = 1.0$ s?
  (b) At what times, closest to $t = 0$ s, do the two oscillators have zero velocities?
  (c) At what times, closest to $t = 0$ s, are the two oscillators moving with maximum velocity toward the negative x-axis?

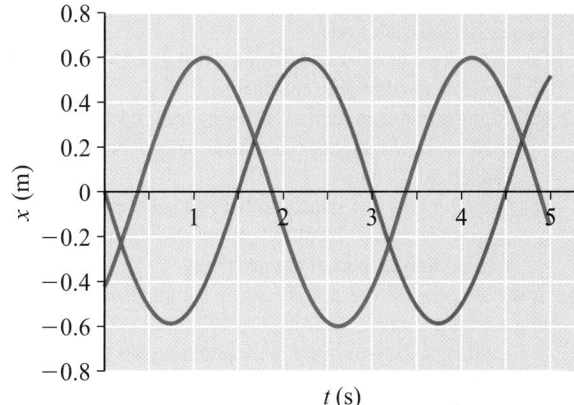

Figure 13-52   Problem 62

63. ✳   Figure 13-53 shows the displacement, $x(t)$, curves for three experiments involving three different masses attached to identical springs oscillating with simple harmonic motion. Rank the curves as indicated.

CHAPTER 13 | **OSCILLATIONS**   377

(a) the system's period
   Greatest ---------- --------------- ------------- Lowest
(b) the total energy of the mass–spring system
   Greatest ---------- --------------- ------------- Lowest
(c) the potential energy of the spring at $t = 0$
   Greatest ---------- --------------- ------------- Lowest
(d) the kinetic energy of the spring at $t = 0$
   Greatest ---------- --------------- ------------- Lowest

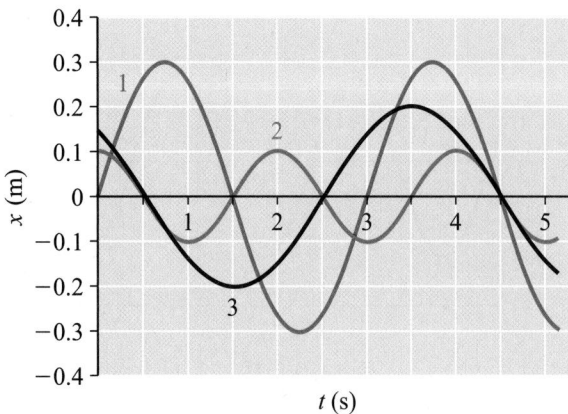

Figure 13-53 Problem 63

Section 13-10 Damped Oscillations

64. ✷✷ A damped harmonic oscillator consists of a block of mass 2.0 kg and a spring with a spring constant $k = 10.0$ N/m. Initially, the system oscillates with an amplitude of 25 cm. Because of the damping, the amplitude decreases by 75% of its initial value at the end of four oscillations.
   (a) What is the value of the damping constant, b?
   (b) What percentage of initial energy has been lost during these four oscillations?

65. ✷✷ A damped mass–spring oscillator loses 5% of its energy during each cycle.
   (a) The period is 2.0 s, and the mass is 0.25 kg. What is the value of the damping constant?
   (b) If the motion is started with an initial amplitude of 0.30 m, how many cycles elapse before the amplitude reduces to 0.20 m?
   (c) Would it take the same number of cycles to reduce the amplitude from 0.20 m to 0.10 m? No calculations are needed for this part of the problem.

66. ✷✷ A car and its shock absorbers behave like a damped mass–spring system, with $m = 1500$ kg, $k = 60$ N/m, and $b = 1500$ kg/s. The car hits a pothole and begins to oscillate. After how much time would the amplitude of oscillations drop to half its initial value?

67. ✷✷ The initial amplitudes and b/m values for three damped oscillators are given in Table 13-5. Rank their amplitudes after 100 s, from smallest to largest, indicating any equivalencies by using the equality sign (=).

   Smallest --- Largest

Table 13-5 For Problem 67

Damped harmonic oscillator	Amplitude (cm)	b/m (s⁻¹)
A	10	0.01
B	20	0.02
C	30	0.03

68. ✷✷ (a) A large pendulum consists of a 50.0 kg mass suspended from a 15.0 m wire. Assuming no damping, how many oscillations are completed by the pendulum in one week?
   (b) Slight damping, of course, exists. At a certain time, the amplitude is measured to be 100.0 cm. Five days later, the amplitude has decreased to 80.0 cm. After how many days would the amplitude drop from 100.0 cm to 20.0 cm?

COMPREHENSIVE PROBLEMS

69. ✷✷ The position of an oscillator is known at two times, $t = 0$ and $t = T/4$, where T is the period. Use this information to determine the amplitude and the phase constant of the oscillator.

70. ✷✷✷ A spherical object of mass m and radius r is placed in a large, hemispherical bowl of radius R. The mass rests in the equilibrium position at the bottom of the bowl. The mass is then displaced from its equilibrium position by a small angle, still touching the inner surface of the bowl, and released from rest. The mass rolls down the inner surface of the bowl without slipping. The mass of the bowl is large enough that the bowl remains stationary as the mass moves back to the equilibrium position.
   (a) Show that, when the friction between the mass and the bowl is ignored, the mass will undergo simple harmonic motion about its equilibrium position.
   (b) What is the frequency of the mass, in terms of the variables m, r, and R?

71. ✷✷✷ The gravitational force acting on a particle that is located inside a solid sphere of uniform density is directly proportional to the distance of the particle from the centre of the sphere. Imagine a narrow, evacuated cylindrical hole along the polar diameter of Earth, passing through its centre. A small object is dropped into this hole at one pole. How much time would the object take to reach the other pole? Does this time depend on the mass of the object? Explain why or why not.

72. ✷✷✷ A simple pendulum consists of a mass m attached to a string of linear mass density μ and length L. Determine the period for the pendulum.

73. ✷✷ An object of uniform cross-sectional area A floats in a liquid. When at rest, a volume V_0 of the object is immersed in the liquid. Show that when pushed slightly farther into the liquid and released, the object will undergo simple harmonic motion with the period $2\pi\sqrt{\dfrac{V_0}{gA}}$, where g is the acceleration due to gravity.

74. ✱✱ A mass attached to a spring has a period of T seconds. The mass is displaced from its equilibrium position and then released from rest. Determine the time (in terms of T) when
(a) the kinetic energy of the mass equals the potential energy of the spring;
(b) the kinetic energy of the mass is twice the potential energy of the spring.

75. ✱✱ A block of mass M, resting on a frictionless surface, is attached to a spring with a spring constant k (Figure 13-54). The other end of the spring is tied to a rigid support. The mass is initially at rest. A bullet of mass m, moving horizontally with speed v, strikes the block and bounces off it, straight back. Determine expressions for
(a) the speed of the block immediately after the collision;
(b) the amplitude of the block–spring oscillations as a result of the collision.

Figure 13-54   Problem 75

76. ✱✱ (a) Show that when the bob of a simple pendulum is displaced by an angle θ from the equilibrium position, the potential energy of the pendulum increases by $mgL(1 - \cos\theta)$.
(b) If the bob is displaced by an angle θ and released from rest, show that the speed of the bob when it passes through the equilibrium position is given by

$$v = \sqrt{2gL(1 - \cos\theta)}$$

See the text online resources at
www.physics1e.nelson.com for Open Problems
and Data-Rich Problems related to this chapter.

Learning Objectives

When you have completed this chapter you should be able to:

1   Describe the nature and classification of waves.

2   Describe how a mechanical wave travels through a medium.

3   Write equations for a travelling waveform moving toward the left or the right from a given algebraic form at a fixed time.

4   Distinguish between the wave speed and the instantaneous speed of a segment of a medium, calculate the instantaneous speed of a segment of a medium, and explain how the wave speed in a string depends on the inertial and elastic properties of the string.

5   Determine the amplitude, wavelength, frequency, and phase constant for a travelling harmonic wave from a given mathematical form.

6   Describe what is meant by position plots and time plots of a wave. Determine the mathematical equation for a wave from its position and time plots.

7   Distinguish between the phase and the phase constant for a harmonic wave. Calculate the phase difference between two points of a medium in the presence of a wave.

8   Explain how energy is carried along a string by a travelling wave and how the kinetic energy and the potential energy of an element of string vary with time.

9   Describe the principle of superposition for waves. Explain the consequence of this principle when two or more waves are simultaneously present in a medium.

10 Determine the waveform of the resultant wave when two or more waves are simultaneously present in a medium.

11 Describe the behaviour of reflected and transmitted waves at the boundary between two media.

12 Explain what standing waves are, and determine the location of nodes and antinodes of a standing wave from its mathematical equation.

13 Calculate the resonant frequencies of a vibrating string that is fixed at both ends.

For additional Making Connections, Examples, and Checkpoints, as well as activities and experiments to help increase your understanding of the chapter's concepts, please go to the text's online resources at www.physics1e.nelson.com.

Chapter 14
Waves

Earthquakes release an enormous amount of energy (Figure 14-1). This energy is carried from the origin of the earthquake, through Earth, by longitudinal and transverse waves over large distances. What are waves? How do waves transport energy? How does a longitudinal wave differ from a transverse wave? What factors determine the speed with which waves travel? Is the motion of a wave different from the motion of the constituents of the medium through which it moves?

© Maciej Dakowicz/Alamy

Figure 14-1   Damage caused by the 2005 earthquake in Islamabad, Pakistan

14-1   The Nature, Properties, and Classification of Waves

Wave motion is very different from the motion of objects. Some waves, such as waves in an ocean, need a medium in which to travel, but other waves, such as light, do not. A wave passes through a medium unaffected by the presence of other waves in that medium. In this chapter, we will explore the laws governing wave motion and derive mathematical relationships for some properties of waves.

What is a wave? Consider a long rope stretched taut between two rigid supports. If you pluck the rope near one end and then release it, a disturbance originates from where you plucked the rope and travels along the length of the rope. The disturbance moves along the rope, but it does not take the rope with it. As the disturbance passes through a given section of the rope, that section is displaced from its equilibrium position and then returns to it after the disturbance has passed through. Therefore, we can define **wave** as the motion of a disturbance. Waves are classified into two major categories: mechanical waves and electromagnetic waves.

Mechanical Waves A **mechanical wave** travels through a physical material, or **medium**, and the speed of the wave depends on the properties of the medium. The particles of the medium must interact such that a disturbance of particles at any given point is transmitted to adjacent particles. The vibration of a violin string, ripples on the surface of water (Figure 14-2), and seismic waves in Earth's crust are all examples of mechanical waves.

Stockbyte/Thinkstock

Figure 14-2   Concentric waves on a water surface

Electromagnetic Waves Light, microwaves, radio waves, and X-rays are all manifestations of **electromagnetic waves**, or **electromagnetic radiation**. Electromagnetic waves are generated by charged particles that accelerate; for example, electrons moving back and

forth in the antenna of a radio transmitter. Unlike mechanical waves, electromagnetic waves do not need a medium in which to travel. The light that we receive from the Sun travels through the vacuum between the Sun and Earth. This chapter deals with mechanical waves; electromagnetic waves are discussed in Chapter 26.

Waveform The shape or pattern of a wave is called its **waveform**. Waveforms can be represented by plotting some property of the medium as the wave moves through it. For example, a sound wave in air creates regions of higher and lower pressure relative to undisturbed air. If we graph the pressure as a function of time at a given location, the resulting pressure versus time graph would show the shape of the sound wave at that location. A waveform can also be displayed by plotting the shape of the disturbance at a fixed time as a function of position in the medium through which the wave is travelling. Figure 14-3 shows the waveform of a sinusoidal wave on a long string, and Figure 14-4 shows the more complex waveform of a note played on a violin string. Figure 14-5 shows an image of a complex waveform for waves on the surface of an ocean at a given instant.

The waveform for a sound wave in the air can also be obtained by graphing either the variation in the air pressure or the displacement from the equilibrium position of a small element of air as a function of time. These are all different ways of visualizing the same wave.

Figure 14-3   The waveform (at a fixed time) generated on a long string by shaking one end of the string in simple harmonic motion

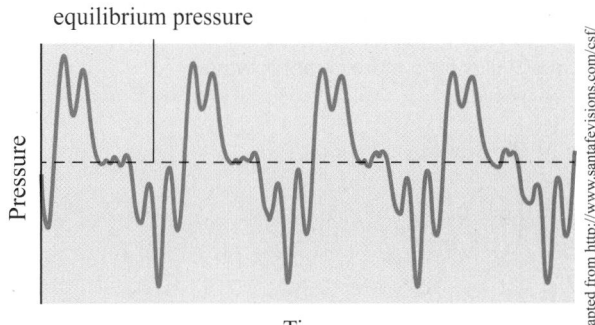

Adapted from http://www.santafevisions.com/csf/images/lectures/violin.gif

Figure 14-4   A waveform of sound waves generated when playing a note on a violin. The plot shows the change in air pressure with respect to the equilibrium pressure (dashed line, corresponding to no sound) at a given distance from the violin.

Figure 14-5 Complex waveforms on the ocean surface

Pulse A **pulse** is a disturbance of short duration. The disturbance generated on a stretched string by a quick flip of the hand, up and back to the original position, is a pulse (Figure 14-6). Similarly, a single handclap generates a sound pulse. The shape of a pulse depends on the motion of the source that generates the pulse.

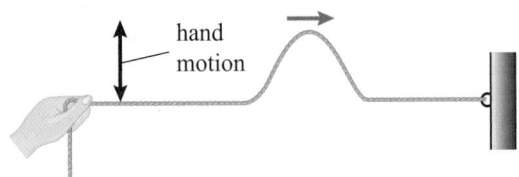

Figure 14-6 A quick flick on one end of a stretched string creates a pulse that travels along the string.

Continuous Wave A **continuous wave** (also called a wave) is a continuous series of pulses. Continuously flipping one end of a taut string up and down generates a continuous wave on the string. A vibrating tuning fork generates continuous sound waves as its tines oscillate in repetitive simple harmonic motion.

Wave Cycle A continuous wave has a waveform that repeats over and over again. The fundamental unit of the repeating pattern is called a **wave cycle**. The duration of the wave cycle is the period of the wave. Figure 14-7 shows the wave cycles for a sine wave, and Figure 14-8 shows the wave cycles for a more complex wave.

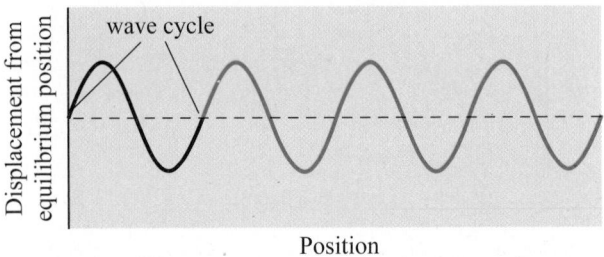

Figure 14-7 The black portion of the wave represents a wave cycle for a sinusoidal wave.

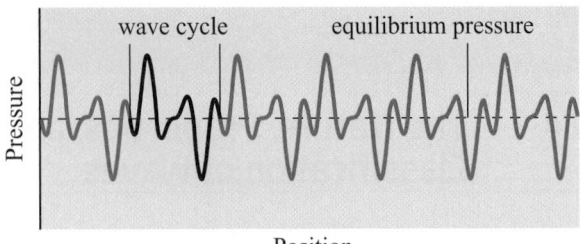

Figure 14-8 The black portion of the wave represents a wave cycle for a more complex wave.

Wave Speed The speed at which a disturbance travels through a medium is called the **wave speed**, usually denoted by the letter v. For example, the speed of sound in dry air at sea level at 20°C is approximately 343 m/s.

Transverse Waves As a wave passes through a medium, the constituents (atoms and molecules) of the medium are temporarily displaced from their equilibrium positions in response to the forces exerted by the wave. In a **transverse wave**, the constituents of the medium move *perpendicular* to the direction of propagation of the wave (Figure 14-9). For example, a pulse generated by moving one end of a taut horizontal string up and down travels horizontally along the length of the string, and the segment of the string that the pulse is passing through is displaced vertically from the rest position.

Transverse Waves

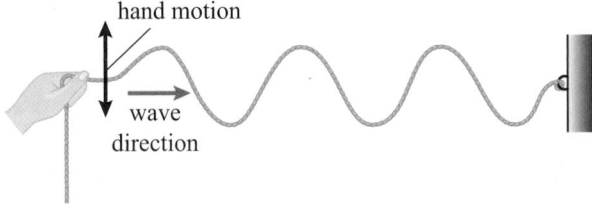

Figure 14-9 A transverse wave on a string

Longitudinal Waves In a **longitudinal wave**, the constituents of the medium move *parallel* to the direction of motion of the wave. For example, alternately pushing and pulling on one end of a long spring generates a wave that moves along the length of the spring (Figure 14-10). As the wave passes through a given section of the spring, the coils of that section move parallel to the direction of motion, being alternately compressed and stretched. Sound waves are examples of three-dimensional longitudinal waves. As a sound wave passes through air, the air molecules oscillate in the direction of wave motion, alternately creating regions of higher and lower pressure.

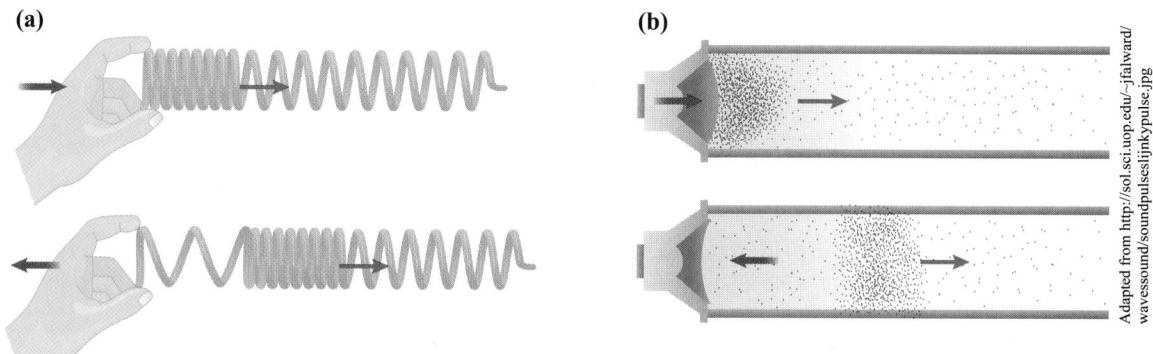

(a)

(b)

Figure 14-10 (a) A longitudinal wave moving along a stretched string. The movement of the coils is in the direction of wave motion, creating regions of higher and lower pressure (compressions and rarefactions, respectively). (b) A sound wave traveling along the length of an air-filled tube. As the diaphragm of the speaker oscillates, it creates periodic pattern of compression and expansion of air.

 MAKING CONNECTIONS

Earthquake Waves

During an earthquake, both transverse and longitudinal waves are generated in the Earth's mantle. The longitudinal waves, called P waves (primary waves), alternately compress and stretch the ground in the direction of wave motion. Transverse waves, called S waves (secondary waves), displace the ground up and down, perpendicular to the direction of wave motion (Figure 14-11). In solids, P waves move approximately twice as fast as S waves. Therefore, during an earthquake, we usually feel two jolts because the P waves arrive ahead of the S waves. The major damage in an earthquake is caused by the up-and-down motion of the ground, that is, by the S waves. Thus, the P waves can give some warning that the more destructive S waves are coming.

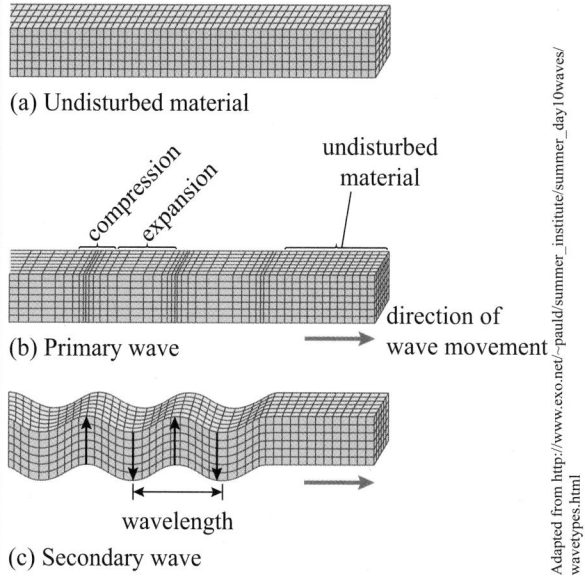

(a) Undisturbed material

(b) Primary wave

(c) Secondary wave

Figure 14-11 Waves generated in an earthquake consist of both longitudinal (P) and transverse (S) waves. The horizontal arrows show the direction of wave propagation.

LO 2

14-2 The Motion of a Disturbance in a String

 ONLINE ACTIVITY

A Travelling Pulse

The e-resource that accompanies every new copy of this textbook contains an Online Activity using the PhET simulation "A Travelling Pulse." Work through the simulation and accompanying questions to gain an understanding of the motion of a travelling pulse.

Consider a long, stretched string with one end tied to a rigid support and the other pulled taut by your hand. If you quickly flip your hand up and down, you will generate a pulse. The pulse travels along the length of string with a constant speed. This example demonstrates two important properties of waves:

■ **Waves carry energy.** In moving the string up and down, your hand did work on the string, thus imparting energy to the string. The pulse carries this energy as it travels along the string. It takes energy to generate a pulse, and the pulse carries this energy as it propagates through the medium. A wave can therefore be described as the motion of energy through a medium.

■ **The medium does not travel with the wave.** A given segment of the string is displaced from its equilibrium position while the pulse passes through it and then returns to the equilibrium position once the pulse has gone by. There is no net movement of the string in the direction of motion of the pulse. Similarly, a sound wave does not carry air along with it; the air molecules only oscillate back and forth in the direction of the motion of the sound wave. Do not confuse the motion of the wave with the motion of the particles of the medium.

What causes a pulse to move through a string? In Figure 14-12, a pulse is moving from left to right along a string that is held under a constant tension of magnitude T. Let us analyze the motion of a segment of the string located at point P as the pulse passes through it. Point E is at the leading edge of the pulse, and point A at the trailing edge.

(a) Before the pulse arrives, point P is at rest and is being pulled with tension T by the string to its left and its right. When point E of the pulse reaches P, a slightly increased tension is exerted on P from the string to its left, causing it to accelerate upward.

(b) As point P accelerates upward, it experiences a downward force due to the tension from the string to its right. At the same time, the increased tension from the left keeps it moving upward.

(c) As P moves farther up, the increased pull from its right starts to decelerate it, until it momentarily comes to a stop at C, the point of maximum displacement on the pulse. At this displacement, P is at rest and is now being pulled downward by the string from both sides. Therefore, it has maximum downward acceleration at its farthest displacement.

(d) As P accelerates toward the equilibrium position, it is being pulled upward by the string to its right and downward by the string to its left.

(e) The point P passes through point A on the pulse, finally coming to rest after the pulse has passed through.

So, a pulse (or a wave) moves through a medium due to the forces exerted on a given segment of the medium in the presence of the pulse, by the elements of the medium around that segment.

Figure 14-13 shows a plot of displacement versus time for point P. Time $t = t_1$ corresponds to the arrival of the leading edge (E) of the pulse, and $t = t_2$ corresponds to the departure of the trailing edge (A).

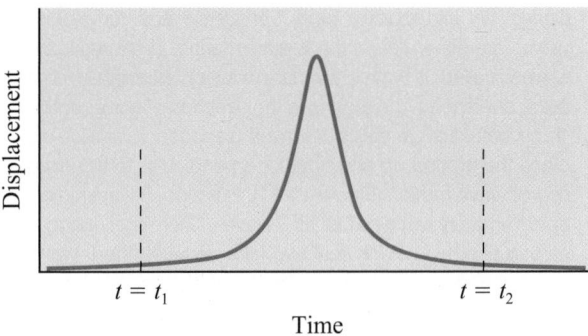

Figure 14-13 A displacement versus time plot for point P (Figure 14-12) as a pulse passes through it

☑ CHECKPOINT

C-14-1 Displacement Graph

The pulse in Figure 14-14 is about to pass through point P. Which of the graphs in Figure 14-15 correctly describes the displacement of point P, as a function of time, when the pulse passes through P?

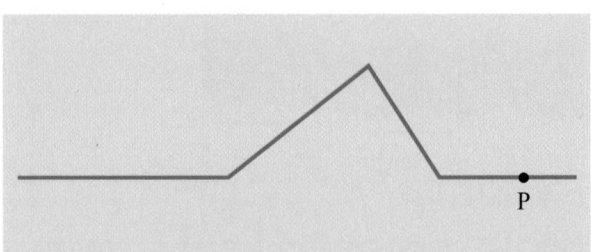

Figure 14-14 C-14-1

C-14-1 (d)

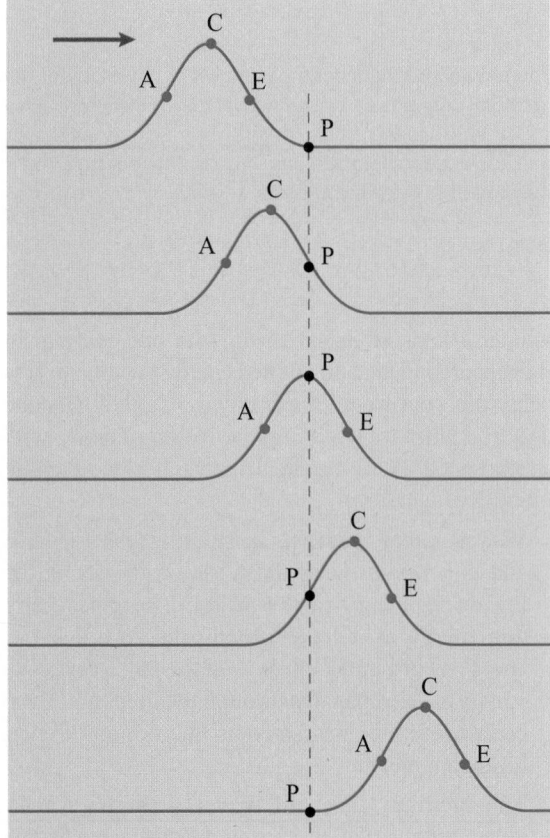

Figure 14-12 The motion of a point (P) on a string as a pulse passes through it. The pulse exerts a net force at P, causing it to be displaced from its equilibrium position.

(a)

(b)

(c)

(d)

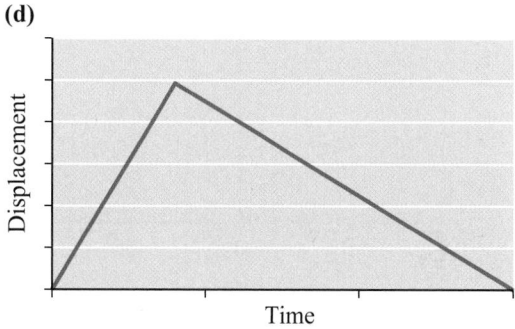

Figure 14-15   C-14-1

LO 3

14-3 Equation for a Pulse Moving in One Dimension

Consider a pulse travelling along a stretched string along the x-axis. The transverse displacement of an element of the string from its equilibrium position is denoted by $D(x, t)$ and is plotted along the y-axis. The displacement is a function of position along the string (x) and time (t). For example, $D(1, 0)$ represents the displacement from the equilibrium position at time $t = 0$ of an element of the string that is positioned at $x = 1$ m. In the absence of any disturbance, every element of the string is at its equilibrium position, and

$$D(x, t) = 0 \quad \text{for all } x \text{ and } t \tag{14-1}$$

A waveform can be described by giving the displacement of each element of the medium at a fixed time. Figure 14-16(a) shows a pulse with a waveform given by the equation (x and D are in meters and t in seconds).

$$D(x, 0) = \frac{4}{x^4 + 2}$$

Let us assume that the pulse is moving in the positive x-direction with a speed of 3.0 m/s. After 1.0 s, the pulse

has moved 3.0 m toward the right, and the peak of the pulse is at $x = 3.0$ m, as shown in Figure 14-16(b). The waveform of the pulse at $t = 1.0$ s is given by the equation

$$D(x, 1.0) = \frac{4}{(x - 3.0)^4 + 2}$$

After 2.0 s, the pulse has moved 3.0 m farther and it's peak is centred 6.0 m from the origin. Its waveform at $t = 2.0$ s is described by the equation

$$D(x, 2.0) = \frac{4}{(x - 6.0)^4 + 2}$$

For any given time t, the peak of the pulse is at $x = 3.0t$, and its waveform is given by

$$D(x, t) = \frac{4}{(x - 3.0t)^4 + 2}$$

If the pulse is moving to the right at an arbitrary speed v (instead of the speed of 3.0 m/s assumed above), we can describe the waveform of the pulse by substituting v for the speed in the preceding equation:

$$D(x, t) = \frac{4}{(x - vt)^4 + 2}$$

The above analysis can be generalized to write an equation for the waveform of any travelling pulse, given

(a)

(b)

(c)

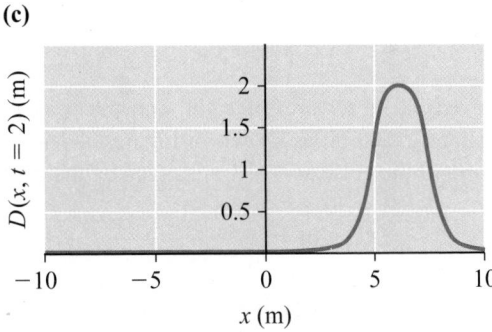

Figure 14-16 Displacement versus position plots of a pulse travelling in the positive *x*-direction with a speed of 3.0 m/s: (a) at $t = 0$, (b) at $t = 1.0$ s, and (c) at $t = 2.0$ s.

its waveform at $t = 0$. If at $t = 0$ the waveform of a pulse is described by a function $D(x)$ and the pulse is moving in the direction of increasing x with speed v, then the waveform at any time t is obtained by the substitution $x \rightarrow x - vt$ in $D(x)$. Similarly, if the pulse is moving in the direction of decreasing x with speed v, the waveform at any time t is obtained by substituting $x \rightarrow x + vt$ in $D(x)$. A function of the form $D(x \mp vt)$ is called a **wave function** or a **displacement function** because it describes a travelling wave by giving the displacement of each point of the medium at any given time. In a wave function for a travelling wave, the variables x and t always occur in the combination $x \mp vt$.

 EXAMPLE 14-1

Pulse on a String

A pulse is travelling from left to right with a speed of 2.0 m/s along a stretched string. Figure 14-17 shows the waveform of the pulse when $t = 1.0$ s.

(a) Find the displacements of the string elements at $x = 0.0$ m and $x = 5.0$ m at this time.
(b) Where was the peak of the pulse at $t = 0.0$ s?
(c) Where will the peak of the pulse be 5.0 s later?

Figure 14-17 A displacement versus position plot of a pulse travelling in the positive x-direction

SOLUTION

(a) From the graph, $D(0.0$ m, 1.0 s$) = 0.25$ m, and $D(5.0$ m, 1.0 s$) = 0.15$ m.
(b) At $t = 1.0$ s the peak is located at $x = 2.0$ m. Since the pulse speed is 2.0 m/s, one second earlier the waveform was shifted towards the left by 2.0 m. The peak was therefore located at $x = 0.0$ m.
(c) In 5.0 s, the pulse will travel a distance of 10.0 m farther right from its location at $t = 1.0$ s, so the peak will be at $x = 12.0$ m.

✓ CHECKPOINT

C-14-2   A Travelling Pulse

Which of the following equations does *not* describe a pulse moving toward the direction of increasing *x*?

(a) $D(x, t) = \dfrac{2}{(5.0t - x)^2 + 5}$

(b) $D(x, t) = \dfrac{2}{(x + 0.5t)^2 + 5}$

(c) $D(x, t) = \dfrac{2}{(x - 5.0t)^2 + 5}$

(d) $D(x, t) = \dfrac{3}{(2.0x - 4.0t)^4 + 6}$

C-14-2 (b)

WAVES AND OSCILLATIONS

EXAMPLE 14-2

Using Waveform Equations

The equation for a pulse traveling through a medium is given by

$$D(x, t) = \frac{3.0}{(x - 2.0t)^2 + 5.0}$$

where x and D are in metres, and t is in seconds.

(a) What is the speed of the pulse, and in which direction is it travelling?
(b) What is the displacement of a point of the medium located at $x = 0.30$ m at $t = 1.0$ s?
(c) What is the maximum displacement of a point as the pulse passes through it? Plot the displacement as a function of time of the point at $x = 15.0$ m as the pulse passes through it.
(d) What is the equation of a pulse that has the same speed and shape as the above waveform but is inverted relative to it?

SOLUTION

(a) Since the x and t variables appear in the form $x - 2.0t$, the pulse speed is 2.0 m/s. The minus sign between x and t indicates that the pulse is moving in the direction of increasing x.

(b) Inserting $x = 0.30$ m and $t = 1.0$ s in the given waveform, we get

$$D(0.30 \text{ m}, 1.0 \text{ s}) = 0.38 \text{ m}$$

(c) The numerator of the given waveform is a constant. Therefore, the displacement is maximum when the denominator is minimum. The term $(x - 2.0\,t)^2$ has a minimum value of zero. Therefore, the minimum value of the denominator is 5.0. The maximum displacement of a point of the medium is therefore 3.0 m/5.0 m = 0.60 m. A plot of $D(15, t)$ as function of t is shown in Figure 14-18.

(d) The displacement of the inverted pulse is the negative of the displacement of the given pulse. We therefore multiply the numerator of the given wave function by -1 to obtain the wave function of the inverted pulse. The wave function of the inverted pulse is

$$D(x, t) = \frac{-3.0}{(x - 2.0t)^2 + 5.0}$$

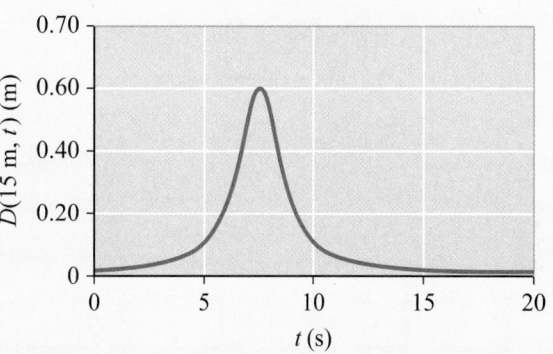

Figure 14-18 The displacement versus time graph at $x = 15.0$ m for the wave function of Example 14-2

Making sense of the result:

Since the wave speed is 2.0 m/s, the peak of the pulse reaches the point at $x = 15.0$ m at $t = 7.5$ s.

LO 4

14-4 Transverse Speed and Wave Speed

In the presence of a wave, the instantaneous speed, $u(x, t)$, of a string element is equal to the rate of change of its displacement, $D(x, t)$, with time. The displacement function depends on both position and time, so the instantaneous speed of an element at x_0 is obtained by taking the derivative of $D(x, t)$ with respect to time while holding the position fixed at $x = x_0$:

$$u(x_0, t) = \left.\frac{\partial D(x, t)}{\partial t}\right|_{x=x_0} \tag{14-2}$$

The symbol ∂ indicates a partial derivative, and the term $\partial/\partial t$ indicates a partial derivative with respect to time. The instantaneous speed can also be obtained by graphing $D(x, t)$ as a function of time for a fixed x. The slope of the graph at a given time is equal to the transverse speed of the segment at that time, and the sign of the slope gives the direction of motion.

EXAMPLE 14-3

Transverse Displacement

A pulse travelling through a string is described by the following wave function:

$$D(x, t) = \frac{2.0}{(x - 0.5t)^2 + 4.0}$$

where x and D are in metres, and t is in seconds.

(a) Find the velocity of the pulse.
(b) At $t = 2.0$ s, what are the speeds of the string elements located at $x = 0.5$ m and $x = 1.5$ m?
(c) Graph the speed of the particle located at $x = 10.0$ m between $t = 0.0$ s and $t = 40.0$ s.

(continued)

SOLUTION

(a) The coefficient of the t term is 0.5, so the pulse speed is 0.5 m/s. The sign between the position x and the time t is negative, so the pulse is travelling in the direction of increasing x.

(b) To determine the speed of a string element, we take the derivative of the wave function with respect to t, while holding x constant:

$$u(x, t) = \frac{\partial}{\partial t}\left(\frac{2.0}{(x - 0.5t)^2 + 4.0}\right)$$

$$= 2.0 \times (-1)\frac{2.0 \times (x - 0.5t) \times (-0.5)}{\left((x - 0.5t)^2 + 4.0\right)^2}$$

$$= \frac{2.0(x - 0.5t)}{\left((x - 0.5t)^2 + 4.0\right)^2}$$

Evaluating $u(x, t)$ at the desired values of x and t gives

$$u(0.5 \text{ m}, 2.0 \text{ s}) = \frac{-1}{(4.25)^2} = -0.055 \text{ m/s}$$

and

$$u(1.5 \text{ m}, 2.0 \text{ s}) = \frac{1}{(4.25)^2} = +0.055 \text{ m/s}$$

At $t = 2.0$ s, the segments at $x = 0.5$ m and $x = 1.5$ m have the same speeds and are moving in opposite directions.

(c) A plot of $u(10.0 \text{ m}, t)$ as a function of time is shown in Figure 14-19.

Figure 14-19 The velocity versus time plot of a segment located at $x = 10.0$ m as the pulse described in Example 14-3 passes through the segment.

Making sense of the result:

From the plot, we see that the string element at $x = 10.0$ m begins to move up (positive speed) slowly, but increases its speed with time until it reaches a maximum speed. The element then starts to slow down and momentarily comes to a stop. The instant where the velocity changes sign corresponds to the centre of the pulse passing through the element. The element then moves downward, back to the equilibrium position, first gaining speed and then slowing down until it returns to its equilibrium position.

✓ CHECKPOINT

C-14-3   Transverse Speed of a String

A waveform is passing through a given element of a string with speed v. What is the speed of the element as a function of time?

(a) v

(b) $-v$

(c) zero

(d) changes with time

C-14-3 (d)

Wave Speed on a String

The speed of a mechanical wave in a medium depends on the elastic and inertial properties of the medium. For a string, the elastic property can be determined from the tension in the string, and the inertial property depends on the linear mass density (mass per unit length), denoted by the Greek letter mu, μ.

If M is the total mass of a string of length L, then

$$\mu = \frac{M}{L} \tag{14-3}$$

The units of linear mass density are kilograms per metre.

As we discussed earlier, in the presence of a wave, a given element of a string is displaced from its equilibrium position by the forces exerted on it by the neighbouring string elements. The greater the tension (the restoring force), the faster a string element moves in response. However, a greater linear mass density means greater inertia and, hence, slower movement. The wave speed is determined by the response time of the medium as a wave passes through it. We should therefore expect the wave speed in a string to be greater for greater tension in the string and less for greater linear mass density.

Consider a pulse moving from left to right with a speed v along a string that has a linear mass density μ and is under a tension of magnitude T_S (Figure 14-20). The pulse speed is measured with respect to a reference frame in which the string is at rest, which we will call the string frame. We assume that the height of the pulse is small compared to the length of the string so that the tension and the linear mass density do not change appreciably in the presence of the pulse. To analyze the motion of the pulse, we choose a frame of reference that is moving with the pulse (call it the pulse frame). In this frame, the pulse is stationary and the string is moving with a speed v from right to left. The string and pulse frames are moving with a constant speed relative to each other, so Newton's laws of motion are identical in both frames. Let us focus on a very small element of string at the top of the pulse at some instant, as shown in Figure 14-21. The string element has length ΔL; hence mass $\Delta m = \mu \Delta L$. Since ΔL is very small, this element can be considered to form an arc of a circle of radius R subtending an angle θ with respect to the centre of a circle such that $\Delta L = R\theta$. In the pulse frame, the mass Δm is moving in a circle of radius R with speed v, so it has a centripetal acceleration of magnitude $a_c = \dfrac{v^2}{R}$. The centripetal acceleration is provided by the tension force $\vec{T}_S$ in the string, acting tangentially at each end of the element.

Figure 14-20 A pulse moving with speed v toward increasing x in a string that is under a tension T_S.

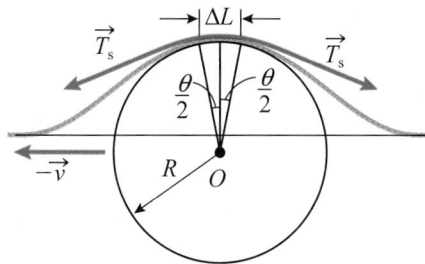

Figure 14-21 The same pulse, viewed from a reference frame in which the pulse is at rest and the string is moving toward the left with speed v.

The two horizontal components of the tension force each have magnitude $T_S \cos(\theta/2)$ and cancel because they are pointing in opposite directions. Each radial component has a magnitude $T_S \sin(\theta/2)$ pointing toward the centre of the circle. Therefore, the net restoring force on the string element is $2T_S \sin(\theta/2)$. According to Newton's second

law, the net restoring force on this element is equal to its mass times the radial acceleration:

$$\Delta m \frac{v^2}{R} = 2T_S \sin\left(\frac{\theta}{2}\right) \qquad (14\text{-}4)$$

Since ΔL and, hence, angle θ are small, we can use the small-angle approximation, $\sin(\theta) \approx \theta$:

$$\Delta m \frac{v^2}{R} \approx 2T_S \left(\frac{\theta}{2}\right) = T_S \theta = T_S \left(\frac{\Delta L}{R}\right)$$

Solving for the wave speed yields

$$v^2 = T_S \left(\frac{\Delta L}{\Delta m}\right) = \frac{T_S}{\mu}$$

$$v = \sqrt{\frac{T_S}{\mu}} \qquad (14\text{-}5)$$

This analysis confirms our assumption that the wave speed increases with greater tension in the string and decreases with greater linear mass density. Since we did not specify a particular waveform, Equation (14-5) is valid for a disturbance of any shape.

✅ **CHECKPOINT**

C-14-4 Wave Speed in a String

A long string of nonzero linear mass density is hanging from a ceiling. A pulse is generated by shaking the string at the bottom end. As this pulse moves upward toward the ceiling, what happens to its speed? Explain your answer.
(a) The speed remains the same.
(b) The speed increases.
(c) The speed decreases.
(d) The speed increases at first, but then decreases closer to the ceiling.

C-14-4 (b) The tension in the string increases with height from the bottom. Therefore, the wave speed increases.

 EXAMPLE 14-4

Pulse Speed on a Hanging String

A string of linear mass density 20.0 g/m and length 10.0 m is hanging vertically from a high ceiling. A mass of 2.00 kg hangs freely from the lower end of the string. A pulse is generated at the lower end of the string and travels upward. What is the speed of the pulse at the lower end of the string, at the middle of the string, and at the top of the string?

SOLUTION

Known quantities:
linear mass density of the string, μ

$$20.0 \text{ g/m} = 20.0 \times 10^{-3} \text{ kg/m}$$

mass of the block, M 2.00 kg
length of the string, L 10.0 m

The tension at any given point along the string is due to the weight of the hanging mass and the weight of the portion of the string below the point. Let x denote the height of a segment of the string, measured from the lower end. The total weight hanging below the point x is $Mg + (\mu x)g$, which is equal to the tension in the string at a height x. Therefore, the wave speed, $v(x)$, at a height x from the lower end of the string is

$$v(x) = \sqrt{\frac{Mg + \mu x g}{\mu}}$$

(continued)

Evaluating the wave speed at $x = 0$, $x = 5.0$ m, and $x = 10.0$ m, we get

$$v(0) = \sqrt{\frac{Mg}{\mu}} = \sqrt{\frac{2.00 \text{ kg} \times 9.81 \text{ m/s}^2}{20.0 \times 10^{-3} \text{ kg/m}}} = 31.3 \text{ m/s}$$

$$v(5.0 \text{ m}) = \sqrt{\frac{Mg + 5.0 \, \mu g}{\mu}}$$

$$= \sqrt{\frac{(2.00 \text{ kg} + 5.0 \text{ m} \times 20.0 \times 10^{-3} \text{ kg/m}) \times 9.81 \text{ m/s}^2}{20.0 \times 10^{-3} \text{ kg/m}}}$$

$$= 32.1 \text{ m/s}$$

$$v(10.0 \text{ m}) = \sqrt{\frac{Mg + 10.0 \, \mu g}{\mu}}$$

$$= \sqrt{\frac{(2.00 \text{ kg} + 10.0 \text{ m} \times 20.0 \times 10^{-3} \text{ kg/m}) \times 9.81 \text{ m/s}^2}{20.0 \times 10^{-3} \text{ kg/m}}}$$

$$= 32.8 \text{ m/s}$$

Making sense of the result:

The tension in the string increases with height, so the pulse speeds up.

14-5 Harmonic Waves

ONLINE ACTIVITY

Harmonic Waves

The e-resource that accompanies every new copy of this textbook contains an Online Activity using the PhET simulation "Harmonic Waves." Work through the simulation and accompanying questions to gain an understanding of harmonic waves.

A wave generated by a source that is undergoing simple harmonic motion is called a **harmonic wave** or a **sinusoidal wave**. Imagine a long string with one end attached to a source. As the source oscillates, it generates a continuous wave that travels along the string in simple harmonic motion (Figure 14-22).

A snapshot of the string at, say, $t = 0$ corresponds to a sine (or a cosine) function (Figure 14-22) and can be represented in the following form:

$$D(x) = A \sin(kx) \quad \text{at } t = 0 \qquad (14\text{-}6)$$

Here, $D(x)$ denotes the transverse displacement of the element of the string located at x. The parameters A and k are described below.

Amplitude As the sine function oscillates between $+1$ and -1, the displacement $D(x)$ oscillates between the values $\pm A$. A is called the **amplitude** of the wave and corresponds to the maximum displacement from the equilibrium position of any element of the string. Amplitude is always a positive quantity. For waves on a string, amplitude has the dimension of length. Figure 14-23 shows two harmonic waves with different amplitudes.

The point of maximum displacement ($D(x) = +A$) in a cycle is called a **crest**, and the point of minimum displacement ($D(x) = -A$) is called a **trough**.

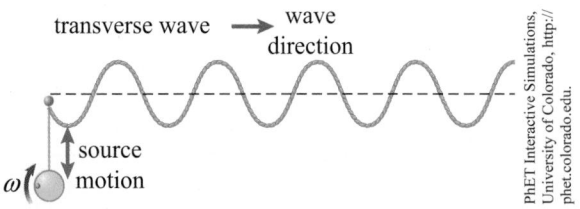

Figure 14-22 A source executing a simple harmonic motion generates a simple harmonic wave.

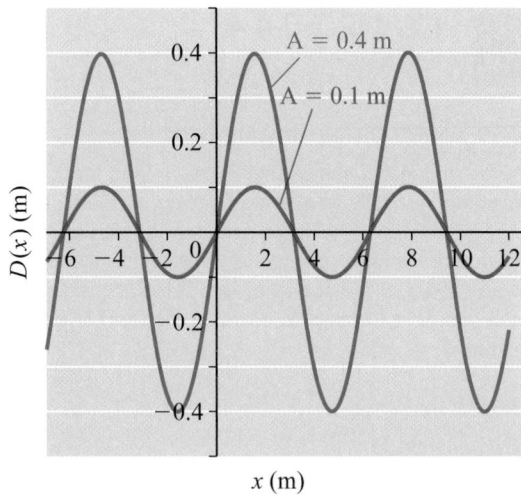

Figure 14-23 A snapshot at a fixed time of two harmonic waves with amplitudes 0.1 m and 0.4 m

CHECKPOINT

C-14-5 Amplitude of a Harmonic Wave

What is the amplitude of the harmonic wave shown in Figure 14-24?

(a) +0.15 m

(b) −0.15 m

(c) +0.30 m

(d) zero

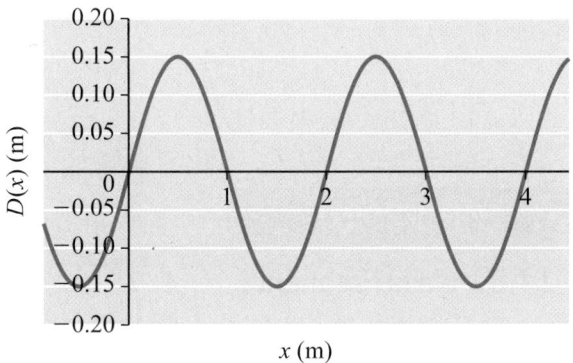

Figure 14-24 C-14-5

C-14-5 (a)

Wavelength The shortest distance over which a wave shape repeats is called the **wavelength**. The usual symbol for wavelength is the Greek letter lambda, λ. Wavelength has dimensions of length. In Figure 14-25, the points at x_1 and x_2 are one wavelength apart and points x_1 and x_3 are two wavelengths apart. Figure 14-26 shows two waves with wavelengths of 2.0 m and 4.0 m.

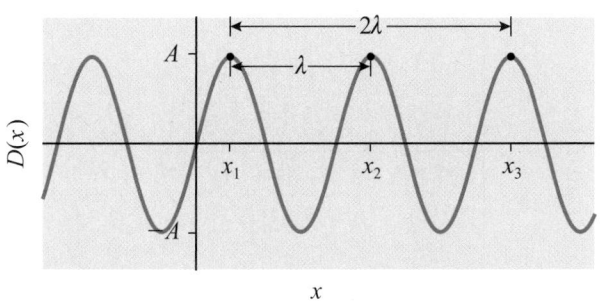

Figure 14-25 The points at $x = x_1$ and $x = x_2$ are one wavelength apart. The points at $x = x_1$ and $x = x_3$ are two wavelengths apart.

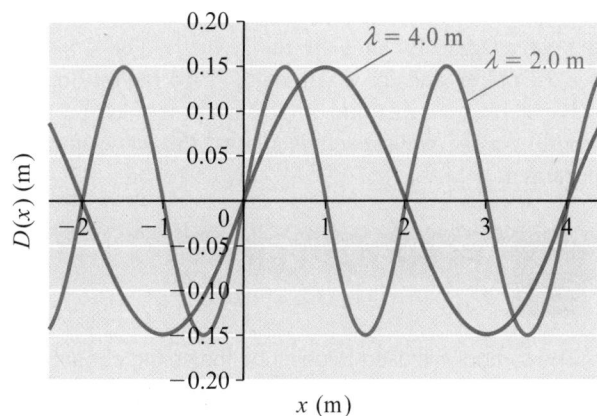

Figure 14-26 Harmonic waves with wavelengths of 2.0 m and 4.0 m

CHECKPOINT

C-14-6 Wavelengths

Figure 14-27 shows a plot of three waves with wavelengths λ_1, λ_2, and λ_3. Which of the following wavelength comparisons is correct?

(a) $\lambda_1 < \lambda_2 < \lambda_3$

(b) $\lambda_1 > \lambda_2 > \lambda_3$

(c) $\lambda_1 < \lambda_3 < \lambda_2$

(d) $\lambda_1 > \lambda_2 < \lambda_3$

C-14-6 (c)

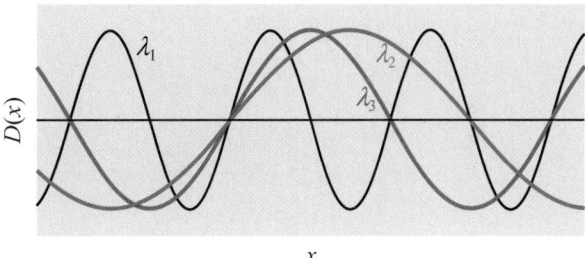

Figure 14-27 C-14-6

Wave Number The parameter k in Equation (14-6) is called the **wave number**. To determine the relationship between k and λ, we use the fact that a waveform repeats over a length λ. Therefore,

$$D(x) = D(x + \lambda) \quad \text{for any given location } x \quad (14\text{-}7)$$

Applying this relationship to the wave function for harmonic waves (Equation (14-6)) gives

$$A \sin(kx) = A \sin(k(x + \lambda)) \quad (14\text{-}8)$$

$$\sin(kx) = \sin(kx + k\lambda) \quad (14\text{-}9)$$

The sine function repeats after 2π rad, so Equation (14-9) is satisfied when $k\lambda = 2\pi$ rad. Therefore,

$$k = \frac{2\pi}{\lambda} \text{ rad/m}$$

Thus, the wave number is 2π times the reciprocal of the wavelength. Its units are radian/metre. For $\lambda = 1$ m, $k = 2\pi$ rad/m, and for $\lambda = 0.5$ m, $k = 4\pi$ rad/m. So, the wave number is a measure of the change of phase per unit length. As the wavelength increases, the wave number decreases.

PEER TO PEER

The symbol k is used to describe the spring constant as well as the wave number. These are two very different quantities; do not confuse one with the other.

Period and Wave Frequency A continuous wave repeats itself in space and time (Figure 14-28). If the period of the source generating a continuous wave is T seconds, one wave cycle is produced in T seconds. As the wave moves through the medium, it takes T seconds for a wave cycle to pass a given point in the medium. The frequency (f) of a mechanical wave is equal to the number of wave cycles passing a fixed point of the medium in one second. Therefore, in one second, an element of the medium will undergo f oscillations.

✓ CHECKPOINT

C-14-7 Wave Numbers

Which of the following relations applies for the waveforms in Figure 14-27?
(a) $k_1 < k_2 < k_3$
(b) $k_1 > k_3 > k_2$
(c) $k_1 < k_2 > k_3$
(d) $k_1 > k_2 < k_3$

C-14-7 (b)

✓ CHECKPOINT

C-14-8 Wave Frequency

Figure 14-29 shows four plots of displacement versus time for four waves passing through a string element at x_0. Order the frequencies of these waves from lowest to highest.

Lowest _____ _____ _____ _____ Highest

C-14-8 (b), (a), (c), (d)

Figure 14-28 The displacement as a function of time as one wave cycle passes through a given point x_0 of the string.

(a)

(b)

(c)

(d)

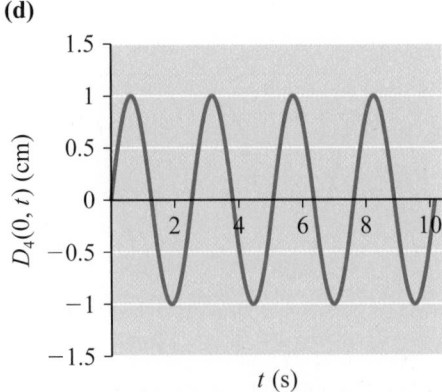

Figure 14-29 C-14-8

Wave Speed Relationships

Imagine a train moving with a constant speed with respect to a stationary observer standing beside the tracks. Each car of the train has a length L, and N cars pass by the observer in one second. Therefore, the speed, v, of the train with respect to the observer is $v = LN$. The same reasoning applies to the speed of a mechanical wave with respect to a medium's frame of reference. Consider a harmonic wave of wavelength λ and frequency f. In one second, f wave cycles, each of length λ, pass through a fixed point of the medium. Therefore, the wave speed with respect to the medium's frame of reference is

KEY EQUATION
$$v = \lambda f \qquad (14\text{-}10)$$

Equation (14-10) is valid for all periodic waves. Using $k = 2\pi/\lambda$ and $f = \omega/2\pi$, we can write the wave speed in terms of the wave number and the angular frequency of a harmonic wave:

$$v = \lambda f = \left(\frac{2\pi}{k}\right)\left(\frac{\omega}{2\pi}\right) = \frac{\omega}{k} \qquad (14\text{-}11)$$

EXAMPLE 14-5

Wave Speed and Wavelength

The speed of sound waves in air is approximately 340 m/s and 1400 m/s in fresh water. What is the change in the wavelength of a sound wave of frequency 400 Hz when it crosses from air into water?

SOLUTION

The frequency of a wave is determined by the frequency of the source and does not change when the wave moves from one medium into another. Using Equation (14-10), the wavelengths of the sound wave in air and in water are as follows:

$$\lambda_{air} = \frac{v_{air}}{f} = \frac{340 \text{ m/s}}{400 \text{ Hz}} = 0.85 \text{ m};$$

$$\lambda_{water} = \frac{v_{water}}{f} = \frac{1400 \text{ m/s}}{400 \text{ Hz}} = 3.5 \text{ m}$$

Making sense of the result:

Since the speed of sound in water is about four times greater than its speed in air, for the same number of wave cycles to pass through a given point in water the wavelength of the sound wave in water must be four times longer than the wavelength in air.

Travelling Harmonic Waves

The displacement of any element of a medium from its equilibrium position changes with time as a wave passes through it. Figure 14-30 shows the displacement of two

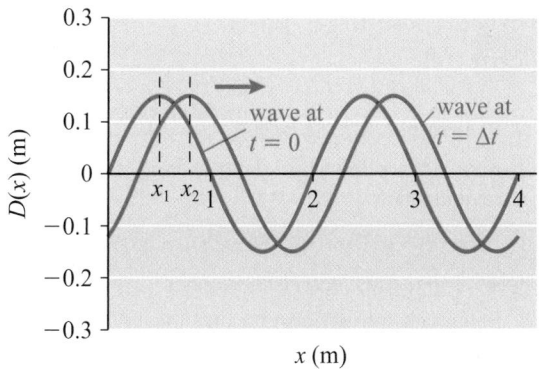

Figure 14-30 The graphs of displacement versus position at time $t = 0$ and $t = \Delta t$ for a sinusoidal wave travelling from left to right

string elements as a wave passes through the string from left to right. At $t = 0$, the point at x_1 is at the crest of the wave and is displaced by 0.15 m, and the point at x_2 is displaced by +0.10 m. A short time later, the wave has moved to the right, and the point at x_1 is displaced by +0.10 m and the point at x_2 by +0.15 m. We can write an equation describing a travelling harmonic wave using the same method that we used to describe a moving pulse. If the equation describing the shape of a wave with a speed of v is known for some fixed time, say $t = 0$, then the wave equation for any arbitrary time t is obtained by

- replacing x by $x - vt$ for a wave travelling in the direction of increasing x
- replacing x by $x + vt$ for a wave travelling in the direction of decreasing x

For a harmonic wave travelling in the direction of increasing x, replacing x by $x - vt$ in Equation (14-6) gives

$$D(x, t) = A \sin(k(x - vt)) \qquad (14\text{-}12)$$

Since $\quad kv = \left(\frac{2\pi}{\lambda}\right)v = 2\pi\left(\frac{v}{\lambda}\right) = 2\pi f = \omega,$

Equation (14-12) can be written as

$$D(x, t) = A \sin(kx - \omega t) \qquad (14\text{-}13)$$

Similarly, the wave function for a harmonic wave travelling in the direction of decreasing x is given by

$$D(x, t) = A \sin(kx + \omega t) \qquad (14\text{-}14)$$

We can also write Equation (14-13) in terms of the wavelength and the time period by using the relationships $k = 2\pi/\lambda$ and $\omega = 2\pi f = 2\pi/T$:

$$D(x, t) = A \sin\left(\frac{2\pi}{\lambda}x - \frac{2\pi}{T}t\right) \qquad (14\text{-}15)$$

Since a harmonic wave is always described by a sine (or a cosine) function, its properties are determined by its amplitude (A), its wavelength(λ), and its period (T).

EXAMPLE 14-6

Travelling Harmonic Wave

A harmonic wave travelling along a string is described by the wave function

$$D(x, t) = (0.10 \text{ m}) \sin[(2.0 \text{ rad/m})x - (3.0 \text{ rad/s})t]$$

where x is in metres, and t is in seconds.

(a) Determine the wavelength and the frequency of the wave.
(b) What is the speed of the wave?
(c) What is the displacement of a segment of the string located at $x = 0.30$ m at $t = 1.0$ s and at $t = 1.2$ s? Determine the average speed of the segment during this time, and compare it to the wave speed.
(d) Plot the displacement of the segment at $x = 0.3$ m between $t = 1.0$ s and $t = 5.0$ s.
(e) Plot the shape of the section of the string between $x = 0.50$ m and $x = 3.0$ m at 0.50 s and 0.60 s.

SOLUTION

(a) By comparing the given equation with the standard equation for a travelling harmonic wave, Equation (14-13), we find that

$$A = 0.10 \text{ m}, k = 2.0 \text{ rad/m}, \text{ and } \omega = 3.0 \text{ rad/s}$$

Therefore, the wavelength and frequency of the wave are

$$\lambda = \frac{2\pi}{k} = \frac{2\pi \text{ rad}}{2.0 \text{ rad/m}} = 3.1 \text{ m}$$

$$f = \frac{\omega}{2\pi} = \frac{3.0 \text{ rad/s}}{2\pi \text{ rad}} = \frac{1.5}{\pi} \text{ Hz} = 0.48 \text{ Hz}$$

(b) The wave speed can be calculated from Equation (14-11):

$$v = \frac{\omega}{k} = \frac{3.0 \text{ rad/s}}{2.0 \text{ rad/m}} = 1.5 \text{ m/s}$$

Since the argument of the sine function is of the form $kx - \omega t$, the wave is travelling in the direction of increasing x.

(c) Substituting the given values for x and t, we get

$$D(0.30 \text{ m}, 1.0 \text{ s}) = (0.10 \text{ m}) \sin(2.0 \times 0.3 - 3.0 \times 1.0)$$

$$= (0.10 \text{ m}) \times (-0.68) = -0.068 \text{ m}$$

$$D(0.30 \text{ m}, 1.2 \text{ s}) = (0.10 \text{ m}) \sin(2.0 \times 0.3 - 3.0 \times 1.2)$$

$$= (0.10 \text{ m}) \times (-0.14) = -0.014 \text{ m}$$

In 0.20 s, this segment moved a distance of 0.054 m, so the average speed was

$$0.054 \text{ m}/0.20 \text{ s} = 0.27 \text{ m/s}.$$

This speed is not the same as the wave speed.

(d) A plot of the displacement of the segment located at $x = 0.30$ m between $t = 1.0$ s and $t = 5.0$ s is obtained by setting $x = 0.30$ m in the wave equation and plotting

the resulting equation between $t = 1.0$ s and 5.0. We thus need to plot

$$D(0.3 \text{ m}, t) = (0.10 \text{ m}) \sin(0.60 - 3.0t) \quad 1.0 \leq t \leq 5.0$$

The resulting plot is shown in Figure 14-31. Note that the above equation is similar to the equation for a simple harmonic oscillator (described by a cosine function) with an angular frequency 3.0 rad/s, amplitude 0.10 m, and phase constant of $\left(\frac{\pi}{2} - 0.60\right)$ rad.

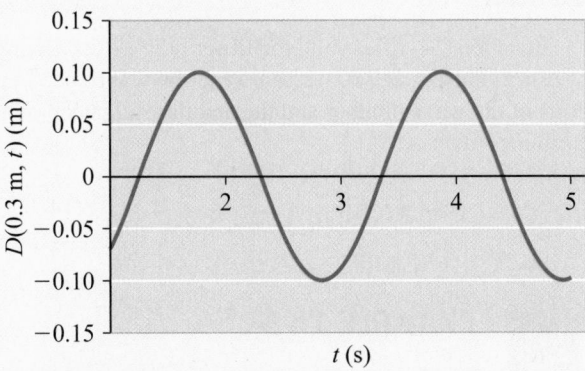

Figure 14-31 The graph of the wave function for Example 14-6 at $x = 0.3$ m for $1.0 \text{ s} \leq t \leq 5.0 \text{ s}$

(e) A plot of the shape of the segment of the string between $x = 0.50$ m and $x = 3.0$ m at 0.50 s and 0.60 s is obtained by setting $t = 0.50$ s and $t = 0.60$ s in the wave equation and plotting the resulting equations between $x = 0.50$ m and $x = 3.0$ m:

$$D(x, 0.50 \text{ s}) = (0.10 \text{ m}) \sin(2.0x - 1.5) \quad 0.5 \leq x \leq 3.0$$

$$D(x, 0.60 \text{ s}) = (0.10 \text{ m}) \sin(2.0x - 1.8) \quad 0.5 \leq x \leq 3.0$$

The resulting graphs are shown in Figure 14-32.

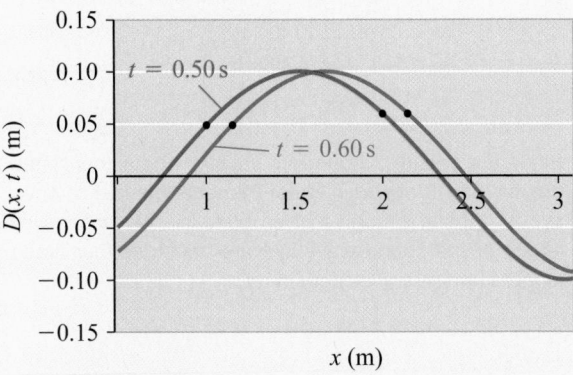

Figure 14-32 Graphs of the wave function for Example 14-6 at $t = 0.50$ s and $t = 0.60$ s

C-14-9 Harmonic Wave Functions

Which of the following wave functions does *not* describe a travelling wave? Explain your reasoning.

(a) $D_1(x, t) = A \sin(x - 4t - 3)$
(b) $D_2(x, t) = A \cos(x - 4t - 3)$
(c) $D_3(x, t) = A \sin(x^2 - 4t)$
(d) $D_4(x, t) = A \sin(x + 4t)$

C-14-9 (c) For a travelling wave, the position x and time t must always occur in the combination $x \mp vt$, where v is the wave speed.

The Phase Constant (ϕ)

Equation (14-13) is not the most general equation for a travelling wave. It restricts the point at $x = 0$ to have zero displacement at $t = 0$. To generalize Equation (14-13), we add a phase constant, ϕ, to the argument of the sine function, as we did for simple harmonic oscillators in Chapter 13. The most general form of the wave function for a harmonic wave travelling in the direction of *increasing x* is

$$D(x, t) = A \sin(kx - \omega t + \phi)$$

$$= A \sin\left(\frac{2\pi}{\lambda} x - \frac{2\pi}{T} t + \phi\right) \quad (14\text{-}16)$$

Similarly, a wave travelling in the direction of *decreasing x* is described by the wave function

$$D(x, t) = A \sin(kx + \omega t + \phi)$$

$$= A \sin\left(\frac{2\pi}{\lambda} x + \frac{2\pi}{T} t + \phi\right) \quad (14\text{-}17)$$

The phase constant ϕ can be positive or negative. The sine function is periodic with a period of 2π rad, so it is sufficient to specify ϕ within a range of 0 to 2π rad. The phase constant is determined by the initial conditions. If the segment at $x = 0$ at $t = 0$ has a displacement D_0, then from Equation (14-16)

$$D_0 = D(0 \text{ m}, 0 \text{ s}) = A \sin(\phi) \quad \text{and} \quad \phi = \sin^{-1}\left(\frac{D_0}{A}\right)$$

We will later see that the phase constant can be uniquely determined from the position or time plot of a harmonic wave.

 EXAMPLE 14-7

Phase Constant of a Harmonic Wave

A sinusoidal wave travelling in the positive x-direction has an amplitude of 0.15 m, a wavelength of 0.40 m and a frequency of 8.0 Hz. The displacement at $t = 0$ of a point located at $x = 0$ is 0.15 m. What is the phase constant of the wave? Write an equation for the wave function of this wave.

SOLUTION

The given wave is travelling along the positive x-axis. For this wave,

$$A = 0.15 \text{ m}, \quad \lambda = 0.40 \text{ m}, \quad \text{and} \quad f = 8.0 \text{ Hz}$$

Inserting these values into Equation (14-17) we obtain

$$D(x, t) = (0.15 \text{ m}) \sin\left(\frac{2\pi}{0.40 \text{ m}} x - 2\pi \times 8.0t + \phi\right)$$

The phase constant is determined by the initial conditions. We are given that the segment at $x = 0$ is displaced by 0.15 m from equilibrium at $t = 0$. Therefore,

$$D(0 \text{ m}, 0 \text{ s}) = 0.15 \text{ m} = (0.15 \text{ m}) \sin\phi$$

$$\sin\phi = 1$$

$$\phi = \frac{\pi}{2} \text{ rad}$$

Therefore, the equation for the wave function is

$$D(x, t)$$

$$= (0.15 \text{ m}) \sin\left((5.0\pi \text{ rad/m})x - (16.0\pi \text{ rad/s})t + \frac{\pi}{2} \text{ rad}\right)$$

In general, ϕ cannot be uniquely determined from a single initial condition because sine is a multi-valued function.

C-14-10 Harmonic Wave Functions

Figure 14-33 shows waveforms (Equation (14-16)) of four waves with the same amplitude and wavelength at $t = 0$. Rank these waveforms in the order of their phase constants, from small to large. Assume that the phase constant is between 0 and 2π rad.

C-14-10 (a), (d), (b), (c)

WAVES AND OSCILLATIONS

(a)

(b)

(c)

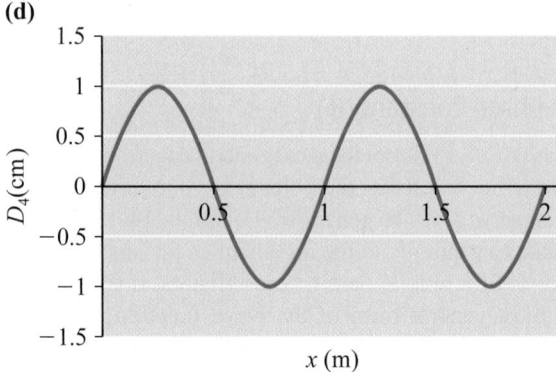

(d)

Figure 14-33 C-14-10

Transverse Velocity and Acceleration for Harmonic Waves

As a harmonic wave travels through a medium, each segment of the medium undergoes periodic motion about its equilibrium position. The instantaneous speed of a segment located at $x = x_0$ is equal to the rate of change of the displacement of the segment. Inserting Equation (14-13) for a harmonic wave into Equation (14-2) and setting $x = x_0$, we obtain the following:

$$u(x_0, t) = \frac{\partial}{\partial t} (A \, \sin(kx - \omega t + \phi))\Big|_{x=x_0}$$

$$= (-\omega A) \cos(kx_0 - \omega t + \phi) \qquad (14\text{-}18)$$

Equation (14-18) shows that the velocity of the segment changes with time, oscillating in the range of $\pm \omega A$. This result is the same as the result for the velocity of a simple harmonic oscillator. The wave speed and the

velocity of a segment of the medium are quite different: a wave moves through a medium with a constant speed, while each segment of the medium oscillates about its equilibrium position with a continuously changing velocity.

The instantaneous acceleration of a segment of a string, $a(x, t)$, is equal to the rate of change of the instantaneous velocity of the segment:

$$a(x_0, t) = \frac{\partial u(x, t)}{\partial t}\Big|_{x=x_0} = \frac{\partial^2 D(x_0, t)}{\partial t^2}$$

$$= -\omega^2 A \, \sin(kx_0 - \omega t + \phi) = -\omega^2 D(x_0, t) \quad (14\text{-}19)$$

The acceleration of a segment is proportional to its displacement and is opposite in sign, a result that also applies to the motion of simple harmonic oscillators. Notice that each segment of the medium has instantaneous acceleration, but the wave itself has no acceleration.

EXAMPLE 14-8

A Harmonic Wave

A harmonic wave on a string is described by the wave function $D(x, t) = (0.20 \text{ m}) \sin(2.0x - 3.0t)$

(a) What is the velocity at $t = 0.0$ s of a segment of the string located at $x = 0.50$ m?

(b) What is the maximum positive velocity of this segment? When is the first time after $t = 0.0$ that the segment attains this velocity?

(c) Plot the segment's velocity for $0.0 \text{ s} \leq t \leq 10.0 \text{ s}$.

SOLUTION

(a) We take the partial derivative of the displacement to find an expression for the velocity:

$$u(x, t) = \frac{\partial}{\partial t}[(0.20 \text{ m}) \sin(2.0x - 3.0t)]$$

$$= (-0.60 \text{ m/s}) \cos(2.0x - 3.0t)$$

Substituting $x = 0.50$ m and $t = 0.0$ s into this expression, we get

$$u(0.50 \text{ m}, 0.0 \text{ s}) = (-0.60 \text{ m/s}) \cos(2.0 \times 0.5 - 3.0 \times 0)$$

$$= -0.32 \text{ m/s}$$

(b) The maximum positive velocity of the segment is $+0.60$ m/s (when the cosine function is -1). To find the time when the segment first attains this velocity, we substitute $+0.60$ m/s in the velocity equation and solve for t:

$$u(0.50 \text{ m}, t) = (-0.60 \text{ m/s}) \cos(1.0 - 3.0t) = +0.60 \text{ m/s}$$

$$(-0.60 \text{ m/s}) \cos(1.0 - 3.0t) = +0.60 \text{ m/s}$$

$$\cos(1.0 - 3.0t) = -1.0$$

$$1.0 - 3.0t = \cos^{-1}(-1.0) = \pm\pi$$

Here, we have retained both signs for π since $\cos(\pm\pi) = -1$. Solving for t, we get

$$t = \frac{1.0 \mp \pi}{3.0}$$

Time must be positive, so

$$t = \frac{1.0 + \pi}{3.0} = 1.4 \text{ s}$$

(c) A plot of the velocity of the segment is shown in Figure 14-34.

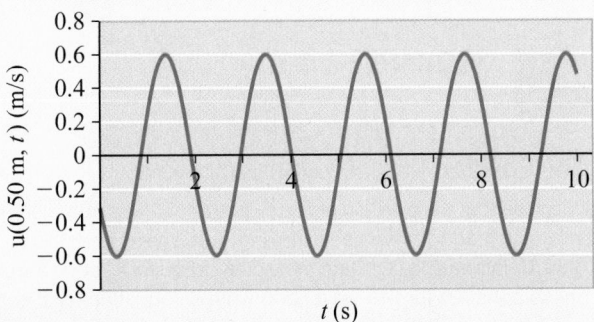

Figure 14-34 A graph of the velocity of the string segment in Example 14-8 at $x = 0.50$ m for $0.0 \text{ s} \le t \le 10.0 \text{ s}$

✓ CHECKPOINT

C-14-11 Transverse Velocity

A displacement versus time graph for a particular segment of a medium is shown in Figure 14-35. List the labelled points on the displacement curve where the segment's velocity is
(a) positive;
(b) negative;
(c) zero.

C-14-11 (a) A, F, K; (b) C, D, I; (c) B, E, G, J

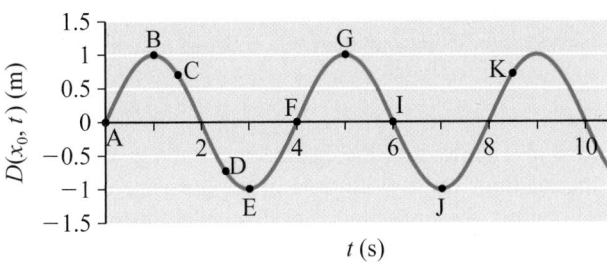

Figure 14-35 C-14-11

LO 6

14-6 Position Plots and Time Plots

A plot of $D(x, t)$ as a function of position x and time t results in a three-dimensional graph. Figure 14-36 shows a plot of $D(x, t) = (1.0 \text{ m}) \sin(2.0\pi x - \pi t)$, with position along the x-axis, time along the y-axis, and the displacement

$D(x, t)$ along the z-axis. Although such a plot is useful in visualizing wave motion, it can be difficult to determine the parameters of a wave (wavelength, frequency, amplitude, and phase constant) from a three-dimensional plot. It is easier to use plots in which one of the variables is kept constant and the wave function is plotted against the other variable.

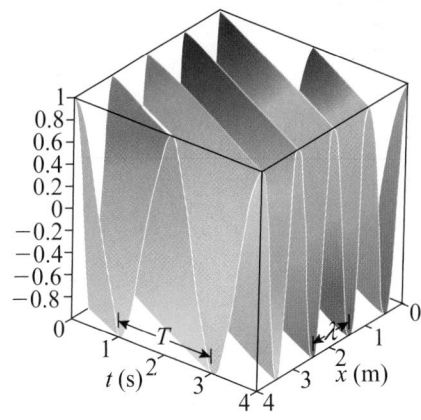

Figure 14-36 A three-dimensional plot of the displacement function $D(x, t) = (1.0 \text{ m}) \sin(2.0\pi x - \pi t)$

Position Plots

A **position plot** for a travelling wave is obtained by keeping the time fixed and plotting the wave function as a function of position (x). If the time is fixed at $t = t_0$, then the position plot is a graph of $D(x, t = t_0)$ as a function of x. A position plot shows the displacement of every section of the medium at a fixed time. It is like taking a picture or a snapshot of the medium, and for this reason, position plots are also called **snapshot graphs**. We can determine the amplitude and the wavelength of the wave from its position plot. Figure 14-37 shows a displacement versus position plot of a harmonic wave. From the graph, notice the following:

■ The maximum displacement of a point is ±1.0 m. Therefore $A = 1.0$ m.

■ The points at $x = 3.0$ m and $x = 6.0$ m are located at two successive crests. Therefore, the wavelength of the wave is $\lambda = 6.0$ m $- 3.0$ m $= 3.0$ m.

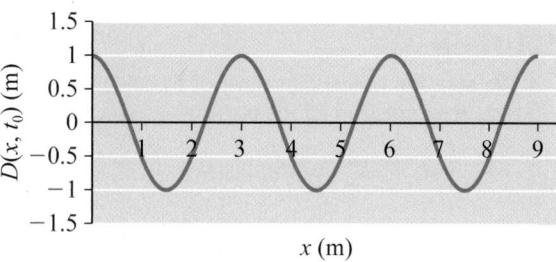

Figure 14-37 The displacement versus position plot of a harmonic wave at a fixed time

EXAMPLE 14-9

The Position Plot of a Harmonic Wave

Suppose that the position plot in Figure 14-37 was obtained at $t = 1.0$ s and the wave is moving at a speed of 1.5 m/s toward increasing x. Determine the frequency and the phase constant for this wave, and write an equation for the wave function of this wave.

SOLUTION

The general equation for a sinusoidal wave travelling toward increasing x is given by

$$D(x, t) = A \sin\left(\frac{2\pi}{\lambda} x - \frac{2\pi}{T} t + \phi\right)$$

Substituting $A = 1.0$ m, $\lambda = 3.0$ m, and $t = 1.0$ s, we get

$$D(x, 1.0\ s) = (1.0\ m) \sin\left(\frac{2\pi}{3.0} x - \frac{2\pi}{T} + \phi\right)$$

The wave speed is 1.5 m/s, and the wavelength is known; therefore, we can determine the period:

$$T = \frac{1}{f} = \frac{\lambda}{v} = \frac{3.0\ m}{1.5\ m/s} = 2.0\ s$$

Insert this value for T into above equation:

$$D(x, 1.0\ s) = (1.0\ m) \sin\left(\frac{2\pi}{3.0} x - \pi + \phi\right)$$

We can determine ϕ by determining the displacement for a given x from the graph and then inserting that information in the wave equation. At $x = 0.0$ m, $D(0.0\ m, 1.0\ s) = 1.0$ m. Inserting these values into the equation yields

$$D(0.0\ m, 1.0\ s) = (1.0\ m) \sin(-\pi + \phi) = 1.0\ m$$

$$\sin(-\pi + \phi) = 1$$

$$-\pi + \phi = \sin^{-1}(1) = \frac{\pi}{2}$$

Therefore, $\phi = \frac{3\pi}{2}$ rad

Therefore, the wave function for the wave is

$$D(x, t) = (1.0\ m) \sin\left(\frac{2\pi}{3.0} x - \pi t + \frac{3\pi}{2}\right) \qquad (14\text{-}20)$$

Making sense of the result:

Evaluating (14-20) for $x = 0.0$ m and $t = 1.0$ s, we get

$$D(0.0\ m, 1.0\ s) = (1.0\ m) \sin\left(-\pi + \frac{3\pi}{2}\right)$$

$$= (1.0\ m) \sin\left(\frac{\pi}{2}\right) = 1.0\ m$$

which is consistent with the value obtained from the graph.

Time Plots

A time plot for a travelling wave is obtained by plotting the wave function $D(x, t)$ as a function of time for a fixed position. If the position is fixed at $x = x_0$, then the time plot is a graph of $D(x = x_0, t)$ as a function of t. A

time plot shows how the displacement of a given point of the medium varies with time in the presence of a wave. Therefore, the time plots are also called history graphs. We can determine the amplitude and the frequency of a wave from its time plot.

EXAMPLE 14-10

The Time Plot of a Harmonic Wave

A sinusoidal wave is travelling with a speed of 20.0 m/s along the positive x-axis on a string. The time plot of the displacement of the section of the string located at $x = 2.0$ m is shown in Figure 14-38. Using the given data and the plot,

(a) determine the wavelength and angular frequency of the wave;
(b) determine the phase constant of the wave;
(c) write an equation for the wave function $D(x, t)$ for the wave.

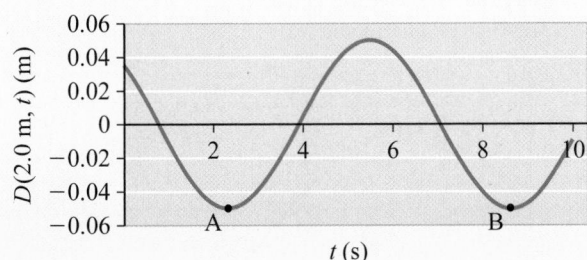

Figure 14-38 The displacement versus time plot of a segment of string located at $x = 2.0$ m as a harmonic wave passes through the point.

SOLUTION

(a) The general equation for a harmonic wave travelling toward increasing x is

$$D(x, t) = A \sin(kx - \omega t + \phi)$$

Inserting $x = 2.0$ m in the above equation we get

$$D(2.0 \text{ m}, t) = A \sin(2.0k - \omega t + \phi)$$

From the graph, $A = 0.05$ m, corresponding to the maximum positive displacement. The period can be calculated from the graph by finding the time difference between two consecutive troughs, for example, points A and B on the graph. Approximately, the period T is:

$$T = t_B - t_A = 6.4 \text{ s}$$

We can now determine the wavelength:

$$\lambda = \frac{v}{f} = vT = (20.0 \text{ m/s})(6.4 \text{ s}) = 128.0 \text{ m}$$

The angular frequency is

$$\omega = \frac{2\pi}{T} = \frac{2\pi \text{ rad}}{6.4 \text{ s}} = 0.98 \text{ rad/s}$$

(b) Since the wave speed is given and we now know the angular frequency, we can determine k:

$$k = \frac{\omega}{v} = \frac{0.98 \text{ rad/s}}{20.0 \text{ m/s}} = 0.05 \text{ rad/m}$$

Inserting the values of A, ω, and k into the wave equation, we get

$$D(2.0 \text{ m}, t) = (0.05 \text{ m}) \sin(2.0 \times 0.05 - 0.98t + \phi)$$

$$= (0.05 \text{ m}) \sin(0.10 - 0.98t + \phi)$$

We can read the displacement at any time on the graph, and then solve the wave equation to determine ϕ. At $t = 0.0$ s, $D(2.0, 0.0) = 0.035$ m, approximately. Therefore,

$$D(2.0, 0.0) = (0.05 \text{ m}) \sin(0.10 + \phi) = 0.035 \text{ m}$$

$$\sin(0.10 + \phi) = \frac{0.035 \text{ m}}{0.05 \text{ m}} = 0.70$$

$$0.10 + \phi = \sin^{-1}(0.70) = 0.77 \text{ rad} \text{ or } (\pi - 0.77) \text{ rad}$$

$$\phi = 0.67 \text{ rad} \text{ or } 2.27 \text{ rad}$$

(c) The value $\pi = 0.67$ rad reproduces the graph shown in figure 14-38. We can now write the general equation for the travelling wave:

$$D(x, t) = (0.05 \text{ m}) \sin(0.05x - 0.98t + 0.67)$$

A common error while doing problems is to omit the phase constant ϕ in the wave function.

LO 7

14-7 Phase and Phase Difference

The argument of the sine (or the cosine) function of a harmonic wave is called the **phase** of the wave. It is measured in radians and is usually denoted by the Greek letter phi, Φ. For the harmonic waves of Equations (14-16) and (14-17), the phase is

$$\Phi(x, t) = kx \mp \omega t + \phi = \left(\frac{2\pi}{\lambda}\right)x \mp 2\pi ft + \phi \quad (14\text{-}21)$$

Do not confuse the phase $\Phi(x, t)$ with the phase constant ϕ. The phase constant is a constant determined by the initial conditions, whereas the phase of a wave is a variable that depends on position and time: Consider two points, x_1 and x_2, that are a distance Δx apart. The difference between the phase at two points, at the same time, is called the **phase difference** and is denoted as $\Delta\Phi$. For harmonic waves,

$$\Delta\Phi = \text{phase of the wave at } x_2 - \text{phase of the wave at } x_1$$
$$= (kx_2 - \omega t + \phi) - (kx_1 - \omega t + \phi)$$
$$= k(x_2 - x_1)$$
$$= k\Delta x$$

$$\Delta\Phi = 2\pi\left(\frac{\Delta x}{\lambda}\right) \quad (14\text{-}22)$$

Table 14-1 shows the phase differences for various distances between two points on a wave.

Two points on a wave that are an integer multiple of wavelengths apart have a phase difference of 2π rad (or an even multiple of π rad). Such points are said to be **in phase** with each other. These points have equal displacements at all times.

Two points that are an odd half-integer multiple (1/2, 3/2, 5/2, ...) of a wavelength apart are π rad (or an odd multiple of π rad) **out of phase** with each other. These points have equal but opposite displacements from the

Table 14-1 Phase Differences for a Periodic Wave

Distance between points, Δx (in multiples of the wavelength)	Phase difference, $\Delta\Phi$ (rad)
0	0
$\lambda/4$	$\pi/2$
$\lambda/2$	π
$3\lambda/4$	$3\pi/2$
λ	2π

equilibrium position at all times. These relationships follow from Equation (14-22). For example, let $\Delta\Phi = \pi$ rad. Then Equation (14-22) can be rearranged as

$$(kx_2 - \omega t + \phi) = (kx_1 - \omega t + \phi) + \pi$$

and

$$D(x_2, t) = A \sin(kx_2 - \omega t + \phi)$$

$$= A \sin(kx_1 - \omega t + \phi + \pi)$$

$$= -A \sin(kx_1 - \omega t + \phi)$$

$$= -D(x_1, t) \qquad (14\text{-}23)$$

Here, we have used the trigonometric identity $\sin(\theta + \pi) = -\sin(\theta)$.

CHECKPOINT

C-14-12   Phase Difference

A harmonic wave is travelling along a string. Two points that were 0.30 m apart when the string was at rest oscillate such that their displacements are always equal and opposite. What is the longest possible wavelength of the wave?

(a)   0.3 m
(b)   0.6 m
(c)   0.9 m
(d)   1.2 m

C-14-12 (b)

LO 8

14-8   Energy and Power in a Travelling Wave

It takes energy to generate a wave, and the wave carries that energy with it as it travels along its medium. Consider a harmonic wave passing through a string of linear mass density μ. A segment of this string has a mass of $\Delta m = \mu \Delta x$, where Δx is the undisturbed length of the segment (i.e., the length of the segment when the string is at rest). In the presence of the wave, the instantaneous kinetic energy of the oscillating segment is

$$\Delta K = \frac{1}{2} \Delta m \big(u(x, t)\big)^2 \qquad (14\text{-}24)$$

where $u(x, t)$ is the instantaneous velocity of the segment.

Using Equation (14-18) for the velocity of the string segment, we can write ΔK as

$$\Delta K = \frac{1}{2} (\mu \Delta x) \big(A^2 \omega^2 \cos^2(kx - \omega t)\big)$$

$$= \frac{1}{2} (\mu \Delta x) \big(A^2 \omega^2 \big(1 - \sin^2(kx - \omega t)\big)\big)$$

$$= \frac{1}{2} (\mu \Delta x) \omega^2 \big(A^2 - D^2(x, t)\big) \qquad (14\text{-}25)$$

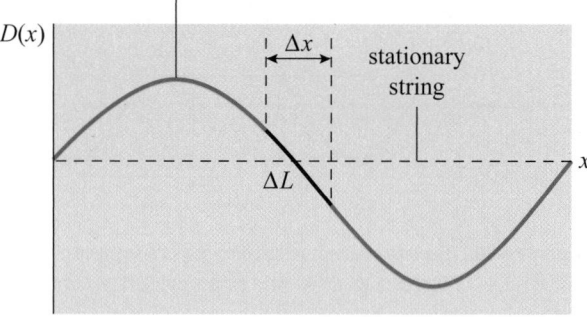

Figure 14-39   The change in the length of an infinitesimal segment of a string in the presence of a harmonic wave. The change is maximum when the segment passes through the origin and is zero at the maximum displacement.

Notice that the vibrating segment has zero kinetic energy at $D(x, t) = \pm A$, the maximum displacement from the equilibrium position where the segment momentarily comes to rest. The segment has a maximum kinetic energy as it passes through its equilibrium position where $D(x, t) = 0$.

As the wave passes through the segment, the length of the segment changes as it is continuously stretched and compressed by the wave. To determine how the change in the length of a segment depends upon the shape of the wave consider a curve in one dimension that is described by a continuous function $f(x)$. The length L of the curve between $x = a$ and $x = b$ is given by the integral

$$L = \int_a^b \sqrt{1 + \left(\frac{df}{dx}\right)^2}\, dx$$

If $\dfrac{df(x)}{dx}$ does not change appreciably in the interval $[a, b]$ then L can be approximated as

$$L = \sqrt{1 + \left(\frac{df}{dx}\right)^2}\,(b - a)$$

Now consider a wave passing through a string. Consider an infinitesimal segment of the string of undisturbed length Δx. In the presence of the wave the string is deformed and let ΔL be the length of the deformed segment (Figure 14-39). Using the previous equation, ΔL can be approximated as,

$$\Delta L = \left(\sqrt{1 + \left(\frac{\partial D(x, t)}{\partial x}\right)^2} \right) \Delta x$$

where we have assumed that the derivative of $D(x, t)$ does not change appreciably within the segment Δx. A general description of wave propagation is quite complicated and we will assume that the amplitude of the wave is small compared to its wavelength. In this approximation the length of the segment in the presence of a wave is not very different from its undisturbed length. Therefore, the second term under the square root is much smaller than 1 and we can expand the square root to write,

$$\Delta L = \left(1 + \frac{1}{2}\left(\frac{\partial D(x, t)}{\partial x}\right)^2\right)\Delta x$$

If Δl is the change in the length of the segment due to the presence of the wave, then $\Delta L = \Delta x + \Delta l$ and Δl is given by,

$$\Delta l = \Delta L - \Delta x = \frac{1}{2}\left(\frac{\partial D(x, t)}{\partial x}\right)^2 \Delta x \qquad (14\text{-}26)$$

For a harmonic wave, $D(x, t) = A \sin(kx - \omega t)$. Therefore,

$$\frac{1}{2}\left(\frac{\partial D(x, t)}{\partial x}\right)^2 = \frac{1}{2}\left(\frac{2\pi A}{\lambda}\right)^2 \cos^2(kx - \omega t)$$

and we can see that in the approximation $2\pi A \ll \lambda$, $\frac{1}{2}\left(\frac{\partial D(x, t)}{\partial x}\right)^2 \ll 1$.

In the absence of a wave a string segment is under a tension T (the tension can vary from segment to segment as is the case for a string of non-negligible linear mass density hanging from a support). As the wave passes through the segment the tension in the segment varies continuously. We now assume that for waves with $A \ll \lambda$ the change in the tension in the presence of a wave is negligible. In this approximation the work done by the wave in changing the length of the segment, W, is given by:

W = Tension in the string × change in the length of the segment

$$= T(\Delta l) = T\left(\frac{1}{2}\left(\frac{\partial D(x, t)}{\partial x}\right)^2 \Delta x\right)$$

It is this work that is stored as the instantaneous elastic potential energy, ΔU of the segment. Therefore,

$$\Delta U = \frac{1}{2}T\left(\frac{\partial D(x, t)}{\partial x}\right)^2 \Delta x \qquad (14\text{-}27)$$

For a harmonic wave, $\dfrac{\partial(A \sin(kx - \omega t))}{\partial x} = Ak \cos(kx - \omega t)$; therefore, the stored potential energy is

$$\Delta U = \frac{1}{2}\left(T\Delta x\right)\left(A^2 k^2 \cos^2(kx - \omega t)\right)$$

$$= \frac{1}{2}\left(Tk^2\Delta x\right)\left(A^2 - A^2 \sin^2(kx - \omega t)\right)$$

$$= \frac{1}{2}\left(Tk^2\Delta x\right)\left(A^2 - D^2(x, t)\right) \qquad (14\text{-}28)$$

Using $v = \sqrt{\dfrac{T}{\mu}}$ for the wave speed along a string and $v = \dfrac{\omega}{k}$, we get $Tk^2 = \mu\omega^2$, and Equation (14-28) can then be written as

$$\Delta U = \frac{1}{2}(\mu\Delta x)\omega^2\left(A^2 - D^2(x, t)\right) \qquad (14\text{-}29)$$

Notice that the equation for stored instantaneous potential energy is exactly the same as the equation for the instantaneous kinetic energy of the segment (Equation (14-25)). As a harmonic wave travels through a string, the instantaneous kinetic energy and the potential energy of any segment of the string are *equal*. When the segment is at its maximum displacement ($D = \pm A$), both the kinetic energy and the potential energy of the segment are zero, so the segment has zero total energy. When the segment is passing through the equilibrium position ($D = 0$), it has maximum potential energy as well as maximum kinetic energy. Thus, the behaviour of a string element in the presence of a travelling wave is different from the behaviour of a simple harmonic oscillator. The kinetic and potential energies of a single oscillator continually transform back and forth into each other. When the kinetic energy of a simple harmonic oscillator is maximum, its potential energy is minimum, and vice versa.

The total energy of the segment (ΔE) is equal to the sum of its instantaneous kinetic and potential energies:

$$\Delta E = \Delta K + \Delta U = (\mu\Delta x)\omega^2 A^2 \cos^2(kx - \omega t) \qquad (14\text{-}30)$$

This energy flows in the direction of the wave motion. The segment receives energy at one end, stores it briefly, and then passes it to the neighbouring segment at the other end. If energy ΔE is transmitted through a segment of length Δx in time Δt, then the energy transmitted across the segment per unit time is

$$\frac{\Delta E}{\Delta t} = \mu\left(\frac{\Delta x}{\Delta t}\right)\omega^2 A^2 \cos^2(kx - \omega t)$$

The ratio $\Delta x/\Delta t$ is the wave speed. Therefore, the instantaneous power, P, delivered by the wave is

$$P = \frac{\Delta E}{\Delta t} = \mu v\omega^2 A^2 \cos^2(kx - \omega t) \qquad (14\text{-}31)$$

The average rate at which the power is transmitted by the wave is obtained by integrating Equation (14-31) over a complete wave cycle. The average value of the square of a cosine (or a sine) function over an integer number of cycles is ½. Thus, the average power transmitted by a harmonic wave, P_{avg}, is

$$P_{avg} = \frac{1}{2}\mu v\omega^2 A^2 = 2\pi^2\mu v f^2 A^2 \qquad (14\text{-}32)$$

The average power delivered by a wave is proportional to the linear mass density of the string, the wave speed, the square of the amplitude of the wave, and the square of its frequency. The dependence of the average power of a wave on the square of its amplitude and frequency is valid for other types of waves as well, including ocean waves, seismic waves, and electromagnetic waves. In real-life situations frictional forces cause a wave to lose its energy as it moves in a medium. As a wave gradually loses its energy it finally disappears. We will neglect effects of frictional forces on wave motion. This means that a wave travels through a uniform medium without a change in its energy.

Note that Equations (14-24) and (14-27) are strictly valid only in the limit when $\Delta x \rightarrow 0$. For a segment of finite length, we must integrate these equations over the length of the segment.

✅ CHECKPOINT

C-14-13   Wave Power

Figure 14-40 shows time plots of four waves passing through four identical strings held under the same tension. Rank these waves, from highest to lowest, on the basis of the power carried by the waves.

Highest _____   _____   _____   _____ Lowest

C-14-13 (b), (d), (c), (a)

(a)

(b)

(c)

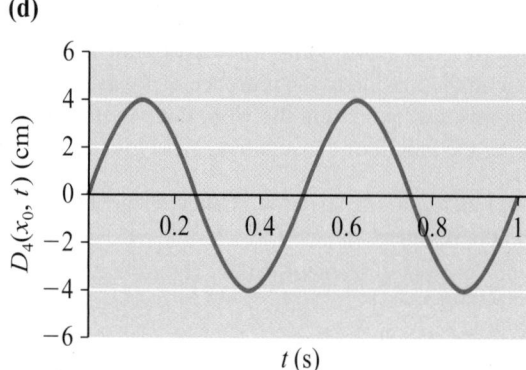

(d)

Figure 14-40   C-14-13

🧩 MAKING CONNECTIONS

Energy from Ocean Waves

Waves in an ocean are generated by the interaction of the wind with the water's surface. If the wind speed just above the water surface is greater than the speed of the water waves, then energy is transferred from the wind to the water, causing the amplitude of the waves to increase. When the following three conditions are satisfied (this happens during ocean storms), the waves are transformed into a series of repetitive swells:

- The wind speed is large (several tens of kilometres per hour).
- The water surface over which the wind is blowing is large (thousands of square kilometres).
- The wind blows in one direction for a sustained period of time (several hours).

A swell is a long, thick wave that has a height of several metres. Swells are characterized by the wave height, H, which is the crest-to-trough height of the swell, and the period T between successive swells. A phrase such as, "a 3 metre swell at 12 seconds" means 3 metre-high swells that are 12 seconds apart.

Swells carry a large amount of energy and travel over large distances from the point of origin. An analysis similar to the analysis that led to the expression for the average power for one-dimensional waves (Equation (14-32)) shows that the average power per unit length carried by swells is given by

$$P_{\text{avg}} = \left(\frac{1}{32\pi}\right)\rho_{\text{ocean}} g^2 T H^2$$

Here, g is the acceleration due to gravity, ρ_{ocean} is the density of ocean water, T is the period of the swells, and H is the wave height. Note that just as for one-dimensional waves, the power carried by swells is proportional to the square of the swell height and the density of the medium.

Daily swell heights for all oceans are published by the NOAA (National Oceanic and Atmospheric Administration). The period between the swells can range from 12 s to 25 s. Let us assume the following data for swells in the deep ocean:

$$\rho_{\text{ocean}} = 1030 \text{ kg/m}^3, H = 3.0 \text{ m, and } T = 14.0 \text{ s}$$

The average power per unit length carried by these swells is

$$P_{\text{avg}} = \left(\frac{1}{32\pi}\right)(1030 \text{ kg/m}^3)(9.81 \text{ m/s}^2)^2 (14.0 \text{ s})(3.0 \text{ m})^2$$

$$\approx 124\,000 \text{ W/m}$$

This is a large amount of freely available, environment-friendly and renewable power, which if efficiently harnessed could greatly help in solving the world's energy problems.

There are a number of efforts underway to harvest this power. One such method is called the oscillating water column (OWC) generator (Figure 14-41). It consists of an air-filled enclosure that has two openings, one underneath and the other at the top. An air turbine (a device that generates electricity when rotated) is connected to the top opening. The enclosure is partially submerged in the water such that the bottom opening is below the waterline and the top opening is above the maximum swell height. When a swell reaches the enclosure, the column of water in the enclosure rises. The air inside the enclosure is compressed and is forced out from the top opening past the turbine. This causes the turbine to rotate and generate electricity. As the swell recedes, the air pressure within the enclosure decreases. This forces the outside air into the enclosure past the turbine, again rotating the turbine and generating electricity.

Located on the island of Islay, off Scotland's west coast, LIMPET (Land Installed Marine Powered Energy Transformer) is the first commercial OWC generator and produces 500 kW of power (Figure 14-42). Scotland is the world leader in producing energy from renewable resources.

(a)

The air turbine rotates when the air is forced out of the enclosure due to the incoming swell.

incoming swell

(b)

The air turbine rotates when the air is forced into the enclosure when swell leaves.

outgoing swell

Figure 14-41 An oscillating water column generator. The incoming swell forces the air out of the enclosure, which rotates the air turbine and generates electricity. The outgoing swell forces the air into the enclosure, again generating electricity.

Figure 14-42 This depiction of LIMPET, located off the west coast of Scotland, is the first commercial OWC generator.

LO 9

14-9 Superposition of Waves

The space around us is filled with electromagnetic waves of various frequencies from radio and TV stations, cell-phones, satellites, and other sources. The electric current induced by this multitude of waves in the antennas of a radio receiver is quite complicated. However, if you tune a radio to a particular frequency, say 106.9 MHz, you clearly hear the broadcast at that frequency. The electromagnetic waves of a particular frequency are completely unaffected by the presence of other waves. The same is true of sound waves in air. While listening to a concert, you can often identify the sound generated by a particular instrument or singer, even though the sound waves from a number of sources reach your ears simultaneously.

Principle of Superposition of Waves When more than one wave is present in a medium at the same time, the resultant wave at any point in the medium is equal to the algebraic sum of the individual waves at that point.

For mechanical waves, the superposition principle states that a medium responds to the effects of each wave individually. The presence of other waves does not change in any way the effect of a particular wave on a medium. When multiple waves are travelling along a string, the net displacement of a segment of the string at any time is equal to the sum of the displacements produced by the individual waves.

For a one-dimensional medium carrying a set of waves with individual waveforms given by $D_1(x, t), D_2(x, t), D_3(x, t), \ldots$, the resultant waveform, $D(x, t)$, of the medium (i.e., the shape of the resultant wave) is equal to the sum of the individual waveforms:

$$D(x, t) = D_1(x, t) + D_2(x, t) + D_3(x, t) + \cdots \quad (14\text{-}33)$$

The wave obtained by adding the component waves is called the **resultant wave**. The process of combining the waves to produce a resultant wave is called the **superposition** of waves. The physical phenomenon of two or more waves combining to produce a resultant wave is called the **interference** of waves.

EXAMPLE 14-11

Superposition of Two Harmonic Waves

Two harmonic waves with different frequencies, moving in opposite directions, are present in a string at the same time. The wave functions of these waves are given by

$$D_1(x, t) = (0.10 \text{ m}) \sin(2.0x - 10.0t)$$

$$D_2(x, t) = (0.15 \text{ m}) \sin(4.0x + 20.0t)$$

Find the displacement of a segment of the string located at $x = 1.5$ m at $t = 2.0$ s. Plot the displacement of the string between $x = 0.0$ and $x = 2.0$ m at $t = 1.0$ s.

SOLUTION

We apply the principle of superposition to obtain the resultant displacement at $x = 1.5$ m and $t = 2.0$ s.

$$D(1.5, 2.0) = D_1(1.5, 2.0) + D_2(1.5, 2.0)$$
$$= (0.10 \text{ m}) \sin(2.0 \times 1.5 - 10.0 \times 2.0)$$
$$\quad + (0.15 \text{ m}) \sin(4.0 \times 1.5 + 20.0 \times 2.0)$$
$$= 9.6 \text{ cm} + 13.5 \text{ cm}$$
$$= 23.1 \text{ cm}$$

Figure 14-43 shows the position plot of the two waves and the resultant wave. Note that displacement of each point on the resultant wave is equal to the sum of the displacements due to the component waves.

Figure 14-43 Position plots of $D_1(x, 1.0 \text{ s})$ (red), $D_2(x, 1.0 \text{ s})$ (blue), and the resultant wave (black)

The principle of superposition of waves is easy to understand for mechanical waves. A wave exerts a tension force at the points of the medium through which it is passing. When two or more waves are present, the net force at a point is equal to the sum of all the forces. Therefore, the resultant displacement of a point is equal to the sum of the displacements due to the individual waves. The fact that two or more waves can simultaneously exist in a medium without changing the shape of each other is quite remarkable.

When particles collide, they scatter off each other and their velocities change in accordance with Newton's laws of motion. Waves, on the other hand, pass through each other unaffected.

Figure 14-44 shows a series of position plots of two pulses moving in opposite direction along a stretched string. When the pulses are far apart, the segment of the string where each pulse happens to be conforms to the shape of the pulse. When the two pulses overlap, the shape of the string at the overlapping positions corresponds to the combined effect of the two pulses, in accordance with the principle of superposition.

Now consider two pulses moving in opposite directions with identical shapes but opposite displacements.

Figure 14-45 shows a series of position plots of the two pulses at different times. At the instant their centres overlap, the individual displacements cancel each other completely and the string is fully straight. Then each pulse reappears, continuing to travel in its original direction of motion with its original shape.

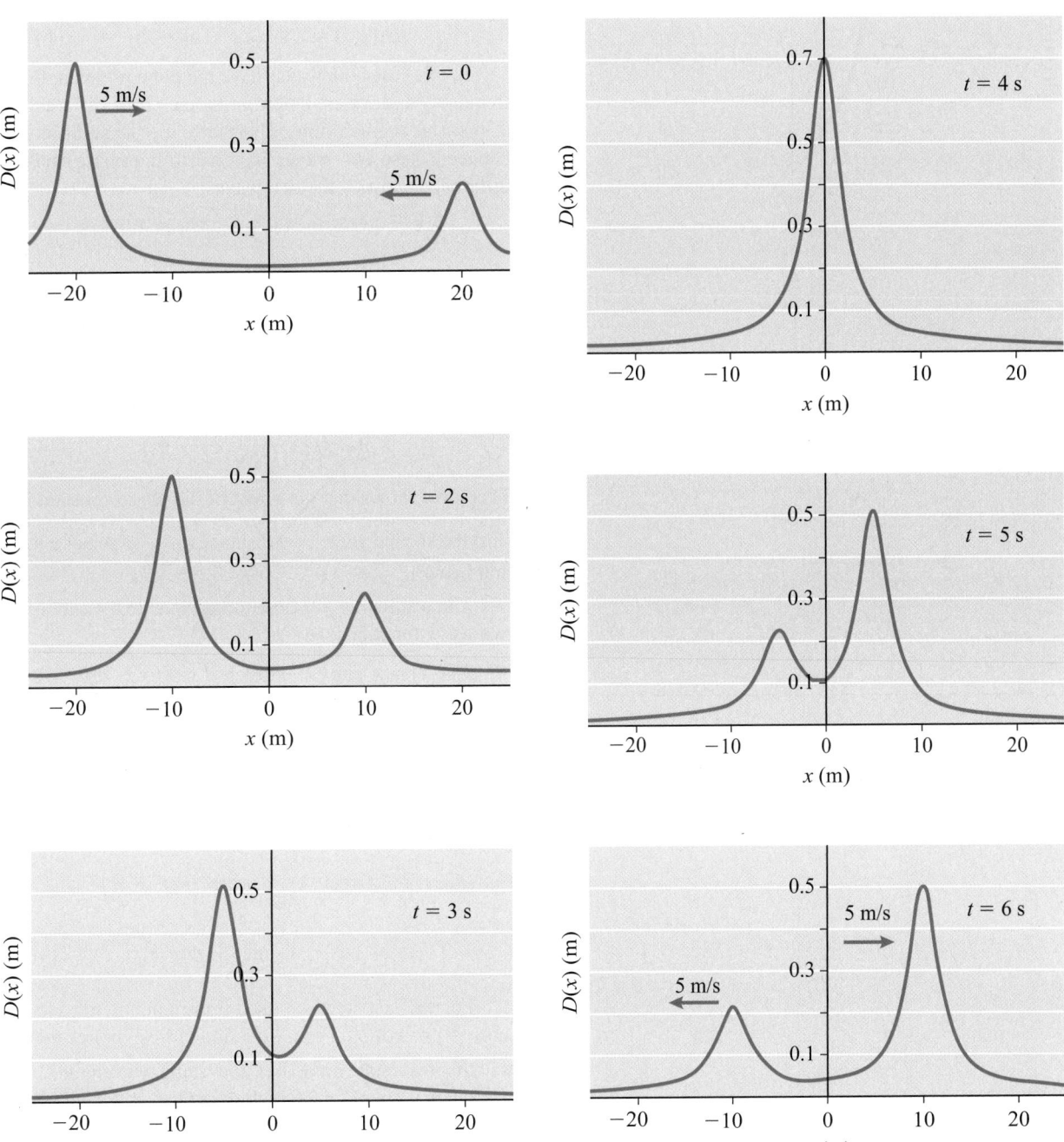

Figure 14-44 Interference of two pulses travelling in opposite direction

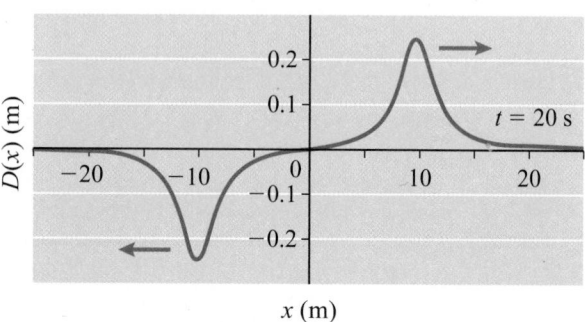

Figure 14-45 Interference of two pulses with identical shapes but opposite amplitudes travelling in opposite directions

14-10 Interference of Waves Travelling in the Same Direction

We now derive a mathematical expression for the resultant wave from the superposition of two waves with the same wavelength, frequency, amplitude, and direction but different phase constants. To simplify the algebra, we will assume that the phase constant of one of the waves is zero. Then, the wave functions of the two waves are

$$D_1(x, t) = A \sin(kx - \omega t)$$
$$D_2(x, t) = A \sin(kx - \omega t + \phi) \quad (14\text{-}34)$$

In this case both waves have the same wavelength and frequency and we are determining the phase difference at a point at a given time. Therefore, the phase difference between these two waves is equal to the phase constant difference between the two waves:

$$\Delta\Phi = \text{ phase of wave 2} - \text{phase of wave 1}$$
$$= (kx - \omega t + \phi) - (kx - \omega t) = \phi \quad (14\text{-}35)$$

The wave function of the second wave can be rewritten as

$$D_2(x, t) = A \sin\left(k\left(x + \frac{\phi}{k}\right) - \omega t\right)$$

$$= A \sin\left(k\left(x + \frac{\phi}{2\pi}\lambda\right) - \omega t\right) \quad (14\text{-}36)$$

This form shows that the effect of the phase constant is to shift the second wave by a distance of $\dfrac{\lambda}{2\pi}\phi$ relative to the first wave.

The resultant wave function is equal to the sum of the two wave functions:

$$D(x, t) = A \sin(kx - \omega t) + A \sin(kx - \omega t + \phi)$$

Using the trigonometric identity

$$\sin(a) + \sin(b) = 2\cos\left(\frac{a - b}{2}\right)\sin\left(\frac{a + b}{2}\right)$$

with $a = kx - \omega t$ and $b = kx - \omega t + \phi$, the equation for the resultant wave function becomes

$$D(x, t) = 2A \cos\left(-\frac{\phi}{2}\right)\sin\left(kx - \omega t + \frac{\phi}{2}\right)$$

$$= 2A \cos\left(\frac{\phi}{2}\right)\sin\left(kx - \omega t + \frac{\phi}{2}\right) \quad (14\text{-}37)$$

For the last step, we used the trigonometric identity $\cos(-\theta) = \cos(\theta)$. From Equation (14-37) notice that

- the resultant wave has the same wavelength, frequency, speed, and direction as the component waves;
- the phase constant of the resultant wave is the mean of the phase constant of the component waves, $\dfrac{0 + \phi}{2} = \dfrac{\phi}{2}$;

- the amplitude of the resultant wave, called the **resultant amplitude**, is $2A \cos\left(\dfrac{\phi}{2}\right)$.

The resultant amplitude depends on the amplitude of the component waves and on the difference between their phase constants. Figure 14-46 shows a plot of the resultant amplitude as a function of ϕ.

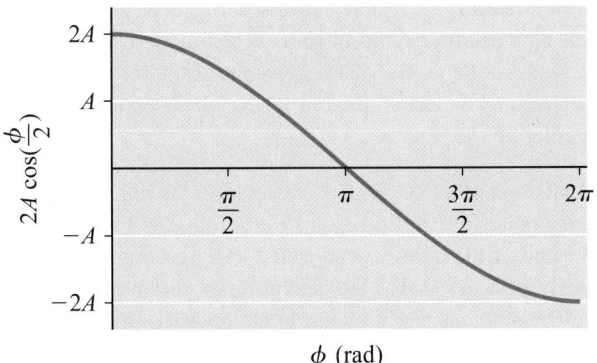

Figure 14-46 The amplitude of the resultant wave as a function of the difference between the phase constants of two waves that have the same wavelength, frequency, amplitude, and direction

When the phase difference between the two waves is zero or an integer multiple of 2π radians, the waves are said to be in phase. In this case the waves reinforce each other, with the crest of one coinciding with the crest of the other, as shown in Figure 14-47. The amplitude of the resultant wave is $2A$. Such an alignment of the waves is called **constructive interference** of waves.

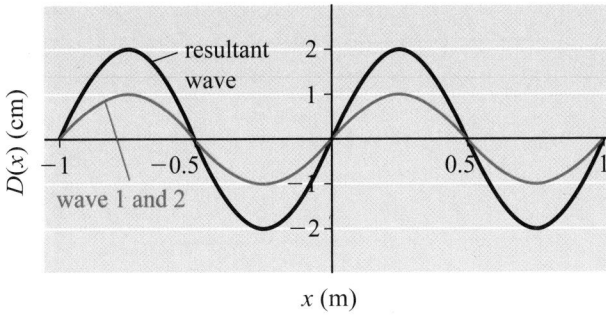

Figure 14-47 Constructive interference of two waves moving in the same direction with equal amplitudes and wavelengths, and a phase constant difference of 0 rad

When the phase difference between the two waves is an odd integer multiple of π radians the waves are said to be out of phase. In this case the two waves completely cancel each other, as shown in Figure 14-48. Now the displacements of the two component waves are exactly equal and opposite at every point, with the crest of one coinciding with the trough of the other, resulting in **destructive interference** of the waves.

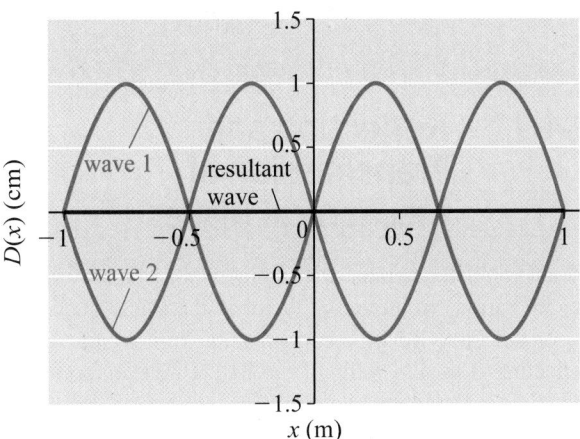

Figure 14-48 Destructive interference of two waves moving in the same direction with equal amplitudes and wavelengths, and a phase constant difference of π rad

For the phase difference that is other than an integer multiple of π rad, the resultant amplitude is intermediate between the constructive ($2A$) and destructive (0) values. As an example, Figure 14-49 shows the interference at time $t = 0$ of two waves with amplitudes of 1.0 m, wavelengths of 1.0 m, and a phase constant difference of $\pi/3$ rad. From Equation (14-37), the resultant wave has an amplitude of $2 \times (1.0 \text{ m}) \times \cos(\pi/6) = 1.73$ m. From the plot, we can see that the crests of the component waves are displaced with respect to each other by

$$\frac{\phi}{2\pi}\lambda = \frac{\pi/3 \text{ rad}}{2\pi \text{ rad}} \times 1.0 \text{ m} = 0.16 \text{ m}.$$

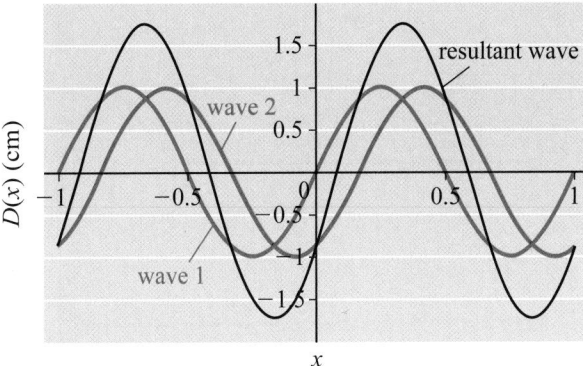

Figure 14-49 The interference of two waves of equal amplitudes and wavelengths and a phase constant difference of $\pi/3$ rad

✓ **CHECKPOINT**

C-14-14 Wave Interference

Two waves with the same amplitude and frequency are $\pi/2$ rad out of phase when they arrive at point P. What is the amplitude of the resultant wave at P?

(a) $2A$

(b) $1.7A$

(c) $1.4A$

(d) $0.7A$

C-14-14 (c)

LO 11

14-11 Reflection and Transmission of Mechanical Waves

When you listen to music in a concert hall, you may notice reverberation, an important factor in the sound quality of a concert hall. Reverberation results from sound waves reflecting from the walls of the hall. However, someone standing in the lobby can also hear the music, albeit considerably muffled. So, a portion of the sound waves have gone through the walls. In this section, we introduce the concepts of reflection and transmission of mechanical waves at boundaries.

Reflection at a Fixed End Consider a transverse pulse travelling along a string toward an end that is tied to a rigid support, as shown in Figure 14-50. The transverse displacement of the **incident pulse** shown is upward, so the string exerts an upward force on the support when the pulse reaches the fixed end. The rigid support exerts an equal and opposite reaction force on the string, in accordance with Newton's third law of motion. The reaction force generates an inverted pulse that moves in the opposite direction. A reflection from a rigid boundary is called a **hard reflection**.

When a wave reflects from a fixed end, the reflected wave is inverted (compared to the incident wave) and moves in the opposite direction. Since the rigid end has zero displacement, the incident and the reflected waves always cancel each other at this end. Therefore, they are out of phase by π rad (180°). When a wave is reflected from a *fixed* end, the phase constant of the reflected wave changes by π rad compared to the phase constant of the incident wave.

Reflection at a Free End Now consider the reflection of the same pulse from an end that is free to move. In Figure 14-51, the end of the string is connected to a light ring that slides freely along a rod. The incident pulse exerts an upward force on the ring, causing it to accelerate upward. Due to inertia, the ring keeps moving upward, overshooting the maximum height of the pulse and pulling the string with it. The reaction force exerted on the string by the ring generates a *backward-moving pulse that is not inverted*. Such a reflection is called

Figure 14-50 Hard reflection: A pulse reflected from a fixed end is inverted. The phase of reflected pulse is changed by π radians.

Figure 14-51 Soft reflection: A pulse reflected from a free end is not inverted. The phase of the reflected pulse does not change.

a **soft reflection**. If the amplitude of the pulse is A, the maximum displacement of the ring from its equilibrium position is $2A$. Therefore, for a soft reflection, the incident and the reflected pulses are in phase with each other.

The same is true for waves. When a wave reflects from a free end, the reflected wave is not inverted (compared to the incident wave) and moves in the opposite direction. In this case, the incident and the reflected waves are in phase with each other.

Often, an end is neither completely rigid nor fully free. Consider a string of linear mass density μ_1 attached to a string of linear mass density μ_2, with $\mu_1 < \mu_2$. When a pulse travelling in the lighter string reaches the junction between the two strings, part of the incident pulse is reflected and part is transmitted into the thicker string. The reflected pulse is inverted, but the transmitted pulse is not (Figure 14-52). For the reverse situation, when an incident pulse travelling in the thicker string reaches the boundary with the lighter string, neither the reflected pulse nor the transmitted pulse is inverted (Figure 14-53).

The total energy of the incident wave is equal to the sum of the energies of the reflected and transmitted waves.

Although our discussion here has dealt only with one-dimensional waves on strings, two- and three-dimensional mechanical waves and electromagnetic waves show the same behaviour at boundaries between media.

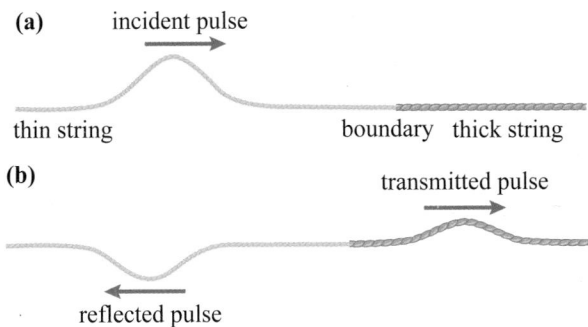

Figure 14-52 When a pulse travels from a less dense medium to a denser medium ($\mu_1 < \mu_2$), the reflected pulse is inverted, but the transmitted pulse is not.

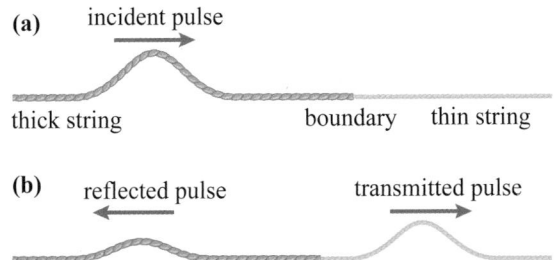

Figure 14-53 When a pulse travels from a dense medium to a less dense medium ($\mu_1 > \mu_2$), neither the reflected pulse nor the transmitted pulse is inverted.

LO 12

14-12   Standing Waves

In Section 14-10, we discussed the interference of two waves that are moving in the same direction. Now we consider two harmonic waves of equal amplitude, wavelength, and frequency that are moving in *opposite* directions. For mathematical simplicity, we assume that the phase constants of both waves are zero; however, in general, they may be unequal. The wave functions for the two waves are

$$D_1(x, t) = A \sin(kx - \omega t)$$

(wave moving in direction of increasing x)

$$D_2(x, t) = A \sin(kx + \omega t)$$

(wave moving in direction of decreasing x) (14-38)

According to the principle of superposition, the resultant wave function is

$$\begin{aligned} D(x, t) &= D_1(x, t) + D_2(x, t) \\ &= A \sin(kx - \omega t) + A \sin(kx + \omega t) \\ &= A[\sin(kx - \omega t) + \sin(kx + \omega t)] \end{aligned} \quad (14\text{-}39)$$

Using the trigonometric identity
$$\sin(a - b) + \sin(a + b) = 2\sin(a) \cos(b),$$
with $a = kx$ and $b = \omega t$, we obtain

$$D(x, t) = 2A \sin(kx) \cos(\omega t) \quad (14\text{-}40)$$

What kind of wave does this wave function describe? A travelling wave must have the position x and time t in the combination $x \mp vt$, where v is the wave speed. In Equation (14-40), the x and t variables appear separately, with x in the argument of the sine function and t in the cosine function. Therefore, the wave described by Equation (14-40) is not a travelling wave. We call it a **standing wave**. Let us define a position-dependent amplitude, $A(x)$, as

$$A(x) = 2A \sin(kx) = 2A \sin\left(2\pi \frac{x}{\lambda}\right) \quad (14\text{-}41)$$

Using this definition, we can rewrite Equation (14-40) as

$$D(x, t) = A(x) \cos(\omega t) \quad (14\text{-}42)$$

When two waves of equal wavelength, frequency, and amplitude but moving in opposite directions combine, each segment of the string oscillates in simple harmonic motion with the frequency of the individual waves and an amplitude that depends on the location of the segment along the string.

Figure 14-54 shows a plot of $A(x)$ as a function of position. Amplitude is a sine function, so certain points on the string have zero amplitude and remain at rest at all times. These points are called **nodes**. The points that move with the maximum possible amplitude of $2A$ are called **antinodes**. All the other points have amplitudes between zero and $2A$. Any two points that are one wavelength apart have the same amplitude because

$$A(x_0 + \lambda) = 2A \sin\left(2\pi \frac{x_0 + \lambda}{\lambda}\right) = 2A \sin\left(2\pi \frac{x_0}{\lambda} + 2\pi\right)$$

$$= 2A \sin\left(2\pi \frac{x_0}{\lambda}\right) = A(x_0) \qquad (14\text{-}43)$$

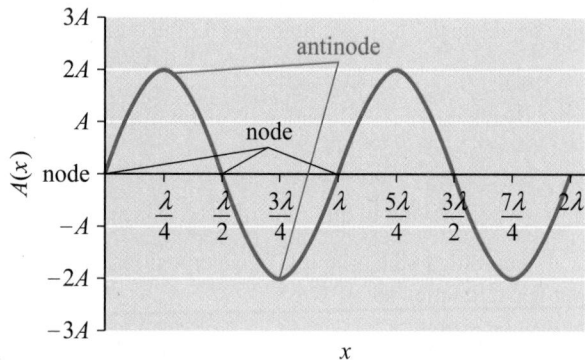

Figure 14-54 The amplitude $A(x)$ of a standing wave as a function of position along a string

Since nodes do not move, no energy flows along a standing wave. As each segment of the medium between two consecutive nodes oscillates in simple harmonic motion its energy continuously transforms between kinetic energy and elastic potential energy, just as for a single harmonic oscillator.

Location of Nodes and Antinodes At the nodes of a standing wave, $A(x) = 0$. Therefore, nodes occur when

$$\sin\left(\frac{2\pi}{\lambda} x\right) = 0$$

$$\frac{2\pi}{\lambda} x = m\pi \qquad m = 0, \pm 1, \pm 2, \dots$$

$$x = m\frac{\lambda}{2} \qquad m = 0, \pm 1, \pm 2, \dots$$

$$x = 0, \pm\frac{\lambda}{2}, \pm\lambda, \pm\frac{3\lambda}{2}, \pm 2\lambda, \dots \quad (14\text{-}44)$$

Thus, the distance between two consecutive nodes is half a wavelength.

At the antinodes of a standing wave, $A(x) = \pm 2A$, which occurs when

$$\sin\left(\frac{2\pi}{\lambda} x\right) = \pm 1$$

$$\frac{2\pi}{\lambda} x = \left(m + \frac{1}{2}\right)\pi \qquad m = 0, \pm 1, \pm 2, \dots$$

$$x = \left(m + \frac{1}{2}\right)\frac{\lambda}{2} \qquad m = 0, \pm 1, \pm 2, \dots$$

$$x = \pm\frac{\lambda}{4}, \pm\frac{3\lambda}{4}, \pm\frac{5\lambda}{4}, \dots \qquad (14\text{-}45)$$

EXAMPLE 14-12

Standing Wave

The amplitude of a standing wave is given by

$$A(x) = (25.0 \text{ cm}) \sin(2.00x)$$

where x is in meters. Determine

(a) the amplitude and the wavelength of the constituent travelling waves;
(b) the location of the first three nodes and the first three antinodes along the positive x-axis;
(c) the location of the first point from the origin where the amplitude is 0.20 m.

SOLUTION

(a) Comparing the given amplitude to Equation (14-41), we find that

$$2A = 25.0 \text{ cm} \quad \text{and} \quad \frac{2\pi}{\lambda} = 2.00 \text{ rad/m}$$

Therefore,

$$A = 12.5 \text{ cm} \quad \text{and} \quad \lambda = \pi \text{ m} = 3.14 \text{ m}$$

(b) The positions of nodes and antinodes can be calculated from Equations (14-44) and (14-45).
The first three nodes occur at

$$x = 0.00 \text{ m}, \quad x = \frac{\lambda}{2} = \frac{3.14 \text{ m}}{2} = 1.57 \text{ m}, \quad x = \lambda = 3.14 \text{ m}$$

The first three antinodes occur at

$$x = \frac{\lambda}{4} = \frac{3.14 \text{ m}}{4} = 0.78 \text{ m}, \quad x = \frac{3\lambda}{4} = 2.36 \text{ m},$$

$$x = \frac{5\lambda}{4} = 3.92 \text{ m}$$

(c) To determine the location of the point that has amplitude of 0.20 m, we substitute this value in the amplitude equation and solve for x:

$$0.20 \text{ m} = (0.25 \text{ m}) \sin(2.00x)$$

$$\sin(2.00x) = 0.80 \Rightarrow 2.00x = \sin^{-1}(0.80) \Rightarrow x = 0.46 \text{ m}$$

The distance between consecutive antinodes is also half a wavelength. An adjacent node and an antinode are a quarter of a wavelength apart.

To see how a string oscillates in a standing wave pattern, we write the wave function in terms of the time period ($T = 1/f$):

$$D(x, t) = 2A \sin\left(\frac{2\pi}{\lambda}x\right) \cos\left(\frac{2\pi}{T}t\right) \qquad (14\text{-}46)$$

Figure 14-55 shows the displacement of a section of the oscillating string at intervals of $T/8$ from $T = 0$ to $t = T/2$. Table 14-2 compares the displacements of the section of the string between the first two nodes during the first half of a cycle. The motion for the next half of the period is in the opposite direction. This oscillatory motion repeats every cycle.

All points between two consecutive nodes oscillate in phase with each other. The mean speed is greatest for the antinode because it has to cover the longest distance ($8A$) in one period. The speed decreases away from the antinode and is zero at the nodes. Note that the motion of the section between the next two nodes is π rad out of phase with the first section. The sine term in the wave function accounts for this property:

$$A\left(x_0 + \frac{\lambda}{2}\right) = 2A \sin\left(\frac{2\pi}{\lambda}\left(x_0 + \frac{\lambda}{2}\right)\right) = 2A \sin\left(\frac{2\pi}{\lambda}x_0 + \pi\right)$$

$$= -2A \sin\left(2\pi \frac{x_0}{\lambda}\right) = -A(x_0) \qquad (14\text{-}47)$$

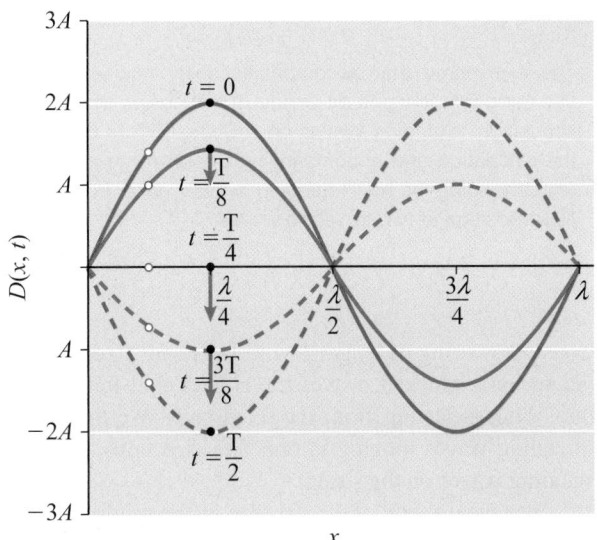

Figure 14-55 Displacement versus position for a standing wave on a string during the first half of a cycle. The position x is expressed in multiples of the wavelength λ.

Table 14-2 Displacements of string elements between two nodes of a standing wave during half a time period

Time	Wave function	Displacement
0	$D(x, 0) = 2A \sin\left(\frac{2\pi}{\lambda}x\right)\cos(0) = 2A \sin\left(\frac{2\pi}{\lambda}x\right)$	each segment is at its maximum displacement
$T/8$	$D\left(x, \frac{T}{8}\right) = 2A \sin\left(\frac{2\pi}{\lambda}x\right)\cos\left(\frac{2\pi}{T}\frac{T}{8}\right) = \sqrt{2}A \sin\left(\frac{2\pi}{\lambda}x\right)$	displacement of each segment decreases by factor of $\sqrt{2}$
$T/4$	$D(x, T/4) = 2A \sin\left(\frac{2\pi}{\lambda}x\right)\cos\left(\frac{2\pi}{T}\frac{T}{4}\right) = 0$	each segment reaches equilibrium position
$3T/8$	$D\left(x, \frac{3T}{8}\right) = 2A \sin\left(\frac{2\pi}{\lambda}x\right)\cos\left(\frac{2\pi}{T}\frac{3T}{8}\right) = -\sqrt{2}A \sin\left(\frac{2\pi}{\lambda}x\right)$	each segment reaches $1/\sqrt{2}$ of maximum displacement at the opposite side of the equilibrium position
$T/2$	$D(x, T/2) = 2A \sin\left(\frac{2\pi}{\lambda}x\right)\cos\left(\frac{2\pi}{T}\frac{T}{2}\right) = -2A \sin\left(\frac{2\pi}{\lambda}x\right)$	each segment is at its maximum displacement at the opposite of the equilibrium position

14-13 Standing Waves on Strings

ONLINE ACTIVITY

Standing Waves on a String

The e-resource that accompanies every new copy of this textbook contains an Online Activity using the PhET simulation "Standing Waves on a String." Work through the simulation and accompanying questions to gain an understanding of how standing waves arise from the superposition of two travelling waves.

Consider a string of length L with both ends fixed. When the string is plucked, waves travel back and forth along the string, reflecting from the fixed ends. We then have travelling waves moving in opposite directions, creating standing waves on the string.

We take one end of the string to be the origin, and the other end to be at $x = L$. Because the string is clamped at both ends the amplitude must be zero at $x = 0$ and at $x = L$. The amplitude in 14-41 is zero at $x = 0$ and for it to be zero at $x = L$ we must have,

$$\sin\left(\frac{2\pi}{\lambda} L\right) = 0 \qquad (14\text{-}48)$$

and

$$\frac{2\pi}{\lambda} L = m\pi \qquad (14\text{-}49)$$

where m is a positive, nonzero integer (1, 2, 3, …).

Rearranging Equation (14-49), we find that a string with both ends fixed can oscillate in a standing wave pattern only with following wavelengths:

$$\lambda_m = \frac{2L}{m} \quad m = 1, 2, 3, 4, \ldots \qquad (14\text{-}50)$$

The longest of these wavelengths is $\lambda_1 = 2L$. The series of wavelengths continues with $\lambda_2 = L$, $\lambda_3 = \frac{2L}{3}$, $\lambda_4 = \frac{L}{2}$, and so on. Notice that each of these wavelengths is such that an integral number of half wavelengths Figures 14-56 and 14-57. These standing waves are called the **normal modes** of vibration of the string.

The frequencies corresponding to the normal modes of vibration are

$$f_m = \frac{v}{\lambda_m} = \frac{m}{2L} v = \frac{m}{2L} \sqrt{\frac{T}{\mu}} \qquad (14\text{-}51)$$

The lowest frequency corresponds to the longest wavelength, $\lambda_1 = 2L$, and is called the **fundamental frequency** or the **first harmonic**:

Fundamental frequency:

$$f_1 = \frac{v}{\lambda_1} = \frac{1}{2L} v = \frac{1}{2L} \sqrt{\frac{T}{\mu}} \qquad (14\text{-}52)$$

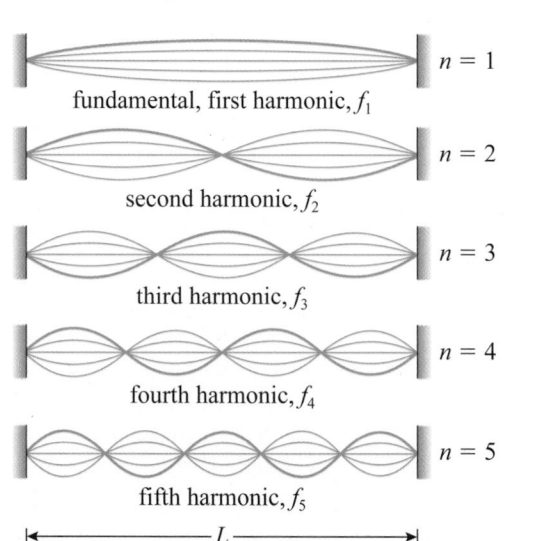

Figure 14-56 Standing waves on a string: the first five harmonics

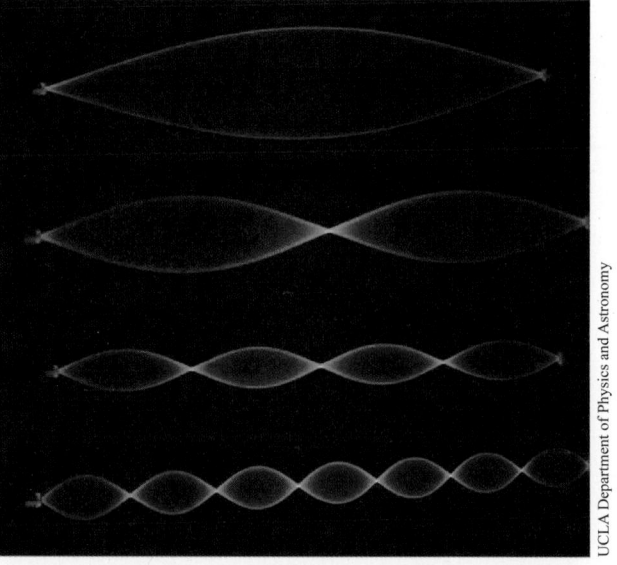

Figure 14-57 The envelopes of a fluorescent string vibrating in various normal modes. The images were taken by shining a strobe light on the string.

Notice that the fundamental frequency of a string is proportional to the square root of the tension in the string and is inversely proportional to its length and to the square root of the linear mass density. Higher frequencies of vibration correspond to higher values of m and are integer multiples of the fundamental frequency, since:

$$f_m = m f_1 \quad m = 1, 2, 3, 4, \ldots \qquad (14\text{-}53)$$

The allowed frequencies are called **harmonics** or **resonant frequencies**. The first few harmonics are

$$f_1 = \frac{1}{2L}\sqrt{\frac{T}{\mu}} \qquad \text{fundamental or first harmonic}$$

$$f_2 = 2f_1 \qquad \text{second harmonic}$$

$$f_3 = 3f_1 \qquad \text{third harmonic}$$

$$f_4 = 4f_1 \qquad \text{fourth harmonic} \qquad (14\text{-}54)$$

When plucked, a string can vibrate in a single normal mode or a combination of several modes. Notice that m is equal to the number of antinodes between the fixed ends of the string.

As we will see in the next chapter, sound waves in a wind instrument also form standing wave patterns.

EXAMPLE 14-13

String Vibrations

A violin string is 0.32 m long and has a linear mass density of 3.83×10^{-4} kg/m. The tension in the string is kept at 70.0 N.

(a) What is the wave speed in the string?

(b) If the string is plucked and allowed to vibrate, approximately how many times would a wave reflect from one end of the string in one second?

(c) What are the wavelengths and frequencies of the first three normal modes of vibration of the string? What is the spacing between two consecutive nodes for each mode? Sketch the vibration pattern of these modes.

(d) By what amount should the tension be changed to decrease the fundamental frequency by 7 Hz?

SOLUTION

When plucked, the string will vibrate in a standing wave pattern with the wavelengths and frequencies of normal modes given by Equations (14-50) and (14-53).

(a) The wave speed in the string is determined by the tension and linear mass density of the string:

$$v = \sqrt{\frac{T}{\mu}} = \sqrt{\frac{70.0\text{ N}}{3.83 \times 10^{-4}\text{ kg/m}}} = 427\text{ m/s}$$

(b) The time that the waves take to travel from one end of the string to the other is

$$\Delta t = \frac{\text{length of the string}}{\text{wave speed}} = \frac{0.32\text{ m}}{427\text{ m/s}} = 7.5 \times 10^{-4}\text{ s}$$

Each reflection from a given end requires a back-and-forth trip. Therefore, the number of times a wave reflects from a given fixed end in one second is

$$\frac{1.0\text{ s}}{2 \times \Delta t} = \frac{1.0\text{ s}}{2 \times 7.5 \times 10^{-4}\text{ s}} = 668$$

(c) For the first harmonic:

$$\text{wavelength: } \lambda_1 = 2L = 2 \times 0.32\text{ m} = 0.64\text{ m}$$

$$\text{frequency: } f_1 = \frac{v}{\lambda_1} = \frac{427\text{ m/s}}{0.64\text{ m}} = 668\text{ Hz}$$

The distance between consecutive nodes: $= \lambda_1/2 = 0.32$ m

For the second harmonic:

$$\text{wavelength: } \lambda_2 = 2L/2 = 0.32\text{ m}$$

$$\text{frequency: } f_2 = 2f_1 = 1336\text{ Hz}$$

The distance between consecutive nodes: $= \lambda_2/2 = 0.16$ m

For the third harmonic:

$$\text{wavelength: } \lambda_3 = 2L/3 = 0.21\text{ m}$$

$$\text{frequency: } f_3 = 3f_1 = 2004\text{ Hz}$$

The distance between consecutive nodes: $= \lambda_3/2 = 0.11$ m

The vibration patterns of these modes are shown in Figure 14-58.

(d) The new fundamental frequency will be $f_1' = 668\text{ Hz} - 7\text{ Hz} = 661\text{ Hz}$. To decrease the fundamental frequency, the tension in the string must be decreased. If T' is the new tension, then

$$f_1' = \frac{1}{2L}\sqrt{\frac{T'}{\mu}}$$

Squaring the above equation, we can write

$$T' = 4L^2\mu(f_1')^2$$

The fractional decrease in tension needed to lower the frequency by 7.000 Hz is therefore,

$$\frac{T - T'}{T} = \frac{(f_1)^2 - (f_1')^2}{(f_1)^2} = \frac{(668\text{ Hz})^2 - (661\text{ Hz})^2}{(668\text{ Hz})^2} = 0.0208$$

Therefore, the tension in the string needs to be decreased by $0.0208 \times 70.0\text{ N} = 1.46\text{ N}$.

Making sense of the result:

The fundamental frequency calculated in part (c) is the same as the number of times the wave reflects from a fixed end of the string (calculated in part (b)).

(a)

(b)

(c)

Figure 14-58 Example 14-13. Envelopes of the displacements for (a) the first harmonic, (b) the second harmonic, and (c) the third harmonic.

CHECKPOINT

C-14-15 Standing Wave Nodes

How many nodes are present between the fixed ends of a string vibrating at its seventh harmonic?

(a) 5
(b) 6
(c) 7
(d) 8

C-14-15 (b)

MAKING CONNECTIONS

Lasers

Standing waves are also formed by electromagnetic waves. For example, there are standing waves in lasers. A gas laser consists of a cylindrical tube that is filled with gases, such as helium and neon. The cavity has a reflecting mirror at one end and a partially reflecting mirror at the other (Figure 14-59). An electric voltage pumps energy into the gases, putting many of the gas atoms into an excited state. The atoms produce light when they release the absorbed energy. This light reflects back and forth between the two mirrors, thus creating standing waves within the cavity. Each pass of the light between the mirrors stimulates more of the excited atoms to emit light. The partially reflecting mirror allows some of the light to pass through. Because of the stimulated emission process, laser light has spatial coherence, allowing it to be very tightly focused.

Figure 14-59 The inner workings of a laser

FUNDAMENTAL CONCEPTS AND RELATIONSHIPS

A mechanical wave is the propagation of a disturbance through a medium. It takes energy to generate a wave, and the wave carries this energy with it as it passes through a medium. In the presence of a wave, the particles of the medium oscillate about their equilibrium positions but are not carried along with the wave. The wave speed in a medium depends on the properties of the medium.

In longitudinal waves, the motion of the particles of the medium is in the direction of propagation of the wave. In transverse waves, the motion of the particles of the medium is perpendicular to the direction of wave propagation.

The shape of a wave is a function of position (x) and time (t). If the shape of a wave is described by a function $D(x)$ at $t = 0$, then its shape at time x is obtained by the substitution $x \rightarrow x \mp vt$, where v is the speed of the wave. $D(x, t)$ is called the wave function.

A sinusoidal wave is generated by a source that is undergoing simple harmonic motion. The wave function of a sinusoidal wave is given by

$$D(x, t) = A \sin(kx - \omega t + \phi)$$

$$= A \sin\left(\frac{2\pi}{\lambda} x - 2\pi ft + \phi\right)$$

where A is the amplitude, λ is the wavelength, f is the frequency, $k = \frac{2\pi}{\lambda}$ is the wave number, $\omega = 2\pi f = \frac{2\pi}{T}$ is the angular frequency, and ϕ is the phase constant.

For a continuous wave, the wave speed is related to the wavelength and frequency by $v = \lambda f$.

A position plot displays the shape of the wave as a function of position for a fixed time. A time plot displays the displacement of a point of the medium as a function of time.

The value of the phase constant can be determined from a position or a time plot.

The average power transported by a wave is

$$P_{avg} = 2\pi^2 \mu v f^2 A^2$$

When two or more waves are simultaneously present in a medium, the resultant wave function is equal to the sum of the individual wave functions.

When two waves with the same wavelength and frequency and travelling in the same direction combine, the amplitude of the resultant wave is maximum when the phase difference between the two waves is an integer multiple of 2π rad, and zero when the difference is an odd multiple of π rad.

Two waves with the same frequency and amplitude travelling in opposite directions produce standing waves. The wave function of a standing wave is

$$D(x, t) = 2A \sin(kx) \cos(\omega t)$$

where A is the amplitude of each wave. In a standing wave, energy is not transported through the medium.

The harmonics for a string that is fixed at both ends are

$$\lambda_m = \frac{2L}{m}; \quad f_m = m\frac{v}{2L}$$

where L is the length of the string, v is the wave speed, and $m = 1, 2, 3, \ldots$.

Key Terms: amplitude, antinodes, constructive interference, continuous wave, crest, destructive interference, displacement function, electromagnetic radiation, electromagnetic wave, first harmonic, fundamental frequency, hard reflection, harmonic wave, harmonics, in phase, incident pulse, interference, longitudinal wave, mechanical wave, medium, nodes, normal modes, out of phase, phase difference, phase, position plot, principle of superposition, pulse, resonant frequencies, resultant amplitude, resultant wave, sinusoidal wave, snapshot graphs, soft reflection, standing wave, superposition, transverse wave, trough, wave cycle, wave function, wave number, wave speed, wave, waveform, wavelength

QUESTIONS

1. Indicate whether each of the following statements is true (T) or false (F). Explain your answer.
   (a) As a wave passes through a medium, it carries the particles of the medium with it.
   (b) The wave speed of a mechanical wave depends on the frequency of the source.
   (c) Waves with different wavelengths, travelling in the same medium, can have different speeds.
   (d) Mechanical waves do not require a medium to travel.
   (e) A wave, travelling in a uniform medium, can have a nonzero acceleration.
   (f) A travelling wave can have zero energy.
   (g) When two waves of amplitudes 1.0 m and 2.0 m interfere, the amplitude of the resultant wave can be greater than 3.0 m.
   (h) Standing waves can be produced by combining two waves that are moving in the same direction.

2. Waves carry energy. Do they carry momentum?

3. The speed of sound waves in air is 340 m/s, and the speed of light is 3×10^8 m/s. Consider a sound wave and a light wave of 20 000 Hz. Which wave has the longer wavelength?

4. In the presence of a travelling wave, are there any particles of the medium that are always at rest?

5. Suppose a longitudinal harmonic wave is moving through a spring with a speed of 10.0 m/s. Does each coil of the spring oscillate about its equilibrium position with a speed of 10.0 m/s?

6. A pulse with total energy E_1 is travelling along a string that is connected to another string. At the boundary of the two strings, the pulse is partly transmitted into the second string and partly reflected into the first string. The energy of the reflected pulse is E_2, and the energy of the transmitted pulse is E_3. How are E_1, E_2, and E_3 related? What assumptions did you make to reach your conclusion?

7. When you throw a stone on the surface of a pond, the wave travels away from the stone and its amplitude decreases. Why?

8. Two wires, stretched under the same tension, have linear mass densities that differ by a factor of 4. The wave speeds in the wires differ by a factor of
(a) ½ (b) 1 (c) 2 (d) 4 (e) 16

9. When the frequency of a wave is doubled, the energy required to generate the wave
(a) remains the same;
(b) increases by a factor of 2;
(c) increases by a factor of 4;
(d) decreases by a factor of 2;
(e) decreases by a factor of 4.

10. A source is generating waves at a certain frequency. When the power of the source is doubled, the amplitude of the wave
(a) remains the same;
(b) increases by a factor of $\sqrt{2}$;
(c) increases by a factor of 2;
(d) decreases by a factor of $\sqrt{2}$;
(e) decreases by a factor of 2.

11. Why do two travelling waves (moving in opposite directions) with the same amplitude and frequency generate a standing wave? Can we produce a standing wave by combining two travelling waves of (a) the same amplitude but different frequencies, and (b) different amplitudes but the same frequency?

12. A travelling wave transports energy between two points. Can a standing wave transport energy between two points?

13. Standing waves with a frequency of 440 Hz are generated on two strings of the same length. Must the two strings have exactly the same linear mass density?

14. Two waves of amplitude A are travelling on a string. The waves interfere to produce a resultant wave of amplitude $A/2$. Is the total energy of the resultant wave equal to the sum of the energies of the constituent waves? Explain your reasoning.

PROBLEMS BY SECTION

For problems, star ratings will be used, (✱, ✱✱, or ✱✱✱), with more stars meaning more challenging problems.

Section 14-1   The Nature, Properties, and Classification of Waves

15. ✱ A wave has a wavelength of 2.00 m and a frequency of 10 Hz. Determine
(a) the wave number;
(b) the angular frequency;
(c) the wave speed.

16. ✱ The speed of sound in air is 340.0 m/s. The angular frequency of a sound wave is 600.0 rad/s. Determine the wave's
(a) frequency;
(b) period;
(c) wavelength.

17. ✱ A wave is travelling along a stretched string with a speed of 150.0 m/s. The wave number of the wave is 6.00 rad/m. Determine the wave's
(a) frequency;
(b) period;
(c) wavelength.

18. ✱ Electromagnetic waves in a microwave oven have a frequency of 4.0×10^9 Hz. What is the wavelength of the waves?

19. ✱ Human ears can detect sounds with frequencies from 20 Hz to 20 000 Hz. The speed of sound in air is approximately 340 m/s. What is the corresponding range of wavelengths that the human ear can detect?

20. ✱ A transverse wave takes 2.0 s to travel the full length of a 3.0 m string. What is the frequency of the wave when its wavelength is 0.50 m?

21. ✱ The speed of sound waves in ocean water is approximately 1500 m/s. Dolphins produce sound waves with frequencies in the range of 250 Hz to 150 kHz. What is the range of wavelengths of the waves in ocean water and in air?

22. ✱ The speed of sound in air is 340 m/s, and the speed of light is 3.0×10^8 m/s. What frequency of light has the same wavelength as a 200.0 Hz sound wave?

Sections 14-2 and 14-3   The Motion of a Disturbance in a String and Equation for a Pulse Moving in One Dimension

23. ✱✱ At time $t = 0$, the transverse displacement of a string due to a pulse travelling through it is given by the equation

$$D(x, 0) = \frac{0.3}{x^2 + 1.2}$$

where x and D are measured in metres. When the pulse is moving at 4.0 m/s toward the positive x-axis, what is the displacement of a segment of the string that is located at $x = 1.5$ m at $t = 2.0$ s? What are the maximum and minimum displacements of this segment ?

24. ✱ Rank the following pulses in order of wave speed and amplitude, from lowest to highest, using an equality sign where needed. Also, indicate the direction of motion of each pulse.

(a) $D(x, t) = \dfrac{3}{4 + (x - 0.5t)^2}$

(b) $D(x, t) = \dfrac{2}{3 + (x + t)^2}$

(c) $D(x, t) = \dfrac{11}{15 + (x - 5t)^2}$

(d) $D(x, t) = \dfrac{7}{5 + (x + 4t)^2}$

25. ✷✷ The displacement function for a pulse passing through a string is given by

$$D(x, t) = \frac{-3.0}{6.0 + (x + 3.0t)^2}$$

where x and D are in metres, and t is in seconds.
(a) What is the speed of the pulse, and in which direction is it travelling?
(b) What is the displacement of a point located at $x = 2.0$ m at $t = 3.0$ s?
(c) What is the maximum displacement of a point on the string?
(d) What is the transverse speed of a point that is located at $x = 2.0$ m at $t = 3.0$ s?
(e) Draw a displacement versus position (x) graph for $t = 0, 1, 2,$ and 3 s for this waveform.

26. ✷ At time $t = 0$, the transverse displacement of a string due to a pulse travelling at a speed of 2.0 m/s is given by

$$D(x, 0) = \frac{3.0}{x^4 + 10.0}$$

where x and D are in metres.
(a) What is the amplitude of the pulse?
(b) What is the displacement of a point located at $x = 1.0$ m at $t = 0.0$ s?
(c) The pulse is moving toward increasing x. Write an equation that describes the displacement of the string as a function of x and t.
(d) What is the transverse speed of a point located at $x = 2.0$ m at $t = 1.0$ s?

27. ✷✷ The displacement function for a travelling pulse is given by the equation

$$D(x, t) = \begin{cases} +2 \text{ m} & \text{if } |x - t| \leq 1 \\ -2 \text{ m} & \text{if } |x - t| > 1 \end{cases}$$

where x is in metres, and t is in seconds.
(a) Draw a displacement versus position graph for the pulse for $t = 1, 2,$ and 3 s.
(b) Determine the wave speed from the graph. In which direction is the disturbance travelling?
(c) What is the displacement of a point located at $x = 0.5$ m at $t = 1.0$ s?
(d) Write the equation for a pulse that has the same shape but is moving in the opposite direction.

28. ✷ The displacement function for a pulse travelling along a string is given by the equation

$$D(x, t) = \frac{4.0}{(x - 2.0t - 10.0)^2 + 6.0}$$

where x is in metres, and t is in seconds.
(a) What is the speed of the pulse, and in which direction is it travelling?
(b) What is the displacement of a point located at $x = 4.0$ m at $t = 1.0$ s?
(c) What is the speed of a point that is located at $x = 5.0$ m at $t = 1.0$ s?
(d) What is the maximum displacement of a point as the pulse passes through it?

29. ✷✷ Two pulses travelling on a string are described by the displacement functions

$$D_1(x, t) = \frac{2}{(x - 5t + 10)^2 + 4}$$

and

$$D_2(x, t) = \frac{-2}{(x + 5t - 10)^2 + 4}$$

(a) Sketch the shape of the pulses at $t = 0$.
(b) Where is the peak of each pulse located when $t = 0$?
(c) At what time do the two pulses cancel each other?
(d) Is there a point along the string that has zero displacement for all t? If yes, where is it located?

Section 14-4 Transverse Speed and Wave Speed

30. ✷ The E string on a violin has a diameter of 0.0039 cm and is made of steel of density (ρ) 7.86 g/cm³. What is the linear mass density of the string?

31. ✷ The density of steel is 7.86 g/cm³. It is observed that transverse waves in a 0.40 mm diameter steel wire propagate at 160 m/s. What is the tension in the wire?

32. ✷✷ A steel wire of cross-sectional area 1.00×10^{-4} m² is under a tension of 1.00×10^3 N. At what speed does a transverse wave move along the wire? The density of steel is 7860 kg/m³.

33. ✷ A 50.0 m string of uniform density has a mass of 0.10 kg. When the tension in the string is 100.0 N, what is the speed of a pulse travelling along the length of the string?

34. ✷✷ The string in problem 33 is hanging from a tall ceiling, and a weight is tied to its lower end to produce a tension of 10.0 N. A pulse generated at the lower end travels upward toward the ceiling. The pulse passes through the midpoint of the string.
(a) What is the speed of the pulse at the midpoint if the mass of the string is ignored?
(b) What is the speed of the pulse at the midpoint if the mass of the string is *not* ignored?

35. ✷ String A is 10.0 m long and has a linear mass density of 2.0 g/m. String B is 20.0 m long and has a linear mass density of 5.0 g/m. The two strings are tied together and kept under a constant tension of 50.0 N. Two identical pulses are generated at the two opposite ends of the strings. Where will the two pulses first meet along the string?

36. ✷✷✷ A thick rope of uniform density hangs vertically from a fixed support.
(a) Show that the wave speed at a distance y from the lower end is given by $v = \sqrt{yg}$.
(b) Show that the time taken by a pulse to travel the entire length L of the rope is given by $t = 2\sqrt{L/g}$.

37. ✷✷ The speed of waves on the surface of a deep ocean is given by $v = \sqrt{\dfrac{g\lambda}{2\pi}}$, where λ is the wavelength of the waves, and g is the acceleration due to gravity.
(a) Show that the expression on the right side of this equation has the dimensions of speed.
(b) What is the period of the waves?
(c) Successive crests pass a stationary oil platform every 12 s. What is the speed of the waves?

Section 14-5   Harmonic Waves

38. ✱   The displacement versus position graphs at a fixed time for four waves, travelling in the same medium, are shown in Figure 14-60. Rank these waves in order of frequency, from highest to lowest.

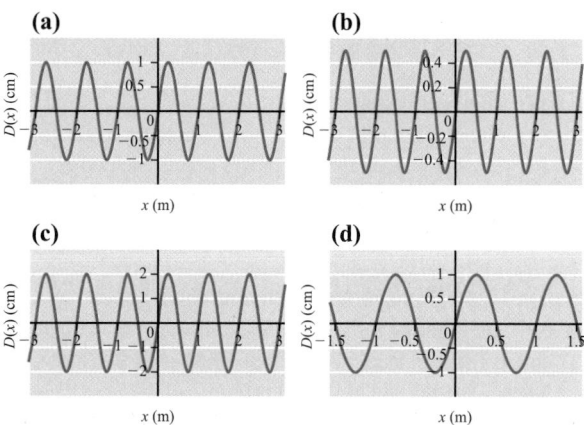

(a)

(b)

(c)

(d)

Figure 14-60   Problem 38

39. ✱   Rank the following waves in order of wavelength and frequency, from lowest to highest. Also, indicate the direction of motion, toward increasing or decreasing x, for each wave. Use an equality sign where needed. Here x is in meters and t in seconds.
(a) $D(x, t) = (0.25 \text{ m}) \sin(x - t)$
(b) $D(x, t) = (1.50 \text{ m}) \sin(2x - t)$
(c) $D(x, t) = (1.25 \text{ m}) \sin(x - 2t)$
(d) $D(x, t) = (0.75 \text{ m}) \sin\left(\dfrac{x}{2} - t\right)$

40. ✱   Wave functions for five waves are given below. Here, x is in metres, and t is in seconds.
(a) $D(x, t) = (0.3 \text{ m}) \sin(x - 3t)$
(b) $D(x, t) = (0.5 \text{ m}) \sin(2x - 6t)$
(c) $D(x, t) = (0.6 \text{ m}) \sin(2x - 3t)$
(d) $D(x, t) = (0.1 \text{ m}) \sin(4x - 7t + \pi)$
(e) $D(x, t) = (0.5 \text{ m}) \sin(7x - 13t + \pi/2)$
Using equality signs where needed, rank the waves from lowest to highest in terms of
(a) amplitude;
(b) wavelength;
(c) frequency;
(d) wave speed.

41. ✱✱   Write a displacement equation for each wave shown in Figure 14-61. The wavelength of D_2 is 1.0 m. Assume that both waves are moving in the same direction in the same medium.

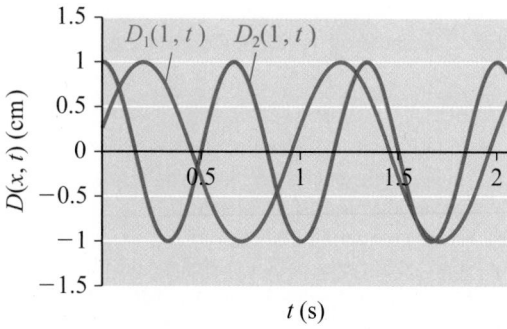

Figure 14-61   Problem 41

42. ✱✱   The displacement function at $x = 2.00$ m for a travelling wave is given by the following equation:

$$D(2.00, t) - (0.75 \text{ m}) \sin(10.0 - 3.50t)$$

where x is in metres, and t is in seconds.
(a) If the phase constant is zero, what are the wavelength, frequency, and speed of the wave?
(b) Write the displacement function for an arbitrary x and t, that is, an equation for $D(x, t)$.

43. ✱✱   The displacement function at $t = 2.00$ s for a travelling wave is given by the following equation:

$$D(x, 2.00) - (0.05 \text{ m}) \sin(10.0x - 15.0)$$

where x is in metres and t is in seconds.
(a) If the phase constant is zero, what are the wavelength, frequency, and speed of the wave?
(b) Write a displacement equation, that is, an equation for $D(x, t)$, for the wave.

44. ✱✱   The displacement function for a travelling wave is given by the following equation:

$$D(x, t) = (0.75 \text{ m}) \sin(10.0x - 3.5t + 0.25)$$

where x is in metres, and t is in seconds.
(a) What are the wavelength, frequency, phase constant, and speed of the wave?
(b) What is the displacement of a point located at $x = 0.10$ m at $t = 2.0$ s?
(c) What is the displacement of a point located at $x = 0.15$ m at $t = 2.0$ s?
(d) At $t = 0.5$ s, what is the phase difference between points located at $x = 1.0$ m and $x = 1.5$ m along the wave?
(e) What is the phase difference between the displacements of a point that is located at $x = 1.0$ m at $t = 1.0$ s and $t = 1.5$ s?
(f) Plot $D(x, 1.0)$ and $D(x, 1.5)$ for the wave. Determine the wave speed from these graphs. Does this speed agree with the speed calculated in part (a)?

45. ✱✱   A wave on a string is described by the displacement equation

$$D(x, t) = (0.02 \text{ m}) \sin\left(\frac{2\pi}{3.0}x + \frac{2\pi}{6.0}t - \frac{\pi}{3}\right)$$

where x is in metres, and t is in seconds.
(a) What are the wavelength, frequency, and speed of the wave?
(b) In which direction is the wave travelling?
(c) Draw a displacement versus time graph for a point located at $x = 0.5$ m for $t = 0$ s to $t = 2.0$ s.
(d) What is the velocity of a segment of the string that is located at $x = 0.5$ m at $t = 2.0$ s?

46. ✱✱✱   A transverse wave on a string is described by the displacement function

$$D(x, t) = (0.20 \text{ m}) \sin(\pi x + 2\pi t)$$

where x is in metres, and t is in seconds.
(a) What is the wave speed, and in which direction is the wave travelling?
(b) What is the transverse velocity of the segment at $x = 0.5$ m at $t = 3.0$ s?
(c) What is the transverse acceleration of the segment at $x = 0.5$ m at $t = 3.0$ s?

Section 14-6 Position Plots and Time Plots

47. ✳ A position plot of a 5.0 Hz wave, moving toward the right, is shown in Figure 14-62.
(a) What is the wavelength of the wave?
(b) What is the speed of the wave?
(c) What is the phase constant of the wave?
(d) Write a displacement equation for the wave.

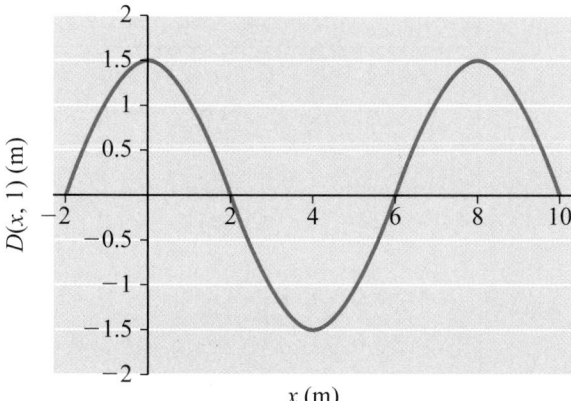

Figure 14-62 Problem 47

48. ✳ The time plot for a wave of wavelength 0.50 m, moving toward the right, is shown at $x = 1.2$ m in Figure 14-63.
(a) What is the angular frequency of the wave?
(b) What is the speed of the wave?
(c) What is the phase constant of the wave?
(d) Write a displacement equation for the wave.

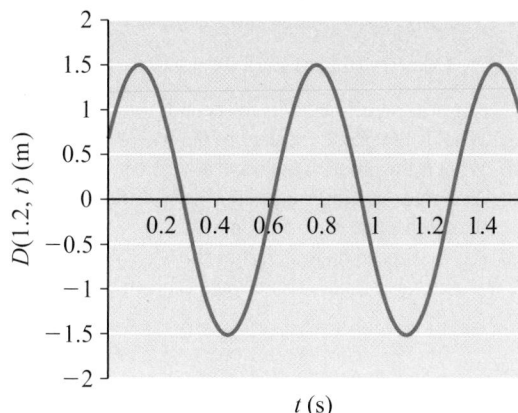

Figure 14-63 Problem 52

49. ✳ A time plot of a wave of at $x = 2.5$ m, moving toward the right with a speed of 4.0 m/s, is shown in Figure 14-64.
(a) What is the frequency of the wave?
(b) What is wavelength of the wave?
(c) What is the phase constant of the wave?
(d) Write a displacement equation for the wave.

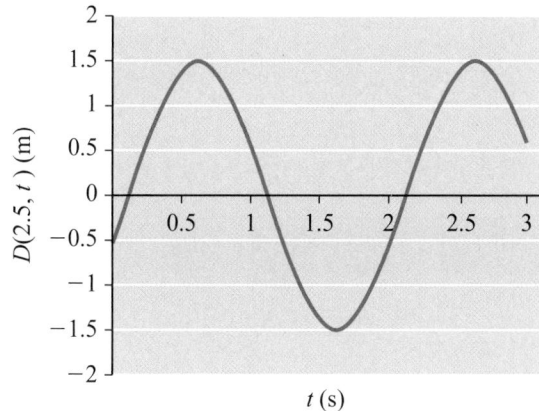

Figure 14-64 Problem 49

50. ✳✳ The position plots for two 50 Hz waves at $t = 1.0$ s are shown in Figure 14-65.
(a) What is the phase difference between the waves if the waves are moving toward the right?
(b) What is the phase difference between the waves if the waves are moving toward the left?

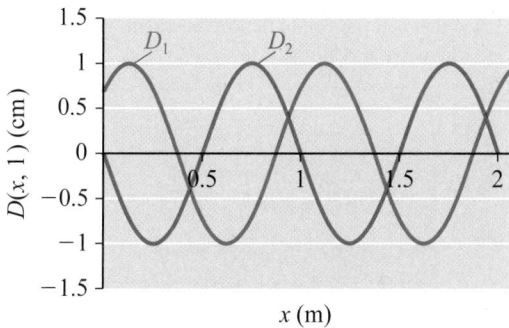

Figure 14-65 Problem 50

51. ✳✳ Figure 14-66 shows a position plot at $t = 1.0$ s and a time plot at $x = 1.2$ m for a travelling wave.
(a) Write a general displacement equation for the travelling wave.
(b) Write an equation for a travelling wave that, when combined with the wave in part (a), produces a standing wave.

(a)

(b)

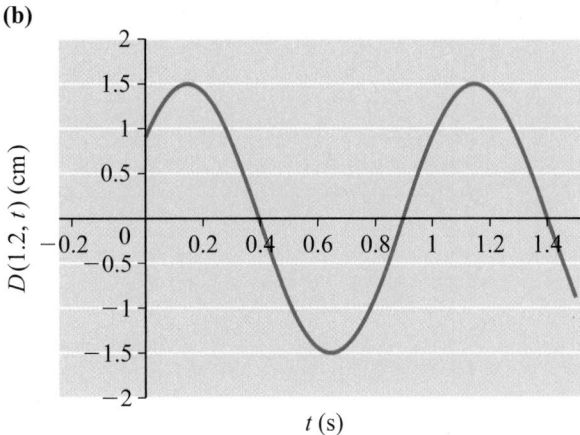

Figure 14-66 Problem 51

52. ✷✷ A one-dimensional wave is travelling along the x-axis in the positive direction. The graph in Figure 14-67(a) shows the displacement versus position for the wave at t = 1.0 s. The graph in Figure 14-67(b) shows the displacement versus time graph of the wave at x = 1.0 m.
   (a) Use the graphs to determine the displacement equation, that is, $D(x, t)$, for the wave.
   (b) What is the speed of the wave, and in which direction is the wave travelling?

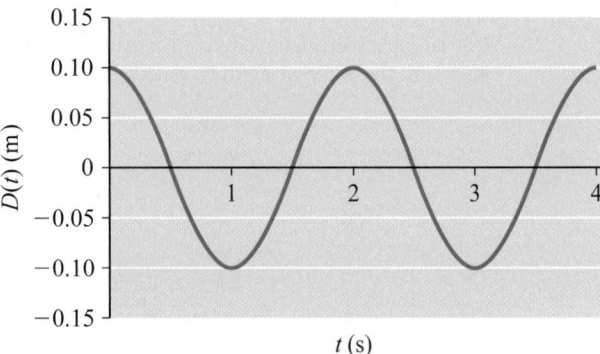

Figure 14-67 Problem 52

53. ✷✷ A one-dimensional wave is travelling along the x-axis in the positive direction. The graph in Figure 14-68(a) shows the displacement versus position for the wave at t = 1.0 s. The graph in Figure 14-68(b) shows the displacement versus time graph of the wave at x = 1.0 m.

(a) Use the graphs to determine the displacement equation, that is, $D(x, t)$, for the wave.
(b) What is the speed of the wave?

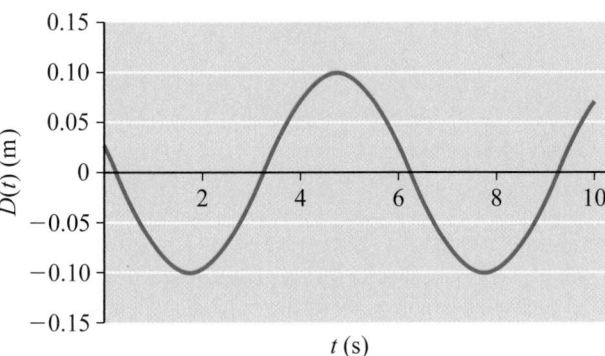

Figure 14-68 Problem 53

Section 14-7 Phase and Phase Difference

54. ✷ The wave function of a wave is given by the equation

$$D(x, t) = (0.2 \text{ m}) \sin(2.0x - 4.0t + \pi)$$

where x is in metres, and t is in seconds.
   (a) What is the phase constant of the wave?
   (b) What is the phase of the wave at t = 1.0 s and x = 0.5 m ?
   (c) At a given instant, what is the phase difference between two points that are 0.5 m apart?
   (d) At what speed does a crest of the wave move?

55. ✷✷ The wave function of a wave is given by the equation

$$D(x, t) = (0.1 \text{ m}) \sin(4.0x - 0.5t + 0.2)$$

where x is in metres, and t is in seconds.
   (a) What is the phase constant of the wave?
   (b) What is the phase of a point located at x = 0.5 m at t = 2.0 s?
   (c) Does the phase of the point x = 0.5 m change with time? Explain your reasoning.
   (d) At t = 1.0 s, what is the phase difference between two points that are 0.1 m apart?
   (e) Is the phase difference between these points the same at t = 2.0 s?
   (f) What is the phase difference between two points that are one wavelength apart?

56. ✷✷ The wave function of a wave at $x = 2.0$ m is given by the equation

$$D(2.0 \text{ m}, t) = (0.3 \text{ m}) \sin(12.0 + 5.0t)$$

where x is in metres, and t is in seconds.
(a) What is the phase of the point at $t = 1.0$ s?
(b) By how much does the phase at this location change between $t = 1.0$ s and $t = 2.0$ s?
(c) Is the phase constant of the given wave 0 rad? Explain your reasoning.

Section 14-8 Energy and Power in a Travelling Wave

57. ✷ A string of linear mass density 100.0 g/m is under 200.0 N of tension. A sinusoidal wave of frequency 10.0 Hz and amplitude 1.0 cm is propagating along the string. What is the average power carried by the wave?

58. ✷ A sinusoidal wave on a string is described by the equation

$$D(x, t) - (0.10 \text{ m}) \sin(4\pi x - 200\pi t)$$

The linear mass density of the string is 10.0 g/m. Determine the direction of motion and the average power transmitted by the wave.

Sections 14-9 and 14-10 Superposition of Waves and Interference of Waves Travelling in the Same Direction

59. ✷ Figure 14-69 shows position versus displacement plots, at a common time, of two harmonic waves that are travelling in the same medium in the same direction.
(a) Plot the resultant wave.
(b) What are the amplitude and the wavelength of the resultant wave?
(c) Is the speed of the resultant wave the same as the speed of the component waves? Explain your reasoning.
(d) Is the resultant wave a standing wave or a travelling wave?

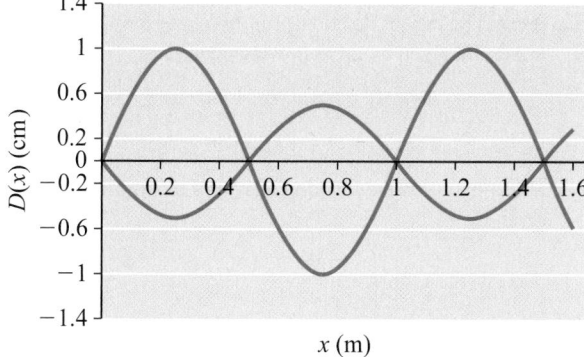

Figure 14-69 Problem 59

60. ✷ Figure 14-70 shows displacement versus position plots for two harmonic waves. Draw the resultant wave obtained by adding these two waves.

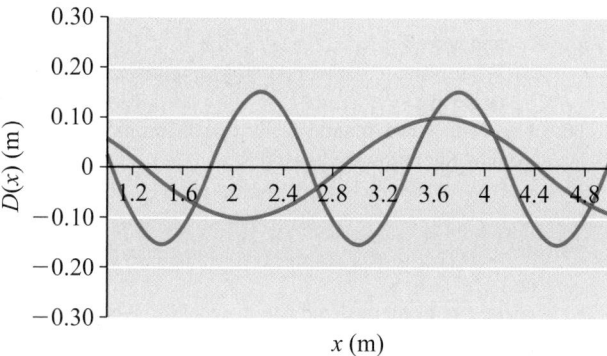

Figure 14-70 Problem 60

61. Figure 14-71 shows plots of two waves at time $t = 0$. Plot the resultant wave on the same graph, and determine the phase constant.

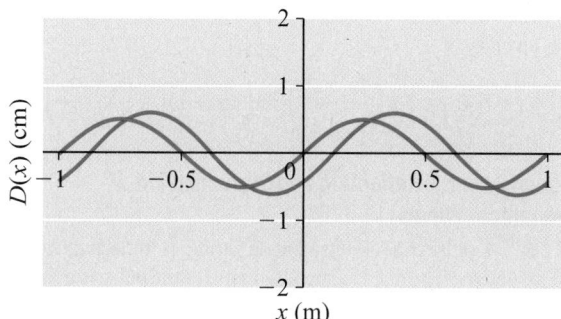

Figure 14-71 Problem 61

62. ✷ Two waves with the same amplitude and frequency arrive at a point P. The waves are $\pi/2$ rad out of phase at this point. What is the resulting amplitude of the waves at P?

63. ✷✷ Two harmonic waves with the same amplitude, wavelength, and frequency are travelling in the same direction. The amplitude of the resultant wave is half the amplitude of the component waves. Find the difference between the phase constants of the two component waves, and the difference between the phase constant of the resultant wave and that of either component wave.

64. ✷ The phase constants of two harmonic waves of same frequency and wavelength, travelling in the same direction, differ by π radians. The amplitude of the waves are A_1 and A_2. Determine the amplitude of the resultant wave.

$$D_1(x, t) = A_1 \sin(kx - \omega t)$$

$$D_2(x, t) = A_2 \sin(kx - \omega t + \pi)$$

Using $\sin(\theta + \pi) = -\sin\theta$, we can write

$$D_2(x, t) = -A_2 \sin(kx - \omega t)$$

The wave function for the resultant wave is,

$$D(x, t) = D_1(x, t) + D_2(x, t) = A_1 \sin(kx - \omega t) - A_2 \sin(kx - \omega t)$$

$$= (A_1 - A_2) \sin(kx - \omega t)$$

Therefore the amplitude of the resultant wave is $A_1 - A_2$

65. ✷ Two waves of amplitudes A_1 and A_2 are simultaneously present in the same medium.

(a) What is the minimum possible amplitude of the resultant wave?

(b) What is the maximum possible amplitude of the resultant wave?

66. ✹✹ The wave functions for two harmonic waves are given by the following equations:

$$D_1(x, t) = (0.2 \text{ m}) \sin(2.0x - 3.0t)$$

$$D_2(x, t) = (0.2 \text{ m}) \sin(2.0x - 3.0t + \pi/4)$$

where x is in metres, and t is in seconds. Write the wave function of the resultant wave when the two waves interfere. What is the amplitude of the resultant wave?

67. ✹✹ The wave functions for two harmonic waves that interfere with each other are given by the following equations:

$$D_1(x, t) = A \sin(kx - \omega t + \phi_1)$$

$$D_2(x, t) = A \sin(kx - \omega t + \phi_2)$$

Show that the wave function $D(x, t)$ of the resultant wave is given by

$$D(x, t) = 2A \, \cos\!\left(\frac{\phi_1 - \phi_2}{2}\right) \sin\!\left(kx - \omega t + \frac{\phi_1 + \phi_2}{2}\right)$$

Section 14-11 Reflection and Transmission of Mechanical Waves

68. ✹ A pulse travelling along a string is reflected from a boundary. The displacement function of the pulse is given by

$$D(x, t) = \frac{1}{(x - 3t)^2 + 4}$$

(a) Write the displacement function of the reflected pulse for a hard reflection.

(b) Write the displacement function of the reflected pulse for a soft reflection.

69. ✹ A harmonic wave travelling along a string is reflected from a boundary. The displacement function of the wave is given by

$$D(x, t) = 0.2 \sin(3x + 4t)$$

(a) Write the displacement function of the reflected wave for a hard reflection.

(b) Write the displacement function of the reflected wave for a soft reflection.

Section 14-12 Standing Waves

70. ✹ A travelling wave is described by the following equation:

$$D(x, t) = (0.5\text{m}) \sin(x - 4t)$$

where x is in metres, and t is in seconds. Which of the following wave(s), when added to the above wave, will result in a standing wave?

(a) $D(x, t) = (-0.5 \text{ m}) \sin(x - 4t)$

(b) $D(x, t) = (0.5 \text{ m}) \sin(x - 4t)$

(c) $D(x, t) = (-0.5 \text{ m}) \sin(x) \cos(4t)$

(d) $D(x, t) = (0.5 \text{ m}) \sin(x + 4t)$

71. ✹ A string oscillates in a standing wave pattern given by the equation (x is in meters and t in seconds)

$$D(x, t) = (1.50 \text{ cm}) \sin((0.6)x) \cos((20\pi)t)$$

(a) Determine the wavelength and frequency of the standing wave.

(b) What is the distance between two consecutive nodes, and between a node and an adjacent antinode?

(c) Write equations for two travelling waves that produce the above standing wave when superimposed.

(d) What is the displacement of a particle located at $x = 0.20$ m when $t = 3.0$ s?

72. ✹ The equation for a sinusoidal wave is given by

$$D(x, t) = (0.50 \text{ cm}) \sin\!\left(2.0x - 20.0t + \frac{\pi}{4}\right)$$

where x is in metres, and t is in seconds.

(a) Write the equation of a wave that produces a standing wave when added to the given wave.

(b) Find the frequency of the resulting standing wave.

(c) What is the amplitude of a point located at an antinode of the standing wave?

(d) Find the distance between two consecutive nodes of the standing wave.

(e) Find the distance between a node and an adjacent antinode in the standing wave.

73. ✹✹ Two sinusoidal waves, simultaneously travelling along a string, are described by the following equations:

$$D_1(x, t) = (1.0 \text{ mm}) \sin(\pi x - 0.5\pi t)$$

$$D_2(x, t) = (1.0 \text{ mm}) \sin(\pi x + 0.5\pi t)$$

where x is in metres, and t is in seconds.

(a) Determine the equation for the resultant wave.

(b) Starting from $x = 0$, what are the values of x corresponding to the first three nodes and the first three antinodes?

(c) What is the distance between two consecutive nodes?

74. ✹✹ Two sinusoidal waves with frequencies of 100 Hz, wavelengths of 0.500 m, amplitudes of 2.0 cm, and phase constants of 0 are moving in opposite directions along a string interfere to produce a standing wave.

(a) Starting from $x = 0$, what are the values of x corresponding to the first three nodes and the first three antinodes of the standing wave?

(b) What is the amplitude of the standing wave at $x = 2.0$ m?

75. ✹✹ The wave functions for two interfering harmonic waves are given by the following equations:

$$D_1(x, t) = A \sin(kx - \omega t + \phi_1)$$

$$D_2(x, t) = A \sin(kx + \omega t + \phi_2)$$

Show that the wave function $D(x, t)$ of the resultant wave is given by

$$D(x, t) = 2A \, \sin\!\left(kx + \frac{\phi_1 + \phi_2}{2}\right) \cos\!\left(\omega t - \frac{\phi_1 - \phi_2}{2}\right)$$

Section 14-13 Standing Waves on Strings

76. ✹ A stretched string of length L that is fixed at both ends is vibrating in its third harmonic. How far from the end of the string can the blade of a screwdriver be placed against the string and not disturb the amplitude of the vibrations?

77. ✹ A string fixed at both ends is 0.500 m long and has a tension that causes the frequency of the first harmonic (fundamental frequency) to be 260.0 Hz. The tension is increased by 4%. What is the new fundamental frequency of the string?

78. ✳✳ A 1.0 m-long string is fixed at both ends.
   (a) What is the longest wavelength standing wave that can exist on this string?
   (b) The wave speed on the string is 300.0 m/s. What is the frequency of the longest standing wave?
   (c) How should the length of the string be changed to increase the lowest frequency of the string by 5% while keeping the tension the same?

79. ✳✳ A 2.0 m-long string is clamped at both ends and is vibrating so that there are two nodes along the length of the string, in addition to the nodes at the fixed ends.
   (a) What is the longest wavelength standing wave possible on this string?
   (b) When the wave speed is 200.0 m/s, what frequency corresponds to the longest wavelength?
   (c) Can a standing wave of frequency 1.5 times the frequency calculated in part (b) be generated on this string without changing its length or tension?
   (d) Determine the frequencies of the second and fourth harmonics, and sketch the corresponding vibrational pattern of the string.

80. ✳✳ A string is clamped at both ends. The string vibrates at 440 Hz and 660 Hz with no other vibrational frequencies in between. What is the lowest frequency at which this string can vibrate in a standing wave pattern?

81. ✳✳ A string of length 100.0 cm is fixed at both ends and is kept under a tension of 25.00 N. The linear mass density of the string is 0.650 g/m.
   (a) What is the lowest resonant frequency of the string?
   (b) What are the frequencies of the second and third harmonics?
   (c) What is the wave speed for this string?
   (d) What is the lowest resonant frequency when the tension is increased to 35.00 N?

82. ✳✳ One end of a string with a linear mass density of 7.60×10^{-4} kg/m is connected to an oscillator with a frequency of 50.0 Hz. The other end is connected to a hanging variable mass, as shown in Figure 14-72. The string passes over a pulley, and the string between the oscillator and the pulley can vibrate freely.
   (a) The length of the vibrating section of the string is 0.500 m. What mass must be attached so that the string vibrates with (i) one antinode between the pulley and the oscillator and (ii) three antinodes between the pulley and the oscillator?
   (b) What is the oscillation frequency of the string in each case in part (a)?

Figure 14-72   Problem 82

See the text online resources at
www.physics1e.nelson.com for Open Problems
and Data-Rich Problems related to this chapter.

Learning Objectives

When you have completed this chapter you should be able to:

1 Describe the physical characteristics of a sound wave.

2 State and use Huygens' principle.

3 Apply Huygens' principle to describe the reflection and refraction of sound waves.

4 Discuss the conditions necessary for standing waves.

5 Sketch a spatial interference pattern.

6 Calculate the sound intensity level from the intensity of a sound, and vice versa.

7 Perform calculations using the Doppler effect equations.

Chapter 15
Interference and Sound

Rock bands use stacks of powerful amplifiers to amplify their music for concerts in large venues (Figure 15-1). Such sound systems typically have a total power of approximately 50 kW. How much of this power reaches your ears when you attend a rock concert? Can such power levels be harmful? In 2009, Ottawa bylaw officers measured a sound intensity level of 136 dB during an outdoor performance by KISS. The officers asked the band to turn the volume down. What is a dB? What are safe sound intensity levels? We will explore the answers to these questions in this chapter.

Figure 15-1   Alberta rock band Nickelback on stage

For additional Making Connections, Examples, and Checkpoints, as well as activities and experiments to help increase your understanding of the chapter's concepts, please go to the text's online resources at www.physics1e.nelson.com.

15-1 Sound Waves

In Chapter 14, we restricted our discussion of wave motion to one dimension. In this chapter, we will examine the properties of sound waves. Sound waves can propagate in three dimensions, and we will devote some attention to those features. We will also consider their propagation in one and two dimensions. Sound waves are a particular type of mechanical wave, so the way they behave is characteristic of many other types of waves. For example, the principles of wave interference described in Section 14-10 apply to all types of waves, including sound waves.

Sound Waves Are Longitudinal Waves

Recall from Chapter 14 that sound is a longitudinal wave: As a sound wave passes through a medium, the molecules of the medium oscillate, alternately creating regions of higher and lower pressure. The molecules are displaced in directions that are either parallel or antiparallel to the direction of propagation of the wave. Figure 15-2 illustrates the direction of propagation, as well as the alternate compressions (darker regions) and rarefactions (lighter regions) in a longitudinal wave.

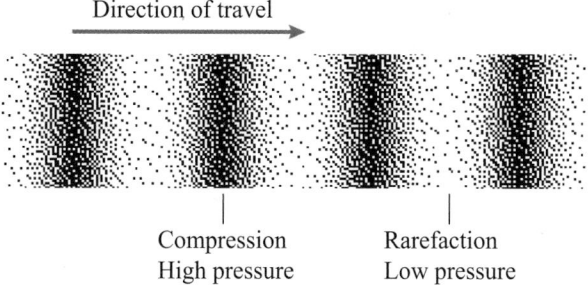

Figure 15-2 Particle positions at an instant of time in a longitudinal wave

Accompanying these particle displacements are pressure variations. In regions in which the medium is compressed, the pressure is elevated above the normal or ambient pressure. Regions in which the medium is "stretched" (rarefaction) correspond to a lowering of the pressure. These pressure variations are illustrated in Figure 15-3. Note that pressure variations are scalar quantities and are not to be confused with the particle displacements, which are either parallel or antiparallel to the direction of propagation.

The Speed of Sound

The speed of a sound wave depends on the properties of the medium through which it is propagating. As you learned in Chapter 14, the speed of waves on strings depends on

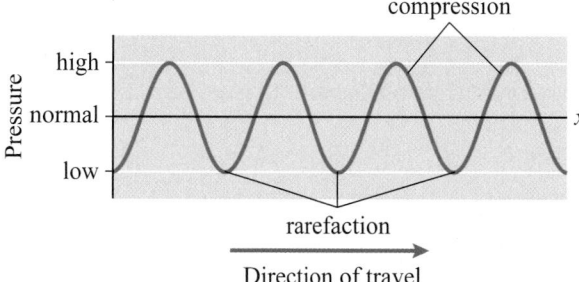

Figure 15-3 Pressure variations in the sound wave illustrated in Figure 15-2.

the linear mass density of the string, μ, and the tension in the string, T_s:

$$v = \sqrt{\frac{T_s}{\mu}} \qquad (14\text{-}5)$$

Note that Equation (14-5) uses a measure of the stiffness of the medium (T_s) and a measure of how much mass is oscillating (μ). The appropriate measure of the string's mass is the linear mass density. For air, and any other medium through which sound can pass, we have to use the density, ρ. What quantity can we use to describe how "stiff" air is?

We know that a spring constant, a measure of the stiffness of a spring, is defined in terms of how much the length of the spring changes when we exert a force on it. For a three-dimensional material, we instead measure by what fraction the volume changes when we change the pressure exerted on the material. The ratio of the change in pressure, Δp, divided by the fractional change in the volume, $\frac{\Delta V}{V}$, is called the **bulk modulus**, B:

$$B = -\frac{\Delta p}{\frac{\Delta V}{V}} = -V\frac{\Delta p}{\Delta V} \qquad (15\text{-}1)$$

The negative sign is included in the definition because the sign of the fractional change in volume is always the opposite of the sign of the pressure change. Paralleling Equation (14-5), the speed of sound is given by

$$v = \sqrt{\frac{B}{\rho}} \qquad (15\text{-}2)$$

This result is quite general and is not restricted to air, or for that matter, even gases. Table 15-1 gives the speed of sound, the bulk modulus, and the density for a variety of media. Note the general trend of increasing sound speed as we move from gases to solids, even though solids tend to be denser than gases. The differences in bulk modulus more than offset the effects of the differences in density.

Table 15-1 Bulk Modulus, Density, and Speed of Sound for Various Media

Medium	Speed of sound (m/s)	Bulk modulus (Pa)	Density (kg/m³)
Air (20°C) (gas)	343	1.01×10^5	1.2
Hydrogen (gas)	1 284	1.51×10^5	0.089
Water (20°C) (liquid)	1 482	2.2×10^9	1 000
Mercury (liquid)	1 450	2.85×10^{10}	13 534
Gold (solid)	3 054	1.8×10^{11}	19 300
Diamond (solid)	11 200	4.42×10^{11}	3 520

As shown in Chapter 14, the relationship between the speed of a sound wave, v, the frequency, f, and the wavelength, λ, is

$$v = f\lambda = \frac{\omega}{k} \qquad (14\text{-}11)$$

EXAMPLE 15-1

Equation for Longitudinal Waves

Write a general expression for the wave function of a longitudinal wave that travels in the z-direction.

SOLUTION

We will still use the variable s to denote the wave function. However, the wave function will depend on the position, z, and the time, t:

$$s(z, t) = s_{\mathrm{m}}\cos(kz - \omega t + \phi)$$

MAKING CONNECTIONS

Wavelength-Dependent Wave Speeds

From Equation (15-2), we can see that the speed of sound in air does not depend on wavelength. However, some types of waves have speeds that are wavelength-dependent. For example, the speed of light in glass varies with the wavelength of the light, which can be demonstrated by the separation of colours of light by a prism. When the wave speed does have wavelength dependence, we say that there is "dispersion" in the medium.

If the speed of sound in air were wavelength-dependent, different musical notes would arrive at slightly different times, and the farther you were from the sound source, the more pronounced this effect would be.

Relationships Between Displacement, Pressure, and Intensity

In addition to having a wave function that describes the displacements of the particles from their equilibrium positions, a sound wave also has pressure variations associated with these particle displacements. The gas that a wave travels through has an ambient pressure. The pressure variations of the wave correspond to increases and decreases in pressure from the ambient pressure. The pressure variations have the same periodicity, in both time and space, as the displacement variations. For a sound wave with a single frequency (a pure tone), the pressure varies sinusoidally. However, the phase relationship between the pressure variations and the displacement is subtle.

We will make our argument in one dimension for clarity and simplicity. One might at first think that the pressure variation is greatest where the displacement is greatest, that is, the two variations are in phase. They are not. If we consider a region of high pressure, it is a region that must have had particles pushed into it from the right and the left. On the left side of a high pressure region, the displacement is positive, while on the right side, it is negative. It must be the case then that the displacement goes through zero right at the point of maximum pressure. A similar argument can be made for the low pressure regions and the point of low pressure is also then a point of zero displacement.

We begin by considering an element of a medium located at the point x, with width Δx and cross-sectional area A. The displacement of this element is in the range $0 < s < s_{\mathrm{m}}$. The volume of medium in the element will

Mathematical Description of the Displacement Amplitude

We can write a mathematical description of a travelling sinusoidal sound wave in one dimension. We will assume that the wave is travelling in the positive x-direction; therefore, the displacement in sound waves is longitudinal in the x-direction. Since we are already using x to denote position, we use s to denote the wave function:

KEY EQUATION $s(x, t) = s_{\mathrm{m}}\cos(kx - \omega t + \phi) \qquad (15\text{-}3)$

In Equation (15-3), s_{m} is the maximum displacement from equilibrium, $k = \dfrac{2\pi}{\lambda}$ is the wave number, $\omega = 2\pi f = \dfrac{2\pi}{T}$ is the angular frequency, and ϕ is the phase constant.

change with x because the displacement of the right side of the element will generally be different from the displacement of the left side. The change in displacement is given by

$$\Delta s = \frac{\partial s}{\partial x} \Delta x \qquad (15\text{-}4)$$

We have used the partial derivative because, in general, s depends on both x and t. To find the change in volume, we multiply Δs by the cross-sectional area A:

$$\Delta V = A \Delta s = A \frac{\partial s}{\partial x} \Delta x \qquad (15\text{-}5)$$

However, the volume of the element is $V = A \Delta x$, to first order, so Equation (15-5) can be rearranged to give

$$\frac{\Delta V}{A \Delta x} = \frac{\Delta V}{V} = \frac{\partial s}{\partial x} \qquad (15\text{-}6)$$

Since we are interested in how the pressure changes, we use the bulk modulus to make the connection between the volume change and pressure change. Solving Equation (15-1) for $\Delta V / V$ and substituting into Equation (15-6), we obtain

$$\frac{\Delta V}{V} = -\frac{\Delta p}{B} = \frac{\partial s}{\partial x} \qquad (15\text{-}7)$$

and

$$\Delta p = -B \frac{\partial s}{\partial x} \qquad (15\text{-}8)$$

We can use Equation (15-3) to calculate the partial derivative of s with respect to x:

$$\frac{\partial s}{\partial x} = \frac{\partial}{\partial x} s_{\mathrm{m}} \cos(kx - \omega t + \phi) = -k s_{\mathrm{m}} \sin(kx - \omega t + \phi)$$

$$(15\text{-}9)$$

Finally, we substitute this result into Equation (15-8) to obtain

KEY EQUATION $\qquad \Delta p = B k s_{\mathrm{m}} \sin(kx - \omega t + \phi) \qquad (15\text{-}10)$

We can identify the amplitude of the pressure variations as the coefficient of the sine factor in Equation (15-10):

$$\Delta p_{\mathrm{m}} = B k s_{\mathrm{m}} \qquad (15\text{-}11)$$

Comparing Equations (15-3) and (15-10), we see that although the wave function has a cosine function and the pressure is a sine function, the arguments are the same in both cases. Thus, they have the same wavelength, period, and wave speed but are $\pi/2$ rad out of phase with each other. These relationships are plotted in Figure 15-4 for a 1000 Hz sound wave with an amplitude that is at the threshold of human hearing—if this sound wave were any quieter, we would not be able to hear it. It is quite incredible that our ear is able to detect displacements on the order of 10^{-11} m, less than the diameter of a hydrogen atom. If our ears were slightly more sensitive, we would hear a constant background noise due to the thermal motion of the air molecules. Our ears have evolved to be as sensitive as possible without subjecting us to constant background noise.

Figure 15-4 Displacement and pressure variations for a 1000 Hz sound wave in air

 EXAMPLE 15-2

Wave Amplitude

Calculate the displacement amplitude of a 1000 Hz sound wave whose pressure amplitude is 20.0 μPa.

SOLUTION

We rearrange Equation (15-11):

$$s_{\mathrm{m}} = \frac{\Delta p}{Bk} = \frac{\Delta p}{B \dfrac{2\pi f}{v}} = \frac{0.000\,020\,0 \text{ Pa}}{1.01 \times 10^5 \text{ Pa} \dfrac{2\pi(1000 \text{ s}^{-1})}{343 \text{ m} \cdot \text{s}^{-1}}}$$

$$= 1.08 \times 10^{-11} \text{ m}$$

☑ CHECKPOINT

C-15-1 Sound Wave Amplitudes

The displacement amplitude and pressure amplitude for a sound wave are
(a) in phase;
(b) out of phase by $\pi/2$ rad;
(c) out of phase by π rad;
(d) none of the above.

C-15-1 (b)

We now examine the energy that a sound wave delivers. We define the **intensity**, I, of a wave as the power delivered per unit area:

$$I = \frac{P}{A} \qquad (15\text{-}12)$$

where P is the rate at which the wave delivers energy, and A is the area that the wave is impinging upon.

As was shown in Chapter 14, the power of a mechanical wave in one dimension is

$$P_{\text{avg}} = \frac{1}{2} \mu v \omega^2 A^2 \qquad (14\text{-}32)$$

where μ is the linear mass density, v is the wave speed, ω is the angular frequency, and A is the amplitude of the wave.

For a sound wave, we replace the linear mass density, which has units of kg/m, with the mass density ρ, which has units of kg/m^3. This substitution gives a quantity with units of W/m^2. As described above, we denote the amplitude of a sound wave as s_{m}. Thus, the intensity of a sound wave is given by

$$I = \frac{1}{2} \rho v \omega^2 s_{\text{m}}^2 \qquad (15\text{-}13)$$

where ρ is the density of the medium, and v is the wave speed.

EXAMPLE 15-3

Intensity and Power

Determine the intensity of the 1000 Hz sound wave in Example 15-2. Assuming that a human eardrum has a radius of approximately 0.75 cm, determine how much power is delivered to the ear by such a wave.

SOLUTION

We use Equation (15-13) to determine the intensity:

$$I = \frac{1}{2}(1.2\ \text{kg/m}^3)(343\ \text{m/s})(2\pi)^2(1000\ \text{s}^{-1})^2(1.08 \times 10^{-11}\ \text{m})^2$$

$$= 9.48 \times 10^{-13}\ \text{W/m}^2$$

To find the power, we solve Equation (15-12) for P:

$$P = IA = (1.5 \times 10^{-13}\ \text{W/m}^2)(\pi)(0.0075\ \text{m})^2$$

$$= 1.67 \times 10^{-16}\ \text{W}$$

LO 2

15-2 Wave Propagation and Huygens' Principle

Spherical Waves

We begin by considering a single point source of sound. This situation is analogous to dropping a stone in a pool of water. In the water, the wave just travels along the surface

of the water (Figure 15-5(b)). However, the sound wave can propagate in three dimensions and it does so uniformly in all directions (Figure 15-5(a)). Each crest of a sound wave forms a **spherical wave** front centred on the point source.

(a)

(b)

Figure 15-5 (a) A spherical wave propagates in three dimensions from a point source. (b) The waves created by dropping a stone into a pond propagate uniformly on the water's surface.

The mathematical description of such a wave is much more complex that the description given in Equation (15-3) for a one-dimensional wave. Rather than being a function of just x and t, a spherical wave is a function of x, y, z, and t because the wave propagates in three dimensions. In addition, the amplitude of a spherical wave decreases as it moves outward from the source. We will consider these amplitude changes in more detail in Section 15-6. Christiaan Huygens, a seventeenth-century Dutch scientist, recognized that he could describe the propagation of a wave by considering any point on a wave front to be a source of spherical waves. The resulting wave is then determined by adding all the waves from the point sources. His description is called **Huygens' principle**. Here is a procedure for applying Huygens' principle:

1. Select a set of equally spaced points on a wave front.

2. Draw a set of circles, all of the same radius, each centred at a different point.

3. Construct the resulting wave front by drawing tangents to the circles.

WAVES AND OSCILLATIONS

Figure 15-6 illustrates this procedure for a spherical wave.

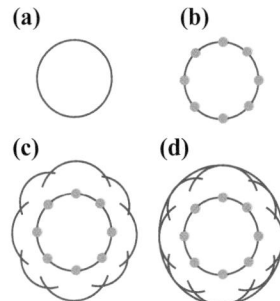

(a)   **(b)**

(c)   **(d)**

Figure 15-6   Applying Huygens' principle. (a) Draw a spherical wave. (b) Draw a set of points along the wave front. (c) Draw a spherical wave from each point. (d) Draw a curve tangent to the individual circular wave fronts.

Plane Waves

We will now consider **plane waves**, which have wave fronts that lie in planes parallel to each other (Figure 15-7). You can create a plane wave in a pond by dropping a long, straight stick horizontally into the pond. You can create a plane sound wave by vibrating a large, flat plate in a direction perpendicular to the surface of the plate. We can use Huygens' principle to show that a plane wave, once created, continues to propagate as a plane wave. In Figure 15-8, notice that the crests and troughs of the wave are parallel to each other.

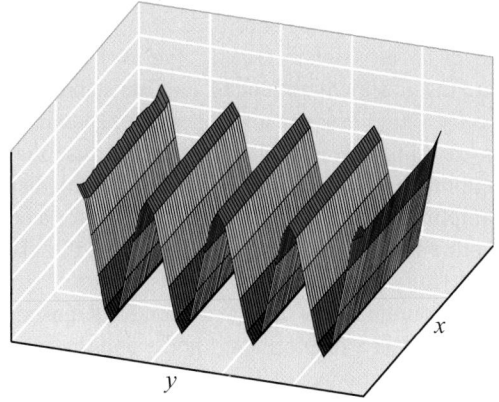

Figure 15-7   A plane wave propagating in the positive x-direction. The wave front extends in the y-direction, and the vertical axis shows the wave displacement amplitude.

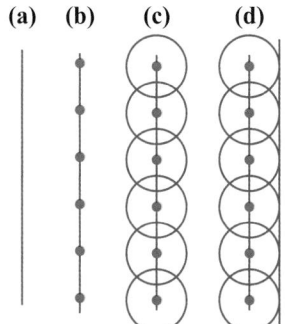

(a)   **(b)**   **(c)**   **(d)**

Figure 15-8   Huygens' principle applied to a plane wave that is travelling from left to right

15-3   Reflection and Refraction

We can use Huygens' principle to help understand what happens to a sound wave when it encounters a boundary between two media where the wave speed changes. At such a boundary, some of the wave can be reflected from the boundary and some transmitted across the boundary into the second medium. The portion of the wave that is reflected travels back into the first medium. As we saw in Chapter 14, the reflected wave may have its phase changed upon reflection. The angle of reflection is the same as the angle of incidence.

The portion of the wave that crosses the boundary into the second medium is called a refracted wave. In general, the direction of propagation of the refracted wave is different from the direction of the incident wave. When the wave speed is slower in the second medium, the wave tends to bend toward the slower medium. When the wave speed is higher in the second medium, the wave tends to bend away from the second medium.

We can use Huygens' principle to understand these effects. Consider a plane wave encountering a planar boundary. To apply Huygens' principle, we place a line of equally spaced dots at the boundary and draw a set of concentric half circles around the dots above and below the boundary, as shown in Figure 15-9. In the figure, the speed of sound in medium 2 is less than the speed of sound in medium 1. We draw only half circles because the speed of the wave is different above and below the boundary. In addition, the circles in the lower half of the figure are drawn closer together because the wave speed is slower in the lower medium.

When we draw our circles, we need to deal with the complication that the wave encounters the boundary at some angle. The incident wave encounters the left-hand points on the boundary before the right-hand points on the boundary. Therefore, the circles we draw around the leftmost points are larger because the wave has had more time to travel from those points. As we move to the right, the circles get successively smaller. Notice that the reflected wave front is perpendicular

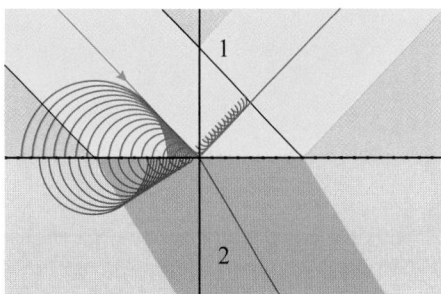

Figure 15-9   Reflection and refraction at a boundary

to the incident wave front. However, the angle of the transmitted wave, measured with respect to the surface normal, is smaller than the incident angle. The wave is bent toward the medium with the lower speed, medium 2. However, if the angle of incidence is 0°, measured with respect to a boundary normal, then both the transmitted wave and the reflected wave leave normal to the boundary.

MAKING CONNECTIONS

Refraction of Water Waves

Have you ever wondered why the orientation of waves reaching a beach is always parallel to the beach? For water waves, the speed of the wave depends on the depth of the water when the depth is comparable to the wavelength. As the depth decreases, so does the wave speed. As we saw in Figure 15-9, waves tend to bend toward the region of slower wave speed. Thus, the water waves tend to bend until they are parallel with the beach, as the image in Figure 15-10 illustrates. Despite the complexity of the coastline, the waves are still coming in parallel to the beach.

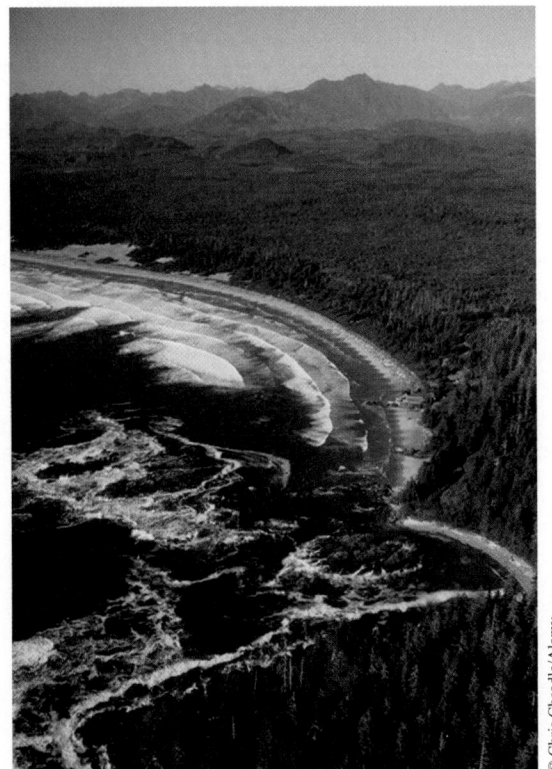

Figure 15-10   Ocean waves coming in parallel to the beach at Pacific Rim National Park in British Columbia

LO 4

15-4 Standing Waves

The formation of standing waves in brass, woodwind and stringed instruments is responsible for producing musical notes. Standing waves are also important in other applications not involving sounds. The resonant cavity of a laser reflects the light waves back and forth and creates standing light waves with a well-defined frequency and wavelength. Piezoelectric crystals used in quartz watches are another example where standing waves are used to create an electrical oscillator with a very precise frequency.

We start by considering a long, narrow column of air, such as might be found in a flute. Sound waves can propagate along the length of the column. When they reach the ends of the column, they are reflected, at least partially, and we have the condition for standing waves that we saw in Chapter 14: waves with the same frequency travelling in opposite directions in the same medium.

Similar to strings, air columns can have open (free) and closed (fixed) ends. The closed end of an air column is a displacement node because the air molecules cannot oscillate across such a fixed boundary. If we have an air column that is closed at both ends, the displacement function must have nodes at both ends. The possible standing waves in such a column are then exactly what we would see for a string with fixed ends.

With an air column that is open at both ends, the ends are displacement antinodes, and the standing waves are exactly what we see in a string with open ends. These results are summarized in Table 15-2. Because the pressure amplitude is $\pi/2$ rad out of phase with the displacement amplitude, the displacement amplitude for a column with both ends open has the same shape as the pressure amplitude for a column with both ends closed.

For a column that is open at one end and closed at the other, one end is a displacement node, and the other is a displacement antinode; see Table 15-3.

Standing waves contribute to the rich variety of tones that musical instruments produce. In Table 15-2, we listed the different types of standing waves that are possible for an air column with both ends open or closed. The rich sound of a musical instrument results from the fact that a string or an air column can simultaneously sustain combinations of the possible allowed standing waves. In fact, it is very difficult to avoid having multiple standing waves. The lowest frequency standing wave is called the fundamental and the higher frequencies are called the harmonics.

Figure 15-11 shows the amplitude as recorded on a computer by blowing into one end of a pipe that is open at the other end. The variations in amplitude are periodic but are clearly not simple sine waves. We can create complex periodic waveforms by adding together sine waves with frequencies that are all integer multiples (harmonics) of a fundamental frequency. For example, the wave in

Text in image: © Chris Cheadle/Alamy

Table 15-2 Standing waves in an air column with either both ends closed or both ends open

Displacement amplitude		Mode	Wavelength	Frequency
Both ends closed	Both ends open			
		$n = 1$	$\lambda_1 = 2L$	$f_1 = \dfrac{v}{2L}$
		$n = 2$	$\lambda_2 = \dfrac{2L}{2}$	$L\,f_2 = \dfrac{2v}{2L}$
		$n = 3$	$\lambda_3 = \dfrac{2L}{3}$	$f_3 = \dfrac{3v}{2L}$
		$n = 4$	$\lambda_4 = \dfrac{2L}{4}$	$f_4 = \dfrac{4v}{2L}$
General case		n	$\lambda_n = \dfrac{2L}{n}$	$f_n = \dfrac{nv}{2L}$

Table 15-3 Displacement amplitudes for a tube that is open at one end and closed at the other end

Displacement amplitude	Mode	Wavelength	Frequency
One end open, one end closed			
	$n = 1$	$\lambda_1 = 4L$	$f_1 = \dfrac{v}{4L}$
	$n = 3$	$\lambda_2 = \dfrac{4L}{3}$	$f_2 = \dfrac{3v}{4L}$
	$n = 5$	$\lambda_3 = \dfrac{4L}{5}$	$f_3 = \dfrac{5v}{4L}$
	$n = 7$	$\lambda_4 = \dfrac{4L}{7}$	$f_4 = \dfrac{7v}{4L}$
General case	only odd n	$\lambda_n = \dfrac{4L}{n}$	$f_n = \dfrac{nv}{4L}$

Figure 15-12 is the sum of sine waves with five different frequencies:

$$s(t) = 0.56\cos(t) + 0.75\cos(3t) + 0.33\cos(5t)$$
$$- 0.16\cos(7t) - 0.05\cos(9t)$$

Notice that the waveform in Figure 15-12 is quite similar to the wave in Figure 15-11. You can also see that all the waves have a frequency that is an integer multiple of the fundamental, $f_0 = \dfrac{1}{2\pi}\ \mathrm{s}^{-1}$. For this particular wave,

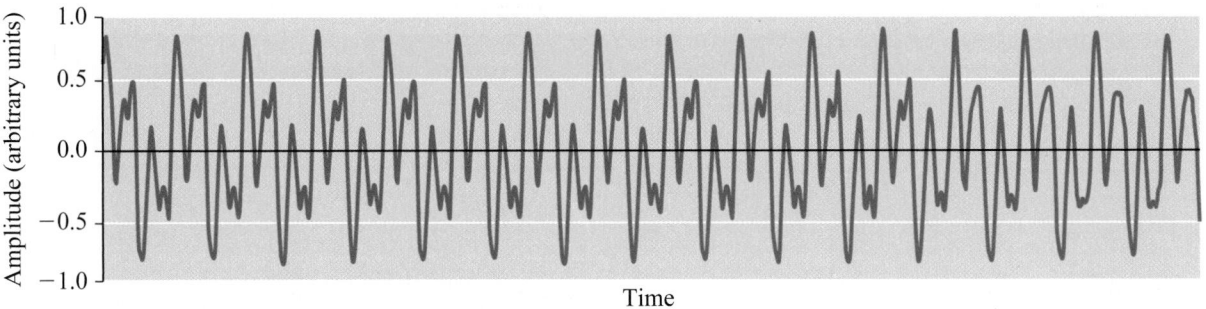

Figure 15-11　Sound wave amplitude for a pipe that was blown into one end, recorded over an interval of approximately 0.1 s

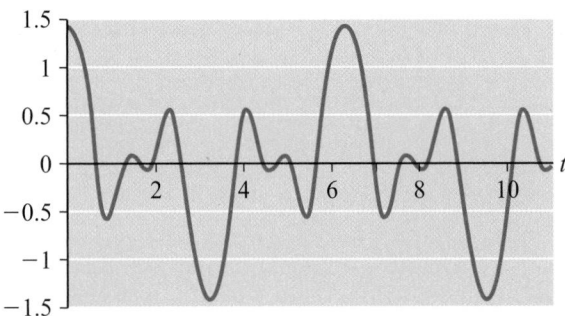

Figure 15-12　A standing wave composed of multiple sine waves with different frequencies and amplitudes

the frequencies are all odd multiples of the fundamental, and some of the terms have a negative sign, which means they are 180° out of phase from a positive term with the same frequency.

Musical instruments produce a sound that is composed of many different frequencies. For instruments that rely on vibrating air columns (e.g., trumpets, flutes, and clarinets), vibrating strings (e.g., violins, pianos, and banjos), and vibrating bars (e.g., xylophones, marimbas, and gamelans), the component frequencies in the sound are integer multiples of the fundamental frequency. However, the sound from instruments based on a vibrating membrane (such as drums) contains multiple frequencies that are not all integer multiples of the fundamental. This is why a drum does not have as distinct a note as, say, a trumpet.

Fourier's Theorem

Figure 15-12 demonstrates that we can generate complicated waveforms by combining sinusoidal waveforms of different wavelengths. In fact, a theorem developed by the French mathematician Jean Baptiste Joseph Fourier (1768–1830) states that *any* periodic waveform can be written as a sum of sine and cosine waves with appropriately chosen frequencies and amplitudes.

According to Fourier's theorem, if a function has a period T, such that $D(t) = D(t + T)$, then

$$D(t) = a_0 + \sum_{n=1}^{\infty}\left[a_n\cos\left(\frac{2\pi nt}{T}\right) + b_n\sin\left(\frac{2\pi nt}{T}\right)\right] \quad (15\text{-}14)$$

Note that the arguments of the sine and cosine functions are all integer multiples of a fundamental frequency that is related to the period by

$$f_0 = \frac{1}{T} \quad (15\text{-}15)$$

In stating the theorem, Fourier also described how to determine the constants a_n and b_n when given $D(t)$. The technique, which involves integral calculus, is beyond the scope of this book.

ONLINE ACTIVITY

Making Waves

The e-resource that accompanies every new copy of this textbook contains an Online Activity using the PhET simulation "Making Waves." Work through the simulation and accompanying questions to gain an understanding of making waves.

LO 5

15-5　Interference

Interference of three-dimensional waves is more complex than interference of one-dimensional waves. In one dimension, waves with the same frequency and wavelength have a fixed phase difference that depends only on the difference between the phase constants of the two waves. The phase difference does not vary with time or position. However, when waves propagate in different directions, their relative phase varies with position.

Figure 15-13 shows two plane waves with the same wavelength and frequency propagating in slightly different directions. We can see that there are points where the two waves are perfectly in phase. At those locations, the amplitudes add, resulting in constructive interference. At the points where the two waves are perfectly out of phase, we observe destructive interference.

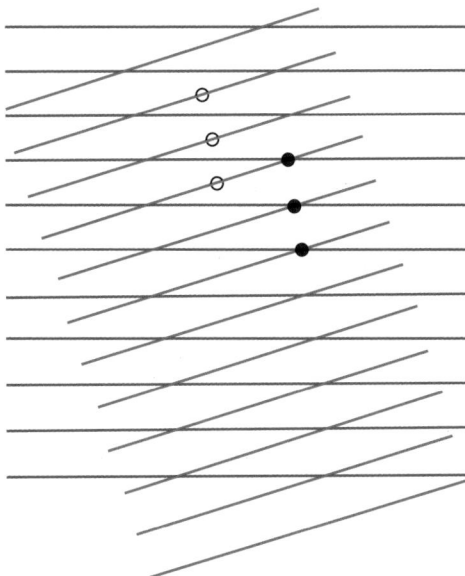

Figure 15-13 Two plane waves with the same wavelength travelling in slightly different directions. The lines represent the crests of the waves, the black circles indicate where the waves are perfectly in phase, and the open circles indicate where the two waves are exactly out of phase

Interference in Space

ONLINE ACTIVITY

Wave Interference

The e-resource that accompanies every new copy of this textbook contains an Online Activity using the PhET simulation "Wave Interference." Work through the simulation and accompanying questions to gain an understanding of interference patterns produced by sound waves.

We will start our exploration of interference in more than one dimension by considering the interference pattern that is produced by two point sources. The two sources are in phase and produce waves with the same frequency and wavelength. Each source produces spherical wave fronts. We can look at the resulting interference pattern in a plane that contains the two sources, as shown in Figure 15-14. This image represents the interference pattern at a particular instant in time.

We can determine mathematically the points where the two waves interfere constructively. At such points, the two waves must be in phase regardless of the elapsed time. Let us consider an instant when both sources are at the peak positive amplitude. If we then locate a point where the waves are interfering constructively, both waves must be at peak positive amplitude at that point. There must be an integer number of wavelengths between each source and the point we are considering. Figure 15-15 shows one such point.

If we consider the instant at which the two sources are at a large negative amplitude, there will still be an integer number of wavelengths between each source and

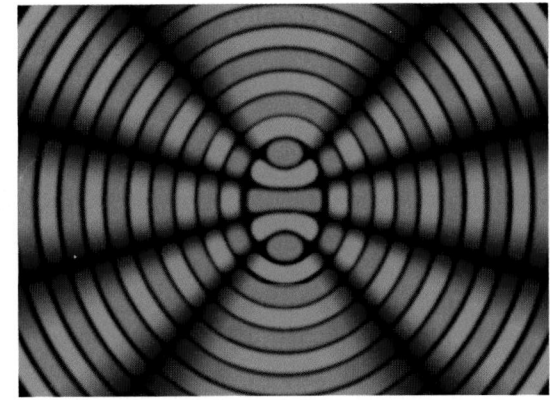

Figure 15-14 Interference from two identical point sources. The red regions represent large positive amplitude, and the green regions represent large negative amplitude.

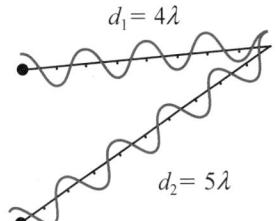

Figure 15-15 The constructive interference of two waves from two point sources

our point; thus, there will still be constructive interference. However, this interference now produces a large negative amplitude. Any point that is an integer multiple of wavelengths from both sources will undergo continuous constructive interference. Defining d_1 as the path length from source 1, and d_2 as the path length from source 2, the condition for constructive interference is

$$d_1 = m\lambda, \quad m = 1, 2, 3, \ldots$$
$$d_2 = n\lambda, \quad n = 1, 2, 3, \ldots \qquad (15\text{-}16)$$

where m and n are integers.

These conditions can be made more general by considering the difference between the distances from the two sources to the point of constructive interference:

$$\Delta d = d_2 - d_1 = (n - m)\lambda = p\lambda,$$
$$p = 0, \pm 1, \pm 2, \pm 3, \ldots \qquad (15\text{-}17)$$

Thus, the condition for constructive interference is that the path difference between the two sources must be an integer multiple of the wavelength.

The approach we have used to arrive at Equation (15-17) demanded that both paths individually be integer multiples of the wavelength. However, that does not necessarily have to be true for Equation (15-17) to be satisfied. Let us see if the path difference condition is sufficient to produce constructive interference.

As a spherical wave travels away from its source, it oscillates in space and time. However, the amplitude is constant over any spherical surface centred on the source. Thus, the spatial variations can be described as a function of r, the distance from the source. We can write the wave function of a spherical wave as

$$s(r, t) = s_m(r)\cos(kr - \omega t + \phi) \qquad (15\text{-}18)$$

This expression looks very similar to the expression for a one-dimensional wave in Equation 15-3. In the argument of the cosine, we have simply replaced x with r. However, the amplitude for a spherical wave depends on r because, unlike a wave in one dimension, the wave front spreads out over a larger area as it propagates outward; consequently, the amplitude decreases as r increases. We will consider this in more detail in Section 15-6.

For two waves to be in phase, the arguments of the cosine function for each wave must differ by an integer multiple of 2π. Since we have assumed that both waves are in phase, we can neglect the phase constant and write

$$(kd_2 - \omega t) - (kd_1 - \omega t) = k(d_2 - d_1) = n2\pi \qquad (15\text{-}19)$$

$$\therefore d_2 - d_1 = n\frac{2\pi}{k} = n\lambda, \quad n = 0,\ \pm1,\ \pm2,\ \pm3$$

From this result, we can see that constructive interference occurs whenever the path difference is an integer multiple of the wavelength. For the special case of $d_2 = d_1 = d$, we can easily add the two waves together to find the resultant wave:

$$\begin{aligned} S(d, t) &= s_m(d)\cos(kd - \omega t + \phi) \\ &\quad + s_m(d)\cos(kd - \omega t + \phi) \\ &= 2s_m(d)\cos(kd - \omega t + \phi) \end{aligned} \qquad (15\text{-}20)$$

We again consider a situation in which both sources are in phase but will turn our attention to determining the

conditions for destructive interference. At an instant when both sources are at a large positive amplitude, we select a point at which the wave from one source is at a large positive amplitude, and the wave from the other source is at a large negative amplitude. One such point is shown in Figure 15-16. At this point, the wave from source 1 has travelled 3 wavelengths, and the wave from source 2 has travelled 3.5 wavelengths. Consequently, the waves are completely out of phase and interfere destructively. The amplitude at this point is quite small.

Destructive interference occurs when one path is an integer number of wavelengths and the other is a half-integer multiple; so, the general condition for destructive interference is that the path difference is a half-integer multiple of the wavelength (i.e., an odd number of half wavelengths):

$$\Delta d = d_2 - d_1 = \frac{(2n + 1)}{2}\lambda = \left(n + \frac{1}{2}\right)\lambda,$$

$$n = 0,\ \pm1,\ \pm2,\ \pm3 \qquad (15\text{-}21)$$

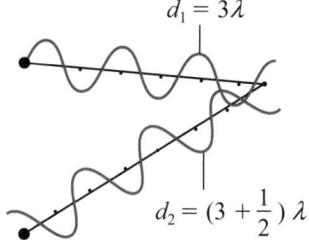

Figure 15-16 The destructive interference of two waves from two point sources

EXAMPLE 15-4

Interference Pattern

Two speakers are placed 2.0 m apart and are emitting sound waves with the same amplitude and phase; the wavelength is 1.0 m. Sketch the interference pattern, and identify points of constructive and destructive interference.

SOLUTION

We make a sketch with concentric circles around the two sources, with each circle having a radius that is one wavelength larger than the previous circle (Figure 15-17).

We can clearly see that there are points where the wave crests overlap—points of constructive interference. However, recognizing that the waves are travelling waves, we should also consider what happens for all times, not just the particular instant shown here.

For all points on the horizontal line that passes midway between the two sources, the path difference is zero.

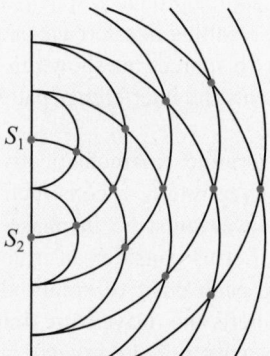

Figure 15-17 Wave fronts for the speakers in Example 15-4

Therefore, that entire line will undergo constructive interference. The other points of constructive interference also appear to lie on lines, and we ask whether these lines are also lines of constructive interference.

Consider the situation in Figure 15-18. To have constructive interference, the path difference must be an integer multiple of the wavelength:

$$d_2 - d_1 = \sqrt{x^2 + (y+1)^2} - \sqrt{x^2 - (y-1)^2} = n\lambda$$

The general solution to this equation is complex. However, for $n\lambda = 1.0$ m, we find that

$$y = \pm \frac{\sqrt{12x^2 + 9}}{6}$$

Figure 15-19 shows the two curves defined by this equation.

We can see that the curves pass through the points of intersection of the circles representing wave fronts 1.0 m apart. Because every point on the two curves satisfies the

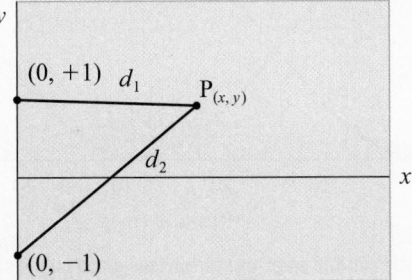

Figure 15-18 The path difference between the two sources and a point (x, y)

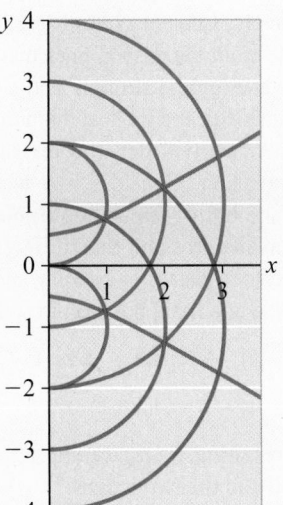

Figure 15-19 The two lines show the solutions corresponding to a path difference of one wavelength.

condition that the path length difference is one wavelength, constructive interference occurs all along the curves. We can infer that there is a curved line of destructive interference somewhere between each of the plotted curves and the x-axis, where the path difference is zero.

 CHECKPOINT

C-15-2 Interference Patterns

Two sources that are in phase produce an interference pattern like the pattern in Figure 15-14. If the two sources are out of phase by π, we will observe

(a) no interference pattern;

(b) the same interference pattern;

(c) an interference pattern similar to Figure 15-14 but with the destructive interference located where the constructive interference was observed, and vice versa;

(d) none of the above.

C-15-2 (c) One source will emit a crest, and $T/2$ s later, the other source will emit a crest. The crest from the first source will have travelled half a wavelength in that time. All the places where the in-phase sources had an integer wavelength path difference will now have a half-integer difference in the path and will be the locations of destructive rather than constructive interference. A similar argument applies to the locations that experienced destructive interference with the sources in phase, and they will be the locations of constructive interference.

Beats

In the patterns described above, regions of constructive and destructive interference are separated by intervals of space. Constructive and destructive interference can also be separated by intervals of time.

When we listen to two waves that have slightly different frequencies, we hear the amplitude increasing and decreasing as a function of time. This type of variation in amplitude is called a **beat**. The rate at which the

amplitude varies is proportional to the frequency difference. However, when the frequency difference is large, we hear the sounds as two distinct tones rather than one tone that varies in intensity.

We can see why this occurs by considering two sources that differ slightly in frequency and imagine we are equidistant from the two. The amplitudes of the two sources and the resultant total amplitude are shown in Figure 15-20.

The sound waves start off interfering constructively. Halfway across the graph, the red wave has completed two full cycles, and the blue wave has only completed one and

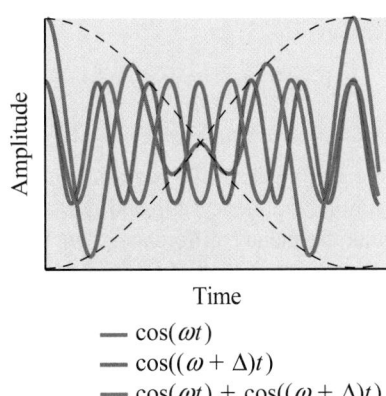

— $\cos(\omega t)$
— $\cos((\omega + \Delta)t)$
— $\cos(\omega t) + \cos((\omega + \Delta)t)$

Figure 15-20 Two waves that differ by Δ Hz (red and blue) and the resultant wave (green). The dashed lines outline the amplitude envelope of the resultant wave.

one-half cycles. Consequently, they are out of phase and interfere destructively. Two red cycles later, they are back in phase. We can see from the dashed lines that the amplitude of the resultant wave (green) strongly and slowly varies in amplitude (is modulated), and we hear this as a beating effect.

For simplicity, we will develop the mathematics of beats in one dimension. Consider two waves that have the same amplitude but different frequencies, ω_1 and ω_2, with different wavelengths and thus different wave numbers. We will set both phases to be zero and consider the resultant amplitude at a fixed position in space, $x = x_o$:

$$s_1(x_o, t) = s_m \cos(k_1 x_o - \omega_1 t)$$
$$s_2(x_o, t) = s_m \cos(k_2 x_o - \omega_2 t)$$

(15-22)

To find the resulting wave, we apply the principle of superposition and add the two waves:

$$
\begin{aligned}
s_{Total}(x_o, t) &= s_1(x_o, t) + s_2(x_o, t) \\
&= s_m \cos(k_1 x_o - \omega_1 t) + s_m \cos(k_2 x_o - \omega_2 t) \\
&= s_m[\cos(k_1 x_o - \omega_1 t) + \cos(k_2 x_o - \omega_2 t)] \\
&= 2 s_m \cos \frac{1}{2}(k_1 x_o - \omega_1 t + k_2 x_o - \omega_2 t) \\
&\quad \times \cos \frac{1}{2}(k_1 x_o - \omega_1 t - k_2 x_o + \omega_2 t) \\
&= 2 s_m \cos\left[\left(\frac{k_1 + k_2}{2}\right)x_o - \left(\frac{\omega_1 + \omega_2}{2}\right)t\right] \\
&\quad \times \cos\left[\left(\frac{k_1 - k_2}{2}\right)x_o - \left(\frac{\omega_1 - \omega_2}{2}\right)t\right]
\end{aligned}
$$

(15-23)

To simplify this expression, we define the following quantities:

$$\varpi = \frac{\omega_1 + \omega_2}{2} \quad \text{Mean angular frequency}$$

$$\Delta\omega = \frac{\omega_1 - \omega_2}{2} \quad \text{Angular frequency difference}$$

(15-24)

With these definitions, and setting $x_0 = 0$, we can rewrite Equation (15-23) as

$$s_{Total}(0, t) = 2 s_m \cos(-\varpi t)\cos(-\Delta\omega t)$$
$$= 2 s_m \cos(\varpi t)\cos(\Delta\omega t)$$

(15-25)

We have a product of two cosines—one oscillates at the mean angular frequency ϖ, and the other oscillates at the angular frequency difference $\Delta\omega$. When the two waves are close to each other in frequency, we hear a tone that has the mean frequency and whose amplitude varies by the difference in angular frequency. Figure 15-21 shows the result of adding waves with frequencies of 244 Hz and 256 Hz. The red line shows the resultant amplitude, and the blue lines show the envelope of the amplitude modulation. From this figure, we can see that the amplitude modulation has a period of approximately

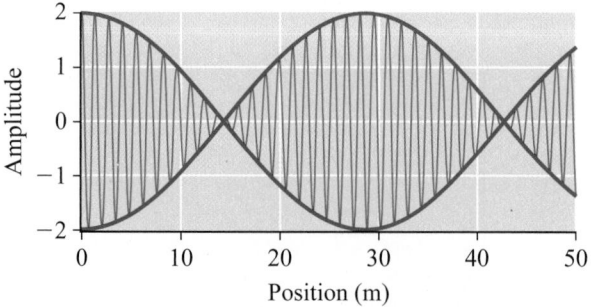

Figure 15-21 (a) The amplitude versus time graph for the sum of a 244 Hz wave and a 256 Hz wave. (b) The amplitude versus position graph for the sum of a 244 Hz wave and a 256 Hz wave. Notice that the spatial modulation of the amplitude has a wavelength of approximately 30 m, which corresponds to a frequency of about 12 Hz.

0.08 s, which corresponds to a frequency of approximately 12 Hz. We hear this modulation as a "beating" with a frequency of 12 Hz. However, when the two waves are not close in frequency, we perceive them as two separate tones (Figure 15-22).

Interestingly, if we listen to two separate tones through headphones, with one frequency in one ear and a different but close frequency in the other, we still hear beats even though the waves do not interfere outside of our head. It is believed that this effect, called binaural beating, results from the way our brain processes signals from our ears rather than from any interference of the sound waves travelling through our skull.

Figure 15-22 Two sound waves with very different frequencies. The black wave has a frequency of 600 Hz, and the blue wave has a frequency of 3000 Hz. In the resultant wave, shown in red, we can clearly see both waves.

CHECKPOINT

C-15-3 Beat Frequency

A 100 Hz tone and a 105 Hz tone are played simultaneously. What is the beat frequency?

(a) 2.5 Hz

(b) 5 Hz

(c) 10 Hz

(d) 205 Hz

C-15-3 (b) Even though the difference frequency is defined as $\Delta\omega = \dfrac{\omega_1 - \omega_2}{2}$, the effect of this is to modulate the amplitude. The amplitude will be large for both the positive and negative crests; therefore, the amplitude modulations have the same frequency as the difference between the two frequencies.

MAKING CONNECTIONS

Using Beat Frequencies

Beats can indicate that an instrument is out of tune. Musicians can tune an instrument by adjusting the frequency of its notes until the beating stops (Figure 15-23). The beating effect is part of the reason why instruments that are badly out of tune sound so disagreeable. However, a Balinese gamelan orchestra uses pairs of instruments that are intentionally slightly out of tune from each other to add a "shimmering" effect to the music (Figure 15-24).

Figure 15-23 Orchestral musician tuning a violin

Figure 15-24 A Balinese gamelan orchestra

EXAMPLE 15-5

Beat Frequency

The advantage of the beat tuning technique described in the Making Connections feature above is readily apparent when you consider that the smallest change in frequency that the human ear can detect for tones that are played one after the other is about 3 Hz for tones around 440 Hz. What beat frequency would you hear if you played a 440 Hz and a 443 Hz tone simultaneously?

SOLUTION

The amplitude of the resultant wave would be modulated at 3 Hz.

Making sense of the result:

A 3 Hz modulation is very easy to detect. In fact, at 440 Hz it is relatively easy to hear beats with frequency differences as low as 1 Hz.

LO 6

15-6 Measuring Sound Levels

As mentioned previously, the quietest sound that a human ear can detect corresponds to an intensity on the order of 10^{-12} W/m². At the other extreme, the threshold of pain at 1000 Hz corresponds to an intensity of about 10 W/m². This intensity is what you would experience if you were standing about 8 m from a jetliner during takeoff. The human ear has a dynamic range of approximately 10^{13}. For such large ranges, it is often convenient to work with a logarithmic scale. In fact, human perception of sound is nonlinear, much like a logarithmic scale.

Decibels

The **decibel** (dB) is a commonly used unit for logarithmic comparisons of power or intensity. Since logarithmic scales are *relative* scales, we must choose a reference level. For sound, we use the quietest intensity level we can hear at 1000 Hz as our reference intensity. We define the zero of our logarithmic sound intensity scale to correspond to this reference level:

The reference sound intensity level of 0 dB corresponds to an intensity $I_0 = 10^{-12}$ W/m².

(15-27)

To express any other **sound intensity level** (β) in decibels, we use

 $\beta(I) = 0 \text{ dB} + 10 \log_{10}\left(\dfrac{I}{I_0}\right)$ (15-28)

The decibel is a convenient unit for β because it is about the smallest change in sound level that a human can typically detect. Table 15-4 lists typical β for some common sources. Since decibels indicate the logarithm of the ratio of two levels, "small" differences in decibel measurements indicate much larger differences in the intensity I.

Table 15-4 Typical sound intensity levels for some common sources

Sound source	Sound intensity level (dB)
Chainsaw, 1 m away	110
Lawn mower, 1 m away	107
Dance club speaker, 1 m away	100
Portable stereo, half-volume	94
Transport truck, 10 m away	90
Side of busy road	80
Vacuum cleaner, 1 m away	70
Normal speech, 1 m away	60
Average home	50
Quiet library	40
Quiet bedroom at night	30
Quiet sound studio	20

 EXAMPLE 15-6

Sound Intensity Levels

The difference between the β for a lawn mower and a chainsaw is 3 dB. By what factor is the sound intensity of a chainsaw greater than the sound intensity of a lawn mower?

SOLUTION

We use Equation (15-29) to write

$$\beta(\text{chainsaw}) - \beta(\text{lawn mower}) = 3 \text{ dB}$$

$$\therefore \log_{10}\left(\frac{I_{\text{chainsaw}}}{I_{\text{lawn mower}}}\right) = 0.3$$

$$I_{\text{chainsaw}} = I_{\text{lawn mower}} 10^{0.3} \approx 2 I_{\text{lawn mower}}$$

So, an increase of 3 dB is actually a doubling of the sound intensity.

Making sense of the result:

This highlights the fact that the decibel scale is a logarithmic scale – the sound in tensity level only increases by 3 dB even though the intensity is doubled.

 EXAMPLE 15-7

Concert Sound Level

You are at a concert where the sound intensity measures 3.00 W/m². What is the sound intensity level in decibels? Should you be concerned about your hearing?

SOLUTION

We use Equation (15-28) and recall that $I_0 = 10^{-12}$ W/m².

$$\beta = 0 \text{ dB} + 10 \log\left(\frac{I}{I_0}\right)$$

$$= 0 \text{ dB} + 10 \log\left(\frac{3 \text{ Wm}^{-2}}{10^{-12} \text{ Wm}^{-2}}\right) = 125 \text{ dB}$$

This level is close to the threshold of pain and is substantially above the level permitted in a workplace. There is cause for concern about hearing loss, especially since hearing damage is cumulative and permanent.

Making sense of the results:

This sound intensity level is not uncommon at rock concerts.

In Example 15-7, the sound intensity was a modest 3 W/m² at what most of us would consider a loud concert. What happens to all the power from the amplifiers? The loudness of a sound from a point source diminishes as you move away from the source. However, as you move around a point source at a constant distance from the source, the loudness does not vary. In practice, it is very difficult to produce an actual point source for sound waves, but the mathematics of such a source is relatively straightforward.

If we think about conservation of energy, it makes sense that the intensity of a source diminishes as we move away from the source. If we enclose the source with a spherical surface of radius r, all the power radiated by the source must pass through the surface of the sphere. If the power is radiated **isotropically** (uniformly in all directions) by the source, then the intensity over the surface of the sphere will be uniform. Consequently, the total power radiated, P, is equal to the intensity, I, multiplied by the surface area of the sphere:

$$P = I4\pi r^2 \qquad (15\text{-}29)$$

We can rearrange this equation to find

KEY EQUATION
$$I = \frac{P}{4\pi r^2} \qquad (15\text{-}30)$$

The intensity of a sound source that radiates equally in all directions falls off as the inverse square of the distance from the source.

Hearing Safety

Exposure to loud noise can cause permanent damage to your hearing. In Canada, provincial Occupational Health and Safety legislation restricts the maximum β in a work environment for an 8 h shift to between 85 and 90 dB. If the β is above this limit, either hearing protection must be worn or the exposure time must be reduced to avoid damage to the workers' hearing.

Musicians are particularly concerned about hearing damage. Interestingly, it appears that the incidence of hearing loss is higher among classical musicians than

among rock musicians. High-quality earplugs that reduce the sound level while still allowing a musician to hear the music adequately have been developed to help prevent such hearing loss (Figure 15-25). Ideally, the earplugs should reduce all frequencies equally.

Figure 15-25 High-quality earplugs provide uniform attenuation at all frequencies.

EXAMPLE 15-8

Speaker Power

Assume that you are 50 m from the stage at a concert where the sound intensity is 0.30 W/m². How much power are the speakers radiating, assuming that they direct their sound output toward the audience such that the sound intensity in that direction is 8 times greater than if the speakers were isotropic point sources?

SOLUTION

The sound intensity is 3 W/m². To find the power for an isotropic source, we rearrange Equation (15-29):

$$P = I4\pi r^2 = (0.3 \text{ W/m}^2)(4\pi 50^2) = 9.42 \times 10^3 \text{ W}$$

Since the speakers direct most of their sound toward the audience, their total sound output is about 8 times less than

an isotropic source that produces the same sound intensity in all directions:

$$P = \frac{9.42 \times 10^3 \text{ W}}{8} = 1.2 \text{ kW}$$

Making sense of the results:

We can see that the point source must be quite powerful to produce this level of sound at 50 m. However, as we noted at the beginning of the chapter, amplifier systems at rock concerts typically are capable of delivering approximately 50 kW of electrical power to the speakers. Speakers themselves typically only have an efficiency of about 1%. Thus, if we connect a 100 W amplifier to a speaker, it only radiates about 1 W in sound energy. The rest of the energy is consumed in heating the wires in the speakers and the output stages of the amplifiers.

Refraction of Sound Waves

If you have ever been out in a canoe on a large lake on a calm day, you may have noticed that it is possible to hear sounds clearly over long distances. The air near the surface of the lake is cooler than the air well above the lake

(Figure 15-26). The speed of sound in air increases as the temperature increases. The sound waves tend to be refracted down to the surface of the lake. Consequently, the power from the sound source is not radiated isotropically.

(continued)

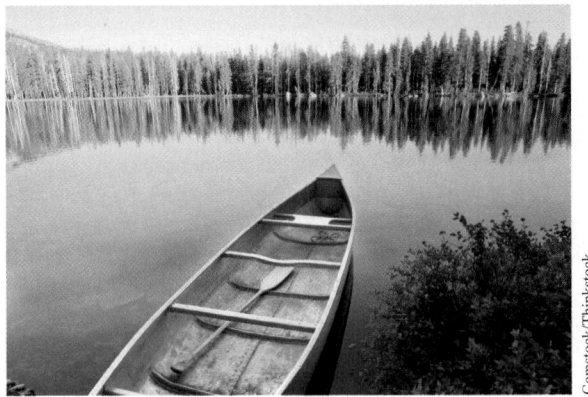

Figure 15-26 On a calm warm day, cool air over the surface of a lake causes sound waves to refract down toward the lake surface, greatly increasing the range of the sound.

Response of the Ear

When we listen to sound, a complex process takes place that converts the sound pressure in the air into our perception of the sound. The outer ear captures the sound waves and directs them to the eardrum, located in the middle ear (Figure 15-27). The sound pressure makes the eardrum vibrate, and three delicate bones connected to the eardrum transmit these vibrations to the inner ear. The inner ear contains the cochlea (shown as a blue coiled object), which is filled with fluid. The vibrations in this fluid stimulate delicate "hairs" on the basilar membrane. Nerve cells in the basilar membrane convert the vibrations of the hairs into electrical impulses that travel to the brain over the auditory nerve.

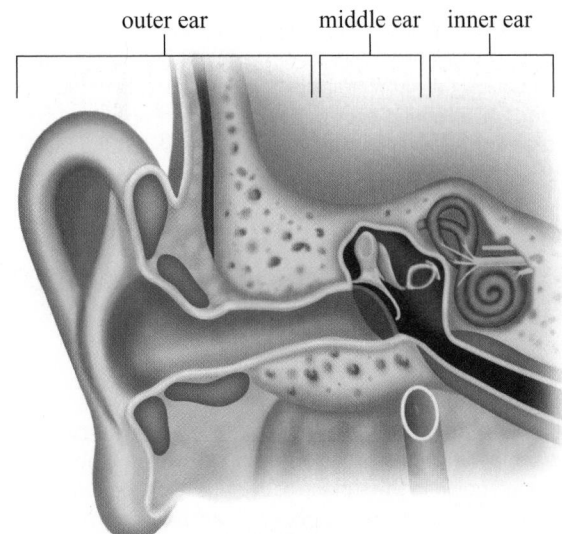

Figure 15-27 A cross section of the human ear

MAKING CONNECTIONS

Ears Can Protect Themselves

The bones in the middle ear are surrounded by muscles that can contract and limit the motion of the bones when the ear is subjected to very loud sounds. Loud percussive noises (Figure 15-28), such as those made when we hammer nails, come too quickly to allow the muscles to respond. Consequently, such sounds can be particularly damaging despite their short duration.

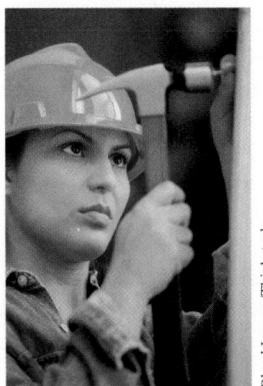

Figure 15-28 Hearing protection can prevent damage from loud percussive noise.

LO 7

15-7 The Doppler Effect

When we stand on the side of a highway and a car goes by, the sound that it generates has a higher frequency when the car is approaching and shifts to a lower frequency as the car passes by. Similarly, if we generate sound as we stand on the side of the road, the people in the car would hear a higher frequency as they approach and a lower frequency as they move away. In general, if there is relative motion between the source of the sound and the receiver of the sound, the frequency at the receiver is different from the frequency that is transmitted. If the two are moving toward each other, the received frequency is higher, and if they are moving apart, the received frequency is lower. This phenomenon is called the **Doppler effect**, named for the Austrian physicist Christian Doppler (1803–53), who first described it.

It is possible to write a single equation for the Doppler effect that covers the three possible scenarios:

- stationary source, moving receiver
- stationary receiver, moving source
- moving source and receiver

We denote the frequency of the source as f_s and the frequency at the receiver as f_r. If the source and receiver are moving along the line that connects them, the general relationship between the two frequencies is

KEY EQUATION
$$f_r = \frac{v \pm v_r}{v \mp v_s} f_s \qquad (15\text{-}31)$$

where v is the speed of sound in air, and v_r and v_s are the speed of the receiver and the source, respectively, relative to the air.

Note that Equation (15-31) has $\pm$ in the numerator but $\mp$ in the denominator. If the receiver is moving toward the source, we observe an increased frequency and use the top sign in the numerator. If the source is moving toward the receiver, we also observe an increase in the frequency and use the top sign in the denominator. In both cases, if the motion is toward, we use the top sign.

We now consider the common special cases, where either $v_r = 0$ or $v_s = 0$.

Moving Source, Stationary Receiver

Figure 15-29(a) shows a stationary source emitting spherical waves, which travel outward at the same speed in all directions. A receiver at any point around this source will detect the same frequency because the wave fronts are all equally spaced and all travel at the same speed. However, if the source is moving to the right, the wave fronts to the right of the source are closer together, and the wave fronts to the left

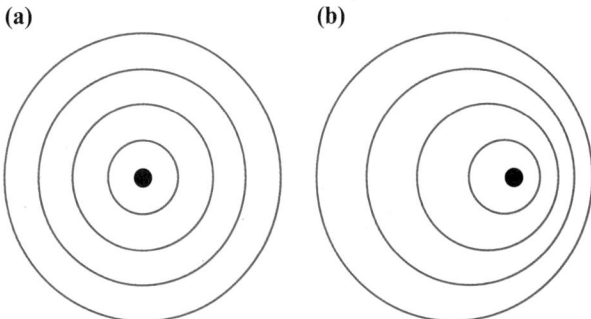

(a) **(b)**

Figure 15-29 (a) A stationary source emitting waves (b) The source moving to the right

of the source are farther apart, as shown in Figure 15-29(b). Since the waves are still all travelling at the same speed, a receiver to the right of the source will detect more waves per second than a receiver to the left of the source will. More waves per second correspond to a higher frequency, and fewer waves per second correspond to a lower frequency.

Another way to understand the Doppler effect is this: Recall that the speed of sound in air at a given temperature is constant. Clearly, the wavelength to the right of the source is shorter, and since $f = v/\lambda$, a shorter wavelength requires a higher frequency.

We can calculate the change in the wavelength. Consider the shortened wavelengths observed to the right of the source. The source emits a wave crest every T seconds, where T is the period of the source. In each period, the wave crest travels a distance $D = vT$, where v is the speed of the wave. In the same time, the source travels a distance $d = v_s T$, where v_s is the speed of the source. Thus, the distance between the source and the last wave crest it emitted is

$$\lambda_r = \frac{v}{f_r} = D - d = (v - v_s)T = \frac{v - v_s}{f_s} \qquad (15\text{-}32)$$

Solving for f_r we find that for a source moving toward a *stationary* receiver:

$$f_r = \frac{v}{v - v_s} f_s \qquad (15\text{-}33)$$

Had we considered the situation with the receiver to the left of the source (i.e., the source moving *away* from a stationary receiver), we would have a plus sign in the denominator of Equation (15-33).

EXAMPLE 15-9

The Doppler Effect

A high-speed train travelling at 250 km/h blows its whistle as it approaches a level crossing. The whistle has a frequency of 660 Hz. What frequency would be perceived by someone standing at the crossing as the train

(a) approaches the crossing;
(b) moves away from the crossing?

SOLUTION

(a) The speed of the source (the train) is 250 km/h = 69.4 m/s, and the speed of the receiver is zero. The speed of sound in air is approximately 343 m/s. Substituting these values into Equation (15-31) gives

$$f_r = \frac{343 \text{ m/s} + 0 \text{ m/s}}{343 \text{ m/s} - 69.4 \text{ m/s}} \, 660 \text{ Hz} = 828 \text{ Hz}$$

Notice that we chose a minus sign in the denominator: since the train was moving toward the receiver, we use the top sign.

(b) When the train is leaving the crossing, we use the bottom sign:

$$f_r = \frac{343 \text{ m/s} - 0 \text{ m/s}}{343 \text{ m/s} + 69.4 \text{ m/s}} \, 660 \text{ Hz} = 549 \text{ Hz}$$

Making sense of the result:

As the train approaches we hear a higher pitch and as it moves away we hear a lower pitch. This is similar to what we observe when we stand near the side of a highway.

MAKING CONNECTIONS

Hubble and the Expanding Universe

The Doppler effect also occurs with electromagnetic radiation, including visible light. American astronomer Edwin Hubble (1889–1953) made a plot of the Doppler shift of distant nebulas versus their distance from Earth. Hubble noted that almost all the objects had a downward shift in frequency and that the magnitude of the shift is approximately proportional to the distance from Earth to the object. Using these data, Hubble concluded that the universe is expanding. His original data are shown in Figure 15-30. The slope of the plot of recession velocity versus distance is called Hubble's constant. The best current estimate for Hubble's constant is 72.6 ± 3.1 (km/s)/Mpc. A parsec (pc) is a commonly used astronomical unit equal to approximately 3.09×10^{16} m, or 3.26 light years.

Figure 15-30   Hubble's plot of nebula speeds versus distance from Earth

Moving Receiver, Stationary Source

When the source is stationary and the receiver is moving away from it, the receiver detects a reduction in the wave speed, and a receiver moving toward the source detects an increase in the wave speed. First, consider a receiver approaching the stationary source. When the receiver is positioned at a wave crest, the next wave crest is a distance λ away and moving toward the receiver at a speed relative to the receiver of $v + v_r$. The time this second wave crest takes to reach the receiver is

$$T_r = \frac{\lambda}{v + v_r} = \frac{v}{v + v_r}\left(\frac{1}{f_s}\right) = \frac{1}{f_r} \qquad (15\text{-}34)$$

Solving for f_r we find

$$f_r = \frac{v + v_r}{v} f_s \qquad (15\text{-}35)$$

This expression is the special case of Equation (15-31) with $v_s = 0$ and the receiver moving toward the source. For a receiver moving away from a stationary source, we use a minus sign in the numerator of Equation (15-35).

With Equations (15-33) and (15-35), we can readily derive the general result we stated in Equation (15-31). If both the source and the receiver are moving, we replace the stationary source frequency in Equation (15-35) with the stationary receiver frequency from Equation (15-33) to find

KEY EQUATION $\quad f_r = \dfrac{v \pm v_r}{v} \dfrac{v}{v \mp v_s} f_s = \dfrac{v \pm v_r}{v \mp v_s} f_s \qquad (15\text{-}31)$

The sign conventions use the upper sign when the motion is toward and the lower sign when the motion is away.

Shock Waves

When the speed of the source reaches the wave speed, many wave crests overlap, producing an intense wave front (Figure 15-31(a)). Such wave fronts are called a shock wave.

You may have heard such a shock wave from a supersonic plane. Shock waves can also occur in water, as shown in Figure 15-31(b).

iStockphoto/Thinkstock

Figure 15-31   (a) Overlapping wave crests produce a shock wave. (b) The wake behind a boat is a type of shock wave.

In a sound wave, the particles of the medium are displaced longitudinally. Associated with the displacement of the medium is a pressure wave that is $90°$ out of phase with the displacement amplitude. Sound waves can travel in three spatial dimensions, and we can describe their propagation qualitatively using Huygens' principle. When a sound wave encounters a boundary, it can reflect and refract. When we confine a sound wave, for example, in a pipe, we get standing waves at specific frequencies. Two sound sources produce a three-dimensional spatial interference pattern. They can also produce a temporal interference pattern if they have different frequencies. Sound intensity levels are expressed in decibels, which are units of a logarithmic scale. The received frequency of a sound wave depends on the relative motion of the source and the receiver.

A sound wave is a longitudinal wave with a wavelength λ and a frequency f, with a displacement amplitude that can be written as

$$s(x, t) = s_{m}\cos(kx - \omega t + \phi) \tag{15-3}$$

where s_{m} is the maximum displacement, $k = \dfrac{2\pi}{\lambda}$, $\omega = 2\pi f$, and ϕ is the phase constant.

Associated with the displacement amplitude is a pressure wave, which is given by

$$\Delta p = Bks_{m}\sin(kx - \omega t + \phi) \tag{15-10}$$

where Δp is the deviation from the ambient pressure.

The speed of a sound wave is related to the frequency and the wavelength:

$$v = f\lambda$$

The speed of a sound wave in a gas is related to the properties of the gas through the mass density, ρ, and the bulk modulus, B, $v = \sqrt{\dfrac{B}{\rho}}$.

We can use Huygens' principle to qualitatively describe how a wave propagates by considering each point on a wave to be a source of spherical waves. Drawing equally spaced concentric circles around each point illustrates how the wave propagates.

When encountering a boundary, a wave can reflect and refract. The reflected portion of the wave leaves the boundary with the angles of incidence and reflection the same. The refracted portion of the wave passes the boundary and bends at an angle that depends on the wave speeds on each side of the boundary.

For speed of sound v and a pipe of length L, which is open at both ends or closed at both ends, the frequency of the standing waves is given by

$$f_{n} = \frac{nv}{2L}, \quad n = 1, 2, 3, \ldots$$

For a pipe that is open at one end and closed at the other,

$$f_{n} = \frac{nv}{4L}, \quad n = 1, 3, 5, \ldots$$

For two sources that are in phase, we get constructive interference when the path difference between the two sources is an integer number of wavelengths. We get destructive interference when the path difference is a half-integer number of wavelengths.

When we have two sources that differ in frequency, the tone we hear has a frequency that is the average of the two frequencies and is amplitude-modulated (beats) at a frequency equal to the difference of the two frequencies.

The intensity of a point sound source a distance r from the source is

$$I = \frac{P}{4\pi r^{2}} \tag{15-30}$$

where P is the total power radiated by the source. The sound intensity level (β) is defined as

$$\beta(I) = 0 \text{ dB} + 10 \log_{10}\left(\frac{I}{I_{0}}\right) \tag{15-28}$$

where the unit is the decibel (dB), and $I_{0} = 10^{-12}$ W/m² is the intensity that corresponds to 0 dB.

When a source and a receiver are in motion relative to each other, the frequency detected by the receiver is not the same as the frequency emitted by the source. For sound waves,

$$f_{r} = \frac{v \pm v_{r}}{v \mp v_{s}}f_{s} \tag{15-31}$$

where f_{s} is the frequency of the source, f_{r} is the frequency at the receiver, v is the speed of sound in air, and v_{r} and v_{s} are the speed of the receiver and the source, respectively, relative to the air. When the motion is toward, we use the top sign, and when the motion is away, we use the bottom sign.

Applications: musical instruments, musical beats, tuning musical instruments, hearing safety, shock waves

Key Terms: beat, bulk modulus, decibel, Doppler effect, Huygens' principle, intensity, isotropic, plane wave, sound intensity level, spherical wave

QUESTIONS

1. A sound wave is a longitudinal wave. True (T) or false (F)?
2. The displacement and pressure amplitudes are
   (a) in phase;
   (b) out of phase by 90°;
   (c) out of phase by 180°.
3. When we double the frequency of a sound wave, by what factor does the wavelength change?
4. When we double the power produced by a source, by what factor does the displacement amplitude of the resultant wave change?
5. The angle of incidence and the angle of reflection are equal when a sound wave reflects from a boundary. True (T) or false (F)?
6. When a sound wave crosses a boundary from a medium of high speed to one of lower speed, the wave
   (a) bends toward the surface normal;
   (b) bends away from the surface normal;
   (c) simply slows without changing direction.
7. We create a standing wave when we confine a wave between boundaries. True (T) or false (F)?
8. The frequencies of standing waves in an air column are all integer multiples of each other. True (T) or false (F)?
9. It is not possible to have multiple standing waves simultaneously in an air column. True (T) or false (F)?
10. We only get an interference pattern when two sources are in phase. True (T) or false (F)?
11. When we double the intensity of a sound source, by what value does the sound intensity level increase?
12. The intensity of a sound source decreases by what factor when we triple the distance from the source?
13. Sound waves can interfere
    (a) spatially;
    (b) temporally;
    (c) both spatially and temporally;
    (d) neither spatially nor temporally.
14. The speed of sound in a gas does not depend on the type of gas. True (T) or false (F)?

PROBLEMS BY SECTION

For problems, star ratings will be used, (✱, ✱✱, or ✱✱✱), with more stars meaning more challenging problems.

Section 15-1   Sound Waves

15. ✱   A wave is observed to have a frequency of 1000 Hz in air. What is the wavelength?
16. ✱✱   When the bulk modulus increases by 10%, by what factor does the speed of sound change?
17. ✱   A depth sounder on a boat sends out a pulse and listens for the echo from the bottom. The water is 30 m deep. How long will it take for the echo to come back?
18. ✱✱   A sound wave is described by $s(x, t) = 2 \times 10^{-9}$ $\cos(37x - 12\,566t)$. Find the wave's
    (a) wavelength;
    (b) frequency;
    (c) speed.
19. ✱   Write an expression for the displacement amplitude of a 60 Hz sound wave.
20. ✱✱   Medical ultrasound machines typically use frequencies in the range of 2 to 18 MHz.

(a) What are the corresponding wavelengths in air?
(b) Given that the human body contains primarily water, estimate the wavelength range in the body.

21. ✱   A 1000 Hz sound wave with a pressure amplitude of 20 N/m² in air is at the threshold of pain for human hearing. It has pressure amplitude of $\Delta p = 20$ N/m². What is the corresponding displacement amplitude?
22. ✱   What is the intensity of a 2000 Hz sound wave in air that has a pressure amplitude of 1.5×10^{-3} N/m²?
23. ✱   When the displacement amplitude of a sound wave doubles, by what factor does the pressure amplitude increase?

Section 15-2   Wave Propagation and Huygens' Principle

24. ✱✱   For the wave front in Figure 15-32, use Huygens' principle to determine the shape of the resulting wave front
    (a) close to the original wave front;
    (b) far from the original wave front.

Figure 15-32   Problem 24

Section 15-3   Reflection and Refraction

25. ✱✱   In Figure 15-33, a plane wave is approaching a rectangular corner. Draw the reflected wave using Huygens' principle.

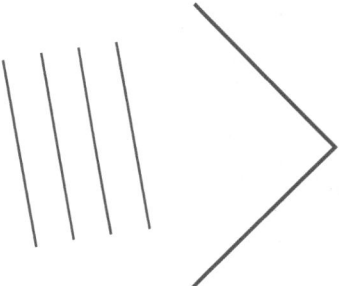

Figure 15-33   Problem 25

Section 15-4   Standing Waves

26. ✱   An air column has a length of 0.75 m and is open at both ends. What is the lowest frequency of a standing wave that can be found in such a column?
27. ✱✱   Standing waves can form in a box. A speaker enclosure has the dimensions $10 \times 15 \times 20$ cm. What are the lowest-frequency standing waves that can form in the three directions?

Section 15-5   Interference

28. ✱✱   Two point sources are 25 cm from each other, and both emit a 2000 Hz tone. Sketch the resulting interference pattern.
29. ✱   Two violins produce frequencies of 294 Hz and 296 Hz. What beat frequency is heard?
30. ✱✱   Two point sources are in phase and emit a 500 Hz tone. What is the phase difference at a point that is 5.0 m from one source and 6.0 m from the other?
31. ✱   In North America, the telephone dial tone consists of a 350 Hz tone and a 440 Hz tone. What is the resulting beat frequency?

Section 15-6   Measuring Sound Levels

32. ✳   A food processor produces a sound intensity level of 87 dB at a distance of 1.0 m. How many watts per square metre does this intensity level correspond to?

33. ✳   A single table saw produces a sound intensity level of 95 dB at a distance of 3.0 m. What would the sound intensity level be for two table saws, both 3.0 m away?

34. ✳   What is the displacement amplitude for an 800 Hz sound wave that has an intensity of 20 mW/m²?

Section 15-7   The Doppler Effect

35. ✳   A police siren emits a frequency of 700.0 Hz. When the police car approaches you at a speed of 80.0 km/h, what frequency do you perceive?

36. ✳   The Doppler effect can be used to measure flow rates in blood. A sound wave is transmitted into a vein or an artery, where it is reflected off the moving red blood cells and then detected by a receiver. A transmitter uses a frequency of 10.0 MHz, and the blood flow rate is 0.50 m/s. What is the expected frequency shift?

37. ✳   Some bats use the Doppler shift to detect the direction of motion of prey. The bats use a frequency of about 61 kHz and can detect shifts as small as 35 Hz. To what velocity does this correspond?

COMPREHENSIVE PROBLEMS

38. ✳   At large concerts, it is sometimes disconcerting to observe the musicians moving apparently out of sync with the music. This results from the time it takes the sound to travel from the stage to you. When the musicians are playing at 100 beats/min, at what distance from the stage will they appear to be one full beat behind?

39. ✳   While camping, you observe a lightning bolt, and 4.0 s later you hear the associated thunder. How far away is the strike?

40. ✳   Two sound sources differ in sound intensity level by 20 dB. Find the ratio of their
    (a) displacement amplitudes;
    (b) pressure amplitudes;
    (c) intensities.

41. ✳   Two sound waves of frequency 200 Hz travel in the x-direction. One has a displacement amplitude of 10 nm, and the other has a displacement amplitude of 20 nm. The phase difference between them is $\pi/4$ rad. What is the resultant amplitude?

42. ✳✳   If we allow the phase difference between the two waves in problem 41 to vary, what are the smallest and largest possible displacement amplitudes?

43. ✳   The sound intensity level a distance of 23 m from an isotropic point source is 65 dB. What is the intensity, and what is the total power radiated by the source?

44. ✳✳✳   B flat played on a tuba has a fundamental frequency of approximately 117 Hz.
    (a) Write equations for this fundamental and for second and third harmonics with half the amplitude of the fundamental.
    (b) On the same graph, plot the waveform of the fundamental, the two harmonics, and the resultant of these three tones.

(c) Compare the period and amplitude of the resultant to those of the fundamental.

45. ✳✳   Police radar uses the Doppler shift to determine whether you are speeding. The radar unit transmits a wave, which is reflected off your car and Doppler shifted as a result. Rather than detecting the frequency shift directly, the radar receiver mixes the transmitted and received beam and measures the resulting beat frequency. For a radar unit that operates at 10 GHz and for a driving speed of 130 km/h, what beat frequency would result?

46. ✳   An interesting demonstration of standing waves can be done by twirling one end of a flexible, corrugated hose that is open at both ends. The twirling motion causes air to be drawn through the hose, and air rushing over the corrugations creates turbulence in the hose. The turbulence excites waves. It is usually possible to get the three lowest tones from the hose. For a 1.50 m hose, what frequencies will be heard?

47. ✳✳   A bugle is an instrument that has an air column with a fixed length. Different notes can be achieved by exciting different harmonics in the column, not an easy task. The bugle actually acts like a tube with both ends open. For a certain length L, a bugle will have a sequence of four consecutive harmonics with frequencies very nearly equal to the frequencies associated with the notes G_4, C_5, E_5, and G_5. Determine the length of the bugle.

48. ✳✳   A plane wave is normally incident on a boundary where the speed of sound changes. The wave travels from a medium in which the speed of sound is 1200 m/s to a medium in which the speed of sound is 400 m/s. The incident wave has a frequency of 440 Hz.
    (a) What is the period of a 440 Hz wave?
    (b) How far will a wave front travel in the second medium during the time from part (a)?
    (c) What is the wavelength in the second medium?
    (d) What is the frequency in the second medium? (Hint: Calculate from the speed and the wavelength from part (c).)

49. ✳✳   The Bay of Fundy has some of the highest tides in the world, with a tidal range as large as 15 m. The bay behaves like a resonator that is open at one end and closed at the other. The Bay of Fundy is approximately 270 km long, and the tidal period is approximately 12.5 h. Assuming that the tide corresponds to the lowest-frequency standing wave, what is the wave speed in the Bay of Fundy?

50. ✳   The air cavity inside stringed instruments is designed to have resonant frequencies that are close to the resonant frequencies that are produced by the instrument. For a 12 cm-deep guitar body cavity, what are the frequencies of the standing waves that could form between the top and back of the guitar?

51. ✳✳   The reverberation time in a performance space is a measure of how long it takes a sound to fade away once it has stopped being produced. The time is defined as the amount of time it takes the sound to drop by 60 dB. For an initial sound intensity of 0.2 W/m², what is the corresponding reduction in intensity after one reverberation time?

52. ✳✳✳   A sound wave loses 10% of its intensity each time it reflects from a particular surface. How many reflections will be required to reduce the sound intensity level by 3 dB?

53. ** To determine the depth of a well, you drop a stone into the well and hear the splash 2.8 s later. How deep is the well?

54. ** A jet produces a sound intensity level of approximately 140 dB at a distance of 30 m. Assuming that your ear has an effective radius of about 2.0 cm, what is the total power received by your ear?

55. *** One way of measuring the frequency of a sound is to make measurements of the interference pattern that is created when the sound is produced by two sources. You find that for two sources, the sound is loudest when you are equidistant from the sources. You then determine that there is a minimum in intensity when you are 2.8 m farther from one source than the other. What is the frequency of the source?

56. ** Standing waves in your music room can cause certain spots in the room to have lower volume levels. For a room with the dimensions 4.00 × 6.00 × 2.40 m, what are the possible standing wave frequencies?

57. ** Standing waves can be set up in solid rods, and this is the basis for a popular demonstration. The rod, clamped in the middle, is stroked (sometimes a cloth is used) and begins to emit a tone. For a 120 cm aluminum rod in which the speed of sound is 6420 m/s, find
(a) the fundamental wavelength on the rod;
(b) the fundamental frequency.

58. ** A typical audio amplifier does not have the same gain at all frequencies. An amplifier is rated 100.0 W at 1.0 kHz. The manufacturer specifies that the output falls by 3.0 dB when the frequency output is 20 kHz. What power can the amplifier produce at 20 kHz?

59. *** Show that we still get beats if we add a phase constant to each of the two waves in Equation (15-22).

60. ** You have been asked to design a set of pipes for an organ. Assuming that the pipes have one open end and one closed end, calculate the lengths you will need to go from 440 Hz (A_4) to 880 Hz (A_5).

DATA-RICH PROBLEM

61. *** You have been hired by an environmental consulting firm to do a noise analysis for a quarry operation. The operation uses two trucks, a drill, and a crusher. The manufacturers of the equipment provided the specifications in Table 15-5 for the sound intensity level of the various pieces of equipment. Local bylaws specify that the maximum sound intensity level at the perimeter of the quarry property not exceed 85 dB. How close to the perimeter can the quarry operate all these devices simultaneously?

Table 15-5 Data for Problem 61

Equipment	Sound intensity level at 10 m
Truck	85 dB
Drill	110 dB
Crusher	110 dB

OPEN-ENDED PROBLEM

62. *** Many of us have heard the effect that can be produced by inhaling helium and speaking. The speaker's voice is shifted to higher frequencies. Discuss the physics behind this effect.

 See the text online resources at www.physics1e.nelson.com for Open Problems and Data-Rich Problems related to this chapter.

Learning Objectives

When you have completed this chapter you should be able to:

1 List the three main states of matter and some of their associated physical characteristics.

2 Discuss the state variables for some systems.

3 Describe the relationship between the average speed of a particle in a gas and the pressure that the gas exerts on the walls of its container.

4 Define thermometric property and do calculations related to the linear expansion of solids.

5 Be familiar with some common temperature scales.

6 State the zeroth law of thermodynamics, and describe the concept of thermal equilibrium.

7 Use the ideal gas law to determine various properties of a gas.

8 Describe how a universal temperature scale can be based on an ideal gas thermometer.

9 Discuss the relationship between the temperature of a gas and the average mechanical energy of the atoms in that gas.

10 Qualitatively interpret a probability distribution.

11 Read a phase diagram.

Chapter 16
Temperature and the Zeroth Law of Thermodynamics

The morning landscape in Figure 16-1 contains all the familiar phases of matter: solids in the trees and bushes, liquid in the tiny droplets of water in the fog, and invisible gases in the air.

On such a morning, you would find the grass and bushes covered with heavy dew. In addition to liquid water, fog also contains a lot of water vapour. So, water is present in two phases: liquid and gas. While this is a relatively common occurrence for water, we might wonder whether other substances exhibit this behaviour, and whether water can be present in all three states simultaneously.

Courtesy of Rebecca Dimock

Figure 16-1 Early morning fog in the Annapolis Valley, Nova Scotia

 For additional Making Connections, Examples, and Checkpoints, as well as activities and experiments to help increase your understanding of the chapter's concepts, please go to the text's online resources at www.physics1e.nelson.com.

16-1 Solids, Liquids, and Gases

On Earth, matter is usually found in **solid**, **liquid**, and **gas** phases, as described in Chapter 12. Under suitable conditions, most materials can exist in all three phases. For example, water exists as a solid (ice), a liquid, and a gas (steam). However, there is a fourth phase, called plasma. Plasma is a highly ionized gas and is the predominant phase in the universe. For example, the Sun, which contains over 99% of the mass in the solar system, is in a plasma phase.

Solids tend to be very rigid and do not compress or stretch readily. The atoms in solids tend to remain at fixed positions, but the atoms in a liquid and in a gas can drift around. However, the atoms in solids do vibrate around their equilibrium positions. In crystalline solids, the atoms are arranged in regular periodic arrays, and each atom is bonded to a set of adjacent atoms (Figure 16-2). Crystalline solids are usually composed of a collection of small crystals. Even microscopic individual crystals still contain a huge number of atoms and extend over a very large number of interatomic distances. Such materials are described as having long-range order. In some materials, such as diamond, the individual crystals are often large enough to be seen without magnification (Figure 16-3).

Figure 16-3 A large single crystal of diamond. The faces of the diamond correspond to planes in the crystal structure

solid

Figure 16-2 A periodic lattice from a solid. The atoms are closely spaced and form a periodic array.

liquid

Figure 16-4 A liquid phase does not have long-range order, and the atoms are somewhat more loosely packed than atoms in the solid phase.

Liquid phases are usually somewhat less dense and more compressible than the solid phase of the same material. Liquids have no long-range order (Figure 16-4). Over short distances, the positions—and sometimes the relative orientations—of adjacent particles in liquids are correlated. As a liquid cools, the extent of this ordering increases, becoming long-range order at the freezing point. Liquids consist of either atoms or molecules.

Gases are much less dense and much more compressible than the liquid and solid phases of the same material. Gases have no long- or short-range order, so the particles are completely free to move around (Figure 16-5). Like liquids, gases are comprised of either atoms or molecules.

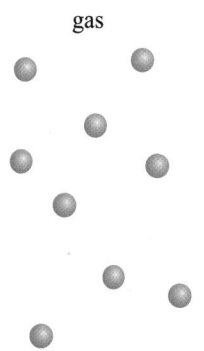

gas

Figure 16-5 The atoms in a gas are relatively far apart, and their positions are not related to each other.

THERMODYNAMICS

16-2 State Variables

The goal of **thermodynamics** is to describe energy transfers in the form of work and thermal energy (heat) from a macroscopic point of view using variables such as temperature and **pressure**. A key concept in thermodynamics is the state of a system. If we know all the forces that act on a particle, and we know the position and velocity of the particle at some instant, then we can, at least in principle, describe the motion of the particle for all time. This description would then specify the state of the particle.

Consider a flask of gas sitting on a lab bench. This system consists of a vast number of individual particles, so single-particle descriptions are not suitable. We can describe the state of a pure gas in terms of the number of particles in the system, the **volume** of space occupied, the pressure that the gas exerts on the walls of the container, and the temperature of the gas. We will discuss these **state variables** for gases in Section 16-6. Other types of systems use different variables. Table 16-1 gives a few examples. Notice that all the variables in Table 16-1 refer to macroscopic properties of the systems. None of the variables relate to properties of specific individual particles.

 CHECKPOINT

C-16-1 Are the Details in the State Variables?

Do state variables depend on all the details of the particles in a system?

C-16-1 Yes. For example, we will see that the pressure in a gas depends on the average speed squared of the particles in the gas. To determine the average, you need to know the speeds of all the particles.

 CHECKPOINT

C-16-2 Can the Details be Extracted from the State Variables?

Can information about specific particles in a system be determined from the state variables?

C-16-2 No. The state variables usually allow us to determine some average values. However, it is not possible to extract information about single particles from the average values.

 PEER TO PEER

State Variables

You can think of state variables as the values you must know in order to recreate a particular thermodynamic system. For any ideal gas system, you only need to know three state variables to know everything about the system. This means that you can calculate any state variable from any three other state variables. For example, if you know the volume, temperature, and number of particles in a system, you can use the ideal gas law to determine the pressure. More complex systems are more difficult to describe and therefore require more state variables to do so.

The fundamentals of thermodynamics were developed during the Industrial Revolution in the late eighteenth and early nineteenth centuries. In fact, one of the first applications of thermodynamics was to improve the efficiency of steam engines (Figure 16-6). The original focus of thermodynamics was the behaviour of gases (steam in particular) and the conversion of thermal energy into mechanical energy.

The four thermodynamic state variables for a gas are the number of gas particles, the volume, the pressure, and the temperature. The number of particles in a gas is a straightforward quantity. Similarly, the volume occupied by a gas is relatively straightforward to understand and is usually simply determined by the size of the container holding the gas. However, the remaining two variables—pressure and temperature—are somewhat more subtle. We will first consider pressure.

16-3 Pressure

The pressure that a gas exerts on the walls of its container results from collisions of individual particles with the wall of the container (Figure 16-7). When a particle collides with the walls of its container, it exerts a force on the wall, and the wall exerts a force on the particle.

Table 16-1 State variables for various types of systems

System	State variables			
Gas	Temperature	Volume	Pressure	Number of particles
Magnet	Temperature	Magnetization	Applied magnetic field	Number of magnetic dipoles*
Dielectric	Temperature	Polarization	Applied electric field	Number of electric dipoles*
*Chapters 19 and 23 describe electric and magnetic dipoles, respectively.				

Figure 16-6 Improving the efficiency of steam engines provided much of the impetus for the early work in thermodynamics.

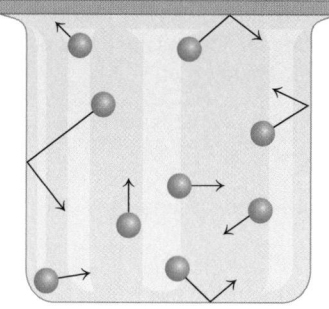

Figure 16-7 Gas atoms collide with the walls of a container.

Consider a gas particle of mass m that is moving with velocity $\vec{v}_{\text{initial}} = v_x\hat{i} + v_y\hat{j} + v_z\hat{k}$ in a cubic container with sides of length L and volume $V = L^3$. We assume that the collisions of the gas particles with the walls of the container are elastic. Then, the particle has the same speed before and after the collision, but one component of the particle's velocity has been reversed. For example, if a particle collides with a vertical wall (perpendicular to the x-axis), the x-component of the velocity is reversed, and $\vec{v}_{\text{final}} = -v_x\hat{i} + v_y\hat{j} + v_z\hat{k}$, the change in momentum of the particle is then

$$\Delta\vec{p}_{\text{particle}} = \vec{p}_{\text{f}} - \vec{p}_{\text{i}} = m\vec{v}_{\text{final}} - m\vec{v}_{\text{initial}} = -2mv_x\hat{i} \quad (16\text{-}1)$$

Since momentum is conserved, the change in momentum of the container wall must be exactly equal but opposite to the change in momentum of the particle:

$$\Delta\vec{p}_{\text{wall}} = -\Delta\vec{p}_{\text{particle}} = 2mv_x\hat{i} \quad (16\text{-}2)$$

Equation (16-2) gives us the momentum imparted to the container wall due to a single collision with a single particle. To determine the pressure exerted by the gas, we also need to know the rate at which collisions are occurring.

The time between collisions between a given particle and the vertical wall is the time it takes the particle to cross the length of the container twice:

$$\Delta t = \frac{2L}{v_x} \quad (16\text{-}3)$$

Dividing Equation (16-2) by Equation (16-3), we obtain the rate of change of momentum of the wall:

$$\frac{\Delta p_{\text{wall}}}{\Delta t} = \frac{2mv_x}{\dfrac{2L}{v_x}} = \frac{mv_x^2}{L} \quad (16\text{-}4)$$

As we saw in Chapter 7, we can write Newton's second law as $F = \dfrac{\Delta p}{\Delta t}$, so

$$F_x = \frac{\Delta p}{\Delta t} = \frac{mv_x^2}{L} \quad (16\text{-}5)$$

To determine the total force on the wall, we sum this expression over all the particles to get

$$
\begin{aligned}
F_{x,\text{Total}} &= \sum_{\text{i}} F_{x,\text{i}} \\
&= \sum_{\text{i}} \frac{mv_{x,\text{i}}^2}{L} \\
&= \left(\frac{mv_{x,1}^2}{L} + \frac{mv_{x,2}^2}{L} + \frac{mv_{x,3}^2}{L} + \cdots + \frac{mv_{x,N}^2}{L}\right) \\
&= \frac{m}{L}\left(v_{x,1}^2 + v_{x,2}^2 + v_{x,3}^2 + \cdots + v_{x,N}^2\right) \quad (16\text{-}6)
\end{aligned}
$$

where N is the total number of particles in the box.

We note that the average values of the square of the x-components of the velocities is given by

$$\left(v_x^2\right)_{\text{avg}} = \frac{1}{N}\left(v_{x,1}^2 + v_{x,2}^2 + v_{x,3}^2 + \cdots + v_{x,N}^2\right) \quad (16\text{-}7)$$

Rearranging this equation gives

$$N(v_x^2)_{\text{avg}} = (v_{x,1}^2 + v_{x,2}^2 + v_{x,3}^2 + \cdots + v_{x,N}^2) \quad (16\text{-}8)$$

Thus, the total force exerted on the wall of the container is

$$F_{x,\text{Total}} = \frac{m}{L} N(v_x^2)_{\text{avg}} \quad (16\text{-}9)$$

To determine the pressure that the gas exerts on the wall, we divide the total force by the area of the wall to obtain

THERMODYNAMICS

AbleStock.com/Thinkstock

$$P = \frac{\dfrac{m}{L} N(v_x^2)_{avg}}{L^2} = \frac{mN(v_x^2)_{avg}}{L^3} \qquad (16\text{-}10)$$

We can generalize this expression to three dimensions. For any set of velocities,

$$(v^2)_{avg} = (v_x^2)_{avg} + (v_y^2)_{avg} + (v_z^2)_{avg} \qquad (16\text{-}11)$$

Because the container is cubical and there is nothing special about either the x-, y-, or z-directions (if we assume there are no external forces acting on the gas), it must be true that

$$(v_x^2)_{avg} = (v_y^2)_{avg} = (v_z^2)_{avg} = \frac{(v^2)_{avg}}{3} \qquad (16\text{-}12)$$

Finally, recalling that $V = L^3$, we can write

KEY EQUATION
$$P = \frac{1}{3} \frac{mN(v^2)_{avg}}{V} \qquad (16\text{-}13)$$

 CHECKPOINT

C-16-3 Pressure-Speed Connection in a Gas

If we observe that the pressure in a sample of gas changes, can we conclude that the average speed squared has changed correspondingly?

C-16-3 No. Either the number of particles or the volume that the gas occupies could have changed instead.

You may have questioned the assumption that particles colliding with the container emerge from the collision with their speeds unchanged. Implicit in this assumption is that the velocity of the container did not change as a result of the collision. This situation might appear to violate conservation of momentum. However, gas molecules collide with all the walls of the container. For every particle that collides with the wall on the right, there is a particle with the same speed that collides with the wall on the left. If this were not the case, the total momentum of the gas would be changing, and the centre of mass of the gas would move. In other words, the net momentum transfer resulting from the random motion of the huge number of gas particles averages to zero.

You might also wonder about the assumption implicit in Equation (16-3), which states that the time for the particle to go from one side of the container to the other is equal to the length of the container divided by the speed of the particle. This relationship is only strictly true for a single particle that does not change its velocity during this time. In fact, a particle in a gas collides with other particles, and its velocity does change. However, our assumption is still valid overall because the gas is in a state of equilibrium, with constant volume, pressure, and temperature. To maintain the state of equilibrium, it must be true that for every collision that causes a particle with speed v_x to change to v_x', there must be a corresponding collision that causes a particle with speed v_x' to change to v_x. Otherwise,

 EXAMPLE 16-1

How Fast Are the Molecules Going?

At standard temperature (0°C) and pressure (101.325 kPa), or STP, a mole of gas occupies a volume of 22.4 L. The molar mass of an atom of nitrogen is approximately 14 g/mol. Nitrogen is diatomic, that is, its atoms bond in pairs to form N_2 molecules. Estimate the average speed of a nitrogen molecule at STP.

SOLUTION

Since each nitrogen molecule contains two atoms, a mole of nitrogen gas has a mass of approximately 28 g. Solving Equation (16-13) for the speed, we get

$$(v^2)_{avg} = \frac{3PV}{mN}$$

$$\therefore \sqrt{(v^2)_{avg}} = v_{rms} = \sqrt{\frac{3PV}{mN}}$$

where v_{rms} is the root-mean-square velocity, that is, the square root of the average (or mean) of the squares of the individual velocities.

Substituting in the known values, and converting litres to cubic metres (1 L = 0.001 m³):

$$v_{rms} = \sqrt{\frac{3 \times 101.325 \times 10^3 \, kg \cdot m^{-1}s^{-2} \times 0.0224 \, m^3}{0.028 \, kg}}$$

$$= \sqrt{2.43 \times 10^5 \, m^2 \, s^{-2}} = 490 \, m/s$$

Making sense of the result:

We can use other data to make a second estimate. It takes 200 kJ of energy to convert 1.0 kg of liquid nitrogen to gaseous nitrogen. If we assume that all of this energy goes into kinetic energy, then

$$v = \sqrt{\frac{2E}{m}} = \sqrt{\frac{2 \times 2 \times 10^5 \, J/kg}{1 \, kg}} = 630 \, m/s$$

Thus our result seems reasonable.

the total momentum of the gas would change, and the gas would not be in a state of equilibrium. *On average*, there is a constant number of particles with a given speed v_x, and that number of particles will, *on average*, cross the container in the time L/v_x.

LO 4

16-4 Temperature and Thermal Expansion

Temperature is something that we all discuss on a daily basis: What is the temperature today? At what temperature should the oven be set to cook a particular dish? In fact, thermometry (the measurement of temperature) was so taken for granted that the law of thermodynamics that deals with it was only developed after the formulation of the other laws. Since these had already been assigned to the first and second laws, thermometry was assigned the zeroth law.

To measure temperature, we use a thermometric property of some object. One common thermometer is the alcohol thermometer, which relies on the thermal expansion of alcohol: as we heat alcohol, it expands. There are many other materials that have thermometric properties that are utilized in thermometers and some of them are listed in Table 16-2.

Table 16-2 Various common temperature-measuring devices and their associated thermometric properties

Temperature-measuring device	Thermometric property
Galinstan*	Thermal expansion
Bimetallic strip	Differential thermal expansion
Thermocouple	Thermoelectric effect

*Galinstan is a liquid alloy used as a substitute for mercury, which is highly toxic.

 CHECKPOINT

C-16-4 Wood or Metal: Which Is "Warmer"?

Consider a metal object and a wooden object sitting on a desk. Typically, the metal object feels cool to the touch, but the wooden object does not. Yet, the objects have been sitting side by side in the same room and should be at the same temperature. Why do they not feel equally warm?

C-16-4 When we touch the wood or the metal, thermal energy is transferred from our hand to the object. Wood has a very poor ability to conduct thermal energy, so the spot we are touching quickly warms up to about the same temperature as our hand. Metal, however, is a very efficient conductor, and the energy from our finger rapidly spreads into the entire object. The spot we are touching does not warm up much and therefore feels cooler than the wood.

Thermal Expansion of Solids

Most solids expand when heated. For a long, thin piece of solid material, the expansion occurs principally along the long axis of the material.

If we do not change the temperature of the material too much, we can make the assumption that the amount of expansion is linear, that is, the expansion is directly proportional to the temperature change. Since the molecular bonds in the material expand equally, the total amount of expansion is also proportional to the length of the material. For a given temperature change, the increase in the length of a piece of metal that is 100 cm long is twice the increase in a 50 cm length of the same material.

 MAKING CONNECTIONS

Using Physics in the Kitchen

When we have a jar that is difficult to open, it can help to run hot water over the lid of the jar. The metal in the lid expands, making it easier to unscrew.

For linear thermal expansion, the coefficient of thermal expansion, α, is defined as

KEY EQUATION
$$\alpha = \frac{\Delta L}{L} \frac{1}{\Delta T} \qquad (16\text{-}14)$$

where ΔL is the observed increase in length, L is the original length of the object, and ΔT is the temperature change.

To calculate the change in length of a sample, we rearrange Equation (16-14):

$$\Delta L = L\alpha\Delta T \qquad (16\text{-}15)$$

Because the expansion depends on the details of the bonds between the atoms in the solid, different materials expand to a greater or lesser extent. Table 16-3 lists some coefficients of linear thermal expansion for various materials.

Table 16-3 Coefficients of linear thermal expansion for various materials

Material at 20°C	Coefficient of thermal expansion (10^{-6}/°C)
Aluminum	23
Brass	19
Carbon steel	10.8
Concrete	12
Copper	17
Diamond	1
Glass	8.5
Gold	14
Iron	11.1
Lead	29
Pine (perpendicular to the grain)	34
Stainless steel	17.3
Water	~69

When designing devices that are made with different materials, it is important to keep in mind these differing coefficients to ensure that the devices function properly across the range of temperatures in which they are expected to operate.

EXAMPLE 16-2

Thermal Expansion of a Rod

A long, thin rod of aluminum is measured to have a length of 20.0°C. What will be the change in the length of the rod at 30.0°C?

SOLUTION

According to Table 16-3, the coefficient of linear thermal expansion for aluminum is 23.0×10^{-6}/°C. We use Equation (16-15) to find the change in length:

$$\Delta L = L \alpha \Delta T$$

$$= (0.250 \text{ m}) (23.0 \times 10^{-6}/°C) (30.0°C - 20.0°C)$$

$$\Delta L = 5.75 \times 10^{-5} \text{ m}$$

Thus, at 30.0°C, the change in length of the rod will be about 0.058 mm.

Note that if we had cooled the rod by 10°C, the change in length would have been negative.

MAKING CONNECTIONS

Sea Level Rise: Melting Ice or Thermal Expansion?

As the temperature of the atmosphere increases, so too does the temperature of the ocean, although it takes some time for the lower levels of the ocean to warm up (Figure 16-8).

As the oceans warm, they also expand. Some researchers predict that this thermal expansion will make the largest contribution to sea level rise caused by global warming.

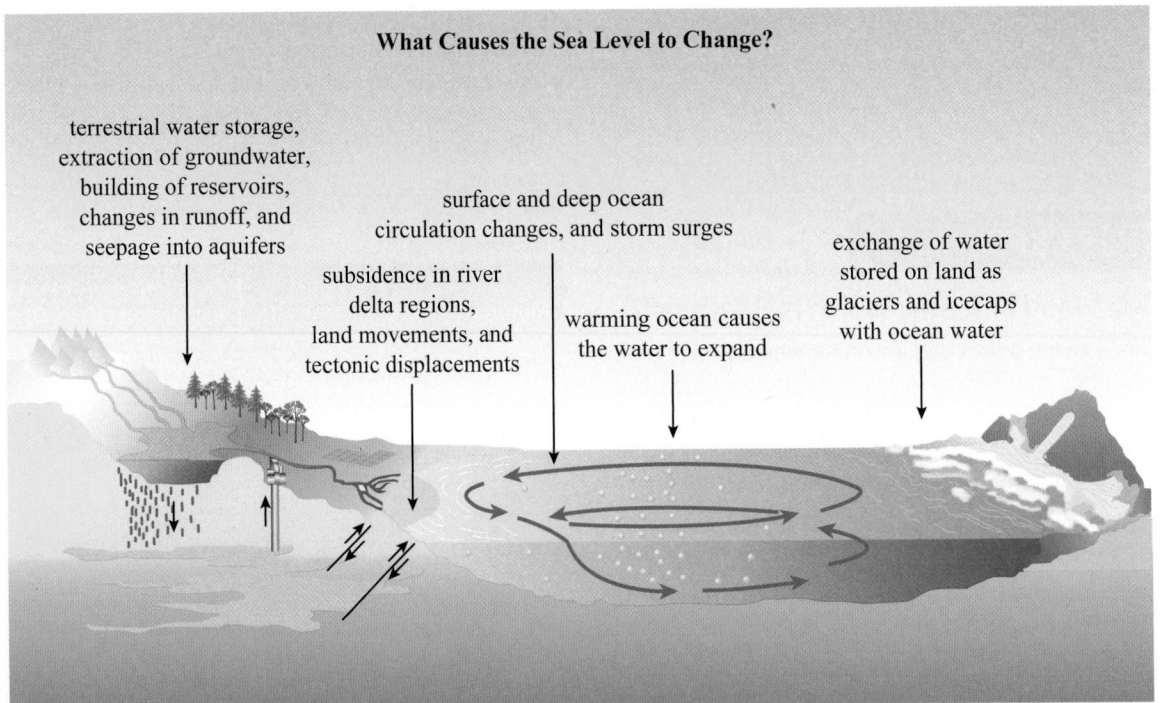

What Causes the Sea Level to Change?

terrestrial water storage, extraction of groundwater, building of reservoirs, changes in runoff, and seepage into aquifers

surface and deep ocean circulation changes, and storm surges

exchange of water stored on land as glaciers and icecaps with ocean water

subsidence in river delta regions, land movements, and tectonic displacements

warming ocean causes the water to expand

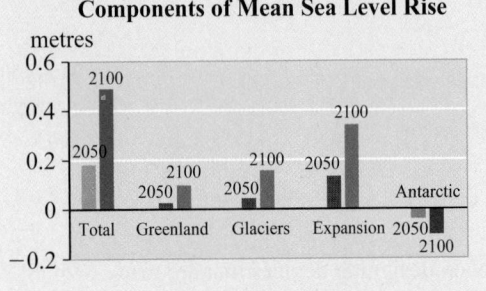

Components of Mean Sea Level Rise

Various assumptions need to be made to model sea level rises. The model depicted to the left assumes rapid economic growth, a global population that peaks in the 21st century and then declines, and the rapid introduction of new and more efficient technologies. The model also assumes a more global economy with a substantial reduction in regional differences in per capita incomes. Another key assumption is what energy sources we rely on, and this model assumes a fossil fuel intensive energy supply.

UNEP/GRID-Arendal

Figure 16-8 Thermal expansion and global warming

LO 5

16-5 Thermometers and Temperature Scales

Temperature-sensitive properties of materials are the basis for various types of thermometers, which convert measurements of quantities such as volume and resistance into temperature readings. In order to quantify the measurements, we must assign a temperature scale to the readings of our thermometer. The most common scale in the world today is the Celsius scale, which assigns a temperature of 0°C to the freezing point of water and 100°C to the boiling point. In practice, we also need to specify the pressure for these calibration points because many of them also depend on the pressure. For example, we need to increase the cooking times for foods boiled at high altitudes because, at higher altitudes, the air pressure is lower and the boiling temperature of water is also lower.

Celsius and Fahrenheit There are several different temperature scales in use today. In addition to the Celsius scale, another common scale is the Fahrenheit scale, which was used in Canada until the 1970s, when the metric system was introduced. Fahrenheit is still used in the United States and Belize. However, even in some countries that have "gone metric," we still see the Fahrenheit scale on some household appliances, most commonly ovens. In the Fahrenheit system, the freezing point of water is 32°F, and the boiling point is 212°F. We can use these reference points to convert temperatures from Celsius to Fahrenheit. In the Celsius system there are 100 increments between freezing and boiling, and in Fahrenheit there are 180 increments. Therefore, the ratio of a Fahrenheit degree to a Celsius degree is 180°F/100°C = 9°F/5°C. We also note that 0°C corresponds to 32°F. Thus,

$$°F = \frac{9°F}{5°C} \times °C + 32°F \quad °C = \frac{5°C}{9°F} \times (°F - 32°F) \quad (16\text{-}16)$$

EXAMPLE 16-3

Temperature Conversion

John, a hungry physics student, wants to cook his favourite recipe, Tandoori chicken. The recipe instructs him to preheat the oven to 175°C. However, the dial on his oven is labelled in Fahrenheit. What setting should John use?

SOLUTION

John needs to find the Fahrenheit equivalent for a Celsius temperature. Using Equation (16-16), we get

$$°F = \frac{9°F}{5°C} \times 175°C + 32°F$$
$$= 315°F + 32°F$$
$$°F = 347°F$$

Kelvin The Kelvin scale was developed by William Thomson (British physicist and engineer, also known as Lord Kelvin, 1824–1907) after he discovered that there is a lower limit for temperature, called absolute zero (about −273.15°C). No object can have a temperature below this limit. The Kelvin scale uses the same degree increments as the Celsius scale, but absolute zero is denoted as 0 K. (Note that SI does not use a degree symbol with kelvin units.) As you will see below, the Kelvin scale is particularly useful for thermodynamic calculations and predicting the behaviour of gases. Table 16-4 lists some common temperatures in Celsius, Kelvin, and Fahrenheit.

Table 16-4 Some common temperatures listed in Celsius, Kelvin, and Fahrenheit

	Celsius	Kelvin	Fahrenheit
Absolute zero	−273.15°C	0 K	−459.67°F
Freezing point of water	0°C	273.15 K	32°F
Typical normal body temperature	37°C	310.15 K	98.6°F
Room temperature	20°C	293.15 K	68°F
Surface temperature of the Sun	~6 000°C	~6 273 K	~10 832°F

LO 6

16-6 The Zeroth Law of Thermodynamics

Suppose we place a thermometer in thermal contact with an object that is at some steady temperature. We will assume that our thermometer is so small that it does not affect the temperature of the object. The thermometer reading will change when we first establish contact with the object, but the reading will eventually reach a steady state. At this point, we say that the thermometer is in thermal equilibrium with the object. Two objects are said to be in **thermal equilibrium** when placing them in thermal contact with each other causes no change in the temperature of either object—they are at the same temperature.

If we now place the thermometer in thermal contact with a second object and find that the thermometer stabilizes at the same reading, we can say that the thermometer is in thermal equilibrium with the second object.

If we then place the first and second objects in thermal contact with each other, the reading on our thermometer will not change, regardless of which object it contacts. This observation demonstrates the **zeroth law** of thermodynamics: If object A is in thermal equilibrium with object B, and object A is also in thermal equilibrium with object C, then object C is also in thermal equilibrium with object B.

In other words, if we get the same temperature reading on a thermometer from two different objects, then those objects are at the same temperature, and if we put them together, their temperatures will not change.

Table 16-5 Densities and interatomic distances for phases of water

Material	Density (g/cm^3)	Typical interatomic distance (nm)
Ice at −10°C	0.998	≈0.3–0.4
Water at 40°C	0.992	≈0.3–0.4
Water vapour at 100°C and 1 atm	0.000 59	≈3–4

✓ CHECKPOINT

C-16-5 Holding Hands

When we hold hands with another person, we notice that the other person's hand feels warm, even if the temperature of the other person is exactly the same as ours. Are we in thermal equilibrium with the other person? If so, why does the person's hand feel warm to us?

C-16-5 Our hand is in thermal equilibrium with the other person's hand if that person is at the same temperature as us. Our skin is usually in contact with air that is cooler than our body temperature. As a result, thermal energy is constantly flowing out of our bodies into our surroundings, and our skin is at some temperature that is lower than our body's core temperature. Further, we have evolved to feel comfortable (i.e., neither hot nor cold) at "room temperature," so when we hold hands, the temperature of the surface of our skin is elevated above its normal temperature, and that feels warm to us.

16-7   Ideal Gases

One difficulty with the system of thermometry we have described so far is getting two thermometers to agree with each other. While identical thermometers are consistent with each other, two different thermometers typically agree precisely only at common calibration points. Much experimentation has been done to develop thermometers that are accurate over their full range. The discrepancies between various types of thermometers are largely due to differences in the materials used to construct them. The particles in solids and liquids are relatively close together and interact strongly with each other. These interactions differ in the various materials and affect how the materials respond to changes in temperature. Consequently, the materials can differ both in the linearity of their response and in the temperature range in which the nonlinear response becomes significant.

Gases, however, have greater interatomic distances and, consequently, much weaker interactions between particles. The behaviour of gases is much more weakly dependent on the type of gas being studied than is the case for solids and liquids. Table 16-5 compares the densities and interatomic distances for ice, liquid water, and water vapour.

Much experimentation has been done with gases to determine the relationships between their pressure, temperature, and volume. The motivation for some of these studies was not to develop better thermometers, but to understand and improve the hot-air balloon!

ONLINE ACTIVITY

The e-resource that accompanies every new copy of this textbook contains an Online Activity using the PhET simulation "Gas Properties." Work through the simulation and accompanying questions to gain an understanding of gas properties.

It was found that all gases exhibit the same behaviour at low densities, regardless of the chemical composition of the gas. This behaviour is described by the **ideal gas law**:

KEY EQUATION
$$PV = NkT \qquad (16\text{-}17)$$

where P is the pressure, V is the volume, N is the number of particles in the gas, T is the temperature in kelvins, and k is Boltzmann's constant, 1.38×10^{-23} J·K^{-1} (named for the Austrian physicist Ludwig Boltzmann, 1844–1906).

MAKING CONNECTIONS

Moles versus Number of Particles

Chemists often write the ideal gas law in terms of the number of moles of gas that are present. The amount (number of moles in the gas) is the number of particles in the gas expressed as a multiple of Avogadro's number, 6.02×10^{23}. Avogadro's number, N_A, is defined as the number of atoms in exactly 12 g of carbon-12. Denoting the amount as n, we can then write

$$PV = nN_A kT = nRT$$

where

$$R = N_A k = 8.314\,472 \text{ J·K}^{-1} \cdot \text{mol}^{-1}$$

EXAMPLE 16-4

Using the Gas Law

The pressure in a 5.00 L flask of gas is 200 kPa, and the temperature is 300 K. Assuming that the gas is ideal, how many gas particles are in the flask?

SOLUTION

We can use the ideal gas law and solve for N:

$$N = \frac{PV}{kT}$$

We can now substitute in the known quantities. However, we must be careful to convert the units properly:

$1\,\text{Pa} = 1\,\text{N/m}^2 = 1\,\text{kg} \cdot \text{m}^{-1} \cdot \text{s}^{-2}$;

$1\,\text{J} = 1\,\text{N} \cdot \text{m} = 1\,\text{kg} \cdot \text{m}^2 \cdot \text{s}^{-2}$; $1\,\text{L} = 1000\,\text{cm}^3 = 10^{-3}\,\text{m}^3$

Using these conversions we find that

$$N = \frac{PV}{kT} = \frac{(200\,000\,\text{kg} \cdot \text{m}^{-1}\text{s}^{-2})\,(5 \times 10^{-3}\,\text{m}^3)}{(1.38 \times 10^{-23}\,\text{kg} \cdot \text{m}^2\text{s}^{-2}\text{K}^{-1})\,(300\,\text{K})}$$

$$= 2.42 \times 10^{23}$$

Making sense of the result:

At first glance, the result seems like a huge number. However, one mole of gas at standard temperature (0°C) and pressure (101.325 kPa) occupies 22.4 L. We have arrived at a result of 2.42×10^{23} particles, which is about one-third of a mole. Our sample is at approximately twice standard pressure but somewhat less than one-quarter of the volume, so we would expect to have approximately 0.5 mol of gas. However, our temperature is higher than STP, so we should have slightly less than 0.5 mol. Thus, our result seems quite reasonable.

Note that real gases behave like ideal gases only at very low densities. Nonetheless, it is somewhat remarkable that all gases do behave the same at low densities, and we will devote some time to understanding why that is so.

Recall that gases are much less dense than liquids and solids and that the typical distance between particles in a gas is large. Consequently, gas particles in a container zoom about with relatively few interactions with the other gas particles. As we reduce the density of the gas, the fraction of the time that a particle spends interacting with other particles becomes smaller and smaller. As a result, the details of those interactions become less and less significant in determining the behaviour of the gas.

Figure 16-9 shows a generic interatomic potential. At large separations, the interaction energy goes to zero. At small separations, the interaction energy is very large and positive: the particles strongly repel each other. The minimum in the potential corresponds to the equilibrium bond length. The particles in a low-density gas spend most

of their time at large interparticle distances. When they do collide, the large repulsive force dominates the interaction. Consequently, a reasonable approximation to this potential, called the hard sphere approximation, is shown in Figure 16-10.

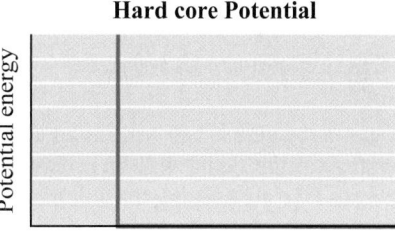

Hard core Potential

Figure 16-10 The hard sphere approximation for the interatomic potential for a dilute gas. The particles experience an infinitely strong repulsive force when they collide but do not interact otherwise.

Even though the approximate potential in Figure 16-10 does not have a minimum, this is not significant for a dilute gas, because the minimum is responsible for the attractive forces that ultimately produce the liquid and solid phases, which are not observed to occur in a dilute gas. Another way of thinking about it is that in a dilute gas, the particles do not spend a lot of time close to each other; thus, the short-range details of the potential are not important.

It is possible to calculate the equation of state for a system of particles that interact through the hard sphere potential, although the details are beyond the scope of this book. The result is the ideal gas law, $PV = NkT$.

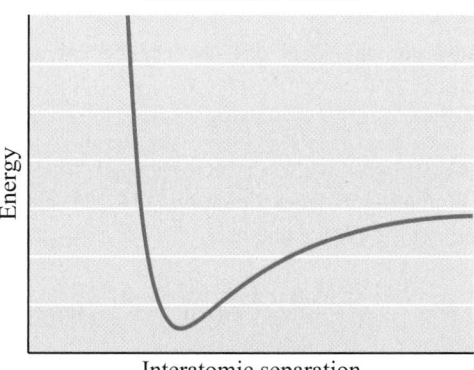

Interatomic Potential

Figure 16-9 A typical interatomic potential

THERMODYNAMICS

16-8 The Constant-Volume Gas Thermometer

The discovery that all gases have the same behaviour in the low-density limit provided the basis for a universal thermometer against which all other thermometers could be calibrated: the **constant-volume gas thermometer**. This thermometer measures the temperature of a gas-filled bulb. In Figure 16-11, this bulb is shown submerged in a liquid of unknown temperature. The bulb on the right is filled with mercury, and its height can be adjusted so that the mercury in the left-hand tube is brought to the zero mark, keeping the volume of gas constant. The difference in height between the left- and right-hand tubes can then be used to determine the pressure in the gas-filled bulb.

Figure 16-11 A constant-volume gas thermometer

If the gas in the bulb is behaving as an ideal gas, then the temperature of the liquid will be proportional to the pressure because we are keeping the volume and quantity of the gas constant:

$$T = CP \tag{16-18}$$

where C is a constant that must be determined.

The pressure in the bulb is given by

$$P = P_{atm} - \rho_{Hg}gh \tag{16-19}$$

where P_{atm} is the atmospheric pressure, ρ_{Hg} is the density of mercury, g is the acceleration due to gravity, and h is the height difference between the two mercury columns.

To determine C, we place the bulb at some known temperature, say the triple point of water (see Section 16-13), and we measure the pressure. At this point we find that

$$T_3 = CP_3 \tag{16-20}$$

Eliminating C from Equations (16-18) and (16-20), we find

$$T = T_3 \frac{P}{P_3} = (273.16 \text{ K}) \frac{P}{P_3} \tag{16-21}$$

Carrying out this procedure with different gases, we find that we get consistent results only at very low densities. By making a series of measurements with successively smaller densities, we can find a limit that gives a precise value for the temperature:

$$T = (273.16 \text{ K}) \lim_{\substack{gas \to 0 \\ density}} \frac{P}{P_3} \tag{16-22}$$

It was discovered that for all gases at the low-density limit, the pressure goes to zero at the same temperature: 0 K. Figure 16-12 shows data confirming this discovery.

Figure 16-12 The three lines represent running the experiment with successively smaller quantities of gas in the system. The lines all go to zero at approximately −273°C.

16-9 Temperature and Mechanical Energy

We now return to the question of the relationship between the temperature—a macroscopic thermodynamic variable—and the molecular-level details of the system under consideration.

Recall that for a gas we showed that

$$P = \frac{1}{3} \frac{mN(v^2)_{avg}}{V} \tag{16-23}$$

However, the ideal gas law can be solved for the pressure:

$$P = \frac{NkT}{V} \tag{16-24}$$

Eliminating P from Equations (16-23) and (16-24) and solving for kT, we obtain

$$kT = \frac{VmN(v^2)_{avg}}{3NV} = \frac{1}{3} m(v^2)_{avg} = \frac{2}{3}\left(\frac{1}{2}m(v^2)_{avg}\right) \tag{16-25}$$

Note that the product kT has units of energy. The last term in parentheses in Equation (16-25) is the average kinetic energy of a particle in the gas. Rearranging Equation (16-25) yields

KEY EQUATION
$$\frac{3}{2}kT = \frac{1}{2}m(v^2)_{avg} \qquad (16\text{-}26)$$

Equation (16-26) provides us with an elegant interpretation of the temperature of a gas: the **temperature** of a gas is a measure of the average kinetic energy of the gas particles.

Equipartition of Energy

We now consider the factor of 3 that appears on the left-hand side of Equation (16-26). It originated in Equation (16-13), which we derived, arguing that

$$\left(v_x^2\right)_{avg} = \left(v_y^2\right)_{avg} = \left(v_y^2\right)_{avg} = \frac{(v^2)_{avg}}{3} \qquad (16\text{-}12)$$

The factor of 3 results from the fact that there are three directions in which the gas particles can move. If the gas were confined to the xy-plane, we would have concluded that

$$\left(v_x^2\right)_{avg} = \left(v_y^2\right)_{avg} = \frac{(v^2)_{avg}}{2} \qquad (16\text{-}27)$$

In both cases, the energy in the system is equally divided among the "degrees of freedom" that the system has. The number of ways in which a system can absorb energy equals the number of degrees of freedom for the system. For an ideal gas in a container, there are three directions in which the gas particles can move, and the kinetic energy associated with those three directions is equal. The energy is "equally partitioned" between the three directions.

Energy considerations are more complex for a gas in which the particles are molecules instead of single atoms. For example, the atmosphere is composed of approximately 78% nitrogen gas, which consists of nitrogen molecules, N_2. A gas molecule has three degrees of freedom associated with its translational kinetic energy. However, the molecule can also absorb kinetic energy in other ways, which give it more degrees of freedom than a single atom.

A nitrogen molecule, or any diatomic molecule, can rotate and vibrate. In Chapter 8, you learned that the kinetic energy associated with a rotating object is

$$K_{rot} = \frac{1}{2}I\omega^2 \qquad (16\text{-}28)$$

where I is the moment of inertia about the axis of rotation, and ω is the angular velocity.

For a diatomic molecule, the rotational inertia about the axis that joins the two atoms is negligible. However, there is a significant moment of inertia about the two axes that are perpendicular to the axis that joins the two atoms. Since neither of these two rotational axes is preferred, the energy is evenly distributed between them. Thus, a diatomic gas molecule has two additional degrees of freedom associated with the rotational kinetic energy.

There is also energy associated with vibrations in a diatomic gas. The chemical bond that joins the two atoms is not infinitely rigid, and it has a potential as shown in Figure 16-13. This potential curve is called the Lennard–Jones potential. The equilibrium separation corresponds to the minimum in the potential. However, the minimum in the potential is a stable equilibrium point. For reasonably small excursions from the equilibrium point, the potential can be modelled as a parabolic well.

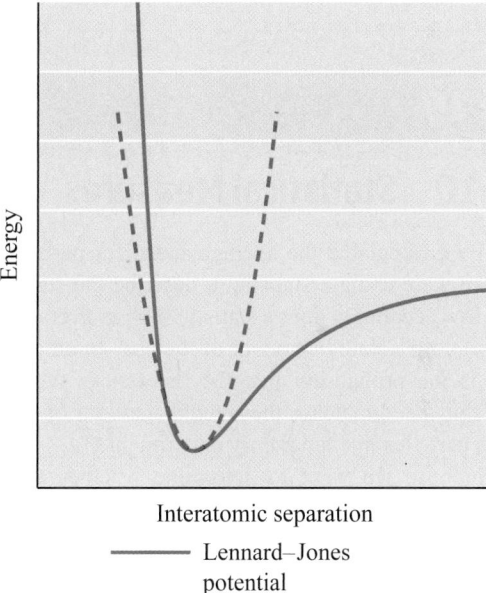

Figure 16-13 The interatomic potential in a diatomic molecule. The solid curve represents the Lennard–Jones potential, and the dashed curve shows a parabolic approximation.

As we know from Chapter 14, a parabolic potential is characteristic of a simple harmonic oscillator. Associated with the oscillator are two energies: the kinetic energy and the potential energy of vibration. Thus, a diatomic gas molecule has two additional degrees of freedom associated with its vibrational motion.

The effects of these additional degrees of freedom are observed in experimental data, but only at sufficiently high temperatures. A full understanding of molecular energy requires quantum theory, which is introduced in Chapter 31.

THERMODYNAMICS

Are there always three degrees of translational kinetic energy in a gas? Can atoms and molecules be confined to an xy-plane? In fact, there are various techniques for making such systems. One involves creating layered structures of semiconducting materials. At the interface between the two appropriately chosen materials, a large potential well can be created that confines the electrons in a very narrow region at the interface, as illustrated in Figure 16-14.

The electrons are free to move parallel to the interface. In the perpendicular direction, the confinement is such that the laws of classical physics no longer apply, and we must use quantum mechanics. (Strictly speaking, we need to use quantum mechanics in the other direction as well, but the quantization effects are very small.) The result is that the energy for motion in the z-direction becomes quantized, with only certain discrete values allowed. If we cool such a device to very low temperatures, the electrons are all forced into the lowest energy level for motion perpendicular to the interface

while remaining relatively free to move in the xy-plane parallel to the interface. Studies of such a system in strong magnetic fields led to the discovery of the integer quantum Hall effect. In 1985, the Nobel Prize in Physics was awarded to German physicist Klaus von Klitzing for this discovery.

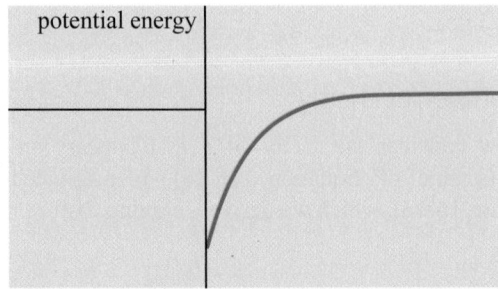

Figure 16-14 Deep potential "well" near the sample surface

LO 10

16-10 Statistical Measures

When we calculated the average speed for particles in a gas, we were using a statistic to describe the state of the gas. How accurately does a statistic such as average speed represent the actual behaviour of the gas? In other words, what is the probability that the theoretical average we have calculated matches the actual behaviour of a sample of gas particles in a laboratory experiment?

Suppose we want to determine whether a coin is "fair," that is, balanced such that there is an equal likelihood of getting either heads or tails when we toss the coin. To quantify the results of the coin toss, we count heads as 1 and tails as 0. If the coin is fair, the average value of a large number of coin tosses should be 0.5. The results of such an experiment are shown in Figure 16-15. We find that the running average gives a reliable description of the coin only when the number of coin tosses is quite large.

Similarly, there must be a large number of particles in a system in order for a statistical thermodynamic description to be applicable. As the number of particles in the sample increases, so does the probability that the

Figure 16-15 Running average value in a coin toss experiment

statistics accurately describes the sample. As the number of particles approaches infinity, the system approaches the **thermodynamic limit**, where the statistical measures are completely accurate. For practical purposes, when the number of particles in a system is of the order of Avogadro's number, the system behaves as though it were at the thermodynamic limit.

However, statistical measures do not give any details about the behaviour of an individual particle. For example, the average speed gives no indication of the proportion of particles that are moving faster or slower than the average at any given moment.

EXAMPLE 16-5

Road Trip Statistics

Suppose that a family drives from Nova Scotia to Manitoba and back for their summer holiday. The total time for the trip is three weeks, with lots of stops along the way to visit and

sightsee. The distance from Halifax to Winnipeg is approximately 3700 km. When they are in the car, they drive at a speed of 120 km/h. What is the average speed? Plot the speed probability distribution.

SOLUTION

To determine the average speed, we divide the total distance travelled, 7400 km, by the total time, 504 h:

$$\text{speed}_{\text{avg}} = \frac{\text{total distance travelled}}{\text{total time for trip}} = \frac{7400 \text{ km}}{504 \text{ h}} = 14.7 \text{ km/h}$$

On this basis alone, you might conclude that this family has very cautious drivers, not to mention being fairly annoying to fellow motorists.

To construct the speed probability distribution, we need to calculate the probability that the car is travelling at a particular speed. We already know that the total time for the trip is 504 h. The time spent driving is

$$\frac{7400 \text{ km}}{120 \text{ km/h}} = 61.7 \text{ h}$$

Therefore, the probability of finding the speed to be 120 km/h at any time on the trip would be

$$\text{Prob}(120 \text{ km/h}) = \frac{\text{time at 120 km/h}}{\text{total time for trip}} = \frac{61.7 \text{ h}}{504 \text{ h}} = 0.122$$

The remainder of the trip, 504 h − 61.7 h = 442.3 h, was spent at a speed of 0 km/h. Thus, the probability of the speed being 0 km/h is

$$\text{Prob}(0 \text{ km/h}) = \frac{442.3}{504} = 0.878$$

The probability of finding any other speed is zero, if we neglect the short times it takes to get up to speed

and slow to a stop. Figure 16-16 shows a plot of these probabilities.

Figure 16-16 Speed probability distribution

Making sense of the result:

Clearly, the speed probability distribution gives us more information about the trip than simply knowing the average speed. For example, we can see that when the family does drive, they travel pretty quickly, but they do not drive that often. Nevertheless, a lot of details are still unknown—how many stops were made and where, for example.

What is the probability that an individual particle in a gas will have a speed within a certain range? An apparatus for measuring the distribution of particle speeds is shown in Figure 16-17(a). The source oven is filled with a gas. The collimating slits ensure that only those gas molecules with horizontal velocity enter the apparatus. The shutter on the left is opened for a brief time, Δt. At some time τ later, the right-hand shutter is also open for a time Δt. The detector produces a signal proportional to the number of particles that impinge upon it. To determine the speeds of the particles that can make it

through the apparatus, we construct a space-time diagram as shown in Figure 16-17(b).

As we can see from Figure 16-17(b), the slowest particles detected travel the distance D in a time $\tau + \Delta t$, and the fastest particles can make it in a time $\tau - \Delta t$. Thus, particles that make it through to the detector while the right-hand shutter is open have a speed, s, that falls in the range of

$$\frac{D}{\tau + \Delta t} \leq s \leq \frac{D}{\tau - \Delta t} \qquad (16\text{-}29)$$

(b)

(a)

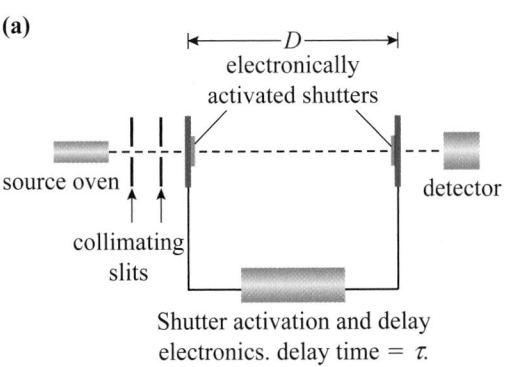

Figure 16-17 (a) An apparatus for measuring molecular speed distributions (b) A space-time diagram for a molecular speed apparatus

where D is the distance between the shutters of the apparatus, τ is the delay between the opening of the left- and right-hand shutters, and Δt is the length of time that the shutters are open.

Only those particles that have a speed in this range will be detected. Particles with faster speeds will hit the right-hand shutter before it opens, and particles with slower speeds will arrive after the shutter has closed. By varying τ, we can select the speed of the particles that reach the detector; thus, we can get data that tells us how many particles there are in a particular speed range. By making Δt small, we can improve the resolution of the measurement, at the cost of increasing the time needed to complete the experiment.

The results of such experiments show that the particle speeds have a well defined distribution that depends upon the temperature of the gas, as shown in Figure 16-18. The experimental results are well modelled by a distribution predicted by a statistical model developed by James Clerk Maxwell (Scottish physicist, 1831–79) and Ludwig Boltzmann (Figure 16-19). The equation they derived for this distribution is

$$f(v) = 4\pi \left(\frac{m}{2\pi kT} \right)^{3/2} v^2 e^{-[mv^2/2kT]} \qquad (16\text{-}30)$$

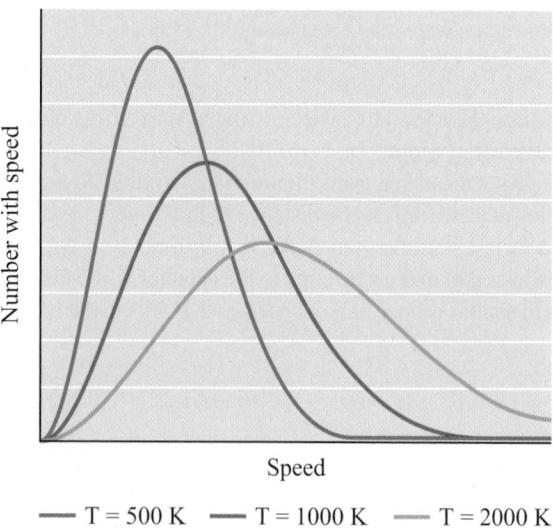

Figure 16-18   The Maxwell-Boltzmann distribution for three different temperatures. It has been shown to be in excellent agreement with experimental data.

The fact that we get a well-defined velocity distribution indicates that the system is near the thermodynamic limit and is in thermal equilibrium. There are many possible ways to distribute the speeds such that the total kinetic energy is the same. However, as the particles collide with each other, they quickly equilibrate, and the speed distribution approaches the Maxwell–Boltzmann distribution.

Figure 16-19   Ludwig Boltzmann's grave in Austria. At the top of the marker you can see Boltzmann's equation for entropy, one of the fundamental concepts of thermodynamics.

LO 11

16-11   Phase Diagrams

Now that we have a set of variables to describe a thermodynamic system, we can begin to discuss the behaviour of a particular system as a function of those variables. We can use a **phase diagram**, which is a diagram that shows the phases of a substance at various temperatures and pressures. A phase diagram for water is shown in Figure 16-20.

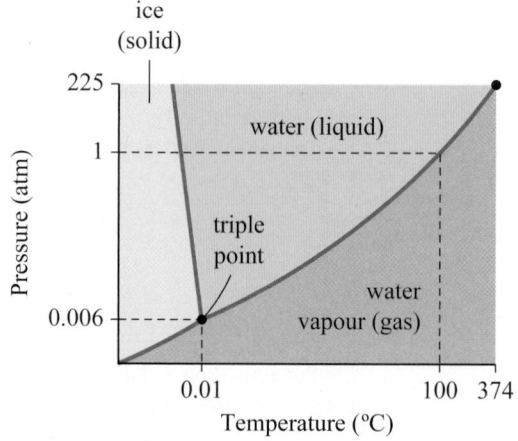

Figure 16-20   A phase diagram for water

We can see that there are distinct regions for each phase. Along the boundaries of each region, different phases can coexist at the same temperature and pressure. In Figure 16-20, there is a broad range of pressures and temperatures where water can coexist in liquid and gas phases along the line extending to the right of the triple point. The conditions for the mist in the image at the beginning of this chapter lie on this line. The triple point, where the three lines intersect in a phase diagram, indicates the pressure and temperature at which the solid, liquid, and gas can all coexist. For water, the triple point occurs at a pressure of 0.006 atm and a temperature of 0.01°C.

Also note that the line separating the solid and liquid phases for water slants upward to the left from the triple point. This slope is a consequence of the fact that water expands when it freezes. If we increase the pressure on solid ice while keeping the temperature constant, we move upward in the phase diagram. Moving upward in the solid region, we cross the boundary into the liquid region, indicating that the ice will melt.

MAKING CONNECTIONS

Water Is Unique

Water is an important exception to the general tendency of liquids and solids to contract as they cool. Water begins to expand as we cool it below about 4°C. The fact that ice floats is a consequence of this expansion.

Lakes cool at the surface due to contact with cold air. If ice were denser than water, bodies of water would fill with ice from the bottom up in freezing weather. Skating on a lake or pond would be possible only after all the water had frozen (Figure 16-21).

Figure 16-21 Outdoor skating

PUTTING IT ALL TOGETHER

In this chapter, we used state variables, such as volume, pressure, and temperature, to describe a macroscopic assembly of particles. We also established a firm basis for the science of thermometry with the zeroth law of thermodynamics.

We used average values to relate state variables to the properties of the individual particles, recognizing that statistical measures are fully valid only for a system with a very large number of particles.

We can relate the pressure that a gas exerts on the walls of its container to the square of the average speed of the gas particles:

$$P = \frac{1}{3}\frac{mN(v^2)_{avg}}{V} \tag{16-13}$$

At the low-density limit, all gases exhibit the same behaviour, which is described by the ideal gas law:

$$PV = NkT \tag{16-17}$$

We can use the universal low-density behaviour of gases to create an ideal gas thermometer:

$$T = T_3\frac{P}{P_3} = (273.16\ \text{K})\frac{P}{P_3} \tag{16-21}$$

The temperature of a gas is a measure of the average kinetic energy of the molecules in the gas:

$$\frac{3}{2}kT = \frac{1}{2}m(v^2)_{avg} \tag{16-26}$$

The zeroth law of thermodynamics is the foundation for the measurement of temperature: If object A is in thermal equilibrium with object B, and object A is also in thermal equilibrium with object C, then object C is also in thermal equilibrium with object B.

A phase diagram captures the complex behaviour of a material as its temperature and pressure are varied.

APPLICATIONS

Applications: thermometry, gas handling, material response to temperature and pressure changes

Key Terms: constant-volume gas thermometer, gas, ideal gas law, liquid, phase, phase diagram, probability distribution, solid, state variable, temperature, thermal equilibrium, thermodynamic limit, thermodynamics, volume, zeroth law of thermodynamics

QUESTIONS

1. Two different thermometers are calibrated by placing them in the same ice-water bath and then in the same pot of boiling water. The respective readings of each thermometer are assigned to be 0°C and 100°C. The scale on each thermometer is divided into 100 equal increments. If we then use these thermometers to measure the temperature of an object that is at some intermediate temperature, will they agree?

2. A piece of sodium metal and a bath of water are at the same temperature. If we place them in thermal contact, their temperatures do not change. However, if we place them in physical contact, we observe a rather spectacular reaction. Discuss.

3. The maximum in a probability distribution indicates the value that is most likely to be observed. Is the most likely value the same as the average value? Use sketches to illustrate your answer.

4. Most materials expand as they are heated. Discuss why this occurs using the shape of the Lennard–Jones potential. Explain why this expansion would be greatly reduced if the potential were parabolic as shown in Figure 16-13.

5. Using the phase diagram of water in Figure 16-20, discuss whether there are limits on the temperatures at which the three phases can be observed. How would your answers change if liquid water did not expand when cooled?

6. A mercury thermometer uses the fact that mercury expands when heated. The coefficient of thermal expansion for mercury is $6 \times 10^{-5}/°C$. This means that if we heat a column of mercury that is 1 m long and cause its temperature to change by 1°C, the column will increase in length by 6×10^{-5} m. However, we are all familiar with mercury thermometers, where the length of the mercury column seems to change by several centimetres with small changes in temperature. How do mercury thermometers work?

7. Two gases are at the same temperature. Do they necessarily exert the same pressure on the walls of their respective containers?

8. Measuring very high temperatures is a challenge. If we try to use a gas, we must measure its pressure and, hence, use a container. However, imagine trying to insert a container of gas into a lava flow. Considering your own experience with the light that is emitted by hot objects, suggest a method of determining the relative temperatures of such objects.

9. Figure 16-22 shows the pressure in a gas sample as a function of temperature, with the volume held constant. Does this gas behave as an ideal gas?

Figure 16-22 Question 9

10. When we derived an expression for the pressure that a gas exerts on the walls of its container, we assumed that the collisions that the gas particles made with the walls were elastic, and the particles emerged from the collisions with the same speed. If that assumption were invalid, what would happen to the gas in the container?

11. Many of us have experienced the phenomenon of a "sea breeze." We wake up on a summer morning near the water and there is no wind. As the Sun rises and the air begins to warm up, we notice that a breeze develops that blows from the water onto the land. Keeping in mind that winds blow from areas of high pressure to areas of low pressure, explain why sea breezes occur.

12. The ideal gas model assumes that there are no attractive interactions between the molecules of a gas, and this is a good approximation at low densities. At densities where the approximation is not valid, would you expect the pressure of a real gas to be higher or lower than the pressure of an ideal gas? Assume the same quantity, volume, and temperature for the real and ideal gases.

13. Many of us have had the experience of pumping up a bicycle tire with a manual pump. You may have observed while doing so that the hose from the pump to the tire becomes warm to the touch, indicating that the pumping process causes the temperature of the air to increase. Clearly, compressing the gas in the pump causes the volume to decrease and the pressure to increase. We might assume, for example, that if we halve the volume we double the pressure. However, if the temperature also increases, is this assumption true? Explain your answer.

PROBLEMS BY SECTION

For problems, star ratings will be used, (✱, ✱✱, or ✱✱✱), with more stars meaning more challenging problems.

Section 16-1   Solids, Liquids, and Gases

14. ✱✱ The density of solid copper is approximately 9.00 g/cm³, and the atomic mass of copper is 63.5 amu.
    (a) How many atoms of copper are in 1.00 cm³ of copper?
    (b) What is the average volume occupied by one copper atom?
    (c) What is the average distance between adjacent copper atoms?

15. ✱ One mole of gas at standard temperature and pressure occupies 22.4 L. What is the average volume occupied by each gas particle? What is the average spacing between particles?

Section 16-2   State Variables

16. ✱ If we know the temperature of a system in Celsius, is the temperature in Kelvin an additional state variable?

17. ✱ The reading of a thermometer in some system fluctuates. Is the system in thermodynamic equilibrium?

Section 16-3   Pressure

18. ✱ The SI unit of pressure is the pascal. Other common units of pressure are the atmosphere, the bar, the torr, and pounds per square inch (psi). Look up the conversions between these units, and express the standard pressure of 101.325 kPa in units of

(a) atmospheres;
(b) torr;
(c) pounds per square inch.

19. ✱✱ Racing bicycles have very narrow tires, which reduce the rolling friction. Assume that such a tire has a diameter of 70.0 cm with a cross-sectional area of 4.00 cm², and that it is inflated to 110 psi.
    (a) Convert the pressure to pascals.
    (b) Determine the volume of the tire.
    (c) Assuming $mv^2 = 1.24 \times 10^{-20}$ J, how many moles of gas are required to inflate the tire?
    (d) What would the mass of this gas be if it were nitrogen? Helium?

Section 16-4   Temperature and Thermal Expansion of Solids

20. ✱ Bridges are usually constructed with expansion joints so that they do not buckle when heated (Figure 16-23). Assume that the expansion of a bridge is dominated by structural steel with a linear thermal expansion coefficient of approximately 1 % 10⁻⁵/°C. Suppose we wish to design expansion gaps in a bridge that will be exposed to temperatures ranging from −40°C to +40°C.
    (a) At what temperature will the gap be the largest?
    (b) How long can a section be if it can expand no more than 3.0 cm?

Figure 16-23 Problem 20

THERMODYNAMICS

Natursports/Shutterstock

21. ✻ A 2.00 m copper rod has its temperature increased from 20.0°C to 40.0°C. By how much does the length of the rod increase?

22. ✻✻ A thin aluminum ring has a circumference of 20.0 cm. The ring is heated from 10.0°C to 50.0°C. By how much does the diameter of the ring change?

Section 16-5 Thermometers and Temperature Scales

23. ✻ Convert −40°F to Celsius.

24. ✻ On February 3, 1947, Weather Service of Canada workers at Snag, Yukon, filed a notch into the glass casing of an alcohol thermometer because the indicator within fell below the lowest number, −80°F. When they later sent the thermometer to Toronto, officials there determined that the temperature had dropped to −81.4°F, the lowest official temperature ever recorded in North America. Convert this temperature to
 (a) Celsius;
 (b) kelvins.

25. ✻ The highest temperature ever officially recorded is 58°C at Azizia, Libya, in the Sahara desert, recorded in 1922. Convert this temperature to
 (a) Fahrenheit;
 (b) kelvins.

Sections 16-7 and 16-8 Ideal Gases and The Constant-Volume Gas Thermometer

26. ✻✻ A hydrogen gas thermometer is found to have a volume of 100.0 cm³ when placed in an ice-water bath at 0°C. When the same thermometer is immersed in boiling liquid bromine, the volume of hydrogen at the same pressure is found to be 121.6 cm³. Assume that the hydrogen gas behaves as an ideal gas.
 (a) Write the ideal gas law for both circumstances.
 (b) Recognizing that the pressure is the same for both cases, solve each equation in part (a) for pressure.
 (c) Set the two equations in part (b) equal to each other, and find the temperature of the bromine.

27. ✻✻ A high-altitude balloon contains helium, whose molar mass is 4.00 g/mol. At the balloon's maximum altitude, its volume is 792 m³ and the outside temperature and pressure are −53.0°C and 5.00 kPa, respectively. Assume that the helium in the balloon is in equilibrium with the outside air temperature and pressure.
 (a) Use the ideal gas law to find the amount of helium in the balloon at maximum altitude.
 (b) Assuming no loss of helium, what would be the volume of the balloon when it was launched from the ground, where the air temperature and pressure are 20.0°C and 101 kPa, respectively?
 (c) If the volume of the balloon when it was launched was 65.0 m³, how many moles of helium were lost during the balloon's ascent?

28. ✻ An 18.0 L container holds 16.0 g of oxygen gas (O_2) at 45.0°C.
 (a) Look up the molar mass of oxygen gas, and use that to determine the amount of gas.
 (b) Use the ideal gas law to find the pressure in the container.

29. ✻ All gases behave as ideal gases at sufficiently low densities. Consider one mole of gas at standard temperature and pressure, and assume that it behaves as an ideal gas.

(a) Calculate the volume occupied by the gas.
(b) If the gas is nitrogen, what is the average speed of a nitrogen molecule?
(c) If the gas is oxygen, what is the average speed of an oxygen molecule?
(d) What is the ratio of the average speeds from parts (b) and (c)?

Section 16-9 Temperature and Mechanical Energy

30. ✻✻ Massive amounts of energy are transferred when the atmosphere close to Earth's surface shifts between daytime and nighttime temperatures.
 (a) Calculate the number of atoms in 1.0 km³ of nitrogen gas at standard temperature and pressure, assuming that nitrogen behaves as an ideal gas.
 (b) Calculate the average kinetic energy of a nitrogen molecule at 25°C.
 (c) Calculate the average kinetic energy of a nitrogen molecule at 15°C.
 (d) How much energy transfers out of 1.0 km³ of nitrogen gas when it cools by 10°C?

31. ✻ What would the average speed of a nitrogen molecule be at the temperatures in problems 24 and 25?

Section 16-10 Statistical Measures

32. ✻ Consider the function $f(t) = at^2$. What is the average value of f between 0 and T seconds? Hint – you will need to do some integration.

33. ✻✻ Use Equation 16-30 to compute the average speed of a molecule in an ideal gas.

Section 16-11 Phase Diagrams

34. Solid carbon dioxide is sometimes called *dry ice* because it goes directly from the solid to the gas phase when warmed at a pressure of one atmosphere. What can you infer about the triple point of carbon dioxide?

35. On a newly discovered planet, the atmospheric pressure is 500 kPa and the temperature is 200 K. What phase(s) will water, oxygen, nitrogen and carbon dioxide be found in? You will need to look up the phase diagrams for these substances.

COMPREHENSIVE PROBLEMS

36. ✻ We have an ideal gas, and we double the quantity of gas. What other quantities would we need to change, and how, for the temperature of the gas to remain constant?

37. ✻ Your body contains approximately 60% water by mass. Estimate the amount of water in your body.

38. ✻✻ Very cold temperatures cannot be measured with a simple mercury thermometer since mercury freezes at −38.72°C. However, solid-state diodes (described in Chapter 32) can be used as sensors for low temperatures. The relationship between the current, I, in a diode and the voltage, V, across the diode is $I = I_0(e^{q_e V/kT} - 1)$, where I_0 is the saturation current, q_e is the fundamental unit of charge (1.602×10^{-19} C), k is Boltzmann's constant, and T is the temperature in kelvins.
 (a) Rearrange this equation to find the relationship between the voltage across the diode and the temperature, assuming a constant current through the diode.

(b) Assuming $I_0 = 1.0 \times 10^{-12}$ A, what voltage will appear across the diode at a temperature of −190°C if a current of 5.0 mA flows through the diode?

39. ✶   The lowest temperatures ever reached in a laboratory are in the order of nanokelvins, 10^{-9} K. Calculate the average speed of a cesium atom in a gas that has been cooled to 1×10^{-9} K.

40. ✶✶   You work for a major theatre chain and have been assigned the job of determining the optimum conditions for running the popcorn machines to get the largest, most uniformly sized popcorn. The popcorn size distributions from two runs of your experiment are shown in Figure 16-24.
(a) Which series has the largest average kernel size?
(b) Which series has the most uniform kernel size?
(c) Which series has the largest number of kernels?
(d) Which series is the more reliable measurement?

series 1   series 2

Figure 16-24   Problem 40

41. ✶   A 30 L container at a pressure of 150 kPa contains three moles of an ideal gas. What is the temperature?

42. ✶✶   A sample of ideal gas is compressed until its volume is one-third of its original volume.
(a) If the compression is carried out at constant temperature, what is the resulting pressure?
(b) If the compression is carried out at constant pressure, what is the resulting temperature?

43. ✶✶✶   We showed that the pressure in a gas is given by $P = \dfrac{1}{3}\dfrac{mNv_{avg}^2}{V}$. Estimate the number of collisions per second per unit area on the walls of the container. Use data for nitrogen at standard temperature and pressure.

44. ✶✶   You are designing storage tanks that will be filled with helium. The tanks will be filled at 20°C to a pressure of 100.0 atm. What maximum and minimum pressures would you expect to observe in a full tank if the ambient temperature will range from −20°C to 30°C?

45. ✶   Gravity pulls the matter in a star toward the centre. When the star eventually collapses, the volume is reduced, and the temperature and pressure increase. The increase in pressure is eventually sufficient to offset the gravity, and the star is said to be in hydrostatic balance. The increase in temperature is enough to allow nuclear fusion to occur, which further increases the temperature, leading to a new equilibrium. The net result is that in the core of a star we observe very high densities, pressure, and temperatures. At the core of the Sun, the density is approximately 150 000 kg·m⁻³, and the temperature is as high as 13.6×10^6 K. In this question, we naively treat the Sun as an ideal gas. Calculate
(a) the pressure at the core of the Sun;
(b) the average speed of a hydrogen atom at the core of the Sun.

46. ✶✶✶   An alternative method of measuring the velocity distribution of particles in a gas is to use a rotating cylinder with a helical groove cut in the side, as shown in Figure 16-25. If the cylinder is rotating at an angular speed of ω, what speed of particle will it be selecting for? Express your result in terms of the length and radius of the cylinder, the helix angle, and the angular speed. Discuss how the width of the groove impacts measurements made with this device.

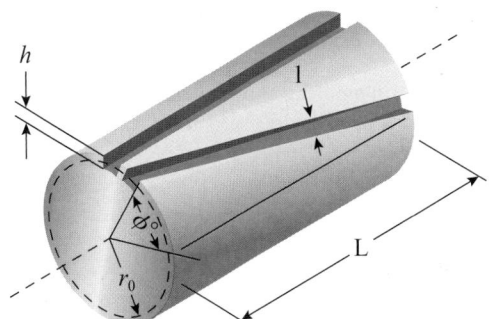

Figure 16-25   Problem 46

47. ✶✶   Table 16-6 contains some observations of the phase that a particular substance is in at various pressures and temperatures. Use the data to construct a rough phase diagram.

Table 16-6   Data for Problem 47

Temperature (°C)	Pressure (atm)	Observed phase
−78	0.5	Gas
−78	1.1	Solid
−78	10	Solid
−57	4	Gas
−57	6	Solid
−57	19	Solid
−50	6	Gas
−50	7	Liquid
−50	15	Solid
−30	7	Gas
−30	8	Liquid
−30	100	Solid

48. ✶✶   Roll a six-sided die, and record the numbers that you roll in a spreadsheet. Calculate the running average value of all of your rolls as you go, and plot your data.

Table 16-7 Data for Problem 49

T (K)	77.0	97.0	117	137	157	177	197	217	237	257	277	297	317
R (Ω)	132.2	134.7	139.5	143.8	146.5	152.0	153.6	157.8	163.7	167.8	168.9	172.2	177.9

49. ✷✷ To construct a thermometer, a physicist decides to use a resistor whose resistance varies as a function of temperature. The physicist and her team construct a circuit that passes a constant current of 1 mA through the resistor; they then measure the voltage across the resistor. Table 16-7 lists the calibration data. Fit the data with a suitable polynomial, and predict the resistance value at a temperature of 400 K.

OPEN PROBLEM

50. ✷✷✷ In problem 43, we found that there is an enormous number of collisions per second with the walls of a container holding a gas. Suppose the pressure sensor in our system has a response time of 1 ms. To what value would we have to reduce the concentration of particles in our system to see fluctuations in the reading of the pressure sensor? State any assumptions you make.

See the text online resources at
**www.physics1e.nelson.com for Open Problems
and Data-Rich Problems related to this chapter.**

Chapter 17
Heat, Work, and the First Law of Thermodynamics

The first step in brewing beer is mixing malted barley and water together to form a "mash" and holding them at a temperature that allows enzymes in the barley to convert starches into fermentable sugars (Figure 17-1). How can the brewer determine what initial temperature of the water will produce the desired temperature for the mixture?

Figure 17-1   Malted barley and water in the mash tun at a brewery

© Cephas Picture Library/Alamy

Learning Objectives

When you have completed this chapter you should be able to:

1 Discuss what is meant by the term "heat."

2 Define specific heat, and perform calculations using heat capacities.

3 Define heats of fusion and vaporization, and use them in calculations.

4 Calculate the work done by a gas during an expansion.

5 State the first law of thermodynamics.

6 Define what is meant by an adiabatic, isothermal, isobaric, and an isochoric process.

7 Describe various heat transfer processes, and perform calculations for radiative and conductive processes.

THERMODYNAMICS

For additional Making Connections, Examples, and Checkpoints, as well as activities and experiments to help increase your understanding of the chapter's concepts, please go to the text's online resources at www.physics1e.nelson.com.

LO 1

17-1   What Is Heat?

The nature of heat puzzled scientists for centuries. Early theories suggested that there was a "caloric fluid" that flowed from a hot object to a cold object. The rather notorious Anglo-American physicist, inventor, and military officer Count Rumford (Benjamin Thompson, 1753–1814) was the first to dispel this notion and make a connection between motion and heat. He performed an experiment in which he bored out the barrel of a brass cannon with a dulled cutting tool in a water bath. He then monitored the temperature of the water to determine the heat produced. He showed that this mechanical process could produce significant heat continuously for a virtually unlimited time. He also noted that the thermal properties of the shavings produced as a result of the boring appeared to be unchanged. On the basis of these results, Rumford claimed that there could not be a caloric fluid because surely it would be depleted and would limit the amount of heat that could be extracted from an object. In addition, if the heat were some type of fluid, you would expect there to be some observable change in the physical properties of a material if the fluid were removed. He suggested that there is a connection between motion and heat.

In Chapter 16, we noted that the temperature of a simple gas is a measure of the average kinetic energy of the particles that make up the gas. Almost every day, we see that adding heat to an object increases its temperature. Based on these observations, it would seem that heating an object somehow transfers energy into the object and, at least in the case of an ideal gas, the energy is stored in the form of kinetic energy. For more complex systems, the energy may also be stored as potential energy. **Heat** is the energy that is transferred from one object to another when a temperature difference exists between the two objects. Because heat is a form of energy, called thermal energy, it can be measured in joules.

✓ CHECKPOINT

C-17-1 Thermal Energy and Temperature

Does an object at a higher temperature always contain more thermal energy than an object at a lower temperature?

C-17-1 No. As we will see in quantitative detail in Section 17-2, in general, different materials exhibit different temperature changes when we add the same amount of thermal energy to them.

We can gain some insight into how this energy transfer occurs by considering what happens when we place a hot object and a cold object in contact with each other. The particles in the hot object have greater kinetic energy than the particles in the cold object. When they collide, energy

is transferred from the more energetic particles to the less energetic particles. Consequently, the average kinetic energy decreases in the hot object and increases in the cold object, and the temperatures of the objects change correspondingly.

LO 2

17-2   Temperature Changes Due to Thermal Energy Transfer

We now consider quantitatively how the temperature of an object changes when it absorbs thermal energy. Since temperature is a measure of the average kinetic energy of the particles (at least for an ideal gas), we would expect that the temperature change due to a thermal energy flow into a material would simply be proportional to the amount of thermal energy put into the system:

$$\Delta T \propto Q \qquad (17\text{-}1)$$

where ΔT is the temperature change, and Q is the thermal energy flow into the system.

Notice that when $Q > 0$, $\Delta T > 0$, and when $Q < 0$, $\Delta T < 0$.

Two other factors affect the change in temperature: the mass of the material present and the type of material. The fact that the temperature change depends on the amount of material makes sense. For example, if we double the number of particles, the quantity of kinetic energy added per particle is halved, and we would expect the change in temperature to be halved as well.

ONLINE ACTIVITY

The e-resource that accompanies every new copy of this textbook contains an Online Activity using the PhET simulation "States of Matter." Work through the simulation and accompanying questions to gain an understanding of how molecular motion changes according to the state of matter.

As demonstrated in the "States of Matter" online activity, the increase in temperature also depends on the number of different ways that particles in the material can move and, hence, have kinetic or potential energy. The atoms of a simple monatomic gas, such as helium, can only have kinetic energy from linear motion. A diatomic gas, such as oxygen, can also have rotational kinetic energy, vibrational kinetic energy, and vibrational potential energy. These different modes are illustrated in Figure 17-2.

As we discussed in Chapter 16, these different ways of absorbing energy are called degrees of freedom, and the absorbed energy is shared equally between all the degrees

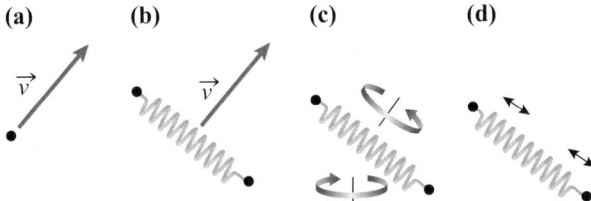

(a)   **(b)**   **(c)**   **(d)**

Figure 17-2 (a) A monatomic particle with translational kinetic energy (b), (c), and (d) A diatomic particle with translational, rotational, and vibrational energy

of freedom. As a result, the absorption of a given amount of thermal energy causes a much smaller increase in translational kinetic energy for a diatomic gas than for an equal mass of monatomic gas.

We can quantify these differences by defining the **heat capacity**, C, of an object as the ratio of the thermal energy added to the object to the resulting temperature change:

$$C = \frac{Q}{\Delta T} \tag{17-2}$$

An object with a large heat capacity undergoes a small temperature change for a given thermal energy input compared to an object with a small heat capacity. The units of heat capacity are J/K.

The amount of thermal energy that a sample can absorb depends on both the type and quantity of material in the sample. We define the **specific heat capacity**, c, as $c = C/m$, which is the quantity of thermal energy required to raise the temperature of a unit mass by a unit degree. This allows us to relate thermal energy flows and temperature changes:

KEY EQUATION $$Q = mc\Delta T \tag{17-3}$$

where Q is the amount of thermal energy, m is the mass of the sample, c is the specific heat capacity, and ΔT is the temperature change, $T_{\text{finial}} - T_{\text{initial}}$.

Table 17-1 lists the specific heat capacities of some common substances.

Table 17-1 Specific heat capacities of some common materials at constant pressure (101 kPa)

Material	Specific heat capacity, c (J/(kg·K))
Copper (20°C)	390
Lead (20°C)	129
Glass (20°C)	840
Air (20°C)	1000
Animal tissue (20°C)	3500
Water, ice (−5°C)	2100
Water, liquid (15°C)	4186
Water, steam (110°C)	2010

EXAMPLE 17-1

Temperature Change Due to an Input of Thermal Energy

Find the resultant temperature change when 100 g of water absorbs 1000 J of thermal energy.

SOLUTION

We simply rearrange Equation (17-3) to solve for the temperature change:

$$\Delta T = \frac{Q}{mc} = \frac{1000 \text{ J}}{(0.1 \text{ kg})(4186 \text{ J/(kg·K)})} = 2.39 \text{ K}$$

Making sense of the results:

1000 J is the energy consumed by a 100 W lightbulb in 10 seconds so we would not expect a large increase in T.

MAKING CONNECTIONS

Water Helps Regulate Temperature

You will note that the specific heat capacity of water is the highest value listed in Table 17-1. This property of water helps us regulate our body temperature and plays a significant role in moderating the climate of coastal areas. For example, in Halifax, Nova Scotia, average daily temperatures are −4.4°C in January and 18.6°C in July. In contrast, the corresponding averages for Winnipeg, Manitoba, are −17.8°C and 19.5°C, respectively. The climate in Halifax is moderated by the Atlantic Ocean.

LO 3

17-3 The Flow of Thermal Energy Between Objects

We now consider what happens when a hot object is in thermal contact with a cold object. For simplicity, we will assume that the two objects are completely thermally insulated from their surroundings. Thus, there is no thermal energy flow into or out of the surroundings, any thermal energy that leaves the hot object will go into the cold object, and

$$Q_{\text{hot}} = -Q_{\text{cold}} \tag{17-4}$$

We can replace the thermal energy flows for each object with the mass, specific heat, and temperature change from Equation (17-3) to get

$$m_{\text{hot}}c_{\text{hot}}\Delta T_{\text{hot}} = -m_{\text{cold}}c_{\text{cold}}\Delta T_{\text{cold}} \tag{17-5}$$

THERMODYNAMICS

✓ CHECKPOINT

C-17-2 Heat and Temperature

When you bite into a hot, fruit-filled pastry, you can burn your mouth on the filling but not likely on the crust. Does this mean that the filling is at a higher temperature?

C-17-2 No, the entire pastry is at approximately the same temperature. However, the heat capacity of the filling is much higher than the heat capacity of the crust, so there is more energy in the filling. When the pastry contacts your mouth, much more energy is transferred to your mouth from the filling than from the crust; thus, the final temperature of your mouth is increased to the point of pain.

EXAMPLE 17-2

Mashing

A particular recipe for beer calls for 3.00 kg of crushed barley to be mixed with 14.0 L of water. The desired temperature of the mixture is 66.0°C. The crushed barley is at room temperature, and malted barley has a specific heat of 1600 J/(kg·K). What should the initial temperature of the water be?

SOLUTION

Table 17-1 lists a specific heat of 4186 J/(kg·K) for water. Since 1 L of water has a mass of 1 kg, we have 14 kg of water. Applying Equation (17-5), we get

$$m_{water}c_{water}(T_{f,water} - T_{i,water}) = -m_{malt}c_{malt}(T_{f,malt} - T_{i,malt})$$

Since $T_{f,water} = T_{f,malt} = 66°C$,

$$T_{i,water} = \frac{-m_{malt}c_{malt}(66°C - T_{i,malt}) - m_{water}c_{water}(66°C)}{-m_{water}c_{water}}$$

$$= \frac{-3\text{ kg} \cdot 1600\text{ J/(kg·°C)} \cdot (66°C - 20°C) - 14\text{ kg} \cdot 4186\text{ J/(kg·°C)} \cdot 66°C}{-14\text{ kg} \cdot 4186\text{ J/(kg·°C)}}$$

$$= 69.8°C$$

Note that we have kept our temperatures in Celcius and have expressed the specific heats in Celcius. We can do this because the unit in the specific heat is a change in temperature and a change of 1 K is the same as a change of 1°C.

Making sense of the results:

The initial water temperature is only a few degrees higher than the temperature of the final mixture because there is much more water than malt and the specific heat of water is greater than the specific heat of malted barley.

👥 PEER TO PEER

When we do these calculations we are really just keeping track of energy. The energy that is removed from the hot object goes into the cold object.

EXAMPLE 17-3

Using Specific Heat

A 20 g block of copper at a temperature of 100°C is placed in good thermal contact with a 40 g block of copper that is at 10°C. What is the final temperature?

SOLUTION

We once again apply Equation (17-5):

$$m_{hot}c_{hot}\Delta T_{hot} = -m_{cold}c_{cold}\Delta T_{cold}$$

In this case, both blocks are made of the same material, so the specific heat cancels. We also know that $T_{f,hot} = T_{f,cold} = T_f$. Therefore,

$$T_f = \frac{m_{hot}T_{hot} + m_{cold}T_{cold}}{m_{hot} + m_{cold}}$$

$$= \frac{0.02\text{ kg} \times 100°C + 0.04\text{ kg} \times 10°C}{0.02\text{ kg} + 0.04\text{ kg}} = 40°C$$

Making sense of the results:

The final temperature is closer to the temperature of the larger block than to the temperature of the smaller block because the larger block has a larger heat capacity.

LO 4

17-4 Phase Changes and Latent Heat

When you boil water in a pot on a stove, the burner transfers significant energy to the pot, yet the temperature of the liquid water sits at 100°C. The energy being absorbed by the water is being used to break the bonds that exist between the water molecules in the liquid phase so they can move to the gas phase. In some sense, the thermal energy that is being added to the system is "hidden" in that it does not manifest itself as a resulting temperature change. We call these "hidden" heats **latent heats**.

A latent heat called the **heat of vaporization** is associated with the transition from the liquid to the gaseous state. Similarly, a latent heat called the **heat of fusion** is associated with the change from the solid to the liquid state. When a material goes from the solid to the liquid state, the thermal energy responsible for the phase change must be supplied from the surroundings. Conversely, when the material goes from the liquid to the solid state, the thermal energy is liberated to the surroundings. Table 17-2 lists heats of fusion and vaporization for a few common materials. Note that all the values are in J/kg. To determine the total amount of thermal energy required to make a transition, we need to multiply these quantities by the mass of the sample. Thus, the thermal energy required for the phase change is

KEY EQUATION
$$Q_{phase} = mL \tag{17-6}$$

where m is the mass of the sample, and L is the latent heat.

THERMODYNAMICS

Table 17-2 Melting points, heats of fusion, boiling points, and heats of vaporization for some common materials

Substance	Melting point (°C)	Heat of fusion (J/kg)	Boiling point (°C)	Heat of vaporization (J/kg)
Water	0	3.33×10^5	100	2.26×10^6
Nitrogen	−210	2.6×10^4	−195.8	2×10^5
Lead	327	2.5×10^4	1750	8.7×10^5
Tungsten	3410	1.84×10^5	5900	48×10^5

 CHECKPOINT

C-17-3 Does the Temperature Always Change When We Add Thermal Energy?

True or false? When we heat a substance, we always observe an increase in temperature.

C-17-3 No. We only observe a temperature change when the substance is not undergoing a change of phase.

 MAKING CONNECTIONS

Reusable Hand Warmers

Reusable hand warmers rely on heat of fusion (Figure 17-3). They contain a supersaturated solution of a salt, typically sodium acetate. Snapping the small metal circle in the package triggers the crystallization of the solution from the liquid to the solid state. You can see that the pack is crystallizing in the image from right to left. As it crystallizes, the heat of fusion is liberated, and the pack feels quite warm to the touch. Once the pack has liberated all the heat, the pack contents can be returned to the liquid state by immersing in boiling water.

Peter Williams

Figure 17-3 A reusable heat pack

When doing thermal energy flow (calorimetric) calculations, we consider whether a substance goes through a phase change. If so, we modify Equation (17-5) as follows:

$$Q = mc_1(T_{\text{phasechange}} - T_i) + mL + mc_2(T_f - T_{\text{phasechange}})$$
(17-7)

where c_1 and c_2 are the specific heats of the two phases, and L is the latent heat per unit mass.

 EXAMPLE 17-4

Temperature Changes Due to Thermal Energy When There Is Also a Phase Change

How much thermal energy is required to raise the temperature of 250 g of water from 80.0°C to 120°C?

SOLUTION

For clarity, we will calculate each term in Equation (17-7) separately. We will first heat the water to 100°C. We assume that the specific heat of water in this temperature range is a constant 4186 J·kg^{-1}°C^{-1}.

$$Q_{80-100} = (0.25 \text{ kg}) (4186 \text{ J/(kg·K)}) (100°C - 80°C)$$
$$= 20\,930 \text{ J}$$

Now, we will determine how much thermal energy is required to convert the water to steam. We refer to Table 17-2 to find the heat of vaporization for steam to write:

$$Q_{\text{water-steam}} = (0.25 \text{ kg}) (2.26 \times 10^6 \text{ J/kg}) = 5.65 \times 10^5 \text{ J}$$

Finally, we will determine how much thermal energy is required to raise the temperature of the steam from 100°C to 120°C. We will assume that the specific heat of the steam is a constant 2010 J/(kg·K):

$$Q_{100-120} = (0.25 \text{ kg}) (2010 \text{ J/(kg·k)}) (120°C - 100°C)$$
$$= 10\,050 \text{ J}$$

Thus, the total thermal energy required to heat 250 g of water from 80°C to 120°C is

$$Q_{\text{Total}} = Q_{80-100} + Q_{\text{water-steam}} + Q_{100-120}$$

$$= 20\,903 \text{ J} + 565\,000 \text{ J} + 10\,050 \text{ J} = 596\,000 \text{ J}$$

Making sense of the results:

Since the heat of vaporization is quite large compared to the thermal energy required to warm the water and subsequently the steam, most of the energy required goes in to the phase change.

CHAPTER 17 | **HEAT, WORK, AND THE FIRST LAW OF THERMODYNAMICS**

THERMODYNAMICS

EXAMPLE 17-5

What Is the Final Phase?

Calculate the final temperature and phase when 100 g of ice at –5.00°C is mixed with 200 g of water at 20.0°C, as well as how much ice will be melted. Assume that no thermal energy is transferred between the mixture and its surroundings.

SOLUTION

There are three possible outcomes for a mixture of ice and water. First, all the ice could melt, resulting in water at some temperature above 0°C. Second, all the liquid could freeze, producing ice at some temperature below 0°C. Finally, the resultant mixture could be part liquid and part solid at a temperature of 0°C. First, we compare the thermal energy transfers required to bring the water and the ice to 0°C:

$$Q_{cool\ water} = (0.2\ kg)(4186\ J/(kg \cdot K))(0°C - 20°C) = -16\ 744\ J$$

$$Q_{warm\ ice} = (0.1\ kg)(2100\ J/(kg \cdot K))(0°C - (-5°C)) = 1050\ J$$

The minus sign indicates that cooling the water transfers thermal energy *from* the water to the ice. The net transfer of energy to the ice is $16\ 744\ J - 1050\ J = 15\ 694\ J$. We now calculate how much ice will be melted by this energy transfer using the heat of fusion from Table 17-2:

$$m = \frac{Q}{L} = \frac{15\ 694\ J}{3.33 \times 10^5\ J/kg} = 4.71 \times 10^{-2}\ kg = 47.1\ g$$

Thus, 47.1 g of ice at 0°C will be converted to liquid at 0°C. So, the final mixture will be 247 g of liquid water and 52.9 g of solid water, all at a temperature of 0°C.

Making sense of the results:

In this case the final outcome was a solid liquid mixture at 0°C. This is because the energy that was transferred form the water to the ice was only enough to warm the ice to the melting point and then melt some of it. If the ice melted had exceeded 100 g, we would have concluded that the final mixture would have been all water. Conversely, if the energy required to warm the ice had been greater than the energy required to cool the water, some or perhaps all of the water would have frozen.

MAKING CONNECTIONS

Home Ice-Cream Makers

A home ice cream maker (like the one in Figure 17-4) utilizes a mixture of ice and salt water to achieve temperatures below 0°C. Crushed ice and a saturated salt solution are placed in the outer part of the ice cream maker. The saturated salt solution has a freezing temperature of about $-21°C$. Thermal energy flows from the salt solution into the ice. The solution cools and the ice warms up. However, because the ice dominates the mixture, the final temperature of the mixture is closer to the initial temperature of the ice than that of the salt solution.

© Susanne Kischnick/Alamy

Figure 17-4 A home ice-cream maker uses a mixture of ice and salt water to freeze the ice cream.

LO 5

17-5 Work

In the previous sections, we examined thermal energy going into and out of a system. We now consider the effects of doing work on a system and having a system do work on its surroundings. We saw in Chapter 6 that doing work on an object changes its kinetic energy. Recall that when a force $\vec{F}$ is exerted on a particle that moves from $\vec{r}_1$ to $\vec{r}_2$, the work done is

$$W_F = \int_{\vec{r}_1}^{\vec{r}_2} \vec{F} \cdot d\vec{r} \qquad (6\text{-}13a)$$

We also recall the work–kinetic energy theorem, which for a single particle states that

$$W_T = \Delta K \qquad (6\text{-}5)$$

where ΔK is the change in kinetic energy of the particle.

How can we do work on an assembly of particles? Let us consider a gas in a cylinder as shown in Figure 17-5. A force is applied to the piston, causing it to move to the left and compress the gas. As the piston moves, the pressure in the gas increases. To determine the work, we consider an infinitesimal motion of the piston, denoted as dx in the figure. Accompanying this motion is an infinitesimal change in the volume of the gas: $dV = A dx$, where A is the cross-sectional area of the cylinder. The work that the applied force does on the piston is given by

$$dW = \vec{F} \cdot d\vec{r} = F dx \qquad (17\text{-}8)$$

because the force and the displacement are parallel.

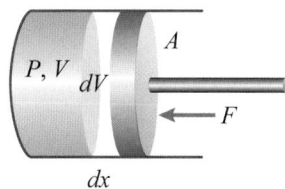

Figure 17-5 A piston compressing a gas

If we assume that the piston either starts and ends at rest, or travels at a constant speed during the process, the kinetic energy of the piston does not change and therefore the net work done on the piston must be zero. Therefore, the gas must do some negative work on the piston to balance off the positive work done by the external force. The pressure of the gas exerts a force to the right on the piston and has a magnitude of $F_{\text{pressure}} = PA$. The work done on the piston by the gas is then

$$dW_{\text{gas on piston}} = -PA dx \qquad (17\text{-}9)$$

The minus sign is included because the force and the displacement are in opposite directions. Since the net work done on the piston is zero,

$$F dx - PA dx = 0 \qquad (17\text{-}10)$$

Thus, the force applied on the piston by the gas has the magnitude PA and is equal but opposite to the applied external force. Newton's third law requires that the piston exert an equal but opposite force on the gas. Consequently, the piston does work on the gas:

$$dW_{\text{piston on gas}} = PA dx = P dV \qquad (17\text{-}11)$$

Consider what happens when the gas particles collide with the walls of the cylinder. If we assume that the collisions are elastic, then a particle emerges from the collision with the same speed and, therefore, the same kinetic energy that it had going into the collision. The only change that occurs is that the component of its velocity vector perpendicular to the face of the piston changes sign. However, in a collision with the moving piston, a gas particle emerges with that component of its velocity increased in addition to having its sign changed. As we saw in Chapter 7, for a perfectly elastic collision in one dimension,

$$u' = \frac{2mv + (M - m)u}{M + m} \qquad (7\text{-}21)$$

and

$$v' = \frac{-2Mu + (M - m)v}{M + m} \qquad (7\text{-}22)$$

where M and m are the masses of the two objects, and u and v are the velocities, with the primes denoting final velocities. In our case, the mass of the piston is much larger than the mass of the gas particle, $M \gg m$. In that limit, Equations (7-21) and (7-22) reduce to

$$V_{\text{f}} = V_{\text{i}} \quad \text{and} \quad v_{\text{f}} = -v_{\text{i}} + 2V_{\text{i}} \qquad (17\text{-}12)$$

Thus, the increase in that component is twice the speed of the piston.

The general expression for calculating the work done in a finite volume change is

KEY EQUATION $$W = \int_{V_1}^{V_2} P dV \qquad (17\text{-}13)$$

The magnitude of this quantity corresponds to the area under the curve on a **P-V diagram**, as shown in Figure 17-6. Note that the amount work done in going from one point to another depends on the details of the process used.

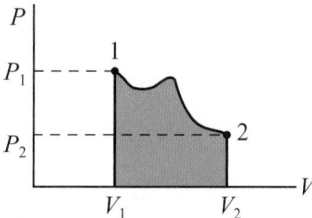

Figure 17-6 A P-V diagram. The shaded area represents the work done in the process of going from point 1 to point 2.

 CHECKPOINT

C-17-4 Expansions and Work

Is the following statement true or false: If we carry a gas through an expansion and then compress it so that it returns to its original volume, the net work done by the gas is always zero because there is no net volume change.

C-17-4 The work done depends on the way we carry out the process. The net work is zero only if we carry out the expansion and compression along the same path on the P-V diagram. Otherwise, the area under the two paths will be different and some net work will have been done.

LO 6

17-6 The First Law of Thermodynamics

We saw in the previous two sections that we can add energy to a thermodynamic system in two ways: heating the system and doing work on the system. Applying conservation of energy to a thermodynamic system suggests that the change in energy of the system is simply the sum of the thermal energy put into the system and the work done by the system:

KEY EQUATION $$\Delta U = Q - W \qquad (17\text{-}14)$$

THERMODYNAMICS

where U is the internal energy of the system, Q is the thermal energy flow *into* the system, and W is the work done by the system.

Equation (17-14) is the **first law of thermodynamics**, which states that the change in the internal energy of a system is equal to the thermal energy that flows into the system less the work done by the system.

What do we mean by "a system" in this context? In general, a system is simply the set of objects we wish to consider, and we are usually free to define our system in any way that is convenient for us. For example, if we wished to consider what happens when we place a quantity of ice cubes in a glass of water, we could define our system as the water and the ice. This choice is a **closed system** because the total amount of material in the system does not change—no mass enters or leaves the system. Alternatively, we could have chosen just the glass of water as our system. This choice defines an **open system**, which is a system in which mass is added or removed.

Another important distinction is deciding whether or not the system is isolated. In an isolated system, we assume that no thermal energy flows into or out of the system. In general, it is very difficult to make an isolated system. A reasonable approximation might be an insulated cup, which can be treated as an isolated system over short periods of time.

We should also pay some attention to the sign conventions that we have adopted in writing the First Law this way. When thermal energy flows into the system, which raises the internal energy, we consider that to be a positive flow and Q is positive. Conversely, when thermal energy flows out of the system, the internal energy decreases and Q is negative.

 CHECKPOINT

C-17-5 Sign of the Thermal Energy Flow

Ice cubes are placed in a glass of water. When we consider the ice and the liquid to be separate systems, which of the following statements is (are) correct?

(a) The thermal energy flow from the ice is negative.

(b) The thermal energy flow from the liquid is negative.

(c) The thermal energy flow from the ice is equal in magnitude but opposite in sign to the thermal energy flow from the liquid.

C-17-5 (b) and (c) Thermal energy flows from the warm liquid to the cooler ice. Thus, for the liquid system, the thermal energy flow is negative (out of the liquid); for the ice system, the thermal energy flow is into the liquid and is positive. The magnitude of the two flows is the same, assuming that the two systems are isolated from all other systems except each other. All the thermal energy that flows from the liquid goes into the ice.

The negative sign in front of the work term in Equation (17-14) is necessary because we are considering the work done *by* the system. When a system does work, it loses energy.

The variables in Equation (17-14) can have quite different properties. For example, in the case of an ideal gas, the internal energy is a function of the temperature of the

EXAMPLE 17-6

Internal Energy and Thermal Energy

Calculate the change in internal energy of an isolated closed system when

(a) 3000 J of thermal energy is added to it, and the system does 4000 J of work;

(b) 3000 J of thermal energy is added to it, and 4000 J of work is done on the system.

SOLUTION

The calculation is a straightforward application of Equation (17-14), but we need to pay careful attention to the sign conventions.

(a) When thermal energy is added to the system, it increases the internal energy of the system and is therefore positive. When the system does work, energy is transferred out of the system. Thus, the net change in the internal energy is

$$\Delta U = Q - W = 3000 \text{ J} - 4000 \text{ J} = -1000 \text{ J}$$

(b) When the system has 4000 J of work done on it, the work done by the system is –4000 J. Then the net change in internal energy is

$$\Delta U = Q - W = 3000 \text{ J} - (-4000 \text{ J}) = 7000 \text{ J}$$

Making sense of the results:

In the first case the system loses more energy by doing work than it received in the form of thermal energy so the net change in internal energy is negative. In the second, work is done ON the system which raises its internal energy.

gas and has a well-defined value at a given temperature. However, as we saw in Section 17-5, the work done in a process that carries a system from one state to another depends on how the process is carried out. Consequently, we cannot associate a quantity of work with the state of a system. In addition, the thermal energy that flows into or out of a system must depend on the way the process is carried out, such that variations in the work done are offset by equal but opposite variations in the thermal energy flow. Otherwise, the first law would not hold for all processes. We will discuss different types of processes in the next section.

 CHECKPOINT

C-17-6 Thermal Energy and Work

Is the following statement true or false: The thermal energy that flows and the work that is done in a process depend on the details of the process, but the difference between the thermal energy and the work, $Q - W$, does not, and only depends on the start and end points of the process.

C-17-6 True. By the first law of thermodynamics, $Q - W = \Delta U$, and the internal energy is a state variable. Therefore, the change in the internal energy in any process that has the same start and end points must be the same, regardless of the details of the process.

17-7 Different Types of Processes

When an isolated system undergoes a process, there is no thermal energy flow into or out of the system. We call such a process **adiabatic**. A process carried out at constant temperature is called an **isothermal** process. Similarly, an **isobaric** process is carried out at constant pressure, and an **isochoric** process is carried out at constant volume.

Isothermal Processes

For an ideal gas,

$$PV = nRT \qquad (17\text{-}15)$$

where P is the pressure, V is the volume, n is the number of moles of the gas, R is the gas constant, and T is the temperature in kelvins. When we carry out an isothermal process on a closed system, the product PV must be a constant. Since the internal energy for an ideal gas depends only on the temperature, in an isothermal process there is no change in the internal energy of the system. Thus,

$$\Delta U = W - Q = 0$$

$$\therefore W = Q \qquad (17\text{-}16)$$

Consider the work done in an isothermal expansion of an ideal gas in a cylinder with a movable piston. We immerse the cylinder in a bath of water large enough that any thermal energy flow between the cylinder and the bath has a negligible effect on the temperature of the bath. We also carry out the process slowly enough that the temperature of the gas in the piston is always the same as the temperature of the surrounding water bath. In fact, this process is only **quasistatic** because the piston would have to move infinitesimally slowly to keep the temperature completely constant. We can apply Equation (17-13) to calculate the work done by the gas:

$$W = \int_{V_1}^{V_2} P\,dV = \int_{V_1}^{V_2} \frac{nRT}{V}\,dV = nRT\,\ln\left(\frac{V_2}{V_1}\right) \quad (17\text{-}17)$$

Here, we have used the ideal gas law to replace P in the integral with $\dfrac{nRT}{V}$. The work in such a process is illustrated in Figure 17-7. The solid black line is an **isotherm** (line of constant temperature), and the work done is the shaded area under the curve.

Isobaric Processes and the Constant Pressure Heat Capacity

When we hold the pressure constant while compressing an ideal gas,

$$P = \frac{nRT}{V} = \text{constant} = P_c \qquad (17\text{-}18)$$

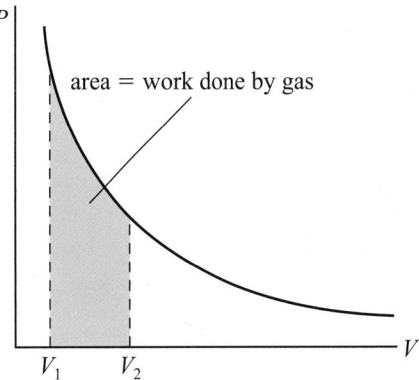

Figure 17-7 A P-V diagram for an isothermal process in an ideal gas

EXAMPLE 17-7

Work and Thermal Energy Flow

One mole of an ideal gas is compressed in an isothermal process at a temperature of 293 K. The initial volume is 22.0 L, and the final volume is 10.0 L. Calculate the work done and the thermal energy flow into the system.

SOLUTION

We use Equation (17-17) to find the work:

$$W = nRT \ln\left(\frac{V_2}{V_1}\right)$$

$$= 1\text{ mol} \times 8.314\text{ J/(K}\cdot\text{mol)} \times 293\text{ K} \times \ln\left(\frac{10\text{ L}}{22\text{ L}}\right)$$

$$= -1920\text{ J}$$

We use Equation (17-16) to find the thermal energy flow during the compression:

$$W = Q = -1920\text{ J}$$

The negative sign indicates that thermal energy flowed out of the system.

Making sense of the results:

The gas exerts an outward pressure on the piston, which moves inward to reduce the volume; thus, the work done by the gas is negative.

$$\therefore T = \frac{P_c V}{nR}$$

Consequently, as we change the volume, we must also change the temperature by allowing thermal energy to flow into or out of the system. In Chapter 16, we showed that the average kinetic energy of the particles in an ideal monatomic gas is related to the temperature of the gas. For a monatomic ideal gas, all of the energy is the translational kinetic energy of the gas particles. It therefore follows that the internal energy of an ideal gas is given by

$$U = \frac{3}{2} nRT \qquad (17\text{-}19)$$

Therefore, this isobaric process involves thermal energy flow, work, and a change in the internal energy of

THERMODYNAMICS

the system. Since the pressure is constant, the integral for work in Equation (17-13) becomes

$$W = \int_{V_1}^{V_2} P dV = P_c \int_{V_1}^{V_2} dV = P_c(V_2 - V_1) \quad (17\text{-}20)$$

The shaded area in Figure 17-8 corresponds to the work done in this process.

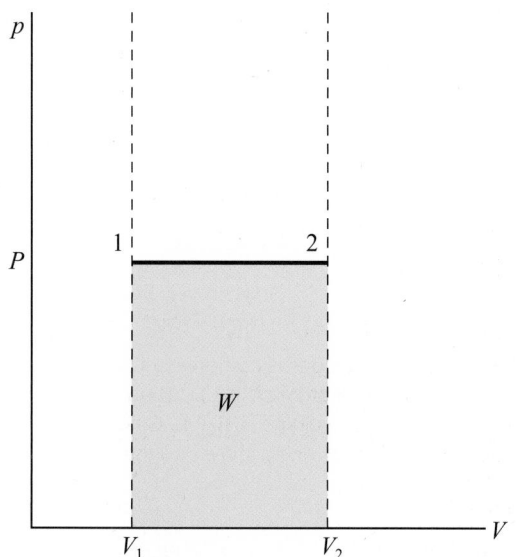

Figure 17-8 An isobaric process

The temperature change can be found using Equation (17-15):

$$\Delta T = T_2 - T_1 = \frac{P_c V_2}{nR} - \frac{P_c V_1}{nR} = \frac{P_c}{nR}(V_2 - V_1) \quad (17\text{-}21)$$

Using this expression for the temperature change with Equation (17-19) gives an equation for the change in internal energy:

$$\Delta U = \frac{3}{2} nR\Delta T = \frac{3}{2} P_c(V_2 - V_1) \quad (17\text{-}22)$$

Rearranging Equation (17-14) to solve for the thermal energy flow, we get

$$Q = \Delta U + W = \frac{3}{2} P_c(V_2 - V_1) + P_c(V_2 - V_1)$$

$$= \frac{5}{2} P_c(V_2 - V_1) \quad (17\text{-}23)$$

When the ideal gas is compressed, all the terms on the right of Equation (17-23) are negative, and thermal energy must flow out of the system. However, if the ideal gas expands, all the terms on the right of Equation (17-23) are positive, and thermal energy must flow into the system. From Equation (17-21), we can see that in the case of an expansion ($V_2 > V_1$), the work done by the gas is positive and, hence, comes at a cost to the internal energy of the gas; for a compression the converse is true.

We can use Equation (17-23) to determine the heat capacity for an ideal gas in an isobaric process. The heat

EXAMPLE 17-8

Work, Thermal Energy, and Internal Energy

One mole of an ideal gas is at a temperature of 293 K and a volume of 22.0 L. It is compressed in an isobaric process to a volume of 10.0 L. Find the work done, the thermal energy flow, and the change in internal energy.

SOLUTION

We can use Equation (17-20) to calculate the work, but first we need to determine the pressure. To do that, we use the ideal gas law and solve for P:

$$P = \frac{nRT}{V} = \frac{1 \text{ mol} \times 8.314 \text{ J/(mol·K)} \times 293 \text{ K}}{0.0220 \text{ m}^3} = 110727 \text{ Pa}$$

We can then determine the work:

$$W = P(V_2 - V_1) = (110\,727 \text{ Pa})(0.0100 \text{ m}^3 - 0.0220 \text{ m}^3)$$

$$= -1330 \text{ J}$$

The change in internal energy is

$$\Delta U = \frac{3}{2} P(V_2 - V_1) = \frac{3}{2} W = -1990 \text{ J}$$

and the thermal energy flow is

$$Q = \frac{5}{2} P(V_2 - V_1) = \frac{5}{2} W = -3320 \text{ J}$$

Making sense of the results:

The work done by the gas is negative, as it always is in a compression.

capacity is the ratio of the thermal energy that flows into a system to the resulting temperature change, so

KEY EQUATION
$$C_p = \frac{Q}{\Delta T} = \frac{\frac{5}{2} P(V_2 - V_1)}{T_2 - T_1}$$

$$= \frac{5}{2} \frac{nR(T_2 - T_1)}{T_2 - T_1} = \frac{5}{2} nR \quad (17\text{-}24)$$

The subscript P denotes this is a constant pressure heat capacity.

Isochoric Processes

In an isochoric process, we keep the volume constant and vary the pressure. We can carry out such a process by placing the gas in thermal contact with a thermal energy reservoir and allow thermal energy to flow into or out of the gas. Because the volume is constant, no work is done in such a process, and the first law becomes

$$\Delta U = Q \quad (17\text{-}25)$$

Figure 17-9 shows a P-V graph for an isochoric process. Note that since the volume does not change there is no area under the curve and no work is done.

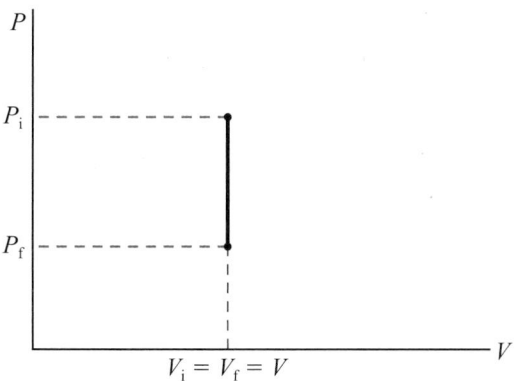

Figure 17-9 In an isochoric process, the pressure and temperature are varied and the volume remains constant.

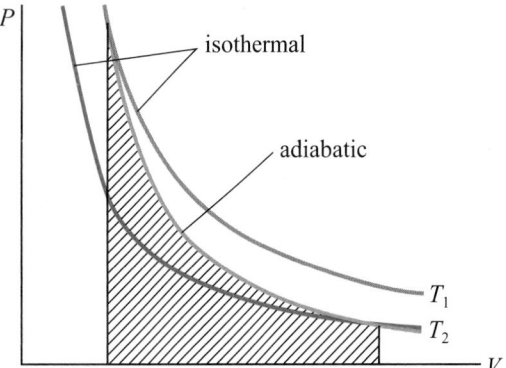

Figure 17-10 An adiabatic expansion

There is a change in the temperature of the gas during this process, and the change in internal energy is

$$\Delta U = \frac{3}{2} nR(T_2 - T_1) = Q \qquad (17\text{-}26)$$

The heat capacity for an isochoric process can be found from Equation (17-26):

$$C_V = \frac{Q}{\Delta T} = \frac{3}{2} nR \qquad (17\text{-}27)$$

The heat capacity for an isochoric process is less than the heat capacity for an isobaric process. For a given input of thermal energy, we get a larger temperature change in an isochoric process than we get in an isobaric process. This is because in an isochoric process, no work is done by the gas, whereas in an isobaric process, the gas does work and some of the thermal energy is converted into work done by the gas.

Adiabatic Processes

In an adiabatic process, there is no thermal energy flow into or out of the system. Such a process can be accomplished by carefully insulating the system from the surroundings. Carrying out a process very quickly also reduces thermal energy flow to and from the system. In an adiabatic process, the first law of thermodynamics becomes

$$\Delta U = -W \qquad (17\text{-}28)$$

In an adiabatic expansion, work is done and the internal energy of the system changes. For an ideal gas, this change means there is also a change in the temperature of the system and, hence, changes in the pressure and volume of the gas. Since the gas does work as it expands, the change in the internal energy is negative. If the process starts at some temperature T_1 and ends at a lower temperature T_2, the P-V diagram for the process would look like Figure 17-10.

The online supplement for this textbook shows that for adiabatic processes in ideal gases,

$$PV^\gamma = \text{constant} \qquad (17\text{-}29)$$

where $\gamma = \dfrac{f+2}{f}$, and f is the number of degrees of freedom, as discussed in Chapter 16.

A monatomic gas only has three translation degrees of freedom and therefore $f = 3$. Therefore, the work done in an adiabatic expansion of a monatomic ideal gas is

$$W = \int_{V_1}^{V_2} P\,dV = \int_{V_1}^{V_2} \frac{C}{V^\gamma}\,dV = \frac{1}{1-\gamma} \frac{C}{V^{\gamma-1}} \bigg|_{V_1}^{V_2} \qquad (17\text{-}30)$$

The constant C can be determined from the starting point using the Ideal Gas Law.

State Variables

We now investigate the idea that the internal energy is a state variable, but neither the work nor the thermal energy is. We will compare two different ways of getting from an initial state to a final state for an ideal gas: an isothermal process that takes an ideal gas from (P_1, V_1, T) to (P_2, V_2, T) versus a two-step process with an isochoric stage from (P_1, V_1, T) to (P_2, V_1, T_i), followed by an isobaric stage from (P_2, V_1, T_i) to (P_2, V_2, T). These two different processes are illustrated in Figure 17-11.

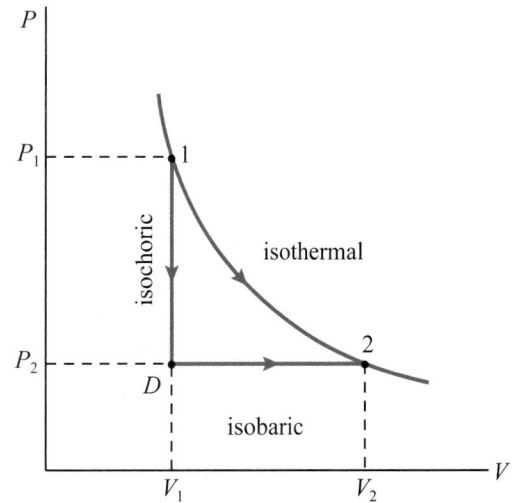

Figure 17-11 An isothermal process and an isochoric/isobaric process

For the isothermal process, there is no change in the temperature and, therefore, no change in the internal energy. We know from Equation (17-17) that the work done is

$$W_{\text{isothermal}} = nRT \ln \left(\frac{V_2}{V_1} \right) = Q \qquad (17\text{-}31)$$

For the isobaric process,

$$W_{\text{isobaric}} = P_2(V_2 - V_1) \qquad (17\text{-}32)$$

$$Q_{\text{isobaric}} = \frac{5}{2} P_2(V_2 - V_1) \qquad (17\text{-}33)$$

and

$$\Delta U_{\text{isobaric}} = Q_{\text{isobaric}} - W_{\text{isobaric}} = \frac{3}{2} P_2(V_2 - V_1) \qquad (17\text{-}22)$$

For the isochoric process, there is no work done but there is a change in temperature and, hence, a change in the internal energy:

$$\Delta U_{\text{isochoric}} = \frac{3}{2} P_2(V_1 - V_2) = Q_{\text{isochoric}} \qquad (17\text{-}34)$$

The net change in internal energy for the two-step isobaric–isochoric process is

$$\Delta U_{\text{isochoric+isobaric}} = \frac{3}{2} P_2(V_2 - V_1) + \frac{3}{2} P_2(V_1 - V_2)$$

$$= 0 = \Delta U_{\text{isothermal}} \qquad (17\text{-}35)$$

However, we can see that the net work in the isothermal process and the two-step process is not the same, and the net thermal energy flows also differ.

LO 8

17-8 Thermal Energy Flow

We will conclude this chapter with a discussion of three different processes by which thermal energy can move from one body to another.

Radiation

All objects whose temperature is not zero emit electromagnetic radiation, which carries energy. An object radiates energy at a rate given by

 KEY EQUATION
$$P_{\text{radiate}} = \sigma \varepsilon A T_{\text{object}}^4 \qquad (17\text{-}36)$$

where $\sigma = 5.67 \times 10^{-8}$ W/(m$^2 \cdot$K^4) is the **Stefan–Boltzmann constant**, ε is the emissivity coefficient of the object, A is the surface area of the object, and T is the temperature of the objects in kelvins. The emissivity coefficient is a characteristic of the surface of the object and has a value between 0 and 1. A surface with an emissivity of 1 is called a **blackbody radiator**. The details of blackbody radiation are discussed in Chapters 30 and 31. Table 17-3 lists emissivity coefficients for some common materials.

Table 17-3 Emissivity coefficients for various materials

Surface material	Emissivity coefficient
Aluminum, heavily oxidized	0.2–0.31
Aluminum, highly polished	0.039–0.057
Asphalt	0.93
Brick, red, rough	0.93
Concrete	0.85
Cotton, cloth	0.77
Glass, smooth	0.92–0.94
Gold, polished	0.025
Ice, rough	0.985
Oil paints, all colours	0.92–0.96
Porcelain, glazed	0.92
Paper	0.93
Plastics	0.91
Water	0.95–0.963

If we were to put some typical values into Equation (17-36), such as an emissivity on the order of 1 and temperature, T, on the order of 300 K, we would see that many everyday objects at room temperature radiate energy at a rate of approximately 400 W/m^2. You might think that these objects would cool down rapidly. However, each object is also receiving radiation from its surroundings at a rate that depends on the ambient temperature:

$$P_{\text{absorbed}} = \sigma \varepsilon A T_{\text{ambient}}^4 \qquad (17\text{-}37)$$

The net power radiated by an object is

$$P_{\text{net}} = P_{\text{radiate}} - P_{\text{absorbed}} = \sigma \varepsilon A \left(T_{\text{object}}^4 - T_{\text{ambient}}^4 \right) \qquad (17\text{-}38)$$

We have used the same constant and parameters in Equations (17-37) and (17-38). If these values were not the same, there would be a net flow of energy between an object and its surroundings when they are at the same temperature. Since such flows do not occur, we know that objects are equally efficient at radiating as they are at absorbing.

Conduction

When two objects are in contact with each other, the molecules at the contact surface can collide and transfer energy between the objects. Within a single object that has one end at a higher temperature than the other, the particles that make up the object itself can transfer energy between each other, and this leads to a flow of thermal energy from the hot end to the cold end.

EXAMPLE 17-9

Hot Coffee

A ceramic cup of hot coffee is placed on the counter. The liquid is at a temperature of 80°C. Estimate the rate at which the cup and coffee are radiating energy.

SOLUTION

We will use Equation (17-38), but we will have to make some assumptions about the surface area of the cup, the emissivity of the cup, and the ambient temperature. We can model the cup as a cylinder of radius 5.0 cm and height 10 cm. Then we get a surface area of $A = \pi r^2 + \pi r^2 + 2\pi rh = 0.047 \text{ m}^2$. From Table 17-3, we see that that the emissivity of glazed porcelain is 0.92 and the emissivity of water is approximately 0.96. So, we can use 0.92 as a reasonable approximation for the cup and coffee system. Assuming that the ambient temperature is 21°C, and converting to kelvins, we get

$$P = (5.67 \times 10^{-8} \text{ Wm}^{-2}\text{K}^4) \, (0.92) \, (0.047 \text{ m}^2)$$
$$\times \, [(353.16 \text{ K})^4 - (294.16 \text{ K})^4] = 20 \text{ W}$$

Making sense of the results:

Putting your hand near a 20 W lightbulb is similar to grabbing a hot cup of coffee. Putting a lid on the cup and insulating it can significantly reduce the surface temperature of the container and reduce the cooling rate.

It is found experimentally that when we place a layer of material between a hot reservoir and a cold reservoir, the rate at which thermal energy is transported from the hot side of the material to the cold side is given by

KEY EQUATION
$$P_{\text{conduction}} = \frac{\kappa A(T_{\text{hot}} - T_{\text{cold}})}{x} \qquad (17\text{-}39)$$

where κ is the thermal conductivity, in units of W/(m·K), A is the area, and x is the thickness of the material.

Table 17-4 lists the thermal conductivities of various materials. In general, gases tend to have low thermal conductivities, insulating solids have somewhat higher conductivities, and metals have the highest thermal conductivities (metals are also good conductors of electricity). In metals, both the atoms and some of the electrons contribute to the thermal conductivity. For materials that do not conduct electricity, only the atoms in the material can transport the energy.

Table 17-4 Thermal conductivities of various materials

Material/Substance	Thermal conductivity at 25°C W/(m·K)
Argon (gas)	0.016
Brass	109
Brickwork, common	0.6–1.0
Carbon dioxide (gas)	0.0146
Concrete, stone	1.7
Copper	401
Glass	1.05
Gold	310
Hardwoods (oak, maple, etc.)	0.16
Insulation materials	0.035–0.16
Mica	0.71
Nitrogen (gas)	0.024
Paper	0.05
Snow (temperature < 0°C)	0.05–0.25
Water	0.58

EXAMPLE 17-10

Efficient Windows

A 2.0 m² glass window is at a temperature of 20°C on the inside surface and –20°C on the outside. The glass is 5.0 mm thick. At what rate does energy flow out through the window glass? Compare this rate to thermal energy flow through a double-paned window that has a 2.0 cm gap between the panes filled with argon gas.

SOLUTION

For the single-paned window, we use Equation (17-39) and the thermal conductivity of glass from Table 17-4:

$$P_{conduction} = \frac{1.05 \text{ W/(m} \cdot \text{K) } 2.0 \text{ m}^2(+20°C - (-20°C))}{0.005 \text{ m}}$$

$$= 16\ 800 \text{ W}$$

For the double-paned window, we will assume that both surfaces of the inner pane are at 20°C and both surfaces of the outer pane are at –20°C. Then, the rate of thermal energy conduction through the argon between the panes is

$$P_{conduction} = \frac{0.016 \text{ W/(m} \cdot \text{K) } 2 \text{ m}^2(+20°C - (-20°C))}{0.02 \text{ m}} = 64 \text{ W}$$

Making sense of the results:

Clearly, investing in double-paned windows can result in a significant reduction in energy consumption. We should also remember that windows allow energy to flow into a house in the form of sunlight and are important elements in a passive solar design.

Another measure of thermal energy flow through a material that is commonly used when discussing insulation is the R value, a measure of the thermal resistance of a material. The R value combines the thermal conductivity with the thickness of the material, l, and is defined as

$$R = \frac{l}{\kappa} \qquad (17\text{-}40)$$

In the SI system, the units for R are m²/(W · K), often called RSI values. R values simplify calculations for building insulation. If you use a multilayered insulation system, you simply add the R values for the individual layers to find the total R value. To calculate the thermal energy loss of a particular building element, you divide the area of the element by the total R value and multiply by the temperature difference across the element.

Convection

The third thermal energy transport mechanism is convection, which occurs in gases and liquids that are in a gravitational field. When we heat a portion of a gas or liquid, it expands and the density decreases. If there is a cooler, higher-density region above it, there is a buoyant force on

 EXAMPLE 17-11

Insulate the Roof

A cathedral (sloped) ceiling in a house has a total area of 100 m². The ceiling is covered with drywall (RSI 0.45), and the rafters are filled with insulation (RSI 40). The roof is sheathed with ½-in. plywood (RSI 0.63) and asphalt shingles (RSI 0.44). Calculate the rate at which thermal energy is lost, assuming that the interior temperature is 20°C and the exterior temperature is 2°C. Assume that thermal energy losses from conduction through the wood rafters and from ventilation of the roof cavity are negligible.

SOLUTION

$$P_{conduction} = \frac{A}{R}(T_{inside} - T_{outside})$$

$$= \frac{100 \text{ m}^2}{[0.45 + 40 + 0.63 + 0.44] \text{ m}^2/(\text{W} \cdot \text{K})}(20°C - 2°C)$$

$$= 43 \text{ W}$$

Making sense of the results:

While this may seem like a modest amount of power, we must remember that this is lost continuously and over time can be quite a loss of energy. Also note that we did not convert the temperatures to kelvins since they appear as a difference and temperature differences are the same in both scales.

---

the warmer portion, so the warmer portion rises and the cooler region falls.

You may have experienced getting into a swimming pool or a lake on a hot summer day and noticed that the top layers of the water are nice and warm, but the water farther down is much cooler. Sunlight striking the surface heats the water from the top. The less dense warm water stays at the top, and the cooler water stays below it. There is no circulation; thus, no thermal energy is transported by the water other than by conduction from the top to the bottom of the lake. Figure 17-13 shows temperature

Figure 17-13 Thermal stratification in a lake

measurements in a lake on a sunny summer day. Near the surface, the temperature approaches 25°C, and at the deepest point it is only about 6°C. The lake is **thermally stratified**.

However, when a fluid is heated from the bottom, the heated material rises and the cooler material at the top falls. When the cooler material gets to the bottom, it is heated and rises. As long as the rising material loses thermal energy by some process, there will be a continuous **convection cycle**.

Hot water heating systems use both radiation and convection to distribute the thermal energy throughout a room. The radiator through which the hot water is circulated is usually placed on or near the floor close to a wall. The radiator is shaped to have a large surface area to radiate thermal energy, which is absorbed by the air surrounding the radiator. The heated air rises, and the cooler air falls. Because the radiator is placed against one wall, the hot air tends to spread out across the ceiling away from that wall, pushing the cooler air toward the opposite wall where it falls, as shown in Figure 17-14. This cooler air is then drawn across the floor to the radiator, where it is heated. It then rises to the ceiling, continuing the convection cycle.

Figure 17-14   Thermal convection in a home heating system

 MAKING CONNECTIONS

Interesting Convection Patterns

A Rayleigh–Bénard cell consists of two parallel plates that are sealed together around the edges with the space between the plates filled with liquid. The bottom plate is heated, and the top plate is cooled. When the temperature difference between the top and bottom plates is quite small, there is simply thermal conduction through the liquid. As the temperature difference increases, however, convection occurs, creating beautiful patterns like the patterns in Figure 17-15. Different patterns emerge as the difference in temperature between plates changes. When the temperature difference is great enough, the convection becomes completely turbulent. This sequence of transitions from a well-ordered system to chaotic turbulence is characteristic of nonlinear dynamical systems.

Convection is also very important in global climate. The Gulf Stream is a large convection current that transports warm water from the Gulf of Mexico to the North Atlantic. The Gulf Stream is responsible for keeping Europe warm. Palm trees can grow in southern England despite it being at about the same latitude as Winnipeg, Manitoba.

Scott Camazine/Photo Researchers, Inc.

Figure 17-15   A time exposure of convection in hexagonal Rayleigh–Bénard cells. The warm fluid is up-welling from the centre of each cell and sinking around the edges. The flow has been made visible by adding small aluminum flakes to the liquid.

The internal energy of a system can be changed by adding thermal energy to the system and by the system doing work on its surroundings. In various processes with the same start and end points, quantities of thermal energy and work may differ, but the net energy change of the system is always the same, which is the basis of the first law of thermodynamics. The ability of a material to absorb thermal energy is characterized by the material's specific heat capacity. When a system undergoes a change of phase, the latent heat of the phase change must be either added or removed. During a phase change, the temperature does not change.

When we heat an object, we transfer energy to the object. The thermal energy required to take a system from one state to another depends on how the process is carried out.

When we heat a system and there is no phase change, the relationship between the thermal energy added and the resulting temperature change is

$$Q = mc\Delta T \qquad (17\text{-}3)$$

where Q is the thermal energy, m is the mass, c is the specific heat, and ΔT is the temperature change.

When we heat a system and it undergoes a phase change, the temperature of the system does not change. Instead, the thermal energy is used to change the phase, and the thermal energy required is given by

$$Q_{\text{phase}} = mL \qquad (17\text{-}6)$$

where m is the mass of the system, and L is the latent heat of the transition.

When a system undergoes a volume change, the system does work that is given by

$$W = \int_{V_1}^{V_2} P\,dV \qquad (17\text{-}13)$$

The work done depends on the path taken between the start and end points.

The first law of thermodynamics states that the difference between the thermal energy that is added to a system and the work done by the system is independent of the path taken to carry out the process. This difference is called the internal energy, U:

$$\Delta U = Q - W \qquad (17\text{-}14)$$

There are countless ways that a system can be taken from one point on a P-V diagram to another. Some common processes are the following:

- Isobaric: constant pressure
- Isochoric: constant volume
- Isothermal: constant temperature
- Adiabatic: no thermal energy flow

Thermal energy transfer mechanisms include radiation, conduction, and convection:

- Radiation: $P_{\text{net}} = \sigma \varepsilon A(T_{\text{object}}^4 - T_{\text{ambient}}^4)$ (17-38)

where σ is the Stefan–Boltzmann constant, ε is the emissivity, and A is the surface area.

- Conduction: $P = \dfrac{\kappa A \Delta T}{x}$ (17-39)

where κ is the thermal conductivity, A is the surface area, x is the thickness of the material, and ΔT is the temperature difference.

- Convection: a gas or fluid that is heated tends to expand and rise.

APPLICATIONS

Applications: heating, cooling, moving thermal energy, energy-efficient windows and buildings, temperature control of mixtures

Key Terms: P-V diagram, adiabatic, blackbody radiation, closed system, convection cycle, first law of thermodynamics, heat capacity, heat of fusion, heat of vaporization, heat, isobaric, isochoric, isotherm, isothermal, latent heats, open system, quasistatic, specific heat capacity, Stefan–Boltzmann constant, thermally stratified

QUESTIONS

1. You are asked to heat a sample of gas using the least amount of thermal energy possible. Should you carry out a constant volume process or a constant pressure process?
2. Two boards of insulation are made of the same material. One board is twice as thick as the other.
   (a) Which board has the greater thermal conductivity?
   (b) Which board has the greater R value?
3. Figure 17-16 shows a temperature versus time graph for a sample that is heated from the solid phase to the gas phase. What phases are present in each section?

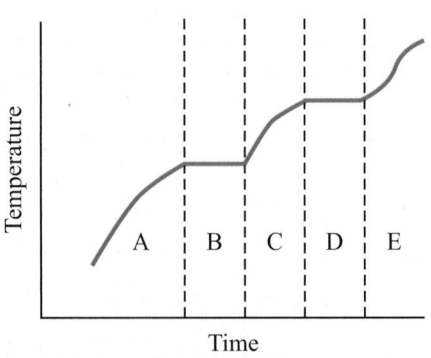

Figure 17-16   Question 3

4. Two objects with the same mass are heated to the same temperature. They are placed into identical insulated water baths that are initially at room temperature. The final temperatures of the water baths with the objects in them are different. Would the bath with the higher final temperature contain the object with the greater or the lesser specific heat?

5. In the winter, even though the air temperature in our homes is 20°C, we usually wear warm clothes. Conversely, on a summer evening, when the inside temperature of the house drops to 20°C, we tend to be quite comfortable in shorts and a light shirt. Why? (Hint: Consider differences in radiant energy absorption by your skin.)

6. Figure 17-17 shows a temperature versus time graph for a sample of ice and a sample of water that are mixed together. What is the final phase?

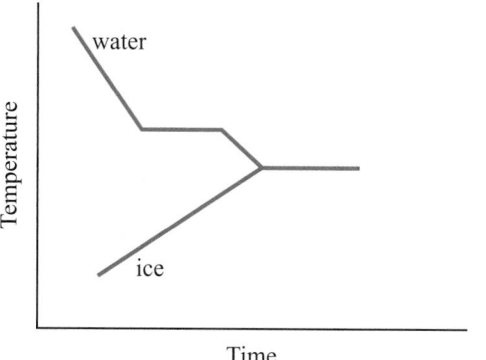

Figure 17-17   Question 6

7. For the *P-V* process for an ideal gas shown in Figure 17-18, the system starts at A and goes to B. Is the work done positive or negative?

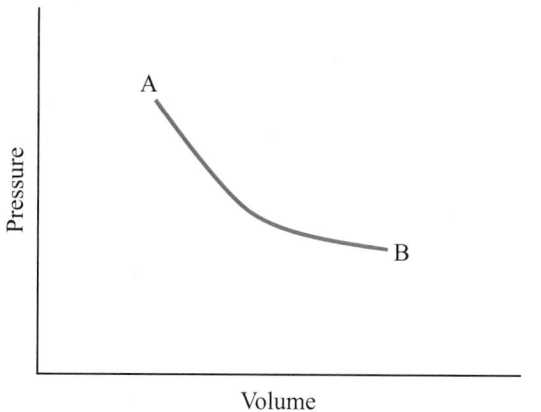

Figure 17-18   Question 7

8. When we double the temperature of a hot object, by what factor does the rate at which it radiates energy increase?

9. When we double the temperature difference across an insulator, by what factor does the thermal energy flow increase?

10. Many animals sweat or perspire in order to keep cool in hot weather. How does perspiring help remove thermal energy from your body?

11. In the cooking technique called *sous-vide*, raw food is sealed in an airtight plastic bag and then immersed in a hot water bath; the water temperature is maintained at the desired final temperature for the food. The types of changes in the food as it cooks are determined primarily by the temperature of the water bath, so the flavour and texture of food

cooked *sous-vide* differ from those of the same food cooked by other methods that use higher temperatures. Many types of food can be held in a *sous-vide* water bath for extended periods without overcooking. Explain why this is so.

12. Pressure cookers are used to shorten cooking times for foods that are boiled. How do pressure cookers work?

13. A hot object and a cold object are placed into thermal contact. Thermal energy flows from the hot object to the cool object, and eventually they reach the same temperature. Under what circumstances is the final temperature halfway between the two initial temperatures?

PROBLEMS BY SECTION

For problems, star ratings will be used, (✻, ✻✻, or ✻✻✻), with more stars meaning more challenging problems.

Section 17-2   Temperature Changes Due to Thermal Energy Transfer

14. ✻   A 50.0 g block of copper is at a temperature of 20°C. When 2.00 kJ of thermal energy is added to the block, what temperature will it reach?

15. ✻✻   You wish to make tea, and you put 1.0 kg of water at 5°C into your 500.0 W kettle. How long will it take for the water to boil? Assume no thermal energy loss.

16. ✻✻   Alberta experiences chinooks in the winter, which are winds that can cause the air temperature to rise dramatically. In one instance, the temperature rose from –30°C to 5°C. We assume that the atmosphere is heated by this amount to an altitude of 5.0 km. How much thermal energy per square kilometre is required?

Section 17-3   Thermal Energy Flow Between Objects

17. ✻   A 100.0 g block of aluminum at a temperature of 100°C is placed in a well-insulated water bath at a temperature of 10°C. If the bath holds 500.0 g of water, what is the temperature of the system when it reaches equilibrium?

18. ✻   Twenty grams of milk at a temperature of 2°C is added to 300.0 g of coffee at a temperature of 170°C. The coffee cup has a heat capacity of 109 J/K. What is the equilibrium temperature of the mixture?

19. ✻✻   You determine in an experiment that a 100.0 g sample of an unknown metal experiences a 5.0°C temperature increase when 65 J of thermal energy is added to the sample. What is the specific heat of the metal?

Section 17-4   Phase Changes and Latent Heat

20. ✻✻   A 500.0 W kettle has 1.00 kg of water in it at a temperature of 100°C. How long will it take for the kettle to boil dry?

21. ✻✻   Twenty grams of ice at a temperature of –4°C is placed into 100.0 g of water at a temperature of 30°C. Find the final temperature of the mixture.

Section 17-5   Work

22. ✻✻   One mole of an ideal gas is compressed isobarically from a volume of 40.0 L to a volume of 30.0 L at a pressure of 110.0 kPa. How much work is done?

23. ✻   Three moles of an ideal gas is expanded isothermally at a temperature of 100°C. During the expansion, the volume increases by 30%. How much work is done?

24. ✻   A gas undergoes an isochoric process in which the pressure increases from 50 kPa to 100 kPa. How much work is done?

Section 17-6 The First Law of Thermodynamics

25. ✳ An ideal gas undergoes an isothermal process in which the gas does 100 J of work. How much thermal energy flows into the gas?

26. ✳✳ A gas undergoes an isothermal expansion from a certain pressure, temperature, and volume. The volume of the gas doubles. The same gas undergoes an adiabatic expansion from the same starting point, and again the gas doubles in volume. In which process is more work done? In which process is there a thermal energy flow?

Section 17-8 Thermal Energy Flow

27. ✳ Comet Ikeya-Seki was observed to reach a temperature of approximately 650°C at its closest approach to the Sun. At what rate does the comet radiate energy per square metre? Assume an emissivity of 0.50.

28. ✳ A particular insulating material is 10 cm thick and has an RSI value of 40. What is the thermal conductivity of the material?

29. ✳✳ A 0.50 cm-thick piece of copper has an area of 4.0 cm². One side of the copper is at a temperature of 100°C, and the other side is at 400°C. At what rate does thermal energy flow through the copper?

COMPREHENSIVE PROBLEMS

30. ✳✳ The Sun delivers radiant energy to Earth at a rate that averages approximately 1.3 kW per square metre of the cross-sectional area of Earth. If we assume that all of this energy is absorbed by Earth (i.e., that Earth has an emissivity of 1), at what rate must Earth radiate energy if Earth's temperature is to remain constant? What would that constant temperature be? (Note: The actual temperature of Earth is somewhat higher due to the greenhouse effect of the atmosphere.)

31. ✳ A researcher heats a 100.0 g sample using a 400.0 W resistive heater. The temperature of the sample increases at a rate of 3.0°C/min. Find the heat capacity and the specific heat of the sample.

32. ✳✳ (a) The Sun has a surface temperature of approximately 6000 K and a diameter of 14×10^5 km. What is the total power that is radiated by the Sun?
(b) The Sun derives its energy from fusion reactions in which $E = mc^2$, where E is the energy released, m is the mass converted to energy, and c is the speed of light. At what rate does the Sun lose mass as radiated energy?

33. ✳✳ Athletes can lose significant amounts of water from their bodies as a result of perspiring during a competition. During an intense 2.0 h tennis match, a player lost 4.0 kg. Estimate the rate at which her body was using energy to evaporate all of that perspiration. List any assumptions you make for the estimate.

34. ✳✳ Three moles of an ideal gas are at a temperature of 273 K and a pressure of 100.0 kPa. The gas is compressed isothermally to half of its original volume. It is then expanded adiabatically to its original state. How much work is done in the compression? In the expansion? What is the highest temperature reached by the gas? The lowest?

35. ✳✳ One and a half moles of an ideal gas is at standard temperature and pressure. The gas is isothermally compressed to one-third of its original volume. How much work is done?

36. ✳✳ The compression in problem 35 is carried out adiabatically. How much work is done in that case?

37. ✳✳ The Etruscan shrew is a small rodent that has a very high metabolic rate: its heart beats over 1000 times/min. The animal is approximately 4 cm long and has a body temperature of approximately 36°C. Estimate the rate at which it loses thermal energy due to radiation.

38. ✳✳✳ The ice on the surface of a lake in winter helps insulate the lake. For these questions, assume that the air temperature is –10°C and the ice is 10 cm thick.
(a) How much thermal energy per square metre is lost from the water to the atmosphere?
(b) Using the rate in part (a), determine how long it will take another centimetre of ice to form.

39. ✳✳ An ideal gas is in a cylinder that has a light frictionless piston. The pressure outside the cylinder is 101 kPa. You slowly add 1000.0 kJ of thermal energy to the gas and observe the volume to increase from 10.0 m³ to 15.0 m³. Determine the work done by the gas and the change in internal energy of the gas.

40. ✳✳ An ideal gas is at an initial pressure of 3000.0 kPa and a volume of 7.0 L. The gas is expanded at constant pressure to a volume of 11.0 L. It is then cooled at constant volume until the temperature returns to its original value. Calculate the work done, the change in internal energy, and the thermal energy flow into or out of the gas.

41. ✳✳ Consider an ideal gas that initially has a pressure of 1909 kPa and a volume of 11.0 L. The gas is compressed at constant pressure to a volume of 7.0 L. Thermal energy is then added at constant volume until the pressure reaches 3000.0 kPa. Calculate the work done, the change in internal energy, and the thermal energy flow into or out of the gas.

42. ✳✳ An ideal monatomic gas undergoes the increases in pressure and volume shown in the graph in Figure 17-19. The initial temperature of the gas is 297 K.
(a) Calculate the work done.
(b) Calculate the temperature of the gas at the end of the process.
(c) Calculate the change in internal energy in this process.
(d) Calculate the quantity of thermal energy added to or removed from the gas during this process.

Figure 17-19 Problem 42

43. An ideal gas is carried through a thermodynamic cycle consisting of two isobaric and two isothermal processes, as shown in Figure 17-20. Calculate the total work done in the entire cycle when $V_1 = 0.50$ m³, $V_2 = 2.0$ m³, $P_1 = 1.00 \times 10^5$ Pa, and $P_2 = 1.50 \times 10^5$ Pa.

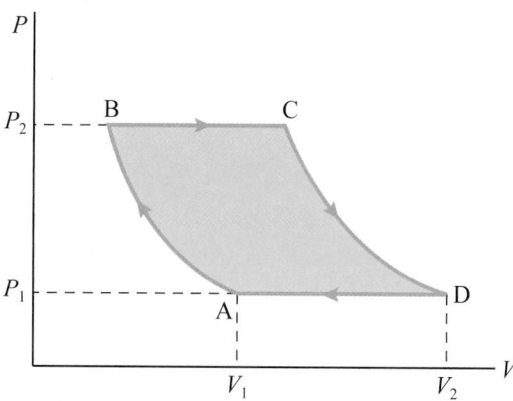

Figure 17-20 Problem 43

44. ✲✲ One mole of an ideal gas is at a temperature of 50°C and a pressure of 100.0 kPa. The gas is expanded isothermally to three times its initial volume. Find the final temperatures, the thermal energy input to the gas, the work done by the gas, and the change in the internal energy of the gas. Draw the process on a *P-V* diagram.

45. ✲✲ One mole of an ideal gas is at a temperature of 50°C and a pressure of 100.0 kPa. The gas is expanded isobarically to three times its initial volume. Find the final temperature, the thermal energy input to the gas, the work done by the gas, and the change in the internal energy of the gas. Draw the process on a *P-V* diagram.

46. ✲✲ One mole of an ideal gas is at a temperature of 50°C and a pressure of 100.0 kPa. The gas is expanded adiabatically to three times its initial volume. Find the final temperatures, the thermal energy input to the gas, the work done by the gas, and the change in the internal energy of the gas. Draw the process on a *P-V* diagram.

DATA-RICH PROBLEM

47. Table 17-5 lists temperatures recorded as a sample of water was heated from about 5°C to 30°C. The ambient temperature in the room was 20.0°C. The thermal energy was supplied by a 10.0 W heater. Determine the mass of the water being heated.

OPEN PROBLEM

48. ✲✲✲ We can reduce the cost of heating our homes by increasing the amount of insulation we put into them. Of course, there is a cost to installing the insulation. Suppose that insulation costs $X per square metre per centimetre and that thermal energy costs $Y per joule. Determine the thickness of insulation one should use when building a house that minimizes the total cost of insulating and heating the house. You will have to make some assumptions about how long the house will be lived in, the thermal conductivity of the insulation, and the average inside–outside temperature difference.

Table 17-5 Data for Problem 47

Time (s)	Temperature (°C)
0	4.79
40	6.53
80	8.04
120	9.86
160	11.05
200	12.50
240	13.56
280	14.94
320	15.80
360	17.04
400	17.83
440	18.90
480	19.98
520	20.90
560	21.92
600	22.51
640	23.63
680	24.28
720	24.94
760	26.00
800	26.31
840	27.03
880	27.71
920	28.62
960	28.98
1000	29.88
1030	30.06

 See the text online resources at www.physics1e.nelson.com for Open Problems and Data-Rich Problems related to this chapter.

Learning Objectives

When you have completed this chapter you should be able to:

1. Describe the operation of a refrigerator and a heat engine.

2. Sketch a Carnot cycle on a *PV* diagram.

3. Describe entropy in terms of macroscopic variables.

4. State the second law of thermodynamics in several ways.

5. Discuss the circumstances under which thermodynamics is applicable.

6. Discuss some of the consequences of the second law of thermodynamics.

7. Describe entropy from a microscopic perspective.

For additional **Making Connections, Examples, and Checkpoints**, as well as activities and experiments to help increase your understanding of the chapter's concepts, please go to the text's online resources at **www.physics1e.nelson.com**.

Chapter 18
Heat Engines and the Second Law of Thermodynamics

What do the refrigerator and car engine in Figure 18-1 have in common? In the refrigerator, we supply electrical energy to power a motor, which does work on a gas. The gas is circulated in such a way that thermal energy is transferred from the interior of the refrigerator to the coils on the back. Work is done on the system, and thermal energy is moved from a lower to a higher temperature.

In the car engine, fuel is burned at a high temperature inside cylinders with pistons. The hot gases do work on the pistons, pushing them outward. The gases cool somewhat during this expansion, and are then ejected through the exhaust pipes. The moving pistons rotate the crankshaft, which is connected to a gearing system that drives the wheels. In this engine, thermal energy is moved from a higher temperature to a lower temperature, and the system does work on the car.

One of the goals of this chapter is to understand how heat pumps and heat engines work. As you will see, the study of these devices was a key step in the formulation of the second law of thermodynamics. The second law of thermodynamics introduces the concept of entropy. We will first describe entropy in terms of macroscopic quantities: thermal energy and temperature. Later in the chapter, we will look at a microscopic description. We begin our approach to the second law by examining in more detail how heat engines and refrigerators (and heat pumps) work.

Figure 18-1 A refrigerator and an internal combustion engine

18-1 Heat Engines and Heat Pumps

Heat Engines

A **heat engine** is a device that performs work while allowing thermal energy to flow from a high-temperature reservoir to a low-temperature reservoir, as shown in Figure 18-2. A **reservoir** is an object whose heat capacity is so large that it can be considered to be at a constant temperature. The reservoir from which we are removing thermal energy is called the source, and the reservoir that we are putting thermal energy into is called the **sink**.

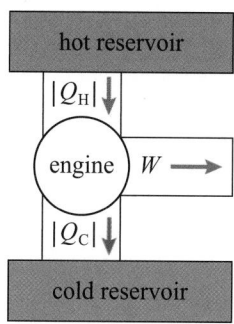

Figure 18-2   Schematic illustration of a heat engine

A high-temperature source provides thermal energy to the engine. For a heat engine, the temperature of the source of thermal energy is always greater than the temperature of the heat sink. The engine does some work and also delivers a quantity of thermal energy to the low-temperature sink. Since energy must be conserved, for a heat engine,

KEY EQUATION
$$|Q_{\text{H}}| = |Q_{\text{C}}| + |W| \qquad (18\text{-}1)$$

In Equation (18-1), we have used absolute value signs to avoid any ambiguity that could arise from the fact that the thermal energy flows can be either positive or negative, depending on our reference point. For example, Q_{H} is a negative thermal energy flow from the perspective of the high-temperature source but positive from the perspective of the engine.

Steam engines are a common type of heat engine. While obsolete for railway locomotives, a steam engine is the mechanism that converts thermal energy into mechanical energy in thermal power plants (both fossil fuel-fired and nuclear). In a steam engine, water is heated to produce steam at a high temperature. In a fossil fuel-fired plant, hot combustion gases from burning fuel heat the water flowing through tubes in the boiler. In a nuclear

C-18-1 Sign Conventions for the Flow of Thermal Energy

The flow of thermal energy is

(a) positive from the perspective of the engine and negative from the perspective of the low-temperature sink;

(b) positive from the perspective of both the engine and the low-temperature sink;

(c) negative from the perspective of the engine and positive from the perspective of the low-temperature sink;

(d) negative from the perspective of both the engine and the low-temperature sink.

C-18-1 (c) Thermal energy flows out of the engine into the low-temperature sink.

plant, the reactor core has coolant pumped through it to capture the thermal energy from the fission reaction, and that thermal energy is used to generate the steam. Such plants use a steam turbine to drive an electrical generator, as shown in Figure 18-3.

Figure 18-3   A steam turbine in a nuclear power plant

Often, pressurized water is circulated through the reactor core to both control the temperature of the core and capture the thermal energy that is generated by the nuclear reactions in the core. Since the water is under high pressure, it can be heated to a high temperature (well above 100°C) without boiling. The pressurized water circulates through pipes in the steam generator, where thermal energy from these pipes vaporizes water in the turbine loop to form high-pressure steam. The reactor coolant, which has been exposed to intense radiation in the core, circulates in a closed loop of piping and does not mix with the water in the turbine loop. The steam is fed to the turbine, causing it to turn. The turbine drives the generator, producing electrical power. The steam expands and cools as it passes through the turbine into the condenser. However, not all the thermal energy in the high-temperature steam is converted into work in the turbine. To increase the efficiency of the engine, we cool the steam in the condenser so that the thermal energy that is ejected from the engine is at as low a temperature as possible. We will examine heat engine efficiencies in Section 18-2.

A heat engine always generates some waste thermal energy. In a car, that thermal energy is dissipated through the radiator, although some of this thermal energy can be used to warm the passenger compartment on cold days. The fact that waste thermal energy is unavoidable leads to the **Kelvin–Planck statement** of the **second law of thermodynamics**: No heat engine can convert thermal energy completely into work.

The device shown in Figure 18-4 is not possible because the sole effect of the device is to convert thermal energy completely into work.

Figure 18-4 This heat engine process is impossible, a fact summarized by the Kelvin–Planck statement of the second law of thermodynamics.

Heat Pumps and Refrigerators

Heat pumps and refrigerators are both devices that use energy to transfer thermal energy from a low-temperature source to a high-temperature sink, as shown in Figure 18-5.

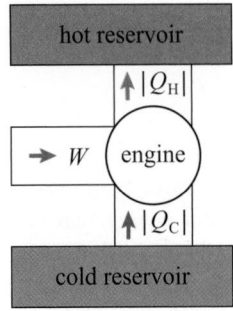

Figure 18-5 A heat pump and a refrigerator

When we apply conservation of energy to a heat pump, we find that the thermal energy expelled to the high-temperature reservoir must equal the thermal energy

extracted from the low-temperature reservoir plus the work input:

$$|Q_H| = |W| + |Q_C| \qquad (18\text{-}2)$$

We are all familiar with household refrigerators, which transfer thermal energy from a cool interior compartment to the warmer surroundings. With heat pumps, the focus is on the thermal energy that is delivered to the higher-temperature area. For example, a heat pump for home heating extracts thermal energy from the cool ground outside and delivers it to the warmer interior of the house. A heat pump operates on the same physical principles as a refrigerator, and both transfer thermal energy from a cooler body to a warmer body. The distinction between the two is based on their purpose.

Figure 18-6 shows the basic operation of a refrigerator. The pressure in the internal heat exchange coil (on the left side of the diagram) is lower than in the external heat exchange coil (right side). When gas passes through the expansion valve, the gas expands and cools to a temperature that is lower than the temperature of the interior of the refrigerator. Thermal energy in the air in the refrigerator is then absorbed by the cool gas. The gas is then compressed as it is pumped into the external heat exchange coil. This compression raises the temperature of the gas above the ambient temperature in the room, so thermal energy transfers from the coil on the back of the refrigerator to the room. Thus, the refrigerator transfers thermal energy from its cool interior to the warmer room. To accomplish this transfer, energy must be supplied to the compressor, which does work on the gas.

Figure 18-6 Schematic of a refrigerator

We find that thermal energy does not flow from a cooler body to a warmer body unless we do work on the system. This observation leads to the **Clausius statement** of the second law of thermodynamics: No device can transfer thermal energy from a low-temperature reservoir to a high-temperature reservoir without work input. The heat pump shown in Figure 18-7 violates the Clausius statement because it has no work input.

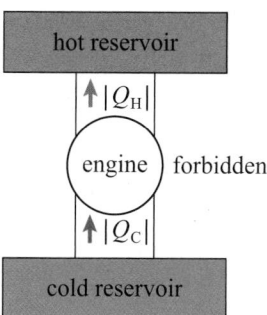

Figure 18-7 This heat pump process is impossible, a fact summarized by the Clausius statement of the second law of thermodynamics.

18-2 Efficiency and the Carnot Cycle

We have seen that heat engines cannot completely convert thermal energy into work and that heat pumps (e.g., refrigerators) cannot simply move thermal energy from a low temperature to a high temperature without work being done. Of course, when designing a heat engine, we would like to extract as much work as possible from a given quantity of thermal energy. Similarly, we would like to minimize the energy that a heat pump requires to transfer a given quantity of thermal energy. Now, we consider the efficiencies of these devices.

Heat Engine Efficiency

For a heat engine, we are primarily interested in maximizing the amount of work done for a given thermal energy input. In so doing, we are also minimizing the thermal energy output, which is typically lost energy. We define the **efficiency**, η, of a heat engine as the ratio of the work done, $|W|$, to the thermal energy input, $|Q_H|$:

KEY EQUATION
$$\eta \equiv \frac{|W|}{|Q_H|} = \frac{|Q_H| - |Q_C|}{|Q_H|} \qquad (18\text{-}3)$$

We used Equation (18-1) to replace W in the numerator with $|W| = |Q_H| - |Q_C|$.

CHECKPOINT

C-18-2 Limits of Efficiency

A manufacturer of a heat engine claims that the engine has an efficiency of 110%. Is this claim reasonable?

EXAMPLE 18-1

Work, Thermal Energy, and Efficiency

An automobile engine has an efficiency of 20.0%. When the engine radiates thermal energy at a rate of 100 kJ/s, how much mechanical work is being extracted?

SOLUTION

Looking at Equation (18-3), we can see that to determine W, we need to know Q_H. We are given Q_C, so we start by solving for Q_H:

$$|Q_C| = |Q_H| - \eta|Q_H|$$

$$\therefore |Q_H| = \frac{|Q_C|}{1 - \eta} = \frac{100 \text{ kJ/s}}{1 - 0.2} = 125 \text{ kJ/s}$$

We can now determine the work to be

$$W = |Q_H| - |Q_C| = 125 \text{ kJ/s} - 100 \text{ kJ/s} = 25 \text{ kJ/s}$$

Making sense of the results:

The work done is certainly less than the total thermal energy input, so we know that the engine has not violated the Kelvin–Planck statement of the second law of thermodynamics.

Efficiency of Heat Pumps and Refrigerators

For heat pumps and refrigerators, the definition of efficiency centres on the amount of work done per unit of thermal energy transferred. For a refrigerator, we are primarily interested in the thermal energy transferred from the cool interior. Therefore, we define the **coefficient of performance of a refrigerator**, CP_R, as

KEY EQUATION
$$CP_R = \frac{|Q_C|}{|W|} \qquad (18\text{-}4)$$

In contrast, our primary focus with a heat pump is the thermal energy delivered to the higher temperature. We define the **coefficient of performance of a heat pump**, CP_H, as

KEY EQUATION
$$CP_H = \frac{|Q_H|}{|W|} \qquad (18\text{-}5)$$

CHECKPOINT

C-18-3 Too Good to Be True?

A manufacturer of home-heating systems claims that its heat pump can deliver up to 3 J of thermal energy for every joule of electrical energy it consumes. Is this even possible?

C-18-3 Yes. An examination of Figure 18-5 shows that the thermal energy delivered to the high-temperature reservoir is, in general, greater than the energy supplied in the form of work. As we will see later in this section, the actual ratio depends on the difference between the inside and outside temperatures.

C-18-2 No. Getting more work out than energy you put in would violate conservation of energy. In addition, we can see from 18-3 that the only way to get an efficiency greater than 1 is to have $Q_C < 0$ which is not possible.

Getting More Than You Pay For

Heat pumps can provide more energy than they consume, but most other forms of home heating cannot (Figure 18-8). Electrical heating systems convert the electrical energy directly into thermal energy, producing a thermal energy output equal to the electrical energy input. Home-heating systems that burn a fuel, such as natural gas, oil, or wood, always lose some thermal energy with exhaust gases that go up the chimney. These losses can be significant—around 20% for a typical midefficiency furnace. However, the latest high-efficiency gas furnaces can reduce the waste thermal energy to as little as 4%.

Figure 18-8   In this heat pump installation, the coils that absorb the thermal energy from outside of the house are installed underground. Heat pump efficiencies decrease substantially when the source reservoir temperature is low compared to the interior temperature. The underground installation helps guard against low air temperatures.

The Carnot Cycle

French scientist Nicolas Léonard Sadi Carnot (1796–1832) devoted considerable attention to maximizing the efficiencies of heat engines. His studies introduced some of the fundamental concepts of thermodynamics.

As is often done in physics, Carnot chose to study a simple model system under ideal conditions. He considered a heat engine that used an ideal gas in a reversible process. A **reversible process** is a process that is carried out so slowly that the system is always in equilibrium. A system is in equilibrium when none of the state variables vary with time or location within the system. An ideal gas is in equilibrium when the pressure and temperature are the same everywhere in the gas, the number of particles in the gas is constant, and the volume of the gas is constant.

All real processes are *irreversible*. For example, if we suddenly compress a gas, we create a pressure wave that travels throughout the gas. The wave is a region of high pressure, and the compression reduces the volume of the gas, so clearly the gas is no longer in equilibrium. As the wave propagates, the molecules in the gas collide with each other. The wave spreads out, and eventually the pressure becomes uniform. At this point, the gas has returned to equilibrium.

Imagine breaking up the sudden compression into smaller, sudden steps. At each step, the departure from equilibrium is smaller than for the entire compression, and the time taken to return to equilibrium is shorter. If we break up the process into a series of infinitesimal steps, the departure from equilibrium becomes infinitesimally small, and the time away from equilibrium becomes infinitesimally short. In the limit, the process maintains equilibrium throughout. However, it will take an infinite amount of time to carry out the process. Nonetheless, this infinite process can be a useful theoretical construct.

Figure 18-9 illustrates the **Carnot cycle** for a heat engine. As indicated by the arrows, we proceed clockwise around the cycle. We could also proceed counterclockwise to describe a cycle for a heat pump or refrigerator.

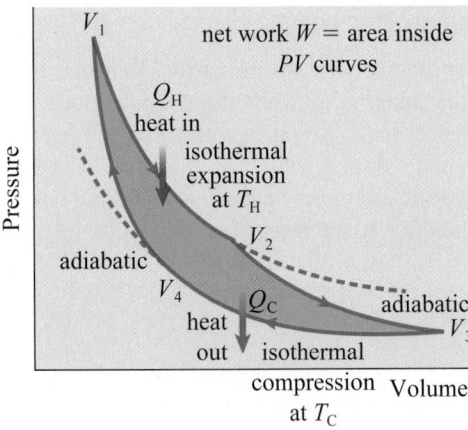

Figure 18-9   A *PV* diagram for a heat engine operating on a Carnot Cycle

The Carnot cycle consists of four reversible processes. Starting in the upper-left-hand corner of the cycle, we first carry out an isothermal expansion at T_H. During this process, thermal energy $|Q_H|$ is absorbed by the gas. We then carry out an adiabatic expansion where no thermal energy flows into or out of the gas, but the temperature is lowered to T_C. Once we reach that temperature, we carry out an isothermal compression, and thermal energy $|Q_C|$ is removed from the system. The final step is an adiabatic compression that raises the temperature back to T_H. During the two

THERMODYNAMICS

expansions, the gas does work on its surroundings. During the two compressions, the surroundings do work on the gas. The net work done on the surroundings is represented by the pink shaded area in Figure 18-9.

We can determine how much work is done in this cycle by calculating all the thermal energy flows. In Chapter 17, we showed that for isothermal processes

$$W = \int_{V_1}^{V_2} P\,dV = \int_{V_1}^{V_2} \frac{nRT}{V}\,dV = nRT \ln\left(\frac{V_2}{V_1}\right) \quad (18\text{-}6)$$

Thus, for the Carnot cycle we have

$$W_{1-2} = nRT_H \ln\left(\frac{V_2}{V_1}\right) \quad (18\text{-}7)$$

Since this is an isothermal process, there is no change in the internal energy of the gas. The first law of thermodynamics then reduces to $W = Q$ (Equation (17-16)) and we can write:

$$|Q_H| = nRT_H \ln\left(\frac{V_2}{V_1}\right) \quad (18\text{-}8)$$

All the same arguments apply to the isothermal compression. Therefore,

$$|Q_C| = nRT_C \ln\left(\frac{V_3}{V_4}\right) \quad (18\text{-}9)$$

Recall that $|Q_C|$ is the magnitude of the thermal energy absorbed by the low-temperature reservoir. To make that quantity positive, we used $\frac{V_3}{V_4}$ as the argument of the logarithm.

For the adiabatic portions of the cycle, we can eliminate the volumes and express the thermal energy flows simply in terms of the temperatures. In Chapter 17, we showed that for an ideal gas undergoing an adiabatic process,

$$PV^\gamma = \text{Constant} \quad (18\text{-}10)$$

where $\gamma = \dfrac{f+2}{f}$, and f is the number of degrees of freedom.

Thus we can write

$$P_2 V_2^\gamma = P_3 V_3^\gamma$$
$$P_4 V_4^\gamma = P_1 V_1^\gamma \quad (18\text{-}11)$$

Since we are using an ideal gas, we can apply the ideal gas law to give

$$\frac{P_2 V_2}{T_H} = \frac{P_3 V_3}{T_C} \quad (18\text{-}12)$$
$$\frac{P_4 V_4}{T_C} = \frac{P_1 V_1}{T_H}$$

In Equation (18-12), all four ratios are equal, but we have written the pairings to mirror those in Equation (18-11). Dividing the corresponding pairs in Equation (18-11) by those in Equation (18-12), we obtain

$$T_H V_2^{\gamma-1} = T_C V_3^{\gamma-1}$$
$$T_H V_1^{\gamma-1} = T_C V_4^{\gamma-1} \quad (18\text{-}13)$$

Dividing the upper row by the lower row in Equation (18-13) we find

$$\left(\frac{V_2}{V_1}\right)^{\gamma-1} = \left(\frac{V_3}{V_4}\right)^{\gamma-1} \quad (18\text{-}14)$$

Therefore,

$$\ln\left(\frac{V_2}{V_1}\right) = \ln\left(\frac{V_3}{V_4}\right) \quad (18\text{-}15)$$

Using this result in Equations (18-8) and (18-9), and combining those two equations, we find (finally!)

$$\frac{|Q_H|}{|Q_C|} = \frac{T_H}{T_C} \quad (18\text{-}16)$$

Rearranging gives

$$\frac{|Q_H|}{T_H} = \frac{|Q_C|}{T_C} \quad (18\text{-}17)$$

We can use this result to determine the efficiency of a heat engine operating on a Carnot cycle:

KEY EQUATION
$$\eta = \frac{|Q_H| - |Q_C|}{|Q_H|} = 1 - \frac{|Q_C|}{|Q_H|} = 1 - \frac{T_C}{T_H} \quad (18\text{-}18)$$

It is important to remember that these temperatures are expressed using the Kelvin scale. Note that the efficiency of such an engine is always less than 1 for an engine operated at finite temperatures. Since the Kelvin–Planck statement of the second law indicates that a heat engine cannot have an efficiency of 100%, T_C must be greater than 0 K.

Carnot was able to prove that this efficiency applies for any reversible engine operating between two given temperatures. He was also able to show that any irreversible engine has a lesser efficiency. This result is called **Carnot's theorem** and can stated as follows: The efficiency of any reversible engine operating between two fixed temperatures is the same, and any irreversible engine operating between the same two temperatures is not as efficient.

Thus, Carnot's theorem establishes the benchmark against which all real (irreversible) engines are measured.

As mentioned earlier, we can run the Carnot cycle in reverse. The coefficients of performance for a heat pump and for a refrigerator operating on a Carnot cycle are as follows:

Heat pump: $\text{CP}_H = \dfrac{|Q_H|}{|W|} = \dfrac{|Q_H|}{|Q_H| - |Q_C|}$

KEY EQUATION
$$= \frac{1}{1 - \dfrac{|Q_C|}{|Q_H|}} = \frac{1}{1 - \dfrac{T_C}{T_H}} = \frac{T_H}{T_H - T_C}$$
$$(18\text{-}19)$$

Refrigerator: $CP_R = \dfrac{|Q_C|}{|W|} = \dfrac{|Q_C|}{|Q_H| - |Q_C|} = \dfrac{1}{\dfrac{|Q_H|}{|Q_C|} - 1}$

 $= \dfrac{1}{\dfrac{|T_H|}{|T_C|} - 1} = \dfrac{T_C}{T_H - T_C}$ (18-20)

Note that in both cases the temperature difference $T_H - T_C$ appears in the denominator. As a consequence, the coefficients of performance of these devices decrease as the temperature difference increases. Thus, we can reduce the energy consumption of a refrigerator by placing it in a cool place. Similarly, heat pumps work more effectively when the low-temperature reservoir is not too cold.

 CHECKPOINT

C-18-4 Practical Limits to Efficiency

Can a heat engine operating between 400 K and 300 K have an efficiency of 24%?

imum possible efficiency.

unlikely because the best practical heat engines only approach approximately 75% of the max-

Thus, an efficiency of 24% is theoretically possible. However, an efficiency this high is very

$$u = 1 - \dfrac{T_C}{T_H} = 1 - \dfrac{300\ \text{K}}{400\ \text{K}} = 0.25 = 25\%$$

temperatures is

C-18-4 The maximum possible efficiency for a reversible engine operating between the given

LO 3

18-3 Entropy

Now, we consider the direction of the thermal energy flow during a Carnot cycle in a heat engine. Q_H is the thermal energy absorbed by the engine at the high temperature, and our convention is that a heat flow into the engine is a positive quantity. Therefore, $|Q_H| = Q_H$. However, Q_C is the thermal energy transferred from the engine to the low-temperature reservoir and is therefore a negative quantity. Thus $|Q_C| = -Q_C$. Therefore, we can write Equation (18-17) as

$$\dfrac{Q_H}{T_H} = -\dfrac{Q_C}{T_C} \qquad (18\text{-}21a)$$

or

$$\dfrac{Q_H}{T_H} + \dfrac{Q_C}{T_C} = 0 \qquad (18\text{-}21b)$$

CHECKPOINT

C-18-5 Sign Conventions

How would Equation (18-21a) change if we considered a heat pump instead of a heat engine?

equation identical to Equation (18-21a).

Thus, Equation (18-17) would become $-\dfrac{Q_H}{T_H} = \dfrac{Q_C}{T_C}$, and multiplying both sides by −1 gives an

C-18-5 There would be no change. We would replace $|Q_H|$ with $-|Q_H|$ and $|Q_C|$ with Q_C.

So far, we have demonstrated only that Equation (18-21a) applies to a Carnot cycle. We now consider whether this relationship applies to any reversible process for an ideal gas. Any such process takes us from one point on a PV diagram to another, and can be accomplished with a combination of a reversible adiabatic process and a reversible isothermal process. The adiabatic process gets us from the starting temperature to the isotherm that the end point lies on. Then the isothermal process takes us to the end point, as shown in Figure 18-10. We could also start with an isothermal process and then carry out an adiabatic process.

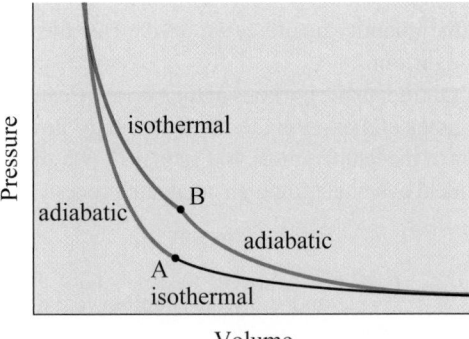

Figure 18-10 We can get from point A to point B by either an adiabatic process (green) followed by an isothermal process (red) or an isothermal process (black) followed by an adiabatic process (blue).

We could break up any arbitrary reversible process that takes the system from A to B into a large number of reversible adiabatic and isothermal steps. The greater the number of steps, the better the approximation. Similarly, we could approximate any arbitrary reversible process that returns the system to point A from point B with a large number of reversible adiabatic and isothermal steps. Thus, any reversible process over a closed path (a path that returns the system to its starting point) can be approximated as a series of small adiabatic and isothermal steps. In the limit of an infinite number of steps, we would have a perfect approximation. Thus, any closed path on a *PV* diagram can be approximated with arbitrary accuracy with a series of Carnot cycles. Such a process is illustrated in Figure 18-11.

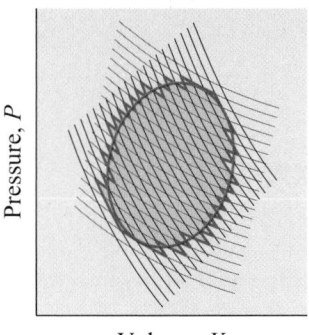

Figure 18-11   Approximating an arbitrary process with a series of Carnot cycles

We now turn our attention to the thermal energy flows that occur in such a step-wise process. We can generalize Equation (18-21b) to

$$\sum_{\text{reversible}} \frac{Q_i}{T_i} = 0 \qquad (18\text{-}22)$$

where Q_i denotes the thermal energy flow for each successive step, and the temperature at which it occurs is T_i.

In the limit of an infinite number of steps, the summation becomes an integral:

$$\oint_{\text{reversible}} \frac{dQ}{T} = 0 \qquad (18\text{-}23)$$

Since the integral sums to zero over a closed path, regardless of the shape of the path, there must be a state variable associated with this integral. Recall that a state variable has a well-defined value that is unique to the state of the system. For example, when we specify the temperature, pressure, and volume for an ideal gas with a fixed number of particles, we have specified the state of the gas. We saw in Chapter 17 that for an ideal gas the internal energy is a state variable, but the work and the thermal energy are not.

We call this new state variable **entropy** and denote it with the symbol S. For a reversible process going from *a* to *b*,

$$S_b - S_a = \Delta S = \int_a^b \frac{dQ}{T} \qquad (18\text{-}24)$$

From Equation (18-24) we can identify

KEY EQUATION
$$dS = \frac{dQ}{T} \qquad (18\text{-}25)$$

Entropy has units of J/K.

18-4   Entropy and the Second Law of Thermodynamics

We have several statements of the second law of thermodynamics so far, and it is not immediately apparent how they are related. As you will see, these statements are all linked by the concept of entropy.

Clausius Statement Violated: The Clausius statement says that thermal energy cannot flow from a cold temperature to a hot temperature without work being done. If such a process were to occur, the net effect would be a flow of thermal energy out of the low-temperature reservoir while the same amount of thermal energy flows into the high-temperature reservoir. The net change in entropy of the system comprised of the two reservoirs is:

$$\Delta S = -\frac{Q}{T_C} + \frac{Q}{T_H} < 0 \qquad (18\text{-}26)$$

Thus, if the Clausius statement of the second law is violated, the entropy of the system will decrease.

Kelvin–Planck Statement Violated: The Kelvin–Planck statement of the second law says that no process can completely convert thermal energy into work. If such a process were possible, the net effect would be a removal of thermal energy from a high-temperature reservoir. Thus, the change in entropy would be:

$$\Delta S = \frac{-Q}{T_H} < 0 \qquad (18\text{-}27)$$

Thus, if the Kelvin Planck statement of the second law is violated, the entropy will decrease.

Carnot Theorem Violated: Suppose we have a heat engine that is more efficient than a Carnot engine. For example, the efficiency would be greater if the thermal energy flow at the lower temperature were somewhat reduced with a corresponding increase in the amount of work done by the engine. We denote the reduced low-temperature thermal energy flow as Q'_C. We are assuming that $Q'_C < Q_C$, where Q_C is the heat flow for the engine that does not violate Carnot's Theorem. It must also be true that $\frac{Q'_C}{T_C} < \frac{Q_C}{T_C} = \frac{Q_H}{T_H}$. The change in entropy will then be:

$$\Delta S = \frac{Q'_C}{T_C} - \frac{Q_H}{T_H} < \frac{Q_C}{T_C} - \frac{Q_H}{T_H} = 0 \qquad (18\text{-}28)$$

Thus, there would also be a decrease in entropy if a heat engine was more efficient than a Carnot engine.

Summary: We can see that in each of these processes that violate one of the statements of the second law, the entropy of the system *decreases*. However, we have restricted ourselves to considering reversible processes so far, and all real processes are irreversible. For a real heat engine, we find that we are able to extract less work than with a reversible process. Since energy is conserved, the thermal energy delivered to the low-temperature reservoir must be greater than for a reversible process. The change in entropy for an irreversible process is

Efficiency less than Carnot efficiency:

$$\Delta S = \frac{-Q_H}{T_H} + \frac{Q_C''}{T_C} > 0 \qquad (18\text{-}29)$$

where $Q_C'' > Q_C$ is the increased amount of thermal energy.

Thus, the entropy either remains constant (reversible processes) or increases (all real processes). The most general statement of the **second law of thermodynamics** is as follows: **The total entropy never decreases. The result of all real processes is that the total entropy increases**.

EXAMPLE 18-2

Entropy Change in a Reversible Process

Find the change in total entropy when a sample of liquid water at 0°C is placed in thermal contact with a large reservoir that has a temperature infinitesimally lower than 0°C.

SOLUTION

Thermal energy will slowly flow out of the liquid into the reservoir, and the liquid will freeze. The change in entropy of the liquid is

$$\Delta S_{\text{liquid}} = \frac{-mL}{273.15 \text{ K}}$$

where m is the mass of the liquid, and L is the latent heat associated with the liquid–solid transition.

We can see that the entropy of the liquid decreases. However, we must also consider the change in the entropy of the reservoir. The thermal energy that comes out of the liquid transfers to the reservoir. Since the reservoir is large and its temperature is only infinitesimally lower than the temperature of the liquid water, the amount of thermal energy transferred to the reservoir is so small that the temperature of the reservoir does not change. The change in the entropy of the reservoir is

$$\Delta S_{\text{reservoir}} = \frac{mL}{273.15 \text{ K}}$$

Thus, the total change in the entropy of the system consisting of the liquid sample and the reservoir is zero.

Making sense of the result:

The process is reversible, so we would expect the change in the entropy to be zero.

EXAMPLE 18-3

Entropy Change in an Irreversible Process

Find the change in the total entropy when a small sample of liquid water at 10.00°C is placed in thermal contact with a large thermal reservoir at exactly 0°C.

SOLUTION

Thermal energy flows from the warmer sample into the cooler reservoir until the sample cools to 0.000°C. The thermal energy that flows out of the water sample is

$$Q_{\text{water}} = -mc\Delta T$$

where m is the mass of the liquid, c is the specific heat, and $\Delta T = 283.15 \text{ K} - 273.15 \text{ K} = 10.00 \text{ K}$.

The same amount of thermal energy flows into the reservoir. Since the reservoir is large, its temperature does not change, and the change in entropy of the reservoir is

$$\Delta S_{\text{reservoir}} = \frac{mc\Delta T}{273.15 \text{ K}} = 0.0366 \, mc > 0$$

However, the temperature of the liquid is not constant during the thermal energy transfer. Recall from Chapter 17 that the thermal energy flow in a process where there is a finite temperature change is

$$Q = mc\Delta T \qquad (17\text{-}3)$$

For an infinitesimal temperature change, Equation (17-3) becomes

$$dQ = mc\,dT$$

We can use this result to determine the change in entropy for the liquid:

$$\Delta S_{\text{liquid}} = \int \frac{dQ}{T} = \int_{283.15 \text{ K}}^{273.15 \text{ K}} \frac{mc\,dT}{T}$$

$$= mc \ln\left(\frac{273.15 \text{ K}}{283.15 \text{ K}}\right) = -0.0360 \, mc$$

The decrease in the entropy of the liquid is less than the increase in the entropy of the reservoir. Thus, the total entropy has increased by $0.0007 \, mc$.

Making sense of the result:

The change in the total entropy is positive, as we would expect for an irreversible process.

18-5 The Domain of the Second Law of Thermodynamics

The second law of thermodynamics establishes a direction for all processes. Only those processes that lead to a net increase in entropy are observed to occur.

In Example 18-3, we examined a situation in which an object was placed in contact with a cooler object. Thermal energy flowed from the warmer object to the cooler object, and the total entropy increased. Imagine two bodies at the same temperature—placed in contact—and having thermal energy flow from one to the other so that a temperature difference is created. Such a process would not violate conservation of energy because the exact amount of thermal energy that left one body would move to the other body. However, this flow of thermal energy would result in a net decrease in entropy and, hence, violate the second law. Such processes are never observed.

Let us consider some other natural processes for which the reverse process never occurs. Place a full glass of water on the counter in a sealed room. Over the period of a few days, the water evaporates from the glass, and the room becomes somewhat more humid. However, placing an empty glass on the counter never results in the room becoming less humid and the glass filling with water. Similarly, when we open a bottle of perfume in a closed room, the scent spreads throughout the room and remains uniformly distributed after all the perfume has evaporated from the bottle. We do not observe the perfume refilling the bottle, and the scent does not even concentrate in any one part of the room. In both of these examples, the water in the glass and the perfume in the bottle are at the same temperature as the room. It is not obvious how thermal energy has flowed from the room into the perfume or the water to cause it to evaporate. However, we can see that the degree of organization or "order" of the system has changed. Having all the water in the glass is a state that is more ordered than having the water molecules dispersed throughout the room. Conversely, there is more disorder when the water or perfume molecules are dispersed throughout the room than there is when they are all in a container. Such examples lead us to another statement of the second law: All natural processes tend toward states of greater disorder.

So, we can interpret the entropy of a system as a measure of the disorder in the system. Increasing disorder equates to increasing entropy.

We can think of many examples of things that we see happen on a regular basis that are never observed to occur in reverse. If a glass falls off a counter and shatters on the floor, we would be startled to see the shards come back together and the reformed glass then leap back up onto the counter. In fact, we usually find it amusing to watch motion pictures run backward because of the absurdity of seeing such processes occur in reverse.

However, we can think of certain events that are readily reversible. For example, consider a perfectly elastic collision between two particles, as shown in Figure 18-12. In the collision, both momentum and energy are conserved. We can see that the collision in Figure 18-12(b) is simply the reverse of the collision in Figure 18-12(a). If we were to observe either of these collisions, we would not in any way be able to tell if the "clock" was running backward or forward.

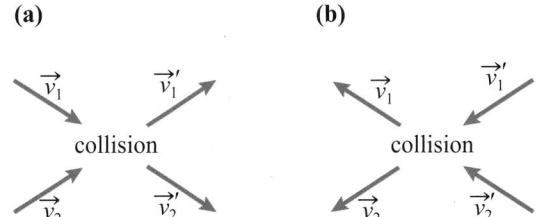

Figure 18-12 Two perfectly elastic collisions between two point particles

What about entropy? Doesn't the second law of thermodynamics apply to all processes? One important difference between the processes illustrated in Figure 18-12 and the processes that we discussed previously is the number of particles involved.

Consider a vessel full of some ideal gas and equipped with pressure and temperature sensors. As we saw in Chapter 16, the pressure is created by collisions between the gas molecules and the walls of the container. Similarly, the temperature is measured by allowing the thermometer to experience collisions with the gas molecules. These collisions transfer energy back and forth between the gas molecules and the material the thermometer is made of. Eventually, the net transfer of energy back and forth is zero, and the gas and the thermometer are in thermal equilibrium. At this point, the thermometric property of the thermometer stops changing and we can read the temperature. For nitrogen gas at standard temperature and pressure, there are approximately 10^{25} molecular collisions with the walls of the container per square metre per second. As a result, the readings on the pressure and temperature sensors appear constant.

Now consider what happens when we remove some of the gas molecules. We would expect the reading on the pressure sensor to decrease because there would be fewer gas molecules colliding with the sensor. If we are careful about how we remove molecules, it is possible to keep the temperature constant. We would need to ensure that the average kinetic energy of the molecules we remove is equal to the average kinetic energy of the molecules in the gas. However, as we reduce the number of molecules further and further, we would find that the frequency of collisions with the pressure sensor would become so low that instead of a steady reading, we would see spikes in the readings as individual molecules hit the sensor.

Our thermometer readings would also start to become peculiar as the number of gas molecules got very low. As we saw in Chapter 16, not all the molecules have the same speed in a gas; hence, there is a distribution of kinetic energies, as well. As the frequency of collisions becomes very small, we would expect fluctuations in the temperature reading, assuming we have an "ideal" thermometer that responds instantaneously to each collision. When a very energetic molecule hits the sensor, it deposits a lot of energy and makes the sensor reading spike up. Conversely, when a low-speed molecule collides with the sensor, we would detect a transfer of energy from the sensor to the molecule, and the reading would spike down.

Once the number of molecules gets so low that the readings of our pressure sensor and thermometer are no longer constant, these variables are no longer well-defined, and we begin to question whether a thermodynamic description of the system is appropriate.

How would we approach thinking about the entropy of a system that contained a single molecule? It would be very difficult to discuss, for example, the degree of disorder of a single gas molecule. Suppose we had two molecules. As they moved around the container we would not be terribly surprised if at some point they passed very close to each other. However, if we have several moles of gas in the container, we would be completely surprised if all of those molecules congregated in the same place at the same time.

All of these state variables—temperature, pressure, and entropy—are based on averages, and these averages are meaningful only when taken over a large number of particles. The number of particles needed to make a thermodynamic approach valid is not precisely known. Strictly speaking, thermodynamics only applies perfectly in the limit of an infinite number of particles—the **thermodynamic limit**. As we reduce the number of particles, we begin to see fluctuations in the readings of thermometers and pressure sensors. Once the fluctuations become comparable to the readings themselves, the variables are no longer well-defined. Then we are no longer in the thermodynamic limit, and the concept of entropy no longer applies.

✓ CHECKPOINT

C-18-6 When Does Thermodynamics Apply?

Which of the following systems can we describe using thermodynamics?

(a) a gas sample containing several moles of gas
(b) a hydrogen atom
(c) a transistor
(d) a galaxy

C-18-6 All but (b) should be amenable to a thermodynamic analysis, because these systems contain a very large number of particles.

18-6 Consequences of the Second Law of Thermodynamics

Absolute Zero

In Section 18-2, we found that the efficiency of a heat engine operating on a Carnot cycle is

$$\eta = \frac{|Q_H| - |Q_C|}{|Q_H|} = 1 - \frac{|Q_C|}{|Q_H|} = 1 - \frac{T_C}{T_H} \quad (18\text{-}18)$$

The Kelvin–Planck statement of the second law asserts that no heat engine can convert all the thermal energy into work. Consequently, $|Q_C| > 0$ and $\eta < 1$. It follows that, for any finite T_H, $T_C > 0$. Thus, there is a lower limit on temperature below which one cannot go. That limit temperature is called **absolute zero**. Absolute zero is 0 K on the Kelvin scale and –273.15°C on the Celsius scale. The lowest temperatures that have been produced in laboratories are on the order of 10^{-10} K.

This lower limit rests on the assumption that T_H is finite. Since temperature is a measure of the average kinetic energy of the particles in a system, we could have an infinite temperature only if we had an infinite amount of energy.

MAKING CONNECTIONS

Additional Limits from Quantum Mechanics

As we will see in Chapter 31, one of the foundations of quantum physics is the uncertainty principle. One of the uncertainty relations states that the precision with which we can simultaneously measure the position, x, and the momentum, p, of a particle are limited such that

$$\Delta x \Delta p \geq \frac{h}{4\pi} \quad (18\text{-}30)$$

where $h \cong 6.62 \times 10^{-34}$ J·s is Planck's constant.

Because of this limit, confining a particle to a region of space results in some uncertainty about the momentum of the particle and, hence, about its kinetic energy. Therefore, we cannot reduce the kinetic energy of such a particle to zero.

Heat Death

The second law of thermodynamics provides all real processes with a direction, which some call the "arrow of time." We now consider in which direction the arrow points and the ultimate fate of the universe.

All processes either move thermal energy from a high temperature to a lower temperature and do some work, or require some work input to move thermal energy from a lower temperature to a higher temperature. The second law establishes that to perform work, we must have temperature differences because there is no heat engine that can solely convert thermal energy into work. Consider a simple model universe that consists of a high-temperature reservoir and a low-temperature reservoir. If we run a heat engine between these two reservoirs, the engine will perform work *and* transfer thermal energy from the high-temperature reservoir to the low-temperature reservoir. The temperature of the high-temperature reservoir

decreases as we remove thermal energy from it, and the temperature of the low-temperature reservoir increases as thermal energy is deposited into it. At some point, the two reservoirs reach the same temperature, making it impossible for the heat engine to produce any more work. This universe will have reached a state where the energy is uniformly distributed throughout the universe, and no further energy flow is possible. Such a universe will be "dead" in the sense that all matter will be uniformly distributed and no life possible.

MAKING CONNECTIONS

You Are a Thermodynamic Fluctuation

The fact that the universe is not in heat death tells us something about the structure of the universe at its beginning. Because there is structure in the universe today, there must have been structure in the universe when it began. This structure is responsible for our existence. Evidence of this structure can be seen in two ways. The way that matter is distributed in the universe on a very large scale—how galaxies are distributed, for example—reflects the structure in the early universe. Other signs of structure are the fluctuations in the cosmic microwave background (CMB) radiation throughout the universe.

As we saw in Chapter 15, the Doppler shift of the light emitted from distant galaxies is evidence that the universe is expanding. Observations suggest that all matter in the universe was primarily hydrogen, and most of the other elements were formed by fusion processes in stars that coalesced from the hydrogen gas. If we extrapolate the expansion backward to before the formation of stars, the universe becomes denser, and it becomes hotter if we model the hydrogen as a simple ideal gas (a reasonable approximation). At some earlier time, the temperature would have been hot enough that all the hydrogen atoms

would have been ionized into electrons and protons. Light scatters from charged particles, and any radiation in the universe at that point would have been interacting significantly with the matter in the universe.

When we have light that is in thermal equilibrium with matter at a fixed temperature, the light has a characteristic pattern of frequencies called a blackbody spectrum. Analysis of such a spectrum reveals the temperature of the object. The CMB radiation is microwave radiation that appears almost uniformly throughout the sky. A detailed analysis of this radiation shows that it is consistent with an object that was in thermodynamic equilibrium with the radiation, and the characteristic temperature of the radiation is now 2.725 K. The "photon gas" that makes up the CMB radiation has also expanded and has been shifted to longer wavelengths and lower frequencies as a result. Consequently, it appears to be at this low temperature from a temperature of approximately 3000 K at which the electrons and protons combined. Although the radiation is remakably uniform, it is not perfectly so, and these fluctuations are believed to result from small deviations from equilibrium that existed very early in the universe (Figure 18-13).

(a)

NASA

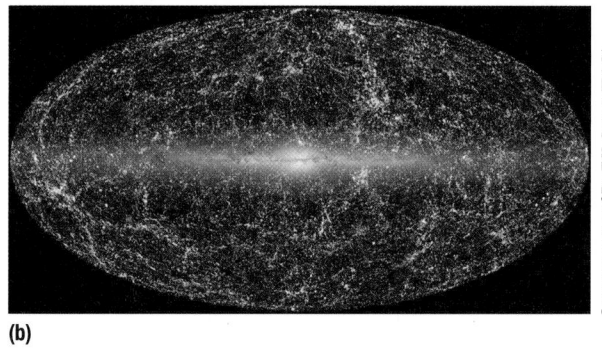

(b)

Image courtesy of Dr. T.H. Jarrett (IPAC/Caltech)

Figure 18-13 (a) Fluctuations in the cosmic microwave background. False-colour images of the temperature inferred from CMB radiation measurements made by the COBE satellite. In the upper image, red corresponds to 2.729 K, and blue corresponds to 2.721 K. The dominant structure is due to the Doppler shift that results from Earth's motion. When that is removed, we are left with the bottom image, where the difference between red and blue is only 0.002 K. (b) Large-scale structure in the universe. A near-infrared image of the sky, which reveals the distribution of galaxies beyond our own. Galaxies that are coloured blue are closest, and galaxies that are coloured red are farthest away. The band across the centre is the Milky Way galaxy.

Such a universe would not necessarily have zero energy. The energy will simply be uniformly distributed. Fortunately, this process will take a very long time to occur. The universe is estimated to be approximately 14 billion years old, and there is still plenty of thermal energy flowing around. In fact, there is even some debate over whether this heat death will be the ultimate fate of our universe, because it depends on some things that we do not know, such as whether the universe will continue to expand or whether the expansion will stop and the universe will collapse back in on itself.

LO 7

18-7 A Microscopic Look at Entropy

We begin our comparison between the macroscopic definition and the microscopic details of a system by examining the concepts of macrostate and microstate.

A "macrostate" of an ideal gas is specified by knowing the pressure, volume, temperature, and number of particles in the system. We have no knowledge of the positions or speeds of any of the individual particles in the system. We have knowledge of the average speeds of the particles because, as we saw in Chapter 16, we can relate the temperature and the pressure of an ideal gas to those quantities. We also know that the average velocity of all the particles is zero if the gas has no net momentum.

A "microstate" of an ideal gas is specified by stating how many particles have a particular speed and the volume of the gas. Knowing this, we can then calculate the total number of particles, the average speed, and, hence, the temperature and pressure.

Let us consider how many different microstates correspond to a given macrostate. The two quantities that must be conserved are the number of particles and the average speed of the particles. One way to create the macrostate is to give each particle the same speed equal to the average speed. There is only one way to do this, however. We say that such a microstate is highly ordered.

Suppose we increase the speed of one particle and decrease the speed of another particle in such a way that we do not change the average speed. There are many ways to do this simple modification because we can change the speed by an arbitrary amount. If we extend this to changing the speeds of three particles, we get an even larger number of ways to do it. These microstates would be less ordered, or more disordered, because the number of particles with changed speeds increases.

The order of a state is a reasonably intuitive concept. For example, we say that a solid material is more ordered than its liquid state. The solid consists of an array of atoms and/or molecules that are in well-defined positions relative to one another. The liquid state does not have well-defined positions for the atoms and/or molecules, and

even the shape of the liquid is less defined than the solid, as illustrated in Figure 18-14.

© lee hacker/Alamy

Figure 18-14   A melting ice cube. The solid is in a more ordered state than the liquid.

Similarly, the gas is less ordered than the liquid, and we expect the water from the melted ice cube to eventually evaporate into the air. The natural tendency is toward states of greater disorder.

☑ CHECKPOINT

C-18-7 Thermal Energy and Order

Which is more ordered, a cold block of metal and a hot block of metal, or the two blocks pressed together at some intermediate equilibrium temperature?

It appears as though systems naturally evolve toward states of higher disorder, and from our consideration of the microstates of a gas, there are many more disordered microstates corresponding to a particular macrostate than there are ordered microstates.

A fundamental concept in statistics is that the probability of an event occurring is proportional to the number of different ways that the event can occur. For example, if we roll a die, there are six possible outcomes, each of which is equally likely. The probability of getting any number on the roll is then 1/6. Suppose we roll two dice, one red and one blue. There are now 36 possible outcomes for the roll because, for each number on the first die, there are 6 possible outcomes for the second die. If we categorize the rolls based on the sum of the two dice, we get the results in Table 18-1. The most probable outcome is that the dice sum to 7 because that result can be achieved in the largest number of ways.

Table 18-1 Outcomes for rolling two dice

Sum of two dice (macrostate)	Possible outcomes (microstates)						Number of outcomes (microstates)
2	1, 1						1
3	1, 2	2, 1					2
4	1, 3	2, 2	3, 1				3
5	1, 4	2, 3	3, 2	4, 1			4
6	1, 5	2, 4	3, 3	4, 2	5, 1		5
7	1, 6	2, 5	3, 4	4, 3	5, 2	6, 1	6
8	2, 6	3, 5	4, 4	5, 3	6, 2		5
9	3, 6	4, 5	5, 4	6, 3			4
10	4, 6	5, 5	6, 4				3
11	5, 6	5, 6					2
12	6, 6						1

We make the analogy between the dice and the thermodynamics systems as follows: Knowing the sum of the two dice is equivalent to specifying the macrostate of the system. Specifying the result of each die is specifying the microstate.

The important point to note is that, in general, there is more than one way to achieve the same outcome. The same is true of a thermodynamic system. In general, there are many microstates that correspond to the same macrostate.

EXAMPLE 18-4

States for Coin Tosses

Enumerate the possible macrostates (number of heads and tails) and their corresponding microstates (head and tail for each coin) when 4 coins are tossed.

SOLUTION

The possible macrostates are 4 tails, 3 tails and 1 head, 2 tails and 2 heads, 1 tail and 3 heads, and 4 heads.

There is only one microstate each corresponding to all heads (H, H, H, H) and all tails (T, T, T, T).

For the case of 3 tails and 1 head, we can have (T, T, T, H), (T, T, H, T), (T, H, T, T), and (H, T, T, T). The case of

1 tail and 3 heads can be arrived at by switching all heads for tails, and vice versa. We then obtain (H, H, H, T), (H, H, T, H), (H, T, H, H), and (T, H, H, H).

The case of 2 tails and 2 heads can be arrived at with (H, H, T, T) and (T, T, H, H), (H, T, T, H) and (T, H, H, T), and (H, T, H, T) and (T, H, T, H). Note that each pair is the complement of each other. We summarize our results in Table 18-2.

Making sense of the result:

We find that the most number of microstates corresponds to the most "disordered" state: 2 heads and 2 tails.

Table 18-2 Results for Example 18-4

Macrostate	Microstates						Number of microstates
4 tails	T, T, T, T						1
3 tails, 1 head	H, T, T, T	T, H, T, T	T, T, H, T	T, T, T, H			4
2 tails, 2 heads	T, T, H, H	H, H, T, T	T, H, H, T	H, T, T, H	T, H, T, H	H, T, H, T	6
1 tail, 3 heads	T, H, H, H	H, T, H, H	H, H, T, H	H, H, H, T			4
4 heads	H, H, H, H						1

The results of a coin toss can be generalized to a large number of coins. For example, if we toss 100 coins, there are on the order of 10^{30} possible microstates. On the one hand, the probability of getting all heads or tails is thus about 10^{-30}, a very small number. On the other hand, approximately 8% of the states are states in which there are the same number of heads and tails. Seventy-three percent of the states lie between 45% and 55% being all heads. As we increase the number of coins, more and more of the probability shifts toward the same number of heads and tails, and in the limit of an infinite number of coins, the probability of getting 50% tails and 50% heads is 1.

The same thing happens in a thermodynamic system. All microstates are occupied with equal probability, but the macrostate that is observed is the one that has the largest number of microstates associated with it. In the thermodynamic limit of an infinite number of particles, this corresponds to a single macrostate. If we prepare a system in another macrostate that has fewer microstates, it will evolve toward the macrostate with the largest number of microstates, and this is also the most disordered state.

THERMODYNAMICS

FUNDAMENTAL CONCEPTS AND RELATIONSHIPS

A study of the efficiency of heat engines and heat pumps led to the development of an idealized, reversible process called the Carnot cycle. The expression obtained for the efficiency of a Carnot cycle suggests the existence of a state variable called entropy. For reversible processes, the entropy remains constant; for all other processes, the entropy increases. This led to the most general statement of the second law of thermodynamics: entropy always increases or stays the same. The idea of entropy also leads to an understanding of the "arrow of time." Processes that are always observed to occur in one direction but never in reverse are processes in which the entropy increases.

A heat engine is a device that moves thermal energy from a high temperature to a low temperature and performs some work:

$$|Q_H| = |Q_C| + |W| \qquad (18\text{-}1)$$

The Kelvin–Planck statement of the second law of thermodynamics: No heat engine can convert thermal energy completely into work.

A heat pump is a device that uses work input to move thermal energy from a low temperature to a high temperature:

$$|Q_H| = |W| + |Q_C| \qquad (18\text{-}2)$$

The Clausius statement of the second law of thermodynamics: No heat pump can move thermal energy from a low temperature to a higher temperature without work input.

A reversible process is an idealized process that occurs so slowly that the system remains in equilibrium throughout the process.

The most efficient heat engine possible is one that uses reversible processes. The processes used are adiabatic and isothermal processes, and such an engine is said to operate on a Carnot cycle. For all Carnot cycles operating between a low and a high temperature, we find that

$$\frac{Q_H}{T_H} + \frac{Q_C}{T_C} = 0 \qquad (18\text{-}21b)$$

Entropy is a state variable defined by

$$dS = \frac{dQ}{T} \qquad (18\text{-}25)$$

The second law of thermodynamics states that the total entropy never decreases. The result of all real processes is that the total entropy increases.

Entropy is a measure of the disorder of a system. The second law of thermodynamics can be stated as all natural processes tend toward states of greater disorder.

Thermodynamics only applies to systems where the number of particles is large enough that statistical measures of their behaviour are valid.

APPLICATIONS

Applications: engine design, home heating, refrigeration, power generation, improving energy efficiency of devices,

Key Terms: absolute zero, Carnot cycle, Carnot's theorem, Clausius statement, coefficient of performance of a heat pump, coefficient of performance of a refrigerator, efficiency, entropy, heat engine, heat pump, Kelvin–Planck statement, reservoir, reversible process, second law of thermodynamics, sink, thermodynamic limit

QUESTIONS

1. Two physics students are trying to stay cool on a hot day. They read in their textbook that a refrigerator and an air conditioner are both heat pumps. They come up with the idea of opening the door of their refrigerator to try to cool their kitchen. Will this idea work? Explain.

2. Discuss how the order of a system changes as it is cooled from the gas phase to the solid phase.

3. An inventor claims to use the waste thermal energy from a heat engine as thermal energy input to the engine and achieve an efficiency greater than the Carnot efficiency. What is wrong with this claim?

4. When a liquid evaporates from a glass into a room, it is not completely clear how the latent heat of evaporation is supplied to the water for it to evaporate, because, if the water is in equilibrium with the room, the water is at the same temperature as the room and thermal energy only flows sponta-

neously from high to low temperatures. Can this truly be an equilibrium process?

5. An inventor has designed a heat pump to recover waste thermal energy from a heat engine and use it to resupply the engine. The inventor claims that adding this heat pump will improve the efficiency of any heat engine. Is this claim valid? Explain.

6. Which is the most effective way to increase the efficiency of a Carnot heat engine: increase the temperature of the high-temperature reservoir or decrease the temperature of the low-temperature reservoir by the same amount?

7. Two keen physics students tell their parents they do not want to clean their rooms because they feel they will be violating the second law of thermodynamics. Are they correct? Explain.

8. Describe a process that respects conservation of energy but violates the second law.

9. In Section 18-5, we considered an elastic collision between two particles and claimed that we are unable to distinguish between the collisions depicted in Figure 18-12 on thermodynamic grounds. Suggest another example.

PROBLEMS BY SECTION

For problems, star ratings will be used, (✷, ✷✷, or ✷✷✷), with more stars meaning more challenging problems.

Section 18-1   Heat Engines and Heat Pumps

10. ✷ A heat engine exhausts 1100 J of thermal energy in the process of doing 400 J of work. What is the thermal energy input to the engine?

11. ✷ A car has a fuel consumption of 5.8 L/100 km while travelling at 100 km/h. The power output of the engine is 40 kW. When we burn gasoline, the thermal energy released is approximately 33 MJ/L
    (a) How many litres of gasoline does the engine burn per second?
    (b) How much thermal energy is released per second as a result?

12. ✷ A heat engine absorbs 20 kJ of energy from a high-temperature reservoir and ejects 15 kJ of energy to a low-temperature reservoir. How much work does the engine perform?

13. ✷✷ A camper has a propane-powered refrigerator. The refrigerator consumes approximately 0.50 kg of propane per day. When propane is burned completely, it releases approximately 50 MJ/kg. The interior of the refrigerator is maintained at an average temperature of 2°C, and the exterior temperature is 22°C. Assume that the refrigerator has a surface area of 2.0 m² and that the insulation has an R factor of 3.2 K·m²/W.
    (a) At what rate does thermal energy flow into the refrigerator? (Hint: See Example 17-9.)
    (b) At what rate must the refrigerator remove thermal energy from its interior?
    (c) How much energy is released per second from the propane combustion?
    (d) How much thermal energy is delivered to the room?

14. ✷ A heat pump operates between the outside and the inside of a house. The pump absorbs 10 kJ of energy from the cool exterior and delivers 30 kJ to the interior. How much work is required to accomplish this?

Section 18-2   Efficiency and the Carnot Cycle

15. ✷ A heat engine operates between a high temperature of 1000.0°C and a low temperature of 500.0°C. Assume it operates on a Carnot cycle.
    (a) What is the efficiency of the engine?
    (b) When the heat engine absorbs 100.0 kJ from the high temperature reservoir, how much work does it do?

16. ✷ A heat pump operates on a Carnot cycle between the cool exterior of a house and the warm interior. When the exterior temperature is 2°C and the interior is 20°C, what is the coefficient of performance for the heat pump?

17. ✷✷ The heat pump in problem 16 delivers 63 MJ of energy to the interior of the house. How much energy input is required in the form of work?

Section 18-3   Entropy

18. ✷✷✷ Suppose we have two identical samples, one at a temperature of T_2 K and the other at a temperature of $-T_1$ K. The two samples are placed in thermal contact.
    (a) Determine the final temperature of the two samples.
    (b) What is the change in entropy of the sample that was at temperature T_2?
    (c) What is the change in entropy of the sample that was at temperature $-T_1$?
    (d) What is the net change in entropy?
    (e) Are negative temperatures possible?

Section 18-4   Entropy and the Second Law of Thermodynamics

19. ✷✷ An ideal gas is placed in good thermal contact with a reservoir, which is held at a constant temperature of 200.0 K. The temperature of the gas is initially 210.0 K. The volume of the gas is held constant.
    (a) How much thermal energy, per mole of gas, flows from the gas to the reservoir to cool the gas to 200.0 K?
    (b) What is the change in the entropy of the reservoir?
    (c) What is the change in the entropy of the gas? (Hint: You will need to do an integral.)
    (d) What is the overall change in the entropy of the reservoir–gas system?
    (e) Is this a reversible process?

20. ✷✷ A 20.0 g piece of copper at an initial temperature of 50.0°C is placed in an insulated container that holds 300.0 g of water at an initial temperature of 20.0°C.
    (a) What is the final temperature of the system?
    (b) How much thermal energy flows out of the copper?
    (c) How much thermal energy flows into the water?
    (d) What is the change in the entropy of the copper?
    (e) What is the change in the entropy of the water?
    (f) What is the total change in the entropy of the copper–water system?

21. ✷✷✷ A heavy-duty restaurant pot is placed on a hot burner on a stove. The temperature of the burner is 500.0°C, and the pot is filled with tap water at a temperature of 10.0°C. The bottom of the pot is steel with a thickness of 1.0 cm and an area of 0.030 m².
    (a) At what rate does thermal energy flow through the bottom of the pot?
    (b) What is the change in entropy per second of the burner?
    (c) What is the change in entropy per second of the water?
    (d) What is the net rate of change of entropy?

Section 18-5   The Domain of the Second Law of Thermodynamics

22. ✷✷ Suppose you toss six coins, which can land either heads or tails. If we treat each coin as being unique, there are 64 unique results for the toss. However, in terms of the number of heads and tails possible, there are only 7 possible outcomes for a toss, ranging from all heads and no tails to all tails and no heads.
    (a) How many ways can you toss all heads?
    (b) How many ways can you toss all tails?
    (c) How many ways can you toss only one head?
    (d) Only two heads?
    (e) Three heads?
    (f) Four heads?

(g) Five heads

(h) Plot a histogram showing the number of ways each result can be achieved.

23. ✱ There is a seemingly never-ending reduction in the size of electronic devices. However, at some point, transistors will become small enough that they will no longer be in the thermodynamic limit. How many atoms would be in a cube of silicon that is 20 nm on edge?

24. ✱✱ In a coin toss experiment, there are only two possible outcomes, and the probability of getting a certain outcome is given by the binomial probability distribution. When we toss N coins, the probability of getting n heads is given by

$$P(n, N) \frac{N!}{n!(N-n)!} 0.5^N$$

(a) Use software (such as a spreadsheet) to calculate $P(n, N)$ for tossing 1000 coins for $0 \le n \le 1000$.

(b) What is the probability of getting 500 heads?

(c) What is the probability of getting between 450 and 550 heads?

Section 18-6 Consequences of the Second Law of Thermodynamics

25. ✱✱ A 200.0 g block of copper is cooled to very low temperatures. Calculate the reduction in the entropy of the block in going from

(a) 100.0 K to 99 K;

(b) 10.0 K to 9.0 K;

(c) 2.0 K to 1.0 K;

(d) 0.020 K to 0.010 K.

For simplicity, assume that the specific heat of copper does not change as the block is cooled. (In fact, the specific heat is proportional to T^3.)

26. ✱ Helium atoms interact very weakly with each other—helium only liquefies at 4.2 K at standard pressure.

(a) What is the kinetic energy of a helium atom at 100 K?

(b) At 10 K?

27. ✱✱ A suggested strategy to avoid heat death is to use the work from a heat engine to drive a heat pump to restore the original temperature difference. Is this possible?

COMPREHENSIVE PROBLEMS

28. ✱✱ Human beings are highly ordered arrangements of molecules. Do we violate the second law of thermodynamics?

29. ✱ You have been asked to review a patent application for a new design of a heat engine. The inventor has supplied the following data on the application:

• The engine takes in 4.50 kW of thermal energy at a temperature of 400 K.

• The engine expels 2.00 kW of thermal energy at a temperature of 200 K.

• The engine performs work at the rate of 2.5 kW.

Do you approve or reject the application? Explain.

30. ✱✱✱ A reasonable approximation to the cycle in an internal combustion engine is the Otto cycle, which is depicted in Figure 18-15. At point 1, the piston is at the bottom of its stroke, and the cylinder is filled with exhaust gases. The exhaust valve was opened at point 4. The exhaust valve remains open, and the piston moves up in the cylinder, pushing the exhaust gases out. This is depicted by the blue horizontal line in the figure. Once the piston reaches the top of the stroke, the exhaust valve closes and the intake valve opens. The piston moves down and draws the fuel–air mixture into the cylinder. This is represented by the green horizontal line, and we have returned to point 1. The intake valve closes and the piston moves up, compressing and raising the temperature of the fuel–air mixture along the yellow line, which we approximate as an adiabatic process ending at point 2. The spark plug then ignites the fuel–air mixture, leading to a sharp increase in temperature and pressure as indicated by the red segment going from 2 to 3, which we approximate as an isochoric process. This increase in pressure forces the piston downward along the red line from 3 to 4, and the pressure and temperature both decrease in this adiabatic process. At point 4, the exhaust valve opens, and the pressure and temperature drop along the blue line from 4 to 1 in another isochoric process. Compare the work done in this process with the work that would be done by a Carnot cycle operating between temperatures T_1 and T_3.

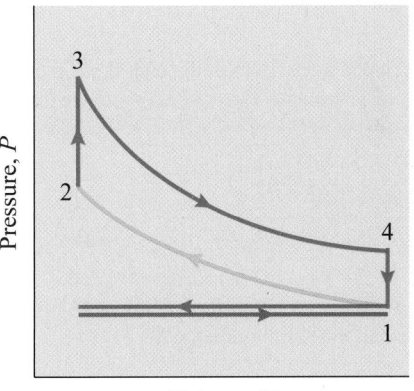

Figure 18-15 Problem 30. The Otto cycle

31. ✱✱✱ We want to investigate how the reading on a sensor in a system becomes more stable as we add more particles to the system. We will consider a very simple one-dimensional system of length L. All the particles in the system travel with the same speed, v m/s. A detector that has the width $\Delta L << L$ is placed in the middle of the system; the detector measures some property of the particle. When a particle is in the detector, the detector reads a value Y_0, and when there is no particle in the detector, it reads 0. If we have a single particle in the system, it will travel from one end to the other in time $T = \dfrac{L}{v}$. During that time, it will pass through the detector once and spend time $\Delta T = \dfrac{L}{v}$ in the detector. As a result, the reading on our sensor will look something like that shown in Figure 18-16. Clearly, the sensor reading is not very stable.

Assuming that the particle enters the sensor at time t_0, the average value of the sensor reading is

$$\bar{Y} = \frac{1}{T} \int_0^T Y(t)\,dt = \frac{1}{T} \int_{t_0}^{t_0 + \frac{\Delta L}{v}} Y_0\,dt = \frac{1}{T} \frac{Y_0 \Delta L}{v}$$

$$= \frac{v}{L} \frac{Y_0 \Delta L}{v} = \frac{Y_0 \Delta L}{L}$$

We will characterize the "wobble" in the sensor with a statistical measure called the standard deviation, σ_Y. It is defined as the positive square root of the following expression

$$\sigma_Y^2 = \frac{1}{T} \int_0^T (Y(t) - \bar{Y})^2 \, dt$$

Notice that this measure adds up the square of the deviations from the average value and is zero when the sensor reading is constant. For our sensor, with a single particle, we calculate this as follows. We will simplify the integrations by assuming $t_0 = 0$:

$$\sigma_Y^2 = \frac{1}{T} \int_0^T \left(Y(t) - \frac{Y_0 \Delta L}{L} \right)^2 dt$$

$$= \frac{1}{T} \left[\int_0^{\frac{\Delta L}{v}} \left(Y_0 - \frac{Y_0 \Delta L}{L} \right)^2 dt + \int_{\frac{\Delta L}{v}}^0 \left(\frac{Y_0 \Delta L}{L} \right)^2 dt \right]$$

$$= \frac{Y_0^2 \Delta L}{L} \left(1 - \frac{\Delta L}{L} \right)$$

We skipped a few steps at the end. If we examine the ratio of σ_Y to $\bar{Y}$, remembering that $\Delta L \ll L$, we find

$$\frac{\sigma_Y}{\bar{Y}} = \frac{\sqrt{\dfrac{Y_0^2 \Delta L}{L} \left(1 - \dfrac{\Delta L}{L} \right)}}{\dfrac{Y_0 \Delta L}{L}} \cong \sqrt{\frac{L}{\Delta L}} \gg 1$$

Generalize this result to N particles, and show that the ratio gets small in the limit of large N.

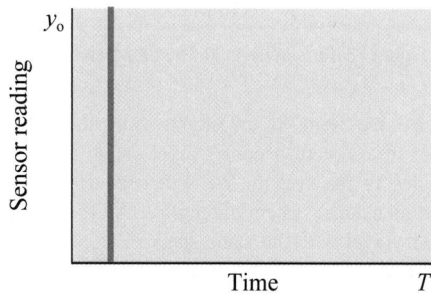

Figure 18-16   Problem 31

32. ✱✱ A block of wood is sliding across a rough surface. The block has a mass of 10 kg and is moving with an initial speed of 10 m/s. It slides across the surface and comes to rest. The kinetic energy of the block has been converted into thermal energy. Assume that the block and the surface are at a constant temperature of 293 K.
(a) What is the kinetic energy of the block?
(b) How much thermal energy has been deposited in the block and the surface?
(c) What is the change in entropy?
(d) Is this a reversible process?

33. ✱✱ A kettle is put onto the stove and brought to a boil. One litre of liquid water is converted to steam. What is the change in the entropy of the litre of water if the initial temperature of the water was 22°C?

34. ✱✱ A typical shower uses approximately 6 L of water per minute. When the temperature of the water is 40°C and the water ends up in the sewer at 5°C, how much does the entropy of the water change during a 5 min shower?

35. ✱✱ Wood burns at a temperature of approximately 500°C. One cord of seasoned (dry) firewood releases approximately 20×10^9 J of energy when it is burned. All the energy is delivered to the room at approximately 20°C. What is the change in the entropy of the wood? The air in the room? The total change in entropy of the wood–air system?

36. ✱✱ An air conditioner removes thermal energy from the interior of a house at 20°C and releases it to the exterior of the house at 35°C. Assume that the system operates on a Carnot cycle. How much energy is removed from the interior per 1000 J of energy supplied to run the device?

37. ✱✱ The temperature of 10.0 mol of an ideal monatomic gas is increased reversibly from 300.0 to 400.0 K. During the process, the volume is held constant. What is the change in the entropy of the gas?

38. ✱✱ Two constant-temperature reservoirs are kept in thermal contact with a 1.00 kg aluminum bar. One reservoir is at 100.0°C, and the other is at 50.0°C. After 1000.0 J of energy has flowed from the hotter reservoir to the cooler reservoir, determine the change in entropy of
(a) the hotter reservoir;
(b) the cooler reservoir;
(c) the bar.

39. ✱✱ Heat pumps for home heating commonly have a coefficient of performance, CP_H, of about 3.5 when the temperature inside the home is 20°C and the temperature outside is 10°C. Estimate the coefficient of performance for such a heat pump when the outside temperature is −10°C, assuming that the change in the coefficient is determined primarily by the difference between the inside and outside temperatures.

40. ✱ Coal releases approximately 30 MJ/kg of thermal energy when it is burned. A power plant produces approximately 1500 MW h while burning 7.0×10^5 kg of coal. How efficient is the plant?

41. ✱✱✱ Ice on the surface of a lake forms an insulating barrier that slows further freezing. Assume that the temperature profile shown in Figure 18-17 is valid for this question.
(a) For a given thickness of ice, at what rate will thermal energy go from the water to the air?
(b) At what rate will the entropy of the air change?
(c) At what rate will the entropy of the water change?
(d) At what rate will the thickness of the ice grow?
(e) Put your answers to (a) and (d) together to show that the thickness of the ice is proportional to the square root of time.

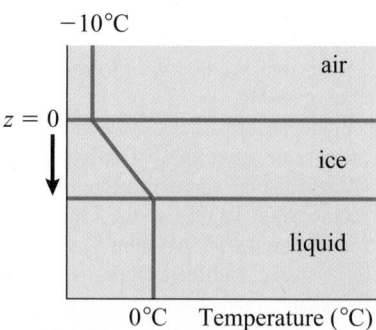

Figure 18-17   Problem 41

42. ✷✷ When dealing four cards from a full deck of cards, determine the probability of dealing
   (a) four cards of the same face value, (e.g., four 3's or four queens);
   (b) four completely different cards (i.e., four cards all with different face values and different suits).
   (Hint: Consider how many different ways (microstates) each outcome (macrostate) can be generated.)

43. ✷✷ An object of mass m falls from a height h and lands on the ground, coming to rest at temperature T. What is the change in the entropy of the system? Discuss the change in entropy in terms of order and disorder.

44. ✷✷ You clean your room and put all of your clothes and books away in an orderly arrangement. The disorder of your room has decreased. Have you violated the second law of thermodynamics? Note: This is *not* a good excuse for refusing to clean your room.

45. ✷✷ As we mentioned in the Making Connections feature called "You Are a Thermodynamic Fluctuation," there was some structure in the early universe. The fluctuations have been amplified by gravity over time to result in the large-scale structure we observe in the universe today: galaxies, etc. Thus, it seems that the universe has evolved from a very nearly uniform state (high disorder) to one of significant structure (low disorder). Does this violate the second law of thermodynamics?

46. ✷✷ The coefficient of performance of a heat pump decreases as the temperature difference increases, so heat pumps become less efficient on very cold days. A clever scientist came up with the idea of using two heat pumps in a home heating system. One transfers heat from the surroundings of the house to a heat reservoir, and runs only when the exterior temperature is high enough for the process to be reasonably efficient. The second transfers heat from the reservoir to the interior of the house, and runs when the house needed to be heated. Discuss the practicality of such an arrangement. What would be a good material to use for the heat reservoir? How large would the reservoir need to be? Clearly state any assumptions.

47. ✷✷✷ Assume that a Carnot heat engine operates between reservoirs T_2 and T_1. The work derived from the heat engine is used to drive an irreversible heat pump that operates between the same two reservoirs. Consider the following three cases:
   (a) The efficiency of the irreversible heat pump is less than the efficiency of the Carnot engine.
   (b) The efficiency of the irreversible heat pump is the same as the efficiency of the Carnot engine.
   (c) The efficiency of the irreversible heat pump is greater than the efficiency of the Carnot engine.
   Prove Carnot's theorem by showing that only one of these cases is possible.

48. ✷✷ Five moles of an ideal monatomic gas is at STP. Calculate the work done and the change in the entropy of the gas when the gas doubles in volume during an
   (a) adiabatic expansion;
   (b) isothermal expansion.

OPEN PROBLEM

49. ✷✷ People have long looked to natural temperature gradients to exploit for driving heat engines. Examples include differences in sea water temperature between the surface and the depths, and differences in the temperature between Earth's surface and core. Estimate the efficiencies of these two examples, and identify any adverse environmental effects that a scheme using these differences may pose.

 See the text online resources at www.physics1e.nelson.com for Open Problems and Data-Rich Problems related to this chapter.

ELECTRICITY,
MAGNETISM, AND OPTICS

Learning Objectives

When you have completed this chapter you should be able to:

1 Explain triboelectric charging, and predict static electricity situations.

2 State the SI unit of charge, and solve problems involving charge and current relationships.

3 Use Coulomb's law to calculate the electric forces exerted by point charges.

4 Calculate the net electric force exerted by multiple point charges using vector superposition.

5 Define electric field, and determine the electric field for various arrangements of point charges.

6 Differentiate between conductors, insulators, and dielectrics.

7 Define electric dipoles, and calculate the electric field for a dipole.

8 Apply electric dipole physics to molecular physics.

9 Define charge density, and use calculus to find the electric field in situations involving uniform charge density.

Chapter 19
Electric Charges and Forces

During a spectacular lightning discharge, there is a massive flow of electric charge (Figure 19-1). Scientists do not completely understand how clouds become electrically charged or how lightning strokes are initiated. In this chapter, we will examine the forces that charged objects exert and apply electric fields to various natural and technological phenomena.

Figure 19-1 A lightning strike results in the flow of a huge electric charge.

For additional Making Connections, Examples, and Checkpoints, as well as activities and experiments to help increase your understanding of the chapter's concepts, please go to the text's online resources at www.physics1e.nelson.com.

19-1 Static Electricity

Static electricity surrounds us in ways that are both obvious and subtle. If on a dry day you scuff your shoes across a carpeted floor, your body will become charged. When you then reach for a metal doorknob, current will suddenly flow to discharge your body.

Static electricity, which is stationary charge on insulated objects, is one of the oldest recorded topics in physics. Thales of Miletus (circa 624–546 BCE) made observations of static electricity. Indeed, the word "electricity" comes from the Greek word for amber, which is one of the first materials found to be easily charged with static electricity.

A key discovery was that electric charges come in two forms—positive and negative. As far as we know, the universe is overall electrically neutral, that is, it has equal amounts of positive and negative charge. Electric charges exert forces on each other: like electric charges (e.g., two negative charges) repel, and unlike charges (one negative and one positive) attract each other.

You probably associate static electricity with rubbing two objects against each other. In fact, charge can be transferred when two different materials come into contact and are then separated. When we rub two materials together, points on the surface of one object repeatedly contact and separate from points on the surface of the other object, causing positive charge to build up on one of the objects and negative charge on the other. This process is called triboelectric charging.

However, rubbing two materials together will result in significant charging only when the two materials have different electrical properties, with one holding on to electrons weakly compared to the other. Qualitatively, we can indicate the relative ability of materials to retain electrons by placing the materials in a triboelectric series, like the one in Table 19-1. For example, if we rub a rubber balloon with hair, the table indicates that the hair will become positively charged and the balloon will become negatively charged. No significant charging will result from rubbing together two materials that are in the centre of the table, near the positive to negative transition, or two materials that are close together in the table and are both in the positive or negative column. For example, rubbing cotton with wood will leave the wood with a tiny amount of positive charge at most.

Metals can be charged by the photoelectric effect, in which incident photons provide the energy needed to strip some electrons from the surface of the metal. Heating can also cause charging in some materials.

How do separated charges behave? When a charged object is placed in contact with an object with the opposite charge sign, or with an object with a much different concentration of charge, a flow of charges results.

Table 19-1 The triboelectric series for some common materials

Positive	Negative
Leather	
Glass	
Hair	
Nylon	
Wool	
Silk	
Paper	
Cotton	
	Wood
	Acrylic
	Rubber
	Polyester
	Polystyrene foam
	Plastic wrap
	Polyethylene
	Polyvinyl chloride (PVC)
	Silicon

The insulating medium between the objects may break down, allowing a sudden flow of charge, often with a visible spark.

Even if there is no actual flow of charges, one charged object can influence the distribution of charges on other objects. When you move a positively charged object near a fluid object, where both positive and negative charge carriers can move, the negative charges (electrons) on the second object are attracted by the positive charge and move closer to the first object. Similarly, positive charges on the second object are repelled and shift away from the positively charged object. In solids, usually only the electrons are free to actually move.

When a flexible, nonconducting object is charged, the charges in the different parts of the object repel each other. This repulsion makes people's hair stand on end when they have either a positive or a negative charge (Figure 19-2). Therefore, objects can be charged by conduction (when charges actually flow between objects) and by induction (no actual contact).

 ONLINE ACTIVITY

The e-resource that accompanies every new copy of this textbook contains an Online Activity using the PhET simulation "Balloons and Static Electricity." Work through the simulation and accompanying questions to gain an understanding of static electrical charging and forces.

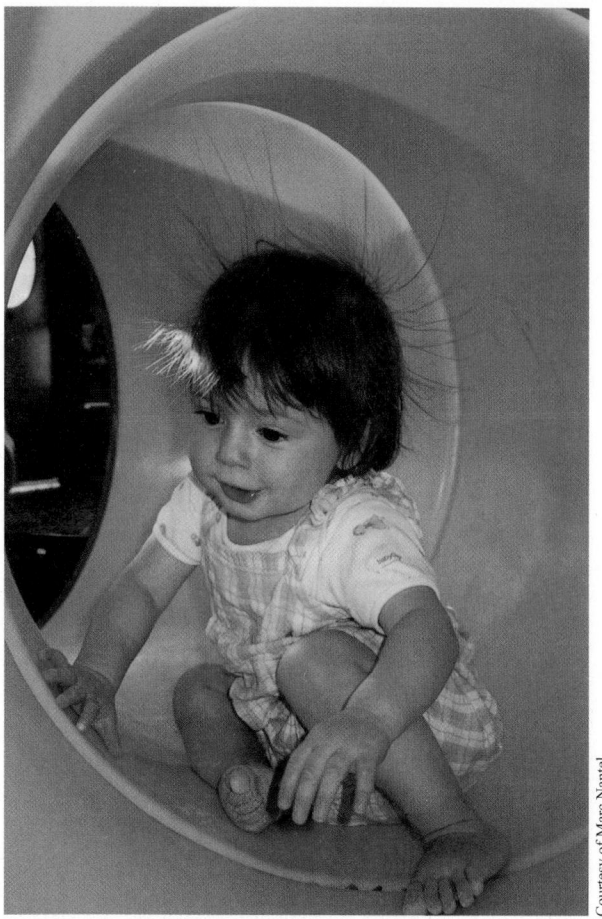

Figure 19-2 The child became charged by triboelectricity as she slid down the plastic tube, causing her hair to stand on end.

Courtesy of Marc Nantel

✓ CHECKPOINT

C-19-1 Clothes Dryer Charges

Some polyester shirts and wool socks have been tumbled together in a clothes dryer. Which of the following statements is correct? Refer to Table 19-1.

(a) The socks are negatively charged, and the shirts are positively charged. The socks stick to each other, and the shirts stick to each other.

(b) The socks are positively charged, and the shirts are negatively charged. The socks stick to each other, and the shirts stick to each other.

(c) The socks are negatively charged, and the shirts are positively charged. The socks stick to the shirts but not to the other socks.

(d) The socks are positively charged, and the shirts are negatively charged. The socks stick to the shirts but not to the other socks.

C-19-1 (d) From Table 19-1, wool is on the positive side, and polyester is on the negative side, which tells us how they are charged when they are rubbed together. Since unlike charges attract (and like repel), we expect the positively charged socks to stick to the negatively charged shirts. You might think that charge would flow to immediately neutralize the clothes in contact with each other, but the rate of flow is actually very low.

19-2 Electric Charge

The SI unit for electric charge is the coulomb (C), named in honour of Charles Augustin de Coulomb (1736–1806). In terms of the SI base units, one coulomb is equal to one ampere times one second: $1\ C = 1\ A \cdot s$. Current is the rate of flow of charges.

KEY EQUATION
$$I = \frac{\Delta q}{\Delta t} = \frac{dq}{dt} \qquad (19\text{-}1)$$

where I is the current (in amperes), and Δq is the amount of charge (in coulombs) that flows past a point in a time Δt (in seconds). We can think of the units of charge and time as infinitesimally small giving the differential form.

The magnitude of the electric charge on a proton or an electron is called the elementary charge, and is usually represented by the symbol e. The coulomb is vastly larger than the elementary charge: $e = 1.60 \times 10^{-19}\ C$.

✓ CHECKPOINT

C-19-2 Charge and Current

Suppose that some dust grains have become charged such that each particle has a charge of 0.0200 C. What current is flowing if 100 of these dust grains pass a point in 0.100 s?

(a) 0.500 A
(b) 2.00 A
(c) 20.0 A
(d) 200. A

C-19-2 (c) With each charge being 0.0200 C, if we have 100 of the charges, we will have a total charge of 2.00 C flowing in 0.100 s. Thus, the current is 2.00 C/0.100 s = 20.0 A.

 EXAMPLE 19-1

How Many Charges per Second?

A current of 0.250 μA is flowing in a circuit. How many elementary charge units are passing a point in this circuit each second?

SOLUTION

First, we convert the current into amperes:

$$\frac{0.250\ \mu A}{1} \times \frac{1 \times 10^{-6}\ A}{1\ \mu A} = 2.50 \times 10^{-7}\ A$$

A current of 1 A means that the amount of charge flowing is 1 C every second. Therefore, the rate of flow of charge is

$$\frac{2.50 \times 10^{-7}\ C}{1\ s} \times \frac{1\ e}{1.60 \times 10^{-19}\ C} = \frac{1.56 \times 10^{12}\ e}{1\ s}$$

Therefore, we would have 1.56×10^{12} elementary charge units (in this case, they would be negatively charged electrons) flowing past a point in the circuit every second.

Making sense of the result:

The elementary unit of charge, e, is truly a tiny charge, so it is reasonable that we would find a huge number flowing every second.

LO 3

19-3 Coulomb's Law

We saw in Chapter 11 that every mass in the universe attracts every other mass with a force directed along a line joining the centres of the two masses, and with a magnitude given by the following equation:

$$|\vec{F}_{\text{G}}| = \frac{GMm}{r^2} \qquad (11\text{-}1)$$

In an analogous way, every electric charge exerts a force on every other electric charge, with the force being attractive when the charges are of opposite sign (i.e., one positive and one negative charge) and repulsive when the two charges are of the same sign.

The electric force, $\vec{F}_{1\to2}$, that a point electric charge q_1 exerts on a second point charge q_2 is given by Coulomb's law:

KEY EQUATION
$$\vec{F}_{1\to2} = \frac{1}{4\pi\varepsilon_0} \frac{q_1 q_2}{r^2} \hat{r}_{1\to2} \qquad (19\text{-}2)$$

Here, q_1 and q_2 are the two electric charges, expressed in coulombs with the appropriate positive or negative sign. The distance between the point charges is r (expressed in metres). A unit vector, $\hat{r}_{1\to2}$, points in the direction from charge 1 toward charge 2. (Recall that a unit vector is a dimensionless vector of magnitude 1.) The constant that plays the role that G does in the universal gravitational force is the term $\dfrac{1}{4\pi\varepsilon_0}$, which is called Coulomb's constant. In Chapter 20, we will show why we prefer to express this constant in terms of ε_0, which is a constant of nature called the permittivity of free space and has a value of 8.85×10^{-12} C²/N·m². Coulomb's law is only applicable to static situations (charges not moving).

EXAMPLE 19-2

Charge on Hanging Masses

Two 100 g balls with equal but unknown charge Q are hanging from massless strings, each making an angle of 15.0° with the vertical, as shown in Figure 19-3. The strings are 40.0 cm long. Find the magnitude of Q. Can you determine the sign of the charges from the information provided?

Note that Equation (19-2) gives us the direction of the force exerted by q_1 on q_2. This force is in the same direction as a unit vector pointing from q_1 toward q_2 when both charges are of the same sign, and is in the opposite direction when one charge is positive and one is negative. We can use Newton's third law to find the force that q_2 exerts on q_1. This force has the same magnitude but the opposite direction as the force that q_1 exerts on q_2.

PEER TO PEER

I find I make fewer mistakes when applying Coulomb's law in situations involving multiple charges if I draw the directions of the arrows for each electric force using the fact that like charges repel and unlike charges attract. I then just use Coulomb's law to find the magnitude of each force, putting in the directions myself.

Coulomb's law applies to point charges. Later in the chapter, we will see how to apply Coulomb's law to situations with multiple point charges, and then in situations with distributions of charges. We will relate electric force and electric field for point charges; one way to consider the inverse square relationship is that the field is spread over a sphere of area $4\pi r^2$ as distance r increases.

 CHECKPOINT

C-19-3 Electric Force Directions

Charge q_1 is a negative charge (say $-Q$) placed in the xy-plane at position ($x = 0$ m, $y = -1$ m), and charge q_2 is a positive charge (say $+Q$) placed at (0 m; $+1$ m). In terms of the ijk unit vector notation, what is the direction of the force acting on charge q_2?

(a) The direction is in the $+\hat{i}$-direction.
(b) The direction is in the $-\hat{i}$-direction.
(c) The direction is in the $+\hat{j}$-direction.
(d) The direction is in the $-\hat{j}$-direction.

C-19-3 (d) First, find the direction of the unit vector pointing from q_1 toward q_2, which must be in the $+\hat{j}$-direction. Since one charge is positive and the other is negative, there is a net negative sign in the equation for Coulomb's law. Thus, the direction of the force on q_2 must be opposite to the direction of the unit vector. This makes sense because unlike charges attract, so the top charge (2) will be pulled downward, in the $-y$-direction.

SOLUTION

As shown in Figure 19-4, there are three forces acting on each ball: a repulsive electric force, a downward gravitational force, and a tension force, $\vec{T}$, directed along the string.

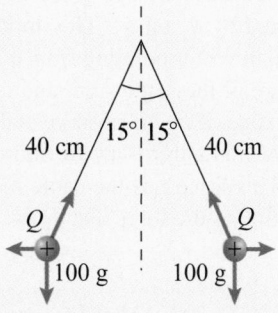

Figure 19-3 The charged balls repel each other. Tension and gravity provide the other forces that together keep the system in static equilibrium.

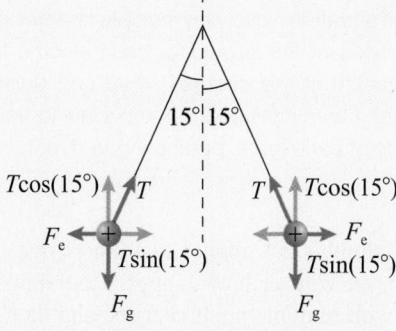

Figure 19-4 Forces acting on the suspended charged balls

The system is in static equilibrium, so these forces must add to zero. Therefore, the vertical component of the tension must balance the gravitational force:

$$T \cos(15.0°) = mg$$

Solving for the magnitude of the tension gives

$$T = \frac{mg}{\cos(15.0°)} = \frac{(0.100 \text{ kg})(9.81 \text{ m/s}^2)}{\cos(15.0°)} = 1.02 \text{ N}$$

Similarly, the horizontal component of the tension must balance the horizontal electric force. This horizontal component is

$$T \sin(15.0°) = (1.02 \text{ N})\sin(15.0°) = 0.263 \text{ N}$$

Applying Coulomb's law for the electric force, we have

$$0.263 \text{ N} = \frac{1}{4\pi\varepsilon_0} \frac{Q^2}{r^2} \quad \text{and} \quad Q^2 = 4\pi\varepsilon_0 r^2 \times 0.263 \text{ N}$$

We can use trigonometry to find the distance, r, between the charges:

$$r = 2 \times 0.400 \text{ m} \times \sin(15.0°) = 0.207 \text{ m}$$

Now we can solve for the unknown charge Q:

$$Q^2 = (4\pi)\, 8.85 \times 10^{-12} \frac{\text{C}^2}{\text{N m}^2} (0.207 \text{ m})^2 \times 0.263 \text{ N}$$

$$Q = 1.25 \times 10^{-12} \text{ C} = 1.25 \text{ pC}$$

However, we cannot determine the sign of the charges from the information provided. The charges could be both negative or both positive because the electric force would be repulsive in either case.

Making sense of the result:

This configuration is used in an electroscope to measure electric charges. If the balls had been separated at a greater angle, then we would find a greater value for Q. Note that in most everyday situations, static charges are a tiny fraction of a coulomb. In this solution, we used the fact that like charges repel to determine the direction of the electric forces. We could have obtained this result more formally using unit vector notation.

LO 4

19-4 Multiple Point Charges

When there are more than two point charges, we find the net force on any one charge by using Coulomb's law to find the force from each of the other charges, and then add these forces as vectors. Calculating the net force in this way is an application of the superposition principle. The superposition principle, widely used in physics, states that in a linear system you can find the response due to individual stimuli (in this case, the electric force due to one charge) and then add the individual contributions to find the total response. You can often use symmetry to simplify the situation because some components will cancel and do not need to actually be calculated.

The following example illustrates the process of combining electric forces for multiple point charge situations. The technique is the same as the technique you used for net forces in the dynamics chapter, except that you now use Coulomb's law to calculate the forces.

✓ CHECKPOINT

C-19-4 Zero Net Electric Force

As shown in Figure 19-5, the charge $-2Q$ is located on the x-axis at a distance D to the right of another charge, $+Q$. Is there a point anywhere on the x-axis where we could place a third positive charge, $+q$, such that it would experience zero net electric force?

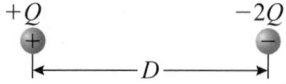

Figure 19-5 C-19-4

(a) No, there is no such point.
(b) Yes, and it would be between the charges.
(c) Yes, and it would be to the left of the $+Q$ charge.
(d) Yes, and it would be to the right of the $-2Q$ charge.

EXAMPLE 19-3

Net Force from Three Charges

Charge q_1 is $+1.00\,\mu C$. Located 0.500 m to the right of q_1 is charge q_2 ($+2.00\,\mu C$). At a distance of 1.50 m in the $+y$-direction from charge q_1 is charge q_3 ($-3.00\,\mu C$). Find the magnitude and direction of the net electric force on charge q_1.

SOLUTION

Let us define a coordinate system with charge q_1 at the origin, as shown in Figure 19-6. We now apply the vector form of Coulomb's law. Since we want to find the net force acting on q_1, we draw unit vectors for direction from q_2 and q_3 to q_1. When we draw the directions of the corresponding forces, we need to keep in mind that the charge product $q_2 q_1$ is positive and the charge product $q_3 q_1$ is negative. Consequently, the direction of the force that q_3 exerts on q_1 will be opposite to that of $\hat{r}_{3\to1}$.

Figure 19-6 Forces acting on q_1

We will first calculate the force that q_2 exerts on q_1:

$$\vec{F}_{2\to1} = \frac{1}{4\pi\varepsilon_0}\frac{q_2 q_1}{r_{21}^2}\hat{r}_{2\to1}$$

$$= \frac{1}{4\pi\varepsilon_0}\frac{(+2.00\times10^{-6}\,\text{C})(+1.00\times10^{-6}\,\text{C})}{(0.500\,\text{m})^2}\hat{r}_{2\to1}$$

$$= 0.0719\,\text{N}\,\hat{r}_{2\to1}$$

Since the unit vector points in the $-x$-direction, we express the force as $-0.0719\,\text{N}\,\hat{i}$.

Similarly, for the force that q_3 exerts on q_1 we have

$$\vec{F}_{3\to1} = \frac{1}{4\pi\varepsilon_0}\frac{q_3 q_1}{r_{31}^2}\hat{r}_{3\to1}$$

$$= \frac{1}{4\pi\varepsilon_0}\frac{(-3.00\times10^{-6}\,\text{C})(+1.00\times10^{-6}\,\text{C})}{(1.50\,\text{m})^2}\hat{r}_{3\to1}$$

$$= -0.0120\,\text{N}\,\hat{r}_{3\to1}$$

The $\hat{r}_{3\to1}$ is in the $-y$-direction, and the force $\vec{F}_{3\to1}$ is in the $+y$-direction:

$$\vec{F}_{3\to1} = +0.0120\,\text{N}\,\hat{j}$$

Thus, the net force acting on q_1 can be written as

$$\vec{F}_{\text{net}\to1} = -0.0719\,\text{N}\,\hat{i} + 0.0120\,\text{N}\,\hat{j}$$

The magnitude of the net force is

$$\sqrt{(-0.0719)^2 + (+0.0120)^2} = 0.0729\,\text{N}.$$

The direction will be above the $-x$-axis at an angle ϕ given by

$$\tan\phi = \frac{0.0120}{0.0719}; \quad \phi = 9.46°$$

Making sense of the result:

Charge 2 will repel charge 1, so charge 1 will experience a force directed to the left. Charge 3 will attract charge 1, so charge 1 will experience an upward force. These directions are consistent with what we found. The separation of charges 1 and 3 is three times the separation for charges 1 and 2, and the same magnitude of charge 3 is 1.5 times the magnitude of charge 2. Thus, the inverse square distance factor, 1/9, will more than offset the greater charge magnitude. When we combine the two forces, we should get a net force that is to the left and upward, with the leftward component dominating, which is consistent with our result.

LO 5

19-5 Electric Field

When a charge experiences an electric force, you could consider that the force is caused by a so-called action-at-a-distance, with the force caused directly by various other charges located at different displacements. Another way to explain the force is that the other charges create an electric field, and the charge responds to that electric field. Both views are valuable frameworks. When dealing with a few charges, it is often easiest to consider a vector superposition

of individual electric forces. However, as we will see in Chapter 21, we can define an energy density present in an electric field, and it is helpful in situations involving distributions of many charges to think in terms of the electric field.

Recall that we can think of the vector of the acceleration due to gravity, $\vec{g}$, as being a gravitational field, as described in Chapter 11. If we place a mass m in this field, we observe a gravitational force $m\vec{g}$. If we want to calculate the gravitational field in a region, we can find the gravitational force on some test mass, divide by that mass, and the result would be the field, $\vec{g}$.

We can define an electric field in an analogous way: place a small positive test charge q in a region, measure the electric force on it, $\vec{F}$, and divide that force by the test charge to obtain the field, $\vec{E}$. In the same way that current is defined in terms of a flow of positive charges by convention only, again, electric field direction is by convention always defined in terms of the force on a positive test charge. Therefore, electric field is simply defined as the electric force per unit charge.

KEY EQUATION

$$\vec{E} = \frac{\vec{F}}{q} \qquad (19\text{-}3)$$

For an electric field at a distance r from a single point charge Q, we can combine Coulomb's law with Equation (19-3):

KEY EQUATION

$$\vec{E} = \frac{1}{4\pi\varepsilon_0}\frac{Q}{r^2}\hat{r} \qquad (19\text{-}4)$$

where $\hat{r}$ is a unit vector pointing away from the charge Q.

The electric field points *away* from a positive charge and *toward* a negative charge. The electric field is strongest near the charge, and decreases as the inverse square of the distance from the centre of the charge. These characteristics are illustrated in Figure 19-7. When we have multiple charges we draw electric field lines that start on positive charges and end on negative charges.

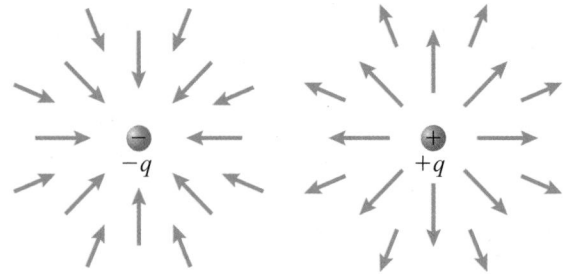

Figure 19-7 The electric field vectors for various points around a negative charge (left) and a positive charge (right)

☑ **CHECKPOINT**

C-19-5 Charged Particle in an Electric Field

A negatively charged particle is moving in the $+x$-direction, but is slowing down. What is the direction of the electric field?
(a) $+x$-direction
(b) $-x$-direction
(c) $+y$-direction
(d) $-y$-direction

C-19-5 (a) The particle is moving in the $+x$-direction but slowing, so the acceleration must be in the $-x$-direction. From Newton's second law, the force must be in the $-x$-direction as well. From $\vec{F} = q\vec{E}$, if the charge (q) is negative, then $\vec{F}$ and $\vec{E}$ must be in opposite directions. Therefore, the electric field is in the $+x$-direction.

🧩 MAKING CONNECTIONS

Inkjet and Laser Printers

Inkjet printers use electric fields to deflect tiny charged drops of ink to form the desired character or image (Figure 19-8). The drops are formed by rapidly heating and vaporizing ink in a chamber or by using a piezoelectric crystal to compress the ink chamber. In both cases, the drops carry charges, and their vertical motion is precisely controlled by electric fields. Modern inkjet printers have ink drops as small as a picolitre, and the ink is typically moving at 10 to 50 m/s when it leaves the print head. The electric fields used for deflection are typically of the order of 100 kN/C. The concept of inkjet printing was first proposed by Lord Kelvin in 1867, but the technology did not become commercially viable until the mid–1970s.

Laser printers use electric charge as well but in a different way. Laser light temporarily changes the conductivity of a photosensitive coating on a drum inside the printer, creating a pattern of static charge on the drum. Toner particles are electrostatically attracted to the charged areas of the drum. When a sheet of paper contacts the drum, most of the toner particles adhere to the paper, transferring the image. A heater then fuses the powdery toner to the paper, making the image indelible.

Figure 19-8 Inkjet printers use electric fields to deflect charged ink drops.

North America's First Electron Microscope

The resolution of an optical microscope is limited according to the wavelength of the light that it uses. The effective wavelength of electrons—the de Broglie wavelength (Chapter 31)—can be much shorter than the wavelength of visible light. In 1938, two University of Toronto graduate students, James Hillier (1915–2007) and Albert Prebus (1913–1997), built the first electron microscope in North America. The instrument, which was capable of magnifications of 20 000×, was a major advance over earlier European attempts and is sometimes credited as the first successful electron microscope. Electron microscopes use electric fields to accelerate the electrons that form the image and to position the beam of electrons when scanning an object. Scanning electron microscopes (SEMs) have many applications in science, engineering, and medicine. Figure 19-9 is an image of a virus taken with a modern electron microscope.

CLIPAREA I Custom media/Shutterstock

Figure 19-9   A modern electron microscope image of a virus

LO 6

19-6 Conductors, Insulators, and Dielectrics

We can classify materials according to how readily charge in the materials moves in response to an electric field. Some of the electrons in a conductor, such as copper, move readily when we apply an electric field, resulting in significant current flow. In contrast, the electrons and protons in insulators are so tightly bound in the atoms that

applying an electric field produces very little, if any, current. Examples of good insulators are glass, ceramic, and paper, although many other materials such as various plastics are relatively good insulators.

Some molecules are polarized, although they have no net charge. In polar molecules, some electrons that are shared in bonds with the atoms in the molecule are pulled more strongly to one or more of the atoms, causing an uneven distribution of charge within the molecule. For example, in a water molecule, the oxygen atom attracts the shared electrons more strongly, so it is negatively charged, and the hydrogen atoms are positively charged. In the presence of an external electric field, the charged atoms in polar molecules shift somewhat. This slight shift of charge causes an additional electric field that opposes the external electric field. Materials that respond to electric fields in this way are called dielectrics. There is very little current in dielectrics: charge displaces within the polar molecules but does not leave the molecules with a net overall charge. In Chapter 21, we will examine an important application of dielectric materials in electronics.

The term "dielectric" was first proposed by the historian and philosopher of science William Whewell (1794–1866). He also introduced a number of other terms, including anode, cathode, physicist, and scientist.

 CHECKPOINT

C-19-6 Fields in Dielectrics

Two parallel horizontal conductors are separated by a liquid dielectric. The top conductor is positively charged, and the bottom conductor is negatively charged. How are the molecules in the dielectric material aligned?

(a) The positive ends of the molecules are toward the top.
(b) The positive ends of the molecules are toward the bottom.
(c) The positive ends of the molecules are toward the right.
(d) There is no net alignment of the molecules.

C-19-6 (b) The dielectric molecules are attracted toward the negative external charges (and the dielectric negative ends of the molecules are attracted to the positive external charges).

LO 7

19-7 Electric Dipoles

A particularly important configuration of charges is an electric dipole, two charges of equal magnitude but opposite sign that are separated by some distance, d. Even though the net charge is zero, a dipole produces an electric field because of the separation of the charges. The superposition principle applies for electric fields. In situations with multiple point charges, we can find the electric field at a point due to each charge separately, and then do a

EXAMPLE 19-4

Electric Field of a Dipole

Consider a dipole lying along the x-axis and centred at the origin. Derive an expression for the electric field at a point on the y-axis.

SOLUTION

Let r represent the distance from the point on the y-axis to the origin (the centre of the dipole). Draw the dipole and the electric field vectors for each charge, as shown in Figure 19-10. The magnitudes of these two electric field vectors are the same because the magnitudes of the charges are identical, as are the distances to the point. Therefore, by symmetry, the downward component of one field vector exactly cancels the upward component of the other, and the net electric field is the sum of

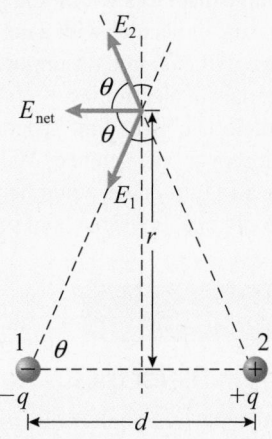

Figure 19-10 An electric dipole with charges $-q$ and $+q$ separated by distance d

the two x-components (both of which point in the $-x$-direction). Symmetry can often be used to simplify applications of the superposition principle for electric forces and fields.

The magnitudes of each of the two electric field vectors are

$$|\vec{E_1}| = |\vec{E_2}| = \frac{1}{4\pi\varepsilon_0}\frac{q}{\left(r^2 + \left(\frac{d}{2}\right)^2\right)} = \frac{1}{4\pi\varepsilon_0}\frac{q}{\left(r^2 + \frac{d^2}{4}\right)}$$

The net electric field is to the left ($-\hat{i}$-direction), and its magnitude is

$$|\vec{E}_{net}| = 2|\vec{E_1}|\cos\theta$$

We can express the angle θ in terms of d and r:

$$\cos\theta = \frac{d/2}{\sqrt{r^2 + \frac{d^2}{4}}}$$

Combining these results we have

$$\vec{E}_{net} = \left(-\frac{1}{4\pi\varepsilon_0}\right)\frac{qd}{\left(r^2 + \frac{d^2}{4}\right)^{3/2}}\hat{i}$$

Making sense of the result:

The direction of the field is the same as the direction from the positive to the negative charge, which makes sense. At large distances from the dipole ($r \gg d$), the electric field is approximately proportional to $1/r^3$, and the field from a single charge is proportional to $1/r^2$. Thus, the electric field from a dipole weakens faster with increasing distance, which makes sense because the fields from the two opposite charges partially cancel each other.

vector addition to obtain the net electric field at the point. In Example 19-4, we demonstrate this process for the electric field of a dipole.

We saw in the above example that the electric field for a dipole goes down as the inverse cube of the distance. It is common to define an electric dipole moment, $\vec{p}$, for charges $+q$ and $-q$ separated by a displacement vector $\vec{d}$ that points from the negative charge to the positive charge:

$$\vec{p} = q\vec{d} \qquad (19\text{-}5)$$

The SI units for electric dipole moments are coulomb-metres. With this definition, we can write the result from Example 19-4 as

$$\vec{E}_{net} \approx \frac{1}{4\pi\varepsilon_0}\frac{\vec{p}}{r^3} \qquad \text{when } r \gg d \qquad (19\text{-}6)$$

✓ CHECKPOINT

C-19-7 Dipole Moments

A proton and an electron are separated by 10 nm. What is the dipole moment?

(a) 1.6×10^{-27} C m directed toward the proton

(b) 1.6×10^{-27} C m directed toward the electron

(c) 3.2×10^{-27} C m directed toward the proton

(d) 3.2×10^{-27} C m directed toward the electron.

C-19-7 (a) The dipole moment is $\vec{p} = q\vec{d}$, where $\vec{d}$ is a displacement vector from the negative to the positive charge, and q is the magnitude of each charge (1.6×10^{-19} C).

19-8 Electric Dipoles in Molecular Physics

Electric dipoles are of crucial importance in molecular physics. The 1936 Nobel Prize in Chemistry was awarded to Petrus Josephus Wilhelmus Debye (1884–1966) for his work on dipole moments. A convenient unit for dipole moments of molecules, the debye (D), was named in his honour (1 D = 3.34×10^{-30} C m).

Electrically, polar molecules can be approximated by a dipole moment. Among the strongest molecular dipoles are simple ionic compounds such as sodium chloride (NaCl), with a dipole moment of 9.0 D, and potassium bromide (KBr), with a dipole moment of about 10.4 D. Molecules of diatomic elements, such as chlorine (Cl_2), have no separation of charge and, hence, zero dipole moment. A number of other nonionic molecules, including CO and CO_2, have near zero dipole moments.

When an external electric field is applied to a substance with a dipole moment, the molecules will align with the field if the molecules are free to rotate. Such movement is the basis for heating in microwave ovens. In a typical household microwave oven, the electric field is applied at a frequency of 2.45 GHz. Water is a polar molecule, with a dipole moment of approximately 1.8 D. As the molecules try to align with the alternating electric field, they rapidly rotate back and forth, and the resulting thermal motion heats the food. Effective heating by this technique depends on a polar molecule (e.g., microwaves heat water more effectively than fats) and an appropriate applied frequency. The technique is called dielectric heating (because it works effectively on dielectrics, materials with polar molecules and electric insulator properties). There are many other applications of electric dipoles in science and technology.

✓ CHECKPOINT

C-19-8 Heating Polar Molecules

Polar molecules experience the most heating in an alternating electric field when their dipole moments are
(a)   0.0 D;
(b)   5.0 D;
(c)   8.0 D;
(d)   11 D.

19-9 Electric Fields from Continuous Charge Distributions

So far, we have calculated electric fields produced by a finite number of discrete point charges. Now, we will consider electric field due to a charge spread uniformly over a linear element (and later over a surface). First, we need to introduce the concept of linear charge density. When a total charge Q is uniformly distributed along a thin rod of length L, we can define a linear charge density, μ, as follows:

$$\mu = \frac{Q}{L} \tag{19-7}$$

Linear charge density has units of C/m.

✓ CHECKPOINT

C-19-9 Linear Charge Density

When a charge of 25 μC is uniformly distributed on a thin rod that is 2.5 cm long, the linear charge density is
(a)   1.0 mC/m;
(b)   10 mC/m;
(c)   10 C/m;
(d)   1.0 kC/m.

To find the net electric field around a continuously distributed charge, we need to use calculus. We write an expression for the field produced by each infinitesimal charge element dq, and then integrate over all of the charge. Often, symmetry allows us to simplify the calculation because components in one or more directions cancel out.

Consider a thin, uniformly charged rod with linear charge density μ. If we define an x-axis along the length of the rod, a length element dx of the rod will have a charge dq given by

$$dq = \mu \, dx \tag{19-8}$$

The contribution from this element of charge to the electric field at a point a distance r from the element is

$$d\vec{E} = \frac{1}{4\pi\varepsilon_0} \frac{dq}{r^2} \hat{r} \tag{19-9}$$

Note that the unit vector $\hat{r}$ points away from the charge element. Therefore, the electric field element points away from the charge element when dq is positive, and toward the charge element when dq is negative.

We then integrate over the entire charged object to find the total electric field:

KEY EQUATION
$$\vec{E}_{net} = \int d\vec{E} = \int_{object} \frac{1}{4\pi\varepsilon_0} \frac{dq}{r^2} \hat{r} \tag{19-10}$$

ELECTRICITY, MAGNETISM, AND OPTICS

EXAMPLE 19-5

Electric Force from a Charged Rod

A thin, horizontal rod of length 2.00 m carries a total positive charge of $98.0\ \mu C$ distributed uniformly along its length. A single small charge, Q, of $+2.00\ \mu C$, is 1.00 m from the right end of the charged rod. Find the total electric force acting on the charge Q.

SOLUTION

As we can see from Figure 19-11, every element of the charge on the rod produces a contribution to the electric field that is in the $+x$-direction, so these contributions all simply add together.

$+98\ \mu C$ $+2\ \mu C$

$\xleftarrow{\hspace{1em}} x \xrightarrow{\hspace{3em}}$

dq dE

$\xleftarrow{\hspace{2em}} 2.0\ \text{m} \xrightarrow{\hspace{2em}}\!\!\xleftarrow{\hspace{1em}} 1.0\ \text{m} \xrightarrow{\hspace{1em}}$

Figure 19-11   To find the net electric field from a charged rod, we integrate the contributions from the various infinitesimal elements of the rod

First, we will express the charge element dq in terms of the linear charge density of the rod and a length element dx:

$$\mu = \frac{\text{charge}}{\text{length}} = \frac{98.0 \times 10^{-6}\ \text{C}}{2.00\ \text{m}} = 4.90 \times 10^{-5}\ \text{C/m}$$

$$dq = \mu\, dx = 4.90 \times 10^{-5}\, dx$$

Note that we have omitted units in the intermediate steps for clarity. Since all elements of charge on the rod produce force elements with the same direction, we can simplify Equation (19-5) to solve for just the magnitude of the electric force:

$$E_{\text{net}} = \frac{1}{4\pi\varepsilon_0} \int_1^3 \frac{\mu\, dx}{x^2}$$

$$= \frac{\mu}{4\pi\varepsilon_0} \int_1^3 \frac{dx}{x^2}\, dx$$

$$= \frac{4.90 \times 10^{-5}}{4\pi\varepsilon_0} \left[-\frac{1}{x} \right]_1^3$$

$$= \frac{4.90 \times 10^{-5}}{4\pi\varepsilon_0} \left[-\frac{1}{3} - \left(-\frac{1}{1} \right) \right]$$

$$= \frac{4.90 \times 10^{-5}\ \text{C/m}}{4\pi[8.85 \times 10^{-12}\ \text{C}^2/(\text{N}\cdot\text{m}^2)]} \left[+\frac{2}{3} \right]$$

$$= 294\ \text{kN/C}$$

To find the force on the charge Q, we use the relationship between electric field and force:

$$F = qE = (2.00 \times 10^{-6}\ \text{C})(2.94 \times 10^5\ \text{N/C}) = 0.588\ \text{N}$$

Since Q is a positive charge, the direction of the force is the same as for the electric field, that is, in the $+x$-direction.

Making sense of the result:

To check the order of magnitude of the force, we can replace the charge along the rod with a point charge of $98\ \mu C$ at a distance of 2 m (average position) from the other charge. Applying Coulomb's law then gives a force of 0.4 N, which indicates that the answer we obtained is reasonable. (Note that this approximation underestimates the actual force because the force depends on the inverse square of the distance between the charges.)

We can use similar calculus techniques to find the net force on a charged object placed near a line of charge, but usually it is easier to solve such problems in two steps: first, calculate the electric field and second, use the relationship between force, charge, and electric field.

We can define surface charge density, σ, as the charge per unit area of a surface. We give both regular and differential forms of the surface charge density relationship below:

$$\sigma = \frac{Q}{A} = \frac{dq}{dA} \qquad (19\text{-}11)$$

In Example 19-6, we demonstrate the use of superposition, integration, symmetry, and surface charge density in finding the electric field of a charged ring and an infinite plane.

EXAMPLE 19-6

Electric Field from a Charged Ring and an Infinite Plane

(a) A horizontal ring of radius r carries a uniform charge and has a total charge of $+Q$. What is the electric field at a point a distance y vertically above the plane of the ring?

(b) Use your result from part (a) to find the electric field a distance y above an infinite plane with surface charge density σ.

SOLUTION

(a) The situation is shown in Figure 19-12, where we have shown the electric field dE from an infinitesimal charge element dq. We can consider the infinitesimal element as approximating a point charge dq, and we obtain the following relationship for the magnitude of the contribution of the electric field from dq:

$$dE = \frac{1}{4\pi\varepsilon_0}\frac{dq}{(y^2 + r^2)}$$

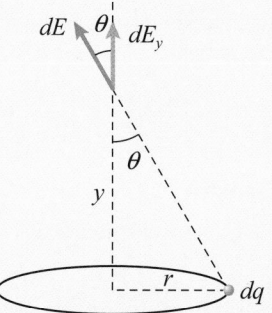

Figure 19-12 Example 19-6. A narrow ring carries a uniform charge.

At this point, it is important to consider the symmetry of the situation. Since the ring is uniformly charged, the components of the electric field in the xz-plane will all cancel, and the net electric field must be vertical. The vertical component is given by

$$dE_y = dE\cos\theta = \frac{1}{4\pi\varepsilon_0}\frac{y\,dq}{(y^2 + r^2)^{3/2}}$$

where we have used

$$\cos\theta = \frac{y}{(y^2 + r^2)^{1/2}}$$

Calculating the integral, and using the facts that both y and r are constant in this situation and that the integral of the charge element around the ring is simply the total charge on the ring Q, we have the following for the net electric field:

$$E = \int dE_y = \frac{1}{4\pi\varepsilon_0}\frac{y\,Q}{(y^2 + r^2)^{3/2}}$$

This expression gives the magnitude of the electric field. Its direction is vertically upward when the ring charge Q is positive and downward when it is negative.

(b) In this situation, we want to consider the electric field at a distance y above an infinite plane that is uniformly charged with surface charge density. The approach we will take is to imagine the plane as being made up of an infinite number of rings of charge of increasing radius. We then integrate over those rings.

If we consider a ring of radius r and width dr, then the charge on just that ring (the entire ring, not just a little piece of it as we did for dq), which we will call dQ, is given by $dQ = 2\pi dr\sigma$.

As in part (a), the symmetry of the situation means that the net electric field must be in a vertical direction. We can use the result from part (a) to find the contribution to this electric field just due to this ring:

$$dE = \frac{1}{4\pi\varepsilon_0}\frac{y\,dQ}{(y^2 + r^2)^{3/2}} = \frac{1}{4\pi\varepsilon_0}\frac{y\,2\pi rdr\sigma}{(y^2 + r^2)^{3/2}}$$

To find the total electric field (due to all the rings), we need to integrate over all r values to infinity:

$$E = \int dE = \frac{2\pi\sigma}{4\pi\varepsilon_0}\int_{r=0}^{r=\infty}\frac{y\,rdr}{(y^2 + r^2)^{3/2}}$$

In calculating the integral, we can make the following substitution:

$$u = y^2 + r^2 \quad \text{and} \quad du = 2rdr$$

This allows us to calculate the integral:

$$E = \frac{\pi\sigma y}{4\pi\varepsilon_0}\int_{u=y^2}^{u=\infty}\frac{du}{u^{3/2}} = \left[-2\frac{\pi\sigma y}{4\pi\varepsilon_0}u^{-1/2}\right]_{y^2}^{\infty}$$

$$= -2\frac{\sigma y}{4\varepsilon_0}\left(0 - \frac{1}{y}\right) = \frac{\sigma}{2\varepsilon_0}$$

The electric field is vertical (upward when the charge is positive) and has the magnitude given.

Making sense of the result:

At first glance, it may seem surprising that the strength of the electric field does not depend on the distance y that you are above the charged plane. Qualitatively, what is happening here is that, although the element of the charged plane nearest to you does produce a stronger electric field, the contributions from more distant parts of the plane are at a larger angle when you are close and have less contribution. The net result is independent of the distance from the charged infinite plane. In Chapter 20, you will see an alternative way to get the same result.

Most everyday forces are electromagnetic in nature. Electric charges are either positive or negative. Like charges repel, and unlike charges attract.

Static electricity is the surface charging of an object. Some materials hold on to their electrons more firmly than other materials, and contact followed by separation causes triboelectric charging. The triboelectric series is used to predict which materials take on positive charges and which take on negative charges when rubbed.

Conductors are materials in which a current flows in response to an applied electric field. Insulators are materials in which virtually no current flows. Dielectrics do not have a net current flow, but they do have charges that can be displaced when an electric field is applied.

Charge and Current

The SI unit of charge is the coulomb, C. Current is the amount of charge that flows past a point in a given time:

$$I = \frac{\Delta q}{\Delta t} \qquad (19\text{-}1)$$

Coulomb's Law

The electric force that one charge exerts on another is given by Coulomb's law:

$$\vec{F}_{1 \to 2} = \frac{1}{4\pi\varepsilon_0} \frac{q_1 q_2}{r^2} \hat{r}_{1 \to 2} \qquad (19\text{-}2)$$

where q_1 and q_2 are the two charges, r is the distance between the centres of the charges, $\hat{r}_{1 \to 2}$ is a unit vector that points from q_1 to q_2, and ε_0 is the permittivity of free space.

Superposition Principle

For systems of point charges, we can find the electric force or electric field separately for each charge, and then vectorially add the results to find the net force or field.

Electric Field

The electric field equals the net electric force on a small positive point charge divided by the magnitude of the point charge:

$$\vec{E} = \frac{\vec{F}}{q} \qquad (19\text{-}3)$$

Dipole Moment

For two opposite charges of equal magnitude $(+q, -q)$ separated by displacement $\vec{d}$, the dipole moment is

$$\vec{p} = q\vec{d} \qquad (19\text{-}5)$$

Linear Charge Density

Linear charge density is the charge per unit length:

$$\mu = \frac{Q}{L} = \frac{dq}{d\ell} \qquad (19\text{-}7)$$

Electric Fields for Continuous Charge Distributions

For continuous charge distributions, the net electric field is given by

$$\vec{E}_{\text{net}} = \int_{\text{object}} \frac{1}{4\pi\varepsilon_0} \frac{dq}{r^2} \hat{r} \qquad (19\text{-}10)$$

Often, we can use symmetry to simplify the calculations.

When dealing with surfaces, a surface charge density is used in the calculations:

$$\sigma = \frac{Q}{A} = \frac{dq}{dA} \qquad (19\text{-}11)$$

Applications: static charging, electrostatic damage to electronics, inkjet and laser printers, electron microscopes, microwave heating

Key Terms: conductor, coulomb, Coulomb's law, current, dielectric, dipole moment, electric dipole, elementary charge, electric field, insulator, linear charge density, permittivity of free space, static electricity, superposition principle, surface charge density, triboelectric charging, triboelectric series

QUESTIONS

1. You remove a piece of polyethylene tape from a glass surface. Which statement best describes what will happen? Table 19-1 may be helpful in answering this question.
   (a) You need rubbing for triboelectric charging, so no electrostatic charge results.
   (b) The tape becomes positively charged, and the glass becomes negatively charged.
   (c) The glass becomes positively charged, and the tape becomes negatively charged.
   (d) Both the tape and the glass take on a positive charge.

2. You rub a balloon through your hair, and it becomes charged. You rub a second similar balloon through your hair. When the two balloons are tied to electrically insulating strings, what happens when you hold the end of the strings away from the balloon?
   (a) The balloons move away from each other.
   (b) The balloons come together and discharge.
   (c) The balloons come together, but each retains its charge.
   (d) The balloons come together, and each charges to a higher value.

3. Use Table 19-1 to determine which of the following combinations will have the greatest triboelectric charge when rubbed together.
   (a) paper and polystyrene
   (b) plastic wrap and glass
   (c) hair and wool
   (d) PVC and polyethylene tape

4. Refer to Table 19-1 when answering this question. Which of the following statements applies when you rub plastic wrap and glass together?
   (a) There is very little charging.
   (b) The glass becomes positively charged, and the plastic wrap becomes negatively charged.
   (c) The glass becomes negatively charged, and the plastic wrap becomes positively charged.
   (d) Both the glass and the plastic wrap become positively charged.

5. A charge of 50 μC flows past a point in 0.05 s. What is the resulting current?
   (a) 10 μA
   (b) 10 mA
   (c) 1 mA
   (d) 10 A

6. A positive charge q is placed near a larger, stationary positive charge $5q$. What happens when you let go of charge q?
   (a) It accelerates toward the larger charge.
   (b) It moves away from the larger charge with constant velocity.
   (c) It moves away from the larger charge with constant acceleration.
   (d) It moves away from the larger charge but with an acceleration that decreases the farther it moves away.

7. You have two charges, q and $4q$. Charge q exerts a force of magnitude F on charge $4q$. What is the magnitude of the force that $4q$ exerts on q?
   (a) $F/4$
   (b) F
   (c) $2F$
   (d) $4F$

8. You have two fixed charges, $+Q$ on the left and $+4Q$ on the right. Where could you place a third small charge, $+q$, along a line joining the two charges such that it would experience zero net electrostatic force?
   (a) at a point beyond the $+Q$ charge to the left
   (b) at a point 1/4 of the way between the charges, closer to the $+Q$ charge
   (c) at a point 1/3 of the way between the charges, closer to the $+Q$ charge
   (d) There is no location where it would experience zero electrostatic force.

9. Two charges, $+Q$ and $+Q$, are located a distance L apart. A second pair of charges, $+2Q$ and $-2Q$, is placed at a distance $2L$ apart. Which situation has the larger electrostatic force?
   (a) The force is the same in the two situations.
   (b) The force is greater in the case with the $+Q$ charges.
   (c) The force is greater in the case with the $+2Q$ charges.
   (d) Since the charges are both positive, the force is zero in both cases.

10. A proton is moving in the $+x$-direction, and its speed is increasing with a constant acceleration. What can you conclude about the external electric field?
    (a) The electric field is in the $-x$-direction and constant.
    (b) The electric field is in the $-x$-direction and increasing.
    (c) The electric field is in the $+x$-direction and constant.
    (d) The electric field is in the $+x$-direction and increasing.

11. A positive charge is located at the origin of an xyz-coordinate system. What is the electric field direction at the point $(x = 0, y = -1, z = 0)$?
    (a) $-\hat{i}$
    (b) $+\hat{i}$
    (c) $-\hat{j}$
    (d) $+\hat{j}$

12. An electron is located at the origin of an xyz-coordinate system. What is the direction of the electric field at the point $(x = 1, y = 0, z = 0)$?
    (a) $-\hat{i}$
    (b) $+\hat{i}$
    (c) $-\hat{j}$
    (d) $+\hat{j}$

13. You are midway between two charges, $+Q$ to the left ($-x$-direction) and $-Q$ to the right ($+x$-direction). What is the direction of the electric field at your location?
    (a) $-\hat{i}$
    (b) $+\hat{i}$
    (c) There is zero net electric field at your location.
    (d) $+\hat{j}$

14. If you slightly increase the distance between a proton and an electron, what happens to the dipole moment?
    (a) The dipole moment is slightly larger.
    (b) The dipole moment is slightly smaller.
    (c) There is no dipole moment before or after.
    (d) The dipole moment has not changed because the charges are still the same.

15. Consider two oppositely charged vertical parallel plates with equal magnitudes of charge. The left plate is positively charged. The space between the conductors is filled with a dielectric. How are the molecules in the dielectric material aligned?
    (a) The positive ends of the molecules are toward the left.
    (b) The positive ends of the molecules are toward the right.
    (c) The positive ends of the molecules are toward the bottom.
    (d) There is no net alignment of the molecules.

16. A radio wave has an electric field of approximately 0.150 N/C. What is the approximate electric force on an electron that is in the path of the radio wave?
    (a) 0.15 N in a direction opposite to the direction of the electric field
    (b) 0.15 N in the same direction as the direction of the radio wave electric field
    (c) 2.4×10^{-20} N in a direction opposite to the direction of the electric field
    (d) 2.4×10^{-20} N in the same direction as the direction of the radio wave electric field

PROBLEMS BY SECTION

For problems, star ratings will be used, (✷, ✷✷, or ✷✷✷), with more stars meaning more challenging problems.

Section 19-2 Electric Charge

17. ✷ If 250 billion elementary charges pass a point in one second, what is the current?

18. ✱✱ When you rub a gold sphere with rabbit's fur, the gold takes on a negative charge (and the rabbit fur positive). Suppose the gold sphere has a mass of 100.0 g and it obtains a net charge of 0.500 μC. Estimate the ratio of the number of electrons added to the gold sphere to the number originally there.

Section 19-3   Coulomb's Law

19. ✱✱ One estimate of the mean separation between the proton and the electron in a hydrogen atom is 53.0 pm (if we picture them as classical particles).
    (a) What is the Coulomb force between the proton and the electron?
    (b) Set this force equal to ma, where the acceleration is a classical centripetal acceleration. What is the speed of the electron?
20. ✱✱ A $+35.0$ nC charge is place at the origin of a coordinate system, and a -25.0 nC charge is placed at the point (0.100 m, -0.0500 m).
    (a) Determine the magnitude and the direction of the force that the positive charge exerts on the negative charge. Use ijk notation for your result.
    (b) Determine the magnitude and direction of the force of the negative charge on the positive charge.

Section 19-4   Multiple Point Charges

21. ✱✱ A -2.00 μC is 3.00 cm to the left of a $+1.00$ μC charge. Where must a $+4.00$ μC be placed to make the net electric force on the $+1.00$ μC zero?
22. ✱✱✱ A square has charge $+Q$ on three of the corners and charge $-Q$ on the bottom-right corner. A charge $+q$ is placed at the exact centre of the square. What is the net force on this charge?

Section 19-5   Electric Field

23. ✱✱ Strong electric fields can be present inside biological cell membranes. The field in a particular cell at a certain location is 5.00 N/μC (note the units). What force magnitude would result on an ion that has a charge of $+2e$?
24. ✱✱ Consider a point midway between two point charges, a $+24.0$ pC charge on the right and a -12.0 pC charge on the left. The charges are separated by 0.240 mm. What is the electric field at this midway point?

Section 19-7   Electric Dipoles

25. ✱✱ An electric dipole consists of one positive and one negative elementary charge. The magnitude of the dipole moment is 1.75 D. What is the separation of the two charges? Express your answer in nanometres.
26. ✱✱ In Example 19-4, we found an expression for the electric field for a dipole when we were at a perpendicular distance from the dipole. Derive a similar expression for the case when you are at a distance in line with the dipole direction. What is the approximate expression for this case when the distance from the dipole is much greater than the dimensions of the dipole?

Section 19-9   Electric Fields from Continuous Charge Distributions

27. ✱✱ A thin, fixed rod is 25.0 mm long and carries a total positive charge of 75.0 nC distributed uniformly along its length. A single small charge $Q = 1.00$ μC is placed 100.0 mm from the left end of the charged rod. What is the total electric force on the charge Q?
28. ✱✱ A thin, fixed rod is 40.0 mm long and carries a total positive charge of $+40.0$ μC. The rod is located on the x-axis, with its centre at the origin of the coordinate system. What is the electric field at a point on the y-axis directly above the centre of the rod and a distance of 120. mm from the rod?

COMPREHENSIVE PROBLEMS

29. ✱ There are a number of different types of lightning with different characteristics. One estimate for a large bolt of positive polarity lightning is that it transfers 350 C of charge and has an average current of 120 kA. Estimate how long the lightning current flows.
30. ✱ Many first-year university textbooks write Coulomb's law with a simplified form for the constant, calling the whole expression k. What is the value of k?
31. ✱✱ Suppose that Earth and the Moon had opposite charges of equal magnitude. What would that magnitude be if the electrostatic force equalled the gravitational attraction between Earth and the Moon?
32. ✱✱ Charges Q and q are stationary at opposite corners of a square with side a as shown in Figure 19-13. Assume that Q is a positive charge. Derive an expression for the value of charge q in terms of the variables of the question so that there is no net force on charge Q. As part of your answer, clearly indicate whether charge q is positive or negative.

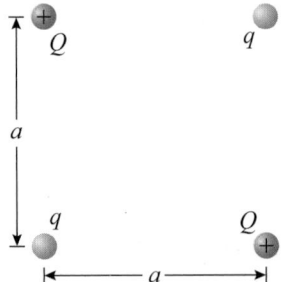

Figure 19-13   Problem 32

33. ✱✱ Three equal charges of $+0.250$ μC are at the vertices of an equilateral triangle with sides 3.25 cm. What is the magnitude of the force on each charge? What is the direction of each force?
34. ✱✱ In Figure 19-14, a small charge with a mass of 1.0 g and a charge of -0.80 μC is attached to the end of a massless string. The charge hangs at an angle of 30° when a uniform electric field, E, is applied in the horizontal direction.
    (a) What is the direction of the electric field?
    (b) Draw an FBD for the small mass.
    (c) Calculate the magnitude of the electric field.

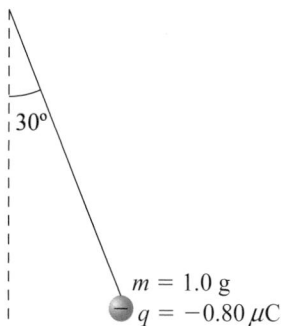

Figure 19-14 Problem 34

35. ✷✷ Four electric charges are arranged as shown in Figure 19-15. Find the electric field at point P.

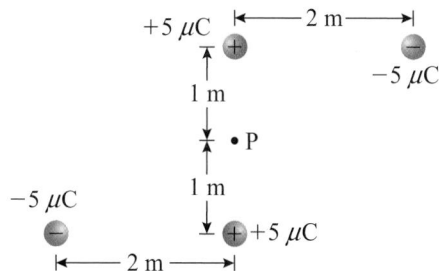

Figure 19-15 Problem 35

36. ✷✷✷ The charges in Figure 19-16 are held rigidly in their positions. Determine the direction and magnitude of the net electric field at point P. Assume the charges and other quantities are known to two significant figures (e.g. $2.0 \, \mu C$). (Hint: Use symmetry to simplify the calculation.)

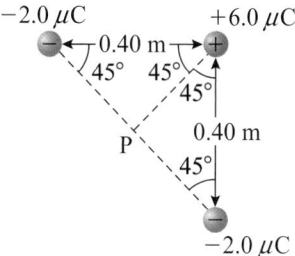

Figure 19-16 Problem 36

37. ✷ A proton has an instantaneous acceleration of 4.0×10^8 m/s² in the +x-direction. What electric field could account for this acceleration?

38. ✷ The electric field near Earth's surface points downward and has a magnitude of approximately 150 N/C. What is the force on an electron due to this electric field? Find the ratio of this force to the gravitational force on the electron.

39. ✷✷ A positive ion with a charge of $+3e$ is fixed at the origin of a coordinate system. An electron is located at $x = 0.20 \, \mu$m on the x-axis. Where is the electric field zero?

40. ✷✷ A cathode ray tube (CRT) display accelerates an electron from rest to 2.50×10^7 m/s within a distance of 0.950 cm in the acceleration region.

(a) What is the electric field in the CRT acceleration chamber?

(b) Does the electric field point in the direction of acceleration of the electron or in the opposite direction? Explain.

41. ✷✷✷ In the deflection area of an inkjet printer there is an electric field of 95.0 kN/C. Assume that one drop has a volume of 2.00 pL (picolitre) and that the density of the ink is the same as the density of water. A drop moves horizontally at a speed of 25.0 m/s and travels between deflecting plates with a length of 1.25 cm. What is the magnitude of the charge on the drop when the vertical deflection is 0.225 mm? Neglect gravitational acceleration, aerodynamic drag, and any effect from Earth's electric field.

42. ✷✷✷ Consider two dipoles, each with the same dipole moment p (from $+q$ and $-q$ charges separated by distance d). The dipoles are both aligned with the x-axis and are located a distance D apart. The negative end of the right dipole faces the positive end of the left dipole.

(a) Derive an expression for the net force between the dipoles.

(b) What does this expression reduce to for the large-distance approximation of $D \gg d$?

43. ✷✷✷ Dipoles placed in an electric field experience a torque unless they are aligned with the field. For an electric field of 30.0 kN/C directed in the +x-direction, find the torque exerted on a dipole that is aligned in the y-axis direction, with a charge of $-e$ at $y = -25$ nm and $+e$ at $y = +25$ nm.

44. ✷✷✷ In 1909, Robert Millikan (1868–1953) used his famous 1909 oil drop experiment to estimate the elementary charge. In this experiment, small drops of oil were charged, and an electric field was adjusted to help the drops stay suspended. From the required field, the charge on various drops was calculated—they were all whole-number ratios of the elementary charge. The actual analysis has to take into account buoyant forces and other details, but we will simplify it for this problem. Assume that the density of the oil used in the experiment is 820 kg/m³, a drop has a radius of 1.10 μm, and a drop has a charge of $+2e$. What applied electric field is necessary to keep the drop suspended against gravity?

45. ✷✷ A charge of 25.0 μC is uniformly distributed around a thin ring of radius 4.00 cm that is in the xy-plane and centred at the origin. What is the electric field at a point 10.0 cm directly above the centre of this ring? (Hint: Use symmetry to simplify the situation.)

46. ✷✷✷ A charge, α, is uniformly distributed over a circular plate of radius R. Derive an expression for the electric field at a distance d above the centre of the plate.

DATA-RICH PROBLEM

47. ✷✷✷ Table 19-2 lists the electric field at various locations along the +x-axis. The negative sign means that the electric field points in the -x-direction. Assuming that the field was created by a single point charge, determine the value and location of this charge.

OPEN PROBLEM

48. ✷✷✷   In this problem, we will consider several aspects of atmospheric electricity and lightning discharges. Look up information or make reasonable estimates as required.

   (a) First, consider how charging might take place. Lightning-producing clouds are typically very high (10 km or more above surface), and during the active phase of these storms, there are extensive updrafts and downdrafts. In addition, the clouds usually have a mix of hail and water droplets. Explain a possible production mechanism for charging by triboelectricity. Do you expect the top of the clouds to be positively or negatively charged?

   (b) As shown in the opening image of this chapter, the lightning strike does not take place in one direct line but in a series of near paths. Explain what you think might be happening.

   (c) Weather services register lightning strikes (e.g., you can look up recent Canadian strikes on the Atmospheric Environment Service website). Try to find similar information for other countries. Where does the highest number of lightning strikes per day occur?

   (d) List some guidelines for protecting human life during a lightning storm. Certain animals are more frequently killed by lightning. Which animals do you think might be at greater risk? Why?

   (e) What are some concerns regarding airplanes and lightning discharges?

See the text online resources at www.physics1e.nelson.com for Open Problems and Data-Rich Problems related to this chapter.

Chapter 20
Electric Potential and Gauss's Law

anada is home to the world's most powerful cyclotron, called TRIUMF (Figure 20-1). TRIUMF is an abbreviation of the original name of the laboratory, the Tri-University Meson Facility, which now has 17 Canadian universities as members. This cyclotron can accelerate protons to almost three-quarters the speed of light. A cyclotron uses a magnetic field to direct charged particles in a circular path inside an acceleration chamber, which is divided into two semicircular sections. Twice during each circuit of the chamber, the charged particles are accelerated by an electric field produced by an electric potential difference between the two sections. The TRIUMF cyclotron accelerates more than a thousand trillion protons each second. While much of the work at TRIUMF focuses on understanding subatomic physics, the cyclotron has a number of other purposes. For example, TRIUMF produces radioisotopes for medical research, and proton beams from the cyclotron have been used to image cancers and to treat eye cancers. Scientists at TRIUMF conduct research in materials science, superconductivity, and hydrogen energy storage mechanisms. Outstanding undergraduate science students are eligible to apply for a TRIUMF summer award.

ELECTRICITY, MAGNETISM, AND OPTICS

Courtesy of TRIUMF

Figure 20-1 A portion of the cyclotron area at TRIUMF, the largest and most powerful cyclotron-based accelerator in the world.

For additional Making Connections, Examples, and Checkpoints, as well as activities and experiments to help increase your understanding of the chapter's concepts, please go to the text's online resources at www.physics1e.nelson.com.

20-1 Electric Potential Energy

Since electrostatic forces are conservative, electric potential energy exists. We can equivalently consider the potential energy as equal to the positive of the work done by an external agent to produce a configuration of charges, or the negative of the work done by the electric field in moving the charges to the configuration. These are equivalent since the external agent must apply a force that is equal in magnitude but opposite in direction to the electric force exerted on a charge by the other charges in the configuration.

Consider the following situation: a positive charge Q is fixed in position, and a second positive charge q is gradually brought toward it. Since like charges repel, a force in the direction of motion is needed to bring the two charges closer together, say, from distance r_1 to r_2 as shown in Figure 20-2. The work done by this external applied force is positive. The force needed varies with the distance between the charges according to Coulomb's law. The Coulomb force on charge q is directed to the left, so the force needed to move the charge is to the right (shown as F in the diagram). The electric potential energy for a charge configuration represents the work that must be done to assemble the distribution.

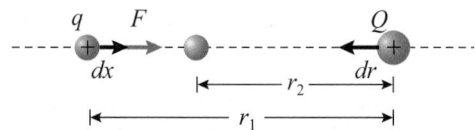

Figure 20-2 A force F must be applied to the right to move a positive charge q toward a fixed positive charge Q.

Since r decreases as the x-coordinate of charge q increases, $dx = -dr$. We can use this relationship when integrating to find the work W_a done by the agent moving the charge.

$$W_a = \int \vec{F} \cdot d\vec{x} = \int_{r_1}^{r_2} \frac{1}{4\pi\varepsilon_0} \frac{Qq}{r^2} (-dr) \qquad (20\text{-}1)$$

When we take the constant terms out of the integral and integrate, we have

$$W_a = -\frac{Qq}{4\pi\varepsilon_0} \int_{r_1}^{r_2} \frac{1}{r^2}\, dr = -\frac{Qq}{4\pi\varepsilon_0}\left[-\frac{1}{r}\right]_{r_1}^{r_2} = \frac{Qq}{4\pi\varepsilon_0}\left[\frac{1}{r_2} - \frac{1}{r_1}\right]$$

$$(20\text{-}2)$$

Since r_2 is less than r_1, the work required to move the charge is positive as expected. Since the external agent did positive work to move charge q, we must have increased the potential energy of the configuration. Equation (20-3) below gives the electric potential energy difference when we move a charge q from a distance r_1 to r_2 from a fixed charge Q. Note that this equation applies for like and

unlike charges as long as the signs of the charges are properly used in the equation:

$$\Delta U_E = \frac{Qq}{4\pi\varepsilon_0}\left[\frac{1}{r_2} - \frac{1}{r_1}\right] \qquad (20\text{-}3)$$

All potential energies are relative. For example, when a mass is on a table, we can use the surface of the table as the zero reference for gravitational potential energy, resulting in the mass having zero potential energy. However, if we use the floor as the zero reference, the mass has positive potential energy. We saw in Chapter 11 that, generally, the most convenient way to define the zero of gravitational potential energy is for the zero level to be when the masses are infinitely far apart. A similar convention is often adopted for electrostatic potential energy—zero when the charges are infinitely far apart. When $r_1 \to \infty$, Equation (20-3) gives the following expression for the potential energy when charges q and Q are a distance r apart:

KEY EQUATION
$$U_E = \frac{Qq}{4\pi\varepsilon_0 r} \qquad (20\text{-}4)$$

Since the existence of potential energy depends on the nature of the field (it being conservative), many physicists prefer to calculate the work W_E done by the electric force when the charge moves (which will be negative in this case), and then set the potential energy $U_E = -W_E$.

☑ CHECKPOINT

C-20-1 Electric Potential Energy

Consider a situation similar to Figure 20-2 but with charge q being negative. The negative charge is moved from a very great distance along the x-axis to a position a short distance from the positive charge. If the potential energy was zero when the charges were infinitely separated, then

(a) positive work must be done by an external agent to move the charge to that position, and the potential energy of the system is positive;

(b) positive work must be done by an external agent to move the charge to that position, and the potential energy of the system is negative;

(c) negative work must be done by an external agent to move the charge to that position, and the potential energy of the system is positive;

(d) negative work must be done by an external agent to move the charge to that position, and the potential energy of the system is negative.

C-20-1 (d) The charges are of opposite sign and therefore attract each other. The applied external force counteracts this force and must be in the $+x$-direction, and the distance moved is along the negative x-direction. Therefore, the net amount of work is negative. This results in a negative potential energy.

Equation (20-4) gives the potential energy for a pair of charges. With three charges, there is a potential energy contribution from each pairing of the charges, where r_{ij} represents the distance between charge i and charge j.

$$U_{total} = \frac{q_1 q_2}{4\pi\varepsilon_0 r_{12}} + \frac{q_1 q_3}{4\pi\varepsilon_0 r_{13}} + \frac{q_2 q_3}{4\pi\varepsilon_0 r_{23}} \qquad (20\text{-}5)$$

We can readily extend this idea to larger numbers of charges, keeping in mind that we must have a term in the potential energy equation for each pairing and that we do not duplicate any pair (i.e., we would not count both a 12 and a 21 pairing). Example 20-1 shows how to apply the technique to four charges.

 EXAMPLE 20-1

Electric Potential Energy for Four Charges

(a) Find the total work required to move four charges (+2.00, +3.00, +4.00, +5.00 nC) from initial positions infinitely far apart to the corners of a square with sides of 0.150 m, as shown in Figure 20-3.

(b) Define the zero of electric potential energy to be when the charges are infinitely far apart. What is the electric potential energy of the charge configuration shown in Figure 20-3?

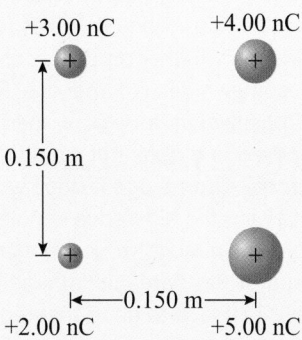

Figure 20-3 Four charges are located at the corners of a square.

SOLUTION

(a) You might at first consider calculating work using force and displacement. However, the forces vary with distance moved, so you would need to integrate, which becomes much more cumbersome once we add the third and fourth charges because the force vectors have different directions. One strength of energy approaches to solving problems is that the potential energy for conservative forces depends only on the actual configuration of the system, and is not affected by how the system got to that configuration. Therefore, we will use electric potential energy, expanding Equation (20-5) to the situation for four charges: We will number the charges 1 to 4 in same order as the charge magnitudes (i.e. the 2 nC charge is number 1).

$$U_{\text{total}} = \frac{q_1 q_2}{4\pi\varepsilon_0 r_{12}} + \frac{q_1 q_3}{4\pi\varepsilon_0 r_{13}} + \frac{q_1 q_4}{4\pi\varepsilon_0 r_{14}} + \frac{q_2 q_3}{4\pi\varepsilon_0 r_{23}} + \frac{q_2 q_4}{4\pi\varepsilon_0 r_{24}} + \frac{q_3 q_4}{4\pi\varepsilon_0 r_{34}}$$

$$= \frac{1}{4\pi\varepsilon_0}\left(\frac{q_1 q_2}{r_{12}} + \frac{q_1 q_3}{r_{13}} + \frac{q_1 q_4}{r_{14}} + \frac{q_2 q_3}{r_{23}} + \frac{q_2 q_4}{r_{24}} + \frac{q_3 q_4}{r_{34}}\right)$$

Note that we have included a term for each charge pair combination, and we have used the r_{ij} notation; for example, r_{34} represents the distance between charges 3 and 4. We use Pythagoras's theorem to obtain the distance between charges on opposite corners of the square. All the terms being summed have identical units, so we can move these units into the common factor for clarity:

$$U_{\text{total}} = \frac{1.00 \times 10^{-18}\ \text{C}^2/\text{m}}{4\pi \times 8.85 \times 10^{-12}\ \text{C}^2/(\text{N} \cdot \text{m}^2)}$$

$$\times\left(\frac{2.00 \times 3.00}{0.150} + \frac{2.00 \times 4.00}{0.212} + \frac{2.00 \times 5.00}{0.150}\right.$$

$$\left. + \frac{3.00 \times 4.00}{0.150} + \frac{3.00 \times 5.00}{0.212} + \frac{4.00 \times 5.00}{0.150}\right)$$

$$= 3.85 \times 10^{-6}\ \text{N} \cdot \text{m}$$

$$= 3.85\ \mu\text{J}$$

Thus, we must do 3.85×10^{-6} J of positive work to bring the charges into this configuration from an initial position of infinite separation.

(b) The electric potential energy of the configuration is simply the work required to create the configuration, which is $+3.85 \times 10^{-6}$ J, or 3.85 μJ.

Making sense of the result:

We expect the work required to be positive because the positive charges repel and it takes positive work to push them closer together. The potential energy is only a few microjoules, which is reasonable because the charges are tiny.

LO 2

20-2 Electric Potential

The **electric potential**, V, is the electric potential energy per unit charge. We can define a zero point for electric potential energy, so the same is true for electric potential. Normally, we define electric potential energy to be zero for infinitely separated charges, but in Chapter 21, when we discuss capacitors, we will invoke alternative definitions for zero potential.

KEY EQUATION
$$V = \frac{U}{q} \qquad (20\text{-}6)$$

The unit of electric potential is the volt (V). We can see from Equation (20-6) that 1 V = 1 J/C. At first, it might seem confusing to use the same symbol V for the quantity electric potential as well as for the unit used to measure it. However, there is a subtle difference in how we write the two symbols: unit symbols should be written as roman (upright) text, and symbols for variables are written as italic (slanted) text.

It is critical to always keep in mind that electric potentials are relative and subject to definition of a zero point.

In electric circuits, this zero point is called ground because in an electrical system it may well be directly connected to a conducting plane placed in the ground. We consider two points, A at +2 V and B at +5 V. Both points are positive with respect to the zero point of potential, but when compared to each other point A has a negative potential difference compared to point B.

We can combine Equation (20-6) with Equation (20-4) to obtain an expression for the electric potential at a distance r from a point charge Q:

 KEY EQUATION

$$V = \frac{Q}{4\pi\varepsilon_0 r} \qquad (20\text{-}7)$$

The electric potential and definition of the zero point expressed in Equation (20-7) are positive in the region near a positive charge, negative near a negative charge, and zero at an infinite distance from the charge Q. It is important to realize that normally electric potential varies with position, and it is often written as $V(r)$ or $V(x)$ to show this dependence. The electric field is often written in a similar way to specifically show its dependence with position.

We apply the superposition principle to situations involving multiple charges, using Equation (20-7) to find the contributions to the potential from each charge. Since these contributions are scalar quantities, it is easy to add them to obtain a net electric potential, as demonstrated in Example 20-2.

ONLINE ACTIVITY

Exploring Electric Potentials

The e-resource that accompanies every new copy of this textbook contains an Online Activity using the PhET simulation "Charges and Fields." Work through the simulation and accompanying questions to gain an understanding of electric potentials.

 PEER TO PEER

I must admit that at first I found the whole idea of electric potentials pretty abstract and confusing. Then I suddenly realized one day that really potentials are telling me whether a positive charge would want to go somewhere. On the one hand, if the potential is positive in a region, then a positive charge is repelled and does not want to go there. It will only go there if some agent does work to move it there. Multiplying the difference in potential by the charge tells you how much work you need to do. On the other hand, if the potential is negative, a positive charge will just love to go to that region and will accelerate to get there.

 ## EXAMPLE 20-2

Electric Potential for Three Charges

(a) Find the electric potential at point P in Figure 20-4.
(b) How much work is required to move a $+1.00\ \mu C$ charge to point P from an infinite distance?

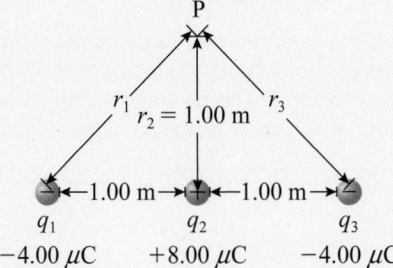

Figure 20-4 Example 20-2. Three charges are in a line; the middle charge is positive.

SOLUTION

(a) According to the superposition principle, we can find the electric potential at point P due to each of the charges separately, and then simply add the three individual potentials algebraically. We use Equation (20-7) with the appropriate charge and the distance from each charge to point P. Distances r_1 and r_3 are both $\sqrt{2}$ m.

$$V = \frac{q_1}{4\pi\varepsilon_0 r_1} + \frac{q_2}{4\pi\varepsilon_0 r_2} + \frac{q_3}{4\pi\varepsilon_0 r_3}$$

$$= \frac{1}{4\pi\varepsilon_0}\left(\frac{-4.0\times10^{-6}\,C}{1.41\ m} + \frac{+8.0\times10^{-6}\,C}{1.00\ m} + \frac{-4.0\times10^{-6}\,C}{1.41\ m}\right)$$

$$= \frac{1\ C/m}{4\pi \times 8.85 \times 10^{-12}\,C^2/(N\cdot m^2)}$$

$$\times (-2.828 \times 10^{-6} + 8.0 \times 10^{-6} - 2.828 \times 10^{-6})$$

$$= +2.11 \times 10^4\,V$$

(b) When the $+1.00\ \mu C$ charge is infinitely far away, it will have an electric potential of zero, so the potential difference between that location and point P is $+21.1$ kV. Since the electric potential is the electric potential energy per unit charge, the electric potential energy when the $1.00\ \mu C$ charge is at point P is

$$(+1.00 \times 10^{-6}\,C)\,(+2.11 \times 10^4\,V) = +0.0211\ J$$

Therefore, the work required to move the charge from infinitely far away to point P is 0.0211 J, or 21.1 mJ.

Making sense of the result:

Although there are equal amounts of positive and negative charge in the initial configuration, the net electric potential is positive because point P is closer to the positive charge than to the negative charges. You may be surprised by how large a potential difference is produced by a small fraction of a coulomb of charge. In part (b), we are moving a positive charge to a region with a more positive electric potential; therefore, the work required is positive.

Example 20-2 demonstrates a useful strategy for finding the work required to move a charge: (1) find the electric potentials at the initial and final points, (2) subtract to determine the electric potential difference, and (3) multiply the potential difference by the charge being moved.

 CHECKPOINT

C-20-2 Work Required to Move a Charge

Two charges, $+Q$ and $-Q$, are located a fixed distance apart along the x-axis, with $+Q$ on the left. A small positive point charge q is moved along the line joining the charges, from a point near the $+Q$ charge to a point near the $-Q$ charge. The work required to move the charge is

(a) positive;
(b) negative;
(c) zero;
(d) either positive or negative depending on where the zero point is for the electric potential.

C-20-2 (b) Since the positive charge q is repelled by the $+Q$ charge and attracted to the $-Q$ charge, a force must be applied in the opposite direction to the motion to keep positive charge from accelerating. In terms of electric potential difference, the electric potential is positive near the $+Q$ charge and negative near the $-Q$ charge because the closer charge dominates in each case. Thus, the potential difference is negative, and when multiplied by a positive charge $+q$, yields a negative value for the work required.

So far, we have only considered a small number of discrete point charges. The same strategies can be used to integrate over continuous distributions of charge to obtain the electric potential, with an infinitesimal charge distribution dq playing the role of the point charge. Since electric potentials are scalar quantities, the integrations are considerably easier than the electric field calculations for charge distributions considered in Chapter 19.

20-3 Equipotential Lines and Electric Field Lines

It is useful to draw lines, called **equipotential lines**, through points of equal electric potential. We see from Equation (20-7) that for a single point charge equipotential lines are concentric circles centred on the point charge. When the charge is positive, the potentials are positive, with increasing values as one approaches the charge. Figure 20-5 shows these equipotential lines along with electric field vector arrows, which point outward from a positive charge. (Recall from Chapter 19 that electric fields point in the direction of the electric force on a positive test charge.)

Note that the electric field arrows are perpendicular to the equipotential lines. In fact, we can easily show that electric field lines are always perpendicular to lines or surfaces made up of points that are at equal electric potentials. Assume for a moment that the electric field $\vec{E}$ is not perpendicular to the equipotential lines. Then there would be a component of the electric field along the equipotential line and, hence, a force acting on any charge moving along the line. But this force does work as the charge moves along the line, which makes the electric potential different at each point, so the line cannot be an equipotential line. This contradiction shows that all electric field lines cross equipotential lines perpendicularly.

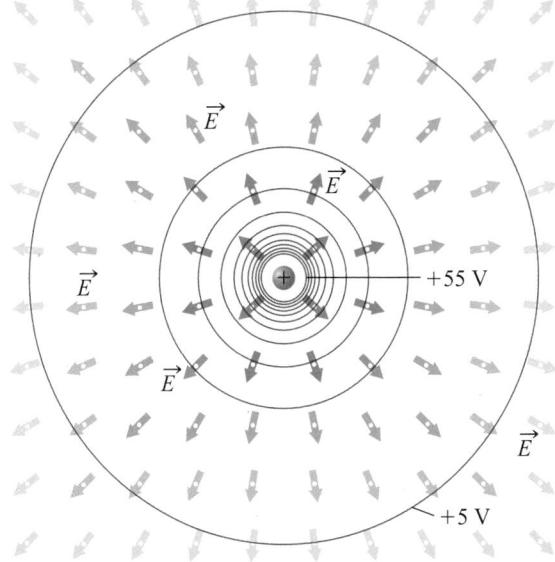

Figure 20-5 Equipotential lines and electric field vectors around a single +1 nC charge. The plotted equipotential lines are for intervals of 5 V starting at +5 V as shown. Note that here the colour intensity of the vector arrows, rather than their length, represents the strength of the electric field.

C-20-3 Zero Net Electric Potential

Two fixed charges have equal magnitude but unknown signs. At the point (2, 0) in the xy-plane, the net electric field points in the $-x$-direction, and the net electric potential is zero. What can you conclude about the signs and locations of the two fixed charges? Draw a diagram of the situation to help you answer this question.

(a) The two charges must both be negative, one to the left and one to the right along the x-axis.

(b) The two charges must have opposite signs and must both be along the x-axis at equal distance from the point where the potential is zero, with the positive charge to the right.

(c) The two charges must have opposite signs and must both be along the x-axis at equal distance from the point where the potential is zero, with the positive charge to the left.

(d) The charges must have opposite signs and be along a vertical line, with the positive charge at the point (2, $+y$) and the negative charge at the point (2, $-y$).

C-20-3 (b) The only way to have zero electric potential is when the charges have opposite signs and are equal distances from the measurement point. For the electric field (which is in the direction of the force on a positive test charge) to point to the left, we must have the positive charge on the right.

Figure 20-6 shows a positive central charge with two lesser negative charges symmetrically placed on

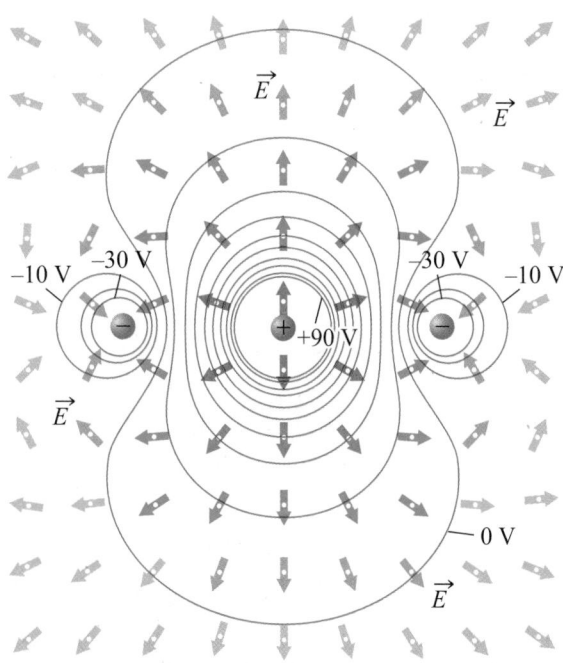

Figure 20-6 Equipotential and electric field vectors for a situation with a positive charge (+4 nC) in the centre and equal negative (−1 nC) charges symmetrically on each side. The plotted equipotential lines are at intervals of 10 V, starting with the equipotential line closest to the positive charge at +90 V. You can produce your own plots like this using the PhET simulation called "Charges and Fields."

each side. The positive charge is +4 nC, and the negative charges are −1 nC. We have plotted only 12 equipotential lines at 10 V intervals. There would be higher and higher potentials near each charge. Note that very near each charge the contribution from that charge dominates, and the equipotential lines approach concentric circles. Farther away, the influence of other charges on the net potential becomes apparent.

 LO 4

20-4 Electric Field from Electric Potential

In Chapter 11, we showed how the gravitational force and the gravitational field can be calculated from the change in gravitational potential energy using partial derivatives. This technique is valid for any conservative force field. The force components can be calculated as follows:

$$F_x = -\frac{\partial U}{\partial x}; \quad F_y = -\frac{\partial U}{\partial y}; \quad F_z = -\frac{\partial U}{\partial z} \quad (20\text{-}8)$$

where U is the potential energy function.

Keep in mind that these partial derivatives are taken with respect to one coordinate variable while holding the others constant. For example, for the potential energy given by $U = 20y - 10$ (in J), the x- and z-components of the corresponding force are zero, and the component in the y-direction is −20 N.

Electric potentials are commonly used in place of electric potential energy when dealing with electric fields. Recall that electric potential is the electric potential energy per unit charge, and electric field is electric force per unit charge. If we divide both sides of Equations (20-8) by charge, we get the following relationships:

KEY EQUATION $\quad E_x = -\frac{\partial V}{\partial x}; \quad E_y = -\frac{\partial V}{\partial y}; \quad E_z = -\frac{\partial V}{\partial z} \quad (20\text{-}9)$

If you know the electric potential at all points, then you can use these relationships to find the electric field at all points. Since electric potentials are scalar quantities, and we can calculate them using the superposition principle, this approach is often the easiest way to calculate electric fields, even in complex situations.

The relationship between the electric field and electric potential suggest that the SI units for electric field are V/m. From the definition of electric field in terms of electric force per unit charge, we had units of N/C. In fact, the definitions of these SI units are such that 1 N/C = 1 V/m.

CHECKPOINT

C-20-4 Electric Potential and Electric Field

The electric potential (in V) in a certain region is given by $V = 3x^2 - 5$. What is the electric field at the origin of the coordinate system?

(a) All components are zero.

(b) The y- and z-components are 0, and the x-component is in the $+x$-direction.

(c) The y- and z-components are zero, and the x-component is in the $-x$ direction.

(d) The x-component is zero, and the y- and z-components are both infinite.

C-20-4 (a) Since y and z do not appear explicitly in the expression for potential, clearly, those components of the electric field are zero everywhere. Taking the partial derivative with respect to the x-coordinate gives $E_x = -6x$, which has a zero value at the origin.

 EXAMPLE 20-3

Electric Field from Electric Potential

In a region of space, the electric potential (in V) is given by

$$V = \frac{25.0}{x^2 + y^2} + 80.0.$$

(a) Is the potential defined everywhere?

(b) Derive expressions for the components of the electric field.

(c) Find the electric field at the point $(1.5 \text{ m}, 0 \text{ m}, 0 \text{ m})$.

(d) Sketch the magnitude of the x-component of the electric field along the x-axis as a function of distance from the origin.

SOLUTION

(a) The potential is defined everywhere except at the origin, where it becomes infinite. The electric potential is positive everywhere and decreases as the distance from the origin in the xy-plane increases. The potential approaches a value of $+80.0$ V for large absolute values of either the x- or y-coordinate.

(b) To obtain the components of the electric field, we use Equation (20-9). Rewriting the electric potential as $V = 25(x^2 + y^2)^{-1} + 80$ makes it easier to take the partial derivatives. For the x-component of the electric field we have

$$E_x = -\frac{\partial V}{\partial x} = -25(-1)(x^2 + y^2)^{-2}(2x) + 0 = +50x(x^2 + y^2)^{-2}$$

Similarly, $E_y = -\dfrac{\partial V}{\partial y} = +50y(x^2 + y^2)^{-2}$

Since z does not appear in the expression for the electric potential, the partial derivative with respect to z is 0, so $E_z = 0$.

(c) Substituting $x = +1.5$ m, $y = 0$ m, and $z = 0$ m into the expressions from part (b), we obtain

$$E_x = 50x(x^2 + y^2)^{-2} = 50(+1.5)(1.5^2 + 0^2)^{-2} = +14.8 \text{ V/m}$$

$$E_y = 0 \text{ V/m}$$

$$E_z = 0 \text{ V/m}$$

Therefore, the electric field at the point $(1.5 \text{ m}, 0 \text{ m}, 0 \text{ m})$ has a magnitude of 14.8 V/m and points in the $+x$-direction.

(d) Figure 20-7 shows plots of the magnitude of the electric potential and the magnitude of the electric field along the x-axis. Note that the magnitudes of both V and E_x are infinite at $x = 0$.

Figure 20-7 (a) A plot of the electric potential (V) as a function of position along the $+x$-axis. (b) A plot of the x-component of the electric field along the $+x$-axis

Making sense of the result:

The first term in the potential function decreases as the square of the distance in the xy-plane from the origin. The constant term does not affect the electric field because the derivative of a constant is zero. Since the electric potential is positive along the x-axis and increases closer to the origin, we expect a positive test charge on the positive x-axis to be repelled with the x-component of the electric field in the positive direction, as obtained above. Careful examination of the two plots shows that the values of E_x correspond to the negative of the slope of the V plot, as expected, because the slope corresponds to the derivative of the potential function.

CHAPTER 20 | ELECTRIC POTENTIAL AND GAUSS'S LAW

You may wonder whether we can reverse the process used above and determine the resultant electric potential for a given electric field. Indeed, you can perform an integral along a path between two points to find the potential difference between the points. In a conservative field, the path taken does not matter, only the end points. We let V_a and V_b represent the electric potentials at points a and b, and $d\vec{r}$ represents a small element of a path from point a to point b. The electric potential difference between the two points is given by the following integral:

KEY EQUATION
$$V_b - V_a = -\int_a^b \vec{E} \cdot d\vec{r} \qquad (20\text{-}10)$$

Of course, absolute electric potential can have an arbitrary constant voltage added; this constant is determined by the definition of the zero point.

Pay careful attention to the signs of the various quantities. You should always check whether the result that you obtain makes sense with what you know about the electric fields and potentials. For example, for the situation in Figure 20-8, an electric field points straight down. By definition, the path element $d\vec{r}$ points along the path going from a to b. Since $\vec{E}$ and $d\vec{r}$ have almost opposite directions, the dot product of the vectors is negative. The right side of Equation (20-10) also has a negative sign, so the net sign is positive, indicating that V_b is greater than V_a.

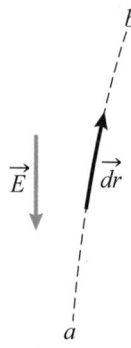

Figure 20-8 In this situation, the electric field points from the top toward the bottom. The element $d\vec{r}$ always points along the integration path going from point a to point b.

Remember that you can take any path in a conservative field, so you are free to choose a path for which the dot product and integral are easy to calculate. For example, *when the electric field is constant, you can choose a straight path from point a to point b.* Then the angle between the field and path vectors is constant, making the dot product easy to integrate.

20-5   The Electron-volt

Rearranging Equation (20-6), we see that, when we move a charge q through an electric potential difference ΔV, the change in electric potential energy, ΔU, is

KEY EQUATION
$$\Delta U = q \Delta V \qquad (20\text{-}11)$$

We have used the notations ΔV and ΔU because we are considering the *changes* in potential and potential energy. When we move a positive charge to a region with higher electric potential, the electric potential energy of the charge increases. However, the electric potential energy of a negative charge decreases when the charge moves to a region of higher electric potential.

When zero work is done by external forces, a loss in electric potential energy must correspond to a gain in some other form of energy, usually kinetic energy. In the TRIUMF cyclotron described at the beginning of this chapter, charged particles are accelerated by changes in electric potential. You can also describe the acceleration in terms of forces exerted on the charged particles in electric fields, but in many situations it is easier to use an energy-potential approach.

In many devices, electrons, protons, and singly charged ions experience changes in electric potential. A small unit of energy called the **electron volt** (eV) is convenient to use in these situations. One electron volt is the energy change when one elementary charge moves through a potential difference of exactly 1 V. For example, cathode ray tubes typically accelerate electrons using a potential difference of approximately 30 000 V, so one electron would have an energy change of 30 000 eV (or 30 keV). Charged particles (often electrons) are accelerated when they pass through a high electric potential difference in a variety of other technological devices, including X-ray machines and electron microscopes.

✓ **CHECKPOINT**

C-20-5 Change in potential

A doubly charged positive ion (charge of $+3.2 \times 10^{-19}$ C) loses 24 keV in going from point A to point B. Which of the following statements describes the potential difference between the points?

(a) Point A is at a higher potential than point B by 24 kV.
(b) Point B is at a higher potential than point A by 24 kV.
(c) Point A is at a higher potential than point B by 12 kV.
(d) Point B is at a higher potential than point A by 12 kV.

C-20-5 (c) Here, a positively charged (positive q) particle loses electric potential energy (ΔU) is negative, so the change in potential (ΔV) must be negative. Therefore, point A is at a higher potential than point B. Since the particle has a charge of $2e$, the voltage difference must just be 12 kV.

It is important to remember that the electron volt is a unit of energy, and the volt is a unit of electric potential difference (energy per unit charge). To convert electron volts to joules, we use the magnitude of the elementary charge, 1.60×10^{-19} C for q:

$$1 \text{ eV} = 1.60 \times 10^{-19} \text{C} \times 1 \text{ V} = 1.60 \times 10^{-19} \text{J}$$

The electron volt is a tiny amount of energy by macroscopic standards, but a convenient amount on the atomic scale. For example, it takes approximately 13.6 eV to ionize

ELECTRICITY, MAGNETISM, AND OPTICS

hydrogen and approximately 5.2 eV to remove the first electron from sodium. Those working with atomic spectra usually state energy level differences in electron volts.

Proton beams, such as those produced at TRIUMF, are used in the treatments of certain types of cancer. Proton beams are preferable to X-rays for cancer treatment for a number of reasons, the main one being that the beam width and the effective depth in the body are better controlled. The proton beam damages the DNA of the cancer cells, thereby reducing the ability of the cancerous cells to reproduce. The proton beam for this type of treatment usually needs to be in the energy range of approximately 75 MeV to 225 MeV, that is, each proton has an energy of between 75 million electron volts and 225 million electron volts.

EXAMPLE 20-4

Proton Beam Therapy

TRIUMF can produce protons with energies between 5 MeV and 120 MeV in the lower energy beam (called BL2C). Consider a beam of 100 MeV protons used for treating cancer.

(a) What is the energy of one of these protons, in joules?
(b) The energy of the protons is built up after many passages through an accelerating potential of 94.0 kV. How many passes are needed to produce 100 MeV protons?
(c) The protons are accelerated by an electric potential difference across a gap between two halves of the cyclotron chamber. Which side of this acceleration region has the higher electric potential?

SOLUTION

(a) Since one electron volt equals 1.60×10^{-19} J,

$$100 \text{ MeV} \times \frac{1.00 \times 10^6 \text{ eV}}{1 \text{ MeV}} \times \frac{1.60 \times 10^{-19} \text{ J}}{1 \text{ eV}} = 1.60 \times 10^{-11} \text{ J}$$

(b) The proton has a charge of one elementary charge $(1.60 \times 10^{-19}$ C). Therefore, the energy gained when it passes through an accelerating potential of 94.0 kV is 94.0 keV. Therefore, the number of passes required is

$$100 \text{ MeV} \times \frac{1.00 \times 10^6 \text{ eV}}{1.00 \text{ MeV}} \times \frac{1 \text{ pass}}{94.0 \text{ keV}} \times \frac{1.00 \text{ keV}}{1.00 \times 10^3 \text{ eV}}$$

$$= 1064 \text{ passes}$$

Therefore, to three significant figures 1060 passages through the accelerating region are needed.

(c) Protons are positively charged; therefore, they are accelerated toward a more negative region. The end of the acceleration region must be at a lower electric potential.

Making sense of the result:

This example demonstrates how convenient the electron volt unit is when dealing with charged particles. Since higher positive electric potentials are more positive regions, it makes sense that positively charged particles lose electric potential energy and gain kinetic energy as they move toward more negative regions.

MAKING CONNECTIONS

Ion Propulsion Rockets

The underlying principle of any rocket engine is conservation of linear momentum: fuel is expelled in one direction with a high relative velocity, and the rocket moves in the opposite direction with the total momentum of the rocket-plus-fuel system unchanged. Most rocket engines, such as those used in space launches, burn fuel for exhaust propulsion. Ion propulsion systems are more environmentally friendly and ultimately more efficient. These systems are relatively simple: the engine ionizes a material, such as xenon gas, using electric energy from solar cells on the spacecraft. The positive ions are then accelerated through a large electric potential difference. High ejection velocities are possible, making ion propulsion quite efficient. Ion propulsion is also safer and more reliable than chemical rockets, which burn fuel in violent oxidation reactions. Ion propulsion systems are in use, but the thrusts available are still tiny compared to chemical propellants. At present, ion propulsion technology is best suited for long-duration voyages where a spacecraft can maintain a tiny thrust for a long time. NASA's *Dawn* mission (Figure 20-9) uses ion propulsion to move the space vehicle into the orbits required to provide close-up views of the two largest asteroids, Ceres and Vesta, from 2007 to 2015.

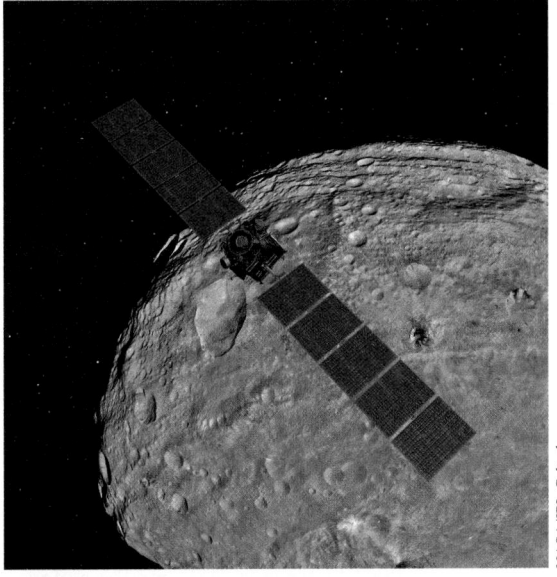

Figure 20-9 NASA's *Dawn* spacecraft uses ion propulsion engines for thrust to put it into orbits around the asteroids Ceres and Vesta.

LO 6

20-6 Electric Fields and Conductors

An ideal conductor allows charges to flow without resistance. An infinitesimally small electric force moves a charge along this conductor. Consequently, when a static charge is on a conductor, there is no electric field component directed along the conductor. Therefore, electric fields are always perpendicular to the surfaces of conductors when no current flows. Note that this orientation applies in static situations, but not when there are moving charges or changing electric fields, which we will consider in Chapter 26. The direction (outward or inward) depends on the charge on the conductor, with positively charged conductors having electric field lines directed outward.

 CHECKPOINT

C-20-6 Electric Field Near Conductors

Which part of Figure 20-10 shows the electric field close to a negatively charged conducting surface?

surface, the answer is (c).
one), so (a) and (d) are excluded. Since the electric field points inward for a negatively charged
C-20-6 (c) The electric field must be perpendicular to the surface of a conductor (at least an ideal

Figure 20-10   C-20-6

The electric potential must be exactly the same everywhere on an ideal conductor. For example, the voltage measured at each end of a conducting wire with zero resistance is the same. Superconductors are the only known materials with zero resistance, but some metals, such as copper and silver, are close enough to being ideal conductors that the potential difference across them can be considered negligible for most situations.

 PEER TO PEER

The idea that the electric field is the rate of change of electric potential is a really powerful and useful idea. It means that if different parts of an object, such as a wire, are at the same voltage (potential), then there is no electric field in that object. Also, when an electric field is constant, you can find a potential difference by simply multiplying the field by the distance. Say the field between two plates is 20 V/m, and the plates are 0.10 m apart. Then the potential difference is 2.0 V.

Consider an ideal conducting spherical shell that has a static charge on it. How is this charge distributed? There must be no electric field inside the material of the conductor; otherwise, the charge carriers in the conductor would move and the charge could not be static. In fact, charges on conductors always stay on the outer surface of the conductor. Later in the chapter, we will see why the charge does not stay on the inner surface of a conducting shell.

LO 7

20-7 Charge Densities

In Chapter 19, we developed the idea of a linear charge density, μ, defined as the charge per unit length of a linear conductor, such as a narrow wire. We similarly define surface charge density, σ, as the charge per unit area of a surface and volume charge density, and ρ, as the charge per unit volume. In a section 20-9, we will use these definitions to determine the electric field near a charged object.

EXAMPLE 20-5

Surface Charge Density on a Conductor

A solid conducting sphere has a radius of 12.0 cm and a total charge of $25.0\ \mu C$. Find the charge density on the surface of this sphere.

SOLUTION

From Section 20-6, we know that all the charge on a conductor goes to its outer surface. The surface area of the sphere is

$$A = 4\pi r^2 = 4\pi(0.120\text{ m})^2 = 0.181\text{ m}^2$$

Therefore, the surface charge density is simply

$$\sigma = \frac{Q}{A} = \frac{25.0\ \mu C}{0.181\text{ m}^2} = 138\ \mu C/m^2 = 1.38 \times 10^{-4}\,C/m^2$$

Making sense of the result:

Remember that μC is a small unit of charge, so the charge density expressed in C/m^2 is expected to be small.

 PEER TO PEER

Whenever you have a charge density question, ask yourself whether it deals with a conductor or an insulator. With an insulator, you must have some assumption (such as evenly distributed charge). With a conductor, all the charge on it goes to the outer surface.

C-20-7 Surface Charge Density

Which of the following has the largest surface charge density, assuming that the total charge on each object is the same?

(a) the surface of a solid conducting sphere of radius r

(b) the surface of a solid insulating sphere of radius r

(c) the surface of a solid conducting cube with all sides of length r

(d) Both (a) and (b) always have exactly the same surface charge density.

C-20-7 (c) For conducting objects, all the charge goes on the outer surface. For nonconducting objects, the charge can be distributed inside as well as on the surface. So, for objects with the same shape, the surface charge density is less on the insulating objects than on the conducting objects. A sphere of radius r has a surface area of $4\pi r^2$, and a cube with sides r has a total surface area of $6r^2$. Therefore, the surface charge density is more on the cube because the same charge is spread over a smaller surface area.

LO 8

20-8 Electric Flux

In Section 20-9, we will introduce Gauss's law, which allows us to calculate (at least in cases with symmetry) the electric field for a known charge distribution. Gauss's law is expressed in terms of a quantity called the electric flux, Φ_E, so before we introduce Gauss's law, we need to introduce flux. When we draw electric field lines closer together, where the magnitude of the electric field is higher, we can consider the electric flux as a measure of how many electric field lines pass through a given area.

A qualitative analogy be helpful. Imagine a heavy rain that is being pushed by a strong wind so that the rain falls at an angle, as shown in Figure 20-11. We are interested in how much rain will be collected by each of the four buckets in a given time. The amounts depend on the rainfall rate, the area of the circular opening of each bucket, and the angle of the falling rain relative to these areas. In Figure 20-11, bucket B will collect the most rainwater (or rain flux). The flux into bucket A is less because the collecting area is less, and the flux into bucket D is less because the rate of rainfall is less. The flux into bucket C is less because the angle of the rainfall results in fewer drops entering the bucket per unit area of the bucket opening.

Figure 20-11 A rainfall analogy to electric flux

Electric flux is a scalar quantity defined as the vector dot product of the electric field, $\vec{E}$, and an area vector, $\vec{A}$, that has a direction perpendicular to the surface (by convention directed outward from the surface) and a magnitude equal to the area over which electric flux is being calculated:

$$\Phi_E = \vec{E} \cdot \vec{A} \tag{20-12}$$

Although we introduced an analogy with a flat surface, electric flux can be calculated over a curved surface as well.

 CHECKPOINT

C-20-8 Electric Flux

The cube in Figure 20-12 is oriented in an electric field pointing in the y-direction. Which of the following statements best describes the electric flux through the faces of the cube?

(a) The only surfaces with nonzero flux are the faces parallel to the yz-plane, and the net flux for the entire cube is zero.

(b) The only surfaces with nonzero flux are the faces parallel to the xy-plane, and the net flux for the entire cube is negative.

(c) The only surfaces with nonzero flux are the faces parallel to the xz-plane, and the net flux for the entire cube is zero.

(d) The only surfaces with nonzero flux are the faces parallel to the xz-plane, and the net flux for the entire cube is negative.

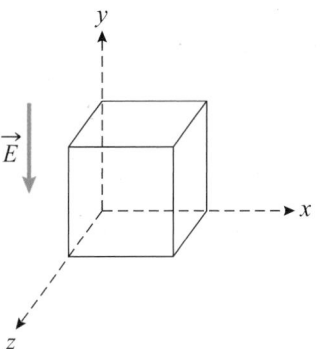

Figure 20-12 C-20-8

C-20-8 (c) Because of the vector dot product, the four faces parallel to the electric field vector all have zero flux. Just the top and the bottom faces of the cube have nonzero electric flux. Since the direction of the area vector is defined as perpendicular to the surface and pointing outward, the electric flux is negative for the top face of the cube (where the vectors $\vec{E}$ and $\vec{A}$ are opposite) and positive for the bottom face (where the vectors are in the same direction). Since the magnitudes of these two fluxes are equal, the net flux through the cube is zero.

In most situations, the electric field varies over the area. We then use an integral over the surface. This integral also accommodates situations in which the angle between the area and the electric field varies (e.g., when the surface is not flat).

KEY EQUATION
$$\Phi_E = \oint \vec{E} \cdot d\vec{A} \tag{20-13}$$

EXAMPLE 20-6

Electric Flux Through a Circle

Derive an expression for the electric flux through a circle of radius r_0 in the xy-plane, centred at the origin. The electric field makes an angle of $30°$ with the z-axis, as shown in Figure 20-13. Assume that the magnitude of the electric field for points in the xy-plane is given by $|\vec{E}| = \dfrac{E_0}{r}$, where r is the distance from the origin.

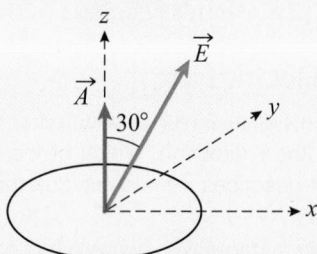

Figure 20-13 Example 20-6

SOLUTION

We will use the definition of electric flux in the derivation of the expression:

$$\Phi_E = \oint \vec{E} \cdot d\vec{A}$$

Since the angle between the electric field and area vectors is a constant $30°$, the dot product gives a factor of $\cos(30°)$. Thus, we can write the electric flux through the circle as

$$\Phi_E = \cos(30°) \oint_{circle} E \, dA = \frac{\sqrt{3}}{2} \oint_{circle} E \, dA$$

There are several options for specifying the area element dA. We will take an infinitesimally thin ring as the differential area element and integrate over concentric rings to make up the circle. As shown in Figure 20-14, each ring has an approximate area of $dA = 2\pi r dr$.

Therefore, the electric flux is

$$\Phi_E = \frac{\sqrt{3}}{2} \oint_{circle} E \, dA = \frac{\sqrt{3}}{2} \int_0^{r_0} \frac{E_0}{r} 2\pi r \, dr$$

$$= \pi\sqrt{3}E_0 \int_0^{r_0} dr = \pi\sqrt{3}E_0 r_0$$

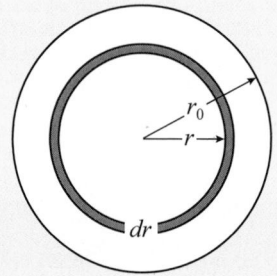

Figure 20-14 We can integrate over the circle by adding rings of length $2\pi r$ and width dr.

Making sense of the result:

You might be surprised that we obtained a finite answer because the electric field is infinite at the origin. However, the flux depends on the angle, area, and electric field, and the region with an infinite electric field has an infinitesimal area, so the total flux through the circle is finite. The electric field goes down with distance from the centre, but the corresponding circular area increases with the square of the radius, so the net flux depends on increases with the radius of the circle, r_0.

LO 9

20-9 Gauss's Law

Gauss's law states that the electric flux through any closed surface is given by the enclosed charge within the region bounded by that surface divided by the permittivity of free space, ε_0, which is a constant of nature:

 KEY EQUATION
$$\oint \vec{E} \cdot d\vec{A} = \frac{q_{enc}}{\varepsilon_0} \qquad (20\text{-}14)$$

where the circle on the integral symbol (called a surface integral) indicates that the integral is taken over a closed surface, and q_{enc} represents the charge *enclosed* by that surface. If both positive and negative charges are enclosed by the surface, it is the net value which appears in Equation 20-14 (e.g. if there was a $+3$ C charge and a -1 C charge we would use $+2$ C).

Although Gauss's law is true for any closed surface, it is only useful for calculating the electric field in situations involving symmetry.

✓ CHECKPOINT

C-20-9 Gauss's Law

Which of the following surfaces will have the greatest net electric flux passing through it if they all enclose the same net electric charge?
(a) a small sphere
(b) a large sphere
(c) a cube
(d) The flux is the same for all three surfaces.

C-20-9 (d) By Gauss's law, the net electric flux through any of the closed surfaces depends only on the enclosed charge, which is the same for all three surfaces.

When applying Gauss's law, you can choose any convenient closed surface, including an irregularly shaped surface. However, choosing a symmetrical surface can greatly simplify the calculations. We will consider spherical symmetry in this section, and in the next section we will apply Gauss's law with several other types of symmetry.

In Chapter 19, we obtained the electric field at a distance r from a point charge Q using Coulomb's law and the definition of the electric field as the force on a small positive test charge divided by that charge. We now apply Gauss's law to the same situation. In Figure 20-15, an electric field is symmetric about a single positive charge, Q. At a given distance r from the charge, the magnitude of the electric field is the same in any direction. Therefore, we choose a sphere centred on the charge Q as the closed surface for application of Gauss's law.

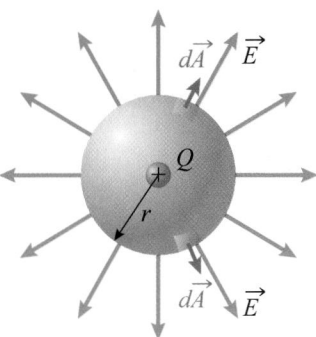

Figure 20-15　A spherical surface surrounding a single positive point charge, Q. The surface area elements, $d\vec{A}$, points radially outward everywhere and are parallel to the electric field at the surface.

Applying Gauss's law to the spherical surface gives

$$\oint \vec{E} \cdot d\vec{A} = \frac{q_{\text{enc}}}{\varepsilon_0} \qquad (20\text{-}15)$$

The directions of the electric field and the surface area element $d\vec{A}$ vary around the sphere, but both are always directed radially outward so they are always parallel to each other. Therefore, the dot product in the equation for Gauss's law produces a constant factor of $\cos(0°)$, which is simply 1. By symmetry, we know that the magnitude of the electric field is the same all over the spherical surface, so E can be taken outside the integral:

$$E \oint dA = \frac{Q}{\varepsilon_0} \qquad (20\text{-}16)$$

Since the integral of dA over the closed surface is simply the surface area of the sphere, we have

$$E(4\pi r^2) = \frac{Q}{\varepsilon_0} \qquad (20\text{-}17)$$

and

$$E = \frac{1}{4\pi \varepsilon_0}\left(\frac{Q}{r^2}\right) = \frac{Q}{4\pi \varepsilon_0 r^2} \qquad (20\text{-}18)$$

Equation (20-18) is the same expression for the magnitude of the electric field at a distance r from a single point charge Q that we obtained using Coulomb's law in Chapter 19. We now see why the constant in Coulomb's law is often written $\dfrac{1}{4\pi \varepsilon_0}$.

Coulomb's law is an inverse square law—the strength of the electric field due to a point charge is inversely proportional to the square of the distance from the charge. We can show that this behaviour is in accordance with Gauss's law. Figure 20-16 shows a single positive point charge Q enclosed by two concentric spheres, one of radius r and one of radius $3r$. The electric field lines that start on the positive charge pass through both spheres. The smaller sphere has more electric field lines per unit area, so the electric field is stronger at its surface than at the surface of the larger sphere. The larger sphere has 9 times the surface area of the smaller sphere, but the number of electric field lines per unit area (a measure of the strength of the electric field) is only 1/9 as much, so the electric flux through each sphere is the same, as predicted by Gauss's law.

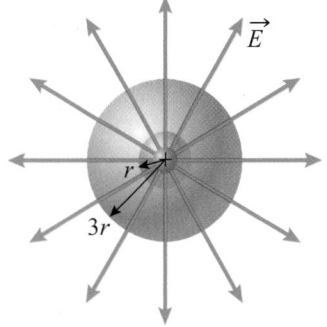

Figure 20-16　Gaussian spheres with different radii do not have the same surface area or electric field strengths, but the electric flux through them is the same.

We can obtain all aspects of Coulomb's law through the combination of the definition of electric field (force per charge) and Gauss's law. In one of the most brilliant contributions to physics, James Clerk Maxwell (1831–79) described all aspects of static and changing electromagnetic fields (and, therefore, electromagnetic waves such as light) in only four powerful equations. Gauss's law is one of these relationships.

20-10 Applications of Gauss's Law

Using symmetry, we can readily apply Gauss's law to calculate electric fields around uniform charge distributions that are spherical, linear, or planar. Often, a calculation for a theoretical infinitely long object will give a good approximation of the electric field, except near the ends, for an actual object, such as a straight wire or a charged plate.

Consider the very long line of positive charge (wire) with uniform linear charge density μ shown in Figure 20-17. We want to find the electric field at a distance r from the line. If we assume that the line of charge is essentially infinite, we know from symmetry that the electric field must point radially away from the line of charge and have the same magnitude E at every point that is at the same radial distance r from the line of charge.

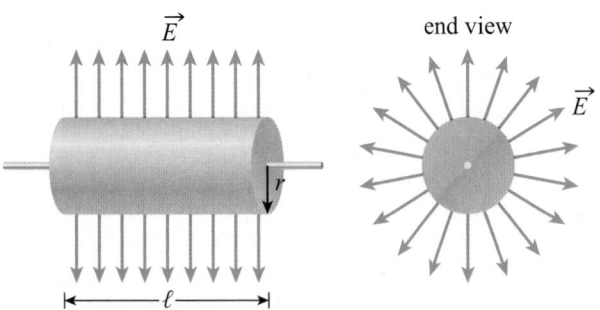

Figure 20-17 A long line of charge with a Gaussian cylinder of radius r and arbitrary length ℓ. The electric field points radially outward from the wire in all directions, but is shown only at the horizontal surfaces of the cylinder here.

Next, we choose a cylinder of arbitrary length ℓ as our closed Gaussian surface. We can find the total electric flux through the cylinder by adding the flux through the flat ends and the curved side of the Gaussian cylinder:

$$\Phi_{\text{left cap}} + \Phi_{\text{curved surface}} + \Phi_{\text{right cap}} = \frac{q_{\text{enc}}}{\varepsilon_0} \quad (20\text{-}19)$$

For each of the parts of the electric flux we have

$$\Phi_E = \oint \vec{E} \cdot d\vec{A} \quad (20\text{-}20)$$

For the left end, $d\vec{A}$ points left, which is perpendicular to the electric field, so the vector dot product yields a flux of zero. Similarly, the electric flux through the right end is also zero. On the curved part of the cylindrical surface, $d\vec{A}$ points radially outward everywhere, the same direction as the local electric field, $\vec{E}$. Therefore, the dot product is simply the product of the magnitudes of the two vectors, and

$$\Phi_E = E \int dA = E(2\pi r \ell) \quad (20\text{-}21)$$

The amount of charge enclosed by the cylindrical Gaussian surface is simply $\mu\ell$, so Gauss's law gives

$$\frac{\mu\ell}{\varepsilon_0} = E(2\pi r \ell) \quad (20\text{-}22)$$

Now we can easily solve for E, the magnitude of the electric field at a distance r from an infinite line with a charge density μ. The direction of the electric field is radially away from the line of charge, as discussed above:

$$E = \frac{\mu}{2\pi r \varepsilon_0} \quad (20\text{-}23)$$

It makes sense that the electric field depends proportionally on the charge per unit length. Note that the electric field is inversely proportional to the radial distance r, not the square of the distance, as it is for a point charge. If the charge were negative, the results would be exactly the same except that the electric field would point toward the line of charge instead of away from it.

✓ CHECKPOINT

C-20-10 Coaxial Cable

Coaxial cables are commonly used for transmitting signals, such as TV and audio. A coaxial cable consists of a central wire, a layer of insulation, a concentric cylindrical second conductor, and an outer layer of insulation. Which of the following statements best describes the electric field between the conductors and outside a length of coaxial cable that has a charge of $+q$ on the inner wire and $-q$ on the outer conductor?

(a) The electric field is directed away from the wire in both regions.

(b) The electric field is directed toward the wire in both regions.

(c) The electric field is zero in the region between the conductors and directed outward outside the cable.

(d) The electric field is directed radially away from the central wire in the space between the conductors and is zero outside the cable.

C-20-10 (d) By Gauss's law, the net electric flux outside the cable must be zero because the total enclosed charge is zero inside a Gaussian box that encloses both the inner wire and the outer conductor. In the space between the conductors, the electric field is directed outward because the inner wire has a positive charge.

Here is a summary of the strategy used to solve problems using Gauss's law:

1. Without some symmetry, Gauss's law is of little use in calculating the electric field, so determine whether the situation has symmetry about a point, line, or plane.

2. Choose an appropriate Gaussian surface according to the symmetry of the situation (spherical or cylindrical). For example, a line of charge has cylindrical symmetry, and a charged sphere or single point charge has spherical symmetry. In the case of planes of charge, we often use cylindrical Gaussian surfaces, although boxes could also be used. The Gaussian surface must be centred around a point or axis of symmetry. The size of the Gaussian surface is determined by the distance to the point where the electric field is to be calculated.

3. Draw a diagram of the situation, including the electric field and the Gaussian surface. Indicate the directions of $d\vec{A}$ on the diagram. Remember that area vectors always point outward perpendicular to the surface.

4. Divide the Gaussian surface into different segments if necessary, ensuring that the entire closed Gaussian surface is included.

5. Note which segments have the electric field perpendicular to the area vector (and hence zero flux) and which have parallel field and area vectors.

6. Calculate the net charge enclosed by the Gaussian surface. Remember that only the charge inside the surface needs to be considered.

7. Use Gauss's law to solve for the magnitude of the electric field. Be sure to state the direction of the field.

8. Check that your answer makes sense.

In Example 20-7, we apply this technique to a planar charge.

EXAMPLE 20-7

Electric Field from an Infinite Plane of Charge

Derive an expression for the electric field a distance d from a very large (assumed infinite) plane of charge that has uniform surface charge density σ.

SOLUTION

We will assume that the charge is positive, although other than reversing the direction of the electric field, the analysis is the same for either sign. To use the planar symmetry of the situation, we draw a cylindrical Gaussian surface that extends an equal distance d above and below the plane of charge, as shown in Figure 20-18. This placement equalizes the flux through the two ends of the cylinder.

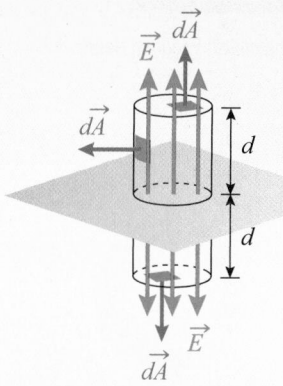

Figure 20-18 A thin, infinite plane of charge with a Gaussian surface extending equal distances above and below the surface. For clarity, the electric field vectors are shown only where they pass through the surface of the Gaussian cylinder.

According to Gauss's law, the net electric flux must equal the enclosed charge divided by the permittivity, ε_0. The cross-sectional area of the Gaussian surface is A, so the enclosed charge is simply σA. Therefore,

$$\Phi_{\text{top cap}} + \Phi_{\text{curved surface}} + \Phi_{\text{bottom cap}} = \frac{q_{\text{enc}}}{\varepsilon_0} = \frac{\sigma A}{\varepsilon_0}$$

In each region, we calculate the electric flux according to the relationship

$$\Phi_E = \oint \vec{E} \cdot d\vec{A}$$

In the figure, we indicate the directions of $\vec{E}$ and $d\vec{A}$ for the three parts of the Gaussian cylinder. For the two ends, $\vec{E}$ and $d\vec{A}$ are parallel, and the magnitude of the electric flux is the same on both. Along the curved side of the cylinder, $\vec{E}$ and $d\vec{A}$ are at right angles, so the electric flux through these curved surfaces of the cylinder is zero.

From the symmetry, we can see that the magnitude of the electric field has a constant value, E, at every point in the ends of the cylinder. Therefore, the flux through each end is simply EA. Applying Gauss's law, we have

$$\Phi_{\text{top cap}} + \Phi_{\text{curved surface}} + \Phi_{\text{bottom cap}} = EA + 0 + EA = \frac{\sigma A}{\varepsilon_0}$$

and

$$E = \frac{\sigma}{2\varepsilon_0}$$

Making sense of the result:

You may be surprised that the strength of the electric field does not depend on the distance from the infinite uniform plane of charge. The contributions to the electric field from distant points almost entirely cancel at points near the plane of charge, but for greater distances from the plane, these contributions cancel less completely, leaving a perpendicular component. This effect exactly offsets the decrease in strength from the point on the plane nearest the measurement point, so the net effect is that the electric field does not depend on distance from the plane. This result is the same result that we obtained in Chapter 19 using direct integration. Gauss's law provides a shorter way to obtain the result for this configuration.

Gauss's law can also be expressed in terms of derivatives, rather than integrals. You will encounter the differential form of Gauss's law if you take advanced physics courses in electricity and magnetism. Gauss's law is not used just to calculate electric fields. Often, we measure the electric field from an object and then use Gauss's law to calculate the charge on the object. Usually, it is easier to measure electric fields (often from the electric potential at two separated points) than to measure electric charge directly.

MAKING CONNECTIONS

Carl Friedrich Gauss

Many consider Carl Friedrich Gauss (1777–1855) to be the greatest mathematician of the last 2000 years (Figure 20-19). He made major contributions to both physics and mathematics, including advances in vector calculus, statistics, electrical theory, astronomy, optics, and geophysics. He showed great promise even as a child and performed his most important work in number theory when he was only 19. Gauss completed his doctorate by the age of 22. He was a perfectionist, often waiting years before publishing results, and then releasing only what he thought was truly important and complete. For example, he developed what we now call Gauss's law in 1835 but did not publish it until 1867. Starting in 1831, Gauss collaborated with the physicist Wilhelm Weber; this collaboration resulted in the application of Gauss's mathematical insights to many branches of physics. One of his first contributions to physics was an accurate calculation of the orbit of the "lost" dwarf planet Ceres by developing new methods for extrapolating the scant data available.

While tremendously successful in his professional life, Gauss faced many personal challenges, including the early death of his first wife and a baby son, the lengthy illness and death of his second wife, and disagreements with his surviving sons. It is said that he struggled with periodic depression, which makes his contributions all the more remarkable. A Canadian national school mathematics prize competition is named in his honour.

Carl Friedrich Gauss (1777–1855), painted by Christian Albrecht Jensen

Figure 20-19   A painting of Carl Friedrich Gauss

LO 11

20-11   Charges, Conductors, and Electric Fields

We argued earlier that in a static situation, an ideal conductor must have zero electric field inside it and the electric field at the surface of the conductor must be exactly perpendicular to the surface. In this section, we will use Gauss's law to examine the charge distribution in conductors and the electric fields just outside conductors.

Consider the positively charged conducting spherical shell shown in Figure 20-20. From the symmetry of the situation, we know that the charge must be distributed symmetrically and the magnitude of the electric field must be the same in all points at a distance r from the centre of the shell. We also know that the electric field inside the conductor must be zero.

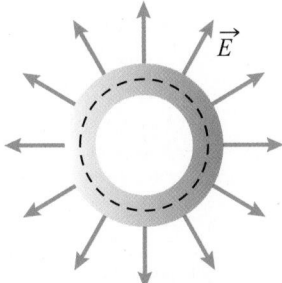

Figure 20-20   Since the electric flux through any spherical Gaussian surface inside the conductor is zero, the enclosed charge must be zero. We can conclude that all charge must reside on the outer surface.

If we draw a spherical Gaussian surface inside the conducting shell, the electric flux through the surface is zero because the electric field in the conductor is zero. Gauss's law then requires that the enclosed charge be zero.

Therefore, the charge on the conducting shell must all reside on the outer surface of the conductor.

In fact, static charges on all conductors always lie on the outer surfaces. Consequently, the electric field lines begin on the outer surface, and there is no static electric field in the cavity inside the conductor, as shown in Figure 20-20.

These charge distributions are examples of electrostatic equilibrium. Electrostatic equilibrium refers to a situation where the charges in the conductor have reached stable equilibrium positions.

C-20-11 Electric Potential of a Charged Conducting Shell

Which statement correctly describes the electric potential around a positively charged spherical conducting shell?

(a) The electric potential is positive and constant inside, and positive and decreasing outside.

(b) The electric potential is zero inside, and positive and decreasing outside.

(c) The electric potential is zero inside, and positive and constant outside.

(d) The electric potential is positive and increasing, both inside and outside.

C-20-11 (a) All the charge goes to the outer surface of the conducting shell. By Gauss's law, the electric field inside the shell is zero. The potential inside the shell is constant but not necessarily zero. Outside the shell, Gauss's law requires an electric field pointing away from the shell, so there must be a positive potential that decreases as the distance from the shell increases.

Although the absence of charges inside a closed conducting surface is strictly true only in electrostatic situations, such surfaces are often effective shields that block electromagnetic waves from penetrating the surface, provided that any holes in the surface are small compared to the wavelength of the electromagnetic waves. Such shielding enclosures are called Faraday cages, after Michael Faraday. In 1816, Faraday used them to demonstrate that no charge was present inside a conducting shell. Actually Benjamin Franklin had made the same discovery earlier while experimenting with charge distributions in 1755.

Suppose we place an uncharged conducting sphere in an external electric field directed downward, as shown in Figure 20-21. The conductor has no net charge, but it does have mobile charge carriers that are free to move within the conductor. We must still have zero net electric field inside the conductor—the mobile charge carriers in the conductor move to positions such that their own electric fields exactly cancel the external electric field. Free electrons move to the top of the sphere, producing a negative charge there and leaving a net positive charge at the bottom of the sphere, a phenomenon called induced charge separation. Note that the electric field lines in Figure 20-21 are distorted by the presence of the conductor: some electric

field lines end at the negative induced charges on the conductor, and an equal number of electric field lines start at the positive induced charges. The total electric field in the conductor is exactly zero in electrostatic equilibrium.

The induced charge can also be explained in terms of electric forces. The external electric field in Figure 20-21 exerts an upward force on negative charges and a downward force on positive charges.

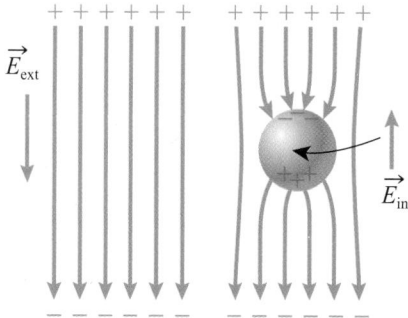

Figure 20-21 Charge separation induced in a conductor by an external electric field

Now consider two conducting spheres with different radii connected with a conducting wire, as shown in Figure 20-22. Recall that the electric field is stronger where the electric field lines are closer together and weaker where the electric field lines are widely separated. Electric charge is placed on one of the spheres. Some of that charge will flow through the conducting wire so that both spheres are charged. In electrostatic equilibrium, the charge must be distributed so that the electric potential is the same on the two spheres (otherwise, charges would flow in the conducting wire and the system would not be in electrostatic equilibrium).

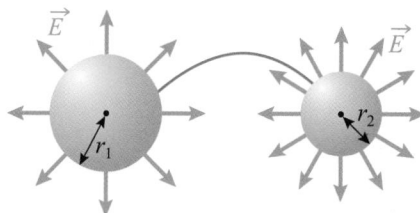

Figure 20-22 Two conducting spheres with different radii connected with a conducting wire

To see how the charge is distributed, first consider the simpler case of a single conducting sphere of radius r_0 that carries charge Q (Figure 20-23). We draw a Gaussian sphere around the conducting sphere at some distance r, with $r > r_0$. By the symmetry of the situation, the electric field has a constant magnitude at any particular distance, and $\vec{E}$ and $d\vec{A}$ are everywhere directed outward and are, therefore, parallel. All the charge is enclosed, so we can apply Gauss's law to find the magnitude of the electric field at a distance r:

$$\oint \vec{E} \cdot d\vec{A} = \frac{q_{enc}}{\varepsilon_0} \qquad (20\text{-}24)$$

$$E(4\pi r^2) = \frac{Q}{\varepsilon_0}$$

$$E = \frac{1}{4\pi\varepsilon_0} \frac{Q}{r^2} \qquad (20\text{-}25)$$

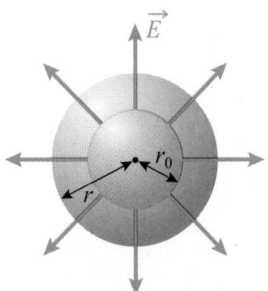

Figure 20-23 A conducting sphere (red) of radius r_0 carries a positive charge Q. The charge is on the outer surface, and the electric field lines start there. We can draw a Gaussian sphere of radius r to find the electric field at that distance.

Equation (20-25) is identical to the expression we obtained in Chapter 19 for the electric field around a point charge Q. Thus, for electric fields and forces outside a uniformly charged sphere, we can treat the sphere as a point charge. We obtain the electric potential V by integrating the electric field over a path from the reference point of zero potential (normally at infinite distance). Since the electric field is the same for a uniformly charged sphere and for a point charge that has the same electric charge Q, the expression we obtained for the electric potential at a distance r from a point charge also applies for the potential from a charged conducting sphere:

$$V = \frac{Q}{4\pi\varepsilon_0 r} \qquad (20\text{-}26)$$

We now return to the two spheres in Figure 20-22. As argued earlier, to be in electrostatic equilibrium the electric potential on each sphere must be the same:

$$\frac{Q_1}{r_1} = \frac{Q_2}{r_2} \qquad (20\text{-}27)$$

where Q_1 and Q_2 are the charges on the spheres of radii r_1, and r_2, respectively.

Comparing the magnitude of the electric field at the surface of the two spheres, we find

$$\frac{E_1}{E_2} = \frac{\dfrac{1}{4\pi\varepsilon_0}\dfrac{Q_1}{r_1^2}}{\dfrac{1}{4\pi\varepsilon_0}\dfrac{Q_2}{r_2^2}} = \frac{Q_1 r_2^2}{Q_2 r_1^2} = \frac{Q_1 r_2^2}{\dfrac{Q_1 r_2}{r_1} r_1^2} = \frac{r_2}{r_1} \qquad (20\text{-}28)$$

We have derived Equation (20-28) for the specific case of two conducting spheres, but the relationship is true in general: The electric field has a larger magnitude when the radius of curvature of a charged conducting surface is smaller. That is, the electric field is stronger near points, and has low values beside flat or nearly flat regions of the surface, as illustrated in Figure 20-24.

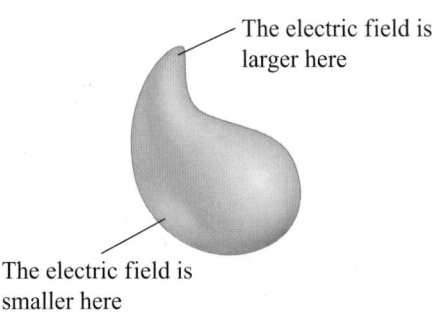

The electric field is larger here

The electric field is smaller here

Figure 20-24 The electric field is stronger near sharply curved points.

Because the electric field is much greater in sharply pointed regions, it may become large enough to ionize material near those points. Corona discharge is electric discharge due to ionization of material in a fluid (such as air) surrounding a conductor. This discharge happens near pointed regions on conductors. This effect is used in ozone generators, in electrostatic precipitators that remove particulates in smokestacks, and in forced-air heating systems (see the Making Connections feature below).

MAKING CONNECTIONS

Electrostatic Precipitators

Undesired by-products of power plant and industrial combustion processes are particulates in the exhaust gases. Electrostatic precipitators can reduce particulate concentrations. The gas containing particulates passes through a chamber region with high voltages applied to a grid with

many sharp points (like a screen with spikes or barbed wire points). Figure 20-25 shows the configuration of a typical electrostatic precipitator. The electric field is high enough near these points to cause corona discharge, which makes some of the particulates electrically charged. When the exhaust then passes near electrode plates with the opposite electric polarity, the charged particles precipitate onto the plates, thus cleaning the exhaust gases. Mechanical or fluid cleaning periodically removes the buildup of particulates from the plates. A similar process can be used in electrostatic cleaners as part of hot-air heating systems in the home and in free-standing air-cleaning units. The idea of charging dust to remove it was devised in the nineteenth century, but the first modern electrostatic precipitator was not patented until 1907, by Dr. Frederick Cottrell. The Research Corporation that Cottrell set up using revenues from these patents continues to fund university scientific research.

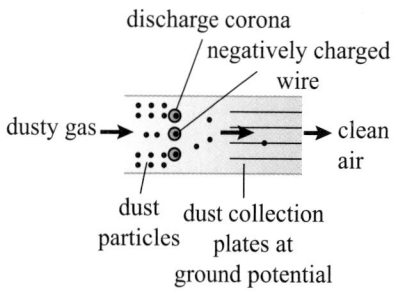

Electrostatic Precipitator

Figure 20-25   An electrostatic precipitator

ELECTRICITY, MAGNETISM, AND OPTICS

When charges of the same sign are moved closer together, positive work is required by an external agent to move them so the electric potential energy increases.

Electric potential, V, is electric potential energy per unit charge. We usually talk of electric potential differences between two points. If the term "electric potential" is used, the zero point should always be specified (usually at an infinite distance).

The electron volt (eV) is a unit of energy that corresponds to the change in electric potential energy when one elementary charge unit moves through a potential difference of one volt.

Electric fields are vector quantities obtained by finding the electric force on a small positive test charge, and then dividing by the charge (units of N/C). These fields also correspond to the negative of the rate of change of electric potential (giving units of V/m).

Equipotential lines and surfaces join points of equal electric potential. Electric field lines show the direction of the electric field at a point and always cross equipotential lines perpendicularly. Electric field lines are drawn more closely together in regions of high electric field intensity.

The surface of an ideal electrical conductor must be at one common electric potential. There is no electric field inside a conductor in electrostatic equilibrium. Electric fields are exactly perpendicular to the surface of a conductor.

Because there is no electric field inside a conductor, all parts of that region must be at the same electric potential.

Charge placed on a conductor distributes itself entirely on the outer surface of the conductor.

For an irregularly shaped conductor, the electric field is stronger in regions where the surface is more pointed.

Electric Potential Energy

The electrostatic potential energy represented by a configuration with two charges, Q and q, a distance r apart is

$$U_{\text{E}} = \frac{Qq}{4\pi\varepsilon_0 r} \qquad (20\text{-}4)$$

We can use the superposition principle to find the total potential energy for configurations of many point charges.

Electric Potential

Electric potential (V) is measured in volts and is defined as electric potential energy per unit charge.

$$V = \frac{U}{q} \qquad (20\text{-}6)$$

We sometimes write ΔV and ΔU to remind ourselves that differences are important and a zero level must be defined.

$$\Delta U = q\Delta V \qquad (20\text{-}11)$$

The electric potential for a point a distance r from a single point charge Q is

$$V = \frac{Q}{4\pi\varepsilon_0 r} \qquad (20\text{-}7)$$

The superposition principle can be used to find the potential due to collections of point charges.

Electric Field from Electric Potential

If we know the electric potential distribution, we can use partial derivatives to find the electric field components:

$$E_x = -\frac{\partial V}{\partial x}; \quad E_y = -\frac{\partial V}{\partial y}; \quad E_z = -\frac{\partial V}{\partial z} \qquad (20\text{-}9)$$

We can integrate the electric field to find the potential difference between two points:

$$V_b - V_a = -\int_a^b \vec{E} \cdot d\vec{r} \qquad (20\text{-}10)$$

Gauss's Law

Gauss's law allows us to calculate the electric field in situations involving spherical, cylindrical, and planar symmetry:

$$\oint \vec{E} \cdot d\vec{A} = \frac{q_{\text{enc}}}{\varepsilon_0} \qquad (20\text{-}14)$$

where the left-hand quantity is the electric flux integrated over some closed surface, and the vector dot product is taken between the electric field and an element of area perpendicular to the surface $d\vec{A}$. The total net electric charge enclosed by the surface is q_{enc}, and ε_0 is a constant of nature called the permittivity of free space.

Applications: atmospheric electricity, radiation treatment using particles from particle accelerators, electron microscopes, electrostatic precipitators for dust control, cathode ray tube devices, ozone production by corona discharge, shielding from electromagnetic fields and waves

Key Terms: corona discharge, electric flux, electric potential, electric potential difference, electric potential energy, electron volt, electrostatic equilibrium, equipotential lines, Gauss's law, induced charge, permittivity of free space, surface integral, surface charge density, volt, volume charge density

QUESTIONS

1. An electron ($-e$ charge) is in orbit about a hydrogen nucleus ($+e$ charge) at a distance r_0. What is the electric potential energy of this system?
   (a) $-\dfrac{1}{4\pi\varepsilon_0}\dfrac{e^2}{r_0}$
   (b) $+\dfrac{1}{4\pi\varepsilon_0}\dfrac{e^2}{r_0}$
   (c) $-\dfrac{1}{4\pi\varepsilon_0}\dfrac{e^2}{r_0^2}$
   (d) $+\dfrac{1}{4\pi\varepsilon_0}\dfrac{e^2}{r_0^2}$

2. You have two unknown point charges, Q_1 and Q_2. You observe that the electric potential is zero exactly halfway between the two charges. What conclusion can you make?
   (a) The two charges have the same magnitude and the same sign.
   (b) The two charges have the same magnitude and the opposite sign.
   (c) There is no way that you can have zero potential at the midpoint.
   (d) The magnitudes of both charges are zero.

3. An electron is moving past a positive ion on the path shown in Figure 20-26. At which of the following points does the electron–ion system have highest speed?
   (a) A
   (b) B
   (c) C
   (d) D

Figure 20-26   Question 3

4. An electron goes from a region at -10 kV to a region at $+30$ kV. What is the electron's change in kinetic energy?
   (a) -40 keV
   (b) $+20$ keV
   (c) $+30$ keV
   (d) $+40$ keV

5. You have two fixed charges, $+Q$ on the left and $-4Q$ on the right, separated by a distance of 3 m. Where is one location with zero electric potential?
   (a) at a point 1 m to the left of the $+Q$ charge
   (b) at a point 1 m to the right of the $+Q$ charge
   (c) at a point 1 m to the left of the $-4Q$ charge
   (d) at a point 1 m to the right of the $-4Q$ charge

6. Which of the following statements is correct?
   (a) Electric field and electric potential are both vectors, and electric flux is a scalar.
   (b) Electric field and electric potential are both scalars, and electric flux is a vector.
   (c) Electric field is a vector, and electric potential and electric flux are scalars.
   (d) Electric field, electric potential, and electric flux are vector quantities.

7. A proton is moving in the $+x$-direction, and its speed is increasing but its acceleration is constant. Which of the following statements is correct?
   (a) The proton is moving into a region with higher electric potential.
   (b) The proton is moving into a region with lower electric potential.
   (c) The acceleration is constant, so there is no change in electric potential.
   (d) The scenario is impossible.

8. A positive charge is located at the origin of an xyz-coordinate system. What can you conclude about the electric potential at the points A$(0, -1, 0)$ and B$(0, +1, 0)$?
   (a) Points A and B are both at a negative electric potential.
   (b) Points A and B are both at a positive electric potential.
   (c) Point A is at a negative electric potential, and point B is at a positive electric potential.
   (d) Point A is at a positive electric potential, and point B is at a negative electric potential.

9. When an electric potential increases with distance in the x-direction according to $V = 3x^2 + 20$, what is the electric field at the point $x = y = z = 1$ m?
   (a) The electric field components are $+20$ N/C in the y- and z-directions and -6 N/C in the x-direction.
   (b) The electric field components are $+20$ N/C in the y- and z-directions and $+6$ N/C in the x-direction.
   (c) The electric field components are zero in the y- and z-directions and $+14$ N/C in the x-direction.
   (d) The electric field components are zero in the y- and z-directions and -6 N/C in the x-direction.

10. When you slightly increase the distance between a proton and an electron in a hydrogen atom, what happens to the electric potential energy of the system?
    (a) The electric potential energy increases.
    (b) The electric potential energy decreases.
    (c) The electric potential energy is negative both before and after but does not change in magnitude.
    (d) The electric potential energy is zero both before and after.

11. A single positive charge is placed inside a small cube. What can be said about the net electric flux over the surfaces of the cube?
    (a) The net electric flux is negative, and the value depends on where in the cube the charge is placed.
    (b) The net electric flux is positive, and the value depends on where in the cube the charge is placed.
    (c) The net electric flux is negative, and the value is independent of the position of the charge inside the cube.
    (d) The net electric flux is positive, and the value is independent of the position of the charge inside the cube.

12. An ion contains a nuclear charge of $+5e$ and has a total electronic charge of $-4e$. What is the electric field at a distance r well outside the atom? Assume spherical symmetry.
    (a) The electric field is $\dfrac{1}{4\pi\varepsilon_0}\dfrac{e}{r^2}$ in a direction radially inward toward the atom.
    (b) The electric field is $\dfrac{1}{4\pi\varepsilon_0}\dfrac{e}{r^2}$ in a direction radially outward away from the atom.
    (c) The electric field is $\dfrac{1}{4\pi\varepsilon_0}\dfrac{5e}{r^2}$ in a direction radially inward toward the atom.
    (d) The electric field is $\dfrac{1}{4\pi\varepsilon_0}\dfrac{9e}{r^2}$ in a direction radially outward away from the atom.

13. You draw a spherical Gaussian surface just outside an electric dipole (equal positive and negative charges separated by a small distance). What can you conclude about the electric flux through that surface?
    (a) As long as the dipole is completely inside the surface, the flux is zero.
    (b) The flux is only zero when the surface is centred on the midpoint of the dipole.
    (c) The flux is approximately but not exactly zero, even when the surface is centred on the midpoint of the dipole.
    (d) None of the above are true.

14. Some charge is placed on a closed (and hollow) tin can. Where is the charge located?
    (a) The charge is located equally throughout the metal of the can.
    (b) All the charge is on the inside surface.
    (c) All the charge is on the outside surface.
    (d) All the charge is on the two ends (none on the curved surface).

15. Two conducting spheres with different radii are joined by a wire, and a negative charge is placed on the system. Which statement correctly describes the electric field and electric potential just outside the surface of each sphere?
    (a) The electric field points toward the surface in both cases and is stronger on the smaller sphere. The electric potential is negative and the same for both spheres.
    (b) The electric field points away from the surface in both cases and is stronger on the smaller sphere. The electric potential is zero for both spheres.
    (c) The electric field points toward the surface in both cases and is stronger on the larger sphere. The electric potential is negative and the same for both spheres.
    (d) The electric field points toward the surface in both cases and is of the same strength. The potential is negative in both cases and higher for the smaller sphere.

16. When the charge on a conducting sphere is doubled, how does the electric potential *inside* the sphere change?
    (a) The electric potential is now four times as large.
    (b) The electric potential is now two times as large.
    (c) The electric potential is now approximately $\sqrt{2}$ times as large.
    (d) The electric potential is zero in both cases.

PROBLEMS BY SECTION

For problems, star ratings will be used, (✷, ✷✷, or ✷✷✷), with more stars meaning more challenging problems.

Section 20-1 Electric Potential Energy

17. ✷ We assume that an electron and a proton are on average separated by 53.0 pm. What is the electric potential energy of a hydrogen atom?

18. ✷✷ How much work is required to move four charges from infinite initial separation into the following configuration along the x-axis: +1.00 μC at +2.00 m, +2.00 μC at +3.00 m, +3.00 μC at +4.00 m, and +4.00 μC at +5.00 m?

Section 20-2 Electric Potential

19. ✷✷ Four +2.00 nC charges are located at the corners of a square that is 10.0 cm on each side. What is the electric potential at the exact centre of the square?

20. ✷✷ In Figure 20-27, a −3.00 μC charge is 0.500 m to the right of a +6.00 μC charge. Point P_1 is located 0.250 m to the right of the negative charge, and point P_2 is located a further 0.250 m to the right.
    (a) What is the electric potential at point P_1?
    (b) What is the electric potential at point P_2?
    (c) How much work is required to move a −1.00 nC charge from point P_2 to P_1?

Figure 20-27 Problem 20

Section 20-3 Equipotential Lines and Electric Field Lines

21. ✷ Figure 20-28 shows plots of some equipotential lines. Sketch in the corresponding electric field lines, and indicate the relative strength of the electric field vectors in different regions.

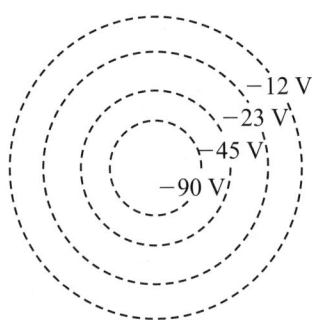

Figure 20-28 Problem 21

22. ✷✷ The central +2 nC charge in Figure 20-29 is flanked on the left and right (at equal distances) by −1 nC charges. Qualitatively sketch the electric field lines and equipotential lines for the system.

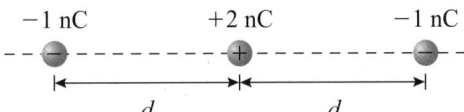

Figure 20-29 Problem 22

Section 20-4 Electric Field from Electric Potential

23. ✷✷ An electric potential in SI units has the form $V = -100.0 + 25.0y$.
    (a) Write expressions for the three components of the electric field.
    (b) What is the electric field at the point (1.00 m, 2.00 m, 0.00)?

24. ✷ In Chapter 19, we noted that high electric fields can be present inside biological cell membranes. The field in a particular cell over a small region is assumed to have a constant value of 5.00 N/μC. What is the potential difference over a 2.00 μm distance?

25. ✷✷ An electric field vector is $\vec{E} = -20\hat{i}$ N/C.
   (a) Describe the equipotential surfaces.
   (b) When $V = +25.0$ V at the origin, where is $V = 75.0$ V?

Section 20-5   The Electron Volt

26. ✷✷ The electron beam in a typical scanning electron microscope produces kinetic energies of 25.0 keV.
   (a) Assuming that relativistic effects can be ignored, what is the speed of the electrons?
   (b) What potential difference is required to produce electrons of this energy?
   (c) This accelerating potential difference is applied over a distance of 2.50 cm. What is the electric field in that region?

27. ✷ A particular chemical reaction requires 2.45 eV to create a new molecule.
   (a) What is this energy per molecule expressed in units of joules?
   (c) What is this energy in thermochemical calories? (1 calorie = 4.184 J)

Section 20-6   Electric Fields and Conductors

28. ✷ A conducting solid oval carries a negative charge (Figure 20-30). Sketch the electric field vectors, if any, (a) inside, (b) on the surface, and (c) slightly away from the surface.

Figure 20-30   Problem 28

Section 20-7   Charge Densities

29. ✷ A charge of $+165\ \mu$C is uniformly distributed around a very thin ring of radius 4.00 cm. What is the linear charge density along the ring?

30. ✷ A thin plate has the dimensions 3.50 cm × 12.0 cm and carries a charge density of $45.0\ \mu$C/m². How many elementary charge units (i.e., 1.60×10^{-19} C) are on the entire plate?

31. ✷ A nonconducting sphere has a radius of 2.75 cm and a total charge of $+3.00\ \mu$C uniformly spread throughout the sphere. What is the volume charge density?

Section 20-8   Electric Flux

32. ✷✷ In a certain region, the electric field points in the $+y$-direction and has a magnitude of 35.0 N/C. What is the electric flux through a circle of radius 1.00 cm that lies in the xz-plane?

33. ✷✷ An electric field $\vec{E}$ is parallel to the axis of a hemispheric shell, as shown in Figure 20-31. What is the electric flux through the surface?

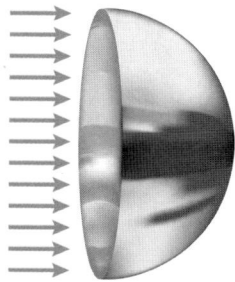

Figure 20-31   Problem 33

Section 20-9   Gauss's Law

34. ✷✷ A +25.0 nC point charge is surrounded by a concentric conducting spherical shell of radius 2.40 cm that carries a charge of −20.0 nC. Derive expressions for the electric field as a function of radial distance both inside and outside the shell.

35. ✷✷ A solid, nonconducting sphere of radius r_0 has a uniform charge density ρ (assume a positive charge). Derive expressions for the electric field as a function of radial distance r both inside and outside the sphere.

Section 20-10   Applications of Gauss's Law

36. ✷✷ A solid, nonconducting very long cylinder of radius r_0 has a uniform charge density ρ. Derive expressions for the electric field as a function of radial distance r both inside and outside the cylinder.

37. ✷✷✷ Two long, concentric conducting cylinders have radii 4.50 cm and 8.50 cm (Figure 20-32). The inner cylinder carries a charge per unit length of $+15.0\ \mu$C/m, and the outer cylinder carries a charge per unit length of $-10.0\ \mu$C/m. Derive and plot expressions for the electric field as a function of radial distance r in all three regions: inside the inner cylinder, the space between the two cylinders, and outside the outer cylinder.

8.50 cm

4.50 cm

Figure 20-32   Problem 37

Section 20-11   Charges, Conductors, and Electric Fields

38. ✷✷✷ A relatively large Van de Graaff generator has a conducting sphere of radius 0.600 m. Assume that a total charge of $94.0\ \mu$C is present on the sphere.
   (a) What is the electric field inside the conducting sphere?
   (b) What is the electric potential at the surface of the conductor?
   (c) What is the electric potential inside the conducting sphere?
   (d) Calculate and plot the electric potential as a function of distance outside the conductor.
   (e) What is the electric field just outside the conductor?

COMPREHENSIVE PROBLEMS

39. ✻ Two protons are 3.00 fm apart. What is the electrostatic energy of this configuration?

40. ✻✻ Three +2.00 nC charges are placed at the vertices of an equilateral triangle with 25.0 cm sides.
    (a) What is the electric potential energy of this charge distribution?
    (b) How much work is needed to bring the charges from an infinite distance into this configuration?

41. ✻✻ Two positive charges in space, each with a charge of +25.0 μC and a mass of 5.00 mg, are initially held in a fixed separation of 35.0 mm. They are released and allowed to separate under mutual repulsion. What speed will they each attain when they are very far apart? (Hint: Use an energy approach. By conservation of linear momentum, you know that the situation must result in symmetric motions relative to the centre of mass.)

42. ✻✻ An electric dipole has two charges, +1.00 nC and −1.00 nC, separated by a distance of 0.250 μm. What is the electric potential energy of the configuration?

43. ✻✻ An electron is accelerated through a constant electric field until it reaches a speed of 3.0×10^5 m/s. What is the potential difference between the starting and ending positions?

44. ✻✻ Consider the distribution of the four point charges shown in Figure 20-33. What is the electric potential, V, at a point halfway between charges A and B?

Figure 20-33 Problem 44

45. ✻✻ Consider the arrangement of four electric charges shown in Figure 20-34. Find the electric potential at point P, assuming that the potential has a zero point at infinite distance.

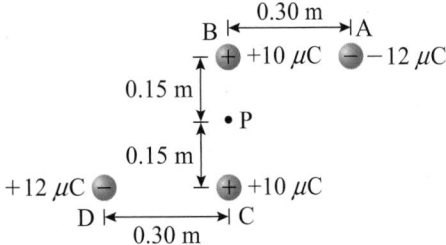

Figure 20-34 Problem 45

46. ✻✻✻ Three identical charges +q are placed at three of the four corners of a square with sides d.
    (a) What is the electric potential at the corner that is missing a charge?
    (b) What is the work required to move a fourth charge +q from an infinite distance to the corner that is missing a charge?

(c) Check your answer to part (b) by calculating the difference between the electrostatic potential energy in the cases of three charges and four charges.

47. ✻✻ In the famous Rutherford scattering experiment, α-particles were fired at a thin gold foil. When the α-particles passed near a nucleus, they were deflected by large angles. For the purpose of this question we are going to consider the situation of an α-particle directed exactly at a gold nucleus. Calculate how close an α-particle could get before it "bounces back" due to electric repulsion. Use energy considerations to solve this problem. Gold has 79 protons in the nucleus. An α-particle is made up of 2 protons and 2 neutrons. The α-particle had a speed of 1.50×10^7 m/s when it was a large distance away. Assume that the gold nucleus is sufficiently massive that it does not move. Treat the gold nucleus as though it were a point charge of the same total charge for purposes of solving this problem.

48. ✻✻ A constant electric field is defined by the following vector:

$$\vec{E} = 4.00\hat{i} + 25.0\hat{k} \text{ N/C}$$

What is the electric flux through a 1.00 m $\times$ 1.00 m area that is in the xy-plane and is centred on the origin?

49. ✻✻ The electric field near Earth's surface points downward and has a magnitude of approximately 150 N/C. If the field were due to a thin layer of charge spread evenly over Earth's surface, what would the magnitude and sign of the electric charge be?

50. ✻✻ (a) Determine the sign and direction of the electric field required for an electron to have an upward electric force that just balances the downward force of gravity.
    (b) With this field, we define the electric potential as 0 V at the surface. What is the potential at a height of 3.00 m? Assume a constant field.

51. ✻✻ A cathode ray tube display accelerates an electron from rest to 2.25×10^7 m/s within a distance of 1.25 cm in the acceleration region.
    (a) What is the magnitude of the electric potential difference in this accelerating region?
    (b) What is the increase in the kinetic energy of the electrons, in eV?

52. ✻✻ (a) A proton is travelling at a speed of 2.50×10^7 m/s. What is the kinetic energy, in eV? Ignore relativistic effects.
    (b) Repeat part (a) with an electron.

53. ✻✻✻ A conducting sphere of radius 12.0 cm has a spherical cavity of radius 5.00 cm in its centre. The sphere carries a total charge of +240 nC. Derive expressions for the electric field magnitude and the electric potential as a function of radial distance, r.

54. ✻✻ A nonconducting plate carries a uniformly distributed charge of 600.0 nC. The plate is 18.00 cm long, 40.00 cm wide, and 0.500 cm thick.
    (a) Use Gauss's law to approximate the electric field at a point 1.00 cm above the surface of the plate.
    (b) What is the electric field at a point 5.00 cm above the plate?

55. ✻✻✻ The diameter of the central wire in a very long triple conductor coaxial cable is 4.00 mm. Concentric around the wire is a middle cylindrical conductor with a diameter of 2.60 cm and an outer conductor with a diameter

of 5.00 cm. At some instant, the wire carries a charge density of $+15.0\ \mu$C/m, the middle conductor carries a charge density of $-7.00\ \mu$C/m, and the outer conductor carries a charge density of $-8.00\ \mu$C/m. Derive expressions for the electric field as a function of radial distance in each of the following regions:

(a) between the wire and the central conductor;
(b) between the middle and outer conductors;
(c) outside the outer conductor.

56. ✶✶ Consider a nonconducting sphere of radius r_0 and uniform charge density ρ.

(a) Derive an expression for the electric field as a function of radial distance for $r < r_0$.
(b) Derive an expression for the electric field as a function of radial distance for $r > r_0$.

57. ✶✶ Figure 20-35 shows a plot of an electric potential as a function of distance along the x-axis. From this information, determine the x-component of the electric field at each of the following displacements:

(a) $x = 0.5$ m;
(b) $x = 1.5$ m;
(c) $x = 3.0$ m;
(d) $x = 4.5$ m.

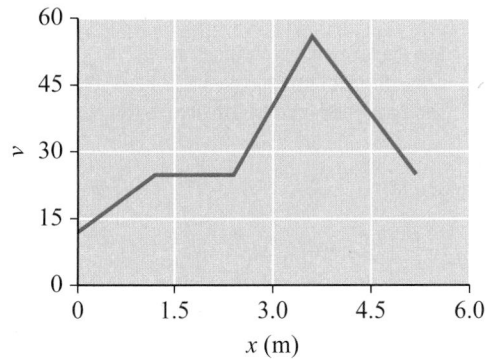

Figure 20-35 Problem 57

58. ✶✶✶ (a) We assume that dry air breaks down electrically when the electric field exceeds 2.50×10^6 V/m. Determine the maximum charge that can be placed on a conducting sphere with each of the following radii: (i) 10.0 cm and (ii) 20.0 cm.

(b) For each case in part (a), determine the surface charge density.

59. ✶✶ As mentioned earlier in the chapter, electrostatic dust precipitators have a section where there are spikes to make high electric fields to ionize the dust grains (followed by high voltages to collect the charged dust later). Assume that the spikes can be approximated by a hemisphere at the top of a rod as shown in Figure 20-36, with a diameter of

Figure 20-36 Problem 59

3.00 mm. When the plate is held at a potential of 800 V, what is the electric field near a spike surface? (Assume for this purpose that it is the same field as for an entire sphere of the same radius.)

60. ✶✶ Imagine that the attractive force between the Sun and Earth were electrostatic rather than gravitational. Assume that Earth has a negative charge and the Sun has a positive charge of equal magnitude.

(a) What charge would produce an electric force equal the actual gravitational force?
(b) Suppose that the charge is distributed evenly over the surface of Earth. What would the surface charge density be?
(c) What would the electric field just above Earth's surface be?
(d) What would the electric potential on Earth's surface be?

61. ✶✶✶ Three drops each have a radius of 250 μm and a charge of 50 nC. Assume that the drops are perfectly spherical (this is a good approximation because of surface tension).

(a) What is the electric potential just outside the surface of one drop?
(b) The three drops coalesce, with the total charge now equally spread over the resultant larger drop. What is the new electric potential?
(c) What is the electric field just outside the drop in each case?

62. ✶✶✶ (a) A positive charge Q is placed at some distance d to the left of an infinite conducting plate that is at zero electric potential (grounded). What sign of charge is induced in the conducting plate at the point nearest Q?
(b) Sketch electric field lines for the situation.
(c) In more advanced electromagnetism courses, a technique called the "method of images" is used. In this case, we can imagine a charge of the opposite sign $-Q$ (an "image charge") placed behind the plate at an equal distance d from the plate. We know this is the correct placement because it will lead to a zero potential along the conducting plate. We can replace the plate with the image charge and consider the two charges when calculating electric potentials. Use this idea to sketch electric equipotential lines.
(d) If $Q = 2.50$ nC and $d = 0.150$ m, what is the electric potential at a point halfway between the charge Q and the plate?

63. ✶✶ Use calculus to find the electric potential at distances $a/2$ and $2a$ from an infinite line of charge having a linear charge density μ (measured in coulombs per metre). Assume that zero electric potential is defined at distance a from the line of charge.

64. ✶✶✶ A thin ring of charge of radius a is in the xy-plane centred on the origin. The ring has linear charge density μ. Derive an expression for the electric potential at the point $(0, 0, z)$. In your final expression, include a constant of integration, representing the zero definition of electric potential.

65. ✶✶✶ Using calculus on differential elements of charge, derive an expression for the electric potential at a point located a distance z above the centre of a disk of radius a. Assume that the disk has a surface charge density of σ (measured in coulombs per square metre).

DATA-RICH PROBLEM

66. ✸✸✸ Table 20-1 lists data on electric potential as a function of radial distance. The object is some sort of a sphere.
(a) Plot the electric potential data.
(b) From the data, determine the approximate electric field, and plot the electric field data. Indicate the direction of the electric field.
(c) Is the object conducting or nonconducting? Justify your answer.
(d) What is the radius of the object?
(e) Calculate the approximate charge distribution for the object: If the charge is all on a surface, calculate the surface charge density. If the charge is uniformly spread throughout the object, calculate the volume charge density, assuming that it is constant. Is the charge positive or negative?
(f) How far from the object would you need to be for the electric potential to be 500 V?

Table 20-1 Data for Problem 66

R (m)	Electric potential (V)
0.0	7000
0.1	7000
0.2	7000
0.3	7000
0.4	7000
0.5	6300
0.6	5250
0.7	4500
0.8	3940
0.9	3500
1.0	3150
1.1	2865
1.2	2625
1.3	2425
1.4	2250
1.5	2100
1.6	1970
1.7	1855
1.8	1750

OPEN PROBLEM

67. ✸✸✸ If you scuff along a carpeted floor on a dry day, and then approach a metal object, such as a doorknob, a spark can jump from your finger. In this problem, you will estimate the amount of charge on your finger as a result of this scenario.
(a) When the electric field reaches a certain value, breakdown will occur. That is what happens when the spark jumps. Use a reliable source to determine a value for breakdown in dry air.
(b) Perform an experiment in which you scuff across a carpet and produce a spark when you point toward a doorknob. From that experiment, estimate the distance that the spark jumps from your finger to the doorknob.
(c) Using your answer from (b) and assuming that the potential on the doorknob is approximately zero, estimate the original electric potential on your finger.
(d) Approximate the end of your finger as a sphere, and estimate its radius. When the charge is equally spread on the surface of the sphere, what is the charge density on the tip of your finger?

 See the text online resources at www.physics1e.nelson.com for Open Problems and Data-Rich Problems related to this chapter.

Chapter 21
Capacitance

Every day defibrillators save many lives. Fibrillation is a series of rapid, irregular muscle contractions. Fibrillation of the heart can be fatal because it prevents proper blood flow and can seriously damage heart tissue. A defibrillator uses an electric shock to momentarily stop the heart and give it a chance to resume beating normally. Figure 21-1 shows a common type of external defibrillator. A miniature internal defibrillator can be implanted in patients who have chronic problems with heart fibrillation. A key part of the defibrillator circuit that delivers a controlled pulse of electric current to the patient is the capacitor, which is a device that can store charge and electrostatic energy. In this chapter, you will learn how capacitors are made, how they operate, and how these relatively simple devices have innumerable applications in modern technologies.

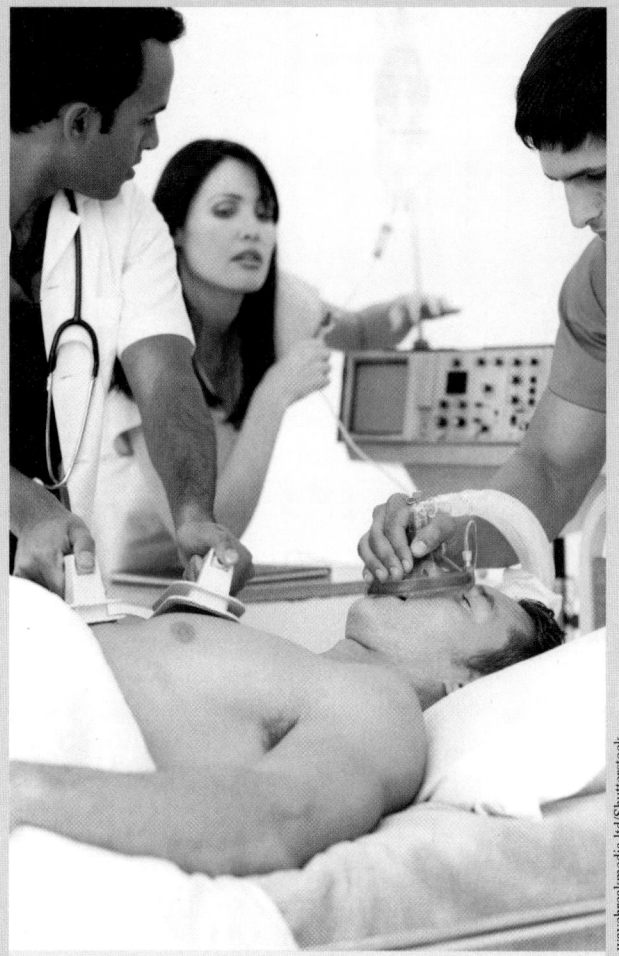

wavebreakmedia ltd/Shutterstock

Figure 21-1 A paddle-type defibrillator

ELECTRICITY, MAGNETISM, AND OPTICS

For additional Making Connections, Examples, and Checkpoints, as well as activities and experiments to help increase your understanding of the chapter's concepts, please go to the text's online resources at www.physics1e.nelson.com.

LO 1

21-1 Capacitors and Capacitance

A capacitor is a device that can store charge and electrostatic energy. You have probably used capacitors numerous times without realizing it. For example, an electronic camera flash uses a capacitor to store the energy that is released during the flash. Turn indicators on some automobiles use a circuit with capacitors and resistors to control the rate at which the signals flash. Electronic cameras and camcorders form images by storing charge on an array of tiny photosensitive capacitors.

Capacitors come in many shapes and sizes and can be made with a variety of materials (Figure 21-2); however, most capacitors consist of two closely spaced parallel plates. When the capacitor is charged, one of these plates carries a positive charge and the other plate carries a negative charge of the same magnitude. The words "capacitor" and "capacity" have the same root, which reflects the key feature of capacitors: their ability to store charge.

Figure 21-2 An assortment of modern capacitors

The first capacitors were invented independently around 1745 by the German scientist Ewald Georg von Kleist (1700–48) and the Dutch professor Pieter van Musschenbroek (1692–1761). The very first models consisted simply of a glass jar of water with a metal wire or chain suspended from an insulating stopper. The water acted as one plate and the experimenter's hand as the other plate. Researchers soon found that coating the inside and outside of the jar with metal foil increased the amount of charge that it could store. These glass jar capacitors were called Leyden jars after the university where Musschenbroek conducted his research. Benjamin Franklin was possibly the first to make a capacitor with flat, parallel plates. Michael Faraday is generally credited with the first clear explanation of the charge storage process in capacitors.

Ideally, the material between the plates of a capacitor is totally nonconducting. Some capacitors simply have air between the plates, but most capacitors have a solid dielectric material (see Chapter 19). Since unlike charges

attract, electrostatic forces hold the charge on the plates of a capacitor when the source of the charge is disconnected from the capacitor. An ideal capacitor would stay charged forever, but an actual capacitor discharges because charge gradually leaks through the dielectric between its plates or through the air between its terminals.

The electric field between the plates of the capacitor in Figure 21-3 is directed from the positively charged plate to the negatively charged plate because that would be the direction of the force on a small positive test charge placed between the plates. As we will prove in the next section, this field is essentially constant: it is the same everywhere between the plates except very near the edge of a plate.

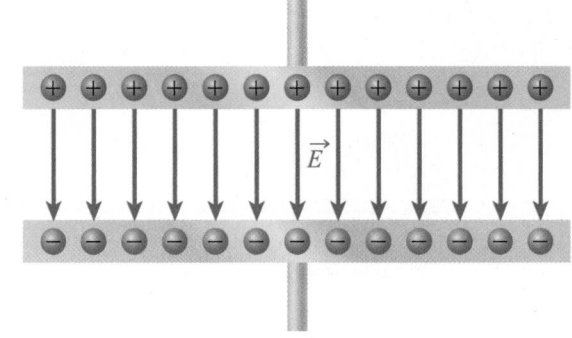

Figure 21-3 An idealized parallel-plate capacitor

Recall from the previous chapter that the electric potential difference between two points can be found by integrating the vector dot product of the electric field along a path joining the points:

$$V_b - V_a = -\int_a^b \vec{E} \cdot d\vec{r} \qquad (20\text{-}10)$$

Since the electric field is constant in both direction and magnitude between the plates of a capacitor, we can take $\vec{E}$ outside the integral, giving a simple expression for the potential difference, V, between the two plates a distance d apart:

KEY EQUATION
$$V = Ed \qquad (21\text{-}1)$$

Keep in mind that it is always the positively charged plate that is at the higher electric potential. If we add more charge to the plates, the electric field must increase in magnitude, and Equation (21-1) indicates that the electric potential difference increases correspondingly. The rate at which changes in charge on a plate, Q, affect the potential

ELECTRICITY, MAGNETISM, AND OPTICS

© David J. Green - electrical/Alamy

difference, V, depends on the physical dimensions of the capacitor (described in the next section) and is called the capacitance, C. Note that Q in the following equation is the magnitude of charge on one of the plates; the sum of the charges on both plates is zero because the plates have opposite charges of equal magnitude.

KEY EQUATION
$$Q = CV \qquad (21\text{-}2)$$

It is important to realize that capacitance is the ratio of charge to potential difference, but the capacitance does not change when you change the potential difference. Capacitance is determined by the dimensions and physical properties of the materials in the capacitor. A useful way to think about capacitance is that it is a measure of how much charge can be stored for a certain voltage. High capacitance means that you can store more charge at the same voltage than in a capacitor with less capacitance.

The SI unit of capacitance is the farad (F), named in honour of Michael Faraday. As you can see from Equation (21-2), one farad is equal to one coulomb per volt. A farad is a huge amount of capacitance for everyday purposes; most capacitors that you will use in the lab or encounter in household devices are measured in pico-farads (pF) or microfarads (μF). By convention, capacitance is rarely measured in nanofarads.

Capacitors are charged by connecting them to a source of electric potential difference, such as the battery in the circuit shown in Figure 21-4. The symbol for a capacitor is simply two parallel lines, representing the plates. Some textbooks show capacitors with one thick or curved line, but in all cases the lines are the same length. The end connected to the positive terminal of the battery is positively charged. We have indicated the polarity in Figure 21-4, but it is not usual to show this on the symbol.

Figure 21-4 When the switch (S) is closed, the capacitor (C) will charge to the same voltage as the ideal battery (V). The end connected to the positive terminal of the battery is the positively charged capacitor side.

 ONLINE ACTIVITY

Capacitor Electric Fields

The e-resource that accompanies every new copy of this textbook contains an Online Activity using the PhET simulation "Capacitor Lab." Work through the simulation and accompanying questions to gain an understanding of capacitor electric fields.

C-21-1 Capacitance

Capacitors A and B are charged by connecting them to identical batteries. Capacitor A has a higher capacitance, even though the distance between the plates is the same for both capacitors. Which of the following statements is correct?

(a) Both capacitors have the same voltage, but A has more charge stored and a larger electric field between the plates.

(b) Both capacitors have the same voltage and electric field, but A has more charge stored.

(c) Both capacitors have the same voltage and electric field, but A has less charge stored.

(d) Capacitor A has a higher voltage, more charge stored, and a larger electric field.

C-21-1 (b) The capacitors are connected to identical batteries, so the potential difference, V, must be the same. Since $Q = CV$, if V is the same and C is greater for capacitor A, clearly it must have more charge. The magnitude of the electric field is given by $E = V/d$. Both V and d are the same, so E must also be the same for both capacitors.

21-2 Electric Fields in Capacitors

In this section, we will show that the electric field is essentially constant in the region between the plates of a capacitor. Consider a parallel-plate capacitor with a positively charged top plate. We will draw a Gaussian cylinder with one end inside the top metal plate and the other end in the space between the plates, as shown in Figure 21-5.

Figure 21-5 Gaussian cylinder in a parallel-plate capacitor

Recall from Chapter 20 that Gauss's law relates the total flux through our Gaussian surface to the enclosed charge:

$$\oint \vec{E} \cdot d\vec{A} = \frac{q_{enc}}{\varepsilon_0} \qquad (20\text{-}15)$$

The total flux equals the sum of the flux through the flat ends and the curved side of the Gaussian cylinder. However, the electric field is zero inside the conductive top plate, so the flux through the top end of the cylinder is zero. Also, the area vector $d\vec{A}$ and the electric field vector $\vec{E}$ are perpendicular to each other for the entire curved side of the cylinder, so the vector dot product makes the net flux on this surface zero. We are left with only the flux through the end that is located between the capacitor plates. Since the area vector for this end is parallel to the electric field, Gauss's law for this situation reduces to

$$EA_e = \frac{\sigma A_e}{\varepsilon_0} \qquad (21\text{-}3)$$

where A_e is the area of the end, and σ is the surface charge density of the plate.

Therefore, the magnitude of the electric field in the space between the plates is simply

$$E = \frac{\sigma}{\varepsilon_0} \qquad (21\text{-}4)$$

This relationship applies as long as our Gaussian surface is located anywhere along the plates and ends anywhere between the plates. Therefore, the electric field between the plates of a parallel-plate capacitor is essentially constant. When you go on to more advanced courses in electromagnetism, you will learn about the variations in the electric field near the ends of the capacitor. However, through most of the capacitor the field is approximately constant, and we will assume that in this textbook.

Most of the time, we do not deal directly with the surface charge density, so Equation (21-4) is rarely used in capacitor problems. If we assume that the area of one plate of the capacitor is A, we can rewrite Equation (21-4) in terms of the total charge, Q, on one plate:

PEER TO PEER

You can solve many capacitor-related problems by applying two simple relationships: $Q = CV$ for all capacitors and $E = V/d$ for the constant electric field between parallel plates, where d is the plate spacing. Remember to make your units consistent (e.g., express the capacitance in farads and the plate spacing in metres).

$$E = \frac{Q}{\varepsilon_0 A} \qquad (21\text{-}5)$$

Multiplying both sides of this equation by the distance between the plates, d, and using the fact that the potential difference between the plates, V, is equal to Ed, we have

$$V = \frac{Qd}{\varepsilon_0 A} \qquad (21\text{-}6)$$

This equation is consistent with our basic capacitor equation, $Q = CV$, when the capacitance of a parallel-plate capacitor is given by

KEY EQUATION
$$C = \frac{\varepsilon_0 A}{d} \qquad (21\text{-}7)$$

This expression for capacitance makes sense because we expect plates with a larger area to be able to store more charge (i.e., have a greater capacitance). Also, with a closer spacing of the plates, the positive and negative charges can more effectively "hold" each other in place; hence the capacitance is greater. Equation (21-7) is exact for a parallel-plate capacitor with a vacuum between the plates, and is a close approximation for capacitors with air between the plates. Later in the chapter, we will see how a dielectric between the plates changes the capacitance.

✓ CHECKPOINT

C-21-2 The Electric Field between Parallel Plates

Two parallel conducting plates are separated by 4.00 mm; one has a voltage of +400 V, and the other is grounded (0 V). Which statement correctly describes the electric field between the plates?

(a) The electric field is 100 kV/m near the midpoint, essentially zero just outside the grounded plate, and essentially infinite just outside the +400 V plate.

(b) The electric field is zero at the midpoint, 100 kV/m pointing away from the +400 V plate, and for the grounded plate the same magnitude but pointing toward the grounded plate.

(c) The electric field is 50 000 V/m at the midpoint, almost 100 kV/m near the +400 V plate, and near zero just outside the grounded plate.

(d) The electric field is constant throughout the entire region at a value of 100 kV/m.

C-21-2 (d) The electric field is essentially constant inside a parallel-plate capacitor. To obtain the value, you divide the potential difference by the plate spacing. Remember that electric potentials (voltages) are always referenced to some zero level, and the answer to this question would not change if we had said the plates were +200 V and −200 V and +100 V and −300 V, or any other combination with the same potential difference.

EXAMPLE 21-1

A Parallel-Plate Capacitor

A capacitor that has two parallel plates of dimensions 8.00 cm $\times$ 10.00 cm separated by 2.00 mm of air is connected to a 24.0 V battery. Find

(a) the capacitance;
(b) the charge on one plate;
(c) the electric field in the capacitor;
(d) the surface charge density on the plates.

SOLUTION

(a) We can find the capacitance from the relationship for a parallel-plate capacitor:

$$C = \frac{\varepsilon_0 A}{d} = \frac{8.85 \times 10^{-12}\,\mathrm{C^2N^{-1}m^{-2}} \times 0.0800\,\mathrm{m} \times 0.100\,\mathrm{m}}{0.002\,00\,\mathrm{m}}$$

$$= 3.54 \times 10^{-11}\,\mathrm{F} = 35.4\,\mathrm{pF}$$

Remember that electric potential is electric potential energy per charge; therefore, $1\,\mathrm{V} = 1\,\mathrm{J/C} = 1\,\mathrm{(N \cdot m)/C}$. Hence, $1\,\mathrm{C^2/(N \cdot m)} = 1\,\mathrm{C/V} = 1\,\mathrm{F}$.

(b) To determine the charge on one plate, we need the capacitance and the applied voltage. The voltage is given, and we now know the capacitance:

$$Q = CV = 3.54 \times 10^{-11}\,\mathrm{F} \times 24.0\,\mathrm{V}$$

$$= 8.50 \times 10^{-10}\,\mathrm{C} = 0.850\,\mathrm{nC}$$

(c) The electric field is essentially constant between the plates of a parallel-plate capacitor. Therefore, we can find the field by dividing the potential difference by the plate spacing:

$$E = \frac{V}{d} = \frac{24.0\,\mathrm{V}}{0.002\,00\,\mathrm{m}} = 1.20 \times 10^4\,\mathrm{V/m} = 12.0\,\mathrm{kV/m}$$

(d) If we rearrange Equation (21-4), we can solve for the surface charge density (remember that all plate charge will go to the surface):

$$\sigma = E\varepsilon_0 = 12\,000\,\mathrm{V/m} \times 8.85 \times 10^{-12}\,\mathrm{C^2N^{-1}m^{-2}}$$

$$= 1.06 \times 10^{-7}\,\mathrm{C/m^2}$$

Making sense of the result:

These results demonstrate that typical plate dimensions and separations for parallel plates in air lead to modest values for the capacitance, usually in the picofarad range. You may be surprised by how large the electric field is, but do not confuse a field of thousands of volts per metre with the potential difference (just 24 V here). Remember that in this capacitor, the field extends only a few millimetres.

LO 3

21-3 Calculating Capacitance

By far, the most common capacitance configuration is that of a parallel plate. We saw in the last section how the capacitance depends on the area and spacing of the plates for a parallel-plate configuration. While most commercial capacitors do not look like parallel plates, they are essentially parallel plates with the conducting plates and insulating layer rolled into a cylindrical shape.

It is important to realize that capacitance is a feature of various configurations of conductors, and not just parallel plates. Capacitance is defined as the ratio of the stored charge, Q, to the potential difference, V, between the plates:

$$C = \frac{Q}{V} \qquad (21\text{-}8)$$

This definition of capacitance is equivalent to the definition in Section 21-1, as we can easily see by comparing Equations (21-2) and (21-8).

✓ CHECKPOINT

C-21-3 Capacitance and Field

Two surfaces (not necessarily planes) have an essentially constant electric field of 1000 V/m between them when a charge of $200\,\mu\mathrm{C}$ is applied to one surface and a charge of $-200\,\mu\mathrm{C}$ is applied to the other surface. The spacing between the surfaces is 5.00 mm. What is the capacitance?

(a) $10\,\mu\mathrm{F}$
(b) $40\,\mu\mathrm{F}$
(c) $250\,\mu\mathrm{F}$
(d) 40 F

C-21-3 (b) Since the electric field is essentially constant, the potential difference must be $Ed = 5$ V. The capacitance is $Q/V = 200\,\mu\mathrm{C}/5$ V $= 40\,\mu\mathrm{F}$.

Here is a procedure for finding the capacitance of any configuration of two plates:

- Assume some arbitrary charge, Q, on the plates, and use Gauss's law to determine the electric field in the space between the plates.

- Integrate the electric field between the two plates to obtain the potential difference, V.

- Take the ratio of the charge Q to the potential difference V to obtain the capacitance.

Example 21-2 demonstrates the application of this procedure.

EXAMPLE 21-2

Capacitance for Coaxial Cylinders

Consider a system of two coaxial cylinders. The inner cylinder is solid and has radius r_1, and the outer cylindrical conductor is thin and has radius r_2. The inner cylinder carries a charge of $+Q$ over a length L, and the outer cylinder carries a charge of $-Q$ over the same length.

(a) What is the capacitance of this configuration?

(b) When $L = 0.500$ m, $r_1 = 4.00$ mm, and $r_2 = 5.00$ mm, what is the capacitance?

SOLUTION

(a) As we saw in the previous chapter, all the charge on the inner solid conductor will be on its outer surface. The outer conductor is thin, and its charge is on its outer surface as well, but we are assuming that the radii of its inner and outer surfaces are essentially the same. As shown in Figure 21-6, the electric field reflects the symmetry of the cylinders and is directed radially outward (from positive to negative charge).

Figure 21-6 Example 21-2. Gaussian cylinder of radius r drawn on coaxial cylindrical conductors

To find the capacitance, we apply the three-stage process of first finding the electric field by Gauss's law, then the potential difference between the surfaces, and finally the charge-to-potential difference ratio.

The system has linear symmetry, so we draw a concentric Gaussian cylinder with length L and radius r intermediate between r_1 and r_2. We can ignore flux through the ends of the Gaussian cylinder because the surface area vector and the electric field are perpendicular, making the vector dot product vanish. Over the curved side of the Gaussian cylinder, the electric field is of constant magnitude (at the same r) and is everywhere parallel to the area vector element, so we simplify the integral as shown below. Note that the enclosed charge is only the charge on the inner conductor, $+Q$.

$$\oint \vec{E} \cdot d\vec{A} = E \oint dA = \frac{q_{enc}}{\varepsilon_0} = \frac{Q}{\varepsilon_0}$$

The area of the curved part of the cylinder is simply the circumference times the length. Thus,

$$E(2\pi rL) = \frac{Q}{\varepsilon_0}$$

$$E = \frac{Q}{\varepsilon_0 2\pi rL}$$

Now we have an expression for the magnitude of the electric field between the two conductors as a function of radial distance. Although this electric field is not constant, we can still integrate it over a path between the cylindrical conductors to find the potential difference. Since $\vec{E}$ and $d\vec{r}$ both point radially outward, their dot product becomes a simple multiplication. Therefore, the potential difference between the plates is

$$V_{r_2} - V_{r_1} = -\int_{r_1}^{r_2} \vec{E} \cdot d\vec{r} = -\int_{r_1}^{r_2} \frac{Q}{\varepsilon_0 2\pi rL}\, dr = -\frac{Q}{2\pi\varepsilon_0 L}\int_{r_1}^{r_2}\frac{1}{r}\, dr$$

Since $\int \frac{1}{r}\, dr = \ln(r)$,

$$V_{r_2} - V_{r_1} = -\frac{Q}{2\pi\varepsilon_0 L}\left[\ln(r_2) - \ln(r_1)\right]$$

$$= -\frac{Q}{2\pi\varepsilon_0 L}\ln\!\left(\frac{r_2}{r_1}\right)$$

To find the capacitance, we simply take the ratio of the magnitude of the charge to magnitude of the potential difference. Note that we ignore the signs of the charge and the potential difference:

$$C = \frac{Q}{V} = \frac{Q}{\dfrac{Q}{2\pi\varepsilon_0 L}\ln\!\left(\dfrac{r_2}{r_1}\right)} = \frac{2\pi\varepsilon_0 L}{\ln\!\left(\dfrac{r_2}{r_1}\right)}$$

(b) Substituting the given quantities yields

$$C = \frac{2\pi\varepsilon_0 L}{\ln\!\left(\dfrac{r_2}{r_1}\right)} = \frac{2\pi \times 8.85 \times 10^{-12}\,\text{C}^2\,\text{N}^{-1}\text{m}^{-2} \times 0.500\,\text{m}}{\ln\!\left(\dfrac{0.005\,00\,\text{m}}{0.004\,00\,\text{m}}\right)}$$

$$= 1.25 \times 10^{-10}\,\text{F} = 125\,\text{pF}$$

Making sense of the result:

Since the electric field in part (a) is directly proportional to the charge and inversely proportional to the length, this field really depends only on the linear charge density (the charge per unit length). It makes sense that the electric field decreases in magnitude as the radial distance increases (to conserve electric flux). In the expression for the potential difference, we found that the potential at the outer conductor is less than the potential at the inner conductor. This makes sense because the inner conductor carries the positive charge. To determine whether the capacitance in part (b) is reasonable, we could approximate the coaxial conductors as a parallel-plate capacitor of area $2\pi rL$ with spacing equal to $r_2 - r_1$:

(continued)

$$C \approx \frac{\varepsilon_0 A}{d} \approx \frac{\varepsilon_0 2\pi r L}{d}$$

$$\approx \frac{8.85 \times 10^{-12}\,\mathrm{C^2\,N^{-1}\,m^{-2}} \times 2\pi \times 0.0045\,\mathrm{m} \times 0.50\,\mathrm{m}}{0.001\,\mathrm{m}}$$

$$\approx 1.25 \times 10^{-10}\,\mathrm{F}$$

Most audio and communications signals are transmitted along coaxial cables with a structure similar to that in Example 21-2. Capacitance is an important consideration for such signal transmissions. Note that we found that the capacitance of coaxial configurations increases linearly with length. Figure 21-7 shows a common type of flexible coaxial cable.

Figure 21-7 The structure of coaxial cables that are frequently used for signal transmission

LO 4

21-4 Combining Capacitors

If you had only, say, 1 μF capacitors and wanted to make a 0.25 μF capacitance, could you do it? What if you needed a 5.0 μF capacitor? In fact, you can indeed combine capacitors in parallel, in series, and in series–parallel combinations to produce different equivalent capacitances. Combinations of capacitors are often used in circuits in which a required value is not available as a standard part.

Capacitors connected in parallel add in the same way that resistors add in series. When you add a capacitor in parallel to another capacitor, you *increase* the net capacitance.

KEY EQUATION
$$C_{\text{parallel}} = C_1 + C_2 \qquad (21\text{-}9)$$

Similarly, capacitors connected in series add like resistors added in parallel. When you insert a capacitor in series with another capacitor, you *decrease* the net capacitance.

KEY EQUATION
$$\frac{1}{C_{\text{series}}} = \frac{1}{C_1} + \frac{1}{C_2} \qquad (21\text{-}10)$$

Recall that components in series are connected such that the current must flow through one component followed by each of the others, and the voltages across the individual components sum to the total applied voltage. When components are connected in parallel, they all share the same voltage; one end of each individual component is connected to the same common terminal, and the other end of each parallel component is connected to a second common terminal.

Consider two capacitors, C_1 and C_2, connected in parallel to a battery, as shown on the left in Figure 21-8. On the right is an equivalent circuit with C_{eq} charged by the same battery. If the circuit really is equivalent, then the same total charge must be redistributed by the battery in both cases:

$$Q_1 + Q_2 = Q_{\text{eq}} \qquad (21\text{-}11)$$

Figure 21-8 A parallel combination of two capacitors charged by battery $\mathcal{E}$ (left) and the equivalent capacitance form of the circuit (right)

The charge on each capacitor is given by $Q = CV$. Since the capacitors are in parallel, the voltage across each capacitor is equal to the battery voltage $\mathcal{E}$, as is the voltage across the equivalent capacitor. Therefore,

$$C_1\mathcal{E} + C_2\mathcal{E} = C_{\text{eq}}\mathcal{E} \qquad (21\text{-}12)$$

We simply divide by $\mathcal{E}$ to get the relationship for capacitors in parallel. Another way to think about capacitors in parallel is this: you are connecting their plates together, thus increasing the total area of the plates. It makes sense that the capacitances will add together in the same way.

When capacitors are combined in series, as in Figure 21-9, all the capacitors have the same charge, and that charge must be the same as on the equivalent single capacitor. The voltages across the individual capacitors must sum to be the applied voltage. See the Peer to Peer box to see how this can be used to derive the relationship for combining capacitors in series.

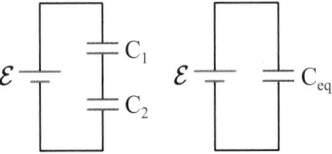

Figure 21-9 A series combination of two capacitors charged by a battery $\mathcal{E}$ (left) and the equivalent capacitance form of the circuit (right)

In this case, the parallel-plate approximation yields the correct result for the capacitor. In fact, when two surfaces are close together compared to their radii, you can often reasonably approximate them as a parallel-plate capacitor.

PEER TO PEER

I had a difficult time remembering that the rules for adding capacitors are opposite to the rules for adding resistors. Finally, the relationship between voltage and capacitance and the relationship between voltage and resistance helped me out. In series, the current is the same and voltages add. So, for resistors $IR_1 + IR_2 = IR_{eq}$; therefore, resistances in series add. But the voltage for a capacitor is given by Q/C, so when adding voltages for capacitors, you have $\dfrac{Q}{C_1} + \dfrac{Q}{C_2} = \dfrac{Q}{C_{eq}}$. Then, you just divide by the constant charge Q to get the formula for capacitors in series.

To understand why the charges on capacitors in series must be the same, consider what a capacitor is—two conducting plates separated by an insulating region. The energy of the battery will cause a flow of charges, charging the left plate of C_1 positively and the right plate of C_3 negatively, as shown in Figure 21-10. The insulating regions of each capacitor prevent the flow of current through the

capacitor, so overall, the regions marked A and B on the figure must remain electrically neutral. The right plate of capacitor C_1 will get some negative charge (electrons) attracted to the positive charges on the other plate of that capacitor. Since region A stays neutral overall, the left end of C_2 must develop an equal magnitude of positive charge. Similar arguments show that all the series capacitors have the same charge magnitudes.

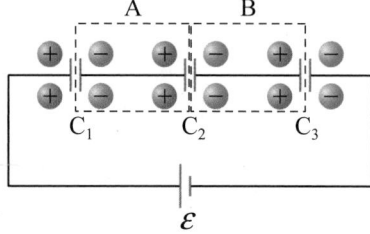

Figure 21-10   Charge on capacitors connected in series

EXAMPLE 21-3

Capacitor Combination

What is the net capacitance between terminals A and B of the circuit in Figure 21-11 when $C_1 = 12.0\ \mu F$, $C_2 = 4.00\ \mu F$, $C_3 = 3.00\ \mu F$, and $C_4 = 6.00\ \mu F$?

Figure 21-11   Example 21-3

SOLUTION

Capacitors C_3 and C_4 are in series because they are connected end to end with no branching to other components at the junction. (Whether they are drawn in a straight line does not matter.) We first find the equivalent capacitance for these two capacitors. For simplicity of notation, we leave out the units until the final step, noting that all the capacitances in this circuit are measured in microfarads:

$$\frac{1}{C_{series}} = \frac{1}{C_3} + \frac{1}{C_4} = \frac{1}{3.00} + \frac{1}{6.00} = \frac{3}{6.00}$$

$$C_{series} = 2.00\ \mu F$$

It is helpful when simplifying combinations of components to redraw the circuit diagram after each step. Figure 21-12 shows the circuit with C_3 and C_4 replaced by the series equivalent of $2.00\ \mu F$.

Figure 21-12   The circuit of Example 21-3 redrawn using the series equivalent for C_3 and C_4

From this revised circuit diagram, it is clear that the next step is to combine the $2.00\ \mu F$ and $4.00\ \mu F$, which are in parallel:

$$C_{parallel} = 2.00\ \mu F + 4.00\ \mu F = 6.00\ \mu F$$

Finally, this equivalent $6.00\ \mu F$ capacitor is in series with the $12\ \mu F$ capacitor, so the net equivalent capacitance is

$$\frac{1}{C_{eq}} = \frac{1}{12.0} + \frac{1}{6.00} = \frac{3}{12.0}$$

$$C_{eq} = 4.00\ \mu F$$

(continued)

Making sense of the result:

When finding equivalents for combinations of capacitors, you should check your results at each stage, remembering that parallel capacitors always add to a greater value than any of the individual capacitors, and series capacitors yield an equivalent that is less than any of the individual capacitors. The calculations satisfy these criteria.

 CHECKPOINT

C-21-4 Combining Capacitors

You have only 6 μF capacitors and need a 2 μF capacitor. How could you get the capacitance you need?

(a) Place three 6 μF capacitors in series with each other.
(b) Place three 6 μF capacitors in parallel with each other.
(c) Place two 6 μF capacitors in series with each other.
(d) It is not possible.

C-21-4 (a) With three 6 μF capacitors in series, $1/C_{eq} = 3/6 \, \mu$F, so the equivalent capacitance, C_{eq}, is 2 μF.

LO 5

21-5 Dielectrics and Capacitors

Chapter 19 introduced dielectrics, which are materials with atoms that align in the presence of an external electric field such that the electric field due to the polar molecules opposes the external field. Although there is no flow of electric charge, the distribution of the charge on these molecules (which are either polar to start with or become polarized in the applied electric field) is slightly changed.

 ONLINE ACTIVITY

Capacitors and Dielectrics

The e-resource that accompanies every new copy of this textbook contains an Online Activity using the PhET simulation "Capacitor Lab." Work through the simulation and accompanying questions to gain an understanding of capacitors and dielectrics.

Figure 21-13 shows a capacitor with a dielectric material inserted between its plates. The top plate is positively charged, so the electric field due to the charge on the plates (the solid arrow) points from top to bottom. The molecules of the dielectric (represented by ovals) orient themselves so that their negative end (shaded blue) is slightly closer to the positively charged top plate on the capacitor. Therefore, the electric field from the polarized dielectric molecules points upward (dashed arrow). The net electric field is the difference between these two electric fields. Thus, the presence of the dielectric reduces the net electric field produced between the plates of a capacitor by a given amount of charge on the plates.

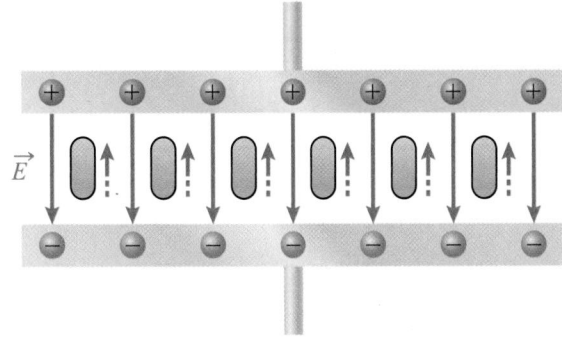

Figure 21-13 A capacitor with a dielectric material between the parallel plates. The internal electric field of the dielectric (dashed red arrows) opposes the field produced by the charges on the capacitor plates (solid red arrows).

If the net electric field is less but the spacing between the plates is unchanged, then the electric potential difference is less for the same amount of charge. Therefore, the capacitance must be greater. The presence of a dielectric between the plates always increases the capacitance. Inserting a dielectric between the plates of a capacitor increases the charge on the plates if they are hooked up to a battery (which holds the voltage constant) or reduces the potential difference if the plates have been disconnected (in which case the charge on the plates is unchanged).

To calculate capacitance in the presence of dielectrics, we simply replace the permittivity of free space, ε_0, with the permittivity of the dielectric material, ε. Permittivity is a measure of the response of a material to an applied electric field depending on the dielectric properties. For example, for a parallel-plate capacitor with a dielectric material between the plates:

$$C = \frac{\varepsilon A}{d} \qquad (21\text{-}13)$$

The dielectric constant (symbol κ) is the dimensionless ratio of the permittivity of the material to the permittivity of free space:

$$\kappa = \frac{\varepsilon}{\varepsilon_0} \qquad (21\text{-}14)$$

Good dielectrics have higher κ values. Table 21-1 lists the dielectric constants for a variety of materials, many of which are used in capacitors. In addition to the dielectric constant, it is important to consider how strong the field can be before electric breakdown occurs and conduction takes place through the dielectric, and this is given in the right-hand column of the table.

Table 21-1 Dielectric constants and maximum electric field strength before electrical breakdown for various materials

Dielectric material	Dielectric constant (κ)	Typical Breakdown Field Strength (MV/m)
Dry air	1.0006	3
Aluminum oxide	7–9	700
Ceramic	10–100	10
Glass	4–10	10
Mica	6.9	150
Mylar	3.1	400
Nylon	3.5	15
Paper	3–4	15
Plexiglas	3.4	40
Polyethylene	2.3	20
Rubber	2–7	10
Skin	30–40	
Tantalum oxide	20–30	500
Water	80	30 (pure water)
Waxed paper	4	50

 CHECKPOINT

C-21-5 A Dielectric in a Capacitor

An air-filled parallel-plate capacitor with capacitance C is charged using a battery of voltage $\mathcal{E}$, which is then disconnected. When a dielectric with dielectric constant 2 is inserted to totally fill the space between the plates,

(a) the capacitance is C, and the voltage is $2\mathcal{E}$;

(b) the capacitance is $2C$, and the voltage is $\mathcal{E}$;

(c) the capacitance is $2C$, and the voltage is $\mathcal{E}/2$;

(d) the capacitance is $2C$, and the voltage is $2\mathcal{E}$.

C-21-5 (c) The dielectric will increase the capacitance by a factor equal to the dielectric constant, so the capacitance becomes $2C$. Since the battery is disconnected, the charge must stay the same on the capacitor. Since $Q = CV$, Q is constant, and C is doubled, the new voltage must be $\mathcal{E}/2$.

21-6 Energy Storage in Capacitors

As mentioned at the beginning of this chapter, one of the main applications of capacitors is storing electrostatic energy. We now develop the key relationships for energy storage in capacitors. The electric potential, V, is defined

as the electrostatic potential energy per unit charge, so you might expect that the energy stored in a capacitor carrying a charge Q would be QV, where V is the voltage across the fully charged capacitor. However, the voltage across the capacitor is zero when charging starts and reaches V only as the last of the charge moves onto the plates.

Consider the electrostatic potential energy of a capacitor of capacitance C as it is charged with Q coulombs of charge. Let V represent the changing voltage on the capacitor. As a small element of charge dq is added, V increases, as does the electric potential energy, U. From the definition of electric potential, we have

$$\int dU = \int_0^Q V \, dq$$

We can express V in terms of the charge q at any given moment:

$$V = \frac{q}{C}$$

Therefore, the electrostatic energy, U, stored in the capacitor when it carries charge Q is given by

$$U = \int_0^U dU = \int_0^Q \frac{q}{C} \, dq = \frac{1}{C} \frac{Q^2}{2}$$

We can use the basic capacitor relationship $Q = CV$ to substitute for either of the variables on the right side of this equation.

Electrostatic potential energy of a charged capacitor

KEY EQUATION $\quad U = \frac{1}{2} \frac{Q^2}{C} = \frac{1}{2} CV^2 = \frac{1}{2} QV \qquad$ (21-15)

 CHECKPOINT

C-21-6 Potential Energy in a Capacitor

A capacitor with air between the plates is charged with a battery, which is then disconnected. If a dielectric (with a dielectric constant greater than that of air) is then slid into the space between the plates, the electrostatic potential energy stored in the capacitor

(a) increases;

(b) decreases;

(c) does not change;

(d) may increase or decrease depending on the sign of the dielectric constant.

C-21-6 (b) Dielectric constants are always positive and have values of 1 and above, so answer (d) does not make sense. When the dielectric is inserted, the capacitance increases. Because the capacitor had previously been disconnected from the battery, no charge can flow and, therefore, charge Q is unchanged. From the equation for the potential energy of a charged capacitor, this means that the energy must have decreased. How is this change of energy possible? The dielectric is pulled into the space; therefore, negative work is done by the person who is guiding the dielectric into the space, and the potential energy of the capacitor decreases. Note that the answer would be different if the capacitor had remained connected to the battery.

The National Ignition Facility

The National Ignition Facility, operated by Lawrence Livermore National Laboratory in the United States, seeks to produce energy by nuclear fusion. In nuclear fusion, light atoms combine into heavier atoms; for example, hydrogen atoms fuse to form helium. Such fusion is the source of energy for the Sun and other stars and could be the basis for power reactors. Fusion reactors are inherently safer than current nuclear fission reactors. However, a key challenge in producing controlled nuclear fusion is creating the incredible pressures and temperatures needed to initiate nuclear fusion. The National Ignition Facility creates these conditions by directing 192 powerful laser beams on a small pellet containing hydrogen. This facility has the world's most powerful laser array. High-energy density capacitors play a crucial role in this facility, storing and then rapidly releasing the energy needed for the powerful laser pulses (Figure 21-14).

Figure 21-14 The target area at the National Ignition Facility. The arm on the right positions the target to a precision of less than the width of a human hair.

EXAMPLE 21-4

Defibrillator

A defibrillator uses a 5000 V power supply to charge a 32.0 μF capacitor. How much energy is stored in this capacitor when fully charged?

SOLUTION

Since we know the voltage and the capacitance, we choose the form of Equation (21-15) that expresses the electrostatic potential energy in terms of these two quantities:

$$U = \frac{1}{2} CV^2 = \frac{1}{2}(32 \times 10^{-6}\,\mathrm{F})\,(5000\,\mathrm{V})^2 = 400\,\mathrm{J}$$

Making sense of the result:

The stored energy is significant—it is approximately equivalent to the energy needed to lift an 40 kg person 0.5 m. Since the energy is not required continuously, storing energy in a capacitor makes it possible to use a power supply that is smaller, lighter, and less expensive than a power supply that supplies energy directly to the defibrillator paddles.

We can think of the energy in a capacitor as being stored in the electric field in the space between the plates of a capacitor. Consider an air-filled parallel-plate capacitor with plate area A and spacing d. For this capacitor, the electric field is related to the potential difference by $V = Ed$. Therefore, the electrostatic potential energy can be written as

$$U = \frac{1}{2} CV^2 = \frac{1}{2} C(Ed)^2 \qquad (21\text{-}16)$$

As we saw earlier in this chapter, the capacitance of this type of capacitor is given by

$$C = \frac{\varepsilon_0 A}{d} \qquad (21\text{-}7)$$

Substituting for C in Equation (20-16) gives

$$U = \frac{1}{2}\left(\frac{\varepsilon_0 A}{d}\right)(Ed)^2 = \frac{1}{2}\,\varepsilon_0 A d E^2 \qquad (21\text{-}17)$$

The area of the plates times the spacing of the plates, Ad, equals the volume between the plates, which is the volume containing the electric field. So, Equation (21-17) shows that the electrostatic energy stored in the capacitor depends on the volume between the plates and the square of the electric field.

It is common, especially in more advanced physics courses, to work in terms of energy density, u, which is defined as energy per unit volume (see Section 26-5). Dividing Equation (21-17) by the volume gives us an expression for the energy density between the capacitor plates:

KEY EQUATION
$$u = \frac{1}{2}\,\varepsilon_0 E^2 \qquad (21\text{-}18)$$

This expression for energy density is generally true for electric fields in a vacuum and can be applied, for example, to the energy density of electromagnetic waves in a vacuum.

ELECTRICITY, MAGNETISM, AND OPTICS

21-7 Applications of Capacitors

Most capacitor applications involve charging or discharging. Let us be clear on what a battery does when it charges a capacitor: The battery provides the energy to move existing charges; it does not create or destroy electric charge. Once the capacitor starts to charge, the flow of more electrons onto the negatively charged plate is opposed by electrostatic repulsion. The battery provides the energy needed to overcome this repulsion and thus provides the electric potential energy stored in the capacitor when it is charged. When a capacitor is discharged by connecting the two plates with an external conductor, electrons flow from the negative plate to the positive plate, eventually resulting in zero net charge on each plate.

Storing Charge

Probably the most direct application of capacitors is the storage of charge. Dynamic random access memory (DRAM), a common type of electronic memory, has arrays of microscopic capacitors, with a high bit or low bit indicated by whether the stored charge is above or below some critical value. In the charge coupled device (CCD) image sensor, tiny capacitors each store charge corresponding to the brightness of a pixel of the image (Figure 21-15).

Figure 21-15 The CCD image sensor is an array of photosensitive capacitive cells.

Storing Energy

A charged capacitor can be used to store electrostatic potential energy, energy that we can get back by allowing partial or full discharge of the capacitor. Capacitors store energy for many electrical devices, including the defibrillator mentioned at the beginning of the chapter. Similarly, camera flash units store the energy needed for a bright flash

in capacitors. A wide variety of "instant on" electronic devices store enough energy to start up in a capacitor even when they are switched off.

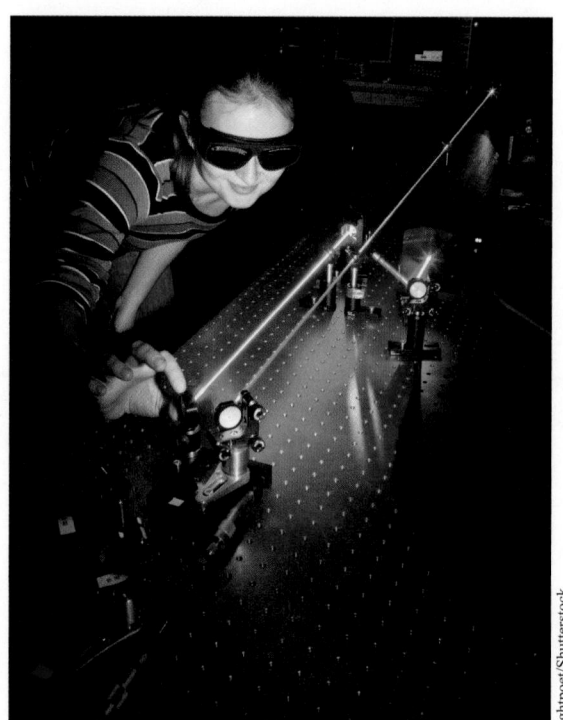
Filtering

In scientific instrumentation, as well as in everyday applications such as audio amplifiers, one needs to filter signals. A filter circuit allows certain frequencies through while blocking others. For example, when you use an audio equalizer or even simple bass and treble controls, you are adjusting the settings of various filters.

Most filters use capacitors, often in combination with resistors. Resistors oppose the flow of current in a circuit: the greater the resistance (symbol R), the less the current that flows for a given applied voltage, as you will see in the discussion of current and resistance in Chapter 22. A low-pass filter lets only low frequencies through. The simplest low-pass filter consists of just a single resistor and a single capacitor, as shown in Figure 21-17. When the frequency of the signal is low, most of the input signal is passed on to the output. When the frequency is high, however, the capacitor cannot rapidly change its voltage as needed for the output voltage to follow the input voltage, and the signal is largely blocked. A cut-off frequency separates the signal frequencies that are largely passed from those that are largely blocked. For the simple RC low-pass filter in Figure 21-17, the cut-off frequency is

$$f_{\text{cutoff}} = \frac{1}{2\pi RC} \qquad (21\text{-}19)$$

Note that the product RC has units of time.

Figure 21-17   The RC low-pass filter

A high-pass filter passes high frequencies and blocks low frequencies. Switching the positions of the capacitor and the resistor in an RC low-pass filter turns it into a high-pass filter. Since the capacitor completely blocks direct current (zero frequency) signals, the majority of electronic devices use capacitors in this high-pass configuration to separate regions of the circuit that have different direct current (DC) voltages.

DC Power Supplies

For efficiency and ease of conversion to different voltages, most power systems transmit electricity as alternating current (AC) with a sinusoidal waveform. In North America, most household circuits have a voltage of 120 V. However, most electronic devices require a lower DC voltage. The power supplies for these devices convert household AC voltage into lower-voltage DC. The essential parts of a power supply are a transformer, which steps down the AC voltage; diodes, which allow the current to flow in only one direction; and a capacitor, which alternately stores and releases charge so that the DC output is approximately constant. Most DC power supplies use electrolytic capacitors, a type of capacitor that allows relatively large capacitances to fit in compact packages (Figure 21-18). Most high-quality power supplies also have a regulator section to help maintain constant voltage under varying conditions.

Uniform Electric Fields

Since the electric field inside a parallel-plate capacitor is essentially constant, this configuration is often used in applications that require uniform electric fields. Such

Figure 21-18   Electrolytic capacitors of the type used in a DC power supply

fields are used to select, direct, and accelerate electrons and other subatomic particles, such as in particle accelerators at major scientific facilities, cathode ray tube (CRT) display devices, X-ray machines, and mass spectrometers.

✓ CHECKPOINT

C-21-7 Particle Physics Experiment

A capacitor in a particle physics experiment provides a uniform electric field of 14 000 V/m directed upward. The plate separation is 1.0 cm. When the bottom plate is at −100 V relative to the ground, what is the potential of the top plate?

(a)   −240 V
(b)   −140 V
(c)   +140 V
(d)   +800 V

C-21-7 (a) The potential difference, ΔV, equals Ed, where d is the plate spacing. Substituting the given values, $d = 0.01$ m and $E = 14\ 000$ V/m, we get $\Delta V = 140$ V. Since the electric field points upward, the bottom plate must be more positive. Therefore, the potential at the top plate must be -100 V $- 140$ V $= -240$ V.

Timing Circuits

Many general-purpose timing circuits use the time required to charge, or discharge, a capacitor to a certain level to control the timing. In these devices, a voltage comparison circuit detects when the voltage across a capacitor reaches the threshold value. Although high-precision timing requires other approaches (such as the use of piezoelectric crystals in resonant circuits), for many technological and scientific applications capacitor timers are sufficient and cost-effective.

Transducers

A transducer is a device that provides an electrical signal that depends on some physical parameter. For example, a microphone is a transducer that transforms sound pressure waves into electrical signals. Many transducers, including

Touch-Sensitive Screens

Touch-sensitive screens on devices such as smart phones and tablets have revolutionized portable electronics. Although several different mechanisms are used for touch screens, including thermal, light, and resistive sensing, the majority of modern touch screens use capacitive effects. Transparent conducting grids lie above the display and the glass layer. In one design, lines for an electrically driven grid are separated by an insulator from sensing lines, forming, in essence, an array of capacitors. When you touch the screen, essentially grounding that point, you alter the charge stored at that point, and the electronics detect which capacitor has been grounded (Figure 21-19).

© Chris Rout/Alamy

Figure 21-19 The iPad and the iPhone use touch screens that depend on capacitive effects.

some types of microphones, pressure sensors, and fluid-level sensors, make use of capacitive effects. Recall that the capacitance depends on the area of the plates, the separation of the plates, and the dielectric properties of any material between the plates:

$$C = \frac{\varepsilon A}{d} \qquad (21\text{-}13)$$

A capacitive-based transducer may work by changing any of these three parameters. For example, in most capacitive pressure sensors, the distance between two plates changes according to the pressure on the transducer. In some position sensors, the effective area of the plates changes as one plate slides relative to the other. The capacitance of a fluid-level sensor can change as the level of a dielectric fluid between the plates of the capacitor changes.

Transistor Capacitors

While passive elements such as resistors and capacitors are fabricated as part of integrated circuits, it is the active semiconductor devices—diodes and transistors—that play the central role. The most important type of transistor for integrated circuits is the metal oxide semiconductor field effect transistor (MOSFET), which is essentially a capacitor with an insulating layer separating two charge regions. Physicist Julius Edgar Lilienfeld (1882–1963) sought the first ever patent for a field effect transistor (FET) in 1925; it was awarded by the Canadian government in 1927. When compared to the other major type of transistor (BJT, or bipolar junction transistor), FET and especially MOSFET devices have significantly lower power requirements. However, it would be many decades before mass production of MOSFETs became practical. Lilienfeld also invented the electrolytic capacitor.

Biological Measurements

Natural systems show capacitive effects, which are widely used in biological research and development. For example, the first accurate measurement of the thickness of the cell wall was done using a capacitive measurement. Samples of different types of cells have capacitances that vary according to the DNA content.

A common type of capacitor consists of two parallel conducting plates separated by a small gap, which may be filled with air or a dielectric. Capacitors are used to store charge and energy, and have many applications, for example, timing and measurement circuits.

Capacitance is a measure of how much charge can be stored for a given applied potential difference and is measured in farads (F).

The Capacitor

The relationship between the stored charge, Q, the applied electric potential difference, V, and the capacitance, C, is

$$Q = CV \qquad (21\text{-}2)$$

Electric Fields

Except for small differences near the ends of the capacitor, the electric field is essentially constant in the area between the plates. The relationship between the potential difference across the plates, V, the plate spacing, d, and the magnitude of the electric field magnitude, E, is

$$V = Ed \qquad (21\text{-}1)$$

Parallel-Plate Capacitors

The capacitance for parallel-plate capacitors depends on the plate area, A, plate spacing, d, and the permittivity of the region between the plates, ε:

$$C = \frac{\varepsilon A}{d} \qquad (21\text{-}13)$$

Calculating Capacitance

The term "capacitance" can be applied to configurations other than parallel plates. For a general configuration, capacitance can be calculated by finding the ratio of charge to potential difference between the conductors. Gauss's law is used to find the electric field between the conductors,

which is then integrated over a path between the conductors to find the potential difference.

Combining Capacitors

When capacitors are connected in parallel, the capacitances add:

$$C_{\text{parallel}} = C_1 + C_2 \qquad (21\text{-}9)$$

When capacitors are connected in series, each of the capacitors has the same charge and

$$\frac{1}{C_{\text{series}}} = \frac{1}{C_1} + \frac{1}{C_2} \qquad (21\text{-}10)$$

Dielectrics and Capacitors

Adding a dielectric between the plates of a capacitor increases its capacitance, and

$$C = \frac{\varepsilon A}{d} \qquad (21\text{-}13)$$

where ε is the permittivity of the dielectric.

The dielectric constant, κ, is a dimensionless ratio of the permittivity of the dielectric to the permittivity of a vacuum:

$$\kappa = \frac{\varepsilon}{\varepsilon_0} \qquad (21\text{-}14)$$

Energy Storage

The energy stored in a capacitor (capacitance C) carrying a charge Q with a voltage V is

$$U = \frac{1}{2}\frac{Q^2}{C} = \frac{1}{2}CV^2 = \frac{1}{2}QV \qquad (21\text{-}15)$$

Energy is stored in the electric field between the plates. The energy density in this field is

$$u = \frac{1}{2}\varepsilon_0 E^2 \qquad (21\text{-}18)$$

Applications: capacitive measurements of biological cells, energy storage in capacitors (e.g., camera flash units and defibrillators), capacitance-based measurements (e.g., fluid levels), touch screens, charge coupled devices (CCDs) for images, timing, filtering of signals

Key Terms: capacitance, capacitor, charging, cut-off frequency, dielectrics, dielectric constant, discharging, energy density, farad, filter, high-pass filter, low-pass filter, permittivity

QUESTIONS

1. A capacitor is connected to a 40 V battery. It is observed that the charge on one of the plates is 160 μC. What is the capacitance?
   (a) 0.25 μF
   (b) 4.0 μF
   (c) 640 μF
   (d) 4.0 F

2. A capacitor has capacitance C and applied voltage V. What is the net total charge on the capacitor?
   (a) 0
   (b) CV
   (c) C/V
   (d) $2CV$

3. The spacing between the plates of a 10 μF capacitor is 1.0 mm. It is connected to a 10 V battery for charging. What is the electric field in the capacitor?
   (a) 10 V/m
   (b) 100 V/m
   (c) 1 000 V/m
   (d) 10 000 V/m

4. Two parallel plates have been charged with equal magnitude and opposite polarity charges. The plates are then disconnected from the charging source, and the spacing between the plates is decreased. Which of the following statements correctly describes what happens?
   (a) The capacitance and the electric potential both decrease.
   (b) The electric potential increases, and the capacitance decreases.
   (c) The capacitance increases, and the potential difference decreases.
   (d) The capacitance and the electric potential both increase.

5. Consider two cylindrical capacitors: A has radii 2.5 cm and 2.6 cm and length 2.0 m; B has radii 1.0 cm and 3.0 cm and length 1.0 m. Rather than using the exact relationship for cylindrical capacitors, you want to approximate them as parallel-plate capacitors. Which of the following statements is best?
   (a) You cannot approximate them at all as a parallel-plate configuration.
   (b) You can approximate A very well, but B will probably be a poor approximation.
   (c) You can approximate B very well, but A will probably be a poor approximation.

6. A coaxial cable is made up of two concentric cylinders. You have two cables, A and B, with the same distance between the cylinders (i.e., from the outside of the inner cylinder to the outer cylinder, which is considered to be thin). Both cables also use the same dielectric material. However, the diameter of cable B is larger than that of cable A. How does the capacitance per length compare with the two cables?
   (a) They both have exactly the same capacitance per length.
   (b) A has a larger capacitance per length.
   (c) B has a larger capacitance per length.
   (d) A has slightly more capacitance per length.

7. You need a slightly smaller capacitance than the value of a capacitor that you are using. How should you add a capacitor to accomplish this?
   (a) Add a much smaller capacitor in series.
   (b) Add a much smaller capacitor in parallel.
   (c) Add a much larger capacitor in series.
   (d) Add a much larger capacitor in parallel.

8. You have a single 1 μF capacitor in a circuit. When you connect a 10 μF capacitor in parallel with the 1 μF capacitor, what is the new effective capacitance?
   (a) 0.1 μF.
   (b) 0.9 μF.
   (c) 1.1 μF.
   (d) 11 μF.

9. An air-filled parallel-plate capacitor initially has some capacitance, C. It is charged using a battery of voltage X, which is left connected. A dielectric with dielectric constant 2 is then inserted, totally filling the space between the plates. Which of the following statements describes the new condition?

(a) The capacitance is C, and the voltage is $2X$.
(b) The capacitance is $2C$, and the voltage is X.
(c) The capacitance is $2C$, and the voltage is $X/2$.
(d) The capacitance is $2C$, and the voltage is $2X$.

10. An air-filled parallel-plate capacitor is charged with a battery, which is kept connected. A dielectric (with a dielectric constant greater than that of air) is then inserted into the space between the plates. What change, if any, is there for the electrostatic potential energy stored in the capacitor?
    (a) The electrostatic potential energy increases.
    (b) The electrostatic potential energy decreases.
    (c) The electrostatic potential energy does not change.
    (d) It depends on whether the dielectric constant is positive or negative.

11. When the voltage applied to the plates of a capacitor is doubled, what happens to the energy stored in the device?
    (a) The energy is one-quarter the previous value.
    (b) The energy is one-half the previous value.
    (c) The energy is double the previous value.
    (d) The energy is four times the previous value.

12. Two capacitors have identical dielectrics and plate spacing, and they are connected to the same battery. However, capacitor A has a plate area that is four times the plate area of capacitor B. How do the electric potential energy values stored in the capacitors compare?
    (a) The electric potential energy of A is 16 times the electric potential energy of B.
    (b) The electric potential energy of A is 4 times the electric potential energy of B.
    (c) The electric potential energy of A is 2 times the electric potential energy of B.
    (d) The electric potential energy of B is 2 times the electric potential energy of A.

13. The power supply you have constructed has too much "ripple," which means that the voltage goes down too much from a constant value during the 1/60 s period until it is refreshed from the transformer–rectifier network. Which statement correctly describes how the situation might be improved?
    (a) You could add a capacitor in series with the existing capacitor.
    (b) You could add a capacitor in parallel with the existing capacitor.
    (c) Changing the capacitor value will not help the situation.
    (d) Replacing the capacitor with a large resistor will help most.

14. In a particle physics experiment, you use the area between capacitor plates to create an electric field for exerting a force on electrons. You find that the force is too strong. What could you do to help the situation?
    (a) Increase the plate area and leave everything else the same.
    (b) Decrease the plate spacing and leave the applied voltage constant.
    (c) Decrease the applied voltage and leave the plate spacing the same.
    (d) Add a second capacitor in parallel, and keep the applied voltage the same.

15. You are using a simple low-pass filter, which consists of a resistor and a capacitor. The cut-off frequency is 500 Hz. However, you need the filter to cut off at about 250 Hz. If you keep the capacitor unchanged, how should you change the resistor value?

(a) Make the resistor value $R/2$.
(b) Make the resistor value $2R$.
(c) Make the resistor value $4R$.
(d) It is impossible to do by changing the resistor. You need to change the capacitor value.

16. You have purchased an old car and find that the turn signals flash too slowly. How can you increase the rate (i.e., the number of flashes per second) of the flasher circuit?
(a) You want to increase the capacitance and can do this by adding another capacitor in parallel with the existing one.
(b) You want to increase the capacitance and can do this by adding another capacitor in series with the existing capacitor.
(c) You want to decrease the capacitance and can do this by adding another capacitor in parallel with the existing capacitor.
(d) You want to decrease the capacitance and can do this by adding another capacitor in series with the existing capacitor.

PROBLEMS BY SECTION

For problems, star ratings will be used, (✽, ✽✽, or ✽✽✽), with more stars meaning more challenging problems.

Section 21-1   Capacitors and Capacitance

17. ✽ You need to store a charge of 6.00 mC when a 12.0 V battery is applied to a capacitor. What must be the capacitance?

18. ✽✽ A capacitor has a charge of 30 μC when a voltage X is applied to it. When the voltage is increased to $X + 10$ V, the new charge on the capacitor is 40.0 μC.
(a) What is the voltage X?
(b) What is the capacitance?

Section 21-2   Electric Fields in Capacitors

19. ✽✽ A capacitor has plates separated by a 2.00 mm air gap. A charge of magnitude 56.0 μC is on each plate. The electric field between the plates is 2400. V/m.
(a) What must be the electric potential difference between the plates?
(b) What is the capacitance?

20. ✽✽ An air-filled parallel-plate capacitor has a plate spacing of 1.50 mm and an electric potential difference of 65.0 V.
(a) What is the electric field between the plates?
(b) Calculate the surface charge density on each plate.
(c) The effective dimensions of the plate are 4.00 cm $\times$ 5.00 cm. What is the total charge on each plate?

Section 21-3   Calculating Capacitance

21. ✽✽ An air-filled parallel-plate capacitor consists of two circular plates of radius 7.50 cm. To achieve a capacitance of 115 pF, what must be the separation of the plates?

22. ✽✽✽ Consider a spherical capacitor that is made up of two spheres of radii 8.00 cm and 9.00 cm. The inner sphere is negatively charged with 450.0 nC, and the outer sphere is positively charged to the same magnitude. Air fills the region between the conductors.
(a) Use Gauss's law to find an expression (as a function of radial distance r) for the electric field in the region between the two spheres. Include the direction of the field.

(b) What is the electric field in the region outside both spheres?
(c) What is the potential difference between the two spheres?
(d) Calculate the capacitance.
(e) Estimate the capacitance by approximating the configuration as a set of parallel plates.

Section 21-4   Combining Capacitors

23. ✽✽ Explain how you could combine an unlimited supply of 500 pF capacitors in order to obtain each of the following:
(a) 0.002 μF;
(b) 0.0001 μF;
(c) 0.001 25 μF.

24. ✽✽ If you had only four capacitors, of values 1.00 μF, 2.00 μF, 3.00 μF, and 4.00 μF, what would be the smallest and the largest capacitances that you could obtain by combinations of these capacitors? In each case, draw the circuit and show the calculation for the equivalent capacitance.

Section 21-5   Dielectrics and Capacitors

25. ✽✽ A parallel-plate capacitor has rectangular plates 6.50 cm $\times$ 8.50 cm, spaced 2.00 mm apart. It has a dielectric material totally filling the space between the plates. The capacitor has a capacitance of 169 pF. Which of the materials in Table 21-1 could be the dielectric?

26. ✽✽ A parallel-plate capacitor has a capacitance of 0.002 00 μF when the space between the plates is filled with air. The plate spacing is 0.200 mm. It is connected throughout to an ideal 24.0 V battery.
(a) What is the charge on each plate?
(b) What is the electric field strength in the space between the plates?
(c) A dielectric with $\kappa = 3.50$ is inserted, totally filling the space between the plates. What is the charge on each plate now?
(d) What is the net electric field between the plates now?

Section 21-6   Energy Storage in Capacitors

27. ✽✽ You have two identical 5.00 μF capacitors and a single 12.0 V ideal battery. What is the total electrostatic energy stored in the system under each of the conditions below?
(a) The two capacitors are placed in series, and that combination is charged with the 12.0 V battery.
(b) The two capacitors are placed in parallel, and that combination is charged with the 12.0 V battery.

28. ✽✽ An air-filled capacitor charged with an ideal battery of voltage V has a total stored electrostatic energy U.
(a) The space between the plates is now filled with a dielectric of dielectric constant κ. The battery is left connected throughout. What is the amount of stored electrostatic energy now?
(b) Answer part (a) for the case in which the battery has been disconnected prior to insertion of the dielectric material.

Section 21-7   Applications of Capacitors

29. ✽✽ A pulsed laser that uses optical flash pumping produces 25.0 mJ per output pulse. Assume that the system is only 20% efficient, so five times as much energy must be stored in a capacitor to fire the flash each time. The capacitance of the capacitor is 50.0 μF. What is the minimum applied voltage needed to charge this capacitor?

30. ✲✲✲ A simple RC high-pass filter is to have a cut-off at 200 Hz.
  (a) Sketch the circuit, with input and output marked.
  (b) Calculate the resistance needed if only a 1.00 μF capacitor is available.

COMPREHENSIVE PROBLEMS

31. ✲ Prove that the permittivity of free space, ε_0, can be expressed in units of F/m.

32. ✲✲ Capacitors can be made on semiconductor integrated circuits essentially as parallel-plate capacitors, although it is difficult to achieve high capacitance values (and also challenging to create precise values). Assume that you want a 13.0 pF capacitor and that the layers of the integrated circuit allow you to have a spacing of 1.50 μm. Assume that the material has a dielectric constant of 3.10. What is the necessary area of the plates on this integrated circuit capacitor?

33. ✲✲ A single cell in a computer memory chip has a capacitance of 90.0 fF (femtofarads) and operates at a voltage of 3.3 V when charged.
  (a) What is the charge on the cell?
  (b) Express this charge in terms of the number of electrons.

34. ✲✲ An air-filled capacitor is to use circular plates of radius 5.00 cm that are separated by 3.00 mm. It is to have a maximum electric field of 1.0 kV/m in the space between the plates. Calculate
  (a) the capacitance;
  (b) the maximum voltage that should be applied;
  (c) the maximum charge that can be stored;
  (d) the maximum energy that can be stored.

35. ✲✲ A capacitor consists of two thin coaxial cylinders. The inner radius is 1.00 cm, the outer radius is 2.50 cm, and the length of both is 2.00 m. They carry equal amounts of charge per unit length but have opposite polarities. What is the capacitance per metre for the system?

36. ✲✲ While it may at first seem strange, you can consider a single charged sphere as having a capacitance.
  (a) Derive an expression for the capacitance of two concentric spheres (assume both are thin) of radii r_1 and r_2. Assume that a vacuum fills the space between the spheres.
  (b) Now, find the limit as we let the second sphere be of infinite radius. This will give you a result for the capacitance of a single sphere.
  (c) What would need to be the radius for a sphere in order for its capacitance to be 100 pF?

37. ✲✲ A parallel-plate capacitor has plates of area A and separation d. Two different dielectrics (κ_1, κ_2) are inserted (Figure 21-20), each filling half the space. Derive an expression for the capacitance. (Hint: Treat the setup as two capacitors in parallel.)

Figure 21-20　Problem 37

38. ✲✲✲ One form of transmission line consists of two long conducting wires (of radius r) separated by a separation distance D. Assume that D is much larger than r and that one wire has a linear charge density of $+\mu$ (measured in coulombs per metre) and the other wire has a linear charge density of $-\mu$.
  (a) Calculate the capacitance per length of the transmission line. In the early days of television the signal from the antenna to the television was carried on this type of transmission line.
  (b) Find a numerical value for the capacitance per unit length for $D = 7.50$ mm and $r = 0.320$ mm.

39. ✲✲✲ Five 2.00 μF capacitors are connected in series, and that combination is charged with a 9.00 V ideal battery.
  (a) What is the charge on one plate of each capacitor?
  (b) What is the voltage across each capacitor?
  (c) How much energy is stored in one capacitor?

40. ✲✲✲ What is the effective capacitance between points A and B in the circuit shown in Figure 21-21?

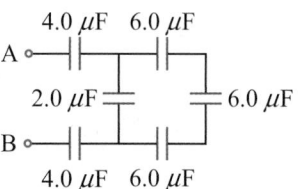

Figure 21-21　Problem 40

41. ✲✲✲ An air-filled parallel-plate capacitor is to be used to create an approximately constant electric field that will accelerate electrons with an acceleration of 7.00×10^{15} m/s^2. Assume that the spacing between the plates is 4.00 mm.
  (a) What is the required electric field strength?
  (b) What must be the applied voltage?
  (c) Will the acceleration of the electrons be in the direction of the higher or lower electric potential plate?
  (d) What time would be required for the electrons to accelerate from rest to a speed of 2.50×10^7 m/s, assuming that there is sufficient space between plates to reach this speed?
  (e) What distance would the electrons travel in reaching this speed? Is it possible to reach this speed with the given plate spacing?

42. ✲✲ A 0.100 μF capacitor is charged to 15.0 V. It is then connected in parallel to a second 0.100 μF capacitor, which was originally uncharged. Calculate the charge and voltage on each capacitor afterward.

43. ✲✲ Consider the circuit shown in Figure 21-22. The voltage is shown at a variety of points in the circuit. You are given the capacitance of the leftmost capacitor. Find the value of each of the other capacitors.

Figure 21-22　Problem 43

44. ✱✱ A 2.00 μF capacitor is charged to 12.0 V. It is then disconnected from the battery and connected across an uncharged 4.00 μF capacitor. Calculate the resulting charge and voltage on each capacitor.

45. ✱✱✱ A 4.00 cm by 5.00 cm parallel plate capacitor uses a dielectric with a dielectric constant of 4.0. The capacitor is designed so that when a potential difference of 60.0 V is applied, the charge stored on each plate has a magnitude of 150. nC. Calculate
(a) the necessary capacitance;
(b) the spacing between the plates;
(c) the energy that will be stored in the capacitor under these conditions;
(d) the electric field due to charge on the plates (not the net field considering the dielectric).

46. ✱✱ An air-filled parallel-plate capacitor has a plate area of 0.100 m², a plate spacing of 2.50 mm, and an applied voltage of 120. V. What is the energy density in the space between the plates of the capacitor?

47. ✱✱ What is the energy density just outside the surface of a conducting sphere of radius 3.50 cm that contains 3.50 nC of charge?

48. ✱✱✱ A charge coupled device (CCD) chip has an active area of 12.0 mm × 9.00 mm and a resolution of 9.0 megapixels; the pixels are square. Note that here we are using the binary definition of megapixel: 1 megapixel = 2^{20} pixels = 1 048 576 pixels.
(a) What are the dimensions of the device, expressed in pixels (i.e., x pixels × y pixels)? Round to the nearest integer number of pixels.
(b) What are the dimensions of each pixel, in micrometres?
(c) Using your answer in (b) as the plate area, and assuming a dielectric constant of 3.1 and a plate spacing of 1.00μm, what is the capacitance of each pixel cell? (Note: This is a very simplified model of a CCD.)

49. ✱✱ The outer layers of biological cells (lipid bilayers) act as capacitors, with the two sides holding opposite charge (the inside of the cell is negatively charged). Cells come in a variety of sizes, but assume a spherical cell of radius 5.00 μm and a dielectric constant of 4.00. Assume an electric potential difference between the layers of 75.0 mV.
(a) The thickness of the bilayer is 20.4 nm. What is the capacitance of the cell? (Hint: Model it as two thin spherical shells.)
(b) What is the charge on each side?
(c) How many elementary charges does this represent?

50. ✱✱ You are camping in a remote location and would like to have power at night. However, you do not want to carry a heavy battery to store the energy from the lightweight solar collectors that you are planning to carry in. The solar array that you have can generate 5.50 W in full sunlight. The array produces a potential difference of 12.0 V.
(a) How much capacitance is needed to store the energy produced by solar array in 6.0 h of full sunlight?
(b) The largest electrolytic capacitors you have available are 1000 μF each. How many would be needed in parallel?
(c) How long could the stored energy power a netbook that uses 8.00 W?

51. ✱✱ Current high-energy capacitor designs (e.g., like those used in the National Ignition Facility) achieve energies of about 2.00 J for every cubic centimetre of the capacitor. If one of these capacitors were in the shape of a cylinder of radius 5.00 cm and length 10.0 cm, and had a maximum voltage rating of 750 V, what would its capacitance be?

DATA-RICH PROBLEM

52. ✱✱✱ Table 21-2 shows a relationship between electric potential (in V) versus radial distance, r (in cm).
(a) Describe a situation that could correspond to these data.
(b) What is the electric field at $r = 6$ cm?
(c) What is the electric field at $r = 12$ cm?
(d) What is the electric field just outside $r = 10$ cm?
(e) If this is some sort of capacitor, what is the voltage between the two plates?
(f) What charge must be on the inner plate to be consistent with the electric field in part (d)?
(g) Use your data from parts (e) and (f) to estimate the capacitance.

Table 21-2 Data for Problem 52

Radial distance (cm)	Voltage (V)
0	27 000
2	27 000
4	27 000
6	27 000
8	27 000
10	27 000
12	22 500
14	19 300
16	16 900
18	15 000
20	13 500
22	13 500
24	13 500
26	13 500
28	13 500

OPEN-ENDED PROBLEM

53. ✱✱✱ In this question, you will make reasonable estimates to determine whether it is feasible to power an electric vehicle with charged capacitors instead of batteries.
(a) Estimate (or look up specifications for) the energy required for an electric vehicle to drive 50 km during one day. List any assumptions that you make for your estimate.
(b) Design an optimally efficient (in terms of energy storage per volume) capacitor. Include the design (e.g., shape and size), materials (e.g., dielectric used), and estimates of the working electric potential and energy storage per capacitor unit. Provide comments on all aspects of your design.

(c) To provide the energy estimated in part (a), what volume and mass of capacitors are required, considering your capacitor design in part (b)? Comment on how reasonable these requirements are.

(d) Current technology uses lithium ion batteries for electric automobile energy storage. Look up, or estimate, the volume and mass requirements using that technology to provide the same amount of energy.

(e) Comment on the general advantages and disadvantages (with an electric automobile context) of capacitor-based energy storage compared to lithium ion batteries or fuel cells.

 See the text online resources at www.physics1e.nelson.com for Open Problems and Data-Rich Problems related to this chapter.

Chapter 22
Electric Current and Fundamentals of DC Circuits

In 1791, an Italian physician and physicist, Luigi Galvani (1737–98), discovered bioelectricity. Shortly thereafter, his associate, Alessandro Volta (1745–1827), invented the first battery. Less than 250 years later, we rely heavily on ubiquitous access to electric power. Terms such as "electricity," "electric current," "electric resistance," "electric power," "energy generation," and "transmission" have entered our everyday vocabulary. Sometimes, we use them interchangeably even though they mean different things. Electric phenomena are the basis of our lives on both microscopic (e.g., physiological phenomena and electron motion) and macroscopic levels (e.g., our use of external electric appliances). To appreciate the importance of electric phenomena in our lives, we only need to recall the havoc in northeastern Canada and the United States on the hot summer day of August 14, 2003. On that day, more than 55 million people in Ontario and in the northeast of the United States were left without electric power due to the failure of a transmission grid as a result of a 3500 MW (3.5×10^9 W) power surge (Figure 22-1). In this chapter, we examine the nature of electric current and its applications in direct current (DC) circuits.

Learning Objectives

When you have completed this chapter you should be able to:

1 Derive and use a microscopic model of electric current to describe its flow through a conductor.

2 Describe the conductivity and resistivity of a metal quantitatively and qualitatively using a microscopic model of electric current.

3 Describe the relationship between electric current, voltage, and resistance in metals using a macroscopic model of electric current: Ohm's law.

4 Analyze simple and compound DC circuits, and use Kirchhoff's laws to predict the circuit power output, current flow, and potential difference across different circuit components.

5 Analyze series RC circuits.

© Reuters/CORBIS

Figure 22-1 The "Northeast Blackout" of 2003 affected more than 55 million people in North America and was the second most widespread blackout in history. A view of the 2003 blackout from space: Ontario and the northeastern United States before (left) and during (right) the blackout.

For additional Making Connections, Examples, and Checkpoints, as well as activities and experiments to help increase your understanding of the chapter's concepts, please go to the text's online resources at www.physics1e.nelson.com.

22-1 Electric Current: The Microscopic Model

In Chapter 20, we saw that electric charges can move through a metal conductor to charge or discharge a capacitor. Various experiments, such as studies of the Hall effect (described in Chapter 23) and the Tolman–Stewart experiment conducted in 1916, have shown that the moving charges in metals are electrons. Thus, we can describe the electric current in metals as an **electron current** or an **electron flow**. In some materials, such as plasmas and ionic solutions, the charge carriers may include positive and negative ions.

We model a metal as a crystal lattice of positively charged ions immersed in a cloud, or sea, of free (delocalized) electrons. In most metals, each atom loses one outer (valence) electron, thus turning the atom into a positive ion (Figure 22-2). The solid metal is held together by the electrostatic attraction forces between the positive lattice and a negative free electron cloud. These forces are called **metallic bonds**. The presence of free electrons in a metal explains why a metal is a good conductor of both thermal energy and electricity. Metal ions are comprised of the metal nuclei and bound electrons. The nuclei of metal atoms consist of protons and neutrons that are approximately 2000 times heavier than electrons ($m_p \approx m_n \approx 2000\, m_e$), so the thermal motion of the metal ions is negligible compared to the thermal motion of the much lighter free electrons. (We will calculate the average speed for thermal motion of electrons in Chapter 32.) Therefore, we consider the heavy lattice ions almost static compared to the fast but randomly moving free electrons. The **free** (or **conduction**) **electron density** (n_e) of a material is defined as the number of free electrons per cubic metre of the material. Table 22-1 lists free electron densities for some familiar metals.

Table 22-1 Conduction electron density in metals

Metal	Conduction electron density, n_e (m^{-3})
Aluminum	6.0×10^{28}
Copper	8.4×10^{28}
Gold	5.9×10^{28}
Iron	8.4×10^{28}
Nickel	9.0×10^{28}
Silver	5.8×10^{28}
Zinc	6.0×10^{28}

EXAMPLE 22-1

Estimating Free Electron Density in Aluminum

Estimate the free electron density in aluminum, given that the density of aluminum is $\rho = 2.7 \times 10^3\, \dfrac{kg}{m^3}$.

SOLUTION

Assuming that every atom of aluminum contributed one free electron to the free electron cloud (Al: $1s^2, 2s^2, 2p^2, 3s^2, 3p^1$), the free electron density is equal to the number of aluminum atoms per cubic metre. That number can be calculated from the molar mass (M_{Al}) and density (ρ_{Al}) of aluminum, and the number of atoms in one mole of aluminum (Avogadro's number N_A):

$$n_e = \frac{\rho_{Al}}{M_{Al}} \cdot N_A = \frac{2.7 \times 10^6\, \dfrac{g}{m^3}}{27\, \dfrac{g}{mol}} \cdot 6.022 \times 10^{23}\, mol^{-1}$$

$$= 6.0 \times 10^{28}\, m^{-3}$$

Making sense of the result:

Our estimated value for the free electron density in aluminum agrees with the observed value listed in Table 22-1. Thus, the free electron cloud model produced a prediction supported by an experiment.

As long as there is no external electric field applied on conduction electrons, their average velocity is zero as they randomly collide with ions of the metal lattice. However, when an external electric field is applied, the resulting net force acting on the electron cloud is directed opposite to the direction of the electric field (Figure 22-3). If the electrons were located in a vacuum, the electric field would have continuously accelerated them to very fast speeds. However, since electrons are located in a metal, they have

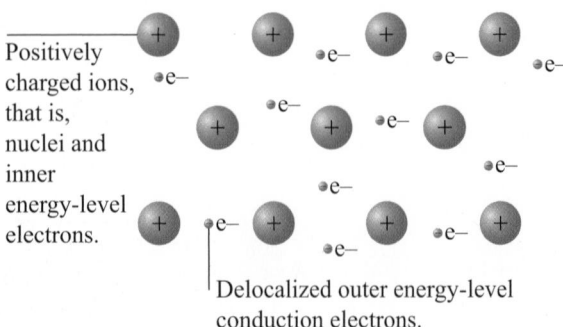

Positively charged ions, that is, nuclei and inner energy-level electrons.

Delocalized outer energy-level conduction electrons.

Figure 22-2 The sea-of-electrons model in a metal. Delocalized conduction electrons are immersed in an ionic lattice of a metal.

to move through the crystal lattice, undergoing multiple scattering, which slows them down, making the average velocity of an electron cloud quite small. This average velocity is called the **drift velocity**, $\vec{v}_d$. For metals, the magnitude of the drift velocity is on the order of 10^{-4} m/s.

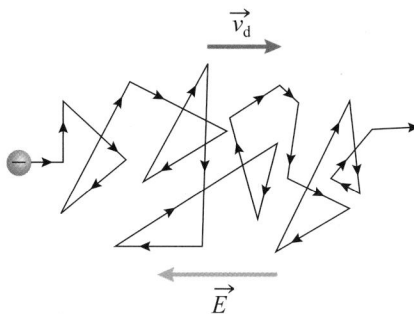

Figure 22-3 An external electric field applies an electric force on electrons in a metal, so they start moving with nonzero average velocity.

Let us estimate the number of electrons, N_e, passing through a cross-sectional area, A, of a cylindrical wire in a time interval, Δt, as a result of an applied external electric field, $\vec{E}$ (Figure 22-4). Note that there is an electric field inside a wire carrying an electric current, in contrast to the electrostatic case, in which the electric field inside a conductor is always zero, as described in Chapter 20.

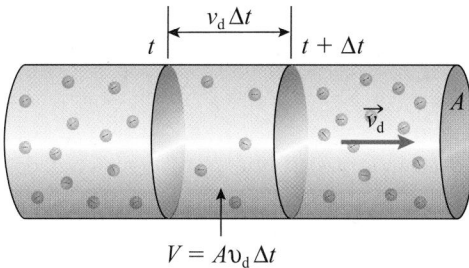

Figure 22-4 The movement of electrons in a conductor under the influence of an external electric field

Since the electron cloud moves a distance $\Delta x = v_d \Delta t$ during the time interval Δt, the number of passing through the cross-sectional area A over this time is

$$N_e = n_e V = n_e A \Delta x = n_e A v_d \Delta t \quad (22\text{-}1)$$

An **electron current** is defined as the number of electrons passing through a cross-sectional area per unit time. Thus,

Key Equation $\qquad I_e = \dfrac{N_e}{\Delta t} = n_e v_d A \quad (22\text{-}2)$

Using the value for the conduction electron density of aluminum from Table 22-1 and a drift velocity of 10^{-4} m/s for electrons, we can estimate the electron current through an aluminum wire with a cross-sectional area of 1 mm^2:

$$I_e = n_e v_d A = \left(6 \times 10^{28} \frac{1}{\text{m}^3}\right)\left(10^{-4} \frac{\text{m}}{\text{s}}\right)(10^{-6} \text{m}^2)$$
$$= 6 \times 10^{18} \text{s}^{-1} \quad (22\text{-}3)$$

It is generally more convenient to measure the current in terms of the amount of electric charge rather than the number of electrons carried through the wire per unit time. The flow of charge is called **electric current, I**. Historically, it was thought that positive charges (protons) and not electrons moved in a wire. For this reason, by convention, electric current is defined as the motion of positively charged particles. Therefore, the directions of electric and electron currents are opposite. Since the electric charge carried by one electron is $q_e = -e = -1.6 \times 10^{-19}$ C, where e is the elementary charge, the relationship between electric current and electron current is as follows:

Key Equation
$$I = I_e \cdot e = (n_e v_d A)e = n_e v_d A(1.6 \times 10^{-19} \text{C}) \quad (22\text{-}4)$$

Electric current is a scalar quantity, even though it describes the motion of charged particles.

The unit of current is the ampere (a base unit of SI), named in honour of the French scientist Andre-Marie Ampère (1775–1836).

We can use Equations (22-3) and (22-4) to estimate the electric current in the aluminum wire in Example 22-1:

$$I = (6.0 \times 10^{18} \text{s}^{-1})(1.6 \times 10^{-19} \text{C}) \approx 1.0 \frac{\text{C}}{\text{s}} = 1.0 \text{ A}$$

The **current density**, J, in a conductor is the amount of electric current flowing per unit of cross-sectional area:

Key Equation
$$J = \frac{I}{A} = \frac{n_e v_d A e}{A} = n_e v_d e$$

$$[J] = \frac{\text{C}}{\text{s} \cdot \text{m}^2} = \frac{\text{A}}{\text{m}^2} \quad (22\text{-}5)$$

☑ **CHECKPOINT**

C-22-1 Relating Current Density in a Wire to its Physical Properties

Which statement must always be true regarding two copper wires carrying currents I_1 and I_2 with current densities J_1 and J_2, respectively? Explain.

(a) If $I_1 = 2I_2$ and $J_1 = J_2$, then $R_1 = \dfrac{1}{2}R_2$.

(b) If $I_1 = 2I_2$ and $J_1 = J_2$, then $R_1 = \sqrt{2}\,R_2$.

(c) If $I_1 = 2I_2$ and $J_1 = J_2$, then $R_1 = 4R_2$.

(d) If $I_1 = 2I_2$ and $J_1 = J_2$, then $R_1 = \dfrac{1}{\sqrt{2}}R_2$.

(e) If $I_1 = I_2$ and $J_1 = J_2$, then $(n_e)_1 = (n_e)_2$.

C-22-1 (b) Using Equation (22-5): $J = \dfrac{I}{A} = \dfrac{I}{\pi R^2} \Rightarrow R = \sqrt{\dfrac{I}{\pi J}} \Rightarrow \dfrac{R_1}{R_2} = \sqrt{\dfrac{I_1 J_2}{I_2 J_1}} = \sqrt{2}$

ELECTRICITY, MAGNETISM, AND OPTICS

Turn on the Light: A Microscopic View of Electric Current

The drift velocity of electrons in metals is of the order of magnitude of 10^{-4} m/s. Why, then, do lights turn on almost instantly when you turn on a light? The answer is that a metal wire is full of conduction electrons even before you turn the light on, and what propagates through the wire is a vibrational wave in the free electron cloud, not the cloud itself. This vibrational wave propagates through the metal lattice at a speed close to the speed of light, thus making the light turn on almost instantaneously.

LO 2

22-2 Electric Conductivity and Resistivity: The Microscopic Model

When a copper wire is connected to a charged capacitor, there is an excess of positive charge at the end of the wire connected to the positive plate of the capacitor (high potential) and an excess of negative charge (electrons) at the end of the wire connected to the negative plate (low potential). The nonuniform charge distribution on the wire's surface creates an internal electric field inside it, $\vec{E}$, directed from high to low external potential. This internal electric field accelerates electrons in the direction opposite to $\vec{E}$:

$$\vec{a} = \frac{\vec{F}}{m_e} = -\frac{e\vec{E}}{m_e} = (-1.6 \times 10^{-19}\,\text{C})\frac{\vec{E}}{m_e} \quad (22\text{-}6)$$

Thus, the collective motion of electrons in the direction opposite to the internal electric field, $\vec{E}$, is superimposed on the random motion of electrons caused by their collisions with the lattice ions (Figure 22-2). The result is the drifting of the electrons in the direction opposite to $\vec{E}$ (Figure 22-3). We can estimate the drift velocity if we assume that, on average, electrons spend τ seconds moving freely between collisions. This value can be measured experimentally and is a property of the metal. As discussed earlier, in the absence of an electric field, the average velocity of the free electrons is zero. During the time between collisions, electrons experience an electrostatic force $\vec{F} = -e\vec{E}$ that causes them to accelerate, gaining drift speed:

$$v_d = \frac{eE}{m}\tau \quad (22\text{-}7)$$

Substituting the above expression into the equations for electron (Equation (22-3)) and electric current (Equation (22-4)), we find the following:

$$I_e = n_e\frac{eE}{m}\tau A = \frac{n_e e \tau A}{m}E$$

KEY EQUATION
$$I = I_e e = \frac{n_e e^2 \tau A}{m}E \quad (22\text{-}8)$$

Equation (22-8) shows that the values of both electron and electric currents are proportional to the electric field. All the other variables in Equation (22-8) are fully determined by the electrons and the physical properties of the metal wire. Let us rewrite the above equation in terms of current density, to exclude the specific dimensions of the wire:

KEY EQUATION
$$J = \frac{I}{A} = \frac{n_e e^2 \tau}{m}E = \sigma E$$

where
$$\sigma = \frac{n_e e^2 \tau}{m} \quad (22\text{-}9)$$

Equation (22-9) describes a causal relationship between the electric field and the current density inside the wire. In this equation, σ is **electric conductivity**, which represents how much electric current flows through the wire in response to a given electric field. Electric conductivity is an internal property of a material. The greater the electric conductivity, the better the material conducts electric current.

Multiplying Equation (22-9) by the cross-sectional area of the wire, A, gives an expression for the electric current:

$$I = JA = \frac{n_e e^2 \tau}{m}AE = \sigma EA = \frac{1}{\rho}EA \quad (22\text{-}10)$$

The **electric resistivity**, ρ, of a material describes how strongly the material opposes the electric current. It is defined as follows:

$$\rho = \frac{1}{\sigma} = \frac{m}{n_e e^2 \tau} \quad (22\text{-}11)$$

The shorter the time (τ) between collisions of electrons within the crystal lattice, the more collisions take place and, consequently, the more kinetic energy of the electrons will be converted into thermal energy. This energy transfer explains why wires heat up when electric current flows through them.

When the temperature rises, the average kinetic energy of the electrons increases. Therefore, the resistivity of a metal also increases with temperature, and the conductivity decreases. As the metals heat up, the time between collisions, τ, decreases, which affects both ρ and σ, as can be seen from Equation (22-11). For most metals, the dependence of resistivity on temperature is linear for a limited range of temperatures:

$$\rho = \rho_0 + \alpha\rho_0(T - T_0) = \rho_0[1 + \alpha(T - T_0)]$$
$$= \rho_0[1 + \alpha\Delta T] \quad (22\text{-}12)$$

ELECTRICITY, MAGNETISM, AND OPTICS

C-22-2 Dimensional Analysis in Action

Which of the following equations has incorrect dimensions?

(a) $[I] = \dfrac{C^2 \cdot s \cdot V}{kg \cdot m^2}$

(b) $[J] = \dfrac{C^2 \cdot s \cdot V}{kg \cdot m^4}$

(c) $[\rho] = \dfrac{kg \cdot m^3}{C^2 \cdot s}$

(d) $[\sigma] = \dfrac{C^2 \cdot s}{kg \cdot m^3}$

(e) $[I_e] = \dfrac{C \cdot s \cdot V}{m^3 \cdot kg}$

C-22-2 (e) $[I_e] = \dfrac{I}{1} = \dfrac{C}{C/s} = \dfrac{A}{C/s} = \dfrac{s}{1} \neq \dfrac{C \cdot s \cdot V}{m^3 \cdot kg}$

where ρ is the resistivity at temperature T, measured in degrees Celsius, ρ_0 is the resistivity at the baseline temperature T_0 (often chosen to be 20°C), and α is the temperature coefficient of resistivity.

The resistivities, conductivities, and temperature coefficients for various materials are given in Table 22-2.

A **superconductor** is a material with a resistivity that abruptly drops to zero below a critical temperature (less than 100 K for all presently known superconductors). We will discuss superconductivity in Chapter 24.

22-3 Ohm's Law and Combinations of Resistors

We want to be able to analyze circuits that contain various combinations of electrical resistors. First, we derive a relationship between electric current, voltage, and resistance.

In Chapter 20, we saw that in the case of a constant and uniform electric field in a wire of length l, the relationship between electric field strength and electric potential is given by $E = V/l$. Substituting for E in Equation (22-10), we obtain

$$I = \sigma EA = \frac{1}{\rho} EA = \frac{1}{\rho} \frac{V}{l} A = \frac{A}{\rho l} V = \frac{V}{R} \quad (22\text{-}13)$$

In this equation, we introduce a new variable, R, for **electric resistance**:

$$R = \rho \frac{l}{A} \quad (22\text{-}14)$$

where A is the cross-sectional area of the material, l is its length, and ρ is its resistivity.

If we know the resistivity of a material, we can calculate the resistance of a wire or cylinder of the material with any arbitrary length or cross-sectional area. From Equation (22-14) we deduce that a wire with a shorter length or a larger cross-sectional area has a lower resistance than a longer or narrower wire.

Table 22-2 Resistivities, conductivities, and temperature coefficients for different metals

Material	$\rho(\Omega \cdot m)$ at 20°C	$\sigma(\Omega \cdot m)^{-1}$ at 20°C	Temperature coefficient α (K^{-1})
Aluminum	2.65×10^{-8}	3.77×10^7	3.8×10^{-3}
Copper	1.72×10^{-8}	5.95×10^7	4.3×10^{-3}
Gold	2.24×10^{-8}	4.10×10^7	3.4×10^{-3}
Iron	9.7×10^{-8}	1.00×10^7	5.0×10^{-3}
Lead	2.06×10^{-7}	4.55×10^6	3.9×10^{-3}
Nichrome	1.10×10^{-6}	9.09×10^5	4.0×10^{-3}
Nickel	6.85×10^{-8}	1.43×10^7	6.41×10^{-3}
Platinum	1.06×10^{-7}	9.43×10^6	3.92×10^{-3}
Silver	1.59×10^{-8}	6.30×10^7	3.8×10^{-3}
Tin	1.09×10^{-7}	9.17×10^6	4.5×10^{-3}
Tungsten	5.60×10^{-8}	1.79×10^7	4.5×10^{-3}
Zinc	5.90×10^{-8}	1.69×10^7	3.7×10^{-3}

Source: The Engineering Toolbox http://www.engineeringtoolbox.com/index.html
http://www.engineeringtoolbox.com/resistivity-conductivity-d_418.html

ELECTRICITY, MAGNETISM, AND OPTICS

Equation (22-13) was derived for a very special case—an electrical wire with a uniform cross-sectional area—but it can be generalized for a large variety of electrical components, such as resistors. The equation shows a very important relationship between electric current flowing through a wire and the voltage across it: the amount of electric current is directly proportional to the voltage. Introducing electric resistance allows us to describe the proportionality between electric current and voltage, and the physical properties of the wire in a very succinct form, called **Ohm's law**: Electric current across an ohmic circuit element, such as a resistor, is directly proportional to the voltage applied across it (Figure 22-5(a)):

Ohm's Law

KEY EQUATION

$$I = \frac{V}{R}$$

$$[I] = \frac{V}{\Omega} = A \tag{22-15}$$

This relationship was discovered by Georg Simon Ohm (1787–1854), a German scientist. The unit of electric resistance, the ohm (Ω), is named in his honour. Ohm's law is widely used in designing and analyzing electric circuits. However, you have to remember that electric resistivity, ρ, and electric resistance, R, for many materials depend on temperature. Thus, the slope of the I-V curve (the common name for the $I(V)$ graph) might change if the material heats up as the current increases (Figure 22-5(b)). Such heating could make the relationship nonlinear. Ohm's law accurately describes how current through the element of a circuit depends on the voltage across it only when the temperature of the element remains constant. For example, the resistance of the filament of an incandescent light bulb increases markedly when current heats it enough to produce visible light; therefore, it is a nonohmic circuit element. It also explains why the light bulbs usually burn out just as they are turned on. A diode is another example of a nonohmic circuit element (Figure 22-5(c)).

The dependence of resistance on temperature can be derived from Equations (22-12) and (22-14):

$$I = \frac{V}{R} = \frac{VA}{\rho_0[1 + \alpha(T - T_0)]l} = \frac{V}{R_0[1 + \alpha(T - T_0)]} \tag{22-16}$$

Components with resistance, such as the ones shown in Figure 22-6(a), are found in all electrical and electronic devices. Often an electric circuit for a particular application will need circuit elements that have constant specified resistances. These circuit elements are called **resistors**. The symbol for resistor used in circuit diagrams is shown in Figure 22-6(b). To mark the value of resistance on small circuit elements, engineers and scientists use colour coding (Figure 22-6(c)).

(a)

edography/Shutterstock

(b)

Figure 22-6 (a) Miniature electrical resistors inside a digital watch. (b) The symbol for a resistor in a circuit diagram. (c) Colour coding scheme of electrical resistors. In order to find the value of the resistor you first look at the left bands (25, 460, 276 in our examples) and then multiply this value by the corresponding power of 10 expressed by the following band. This value can be found in the second column of the diagram above (1000, 1000 and 0.01 in our examples). The value of the resistance is the product of the two: 25 kΩ, 460 kΩ and 2.76 Ω for the resistors in the example.

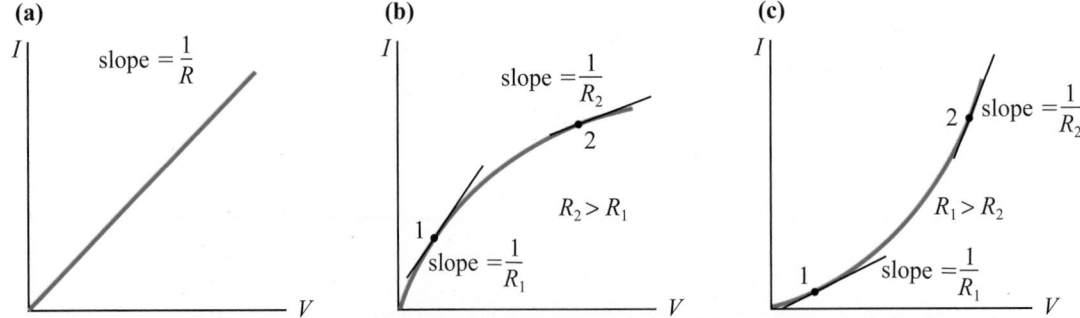

Figure 22-5 The dependence of current on applied voltage across (a) an ohmic circuit element and two nonohmic circuit elements, (b) an incandescent light bulb, and (c) a diode

ELECTRICITY, MAGNETISM, AND OPTICS

EXAMPLE 22-2

Applying Ohm's Law to a Current-Carrying Wire

A potential difference of 100 mV is applied across a 5.0 m-long metal wire that has a diameter of 0.50 mm. The resulting electric current is 66.7 mA.

(a) Find the resistance of the wire.
(b) Find the resistivity of the wire, and suggest what metal it could be made of.

SOLUTION

(a) We assume that the metal wire obeys Ohm's law. Therefore, we can find its resistance, R, using Equation (22-15):

$$R = \frac{V}{I} = \frac{100 \text{ mV}}{66.7 \text{ mA}} = \frac{100 \times 10^{-3} \text{ V}}{66.7 \times 10^{-3} \text{ A}} = 1.50 \ \Omega$$

(b) Rearranging Equation (22-14), we can find the resistivity of the wire:

$$\rho = \frac{RA}{l} = \frac{(1.5 \ \Omega) \cdot \pi (0.25 \times 10^{-3} \text{m})^2}{5.0 \text{ m}} = 5.9 \times 10^{-8} \ \Omega \cdot \text{m}$$

Comparing this value to the resistivities listed in Table 22-2, we find that the wire could be made of zinc.

Making sense of the result:

Resistance and resistivity are different physical quantities. The resistance provides a description of how a particular piece of wire resists the flow of electric current. The resistivity is the resistance of a hypothetical wire that is 1 m long and has a cross-sectional area of 1 m^2.

 CHECKPOINT

C-22-3 Ranking Resistances of Metal Wires

Which of the following statements correctly represents the ranking of the resistances of the five copper wires shown in Figure 22-7? How would you rank the resistivities of these wires?

(a) $R_B > R_A = R_E > R_C > R_D$
(b) $R_B > R_A > R_E > R_C > R_D$
(c) $R_C > R_A = R_E > R_B > R_D$
(d) $R_D > R_C > R_E = R_A > R_B$
(e) $R_D > R_A = R_E > R_C > R_B$

each with resistance R. What is the effective resistance of a combination of these resistors? The answer depends on how the resistors are connected.

Resistors Connected in Series When two or more resistors are connected in series (Figure 22-8(a)), the circuit is called a **series circuit**. In that case, the same electric current flows through each resistor because the current that exits the first resistor must enter the second resistor. Connecting two identical resistors in series affects their total length while leaving the other parameters unchanged. It is convenient to define equivalent resistance of electric circuits that are comprised of two or more resistors. The **equivalent resistance** is the

(a)   (b)   (c)

L, R $2L, R$ $2L, 2R$

(d)   (e)

$L, 2R$ $4L, 2R$

Figure 22-7 C-22-3

Now we are ready to explore how Ohm's law can be applied to a combination of resistors. Let us model a resistor as a wire of length l, cross-sectional area A, and resistivity ρ. Imagine that we have two identical resistors,

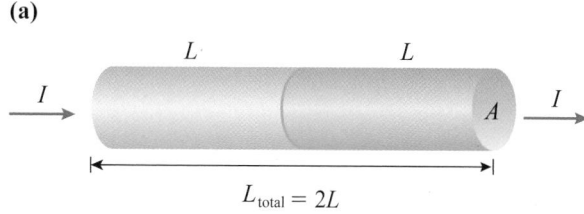

(a)

I → L L A → I

$L_{\text{total}} = 2L$

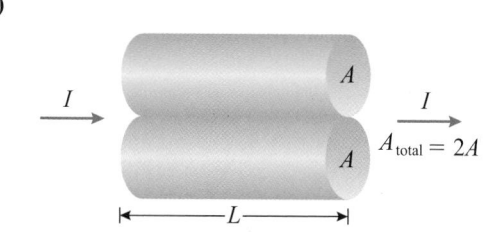

(b)

I → A → I

A $A_{\text{total}} = 2A$

L

Figure 22-8 Electrical resistors connected (a) in series and (b) in parallel resistance of a resistor that, when connected in series to the circuit's battery will produce an electric current equal to the electric current in the original circuit. To find the value of the equivalent resistance in a series circuit, R_{series}, we use Equation (22-14):

$$R_{series} = \rho \frac{2L}{A} = 2\rho \frac{L}{A} = 2R \qquad (22\text{-}17)$$

Although this expression was derived for a special case of two identical resistors, we will show later that it applies to any number of different resistors connected in series:

KEY EQUATION
$$R_{series} = \sum_{i=1}^{N} R_i \qquad (22\text{-}18)$$

Equation (22-18) indicates that connecting resistors in series increases their equivalent resistance. The term "equivalent" here indicates that resistors connected in series behave as one equivalent resistor that has a resistance equal to the sum of all the individual resistances.

Resistors Connected in Parallel A circuit in which the electric current has more than one continuous path for electrons to flow (**electric circuit branch**) is called a **parallel circuit**. A parallel circuit has **junction points**, which are points where electric current can split up and flow through different branches of the circuit. When two resistors are connected in parallel (Figure 22-8(b)), some of the current will flow through the first resistor, and the rest will flow through the second resistor. Connecting identical resistors in parallel increases the effective cross-sectional area while leaving the length of the path for electrons unchanged. To find the value of the equivalent resistance for two identical resistors connected in parallel, $R_{parallel}$, we again use Equation (22-14):

$$R_{parallel} = \rho \frac{L}{2A} = \frac{1}{2} \rho \frac{L}{A} = \frac{R}{2} \qquad (22\text{-}19)$$

Although we used a special case of two identical resistors to derive this equation, we will show later that the equivalent resistance for any number of different resistors connected in parallel is given by

KEY EQUATION
$$\frac{1}{R_{parallel}} = \sum_{i=1}^{N} \frac{1}{R_i} \qquad (22\text{-}20)$$

Equation (22-20) indicates that connecting resistors in parallel reduces the equivalent resistance.

Power in DC Circuits

In Chapter 20, we defined electric potential difference as the potential energy gained by a unit charge when it moves from one point to another: $V = \dfrac{U_{el}}{q}$. In Section 22-2, we saw that a current of free electrons in a material heats

PEER TO PEER

I always thought that in a parallel circuit, the current splits equally between the circuit's parallel branches. Then I realized that this is only true when the branches have equal resistance. The current will split in a way such that the more resistance the branch has, the less current will flow through it. For example, if 6 A of current enters the junction and splits between two branches of 2 Ω and 4 Ω, the branch that has twice the resistance will receive half the current. Thus, the current will split into 4 A and 2 A, respectively. Keeping this in mind always helps me determine whether my circuit solution is reasonable.

the material through collisions with its atoms. We define **power**, P, as the rate of energy transfer. For the thermal energy dissipated as a result of electric charges flowing through a material,

$$P = \frac{Q_{heat}}{\Delta t} \qquad (22\text{-}21)$$

where Q_{heat} is heat dissipated during the time interval, Δt.

When current flows in a material, the electric potential energy of the electrons is converted first to kinetic energy and then into heat, which is dissipated by the material. Therefore,

$$P = \frac{Q_{heat}}{\Delta t} = \frac{qV}{\Delta t} = IV \qquad (22\text{-}22)$$

Using Ohm's law, Equation (22-22) can be rewritten as follows:

KEY EQUATION
$$P = IV = I^2 R = \frac{V^2}{R}$$

$$[P] = \frac{J}{s} = watt\ (W);$$

$$1W = 1A \cdot 1V \qquad (22\text{-}23)$$

Consequently, the heat dissipated by a conductor of resistance R carrying electric current I over the time period Δt can be expressed as

KEY EQUATION
$$Q_{heat} = I^2 R \Delta t \qquad (22\text{-}24)$$

This relationship was discovered independently by two scientists: British physicist James Prescott Joule (1818–89) and Russian-German-Estonian physicist Heinrich Lenz (1804–65), and is often called Joule–Lenz's law.

From this example we conclude that in a parallel electric circuit, the power output of the circuit equals the sum of the nominal power outputs of its components.

Household Electrical Wiring: How Much Is Too Much?

All your household appliances are connected in parallel to the external source of electric energy through the electrical outlets in the walls (Figure 22-9). By connecting more appliances in a parallel circuit, you are reducing the equivalent resistance of the combined load, thus increasing the current drawn from the electrical outlet: $I_{outlet} = I_1 + I_2 + I_3 + \cdots$

$$= \frac{V}{R_1} + \frac{V}{R_2} + \frac{V}{R_3} + \cdots = V\left(\frac{1}{R_1} + \frac{1}{R_2} + \frac{1}{R_3} \cdots\right) = \frac{V}{R_{parallel}}.$$

In North America, common electrical outlets are designed to supply currents of up to 15 A. Larger currents in such circuits can overheat the wires enough to damage the insulation and possibly cause a fire. For this reason, strict safety regulations are in place that require all household circuits to be protected by a fuse or a circuit breaker, which will cut off the current when the circuit is overloaded.

In an electric circuit, the total power output of the circuit equals the sum of the power outputs of all the appliances. Therefore, to avoid reaching the maximum allowable current, the sum of all the power values produced by the appliances should not exceed 1800 W. For a parallel circuit at your home:

$$P_{max} = IV = (I_1 + I_2 + I_3 + \cdots)V$$

$$= I_1V + I_2V + I_3V + \cdots = \frac{V^2}{R_1} + \frac{V^2}{R_2} + \frac{V^2}{R_3} + \cdots$$

$$= P_1 + P_2 + P_3 + \cdots$$

$$P_{max} = IV = (15 \text{ A})(120 \text{ V}) = 1800 \text{ W}$$

$$P_1 + P_2 + P_3 + \cdots \leq P_{max}$$

In a series circuit connected to an energy source V, the power output of the circuit will also be equal to the sum of the power outputs of its elements. However, the power output of each of the elements will be less than the nominal power output they were designed for, as the potential difference across each one of them will be less than V (which was used to calculate the nominal power output).

Figure 22-9 All your household appliances are connected in parallel to the external source of electric energy through the electrical outlets in the walls.

The power output increases when the number of electrical appliances connected in parallel increases:

KEY EQUATION
$$P_{parallel} = \sum_{i=1}^{N} P_i \qquad (22\text{-}25)$$

The electric energy used by a load (such as a toaster or other appliance) in a circuit depends on the power drawn from the source and on the length of time, Δt, that the circuit operates:

$$E_{el} = P\Delta t \qquad (22\text{-}26)$$

Electrical utility companies commonly measure energy consumption in kilowatt hours (kW·h) (Figure 22-10), although some companies now use megajoules (MJ) for their billing. By converting hours to seconds, we can convert a kilowatt hour into joules:

$$1 \text{ kW·h} = (1000 \text{ W})(3600 \text{ s}) = 3.6 \times 10^6 \text{ J} = 3.6 \text{ MJ}$$
$$(22\text{-}27)$$

BC Hydro Electric Charges

Apr 07 to Jun 04 (Residential Conservation Rate 1101)	
Basic charge: 59 days @ $0.13410/day	7.91*
Usage charge:[1]	
Step 1: 935 kW.h @ $0.06270/kW.h	58.62*
Step 2: 0 kW.h @ $0.08780/kW.h	0.00
Rate Rider at 4.0%	2.66*
Innovative Clean Energy Fund Levy at 0.4%	0.28
Regional transit levy: 59 days @ $0.06240/day	3.68*
* GST	3.64
	$76.79

Your total consumption for the billing period is 935 kW.h

Figure 22-10 A residential electricity bill

PEER TO PEER

I knew that on an electrical bill the energy is written as $kW \cdot h$, but only recently did I notice that it stands for the product of the two, not the ratio (i.e., it is *not* kW per hour). When measuring energy consumption in everyday life, the unit J is inconvenient because it is too small. So we use $kW \cdot h$ instead: $1 \ kW \cdot h = 3.6 \times 10^6 \ J$ (Equation 22-27). To show the ratio of two numbers, we use the term "per" (e.g., kilometres per hour), but when we want to show the product of two numbers, we just write them side by side (e.g, kilowatt hours).

MAKING CONNECTIONS

The Invention of the Incandescent Light Bulb

Humphry Davy (1778–1829) built the first electric light bulb in 1809. Over the next century, a number of inventors developed a variety of designs, which led to the mass production of affordable and reasonably reliable incandescent light bulbs. Among these inventors were Sir Joseph Wilson Swan (1828–1914) in England, who invented a bulb using a carbon filament, and Henry Woodward and Mathew Evans in Canada, who ended up selling their incandescent light bulb patent to an American, Thomas Alva Edison (1847–1931). Edison had the financial backing needed to improve and commercialize their design. Today, the filaments in incandescent lamps are made of tungsten, which can withstand higher temperatures and therefore produce more light than carbon filaments. The energy efficiency of tungsten bulbs is 2% to 10% because only a tiny fraction of the light they emit is in the range of visible light. Traditional incandescent light bulbs are gradually being replaced by more efficient bulbs, such as halogen incandescent light bulbs, fluorescent lamps, and light-emitting diodes.

EXAMPLE 22-3

Power Rating of a Light Bulb

The antique 30 W light bulb in Figure 22-11(a) was designed to be connected to a 200 V battery. What is the power output of this light bulb when it is operated by a 100 V battery (Figure 22-11(b))?

SOLUTION

The power output of the light bulb depends on the resistance of its filament and the voltage of the energy

Figure 22-11 This antique incandescent light bulb uses a carbon filament. The grey colour results from sublimated carbon that is gradually deposited on the inner surface of the bulb. This light bulb is powered by (a) a 200 V source and (b) a 100 V source.

source. The nominal resistance of the carbon filament of this light bulb is

$$R = \frac{V_1^2}{P_1} = \frac{(200 \ \text{V})^2}{30 \ \text{W}} = \frac{4.0 \times 10^4 \ \text{V}^2}{30 \ \text{W}} = 1.33 \ k\Omega$$

When the light bulb is connected to a source with a lower voltage, the power output is reduced:

$$P_2 = \frac{V_2^2}{R} = \frac{(100 \ \text{V})^2}{1.33 \ k\Omega} = \frac{1.0 \times 10^4 \ \text{V}^2}{1.33 \times 10^3 \ \Omega} = 7.5 \ \text{W}$$

Making sense of the result:

Since the power depends on the square of the voltage, decreasing the voltage across the light bulb by half reduces the power output of the bulb by a factor of 4.

MAKING CONNECTIONS

You Are Grounded!

You probably have heard the term "electrical grounding" and may have wondered what it meant. **Electrical grounding** was invented as a safety measure to protect people, who, when accidentally touching a faulty electrical appliance, become a path for the electric current and being shocked by it. For example, consider the heating element inside your electric toaster. When the toaster is operating properly, electric current flows through the heating element, thus heating your bagel. If for any reason the heating element comes into contact with the metal case of the toaster, the case will become a part of the circuit. If you accidentally touch the case, your body will become a part of the circuit as well, providing a path for electric current from a high-voltage power source (high potential energy) to the ground (low potential energy): the current will flow through the device, through the metal case, and through you into the ground. This will expose you to high voltage.

While the current through your body might be low due to your high internal resistance, it is still very dangerous. A person will feel a shock if only 5 mA of current flows

ELECTRICITY, MAGNETISM, AND OPTICS

through the body; 100 mA of current could be lethal. The danger of relatively low electric current is very important because a circuit breaker is built to break the circuit when the current levels reach much higher values in order to prevent fires (15 A for Canadian households), while not protecting people from being shocked. By connecting the toaster to the ground, the current will have an alternative (parallel) path to flow from the toaster case to the ground circuit. If the ground circuit has less resistance than the circuit through your skin, you will be safe. This low-resistance return path for electric current (grounding circuit) protects you from being shocked by the toaster. Since the resistance of a ground circuit is much higher than the resistance of the original circuit when the toaster is operating properly, the electric current only flows through the ground circuit when there is a problem.

LO 4

22-4 Analysis of DC Circuits and Kirchhoff's Laws

A **direct current (DC)** circuit is a circuit in which electric current flows in one direction. A simple DC electric circuit consists of circuit elements, such as batteries, resistors, switches, capacitors, and diodes, connected by conducting wires that allow the flow of electric current. We will use the concepts of electric current I, resistance R, power P, and voltage V to describe the operation of these circuits. First, we consider the source of electric potential energy.

Source of Electromotive Force Electric batteries and generators are sources of electric energy. Historically, they were called the **sources of electromotive force** (emf). However, as you will see below, electromotive force is not

a real force but is a measure of electric potential energy per unit charge. To avoid confusion while adhering to the tradition, we will use the abbreviation emf to refer to a constant voltage produced by a source. Figure 22-12 shows representations of an ideal battery and a real battery.

Figure 22-12 Schematic representations of ideal and real batteries. (a) An ideal battery (b) A real battery that has a nonzero internal resistance r

The simple DC circuit in Figure 22-13(a) consists of a battery connected to a light bulb with conducting wires. As a result of the energy stored in the battery, an electric current, I, flows through the circuit. We assume that the resistance of the wires compared to the resistance of the light bulb is negligible, but the emf source (the battery) is real, which means it has an internal resistance, r. We apply Ohm's law (Equation (22-15)) to each segment of this circuit, and summarize the results in Table 22-3.

Figure 22-13 Analysis of a simple DC circuit consisting of a non-ideal emf source (a battery), a light bulb, and connecting wires. (a) A schematic representation of a real-life circuit real-life circuit (b) A schematic representation of the energy use in the circuit

Table 22-3 Analysis of a simple DC circuit containing a nonideal emf source (a battery), a light bulb, and connecting wires for Figure 22-13

Segment	I (A)	R (Ω)	V (V)	Meaning
AB	I	R	$V_{AB} = V_B - V_A = (\varepsilon - Ir)$	The positive terminal, B, has a higher potential than the negative terminal, A; therefore V_{AB} is positive. Moving charged particles (electric current) will use some of the potential energy Ir from the emf source ε to move through it. Thus, the voltage output between the terminals of the emf source V_{AB} will be reduced by Ir as compared to emf ε.
BC	I	0	$V_{BC} = V_B - V_C = I(0\ \Omega) = 0$ V	No energy is used by the current to flow through this segment: zero potential difference.
CD	I	R_{bulb}	$V_{CD} = V_C - V_D = IR_{bulb}$	Electric current converts electric potential energy into heat, thus dissipating electric energy.
DA	I	0	$V_{DA} = V_D - V_A = I(0\ \Omega) = 0$ V	No energy is used by the current to flow through this segment: zero potential difference.
Total: ABCDA	I	$R_{bulb} + r$	$IR_{bulb} - (\varepsilon - Ir) = 0$ $IR_{bulb} - \varepsilon + Ir = 0$ $IR_{bulb} = V_{AD} = \varepsilon - Ir$	Since electric charges complete an entire circuit, their total change in electric potential must be zero: $\Delta V_{total} = 0$.

CHAPTER 22 | ELECTRIC CURRENT AND FUNDAMENTALS OF DC CIRCUITS

The analysis presented in Table 22-3 is an application of the law of conservation of energy to a simple DC circuit. As with gravitational potential energy, if one returns to the starting point, the sum of all the gains and losses of potential energy along the way must be zero.

Since any real (non-ideal) emf source has some internal resistance, we represent it by connecting an internal resistance, r, in series to an ideal source that has zero internal resistance (Figure 22-13(b)).

Figure 22-13(c) shows potential differences across different elements of the circuit. This diagram clearly identifies the segments where the sources of energy are located. When a positive electric charge moves from point A to point B, it gains energy. Since the internal resistance of the source is nonzero, the energy gain per unit charge is $\varepsilon - IR$. On the other hand, there is no change in energy when the charge moves along segments BC and DA; and the electric potential energy of the charge is converted into thermal energy over the segment CD.

The analysis above shows that the emf and the voltage across the terminals of a source will differ by an amount that depends on the current and the internal resistance of the source:

$$V_{AB} = \varepsilon - Ir \qquad (22\text{-}28)$$

Since the light bulb is connected to the emf source with conducting wires that have negligible resistance, the potential difference across the light bulb equals the potential difference across the terminals of the emf source:

$$IR_{bulb} = \varepsilon - Ir$$

$$I = \frac{\varepsilon}{(R_{bulb} + r)} \qquad (22\text{-}29)$$

The thermal energy dissipated by this circuit includes the energy dissipated by the light bulb and by the battery:

$$I\varepsilon = I^2 R_{bulb} + I^2 r \qquad (22\text{-}30)$$

This example illustrates how the law of energy conservation applies to electric circuits.

Meters in an Electric Circuit To confirm that our analysis of a circuit is correct, we can use an **ammeter** to measure the currents through various circuit elements and a **voltmeter** to measure the potential differences across them. An ammeter has to be connected in series with the circuit elements so that the same current flows through the ammeter and through the element, as shown in Figure 22-14(a). An ammeter has a very low internal resistance to minimize its effect on the circuit. However, a voltmeter has to be connected in parallel to a circuit element to measure the voltage across it. A voltmeter (Figure 22-14(b)) has a very high internal resistance, producing a minimal effect on the circuit. Both ammeter and voltmeter measurements have to be made on an operating circuit.

An ohmmeter (Figure 22-14(c)) measures the electrical resistance of a circuit element when the element is disconnected from a circuit. A device that can be set

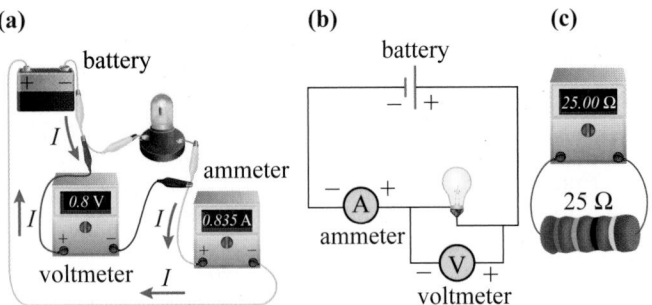

Figure 22-14 (a) An ammeter and a voltmeter connected to an electric circuit (b) A circuit diagram showing the ammeter and the voltmeter (c) An ohmmeter connected to a resistor

to measure current, voltage, or resistance is called a **multimeter**.

You can solve many problems dealing with simple DC circuits. However, to develop a real physics intuition, you have to have some hands-on experience with constructing and analyzing circuits. Today, you can gain such experience—even if you do not have a real-life circuit in front of you—by trying the online activity below.

ONLINE ACTIVITY

Simple and Not So Simple DC Circuits

The e-resource that accompanies every new copy of this textbook contains an Online Activity using the PhET simulation "Circuit Construction Kit." Work through the simulation and accompanying questions to gain an understanding of DC circuits.

Kirchhoff's Laws

Series Circuits Figure 22-15 shows three different light bulbs connected in series with an ideal battery (zero internal resistance) and a switch. When the switch is closed, it has negligible resistance and the potential difference across it is zero. As a consequence of the law of energy conservation, the potential difference across the battery, V_B, must be equal to the sum of potential differences across all circuit elements. Therefore,

$$V_B = V_1 + V_2 + V_3 \qquad (22\text{-}31)$$

Equation (22-31) can be generalized to state that the algebraic sum of all the potential differences (voltages) across all circuit elements around a closed circuit loop must be zero:

KEY EQUATION
$$\sum_{closed\ loop} \Delta V = 0 \qquad (22\text{-}32)$$

This loop rule, called **Kirchhoff's second law**, after German physicist Gustav Kirchhoff (1824–87), is a consequence of the law of energy conservation.

Next, we note that the same current I flows through all the elements of our series circuit. Applying Ohm's law to each light bulb and once again using R_{series} for the equivalent resistance of the entire series circuit, we obtain

$$V_B = IR_1 + IR_2 + IR_3 = I(R_1 + R_2 + R_3)$$

Since $\quad V_B = IR_{series}$,

$$IR_{series} = I(R_1 + R_2 + R_3)$$

$$R_{series} = R_1 + R_2 + R_3$$

We can apply the same analysis to any number of different resistors connected in series, thus proving a general formula for resistors connected in series we discussed earlier:

KEY EQUATION
$$R_{series} = \sum_{i=1}^{N} R_i \qquad (22\text{-}18)$$

✅ **CHECKPOINT**

C-22-4 Three Light Bulbs in a Series Electric Circuit

The three light bulbs in Figure 22-15 are connected in series. The middle light bulb is the brightest, and the bulb on the left is the dimmest. For these light bulbs,

(a) $R_1 > R_3 > R_2$; $I_1 = I_2 = I_3$;
(b) $R_1 = R_2 = R_3$; $I_1 < I_3 < I_2$;
(c) $R_1 > R_3 > R_2$; $I_1 < I_3 < I_2$;
(d) $R_1 = R_2 = R_3$; $I_2 < I_3 < I_1$;
(e) $R_2 > R_3 > R_1$; $I_1 = I_2 = I_3$.

Figure 22-15 Three non-identical light bulbs connected in series

Parallel Circuits Figure 22-16 shows three light bulbs connected in parallel to an ideal battery. Each light bulb is connected directly to the battery and thus forms a separate branch of the circuit. A point where different branches join together is called a **junction**, or a **node**. Figure 22-17 shows currents flowing through two different junctions. As discussed earlier, the current does not necessarily divide equally among the parallel branches, but the amount of current entering any branch must equal the current exiting it, and the current entering any junction must equal the

current exiting it. In the junction rule, we use a plus sign before the currents entering the junction and a minus sign before the currents exiting the junction.

KEY EQUATION
$$\sum_{junction} I_i = 0 \quad \text{or} \quad \sum_{junction} I_{entering} = \sum_{junction} I_{exiting}$$

$$(22\text{-}33)$$

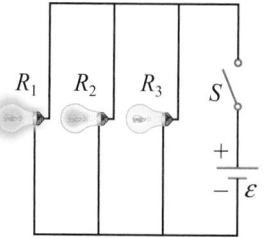

Figure 22-16 Three light bulbs connected in parallel

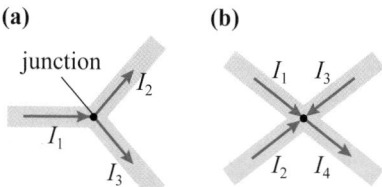

Figure 22-17 The amount of current entering a junction equals the amount of current leaving it.

This junction rule, called **Kirchhoff's first law**, is a consequence of the **law of electric charge conservation**, which states electric change can neither be created nor destroyed.

Applying Kirchhoff's junction rule to our parallel circuit, we see that the current leaving the battery must equal the sum of the currents in the three branches:

$$I_B = I_1 + I_2 + I_3 \qquad (22\text{-}34)$$

Since each light bulb is connected directly to the battery, the potential difference across each bulb must be equal to the potential difference across the battery's terminals. Applying Ohm's law to substitute for the currents in Equation (22-34), we find

$$\frac{V_B}{R_{parallel}} = \frac{V_B}{R_1} + \frac{V_B}{R_2} + \frac{V_B}{R_3}$$

$$\frac{1}{R_{parallel}} = \frac{1}{R_1} + \frac{1}{R_2} + \frac{1}{R_3}$$

We can apply the same argument to any number of different resistors connected in parallel, thus proving a general formula for resistors connected in parallel we discussed earlier:

KEY EQUATION
$$\frac{1}{R_{parallel}} = \sum_{i=1}^{N} \frac{1}{R_i} \qquad (22\text{-}20)$$

ELECTRICITY, MAGNETISM, AND OPTICS

C-22-5 Three Light Bulbs in a Parallel Electric Circuit

The three light bulbs in Figure 22-16 are connected in parallel to a battery. The left light bulb is the brightest, and the light bulb in the middle is the dimmest. For these light bulbs,

(a) $R_1 > R_3 > R_2$; $I_1 = I_2 = I_3$;
(b) $R_1 = R_2 = R_3$; $I_1 > I_3 > I_2$;
(c) $R_1 > R_3 > R_2$; $I_1 < I_3 < I_2$;
(d) $R_1 = R_2 = R_3$; $I_2 < I_3 < I_1$;
(e) $R_2 > R_3 > R_1$; $I_1 > I_3 > I_2$.

C-22-5 (e) The light bulbs are connected in parallel; therefore, they have the same potential difference across them. Since $P = \dfrac{V^2}{R}$ (Equation (22-23)), the lower the resistance of the light bulb, the more current will flow through it and the more power it will produce, thus dissipating more heat.

We can use the properties of series and parallel circuits to find the simplest way to determine the power developed in the components of the circuits. In a series circuit, the same current flows through each component, so it is convenient to use the equation $P = I^2R$ to determine the power dissipated by each component. However, the branches in a parallel circuit are all connected to the same battery; thus, the potential differences across them must be the same and equal to the potential difference across the battery. Therefore, it is more convenient to use $P = \dfrac{V^2}{R}$ to determine the thermal energy (heat) dissipated by each branch.

C-22-6 The Relative Brightness of Two Light Bulbs in Series and Parallel Electric Circuits

A 25 W light bulb and a 40 W light bulb are first connected in series. Then they are connected in parallel. Which statement describes their relative brightness?

(a) The 25 W light bulb is brighter in a series circuit and dimmer in a parallel circuit.
(b) The 25 W light bulb is brighter in a parallel circuit and dimmer in a series circuit.
(c) The 25 W light bulb is always brighter.
(d) The 40 W light bulb is always brighter.

C-22-6 (b) The light bulbs are designed to be used in a parallel circuit; therefore, the higher the nominal power (power rating), the less the resistance of the light bulb (higher-rated light bulbs have thicker filaments).

 PEER TO PEER

I never thought that a power rating for a light bulb (nominal power) is the power it will generate only when connected to the power source for which the light bulb was rated. If the light bulb is designed to generate 40 W while connected to a 110 V source, it will generate less power if it is connected to a lower-voltage source. Thus, the nominal power is not necessarily the power that the light bulb generates.

 ## EXAMPLE 22-4

Resistors in Series and in Parallel

Two different combinations of resistors are shown in Figure 22-18. Each combination then is connected to a 12.0 V battery which is connected to points A and B respectively. The current through the battery in the first circuit is twice the current in the second circuit.

(a) What is the value of R in the second circuit?
(b) Find the currents in each branch of the two circuits.

Figure 22-18 Example 22-4

SOLUTION

(a) In the first circuit diagram, the two resistors in the top branch are connected in series, so their equivalent resistance is

$$(R_{eq})_{top1} = 35.0\ \Omega + 82.0\ \Omega = 117\ \Omega$$

Since the top and the bottom branches are connected in parallel, their equivalent resistance is

$$\frac{1}{R_{eq1}} = \frac{1}{117\ \Omega} + \frac{1}{45.0\ \Omega}$$

$$\frac{1}{R_{eq1}} = 0.0308\ \Omega^{-1}$$

$$R_{eq1} = 32.5\ \Omega$$

In the final answer, three significant figures are kept in accordance with the significant figure rules.

We now use Ohm's law to find the current flowing through the battery (again, we round the final answer to three significant figures):

$$I_{battery} = \frac{V_{battery}}{R_{eq1}} = \frac{12.0 \text{ V}}{32.5 \text{ } \Omega} = 0.36923 \text{ A} \approx 0.369 \text{ A}$$

Since the current in the second circuit is half the current in the first one, $I_2 = 0.185$ A. To produce this current by connecting the combination of resistors to a 12.0 V battery, the equivalent resistance in the second circuit must be

$$R_{eq2} = \frac{12.0 \text{ V}}{0.18462 \text{ A}} = 64.998 \text{ } \Omega \approx 65.0 \text{ } \Omega$$

To find R, we write an equation for the equivalent resistance in the second circuit:

$$R_{eq2} = 64.998 \text{ } \Omega$$

$$64.998 \text{ } \Omega = 15.0 \text{ } \Omega + \left[\frac{1}{55.0 \text{ } \Omega} + \frac{1}{R} \right]^{-1}$$

$$49.998 \text{ } \Omega = \frac{(55.0 \text{ } \Omega)R}{R + 55.0 \text{ } \Omega}$$

$$(49.998 \text{ } \Omega)R + 2749.89 \text{ } \Omega^2 = (55.0 \text{ } \Omega)R$$

$$R = 549.803 \text{ } \Omega \approx 549 \text{ } \Omega$$

(b) To find the currents flowing through each branch, we apply Ohm's law to each one. For diagram (a), both branches have a potential difference of 12.0 V across them. Therefore, the currents through the branches are as follows:

$$I_{top1} = \frac{12.0 \text{ V}}{117 \text{ } \Omega} = 0.103 \text{ A}$$

$$I_{bottom1} = \frac{12.0 \text{ V}}{45.0 \text{ } \Omega} = 0.267 \text{ A}$$

For diagram (b), the potential difference across each resistor is less than 12.0 V because the parallel combination of resistors is connected in series with the 12.0 Ω resistor. Therefore,

$$I_{top2} = \frac{12.0 \text{ V} - (0.185 \text{ A})(15.0 \text{ } \Omega)}{55.0 \text{ } \Omega} = 0.168 \text{ A}$$

$$I_{bottom2} = \frac{12.0 \text{ V} - (0.185 \text{ A})(15.0 \text{ } \Omega)}{549.803 \text{ } \Omega} = 0.0168 \text{ A}$$

Making sense of the result:

To verify our calculations, we can check that the sum of the currents in each circuit equals the current flowing through the battery:

$$I_{battery1} = I_{top1} + I_{bottom1} = 0.369 \text{ A} \quad \text{and}$$

$$I_{battery2} = I_{top2} + I_{bottom2} = 0.185 \text{ A}$$

Example 22-4 demonstrates two important properties of parallel circuits:

■ In a parallel circuit, the ratio of the currents through parallel branches is equal to the inverse ratio of the resistances of these branches. Consequently, a branch with less resistance has greater current. In the example above,

$$\frac{I_{top2}}{I_{bottom2}} = \frac{R_{bottom2}}{R_{top2}}$$

$$\frac{0.168 \text{ A}}{0.0168 \text{ A}} = \frac{549 \text{ } \Omega}{55.0 \text{ } \Omega}$$

■ The equivalent resistance in a parallel circuit is less than the lowest resistance of its branches. In the example above, the equivalent resistance of the parallel components of the second circuit is 50.0 Ω, which is less than either 55.0 Ω or 549 Ω. (See problems 33, 43, and 44 at the end of the chapter.)

Resolving Compound Electric Circuits Most practical circuits that you will deal with have both series and parallel connections of multiple resistors. In this section, we describe a strategy for finding a single equivalent resistance for these circuits. This strategy works for many, but not all, circuits. For circuits that cannot be analyzed in this way, you can use Kirchhoff's rules as described in the next section.

A Strategy for Resolving Compound DC Circuits

1. Draw the circuit diagram, label all the circuit elements, and record all the known quanitites.

2. Identify all the resistors connected in series. Calculate the equivalent resistance of each set of series-connected resistors using Equation (22-18):

$$R_{series} = R_1 + R_2 + R_3 + \cdots R_N = \sum_{i=1}^{N} R_i$$

3. Draw a simplified circuit diagram, and replace each series combination of resistors with its equivalent resistance.

4. Identify all the resistors connected in parallel. Calculate the equivalent resistance of each set of parallel-connected resistors using Equation (22-20):

$$\frac{1}{R_{parallel}} = \frac{1}{R_1} + \frac{1}{R_2} + \frac{1}{R_3} + \cdots \frac{1}{R_N} = \sum_{i=1}^{N} \frac{1}{R_i}$$

5. Draw a further simplified circuit diagram, and replace each parallel combination of resistors with its equivalent resistance.

6. Repeat steps 4 and 5, if necessary, for series and parallel combinations that include the equivalent resistances until you have simplified the DC circuit to one final equivalent resistance.

EXAMPLE 22-5

Analyzing a Compound Electric Circuit

Figure 22-19 shows a compound electric circuit that has nine identical $2.0\ \Omega$ resistors connected to a 12 V ideal battery.

(a) Find the equivalent resistance for the circuit.
(b) Find the total electric current flowing from the battery.
(c) Identify the resistors that have the greatest potential difference across them.

Figure 22-19   Example 22-5

SOLUTION

(a) Although the resistors are identical, the currents flowing through them likely differ. Therefore, we give each resistor a separate label: R_A, R_B, R_C, and so on. Then, we identify the resistors connected in series: R_G, R_H, and R_I. The equivalent resistance of these resistors is

$$R_{GHI} = R_G + R_H + R_I = 6.0\ \Omega$$

We can simplify the circuit using this equivalent resistance (Figure 22-21(b)). Since there are no more resistors connected only in series, we now find the equivalent resistance for the two resistances connected in parallel, R_E and R_{GHI}:

$$\frac{1}{R_{EGHI}} = \frac{1}{R_E} + \frac{1}{R_{GHI}} = \frac{1}{2.0\ \Omega} + \frac{1}{6.0\ \Omega} \Rightarrow \frac{1}{R_{EGHI}} = \frac{2}{3\ \Omega}$$

$$\Rightarrow R_{EGHI} = \frac{3}{2}\ \Omega$$

We can simplify the circuit further using the equivalent resistance R_{EGHI} (Figure 22-21(c)). In this simplified circuit, R_D, R_F, and R_{EGHI} are connected in series. Their equivalent resistance is

$$R_{DEFGHI} = 2.0\ \Omega + 2.0\ \Omega + \frac{3}{2}\ \Omega = \frac{11}{2}\ \Omega = 5\frac{1}{2}\ \Omega$$

Continuing this procedure, we eventually get a circuit that has only one equivalent resistance (Figure 22-19(e)):

$$R_{ABCDEFGHI} = 5\frac{7}{15}\ \Omega$$

(b) We apply Ohm's law to find the current output of the battery:

$$I_{battery} = \frac{\varepsilon}{R_{ABCDEFGHI}} = \frac{12\ V}{5\frac{7}{15}\ \Omega} = \frac{90}{41}\ A$$

(c) To identify the resistors that have maximum potential difference across them, we use Ohm's law once again. Since all the resistors have the same resistance, the one with the largest current flowing through it will have the greatest potential difference across it. Based on the diagram, you can see that all the current from the battery flows through resistors R_A and R_C, and only part of this current flows through the other resistors. Therefore, the potential difference across resistors R_A and R_C is the highest among them all:

$$V_A = V_C = I_{battery}(2.0\ \Omega) = \left(\frac{90}{41}\ A\right)(2\ \Omega) = 4\frac{16}{41}\ V$$

Making sense of the result:

Notice the problem solving strategy: we identify a part of a circuit that can be simplified and continue step by step simplification until we arrive at the final circuit that contains a battery and an equivalent resistance. You can check your calculations using PhET or other circuit simulation software by constructing and measuring the circuit.

When I used to see a resistor-filled circuit diagram, such as the one in Example 22-5, I always cringed. They always look so complicated, and I never knew where to start. Then I learned a strategy to reduce the number of resistors: start in the deepest or most inward branch of the circuit that contains a resistor, then move step by step, simplifying the circuit. It's really important to redraw the circuit after you simplify it, after every step, as shown in Example 22-5. This way, the circuit just keeps simplifying, until it has only one equivalent resistor. Once I discovered this strategy, it made solving circuit problems much easier.

 CHECKPOINT

C-22-7 Ranking the Relative Brightness of Light Bulbs

The circuit in Figure 22-20 has six identical light bulbs and an ideal battery. Initially, switch S is open. Which of the following statements correctly describes what happens to the brightness of bulbs A and D when the switch is closed?

(a) Bulb A gets brighter, and bulb D turns off.
(b) Bulb A gets dimmer, and bulb D gets brighter.
(c) Bulb A turns off, and bulb D gets brighter.
(d) The brightness of bulbs A and D remains the same.
(e) Both bulbs A and D get dimmer.

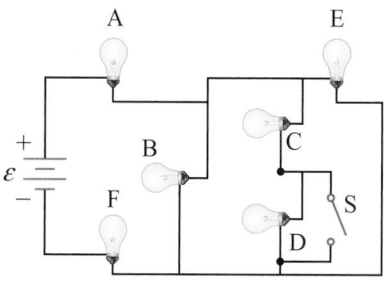

Figure 22-20 C-22-7

through bulb D.
lent resistance means increasing the current through bulb A and decreasing the current to zero
After the switch is open, the equivalent resistance will be $R_{eq_2} = 2\frac{1}{3}R$. Decreasing equiva-
all light bulbs have equal resistance R, the equivalent resistance initially equals $R_{eq_1} = 2\frac{5}{2}R$.
will bypass light bulb D. Thus, the equivalent resistance of the circuit will decrease. Assuming that
C-22-7 (a) By closing switch S, you will create a zero-resistance path for the electric current that

Consider a circuit with two parallel branches, one of which has zero resistance. The equivalent resistance of such a circuit is also zero:

$$\frac{1}{R_{parallel}} = \frac{1}{R_1} + \frac{1}{R_2}$$

$$R_{parallel} = \frac{R_1 \cdot R_2}{R_1 + R_2} = \frac{0 \cdot R_2}{R_2} = 0 \qquad (22\text{-}35)$$

According to Ohm's law, a zero–equivalent resistance electric circuit connected to a real battery produces a very high electric current. The current will travel through the zero-resistance branch, causing overheating. We say that the branches with nonzero resistance are "shorted out" by the branch with no resistance. A **short circuit** can damage power supplies and cause a fire or shock hazard.

The potential differences across the parallel branches of this circuit are also zero:

$$V_{across\ branch1} = R_1 \cdot I_1 = 0 \cdot I = 0\ V$$

$$V_{across\ branch2} = R_2 \cdot I_2 = R_2 \cdot 0 = 0\ V \qquad (22\text{-}36)$$

Applications of Kirchhoff's Circuit Laws

The technique of analyzing a circuit by replacing series and parallel combinations of resistances with equivalent resistances might not work for some circuits, such as those with configurations of components that cannot be separated into series and parallel connections. For such circuits, we apply Kirchhoff's circuit laws, which we introduced earlier in this section (Equations (22-32) and (22-33)):

Kirchhoff's Circuit Laws

KEY EQUATION
$$\sum_{junction} I_i = 0 \quad or \quad \sum_{junction} I_{entering} = \sum_{junction} I_{exiting}$$

$$(22\text{-}33)$$

KEY EQUATION
$$\sum_{closed\ loop} \Delta V = 0 \qquad (22\text{-}32)$$

As we discussed earlier, Kirchhoff's first law is a consequence of the electric charge conservation, and it states that the algebraic sum of the currents at a junction is zero provided the sign notation for the currents mentioned earlier is used. Kirchhoff's second law is a consequence of energy conservation, and it states that the algebraic sum of potential differences (voltages) across the circuit elements around a closed circuit loop is zero. To demonstrate a step by step procedure for applying Kirchhoff's laws, we analyze the circuit shown in Figure 22-21. Our goal is to find the current flowing through each circuit branch and the potential difference across each circuit element.

Circuit Analysis Using Kirchhoff's Laws

1. Identify the loops. For example, the circuit in Figure 22-21 has three loops, two of which are independent loops.

2. Identify and label all the junctions and branches. In Figure 22-21, there are three parallel branches (top, middle, and bottom) connecting the two junctions, A and B.

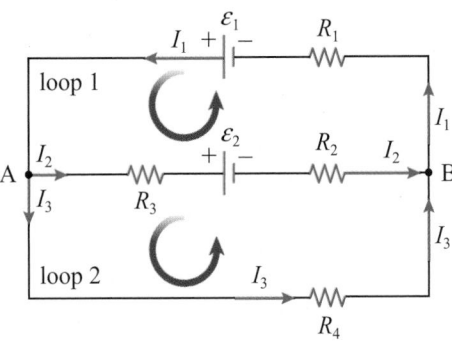

Figure 22-21 Circuit analysis using Kirchhoff's laws

3. Label the currents in each branch. Do not worry about current direction. You can choose it arbitrarily. If your choice happens to be opposite to the actual direction of the current, the value of the current you calculate will be negative. In Figure 22-21, we have three different currents: I_1, I_2, and I_3. Remember that the current through each series component in a branch is the same. For example, the current leaving the battery ε_1 in Figure 22-21 is labelled I_1 because it is the same current as the current entering R_1 from junction B.

4. Apply Kirchhoff's first law (junction rule) to each junction. In general, the number of independent equations you obtain from applying the junction rule is one fewer than the number of junctions. In our example, the equations for junctions A and B are the same: $I_1 = I_2 + I_3$.

5. Apply Kirchhoff's second law (loop rule) to as many loops as needed to include each branch of the circuit at least once. Choose a starting point in each loop, and make sure that you go around the entire loop. Follow the same direction (clockwise or counterclockwise)

in each loop, and use the sign convention listed in Table 22-4. For example, in loop 1, we have chosen the positive terminal of the battery as the starting and ending point and counterclockwise as the direction.

For our example, the application of the Kirchhoff's second law produces the following equations:

Loop 1: $-I_2R_3 - \varepsilon_2 - I_2R_2 - I_1R_1 + \varepsilon_1 = 0$

Loop 2: $I_2R_3 - I_3R_4 + I_2R_2 + \varepsilon_2 = 0$

6. Identify the number of unknowns, and choose an equal number of independent equations. For the example circuit, we have three unknowns (currents I_1, I_2, and I_3), so we need three independent equations. Solve the equations in general form. Do not substitute known values right away.

$$\begin{cases} -I_2R_3 - \varepsilon_2 - I_2R_2 - I_1R_1 + \varepsilon_1 = 0 \\ I_2R_3 - I_3R_4 + I_2R_2 + \varepsilon_2 = 0 \\ I_1 = I_2 + I_3 \end{cases}$$

$$\begin{cases} -I_2(R_3 + R_2) - \varepsilon_2 - I_1R_1 + \varepsilon_1 = 0 \\ I_2(R_3 + R_2) - I_3R_4 + \varepsilon_2 = 0 \\ I_1 = I_2 + I_3 \end{cases}$$

$$\begin{cases} -I_2(R_3 + R_2) - \varepsilon_2 - (I_2 + I_3)R_1 + \varepsilon_1 = 0 \\ I_2(R_3 + R_2) - I_3R_4 + \varepsilon_2 = 0 \\ I_1 = I_2 + I_3 \end{cases}$$

$$\begin{cases} I_2 = \dfrac{\varepsilon_1 - I_3(R_1 + R_4)}{R_1} \\ I_2 = \dfrac{I_3R_4 - \varepsilon_2}{R_2 + R_3} \\ I_1 = I_2 + I_3 \end{cases}$$

Table 22-4 Sign conventions for applying Kirchhoff's laws

Sign convention	Potential difference across the resistor	Potential difference across the source
Positive: 1. The direction of the traverse around the loop is opposite to the direction of the current through the resistor (R). 2. The direction of the traverse around the loop is from the positive terminal to the negative terminal of the emf source.	IR	ε
Negative: 1. The direction of the traverse around the loop coincides with the direction of the current through the resistor (R). 2. The direction of the traverse around the loop is from the negative terminal to the positive terminal of the source.	$-IR$	$-\varepsilon$

$$I_1 = \frac{\left(\dfrac{\varepsilon_1(R_2 + R_3) + \varepsilon_2 R_1}{R_1 R_4 + R_1 R_2 + R_1 R_3 + R_2 R_4 + R_3 R_4}\right) R_4 - \varepsilon_2}{R_2 + R_3}$$

$$+ \frac{\varepsilon_1(R_2 + R_3) + \varepsilon_2 R_1}{R_1 R_4 + R_1 R_2 + R_1 R_3 + R_2 R_4 + R_3 R_4}$$

$$I_2 = \frac{I_3 R_4 - \varepsilon_2}{R_2 + R_3} = \frac{\left(\dfrac{\varepsilon_1(R_2 + R_3) + \varepsilon_2 R_1}{R_1 R_4 + R_1 R_2 + R_1 R_3 + R_2 R_4 + R_3 R_4}\right) R_4 - \varepsilon_2}{R_2 + R_3}$$

$$I_3 = \frac{\varepsilon_1(R_2 + R_3) + \varepsilon_2 R_1}{R_1 R_4 + R_1 R_2 + R_1 R_3 + R_2 R_4 + R_3 R_4}$$

7. Use dimensional analysis to check whether the expressions for the unknowns make sense. In our example, all the final variables (I) should have units of current (amperes or volts/ohm).

8. Use limiting cases to check your final answers. Example 22-6 demonstrates this technique.

9. Use Ohm's law and the calculated values for the currents to find potential differences across the circuit elements.

EXAMPLE 22-6

Analyzing a Multiloop and Multibattery Electric Circuit

Assume that the components in Figure 22-21 have the following values:

$$\varepsilon_1 = 10 \text{ V}; \quad \varepsilon_2 = 15 \text{ V (ideal batteries)}$$

$$R_1 = 5.0 \ \Omega; \quad R_2 = 10 \ \Omega; \quad R_3 = 15 \ \Omega; \quad R_4 = 20 \ \Omega$$

(a) Find the current through each branch of the circuit.
(b) Find the potential differences across each circuit element.
(c) How will your answers to parts (a) and (b) differ if the batteries are real batteries?

SOLUTION

(a) Substituting the given values into the expressions derived for the currents in the three branches, we obtain the following:

$$I_2 = \frac{I_3 R_4 - \varepsilon_2}{R_2 + R_3} = \frac{\left(\dfrac{\varepsilon_1(R_2 + R_3) + \varepsilon_2 R_1}{R_1 R_4 + R_1 R_2 + R_1 R_3 + R_2 R_4 + R_3 R_4}\right) R_4 - \varepsilon_2}{R_2 + R_3}$$

$$= -\frac{7}{29} \text{ A}$$

$$I_3 = \frac{\varepsilon_1(R_2 + R_3) + \varepsilon_2 R_1}{R_1 R_4 + R_1 R_2 + R_1 R_3 + R_2 R_4 + R_3 R_4} = \frac{13}{29} \text{ A}$$

$$I_1 = I_2 + I_3 = \frac{6}{29} \text{ A}$$

These results indicate that our choice of current directions for I_1 and I_3 was correct, and current I_2 flows in the opposite direction to the one chosen in the diagram.

(b) To find the potential differences across each resistor, we use Ohm's law:

$$V_{R_1} = I_1 R_1 = \left(\frac{6}{29} \text{ A}\right)(5.0 \ \Omega) = 1\frac{1}{29} \text{ V}$$

$$V_{R_2} = I_2 R_2 = \left|-\frac{7}{29} \text{ A}\right|(10 \ \Omega) = \frac{70}{29} \text{ V} = 2\frac{12}{29} \text{ V}$$

$$V_{R_3} = I_2 R_3 = \left|-\frac{7}{29} \text{ A}\right|(15 \ \Omega) = \frac{105}{29} \text{ V} = 3\frac{18}{29} \text{ V}$$

$$V_{R_4} = I_3 R_4 = \left(\frac{13}{29} \text{ A}\right)(20 \ \Omega) = \frac{260}{29} \text{ V} = 8\frac{28}{29} \text{ V}$$

(c) If the batteries are real, we have to account for their internal resistance. This can be done by adding their internal resistances in series to each battery. For the circuit, it is equivalent to replacing resistances R_1 and R_2 with R'_1 and R'_2:

$$\begin{cases} R'_1 = R_1 + r_1 \\ R'_2 = R_2 + r_2 \end{cases}$$

Making sense of the result:

To check your calculations, you can use Wolfram Alpha (www.wolframalpha.com) or other software for solving simultaneous equations. However, checking the calculations does not tell you if your initial equations are correct. To check these equations, we can apply dimensional analysis, analysis of limiting cases, and the law of conservation of energy.

Dimensional analysis: The currents should all have dimensions of amperes:

(continued)

$$[I_3] = \frac{\varepsilon_1(R_2 + R_3) + \varepsilon_2 R_1}{R_1 R_4 + R_1 R_2 + R_1 R_3 + R_2 R_4 + R_3 R_4}$$

$$= \frac{V \cdot \Omega + V \cdot \Omega}{\Omega^2} = \frac{V}{\Omega} = A$$

$$[I_2] = \frac{I_3 R_4 - \varepsilon_2}{R_2 + R_3} = \frac{A \cdot \Omega - V}{\Omega} = \frac{V}{\Omega} = A$$

$$[I_1] = I_2 + I_3 = A + A = A$$

Dimensional analysis helps identify errors, but does not guarantee that the solution is correct.

Limiting case analysis: Let us consider the limiting case, where no source of energy is present in the circuit. If we replace the two batteries with conductors, all the currents become zero. The expressions derived above equal zero when the two emfs are zero. So, these equations are valid for this limiting case. If only the first battery, ε_1, is removed from the circuit, then the equivalent resistance of the circuit becomes

$$R_{eq1} = 10\ \Omega + 15\ \Omega + \frac{1}{\dfrac{1}{5.0\ \Omega} + \dfrac{1}{20\ \Omega}} = 29\ \Omega$$

and the current flowing through the battery is

$$I_2 = \frac{\varepsilon_2}{R_{eq1}} = \frac{15\ V}{29\ \Omega} = \frac{15}{29}\ A$$

Let us check that by substituting $\varepsilon_1 = 0$ in the general solution we get the same answer:

$$(I_2)_{\varepsilon_1 = 0} = \frac{\left(\dfrac{\varepsilon_2 R_1}{R_1 R_4 + R_1 R_2 + R_1 R_3 + R_2 R_4 + R_3 R_4}\right)R_4 - \varepsilon_2}{R_2 + R_3} = \frac{15}{29}\ A$$

Energy analysis: You can check that the potential differences across each branch between junctions A and B are equal, as we would expect because the branches are parallel.

Top branch:

$$V_{AB} = I_1 R_1 - \varepsilon_1 = \left(\frac{6}{29}\ A\right)(5.0\ \Omega) - 10\ V$$

$$= \left(\frac{30}{29} - 10\right)V = -8\frac{28}{29}\ V$$

Middle branch:

$$V_{AB} = -I_2 R_3 - \varepsilon_2 - I_2 R_2 = -\left(-\frac{7}{29}\ A\right)(15\ \Omega)$$

$$-15\ V - \left(-\frac{7}{29}\ A\right)(10\ \Omega) = -8\frac{28}{29}\ V$$

Bottom branch:

$$V_{AB} = -I_3 R_1 = \left(-\frac{13}{29}\ A\right)(20\ \Omega) = -8\frac{28}{29}\ V$$

The values for V_{AB} are identical across each branch, which tells us that we solved the problem correctly. Note that when finding potential difference V_{AB} along a particular branch, we traverse the branch from point A to point B.

PEER TO PEER

It is important to remember that you always have to try and make sense of your values, especially when the intermediate calculations are long and tedious. It is easy to make a mathematical error. A careful analysis of your answer will help you catch it.

LO 5

22-5 RC Circuits

Online Activity 22-1 examines a circuit with a capacitor, a battery, a light bulb, and two switches. The brightness of the light bulb in this circuit changes with time, indicating that the current through the circuit is time-dependent. Since the voltage across the capacitor and across the light bulb are both related to the current in the circuit, these voltages also vary with time. We will now use Kirchhoff's laws to derive mathematical expressions describing how the current and voltage change as a capacitor charges and discharges.

Charging a Capacitor

Consider the circuit in Figure 22-22. Let us first close switch S_1 while switch S_2 remains open. By closing this switch, we complete a series circuit that includes an initially uncharged capacitor (C), a resistor (R), and an emf source (ε). Closing switch S_1 starts the process of charging the capacitor. Applying Kirchhoff's loop rule to the circuit, we obtain

$$\varepsilon - IR - V_C = 0 \tag{22-37}$$

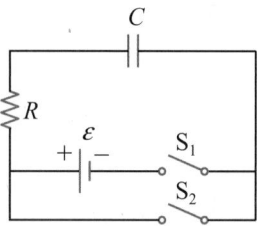

Figure 22-22 A simple RC circuit

In Chapter 21, we discussed how the potential difference across a capacitor, V_C, is related to the charge on each of its plates (q) and to its capacitance (C):

$$V_C = \frac{q}{C}$$

In this section, we consider what happens during the charging and discharging of a capacitor. During these processes the values of electric current in the circuit, the charge on capacitor's plates, and the voltage across the plates are continuously changing. We will use lower case letters for these instantaneous values. Do not confuse I, which stands for a constant value of electric current, and $i(t)$, which represents an instantaneous value of electric current. Since $i(t) = \dfrac{dq}{dt}$, we can substitute for I and V_C in Equation (22-37) to obtain

$$\varepsilon - \frac{dq}{dt}R - \frac{q}{C} = 0 \qquad (22\text{-}38)$$

As the electric charge on the capacitor's plates varies with time, we denote it as $q(t)$. Then we can rearrange the above equation as follows:

$$\varepsilon - \frac{dq(t)}{dt}R - \frac{q(t)}{C} = 0$$

$$\frac{dq(t)}{dt} = \frac{\varepsilon}{R} - \frac{q(t)}{RC}$$

$$\frac{dq(t)}{dt} = -\frac{q(t) - \varepsilon C}{RC}$$

$$\frac{dq(t)}{q(t) - \varepsilon C} = -\frac{dt}{RC} \qquad (22\text{-}39)$$

This is a differential equation. Noticing that the quantity εC is constant, we can use the following substitution:

$$x(t) \equiv q(t) - \varepsilon C \Rightarrow dx(t) \equiv dq(t) \qquad (22\text{-}40)$$

As a result, Equation (22-39) can be rewritten as follows:

$$\frac{dx(t)}{x} = \frac{dt}{RC} \qquad (22\text{-}41)$$

Moreover, at time $t = 0$, the capacitor had no charge on its plates, so $q(t = 0) = 0 \Rightarrow x(t = 0) = -\varepsilon C$, and at time t, the plates had a charge of $\pm q(t)$. Therefore, $x(t) = q(t) - \varepsilon C$. Integrating both sides of Equation (22-41) gives

$$\int_{-\varepsilon C}^{q - \varepsilon C} \frac{dx(t)}{x} = -\int_{0}^{t} \frac{dt}{RC}$$

$$\ln\left(x(t)\right)\Big|_{-\varepsilon C}^{q(t) - \varepsilon C} = -\frac{t}{RC}$$

$$\ln\left(q(t) - \varepsilon C\right) - \ln(-\varepsilon C) = -\frac{t}{RC}$$

$$\ln\frac{q(t) - \varepsilon C}{-\varepsilon C} = -\frac{t}{RC}$$

$$\frac{q(t) - \varepsilon C}{-\varepsilon C} = e^{-\frac{t}{RC}}$$

$$q(t) = (-\varepsilon C)e^{-\frac{t}{RC}} + \varepsilon C = \varepsilon C(1 - e^{-\frac{t}{RC}}) \quad (22\text{-}42)$$

Recall that the maximum amount of charge, q_{max}, on the capacitor's plates is $V_{max}C$. In our circuit, the maximum potential difference across the plates equals the potential difference across the battery, so $V_{max} = \varepsilon$. Therefore, while a capacitor is charging, the charge on the capacitor's plates can be expressed as follows:

KEY EQUATION $\quad q(t) = q_{max}\left(1 - e^{-\frac{t}{RC}}\right) = q_{max}\left(1 - e^{-\frac{t}{\tau}}\right)$

$$(22\text{-}43)$$

Let us check the dimensions of the product RC:

$$[RC] = \Omega \cdot F = \frac{V}{A} \cdot \frac{C}{V} = \frac{C}{A} = \frac{C}{C/s} = s \qquad (22\text{-}44)$$

The product of the capacitance and the resistance of a series RC circuit has dimensions of time. It is called the **time constant** of the circuit, $\tau = RC$. The exponent in Equation (22-43) is therefore dimensionless: $\left[\dfrac{t}{\tau}\right] = 1$.

Differentiating Equation (22-45), we find the dependence of electric current on time:

$$i(t) = \frac{dq(t)}{dt} = \frac{d\left(q_{max}\left(1 - e^{-\frac{t}{\tau}}\right)\right)}{dt} = \frac{\varepsilon C}{RC}e^{-\frac{t}{\tau}} = \frac{\varepsilon}{R}e^{-\frac{t}{\tau}}$$

$$(22\text{-}45)$$

The ratio $\dfrac{\varepsilon}{R}$ represents the maximum current in this RC circuit (the current in the absence of a capacitor). This current flows in the circuit only at the initial moment when the switch is being closed ($t = 0$) and the capacitor begins charging. Thus,

KEY EQUATION $\quad i(t) = I_0 e^{-\frac{t}{RC}} = I_0 e^{-\frac{t}{\tau}} \qquad (22\text{-}46)$

Equation (22-46) describes the dependence of electric current on time during the charging of a capacitor in an RC circuit. In Chapter 21, we found that the energy stored in a charging capacitor depends on the square of the charge stored by it. Substituting for $q(t)$ in Equation (21-15), $U = \dfrac{1}{2}\dfrac{Q^2}{C}$, gives an expression for the energy stored in a capacitor as a function of time:

$$U_C = \frac{(q(t))^2}{2C} = \frac{\left(q_{max}\left(1 - e^{-\frac{t}{RC}}\right)\right)^2}{2C} \qquad (22\text{-}47)$$

Graphs of $q(t)$ and $i(t)$ are shown in Figure 22-23.

The voltage across the capacitor asymptotically approaches its maximum value, $V_{max} = \varepsilon C$. At the same time, the current in this RC circuit decreases exponentially but never reaches zero. For exponential decay or growth, it is common to describe the relationship using the time constant. In the case of charging a capacitor, $\tau = RC$.

CHAPTER 22 | **ELECTRIC CURRENT AND FUNDAMENTALS OF DC CIRCUITS**

ELECTRICITY, MAGNETISM, AND OPTICS

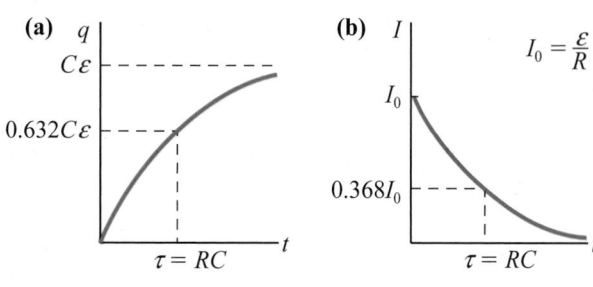

(a) q

$C\varepsilon$ ----

$0.632C\varepsilon$ ---

$\tau = RC$ t

(b) I

$I_0 = \dfrac{\varepsilon}{R}$

I_0

$0.368I_0$ ---

$\tau = RC$ t

Figure 22-23 (a) Graph of charge versus time and (b) graph of current versus time for a charging capacitor in a series RC circuit

The time constant is the time that it takes the capacitor to charge to about 63% of its maximum value, q_{max}:

$$q(\tau) = q_{max}\left(1 - e^{-\frac{\tau}{\tau}}\right) = q_{max}\left(1 - \frac{1}{e}\right) \approx 0.632\, q_{max}$$

$$(22\text{-}48)$$

At the same time ($t = \tau$), the current across the capacitor is a factor of e less than the maximum current, I_0:

$$i(\tau) = I_0 e^{-\frac{\tau}{\tau}} = \frac{I_0}{e} \approx 0.368\, I_0 \qquad (22\text{-}49)$$

The relationships described above explain the observed light bulb brightness in a series RC circuit. When the capacitor begins charging, the current is near its maximum and the light bulb is bright. Then the light bulb gets progressively dimmer, and eventually stops glowing.

Therefore, the charge on the plates of a capacitor when it is charging can be expressed as follows:

KEY EQUATION $q(t) = q_{max}\left(1 - e^{-\frac{t}{RC}}\right) \qquad (22\text{-}50)$

For most practical purposes, a capacitor can be considered fully charged after 5 to 10 time constants have elapsed.

Discharging a Capacitor

Let us now assume that the capacitor in Figure 22-22 is fully charged, and we open switch S_1 and close switch S_2. The capacitor and resistor are now connected in a loop with S_2, leaving out the emf source (ε). With these

---

EXAMPLE 22-7

Charging a Capacitor

A 500 μF capacitor is connected to a 9.00 V battery through a 2.00 kΩ resistor, thus completing a simple RC series circuit.

(a) How long does it take for the capacitor to reach 99% of q_{max}? Express your answer in terms of the time constant, τ.
(b) How much charge will have accumulated on the capacitor's plates when $t_1 = \tau$, $t_2 = 2\tau$, and $t_3 = 3\tau$?
(c) Find the current in the circuit at the times in part (b).

SOLUTION

(a) For this circuit,

$$\tau = RC = (2 \times 10^3\,\Omega)(500 \times 10^{-6}\,F) = 1.00\ s$$

$$0.99\, q_{max} = q_{max}\left(1 - e^{-\frac{t}{RC}}\right)$$

$$0.99 = 1 - e^{-\frac{t}{RC}}$$

$$e^{-\frac{t}{RC}} = 0.01$$

$$t = -RC\ln(0.01) = -\tau\ln(0.01) = 4.61\,\tau$$

$$t = -(2 \times 10^3\,\Omega)(500 \times 10^{-6}\,F)\ln(0.01)$$

$$= 4.61\ s$$

(b)

$$q_{max} = \varepsilon C = (9\ V)(500 \times 10^{-6}\,F) = 4.5 \times 10^{-3}\,C$$

$$q(\tau) = q_{max}\left(1 - e^{-\frac{\tau}{RC}}\right) = q_{max}\left(1 - e^{-\frac{RC}{RC}}\right)$$

$$= q_{max}\left(1 - \frac{1}{e}\right) \approx 0.632\, q_{max} = 2.84 \times 10^{-3}\,C$$

$$q(2\tau) = q_{max}\left(1 - e^{-\frac{2\tau}{RC}}\right) = q_{max}\left(1 - e^{-\frac{2RC}{RC}}\right)$$

$$= q_{max}\left(1 - \frac{1}{e^2}\right) \approx 0.865\, q_{max} = 3.89 \times 10^{-3}\,C$$

$$q(3\tau) = q_{max}\left(1 - e^{-\frac{3\tau}{RC}}\right) = q_{max}\left(1 - e^{-\frac{3RC}{RC}}\right)$$

$$= q_{max}\left(1 - \frac{1}{e^3}\right) \approx -0.950\, q_{max} = 4.28 \times 10^{-3}\,C$$

(c)

$$I_0 = \frac{\varepsilon}{R} = \frac{(9.00\ V)}{2000\ \Omega} = 4.50 \times 10^{-3}\,A$$

$$I(\tau) = I_0 e^{-\frac{\tau}{RC}} = I_0 e^{-\frac{RC}{RC}} = I_0\left(\frac{1}{e}\right) \approx 0.368\, I_0 = 1.66 \times 10^{-3}\,A$$

$$I(2\tau) = I_0 e^{-\frac{2\tau}{RC}} = I_0 e^{-\frac{2RC}{RC}} = I_0\left(\frac{1}{e^2}\right) \approx 0.135\, I_0 = 0.609 \times 10^{-3}\,A$$

$$I(3\tau) = I_0 e^{-\frac{3\tau}{RC}} = I_0 e^{-\frac{3RC}{RC}} = I_0\left(\frac{1}{e^3}\right) \approx 0.050\, I_0 = 0.224 \times 10^{-3}\,A$$

Making sense of the result:

It is often helpful to calculate the time constant of the circuit first ($\tau = 1$ s in this example). You can see that the capacitor acquired 99% of its maximum charge before $t = 5\tau = 5$ s.

connections, the capacitor will start discharging. Applying Kirchhoff's loop rule to the circuit, we obtain

$$-V_C - IR = 0$$

$$-\frac{q(t)}{C} - IR = 0 \qquad (22\text{-}51)$$

We can derive equations for the charge on the capacitor's plates and the current through the circuit by using the same method as for the example above. For the discharging capacitor,

$$-\frac{q(t)}{C} - \frac{dq(t)}{dt}R = 0 \qquad (22\text{-}52)$$

At time $t = 0$, the capacitor was fully charged: $q(0) = C\varepsilon = q_{max}$. Again, we find the solution by rearranging the differential equation and integrating both sides:

$$\frac{dq(t)}{q(t)} = -\frac{dt}{RC}$$

$$\int_{\varepsilon C}^{q(t)} \frac{dq(t)}{q(t)} = -\int_0^t \frac{dt}{RC}$$

$$\ln q(t) - \ln(\varepsilon C) = -\frac{t}{RC}$$

$$\ln \frac{q(t)}{\varepsilon C} = -\frac{t}{RC}$$

$$q(t) = q_{max} e^{-\frac{t}{RC}} = q_{max} e^{-\frac{t}{\tau}} \qquad (22\text{-}53)$$

Therefore, the charge on the plates of a capacitor when it is discharging can be expressed as follows:

KEY EQUATION $\quad q(t) = q_{max} e^{-\frac{t}{RC}} = q_{max} e^{-\frac{t}{\tau}} \qquad (22\text{-}54)$

Consequently, the current is

$$I(t) = \frac{dq(t)}{dt} = \frac{d\left(q_{max} e^{-\frac{t}{RC}}\right)}{dt}$$

$$= -\frac{\varepsilon C}{RC} e^{-\frac{t}{RC}} = -\frac{\varepsilon}{R} e^{-\frac{t}{RC}} = -\frac{\varepsilon}{R} e^{-\frac{t}{\tau}}$$

The ratio $\dfrac{\varepsilon}{R}$ once again represents the maximum current in this RC circuit. This current flows at the instant that switch S_2 is being closed ($t = 0$). Thus,

$$I(t) = -I_0 e^{-\frac{t}{RC}} = -I_0 e^{-\frac{t}{\tau}} \qquad (22\text{-}55)$$

The negative value for current in this equation indicates that it flows in an opposite direction to the current when the capacitor was charging.

Equations (22-54) and (22-55) describe the time dependence of the current and charge during the discharge of a capacitor in an RC circuit. As above, we use these equations to find an expression for the electric potential energy stored in the capacitor at any given moment:

$$U_C = \frac{(q(t))^2}{2C} = \frac{\left(q_{max} e^{-\frac{t}{RC}}\right)^2}{2C} = \frac{(q_{max})^2 e^{-\frac{2t}{RC}}}{2C} \qquad (22\text{-}57)$$

Figure 22-24 shows $q(t)$ and $i(t)$ graphs for a capacitor discharging through a resistor. Comparing the graphs in Figures 22-23 and 22-24, we see that $i(t)$ graphs are the same: the current starts with its maximum value and then diminishes exponentially with time. However, $q(t)$ graphs for charging and discharging a capacitor have opposite trends.

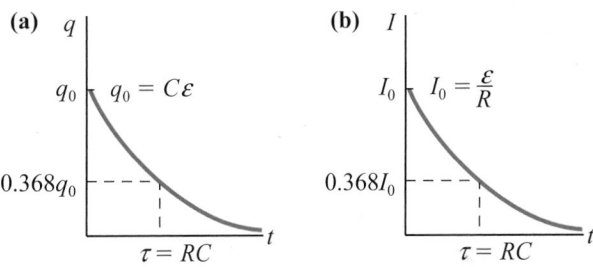

(a) Graph with q on vertical axis: q_0, $q_0 = C\varepsilon$, $0.368q_0$, $\tau = RC$, t

(b) Graph with I on vertical axis: I_0, $I_0 = \dfrac{\varepsilon}{R}$, $0.368I_0$, $\tau = RC$, t

Figure 22-24 (a) Graph of charge versus time and (b) graph of current versus time for a discharging capacitor in a series RC circuit. The absolute value of current is shown in the graph.

The analysis above can be extended to circuits that have a combination of resistors and capacitors by calculating the equivalent resistance and the equivalent capacitance of the combinations.

C-22-8 Discharging a Capacitor

The five circuits in Figure 22-25 consist of identical capacitors and resistors. At $t = 0$ s, the switch is closed and the capacitors start discharging. In which circuit will the capacitors discharge to 5% of their maximum charge in the shortest time? Explain.

(a) circuit 1
(b) circuit 2
(c) circuit 3
(d) circuit 4
(e) circuit 5

Figure 22-25 C-22-8

C-22-8 (d) The circuit time constant $\tau = RC$ shows how fast the capacitor in the circuit is going to be charging or discharging. The time constants of the circuits are $\tau_1 = RC;\ \tau_2 = 2R = \dfrac{2}{C} = RC;\ \tau_3 = 2C = RC;\ \tau_4 = \dfrac{R}{2} = \dfrac{2}{C} = \dfrac{RC}{2};\ \tau_5 = \dfrac{RC}{4} = 2R2C = 4RC$

How Do Touch-Sensitive Lamps Work?

Touch-sensitive lamps commonly have a power-level switch controlled by an RC circuit. Your body has some capacitance. When you touch the metal body of a touch-sensitive lamp, you add capacitance to the control circuit and usually some static charge from your body, as well. These changes in the capacitance trigger the control circuit such that it changes the voltage supplied to the lamp's bulb. Often, touch-sensitive switches cycle through three power levels: off-low-medium-high, then back to off.

ELECTRICITY, MAGNETISM, AND OPTICS

Metals have free electrons that can drift in the direction opposite to the external electric field. Moving electric charges create electric current. Electric current flowing through a conductor carries energy that can be converted into heat, which is dissipated by the conductor. There is a fundamental relationship between electric current through a resistor, its resistance, and the voltage across it. DC circuits have currents that flow in only one direction through the components. Ohm's law, equivalent resistances, and Kirchhoff's laws can be used to analyze circuits that contain a number of components such as resistors, capacitors, and emf sources.

Electric Current: The Microscopic Model

Electron current in a metal wire i is related to electric current in it and electric current density, J. They depend on electron density n_e, drift speed v_d, and cross-sectional area A:

$$I = I_e \cdot e = (n_e v_d A)e = n_e v_d A(1.6 \times 10^{-19} C) \qquad (22\text{-}4)$$

Electric conductivity and resistivity, ρ, describe how well a specific metal conducts current; these properties are temperature-dependent:

$$\sigma = \frac{1}{\rho} = \frac{n_e e^2 \tau}{m} \qquad (22\text{-}9)$$

$$\rho = \rho_0 [1 + \alpha \Delta T] \qquad (22\text{-}12)$$

where α is the temperature coefficient of resistivity, T is temperature, and ρ_0 is the resistivity at a baseline temperature.

Electric Current, Resistance, and Voltage: The Macroscopic Model

Electric current I through a resistor R and the voltage across it V are related by Ohm's law:

$$I = \frac{V}{R} \text{ (for ohmic resistors only)} \qquad (22\text{-}15)$$

The resistance of a conductor depends on its resistivity ρ, length l, and cross-sectional area A:

$$R = \rho \frac{l}{A} \qquad (22\text{-}14)$$

A resistor R that carries electric current I dissipates thermal energy Q_{heat} according to Joule–Lenz's law:

$$Q_{heat} = I^2 R \Delta t \qquad (22\text{-}24)$$

In a series RC circuit, electric current across the capacitor changes as

$$i(t) = I_0 e^{-\frac{t}{RC}} \qquad (22\text{-}46)$$

Electric charge on the capacitor's plates:

$$q(t) = q_{max}\left(1 - e^{-\frac{t}{RC}}\right) \quad \text{charging} \qquad (22\text{-}50)$$

$$q(t) = q_{max}\left(e^{-\frac{t}{RC}}\right) \quad \text{discharging} \qquad (22\text{-}54)$$

Analysis of Compound Electric Circuits: Equivalent Resistance

For resistors connected in series:

$$R_{series} = \sum_{i=1}^{N} R_i \qquad (22\text{-}18)$$

For resistors connected in parallel:

$$\frac{1}{R_{parallel}} = \sum_{i=1}^{N} \frac{1}{R_i} \qquad (22\text{-}20)$$

Kirchhoff's Laws for Analysis of Multiloop DC Circuits

Junction rule (conservation of current):

$$\sum_{junction} I_{entering} = \sum_{junction} I_{exiting} \qquad (22\text{-}33)$$

Loop rule (energy conservation):

$$\sum_{closed\ loop} \Delta V = 0 \qquad (22\text{-}32)$$

Applications: electrical appliances, residential and commercial electrical wiring, physiological applications

Key Terms: ammeter, conduction electron density, current density, direct current (DC) circuit, drift velocity, electric circuit junction, electric current, electric conductivity, electric resistivity, electrical grounding, electron current, free electron density, electron flow, equivalent resistance, junction points, Kirchhoff's first law, Kirchhoff's second law, law of electric charge conservation, metallic bonds, multimeter, node, ohmmeter, Ohm's law, parallel circuit, power, series circuit, short circuit, superconductor, time constant, voltmeter

QUESTIONS

1. Which of the following statements is true? Explain your answer.
   (a) The directions of electron and electric currents in metals are always the same.
   (b) If the resistivities of two wires are the same, then their resistances must also be the same.
   (c) The free electron density in a metal depends on its temperature.
   (d) Resistivity and conductivity of a metal depend on the metal's temperature.
   (e) The electron drift speed depends on the resistivity of the metal.

2. In a box labelled "spare 40 W and 60 W bulbs," there are two clear incandescent light bulbs. One bulb has a thicker filament than the other. Which bulb is likely to be rated at 40 W?
   (a) thicker filament
   (b) thinner filament
   (c) It is impossible to know without trying each one.

3. In circuit I, a light bulb is connected to a battery. In circuit II, 10 light bulbs, all identical to the light bulb in circuit I, are connected in series to an identical battery. In circuit III, 10 light bulbs are connected in parallel to an identical battery. Which statement correctly describes the relative power outputs of these circuits? Explain your answer.
   (a) circuit I > circuit II > circuit III
   (b) circuit III > circuit I > circuit II
   (c) circuit I = circuit II = circuit III
   (d) circuit II > circuit I > circuit III
   (e) circuit III > circuit II > circuit I

4. In circuit I, a light bulb is connected to a battery. In circuit II, 10 light bulbs, all identical to the light bulb in circuit I, are connected in series to the same type of battery. Which circuit in Question 3 will produce the most light? Explain.
   (a) circuit I
   (b) circuit II
   (c) circuit III
   (d) Circuits II and III will both produce 10 times more light than circuit I.

5. The resistance of a long copper wire is R. You cut the wire into N equal pieces and connect the pieces in parallel. What is the resistance of this combination?
   (a) $N^2 R$
   (b) NR
   (c) R
   (d) R/N
   (e) R/N^2

6. How much energy is consumed by a 60 W bulb if you leave it on for 10 h?
   (a) 60 kW·h
   (b) 60 kJ
   (c) 600 J
   (d) 3600 kW·h
   (e) 2.16 MJ

7. How much energy is stored in a fully charged 12 V car battery rated at 100 A·h?
   (a) 12 MW·h
   (b) 1.2 MW·h
   (c) 1.0 MW·h
   (d) 4320 kJ
   (e) 432 kW·h

8. What is the ratio of the filament resistance of a 70 W light bulb designed for North America (110 V) and a 140 W light bulb designed for Europe (220 V)?
   (a) 2:1
   (b) 1:2
   (c) 2:2
   (d) 4:1
   (e) 1:4
   (f) 1:1

9. You have six identical 4.0 Ω resistors and one 3.0 V battery. How can you connect the resistors to the battery such that the current flowing from the battery is 4.5 A?
   (a) Attach all the resistors in series, and connect them to the battery.
   (b) Attach all the resistors in parallel, and connect them to the battery.
   (c) Attach two resistors in series and the rest in parallel, and then connect them to the battery.
   (d) Attach three resistors in series and the rest in parallel, and then connect them to the battery.
   (e) It is not possible.

10. Which of the following statements is false?
    (a) According to Ohm's law, an increase in a circuit's equivalent resistance means a decrease in the circuit's power output.
    (b) A voltmeter is always connected in parallel to the circuit element across which we want to measure a voltage.
    (c) An ammeter is always connected in series to the circuit element through which we want to measure a current.
    (d) A short circuit means a circuit that has a very large electric resistance, thus becoming a fire hazard.
    (e) If a piece of metal wire is heated, its resistivity will increase and its conductivity will decrease.

11. You need a 6.0 Ω resistor, but you only have four 4.0 Ω resistors. How many resistors should you use and how should you connect them to produce a total resistance of 6.0 Ω?
    (a) three 4.0 Ω resistors connected in series
    (b) three 4.0 Ω resistors connected in parallel
    (c) two 4.0 Ω resistors connected in parallel and then connected in series to a 4.0 Ω resistor
    (d) two 4.0 Ω resistors connected in series and then connected in parallel to a 4.0 Ω resistor
    (e) two 4.0 Ω resistors connected in parallel, connected in series to two 4.0 Ω resistors connected in series.

12. What is the operating resistance of a 100 W (110 V) light bulb?
    (a) 1.11 Ω
    (b) 12.1 Ω
    (c) 121 Ω
    (d) 12.1 kΩ
    (e) 121 kΩ

13. Figure 22-26 shows an electric circuit consisting of six identical light bulbs, each with an operating resistance of 10.0 Ω, connected to a 20.0 V battery. What is the equivalent resistance of this circuit?
    (a) 60 Ω
    (b) 22.5 Ω
    (c) 21.7 Ω
    (d) 18.8 Ω
    (e) 18.4 Ω

Figure 22-26 Question 13

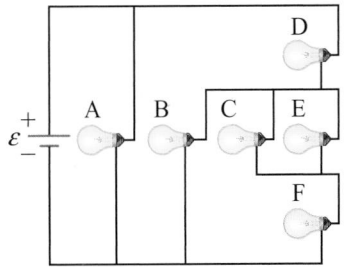

Figure 22-28 Question 18

14. What is the electric current flowing through the battery in Figure 22-26?
    (a) 1.07 A
    (b) 1.0 A
    (c) 0.92 A
    (d) 0.89 A
    (e) 1.09 A

15. In the circuit shown in Figure 22-26, what is the value of the potential difference between points A and B?
    (a) 9.24 V
    (b) 10.0 V
    (c) 18.46 V
    (d) 5.45 V
    (e) 4.62 V

16. Rank the brightness of the light bulbs in Figure 22-26. Use your answers to questions 13, 14, and 15.
    (a) $M > N > O > P = Q > R$
    (b) $M > N = O < P = Q > R$
    (c) $M > R > N = O > P = Q$
    (d) $M = N = O > P = Q < R$
    (e) $M > N > O > R > P = Q$

17. A 12.0 V battery is connected to a combination of nine identical 400.0 Ω resistors shown in Figure 22-27. What is the current flowing through this battery?
    (a) 3.33 mA
    (b) 1.33 A
    (c) 8.92 mA
    (d) 11.0 mA
    (e) 533 A

Figure 22-27 Question 17

18. Six identical 5 Ω light bulbs—A, B, C, D, E, and F—are connected as shown in Figure 22-28; $\varepsilon = 24$ V. Which of the following ranks the amount of current flowing through the light bulbs from the largest to the smallest?
    (a) $I_D, I_A, I_B, I_F, I_C = I_E$
    (b) $I_A, I_D, I_B = I_F, I_C = I_E$
    (c) $I_A, I_D, I_B, I_F, I_C = I_E$
    (d) $I_A, I_D, I_B, I_F = I_C = I_E$
    (e) $I_D, I_A, I_B, I_F = I_C = I_E$

19. A 5 μF capacitor is connected through a 10 kΩ resistor to a 12 V battery. It takes this capacitor 0.002 s to be charged to 95% of its maximum charge in this series circuit. How long would it take the same capacitor to get 95% charged (if initially uncharged) when the 12 V battery is replaced by a 6 V battery?
    (a) 0.004 s
    (b) 0.002 s
    (c) 0.001 s
    (d) less than 0.001 s
    (e) more than 0.001 s

20. An RC electric circuit is shown in Figure 22-29. Initially, switch S is open and the capacitors are completely uncharged. After the switch has been closed for 3 min, how much energy will be stored in the 5 μF capacitor?
    (a) 0.3 mJ
    (b) 0.6 mJ
    (c) 1.5 mJ
    (d) 36 mJ
    (e) 72 mJ

Figure 22-29 Question 20

21. A series RC circuit consists of a resistor, a battery, wires, and a capacitor. You are worried that the wires might not withstand the current flowing through them as the capacitor charges. When are the wires more likely to burn?
    (a) right at the beginning of the experiment, when the capacitor just starts charging
    (b) in the middle of the experiment, when the capacitor is about 32% charged
    (c) in the middle of the experiment, when the capacitor is about 67% charged
    (d) at the end of the experiment, when the capacitor is almost fully charged

PROBLEMS BY SECTION

For problems, star ratings will be used, (*, **, or ***), with more stars meaning more challenging problems.

Section 22-1   Electric Current: The Microscopic Model

22. *   Estimate the conduction (free) electron density in copper given that the atomic mass of copper is 63.546 amu and the density of copper at 293 K is 8.96 g/cm³. Compare your estimate with an experimentally established value.

23. *   An electric current of 2.0 A is flowing through a copper wire that has a diameter of 2.0 mm.
   (a) At what rate are electrons flowing through the wire?
   (b) Find the electric current density in the wire.
   (c) How much electric charge will be carried through the cross section of the wire in 2.0 s?

24. *   Your classmate heard the professor mention that the average drift speed of electrons in metals is about 0.1 mm/s. This did not make sense to your friend: "If this were true," he said, "I would have to keep my key in the ignition switch of my car for more than 4 h in order for the electrons to flow from the negative terminals of the battery to the starter motor and back to the positive terminal of the battery." How will you respond to him?

Section 22-2   Electric Conductivity and Resistivity: The Microscopic Model

25. *   An electric current of 2.0 A flows through a copper wire at 15°C. The wire has a circular cross section and is 2.0 mm in diameter.
   (a) Using the information in Table 22-2, find the conductivity and the resistivity of the wire.
   (b) Using Table 22-1, find the drift velocity of the conduction electrons in the wire.
   (c) Estimate the mean time, τ, that free electrons in the wire spend between collisions.
   (d) Find the electric field inside the wire.
   (e) How would your answers to parts (a) to (d) change if the electric current through the wire doubled (i.e., 4.0 A instead of 2.0 A)?
   (f) Suppose the cross-section of the wire were changed from circular to square with the same area. What effect would this change have on the quantities in parts (a) to (d)?
   (g) How will the quantities in parts (a) to (d) change when the temperature of the wire is raised from 15°C to 35°C while the current remains constant?

Section 22-3   Ohm's Law and Combinations of Resistors

26. *   The resistance of the tungsten filament in a common size of incandescent light bulb is 15 Ω. The diameter of the filament is 0.1 mm. What is the length of the filament wire?

27. *   The mass of the copper wire in a coil is 1.78 g, and its resistance is 34 Ω. Estimate the length and the cross-sectional area of the wire in the coil.

28. *   (a) Compare the relative change of the resistances and resistivities of an aluminum wire and a Nichrome wire that are both heated to 500°C. Neglect the thermal expansion of the wires.
   (b) How would your answers change if the lengths of the wires were doubled?
   (c) How would your answers change if you took the thermal expansion of the wires into account?

29. *   How much current flows through a 1.0 kW electric toaster connected to a 120 V outlet? (The voltage is AC, not DC, but the physics is the same.) How many toasters can you operate at the same time before you blow a fuse? (Assume that the maximum allowed current is 15 A.)

30. *   You have two conductors: Conductor 1 is a 10 m-long tin wire that has a 1 mm² square cross-sectional area. Conductor 2 is a strip of tin foil that you have shaped into a strip with the dimensions 10 mm × 200 mm, and a thickness of 100 μm. Compare the resistance of both conductors, assuming that the current flows along the longest dimension of the tin strip.

31. *   A wire made of a manganese alloy has a length of 56 cm and a diameter of 0.40 mm. The resistivity of the alloy is $\rho = 0.45\ \mu\Omega \cdot m$.
   (a) Find the resistance of the wire.
   (b) Find the current through the wire when the potential difference across it is 3.0 V.

32. *   Will a 3 m-long aluminum wire melt if you heat it enough to triple its resistance? Explain your answer.

33. **   Prove that an equivalent resistance of any number of parallel resistors will never exceed the resistance of the branch that has the least resistance among all parallel branches:

$$R_{parallel} < (R_{min})_{\text{in parallel branch}}$$

34. *   Use Figure 22-9 to determine three possible different combinations of household electrical appliances that can be used simultaneously without exceeding the maximum current of 15 A.

35. **   You decide to decrease your electricity bill by reducing the usage of your toaster from 5 min/day to 3 min/day. Your toaster has a wattage of 1 kW, and you pay 6.5¢ per kW·h.
   (a) How much energy will you save per year?
   (b) How much less will you pay for your electricity bill per year than before?
   (c) Suggest more effective ways of reducing your electricity bill.

Section 22-4   Analysis of DC Circuits and Kirchhoff's Laws

36. *   Ohm's law is sometimes expressed as $I = \dfrac{V}{R}$, $I = \dfrac{\varepsilon}{R}$, or $I = \dfrac{\varepsilon}{R + r}$. Explain when each equation applies and the differences between them.

37. *   You have four identical ideal 1.5 V batteries. How should you connect these batteries to produce
   (a) a 1.5 V emf source that will outlast a single battery;
   (b) a 3.0 V emf source that will outlast a single battery;
   (c) a 4.5 V emf source;
   (d) a 6.0 V emf source?
   Check your answer by using PhET computer simulations used earlier in Online Activities.

38. **   A series electric circuit consists of a real battery ($\varepsilon = 12$ V; $r = 3.0\ \Omega$) and an ohmic resistor with $R = 20\ \Omega$.
   (a) Draw a schematic diagram for the circuit.
   (b) Derive an expression to determine how much current will flow through the battery in the circuit, and calculate the current.

(c) Derive an expression to determine the potential difference across the battery, and calculate its value.
(d) Without doing the calculations, describe what will happen to the answers above if another identical resistor ($R = 20\ \Omega$) were connected
   (i) in parallel with the original resistor;
   (ii) in series with the original resistor.
(e) Modify the expressions derived in parts (b) and (c) to determine whether your predictions in part (d) were correct. Calculate the current through the circuit and the potential difference across the battery.

39. **✶✶** Figure 22-30 shows an electric circuit that consists of nine resistors (all multiples of R = 5.0 Ω) connected to a 12 V ideal emf source.
(a) Calculate the equivalent resistance of the circuit and the electric power dissipated by it.
(b) Find the electric current that flows through each resistor.
(c) Calculate the potential differences across each resistor.
(d) Calculate the power dissipated by each resistor.
(e) Compare your answer in question (d) to the electric power dissipated by the circuit you found in part (a). What does your comparison tell you?

Figure 22-30 Problem 39

40. **✶✶** An ammeter is always connected in series (it has a very small internal resistance), and a voltmeter is always connected in parallel (it has a very large internal resistance). Assume that the resistance of an ammeter is 1 $\mu\Omega$ and the resistance of a voltmeter is 10.0 kΩ. If a series circuit consists of a 12 V battery and a 5.0 Ω resistor, what will be the effect of measuring the electric current in this circuit and the electric voltage across the resistor using the ammeter and the voltmeter? Calculate the percentage change of electric current and voltage before and after the meters are connected.

41. **✶✶** Seven identical resistors, resistance R, are connected as shown in Figure 22-31. What is the equivalent resistance between points A and B?

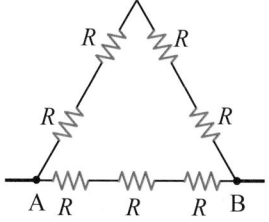

Figure 22-31 Problem 41

42. **✶✶✶** Twelve identical resistors, resistance R, are connected as shown in Figure 22-32. What is the equivalent resistance between points A and B?

Figure 22-32 Problem 42

43. **✶✶✶** Twelve identical 3.0 Ω, resistors are connected as shown in Figure 22-33. What is the equivalent resistance between points A and B?

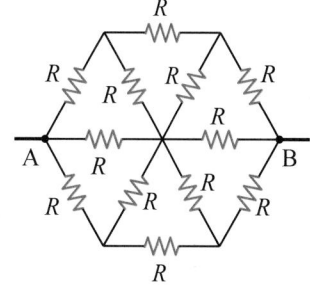

Figure 22-33 Problem 43

44. **✶✶✶** Figure 22-34 shows a multiloop electric circuit. The components have the following values:

$$\varepsilon_1 = 10\ \text{V}; \quad \varepsilon_2 = 20\ \text{V (ideal batteries)}$$

$$R_1 = 10\ \Omega; \quad R_2 = 15\ \Omega; \quad R_3 = 15\ \Omega; \quad R_4 = 20\ \Omega$$

(a) Find the currents through all the branches of the circuit.
(b) Find the potential differences across each circuit element.
(c) How will your answers to parts (a) and (b) differ if the batteries are real batteries?

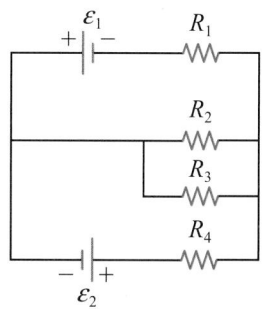

Figure 22-34 Problem 44

Section 22-5 RC Circuits

45. ✶ A series RC circuit consists of a 300.0 μF capacitor, a 12 V battery, and a 5.0 kΩ resistor.
  (a) How long does it take for the capacitor to reach 99.9% of q_{max}? Express your answer in terms of the time constant, τ.
  (b) How much charge will be accumulated on the capacitor's plates after $t_1 = \tau$, $t_2 = 2\tau$, and $t_3 = 3\tau$?
  (c) What will be the value of electric current in the circuit in each of the above cases?

46. ✶ An initially uncharged 500 μF capacitor is being connected in series to a 2500 V emf source with a 10 kΩ resistor.
  (a) How long will it take for the voltage across this capacitor to reach 95% of its maximum value? What is its maximum value?
  (b) When the capacitor reaches 95% of its maximum voltage, the charging circuit is removed and the capacitor is left charged but can slowly discharge through the moist air with a resistance of 1000 MΩ. How long will it take to discharge the capacitor to 20 V?

COMPREHENSIVE PROBLEMS

47. ✶ A high-quality dry-cell D battery (1.5 V) can provide 0.67 A for about 5 h before dying.
  (a) Compare the cost of energy you get from buying a dry-cell D battery to the cost of the energy you get from the electrical outlet in your wall. (Assume that the cost of 1 kW·h is 10¢ and one D dry-cell battery costs $1.)
  (b) How much would an average Canadian household pay per month for their electric bill if they were to use dry cells to satisfy all their electricity needs?

48. ✶ When operating, the resistance of the filament of a 60 W incandescent light bulb is 240 Ω. However, if you were to measure the resistance of the filament using a regular ohmmeter, you would find it to be about 12 times smaller (20 Ω).
  (a) How can you explain this phenomenon? (Hint: Think what happens to the light bulb when it is plugged in.)
  (b) Do appropriate calculations to show that when operating, the tungsten filament does have the appropriate resistance.
  (c) Based on your calculations, estimate the temperature of the incandescent light bulb when operating.

49. ✶ Four resistors connected to a battery have the resistances indicated in Figure 22-35; $R = 5\ \Omega$. The battery is an ideal emf source that maintains a constant potential difference across it of 10 V. A voltmeter is used to measure the following potential differences: V_{AB}, V_{AG}, V_{BH}, V_{HF}, V_{FG}, V_{HD}, V_{EF}, V_{CD}, V_{DE}, V_{BE}, and V_{CF}. Determine the readings of the voltmeter.

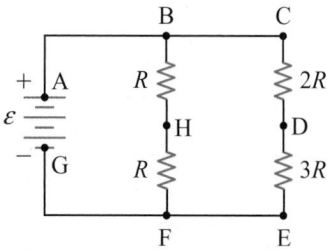

Figure 22-35   Problem 49

50. ✶✶ An electric circuit consists of six resistors and a real battery with $\varepsilon = 9.0$ V and an internal resistance $r = 0.5\ \Omega$ (Figure 22-36). Find the battery current, the voltage across each resistor, and the terminal voltage of the battery.

Figure 22-36   Problem 50

51. ✶ Figure 22-37 shows two different electric circuits.
  (a) Compare the current flowing through the batteries in both circuits.
  (b) Compare the potential differences across points A and B in each circuit. What does your comparison tell you?
  (c) Compare the brightness of the light bulbs in both circuits.

Figure 22-37   Problem 51

52. ✶ In Figure 22-38, a metal net is connected to an emf source through points A and B. The conductivity of the external (solid lines) frame of the net is very high, and the resistance of each internal (dashed) segment is r. What is the resistance of the net between points A and B?

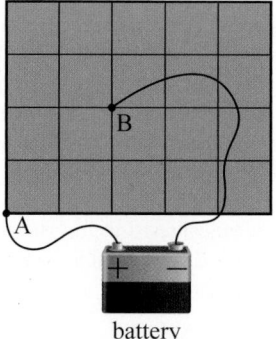

Figure 22-38   Problem 52

53. ✶✶✶ You have an aluminum wire and a tungsten wire that have the same resistance at room temperature. The temperature resistivity coefficient of aluminum is approximately half the temperature resistivity coefficient of tungsten (see Table 22-2). When answering the questions below, assume that when their temperatures are equal the wires lose thermal energy to the surrounding at the same rate. For each part, provide a detailed explanation.

(a) When the wires are connected in parallel to the battery, which wire will have a higher temperature when the system reaches equilibrium?

(b) When the wires are connected in series to the battery, which wire will have a higher temperature when the system reaches equilibrium?

54. ✷✷ Twelve identical 4.0 Ω resistors are connected to form a cube as shown in Figure 22-39. What is the equivalent resistance between points A and B?

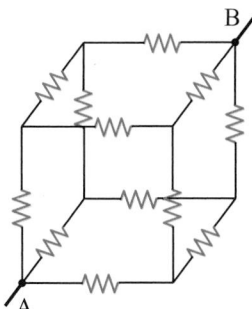

Figure 22-39   Problem 54

55. ✷✷✷ Figure 22-40 shows a multiloop electric circuit. The values of the components are as follows:

$\varepsilon_1 = 15.0$ V; $\varepsilon_2 = 20.0$ V (ideal batteries)

$R_1 = 10.0$ Ω; $R_2 = 10.0$ Ω; $R_3 = 15.0$ Ω; $R_4 = 20.0$ Ω;

$R_5 = 20.0$ Ω; $R_6 = 10.0$ Ω

(a) Find the currents through each branch of the circuit.

(b) Find the potential difference across each circuit element.

(c) How will your answers to parts (a) and (b) differ if the batteries are real batteries?

Figure 22-40   Problem 55

56. ✷✷ Design an electric circuit that can be described by the following system of equations:

$$\begin{cases} I_1 + I_2 + I_3 = 0 \\ -V_1 + I_1 R_1 - I_3 R_3 = 0 \\ I_3 R_3 - I_2 R_2 + V_2 = 0 \end{cases}$$

57. ✷✷✷ Analyze the DC circuit shown in Figure 22-41 by finding the currents flowing through each light bulb and the potential differences across each light bulb. All the emf sources in this circuit are ideal. The values of the emf sources and resistances are as follows:

$\varepsilon_1 = 25.0$ V; $\varepsilon_2 = 10.0$ V; $\varepsilon_3 = 12.0$ V

$R_1 = 10.0$ Ω; $R_2 = 5.0$ Ω; $R_3 = 15.0$ Ω; $R_4 = 20.0$ Ω;

$R_5 = 10.0$ Ω

Figure 22-41   Problem 57

58. ✷✷ A capacitor is connected to a battery through a resistor and allowed to charge fully, such that the voltage across the capacitor is V. The distance between the plates of the capacitor is quadrupled, and the capacitor remains connected to the battery. Describe what happens to the charge (Q) of the capacitor, its electric field (E), and the voltage across it (V) as a result of the increase in distance. Explain why this happens. Include appropriate diagrams to support your explanation.

59. ✷✷ The switch in the circuit shown in Figure 22-42 has been closed for a long time.

(a) Find the potential difference across each circuit element.

(b) When the switch is reopened, what is the time constant of the circuit?

Figure 22-42   Problem 59

60. ✷✷✷ A pacemaker uses an RC circuit for timing.

(a) If the pacemaker is to operate at 72 beats/min, and if the circuit "fires" when the capacitor is 65% charged, what must the time constant be?

(b) When $C = 7.50$ μF, what value of resistance should be used?

61. ✷✷✷ Assume that a capacitor used for smoothing in a DC power supply is to discharge by no more than 0.100 V (from an initial value of 10.0 V) during 1/60 s. The power supply has specifications of output voltage of 10.0 V and a supplied current of 250.0 mA.

(a) What is the effective "resistance" of the circuit connected to the power supply?

(b) What must the minimum value for the smoothing capacitor be?

CHAPTER 22 | **ELECTRIC CURRENT AND FUNDAMENTALS OF DC CIRCUITS**

ELECTRICITY, MAGNETISM, AND OPTICS

(c) How much energy is stored in the capacitor when it is fully charged?

(d) How much energy has been drawn from the capacitor when it has had its voltage reduced by 0.100 V?

DATA-RICH PROBLEM

62. ★★★ Figure 22-43 is a voltage versus time graph for a circuit using a 5.00 μF capacitor and an unknown resistor.
(a) What is the time constant?
(b) What is the approximate value of the resistor?
(c) What is the approximate voltage of the charging source?

Figure 22-43 Problem 62

See the text online resources at www.physics1e.nelson.com for Open Problems and Data-Rich Problems related to this chapter.

Chapter 23
Magnetic Fields and Magnetic Forces

A Magnetic Stripe That Changed Our Lives

You have probably used a plastic card bearing a magnetic stripe, such as a credit card, a debit card, a student identification card, a driver's licence, or a transit pass (Figure 23-1). Have you ever wondered how these cards work? Magnetic stripes are made of iron-based magnetic particles approximately 5 μm across embedded in a thin film. To record information on the magnetic stripe, groups of these particles are magnetized in a specific direction. The card is read by measuring the magnetic field along the stripe.

The process of attaching a magnetic stripe to a plastic card was invented in the 1960s by the IBM engineer Forrest Parry, who was working on developing a security system for the U.S. government. Parry found that it was quite difficult to attach a magnetic stripe to a plastic card without warping the stripe or having its surface damaged or contaminated by the adhesives used to glue it in place. The solution came from Parry's wife, Dorothea Tillia, who suggested using an iron to bond the stripe to the card.

The concept of the modern credit card was invented by Frank McNamara in 1949. It was intended to be used by salesmen who had to travel a lot and did not want to carry large amounts of cash to pay for their meals and entertainment. Hence, the first credit card was the "Diners' Club" card.

Figure 23-1 The magnetic stripe on the back of a credit card

When you have completed this chapter you should be able to:

1. Describe magnetic fields using the concept of magnetic field lines.
2. Calculate the magnitude and direction of the magnetic force acting on a charged particle moving inside a uniform magnetic field.
3. Describe the trajectory of a charged particle moving inside crossed electric and magnetic fields.
4. Calculate the magnitude and direction of a magnetic force and a magnetic torque applied by a magnetic field on current-carrying wires and current-carrying loops.
5. Calculate the magnitude and direction of a magnetic field created by moving electric charges using the Biot–Savart law.
6. Calculate the magnitude and direction of a magnetic field created by moving electric charges using Ampère's law.
7. Describe the magnetic forces between two current-carrying wires quantitatively and qualitatively.
8. Describe the magnetic properties of materials and their applications.
9. Describe some everyday applications of electromagnetic phenomena.

ELECTRICITY, MAGNETISM, AND OPTICS

For additional Making Connections, Examples, and Checkpoints, as well as activities and experiments to help increase your understanding of the chapter's concepts, please go to the text's online resources at www.physics1e.nelson.com.

23-1 Magnetic Field and Magnetic Force

Earlier chapters describe electric and gravitational fields and illustrate how field lines can be used to represent the direction and strength of a field. The concepts of fields and field lines were originally developed by the British scientist and engineer Michael Faraday (1791–1867) to model a magnetic field. In 1845, Faraday used **magnetic field lines** (which he called "lines of force") to describe the space surrounding a permanent magnet.

You can map the **magnetic field** created by a small permanent magnet by moving a compass around it. The needle of the compass is itself a small magnet, and it deflects differently, depending on its location relative to the magnet.

ONLINE ACTIVITY

Magnetic Field Lines Created by a Permanent Magnet

The e-resource that accompanies every new copy of this textbook contains an Online Activity using the PhET simulation "Magnet and Compass." Work through the simulation and accompanying questions to gain an understanding of magnetic field lines created by a permanent magnet.

The end of the compass needle pointing toward Earth's North Pole is called a north-seeking pole, or simply a north pole. Similarly, the end of the compass needle pointing toward Earth's South Pole is called a south-seeking pole, or a south pole (Figure 23-2(a)). Earth's magnetic field is likely caused by electric currents flowing in its molten outer core. This magnetic field is similar to the field that would be produced by a giant bar magnet located inside Earth's core. Earth's magnetic north pole would be located near the south pole of this imaginary bar magnet, and Earth's magnetic south pole would be near the magnet's north pole. The magnetic and geographical poles do not coincide. In fact, the magnetic poles are gradually moving relative to the geographic poles.

Another way of displaying the magnetic field patterns is to use iron filings. In a magnetic field, suspended iron filings behave as tiny magnets and orient themselves in the direction of the field lines. To visualize a 3-D magnetic field, we can use iron filings suspended in oil (Figure 23-2(b)). Figures 23-2(c)–(e) show magnetic field lines created by different magnet configurations.

Like an electric field, a magnetic field is a vector physical quantity. We defined the direction and magnitude of an electric field in terms the electrostatic force that it exerts on a test charge. The electric field and the electrostatic force are always collinear, and $\vec{F}_E = q\vec{E}$,

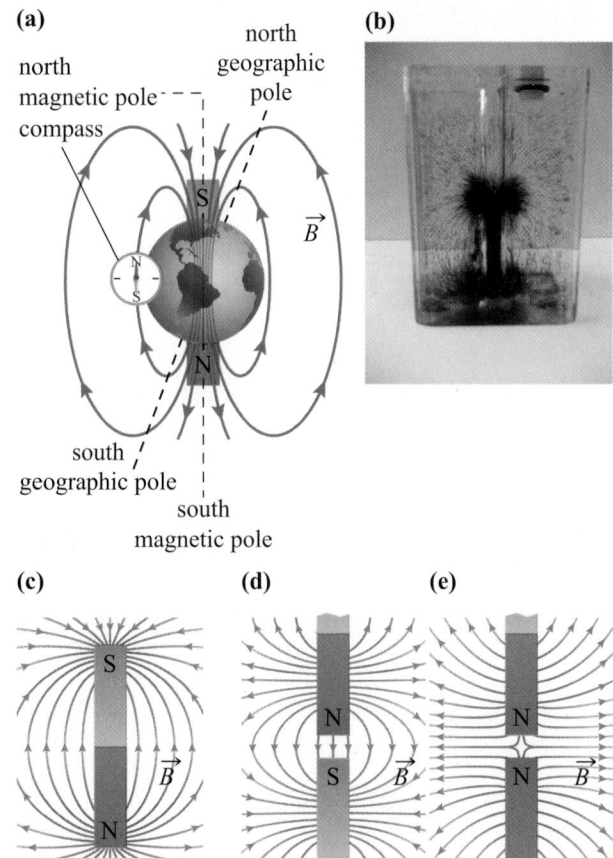

Figure 23-2 (a) Earth's magnetic field can be visualized as the magnetic field of a giant bar magnet. (b) Magnetic field lines created by iron fillings suspended in oil. (c), (d), and (e) Magnetic field patterns indicated by iron fillings.

where $\vec{F}_E$ is an electrostatic force, q is an electric charge, and $\vec{E}$ is the electric field. Although the electrostatic force $\vec{F}_E$ and the magnetic force $\vec{F}_B$ on a charged particle are both proportional to the particle's charge, the **magnetic force** also depends on the velocity of the particle and is *not* collinear with the magnetic field. The forces exerted on charged particles by electric and magnetic fields thus differ significantly:

■ The magnitude of the magnetic force $\vec{F}_B$ on a charged particle is proportional to particle's speed. A faster-moving particle experiences a greater magnetic force, and a particle at rest does not experience any magnetic force.

■ Unlike the electrostatic force, the magnetic force depends on the direction of the motion of a charged particle relative to the direction of the magnetic field. For a given speed, a charged particle moving perpendicular to the magnetic field lines experiences the greatest magnetic force, and a charged particle moving parallel to the magnetic field lines experiences no magnetic force. When the particle moves at an angle θ to the magnetic field lines, the force exerted by the magnetic field is proportional to $\sin\theta$ (Figure 23-3).

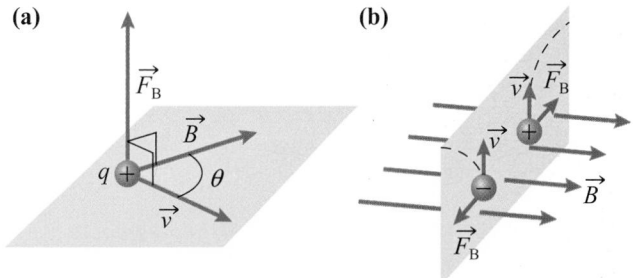

Figure 23-3 The magnetic force acting on a charged particle moving inside a uniform magnetic field at different angles relative to the magnetic field lines

Figure 23-4 (a) The direction of the magnetic force on a charged particle q moving with velocity $\vec{v}$ in a magnetic field $\vec{B}$. (b) Magnetic forces on positively and negatively charged particles moving inside a uniform magnetic field $\vec{B}$. (We indicate magnetic field vectors with green arrows, magnetic forces with blue arrows, and velocities with red arrows.)

■ The magnetic force is always perpendicular to a particle's velocity $\vec{v}$ and to the magnetic field $\vec{B}$: $\vec{F}_B \perp \vec{v}$ and $\vec{F}_B \perp \vec{B}$. In Figure 23-3 it is directed in or out of the page. Consequently, the magnetic force does zero work on the particle:

$$dW = (\vec{F}_B \cdot d\vec{r}) = (\vec{F}_B \cdot \vec{v}dt) = (\vec{F}_B \cdot \vec{v})dt = 0$$

Thus, the magnetic force cannot change the particle's kinetic energy (just as the gravitational force acting on an object in a circular orbit does not change its kinetic energy, Chapter 11). However, the magnetic force can affect the *direction* of the particle's motion.

Can you think of any case in which a force applied on an object does not change its speed, but only affects the direction of its motion? (See Section 4-3)

Calculating the magnitude and direction of the magnetic force requires knowing not only the magnetic field $\vec{B}$ and the particle's charge q, but also its velocity $\vec{v}$. Moreover, the three vectors $\vec{v}$, $\vec{B}$, and $\vec{F}_B$ are never located all in the same plane. Therefore, problems involving magnetic forces are always three-dimensional. The magnetic force acting on a charged particle can be described succinctly using a cross product (Figure 23-4):

KEY EQUATION
$$\vec{F}_B = q\vec{v} \times \vec{B} \qquad (23\text{-}1)$$

which means $\vec{F}_B = qvB\sin\theta$

Equation (23-1) indicates that the magnetic forces acting on oppositely charged particles moving in the same direction in a uniform magnetic field are indeed opposite. Remember that cross products are anticommutative: $\vec{v} \times \vec{B} = -\vec{B} \times \vec{v}$.

When a charged particle is moving inside crossed magnetic and electric fields, it experiences both electrostatic and magnetic forces.

KEY EQUATION
$$\vec{F} = \vec{F}_E + \vec{F}_B = q\vec{E} + q\vec{v} \times \vec{B} \qquad (23\text{-}2)$$

The expression for the combined force is called the **Lorentz force**, in honour of the Dutch physicist Hendrik Lorentz (1853–1928), who formulated it in 1892. However, Equation (23-2), albeit not in the present form, appeared 31 years earlier in the works of the Scottish mathematician and theoretical physicist James Clerk Maxwell (1831–79).

The SI unit of magnetic field is the **tesla** (T), named for the Serbian-American inventor, physicist, and engineer Nikola Tesla (1856–1943). A particle carrying a charge of 1 C moving with a speed of 1 m/s perpendicular to a magnetic field of 1 T will experience a magnetic force of 1 N:

$$1\,\text{N} = 1\,\text{C} \cdot 1\,\frac{\text{m}}{\text{s}} \cdot 1\,\text{T} \quad \text{or} \quad 1\,\text{T} = 1\,\frac{\text{N} \cdot \text{s}}{\text{C} \cdot \text{m}}$$

The tesla can be also expressed as follows:

$$1\,\text{T} = 1\,\frac{\text{N} \cdot \text{s}}{\text{C} \cdot \text{m}} = 1\,\frac{\text{N}}{\text{C/s} \cdot \text{m}} = 1\,\frac{\text{N}}{\text{A} \cdot \text{m}} \qquad (23\text{-}3)$$

A magnetic field of 1 T is quite strong. The magnetic fields that we typically encounter have strengths of less than 1 mT.

Another commonly used unit for magnetic fields is the gauss (G), named after the German mathematician and physical scientist Carl Friedrich Gauss (1777–1855): $10^4\,\text{G} = 1\,\text{T}$ or $1\,\text{G} = 10^{-4}\,\text{T}$

 MAKING CONNECTIONS

Strengths of Magnetic Fields

Earth: The strength of the magnetic field at Earth's surface ranges from about 30 μT (0.3 G) in an area including most of South America and South Africa to over 60 μT (0.6 G) in areas near the magnetic poles (in northern Canada and southern Australia).

The solar wind: The solar wind is a stream of plasma that flows away from the Sun. The magnetic field of the solar wind ranges between 5 and 10 nT (approximately 1000 times weaker than Earth's magnetic field).

(continued)

Magnetic resonance imaging (MRI): MRI is a technique for producing diagnostic images of tissue inside the body. MRI scanners generally use powerful magnetic fields in the range of 0.5 to 3.0 T.

Atoms: The effective magnetic field felt by an electron inside an alkali atom can exceed 1000 T. These atomic magnetic fields are responsible for fine structure splitting of atomic spectra measured in laboratories.

Earth's magnetosphere: Earth's magnetosphere, a region in space dominated by the magnetic field, was detected in 1958 during the *Explorer 1* mission. The magnetosphere is located above the ionosphere and extends as far as 25 Earth radii. Two main phenomena affect Earth's magnetosphere: Earth's internal magnetic field (30 to 60 μT) and the solar wind (5 to 10 nT).

Laboratory experiments: To date, the strongest continuous magnetic field created in a lab is a 45 T field at the National High Magnetic Field Laboratory of Florida State University. Pulsed electromagnets have created fields of over 100 T for a few milliseconds.

Tokamaks: These experimental fusion reactors use magnetic fields of 8 to 10 T to confine extremely hot plasma. The word "tokamak" is a Russian acronym for toroidal chamber with magnetic coils.

Neutron stars: The magnetic field on the surface of a neutron star is calculated to be of the order of magnitude of 10^8 T, making neutron stars the strongest known naturally occurring magnetic fields.

✅ CHECKPOINT

C-23-1 Ranking the Magnitudes of Magnetic Forces

Four charged particles, $q_1 = q$, $q_2 = -q$, $q_3 = 2q$, and $q_4 = -2q$, moving with velocities $\vec{v}_1 = 3v\hat{i}$, $\vec{v}_2 = 3v\hat{j}$, $\vec{v}_3 = -2v\hat{i}$, and $\vec{v}_4 = 3v\hat{j}$, enter a uniform magnetic field $\vec{B} = -2B_0\hat{i}$. Which of the following correctly ranks the magnitudes of the magnetic forces acting on these particles?

(a) $F_{B_1} < F_{B_2} < F_{B_3} < F_{B_4}$
(b) $F_{B_1} > F_{B_2} > F_{B_3} > F_{B_4}$
(c) $F_{B_1} = F_{B_2} > F_{B_3} = F_{B_4}$
(d) $F_{B_1} > F_{B_2} = F_{B_3} > F_{B_4}$
(e) $F_{B_1} = F_{B_3} < F_{B_2} < F_{B_4}$

C-23-1 (e) The magnetic force can be calculated using Equation (23-1).

"Hand" rules are useful shortcuts for determining the direction of a magnetic force without calculating a cross product. Here are two commonly used versions of the **right-hand rule** (Figure 23-5).

Right-hand rule 1: Point your four right-hand fingers in the direction of motion of a positively charged particle, and curl them in the direction of the magnetic field. Your stretched-out thumb points in the direction of the magnetic force acting on the particle (Figure 23-5(a)).

Right-hand rule 2: Position your right-hand so that your fingers point in the direction of the magnetic field and your thumb points in the direction of the velocity of a positively charged particle. When you then bend your fingers 90°, they point in the direction of the magnetic force (Figure 23-5(b)).

Check for yourself that both versions of the right-hand rule are equivalent. For a negatively charged particle, the direction of the magnetic force is opposite to the direction indicated by the right-hand rule. (Alternatively, you can use corresponding left-hand rules for negative charges.)

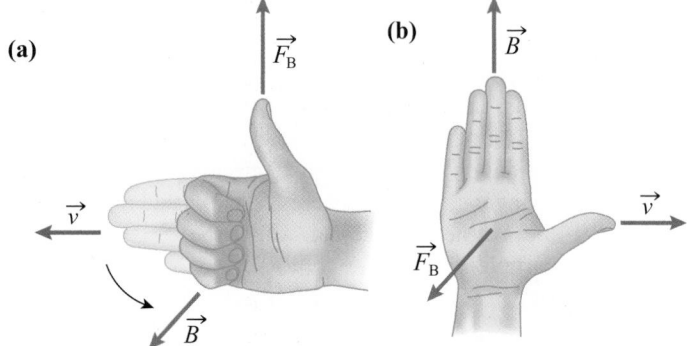

(a)
(b)

Figure 23-5 Right-hand rules for the direction of a magnetic force acting on a charged particle. (a) Right-hand rule 1 (b) Right-hand rule 2

👥 PEER TO PEER

Don't be intimidated if your professor uses a right-hand rule that you're not aware of. There are a number of different ways to use the right-hand rule. Make sure that you use a right-hand rule correctly, and find the one that works for you. All produce the same result as long as you use them properly. Remember that a right-hand rule applies to a positively charged particle. A negatively charged particle will produce the opposite result.

Since problems involving the calculation of magnetic forces involve three dimensions, it is useful to have conventions to represent 3-D situations as 2-D diagrams. We represent a vector directed away from us into the page as a cross (like the end view of the tail of an arrow). Similarly, we represent a vector coming out of the page toward us as a dot (the front view of the tip of an arrow) (Figure 23-7).

ELECTRICITY, MAGNETISM, AND OPTICS

EXAMPLE 23-1

The Right-Hand Rule and the Cross Product

A positively charged particle q enters a region with a uniform magnetic field $\vec{B} = B_0\hat{i}$. The particle's initial velocity is $\vec{v} = 3v_0\hat{j}$ (Figure 23-6). Determine the magnetic force acting on the particle just as it enters the magnetic field.

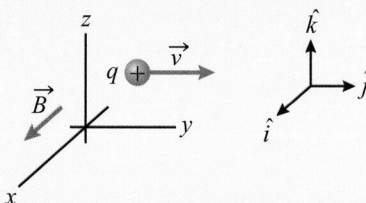

Figure 23-6 A positively charged particle moves inside a uniform magnetic field.

SOLUTION

Method 1: Using the Right-Hand Rule

Applying the right-hand rule, we find that the magnetic force points in the negative z-direction:

Its magnitude is given by:

$$F_B = qvB\sin\theta = q3v_0B_0\sin(90°) = 3qv_0B_0$$

In vector form, the magnetic force can be represented as $\vec{F}_B = -3qv_0B_0\hat{k}$.

Method 2: Using a Cross Product

$$\vec{F}_B = q\vec{v} \times \vec{B} = q3v_0\hat{j} \times B_0\hat{i} = q3v_0B_0 \begin{pmatrix} \hat{i} & \hat{j} & \hat{k} \\ 0 & 1 & 0 \\ 1 & 0 & 0 \end{pmatrix}$$

$$= q3v_0B_0(\hat{i}(1\cdot0 - 0\cdot0) - \hat{j}(0\cdot0 - 1\cdot0) + \hat{k}(0\cdot0 - 1\cdot1))$$

$$= -3qv_0B_0\hat{k}$$

Making sense of the result:

As expected, both solutions produce the same result. Method 1 appears to be more straightforward; however, in general, an algebraic approach is more versatile.

(a) $\vec{B}$ out of page:

(b) $\vec{B}$ into page:

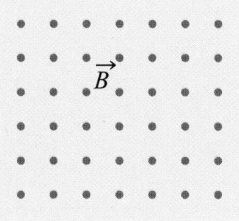

(c) $\vec{B}$ out of page:

(d) $\vec{B}$ into page:

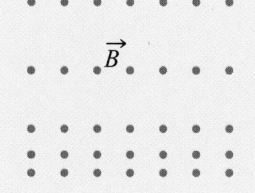

Figure 23-7 Representations of magnetic fields perpendicular to the page. (a) A uniform magnetic field is directed out of the page. (b) A uniform magnetic field is directed into the page. (c) A nonuniform magnetic field is directed out of the page. (d) A nonuniform magnetic field is directed into the page.

A **uniform magnetic field** is a magnetic field that is not changing in space (Figures 23-7(a) and (b)), and a **constant magnetic field** is a magnetic field that is not changing in time. A uniform field is not necessarily constant nor is a constant field necessarily uniform.

PEER TO PEER

Students are often confused over what the dots and crosses mean. They are just ways of representing vectors perpendicular to the page (magnetic fields in this case). I like to think of them as the head and tip of an arrow. The dot is like the arrowhead facing you, and this means that the field is coming out of the page. The cross is like the feather end of the arrow, and this means that the field is going into the page, away from you.

CHECKPOINT

C-23-2 Determining the Direction of the Magnetic Force

An electron is moving at speed v in the positive x-direction. The magnetic field $\vec{B}$ points in the negative x-direction. What is the magnetic force experienced by the electron?

(a) $F_B = qvB$ points in the positive y-direction.
(b) $F_B = qvB$ points in the negative y-direction.
(c) $F_B = qvB$ points in the positive z-direction.
(d) $F_B = qvB$ points in the negative z-direction.
(e) $F_B = 0$.

C-23-2 (e) A magnetic field that is parallel or antiparallel to the charged particle's velocity exerts no force on the particle (Equation 23-1).

LO 2

23-2 The Motion of a Charged Particle in a Uniform Magnetic Field

As noted earlier, the magnetic force acting on a charged particle can change the *direction* of its motion but not its speed. In this section, we use Newton's laws and the description of the magnetic force (Equation (23-2)) to investigate the trajectory of a charged particle moving in a uniform magnetic field.

A Charged Particle Moving Perpendicular to a Uniform Magnetic Field

Consider an electron moving with velocity $\vec{v}$ entering a uniform magnetic field $\vec{B}$, directed *perpendicular* to the electron's velocity (Figure 23-8). Since the magnetic force acting on the electron is perpendicular to its velocity (Equation (23-1)), the electron moves along a circular path (Section 4-3), accelerating at the rate of $a_R = \dfrac{v^2}{R}$. Applying Newton's second law to this circular motion, we get

$$\sum \vec{F} = m\vec{a}$$

$$qvB\sin\theta = ma_R$$

$$qvB\sin(90°) = m\frac{v^2}{R}$$

$$qvB = m\frac{v^2}{R}$$

Therefore, the radius of the electron's trajectory can be expressed as follows:

KEY EQUATION
$$R = \frac{mv}{qB} = \frac{p}{qB} \tag{23-4}$$

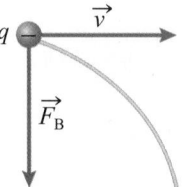

Figure 23-8 An electron moving in a uniform magnetic field directed perpendicular to its velocity has a circular trajectory.

Equation (23-4) indicates that the radius of the electron's trajectory is proportional to the magnitude of its momentum, $\vec{p}$. The radius of the motion is inversely proportional to the particle's charge and the magnitude of the magnetic field.

We now consider a stream of identical particles all moving in the same direction but with different speeds. For these identical particles, the mass-to-charge ratio is the same. Therefore, when the stream of charged particles enters a uniform magnetic field perpendicular to its direction, the particles move along circles of different radii (Equation (23-4)), defined by the particle's momentum and charge, and the magnitude of the magnetic field.

For circular trajectories of a charged particle, we can define the period, T, as the time required for a particle to complete one revolution:

$$T = \frac{2\pi R}{v} \tag{23-5}$$

Combining Equations (23-4) and (23-5) yields

KEY EQUATION
$$T = \frac{2\pi R}{v} = \frac{2\pi mv}{vqB} = 2\pi \frac{m}{qB} \tag{23-6}$$

The period, T, depends only on the ratio m/qB and is not affected by the particle's speed. Since the relationship between magnitudes of angular and linear velocities is $v = \omega R$,

$$\omega = \frac{v}{R} = \frac{qBR}{mR} = \frac{qB}{m}$$

KEY EQUATION
$$f_{cyc} = \frac{\omega}{2\pi} = \frac{qB}{2\pi m} \tag{23-7}$$

Thus, the angular velocity ω of a charged particle moving in a uniform magnetic field $\vec{B}$ directed perpendicular to its velocity depends only on the particle's charge and mass, and the magnitude of the magnetic field. The final expression in Equation (23-7) does not include the particle's linear velocity or the radius of its trajectory. The angular velocity of the particle, ω, divided by 2π has the unit of 1/s or Hz: $[\omega] = \dfrac{1}{s} \equiv$ Hz, and is called the **cyclotron frequency**, f_{cyc} (recall that $\omega = 2\pi f$ (Equation (13-5))). A **cyclotron** is a type of particle accelerator that accelerates charged particles moving along circular trajectories (see Section 23-3).

A Charged Particle Moving at an Angle to a Magnetic Field

Let us consider what happens when a charged particle enters a uniform magnetic field at an arbitrary angle θ to the field lines. We can resolve the particle's velocity into components perpendicular ($\vec{v}_\perp$) and parallel ($\vec{v}_\parallel$) to the magnetic field (Figure 23-9):

$$\vec{v} = \vec{v}_\parallel + \vec{v}_\perp, \quad v = \sqrt{v_\parallel^2 + v_\perp^2} \tag{23-8}$$

where $v_\parallel = v\cos\theta$ and $v_\perp = v\sin\theta$.

The magnetic force acting on the particle due to the perpendicular component of the particle's velocity causes

608   SECTION FOUR | ELECTRICITY, MAGNETISM, AND OPTICS   NEL

ELECTRICITY, MAGNETISM, AND OPTICS

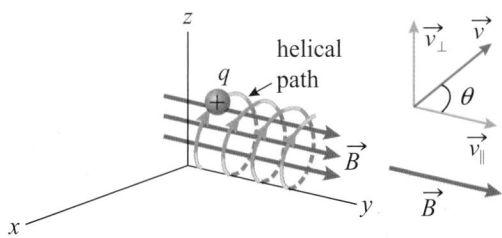

Figure 23-9 A positively charged particle moving at angle θ to a uniform magnetic field follows a helical trajectory.

the particle to move along a circular path with a radius given by Equation (23-4). The velocity component parallel to the magnetic field lines is not affected by the magnetic field. The magnetic force can be expressed as

$$\vec{F}_B = q\vec{v} \times \vec{B} = q(\vec{v}_\parallel + \vec{v}_\perp) \times \vec{B}$$

$$= q\vec{v}_\parallel \times \vec{B} + q\vec{v}_\perp \times \vec{B} = q\vec{v}_\perp \times \vec{B} \qquad (23\text{-}9)$$

The charged particle follows a helical path, a combination of circular motion in a plane perpendicular to the magnetic field and uniform linear motion parallel to it:

> In the plane perpendicular to the magnetic field—circular motion
>
> $$R = \frac{mv\sin\theta}{qB}$$
>
> In the direction parallel to the magnetic field—motion with constant velocity:
>
> $$\vec{v}_\parallel = \text{const}$$
>
> The distance along the magnetic field travelled by a charged particle during one revolution (pitch):
>
> $$d = v_\parallel T = v\cos\theta\, \frac{2\pi m}{qB} = \frac{2\pi mv\cos\theta}{qB} \qquad (23\text{-}10)$$

MAKING CONNECTIONS

Auroras

Spectacular auroras occur in regions 10° to 20° of latitude from the magnetic poles: the aurora borealis (Northern Lights) in the northern hemisphere (Figure 23-10(a)) and the aurora australis in the southern hemisphere. Auroras happen when fast-moving electrons and protons enter the magnetosphere, which is the region of strong magnetic fields with a boundary at approximately 15 to 25 Earth radii above Earth's surface. These charged particles follow magnetic field lines moving along helical paths (Figure 23-10(b)), as described above. When they collide with oxygen and nitrogen atoms at altitudes of 30 to 300 km above Earth's surface, they transfer part of their energy to atmospheric atoms and molecules, which then emit visible light. The colour of the aurora depends on the atoms that were struck and on the altitude of the collision.

(a)

(b)

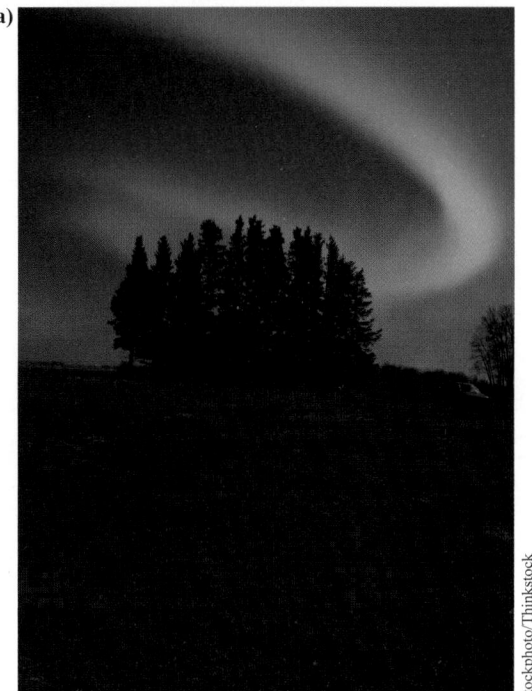

Figure 23-10 (a) The aurora borealis in Northern Canada (b) Schematic representation of auroras

iStockphoto/Thinkstock

EXAMPLE 23-2

A Charged Particle Moving in a Uniform Magnetic Field

A particle with a charge of 1.3 nC and a mass of 1.4×10^{-9} kg moves with velocity $\vec{v} = (1.5 \times 10^4\,\hat{i} + 1.2 \times 10^4\,\hat{j})$ m/s as it enters a uniform magnetic field $\vec{B} = (1.4\hat{i})$ T.

(a) Determine the magnetic force acting on the particle as it enters the field.
(b) Describe the path of the particle, and determine its radius.
(c) Find the time it takes for the particle to complete one revolution.
(d) Find the distance the particle travels along the magnetic field lines as it completes one revolution.

SOLUTION

(a) Part (a) can be solved in a way similar to the previous example:

$$\vec{F}_B = q\vec{v} \times \vec{B} = (1.3 \times 10^{-9})\begin{pmatrix} \hat{i} & \hat{j} & \hat{k} \\ 1.5 \times 10^4 & 1.2 \times 10^4 & 0 \\ 1.4 & 0 & 0 \end{pmatrix} \text{N}$$

$$= [(1.3 \times 10^{-9})10^4\,(\hat{i}(1.2 \cdot 0 - 0 \cdot 0) - \hat{j}(1.5 \cdot 0 - 0 \cdot 1.4)$$

$$+\,\hat{k}(1.5 \cdot 0 - 1.4 \cdot 1.2))] \text{ N}$$

$$= 1.3 \times 10^{-5}(-1.68\,\hat{k})\text{ N} = -2.184 \times 10^{-5}\,\hat{k}\text{ N}$$

(b) The particle moves along a helical path because the $\hat{i}$ component of the particle's velocity is parallel to the magnetic field and therefore unaffected by the field. So, the particle moves along a circular trajectory in the yz-plane while moving with a constant velocity in the x-direction. The radius of the helix can be found using Equation (23-4):

$$R = \frac{mv}{qB} = \frac{(1.4 \times 10^{-9}\text{ kg})(1.2 \times 10^4\text{ m/s})}{(1.3 \times 10^{-9}\text{ C})(1.4\text{ T})} = 9.2 \times 10^3\text{ m}$$

(c) We can use Equation (23-6) to find the period of the particle's motion:

$$T = \frac{2\pi m}{qB} = \frac{2(3.14)(1.4 \times 10^{-9}\text{ kg})}{(1.3 \times 10^{-9}\text{ C})(1.4\text{ T})} = 4.8\text{ s}$$

(d) It will take the particle 4.8 s to complete one revolution. During this time, the particle will move a distance d along the x-axis. This distance is called the **pitch of the helix**:

$$d = Tv_x = (4.8\text{ s})(1.5 \times 10^4\text{ m/s}) = 7.2 \times 10^4\text{ m}$$

✓ CHECKPOINT

C-23-3 Location of Auroras

Auroras do not occur near the equator because
(a) charged particles never reach the equator;
(b) the magnetic field lines are strongest near the equator;
(c) the magnetic field lines near the equator are directed toward Earth;
(d) there is no magnetosphere near the equator.
(e) None of the above are true.

C-23-3 (a): Charged particles emitted by the Sun follow a helical path around B lines reaching north and south poles, but not the equator.

LO 3

23-3 Applications: Charged Particles Moving in a Uniform Magnetic Field

The behaviour of charged particles in a magnetic field has a number of applications in medicine, industry, and scientific research. In this section, we see how the physics of this behaviour is used in several important devices.

Velocity Selector

Charged particles produced in a lab or arriving from the Sun often have a range of velocities. A **velocity selector** uses perpendicular electric and magnetic fields to filter a beam of charged particles, leaving only those with a particular desired velocity.

The velocity selector in Figure 23-11 has a chamber with crossed uniform electric and magnetic fields. When a positively charged particle enters this chamber, the electric force

(a)

(b)

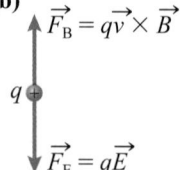

Figure 23-11 (a) An operation of a velocity selector (b) A free body diagram for a positively charged particle moving inside a velocity selector

ELECTRICITY, MAGNETISM, AND OPTICS

acting on the particle is directed along the electric field lines (downward). However, the magnetic force is directed upward, perpendicular to the magnetic field lines. Particles with a velocity such that the magnetic force balances the electric force pass through the chamber without changing their velocity, as indicated by the dashed line. All other charged particles are deflected either upward or downward, depending on which force is stronger for each particular particle. The magnetic force can be opposite to the electric force only for the particles entering the region with the velocity perpendicular to both $\vec{B}$ and $\vec{E}$. When the electric and magnetic forces balance,

$$F_E = F_B$$

$$qE = qvB \qquad (23\text{-}11)$$

$$E = vB$$

and

 KEY EQUATION
$$v = \frac{E}{B} \qquad (23\text{-}12)$$

Thus, only the particles whose velocity satisfies Equation (23-12) pass straight through the velocity selector.

☑ CHECKPOINT

C-23-4 Operation of the Velocity Selector

Which of the following statements represents the operation of a velocity selector?

(a) The polarity of the electric and magnetic fields in the velocity selector depends on the charge of the selected particles.

(b) The ratio $\frac{E}{B}$ of the velocity selector depends on the mass of the charged particles to be selected.

(c) If in the velocity selector $E \neq vB$, the particle moves in a circular trajectory.

(d) If a positive particle is moving faster than the speed chosen in the velocity selector, the particle is deflected in the direction of the electric field.

(e) If a negative particle is moving faster than the speed chosen in the velocity selector, the particle is deflected in the direction of the electric field.

C-23-4 (e) The speed of the particle has no effect on the electric force acting on it, but it has an effect on the magnetic force. Since the particle in the velocity selector is negatively charged, the magnetic force acting on it will be larger than the electric force. The magnetic force in that case is directed in the direction of the electric field.

Mass Spectrometers

The charged particles that pass through a velocity selector all have the same velocity, but they do not necessarily have the same masses or charges. A **mass spectrometer** sorts particles by their mass-to-charge ratios, making it a powerful tool for chemical analysis.

As shown in Figure 23-12, a mass spectrometer has a velocity selector connected to a chamber with a uniform magnetic field. Charged particles leaving a velocity selector have a known speed. When they enter the second chamber, the uniform magnetic field causes them to move along

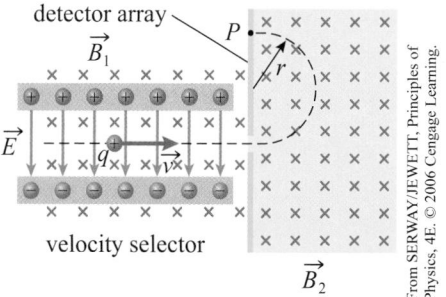

Figure 23-12 An operation of a mass spectrometer

From SERWAY/JEWETT, Principles of Physics, 4E. © 2006 Cengage Learning.

semicircular trajectories with radii that depend on their mass and charge. Combining Equations (23-4) and (23-12):

$$R = \frac{mv}{qB_2} = \frac{mE}{qB_1 B_2} \qquad (23\text{-}13)$$

where B_1 represents the magnetic field inside the velocity selector chamber 1, and B_2 represents the magnetic field inside chamber 2.

Thus, by measuring the radius of a particle's trajectory, we can determine its mass-to-charge ratio:

$$\frac{m}{q} = \frac{RB_1 B_2}{E} \qquad (23\text{-}14)$$

If the magnetic fields in the two chambers are equal which is not always the case, $B_1 = B_2 = B$,

$$\frac{m}{q} = \frac{RB^2}{E} \qquad (23\text{-}15)$$

Some mass spectrometers might not have a separate velocity selector chamber as shown in Figure 23-12. These devices (Figures 23-12 and 23-73) have an advantage of being lighter and lower in cost. We will consider this model of a mass spectrometer in the comprehensive problems at the end of the Chapter, Problem 74.

☑ CHECKPOINT

C-23-5 The Motion of Charged Particles Inside a Mass Spectrometer

An electron beam travels through a mass spectrometer of the type shown in Figure 23-12. Then a beam of protons travels through the same mass spectrometer. What is true about the trajectories of the electrons and protons inside the mass spectrometer?

(a) The electron beam has a smaller radius of curvature than the proton beam. The beams bend in opposite directions.

(b) The electron beam has a larger radius of curvature than the proton beam. The beams bend in the same direction.

(c) The electron beam has a smaller radius of curvature than the proton beam. The beams bend in the same direction.

(d) The electron beam has a larger radius of curvature than the proton beam. The beams bend in opposite directions.

C-23-5 (a) Electrons and protons have the opposite charges but very different masses: the mass of a proton is about 2000 times larger than the mass of an electron. Thus, a proton beam will have a radius 2000 times smaller than an electron beam. The beams will also bend in different directions.

ELECTRICITY, MAGNETISM, AND OPTICS

Determining the Electron's Charge-to-Mass to Ratio

Experiments by the British scientist J.J. Thomson (1856–1940) culminated in 1897 with the discovery of the electron. Thomson was able to measure the electron's charge-to-mass ratio using a cathode ray tube (CRT), like the one shown in Figure 23-13. Thomson's apparatus consisted of a vacuum tube containing an electrode that emits electrons (a cathode), a fluorescent screen, plates to apply an electric field, and coils to apply a magnetic field. Electrons emitted by the cathode passed through a slit into a region of crossed magnetic and electric fields; then they struck a fluorescent screen, causing it to glow at the point of impact. Thomson determined the radius of the path of the electrons from the position of the glowing point. Then he calculated the $\frac{m_e}{e}$ ratio for the electron:

$$\frac{m_e}{e} = \frac{RB}{v} \qquad (23\text{-}16)$$

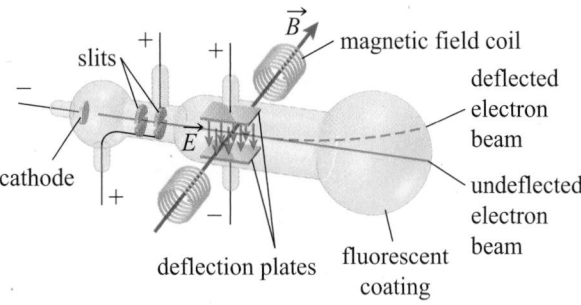

Figure 23-13 Thomson used a cathode ray tube, similar to the one shown here, to determine the charge-to-mass ratio for the electron.

Although Thomson's equipment produced only an approximate value for the ratio $\frac{m_e}{e}$, it was nonetheless clear evidence that "cathode rays" are tiny negatively charged particles. Almost 15 years later, American physicist Robert A. Millikan (1868–1953) devised an ingenious method to use electric fields to determine the mass and charge of the electron. Millikan's experiment is described in Chapter 30.

✓ CHECKPOINT

C-23-6 The Cathode Ray Tube

Which particle *cannot* be measured by using the crossed magnetic and electric field method suggested by Thomson? Explain.

(a) electron
(b) proton
(c) α-particle
(d) neutron
(e) hydrogen ion

C-23-6 (d) Neutrons have no electric charge, so they will move along the cathode ray tube undeflected.

MAKING CONNECTIONS

Uranium Enrichment During World War II

The mass spectrometer played a crucial role in the Manhattan Project, which is the name given to the secret project to develop the atomic bomb during World War II. Uranium-235 is the only naturally occurring isotope that undergoes nuclear fission when bombarded with thermal (low-energy) neutrons. However, only 0.711% of the uranium in ore is uranium-235 (most of the remainder is the much more stable isotope uranium-238). The Manhattan Project had to develop ways to separate the isotopes. One of the methods used was a form of multistage mass spectrometer, called a calutron (Figure 23-14). Today, more effective methods of isotope separation are used. Canada produces approximately 20% of the world's uranium. Approximately 50% of the electrical power in Ontario and 15% of the overall power generation in Canada are generated using unenriched uranium.

Figure 23-14 (a) The Oakridge Nuclear Plant (b) An original drawing of a calutron used to separate uranium isotopes in the Manhattan Project

Cyclotrons

Cyclotrons use electric and magnetic fields to accelerate charged particles to very high kinetic energies. A uniform electric field acts on a charged particle that moves a distance d along the electric field lines, and changes the particle's kinetic energy:

$$\Delta KE = W$$
$$W = (\vec{F}_E \cdot \vec{d}) = qEd = q\Delta V$$
$$\Delta KE = qEd = q\Delta V \qquad (23\text{-}17)$$

Increasing the distance d and strengthening the electric field E increase the gain in the particle's kinetic energy. However, building very long accelerators and using extremely strong electric fields is costly and technologically difficult. The cyclotron reduces some of these challenges by using crossed electric and magnetic fields to accelerate charged particles as they follow a spiral path (Figure 23-15).

A cyclotron consists of two hollow semicircular chambers called "dees," labelled D_1 and D_2 in Figure 23-15(a). A high-frequency voltage is applied across the gap between the dees, and a uniform magnetic field is directed perpendicular to the plane of the dees. An ion source feeds ions to the centre of the cyclotron (point P). When an ion enters a region of the uniform magnetic field, it starts moving along a circular path, repeatedly crossing the gap between the dees. The alternating voltage source across the gap ensures that the ion experiences an electric field in the direction of its motion, thus increasing its kinetic energy. Example 23-3 discusses how the frequency of the alternating voltage across the gap is calculated. After exiting the gap between the dees, the ion is subjected to a uniform magnetic field only, perpendicular to its velocity. As a result, the ion continues to move along a circular path. However, since the speed of the ion is higher than it was before the ion entered the gap, the radius of the ion's trajectory increases. When the ion enters the second gap between the dees, the potential difference between the dees has reversed, so the electric field once again points in the direction of the ion's motion and the ion gains additional kinetic energy. After the ion exits the gap between the dees, it is moving even faster, causing an additional increase in the radius of its trajectory. Eventually, the trajectory radius increases enough that the ion travels along the edge of the dee and passes through the exit slit.

The exit kinetic energy of the ion is

$$KE = \frac{mv^2}{2} = \frac{m}{2}\left(\frac{RqB}{m}\right)^2 = \frac{R^2 q^2 B^2}{2m} \qquad (23\text{-}18)$$

Cyclotrons can accelerate charged particles to speeds approaching the speed of light. For example, the giant TRIUMF cyclotron (www.triumf.info/), located in Vancouver, British Columbia, accelerates negatively charged hydrogen ions (called H^- ions) to 224 000 km/s, approximately 75% of the speed of light (Figure 23-16). These high-energy particles can be used to produce beams of neutrons and short-lived particles called mesons. TRIUMF is so efficient at creating pi-mesons that it is often called a meson factory. The high-energy particles are also used to treat several forms of cancer.

(a)

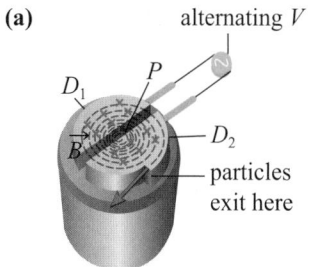

alternating V

D_1

P

B

D_2

particles exit here

(b)

LAWRENCE BERKELEY LABORATORY/SCIENCE PHOTO LIBRARY

Figure 23-15 (a) A schematic representation of a cyclotron (b) The first cyclotron, which was invented by E.O. Lawrence and M.S. Livingston in 1934

Courtesy of TRIUMF

Figure 23-16 The cyclotron at TRIUMF is the largest cyclotron in the world. It accelerates 1000 trillion protons per second to speeds of 75% of the speed of light. TRIUMF's cyclotron accelerates H^- ions: it can extract multiple beams with varying energy levels at very high efficiency, making TRIUMF's accelerator one of the most versatile accelerators in the world.

EXAMPLE 23-3

The Cyclotron Frequency of the TRIUMF Cyclotron

The magnetic field used in the cyclotron at TRIUMF to accelerate H⁻ ions for the purpose of medical research has a strength of 0.3 T.
(a) Calculate the cyclotron frequency of the TRIUMF cyclotron.
(b) Calculate the period of one revolution.
(c) How will increasing the magnetic field approximately 1.5 times, to 0.46 T, affect the cyclotron frequency?

SOLUTION

(a) As per Equation (23-7), the cyclotron frequency is defined as the inverse of the time it takes a charged particle to complete one revolution inside a cyclotron (a circular trajectory of radius R). In the case of the TRIUMF cyclotron, the charged particles are H⁻ ions. Therefore, their charge is -1.6×10^{-19} C, and their mass is approximately equal to the mass of one proton (the mass of two electrons is negligible):

$$m_{\text{p}} \sim 1800\, m_{\text{e}} \rightarrow m_{\text{H}^-} = m_{\text{p}} + 2\, m_{\text{e}} \sim 1.67 \times 10^{-27} \text{kg}$$

Therefore, the cyclotron frequency, f_{cyc}, is

$$f_{\text{cyc}} = \frac{qB}{2\pi m} = \frac{|-1.6 \times 10^{-19}\ \text{C}|(0.3\ \text{T})}{2\pi(1.67 \times 10^{-27}\ \text{kg})} = 4.6 \times 10^{6}\ \text{s}^{-1}$$

$$f_{\text{cyc}} = 4.6 \times 10^{6}\ \text{Hz}$$

The cyclotron frequency is always positive, thus we use the absolute value of the charge in calculations.
(b) The period of one revolution is the inverse of the cyclotron frequency:

$$T = \frac{1}{f_{\text{cyc}}} = (4.6 \times 10^{6}\ \text{s}^{-1})^{-1} = 2.2 \times 10^{-7}\ \text{s}$$

(c) As discussed in part (a), the cyclotron frequency is proportional to the magnetic field and to the charge-to-mass ratio of the accelerated particles (Equation (23-7)). Therefore, if the magnetic field were to increase 1.5 times, the cyclotron frequency would increase in the same proportion.

Making sense of the result:

The cyclotron frequency does not depend on the radius of the particle's motion. Thus, the oscillating voltage between the dees can have a constant frequency, equal to the cyclotron frequency.

MAKING CONNECTIONS

Cyclotron Use in Medicine

Canada is one of the world leaders in the production of medical isotopes. For example, MDS Nordion, a leading Canadian company that produces radioisotopes for cancer treatment, has recently started building its fourth cyclotron. The new cyclotron will help meet a growing demand for the radioisotopes iodine-123 and palladium-103. Palladium-103 is used to treat prostate cancer, and iodine-123 is primarily used in the diagnosis of thyroid cancer and neurological disorders (such as Alzheimer's and Parkinson's disease). MDS Nordion supplies over two-thirds of the world's reactor-produced isotopes. It is estimated that nearly 20 million nuclear medicine procedures are conducted worldwide each year. MDS Nordion currently operates two commercial cyclotrons at the TRIUMF site located on the University of British Columbia campus, and a third in Fleurus, Belgium.

The Hall Effect

Magnetic fields can also be used to determine the sign of charge carriers. Let us consider a wide conducting strip carrying electric current placed in a uniform magnetic field B perpendicular to it (Figure 23-17). As discussed in Section 22-1, charge carriers can be considered to move with an average drift speed, v_{drift}. Let us assume that an electric current flows clockwise (Figure 23-17(a)), which can be represented by negative charges moving counter-clockwise or by positive charges moving clockwise. When positive charge carriers are moving clockwise, the magnetic field pointing perpendicular to the conducting strip (directed into the strip) exerts an upward magnetic force on the positive charges, so they move toward the top of the conducting plate. The positive charge carriers leave an excess of negative charges behind and create a potential difference across the conductor. The top of the conductor

Figure 23-17 The Hall effect. A magnetic field perpendicular to the direction of an electric current creates a potential difference in the direction perpendicular to both the electric current and the magnetic field. (a) The Hall effect in the case of positive charge carriers. (b) The Hall effect in the case of negative charge carriers. The Hall potential difference is indicated with red for the higher potential and blue for the lower potential.

has a *higher* potential than the bottom. When negative particles are moving counterclockwise, the same magnetic field exerts an upward magnetic force on them, shifting them toward the top of the conductor, leaving an excess of positive charges behind. Then the top of the conductor becomes negatively charged and has a *lower* potential than the bottom. This effect, called the **Hall effect**, is named after Edwin Hall (1855–1938), who discovered it in 1879.

However, almost 50 years before Hall, Faraday realized that moving electric charges in a uniform magnetic field create a potential difference in the direction perpendicular to the magnetic field and the charge flow. In the 1830s, Faraday made an unsuccessful attempt to measure the electric voltage between the sides of the Waterloo Bridge across the Thames River in London, England. Faraday expected that salt water flowing through Earth's magnetic field as the tide ebbed would produce a potential difference across the river (Figure 23-18).

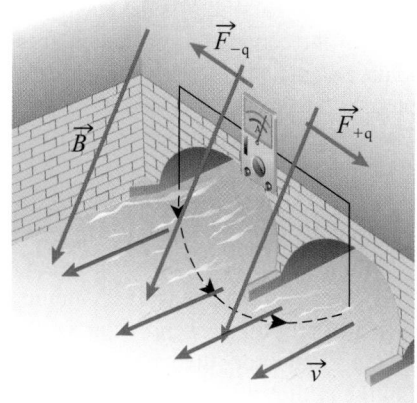

Figure 23-18 In 1831, Faraday tried to measure the potential difference across the Waterloo Bridge over the Thames River due to the motion of ions in Earth's magnetic field. This schematic representation of Faraday's original experiment shows the magnetic field, velocity, and magnetic forces acting on positive and negative ions.

 ## EXAMPLE 23-4

The Thames River as a Natural Dynamo

Estimate the maximum potential difference across the Waterloo Bridge in London, England. The width of the Thames River at the bridge is approximately 265 m. The average flow rate of the Thames at London is approximately 65 m³/s. Assume that the average depth of the river near the bridge is 2.5 m. London is close enough to the sea that the tides affect the flow of the river at Waterloo Bridge and the salinity of the water—the river here is brackish.

SOLUTION

Let us consider a charged particle moving in brackish Thames River waters, and assume that the vertical component of the Earth's magnetic field in this area equals 45 μT. The magnitude of the magnetic force acting on such a particle is given by the magnetic component of the Lorentz force, Equation (23-2): $F_B = qvB$.

Since the magnetic force is perpendicular to both the magnetic field and the direction of water flow, the force is parallel to the bridge. This magnetic force acts in opposite directions on positive and negative ions, creating a potential difference of $\Delta V = Ed$. In a state of equilibrium, the electrostatic force attracts oppositely charged ions and is equal to the magnetic force pulling them apart:

$$qE = qvB$$

$$\text{therefore,} \quad \Delta V = vBd$$

$$E = \frac{\Delta V}{d} \qquad (23\text{-}19)$$

The average speed of the water flowing past the bridge equals the flow rate divided by the cross-sectional area of the river:

$$v \approx \frac{65 \text{ m}^3/\text{s}}{(2.5 \text{ m})(265 \text{ m})} \approx 0.10 \text{ m/s}$$

The flow rate is greatest when the tide is ebbing. As an approximation, we assume that the maximum speed of the water is approximately 0.3 m/s (three times the average value). Then the potential across the bridge is

$$\Delta V = vBd \approx \left(0.3 \, \frac{\text{m}}{\text{s}}\right)(0.045 \text{ mT})(265 \text{ m}) = 0.0036 \text{ V} = 3.6 \text{ mV}$$

Notice that ion concentration in the Thames River does not affect the potential difference caused by the Hall effect.

Making sense of the result:

The potential difference across the river is quite small. To detect this voltage, Faraday would have needed a voltmeter sensitive enough to measure a few millivolts.

A potential difference across a conductor is called the **Hall voltage**. A metal conductor has only one type of charge carrier. Therefore, the Hall voltage can be calculated using the fact that in a steady state, the electric field across the conductor creates an electric force equal and opposite to the magnetic force acting on the charge carriers. In a metal conductor with cross-sectional area $w \times h$ (as in Figure 23-17),

$$F_B = qv_{\text{drift}}B$$

$$F_E = qE_{\text{Hall}} = q\frac{\Delta V_{\text{Hall}}}{w}$$

$$F_E = F_B, \quad \text{thus} \quad qv_{\text{drift}}B = q\frac{\Delta V_{\text{Hall}}}{w}$$

$$v_{\text{drift}} = \frac{\Delta V_{\text{Hall}}}{wB}$$

$$\Delta V_{\text{Hall}} = wv_{\text{drift}}B = w\frac{J}{ne}B = w\frac{\text{I/A}}{ne}B = w\frac{\text{I}}{whne}B$$

KEY EQUATION $$\Delta V_{\text{Hall}} = \frac{IB}{hne} \qquad (23\text{-}20)$$

where I is the electric current, B is the magnetic field, h is the dimension of the conductor perpendicular to both the current and the magnetic field, n is the charge density, and e is the charge of the charge carrier.

The Hall effect is used in magnetic field sensors, proximity switches, position and speed detectors, and current sensors. A Hall effect sensor is a transducer that varies its output voltage in response to changes in a magnetic field. To determine the strength of a magnetic field, a Hall effect sensor measures voltage across a conductor in a direction perpendicular to the current flowing in the conductor, then uses this voltage, the current, and the width of the conductor to calculate the magnetic field.

 EXAMPLE 23-5

Using the Hall Effect to Describe Electric Properties

A Hall probe containing a strip of copper 1.5 mm thick and 1.0 cm wide is placed in a uniform magnetic field of 1.5 T. The probe measures a Hall voltage of 0.108 μV when the current through the probe is 10.0 A. Determine

(a) the drift speed of the electrons in the copper strip;
(b) the strength of the electric field across the strip;
(c) the density of the charge carriers in the strip.

SOLUTION

(a) Equation (23-20) can be used to find the drift speed of the electrons inside the probe:

$$v_{drift} = \frac{\Delta V_{Hall}}{wB} = \frac{0.108 \times 10^{-6}\,V}{(1.5 \times 10^{-3}\,m)(1.5\,T)} = 4.8 \times 10^{-5}\,m/s$$

(b) The strength of the electric field inside the wire due to the Hall effect is

$$E = \frac{\Delta V_{Hall}}{w} = \frac{(0.108\,\mu V)}{(1.5 \times 10^{-3}\,m)} = 72\,\mu V/m$$

(c) Then the density of the charge carriers can be expressed as follows:

$$n = \frac{I}{ev_{drift}A}$$

$$= \frac{10\,A}{(1.6 \times 10^{-19}\,C)(4.8 \times 10^{-5}\,m/s)(1.5 \times 10^{-3}\,m)(1.0 \times 10^{-2}\,m)}$$

$$= 8.7 \times 10^{28}\,m^{-3}$$

Making sense of the result:

The values of the drift speed (4.8×10^{-5} m/s) of the charge carriers in copper and of the charge density in copper (8.7×10^{28} m^{-3}) agree with the experimentally measured values listed in Table 22-1 in Chapter 22.

23-4 The Magnetic Force on a Current-Carrying Wire

As discussed earlier, a charged particle moving inside a magnetic field experiences a magnetic force. Therefore, a magnetic field exerts a force on each charged particle moving inside a current-carrying wire. The net magnetic force acting on the wire depends on the net motion of the electrons and, hence, on the drift velocity of the electrons.

Figure 23-19 shows a series of experiments to study the effect of a magnetic field on a current-carrying wire placed between the poles of a permanent magnet. As expected, a wire carrying no current is unaffected by the magnetic field. Since free electrons inside the wire move randomly, the effect of the magnetic field on the electrons averages to zero. However, when electric current flows in the wire, the effect of a magnetic field becomes pronounced. The direction of conventional current is defined in terms of the flow of positively charged carriers. Thus, to find the direction of the magnetic force on the wire, we use the same right-hand rule that we used to find the direction of a magnetic force acting on a positively charged carrier. Note that a

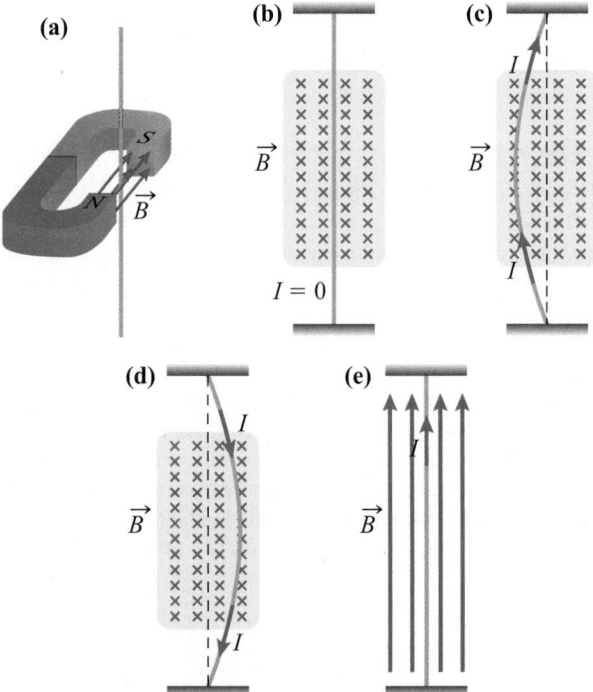

Figure 23-19 The effect of a magnetic field on a current-carrying wire. (a) A wire suspended vertically between the poles of a permanent magnet. (b) A wire carrying no current ($I = 0$) in a uniform magnetic field perpendicular to the wire. (c) A wire carrying an upward nonzero current in a uniform magnetic field perpendicular to the wire. (d) A wire carrying a downward nonzero current in a uniform magnetic field perpendicular to the wire. (e) A wire carrying a nonzero current in a uniform magnetic field parallel to the wire.

ELECTRICITY, MAGNETISM, AND OPTICS

magnetic field directed parallel or antiparallel to the current does not exert any force on the wire.

We now derive an expression for the magnitude of the magnetic force on a current-carrying wire of length l and cross-sectional area A (Figure 23-20). Let us assume that all charge carriers have the same charge e and are moving with the same drift velocity $\vec{v}_{\text{drift}}$. Then the magnetic force acting on each charge carrier is

$$\vec{F}_{\text{B}} = e\vec{v}_{\text{drift}} \times \vec{B}$$

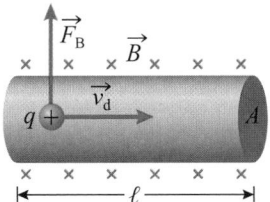

Figure 23-20 Calculating the effect of the magnetic field on a straight current-carrying wire. To calculate the net force on a segment of the wire, we calculate the magnetic force on each moving charge (considering that the charges move with an average drift velocity $\vec{v}_{\text{drift}}$) and then add the forces acting on all the charges in this segment.

Since the volume of the wire is Al, the total charge on the charge carriers in the wire is $q = nAl$, where n is the charge density in the wire.

The net magnetic force acting on the wire is

$$\vec{F} = nlA\vec{v}_{\text{drift}} \times \vec{B} \qquad (23\text{-}21)$$

Substituting $I = nv_{\text{drift}}$, and recalling that the direction of the drift velocity of positive charge carriers coincides with the direction of the electric current, we get

$$\vec{F} = nAv_{\text{drift}}\vec{l} \times \vec{B} = I\vec{l} \times \vec{B}$$

KEY EQUATION $\qquad \vec{F} = I\vec{l} \times \vec{B} \qquad (23\text{-}22)$

where $\vec{l}$ is a small segment of the wire in the direction of the electric current I.

For the above calculation we assumed that the wire is straight.

We now consider a wire of arbitrary shape (Figure 23-21), which consists of an infinite number of small segments, $d\vec{l}$. The magnetic force acting on each

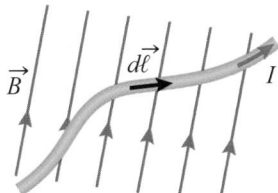

Figure 23-21 Calculating the effect of the magnetic field on a current-carrying wire of an arbitrary shape. The wire can be split into small straight segments $d\vec{l}$, and the effects of the magnetic field on each segment can be added to find the net force on the wire.

individual segment is expressed by Equation (23-22). To find the net magnetic force on the wire, we integrate the forces on the infinitesimal segments:

KEY EQUATION $\qquad \vec{F} = \int_a^b Id\vec{l} \times \vec{B} = -I\int_a^b \vec{B} \times d\vec{l} \qquad (23\text{-}23)$

Note that *the electric current flowing through all the wire segments is the same.* The negative sign on the right-hand side of the equation results from the noncommutative nature of the cross product: $d\vec{l} \times \vec{B} = -\vec{B} \times d\vec{l}$. Equation (23-23) provides a way to define the ampere: a uniform magnetic field of 1 T perpendicular to the wire carrying an electric current of 1 A exerts a force of 1 N on a 1 m length of wire:

$$1\text{ A} = \frac{1\text{ N}}{1\text{ m} \cdot 1\text{ T}} \qquad (23\text{-}24)$$

☑ **CHECKPOINT**

C-23-7 The Magnetic Force Acting on a Current-Carrying Wire

An electric current of 1 A flows through a long, straight vertical conductor located near Earth's surface. The horizontal component of Earth's magnetic field in this location is 0.4 mT. What is the magnetic force per metre of this conductor?

(a) 0

(b) 0.0004 N

(c) 4 N

(d) 2500 N

PEER TO PEER

At first, I was stressed out by the electricity and magnetism equations. However, later I realized that I should start with understanding the problem and finding the approximate answer conceptually before jumping into the calculations. Understanding the problem first made it much easier for me.

LO 5

23-5 The Torque on a Current-Carrying Loop in a Magnetic Field

Let us consider a rectangular loop of wire, with the dimensions a and b, carrying a current in a clockwise direction (Figure 23-22(a)). This loop is placed in a uniform magnetic field of magnitude B directed to the right.

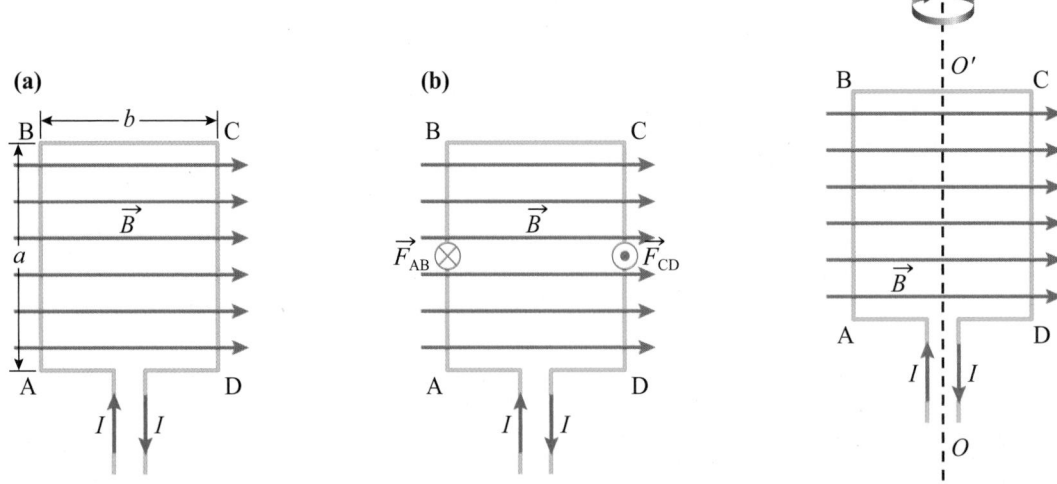

Figure 23-22 (a) A rectangular current-carrying loop in a uniform magnetic field (b) Magnetic forces acting on opposite sides of the carrying-current loop (c) A couple moment M (torque) acting on a current-carrying loop in a uniform magnetic field

The magnetic field is parallel to the sides BC and DA of the loop; therefore $\vec{F}_{BC} = \vec{F}_{DA} = 0$. Sides AB and CD are perpendicular to the magnetic field. Since an electric current in AB flows in the opposite direction to the electric current in CD, the magnetic forces on the two opposite sides of the loop are opposite to each other: the magnetic force on AB is directed into the page, and the magnetic force on CD is directed out of the page (Figure 23-22(b)). Mathematically, this result can be expressed as follows:

$$\vec{F}_{ABCD} = \vec{F}_{AB} + \vec{F}_{BC} + \vec{F}_{CD} + \vec{F}_{DA}$$

$$= I\vec{l}_{AB} \times \vec{B} + I\vec{l}_{BC} \times \vec{B} + I\vec{l}_{CD} \times \vec{B} + I\vec{l}_{DA} \times \vec{B}$$

$$= \vec{F}_{AB} + \vec{F}_{CD}$$

Forces $\vec{F}_{AB}$ and $\vec{F}_{CD}$ are equal in magnitude and opposite in direction:

$$F_{AB} = F_{CD} = IaB \qquad (23\text{-}25)$$

Since forces $\vec{F}_{BC}$ and $\vec{F}_{DA}$ are applied to opposite sides of the loop and are perpendicular to its plane, they create a torque on the loop, causing it to rotate about the OO'- axis (Figure 23-22(c)). The magnitude of this torque is

$$\tau_{ABCD} = \tau_{AB} + \tau_{CD}$$

$$= \frac{b}{2}F_{AB} + \frac{b}{2}F_{CD} = \frac{b}{2}IaB + \frac{b}{2}IaB = abIB = AIB$$

KEY EQUATION $\qquad \tau_{ABCD} = IAB \qquad (23\text{-}26)$

where A is the area of the loop: $A = ab$.

The torque exerted on the loop by the magnetic field is proportional to the area of the loop, the electric current and the magnitude of the magnetic field. It is common to describe the orientation of the loop by a vector $\vec{A}$,

such that its magnitude represents the area of the loop and its direction is defined by a right-hand rule: when the fingers of your right-hand are curled in the direction of the electric current, the direction of the thumb indicates the direction of the vector $\vec{A}$ (Figure 23-23).

Figure 23-23 A right-hand rule for determining the orientation of a current-carrying loop using vector $\vec{A}$ and magnetic dipole moment $\vec{\mu}$

As discussed in Chapter 8, torque is a vector quantity that points along the object's axis of rotation, indicating clockwise or counterclockwise rotation. The direction of rotation is called the "sense of the torque vector." For the loop in Figure 23-24, for example, the direction of $\vec{\tau}$ is perpendicular to both vectors $\vec{A}$ and $\vec{B}$:

KEY EQUATION $\qquad \vec{\tau}_{ABCD} = I\vec{A} \times \vec{B} \qquad (23\text{-}27)$

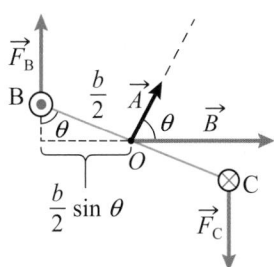

Figure 23-24 The top view of a vertical rectangular loop carrying a current submerged in a uniform horizontal magnetic field. The loop is free to rotate about its vertical axis of symmetry.

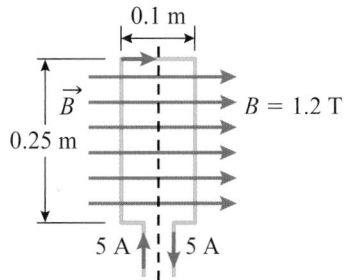

Figure 23-25 A rectangular loop carrying current in a uniform magnetic field. The maximum torque on this current-carrying loop is 0.15 N·m.

The magnitude of the torque is determined by the properties of the loop and the external magnetic field. To describe these properties, we introduce the concept of a **magnetic dipole moment**, $\vec{\mu}$, (Figure 23-24) defined as follows:

$$\vec{\mu} = I\vec{A} \qquad (23-28)$$

The concept of a magnetic dipole helps clarify a key difference between electric and magnetic fields. An electric dipole, a combination of two equally but oppositely charged objects, in a uniform electric field experiences a torque that acts to align the dipole with the electric field lines: $\vec{\tau}_E = \vec{p} \times \vec{E}$ (Chapter 20), where $\vec{p}$ is the electric dipole moment. However, there is no evidence that magnetic charge or magnetic monopoles exist. A **magnetic dipole** is a current-carrying loop or a permanent magnet, not a combination of two magnetic monopoles. A magnetic field exerts a torque on a magnetic dipole, directed such that the magnetic dipole moment aligns with the magnetic field:

KEY EQUATION $$\vec{\tau}_B = \vec{\mu} \times \vec{B} \qquad (23-29)$$

You may have noticed that a current-carrying loop behaves like a permanent magnet, with the polarity of the loop depending on the direction of the current. Only the magnetic field component parallel to the plane of the loop results in a torque on it. When a loop positioned parallel to a uniform magnetic field starts rotating, the effect of the magnetic field on the loop changes (Figure 23-24).

The component of the magnetic field parallel to the surface of the rotating loop can be expressed as $B\sin\theta$, where θ is an angle between the normal to the loop and the magnetic field (Figure 23-25). Then the net torque (a couple moment) produced by the magnetic force is

$$\tau = blaB\sin\theta = AIB\sin\theta \qquad (23-30)$$

Consequently, the net torque varies as the loop rotates, reaching a maximum value when the magnetic field is parallel to the plane of the loop ($\theta = 90°$) and a minimum value when the magnetic field is perpendicular to it ($\theta = 0°$):

$$\tau_{max} = blaB, \quad \text{when } \theta = 90°$$

$$\tau_{min} = blaB\sin(0°) = 0, \quad \text{when } \theta = 0° \qquad (23-31)$$

Although we derived this expression for a rectangular loop, we can show that it holds true for any loop.

✓ **CHECKPOINT**

C-23-8 Interactions of a Current-Carrying Loop and a Magnetic Field

A current-carrying loop hangs from a light thread (Figure 23-26). A permanent magnet is brought close to the loop so that the north pole of the magnet faces the loop. Which statement best describes the interaction between the magnet and the loop?

Figure 23-26   C-23-8

(a) There is no interaction because electric currents do not repel or attract magnets.
(b) They will attract.
(c) They will repel.
(d) There is not enough information to determine.

C-23-8 (b) The loop is a magnetic dipole oriented such that it is attracted to the magnet.

CHAPTER 23 | **MAGNETIC FIELDS AND MAGNETIC FORCES**

MAKING CONNECTIONS

Can You Build an Electric Motor?

Many everyday appliances use electric motors, for example, microwave ovens, food processors, fans, DVD players, and washing machines. These motors all apply the concept of a magnetic torque on a current-carrying loop. Although some motors, such as those in DVD players and computer hard drives, involve complex precision technology, you can build a simple DC motor with a flashlight, a battery, a few pieces of wire, a small permanent magnet, a pair of scissors, a few paper clips, tape, and any convenient cylinder, such as a thick marker, as shown in Figure 23-27.

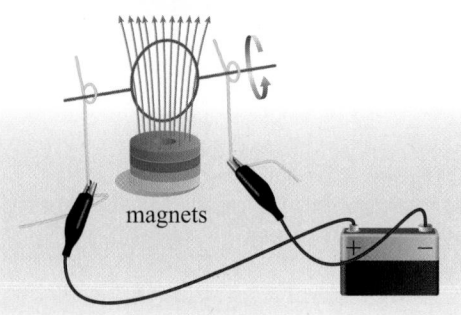

Figure 23-27   A simple DC motor

EXAMPLE 23-6

The Magnetic Torque on a Simple Electric Motor

Estimate the maximum value of the magnetic torque in the simple electric motor shown in Figure 23-27, given that the magnetic field created by the magnets is 0.4 T, the battery has a voltage of 3 V, the number of copper wire loops is 2, and the area of each loop is 5 cm². Assume that the resistance of all the wire used in the motor is 1 Ω.

SOLUTION

The magnetic torque acting on a current-carrying loop in a magnetic field can be found using Equation (23-26):

$$\tau_{max} = 2IAB = 2\frac{V}{R}AB = 2\frac{3\text{ V}}{1\text{ }\Omega}(5 \times 10^{-4}\text{ m}^2)(0.4\text{ T})$$

$$= 2 \times 6 \times 10^{-4}\frac{\text{V} \cdot \text{m}^2 \cdot \text{T}}{\Omega} = 1 \times 10^{-3}\text{ N} \cdot \text{m}$$

Making sense of the result:

This simple experiment helps us make two interesting observations: First, the low resistance of the wire allows a relatively large current to flow from the battery. Nevertheless, the magnetic torque is very small. Therefore, in order for this motor to work, the loop has to be very light. The second observation is more profound. We discussed earlier that magnetic force cannot do work on a moving charge because it is directed perpendicular to the charge's velocity (Equation (23-1)). In the case of an electric motor, it appears to do work on a moving current-carrying wire. How can this happen? The solution to this apparent paradox is that the physics involved is really more complex than finding the magnetic force on the electrons flowing in the wire, although that standard derivation does give the correct answer. In reality, the magnetic force on the electrons exerts a transverse force on them as in the Hall effect (Equation (23-20)). This charge separation creates an electric force across the wire, which exerts a force on the positively charged lattice atoms, doing the work.

LO 6

23-6   The Biot-Savart Law

Electric charges create electric fields, but we have yet to discuss the sources of magnetic fields. The nature of the sources of magnetic fields was a mystery for centuries. The equations we have derived so far show how magnetic fields affect moving charges, but not vice versa. Is it reasonable to assume that a magnetic field can be affected by an electric current? Could electric current be a source

of a magnetic field? Although the answers to these questions might seem obvious today, two centuries ago they were at the forefront of scientific inquiry. The first step in solving this mystery was made by the Danish scientist Hans Christian Oersted (1777–1851).

In 1820, Oersted was able to show that a compass needle deflects from magnetic north when an electric current is switched on or off in a nearby wire. This discovery motivated the French scientists Jean-Baptiste Biot (1774–1862) and Félix Savart (1791–1841) to conduct experiments not only to confirm that electric current is the source of magnetic fields, but to find a mathematical expression for the relationship between them. The Biot–Savart law describes how to calculate a magnetic field generated by an infinitesimally small element of a steady electric current (Figure 23-28):

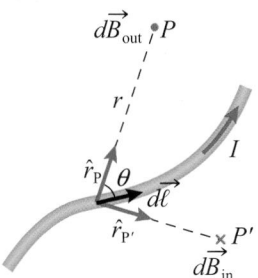

Figure 23-28 Using the Biot–Savart law to find a magnetic field created by a long current-carrying wire

KEY EQUATION $$d\vec{B} = \frac{\mu_0}{4\pi} \frac{I d\vec{l} \times \hat{r}}{r^2} = \frac{\mu_0}{4\pi} \frac{I d\vec{l} \times \vec{r}}{r^3} \quad (23\text{-}32)$$

In this equation, $d\vec{l}$ is an infinitesimally small element of the wire carrying current I, so $d\vec{l}$ is directed in the direction of the conventional current; r is the distance from the element $d\vec{l}$ to an arbitrary point at which the magnetic field is calculated; $\hat{r}$ is the unit vector in the direction of $\vec{r}$; and μ_0 is a constant called the **permeability constant of free space** and has a value of $4\pi \times 10^{-7}$ T·m/A. The term 4π was introduced in the definition in order to make Maxwell equations of electromagnetism appear more elegant.

Let us examine the Biot–Savart law more closely:

■ Since the expression for $d\vec{B}$ includes the cross product of $d\vec{l}$ and $\vec{r}$, $d\vec{B}$ is always perpendicular to the element of electric current that created it, as well as to the position vector $\vec{r}$ connecting $d\vec{l}$ and the point at which the magnetic field is calculated. The direction of the magnetic field can be observed experimentally by spreading iron filings near a long, straight current-carrying wire (Figure 23-29).

■ $d\vec{B}$ is inversely proportional to the square of the distance from the point for which the field is calculated to the infinitesimal element of electric current creating the field.

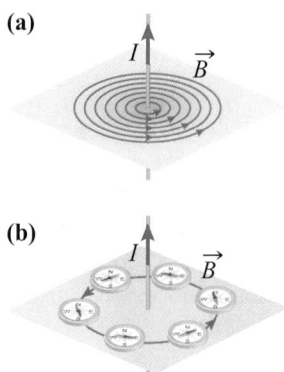

Figure 23-29 (a) A magnetic field created by a long, current-carrying wire. Iron fillings spread near the current-carrying wire. (b) Compass needles indicate the direction of the magnetic field surrounding the current-carrying wire.

■ Both $d\vec{B}$ and the resultant magnetic field are proportional to the current.

■ Since $d\vec{B}$ is proportional to the cross product of $d\vec{l}$ and $\vec{r}$, the magnetic field depends on the sine of the angle between the infinitesimally small element of the wire $d\vec{l}$ and the position vector $\vec{r}$.

The Biot–Savart law for magnetism is analogous to Coulomb's law for electric fields. Both laws show that the strength of the field is proportional to its source: q for an electric field and the electric current element $d\vec{l} = I d\hat{l}$, where $\hat{l}$ is a unit vector in the $\vec{l}$ direction for a magnetic field. In both laws, the field strength is inversely proportional to the square of the distance from the source of the field.

However, there are also some important differences: electric field lines always start on positive charges and end on negative charges. Sources of electric field (positive and negative charges) can be separated, with a positive charge defined as a source and a negative charge as a sink. We cannot, however, separate the south and north poles of a magnet because, as far as we know, there are no magnetic monopoles. Since there is no magnetic monopole, magnetic field lines are continuous and closed: they have neither a beginning nor an end, and they surround the electric current that created them.

EXAMPLE 23-7

The Magnetic Field Created by an Infinitely Long, Straight Current-Carrying Wire

Find the magnetic field created by the infinitely long, straight current-carrying wire at point P located a distance R from the wire (Figure 23-30).

SOLUTION

We use the Biot–Savart law to calculate the magnetic field at point P. Let us denote the current flowing through the wire I, and the shortest distance from the wire to the point where the magnetic field is to be calculated R. To use the Biot–Savart law (Equation (23-32)), we have to calculate the contribution of the current in the wire segment $d\vec{l}$ ($d\vec{l} = I d\hat{l}$)

(continued)

ELECTRICITY, MAGNETISM, AND OPTICS

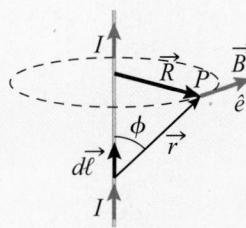

Figure 23-30 Calculating the magnetic field created by an infinitely long, straight current-carrying wire at point P located a distance R from the wire

to the magnetic field at point P. Then we add the effects of all the segments along the wire. The magnetic field produced by the wire segment $\vec{dl}$ has the direction $\vec{dl} \times \vec{r}$. The resultant vector is perpendicular to both $\vec{dl}$ and $\vec{r}$, and we denote its direction using a unit vector, $\hat{e}$:

$$\vec{dl} \times \vec{r} = \hat{e} r \sin \theta \, dl = \hat{e} R \, dl$$

Now we are ready to apply the Biot–Savart law:

$$\vec{dB} = \frac{\mu_0}{4\pi} \frac{I \vec{dl} \times \vec{r}}{r^3} = \frac{\mu_0 I}{4\pi} \frac{\hat{e} R \, dl}{(l^2 + R^2)^{3/2}} = \hat{e} \frac{\mu_0 I R}{4\pi} \frac{dl}{(l^2 + R^2)^{3/2}}$$

where $r = \sqrt{l^2 + R^2}$.

The magnitude of the magnetic field at point P can be expressed as follows:

$$\vec{B} = \int_{-\infty}^{\infty} \hat{e} \frac{\mu_0 I R}{4\pi} \frac{dl}{(l^2 + R^2)^{3/2}} = \hat{e} \frac{\mu_0 I R}{4\pi} \int_{-\infty}^{\infty} \frac{dl}{(l^2 + R^2)^{3/2}}$$

This integral can be calculated using two substitutions: $x = el/R$ and then $x = \cot \phi = \cos \phi / \sin \phi$:

First substitution: $\dfrac{l}{R} \equiv x; \; dl = R \, dx$

$$\vec{B} = \hat{e} \frac{\mu_0 I R}{4\pi} \int_{-\infty}^{\infty} \frac{dl}{(l^2 + R^2)^{3/2}} = \hat{e} \frac{\mu_0 I R}{4\pi R^3} \int_{-\infty}^{\infty} \frac{R \, dx}{(x^2 + 1)^{3/2}}$$

$$= \hat{e} \frac{\mu_0 I R^2}{4\pi R^3} \int_{-\infty}^{\infty} \frac{dx}{(x^2 + 1)^{3/2}} = \hat{e} \frac{\mu_0 I}{4\pi R} \int_{-\infty}^{\infty} \frac{dx}{(x^2 + 1)^{3/2}}$$

Second substitution: $x \equiv \cot \phi = \dfrac{\cos \phi}{\sin \phi}; \; dx = -\dfrac{d\phi}{\sin^2 \phi}$

$$\vec{B} = \hat{e} \frac{\mu_0 I}{4\pi R} \int_{-\infty}^{\infty} \frac{dx}{(x^2 + 1)^{3/2}} = -\hat{e} \frac{2\mu_0 I}{4\pi R} \int_{0}^{\pi/2} \sin \phi \, d\phi = \hat{e} \frac{\mu_0 I}{2\pi R}$$

Making sense of the result:

It has been confirmed experimentally that the magnetic field created by current flowing through a long, straight wire is directed along concentric circles surrounding the wire (Figure 23-29). The magnitude of this field is described by the following equation:

$$B_{\text{outside straight wire}} = \frac{\mu_0 I}{2\pi R} \qquad (23\text{-}33)$$

The direction of the magnetic field due to the electric current can be conveniently established using the following right-hand rule for the magnetic field (Figure 23-31): When you point your right thumb in the direction of the electric current and curl your fingers, the direction of the curled fingers is the direction of the magnetic field lines that form concentric circles around the current-carrying wire.

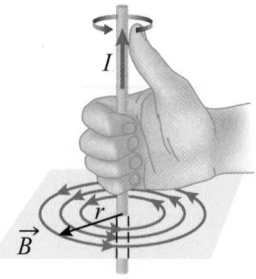

Figure 23-31 The right-hand rule for determining the direction of the magnetic field surrounding a long, straight current-carrying wire. The magnetic field lines create closed circular concentric loops. The magnitude of the magnetic field can be found using Equation (23-34).

EXAMPLE 23-8

The Magnetic Field Created by a Circular Current-Carrying Loop

Derive an expression for the magnetic field along the axis of symmetry of a circular loop of radius R that carries an electric current I.

SOLUTION

For convenience, we choose a coordinate system with the current-carrying loop in the yz-plane (Figure 23-32). We divide the loop into an infinitely large number of infinitesimally small current elements: $\vec{dl} = I \vec{dl}$. The Biot–Savart law

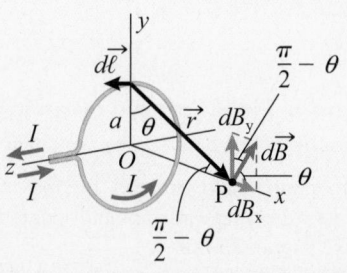

Figure 23-32 Using the Biot–Savart law to calculate the magnetic field created by a current-carrying loop along its axis of symmetry

can be used to calculate the contribution of each element to the magnetic field at point P on the axis of symmetry of the loop. Let x be the distance of this point from the centre of the loop, O. Then,

$$d\vec{B} = \frac{\mu_0}{4\pi} \frac{Id\vec{l} \times \hat{r}}{r^2} = \frac{\mu_0 I}{4\pi} \frac{d\vec{l} \times \hat{r}}{(x^2 + R^2)}$$

First, consider the cross product of $d\vec{l}$ and $\vec{r}$. Since point P is on the loop's axis of symmetry, the unit vector $\hat{r}$ is perpendicular to any loop element $d\vec{l}$. Therefore, $|d\vec{l} \times \hat{r}| = dl$. The direction of the cross product, shown in Figure 23-32, is perpendicular to both $d\vec{l}$ and $\vec{r}$. The magnetic field created by this loop element has two components, B_y and B_z. Now consider another element of the loop $d\vec{l}''$ symmetrical to the element $d\vec{l}$ about the centre of the loop. Since the elements $d\vec{l}$ and $d\vec{l}''$ are antiparallel, the magnetic fields created by them have x-components that are equal and y-components that are equal in magnitude but opposite in direction. The same consideration is applicable to any pair of symmetrical loop elements: the magnetic field components perpendicular to the loop's axis of symmetry cancel out, and the components parallel to the axis add up. The dB_x components can be calculated as follows:

$$dB_x = \frac{\mu_0}{4\pi} \frac{I\cos\theta\, dl}{(x^2 + R^2)} = \frac{\mu_0}{4\pi} \frac{Idl}{(x^2 + R^2)} \cos\theta$$

$$= \frac{\mu_0}{4\pi} \frac{Idl}{(x^2 + R^2)} \frac{R}{(x^2 + R^2)^{1/2}}$$

$$= \frac{\mu_0}{4\pi} \frac{IRdl}{(x^2 + R^2)^{3/2}}$$

To calculate the net magnetic field created by the entire loop, we integrate the contributions of all the loop elements:

$$\vec{B}_{loop} = \int_0^{2\pi R} \frac{\mu_0}{4\pi} \frac{IRdl}{(x^2 + R^2)^{3/2}}$$

$$= \frac{\mu_0 IR}{4\pi(x^2 + R^2)^{3/2}} \int_0^{2\pi R} dl = \frac{\mu_0 IR 2\pi R}{4\pi(x^2 + R^2)^{3/2}}$$

$$= \frac{\mu_0}{2} \frac{IR^2}{(x^2 + R^2)^{3/2}}$$

Thus, the magnetic field created by a circular current-carrying loop along its axis of symmetry is collinear with this axis. The direction of the magnetic field can be determined using the right-hand rule for current, and its magnitude is

KEY EQUATION
$$B_{loop} = \frac{\mu_0}{2} \frac{IR^2}{(x^2 + R^2)^{3/2}} \qquad (23\text{-}34)$$

Making sense of the result:

The magnetic field of a circular current-carrying loop along its axis of symmetry diminishes to zero when the distance from the centre of the loop is much larger than the size of the loop: $x \gg R$. The magnetic field in the centre of the loop ($x = 0$) is

KEY EQUATION
$$B_{loop\text{-}centre} = \frac{\mu_0}{2} \frac{IR^2}{R^3} = \frac{\mu_0}{2} \frac{I}{R} \qquad (23\text{-}35)$$

✓ CHECKPOINT

C-23-9 Ranking Magnetic Field Strengths Produced by Current-carrying Circular Loops

For the six circular loops in Figure 23-33, which of the following correctly ranks the magnetic fields in the centre of each loop?

(a) $B_D > B_B > B_A > B_C > B_F > B_E$
(b) $B_E > B_F = B_A > B_C = B_B > B_D$
(c) $B_D > B_B = B_C > B_A = B_F > B_E$
(d) $B_B > B_C = B_A > B_E = B_F > B_D$
(e) $B_B = B_D > B_C > B_A > B_E > B_F$

Figure 23-33 C-23-9

23-7 Ampère's Law

Although the Biot–Savart law provides a technique for determining the magnetic field around a wire of any arbitrary shape, the calculations can be somewhat daunting even for relatively simple shapes, as we saw in Examples 23-7 and 23-8. In this respect, the Biot–Savart law resembles Coulomb's law, which can be used to determine the electric field around any distribution of charge, but involves difficult calculations unless the charge distribution is highly symmetric. When a continuous charge distribution has certain symmetries, Gauss's law significantly simplifies the electric field calculations. Similarly, Ampère's law allows much easier calculations of magnetic fields when the path of the current is symmetric. Before we define Ampère's law, we need to introduce line integrals.

Overview of Line Integrals

In the previous chapters, we saw that the integral of a function $\int f(x)dx$ is an infinite sum of the products of the values of the function $f(x)$ and an infinitesimally small element dx along a path of integration from $x = a$ to $x = b$, where a and b are the limits of integration. Thus, dx is the lower limit of small steps, Δx, along the path of integration, and

$$\lim_{N \to \infty} \sum_{i=1}^{N} f(x)\Delta x_i = \int_a^b f(x)dx \qquad (23\text{-}36)$$

Consequently, an integral represents the area under the graph of $f(x)$ (Figure 23-34).

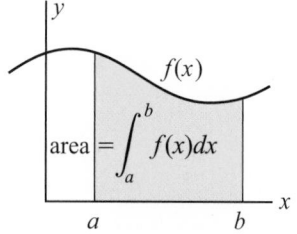

Figure 23-34 The area under the graph represents integration geometrically.

Sometimes, instead of varying an independent variable along a linear coordinate axis, it is useful to vary it along a 2-D or a 3-D curve (Figure 23-35). Such line integrals are defined as

$$\lim_{N \to \infty} \sum_{i=1}^{N} f(\vec{r}) \cdot \Delta \vec{l}_i = \int_a^b f(\vec{r}) \cdot d\vec{l} \qquad (23\text{-}37)$$

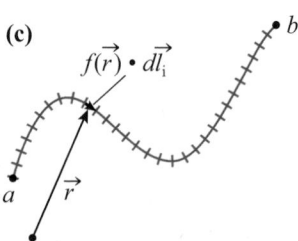

Figure 23-35 (a) An arbitrary line connecting two points a and b along the path of integration. (b) The line can be divided into infinitesimally small segments, $d\vec{l}_i$. (c) The integral along the path of integration can be considered as a sum of the products of $f(\vec{r})$ and $d\vec{l}_i$.

An integral defined by Equation (23-37) is a general case of a line integral along a straight line, such as the x-coordinate axis. The function $f(x)$ that we want to integrate can be either a scalar or a vector function. Each infinitesimally small element of the curve is considered an infinitesimally small vector element $d\vec{l}$ because it has both magnitude and direction. Consider a line integral where the magnetic field has a constant magnitude and is parallel to the line of integration. In this case, the integral of the scalar product of the magnetic field and the curve element can be simplified. Notice that although the magnetic field depends on $\vec{r}$, for simplicity, we will omit $\vec{r}$ from the notation of the magnetic field and will denote it as $\vec{B}$.

$$\int_a^b \vec{B} \cdot d\vec{l} = \int_a^b B \cdot dl = B\int_a^b dl = BL \qquad (23\text{-}38)$$

where L is the total displacement along the curve of integration.

Closed loop path integrals are integrals around a closed loop, so the initial and the final points of integration coincide. Such integrals are denoted by a small circle over the conventional integral sign, for example, $\oint \vec{B} \cdot d\vec{l}$. Closed loop path integrals are used to calculate the magnetic fields created by electric currents.

Ampère's Law

French physicist and mathematician André-Marie Ampère (1775–1836), for whom the unit of electric current is named, studied magnetic fields created by current-carrying

ELECTRICITY, MAGNETISM, AND OPTICS

wires. He noticed that the direction of a magnetic field is always perpendicular to the direction of the wire. He also realized that magnetic field lines should have a symmetry corresponding to the symmetry of the wire. The only way this symmetry can occur is if the magnetic field lines form concentric circles around the wire in the planes perpendicular to it (Figure 23-36).

Figure 23-36 The magnetic field created by a straight current-carrying wire

path of integration

Ampère's law states:

1. The magnetic field in the space around an electric current is proportional to the electric current that serves as its source, just as an electric field is proportional to the charge that serves as its source.

2. For any closed loop path (Figure 23-36), the sum of the scalar products of the length element $d\vec{l}$ and of the magnetic field $\vec{B}$ (which can be expressed as the line integral along the closed path) is equal to the permeability of free space (or magnetic constant) μ_0 multiplied by the electric current enclosed by the loop:

KEY EQUATION
$$\oint \vec{B} \cdot d\vec{l} = \mu_0 \Sigma I_{enc} \qquad (23\text{-}39)$$

When electric currents enclosed by an integral path flow in opposite directions, they have opposite signs. The direction of the magnetic field can be determined using a right-hand rule: when you orient your right-hand such that the thumb points in the direction of the electric current, the curled fingers point in the direction of the magnetic field. It is especially convenient to apply Ampère's law when one can choose a path of integration along the magnetic field lines. Notice that only electric currents *enclosed* by the path of integration contribute to the magnetic field (similar to Gauss's law (Equation (20-15), where only electric charges enclosed

by the surface are used for calculating the electric flux contribute to the electric field).

Applications of Ampère's law

We now look at three common applications of Ampère's law.

An Infinitely Long, Straight Current-carrying Wire
For a path of integration, we choose a circle centred on the wire and perpendicular to it (Figure 23-36). The magnetic field at any point along this path is tangent to the integration path element $d\vec{l}$ Moreover, due to the symmetry of the problem, the magnitude of the magnetic field is constant along the path of integration. Thus, we can apply Ampère's law (Equation (23-39)) as follows:

$$\oint \vec{B}(\vec{r}) \cdot d\vec{l} = \mu_0 \Sigma I_{enc}$$

$$\oint B dl = 2\pi R B = \mu_0 I$$

$$B = \frac{\mu_0 I}{2\pi R}$$

The same result was obtained earlier using the Biot–Savart law (Equation (23-33)). However, the application of Ampère's law greatly simplifies the derivation.

EXAMPLE 23-9

The Magnetic Field Around a Long, Straight Current-Carrying Wire

Find the magnetic field at a distance of 1 m from a long, straight wire carrying a current of 1 A.

SOLUTION

This problem can be solved using Equation (23-33):

$$B = \frac{\mu_0 I}{2\pi R} = \frac{\left(4\pi \times 10^{-7} \dfrac{\text{T} \cdot \text{m}}{\text{A}}\right)(1\ \text{A})}{2\pi(1\ \text{m})}$$

$$= 2 \times 10^{-7}\,\text{T} = 2 \times 10^{-4}\,\text{mT}$$

Making sense of the result:

It is worth noting that Earth's magnetic field is in the range of 30 to 60 mT, which is approximately 10^5 times stronger than the magnetic field 1 m from a long, straight wire carrying 1 A of electric current.

C-23-10 Ranking the Line Integrals

Figure 23-37 shows six different configurations of electric currents going into or coming out of the plane of the page. The closed loop path of integration is indicated by a dashed line in each diagram. Which of the following correctly ranks the magnitudes of the $(\vec{r} \cdot \vec{dl})$ integrals for these configurations?

(a) (c) > (e) > (a) = (b) = (d) > (f)
(b) (f) > (a) = (b) = (d) = (c) > (e)
(c) (c) > (e) > (a) > (b) > (f) > (d)
(d) (f) > (e) > (d) = (b) = (a) > (c)
(e) (a) > (d) > (b) = (c) = (d) > (f)

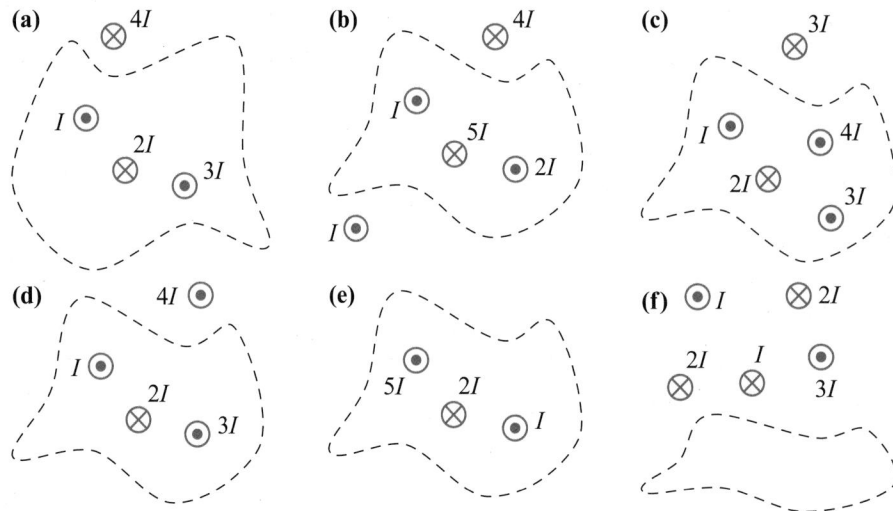

Figure 23-37   Checkpoint C-23-10

C-23-10 (a) According to Ampère's law, the path that encloses the most electric current has the highest value of the $(\overleftarrow{r}) \cdot (\overleftarrow{dl})$ loop integral along it.

 EXAMPLE 23-10

The Magnetic Field Inside and Outside a Long, Current-Carrying Wire

Determine the magnetic field inside and outside a long wire of radius R carrying an electric current I.

SOLUTION

Figure 23-38(a) shows an electric wire of radius R carrying an electric current I. To find the magnetic field inside the wire, let us assume a constant current density across the wire. Let us choose an arbitrary point P a distance r from the centre of the wire, such that $r < R$. Now let us draw a circular loop of radius r with the wire at its centre. This loop goes through point P. To find how much current is enclosed by the loop, we calculate the current density inside the wire and multiply it by the cross-sectional area of the wire enclosed by it:

$$\sigma = \frac{I}{\pi R^2} \qquad (23\text{-}40)$$

$$B_{\text{inside wire}} = \frac{\mu_0 I}{2\pi R} = \frac{\mu_0}{2\pi r}(\sigma \pi r^2)$$

$$= \frac{\mu_0}{2\pi r}\frac{I}{\pi R^2}\pi r^2 = \frac{\mu_0 I}{2\pi R^2}r, \; r \leq R$$

Making sense of the result:

The magnetic field *inside* the current-carrying wire is proportional to the distance from the centre of the wire. This relationship is shown in Figure 23-38(b). The magnetic field *outside* the wire is inversely proportional to the distance from its centre, as indicated by Equation (23-33):

$$B = \frac{\mu_0 I}{2\pi r}, \; r \geq R$$

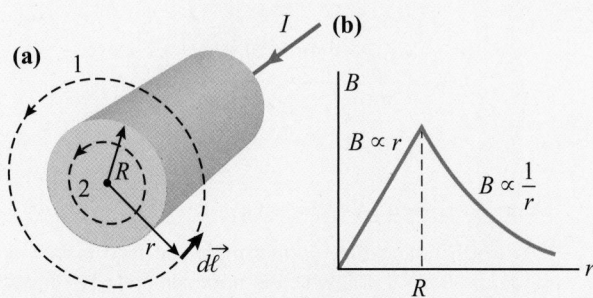

Figure 23-38   (a) The cross section of a current-carrying wire (b) A plot of $B(r)$, the dependence of the magnitude of the magnetic field created by the current-carrying wire on distance from the centre of the wire

Coaxial Cables A **coaxial cable** is an electrical cable that consists of an inner conductor surrounded by an outer cylindrical conductor. The two conductors are separated by an insulating spacer of constant thickness, and the cable usually has an insulating jacket over the outer conductor (Figure 23-39). The term "coaxial" is used because the inner and outer conductors have the same axis of symmetry. Coaxial cables have many applications, in particular, for carrying high-frequency voltages such as for TV and radio signals. Coaxial cables have electrical characteristics that help reduce signal losses in long cable runs and shield against electromagnetic interference.

Figure 23-39 A cross section of a coaxial cable

Ampère's law can be used to show that the electromagnetic field carrying the signal exists only in the space between the inner and the outer conductors of the coaxial cable when the currents passed in the two conductors have equal magnitude but opposite direction. Outside the coaxial cable, the magnetic field created by the electric current flowing in one direction is equal and opposite to the magnetic field created by the opposite current (see problem 68).

Solenoids In Chapter 20, we found that two large, oppositely charged parallel metal plates create a uniform electric field in the space between them. A solenoid is a device that produces a nearly uniform magnetic field (Figure 23-40). Solenoids are widely used because the uniform magnetic field inside them is determined solely by their structure and the current flowing through them.

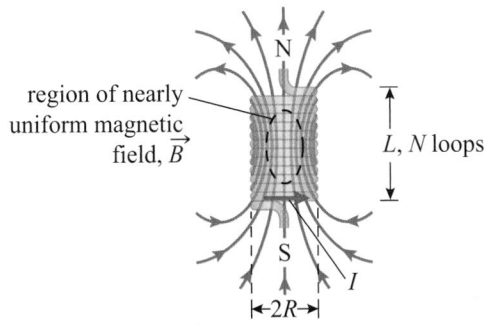

Figure 23-40 A solenoid of length L and radius r consists of N current-carrying loops.

Recall the expression for the magnetic field created in the centre ($x = 0$) of a current-carrying loop (Equation (23-35)):

$$B_{\text{loop-centre}} = \frac{\mu_0}{2} \frac{I}{R}$$

Now we draw magnetic field lines created by a single loop, by three neighbouring loops, and by four neighbouring loops (Figure 23-41).

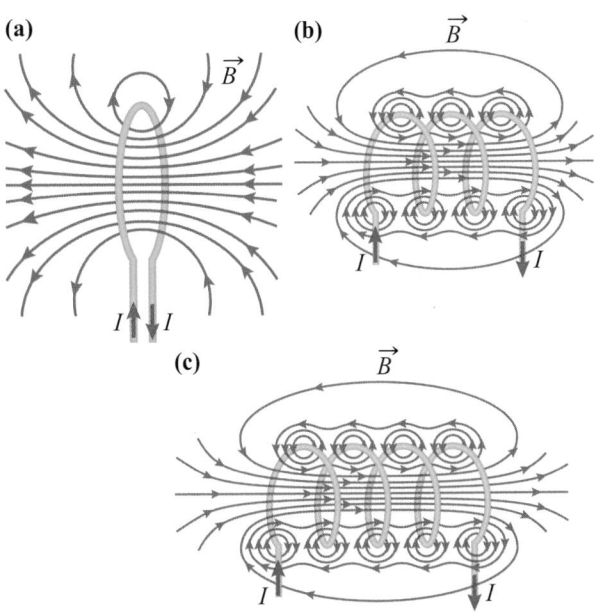

Figure 23-41 Calculating the magnetic field of a solenoid (a) The magnetic field created by a single loop (b) The magnetic field created by three adjacent loops (c) The magnetic field created by four adjacent loops

Magnetic field lines inside the loops (Figure 23-41) reinforce each other, and the magnetic field lines outside the loops nearly cancel each other. The more loops that are placed next to each other, the more their magnetic fields are reinforced inside the loops and cancel each other outside of them. A cross section of a 10-loop solenoid is shown in Figures 23-42a and 23-42b. The magnetic field is almost uniform in the centre of the solenoid and is relatively weak outside of it. Increasing the number of loops increases the magnetic field inside the solenoid and makes it more uniform. A solenoid is also called an **electromagnet** because it uses an electric current to produce a magnetic field. However, unlike permanent magnets, the magnetic field of a solenoid depends on its structure and on the amount and direction of the electric current that flows through it.

Figure 23-42(a) illustrates how Ampère's law can be used to calculate the magnetic field inside a long solenoid of length L and radius R, which has N loops and carries current I. The axis of a solenoid is directed along the x-axis. It is convenient to choose a rectangular path of integration with length a and width b. The magnetic field

(a)

(b)

(c)

$\vec{B}$

$B = \mu n I$

$\vec{B}$

Figure 23-42 (a) The magnetic field created by a 10-loop solenoid, (b) a cross section of a very long solenoid, and (c) a very long solenoid consisting of *N* loops.

inside the solenoid is directed along its axis of symmetry, so the field is parallel to the long side of the rectangle, *a*, and perpendicular to the short side, *b*. The magnetic field outside the solenoid is negligible. Since Ampère's law is defined in terms of the dot product of the magnetic field and the path element $d\vec{l}$, this product is nonzero only for one side of the rectangle: the long side located inside the solenoid. Applying Ampère's law,

$$\oint \vec{B} \cdot d\vec{l} = \mu_0 \sum I_{enc}$$

$$\oint B\hat{i} \cdot d\vec{l} = \mu_0 \frac{INa}{L}$$

$$B \oint \hat{i} \cdot d\vec{l} = \mu_0 \frac{INa}{L}$$

$$Ba = \mu_0 \frac{INa}{L}$$

KEY EQUATION
$$B_{solenoid} = \mu_0 \frac{IN}{L} = \mu_0 In \qquad (23\text{-}41)$$

Therefore, the magnetic field inside the solenoid depends on the density of the solenoid's loops ($n = N/L$) and the value of the electric current *I*. Unlike the magnetic field created by a current-carrying loop, the magnetic field of a solenoid does not depend on the radius of the coils.

The Magnetic Field of a Solenoid

The e-resource that accompanies every new copy of this textbook contains an Online Activity using the PhET simulation "Magnets and Electromagnets." Work through the simulation and accompanying questions to gain an understanding of how the current and the number of loops affect the magnetic field of a solenoid.

MAKING CONNECTIONS

Magnetic Resonance Imagining

Although magnetic resonance imaging (MRI) was invented less than 40 years ago, it is difficult to imagine a modern hospital without it. MRI technology produces images of body tissue and can show details of soft tissue that are not visible using other imaging technologies, such as X-rays and CT scans (Figure 23-43). In addition to the advantage of being good resolution for soft tissue we should mention that it does not involve ionizing radiation so does not carry the potential risks of X-rays. MRI machines generate these images by measuring changes in the magnetic fields of hydrogen nuclei in water molecules in the body (human bodies are 70% water). These changes are caused by nuclear magnetic resonance of the hydrogen nuclei: when a person is inside the scanner, the hydrogen nuclei (i.e., protons) align with a strong magnetic field in the scanning chamber. A radio wave at just the right frequency (the resonance frequency) for the protons to absorb energy pushes some of the protons out of alignment. The protons then snap back to alignment, producing a detectable rotating magnetic field as they do so. Since protons in different types of tissue (e.g., fat, muscle, or a tumour) realign at different speeds, these tissues can be distinguished in the MRI images. Most MRI systems use a superconducting magnet to generate the very strong magnetic fields required, typically 0.5 to 3.0 T (approximately 10 000 times greater than Earth's magnetic field).

iStockphoto/Thinkstock

Figure 23-43 An MRI scanner

ELECTRICITY, MAGNETISM, AND OPTICS

EXAMPLE 23-11

Coils in an MRI Magnet

Some MRI scanners need a magnetic field of 3.0 T. How many loops per metre does an electromagnet require to produce this magnetic field with a current of 15 A?

SOLUTION

We can use Equation (23-41) to calculate the turn density of the coil:

$$B = \mu_0 \frac{IN}{L} = \mu_0 In$$

Therefore,

$$n = \frac{B}{\mu_0 I} = \frac{3.0 \text{ T}}{(4\pi \times 10^{-7} \text{ T·m/A})(15 \text{ A})} = 1.6 \times 10^5 \text{ loops/m}$$

Making sense of the result:

The number of loops per metre is very high. To have fewer turns and to reduce the size of the electromagnets, MRI scanners use a superconducting wire that allows significantly higher electric currents.

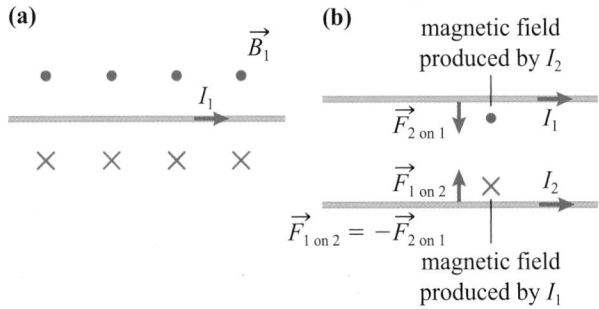

Figure 23-44 (a) Magnetic field produced by I_1 (b) Two parallel wires carrying currents that flow in the same direction attract each other (c) Two parallel current-carrying wires carrying opposite currents repel each other

23-8 The Magnetic Force Between Two Parallel Conductors

Imagine two straight, parallel wires that are separated by a distance R and carry electric currents I_1 and I_2 in the same direction, as shown in Figure 23-44(b). We first consider the magnetic field created by the top wire. Using the right-hand rule (Figure 23-44(a)), we see that directly below the wire, the magnetic field created by I_1 is directed into the page and, hence, is perpendicular to the bottom wire. The magnitude of this field is given by Equation (23-33): $B = \dfrac{\mu_0 I_1}{2\pi R}$.

Applying the right-hand rule for the magnetic force exerted on a moving charge shows that the magnetic field from the top wire exerts an upward force on the bottom current-carrying wire. Consequently, the magnetic force experienced by I_2 due to the magnetic field created by I_1 is a force of attraction. The magnitude of this force can be calculated using Equation (23-22):

$$\vec{F}_{1 \text{ on } 2} = I_2 \vec{l} \times \vec{B}$$

$$F_{1 \text{ on } 2} = I_2 l \frac{\mu_0 I_1}{2\pi R} = \frac{\mu_0 I_1 I_2 l}{2\pi R} = \frac{\mu_0 I_1 I_2}{2\pi R} l$$

Similarly, the magnetic force experienced by I_1 due to the magnetic field created by I_2 is directed toward I_2 (force of attraction) and has a magnitude of

$$F_{2 \text{ on } 1} = I_1 l \frac{\mu_0 I_2}{2\pi R} = \frac{\mu_0 I_1 I_2 l}{2\pi R} = \frac{\mu_0 I_1 I_2}{2\pi R} l$$

The fact that these two forces are equal in magnitude and opposite in direction is not surprising because the interacting wires should obey Newton's third law. The magnitude of force per unit length is also equal for the two wires:

KEY EQUATION
$$\frac{F_{1 \text{ on } 2}}{l} = \frac{F_{2 \text{ on } 1}}{l} = \frac{\mu_0 I_1 I_2}{2\pi R} \qquad (23\text{-}42)$$

We also use the right-hand rule to show that whenever two parallel wires carry electric currents in opposite directions, they repel each other (Figure 23-44(c)).

The force between two straight, parallel wires provides a way to define the SI unit of electric current, the ampere: When two parallel wires each carry an electric current of 1 A and are a distance of 1 m from each other, the force between the wires per metre of wire is

$$\frac{F_{1 \text{ on } 2}}{l} = \frac{F_{2 \text{ on } 1}}{l} = \frac{\mu_0 I_1 I_2}{2\pi R} = \frac{\mu_0 (1 \text{ A})(1 \text{ A})}{2\pi (1 \text{ m})}$$

$$= \frac{4\pi \times 10^{-7}}{2\pi} \text{ N/m} = 2 \times 10^{-7} \text{ N/m}$$

ELECTRICITY, MAGNETISM, AND OPTICS

C-23-11 The Magnetic Force Between Two Current-carrying Wires

Two straight, parallel wires carry electric currents of $I_1 = I$ and $I_2 = 2I$. Which of the following equations correctly describes the magnetic forces that the wires exert on each other?

(a) $F_{1\,\text{on}\,2} = 4F_{2\,\text{on}\,1}$

(b) $F_{1\,\text{on}\,2} = 2F_{2\,\text{on}\,1}$

(c) $F_{1\,\text{on}\,2} = F_{2\,\text{on}\,1}$

(d) $2F_{1\,\text{on}\,2} = F_{2\,\text{on}\,1}$

(e) $4F_{1\,\text{on}\,2} = F_{2\,\text{on}\,1}$

C-23-11 (c) Think of Newton's third law.

LO 9

23-9 The Magnetic Properties of Materials

We began this chapter with a discussion of how credit cards work. To store information on a credit card, an array of microscopic magnetic particles is arranged. Credit cards, magnetic tapes, computer hard drives, and refrigerator magnets all depend on the magnetic properties of matter. Materials can be classified into three broad groups based on their magnetic properties: paramagnetic, diamagnetic, and ferromagnetic. The behaviour of these three classes of materials arises from interactions at the atomic level, which we look at first.

The Bohr Magneton

At the beginning of the 20th century, the Danish physicist Niels Bohr (1885–1962) proposed a model for the structure of atoms (which will be discussed in more detail in later chapters). In Bohr's model, atoms consist of a nucleus made of protons and neutrons, with electrons orbiting around it. These orbiting electrons behave like microscopic current loops inside the atoms and give the atom a magnetic moment. In some materials, the atomic magnetic moments orient themselves along the external magnetic field, increasing its strength. We call this process magnetization and say that some materials can be magnetized by an external magnetic field.

To understand how atomic magnetic moments contribute to the magnetic properties of matter, we look at hydrogen, the simplest atom. A semiclassical model of the hydrogen atom consists of one electron orbiting a nucleus that consists of just one proton. As discussed in Section 23-5, the magnetic dipole moment of the current loop can be expressed as $\vec{\mu} = I\vec{A}$, where I is the electric current and A is the area of the loop (Equation (23-28)). Applying this definition to an electron orbiting the nucleus in the atom of hydrogen, we obtain the following:

$$\mu = IA = \frac{e}{T}A = \frac{e}{2\pi r/v}\pi r^2 = \frac{evr}{2} \qquad (23\text{-}43)$$

It is often convenient to express the dipole moment of the hydrogen atom in terms of the angular momentum of the electron moving in a circular orbit (Section 8-7), $L = rp = rmv$:

KEY EQUATION
$$\mu = \frac{evr}{2} = \frac{e}{2m}L \qquad (23\text{-}44)$$

Expression (23-44) shows that the magnetic dipole moment of a hydrogen atom is proportional to the angular momentum of its electron. Bohr suggested that this angular momentum must always be an integer multiple of $h/2\pi$, where h is Planck's constant, a fundamental physical constant with a value of 6.626×10^{-34} J·s. Bohr's postulate of the quantization of angular momentum is a cornerstone of quantum mechanics (Chapter 31). Until now, we have discussed angular momentum as being a continuous physical quantity. However, due to the extremely small value of Planck's constant, the quantum properties of matter only become relevant at the atomic scale.

Using the quantization of angular momentum, we can rewrite Equation (23-44) as

KEY EQUATION
$$\mu_n = \frac{e}{2m}L_n = \frac{e}{2m}\frac{h}{2\pi}n = \frac{eh}{4\pi m}n = \mu_B n$$

$$where\ n = 1, 2, 3, \ldots \qquad (23\text{-}45)$$

The quantity $\mu_B = \frac{eh}{4\pi m} = 9.274 \times 10^{-24}\ \frac{J}{T} = 9.274 \times 10^{-24}\ A\cdot m^2$ represents the fundamental unit of magnetic moment and is called the **Bohr magneton**. It is convenient to express magnetic moments of charged particles in terms of the Bohr magneton. Equation (23-45) can also be used to calculate the magnetic potential energy for a magnetic moment in a magnetic field:

$$U = -\vec{\mu} \cdot \vec{B} \qquad (23\text{-}46)$$

Electron Spin

In 1925, three young doctoral students—George Uhlenbeck, Samuel Goudsmit, and Ralph Kronig—independently came up with the then radical idea that an electron has an inherent property (similar to its mass or charge) that interacts with a magnetic field. This inherent property of an electron is called **electron spin**, and its magnitude is slightly greater than the Bohr magneton, $1.001\ \mu_B$. The net magnetic moment of an atom is the sum of the magnetic moments of the electrons orbiting the nucleus and the spins of these electrons.

Paramagnetism

In most materials, the net magnetic moments in the atoms are zero. However, in certain materials, some atoms have nonzero magnetic moments on the order of magnitude of

ELECTRICITY, MAGNETISM, AND OPTICS

NEL

the Bohr magneton, μ_B. When these materials—called paramagnetic materials—are placed in an external magnetic field, the unbalanced atomic magnetic moments experience a magnetic torque of $\vec{\tau} = \vec{\mu} \times \vec{B}$. The unbalanced torques will try to align with the magnetic field to decrease the potential energy of the system, thus increasing the net magnetic field. This phenomenon is called **paramagnetism**.

Since the net magnetic dipole moment of the material $\vec{\mu}_{net}$, depends on the amount of material, it is more meaningful to discuss the net magnetic dipole moment of the material per unit volume. This quantity is called the **magnetization** of the material and is denoted by $\vec{M}$:

$$\vec{M} = \frac{\vec{\mu}_{net}}{V} \qquad (23\text{-}47)$$

This additional magnetic dipole moment produces an additional magnetic field, which can be calculated in the

same way as the magnetic field produced by a current-carrying loop: $\vec{B}_{add} = \mu_0 \vec{M}$. Then the total magnetic field inside such a material is given by

$$\vec{B} = \vec{B}_0 + \mu_0 \vec{M}$$

where $\vec{B}_0$ is an external magnetic field.

It is common to describe the behaviour of paramagnetic materials using the **relative permeability** of the material, K_m, which indicates how much greater the magnetic field inside a paramagnetic material in an external magnetic field is relative to the external magnetic field:

KEY EQUATION $$\vec{B} = \vec{B}_0 + \mu_0 \vec{M} = K_m \vec{B}_0 \qquad (23\text{-}48)$$

Table 23-1 shows the relative magnetic permeability K_m of different paramagnetic and diamagnetic materials K_m at room temperature (20°C). The relative magnetic

Table 23-1 Relative magnetic permeability K_m and magnetic susceptibility of paramagnetic and diamagnetic materials, χ_m, at room temperature (20°C)

Material	Relative magnetic permeability, K_m	Magnetic susceptibility, $\chi_m = K_m - 1$
Paramagnetic materials		
Iron oxide	1.007 20	0.007 20
Ammonium iron (III) sulfate (iron alum)	1.000 66	0.000 66
Uranium	1.000 40	0.000 40
Platinum	1.000 26	0.000 26
Tungsten	1.000 068	0.000 068
Cesium	1.000 051	0.000 051
Aluminum	1.000 022	0.000 022
Lithium	1.000 014	0.000 014
Magnesium	1.000 012	0.000 012
Sodium	1.000 007 2	0.000 007 2
Oxygen gas	1.000 001 9	0.000 001 9
Diamagnetic materials		
Bismuth	0.999 834	−0.000 166
Mercury	0.999 971	−0.000 029
Silver	0.999 974	−0.000 026
Carbon (diamond)	0.999 979	−0.000 021
Carbon (graphite)	0.999 984	−0.000 016
Lead	0.999 982	−0.000 018
Sodium chloride	0.999 986	−0.000 014
Copper	0.999 99	−0.000 010
Water	0.999 991	−0.000 009 1

Source: Adapted from HyperPhysics, http://hyperphysics.phy-astr.gsu.edu/hbase/tables/magprop.html

ELECTRICITY, MAGNETISM, AND OPTICS

permeabilities of these materials typically range from 1.000 01 to 1.003 00. Another way of thinking of the relative permeability is to use a coefficient that compares the magnetic field inside the paramagnetic material to the magnetic field in a vacuum immersed in the same external magnetic field. This view of relative permeability allows us to extend all the magnetic field derivations to current-carrying conductors embedded in paramagnetic materials. For such derivations, μ_0 should be replaced by the quantity **magnetic permeability**, μ:

$$\mu = \mu_0 K_m \qquad (23\text{-}49)$$

Be careful not to confuse the magnetic permeability of a material with the magnetic dipole moment, which is also denoted by μ.

Another convenient way of describing paramagnetic materials is to use **magnetic susceptibility**, χ_m, which shows by how much the magnetic permeability differs from 1. For paramagnetic materials, the magnetic susceptibility is positive (Table 23-1):

$$\chi_m = K_m - 1 \qquad (23\text{-}50)$$

For paramagnetic materials, the magnetization of the material depends on the absolute temperature. In many cases, this dependence can be described by Curie's law, named after the renowned French scientist Pierre Curie (1859–1906):

$$M = C\,\frac{B}{T} \qquad (23\text{-}51)$$

where C is the Curie constant, T is the absolute temperature, and B is the magnetic field.

Different materials have different Curie constants. Curie's law can be understood when we consider how the variables affect atomic magnetic moments: thermal motion tends to randomize atomic magnetic moments, and an external magnetic field tends to align the atomic magnetic moments along its direction.

Diamagnetism

Unlike paramagnetic materials, the net magnetic moments of atoms in diamagnetic materials are zero when there is no external magnetic field present. However, an external magnetic field affects the motion of the electrons within the atoms, creating magnetic dipole moments within the atoms. These induced magnetic dipole moments are always directed opposite to the external magnetic field. We explain the reason for this orientation when we talk about Faraday's law of electromagnetic induction in Chapter 24. **Diamagnetism** can be compared to the effect of polarization by an electric field (Section 19-6). For example, a neutral water molecule can become polarized in the presence of an electric field and act as an electric dipole.

Diamagnetic materials always have a negative susceptibility, as shown in Table 23-1. The relative permeability of diamagnetic materials is slightly less than 1. In contrast to paramagnetic materials, the susceptibility

of diamagnetic materials is almost independent of temperature.

Ferromagnetism

Note that Table 23-1 does not include the metals that come to mind when we talk about magnetic materials: iron, nickel, cobalt, and their alloys. These materials constitute a third group of magnetic materials, called ferromagnetic materials. In ferromagnetic materials, strong magnetic interactions between atomic magnetic dipole moments make them align parallel to each other within small regions even when there is no external magnetic field, a phenomenon called **ferromagnetism**. These regions of aligned dipole moments are called **magnetic domains**. The magnetic moment of a magnetic domain can be of the order of magnitude of 1000 μ_B because most of the individual atomic magnetic moments within the domain are parallel.

When there is no external magnetic field $\vec{B}_0$, as shown in Figure 23-45, the magnetic domains are randomly oriented, and the net magnetic moment of the material is almost zero. However, in the presence of an external magnetic field $\vec{B}_0$, the domains orient themselves in the direction of the field, creating a strong magnetic field within the material.

The boundaries of magnetic domains in ferromagnetic materials are also affected by the external magnetic field: domains oriented in the direction of an external magnetic field grow as the external magnetic field increases. As a result, the relative magnetic permeability

Figure 23-45 (a) Magnetic domains inside a ferromagnetic material when there is no external magnetic field present. (b) Magnetic domains in the presence of a weak external magnetic field. (c) Magnetic domains in the presence of a strong magnetic field.

Table 23-2 Initial and maximum relative magnetic permeability K_m of some ferromagnetic materials at room temperature (20°C)

Material	Treatment	Initial relative magnetic permeability, K_m	Maximum relative magnetic permeability, K_m
Iron, 99.8% pure	Annealed	150	5 000
Iron, 99.95% pure	Annealed in hydrogen	10 000	200 000
78 Permalloy	Annealed, quenched	8 000	100 000
Superpermalloy	Annealed in hydrogen, controlled cooling	100 000	1 000 000
Cobalt, 99% pure	Annealed	70	250
Nickel, 99% pure	Annealed	110	600

Source: Adapted from HyperPhysics, http://hyperphysics.phy-astr.gsu.edu/hbase/tables/magprop.html

of a ferromagnetic material, K_m is 1000 to 1 000 000 times greater than the relative permeability of a paramagnetic material (Table 23-2). Consequently, ferromagnetic materials, such as iron, are strongly attracted to a permanent magnet, and a paramagnetic material, such as aluminum, is not. Aluminum does experience some magnetic attraction, but it is thousands of times weaker than for iron in the same magnetic field.

When an external magnetic field increases, more and more domains in the ferromagnetic material align with the external magnetic field. As a result, the magnetization and the relative magnetic permeability of the ferromagnetic material increase. When all the magnetic domains have aligned with the external magnetic field, increasing the external field will not produce any further increase in relative permeability. At this point, the ferromagnetic sample has reached **saturation of magnetization**.

Figure 23-46 shows how the magnetization of a ferromagnetic material varies as the external magnetic field changes. This nonlinear behaviour is called magnetic **hysteresis**. As the external magnetic field increases, the magnetization of a ferromagnetic sample increases until it reaches its maximum value (saturation). However, when the external magnetic field is removed, the magnetization of the sample does not completely disappear. To decrease magnetization of the sample to zero, one has to apply a magnetic field in the opposite direction. The area inside the hysteresis curve indicates the relative amount of energy required to magnetize and demagnetize a sample: A ferromagnetic material that has a large area inside the hysteresis loop requires more energy to remagnetize it than a ferromagnetic sample with a smaller area. The more energy required to demagnetize a material, the better this material can serve as a permanent magnet.

Figure 23-46 Magnetic hysteresis loop for a ferromagnetic material

FUNDAMENTAL CONCEPTS AND RELATIONSHIPS

Electric charges *moving* inside a magnetic field experience a magnetic force that is perpendicular to both the magnetic field and the velocity of the moving charge. Moving electric charges (electric currents) are the sources of magnetic fields.

The Magnetic Force

Charged particles moving inside a magnetic field experience a magnetic force that is perpendicular to both the particle's velocity and the direction of the magnetic field:

$$\vec{F}_B = q\vec{v} \times \vec{B} \qquad (23\text{-}1)$$

The Lorentz Force

Charged particles moving in electric and magnetic fields experience a Lorentz force:

$$\vec{F}_B = q\vec{E} + q\vec{v} \times \vec{B} \qquad (23\text{-}2)$$

Magnetic Fields

Moving electric charges (electric currents) are the sources of magnetic fields. A magnetic field produced by a small element of wire carrying a current can be calculated using the Biot–Savart law:

$$d\vec{B} = \frac{\mu_0}{4\pi} \frac{I d\vec{l} \times \hat{r}}{r^2} \qquad (23\text{-}32)$$

Ampère's Law

Ampère's law provides another way of calculating the magnetic field created by a wire carrying a current:

$$\oint \vec{B} \cdot d\vec{l} = \mu_0 \sum I_{enc} \qquad (23\text{-}39)$$

APPLICATIONS

Applications: mass spectrometers, electromotors, magnetic field sensors, proximity switches, position and speed detectors, current sensors, coaxial cables, solenoids

Key Terms: Ampère's law, Bohr magneton, constant magnetic field, coaxial cable, cyclotron, cyclotron frequency, diamagnetism, electromagnet, electron spin, ferromagnetism, Hall effect, Hall voltage, hysteresis, Lorentz force, magnetic dipole, magnetic dipole moment, magnetic domains, magnetic field, magnetic field lines, magnetic force, magnetic permeability, magnetic susceptibility, magnetization, mass spectrometer, paramagnetism, permeability constant of free space, pitch of a helix, relative permeability, right-hand rule, saturation of magnetization, tesla, uniform magnetic field, velocity selector

QUESTIONS

1. A charged particle moves in a straight line through a particular region of space. Could there be a nonzero magnetic field in this region? Explain.

2. An electron is travelling above and parallel to a conducting wire when a current is turned on in the wire. The electron curves upward and away from the wire. Initially, the current in the wire
   (a) is running parallel to the electron;
   (b) is running perpendicular to the electron;
   (c) is running opposite the electron;
   (d) has no effect on the electron.
   (e) None of the above are true.

3. Find the direction of the force on the negative charge for each diagram in Figure 23-47. Copy the diagram into your notebook, and indicate the direction of the force on your diagram.

Figure 23-47 Question 3

4. A charged particle is moving along a circular path under the influence of a uniform magnetic field. An electric field that points in the same direction as the magnetic field is turned on. Describe the path that the charged particle will take.

5. A rectangular piece of a semiconductor is inserted in a magnetic field, and a battery is connected to its ends, as shown in Figure 23-48. When a sensitive voltmeter is connected between points D (back face) and C (front face), it is found that point D is at a higher potential than point C. What is the sign of the charge carriers in this semiconductor material? Explain.

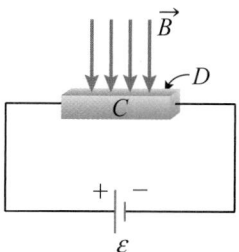

Figure 23-48   Question 5

6. Figure 23-49 shows several situations with two wires with currents of 2 A flowing out of or into the page. Rank the magnetic field at point P from greatest to least. Denote the downward-pointing field as negative and the upward-pointing field as positive. Assume that all the adjacent lines on the grid are separated by the same distance and that each situation is independent of the others.

Greatest 1____   2____   3____   4____   5____   6____
7 ____ Least
Or, the magnetic field is the same in all cases. _____
Or, the magnetic field is zero in all cases. _____
If the magnetic field is the same for two or more cases, clearly indicate it on the ranking scheme. Explain your reasoning.

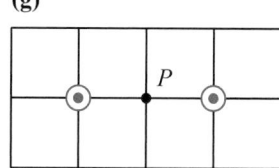

Figure 23-49   Question 6

7. The coil shown in Figure 23-50 attracts a permanent bar magnet. Which of the following statements correctly describes the polarity of the permanent bar magnet? Draw a diagram for this problem, and show your solution on the diagram.
  (a) The end of the bar magnet that is close to the coil is north; the opposite end is south.
  (b) The end of the bar magnet that is close to the coil is south; the opposite end is north.
  (c) The upper side of the bar magnet is north; the bottom side is south.
  (d) The upper side of the bar magnet is south; the bottom side is north.
  (e) Since we have a permanent magnet, we cannot determine its polarity in this example.
  (f) A permanent magnet (unlike a temporarily magnet) has zero polarity; it is neutral.

Figure 23-50   Question 7

8. The coil in Figure 23-51 repels the iron bar on the left by causing it to become a temporary magnet. Which of the following statements correctly describes the polarity of the temporary bar magnet in this situation? Draw a diagram for this problem, and show your solution on the diagram.
  (a) The end of the bar magnet that is close to the coil is north; the opposite end is south.
  (b) The end of the bar magnet that is close to the coil is south; the opposite end is north.
  (c) The upper side of the bar magnet is north; the bottom side is south.
  (d) The upper side of the bar magnet is south; the bottom side is north.
  (e) A temporary magnet cannot repel the coil, so this example does not make sense.

Figure 23-51   Question 8

9. A proton moving upward enters the magnetic field created by two bar magnets, as shown in Figure 23-52. The direction of the force exerted on the proton by the magnetic field is
   (a) in the direction of the proton motion (upward);
   (b) in the direction opposite to the proton motion (downward);
   (c) left (i.e., toward the south pole of the left magnet);
   (d) right (i.e., toward the north pole of the right magnet);
   (e) into the page;
   (f) out of the page.

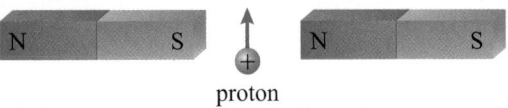

proton

Figure 23-52   Question 9

10. What would your answer be to question 9 if there was an electron instead of a proton?

11. An electric current in a long, vertical straight wire is flowing straight up (Figure 23-53). What is the direction of the magnetic field at point P?
   (a) up (vertically)
   (b) down (vertically)
   (c) into the page
   (d) out of the page
   (e) to the right (horizontally)
   (f) to the left (horizontally)

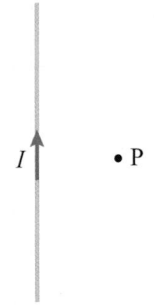

I • P

Figure 23-53   Question 11

12. Two long, parallel vertical wires carry electric currents of I and $3I$ in the same direction, as shown in Figure 23-54. The wires are separated by a distance d. At what point will the net magnetic field created by these wires equal zero?
   (a) between the wires, a distance $d/3$ from the wire carrying current I and $2d/3$ from the wire carrying current $3I$
   (b) between the wires, a distance $2d/3$ from the wire carrying current I and $d/3$ from the wire carrying current $3I$
   (c) outside the wire carrying current I, a distance $d/3$ from this wire and $4d/3$ from the wire carrying current $3I$
   (d) between the wires, a distance $d/4$ from the wire carrying current I and $3d/4$ from the wire carrying current $3I$
   (e) between the wires, a distance $3d/4$ from the wire carrying current I and $d/4$ from the wire carrying current $3I$

I $3I$

$\leftarrow$————— d —————$\rightarrow$

Figure 23-54   Question 12

13. A negatively charged ion moving with speed v (Figure 23-55) enters three adjacent regions of uniform magnetic fields but of varying magnitudes. The magnetic field in each region is directed perpendicular to the plane of the paper. Which of the following correctly ranks the magnitudes of these magnetic fields?
   (a) $B_3 > B_1 > B_2$
   (b) $B_3 > B_2 > B_1$
   (c) $B_2 > B_1 > B_3$
   (d) $B_2 > B_3 > B_1$
   (e) $B_1 > B_3 > B_2$
   (f) $B_1 > B_2 > B_3$

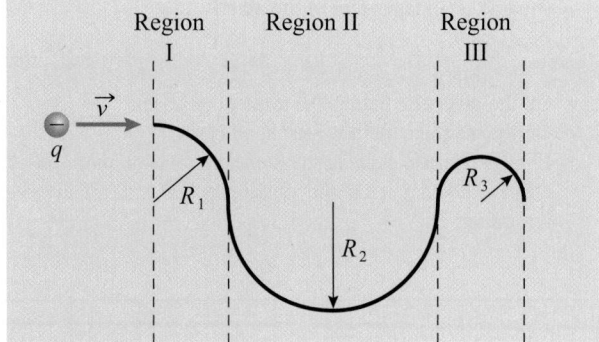

Figure 23-55   Question 13

14. What is true about the directions of the magnetic fields in question 13?
   (a) Magnetic fields B_1 and B_3 are into the page, and magnetic field B_2 is out of the page.
   (b) Magnetic fields B_1 and B_3 are out of the page, and magnetic field B_2 is into the page.
   (c) All the magnetic fields are directed into the page.
   (d) All the magnetic fields are directed out of the page.
   (e) There is not enough information to determine the directions.

15. An electron is moving inside the cathode ray tube shown in Figure 23-13. What happens to the deflection of an electron when the magnetic field decreases by a factor of 2?
   (a) The deflection remains the same.
   (b) The deflection decreases by a factor of 2.
   (c) The deflection decreases but not by a factor of 2.
   (b) The deflection increases by a factor of 2.
   (c) The deflection increases but not by a factor of 2.

16. Rank the magnitudes of the magnetic fields produced at the centres of curvature (points O) of the wires shown in Figure 23-56 from highest to lowest. All the wires carry an identical electric current. The radii of the circular segments are shown in the figure.

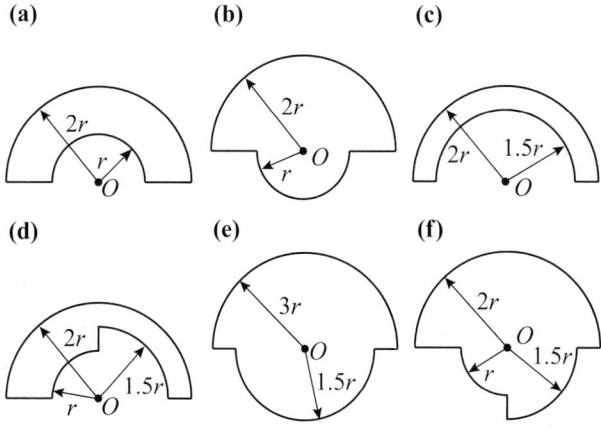

Figure 23-56 Question 16

17. Rank the magnitudes of the magnetic fields in the centres of the six squares shown in Figure 23-57. The magnetic fields are created by four different parallel (or antiparallel) currents I flowing through the vertices of each square.

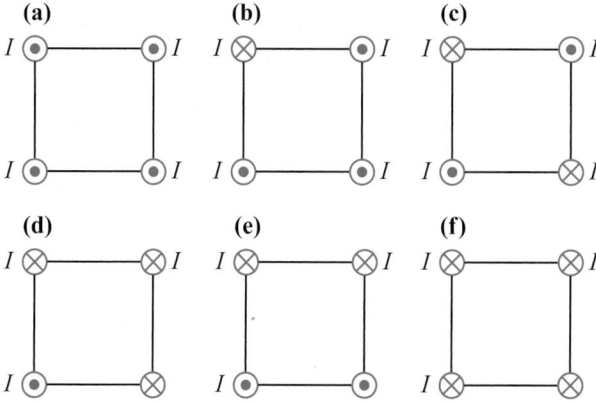

Figure 23-57 Question 17

18. Each wire in Figure 23-58 carries an electric current of 1 A into or out of the page. Each diagram indicates the path of a line integral $\oint \vec{B} \cdot \vec{dl}$ to be used while applying Ampère's law for calculating the magnitude of the resultant magnetic field. Rank the values of each line integral.

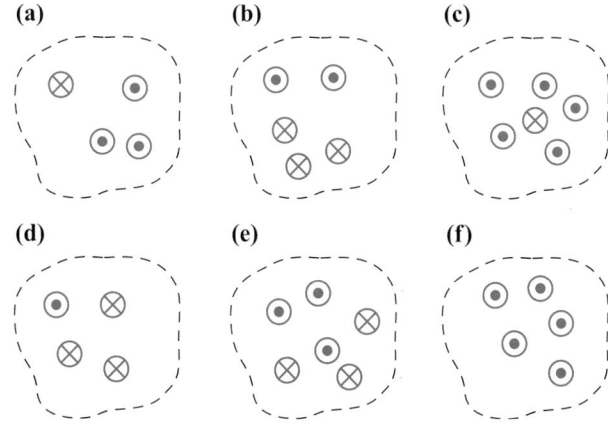

Figure 23-58 Question 18

PROBLEMS BY SECTION

For problems, star ratings will be used, (✶, ✶✶, or ✶✶✶), with more stars meaning more challenging problems.

Section 23-1 Magnetic Field and Magnetic Force

19. ✶ An electron moving with velocity $\vec{v} = (4.0\hat{i} - 8.0\hat{k}) \times 10^4$ m/s enters a region where a uniform magnetic field $\vec{B} = (0.80\hat{i} - 0.60\hat{j})$ T is present. Determine the force on the electron in algebraic notation, and find its magnitude.

20. ✶ An electron, accelerated to an energy of 100 eV, enters a region with a uniform magnetic field with a magnitude of 1.5×10^{-2} T. When a magnetic field is directed in the positive x-direction and an electron is moving in the positive z-direction, what are the direction and magnitude of the force acting on the electron? How would your answer change if instead of an electron, a proton with the same energy entered this region of the magnetic field? Explain.

21. ✶ A particle carrying a charge of 10 nC is moving at $0.001c$ along the positive x-direction in the region of a uniform magnetic field directed in the negative z-direction. The magnitude of the magnetic field is 1.5 T. Calculate the magnetic force on the particle.

22. ✶✶ A droplet has a mass of 2×10^{-4} g, a charge of 35 nC, and an initial horizontal velocity of $6 \times 10^5 \hat{j}$ m/s. The electric field in the vicinity of Earth is approximately 100 N/C directed downward, and the magnetic field in this area was measured to be approximately 45 μT directed parallel to Earth's surface. Find the magnitude and direction of a magnetic field that will keep the droplet moving in this direction. What assumptions did you make to solve the problem? (Hint: Do you need to take the gravitational force into account?)

Section 23-2 The Motion of a Charged Particle in a Uniform Magnetic Field

23. ✶ An electron, a proton, and an α-particle are moving with the same velocity. At some instant, they enter a region of a uniform magnetic field B pointing perpendicular to the direction of their motion.
    (a) Compare the radii of the orbits of the particles upon entering this region.
    (b) Compare the time it takes for each particle to complete one revolution (period of motion).

24. ✶ An electron moving perpendicular to a magnetic field of magnitude 2.4×10^{-2} T has a circular trajectory of radius r. The speed of the electron is 5% of the speed of light.
    (a) Determine the radius of the electron's trajectory.
    (b) Determine the period of revolution.

25. ✶ A proton enters a uniform magnetic field that causes it to follow a circular path of radius 0.1 mm. The field strength is 0.9 T.
    (a) Calculate the proton's speed.
    (b) Calculate the proton's angular frequency.

26. ✶✶ An electron of mass m and charge $-e$ moving with an initial velocity of $\vec{v} = v_{ox}\hat{i} + v_{oy}\hat{j}$ m/s enters a uniform magnetic field, $\vec{B} = B\hat{j}$. Derive an expression, in algebraic notation, for the velocity of the electron at any later instant, t.

27. ✶✶ A proton (mass m_p), a deuteron ($m_d = 2m_p$, $Q = e$), and an α-particle ($m_\alpha = 4m_p$, $Q = 2e$) are accelerated by the same potential difference, V, and then enter a uniform magnetic field, B, where they move in circular paths

perpendicular to B. Express the radii of the paths for the deuteron and α-particle in terms of the proton's radius.

28. ✱✱ An electron is travelling at 100 km/s parallel to a long, straight horizontal conductor a distance of 3 cm from the conductor. A current of 12 A runs through the wire as the electron travels parallel to it, in the same direction as the electron's velocity. Find the strength and direction of the external electric field that will prevent the electron from deviating from its original path. Express your answer in V/m.

Section 23-3   Applications: Charged Particles Moving in a Uniform Magnetic Field

29. ✱ A magnetic field at the surface of a neutron star can reach 3.0×10^7 T. What would be the radii of curvature for the paths of three electrons moving in the vicinity of a neutron star with speeds of 5%, 10%, and 20% of the speed of light? What are the magnitudes of the magnetic forces acting on these electrons?

30. ✱ Charged particles, such as protons, found in cosmic rays, enter the Van Allen belt as they approach Earth. The average strength of the magnetic field of the lower Van Allen belt at a height of 3000 km is approximately 10^{-5} T. What is the radius of curvature and the cyclotron frequency of a proton entering the lower Van Allen belt at an angle of 30°? Describe the trajectory of the proton. Why do scientists say that a "proton was captured by the Van Allen belt"?

31. ✱ Cathode ray tube TVs used crossed electric and magnetic fields to deflect electrons before they struck a fluorescent screen. A beam of electrons moving with speed 6.0×10^7 m/s strikes a fluorescent screen located 0.5 m away. However, even when the magnetic field of the cathode ray is turned off, Earth's magnetic field is still present in that region. When an electron beam is moving along the x-axis, and Earth's magnetic field in the region has the components ($15\ \mu$T, $20\ \mu$T, $18\ \mu$T), what is the deflection, in mm, of the electron beam due to Earth's magnetic field?

32. ✱✱ The core of any microwave oven is a high-voltage system called a magnetron tube. The magnetron is a diode-type electron tube, which is used to produce electromagnetic waves with a frequency of 2450 MHz which are strongly absorbed by water. These waves are microwaves and have a wavelength of approximately 0.12 m. Determine the strength of the magnet needed to move the electrons in the oven at that frequency.

Section 23-4   The Magnetic Force on a Current-Carrying Wire

33. ✱ (a) Find the magnetic force per unit length on a long, straight wire carrying 10 A of electric current in the positive x-direction in a uniform magnetic field of 1.8 T directed in the negative y-direction.
(b) What happens to this force when the magnitude of the magnetic field is halved and the electric current in the wire doubles?

Section 23-5   The Torque on a Current-Carrying Loop in a Magnetic Field

34. ✱✱ A copper loop in the shape of an equilateral triangle ABC (AB = AC = BC = a), carrying an electric current I, is placed in a uniform magnetic field of magnitude B directed parallel to the side AC of the triangle (Figure 23-59).

(a) Find the magnitude of the magnetic force exerted on each side of the triangle.
(b) Find the net magnetic force exerted on this triangle by the magnetic field.
(c) Find the magnetic torque exerted on this triangle by this magnetic field.

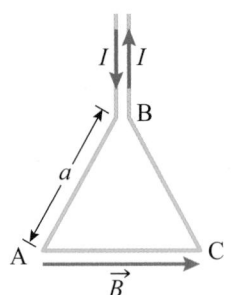

Figure 23-59   Problem 34

Section 23-6   The Biot–Savart Law

35. ✱✱ A long, straight conductor carrying a current I splits into two identical semicircular arcs of 5 cm radius (Figure 23-60). Calculate the magnitude and the direction of the magnetic field at the centre of the circle (point O).

Figure 23-60   Problem 35

36. ✱✱ The magnitude of a magnetic field created by a long, straight wire carrying an electric current 25 cm from the wire is 10.0 μT.
(a) What is the current through the wire?
(b) What will happen to the magnitude of the magnetic field when the current in the wire doubles?

Section 23-7   Ampère's Law

37. ✱ A lightning bolt can generate electric currents of magnitude on the order of 10^4 A. What are the magnitudes of the magnetic fields generated 10 m, 100 m, and 1 km from a lightning bolt that generated a 10^4 A current? Compare these magnitudes to the magnitude of Earth's magnetic field.

38. ✱ A straight, vertical wire carries 10 A of electric current. Find the direction and magnitude of the magnetic fields created by this wire at a distance of (a) 10 cm, (b) 20 cm, and (c) 50 cm from the wire.

39. ✱✱ An electric cable has a diameter of 10 mm and carries an electric current of 25 A.
(a) Find the strength of the magnetic field at distances of 2 mm and 7 mm from the centre of the cable.
(b) Plot a graph of the magnitude of the magnetic field as a function of the distance from the centre of the cable for each distance.
(c) In Examples 23-9 and 23-10, we found expressions for the magnitude of the magnetic field inside and outside a long wire. Compare these expressions with the expressions for the magnitude of the electric field inside and outside of a uniformly charged insulating sphere. What does this comparison tell you?

Section 23-8 The Magnetic Force Between Two Parallel Conductors

40. ✶ Two straight, parallel superconducting cables can withstand a maximum tension of 10 kN/m without breaking. An engineer wants to use these cables in a device in which each cable is securely fastened and carries 20 kA of electric current. The currents will be flowing in opposite directions. What is the minimum separation required to keep the cables from breaking? Explain.

41. ✶ A rectangular loop of dimensions of 30 cm × 20 cm carries 10 A of electric current. Find the magnitudes and directions of the magnetic forces experienced by the sides of the loop.

Section 23-9 The Magnetic Properties of Matter

42. ✶ The magnetic field inside an air-filled solenoid is measured to be 1.2 T. Will the field inside the solenoid change if the air is pumped out of the core of the solenoid? If so, by how much will the magnetic field change? Explain.

COMPREHENSIVE PROBLEMS

43. ✶✶ A circular 100-loop coil with a radius of 0.10 m is located in a horizontal (*xy*) plane as shown in Figure 23-61. A 10 A current flows clockwise through the coil (as seen from above). The coil is in a uniform magnetic field of 1.0 T, directed in the positive *y*-direction with $\vec{B} = 1.0\hat{j}$.
(a) Calculate the magnetic moment of each loop of the coil.
(b) Calculate the magnetic moment of the entire coil.
(c) Calculate the magnetic torque experienced by the coil. Clearly identify the direction of the torque.
(d) How will your answers to parts (a) to (c) change if the direction of the current in the coil is reversed?
(e) How will your answers to parts (a) to (c) change if the direction of the magnetic field is changed to $\vec{B} = 1.0\hat{i}$ or $\vec{B} = 1.0\hat{k}$? Explain.
(f) The coil rotates from the position where its magnetic moment is parallel to the direction of the magnetic field to the position where its magnetic moment is perpendicular to it. What is the change in the potential energy of the coil? Explain.

Figure 23-61 Problem 43

44. ✶✶ A current-carrying wire is located along the *y*-axis. An electric current of 5.0 A flows through this wire in the positive *y*-direction. Calculate the magnetic force per metre experienced by this wire when it is placed in the following magnetic fields:

(a) $\vec{B} = (0.55 \text{ T})\hat{i}$
(b) $\vec{B} = (0.55 \text{ T})\hat{j}$
(c) $\vec{B} = (0.55 \text{ T})\hat{k}$
(d) $\vec{B} = (0.55 \text{ T})\hat{i} - (0.55 \text{ T})\hat{k}$

45. ✶✶ A long wire carries a 5 A current toward the top of the page, as shown in Figure 23-62. An electron is travelling parallel to the wire with an initial separation of 2 cm and an initial velocity of 1.5×10^7 m/s toward the top of the page.
(a) What are the magnitude and direction of the magnetic force on the electron?
(b) What would your answers to part (a) be if instead of an electron, there was a proton moving in the same direction at the same speed?
(c) Draw the trajectories of the electron and the proton.

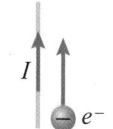

Figure 23-62 Problem 45

46. ✶✶✶ A charged particle carrying charge *q* and moving with a speed *v* enters a region of a uniform magnetic field $\vec{B}$ directed perpendicular to the particle's velocity. The particle spends time Δt in this region.
(a) Assume that Δt is small. Estimate the angle of deflection θ of the particle. When we assume that Δt is small, to what are we comparing Δt? Explain your answer.
(b) Suppose the charged particle is an electron moving with a speed of 1.5×10^7 m/s. What should be the size (length along the particle's original path) of the region of a uniform magnetic field of 2.0×10^{-2} T to produce a deflection of 0.15 rad?

47. ✶✶ A long, straight conductor carrying current *I* splits into two semicircular arcs. The resistance of the bottom arc is twice the resistance of the upper arc (Figure 23-63). Calculate the magnitude and the direction of the magnetic field in the centre of the circle (point O).

Figure 23-63 Problem 47

48. ✶✶ In Figure 23-64, three parallel, current-carrying wires pass through the vertices of an equilateral triangle. Two wires carry an electric current that is going into the page, and one wire carries an electric current that is coming out of the page. The magnitudes of the currents are 5 A, and the side of the equilateral triangle is 5 cm. Find the magnitude and direction of the magnetic field in the centre of the triangle (point O).

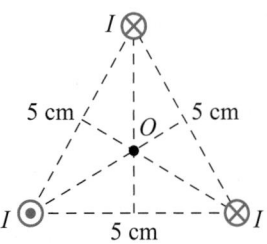

Figure 23-64   Problem 48

49. ✹✹✹   A 40-cm solenoid has a radius of 0.05 m and consists of 50 circular loops. An electric current of 1 A flows through the solenoid.
   (a) Determine the magnetic field in the centre of the solenoid.
   (b) Derive the formula for magnetic field along the axis of symmetry of the solenoid (Hint: you can check if your answer makes sense by using it to answer part (a)). Use this formula to find the magnetic field along the axis of the solenoid 15 cm off its centre.
   (c) Plot the graph of the magnetic field along the axis of symmetry of the solenoid as a function of the distance from its centre (use graphing software).
   (d) From the plot in (c) estimate at what distance from the centre of the magnetic field will decrease by 5%.

50. ✹✹   Two parallel, current-carrying wires are separated by a distance d. The wires carry currents of equal magnitude. In Figure 23-65(a), both currents are going into the page, and in Figure 23-65(b), one current is going into the page and the other is coming out of the page. At what location(s) does the resultant magnetic field equal zero for each one of the cases? Explain.

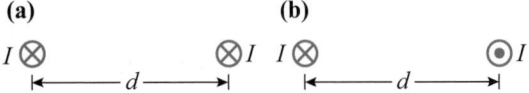

Figure 23-65   Problem 50

51. ✹✹   Two long, parallel vertical wires carry electric currents of I and $2I$ in opposite directions (Figure 23-66). Find the magnitude and direction of the magnetic fields at points A, B, C, and D when $I = 10$ A and the distance between the wires is 1 m.

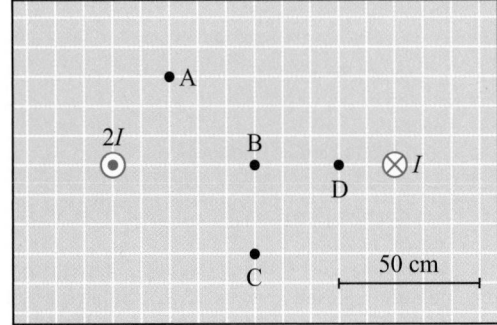

Figure 23-66   Problem 51

52. ✹✹   Two parallel wires separated by a distance of 50 cm carry currents in opposite directions (Figure 23-67). The left wire carries a current of 10 A. Point A is the midpoint

between the wires, and point B is 5 cm to the left of the 10 A current. The current I is adjusted so that the magnetic field at B is zero.
   (a) Find the strength of the magnetic field 10 cm to the left of A.
   (b) Find the direction and magnitude of the force per unit length on the left wire. Ignore gravity.

Figure 23-67   Problem 52

53. ✹✹✹   A rectangular loop of dimensions 30 cm × 20 cm carries 10 A of electric current (Figure 23-68). The loop is oriented in the xz-plane. Sixty centimetres from the centre of the loop is an infinitely long, straight vertical wire carrying 20 A of electric current in the positive z-direction. Find the resultant force acting on the loop.

Figure 23-68   Problem 53

54. ✹✹✹   A long copper wire of length L carries an electric current I. This wire can be formed into a coil consisting of multiple loops while preserving the total length of the wire. Find the optimal number of the loops in such a coil to maximize the torque on the coil when it carries a current I in a uniform magnetic field B. Prove that the maximum value of this torque is $\tau = (1/4\pi)L^2IB$.

55. ✹✹   Two very long, straight, parallel wires carry 10 A (wire A) and 15 A (wire B) of electric current. The currents in the wires flow in the same direction. The wires are located 1 m from each other.
   (a) What is the magnitude of the magnetic force that wire A exerts on wire B?
   (b) What is the magnitude of the magnetic force that wire B exerts on wire A?
   (c) Compare your answers to parts (a) and (b) and explain them.

56. ✹✹   A straight, vertical wire carries a downward current of 1.5 A. The wire is placed in the area of a uniform magnetic field created by a strong superconducting electromagnet. The magnitude of this magnetic field reaches 0.65 T. Calculate the magnetic force per metre of the wire applied on it by the magnetic field when
   (a) the direction of the magnetic field is perpendicular to the direction of the wire;
   (b) the direction of the magnetic field is parallel to the wire;
   (c) the magnetic field is directed at 45° to the direction of the wire.
   Draw a diagram for each scenario.

57. ✷✷ A proton, moving (at $t = 0$) with the velocity $v_x = 2.0 \times 10^5$ m/s, $v_y = 0$, $v_z = 1.5 \times 10^5$ m/s, enters a uniform magnetic field of 0.45 T directed along the z-axis. Assume that the only force acting on the proton is a magnetic force.
(a) Find the magnetic force acting on the proton as it enters the region of the magnetic field, and compare its magnitude with the gravitational force. Use your findings to justify the above assumption.
(b) Describe the proton's trajectory as it travels inside the magnetic field region.
(c) Find the proton's angular speed and its period of circular motion.
(d) Find the radius of the proton's trajectory and the pitch of the helix.
(e) How would your answers to parts (a) to (d) change if instead of a proton, an electron entered this region at the same velocity?

58. ✷✷✷ An α-particle travels in a circular path of radius 15 cm in a uniform magnetic field of 2.0 T.
(a) Find the speed of the α-particle. Express the speed of the α-particle as a fraction of the speed of light. Is it justified to use classical physics to express the kinetic energy of this particle, or should relativity be used?
(b) What is the kinetic energy of the α-particle in eV?
(c) What is the potential difference that the α-particle should be accelerated through in order to achieve such a speed?

59. ✷✷ At a certain instant, an electron with kinetic energy of 20 eV is moving in the westward direction. The horizontal component of Earth's magnetic field in this region is 18 μT north, and its vertical component is 48 μT down.
(a) What is the trajectory of the electron?
(b) What is the radius of curvature of the electron's path?
(c) Should the gravitational force on the electron be taken into account? Justify your answer.

60. ✷✷ A proton is moving in a circular orbit of radius 14 cm in a uniform 0.35 T magnetic field perpendicular to the velocity of the proton.
(a) Find the linear speed of the proton.
(b) Find the time it takes for the proton to complete one revolution.
(c) Suddenly a uniform electric field is turned on in the direction opposite to the direction of the magnetic field. The magnitude of the electric field is 400 kV/m.
(i) Draw a Free Body Diagram for the proton.
(ii) Draw the trajectory of the proton when both fields are on. Explain your drawing.
(iii) How would your answers to parts (i) and (ii) be different if instead of a proton an electron was moving in a uniform magnetic field of 0.35 T and the electric field mentioned above was turned on?

61. ✷✷ Due to health regulations, the magnetic field produced by a current-carrying wire at a distance of 30 cm is not allowed to exceed Earth's magnetic field (5.5×10^{-5} T).
(a) What is the maximum current that this wire can carry?
(b) What would the maximum current be if the regulations require the same limit at half the original distance (15 cm)?

62. ✷✷ A jumper cable used to start a vehicle with a dead battery carries a 65 A current. What is the magnitude of the magnetic field created by the cable (a) 10 cm away and (b) 25 cm away? Compare the values of the magnetic field to the magnitude of Earth's magnetic field.

63. ✷✷ A long, straight wire carrying an electric current of 50 A is placed in a uniform magnetic field of 5 mT directed toward the north magnetic pole. The wire is oriented such that the electric current flows from east to west. Find the points at which the net magnetic field is zero. (Hint: Think of the influence of Earth's magnetic field).

64. ✷✷ Figure 23-69 shows six sets of five equally spaced long, parallel wires carrying electric currents of equal magnitudes into or out of the page. Rank the magnitude of the magnetic forces acting on the central wire in each scenario, due to the currents in the other four wires, from greatest to smallest. Support your ranking with the appropriate calculations.

Figure 23-69 Problem 64

65. ✷✷✷ What is the magnetic field inside a 2 m-long solenoid of radius $r = 0.25$ m that has 1000 turns per metre and 10 A of current flowing through it? How does the magnetic field change if an iron core with $\chi_{iron} = 4.0 \times 10^3$ is inserted inside this solenoid?

66. ✷✷✷ Two charged particles moving with the same speed but in opposite directions pass each other at a distance d. The speed of the particles, v, is much less than the speed of light, $v \ll c$. Compare the magnetic and electric forces that the particles exert on each other at the instant when the particles are closest to each other. Perform the calculations for the three cases listed below.
(a) The charged particles are two protons.
(b) The charged particles are two electrons.
(c) One charged particle is a proton, and the other charged particle is an electron.

67. ✷✷ A 15 A current flows through a 50 cm-long solenoid that has 1000 turns. By how much will the magnetic field inside this solenoid change if the air in its core is replaced by
(a) water at 20°C;
(b) oxygen at 20°C;
(c) a vacuum?

68. ✷✷ A coaxial cable has an external radius of 5 mm and an internal radius of 2 mm, and carries an electric current of 5 A in each direction. Derive an equation to determine the magnitude and direction of the magnetic field generated by this cable at an arbitrary distance from its axis. Then use your equation to find the magnetic fields at the following distances from its axis:
(a) 1 mm
(b) 3 mm
(c) 5 mm

69. ★★★ Use the Biot–Savart law to prove that the magnetic field at the centre of a wire shaped into a circular arc (Figure 23-70) is given by $B = \dfrac{\mu_0 I \phi}{4\pi R}$, where I is the electric current flowing through the wire, R is the radius of curvature, and ϕ is the angle subtended by the circular arc.

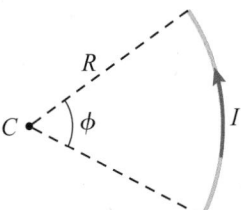

Figure 23-70   Problem 69

70. ★★★ Each wire in Figure 23-71 carries an electric current of 5 A.
   (a) In Figure 23-71(a), the radius of the arc is 10 cm. Find the strength of the magnetic field in the centre of the wire.
   (b) In Figure 23-71(b), the radius of the arc is 20 cm. Find the strength of the magnetic field in the centre of the wire. (Hint: Use your result from part (a).)

(b)

(a)

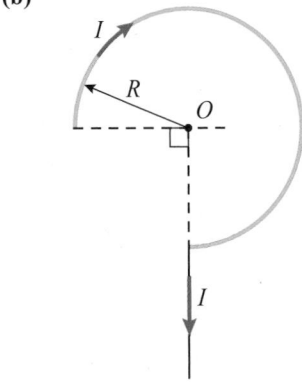

Figure 23-71   Problem 70

71. ★★★ A straight wire of length L carries an electric current of magnitude I.
   (a) Use the Biot–Savart law to prove that the magnetic field at point P_1, located a distance R from the centre of the wire along the line perpendicular to it, can be expressed as

$$B_{P_1} = \frac{\mu_0 I}{2\pi R} \frac{L}{(L^2 + 4R^2)^{1/2}}$$

   (b) What result would you expect if the length of the wire were to become infinitely long? Show that the result you proved in part (a) satisfies the limiting case of an infinitely long wire.

   (c) Use the Biot–Savart law to prove that the magnetic field at point P_2, located a distance R from the end of the wire along the line perpendicular to it, can be expressed as

$$B_{P_2} = \frac{\mu_0 I}{4\pi R} \frac{L}{(L^2 + 4R^2)^{1/2}}$$

   (d) Show that the results you proved in parts (a) and (c) are consistent.

72. ★★★ A copper wire is bent to create a square loop with sides of length d. An electric current of magnitude I flows counterclockwise through this loop. Find the magnitude and direction of the magnetic field created in the centre of this loop.

73. ★★★ A toroidal (donut-shaped) coil is closely wound with one continuous wire as shown in Figure 23-72. Assume that the current in the wire is I, the number of turns in the coil is N, and the radius of the toroid (the distance between the inner and outer radii of the toroid) is R.
   (a) Derive an expression for the magnetic field both inside and outside the toroid.
   (b) Compare your answer for the magnetic field of a toroid to the magnetic field of a solenoid.
   (c) What does your comparison in part (b) tell you?
   (d) Calculate the magnetic field of a 1000-turn toroid of radius 0.5 m when 15 A of electric current flows through it.
   (e) Compare your result in part (d) to the magnetic field inside a 1 m-long solenoid that has 1000 turns and a 15 A current flowing through it. What does your comparison tell you?

Figure 23-72   Problem 73

74. ★★★ As discussed earlier, there are different versions of mass spectrometers. Figure 23-74 shows a mass spectrometer that does not have a velocity selection stage (compare it to the mass spectrometer shown in Figure 23-12). Derive an expression that shows how the radius of the charged particle in the mass spectrometer depends on the mass of the particle, its charge, magnetic field $\vec{B}$ and the potential difference V. What assumption have you made for your derivation? Compare the operation of two mass spectrometer models (Figure 23-12 and Figure 23-73) and discuss what are their advantages and disadvantages.

Gas is ionized by a beam of electrons. Beam of positive ions leaves ionization chamber

In acceleration chamber, positively charged ions are accelerated by a strong electric field.

Charged ions of different masses strike a photo-detector's screen at different locations.

Source of gaseous sample

Magnetic and electric fields adjusted to permit only ions traveling at a particular speed to pass through entrance slit

Windings of an electromagnet that creates a constant and uniform magnetic field inside the second chamber

Figure 23-73 Problem 74

DATA-RICH PROBLEM

75. ✱✱✱ Could using a device such as an iPod be hazardous for a person who has a pacemaker? A paper called "Low frequency magnetic emissions and resulting induced voltages in a pacemaker by iPod portable music players," by Howard Bassen, attempts to provide an experimental answer to this question (www.pubmedcentral.nih.gov/articlerender.

fcgi?artid=2265271). Use the information provided in the paper to give scientifically warranted recommendations regarding the use of portable music devices by people who have pacemakers.

See the text online resources at www.physics1e.nelson.com for Open Problems and Data-Rich Problems related to this chapter.

ELECTRICITY, MAGNETISM, AND OPTICS

Learning Objectives

When you have competed this chapter you should be able to:

1 Describe how to induce an electric current using a magnetic field.

2 Define and calculate magnetic flux and its rate of change.

3 Describe Faraday's law of electromagnetic induction quantitatively and qualitatively.

4 Use Lenz's law and Faraday's law of electromagnetic induction to determine the direction and magnitude of an induced emf and an induced electric current.

5 Describe the concepts of self-inductance of a solenoid and mutual inductance of two adjacent solenoids.

6 Calculate the energy stored in a magnetic field.

7 Describe the behaviour of RL circuits qualitatively and quantitatively.

8 Apply Faraday's law of electromagnetic induction to power generation, transmission, and other everyday applications.

For additional Making Connections, Examples, and Checkpoints, as well as activities and experiments to help increase your understanding of the chapter's concepts, please go to the text's online resources at www.physics1e.nelson.com.

Chapter 24
Electromagnetic Induction

Canada is the second-largest country in the world, and more than one-third of its CO_2 emissions come from transportation. For example, a round-trip flight from Vancouver to Toronto (6717 km) produces 1.536 tonnes of CO_2 emissions.[1] Consequently, the development of energy-efficient transportation is an important issue for Canadians. One promising technology is magnetic levitation (maglev) trains, which are suspended over conductive rails by electromagnetic forces (Figure 24-1). Maglev trains are powered by electric energy. In addition to reducing CO_2 emissions, (assuming the electricity is not fossil fuel generated) maglev trains are fast—they reach speeds comparable to modern jets: up to 590 km/h. They can travel at high speeds because levitating the train eliminates friction and vibrations from contact between wheels and the rails, thus reducing waste energy.

The first maglev vehicle patent was obtained in 1902 by an American, A. Zehden. However, this technology required very strong magnets, so practical maglev vehicles were not developed until the early 1960s. The world's first commercial maglev train was introduced in Birmingham, England, in 1984, but it was discontinued in 1995 due to technical difficulties. In 2002, German scientists and engineers from Transrapid International developed the first maglev system still in operation today: a maglev line between Shanghai's Pudong airport and its financial district (30.5 km). Other commercial maglev trains operate in Japan. Currently, there are a few maglev projects under construction in Europe, China, Japan, and the United States. These maglev trains use the law of electromagnetic induction discovered by Michael Faraday almost two centuries ago, and they might be a solution to the energy-efficient transportation problem.

Figure 24-1 Magnetic levitation (maglev) trains are high-speed trains that use the law of electromagnetic induction for their operation.

[1]www.less.ca

24-1   In Faraday's Lab

As described in Section 23-6, the first solid evidence for a connection between electric and magnetic phenomena was the discovery by Danish physicist and chemist Hans Christian Oersted. In 1821, while showing an experiment during a lecture, Oersted noticed that an electric current deflects the needle of a nearby compass, suggesting that the current generates a magnetic field. Oersted's findings intrigued a British scientist, Michael Faraday (1791–1867), who set out in the late 1820s to answer one of the most important questions in the history of physics: *If electric current is capable of generating a magnetic field, can a magnetic field generate an electric current?*

You can recreate Faraday's experiments either with available equipment or with the computer simulation in the Online Activity below.

ONLINE ACTIVITY

In Faraday's Lab

The e-resource that accompanies every new copy of this textbook contains an Online Activity using the PhET simulation "Faraday's electromagnetic lab: pickup coil." Work through the simulation and accompanying questions to gain an understanding of Faraday's experiments.

Faraday observed that moving a magnet toward a coil generated an electric current in one direction, and moving the same magnet away from the coil generated an electric current in the opposite direction. When the magnet was not moving relative to the coil, no current flowed in the coil. Faraday made several important observations from this series of experiments:

- A magnetic field can induce (generate) an electric current in a coil; thus, the phenomenon is called **electromagnetic induction** (Figure 24-2).

- The magnitude and the direction of the induced current depend on the rate of change of the magnetic field and the rate of change of the effective area of the loops in the coil. As soon as the change ceases the generation of electric current stops.

 CHECKPOINT

C-24-1 In Michael Faraday's Lab

Which of the following actions in Faraday's experiment will NOT induce electric current in the pickup coil, provided other parameters remain unchanged?

(a)

(b)

(c)

Figure 24-2   The effects of moving a magnet through a conducting loop. (a) When the magnet moves into the loop, an electric current is generated in the loop that is registered by an ammeter. (b) When the magnet is stationary relative to the loop, no current in the loop is detected. (c) When the magnet moves away from the loop, a current is generated again. The direction of the current depends on the relative motion of the magnet and the loop.

Faraday's original experiment (Figure 24-3) used two adjacent electric circuits called the primary and secondary circuits. The primary circuit consisted of a coil connected to a battery, while the secondary circuit consisted of a coil connected to an ammeter. The coils were made of insulated wire wound around opposite sides of an iron ring, which kept most of the magnetic field inside the coils. Faraday noticed that, when a steady electric current flowed through the primary circuit, the ammeter in the secondary circuit indicated zero current. However, when the circuit was switched on or off, the needle of the ammeter deflected for an instant, indicating the generation of an electric current in the secondary circuit. Faraday hypothesized

(a)   gradually changing the number of loops
(b)   gradually changing the strength of the magnet
(c)   gradually changing the area of the loops
(d)   using the most powerful stationary magnet available

C-24-1 (d) To generate an induced current, a magnetic flux has to be changing with time.

ELECTRICITY, MAGNETISM, AND OPTICS

Figure 24-3 A modern version of Faraday's original experiment

that the changing electric current in the primary circuit induced a changing magnetic field that, in turn, induced a changing electric current in the secondary circuit. The secondary coil is sometimes called a pick-up coil because it is picking up the signal of the varying magnetic field.

LO 2

24-2 Magnetic Flux and Its Rate of Change

Faraday introduced the concept of **magnetic flux** to explain how a varying magnetic field generates electric current. You learned about electric flux in Chapter 20. To distinguish between electric and magnetic flux, we will use the subscripts E and B, respectively. In Section 23-5, we defined the orientation of a loop of wire with area vector, $\vec{A}$, where the magnitude of the vector represents the loop's area and the vector's direction is perpendicular to the plane of the loop. If a uniform magnetic field is directed at an arbitrary angle θ to the surface of the loop (Figure 24-4(a)), the magnetic flux through the loop is defined as follows:

$$\Phi_B = BA \cos\theta = BA_{\text{eff}} \qquad (24\text{-}1)$$

The product of the area of the loop and the cosine of the angle between the normal to the loop and the direction of the magnetic field ($A \cos\theta$) is called the **effective surface area** of the loop. The magnetic flux passing through the loop is a scalar quantity. It can be described more concisely using a scalar product:

KEY EQUATION $\qquad \Phi_B = BA \cos\theta = \vec{B} \cdot \vec{A} \qquad (24\text{-}2)$

When a uniform magnetic field is directed perpendicular to the plane of the loop (Figure 24-4(b)), the angle between the vectors $\vec{A}$ and $\vec{B}$ is 0°; thus, the magnetic flux Equation (24-2) becomes

$$\Phi_{B(\vec{B}\perp\vec{A})} = BA \cos(0°) = BA \qquad (24\text{-}3)$$

When magnetic field lines are parallel to the plane of the loop, the angle between $\vec{A}$ and $\vec{B}$ is 90°, and the magnetic flux through the loop is zero (Figure 24-4(c)):

$$\Phi_{B(\vec{B}\parallel\vec{A})} = BA \cos(90°) = 0 \qquad (24\text{-}4)$$

The SI unit for magnetic flux is the weber (Wb):

$$1\,\text{Wb} = 1\,\text{T}\cdot\text{m}^2 \qquad (24\text{-}5)$$

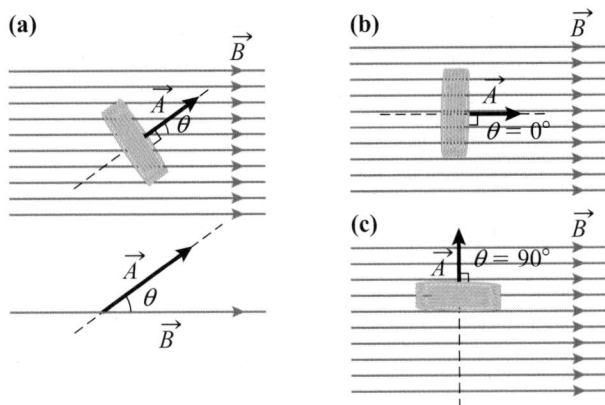

Figure 24-4 Magnetic flux through a closed loop can be expressed as $\Phi_B = BA \cos\theta$. (a) A magnetic field is directed at an angle θ to the normal to the loop. (b) The magnetic field is perpendicular to the plane of the loop $\Phi_B = BA \cos(0°) = BA$. (c) The magnetic field is parallel to the plane of the loop, $\Phi_B = BA \cos(90°) = 0$.

Thus, 1 Wb of magnetic flux passes through each square metre of a surface perpendicular to a magnetic field of 1 T.

For a nonuniform magnetic field, magnetic flux is defined as follows:

$$\Phi_B = \vec{B}_{\text{avg}} \cdot \vec{A} = B_{\text{avg}} A \cos\theta = B_{\text{avg}} A_{\text{eff}} \qquad (24\text{-}6)$$

We can divide an arbitrary surface into infinitesimally small area elements, $d\vec{A}_i$, calculate magnetic flux $d\Phi_i$ through each element, and then integrate over the surface to determine the net flux:

$$\Phi_B = \int \vec{B} \cdot d\vec{A} = \int d\Phi \qquad (24\text{-}7)$$

For a three-dimensional surface (Figure 24-5), the direction of the normal to the surface element is always chosen to point outward. Therefore, a vector field directed inward through the surface generates a negative flux, and a field directed outward generates a positive flux. For a flat surface, the direction of the normal to the surface (and, hence, the sign of the magnetic flux) is chosen arbitrarily.

The breakthrough idea in Faraday's discovery of electromagnetic induction was his realization of the importance of the **rate of change of magnetic flux**. Since the coil can be represented as a stack of closed loops, we first

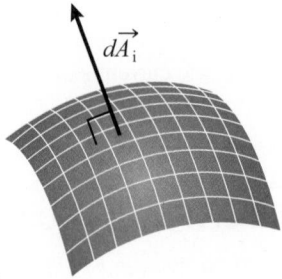

Figure 24-5 A three-dimensional surface can be divided into small surface elements $d\vec{A}_i$. The direction of each $d\vec{A}_i$ element is normal to the surface and points outward.

learn how to calculate the rate of change of the magnetic flux through a single loop:

$$\frac{d\Phi_B}{dt} = \frac{d(\vec{B} \cdot \vec{A})}{dt}$$

$$\frac{d\Phi_B}{dt} = \frac{d[B(t)A(t)\cos(\theta(t))]}{dt}$$

$$\frac{d\Phi_B}{dt} = \underbrace{\frac{dB(t)}{dt}A(t)\cos(\theta(t))}_{B \text{ changes with time}} + \underbrace{\frac{dA(t)}{dt}B(t)\cos(\theta(t))}_{\text{Area changes with time}}$$

$$+ \underbrace{\frac{d\cos(\theta(t))}{dt}B(t)A(t)}_{\text{Angle } \theta \text{ changes with time}} \qquad (24\text{-}8)$$

CHECKPOINT

C-24-2 Magnetic Flux Through the Surface of a Closed Loop

The magnetic flux through the plane of a closed loop is zero ($\Phi_B = 0$). What can you conclude?

(a) There is no magnetic field in that space.
(b) Magnetic field lines are perpendicular to the plane of the loop.
(c) Magnetic field lines are parallel to the plane of the loop.
(d) The effective area of the loop is zero.
(e) Statements (a), (c), and (d) may be true.
(f) Statements (a), (b), (c), and (d) may be true.

C-24-2 (e) Use the definition of magnetic flux.

EXAMPLE 24-1

Calculating the Rate of Change of a Magnetic Flux

A rectangular copper loop with the dimensions $l = 60$ cm and $w = 20$ cm is moving at a constant speed of 1 m/s out of the region of a uniform time-varying magnetic field, $B(t) = (0.001t + 0.005)$ T. The magnetic field is directed perpendicular to the loop (Figure 24-6).

(a) Derive an expression that describes how the rate of change of the magnetic flux through the loop varies with time.
(b) Find the rate of change of the magnetic flux through the loop at the instant when the loop is halfway out of the magnetic field region.
(c) Answer part (b) when the loop has moved completely out of the magnetic field region.
(d) Find the rate of change of the magnetic flux when the loop is stationary, while located within the magnetic field.
(e) Recalculate parts (a) to (d) for a coil consisting of three loops, all identical to the loop described above.

Figure 24-6 A rectangular copper loop is moving at a constant speed out of a uniform magnetic field region.

SOLUTION

(a) We use Equation (24-8) to calculate the rate of change of the magnetic flux. The plane of the loop is always perpendicular to the magnetic field, $\cos(0°) = 1$, so Equation (24-8) simplifies to

$$\frac{d\Phi_B}{dt} = \frac{dB(t)}{dt}A(t)\cos\theta + \frac{dA(t)}{dt}B(t)\cos\theta$$

Since the loop is pulled out of the magnetic field region with constant speed v, the effective area of the loop changes with time. If $t = 0$ s was the last instant when the loop was entirely inside the magnetic field region, at time t, its effective area is $A(t) = w(l - vt)$, provided $t \leq \frac{l}{v}$. Therefore,

$$\frac{dA(t)}{dt} = \frac{d[w(l - vt)]}{dt} = -wv$$

Consequently, the rate of change of the magnetic field is

$$\frac{d\Phi_B}{dt} = \frac{dB(t)}{dt}A(t) - wvB(t) = \frac{dB(t)}{dt}w(l - vt) - wvB(t)$$

$$\frac{d\Phi_B}{dt} = \left(0.001\,\frac{\text{T}}{\text{s}}\right)(0.2\text{ m})\left(0.6\text{ m} - \left(1\,\frac{\text{m}}{\text{s}}\right)t\right)$$

$$- (0.2\text{ m})\left(1\,\frac{\text{m}}{\text{s}}\right)(0.001t + 0.005)\text{ T}$$

$$\frac{d\Phi_B}{dt} = \left(0.0002\,\frac{\text{T}\cdot\text{m}}{\text{s}}\right)\left(0.6\text{ m} - \left(1\,\frac{\text{m}}{\text{s}}\right)t\right)$$

$$- \left(0.2\,\frac{\text{m}^2}{\text{s}}\right)(0.001t + 0.005)\text{ T}$$

while $t \leq \dfrac{l}{v} = 0.6$ s

This result applies only during the time when the loop is partially submerged in the magnetic field region. After 0.6 s, the loop is clear of the magnetic field and is no longer affected by it. As expected, the rate of change of the magnetic field is time-dependent.

(continued)

ELECTRICITY, MAGNETISM, AND OPTICS

(b) The loop is halfway out of the magnetic field at $t = 0.3$ s, and

$$\left.\frac{d\Phi_B}{dt}\right|_{t=0.3\,\text{s}} = \left|\left(0.0002\ \frac{\text{T}\cdot\text{m}}{\text{s}}\right)\left(0.6\ \text{m} - \left(1\ \frac{\text{m}}{\text{s}}\right)(0.3\ \text{s})\right)\right.$$

$$\left. - \left(0.2\ \frac{\text{m}^2}{\text{s}}\right)(0.001(0.3\ \text{s}) + 0.005)\ \text{T}\right|$$

$$\left.\frac{d\Phi_B}{dt}\right|_{t=0.3\,\text{s}} = \left|\left(6\times10^{-5}\ \frac{\text{T}\cdot\text{m}^2}{\text{s}}\right) - \left(106\times10^{-5}\ \frac{\text{T}\cdot\text{m}^2}{\text{s}}\right)\right|$$

$$= \left(100\times10^{-5}\ \frac{\text{Wb}}{\text{s}}\right) = 1.0\ \frac{\text{mWb}}{\text{s}}$$

(c) When the loop has moved completely outside the magnetic field region, the magnetic flux through the loop is $\Phi_B \equiv \text{const} \equiv 0$, and the rate of change is $\dfrac{d\Phi_B}{dt} \equiv 0$.

(d) When the entire loop is held stationary in the uniform magnetic field, the effective area of the loop remains constant. However, the magnetic flux through the loop is changing because the magnetic field depends on time. Consequently, Equation (24-8) simplifies to

$$\frac{d\Phi_B}{dt} = \frac{dB(t)}{dt}A(t) = \left(0.001\ \frac{\text{T}}{\text{s}}\right)(0.6\ \text{m})(0.2\ \text{m}) = 0.12\ \frac{\text{mWb}}{\text{s}}$$

(e) When the number of loops is tripled, the effective area of the loop and thus the rate of change of the magnetic

flux also triples. Therefore, the answers to the previous parts triple as well.

Making sense of the result:

Since the solution to the problem depends on the expression derived in part (a), we can check it using dimensional analysis:

$$\left[\frac{d\Phi_B}{dt}\right] = \frac{\text{Wb}}{\text{s}}$$

$$\left[\left(0.0002\ \frac{\text{T}\cdot\text{m}}{\text{s}}\right)\left(0.6\ \text{m} - \left(1\ \frac{\text{m}}{\text{s}}\right)t\right) + \left(0.2\ \frac{\text{m}^2}{\text{s}}\right)(0.001t + 0.005)\ \text{T}\right]$$

$$= \frac{\text{T}\cdot\text{m}^2}{\text{s}} = \frac{\text{Wb}}{\text{s}}$$

The answer to part (c) is zero because the loop in (c) is unaffected by the magnetic field. The answers to parts (b) and (d) are nonzero because the rate of change of the magnetic flux in (b) is larger than in part (d). In part (d), the only cause of the change of the magnetic flux is the change of the magnetic field. Yet in part (b) the changes in both the magnetic field and in the effective area of the loop contribute to the change in the magnetic flux. Therefore, the final result in part (b) is larger than the result in part (a).

In general, when the number of loops in a coil is N,

KEY EQUATION
$$\left(\frac{d\Phi_B}{dt}\right)_{\text{coil}} = N\left(\frac{d\Phi_B}{dt}\right)_{\text{loop}} \tag{24-9}$$

LO 3, LO 4, LO 8

24-3 Faraday's Law of Electromagnetic Induction

In 1831, Faraday formulated a theory of electromagnetic induction to explain the results of his experiments: a time-varying magnetic flux through a conducting loop induces an emf ε across the loop. This emf has a magnitude equal to the rate of change of the magnetic flux through this loop and is directed such that it opposes the change of the magnetic flux that causes it.

The part of this law pertaining to the direction of induced current was rediscovered in 1834 by the Russian physicist Heinrich Lenz (1804–65) and is called **Lenz's law**.

The law of electromagnetic induction

KEY EQUATION
$$\varepsilon_{\text{ind}} = -\left(\frac{d\Phi_B}{dt}\right) \tag{24-10}$$

The minus sign in the law of electromagnetic induction emphasizes that the direction of the induced current is such that the magnetic field induced by it ($\vec{B}_{\text{ind}}$) *opposes* the change of the magnetic flux. It is important to note that if the direction of the induced emf were not opposed to the change of the magnetic flux, one would get a perpetual motion machine arising from the positive feedback. This "runaway" scenario contradicts the law of energy conservation. You might want to revisit the Online Activity on page 645 now, paying special attention to the direction and magnitude of the induced current.

To determine the units for the rate of the change of the magnetic flux, we first use the Lorentz force expression $\vec{F}_B = q\vec{v} \times \vec{B}$ (Equation (23-2)) to link the unit of magnetic field, the tesla, to other SI units:

$$[\vec{B}] = \text{T} = \frac{\text{N}}{\text{C}\cdot\dfrac{\text{m}}{\text{s}}} = \frac{\text{kg}\cdot\dfrac{\text{m}}{\text{s}^2}}{\text{C}\cdot\dfrac{\text{m}}{\text{s}}} = \frac{\text{kg}}{\text{s}\cdot\text{C}} \Rightarrow \text{T} \equiv \frac{\text{kg}}{\text{s}\cdot\text{C}}$$

Therefore,
and

$$\left[\frac{d\Phi}{dt}\right] = \frac{\text{Wb}}{\text{s}} = \frac{\text{T}\cdot\text{m}^2}{\text{s}} = \frac{\text{kg}}{\text{s}\cdot\text{C}}\cdot\frac{\text{m}^2}{\text{s}} = \frac{\text{kg}\cdot\text{m}^2}{\text{s}^2\cdot\text{C}} = \frac{\text{J}}{\text{C}} = \text{V}$$

$$\tag{24-11}$$

Thus, dimensional analysis confirms that a rate of change of magnetic flux corresponds to a potential difference. A conducting loop in a magnetic field acts as an emf source, as long as the magnetic flux keeps changing in time.

When a coil consists of N identical loops, the law of electromagnetic induction (Equation (24-10)) can be rewritten as follows:

KEY EQUATION
$$\varepsilon_{ind} = -\left(\frac{d\Phi_B}{dt}\right)_{coil} = -N\left(\frac{d\Phi_B}{dt}\right)_{loop} \quad (24\text{-}12)$$

Expressions (24-8) and (24-12) can be combined to obtain the following:

$$\varepsilon_{ind} = -N\left(\underbrace{\frac{dB(t)}{dt}A(t)\cos(\theta(t))}_{B \text{ changes with time}} + \underbrace{\frac{dA(t)}{dt}B(t)\cos(\theta(t))}_{Area \text{ changes with time}}\right.$$
$$\left. + \underbrace{\frac{d\cos(\theta(t))}{dt}B(t)A(t)}_{Angle\ \theta \text{ changes with time}}\right) \quad (24\text{-}13)$$

✅ CHECKPOINT

C-24-3 Induced Current in a Metal Ring

In which direction should the bar magnet in Figure 24-7 move to induce the electric current in the ring shown in the diagram?

(a) horizontally toward the ring
(b) horizontally away from the ring
(c) upward
(d) downward
(e) None of the above will generate an electric current in the ring.

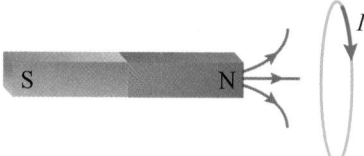

Figure 24-7 C-24-3

C-24-3 (b) Apply Lenz's law.

Eddy Currents

In 1851, almost 20 years after Faraday's original discovery, French physicist Léon Foucault (1819–68) noticed that when the magnetic flux through a large piece of conductor varies with time, the flux induces electric current loops inside the conductor (Figure 24-8(a)). Such currents are called **eddies** or **eddy currents**. According to Faraday's law, these induced currents oppose the change of the original magnetic flux. The faster the magnetic flux changes, the stronger the induced eddy currents. The direction of eddies can be determined using Lenz's law. Since all conductors (except for superconductors) have some resistance, eddy currents generated in them by the changing magnetic

flux will heat the conductors up. This heating means generation of thermal energy. Therefore, eddy currents can convert mechanical energy (e.g. motion of a conductor through the non-uniform magnetic field) into thermal energy (heat dissipated by the current-carrying conductors). (In some applications such as hand-crank generator flashlight, http://www.youtube.com/watch?v=Im14gk4L7XA, the electrical

(a)

(b)

(c)

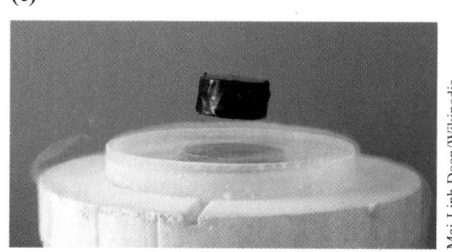

Mai-Linh Doan/Wikipedia

Figure 24-8 Eddy currents. (a) One of the eddy current loops inside a metal sheet moving to the right through a nonuniform magnetic field (b) Two metal pendulums are placed between the poles of two permanent magnets. As the pendulums swing, eddy currents generated in the pendulum that consists of a solid metal sheet (no slits) generate eddy currents that slow the motion of the pendulum (a consequence of Lenz's law). On the other hand, the pendulum that has slits in it does not slow down, since slits prevent generation of eddy currents. (c) A magnet levitates above a superconductor as a result of the Meissner effect.

energy generated by the changing magnetic flux can be stored). However, in other applications, the energy dissipation from eddy currents is undesirable. To reduce energy losses, conductors are made with slits (Figure 24-8(b)) or are constructed with thin layers (laminations) that are insulated from each other by a nonconductive varnish. These techniques reduce undesirable eddy currents by making the area of the closed loops much smaller. Eddy currents are used in magnetic brakes and magnetic dampers, as discussed in the following section.

Since superconductors have zero resistance (and, hence, no thermal energy dissipation), eddy currents generated on the surface of a superconductor will flow indefinitely. Consequently, such eddy currents induce a magnetic field that *completely* opposes the external magnetic flux that induced them. Consequently, the superconductor expels the magnetic field from its interior. This phenomenon is called the Meissner effect (Figure 24-8(c)), named after the German physicist Walther Meissner (1882–1974), who discovered it in 1933.

EXAMPLE 24-2

Emf Induced by a Time-dependent Magnetic Field

A rectangular stationary loop with dimensions $l = 60$ cm and $w = 20$ cm (Figure 24-9) and a resistance of 0.1 Ω is placed perpendicular to a uniform time-dependent magnetic field with $B(t) = (0.04t + 0.002)$ T. Calculate the direction and the magnitude of the electric current induced in the loop.

$$B(t) = (0.04t + 0.002) \text{ T}$$

Figure 24-9 A rectangular loop in a uniform magnetic field that changes with time

SOLUTION

First, we use expression (24-8) with $N = 1$ to find the magnitude of the emf induced in the loop:

$$|\varepsilon_{ind}| = N \left| \underbrace{\frac{dB(t)}{dt}}_{B \text{ changes with time}} A(t) \cos(\theta(t)) \right|$$

$$= \left(\underbrace{\frac{d(0.04t + 0.02) \text{ T}}{dt}}_{B \text{ changes with time}} (0.2 \text{ m} \cdot 0.6 \text{ m}) \cos(0°) \right)$$

$$|\varepsilon_{ind}| = 0.04 \frac{\text{T}}{\text{s}} \cdot 0.12 \text{ m}^2 = 0.0048 \text{ V} = 4.8 \text{ mV}$$

To calculate the magnitude of the induced current, we use Ohm's law:

$$I_{ind} = \frac{\varepsilon_{ind}}{R} = \frac{4.8 \text{ mV}}{0.1 \text{ Ω}} = 48 \text{ mA}$$

Since the magnetic field increases with time, the magnetic flux increases with time. According to Lenz's law, the direction of the induced current is such that the induced magnetic field $\vec{B}_{ind}$ opposes the magnetic flux change. The direction of the original magnetic field $\vec{B}(t)$ is into the page in Figure 24-9, so the direction of the induced magnetic field $\vec{B}_{ind}$ will be opposite: out of the page. Therefore, the induced current flows counterclockwise.

Making sense of the result:

Since the original magnetic field $\vec{B}(t)$ changes with time linearly, its rate of change is constant. Consequently, the induced emf and induced current are also constant.

Additional examples in the online supplement to this textbook consider more complex time-dependent magnetic fields. Next, we consider **motional emf**, which is emf induced by the motion of a rectangular coil (consisting of N rectangular loops) inside or outside a region of a constant and uniform magnetic field. Let us first assume that each rectangular loop has a width w and length l. Let us calculate the induced emf in this rectangular coil when it enters or leaves a region of a uniform and constant magnetic field, $\vec{B} = $ const, that is directed perpendicular to the coil $\theta = 0°$. (See Figure 24-10(a) in Example 24-3 below.) Moreover, the coil is moving at a constant velocity $\vec{v}$ such that its plane is perpendicular to the direction of the magnetic field. We

can use Equation (24-13) to derive the expression for the induced emf:

$$\varepsilon_{ind} = -\left(\frac{d\Phi_B}{dt}\right)_{coil} = -N\left(\frac{d\Phi_B}{dt}\right)_{loop}$$

KEY EQUATION (24-14)

$$= -NB\left(\underbrace{\frac{dA(t)}{dt}}_{\substack{\text{Area changes} \\ \text{with time}}}\right) = -NBvl$$

where N is the number of loops in the coil, and l is the length of the side of the coil that is in the field and perpendicular to both the field and the velocity.

EXAMPLE 24-3

Motional emf in a Rectangular Coil

A horizontal, rectangular (5 cm × 10 cm) coil has 100 turns and a total resistance of 10 Ω. Initially, it is sitting with the right side of the coil at the edge of a vertical uniform magnetic field with $B = 0.4$ T as shown in Figure 24-10(a). The coil is then pulled to the right, out of the field.

(a) Determine the velocity of the coil needed to generate an emf of 0.5 V.
(b) Draw a graph that shows the dependence of the induced emf on time while the coil is leaving the magnetic field at the constant velocity $\vec{v}$ found in part (a).
(c) Find the direction and the magnitude of the electric current induced in the coil during this process.
(d) Find the magnitude of the external force $\vec{F}_{ext}$ needed to pull the coil with the constant velocity $\vec{v}$. Assume that friction is negligible.
(e) Sketch a graph that represents the dependence of the external force $\vec{F}_{ext}$ on time. Compare this graph with the emf graph in part (b).

SOLUTION

(a) Applying expression (24-2), we find the speed of the coil that will generate 0.5 V of induced emf:

$$v = \frac{\mathcal{E}_{ind}}{NBl} = \frac{0.5 \text{ V}}{(100)(0.4 \text{ T})(0.05 \text{ m})} = 0.25 \text{ m/s} = 25 \text{ cm/s}$$

(b) Moving at a constant speed of 25 cm/s, it takes 0.4 s for the coil to move 10 cm to get completely outside of the magnetic field region. During this time, the induced emf is constant (Equation (24-14)). However, as soon as the coil is clear of the magnetic field, the induced emf drops to zero (Figure 24-10(b)).
(c) To find the magnitude of the electric current induced in the coil while it is being pulled out of the field, we use Ohm's law:

$$I_{ind} = \frac{\mathcal{E}_{ind}}{R} = \frac{NBvl}{R} = \frac{0.5 \text{ V}}{10 \text{ Ω}} = 0.05 \text{ A} = 50 \text{ mA}$$

Since the magnetic flux through the coil decreases as the coil leaves the field, Lenz's law requires the induced current to move in the clockwise direction to induce a magnetic field $\vec{B}_{ind}$ in the same direction as the original field (into the page).
(d) Since the coil is moving with a constant velocity, the net force acting on it must be zero. It might seem that there is no need to apply an external force $\vec{F}_{ext}$ to pull the coil out of the magnetic field. However, let us consider every side of the coil more carefully (Figure 24-10(c)). Side AB

Figure 24-10 A rectangular coil leaves a uniform magnetic field

is never inside the magnetic field, so there is no magnetic force acting on it. However, there are magnetic forces acting on the other three sides. Using the right-hand rule for magnetic force (Section 23-1) and the direction of the current in the coil, we find that the magnetic forces acting on sides AD and BC are directed upward and downward, respectively (Figure 24-10(c)), and the magnetic force acting on side CD is directed to the left, resisting the coil's motion. The magnitude of the resultant horizontal magnetic force can be found using Equation (23-22):

$$F_B = N(IlB) = N\frac{NBvl}{R}lB = \frac{N^2B^2l^2v}{R} \quad (24\text{-}15)$$

Since the net force acting on the coil is zero, the external force pulling it must be directed right with a magnitude of

$$F_{ext} = \frac{NB^2l^2v}{R} = \frac{(100)^2(0.4 \text{ T})^2(0.05 \text{ m})^2(0.25 \text{ m/s})}{10 \text{ Ω}}$$
$$= 0.1 \text{ N}$$

(e) Figure 24-10(d) shows how the magnitude of the pulling force $\vec{F}_{ext}$ depends on time.

Making sense of the result:

The induced emf is generated as long as the magnetic flux through the coil is changing, that is, while the coil is moving out of the region of a constant and uniform magnetic field. During this process, the magnetic force acting on the coil resists its motion.

Let us consider the energy transfer in Example 24-3. We generated an electric current by pulling the coil out of the region of a constant and uniform magnetic field. To pull the coil with constant velocity $\vec{v}$, we had to apply an external force $\vec{F}_{ext}$ (Figure 24-11), which did work on the coil (Equation (24-15)). The rate at which this work was done equals the generated power, P:

$$P = F_{ext}v = \frac{N^2B^2l^2v}{R}v = \left(\frac{NBlv}{R}\right)^2 R = \left(\frac{\mathcal{E}_{ind}}{R}\right)^2 R = I_{ind}^2 R$$

$$(24\text{-}16)$$

Expression (24-16) shows that the power delivered by the external force dissipates as thermal energy.

ELECTRICITY, MAGNETISM, AND OPTICS

Figure 24-11 (a) A conducting bar is in a constant and uniform magnetic field. An applied force $\vec{F}_{app}$ slides the bar with constant velocity along the frictionless rails. (b) The equivalent circuit diagram for the conducting bar sliding with constant velocity

We now consider motional emf at the microscopic level. When a conductor is moving with a constant velocity $\vec{v}$ inside a uniform magnetic field $\vec{B}$ directed perpendicular to its velocity, the charged particles inside the conductor move with the same average velocity $\vec{v}$. Therefore, positive and negative charges inside this moving conductor experience magnetic forces acting in opposite directions (Figure 24-12). Consequently, the charges separate. Electrons are free to move inside a conductor, so they will be affected by this magnetic force. This will generate a potential difference across the conductor. The separation of charges continues until the magnitudes of the electric and magnetic forces equalize:

$$F_E = F_B \Rightarrow E_{ind}q = qvB_\perp \text{ Thus, } E_{ind} = vB_\perp$$

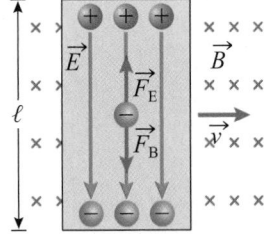

Figure 24-12 The separation of charges inside a rectangular conductor of length *l* moving inside a region of a uniform magnetic field $\vec{B}$, directed perpendicular to its velocity $\vec{v}$.

In a conductor of length *l*, an induced electric field $\vec{E}_{ind}$ causes an induced potential difference across the conductor:

KEY EQUATION
$$\Delta V_{ind} = E_{ind}l = vB_\perp l \qquad (24\text{-}17)$$

Motional emf is induced as long as the velocity of a conductor moving in a magnetic field is directed at a nonzero angle to the field. The velocity of the conductor does not need to be constant or perpendicular to the magnetic field.

C-24-4 Charge Distribution on a Plate Moving Inside a Magnetic Field

A rectangular metal plate initially located outside of a constant and uniform magnetic field is given an initial push toward the magnetic field region and then let go. Which diagram in Figure 24-13 correctly represents the charge distribution on the plate as it moves into the magnetic field?

(a) diagram (a)
(b) diagram (b)
(c) diagram (c)
(d) diagram (d)
(e) diagram (e)
(f) diagram (f)

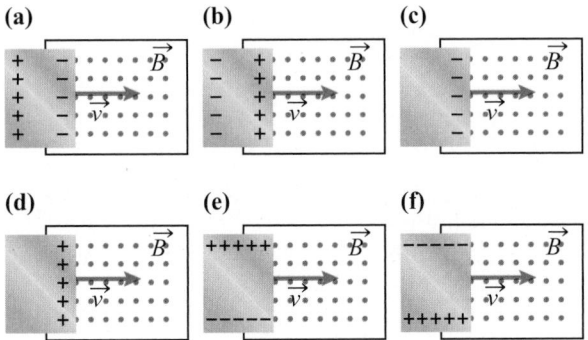

Figure 24-13 C-24-4

C-24-4 Diagram (f): Use the right-hand rule to determine the direction of the magnetic force on charged particles moving inside a magnetic field.

LO 4, LO 8

24-4 Applications of Faraday's Law of Electromagnetic Induction

Few scientific discoveries have had a more significant impact on our lives than Faraday's law of electromagnetic induction. In this section, we look at a few examples of how this law is applied in the multitude of devices used in homes, businesses, and industry every day.

AC Power Generators

Most of the electric power generated in the world is produced using generators that convert mechanical energy into electric energy using the same principle as the simple generator shown in Figure 24-14.

The core of an electric generator is a system of conductive loops that rotate in a uniform and constant magnetic field. As each loop rotates, the magnetic flux through it changes, inducing an emf in it. Consider a single loop of area *A* and resistance *R* rotating at an angular velocity ω

(a)

electrical output

rotating coil

$\vec{B}$

N

S

(b)

— magnetic flux — induced emf

Figure 24-14 (a) The operation of an electric generator (b) The dependence of the magnetic flux through the rotating coil and of the induced emf on time

in a uniform magnetic field (Figure 24-14(a)). The angle between the magnetic field and the normal to the plane of the loop can be described as $\theta = \omega t$. Then the rate of change of the magnetic flux through the loop is

$$\frac{d\Phi}{dt} = \frac{dBA\,\cos(\theta)}{dt} = \frac{dBA\,\cos(\omega t)}{dt} = -BA\,\omega\sin(\omega t)$$

(24-18)

We use Faraday's law to find the magnitude of the induced emf in the loop:

$$\varepsilon_{ind} = BA\omega\sin(\omega t) = V_{max}\sin(\omega t)$$

$$V_{max} = BA\omega = 2\pi BAf$$

$$I = I_{max}\sin(\omega t)$$

(24-19)

where

$$I_{max} = \frac{BA\omega}{R} = \frac{2\pi BAf}{R}$$

Thus, such a generator produces an emf that varies sinusoidally with time, which is the most common type of **alternating emf**. An electric current generated by such a generator is called **alternating current** (AC). For a rotating coil with N loops, the expressions for emf and current become the following:

KEY EQUATION

$$\varepsilon_{ind} = NBA\,\omega\sin(\omega t)$$

$$I = I_{max}\sin(\omega t)$$

(24-20)

where

$$I_{max} = \frac{NBA\omega}{R} = \frac{2\pi NBAf}{R}$$

and the frequency of the alternating current (f) depends on the angular speed of the coil (ω), $f = \dfrac{\omega}{2\pi}$.

 MAKING CONNECTIONS

Sources of Electrical Energy

In North America, the standard frequency for AC current is 60 Hz; in Europe and some other countries it is 50 Hz. The most common source of mechanical energy for electric generators worldwide is steam from burning fossil fuels or from nuclear fission. However, in Canada, the most common source of this mechanical energy is hydroelectric—kinetic energy from falling water (approximately 60%). Coal supplies slightly less than 20%, and nuclear power supplies approximately 12%. The proportions vary considerably by province: in 2007, more than half of Ontario's energy came from nuclear power, and only 21% was hydroelectric. However, Alberta generates almost 90% of its electricity with coal and natural gas. More than 70% of the energy for generators in the United States is produced using fossil fuels and nuclear power.

The same principles of electromagnetic induction that underlie giant generators (Figure 24-15) feeding national power grids also apply to the diesel-powered backup generators for hospitals, gas-powered portable generators on

Figure 24-15 The Churchill River Underground Power Plant in Newfoundland and Labrador is the second largest underground power plant in the world.

(continued)

ELECTRICITY, MAGNETISM, AND OPTICS

construction sites, and even hand-powered flashlights, like those shown in Figure 24-16. The flashlight on the bottom right uses a stationary coil and a movable magnet. Shaking this flashlight makes the magnet slide back and forth through the coil, changing the magnetic flux and, hence, inducing current. On the hand-crank flashlights, turning the crank either rotates a coil inside a stationary magnet directly or winds a spring, which powers clockwork, which rotates the coil until the spring runs down. All of these flashlights use electromagnetic induction to convert human mechanical energy into electrical energy.

Dennis Steen/Shutterstock

By Richard Wheeler (Zephyris) 2007.

ktynzq/Shutterstock

Figure 24-16   Several types of hand-powered generator flashlights

EXAMPLE 24-4

Exploring How an Electric Generator Works

An electric generator has a circular coil with 20 turns, each 60 cm in diameter. The coil makes 60 revolutions per second. What strength of magnetic field is required to produce a peak output voltage of 170 V?

SOLUTION

Using expression (24-20) we find

$$V_{max} = NBA\omega = 2\pi NBAf$$

$$B = \frac{V_{max}}{2\pi NAf} = \frac{V_{max}}{2\pi N\left(\pi \dfrac{d^2}{4}\right)(60 \text{ Hz})}$$

$$= \frac{170 \text{ V}}{2\pi(20 \text{ turns})\left(\pi \dfrac{(0.6 \text{ m})^2}{4}\right)(60 \text{ Hz})} = 0.08 \text{ T}$$

Making sense of the result:

A strong permanent magnet can produce a magnetic field strength in the order of 0.1 T, so a value of 0.08 T seems reasonable.

Transformers

Have you ever wondered why electric power delivered to our homes comes in the form of alternating current (AC) rather than direct current (DC)? At the end of the 19th century, the struggle between the proponents of AC and DC power generation was so fierce that it was dubbed the "war of currents." Thomas Alva Edison (1847–1931) advocated for DC power, and Nikola Tesla (1856–1943, Figure 24-17) and George Westinghouse (1846–1914) promoted AC power generation. The key issue that swayed the argument in favour of AC power was power transmission. For various practical reasons, large electric power plants are often located away from densely populated areas, so the electric current has to travel through wires over long distances to reach consumers. The amount of power dissipated as thermal energy in the transmission lines depends on the current and the resistance of the wires (Equation (22-23)):

$$P = I^2R$$

The longer the transmission line, the higher its resistance and the greater the proportion of electrical energy lost as thermal energy. Edison tried to minimize such losses by building electric power plants very close to the consumers, for example, in New York City. However, this approach turned out to be impractical.

Figure 24-17 The Nikola Tesla monument in Queen Victoria Park at Niagara Falls, Ontario, was unveiled on July 9, 2006. Tesla is standing on top of an AC motor, one of the 700 inventions he patented.

© Kevin Woodrow/Alamy

Tesla realized that AC voltages can be stepped up or down using a **transformer**. By using high voltages and much smaller currents, the thermal energy dissipation in transmission lines can be greatly reduced. In Edison's time, there was no practical way of changing the voltage of large amounts of DC. Transformers use changing magnetic flux, so they do not work with DC.

ONLINE ACTIVITY

Operation of a Transformer

The e-resource that accompanies every new copy of this textbook contains an Online Activity using the PhET simulation "Faraday's Electromagnetic Lab: Transformer." Work through the simulation and accompanying questions to gain an understanding of the operation of transformers.

A transformer consists of two adjacent coils: a primary coil, which is connected to the power supply, and a secondary coil, which is connected to the load (Figure 24-18). Usually, the coils are wound on the same iron core. A transformer is based on the principle of **mutual inductance**: a changing magnetic field through the primary coil generates a time-varying magnetic field through the secondary coil. In the primary coil of a transformer, an AC current produces a time-varying magnetic field. This field creates a time-varying magnetic flux through the secondary coil, which induces a time-varying

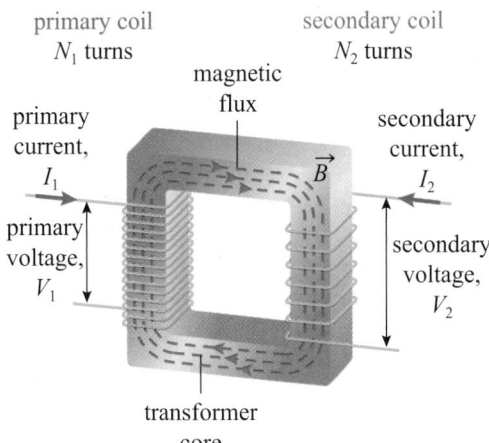

Figure 24-18 A step-down transformer

emf in the secondary coil. Let us denote the potential difference across the primary coil as V_1 and the potential difference across the secondary coil as V_2. Sinusoidal alternating currents continuously change their magnitude and direction, but due to symmetry considerations, the average values of V_1 and V_2 are zero. For power calculations, we need to use root-mean-square values for AC voltages and currents. You will learn more about root-mean-square values in Chapter 25.

Transformers transform lower voltages into higher voltages and back by converting electric energy into magnetic energy and vice versa. The law of energy conservation states that if there are no energy losses (such as dissipated thermal energy), the power delivered to the primary coil equals the power generated in the secondary coil. Using Equation (22-23),

$$P_1 = P_2$$
$$I_1 V_1 = I_2 V_2 \qquad (24\text{-}21)$$

Since power dissipated through thermal energy is proportional to the square of the current ($P = I^2 R$), to reduce energy transmission losses, the current transmitted through the wires should be as low as possible. Therefore, power stations have transformers that step up the voltage (transforming low voltage into high voltage) and, consequently, lower the current in the transmission lines. Transmission voltages generally range from 138 kV to 765 kV. At the other end of the transmission line, the voltage is stepped down (transformed from high voltage to lower voltage) in several stages to the standard 120 V/240 V for most residential customers.

In the next section, we will discuss the phenomenon of mutual inductance in more detail and will prove that the function of a transformer (step-up or step-down) is defined by the ratio of the number of loops in its primary and secondary coils:

$$\frac{V_2}{V_1} = \frac{N_2}{N_1} \qquad (24\text{-}22)$$

C-24-5 The Operation of a Transformer

Figure 24-19 shows a simple transformer: two coils wound around an iron core. The primary coil is connected to a DC power supply. The voltmeter shows the voltage induced in the secondary coil of the transformer. When was the voltmeter reading taken?

(a) The voltmeter reading was taken during the opening of a circuit, when the current in the circuit was decreasing.

(b) The voltmeter reading was taken during the closing of a circuit, when the current in the circuit was increasing.

(c) The voltmeter reading was taken when the current in the circuit was steady: the circuit has been closed for a while.

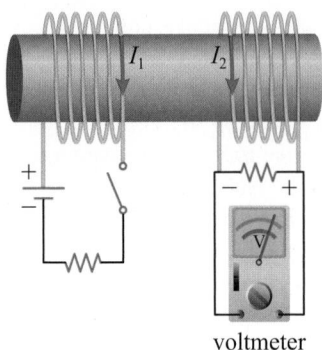

voltmeter

Figure 24-19   C-24-5

C-24-5 Use Lenz's law to verify that (a) is correct.

EXAMPLE 24-5

Transformer Ratio

A transformer is used to change a 120 V, 3 A current to 12 V in a power supply for a laptop computer.

(a) Is this a step-up or a step-down transformer? Explain.

(b) What is the ratio of the number of secondary coil turns to the number of primary coil turns in this transformer?

(c) What current would be induced in the secondary coil?

SOLUTION

(a) This is a step-down transformer because it transforms higher voltage into lower voltage.

(b) To find the ratio of the number of secondary coil turns to the number of primary coil turns, we use Equation (24-22):

$$\frac{V_2}{V_1} = \frac{N_2}{N_1} \Rightarrow \frac{N_2}{N_1} = \frac{12\text{ V}}{120\text{ V}} = 1{:}10$$

(c) The law of energy conservation can be used to find the current in the secondary coil (Equation (24-21)):

$$I_1 V_1 = I_2 V_2 \Rightarrow I_2 = \frac{I_1 V_1}{V_2} = \frac{(3\text{ A})(120\text{ V})}{12\text{ V}} = 30\text{ A}$$

Making sense of the result:

The secondary current is 10 times higher than the primary current, which is reasonable because the secondary voltage is 10 times lower than the primary voltage.

Electromagnetic Damping and Electromagnetic Braking

Electromagnetic dampers consist of a strong electromagnet or permanent magnet and a moving nonmagnetic metal part whose vibrations are to be damped. In the presence of a strong magnetic field, eddy currents are created inside the moving part. These eddy currents oppose the motion that induces them, and thus act to damp vibration in the moving part. Electromagnetic dampers are clean, easy to adjust, and less sensitive to temperature than hydraulic dampers. For example, when sheet metal is fed along a production line for coating, transverse vibrations in the sheets can make the coating uneven. Placing a powerful electromagnet near the sheets induces eddy currents that dampen the vibrations significantly.

Electromagnetic braking uses the same principle, slowing an object by inducing eddy currents that convert mechanical energy into thermal energy or electrical energy. The first electromagnetic brakes were used on mountain trains. Today, electromagnetic brakes are used on vehicles such as roller coasters, conventional and magnetic levitation trains, and hybrid vehicles. To slow the vehicle, the electric drive motor is switched to act as a generator. The motor then converts kinetic energy into electric energy and thermal energy. As a coil on the shaft of the motor rotates in a magnetic field inside the motor, a **back emf** is induced on the coil and fed to the battery, recharging it while slowing the rotation of the motor shaft.

The Electric Guitar

In both acoustic and electric guitars, sound is produced by the vibration of the strings. In an acoustic guitar, the sound of the strings is boosted by a soundboard and a resonant cavity (Chapter 15). In an electric guitar, electromagnetic induction produces an electric current that corresponds to the vibration of the strings. The signal is then amplified electronically to drive a speaker. The strings of an electric guitar are made of a ferromagnetic material (often steel or nickel). Electric guitars typically have two or three sets of pickups positioned along the strings (Figure 24-20(a)). Each set has six coils, one aligned directly underneath each of the six strings on

(b)

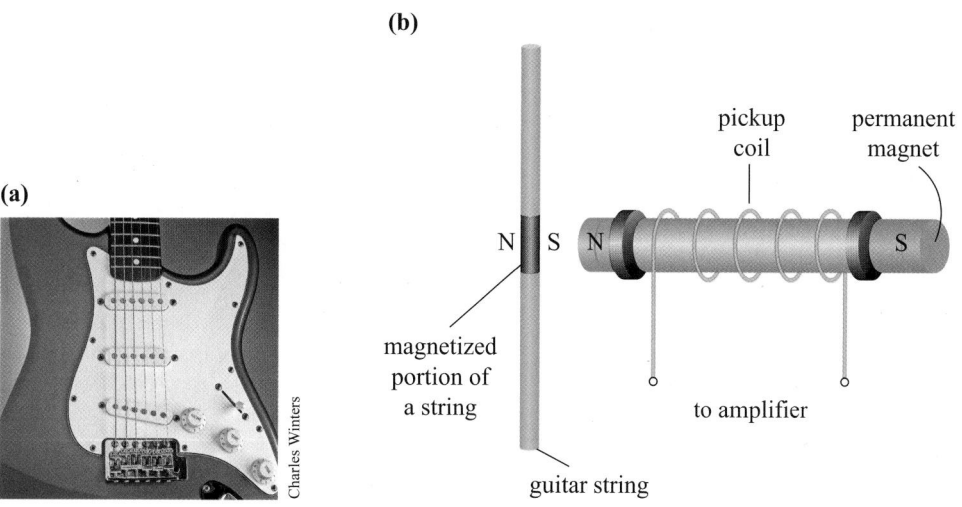

Figure 24-20   The operation of an electric guitar. (a) Pickup coils (metal circles) are located under the string of an electric guitar. (b) The moving steel string changes the flux from the permanent magnet, inducing a current in the pickup coil.

the guitar. The design of the coils and the position of the pickups affect the signal they produce, so the guitarist can change the sound of the guitar by choosing which pickups are connected to the amplifier. The vibrating ferromagnetic strings create time-varying magnetic fields through the pickup coils (Figure 24-20(b)), which then induce a changing emf and a changing electric current, which control the sound created by the speakers.

Metal Detectors

There are several types of metal detectors, and they all use electromagnetic induction. In the single-coil metal detector shown in Figure 24-21(a), a pulsing current is applied to the coil. This current induces a time-varying magnetic field and, hence, a time-varying magnetic flux through nearby objects. In metal objects, the varying flux induces eddy currents, which generate their own magnetic fields, shown in red in Figure 24-21(b). As a result, an electric current opposite in direction to the original electric current is generated in the coil. This change of current in the coil is used to detect the presence of metal.

In a two-coil metal detector (Figure 24-21(c)), the transmitter and receiver coils act as a transformer. The presence of metal near the receiver coil induces eddy currents that change the current in the receiver coil.

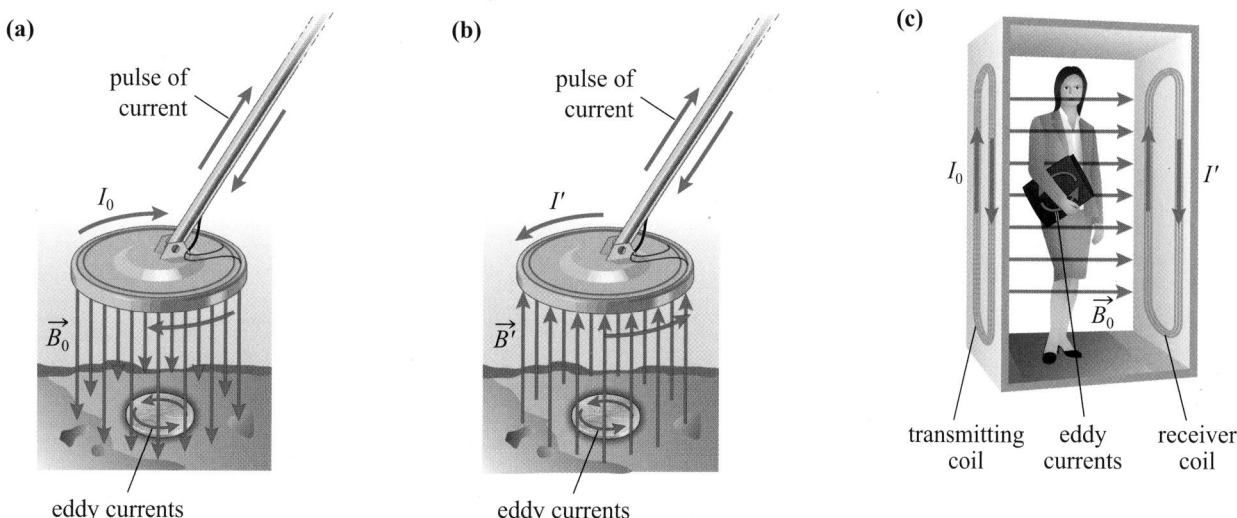

Figure 24-21   Metal detectors. (a) A single-coil metal detector consists of a transmitter coil that generates a pulsing magnetic field. (b) In the presence of a coin, for example, in the vicinity of the coil, the eddy currents induced in the coin affect the current in the transmitter coil and indicate the presence of metal. (c) A two-coil metal detector

24-5 Induced emf and Induced Electric Fields

We now take a closer look at the current induced in a circular conducting loop by a time-varying magnetic flux (Figure 24-22). As the magnetic field in the shaded area changes with time, it generates a changing magnetic flux through the loop. This changing magnetic flux induces an electric current in the loop. A galvanometer (G) connected in series with the loop can measure this current. The presence of an electric current indicates the presence of an electric field that applies an electric force on charge carriers. Since the loop is circular, the direction of the electric field must be tangent to the loop to force the charged particles to move along the circular loop (Figure 24-22(a)). An electric field is induced in the space surrounding the time-varying magnetic flux (Figure 24-22(b)). Note that this **induced electric field**, $\vec{E}_{\text{ind}}$, is independent of the presence of free electric charges in the space. The properties of an induced electric field differ substantially from the properties of an electrostatic electric field.

Table 24-1 compares the properties of the electrostatic field $\vec{E}$ created by a stationary electric charge and with the properties of an induced electric field $\vec{E}_{\text{ind}}$.

Say there is a positively charged particle q moving along a circular loop in a counterclockwise direction in Figure 24-22(a). Let us calculate the work done by the induced electric field on the particle. Both the force acting on the charged particle and the induced electric field lines are always tangent to the loop. From symmetry considerations, we deduce that at any given instant the magnitude of the induced electric field $\vec{E}_{\text{ind}}$ is the same at every

Figure 24-22 (a) A galvanometer, G, records a presence of electric current in the coil, induced by a magnetic field through the coil, $\vec{B}$, that changes with time. (b) Induced electric field lines surrounding the area of the increasing magnetic field

point along the circular loop. When the electric charge moves a distance ds along the loop, the work done on it is $dW = \vec{F} \cdot d\vec{s} = q\vec{E} \cdot d\vec{s}$. The work done on a charged particle is $W = q\varepsilon_{\text{ind}}$. Therefore,

$$\begin{cases} W_{\text{along closed loop}} = \oint dW = \oint q\vec{E}_{\text{ind}} \cdot d\vec{s} \\ \qquad\qquad = qE_{\text{ind}} \oint d\vec{s} = qE_{\text{ind}} 2\pi r \\ W_{\text{along the loop}} = q\varepsilon_{\text{ind}} \end{cases} \qquad (24\text{-}23)$$

Since both expressions represent the work done on a charged particle moving along a closed circular loop, they must be equal, and

$$E_{\text{ind}} = \frac{\varepsilon_{\text{ind}}}{2\pi r} = \frac{1}{2\pi r}\left(-\frac{d\Phi_B}{dt}\right)$$

$$= -\frac{1}{2\pi r}\left(\frac{dB(t)}{dt}\right)\pi r^2 = -\frac{r}{2}\left(\frac{dB(t)}{dt}\right) \qquad (24\text{-}24)$$

Table 24-1 Comparison of electrostatic field $\vec{E}$ and induced electric field $\vec{E}_{\text{ind}}$

	Electrostatic field $\vec{E}$ (Figure 19-7)	Induced electric field $\vec{E}_{\text{ind}}$ (Figure 24-24(b))
Definition	The electrostatic field $\vec{E}$ at a point represents the **electrostatic force** per unit charge acting on a test charge at that point: $\vec{E} = \dfrac{\vec{F}_{q_0}}{q_0}$. The electrostatic field is a vector field directed parallel to the electrostatic force acting on a positively charged test particle.	An induced electric field $\vec{E}_{\text{ind}}$ at a certain point in space represents the **electromagnetic force** per unit charge acting on a test charge placed at that point: $\vec{E}_{\text{ind}} = \dfrac{\vec{F}_{q_0}}{q_0}$. The induced electric field is a vector field directed parallel to the electromagnetic force acting on a positively charged test particle.
Source	Electric charges	Time-varying magnetic fields
Field lines	Emanate from positive charges and point toward negative charges	Are closed loops without a beginning or an end point
Work done on a charged particle	Depends on initial and final position of the particle, but not on the path between them; hence, electrostatic fields are **conservative fields**.	Done on a charged particle moving along a closed loop and is nonzero; hence, induced electric fields are **nonconservative fields**.
Potential energy	The electrostatic potential energy is defined as $U_E = \dfrac{Qq}{4\pi\varepsilon_0 r}$.	We cannot associate potential energy with an induced electric field because it is a nonconservative field.

In this derivation, we expressed the magnetic flux through the loop as $\Phi_B = B(t)\pi r^2$. Equation (24-24) indicates that as long as the magnetic field varies with time, an induced electric field is created. The minus sign indicates that the induced electric field opposes the change in the magnetic flux. Since the change in the magnetic flux is proportional to the area of the loop (provided the loop is situated within the magnetic field region), a bigger loop means a greater induced emf and induced electric field. Therefore, the intensity of the induced field increases as one moves away from the centre of the loop, as long as the loop is located inside the region of changing magnetic field. (Notice how the density of electric field lines changes in Figure 24-2(b).)

Equation (24-23) can be extended to the general case of a closed path of an arbitrary shape:

$$\begin{cases} W_{\text{along the loop}} = \oint \vec{F} \cdot d\vec{s} = \oint q\vec{E}_{\text{ind}} \cdot d\vec{s} \Rightarrow \varepsilon = \oint \vec{E}_{\text{ind}} \cdot d\vec{s} \\ W_{\text{along the loop}} = q\varepsilon_{\text{ind}} \end{cases}$$

$$\oint \vec{E} \cdot d\vec{s} = -\frac{d\Phi_B}{dt}$$

This equation leads to a reformulation of **Faraday's law of electromagnetic induction**: A time-varying magnetic flux in a certain region of space always induces a *nonconservative* electric field $\vec{E}_{\text{ind}}$ in that region, even in the absence of electric charges. The magnitude of this induced electric field can be expressed as follows:

KEY EQUATION $\quad \varepsilon_{\text{ind}} = \oint \vec{E}_{\text{ind}} \cdot d\vec{s} = -\frac{d\Phi_B}{dt} \quad$ (24-25)

The direction of the induced electric field is such that the electric current generated by it induces a magnetic field that *opposes* the change in the magnetic flux.

✅ **CHECKPOINT**

C-24-6 Exploring Induced Electric Fields

A uniform but time-varying magnetic field $\vec{B}(t)$ induces an electric field that is constant in magnitude but has a varying direction. What must be true about the dependency of the magnetic field $\vec{B}$ on time?

(a) The magnetic field does not depend on time; it is constant.

(b) The magnetic field changes with time exponentially: $B(t) = Be^{-Ct}$, where C is a constant.

(c) The magnetic field changes with time at a constant rate.

(d) The magnetic field B changes with time as $B(t) = At^2 + Bt + C$, where A, B, and C are constants.

C-24-6 (c) Use Faraday's law of electromagnetic induction: An induced electric field is proportional to the rate of change of the magnetic field. Therefore, a magnetic field that changes with time at a constant rate induces a constant electric field.

24-6 Self-Inductance and Mutual Inductance

Understanding the concepts of induced emf and induced electric field leads us to another question: What effect does the presence of a loop have on a circuit consisting of a battery, a resistor, a coil, and a switch? To distinguish between the emf from a battery and an induced emf, we denote the battery emf as ε and the emf induced in a coil as ε_{ind}. Induced currents can be generated in any closed loop. However, they become especially important in circuits containing various coils, significantly affecting the behaviour of the circuit. Such elements of electric circuits are called **inductors** and are denoted in circuit diagrams by a helical symbol and the letter L (in honour of Heinrich Lenz), as shown in Figure 24-23. Let us consider an inductor that has a very low internal resistance ($R \approx 0\ \Omega$) and is connected in series to resistor R and a battery ε. This circuit is called a **series RL circuit**.

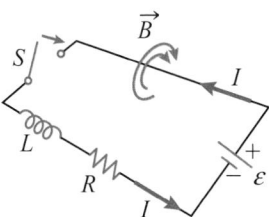

Figure 24-23 An RL circuit consists of a battery, a switch, a resistor, and a coil (also called an inductor and denoted as L).

First, we consider the circuit when the switch has been in position a for a long time, and a steady electric current flows through the RL circuit (Figure 24-24). Let us use a voltmeter (V_L) to measure the potential difference across

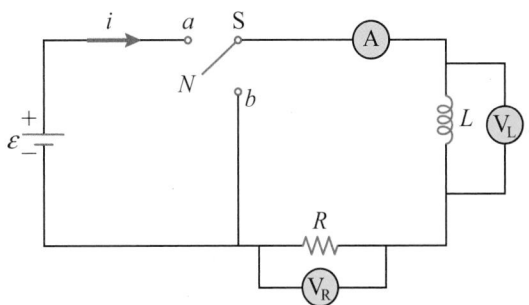

Figure 24-24 The RL circuit in this figure is designed such that when the switch is in position a, the battery, the resistor, and the inductor are part of the series circuit. When the switch is in position b, only the resistor and the inductor are part of the series circuit. When the switch is in position a, the potential difference across the inductor is zero: $\varepsilon_{\text{ind}} = 0$. When the current across the inductor is changing (the switch is being flipped from a to b or vice versa), the changing current in the circuit induces a nonzero emf across the inductor: $\varepsilon_{\text{ind}} \neq 0$.

ELECTRICITY, MAGNETISM, AND OPTICS

the inductor ε_{ind} and an ammeter to measure the current I through it. Since a constant current flows through the inductor, it generates a constant magnetic field, so there is no change in the magnetic flux through the inductor. In that case, the inductor behaves like an ordinary wire with a very low resistance R. Therefore, the potential difference across this inductor is zero: $\varepsilon_{ind} \approx 0$.

Now let us examine what happens immediately after the switch has been flipped from position a to position b. Initially, the current in the circuit was I. However, as the battery was effectively removed from the circuit, the current in the circuit will start decreasing. The changing electric current in the circuit induces emf ε_{ind} in the coil. We apply Faraday's law to find the direction of the induced emf. Since the induced emf is directed such that it reduces the change in the magnetic flux through the coil, this emf induces electric current I_{ind} in the direction of the original current I. Thus, the magnetic field, $\vec{B}_{ind}$, by this current, is directed in the direction of the original magnetic field created by the decreasing current. If we flip the switch once again from position b to position a, induced emf will have an opposite direction. In this case, the induced electric current I_{ind} will be in the opposite direction to the current I in the circuit. The Online Activity below provides an opportunity to take a closer look at the behaviour of an RL circuit and build a deeper understanding of this phenomenon.

ONLINE ACTIVITY

An Induced emf Across an Inductor in an RL Circuit

The e-resource that accompanies every new copy of this textbook contains an Online Activity using the PhET simulation "Circuit construction kit (AC + DC)". Work through the simulation and accompanying questions to gain an understanding of the behaviour of RL circuits.

An induced emf across an inductor, ε_{ind}, is often called a "back emf" because it opposes the change in the magnetic flux. A back emf has a nature different from the emf created by the battery. In a coil consisting of many loops, the magnetic field of each loop creates a back emf, which affects each of the other loops, multiplying the effect. Every loop affects itself as well as its neighbours. Therefore, the process of creating a back emf is called **self-induction**.

Recall from Chapter 23 that a solenoid is a coil with evenly spaced turns. Consider a solenoid of radius r and length l, consisting of N closely spaced loops. Using Faraday's law (Equation (24-10)) and the expression for the magnetic field of a solenoid (Equation (23-47)), the back emf can be expressed as follows:

$$\varepsilon_{ind} = -N\left(\frac{d\Phi_B}{dt}\right) = -N\frac{d}{dt}(BA) = -NA\frac{dB(t)}{dt}$$

$$B(t) = \mu_0 nI = \mu_0\frac{N}{l}I \Rightarrow \frac{dB(t)}{dt} = \mu_0\frac{N}{l}\frac{dI}{dt}$$

$$\varepsilon_{ind} = -N\pi r^2\mu_0\frac{N}{l}\frac{dI}{dt} = -\mu_0\frac{N^2\pi r^2}{l}\frac{dI}{dt} = -L\frac{dI}{dt} \quad (24\text{-}26)$$

The back emf depends on the rate of change of the electric current through the solenoid and on its **inductance**, L:

$$\varepsilon_{ind} \equiv \varepsilon_L = -L\frac{dI}{dt} \quad (24\text{-}27)$$

While expression (24-27) is true for any inductor, the following derivation for inductance L only applies to a solenoid with circular coils:

$$L_{solenoid} = \mu_0\frac{N^2\pi r^2}{l} = \mu_0\frac{N^2 l\pi r^2}{l^2} = \mu_0 n^2 V \quad (24\text{-}28)$$

where $V = l\pi r^2$.

The inductance of a solenoid depends solely on its geometric and physical properties. The inductance is proportional to the volume V and to the square of the density of the loops. A solenoid with an iron core often has an inductance hundreds of times greater than an air-core solenoid of the same size. Using the expression for magnetic permeability (Equation (23-57)), the inductance of a circular solenoid can be expressed as

$$L_{solenoid} = K_m\mu_0\frac{N^2 l\pi r^2}{l^2} = \mu n^2 V \quad (24\text{-}29)$$

where μ is the magnetic permeability of the core.

We can also calculate the inductance of a solenoid using the expression for its magnetic field (Equation (23-41)) and the law of electromagnetic induction:

$$\varepsilon_L = -N\frac{d\Phi_B}{dt} = -N\frac{d(\mu_0 nIA)}{dt} = -N\mu_0 nA\frac{dI}{dt} = -L\frac{dI}{dt}$$

$$L \equiv N\mu_0 nA = \mu_0 n^2 Al \quad (24\text{-}30)$$

Therefore, considering that for a circular coil $\Phi_B = \mu_0 nAI$, the inductance of a solenoid can be expressed as

$$L = N\mu_0 nA = \frac{N\Phi_B}{I} \quad (24\text{-}31)$$

Let us compare the functions of resistors and inductors in electric circuits. Since $I = \frac{\Delta V}{R}$, increasing the resistance in a circuit reduces the current that converts electric energy to thermal energy. The potential difference across a resistor is proportional to both the current and the resistance: $\Delta V = IR$ (Figure 24-25(a)). Current flows through a resistor from a higher potential to a lower potential. Inductors, however, act to oppose or impede the *change* in electric current. The potential difference across an inductor is proportional to its inductance and the rate of change of the electric current: $\Delta V = -L\dfrac{dI}{dt}$. Notice the direction of the induced potential difference in the inductor (Figure 24-25(b)). Thus, inductors behave somewhat like resistors to AC current. However, when a steady (DC) current flows through the inductor, the potential difference across the inductor is zero because $\dfrac{dI}{dt} = 0$. The inductor provides no resistance in a DC circuit. This explains why a transformer can burn out if you try to use DC instead of AC current.

(a)

a

i R

b

$V_{ab} = iR$

(b)

a

i L

b

$V_{ab} = \dfrac{di}{dt} L$

Figure 24-25 Comparison of the resistive and inductive elements of an electric circuit. (a) In the case of a resistor, electric current flows from a higher potential to a lower potential: $V_a > V_b$. (b) In the case of an inductor, the induced current in the coil generates a potential difference V_{ab} such that $V_b > V_a$.

The SI unit for inductance is the henry (symbol H), named after renowned American physicist Joseph Henry (1797–1878). An inductance of 1 H corresponds to an inductor that self-induces an emf of 1 V when an electric current in it changes at a rate of 1 A/s:

$$1\ \text{H} = \frac{1\ \text{V}}{1\ \text{A/s}} \qquad (24\text{-}32)$$

☑ **CHECKPOINT**

C-24-7 The Direction of an Induced emf in an Ideal Inductor

Figure 24-26 shows an ideal inductor ($R = 0$). The potential at point a of the inductor is measured to be higher than the potential at point b. Which of the following current flow(s) could account for this potential difference?

(a) A steady current flows from a to b.
(b) A steady current flows from b to a.
(c) A decreasing current flows from a to b.
(d) An increasing current flows from a to b.
(e) A decreasing current flows from b to a.
(f) An increasing current flows from b to a.

a $V_a > V_b$ b

Figure 24-26   C-24-7

C-24-7 (d) and (e) according to Faraday's or Lenz's law

The transformer discussed earlier is an example of two **coupled inductors**. Consider two solenoids wound around an iron core such that the magnetic fluxes through them are equal, as shown in Figure 24-27. Let us assume that the solenoids have N_1 and N_2 loops, respectively, the average circumference of the iron core is l, and its cross-sectional area is A. Faraday's law of electromagnetic induction states that a time-varying electric current in the first solenoid $I_1(t)$ generates a time-varying magnetic field $\vec{B}(t)$ through the second solenoid, which induces an electric current in it. Let us calculate this induced current.

To calculate the magnetic field induced by the first solenoid, we apply Ampère's law (Equation (23-39)) to a closed contour L located inside the iron core:

$$\oint_L \vec{B} \cdot d\vec{l} = \mu_0 \sum I_{\text{enc}}$$

Since the magnetic field has the same magnitude at all points inside the iron core and is always directed along the tangent to the contour L, we rewrite the above equation as

$$Bl = \mu N_1 I$$

$$B = \frac{\mu N_1 I}{l}$$

where μ is the magnetic permeability of iron.

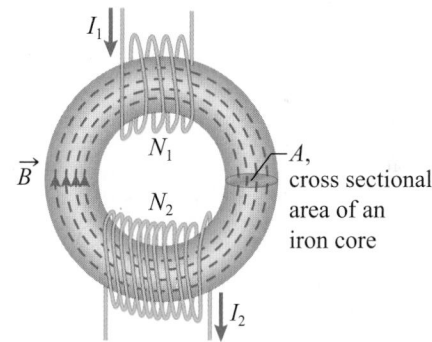

Figure 24-27   Two coils coupled through an iron core of average circumference l and cross-sectional area A

Using Faraday's law, the induced emf in the second solenoid can be expressed as

$$(\varepsilon_{\text{ind}})_2 = -N_2\left(\frac{d\Phi_B}{dt}\right)_2 = -N_2\frac{d(AB)}{dt} = -N_2 A \frac{dB}{dt}$$

$$= -N_2 A \frac{d}{dt}\frac{\mu N_1 I_1}{l} = -\frac{\mu A N_2 N_1}{l}\frac{dI_1}{dt}$$

Then, the emf induced in the second solenoid can be expressed as

$$(\varepsilon_{\text{ind}})_2 = -M_{12}\frac{dI_1}{dt} \qquad (24\text{-}33)$$

where

$$M_{12} = \mu A \frac{N_1 N_2}{l}$$

The coefficient M_{12}, called the **coefficient of mutual inductance**, indicates the strength of coupling between the two coils. Similarly, if we consider the emf induced in the first solenoid by the varying current in the second, we find that

$$\varepsilon_1 = -M_{21}\frac{dI_2}{dt}; \quad M_{21} = \mu A \frac{N_2 N_1}{l} \qquad (24\text{-}34)$$

Comparing Equations (24-33) and (24-34) we see that

$$M_{12} = M_{21} = M \qquad (24\text{-}35)$$

Thus, the coefficient of mutual inductance between two coils is denoted as M.

Now we consider two long, thin solenoids of length l wound over top of each other (Figure 24-28). The inner and outer solenoids have N_1 and N_2 coils, respectively, and their cross-sectional area is A. Once again, if a time-varying electric current flows through the inner solenoid I_1, it will induce an emf ε_2 in the outer coil:

$$(\varepsilon_{\text{ind}})_2 = -N_2 \frac{d\Phi_B}{dt} = -N_2 \frac{d(AB)}{dt} = -N_2 A \frac{dB}{dt}$$

$$= -N_2 A \frac{d}{dt}\frac{\mu N_1 I_1}{l} = -\frac{\mu A N_2 N_1}{l}\frac{dI_1}{dt}$$

and

$$(\varepsilon_{\text{ind}})_2 = -M\frac{dI_1}{dt}$$

where

$$M = \frac{\mu A N_2 N_1}{l} \qquad (24\text{-}36)$$

Figure 24-28 Two long, thin concentric solenoids wound on top of each other

Once again, we can show that the mutual inductance $M_{12} = M_{21} = M$. Comparing expression (24-36)

with the expression for the self-inductance of a solenoid (Equation (24-30)), we find

$$M = \frac{\mu A N_1 N_2}{l} = \sqrt{L_1 L_2} \qquad (24\text{-}37)$$

In the derivation above we assumed that all the magnetic flux created by the first solenoid passes through the second solenoid. This is rarely true; therefore, formula (24-37) should be adjusted to include the coefficient of coupling between the inductors, k that varies from 0 to 1. Perfectly coupled coils have a coefficient of coupling of 1, and perfectly decoupled coils have a coefficient of coupling of zero:

$$M = k\sqrt{L_1 L_2} \qquad (24\text{-}38)$$

MAKING CONNECTIONS

How Does an Electric Toothbrush Work?

Have you ever wondered how an electric toothbrush can be recharged when the charging stand and the battery-powered toothbrush are both sealed and have no exposed metal contacts to make a circuit (Figure 24-29)? The answer is **inductive charging**. The stand and toothbrush essentially operate like a transformer. The base connected to the AC power outlet serves as the primary coil, and a coil inside the toothbrush acts as the secondary coil. Every time you slide your toothbrush into the base, you couple these coils, allowing current to be induced in the toothbrush. The induced current recharges the toothbrush battery and also makes the handle of the toothbrush feel warm. So, the next time you brush your teeth, think of Michael Faraday, whose discovery of electromagnetic induction made the operation of your toothbrush possible.

iStockphoto/Thinkstock

Figure 24-29 An electric toothbrush uses an inductive charger, which operates like a transformer.

EXAMPLE 24-6

Calculating the Mutual Inductance of Two Coupled Solenoids

Two long, thin solenoids of equal length l and equal radius r are wound over top of each other (Figure 24-28). The inner and outer solenoids have N_1 and N_2 turns, respectively. An alternating electric current $I_1 = (I_1)_{max} \sin(\omega t)$ flows through the inner solenoid, where $(I_1)_{max}$ represents the maximum value of the current in the inner solenoid.

(a) Calculate the mutual inductance of the solenoids, and prove that it satisfies Equations (24-36) and (24-38).
(b) Find an expression for the emf induced in the outer solenoid.

SOLUTION

(a) Using Equation (24-31) for the inductance of a solenoid and substituting it into Equation (24-36) for the mutual inductance of two solenoids gives

$$M = \mu_0 A \frac{N_1 N_2}{l} = \sqrt{\left(\mu_0 A N_1^2 \frac{N_1}{l}\right)\left(\mu_0 N_2 A \frac{N_2}{l}\right)} = \sqrt{L_1 L_2}$$

In this derivation, we assumed that the cross-sectional area of each solenoid is A. Since the outer solenoid surrounds the inner solenoid, all the magnetic flux produced by the inner solenoid passes through the outer solenoid. Therefore $k = 1$, and the solution satisfies Equation (24-38).

(b) To find the emf induced in the outer solenoid, we use Faraday's law (Equation (24-33)) for a solenoid, taking into account that the cross-sectional area of a solenoid is πr^2:

$$(\varepsilon_{ind})_2 = -M \frac{dI_1}{dt} = -M \frac{d}{dt}(I_1)_{max} \cos(\omega t)$$

$$= (I_1)_{max} \omega \mu_0 \pi r^2 \frac{N_1 N_2}{l} \sin(\omega t)$$

Making sense of the result:

As expected, a sinusoidal emf is generated in the outer coil because it is being induced by a sinusoidal current, $I = (I_1)_{max} \sin(\omega t)$.

LO 7

24-7 RL Circuits

We now consider a circuit in which an inductor and a resistor are connected in series to a battery (Figure 24-30). Since an ideal inductor influences the circuit behaviour only when the current through it varies in time, the inductor shown in Figure 24-30 affects the current only for a relatively short time immediately after the switch is

turned on or off. (See the Online Activity box on page 660 for simulations of such transient currents.)

Let us assume that the switch is being flipped to position a (from a neutral position, N) at time $t = 0$. Since all the circuit components are connected in series, the law of energy conservation and Kirchhoff's second law require that

$$\varepsilon + V_R + V_L = 0 \Rightarrow \varepsilon - IR + \varepsilon_{ind} = 0 \qquad (24\text{-}39)$$

where

$$V_R = -IR \quad \text{and} \quad V_L = \varepsilon_{ind} = -L \frac{dI}{dt}$$

These expressions can be combined:

$$\varepsilon - IR - L \frac{dI}{dt} = 0$$

$$\frac{dI}{dt} = \frac{\varepsilon - IR}{L} \qquad (24\text{-}40)$$

To solve this differential equation, we set $x = \varepsilon - IR$. Then $dx = d(\varepsilon - IR) = -R dI$, and Equation (24-40) can be rewritten as

$$x + \frac{L}{R} \frac{dx}{dt} = 0 \Rightarrow \frac{dx}{x} = -\frac{R}{L} dt \qquad (24\text{-}41)$$

We solve this equation by integrating both sides. Since we are interested in knowing the behaviour of the current from time $t_i = 0$ to $t_f = t$, we choose these values as the limits of integration. The corresponding limits of integration for the variable x are denoted as x_i and x:

Figure 24-30 A series RL circuit consists of at least a resistor (R) and an inductor (L). Such a circuit might (when the switch is in position a) or might not (when the switch is in position b) include a battery. The ammeter registers electric current flowing through the resistor and an inductor, while the voltmeters (V_L and V_R) measure potential differences across the inductor and the resistor respectively.

$$\int_{x_i}^{x}\frac{dx}{x} = -\int_{0}^{t}\frac{R}{L}dt \;\Rightarrow\; \int_{x_i}^{x}\frac{dx}{x} = -\frac{R}{L}\int_{0}^{t}dt$$

$$\ln\frac{x}{x_i} = -\frac{R}{L}t \;\Rightarrow\; \frac{x}{x_i} = e^{-\frac{R}{L}t} \qquad (24\text{-}42)$$

Let us verify that Equation (24-42) has the proper units. The exponent should be a pure number, so the units of the ratio L/R should be time (seconds). Substituting for the inductance gives

$$\left[\frac{L}{R}\right] = \frac{\dfrac{\text{V}\cdot\text{s}}{\text{A}}}{\Omega} = \frac{\Omega\cdot\text{s}}{\Omega} = \text{s}$$

Using $x = \varepsilon - IR$ and taking into account that the current at $t = 0$ is zero, Equation (24-42) can be rewritten as

$$x = x_i e^{-\frac{R}{L}t} \;\Rightarrow\; \varepsilon - IR = (\varepsilon - I_iR)e^{-\frac{R}{L}t}$$

$$\varepsilon - IR = \varepsilon e^{-\frac{R}{L}t}$$

$$\varepsilon\left(1 - e^{-\frac{R}{L}t}\right) = IR$$

$$I(t) = \frac{\varepsilon}{R}\left(1 - e^{-\frac{R}{L}t}\right)$$

The ratio L/R has the units of time and is called the **time constant**, τ, of the RL circuit. Substituting for L/R in the above expression gives

$$I(t) = \frac{\varepsilon}{R}\left(1 - e^{-\frac{t}{\frac{L}{R}}}\right) = \frac{\varepsilon}{R}\left(1 - e^{-\frac{t}{\tau}}\right)$$

where $\tau = L/R$.

Figure 24-31(a) shows a graph of $I(t)$ versus time.

The $I(t)$ graph starts from zero, then the current increases exponentially with time, asymptotically approaching its maximum value of $\dfrac{\varepsilon}{R}$. This is the value of the current that would have been observed if there were no inductor in the circuit. This makes sense because as the time increases, the rate of change of current diminishes (the current approaches the steady state as the $I(t)$ slope becomes more horizontal); as a result, ε_L approaches zero. Another important observation is that at time $t = \tau$, the current reaches approximately 63% of its maximum value:

$$I(\tau) = \frac{\varepsilon}{R}\left(1 - e^{-\frac{\tau}{\tau}}\right) = \frac{\varepsilon}{R}(1 - e^{-1}) \approx \frac{\varepsilon}{R}(1 - 0.368)$$

$$= 0.632\frac{\varepsilon}{R} \qquad (24\text{-}43)$$

To examine how the potential difference across the inductor ε_L depends on time, we find the derivative of current with respect to time $\dfrac{dI(t)}{dt}$ and again use Equation (24-27):

(a)

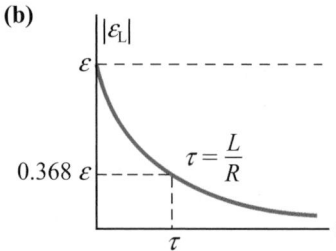

(b)

Figure 24-31 (a) The dependence of the electric current through the series RL circuit on time when the switch is flipped from position N to position a. (b) The dependence of the induced emf across the inductor on time when the closed switch is flipped from position N to position a.

$$\varepsilon_L = -L\frac{dI}{dt} = -L\frac{d}{dt}\left(\frac{\varepsilon}{R}\left(1 - e^{-\frac{t}{\tau}}\right)\right) = -\frac{L\varepsilon}{R\tau}e^{-\frac{t}{\tau}}$$

$$= -\frac{L\varepsilon}{R\dfrac{L}{R}}e^{-\frac{t}{\tau}} = -\varepsilon e^{-\frac{t}{\tau}}$$

Combining the results of the above equations, we obtain

$$I(t) = \frac{\varepsilon}{R}\left(1 - e^{-\frac{t}{\tau}}\right)$$

KEY EQUATION

$$\varepsilon_L = -\varepsilon e^{-\frac{t}{\tau}} \qquad (24\text{-}44)$$

where $\tau = L/R$.

From a graph of $\varepsilon_L(t)$, we see that at $t = 0$, $\varepsilon_L(0) = \varepsilon$, and $\varepsilon_L(t)$ diminishes exponentially with time (Figure 24-34(b)). Therefore, at $t = 0$, the emf across the inductor equals the value of emf supplied by the battery. Then, after τ seconds have passed, the value of ε_L is approximately equal to 37% of ε:

$$\varepsilon(\tau) = \varepsilon e^{-\frac{\tau}{\tau}} = \varepsilon e^{-1} \approx 0.368\varepsilon \qquad (24\text{-}45)$$

Both graphs are asymptotic: it takes an infinitely long time for the current to reach its steady-state value of ε/R and for the voltage across the inductor ε_L to reach zero. However, the time constants for most practical inductive circuits are short enough that the difference between the actual value and the final steady-state value is negligible after a few seconds.

We can use the same method to determine what happens to the circuit when the switch being moved to position b, after it has been in position a for a long time.

C-24-8 Exploring RL Circuits

How long does it take for the current in the RL circuit in Figure 24-30 to reach 74% of its maximum value (Figure 24-31(a))?

(a) τ

(b) 1.34τ

(c) 1.50τ

(d) 2.00τ

(e) 2.67τ

C-24-8 (b) Use Equation (24-44) to check the answer.

(a)

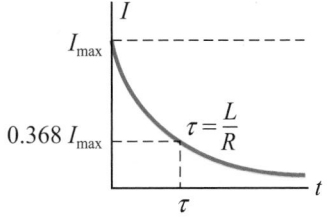

Taking the instant when the switch is opened as $t = 0$, we have $I(0) = I_{max}$ and $I(\infty) = 0$. Then the expressions for the current in the circuit and the emf across the inductor can be rewritten as follows:

$$I(t) = \frac{\varepsilon}{R}e^{-\frac{t}{\tau}}$$

$$\varepsilon_L = \varepsilon e^{-\frac{t}{\tau}} \qquad (24\text{-}46)$$

Figure 24-32 shows how $I(t)$ and $\varepsilon_L(t)$ change immediately after the switch in a series RL circuit is open.

The time constant of the circuit τ determines how fast the current in the circuit reaches its steady state when the

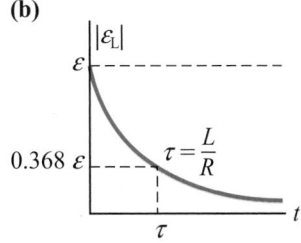
(b)

Figure 24-32 A switch in a series RL circuit is being moved from position a to position b. (a) The dependence of the electric current through this circuit on time, $I(t)$. (b) The dependence of the potential difference across the inductor on time, $\varepsilon_{ind}(t)$.

EXAMPLE 24-7

Exploring RL Circuits

In the RL circuit shown in Figure 24-33, the battery has an emf of 20 V, the resistor has a resistance of 10 Ω, and the inductor has an inductance of 40 mH.

(a) How long after the switch is set to position a does it take the current to reach 99% of its maximum value? Express your answer in seconds and in terms of the time constant.

(b) What is the current at this time?

(c) When the current reaches 99% of its maximum value, what is the emf across the inductor?

(d) After 20 s, the switch is moved to position b. How long will it take for the current to decrease to 10% of its maximum value?

Figure 24-33 An RL circuit

SOLUTION

(a) To find how long it will take the current to reach 99% of its maximum value, we use Equation (24-44):

$$I(t) = \frac{\varepsilon}{R}\left(1 - e^{-\frac{t}{\tau}}\right)$$

$$0.99\frac{\varepsilon}{R} = \frac{\varepsilon}{R}\left(1 - e^{-\frac{t}{\tau}}\right)$$

$$e^{-\frac{t}{\tau}} = 0.01 \Rightarrow t = -\tau\ln(0.01)$$

$$t = 4.6\,\tau = 4.6\frac{L}{R} = 4.6\frac{40 \times 10^{-3}\,\text{H}}{10\,\Omega} = 4.6(4\,\text{ms}) = 18.4\,\text{ms}$$

The current will reach 99% of its maximum value after 4.6τ or 18.4 ms.

(b) The maximum value of the current is

$$I_{max} = \frac{\varepsilon}{R} = \frac{20\,\text{V}}{10\,\Omega} = 2\,\text{A}$$

and 99% of this value is 1.98 A.

(c) When the current reaches 99% of its maximum value, the emf across the inductor can be calculated using Equation (24-46):

$$\varepsilon_L = \varepsilon e^{-\frac{t}{\tau}} = (20\,\text{V})e^{-\frac{18.4\,\text{ms}}{4\,\text{ms}}} = (20\,\text{V})e^{-4.6} = 0.20\,\text{V}$$

(continued)

(d) The switch is moved to position b after 20 s. Since this time interval is over 4000 times longer than the circuit time constant (4.6 ms), we can assume that at the time the switch was moved to position b, the current has reached its maximum value. Let t_{10} represent the time when the current reaches 10% of its initial value. Then, $I(t_{10}) = 0.1\,I_{max} = 0.1\dfrac{\varepsilon}{R}$. Substituting into Equation (24-46) gives

$$0.1\,I_{max} = \frac{\varepsilon}{R}e^{-\frac{t}{\tau}} \Rightarrow 0.1\frac{\varepsilon}{R} = \frac{\varepsilon}{R}e^{-\frac{t}{\tau}}$$

$$0.1 = e^{-\frac{t}{\tau}} \Rightarrow t = -\tau\ln(0.1) = 2.3\,\tau = 2.3(4.6\text{ ms}) = 11\text{ ms}$$

Making sense of the result:

Due to the presence of an inductor in an electric circuit, the current cannot reach its maximum value or diminish to zero instantaneously.

circuit is being connected to a battery and how quickly it diminishes to zero when a battery is being switched off. Since the time constant is directly proportional to L and inversely proportional to R $\left(\tau = \dfrac{L}{R}\right)$, when $\dfrac{L}{R} \gg 1$ the inductor dominates the circuit behaviour (inductive circuit), and when $\dfrac{L}{R} \ll 1$, the resistance dominates the circuit behaviour (resistive circuit).

LO 6, LO 8

24-8 Energy Stored in a Magnetic Field

Example 24-7 demonstrated that the current in an RL circuit continues to flow for some time after the battery is disconnected. This tells us that an inductor is supplying energy to the circuit, somewhat like a battery. In fact, the energy is stored in the magnetic field of the inductor.

Let us once again consider the RL circuit shown in Figure 24-36. When the switch is in position a, the law of energy conservation demands that Equation (24-40) holds true. Multiplying both sides of this equation by I, we obtain

$$\varepsilon I - I^2R - LI\frac{dI}{dt} = 0$$

$$\varepsilon I = I^2R + LI\frac{dI}{dt} \tag{24-47}$$

This equation describes the law of energy conservation for an RL circuit. The εI component represents the rate at which the energy is supplied by the battery to the circuit, that is, the electric power delivered to circuit. Some of this energy is delivered to the resistor at the rate of I^2R, and the rest is delivered to the inductor at the rate of $LI\dfrac{dI}{dt}$. Since the latter expression represents the rate of energy transfer to the inductor, we define the energy stored in the magnetic field, U_B, as

$$\frac{dU_B}{dt} = LI\frac{dI}{dt}$$

$$dU_B = LIdI \tag{24-48}$$

Knowing the rate of energy transfer to the inductor, we calculate the energy stored in the inductor as the current in the circuit changes from zero to its maximum (steady-state) value:

$$\int_0^{U_B} dU_B = \int_0^I LIdI$$

 KEY EQUATION
$$U_B = \frac{LI^2}{2} \tag{24-49}$$

This important relationship indicates that the energy stored in the inductor is proportional to the inductance L and to the square of the current flowing through the inductor. This energy is stored in the magnetic field of the inductor when current flows through it. Notice that U_B is present only when charge is moving through the inductor. This magnetic energy expression is somewhat analogous to the kinetic energy, $KE = \dfrac{mv^2}{2}$, and to the potential energy stored in the electric field of a capacitor, $U_E = \dfrac{CV^2}{2}$ (Equation (21-15)).

Let us now examine Faraday's law of electromagnetic induction from the energy perspective. When electric current starts flowing through a solenoid, the current generates a magnetic field inside the solenoid. We determined (Equation (24-49)) that magnetic energy is stored within this magnetic field. This energy must have come from somewhere. As the current builds up in the circuit, the inductor acts to oppose that current increase by converting kinetic energy carried by the moving electric charges into magnetic energy stored in the inductor's magnetic field.

When the switch disconnects the battery, moving into position b (Figure 24-33), the current in the solenoid, already at its maximum value, starts to drop off. The energy stored in the inductor also begins to drop off according to Equation (24-49). Since the inductor acts to oppose the resulting decrease in current, the energy stored in the magnetic field is used to reduce the rate of decrease of electric current. The inductor uses the energy stored in its magnetic field to prevent the abrupt decrease of the current in the circuit. Therefore, the current induced by the inductor is directed such that the induced emf has the same direction as the emf from the original battery. As when the circuit was first switched on, the inductor opposes the *change* in electric current.

Let us calculate the energy stored in the magnetic field of a long solenoid with loop density n, radius R, and length L:

$$U_B = \frac{LI^2}{2} = \frac{1}{2}(\mu_0 n^2 lA)I^2 = \frac{1}{2}\mu_0 n^2 l(\pi R^2)\left(\frac{B}{\mu_0 n}\right)^2 = \frac{B^2}{2\mu_0}l(\pi R^2)$$

$$U_B = \frac{B^2}{2\mu_0}V \qquad (24\text{-}50)$$

where $V = \pi R^2 l$ is the volume of a solenoid.

Equation (24-50) shows that the energy stored in a solenoid is determined by its volume and the strength of the magnetic field inside it. For many practical applications, it is useful to know the energy density (amount of energy per unit volume):

KEY EQUATION
$$u_B = \frac{B^2}{2\mu_0} \qquad (24\text{-}51)$$

Although we derived Equation (24-51) for a particular type of solenoid, the equation is valid for any region of space that contains a magnetic field. Compare this equation to the equation describing the energy density of the electric field, $u = \frac{1}{2}\varepsilon_0 E^2$ (Equation (21-18)). Both types of energy density depend solely on the square of the magnitude of the respective field. We will apply this crucial observation when discussing the transfer of energy by electromagnetic waves in Chapter 25.

It will be instructive to compare what you have learned about RL circuits in this chapter with what you learned about RC circuits in Chapter 22, focusing on the functions of the resistors, capacitors, and inductors in the circuits. We will discuss circuits combining resistors, capacitors, and inductors in Chapter 25 in detail.

MAKING CONNECTIONS

Maglev Trains

There are three major technologies currently in use for magnetic levitation in maglev trains. The electromagnetic system (EMS) uses the attractive force between a support magnet located under the vehicle and a rail that is made of a ferromagnetic material. The attractive force between the steel rail and the supporting magnet (Figure 24-34(a)) lifts the vehicle upward. EMS technology is used in German maglev trains. The distance between the electromagnets and the rail is critical, so this system requires a control mechanism that continuously measures this distance and adjusts it by varying the current in the guidance magnet.

The electrodynamic system (EDS) is used in the maglev trains in Japan (Figure 24-34(b)). These trains carry magnets, and a large metal plate runs along the centre of the maglev track. When a train's magnet passes over the plate, eddy currents are created in the metal plate, generating a repulsive force between the track and train. The EDS maglev trains are naturally stable: if the train drops toward the rail, the repulsive force increases; if the train rises too high above the rail, the repulsive force decreases. However, the electromagnetic repulsion is present only when the vehicle is moving, and the vehicle must have wheels for takeoff and landing (unlike EMS trains). Moreover, the electric currents induced in the metal plates dissipate energy and create a significant drag force on the vehicle.

Inductrack, a variation of the EDS technology, uses permanent room-temperature magnets attached to the vehicle (instead of electromagnets or superconducting magnets) to produce a magnetic field that levitates the train over passive coils. Like EDS trains, inductrack maglev trains do not levitate until they are in motion (Figure 24-34(c)).

(a)

Electromagnetic Suspension

Electromagnets are located on the train cars, and they lift the train cars.

(b)

Electrodynamic Suspension

Electromagnets are located on the guideway, and they lift the train cars.

(c)

Inductrack

Permanent magnets levitate over passive coils.

Figure 24-34 (a) The electromagnetic suspension system is used in German maglev trains. (b) The electrodynamic suspension system is used in Japanese maglev trains. (c) Inductrack uses permanent magnets to levitate the train over passive coils.

FUNDAMENTAL CONCEPTS AND RELATIONSHIPS

A time-varying magnetic flux through a conducting loop induces an emf in it. To produce a magnetic flux that changes with time, we must either have a magnetic field that changes with time or a constant magnetic field but a loop whose effective area changes with time. The latter can be achieved, for example, by rotating the loop in the constant magnetic field.

Magnetic Flux Through a Surface

Magnetic flux is a scalar quantity (can be positive, negative, or zero) and is defined as an integral over the surface:

$$\Phi_B = \int \vec{B} \cdot d\vec{A} = \int d\Phi \qquad (24\text{-}7)$$

Magnetic flux changes with time when the magnetic field, the area of the loop, and/or the angle between the normal to the loop and the magnetic field vary in time (Figure 24-3).

Faraday's Law of Electromagnetic Induction

A time-varying magnetic flux through a conducting loop induces an emf in it:

$$\varepsilon = -\left(\frac{d\Phi_B}{dt}\right) \qquad (24\text{-}10)$$

Lenz's law: The directions of the induced emf and induced current always oppose (reduce) the change of the magnetic flux that caused it.

Energy Stored in a Magnetic Field

1. The energy stored in a magnetic field can be expressed as

$$U_B = \frac{LI^2}{2} \qquad (24\text{-}49)$$

2. The energy stored in a magnetic field per unit volume is

$$u_B = \frac{B^2}{2\mu_0} \qquad (24\text{-}51)$$

APPLICATIONS

Applications: AC current generator, AC circuits (RL, RC, and RLC circuits), electromagnetic brakes and dampers, transformers, metal detectors, electric guitars, magnetic levitation trains

Key Terms: alternating current, alternating emf, back emf, coefficient of mutual inductance, conservative fields, coupled inductors, eddies, eddy currents, effective surface area, electromagnetic force, electromagnetic induction, electrostatic force, Faraday's law of electromagnetic induction, induced electric field, inductance, inductive charging, inductors, Lenz's law, magnetic flux, motional emf, mutual inductance, nonconservative fields, rate of change of the magnetic flux, self-induction, series RL circuit, transformer

QUESTIONS

1. A rectangular metal loop is located in the region of a uniform magnetic field directed perpendicular to the plane of the loop (Figure 24-35) An induced electric current in the clockwise direction is detected in the loop. What must be true about this magnetic field?
   (a) The magnetic field is directed into the page and is steady.
   (b) The magnetic field is directed into the page and is increasing.
   (c) The magnetic field is directed into the page and is decreasing.
   (d) The magnetic field is directed out of the page and is steady.
   (e) The magnetic field is directed out of the page and is increasing.
   (f) The magnetic field is directed out of the page and is decreasing.
   (g) Only (a) and (b) are possible correct answers.

   (h) Only (c) and (e) are possible correct answers.
   (i) Only (d) and (f) are possible correct answers.
   (j) All the answers can be correct.

Figure 24-35 Question 1

2. Which of the following descriptions accurately describes a magnetic field that can produce a constantly increasing emf in a loop oriented perpendicular to its direction?
   (a) $B(t) = (3t + 278)$ mT
   (b) $B(t) = (5\sin(2\pi t))$ mT
   (c) $B(t) = (5\sin(3\pi t) + 10)$ mT
   (d) $B(t) = (5\cos(3\pi t) + 30)$ mT
   (e) $B(t) = (5t^2 + 15)$ mT

3. Which of the following scenarios describes a magnetic flux through a loop that does not change with time?
   (a) A uniform magnetic field changes with time, and the area of the loop and its orientation remain constant.
   (b) A loop is moving with constant velocity through a non-uniform and constant magnetic field.
   (c) A square loop is rotating about its diagonal in a uniform and constant magnetic field directed perpendicular to the loop's diagonal.
   (d) A loop is moving at a constant velocity (its orientation is not changing) in a uniform and constant magnetic field.
   (e) The area of a loop is increasing at a constant rate while it is in a uniform magnetic field whose magnitude is also increasing and whose direction is always parallel to the surface of the loop.

4. In which one of the following cases will there be an electric current induced in the ring in Figure 24-36?
   (a) The ring is moving in a horizontal direction in a uniform and constant magnetic field.
   (b) The ring is moving vertically in a uniform and constant magnetic field.
   (c) The ring is stationary, but the magnetic field is increasing.
   (d) The ring is stationary, but the magnetic field is decreasing.
   (e) The ring is spinning about a horizontal axis parallel to the B lines and going through its centre.
   (f) The ring is spinning about a horizontal axis perpendicular to the B lines and going through its centre.
   (g) The ring is spinning about a vertical axis perpendicular to the B lines and going through its centre.
   (h) The ring is moving diagonally (without spinning) at a constant velocity.
   (i) The ring is moving diagonally (without spinning) but at an increasing speed.
   (j) parts (b), (c), (d), (e), (f), (h), and (i)
   (k) parts (c), (d), (f), (g), and (i)
   (l) parts (c), (d), (f), and (g)
   (m) parts (a), (c), (d), (f), (g), and (i)
   (n) All of the above.

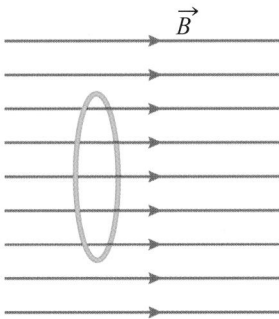

Figure 24-36   Questions 4 and 5

5. What will the direction of the current in the metal ring in Figure 24-36 be if the magnetic field starts increasing?
   (a) counterclockwise (if seen from the left)
   (b) clockwise (if seen from the left)
   (c) There will be no current generated in the ring, and the ring is stationary.

6. A metal ring is dropped between the poles of a horseshoe magnet (Figure 24-37). What is true about the electric current in the ring as observed if looking from the right?
   (a) The electric current is clockwise while entering the loop and counterclockwise while exiting.
   (b) The electric current is counterclockwise when entering the loop and clockwise while exiting.
   (c) The electric current is counterclockwise all the time.
   (d) The electric current is clockwise all the time.
   (e) There is no current in the ring.

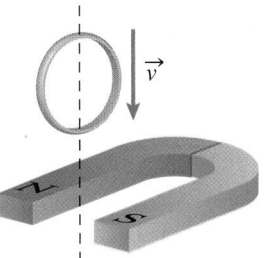

Figure 24-37   Question 6

7. A very light, 1 m-long metal rod is placed in a uniform and constant horizontal magnetic field $B = 1$ T, directed as shown in Figure 24-38. The rod is free to move along frictionless metal rails that have negligibly small resistance. A light bulb, requiring a potential difference of 10 V to light up, is attached to the metal rails as shown. What should the minimal speed of the rod be for the bulb to light up?
   (a) 0.01 m/s
   (b) 0.1 m/s
   (c) 1 m/s
   (d) 10 m/s
   (e) 100 m/s

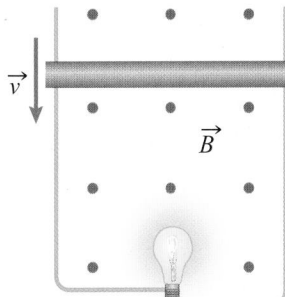

Figure 24-38   Question 7

8. Three light rectangular blocks with the same shape and approximately the same mass are placed on top of an aluminum inclined plane relatively far from each other (Figure 24-39). Block A is made of wood, block B is made of a magnetic material, and block C is made of copper. The coefficients of friction between the blocks and the incline are negligible. The blocks, initially at rest, are let go from the top of the incline. Which of the following statements most accurately describes what happens to the blocks?

ELECTRICITY, MAGNETISM, AND OPTICS

(a) The copper block will come down last because copper is attracted to aluminum.

(b) The magnetic block will come down last because, during its motion, eddy currents will be created in the incline, which will slow it down significantly.

(c) The magnetic block will come down first because, during its motion, eddy currents will be created in the incline, speeding it up.

(d) The wooden block will come down last because wood will experience significant friction that will slow it down.

(e) Since friction is negligible, all the blocks will have the same acceleration. As a result, they will reach the bottom simultaneously.

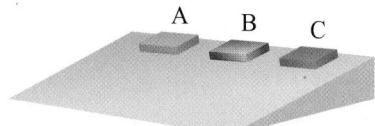

Figure 24-39   Question 8

9. A step-up transformer has $N_1 = 25$ loops and $N_2 = 25\ 000$ loops. The output voltage, V_2, has to be 120 000 V, and the output I_2 has to be 10 mA. What are the input V_1 and I_1?
   (a) $V_1 = 12$ V; $I_1 = 10$ mA
   (b) $V_1 = 12$ V; $I_1 = 1$ A
   (c) $V_1 = 12$ V; $I_1 = 10$ A
   (d) $V_1 = 120$ V; $I_1 = 10$ A
   (e) $V_1 = 120$ V; $I_1 = 1$ A
   (f) $V_1 = 120$ V; $I_1 = 10$ mA
   (h) $V_1 = 120$ V; $I_1 = 0.01$ mA

10. A current through an inductor starts increasing at a steady rate as shown in Figure 24-40. Which of the following is true about the potentials at points a and b?
   (a) $V_a = V_b$
   (b) $V_a < V_b$
   (c) $V_a > V_b$

Figure 24-40   Question 10

11. A circular copper loop is moving away from a long, straight wire-carrying current as shown in Figure 24-41. Which of the following statements correctly describes what is happening to the loop as a result of its motion?
   (a) A circular loop is attracted to the current-carrying wire because copper is magnetic and it is attracted to the current.
   (b) A current in the clockwise direction is induced in the loop, which makes the loop attracted to the current-carrying wire.
   (c) A current in the counterclockwise direction is induced in the loop, which makes the loop attracted to the current-carrying wire.

(d) A current in the clockwise direction is induced in the loop, which makes the loop repelled by the current-carrying wire.

(e) A current in the counterclockwise direction is induced in the loop, which makes the loop repelled by the current-carrying wire.

(f) There are no forces of attraction or repulsion between the loop and the wire because copper is a nonmagnetic material.

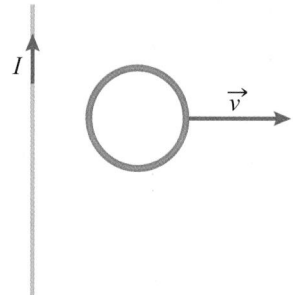

Figure 24-41   Question 11

12. A circular copper loop is moving along a long straight wire carrying current as shown in Figure 24-42. The speed of the loop is steadily increasing. Which of the following statements correctly describes what is happening to the loop as a result of its motion?
   (a) A counterclockwise current is induced in the loop, which causes the loop to be attracted to the current-carrying wire.
   (b) A clockwise current is induced in the loop, which makes the loop repelled by the current-carrying wire.
   (c) A current in the counterclockwise direction is induced in the loop, which makes the loop repelled by the current-carrying wire.
   (d) A current in the clockwise direction is induced in the loop, which makes the loop attracted by the current-carrying wire.
   (e) There are no forces of attraction or repulsion between the loop and the wire because the magnetic flux is constant.

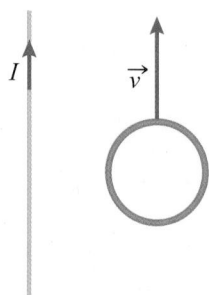

Figure 24-42   Question 12

13. A metal ring is moved over a permanent magnet, as shown in Figure 24-43(a). Which of the following statements correctly describes what happens to the ring and the direction of the current as seen from above?
   (a) A clockwise current is induced in the ring before it reaches the middle of the magnet, then the induced current flows in the counterclockwise direction.

(b) A counterclockwise current is induced in the ring before it reaches the middle of the magnet, then the current is induced in the clockwise direction.

(c) A clockwise current is induced in the ring only when the magnet is inside the ring.

(d) A counterclockwise current is induced in the ring before it reaches the magnet, then the current stops.

(e) A current is induced in the ring in the clockwise direction only when the ring leaves the magnet. (The magnet is no longer inside the ring.)

Figure 24-43 Questions 13 and 14

14. A metal ring is located near a permanent magnet, as shown in Figure 24-43(b). The ring is moved away from the magnet. Which of the following statements correctly describes the direction of the current in the ring as seen from above?

(a) There is an induced current in the clockwise direction.

(b) There is an induced current in the counterclockwise direction.

(c) There is no current induced in the ring.

15. Rank the inductances of the five solenoids shown in Figure 24-44 from the largest to the smallest.

Largest 1_____ 2_____ 3_____ 4_____ 5_____ Smallest

Or, the inductances are the same in all cases. _____

If the inductance is the same for two or more cases, clearly indicate it on the ranking scheme. Explain your reasoning.

Figure 24-44 Questions 15, 16, and 17

16. The same electric current flows through the five solenoids shown in Figure 24-44. Rank the energy stored inside the solenoids from largest to smallest if a 1 A electric current flows through each solenoid.

Largest 1_____ 2_____ 3_____ 4_____ 5_____ Smallest

Or, the stored energy is the same in all cases. _____

If the energy is the same for two or more cases, clearly indicate it on the ranking scheme. Explain your reasoning.

17. Rank the energy density inside the five solenoids in Figure 24-44 from largest to smallest when a 1 A electric current flows through each solenoid.

Largest 1_____ 2_____ 3_____ 4_____ 5_____ Smallest

Or, the energy density is the same in all cases. _____

If the energy density is the same for two or more cases, clearly indicate it on the ranking scheme. Explain your reasoning.

18. Two copper coils are facing each other as shown in Figure 24-45. Coil 1 is connected to a battery, so a constant current I_1 flows through it. At time $t = 0$, coil 2 starts moving to the right with a velocity $\vec{v}$. Which of the following statements correctly describes what happens as soon as coil 2 starts moving away from coil 1?

(a) There is an induced current in coil 2 in the clockwise direction if looking from the right.

(b) There is an induced current in coil 2 in the counterclockwise direction if looking from the right.

(c) There is no current induced in coil 2 because current I_1 is constant.

(d) All the above are possible, depending on the velocity of coil 2 relative to coil 1.

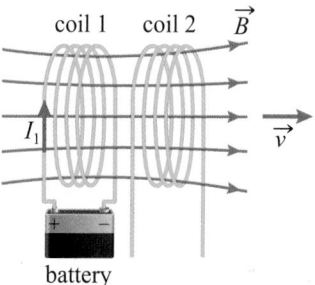

Figure 24-45 Question 18

19. Five series RL circuits (like the RL circuits in Figure 24-30) have different resistors, inductors, and batteries. The information for each circuit is shown in Table 24-2. The circuits were initially off; at time $t = 0$ s, they were turned on. Rank the times it takes for the current in each circuit to reach 75% of its maximum value from largest to smallest.

Largest 1_____ 2_____ 3_____ 4_____ 5_____ Smallest

Or, the time is the same for each circuit. _____

If the time is the same for two or more circuits, clearly indicate that on the ranking scheme. Explain your reasoning.

Table 24-2 Data for Question 19

	R	L	ε
Circuit 1	10 Ω	10 mH	10 V
Circuit 2	20 Ω	20 mH	10 V
Circuit 3	20 Ω	20 mH	20 V
Circuit 4	50 Ω	30 mH	25 V
Circuit 5	50 Ω	30 mH	30 V

20. Five different series RL circuits (like the RL circuits in Figure 24-30) have the same batteries (20 V each), the same resistors (20 Ω each), but different inductors: $L_1 = 10$ mH, $L_2 = 20$ mH, $L_3 = 30$ mH, $L_4 = 40$ mH, and $L_5 = 50$ mH. The circuits are initially turned off, then turned on. What is true about the values of the maximum current in these circuits?

(a) $I_1 > I_2 > I_3 > I_4 > I_5$
(b) $I_1 < I_2 < I_3 < I_4 < I_5$
(c) $I_1 = I_2 = I_3 = I_4 = I_5$

21. The light bulb in Figure 24-46 is a part of a series circuit powered by an AC source and containing an air-filled inductor. When an iron bar is moved inside the inductor's core, what happens to the brightness of the light bulb? Explain.

(a) The brightness remains the same.
(b) The brightness increases.
(c) The brightness decreases.

Figure 24-46 Question 21

PROBLEMS BY SECTION

For problems, star ratings will be used, (✷, ✷✷, or ✷✷✷), with more stars meaning more challenging problems.

Sections 24-1 and 24-2   In Faraday's Lab and Magnetic Flux and Its Rate of Change

22. ✷ A circular loop of area $A = 3$ cm² is in a region where there is a uniform and steadily decreasing magnetic field. The direction of the magnetic field is at a 30° angle to the loop's axis of symmetry. At time $t = 2$ s, the magnetic field has a magnitude of 10 mT, and at time $t = 6$ s, the magnetic field has a magnitude of 2 mT. Find the rate of change of the magnetic flux through the loop.

23. ✷ A rectangular copper loop with the dimensions $l = 50$ cm and $w = 25$ cm is moving at a constant velocity of 1.5 m/s out of a uniform magnetic field directed into the page. The magnetic field varies with time as $B = (0.002t + 0.001)$ T.

(a) Find the rate of change of the magnetic flux through this loop at the instant when the loop is three-quarters of its length out of the magnetic field.
(b) Answer part (a) when the loop has moved completely out of the magnetic field.
(c) How will your answers to parts (a) and (b) change if, instead of a single loop, there is a coil consisting of 20 loops?

24. ✷ A square copper loop 5 cm × 5 cm is located in an area of a uniform, time-varying magnetic field. The loop is directed perpendicular to the magnetic field. The magnetic field varies with time, as shown in Figure 24-47. Find the emf induced in the loop during the time intervals I, II, and III.

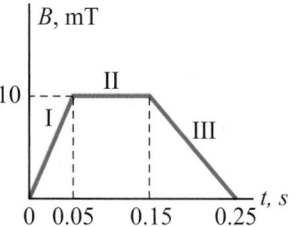

Figure 24-47   Problem 24

25. ✷ Two long, concentric (coaxial) copper solenoids have the same length $l = 50$ cm. Their radii are $R_1 = 2.0$ cm and $R_2 = 3.0$ cm, and their numbers of turns are 100 and 200, respectively. When the electric current in the inner solenoid changes at a rate of 0.5 A/s, what is the rate of change of the magnetic field through the outer solenoid?

26. ✷✷ A circular coil with 10 loops and the radius of 5 cm rotates about its diameter in a uniform and constant magnetic field with a magnitude of 10 mT. The axis of rotation of a coil is always perpendicular to the magnetic field. When the angular velocity of rotation is 4 rpm, what is the maximum rate of change of the magnetic flux through the coil?

Section 24-3   Faraday's Law of Electromagnetic Induction

27. ✷ The magnetic flux through a solenoid with 100 turns changes from 6×10^{-2} Wb to 8×10^{-4} Wb in 0.1 s. Find the induced emf in the coil.

28. ✷ A rectangular metal loop with the dimensions 5 cm × 10 cm is inserted in a uniform but changing magnetic field. The resistance of the loop is 3 Ω. Find the rate of change of the magnetic field so that 5 A of electric current is induced in the loop when

(a) the magnetic field is perpendicular to the surface of the loop;
(b) the magnetic field is directed at a 30° angle to the loop's normal;
(c) the magnetic field is directed parallel to the loop's surface.

29. ✷ To find the magnitude of the magnetic field between the poles of a large horseshoe magnet, a coil of wire consisting of 20 turns and with a cross-sectional area of 6 cm² is inserted between the poles of the magnet perpendicular to the magnetic field lines. When the coil is pulled out of the magnetic field in 0.02 s, an induced emf of 45 mV was measured in the coil. What is the magnitude of the magnetic field between the poles of the magnet?

30. ✷✷ A rectangular 4 cm × 8 cm metal coil has 50 turns and a resistance of 10 Ω. The coil is placed on a horizontal frictionless table and is completely submerged in a vertical uniform magnetic field, $B = 0.5$ T.

(a) How fast should the coil be pulled out of this uniform magnetic field so that the emf generated in the coil reaches 0.5 V? The velocity of the coil $\vec{v}$ is directed perpendicular to one of its sides (Figure 24-48).
(b) Draw a graph that shows the dependence of the induced emf on time while the coil is pulled out of the magnetic field at the constant velocity $\vec{v}$ found above.

(c) Find the direction and the magnitude of the electric current induced in the coil while it is pulled out of the magnetic field at a constant velocity $\vec{v}$.

(d) Find the force F that needs to be applied to pull the coil with this constant velocity $\vec{v}$.

(e) Draw the graph representing the dependence of F on time. Draw this graph under the graph you made for part (b) to see the correspondence.

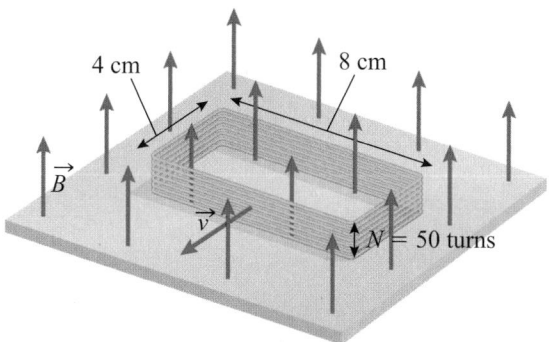

Figure 24-48   Problem 30

31. ✷✷   Another way of looking at Faraday's law of electromagnetic induction is to consider the induced charge in a wire placed in a magnetic field (due to the Lorentz force acting on moving electric charges submerged in magnetic fields). An induced charge in a coil with N loops and resistance R can be expressed as

$$\Delta Q = \frac{-N\Delta\Phi_B}{R}$$

Show that this relationship is a direct consequence of the law of the electromagnetic induction.

32. ✷✷✷   A conducting bar of length $l = 0.2$ m and mass $m = 0.1$ kg is submerged in a uniform and constant magnetic field $B = 0.5$ T (Figure 24-49). The bar is free to slide down two frictionless metal rails. At time $t = 0$ s, the bar is given an initial velocity $\vec{v}_i = 2$ m/s directed to the right. The resistance $R = 4\ \Omega$ comprises the resistance of the entire circuit.

(a) Find the velocity of the bar 1.0 s after it was pushed.

(b) Find the acceleration of the bar at $t = 1.0$ s

(c) Determine the distance travelled by the bar during the first second.

(d) Calculate the distance travelled by the bar before its speed decreases to $v_i/4$.

Figure 24-49   Problem 32

33. ✷✷   A metal rod of mass $m = 250$ g and length $L = 50$ cm is free to slide on two parallel, frictionless horizontal metal rails (Figure 24-50). The rails are connected at one end so that they

and the rod form a closed circuit. The rod and the rails have a resistance of $2\ \Omega$. A uniform magnetic field B, perpendicular to the plane of this circuit and directed vertically upward, is decreasing at a constant rate of 0.3 mT/s. At time $t = 0$ s, the magnetic field has a strength of 15 mT, and the rod is at rest at a distance of 150 cm from the connected end of the rails.

(a) Describe the forces acting on the rod at $t = 0$ s and thereafter.

(b) Derive an expression for the acceleration of the rod at $t = 0$ s in terms of the given quantities.

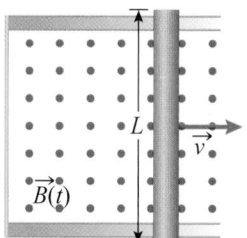

Figure 24-50   Problem 33

Section 24-4   Applications of Faraday's Law of Electromagnetic Induction

34. ✷✷   An engineering student proposes to use the law of electromagnetic induction to power aircraft equipment. The proposal uses the Boeing 757-200 as an example. The aircraft has a maximum cruising speed of 914 km/h and a wing span of 38.05 m. Suppose the aircraft is flying in the region where the vertical component of the magnetic field is 3.5×10^{-5} T. What is the potential difference generated across the wings? Is it sufficient to power the onboard equipment?

35. ✷✷   A simple AC generator, like the one described in Figure 24-14, consists of a single loop with an area of 100 cm^2 that rotates with a frequency of 60 Hz in a uniform magnetic field of 3.0×10^{-2} T.

(a) Find the maximum emf generated by this generator.

(b) Find the maximum emf that would be generated by this generator if instead of a single loop it had 100 loops.

(c) Describe mathematically how the output voltage depends on time. The output voltage is maximum when $t = 0$ s.

36. ✷✷   A horizontal copper disk 40 cm in diameter is placed in a constant and uniform magnetic field of 8.0×10^{-2} T directed vertically upward. The disk spins about a vertical axis through its centre, with an angular velocity of 10 rad/s.

(a) What is the magnitude of the emf induced between the centre of the disk and its rim?

(b) What is the magnitude of the induced emf between diametrically opposite points on the disk?

37. ✷✷   To reduce the 120 V AC voltage to 12 V AC to operate a standard doorbell, a step-down transformer is used.

(a) This transformer has 1000 turns in the primary coil. How many turns does it have in the secondary coil?

(b) When the doorbell button is pushed, 15 Ω of resistance is introduced in the secondary circuit. What are the values of the maximum and rms currents in the primary and in the secondary coils? Assume that the transformer used in the doorbell has 100% efficiency.

(c) How would your answers to part (b) change if the efficiency of the transformer were 90%?

Section 24-5 Induced emf and Induced Electric Fields

38. ✷✷ A circular loop of radius $R = 10$ cm is coaxial with a solenoid ($r = 0.05$ m, $l = 1.00$ m, and 750 turns), as shown in Figure 24-51. The current in the circuit decreases linearly from 1.0 A to 4.6 A over 0.5 s.
   (a) Calculate the induced emf in the loop.
   (b) Calculate the induced electric field in the loop.
   (c) When the resistance of the loop is 20 Ω, how much energy will be dissipated by the loop in 10 s?

Figure 24-51   Problem 38

39. ✷✷ A solenoid with length L and radius R has N turns. The solenoid carries an electric current that varies with time as $I(t) = I_{max}\sin(2\pi f t)$, where I_{max} represents the maximum current, and f is the frequency of the AC source.
   (a) Derive an expression that describes the magnitude of the electric field outside the solenoid ($r > R$).
   (b) Derive an expression that describes the magnitude of the electric field inside the solenoid ($r < R$).
   (c) For parts (a) and (b), draw graphs showing the dependence of the magnitude of the electric field on r: $E(r)$ at a certain instant of time t. Explain your results.

Section 24-6 Self-Inductance and Mutual Inductance

40. ✷ The current through an air-filled coil is changing at a rate of 200 A/s. As a result of this change, the induced emf in it is measured to be 250 V.
   (a) What is the self-inductance of the coil?
   (b) How will the value of the self-inductance change if an iron core is inserted in this coil?
   (c) What is the rate of change of the magnetic flux through this coil?

41. ✷✷ Two electric circuits are located near each other. When the electric current in the first circuit changes at a rate of 200 A/s, an induced emf of 20 V is measured in the second circuit. What is the mutual inductance of the circuits?

42. ✷✷ Two coaxial air-filled solenoids of equal length ($L = 20$ cm) and equal radius ($r = 5$ cm) have 100 turns and 400 turns.
   (a) Find the mutual inductance of these solenoids.
   (b) How will your answer to part (a) change if iron cores are inserted in the solenoids?

43. ✷✷ A 2 m-long air-filled solenoid has a radius of 5 cm and 25 000 turns. A 500-turn coil is tightly wound over the outside of the solenoid, as shown in Figure 24-52.
   (a) Find the mutual inductance of the solenoid and the coil.
   (b) The current in the solenoid is changing according to $I_{solenoid}(t) = \left(10\dfrac{\text{A}}{\text{s}}\right)t$. Find the emf induced in the coil.

Figure 24-52   Problem 43

Section 24-7 RL Circuits

44. ✷✷ A series electric circuit consists of a 24 V battery, a switch, a 20 Ω resistor, and an inductor. The inductor is a 30 cm-long solenoid that has 1000 turns and a radius of 2.5 cm. After the switch is closed, it takes 0.28 ms for the current in the circuit to reach half of its maximum value.
   (a) Find the inductance of the solenoid using two different methods.
   (b) Derive an expression for the induced emf in the solenoid during the time when the current is changing from zero to its maximum value.
   (c) Find the value of the maximum current in the circuit.
   (d) Find the value of the voltage across the resistor when the induced emf in the solenoid is at one-half of its maximum value.
   (e) The switch is left on for a long time. What is the value of the induced emf in the solenoid at $t = 0.1$ s? What is the value of the voltage across the resistor at that time?

Section 24-8 Energy Stored in a Magnetic Field

45. ✷ Find the energy stored in the solenoid described in problem 44 when the current in the circuit
   (a) is at its maximum value;
   (b) is at one-half of its maximum value;
   (c) is zero.

COMPREHENSIVE PROBLEMS

46. ✷✷ A square loop of wire with a side length of 5 cm is bent along its diagonal such that its two triangular halves are located in perpendicular planes (Figure 24-53). The loop is then submerged in a uniform magnetic field B, directed perpendicular to the diagonal of the square loop and at a 45° angle to the planes of the bent square. The resistance of the loop is 3 Ω, and the magnetic field in the region changes with time according to $B(t) = (50t + 3)$ mT. Find the magnitude and the direction of the induced emf and the induced electric current in the loop during the time interval 0 s < $t < 5$ s.

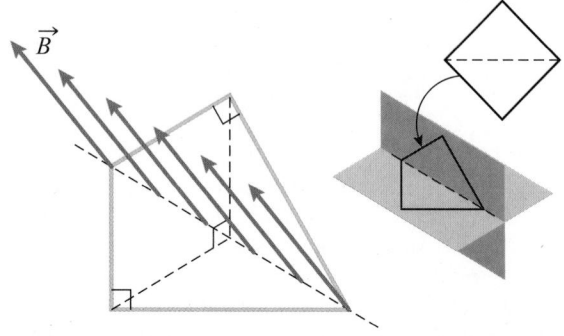

Figure 24-53   Problem 46

47. ✷✷ A horizontal, 10 cm-long aluminum rod is located inside a vertical, uniform, and constant magnetic field of 35 mT. One end of the rod is secured to a hinge, such that the rod can rotate around it in a horizontal plane (Figure 24-54). The rod makes five revolutions per second. What is the induced emf along its length?

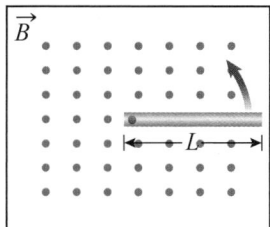

Figure 24-54 Problem 47

48. ✷✷ Two concentric aluminum loops have radii $R_1 = 2.5$ cm and $R_2 = 5$ cm.
(a) The current in the outside loop is changing according to $I(t) = [15\sin(30(\pi t))]$ A, and the resistance of the inside loop is 10 Ω. Derive expressions for the induced emf and the induced electric current in the inside loop.
(b) Compare the frequency of the induced emf and the induced current with the frequency of the current in the outside loop.

49. ✷✷ A copper loop in the shape of an equilateral triangle with sides of 10 cm is placed in a uniform and constant magnetic field, $B = 45.0$ mT, directed perpendicular to the surface of the loop (Figure 24-55). The resistance of the loop is 1 Ω. The loop is pulled out of the magnetic field with a constant velocity of 2 m/s.
(a) What is the magnitude of the emf induced in the loop as it moves out of the magnetic field?
(b) What is the direction of the induced emf?
(c) Find the force that needs to be exerted on the loop to pull it out with the speed of 4 m/s.

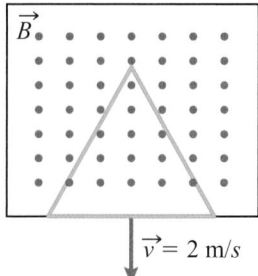

Figure 24-55 Problem 49

50. ✷✷ A rectangular metal loop is located a distance d from a long, straight wire carrying electric current I (Figure 24-56). The short side of the loop is parallel to the wire, and the loop and the wire are located in the same plane. The electric current in the wire changes according to $I(t) = [10\sin(120\pi t)]$ A. Derive an expression that describes the rate of change of the magnetic flux through the loop.

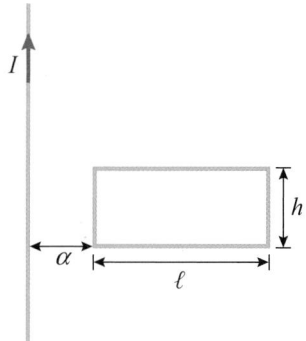

Figure 24-56 Problem 50

51. ✷✷ A circular elastic band is made out of a conducting material. Its initial radius is 5 cm, and it is stretched in a horizontal plane. The band is located in a uniform and constant magnetic field of $B = 50.0$ mT. The band begins expanding uniformly in all directions such that its radius changes at a rate of 0.025 cm/s. What is the induced emf in the band?

52. ✷ A square metal loop 5 cm × 5 cm is located in a region of a nonuniform but constant magnetic field directed perpendicular to the surface of the loop (Figure 24-57). The magnetic field is uniformly decreasing along the direction of motion of the loop at a rate of 5 mT/cm. When the loop is pulled with a velocity of 1 m/s as shown in the figure, what is the emf induced in it?

Figure 24-57 Problem 52

53. ✷✷✷ A metal rod with length $L = 0.2$ m, mass 300 g, and cross-sectional area $A = 0.01$ m² slides down frictionless, tilted aluminum rails (Figure 24-58). The rod starts from rest at the top of the rails tilted at a 30° angle to the horizontal. The apparatus is placed in a uniform and constant vertical magnetic field $B = 0.5$ T, and the resistance of the rails and the rod is 2 Ω.
(a) Describe the motion of the rod as it slides down the incline. Indicate all the forces acting on the rod.
(b) Derive an expression for the induced emf across the rod as a function of the rod's velocity.
(c) Derive an expression for the velocity of the rod as a function of time.
(d) Derive an expression of the emf induced across the rod as a function of time.
(e) Find the terminal velocity of the rod as it slides down the incline.

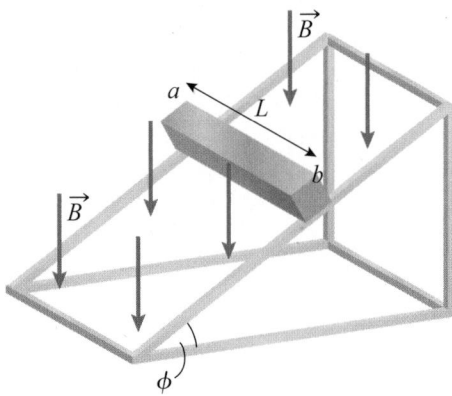

Figure 24-58 Problem 53

54. ✷✷ Two coaxial air-filled solenoids A and B of equal length ($L = 20$ cm) and radii $R_A = 2.5$ cm and $R_B = 5.0$ cm have $N_A = 100$ turns and $N_B = 400$ turns.
   (a) When the current in the inner solenoid increases from 0 to 15 A in 0.01 s, what is the induced emf in the outer solenoid during that time?
   (b) After 0.01 s, the current in the inner solenoid becomes constant. What is the induced emf in the outer solenoid at that time?
   (c) How will your answer to part (b) change if an iron core is inserted into the solenoids?

55. ✷✷ Toroidal inductors (Figure 24-59) and transformers are electronic components that consist of a ring-shaped magnetic core around which a wire is wound to make an inductor. A toroidal inductor is built by winding 10 000 turns of wire around a soft iron ring with a rectangular cross section of 10 mm × 15 mm. The mean radius of the inductor is 30 cm, and the magnetic permeability of iron used in this particular inductor is $\mu = 1.2 \times 10^{-3}$ H/m.
   (a) Calculate the magnetic field inside the toroid when the electric current through it is 10 A.
   (b) Calculate the inductance of the toroid.
   (c) How will your answers to parts (a) and (b) change if the electric current through the inductor were to double?

core

Courtesy of Digi-Key

Figure 24-59 Problem 55

56. ✷✷ Two coaxial coils oriented in the same plane are arranged as shown in Figure 24-60. The larger coil has a radius R_1 and N_1 turns, and the smaller coil has a radius R_2 and N_2 turns, where $R_2 \ll R_1$.
   (a) Derive an expression for the mutual inductance of these coils.
   (b) When the current in the outer coil is changing at the rate of 150 A/s, what is the value of the induced emf in the inner coil?

(c) If the inner coil is connected to the emf source, and the outer coil is not, how will the value of mutual inductance of these coils change? How can you make sense of your answer?

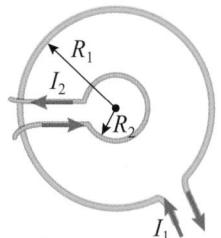

Figure 24-60 Problem 56

57. ✷✷ The starter motor in your car has a resistance of 0.40 Ω in its armature windings. (An armature is a conductive coil used in a motor.) The car battery has a voltage of 12 V. When the motor operates at its normal speed, it has a back emf of approximately 10 V. Estimate the amount of current this car motor draws
   (a) when running at its operating speed;
   (b) during start-up.

58. ✷✷ The switch in the circuit shown in Figure 24-61 has been open for a long time. Then, at $t = 0$ s, the switch is closed.
   (a) Derive a relationship that shows how the current through the inductor depends on time.
   (b) Derive an expression that shows how the induced emf across the inductor depends on time.
   (c) Plot the graphs of the quantities found in parts (a) and (b) as function of time. Assume the inductor is ideal.

Figure 24-61 Problem 58

59. ✷✷ Faraday's law of electromagnetic induction can be used to measure the speed of a projectile. Imagine a small magnet imbedded into a projectile as shown in Figure 24-62. The apparatus consists of two coils separated by a distance d; each coil is connected to an oscilloscope. When the magnet-carrying projectile passes through a coil, it induces an emf in the coil that is recorded by the oscilloscope. Since the time of the emf pulses can be recorded accurately, the speed of the projectile can be calculated.
   (a) Draw what you would see in an oscilloscope connected across the inductor when the projectile is approaching the coils from the left. Denote a current flowing in the clockwise direction as seen by the observer launching the projectile as positive. Clearly indicate the pulse induced in coil 1 and in coil 2.
   (b) How will your graphs change if the projectile were to move twice as fast?

(c) How will your graphs change if the number of loops in the left coil were to double, and the number of loops in the right coil remained the same?

(d) The time separation between the pulses is 3.5 ms, and the distance between the coils is 2.0 m. What is the speed of the projectile?

(e) The method described above is used to measure the speed of a projectile up to tens of kilometres per second. What are some limitations of this method? Why do you think it is not used to measure speeds of faster projectiles?

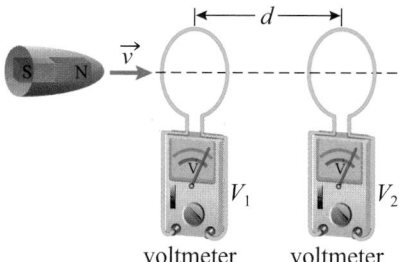

Figure 24-62 Problem 59

60. ✹✹ A loop of wire, shown in Figure 24-63, is placed in a uniform but nonconstant magnetic field. The magnitude and the direction of the magnetic field change continuously with time, such that at times t_1 and t_2 the magnetic flux values are $\Phi_B(t_1)$ and $\Phi_B(t_2)$, respectively.

(a) Prove that the net electric charge that flows through the resistor R during the time $t = t_2 - t_1$ can be expressed as

$$q(t) = \frac{\Phi(t_2) - \Phi(t_1)}{R}$$

(b) Analyze the expression for electric charge for the limiting cases: constant magnetic field, very low resistance, and very high resistance. What does the analysis of these limiting cases tell you?

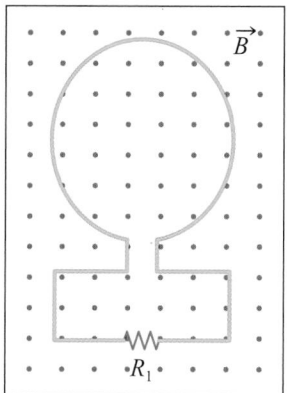

Figure 24-63 Problem 60

61. ✹✹ A piece of insulated conducting wire is shaped as shown in Figure 24-64. The radius of the large circle is twice as large as the radius of the small circle (4 cm and 2 cm, respectively), and the resistance of the small circle is 0.5 Ω. The wire is placed in a uniform but time-varying magnetic field. The rate of change of the magnetic field is 0.5 mT/s. Find the direction and magnitude of the electric current induced in the wires in each case.

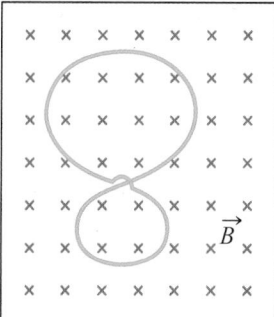

Figure 24-64 Problem 61

62. ✹✹ A series RL circuit consists of a 9 V battery, a 5 Ω resistor, and a 4 H inductor. After the circuit has been turned on, how long will it take for the current in this circuit to reach

(a) 50% of its maximum value;
(b) 75% of its maximum value;
(c) 90% of its maximum value;
(d) 100% of its maximum value?

Supplement your algebraic solution by drawing an $I(t)$ graph for this circuit.

63. ✹✹ A series RL circuit consists of a 12 V battery, a 15 Ω resistor, and a 3 H inductor.

(a) Complete Table 24-3 by calculating the values of the voltage across the resistor, the voltage across the inductor V_L, the current through the battery, and the voltage across it.

(b) How do you know if the values you found are reasonable? Assume the battery is ideal.

(c) How would the answers to the problem change if the battery were not ideal?

Table 24-3 Table for Problem 63

Time	V_R (V)	V_L (V)	I (A)	ε battery (V)
0.001 s				
0.01 s				
0.1 s				
0.2 s				
0.4 s				

64. ✹✹✹ You and your friend decide to design a simple magnetometer. Your device is mounted on your car and consists of a 1 m-long metal pole (a radio antenna, e.g.) and a voltmeter that measures the potential difference across it. When the car is moving due north along a horizontal road at a speed of 60 km/h, your magnetometer measures the potential difference across the antenna to be 1 mV. You know that in the area where you were driving the direction of the magnetic field is approximately 63° below the horizontal. What is the magnitude of the magnetic field there?

65. ✱✱✱ A 60 cm-long bar can slide without friction along the long horizontal rails shown Figure 24-65. The bar and the rails have a resistance of 5 Ω, and the entire setup is placed in a uniform magnetic field of 3.0 mT.
(a) Derive an expression for the force needed to move the bar to the left at a constant speed.
(b) Use the expression you derived in part (a) to find the force required to pull the bar with a constant speed of 1.5 m/s.
(c) Derive the expression for the energy delivered to the resistor in t seconds when the bar is moving to the right with constant speed $\vec{v}$.
(d) How much energy was delivered to the resistor when the bar was moving with a speed of 1.5 m/s during 3 s?

Figure 24-65   Problem 65

66. ✱✱✱ The 60 cm-long metal bar in Figure 24-58 (see problem 53) is given an initial velocity of 2.5 m/s up the incline (up and right) and then let go.
(a) How long will it take the bar to stop moving, and how far will the bar move along the metal rails provided the friction between the rails and the bar is negligible?
(b) Provide at least two ways of solving the problem: one solution based on Newton's laws and another solution based on energy considerations.

67. ✱✱ The RL circuit shown in Figure 24-33 is designed such that the switch S is used to choose between having the battery as part of the circuit (the switch is in position a) or having the battery out of the circuit (the switch is in position b). The battery has an emf of 12 V, the resistance of the resistor is $R = 5$ Ω, and the inductance of the inductor is $L = 40$ mH. At first, the switch is moved to position a.
(a) How long does it take the current to reach 99% of its maximum value? What is this value?
(b) When the current reaches 99% of its maximum value, what is the emf across the inductor?
(c) After 25 s, the switch is moved to position b. How long will it take for the current to decrease by 50% of its maximum value?

68. ✱✱ A medium-sized electric generating plant produces electric energy at 50 A and 20 kV. The electricity produced by the power plant has to be transmitted 50 km over transmission lines, which have a resistance of 10.0 Ω/km.
(a) Compare the power output of this electric generating plant with the electric power output of the Sir Adam Beck #2 Power Plant at Niagara Falls, Ontario (1.29 GW).
(b) What is the power loss during the transmission when the energy is transmitted at 20 kV?
(c) To what value should the voltage output value of the generator be stepped up to reduce the energy loss by a factor of 20?
(d) How many average families can this power plant support? List and justify your assumptions.

69. ✱✱✱ An electric generator consists of a 1000-loop coil formed into a rectangle with sides of 30 cm and 60 cm. The coil is placed in a uniform magnetic field of magnitude $B = 2.5$ T.
(a) Derive an expression for the emf induced by this generator, provided the coil is spun about the axis perpendicular to the magnetic field with angular velocity ω.
(b) Find the maximum value of the emf produced by this generator when the coil is rotated about an axis perpendicular to $\vec{B}$ with $f = 1200$ rev/min.
(c) Use the expression for the induced emf to explain what can be done to the generator to change the frequency of the induced emf and/or its maximum value.

70. ✱✱ The RL circuit in Figure 24-66 consists of an ideal battery, a resistor R, an inductor L, and a switch S that has three different positions: a, b, and N. Initially, the switch is in position N (the circuit is open). After the circuit switch is moved to position a and time Δt elapses, the current in the circuit reaches 50% of its maximum value. The switch remains in position a for the time interval $10\,\tau$. After that, the switch is moved into position b.
(a) What is the time constant of this RL circuit?
(b) What are the values of the electric current in the circuit, the energy density in the coil, and the emf across the coil when
  (i) the switch is at position a for a long time;
  (ii) after the switch moved to position b and τ seconds elapsed;
  (iii) after the switch moved to position b and $3\,\tau$ seconds elapsed?
Support your answers with a graph of electric current in the circuit versus time.

Figure 24-66   Problem 70

71. ✱✱ The value of the electric field at Earth's surface depends on the location on Earth, the time of the day, and the weather, and ranges from 40 V/m to 150 V/m. The magnetic field values at Earth's surface vary from 25 000 nT to 57 000 nT (being the highest at the poles). Compare the energy densities for the electric and magnetic fields near Earth's surface. What does your comparison mean?

72. ✱✱✱ Magnetic resonance imaging (MRI) is a medical technique used widely in modern hospitals to produce images of the interior of the body. During an MRI scan, a patient is placed in a region of a constant magnetic field of up to 5.0 T. While operating properly, the magnetic field is kept constant. However, in case of equipment failure, the magnetic field might be shut off suddenly. This rapidly decreasing magnetic field will generate an induced emf, potentially affecting the charged particles in the body fluids and producing life-threatening electric currents. To prevent this from happening, the engineers make sure that it takes at least the time τ to shut the MRI magnetic field off.

(a) Taking into account that for safety reasons, the maximum induced emf should be kept below 0.010 V and the cross-sectional area of the part of the human body exposed to the magnetic field does not exceed 0.04 m², calculate the value of τ.

(b) One of the important safety regulations for an MRI scan is that a patient or a technician never wear any metal objects. Explain why.

73. ✷✷✷ Baby breathing monitors became popular in the mid–1990s and today are widely used in homes and in hospitals. Although not a life-saving device by itself, the monitor is sensitive to the slightest movement and will sound an alarm if the baby stops breathing for more than 20 s. There are two main types of baby monitor sensors: the mattress pad sensor, placed under the baby's mattress, and the body sensor, which is a thin belt wrapped around the baby's chest. Both detect the changes in Earth's magnetic field flux through a coil as a result of the movement of a patient. The simplest baby body sensor monitor consists of a 300-turn coil wrapped around the baby's chest and connected to a sensitive voltmeter. Estimate the sensitivity of this voltmeter (how accurately it should be able to measure an induced emf) to measure the emf induced by a baby's breathing. Assume that the magnetic field component perpendicular to the coils is 30 μT, the newborn's normal breathing rate is approximately 40 times a minute (much higher than an adult rate), and the smallest increase in the area of a belt around the baby's chest is 5 cm².

74. ✷✷✷ An electric guitar uses a circular pickup coil that detects the vibrations of a steel guitar string. The pickup coil consists of a magnet and a coil with as many as 7000 turns of wire wound around it. A vibrating steel string modulates a changing magnetic flux through the pickup coil. The component of the magnetic field perpendicular to the pickup coil can be described as $B(t) = [45.0 + 3.5\sin(2\pi622t)]$ mT. The diameter of a pickup coil is 5.4 mm. Find the time dependence of the emf induced in the coil.

75. ✷✷✷ Engineers proposed using superconducting materials in the coaxial cables for power transmission lines. A diagram of such a cable is shown in Figure 24-67. The inner and outer wires of such a superconducting cable carry currents in opposite directions. Such a superconducting coaxial cable can carry up to 1 GW of power (1×10^9 W) at 200 kV DC over a distance of 1500 km, almost without a loss in power strength. The radius of an inner wire is 2.5 cm, and the radius of an outer wire if 5.0 cm. Calculate the value of
(a) the magnetic field at the surface of the inner conductor;
(b) the magnetic field at the inner surface of the outer conductor;

(c) the magnetic field at the outer surface of the outer conductor;
(d) the energy density between the conductors;
(e) the energy stored in the space between the conductor over the 1500 km distance;
(f) the pressure exerted on the outer conductor as a result of the power transmission.

76. ✷✷✷ The switch in the circuit shown in Figure 24-68 has been open for a long time. Then, at $t = 0$ s, the switch is closed.
(a) Derive an expression that shows how the current in the inductor, the current in the battery, and the current in the switch depend on time.
(b) Derive an expression that shows how the induced emf across the inductor depends on time.
(c) Plot the graphs for the quantities found in parts (a) and (b) as a function of time. Assume it is an ideal inductor.

Figure 24-68 Problem 76

77. ✷✷✷ Show that the equivalent inductance of two or more inductors can be calculated as follows:
(a) For inductors connected in series:

$$L_{eq} = L_1 + L_2 + \cdots$$

(b) For inductors connected in parallel:

$$\frac{1}{L_{eq}} = \frac{1}{L_2} + \frac{1}{L_2} + \cdots$$

(c) Use the PhET simulations from the online activities to verify whether these formulas hold.
(d) Design more complex RL circuits, calculate their parameters, and then test your prediction using the PhET simulations.

78. ✷✷✷ A student in a physics lab investigates the operation of an AC generator. He collects data on the emf of the generator as a function of time (Figure 24-69). The generator consists of a 200-turn coil that is spun in a strong magnetic field. The coil has a cross-sectional area of 25 cm².
(a) Determine the frequency of the generated current f (Hz).
(b) Determine the frequency of the rotation of the coil f (Hz) and its angular frequency ω (rad/s).
(c) Find the magnitude of the magnetic field.

Figure 24-69 Problem 78

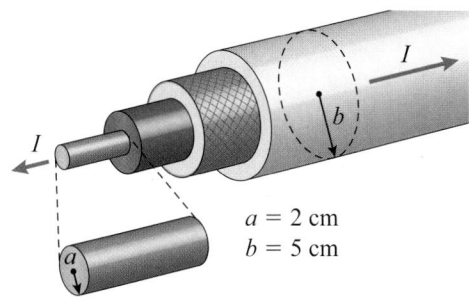

Figure 24-67 Problem 75

ELECTRICITY, MAGNETISM, AND OPTICS

79. ✱✱✱ A student in a physics lab investigates the operation of a series RL circuit. She uses a 2 Ω resistor, a battery, an inductor with an unknown value, and a voltmeter. After the inductor is connected to the battery (the switch is in position *a*), she measures the voltage across the inductor at different times. Her observations are recorded in Table 24-4.

(a) Find the inductance value of the inductor.

(b) Find the time constant of the circuit.

(c) Find the values of electric current at each time. (Hint: Use a spreadsheet program to answer the questions.)

Table 24-4 Data for Problem 79

Time (ms)	V_L (V)	Time (ms)	V_L (V)	Time (ms)	V_L (V)
1.0	8.598	3.5	3.737	6.0	1.624
1.5	7.278	4.0	3.163	6.5	1.375
2.0	6.161	4.5	2.678	7.0	1.164
2.5	5.215	5.0	2.267	7.5	0.985
3.0	4.414	5.5	1.919	8.0	0.834

OPEN PROBLEM

80. ✱✱✱ An 8 cm-long bar magnet (Figure 24-70) is dropped into a copper coil from a height of 50 cm above the coil. The coil is 5 cm in radius, consists of 400 turns, is 10 cm long and the bar magnet has a strength of 0.01 T.

(a) Estimate the value of the maximum emf induced in the coil. Clearly list your assumptions.

(b) Draw the graphs of the induced emf in the coil as a function of time and distance.

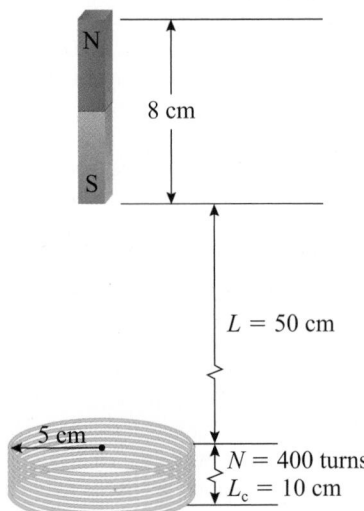

Figure 24-70 Problem 80. Not to scale

ELECTRICITY, MAGNETISM, AND OPTICS

Chapter 25
Alternating Current Circuits

LOOK AHEAD

According to Natural Resources Canada, the annual energy consumption of an average Canadian household in 2007, including all energy sources except fuel for transportation, was 106×10^9 J $\approx$ 30 000 kW·h (distributed as shown in Figure 25-1). This annual consumption is equivalent to 3.35 kJ of energy per second year-round. Approximately 38% of this energy is electricity supplied in the form of alternating current (AC) through a nation-wide power grid, a gigantic network of AC circuits. This chapter introduces the basic principles of AC circuits.

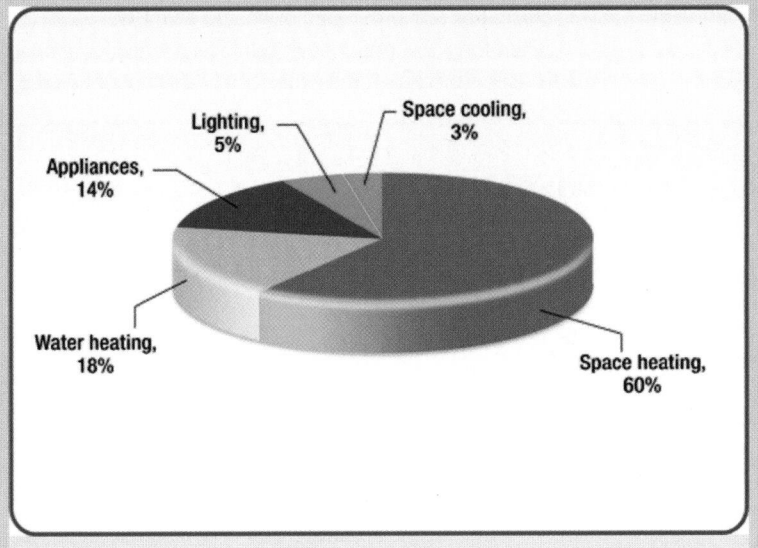

Energy Efficiency Trends in Canada 1990-2005, page numbers, or URL, Office of Energy Efficiency, Natural Resources Canada. Reproduced with the permission of the Minister of Public Works and Government Services Canada, 2012.

Figure 25-1 Domestic energy usage in Canada

Learning Objectives

When you have completed this chapter you should be able to:

1. Derive expressions for the current in alternating current (AC) circuits with resistive, inductive, or capacitive loads.

2. Analyze an LC circuit.

3. Use phasors and phasor diagrams to represent electric current and voltage in AC circuits.

4. Analyze series RLC circuits.

5. Determine the resonant frequency of a series RLC circuit, and describe how the current and component voltages vary with the frequency of the emf.

6. Calculate power dissipated in AC circuits.

ELECTRICITY, MAGNETISM, AND OPTICS

For additional Making Connections, Examples, and Checkpoints, as well as activities and experiments to help increase your understanding of the chapter's concepts, please go to the text's online resources at www.physics1e.nelson.com.

25-1   Simple Loads in AC Circuits

In Chapter 24, we discussed the generation of electrical energy. Most modern power plants generate **alternating current** (AC), which is electric current that changes direction and magnitude many times a second. In North America, the frequency of AC oscillation is 60 Hz; in the European Union and many other countries, the AC frequency is 50 Hz.

Due to the dependence of AC current and voltage on time, we cannot describe these quantities the same way we did for DC circuits. However, we can apply the major concepts and relationships of DC circuits, such as potential difference, electric current, resistance, and Ohm's law, to AC circuits by using some additional techniques and terminology. Unlike a DC circuit, an AC power supply produces an oscillating emf (voltage) that continually changes its magnitude and direction. Therefore, the voltage across the power supply changes with time. For AC circuits, we denote instantaneous values of current and voltage using lower case letters: $i(t)$ and $v(t)$, respectively.

ONLINE ACTIVITY

Exploring AC Circuits

The e-resource that accompanies every new copy of this textbook contains an Online Activity using the PhET simulation "Exploring AC circuits." Work through the simulation and accompanying questions to gain an understanding of more types of AC circuits.

In this chapter, we discuss AC power supplies with an emf that can be described using a sine function:

$$\boxed{\text{KEY EQUATION}} \quad \varepsilon(t) = \varepsilon_{max}\sin(\omega t) = \varepsilon_{max}\sin(2\pi f t) \quad (25\text{-}1)$$

where ε_{max} is the maximum value of the emf, and ω and f are the angular frequency and frequency, respectively.

This emf oscillates between $\pm\varepsilon_{max}$ and has an average value of zero. As you will see below, the current in a sinusoidal AC circuit oscillates between $\pm i_{max}$ and also has an average value of zero. Generally, the **root-mean-square** (rms) values of AC emfs and currents are much more useful for circuit calculations. An rms value of a physical quantity is the square root of the mean value of the square of the mathematical function describing the quantity. For sinusoidal quantities we can take the root-mean over one period, or cycle. If emf $\varepsilon = \varepsilon_{max}\sin(\omega t)$, the period of oscillation is $T = \dfrac{1}{f} = \dfrac{1}{\omega/2\pi} = \dfrac{2\pi}{\omega}$. So, we can substitute the sine function into the expression for an rms value and then integrate over one period:

$$\varepsilon_{rms} = \sqrt{(\varepsilon^2)_{avg}}$$

$$= \sqrt{(\varepsilon_{max}^2\sin^2(\omega t))_{avg}}$$

$$= \varepsilon_{max}\sqrt{(\sin^2(\omega t))_{avg}}$$

$$= \frac{\varepsilon_{max}}{\sqrt{2}}$$

$$(\sin^2(\omega t))_{avg} = \frac{\displaystyle\int_0^T \sin^2(\omega t)\, dt}{T}$$

$$= \frac{1}{T}\int_0^T \frac{1}{2}(1 - \cos(2\omega t))\, dt$$

$$= \frac{1}{T}\frac{T}{2}$$

$$= \frac{1}{2} \qquad (25\text{-}2)$$

The same method can be used to find the rms value of AC current or any other sine or cosine quantity.

Therefore, the rms values for the emf and current in an AC circuit, with a sinusoidal emf, can be expressed as follows:

$$\boxed{\text{KEY EQUATION}} \quad \varepsilon_{rms} = \frac{\varepsilon_{max}}{\sqrt{2}} \qquad (25\text{-}3)$$

$$\boxed{\text{KEY EQUATION}} \quad i_{rms} = \frac{i_{max}}{\sqrt{2}}$$

☑ CHECKPOINT

C-25-1 The rms Value of a Square Wave

Is the rms value of a square wave greater or less than the rms value of a sine wave?

C-25-1 Greater. The rms is defined as the square root of the area under one full waveform squared, divided by the period of the waveform. Since a square wave has an amplitude that is either $\pm\varepsilon_{max}$, the square of that is ε_{max}^2, and the area is $\varepsilon_{max}^2 T$. When we divide by the period and take the square root, we obtain $\varepsilon_{rms} = \varepsilon_{max}$ for a square wave.

Resistive Load

Let us start with an AC circuit that consists simply of an AC power supply connected to a resistor, as shown in Figure 25-2.

Figure 25-2   An AC circuit with a resistive load

Just as with DC circuits, we can apply Kirchhoff's loop rule to obtain

$$\varepsilon(t) - iR = 0 \qquad (25\text{-}4)$$

Solving for i and using Equation 25-1 for the emf we find

$$i(t) = \frac{\varepsilon(t)}{R} = \frac{\varepsilon_{max}}{R}\sin(\omega t) = i_{max}\sin(\omega t) \qquad (25\text{-}5)$$

Not surprisingly, the current in this circuit is also sinusoidal, and the amplitude of the current is related to the amplitude of the emf:

$$i_{max} = \frac{\varepsilon_{max}}{R} \qquad (25\text{-}6)$$

Equation (25-6) has the same form as Ohm's law, $I = \dfrac{V}{R}$: current equals voltage divided by resistance. Figure 25-3 shows a plot of the emf and the resulting current versus time.

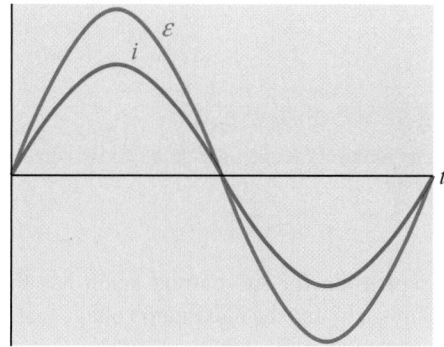

Figure 25-3 The emf and current in an AC circuit with only a resistor. Note that they are in phase.

The voltage that appears across the resistor is given by

$$V_R(t) = i(t) = i_{max}R\sin(\omega t) \qquad (25\text{-}7)$$

From this equation we can identify the maximum value of the voltage across the resistor:

KEY EQUATION
$$V_{R,\,max} = i_{max}R \qquad (25\text{-}8)$$

EXAMPLE 25-1

Current in a Resistor

A 1.0 kΩ resistor is connected across an emf where $\varepsilon(t) = 20\sin(50t)$. Find the maximum current.

SOLUTION

From the given emf, we identify $\varepsilon_{max} = 20$ V. We can then use Equation (25-6) to find the maximum current:

$$i_{max} = \frac{\varepsilon_{max}}{R} = \frac{20\text{ V}}{1000\ \Omega} = 20\text{ mA}$$

Making sense of the result:

We have a 1000 Ω resistor. With a 1 volt source, we would expect 1 mA.

Inductive Load

We now consider an AC emf connected to just an inductor, as shown in Figure 25-4.

Figure 25-4 An AC circuit with an inductive load

Again applying Kirchhoff's loop rule, we obtain

$$\varepsilon(t) - L\frac{di}{dt} = 0 \qquad (25\text{-}9)$$

Solving for the derivative of the current, we find

$$\frac{di}{dt} = \frac{\varepsilon(t)}{L} = \frac{\varepsilon_{max}}{L}\sin(\omega t) \qquad (25\text{-}10)$$

A solution to this differential equation is

$$i(t) = -\frac{\varepsilon_{max}}{\omega L}\cos(\omega t) = -i_{max}\cos(\omega t) \qquad (25\text{-}11)$$

We note that $i_{max} = \dfrac{\varepsilon_{max}}{\omega L}$, has the form of Ohms's Law—a current is given by a voltage divided by another quantity. The quantity ωL does indeed have units of ohms and we define the **inductive reactance** as

KEY EQUATION
$$X_L = \omega L \qquad (25\text{-}12)$$

If we plot the emf and the current through the inductor, as shown in Figure 25-5, we can see the current peaks a quarter cycle after the emf. Thus, the current *lags* the emf.

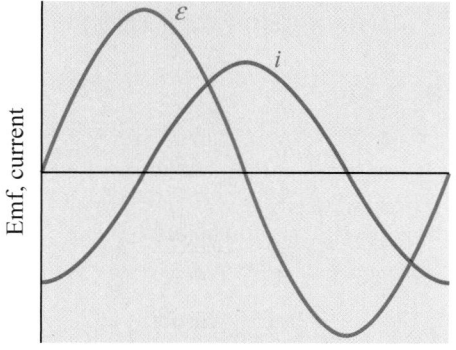

Figure 25-5 Plot of the emf and the current for an AC circuit with an inductive load. Note that the voltage leads the current.

Since $\sin\!\left(\omega t - \dfrac{\pi}{2}\right) = -\cos(\omega t)$, we can rewrite the current as

$$i(t) = i_{max}\sin\!\left(\omega t - \frac{\pi}{2}\right) \qquad (25\text{-}13)$$

The maximum voltage that appears across the inductor is given by

$$V_{L,\,max} = i_{max}X_L \qquad (25\text{-}14)$$

 CHECKPOINT

C-25-2 Inductor Current

How can you verify that the inductor current really is a cosine function of time?

C-25-2 Differentiate Equation (25-11) and substitute it into Equation (25-10).

Capacitive Load

We now consider an AC voltage source connected to just a capacitor, as shown in Figure 25-6.

Figure 25-6 An AC circuit with a capacitive load

Applying Kirchhoff's loop rule, we obtain

$$\varepsilon(t) - \frac{q}{C} = 0 \qquad (25\text{-}15)$$

Solving for q, the charge on the capacitor, gives

$$q(t) = C\varepsilon(t) = C\varepsilon_{max}\sin(\omega t) \qquad (25\text{-}16)$$

Since $\quad i(t) = \dfrac{dq}{dt},$

$$i(t) = \frac{d}{dt}\left(C\varepsilon_{max}\sin(\omega t)\right)$$

$$= C\varepsilon_{max}\frac{d\sin(\omega t)}{dt}$$

$$= \omega C\varepsilon_{max}\cos(\omega t) \qquad (25\text{-}17)$$

We now define **capacitive reactance** as

$$X_C = \frac{1}{\omega C} \qquad (25\text{-}18)$$

We can then write

$$i_{max} = \frac{\varepsilon_{max}}{X_C} \qquad (25\text{-}19)$$

 EXAMPLE 25-2

Current in an Inductor

A 10 mH inductor is connected across an emf $\varepsilon(t) = 50\sin(20t)$. Find

(a) the inductive reactance;
(b) the maximum current through the inductor.

SOLUTION

(a) We will use Equation (25-12) for the inductive reactance. From the given emf, we see that $\omega = 20$ rad/s; thus,

$$X_L = \omega L = (20\text{ rad/s})(10\text{ mH}) = 0.20\ \Omega$$

(b) Noting that $\varepsilon_{max} = 50$ V, we can find the maximum current:

$$i_{max} = \frac{\varepsilon_{max}}{X_L} = \frac{50\text{ V}}{0.20\ \Omega} = 250\text{ A}$$

Making sense of the result:

The inductive reactance is very small so we have a very large current.

The expression for AC current again has the same form as Ohm's law, and the maximum voltage that appears across the capacitor is given by

$$V_{C,\,max} = i_{max}X_C \qquad (25\text{-}20)$$

When we plot the emf and the current through the capacitor, as shown in Figure 25-7, we again see that the emf and the current are not in phase. However, now the current *leads* the emf by a quarter cycle.

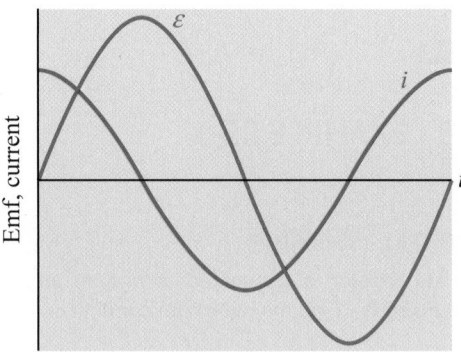

Figure 25-7 Plot of the emf and the current for an AC circuit with a capacitive load. Note that the current leads the emf.

Again, we can rewrite the current in terms of a sine function. Since $\sin\left(\omega t + \dfrac{\pi}{2}\right) = \cos(\omega t)$,

$$i(t) = i_{max}\sin\left(\omega t + \frac{\pi}{2}\right) \qquad (25\text{-}21)$$

In contrast to the inductor current, the phase difference in the argument of the sine function of the current is positive.

EXAMPLE 25-3

Current in a Capacitor

A 1 nF capacitor is connected across an emf $\varepsilon(t) = 30\sin(40t)$. Find

(a) the capacitive reactance;
(b) the maximum current through the capacitor.

SOLUTION

(a) We will use Equation (25-18) for the capacitive reactance. From the emf, we can see that $\omega = 40$ rad/s; thus,

$$X_C = \frac{1}{\omega C} = \frac{1}{(40 \text{ rad/s})(1 \text{ nF})} = 25 \text{ M}\Omega$$

(b) Noting from the emf that $\varepsilon_{max} = 30$ V, we can find the maximum current:

$$i_{max} = \frac{\varepsilon_{max}}{X_C} = \frac{30 \text{ V}}{25 \text{ M}\Omega} = 1.2 \ \mu A$$

Making sense of the result:

In this case the capacitive reactance is quite large as we therefore have a small current.

The introduction of either an inductor or a capacitor into an AC circuit produces the interesting result that the current is not in phase with the emf. Such components are called **reactive circuit elements**, and the phase difference between the emf and the current is called the **phase shift**.

LO 2

25-2 The LC Circuit

We will now consider the simple combination of a capacitor and an inductor, called an **LC circuit**, shown in Figure 25-8. Instead of using an AC emf, we will charge the capacitor with a battery. We will remove it

EXAMPLE 25-4

An LC Circuit

A 1.0 μF capacitor charged with 10 μC is connected across a 10 mH inductor.

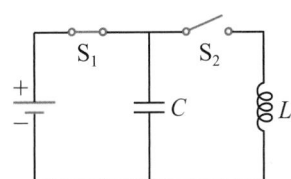

Figure 25-8 An LC circuit with the capacitor fully charged

from the battery and connect it to an inductor. We will assume that the conductors and components in this circuit have no resistance. Once the capacitor has been fully charged to some initial charge q_0 by the battery, we open switch S_1 to disconnect the battery. At time $t = 0$ we then close switch S_2 to connect the inductor across the capacitor.

We apply Kirchhoff's law around the loop containing the capacitor and the inductor to find

$$\frac{q}{C} - L\frac{di}{dt} = 0 \qquad (25\text{-}22)$$

Since $i = -\dfrac{dq}{dt}$, $\dfrac{di}{dt} = -\dfrac{d^2q}{dt^2}$, which we substitute into Equation (25-22):

$$\frac{q}{C} + L\frac{d^2q}{dt^2} = 0 \qquad (25\text{-}23)$$

$$\frac{d^2q}{dt^2} = -\frac{1}{LC}q$$

This equation has the same form as a simple harmonic oscillator:

$$a(t) = -\omega^2(A\cos(\omega t + \phi)) = -\omega^2 x(t) \qquad (13\text{-}18)$$

Just as for simple harmonic oscillators (Chapter 13), Equation (25-23) has a solution of the form

$$q(t) = q_0\cos(\omega_0 t) \qquad (25\text{-}24)$$

where $\omega_0 = \dfrac{1}{\sqrt{LC}}$

So, the charge on the capacitor, and hence the current in the circuit, oscillates with a frequency that is determined by the value of the inductor and the capacitor.

The LC circuit is analogous to the oscillating mass–spring system in Chapter 13. Inductance plays the role

(a) What is the initial voltage across the capacitor?
(b) What is the angular frequency of the oscillation?
(c) Plot the voltage across the capacitor and the voltage across the inductor versus time.

(continued)

SOLUTION

(a) We use $CV = q$ for a capacitor to solve for V:

$$V = \frac{q}{C} = \frac{10\ \mu C}{1\ \mu F} = 10\ V$$

(b) We calculate the angular frequency of oscillation using Equation (25-24):

$$\omega_0 = \frac{1}{\sqrt{LC}} = \frac{1}{\sqrt{10\ mH \times 1\ \mu F}} = 1.0 \times 10^4\ s^{-1}$$

(c) The voltage across the capacitor is found from

$$v_C(t) = \frac{q(t)}{C} = \frac{10\ \mu C \cos(\omega_0 t)}{1\ \mu F} = 10 \cos(1.0 \times 10^4\ s^{-1} \cdot t)\ V$$

The voltage across the inductor is given by

$$v_L(t) = L \frac{d^2 q}{dt^2}$$

$$= -L q_0 \omega_0^2 \cos(\omega_0 t)$$

$$= -10\ mH \cdot 10\ \mu C \cdot (1.0 \times 10^4\ s^{-1})^2 \cos(1.0 \times 10^4\ s^{-1}\ t)$$

$$= -10 \cos(1.0 \times 10^4\ s^{-1}\ t)$$

The plot of these voltages in Figure 25-9 shows that they are exactly out of phase.

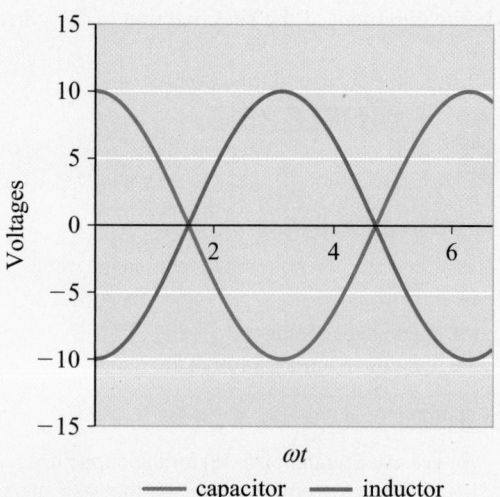

Figure 25-9 Capacitor and inductor voltages versus time

Making sense of the result:

At any instant in time, the voltage across the capacitor is exactly opposite to the voltage across the inductor. This must be the case in order for Kirchoff's Loop rule to be satisfied.

---

of the mass, and the reciprocal of the capacitance plays the role of the spring constant. We can see the similarity between the two systems by comparing Equations 25-23 and 13-18.

✓ CHECKPOINT

C-25-3 How to Change the Resonant Frequency

How does the resonant frequency of an LC circuit change when we increase the capacitance?

C-25-3 The resonant frequency decreases when we increase the capacitance because the capacitance is in the denominator of the expression.

LO 3

25-3 Phasors

As we saw in Section 25-1, the phases of the AC current and voltage sine functions can shift relative to each other depending on the nature of the load in the circuit. We will use phasors to represent and manipulate the various quantities in an AC circuit. Phasors are well suited to this task because they can describe the amplitude and phases of these quantities as well as the phase relationships between them.

A phasor is a vector in two dimensions. The tail of the vector is located at the origin of the coordinate system and the vector has a length that is equal to the amplitude of the quantity we wish to represent. The angle that the phasor makes with the horizontal axis is time dependent and as a consequence, the phasor rotates in a counterclockwise fashion about the origin. The rate at which it rotates is determined by the angular frequency, ω. We can determine the instantaneous value of the quantity the phasor is being used to represent by taking the projection of the phasor onto the vertical axis. For example, the phasor representing the emf in an AC circuit would have a length of ε_{max} and would make an angle of $\theta = \omega t$ with the x-axis. The projection of the phasor on the vertical axis is then $\varepsilon_{max}\sin(\omega t)$ which is $\varepsilon(t)$, as given by Equation (25-1).

When we examined an AC circuit with just an inductor, we found that the current in the circuit was given by $i_{max}\sin\left(\omega t - \frac{\pi}{2}\right)$. We would represent this with a phasor with length of i_{max} that makes an angle of $\theta = \omega t - \frac{\pi}{2}$ with the positive x-axis.

In general, the values of voltage across and current through a circuit element can be represented mathematically as follows:

KEY EQUATION
$$\begin{cases} \varepsilon(t) = \varepsilon_{max}\sin(\omega t) \\ i(t) = i_{max}\sin(\omega t - \phi) \end{cases} \quad (25\text{-}25)$$

where ϕ is the phase shift.

CHECKPOINT

C-25-4 Phase Shift

Which statement correctly describes the electric circuit represented in Figure 25-10(b)?

(a) The phase shift between the current and the voltage is π (current lags voltage).

(b) The phase shift between the voltage and the current is $\dfrac{\pi}{2}$ (current lags voltage).

(c) The phase shift between the current and the voltage is $\dfrac{\pi}{4}$ (current lags voltage).

(d) The phase shift between the current and the voltage is $-\dfrac{\pi}{2}$ (voltage lags current).

(a)

(b)

(c)

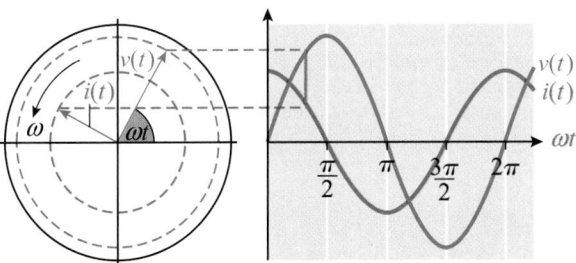

Figure 25-10 Phasor diagrams representing the instantaneous values of voltage and current with (a) zero phase shift (resistor), (b) current lagging voltage (inductor), and (c) voltage lagging current (capacitor).

LO 4

25-4 Series RLC Circuits

We now consider a circuit with an AC emf, a resistor, an inductor, and a capacitor in series, as shown in Figure 25-11. This type of circuit is called a **series RLC circuit**.

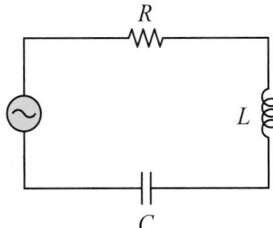

Figure 25-11 A series RLC circuit

Kirchhoff's loop rule (Chapter 22, Section 22-4) requires that the instantaneous voltages around the circuit loop add to zero. We can use phasors in this process to determine the magnitude and phase of the current in the circuit. We know the phase of the voltages across the resistor, inductor, and capacitor relative to the current. Specifically, the resistor voltage is in phase with the current, the inductor voltage leads the current, and the capacitor voltage lags the current. Because all the circuit elements are connected in series, the same current flows through them. Consequently, we can draw a phasor diagram for the circuit by first placing a phasor that represents the current on the circuit. This phasor makes an angle of $\omega t - \phi$ with respect to the positive x-axis, as shown in Figure 25-12.

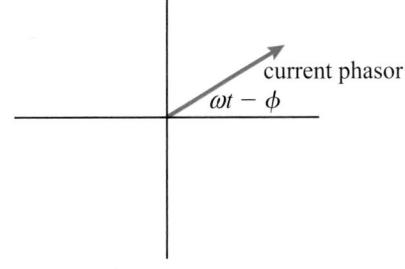

Figure 25-12 A current phasor for the series RLC circuit

Next, we use the voltage–current phase relationships to place phasors for the resistor, inductor, and capacitor voltages, as shown in Figure 25-13.

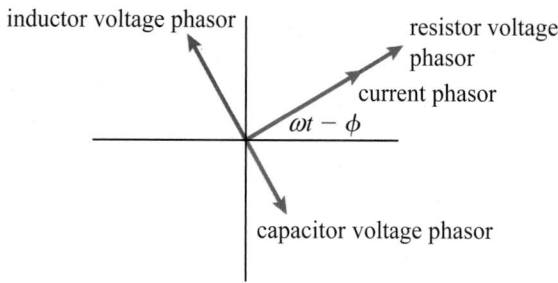

Figure 25-13   Current and voltage phasors diagrams for the RLC circuit

By Kirchhoff's loop rule, the vector sum of the voltage phasors must equal the AC emf phasor, as shown in Figure 25-14.

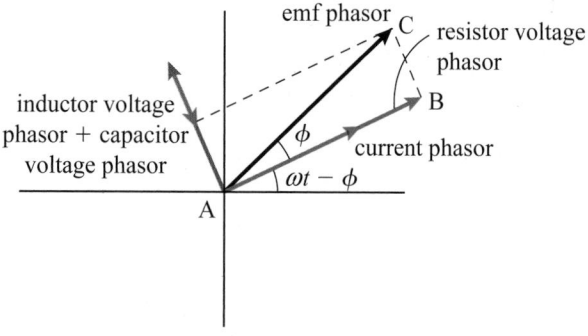

Figure 25-14   The sum of the phasors for the resistor, inductor, and capacitor voltages gives the AC emf phasor.

In the right-angled triangle ABC in Figure 25-14, the emf phasor is the hypotenuse. The other two sides are given by the magnitude of the resistor voltage phasor and the difference in magnitude of the inductor voltage phasor and the capacitor voltage phasor. Applying Pythagoras' theorem gives

$$\varepsilon_{max}^2 = v_{R,max}^2 + (v_{L,max} - v_{C,max})^2 \qquad (25\text{-}26)$$

Substituting from Equations (25-8), (25-14), and (25-20), we get

$$\varepsilon_{max}^2 = (i_{max}R)^2 + (i_{max}X_L - i_{max}X_C)^2 \qquad (25\text{-}27)$$

Solving for i_{max} gives

$$i_{max} = \frac{\varepsilon_{max}}{\sqrt{R^2 + (X_L - X_C)^2}} \qquad (25\text{-}28)$$

Equation (25-28) gives us the magnitude of the maximum current in the circuit as a function of the magnitude of the maximum AC emf, the resistance, the capacitive reactance, and the inductive reactance. Note that both the capacitive and inductive reactance depend on the frequency of the AC emf; consequently, so does the magnitude of the maximum current.

We can see that Equation (25-28) looks similar to Ohm's law in that a current is given by a ratio of a voltage to a quantity that has units of ohms. We define **impedance**, Z, as

KEY EQUATION $$Z = \sqrt{R^2 + (X_L - X_C)^2} \qquad (25\text{-}29)$$

Now that we have found an expression for the magnitude of the maximum current, we turn our attention to the phase angle. From Figure 25-14 we can see that the tangent of the angle ϕ is given by

KEY EQUATION $$\tan\phi = \frac{i_{max}(X_L - X_C)}{i_{max}R} = \frac{X_L - X_C}{R} \qquad (25\text{-}30)$$

✓ CHECKPOINT

C-25-5 The Sign of the Phase Angle

In Figure 25-15, is the phase angle positive or negative? Explain.

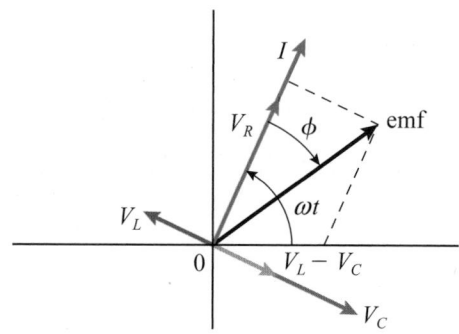

Figure 25-15   C-25-5

C-25-5 Negative. The current is leading the emf in the circuit, that is, it is farther counterclockwise than the emf. The emf makes an angle of ωt with the x-axis, and the current makes an angle of $\omega t - \phi$ with the x-axis. The only way for $\omega t - \phi > \omega t$ is if $\phi < 0$. Another way of telling is by noting that $V_C > V_L$, means that $X_C > X_L$. Then, from Equation (25-30) the phase angle must be negative.

EXAMPLE 25-5

Analyzing a Series RLC Circuit

A series RLC circuit consists of an emf $\varepsilon = 100\sin(8000t)$, a 100 Ω resistor, 1.00 mH inductor, and a 1.00 μF capacitor. Determine

(a) the maximum current flow;
(b) the phase shift;
(c) the maximum voltages across the three components.

SOLUTION

(a) We will first determine the impedance using Equation (25-29):

$$Z = \sqrt{R^2 + (X_L - X_C)^2}$$

$$= \sqrt{100^2 + \left(8000 \text{ s}^{-1} \cdot 1 \text{ mH} - \frac{1}{8000 \text{ s}^{-1} \cdot 1\,\mu\text{F}}\right)^2} = 154 \text{ Ω}$$

Using this result, we can determine i_{max}:

$$i_{max} = \frac{\varepsilon_{max}}{Z} = \frac{100}{154\ \Omega} = 0.650\ \text{A}$$

(b) The phase shift can be determined using Equation (25-30):

$$\tan\phi = \frac{X_L - X_C}{R} = \frac{(8000\ \text{s}^{-1})(1\ \text{mH}) - \dfrac{1}{(8000\ \text{s}^{-1})(1\ \mu\text{F})}}{100} = -1.17$$

Therefore $\phi = \tan^{-1}(-1.17) = -49.5°$

(c) We find the maximum voltages across the resistor, the inductor, and the capacitor from Equations (25-8), (25-14), and (25-20), respectively:

$$v_{R,\,max} = i_{max}\,R = (0.650\ \text{A})(100\ \Omega) = 65.0\ \text{V}$$

$$v_{L,\,max} = i_{max}\,X_L = (0.650\ \text{A})(8000\ \text{s}^{-1})(1\ \text{mH}) = 5.20\ \text{V}$$

$$v_{C,\,max} = i_{max}\,X_C = (0.649\ \text{A})\frac{1}{(8000\ \text{s}^{-1})(1\ \mu\text{F})} = 81.2\ \text{V}$$

Making sense of the result:

We can get a sense of what is going on in the circuit from the reactances:

$$X_L = \omega L = (8000\ \text{s}^{-1})(1\ \text{mH}) = 8.00\ \Omega$$

$$X_C = \frac{1}{\omega C} = \frac{1}{(8000\ \text{s}^{-1})(1\ \mu\text{F})} = 125\ \Omega$$

The capacitive reactance is much larger than the inductive reactance but comparable to the resistance. As such, we would expect the circuit to behave somewhere in between a purely resistive circuit and a purely capacitive circuit. In a purely resistive circuit, the current is in phase with the applied emf. In a purely capacitive circuit, the current leads the applied emf. For this particular circuit, we found a phase angle of −49.5°. The minus sign indicates that the current is leading the emf, and the 49.5° angle indicates that the circuit is somewhere between a purely resistive circuit ($\phi = 0$) and a purely capacitive circuit ($\phi = -90°$), as we predicted from the reactances.

The maximum voltages that appear across the components in an AC circuit reflect their reactances. The maximum voltage across the inductor in Example 25-5 is quite small, and the maximum voltage that appears across the capacitor is relatively large. When we add the maximum voltages, we get

$$v_{R,\,max} + v_{L,\,max} + v_{C,\,max} = 65.0\ \text{V} + 5.20\ \text{V} + 81.2\ \text{V}$$

$$= 151\ \text{V}$$

This total is much greater than the maximum applied voltage of 100 V. How can this be? Remember that the voltages that appear across these elements are time-dependent and are not in phase. The three voltages do not all reach their maximum voltage simultaneously, as we can see from a plot of the voltages versus time. The time-dependent voltages across the RLC circuit elements are given by

KEY EQUATION $\qquad v_R(t) = Ri(t) = Ri_{max}\sin(\omega t - \phi)$

KEY EQUATION
$$v_L(t) = L\frac{di}{dt} = \omega L i_{max}\cos(\omega t - \phi)$$
$$= X_L i_{max}\cos(\omega t - \phi) \qquad (25\text{-}31)$$

KEY EQUATION
$$v_C(t) = \frac{q(t)}{C} = -\frac{1}{\omega C}i_{max}\cos(\omega t - \phi)$$
$$= -X_C i_{max}\cos(\omega t - \phi)$$

In Figure 25-16, we plot the voltages versus time along with the AC emf. The voltages are indeed out of phase, and the component voltages do sum to the AC emf at any given instant.

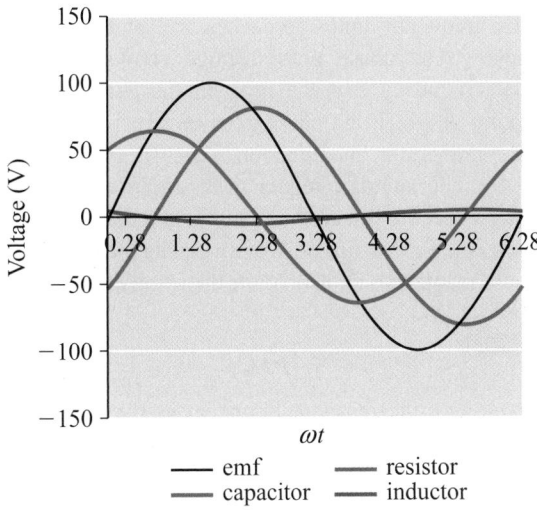

Figure 25-16 Voltages versus time for the RLC circuit of Example 25-5

LO 5

25-5 Resonance

We will now consider how the maximum current and the phase shift vary in a series RLC circuit as we change the frequency of the emf. The inductive reactance increases as we increase the frequency. On the other hand, the capacitive reactance increases when we decrease the frequency. At low frequencies, we would expect the capacitor to dominate the behaviour of the circuit, while at high frequencies, we would expect the inductor to dominate. In Figure 25-17, we plot i_{max} and ϕ versus angular frequency for a series RLC circuit with

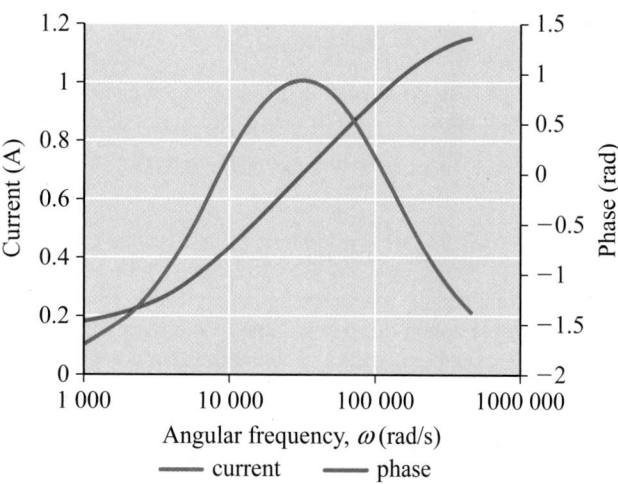

Figure 25-17 Current (left axis) and phase (right axis) versus angular frequency for an RLC circuit

$R = 100 \ \Omega$, $L = 1$ mH, and $C = 1 \ \mu$F. The applied emf has an amplitude of 100 V.

We can see that the current tends to small values at low and high angular frequencies and peaks at some intermediate frequency. The phase starts off at $-\pi/2$ at low angular frequency and approaches $\pi/2$ at high angular frequency. The phase goes through zero at the same angular frequency that the peak in the current occurs. The point at which the phase goes to zero is determined by the numerator in Equation (25-30), that is when $X_L = X_C$. The angular frequency at which the inductive and capacitive reactances are equal is the **resonance frequency**, ω_0, for this circuit. We can equate the expressions for these reactances and solve for the resonant frequency:

$$\boxed{\text{KEY EQUATION}} \qquad \omega_0 = \frac{1}{\sqrt{LC}} \qquad (25\text{-}32)$$

Note that the resonant frequency is the same as the oscillation frequency of the LC circuit with no resistance in Section 25-2. The driven RLC circuit is analogous to a driven damped harmonic oscillator, with the resistance acting as the damping term. In Figure 25-18, we plot the

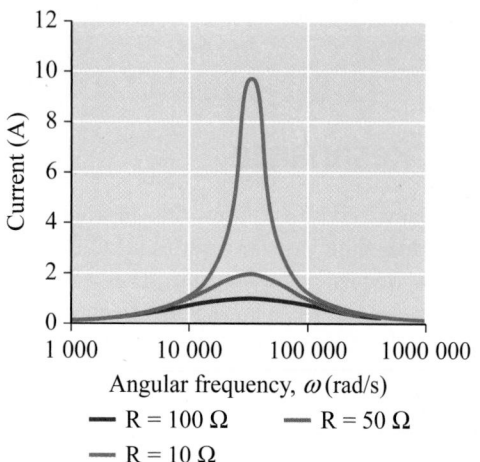

Figure 25-18 Current versus angular frequency in a driven RLC circuit. The different curves correspond to different values of the resistance.

current in a driven RLC circuit with various values of resistance while holding the inductance and the capacitance constant. As the resistance decreases, the resonance peak gets taller and narrower.

Now consider what happens to the voltages that appear across the various circuit elements as we vary the frequency. We plot the maximum resistor, capacitor, and inductor voltages versus frequency for $R = 50 \ \Omega$, $L = 1$ mH, and $C = 1 \ \mu$F in Figure 25-19. We can see that the voltage across the resistor has the same shape as the current, which we would expect from Ohm's law. At low frequencies, the voltage across the capacitor becomes large, which is reasonable because the capacitive reactance gets large at low frequencies. Conversely, at high frequencies, the voltage across the inductor becomes relatively large because the inductive reactance gets large at high frequencies.

Remember that Figure 25-19 shows voltage *amplitudes*, which do not generally add to the applied emf of 100 V. However, at resonance, the voltage across the resistor is exactly 100 V, and the voltages across the inductor and the capacitor add to zero because they have equal magnitudes and opposite phases.

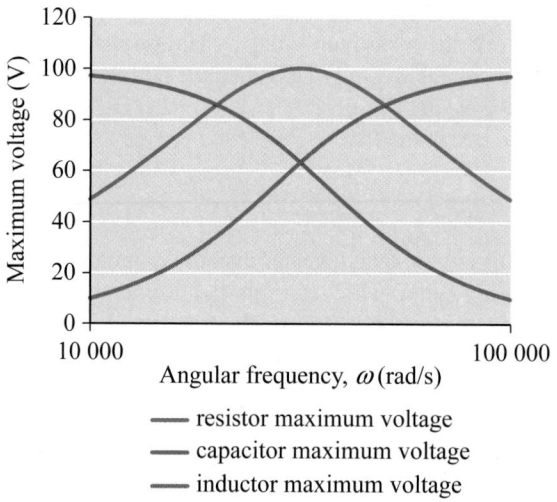

— resistor maximum voltage
— capacitor maximum voltage
— inductor maximum voltage

Figure 25-19 Voltage amplitudes versus angular frequency in an RLC circuit with $R = 50 \ \Omega$

If we reduce the value of the resistor, the resonance sharpens. We plot the maximum resistor, capacitor, and inductor voltage versus frequency for $R = 10 \ \Omega$, $L = 1$ mH, and $C = 1 \ \mu$F in Figure 25-20. The maximum resistor voltage remains unchanged. At low frequency, the capacitor voltage is approximately 100 V, and the inductor voltage is small. At high frequencies, the inductor voltage is approximately 100 V, and the capacitor voltage is small. In the resonance region, the capacitor and inductor voltages are quite large and substantially exceed the applied voltage. We must again remind ourselves that the graphs shows amplitudes and the voltages are not in phase.

Figure 25-20　Maximum voltages in an RLC circuit for $R = 10\ \Omega$

— resistor maximum voltage
— capacitor maximum voltage
— inductor maximum voltage

We characterize the width of a resonance by a **quality factor**, or **Q-factor**, defined as

$$Q = \frac{f_0}{\Delta f} \qquad (25\text{-}33)$$

where f_0 is the centre frequency of the peak, and Δf is the half-power bandwidth of the peak.

The **half-power bandwidth** is the difference between the two frequencies at which the power in the circuit is equal to half the power at the centre frequency (Figure 25-21). A circuit with a high Q-factor has a narrow, tall resonance peak. The Q-factor can be determined from the power dissipated in the resistor of the circuit, which we will examine more closely in the next section.

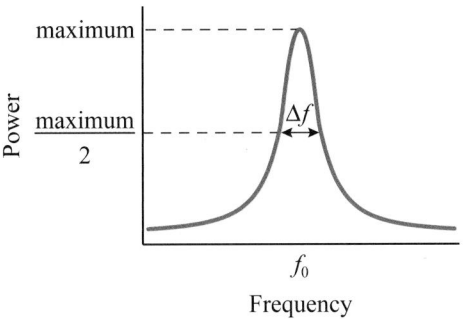

Figure 25-21　f_0 and Δf on a resonance peak

MAKING CONNECTIONS

Radio Resonance

A tuner in a radio or a TV is an RLC circuit. Typically, the resistance in the circuit is made as small as possible to give as sharp a resonance as possible. Figure 25-22 shows a type of tuning capacitor commonly used in radios with manual tuning. Turning the knob on the capacitor changes the overlap between the interleaved stationary bottom plates and the moving top plates, changing the effective area of the capacitor plates.

Figure 25-22　A variable tuning capacitor

LO 6

25-6　Power in AC Circuits

The individual voltages and currents in an AC circuit all have an average value of zero, because they oscillate symmetrically about zero. However, the power dissipated in a circuit element is the time average of the product of the voltage and the current, and, as we will see, this product generally does not average to zero.

We first consider the power dissipated by a resistor in an RLC circuit. The current through the resistor is given by the following equation:

$$i(t) = i_{\max}\sin(\omega t - \phi) \qquad (25\text{-}34)$$

where $i_{\max}$ is given by Equation (25-28), and ϕ is given by Equation (25-30).

The time-dependent voltage across the resistor is given by

$$v_R(t) = Ri(t) = Ri_{\max}\sin(\omega t - \phi) \qquad (25\text{-}35)$$

The instantaneous power dissipated by the resistor is thus given by

$$P_R(t) = v_R(t)i(t) = Ri_{\max}\sin(\omega t - \phi)i_{\max}\sin(\omega t - \phi)$$

$$= i_{\max}^2 R\sin^2(\omega t - \phi) \qquad (25\text{-}36)$$

Recall from Section 25-1 that the average value of $\sin^2(x) = \dfrac{1}{2}$. So, averaging the instantaneous power over one cycle gives

KEY EQUATION　$P_{R,\,\text{avg}} = \dfrac{i_{\max}^2}{2}R = \left(\dfrac{i_{\max}}{\sqrt{2}}\right)^2 R = i_{\text{rms}}^2 R \qquad (25\text{-}37)$

MAKING CONNECTIONS

Heating with Electricity

A heater element toaster (Figure 25-23) is a simple example of resistive power dissipation in an AC circuit. Similar elements are found in electric water heaters, electric clothes dryers, blow dryers, and electric baseboard heaters. Heating elements are made of metals with low resistance and high melting points. The low resistance of heating elements enables them to heat rapidly and efficiently.

Yiargo/Shutterstock

Figure 25-23   Heater elements in a toaster

For the inductor, we also calculate the power as the product of the voltage and the current. We get the voltage across the inductor from Equation (25-31):

$$P_L(t) = v_L(t)i(t) = X_L i_{max}\cos(\omega t - \phi)i_{max}\sin(\omega t - \phi)$$

$$= X_L i_{max}^2 \cos(\omega t - \phi)\sin(\omega t - \phi) \qquad (25\text{-}38)$$

When we average this power over one complete cycle, we can use the trigonometric identity $\sin(2\theta) = 2\sin(\theta)\cos(\theta)$ to get

$$P_{L,avg} = \frac{1}{T}\int_0^T X_L i_{max}^2 \cos(\omega t - \phi)\sin(\omega t - \phi)dt$$

$$= \frac{X_L i_{max}^2}{T}\int_0^T \cos(\omega t - \phi)\sin(\omega t - \phi)dt$$

$$= \frac{X_L i_{max}^2}{T}\int_0^T \frac{1}{2}\sin(2(\omega t - \phi))dt \qquad (25\text{-}39)$$

$$= 0$$

A plot of the instantaneous power versus time verifies that the average power is zero. For each complete cycle, the area enclosed by the curve below the x-axis exactly cancels the area enclosed by the curve above the x-axis, as can be seen in Figure 25-24.

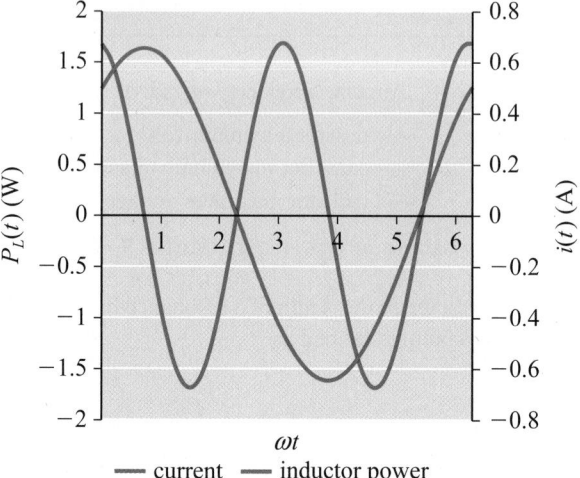

— current — inductor power

Figure 25-24   Power (red) and current (blue) in an inductor in an RLC circuit with $R = 100\ \Omega$, $L = 1$ mH, $C = 1\ \mu$F and $\varepsilon_{max} = 100$ volts. Note that as the magnitude of the current increases, the inductor absorbs energy. As the current returns to zero, the inductor delivers energy back to the circuit.

When the power is positive the inductor is absorbing energy. Unlike the resistor, this energy is not dissipated, but is stored in the magnetic field of the inductor. While the current is increasing, the inductor absorbs energy. When the current starts to decrease, the inductor returns the stored energy to the circuit, and the power is negative.

 EXAMPLE 25-6

Power and the Capacitor

Find the average power in the capacitor in an RLC circuit.

SOLUTION

The instantaneous power in the capacitor is given by

$$P_C(t) = v_C(t)i(t) = -X_C i_{max}\cos(\omega t - \phi)i_{max}\sin(\omega t - \phi)$$

$$= -X_C i_{max}^2 \cos(\omega t - \phi)\sin(\omega t - \phi)$$

Averaging this power over one complete cycle, we find

$$P_{C,avg} = \frac{1}{T}\int_0^T -X_C i_{max}^2 \cos(\omega t - \phi)\sin(\omega t - \phi)dt$$

$$= \frac{-X_C i_{max}^2}{T}\int_0^T \cos(\omega t - \phi)\sin(\omega t - \phi)dt = 0$$

Making sense of the result:

The instantaneous power in the capacitor is exactly out of phase with the instantaneous power in the inductor. When the inductor is absorbing energy, the capacitor is releasing energy and vice versa.

Thus, we find that the only element in the driven RLC circuit that dissipates power is the resistor. Total energy must be conserved, so all the power dissipated by the resistor comes from the emf source. The amount of power that the resistor absorbs depends on the frequency. Figure 25-25 shows a plot of power transfer versus frequency for an RLC circuit with $R = 10\ \Omega$, $C = 1\ \mu F$, and $L = 10$ mH.

Frequency dependence is an important general result for all oscillating systems. The maximum energy transfer between the driving emf and the oscillator occurs when the frequency of the driver matches the resonance frequency of the oscillator.

Figure 25-25 Frequency dependence of power dissipated in the resistor of an RLC circuit

Now, we will examine how the quality factor is related to the power curve and the parameters of the RLC circuit. We can use Equation (25-37) to find the power in the resistor:

$$P_{R,\ avg} = i_{rms}^2 R = \frac{\mathcal{E}_{rms}^2 R}{Z^2} = \frac{\mathcal{E}_{rms}^2 R}{R^2 + (X_L - X_C)^2} \quad (25\text{-}40a)$$

Since the Q factor is defined by the half-power points, we need to find the frequencies at which the power is half the peak value. The peak occurs at resonance when $Z = R$; thus, $P_{R,\ avg,\ max} = \dfrac{\mathcal{E}_{rms}^2}{R}$. At the half-power points, the denominator in Equation (25-40a) must be $2R^2$, which requires $(X_L - X_C)^2 = R^2$. Substituting expressions for the reactances, we find

$$\left(\omega L - \frac{1}{\omega C}\right)^2 = R^2$$

or

$$\left(\omega L - \frac{1}{\omega C}\right) = \pm R \quad (25\text{-}40b)$$

This equation yields a quadratic for the angular frequency:

$$L\omega^2 \mp R\omega - \frac{1}{C} = 0 \quad (25\text{-}40c)$$

There are two possible solutions for $+R$ and two for $-R$:

$$\omega = -\frac{\sqrt{C}\sqrt{CR^2 + 4L} - CR}{2CL}$$

$$\omega = \frac{-\sqrt{C}\sqrt{CR^2 + 4L} - CR}{2CL}$$

$$\omega = \frac{\sqrt{C}\sqrt{CR^2 + 4L} + CR}{2CL}$$

$$\omega = \frac{\sqrt{C}\sqrt{CR^2 + 4L} - CR}{2CL} \quad (25\text{-}40d)$$

Since the angular frequency must have a positive value, we reject the two negative roots. The half-power bandwidth is equal to the difference between the two positive roots:

$$\Delta\omega = \frac{\sqrt{C}\sqrt{CR^2 + 4L} + CR}{2CL} - \frac{\sqrt{C}\sqrt{CR^2 + 4L} - CR}{2CL}$$

$$= \frac{R}{L} \quad (25\text{-}40e)$$

Therefore, for a series RCL circuit, the quality factor can be expressed as follows:

KEY EQUATION $\quad Q = \dfrac{f_0}{\Delta f} = \dfrac{\omega_0}{\Delta\omega} = \dfrac{\sqrt{\dfrac{1}{LC}}}{\dfrac{R}{L}} = \dfrac{1}{R}\sqrt{\dfrac{L}{C}} \quad (25\text{-}41)$

EXAMPLE 25-7

The Q-factor from the Resistor Power

Find the quality factor for the RLC circuit with the power curve shown in Figure 25-25.

SOLUTION

The component values are R = 10 Ω, C = 1 μF, and L = 10 mH. Therefore,

$$Q = \frac{1}{R}\sqrt{\frac{L}{C}} = \frac{1}{10\ \Omega}\sqrt{\frac{10\ \text{mH}}{1\ \mu F}} = 10$$

Making sense of the result:

We can compare this value to what we would expect for a radio receiver. FM radios receive signals with a frequency on the order of 100 MHz and the channels are separated by 0.2 MHz. The Q-factor for the tuner must therefore be on the order of 100 MHz/0.2 MHz = 500. So, a value of 10 is not a particularly high Q value.

ELECTRICITY, MAGNETISM, AND OPTICS

A driven series RLC circuit has a response that depends strongly on the frequency of the emf. In general, the current in the circuit is not in phase with the emf. If the frequency is near the resonance frequency of the circuit, there is a large power transfer from the emf to the resistor in the circuit.

In an AC circuit, the emf and the current have the general forms

$$\varepsilon(t) = \varepsilon_{max} \sin(\omega t) \tag{25-1}$$

and

$$i(t) = i_{max} \sin(\omega t - \phi) \tag{25-34}$$

We express their effective values with rms quantities, which for sine and cosine functions are

$$\varepsilon_{rms} = \frac{\varepsilon_{max}}{\sqrt{2}} \quad \text{and} \quad i_{rms} = \frac{i_{max}}{\sqrt{2}} \tag{25-3}$$

In a purely resistive circuit, the current is in phase with the emf and is given by

$$i(t) = \frac{\varepsilon_{max}}{R} \sin(\omega t) = i_{max} \sin(\omega t) \tag{25-5}$$

In a purely inductive circuit, the voltage leads the current by 90°, and the current is given by

$$i(t) = -\frac{\varepsilon_{max}}{\omega L} \cos(\omega t) = \frac{\varepsilon_{max}}{X_L} \sin\left(\omega t + \frac{\pi}{2}\right)$$

We define the inductive reactance as follows:

$$X_L = \omega L \tag{25-12}$$

In a purely capacitive circuit, the voltage lags the current by 90°, and the current is given by

$$i(t) = \varepsilon_{max} \omega C \sin\left(\omega t - \frac{\pi}{2}\right) = \frac{\varepsilon_{max}}{X_C} \sin\left(\omega t - \frac{\pi}{2}\right)$$

The capacitive reactance is defined as

$$X_C = \frac{1}{\omega C} \tag{25-18}$$

In an LC circuit, with the capacitor initially charged we see oscillations, and the charge on the capacitor is given by

$$q(t) = q_0 \cos(\omega_0 t) \tag{25-24}$$

where

$$\omega_0 = \frac{1}{\sqrt{LC}}$$

Phasors can be used to represent voltages in AC circuits. The magnitude of a phasor is given by the magnitude of the signal we are representing. The phasor makes an angle with the x-axis that varies with time. The instantaneous value of the signal can be found by taking the projection of the phasor onto the y-axis.

In an AC series RLC circuit, the current has the form

$$i(t) = i_{max} \sin(\omega t - \phi) \tag{25-34}$$

where

$$i_{max} = \frac{\varepsilon_{max}}{\sqrt{R^2 + (X_L - X_C)^2}} \tag{25-28}$$

and

$$\tan \phi = \frac{X_L - X_C}{R} \tag{25-30}$$

The time-dependent voltages across the circuit elements are given by

$$v_R(t) = R i_{max} \sin(\omega t - \phi)$$
$$v_L(t) = X_L i_{max} \cos(\omega t - \phi) \tag{25-31}$$
$$v_C(t) = X_C i_{max} \cos(\omega t - \phi)$$

The quality factor of an RLC circuit is defined as

$$Q = \frac{f_0}{\Delta f} \tag{25-33}$$

where Δf is the half-power width and f_0 is the centre frequency.

In a series RLC circuit, the only element that dissipates power is the resistor. The instantaneous power dissipated is

$$P_R(t) = i_{max}^2 R \sin^2(\omega t - \phi) \tag{25-36}$$

and the average power dissipated is

$$P_{R,avg} = i_{rms}^2 R \tag{25-37}$$

APPLICATIONS

Applications: energy consumption, power plants, electric circuits, radio and TV tuners, home heater elements

Key Terms: alternating current, capacitive reactance, half-power bandwidth, impedance, inductive reactance, LC circuit, phase shift, phasors, quality factor, Q-factor, reactive circuit elements, resonance frequency, root-mean-square, series RLC circuit

QUESTIONS

1. An RLC circuit is driven at a certain frequency. When the frequency is increased, the maximum current decreases. Is the circuit above or below resonance?
2. What is the phase angle of a series RLC circuit at resonance?
3. At very low frequencies, an inductor has little effect on a series RLC circuit. Why?
4. Two circuits with identical inductors and capacitors have quality factors of 10 and 100. Which has the smaller resistor?
5. Does capacitive reactance increase or decrease with frequency?
6. When we add the maximum voltages that appear across the elements in a series RLC circuit, we usually get a voltage that is greater than the applied emf. Does this violate Kirchhoff's loop rule?
7. Can the rms values of a voltage ever exceed the maximum value? Equal it?
8. What is the limit on the maximum voltage that can appear across the capacitor (or inductor) in a series RLC circuit?
9. What does the phase angle in a series RLC circuit tend toward as the frequency increases significantly?
10. For fixed L and C, how does one increase the quality factor of a series RLC circuit?

PROBLEMS BY SECTION

For problems, star ratings will be used, (✶, ✶✶, or ✶✶✶), with more stars meaning more challenging problems.

Section 25-1   Simple Loads in AC Circuits

11. ✶ Show that the rms value for a triangular wave is $\frac{1}{\sqrt{3}}\varepsilon_{max}$.

12. ✶ The wall outlets in many North American homes provide 110 V rms. What is the maximum amplitude of this emf? In Europe, it is 220 V rms. What is this maximum amplitude?

13. ✶ A 60 W light bulb is connected to 120 V rms.
    (a) What is the rms current?
    (b) What is the resistance of the bulb filament?

14. ✶ A 200 Ω resistor is connected across a 20 V emf with a frequency of 440 Hz.
    (a) What is the amplitude of the current?
    (b) What is the angular frequency of the emf?
    (c) Write an expression for the current through the resistor as a function of time.

15. ✶ A 1 mH inductor is connected across a 20 V rms emf with an angular frequency of 1000 rad/s.
    (a) What is the inductive reactance?
    (b) What is the amplitude of the resulting current?
    (c) Write an expression for the current through the inductor as a function of time.

16. ✶ An inductor is connected across a 100 V emf (RMS) with an angular frequency of 2000 rad/s. The resulting current is given by $i(t) = 0.001 \sin\left(\omega t - \frac{\pi}{2}\right)$.
    (a) What is the amplitude of the current?
    (b) What is the inductive reactance?
    (c) What is the inductance?

17. ✶ A 1.0 nF capacitor is connected across a 20 V rms emf with an angular frequency of 100 rad/s.
    (a) What is the capacitive reactance?
    (b) What is the amplitude of the current in the circuit?
    (c) Write an expression for the current in the circuit as a function of time.

18. ✶ A 1 nF capacitor is connected across an emf with a frequency of 200 Hz. A current with an amplitude of 10 μA results.
    (a) What is the capacitive reactance?
    (b) What is the amplitude of the emf?

Section 25-2   The LC Circuit

19. ✶✶ A 10 μF capacitor is charged and connected across an inductor. The resulting current has a frequency of 10 kHz. What is the value of the inductor?

20. ✶ A 1 nF capacitor is charged and connected across a 0.1 mH inductor. What is the angular frequency of the resulting oscillation?

21. ✶ By what factor does the resonance frequency increase when the capacitance of a capacitor is doubled in value?

Section 25-3   Phasors

22. ✶ The voltage across a circuit element is represented by a phasor with an amplitude of 100 V that makes an angle of $\theta = 200t$ with respect to the positive x-axis. What is the value of the voltage across the element at $t = 10$ s?

23. ✶✶ The current in a circuit is represented by a phasor with an amplitude of 20 mA that makes an angle of $\theta = 5000t$ with respect to the positive x-axis.
    (a) At what time(s) is the current zero?
    (b) At what times is the current at its maximum positive value?
    (c) At what times is the current at its maximum negative value?

24. ✶✶ A circuit has three elements and an emf. The phasors representing the voltages across the elements are given by $v_1 = 20 \sin(\omega t)$, $v_2 = 40\cos(\omega t)$, and $v_3 = -60\cos(\omega t)$. Add the three phasors to find
    (a) the amplitude of the emf;
    (b) the angle between the emf and v_1.

Section 25-4   Series RLC Circuits

25. ✶✶ A 10 Ω resistor, a 5.0 nF capacitor, and a 20 mH inductor are connected in series across an emf $\varepsilon(t) = 25 \sin(33\,000t)$.
    (a) Find the capacitive reactance.
    (b) Find the inductive reactance.
    (c) Find the impedance.
    (d) Find the phase angle.
    (e) Find the amplitude of the current, i_{max}.

26. ✶✶ A 25 nF capacitor and a 30 mH inductor are connected in series with an unknown resistor across an emf $\varepsilon(t) = 25 \sin(33\,000t)$. The current has amplitude of 0.10 A.
    (a) What is the value of the resistor?
    (b) What is the phase angle?

27. ✶✶ A 100 Ω resistor and a 5.0 nF capacitor are connected in series with an unknown inductor across an emf $\varepsilon(t) = 250 \sin(9000t)$. The phase angle is -0.79 rad.
    (a) What is the inductive reactance?
    (b) What is the inductance?
    (c) What is the amplitude of the current?

Section 25-5 Resonance

28. ✻ An RLC circuit has an inductor of 23 mH, a capacitor of 1.0 μF, and a resistor of 20 Ω. What is the resonance frequency?

29. ✻ An RLC circuit has a quality factor of 15, a capacitor of 0.10 μF, and a resistor of 100 Ω. What is the value of the inductor?

30. ✻✻✻ A circuit has a quality factor of 200. By what factor, compared to the value at resonance, does the maximum current change when the frequency is changed to 90% of the resonance frequency?

Section 25-6 Power in AC Circuits

31. ✻ A 110 V rms house circuit is fused for 15 A. How many 100 W light bulbs can go on the circuit?

32. ✻ How much power is dissipated in a 100 Ω resistor connected across an emf with an amplitude of 10 V?

33. ✻✻ A series LC circuit is connected across an emf. At $t = 0$, the capacitor is uncharged and there is no current flowing in the circuit. The emf is then turned on. Does the emf deliver any energy to the circuit?

COMPREHENSIVE PROBLEMS

34. ✻✻ A 20 Ω resistor, a 1.0 nF capacitor, and a 20 mH inductor are in series with an emf $\varepsilon(t) = 50\sin(6000)t$. Find the maximum current and the phase angle.

35. ✻✻ A 40 Ω resistor, a 200 μF capacitor, and an inductor are in series with an emf $\varepsilon(t) = 25\sin(4000)t$. The current is observed to have an amplitude of 45 mA. Find the inductance and the phase angle.

36. ✻✻✻ RLC circuits can be used as filters. Plot the amplitude of the maximum voltage versus frequency across the resistor, capacitor, and inductor for a series RLC circuit with a 500 Ω resistor, 1.0 μF capacitor, and a 100 mH inductor. Assume that the emf has an amplitude of 1.0 V, and plot in the range $0 < \omega < 20\,000$ rad/s.

37. ✻ At an angular frequency of 1000 rad/s, a capacitor has a reactance of 300 Ω. What is its capacitance value?

38. ✻✻ A series RLC circuit is observed to have a phase angle of $-45°$. The amplitude of the voltage across the capacitor is twice the amplitude of the voltage across the resistor. Relative to the capacitor, what is the amplitude of the voltage across the inductor?

39. ✻✻ The windings in electric motors have an electrical resistance from the wire as well as an inductance from the fact that they are wound in coils. A motor that is operated at 110 V rms, 60 Hz, draws 2.0 A of current. The windings have a resistance of 30 Ω. What is the inductance of the motor?

40. ✻✻ A 0.10 μF capacitor and a 10 mH inductor are connected together. The current in the circuit is observed to have a maximum amplitude of 10 mA. What is the frequency of oscillation, and what is the maximum charge on the capacitor?

41. ✻✻ A 0.10 μF capacitor is in an LC circuit that oscillates at 200 kHz. The capacitor is initially charged to 200 nC. What is the value of the inductor, and what is the amplitude of the current?

42. ✻✻ In an RLC circuit, the resistor, inductor, and capacitor have a voltage amplitude of 10 V, 20 V, and 30 V, respectively. What are the phase angle and the amplitude of the emf generator?

43. ✻✻ In an RLC circuit, the voltage amplitude across the resistor is 20 V, the frequency is 2000 Hz, and the phase angle is $-45°$. What is the instantaneous voltage across the resistor at $t = 10$ s?

44. ✻✻ A series circuit contains a resistor and an inductor. The inductance is 10 mH, and the emf has a frequency of 2.0 kHz. The phase angle is 45°. What is the value of the resistance?

45. ✻ A series circuit contains a resistor and a capacitor. The emf is $\varepsilon(t) = 45\sin(5000t)$, and the current is $i(t) = 0.001\sin(5000t + 0.7$ rad$)$. Find the resistance and the capacitance.

46. ✻ A 20 Ω resistor, a 1.0 nF capacitor, and a 20 mH inductor are in series with an emf $\varepsilon(t) = 50\sin(6000t)$. How much power does the resistor dissipate?

47. ✻✻ What should the phase angle be in a circuit to deliver maximum power from the emf (source) to the circuit (load)?

48. ✻✻ The emf and current in an RLC circuit are shown in Figure 25-26. What are the phase angle and resistance?

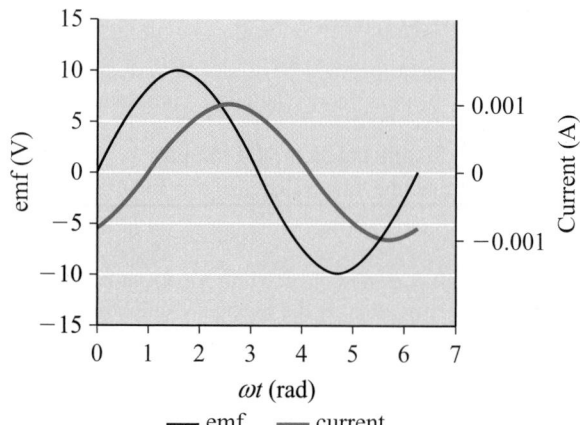

Figure 25-26 Problem 48

49. ✻✻ A filter motor on a swimming pool draws 5.0 A from a 110 V, 60 Hz wall socket and dissipates 500 W rms of power. Find the resistance of the motor and the phase angle.

50. ✻ The receiver in an FM radio has to be tuned from a low frequency of 88 MHz up to 108 MHz. A fixed inductor of 3.0 nH is used. What range must the variable tuning capacitor have to tune across this frequency band?

51. ✻✻✻ FM radio stations are typically separated by 200 kHz in frequency. Each individual station transmits energy in a frequency range from approximately 20 kHz below to 20 kHz above its centre frequency. Design a tuner circuit for an FM radio receiver, using 101.5 MHz as the centre frequency. Your tuning circuit should have a half-power bandwidth of 40 kHz and use a 10 nH inductor.

DATA-RICH PROBLEM

52. ✱✱✱ The markings have worn off the resistor in a series RLC circuit. The circuit has a 1.0 μF capacitor and a 100 mH inductor. You measure the voltage across the resistor as you vary the frequency of the emf and obtain the results in Table 25-1. Use some fitting software to determine the amplitude of the emf and the unknown resistance.

Table 25-1   Voltage versus frequency Data for Problem 52

Angular frequency (rad/s)	Resistor voltage (V)
1 000	11.0
2 000	31.6
3 000	94.9
4 000	55.5
5 000	31.6
6 000	22.5
7 000	17.7
8 000	14.7
9 000	12.6
10 000	11.0
11 000	9.9
12 000	8.9
13 000	8.1
14 000	7.5
15 000	7.0
16 000	6.5
17 000	6.5

OPEN PROBLEM

53. ✱✱✱ You wish to design a very sensitive thermometer that is based on a capacitor whose capacitance changes with temperature. Consider using a driven RLC circuit, and outline the strategy you would take.

See the text online resources at www.physics1e.nelson.com for Open Problems and Data-Rich Problems related to this chapter.

ELECTRICITY, MAGNETISM, AND OPTICS

Chapter 26

Electromagnetic Waves and Maxwell's Equations

The radio waves transmitted by local radio stations, the red laser light used to scan a product's bar code, the radiation that cooks food in a microwave oven, the sunlight that makes life possible on Earth, the X-rays used to detect hidden cavities in your teeth, and the gamma rays emitted by quasars are all electromagnetic waves with different frequencies.

Along with Newton's laws of motion, relativity theory, and quantum mechanics, the theory of electromagnetic waves forms the backbone of modern science and our understanding of the universe. In this chapter, we explore some of the important properties of electromagnetic waves.

Figure 26-1 The galaxy cluster Abell 2744 emits visible light and X-rays.

NASA, ESA, ESO, CXC & D. Coe (STScI/J. Merten (Heidelberg/Bologna)

For additional Making Connections, Examples, and Checkpoints, as well as activities and experiments to help increase your understanding of the chapter's concepts, please go to the text's online resources at www.physics1e.nelson.com.

26-1 The Laws of Electric and Magnetic Fields

So far, we have used electric and magnetic fields to describe the forces that act on stationary and moving charges. In 1862, Scottish physicist James Clerk Maxwell (1831–1879) showed that electric and magnetic fields are two components of a single field, which he called the **electromagnetic field**. Maxwell had noted the strong similarities between the key laws that describe electric and magnetic fields, and developed a unified model of the fields. Maxwell's theory also predicts that oscillating electromagnetic waves can exist in free space and can travel without any medium.

Chapters 20, 23, and 24 discussed two laws that govern electric fields and two similar laws for magnetic fields. Table 26-1 lists the equations for these four laws, and the laws themselves are summarized below.

Table 26-1 Four laws of electric and magnetic fields (for steady currents)

Gauss's law for electric fields	Faraday's law for electric fields
$\oint \vec{E} \cdot d\vec{A} = \dfrac{q_{enc}}{\varepsilon_0}$ (20-14)	$\oint \vec{E} \cdot d\vec{l} = -\dfrac{d\Phi_B}{dt}$
Gauss's law for magnetic fields	**Ampère's law for magnetic fields**
$\oint \vec{B} \cdot d\vec{A} = 0$	$\oint \vec{B} \cdot d\vec{l} = \mu_0 \sum I_{enc}$ (23-39)

Gauss's Law for Electric Fields

Gauss's law for electric fields relates the net flux of an electric field $\vec{E}$ through a closed surface, S, to the net electric charge, q_{enc}, enclosed within that surface. The quantity $\oint \vec{E} \cdot d\vec{A}$ is the electric flux through the surface S and is denoted Φ_E. The area vector element $d\vec{A}$ is normal to the surface and directed outward (Figure 26-2).

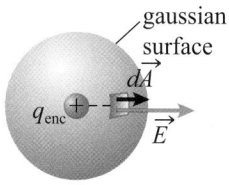

Figure 26-2 Gauss's law for electric fields. The total electric flux through a closed surface is proportional to the total charge enclosed within the surface.

Gauss's Law for Magnetic Fields

Gauss's law for magnetic fields states that the net flux of a magnetic field $\vec{B}$ through a closed surface, S, is zero. The quantity $\oint \vec{B} \cdot d\vec{A}$ is the magnetic flux through the surface S and is denoted by the symbol Φ_B.

Faraday's Law for Electric Fields

Faraday's law for electric fields states that a changing magnetic flux produces an electric field and therefore induces an electromotive force (emf) in a closed circuit. The magnitude of the induced emf is equal to the rate of change of the magnetic flux through the circuit bounded by a closed loop, C.

Ampère's Law for Magnetic Fields

A steady electric current generates a magnetic field that encircles the current. Ampère's law for magnetic fields states that the line integral of a magnetic field $\vec{B}$ around a closed loop that encloses the current is equal to the permeability times the electric current passing through the loop.

As you can see from Table 26-1, the equations for these four laws are quite similar. The left sides of the equations for both of Gauss's laws describe the flux of the field through a closed surface. For the electric field, the flux depends on the quantity of the charge enclosed by that surface. When the net charge within the surface is positive, a net electric flux *emerges* through the surface. When the net charge is negative, a net flux *enters* the surface. However, the net magnetic flux through a closed surface is zero because sources that produce magnetic fields always have both a north pole and a south pole. An isolated magnetic monopole (a single north pole or a single south pole) has never been observed. Therefore, magnetic field lines that originate from the north pole of a magnet eventually have to reenter the magnet's south pole. Consequently, the magnetic flux leaving a closed surface equals the flux reentering the surface. If magnetic monopoles are ever discovered, a nonzero term will have to be added to the right side of the equation for Gauss's law for magnetic fields, making it completely symmetric with Gauss's law for electric fields.

The left sides of the equations for Faraday's law for electric fields and Ampère's law for magnetic fields are both line integrals of a field around a closed loop. However, the right sides of these two equations look quite different. The equation for Faraday's law has no term proportional to a current because there are no magnetic monopoles. (If monopoles did exist, then the right side of Faraday's law would have a term describing a magnetic current due to the flow of magnetic monopoles.) The right side of the equation for Faraday's law indicates that the induced electric field is proportional to the rate of change of the magnetic flux, but Ampère's law has no corresponding term. Does a changing electric flux produce a magnetic field?

Maxwell realized that Ampère's law as stated in Table 26-1 holds only when the electric current I is constant. When a current changes with time, the equation for Ampère's law needs an additional term.

✓ **CHECKPOINT**

C-26-1 Gauss's Law for Electric Fields

Rank each charge configuration in Figure 26-3 in decreasing order of the amount of net electric flux passing through each closed spherical surface. Use the equality sign if needed.

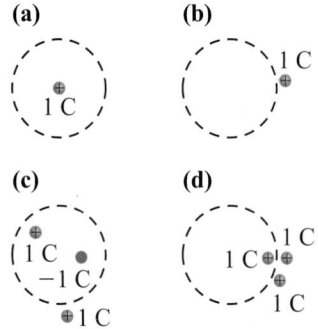

Figure 26-3 C-26-1

C-26-1 (a) = (b) < (d) = (c)

LO 2

26-2 Displacement Current and Maxwell's Equations

Consider the process of charging a parallel plate capacitor by passing a current $I(t)$ through a connecting wire. As the capacitor charges, a charge $q(t)$ accumulates on its plates and an instantaneous electric field $\vec{E}(t)$ begins to build up between the two plates of the capacitor. We know that a magnetic field exists around a current-carrying wire. Let us apply Ampère's law to the circular loop a of radius R

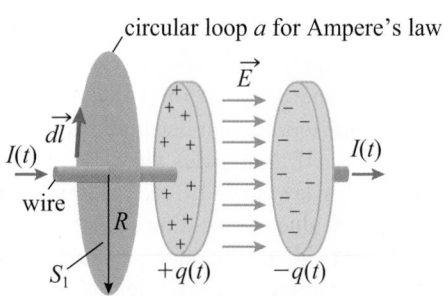

Figure 26-4 A parallel-plate capacitor being charged. The conduction current $I(t)$ induces a magnetic field around the wire, which can be calculated by applying Ampère's law to the flat surface S_1.

that is centered on the wire, as shown in Figure 26-4. The loop a encloses the planar surface S_1 that is being penetrated by the current $I(t)$. So the magnitude of the magnetic field around the path is $B(t) = (\mu_0/2\pi R)\,I(t)$. Now let us distort the surface S_1 (still bounded by the same circular loop a) to a surface S_2 such that the surface S_2 passes through the space between the plates of the capacitor, as shown in Figure 26-5. In this case, no current is penetrating through the surface S_2 (because there is no flow of current between the two plates) and hence the line integral of $\vec{B}$ around the circular path vanishes. So Ampère's law, as given by (23-39) is ambiguous. It gives different results depending upon how the surface bounding the circular loop is chosen. Notice that this ambiguity arises only when the capacitor is being charged (or discharged). When the capacitor is fully charged, current flow through the wire stops and the line integral of $\vec{B}$ is zero for any choice of surface bounding the circular path.

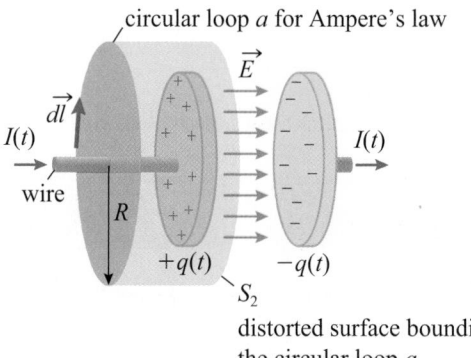

distorted surface bounding the circular loop a

Figure 26-5 There is no conduction current through the surface S_2, however, there is a displacement current through this surface due to a changing electric flux as the charge builds up on the capacitor plates.

Maxwell noted that as the capacitor is being charged, the flux of the electric field between the capacitor plates increases with time. So the *changing* electric field must generate a magnetic field that has exactly the same magnitude as the magnetic field generated when the electric current passes through surface S_1. Using this insight, Maxwell extended Ampère's law:

KEY EQUATION
$$\oint \vec{B} \cdot d\vec{l} = \mu_0 I + \mu_0 \left(\varepsilon_0 \frac{d\Phi_E}{dt} \right) \qquad (26\text{-}1)$$

where Φ_E denotes electric flux, $\dfrac{d\Phi_E}{dt}$ is the rate of change of electric flux with time, and $I = \sum I_{\text{enc}}$. Equation (26-1) is called the Ampere-Maxwell equation. The quantity $\varepsilon_0 \dfrac{d\Phi_E}{dt}$ has units of current as shown below and is called the **displacement current**, I_{disp}:

$$[I_{\text{disp}}] = \frac{C^2}{N \cdot m^2} \times \frac{1}{s} \times \frac{N \cdot m^2}{C} = \frac{C}{s}$$

Note that no current is actually displaced; the name simply distinguishes the quantity $\varepsilon_0 \dfrac{d\Phi_E}{dt}$ from the current I that is due to a flow of charges. With the addition of the displacement current term, Ampère's law gives an unambiguous result for the magnetic field generated around any closed loop when the current is not steady. This is how it works for the two surfaces of Figures 26-4 and 26-5.

- Surface S_1: A current I flows through the surface, but there is no electric flux through S_1. Hence, the displacement current is zero, and the magnetic field is generated by the conduction current I.

- Surface S_2: There is no conduction current through the surface, but there is a changing electric flux through it. The magnetic field is generated by the displacement current (i.e., by the changing electric flux). As will be shown below, this magnetic field is exactly the same as the magnetic field generated by the conduction current I.

- When the capacitor is fully charged, the conduction current stops. There is an electric flux through the surface S_2, but it is now constant. So the displacement current is zero, and there is no magnetic field in the circuit.

We showed that the dimension of the displacement current is the same as that of the conduction current (due to the flow of electrons). Now we will show that it has the same magnitude. Let $q(t)$ be the charge on a capacitor plate at time t. The charge density (charge per unit area) on a plate of cross-sectional area A is

$$\sigma(t) = \frac{q(t)}{A}$$

Therefore, the magnitude of the electric field between the two plates of the capacitor, ignoring edge effects, is

$$E(t) = \frac{\sigma(t)}{\varepsilon_0} = \frac{q(t)}{A\varepsilon_0}$$

Assuming that the electric field between the plates is uniform, the electric flux, Φ_E, between the plates is the product of the electric field and the area of a single plate:

$$\Phi_E(t) = E(t) \times A = \frac{q(t)}{\varepsilon_0}$$

The rate of change of the electric flux is

$$\frac{d\Phi_E(t)}{dt} = \frac{1}{\varepsilon_0}\frac{dq(t)}{dt}$$

Since the rate of change of charge on the plates is equal to the conduction current,

$$I_{\text{disp}}(t) = \varepsilon_0 \frac{d\Phi_E(t)}{dt} = \frac{dq(t)}{dt} = I(t)$$

Thus, *the magnitude of the displacement current is the same as the magnitude of the conduction current.*

We emphasize that no electrons are flowing across the gap between the capacitor plates. The changing electric flux between the plates generates a magnetic field that is exactly the same as the magnetic field generated by the conduction current in the wires (Figure 26-6). If the circuit were enclosed in a sealed box, you would detect a magnetic field around the box, but you would not be able to do any experiment to determine whether the magnetic field is due to a conduction current or a displacement current.

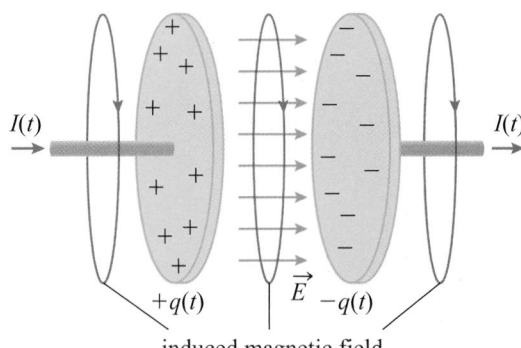

Figure 26-6 As a capacitor is charged, the induced magnetic field around the wire is due to the conduction current, and the field induced between the plates is due to the displacement current.

✓ CHECKPOINT

C-26-2 Displacement Currents

The graph in Figure 26-7 shows the magnitudes of four changing uniform electric fields, all with the same direction. Consider a square loop of area 5 cm² perpendicular to the electric fields. Rank the displacement currents from the changing fields from least to greatest.

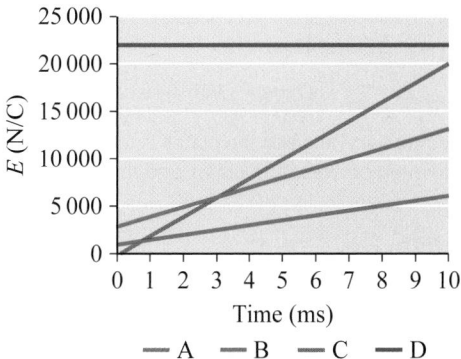

Figure 26-7 A graph of magnitudes of four electric fields as a function of time.

C-26-2 C < A < B < D

ELECTRICITY, MAGNETISM, AND OPTICS

EXAMPLE 26-1

Magnetic Field Due to the Displacement Current

D.F. Bartlett and T.R. Cole measured the induced magnetic field inside a charging capacitor in 1984 using a very sensitive magnetometer. They used a capacitor with two circular plates that were 1.22 cm apart. Instead of using direct current to charge the capacitor, they charged the capacitor using a generator that produced a 1.25 kHz sinusoidal voltage with an amplitude of 340 V. The arrangement is similar to the one shown in Figure 26-6. Calculate the peak strength of the magnetic field around a loop of radius 2.00 cm centred on the central axis of the capacitor. Assume that the radius of the capacitor plates is 6.00 cm so that the edge effects can be ignored.

SOLUTION

As the capacitor is being charged, the electric flux between the plates increases and generates a magnetic field. The magnitude and direction of the induced magnetic field are given by the Ampère–Maxwell Equation (26-1). Since there is no conduction current in the region between the plates,

$$\oint \vec{B} \cdot d\vec{l} = \mu_0 \left(\varepsilon_0 \frac{d\Phi_E}{dt} \right)$$

We now evaluate each side of this equation.

Right side:

The electric field is perpendicular to the surface of the plates and has the same magnitude, $E(t)$, anywhere between the plates at a given time t. The electric flux, Φ_E, through a disk of radius r centred on the central axis of the plates is given by

$$\Phi_E(t) = E(t) \times \pi r^2$$

The rate of change of this flux is

$$\frac{d\Phi_E(t)}{dt} = \pi r^2 \frac{dE(t)}{dt} \tag{1}$$

The potential difference, V, between the two plates has a maximum value of 340 V and varies sinusoidally with a frequency of 1.25 kHz. Therefore,

$$V(t) = V_{max} \sin(2\pi ft) = (340 \text{ V}) \sin(2\pi \times 1250t)$$

The electric field between the plates is related to the potential difference between the plates and their separation, L:

$$E(t) = \frac{V(t)}{L} = \frac{(340 \text{ V}) \sin(2\pi \times 1250t)}{L}$$

Inserting this expression for $E(t)$ into the right side of Equation (1), we get

$$\mu_0 \varepsilon_0 \frac{d\Phi_E(t)}{dt} = \mu_0 \varepsilon_0 \frac{\pi r^2}{L} (340 \text{ V}) \frac{d\sin(2500\pi t)}{dt}$$

$$= 85 \times 10^4 \left(\frac{\pi^2 r^2 \mu_0 \varepsilon_0}{L} \right) \cos(2500\pi t) \, \frac{\text{C}}{\text{s}}$$

Left side:

We need to calculate the line integral around the circular amperian loop of radius r shown in Figure 26-6. The magnetic field is tangent to the loop, and the loop is centred symmetrically on the central axis. Therefore, the magnetic field has the same magnitude at every point around the loop. Taking $\vec{B}$ and the path element $d\vec{l}$ parallel to each other, we get

$$\oint \vec{B} \cdot d\vec{l} = \oint B \cos(0) \, dl = B \oint dl = B(2\pi r)$$

Equating the left and the right sides, we get the magnitude of the induced magnetic field around a circular loop of radius r that lies inside the two plates:

$$B(t) = 85 \times 10^4 \left(\frac{\pi r \mu_0 \varepsilon_0}{2L} \right) \cos(2500\pi t) \text{ T}$$

Notice that both the electric field and the induced magnetic field oscillate with the same frequency. The amplitude of the magnetic field is equal to the factor multiplying the cosine term. Substituting the known values $r = 2.00 \times 10^{-2}$ m, $L = 1.22 \times 10^{-2}$ m, $\mu_0 = 4\pi \times 10^{-7}$ T·m/A, and $\varepsilon_0 = 8.85 \times 10^{-12}$ C²/N·m², we get

$$B(t) = (2.5 \times 10^{-11}) \cos(2500\pi t) \text{ T}$$

The peak strength of the magnetic field occurs when $\cos(2500\pi t) = \pm 1$ and is therefore $\pm 2.5 \times 10^{-11}$ T.

Making sense of the results:

The magnitude of Earth's magnetic field ranges between 3×10^{-5} T and 6×10^{-5} T, which is much larger than the induced magnetic field in even a large capacitor. Physicists in the time of Faraday and Ampère were unable to detect such small magnetic fields due to changing electric flux in their experiments.

The four equations in Table 26-1, with Ampère's law replaced by Equation (26-1), are collectively called **Maxwell's equations**. **Maxwell's equations** describe the physics of all phenomena that involve electric and magnetic fields, including the attraction between nuclei and electrons, van der Waals forces between molecules, friction, viscosity, and the propagation of electromagnetic waves. Maxwell's equations are to electromagnetism what Newton's laws are to mechanics.

Maxwell's Equations in a Vacuum

There is no matter in a vacuum and therefore there are no charges or currents. Consequently, Maxwell's equations become much simpler in a vacuum:

$$\oint \vec{E} \cdot d\vec{A} = 0 \quad \text{(Gauss's law for electric fields)} \quad (26\text{-}2)$$

$$\oint \vec{B} \cdot d\vec{A} = 0 \quad \text{(Gauss's law for magnetic fields)} \quad (26\text{-}3)$$

$$\oint \vec{E} \cdot d\vec{l} = -\frac{d\Phi_B}{dt} \quad \text{(Faraday's law)} \quad (26\text{-}4)$$

$$\oint \vec{B} \cdot d\vec{l} = \mu_0 \varepsilon_0 \frac{d\Phi_E}{dt} \quad \text{(Ampère's law)} \quad (26\text{-}5)$$

In a vacuum, Gauss's laws for electric and magnetic fields are identical in form. There are no charges in a vacuum, so the net flux of the electric field through any closed surface is zero. Similarly, the net magnetic flux through a closed surface is zero because there are no magnetic monopoles.

Other than a multiplying factor, Faraday's and Ampère's laws also have identical forms in a vacuum. A changing magnetic flux produces an electric field, and any change in electric flux produces a magnetic field. The minus sign on the right side of Faraday's law is equivalent to Lenz's law, which states that the direction of the induced electric field is such that the induced current produces a magnetic field that opposes the original change in the magnetic flux.

Equations (26-4) and (26-5) indicate that the electric and magnetic fields in a vacuum are coupled. If the strength of the electric field changes with time, the changing electric flux produces a magnetic field that also changes with time. A changing magnetic field means that magnetic flux is changing; hence, a changing electric field is produced. This electric field, in turn, produces a changing magnetic field, and so on, as shown in Figure 26-8. The coupling of the electric and the magnetic fields continuously produces electric and magnetic fields and sets up an electromagnetic wave that propagates in the vacuum. We will study such propagation in the next section.

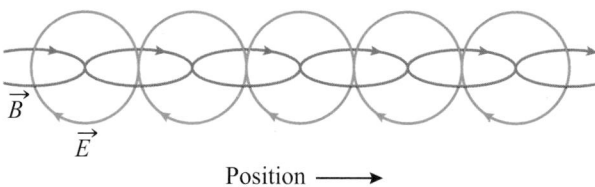

Figure 26-8 An electric field that changes with time produces a magnetic field that also changes with time, and vice versa.

Equations (26-2) to (26-5) contain integrals of electric and magnetic fields and are called "Maxwell's equations in integral form." We can also write Maxwell's equations in terms of derivatives of electric and magnetic fields, and those equations are called "Maxwell's equations in differential form."

Maxwell published his work in 1873. Although relativity and quantum theory were developed in the next century, Maxwell's equations are consistent with both of these theories. Maxwell's careful synthesis of a large amount of seemingly unrelated experimental data produced one of the most elegant theories in physics.

LO 3

26-3 Electromagnetic Waves

An **electromagnetic wave** consists of coupled electric and magnetic fields that vary with time. Electromagnetic waves are generated by accelerating charges and by quantum-mechanical processes (see Section 26-7). Once an electromagnetic wave is generated, it propagates on its own and does not require any physical medium to continue its propagation. The wave generates itself because a time-varying electric field generates a time-varying magnetic field, which in turn generates a time-varying electric field. This process is continuously repeated.

Electromagnetic waves have the following properties:

■ They are *three-dimensional* because the electric field, the magnetic field, and the propagation direction of the wave are all perpendicular to each other.

■ Electromagnetic waves are *transverse*. The direction of motion of the wave is perpendicular to the direction of oscillation of the wave's electric and magnetic fields.

■ In a vacuum, all electromagnetic waves travel at the speed of light.

The existence of electromagnetic waves can be derived from Maxwell's equations using vector calculus. However, here we will simply show that a sinusoidal plane electromagnetic wave in a vacuum satisfies Maxwell's equations.

Recall from Chapter 15 that a plane wave is a wave with a fixed frequency whose wave fronts are parallel planes of constant amplitude and propagate perpendicular to the direction of motion of the wave. Therefore, the properties of a plane wave are the same when observed at any point on a given plane perpendicular to the direction of motion of the wave (Figure 26-9). A plane wave

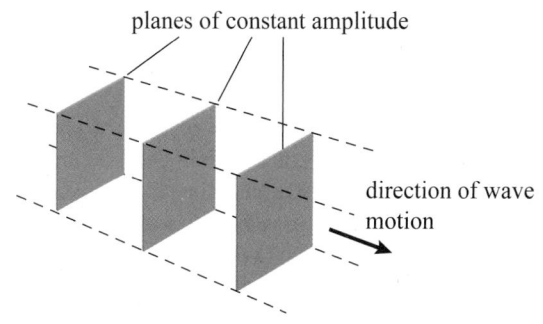

Figure 26-9 A three-dimensional plane wave

is an approximation of a spherical wave that propagates outward from a source. At distances that are much larger than the wavelength of the emitted wave, a spherical wave can be accurately approximated as a plane wave.

Figure 26-10 shows a sinusoidal plane electromagnetic wave travelling in the positive x-direction. The directions of oscillation of the electric field and the magnetic field are chosen as the y-axis and z-axis, respectively. Recall from Chapter 14 that a sinusoidal wave travelling in the positive x-direction is described by the function $\sin(kx - \omega t)$. Here, $k = \dfrac{2\pi}{\lambda}$ is the wave number, and $\omega = 2\pi f$ is the angular frequency of the wave. The components of the electric and magnetic fields of the plane electromagnetic wave in Figure 26-10 are as follows:

$$E_x(x,t) = 0; \; E_y(x,t) = E_0 \sin(kx - \omega t); \; E_z(x,t) = 0 \quad (26\text{-}6)$$

$$B_x(x,t) = 0; \; B_y(x,t) = 0; \; B_z(x,t) = B_0 \sin(kx - \omega t) \quad (26\text{-}7)$$

where E_0 is the amplitude of the electric field (measured in V/m), and B_0 is the magnitude of the magnetic field (measured in T).

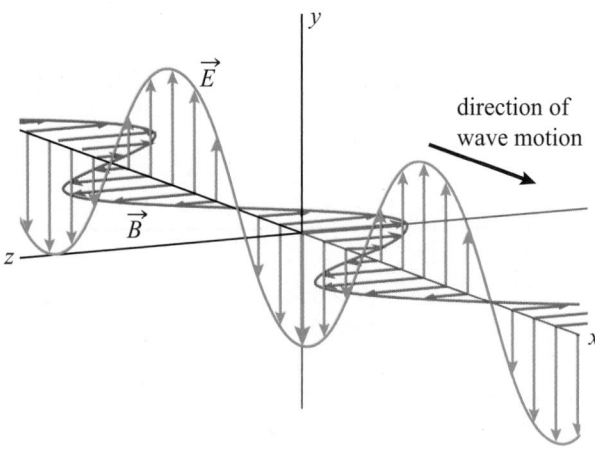

Figure 26-10 A sinusoidal travelling electromagnetic wave at a moment when the wave crests are at the origin of the coordinate system

toward the reader, then at any given instant the electric field has the same magnitude and direction at all points in the plane of the page of the paper. The same is true for the magnetic field. The magnitudes of the electric and magnetic fields oscillate with time, reaching maximum and minimum values in unison (Figure 26-11).

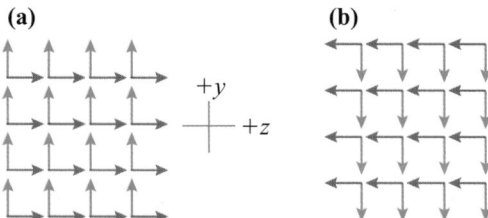

Figure 26-11 (a) A sinusoidal plane electromagnetic wave at an instant when the electric fields (red arrows) and the magnetic fields (green arrows) reach their maximum strengths in the positive y- and z-directions, respectively. The electric and magnetic fields are in the plane of the paper, and the wave is travelling toward the reader. (b) The electric and magnetic fields of the wave half a time period later

✓ CHECKPOINT

C-26-3 Field Orientation of Electromagnetic Waves

Figure 26-12 shows four orientations of electric (red) and magnetic (green) fields in a vacuum at a given instant of time. Which orientation could represent an electromagnetic wave? Explain why, and determine the direction in which the wave is travelling.

Figure 26-12 C-26-3

C-26-3 (a) and (d) The electric and magnetic fields are perpendicular to each other. These waves are travelling in a direction that is perpendicular to the plane of the page.

Note that the electric and magnetic fields have the same wavelength, frequency, and phase constant (which we have chosen to be zero). According to Equation (26-6), the magnitude of the electric field at $x = 0$ varies with time as $E_0 \sin(-\omega t) = -E_0 \sin(\omega t)$, and the direction of the electric field parallels the y-axis. A charge $+q$ at $x = 0$ would experience a force of magnitude $qE_0 \sin(-\omega t)$ directed along the y-axis. Similarly, the magnitude of the magnetic field at $x = 0$ varies with time as $B_0 \sin(-\omega t)$, and the direction of the magnetic field parallels the z-axis.

For a plane electromagnetic wave travelling along the x-axis, the magnitudes of the electric and magnetic fields are the same everywhere on a given yz-plane. If the yz-plane is the plane of the page with the x-axis pointing

We will now consider each of Maxwell's equations as they apply to a plane electromagnetic wave travelling in a vacuum.

Gauss's Law for Electric Fields

Since there are no charges in a vacuum, the net flux of the electric field through any closed surface must be zero. Imagine a cubical Gaussian surface of length h, like the one in Figure 26-13(a), in the presence of a plane electromagnetic wave described by Equations (26-6) and (26-7). Since the electric field does not depend on the y-coordinate, the magnitude of the electric field entering the cube through the bottom surface is equal to the magnitude of the field that leaves the top surface of the cube.

Therefore, the electric flux entering the bottom surface is equal to the flux leaving the top surface. The electric flux through the surfaces in the *xy*- and *yz*-planes of the cube is zero because the direction of the electric field is parallel to these surfaces. Therefore, the net electric flux through the box is zero, and Gauss's law for electric fields is satisfied.

(a)   The net electric flux through the Gaussian box is zero.

(b)   The net magnetic flux through the Gaussian box is zero.

Figure 26-13   (a) The electric flux entering the bottom surface in the *xz*-plane of the cube is equal to the flux leaving through the corresponding top surface. (b) Similarly, the magnetic flux entering from the back *xy*-surface of the cube is equal to the flux leaving from the corresponding front surface.

Gauss's Law for Magnetic Fields

Consider the magnetic flux through the cubical Gaussian surface in Figure 26-13(b). The magnetic flux through the surfaces in the *xz*- and *yz*-planes of the cube is zero because the direction of the magnetic field is parallel to these surfaces. The magnetic field does not depend on the *z*-coordinate, so the magnetic flux entering the back surface in the *xy*-plane is equal to the flux leaving through the parallel front surface. Thus, the net magnetic flux through the Gaussian surface is zero, and Gauss's law for magnetic fields is satisfied.

Faraday's Law

According to Faraday's law, a changing magnetic flux through a closed loop induces an electric field around that loop. The loop may be an imaginary surface or a part of a physical circuit, like a loop of wire. Consider an infinitesimally narrow, rectangular strip in the *xy*-plane, with height *h* and width Δx (Figure 26-14) in the presence of a plane electromagnetic wave described by Equations (26-6) and (26-7). The left edge of the strip is at *x*, and the right edge is at $x + \Delta x$. The width is sufficiently small that the magnitude of the magnetic field passing through the strip can be assumed to remain constant over the entire width. Both the magnetic field and the normal to the strip are pointing along the *z*-axis. Therefore, the magnetic flux

through the strip is the product of the magnitude of the magnetic field $B_z(x, t)$ and the area of the strip:

$$\Phi_B(x, t) = B_z(x, t)(h\Delta x)$$

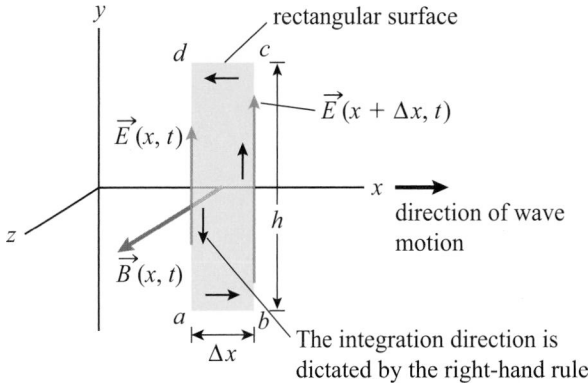

Figure 26-14   The closed rectangular surface in the *xy*-plane has an infinitesimally small width. The surface is placed in the path of an electromagnetic wave that has $\vec{E}$ along the *y*-axis and $\vec{B}$ along the *z*-axis.

The magnetic field is a function of time, so the magnetic flux passing through the strip changes with time, and

$$\frac{\partial}{\partial t}\Phi_B(x, t) = \frac{\partial}{\partial t}[B_z(x, t)(h\Delta x)] = (h\Delta x)\frac{\partial B_z(x, t)}{\partial t} \quad (26\text{-}8)$$

The changing magnetic flux induces an electric field around the strip. *This induced electric field is the electric component of the electromagnetic wave.* The magnetic field points along the positive *z*-axis; therefore, according to the right-hand rule, the circulation of the electric field is in the counterclockwise direction, as shown in Figure 26-14:

$$\oint \vec{E} \cdot d\vec{l} = \int_a^b \vec{E} \cdot d\vec{l} + \int_b^c \vec{E} \cdot d\vec{l}$$

$$+ \int_c^d \vec{E} \cdot d\vec{l} + \int_d^a \vec{E} \cdot d\vec{l} \quad (26\text{-}9)$$

The electric field has only a *y*-component. The first and third line integrals therefore vanish because the line elements $d\vec{l}$ for these sections point along the *x*-axis, perpendicular to the direction of the electric field. The line integral along the path *bc* is

$$\int_b^c \vec{E} \cdot d\vec{l} = \left(E_y(x + \Delta x, t)\hat{y}\right) \cdot (h\hat{y}) = hE_y(x + \Delta x, t)$$

Here, the electric field is evaluated at $x + \Delta x$, the location of the *bc* edge of the strip. The path *da* is traversed in the negative *y*-direction, from top to bottom (counterclockwise). Therefore,

$$\int_d^a \vec{E} \cdot d\vec{l} = \left(E_y(x, t)\hat{y}\right) \cdot \left((h)(-\hat{y})\right) = -hE_y(x, t)$$

Combining contributions from all the sides, the line integral of the electric field around the rectangular strip is

$$\oint \vec{E} \cdot d\vec{l} = E_y(x + \Delta x, t)h - E_y(x, t)h \quad (26\text{-}10)$$

The expression on the right side is related to the derivative of $E_y(x, t)$ with respect to x. By definition, the derivative of a function $f(x)$ of a variable x is related to its values at two nearby points by

$$\frac{df(x)}{dx} = \lim_{\Delta x \to 0} \left(\frac{f(x + \Delta x) - f(x)}{\Delta x} \right)$$

Therefore, as the width $\Delta x \to 0$,

$$E_y(x + \Delta x, t) - E_y(x, t) \Rightarrow \Delta x \frac{dE_y(x, t)}{dx}$$

and

$$\oint \vec{E} \cdot d\vec{l} = (h\Delta x)\frac{dE_y(x, t)}{dx} \quad (26\text{-}11)$$

Inserting Equations (26-11) and (26-8) into Equation (26-4), we get

$$\frac{\partial E_y(x, t)}{\partial x} = -\frac{\partial B_z(x, t)}{\partial t} \quad (26\text{-}12)$$

In Equation (26-12), we have replaced the total derivative of the electric field by the partial derivative because the electric field depends on both position and time. Equation (26-12) emphasizes that while taking the derivative of $E_y(x, t)$ with respect to x, we must keep t constant. Similarly, while taking the derivative of $B_z(x, t)$ with respect to t, we must keep x constant.

When applied to plane electromagnetic waves, Faraday's law requires that *the rate of change of the electric field with position in the neighbourhood of a given point be equal and opposite to the rate at which the magnetic field changes with time at that point*. Therefore, if the electric field changes with position, then the magnetic field must change with time.

So far, we have not used the fact that the plane electromagnetic wave is a sinusoidal travelling wave. We can evaluate both sides of Equation (26-12) using Equations (26-6) and (26-7):

$$\frac{\partial E_y(x, t)}{\partial x} = \frac{\partial}{\partial x}(E_0 \sin(kx - \omega t)) = kE_0\cos(kx - \omega t)$$

and

$$-\frac{\partial B_z(x, t)}{\partial t} = -\frac{\partial}{\partial t}(B_0 \sin(kx - \omega t)) = \omega B_0\cos(kx - \omega t)$$

Dividing out the cosine term, we get

$$kE_0 = \omega B_0 \quad (26\text{-}13)$$

Thus, Faraday's law requires that the amplitude of the magnetic field be equal to the amplitude of the electric field divided by the speed of the electromagnetic waves:

$$B_0 = \frac{k}{\omega}E_0 = \frac{E_0}{v_{em}} \quad (26\text{-}14)$$

where $v_{em} = \omega/k$ is the speed of the electromagnetic waves in a vacuum.

Ampère's Law

The magnetic field of the plane electromagnetic wave points along the z-axis (Equation (26-7)). We therefore choose a closed rectangular strip in the xz-plane, as shown in Figure 26-15.

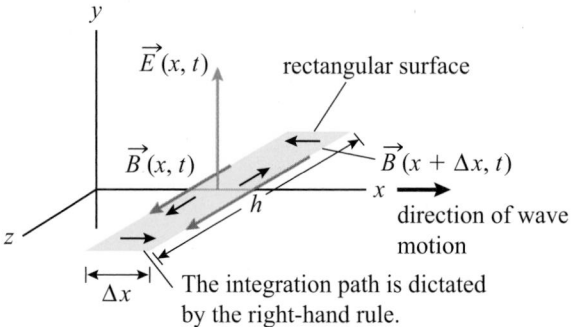

Figure 26-15 The closed rectangular surface has an infinitesimally small width in the xz-plane. The electromagnetic wave has $\vec{E}$ along the y-axis and $\vec{B}$ along the z-axis.

The electric field is normal to the plane of the strip, and the electric flux passing through the strip is the product of the magnitude of the electric field $E_y(x, t)$ and the area of the strip:

$$\Phi_E(x, t) = E_y(x, t)(h\Delta x)$$

The rate at which the electric flux passing through the strip varies with time is given by

$$\frac{\partial}{\partial t}\Phi_E(x, t) = \frac{\partial}{\partial t}[E_y(x, t)(h\Delta x)] = (h\Delta x)\frac{\partial E_y(x, t)}{\partial t} \quad (26\text{-}15)$$

The changing electric flux induces a magnetic field around the rectangular strip. *The induced magnetic field forms the magnetic field component of the electromagnetic wave.* The calculation of the line integral of the magnetic field around the strip is similar to the calculation for the electric field. The result is

$$\oint \vec{B} \cdot d\vec{l} = (h\Delta x)\frac{\partial B_z(x, t)}{\partial x} \quad (26\text{-}16)$$

Inserting Equations (26-15) and (26-16) into Equation (26-5) for Ampère's law we get

$$\frac{\partial B_z(x, t)}{\partial x} = -\varepsilon_0 \mu_0 \frac{\partial E_y(x, t)}{\partial t} \quad (26\text{-}17)$$

When applied to plane electromagnetic waves, Ampère's law requires *that the rate of change of the magnetic field with position in the neighbourhood of a given point be proportional and opposite to the rate at which the electric field changes with time at that point.*

Inserting Equations (26-6) and (26-7) and evaluating the partial derivatives with respect to x and t, we get the following constraint imposed by Ampère's law on electric and magnetic fields of a plane electromagnetic wave:

$$kB_0 = \varepsilon_0 \mu_0 (\omega E_0) \qquad (26\text{-}18)$$

The Speed of Electromagnetic Waves

When we combine the constraints of Equations (26-14) and (26-18), we get a remarkable result. Equation (26-18) can be written as

$$B_0 = \varepsilon_0 \mu_0 \left(\frac{\omega}{k}\right) E_0 = \varepsilon_0 \mu_0 v_{em} E_0 \qquad (26\text{-}19)$$

Combining Equations (26-19) and (26-14), we get

$$v_{em} = \frac{1}{\sqrt{\varepsilon_0 \mu_0}} \qquad (26\text{-}20)$$

This equation shows that the speed of electromagnetic waves in a vacuum depends *only* on the electric permittivity and magnetic permeability of the vacuum. The speed does not depend on the frequency, wavelength, or amplitude of the waves. Substituting the known values for ε_0 and μ_0 into Equation (26-20) gives

$$v_{em} = \frac{1}{\sqrt{\varepsilon_0 \mu_0}}$$

$$= \frac{1}{\sqrt{(8.85 \times 10^{-12}\, \text{C}^2/\text{N}\cdot\text{m}^2)(4\pi \times 10^{-7}\, \text{T}\cdot\text{m}/\text{A})}}$$

$$= 3.00 \times 10^8\, \text{m/s} \qquad (26\text{-}21)$$

which is precisely the speed of light, c. Therefore,

KEY EQUATION $$v_{em} = c$$

So, Maxwell's equations predict that *all electromagnetic waves in a vacuum travel at the speed of light.* From Equation (26-19), using $k = 2\pi/\lambda$ and $\omega = 2\pi f$, we get

$$c = \frac{\omega}{k} = \lambda f \qquad (26\text{-}22)$$

To summarize, Maxwell's equations predict the existence of electromagnetic waves with the following properties:

1. Electromagnetic waves travel in a vacuum at the speed of light. The wavelength and frequency of an electromagnetic wave are related to its speed by $\lambda f = c$.

2. The electric and magnetic fields are perpendicular to the direction of propagation of the wave. Electromagnetic waves are therefore transverse waves.

3. The electric and magnetic fields are perpendicular to each other.

4. The amplitudes of electric and magnetic fields of an electromagnetic wave are related such that

$$B_0 = \frac{E_0}{v_{em}} = \frac{E_0}{c}$$

We have shown that sinusoidal plane electromagnetic waves satisfy Maxwell's equations. Electromagnetic waves with complex waveforms can be produced by the superposition of sinusoidal waves of various frequencies and amplitudes, similar to the way that complex sound waves can be formed by a combination of harmonics. Since Maxwell's equations are linear in electric and magnetic fields, such complex electromagnetic waves also satisfy Maxwell's equations.

✓ CHECKPOINT

C-26-4 Plane Electromagnetic Waves

The four plots in Figure 26-16 show the directions of electric (red) and magnetic (green) fields at a given instant. Determine whether the electric and magnetic fields in each plot correspond to those of a plane electromagnetic wave. If yes, determine the direction of motion of the wave.

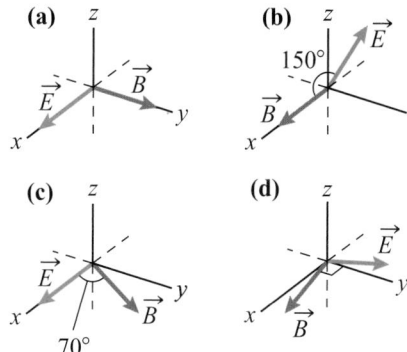

Figure 26-16 C-26-4

C-26-4 (a) Yes, z; (b) no; (c) no; (d) yes, z.

EXAMPLE 26-2

A Helium–Neon Laser

A helium–neon laser produces a beam of light with a wavelength of 633 nm and an electric field strength of 1000 V/m.

The direction of motion of the beam is taken to be along the z-axis, and the electric field is parallel to the x-axis. A laser beam is an electromagnetic plane wave.

(continued)

(a) Find the frequency of the laser beam.

(b) What are the amplitude and the direction of the magnetic field of the beam?

SOLUTION

(a) The wavelength and the frequency of this wave are related by Equation (26-22). Rearranging and solving for f, we get

$$f = \frac{c}{\lambda} = \frac{3.00 \times 10^8 \, \text{m/s}}{633 \times 10^{-9} \, \text{m}} = 4.74 \times 10^{14} \, \text{Hz}$$

(b) The amplitude of the magnetic field is related to the amplitude of the electric field by Equation (26-14). Solving for B_0 gives

$$B_0 = \frac{E_0}{c} = \frac{1000 \, \text{V/m}}{3.00 \times 10^8 \, \text{m/s}} = 3.33 \times 10^{-6} \, \text{T}$$

The direction of the magnetic field is perpendicular to the direction of motion of the laser beam and that of the electric field; therefore, the direction of the magnetic field must be parallel to the y-axis.

LO 4

26-4 The Electromagnetic Spectrum

The wavelength (and, hence, the frequency) of an electromagnetic wave can have any positive value. The largest possible wavelength is limited only by the size of the universe. The shortest wavelength can tend to zero. This range of allowed wavelengths is called the **electromagnetic spectrum**. As shown in Figure 26-17, regions of the electromagnetic spectrum have their own names. We outline below the properties and some applications of electromagnetic waves in each region. Note, however, that the boundaries of these regions are not precisely defined.

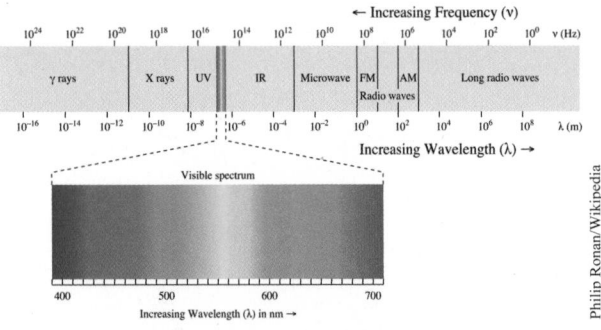

Figure 26-17 The electromagnetic spectrum

Radio Waves

Radio waves ($\lambda \approx 100$ km to 0.3 m, $f \approx 3 \times 10^3$ Hz to 1×10^9 Hz) have the longest wavelengths in the electromagnetic spectrum. Radio waves are used for radio and TV signals, radar, and numerous other applications. Some properties of radio waves depend on the wavelength. For example, the shorter-wavelength radio waves can reflect off the ionosphere and thus travel beyond the horizon. Consequently, shortwave radio broadcasts (1.8 to 30 MHz) can be received at long distances from the transmitter.

Many galaxies, including the Milky Way, emit radio waves as well as visible light. The radio emissions from galaxies are generated by fast-moving electrons that spiral around the magnetic field lines of the galaxy. The branch of astronomy that studies radio waves from space is called **radio astronomy**. The study of radio wave emission from galaxies provides extremely useful information that cannot be obtained by observing only visible light. Figure 26-18 shows a radio image of the central region of the Milky Way. The arrow in the image indicates a supernova remnant at the location of a newly discovered source of low-frequency radio waves.

Figure 26-18 A radio frequency image of the central region of the Milky Way

Microwaves

Applications of microwaves ($\lambda \approx 0.3$ m to 0.3 mm, $f \approx 1 \times 10^9$ Hz to 1×10^{12} Hz) include wireless communications (Bluetooth and IEEE 02.11 devices use 2.4 GHz microwaves), cellphone networks, satellite radio (2.3 GHz), garage door openers, the Global Positioning System (GPS), and microwave ovens. Microwaves are also generated by stars, including the Sun.

Infrared Radiation

The infrared region ($\lambda \approx 0.3$ mm to 700 nm, $f \approx 1 \times 10^{12}$ Hz to 4.3×10^{14} Hz) of the electromagnetic spectrum lies between microwaves and visible light. The part of the region closest to microwaves is called **far infrared**, and the part closest to visible light is called **near infrared**. All objects emit some infrared radiation. Even cold objects, like snow, emit infrared radiation. The hotter an object, the more infrared radiation it emits. The heat from the Sun and from a fire is transmitted largely in the form of infrared radiation. Although human eyes cannot detect infrared radiation, many organisms have infrared receptors that allow them to see warm-blooded animals in the dark. Images taken with sensors and photographic emulsions that respond to infrared radiation are called **thermograms**. Figure 26-19 shows a thermogram that has been processed so that the red colour corresponds to higher temperatures and the blue colour to lower temperatures. The red colour on the roof and the middle window indicates heat escaping from the house, a sign of poor insulation in these areas.

Near-infrared radiation is used in TV remote controls and similar devices. A near-infrared wave causes much less heating than a far-infrared wave.

Figure 26-19 A thermogram of a house at night. The scale at the right shows the temperatures corresponding to the various colours.

Visible Light

The narrow band of the electromagnetic spectrum that our eyes can detect is called the **visible spectrum** ($\lambda \approx 750$ nm to 400 nm, $f \approx 4.0 \times 10^{14}$ Hz to 7.5×10^{14} Hz). Although we divide the visible spectrum into colours (often listed as red, orange, yellow, green, blue, indigo, and violet), the wavelength changes continuously within the region, as shown in the enlarged section of Figure 26-17. Human eyes are most sensitive to light with a wavelength of 555 nm, which has a green colour.

Ultraviolet Light

The ultraviolet (UV) region ($\lambda \approx 400$ nm to 10 nm, $f \approx 7.5 \times 10^{14}$ Hz to 3×10^{16} Hz) lies between visible light and X-rays. Ultraviolet radiation has sufficient energy to ionize atoms and cause chemical reactions. Many minerals absorb ultraviolet light and then emit radiation of longer wavelengths, often in the visible spectrum (Figure 26-20). This process is called **fluorescence**. UV lamps are commonly used in suntanning salons.

Figure 26-20 Various coloured light emitted by different minerals when exposed to ultraviolet radiation

The Sun is a major source of UV radiation, which is divided into three bands: UV-A, UV-B, and UV-C.

■ The UV-A band contains wavelengths between 320 nm and 400 nm. Radiation in the UV-A band is not absorbed by the atmosphere and is not particularly harmful to living organisms.

■ The UV-B band contains wavelengths between 320 nm and 280 nm. Radiation in this band is partly absorbed by the atmosphere. UV-B radiation is harmful to human skin, and prolonged exposure can cause skin cancer.

■ The UV-C band contains wavelengths between 280 nm and 100 nm. Radiation in this band is mostly absorbed by the atmosphere. UV-C radiation is more energetic and harmful to human skin. Prolonged exposure can cause sunburn and eventually skin cancer.

 MAKING CONNECTIONS

What Is the UV Index?

The UV index is a numbering system that indicates the level of harm to human skin due to the intensity of UV radiation (Table 26-2). This index was first developed by Environment Canada, and Canada was the first country to issue UV level forecasts for the public. The UV index is now standardized by the World Health Organization and used worldwide. The higher the category, the more harmful the effect of the UV radiation. A color is associated with each risk category.

Table 26-2 The UV Index

UV index	Risk category	Associated colour
0–2	Low	Green
3–5	Moderate	Yellow
6–7	High	Orange
8–10	Very high	Red
11 and higher	Extreme	Violet

X-rays

Electromagnetic waves with wavelengths between 3 nm and 0.003 nm are called X-rays ($\lambda \approx 3$ nm to 0.003 nm, $f \approx 1 \times 10^{17}$ Hz to 1×10^{20} Hz). Wavelengths between 10 nm and 0.10 nm are classified as **soft X-rays**, and wavelengths between 0.10 nm and 0.010 nm are classified as **hard X-rays**. Hard X-rays easily penetrate skin and soft body tissue but are absorbed by dense material such as bones and teeth. Hard X-rays are used for making diagnostic images of the interior of the human body, studying the structure of crystals, and nondestructive testing of objects such as structural materials. The X-ray image is either recorded on a film (a **radiograph**) or stored as an electronic image. Different materials appear light or dark depending on the proportion of the X-rays the materials absorb. Calcium in bones absorbs large amounts of X-rays; therefore, bones look white on the radiograph, which is viewed as a negative image. Fat and other soft tissue absorbs less and look grey. Air absorbs very few X-rays, so lungs look black. X-rays are high-energy radiation, and overexposure to them can cause severe damage to human tissue.

Gamma Rays

Gamma rays ($\lambda \approx 0.003$ nm and smaller, $f \approx 1 \times 10^{20}$ Hz and greater) are the most energetic electromagnetic waves. Gamma rays are emitted by nuclei and can also be produced by collisions of particles and antiparticles. Gamma rays are highly penetrating radiation, and in recent years they have been used in different industries, for example: cancer therapy (destroying cancerous cells), food preservation (killing bacteria and other microorganisms in food), sterilizing medical equipment (killing bacteria), and testing structural materials. Gamma ray detectors are also used in seaports and border crossings to image the insides of containers (Figure 26-21).

United States Department of Homeland Security

Figure 26-21 A gamma ray image of the inside of a truck. The image clearly shows the boxes and human figures (dummies) inside the truck.

LO 5

26-5 The Energy and Momentum of Electromagnetic Waves

Like all waves, electromagnetic waves carry energy and momentum. The energy and momentum flow in the direction of motion of the wave. In this section, we will explore how the energy and momentum of an electromagnetic wave depend on the amplitudes of the wave's electric and magnetic fields.

We know from electrostatics that electric and magnetic fields store energy. Recall that the magnitude of the *electric energy density* (the electric potential energy per unit volume) stored in any electric field $\vec{E}$ is

$$u_E = \frac{1}{2}\varepsilon_0 E^2 \qquad (21\text{-}18)$$

Similarly, the magnitude of the *magnetic energy density* of a magnetic field $\vec{B}$ is

$$u_B = \frac{1}{2\mu_0}B^2 \qquad (24\text{-}51)$$

Consider a sinusoidal electromagnetic wave described by Equations (26-6) and (26-7) travelling in a vacuum in the positive *x*-direction. Substituting Equations (26-6) and (26-7) into the equations for the energy densities gives

$$u_E(x, t) = \frac{1}{2}\varepsilon_0 E_y^2 = \frac{1}{2}\varepsilon_0 E_0^2 \sin^2(kx - \omega t)$$

$$u_B(x, t) = \frac{1}{2\mu_0}B_z^2 = \frac{1}{2\mu_0}B_0^2 \sin^2(kx - \omega t) \qquad (26\text{-}23)$$

Note that the energy densities for a travelling wave are position- and time-dependent. Since $B_0 = E_0/c$ for an electromagnetic wave and $c = \dfrac{1}{\sqrt{\varepsilon_0 \mu_0}}$, the magnetic energy intensity can be written as follows:

$$u_B(x, t) = \frac{1}{2\mu_0}\left(\frac{E_0}{c}\right)^2 \sin^2(kx - \omega t)$$

$$= \frac{1}{2}\varepsilon_0 E_0^2 \sin^2(kx - \omega t) \qquad (26\text{-}24)$$

Thus, *in a travelling electromagnetic wave, the energy density of the magnetic field is equal to the energy density of the electric field.* In any given region of space, the total energy of the electromagnetic field is equally shared between the electric and the magnetic fields. This result is a consequence of the symmetry of Maxwell's equations with respect to electric and magnetic fields in a vacuum.

The total energy density of the wave, $u(x, t)$, is the sum of the electric and magnetic energy densities and can be written in several equivalent ways:

$$u(x, t) = u_E(x, t) + u_B(x, t)$$

$$= 2u_E(x, t) = \varepsilon_0 E_0^2 \sin^2(kx - \omega t)$$

$$= \varepsilon_0 c E_0 B_0 \sin^2(kx - \omega t) \qquad (26\text{-}25)$$

The amount of energy flowing per unit time per unit area perpendicular to the direction of the wave propagation is called the **energy flux** (S) or the **intensity** of the wave. Imagine a 1.0 m² rectangular area perpendicular to the path of an electromagnetic wave in a vacuum. Since an electromagnetic wave travels with the speed of light, c, the energy that passes through the rectangular area in one second is equal to the energy contained in a cube of volume $1.0 \text{ m}^2 \times c \times 1.0 \text{ s}$ (Figure 26-22). Therefore,

$$S(x, t) = u(x, t)c = \varepsilon_0 c^2 E_0 B_0 \sin^2(kx - \omega t)$$

$$= \frac{E_0 B_0}{\mu_0} \sin^2(kx - \omega t) \qquad (26\text{-}26)$$

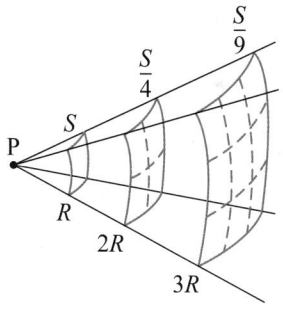

Figure 26-22 The energy flux, S, is defined as the amount of electromagnetic energy that passes through a unit cross-sectional area perpendicular to the path of the incident wave, in one second.

The units of energy flux are W/m^2, which we can confirm by considering the dimensions of the quantities in Equation (26-26):

$$[\mu_0] = \frac{\text{kg} \cdot \text{m}}{\text{C}^2}; \qquad [E_0] = \frac{\text{N}}{\text{C}}; \qquad [B_0] = \text{T} = \frac{\text{kg}}{\text{C} \cdot \text{s}}$$

So,

$$[S] = \frac{\text{C}^2}{\text{m} \cdot \text{kg}} \times \frac{\text{N}}{\text{C}} \times \frac{\text{kg}}{\text{C} \cdot \text{s}} = \frac{\text{N}}{\text{m} \cdot \text{s}} = \frac{\text{N} \cdot \text{m}}{\text{m}^2 \cdot \text{s}} = \frac{\text{W}}{\text{m}^2}$$

The energy flow varies with the frequency of the electric and magnetic fields. We do not notice these fluctuations in visible light because the frequency ($\sim 10^{14}$ Hz) is far too fast for us to perceive. The time and position average of the function $\sin^2(kx - \omega t)$ over one oscillation is $\frac{1}{2}$. Therefore, we can define the **average intensity** ($\bar{S}$), or **irradiance**, as

$$\bar{S} = \frac{E_0 B_0}{2\mu_0} = \frac{E_0^2}{2\mu_0 c} = \frac{\varepsilon_0 c E_0^2}{2} \qquad (26\text{-}27)$$

From Equations (26-26) and (26-27) the total average energy density, $\bar{u}$ of an electromagnetic wave is related to the average intensity of the wave by

$$\bar{S} = c\bar{u} \qquad (26\text{-}28)$$

Notice that as for mechanical waves, the energy carried by an electromagnetic wave is proportional to the square of its amplitude.

Consider a source of power that is emitting spherical electromagnetic waves, for example, a light bulb or an omnidirectional radio antenna. The waves emerging from the source spread outward at the speed of light. At a distance r from the source, the power, P, is distributed over a surface area, $4\pi r^2$. Therefore, the average intensity of the waves, $\bar{S}$, at a distance r from the source is

$$\bar{S} = \frac{P}{4\pi r^2} \qquad (26\text{-}29)$$

The intensity of electromagnetic waves falls off as the square of the distance from the source (Figure 26-23).

Figure 26-23 The intensity of the three-dimensional spherical waves falls off as the square of the distance from the point source producing the waves.

 EXAMPLE 26-3

Solar Power Reaching Earth

The average intensity of sunlight, $\bar{S}$, at the upper atmosphere is approximately 1400 W/m^2. The average distance between Earth and the Sun is 150 million kilometres, and the mean radius of Earth is 6370 km.

(a) What are the amplitudes of the electric and magnetic fields in the sunlight?

(b) What is the average power generated by the Sun?

(c) What is the total power intercepted by Earth?

SOLUTION

(a) From Equation (26-27), the amplitude of the electric field is related to the average intensity by

$$E_0 = \sqrt{2\mu_0 c\bar{S}}$$

$$= [2(4\pi \times 10^{-7} \text{ T} \cdot \text{m/A})(3.00 \times 10^8 \text{ m/s})(1400 \text{ W/m}^2)]^{1/2}$$

$$= 1030 \text{ V/m}$$

(continued)

The amplitude of the magnetic field is

$$B_0 = \frac{E_0}{c} = \frac{1030 \text{ V/m}}{3.00 \times 10^8 \text{ m/s}} = 3.4 \times 10^{-6} \text{ T}$$

(b) The average power of the Sun can be calculated from Equation (26-29):

$$P = 4\pi r^2 \bar{S} = 4\pi (150 \times 10^9 \text{ m})^2 (1400 \text{ W/m}^2) = 4.0 \times 10^{26} \text{ W}$$

(c) The total power intercepted by Earth is the product of the incident intensity and the cross-sectional area of Earth. At such a large distance from the Sun, electromagnetic waves reaching Earth are plane waves.

Earth projects a disk with a radius of 6370 km to these waves. Therefore, the power intercepted by Earth is

$$P_{\text{Earth}} = (\pi r_{\text{Earth}}^2)\bar{S} = \pi(6370 \times 10^3 \text{ m})^2 \times (1400 \text{ W/m}^2)$$

$$= 1.8 \times 10^{17} \text{ W}$$

Making sense of the results:

In 2008, the average power consumption of the world was approximately 1.5×10^{13} W. So, the Sun could provide all of this power if sufficiently efficient and cost-effective ways to harness Solar power can be developed.

MAKING CONNECTIONS

How Do Microwave Ovens Heat Food?

A microwave oven produces electromagnetic radiation with a frequency of approximately 2.45 GHz ($\lambda \cong 0.12$ m). This radiation is absorbed by water, fats, and sugar molecules contained in the food, thus heating and cooking the food. A water molecule consists of one oxygen and two hydrogen atoms and is electrically neutral overall. However, within the water molecule the oxygen atom is slightly negatively charged, and the hydrogen atoms are slightly positively charged. Therefore, the water molecules act like tiny dipoles. In the presence of an electric field, the positive end of a dipole experiences a force in the direction of the electric field, and the negative end experiences a force in the opposite direction, thus twisting the dipole. When an alternating electromagnetic field is applied to a water molecule, as in a microwave oven, the molecule continuously twists about its centre of charge (Figure 26-24). In a microwave oven, water molecules absorb electromagnetic energy from the microwaves and convert it to thermal energy (molecular motion). The molecules then transfer this thermal energy to adjacent nonpolar molecules by colliding with them and this heating the food. Microwave radiation heats food quickly and effectively because it is absorbed throughout the food. Longer wavelengths would pass through the food, and shorter wavelengths would be absorbed, mainly at the surface. Microwave radiation does not have sufficient energy to remove electrons from the atoms, so it cannot chemically change the food.

(a)

no electric field

(b)

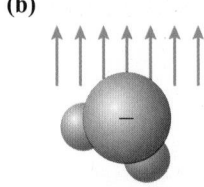

electric field pointing upwards

(c)

electric field pointing downwards

Figure 26-24 Orientations of a water molecule with (a) no electric field, (b) the electric field directed upward, and (c) the electric field directed downward

ONLINE ACTIVITY

Microwave Ovens

The e-resource that accompanies every new copy of this textbook contains an Online Activity using the PhET simulation "Microwaves." Work through the simulation and accompanying questions to gain an understanding of how microwave ovens work.

The Poynting Vector and Wave Momentum

Recall that the cross product of vectors $\vec{a}$ and $\vec{b}$ is a vector that is perpendicular to the plane containing the two vectors. Since the direction of energy flow of an electromagnetic wave is perpendicular to the direction of the electric and magnetic fields, we define a vector, $\vec{S}$, called the **Poynting vector**, as

$$\vec{S} = \frac{1}{\mu_0} \vec{E} \times \vec{B} \tag{26-30}$$

The magnitude of $\vec{S}$ is the energy flux S as defined by Equation (26-26), and the direction of $\vec{S}$ is that of the wave propagation, as shown in Figure 26-25. The magnitude of the Poynting vector oscillates with the frequency of the wave. The Poynting vector was introduced by the English physicist John Henry Poynting (1852–1914).

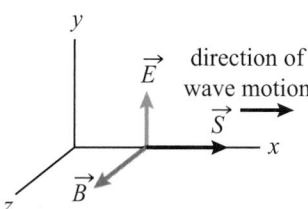

The electric and magnetic fields are in the yz-plane.

Figure 26-25 The Poynting vector $\vec{S}$ of an electromagnetic wave

Electromagnetic Wave momentum

Consider an electromagnetic wave incident on a collection of charges. Even if all the charges are initially at rest the electromagnetic wave exerts a force on the charges causing them to move and hence gain momentum. By the law of conservation of momentum an increase in the momentum of the charges is equal to the loss in the wave's momentum. Therefore, in addition to energy an electromagnetic wave must also carry momentum.

According to quantum theory, electromagnetic waves are made of massless particles, called photons (see Chapter 33). The relationship between the energy and momentum of a fast moving particle is given by Einstein's relativistic energy-momentum relationship (Chapter 27),

$$E = \sqrt{m^2c^4 + p^2c^2}$$

where E is the energy, p the momentum, m is the rest-mass of the particle and c is the speed of light in vacuum. The energy E and the magnitude of momentum p of a photon ($m = 0$) are therefore related by,

$$p = \frac{E}{c} \qquad (26\text{-}31)$$

Because electromagnetic waves are made of photons the relationship (26-31) also applies to these waves. The momentum carried by an electromagnetic wave is equal to the energy of the wave divided by the speed of light. It follows from the above equation that the magnitude of time-averaged **radiation momentum density** (momentum per unit volume) $\bar{g}_{\text{rad}}$ of an electromagnetic wave equals its time-averaged energy density $\bar{u}$ (Equation 26-28), divided by c.

$$\bar{g}_{\text{rad}} = \frac{\bar{u}}{c} = \frac{1}{c}\left(\frac{\varepsilon_0 E_0^2}{2}\right) \qquad (26\text{-}32)$$

As the direction of energy and momentum flow of a particle is along its direction of motion, therefore the momentum of an electromagnetic wave points along the direction of the energy flow. As defined in Equation (26-30) the Poynting vector $\vec{S}$ describes the direction of energy flow of an electromagnetic wave. Therefore the radiation momentum density $\vec{g}_{\text{rad}}$ (a vector quantity that describes both the magnitude of the momentum density as well as the direction of the momentum flow) of an electromagnetic wave is related to $\vec{S}$ by

$$\vec{g}_{\text{rad}} = \frac{\vec{S}}{c^2} \qquad (26\text{-}33)$$

The momentum density equals the energy flux divided by c^2. This relationship between the energy flux of a wave and its momentum density is also true for other types of waves, for example sound waves.

The relationship (26-33) can also be derived using Maxwell's equations and was known before the advent of quantum theory. However, that derivation involves the use of vector calculus, a branch of mathematics that you may not have yet studied.

LO 6

26-6 Radiation Pressure

Radiation pressure is defined as the force per unit area exerted by electromagnetic waves on an object. Light falling on the surface of an object, such as black paper, is absorbed by the surface. As the light is absorbed, it transfers its momentum to the object, thus exerting a force on the object. By definition, the force per unit area on a surface is the pressure exerted on that surface. Therefore, light exerts pressure on an absorbing surface.

Now consider electromagnetic radiation with average intensity $\bar{S}$ incident on a surface of area A that absorbs all the incident radiation. The energy Δu absorbed by the surface in time Δt is given by,

energy absorbed = average intensity $\times$ surface area $\times \Delta t$

$$\Delta \bar{u} = \bar{S} A \Delta t$$

Since all the radiation is absorbed, the total momentum transferred to the surface, Δp, is equal to the time-averaged **radiation momentum**, $\bar{g}_{\text{rad}}$, given by Equation (26-32):

$$\Delta p = \frac{\Delta \bar{u}}{c} = \frac{1}{c}\left(\bar{S} A \Delta t\right)$$

The rate of change of momentum per unit area of the surface is therefore,

$$\frac{1}{A}\left(\frac{\Delta p}{\Delta t}\right) = \frac{\bar{S}}{c}$$

The quantity on the left side is the pressure exerted on the surface by the radiation, P_{rad}. Therefore, radiation pressure on a perfectly absorbing surface is

ELECTRICITY, MAGNETISM, AND OPTICS

$$P_{rad} = \frac{\bar{S}}{c} \text{ (for a perfectly absorbing surface) } (26\text{-}34)$$

If the surface is a perfect reflector, then the change in the momentum of the reflected waves is $2\Delta p$ for normal incidence (just like the change in the momentum of a ball

when it rebounds off a perfectly elastic surface is twice the incident momentum). The pressure exerted on a perfectly reflecting surface is

$$P_{rad} = 2\frac{\bar{S}}{c} \text{ (for a perfectly reflecting surface) } (26\text{-}35)$$

EXAMPLE 26-4

Radiation Pressure at Earth's Surface

Estimate the radiation pressure at Earth's surface. Is this pressure uniformly distributed over Earth's entire surface at any given time?

SOLUTION

The intensity of solar radiation at Earth's surface is approximately 1000 W/m². It is less than the intensity at the top of Earth's atmosphere due to absorption in the atmosphere. Assuming that most of the sunlight is reflected from Earth's surface, the radiation pressure exerted on Earth's surface is approximately

$$P_{rad} = 2\frac{1000 \text{ W/m}^2}{3.00 \times 10^8 \text{ m/s}} = 6.7 \times 10^{-6} \text{ Pa}$$

The amount of sunlight reflected by Earth's surface varies considerably across the surface and depends on the presence of cloud cover, snow, and other variables. Our assumption that all incident sunlight is reflected overestimates the radiation pressure. The radiation pressure is only exerted on the side of Earth that is facing the Sun.

Making sense of the result:

The atmospheric pressure at sea level is 1.01×10^5 Pa, so the radiation pressure is approximately 7×10^{-11} times smaller than the atmospheric pressure. That is why we do not feel radiation pressure.

EXAMPLE 26-5

Solar Sailing: IKAROS

On May 21, 2010, the Japan Aerospace Exploration Agency launched a spacecraft, called IKAROS (Interplanetary Kitecraft Accelerated by Radiation of the Sun), toward Venus to demonstrate propulsion by a solar sail. The sail of the spacecraft (Figure 26-26(a)) consists of 200 m² of a reflective polyimide resin membrane as thin as 0.0075 mm. The mass of the solar sail is 15 kg, and the total mass of the spacecraft is 315 kg. Once clear of Earth, the spacecraft is propelled toward Venus by pressure from solar radiation. Estimate the magnitude of acceleration generated by radiation pressure on the sail.

SOLUTION

Let $\bar{S}_E$ be the solar radiation intensity just above Earth's atmosphere. When the spacecraft is at a distance r from the Sun, the solar intensity at this distance, $\bar{S}(r)$, can be related to $\bar{S}_E$ using Equation (26-29):

$$\bar{S}(r) = \bar{S}_E\left(\frac{r_E^2}{r^2}\right)$$

where r_E is the distance between Earth and the Sun. Assuming that all incident radiation is reflected, the radiation pressure at the distance r from the Sun is

$$P_{rad}(r) = \frac{2\bar{S}(r)}{c}$$

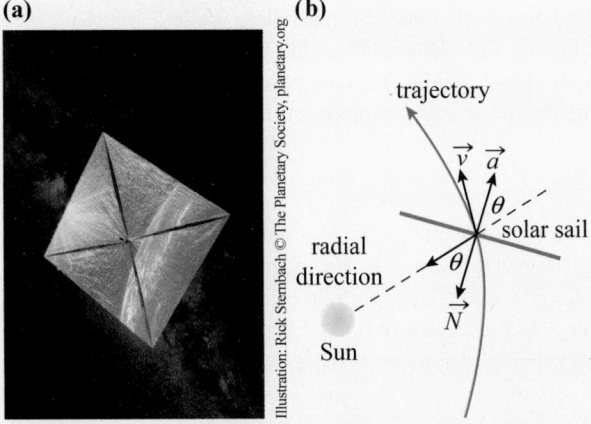

Figure 26-26 (a) The IKAROS solar sail (b) The orientation of the solar sail with respect to the Sun. The normal to the sail ($\vec{N}$) makes an angle θ with respect to the radial direction. The red curve shows the trajectory of the spacecraft.

For a solar sail with cross-sectional area A perpendicular to the incident radiation, the magnitude of the force exerted by the radiation is

$$F_{rad}(r) = A \times P_{rad}(r) = \frac{2A\bar{S}(r)}{c}$$

For a general trajectory (Figure 26-26(b)), the sail makes an angle θ with respect to the radial direction between the Sun and the spacecraft. So, the effective area perpendicular to the direction of the incident radiation is $A \cos\theta$ (normal incidence corresponds to $\theta = 0°$). The magnitude of acceleration generated by the radiation pressure is

$$a(r) = \frac{F_{\text{rad}}(r)}{m} = \bar{S}_\text{E}\left(\frac{r_\text{E}^2}{r^2}\right)\frac{2(A\cos\theta)}{mc}\cos\theta$$

where m is the total mass of the spacecraft.

The extra factor of $\cos\theta$ is due to the fact that the acceleration is directed perpendicular to the plane of the sail, at an angle θ with respect to the radial direction.

Let us assume that $r = r_\text{E}$, $\theta = 0°$ (normal incidence), and $\bar{S}_\text{E} = 1400$ W/m². Then the acceleration due to

radiation pressure when the spacecraft is close to Earth is approximately

$$a(r_\text{E}) = \frac{(1400 \text{ W/m}^2)2(200 \text{ m}^2)}{(315 \text{ kg})(3.00 \times 10^8 \text{ m/s})} = 5.9 \times 10^{-6} \text{ m/s}^2$$

Making sense of the result:

The acceleration from solar radiation is very small. The effects of gravity from Earth and the Sun would have to be included when calculating the trajectory of the spacecraft.

 CHECKPOINT

C-26-5 Solar Sail

Assume that the speed of the IKAROS spacecraft when it leaves the atmosphere is 7 km/s and that the acceleration due to solar radiation pressure remains constant at 6×10^{-6} m/s². How long would it take the spacecraft to double its speed?

(a) approximately 6 months
(b) between 5 and 10 years
(c) between 10 and 50 years
(d) over 100 years

C-26-5 (c)

LO 7

26-7 How Are Electromagnetic Waves Generated?

To generate electromagnetic waves, a source must lose energy. The waves carry this energy away from the source. Electromagnetic waves are generated during two types of processes: quantum-mechanical processes and acceleration of a charged particle.

Quantum-Mechanical Processes

When an electron in an atom makes a transition from a state of higher energy to a state of lower energy, it radiates the energy difference as electromagnetic waves. Infrared light, visible light, ultraviolet light, and X-rays can be generated by this process. We will discuss the process of emission of light from atoms in Chapter 31.

Protons and neutrons in a nucleus can also make transitions from a state of higher energy to a state of lower energy, emitting gamma rays in the process. We will discuss the process of emission of gamma rays from a nucleus in Chapter 33.

In matter–antimatter annihilation, a charged particle can combine with its antiparticle (which has an equal

but opposite charge), and the mass of both particles can be converted to energy in the form of electromagnetic radiation. Conversion of mass to energy is discussed in Chapter 29, but the details of annihilation processes are beyond the scope of this book.

Acceleration of a Charged Particle

Electromagnetic waves cannot be generated by stationary charges or charges that are moving with a constant velocity. A stationary charge has an electric field but no magnetic field. Therefore, it cannot generate an electromagnetic wave. The electric field of a stationary charge remains *attached* to the charge, never leaving it.

Now consider a charge moving with a constant velocity. In a frame of reference in which the charge is moving with a uniform velocity, it generates both electric and magnetic fields. However, in the frame of reference of the charge, the charge is at rest and therefore has only an electric field. So, in its own frame of reference, a charge cannot emit an electromagnetic wave. The laws of physics must be the same in all frames that move with a uniform velocity with respect to each other. So, if a charge cannot emit electromagnetic radiation in its own reference frame, it cannot emit electromagnetic radiation in any frame moving with uniform velocity with respect to the charge.

Thus, a charge can emit electromagnetic radiation in free space *only when accelerating.* Note the qualifier "in free space." A charge moving with a uniform velocity through a dielectric medium can emit electromagnetic radiation. Such radiation is called **Cerenkov radiation**.

We briefly discuss a few examples of electromagnetic radiation emitted during accelerated motion of charged particles.

A Charge (or an Electric Current) Undergoing Simple Harmonic Motion Simple harmonic motion is accelerated motion. Therefore, a charged particle or an electric current undergoing oscillatory motion emits electromagnetic waves. The frequency of the radiated wave is equal to the oscillation frequency of the motion. Harmonic motion of charges underlies the generation of electromagnetic waves for radio, TV, and other types of communication signals.

A Charged Particle Moving in a Curved Orbit Motion along a curved path is accelerated motion. Therefore, a charged particle moving in a curved trajectory radiates electromagnetic waves, called **synchrotron radiation** or **cyclotron radiation**. The term "synchrotron radiation" is generally used for radiation emitted by a charged particle rapidly moving in a magnetic field. In the presence of a magnetic field, a charged particle experiences a force that is perpendicular to both the particle's velocity and the direction of the magnetic field. This force causes the particle to spiral in the direction of the magnetic field (Figure 26-27), continuously emitting electromagnetic radiation. Synchrotron radiation is also generated when fast-moving electrons spiral through the intense magnetic fields of astronomical objects.

A Charged Particle Stopping Abruptly When a fast-moving charged particle stops abruptly, it undergoes large deceleration and emits electromagnetic radiation. Such radiation is called **bremsstrahlung**, a German word meaning "braking radiation." The X-ray machine in a dentist's office generates X-rays by firing a beam of fast-moving electrons at a metal plate. When electrons collide with the plate, they come to an abrupt stop, emitting X-rays. The electrons actually follow a curved path as they decelerate, as shown in Figure 26-28. The electric fields of nuclei of the metal atoms deflect the electrons.

Figure 26-27 The spiral motion of a charged particle in a magnetic field generates synchrotron radiation.

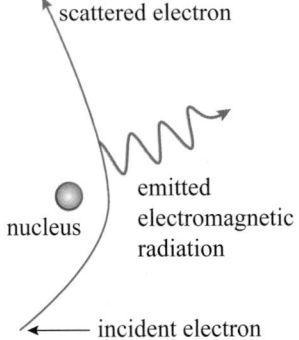

Figure 26-28 An electron moving within the vicinity of a nucleus is deflected by the electric and magnetic forces between the two.

MAKING CONNECTIONS

Half-Wave Dipole Antennas

A dipole antenna consists of two conducting metal wires (or rods) of equal length, separated by a narrow insulating gap. Dipole antennas can be used both for generating and receiving electromagnetic waves. To generate waves, the two wires are powered, often by connecting them to an RLC circuit as shown in Figure 26-29. (An RLC circuit is an electrical circuit consisting of a resistor, an inductor, and a capacitor.) The current and voltage in the wires vary sinusoidally with frequency f_0, the resonant frequency of the RLC circuit. The sinusoidal output of the RLC circuit forces electrons in the antenna to undergo harmonic motion, and the accelerated electrons emit electromagnetic waves at frequency f_0, which then propagate away from the antenna. The frequency of the emitted waves can be changed by changing the inductance or the capacitance in the RLC circuit.

For maximum efficiency, the total length of the antenna should be approximately half the wavelength of the electromagnetic waves emitted. So, for example, if the frequency of the RLC circuit is 98 MHz (which is approximately in the middle of the FM radio band in North America), the wavelength of the emitted waves is

$$\lambda_0 = \frac{c}{f_0} = \frac{3 \times 10^8 \text{ m/s}}{98 \times 10^6 \text{ Hz}} = 306 \text{ cm}$$

Therefore, the optimal total length of the antenna is 153 cm, and each wire would be approximately 75 cm long (depending on the length of the gap between the two wires). The exact length for the antenna is also affected by other factors, such as the thickness of the wires and the height of the antenna above the ground.

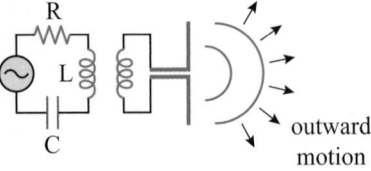

Figure 26-29 A dipole antenna connected to an RLC circuit emits electromagnetic waves at the resonant frequency of the RLC circuit.

26-8 Polarization

The **polarization** of a transverse wave at a given point is defined as the direction of oscillation of the wave at that point. For example, for the transverse wave moving along the string in Figure 26-30, the motion of the string is taken to be along the y-axis, so the wave is said to be polarized along the y-axis.

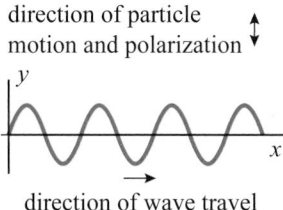

direction of particle motion and polarization

direction of wave travel

Figure 26-30 A wave on a string is polarized perpendicular to its direction of motion.

By convention, the polarization of electromagnetic waves is described by specifying the orientation of the wave's *electric* field at a given point over one time period. The electric field of the plane electromagnetic wave in Figure 26-31(a) is oriented along the y-axis, so the wave is polarized along the y-axis. When the electric field is oriented along a single line, the polarization is called **linear polarization**, and the wave is **linearly polarized**. The plane containing the electric field and the direction of motion of the wave is called the **plane of polarization**. The wave shown in Figure 26-31(a) is linearly polarized along the y-axis, and its plane of polarization is the yz-plane.

Since the electric and magnetic fields of electromagnetic waves are always perpendicular to the direction of motion, these waves are always polarized perpendicular to their direction of motion. Figure 26-31(b) shows a wave linearly polarized at an angle of 45° in the xy-plane.

✓ CHECKPOINT

C-26-6 Polarization

A linearly polarized electromagnetic wave is travelling along the x-axis, and its magnetic field is oscillating along the y-axis. Which of the following statements is correct?

(a) The wave is polarized along the y-axis, and its plane of polarization is the xy-plane.

(b) The wave is polarized along the x-axis, and its plane of polarization is the xy-plane.

(c) The wave is polarized along the z-axis, and its plane of polarization is the xz-plane.

(d) The wave is polarized along the y-axis, and its plane of polarization is the xz-plane.

C-26-6 (c) As the wave is moving along the x-axis the electric and magnetic fields are in y-z plane. Since the magnetic field is along the y-axis, the electric field must be along the z-axis. Therefore, the wave is polarized along the z-axis and its plane of polarization is the xz plane.

(a)

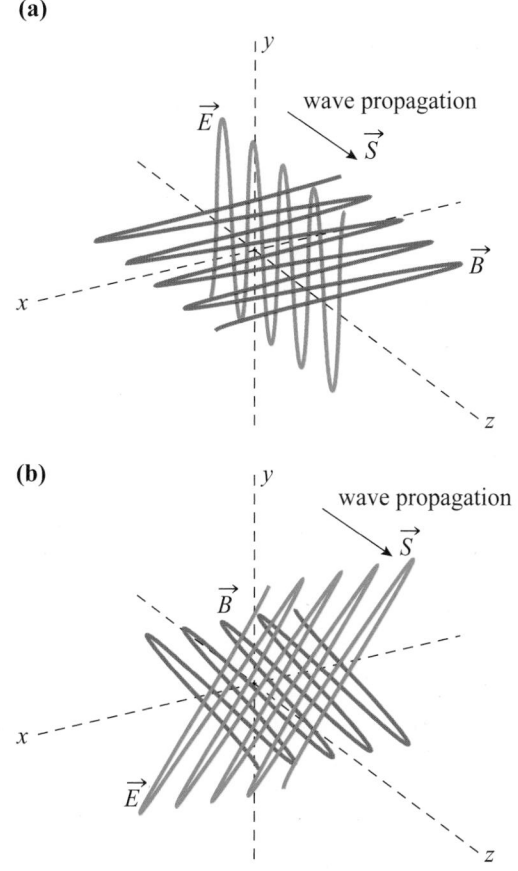

(b)

Figure 26-31 (a) A linearly polarized electromagnetic wave with its electric field along the y-axis (b) An electromagnetic wave linearly polarized at an angle of 45° in the xy-plane

Unpolarized Light

Light from the Sun and from a common light bulb consists of a large number of waves emitted by individual atoms. In general, an electromagnetic wave emitted by one atom has no correlation with the waves emitted by other atoms, and the electric fields of individual waves point in random directions. For example, in a 50 W light bulb, approximately 10^{19} atoms emit individual electromagnetic waves with randomly oriented electric fields every second. Such light is **unpolarized**. The electric field vector of unpolarized light consists of a random superposition of a very large number of linearly polarized waves (Figure 26-32).

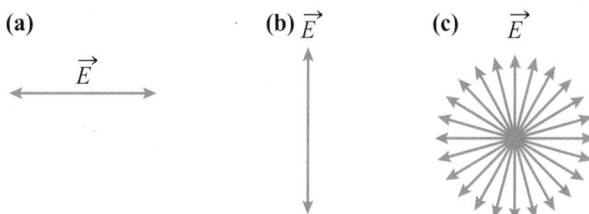

Figure 26-32 (a) An electromagnetic wave polarized along the x-axis (b) An electromagnetic wave polarized along the y-axis (c) An unpolarized electromagnetic wave consisting of random superposition of linearly polarized waves. The waves are travelling perpendicular to the page.

Polarization by Absorption

We can produce polarized light from a beam of unpolarized light by filtering out all but one of the polarizations. **Polarizers** are materials that allow the electric field only with a particular orientation to pass through. The electric field of the transmitted light is oriented in that direction and is therefore polarized.

A commonly used polarizer is Polaroid film, which is a flexible, plastic sheet made of polyvinyl alcohol (PVA). The sheet is heated and stretched to align the PVA molecules into long, parallel rows. The sheet is then dipped into an iodine dye. The iodine atoms, which have loosely bound outer electrons, bond to the long chains of PVA molecules, making them conductive. The PVA chains can then act as long, extremely thin, closely spaced metal wires. When unpolarized light falls on the plastic film, the electric field of the light exerts a force on the free-moving iodine electrons. The electrons can move along the length of the stretched chains of molecules, so the component of the electric field that is parallel to the direction of these chains is easily absorbed by the electrons. However, the iodine electrons cannot readily move perpendicular to the stretched chains and therefore cannot absorb the component of the electric field in this direction. Consequently, this perpendicular component of the electric field passes through the material. The direction along which the electric field is transmitted is called the **transmission axis**, or **polarizing axis**. Note that the transmission axis of Polaroid film is perpendicular to the axis of long-chained molecules.

Figure 26-33 shows unpolarized light incident on a polarizer that lies in the plane of the page. The red arrow shows the direction of the electric field, $\vec{E}$, of one of the randomly polarized light waves. The transmission axis of the polarizer is along the y-axis, and the electric field makes an angle θ with respect to the transmission axis. The electric field can be resolved into a component along the transmission axis ($E_y = E\cos\theta$) and a component perpendicular to the transmission axis ($E_x = E\sin\theta$). Only the E_y component passes through the polarizer.

When a polarized beam of light passes through a polarizer, the intensity of the transmitted light depends on the angle between the light's initial polarization direction and the transmission axis of the polarizer. Let I_0 be the intensity of a polarized beam of light that falls on a polarizer. We denote the angle between the polarization direction of the light and the transmission axis of the polarizer as θ and the intensity of light that is transmitted through the polarizer as $I(\theta)$. Then

KEY EQUATION
$$I(\theta) = I_0 \cos^2\theta \qquad (26\text{-}36)$$

This relationship was discovered experimentally by Étienne-Louis Malus in 1809 and is called the **Malus' Law**. We can derive Malus' law from the properties of electromagnetic waves.

Consider the arrangement shown in Figure 26-34. A beam of unpolarized light is incident on polarizer P_1, which polarizes the light along its transmission axis. Let the amplitude of the electric field that is transmitted through P_1 be E_0, and the intensity of the transmitted light be I_0. A second polarizer, P_2 (called an **analyzer**), is placed in front of the polarized beam with its polarization axis at an angle θ with respect to the transmission axis of P_1. The amplitude of the electric field transmitted through P_2 is $E_0\cos\theta$. The intensity of light is proportional to the square of the amplitude of the electric field (Equation (26-27)); therefore, the intensity of the light transmitted through P_2 is

$$I(\theta) = \frac{1}{2}c\varepsilon_0(E_0\cos\theta)^2 = \left(\frac{1}{2}c\varepsilon_0 E_0^2\right)\cos^2\theta$$

$$= I_0 \cos^2\theta \qquad (26\text{-}37)$$

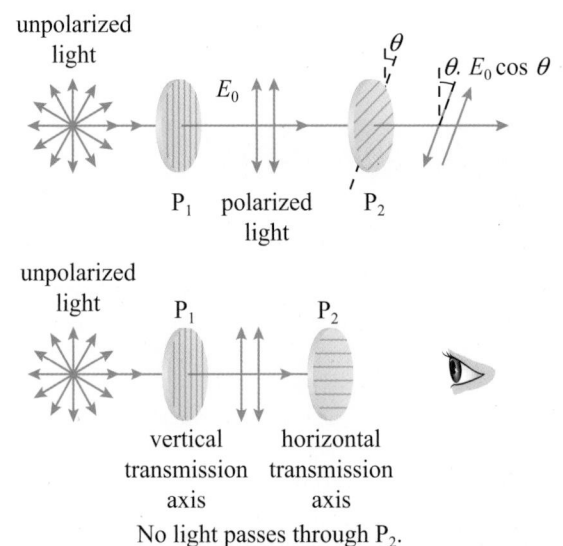

Figure 26-34 No light passes through P_2.

Figure 26-34 When an electromagnetic wave passes through a polarizer, only the component of the electric field that is parallel to the transmission axis of the polarizer is transmitted. The component that is perpendicular to the transmission axis is blocked.

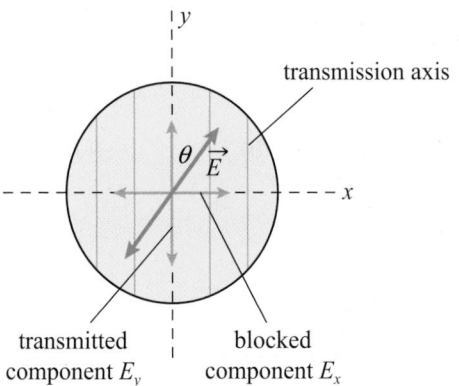

Figure 26-33 Only the component of the electric field that is parallel to the transmission axis of the polarizer is transmitted. The component that is perpendicular to the transmission axis is blocked.

Note these consequences of Malus' law (see Figure 26-34):

- When the polarizing axes of two polarizers are parallel to each other ($\theta = 0°$ or $\theta = 180°$), the intensity of the light passing through the analyzer remains I_0. The analyzer does not block any light that is polarized along its transmission axis.

- When the polarizing axes of two polarizers are perpendicular to each other ($\theta = 90°$), the intensity of the light passing through the analyzer is zero. The analyzer completely blocks the light that is polarized perpendicular to its transmission axis.

 CHECKPOINT

C-26-7 Malus' Law

A beam of light is polarized along the vertical axis and passes through two polarizers. The transmission axis of the first polarizer (P_1) is inclined at 45° with respect to the light's polarization direction. The transmission axis of the second polarizer (P_2) is inclined at 45° with respect to P_1 so that it is perpendicular to the polarization direction of the incident beam. The intensity of the initial beam is I_0. What is the intensity of the beam after it passes through P_2?

(a) zero (b) $\dfrac{I_0}{4}$ (c) $\dfrac{I_0}{2}$ (d) I_0

C-26-7 (c) The transmission axes of the two polarizers are inclined at 45° and cos(45°) = 1/√2.

Polarization by Reflection

A wave travelling in a medium divides into two parts when it encounters the boundary of another medium. The part that is reflected into the original medium is called the **reflected wave**. The part that is transmitted into the second medium is the **transmitted wave**. The plane containing the direction of the incident wave and the normal to the surface between the two media is called the **plane of incidence** (Figure 26-35).

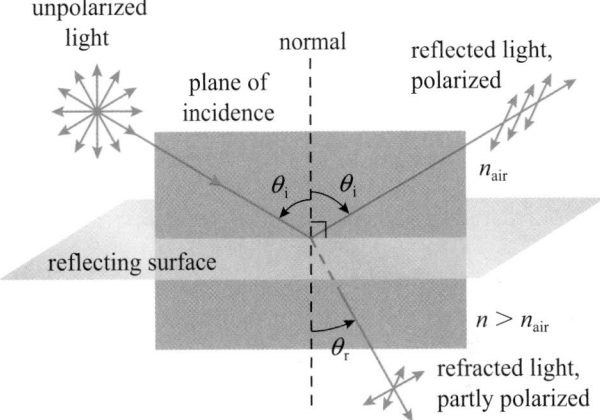

Figure 26-35 The incident, reflected, and transmitted waves at the boundary between two media. Here θ_i is the angle of incidence with respect to the normal to the reflecting surface and θ_r is the angle of refraction.

Consider a beam of unpolarized light incident at an angle to the surface of a flat glass plate in air. The electric field of the incident beam can be resolved into two components: one in the plane of incidence and the other

EXAMPLE 26-6

Beam Intensity Through Two Polarizers

A beam of unpolarized light, with intensity I_0, is incident on a polarizer with a vertical transmission axis. The beam then passes through a second polarizer, the transmission axis of which is at an angle of 60° to the vertical.

(a) What is the intensity of the light after it passes through the first polarizer?
(b) What is the intensity of the beam that emerges from the second polarizer in terms of I_0?

SOLUTION

(a) We can resolve the electric field of the unpolarized light into two perpendicular components. We choose one component to be in the vertical direction and the other component to be in the horizontal direction. For unpolarized light, these two components have equal magnitudes. When the light passes through the first polarizer, the vertical component is transmitted and the horizontal component is absorbed. Therefore, half the incident intensity passes through the first polarizer:

$$I_1 = \frac{I_0}{2}$$

(b) The light that passes through the first polarizer is linearly polarized. We can now apply Malus' law to find its intensity after transmission through the second polarizer:

$$I_2 = I_1(\cos(60°))^2 = \frac{I_0}{2} \times \left(\frac{1}{2}\right)^2 = \frac{I_0}{8}$$

Making sense of the result:

Only 12.5% of the initial intensity emerges from the second polarizer.

perpendicular to it, as shown in Figure 26-36. For unpolarized light, both components have the same amplitude.

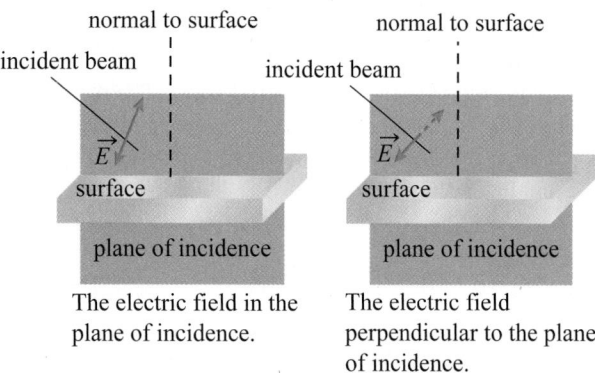

The electric field in the plane of incidence.

The electric field perpendicular to the plane of incidence.

Figure 26-36 The electric field of the incident beam of light can be resolved into a component in the plane of incidence (the plane of the page) and a component perpendicular to the plane of incidence.

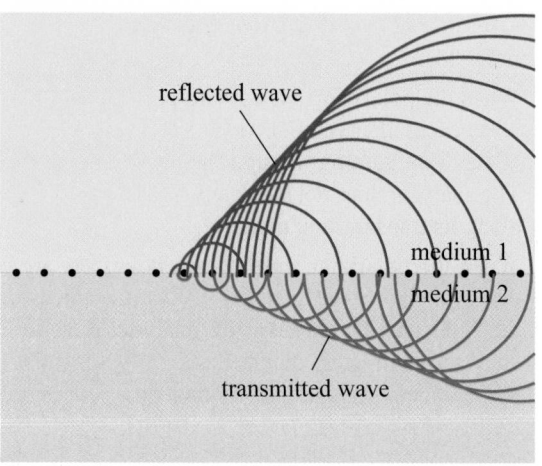

Figure 26-37 When subjected to an incident electromagnetic wave the atomic electrons at the interface of two surfaces act as sources of spherical electromagnetic waves that form the reflected and the transmitted waves.

The reflected and transmitted waves are generated by electrons on the surface of the glass plate. The incident electric field exerts an oscillatory force on the electrons, causing them to oscillate. Each oscillating electron becomes a source of electromagnetic waves that propagate outward, forming the reflected and transmitted waves (Figure 26-37).

When an incident wave hits the surface of the glass plate, the component of the electric field perpendicular to the plane of incidence lies in the plane of the glass plate. Electrons oscillating in response to the perpendicular component do so in the plane of the plate; they therefore radiate in a direction perpendicular to the surface of the plate. Hence, the electric field of this part of the reflected and transmitted waves is polarized perpendicular to the plane of incidence.

The component of the electric field that is in the plane of incidence is perpendicular to the surface of the glass plate. Electrons responding to this component oscillate perpendicular to the surface of the glass plate and therefore emit radiation in the plane of the plate (like waves propagating on the surface of a pond when a stone is thrown into it). Virtually all of this radiation transmits into the body of the glass plate. Therefore, the transmitted wave has an electric field with substantial components parallel and perpendicular to the plane of incidence, and the reflected wave is predominantly polarized perpendicular to the plane of incidence (Figure 26-36).

Maxwell's Equations

Maxwell discovered that a varying electric flux produces a magnetic field. To incorporate this phenomenon, he added a displacement current term, $\varepsilon_0 \dfrac{d\Phi_E}{dt}$, to Ampère's law.

Maxwell's equations describe how electric charges, electric currents, electric fields, and magnetic fields interact with each other; the equations also predict the existence of travelling electromagnetic waves.

Properties of Electromagnetic Waves

Electromagnetic waves sustain themselves once generated, travel in a vacuum at the speed of light, and have electric and magnetic fields in phase with each other with amplitudes related by $E_0 = cB_0$.

In a travelling electromagnetic wave, the electric field, the magnetic field, and the direction of motion of the wave are all perpendicular to each other.

Electromagnetic waves span a broad spectrum, from very large to extremely small wavelengths. Radio waves, TV waves, microwaves, infrared radiation, visible light, ultraviolet radiation, X-rays, and gamma rays are all electromagnetic waves.

Electromagnetic waves are generated by quantum-mechanical processes and by accelerating charged particles.

Electromagnetic waves carry energy and momentum. The rate at which energy is transported per unit area per unit time is called the energy flux or intensity. The Poynting vector, $\vec{S}$, describes the magnitude of energy flux and the direction of the energy flow in electromagnetic waves:

$$\vec{S} = \frac{1}{\mu_0}\vec{E} \times \vec{B} \tag{26-29}$$

The average intensity flow in one cycle is given by the time-averaged energy flux, $\bar{S}$:

$$\bar{S} = \frac{E_0 B_0}{2\mu_0}$$

In an electromagnetic wave, the energy density of the electric field is equal to the energy density of the magnetic field.

The intensity of a three-dimensional spherical wave varies inversely as the square of the distance from the source.

Electromagnetic waves exert radiation pressure on objects:

$$P_{\text{rad}} = \frac{\bar{S}}{c} \text{ (when absorbed)} \tag{26-34}$$

$$P_{\text{rad}} = 2\frac{\bar{S}}{c} \text{ (when reflected)} \tag{26-35}$$

Polarization describes the orientation of the electric field of an electromagnetic wave. The plane of polarization is a plane that contains the direction of motion of the wave and the direction of the electric field of the wave.

When polarized light with intensity I_0 falls on a polarizer, which has its transmission axis inclined at an angle θ with respect to the direction of polarization of the incident light, the intensity of the emitted light, I_1, is related to I_0 by Malus' law:

$$I_1 = I_0 \cos^2\theta \tag{26-36}$$

APPLICATIONS

Applications: wireless communications, astronomical research, food preservation, security, GPS, UV index, medical applications

Key Terms: average intensity, bremsstrahlung, Cerenkov radiation, cyclotron radiation, displacement current, electromagnetic field, electromagnetic spectrum, electromagnetic wave, energy flux, far infrared, fluorescence, hard X-rays, intensity, irradiance, linear polarizer, linearly polarized, Malus' law, Maxwell's equations, momentum density, near infrared, plane of incidence, plane of polarization, polarization axis, polarization, polarizer, Poynting vector, radiation momentum, radiation pressure, radio astronomy, radiograph, reflected wave, soft X-rays, synchrotron radiation, thermograms, transmission axis, transmitted wave, unpolarized, visible spectrum

QUESTIONS

1. Explain why a displacement current is needed to explain the propagation of electromagnetic waves.

2. An electromagnetic wave is travelling along the positive z-axis. If at a given instant and position the magnetic field is pointing along the negative y-axis, the electric field at that instant and position points along the
   (a) negative y-axis;
   (b) positive y-axis;
   (c) negative x-axis;
   (d) positive x-axis.

3. Which of the following sets of equations for electric and magnetic fields do *not* represent a travelling electromagnetic wave? Why not?

   (a) $\vec{E}(x, t) = E_0 \sin(x - ct)\vec{z}$; $\vec{B}(x, t) = \dfrac{E_0}{c}\sin(x - ct)\vec{z}$

   (b) $\vec{E}(x, t) = E_0 \sin(x - ct)\vec{z}$; $\vec{B}(x, t) = \dfrac{E_0}{c}\sin(x - ct)\vec{y}$

   (c) $\vec{E}(x, t) = E_0 \sin(x - ct)\vec{y}$; $\vec{B}(x, t) = \dfrac{E_0}{c}\cos(x - ct)\vec{z}$

   (d) $\vec{E}(x, t) = E_0 \sin(x - ct)\vec{y}$; $\vec{B}(x, t) = E_0 \sin(x - ct)\vec{z}$

4. Give several examples of electromagnetic waves from very large to very small frequencies. Compare and contrast your examples.

5. Compare and contrast sound and electromagnetic waves.

6. (a) A dipole antenna transmits electromagnetic waves at a frequency of 1130 kHz. Determine the length of the antenna.
   (b) A dipole antenna transmits radio waves at a frequency of 98.8 MHz. Determine the length of the antenna.

7. Describe various ways that a charged particle can emit electromagnetic radiation.

8. Unpolarized light is incident on two polarizers (Figure 26-34). Their transmission axes are oriented so that no light is transmitted through the second polarizer. A third polarizer with its transmission axis oriented at 45° with respect to that of the first polarizer is inserted between the two. Can the light now pass through the second polarizer? Explain your reasoning.

9. Can an isolated charged particle move in a circular orbit of constant radius while emitting electromagnetic radiation? Explain your reasoning.

10. Can a laser beam be used to levitate a small and perfectly reflecting object? Explain your reasoning.

11. Is the radiation pressure greater on a surface that absorbs all incident radiation or a surface that is a perfect reflector? Explain your reasoning.

PROBLEMS BY SECTION

For problems, star ratings will be used, (∗, ∗∗, or ∗∗∗), with more stars meaning more challenging problems.

Section 26-2   The Displacement Current and Maxwell's Equations

12. ∗ A uniform electric field is increasing at a rate of 2.0×10^6 V/m·s. What is the displacement current through an area of 5.0×10^{-4} m² that is perpendicular to the direction of the field?

13. ∗ A parallel-plate capacitor has circular plates of radius 0.01 m each that are 0.10 m apart. The capacitor is being charged at a rate of 100 V/m·s. What is the displacement current in the capacitor?

14. ∗∗ Circular plates of a capacitor are 5.0 cm in radius, 2.0 mm apart, and have air between them. The voltage across the plates is changing at a rate of 60.0 V/s. Determine
   (a) the rate of change of the electric field between the plates;
   (b) the displacement current between the plates.

15. ∗ A uniform displacement current of magnitude 1.00 A, directed out of the plane of the paper, passes through a circular region of radius 5.00 cm. Determine the magnitude of the magnetic field due to the displacement current at the radial distances (a) 2.00 cm and (b) 10.00 cm.

16. ∗∗ A parallel-plate capacitor with circular plates of radius 10 cm is being charged. Consider a circular loop of radius 20 cm centred at the central axis of the plates. The displacement current through the loop is 1.0 A. What is the rate of change of the magnitude of the electric field between the plates?

17. ∗∗ A 100 pF capacitor has circular plates of 10.0 cm radius that are 5.0 mm apart and have air between them. The capacitor is charged by connecting it to a 12.0 V battery through a 1.0 Ω resistor.
   (a) Determine the current through the plates at $t = 0$ (when the battery is connected).
   (b) What is the current through the plates at $t = 60$ s?
   (c) Determine the rate at which the electric field between the plates changes at $t = 0$ and at $t = 60$ s.
   (d) Determine the magnetic field between the plates at $t = 0$ and at $t = 60$ s.

18. ∗∗ The rate of change of a uniform electric field through a region of space is shown in Figure 26-38. Determine the displacement current through a 0.25 m² area perpendicular to the direction of the electric field during the time intervals $0 < t < 12$, $12 < t < 22$, and $22 < t < 30$, where t is in milliseconds.

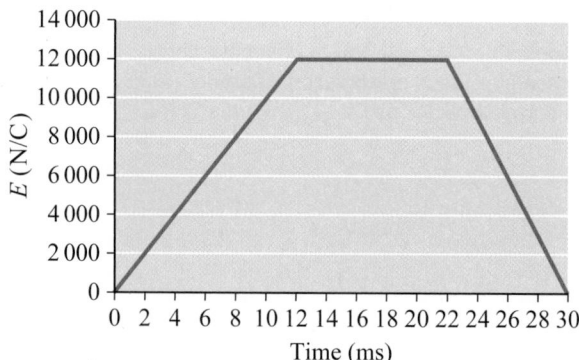

Figure 26-38   Problem 18

Section 26-3   Electromagnetic Waves

19. ∗ Determine the direction of propagation of the electromagnetic wave for each of the following cases.
   (a) $\vec{E}$ is in the z-direction; $\vec{B}$ is in the x-direction.
   (b) $\vec{E}$ is in the y-direction; $\vec{B}$ is in the x-direction.
   (c) $\vec{E}$ is in the x-direction; $\vec{B}$ is in the y-direction.
   (d) $\vec{E}$ is in the z-direction; $\vec{B}$ is in the x-direction.

20. ✳ The amplitude of the magnetic field of an electromagnetic wave is 2.0×10^{-2} T. What is the amplitude of the electric field of the electromagnetic wave? If the magnetic field is pointing along the positive z-axis, what are the possible directions of the electric field?

21. ✳ What is the amplitude of the magnetic field of an electromagnetic wave that has a electric field of amplitude 5.0×10^{-3} V/m? When the magnetic field is pointing along the positive x-axis, what are the possible directions of the electric field?

22. ✳ The electric field of an electromagnetic wave in a vacuum is given by

$$\vec{E}(x, t) = (100 \text{ V/m}) \sin(2.00 \times 10^{9}x - \omega t)\hat{z}$$

(a) What are the wavelength and frequency of this wave?
(b) Write an equation for $\vec{B}(x, t)$.
(c) In which direction is the wave travelling?

23. ✳ The magnetic field of an electromagnetic wave is given by

$$\vec{B}(x, t) = (5.0 \times 10^{-6} \text{ T}) \sin(4.00 \times 10^{9}x - \omega t)\hat{k}$$

(a) Determine the wavelength and the frequency of the wave.
(b) What is the amplitude of the wave's electric field?
(c) Write an the equation for $\vec{E}(x, t)$.
(d) What is the direction of propagation of the wave?

24. ✳ A plane electromagnetic wave with a wavelength of 500 nm is travelling in the positive x-direction, and its electric field points along the z-axis. The amplitude of the electric field is 2.0 V/m. Write the equations for the electric and magnetic fields of this wave as a function of position and time.

25. ✳ A sinusoidal electromagnetic field is travelling in the $+y$-direction. At a certain position and time its electric field has a magnitude of 1.0 V/m and is pointing along the $-x$-direction. What are the magnitude and direction of the wave's magnetic field at that position and location?

26. ✳ The electric field of an electromagnetic wave is given by

$$\vec{E}(y, t) = (5.0 \text{ V/m}) \sin\left(\frac{2\pi}{10^{6}}y - 2\pi f t\right)\hat{i}$$

(a) Determine the frequency of the wave.
(b) In which direction is the wave travelling?
(c) Write an equation for $\vec{B}(y, t)$.

Section 26-4 The Electromagnetic Spectrum

27. ✳ Rank the following electromagnetic waves in the order of their frequencies from smallest to largest: infrared light, gamma rays, yellow light, blue light, microwaves, and X-rays.

28. ✳ What is the wavelength of gamma rays that have a frequency of 10^{20} Hz?

29. ✳ Determine the frequency of light of wavelengths (a) 450 nm, (b) 533 nm, and (c) 700 nm.

30. ✳ Which of the following combinations of wavelength and frequency are not allowed for electromagnetic waves in a vacuum? Why? For the speed of light, use $c = 3.0 \times 10^{8}$ m/s.
(a) $\lambda = 100.0$ m; $f = 3.0 \times 10^{6}$ Hz
(b) $\lambda = 0.1$ m; $f = 3.0 \times 10^{8}$ Hz
(c) $\lambda = 6.0 \times 10^{-7}$ m; $f = 5.0 \times 10^{14}$ Hz
(d) $\lambda = 2.0 \times 10^{-15}$ m; $f = 3.0 \times 10^{-23}$ Hz

Sections 26-5 and 26-6 Energy and Momentum of Electromagnetic Waves, and Radiation Pressure

31. ✳ A 100 W point source produces light with a wavelength of 550 nm. Determine the amplitudes of the electric and magnetic fields of the light at (a) 10 m, (b) 100 m, and (c) 10^{5} m from the source. What is the intensity of the electromagnetic wave at the three distances?

32. ✳ The intensity of the sunlight on top of Earth's atmosphere is approximately 1400 W/m^2. Calculate the amplitudes of the electric and magnetic fields corresponding to this intensity.

33. ✳ A laser beam delivers 0.10 J/m^2 of energy in 5.0 s to a surface that is placed normal to the direction of motion of the beam. What is the irradiance of the beam?

34. ✳✳ A laser pointer has a radius of 2.0 mm and delivers 3.2×10^{-3} W of power.
(a) What is the intensity of the laser light?
(b) For a beam length of 5.0 m, how much energy is contained in the beam?

35. ✳ The cosmic microwave background (CMB) radiation is the electromagnetic radiation that is left over from the Big Bang and fills the universe. It has an average energy density of 4×10^{-14} J/m^3. What are the peak values (i.e., the amplitudes) of the electric and magnetic fields of the CMB radiation?

36. ✳✳ The National Ignition Facility in California fired the world's most powerful laser in March 2012. The pulse from the laser lasted for 23×10^{-9} s and delivered 4.11×10^{14} W of power.
(a) What was the total energy delivered by the laser?
(b) What were the amplitudes of the electric and magnetic fields?
(c) If the laser beam was focused on an area of 1.0 mm^2, what was the radiation pressure on this area?

37. ✳ The irradiance at 1.0 m from a light bulb is 2.0 W/m^2. What is the irradiance from this bulb at a distance of 5.0 m?

38. ✳ A point source emits lights uniformly in all directions. At 10.0 m from the source, the electric field strength of the light is 50 N/C. Determine the average power emitted by the source.

39. ✳ The average solar power that reaches Earth's surface is 1000 W/m^2. What is the magnitude of the force exerted by the sunlight on a free electron that is moving with a speed of 1.1×10^{5} m/s?

40. ✳✳✳ Radiation pressure of a laser beam can be used to levitate a small particle. What should be the minimum power of a laser beam to levitate a particle of mass 1.0×10^{-6} kg? Assume that the beam is totally reflected from the particle and that the cross-sectional area of the particle is greater than the cross-sectional area of the beam.

41. ✳✳ Photovoltaic cells convert electromagnetic radiation into electricity. A solar panel consists of an assembly of photovoltaic cells (Figure 26-39). A typical efficiency of these cells is approximately 15%, which means that 15% of the incident radiation is converted into electricity. For solar panels that cover an area of 16 m^2 on the roof of a house, how many kilowatt hours of energy will the panels generate in one year? Assume that the average solar intensity at Earth's surface is approximately 200 W/m^2. This takes into account the fact that Earth represents a disk of radius 6370 km to the incoming sunlight and that a fraction of the incident energy is either absorbed or reflected by the atmosphere.

Figure 26-39 Problem 41. Solar panels on a rooftop

42. ✳✳ A sinusoidal electromagnetic wave from a satellite passes perpendicularly through a window with a cross-sectional area of 0.75 m². The amplitude of the electric field of the wave entering the window is 4.0×10^{-3} V/m. How much energy passes through the window in one hour?

43. ✳✳ A laser has a power of 5.00 mW and produces a spot with a radius of 1.00 mm on a white screen.
   (a) What is the intensity of the laser light on the spot?
   (b) When all the light reflects from the white screen, what are the radiation pressure and the magnitude of the force exerted by the light on the screen?

44. ✳✳ The power emitted by the Sun is approximately 3.8×10^{26} W.
   (a) What is the average intensity of the Sun's electromagnetic waves at Earth's location? (The average distance between the Sun and Earth is approximately 1.5×10^{11} m.)
   (b) What is the radiation pressure exerted on Earth by the sunlight?

Section 26-8 Polarization

45. ✳ An electromagnetic wave is moving in the direction of the positive x-axis. At a given instant the magnetic field of the wave is pointing toward the negative z-axis. What is the direction of polarization of the wave?

46. ✳ In a certain setup, light is polarized in the vertical direction, and only 75% of the incident light is transmitted through the polarizer. What angle does the transmission axis of the polarizer make with the vertical?

47. ✳ In a certain setup, light is polarized in the horizontal direction, and only 25% of the incident light is transmitted through the polarizer. What angle does the transmission axis of the polarizer make with the vertical?

48. ✳ Linearly polarized light is incident on a polarizer (see Figure 26-33). The intensity of the incident light is 1.0 W/m². Calculate the intensity of the transmitted light when the angle between the transmission axis and the direction of the electric field is (a) 30°, (b) 60°, and (c) 90°.

49. ✳ Unpolarized light passes through two polarizers (see Figure 26-34), and the transmission axis is inclined at an angle of 45° with respect to each other. The initial light has an intensity of 1.0 W/m². What is the intensity of the light that emerges from the second polarizer?

50. ✳✳ Linearly polarized light with an intensity of 5.0 W/m² passes through two polarizers. The transmission axis of the first polarizer is inclined at 45°, and the transmission axis of the second polarizer is inclined at 90° with respect to the polarization direction. What is the intensity of the transmitted light?

51. ✳✳ The transmission axes of two polarizers are inclined at 90° with respect to each other. For the following cases, determine the ratio of incident to transmitted intensity of the light that passes through the two polarizers.
   (a) The initial light is unpolarized.
   (b) The initial light is linearly polarized with its polarization axis along the transmission axis of the first polarizer it encounters.
   (c) The initial light is linearly polarized with its polarization axis perpendicular to the transmission axis of the first polarizer it encounters.
   (d) The initial light is linearly polarized with its polarization axis inclined at 45° with respect to the transmission axis of the first polarizer it encounters.

52. ✳✳ Vertically polarized light is incident on two polarizers, as shown in Figure 26-40. The transmission axis is indicated by the solid black line. The incident intensity is 5.0 W/m². Calculate the transmitted intensity for each configuration.

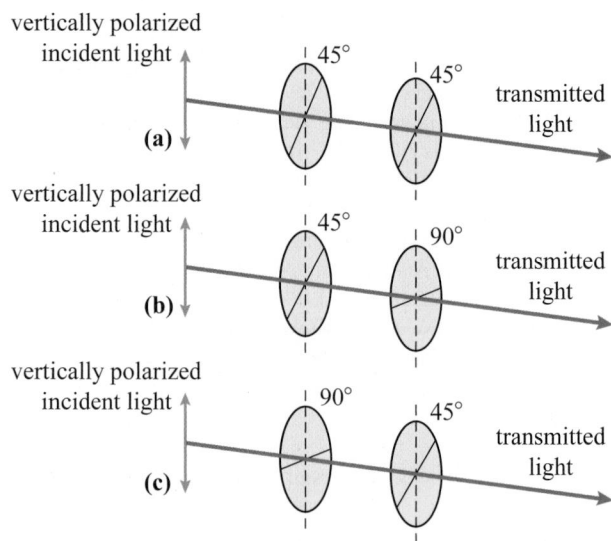

Figure 26-40 Problem 52

53. ✳✳ Unpolarized light from a laser is incident on three polarizers, as shown in Figure 26-41. The intensity of the incident light is I_0. Determine the intensity of the light at points A, B, and C.

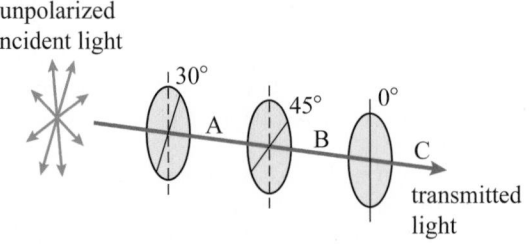

Figure 26-41 Problems 53 and 54

54. ✸✸ Consider the transmission of unpolarized light through three polarizers (Figure 26-41). Determine the ratio of incident to transmitted intensity for the following combinations of the directions of the transmission axes of the polarizers. The angles are with respect to the horizontal direction.
(a) 90°; 45°; 0°
(b) 90°; 45°; 90°
(c) 90°; 0°; 45°
(d) 0°; 45°; 45°

55. ✸✸✸ Five identical polarizers are placed in a straight line next to each other with their transmission axes inclined at angles θ, 2θ, 3θ, 4θ, and 5θ from left to right, with respect to the vertical direction. Unpolarized light is incident from the left. What must the angle θ be so that only 10% of the incident light is transmitted through the polarizers?

COMPREHENSIVE PROBLEMS

56. ✸✸✸ Two identical circular disks of radius 1.00 cm and mass 2.00 g each are attached to the opposite ends of a thin wire of length 10.0 cm and mass 1.00 g. The wire is free to rotate about an axis that passes through its centre (Figure 26-42). One of the disks is coloured black to absorb all the incident radiation, and the other disk is a perfect reflector. The arrangement is placed in the sunlight, which falls perpendicular on one face of each disk.
(a) Will the wire rotate in the presence of the sunlight? Explain your reasoning.

Figure 26-42   Problem 56

(b) If yes, determine the angular acceleration of the wire.
(c) Will the wire rotate if both disks are perfect reflectors or coloured black? Ignore the effect of air drag in this problem. The intensity of the sunlight is 1000 W/m².

57. ✸✸✸ As a comet approaches the Sun, the solar radiation causes some materials within the comet to vaporize and form an atmosphere around the comet. The force due to solar radiation pressure pushes lighter particles in the atmosphere away from the Sun, forming a tail (Figure 26-43). Assume that the particles in the comet's atmosphere are spherical and have a density of 2500 kg/m³.
(a) For what size would the force due to radiation pressure be equal to the gravitational force of the Sun on a particle? Assume that a particle reflects all incident radiation.
(b) What would happen to the particles smaller than the size you calculated in part (a)? Would you expect the comet's tail to become longer as the comet approaches the Sun?

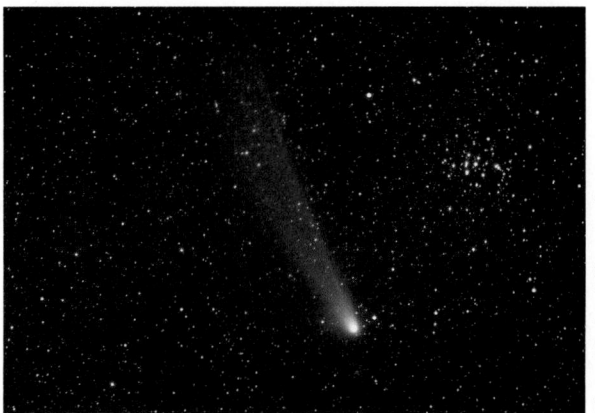

Figure 26-43   Problem 57. Comet Hyakutake

 See the text online resources at www.physics1e.nelson.com for Open Problems and Data-Rich Problems related to this chapter.

Learning Objectives

When you have completed this chapter you should be able to:

1. Provide evidence for the geometric optics approach to describing optical phenomena and solving problems.

2. Derive the law of reflection from Fermat's principle, and use the law of reflection to describe image formation in a plane mirror.

3. Use the law of reflection to predict the properties of optical images formed by plane and spherical mirrors.

4. Derive the law of refraction from Fermat's principle, and use the law of refraction to describe light propagation through various media.

5. Use the law of refraction to predict the properties of optical images formed by thin lenses.

6. Use the laws of geometric optics to describe how the human eye and corrective lenses function.

7. Use the concept of Brewster's angle to describe the polarization of refracted and reflected light.

For additional Making Connections, Examples, and Checkpoints, as well as activities and experiments to help increase your understanding of the chapter's concepts, please go to the text's online resources at www.physics1e.nelson.com.

Chapter 27
Geometric Optics

Cloud Gate: Where Art and Science Converge

The 110-tonne elliptical sculpture shown in Figure 27-1 stands in Chicago's Millennium Park. This sculpture was designed by the British artist Anish Kapoor, who forged highly polished, curved steel plates to form a huge bean-shaped sculpture that reflects Chicago's famous skyline. The sculpture is called "Cloud Gate" because the shape of the sculpture resembles the shape of a cloud. The visitors are invited to walk through the sculpture and view their images on the curved mirror surfaces. The sculpture is warmly referred to by the locals as "the bean." This unforgettable sculpture, a convergence of science and art, makes us wonder about the formation of optical images, and prompts us to ask scientific questions that can be answered using the concepts of geometric optics, the topic of this chapter.

Courtesy of Tetyana Antimirova

Figure 27-1 The Cloud Gate sculpture in Chicago's Millennium Park

27-1 Evidence for the Geometric Optics Approach to Describing Optical Phenomena

Although light is an electromagnetic wave, it is often possible to describe the propagation of light using a simplified model in which light is treated as a collection of geometric rays. The geometric optics approach to describing light is applicable only when the dimensions of the objects that light interacts with are much larger than the wavelength of the light (400–700 nm for visible light (Chapter 26)). The branch of physics that uses ray approximations is called **geometric** or **ray optics**. The ray approximation is a straightforward geometric way of describing optical phenomena such as reflection, refraction, eclipses, rainbows, mirages, and images created by mirrors and lenses. However, some light phenomena, including interference and diffraction, can be explained only by considering the wave nature of light, as you will learn in Chapter 28.

In geometric optics, we treat light as a collection of rays that propagate in straight lines through a uniform medium. In physics, a **medium** (plural form media) is any substance through which something (such as waves or energy) is transmitted. For example, water, air, and metals can be media for both mechanical and electromagnetic waves.

Let us consider a few of the natural phenomena that led to the development of geometric optics. A shadow is created when an opaque object is located between a light source and a screen. An **opaque** material does not transmit any light. In contrast, **transparent** materials transmit all light, and **translucent** materials transmit some light. The location and size of the shadows can be predicted by noting that light in a homogeneous medium propagates in straight lines, a property called the rectilinear propagation of light. If a light source produces a beam of parallel rays (also called collimated light), it produces a shadow, as shown in Figure 27-2(a). A beam of parallel rays results when a light source is located very far from an observer, such as the Sun's rays observed from Earth. Some light sources, such as lasers, can produce a highly collimated beam of light with rays that are almost parallel. A noncollimated light source, such as an incandescent light bulb, produces light rays propagating in many different directions. When such a source is relatively close to the object casting a shadow, or when a light source is much bigger than an object casting a shadow, both a **full shadow (umbra)** and a **partial shadow (penumbra)** result (Figure 27-2(b)).

Observations of shadows cast by terrestrial and celestial bodies support the geometric optics approach: light appears to propagate in straight lines and can therefore be considered as a collection of rectilinear rays. As shown in Figure 27-3, we can use light rays to help understand how eclipses occur.

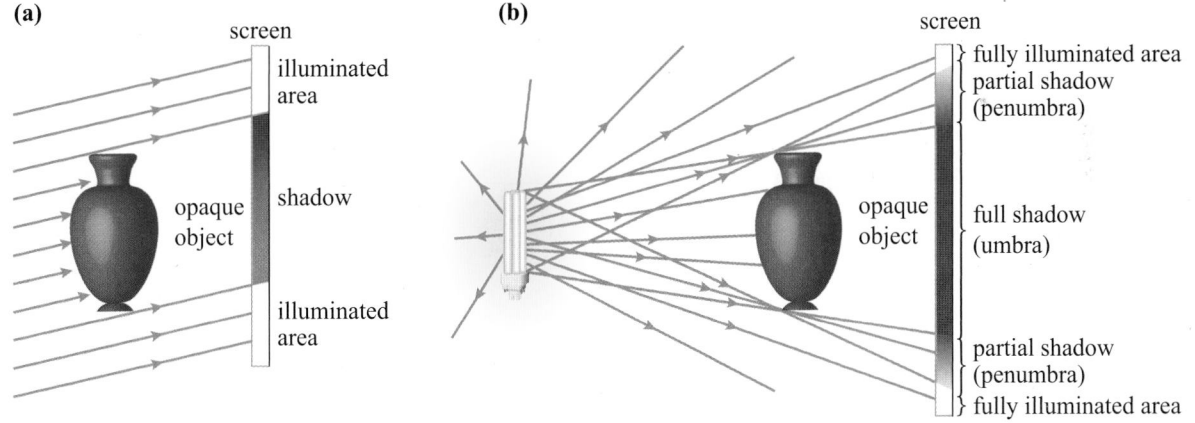

Figure 27-2 Opaque objects casting shadows while illuminated by (a) a collimated beam of light and (b) a noncollimated beam of light.

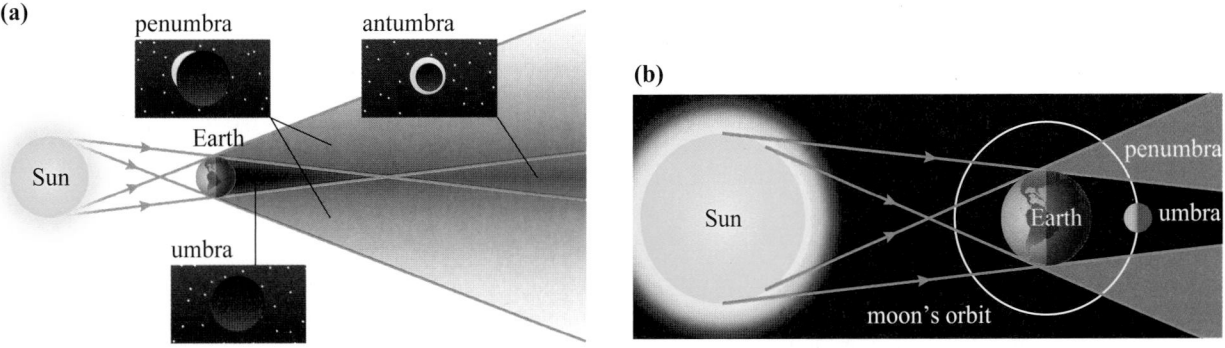

Figure 27-3 (a) Shadows cast by Earth (b) A total lunar eclipse (both figures are not drawn to scale)

✓ CHECKPOINT

C-27-1 Partial Shadows

To produce a partial shadow,

(a) the light rays illuminating an opaque object must diverge (spread apart) or converge (come together);

(b) the light rays illuminating an opaque object must come from a distant light source;

(c) the light rays illuminating an opaque object must be parallel;

(d) both the light source and the opaque object must be of comparable size;

(e) Both (a) and (d) are true.

C-27-1 (e) See Figure 27-3.

ONLINE ACTIVITY

Measuring Scaling the Earth

The e-resource that accompanies every new copy of this textbook contains an Online Activity using the "Measuring Scaling the Earth" simulation. This online activity will help you explore in more detail how Eratosthenes was able to measure the circumference of the earth without leaving Egypt more than 2000 years ago.

➕➖ EXAMPLE 27-1
➗✖

Eratosthenes' Estimate of Earth's Radius

Ancient Greek mathematician, geographer, and astronomer Eratosthenes (276–194 BCE) estimated the distance between Syene and Alexandria, Egypt, to be approximately 4400 stadia (singular form stadium) by averaging the speed of camel caravans. The stadium is an ancient unit of length, but it is not known whether Eratosthenes' calculations used the Greek stadium (157 m) or the Egyptian stadium (166 m). Eratosthenes measured the angle between the two cities to be $7°12'$, as shown in Figure 27-4. Use Eratosthenes' data to estimate Earth's radius in stadia and metres (using both conversion factors), and compare the results to the current accepted value of 6370 km.

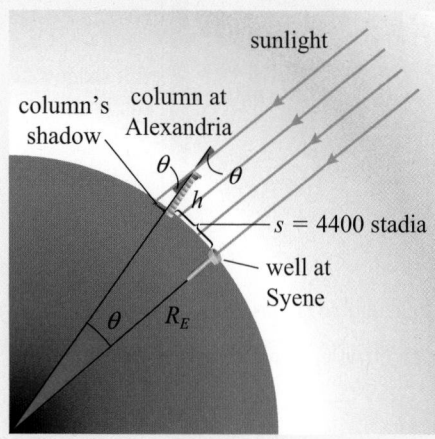

Figure 27-4 Example 27-1

SOLUTION

From the diagram of Eratosthenes' calculations, we can see that the angle between radii passing through Syene

and Alexandria equals the angle between the Sun's rays and the column in Alexandria. Therefore, the ratio of this angle to $360°$ (a full circle) equals the ratio of the distance between the cities (length of the arc) to Earth's circumference:

$$\frac{7°12'}{360°} = \frac{d_{\text{Alexandria-Syene}}}{2\pi R_{\text{E}}}$$

$$R_{\text{E}} = \frac{360°}{7°12'} \frac{d_{\text{Alexandria-Syene}}}{2\pi} = \left(\frac{50.0}{2\pi}\right)(4400 \text{ st}) = 35\ 014 \text{ st}$$

Assuming Greek stadia:

$$R_{\text{E}_1} = 35\ 014 \text{ st} \cdot 157 \frac{\text{m}}{\text{st}} = 5.50 \times 10^6 \text{ m} = 5500 \text{ km}$$

Assuming Egyptian stadia:

$$R_{\text{E}_2} = 35\ 014 \text{ st} \cdot 166 \frac{\text{m}}{\text{st}} = 5.82 \times 10^6 \text{ m} = 5810 \text{ km}$$

The current accepted value for the mean radius of Earth is 6370 km. Depending on which stadium unit Eratosthenes used, his estimate was either 9% or 14% off the actual value: $R_{\text{E}} = 6400$ km.

Making sense of the result:

Eratosthenes' estimate was based on basic geometry, the observation that the Sun's rays are straight and parallel, and the assumption that Earth is spherical (or nearly so). Since the actual angle between Syene and Alexandria is approximately $7°5'$, Eratosthenes' estimate of $7°12'$ gives a slightly low value for Earth's radius.

We finish this section with a brief discussion of optical image formation. In geometric optics, an **object** can be any material that either emits or reflects light. Whatever the object is, light rays emanate from it. Although all real objects have some physical size, objects can often be modelled as a point. For example, a small light-emitting diode viewed from a distance of 5 m can be considered a point object (a point light source), but when viewed from a distance of 5 cm, we usually need to treat it as an extended object (an extended light source). Extended objects (like the Sun) can be visualized as a combination of an infinite number of point objects.

An **image** is an apparent reproduction of an object formed by a lens, mirror, and/or other optical apparatus through the one-to-one correspondence between each point on an object and each image point. If the light rays emanating from an object converge after propagating through the optical apparatus, they form a **real image**, which can be observed using a screen placed where the rays converge. If the light rays diverge and only their imaginary extensions converge, they form a **virtual image**, which cannot be captured on a screen. Although our brains perceive a virtual image to be in a particular location, light does not actually emanate from that location. Light rays that form virtual images diverge.

We explore the formation of images by optical apparatus in the following sections. However, an image can be formed without mirrors or lenses using a simple technique first recorded approximately 2400 years ago by both Greek and Chinese scholars. A **pinhole camera** (or *camera obscura*) is a closed box, or a darkened room, with a screen on one side and a small hole (a pinhole aperture) in the middle of the opposite side. Some of the light emitted by an object passes through the pinhole to reach the screen. Since the pinhole is very small and the light propagates in straight lines, one can imagine that only one ray emitted by every point of an object will be able to reach the screen, thus creating a real image that corresponds point by point to the object (Figure 27-5(a)). The images created by pinhole cameras are reversed left to right and inverted (upside

Figure 27-5 (a) The formation of an image of a person in a pinhole camera. (b) On January 24, 1544, Reinerus Gemma-Frisius observed an eclipse of the Sun at Louvain, Belgium. Later, he used this illustration of the event in his book *De Radio Astronomica et Geometrico*, 1545. It is thought to be the first published illustration of a *camera obscura*.

down). In the mid-1820s, the first photographs were made by substituting a plate of photosensitive material for the screen in a *camera obscura*.

MAKING CONNECTIONS

Pinhole Cameras in Nature, Art, and Science

Walking through the shady trees of the beautiful University of British Columbia campus, one sometimes notices projected images of the Sun on the ground under the trees. These images are formed by the small openings in the leaf canopy. Thousands of years ago, ancient Chinese, Greeks, and Arabs used pinhole cameras to study the nature of light. By the Renaissance, many people were fitting darkened rooms with an outside wall with a pinhole so that the outdoor scene could be projected onto an opposite interior wall. Newton conducted optics experiments with such a room. Some famous Renaissance painters used images from pinhole cameras to help them draw small details in their paintings, and the military used pinhole cameras for surveillance. Later on, rich nobility used darkened carriages for entertainment, like the one sculpted by Rodney Graham and exhibited on the University of British Columbia campus shown in Figure 27-6. Pinhole cameras are still used in artists' studios and in science labs today. Scientists use pinhole cameras to make images using types of radiation that do not pass through lenses readily, such as α-particles and gamma rays.

Figure 27-6 "The Millennial Time Machine" is a sculpture by Rodney Graham. The sculpture is an artistic rendition of a pinhole camera used for entertainment during the Renaissance. Rodney Graham, Millennial Time Machine, 2003. Landau carriage converted to a mobile camera obscura. Collection of the Morris and Helen Belkin Art Gallery, The University of British Columbia. Gift of the artist with support from the Canada Council for the Arts Millennium Fund, the Morris and Helen Belkin Foundation, British Columbia 2000 Recognition Plan, and The University of British Columbia, 2003. Photo: Tim Newton

Camera Obscura, Reinerus Gemma-Frisius, 1544

27-2 Reflection of Light

The rectilinear propagation of light in a homogenous medium follows from the "principle of least time" postulated by the French mathematician Pierre de Fermat (1601–65). The principle of least time, or **Fermat's principle**, states that a light ray propagating between any two points always takes the path that it can traverse (travel across) in the shortest time. During the 20th century, Fermat's principle was modified to describe newly discovered phenomena, such as the curved paths of light near supermassive objects, a result of gravitational lensing. For our purposes in this chapter we can use Fermat's original principle.

When light rays encounter a surface or a boundary between two media, some of them change their direction and bounce back. This process is called **reflection**. A very smooth surface (such as a mirror) produces specular reflection, and a rough surface (such as a painted wall) produces diffuse reflection (Figure 27-7). Note that the roughness or the smoothness of the surface is measured compared to the wavelength of the light reflecting from the surface.

A ray travelling toward a surface is called an **incident ray**, and a ray leaving the surface is called a **reflected ray**. By convention, the **angle of incidence** and the **angle of reflection** are measured relative to a normal to the surface at the point of incidence (Figure 27-8). Therefore, a light ray perpendicular to the surface has an angle of incidence of 0°. All reflected rays have a direction determined by the **law of reflection**: the incident ray, the reflected ray, and the normal to the surface at the point of incidence lie in

(a) Specular Reflection

(b) Diffuse Reflection

Figure 27-7 Specular and diffuse reflection

the same plane, and the angles of incidence and reflection are always equal:

$$\theta_1 = \theta_1' \qquad (27\text{-}1)$$

KEY EQUATION

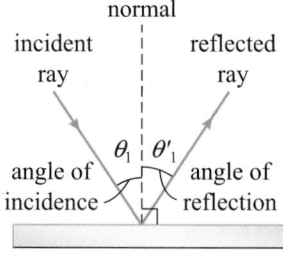

Figure 27-8 The law of reflection

EXAMPLE 27-2

The Law of Reflection

Show that the law of reflection is a direct consequence of Fermat's principle.

SOLUTION

We need to prove that the shortest path for a ray that leaves point A and reflects from a surface to pass through point B occurs when the angles of incidence and reflection are equal. Let us choose the two possible optical paths shown in Figure 27-9. For path AOB, the angles of incidence and reflection are equal ($\theta_1 = \theta_1'$), but they are not equal for path AO′B ($\theta_2 \neq \theta_2'$).

We draw a normal to the mirror from point B and extend it an equal distance beyond the mirror to point B′. To emphasize that the rays through the mirror are imaginary, we show them as dashed lines. We now extend ray AO beyond the mirror to point B′. Right-angled triangles BOC and B′OC are congruent because they have a common side (OC) and their corresponding angles are equal. Therefore,

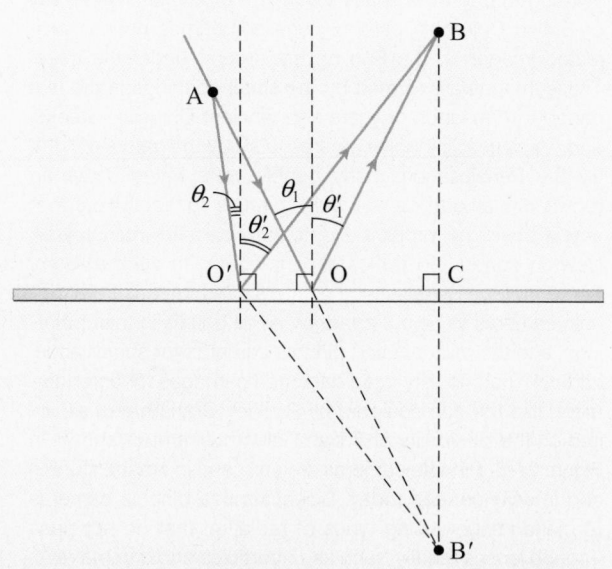

Figure 27-9 Proof of the law of reflection using Fermat's principle

their corresponding sides must be equal: OB = OB'. By the same reasoning, O'B = O'B'. Now we compare the optical paths AOB and AO'B:

$$AOB = AO + OB = AO + OB' = AOB'$$

$$AO'B = AO' + O'B = AO' + O'B' = AO'B'$$

$$AOB' < AO'B' \Rightarrow AOB < AO'B$$

The same argument applies for any location of point O' on the mirror. Therefore, the path with $\theta_i = \theta_i'$ is shorter than any path where the angles of incidence and reflection differ, such as path AO'B.

In Example 27-2, the speed of light along both paths is the same, so the fastest optical path coincides with the shortest geometric path. However, when light travels through different media, the fastest optical path and the shortest geometric paths could differ. Moreover, if the light were to travel in the opposite direction along the path (from point B to point A), our conclusions would not change: thus, the optical path between two points is the shortest path independent of the direction of the light travel. In other words, the *optical path is reversible*. We will rely on *reversibility of light* when solving problems and analyzing more complex optical apparatus.

 CHECKPOINT

C-27-2 Photographing a Mirror Image

You are photographing your friend looking at herself in a wall mirror. You want to make sure your photo captures the reflection of her face. While you are photographing her, what must she see in the mirror?

(a) her face
(b) her entire body
(c) your entire body
(d) your camera

C-27-2 (d) Due to the reversibility of light, she must see your camera for the camera to "see" her eyes and her face. However, she might not be able to see herself in the mirror.

 PEER TO PEER

I remember drawing ray diagrams in high school. However, I never paid too much attention to them then because I didn't realize their value. Now I can see that they help me visualize how images are formed. There is also value in making a scale diagram. I especially like to experiment with computer simulations and vary the parameters of the mirrors to see how these changes affect the formation of an image. I think computer simulations are especially helpful for geometric optics.

27-3 Mirrors

In this section, we use the law of reflection to understand the formation of images by mirrors.

Images in Plane Mirrors

A **plane mirror** is a highly reflective flat surface, such as a flat sheet of glass with a reflective coating or the surface of a calm lake. Consider an arbitrary point P on a coffee mug sitting in front of a mirror. When illuminated by incident rays with a range of angles of incidence, which is the usual case when the source of illumination is an extended source (e.g., lighting in a room), the optically rough surface of the mug reflects incident light in all outward directions (diffuse reflection). Some of the light rays emanating from point P reach the mirror and reflect off it in accordance with the law of reflection, as shown in Figure 27-10(a).

To an observer standing in front of the mirror, it appears as if the light rays originated from behind the mirror, from point P', which is a virtual image of point P. The image is a virtual image because the light rays diverge after reflecting from the mirror (no light rays actually penetrate the mirror's reflective surface). Rays from other points on the surface of the mug reflect in the same way. The region where the image can be seen is called the **field of view**. This region is bounded by rays connecting the end points of an extended object to the adjacent ends of the mirror as shown in Figure 27-10(b).

Figure 27-11 is an example of a ray diagram. In a **ray diagram**, an object is represented by an arrow to make it clear when the image is inverted. We call the distances between the object and the mirror and the image and the mirror d_o and d_i, respectively. We also call an imaginary line perpendicular to the mirror and passing through its centre the **principal optical axis** of the mirror. In the case of a plane mirror, there is no special point that is considered to be the centre of the mirror. (Thus, a plane mirror can have an infinite number of principal optical axes.) However, the concept of a principal optical axis will be particularly significant for curved mirrors, as you see later in the chapter.

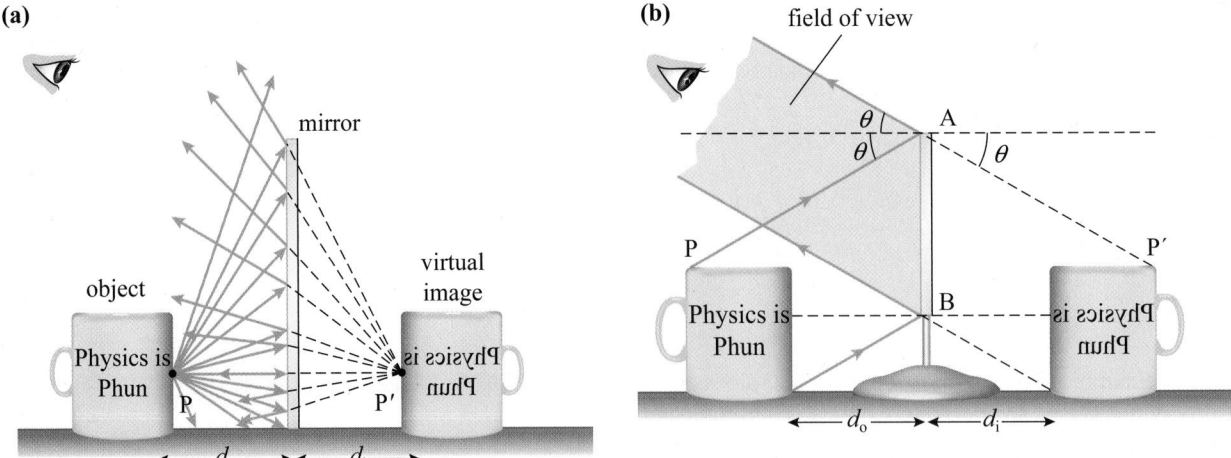

(a)

mirror

object

virtual image

Physics is Phun

Physics is Phun

R

P'

d_o

d_i

(b)

field of view

θ

θ

A

θ

P

Physics is Phun

B

Physics is Phun

P'

d_o

d_i

Figure 27-10 (a) Some of the light rays from point P reflect off the mirror in accordance with the law of reflection. (b) Only the observers located within the shaded region can see the reflection of the entire coffee mug in the mirror.

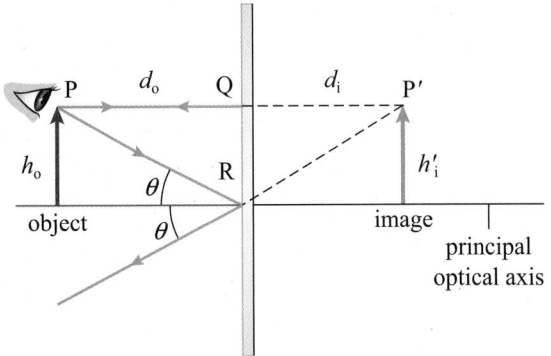

P

d_o

Q

d_i

P'

h_o

θ

R

h'_i

object

θ

image

principal optical axis

Figure 27-11 A ray diagram representing the formation of an image in a plane mirror. An object (represented by an upright arrow) is located along the mirror's principal axis. Ray diagrams often show a number of light rays emanating from the object and reflected off a mirror. It is convenient to use special rays: a ray parallel to the principal axis and a ray reflected off the mirror at the location where the principal axis meets the mirror (point R). The intersection of these rays helps identify the location of the image.

To find the location of the image of the top point of the arrow, we need to use at least two rays emanating from it. It is convenient to use a ray parallel to the principal

optical axis of the mirror, such as PQ in Figure 27-11. For the second ray, we use a light ray reflected off the mirror at the point where the principal axis meets the mirror, ray PR. We can easily draw its reflection, remembering that the angles of incidence and reflection are equal. The intersection of the continuations of these two rays gives the location of the image of point P (point P'). Extensions of any additional reflected rays pass through the same image point. The same procedure can be repeated for any other point of the object.

Right-angled triangles PQR and P'QR are congruent because they have a common side and equal angles. As a result, their corresponding sides are equal: PQ = QP'. Therefore, the distances from the object to the mirror and from the mirror to the image are equal: $d_o = d_i$. Thus, the sizes of the image and the object are equal: $h_i = h_o$, and the magnification, M, of a plane mirror is 1:

$$M = \frac{h_i}{h_o} = 1 \qquad (27\text{-}2)$$

To summarize, a plane mirror forms a virtual, upright image that is the same size as the object ($M = 1$).

MAKING CONNECTIONS

Why Are Images in a Plane Mirror Laterally Inverted?

Some people believe that a mirror exchanges left and right. However, this is not so.

You can test this by standing in front of a mirror. Move your right arm about, and it seems that the left arm in the mirror is moving. However, this is is not so. It only appears to be so because if your identical twin were to stand in place of the image, and then the mirror were removed, it would indeed be the twin's left arm moving. If you hold something in your right hand, you will see that the mirror image arm that is moving also holds something; that is, it is the mirror image of the right hand moving, not a left hand.

The moving hand in the mirror appears to be a left hand, because we are used to seeing real people standing in front of us, and if it were a real person, then it would be a left hand moving, not a right hand. The mirror fools us into thinking of the image as a real person standing before us.

What the mirror actually does is to invert in a direction perpendicular to the mirror; that is, it inverts "depth." You can see this by having several people stand in a line that is perpendicular to the mirror, in the order A, B, and C. The mirror images are in the order C, B, and A.

A further illustration results from placing a mirror on a floor and standing on it, or looking up at a mirror on a ceiling. This should clarify that a mirror inverts "depth" and not left and right.

Someone may counter-argue by saying, "Suppose that we write the word "PHYSICS" on translucent paper, and then turn it over and place it in front of a mirror. It's clear that the mirror has inverted the word laterally; that is, exchanged left and right."

PHYSICS (mirror-reversed)

Note carefully that it is the person that has inverted the piece of paper laterally, not the mirror! Mirrors invert depth, not laterally.

EXAMPLE 27-3

Mirror, Mirror on the Wall, I Want to See My Full-Length Reflection or Nothing at All!

(a) If you are 1.80 m tall, what is the minimum mirror length in which you can see your full-length reflection from the top of your head to your toes?
(b) How high above the floor should you mount the mirror?
(c) Will you see more of yourself in a smaller mirror if you back away from it?

SOLUTION

(a) First, we draw a ray diagram of your reflection in the mirror, as shown in Figure 27-12. Point S represents the location of your shoes, and point E is at the height of your eyes. For you to see the image of your shoes, ray OE must enter your eyes. If the mirror is the minimum length that allows you to see your full-length reflection, point O will be at the bottom edge of the mirror. In right-angle triangle $EE'S'$, the point N bisects side EE', so $NO = \frac{1}{2}E'S'$. By similar reasoning, $NM = \frac{1}{2}E'H'$, and

$$OM = (ON + NM) = \frac{1}{2}S'E' + \frac{1}{2}E'H'$$

$$= \frac{1}{2}(S'E' + E'H') = \frac{1}{2}S'H' = \frac{1}{2}SH$$

Since SH is your height (1.80 m), the minimum length for the mirror is 90 cm (half your height).

(b) Assuming that the height of your eyes (point E) is approximately 1.66 m, then $EH = 14$ cm and $NM = 7$ cm. Therefore, the top edge of the mirror should be 173 cm above the floor.

(c) The calculation of the minimum mirror length for a complete image did not involve the object-mirror distance d_o at all. Therefore, you cannot see a full image of yourself in a mirror less than half your height, regardless of your distance from the mirror. You can confirm this result by redrawing Figure 27-12 with the object distance doubled.

Figure 27-12 To see your entire body in the mirror, the length of the mirror must be at least half your height.

Making sense of the result:

The reasoning in part (a) does not depend on your height. Therefore, everyone who wants to see their full-length image needs a mirror no shorter than half their height. In addition, it is important how high off the floor you mount the mirror. The height of mounting of the mirror must be exactly right for the person for this to work. Since people who have different heights will be using the same mirror, full length mirrors are in practice much longer than the minimum. Contrary to what one might assume, moving away from a shorter mirror or coming closer to it does not let you see a complete image of yourself.

Images Produced by Spherical Mirrors

Spherical mirrors are segments of a spherical reflective surface (Figure 27-13). The inner face of the segment is a **concave mirror** because it curves away from objects reflected in it, and the outer face of the segment is a **convex mirror** because it curves toward objects reflected in it.

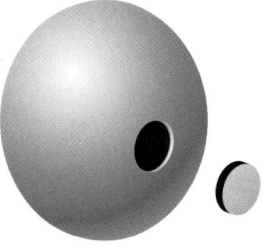

Figure 27-13 Spherical mirrors

Images Produced by Concave Spherical Mirrors The **centre of curvature** of the mirror shown in Figure 27-14 is the centre of the sphere. We denote the centre of curvature as C and the **radius of curvature** as R. Note, light rays passing through the centre of curvature of the mirror (point C) are perpendicular to the mirror. The line passing through the centre point of the mirror, V, and the centre of curvature is the principal optical axis. If we place a light source beyond the centre of curvature of this concave mirror (Figure 27-14(b)), some of the light rays will reach the mirror, reflect from it, and converge at point I, which denotes the location of a real image of the light source. To an observer, the light rays appear to originate from this point.

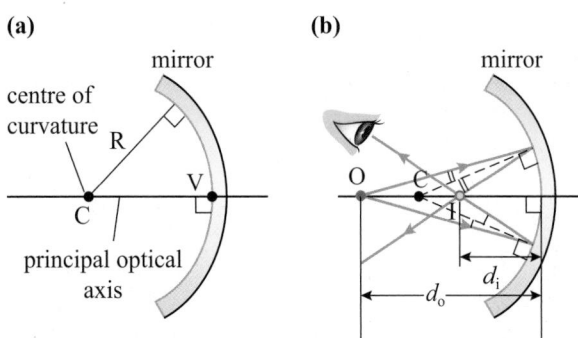

Figure 27-14   (a) A concave spherical mirror (b) Image formation by a concave mirror

The light rays from a point on an object do not all pass through a single point when they reflect from a spherical mirror. Spherical mirrors produce sharp images only for paraxial rays, which are incident rays that are close to the principal axis and form small angles with it (Figure 27-15). To form a sharp image in a spherical mirror, the size of the object must be small compared to the radius of the mirror.

Figure 27-15   Spherical mirrors create sharp images only when the rays coming from an object are close to the principal optical axis and are nearly parallel to it.

Consider the small candle positioned at point O along the principal axis of a concave spherical mirror in Figure 27-16. If the candle is very distant from the mirror, the rays that reach the mirror are essentially parallel to the principal optical axis (these rays are called P-rays).

Since the candle is small and positioned along the principal optical axis, the rays are close to the axis. After reflecting off the mirror, the rays converge at the point F, the **focal point** of the mirror. The distance from the focal point to the mirror is called the **focal length** of the mirror, f. Concave mirrors have a *real* focal point and a *positive* focal length.

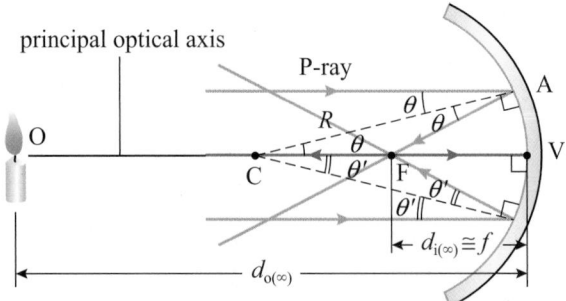

Figure 27-16   The formation of an image of a distant object by a concave spherical mirror

In Figure 27-16, ray OA has an angle of incidence θ. Since the angles of reflection and incidence are equal, triangle CFA is an isosceles triangle with $\sphericalangle ACF = \sphericalangle CAF = \theta$ and CF = AF. Moreover, AF ≈ FV because we are dealing with paraxial rays. Therefore,

$$CF = VF \approx \frac{R}{2} \text{ and}$$

for concave spherical mirrors:

$$f = \frac{R}{2} \tag{27-3}$$

We can use a ray diagram to determine the properties of an image formed by a mirror. Consider a luminous object of height h_o located a distance d_o from a concave mirror (Figure 27-17). The image of a point at the top of the object is formed by reflected rays converging to a point. The location of this image point, O′, is given by the intersection of any two of these rays. We can confirm this location by drawing a third ray. As shown in Figure 27-17, a ray that reflects from the centre of the mirror (V) has a path that is symmetrical about the principal axis. It is often convenient to use three special rays:

- P-ray: a ray parallel to the principal axis. After being reflected, P-rays pass through the focal point of the mirror.

- C-ray: a ray that passes through the mirror's centre of curvature. Since incident C-rays are perpendicular to the mirror's surface ($\theta = 0°$), the reflected C-ray coincides with the incident ray.

- F-ray: a ray that passes through the focal point. After reflecting from the mirror, F-rays are parallel to the principal optical axis. The path of an F-ray is the reverse of the path of a P-ray due to reversibility of light.

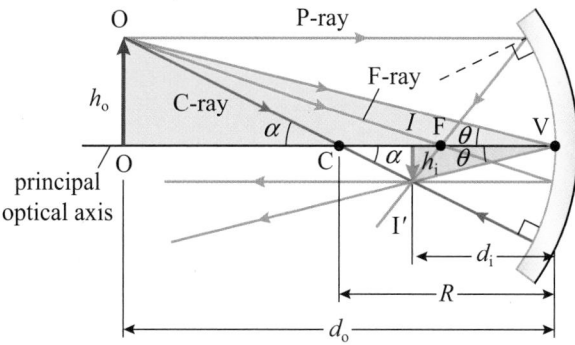

Figure 27-17 We can use a ray diagram to help determine the properties of an image formed by a mirror.

We follow the same procedure to locate the image of any point of the object. Since the image is inverted, its height h_i is represented by a negative quantity. We define the magnification of a concave spherical mirror by considering the similar right-angled triangles O'OC and CII':

$$M = \frac{h_i}{h_o} = -\frac{R - d_i}{d_o - R}$$

Since triangles O'OV and I'IV are also similar, the ratios of their corresponding sides are equal. This allows us to derive the magnification equation for a spherical mirror:

$$M = \frac{h_i}{h_o} = -\frac{d_i}{d_o}$$

Equating the two previous expressions yields

$$\frac{R - d_i}{d_o - R} = \frac{d_i}{d_o}$$

$$\frac{1}{d_o} + \frac{1}{d_i} = \frac{2}{R} \qquad (27\text{-}4)$$

Using Equation (27-3) to substitute for the focal length of a concave mirror allows us to derive a complete algebraic description of image formation by spherical mirrors:

KEY EQUATION
$$\frac{1}{d_o} + \frac{1}{d_i} = \frac{1}{f} \qquad (27\text{-}5a)$$

KEY EQUATION
$$M = \frac{h_i}{h_o} = -\frac{d_i}{d_o} \qquad (27\text{-}5b)$$

Equations (27-5) are used to describe images formed by concave spherical mirrors. A plane mirror can be considered to be a part of an infinitely large concave mirror—a concave mirror with an infinitely large focal length, $f = \infty$. Substituting this information into Equations (27-5), we find that this description also applies to plane mirrors. We will soon discover that these equations apply to convex mirrors, as well. Thus, Equations (27-5) are called the mirror equations.

You can use these expressions to determine the location of an image when a ray diagram approach might be too cumbersome or imprecise. You can also verify that the ray diagrams you construct represent the images correctly.

✔ CHECKPOINT

C-27-3 Concave Spherical Mirrors

Which of the following statements correctly describes images formed by a concave spherical mirror?
(a) When $d_o \gg f$, $d_i = 2f$.
(b) It is impossible to have a negative image–mirror distance ($d_i < 0$) in a spherical mirror.
(c) Spherical mirrors always produce virtual, inverted, and magnified images.
(d) When $0 < d_o < f$, the image is virtual, magnified, and upright.
(e) When $d_o = f$, the image is virtual and located at $d_i = -2f$.
(f) None of the above are correct.

C-27-3 (d) Draw a ray diagram or use Equation (27-5a) to verify that this answer is correct.

 EXAMPLE 27-4

Touching an Image

While waving your finger in front of a concave surface of a shiny metal soup ladle, you notice that at one point your finger appears to touch its own reflection. The ladle has a spherical shape with a diameter of 10 cm. How far is your finger from the ladle when this happens?

SOLUTION

We consider the ladle to be a spherical concave mirror, so its focal length is equal to half its radius (Equation (27-3)):

$$f = \frac{R}{2} = \frac{D/2}{2} = 2.5 \text{ cm}$$

Since the locations of the image and the object (your finger) coincide, $d_i = d_o$. Using the spherical mirror Equation (27-5a) and remembering that a concave mirror has a positive focal length:

$$\frac{1}{d_o} + \frac{1}{d_i} = \frac{1}{f} \Rightarrow \frac{2}{d_o} = \frac{1}{f} \Rightarrow d_o = 2f = 5 \text{ cm}$$

$$d_o = d_i = 5 \text{ cm} \Rightarrow M = \frac{h_i}{h_o} = -\frac{d_i}{d_o} = -1$$

Making sense of the result:

Looking at the calculations in this example, we can see that the result can be generalized to any concave spherical mirror: when the object–mirror distance equals twice the focal length ($2f$), the formed image is a real inverted image that is the same size as the object ($M = -1$).

Images in Convex Spherical Mirrors Convex mirrors are diverging mirrors. A collimated light beam parallel to the mirror's principal optical axis is reflected in such a way that the rays appear to diverge from a point behind the mirror (Figure 27-18(a)). This point is called a **virtual focal point** and is denoted as F. Since convex mirrors always form virtual images, their focal length f is always negative. Using similar reasoning as for concave mirrors, we can show that Equation (23-3) holds for convex mirrors if we assign the radius of a convex mirror a negative value. Thus, $f_{convex} = \dfrac{R}{2} < 0$. This sign convention, $+R$ for concave surfaces and $-R$ for convex surfaces, is accepted in mathematics and science. The mirror equations discussed earlier (Equations (27-5)) also hold for convex mirrors. For a convex mirror, both the focal point F and the centre of curvature C are located behind the mirror, as shown in Figure 27-18.

In conclusion, Equations (27-5) can be applied to any plane or spherical mirror provided we only deal with paraxial rays.

(a)

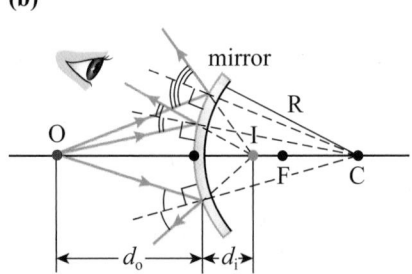

(b)

Figure 27-18 (a) A convex spherical mirror has a virtual focal point. (b) Image formation by a convex mirror

EXAMPLE 27-5

Comparing Convex and Concave Mirrors

A shiny, metal hemispherical bowl has a diameter of 32 cm. You notice that when you place a small object along its axis of symmetry on the concave side, the image of the object appears in front of the mirror, inverted and approximately half the height of the object. Describe the properties of the image formed if you place the object at the same distance from the bowl on the axis of symmetry on the convex side.

SOLUTION

Since the reflective surfaces are spherical, their focal length values are as follows:

$$f_{concave} = \frac{R}{2} = \frac{16\ cm}{2} = 8.0\ cm \quad \text{and}$$

$$f_{convex} = \frac{R}{2} = \frac{-16\ cm}{2} = -8.0\ cm$$

We use the spherical mirror Equations (27-5) to find the object–mirror distance for the real image produced by the concave surface:

$$M = \frac{h_i}{h_o} = -\frac{d_i}{d_o} = -\frac{1}{2} \Rightarrow d_i = \frac{d_o}{2}$$

$$\frac{1}{d_o} + \frac{1}{d_i} = \frac{1}{f} \Rightarrow \frac{1}{d_o} + \frac{1}{\dfrac{d_o}{2}} = \frac{1}{f} \Rightarrow \frac{3}{d_o} = \frac{1}{f}$$

$$d_o = 3f = 3(8.0\ cm) = 24\ cm$$

Now we apply the same spherical mirror equations to find the properties of the image produced by the convex surface:

$$\frac{1}{d_o} + \frac{1}{d_i} = \frac{1}{f} \Rightarrow \frac{1}{24\ cm} + \frac{1}{d_i} = \frac{1}{-8.0\ cm} \Rightarrow \frac{1}{d_i} = -\frac{1}{6\ cm}$$

$$d_i = -6.0\ cm$$

$$M = \frac{h_i}{h_o} = -\frac{d_i}{d_o} = -\frac{-6.0\ cm}{24\ cm} = \frac{1}{4} \Rightarrow h_i = \frac{1}{4}h_o$$

Since d_i is negative and M is positive, the image produced by the convex surface is virtual, four times shorter than the object, and upright.

Making sense of the result:

For an object located $3f$ from a convex mirror, the image is virtual, upright, and one-quarter the height of the object ($M = 0.25$). Concave and convex mirrors with the same curvature produce markedly different images of an object located at the same distance from them.

CHECKPOINT

C-27-4 Spherical Mirrors

A spherical mirror forms an image equal in size to the object. Which of the following statements is correct?

(a) The mirror is concave, the image is real and inverted, and $d_o = f$ and $d_i = 2f$.

(b) The mirror is concave, the image is real and inverted, and $d_o = d_i = 2f$.

(c) The mirror is concave, the image is virtual and upright, and $d_o = -d_i = 2f$.

(d) The mirror is convex, the image is virtual and inverted, and $d_o = -d_i = -2f$.

(e) The mirror is convex, the image is real and upright, and $d_o = d_i = 2f$.

C-27-4 (b) You can verify the correct answer using Equation (27-5a) or by drawing a ray diagram.

Variation of Images with Object Location The ray diagrams and images in Figure 27-19 illustrate that the properties of the image produced by a spherical mirror change as the location of the object relative to the focal point changes. For example, the object in Figure 27-19(a) is to the far left of C, and the image is real and inverted. In Figure 27-19(b), the same object is located between F and the mirror, and the image is virtual and upright.

Table 27-1 on page 738 summarizes the properties of images in spherical mirrors. The absolute value of magnification M shows whether the image is magnified ($|M| > 1$) or reduced ($|M| < 1$), and its sign indicates whether the image is upright (positive) or inverted (negative). Convex mirrors always produce reduced images. Remember that

$$|f| = \frac{R}{2} \text{ for spherical mirrors.}$$

Focal Plane The focal plane of a spherical mirror passes through its focal point and is perpendicular to the principal optical axis. A parallel beam of light incident on a spherical concave mirror always converges at a point located in its focal plane. Only rays parallel to the principal axis converge at the mirror's focal point. A beam of parallel rays that are incident at an angle to the principal axis (Figure 27-20) converge at a **secondary focal point** that is located at the intersection of the focal plane of the mirror and the secondary optical axis. A secondary optical axis passes through the centre of the mirror and is parallel to the beam. Thus, a concave mirror has one primary focal point and an infinite number of secondary focal points. All the focal points of a concave mirror are real. Similar reasoning can be applied to convex mirrors, but all its focal points are virtual.

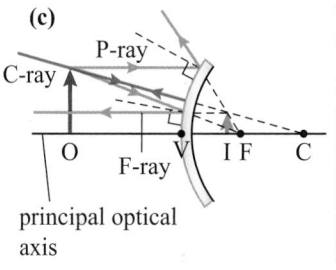

Figure 27-19 Ray diagrams for image formation for an object at various locations near a spherical mirror

From SERWAY/JEWETT, Principles of Physics, 4E. © 2006 Cengage Learning.

ELECTRICITY, MAGNETISM, AND OPTICS

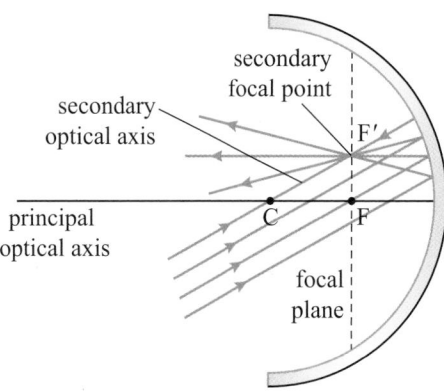

Figure 27-20 The focal plane of a spherical mirror passes through the focal point and is perpendicular to the principal optical axis.

Table 27-1 Properties of images produced by spherical mirrors

Object–mirror distance	Concave mirror: $R > 0, f = R/2 > 0$		Convex mirror: $R < 0, f = R/2 < 0$	
	Mathematical description	Image properties	Mathematical description	Image properties
$0 < d_o < f$	$d_i < 0$	Virtual, appears behind the mirror	$d_i < 0$	Virtual, appears behind the mirror
	$h_i > 0$	Upright	$h_i > 0$	Upright
	$M = \dfrac{h_i}{h_o} = -\dfrac{d_i}{d_o} > 1$	Magnified	$M = \dfrac{h_i}{h_o} = -\dfrac{d_i}{d_o} < 1$	Reduced
	$\lvert d_i \rvert > \lvert d_o \rvert$	Located farther from the mirror than the object	$(0 < M < 1)$ $\lvert d_i \rvert < \lvert d_o \rvert$	Located closer to the mirror than the object
$d_o = \lvert f \rvert$	No image is created		$d_i = \dfrac{f}{2} < 0$	Virtual, appears behind the mirror
			$h_i > 0$	Upright
			$M = -\dfrac{d_i}{d_o} = \dfrac{1}{2}$	Reduced by half
			$\lvert d_i \rvert = \dfrac{\lvert f \rvert}{2} < \lvert d_o \rvert$	Closer to the mirror than the object
$f < d_o < 2f$	$d_i > 0$	Real, appears in front of the mirror	$d_i < 0$ $\lvert d_i \rvert < \lvert f \rvert < \lvert d_o \rvert$	Virtual, appears behind the mirror
	$h_i < 0$	Inverted	$h_i > 0$	Upright
	$M = \dfrac{h_i}{h_o} = -\dfrac{d_i}{d_o} < -1$	Magnified	$M = \dfrac{h_i}{h_o} = -\dfrac{d_i}{d_o} < 1$	Reduced
	$\lvert d_i \rvert > \lvert d_o \rvert$	Farther from the mirror than the object	$\lvert d_i \rvert < \lvert d_o \rvert$	Closer to the mirror than the object
$d_o = 2\lvert f \rvert$	$d_i > 0$ $d_i = -d_o$	Real, appears in front of the mirror	$d_i = \dfrac{2}{3} f < 0$ $0 < \lvert d_i \rvert < \lvert f \rvert$	Virtual, appears behind the mirror
	$h_i < 0$	Inverted	$h_i > 0$	Upright
	$M = \dfrac{h_i}{h_o} = -\dfrac{d_i}{d_o} = -1$	Same size as the object	$M = \dfrac{h_i}{h_o} = -\dfrac{d_i}{d_o} < 1$	Reduced, closer to the mirror than the object
$d_o > 2f$	$d_i > 0$	Real ($d_i > 0$), appears in front of the mirror	$d_i < 0$	Virtual, appears behind the mirror
	$h_i < 0$	Inverted	$h_i > 0$	Upright
	$-1 < M < 0$	Reduced, closer to the mirror than the object	$M = \dfrac{h_i}{h_o} = -\dfrac{d_i}{d_o} < 1$ $\lvert d_i \rvert < \lvert d_o \rvert$	Reduced, closer to the mirror than the object

C-27-5 Focal Points

A parallel beam of light strikes a concave spherical mirror. Which ray diagram in Figure 27-21 correctly represents the reflected beam?

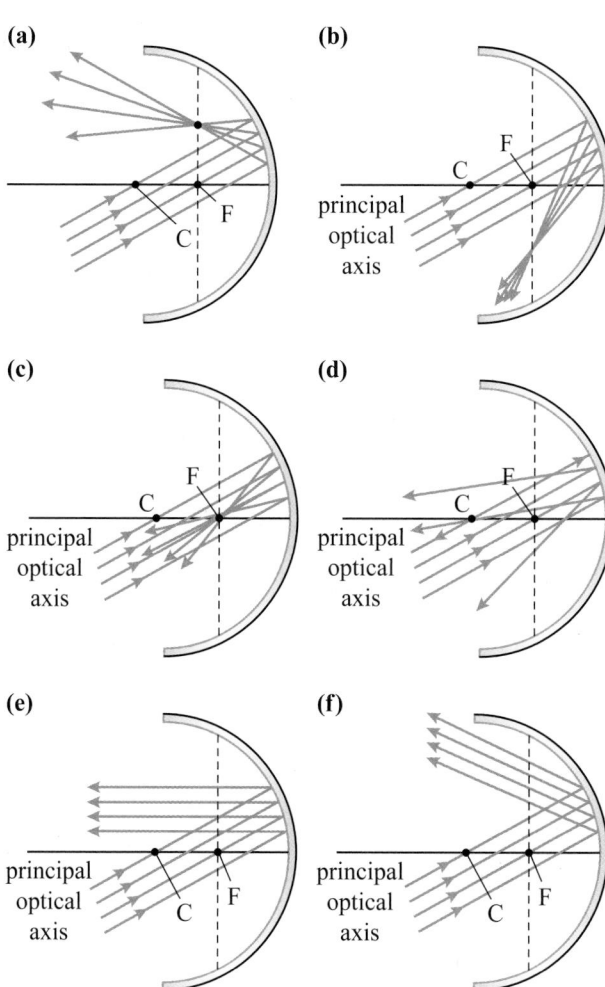

(a)

(b)

principal optical axis

(c)

principal optical axis

(d)

principal optical axis

(e)

principal optical axis

(f)

principal optical axis

Figure 27-21   C-27-5

C-27-5 (a) The parallel beam intersects at the secondary focal point F'. A secondary focal point for a spherical mirror is located at the intersection of the mirror's focal plane and its secondary optical axis (Figure 27-20).

LO 4

27-4   Refraction of Light

Let us consider what happens when a ray of light encounters a boundary between two transparent media, for example, when light travelling through air enters glass. As shown in Figure 27-22, the direction of the light changes. This process is called **refraction**. In the figure, θ_1 denotes the angle of incidence, and θ_2 denotes the **angle of refraction**.

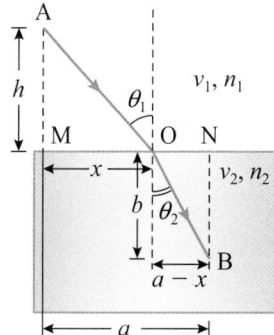

Figure 27-22   Fermat's principle applied to the refraction of light

Fermat's principle provides an explanation for why refraction occurs. The speed of light depends on the medium, so we denote the speed of light in the medium of the incident ray as v_1 and in the medium where the light is refracted as v_2. Consider the ray of light that is travelling from point A to point B in Figure 27-22. The time, t, the light takes to travel between the points is the sum of the time taken along segment AO in the first medium and OB in the second medium. Applying Pythagoras' theorem to the two right-angled triangles AMO and BNO, we obtain the following:

$$t = \frac{\sqrt{h^2 + x^2}}{v_1} + \frac{\sqrt{b^2 + (a-x)^2}}{v_2}$$

According to Fermat's principle, the light follows the path for which this expression has a minimum value. Therefore,

$$\frac{dt(x)}{dx} = 0$$

$$\frac{d}{dt}\left(\frac{\sqrt{h^2 + x^2}}{v_1} + \frac{\sqrt{b^2 + (a-x)^2}}{v_2} \right) = 0$$

$$\frac{x}{v_1\sqrt{h^2 + x^2}} + \frac{-(a-x)}{v_2\sqrt{b^2 + (a-x)^2}} = 0$$

$$\frac{\sin\theta_1}{v_1} = \frac{\sin\theta_2}{v_2} \qquad (27\text{-}6)$$

The **index of refraction** of a medium is defined as

KEY EQUATION
$$n = \frac{c}{v} \qquad (27\text{-}7)$$

where c is the speed of light in a vacuum, and v is the speed of light in the medium.

Therefore, the index of refraction of a vacuum is 1, and $n > 1$ for all other media (Table 27-2). The term **optical density** is often used to describe the refractive properties of materials; the greater a medium's index of refraction, the greater its optical density.

ELECTRICITY, MAGNETISM, AND OPTICS

Table 27-2 Indexes of Refraction for Various Substances

Substance (Solid at 20°C)	Index of refraction	Substance (Liquid at 20°C)	Index of refraction
Cubic zirconium	2.20	Benzene	1.501
Diamond (C)	2.419	Carbon disulfide	1.628
Fluorite (CaF_2)	1.434	Carbon tetrachloride	1.461
Fused quartz (SiO_2)	1.458	Corn syrup	2.21
Gallium phosphide	3.50	Ethyl alcohol	1.361
Glass, crown	1.52	Glycerine	1.473
Glass, flint	1.66	Vegetable oil	1.47
Ice (H_2O)	1.309	Water	1.333
Polystyrene	1.49	**Gases at 0°C, 1 atm**	
Sodium chloride (NaCl)	1.54	Air	1.000 293
		Carbon dioxide	1.000 45

Equation (27-6) is called the **law of refraction**, and it can be rewritten in terms of the indexes of refraction of the two media:

KEY EQUATION $$n_1 \sin\theta_1 = n_2 \sin\theta_2 \qquad (27\text{-}8)$$

This formula was discovered experimentally by the Dutch mathematician Willebrord Snellius (1580–1626) and is often called **Snell's law**. The word "refraction" stems from a Latin word "refract" meaning "to break up." So this phenomenon is called "light refraction" because a light ray "breaks" at the border between two media. Consequently, the objects observed at the interface of different media appear broken or fractured. For example, a pencil placed in a glass filled with water appears to an observer to be broken because the image of the bottom part of the pencil is closer to the water–air boundary than the real pencil is (Figure 27-23).

Applying Snell's law, we conclude the following:

■ When an incident light ray is normal to the boundary between two media ($\theta_1 = 0°$), the **refracted ray** is also going to be normal to the boundary, $\theta_1 = \theta_2 = 0°$.

■ When light propagates from a less optically dense medium to a more optically dense medium ($n_1 < n_2$), the light ray bends toward the normal: $\theta_2 < \theta_1$.

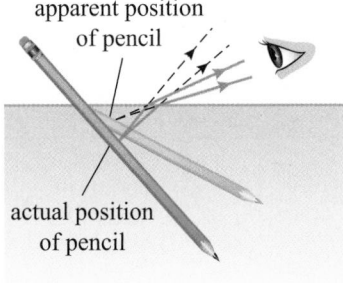

Figure 27-23 A pencil partially submerged in water appears to be broken at the air–water border.

■ When light propagates from a more optically dense medium to a less optically dense medium ($n_1 > n_2$), the light ray bends away from the normal: $\theta_2 > \theta_1$.

■ A beam of polychromatic (light with more than one frequency) light disperses when it crosses the boundary between media because the speed of light in most media varies with the frequency of the light. **Dispersion** is the reason we see a range of colours when a beam of white light enters a prism at an angle (Figure 27-24).

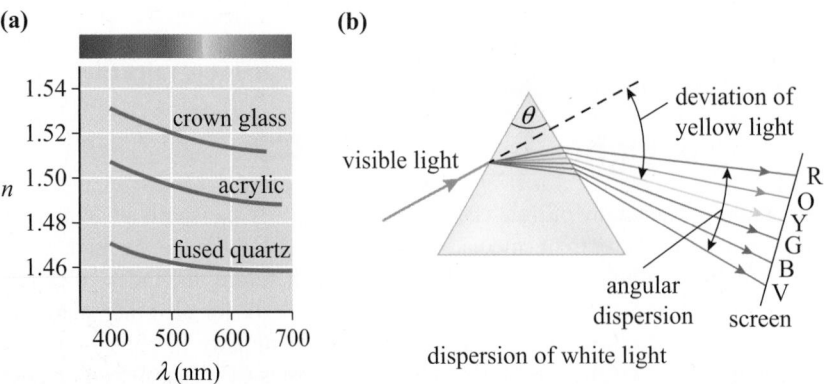

Figure 27-24 (a) Variation of the index of refraction with frequency (b) Dispersion of sunlight in a glass prism

In summary, when a beam of light propagating through air is incident on a glass plate (Figure 27-25), some of the light reflects (1–2 and 3–4), and some of the light refracts (1–3 and 4–5).

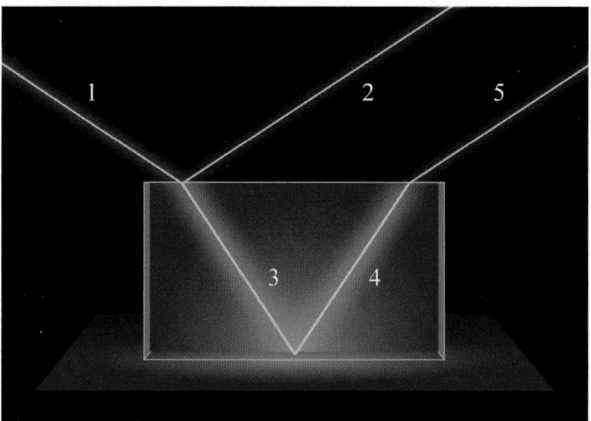

Figure 27-25 A beam of light incident on a glass plate in air. Some of the beam undergoes reflection (paths 1–2 and 3–4), and some undergoes refraction (paths 1–3 and 4–5).

MAKING CONNECTIONS

Mirages

The indexes of refraction of many optical media (e.g., air) vary with temperature. Such variations in the index of refraction of air can cause refractions that produce mirages. For example, when driving on a hot summer day, you may have observed what appear to be puddles on the road that disappear as you approach them. Similarly, in a hot desert people sometimes see what appear to be pools of water. Mirages occur because the air closest to the hot ground or pavement is much hotter than the air above (Figure 27-27). Consequently, the index of refraction of air gradually increases as you move away from the ground. This index of refraction gradient refracts Sun rays that are incident on the pavement at very large angles (remember, the angles are measured from the normal to the surface), making it look as if this light were reflected from water in a pool ahead of the observer. Invoking Fermat's principle once again, we can say that as the speed of light in the air increases with air temperature raising, the light rays bend down where they can travel faster, to make the time from the tree to your eye a minimum.

Although less common, mirages also occur when there is cooler air near the ground with a layer of warmer air above it. If there is a sufficient difference in temperature, light from objects below the horizon is refracted downward enough to offset the curvature of Earth and make the objects visible to a distant observer. This type of mirage can make a ship that is beyond the horizon appear to be on the horizon or even floating above it.

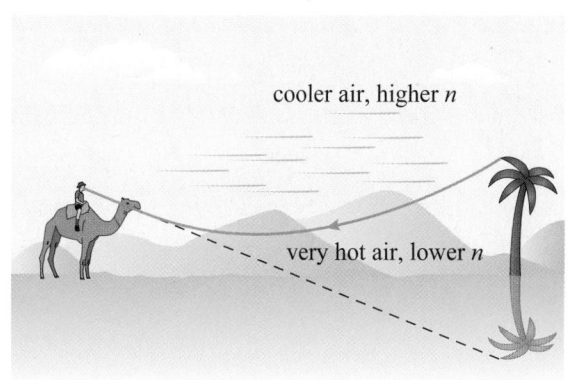

cooler air, higher n

very hot air, lower n

Figure 27-27 The formation of a mirage

Total Internal Reflection

We can rewrite Snell's law (Equation (27-8)) as

$$\sin\theta_2 = \frac{n_1}{n_2}\sin\theta_1 \qquad (27\text{-}9)$$

In Figure 27-28, light is propagating from a more optically dense medium to a less optically dense medium ($n_1 > n_2$) (e.g., from water to air). At some angle of incidence the right side of Equation (27-9) will equal 1, making the angle of refraction, θ_2, equal 90° (Figure 27-28(c)). This angle of incidence is called the **critical angle**, θ_{cr}, and

$$\sin(90°) = \frac{n_1}{n_2}\sin\theta_{cr}$$

$$\frac{n_1}{n_2}\sin\theta_{cr} = 1$$

Therefore,

KEY EQUATION
$$\theta_{cr} = \sin^{-1}\left(\frac{n_2}{n_1}\right) \qquad (27\text{-}10)$$

For example, at a water–air boundary,

$$\theta_{cr} = \sin^{-1}\left(\frac{1}{1.33}\right) \approx 48.6°$$

When the angle of incidence exceeds the critical angle, the light cannot propagate through the boundary between the media and is reflected off it (Figure 27-28(d)). This phenomenon is called **total internal reflection**.

Total internal reflection is essential for the operation of the optical fibres that are widely used in communications and in medicine. The light travelling inside a fibre optic cable undergoes multiple total internal reflections, so the light does not leave the fibre. Consequently, by bending the optical fiber, the beam of light can be forced to bend. Total internal reflection is also used in many other devices, for example, LEDs.

(a)

(b)

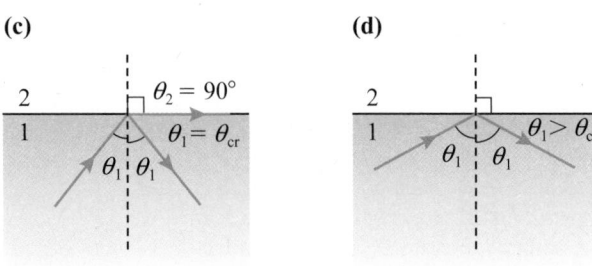

(c)

(d)

Figure 27-28 (a), (b) The propagation of light from a more optically dense medium to a less optically dense medium, $n_1 > n_2$. (c) When the angle of refraction is 90°, the refracted beam propagates along the border between the media. (d) Total internal reflection

Dispersion and total internal reflection are also common in nature. For example, the total internal reflection of light and light dispersion in raindrops produce rainbows. As shown in Figure 27-29, we can approximate raindrops as small spheres. Sunlight entering a raindrop is refracted at the air–water boundary as it enters the raindrop. At the water–air boundary, some of the Sun's rays then undergo total internal reflection.

Refraction of Light in a Triangular Prism

We now consider the refraction of monochromatic light in a triangular prism. Let the index of refraction of the prism, n_p, be higher than the index of refraction of

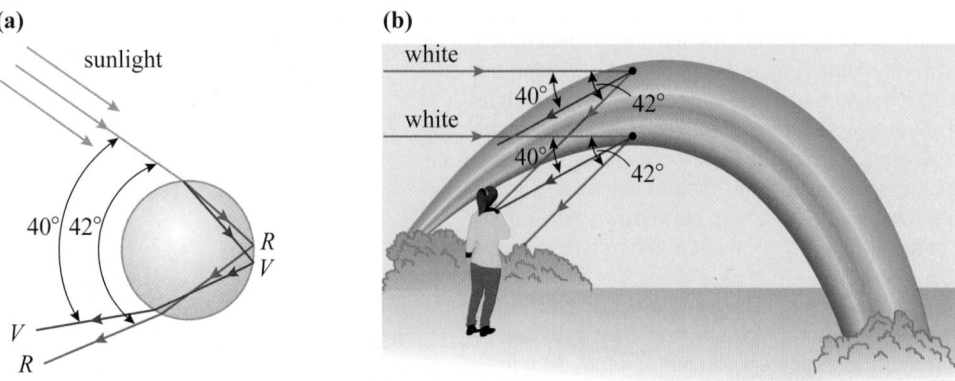

(a)

(b)

Figure 27-29 (a) Dispersion and total internal reflection of light in a spherical raindrop (b) The formation of a rainbow

MAKING CONNECTIONS

Sparkling Diamonds

With an index of refraction of 2.419 for a diamond (Table 27-2), a diamond–air boundary has one of the largest differences in indexes of refraction found in nature. Consequently, the critical angle for a diamond–air boundary is relatively small:

$$\theta_{cr} = \sin^{-1}\left(\frac{1}{2.419}\right) \approx 24.42°$$

This property plays an important role in the brilliance of diamond gemstones. Since the critical angle is so small, light has a tendency to become "trapped" inside a diamond. A light ray will typically undergo total internal reflection several times before finally refracting out of the diamond, often through the same facet that it entered. As a result, the gemstone facets behave like small mirrors and

make the diamond sparkle. The effect can be enhanced by cutting a diamond so that its facets are oriented at angles that will provide paths for multiple internal reflections rather than refraction, as shown in Figure 27-30.

Figure 27-30 (a) A diamond with facets oriented for the total internal reflection of light (b) A diamond with fewer facets oriented such that light entering the top facet passes through the gemstone

---

the ambient medium, n_{amb}, as, for example, in a glass prism surrounded by air. In the prism, the refracted ray is closer to the normal at the air–glass boundary than the incident ray is. The process is reversed at the glass–air boundary. As a result, the light ray exiting the prism is deflected toward its base by an angle δ (Figure 27-31(a)). When white light enters the prism, the deflection of rays of different colours varies with the wavelength of the light. Consequently, the light exiting the prism spreads out with the highest frequencies (shortest wavelengths) being deflected the most (Figure 27-31(b)).

(a)

$$n_p > n_{amb}$$

(b)

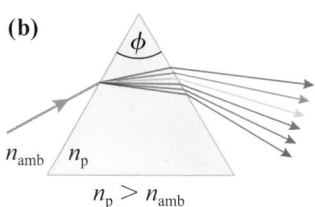

$$n_p > n_{amb}$$

Figure 27-31 (a) The refraction of monochromic light in a triangular prism (b) The dispersion of white light by a triangular prism

CHECKPOINT

C-27-7 The Refraction of Light in a Prism

A horizontal, monochromatic light beam travelling in air enters a glass prism. Which of the diagrams in Figure 27-32 correctly represents the beam that exits the prism?

(a)   **(b)**

(c)   **(d)**

(e)

Figure 27-32 C-27-7

C-27-7 (e) See Figure 27-31.

27-5 Images Formed by Thin Lenses

As shown in the previous section, a triangular prism with a higher index of refraction than the ambient medium deflects light toward its base. In contrast, a parallel beam of light passing through a rectangular prism is not deflected (it will shift up or down a prism parallel to itself). We can understand how images are formed by a curved lens by treating the lens as a stack of prisms (or parts of prisms) with different apex angles, which is the angle formed by the pointed tip of a prism (Figure 27-33).

Figure 27-33 A representation of a biconvex lens (a lens that has convex faces on both sides) as a stack of prisms or parts of prisms with different apex angles

 ONLINE ACTIVITY

Introducing Lenses

The e-resource that accompanies every new copy of this textbook contains an Online Activity using the PhET simulation "Introducing Lenses". Work through the simulation and accompanying questions to gain an understanding of thin lenses and geometric optics.

A **thin lens** is a lens whose thickness is small relative to its other dimensions. Figure 27-34 shows

six basic shapes of thin lenses. The **converging lenses** are biconvex, convex–concave, and plano–convex. The **diverging lenses** are biconcave, convex–concave, and plano–concave.

The distance from an object to a lens, d_o, is measured from the centre plane of the lens. A beam of light parallel to the major optical axis of a converging thin lens (Figure 27-35(a)) converges at the focal point F behind the lens. Such a lens has a positive focal length, $f > 0$. A beam of light parallel to the principal optical axis of a diverging thin lens (Figure 27-35(b)) diverges, such that the rays appear to have originated from an imaginary focal point F in front of the lens. The focal length of a diverging lens is negative, $f < 0$. Every lens has two equidistant focal points, one is located in front of the lens, and the other behind it. We can easily see that symmetrical lenses, like those in Figure 27-35, have this property. Using ray diagrams (or more advanced methods of analysis), we can show that asymmetrical lenses also have two equidistant focal points, one on each side. Consequently, all lenses have two focal planes, each perpendicular to the principal optical axis crossing it at one of the focal points.

(a)

(b)

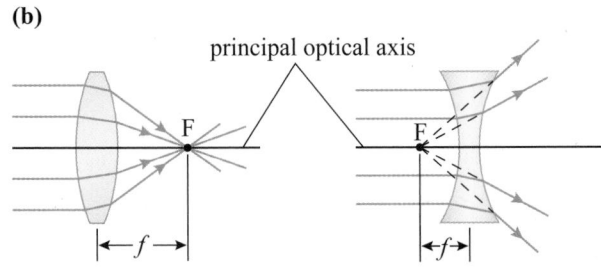

Figure 27-35 Parallel beams of light passing through (a) two stacked prisms, and (b) converging and diverging lenses.

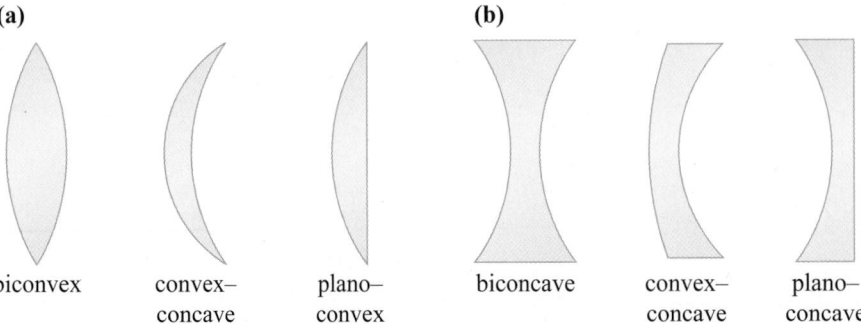

Figure 27-34 Various shapes of lenses (a) Converging lenses have a positive focal length. (b) Diverging lenses have a negative focal length.

PEER TO PEER

It took me a while to convince myself that asymmetrical lenses (e.g., concave–convex lenses and plano–convex lenses) have two symmetrical focal points. Computer simulations that allow me to reorient the lens and change its form helped me. In addition, it is important to remember that you can always reverse the direction of light. So, if a light ray travels from point A to B along a certain path, it will travel along the same path when it travels from point B to point A.

Images Produced by Thin Lenses

We denote the centre point of a converging (biconvex) lens (Figure 27-36) as C. The axis of symmetry perpendicular to the curved surfaces of a lens is the principal optical axis of the lens. Consider a small light source (object) at point O, a distance d_o from a lens. Every ray of light that emanates from point A of the source and reaches the lens refracts according to the law of refraction (Equation (27-8)). If the light source is small and the lens is thin, refracted rays converge at point A′ behind the lens, forming a real image of point A. The same is true for every point of the light source. To an observer standing behind the lens, the light rays appear as if they originated from point A′. The same applies for any other point of the light source. Thin lenses create sharp images when the incident rays are paraxial.

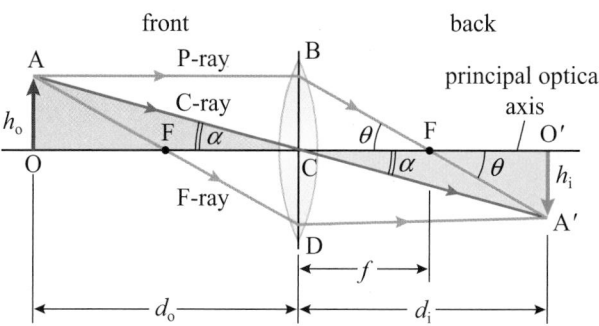

Figure 27-36 A ray diagram for image formation by a converging lens

Now consider an object on the principal axis of a converging lens, as shown in Figure 27-36. Incident light ray AB (a P-ray) passes through the focal point F after refracting through the lens. Ray AD, an F-ray passing through the front focal point of the lens, exits parallel to the principal axis behind the lens. Incident ray AC, a C-ray passing through the centre of the lens, however, does not deflect noticeably because the lens is thin. These rays intersect at point A′, which is the location of the image of point A.

Since the right-angled triangles AOC and CO′A′ are similar, the ratios of the corresponding sides must be equal. This leads to a lens magnification equation, which is identical to the magnification equation for a spherical mirror (Equation (27-5b)):

$$M = \frac{h_i}{h_o} = -\frac{d_i}{d_o}$$

As with mirror magnification, the minus sign indicates that the image is inverted. Right-angled triangles BCF and O′FA′ are also similar, and CB = h_o. Therefore,

$$\begin{cases} \tan \theta = \dfrac{h_o}{f} & \text{from } \Delta \text{BCF} \\ \tan \theta = \dfrac{-h_i}{d_i - f} & \text{from } \Delta A'O'F \end{cases} \Rightarrow$$

$$\frac{h_o}{f} = \frac{-h_i}{d_i - f} \Rightarrow \frac{h_o}{f} = \frac{h_i}{f - d_i} \Rightarrow \frac{h_o}{h_i} = \frac{f}{f - d_i}$$

Combining the two previous equations gives the following equation:

$$-\frac{d_o}{d_i} = \frac{f}{f - d_i}$$

Rearranging the above equation leads to the thin lens equation, which, in combination with the lens magnification equation, provides an algebraic description of image formation by thin lenses:

KEY EQUATION
$$\frac{1}{d_o} + \frac{1}{d_i} = \frac{1}{f} \qquad (27\text{-}11\text{a})$$

KEY EQUATION
$$M = \frac{h_i}{h_o} = -\frac{d_i}{d_o} \qquad (27\text{-}11\text{b})$$

You can use Equations (27-11) to predict the properties of an image formed by a thin lens. You can also use it to check that your ray diagrams correctly represent the images formed by lenses. Noticing the similarities in the derivations of the thin lens and the mirror equations, it should not be surprising that the thin lens equation is the same as the spherical mirror equations, Equations (27-5). However, the relationship between the focal length and the geometric and physical characteristics of the lens is far from obvious. The relationship can be derived theoretically by examining image formation at both refracting surfaces. However, it was first discovered experimentally by lens makers and is thus called the lens maker's equation for a thin lens:

KEY EQUATION
$$\frac{1}{f} = \left(\frac{n_{lens} - n_{amb}}{n_{amb}}\right)\left(\frac{1}{R_1} - \frac{1}{R_2}\right) \qquad (27\text{-}12)$$

where n_{lens} is the index of refraction of the lens, n_{amb} is the index of refraction of the ambient medium, and R_1 and R_2 are the radii of curvature of the front (closest to the light source) and back surfaces (farthest from the light source) of the lens, respectively.

The radius of curvature of a lens is considered to be positive when the surface is convex as observed from the side of the light source, and negative when the surface is concave. For example, R_1 is positive and R_2 is negative for a biconvex lens. The bigger the difference in the indexes of refraction, the shorter the focal length and the more the lens refracts light. Unless indicated otherwise, we assume that the ambient medium is air. Since n_{air} is very close to 1, the lens maker's equation for lenses in air reduces to

$$\frac{1}{f} = \left(\frac{n_{lens} - 1}{1}\right)\left(\frac{1}{R_1} - \frac{1}{R_2}\right) = (n - 1)\left(\frac{1}{R_1} - \frac{1}{R_2}\right) \quad (27\text{-}13)$$

Although we used converging lenses to derive Equations (27-11), the equations hold for diverging lenses, as well. Thus, we derived an algebraic description of images produced by thin lenses and stated the formula for describing the focal length of a lens based on its physical properties (Equation (27-13)). Table 27-3 summarizes the sign conventions for the variables used in these equations. These conventions match the conventions used for spherical mirrors. Notice that the term in front of the

lens indicates that it is the side of the lens that faces the object. Thus, the light from the object enters the lens from its front side and exits from the back side of the lens.

 CHECKPOINT

C-27-8 The Focal Point of a Lens

A light source is placed in front of a biconvex glass lens at its focal point. Which of the following statements correctly describes this situation?

(a) The light beam emerging from behind the lens consists of parallel rays.
(b) The real image of the light source is located at the focal point behind the lens.
(c) The virtual image of the light source is located at the focal point behind the lens.
(d) The virtual image of the light source is located at the focal point in front of the lens.
(e) The real image of the light source is located at the focal point in front of the lens.

C-27-8 (a) Use the reversibility of light to verify the answer.

Table 27-3 Sign conventions for lenses

Quantity	Positive	Negative	Comments
d_o, object–lens distance	The object is in front of the lens (real object).	The object is behind the lens (virtual object).	A negative object–lens distance is often found in systems of multiple lenses.
d_i, object–image distance	The image is behind the lens (real image).	The image is in front of the lens (virtual image).	
f, the focal length	Converging lens	Diverging lens	The focal length depends on the geometry of the lens, the properties of the materials it is made of, and the surrounding medium.
h_o, the height of an object	Most cases: when an object is located above the principal optical axis.	Special cases: when an object is located above the principal optical axis.	Whatever the object orientation is, it is most often considered to be positive.
h_i, the height of an image	An image is upright compared to the object.	An image is inverted compared to the object.	
n_{lens}, the index of refraction of the lens	Always positive		Important for calculating the focal length of a lens using the lens maker's equation, Equation (27-12)
n_{amb}, the index of refraction of the surrounding medium	Always positive		
R_1, the radius of curvature of the lens surface closest to the light source	The surface is convex (from the side of the light source).	The surface is concave (from the side of the light source).	Important for calculating the focal length of a lens using the lens maker's equation, Equation (27-12)
R_2, the radius of curvature of the lens surface farthest from the light source	The surface is convex (from the side of the light source).	The surface is concave (from the side of the light source).	

EXAMPLE 27-6

Is a Biconvex Lens a Converging Lens After All?

A biconvex lens shell made of a thin transparent plastic is filled with water. A 5 cm–tall plastic toy placed 25 cm in front of the lens produces a sharp image that is 10 cm tall on a screen. The lens shell is then emptied and filled with vegetable oil ($n_{oil} = 1.47$). Ignore the small amount of refraction in the plastic holder in solving the problem.

(a) What is the focal length of the lens when it is filled with water?

(b) What are the characteristics of the image of the toy when the lens is filled with water?

(c) What is the focal length of the lens when it is filled with oil?

(d) What are the characteristics of the image of the toy on the screen when the lens is filled with oil?

(e) The same demonstrations were performed when the lens was filled with air and immersed in water. What would the properties of the toy's image be in that case?

SOLUTION

(a) From Table 27-1, we have $n_{water} = 1.33$. Since the image is real and it is a one-lens system, it must be inverted. Therefore, $h_i = -10$ cm, and the magnification is $M = \dfrac{h_i}{h_o} = \dfrac{-10 \text{ cm}}{5 \text{ cm}} = -2$.

From Equation (27-11a), we can see that the image–lens distance must be twice as long as the object–lens distance: $d_i = (25 \text{ cm}) \cdot 2 = 50$ cm.

Now we use the thin lens equation, Equation (27-11a), to find the focal length of the water-filled lens in air:

$$\frac{1}{d_o} + \frac{1}{d_i} = \frac{1}{f_{w\text{-}a}}$$

$$\frac{1}{25 \text{ cm}} + \frac{1}{50 \text{ cm}} = \frac{1}{f_{w\text{-}a}} \Rightarrow f_{w\text{-}a} = 16.67 \text{ cm} \approx 17 \text{ cm}$$

(b) The height and magnification found in part (a) indicate that the image is real, inverted, and magnified ($M = -2$).

(c) When the lens is filled with oil instead of water, the index of refraction of the lens changes but R_1 and R_2 do not change. We can use the lens maker's equation, Equation (27-13), to find the focal length of the lens filled with oil and surrounded by air, f_{o-a}:

$$\frac{1}{f_{w\text{-}a}} = (n_w - 1)\left(\frac{1}{R_1} - \frac{1}{R_2}\right)$$

$$\left(\frac{1}{R_1} - \frac{1}{R_2}\right) = \frac{1}{f_{w\text{-}a}(n_w - 1)} = \frac{1}{(16.67 \text{ cm}) \times 0.33} = 0.182 \text{ cm}^{-1}$$

$$\frac{1}{f_{o\text{-}a}} = (n_o - 1)\left(\frac{1}{R_1} - \frac{1}{R_2}\right) = (1.47 - 1)(0.182 \text{ cm}^{-1}) = 0.085$$

$$f_{o\text{-}a} = 11.702 \text{ cm} \approx 12 \text{ cm}$$

The focal length of the oil-filled lens is shorter than the focal length of the water-filled lens.

(d) The object is 25 cm from the oil-filled lens:

$$\frac{1}{d_o} + \frac{1}{d_i} = \frac{1}{f_{w\text{-}a}}$$

$$\frac{1}{25 \text{ cm}} + \frac{1}{d_i} = \frac{1}{11.702 \text{ cm}} \Rightarrow \frac{1}{d_i} = 0.045$$

$$d_i = 22.0 \text{ cm}$$

Since the image–lens distance is positive, the image is real. We find the image height and the magnification using Equation (27-11b):

$$M = -\frac{d_i}{d_o} = -\frac{22 \text{ cm}}{25 \text{ cm}} = -0.88$$

$$M = \frac{h_i}{h_o} \Rightarrow h_i = h_o \cdot M = -8.8 \text{ cm}$$

Thus, the image is real, inverted, and reduced.

(e) From Equation (27-12), we see that when the lens is filled with air and surrounded by water, it becomes a diverging lens. Thus, the lens produces images that are virtual and upright. These images cannot be viewed on a screen.

Making sense of the result:

This example illustrates that the focal length of a lens depends on its geometry, the material it is made of, and the optical properties of the ambient medium. A lens can be either converging or diverging, depending on the relative indexes of refraction of the lens and the ambient medium.

Ray Diagrams for Thin Lenses

In Figure 27-36, we constructed a basic ray diagram for a thin lens. We now look at ray diagrams for thin lenses in more detail. As with spherical mirrors, we find the image locations first by finding the location of the image of a point at the top of the object. Again, it is convenient to use three special rays:

■ The P-ray is parallel to the principal optical axis. For a *converging* lens, the P-ray passes through the focal

point *behind* the lens (Figure 27-37(a), (b)). However, for a *diverging* lens, the P-ray refracts such that it appears to originate from a focal point *in front* of the lens (Figure 27-37(c)).

■ The C-ray passes through the centre of the lens. The displacement of this ray is negligible for thin lenses.

■ The F-ray passes through the focal point in front of the lens. This ray refracts such that the ray emerges from the lens parallel to its principal axis.

(a)

(b)

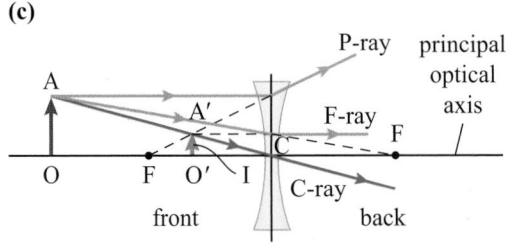

(c)

Figure 27-37 Ray diagrams for locating images created by thin lenses. (a) An object that is farther than twice the focal length from a converging lens creates a real, inverted, and reduced image. (b) An object between the mirror and the focal length of a converging lens creates a virtual, upright, and reduced image. (c) An object in front of a diverging lens creates a virtual, upright, and reduced image.

The point where the rays actually intersect indicates the location of a real image, and the point from which the diverging rays appear to originate indicates the location of a virtual image. With thin lenses and small objects, every point of an object is located approximately the same distance from the lens, and every point of the image is also located approximately the same distance from the lens. Thus, the image location is determined by drawing a normal to the principal axis from the point where the image of point A is located.

Table 27-4 summarizes the properties of images formed by thin lenses. The absolute value of magnification indicates whether the image is magnified ($|M| > 1$) or reduced ($|M| < 1$), and its sign indicates whether the image is upright (positive) or inverted (negative). Note that diverging lenses *always* produce reduced virtual images.

ONLINE ACTIVITY

Playing with Lenses

The e-resource that accompanies every new copy of this textbook contains an Online Activity using the "Playing with Lenses" simulation. Use this simulation and Table 27-4 to explore the formation of images by thin lenses.

 CHECKPOINT

C-27-9 The Image in a Converging Lens

A lit candle is 20 cm from a converging lens ($f = 10$ cm). What happens to the candle's image if you cover the bottom part of the lens?

(a) Only the bottom half of the original image is visible; the image is dimmer.

(b) Only the top half of the original image is visible; the image is dimmer.

(c) The image is dimmer and twice as small than before, but the entire candle is visible.

(d) The image is the same size as before but dimmer; the entire candle is visible.

(e) The image does not change in brightness but is twice as big; the entire candle is visible.

C-27-9 (d) Draw a ray diagram to verify the answer. Every point of an object emits light in all directions, so rays from every point of an object go through different parts of the lens. When some of the lens is blocked, fewer rays converge at the location of an image, thus creating a dimmer image.

Table 27-4 Properties of images created by thin lenses

| Object–mirror distance | Converging lens: $f > 0$ | | Diverging lens: $f < 0$ | |
	Mathematical description	Image properties	Mathematical description	Image properties
$0 < d_o < f$	$d_i < 0$	Virtual, appears in front of the lens	$d_i < 0$	Virtual, appears in front of the lens between the lens and its front focal point
	$h_i > 0$	Upright	$h_i > 0$	Upright
	$M = \dfrac{h_i}{h_o} = -\dfrac{d_i}{d_o} > 1$	Magnified	$M = \dfrac{h_i}{h_o} = -\dfrac{d_i}{d_o} < 1$	Reduced
	$|d_i| > |d_o|$	Located farther from the lens than the object	$(0 < M < 1)$ $|d_i| < |d_o|$	Located closer to the lens than the object
$d_o = |f|$	No image is created		$d_i = \dfrac{f}{2} < 0$	Virtual, appears behind the lens
			$h_i > 0$	Upright
			$M = -\dfrac{d_i}{d_o} = \dfrac{1}{2}$	Reduced by half
			$|d_i| = \dfrac{|f|}{2} < |d_o|$	Closer to the lens than the object
$f < d_o < 2f$	$d_i > 0$	Real, appears on the other side of the lens	$d_i < 0$ $|d_i| < |f| < |d_o|$	Virtual, appears in front of the lens between the lens and its front focal point
	$h_i < 0$	Inverted	$h_i > 0$	Upright
	$M = \dfrac{h_i}{h_o} = -\dfrac{d_i}{d_o} < -1$	Magnified	$M = \dfrac{h_i}{h_o} = -\dfrac{d_i}{d_o} < 1$	Reduced
	$|d_i| > |d_o|$	Farther from the lens than the object	$|d_i| < |d_o|$	Closer to the lens than the object
$d_o = 2|f|$	$d_i > 0$ $d_i = -d_o$	Real, appears on the other side of the lens	$d_i = \dfrac{2}{3}f < 0$ $0 < |d_i| < |f|$	Virtual, appears in front of the lens between the lens and its front focal point
	$h_i < 0$	Inverted	$h_i > 0$	Upright
	$M = \dfrac{h_i}{h_o} = -\dfrac{d_i}{d_o} = -1$	Same size as the object	$M = \dfrac{h_i}{h_o} = -\dfrac{d_i}{d_o} < 1$	Reduced
$d_o > 2f$	$d_i > 0$	Real, appears on the other side of the lens	$d_i < 0$	Virtual, appears in front of the lens between the lens and its front focal point
	$h_i < 0$	Inverted	$h_i > 0$	Upright
	$-1 < M < 0$	Reduced	$M = \dfrac{h_i}{h_o} = -\dfrac{d_i}{d_o} < 1$	Reduced
	$|d_i| < |d_o|$	Closer to the mirror than the object	$|d_i| < |d_o|$	Closer to the lens than the object

ELECTRICITY, MAGNETISM, AND OPTICS

EXAMPLE 27-7

Combination of Two Thin Lenses

Two converging lenses with focal lengths of $f_1 = 20$ cm and $f_2 = 25$ cm are placed parallel to each other such that their centres are aligned along the principal optical axis (Figure 27-38). The lenses are separated by a distance $d = 10$ cm. A 30 cm–tall object is located 100 cm to the left of lens 1 (L_1). Describe the properties of the image produced by this combination of lenses.

Figure 27-38 Image formed by a combination of two lenses. The image formed by lens L_1 is shown in orange; the final image, formed by L_2, is shown in green. Dashed lines show the location of an image that would have been formed by L_1 if L_2 were not present.

SOLUTION

Let us consider the image formed by the first lens, L_1. The distance between the object and L_1 is 100 cm. We can use the thin lens equation (Equation (27-11a)) to find the lens–image distance:

$$\frac{1}{d_{o1}} + \frac{1}{d_{i1}} = \frac{1}{f_1}$$

$$\frac{1}{100 \text{ cm}} + \frac{1}{d_{i1}} = \frac{1}{20 \text{ cm}} \Rightarrow d_{i1} = 25.0 \text{ cm}$$

This distance is shown in red in the diagram. Since the lens–image distance is positive, the image is real and inverted.

We find its height by using the lens magnification equation (Equation (27-11b)):

$$M_1 = \frac{h_{i1}}{h_{o1}} = -\frac{d_{i1}}{d_{o1}} = -\frac{25 \text{ cm}}{100 \text{ cm}} = -\frac{1}{4}$$

$$h_{i1} = -\frac{1}{4} h_{o1} = -\frac{30 \text{ cm}}{4} = -7.5 \text{ cm}$$

The negative height of the image indicates that it is inverted. We now consider the second lens, L_2. The image formed by L_1 becomes an object for L_2. However, since L_2 is located to the left of the image created by L_1, we are dealing with a virtual object and $d_{o2} < 0$. From Figure 27-38 we see that the distance between the second lens and the virtual object is 15 cm. Now we use the thin lens equation (Equation (27-11a)) to find the location of the image formed by L_2:

$$\frac{1}{d_{o2}} + \frac{1}{d_{i2}} = \frac{1}{f_2}$$

$$\frac{1}{-15 \text{ cm}} + \frac{1}{d_{i2}} = \frac{1}{25 \text{ cm}} \Rightarrow d_{i2} = 9.4 \text{ cm}$$

To find the height of the final image we use the lens magnification equation (Equation (27-11b)):

$$M_2 = \frac{h_{i2}}{h_{o2}} = -\frac{d_{i2}}{d_{o2}} = -\frac{9.4 \text{ cm}}{-15.0 \text{ cm}} = 0.63$$

$$h_{i2} = 0.6 \, h_{o1} = 0.63(-7.5 \text{ cm}) = -4.7 \text{ cm}$$

The overall magnification is

$$M_{\text{final}} = M_1 \cdot M_2 = \left(-\frac{1}{4}\right) \cdot (0.63) = -0.16$$

$$h_{i2} = -\frac{1}{8} h_{o1} = (-0.16)(30 \text{ cm}) = -4.7 \text{ cm}$$

Thus, the final image is inverted (relative to the object) and reduced such that the height and thickness of the final image are 16% of the original object's height and thickness, respectively.

Example 27-7 illustrates how to find an image formed by a combination of thin lenses. In general, a combination of thin lenses can be replaced by a single equivalent lens. The optical properties of the lenses and the distance between them define the properties of the image. Lens combinations are used in telescopes, microscopes, cameras, and other optical devices. Interestingly, when you wear prescription eyeglasses, each of your eyes sees an image formed by a combination of the lens in the glasses and the lens in your eye.

LO 6

27-6 The Human Eye and Vision Correction

We can use the principles of geometric optics to understand how the human eye functions. Indeed, we can use the same principles to understand the functioning of the eyes of insects, fish, birds, and other animals, even though

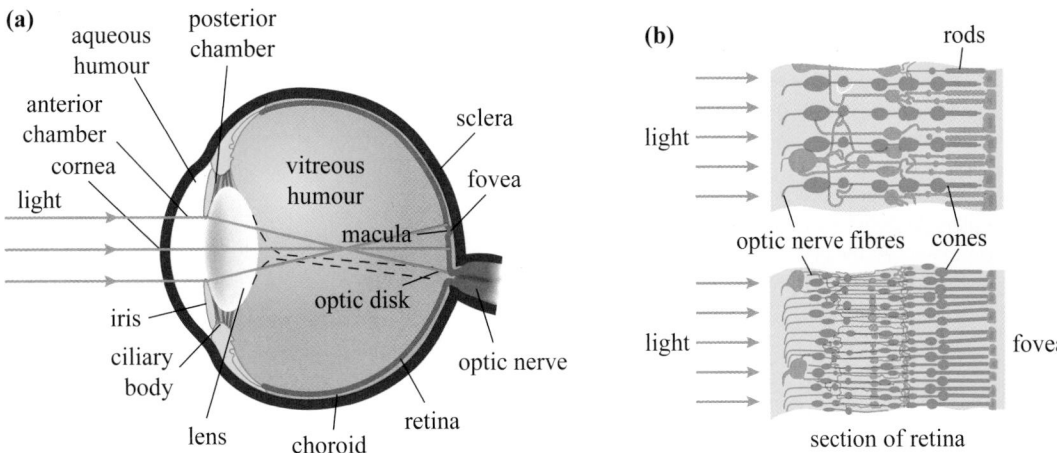

(a)

posterior chamber

aqueous humour

anterior chamber

cornea

light

iris

ciliary body

lens

choroid

vitreous humour

sclera

fovea

macula

optic disk

optic nerve

retina

(b)

rods

light

optic nerve fibres   cones

light

fovea

section of retina

Figure 27-39   (a) A schematic representation of a human eye (b) Details of the retina

the structure of their eyes is often radically different from the structure of human eyes.

The human eye is a complex optical device (Figure 27-39). Optically, the key parts are as follows: The iris adjusts the size of the opening (the pupil) through which light passes to the lens, which focuses the light onto the retina. The cornea and a chamber of fluid (called the aqueous humour) are in front of the lens, and a thicker fluid (called the vitreous humour) is in a chamber between the lens and the retina. The retina contains cone cells, which provide black and white vision in dim light, and rod cells, which provide colour vision in daylight. The optic nerve carries electrochemical signals from the retina to the brain.

For us to perceive an object clearly, the light emitted by the object has to be focused on the retina. Since the lens–retina distance is essentially constant, muscles in the human eye change the shape and, hence, the focal length of the lens to focus images of the objects located at different distances from the eye. (In contrast, a fish eye can adjust the distance between its lens and retina.) If the eye is too long or the combination of cornea and lens too powerful optically, the eye muscles may not be able to bring distant objects completely into focus, resulting in nearsightedness (myopia). Similarly, a short eye or a weaker lens can cause farsightedness (hyperopia).

Optometrists use a unit called a dioptre (symbol D) to measure the optical power of a lens. **Optical power** is the reciprocal of the focal length:

$$P = \frac{1}{f} \Rightarrow [P] \equiv D = \frac{1}{m} \qquad (27\text{-}14)$$

For example, a lens that has a focal length of 10 cm has an optical power of 10 D. Note that "optical power" has nothing to do with the rate at which energy is transferred.

The optical power of a combination of two or more thin lenses that are close together (essentially at the same location) is approximately equal to the sum of optical powers of the individual lenses:

$$D_{net} = \sum_{i=1}^{n} D_i \qquad (27\text{-}15)$$

ONLINE ACTIVITY

Exploring Human Vision

The e-resource that accompanies every new copy of this textbook contains an Online Activity using the "Exploring Human Vision" simulation. Use this activity to explore the human eye.

 CHECKPOINT

C-27-10 Exploring Lenses

You are given two lenses: one has an optical power of –3 D, and the other one has an optical power of –6 D. How can you tell which lens is which?

C-27-10 Since both powers are negative, both focal lengths are negative and both lenses are diverging. The lenses can be plano-concave or biconcave. According to Equation (27-14), the lens that has the shorter focal length has the greater optical power. We can measure the focal length by diverging a parallel beam of sunlight and tracing where the divergent beam will appear to emerge. The lens that diverges the beam more has a higher optical power (in absolute value).

LO 7

27-7 Brewster's Angle

In Chapter 26, we considered the physical properties of light as an electromagnetic wave. Using Maxwell's equations, it can be shown that reflected light is completely polarized when the reflected and the transmitted waves form a right angle (Figure 27-41). This fact was first discovered by Scottish scientist David Brewster (1781–1868) in 1811, long before Maxwell's equations were developed. The angle of incidence for this complete polarization is called **Brewster's angle**, θ_B.

Figure 27-41 When the incident angle is equal to Brewster's angle θ_B, the reflected light is completely polarized perpendicular to the plane of incidence. The transmitted light is partially polarized parallel to the plane of incidence.

Let the index of refraction of air be n_1 and the index of refraction of the crown glass in Figure 27-41 be n_2. Let θ_B be the angle that the incident beam makes with respect to the normal to the plate and θ_2 the angle that the transmitted beam makes with respect to the normal.

(a) The angle between the reflected beam and the normal is θ_B.

(b) The angle of incidence and the angle of transmission are related by Snell's law:

$$n_1 \sin\theta_B = n_2 \sin\theta_2$$

We require that the reflected and the transmitted beams make a right angle. From the geometry of Figure 27-37, we have

$$\theta_B + 90° + \theta_2 = 180°$$

$$\theta_2 = 90° - \theta_B$$

Inserting θ_2 into Snell's law, we get

$$n_1 \sin\theta_B = n_2 \sin(90° - \theta_B) = n_2 \cos\theta_B$$

which gives

$$\tan\theta_B = \frac{n_2}{n_1}$$

$$\theta_B = \tan^{-1}\left(\frac{n_2}{n_1}\right) \quad \text{(Brewster's angle)} \quad (27\text{-}16)$$

For the air–glass interface, $n_1 = 1$ (air) and $n_2 = 1.52$ (glass). The value of Brewster's angle in this case is

$$\theta_B = \tan^{-1}\left(\frac{1.52}{1.00}\right) = 57°$$

(1) When unpolarized light from air strikes a glass surface at 57° (with respect to the normal to the surface), the reflected light is completely polarized perpendicular to the plane of incidence.

(2) When the incident light is polarized in the plane of incidence and strikes the glass surface at 57°, there is no reflected beam. The incident beam is completely transmitted into the glass.

 MAKING CONNECTIONS

Polarizing Sunglasses

Sunlight is unpolarized. Light reflected from a horizontal surface is partially polarized in the horizontal plane. Polarizing sunglasses contain a coating that absorbs light that is polarized in the horizontal direction (Figure 27-42), greatly reducing the glare from road and water surfaces.

vision without Polaroid sunglasses

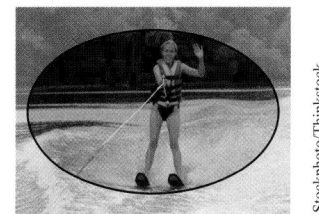

vision with Polaroid sunglasses

Figure 27-42 Polaroid sunglasses reduce glare by absorbing horizontally polarized sunlight.

Fermat's Principle

In a homogeneous medium, light propagating between any two points always takes the path that it can traverse in the least time.

Optical Images

For us to see an object or an image, the light reflecting from the object must enter our eye. A real image of an object is created when the rays emitted by the object converge after reflecting from a mirror or refracting in a lens. A virtual image appears where the imaginary continuations of the rays emitted by the object meet after refraction or reflection. Such images can be analyzed using the mirror and lens equations or ray diagrams.

The Law of Reflection

The angle of incidence equals the angle of reflection:

$$\theta_1 = \theta'_1 \qquad (27\text{-}1)$$

The incident ray, the reflected ray, and the normal to the surface at the point of incidence are always located in the same plane.

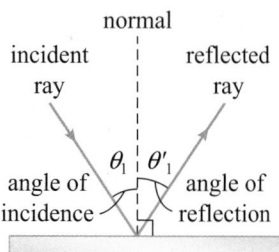

The Law of Refraction

When a ray of light propagates through the boundary between two media, it changes direction such that

$$n_1 \sin\theta_1 = n_2 \sin\theta_2 \qquad (27\text{-}8)$$

where n_1 and n_2 are the indexes of refraction of the two media, and θ_1 and θ_2 are the angles of incidence and refraction, respectively.

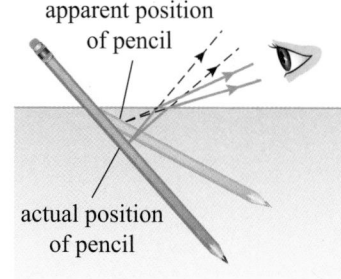

The index of refraction can be expressed as follows:

$$n = \frac{c}{v} \qquad (27\text{-}7)$$

where c and v are the speeds of light in a vacuum and the medium, respectively.

When light propagates from a more optically dense medium to a less optically dense medium, total internal reflection occurs when the angle of incidence exceeds the critical angle.

(a)

(b)

The Formation of Images by Mirrors and Lenses

The locations of an object and the image formed by a spherical mirror or a thin lens are related as follows:

$$\frac{1}{d_o} + \frac{1}{d_i} = \frac{1}{f} \qquad (27\text{-}5a)$$

where f is the focal length. The focal length is positive for concave mirrors and converging lenses and negative for concave mirrors and diverging lenses.

The magnification of an image is determined using the following equation:

$$M = \frac{h_i}{h_o} = -\frac{d_i}{d_o} \qquad (27\text{-}5b)$$

where d_o and d_i are the object–mirror/lens and the image–mirror/lens distances, respectively.

Brewster's Angle

Brewster's angle is the angle of incidence at which light reflected from a surface is completely polarized:

$$\theta_B = \tan^{-1}\left(\frac{n_2}{n_1}\right) \qquad (27\text{-}16)$$

APPLICATIONS

Applications: optical instruments, natural phenomena such as shadows, eclipses, rainbows, and mirages; fibre optics cables; polarized sunglasses

Key Terms: angle of incidence, angle of reflection, angle of refraction, Brewster's angle, centre of curvature, concave mirror, converging lens, convex mirror, critical angle, dispersion, diverging lens, Fermat's principle, field of view, focal length, focal point, full shadow, geometric optics, image, incident ray, index of refraction, law of reflection, law of refraction, medium, object, opaque, optical density, optical power, partial shadow, penumbra, pinhole camera, plane mirror, principal optical axis, radius of curvature, ray diagram, ray optics, real image, reflected ray, reflection, refracted ray, refraction, secondary focal point, Snell's law, thin lens, total internal reflection, translucent, transparent, umbra, virtual focal point, virtual image

QUESTIONS

1. You are standing 2 m in front of a plane mirror, then move 0.5 m away from the mirror. By how much will the distance between you and your image change?

2. You are moving toward a mirror at the speed of 0.5 m/s. How fast are you moving toward your own image?

3. Why does an image in a plane mirror undergo depth inversion? Is it correct to say that a mirror interchanges left and right? Why is there no up–down inversion? Use a plane mirror, a piece of paper with a word written on it, and a ray diagram to test your explanation. How is this phenomenon used in everyday life? (Hint: Think of ambulances.)

4. Ordinary window glass reflects approximately 4% of incident light and transmits the remaining 96%. Use this information to explain why ordinary window glass works like a mirror for a person inside the house at night but looks transparent during the day. Remember, the light entering the glass encounters two boundaries: an air–glass boundary upon entering the glass pane and a glass–air boundary at the exit.

5. Which of the following statements correctly describes the images formed by a plane mirror?
   (a) The images are always real, inverted, and magnified.
   (b) The images are always virtual, inverted, and reduced.
   (c) The images are always virtual, upright, and the same size as the objects.
   (d) The images are always virtual and upright, and they appear closer than the objects.
   (e) The images in the mirror can be viewed from any location in front of the mirror.

6. A friend of yours claims that the image formed in a plane mirror is formed on the surface of the mirror, so the image does not have depth. What would you do to convince your friend that the location of the image is behind the mirror?

7. On a car's side-view mirror you notice a warning: "Objects in mirror are closer than they appear." Explain the physics behind this statement. What kind of mirror is it, and why is it used as a side-view mirror?

8. While surfing the Internet, you noticed an advertisement for a car rear-view mirror that claims to solve the blind spot problem (also often called the dead zone problem). The problem arises because the driver's field of view in a regular rear-view mirror is limited. Thus, the driver must move her head to be able to see all the cars in the vicinity of her own car. Alternatively, she must use additional mirrors. Draw a ray diagram to illustrate the problem, and propose a solution using optical devices discussed in this chapter. See whether your proposal makes sense the next time you have a chance to take a closer look in a rear-view mirror.

9. Figure 27-43 shows a rose that is located off to one side of a plane mirror. Use a ray diagram to answer the following questions.
   (a) Will an image of the rose be formed?
   (b) If you answered "yes" to part (a), where will the image be formed, and where should an observer be located to be able to view the image? (That is, determine the field of view for this image.)

Figure 27-43 Questions 9 and 10

10. Replace the plane mirror in Figure 27-43 with a concave mirror and then with a convex mirror, and answer the questions asked in question 9. Compare your answers for the three different mirrors, and draw conclusions.

ELECTRICITY, MAGNETISM, AND OPTICS

11. Two plane mirrors form the legs of a 60° angle. An object is placed between the two mirrors. What is the maximum number of images that could be viewed while looking into the mirrors? Explain.

12. If you want to see a full-length image of yourself (from the top of your head to your toes) in a plane mirror, what should the minimum size of the mirror be?
    (a) slightly more than my height
    (b) exactly my height
    (c) about three-quarters of my height
    (d) about one-half of my height
    (e) about one-quarter of my height
    (f) It depends on my distance from the mirror.

13. Explain your reasoning for your answer to question 12. How would your answer change if instead of a plane mirror you used a spherical mirror?

14. The Mirage is a classic toy that consists of two parabolic mirrors (Figure 27-44). The top mirror has a small circular opening on the top. When assembled, the mirrors produce a lifelike three-dimensional image that looks real but is an illusion. Draw a ray diagram showing how the image of the dials in Figure 27-44 is created. Why should the object be small? How many images will be created by this toy? What is their relative orientation?

Courtesy of Marina Milner-Bolotin

Figure 27-44 Question 14

15. You decide to buy yourself a magnifying makeup mirror. What kind of a mirror is it? Use a ray diagram to explain your answer.

16. Use the mirror equation to plot a graph of the magnification of (a) a concave mirror, (b) a convex mirror, and (c) a plane mirror as a function of the object–mirror distance. Use Tables 27-1 and 27-3 to make sense of your graphs. Do the two representations (the table and your graph) match?

17. Three light rays simultaneously emitted by a candle pass through a converging lens as shown in Figure 27-45. Which ray arrives at the screen first? Explain.

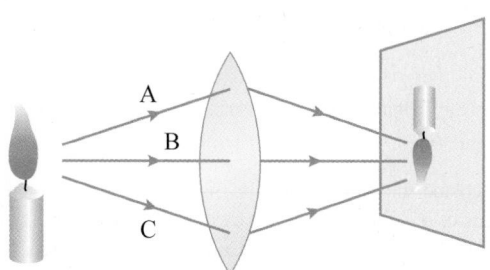

Figure 27-45 Question 17

18. Use the thin lens equation to plot a graph of the magnification of (a) converging lenses and (b) diverging lenses as a function of the object–lens distance. Use Table 27-4 to make sense of your graphs. Do the two representations (the table and your graph) match? Compare the graphs you plotted for the mirrors in question 16 and for the lenses. What do they have in common, and how they differ?

19. An object is placed in front of a converging lens such that its image is formed on a screen. What happens to the image when the bottom half of the lens is covered? Explain.

20. Give an example of where one lens can be a converging lens under some conditions and diverging under other conditions. Where might this be important? (Hint: Think of seeing under water.)

21. A parallel beam of light is incident on a converging lens at an angle to the principal optical axis. Draw a ray diagram illustrating the propagation of the beam through the lens. Will the beam converge after exiting the lens? Where will it happen? Compare your result to the case of a spherical concave mirror. (Hint: Think of the lens's focal plane.)

22. Explain the difference between nearsighted vision and farsighted vision, and use a ray diagram to show what lenses can be used to correct these problems.

PROBLEMS BY SECTION

For problems, star ratings will be used, (✷, ✷✷, or ✷✷✷), with more stars meaning more challenging problems.

Section 27-1 Evidence for the Geometric Optics Approach to Describing Optical Phenomena

23. ✷ You want to figure out the height of a tree on your street. You use a 1 m–long stick as a gnomon (Figure 27-46). (A gnomon is an object that is positioned perpendicular to the ground, and the length of its shadow is used to determine other lengths or distances.) On a bright sunny day you find that at a certain time, the shadow cast by a gnomon is 3 m long. The gnomon is 100 m behind the tree, and you measure the shadow just before the Sun disappears behind the tree. What is the height of the tree? What assumptions did you make?

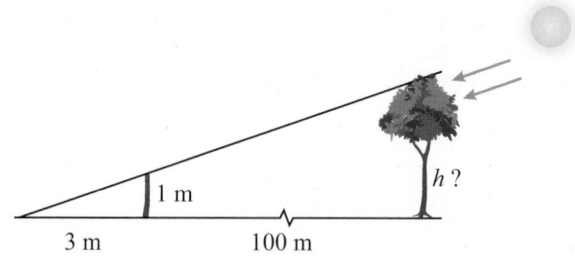

Figure 27-46 Problem 23

24. ✷ A pinhole camera like the one in Figure 27-4(a) is 20 cm deep and has a hole 1 mm in diameter. A 20 m–tall tree is 50 m from the camera. What is the size of the image of the tree on the camera screen? How does the size of the image change when the diameter of the pinhole is doubled? Does anything else change as a result? Explain.

ELECTRICITY, MAGNETISM, AND OPTICS

25. ✲✲ An eclipse of the Sun can be total or partial, depending on where on Earth one is located, as shown in Figure 27-47.
   (a) Explain what observers A, B, and C will see during the eclipse.
   (b) The figure shows an instant of a solar eclipse. Show how the area of the total and partial solar eclipse moves on Earth's surface during the eclipse.

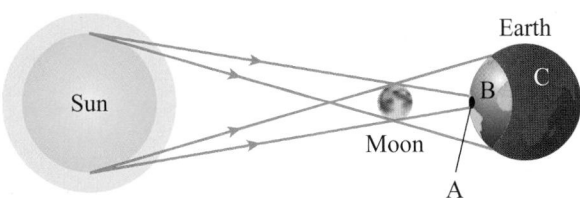

Figure 27-47 Problem 25. A total solar eclipse (not to scale)

26. ✲✲ The phases of the Moon are shown in Figure 27-48.
   (a) Use Figure 27-48 to explain what phase the Moon is in during a total solar eclipse. Why is a total solar eclipse such a rare phenomenon?
   (b) At some point, one can see the Moon and the Sun. When does this happen in terms of the phases of the Moon?
   (c) The popular phrase "the dark side of the Moon" reflects the fact that we only can see one side of the Moon. Explain why.
   (d) Draw a diagram that shows how lunar eclipses happen and why at different locations around the globe people can see a total or a partial lunar eclipse.

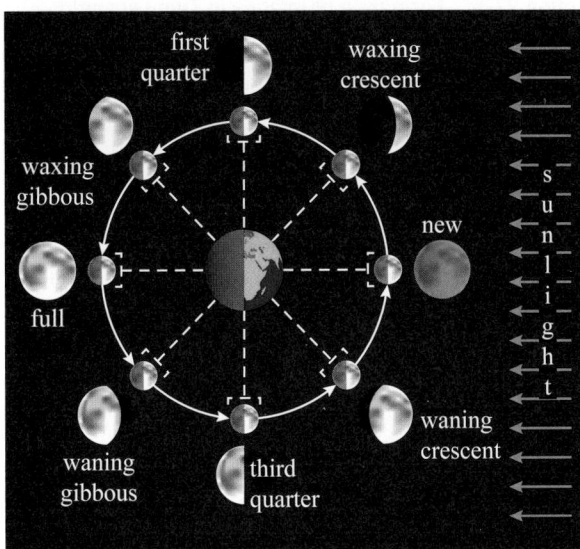

Figure 27-48 Problem 26

Section 27-2   Reflection of Light

27. ✲ A laser beam is reflected by a plane mirror. It is observed that the angle between the incident and reflected beams is 45°. The mirror is then rotated so that the angle of incidence decreases by 7°. What is the new angle between the incident and reflected beams?

28. ✲ A laser beam enters a 10 m–long optical fibre at an angle of incidence of 60°, which far exceeds the critical angle of the fibre. The core of the fibre is 60 mm in diameter, and the cladding is 140 mm (Figure 27-49). How many times will the laser light be reflected inside the fibre prior to exiting it?

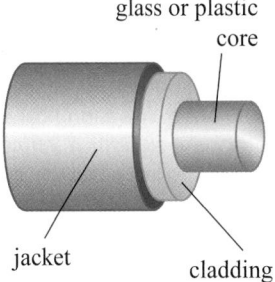

Figure 27-49 Problem 28

29. ✲ Sunlight enters a room at an angle of 30° above the horizontal and reflects from a small mirror lying flat on the floor. The reflected light forms a spot on a wall that is 2.5 m behind the mirror (Figure 27-50). You place a notebook under the edge of the mirror nearest to the wall, tilting it upward by 4°. How much higher on the wall (Δh) is the spot?

Figure 27-50 Problem 29

Section 27-3   Mirrors

30. ✲ Two plane mirrors meet at an angle θ, such that $\dfrac{360°}{\theta}$ is a whole number.
   (a) Use ray diagrams to help you derive an equation that shows the number of images that will be created by this configuration of mirrors.
   (b) Apply the equation you derived in part (a) to the case where two mirrors are parallel to each other. What conclusion can you make? How can you verify your conclusion?

31. ✲ A periscope is an optical device that allows you to see from a concealed position. The simplest periscope consists of a tube at each end of which are tilted mirrors. Draw a diagram of a periscope, and use your diagram to answer the following questions.
   (a) Is the image of an object seen in your periscope inverted? Explain why or why not.
   (b) Sometimes two simple lenses are added to a periscope to focus the images of the objects observed. Use a ray diagram to show how this works.
   (c) Now suppose that you use curved mirrors in your periscope. Would this be a viable design? Explain why or why not.

32. ✲ A 1.5 m–tall child is standing in front of a plane mirror. His eyes are 9 cm below the top of his head.

(a) What is the shortest mirror height for the child to see his entire body?

(b) How high from the floor should the mirror hang so that the child can see his entire body?

(c) How will your answers to parts (a) and (b) change if instead of a plane mirror a spherical mirror is used?

(d) If the child moves back from the mirror, will he be able to use a mirror smaller than the size you found in part (a) and still see his entire body?

33. ✱ While playing with a metal salad spoon, you notice that when looking at the concave part of the spoon, your finger appears to touch its image when 7.0 cm from the inner surface of the spoon. Assume that the spoon is a spherical mirror surface.

(a) What is the focal length?

(b) What are the properties of the image of your finger?

34. ✱✱ A 15 cm–tall object is placed 0.50 m in front of a concave spherical mirror that is 60 cm in diameter. Justify your answers to the following questions.

(a) Where will the image of the object be formed?

(b) Describe the properties of the image (type, orientation, and magnification).

(c) Instead of a concave mirror, you use a convex mirror. How do your answers to parts (b) and (c) change?

35. ✱✱ To increase security in a store, a storekeeper decides to replace an old plane mirror with a new convex mirror with the same diameter. He finds a 12-in. heavy-duty, glass, indoor, round spherical mirror. The seller claims that the mirror has a 160° field of view area that is ideal for the store.

(a) What should the focal length of the mirror be for the specifications to be true?

(b) If a person stands 2.0 m away from this mirror, how large will his image be in the mirror? What are the properties of the image?

Section 27-4 Refraction of Light

36. ✱ A laser beam is shone on a horizontal slab of 3 cm–thick glass ($n = 1.5$) at a 30-degree angle from the normal. By how much will the light ray be shifted from along the glass slab upon exiting it? (How far is the distance between the point where the laser beam leaves the glass slab and where it would have left it if the slab had an index of refraction $n = 1$)?

37. ✱ An underwater flashlight is pointed upward at the bottom of a 2 m–deep pool filled with fresh water. Estimate the area of the light spot at the surface of the pool. List any assumptions you make for your estimate.

38. ✱ (a) Find the index of refraction of a substance in which light travels 0.620 m in 3.00 ns.

(b) How far does light travel in air in 3.00 ns?

(c) How long does it take light to travel 1.00 m in oil?

39. ✱✱ A coin is placed at the bottom of an 80 cm–deep aquarium filled with water. You are looking at the coin from above the aquarium. How deep will the coin appear to you? Explain using ray diagrams and the necessary calculations.

40. ✱ While scuba diving, you look up and notice a beautiful sunset. To you, the Sun appears to be 50° above the horizon, yet it was a sunset. What is the actual angle of the Sun above the horizon? Explain.

41. ✱ A ray of light enters the long side of a 45°-90°-45° prism and undergoes two total internal reflections (Figure 27-51). As a result, the ray reverses its direction.

(a) What is the minimum value of the prism's index of refraction for these internal reflections to be total?

(b) How will your answer to part (a) change if instead of air the prism is submerged in oil?

Figure 27-51 Problem 41 and problem 57

42. ✱ A ray of He-Ne laser light is incident on a 45°-90°-45° glass prism at an angle of 60°, 5.0 cm above its base, as shown in Figure 27-52.

(a) What is the angle of deflection (the angle between the incident and refracted beams) for this ray as it exits the prism?

(b) At what point will the ray exit the prism?

(c) How will your answers to parts (a) and (b) change if instead of a He-Ne laser (that produces a monochromatic light), white light were used for this question?

Figure 27-52 Problem 42

Section 27-5 Formation of Images in Thin Lenses

43. ✱ A lens for an old-fashioned 35 mm camera has a focal length of 45 mm.

(a) How close to the film should the lens be placed to form a sharp image of an object that is 6.0 m away?

(b) What is the magnification of the image on the film?

44. ✱ A plastic magnifying lens ($n = 1.45$) has a magnification of 2.5 when the object–lens distance is 4.0 cm. The lens is placed in a large container filled with water, and its magnification is measured when the object–lens distance is once again 4 cm.

(a) What is the magnification?

(b) Justify your answer to part (a) using (i) the lens maker's equation and (ii) the law of refraction.

45. ✱ A thin lens has a focal length of −40 cm. Justify your answers to the following questions using equations and ray diagrams.

(a) Find the image distance and magnification that result when an object is placed 25 cm in front of the lens.

(b) What type of lens is it?

(c) What type of image is it?

46. ✱✱ Two converging thin lenses ($f_1 = 15$ cm and $f_2 = 20$ cm) are separated by 25 cm. A candle is placed 40 cm in front of the first lens.

(a) Where will the image of the candle be created by the two-lens system?

(b) Describe the properties of the image.

(c) Describe how the image changes as the lenses are moved closer together.

(d) The lenses are now moved right next to each other. What is the equivalent optical power of the system? What conclusion can you draw?

47. ✱✱ A glass concave–convex lens has the radii of 12 cm and 14 cm, respectively. A 5 cm–tall object is placed 12 cm in front of the lens. Justify your answers to the following questions using equations and ray diagrams.
   (a) The lens and the object are submerged in water. What is the magnification of the lens?
   (b) The lens is now submerged in air. What is the magnification?

COMPREHENSIVE PROBLEMS

48. ✱ An observer is positioned so that the far edge of the bottom of an empty glass is just visible (Figure 27-53). When the glass is filled to the top with water, the centre of the bottom of the glass is just visible to the observer. The width of the glass is 6.8 cm.
   (a) What is the height h of the glass?
   (b) What does the observer see when the glass is filled with oil instead of the water?

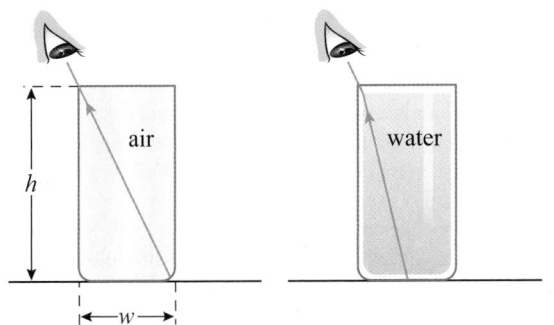

Figure 27-53   Problem 48

49. ✱ A small light source is located at the bottom of a large aquarium filled with water. A circular disk is floating on the surface of the water right above the light source. The height of the water in the aquarium is H. What should the smallest radius of the disk be such that no light exits the surface of the water?

50. ✱ A light ray is incident on a horizontal glass slab like the one in Figure 27-25 such that the angle between the reflected and refracted rays is 90°. What is the angle of incidence of this ray?

51. ✱ A narrow beam of light is incident on an air–glass boundary at an angle of 30°. Calculate the angle of deflection (the angle between the incident and refracted beams) of this beam. Will the angle of deflection increase or decrease when the angle of incidence increases to 45°? Justify your answer with calculations and a diagram.

52. ✱ While visiting the Vancouver Aquarium, you watch beluga whales through the glass wall of their water tank. When a beluga is swimming toward you at a speed of 1 m/s, what is the apparent speed of the whale as observed by you? Should these two speeds be different? Explain.

53. ✱ A small object is 0.15 m from a diverging lens, which has a focal length of −30 cm. At what distance from the lens is the image of this object located? Will you be able to see this image on a screen? Justify your answer using calculations and a diagram.

54. ✱ An object is 4.2 m from a screen.
   (a) Where should you place a converging lens to see a focused image of the object on the screen with 20× magnification?
   (b) Determine the focal length and the optical power of the lens.

55. ✱ A camera lens has a focal length of 5.0 cm and a CCD sensor 6.0 cm × 8.0 cm. You want to take an image of an object that is 60 cm × 60 cm.
   (a) How far from the object should the lens of the camera be to make the image as big as possible?
   (b) How far from the lens should the CCD screen of the camera be?

56. ✱ A monochromatic parallel beam of light emitted by a He-Ne laser strikes a triangular 30°−60°−90° prism surrounded by air 5 cm above the base of the prism. The incident beam is perpendicular to one of the prism's faces, as shown in Figure 27-54. After undergoing refraction in the prism the light comes out of the bottom right vertex of the prism. Calculate the index of refraction of the prism for this beam of light. Draw a ray diagram of this scenario.

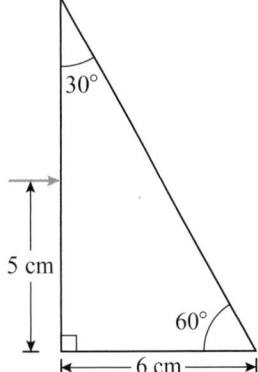

Figure 27-54   Problem 56

57. ✱✱ The 45°-90°-45° triangular glass prism in Figure 27-51 is surrounded by air. A ray of light is incident on the left face at an angle of 60.0°. The point of incidence is high enough that the refracted ray hits the opposite sloping side.
   (a) Through which face of the prism does the light exit?
   (b) What is the angle that the exiting beam of light makes with the horizontal?
   (c) The length of the hypotenuse of the cross-sectional area of the prism is 10 cm. How high from the bottom of the prism will the light ray exit it?

58. ✱ A 5.0 cm–tall vertical rod is 60 cm from a convex mirror, which has a radius of curvature of 40 cm.
   (a) Where is the image of the rod located?
   (b) What are the properties of the image?

59. ✱✱ Two thin lenses have an optical power of 4.0 D and 5.0 D, respectively. They are 0.90 m from each other.
   (a) Find the focal length of the lenses.
   (b) Find the location of the image of an object that is 0.50 m in front of the first lens.
   (c) Can you use the equation for the combined power of the lens to solve this problem? Justify your answer.

60. ✷✷ During an eye exam, you notice that to be able to read a book you have to hold the book 16 cm in front of your eyes. For a healthy eye, this distance should be approximately 25 cm.
(a) What vision problem does this indicate?
(b) What further information do you need to determine the optical power of eyeglasses to correct your vision?

61. ✷✷✷ Derive an equation for the angle of deflection of a light beam propagating through an isosceles triangular prism as a function of the angle of the apex of the prism (ϕ), its index of refraction n, and the angle of incidence θ_i. Assume that the base of the prism is its largest face and that the light beam encounters the prism high enough that the refracted ray meets the opposite sloping side.

62. ✷✷✷ A horizontal monochromatic circular beam of light (3.0 mm in diameter) is incident on an equilateral triangular prism made of glass ($n = 1.52$) which sits on a table. The point of incidence is 8 cm above the base of the prism. The side of the equilateral triangle is 15 cm. What is the area of the spot formed by the emergent beam on the table?

63. ✷✷ A parallel beam of light is incident on a diverging lens that has a focal length of F_1.
(a) How far from the diverging lens should a converging lens ($F_2 = -3F_1$) be located so that the beam exiting the diverging lens is also a parallel beam of light?
(b) How will your answer to part (a) change if the light beam enters first the converging lens and then the diverging lens? Explain.

64. ✷✷ A point light source is located at the bottom of a 20 cm–deep aquarium filled with glycerine ($n = 1.47$), as shown in Figure 27-55. A screen is placed 10 cm above the surface of the glycerine. A converging lens with a focal length of 1.6 cm is placed between the surface of the glycerine and the screen. The lens is used to create a focused image of the source on the screen. What is the size of the image of the light source on the screen as compared to the size of the light source?

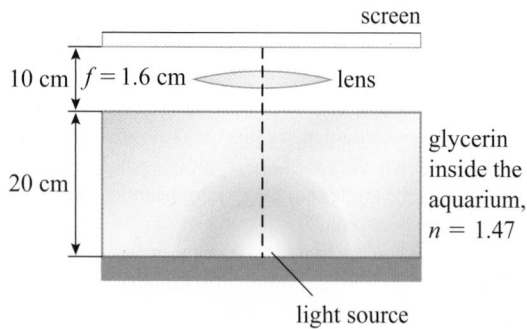

10 cm | $f = 1.6$ cm — lens
screen
20 cm
glycerin inside the aquarium, $n = 1.47$
light source

Figure 27-55 Problem 64

65. ✷✷ Water is poured inside a concave mirror, which has a radius of 50 cm.
(a) Draw a diagram of this optical system.
(b) What are the focal lengths and optical powers of the mirror and the water lens?
(c) What is the optical power of the system?

66. ✷✷✷ The inside surface of a convex part of a plano–convex plastic lens with radius $R = 40$ cm and index of refraction $n = 1.52$ is coated with silver, so the inside of the convex surface acts as a mirror (Figure 27-56). A collimated beam of light parallel to the principal optical axis of the lens is incident on the mirror surface.
(a) Describe what happens to the beam of light in the figure.
(b) Where is the focal point of the system?
(c) An object is placed 15 cm in front of the plane side of the lens. Where is its image located? Describe the properties of the image.

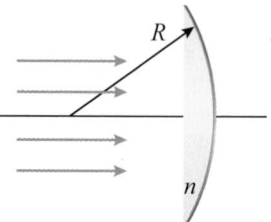

R
n

Figure 27-56 Problem 66

67. ✷✷✷ A small 50 g marble is moving at 6.0 m/s along the principal optical axis of a converging plano–convex lens as shown in Figure 27-57. The focal length of the lens is 10 cm. The lens is securely mounted on a stand, and the friction between the stand and the table is negligible. The combined mass of the lens and the stand is 0.4 kg. After an elastic collision between the marble and the lens, the marble bounces back. Find the time interval during which the lens produces a virtual image of the marble. Neglect the gravitational force.

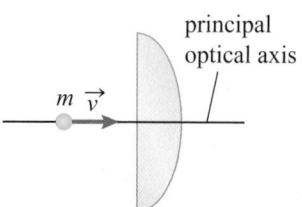

principal optical axis
$m\ \vec{v}$

Figure 27-57 Problem 67

68. ✷✷ A converging beam of light is incident on a screen as shown in Figure 27-58. The screen has an opening 10 cm in diameter exactly at the location of the beam. The angle between the envelope of the beam and the axis of symmetry of the opening is 30°. Support your answers to the following questions with ray diagrams.
(a) How far from the screen does the beam converge after going through the opening?
(b) A converging lens of $P_1 = 10$ D is placed in the opening. How far from the screen does the beam converge?
(c) If instead of a converging lens a diverging lens of $P_2 = -10$ D is placed in the opening, how far from the screen will the beam converge?

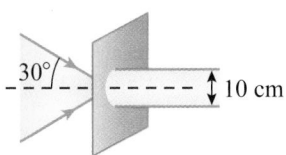

Figure 27-58 Problem 68

69. ✱✱✱ Prove that a pair of thin lenses located very close to each other (essentially at the same location) act as a single lens with a power P_{net} equal to the sum of powers of the two lenses making up the combination, as shown in Equation (27-15).

70. ✱✱✱ Prove that the combination of a thin lens and a spherical mirror placed very close to each other (essentially at the same location) acts as a single mirror with a power P_{net} equal to the sum of powers of the lens and mirror making up the combination, as shown in Equation (27-15).

71. ✱✱✱ A thin converging lens is connected with a string to the ceiling, as shown in Figure 27-59. The lens is moving along a circle such that its optical axis always remains perpendicular to the floor. The lens produces an image of the hook on the ceiling that the string is attached to. At what angular velocity of the lens is the image always located on the floor? The height of the ceiling is 3 m.

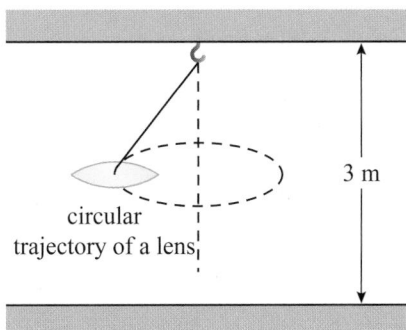

Figure 27-59 Problem 71

72. ✱✱✱ A magnifying glass (either a biconvex lens or a combination of two lenses) is a converging lens that has a short focal length, approximately 1 cm to 10 cm, depending on the required magnification. When an object is located between the lens and its focal point, its virtual magnified image should be located at a distance of approximately 25 cm, which is the optimal distance for a healthy eye.
(a) Prove that the magnification of a magnifying glass is given by:

$$M = \frac{f}{f - d_o}$$

(b) Prove that the magnification of a magnifying glass with the above focal length that varies from 1 cm to 10 cm, varies between 3.5 and 26.

73. ✱✱✱ A compound microscope is an optical instrument that uses at least two converging lenses separated by a distance L. Two converging lenses for a compound microscope have focal lengths of 5.0 cm and 30.0 cm and are positioned such that they are 40.0 cm apart (Figure 27-60).
(a) How far from the left lens should an object be located so that the final image is located exactly at the midpoint between the two lenses?

(b) What is the overall magnification of the compound microscope?
(c) The final image in this case is inverted. Is it possible to produce an upright final image with this microscope?
(d) Use a graphing software, the equations you derived and/or a virtual microscope simulation (such as http://www.udel.edu/biology/ketcham/microscope/scope.html) to explore how the change in each one of the parameters of the compound microscope (distance between the lenses, power of each one of the lenses, and the distance between the object and the objective lens) will affect the magnification.

Figure 27-60 Problem 73

74. ✱✱✱ Galileo's telescope used two lenses: a diverging lens as an eyepiece and a large converging lens as an objective. Kepler's telescope used two converging lenses. Kepler's telescope produced inverted images, and Galileo's telescope produced upright images. Galileo's design has a serious disadvantage that makes it almost useless (except as opera glasses). Draw a ray diagram for each telescope, and explain the problem with Galileo's telescope.

75. ✱✱✱ A friend of yours claims that he can design an optical system that can project all the light from a large light source into a significantly reduced image of the light source.
(a) Is this optical system possible? Will it violate any law of physics? (Hint: Think thermodynamics.)
(b) Would a system that focused all the light from a small light source into a magnified image be possible? Explain why or why not.

DATA-RICH PROBLEM

76. ✱✱✱ Use Online Activities described on pages 744, 749, or 751 to collect quantitative data on the relationship between the parameters of spherical mirrors and lenses. Use the simulations to illustrate how the images are formed and how their parameters change as a function of the object–optical mirror/lens distance. Prove that the same equation can be used to describe spherical mirrors and thin lenses.

OPEN PROBLEM

77. ✷✷✷ In his paper titled "Optics Demonstration with Student Eyeglasses Using the Inquiry Method,"[1] Dr. Mark C. James explains how he can "magically" guess his students' eyeglass prescriptions by looking through their eyeglasses. Explore the paper, and develop a short presentation to share this method with other students. Will this technique work for both positive and negative prescription eyeglasses? Why or why not?

---

[1] Mark C. James, "Optics Demonstration with Student Eyeglasses Using the Inquiry Method," *The Physics Teacher*, 49, September, 357–359.

See the text online resources at **www.physics1e.nelson.com for Open Problems and Data-Rich Problems related to this chapter.**

Chapter 28
Physical Optics

igure 28-1 shows a radio telescope near the town of Penticton in southern British Columbia. Why do radio telescopes need such large diameters to have even modest resolution? How do you think the resolution of this telescope compares to the resolution of a small optical telescope or a pair of binoculars? The surface of this radio telescope is not even solid, whereas the surface of a mirror in an optical telescope needs to be very precise. Why the difference?

Photo of The 26-Metre Telescope in Winter, http://www.nrc-cnrc.gc.ca/obj/hia-iha/doc/26-m-Telescope-in-the-Snow.pdf National Research Council Canada. Reproduced with the permission of the Minister of Public Works and Government Services, 2012.

Figure 28-1 The radio telescope operated by the National Research Council near Penticton, British Columbia

Learning Objectives

When you have completed this chapter you should be able to:

1 Differentiate between physical optics and geometric optics, and recognize when a physical optics approach must be used when solving problems.

2 Explain conditions under which the interference of light waves can be observed.

3 Calculate conditions for constructive and destructive interference of light in double-slit situations.

4 Explain how a diffraction grating works and the advantages over two slits, and perform related calculations.

5 Perform calculations related to thin film interference, and describe applications of thin films.

6 Perform calculations related to diffraction from single slits.

7 Qualitatively describe the intensity pattern from a real double slit.

8 Explain resolution limits, and perform related calculations.

ELECTRICITY, MAGNETISM, AND OPTICS

For additional Making Connections, Examples, and Checkpoints, as well as activities and experiments to help increase your understanding of the chapter's concepts, please go to the text's online resources at www.physics1e.nelson.com.

LO 1

28-1 Physical and Geometric Optics

In Chapter 27, we saw that light rays can be applied to solve problems in reflection and refraction, including how images are formed. That branch of optics is called **geometric (ray) optics** because we apply the geometry of light rays to solve the associated problems. The geometric optics approach is accurate as long as the wavelength of the light is considerably smaller than the dimensions of the objects being studied. When that condition is not true, and the wavelengths are comparable to the object dimensions, we can no longer accurately use ray optics, and we must consider interference between waves travelling different paths. This branch of optics, which forms the content of this chapter, is called **wave** or **physical optics**.

While we will primarily discuss phenomena involving visible light, the concepts can be applied to any type of electromagnetic radiation.

Recall that the speed of light, c, the wave frequency, f, and the wavelength, λ, for all types of electromagnetic waves are related by the following equation:

$$c = f\lambda \qquad (28\text{-}1)$$

Physical optics effects are more pronounced for electromagnetic radiation with longer wavelengths because more objects have dimensions similar to the waves. Thus, the wave behaviour of radio waves is more readily evident than the wave behaviour of visible light, and much more apparent than the wave behaviour of gamma rays.

✓ CHECKPOINT

C-28-1 Physical or Geometric Optics?

When analyzing the following situations, which is (are) likely to require a physical optics approach?
(a) Visible light is used to illuminate a coin.
(b) FM radio waves with a frequency of 100 MHz come through the window of your home.
(c) X-rays are used to take an image of a bone that is approximately 3×8 cm.
(d) Infrared light shines on a red blood cell, which has a width of approximately 8 μm.

C-28-1 (b) and (d) The wavelength of visible light (400 to 700 nm) is much smaller than a coin, and the wavelength of X-rays (0.01 to 10 nm) is much smaller than the bone, so geometric optics approximations are valid for these two situations. However, the wavelength of FM radio waves (approximately 3 m) is comparable to the dimensions of a window, and the wavelength of infrared light (approximately 0.7 μm to 300 μm) is comparable to or larger than the width of a red blood cell.

LO 2

28-2 Interference

At the core of physical optics is **interference** between waves. Recall from Chapters 14 and 15 that two mechanical waves with the same frequency will interfere constructively when they are in phase and destructively when they are out of phase. Electromagnetic waves interfere in exactly the same way. Of course, waves usually do not have precisely the same amplitudes and phase difference to completely reinforce or completely cancel each other.

ONLINE ACTIVITY

Interference of Light Waves

The e-resource that accompanies every new copy of this textbook contains an Online Activity using the PhET simulation "Wave Interference." Work through the simulation and accompanying questions to gain an understanding of how light waves interfere.

In interference experiments and calculations, we need to consider both the wavelength and the relative phase of the waves. Ordinary light bulbs produce light with mixed wavelengths and random phases. **Lasers**, however, produce light that is both **monochromatic** (with exactly a single wavelength and, hence, colour) and **coherent** (with the same phase). (Gas discharge tubes with appropriate filters can produce monochromatic light, but the light lacks coherence over significant time intervals.)

Complications also arise from the fact that the planes of the electric and magnetic fields are important when we add two electromagnetic waves. However, the addition of two waves with the same frequency and polarization is fairly straightforward. The intensity of the resulting combination is given by

$$I = I_1 + I_2 + 2\sqrt{I_1 I_2}\,\cos(\Delta\phi) \qquad (28\text{-}2)$$

where the intensities of the waves are I_1 and I_2, and the phase difference between them is $\Delta\phi$.

When the intensities of the two initial waves are equal ($I_1 = I_2$) and the phase difference $\Delta\phi$ is 0 (or any integer multiple of 2π), the total intensity is four times the intensity of either wave alone. When $\Delta\phi$ is π rad (or 3π, 5π, and so on), the resultant intensity I is zero.

The phase difference for complete destructive interference of two waves with the same intensity, frequency, and polarization can be expressed as a fraction of a wavelength, as an angle, or in radians. For example, phase difference for destructive interference:

$$\frac{\lambda}{2} \equiv 180° \equiv \pi \text{ rad} \qquad (28\text{-}3)$$

Thinking about phase differences in terms of distances expressed in wavelengths can be particularly helpful when analyzing situations involving interference. If the wave is travelling in a medium, we must use the wavelength of the electromagnetic wave in that medium.

One way to observe interference is to use a **Michelson interferometer**: split a beam of light, send the two parts along different paths of different lengths (causing a phase difference), and then recombine them. The key elements of a Michelson interferometer are shown in Figure 28-2. This instrument can measure tiny shifts in distance or wave speed. The light beam from the source (usually a laser) is split into two beams by a partially silvered mirror. One beam reflects off the silvered edge of the beam splitter (which is assumed to be farther from the source) and bounces off the movable mirror before passing through the splitter again to the detector. The other beam passes through the splitter, bounces off a fixed mirror back to the splitter, and then reflects from the splitter to the detector. The glass compensator plate makes the total distance that each beam travels in glass equal. If the beam splitter is silvered on the side away from the source, the compensator is placed as shown in Figure 28-2.

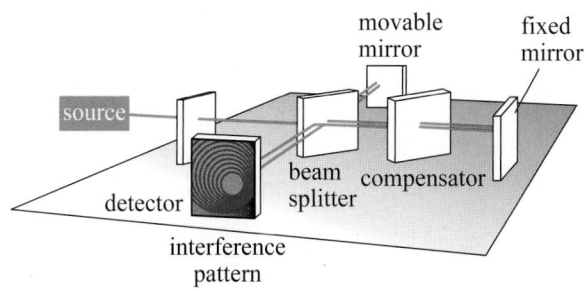

Figure 28-2 In a Michelson interferometer, a light beam is split by a half-silvered mirror, the two parts travel different paths, and then combine to produce an interference pattern.

When a Michelson interferometer is placed in a vacuum, the speed of light in both arms is c. If we start with the two paths equal in length and gradually adjust the position of the movable mirror, the image at the detector goes from bright (zero path difference) to dim (a path difference of one-half wavelength) to bright again (one wavelength difference), to dim again (one and one-half wavelength difference), and so on. Note that the change in the path difference distance is twice the distance d that the mirror moves because the light travels to the movable mirror and then back to the splitter. For example, moving the mirror a quarter wavelength changes the path length by a half wavelength. The condition for an intensity maximum at the detector (constructive interference) is

$$m\lambda = 2d \qquad (28\text{-}4)$$

where m is any integer, and λ is the wavelength of the light.

When the path difference is 0, constructive interference produces a bright disk of light at the detector; a total path difference of 0.5λ wavelength produces complete destructive interference. As we continue to increase the path difference, the centre of the output is bright for integer multiples of the wavelength and dark for $m + \frac{1}{2}$ multiples, but there are also alternating light and dark rings around the centre, as shown in Figure 28-3. These interference patterns result from the fact that the path lengths increase slightly with distance from the centre of the detector.

Figure 28-3 The output of a Michelson interferometer for various different path length differences (labelled as multiples of the wavelength)

The Michelson interferometer is named in honour of its inventor, physicist Albert Abraham Michelson (1852–1931). He worked with chemist Edward Williams Morley (1838–1923) to improve the precision of his original model. In 1887, Michelson and Morley used the improved version shown in Figure 28-4 to carry out a key experiment in the development of modern physics. Their results revealed that the speed of light in a vacuum is the same for all observers, regardless of their relative motion. As you will see in Chapter 29, this puzzling observation became one of the postulates of special relativity.

Figure 28-4 The original Michelson–Morley interferometer

Interferometers continue to play central roles in modern physics. They provide ways to detect tiny changes in distance and to measure indexes of refraction with outstanding precision. An interferometer is a key instrument in the search for evidence of gravitational waves.

ELECTRICITY, MAGNETISM, AND OPTICS

C-28-2 Interferometer

A Michelson interferometer is operating in a vacuum chamber using light from a laser with a wavelength of 632 nm. The movable mirror is adjusted to give a maximum bright output at the detector. How far must the movable mirror be moved to produce the next output maximum?

(a) 1264 nm

(b) 948 nm

(c) 632 nm

(d) 316 nm

C-28-2 (d) The path difference must be increased by one wavelength for the next maximum. Since the change in path length is double the distance that the mirror moves, the mirror needs to move by only 316 nm.

 MAKING CONNECTIONS

Laser Interferometer Gravitational-Wave Observatory

Laser Interferometer Gravitational-Wave Observatory (LIGO) is an experiment designed by the California Institute of Technology and the Massachusetts Institute of Technology with the aim of detecting gravitational waves. Gravitational waves have been called "ripples in spacetime." Some scientists believe that events including the coalescence of neutron stars and black holes produce detectable gravitational radiation. Essentially, LIGO is a perpendicular arm interferometer with incredible precision. The interferometer arms are each approximately 4 km long. When fully operational, the system will be able to detect length changes of approximately one hundred-millionth of the diameter of a hydrogen atom. To help rule out local "noise," the system has LIGO detectors at Livingston, Louisiana, and at Hanford, Washington (Figure 28-5). A real gravitational wave would be detected at both sites, whereas many noise sources that might mimic gravitational waves would appear only at one site or another. The construction began in 1994, first monitoring started in 2001, and the system is expected to be fully operational in 2014.

LIGO Laboratory

Figure 28-5 A portion of the Hanford, Washington, station of the LIGO interferometer

---

28-3 Double-Slit Interference

At the beginning of the nineteenth century, Thomas Young (1773–1829) performed a crucial experiment that showed that light has wave properties. Young had worked previously with sound waves, and in 1799 he proposed that light is also a wave. This theory contradicted the view expressed by many important scientists, who shared Isaac Newton's firm belief that light was composed only of tiny particles. General acceptance of Young's double-slit experiment as evidence for the wave nature of light did not come for some time.

Figure 28-6 shows the basics of Young's experiment. A source of light illuminates two narrow, closely spaced slits, projecting an image on a screen some distance away. If light were particles, one would expect two maxima on the screen corresponding to straight-line motion of the particles through the two narrow slits. However, if light acted as a wave, the contributions from the two slits would interfere with each other. Young observed an alternating series of light and dark bars on the screen, confirming the wave nature of light. Figure 28-7 shows results from a modern re-enactment of the double-slit experiment using laser light.

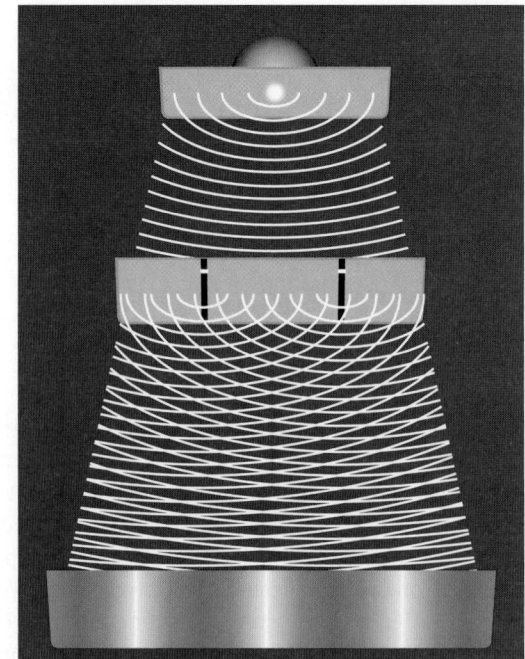

Figure 28-6 Young's double-slit experiment. Light illuminates two narrow slits, producing an interference pattern of alternating bright and dark patches on a distant screen.

To derive a mathematical relationship for the angles at which maxima in light intensity from a double slit occur, we begin with the geometry shown in Figure 28-8. Light passing through slit S_1 travels a shorter distance to Point P than does light passing through slit S_2. We draw a right-angled triangle

Figure 28-7 Output on a screen from an actual double-slit experiment performed using red (wavelength 632 nm) and green (543 nm) laser light with two slits 75 μm apart. The interference pattern is more spread out for the longer wavelength.

to divide the distance from S_2 to P into two parts, as shown. The distance from S_1 to P is approximately equal to the distance from point Q to P as long as the distance D from the slits to the screen is much greater than the slit spacing d. This restriction is called the **Fraunhofer condition** (named for the German optician and physicist Joseph von Fraunhofer (1787–1826)), and is true in most important cases. We can readily show that the upper angle in the right-angled triangle equals the deflection angle θ to point P (measured from a normal to the slit plate). Therefore, the extra distance that the light from S_2 travels to reach point P is $d\sin\theta$.

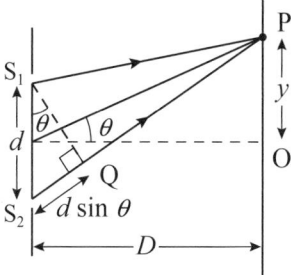

Figure 28-8 Geometry of the double-slit experiment. The distance between the two slits is d, and the distance to the screen is D.

When the extra distance corresponds to exactly a whole number of wavelengths, then constructive interference occurs, and a bright region appears on the screen at angle θ. Thus, constructive interference occurs when

KEY EQUATION $$d\sin\theta = m\lambda \qquad (28\text{-}5)$$

where m is an integer, often called the spectral order.

When $m = 0$, we get an angle of $0°$ for any wavelength. Therefore, when we shine white light at a pair of slits, the beam in the centre of the interference pattern is white.

For all other values of m, the maxima for different wavelengths occur at different angles, with longer wavelengths having larger deflection angles. The direct central maximum is called the zero order, and the spectral orders on one side are normally labelled $m = +1, +2$, and so on, and the spectral orders on the other side are labelled $m = -1, -2$, and so on.

 ONLINE ACTIVITY

The Behaviour of Light

The e-resource that accompanies every new copy of this textbook contains an Online Activity using the PhET simulation "Wave Interference." Work through the simulation and accompanying questions to gain an understanding of the behaviour of light in a double slit experiment.

 CHECKPOINT

C-28-3 Double-Slit Spacing

When analyzing the interference pattern from two slits, you see that the $m = 1$ and $m = 2$ fringes are too close together to measure easily. Which of the following statements correctly describes the ways that the interference pattern spacing could be increased?

(a) Move the slits farther from the screen, reduce the wavelength of the light, or increase the spacing of the slits.

(b) Move the slits farther from the screen, increase the wavelength of the light, or increase the spacing of the slits.

(c) Move the slits farther from the screen, reduce the wavelength of the light, or decrease the spacing of the slits.

(d) Move the slits farther from the screen, increase the wavelength of the light, or decrease the spacing of the slits.

C-28-3 (d) To increase the spacing between the fringes, we could simply keep the angle same and move the screen farther away. Otherwise, we have to increase the angle. Rearranging Equation (28-5), we have $\sin\theta = \dfrac{m\lambda}{d}$, which shows that we get a larger angle when d is less or when λ is more.

In interferometry experiments, we often want to know the displacement distance y and not the angle, (see Figure 28-8). The displacement is readily determined using trigonometry (see Example 28-1 below). Other times, we want to determine the maximum order (m value) that can occur. In that case, we set $\sin\theta$ to 1 (because just under $90°$ is the maximum angle possible), solve for a value of m, then truncate this value down to the nearest integer. For example, if setting $\sin\theta = 1$ gives $m = 3.8$, then 3 spectral orders are visible.

Although we are usually primarily concerned with the bright fringes in an interference pattern, we sometimes want to find the angles for the dark fringes. The condition for destructive interference is simply a path difference equal to an odd number of half wavelengths:

$$d \sin\theta = \left(m + \frac{1}{2}\right)\lambda \qquad (28\text{-}6)$$

EXAMPLE 28-1

Double-Slit Interference Pattern

A laser beam with a wavelength of 532 nm illuminates a double slit and produces an interference pattern on a screen 9.00 cm away. The distance between the $m = 0$ fringe and one of the $m = 1$ fringes is 0.850 cm on the screen. How many bright fringes are on the screen? Assume that the screen extends to a large distance in each direction and that even the faint fringes are visible.

SOLUTION

Here, we have the geometry shown in Figure 28-8 with values of $D = 9.00$ cm, $\lambda = 532$ nm, and $y = 0.850$ cm for $m = 1$. We do not have values for d and θ.

Our strategy will be to first use trigonometry to find the angle θ, and then solve for the slit spacing d. Then we can calculate the maximum order m_{max}.

We need to use consistent units, so we will express all distances in metres:

$$\tan\theta = \frac{8.50 \times 10^{-3} \text{ m}}{9.00 \times 10^{-2} \text{ m}} = 0.0944$$

$$\theta = 5.40°$$

Now we can use Equation (28-5) to solve for the slit spacing d:

$$d = \frac{m\lambda}{\sin\theta} = \frac{1 \times 5.32 \times 10^{-7}}{\sin(5.40°)} = 5.66 \times 10^{-6} \text{ m}$$

Now we find the maximum possible value for m by setting $\theta = 90°$:

$$m = \frac{d \sin\theta}{\lambda} = \frac{5.65 \times 10^{-6} \text{ m} \times \sin(90°)}{5.32 \times 10^{-7} \text{ m}} = 10.6$$

Truncating down to the nearest integer gives $m = 10$. Therefore, there are 10 bright fringes on each side of the central zero-order maximum, making a total of 21 bright fringes.

Making sense of the result:

The angle of approximately 5° that we found for $m = 1$ seems reasonable for the given dimensions. Since the bright fringes become farther apart with each order, it seems reasonable that there are 10 fringes in the 90° span on each side of the zero-order maximum.

The analysis above holds only if the slits themselves are narrow compared to the distance separating them. We will see later in the chapter that there are diffraction effects for light from a single slit that need to be taken into account for precision work. In addition, we have only considered the angles corresponding to the maxima (or minima) and the intensities of the various fringes. Generally, the higher the order, the dimmer each successive maxima. Finally, we mention again that our derivation depends on the Fraunhofer condition, $D \gg d$.

LO 4

28-4 Diffraction Gratings

You saw in Chapter 27 that a glass prism can be used to split light into different colours in order by wavelength. Different wavelengths of light have slightly different speeds in glass (and, therefore, different indexes of refraction), resulting in varying angles of refraction at the prism's air–glass interfaces. However, most scientific instruments that divide light into constituent wavelengths use **diffraction gratings**, which are plates with typically a thousand or more evenly spaced slits close together.

The condition for constructive interference with many slits with constant spacing d between adjacent slits is exactly the same as for one pair of slits. This is because if one pair of slits meets the condition for constructive interference, then all pairs will. For example, if distance b–c for the diffraction grating in Figure 28-9 is one wavelength, then distance d–e is two wavelengths, and distance f–g is three wavelengths. Consequently, the angles for constructive interference for light of wavelength λ passing through the diffraction grating are given by

$$d \sin\theta = m\lambda \qquad (28\text{-}7)$$

where m is the spectral order.

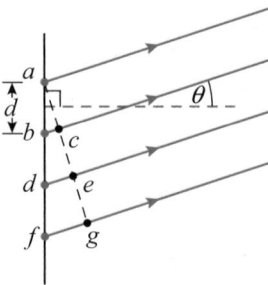

Figure 28-9 Geometry of the diffraction grating. The distance between each pair of slits is d, so the relative distance to the screen is one wavelength longer (for the correct angle) for light from each subsequent slit.

For calculations involving first-order angles, it is often possible to use the small-angle approximations: when an angle is small, then the sine, the tangent, and the angle itself (expressed in radians) are all approximately equal:

$$\theta \approx \sin\theta \approx \tan\theta \qquad (28\text{-}8)$$

Why bother with a diffraction grating when the angles are exactly the same as with just two slits? There are two reasons. First, with many slits, more light passes through the plate, and the interference pattern is much brighter. Second, with more slits the diffraction grating better resolves different wavelengths. Figure 28-10 illustrates how the constructive interference peaks for monochromatic light become markedly sharper as the number of slits increases. Thus, diffraction gratings are able to produce separate peaks for wavelengths of light that are only slightly different. Figure 28-11 shows a high-resolution spectrum of light from the Sun, made with a series of diffraction gratings with different slit spacings.

Figure 28-11 Spectrum of visible light from the Sun, compiled from a series of detailed spectra of regions of the visible spectrum. The dark lines represent absorption by different atoms and molecules in the outer regions of the Sun.

Two-slit Diffraction Pattern

Four-slit Diffraction Pattern

Eight-slit Diffraction Pattern

Figure 28-10 Interference patterns for monochromatic light passing through gratings with 2, 4, and 8 slits

Usually, diffraction grating specifications are given in terms of the number of lines per millimetre, N, rather than the distance, d, between adjacent slits. Clearly,

$$d = \frac{1}{N} \qquad (28\text{-}9)$$

For example, if there are 200 lines/mm, then $d = 1/200$ of a millimetre, or 5×10^{-6} m.

Throughout this section we have assumed a narrow, parallel beam of light. If the light source is not a laser, a slit should be placed between the source and the diffraction grating so that the incident light that reaches the grating is essentially a narrow, parallel beam.

The diffraction gratings we have described so far are all **transmission gratings**, which have evenly spaced slits. There are also **reflection gratings**, which have evenly spaced lines that reflect light, producing interference patterns similar to those from transmission gratings.

☑ **CHECKPOINT**

C-28-4 Two Diffraction Gratings

You have two diffraction gratings. Grating A is known to have 100 lines/mm, but the specifications sheet has been lost for grating B. The first-order fringe produced by monochromatic light shone on the two gratings appears at an angle of 4° for grating A and an angle of 8° for grating B. What is the spacing of the slits in grating B?

(a) 25 lines/mm

(b) 50 lines/mm

(c) 200 lines/mm

(d) 400 lines/mm

C-28-4 (c) Rearranging the diffraction grating equation to $d = \frac{m\lambda}{\sin\theta}$, we see that a larger angle corresponds to a smaller slit spacing d. Evaluating the sines (or using the small-angle approximation) shows that the spacing d for grating B is half that of grating A. Therefore, grating B has twice as many lines per millimetre, which is 200 lines/mm.

EXAMPLE 28-2

Design a Spectrometer

You want to design a spectrometer to cover the visible spectrum from violet (400 nm) to red (700 nm) using first-order images from a diffraction grating with 400 lines/mm. The spacing from the grating to the electronic image sensor used to record the spectrum is 6.00 cm. What is the minimum dimension for the image sensor?

SOLUTION

The design of the spectrometer is shown in Figure 28-12. The sensor needs to cover the first-order image on only one side. From the relationship $d \sin\theta = m\lambda$, we know that the longer wavelength corresponds to a larger deflection angle. Therefore, B corresponds to the red light, and A corresponds to the violet light. We will find the angles for these limiting wavelengths and use the angles and the distance to the screen to find the two displacements y_A and y_B. The sensor must cover the ranges of these two displacements.

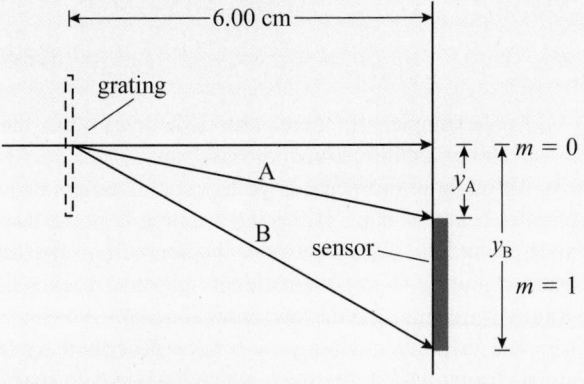

Figure 28-12   Example 28-2

First, we use the number of lines per millimetre to find the slit spacing d: $d = \dfrac{1}{400}$ mm $= 2.50 \times 10^{-6}$ m

Next, we solve for the angle to the A line in the diagram (400 nm wavelength):

$$\sin\theta_A = \frac{m\lambda}{d} = \frac{1 \times 4.00 \times 10^{-7}\,\text{m}}{2.50 \times 10^{-6}\,\text{m}} = 0.160$$

$$\theta_A = 9.21°$$

We use the same method to find the angle to the B line (700 nm wavelength):

$$\sin\theta_B = \frac{m\lambda}{d} = \frac{1 \times 7.00 \times 10^{-7}\,\text{m}}{2.50 \times 10^{-6}\,\text{m}} = 0.280$$

$$\theta_B = 16.26°$$

We use trigonometry to solve for the distances y_A and y_B:

$$y_A = 6.00\,\text{cm} \times \tan(9.21°) = 0.973\,\text{cm}$$

$$y_B = 6.00\,\text{cm} \times \tan(16.26°) = 1.750\,\text{cm}$$

Therefore, the minimum length of the light sensor must be

$$1.750 - 0.973\,\text{cm} = 0.777\,\text{cm}$$

Making sense of the result:

The detector size of just under 1 cm seems reasonable. Most commercial spectrometers are designed not to include the zero-order maximum because it does not provide any spectral information. If we had wanted to detect the zero-order maximum as well as one set of first-order images, the detector would have to be twice as wide. We should also comment on significant digits here. When you are near the boundary above 9, the simplified significant digits rule breaks down. Clearly, when a number such as 9.98 increases slightly to 10.02, it should not lose precision. Here, we have kept one additional significant figure in the intermediate steps for the angle just over 16° for this reason.

MAKING CONNECTIONS

Quantum Control and Ultrafast Physics

One of the top laboratories in the world for studying very fast physics is the Joint Attosecond Science Laboratory (JASLab), a laser facility shared by the National Research Council of Canada and the University of Ottawa. In 2010, researchers at JASLab were able to study how a bromine molecule breaks apart, including how the electron clouds shifted during the breakup. Researchers use reflection gratings with a masking system to shape electromagnetic pulses that apply the optimum electromagnetic forces to break apart the molecules. The mask lets the researcher control the amounts of different wavelengths in the pulse, somewhat like the way an audio equalizer controls the various frequencies in the input to an amplifier (see Molecular Physics online activity).

Dr. Paul Corkum of JASLab has won numerous honours and awards and is considered one of the pioneers of attosecond physics.

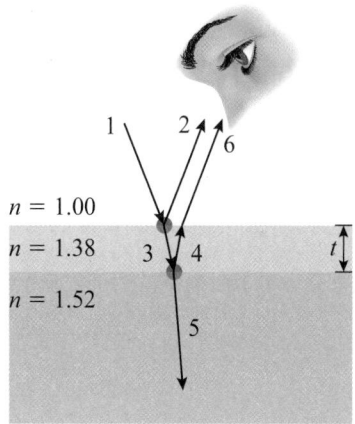

Figure 28-13 Lasers can be used in high-speed molecular observation and control techniques.

ONLINE ACTIVITY

Molecular Physics

The e-resource that accompanies every new copy of this textbook contains an Online Activity using the PhET simulation "Quantum Control." Work through the simulation and accompanying questions to gain an understanding of how electromagnetic pulses are used to research molecular physics.

LO 5

28-5 Thin Film Interference

In his book on optics, Isaac Newton wrote, "It has been observed by others that transparent substances such as glass, water, air, etc. when made very thin by being blown into bubbles, or otherwise formed into plates, do exhibit various colors according to their various thinness, altho' at a greater thickness they appear very clear and colourless." In this section, we consider **thin film interference**, which produces the colours of soap bubbles, oil films, peacock feathers, and some types of butterflies. Thin films are also important in antireflection coatings in quality optics and in energy-saving technologies for windows.

Constructive or destructive interference can occur at the surface of a thin film, depending on the thickness of the film and any phase changes that occur during reflection at the surfaces of the film.

Let us review what happens when waves encounter a boundary between media with different optical properties. As we have seen with mechanical waves, some of the incident wave is reflected from the boundary surface, and some is transmitted through it. Light will reflect, at least partially, from the boundary surface when there is a difference in the index of refraction (which corresponds to the speed of light in the medium).

Figure 28-14 shows a thin film designed to reduce reflection from a lens. Here, we have air ($n = 1.00$) and glass (assume $n = 1.52$) separated by a thin coating of uniform thickness, t. For a variety of practical reasons, magnesium fluoride (MgF_2) is one of the most common materials for antireflection coatings on optical lenses and glasses. Magnesium fluoride has an index of refraction of 1.38.

Figure 28-14 Reflection and refraction of light at the boundary surfaces of an antireflection coating. The small circles indicate a hard reflection (phase change).

When a light ray (1) strikes the air–coating boundary, it is partly reflected (2) and partly transmitted (3). The transmitted light is refracted because the index of refraction of the coating is greater than the index of refraction of air. When the refracted light (3) encounters the coating–glass boundary, the light is again partly reflected (4) and partly transmitted with further refraction (5). When (4) strikes the coating–air boundary, part will be transmitted back into the air (6). The net reflection depends on the interaction between (2) and (6). If they are in phase, constructive interference will make the net reflected ray relatively bright. If (2) and (6) are out of phase, destructive interference will result and there will be essentially no net reflection. The function of an antireflection coating is to use destructive interference to minimize the amount of light reflected from the lens.

The relative phase of (2) and (6) depends, in part, on the extra distance travelled by the light (6). If for now we assume that the initial ray (1) is near vertical, then the distance travelled by light after leaving the coating is essentially equal and constant, and the extra distance travelled by (6) is approximately $2t$, twice the thickness of the coating.

Note that the extra distance is travelled in the coating, not in the air. When an electromagnetic wave is in a medium, the wavelength is shorter than it would be in a vacuum. The frequency does not change when the wave enters a different medium, but the speed decreases compared to the speed in a vacuum; therefore, the wavelength decreases correspondingly. The wavelength λ_n of the wave in a medium with an index of refraction n is given by

$$\lambda_n = \frac{\lambda}{n} \qquad (28\text{-}10)$$

Now we must take into account the phase changes that can occur when a wave reflects. Reflected light will experience a 180° phase change (a **hard reflection**) when it reflects from a medium with a greater index of refraction, and no **phase change** (a soft reflection) when the reflection is from a medium with a smaller index of refraction. For the coating we are considering, there is a phase change for incident light reflected from both the outer and inner surfaces.

Combining these ideas, we find that a condition of destructive interference (minimal reflection) for near vertical incident light on the thin coating on glass is

$$2t = \frac{\lambda_n}{2} \text{ (destructive interference)} \qquad (28\text{-}11)$$

Equation (28-11) gives the thinnest film that will produce destructive interference. We will also have destructive interference when the difference in path length is $3\lambda_n/2$, $5\lambda_n/2$, and so on. The general relationship is

$$t = \frac{(2m + 1)\lambda_n}{4} \text{ (destructive interference)} \qquad (28\text{-}12)$$

where m is an integer ≥ 0.

Note that this relationship holds only for media that cause a phase shift at each of the two reflections. Also, a different condition is required for constructive interference, as demonstrated in the next example.

EXAMPLE 28-3

Soap Bubble

When a narrow beam of white light shines almost perpendicular to the surface of a soap bubble, the bubble appears blue (480 nm in air). The index of refraction of the soapy water is approximately 1.33. What are three possible thicknesses for the wall of the soap bubble?

SOLUTION

The situation is pictured in Figure 28-15. Here, we have a 180° phase change at the first reflection but not at the second. To have *constructive* interference (as required for the blue appearance of the outer surface), the extra distance travelled by the light that reflects from the inner surface must correspond to another 180° phase change. Therefore,

$n = 1.33$

$n = 1.00 \qquad n = 1.00$

t

soap
film

Figure 28-15 Example 28-3. Reflection and refraction at the surfaces of a soap bubble. The small circle indicates a hard reflection.

$$2t = \frac{\lambda_n}{2}$$

We need to find the wavelength of the blue light in the soapy water:

$$\lambda_n = \frac{\lambda}{n} = \frac{480 \text{ nm}}{1.33} = 361 \text{ nm}$$

Therefore, the thinnest possible wall for a blue colour is

$$t = \frac{\lambda_n}{4} = \frac{361 \text{ nm}}{4} = 90.2 \text{ nm}$$

The next two possible conditions for constructive interference of blue light are

$$2t = \frac{3\lambda_n}{2}; \text{ therefore } t = 271 \text{ nm}$$

$$\text{and} \quad 2t = \frac{5\lambda_n}{2}; \text{ therefore } t = 451 \text{ nm}$$

Making sense of the result:

As expected, the wavelength of the light in the soapy water is less than the wavelength in air. We need an extra 180° phase shift added to the phase shift of the reflected wave to produce constructive interference, so we need wall thicknesses of $\lambda_n/4$, $3\lambda_n/4$, $5\lambda_n/4$, and so on. We have calculated wall thicknesses for the maximum reflection of blue light. However, the incident light is white, so there will also be partial reflection of other colours.

Figure 28-16 shows a soap film exposed to white light. You can see that the film of soapy water has a varying thickness; we would expect the higher parts of the film to be thinner due to the effect of gravity. The variations in thickness result in different colours being dominant in constructive interference in various regions of the soap film.

Thin films are critical in modern science and technology. (See the Making Connections feature on applications to energy-efficient windows, called "Smart Windows.")

Figure 28-16 The varying thicknesses of the soap bubble result in constructive interference for different colours in different regions.

 CHECKPOINT

C-28-5 Antireflection Coatings

Laser light with a wavelength (in a vacuum) of 632 nm is incident approximately normal to the surface of a thin coating ($n = 1.5$) on plastic ($n = 1.3$). Which thickness results in the minimum amount of reflected laser light?

(a) 105 nm
(b) 122 nm
(c) 211 nm
(d) 243 nm

C-28-5 (c) There is a hard reflection at the first surface, but not at the second. We want no reflection, so there must be a total phase difference of an odd number of half wavelengths. Since the hard reflection is equivalent to one half wavelength, the extra distance must be a whole number of wavelengths: $2t = \lambda_n = \frac{\lambda}{n} = \frac{632 \text{ nm}}{1.5}$. Therefore, $t = 211$ nm. Note that zero thickness would also give minimum reflection.

We have not formally done the case where the incident light is not normal to the surface, but exactly the same principles apply. The only difference is that you need to take into account that the difference travelled is not the thickness (in each direction) but rather $\frac{t}{\cos\theta}$, where θ is the angle of the light relative to a perpendicular to the surface.

 LO 6

28-6 Single-Slit Diffraction

Two related but distinct phenomena—interference and diffraction—form the core of physical optics. Both depend on the wave nature of light. Interference allows us to understand how waves that travel different paths in the same medium interact. When analyzing diffraction, we consider each point on a wave front to be the source of new waves, which then allows us to analyze the interference between those waves. Diffraction gives us a model to understand how light spreads out and bends around a small object. Diffraction is also crucial in understanding the resolution limit (Section 28-8) imposed by the wave nature of light.

Young's double-slit experiment was a crucial proof of the wave nature of light, but another experiment, this one by Dominique Arago involving diffraction, played a critical historical role. To understand diffraction, we will use **Huygens' principle** (sometimes called the Huygens–Fresnel principle), which was described in Section 15-2. According to Huygens' principle, each point along a wave front acts as a new source of waves that spread out spherically from that point. The wave front at a later time is defined by the unique tangent to the spherical waves at that time.

In 1818, Augustine Fresnel presented a paper on diffraction as part of a scientific competition on the nature of light. One of the judges, Siméon Poisson, a critic of the wave view of light, noted that Fresnel's theory predicts that the image produced by a circular disk that is illuminated by a light source will have a bright spot in the middle, right where intuition suggests that the shadow should be darkest. Huygens' waves from all around the edges of the disk would travel an equal distance to the centre of the shadow, so constructive interference would produce a bright spot at that point. Poisson argued that this apparently absurd prediction disproved Fresnel's wave theory of light.

The head judge in the competition, Dominique François Jean Arago (1786–1853), decided to try the experiment, and it produced a bright spot in the middle of the shadow exactly as predicted by Fresnel's wave theory. Ironically, this phenomenon of a bright spot in the centre of a shadow cast by a circular object has become known as a Poisson spot, even though Poisson's goal was to discount, not to prove, the wave nature of light. Two other scientists had discovered the phenomenon of the spot almost a century earlier, but their work had been forgotten. A modern version of this experiment is shown in Figure 28-18.

Courtesy of Robert Vanderbei

Figure 28-18 A laser beam and a telescope with a central obstruction is used to demonstrate the Poisson spot. The Poisson spot in the centre is clearly visible in the expanded view of the image, along with an interference pattern of alternating light and dark rings.

Consider a single slit illuminated by a source of parallel monochromatic light, which is also coherent, at least over a short distance. At first, you might think that a single slit would not produce any interference effects, but when we apply Huygens' principle to the area of the slit, we have a whole series of point sources of waves that can interfere with each other. Figure 28-19 shows a row of such sources spanning the width, w, of a slit. We can pair each source in the upper half of the slit with a corresponding source in the lower half. For sources shown in Figure 28-19, we would pair the first source (counting from the top) with the fifth source, the second with the sixth, the third with the seventh, and so on. For a distant screen, the path difference for each pair of sources is approximately $\frac{w}{2}\sin\theta$.

If this path difference equals an odd number of half wavelengths, destructive interference will produce a dark region on the screen:

$$\frac{w}{2}\sin\theta = \frac{2m+1}{2}\lambda \quad \text{or} \quad w\sin\theta = (2m+1)\lambda \quad (28\text{-}13)$$

where m is an integer ≥ 0.

Thus, destructive interference will occur when $w\sin\theta = \lambda, 3\lambda, 5\lambda, 7\lambda \ldots$

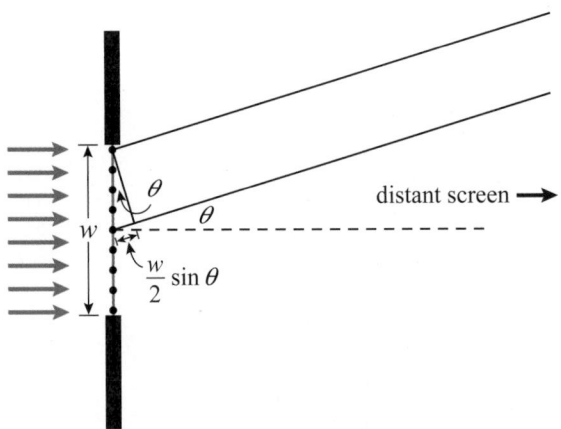

Figure 28-19 Huygens' secondary sources (marked as circles) in a row across the width of a single slit

However, we could divide the secondary sources into four regions instead if two, and pair each source in the first (uppermost) region with a corresponding source in the next region, and similarly pair each source in the third region with a corresponding source in the fourth region. Now the condition for a half-wavelength path difference is

$$\frac{w}{4}\sin\theta = \frac{2m+1}{2}\lambda \quad \text{or} \quad w\sin\theta = 2(2m+1)\lambda \quad (28\text{-}14)$$

This will result in destructive interference when $w\sin\theta = 2\lambda, 6\lambda, 10\lambda, \text{K.}$

We can apply the same reasoning when we divide the secondary sources into any even number of groups. Combining all these conditions for destructive interference gives a simple relationship for the destructive interference cases:

KEY EQUATION
$$w\sin\theta = m\lambda \quad (28\text{-}15)$$

where m is an integer > 0.

This relationship is almost the same as the relationship for *constructive* interference for a double slit, where the width w of the single slit replaces the spacing d between the double slits. However, it is important to remember for the single slit that the equation gives the angle for dark fringes (destructive interference), and the similar equation for the double slit gives the angles for bright fringes (constructive interference). Also, the double-slit equation can be applied with $m = 0$, but the single-slit equation cannot.

Figure 28-20 shows the interference pattern on a screen one meter away for a single slit 0.40 mm wide illuminated with 633 nm (red) laser light. The pattern has a bright central band, becomes dark at a distance of about 1.6 mm on each side, and then alternates light fringes (with decreasing intensity) and dark fringes as the distance from the centre increases. Note that most of the intensity is contained within the **central maximum**, which we define as the region between the first minimum on either side.

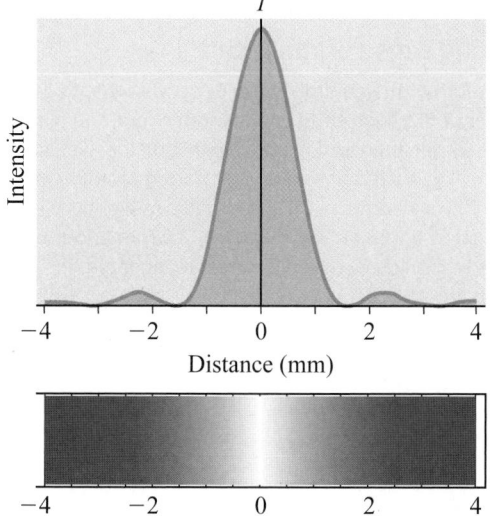

Figure 28-20 The interference pattern and relative intensity plot for coherent light with a wavelength of 633 nm falling on a single slit of width 0.40 mm. The plot gives distance in mm on a screen that is one meter away from the slit.

What will happen if we make the single slit wider? Figure 28-21 shows the results of using the same light source (633 nm) with a slit 0.80 mm wide. Now the

Figure 28-21 The interference pattern for a situation like Figure 28-20 but for a single slit of width 0.80 mm

central band extends only to approximately 0.8 mm on each side. Making the slit wider has made the image on the screen narrower.

From Equation (28-15), it is clear why the size of the central maximum of the interference pattern for a given wavelength decreases when we make the slit width w larger. Equation (28-15) also indicates that a smaller slit width will produce larger angles for the same wavelength, resulting in a broader pattern, as shown in Figure 28-22. Note that in this diffraction pattern the maxima are not located halfway between the minima.

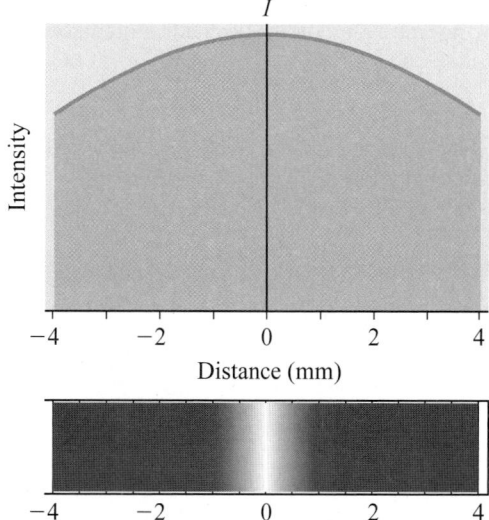

Figure 28-22 The interference pattern and relative intensity plot for a narrower single slit of width 0.05 mm

✓ **CHECKPOINT**

C-28-6 Single-slit Diffraction with Blue and Red Filters

White light passes through a blue filter before reaching a narrow single slit. Which statement correctly describes the positions of the first bright band away from the centre when a red filter replaces the blue filter?

(a) The position is exactly the same when either filter is used.

(b) The bright band is slightly closer to the centre when the blue filter is used.

(c) The bright band is slightly farther from the centre when the blue filter is used.

(d) There is only the central bright band because you need white light to form a complete interference pattern.

C-28-6 (b) Red light has a longer wavelength than blue light (and passing the white light through a coloured filter leaves light predominantly of that colour). When the wavelength is longer, the angle θ for both dark and light fringes is greater. Therefore, we expect the positions of the next bright fringe to be more distant for red and closer for blue.

ELECTRICITY, MAGNETISM, AND OPTICS

EXAMPLE 28-4

Single-Slit Image Width

A single slit of width 0.750 mm is illuminated by red laser light with a wavelength of 656 nm. Find the width of the central (bright) band on a screen 1.00 m from the slit.

SOLUTION

The width of the central band is usually defined as the distance from the first dark region on one side of the central band to the first dark region on the other side. Therefore, we will find the angle of the first dark region on either side and then use trigonometry to calculate the distance on the screen.

The first angle at which destructive interference occurs has $m = 1$, so

$$w \sin\theta = \lambda$$

We need to use consistent units, so we will express all distances in metres:

$$\sin\theta = \frac{\lambda}{w} = \frac{6.56 \times 10^{-7}\ \text{m}}{7.50 \times 10^{-4}\ \text{m}} = 8.75 \times 10^{-4}$$

$$\theta = 0.0501°$$

The distance from the centre to this point is

$$y = 1.00\ \text{m} \times \tan(0.0501°) = 8.74 \times 10^{-4}\ \text{m}$$

The width of the central band is double this distance: 1.75×10^{-3} m, or 1.75 mm.

Making sense of the result:

The width of the central image depends on both the wavelength of the light and the slit width. Here, the slit is three times wider than in Figure 28-20, but the wavelength is longer. The wider slit makes the pattern narrower, and the longer wavelength makes the pattern slightly wider. Our calculated angle of approximately 0.05° is about twice the angle in the figure, which appears reasonable.

LO 7

28-7 Actual Intensity Pattern for Double Slits

The equations we derived in Section 28-3 do not, in fact, fully describe double-slit interference. When we want the precise intensity pattern, we must consider both the double-slit interference and the diffraction effects due to the finite width of the slits. The exact mathematics of this combination can be daunting, but we can use a simple approximation that is reasonably accurate.

Let us first briefly compare what we have found for single-slit and double-slit interference patterns (Figures 28-23 and 28-24):

single slit: for destructive interference, $w \sin\theta = m\lambda$

double slit: for constructive interference, $d \sin\theta = m\lambda$

Here, λ is the wavelength of the light, θ is the angle from the centre of the pattern, m is the spectral order, w is the slit width, and d is the spacing between the slits.

Both patterns have most of the intensity in the central bright band, and the intensity of the other bright fringes decreases farther from the central point.

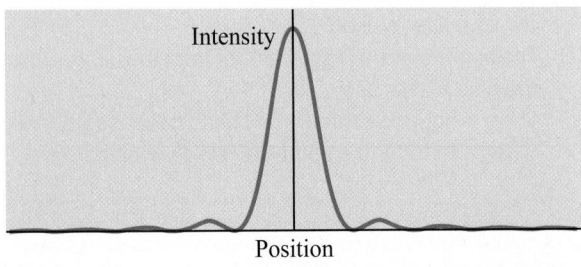

Figure 28-23 A simplified intensity plot for interference from a single slit

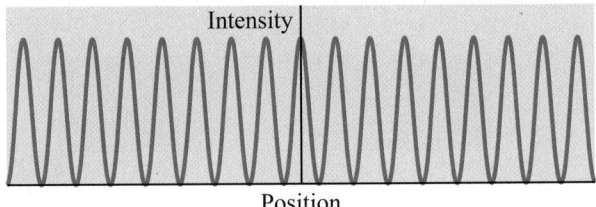

Figure 28-24 A simplified intensity plot for interference from an idealized double slit

We can approximate the actual interference pattern for a double slit by multiplying the intensity pattern for the idealized double slit by the intensity pattern for single-slit diffraction, giving the intensity pattern shown in Figure 28-25. Compare this prediction with the actual patterns for two slits in Figure 28-7. Note in Figure 28-25 that some of the double-slit orders are in essence removed by corresponding to a minimum of the single-slit pattern.

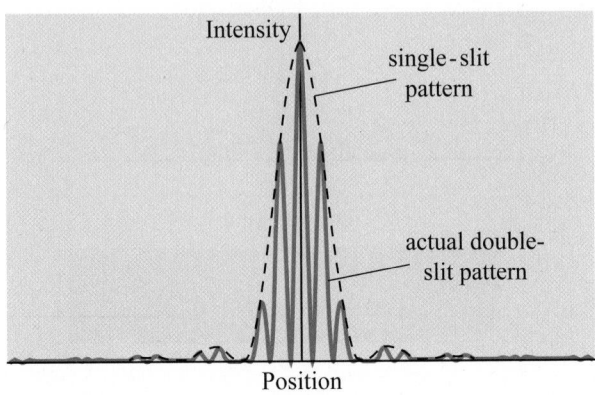

Figure 28-25 The approximate intensity plot for a double slit with diffraction from the individual slits factored in

C-28-7 A Real Double Slit with Different Slit Widths

When you shine monochromatic light through two slits, you are able to distinguish the central bright fringe and three more light fringes on each side. You make the slits narrower and increase the intensity of the incident light to keep the amount of light that passes through the slits constant. What will you observe?

(a) the same number of light fringes in the same positions

(b) the same number of light fringes, but closer to the centre

(c) fewer light fringes

(d) more light fringes

C-28-7 (d) We multiply the single-slit pattern by the theoretical double-slit interference pattern. If we make the slits narrower, the single-slit image becomes broader, and we are able to see a higher order (i.e., more) of the double-slit images. The narrowing of the slits could make the higher-order fringes too faint to see if we had not increased the incident intensity.

 EXAMPLE 28-5

How Many Fringes?

A laser emitting green coherent light with a wavelength of 532 nm illuminates two slits that are 0.150 mm apart. The width of each slit is 0.0200 mm. How many fringes are visible on a screen? Assume that only the fringes within the central maximum are visible.

SOLUTION

We need to determine how many fringes are within the single-slit envelope at the centre of the interference pattern, which has a shape like that shown Figure 28-25. We will write an expression for the angle corresponding to the first zero for the single-slit pattern, set that expression equal to the expression for double-slit interference, and solve for the order m.

The $m = 1$ zero for single-slit diffraction occurs when $\sin\theta = \dfrac{1\lambda}{w}$.

For double slit-interference we have $\sin\theta = \dfrac{m\lambda}{d}$.

Equating these two expressions and solving for m, we have

$$m = \frac{d}{w} = \frac{1.50 \times 10^{-4}\,\text{m}}{2.00 \times 10^{-5}\,\text{m}} = 7.5$$

Thus, double-slit fringes up to and including the 7th order are visible, but the 8th and higher fringes are outside the single-slit envelope; they are assumed to be too faint to be seen. Since the double-slit pattern has a bright zero-order central band, a total of 15 fringes are visible (-7, -6, -5, -4, -3, -2, -1, 0, 1, 2, 3, 4, 5, 6, and 7).

Making sense of the result:

Although 15 light fringes are visible, they are not equally bright: the higher-order fringes are much fainter, as we would expect from shape of the combined pattern in Figure 28-25.

 LO 8

28-8 Resolution Limit

We have seen that the wave nature of light can produce diffraction and interference. These effects can mask the true nature of objects. We call the smallest dimensions that we can see the **resolution limit**.

For example, suppose you are on the Prairies on a clear night and see an automobile far away. The automobile has two headlights, but if it is too far away, you will not be able to clearly resolve them as two separate sources of light; instead you see just see one bright light. In this situation, the distance separating the two headlights is beyond the resolution limit. Figure 28-26 shows two adjacent light sources photographed such that they are not resolved (left), marginally resolved (centre), and resolved (right).

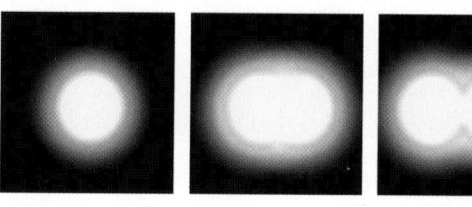

Figure 28-26 Improving the resolution reveals that the bright spot in the left photograph is actually two separate sources. OpenStax College. (2012, July 19). Limits of Resolution: The Rayleigh Criterion. Retrieved from the Connexions Web site: http://cnx.org/content/m42517/1.4/

The resolution limit depends on the wavelength of the light and the diameter of the light-gathering area of the detection instrument. We can see the origin of this relationship by considering the effect of a single slit: the narrower the slit relative to the wavelength, the wider the central image. Therefore, the smaller the aperture (opening) of the observing instrument, the more spread out the image, making two adjacent point objects more difficult, or even impossible, to resolve.

 PEER TO PEER

When solving double- and single-slit problems, it is easy to forget to make the units consistent. Usually, widths and spacing of slits are given in millimetres and wavelengths in nanometres, but for calculations, they have to be in the same units. Similarly, in resolution problems, the diameter and the wavelength need to be in the same units.

ELECTRICITY, MAGNETISM, AND OPTICS

Figure 28-27 shows the diffraction pattern for coherent light with a wavelength of 532 nm passing through an aperture with a diameter of 0.1 mm. The central image has a radius of approximately 0.19°, surrounded by alternating light and dark rings, called an **Airy pattern** (named for the English astronomer George Biddell Airy, 1801–92). If we make the aperture much larger, the pattern has essentially the same form, but there is much better resolution. With an aperture diameter of 15 mm, the central image is only approximately 0.0012° across.

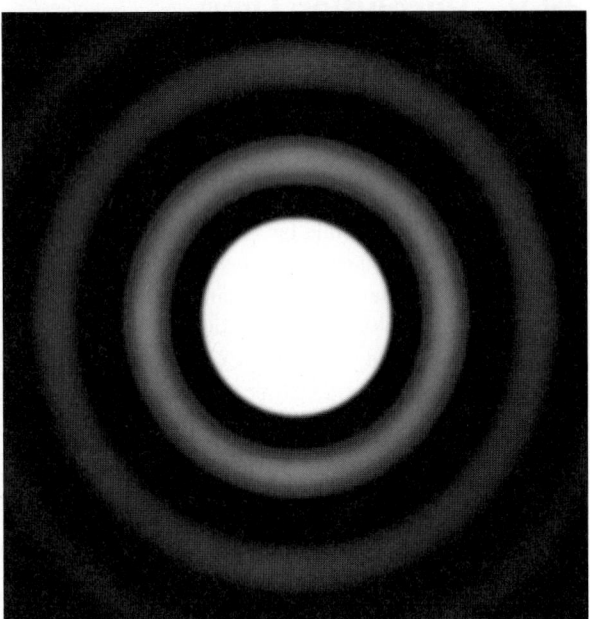

Figure 28-27 With a small aperture, the effects of diffraction are apparent: the image is broad and surrounded by an Airy pattern of light and dark rings.

The Rayleigh criterion is widely used to decide whether two sources are too close to be clearly distinguished. Developed by the English physicist Baron Rayleigh (John William Strutt, 1842–1919), the **Rayleigh criterion** states that two objects can be resolved if they are separated by an angle at least large enough for the central maximum of one object to correspond to the first minimum in the diffraction pattern of the other.

If we apply the equation for single-slit diffraction to the Rayleigh criterion, we have $m = 1$ because the criterion involves the first minimum of the diffraction pattern. It follows that for light with a wavelength of λ

and an optical instrument whose aperture has an effective diameter D, the minimum resolved angle, θ_r, is given by

$$\sin\theta_r = \frac{\lambda}{D} \qquad (28\text{-}16)$$

When the angular separation of two objects is less than θ_r, they cannot be resolved. However, most apertures in optical instruments are circular. The mathematics of the fringes around images formed from small circular apertures (Figure 28-27) is somewhat more complex. The Rayleigh criterion for circular apertures can be shown to be

 $$\sin\theta_r = 1.22\frac{\lambda}{D} \qquad (28\text{-}17)$$

When applying this criterion, we must ensure that the wavelength and diameter are expressed in the same units. Optical telescopes with a large diameter collect more light, which helps us detect faint objects; the large diameter also improves resolution. Equation (28-17) also applies to the resolution limit for the human eye, which depends on the diameter of the pupil.

The resolution criterion can be applied to any wavelength, not just to the visible region of the electromagnetic spectrum. We now see that radio telescopes (like the one shown at the beginning of the chapter) must have large diameters to have even modest angular resolution because radio wavelengths are relatively long. While radio telescopes must be large in diameter, because the wavelength is so large the requirements on surface smoothness is much less stringent than for optical telescopes.

☑ CHECKPOINT

C-28-8 Binocular Resolution

For binoculars, the specification 10×50 means a magnification of $10\times$ and an objective lens diameter of 50 mm. (The objective lens is the lens that is closest to the object being viewed. In the case of binoculars it is the main lens.) Which of the following specifications has the best resolution?

(a) 7×35
(b) 8×40
(c) 12×40
(d) 10×50

C-28-8 (d) Resolution depends on the effective diameter of the aperture, not the magnification.

EXAMPLE 28-6

Resolving Binary Stars

A pair of stars in a binary system is separated by 0.200 light years. The system is 7800 light years from us, and their orbits are oriented perpendicular to our line of sight.

(a) What minimum aperture diameter for your telescope do you need to resolve them? Assume light from the stars has a wavelength of 525 nm.

(b) If the stars are bright enough, can they be resolved by an unaided human eye? Explain.

SOLUTION

(a) We first need to find the angle subtended by the two stars as seen from Earth. Then we will use the Rayleigh criterion with this angular separation to calculate the necessary diameter:

$$\tan\theta = \frac{0.200 \text{ ly}}{7800 \text{ ly}} = 2.56 \times 10^{-5}$$

$$\theta = \tan^{-1}(2.56 \times 10^{-5}) = 1.47 \times 10^{-3\circ}$$

Now we use the Rayleigh criterion and solve for the diameter D:

$$D = 1.22\,\frac{\lambda}{\sin\theta_r} = 1.22\,\frac{5.25 \times 10^{-7} \text{ m}}{\sin(1.47 \times 10^{-3\circ})} = 0.025 \text{ m}$$

A telescope with an aperture diameter of 2.5 cm is the minimum size needed.

(b) The diameter of the pupil of the human eye, even under dark-adjusted conditions, is only approximately 5 mm, much less than the required aperture diameter. Therefore, the binary stars cannot be resolved by an unaided human eye.

Making sense of the result:

It makes sense that an optical instrument with an aperture diameter of only 2.5 cm can resolve closely spaced objects better than the human eye can because the aperture diameter of the eye is only approximately one-fifth as large. The Rayleigh criterion only gives us the resolution imposed by the wave nature of light. Other effects, such as imperfections in the optics and motion of air with varying temperatures and humidity, can increase the resolution limit.

According to Equation (28-17), there are two ways to improve the resolution for circular apertures: increase the effective diameter of the optical instrument and/or use light with a shorter wavelength. For example, other things being equal, a system using ultraviolet light will resolve features better than a system that uses infrared or visible light.

Electron Microscopes

In 1938, two students at the University of Toronto, James Hillier and Albert Prebus, completed the first workable electron microscope. Their instrument achieved magnifications of 7000×, several times greater than the best optical microscopes. Hillier and Prebus based their design on the work of the German physicist Ernst August Friedrich Ruska (1906–1988), who showed that electrons can be used in a microscope instead of light to greatly increase the resolution of the images. As you will see in later chapters, electrons have both wave and particle properties, including an effective wavelength that depends on the momentum of the electrons. Electron wavelengths are tiny compared to even ultraviolet light, so the most modern electron microscopes can have resolutions less than 50 pm. Figure 28-28 is an electron microscope image of part of a butterfly wing.

The wave nature of electrons means that a small aperture for an electron beam will produce an interference pattern very similar to that seen in Figure 28-27 for light. Figure 28-29 shows the diffraction pattern of a beam of electrons passing through a graphite block.

Figure 28-28 An electron microscope image of part of a butterfly wing reveals a pattern in the wing's structure. This pattern acts like a diffraction grating, and the colour of the wing results primarily from interference effects rather than pigments.

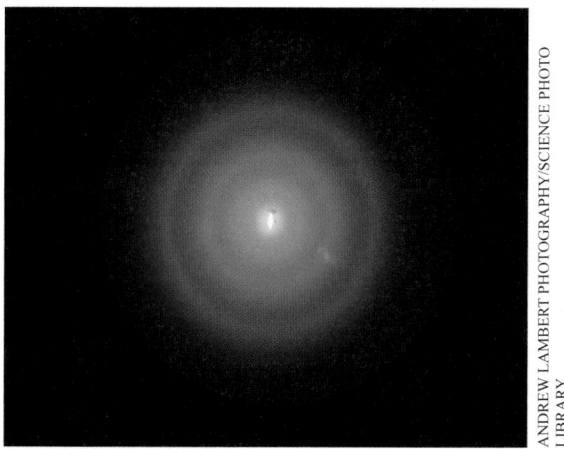

Figure 28-29 Diffraction of an electron beam passing through graphite

Geometric and Physical Optics

When solving problems in which the size of an object is comparable to the wavelength of the electromagnetic radiation illuminating it, we must consider explicitly the wave nature of light including interference effects. We can consider light to travel in straight lines (rays) in geometric optics only when the wavelength is much shorter than the dimensions of the object being studied.

Conditions for Interference

Electromagnetic waves, including light, interfere constructively when they are in phase and destructively when they are out of phase. Phase differences between two waves are often due to a greater distance travelled by one of the waves.

Ideal Double Slit

Ignoring diffraction from the individual slits, constructive interference of light of wavelength λ passing through two narrow slits separated by a distance d occurs at any angle θ for which

$$d\sin\theta = m\lambda \qquad (28\text{-}5)$$

where m, the spectral order, is an integer ≥ 0.

Thin Film Interference

Antireflection optical coatings are designed so that that reflected waves from the top and the bottom of the coating are out of phase, resulting in destructive interference and little net reflected wave. Hard reflections occur at a boundary with a medium with a higher index of refraction and cause a $180°$ phase shift. When considering the phase difference due to distance travelled, we must take into account the wavelength in the medium:

$$\lambda_n = \frac{\lambda}{n} \qquad (28\text{-}10)$$

Single-Slit Diffraction

By Huygens' principle, we can regard the region in a single slit to be composed of a number of secondary sources. *Destructive* interference between waves from these sources in a slit of width w occurs at any angle θ for which

$$w\sin\theta = m\lambda \qquad (28\text{-}15)$$

where m is an integer > 0.
The narrower the slit, the wider the region over which the light is spread. Most of the intensity falls between the $m = 1$ dark fringes, and we define the size of the central maximum as the distance between these fringes.

Actual Intensity Pattern of Double Slits

The interference pattern for a double slit is approximately equal to the product of the ideal intensity pattern for single-slit diffraction times the intensity pattern for an ideal double slit.

Resolution Limit

The wave nature of light limits the detail that can be seen with a given wavelength. The Rayleigh criterion states that two objects can be resolved if they are separated by an angle at least large enough for the central maximum of one object to correspond to the first minimum in the diffraction pattern of the other.

For a circular aperture of diameter D, the smallest subtended angle θ_r that can be resolved with light (or any other electromagnetic wave) of wavelength λ in accordance with the Rayleigh criterion is

$$\sin\theta_r = 1.22\frac{\lambda}{D} \qquad (28\text{-}17)$$

APPLICATIONS

Applications: interferometry, diffraction gratings, spectrometers, quantum control and ultrafast physics, antireflection optical coatings, energy-efficient windows, resolution of optical instruments, electron microscopes

Key Terms: Airy pattern, aperture, central maximum, coherent, diffraction grating, Fraunhofer condition, geometric optics, hard reflection, Huygens' principle, interference, laser, Michelson interferometer, monochromatic, phase change, physical optics, Poisson spot, ray optics, Rayleigh criterion, reflection gratings, resolution limit, spectral order, thin film interference, transmission gratings, wave optics

QUESTIONS

1. Which of the following electromagnetic waves is most accurately approximated by geometric (ray) optics when used to view print on a page?
   (a) infrared
   (b) radio
   (c) visible light
   (d) X-ray

2. A thermal imaging camera can work in two modes, one with a wavelength of 2 μm and the other with a wavelength of 10 μm. The camera is being used to study emissions from an organism that is approximately spherical with a diameter of 5 mm. We want to solve problems related to the thermal imaging camera. Which of the following statements is true?
   (a) We can use geometric optics to solve problems, and the approximation is better at the 2 μm wavelength.
   (b) We can use geometric optics to solve problems, and the approximation is better at the 10 μm wavelength.
   (c) We cannot use geometric optics to solve problems in this situation.
   (d) We can use geometric optics to solve problems, and the approach is equally valid at both wavelengths.

3. A Michelson interferometer apparatus is used with two different monochromatic sources, one with a wavelength of 430 nm and the other with a wavelength of 645 nm. When the light with the 645 nm wavelength is used, 64 fringes are counted for a certain movement of the mirror. What is the number of fringes counted for the same mirror movement when the 430 nm wavelength is used?
   (a) 43
   (b) 64
   (c) 96
   (d) 144

4. Monochromatic laser light with a wavelength of 532 nm illuminates two closely spaced slits to form an interference pattern on a screen. Bright interference dots form at a spacing of approximately 2 cm on the screen. If nothing else is changed when an infrared laser of wavelength 1064 nm is substituted, what is the spacing of the bright interference dots on the screen?
   (a) 0.5 cm
   (b) 1 cm
   (c) 4 cm
   (d) 8 cm

5. One of the wavelengths produced by an argon laser is 488 nm. Light from this laser is used to illuminate two slits, and the pattern on a screen has bright fringes approximately 1.2° on each side of the central bright band. With a different two-slit pattern, the first-order bright fringes are now 2.4° from the central image. The original slits were separated by a distance s. What is the spacing of the second pair of slits?
   (a) $s/4$
   (b) $s/2$
   (c) $2s$
   (d) $4s$

6. Spectrometer A has a diffraction grating with 9000 lines/cm, and spectrometer B (otherwise similar) has a diffraction grating with 3000 lines/cm. Which statement correctly describes the images produced by these spectrometers?
   (a) The first-order image for spectrometer B starts at a larger angle and is more spread out for spectrometer A.
   (b) The first-order image for spectrometer A starts at a larger angle and is more spread out for spectrometer B.

   (c) The first-order images start at the same angle for both spectrometers but are more spread out for spectrometer A.
   (d) The first-order image starts at the same angle for both spectrometers but is more spread out for spectrometer B.

7. A diffraction grating is held in front of the lens of a camera pointed at stars in the night sky. Which of the following statements correctly describes the image produced?
   (a) There is no slit, so there is no interference pattern. Nothing will be seen.
   (b) On each side of each star the light spreads out into colours, with violet toward the centre and red farther out (but no spot directly in centre for each star).
   (c) There is a white dot for each star, and on each side of that dot, the light spreads out into colours, with red nearer the white dot and violet farther away.
   (d) There is a white dot for each star, and on each side of that dot, the light spreads out into colours, with violet nearer the white dot and red farther away.

8. When a slit is narrow in both directions, there are diffraction patterns in both directions. The image in Figure 28-30 was produced with monochromatic blue light and a slit that is narrow in both directions. Which of the following most accurately describes the dimensions of the slit used to produce the image?
   (a) The slit is small in both dimensions but is narrower along the x-axis.
   (b) The slit is small in both dimensions but is narrower along the y-axis.
   (c) The slit is of equal width in both dimensions.
   (d) The relative size cannot be determined from the information provided.

Figure 28-30 Questions 8 and 9

9. If instead of monochromatic blue light one used monochromatic red light to produce an image like the one in Figure 28-30, how will the diffraction pattern change?
   (a) The pattern will not change at all.
   (b) The pattern will be more spread out in both directions.
   (c) The pattern will be more compressed in both directions.
   (d) The pattern will be more compressed in the x-direction but more spread out in the y-direction.

10. A circular diffraction pattern is observed by shining monochromatic light through a small circular opening. Which statement correctly describes how the pattern changes with wavelength and the aperture size?
    (a) The size of the diffraction pattern increases as the size of the opening increases or the wavelength increases.
    (b) The size of the diffraction pattern increases as the size of the opening decreases or the wavelength increases.

(c) The size of the diffraction pattern increases as the size of the opening increases or the wavelength decreases.

(d) The size of the diffraction pattern increases as the size of the opening decreases or the wavelength decreases.

11. When white light is shining perpendicular to the surface of a soap bubble, the wall of the bubble initially appears to be green. What happens to the colour of that region as the soap film begins to evaporate?

(a) The soap film stays the same colour.

(b) The soap film becomes blue.

(c) The soap film becomes yellow.

(d) The soap film becomes white.

12. A test thickness, t, of an antireflection coating works excellently for red light with a wavelength (in a vacuum) of 640 nm. What thickness should be used to prevent reflection of ultraviolet radiation with a wavelength of 160 nm?

(a) $t/4$

(b) $t/2$

(c) $2t$

(d) $4t$

13. A certain microscope is unable to resolve details of an object when visible light is used. Assuming appropriate detectors, how could you improve the resolution?

(a) Use infrared light instead of visible light.

(b) Use ultraviolet light instead of visible light.

(c) Use microwaves instead of visible light.

(d) None of the above will help because the resolution depends only on the diameter of the objective lens.

14. The radio telescope pictured in Figure 28-1 is used to study both atomic hydrogen at a frequency of 1.42 GHz and methanol at a frequency of 6.67 GHz. Compare the angular resolution at these two frequencies.

(a) The angular resolution is the same for both.

(b) The angular resolution is approximately 4.7 times better for the atomic hydrogen observations.

(c) The angular resolution is approximately 4.7 times better for the methanol observations.

(d) The angular resolution is approximately 22 times better for the methanol observations.

PROBLEMS BY SECTION

For problems, star ratings will be used, (✶, ✶✶, or ✶✶✶), with more stars meaning more challenging problems.

Section 28-1 Physical and Geometric Optics

15. ✶ A single wavelength electromagnetic wave passes through an opening that has dimensions of 1 mm × 10 cm. Assess whether geometric optics approximations can be used in each of the following cases. Justify your answer.

(a) visible green light with a wavelength of 532 nm

(b) infrared radiation with a wavelength of 5.00 μm

(c) ultraviolet radiation with a wavelength of 180 nm

(d) radio waves with a frequency of 88.0 MHz

16. ✶✶ The brightness of a surface varies in the form of a sine wave with a wavelength of 2.5 mm (i.e., the distance between one brightness maximum and the next is 2.5 mm). Assume that the wavelength of the radiation must be one-tenth as large as the object being viewed for a geometric optics approach to be valid. What are the wavelength and frequency of the maximum wavelength of radiation that meets this condition, and what type of radiation is this?

Section 28-2 Interference

17. ✶✶ Two radio antennas are being driven in phase by the same 1500 kHz signal. The antennas are 2.00 km from each other.

(a) Would constructive interference or destructive interference occur at a detector located exactly halfway between the two transmitters?

(b) The detector is moved to a position of destructive interference. Determine the minimum distance the detector needs to be moved to this position.

18. ✶ Laser light with a wavelength of 633 nm is used to illuminate a Michelson interferometer in a vacuum chamber. You count 90 fringes as one mirror is moved. Through what distance must the mirror have moved?

Section 28-3 Double-Slit Interference

19. ✶✶ Two narrow slits are 0.250 mm apart. They are illuminated by light from a helium–neon laser with a wavelength of 633 nm, and the interference pattern is observed on a screen 45.0 cm from the slits.

(a) What is the angle between the zero-order maximum and one of the first-order images?

(b) What is the distance between the two first-order images?

20. ✶✶ In a double-slit experiment, you want the separation between the $m = 1$ and $m = 2$ images on the same side of the zero-order maximum to be 4.00 mm when the screen is 0.750 m from the slits. Green laser light with a wavelength of 532 nm is being used for illumination.

(a) What is the required spacing between the slits?

(b) What would the spacing between the $m = 1$ and $m = 2$ images be if a red laser with a wavelength of 633 nm were used for illumination?

Section 28-4 Diffraction Gratings

21. ✶✶ You are asked to design a spectrometer such that on a screen positioned 0.750 m from the diffraction grating the $m = 1$ image for red light with a wavelength of 700 nm is at a distance (from the centre of the $m = 0$ image) of 2.00 cm.

(a) How many lines per centimetre should the diffraction grating have?

(b) What is the distance to the $m = 1$ image for light with a wavelength of 400 nm?

22. ✶✶✶ You want to design a spectrometer to cover the first-order spectrum completely from 360 nm to 675 nm. The image sensor that will detect the first-order image on one side is 22.5 mm across. The spacing from the diffraction grating to the detector is 5.00 cm. What grating spacing (in lines/cm) should be used?

Section 28-5 Thin Film Interference

23. ✶✶ You need to design an antireflection coating for infrared light with a wavelength of 1250 nm. The coating has an index of refraction of 1.38, and it is to be placed on glass with an index of refraction of 1.52.

(a) What is the thinnest coating that you could use? Assume that the infrared beam is perpendicular to the coating.

(b) What wavelength shorter than 1250 nm would produce constructive interference with the coating you designed in part (a)?

24. ✶✶ What is the minimum thickness of a soap bubble wall that will result in constructive interference of yellow light with a wavelength (in a vacuum) of 580 nm. Assume

that the beam is approximately perpendicular to the soap bubble and that the index of refraction of the soapy water is $n = 1.33$.

Section 28-6   Single-Slit Diffraction

25. ✶   Monochromatic light with a wavelength of 589 nm illuminates a single slit. Destructive interference appears at angles of 1.15° on either side of the normal. What is the width of the slit?

26. ✶✶   During a single-slit diffraction experiment, you want the first minimum on either side to be at an angle of 60°. Calculate the required ratio of the slit width to the wavelength of light used.

Section 28-7   Actual Intensity Pattern for Double Slits

27. ✶✶✶   Laser light with a wavelength of 633 nm is used to illuminate two slits separated by 0.125 mm. The width of each slit is 0.0150 mm. Assuming that only fringes between the first minima in the pattern are counted, how many bright fringes are visible?

Section 28-8   Resolution Limit

28. ✶✶   (a) If limited only by the wave nature of light, what would be the minimum angle that could be resolved (assume Rayleigh criterion) by a telescope that has an aperture diameter of 15.0 cm? Assume light with a wavelength of 550 nm.
(b)   If the telescope is used to observe a binary star system that is 16.5 ly away, what is the minimum separation of the stars (in km) that can be resolved (assuming optimum orientation)?

29. ✶✶   You are driving at night when the pupil of your eye is 2.85 mm in diameter. A car's headlights are 1.45 m apart. How close must a distant car be for you to see two separate headlights? Assume a wavelength of 550.0 nm.

COMPREHENSIVE PROBLEMS

30. ✶✶   (a) You are listening to an FM radio station with a frequency of 92.0 MHz, and the combination of the direct signal and a signal reflected from a wall produces destructive interference. What is the closest that you can be to the wall for this condition to be possible?
(b)   How much farther should you move away from the wall to reach a point of constructive interference?

31. ✶✶✶   A Michelson interferometer is constructed such that one of the equal arms can hold gas or be evacuated. Each arm is 0.365 000 m long. An interference pattern is observed as a gas is added slowly, and it is observed that in going from evacuated to containing the gas a total of 205 fringes are observed. The interferometer uses a helium–neon laser with a wavelength of 633.000 nm. Find the index of refraction of the gas.

32. ✶✶   A sodium discharge tube produces results in a doublet of two relatively closely spaced spectral lines with wavelengths of 588.995 and 589.592 nm. You want these lines to be separated by 2.40 mm on a screen that is 50.0 cm from a diffraction grating. How many lines/cm should the diffraction grating have?

33. ✶✶   An argon laser produces ultraviolet light with a wavelength of 264 nm. Light from that laser is used to illuminate two slits separated by exactly 0.100 mm. What is the angular separation between the $m = 0$ spot and the $m = +1$ spot?

34. ✶   Two slits are separated by 0.100 mm. Light with a wavelength of 500 nm is used to illuminate the slits. It is observed that one of the bright fringes is at an angle of 1.43°. Find the spectral order of this band.

35. ✶✶   The angle between the two $m = 2$ maxima is 1.30° when two slits are illuminated with light with a wavelength of 480 nm. Find the spacing between the slits.

36. ✶✶   A laser with a wavelength of 633 nm is used to illuminate a diffraction grating with 7500 lines/cm. What is the total number of bright spots that appear on a screen? (Hint: Remember that the maximum possible value of angle is 90° and that spots appear on both sides of the zero-order maximum.)

37. ✶✶✶   The Ocean Optics company makes a variety of miniature spectrometers, which are widely used in scientific labs. One version of this particular spectrometer uses a diffraction grating with 600 lines/mm and is capable of detecting light from 350 to 1000 nm in a first-order image.
(a)   The grating has a simple design (without mirror stages). What is the deflection angle for light with a wavelength of 350 nm? For a 1000 nm wavelength?
(b)   The detector used in this spectrometer has a row of 2048 pixels, each of which is 14 μm wide. What is the effective width of the detector, in millimetres?
(c)   Considering the angles determined in part (a) and the width of the detector, what is the effective distance from grating to the sensor if the spectrometer does not include mirrors?

38. ✶✶   A thin liquid layer with an index of refraction $n = 1.44$ floats on top of water ($n = 1.33$). What is the minimum thickness possible when light incident perpendicular to the surface is most strongly reflected at a wavelength of 540 nm?

39. ✶✶   Two glass plates with very accurate, flat surfaces are slightly inclined relative to each other as shown in Figure 28-31. The separation, s, between the plates is 75.0 μm at one side and 0 at the opposite side. Monochromatic light with a wavelength of 532 nm in air is vertically incident. Due to interference between the light that reflects at the top and bottom of the varying air gap between the plates, you observe a series of light and dark fringes. How many bright regions do you see when viewing the plates from above?

Figure 28-31   Problems 39 and 40

40. ✶✶   For the two glass wedged plates in problem 39, the space between the plates is now filled with water ($n = 1.33$). How many bright fringes are there?

41. ✶✶   The Canada-France-Hawaii Telescope (CFHT) has an objective mirror 3.60 m in diameter. Assume a wavelength of 550.0 nm.

(a) If only the wave nature of light limited resolution, what would be the limiting angular resolution (assuming the Rayleigh criterion)?

(b) Astronomers often express angles in units of arcseconds (symbol "), where 1" = 1°/3600. Convert your answer from (a) into arcseconds.

42. ✳✳ The Penticton radio telescope in Figure 28-1 has a diameter of 26.0 m. One of the main uses is to study electromagnetic emissions from atomic hydrogen at a frequency of 1.42 GHz. Calculate the angular resolution limit according to the Rayleigh criterion of the telescope at this frequency.

43. ✳✳ To improve effective resolution, separate radio telescopes are connected in interferometry arrays that approximate a single very large instrument. In 1967, this was done for synchronized signals recorded separately at the radio telescope in Penticton, British Columbia, and another radio telescope in Algonquin Park, Ontario, a spacing of approximately 3000 km. What is the approximate angular resolution achieved with this array? Assume the same radio frequency as for the previous question.

44. ✳✳ The Very Long Baseline Array (VLBA) uses 10 telescopes with a baseline extending 8611 km from Mauna Kea, Hawaii, to St. Croix, U.S. Virgin Islands. Find the angular resolution of this array at the extremes of the range of frequencies used: 1.20 GHz to 96.0 GHz.

45. ✳✳ Atmospheric scintillation, which make stars appear to twinkle, limits the resolution of observations made through the atmosphere. A typical resolution limit from atmospheric scintillation is approximately 1.5". What size of instrument aperture would have an equal resolution limit due to the wave nature of light? Assume a wavelength of 500 nm.

46. ✳✳ The Canadian NEOSSat mission utilizes a small satellite (about the size of a suitcase) in Earth orbit to detect asteroids near Earth. NEOSSat uses a telescope with a diameter of 15.0 cm. If there were no other limitations, what is the minimum separation of two sources that this telescope can resolve at a wavelength of 560 nm? Express your answer in units of arcseconds.

47. ✳✳✳ It is desired to have an actual double slit with a spacing such that the third-order image does not appear (because it corresponds to a diffraction minimum from the width of the slits). If the separation between the centres of the slits is 0.125 mm and if light with a wavelength of 532 nm is used, what should the width of the slits be?

DATA-RICH PROBLEM

48. In this question you will plot the relative intensity at different angles θ due to the illumination of a single slit of width w with monochromatic light of wavelength λ. The intensity function is given by the following:

$$I_0\left\{\frac{\sin\left(\frac{\pi w \sin\theta}{\lambda}\right)}{\left(\frac{\pi w \sin\theta}{\lambda}\right)}\right\}^2$$

where I_0 is the intensity at the centre, and the angles are expressed in radians. (The function $\frac{\sin(\pi x)}{\pi x}$ is called the sinc function. It often appears in advanced physics, especially in signal processing.)

(a) Use a computer program or spreadsheet to calculate the intensity for angles from $-60°$ to $+60°$. First, use $w/\lambda = 4$. Plot your output and record the plot.

(b) Determine where the intensity approaches zero. Do these angles correspond to the angles predicted by the single-slit equation derived in Section 28-6?

(c) Now repeat part (a) for $w/\lambda = 8$ and then $w/\lambda = 2$. Compare the resulting plots with the one from part (a). How does the slit width affect the intensity pattern?

(d) Set $w/\lambda = 1$. What do you now observe?

OPEN PROBLEM

49. ✳✳✳ You need to estimate the volume of an oil spill in tropical waters by observing it from the air. When the Sun is approximately overhead and you are nearly directly above the spill, you observe that the most strongly reflected light has a wavelength of approximately 500 nm. When viewed from 1200 m above the water, the spill subtends an angle of approximately 20° side to side. Estimate the volume of oil, clearly stating any assumptions.

See the text online resources at www.physics1e.nelson.com for Open Problems and Data-Rich Problems related to this chapter.

Chapter 29
Relativity

Global Positioning System (GPS) receivers are more than just a convenience for drivers (Figure 29-1). These devices are the core of modern navigation systems for planes and ships and of emergency beacons that broadcast their exact location. A GPS receiver system uses the time delays of signals from a series of satellites orbiting Earth to calculate the position of the receiver. Without corrections for relativistic effects, the GPS system would be hopelessly inaccurate. Relativity theory also underlies research and technology ranging from astrophysics to nuclear medicine to power plants.

kaczor58/Shutterstock

Cristi Matei/Shutterstock

Figure 29-1 A GPS system requires calculations using relativity theory to determine your precise location.

Learning Objectives

When you have completed this chapter you should be able to:

1 Differentiate between special and general relativity, and cite applications of each.

2 Describe the Michelson–Morley experiment, and explain its significance.

3 Use the postulates of special relativity to derive the time dilation relationship.

4 Solve problems related to length contraction and time dilation.

5 Explain mass–energy equivalence, and calculate relativistic momentum and kinetic energy.

6 Add relativistic speeds in special relativity situations.

7 Describe spacetime intervals, and apply them to solve relativity problems.

8 Explain and apply concepts of gravitational time dilation.

9 Explain how special and general relativity considerations affect the GPS system.

 For additional Making Connections, Examples, and Checkpoints, as well as activities and experiments to help increase your understanding of the chapter's concepts, please go to the text's online resources at www.physics1e.nelson.com.

MODERN PHYSICS

29-1 Special and General Relativity

Relativity is divided into two branches: special relativity and general relativity. Most of this chapter deals with special relativity, but we will introduce the basic concepts of general relativity near the end.

Special relativity describes how observers moving relative to each other measure different values for the same physical quantities. These differences are substantial when the relative speed is significant compared to the speed of light (3.0×10^8 m/s) but negligible for moderate relative speeds. Perhaps the most surprising consequence of special relativity is that time slows for moving observers, an effect called **time dilation**. Length measurements also depend on the frame of reference of the observer. Consequently the concepts of kinetic energy and linear momentum developed earlier in this book need to be modified. Relativity theory treats space and time as components of a single four-dimensional entity called **spacetime**.

The speed of light appears to be the upper limit for the speeds of objects in our universe. The relativistic effects on time, length, momentum, and kinetic energy become apparent when the speed of an object approaches the speed of light. Another important consequence of relativity is that energy and mass are viewed as different aspects of a unified entity, and objects possess energy by virtue of their mass.

As the name implies, **general relativity** is a more general form of relativity, but it is also a radically new theory of gravitation. General relativity does not treat gravitation as a force, but rather proposes that masses bend spacetime and that objects move according to the resulting curvature of spacetime. The curvature caused by Earth's mass is small, but its effects are measurable. The effects of the curvature near very dense objects, such as stellar cores and black holes, are extreme and rather surprising.

The concepts of relativity can be challenging because they seem contrary to common sense. However, countless experiments on Earth and observations of objects in space have validated the ideas of both special and general relativity. For example, measurements of the decay of unstable subatomic particles moving at speeds near to the speed of light have confirmed that time dilation occurs exactly as predicted by special relativity.

MAKING CONNECTIONS

Albert Einstein

Figure 29-2   Albert Einstein at about the time he developed special relativity

The name Albert Einstein (1879–1955) is almost synonymous with scientific creativity and intellectual brilliance. Like Newton and many other famous scientists, Einstein

did his most important work as a young scientist. In 1905, Einstein published five works including key papers on relativity, Brownian motion, and the photoelectric effect. Einstein was then only 26 years old, and his ideas were developed when he was even younger (Figure 29-2).

Einstein's contribution to our understanding of the photoelectric effect, for which he later received a Nobel Prize, helped set the foundation for quantum mechanics. However, Einstein is best known for the development of relativity theory. He published the main elements of special relativity in 1905, including a paper that developed the famous $E = mc^2$ mass–energy equivalence, although the term "special relativity" was not introduced until later. Approximately 10 years later, he finished work on a generalization of relativity that incorporated gravitation. This version is what we now call general relativity.

✓ CHECKPOINT

C-29-1 Special and General Relativity

Identify the branch of relativity theory associated with the following concepts:

(a) Gravitational effects result from the curvature of spacetime.

(b) Time passes at a different rate for observers in relative motion at speeds near the speed of light.

(c) Objects cannot travel faster than the speed of light.

(d) Mass has an energy equivalent.

C-29-1 (a): general relativity; (b), (c), and (d): special relativity.

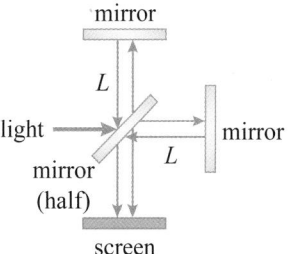

Figure 29-3 Simplified version of the Michelson–Morley experiment

29-2 Reference Frames and the Michelson–Morley Experiment

The story of relativity begins with the speed of light. As you saw in Chapter 26, Maxwell's elegant set of equations summarizing electricity and magnetism led directly to a relationship between the speed of electromagnetic waves in a vacuum (*c*) and the product of the permeability of free space (μ_0) and the permittivity of free space (ε_0):

$$c = \frac{1}{\sqrt{\varepsilon_0 \mu_0}} \qquad (26\text{-}29)$$

Thus, the fields of electricity, magnetism, and optics had been unified. In this chapter, we will see that the speed of light has a key role in mechanics as well.

In the second half of the 19th century, physicists puzzled over the reference frame for measuring the speed of light. Any speed must be measured relative to something. Since electromagnetic waves propagate in a vacuum (for example we get energy radiated from the sun through the largely vacuum of space), the reference frame was not obvious. One proposed solution was the theory that electromagnetic waves travelled relative to an **ether**, a massless, invisible medium present even in a vacuum. So the search was on for the reference frame of this hypothetical ether. The most obvious way to establish this reference frame was to carefully measure the speeds of electromagnetic waves transmitted when moving in different directions relative to the ether. Differences in these speeds would indicate the speed of the ether relative to the observer.

Several attempts were made to detect the flow of the ether, the most famous being the interferometer experiments conducted by Albert Michelson and Edward Morley, working at what is now Case Western Reserve University in Cleveland, Ohio. Interferometers can measure very small changes in distance (comparable to the wavelength of light) and, if the distances are held fixed, correspondingly small differences in the speed of light. Section 28-2 described the principles of interferometers and the basics of the design developed by Michelson and Morley.

Figure 29-3 shows a simplified version of the experiment. As described in Chapter 28, the Michelson interferometer uses a half-silvered mirror to split the light from a source into two beams that travel different paths, and then recombines the beams to produce an interference pattern. In this experiment, the paths in the perpendicular arms of the interferometer have exactly the same length. By varying the orientation of the apparatus relative to the motion of Earth around the Sun, Michelson and Morley hoped to determine the ether reference frame. Thus, the **Michelson–Morley experiment** is an interferometer-based attempt to measure the reference frame for electromagnetic waves such as light.

Assume that the ether frame moves at speed *v* from right to left, as shown in Figure 29-4. Then the beam that travels in the arm *parallel* to the direction of the ether would move at speed *c* + *v* when moving to the right, and at speed *c* − *v* when moving to the left (against the ether). The time for the two-way trip is then

$$t_\parallel = \frac{d}{c-v} + \frac{d}{c+v} \qquad (29\text{-}1)$$

where *d* is the distance travelled.

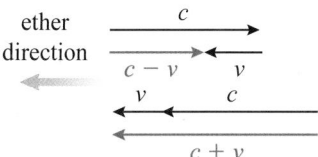

Figure 29-4 Light moving parallel and antiparallel to an assumed ether reference frame

✓ CHECKPOINT

C-29-2 Movement in the Ether

If the ether were moving horizontally at a speed of $0.25c$, what would be the total time required to travel a horizontal distance Δx and back?

(a) $\dfrac{7\Delta x}{4c}$

(b) $\dfrac{2\Delta x}{c}$

(c) $\dfrac{32\Delta x}{15c}$

(d) $\dfrac{9\Delta x}{4c}$

C-29-2 (c) When travel is in the direction of the ether flow, the time is $\dfrac{\Delta x}{0.75c}$. For a direction opposite to the ether flow, the time is $\dfrac{\Delta x}{1.25c}$. Adding these times gives a total travel time of $\dfrac{32\Delta x}{15c}$.

Now consider the movement perpendicular to the ether reference frame (Figure 29-5).

Analogous to rowing a boat across a slow-moving river: if you try to row directly across the river, the current

MODERN PHYSICS

will push you slightly upstream or downstream, and you will not end up exactly opposite where you started. Similarly, the beam in the arm perpendicular to the ether flow would have to be aimed at an angle as shown in Figure 29-5 to compensate for the displacement caused by the ether flow. Applying the Pythagorean theorem gives a travel distance of $\sqrt{c^2 - v^2}$ in each direction. Therefore, the total travel time in the perpendicular arm is

$$t_\perp = \frac{2d}{\sqrt{c^2 - v^2}} \qquad (29\text{-}2)$$

Figure 29-5 Movement perpendicular to an ether reference frame

We can show that the travel times in Equations (29-1) and (29-2) are not equal (see problem 16). Therefore, a precise enough experiment would show the direction and speed of the flow of ether. Since Earth moves at approximately 30 km/s in its orbit around the Sun, Michelson and Morley hoped that orienting their interferometer with one arm parallel to the direction of this motion would result in a detectable difference between the speeds of the two light beams.

Michelson detected no difference when he first tried the experiment in 1881 while working in Germany. After he moved to the United States, Michelson worked with Morley to substantially increase the precision of the equipment, but their experiments in 1887 still found no difference in light speed for the two paths. Einstein, who became interested in the ether question while still in high school, wanted to try the same experiment when he was an undergraduate at university in Zurich, Switzerland, but his professor would not let him use the university laboratory for the experiment. Modern experiments with extremely high precision have confirmed Michelson and Morley's null result.

Therefore experiments suggested that there was no ether, and the speed of light did not depend on the observer's reference frame. As we will see in the next section, the idea that the speed of light does not depend on the reference frame is the starting point for the special theory of relativity.

LO 3

29-3 Postulates of Special Relativity and Time Dilation

Special relativity applies for **inertial reference frames**, which are reference frames in which Newton's laws are valid. We can in general consider these to be unaccelerated

reference frames. To derive special relativity, Einstein made two assumptions:

- **Observers in different inertial reference frames agree on the *laws* of physics**, even though they may measure different *values* for physical quantities. Although we cannot prove that physical laws are the same throughout the universe, so far, there is no evidence to contradict this assumption.

- **All inertial observers measure the same value for the speed of light in a vacuum.** For example, if you were on a spaceship passing a planet at half the speed of light and shine a laser beam in front of you, you and an observer on the planet would both measure the same speed for the laser light. This assumption is contrary to common sense but is consistent with the results of the Michelson–Morley experiments.

We will use the term "event" frequently in this chapter. In relativity theory, an event is something that happens independent of the space and time reference frame used to describe it. Different reference frames assign different space and time values to the same event. Some examples of events are an atom emitting a photon, a radioactive nucleus emitting an α-particle, a flare being fired, and two particles colliding with each other.

MAKING CONNECTIONS

Olympia Academy

Einstein developed special relativity while he was working at the patent office in Bern, Switzerland. He became lifelong friends with two of his co-workers, Maurice Solovine and Conrad Habicht. The three men spent many hours discussing ideas of physics and philosophy, and they called their discussion group the Olympia Academy. Einstein's first wife, Mileva Marić, sometimes took part in the discussions of the Olympia Academy. Marić and Einstein studied physics together at the Zurich Polytechnic.

Time Dilation

Now, we shall see how Einstein's assumptions led to an amazing result: time dilation. We start with a thought experiment. In a **thought experiment**, you use the results of practical experiments to predict how a hypothetical experiment would turn out, especially when you cannot actually perform the hypothetical experiment. Einstein liked thought experiments. Newton used thought experiments in some of his work, as did some of the founders of quantum mechanics.

Consider a clock based on the time that light takes to travel some fixed distance, as shown in Figure 29-6. In the figure, we have a light pulse that starts at the bottom, travels a distance d to the top where it is reflected from a mirror, and then is detected when it reaches the bottom again. We will assume that the light travels in a vacuum inside the clock and that the clock is stationary relative to the reference frame being used for measurements. We can treat the time that the light takes to travel from the bottom to the mirror as a unit of time. Then, for a clock at rest in our reference frame,

$$t_0 = \frac{d}{c} \qquad (29\text{-}3)$$

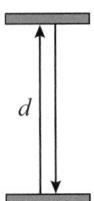

Figure 29-6 A clock based on the time for light to travel a fixed distance. In this case, the clock is not moving relative to an observer.

Now imagine the same clock moving at speed v to the right. Since the entire clock is moving at the same constant speed, the light beam will still reflect from the same point on the mirror, but the distance, s, travelled by the light will be somewhat longer relative to the observer, as shown in Figure 29-7. Using the Pythagorean theorem, we have

$$s = \sqrt{d^2 + v^2 t^2} \qquad (29\text{-}4)$$

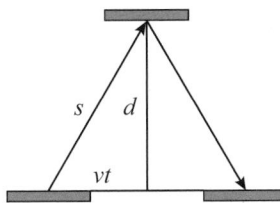

Figure 29-7 A clock moving to the right at speed v relative to an observer

According to the second assumption of special relativity, an observer in the reference frame of the moving clock measures the same speed of light, c. Therefore, the travel time in the moving clock is given by

$$t = \frac{s}{c} = \frac{\sqrt{d^2 + v^2 t^2}}{c} \qquad (29\text{-}5)$$

and

$$t^2 = \frac{d^2}{c^2} + \frac{v^2 t^2}{c^2} \qquad (29\text{-}6)$$

If we substitute $t_0 = d/c$ (from Equation (29-3)) and solve for t, we obtain

$$t = \frac{t_0}{\sqrt{1 - v^2/c^2}} \qquad (29\text{-}7)$$

Usually, it is the intervals between events that are important, for example, the time between a spaceship starting and ending a trip or between the formation and decay of a subatomic particle. Therefore, the time dilation relationship is normally written in terms of time intervals:

KEY EQUATION $$\Delta t = \frac{\Delta t_0}{\sqrt{1 - v^2/c^2}} \qquad (29\text{-}8)$$

The expression Δt_0 is called the **proper time interval**. The **proper time** is the time measured by the observer who records the two events taking place at the *same* spatial coordinates. Although an observer in any other reference frame does not measure the proper time, measurements in that observer's reference frame are still self-consistent and equally valid. The light emission and light detection events occur at the same position in our reference frame in the stationary light clock in Figure 29-6, but not for the moving clock in Figure 29-7.

Note that the denominator in Equation (29-8) is less than 1 since $v < c$. Therefore,

$$\Delta t > \Delta t_0 \qquad (29\text{-}9)$$

Thus, the proper time interval for any pair of events is less than the time interval measured in any other reference frame. In other words, a *moving clock always runs more slowly than the proper time clock*. Strictly speaking, this conclusion applies only to time intervals measured using the passage of light beams because our derivation is based on a clock that uses that process. However, repeated experiments have shown that time dilation affects every process that physicists have been able to test. It appears that moving clocks of any type—mechanical, electronic, biological, chemical, and atomic—run slower. In other words, time itself runs slower and affects all time-related processes. Consequently, an observer cannot detect this slowing by measuring only objects in her own reference frame.

✓ **CHECKPOINT**

C-29-3 Proper Time

If a spaceship travels from Earth to a planet in the Alpha Centauri star system, which observer measures the proper time for the trip?

(a) an observer on Earth
(b) an observer on the planet in the Alpha Centauri star system
(c) an observer on the spaceship
(d) an observer deep in space far from the planet and Earth

C-29-3 (c) For an observer on the spaceship, the trip starts and ends at the same spatial coordinates; therefore, the spaceship reference frame gives the proper time measurement.

EXAMPLE 29-1

Time Dilation of a Muon

A muon is an unstable subatomic particle with the same charge as the electron, but its mass is approximately 207 times the mass of the electron. Muons at rest decay in 2.20 μs on average. Cosmic rays hitting Earth's atmosphere can create muons travelling near the speed of light. Assume that one of these muons has a speed of $0.990c$. Note: the reference frames are not strictly inertial due to the rotation of Earth, but we can neglect that in this question.

(a) What lifetime does an observer on the ground measure for the muon?

(b) How far does the muon travel according to the observer?

(c) How far does the muon travel in its own reference frame?

SOLUTION

(a) The event is the lifetime of the muon. Clearly, the beginning and end of this lifetime have the same location only in a reference frame fixed to the muon. Therefore, an observer in the muon's reference frame measures the proper time, and

$$\Delta t_0 = 2.20 \times 10^{-6} \text{ s}$$

We want to find the time interval Δt for the ground-based observer. The speed of the muon relative to the

ground-based observer is $v = 0.990c$. Applying the time dilation equation gives

$$\Delta t = \frac{\Delta t_0}{\sqrt{1 - v^2/c^2}} = \frac{2.20 \times 10^{-6} \text{ s}}{\sqrt{1 - (0.990)^2}} = 1.56 \times 10^{-5} \text{ s}$$

(b) The ground-based observer sees a muon with the lifetime calculated in part (a) travelling at a speed of $0.990c$. Therefore, the observer measures the distance travelled to be

$$\Delta x = 0.990 \cdot 3.00 \times 10^8 \text{ m/s} \cdot 1.56 \times 10^{-5} \text{ s}$$
$$= 4632 \text{ m} = 4630 \text{ m}$$

(c) From its own reference frame, the motion is only about 653 m in its average lifetime (obtained by multiplying the speed by the time it measures for its lifetime).

Making sense of the result:

The muon is moving very fast relative to the observer, so the observer sees time for the muon substantially dilated. So it is reasonable that an observer in Earth's reference frame measures a lifetime that is almost an order of magnitude longer than that of the muon reference frame. Observations of high-energy muons in the upper atmosphere do indeed confirm the time dilation predicted by special relativity. The next section describes an alternative way to find the distance travelled.

The Twin "Paradox"

Consider identical twins who separate: one stays on Earth, and the other leaves Earth in a spaceship travelling at near the speed of light, then reverses direction to return to Earth at the same speed. According to special relativity, when the twins meet again, the one who travelled on the spaceship expects to be younger than the one who stayed on Earth. Let's view the situation from the reference frame of the travelling twin: Earth moves away from the travelling twin at a high speed, so the twin who stays on Earth should have the slower clock and expects to be younger when they meet. At first glance, it seems paradoxical that both twins expect the other to be younger.

However, the resolution of the paradox lies in the fact that the motion of the twins is not symmetrical. The twin who stays on Earth is always in the same reference frame, and the travelling twin undergoes accelerations: when leaving Earth, when turning around, and when arriving back at Earth. She is on one reference frame during the trip out and a different reference frame on the return, with periods of acceleration in-between. Special relativity applies only to inertial reference frames, and the travelling twin is not in an inertial reference frame when accelerating. It is the twin who experiences the accelerations who will be younger. One way to look at the twin paradox is this: according

to general relativity, we cannot differentiate between an accelerated frame of reference and a frame of reference in a gravitational field. Later in the chapter you will see that time passes more slowly in strong gravitational fields.

ONLINE ACTIVITY

Twin Paradox Simulations

The e-resource that accompanies every new copy of this textbook contains an Online Activity using the "Twin Paradox" simulation. Work through the simulation and accompanying questions to gain an understanding of time dilation.

A similar puzzle involves two spaceships passing each other at constant velocity. Special relativity says that an observer in each spaceship thinks that the other's clock runs slow. Although these observations may appear paradoxical, they are not actually contradictory because the two observers are in different inertial frames of reference. Each observer's measurements are correct and consistent for that observer's reference frame. The two spaceships are at the same position on the axis of the direction of motion only at the instant they pass each other. Thus, they can directly compare their clocks only at that one moment.

29-4 Length Contraction

In this section, we will see that special relativity also predicts that when an object is moving relative to an observer, the length of the object in the direction of motion contracts. To understand length contraction, we first define **proper length** in a manner analogous to proper time: An observer measures the proper length, L_0, of an object only if the object is *not* in motion relative to the observer's reference frame. An observer in any other inertial frame measures a different length, L.

C-29-4 Proper Length

Suppose a spaceship travels between two interstellar beacons that are in the same inertial reference frame. Which observer measures the proper length of the trip?

(a) the observer on the spaceship
(b) the observers at each beacon
(c) both the spaceship observer and the observers at each beacon
(d) none of these observers

C-29-4 (b) The end points of the trip (the spatial interval between the beacons) are not moving relative to the observers at the beacons; therefore, the observers at each beacon measure the proper length.

Consider a hypothetical trip by a spaceship travelling at a constant speed v between exoplanets A and B, which are not moving relative to each other. (We will approximate the situation as inertial reference frames, although actual exoplanets do not have inertial frames of reference due to motion about their parent stars, galactic rotation, and expansion of the universe.) An observer on each exoplanet measures the proper length for the distance between the planets, L_0, but does not measure the proper time, because the beginning and the end of the trip occur at different spatial coordinates in the reference frame of the planet (top of Figure 29-8). Instead, a planetary observer measures a dilated time, Δt. An observer on the spaceship (bottom of Figure 29-8) measures the proper time, Δt_0, because the start and end of the trip do occur at the same spatial coordinates in this reference frame. However, this observer does not measure the proper length because the spaceship moves relative to A and B.

According to the planet-based observer, the velocity of the spaceship is

$$v = \frac{L_0}{\Delta t} \qquad (29\text{-}10)$$

Figure 29-8 From a reference frame on one planet (top) a trip between the planets will take time Δt and be of distance L_0. A spaceship observer will measure time Δt_0 and distance L.

Similarly, from the point of view of the spaceship observer, the velocity for the trip is

$$v = \frac{L}{\Delta t_0} \qquad (29\text{-}11)$$

If we assume that the two observers must agree on the velocity, then we can use Equations (29-10) and (29-11) to obtain the following relationship:

$$L = \frac{L_0 \Delta t_0}{\Delta t} \qquad (29\text{-}12)$$

When we substitute the time dilation result (Equation (29-8)) into Equation (29-12), we obtain the following **length contraction** relationship:

KEY EQUATION $\qquad L = L_0 \sqrt{1 - v^2/c^2} \qquad (29\text{-}13)$

 EXAMPLE 29-2

Spaceship Length Contraction

A spaceship is 25.0 m long as measured in its own reference frame. What length do you measure as it moves past you at 90% of the speed of light?

SOLUTION

First, ask yourself, Who measures the proper length? It is an observer on the spaceship. So, $L_0 = 25.0$ m and $v = 0.900c$. Using the length contraction relationship, we have

$$L = L_0 \sqrt{1 - v^2/c^2} = (25.0 \text{ m})\sqrt{1 - 0.900^2} = 10.9 \text{ m}$$

Making sense of the result:

As expected, the spaceship moving past you is contracted in length in the direction of motion. The value you measure is less than the proper length of the spaceship.

MODERN PHYSICS

PEER TO PEER

I often have trouble figuring out which observer measures the proper length or the proper time. I find it easiest to picture myself sitting in each frame of reference and asking, "What do I experience? From where I am sitting, am I moving along the given distance? Can I time the event without changing where I am looking?" For example, if an electron is accelerated along a linear accelerator before colliding with another particle and I am sitting on the electron, I can see the beginning of my trip and my painful ending (ouch!) without looking away, so I must be measuring the proper time of the trip and therefore the improper, or contracted, length (top of Figure 29-9).

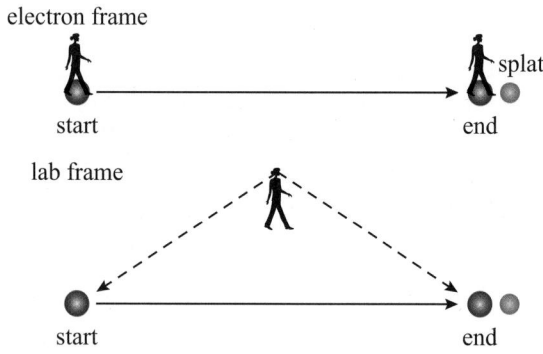

electron frame

start   end

lab frame

start   end

Figure 29-9 A trip in a linear accelerator from an electron's view (top) and an observer watching from the lab (bottom)

If I am an observer in the lab watching the electron whiz by, I need to change where I am looking to see the start and end of the electron's journey, and I don't move relative to the linear accelerator, so I must be measuring the improper time and the proper distance for the trip, that is, the length of the accelerator (bottom of Figure 29-9).

Remember that length contraction takes place only in the direction corresponding to the motion of the object. Various experiments have shown the validity of this length contraction result.

Example 29-3 demonstrates the important result that the observer who measures the proper time for a trip does *not* measure the proper length. In the reference frame of the muon, it has a short lifetime and travels a short distance, while from our perspective the muon has a longer lifetime and travels a correspondingly longer distance. Both views agree on the speed of the muon during its short relativistic flight.

 EXAMPLE 29-3

Muon Length Contraction

Consider the muon in Example 29-1. Show that the length contraction relationship correctly relates the distances measured by an observer on the ground and an observer moving with the muon.

SOLUTION

The muon travels from one point to another during its lifetime. From the muon's viewpoint, it is not the muon but the end points of the trip path that move. Therefore, an observer in the reference frame of the muon measures the proper time, but not the proper length. It is the Earth-based observer who measures the proper length. The starting and ending point of the muon's trip in the atmosphere does not move relative to Earth. (Note that again we are neglecting the noninertial nature of the reference frame for a rotating Earth.)

Using the length contraction relationship,

$$L = L_0\sqrt{1 - v^2/c^2} = (4632 \text{ m})\sqrt{1 - 0.990^2}$$
$$= 653.4 \text{ m} = 653 \text{ m}$$

Making sense of the result:

The results here are consistent with the results we obtained in Example 29-1, demonstrating that time dilation and length contraction provide consistent views of what happens. Different frames of reference agree on the relative velocity but not on the length or the time interval.

Special relativity makes it possible (theoretically at least) to travel to distant exoplanets within a human lifetime. For example, if you travel at 99% of the speed of light, a trip that is 10 y (years) in duration for an Earth-based reference frame will only be approximately 1.4 y to an observer on the spaceship. However, you will learn in the next section that the energy that a spaceship requires to reach speeds close to the speed of light is huge. There is an additional issue related to the physiological effects of acceleration to get to the high relativistic speeds, as well as the time required for reasonable accelerations.

For problems involving space travel, it is often convenient to work in distance units of light years and instead of writing it as ly, to express it as *c* y. This allows us to divide out, for example, the speed of light *c*. We also express all speeds as a fraction of the speed of light and use years as the time unit. This approach to units is demonstrated in Example 29-4. Of course, you can equally well convert all quantities to SI units.

MODERN PHYSICS

EXAMPLE 29-4

Space Travel

A spaceship travels at 98.0% of the speed to light to an exoplanet 45.0 ly away.

(a) What distances do observers on Earth and on the spaceship measure for the trip?

(b) What time does an observer on Earth measure for the one-way trip?

(c) What time do the occupants of the spaceship measure for the one-way trip?

Note that we are ignoring the times required for the acceleration stages.

SOLUTION

The spaceship traveller measures the proper time for the trip because the start and end occur at the same place in the reference frame of the spaceship.

(a) The Earth-based observer measures the proper length for the trip, 45.0 ly. We use the length contraction relationship to find the length for the trip as measured by an observer on the spaceship. For convenience, we will use light years as the unit for length:

$$L = L_0 \sqrt{1 - v^2/c^2} = (45.0 \text{ ly}) \sqrt{1 - 0.980^2} = 8.95 \text{ ly}$$

(b) We now use the notation suggested earlier: an observer on Earth says the distance is 45.0 ly = 45.0c y, and we know the speed is 0.980c, so the time required is

$$\frac{45.0c \text{ y}}{0.980c} = 45.9 \text{ y}$$

(c) To find the time as measured by occupants of the spaceship, we can use the distance result from part (a):

$$\Delta t = \frac{8.95c \text{ y}}{0.98c} = 9.14 \text{ y}$$

Alternatively, we can apply the time dilation relationship:

$$\Delta t = \frac{\Delta t_0}{\sqrt{1 - v^2/c^2}}$$

$$\Delta t_0 = \Delta t \sqrt{1 - v^2/c^2} = (45.9 \text{ y}) \sqrt{1 - 0.98^2} = 9.14 \text{ y}$$

Making sense of the result:

The length contraction and time dilation approaches for finding the trip duration in part (c) yield the same value, as expected.

LO 5

29-5 Relativistic Momentum and Energy

In special relativity, linear momentum can still be expressed in the form $m\vec{v}$, but a velocity-dependent factor appears when we consider the **relativistic momentum**, that is, the momentum in terms of the rest mass, m_0, of an object. **Rest mass** is the mass as measured in a reference frame in which the object is not moving.

KEY EQUATION
$$\vec{p} = m\vec{v} = \frac{m_0 \vec{v}}{\sqrt{1 - v^2/c^2}} \qquad (29\text{-}14)$$

Note that the relativistic factor for momentum has the same form as the factors for time dilation and length contraction. Some experts in relativity prefer to think of this factor as part of the definition of linear momentum. However, we can also consider the mass of an object as increasing with increasing velocity:

KEY EQUATION
$$m = \frac{m_0}{\sqrt{1 - v^2/c^2}} \qquad (29\text{-}15)$$

Probably the most famous equation in all of science is the mass–energy equivalence formulated by Einstein in 1905 in one of his "miracle year" papers:

KEY EQUATION $$E = mc^2 \qquad (29\text{-}16)$$

Actually, the equation $E = mc^2$ never appeared in Einstein's paper. He did argue however that if a body lost energy that its mass would change by that energy loss divided by the square of the speed of light. According to this relationship, it is possible to convert mass into energy, or energy into mass.

Einstein's justification for his derivation of mass–energy equivalence was based on the fact that electromagnetic waves carry momentum. For example, the momentum of sunlight is the main cause of comet tails pointing away from the Sun. In the 19th century, physicists established that a packet of electromagnetic waves with energy E carries linear momentum p of value E/c.

Although Einstein's complete derivation is too complex to include here, we can demonstrate the underlying principle. If the energy of a light wave does have a mass equivalent, m, its momentum would be $p = mv = mc$ because the energy is moving at the speed of light. As described above, the momentum of the light wave also equals E/c, so we can readily obtain $E = mc^2$. However, the real proof of $E = mc^2$ and the relativistic momentum relationship consists of the countless experiments that have proven that the universe operates in a way that is consistent with this relationship.

For example, mass-to-energy conversion occurs in nuclear power reactors, where large nuclei are broken

Mass-Energy Conversion in the Sun

In the dominant reaction that produces energy in the core of the Sun, four hydrogen nuclei (protons) fuse into one helium nucleus (α-particle) and emit two positrons and two neutrinos of negligible mass. The rest mass of a proton is 1.007 276 467 u, and the rest mass of an α-particle is 4.001 506 177 u, where u is the unified atomic mass unit (1 u = $1.660\ 539 \times 10^{-27}$ kg). Material in the interior of stars such as the Sun is essentially totally ionized, so we use the mass of the nuclei, not the neutral atoms, in the calculation. The rest mass of a positron is $9.109\ 38 \times 10^{-31}$ kg.

(a) What is the mass loss in this fusion reaction?
(b) What percentage of the original mass is lost?
(c) How much energy is released by one reaction? Express the energy in joules and electron volts.
(d) The power output of the Sun is 3.85×10^{26} W. How many hydrogen fusion reactions per second are needed to provide this energy output?

SOLUTION

(a) The net reaction is 4 protons → 1 α-particle + 2 positrons + energy. We need to carry several significant figures because the difference in mass is a very small percentage.

Initial mass:

4 protons: $4 \cdot 1.007\ 276\ 467$ u $\cdot 1.660\ 539 \times 10^{-27}$ kg/1u

$= 6.690\ 487\ 428\ 9 \times 10^{-27}$ kg

Products:

α − particle: $1 \cdot 4.001\ 506\ 174\ 7$ u $\cdot 1.660\ 539 \times 10^{-27}$ kg/1u

$= 6.644\ 657\ 065 \times 10^{-27}$ kg

2 positrons: $2 \cdot 9.109\ 38 \times 10^{-31}$ kg $= 1.821\ 876 \times 10^{-30}$ kg

Therefore, the overall mass loss is $4.400\ 849 \times 10^{-29}$ kg.

(b) The percentage of the original mass lost is

$(4.400\ 849 \times 10^{-29}$ kg$/6.690\ 49 \times 10^{-27}$ kg$) \times 100\% = 0.66\%$

(c) To calculate the energy released, we use the mass loss from (a) for m in $E = mc^2$. Note that only the mass difference is converted into energy, not the entire original mass. Putting in the numbers we obtain 3.96×10^{-12} J. Using the conversion 1 eV = 1.60×10^{-19} J, we have 2.48×10^7 eV, or 24.8 MeV.

(d) To determine the number of fusion reactions, we divide the power output of the Sun by the energy produced by each fusion reaction:

$$\frac{3.85 \times 10^{26} \text{ J/s}}{3.96 \times 10^{-12} \text{ J/reaction}} = 9.72 \times 10^{37} \text{ reactions/s}$$

Making sense of the result:

A tiny amount of mass is lost in each hydrogen fusion reaction. In most nuclear reactions only a small part of the original mass is lost. We see that nuclear processes create, per reaction, much more energy than chemical processes (although all energy does ultimately come from the same mass–energy equivalence). The Sun actually has several different nuclear reactions taking place, but the result for the number of reactions per second is a good first approximation. The number of hydrogen fusion reactions is huge, and each one produces two neutrinos. Therefore, the Sun produces an incredible number of neutrinos each second. It has been estimated that approximately 100 trillion neutrinos from the Sun pass through your body each second.

into two or more lighter nuclei. During this process, a small amount of the mass of each original nucleus is converted into energy. The Sun and similar stars obtain their energy by nuclear fusion in their core, where temperatures and pressures are immense. In **nuclear fusion** processes, light elements fuse together to form heavier elements. Each fusion reaction has a small net mass loss; the "lost" mass corresponds to the energy produced. In fact, at the fundamental level, all chemical reactions also derive their energy from slight changes in mass.

Relativistic Kinetic Energy

The classical expression for kinetic energy is not valid at relativistic speeds. The **relativistic kinetic energy** is the difference between the total energy and the rest mass energy:

KEY EQUATION $KE = E - m_0 c^2$ (29-17)

Substituting Equation (29-15), we obtain the following expression for relativistic kinetic energy:

$$KE = mc^2 - m_0 c^2 = m_0 c^2 \left(\frac{1}{\sqrt{1 - v^2/c^2}} - 1 \right) \quad (29\text{-}18)$$

You may wonder how this equation relates to the classical result, $E = \frac{1}{2} mv^2$. If we use a series expansion for the square root function in Equation (29-18), we have

$$KE = m_0 c^2 \left[\left(1 - v^2/c^2 \right)^{-1/2} - 1 \right]$$

$$= m_0 c^2 \left[1 + \frac{1}{2} \frac{v^2}{c^2} - \frac{1 \cdot 3}{2 \cdot 4} \left(\frac{v^2}{c^2} \right)^2 + \frac{1 \cdot 3 \cdot 5}{2 \cdot 4 \cdot 6} \left(\frac{v^2}{c^2} \right)^3 - \cdots - 1 \right]$$

$$= m_0 c^2 \left[\frac{1}{2} \frac{v^2}{c^2} - \frac{3}{8} \left(\frac{v^2}{c^2} \right)^2 + \frac{15}{48} \left(\frac{v^2}{c^2} \right)^3 - \cdots \right]$$

When $v \ll c$, we can ignore all but the first term in the series, leaving an expression that simplifies to the classical result for kinetic energy.

✓ CHECKPOINT

C-29-5 Relativistic Kinetic Energy

If an object has a total relativistic energy that is three times its rest mass energy, what fraction of the total energy is the relativistic kinetic energy?

(a) 1/9

(b) 1/6

(c) 1/3

(d) 2/3

C-29-5 (d) If $E = 3m_0c^2$ and $KE = E - m_0c^2$, then KE must be 2/3 of the total energy, E.

EXAMPLE 29-6

Relativistic Kinetic Energy and Momentum

A spaceship has a rest mass of 2000 kg and travels at 90.0% of the speed of light.

(a) What is the relativistic kinetic energy of the spaceship?

(b) What is the kinetic energy according to classical physics?

(c) What is the relativistic momentum of the spaceship?

SOLUTION

(a) We can use the relationship for relativistic kinetic energy (Equation (29-18)):

$$KE_{rel} = m_0c^2\left(\frac{1}{\sqrt{1 - v^2/c^2}} - 1\right)$$

$$= (2000 \text{ kg})(3.00 \times 10^8 \text{ m/s})^2\left(\frac{1}{\sqrt{1 - 0.900^2}} - 1\right)$$

$$KE_{rel} = 2.33 \times 10^{20} \text{ J}$$

(b) Using the classical mechanics expression for kinetic energy:

$$KE_{cl} = \frac{1}{2}mv^2 = \frac{1}{2}(2000 \text{ kg})((0.900)(3.00 \times 10^8 \text{ m/s}))^2$$

$$= 7.29 \times 10^{19} \text{ J}$$

(c) The relativistic momentum has the same direction as the velocity of the spaceship and has a magnitude of

$$p = mv = \frac{m_0v}{\sqrt{1 - v^2/c^2}} = \frac{(2000 \text{ kg})(2.70 \times 10^8 \text{ m/s})}{\sqrt{1 - 0.900^2}}$$

$$= 1.24 \times 10^{12} \text{ kg} \cdot \text{m/s}$$

Making sense of the result:

We see that the relativistic kinetic energy is significantly greater than the classical kinetic energy, as expected at a speed of 90% of the speed of light. In part (c), we see that the relativistic momentum is similarly much greater than the classical momentum.

We can see that relativistic effects become extreme as v approaches c. For example, if an object were moving at the speed of light, the length of the object would contract to zero, and its kinetic energy would become infinite. The speed of light appears to be a universal upper limit for speeds. Some theoretical models suggest that particles that always travel at speeds greater than the speed of light may exist. To date, these particles, called tachyons, have never been observed.

LO 6

29-6 Relative Speeds

In Figure 29-10, an object moves at some speed v with respect to reference frame B. However, reference frame B is itself moving at speed u with respect to another reference frame, A. What is the speed, v', of the object, as measured by reference frame A? By common sense, if the speeds are much less than the speed of light, the result should be approximately $u + v$. However, all measured speeds must not exceed c, even if the individual speeds u and v are both comparable to the speed of light. The relationship for relativistic speed addition below meets these criteria:

KEY EQUATION
$$v' = \frac{u + v}{1 + \dfrac{uv}{c^2}} \qquad (29-19)$$

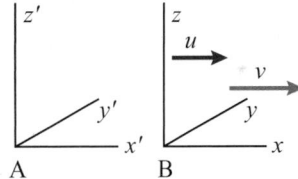

Figure 29-10 An object moves at speed v relative to a reference frame that itself is moving at speed u relative to another observer. What is the speed of the object relative to that observer?

✓ CHECKPOINT

C-29-6 Relative Speeds

An observer on a planet sees two spaceships, each travelling at half the speed of light, approaching from opposite directions. What is the relative speed of one spaceship as measured by the other?

(a) c

(b) $\frac{4}{5}c$

(c) $\frac{3}{4}c$

(d) $\frac{1}{2}c$

C-29-6 (b) If we take the left spaceship as the reference, both u and v are $\frac{1}{2}c$, and

$$v' = \frac{c/2 + c/2}{1 + \frac{(c/2)(c/2)}{c^2}} = \frac{c}{\frac{5}{4}} = \frac{4}{5}c.$$

MODERN PHYSICS

EXAMPLE 29-7

Relative Speeds

An approaching spaceship launches a survey probe toward you. Observers on the spaceship see the probe move at 90% of the speed of light. However, you measure the speed of the probe as 95% of the speed of light. How fast is the spaceship moving in your reference frame?

SOLUTION

Using the notation of Equation (29-19), $v = 0.90c$, $v' = 0.95c$, and u is unknown. Rearranging Equation (29-16) we get

$$v' - v = u - \frac{uvv'}{c^2}$$

Solving for u we have

$$u = \frac{v' - v}{1 - \frac{vv'}{c^2}} = 0.34c$$

Making sense of the result:

Although classical physics predicts that the spaceship is travelling toward you at 5% of the speed of light, relativistic velocity addition shows that the speed of the spaceship is 34% of the speed of light.

LO 7

29-7 Spacetime

When solving problems related to relativity, it is important to consider spatial and time coordinates as integrated into one spacetime system. In spacetime, we view each event as being specified by four spacetime coordinates, such as (ct, x, y, z). The time coordinate is multiplied by the speed of light so that all four coordinates have units of distance. Another observer could associate a different set of spacetime coordinates, (ct', x', y', z'), to the same event, similar to the way two observers could assign different coordinates to the same point in three-dimensional space. For example, if two spots are marked on a blackboard, students could draw different x- and y-axes and assign different coordinates to the same spot, as shown in Figure 29-11. However, the students would agree on the length of the line joining the two spots.

A similar situation applies in spacetime, where different observers agree on the **spacetime interval** described in Equation (29-20) below. We demonstrate in Example 29-8 how the constancy, or invariance, of this spacetime interval can be used to solve special relativity problems. Four-dimensional spacetime diagrams are called Minkowski diagrams, named for the German mathematician Hermann Minkowski (see the Making

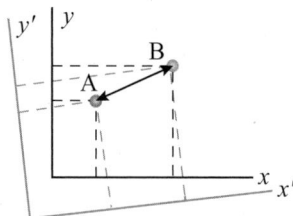

Figure 29-11 In an analogy to spacetime intervals in two dimensional space two reference frames do not agree on the coordinates, but they do agree on the length between the points.

Connections feature below called "The Development of Spacetime Ideas").

KEY EQUATION $\qquad (\Delta s)^2 = (c\Delta t)^2 - (\Delta x)^2 - (\Delta y)^2 - (\Delta z)^2$

$$(29\text{-}20)$$

ONLINE ACTIVITY

Spacetime Interval

The e-resource that accompanies every new copy of this textbook contains an Online Activity using the "Spacetime Interval" simulation. Work through the simulation and accompanying questions to gain an understanding of spacetime intervals.

MAKING CONNECTIONS

The Development of Spacetime Ideas

While the core concepts of spacetime were present in Einstein's 1905 relativity papers, it was Hermann Minkowski (1864–1909) who developed much of the mathematics we now associate with spacetime. Minkowski was Einstein's mathematics teacher at Zurich Polytechnic. The class Minkowski taught Einstein had only five students, including only two physics majors, Einstein and Mileva Marić, who would become Einstein's first wife. Einstein liked to concentrate on what interested him and not always on what was being taught, and in that sense he was a difficult student. It is said that Minkowski was surprised at the brilliance of Einstein's 1905 paper and the extent to which it was influenced by the mathematics Minkowski had taught.

EXAMPLE 29-8

Spacetime Intervals

Karen sees two events occur at the same place but separated by $2.50\ \mu s$ in time. Another observer, Tasha, sees the two events separated by $3.50\ \mu s$ in time. According to Tasha, what is the spatial separation between the two events?

SOLUTION

For Karen, we have $\Delta x = \Delta y = \Delta z = 0$, so the square of the spacetime interval is

$$(\Delta s)^2 = (c\Delta t)^2 - (\Delta x)^2 - (\Delta y)^2 - (\Delta z)^2$$

$$= (3.0 \times 10^8 \text{ m/s} \cdot 2.5 \times 10^{-6} \text{ s})^2 = 5.63 \times 10^5 \text{ m}^2$$

Now, both observers must have the same value for the spacetime interval. Therefore, we have the following, where we have let $\Delta x'$ represent the spatial separation of the events for Tasha:

$$5.63 \times 10^5 \text{ m}^2 = (c\Delta t')^2 - (\Delta x')^2$$

$$= (3.0 \times 10^8 \times 3.5 \times 10^{-6})^2 - (\Delta x')^2$$

$$(\Delta x')^2 = 1.103 \times 10^6 \text{ m}^2 - 5.63 \times 10^5 \text{ m}^2$$

$$\Delta x' = \pm 735 \text{ m}$$

Thus, in Tasha's reference frame the two events are separated by 735 m.

Making sense of the result:

In calculations of spacetime intervals, the squared spatial coordinates are subtracted from the squared time coordinate, so if the time component of a given spacetime interval is less time, the spatial components must also be less, and vice versa. Therefore, it makes sense that Karen measures a shorter time with the events closer together, while Tasha measures a longer time interval and a greater spatial separation.

✓ CHECKPOINT

C-29-7 Two Perspectives

John sees two events occur with a shorter time separation than Jillian, who observes the same two events. What can we conclude about their perspectives on the spatial separation of the events?

(a) John thinks they are closer together spatially.
(b) Jillian thinks they are closer together spatially.
(c) They both agree on exactly the same spatial separation.
(d) The situation of the question is impossible: they cannot measure different time separations between the same events.

C-29-7 (a) If the time separation is less, the spatial separation must also be less for the spacetime interval to be consistent for observers.

LO 8

29-8 Gravitational Time Dilation in General Relativity

Einstein's **mass–energy equivalence** equation, $E = mc^2$, indicates that mass can be created from energy as long as all conservation laws are obeyed. According to classical Newtonian physics, any existing mass would instantly experience the gravitational force exerted by the newly created mass. But this instantaneous effect requires information to be transmitted faster than the speed of light, contrary to special relativity. Einstein saw this problem while developing special relativity and spent about a decade working on a solution: general relativity.

The first step was the equivalence principle. The **equivalence principle** states that the acceleration of a reference frame and the effect of gravity cannot be distinguished. For example, if you were in a small, windowless box, you could not distinguish the presence of a gravitational field from being in an accelerating spaceship far from any masses. In what Einstein later described as his happiest (some prefer the translation "most fortunate") thought, he realized that the way to remove the effect of gravity was to consider reference frames in free fall.

If gravitation is not a force, what does it become in general relativity? The answer is that masses curve spacetime. Imagine first that we had no masses in the universe. The spacetime would be flat and would not vary from location to location. If we then add a mass, say a dense star, the spacetime curves, or warps, in the region near the star, as shown in Figure 29-12. The motion of objects follows this curved spacetime, somewhat like the way your motion is dictated by the contour of the landscape when you drive in hilly country. The renowned American theoretical physicist John Archibald Wheeler (1911–2008) concisely summarized general relativity as this: matter tells spacetime how to curve, and curved spacetime tells matter how to move.

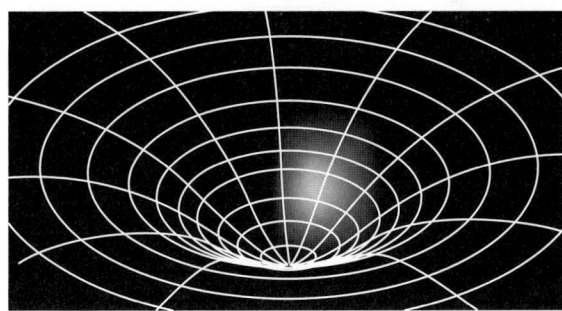

Figure 29-12 Spacetime is curved in the region around a mass

The field equations that describe general relativity are mathematically complex even for symmetrical situations. Nonetheless, general relativity does provide an alternative to Newtonian gravity for explaining the motion of planets and other objects in elliptical orbits around the Sun. The speed of an object in an elliptical orbit varies as its distance from the Sun varies because the curvature of spacetime is greater closer to the Sun.

The presence of masses does not just curve spacetime; it also causes time to slow down. The greater the curvature of spacetime, the more slowly time moves. This effect is called **gravitational time dilation**. To understand why general relativity predicts time dilation, we consider a related effect, gravitational redshift. **Gravitational redshift** is related to the curvature of spacetime and is not the same as the velocity-based redshift of light emitted by sources moving away from you. Velocity-based redshift can be expressed by the following relationship:

$$\frac{\lambda_{obs} - \lambda_{em}}{\lambda_{em}} \approx \frac{v_r}{c} \quad \text{for } v_r \ll c \qquad (29\text{-}21)$$

Here, λ_{em} is the wavelength of light as measured by an observer in the reference frame of the object emitting the light, and λ_{obs} is the wavelength measured by an observer who sees the object moving away with a radial velocity component of v_r. When a light source is moving away from us, it is redshifted (has longer wavelengths), and a light source moving toward us is blueshifted. The redshift and blueshift for light is an analogous wave phenomenon to the Doppler effect for sound.

We will use the thought experiment shown in Figure 29-13 to introduce gravitational redshift. An elevator car has an apparatus on the floor that emits a photon of light toward a detector on the ceiling. At exactly the instant that the photon is emitted, the cable on the elevator is released, allowing the elevator car to fall freely toward Earth. (We neglect Earth's rotation and orbital motion in this thought experiment.)

Figure 29-13 A source emits a photon from the bottom at the instant the elevator goes into free fall. A detector is at the top of the elevator box.

An observer watching from Earth's surface would expect the light to be blueshifted when it reaches the detector because the detector is falling toward the photon. However, Einstein proposed that any reference frame in gravitational free fall is just as valid as a reference frame at rest for general relativity. Therefore, there should not be any unusual results in the free-falling frame of the elevator car; hence, the photon should have the same wavelength when it is detected as when it was emitted, as would be observed in the elevator frame of reference.

We can reconcile the wavelength measurements in the two reference frames only if the curved spacetime of Earth's gravitational field causes a redshift that exactly

cancels the blueshift in the free-fall reference frame. Numerous observations have demonstrated that curved spacetime does indeed cause a gravitational redshift.

Combining the inverse relationship between period and frequency with the wave equation for light, we obtain

$$c = \frac{\lambda}{T} \qquad (29\text{-}22)$$

Since c has a constant value, Equation (29-22) requires a corresponding inverse gravitational time dilation for any gravitational redshift. This gravitational time dilation means that a clock in steeply curved spacetime (near a large mass) will run slower than an identical clock in flat spacetime (far from any significant masses).

It can be shown (see problem 57) that gravitational time dilation is described by the following relationship:

$$\frac{\Delta t_0}{\Delta t_\infty} = \left(1 - \frac{2GM}{R_0 c^2}\right)^{1/2} \qquad (29\text{-}23)$$

where M is the mass (assumed to be spherically symmetric), R_0 is the distance of the observer who measures Δt_0 from the centre of mass M, and Δt_∞ is the time measured by an observer at an infinite distance, where spacetime is assumed to be flat. Since the quantity in brackets is always less than 1, $\Delta t_0 < \Delta t_\infty$.

 CHECKPOINT

C-29-8 Gravitational Time Dilation

Assume that observers in each of the following locations observe the time between the same two distant events (e.g., two supernova explosions in a distant galaxy). Rank, from the shortest to the longest, the time intervals measured by observers at the following locations:

(a) Earth's surface
(b) the surface of a small asteroid at the same distance from the Sun as Earth
(c) the surface of a more massive and higher-density planet
(d) just outside a very dense stellar remnant

C-29-8 (d), (c), (a), (b) The more sharply curved spacetime is, the more time is slowed for that observer. The gravitational field is weakest for the asteroid and strongest for the stellar remnant. Where time passes more slowly, observers will measure a shorter time period for the distant events.

 ONLINE ACTIVITY

Time in Curved Spacetime

The e-resource that accompanies every new copy of this textbook contains an Online Activity using the "Time in Curved Spacetime" simulation. Work through the simulation and accompanying questions to gain an understanding of curved spacetime.

EXAMPLE 29-9

Gravitational Time Dilation Near a Dwarf Star

Say 100.000 s passes on a clock that is in flat spacetime, far away from any masses. What time passes on a clock near the surface of a white dwarf star with a mass of 1.800×10^{30} kg and a radius of 6.100×10^6 m?

SOLUTION

Applying the relationship for gravitational time dilation (Equation (29-23)),

$$\frac{\Delta t_0}{\Delta t_\infty} = \left(1 - \frac{2GM}{R_0 c^2}\right)^{1/2}$$

$$= \left(1 - \frac{2(6.67 \times 10^{-11})(1.80 \times 10^{30})}{(6.10 \times 10^6)(3.00 \times 10^8)^2}\right)^{1/2} = 0.999\,781$$

$$\Delta t_0 = 0.999\,781 \times 100.000 \text{ s} = 99.978\,1 \text{ s}$$

Making sense of the result:

Even near massive high-density objects such as a white dwarf star, the effect of gravitational time dilation is slight.

Black Holes

The right side of Equation (29-23) is zero when

$$R_0 = \frac{2GM}{c^2} \qquad (29\text{-}24)$$

An observer at distance R_0 would record zero local time change even when a distant observer in flat spacetime notes a very long time interval. For this reason, we call this value of R_0 the **event horizon**. The stopping of time at this distance from the mass justifies the expression, "nothing ever happens at the event horizon." We will see below that Equation (29-24) is also an expression for the dimension of a black hole.

The idea of a **black hole**—a region that does not let even light escape—may seem far-fetched, but there is excellent evidence for the existence of several types of black holes in the universe. First, we will use a classical argument to derive the size of the boundary of the black hole. It is a boundary in the sense that if you pass within, you will no longer be able to communicate with the rest of the universe.

Recall from Chapter 11 that escape from near the surface of a spherical object of mass M and radius R requires the total of the kinetic energy and the gravitational potential energy to be at least zero:

$$\frac{1}{2}mv_{esc}^2 - \frac{GMm}{R^2} \geq 0 \qquad (11\text{-}10)$$

When we divide both sides by m, set the expression to the minimum value of 0, use c for the escape speed, and solve for this special value of R, which we will call R_S, we get the following result:

KEY EQUATION
$$R_S = \frac{2GM}{c^2} \qquad (29\text{-}25)$$

While this classical approach to the escape of light is not strictly valid, the field equations of general relativity yield exactly the same result when solved for a single, radially symmetric mass (like a black hole). The escape limit for light, R_S, is called the **Schwarzschild radius** in honour of the German scientist Karl Schwarzschild (1873–1916),

MAKING CONNECTIONS

Black Holes

The first to consider the idea of a dense star trapping light was probably the British clergyman Rev. John Michell (1724–1793), who proposed the idea in 1783. In the view of general relativity, a black hole is an object that is completely collapsed in terms of spacetime curvature. The term "black hole" was coined by American physicist John Wheeler because it was awkward to keep saying "completely gravitationally collapsed object." Wheeler was passionate about the importance of teaching and learning physics and continued to teach first-year physics through most of his career. Together with Edward Taylor he wrote one of the best books on relativity called "Spacetime Physics." Theoretical calculations of the collapse of large stars led to predictions of the existence of stellar black holes. Observations of an object near the X-ray source Cygnus X-1 by Canadian astrophysicists including Dr. Tom Bolton of the University of Toronto established the first (and some of the most conclusive) evidence for the existence of a stellar black hole. Canadian rock band Rush wrote and performed a song called "Cygnus X-1" about this discovery and the strange nature of black holes.

More recent data suggest the existence of much larger black holes at the centres of most galaxies, including our own Milky Way galaxy. The LIGO experiment (see Chapter 28) will provide a means for testing general relativity, including the detection of interacting black holes. The famous physicist Dr. Stephen Hawking (b. 1942) has proposed the possible existence of tiny primordial black holes formed in regions of extremely dense matter during the initial stage of the expansion of the universe.

MAKING CONNECTIONS

Solving for Black Holes

Figure 29-14 Dr. Karl Schwarzschild

Einstein had thought that it would be many years before the field equations for general relativity would be solved, but only months after receiving the equations in a letter from Einstein, the brilliant German physicist and astrophysicist Karl Schwarzschild (Figure 29-14) completed an exact solution for a single, spherical nonrotating mass, a solution that could be applied to black holes. Soon afterward, Schwarzschild produced a more general solution with much wider applications. At that time, Schwarzschild was literally in the trenches of the Russian front during World War I. Tragically, he contracted a rare skin disease and died at age 42, shortly after publishing the second solution.

Schwarzschild published two papers on celestial mechanics before he was 16, and he speculated on the possibility of curved spacetime well before Einstein developed the concept. Although Schwarzschild's best known work is related to general relativity, he also made significant contributions to astronomical research and quantum mechanics.

the first person to solve Einstein's field equations for the case of a black hole.

What are black holes in terms of general relativity? The mass of a black hole is so concentrated that it causes the extreme curvature of spacetime. As described above, gravitational time dilation at the Schwarzschild radius

(event horizon) essentially slows time to a stop. If one is inside the event horizon of a black hole, it is impossible to ever escape from the black hole, or even to send information past the event horizon of the black hole. Thus, the Schwarzschild radius is a measure of the size of a black hole. If we shrink the physical size of an object to less than this limit, the object will become a black hole. For example, an object with Earth's mass would have to shrink to a radius of 8.8 mm to become a black hole. Some very massive stars may become black holes when they exhaust their supply of nuclear fuel and collapse. It appears that natural processes also form much larger black holes at the centres of galaxies.

The concepts of general relativity affect many aspects of astrophysics. For example, the curved spacetime near massive objects bends light in a pattern somewhat like that of a convex lens. This phenomenon is called **gravitational lensing**. However, the spacetime lenses of galaxies are much more irregular than the glass lenses you use in optics experiments. Consequently, gravitational lensing usually produces distorted images, somewhat like viewing a street lamp through a wine glass. The distorted arcs in Figure 29-15 are the lights produced by gravitational lenses.

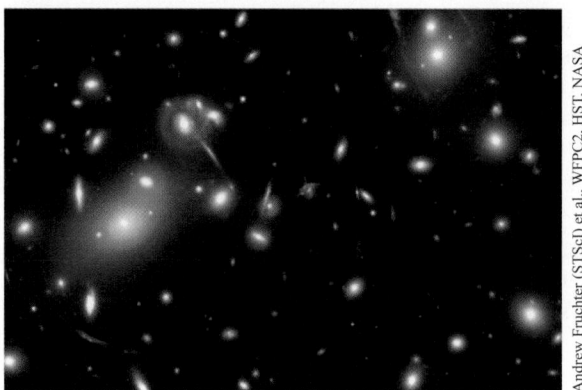

Figure 29-15 This astronomical image is proof that mass really does bend spacetime. The dense galaxy cluster called Abell 2218 bends light from more distant galaxies, stretching their images into arcs.

LO 9

29-9 Relativity and the Global Positioning System

The **Global Positioning System** (GPS) measures time delays in signals received from four different satellites. With this information, we can calculate position coordinates for the receiver and also obtain a precise time. Consider this two-dimensional analogy: If you are on the surface of a flat Earth and know that you are a certain distance, d_1, from a city, then you could draw a circle of radius d_1 around that city. You would then know that your position is somewhere on that circle. If you also know that you are at a distance d_2 from a second city, you could draw a circle of radius d_2 around that city, and you would know that you must be at one of the two intersections of the two circles. For example, Figure 29-16 shows one point, Thompson, Manitoba: the point is 1188 km from Calgary and 2211 km from Montréal. Knowing your distance from a third city lets you uniquely determine which intersection point is correct. The GPS system calculates a position in three dimensional space in a similar manner.

Figure 29-16 You can use known distances from two cities to find two possible locations on a flat Earth (in this case one near Thompson, Manitoba, and the other in the northern United States). The GPS system works in three dimensions with more points.

 CHECKPOINT

C-29-9 GPS Corrections

Which statement correctly summarizes the relativistic effects for a clock on a GPS satellite orbiting Earth as compared to an identical clock on the surface of Earth?

(a) The GPS clock runs faster due to special relativity and more slowly due to general relativity.

(b) The GPS clock runs faster due to general relativity and more slowly due to special relativity.

(c) The GPS clock runs more slowly due to both effects.

(d) The GPS clock runs faster due to both effects.

C-29-9 (b) The clock on the GPS satellite is moving relative to the clock on Earth, so the GPS clock runs more slowly due to special relativistic time dilation. However, the spacetime around the satellite is less curved than the spacetime at Earth's surface, so the satellite clock runs faster because the gravitational time dilation is nearer to Earth's centre.

A consumer-level GPS receiver has a precision of approximately one metre. (More precise receivers are available for military and government operations.) At present, there are more than 30 active GPS satellites. These satellites orbit at a distance of approximately four Earth radii in six planes inclined at about 55° with respect

Figure 29-17 The configuration of GPS satellites

to the equator, as shown in Figure 29-17. Each satellite circles Earth approximately twice every day.

If the desired precision is of the order of one metre, then the time must be accurate to that distance divided by the speed of light or approximately 3.3 ns. The atomic (cesium) clocks that are on the satellites have sufficient precision (a typical cesium atomic clock has a precision of 1 ns or better). However, the satellite clocks are all moving relative to the receiver on Earth, and they are several times farther from Earth's centre.

Even though the curvature of spacetime is not severe near Earth, there is still a small effect due to gravitational time dilation. Time runs more slowly in the more sharply curved spacetime near Earth's surface. Therefore, the satellite clocks will, by general relativity, tick faster than one on Earth. The difference is approximately 46 μs a day (see Example 29-10), great enough that the needed 3.3 ns precision would be lost in less than 10 s.

However, we also must consider special relativity time dilation because the satellites in orbit are travelling at a significant speed compared to the ground-based receiver. Even though the speed is not large compared to the speed of light, the precision needed means that relativistic effects are still important. Note that time dilation causes the satellite clocks to run more slowly than clocks on the ground by approximately 7 μs per day (see Example 29-10). Since the two time dilations have opposite effects, the special relativity dilation partly cancels the general relativity dilation, leaving a net time error of approximately 39 μs per day.

The designers of the GPS system understood both special and general relativity effects and built the corrections into the system's calculations. With these corrections, based directly on Einstein's discoveries about a century ago, we have a remarkably precise system for navigation worldwide.

EXAMPLE 29-10

GPS Relativistic Corrections

(a) The GPS satellites make one orbit around Earth in one-half of a sidereal day. A sidereal (star-based) day is approximately 23 h and 56 min. Find the radius of the orbit of a GPS satellite, assuming that the orbit is circular.

(b) What is the orbital speed of a GPS satellite relative to the centre of Earth?

(c) Calculate the amount per solar day by which a GPS satellite clock differs from a clock on Earth due to special relativistic time dilation. State any assumptions you make for this calculation.

(d) Calculate the amount per solar day by which a GPS satellite clock differs from a clock on Earth due to gravitational time dilation.

SOLUTION

(a) We will first use the Newtonian form of Kepler's third law to solve for the length of the semi-major axis, a. Since we are assuming a circular orbit, this axis will be the same as the radius of the orbit:

$$T^2 = \frac{4\pi^2}{G(m_1 + m_2)} a^3$$

The satellite period T is 11 h and 58 m, or 43 080 s. The mass of the satellite is negligible compared to the mass of Earth, so for the mass term in the denominator we use Earth's mass, 5.974×10^{24} kg. Substituting these values and solving for a gives

$$a = 2.656 \times 10^7 \text{ m}$$

(b) For a circular orbit, the orbital speed is given by $v = \frac{2\pi a}{T}$. Substituting $a = 2.656 \times 10^7$ m and T = 43 080 s, we obtain $v = 3.873 \times 10^3$ m/s.

(c) The clock on Earth moves as Earth rotates, so the relative speed of the satellite varies because its orbit is not directly above the equator. We will assume that the mean speed of the satellite relative to the clock on Earth is equal to the orbital speed of the satellite (which is a reasonable approximation). Both clocks are actually accelerating constantly, so neither is in a truly inertial reference frame. We will assume that the effects of these accelerations can be ignored.

We saw in the special relativity time dilation section that

$$\Delta t = \frac{\Delta t_0}{\sqrt{1 - v^2/c^2}}$$

Using the orbital speed from part (b), we obtain

$$\frac{\Delta t}{\Delta t_0} = 1.000\ 000\ 000\ 0833$$

To find the time dilation per day, multiply the difference between this ratio and 1 by the number of seconds in a solar day:

$$0.000\ 000\ 000\ 0833 \times (24 \text{ h} \times 3600 \text{ s/h}) = 7.20 \times 10^{-6} \text{ s}$$
$$= 7.2\ \mu s$$

(d) The general relativity time dilation relationship relates time measured by an observer at a distance R_0 from the centre of a spherical mass M to the time measured by an observer in flat spacetime, infinitely far from any masses:

$$\frac{\Delta t_0}{\Delta t_\infty} = \left(1 - \frac{2GM}{R_0 c^2}\right)^{1/2}$$

However, we want to compare two clocks that are both in curved spacetime. We use the subscript E to denote quantities related to the Earth clock and the subscript S for quantities related to the satellite clock:

$$\frac{\Delta t_S}{\Delta t_E} = \frac{\left(\frac{\Delta t_S}{\Delta t_\infty}\right)}{\left(\frac{\Delta t_E}{\Delta t_\infty}\right)} = \frac{\left(1 - \frac{2GM}{R_S c^2}\right)^{1/2}}{\left(1 - \frac{2GM}{R_E c^2}\right)^{1/2}}$$

We know from part (a) that $R_S = 2.656 \times 10^7$ m. The radius of Earth is 6.378×10^6 m, and the mass of Earth is 5.974×10^{24} kg. Substituting these values, we obtain

$$\frac{\Delta t_S}{\Delta t_E} = 1.000\ 000\ 000\ 528$$

Again, we multiply the difference between this ratio and 1 by the number of seconds in a solar day to find the time dilation per day:

$$0.000\ 000\ 000\ 528 \times (24 \text{ h} \times 3600 \text{ s/h}) = 4.56 \times 10^{-5} \text{ s}$$
$$= 46\ \mu s$$

Making sense of the result:

In part (a), we found that the GPS satellites were at a radial orbital distance of about four Earth radii, a reasonable result. Because the satellites orbit twice each day, we expect them to be at a lower altitude than the geosynchronous satellites studied in Chapter 11.

In part (b), we obtained an orbital speed of approximately 3.9 km/s, which is somewhat less than the escape speed (11.2 km/s) and therefore a reasonable result. In parts (c) and (d), we found that the correction due to general relativity is greater than the correction due to special relativity, and both corrections are essential for determining accurate locations using the GPS system.

SPECIAL RELATIVITY

Special relativity deals with objects moving near the speed of light as seen from two inertial (nonaccelerating) reference frames. As we approach speeds that are comparable to the speed of light, there is a slowing of time, an increase in linear momentum and energy, and a contraction of the lengths of objects in the direction of motion.

Spacetime Intervals

According to relativity, we must unify space and time coordinates into spacetime. Different observers no longer agree on various physical measurements, although they do agree on spacetime intervals between pairs of events:

$$(\Delta s)^2 = (c\Delta t)^2 - (\Delta x)^2 - (\Delta y)^2 - (\Delta z)^2 \quad (29\text{-}20)$$

Time Dilation

Proper time is measured by an observer for whom the interval events begin and end at the same spatial coordinates. The time interval measured by any other observer is

$$\Delta t = \frac{\Delta t_0}{\sqrt{1 - v^2/c^2}} \quad (29\text{-}8)$$

where Δt_0 is the proper time interval, v is the relative speed, and c is the speed of light.

Length Contraction

In special relativity, the observer for whom an object is not moving measures the proper length, ΔL_0, for that object. Lengths are contracted in the direction of motion such that

$$L = L_0 \sqrt{1 - v^2/c^2} \quad (29\text{-}13)$$

The same observer does not measure both proper time and proper length.

Relativistic Momentum

Relativistic momentum is given by

$$\vec{p} = m\vec{v} = \frac{m_0 \vec{v}}{\sqrt{1 - v^2/c^2}} \quad (29\text{-}14)$$

where m_0 is the rest mass of the object.

Relativistic Kinetic Energy

Mass and energy are different forms of a unified quantity, with mass–energy equivalence given by $E = mc^2$.

Relativistic kinetic energy is defined as the difference between the total relativistic energy, E, and rest mass energy:

$$KE = mc^2 - m_0 c^2 = m_0 c^2 \left(\frac{1}{\sqrt{1 - v^2/c^2}} - 1 \right) \quad (29\text{-}18)$$

Relative Speeds

Special relativity requires that speeds not exceed c. If an object is moving at speed v relative to an inertial reference frame itself moving at speed u relative to another frame, then the speed of the object relative to that second reference frame, v', is given by

$$v' = \frac{u + v}{1 + \dfrac{uv}{c^2}} \quad (29\text{-}19)$$

GENERAL RELATIVITY

General relativity expands special relativity to include the gravitational effects of masses by treating gravitation as a curvature of spacetime.

Gravitational Time Dilation

Time passes more slowly in regions of curved spacetime. The time interval Δt_0 measured by an observer at a distance R_0 from a mass m is

$$\frac{\Delta t_0}{\Delta t_\infty} = \left(1 - \frac{2GM}{R_0 c^2} \right)^{1/2} \quad (29\text{-}23)$$

where Δt is the time interval as measured by an observer far from any masses.

Applications: nuclear fusion power, gravitational lensing, Global Positioning System (GPS), dynamics of particles in accelerators and in cosmic ray physics, black hole physics, space travel

Key Terms: black hole, equivalence principle, ether, event horizon, general relativity, Global Positioning System, gravitational lensing, gravitational redshift, gravitational time dilation, inertial reference frame, length contraction, mass–energy equivalence, Michelson–Morley experiment, nuclear fusion, proper length, proper time interval, proper time, relativistic kinetic energy, relativistic momentum, rest mass, Schwarzschild radius, spacetime interval, spacetime, special relativity, thought experiment, time dilation

QUESTIONS

1. Which of the following statements is correct with respect to the laws of physics in different inertial reference frames?
   (a) The laws of physics are different in different inertial reference frames.
   (b) The laws of physics vary by a constant in different inertial frames.
   (c) The laws of physics are the same in different inertial frames.
   (d) The laws of physics are scaled by some factor in different inertial reference frames.

2. Parvati moves past you at a speed comparable to the speed of light. According to you, when 100 s passes on your watch, how much time passes on Parvati's watch?
   (a) more than 100 s
   (b) exactly 100 s
   (c) less than 100 s but more than 50 s
   (d) exactly 50 s

3. On what do different inertial observers agree?
   (a) only the laws of physics
   (b) only the laws of physics and the speed of light
   (c) the laws of physics, the speed of light, and the spacetime interval between two events
   (d) the laws of physics, the speed of light, the times of events, and the spatial coordinates of events

4. According to special relativity, for observers moving past you at a high speed, you observe their time running slowly, their lengths contracted in the direction of motion, and an increase in their mass. What will these observers say about you?
   (a) My time runs fast, my mass decreases, and my length increases.
   (b) My time runs fast, but my mass and length do not change.
   (c) My time runs slowly, my mass decreases, and my length increases.
   (d) My time runs slowly, my mass increases, and my length contracts.

5. A black hole has a mass M, and its Schwarzschild radius is R_S. We add mass so that its mass is now $2M$. What is the new Schwarzschild radius?
   (a) It is no longer a black hole.
   (b) $R_S/2$
   (c) $1.4R_S$
   (d) $2R_S$

6. If the Sun were to become a black hole (astrophysics tells us that this will not happen for stars the size of the Sun), what would happen to Earth?
   (a) Earth would fall into the black hole Sun.
   (b) Earth would stay in exactly its current orbit.
   (c) Earth would gradually spiral in until it fell within the Schwarzschild radius of the black hole Sun.
   (d) Earth would break free from the gravity of the Sun and move into interstellar space.

7. On which of the following objects would time be expected to run most slowly?
   (a) a planet of mass M and radius R
   (b) a planet of mass M and radius $2R$
   (c) a planet of mass M and radius $R/2$
   (d) a planet of mass $2M$ and radius $2R$

8. Consider three particles. According to some inertial reference frame,

A has a rest mass energy of E_0 and a total energy of $2E_0$;
B has a rest mass energy of $2E_0$ and a total energy of $3E_0$;
C has a rest mass energy of $3E_0$ and a total energy of $4E_0$.
Which statement about the relative rest masses is correct?
   (a) A has the least rest mass, and C has the greatest rest mass.
   (b) A, B, and C all have exactly the same rest mass values.
   (c) C has the least rest mass, and A has the greatest rest mass.
   (d) A and C have a larger rest mass than B.

9. Considering the data in question 8, which statement about the relative kinetic energies of the objects is correct?
   (a) A has the least and C has the greatest kinetic energy.
   (b) C has the least and A has the greatest kinetic energy.
   (c) B has the least and A the greatest kinetic energy.
   (d) All three have exactly the same kinetic energy values.

10. Considering the data from questions 8 and 9, which particle has the greatest speed?
    (a) A
    (b) B
    (c) C
    (d) They all have exactly the same speed.

11. You are travelling in a spaceship at a speed in the $+x$-direction of $0.5c$ relative to an Earth-based reference frame. Within the spaceship you are conducting an experiment in which subatomic particles are moving at a speed of $0.5c$ relative to you (in the $+x$-direction as well). What is the approximate speed of the particles relative to Earth?
    (a) $0.3c$
    (b) $0.5c$
    (c) $0.8c$
    (d) $1.0c$

12. You observe light produced by atoms located just outside the surface of a very dense white dwarf star. A spectral line observed from these atoms compared to the same spectral line on Earth has
    (a) exactly the same wavelength;
    (b) a slightly longer wavelength;
    (c) a slightly shorter wavelength;
    (d) a much longer wavelength.

13. Which of the following statements is *false*?
    (a) The speed of light (in a vacuum) is constant.
    (b) Observers in different inertial reference frames measure the same speed of light.
    (c) The speed of light slows down in the region near (but outside of) a black hole.
    (d) As an object approaches the speed of light, the relativistic kinetic energy of the object becomes very large.

PROBLEMS BY SECTION

For problems, star ratings will be used, (✷, ✷✷, or ✷✷✷), with more stars meaning more challenging problems.

Section 29-1   Special and General Relativity

14. ✷ You are writing an introduction to relativity for non-specialist readers. Select one of the following as the best description of the key idea of *general* relativity. Support your choice through an analysis of all four statements.

(a) No object can travel faster than the speed of light.

(b) When objects travel near the speed of light, time goes more slowly.

(c) Gravity is not a force but rather a curvature of spacetime near massive objects.

(d) The laws of physics are the same for observers in relative motion.

Section 29-2 Reference Frames and the Michelson–Morley Experiment

15. ✱ You learned in Chapter 21 that the permittivity of free space, ε_0, has the value 8.85×10^{-12} C^2/(N·m^2). Use the speed of light $c = 3.00 \times 10^8$ m/s to find a value for the permeability of free space, μ_0.

16. ✱✱ In their 1887 experiment, Michelson and Morley used an effective length of 1.10 m for the distance d in the interferometer. One arm of an interferometer is parallel to the motion of Earth in its orbit at a speed of 29 800 m/s. Calculate the time *difference* for light travel along the two paths, assuming that the standard value for the speed of light is used for the other arm and that speeds add in the Galilean way expected prior to relativity. Use Michelson and Morley's value for d.

Section 29-3 Postulates of Special Relativity and Time Dilation

17. ✱✱ (a) How long, from the perspective of an Earth-bound observer, would a flight around Earth's circumference at an altitude of 1000 km above Earth's surface take if you could travel at 50% of the speed of light? For this question, treat the reference frames as inertial.

(b) What time would be required from the perspective of someone who was in the capsule taking the flight?

18. ✱✱ A fundamental particle called a muon has a lifetime, in its own reference frame, of 2.2 μs. Muons from a cosmic ray atmospheric burst have an apparent lifetime of 15 μs for a ground-based observer. What must be the speed of the muon relative to the ground-based observer?

Section 29-4 Length Contraction

19. ✱✱ The nearest star similar to the Sun is Alpha Centauri, which is at a distance of 4.40 ly. If one travels at 75% of the speed of light, how long would a one-way trip to the star from Earth require from the viewpoint of

(a) an Earth-based observer;

(b) an occupant on the spacecraft?

Assume that the initial acceleration is sufficient that the entire trip is essentially at full speed.

20. ✱✱ (a) If you make a round trip to a star that is 400 ly away (as measured by an Earth-based observer) in a total time of 40 y (as measured by people on the spacecraft), what speed is needed? (Assume, somewhat unrealistically, that the time that the spacecraft takes to accelerate to its cruising speed is negligible.)

(b) What time would the Earth-based observer measure for the trip?

(c) What distance (one way) would the spaceship determine for the trip?

Section 29-5 Relativistic Momentum and Energy

21. ✱✱ Assume that a high-energy cosmic ray has a relativistic kinetic energy of 75 J. Assume that the cosmic ray is a proton. Determine the speed at which it must be travelling relative to Earth.

22. ✱✱ The maximum speed that a meteor can have (relative to Earth) when it is in orbit around the Sun is approximately 73 km/s. Assume that the mass of the meteor is 0.050 kg.

(a) What is the relativistic linear momentum of the meteor, and what is its momentum according to classical Newtonian mechanics? Determine the percentage difference between the two values.

(b) Calculate the relativistic and Newtonian kinetic energies, and the percentage difference between the two.

23. ✱✱ A free neutron can decay into a proton as long as an electron and an electron neutrino are also produced (to conserve electrical charge and lepton characteristics in the reaction). The rest mass of a neutron is $1.674\ 927 \times 10^{-27}$ kg, the proton rest mass is $1.672\ 622 \times 10^{-27}$ kg, and the rest mass of an electron is $9.109\ 38 \times 10^{-31}$ kg. What is the kinetic energy that must be released in the decay? Express your answer in both J and MeV. Assume that no rest mass is associated with the neutrino.

Section 29-6 Relative Speed

24. ✱✱ We are being visited by two alien civilizations, each approaching us from opposite directions. Alien civilization A has a spacecraft approaching us at a relative speed of 0.80c, and civilization B is approaching us at a speed of 0.90c. What speed does civilization A measure for the spacecraft of civilization B?

Section 29-7 Spacetime

25. ✱ Observer Mei sees two events as being simultaneous in time but separated by 300 m in space. Li sees the two events as taking place with a spatial separation of 500 m. What will Li say is the time separation between the two events?

Section 29-8 Gravitational Time Dilation in General Relativity

26. ✱✱ Consider two locations, one at sea level and one high in the Rockies at an elevation of 3400 m above sea level.

(a) Would time move more slowly at sea level or high in the Rockies?

(b) Over a 70 y lifetime, what would be the difference in time, due to gravitational time dilation, between the two locations?

27. ✱ A black hole has a mass that is 12 times the mass of the Sun. What is the radius of the event horizon around the black hole?

Section 29-9 Relativity and the Global Positioning System

28. ✱✱✱ An exoplanet has a mass exactly double Earth's mass, and a radius that is 85% of Earth's radius. The planet has a sidereal (relative to distant stars) rotation period of 7.45×10^4 s. Assume that the exoplanet has a GPS system similar to ours, and that the periods of the GPS satellites are one-half of the planet's sidereal period.

(a) What is the radius (from the centre of the exoplanet) of the satellite orbits?

(b) What is the special relativity time correction between the satellite and the surface of the exoplanet? Express your answer in terms of difference in time over a sidereal "day" on the exoplanet.

(c) What is the general relativity time correction per sidereal day?

(d) What is the net correction per day?

COMPREHENSIVE PROBLEMS

29. ✷ Calculate the length of one light year in metres. Assume that the duration of the year is 365.25 days.

30. ✷✷ The Sun is located at a distance of approximately 25 500 ly from the centre of our Milky Way galaxy. If one wanted to travel from Earth to the centre of the galaxy in 45 y, at what speed should the spacecraft move?

31. ✷✷ The bright star Vega is 25.3 ly from Earth. A planet around that star is featured in the movie *Contact*.

(a) What time is required to send a radio signal from Earth to the planet orbiting Vega and back again?

(b) If a spacecraft were to travel at a speed of $0.985c$ from Earth to the planet and back again, what time does an Earth-based observer measure for the return trip? (Assume that the acceleration time is negligible.)

(c) What time does an observer on the spacecraft measure for the round trip?

32. ✷✷ A friend passes you in her spaceship travelling at a relative speed of $0.80c$. As she passes you, you measure her spaceship to be 5.2 m long and 1.5 m high. It is moving in the direction of the length.

(a) What are the length and height of the spaceship as measured in a reference frame in which the spaceship is at rest?

(b) The rest mass of your friend's spaceship is 5.0×10^3 kg. What is the relativistic kinetic energy of the spaceship as it travels at $0.80c$?

33. ✷✷ A hypothetical high-speed transit line advertises that passengers will be carried from Halifax, Nova Scotia, to Vancouver, British Columbia, a distance of approximately 6000 km, at such a high speed that a passenger's watch only registers half the time for the trip that synchronized clocks on the stations will read.

(a) What is the speed of the transit vehicles?

(b) What is the time for a one-way trip, as measured by the clocks at the stations?

34. ✷ In the electromagnetism chapters, you learned about the electron volt as a unit of energy. What is the speed of an electron that has a total energy, E, of 1.2 MeV? Express your answer in terms of c.

35. ✷✷✷ Older televisions and monitors used CRT (cathode ray tube) displays. Assume that the electrons in a CRT display are accelerated by a potential difference of 32 kV.

(a) Use the classical expression for kinetic energy to determine the speed of the electrons.

(b) Now use relativistic kinetic energy to determine the speed of the electrons.

(c) Are relativistic considerations important in the design of CRT electronics?

36. ✷✷ An electron is accelerated through a potential difference of 18 MV (million volts). Calculate the speed of the electron using classical and relativistic kinetic energy equations. Are relativistic effects important in this situation?

37. ✷ When a star goes supernova, the light from the star might reach Earth a few hours after a burst of neutrinos. Is this in conflict with special relativity? Explain.

38. ✷✷ (a) An interstellar probe travelling at $0.85c$ relative to a local planetary system hits a meteoroid in that planetary system (i.e., the relative speed of the meteoroid is $0.85c$). The meteoroid has a mass of 1.6 g (about half the mass of a penny). Find the relativistic kinetic energy of the meteoroid.

(b) Find the TNT equivalent to the kinetic energy of the meteoroid, given that 1 kg of TNT explosive releases approximately 4.2 MJ.

39. ✷✷✷ The Sun has a luminosity of 3.85×10^{26} W. This is the total of the power that it radiates in all directions.

(a) According to $E = mc^2$, how much mass per second must the Sun convert to create this much power?

(b) The most important nuclear fusion reaction sequence in the Sun is one in which four hydrogen nuclei (protons) combine, forming one helium nucleus (an α-particle), plus some low-mass products such as neutrinos. In this reaction, 0.66% of the initial mass of the hydrogen is converted to energy. What mass of hydrogen must the Sun convert to helium each year?

(c) The current mass of the Sun is about 2.0×10^{30} kg. Approximately 10% of this mass is hydrogen available for nuclear fusion reactions (not all the hydrogen can be brought to core for nuclear reactions). For how long can the Sun produce its current power output?

40. ✷✷ One of the exoplanets recently discovered is around a star 49.8 ly from Earth. You are proposing that a spaceship travel at a speed of $0.980c$ to this exoplanet. Assume this speed is reached sufficiently quickly that we can consider the spaceship to make the entire voyage at this speed.

(a) How long (in years) will a one-way trip take, according to clocks on the spaceship?

(b) What is the time (in years) for the one-way trip, according to Earth-based clocks?

(c) Assume that the spaceship has a mass of 12 million tonnes (1.2×10^{10} kg), which is the minimum suggested by considerations of rocket design and closed systems for long voyages, even with propulsion systems. Find the relativistic kinetic energy of the spaceship when it reaches cruising speed.

41. ✷✷ One observer sees two spacetime events happen separated by 10 ns in time and by 3.0 m in space. Another inertial observer sees the same events as happening at the same time. What will the inertial observer say is the spatial separation of the events?

42. ✷✷ (a) Danielle sees two events occur at the same place but separated by 2.00 μs in time. Another observer, Wei, sees the two events separated by 100.0 m in space. What does Wei say is the time between the two events?

(b) What is the relative speed of their reference frames?

43. ✷✷ In one reference frame two events occur with a spatial separation of 4.0 m and separated in time by 20.0 ns. Another inertial observer travels in such a way as both events occur at the same spatial location.

(a) What approximate time interval does a clock with that observer measure between the two events?

(b) What is the speed of one reference frame relative to the other?

44. ✹✹ The Sun has a mass of approximately 2.0×10^{30} kg and a radius of 6.96×10^8 m.

(a) Assuming that the Sun is spherical, what is its mass density?

(b) If a star has the same density as the Sun but greater mass, could it ever become a black hole? What would the radius need to be?

(c) Does an object need to be very dense to become a black hole? (In 1792, the great French scientist Pierre-Simon de Laplace commented on how a star with such dimensions could trap light.)

45. ✹✹ An interstellar spacecraft of length 250 m is moving at a speed of $0.75c$ with respect to its home planet reference frame. An observer on the planet observes a tiny meteoroid moving in a direction opposite to the motion of the spacecraft and at a speed of $0.50c$ relative to the planet. What is the time, from the viewpoint of an observer on the spacecraft, for the meteoroid to move past the length of the spacecraft?

46. ✹✹ A particle has a total energy E that is exactly three times its rest mass energy. What is the speed of the particle?

47. ✹✹ Consider three inertial reference frames, A, B, and C. B and C are in motion in the same direction (along the $+x$-axis) with respect to A. Assume that reference frame B moves at a relative speed of $0.50c$ with respect to A, and reference frame C moves at a speed of $0.40c$ with respect to B. An object moves, also along the $+x$-axis, at a speed of $0.30c$ with respect to reference frame C. What is the speed of the object as measured by reference frame A?

48. ✹ Look up the lyrics to the song "Cygnus X-1" by the Canadian band Rush. Analyze the lyrics in terms of what we know about black holes and the candidate black hole near the X-ray source Cygnus X-1.

49. ✹✹ Dr. Stephen Hawking and others have proposed the possibility that small black holes, called primordial black holes, might have formed shortly after the Big Bang. It is probable that these primordial black holes were formed with masses ranging from perhaps 10^{-8} to 10^{35} kg. Calculate the Schwarzschild radius for the largest and the smallest of these proposed primordial black holes, and find the mass density of each.

50. ✹✹ If an object has the density of water, what would its radius need to be to become a black hole?

51. ✹✹✹ The world's most powerful particle accelerator, the Large Hadron Collider (LHC) at CERN, became operational in 2008–2009. ATLAS, one of the two main experiments, involves the collaboration of more than 1000 scientists, including many from Canada. This experiment sends protons in opposite directions around a circular path with a 27 km circumference. After acceleration to very high speeds, some of the protons collide head-on, liberating enough energy to create a number of different particles.

(a) The kinetic energy of each proton is up to 7.0 TeV. What is the speed of a proton in terms of c?

(b) How many times per second will a proton go around the 27 km ring?

(c) What is the relativistic momentum of a proton at the kinetic energy stated in part (a)?

52. ✹✹✹ The Alberta Large-area Time-coincidence Array (ALTA) experiment uses an array of detectors, many on roofs of schools, to detect very high-energy cosmic rays by searching for time-coincident events over large areas. Such arrays sometimes detect cosmic rays with energies as high as 10^{19} eV or more. Interestingly, the origin of these high-energy cosmic rays is not definitively known.

(a) A cosmic ray proton has an energy of 10^{19} eV. Which produces higher-energy particles: the CERN LHC or cosmic rays? By what factor?

(b) What is the speed of a proton that has a relativistic kinetic energy of 10^{19} eV?

(c) What is its relativistic momentum?

(d) What is the time we measure for the proton to cross our galaxy, a distance of about 280 000 ly (if we include the halo region)?

(e) What is the time as measured by a clock that travels with the proton?

53. ✹✹✹ One observer sees two particles, each of rest mass m_0, collide head-on, with one particle travelling at 80% of the speed of light in the negative x-direction and the other at 80% of the speed of light in the positive x-direction. A second observer sees the same collision but is in an inertial reference frame in which one of the particles is at rest.

(a) At what speed is the other particle travelling in that reference frame?

(b) Find the kinetic energy of the particles in the first reference frame.

(c) Find the kinetic energy of the moving particle in the second reference frame.

54. ✹✹✹ The rest mass energy, E_0, the total energy, E, and the relativistic momentum, p, are related as follows:

$$E^2 = p^2c^2 + E_0^2$$

Derive this relationship. (Hint: Start with an expression for the relativistic momentum, square that, and multiply by c^2.)

55. ✹✹✹ A particle of rest mass m_0 moves (according to some inertial reference frame) in the $+x$-direction with a speed V. The particle then has a totally inelastic collision with a particle of rest mass $m_0/2$ moving with speed V but in the negative x-direction. What is the rest mass of the particle produced in the collision? (Hint: Both total relativistic energy and total relativistic momentum must be conserved in the collision.)

56. ✹✹✹ Electromagnetic waves carry momentum. The momentum carried by a wave packet of energy E is E/c. The effect of momentum transfer from electromagnetic waves is called radiation pressure.

(a) Derive an expression for the largest radius of a dust grain for which radiation pressure just balances the gravitational pull of the Sun. Assume that dust grains are completely reflecting spheres of radius r and mass density ρ. Denote the luminosity (power radiated) by the Sun as L and the distance of the grain from the Sun as d. (Hint: Think about the momentum transfer from the wave to the dust grain.)

(b) Does the distance from the Sun really matter in this calculation?

(c) Substitute reasonable values, and calculate the maximum radius. The Sun has a luminosity of 3.85×10^{26} W.

57. ✷✷✷ Use the gravitational time dilation thought experiment in the text (section 29-8) to obtain the following expression for gravitational time dilation at a point a distance R_0 from a spherically symmetric mass M:

$$\frac{\Delta t_0}{\Delta t_\infty} = \left(1 - \frac{2GM}{R_0 c^2} \right)^{1/2}$$

DATA-RICH PROBLEM

58. ✷✷✷ Table 29-1 gives the magnitude of the relativistic momentum, p, and the relativistic kinetic energy, KE, for a series of unknown velocities. Use numerical or analytical approaches to use the data to answer the following questions.

(a) What is the rest mass of the object?

(b) What is the range of velocities represented in the table, expressed in terms of the speed of light?

(c) What is the value for relativistic linear momentum in the blank cell in the left column?

(d) What is the value for kinetic energy in the blank cell in the right column?

(e) What total energy, E, corresponds to the data in the last row of the table?

Table 29-1 Data for Problem 58

ρ (kg·m/s)	KE (J)
7.54×10^7	1.13×10^{15}
1.53×10^8	4.64×10^{15}
2.36×10^8	1.09×10^{16}
3.27×10^8	2.05×10^{16}
4.33×10^8	3.48×10^{16}
5.63×10^8	5.63×10^{16}
	9.01×10^{16}
1.00×10^9	

OPEN PROBLEM

59. ✷✷✷ You have asked to plan the first interstellar space mission to carry people to another planetary system.

(a) Look up recent data on exoplanet discoveries, and select a specific target. Explain why you chose that target planet (in terms of habitable zone and other features). Note the distance to the exoplanet system.

(b) What would you consider a reasonable round-trip time for the visit (time as measured by the travellers)?

(c) Use relativity to calculate the speed required, assuming that most of the trip is done at a constant speed and the time for acceleration and deceleration is not significant compared to the duration of the voyage.

(d) How many people, minimum, do you feel that such a mission should carry?

(e) Considering your answers to (b) and (c), what would you suggest as the minimum mass of the spacecraft? Provide the analysis upon which your answer is based (e.g., the amount of mass needed for fuel, food production, protective shell). Canadian scientist Dr. Mike Dixon and his research team at the University of Guelph are working on closed environmental systems to be used to process wastes and grow food for long-duration missions to Mars.

(f) How much energy would be required to accelerate the spacecraft to the speed you calculated in part (c)? Express this energy in joules and in terms of the annual energy use in Canada.

See the text online resources at www.physics1e.nelson.com for Open Problems and Data-Rich Problems related to this chapter.

Chapter 30
Fundamental Discoveries of Modern Physics

The ATLAS detector is part of the Large Hadron Collider at CERN (Figure 30-1). The detector was built as a collaboration among 3000 scientists from more than 169 universities and laboratories and 37 nations—one of the largest collaborations in the physical sciences. Of the 3000 scientists, approximately 1000 are students. The Large Hadron Collider is the most powerful particle accelerator built to date. In 2012, it was producing beams of protons with energies of 4 TeV. An upgrade scheduled for completion in 2014 will enable the collider to accelerate particles to 7 TeV. In this chapter, we will see how particle collisions helped physicists realize that there is a realm of physics beyond what is now called classical physics.

The genesis of classical physics was the publication of Newton's *Philosophiæ Naturalis Principia Mathematica* in 1687, which included his three laws of motion and law of universal gravitation. This view of the world formed the basis of a great deal of scientific advance and reigned well into the 1800s. At that time, experimental results began to emerge that could not be reconciled with some aspects of Newtonian physics. This period coincided with the Industrial Revolution, when scientific advances both enabled and were enabled by technological advances. In this chapter, we will examine some of the experimental evidence and early theoretical efforts that ultimately led to the development of quantum physics.

Much of the progress occurred during a time of scientific confusion and controversy. The emergence of quantum theory was hotly debated because it required discarding some of the concepts that were the basis of extremely successful models for describing nature. We will not present the material in a strictly chronological order but in a sequence that allows for the logical development of the topics. Keep in mind that this sequence does not reflect the considerable difficulties that physicists faced in developing and accepting radical new theories during the first half of the 20th century.

Figure 30-1 The ATLAS detector at CERN

MAXIMILIEN BRICE, CERN/SCIENCE PHOTO LIBRARY

MODERN PHYSICS

For additional Making Connections, Examples, and Checkpoints, as well as activities and experiments to help increase your understanding of the chapter's concepts, please go to the text's online resources at www.physics1e.nelson.com.

30-1 The Connection Between Matter and Electricity

In 1834, Faraday published the results of a series of experiments that led to an understanding of the electrical properties of matter. Faraday passed electric currents through solutions and observed that gas evolves at the electrodes. This process is called electrolysis. A typical setup for such an experiment is shown in Figure 30-2.

Faraday noted that the amount of gas evolved at each electrode is always proportional to the total charge that passes through the cell. He also observed that the relative amounts of gas produced are in the same proportion as

$$2\,H_2O\,(l) \rightarrow O_2\,(g) + 4\,H^+\,(aq) + 4e^-$$

$$2\,H_2O\,(l) + 2\,e^- \rightarrow H_2\,(g) + 2\,OH^-\,(aq)$$

Figure 30-2   An apparatus for the electrolysis of water. A weak acid such as vinegar is added to the water to improve conduction. As current flows through the cell, hydrogen gas and oxygen gas are evolved at the electrodes.

they are known to be in the substance being electrolyzed. For example, in the electrolysis of water, twice as much hydrogen gas (by volume) is evolved as oxygen. Faraday found that when he passed the same amount of electricity through different cells, the mass of each element produced at the electrodes is proportional to the atomic weight of the element divided by its **valence**, which is a measure of the number of bonds formed by each atom of the element. The constant of proportionality, now called Faraday's constant, F, has a value of 96 485 C/mol.

✓ CHECKPOINT

C-30-1 Charge and Mass

Which of the following statements is true?

(a) Faraday demonstrated that the charge-to-mass ratio is the same for all elements.

(b) Faraday demonstrated that each element has a unique charge-to-mass ratio.

(c) Faraday demonstrated that there is no relationship between mass and charge.

(d) Faraday demonstrated that all elements have the same amount of charge but differ in their mass.

C-30-1 (b) Faraday showed that for a given element, the gas evolved at an electrode is always proportional to the total charge passed through the cell and that each element has a unique constant of proportionality.

Faraday's observations established a clear link between matter and electricity. Faraday observed that the ratio of the charge passed through the system to the mass of gas evolved is the same from experiment to experiment for a particular substance, but a different ratio is obtained for different substances. It seemed that each substance had a well-defined charge-to-mass ratio. However, his results gave no insight into whether the variations in the ratio from substance to substance were due to differences in mass, charge, or both.

EXAMPLE 30-1

The Electrolysis of Water

During a water electrolysis experiment, a current of 36 mA is passed through the cell for 60 seconds. How many hydrogen molecules are liberated from the cell during that time?

SOLUTION

The total charge that passes through the cell is given by

$$Q = \int_0^{60} I\,dt = It\Big|_0^{60} = 2.16\text{ C}$$

In Figure 30-2 we see that one hydrogen molecule is liberated for every two electrons that pass through the cell. Therefore, the number of hydrogen molecules produced is

$$N_{H_2} = \frac{Q}{2e} = \frac{2.16\text{ C}}{(2)(1.602 \times 10^{-19}\text{ C})} = 6.74 \times 10^{18}$$

In one mole of material, there are 6.022×10^{23} particles. Thus, the quantity liberated in this example is 1.12×10^{-5}mol.

Making sense of the result:

One mole of ideal gas at standard temperature and pressure occupies 22.4 L, so the volume of gas produced in 1 min occupies approximately 0.25 mL. The current is relatively small, so this amount seems reasonable.

LO 2

30-2 Temperature and the Emission of Light

Some of the early investigations into how the temperature of an object affects the light that it emits played a fundamental role in the development of modern physics.

MAKING CONNECTIONS

Spectrometers

How do we characterize the light that an object emits? One method is to measure the intensity of the light as a function of wavelength. In a **spectrometer**, light passes through a dispersive element, such as a prism or a diffraction grating. The intensity of the light at the different wavelengths is then measured at a detector, as shown in Figure 30-3. Adjusting the angle of the grating determines which wavelengths strike the detector, thus allowing a measurement of intensity as a function of wavelength.

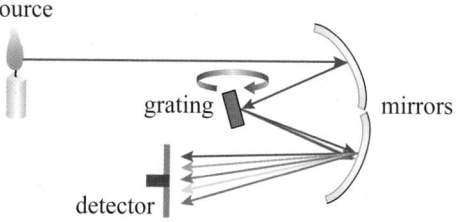

Figure 30-3   A schematic diagram of a spectrometer

ONLINE ACTIVITY

Blackbody Spectra

The e-resource that accompanies every new copy of this textbook contains an Online Activity using the PhET simulation "Blackbody Spectra." Work through the simulation and accompanying questions to gain an understanding of black bodies.

 CHECKPOINT

C-30-2 Blackbodies

As we increase the temperature of a blackbody, the
(a)   intensity changes;
(b)   the peak intensity wavelength changes;
(c)   the general shape of the curve changes.
(d)   None of the above.

C-30-2 (a) and (b) The important feature of blackbody radiators is that they all have the same emission spectra if they are all at the same temperature.

If we heat an object to a sufficiently high temperature, the object will visibly glow. Thus, hot objects, such as the stove element in Figure 30-4, emit visible light. In fact, an object can emit all types of electromagnetic radiation, and the spectrum the object emits is characteristic of the temperature of the object. Detailed experiments show that the behaviour of most hot objects approximates the spectrum of a blackbody. A **blackbody** is an object for which the character of the light emitted depends only on the temperature of the body. The spectrum of a blackbody is continuous; that is, a blackbody emits electromagnetic waves at all wavelengths, although the intensity varies considerably with wavelength.

Figure 30-4   A stove element glows red-hot

Numerous efforts to describe blackbody radiation were made. As we saw in Chapter 17, the total power radiated by a hot object was shown to be

$$P = \sigma \varepsilon A T^4 \qquad (17\text{-}36)$$

where $\sigma = 5.67 \times 10^{-8}\,\text{W/m}^2 \cdot \text{K}^4$ is the Stefan-Boltzmann constant, ε is the emissivity of the object, A is the surface area and T is the temperature of the object in kelvins.

The wavelength at which the greatest intensity of radiation is emitted had been shown to be given by **Wien's Law**

$$\lambda_{\text{max}} = \frac{2.898 \times 10^{-3}\,\text{m} \cdot \text{K}}{T} \qquad (30\text{-}1)$$

 EXAMPLE 30-2

The Power of the Sun

The sun has a surface temperature of about 6000 K and has a surface area of about $6 \times 10^{18}\,\text{m}^2$. How much power does the sun radiate and what is the peak wavelength in the spectrum, assuming it radiates as a blackbody with an emissivity of 1?

(continued)

MODERN PHYSICS

Answer: We can use the two relationships above to find these answers. We can determine the total power radiated using Equation (17-36)

$$P_{sun} = 5.67 \times 10^{-8}\,\frac{W}{m^2 \cdot K^4} \times 1.6 \times 10^{18}\,m^2 \times (6000\,K)^4$$

$$= 1.2 \times 10^{26}\,W$$

The peak wavelength in the spectrum is obtained from Equation (30-1)

$$\lambda_{max} = \frac{2.898 \times 10^{-3}\,m \cdot K}{6000\,K} = 483\,nm$$

Making sense of the results:

The peak wavelength falls right into the visible spectrum of light. In fact, this is why the visible spectrum falls where it does—our eyes have evolved to be most sensitive in the part of the spectrum where there is the most light. The power result is huge. If we divide this result by the surface of a sphere whose radius is the orbit of the earth, we get a power density of about 1.5 kW/m², which agrees reasonably well with the measured solar flux.

These two results were based on thermodynamic reasoning and gave some insight into the nature of black-body radiation. A full description of the spectrum did not come until 1900 when Max Planck (1858–1947) made a bold assumption. He assumed that the atoms in the heated material oscillated with a frequency *f* and emitted radiation when they did so. However, he then postulated that the radiation could only be emitted in discrete amounts with an energy *E* given by

$$E = hf \qquad (30\text{-}2)$$

where $h = 6.626 \times 10^{-34}$ J·s is a constant known as **Planck's constant**. Based on this, he was able to show that the spectral radiance of the emitted radiation depended upon the frequency and the temperature through the following relationship

$$B(f,T) = \frac{8\pi hf^3}{c^2}\frac{1}{e^{\frac{hf}{kT}} - 1} \qquad (30\text{-}3)$$

The units of *B* are J·m⁻². To get the total power radiated by the object, we would need to integrate *B* with respect to *f*. This result showed excellent agreement with experiment. Interestingly, the key assumption of quantizing the energy was not immediately recognized as being of great significance—Planck himself did not place great emphasis on the assumption in his initial work. However, it turned out to be the birth of quantum physics.

MAKING CONNECTIONS

Set the Controls for the Heart of the Sun

The surface temperature of the Sun is approximately 6000 K. Have you ever wondered how this temperature was determined? Obviously, no one has been to the Sun with a thermometer. Astronomers analyze the light emitted by the Sun and compare it to a blackbody spectrum. As shown in Figure 30-5, a blackbody temperature of 5777 K gives a reasonably good fit to the solar spectrum.

extraterrestrial solar spectral irradiance total area: 1367 W/m²

blackbody spectrum for $T = 5777$ K total area: 1367 W/m²

Figure 30-5   A spectrum of the Sun with a blackbody fit

LO 3

30-3   Gas Discharge Spectra

In this section, we will concentrate on the light emitted from excited gases. A familiar source of this type of light is a neon sign, as shown in Figure 30-6. A neon sign is an example of a discharge tube. A **discharge tube** contains a gas at a relatively low pressure. Electrodes at each end of the tube allow an electric potential difference to be applied across the gas. When a sufficiently high potential difference is applied, a current flows through the gas, which then emits light. The spectrum of this light is not continuous—the light is emitted at discrete wavelengths that depend on the type of gas in the tube.

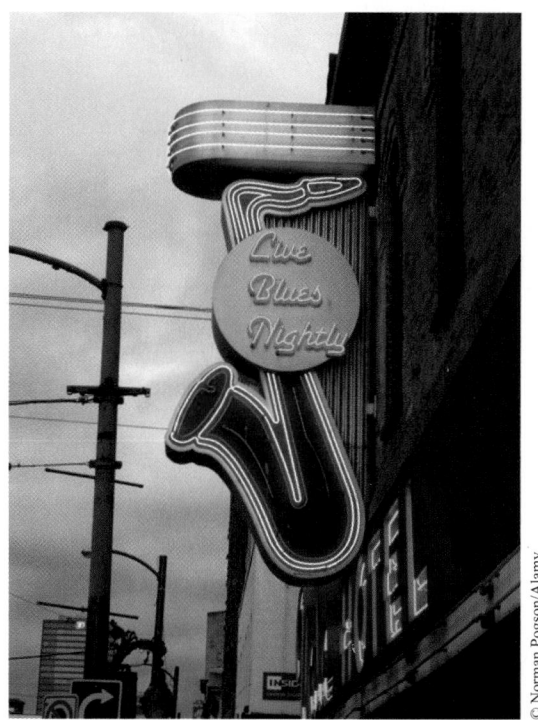

Figure 30-6 A neon sign emits light with a characteristic reddish-orange colour

ONLINE ACTIVITY

Gas Discharge Spectra

The e-resource that accompanies every new copy of this textbook contains an Online Activity using the PhET simulation "Gas Discharge Spectra." Work through the simulation and accompanying questions to gain an understanding of discharge spectra.

Figure 30-7 shows spectra obtained with a spectrometer for a hydrogen discharge tube and a helium discharge tube. You can see that there is only significant intensity at certain wavelengths. Further, the two gases have very different spectra.

These observations presented a formidable challenge. Maxwell had demonstrated in 1846 that light is a form of electromagnetic radiation. In 1887, Hertz had shown that electromagnetic radiation is produced when charged particles are accelerated. The fact that a gas discharge spectrum has discrete frequencies suggests that charged particles in the gas oscillate at only those frequencies. Thus, any theory that tried to account for these spectra would have to include oscillating charged particles that were somehow constrained to oscillate only at certain frequencies.

Figure 30-7 Intensity versus wavelength for a hydrogen discharge tube (left) and a helium discharge tube (right). The insets show the tubes themselves.

Much effort was expended to determine whether these spectral lines could be described by mathematical formulas. The Swedish physicist Johannes Rydberg (1854–1919) is credited with developing a general formula applicable to the spectra of all elements. His formula describes the spectral lines in terms of wave numbers instead of wavelengths. The **wave number**, $\tilde{\nu}$, is simply the reciprocal of the wavelength, λ, and therefore equals the number of wavelengths per unit distance:

KEY EQUATION
$$\tilde{\nu} = \frac{1}{\lambda} \tag{30-4}$$

Rydberg noted that the wave numbers of spectral lines are not uniformly distributed and tend to fall into groups or series. If we assign an integer to each line within a series and plot the wave number versus the series number, we see that each series follows a smooth curve with a similar shape, as shown in Figure 30-8 for hydrogen. This similarity suggested to Rydberg that a single formula could be used to describe all the spectral lines.

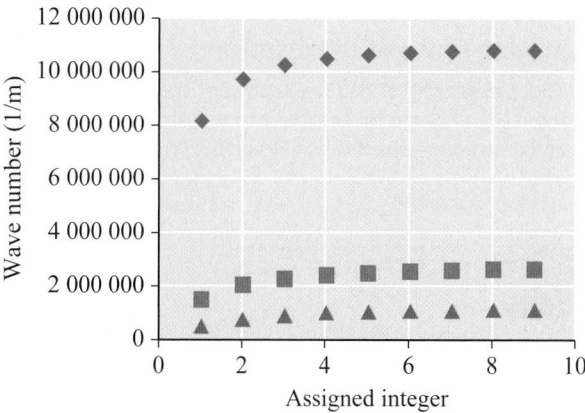

Figure 30-8 Wave number series for hydrogen. These plots are generated by assigning successive integers to each wave number in a series, starting with 1 for the lowest wave number in each series.

CHAPTER 30 | **FUNDAMENTAL DISCOVERIES OF MODERN PHYSICS**

In 1885, the Swiss mathematician Johann Jakob Balmer (1823–98) derived an equation that matches the pattern of the visible spectral lines of hydrogen. When Rydberg saw Balmer's equation, he was able to generalize it, producing an equation that predicts all the spectral lines of hydrogen:

 KEY EQUATION

$$\tilde{\nu} = R_{\mathrm{H}}\left(\frac{1}{n_2^2} - \frac{1}{n_1^2}\right) \qquad (30\text{-}5)$$

where $R_{\mathrm{H}} = 1.097 \times 10^7/\mathrm{m}$ is a constant that Rydberg determined empirically, so it is called Rydberg's constant.

The Balmer series corresponds to $n_2 = 2$, the only hydrogen series to have been observed at the time. Rydberg's equation accurately predicts other series of infrared and ultraviolet lines. Any successful atomic theory would have to produce the Rydberg formula.

 EXAMPLE 30-3

Hydrogen Spectral Lines

Find the wave number, wavelength, and frequency of the hydrogen spectral line that corresponds to $n_2 = 2$, $n_1 = 3$.

SOLUTION

We can calculate the wave number using Equation (30-5):

$$\tilde{\nu} = R_{\mathrm{H}}\left(\frac{1}{n_2^2} - \frac{1}{n_1^2}\right) = (1.097 \times 10^7/\mathrm{m})\left(\frac{1}{2^2} - \frac{1}{3^2}\right)$$
$$= 1.52 \times 10^6/\mathrm{m}$$

The wavelength is the reciprocal of the wave number:

$$\lambda = \frac{1}{\tilde{\nu}} = 6.56 \times 10^{-7}\,\mathrm{m} = 656\,\mathrm{nm}$$

To find the frequency, we recall that $v = f\lambda$; therefore,

$$f = \frac{v}{\lambda} = \frac{3 \times 10^8\,\mathrm{m/s}}{6.56 \times 10^{-7}\,\mathrm{m}} = 4.57 \times 10^{14}\,\mathrm{Hz}$$

Making sense of the result:

The wavelength of 656 nm is toward the violet end of the visible spectrum and corresponds to the blue line in the hydrogen spectrum in Figure 30-7.

PEER TO PEER

Both the wavelength and the frequency are difficult numbers to understand because they are very small and very large. It is useful to remember that the visible part of the electromagnetic spectrum extends from approximately 380 nm (violet) to 750 nm (red), or from 8×10^{12} Hz to 4×10^{12} Hz, respectively.

30-4 Cathode Rays

Experimenters working with gas discharge tubes at extremely low pressures noticed that the glass of the tubes would glow at the end opposite to the cathode, the negative electrode. Further experimentation revealed that an object placed in the tube casts a shadow on the end of the tube away from the cathode (Figure 30-9). This result suggested that whatever was causing the tube to glow was emanating from the cathode. Consequently, these rays were called cathode rays.

ANDREW LAMBERT PHOTOGRAPHY/SCIENCE PHOTO LIBRARY

Figure 30-9 Cathode rays in the tube emanate from the cathode, and when they strike the glass walls of the tube, they cause the tube to glow. When the metal cross is placed in the path, it creates a shadow on the glass.

In a related experiment, a paddle wheel placed between the electrodes rotated when the tube was energized, indicating that whatever was moving through the tube carried both energy and momentum. Further experimentation revealed that cathode rays can be deflected by both electric and magnetic fields, providing evidence that the rays could be moving charged particles.

These experiments culminated in the discovery of the electron by J.J. Thomson in 1897. As described briefly in Section 23-3, Thomson measured the charge-to-mass ratio of cathode rays, thus showing them to be tiny charged particles with mass. We now look at Thomson's experiment in more detail.

Thomson's apparatus is illustrated schematically in Figure 30-10. The filament serves as the cathode in this apparatus. The rays are attracted to the **anode**, the positive electrode of the tube. The anode has a small hole in the middle, which allows a narrow beam of cathode rays to reach the fluorescent zinc sulfide screen at the end of the tube. The beam passes between the deflection plates,

MODERN PHYSICS

J.J. Thomson's Experiment

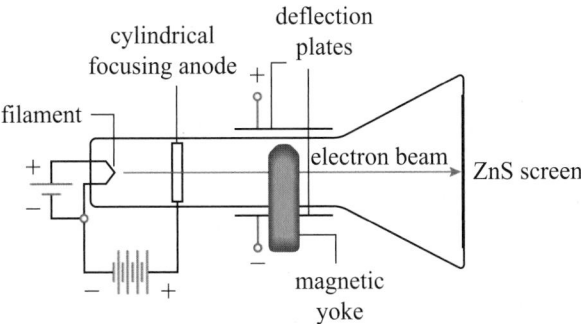

Figure 30-10 A schematic diagram of the apparatus used by J.J. Thomson to measure the charge-to-mass ratio of cathode rays.

which have a potential difference across them that creates an electric field directed downward. In the same region, an electromagnet creates a uniform magnetic field directed into the page.

Thomson made his measurement in two steps. First, he carefully adjusted the electric and magnetic fields until the beam of cathode rays passed straight through the deflection plate area. As described in Chapter 23, Thomson could determine the speed of the particles from the strengths of the electric and magnetic fields. He then turned off the electric field and measured the deflection caused by the magnetic field, which he used to calculate the charge-to-mass ratio of the particles in the cathode rays.

Consider the forces that a charged particle moving with velocity $\vec{v}$ experiences in electric and magnetic fields. These fields, and the associated forces that the particle experiences, are shown in Figure 30-11. The negatively charged particle experiences an upward force due to the electric field and a downward force due to the magnetic field.

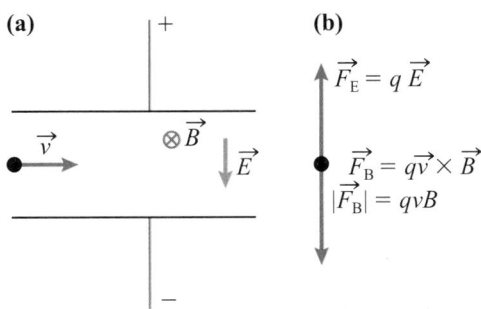

Figure 30-11 Field configuration (a) and free body diagram (b) for Thomson's crossed-field apparatus

When Thomson adjusted the fields so that the beam was not deflected, the force caused by the electric field was equal but opposite to the force caused by the magnetic field. Consequently, as we saw in Chapter 23,

$$qvB = qE$$

and

$$v = \frac{E}{B} \qquad (30\text{-}6)$$

Figure 30-12 shows the geometry of the vertical deflection, δ, when the electric field is switched off. We can see from Figure 30-12 that $r^2 = (r - \delta)^2 + L^2$, where L is the length of the plates. Solving for the path radius r gives

$$r = \frac{\delta^2 + L^2}{2\delta} \qquad (30\text{-}7)$$

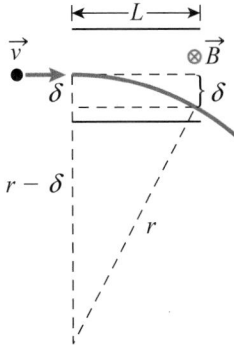

Figure 30-12 Cathode rays deflected by a magnetic field

Thus, by measuring δ, we can determine the radius of curvature of the particle trajectory in the magnetic field. A charged particle in a magnetic field experiences a force that is perpendicular to the velocity of the particle. This force causes the particle to travel along a curved arc of a circle. Since the particle is travelling along a circular path with constant velocity, the magnitude of the acceleration must be

$$a_{\mathrm{c}} = \frac{v^2}{r} = \frac{qvB}{m} \qquad (30\text{-}8)$$

Solving for $\frac{q}{m}$ and using Equation (30-6), we find

$$\frac{q}{m} = \frac{v}{rB} = \frac{E}{rB^2} \qquad (30\text{-}9)$$

With this result, Thomson was able to determine that the negatively charged particles in the cathode rays have a charge-to-mass ratio that is over 1000 times larger than the ratio determined by Faraday for a hydrogen ion. Thomson repeated his experiments with different cathode materials and found the same charge-to-mass ratio. These results led Thomson to postulate that cathode rays contain a new type of particle: the electron. Because Thomson was able to measure only the charge-to-mass ratio, it was unclear whether either the mass or the charge of this new particle were comparable to that of a hydrogen ion.

EXAMPLE 30-4

Determining the Speed of a Charged Particle

In a crossed-field experiment with a magnetic field of 3.00 mT, a potential difference of 200 V is required for a beam of electrons to pass undeflected through the plates, which are separated by 2.00 cm.

(a) How fast are the electrons moving? (Ignore relativistic effects.)

(b) What potential difference is required for the electrons to be accelerated to the speed you found in part (a)?

SOLUTION

(a) We can use Equation (30-6) to calculate the velocity if we know the strength of both fields. Since the electric field is produced by parallel plates, we know that

$$E = \frac{V}{d} = \frac{200 \text{ V}}{0.02 \text{ m}} = 10000 \text{ V/m}$$

Therefore,

$$v = \frac{E}{B} = \frac{10000 \text{ V/m}}{0.003 \text{ T}} = 3.33 \times 10^6 \text{ m/s}$$

(b) If we accelerate an electron at rest through a potential difference of ΔV, the electron will have a kinetic energy of approximately $K = \frac{1}{2}mv^2 = q\Delta V$, because the speed is low enough to ignore relativistic effects. We can solve for ΔV to find

$$\Delta V = \frac{mv^2}{2q} = 31.6 \text{ V}$$

MAKING CONNECTIONS

Display Technology

The Thomson apparatus is the precursor of the cathode ray tube (CRT) display, which has only recently been supplanted by liquid crystal display (LCD) technology. Figure 30-13 shows a CRT display on the left and an LCD display on the right. In a CRT display, an electron beam is swept across the phosphorescent screen line by line, in what is called a raster pattern. Modulating the intensity of the beam as a function of screen position varies the brightness and thus creates an image.

Figure 30-13 Cathode ray tube technology (left) has been replaced by LCD technology (right).

LO 5

30-5 The Millikan Oil Drop Experiment

American physicist Robert Andrews Millikan (1868–1953) conducted a series of experiments that established that there is a fundamental unit of charge. The apparatus that Millikan used is illustrated in Figure 30-14. He used a spray device to produce a mist of tiny oil drops. Some of the oil drops fall into the region between the charged plates, where the drops can be observed with a telescope and manipulated by an electric field between the plates. Millikan observed that these droplets experience a force when placed in an electric field, indicating that they must

Figure 30-14 The Millikan oil drop experiment

be charged. The drops acquire a charge through friction with the nozzle, in much the same way that we acquire a static charge when we rub against a carpet. (Ionizing radiation can also be used to charge the droplets.)

When there is no field, an oil droplet falls under the influence of gravity, but quickly reaches a terminal velocity due to the drag force of the surrounding air. Millikan knew that the magnitude of the drag force is given by Stokes' law, an equation derived by the mathematician and physicist George Gabriel Stokes (1819–1903):

$$F_D = 6\pi r \eta v_d \qquad (30\text{-}10)$$

where r is the radius of the oil drop, η is the viscosity of the air, and v_d is the terminal velocity of the falling droplet.

The weight of the oil drop is partially offset by the buoyant force of the air surrounding it. So, the magnitude of the effective weight of the droplet is

$$W = mg - m_{air}g = \frac{4}{3}\pi r^3 (\rho - \rho_{air})g \qquad (30\text{-}11)$$

where g is the acceleration due to gravity, m is the mass of the droplet, m_{air} is the mass of the same volume of air, and ρ and ρ_{air} are the densities of the oil and the air, respectively.

When the drop reaches terminal velocity, the drag force is equal to the effective weight of the droplet. Therefore,

$$6\pi r \eta v_d = \frac{4}{3}\pi r^3 (\rho - \rho_{air})g$$

and

$$r^2 = \frac{9\eta v_1}{2g(\rho - \rho_{air})} \qquad (30\text{-}12)$$

Once we have a value for r, we can readily calculate the effective weight W using Equation (30-11).

When the electric field E is turned on, the drop begins to rise and quickly reaches a new terminal velocity, v_2. Now the droplet experiences an upward force of magnitude qE, where q is the charge on the droplet, and the downward forces are the effective weight and the drag force, F_D. Thus,

$$qE = W + F_D$$
$$= W + 6\pi r \eta v_2$$
$$= W + \frac{W}{v_1}v_2 \qquad (30\text{-}13)$$

Since the electrodes are parallel plates, the magnitude of the electric field can be readily calculated from the potential between the plates and the plate separation. Thus, by measuring the terminal velocities with the field off and the field on, it is possible to calculate the charge on a drop.

Millikan made the necessary measurements on a number of droplets and calculated the charge on each droplet. He was able to show that all the droplets had charges that are integer multiples of a fundamental unit of charge. In his initial results, published in 1910, he reported a value of 1.63×10^{-19} C for the fundamental

charge. After modifying his equipment and measuring a larger number of droplets, he published a revised value of 1.59×10^{-19} C in 1913. The currently accepted value for the **fundamental** (or **elementary**) **charge**, e, is approximately 1.602×10^{-19} C.

✓ **CHECKPOINT**

C-30-3 The Millikan Experiment

Millikan showed that the charge on an oil drop only comes in discrete units of 1.602×10^{-19} C. True or false?

C-30-3 True. Millikan showed that the fundamental unit of charge is 1.602×10^{-19} C.

  EXAMPLE 30-5

An Oil Drop Experiment

Students replicating Millikan's experiment measure a free-fall terminal velocity of $v_1 = 3.30 \times 10^{-5}$ m/s for an oil droplet. When they apply a potential difference of 10 V to the plates, which are 5 mm apart, the students observe that $v_2 = 2.59 \times 10^{-5}$ m/s. Find the charge on the droplet given that $\rho_{oil} = 1.05 \times 10^3$ g/cm³, $\rho_{air} = 0.129 \times 10^3$ g/cm³, and $\eta = 1.73 \times 10^{-5}$ N·s/m².

SOLUTION

We use Equation (30-12) to determine the radius of the drop from the free-fall velocity v_1:

$$r = \sqrt{\frac{(9)(1.73 \times 10^{-5}\,\text{N·s/m}^2)(3.30 \times 10^{-5}\,\text{m/s})}{(2)(9.81\,\text{m/s}^2)(1.05 \times 10^6\,\text{kg/m}^3 - 0.129 \times 10^6\,\text{kg/m}^3)}}$$

$$= 1.69 \times 10^{-8}\,\text{m}$$

With this radius, we then calculate the effective weight W using Equation (30-11):

$$W = \frac{4}{3}\pi r^3 (\rho - \rho_{air})g = 1.81 \times 10^{-16}\,\text{N}$$

We then use Equation (30-13) and solve for q:

$$q = \frac{W}{E}\left(1 + \frac{v_2}{v_1}\right) = \frac{W}{\frac{V}{d}}\left(1 + \frac{v_2}{v_1}\right) = 1.62 \times 10^{-19}\,\text{C}$$

Dividing this charge by 1.602×10^{-19} C, we obtain e as the charge on the oil drop.

Millikan's results suggest that all particles carry some integer multiple of the same fundamental unit of charge. It appeared that positively and negatively charged particles have the same magnitude of the charge. Thus, the charge-to-mass ratio for cathode ray particles indicates that these particles must be much lighter than atomic masses: that is, electrons are subatomic particles. These ideas set the stage for the development of atomic models.

LO6

30-6 Thomson's Model of the Atom

Atoms are neutral objects in bulk substances, yet Faraday's experiments clearly established that there is electrical charge associated with atoms. Taken together, Thomson's and Millikan's results show that there are negatively charged particles that are much less massive than atoms. So, it seemed reasonable to postulate that atoms consist of a collection of rather massive positive charge that is balanced by the negative charge of the less massive electrons. However, the structure of the atom was unknown.

ONLINE ACTIVITY

Plum Pudding Spectrum

The e-resource that accompanies every new copy of this textbook contains an Online Activity using the PhET simulation "Hydrogen Atoms." Work through the simulation and accompanying questions to gain an understanding of the spectrum predicted by the plum pudding model.

Thomson proposed that the atom was a sphere of positive charge in which were embedded enough electrons to make the atom electrically neutral. This model explained the incompressibility of solids because there would be strong repulsive forces between the spheres of positive charge if they were forced to overlap. Without any such overlap, the electrostatic forces are quite weak because the atoms are neutral overall. Consequently, one would expect these atoms to have strong repulsive interactions when packed together in solids but very weak long-range interactions in a gas state.

In this model, the electrons experience a linear restoring force inside the sphere of positive charge. If we assume this sphere is a cloud of uniform charge density with a radius of r_0 and a total charge of Q, we can use Gauss's law (Chapter 20) to calculate the electric field, E, inside the sphere:

$$E = \frac{Q}{4\pi\epsilon_0}\frac{r}{r_0^3} \qquad (30\text{-}14)$$

The cloud has a positive charge, so the electric field points radially outward. An electron placed in the cloud therefore experiences a force eE that is directed radially inward. Due to spherical symmetry of the positive charge, the net force acting on an electron at any distance r from the centre of the sphere is entirely radial with no transverse components.

Now we consider the motion of an electron inside the sphere of positive charge. We will assume that this motion

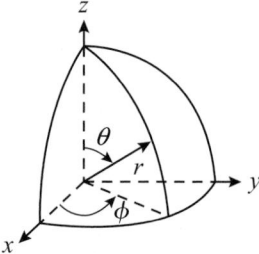
is radial, so the two angular coordinates do not change with time. Using Newton's second law we can then write

$$ma_r = -qE = -\frac{qQ}{4\pi\epsilon_0 r_0^3}r \qquad (30\text{-}15)$$

where a_r is the acceleration in the radial direction.

Equation (30-15) is an equation for simple harmonic motion with an angular frequency given by

$$\omega = \sqrt{\frac{qQ}{4\pi\epsilon_0 r_0^3 m}} \qquad (30\text{-}16)$$

Therefore, electrons in the Thomson model undergo simple harmonic motion inside the atom. Particles undergoing simple harmonic motion are accelerating, and charged particles emit radiation when they accelerate. Thus, we would expect Thomson atoms to radiate light at the frequency of oscillation given by Equation (30-16). This frequency depends on Q, the charge on the cloud. For $q = Q = e$, $m_e = 9.1 \times 10^{-31}$ kg, and $r_0 = 10^{-9}$ m, we find

$$\omega \approx 5 \times 10^{14} \text{ s}^{-1}$$
$$\therefore\ f = 2\pi\omega \approx 3 \times 10^{15} \text{ s}^{-1}$$
$$\therefore\ \lambda = \frac{c}{f} \approx 100 \text{ nm} \qquad (30\text{-}17)$$

This wavelength has the correct order of magnitude, but the model does not explain how an atom can produce many different spectral lines for a given value of Q.

MODERN PHYSICS

30-7   Rutherford Scattering

In 1909, the New Zealand physicist Ernest Rutherford (1871–1937), working with Hans Geiger and Ernest Marsden at the University of Manchester, performed experiments that revealed the structure of the atom. The researchers directed a beam of α-particles produced by the decay of radium at a thin gold foil. Rutherford had made reasonable estimates for the speed and charge-to-mass ratio of α-particles using a crossed-field apparatus similar to Thomson's. The particles scattered from the gold foil were counted using a detector developed by Rutherford and Geiger. A zinc sulfide screen in the detector emitted a flash of light when struck by an α-particle.

ONLINE ACTIVITY

Rutherford Scattering

The e-resource that accompanies every new copy of this textbook contains an Online Activity using the PhET simulation "Rutherford Scattering." Work through the simulation and accompanying questions to gain an understanding of the scattering predicted by the Rutherford model.

MAKING CONNECTIONS

Rutherford's Influence on the Next Generation

Rutherford was born in New Zealand, where he completed his undergraduate studies and a master's degree. After two years of research in electrical technology, he moved to England for more postgraduate studies at the University of Cambridge, where he worked with J.J. Thomson. In 1898, Rutherford was the Chair of Physics at McGill University in Montréal, Québec. While there, he conducted research on radioactivity and the transmutation of elements, for which he was awarded the Nobel Prize in Chemistry in 1908. Rutherford returned to England in 1907 to take the Chair of Physics at the University of Manchester, where he devised his famous scattering experiment and supported the early work of Niels Bohr (whose quantum model of the atom is described in Section 30-9). In 1919, Rutherford became director of the Cavendish Laboratory at the University of Cambridge. A number of the researchers working under his direction received Nobel prizes, including James Chadwick (discovery of the neutron), Edward Appleton (discovery of the ionosphere), and John Cockcroft and Ernest Walton (splitting the atom).

Based on the Thomson model with electrons embedded in a cloud of positive charge, Rutherford's group thought that the α-particles would be deflected by at most a few degrees. Since α-particles are much heavier than electrons, the researchers expected the electrons to have very little effect on the trajectory of the α-particles, and any measurable deflection would be caused by the positive charge. For an α-particle travelling along a line passing through the centre of the cloud, the electric field is given by

$$E = \begin{cases} \dfrac{1}{4\pi\epsilon_0}\dfrac{Q}{r^2} & r > r_0 \\[2ex] \dfrac{1}{4\pi\epsilon_0}\dfrac{Qr}{r_0^2} & r \le r_0 \end{cases} \qquad (30\text{-}18)$$

where r_0 is the radius of the cloud of positive charge.

Integrating this electric field gives the electric potential at the centre of the cloud of positive charge:

$$V_{r=0} = \frac{3Q}{8\pi\epsilon_0 r_0} \qquad (30\text{-}19)$$

The radius r_0 was assumed to be of the order of interatomic spacing. Gold has a molar mass of approximately 197 g/mol and a density of 19.3 g/cm³. Using these data with Avogadro's number, N_A, we can estimate the volume associated with a single gold atom and, hence, its radius:

$$\frac{4}{3}\pi r_0^3 = \frac{197 \text{ g/mol}}{19.3 \text{ g/cm}^3}\frac{1}{N_A} = 1.69 \times 10^{-23} \text{ cm}^3 \qquad (30\text{-}20)$$

$$r_0 = 1.6 \times 10^{-10} \text{ m} \qquad (30\text{-}21)$$

Rutherford's measurements showed that the speeds of the α-particles emitted by radium are in the range of 2 to 3×10^7 m/s. Thus, the α-particles have a kinetic energy of approximately

$$K = \frac{1}{2}mv^2 = 5 \times 10^{-13} \text{ J} \qquad (30\text{-}22)$$

An α-particle has a charge of $+2e$, and a gold nucleus has a charge of $+79e$. We can use Equation (30-19) to find the potential energy of an α-particle at the centre of a cloud of charge of $+79e$ and with a radius of 1.6×10^{-10} m:

$$U = qV = \frac{474e^2}{8\pi\epsilon_0 r_0} = 3.4 \times 10^{-16} \text{ J} \qquad (30\text{-}23)$$

This potential energy is approximately one-thousandth of the kinetic energy of an α-particle, so we would expect the particle to pass right through the cloud of positive charge with very little deflection.

If the path of the α-particle is offset from the centre of the cloud, the calculation of the trajectory is substantially more complex. Suppose the trajectory of the α-particle is off-centre by r_0, the radius of the cloud. We can make a rough estimate of the deflection by assuming that the particle experiences a force that is perpendicular

MODERN PHYSICS

to its trajectory for a time that is equal to the time it takes for the particle to travel the diameter of the atom. The time for the α-particle to travel this distance is

$$\Delta t = \frac{2r_0}{v} = 1.28 \times 10^{-17} \text{ s} \qquad (30\text{-}24)$$

The magnitude of the force varies as the distance between the cloud and the α-particle changes, but we will approximate the force as constant with a magnitude equal to the force that the α-particle experiences at the surface of the atom. The force at the surface of the cloud is given by the product of the charge on the α-particle, $+2e$, and the electric field due to the cloud, which we can get from Equation (30-18):

$$F = qE = (2e)\frac{1}{4\pi\epsilon_0}\frac{79e}{r_0^2} = 1.4 \times 10^{-6} \text{ N} \quad (30\text{-}25)$$

If the particle experiences this force for the entire time it crosses the diameter of the nucleus, the change in momentum perpendicular to the incident direction is given by

$$\Delta p_\perp = F_{\text{avg}}\Delta t = 1.8 \times 10^{-23} \text{ kg·m/s} \qquad (30\text{-}26)$$

Since the mass of the α-particle is approximately four times the mass of a proton, the change in velocity perpendicular to the direction of the incident beam is approximately

$$\Delta v_\perp = \frac{\Delta p}{m} = 2.7 \times 10^3 \text{ m/s} \qquad (30\text{-}27)$$

Since the incident velocity is of the order of 10^7 m/s, the deflection from the incident direction is quite small.

Rutherford and his co-workers observed that the majority of the α-particles did simply pass through the foil; but some were scattered through quite large angles, and a few particles rebounded almost straight back from the foil. Rutherford commented: "It was almost as incredible as if you fired a 15-inch shell at a piece of tissue paper and it came back and hit you."

Rutherford realized that the scattering pattern showed that the positive charge could not be distributed in a cloud throughout the gold atoms. For the α-particles to be scattered backward, the electric fields produced by the positive charges inside the atom must be much greater than the fields predicted by the Thomson model. If we examine Equation (30-18) closely, we can see that the largest value of the electric field that a cloud of charge can produce occurs at the surface of the charge and is given by

$$E_{\text{max}} = \frac{1}{4\pi\epsilon_0}\frac{Q}{r_0^2} \qquad (30\text{-}28)$$

Thus, if the strength of the electric field is much greater, the radius of the cloud of charge must be very small. Rutherford estimated the maximum possible radius of the cloud of positive charge by assuming that

an α-particle travelling straight toward the centre of the cloud of positive charge comes to a stop just at the surface of the positive charge before being accelerated back the way it came. If an α-particle stops in this way, all the kinetic energy of the particle must be converted to potential energy. At the surface of a charged sphere, the electric potential is equal to the electric potential due to a point charge:

$$V = \frac{1}{4\pi\epsilon_0}\frac{Q}{r} \qquad (20\text{-}7)$$

Multiplying the electric potential by the charge of the α-particle gives us the potential energy of the α-particle at the surface of the cloud. The α-particle stops just at the surface of the charge if this electric potential energy equals the initial kinetic energy of the α-particle:

$$qV = \frac{1}{2}mv^2$$

$$\frac{1}{4\pi\epsilon_0}\frac{(2e)(79e)}{r_0} = \frac{1}{2}mv^2 \qquad (30\text{-}29)$$

Solving for r_0 we find

$$r_0 = \frac{79e^2}{\pi\epsilon_0 mv^2} = 1.2 \times 10^{-14} \text{ m} \qquad (30\text{-}30)$$

This result suggests that for the α-particles to be backscattered, all the positive charge in the atom must be concentrated in a sphere with a radius no larger than approximately 10^{-14} m.

Rutherford realized the rather startling implications of this result. If all the positive charge were confined to such a small region, then most of the atom is empty space. If atoms are mostly empty space, what prevents them from collapsing in on each other? The small radius of the charge also indicates that the material carrying the charge has a density of at least 10^{15} kg/m³, many orders of magnitude greater than any previously measured (e.g., lead has a density of approximately 1×10^4 kg/m³).

✓ CHECKPOINT

C-30-4 The Rutherford Experiment

Rutherford's experiments demonstrated that
(a) the Thomson model of the atom is correct;
(b) all the positive charge in an atom is contained in a very small region;
(c) all the negative charge in an atom is confined to a small region.
(d) None of the above.

C-30-4 (b) The fact that some α-particles were scattered completely backward is only consistent with the positive charge concentrated in a very small volume of space.

MAKING CONNECTIONS

The Density of a Black Hole

Is the density of atomic nuclei the greatest in nature? For example, could the density of a black hole be greater? To estimate the density of a black hole, we use the concept of escape velocity introduced in Chapter 11. The minimum launch speed required to escape the gravitational field of an object of mass M and radius r is given by

$$v_{esc}^2 = \frac{2GM}{r}$$

where G is the gravitational constant.

If we use the speed of light as the escape velocity and solve for r, we get

$$r = \frac{2GM}{c^2}$$

The density of such an object is

$$\rho = \frac{M}{\frac{4}{3}\pi r^3} = \frac{M}{\frac{4}{3}\pi \left(\frac{2GM}{c^2}\right)^3} = \frac{3c^6}{32\pi G^3 M^2}$$

For a black hole with the same mass as the Sun, 1.98×10^{30} kg, we get a density of 1.87×10^{19} kg/m³.

Rutherford suggested that his results supported a **planetary model** of the atom, in which the electrons could perhaps be orbiting a central positive charge. However, orbiting electrons would be accelerating and thus should be emitting light continuously. Such electrons would be losing energy and should therefore simply spiral into the positive charge. Nevertheless, the planetary model was a key step in understanding atomic structure.

30-8   The Photoelectric Effect

Some materials can eject electrons when they are exposed to light, a phenomenon called the **photoelectric effect**. Early observations of the photoelectric effect were made by the French physicist Antoine Henri Becquerel (1852–1908) with electrodes immersed in conducting solutions. Heinrich Hertz reported the first observations of a similar effect in air. During his experiments with spark gaps, Hertz found that when he placed a spark gap in the dark to observe the sparks more easily, he had to narrow the gap considerably to get a spark to jump.

ONLINE ACTIVITY

Photoelectric Effect

The e-resource that accompanies every new copy of this textbook contains an Online Activity using the PhET simulation "Photoelectric Effect." Work through the simulation and accompanying questions to gain an understanding of the photoelectric effect.

Systematic investigations of the photoelectric effect revealed the following:

■   Every photoelectric material has a **threshold frequency**, which is the minimum frequency of

MAKING CONNECTIONS

Scattering Experiments

The Rutherford gold-leaf experiment illustrates how information about the structure of matter can be inferred from a scattering experiment. Scattering techniques have been applied in many areas of research, including investigations of the distribution of electrons in atoms and the magnetic properties of nuclei. For example, deep inelastic electron–proton scattering revealed that the proton is not a point particle and led to the discovery of quarks, which we now believe are among the fundamental particles of nature. Using beams of neutrons from research reactors at Chalk River, Ontario, Canadian physicist Bertram Neville Brockhouse (1918–2003) developed the field of neutron scattering (Figure 30-16). Brockhouse initially used those techniques to study the vibrations of atoms in solids.

Figure 30-16   Nobel Prize–winning physicist Bertram Brockhouse

MODERN PHYSICS

incident radiation below which no electrons are emitted, regardless of the intensity of the incident radiation.

■ Above the threshold frequency, the number of electrons emitted is proportional to the intensity of the incident radiation.

■ Above the threshold frequency, the energy of the electrons is not proportional to the intensity of the incident radiation,

The relationship between the number of electrons emitted and the radiation intensity made sense because Maxwell's description of electromagnetic radiation indicates that the total energy in an electromagnetic wave is proportional to the intensity of the wave. However, the existence of threshold frequencies was baffling. Why could one not simply increase the intensity to eject electrons?

In 1900, German physicist Max Planck (1858–1947) developed a theory that describes the blackbody spectra (like those in Figure 30-5). Planck had postulated that electromagnetic radiation comes in discrete parcels, which we now call **photons**. Planck asserted that the energy, E, of a photon is related to its frequency f through the relation

$$E = hf \qquad (30\text{-}31)$$

where $h = 6.626 \times 10^{-34}\ \text{J} \cdot \text{s}$ is Planck's constant.

EXAMPLE 30-6

Photon Energy

Calculate the energy associated with a photon that has a wavelength of 665 nm.

SOLUTION

We can use Equation (30-31) in combination with the relation $f = \dfrac{c}{\lambda}$:

$$E = hf = h\frac{c}{\lambda} = \frac{(6.626 \times 10^{-34}\ \text{J} \cdot \text{s})(3 \times 10^{8}\ \text{m/s})}{6.65 \times 10^{-7}\ \text{m}}$$

$$= 2.99 \times 10^{-19}\ \text{J}$$

Making sense of the result:

This amount of energy is very small, but it is on the scale of energies typically associated with atoms.

In 1905, Einstein applied Planck's concept of quantized radiation energy to the photoelectric effect. He assumed that an electron would absorb only one photon at a time (the probability of absorbing two at the same time is very small). The electrons are bound to the material, so a certain minimum amount of energy is required to liberate an electron. This threshold energy is called the **work function** and is usually denoted E_0. Applying conservation

of energy to the process of an electron absorbing a photon, we find that

$$hf - E_0 = K_{\max} \qquad (30\text{-}32)$$

where $K_{\max}$ is the maximum kinetic energy of the ejected electron. If $K_{\max} > 0$, then $hf > E_0$. This condition requires the photon to have a frequency greater than the threshold frequency, f_0, where

$$f_0 = \frac{E_0}{h} \qquad (30\text{-}33)$$

Radiation with a frequency less than f_0 will not be able to eject an electron from the material because the individual photons do not have sufficient energy. Increasing the intensity of the radiation simply increases the *number* of photons that hit the material, but each individual photon still has insufficient energy to eject an electron from an atom in the material. Once the frequency threshold is exceeded, increasing the intensity increases the number of electrons ejected because each individual photon now has enough energy to free an electron from the material.

Einstein's explanation of the photoelectric effect was an important analysis because it suggested that light, which had been thought to be a wave, has some particle properties. Further, Einstein extended Planck's idea that physical quantities could be quantized. Einstein's 1921 Nobel Prize in Physics was awarded largely for his work on the photoelectric effect.

LO 9

30-9 The Bohr Model of the Atom

To develop an atomic model, the Danish physicist Niels Henrik David Bohr (1885–1962) first considered the simplest possible atom, hydrogen, which has just one electron and one proton. Hydrogen also has the simplest emission spectrum of all the elements. Bohr's model, published in 1913, accurately predicts the emission spectra for hydrogen.

ONLINE ACTIVITY

Bohr Model

The e-resource that accompanies every new copy of this textbook contains an Online Activity using the PhET simulation "The Bohr Model of the Atom." Work through the simulation and accompanying questions to gain an understanding of the spectrum predicted by the Bohr Model.

MODERN PHYSICS

To explain why orbiting electrons do not continuously emit electromagnetic radiation, Bohr postulated that there are stable states for the electrons and that the lowest energy state did not decay. An electron could go from a high-energy state to a low-energy state and vice versa. For energy to be conserved during such transitions, the atom would either emit or absorb energy equal to the difference in energy between the two states of the atom. The energy that was emitted or absorbed is in the form of electromagnetic radiation. This radiation should have frequencies that match the observed spectral lines for the atom.

Planck and Einstein had suggested that energy in electromagnetic radiation is quantized in terms of Planck's constant, so Bohr conjectured that some quantity in the atom is correspondingly quantized. As you saw in Chapter 8, an object of mass m travelling with constant speed v in a circular orbit of radius r has an angular momentum of $L = mvr$. Bohr suggested that the electron in a hydrogen atom has a quantized angular momentum given by

$$L = mvr = n\frac{h}{2\pi} = n\hbar \qquad (30\text{-}34)$$

where n is a positive integer, and $\hbar = \dfrac{h}{2\pi}$.

Each value of the quantum number n is associated with a unique angular momentum and a unique energy, E_n. Bohr assumed that the electron was kept in a circular orbit around the nucleus by the Coulomb force between the positive nucleus and the negative electron. Thus,

$$\frac{ke^2}{r^2} = \frac{mv^2}{r} \qquad (30\text{-}35)$$

where k is Coulomb's constant.

If we solve Equation (30-28) for v, we get

$$v = \frac{n\hbar}{mr} \qquad (30\text{-}36)$$

Substituting this result into Equation (30-35) gives

$$\frac{ke^2}{r^2} = \frac{m}{r}\left(\frac{n\hbar}{mr}\right)^2 \qquad (30\text{-}37)$$

and

$$r = \frac{\hbar^2}{kme^2}n^2 = \frac{L^2}{kme^2} \qquad (30\text{-}38)$$

Thus, only certain orbital radii are permitted. The smallest allowed value of r, corresponding to $n = 1$, is called the **Bohr radius** and has a value of

$$r_0 = \frac{\hbar^2}{kme^2} = 0.529 \times 10^{-10} \text{ m} \qquad (30\text{-}39)$$

Note that the standard notation for the Bohr radius is r_0, even though it is the radius for $n = 1$. The Bohr radius is a reasonable value for the size of a hydrogen atom based on the density of liquid hydrogen. However, the radius of the electron's orbit grows very rapidly with increasing n.

The total energy of an orbiting electron is the sum of its kinetic energy and its electric potential energy. Thus, the energy for any orbit (or state) is given by

$$E_n = \frac{1}{2}mv^2 - \frac{ke^2}{r} \qquad (30\text{-}40)$$

Substituting $mv^2 = \dfrac{ke^2}{r}$ from Equation (30-35), we obtain

$$E_n = \frac{1}{2}\frac{ke^2}{r} - \frac{ke^2}{r} = -\frac{1}{2}\frac{ke^2}{r} \qquad (30\text{-}41)$$

Next, substitute the expression for r from Equation (30-38):

$$E_n = -\frac{1}{2}\frac{ke^2}{\frac{\hbar^2 n^2}{kme^2}} = -\frac{1}{2}\frac{k^2me^4}{\hbar^2 n^2} = -\frac{1}{2}\frac{k^2me^4}{L^2} \qquad (30\text{-}42)$$

Equation (30-42) shows that the quantized angular momentum results in quantized energy levels. Thus, the Bohr model allows only quantized values of the energy. Note that these energy levels are negative because we have taken the potential energy to be zero at infinite separation. For the electron to remain bound to the nucleus, the kinetic energy of the electron must be less than the potential energy. The state with $n = 1$ has the lowest energy and is called the **ground state**.

The change in the energy of the electron when it moves from some state n_1 to a different state n_2 is given by

$$\Delta E_{\text{atom}} = E_{n_2} - E_{n_1} = -\frac{1}{2}\frac{k^2me^4}{\hbar^2}\left(\frac{1}{n_2^2} - \frac{1}{n_1^2}\right) \qquad (30\text{-}43)$$

The change in the energy of the atom is negative when $n_2 < n_1$, that is, when the electron moves to a lower state. To conserve energy, the photon emitted during this transition must have a positive energy equal to the decrease in the energy of the atom:

$$E_{\text{photon}} = hf = -\Delta E_{\text{atom}} \qquad (30\text{-}44)$$

To facilitate comparison with Rydberg's formula, we now derive an expression for the wave number of the photon. Since $\tilde{\nu} = \dfrac{1}{\lambda} = \dfrac{v}{c}$,

$$\tilde{\nu} = \frac{4\pi^2 k^2 me^4}{2h^3 c}\left(\frac{1}{n_2^2} - \frac{1}{n_1^2}\right) \qquad (30\text{-}45)$$

The form of Equation (30-45) is the same as Rydberg's formula (Equation (30-5)). When we evaluate the constant in Equation (30-45), we find $\dfrac{4\pi^2 k^2 me^4}{2h^3 c} = 1.09 \times 10^7 /\text{m}$, which is in excellent agreement with Rydberg's empirical value.

Although Bohr's model was soon supplanted by a more complete description, it was a significant breakthrough in the development of modern physics. While Planck had previously hypothesized that radiation could

CHAPTER 30 | **FUNDAMENTAL DISCOVERIES OF MODERN PHYSICS**

MODERN PHYSICS

EXAMPLE 30-7

The Ground State of Hydrogen

Calculate the energy for the ground state of the hydrogen atom using the Bohr model. Express this energy in J and eV.

SOLUTION

Setting $n = 1$ in Equation (30-42) gives the ground-state energy for hydrogen:

$$E_1 = -\frac{1}{2}\frac{k^2 me^4}{\hbar^2 1^2} = -2.18 \times 10^{-18}\ \text{J}$$

Since $1\ \text{eV} = 1.602 \times 10^{-19}\ \text{J}$,

$$E_1 = -\frac{2.18 \times 10^{-18}\ \text{J}}{1.602 \times 10^{-19}\ \text{J/eV}} = -13.6\ \text{eV}$$

MAKING CONNECTIONS

Stellar Spectra

Have you ever wondered how we know the chemical composition of the Sun? The surface of the Sun is hot and emits a spectrum like a blackbody. That surface is enclosed in an "atmosphere" of hot gases surrounding the Sun. As light from the surface passes through the solar atmosphere, some frequencies are absorbed by the atoms in the atmosphere, causing dark absorption lines in the spectrum, as shown in Figure 30-17. By carefully examining the absorption lines, we can determine which elements are in the atmosphere. Absorption spectroscopy is also a common technique used in chemistry. Light is passed through a sample of interest, and the location and depth of absorption lines give information about the atoms in the sample and their relative abundances.

Data from NREL, Solar Spectra: Air Mass Zero, http://rredc.nrel.gov/solar/spectra/am0/

Figure 30-17 We can see that there is an overall blackbody shape to the spectra. In addition, there are narrow absorption lines due to atoms in the solar atmosphere absorbing some of the light as it passes through the atmosphere.

be quantized, the Bohr model was the first to predict that matter particles can have quantized physical quantities such as angular momentum and energy.

LO 10

30-10 Compton Scattering

Experimental confirmation that light consists of photons was provided in 1923 by American physicist Arthur Holly Compton (1892–1962). He scattered X-rays of well-defined energy from electrons bound to carbon atoms in a graphite sample.

Compton's experimental setup is shown in Figure 30-18. Compton produced X-rays by bombarding a metal target with electrons and then passed the X-rays through a series of collimating slits to produce a narrow beam with a known wavelength. He directed the beam at a carbon target and measured the intensity and energy (and, hence, the wavelength) of the X-rays scattered at various angles.

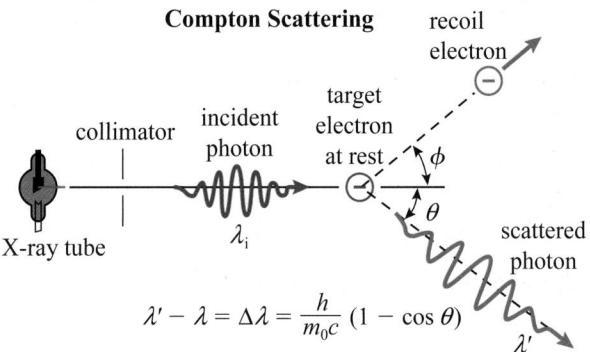

Figure 30-18 A schematic diagram of Compton's scattering experiment

According to the classical theory of scattering, electromagnetic waves passing through an atom force the atom's electrons into harmonic motion. The accelerating electrons then emit radiation with the same frequency as the radiation that caused the electrons to accelerate. Let I_{inc} be the intensity of the X-rays incident on the target and $I_{\text{sca}}(\theta)$ be the intensity of the scattered X-rays at an angle θ relative to the incident beam direction and at a distance R from the scattering target. Then, according to Maxwell's equations,

$$I_{\text{sca}}(\theta) = I_{\text{inc}}\frac{Ne^4}{2m_e^2 c^4 R^2}(1 + \cos^2\theta) \qquad (30\text{-}46)$$

where N is the number of electrons per unit volume of the target, m_e is the electron mass, e is the magnitude of the charge of the electron, and c is the speed of light.

From Equation (30-46) we see that the intensity of the scattered X-rays does not depend on the wavelength

MODERN PHYSICS

of the incident beam. The $\cos^2\theta$ dependence of the scattered intensity is due to the fact that only a fraction of the incident beam is scattered and the rest of the beam passes through the target undeflected.

Earlier experiments by other researchers showed that Equation (30-46) agrees with observations for incident X-rays of low energy at small scattering angles. However, when Compton used high-energy X-rays (of the order of 10^4 eV) and measured the scattered intensity for large angles, he observed that the scattered X-rays have intensity peaks at two different wavelengths (Figure 30-19). One of the wavelengths is the incident wavelength λ. The other wavelength, λ', is larger than λ and increases as the scattering angle θ increases.

Figure 30-19 The intensity of scattered X-rays for various scattering angles

How does the scattering process generate X-rays of different wavelengths? Compton found that all explanations based on classical physics failed to account for the increase in the wavelength of scattered X-rays. For example, the Doppler effect could lower the frequency (and, hence, increase the wavelength) of a scattered X-ray if the target electron were moving away from the incident beam of X-rays. However, a simple calculation showed that for the Doppler effect to account for the observed increase in the wavelength, the electrons in the target would have to be moving away from the direction of incident X-rays at half the speed of light, which is clearly not possible for electrons bound in a carbon atom.

Next, Compton considered the hypothesis proposed by Planck and Einstein that X-rays are photons, each carrying energy $E = hf = \dfrac{hc}{\lambda}$. Then the X-rays are scattered by elastic collisions of an electron and a single photon. The energy of the incident photons is so much greater than the binding energy of an electron in graphite (carbon) that the target electron can be assumed to be initially at rest and free. Figure 30-18 shows the geometry of a collision

between a stationary electron and a photon of wavelength λ. During the collision, the photon transfers some of its energy and momentum to the electron and scatters at an angle θ. The electron, having received energy and momentum, scatters at an angle ϕ. Compton applied the laws of energy and momentum conservation to this scattering process.

Let $\vec{p_\lambda}$, $\vec{p_{\lambda'}}$, $\vec{p_e}$, and $\vec{p_{e'}}$ denote the momentum of the incident photon, the scattered photon, the stationary electron, and the scattered electron, respectively, and let E_λ, $E_{\lambda'}$, E_e, and $E_{e'}$ denote the corresponding energies. Since momentum must be conserved in the scattering process,

$$\vec{p_\lambda} + \vec{p_e} = \vec{p_{\lambda'}} + \vec{p_{e'}} \qquad (30\text{-}47)$$

Similarly, energy conservation requires that

$$E_\lambda + E_e = E_{\lambda'} + E_{e'} \qquad (30\text{-}48)$$

Using relativistic kinematics, the energy and momentum of an electron are related by $E = \sqrt{m_e^2 c^4 + p^2 c^2}$

Assuming that a photon has zero mass, its energy and momentum are related by $E = pc$. Since the target electron is assumed to be at rest, $\vec{p_e} = 0$ and Equation (30-47) can be written as

$$\vec{p_{e'}} = \vec{p_{\lambda'}} - \vec{p_\lambda} \qquad (30\text{-}49)$$

Using the identity $\vec{A} \cdot \vec{A} = A^2$, where A denotes the magnitude of vector $\vec{A}$, we can write

$$\begin{aligned} p_{e'}^2 &= \vec{p_{e'}} \cdot \vec{p_{e'}} \\ &= p_\lambda^2 + p_{\lambda'}^2 - 2\vec{p_\lambda} \cdot \vec{p_{\lambda'}} \\ &= p_\lambda^2 + p_{\lambda'}^2 - 2 p_\lambda p_{\lambda'} \cos\theta \qquad (30\text{-}50) \end{aligned}$$

Equation (30-48) for energy conservation can be written as follows:

$$p_\lambda c + m_e c^2 = p_{\lambda'} c + \sqrt{m_e^2 c^4 + p_{e'}^2 c^2}$$

Bringing the $p_{\lambda'} c$ term to the left and squaring both sides gives

$$p_\lambda^2 + p_{\lambda'}^2 - 2 p_\lambda p_{\lambda'} + 2 m_e c (p_\lambda - p_{\lambda'}) = p_{e'}^2 \quad (30\text{-}51)$$

Inserting Equation (30-50) into Equation (30-51), we obtain

$$m_e c (p_\lambda - p_{\lambda'}) = p_\lambda p_{\lambda'} \cos\theta \qquad (30\text{-}52)$$

The energy and the wavelength of a photon are related by

$$E = pc = hf = \frac{hc}{\lambda}$$

Therefore, the momentum of a photon is given by

$$p = \frac{h}{\lambda} \qquad (30\text{-}53)$$

Writing Equation (30-52) in terms of wavelength, we get

$$\lambda' - \lambda = \frac{h}{m_e c}(1 - \cos\theta) \qquad (30\text{-}54)$$

MODERN PHYSICS

Equation (30-54) relates the wavelength of the scattered photon to the scattering angle and accurately predicts the change in the wavelength of scattered X-rays for a given scattering angle. Notice that the change in the wavelength ($\lambda' - \lambda$) is independent of the original wavelength. The quantity $h/m_e c$ is called the **Compton wavelength**, λ_C, and its value is

$$\lambda_C = \frac{h}{m_e c} = \frac{6.62 \times 10^{-34}\ \text{J} \cdot \text{s}}{(9.11 \times 10^{-31}\ \text{kg})(3.0 \times 10^8\ \text{m/s})}$$

$$= 2.42 \times 10^{-12}\ \text{m} \qquad (30\text{-}55)$$

The small value of the Compton wavelength explains why the classical description of the scattering experiment (Equation (30-46)) agrees with experiments for low-energy (long-wavelength) X-rays. The shift in the wavelength is very small and becomes noticeable only when the incident wavelength is small.

Compton's explanation of his observations proved without a doubt that the scattering of X-rays from electrons is a quantum phenomenon. His experiment showed that single photons transfer energy and momentum to electrons during X-ray scattering; therefore, X-rays and, by extension, other electromagnetic radiation, consist of photons. The scattering of photons from electrons is now called **Compton scattering**, and the increase in the wavelength of the scattered photon is called the **Compton shift**.

EXAMPLE 30-8

Wavelength Change in Compton Scattering

In one of his experiments, Compton scattered 17.4 keV X-rays from a graphite target.

(a) What is the wavelength of photons that are scattered at an angle of 135° with respect to the incident beam?

(b) What is the percentage change in the observed wavelength?

SOLUTION

(a) The wavelength of the incident photons is given by the relationship $E = hc/\lambda$. Therefore,

$$\lambda = \frac{hc}{E} = \frac{(6.62 \times 10^{-34}\ \text{J} \cdot \text{s})(3.0 \times 10^8\ \text{m/s})}{(17.4 \times 10^3\ \text{eV})(1.60 \times 10^{-19}\ \text{J/eV})}$$

$$= 0.71 \times 10^{-10}\ \text{m}$$

The wavelength of the photon scattered at 135° is given by Equation (30-54):

$$\lambda' = \lambda + \frac{h}{m_e c}(1 - \cos\theta)$$

$$= 0.71 \times 10^{-10}\ \text{m} + (2.42 \times 10^{-12}\ \text{m})(1 - \cos(135°))$$

$$= 0.75 \times 10^{-10}\ \text{m}$$

(b) The percentage change in the wavelength is

$$\frac{\lambda' - \lambda}{\lambda} = \frac{0.04 \times 10^{-10}\ \text{m}}{0.71 \times 10^{-10}\ \text{m}} \times 100\% = 5.6\%$$

✓ CHECKPOINT

C-30-5 Compton Scattering

In a Compton scattering experiment, energies of scattered electrons are measured at 0°, 90°, 135°, and 180°. At which angle does the electron carry maximum energy? Assume that initially the target electron is at rest.

(a) 0°
(b) 90°
(c) 135°
(d) 180°

C-30-5 (d)

MODERN PHYSICS

FUNDAMENTAL CONCEPTS AND RELATIONSHIPS

The early developments in modern physics were centred on gaining an understanding of the electrical properties of matter and the generation of light by matter.

Faraday's electrolysis experiments established a clear connection between electric charge and matter. He was able to show that the amount of gas evolved in an electrolysis experiment is proportional to the amount of charge passed through the cell, and he was able to determine charge-to-mass ratios for various elements.

Experiments with light emission established that hot objects emit light at all wavelengths, and the spectrum is characteristic of the temperature of the object. Max Planck was able to accurately describe the spectrum by assuming that the light is emitted by the object in discrete quanta with energy

$$E = hf \qquad (30\text{-}2)$$

Using this assumption, Planck derived the following expression for the intensity of blackbody radiation

$$B(f,T) = \frac{8\pi h f^3}{c^2} \frac{1}{e^{\frac{hf}{kT}} - 1} \qquad (30\text{-}3)$$

which agrees very well with experimental data.

In addition to emission spectra generated by hot objects, gas discharge spectra were also being studied. In contrast with a blackbody, a gas discharge spectrum only contains intensity at certain wavelengths, called spectral lines. In addition, each different gas has a unique set of spectral lines. Balmer analyzed the light from the hydrogen spectrum and showed that it could be described by

$$\tilde{v} = R_H \left(\frac{1}{n_2^2} - \frac{1}{n_1^2} \right) \qquad (30\text{-}5)$$

Experiments with gas discharge tubes at very low pressures led to Thomson's discovery of negatively charged particles (electrons) with a charge-to-mass ratio at least 1000 times larger than any previously observed.

Millikan's oil drop experiments showed that charges come in integer multiples of a fundamental unit of charge e, now known to be approximately $e = 1.602 \times 10^{-19}$ C. This result helped determine that an electron carries a single unit of negative charge and, consequently, must be very light.

Thomson postulated a model of an atom that consisted of electrons embedded in a spherical cloud of positive charge, which contained almost all the mass of the atom. This model predicted that the electrons oscillate, and thus radiate, but only at one frequency for a given element.

Rutherford and his co-workers scattered α-particles off thin gold foils and found that some of the particles were scattered through very large angles. Rutherford showed that the positive charge in the atom must be concentrated in a very small volume, and he proposed an atomic structure with the electrons in orbits about a positive nucleus.

Bohr introduced the notion of stationary orbits that had quantized angular momentum. Bohr's model, which correctly predicts the wavelengths of the spectral lines for hydrogen, was an important first step in the development of quantum theory. He found that the wave numbers could be described by

$$\tilde{v} = \frac{4\pi^2 k^2 m e^4}{2h^3 c} \left(\frac{1}{n_2^2} - \frac{1}{n_1^2} \right) \qquad (30\text{-}45)$$

Compton's X-ray scattering experiments provided proof that Einstein and Planck's idea that light consists of discrete particles called photons is correct.

APPLICATIONS

Applications: particle accelerators, spectrometers

Key Terms: anode, blackbody, Bohr radius, Compton scattering, Compton shift, Compton wavelength, discharge tube, elementary charge, fundamental charge, ground state, photoelectric effect, photons, planetary model, spectral lines, spectrometer, threshold, valence, wave number, work function

QUESTIONS

1. Which experiments led to the postulation of the existence of the electron?
2. When an atom emits light, does the atom make a transition to a higher or a lower electronic energy state?
3. What principal facts need to be explained by a successful atomic model?
4. What is the key property required for a nucleus to backscatter α-particles?
5. What physical quantity did Bohr quantize to develop his model of the atom?
6. How can you distinguish between different gases by examining their discharge spectra?
7. What are the essential differences between a discharge spectrum and a blackbody spectrum?
8. What important question did Millikan resolve with his oil drop experiment?
9. Some electronic medical thermometers measure body temperature with a probe placed in the ear for a few seconds. How might such a device work?
10. What idea did Planck and Einstein introduce that was key to the development of the quantum theory?

MODERN PHYSICS

PUTTING IT ALL TOGETHER

PROBLEMS BY SECTION

For problems, star ratings will be used, (✱, ✱✱, or ✱✱✱), with more stars meaning more challenging problems.

Section 30-1   The Connection Between Matter and Electricity

11. ✱   Calculate the charge-to-mass ratio for hydrogen and oxygen.

12. ✱   During an electrolysis experiment, 3.0 mol of hydrogen gas is evolved.
   (a) How much charge has passed through the cell?
   (b) The current was 10 mA. How long did the experiment take?

Section 30-2   Temperature and the Emission of Light

13. ✱   Figure 30-20 shows blackbody spectra for two objects.
   (a) Which object is at the higher temperature?
   (b) Is the light emitted from these objects visible to the human eye?

Intensity v. Wavelength

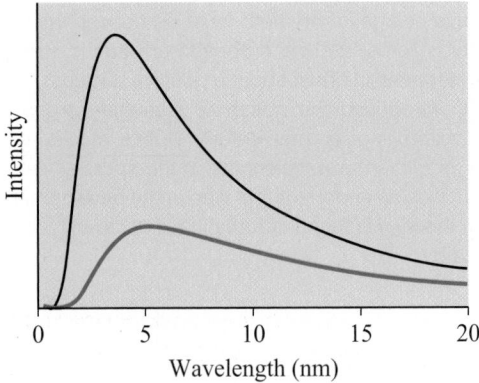

Figure 30-20   Problem 13

Section 30-3   Gas Discharge Spectra

14. ✱✱   Calculate the wave numbers for the Balmer series of hydrogen, which corresponds to $n_2 = 2$.

15. ✱   A spectral line in a hydrogen discharge spectrum is observed to have a wavelength of 1875 nm.
   (a) What wave number does this correspond to?
   (b) What values of n_1 and n_2 does this correspond to?
   (c) What part of the electromagnetic spectrum does this radiation appear in?

Section 30-4   Cathode Rays

16. ✱✱   An electron is accelerated from rest through 200 V. It then enters a crossed-field apparatus whose plates are separated by 1 cm with a potential difference of 100 V across them.
   (a) How fast is the electron moving when it enters the crossed fields?
   (b) What is the electric field between the plates?
   (c) What magnetic field is required to have zero deflection?

17. ✱✱   The path of an electron is observed to have a radius of curvature of 20.3 cm in a magnetic field of 1.8 mT.
   (a) How fast is the electron moving?
   (b) What electric field is required to cause the electron to move in a straight line?

   (c) The field is produced by a potential of 150 V. What is the separation of the plates?

18. ✱✱   An α-particle consists of two protons and two neutrons.
   (a) What is the mass of an α-particle?
   (b) What is the charge of an α-particle?
   (c) We accelerate an α-particle through 200 V. How fast is it moving?
   (d) An electric field of 10 000 V/m causes the α-particle to pass straight through the crossed-field region. What is the magnetic field?
   (e) If we turned off the electric field, what would the radius of curvature of the path of the α-particles be?

19. ✱✱   In our analysis of crossed-field experiments, we did not consider the effects of gravity on the electrons. Suppose we have a plate separation of 2 cm and a voltage of 300 V.
   (a) What is the electric field between the plates?
   (b) What is the force on the electron due to the magnetic field?
   (c) What is the force on the electron due to gravity?
   (d) Was it reasonable to neglect gravity in our analysis? Explain.

Section 30-5   The Millikan Oil Drop Experiment

20. ✱✱   A latex sphere with a radius of 1 μm is between plates that are separated by 15 mm. A potential difference of 38 V is just enough to keep the sphere suspended.
   (a) What is the gravitational force on the sphere?
   (b) What is the electric field between the plates?
   (c) What is the magnitude of the sphere's charge?
   (d) The top plate is positive relative to the bottom plate. What is the sign of the charge on the sphere?

21. ✱✱✱   If you do not account for the buoyancy force on spheres in an oil drop experiment, what percentage error do you make in the weight force for the following? (Hint: You will need to look up the densities of oil and latex.)
   (a) oil drops
   (b) latex spheres

22. ✱✱   Calculate the free-fall speed of a 0.8 μm oil drop. Assume that the density of the oil is 821 kg/m³.

Section 30-6   Thomson's Model of the Atom

23. ✱✱✱   In this question, you will calculate the energy involved in assembling the positive charge in a Thomson atom. Treat the positive charge as a sphere with charge uniformly distributed throughout. The potential at a distance r from the centre is given by

$$V(r) = \frac{kQ_{enc}}{r} = \frac{kQr^2}{R^3}$$

where $+Q$ is the atom's total charge, and R is its radius. Consider a thin shell of charge, with radius r and thickness dr. Such a shell carries the charge

$$dq = 4\pi r^2 dr\rho = \frac{3Qr^2 dr}{R^3}$$

   (a) Put the above expressions together to derive $dU = Vdq$.
   (b) Integrate the resultant expression from $r = 0$ to $r = R$ to get the total potential energy of the sphere of charge.
   (c) Evaluate the resulting expression for $Q = 79e$ and $R = 1$ nm.

24. ✱✱   Use Equation (30-16) to plot ω versus Q for $e < Q < 100e$. Set $q = e$, and use m as the mass of the electron.

MODERN PHYSICS

Section 30-7 Rutherford Scattering

25. ✸✸ Equation (30-30) gives the radius of a nucleus required to stop an α-particle with an initial kinetic energy of energy of $K = \frac{1}{2}mv^2 = 5 \times 10^{-13}$ J at the surface of a nucleus with a radius of 1.2×10^{-14} m.
    (a) Suppose the α-particle has one-half that initial energy. At what distance from the centre of the nucleus will it stop?
    (b) Suppose the α-particle has twice that initial energy. How far into the nucleus would it penetrate?

26. ✸✸ Evaluate the expression you obtained in problem 23(b) for $Q = 79e$ and $R = 1 \times 10^{-14}$ m.

27. ✸ We accelerate an electron from rest through a potential difference of 1000 V.
    (a) What is the kinetic energy of the electron in J?
    (b) What is the speed of the electron?
    (c) What is the kinetic energy of the electron in eV?

28. ✸ Some theorists have speculated that the mass of particles may be due to their "self-energy"; that is, the energy required to assemble them is given by $E = mc^2$. Does the result of problem 26 give a reasonable estimate for the mass of the gold nucleus?

Section 30-8 The Photoelectric Effect

29. ✸ Sodium has a work function of 2.28 eV.
    (a) Convert this energy into J.
    (b) What is the threshold frequency for the photoelectric effect in sodium?
    (c) To what wavelength does this frequency correspond?

30. ✸ A certain material is observed to have a threshold wavelength of 263 nm.
    (a) To what frequency does this wavelength correspond?
    (b) What energy is carried by each photon at this frequency?
    (c) What is the work function of the material?
    (d) Express your result in eV.

31. ✸✸ A student collects the data in Table 30-1 during a photoelectric effect experiment. Recall that $hf - E_0 = K_{max}$, and K_{max} is related to the stopping potential, ΔV, through $K_{max} = e\Delta V$, where $e = 1.602 \times 10^{-19}$ C.
    (a) Convert the wavelengths in Table 30-1 to frequencies.
    (b) Convert the stopping potentials into values of K_{max}.
    (c) Plot K_{max} versus frequency, and fit a straight line.
    (d) How does the value of your slope compare to Planck's constant?
    (e) Divide the intercept by e to get the work function in eV.

Table 30-1 Data for Problem 31

Wavelength (nm)	Stopping potential (V)
260	0.46
250	0.66
240	0.87
230	1.09

Section 30-9 The Bohr Model of the Atom

32. ✸ Consider a spectral line corresponding to a transition from $n_1 = 2$ to $n_2 = 1$.

    (a) Calculate the change in energy of the atom.
    (b) Calculate the frequency of the emitted light.
    (c) Calculate the wavelength of the emitted light.
    (d) Calculate the wave number of the emitted light.

33. ✸ What is the speed of an electron in a Bohr orbit corresponding to $n = 1$?

34. ✸✸ What is the radius of the $n = 7$ orbit of a Bohr atom?

35. ✸ What is the period of the $n = 1$ Bohr orbit?

36. ✸✸ In 1924, Louis de Broglie postulated that particles have a wavelength $\lambda = \frac{h}{p}$, where h is Planck's constant and $p = mv$.
    (a) What is the momentum of an electron in the $n = 1$ Bohr orbit?
    (b) What is the wavelength of an electron in the $n = 1$ orbit?
    (c) What is the circumference of the $n = 1$ orbit?

Section 30-10 Compton Scattering

37. ✸ A Compton scattering experiment uses 350 keV X-rays. The detector is placed at an angle of 40° relative to the incident beam. Find
    (a) the energy of the scattered X-ray;
    (b) the energy of the scattered electron.

38. ✸✸ A photon with a wavelength of 750 pm causes an electron to recoil with a speed of 1.35×10^6 m/s in a Compton experiment. Find
    (a) the change in the photon's wavelength;
    (b) the change in the photon's frequency;
    (c) the angle through which the photon is scattered.

COMPREHENSIVE PROBLEMS

39. ✸ A 5.0 MeV α-particle approaches a nucleus head-on and stops at a distance of 7.0 fm from the centre of the nucleus. What is the charge on the nucleus?

40. ✸ What element is in the sample with the absorption spectra given in Figure 30-21?

400 nm   700 nm

Figure 30-21 Absorption spectra for problem 40

41. ✸✸✸ In the Sun, two protons can fuse together to form deuterium. In this process, a positron, an electron neutrino, and some energy are released. For the reaction to take place, the protons must come close enough together for the weak interaction forces to overcome the repulsive Coulomb interaction. Make a plot of the Coulomb potential as a function of proton–proton separation, assuming that the protons are point charges. It is generally thought that this reaction can proceed at temperatures of approximately 1.5×10^7 K. What separation distance does such a temperature allow?

42. ✸ The diameter of a hydrogen nucleus is taken to be 1×10^{-14} m, and we assume that the volume of a hydrogen atom is defined by the Bohr radius. What is the ratio of the volume of the nucleus to the volume of the atom?

43. ✷✷ In a mass spectrometer, a magnetic field is used to cause charged particles to move in circular paths, with radii $r = \dfrac{mv}{qB}$. Determining r then allows us to measure m, assuming that v, q, and B are known. To simplify the experiment, a velocity selector is used to make sure that all the particles have the same speed and a crossed-field configuration is used. If the radius of the spectrometer magnet is 0.50 m and it has a field strength of 300 mT, what is the maximum speed that incoming particles can have if we wish to measure masses up to the mass of an argon atom? Assume that all particles are singly ionized.

44. ✷✷ Suppose that electrons are accelerated through a 1000 V potential difference and then strike a 5 mg aluminum target. It is observed that the temperature of the target rises by 4° C in 8 s. Assume that all the energy of the electrons is converted into thermal energy. How many electrons have struck the target during the 8 s? Find the beam current.

45. ✷✷ An electron in a cathode ray beam enters a crossed-field region. The plates are 1.0 cm apart and have a potential of 200 V across them. The magnetic field has a strength of 0.80 mT when the electrons pass straight through. When the potential difference is removed, what is the radius of curvature of the electron beam? What is the maximum length of the plates that will allow the beam to exit without hitting the plates? Assume that the beam enters the plates halfway between them.

46. ✷✷ Suppose that an electron orbits a proton and radiates light at the orbital frequency. What does the radius of the orbit need to be for the atom to emit light with a wavelength of 656 nm?

47. ✷✷✷ In Chapter 31, you will see that a blackbody spectrum can be described by $I(f, T) = \dfrac{2hf^3}{c^2} \dfrac{1}{e^{\frac{hf}{kT}} - 1}$ where I is the intensity of the radiation, f is the frequency of the radiation, h is Planck's constant, c is the speed of light, k is Boltzmann's constant, and T is the temperature in K. Differentiate the intensity to find the frequency at which the intensity is maximum.

48. ✷✷ The following series of wave numbers are measured in some other universe that has a different Rydberg constant; 3472, 4688, 5250, 5556, 5740, and 5859/m. Perform a fit to the data to determine n_2 and the value of the Rydberg constant in that universe.

49. ✷✷ It is not known whether the electron is a point particle. If you want to do electron–electron scattering to probe the structure of the electron, what energies are required? Assume that the radius of the electron is 10^{-17} m and that the electrons must be able to get at least this close to each other.

50. ✷✷ An α-particle scatters from an oxygen nucleus. The initial speed of the α-particle is $v_i = 1 \times 10^7$ m/s. The α-particle is deviated by an angle of 90°, and it emerges from the collision with the speed $v_f = 0.317 \times 10^7$ m/s. Determine the magnitude and direction of the recoil momentum of the nucleus, and discuss whether the collision was elastic.

51. ✷✷✷ Suppose you wanted to use a crossed-field apparatus built for studying electrons to measure the charge-to-mass ratio of the proton. What changes would you need to make to the apparatus to make a successful measurement?

52. ✷✷✷ An alternative method of doing the Millikan experiment is to reverse the direction of the applied field and observe the terminal velocity reached in both directions. The radii of the drops must be known. A student is doing such an experiment with 1 μm latex spheres and a voltage of 100 V between plates, which are separated by 5 mm. What velocities would be observed for a drop with a charge of $+e$? $-2e$?

53. ✷✷ The radiant flux at Earth due to the Sun is approximately 1300 W/m². This energy is spread across the visible spectrum and extends both down into the infrared and up into the ultraviolet. For the purpose of this problem, assume that all of this intensity is at 500 nm. How many photons per square metre does such a flux correspond to?

DATA-RICH PROBLEM

54. ✷✷ Go to the text's online resource and download the hydrogen data file, which contains all the wave numbers that are plotted in Figure 30-7. Download the data file, plot the data, and try to fit the data to the form of Equation (30-5). In the process you should arrive at a value for the Rydberg constant.

OPEN PROBLEM

55. ✷✷✷ Hydrogen is attractive as an alternative fuel because it is plentiful and its combustion product is simply water. Estimate how much water would have to be electrolyzed to meet your annual energy needs.

See the text online resources at www.physics1e.nelson.com for Open Problems and Data-Rich Problems related to this chapter.

Chapter 31
Introduction to Quantum Mechanics

Quantum mechanics predicts the behaviour of particles and waves at the atomic and subatomic scale. This behaviour can appear strange and even baffling because it often differs radically from behaviour at the scale of the everyday objects we have handled all of our lives. In this chapter, you will learn how to apply the equations that describe the quantum-mechanical behaviour of particles and the hydrogen atom. Many modern instruments and devices, such as laser scanners, smoke detectors, flash memory, tunnelling electron microscopes, diode lasers, and nuclear magnetic resonance (NMR) machines, are based on the theory of quantum mechanics (Figure 31-1). New theoretical and experimental research involving quantum computers and quantum teleportation has the potential for a new generation of practical applications.

Some derivations in this chapter require a basic understanding of complex numbers and partial derivatives.

Figure 31-1 The principles of quantum mechanics underlie the technology of the memory chips in this music player.

Learning Objectives

When you have completed this chapter you should be able to:

1. Explain de Broglie's hypothesis and the Davisson–Germer experiment, and calculate an object's wavelength given its momentum.

2. Explain how the width of a wave packet is related to the uncertainty in the position of the particle described by the wave packet, and how uncertainties in position and momentum are related by Heisenberg's uncertainty principle.

3. State the four assumptions that Schrödinger made to derive his wave equation, and write the Schrödinger equation in one dimension for a given potential.

4. Write solutions of the Schrödinger equation for $V(x) = V_0$. Calculate the probability density and the probability of finding a particle in a specified region.

5. Calculate the energy and wavelength of the emitted photon when a charged particle makes a transition between two bound states of an infinite square well potential.

6. Describe the behaviour of a wave function in a region where it is forbidden by classical mechanics.

7. Describe the quantization of angular momentum. Determine the values of the total angular momentum quantum number of two or more angular momenta. Calculate the magnitude and the z-component of the total angular momentum quantum number.

8. Explain the similarities and differences between the Bohr model and the Schrödinger equation description of a hydrogen atom.

9. Describe the Stern–Gerlach experiment, and explain how it led to the discovery of quantized spin angular momentum.

 For additional Making Connections, Examples, and Checkpoints, as well as activities and experiments to help increase your understanding of the chapter's concepts, please go to the text's online resources at www.physics1e.nelson.com.

MODERN PHYSICS

31-1 Matter Waves and de Broglie's Hypothesis

Quantum mechanics is the branch of physics that deals with the motion and properties of objects of extremely small masses, such as electrons, protons, neutrons, nuclei, atoms, and molecules. The following comparison will help provide an idea of the scale of subatomic particles: the mass of the smallest known virus is of the order of 10^{-20} kg, which is approximately 10 million times the mass of a proton and 40 000 times the mass of the heaviest nucleus. For subatomic masses Newton's laws of motion break down, and quantum mechanics provides a much more accurate description of the motion of these objects. Numerous experiments have verified the principles of quantum theory and have shown that its strange and far-reaching predictions are correct.

As described in Chapter 30, Planck had to assume that electromagnetic radiation is emitted in discrete parcels, called quanta, to explain the radiation spectrum of a blackbody, and Einstein had to make the same assumption to explain the photoelectric effect. Not only is the radiation *emitted* in quanta, but it also *travels* in quanta. So, by the early 20th century, scientists had realized that electromagnetic waves have particle-like properties. In 1924, French physicist Louis de Broglie (1892–1987) argued that in a symmetric universe particles should also possess wave-like properties. He proposed that *moving objects behave like a wave and therefore have a wavelength and a frequency.* For a particle with energy E and momentum p, the frequency f and the wavelength λ of the associated wave are given by

$$f = \frac{E}{h} \qquad (31\text{-}1)$$

$$\lambda = \frac{h}{p} \qquad (31\text{-}2)$$

Equations (31-1) and (31-2) are the same equations that Planck and Einstein proposed for the energy and momentum of a photon. A wave associated with a moving particle is called a **de Broglie wave** or **matter wave**, and its wavelength is called the **de Broglie wavelength**.

EXAMPLE 31-1

de Broglie Wavelengths

Determine whether the de Broglie effect (wave–particle) is detectable for the following objects:

(a) a space shuttle entering orbit;
(b) a fastball thrown by a major-league pitcher;
(c) an electron accelerated through the potential difference of 1.5 volt.

SOLUTION

(a) The approximate mass of a space shuttle is 22 000 kg, and its speed just before it enters its orbit is approximately 8000 m/s. The de Broglie wavelength of a shuttle at this speed is

$$\lambda = \frac{h}{p} = \frac{6.63 \times 10^{-34}\ \text{J} \cdot \text{s}}{(22\ 000\ \text{kg})(8\ 000\ \text{m/s})} = 3.8 \times 10^{-42}\ \text{m}$$

This wavelength is approximately 10^{-29} times smaller than the size of a proton. No existing instrument or method can measure a wavelength this small. Note that we can use the nonrelativistic formula for momentum ($p = mv$) because the speed of the shuttle is much less than the speed of light.

(b) The mass of a baseball is approximately 0.14 kg, and the speed of the fastest thrown baseball was recorded at 103 mph (46.0 m/s). The de Broglie wavelength of the baseball is

$$\lambda = \frac{6.63 \times 10^{-34}\ \text{J} \cdot \text{s}}{(0.14\ \text{kg})(46.0\ \text{m/s})} = 1.0 \times 10^{-34}\ \text{m}$$

This wavelength is still beyond the measurement range of any instrument.

(c) An electron has a mass of 9.11×10^{-31} kg. So when accelerated through a 1.5 V potential difference it acquires 1.5 eV of kinetic energy. The speed can be calculated using kinematic relationships:

$$K = \text{kinetic energy of the electron} = 1.5\ \text{eV}$$

$$= 1.5 \times 1.60 \times 10^{-19}\ \text{J} = 2.40 \times 10^{-19}\ \text{J}$$

$$v = \sqrt{\frac{2K}{m_e}} = \sqrt{\frac{(2)(2.40 \times 10^{-19}\ \text{J})}{9.11 \times 10^{-31}\ \text{kg}}} = 7.3 \times 10^5\ \text{m/s}$$

Again, the speed is low enough that we can ignore relativistic effects. The de Broglie wavelength of the electron is

$$\lambda = \frac{6.63 \times 10^{-34}\ \text{J} \cdot \text{s}}{(9.11 \times 10^{-31}\ \text{kg})(7.3 \times 10^5\ \text{m/s})} = 1.0 \times 10^{-9}\ \text{m}$$

This wavelength is of the order of the wavelength of high-energy X-rays and the size of the atoms. Such wavelengths are easily measured with modern instruments.

MODERN PHYSICS

C-31-1 The de Broglie Wavelength

An electron, a proton, and an iron nucleus are moving at speeds such that they have the same momentum. Which has the shortest de Broglie wavelength?

(a) the electron

(b) the proton

(c) the iron nucleus

(d) They all have the same wavelength.

C-31-1 (d) The de Broglie wavelength of a particle depends on its momentum only.

Bragg's Law

As described in Chapter 28, the wavelength of light can be measured by observing the diffraction pattern formed when the light passes through a diffraction grating. For accurate measurements, the spacing between the slits of the grating must be of the same order of magnitude as the wavelength of the light. So, the slit separation of an optical diffraction grating is too large for measuring wavelengths of X-rays, for example.

The German physicist Max Theodor Felix von Laue (1879–1960) showed that crystals, whose interatomic spacing is of the order of 10^{-10} m, can be used as a diffraction grating for measuring the wavelengths of X-rays. Figure 31-2 shows X-rays scattering from individual atoms of successive layers of a crystal. The path difference between two rays that scatter at an angle θ from two adjacent planes is $2d\sin\theta$. The condition for constructive interference between these two waves is

$$2d\sin\theta = m\lambda \qquad \text{(Bragg's law)} \qquad (31\text{-}3)$$

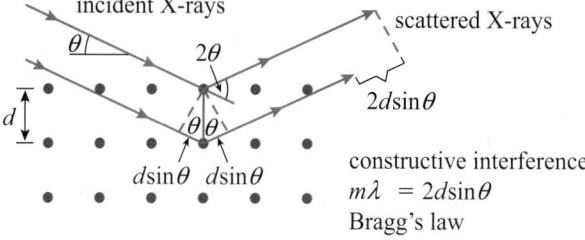

Figure 31-2 The scattering of X-rays from adjacent planes in a crystal. The path difference for waves scattered from two successive planes is $2d\sin\theta$.

where m is a nonzero positive integer. This condition for constructive interference is called **Bragg's law**. Figure 31-3 shows the diffraction pattern produced by X-rays scattering from an aluminum foil.

The Davisson–Germer Experiment

In 1926, the German-born American physicist Walter Maurice Elsasser (1904–91) suggested that if electrons have wavelengths comparable to X-ray wavelengths, then electron scattering from crystals should produce the same type of diffraction pattern as observed in X-ray scattering.

Figure 31-3 A diffraction pattern formed by X-rays scattering from an yttrium aluminum oxide target.

In 1927, American physicists Clinton Davisson (1881–1958) and Lester Germer (1896–1971) tested Elsasser's suggestion and performed electron diffraction experiments using the setup shown in Figure 31-4. Electrons emitted from a heated metal filament are accelerated through a potential difference and collimated into a narrow beam. Davisson and Germer directed the electron beam at the face of a crystal, and they measured the angles at which the reflected beam had maximum intensity. They then used Bragg's law to calculate the wavelength corresponding to these angles—they found that it matched the wavelength predicted by de Broglie. Davisson and Germer varied the potential difference to change the energy of the electrons in the beam, producing maxima and minima at different angles in the diffraction patterns. In each case, the wavelength calculated using the Bragg equation agreed with the de Broglie wavelength. This series of experiments proved conclusively that electrons do have a wavelength. Similar experiments with other particles have confirmed the wave nature of matter.

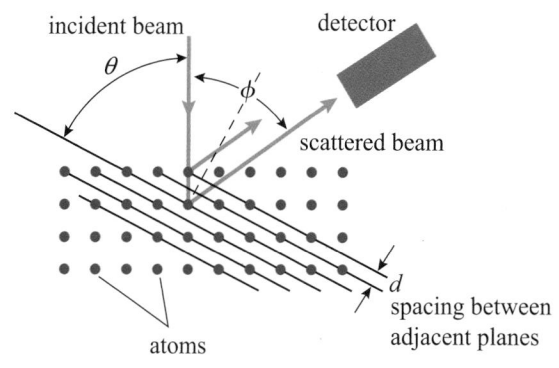

Figure 31-4 The Davisson–Germer experiment, showing the relationship between the incident angle θ and scattering angle ϕ

EXAMPLE 31-2

The de Broglie Wavelength of an Electron

In one experiment, Davisson and Germer accelerated electrons through a potential difference of 54 V and scattered the electrons from a nickel crystal. They observed a maximum in the intensity of reflected electrons at an angle of $\phi = 50°$. The spacing between the scattering planes of the crystal was calculated from X-ray diffraction to be $d = 0.091$ nm. Show that these results agree with de Broglie's hypothesis for particle wavelengths.

SOLUTION

According to Bragg's law, when waves of wavelength λ are reflected from the planes of a crystal that are separated by a distance d, the scattering angle at which maximum constructive interference occurs, θ, is given by Equation (31-3). For scattering from adjacent planes, $m = 1$, the angle θ is the angle of incidence with respect to a particular scattering plane of the crystal (Figure 31-4). By conservation of momentum, the reflected beam must also make an angle θ with respect to the same scattering plane. Therefore, the angle between the incident and the reflected beams, ϕ, is

$$\phi = \pi - 2\theta$$

The location of the detector with respect to the incident beam corresponds to this angle, ϕ. For scattering from two adjacent planes, Bragg's law gives

$$\lambda = 2d \cos\left(\frac{\phi}{2}\right)$$

Inserting the given data we get

$$\lambda = (2)(0.091 \text{ nm}) \cos(25°) = 0.165 \text{ nm}$$

For nonrelativistic speeds, the kinetic energy of an electron accelerated through a potential V is given by

$$\frac{1}{2} m_e v^2 = eV$$

Therefore, the de Broglie wavelength of this electron is

$$\lambda = \frac{h}{m_e v} = \frac{h}{m_e \sqrt{\dfrac{2Ve}{m_e}}} = \frac{h}{\sqrt{2Vem_e}}$$

Substituting $V = 54$ V gives a de Broglie wavelength of

$$\lambda = \frac{6.63 \times 10^{-34} \text{ J·s}}{\sqrt{(2)(54 \text{ eV})(1.60 \times 10^{-19} \text{ J/eV})(9.11 \times 10^{-31} \text{ kg})}}$$

$$= 0.167 \text{ nm}$$

This value is in excellent agreement with the experimentally determined wavelength.

 ## CHECKPOINT

C-31-2 Proton Diffraction

If the electron beam in Example 31-2 is replaced by a proton beam ($m_p \approx 1836 m_e$) of the same de Broglie wavelength, the angle at which the diffraction peak occurs

(a) is the same as for the electron beam;

(b) increases;

(c) decreases;

(d) is too small to measure because protons are much more massive than electrons.

You might think that the wave nature of a beam of electrons is a collective effect rather than a property of individual electrons. However, when a diffraction experiment is performed over an extended period with a low-intensity beam such that one electron at a time scatters from the crystal, the diffraction pattern recorded on a photographic plate is the same as the pattern produced by a short-duration high-intensity beam with the same number of electrons. So, each electron independently displays wave properties.

Diffraction experiments have been performed with a variety of particles. Figure 31-5 shows diffraction patterns of X-ray beams and electron beams passing through

1927 Public Domain

Figure 31-5 Diffraction patterns formed by the scattering of X-ray beams and electron beams from aluminum foil

an aluminum foil. The diffraction of neutron beams from crystal surfaces displays similar diffraction effects.

 ## ONLINE ACTIVITY

Electron Diffraction from a Crystal

The e-resource that accompanies every new copy of this textbook contains an Online Activity using the PhET simulation "Electron Diffraction from a Crystal." Work through the simulation and accompanying questions to learn more about the diffraction of electrons from crystals.

MODERN PHYSICS

31-2 Heisenberg's Uncertainty Principle

What kind of wave is associated with the de Broglie wavelength? In Chapter 14, you learned that a travelling harmonic wave on a string can be described by a sinusoidal wave function of the form $D(x, t) = A \sin(kx - \omega t)$. Sound and electromagnetic waves can also be represented by harmonic waves. The square of the wave function at a given position x and time t is proportional to the *intensity* of the wave at that position and time. If a de Broglie wave is represented by a harmonic wave, then the square of the wave function at a given position and time is related to the *probability* of the object being at that position at that time. However, the sinusoidal wave function has an infinite extent with a constant amplitude, so the probability of finding the particle would be the same for all locations. Therefore, we cannot represent a de Broglie wave with a sinusoidal wave of the form $\sin(kx - \omega t)$. Such a wave function accurately describes the wavelength of the object and, hence, its momentum (because $p = h/\lambda$) but leaves its location undetermined.

We need to represent a moving particle by a wave that is localized in space. Such a wave is called a **wave packet** or a **wave group**. We can construct a wave group by combining a large number of harmonic waves with slightly different wavelengths. Figure 31-6 shows a single harmonic wave with $k = 10$ rad/m and a wave group made up of 10 harmonic waves with the wave number increasing by 0.1 rad/m, starting at 10.0 rad/m. With a localized wave group the position of an object is represented more accurately. However, the wavelength (and, hence, momentum) of the object becomes less certain because the wave group consists of a number of waves with closely spaced wave numbers. In fact, there is a *fundamental* limit to the accuracy with which we can measure both the position and the momentum of a moving object.

Consider a wave group that results from combining two waves with wave numbers k and $k + \Delta k$ and angular

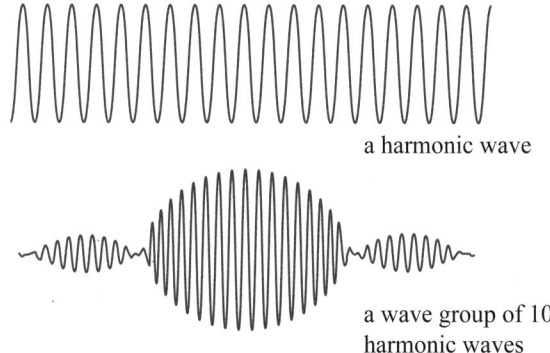

a harmonic wave

a wave group of 10 harmonic waves

Figure 31-6 A harmonic wave and a wave group consisting of 10 harmonic waves with closely spaced wave numbers

frequencies ω and $\omega + \Delta \omega$. The wave function of this wave group is given by,

$$D(x, t) = A \sin((k + \Delta k)x - (\omega + \Delta \omega)t) + A \sin(kx - \omega t)$$

Using the trigonometric identity

$$\sin(A) + \sin(B) = 2 \cos\left(\frac{A - B}{2}\right) \sin\left(\frac{A + B}{2}\right),$$

we get

$$D(x, t) = 2A \cos\left(\frac{\Delta k}{2} x - \frac{\Delta \omega}{2} t\right)$$

$$\sin\left(\frac{2k + \Delta k}{2} x - \frac{2\omega + \Delta \omega}{2} t\right) \quad (31\text{-}4)$$

A graph of this equation at $t = 0$ for $A = 1$ m, $k = 6$ rad/m, and $\Delta k = 1$ rad/m is shown in Figure 31-7. This situation is similar to the situation that you encountered while studying the beat phenomenon for sound waves.

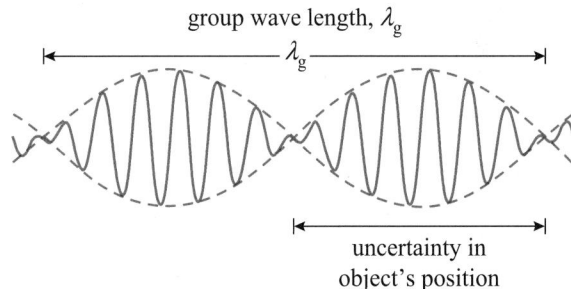

group wave length, λ_g

uncertainty in object's position

Figure 31-7 A wave group. The dashed curve represents the amplitude of the wave group. The uncertainty in the object's position is equal to half the wavelength λ_g of the wave group.

The cosine term in Equation (31-4), called the modulation, has a wave number $\Delta k/2$, and its wavelength is $\frac{2\pi}{\Delta k/2} = \frac{4\pi}{\Delta k}$. From Figure 31-7, we notice that the width Δx of the wave group is equal to half the wavelength of the modulation. Therefore,

$$\Delta x = \frac{1}{2}\left(\frac{4\pi}{\Delta k}\right) = \frac{2\pi}{\Delta k} \quad (31\text{-}5)$$

The uncertainty in the position of the wave group is of the same order of magnitude as its width. Therefore, it is reasonable to assume that the uncertainty in the wave number of the wave group is related to the uncertainty in its position by the following relationship:

$$\Delta k = \frac{2\pi}{\Delta x}$$

From Equation (31-2), we know that the wave number k of the de Broglie wave that is associated with a particle of momentum p is given by

$$p = \frac{h}{\lambda} = \frac{h}{2\pi/k} = \frac{h}{2\pi} k$$

Therefore, an uncertainty Δk in the wave number of the wave group corresponds to an uncertainty Δp in the particle's momentum such that

$$\Delta p = \frac{h}{2\pi} \Delta k$$

Using Equation (31-5), we can now relate the uncertainty in the particle's momentum to the uncertainty in its position (i.e., the uncertainty in the position of the wave group that represents the particle) as

$$\Delta p = \frac{h}{2\pi} \left(\frac{2\pi}{\Delta x} \right) = \frac{h}{\Delta x}$$

The product of the uncertainty in the particle's position (Δx) and the uncertainty in its momentum (Δp) satisfy the relationship

$$\Delta x \, \Delta p = h$$

Note that Δx and Δp are the *minimum* uncertainties due to the wave nature of the object. Any simultaneous measurement of the object's position and momentum will result in uncertainties such that

$$\Delta x \, \Delta p_x \geq h$$

In the above equation we have replaced Δp with Δp_x to emphasize that the uncertainty is in that component of the momentum that is related to the uncertainty in a given component of the position vector. Our approach did not rigorously define what we mean by uncertainty. A more careful analysis yields the relationship called **Heisenberg's uncertainty principle**:

$$\Delta x \, \Delta p_x \geq \frac{\hbar}{2} \qquad (31\text{-}6)$$

The German physicist Werner Karl Heisenberg (1901–76) derived this relationship in 1927 from the general principles of quantum physics. He found that the product of the uncertainties in the simultaneous measurement of the position and momentum of a particle cannot be less than $\hbar/2$, and therefore the position and the momentum cannot be simultaneously measured with infinite accuracy. In very precise measurements, any technique that reduces the uncertainty in the position measurement increases the uncertainty in the momentum measurement, and vice versa.

Since $p_x = mv_x$ where m is the mass of the object, Equation (31-6) can be written in terms of uncertainties in position and velocity:

$$\Delta x \, \Delta v_x \geq \frac{\hbar}{2m} \qquad (31\text{-}7)$$

Let us consider another physical basis of the uncertainty principle. We determine the position of an object by illuminating the object with light. The light scatters from the object, enters the field of view of an eye or a microscope, and forms an image of the object. From Chapter 28 we know that to form a clear image, the wavelength of the light must be less than the dimensions of the object it illuminates. If the wavelength is of the same order of magnitude as the size of the object, the diffraction effects produce a blurred image. Let us use this mechanism to determine the position and the velocity of a small classical object and then an electron.

Consider a carbon particle with a radius of one micron (10^{-6} m) moving with a speed of 0.1 m/s. The density of carbon is 2.25×10^3 kg/m³. Therefore, the mass and momentum of this particle are

$$m = \frac{4\pi}{3} R^3 \times \rho = \frac{4\pi}{3} (10^{-18} \text{ m}^3)(2.25 \times 10^3 \text{ kg/m}^3)$$
$$\approx 10^{-14} \text{ kg}$$

and

$$p = (10^{-14} \text{ kg})(0.1 \text{ m/s}) = 10^{-15} \text{ kg} \cdot \text{m/s}$$

We can form a clear image of such a particle by using green light with a wavelength of 550 nm. (The wavelength of the green light is approximately four times smaller than the diameter of the particle.) To form the image, the light must scatter from the particle. The momentum of a photon with a wavelength of 550 nm is

$$p = \frac{h}{\lambda} = \frac{6.63 \times 10^{-34} \text{ kg} \cdot \text{m}^2/\text{s}}{550 \times 10^{-9} \text{ m}} = 1.2 \times 10^{-27} \text{ kg} \cdot \text{m/s}$$

This is a factor of 10^{-12} smaller than the momentum of the particle. Therefore, when a photon hits a carbon particle, there is no noticeable change in the momentum of the particle.

Now we replace the carbon particle with an electron. Let us assume that the electron is moving with a speed of 10^8 m/s. The momentum of the electron is

$$(9.11 \times 10^{-31} \text{ kg})(10^8 \text{ m/s}) = 9.1 \times 10^{-23} \text{ kg} \cdot \text{m/s}$$

(We have used nonrelativistic kinematics because we are interested in the order-of-magnitude calculations.) The current experimental data suggest that the radius of an electron must be smaller than 10^{-18} m. Let us assume that the radius of an electron is 10^{-18} m. To form a sharp image of an electron, the wavelength of the photon must be at least 10^{-18} m. Suppose that we scatter photons with a wavelength of 10^{-18} m from the electron. The momentum of such a photon is

$$p = \frac{h}{\lambda} = \frac{6.63 \times 10^{-34} \text{ kg} \cdot \text{m}^2/\text{s}}{10^{-18} \text{ m}} = 6.63 \times 10^{-16} \text{ kg} \cdot \text{m/s}$$

This momentum is a factor of 7×10^6 times larger than the momentum of the electron. Such a photon striking an electron would be like a freight train smashing into a bicycle. If we illuminate an electron

MODERN PHYSICS

with a photon of small enough wavelength to form a sharp image, we knock the electron out of its position and do not know its position or momentum. If we try using a photon with a much longer wavelength (and, hence, much smaller momentum) so that the electron is not disturbed, diffraction effects blur its image and we cannot accurately determine the position of the electron.

Our inability to accurately measure both the position and the momentum of a particle is not simply a matter of needing more precise instruments. The uncertainty limit determined by Heisenberg is part of the nature of the microscopic world. At this scale, we have to deal with probabilities rather than definite values. A statement such as "at time $t = 0$, the electron is at $x = 0.2$ m and moving with a speed of 10^4 m/s" gets replaced by "at time $t = 0$, the probability of finding the electron around $x = 0.2$ m is 0.9, and the probability that it is moving with a speed of 10^4 m/s is 0.6." This lack of certainty is one of the more confusing aspects of quantum theory. Even Einstein was perplexed about why nature would have a built-in mechanism that prevents precise measurements.

EXAMPLE 31-3

Uncertainty in the Momentum of a Particle Confined Within a Box

An object with a mass of 1.0×10^{-3} kg is confined within a box of length 1.0 m. What is the minimum uncertainty in the momentum and the speed of the object?

SOLUTION

The object is confined within a box of length 1.0 m, so the maximum uncertainty in the position of the object is 1.0 m. Using $\Delta x = 1.0$ m in Equation (31-6), we get

$$\Delta p_x \geq \frac{\hbar}{2 \times \Delta x} = \frac{1.05 \times 10^{-34}\,\text{J·s}}{2 \times 1.0\,\text{m}} = 0.53 \times 10^{-34}\,\text{kg·m/s}$$

for the minimum uncertainty in the momentum of the particle. Therefore, the speed of the particle cannot be determined to an accuracy better than

$$\Delta v_x = \frac{\Delta p_x}{m} = \frac{0.53 \times 10^{-34}\,\text{kg·m/s}}{1.0 \times 10^{-3}\,\text{kg}} = 5.3 \times 10^{-30}\,\text{m/s}$$

Making sense of the result:

Since the uncertainty in position is very large ($\Delta x \sim 1.0$ m) relative to a microscopic scale, the uncertainty in the speed is very small. Over a billion years, a difference in speed of 5.3×10^{-30} m/s would cause a difference in location of only of 1.7×10^{-13} m.

EXAMPLE 31-4

Uncertainty in the Momentum of a Proton in a Nucleus

Find the minimum uncertainty in the momentum and the speed of a proton in the nucleus of an iron atom. The mass of a proton is 1.67×10^{-27} kg, and the radius of an iron nucleus is approximately 4.6×10^{-15} m.

SOLUTION

The proton is confined within the iron nucleus, so it is reasonable to expect that the uncertainty in the position of the proton is equal to the diameter of the iron nucleus:

$$\Delta x = 2 \times 4.6 \times 10^{-15}\,\text{m} = 9.2 \times 10^{-15}\,\text{m}$$

The uncertainty in the momentum of the proton is then given by

$$\Delta p_x \geq \frac{\hbar}{2 \times \Delta x} = \frac{1.05 \times 10^{-34}\,\text{J·s}}{2 \times 9.2 \times 10^{-15}\,\text{m}} = 5.7 \times 10^{-21}\,\text{kg·m/s}$$

The speed of the proton therefore cannot be determined to an accuracy better than

$$\Delta v_x = \frac{\Delta p_x}{m} = \frac{5.7 \times 10^{-21}\,\text{kg·m/s}}{1.67 \times 10^{-27}\,\text{kg}} = 3.4 \times 10^{6}\,\text{m/s}$$

Making sense of the result:

A proton is localized in a very small volume, so the uncertainty in the location of the proton is very small (Δx: 9.0×10^{-15} m). Therefore, the uncertainty in its speed, and hence momentum, is large.

 CHECKPOINT

C-31-3 Wave Groups

The four wave groups in Figure 31-8 represent the de Broglie wave associated with a moving particle. Rank the wave groups in order of the uncertainty in the particle's speed from smallest to largest.

(a)

(c)

(b)

(d)

Figure 31-8 C-31-3

Other Uncertainty Relationships

The position and the momentum of a particle are called **observables** because these quantities can be measured in an experiment. The uncertainty relationship is also valid for other combinations of observables, such as time, energy, angular momentum, and acceleration. The time–energy uncertainty relationship is

$$\Delta t \, \Delta E \geq \frac{\hbar}{2}$$

where Δt is the time taken to measure the energy of a system (e.g., a hydrogen atom in an excited state), and ΔE is the accuracy to which the energy of the system can be measured in that time. The time–energy uncertainty principle states that the time required to measure the energy of a system to an accuracy of ΔE must be greater than $\frac{\hbar}{2\Delta E}$. Thus, the law of conservation of energy can only be verified to an accuracy of $\frac{\hbar}{2\Delta t}$, where Δt is the time taken to measure the energy.

LO 3

31-3 The Schrödinger Equation

Macroscopic objects obey Newton's laws of motion. Harmonic travelling waves obey the wave equation (see below), and electromagnetic waves are described by Maxwell's equations. What is the nature of the laws that describe matter waves? In 1926, Heisenberg formulated a description of matter waves using matrices. In the same year, German physicist Erwin Schrödinger (1887–1961) used a wave equation to accurately describe the wave behaviour of matter. The two approaches are equivalent. Here, we will follow Schrödinger's approach.

First, we will show that the wave function $D(x, t)$ for a one-dimensional travelling harmonic wave satisfies the following differential equation:

$$\frac{\partial^2 D(x, t)}{\partial x^2} = \frac{1}{v^2} \frac{\partial^2 D(x, t)}{\partial t^2} \qquad (31\text{-}8)$$

where v is the wave speed.

For the left side of Equation (31-8), we take the second derivative of $D(x, t)$ with respect to x while keeping the time t constant:

$$\frac{\partial}{\partial x}(A \sin(kx - \omega t)) = A \frac{\partial}{\partial x} \sin(kx - \omega t)$$

$$= kA \cos(kx - \omega t)$$

$$\frac{\partial^2}{\partial x^2}(A \sin(kx - \omega t)) = \frac{\partial}{\partial x}(kA \cos(kx - \omega t))$$

$$= -k^2 A \sin(kx - \omega t)$$

Similarly, for the right side, we calculate the second derivative of $D(x, t)$ with respect to t while keeping the position x constant:

$$\frac{\partial}{\partial t}(A \sin(kx - \omega t)) = A \frac{\partial}{\partial t} \sin(kx - \omega t)$$

$$= -\omega A \cos(kx - \omega t)$$

$$\frac{\partial^2}{\partial t^2}(A \sin(kx - \omega t)) = \frac{\partial}{\partial t}(-A\omega \cos(kx - \omega t))$$

$$= -\omega^2 A \sin(kx - \omega t)$$

Since $v^2 = \dfrac{\omega^2}{k^2}$,

$$\frac{1}{v^2} \frac{\partial^2}{\partial t^2}(A \sin(kx - \omega t)) = -\left(\frac{k^2}{\omega^2}\right)\omega^2 A \sin(kx - \omega t)$$

$$= -k^2 A \sin(kx - \omega t)$$

Therefore, $D(x, t) = A \sin(kx - \omega t)$ is a solution of Equation (31-8).

To derive a similar differential equation that describes the motion of a matter wave, Schrödinger made the following four assumptions:

1. Matter waves satisfy the de Broglie relation $\lambda = \dfrac{h}{p}$. Therefore, the wave number, k, of a matter wave is related to corresponding particle's momentum:

$$k = \frac{2\pi}{\lambda} = 2\pi\left(\frac{p}{h}\right) = \frac{p}{\hbar} \qquad (31\text{-}9)$$

2. Einstein's equation $f = \dfrac{E}{h}$ relates a particle's energy to the frequency of the corresponding matter wave. Therefore, the angular frequency of a matter wave is related to a particle's energy:

$$\omega = 2\pi f = 2\pi\left(\frac{E}{h}\right) = \frac{E}{\hbar} \qquad (31\text{-}10)$$

3. The phase of a travelling matter wave has the same form as the phase for a mechanical wave. Therefore, the relationships in the first two assumptions can be used to derive an expression for the phase:

$$kx - \omega t = \frac{p}{\hbar}x - \frac{E}{\hbar}t = \frac{1}{\hbar}(px - Et) \qquad (31\text{-}11)$$

4. A wave equation that describes the motion of a particle must satisfy the classical energy–momentum relationship for that particle.

Schrödinger first considered a free particle of mass m that is moving in one dimension (along the x-axis) with momentum p. Assuming that the particle has no potential energy and that relativistic effects are negligible, the total energy E of the particle is equal to its kinetic energy ($p^2/2m$); therefore,

$$\frac{p^2}{2m} = E \qquad (31\text{-}12)$$

Schrödinger realized that to satisfy the energy and momentum relation in Equation (31-12), the wave

MODERN PHYSICS

function of the matter wave must contain a combination of sine and cosine functions of the form

$$\cos\theta + i\sin\theta \qquad (31\text{-}13)$$

Here, i denotes the imaginary number defined by $i^2 = -1$. Therefore, $i = \sqrt{-1}$. Equation (31-13) can be written as an exponential of an imaginary number (problem 35):

$$e^{i\theta} = \cos\theta + i\sin\theta \qquad (31\text{-}14)$$

Using the Greek letter ψ (psi, pronounced "sigh") to denote the wave function for matter waves, Schrödinger wrote the wave function for a particle as a sum of a cosine and a sine travelling wave:

$$\psi(x, t) = \psi_{\text{M}}\left[\cos\left(\frac{p}{\hbar}x - \frac{E}{\hbar}t\right) + i\sin\left(\frac{p}{\hbar}x - \frac{E}{\hbar}t\right)\right] \quad (31\text{-}15)$$

where the constant ψ_{M} is the amplitude of the wave function. Using Equation (31-14), $\psi(x, t)$ can be expressed as a compact exponential function:

$$\psi(x, t) = \psi_{\text{M}}\, e^{\frac{i}{\hbar}(px - Et)} \qquad (31\text{-}16)$$

To write an equation for $\psi(x, t)$ that satisfies the energy relationship (31-12), Schrödinger multiplied both sides of Equation (31-12) by $\psi(x, t)$:

$$\left(\frac{p^2}{2m}\right)\psi(x, t) = E\psi(x, t) \qquad (31\text{-}17)$$

Note that

$$\frac{\partial}{\partial t}\psi(x, t) = \frac{\partial}{\partial t}\left(\psi_{\text{M}}\, e^{\frac{i}{\hbar}(px - Et)}\right) = -\frac{iE}{\hbar}\,\psi_{\text{M}}\, e^{\frac{i}{\hbar}(px - Et)}$$

$$= -\frac{iE}{\hbar}\,\psi(x, t) \qquad (31\text{-}18)$$

Here, we have used the fact that $\dfrac{d}{dx}e^{ax} = ae^{ax}$. Multiplying both sides by $i\hbar$ and using the identity $i \times i = -1$, we get

$$\left(i\hbar\,\frac{\partial}{\partial t}\right)\psi(x, t) = E\psi(x, t) \qquad (31\text{-}19)$$

Similarly, taking the partial derivative of $\psi(x, t)$ with respect to position x gives

$$\frac{\partial}{\partial x}\psi(x, t) = \frac{\partial}{\partial x}\left(\psi_{\text{M}}\, e^{\frac{i}{\hbar}(px - Et)}\right)$$

$$= +\frac{ip}{\hbar}\,\psi_{\text{M}}\, e^{\frac{i}{\hbar}(px - Et)} = +\frac{ip}{\hbar}\,\psi(x, t) \qquad (31\text{-}20)$$

Multiplying both sides by $-i\hbar$, we get

$$\left(-i\hbar\,\frac{\partial}{\partial x}\right)\psi(x, t) = p\psi(x, t) \qquad (31\text{-}21)$$

Schrödinger then took the second derivative of $\psi(x, t)$ with respect to x (since Equation (31-17) contains a p^2):

$$\frac{\partial}{\partial x}\left(\frac{\partial}{\partial x}\psi(x, t)\right) = \frac{ip}{\hbar}\frac{\partial}{\partial x}\left(\psi_{\text{M}}\, e^{\frac{i}{\hbar}(px - Et)}\right)$$

$$= -\frac{p^2}{\hbar^2}\,\psi(x, t) \qquad (31\text{-}22)$$

Multiplying both sides of Equation (31-22) by $-\dfrac{\hbar^2}{2m}$ gives

$$\left(-\frac{\hbar^2}{2m}\frac{\partial^2}{\partial x^2}\right)\psi(x, t) = \frac{p^2}{2m}\,\psi(x, t) \qquad (31\text{-}23)$$

Inserting Equations (31-23) and (31-19) into Equation (31-17), we get the following equation for the wave function:

$$-\frac{\hbar^2}{2m}\frac{\partial^2\psi(x, t)}{\partial x^2} = i\hbar\,\frac{\partial\psi(x, t)}{\partial t} \qquad (31\text{-}24)$$

Note that in obtaining this differential equation,

1. Schrödinger started with the equation that relates the momentum and total energy of the particle;
2. he multiplied the equation by a wave function that uses de Broglie's relation for the wavelength and Einstein's relation for the frequency of the wave;
3. then he derived a wave equation involving derivatives of the wave function by making the following substitutions:

$$E \rightarrow i\hbar\,\frac{\partial}{\partial t} \quad \text{and} \quad p \rightarrow -i\hbar\,\frac{\partial}{\partial x} \qquad (31\text{-}25)$$

Equations (31-19) and (31-21) are called **eigenvalue** equations, and the derivatives $i\hbar\,\dfrac{\partial}{\partial t}$ and $-i\hbar\,\dfrac{\partial}{\partial x}$ are called operators because they operate on the wave function. When the operator $i\hbar\,\dfrac{\partial}{\partial t}$ acts on the wave function $\psi(x, t)$, it gives the energy eigenvalue E. Similarly, the operator $-i\hbar\,\dfrac{\partial}{\partial x}$ gives the momentum eigenvalue when it acts on the wave function. Note that operators in equations are placed to the left of the wave function on which they act.

What happens if the particle also has a potential energy $V(x)$? In classical mechanics, the total energy of the particle is the sum of its kinetic and potential energies:

$$\frac{p^2}{2m} + V(x) = E \qquad (31\text{-}26)$$

Schrödinger applied the procedure outlined above to Equation (31-26) to obtain a differential equation for the wave function when the potential energy is not zero. The resulting equation is called the **time-dependent Schrödinger wave equation**:

KEY EQUATION

$$-\frac{\hbar^2}{2m}\frac{\partial^2\psi(x, t)}{\partial x^2} + V(x)\psi(x, t) = i\hbar\,\frac{\partial\psi(x, t)}{\partial t} \qquad (31\text{-}27)$$

Equation (31-27) describes the motion of a particle with rest mass m that is moving in one dimension and has potential energy $V(x)$. Note that the potential energy term is accounted for by simply multiplying $V(x)$ by the wave function and adding this term to Equation (31-24).

Schrödinger's equation is to quantum physics what Newton's laws are to classical physics. The above procedure of obtaining the Schrödinger equation is a prescription that cannot be derived from more fundamental principles, just as Newton's laws of motion cannot be derived from more fundamental principles. As with Newton's laws of motion and relativity theory, experimental results have consistently agreed with predictions based on the Schrödinger equation.

The Time-Independent Schrödinger Equation

When a particle is in a state with definite energy E (such as an electron moving in a given energy state around a nucleus), the time variation of the wave function of the particle can be written in the simple form $e^{-iEt/\hbar}$. We can then separate the wave function $\psi(x, t)$ into two parts: $\psi(x)$, which only depends on the position x, and $e^{-iEt/\hbar}$, the time-dependent part. Thus,

$$\psi(x, t) = \psi(x)\, e^{-iEt/\hbar}$$

Substituting this wave function into the time-dependent equation, we get

$$-\frac{\hbar^2}{2m}\left(\frac{d^2\psi(x)}{dx^2}\right)e^{-iEt/\hbar} + V(x)\psi(x)e^{-iEt/\hbar} = i\hbar\,\psi(x)\left(\frac{de^{-iEt/\hbar}}{dt}\right)$$

$$= E\psi(x)e^{-iEt/\hbar}$$

Here, we have replaced the partial derivatives with total derivatives because $\psi(x)$ is a function of x only, and the $e^{-iEt/\hbar}$ term only depends on t. Dividing out the factor $e^{-iEt/\hbar}$ gives the **time-independent Schrödinger equation** in one dimension:

KEY EQUATION
$$-\frac{\hbar^2}{2m}\frac{d^2\psi(x)}{dx^2} + V(x)\psi(x) = E\psi(x)$$
(31-28)

For a given $V(x)$, Equation (31-28) is satisfied for only certain values of energy E. These values of energy are called eigenvalues. For a given eigenvalue, the corresponding wave function that satisfies the Schrödinger equation is called the **eigen wave function**, or simply the **eigenfunction**. The first successful application of the Schrödinger equation was in calculating energy eigenvalues and eigenfunctions for the electron in a hydrogen atom.

EXAMPLE 31-5

The Schrödinger Equation for a Harmonic Oscillator Potential

Consider a particle of mass m undergoing simple harmonic motion along the x-axis under the influence of an elastic restoring force, $F(x) = -kx$. Write the Schrödinger equation for this particle.

SOLUTION

As discussed in Chapter 13, the elastic potential energy of a particle undergoing simple harmonic motion is $\frac{1}{2}kx^2$ (Figure 31-9), and the total energy of the particle is

$$\frac{1}{2}mv^2 + \frac{1}{2}kx^2 = \frac{p^2}{2m} + \frac{1}{2}kx^2 = E$$

To obtain the Schrödinger equation for this simple harmonic oscillator, we start with the equation that relates its kinetic energy, potential energy, and total energy. Multiplying the energy equation by the wave function $\psi(x, t)$ gives

$$\left(\frac{p^2}{2m} + \frac{1}{2}kx^2\right)\psi(x, t) = E\psi(x, t)$$

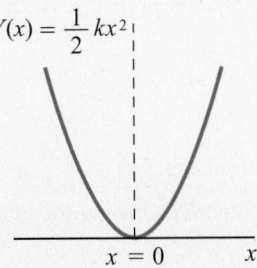

Figure 31-9 A harmonic oscillator potential, $V(x) = \frac{1}{2}kx^2$, in one dimension

Next, we replace the total energy E and the momentum p by the derivative operators from Equation (31-25):

$$\frac{-\hbar^2}{2m}\frac{\partial^2}{\partial x^2}\psi(x, t) + \frac{1}{2}kx^2\psi(x, t) = i\hbar\frac{\partial}{\partial t}\psi(x, t)$$

Making sense of the result:

Here, we have used Schrödinger's procedure to turn the classical energy equation into the corresponding Schrödinger equation. To determine the wave function $\psi(x, t)$, we need to solve this differential equation.

MODERN PHYSICS

The Schrödinger Equation in Three Dimensions

The total energy E of a particle moving in three dimensions is given by

$$\frac{p_x^2}{2m} + \frac{p_y^2}{2m} + \frac{p_z^2}{2m} + V(x, y, z) = E \qquad (31\text{-}29)$$

where p_x, p_y, and p_z are the Cartesian components of the particle's momentum and $V(x, y, z)$ represents the potential energy in three dimensions.

To obtain the corresponding Schrödinger equation, we replace the momentum components and the energy variable with derivatives that act on a wave function $\psi(x, y, z, t)$ that depends on all three space coordinates and time:

$$p_x \rightarrow -i\hbar\frac{\partial}{\partial x}, \quad p_y \rightarrow -i\hbar\frac{\partial}{\partial y}, \quad p_z \rightarrow -i\hbar\frac{\partial}{\partial z},$$

$$E \rightarrow i\hbar\frac{\partial}{\partial t} \qquad (31\text{-}30)$$

The resulting time-dependent Schrödinger equation is

$$-\frac{\hbar^2}{2m}\left(\frac{\partial^2}{\partial x^2} + \frac{\partial^2}{\partial y^2} + \frac{\partial^2}{\partial z^2}\right)\psi(x, y, z, t)$$

$$+ V(x, y, z)\psi(x, y, z, t) = i\hbar\frac{\partial \psi(x, y, z, t)}{\partial t} \qquad (31\text{-}31)$$

As for the one-dimensional case, the right side of Equation (31-31) is equal to $E\psi(x, y, z, t)$ for stationary states. The corresponding time-independent Schrödinger equation is

$$-\frac{\hbar^2}{2m}\left(\frac{\partial^2}{\partial x^2} + \frac{\partial^2}{\partial y^2} + \frac{\partial^2}{\partial z^2}\right)\psi(x, y, z)$$

$$+ V(x, y, z)\,\psi(x, y, z) = E\psi(x, y, z) \qquad (31\text{-}32)$$

The Physical Meaning of the Wave Function

For mechanical waves, $D(x, t)$ denotes the displacement of a particle from its equilibrium position. However, the quantity represented by the wave function $\psi(x, y, z, t)$ is far from obvious. The physical interpretation of $\psi(x, y, z, t)$ was provided by Max Born (1882–1970), a German British physicist and mathematician. Born noted that the square of the absolute value of the wave function is a real number given by

$$\left|\psi(x, y, z, t)\right|^2 = \psi^*(x, y, z, t) \times \psi(x, y, z, t)$$

where $\psi^*(x, y, z, t)$ represents the complex conjugate of the wave function $\psi(x, y, z, t)$. Born argued that this number is equal to the probability per unit volume of finding a particle at position (x, y, z) at time t. For a small volume element $dV = dx\,dy\,dz$ centred around a point (x, y, z), $\left|\psi(x, y, z, t)\right|^2 dV$ is equal to the probability of finding the particle in that volume at time t. Since the probability of finding the particle somewhere in the universe must be one (a certainty),

$$\int_{\text{all space}} \left|\psi(x, y, z, t)\right|^2 dV = 1 \qquad (31\text{-}33)$$

Equation (31-33) is a **normalization condition**, and a wave function that satisfies this condition is called a **normalized wave function**. For the one-dimensional case, dV is replaced by dx.

31-4 Solving the Time-Independent Schrödinger Equation

The Schrödinger equation is a second-order differential equation. Generally, wave functions that satisfy this equation for a given potential energy term $V(x)$ can be complex (i.e., have both a real and an imaginary part). A complete solution of the Schrödinger equation requires that we either know the value of the wave function at two different values of x, or we know the wave function and its derivative at a given x.

Let us consider solutions of two second-order differential equations. These solutions will help us understand how the Schrödinger equation is solved for simple cases.
Case 1:
Consider a second-order differential equation of the form

$$\frac{d^2 f(x)}{dx^2} + a^2 f(x) = 0 \qquad (31\text{-}34)$$

where a is a real number. The two conditions to be satisfied by the function are $f(0) = 0$ and $f(L) = 0$. These conditions require the solution of Equation (31-34) to vanish at $x = 0$ and $x = L$.

A solution of Equation (31-34) can be written in terms of either

(a) sine and cosine functions:

$$f(x) = A_1 \sin(ax) + B_1 \cos(ax)$$

where A_1 and B_1 are arbitrary constants that can be determined by applying the given conditions on $f(x)$

or

(b) exponential functions e^{iax} and e^{-iax}:

$$f(x) = A_2 e^{iax} + B_2 e^{-iax}$$

Using the Euler identity, $e^{i\theta} = \cos\theta + i\sin\theta$, we can see that these forms are equivalent. Let us choose a solution of Equation (31-34) in terms of sine and cosine functions:

$$f(x) = A_1 \sin(ax) + B_1 \cos(ax) \qquad (31\text{-}35)$$

and impose the two conditions on $f(x)$:

$$f(0) = 0 \text{ implies}$$

$$0 = A_1 \sin(0) + B_1 \cos(0) = B_1 \qquad (31\text{-}36)$$

$f(L) = 0$ implies

$$0 = A_1 \sin(aL) + B_1 \cos(aL) \qquad (31\text{-}37)$$

From Equation (31-36), $B_1 = 0$; and inserting this relation into Equation (31-37) we get

$$0 = A_1 \sin(aL)$$

To satisfy Equation (31-37), the sine function must vanish. Therefore,

$$\sin(aL) = 0$$

Therefore,

$$aL = n\pi \quad \text{or} \quad a = n\frac{\pi}{L} \quad \text{where } n = 0, 1, 2, 3, \ldots \quad (31\text{-}38)$$

We see that the conditions have constrained the constant a to only a certain set of values. Thus, a has become "quantized," and the allowed values of the function $f(x)$ are

$$f(x) = A_1 \sin\left(n\frac{\pi x}{L}\right) \quad n = 0, 1, 2, 3, \ldots \quad (31\text{-}39)$$

Case 2:

Consider a second-order differential equation of the form

$$\frac{d^2 f(x)}{dx^2} - a^2 f(x) = 0 \qquad (31\text{-}40)$$

where a is a real number. Let us require that the function $f(x)$ satisfies the following conditions:

$$f(0) = 0 \quad \text{and} \quad \left.\frac{df(x)}{dx}\right|_{x=0} = 1$$

Because of the minus sign in front of the second term in Equation (31-40), its solution can be written in terms of exponential functions e^{ax} and e^{-ax}:

$$f(x) = A_1 e^{-ax} + B_1 e^{ax} \qquad (31\text{-}41)$$

The first condition, $f(0) = 0$, requires $A_1 + B_1 = 0$.
The second condition requires $-aA_1 + aB_1 = 1$.
Solving for A_1 and B_1 we get

$$A_1 = -\frac{1}{2a}, \quad B_1 = \frac{1}{2a}$$

Therefore, the solution is

$$f(x) = \frac{1}{2a}(-e^{-ax} + e^{ax})$$

Initial Conditions and Boundary Values

An **initial value problem** for second-order differential equations consists of finding a solution for $f(x)$ that satisfies conditions of the form

$$f(0) = a \quad \text{and} \quad \left.\frac{df(x)}{dx}\right|_{x=0} = b$$

for given constants a and b.

A **boundary value problem** for a second-order differential equation consists of finding a solution for $f(x)$ that satisfies conditions of the form

$$f(x_1) = a \quad \text{and} \quad f(x_2) = b$$

Unlike an initial value problem, a boundary value problem does not always have a solution.

In the following section, we solve the time-independent Schrödinger equation for two simple cases. In general, solutions of the Schrödinger equation can be quite complicated and we use computers to solve the equation. Examples 31-6 and 31-7 below, although simple, clearly explain the peculiar nature of quantum mechanics.

31-5 A Particle in a One-Dimensional Box

Consider a particle of mass m that is confined within a box of width L. The particle moves freely within the box and experiences elastic collisions with the walls of the box. According to classical physics, the energy and momentum of the particle can vary continuously.

Let us see how the Schrödinger equation describes the motion of a particle confined within a box. In quantum mechanics, an impenetrable wall is replaced by an infinite potential. The probability of finding a particle on the other side of an infinite potential is zero. Thus, an infinite potential acts like a rigid classical wall. We will consider a particle constrained to move along a single axis.

For the particle to be confined within the box, the potential energy outside the region $0 \leq x \leq L$ must be infinite. The potential $V(x)$ is therefore given by

$$V(x) = 0 \quad 0 \leq x \leq L$$

$$V(x) = \infty \quad x < 0 \text{ and } x > L$$

The x-axis is then divided into three regions, as shown in Figure 31-10, and we need to solve the Schrödinger equation in each region. The Schrödinger equations for each region are given in Table 31-1.

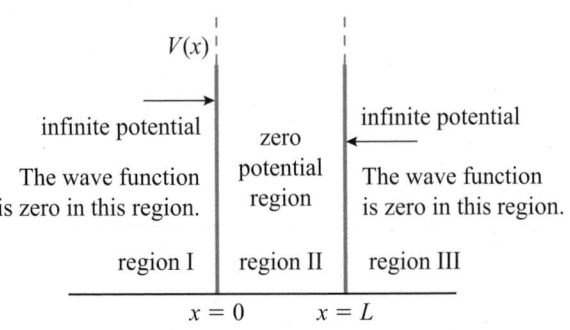

Figure 31-10 An infinite square well potential in one dimension

Table 31-1 The Schrödinger Equation for Regions I, II, and III

Regions I and III ($x < 0$ and $x > L$)	Region II ($0 \leq x \leq L$)
$-\dfrac{\hbar^2}{2m}\dfrac{d^2\psi(x)}{dx^2} + V(x)\,\psi(x)$	$-\dfrac{\hbar^2}{2m}\dfrac{d^2\psi(x)}{dx^2} + V(x)\,\psi(x)$
$= E\,\psi(x)$	$= E\,\psi(x)$
$V(x) = \infty$	Since $V(x) = 0$,
	$-\dfrac{\hbar^2}{2m}\dfrac{d^2\psi(x)}{dx^2} = E\,\psi(x)$
	$\dfrac{d^2\psi(x)}{dx^2} + \dfrac{2mE}{\hbar^2}\,\psi(x) = 0$

In regions I and III, the only possible solution is

$$\psi(x) = 0 \quad \text{for} \quad x < 0 \quad \text{and} \quad x > L$$

For region II, we define $k^2 = \dfrac{2mE}{\hbar^2}$ and write the Schrödinger equation as

$$\frac{d^2\psi(x)}{dx^2} + k^2\psi(x) = 0 \qquad (31\text{-}42)$$

Notice that k is the wave number of the particle and is related to its de Broglie wavelength by $k = 2\pi/\lambda$. As discussed earlier, the solution for Equation (31-42) can be written as

$$\psi(x) = A\sin(kx) + B\cos(kx) \qquad (31\text{-}43)$$

The wave function must vanish in regions I and III, so constants A and B are determined by imposing the following boundary conditions on $\psi(x)$:

(a) $\psi(x) = 0$ at $x = 0$,

which requires $0 = A \times 0 + B$; hence $B = 0$;

(b) $\psi(x) = 0$ at $x = L$,

which requires $0 = A\sin(kL)$.

The constant A cannot be zero (otherwise, the wave function will be zero everywhere), so the second boundary condition can only be satisfied if $\sin(kL) = 0$. Therefore,

$$kL = n\pi \quad n = 1, 2, 3, \ldots \qquad (31\text{-}44)$$

Thus, k is restricted to a discrete set of values, k_n, given by

$$k_n = \frac{n\pi}{L}, \quad n = 1, 2, 3, \ldots \qquad (31\text{-}45)$$

The integer n is called a **quantum number**. Using $k^2 = \dfrac{2mE}{\hbar^2}$, we can see that a particle confined between $x = 0$ and $x = L$ can have only certain discrete values of energy:

$$E_n = \frac{\hbar^2 k_n^2}{2m} = \frac{\hbar^2(n\pi/L)^2}{2m} = n^2\frac{\hbar^2\pi^2}{2mL^2} \qquad (31\text{-}46)$$

The lowest value of energy corresponds to $n = 1$ and is called the **ground-state energy**. For the particle confined within a one-dimensional box, the ground-state energy is

$$E_1 = \frac{\hbar^2\pi^2}{2mL^2} = \frac{h^2}{8mL^2} \qquad (31\text{-}47)$$

States with higher energy are called **excited states**. For an infinite square well potential, the energies of excited states are related to the ground-state energy by

$$E_n = n^2 E_1 \qquad (31\text{-}48)$$

For example, the energy of the state with $n = 2$ is $E_2 = 4E_1$.

The energy of the particle is thus quantized. Figure 31-11 shows allowed energy levels for an infinite square well potential. Such a diagram is called an **energy-level diagram**. Notice that quantum mechanics predicts that the lowest possible energy of a particle confined within a one-dimensional region is nonzero. In fact, quantum mechanics requires that *any* particle confined within some region of space have nonzero kinetic energy. This minimum energy is the ground-state energy (sometimes called the **zero-point energy**). *The smaller the confining space, the greater the zero point energy*, as we can see from Equation (31-47).

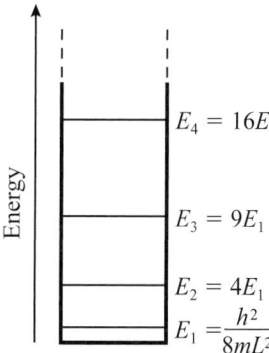

Figure 31-11 Energy levels of the infinite square well potential. The ground state energy is E_1. The energies of the higher levels are given by $E_n = n^2 E_1$.

Wave Functions for an Infinite Square Well Potential

From Equations (31-43) and (31-45) we see that the wave function for quantum state n of a particle confined within an infinite square well potential is

$$\psi_n(x) = A\sin(k_n x) = A\sin\left(\frac{n\pi}{L}x\right) \qquad (31\text{-}49)$$

The constant A can be determined from the normalization condition that the probability of finding the particle somewhere along the x-axis must be one. Since the wave function is zero outside region II, the normalization condition can be obtained by integrating the probability density over the region $0 \leq x \leq L$ for a given quantum state n:

$$\int_{-\infty}^{+\infty} \psi_n^*(x)\,\psi_n(x)\,dx = \int_0^L \psi_n^*(x)\,\psi_n(x)\,dx = 1 \ (31\text{-}50)$$

where $\psi^*(x)$ is the complex conjugate of the wave function $\psi(x)$.

Equation (31-49) shows that $\psi_n(x)$ is real and therefore $\psi_n^*(x) = \psi_n(x)$. Inserting Equation (31-43) into the above integral, we get

$$\int_0^L A^2 \sin^2\left(\frac{n\pi}{L}x\right)dx = 1$$

Since $\int_0^L \sin^2\left(\frac{n\pi}{L}x\right)dx = \frac{L}{2}$ for all n, the normalization condition gives

$$A^2\frac{L}{2} = 1 \implies A = \sqrt{\frac{2}{L}}$$

A is called the **normalization constant**. Notice that the normalization constant does not depend on the quantum number n. Thus, the normalized wave functions for the infinite square well potential are

$$\psi_n(x) = \sqrt{\frac{2}{L}} \sin\left(\frac{n\pi}{L}x\right) \quad n = 1, 2, 3, \ldots \quad (31\text{-}51)$$

The corresponding probability density, $\psi_n^*(x)\,\psi_n(x)$, is

$$\psi_n^*(x)\,\psi_n(x) = \frac{2}{L} \sin^2\left(\frac{n\pi}{L}x\right) \quad (31\text{-}52)$$

Figure 31-12 shows plots of $\psi_n^*(x)\psi_n(x)$ for $n = 1, 2, 3$, and 100. We see that in the ground state, the particle is most likely to be found at the middle of the $0 \le x \le L$ region, that is, at $x = L/2$. For the first excited state ($n = 2$), the particle is most likely to be found at $x = L/4$ and $x = 3L/4$, and the probability of finding the particle at $x = L/2$ is zero. Both of these probability distributions differ radically from classical mechanics, which predicts that the probability of finding the particle is the same everywhere within the region $0 \le x \le L$ and is therefore equal to $1/L$. As the quantum number n increases, the predictions of quantum mechanics approach the classical prediction. As shown by the plot of the probability density for $n = 100$, for large values of n, the maxima and minima of the probability density are very closely spaced, and the probability density is indistinguishable from its average value of $1/L$, which is the classical value. Observation that for large quantum numbers the predictions of quantum mechanics agree with those of classical mechanics is a general property of quantum mechanics. This general property was first stated by Niels Bohr in the form of the **correspondence principle**: "*the results of a quantum theory analysis of a system that involves the use of very large quantum numbers must agree with the results of a classical physics analysis.*"

(a)

(b)

(c)

(d)

Figure 31-12 The probability density $\psi_n^*(x)\,\psi_n(x)$ versus x for a particle in a one-dimensional infinite square well potential of length L, for (a) the ground state, $n = 1$; (b) the first excited state, $n = 2$; (c) the second excited state, $n = 3$; and (d) the state $n = 100$.

EXAMPLE 31-6

An Electron in an Infinite Square Well Potential

Consider an electron confined within a one-dimensional infinite square potential of length 0.50×10^{-10} m, the approximate radius of a hydrogen atom.

(a) Find the ground-state energy of the electron.

(b) Calculate the energies corresponding to the $n = 2$, $n = 3$, and $n = 4$ states.

MODERN PHYSICS

(c) An electron in an excited state can move to a state of lower energy by emitting a photon. Calculate the energies and wavelengths of emitted photons for all transitions from the $n = 4$ state to lower states.

SOLUTION

(a) The ground-state energy for an infinite square well potential is given by Equation (31-47). Using the electron mass for m, 9.11×10^{-31} kg, we get

$$E_1 = \frac{h^2}{8mL^2} = \frac{(6.63 \times 10^{-34} \text{ J} \cdot \text{s})^2}{(8)(9.11 \times 10^{-31} \text{ kg})(0.50 \times 10^{-10} \text{ m})^2}$$

$$= 2.41 \times 10^{-17} \text{ J} = 150.6 \text{ eV}$$

Since atomic energies are commonly given in units of eV, we converted from J using $1 \text{ eV} = 1.60 \times 10^{-19}$ J.

(b) Energies of the excited states corresponding to $n = 2$, $n = 3$ and $n = 4$ are related to the ground-state energy by Equation (31-48). Therefore,

$$E_2 = 4E_1 = 602.4 \text{ eV}$$

$$E_3 = 9E_1 = 1355.4 \text{ eV}$$

$$E_4 = 16E_1 = 2409.7 \text{ eV}$$

(c) An electron in the $n = 4$ state can jump to a state of lower energy by emitting a photon (Figure 31-13). By energy conservation,

$$E_4 = E_m + E_{\text{photon}}$$

where m denotes a state of lower energy ($m < 4$), and E_{photon} is the energy of the photon. The wavelength of a photon is related to its energy E by

$$\lambda = \frac{c}{f} = \frac{hc}{E_{\text{photon}}}$$

where

$$hc = (6.63 \times 10^{-34} \text{ J} \cdot \text{s})(3.0 \times 10^8 \text{ m/s})$$

$$= \frac{1.99 \times 10^{-25} \text{ J} \cdot \text{m}}{1.60 \times 10^{-19} \text{ J/eV}} = 1.24 \times 10^{-6} \text{ eV} \cdot \text{m}$$

Photon energies and wavelengths for all transitions from the $n = 4$ state to lower states are as follows:

(i) $4 \rightarrow 3$ transition

$$E_{\text{photon}} = E_4 - E_3 = 1054 \text{ eV}$$

$$\lambda = \frac{1.24 \times 10^{-6} \text{ eV} \cdot \text{m}}{1054 \text{ eV}} = 1.2 \times 10^{-9} \text{ m}$$

(ii) $4 \rightarrow 2$ transition

$$E_{\text{photon}} = E_4 - E_2 = 1807 \text{ eV}$$

$$\lambda = \frac{1.24 \times 10^{-6} \text{ eV} \cdot \text{m}}{1807 \text{ eV}} = 6.9 \times 10^{-10} \text{ m}$$

(iii) $4 \rightarrow 1$ transition

$$E_{\text{photon}} = E_4 - E_1 = 2259 \text{ eV}$$

$$\lambda = \frac{1.24 \times 10^{-6} \text{ eV} \cdot \text{m}}{2259 \text{ eV}} = 5.5 \times 10^{-10} \text{ m}$$

Figure 31-13 When an electron jumps from a state of higher energy to that of a lower energy it emits a photon. The energy of the emitted photon is equal to the difference of the energies of the two levels.

Example 31-6 demonstrates some of the fundamental features of quantum mechanics:

1. The energy of a particle confined within a finite region of space is quantized. The energy and momentum of the particle can only take on a finite set of values, which are determined by the shape of the potential and the boundary conditions. This quantization is a radical departure from classical mechanics, where energy and momentum have continuous values.

2. A charged, confined particle can only emit a photon when it jumps from a higher energy level to a lower energy level. A particle that is in the ground state of a confining potential cannot emit a photon. Thus, quantum mechanics explains why atoms are stable.

3. The energy of a confined particle cannot be zero. The minimum energy that a particle must have is the ground-state energy (also called the zero-point energy).

4. The smaller the confining space, the greater the ground-state energy. This prediction is experimentally verified by the fact that the energies of protons and neutrons confined within a nucleus are much larger than the energies of the electrons that are bound to atoms The volume of a nucleus is approximately a billion times smaller than the volume of an atom.

C-31-4 A Proton in an Infinite Square Well Potential

Replace the electron in Example 31-6 with a proton. (The mass of a proton is approximately 1836 times the mass of an electron.) The ground-state energy of the confined proton is

(a) less than the ground-state energy of the electron;
(b) the same as the ground-state energy of the electron;
(c) greater than the ground-state energy of the electron;
(d) zero, because protons remain stationary when confined to an infinite square well potential.

C-31-4 (a) Energy levels of a particle confined in an infinite square well potential are inversely proportional to the mass of the particle.

 ## ONLINE ACTIVITY

Quantum Bound States

The e-resource that accompanies every new copy of this textbook contains an Online Activity using the PhET simulation "Quantum Bound States." Work through the simulation and accompanying questions to learn more about the bound states of a finite square well potential.

 ## EXAMPLE 31-7

The Probability of Finding an Electron

A particle is in the ground state of an infinite square well potential. Calculate the probability of finding the particle in the region $0 \le x \le \dfrac{L}{3}$ of the well.

SOLUTION

The probability, P, of finding the particle in the region $0 \le x \le \dfrac{L}{3}$ is obtained by integrating the square of the wave function over the region:

$$P = \int_0^{L/3} \psi^2(x)\,dx$$

For the ground state of an infinite square well potential,

$\psi(x) = \sqrt{\dfrac{2}{L}} \sin\left(\dfrac{\pi x}{L}\right)$. Substituting gives

$$P = \frac{2}{L} \int_0^{L/3} \sin^2\left(\frac{\pi x}{L}\right)^2 dx$$

Let $z = \dfrac{\pi x}{L}$. Then $dx = \dfrac{L}{\pi} dz$ and

$$P = \frac{2}{\pi} \int_0^{\pi/3} \sin^2 z\, dz = \frac{2}{\pi}\left(-\frac{\sqrt{3}}{8} + \frac{\pi}{6}\right) = 0.195$$

The probability of finding the particle in the region $0 \le x \le \dfrac{L}{3}$ is therefore 0.195, or 19.5%.

LO 6

31-6 Quantum Tunnelling

Consider a ball of mass m and kinetic energy K, moving on a frictionless surface toward a hill of height h, as shown in Figure 31-14. The minimum energy needed for the ball to reach the top of the hill is mgh. If $K < mgh$, the ball will not reach the top, and the probability of finding the ball at the right side of the hill is zero. If $K > mgh$, the ball will reach the top and then roll down the right side of the hill. After the ball has crossed the top of the hill, the probability of finding the ball anywhere to the left of the hill is zero.

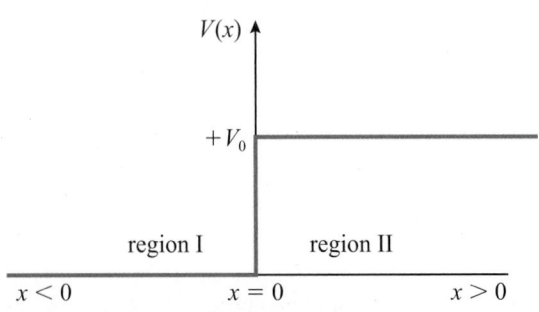

Figure 31-14 A ball moving toward a barrier (hill)

However, quantum mechanics makes a startling prediction: *Even if the kinetic energy of the ball is less than its potential energy required to reach the top of the barrier, the probability of the ball penetrating the barrier is not zero.*

To study the quantum-mechanical analog of the situation in Figure 31-14, we replace the physical barrier by a potential well. We start with a simple **step-function potential**, where the potential energy is zero in one region and has a positive and constant value in the other region. Such potential wells provide a simple model of the potential energy encountered by an electron at the interface of two homogeneous media. We take the potential energy to be zero for $x < 0$ and to have a positive value V_0 for $x \ge 0$, thus dividing the x-axis into two regions, as shown in Figure 31-15.

Figure 31-15 A repulsive finite step potential

MODERN PHYSICS

Now consider a steady beam of particles, each with a mass m and energy $E < V_0$, travelling from left to right toward $x = 0$. Let $\psi_1(x)$ and $\psi_2(x)$ denote a particle's wave function in regions 1 and 2, respectively. For $x < 0$, the Schrödinger equation has the form

$$-\frac{\hbar^2}{2m}\frac{d^2\psi_1(x)}{dx^2} + 0 \times \psi_1(x) = E\,\psi_1(x)$$

$$\frac{d^2\psi_1(x)}{dx^2} + \frac{2mE}{\hbar^2}\psi_1(x) = 0 \qquad (31\text{-}53)$$

In region 2, $V(x) = V_0$; therefore,

$$-\frac{\hbar^2}{2m}\frac{d^2\psi_2(x)}{dx^2} + V_0\psi_2(x) = E\,\psi_2(x)$$

$$\frac{d^2\psi_2(x)}{dx^2} + \frac{2m(E - V_0)}{\hbar^2}\psi_2(x) = 0 \quad (31\text{-}54)$$

To simplify the calculation, we make the following substitutions:

$$k_1^2 = \frac{2mE}{\hbar^2}$$

$$k_2^2 = \frac{2m(V_0 - E)}{\hbar^2} \qquad (31\text{-}55)$$

where k_1 and k_2 are positive, and $\hbar k_1$ and $\hbar k_2$ are the magnitudes of the particle's momenta in regions 1 and 2, respectively. The Schrödinger equations in the two regions can be written as

$$\frac{d^2\psi_1(x)}{dx^2} + k_1^2\psi_1(x) = 0 \quad x < 0$$

$$\frac{d^2\psi_2(x)}{dx^2} - k_2^2\psi_2(x) = 0 \quad x \geq 0 \qquad (31\text{-}56)$$

The solutions of the above two equations are

$$\psi_1(x) = A_1 e^{ik_1 x} + B_1 e^{-ik_1 x} \quad x < 0 \qquad (31\text{-}57)$$

$$\psi_2(x) = A_2 e^{-k_2 x} + B_2 e^{k_2 x} \quad x \geq 0 \qquad (31\text{-}58)$$

The exponential functions e^{ikx} and e^{-ikx} describe travelling waves. To determine the direction of motion of the wave described by each function, we need to examine the time dependence of the wave functions. For a state of energy E, the time dependence of the wave function is of the form $e^{-iEt/\hbar}$. Therefore, the time-dependent wave function in region 1 is

$$\psi_1(x, t) = A_1 e^{ik_1 x} e^{-iEt/\hbar} + B_1 e^{-ik_1 x} e^{-iEt/\hbar} \quad x \leq 0$$

$$= A_1 e^{i(k_1 x - Et/\hbar)} + B_1 e^{-i(k_1 x + Et/\hbar)} \quad x \leq 0 \;(31\text{-}59)$$

The A_1 term represents a particle moving toward the right (toward the origin), and the B_1 term represents a particle moving toward the left (away from the origin). Therefore, the e^{ikx} term represents the incident particle, and the e^{-ikx} term represents the reflected particle.

The wave function and its derivative are continuous at the boundary between the two regions. We therefore require that

$$\psi_1(0) = \psi_2(0)$$

$$\left.\frac{d\psi_1(x)}{dx}\right|_{x=0} = \left.\frac{d\psi_2(x)}{dx}\right|_{x=0} \qquad (31\text{-}60)$$

Since x is positive in region 2, the $e^{k_2 x}$ term becomes infinite as $x \to \infty$. The wave function must remain finite for *all* x, so B_2 must be zero. Therefore, the wave function in region 2 is

$$\psi_2(x) = A_2 e^{-k_2 x} \quad x \geq 0 \qquad (31\text{-}61)$$

Let us now apply the boundary conditions at $x = 0$. The requirement that the wave functions must have the same value at $x = 0$ gives

$$A_1 + B_1 = A_2$$

The requirement that the derivatives of the wave functions must be equal at $x = 0$ gives

$$ik_1 A_1 - ik_1 B_1 = -k_2 A_2$$

Using the above two equations, we can write B_1 and A_2 in terms of A_1:

$$B_1 = \left(\frac{k_1 - ik_2}{k_1 + ik_2}\right)A_1 \quad \text{and} \quad A_2 = \left(\frac{2k_1}{k_1 + ik_2}\right)A_1 \quad (31\text{-}62)$$

Therefore, the wave functions that satisfy the boundary conditions at the interface of the two regions are

$$\psi_1(x) = A_1 e^{-ik_1 x} + \left(\frac{k_1 - ik_2}{k_1 + ik_2}\right)A_1 e^{ik_1 x} \quad x < 0$$

$$\psi_2(x) = \left(\frac{2k_1}{k_1 + ik_2}\right)A_1 e^{-k_2 x} \quad x \geq 0 \qquad (31\text{-}63)$$

Without any loss of generality, we can take $A_1 = 1$; this value corresponds to a probability density where the incident beam contains one particle per unit length. Using the expressions for k_1 and k_2 from Equation (31-55), the wave functions in the two regions are

$$\psi_1(x) = e^{-i\frac{\sqrt{2mE}}{\hbar}x} + \left(\frac{\sqrt{E} - i\sqrt{V_0 - E}}{\sqrt{E} + i\sqrt{V_0 - E}}\right)e^{i\frac{\sqrt{2mE}}{\hbar}x} \quad x < 0$$

$$\psi_2(x) = \left(\frac{2\sqrt{E}}{\sqrt{E} + i\sqrt{V_0 - E}}\right)e^{-\frac{\sqrt{2m(V_0 - E)}}{\hbar}x} \quad x \geq 0 \;(31\text{-}64)$$

Notice the following features of these wave functions:

1. In region 1, the first term represents the wave function of the incident beam, and the second term represents

the wave function of the reflected beam. The term $\left(\dfrac{\sqrt{E} - i\sqrt{V_0 - E}}{\sqrt{E} + i\sqrt{V_0 - E}}\right)$ represents the amplitude of the reflected beam. This is similar to the situation in classical physics where a beam of particles with energy less than the height of the barrier reflects from the barrier.

2. In region 2 the wave function decays exponentially and is not zero. Therefore, a particle can penetrate into a classically forbidden region. Such penetration of a potential barrier is called **tunnelling**. It is a quantum effect and has no analog in classical physics. In classical physics, for $E < V_0$, the entire incident beam must reflect from the barrier, and the probability of finding a particle in region 2 is zero.

A simple but slightly tedious calculation (problem 48) lets us compare the probability densities in the two regions (note that probability densities here are not normalized):

$$\psi_1(x)\,\psi_1^*(x) = 2 + \left(\frac{4E - 2V_0}{V_0}\right)\cos\left(2\,\frac{\sqrt{2mE}}{\hbar}x\right)$$

$$- \left(\frac{4\sqrt{E(E - V_0)}}{V_0}\right)\sin\left(2\,\frac{\sqrt{2mE}}{\hbar}x\right) \quad x < 0$$

$$\psi_2(x)\,\psi_2^*(x) = \left(\frac{4E}{V_0}\right)e^{-\frac{2\sqrt{2m(V_0 - E)}}{\hbar}x} \quad x \geq 0 \quad (31\text{-}65)$$

Figure 31-16 shows plots of these probability densities for a barrier of $V_0 = 5.0$ eV and electrons with $E = 1.0$ eV and $E = 4.9$ eV. Notice that as E approaches V_0,

the probability that a particle can penetrate into the classically forbidden region increases. In our example, the potential extends all the way to infinite x. With real-life potentials of finite width, a detector on the right side of the potential can detect particles that pass through the potential. Such experiments provided a clear verification of quantum mechanics.

If the potential well has a finite width, then the probability of finding the particle at the other side of the potential is nonzero. Consider a potential of width a as shown in Figure 31-17. A steady beam of particles of energy E and mass m is incident on the left side of the potential. The wave function at $x = a$ is given by evaluating $\psi_2(a)$ in Equation (31-64). The probability, P, that the particle reaches the other side of the potential is proportional to the square of $\psi_2(a)$. This probability depends on the incident energy of the particle and on the height and width of the potential barrier:

$$P(E, V_0, a) \propto e^{\frac{-2a}{\hbar}\sqrt{2m(V_0 - E)}} \quad (31\text{-}66)$$

The proportionality sign is necessary because we have not normalized the wave function. We can calculate the normalization factor by solving the Schrödinger equation, with the boundary conditions, for all three regions of the potential in Figure 31-17, and then integrating the square of the wave function. However, the shape of the probability distribution for the tunnelling process is determined by the exponential term. Since an exponential function varies rapidly with its argument, the tunnelling probability is a sensitive function of mass and energy of the particle, and of the height and the width of the potential, as shown in the following example.

(a)

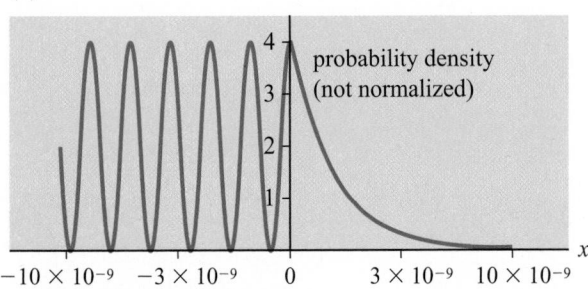

(b)

Figure 31-16 The probability density for scattering from a step potential (Figure 31-15) of $V_0 = 5.0$ eV for electrons incident from left with kinetic energy of (a) 1.0 eV and (b) 4.9 eV.

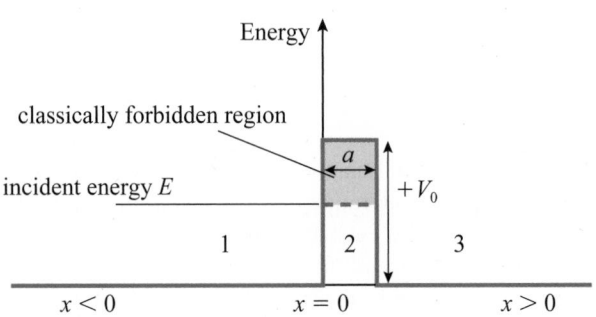

Figure 31-17 A potential barrier of height V_0 and width a

From Equation (31-66) we see that the tunnelling probability decreases exponentially as the width of the barrier increases. Figure 31-18 is a graph of the tunnelling probability of the electrons as a function of potential width.

MODERN PHYSICS

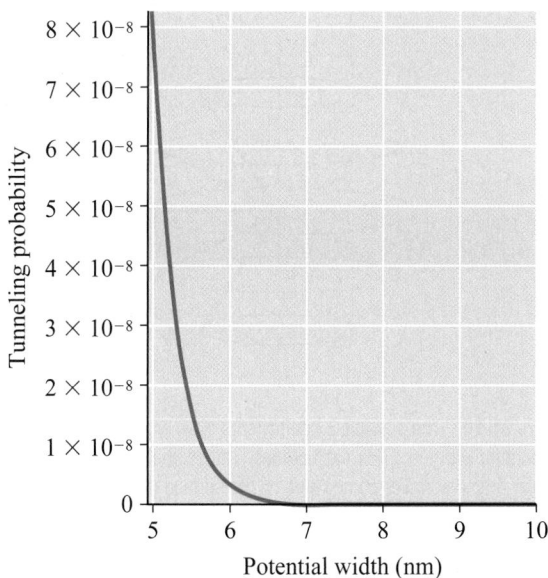

Figure 31-18 The relative tunnelling probability for 1.0 eV electrons as a function of the width of a potential with $V_0 = 5.0$ eV.

Therefore, the probability, P, of the electron tunnelling through the barrier is

$$P \propto e^{-16.3} = 8 \times 10^{-8}$$

The constant of proportionality is of the order of unity. Therefore, P is of the order of 10^{-7}.

(b) The mass of a proton is approximately 1836 times the mass of an electron. Keeping other quantities the same, the tunnelling probability for a proton is

$$P \propto e^{-700} \sim 10^{-300}$$

Therefore, P is of the order of 10^{-300}.

Making sense of the result:

If 10^{10} electrons hit the barrier per second, approximately 1000 will pass through. Such a current can be easily detected with a sensitive instrument. However, if 10^{10} protons per second are incident on the barrier, on average, one proton would tunnel through the barrier every 10^{290} s. No wonder we do not observe objects like baseballs tunnelling through baseball bats or gloves.

 EXAMPLE 31-8

Tunnelling Through a Barrier

Consider an electron of 1.0 eV kinetic energy incident on a potential barrier of width 5.0 nm with $V_0 = 5.0$ eV.

(a) Estimate the probability that the electron will tunnel through the potential.
(b) Estimate the tunnelling probability for a proton with the same initial kinetic energy.

SOLUTION

(a) The tunnelling probability can be determined using Equation (31-66). The values for the variables in the exponent are

$$\hbar = 6.63 \times 10^{-34} \text{ J} \cdot \text{s}$$

$$m = \text{electron mass} = 9.11 \times 10^{-31} \text{ kg}$$

$$E = 1.0 \text{ eV} = 1.0 \times 1.60 \times 10^{-19} \text{ J}$$

$$V_0 = 5.0 \text{ eV} = 5.0 \times 1.60 \times 10^{-19} \text{ J}$$

$$a = 5.0 \text{ nm} = 5.0 \times 10^{-9} \text{ m}$$

Let us first confirm that the exponent is dimensionless:

$$\frac{1}{\text{J} \cdot \text{s}} [\text{kg} \cdot \text{J}]^{1/2} \text{ m} = \frac{[\text{kg}]^{1/2} \cdot \text{m}}{[\text{J}]^{1/2} \cdot \text{s}} = \frac{[\text{kg}]^{1/2} \cdot \text{m}}{[\text{kg} \cdot \text{m}^2 \cdot \text{s}^{-2}]^{1/2} \cdot \text{s}} = 1$$

Evaluating the exponent,

$$\frac{2}{\hbar} \sqrt{2m(V_0 - E)} \, a = \frac{2(5.0 \times 10^{-9} \text{ m})}{6.63 \times 10^{-34} \text{ J} \cdot \text{s}}$$

$$\times \sqrt{2(9.11 \times 10^{-31} \text{ kg})(5.0 - 1.0)(1.60 \times 10^{-19} \text{ J})}$$

$$= 16.3$$

 ONLINE ACTIVITY

Quantum Tunnelling

The e-resource that accompanies every new copy of this textbook contains an Online Activity using the PhET simulation "Quantum Tunnelling." Work through the simulation and accompanying questions to learn more about the tunnelling of electrons through a barrier.

 CHECKPOINT

C-31-5 Tunnelling Through a Barrier

A beam of 2 eV electrons is incident on a square well potential of width 1 nm and height 4 eV. Which of the following statements is correct when the electron beam scatters from the potential barrier?

(a) All electrons transmit through the potential barrier, and there is no reflected beam.
(b) All electrons reflect from the potential barrier, and there is no transmitted beam.
(c) Exactly half the electrons transmit through and the other half reflect.
(d) Some electrons transmit through and the others reflect.

C-31-5 (d)

CHAPTER 31 | INTRODUCTION TO QUANTUM MECHANICS

MODERN PHYSICS

31-7 The Quantization of Angular Momentum

In Section 31-5, you saw that the energy and the momentum of a confined particle are quantized. Is the angular momentum of a confined particle also quantized, as Bohr assumed for his model of the atom? In fact, the Schrödinger wave equation does predict such quantization.

Consider a particle with angular momentum $\overrightarrow{J}$ and magnitude $J = \sqrt{J_x^2 + J_y^2 + J_z^2}$. A given component of $\overrightarrow{J}$ could have values in the range from $-J$ to $+J$. For example, the z-component of angular momentum is given by $J_z = \overrightarrow{J} \cdot \hat{z} = J\cos\theta$, which varies between $-J$ and $+J$ as θ, the angle between the momentum vector and the z-axis, varies between 0 and π rad.

In quantum mechanics, the z-component of angular momentum can have only *certain discrete values*, which differ by multiples of $\hbar$. Note that the units of $\hbar$, $m^2 \cdot kg/s$, have dimensions of angular momentum.

Any quantum-mechanical system, (a free particle, a bound particle or a collection of particles), is assigned a characteristic **total angular momentum quantum number**, j. The z-component of the total angular momentum of the system must be one of the following values in the range of $-j\hbar$ to $+j\hbar$:

$$-j\hbar, \ (-j + 1)\hbar, \ (-j + 2)\hbar, \ \ldots.$$

$$(+j - 2)\hbar, (+j - 1)\hbar, + j\hbar \qquad (31\text{-}67)$$

The quantum number j can only have integer or half-integer values. For a given value of j, there are $2j + 1$ possible values for the z-component of angular momentum. For example, a free electron has $j = \frac{1}{2}$, so there are only two possible values for the z-component of the angular momentum of a free electron: $+\frac{1}{2}\hbar$ and $-\frac{1}{2}\hbar$. Similarly, if $j = 2$ for an atom, the z-component of the atom's angular momentum must be $-2\hbar, -\hbar, 0\hbar, \hbar$, or $2\hbar$ (a total of five possible values, Figure 31-19).

Since the z-axis can be chosen along any direction, the component of angular momentum along *any* axis is quantized in multiples of $\hbar$ in a range determined by the total angular momentum quantum number, j.

The magnitude J of the angular momentum is related to the angular momentum quantum number j as follows:

$$J = \sqrt{J_x^2 + J_y^2 + J_z^2} = \sqrt{j(j + 1)}\,\hbar \qquad (31\text{-}68)$$

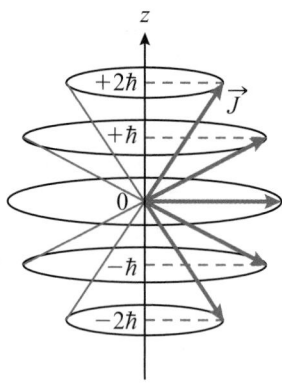

Figure 31-19 The possible z-components of an atom's angular momentum with $j = 2$. For each of these values, the magnitude of the angular momentum lies somewhere on the surface of a cone.

The above relationship can be proven rigorously by treating angular momentum as an operator. Here, we will use a less rigorous derivation developed by the American physicist Richard Feynman (1918–88). Since $J^2 = J_x^2 + J_y^2 + J_z^2$, the average value of J^2 is equal to the sum of the average values of J_x^2, J_y^2, and J_z^2. Since J_x^2, J_y^2, and J_z^2 are scalars, they have the same value for any orientation of the coordinate system. Therefore, the average values of J_x^2, J_y^2, and J_z^2 are all equal, and

$$\bar{J}^2 = 3(\bar{J_z^2}) \qquad (31\text{-}69)$$

where $\bar{J}$ denotes the mean value of J.

Since $J^2 = \overrightarrow{J} \cdot \overrightarrow{J}$ is a scalar, it is independent of the orientation of $\overrightarrow{J}$. Therefore, its average is equal to its value, that is,

$$J^2 = 3(\bar{J_z^2}) \qquad (31\text{-}70)$$

The $2j + 1$ allowed values of J_z are given by Equation (31-67). Therefore,

$$\bar{J_z^2} = \frac{(-j)^2 + (-j+1)^2 + (-j+2)^2 + \cdots + (j-2)^2 + (j-1)^2 + (j)^2}{2j + 1}\hbar^2$$

$$= \frac{(2j^3 + 3j^2 + j)/3}{2j + 1}$$

$$= \frac{j(2j + 1)(j + 1)}{3(2j + 1)}$$

$$= \frac{j(j + 1)}{3}\hbar^2 \qquad (31\text{-}71)$$

Substituting into Equation (31-69) gives

$$J = \sqrt{j(j + 1)}\,\hbar \qquad (31\text{-}72)$$

If the particle is in a state with angular momentum quantum number j, the eigenvalue of the magnitude of the angular momentum operator of the particle is $\sqrt{j(j + 1)}\,\hbar$.

EXAMPLE 31-9

The Angular Momentum of a Bound Electron

The angular momentum quantum number j of an electron in an excited state of a hydrogen atom is 5/2.

(a) What is the magnitude of the electron's angular momentum?

(b) What are the angles between the angular momentum vector and the z-axis?

SOLUTION

(a) The magnitude J of the angular momentum vector for the quantum number $j = 5/2$ is

$$J = \sqrt{j(j+1)}\hbar = \sqrt{\frac{5}{2} \times \frac{7}{2}}\,\hbar = \frac{\sqrt{35}}{2}\,\hbar$$

(b) Since the z-component of the angular momentum is quantized,

$$J_z = m\hbar \quad m = -j, (-j+1), \ldots, (j-1), j$$

Therefore,

$$\cos\theta = \frac{m\hbar}{\sqrt{j(j+1)}\,\hbar} = \frac{m}{\sqrt{j(j+1)}}$$

For $j = \frac{5}{2}$, $m = -\frac{5}{2}, -\frac{3}{2}, -\frac{1}{2}, \frac{1}{2}, \frac{3}{2}, \frac{5}{2}$.

Since, $\cos^{-1}(-x) = \pi - \cos^{-1}(x)$, we only need to evaluate θ for either positive or negative values of m. For $m = \frac{5}{2}$,

$$\cos\theta = \left(\frac{2}{\sqrt{35}}\right)\left(\frac{5}{2}\right) = \frac{5}{\sqrt{35}}$$

$$\theta = \cos^{-1}\left(\frac{5}{\sqrt{35}}\right) = 0.56 \text{ rad} = 32.3°$$

Similarly, we can calculate θ for other values of m:

$m = \frac{5}{2}$ $\theta = 32.3°$; $\qquad m = -\frac{5}{2}$ $\theta = 147.7°$

$m = \frac{3}{2}$ $\theta = 59.5°$; $\qquad m = -\frac{3}{2}$ $\theta = 120.5°$

$m = \frac{1}{2}$ $\theta = 80.3°$; $\qquad m = -\frac{1}{2}$ $\theta = 99.7°$

 CHECKPOINT

C-31-6 Projection of Angular Momentum

For an electron bound in an atom in a state with angular momentum quantum number $j = 3/2$, the possible values of the angle between the angular momentum vector of the electron and the z-axis are

(a) all possible values between 0° and 180°;
(b) 39°, 75°, 141°,105°;
(c) 30°, 60°, 120°, 150°;
(d) 0°, 45°, 90°,135°.

C-31-6 (b)

LO 8

31-8 The Schrödinger Equation for a Hydrogen Atom

The Bohr model of the hydrogen atom (Section 30-9) explains the origin of the hydrogen spectrum but has a number of shortcomings. The only justification that the Bohr model gives for the quantization of the angular momentum of a bound electron is that quantization gives rise to stationary orbits and, hence, to quantized energies of the electron.

The Schrödinger equation provides a better description of the hydrogen atom. You saw in Section 31-4 that the energy of a particle confined within a region of space is quantized. In a hydrogen atom, an electron is bound to a proton by electrostatic attraction. Therefore, we can expect the energy of the electron to be quantized. Now, we will examine the solution for the Schrödinger equation for a hydrogen atom and see how well the resulting energy levels match the hydrogen spectrum.

If the electron in the hydrogen atom is at a distance r from the proton, the electrostatic potential energy between the proton and the electron is

$$V(r) = k\frac{(+e)(-e)}{r} = -k\frac{e^2}{r} \qquad (31\text{-}73)$$

where $k = 1/4\pi\varepsilon_0$ is Coulomb's constant.

The time-independent Schrödinger equation for a particle of mass m, moving in three dimensions in the presence of a potential, is given by Equation (31-32). Since the potential energy $V(r)$ depends only on the radial distance $r = \sqrt{x^2 + y^2 + z^2}$ between the proton and the electron, it is convenient to write the Schrödinger equation in terms of spherical coordinates: r, θ, and ϕ (the radial, polar, and azimuthal coordinates, respectively). As shown in Section 30-6, spherical coordinates are related to the Cartesian coordinates as follows:

$$x = r \sin\theta \cos\phi$$
$$y = r \sin\theta \sin\phi$$
$$z = r \cos\theta \qquad (31\text{-}74)$$

The potential energy does not depend on the θ and ϕ coordinates, so the wave function can be written as a product of three wave functions, each depending on a single spherical coordinate:

$$\psi(r, \theta, \phi) = \psi_1(r)\,\psi_2(\theta)\,\psi_3(\phi) \qquad (31\text{-}75)$$

In Section 31-5, a solution for a one-dimensional Schrödinger equation for a particle bound in an infinite square well potential resulted in a single quantum number, n. The allowed energies and wave functions of the particle are indexed by n. When the Schrödinger equation in three dimensions is solved for the wave functions $\psi_1(r)$, $\psi_2(\theta)$, and $\psi_3(\phi)$, three quantum numbers arise, one for each dimension. These three quantum numbers are not independent. Choosing a specific value for one quantum number restricts the allowed values of the other two. Although solving these differential equations is not very difficult, the procedure is quite lengthy. So we will just summarize the results here.

The Principal Quantum Number, n

The radial wave function $\psi_1(r)$ exists only for positive integer values of a quantum number called the principal quantum number. The **principal quantum number**, n, describes the *size* of the **orbital**, the wave function that indicates the probable location of the bound electron. The energies of the electron are quantized in terms of n:

$$E_n = -\frac{1}{n^2}\left(\frac{m_e k^2 e^4}{2\hbar^2}\right)$$
$$= -\frac{13.6\text{ eV}}{n^2} \quad n = 1, 2, 3, \ldots \quad (31\text{-}76)$$

This expression for the energy levels is exactly the same as the expression for the Bohr model of the hydrogen atom (Equation (30-42)), and the principal quantum number in Equation (31-76) is the same as the number in the Bohr model. Thus, Schrödinger's wave equation accurately predicts the hydrogen energy levels and the resulting spectral lines.

The Orbital Quantum Number, l

The solution of the differential equation for $\psi_2(\theta)$ gives rise to the orbital quantum number. The **orbital quantum number**, l, describes the *shape* of the orbital and is restricted to values from 0 to $n - 1$:

$$l = 0, 1, 2, 3, \ldots, n - 1 \qquad (31\text{-}77)$$

For example, if $n = 3$, the allowed values for l are 0, 1, and 2. The magnitude of the orbital angular momentum, L, of the electron, is related to the orbital quantum number:

$$L = \sqrt{l(l + 1)}\,\hbar \qquad (31\text{-}78)$$

Therefore, the angular momentum of the electron is quantized as multiples of Planck's constant. The quantization of the orbital angular momentum appears naturally when we solve the Schrödinger equation. There is no need to assume, as Bohr did, that the orbital angular momentum is quantized. Note that values of orbital angular momentum given by Equation (31-78) do not agree with the values suggested by Bohr (Equation (30-34)), as can be seen in the examples listed in Table 31-2.

Table 31-2 The Bohr model and the Schrödinger equation predictions of the magnitude of the orbital angular momentum of an electron in a hydrogen atom

	The Bohr model	The Schrödinger equation	
Principal quantum number, n	Orbital angular momentum, L	Orbital quantum number, l	Orbital angular momentum, L
1	$\hbar$	0	0
2	$2\hbar$	0	0
		1	$\sqrt{2}\,\hbar$
3	$3\hbar$	0	0
		1	$\sqrt{2}\,\hbar$
		2	$\sqrt{6}\,\hbar$

For large values of l, $\sqrt{l(l + 1)} \approx l$ and the Bohr model agrees reasonably closely with the orbital angular momentum predicted by the Schrödinger equation. Experiments have confirmed that the values of L predicted by the Schrödinger equation are correct. As shown in Table 31-3, various values of l are commonly designated with lower case letters.

Table 31-3 Letter designations for the orbital quantum number

Orbital quantum number, l	Letter designation
0	s
1	p
2	d
3	f
4	g

The Magnetic Quantum Number, m_l

The solution of the differential equation for the azimuthal angle ϕ gives rise to a quantum number called the magnetic quantum number. The **magnetic quantum number**, m_l, describes the *orientation* of a particular orbital and is restricted to integer values between $-l$ and $+l$:

$$L_z = m_l\hbar \quad m_l = -l, (-l + 1), \ldots, (l - 1), l \quad (31\text{-}79)$$

Shells and Subshells

The orbitals that have the same value of principal quantum number n form a **shell**. Within a shell, orbitals with different orbital quantum numbers form **subshells**. Since for a given n the orbital quantum number l has values from 0 to $n - 1$, there are n subshells within a shell of principal quantum number n. It is customary to denote a subshell by giving the principal quantum number with the letter designation for the orbital quantum number (e.g., $2p$ ($n = 2, l = 1$) and $4d(n = 4, l = 3)$). Table 31-4 shows the possible quantum numbers for the wave function for a hydrogen atom.

Table 31-4 Quantum numbers for a hydrogen atom wave function

Principal, n	Orbital, l	Magnetic, m_l
1	0	0
2	0	0
	1	+1, 0, −1
3	0	0
	1	+1, 0, −1
	2	+2, +1, 0, −1, −2
...	...	...
n	1, 2, 3, 4, ..., $n - 1$	$-l, -l + 1, ... 0, ...$ $l - 1, l$

The Ground State of a Hydrogen Atom

For the ground state, $n = 1$ and the only possible value of l is 0. Therefore, the ground state of a hydrogen atom has zero angular momentum. For $l = 0$, the only possible value of m_l is 0. Thus,

Ground state: $n = 1, l = 0, m_l = 0$, orbital: $1s$

The ground state is the most tightly bound state and has a binding energy of

$$E_1 = -\frac{13.6 \text{ eV}}{1^2} = -13.6 \text{ eV}$$

It takes 13.6 eV of energy to liberate an electron from the ground state of a hydrogen atom, producing a free electron and a positive hydrogen ion (i.e., a proton). The shape of the ground-state orbital is spherical, as shown in Figure 31-20.

Figure 31-20 A plot of the probability density for an electron in an orbital with $l = 0$

The First Excited State The first excited state of the hydrogen atom corresponds to $n = 2$. The energy of the first excited state is

$$E_2 = -\frac{13.6 \text{ eV}}{2^2} = -3.4 \text{ eV}$$

There are two possible values of orbital quantum number, $l = 0$ and $l = 1$. If $l = 0$, then $m_l = 0$ and the orbital angular momentum of the electron is zero. If $l = 1$, then m_l has three possible values: +1, 0, and −1. For $l = 1$, the magnitude of the orbital angular momentum is $\sqrt{2}\,\hbar$, and there are three possible orientations of the orbital, as shown in Figure 31-21. Thus, the first excited state has four subshells, all with the same binding energy. We say that the first excited state of the hydrogen atom is *four-fold degenerate*.

First excited state: $n = 2$ $l = 0$ $m_l = 0$} $2s$

$$\left.\begin{array}{ccc} n = 2 & l = 1 & m_l = +1 \\ n = 2 & l = 1 & m_l = 0 \\ n = 2 & l = 1 & m_l = -1 \end{array}\right\} 2p$$

The $l = 0$ orbital is spherical (Figure 31-20), and the $l = 1$ orbitals have the double-lobe shape shown in Figure 31-21.

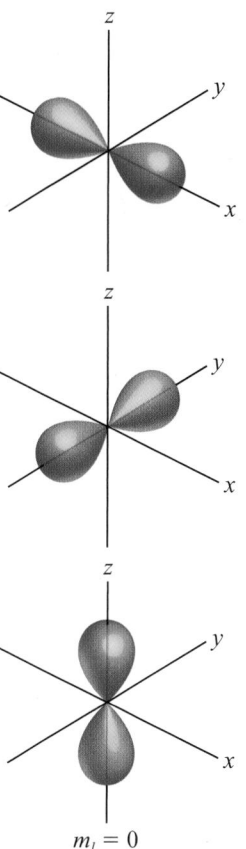

Figure 31-21 A plot of the $2p$ orbitals. Note that the orbital aligned along the z-axis has $m_l = 0$, and the orbitals aligned along the x-axis and y-axis are certain mixtures of quantum states with $m_l = -1$ and $m_l = 1$.

Figure 31-22 shows the energy-level diagram for the hydrogen atom, as predicted by the Schrödinger equation.

Energy Levels of Hydrogen Atom

Figure 31-22 The energy-level diagram for a hydrogen atom

✓ CHECKPOINT

C-31-7 $n = 3$ State of Hydrogen Atom

For a hydrogen atom, how many combinations of l and m_l are possible for the $n = 3$ state?

(a) 6
(b) 9
(c) 12
(d) 24

C-31-7 (b) For $n = 3$ there are three possible values of l. For $l = 2$ there are 5 possible values of m_l, for $l = 1$, 3 values and $l = 0$ one value, for a total of 9 possible combinations.

The Radial Wave Function

The radial wave function is determined by solving the differential equation for the radial coordinate r. The radial wave functions depend on the principal and the orbital quantum numbers and are usually denoted as $R_{nl}(r)$.

The Ground State The ground state has $n = 1$ and $l = 0$; the radial wave function for the ground state is

$$R_{1s}(r) = \left(\frac{1}{\pi a_B^3}\right)^{1/2} e^{-\left(\frac{r}{a_B}\right)} \quad (31\text{-}80)$$

where $a_B = \dfrac{k\hbar^2}{m_e e^2}$ is the Bohr radius.

The radial wave function does not depend on θ and the ϕ coordinates and therefore has the same value everywhere on a spherical shell of radius r. The probability $P_{nl}(r)$ of finding the electron in a spherical shell between r and $r + dr$ is given by

$$P_{nl}(r) = 4\pi r^2 R_{nl}^2(r) \quad (31\text{-}81)$$

The area of a spherical shell is $4\pi r^2$, and the factor of 4π arises because of integration over angles θ and ϕ. The radial probability density of the ground state is obtained by inserting Equation (31-80) into the above equation:

$$P_{1s}(r) = \frac{4}{a_B^3} r^2 e^{-\frac{2r}{a_B}} \quad (31\text{-}82)$$

Figures 31-23 and 31-24 show the wave function and the radial probability density for the ground state. The probability density is zero at the centre of the atom, reaches a maximum at $r = a_B$, and then falls off exponentially. Unlike in the Bohr model, the electron does not move in a well-defined orbit with $r = a_B$, and it is possible to find the electron everywhere away from the origin. The most probable distance is at the Bohr radius.

The First Excited State The first excited state has $n = 2$ with $l = 0$ or $l = 1$. The radial wave functions for these states are

$$R_{2s}(r) = \left(\frac{1}{32\pi a_B^3}\right)^{1/2}\left(2 - \frac{r}{a_B}\right) e^{-\left(\frac{r}{2a_B}\right)} \quad (31\text{-}83)$$

$$R_{2p}(r) = \left(\frac{1}{96\pi a_B^3}\right)^{1/2}\left(\frac{r}{a_B}\right) e^{-\left(\frac{r}{2a_B}\right)} \quad (31\text{-}84)$$

Figure 31-23 The radial wave functions for the ground state ($1s$) and the first excited state ($2s$) of hydrogen as a function of r. The vertical axis is in units of $(1/a_B)^{3/2}$.

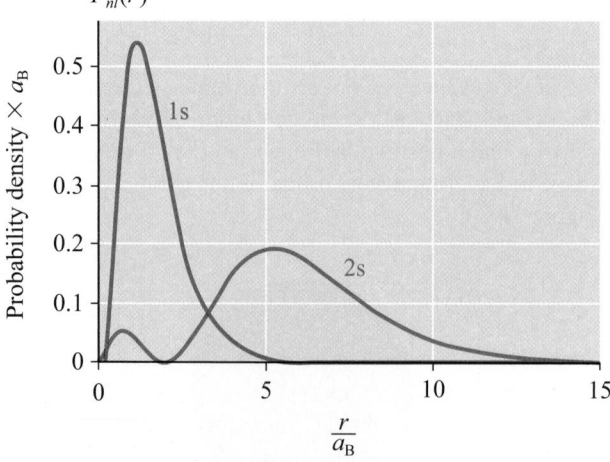

Figure 31-24 The radial probability density for the ground state ($1s$) and the first excited state ($2s$) of hydrogen as a function of r. The probability density is in units of $(1/a_B)$.

MODERN PHYSICS

The wave function and radial probability density for the 2s state are shown in Figures 31-23 and 31-24. For the 2s state, the radial probability density is maximum at $r \approx 5.2a_B$. (The radius of the second energy level in the Bohr model is $4a_B$.) The existence of the smaller secondary maximum at $r = 0.76a_B$ indicates that there is a small probability of finding the 2s electron closer to the nucleus. This feature cannot be explained by Bohr's model.

Although the energies of the hydrogen atom predicted by Schrödinger's equation and the Bohr model agree, the wave function predicted by Schrodinger's equation is very different from the Bohr model. In the Bohr model, an electron travels around the nucleus in distinct circular orbits, where the circumference of each possible orbit equals an integral number of de Broglie wavelengths. In the more accurate quantum-mechanical model based on Schrödinger's equation, there are no orbits. The electron does not follow a well-defined trajectory around the nucleus. A statement such as, "the electron is at a distance of one Bohr radius from the proton" is meaningless in quantum mechanics. We can only talk about the probability of finding the electron at a certain distance from the proton. As shown by the graphs of the radial probability density, the probability of finding the electron at any distance from the proton is nonzero. The Bohr radius for an energy level corresponds approximately to the average of the most probable distances of the electron from the nucleus.

✓ CHECKPOINT

C-31-8 Electron Energy in a Hydrogen Atom

According to the Schrödinger equation, the probability of finding an electron in its ground state at radii other than $r = a_B$ is nonzero. From Figure 31-24, the probability of finding an electron in a hydrogen atom at $r = 2a_B$ is about half the probability at $r = a_B$. At $r = 2a_B$, what is the binding energy of the electron in its ground state?

(a) less than -13.6 eV
(b) equal to -13.6 eV
(c) greater than -13.6 eV
(d) zero

C-31-8 (b) The binding energy of a given state is independent of r.

LO 9

31-9 Intrinsic Angular Momentum—Spin

Because of its success in explaining the hydrogen spectrum and other phenomena, the Schrödinger equation is one of the foundations of modern quantum mechanics. However, by 1925, two experimental results had been confirmed that could not be explained by the Schrödinger wave equation we have discussed so far.

First, hydrogen spectral lines observed at very high resolution were found to each consist of two closely spaced lines. For example, the spectral line corresponding to the 2p state is found to consist of two spectral lines with an energy difference of 4.5×10^{-5} eV (Figure 31-25). Such splitting of spectral lines is called the **fine structure**. The same splitting occurs in the spectra of other elements that have a single electron in the outer shell.

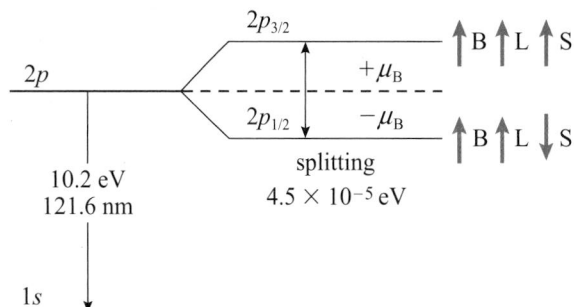

Figure 31-25 Fine structure in hydrogen. The spectral line corresponding to the 2p state consists of two closely spaced spectral lines. States with S parallel to L have a slightly higher energy compared to the states for which S and L are anti-parallel.

Second, when a beam of neutral atoms that have a single electron in the outermost shell passes through a non-uniform magnetic field, the beam splits into two or more beams. This splitting can be explained if the single electron has a nonzero orbital angular momentum. However, this splitting occurs with a beam of hydrogen atoms in their ground state where the electron has no orbital angular momentum. Both of these splitting effects have the same origin in that the charged particles have a magnetic moment.

Magnetic Moment and Orbital Angular Momentum

Consider a particle of charge q and mass m that is moving in a circular orbit of radius R with speed v. The current produced by this particle is equal to the rate of rotation of the charge:

$$I = \frac{q}{(2\pi R/v)} = \frac{qv}{2\pi R} \qquad (31\text{-}85)$$

The magnitude of the magnetic moment produced by this loop of current is equal to the product of the current and the area of the circular orbit:

$$\mu = IA = \left(\frac{qv}{2\pi R}\right)(\pi R^2) = \frac{qvR}{2} \qquad (31\text{-}86)$$

The magnitude of the classical angular momentum of the rotating particle is

$$L = mvR \qquad (31\text{-}87)$$

Therefore, the magnitudes of the magnetic moment and the classical angular momentum of the particle are related as follows:

$$\mu = \frac{q}{2m}L \qquad (31\text{-}88)$$

MODERN PHYSICS

Since the magnetic moment and angular momentum are both vector quantities, we can write Equation (31-88) in vector form:

$$\vec{\mu} = \frac{q}{2m}\vec{L} \qquad (31\text{-}89)$$

For an electron, $q = -e$ and $m = m_e$, so the magnetic moment due to its orbital angular momentum is

$$\vec{\mu}_l = g_l \frac{-e}{2m_e}\vec{L} = -g_l \frac{\mu_B}{\hbar}\vec{L} \qquad (31\text{-}90)$$

where we have introduced the factor $g_l = 1$ to keep the definition of the magnetic moment general. (As we will see later in this section, the g-factor for the quantum-mechanical magnetic moments is not always 1.) The quantity $\dfrac{e\hbar}{2m_e}$ is called the **Bohr magneton**, μ_B, and has a value of

$$\mu_B = \frac{e\hbar}{2m_e} = 9.27 \times 10^{-24}\ \text{A}\cdot\text{m}^2 \qquad (31\text{-}91)$$

When a particle with a magnetic moment is placed in a magnetic field, it experiences a torque that tends to align its magnetic moment along the direction of the magnetic field. The particle, therefore, acquires a magnetic potential energy, U_{mag}, that is given by

$$U_{mag} = -\vec{\mu}_l \cdot \vec{B} \qquad (31\text{-}92)$$

If we choose a z-axis along the direction of the magnetic field, then

$$U_{mag} = -\vec{\mu}_l \cdot \vec{B} = g_l \frac{e}{2m_e}\vec{L}\cdot\vec{B}$$
$$= g_l \frac{e}{2m_e}L_z B = g_l \frac{\mu_B}{\hbar}L_z B \qquad (31\text{-}93)$$

where L_z is the component of the orbital angular momentum along the z-axis. If the angular momentum makes an angle θ with respect to the z-axis, then $L_z B = LB \cos\theta$. As the angle θ varies between 0 and π rad, the magnetic potential energy varies between $+g_l\dfrac{\mu_B}{\hbar}LB$ and $-g_l\dfrac{\mu_B}{\hbar}LB$.

The relationship between angular momentum and the magnetic moment (Equation (31-90)) also holds in quantum mechanics. However, in quantum mechanics angular momentum is quantized; therefore, the magnetic moment is also quantized. Since the allowed values of L_z are $m_l\hbar$, the magnetic potential energy of an electron in a magnetic field (given by Equation (31-93)) is quantized:

$$U_{mag} = m_l(g_l\mu_B B) \qquad (31\text{-}94)$$

Thus, quantum mechanics predicts that the energy of an electron in the presence of a magnetic field changes by discrete amounts.

The Stern–Gerlach Experiment

To determine whether the magnetic moment is quantized, as predicted by quantum mechanics, German physicists Otto Stern (1888–1969) and Walter Gerlach (1889–1979)

performed an ingenious experiment. They prepared a beam of silver atoms by boiling silver in an oven and passed the evaporated atoms through a series of slits to form a narrow beam. The beam was then directed between poles of magnets shaped to produce a nonuniform magnetic field that varied considerably along a line joining the two poles, as shown in Figure 31-26. After passing through the magnetic field, the beam hit a photographic plate and created a mark on the plate.

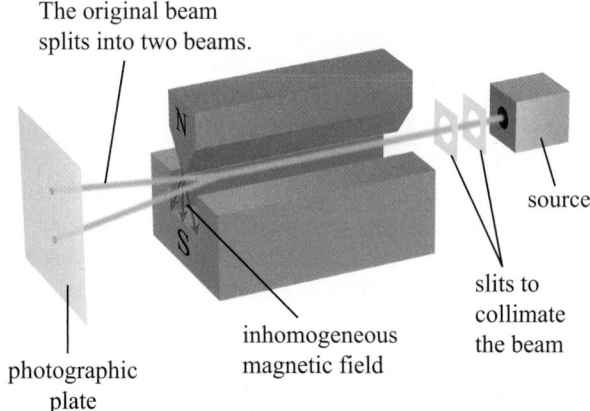

The original beam splits into two beams.

source

slits to collimate the beam

inhomogeneous magnetic field

photographic plate

Figure 31-26 The Stern–Gerlach experiment

Stern and Gerlach argued that if a silver atom *does* have a magnetic moment, then in a magnetic field that is pointing along the z-direction the atom will gain a magnetic potential energy of $-\vec{\mu}\cdot\vec{B} = -\mu B\cos\theta$, where θ is the angle between the directions of the magnetic field and the magnetic moment. Since the magnetic field varies along the z-direction, there will be a net force on the atom given by

$$F_z = -\frac{dU_{mag}}{dz} = (\mu\cos\theta)\frac{dB}{dz} \qquad (31\text{-}95)$$

Silver atoms with their magnetic moments oriented perpendicular to the direction of the magnetic field ($\theta = 90°$) experience no force and pass through the magnetic field undeflected. Atoms with magnetic moments in the direction of the magnetic field ($\theta = 0°$) experience an upward force and are deflected upward. Atoms with magnetic moments directed opposite to the magnetic field ($\theta = 180°$) experience a downward force and are deflected downward.

A silver atom has 47 electrons. In its ground state, the first 46 electrons couple together in pairs that have zero net orbital angular momentum (Figure 31-27). The lone uncoupled electron is in the $l = 0$ state. Thus, a silver atom has zero orbital angular momentum and should have zero magnetic moment. Consequently, a beam of silver atoms should not be deflected by the inhomogeneous field in Stern and Gerlach's experiment.

However, Stern and Gerlach observed that silver atoms formed two spots on the plate, spaced at equal distances above and below the original beam direction. The

MODERN PHYSICS

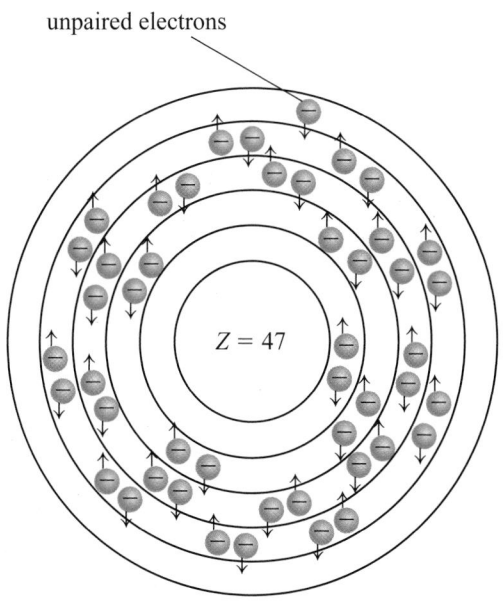

unpaired electrons

Z = 47

Figure 31-27 A Bohr model diagram of a silver atom in its ground state

splitting of the incident beam showed that silver atoms *do* have a magnetic moment and that the component of the magnetic moment along the direction of the magnetic field has only two possible values: either along the direction of the magnetic field or opposite to it.

But a silver atom has zero orbital angular momentum, so how can it have a magnetic moment? In 1925, Dutch American physicists George Uhlenbeck (1900–1988) and Samuel Goudsmit (1902–1978) suggested that electrons possess an intrinsic angular momentum, which they called **spin**. The situation is somewhat analogous to Earth spinning on its axis and thus possessing angular momentum because of this spin (although our current understanding is that an electron is a point particle with zero radius). The **spin quantum number**, s, is a fundamental characteristic of all subatomic particles. The spin angular momentum is denoted by the symbol $\vec{S}$. The magnitude of the spin angular momentum and its projection along the z-axis are given by formulas similar to Equations (31-78) and (31-79):

$$S = \sqrt{s(s+1)}\,\hbar$$

$$S_z = m_s\hbar \quad m_s = -s, (-s+1), \ldots, (s-1), s \quad (31\text{-}96)$$

where m_s is called the spin magnetic quantum number.

Because of its spin, an electron possesses a magnetic moment of a form analogous to Equation (31-93):

$$\vec{\mu}_s = -g_s \frac{\mu_{\text{B}}}{\hbar}\,\vec{S} \quad (31\text{-}97)$$

Experiments have shown that the g-factor, g_s, is approximately equal to 2. Because of its spin, an electron acquires an additional magnetic energy in the presence of a magnetic field:

$$U_{\text{mag,s}} = g_s\mu_{\text{B}}\left(\frac{S_z}{\hbar}\right)B \quad (31\text{-}98)$$

The total magnetic potential energy of an electron is therefore equal to the sum of the magnetic potential energies due to its orbital and spin angular momenta.

The splitting of the beam in the Stern–Gerlach experiment into two beams indicates that S_z for the silver atoms can have two values: $+\dfrac{\hbar}{2}$ and $-\dfrac{\hbar}{2}$. Therefore, $s = \dfrac{1}{2}$ for the silver atom. Since the spin of the silver atom is determined by the spin of the unpaired electron, s must be $\dfrac{1}{2}$ for an electron. The two possible values of m_s for an electron are therefore $+\dfrac{1}{2}$ and $-\dfrac{1}{2}$. An electron with $m_s = +\dfrac{1}{2}$ is called a "spin-up" electron and is represented with an up arrow (↑). An electron with $m_s = -\dfrac{1}{2}$ is called a "spin-down" electron and is represented with a down arrow (↓).

In 1927, American physicists T.E. Phipps and J.B. Taylor used a beam of hydrogen gas in an experiment similar to the Stern–Gerlach experiment. In the Phipps–Taylor experiment, the beam split into two beams. Since the electron in the ground state of hydrogen has zero orbital angular momentum, this experiment again confirmed that an electron has an intrinsic spin, $s = \dfrac{1}{2}$.

Just like mass and charge, spin is an intrinsic property of a particle. When the theory of relativity is combined with quantum mechanics, spin naturally appears as one of the intrinsic properties of a particle. All elementary particles have an intrinsic spin. Particles that have a half-integer spin (such as protons, neutrons, and electrons) are called **fermions**. Particles that have integer spin (such as photons and pions) are called **bosons**.

 EXAMPLE 31-10

The Spin of a Rho Meson (ρ-meson)

A ρ-meson is an elementary particle with spin quantum number $s = 1$.

(a) What is the magnitude of the spin angular momentum of a ρ-meson, and what are the possible values of the projection of its spin along the z-axis?

(b) If a beam of ρ-mesons is used in the Stern–Gerlach experiment into how many beams would the original beam split?

SOLUTION

(a) Using Equation (31-96) with $s = 1$, we find

$$S = \sqrt{s(s+1)}\,\hbar = \sqrt{2}\,\hbar$$

$$S_z = m_s\hbar = -\hbar, 0, \hbar$$

(b) Since S_z has three possible values, there will be three beams emerging from the magnetic field. One beam is along the original direction ($S_z = 0$), one above the original beam ($S_z = +\hbar$), and one below it ($S_z = -\hbar$).

MODERN PHYSICS

CHAPTER 31 | **INTRODUCTION TO QUANTUM MECHANICS**

✓ CHECKPOINT

C-31-9 Particle Spin

How many different values of S_z are possible for a sub-atomic particle with a spin quantum number $s = \frac{3}{2}$?

(a) 2
(b) 4
(c) 6
(d) 8

C-31-9 (b)

Adding Angular Momenta in Quantum Mechanics

In classical mechanics, if particle 1 has angular momentum $\overrightarrow{J_1}$ and particle 2 has angular momentum $\overrightarrow{J_2}$, then the total angular momentum $\overrightarrow{J}$ of the two particles is equal to the vector sum of $\overrightarrow{J_1}$ and $\overrightarrow{J_2}$: $\overrightarrow{J} = \overrightarrow{J_1} + \overrightarrow{J_2}$. The magnitude of $\overrightarrow{J}$ can have any value between $|J_1 - J_2|$ and $J_1 + J_2$.

In quantum mechanics, angular momentum is an operator, and angular momenta are added according to the following rule: Let j_1 and j_2 be the quantum numbers corresponding to angular momenta $\overrightarrow{J_1}$ and $\overrightarrow{J_2}$, respectively. Let j be the quantum number of the total angular momentum $\overrightarrow{J}$ such that $\overrightarrow{J} = \overrightarrow{J_1} + \overrightarrow{J_2}$. Then j can have any of the following values: $j_1 + j_2, j_1 + j_2 - 1, j_1 + j_2 - 2, \ldots, |j_1 - j_2|$.

For a given value of j, the magnitude J of the total angular momentum is $\sqrt{j(j+1)}\,\hbar$, and the z-component of $\overrightarrow{J}$ can have any of the following two $(2j + 1)$ values:

$$j\hbar, (j-1)\hbar, (j-2)\hbar, \ldots, (-j+1)\hbar, -j\hbar$$

This rule for adding angular momenta applies whether $\overrightarrow{J_1}$ and $\overrightarrow{J_2}$ represent angular momenta of two particles or two angular momenta of the same particle.

✓ CHECKPOINT

C-31-10 The Angular Momentum of Two Electrons

Two electrons in an atom each have an angular momentum quantum number $\frac{3}{2}$. What are the allowed values of the quantum number of the total angular momentum of the two electrons?

(a) 0
(b) 3
(c) 3, 2, 1, 0
(d) 3, 2, 1, 0, −1, −2, −3

C-31-10 (c)

➕➖➗✖ EXAMPLE 31-11

Adding Angular Momentum in Quantum Mechanics

What are the possible values of the quantum number j of the total angular momentum $\overrightarrow{J}$ of an electron in a hydrogen atom with orbital angular momentum characterized by the quantum number $l = 1$? Determine the magnitude and allowed values of the z-component of $\overrightarrow{J}$ for each value of j.

SOLUTION

Let $\overrightarrow{L}$ be the orbital angular momentum and $\overrightarrow{S}$ be the spin of the electron. The total angular momentum $\overrightarrow{J}$ of the electron is

$$\overrightarrow{J} = \overrightarrow{L} + \overrightarrow{S}$$

Since $s = \frac{1}{2}$ and $l = 1$, the possible values of j are

$$j = l + \frac{1}{2} = \frac{3}{2} \quad \text{and} \quad j = l - \frac{1}{2} = \frac{1}{2}$$

For $j = \frac{3}{2}$:

$$J = \sqrt{\frac{3}{2}\left(\frac{3}{2} + 1\right)}\,\hbar = \sqrt{\frac{15}{4}}\,\hbar$$

and J_z can have four values: $\frac{3}{2}\hbar, \frac{1}{2}\hbar, -\frac{1}{2}\hbar, -\frac{3}{2}\hbar$.

For $j = \frac{1}{2}$:

$$J = \sqrt{\frac{1}{2}\left(\frac{1}{2} + 1\right)}\,\hbar = \sqrt{\frac{3}{4}}\,\hbar,$$

and J_z can have two values: $\frac{1}{2}\hbar, -\frac{1}{2}\hbar$.

Making sense of the result:

For $l = 1$ there are three possible values of m_l $(-1, 0, +1)$. As the spin projection of an electron can be up or down, an electron with $l = 1$ can exist in six possible states. The number of allowed states remains the same when the spin and the orbital angular momenta are combined. There are four allowed states for $j = 3/2$ and two for $j = 1/2$, adding to a total of six possible states.

MODERN PHYSICS

The Pauli Exclusion Principle

A hydrogen atom has one electron. In its ground state, the electron is in the $n = 1$, $l = 0$ orbital. A helium atom has two electrons, and in the lowest energy state both electrons are in the $n = 1$, $l = 0$ orbital with their spins pointing in opposite directions. A lithium atom has three electrons. In its lowest energy state, the first two electrons are in the $n = 1$, $l = 0$ orbital with their spins pointing in opposite directions, but the third electron is in the higher energy orbital, $n = 2$, $l = 0$. Similarly, a sodium atom has 11 electrons, with 2 electrons in orbitals with $n = 1$, 8 electrons in $n = 2$ orbitals, and one electron in an $n = 3$ orbital. Why are all the electrons of an atom that is in its ground state not in the lowest energy $n = 1$, $l = 0$ orbital? To explain this, Austrian physicist Wolfgang Pauli (1900–58) proposed that *no two electrons within an atom can occupy the same quantum state simultaneously.* This principle is called the **Pauli exclusion principle**.

With the inclusion of spin, the quantum state of an electron in an atom is characterized by four quantum numbers: n, l, m_l, and m_s. Since electrons are added to build heavier atoms, the lowest energy state configuration of electrons must be consistent with the Pauli exclusion principle. Table 31-5 shows how the electrons are distributed in $n = 1$, 2 orbitals.

The Pauli exclusion principle applies to all fermions. There is no limit to the number of bosons that can occupy the same quantum state simultaneously.

Table 31-5 Quantum states of an atom for principal quantum numbers 1 and 2

n	l	m_l	m_s	Capacity of shell or subshell	
1	0	0	+1/2	2 electrons in the 1s shell	A maximum of two electrons in the $n = 1$ shell
		0	−1/2		
2	0	0	+1/2	2 electrons in the 2s subshell	A maximum of 8 electrons in the $n = 2$ shell
		0	−1/2		
	1	+1	+1/2	6 electrons in the 2p subshell	
		+1	−1/2		
		0	+1/2		
		0	−1/2		
		−1	+1/2		
		−1	−1/2		

 CHECKPOINT

C-31-11   Electrons in an $n = 3$ Shell

What is the maximum number of electrons allowed in an $n = 3$ shell?

(a) 3
(b) 6
(c) 12
(d) 18

C-31-11 (d)

MODERN PHYSICS

Quantum mechanics is the branch of physics that deals with the motion of particles at the subatomic scale. The Schrödinger equation describes the behaviour of subatomic matter, predicts that the energy and momentum of a confined particle are quantized, and predicts that a particle can tunnel into a classically forbidden region.

Matter Waves and de Broglie's Hypothesis

De Broglie postulated that a particle of momentum p has a wavelength $\lambda = \dfrac{h}{p}$, where $h = 6.63 \times 10^{-34}$ J·s $= 4.14 \times 10^{-15}$ eV·s is Planck's constant. Since h is extremely small, the wave nature of matter is not observed at a macroscopic level.

Heisenberg's Uncertainty Principle

The product of the uncertainty in a measurement of the position of a particle and the uncertainty in the measurement of its momentum must be greater than $\hbar/2$:

$$\Delta x \, \Delta p \geq \frac{\hbar}{2} \qquad (31\text{-}6)$$

Similarly, a time Δt is needed to measure the energy of a system to uncertainty ΔE, and

$$\Delta E \, \Delta t \geq \frac{\hbar}{2}$$

The Schrödinger Equation

The wave function $\psi(x, t)$ of a particle of mass m satisfies the time-dependent Schrödinger equation:

$$-\frac{\hbar^2}{2m} \frac{\partial^2 \psi(x, t)}{\partial x^2} + V(x) \, \psi(x, t) = i\hbar \frac{\partial \psi(x, t)}{\partial t} \quad (31\text{-}27)$$

where $V(x)$ is the potential energy term.

When the particle is in a state of definite energy E, the wave function can be written as $\psi(x, t) = \psi(x)e^{-iEt}$, and the Schrödinger equation can be written in a time-independent form:

$$-\frac{\hbar^2}{2m} \frac{d^2 \psi(x)}{dx^2} + V(x)\psi(x) = E\psi(x) \qquad (31\text{-}28)$$

where $\psi(x)$ is a function of position.

The energy E is called the energy eigenvalue. $|\psi(x)|^2 dx$ represents the probability of finding the particle in a region of length dx that is centred around the point x. The wave function is normalized such that the probability of finding the particle somewhere in space is one. The normalization condition for the Schrödinger equation in one dimension is

$$\int_{-\infty}^{+\infty} |\psi(x)|^2 \, dx = 1 \qquad (31\text{-}50)$$

Particle in an Infinite Square Well Potential

The quantized energy values for a particle in an infinite square well potential are

$$E_n = n^2 \frac{\hbar^2 \pi^2}{2mL^2} \qquad (31\text{-}46)$$

The principal quantum number n is a positive integer. The wave function of a particle in an infinite square well of width L and energy E_n is

$$\psi_n(x) = \sqrt{\frac{2}{L}} \, \sin\left(\frac{n\pi}{L}x\right) \quad n = 1, 2, 3, \ldots \quad (31\text{-}51)$$

Quantum Tunnelling

The probability of penetration through a barrier of height V_0 and width a by a particle of energy $E < V_0$ is proportional to

$$P \propto e^{-\frac{2}{\hbar}\left(\sqrt{2m(V_0 - E)}\right)a} \qquad (31\text{-}66)$$

Angular Momentum in Quantum Mechanics

Angular momentum is quantized and is characterized by a quantum number j, which is a positive integer or a half-integer. For a given j the magnitude of the angular momentum J is

$$J = \sqrt{j(j + 1)} \, \hbar \qquad (31\text{-}68)$$

The z-component of the angular momentum J_z has the $2j + 1$ allowed values:

$$J_z = m_j \hbar \quad m_j = -j, (-j + 1), \ldots, (j - 1), j$$

The Schrödinger Equation for a Hydrogen Atom

The hydrogen atom is described by a wave function that is characterized by four quantum numbers:

- The principal quantum number n determines the size of the orbital and has the allowed values of $n = 1, 2, 3, \ldots$.
- The angular momentum quantum number l determines the shape of the orbital and has the allowed values of $l = 0, 1, 2, \ldots, (n - 1)$.
- The magnetic quantum number m_l determines the orientation of the orbital for a given l and has the allowed values of $m_l = -l, (-l + 1), \ldots, (l - 1), l$.
- The spin quantum number m_s describes the intrinsic angular momentum of the electron and has the allowed values of $m_s = -\dfrac{1}{2}, \dfrac{1}{2}$.

The energy levels predicted by solving the Schrödinger equation for an electron confined within the Coulomb potential of a proton match the spectral lines of hydrogen.

Spin

Every particle has an intrinsic spin quantum number, s. Particles with half-integer spin are called fermions, and particles with integer spin are called bosons. For electrons, $s = \dfrac{1}{2}$. According to the Pauli exclusion principle, no two fermions within the same atom can have the same set of quantum numbers simultaneously.

QUESTIONS

1. Can you ever say that an electron is at rest? Explain your reasoning.
2. Evaluate the following statement: "There is a nonzero probability that a tennis ball striking a racquet will pass through the racquet without breaking the strings."
3. Could an electron in your body have a nonzero probability of being on the Moon?
4. Say an electron and a proton have the same kinetic energy. Which has the shortest de Broglie wavelength?
5. What is the de Broglie wavelength of a particle at rest?
6. If an electron and a proton have the same total energy (including the rest mass energy), which has the shortest de Broglie wavelength?
7. A high-intensity beam of photons is incident on a gas of hydrogen atoms. The atoms are in their ground states. Would you expect to see any electrons dislodged from the gas when the energy of the photons is
   (a) 10 eV;
   (b) 20 eV;
   (c) 30 eV? Explain your reasoning.
8. What is the maximum number of electrons that can be contained in an $l = 4$ shell?
   (a) 3
   (b) 6
   (c) 9
   (d) 18
   (e) 36
9. Rank the following transitions in a hydrogen atom in order of increasing energy:
   (a) $n = 4 \rightarrow n = 2$
   (b) $n = 6 \rightarrow n = 3$
   (c) $n = 3 \rightarrow n = 1$
   (d) $n = 5 \rightarrow n = 3$
10. An electron in an atom is in the $n = 3$ state. Which of the following values of l are *not* allowed for this electron?
   (a) 0
   (b) 1
   (c) 2

   (d) 3
   (e) 4
11. Two angular momenta with quantum numbers $j_1 = 2$ and $j_2 = 2$ are added. Which of the following quantum numbers for the total angular momentum is *not* possible?
   (a) 0
   (b) 1
   (c) 2
   (d) 3
   (e) 4

PROBLEMS BY SECTION

For problems, star ratings will be used, (✷, ✷✷, or ✷✷✷), with more stars meaning more challenging problems.

Section 31-1   Matter Waves and de Broglie's Hypothesis

12. ✷ Using the relativistic energy momentum relationship $E^2 = p^2c^2 + m_0^2c^4$, where m_0 is the rest mass, p is the momentum, and c is the speed of light, find the de Broglie wavelength of the following:
   (a) a 10 keV electron;
   (b) a 10 keV proton;
   (c) a 500 MeV electron;
   (d) a 500 MeV proton.
13. ✷✷ Calculate the de Broglie wavelength of an electron with a speed of
   (a) $0.9c$;
   (b) $0.99c$.
14. ✷ What is the de Broglie wavelength of a 145 g baseball thrown at a speed of 150 km/h?
15. ✷✷ Calculate the kinetic energy of an electron that has a de Broglie wavelength of 600 nm. Through what electric potential would the electron have to be accelerated to acquire this kinetic energy?
16. ✷✷ What is the speed of an electron with a de Broglie wavelength of
   (a) 10^{-1} m;
   (b) 10^{-5} m;
   (c) 10^{-10} m?

CHAPTER 31 | INTRODUCTION TO QUANTUM MECHANICS

MODERN PHYSICS

Section 31-2 Heisenberg's Uncertainty Principle

17. ✳ What is the uncertainty in the momentum of an electron that is confined within a cube of length 1.0×10^{-10} m? What is the minimum kinetic energy of the electron?

18. ✳✳ The radius of a carbon nucleus is approximately 3×10^{-15} m. What is the uncertainty in the momentum of a proton that is confined within the nucleus? What is the minimum kinetic energy of a proton in this nucleus?

19. ✳ What is the minimum possible uncertainty in the position of a 1 g object with a speed of 100 ± 0.1 m/s?

20. ✳✳ Suppose that the electrons in an atom are confined within the nucleus.
    (a) What is the minimum kinetic energy of an electron that is confined within a silver nucleus that has a radius of 5.7×10^{-15} m?
    (b) Explain why the minimum energy indicates that electrons are not actually confined within the nuclei of atoms.

21. ✳ What is the uncertainty in the energy of a particle that has a half-life of
    (a) 10^{-12} s;
    (b) 10^{-23} s?

22. ✳ An electron has a kinetic energy of 10 eV. If its momentum is uncertain by 0.1%, what is the minimum uncertainty in its position?

23. ✳ An electron has a speed of 10^5 m/s. What is the minimum time needed to measure its energy to an accuracy of $\pm 0.001\%$?

24. ✳✳ A charged pion is an unstable particle that has a rest mass energy of 139.5 MeV and a half-life of 2.6×10^{-8} s. The maximum time available to measure the energy of a charged pion is equal to its half-life. Find the uncertainty in its rest mass energy. The uncertainty in rest mass energy of unstable particles is called the "decay width" of the particle.

Section 31-3 The Schrödinger Equation

25. ✳ What are the units of the one-dimensional Schrödinger wave function $\psi(x)$?

26. ✳ What are the units of the three-dimensional Schrödinger wave function $\psi(x, y, z)$?

Section 31-5 A Particle in a One-Dimensional Box

27. ✳✳ Calculate the lowest three energies for an electron that is confined within a one-dimensional box of length 10^{-10} m (the size of an atom). Sketch the probability density of the electron in each state.

28. ✳✳ Consider a baseball with a mass of 145 g that is confined within a one-dimensional infinite square well potential of width 1.0 m. How large would Planck's constant have to be if the minimum possible energy of the baseball is 10.0 J?

29. ✳✳ Calculate the energies of the photons emitted when an electron in an infinite square well potential of width 50 nm makes the following transitions:
    (a) $n = 2$ to $n = 1$;
    (b) $n = 3$ to $n = 1$;
    (c) $n = 3$ to $n = 2$;
    (d) $n = 1000$ to $n = 1$.

30. ✳✳ A charged particle of unknown rest mass is confined within a one-dimensional square well potential. When the particle drops from the $n = 2$ excited state to the ground state, it emits a photon of 0.12 eV.

The width of the confining potential is 0.1 nm. What is the rest mass of the particle?
Suppose the rest mass of the particle is 105 MeV/c^2. Determine the width of the potential.

31. ✳✳✳ Consider a particle in the ground state of an infinite square well potential of width L. What is the probability of finding the particle between $x = 0$ and $x = L/2$?

32. ✳ A dust particle of mass 10^{-8} kg is confined within a box of width 10^{-3} m and has a speed of 10^{-3} m/s. This particle is considered to be a quantum-mechanical system. What excited state is it in?

33. ✳ An electron is confined within a box of length 0.1 nm. Determine the energy and the wavelength of the emitted photon when the electron makes a transition from an $n = 2$ to an $n = 1$ state.

34. ✳ Let $a = x + iy$ be a complex number, where x and y are real numbers. The complex conjugate of a is $a^* = x - iy$. Show that the product aa^* is always positive.

35. ✳✳ The sine, cosine, and exponential functions can be written as a sum of powers of their arguments. Such a sum is called a series expansion. Series expansions of these three functions are as follows:

$$\sin(x) = x - \frac{x^3}{3!} + \frac{x^5}{5!} - \frac{x^7}{7!} + \frac{x^9}{9!} - \cdots$$

$$\cos(x) = 1 - \frac{x^2}{2!} + \frac{x^4}{4!} - \frac{x^6}{6!} + \frac{x^8}{8!} - \cdots$$

$$e^x = 1 + x + \frac{x^2}{2!} + \frac{x^3}{3!} + \frac{x^4}{4!} + \frac{x^5}{5!} + \frac{x^6}{6!} + \cdots$$

Although each series has an infinite number of terms, for small arguments of the function only a few terms are sufficient to calculate the function to a high enough accuracy.
    (a) For $x = 0.1, 1.0, 10.0$, calculate $\sin(x)$, $\cos(x)$, and e_x using a calculator. Now calculate the functions using the series expansions. How many terms of the series did you have to include in each case to obtain an answer that agrees with the calculator values to three decimal places?
    (b) Use a series expansion of $e^{i\theta}$ to show that $e^{i\theta} = \cos\theta + i\sin\theta$.

36. ✳✳✳ Let $\psi(x, t)$ be the wave function of the particle that satisfies

$$\left(i\hbar \frac{\partial}{\partial t}\right)\psi(x, t) = E\psi(x, t) \quad \text{and} \quad \left(-i\hbar \frac{\partial}{\partial x}\right)\psi(x, t) = p\psi(x, t)$$

Derive an equation for the wave function that satisfies the equation relating the relativistic total energy and the momentum of a particle:

$$p^2c^2 + m_0^2c^4 = E^2$$

where m_0 is the rest mass of the particle.

37. ✳✳ The time-independent wave function of a particle that is confined to move along the x-axis is given by

$$\psi(x) = \begin{cases} A(4 - x^2) & \text{for } -2 \leq x \leq 2 \\ 0 & \text{everywhere else} \end{cases}$$

where A is the normalization constant. Determine A and write the normalized wave function.

38. ✱✱ For the normalized wave function in problem 37, what is the probability of finding the particle (a) in the region $0 \leq x \leq 1$, (b) in the region $1 \leq x \leq 2$

39. ✱ In problem 37 if the mass of the particle is m and its momentum is p write the time dependent normalized wave function of the particle.

40. ✱ Graph the wave functions and the probability densities of a particle in an infinite square well potential with $L = 1$ nm for states with $n = 4$, 5, and 6.

41. ✱✱ A particle is confined within a one-dimensional infinite square well potential of width L. What is the probability of finding the particle in the region $\Delta x = 0.1L$ centred around $x = 0.5L$?

42. ✱✱ Assume that a proton in a carbon nucleus can be modelled as a particle confined within a one-dimensional infinite square well potential. The radius, R, of the carbon nucleus is approximately 2.7×10^{-15} m.
    (a) The width of the potential is $2R$. Estimate the ground-state energy of the proton in MeV.
    (b) Calculate the wavelength of the photon emitted for $n = 2$ to $n = 1$ and $n = 3$ to $n = 2$ transitions.

43. ✱✱ The ground-state energy of an electron in an infinite square well potential is 20 eV.
    (a) What are the energies of the first three excited states of the electron?
    (b) What is the width of the potential well?
    (c) What is the longest wavelength photon that an electron in this potential well can emit?

44. ✱✱ Let E_n be the energy of a particle in a state n of an infinite square well potential. Show that in the limit of large n the fractional difference in the energies of two adjacent states can be approximated as

$$\frac{E_{n+1} - E_n}{E_n} \approx \frac{2}{n}$$

How is this result related to Bohr's correspondence principle?

Section 31-6 Quantum Tunnelling

45. ✱✱ An electron with energy 2 eV is incident on a barrier of height 3 eV and width 0.1 nm. What is the probability that the electron will tunnel through the barrier?

46. ✱✱ An electron beam approaches a potential barrier of height 2 eV and 1 nm width. What must be the energy of the beam if 10% of the electrons are to tunnel through the barrier?

47. ✱✱ A beam of protons with energy 10.0 eV approaches a potential barrier of height 20.0 eV and width 0.5 nm. Approximately 10^{20} protons per second hit the barrier. How long will it take for 1000 protons to pass through the barrier?

48. ✱✱ Derive Equation (31-62) from the wave functions given in Equation (31-61).

Section 31-8 The Schrödinger Equation for a Hydrogen Atom

49. ✱ Calculate the energy of the $n = 5$ state of a hydrogen atom.

50. ✱✱ Calculate the energies and wavelengths of photons emitted for electron transitions from $n = 5$, 6, and 7 states to $n = 3$ states.

51. ✱✱ For an electron in an $n = 4$ state, what are the possible values of the orbital angular momentum quantum

number l? For each value of l, calculate the magnitude of the orbital angular momentum and the possible values of its projection along the z-axis.

52. ✱ How many subshells are possible for the $n = 3$ state of an atom?

53. ✱ What are the lowest possible values of n and l for
    (a) $m_l = -3$;
    (b) $m_l = 0$;
    (c) $m_l = +4$?

54. ✱ An electron in a hydrogen atom is in the $l = 4$ state.
    (a) What are the allowed values of n and m_l for this electron?
    (b) Rank these values in order of the binding energy of the electron, from lowest to highest.

55. ✱✱✱ Use Equation (31-82) to show that the most probable distance for an electron in the ground state of the hydrogen is equal to the Bohr radius.

56. ✱✱✱ Determine the probability of finding an electron in the ground state of the hydrogen atom for $r \leq 4r_0$.

57. ✱✱✱ A muon is an elementary particle with a rest mass energy of 105.7 MeV/c^2 and charge $-e$ (the same as the charge for an electron). A muonic hydrogen atom is formed by replacing an electron in a hydrogen atom with a muon.
    (a) Calculate the energy levels of the muonic hydrogen atom.
    (b) What is the energy of the emitted photon when a muon transitions from an $n = 2$ state to an $n = 1$ state of a muonic hydrogen atom?
    (c) What is the Bohr radius of a muon in the $n = 1$ state?

Section 31-9 The Intrinsic Angular Momentum—Spin

58. ✱✱ What are the possible angles between the spin angular momentum $\vec{S}$ of an electron and the z-axis?

59. ✱✱✱ Assume that an electron is a sphere of radius 10^{-18} m. Using nonrelativistic kinematics, find the speed at its surface when its rotational angular momentum is $\frac{\sqrt{3}}{2}\hbar$.

60. ✱✱ Two electrons in an atom have angular momentum quantum numbers $j_1 = \frac{1}{2}$ and $j_2 = \frac{3}{2}$.
    (a) Calculate all possible values of the quantum number j of the total angular momentum of the two electrons.
    (b) Calculate the magnitude and the z-component of the total quantum number for each value of j.

61. ✱✱ Repeat problem 60 for $j_1 = \frac{3}{2}$ and $j_2 = \frac{3}{2}$.

62. ✱✱ The total angular momentum quantum number for two electrons is 4. The angular momentum quantum number for one of the electrons is $\frac{5}{2}$. What are the possible values of the angular momentum quantum number of the second electron?

63. ✱✱ Show that the maximum number of electrons allowed in a hydrogen shell with principal quantum number n is $2n^2$.

 See the text online resources at www.physics1e.nelson.com for Open Problems and Data-Rich Problems related to this chapter.

MODERN PHYSICS

Learning Objectives

When you have completed this chapter you should be able to:

1 Draw simple crystal structures.

2 Discuss the density of states for the free electrons in a box model.

3 Sketch the electron bands in a periodic potential.

4 Describe the difference between a conductor and an insulator in terms of energy bands.

5 Discuss the basic characteristics of a pure semiconductor.

6 Qualitatively describe how doping a semiconductor changes its conduction properties.

7 Sketch the current-voltage characteristic of a pn junction diode.

8 Name some semiconductor devices.

9 Describe some of the tools used in nanotechnology.

 For additional Making Connections, Examples, and Checkpoints, as well as activities and experiments to help increase your understanding of the chapter's concepts, please go to the text's online resources at www.physics1e.nelson.com.

Chapter 32
Introduction to Solid-State Physics

In a modern circuit board (Figure 32-1), we can see examples of three different types of electrical behaviour exhibited by solids. The fine lines connecting the components are metallic conductors chosen for their ability to conduct electricity efficiently. The board itself is made of a material that conducts electricity very poorly, an electrical insulator. Some of the components on the board are semiconductor devices, which have conductivities between those of a conductor and an insulator. How is it that these solid materials, which have many similar properties, such as density and hardness, can have resistivities ranging over 32 orders of magnitude? In this chapter, we will use a quantum-mechanical approach to understand the behaviour of electrons in solids. For simplicity, we will restrict ourselves to considering crystalline solids.

© Phil Degginger/Alamy

Figure 32-1 A circuit board

32-1 Crystal Structures

In this chapter, we will restrict our discussion to materials that are in crystalline form. In a crystal, the atoms are arranged in periodic arrays that repeat themselves over and over in all three dimensions. Simple examples are the **cubic**, **hexagonal**, and **tetragonal structures** shown in Figure 32-2. The **cubic face-centred structure** has atoms in the middle of each face of the cube as well as at each corner of the cube, and the **cubic body-centred structure** has just one extra atom, which is at the centre of the cube. Note that a full crystal can be created by translating each structure through an appropriate direction and distance. For example, translating the cubic cell repeatedly by the length of one side creates a lattice with no gaps.

Figure 32-3 shows a bulk crystal that has a cubic structure. We can see that the bulk crystal has the same symmetry as the individual cells of the structure. Table 32-1 lists a few materials with their associated crystal structures and typical interatomic spacings, which is called the lattice parameter of the crystal. For all cubic structures, the **lattice parameter**, a, is simply the length

Figure 32-3 A sample of a material with a cubic crystal structure

of one side of the cube. For a tetragonal lattice, a is the length of the sides of the square base, and c is the length of the other edges.

Table 32-1 Some basic structural properties of a few elements

Material	Crystal lattice	Lattice parameter (nm)
Copper	Cubic: face-centred	0.361
Tin	Tetragonal	$a = 0.37, c = 0.337$
Silicon	Cubic: face-centred	0.543

    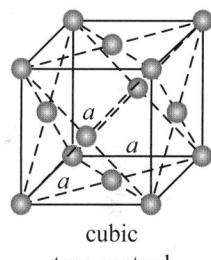

| cubic | hexagonal | tetragonal | cubic body centred | cubic tace centred |

Figure 32-2 Some simple crystal lattices

EXAMPLE 32-1

The Lattice Parameter of Sodium

Sodium has a cubic body-centred structure at 300 K. Determine the lattice parameter.

SOLUTION

Each cubic cell of sodium has an atom at each of its eight corners, and one-eighth of each of those atoms is inside the cube. In addition, one atom sits at the centre of the cubic body-centred structure. There are thus two atoms per cube in such a structure. To find the lattice parameter, we determine the volume per atom, and then find the side length of a cube with the volume of two atoms.

First, we look up the mass density and the atomic mass of sodium: 971 kg/m³ and 0.023 kg/mol, respectively. Dividing

the atomic mass by the density gives the volume per mole, and dividing this quantity by Avogadro's number gives the volume per atom:

$$V = \frac{0.023 \text{ kg/mol}}{971 \text{ kg/m}^{-3}} \frac{1}{6.022 \times 10^{23} \text{ atom/mol}}$$

$$= 3.93 \times 10^{-29} \text{ m}^3/\text{atom}$$

There are two atoms in the unit cube, so the volume of the unit cube is 7.87×10^{-29} m³. Since the volume of the unit cube is $V = a^3$, the lattice parameter is

$$a = \sqrt[3]{V} = \sqrt[3]{7.87 \times 10^{-29} \text{ m}^3} = 4.28 \times 10^{-10} \text{ m}$$

MODERN PHYSICS

Table 32-2 lists some ranges of typical values for crystalline materials.

Table 32-2 Typical values for crystalline materials

Parameter	Typical value
Atomic spacing	On the order of 10^{-9} to 10^{-10} m
Mass density	On the order of 1 to 10 g/cm^3
Number of atoms/cm^3	On the order of 10^{21} to 10^{24}

When a solid is formed, the atoms that compose the solid make chemical bonds with their neighbours. Some of the electrons associated with each atom participate in these chemical bonds and are called **valence electrons**. Valence electrons remain in the vicinity of the atoms with which they are associated. However, some of the outer electrons may *not* participate in the bonding process and move throughout the crystal. In quantum mechanics, electrons are described as waves.

LO 2

32-2 Electrons in a Box

A simple model for the distributed electrons in a material is to treat them as being confined within the material but free to move anywhere inside it. Since all the electrons are negatively charged, they repel each other. For this reason, the free electrons are relatively uniformly distributed throughout the material, particularly in metals. When an electron is liberated from an atom, the atom has a corresponding net positive charge. So, one might expect that the free electrons would be strongly attracted to the positively charged atoms. However, the distributed negative charge of the free electrons offsets this attraction. At a distance of only a few times the lattice parameter, a unit cell of the crystal appears to have virtually no net charge, and the net force on an electron from almost all the other electrons and atoms is very small. This effect is called **screening**.

For simplicity, we will first consider free electrons in one dimension. Assume that the electrons are confined in a length $L = N^*a$, where N^* is the number of atoms in the length and a is the interatomic separation. To determine the wave functions of the free electrons, we need to solve the Schrödinger equation:

KEY EQUATION
$$-\frac{\hbar^2}{2m}\frac{\partial^2 \psi(x)}{\partial x^2} + V(x)\psi(x) = E\psi(x) \quad (31\text{-}28)$$

where $\psi(x)$ is the electronic wave function, m is the mass of the electron, and $V(x)$ is the potential energy.

In our one-dimensional model, $V(x) = 0$ for $0 \le x \le L$ and $V(x) = \infty$ elsewhere. This potential function leaves the electrons free to move within the crystal but confines them to it. So, we have the boundary conditions $\psi(0) = 0$ and $\psi(L) = 0$. We try a solution of the form

$$\psi(x) = A\sin(kx) \quad (32\text{-}1)$$

where A is a normalization constant and k is the wave number.

Since the total probability of finding the electron somewhere along the length L is 1,

$$\int_0^L \psi^2(x)\,dx = 1 \quad \text{and} \quad A = 1/\sqrt{L}$$

Inserting the sinusoidal wave function into Equation (31-28) gives

$$\frac{\hbar^2 k^2}{2m}A\sin(kx) = EA\sin(kx) \quad (32\text{-}2)$$

Therefore,

$$E = \frac{\hbar^2 k^2}{2m} \quad (32\text{-}3)$$

Since the boundary conditions require that $\psi(L) = A\sin(kL) = 0$, $kL = \pi n$

$$\therefore k_n = \frac{\pi}{L}n = \frac{\pi}{N^*a}n \quad (32\text{-}4)$$

Note that we have added a subscript to k corresponding to the integer n. We can write the energy eigenvalues as

KEY EQUATION
$$E_n = \frac{\hbar^2 k_n^2}{2m} \quad (32\text{-}5)$$

So, the energies of the free electrons are quantized. However, the change in k and E_n from one value of n to the next is very small, because for a macroscopic sample, N^* is of the order of at least 10^{23} and a is of the order of 10^{-10} m:

$$\Delta k = k_{n+1} - k_n = \frac{\pi}{N^*a} \sim 10^{-13}/\text{m} \quad (32\text{-}6)$$

$$E_{n+1} - E_n = \frac{\hbar^2 k_{n+1}^2}{2m} - \frac{\hbar^2 k_n^2}{2m}$$

$$= \frac{\hbar^2 n^2}{2mN^2a^2}(2n+1) \sim (2n+1) \times 10^{-81} \quad (32\text{-}7)$$

Thus, the energy eigenvalues are extremely closely spaced, and to a good approximation they may be treated as continuously distributed. A sketch of E_n versus k_n is shown in Figure 32-4. Figure 32-5 shows the wave functions for the lowest three energy levels for this system. Notice the similarity between these wave functions and the standing waves that you studied in Chapter 15.

MODERN PHYSICS

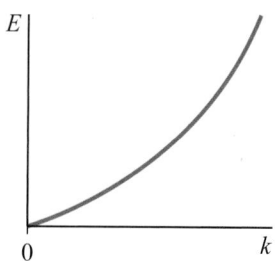

Figure 32-4 Energy versus wave number for free electrons in one dimension

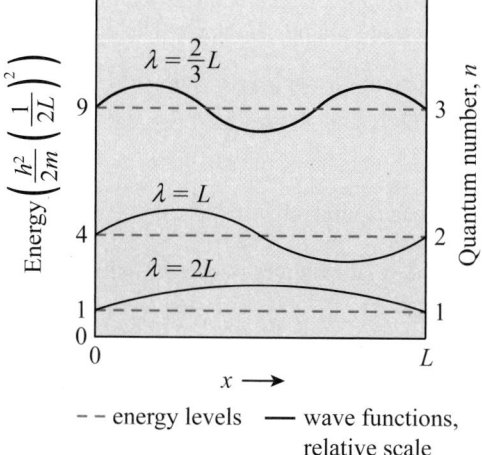

– – energy levels —— wave functions, relative scale

Figure 32-5 Wave functions for the three lowest energy states for free electrons in a box of length L

For the moment, we will only consider the situation that corresponds to a temperature of 0 K. At absolute zero, the electrons occupy the energy states that give the lowest total energy possible. Since electrons are fermions and therefore obey the Pauli exclusion principle, only one electron can occupy each quantum state. Thus, two electrons can be in each energy level, one with spin-up and the other with spin-down. From Equation (32-7) we can see that when $n = N^*$, $k = \dfrac{\pi}{a}$. Thus, there are N^* energy states between $k = 0$ and $k = \dfrac{\pi}{a}$. If each atom contributes a single free electron, then all the states up to $n = \dfrac{N^*}{2}$ will be occupied. In general, if there are N electrons, then the highest occupied state will have a quantum number $n_f = \dfrac{N}{2}$.

We call the highest occupied energy level in a solid the **Fermi level**. Associated with this energy level are the Fermi wave number, k_f, and the **Fermi energy**, E_f:

KEY EQUATION $k_f = \dfrac{\pi N}{aN^*}$ $E_f = \dfrac{\hbar^2 k_f^2}{2m} = \dfrac{\hbar^2 \pi^2}{8ma^2}\left(\dfrac{N}{N^*}\right)^2$ (32-8)

Note that the ratio $\dfrac{N}{N^*}$ is the number of electrons that each atom contributes to conduction. For a typical solid

with $\dfrac{N}{N^*} = 1$ and $a \sim 10^{-10}$ m, $E_f \sim 10^{-18}$ J ~ 10 eV. Since the energy associated with thermal excitation of an electron at room temperature is approximately 0.025 eV, the Fermi energy is relatively high. The speed of electrons at the Fermi level is called the **Fermi speed**.

 CHECKPOINT

C-32-1 Fermi Energy

Why is the Fermi energy large?
(a) The electrons have high energies in the atoms.
(b) Only two electrons can go into any state, so many states are needed.
(c) There are a lot of electrons.
(d) The temperature is large.

C-32-1 (b)

EXAMPLE 32-2

The Fermi Speed

How fast would an electron at the Fermi level be moving in a typical solid?

SOLUTION

If we assume the electron only has kinetic energy and can be treated non-relativistically, then

$$v_f = \sqrt{\dfrac{2E_f}{m}} = \sqrt{\dfrac{2 \times 10^{-18}\,\text{J}}{9.1 \times 10^{-31}\,\text{kg}}} = 1.5 \times 10^6\,\text{m/s}$$

Making sense of the results:

These free electrons are moving at approximately 0.5% of the speed of light.

If we extend our model to a three-dimensional crystal of length L on each side, the wave vectors also become three dimensional with the same quantization along each axis. So, the wave number becomes a wave vector:

$$\vec{k} = \dfrac{\pi}{N_x a}n_x\hat{i} + \dfrac{\pi}{N_y a}n_y\hat{j} + \dfrac{\pi}{N_z a}n_z\hat{k} \qquad (32\text{-}9)$$

where n_x, n_y, and n_z are all positive integers.

At 0 K, the filled levels occupy the first octant of a sphere in "k-space." All states below the Fermi energy are occupied with a probability of one, and all states above the Fermi energy are occupied with a probability of zero—they are not occupied.

As the temperature increases from 0 K, we would expect the electrons to become thermally excited and their energies to increase. However, the electrons are fermions, and they only go into a higher energy state if it is unoccupied. As you saw in Chapter 16, the thermal energy is given by $k_B T$, where k_B is Boltzmann's constant and T is the temperature in kelvins. Only those electrons that are within $k_B T$ of E_f will be able to make a transition to a higher energy level. Because this thermal energy is so small compared to the Fermi energy, only a very small fraction of the electrons are able to make the transition. Thus, the probability of a state being occupied just below the Fermi energy is slightly less than one because some electrons will have been excited out of those states, and the probability of occupation for states slightly above E_f will be slightly greater than zero because some electrons will have been excited into these states.

Figure 32-6 shows how temperature affects $f(E)$, the probability that a state with energy E is occupied. We can see that only those electrons near the Fermi energy are affected by the increase in temperature. For example, $f(E)$ drops to approximately 0.9 near 6.8 eV, just 0.2 eV below E_f for $T = 1000$ K. Even when the temperature increases substantially, only a small fraction of the electrons are able to absorb any of the thermal energy.

Figure 32-6 The probability of a state being occupied by an electron, $f(E)$. The blue curve shows the probability at $T = 100$ K, and the red curve shows the probability at $T = 1000$ K, with a Fermi energy of 7 eV, typical of that found in metals.

✅ CHECKPOINT

C-32-2 Absorption of Thermal Energy in a Crystal

When we put energy into the electrons in a solid by raising the temperature,

(a) all the electrons experience an increase in energy;
(b) only those electrons that have a low energy are affected;
(c) only those electrons near the Fermi level are affected.

C-32-2 (c) Only those electrons near the Fermi level have empty states into which they can be excited.

What will happen if we apply an electric field to the crystal? In the absence of the field, an equal number of electrons travel in any direction, and the net flow of electrons is zero. If we apply an electric field directed from right to left, some of the electrons travel from the left into unoccupied right-travelling states. However, only electrons near the Fermi level can make these transitions because these electrons are the only ones in energy levels close to those of the unoccupied states. Now, we will consider the various factors that determine how many electrons are at the Fermi level.

The total number of electrons, N, in the system is equal to the volume of occupied k-space times the number of states per unit volume in k-space. The occupied volume is the first octant of a sphere of radius k_f, $\frac{1}{8}\frac{4}{3}\pi k_f^3$. From Equation (32-7) and Figure 32-7 we can see that, for a cubic crystal, the sides of the cube in k-space associated with each point are of "length" $\frac{\pi}{N_x^* a}$, $\frac{\pi}{N_y^* a}$, $\frac{\pi}{N_z^* a}$, and thus the total volume associated with each point is $\frac{\pi}{N_x^* a}\frac{\pi}{N_y^* a}\frac{\pi}{N_z^* a} = \frac{\pi^3}{N^* a^3}$, where $N^* = N_x^* N_y^* N_z^*$.

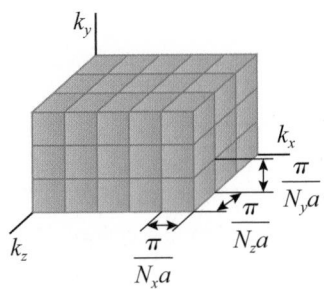

Figure 32-7 Volume associated with points in k-space

For a cubic crystal, $N^* a^3 = V$, where V is the total volume of the crystal. Finally, we must remember that each state can hold two electrons, one spin-up and the other spin-down. Therefore,

$$N = 2\frac{1}{8}\frac{4}{3}\pi k_f^3 \frac{V}{\pi^3} = \frac{1}{3}\frac{k_f^3 V}{\pi^2}$$

Thus,

KEY EQUATION
$$k_f = \left(\frac{3\pi^2 N}{V}\right)^{\frac{1}{3}} \text{ and } E_f = \frac{\hbar^2}{2m}\left(\frac{3\pi^2 N}{V}\right)^{\frac{2}{3}}$$

$$(32\text{-}10)$$

Note that in Equation (32-10) the Fermi energy is completely determined by $\frac{N}{V}$, the number of electrons divided by the volume of the sample. This quantity is the **conduction electron density**.

EXAMPLE 32-3

The Fermi Energy of Aluminum

Aluminum has a density of 2.69 g/cm³ and a molar mass of 26.98 g/mol. Each aluminum atom contributes three conduction electrons. Find the Fermi energy.

SOLUTION

We can determine the conduction electron density, $\frac{N}{V}$, and then use Equation (32-10) to find the Fermi energy. The conduction electron density equals the number of conduction electrons per atom times the number of atoms per m³:

$$\frac{N}{V} = \frac{3 \text{ electrons}}{\text{atom}} \times \frac{2.69 \frac{\text{g}}{\text{cm}^3}}{26.98 \frac{\text{g}}{\text{mol}}} \times 6.022 \times \frac{10^{23} \text{ atoms}}{\text{mol}}$$

$$= 1.80 \times \frac{10^{23} \text{ electrons}}{\text{cm}^3}$$

$$= 1.80 \times 10^{29} \text{ electrons/m}^3$$

Using this value in Equation (32-10) gives

$$E_f = \frac{\hbar^2}{2m} \left(\frac{3\pi^2 N}{V} \right)^{\frac{2}{3}} = 1.85 \times 10^{-18} \text{ J} = 11.5 \text{ eV}$$

We can rearrange Equation (32-10) to show the relationship between the conduction electron density, n, and the Fermi energy more explicitly:

$$N = \frac{1}{3} \frac{V}{\pi^2} \left(\frac{\sqrt{2mE_f}}{\hbar} \right)^3 \tag{32-11}$$

or

KEY EQUATION
$$n = \frac{N}{V} = \frac{1}{3\pi^2} \left(\frac{\sqrt{2mE_f}}{\hbar} \right)^3$$

We could instead ask how many electrons per unit volume have energies *less* than E. We denote this quantity $n(E)$, and it is given by

$$n(E) = \frac{1}{3\pi^2} \left(\frac{\sqrt{2mE}}{\hbar} \right)^3 \tag{32-12}$$

We define the **density of states**, $D(E)$, as the number of states per unit energy per unit volume and get

$$n = \int_0^{E_f} D(E) dE = \int_0^{E_f} \frac{dn(E)}{dE} dE \tag{32-13}$$

and

KEY EQUATION $\quad D(E) = \frac{dn(E)}{dE} = \frac{1}{2\pi^2} \left(\frac{2m}{\hbar^2} \right)^{\frac{3}{2}} E^{\frac{1}{2}} \tag{32-14}$

Figure 32-8 shows a plot of the density of states versus energy.

Figure 32-8　The density of states versus energy

To determine the number of occupied states in an energy interval, we multiply the number of states in that interval, $D(E)$, by the probability that the state is occupied, $f(E)$. It is beyond the scope of this book to derive the probability distribution so we simply give it as

$$f(E) = \frac{1}{e^{\frac{E - E_f}{k_s T}} + 1} \tag{32-15a}$$

With this result and the density of states expression from Equation 32-14 we can write the number of electrons with energy E per unit energy interval, $N(E)$, as

$$N(E) = D(E) f(E)$$

$$= \frac{1}{2\pi^2} \left(\frac{2m}{\hbar^2} \right)^{\frac{3}{2}} E^{\frac{1}{2}} \frac{1}{e^{\frac{E - E_f}{k_s T}} + 1} \tag{32-15b}$$

We can now determine how many electrons participate in conduction. From Figure 32-9 we can see that only

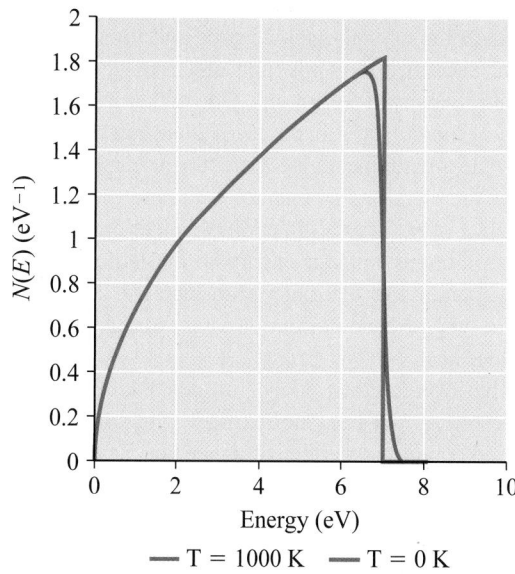

Figure 32-9　The number of electrons per energy interval with a Fermi energy of 7 eV. The blue line corresponds to T = 0 K and the red corresponds to T = 300 K.

those electrons within a small range of the Fermi energy have empty energy states near enough for them to move into in response to an electric field. The width of this range is $\Delta E \sim k_B T$. At room temperature (approximately 300 K), ΔE is only about 0.02 eV, and

$$\text{Number of conduction electrons} \sim D(E)\,f(E)\,\Delta E \sim 1$$
$$\times\ 10^{28}\ \text{eV/m}^3\ 0.02\ \text{eV} \sim 10^{26}\ \text{m}^{-3} \qquad (32\text{-}16)$$

We used a Fermi energy of 7 eV and the approximation that $D(E)\,f(E) \approx D(E_f)\,f(E_f)$ over the narrow range of ΔE. Since the total density of free electrons is of the order $10^{28}\ \text{m}^{-3}$, we can see that only about 1% of those electrons actually participate in conduction.

In our model, there is nothing to stop the motion of conduction electrons within the crystal. Thus, our model represents an ideal **conductor** with no electrical resistance. In actual crystals, the conduction electrons interact with atoms in the lattice in two ways that impede the flow of current: First, the conduction electrons scatter off defects in the crystal, such as gaps and impurities in the lattice. Second, the conduction electrons scatter off lattice vibrations called **phonons**. Due to thermal energy, the atoms in the crystal vibrate about their equilibrium positions. The greater the temperature, the more phonons there are for the electrons to interact with, and the greater the resistance of the crystal. In addition, electrons that scatter off the phonons transfer energy to the lattice and cause the temperature of the material to increase. These effects explain why electric current passing through a resistance produces heat.

The free-electron model allows us to understand simple metals but does not explain why some materials are good conductors while others are not. To explain these properties, our model has to include the electron–lattice interactions.

The key feature of the free electrons in a box model is that because electrons are fermions, only those electrons with energies near Fermi energy are able to participate in conduction and this is a small fraction of the total number of electrons in the material.

32-3 Periodic Potential

In a crystal, the electrons experience a potential due to interactions with the lattice. Because the lattice is composed of repeating units, the potential has a pattern with the same spacing, or periodicity, as the crystal; this potential is called the **periodic potential**.

ONLINE ACTIVITY

Periodic Potentials

The e-resource that accompanies every new copy of this textbook contains an Online Activity using the PhET simulation "Periodic Potentials." Work through the simulation and accompanying questions to gain an understanding of electrons in a periodic potential.

The wave functions for electrons in a periodic potential are not the same as the free-electron solutions we found in the previous section. We will assume that the potential is relatively weak. Although the details of calculating wave functions for a periodic potential are beyond the scope of this book, we can see that the wave functions must have the same periodicity as the potential. The periodicity also appears in the wave vector, $\vec{k}$. A key effect occurs when $|\vec{k}|$ equals integer multiples of $\dfrac{\pi}{a}$, as shown in Figure 32-10.

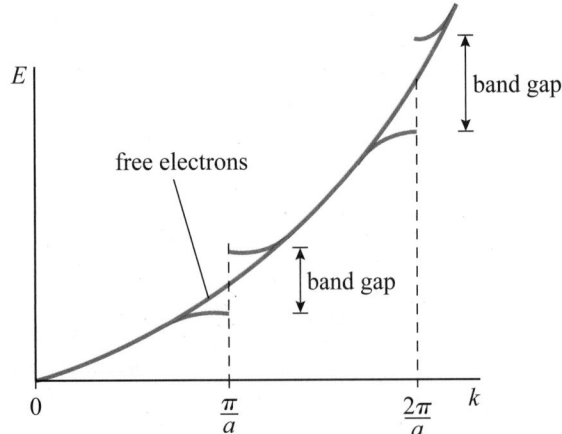

Figure 32-10 E versus k for electrons in a one-dimensional potential of period a

Well away from $|\vec{k}| = \dfrac{\pi}{a}$, the wave functions are essentially the same as the wave functions for free electrons. However, as $|\vec{k}|$ approaches $\dfrac{\pi}{a}$, the energy curve bends away from the parabolic shape we found for free electrons (Figure 32-4) and actually becomes horizontal at $\dfrac{\pi}{a}$, leaving a gap in the energy spectrum. Similar gaps occur when $|\vec{k}| = \dfrac{2\pi}{a}, \dfrac{3\pi}{a}, \dfrac{4\pi}{a}$, and so on. The regions of allowed states are called **energy bands**, and the regions where no solutions are possible are called **band gaps**. The energy bands together with the energy gaps provide a description of the electronic structure of a material and are called the **band structure**. Some typical band gaps for various materials are listed in Table 32-3. Silicon and germanium are **semiconductors**, which are materials whose conductivities lie between insulators and conductors. Silicon and germanium have small band gaps, and diamond, an excellent **insulator**, has a large band gap. In general, as the band gap increases, so does the material's resistivity.

Table 32-3 Band gaps for various materials

Material	Band gap (eV at 300 K)	Resistivity at 20°C ($\Omega \cdot m$)
Silicon	1.11	6.40×10^2
Diamond	5.5	10^{14}
Germanium	0.67	0.46

 CHECKPOINT

C-32-4 The Role of the Periodic Potential

A periodic potential in a crystal
(a) creates an energy gap at the Fermi level;
(b) creates energy gaps at $|\vec{k}| = \dfrac{n\pi}{a}$;
(c) leaves the electronic structure essentially unchanged.

C-32-4 (b) The periodic potential creates periodic gaps in the energy levels, with a periodicity in the wave vector of $|\vec{k}| = \dfrac{n\pi}{a}$.

LO 4

32-4 Metals and Insulators

The band structure of a material indicates the ranges of states for electrons in the material. How these states are filled determines whether a material is a metal or an insulator. We can represent the band structure graphically as shown in Figure 32-11.

Figure 32-11 Schematic representation of the band structure of a material

We will first consider a material with a partially filled band, as illustrated in Figure 32-12. The electrons in the completely filled lower bands cannot participate in conduction because there are no empty states with energies close to the energies in the full bands. The electrons at the top of the filled portion of the partially filled band do have empty states adjacent to them. Therefore, if we apply an electric field, these electrons can move into higher energy states and contribute to an electric current. These excited electrons still scatter off the lattice defects and vibrations. Therefore, the material has some resistance, but it is a good conductor. An electronic structure with a partially filled band is characteristic of a metal.

Figure 32-12 A material with a partially filled band

 CHECKPOINT

C-32-5 Energy Levels in Metals

A metal is able to conduct electricity readily because the electrons in filled states at the Fermi level can be excited into empty states that are
(a) immediately adjacent to the Fermi level;
(b) across a very narrow energy gap.

C-32-5 (a) The low resistivity of a metal is due to the fact that there are empty energy levels infinitesimally close to the filled levels at the Fermi level.

We now consider a material in which the uppermost occupied band is completely filled (Figure 32-13). The empty states are not close in energy to any of the occupied states. If we place such a material in an electric field, the electrons in the occupied states cannot move to higher-energy states unless the electric field is strong enough to give an electron enough energy to cross the gap into the

MODERN PHYSICS

next highest band A material that has only completely filled bands and a fairly large energy gap between the top of the highest filled band and the bottom of the nearest empty band is an insulator. However, if the field strength is sufficiently great, even a very good insulator will conduct, a process called **electrical breakdown**. The maximum field that a material can withstand without breakdown conduction is called the **dielectric strength** of the material. Table 32-4 lists values for some common materials.

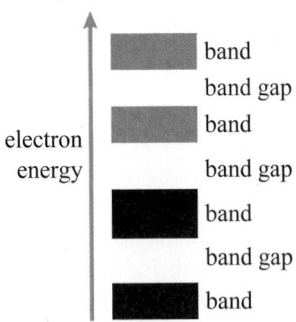

Figure 32-13 In this material, the uppermost occupied band is completely filled with electrons.

Table 32-4 Dielectric strengths for various insulating materials

Material	Dielectric strength (MV/m)
Vacuum	0.8
Air	0.8–3.0
Porcelain	1.6–7.9
Paraffin wax	7.9–11.8
Transformer oil	15.7
Bakelite	11.8–21.7
Rubber	17.7–27.6
Shellac	35.4
Paper	49.2
Teflon	59.1
Glass	79–118
Mica	197

C-32-6 Band Gap in Insulators

The Fermi energy of an insulator lies at the top of a band that is immediately below

(a) a large energy gap;

(b) a small energy gap;

(c) an energy gap wider than the Fermi energy;

(d) an energy gap of any width.

C-32-6 (a) In an insulator, the size of the energy gap is large compared to thermal energy, k_BT. There is insufficient thermal energy to excite electrons across the gap, so there are no electrons available for conduction.

LO 5

32-5 Semiconductors

We now consider what happens if the energy gap between the highest filled band (called the **valence band**) and the first empty band (called the **conduction band**) is not too large, say, less than an order of magnitude greater than the thermal energy. The material has a large number of electrons at the top of the valence band, but they are separated in energy from the empty states in the conduction band by an energy gap, ΔE. Since the electrons can absorb thermal energy from the lattice vibrations, there is a probability that some of the electrons will move to the higher-energy upper band, even though ΔE is greater than the thermal energy, k_BT. Suppose that at zero temperature we have N_0 electrons at the top of the filled band. Some of those electrons can be thermally excited into the bottom of the empty band. The number of thermally excited electrons, N_x, can be approximated by an exponential relationship that depends on the ratio of energy gap to the thermal energy:

$$N_x \approx N_0 e^{-\frac{\Delta E}{k_B T}} \qquad (32\text{-}17)$$

 ## EXAMPLE 32-4

The Effect of Energy Gap on Conduction

Plot $\dfrac{N_x}{N_0}$ versus ΔE for a temperature of 300 K. What is the value of the ratio for $\Delta E = 0.25$ eV and $\Delta E = 1.0$ eV?

SOLUTION

We can rearrange Equation (32-17) to determine $\dfrac{N_x}{N_0}$:

$$\frac{N_x}{N_0} \approx e^{-\frac{\Delta E}{k_B T}}$$

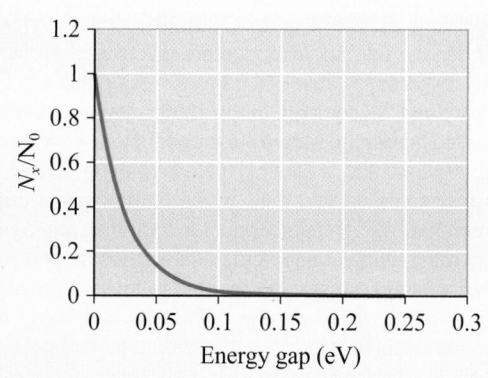

Figure 32-14 A fraction of excited carriers versus band gap. Notice that the ratio drops off very quickly as ΔE increases.

At 300 K, $k_B T = 0.0258$ eV. Figure 32-14 shows a plot of $\frac{N_x}{N_0}$ with ΔE expressed in eV.

For $\Delta E = 0.25$ eV, $\frac{N_x}{N_0} \approx e^{-\frac{0.25 \text{ eV}}{0.0258 \text{ eV}}} \approx 6 \times 10^{-5}$. Once we get out to a band gap of 1 eV, the ratio drops to approximately 10^{-17}.

Making sense of the result:

The ratio drops off very quickly as ΔE increases due to the exponential dependence.

EXAMPLE 32-5

Temperature Dependence of Conduction

For a band gap of 0.25 eV, plot $\frac{N_x}{N_0}$ versus T.

SOLUTION

We use the equation from Example 32-4 to calculate values for the plot (Figure 32-15).

Making sense of the result:

The ratio is essentially zero from 0 K to approximately 200 K. The ratio then begins to increase as a very steep function of T.

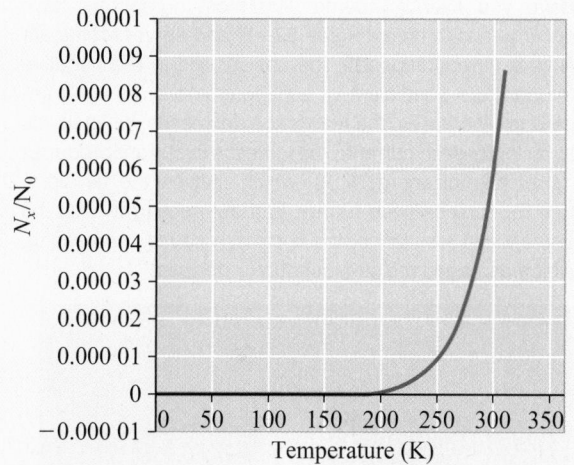

Figure 32-15 Fraction of carriers thermally excited versus temperature

The two previous examples show that for relatively "small" band gaps, some of the electrons from the top fully filled band can acquire enough thermal energy to cross the band gap up into the empty band above. In addition, the number of electrons that are excited across the gap is a very strong function of the temperature, increasing sharply as the temperature increases. Electrons that are excited across the gap are in energy states that are adjacent to empty states. In the presence of an electric field, the electrons can move into the empty states and contribute to conduction in the previously empty band. This type of conduction occurs in semiconductors.

The electrons that are excited across the energy gap leave some empty states, or **holes**, in the previously filled band. The holes are adjacent to filled electronic states, and the electrons from the filled states can move into the

✓ **CHECKPOINT**

C-32-7 Semiconductor Energy Gaps

A semiconductor has a smaller energy gap than an insulator. True or false?

C-32-7 True. In a semiconductor, the energy gap is small enough that some electrons can be excited across the gap and become available for conduction. An insulator has a larger energy gap that prevents conduction.

empty states. So, there is conduction in the previously filled band as well in the previously empty band. The net effect is that the holes appear to move. Because the holes represent the *absence* of an electron, **hole motion** appears like the motion of a positively charged particle, that is, a hole can be treated as a **charge carrier** like a conduction electron.

CHAPTER 32 | INTRODUCTION TO SOLID-STATE PHYSICS

In semiconductors, only a small fraction of electrons in the highest energy states are available to participate in conduction. Consequently, the resistivities of semiconductors are higher than the resistivities of metals. The conduction electrons in semiconductors scatter off the lattice defects and vibrations, like the conduction electrons in metals. Thus, one might expect that the resistance of a **semiconductor** increases with temperature, similar to the increase for metals. However, this increase is overwhelmed by the sharp increase in the number of conduction electrons that are available as the temperature increases (see Figure 32-15). The net result is that the resistance of semiconductors actually decreases as the temperature increases.

MAKING CONNECTIONS

Semiconductor Thermal Runaway

Have you ever wondered why most computers have cooling fans in them (Figure 32-16) and server rooms are always air-conditioned? The answer lies in the negative temperature coefficient of semiconductors. As the temperature increases, the resistance decreases, which leads to an increase in current. The increase in current produces more thermal energy (I^2R), which causes the temperature to increase even further. Without cooling to limit the decrease in resistance, the cycle can cause overheating, which damages the semiconductor devices.

Figure 32-16 This cooling fan is mounted directly on the finned aluminum heat sink that is attached to the central processing unit (CPU) of a computer. The heat sink increases the surface area and, hence, the heat dissipation.

MAKING CONNECTIONS

Superconductors

The most dramatic quantum effect involving resistivity is *superconductivity*, discovered in 1911 by the Dutch physicist Heike Kamerlingh Onnes (1853–1926). Some materials lose *all* of their electrical resistance when cooled below a critical temperature (Figure 32-17). The critical temperatures for the first superconductors discovered are within a few degrees of absolute zero, for example, 4.2 K for mercury and 7 K for lead. In the 1980s, several classes of so-called high-temperature superconductors were discovered. Around 1993, a ceramic material was developed that superconducts up to 138 K, which is the highest critical temperature for a superconductor found to date. High-temperature superconductors generated so much excitement that a special session devoted to them at the March 1987 meeting of the American Physical Society in New York has become known as the "Woodstock of physics." Canadian scientists have made significant contributions to the study of these materials through a national research network of the Canadian Institute for Advanced Research.

Figure 32-17 A magnetic cube levitated by a superconducting disk

LO 6

32-6 Doped Semiconductors

We have seen that the **carrier concentration**, the number of conduction electrons (or holes) per unit volume, has a profound influence on the electrical characteristics of semiconductors. This effect led researchers to look for ways to alter carrier concentrations other than by changing the temperature. The answer lay in altering the chemical composition of semiconductors by adding small amounts of other elements.

Silicon and germanium are the semiconductors used in most solid-state electronic devices. These two elements are both in group IV of the periodic table. All elements in group IV have four valence electrons. Adding tiny amounts of elements from either group III or group V dramatically changes the carrier concentration of a pure group IV semiconductor. The added materials are called **dopants**, and the process of adding them to a semiconductor is called **doping**.

We will first consider the effect of adding group V dopants, which have five valence electrons. For silicon, the dopants are usually phosphorous or arsenic. We will assume

MODERN PHYSICS

that the dopant has not changed the band structure, which is a reasonable approximation if the dopant concentration is low. Four of the valence electrons from each dopant atom contribute to filling the completely filled band, like the valance electrons from an atom of the semiconductor element. However, the fifth valance electron goes into the unfilled band above the filled band. Thus, a group V dopant increases the number of conduction electrons. For example, at room temperature, 1 cm³ of pure germanium contains approximately 2.5×10^{13} electrons that have been thermally excited across the band gap. These conduction electrons have left behind the same number of unoccupied states (holes) below the gap. The addition of 0.001% of arsenic contributes an extra 10^{17} free electrons per cm³, and the resistivity of the semiconductor decreases by a factor of approximately 10 000. Group V dopants are called **donors** because they "donate" electrons to the material. The resulting material is called an **n-type semiconductor** because it has extra carriers for negative charge.

Now let us consider the effect of dopants from group III of the periodic table. These elements have only three valence electrons. Consequently, each dopant atom creates an unfilled state in the highest filled band. Group III dopants are called **acceptors** because they accept electrons out of the band to create holes. Since these holes behave like positively charged carriers, the resulting material is called a **p-type semiconductor**. Boron is a commonly used acceptor.

Thus, dopants can be used to determine the type and concentration of charge carriers in a semiconductor and, hence, the resistivity of the material.

LO 7

32-7 The pn Junction Diode

A **pn junction diode** consists of a piece of p-type semiconductor attached to an n-type semiconductor. Junction diodes are manufactured by diffusing dopants into a single piece of pure semiconductor to create a p-type region and an n-type region. The circuit symbol and the structure for a junction diode are shown in Figure 32-18. The triangular arrowhead in the diode symbol indicates the direction in which conventional current can flow. The p-type material is the anode, and the n-type material is the cathode.

Suppose that the holes and conduction electrons in the diode are initially immobilized. Associated with each conduction electron in the n-type region is a donor atom

Figure 32-18 The circuit symbol for a junction diode (left) and the configuration of a diode semiconductor material (right)

that has a net positive charge. The net charge on the region is still zero because the electron has an equal but opposite charge. Similarly, in the p-type region, there is a negatively charged acceptor atom associated with each hole. As shown in Figure 32-19, the region near the junction has positively charged donors in the n-type material and negatively charged acceptors in the p-type material.

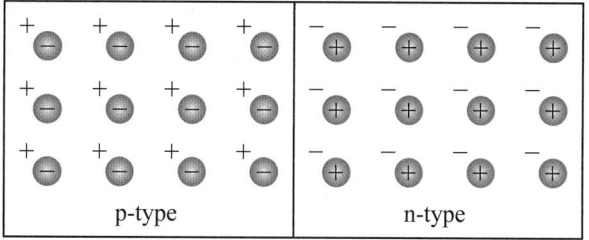

Figure 32-19 A pn junction before electrons and holes are allowed to diffuse across the junction. The circles with positive signs denote acceptor atoms in the crystal lattice, and the circles with negative signs denote donor atoms.

Now consider what happens when the electrons and holes begin to diffuse. Well away from the junction, nothing particularly interesting happens. Near the junction, an electron from the n-type material can move to a lower-energy state by crossing the junction to fill a hole. Both the electron and the hole "disappear," leaving a positively charged donor in the n-type region and a negatively charged acceptor in the p-type region. These charges are fixed in position in the lattice and produce an electric field directed from the n-type material to the p-type material. The field tends to oppose the further diffusion of electrons and holes across the junction. When the field becomes strong enough to stop any further diffusion, there is a **depletion region** with no mobile charge carriers around the junction (Figure 32-20).

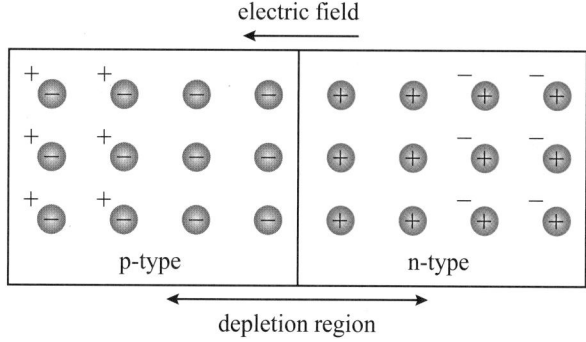

Figure 32-20 Electron–hole recombination at a pn junction

We will now consider what happens if we apply an external potential difference to the diode. Connecting a positive voltage to the cathode and a negative voltage to the anode establishes an electric field with the same direction as the electric field in the depletion region. Consequently, almost no current flows in the device because the effect of

MODERN PHYSICS

the applied voltage widens the depletion region by moving charge carriers farther from the junction. This condition is called **reverse bias** of the diode. There is a small **leakage current** during reverse bias, typically less than a mA for a few volts of bias. However, if the reverse-bias voltage is high enough, electrons will cross the band gap and a large **avalanche current** will flow.

If we apply a positive voltage to the anode and a negative voltage to the cathode, the depletion region narrows. Once we apply sufficient voltage, the depletion region disappears and the diode readily conducts current. This condition is called **forward bias**. For a silicon diode, significant current flows once the forward bias exceeds 0.7 V; for germanium, this voltage is approximately 0.2 V. A typical I–V characteristic for a junction diode is shown in Figure 32-21.

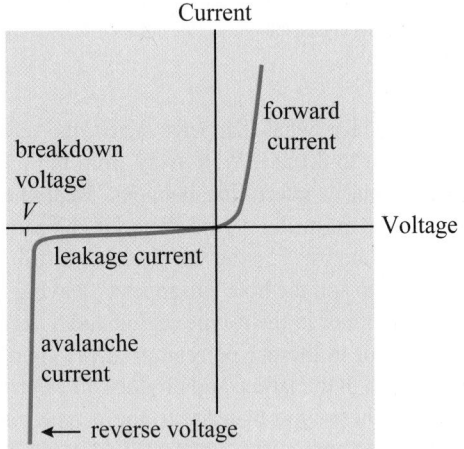

Figure 32-21 A typical I–V characteristic for a pn junction diode

Diodes are commonly used as **rectifiers**, which are circuits that convert alternating current (AC) into direct current (DC). Rectifiers are found on most electronic devices that run from household outlets and in the adapters for charging portable devices.

Figure 32-22 shows a rectifier circuit. When the upper terminal of the sinusoidal AC voltage source is positive,

the diode is forward-biased. Consequently, current flows freely in the circuit, and a voltage appears across the resistor. When the upper terminal of the source is negative, the diode is reverse-biased, no current flows, and no voltage appears across the resistor. Thus, only positive voltages appear across the resistor. The capacitor in parallel with the resistor smooths out much of the fluctuations in the pulsating voltage.

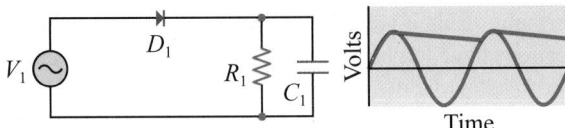

Figure 32-22 A half-wave rectifier with a filter capacitor. The blue curve is the input voltage, and the red curve is the voltage that appears across the resistor.

MAKING CONNECTIONS

Energy Consumption

Rectifier circuits, like the ones found in your mobile device chargers, usually have a voltage regulator attached to the output to ensure that the output voltage is constant. However, these regulator circuits use some power regardless of whether a device is drawing power from the rectifiers. Unplug them when you are not charging. Some regulators have been designed to switch themselves off under no-load conditions to reduce the energy wasted.

LO 8

32-8 Other Semiconductor Devices

A wide variety of devices can be made using various configurations of p-type and n-type semiconductors. For example, the bipolar junction **transistor** (BJT) and the field effect transistor (FET) are three-terminal devices where the current through two of the terminals is controlled by either a small current (BJT) or a small voltage FET to the third terminal. Junction transistors are commonly used in amplifiers, and FETs are the basis of computer logic circuits, processors, and memory. The mass production of transistors began in the 1950s, and semiconductor devices have now almost completely replaced the bulky (and somewhat fragile) vacuum tubes as the active elements in electronic circuits (Figure 32-23).

Significant advances have been made in the fabrication of semiconductor devices. Many low-power applications can be accomplished using **integrated circuits**

Figure 32-23 Vacuum tubes and individual transistors. The transistors are approximately 1 cm high.

rather than individual discrete devices. Integrated circuits, like those in Figure 32-24, are manufactured on a single piece of semiconductor. The process starts with the fabrication of a long, cylindrical crystal of ultrapure silicon. The cylinder is then sliced into circular wafers slightly less than 1 mm thick and typically 10 to 30 cm in diameter. The wafers are then polished to make them very flat.

Once the wafers are polished, the pure semiconductor is doped using a lithographic process. The wafer is coated with photoresists, which are materials that when exposed to light change their chemical properties: they either become soluble in a solvent if they were previously insoluble (positive photoresist), or they become insoluble if they were previously soluble (negative photoresist). A mask is placed on the wafer, and it is exposed to light. The wafer is then washed with a solvent, removing either the exposed positive photoresist or the unexposed negative photoresist. Both methods leave some portions of the wafer coated in photoresist and other portions bare. The wafers are then exposed to the dopant impurities. The dopants can diffuse into the bare areas but are prevented from diffusing into the coated areas by the photoresist. Several cycles of this process are required to produce the desired configuration of n-type and p-type regions in the wafer.

A similar process is used to deposit metal traces that connect various regions of the wafer. Many complete circuits are produced on a single wafer, which is then cut into separate integrated circuits. Each circuit is sealed in

Figure 32-24 An integrated circuit. The eight-pin chip at the top is a 741 operational amplifier that contains approximately 20 transistors. Most of the space is actually taken up with the packaging.

an insulating protective package with connections from the circuit to external metal pins or soldering terminals on the base of the package, allowing for connection on a circuit board.

The development of these techniques for manufacturing integrated circuits has led to the incredible miniaturization of modern electronics. For example, the CPU of a modern personal computer contains approximately 2 billion transistors built on a 21.5×32.5 mm chip. As this miniaturization increases, the transistors become smaller and are placed closer to each other. The closer spacing leads to increased speeds because it takes less time for the signals to travel from one device to the next. However, the transistors rely on the bulk properties of the materials of which they are made. Consequently, there is a limit to how small a transistor can be. Beyond this limit, the extremely small regions of the semiconductor will not display these bulk properties.

EXAMPLE 32-6

Integrated Circuit Transistor

Estimate the area of a typical single transistor in a modern CPU for a personal computer. How many atoms are in such a transistor?

SOLUTION

As described above, the CPU contains 2 billion transistors built on a chip that is 21.5×32.5 mm. We will assume that the transistors cover the entire area of the CPU. Then the area available for each transistor is approximately

(continued)

CHAPTER 32 | **INTRODUCTION TO SOLID-STATE PHYSICS**

MODERN PHYSICS

$$A = \frac{21.5 \text{ mm} \times 32.5 \text{ mm}}{2.0 \times 10^9} = 3.5 \times 10^{-7} \text{mm}^2$$

So, each transistor occupies a square with sides of slightly more than half a micrometre. If we assume that the dopants are diffused to a similar depth in the material, this produces a volume for each device of approximately

$$V = (0.59 \times 10^{-6} \text{m})^3 = 2 \times 10^{-19} \text{m}^3$$

The density of silicon is 2330 kg/m³, and silicon has an atomic weight of 0.028 kg/mol. Therefore, each transistor contains

$$N = \frac{(2.07 \times 10^{-19} \text{m}^3)(2330 \text{ kg/m}^3)}{0.028 \text{ kg/mol}} \sim 2 \times 10^{-14} \text{ mol} \sim 10^{10} \text{ atoms}$$

Making sense of the result:

The CPU does contain other components, such as resistors and diodes. Some area is also used for connections and insulation between the components. Therefore, our estimates of the area and the number of atoms per transistor is likely somewhat high.

MAKING CONNECTIONS

Spin-offs from the Semiconductor Industry

The techniques used to produce integrated circuits are also used to produce solar cells, which use the photovoltaic effect to generate electrical power. To date, the largest photovoltaic installation in Canada is a 100 kW peak output array at the Horse Palace at Exhibition Place in Toronto, Ontario (Figure 32-25). This array has 536 solar modules with a total area of 1428 m². The cost of the project was $1.1 million, and it is expected to pay for itself in 22 years. In addition, it reduces greenhouse gas emissions by 115 tonnes annually.

Figure 32-25 The photovoltaic array at Exhibition Place, Toronto, Ontario

LO 9

32-9 Nanotechnology

In 1959, Richard Feynman gave a talk at a meeting of the American Physical Society called "There's Plenty of Room at the Bottom." Feynman envisioned that it might be possible to manipulate individual atoms and molecules by building a set of precise microscopic tools to build and operate a proportionally smaller set, which would build a still smaller set, and so on down to the smallest possible scale—individual atoms and molecules. If by doing so one could build the analog of transistors, perhaps from single molecules, one could replace the bulk semiconductor technology that we currently use with these new devices.

These ideas spawned the field of study called nanotechnology. **Nanotechnology** generally refers to studies at lengths of less than 100 nm. Research in this field involves different scientists, such as physicists, chemists, engineers, and biologists, to name a few. Investigations include extending existing technology to smaller scales and attempts to control matter on an atomic scale.

Biological systems appear to be excellent prototypes for nanotechnology. Consider the human body: It is able to respond to light, sound, touch, scent, taste, thermal energy, and motion. It is able to use fuel and convert that fuel into energy, motion, and thought. All of this is accomplished by the cells within our bodies, each of which is capable of replicating or building itself as well as all the internal mechanisms that the cells themselves contain to function. Each cell is capable of interacting with other cells so that the whole organism can function. Clearly, the body is built in a bottom-up fashion, starting from a single fertilized ovum. The body does not start from a large, single, undifferentiated mass of material, as we do when we build an integrated circuit.

Specialized tools have been developed to image and manipulate individual atoms. Optical microscopes were the first devices that allowed us a glimpse of details approaching the nanoscale. However, the best resolving

MODERN PHYSICS

MAKING CONNECTIONS

The National Institute for Nanotechnology

The National Research Council of Canada has particular strength in nanotechnology. The National Institute for Nanotechnology (NINT) is an integrated, multidisciplinary institution with researchers in physics, chemistry, engineering, biology, informatics, pharmacy, and medicine (Figure 32-26). Established in 2001, NINT is operated as a partnership between the National Research Council and the University of Alberta, and is jointly funded by the Government of Canada, the Government of Alberta, and the university. NINT describes the focus of its research as "the integration, at the molecular level, of nature's most powerful nano-devices such as proteins, lipids, and other biological structures made from 'soft' organic material, with crystalline semiconductors, metals, and catalysts made from inorganic 'hard' materials." This research could lead to applications in the life sciences, medicine, materials science, electronics, and computation.

Figure 32-26 The offices of the National Institute for Nanotechnology on the north campus of the University of Alberta, Edmonton, Alberta.

EXAMPLE 32-7

Determining the Wavelength of an Electron

Find the wavelength of an electron that has been accelerated through a potential difference of 200 kV.

SOLUTION

Since the electron has been accelerated through a relatively large potential difference, we will use a relativistic approach. The wavelength of a particle moving at a relativistic speed is given by

$$\lambda \approx \frac{h}{\sqrt{2m_0 E\left(1 + \dfrac{E}{2m_0 c^2}\right)}}$$

$$= \frac{6.626 \times 10^{-34}\,\text{J} \cdot \text{s}}{\sqrt{2 \times 9.11 \times 10^{-31}\,\text{kg} \times 3.2 \times 10^{-14}\,\text{J}\left(1 + \dfrac{3.2 \times 10^{-14}\,\text{J}}{2 \times 9.11 \times 10^{-31}\,\text{kg} \times 9 \times 10^{16}\,\text{m}^2\,\text{s}^{-2}}\right)}}$$

$$= 2.5 \times 10^{-12}\,\text{m}$$

Making sense of the result:

This wavelength is substantially smaller than the wavelength of blue light (approximately 400 nm); thus, electron microscopes have a much higher resolving power than optical microscopes.

power of optical microscopes is about 200 nm because their resolution is limited by the wavelength of light. Electron microscopes have greater resolving power because they use a beam of electrons instead of a beam of light. As you saw in earlier chapters, electrons have a shorter wavelength than visible light. In an electron microscope, electrons are accelerated through a potential difference, focused into a small beam, and directed at the sample. We can study either the reflected beam, as is done in a scanning electron microscope (SEM), or the

transmitted beam, as is done in a **transmission electron microscope** (TEM).

The best TEMs have a resolution on the order of 0.05 nm. An example of a high-resolution TEM image is shown in Figure 32-27.

The **scanning probe microscope** (SPM) also has very high resolution. In SPMs, some type of probe is brought very close to the surface of the sample. The probe must interact in some way with the surface, and it must be possible to measure the strength of the interaction. As the

MODERN PHYSICS

Figure 32-27   A high-resolution electron microscopy image of a manganese silicide (MnSi) layer grown on silicon (Si). The uppermost, random-looking layer is a layer of amorphous Si deposited to protect the MnSi layer from the atmosphere.

Figure 32-28   A quantum corral made of 48 iron atoms placed in a circle on a very clean copper surface using an STM.

probe scans across the sample, the strength of the interaction is monitored; plotting the interaction strength versus the probe position gives an "image" of the surface property to which the probe is sensitive. Examples of these types of microscopes are the **scanning tunnelling microscope** (STM), the atomic force microscope (AFM), and the magnetic force microscope (MFM).

The probe in an STM has a very sharp metal tip and a bias voltage is applied to the tip relative to the conducting sample. The tip scans close enough to the sample that electrons can tunnel between the two. Depending on the sign of the bias, electrons either move from full states in the tip to empty states in the sample, or vice versa. The magnitude of the tunnelling current, I_t, depends on both the height, a, and width, Δz, of the potential barrier:

KEY EQUATION

$$I_t \propto e^{-2\kappa \Delta z}$$
$$\kappa = 0.51\sqrt{a} \times 10^{10}/m$$

As a result, the tunnelling current is a very sensitive measure of the height of the probe above the sample. This exponential dependence also localizes the current directly between the point of the metal tip and the sample. As a result, the STM has a resolution in the vertical direction on the order of 0.01 nm and in the horizontal direction on the order of 0.1 nm. Researchers have also discovered that by manipulating the bias voltage, it is possible to pick up and place single atoms on surfaces. Figure 32-28 shows a ring (corral) of iron atoms assembled using this technique. The corral has a diameter of 14.3 nm, and the manipulation and imaging took place at a temperature of 4 K. The circular waves inside the corral are standing waves created by the surface electronic states confined in this circular barrier.

STMs have also been used to study self-assembled monolayers (SAMs). SAMs are typically grown on metal surfaces from molecules that have a long carbon chain. A head group at one end of the chain bonds strongly to the metal surface. For example, alkanethiols have a hydrophilic head group (the thiol), which likes to absorb onto a metal surface. Attached is a hydrophobic carbon chain with a length that can vary. When these molecules are deposited onto a metal surface, typically gold, they form various types of periodic structures on the surface, as shown in the STM image in Figure 32-29. The labels indicate regions of three different molecular structures. Studying this type of self-assembly may lead to the development of techniques for growing nanodevices and structures.

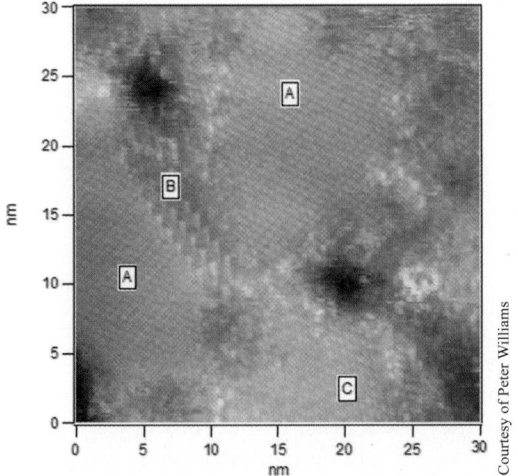

Figure 32-29   An STM image of a 30 × 30 nm square of decanethiol absorbed on a gold surface.

FUNDAMENTAL CONCEPTS AND RELATIONSHIPS

We can use quantum mechanics to explain the resistivity of materials in terms of their electronic structure. Good conductors have partially filled energy bands, insulators have completely filled bands with large band gaps, and semiconductors have filled bands with small band gaps.

Solving the Schrödinger equation for free electrons tells us what quantum states are available.

In one dimension, for a crystal of length L,

$$E_n = \frac{\hbar^2 k_n^2}{2m} \qquad (32\text{-}5)$$

where $k_n = \frac{\pi}{L} n$, $n = 1, 2, 3 \ldots$

In three dimensions, we have the wave vector

$$\vec{k} = \frac{\pi}{N_x a} n_x \hat{i} + \frac{\pi}{N_y a} n_y \hat{j} + \frac{\pi}{N_z a} n_z \hat{k} \qquad (32\text{-}9)$$

where N_x, N_y, and N_z are the numbers of atoms in the x-, y-, and z-directions; n_x, n_y, and n_z are quantum numbers; and a is the lattice spacing for a cubic lattice.

Electrons occupy states up to the Fermi energy:

$$k_f = \frac{\pi N}{aN^*} \qquad E_f = \frac{\hbar^2 k_f^2}{2m} = \frac{\hbar^2 \pi^2}{8m a^2}\left(\frac{N}{N^*}\right)^2 \qquad (32\text{-}8)$$

where N is the total number of electrons, and V is the sample volume.

Only those electrons in states near the Fermi energy can participate in conduction.

The density of states is given by

$$D(E) = \frac{1}{2\pi^2}\left(\frac{2m}{\hbar^2}\right)^{\frac{3}{2}} E^{\frac{1}{2}} \qquad (32\text{-}14)$$

To find the total number of electrons available, we evaluate $D(E)\,f(E)\Delta E$ at E_f, where $f(E)$ is the probability of a state being occupied and $\Delta E \sim k_B T$.

A periodic potential creates bands of energy where there are allowed states, separated by band gaps where there are no allowed states.

Metals have partially filled bands, and insulators have completely filled bands with large energy gaps.

A semiconductor has a completely filled band with a small energy gap.

The number of charge carriers, N_x, thermally excited across the gap, ΔE_{gap}, is approximately

$$N_x = N_0 e^{-\frac{\Delta E_{gap}}{k_B T}} \qquad (32\text{-}17)$$

where N_0 is the number of electrons at the top of the filled band.

We can control the number of charge carriers in a semiconductor by doping to add conduction electrons (n-type) or to remove conduction electrons (p-type).

A pn junction acts as a diode that conducts well under forward bias conditions but does not conduct well under reverse bias.

APPLICATIONS

Applications: pn junction diodes; rectifier circuits to convert AC voltages to DC; integrated circuit technology; nanotechnology tools, such as high-resolution transmission electron microscopes and scanning probe microscopes

Key Terms: acceptors, avalanche current, band structure, carrier concentration, charge carriers, conduction band, conduction electron density, conductor, cubic body-centred structure, cubic face-centred structure, density of states, depletion region, dielectric strength, donors, dopants, doping, electrical breakdown, energy band, energy gap, face-centred cubic structure, Fermi energy, Fermi level, Fermi speed, forward bias, hexagonal structure, hole motion, holes, insulator, integrated circuits, lattice parameter, leakage current, n-type semiconductor, nanotechnology, p-type semiconductor, periodic potential, phonons, pn junction (diode), rectifiers, reverse bias, scanning probe microscope, scanning tunnelling microscope, screening, semiconductor, tetragonal structure, transistor, transmission electron microscope, valence band, valence electrons

QUESTIONS

1. If the temperature of a piece of a metal is increased, does the probability of occupancy 0.1 eV below the Fermi level increase, decrease, or remain the same?

2. All the electrons in a metal are able to participate in the conduction process. True or false?

3. A material with a partially filled band is a(n)
   (a) semiconductor;
   (b) insulator;
   (c) good conductor;
   (d) cannot be determined without knowing the width of the band gaps.

4. Why does a metal have a positive temperature coefficient of resistivity?
  (a) Fewer electrons are available for conduction at higher temperatures.
  (b) There are a greater number of lattice vibrations for the electrons to scatter from at higher temperatures.
  (c) More electrons are excited across the energy gap at higher temperatures.

5. Dopants from group III of the periodic table added to silicon
  (a) donate electrons to make an n-type material;
  (b) donate electrons to make a p-type material;
  (c) accept electrons to make an n-type material;
  (d) accept electrons to make a p-type material.

6. In a p-type semiconductor, the charge carriers are
  (a) electrons;
  (b) holes.

7. The increased resolution of a transmission electron microscope over an optical microscope is due to
  (a) a smaller beam size;
  (b) the shorter wavelength of the electrons;
  (c) the longer wavelength of the electrons;
  (d) none of the above;

8. When a pn junction is reverse-biased, is the electric field in the depletion zone larger or smaller than the electric field found at zero bias?

9. Under forward-bias conditions, the voltage across a pn junction is approximately 0.7 V. Is the diode silicon or germanium?

PROBLEMS BY SECTION

For problems, star ratings will be used, (✶, ✶✶, or ✶✶✶), with more stars meaning more challenging problems.

Section 32-1 Crystal Structures

10. ✶ Associated with a crystal structure is a set of lattice vectors: $\vec{a}$, $\vec{b}$, and $\vec{c}$. For a simple cubic lattice, the lattice vectors are all of the same length a and point in the $\hat{i}$-, $\hat{j}$-, and $\hat{k}$-directions. Thus, $\vec{a} = a\hat{i}$, $\vec{b} = a\hat{j}$, and $\vec{c} = a\hat{k}$. Take one corner of the cube to be at the origin.
  (a) What combination of lattice vectors connects the origin with the point diagonally across the cube?
  (b) What is the length of the vector you found in part (a)?

11. ✶✶ The volume of a unit cell is determined by $(\vec{a} \times \vec{b}) \cdot \vec{c}$.
  (a) What is the volume of a cubic unit cell with side a?
  (b) What is the volume of a rhombohedral unit cell with $\alpha = \beta = \gamma = 60°$ (Figure 32-30)?

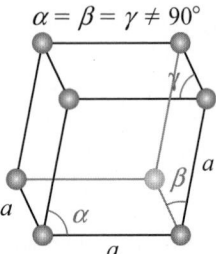

$$\alpha = \beta = \gamma \neq 90°$$

Figure 32-30   Problem 11

12. ✶✶ When planes of atoms in a material satisfy Bragg's law, $n\lambda = 2d \sin\theta$, a diffraction line is observed. In this equation, n is an integer that corresponds to the order of the diffraction line, λ is the wavelength of the X-rays, d is the distance between adjacent planes of atoms, and θ is the angle of incidence.
  (a) For a simple cubic crystal, what is the longest distance between adjacent planes in the crystal?
  (b) When $a = 0.36$ nm and $\lambda = 1$ nm, what angle of incidence satisfies Bragg's law?

Section 32-2 Electrons in a Box

13. ✶ Assume that the total volume of a solid is the sum of the volume occupied by the ions and the volume occupied by the conduction electrons. The density and molar mass of a certain metal are 856 kg/m³ and 16.8 g/mol, respectively. Assume that the radius of an ion is 73.4 pm. What percent of the volume of a sample of this metal is occupied by its conduction electrons?

14. ✶ A certain monovalent metal has a density of 12.75 g/cm³ and a molar mass of 109 g/mol.
  (a) Calculate the number density of conduction electrons.
  (b) Calculate the Fermi energy (in eV).
  (c) What is the Fermi speed?

15. ✶✶ The Fermi energy of a certain metal is 9.69 eV; its density and molar mass are 2.44 g/cm³ and 33.0 g/mol, respectively. Determine the number of conduction electrons per atom.

Section 32-3 Periodic Potential

16. ✶ In quantum mechanics, the momentum of a particle is given by $p = \hbar k$. For a metal with a Fermi energy of 4.0 eV, determine
  (a) the Fermi speed;
  (b) the Fermi wave vector, k_f;
  (c) the Fermi momentum.

17. ✶✶ It is sometimes convenient to assign a particle an "effective mass" that reflects how the particle interacts with its surroundings. In classical physics, for a free particle,

$$E = \frac{p^2}{2m} \quad \text{and} \quad \frac{\partial^2 E}{\partial p^2} = \frac{1}{m}$$

If we extend these relationships to the quantum-mechanical case, we find

$$E = \frac{\hbar^2 k^2}{2m} \quad \text{and} \quad \frac{1}{\hbar^2} \frac{\partial^2 E}{\partial k^2} = \frac{1}{m}$$

Using these ideas, we can define the effective mass, m^*, through

$$\frac{1}{m^*} = \frac{1}{\hbar^2} \frac{\partial^2 E}{\partial k^2}$$

Consider an electron with the E versus k relation in a periodic potential as shown in Figure 32-10 in Section 32-3.
  (a) What is the sign of m^* well away from the band gap?
  (b) What is the sign of m^* just below the band gap?
  (c) What is the sign of m^* just above the band gap?

Section 32-4 Metals and Insulators

18. ✶✶ Insulators can be made to conduct by shining light on them if the energy of the photons is sufficient to excite an electron across the band gap. Recall that for a photon, $E = hf = \frac{hc}{\lambda}$, where h is Planck's constant, f is the frequency, c is the speed of light, and λ is the wavelength. A material has a band gap of 6.53 eV.

(a) What is the band gap in J?

(b) What is the minimum frequency of light that can excite an electron across the gap?

(c) What is the maximum wavelength of the light that will excite an electron across the gap?

19. ✱✱ In a current-carrying conductor, energy is supplied to the conduction electrons by an electric field. When the electrons collide with the atoms in the lattice, they lose some of this energy to the lattice, causing the lattice to vibrate. This process increases the temperature of the lattice. An equilibrium is established when the lattice radiates energy at the same rate that the electrons are transferring energy to the lattice. Consider a conductor that is dissipating energy at the rate of 1.00 W. Assume that the conduction electron density is 1.00×10^{22}/m^3 and that the volume of the conductor is 1.00×10^{-9} m^{-3}.

(a) How many conduction electrons are in the conductor?

(b) At what rate does each individual electron transfer energy to the lattice?

Section 32-5   Semiconductors

20. ✱✱ An insulating material has a band gap of 4.8 eV between the highest filled and lowest empty energy levels. What is the probability that an electron will jump the gap at $T = 300$ K?

21. ✱ A semiconductor has a band gap of 2.0 eV. Using Equation (32-17), determine what fraction of the charge carriers will be excited across the gap at a temperature of 500 K.

22. ✱✱ A simple model for the high temperature resistance of a semiconducting material supposes that the resistance is inversely proportional to the carrier concentration, $R \propto \dfrac{1}{n}$, where $n = \dfrac{N}{V}$ is the total number of carriers, N, divided by the volume of the sample, V. For a semiconductor with a band gap of 0.2 eV, by what factor would the resistance decrease in going from 400 K to 500 K?

Section 32-6   Doped Semiconductors

23. ✱ The concentration of conduction electrons in pure silicon is approximately 10^{16} m^{-3}. What mass of phosphorus is needed to dope 1.0 g of silicon so that the concentration of conduction electrons is increased by a factor of 1.0×10^6? The density of silicon is 2.33 g/cm^3, the molar mass of silicon is 28.1 g/mol, and the molar mass of phosphorus is 30.9758 g/mol.

24. ✱✱ Pure silicon at room temperature has an electron concentration of approximately 5×10^{15} m^{-3} and an equal concentration of holes in the valence band. Suppose that one of every 10^6 silicon atoms is replaced by a phosphorus atom.

(a) What charge carrier concentration will the phosphorus add (in terms of m^{-3})?

(b) What is the ratio of the charge carrier concentration (electrons in the conduction band and holes in the valence band) in the doped silicon to that in pure silicon? The density of silicon is 2.33 g/cm^3, and its molar mass is 28.1 g/mol.

Section 32-7   The pn Junction Diode

25. ✱✱ For an ideal pn junction diode with a sharp boundary between its two semiconducting sides, the current I is related to the potential difference V across the diode as follows:

$$I = I_0(e^{eV/kT} - 1)$$

where I_0, which depends on the materials but not on I or V, is called the reverse saturation current. The potential difference V is positive if the rectifier is forward-biased and negative if it is reverse-biased. For $T = 270$ K, calculate the ratio of the current for a 0.30 V forward bias to the current for a -0.30 V reverse bias.

26. ✱✱✱ When a photon enters the depletion zone of a pn junction, the photon can scatter from the valence electrons there, transferring part of its energy to several electrons, which then jump to the conduction band. Thus, the photon creates electron–hole pairs. For this reason, pn junctions are often used as photon detectors, especially in the X-ray and gamma-ray regions of the electromagnetic spectrum. Suppose a single 699 keV photon transfers its energy to electrons in multiple scattering events inside a semiconductor with an energy gap of 0.90 eV, until all the energy is transferred. Assuming that each electron jumps the gap from the top of the valence band to the bottom of the conduction band, find the number of electron–hole pairs created by the photon.

27. ✱✱ While a diode does conduct quite freely in the forward bias direction, there is still a small voltage drop across the junction, called the knee voltage. This is typically 0.7 V for a silicon diode. Suppose such a diode is used in a rectifier that is intended to supply a current of 100 mA. What must the minimum power rating of the diode be?

Section 32-8   Other Semiconductor Devices

28. ✱ A certain computer chip that has the dimensions 3.79×3.03 cm contains 3.9 million square transistors. What must be their *maximum* dimensions (in terms of microns, or μm)? (Note: Devices other than transistors are also on the chip, and there must be room for the interconnections between the circuit elements.)

29. ✱✱ A silicon-based device consists of a heavily doped region of semiconductor upon which a thin oxide layer is allowed to grow. A metal layer is then deposited on top of the oxide layer. Suppose that the metal layer is 0.62 μm square. The insulating silicon oxide layer that separates the metal layer from the p-type substrate is 0.40 μm thick and has a dielectric constant of 4.0.

(a) What is the equivalent metal–substrate capacitance (treating the metal as one plate and the substrate as the other plate)?

(b) Approximately how many elementary charges appear in the metal when there is a metal–source potential difference of 2.4 V?

30. ✱✱ As miniaturization continues, we will reach the limit where devices get so small that their behaviour deviates significantly from the behaviour of a "bulk" sample. The exact size at which this limit occurs is not well known, but we could estimate it by looking at the number of dopants in a volume. Consider a silicon device that is doped at 0.001%, that is, for every dopant atom, there are 100 000 silicon atoms. How many dopant atoms are in a cube of silicon whose edge is

(a) 1 μm;

(b) 0.1 μm;

(c) 0.001 μm?

MODERN PHYSICS

Section 32-9 Nanotechnology

31. ✲ The current in a scanning tunnelling microscope (STM) is given by

$$I = I_{\text{tunneling}} \propto e^{-2\kappa \Delta z}$$

$$\kappa = 0.51\sqrt{\phi} \times 10^{10}/\text{m}$$

where ϕ the barrier height (in eV), usually taken to be the work function of the metal, and Δz is the thickness of the barrier. Assuming that the work function is 4.83 eV, what percent change in the tunnelling current occurs if Δz increases by 0.1 nm?

32. ✲ A transmission electron microscope (TEM) is operated at a potential difference of 150 000 V.
 (a) What is the wavelength of the electrons that have been accelerated through this potential difference?
 (b) The resolving power of the TEM is taken to be $\Delta l = 1.22 \dfrac{f\lambda}{D}$, where f is the focal length and D is the aperture. Determine Δl, assuming $f = 10$ cm and $D = 2$ cm.

COMPREHENSIVE PROBLEMS

33. ✲✲✲ In problem 19, we examined the rate at which an individual electron transfers energy to the lattice. This energy is provided by the electric field. In that problem, we found that for a 1 mm³ conductor with a carrier density of 10^{22} m⁻³ dissipating 1 W of power, each electron delivers energy to the lattice at a rate of 1×10^{-13} W $\approx 6 \times 10^{5}$ eV/s. Assume that the electron travels between collisions only under the influence of the constant applied electric field, and loses all of its energy to the lattice when it collides. The rate at which the electron loses energy to the lattice must be equal to the rate at which the electric field supplies energy to the electron. Suppose that the conductor is a cube and that a potential difference of 1.0 V is applied.
 (a) What is the strength of the electric field in the sample?
 (b) What acceleration does an electron experience due to that field?
 (c) Write an expression for the speed of the electron (it will depend on time).
 (d) Write an expression for the kinetic energy of the electron.
 (e) Find the rate at which kinetic energy is delivered to the electron.
 (f) Set the result of part (e) equal to 1×10^{-13} W and solve for t.
 (g) The time found in part (e) is half the time between collisions. How far will the electron have moved between collisions? (Hint: Use $x = \dfrac{1}{2}at^2$.)

34. ✲✲✲ The lateral resolution of an STM is related to the exponential dependence of the tunnelling current on distance. We assume that the tip of the STM probe is spherical, and we use the geometry shown in Figure 32-31.
 (a) Show that $z(x) = z_o + R - \sqrt{R^2 - x^2}$.
 (b) Assume that $x \ll R$, and expand $z(x)$ in a Taylor series about $x = 0$ to first order.

(c) Show that $I_t(x) = I_0 e^{-2\kappa\left(z_o + \frac{x^2}{2R}\right)} = I_o(z)e^{-\kappa \frac{x^2}{R}}$, where $I_t = I_0 e^{-2\kappa\Delta z}$.

(d) Recall that $\kappa = 0.51\sqrt{\phi} \times 10^{10}/\text{m}$. Determine the full width at half maximum for $I_t(x)$ for a metal with a work function of 5 eV.

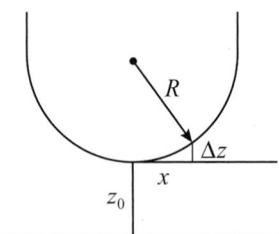

Figure 32-31 Problem 34

35. ✲ Calculate the density of states at an energy of 6 eV.

36. ✲ A fictitious trivalent metal has a molar mass of 59 g/mol and a density of 6.32 g/cm³. What is the number density n of conduction electrons?

37. ✲✲ What is the probability that a state 0.30 eV above the Fermi level will be occupied at (a) $T = 0$ K and (b) $T = 300$ K?

38. ✲✲ At $T = 200$ K, determine the probability of occupancy, $f(E)$, for a state
 (a) 60 meV above the Fermi level;
 (b) 60 meV below the Fermi level.

39. ✲✲ A metal has a Fermi energy of 8.6 eV. At a temperature of 400 K, what is the energy of a state whose probability of being occupied is
 (a) 0.38;
 (b) 0.84?

40. ✲ A state has an energy that is 0.50 eV above the Fermi energy. At what temperature will this state be occupied with a probability of 0.1?

41. ✲ A simple model of the resistance in a metal involves assuming that the electrons scatter every τ seconds. In this model, the resistivity is given by

$$\rho = \frac{m}{ne^2\tau}$$

where n is the conduction electron density, m is the mass of the electron, and e is the charge of the electron. For copper, $n = 8.46 \times 10^{28}$ m⁻³ and $\rho = 1.68 \times 10^{-8}$ Ω·m. Find the scattering time τ.

42. ✲✲ (a) Assuming that the electrons in problem 41 travel at a drift speed of approximately 3.00×10^{-4} m/s, how far do they travel between collisions?
 (b) Assuming that the electrons travel at the Fermi speed of approximately 1.00×10^{6} m/s, how far do they travel between collisions?
 (c) Given that the lattice spacing in copper is 3.6×10^{-10} m, does the result in part (a) make any sense?

43. Capacitors can be fabricated in an integrated circuit by heavily doping a region of the chip, allowing an insulating oxide layer to grow over that region, and then depositing a metal layer on the oxide. The metal layer and the heavily doped region in the wafer act as the plates of the capacitor. What oxide thickness is required for a doped region with an area of 10^{-10} m² if we want a capacitance of 1 pF? Assume silicon oxide, which has a dielectric constant $\kappa \approx 5$.

MODERN PHYSICS

44. ✷✷ A capacitor is fabricated in a silicon integrated circuit. The oxide layer separating the two doped "plates" is 1 nm thick and has a cross-sectional area of 10^{-10} m². The oxide has a resistivity of 10^{13} $\Omega \cdot$m. What is the resistance associated with the oxide layer?

45. ✷✷ Increases in computing speed are largely due to the decreases in size of the CPU units. Making the components closer together decreases the time required for a signal to propagate from one circuit to the next on the chip. Current CPU clock speeds are of the order of 10^9 Hz. If we assume that all the time between clock pulses, that is, 10^{-9} s, is required for the signal to propagate, how far apart are the circuits on such a chip? Assume that the signals propagate at 90% of the speed of light. Is this result reasonable?

46. ✷✷ Resistors are fabricated in integrated circuits by doping regions of the pure semiconductor wafer to decrease their resistance. Suppose we dope an area of 10^{-10} m² that is 1 μm thick. What resistivity is needed to make a 10 kΩ resistor?

47. ✷ A capacitor is built on an integrated circuit. The capacitance is 1 pF, and the resistance of the oxide layer is 10^{14} Ω. What is the time constant associated with this RC combination?

48. ✷✷ The circuit shown in Figure 32-32 is a bridge rectifier. V_1 is a sinusoidal voltage source.
(a) During the first half of the sine cycle, point A is positive with respect to point B. Trace the current path from A to B.
(b) During the second half of the sine cycle, A is negative with respect to B. Trace the current path for this situation.
(c) Based on your findings from parts (a) and (b), sketch V_1 and the voltage across R_1 for one complete cycle.

Figure 32-32 Problem 48

49. ✷✷ Equation (32-8) for the Fermi wave number, $k_f = \dfrac{\pi N}{aN^*}$, suggests that the Fermi wave vector depends on the total sample size, $L = aN^*$. Why is this not the case?

50. ✷ A metal has a Fermi speed of 3.8×10^6 m/s. Find the Fermi energy.

51. ✷✷ Calculate the wavelength for an electron with a wave vector magnitude of $k = \dfrac{\pi}{a}$, where a is the lattice spacing.

52. ✷✷ The bond length in diamond is 0.154 nm, which is the longest bond length that exists for ordinary carbon covalent bonds. Why would you expect the bond length in a diamond crystal to be larger than the bond length in carbon molecules?

53. ✷✷ From classical thermodynamics, we know that the number of degrees of freedom, f, in a system determines the constant volume heat capacity: $C_V = \dfrac{fR}{2}$, where R is the gas constant. In an insulator, the electrons cannot move, whereas in a metal they can. One would therefore expect that there are more degrees of freedom in a metal and that it would have a larger heat capacity than an insulator by an amount of $3R/2 \sim 12.5$ J/mol$\cdot$K. Table 32-5 lists heat capacity data for a number of elements, including metals, a semiconductor, and an insulator. Clearly, the metals do not have heat capacities that are larger by $3R/2$.
(a) When we heat a metal, do all the electrons have their energies increased?
(b) What fraction of the electrons do have their energy increased?
(c) Would you expect the electronic contribution to the heat capacity to be large?

Table 32-5 Data for Problem 53

Element	Heat capacity (J/mol · K)
Sodium	28.277 7
Magnesium	24.791 1
Aluminum	24.389 92
Silicon	19.943 9
Phosphorus	23.837 61
Sulfur	22.609 35

See the text online resources at
www.physics1e.nelson.com for Open Problems
and Data-Rich Problems related to this chapter.

MECHANICS

MODERN PHYSICS

Learning Objectives

When you have completed this chapter you should be able to:

1 Understand nuclear terminology, including atomic mass
 number, atomic number,
 nucleon, nuclide, isotopes,
 isobars, isotones, and atomic
 mass.

2 Calculate the size and density
 of nuclei, and describe the
 properties of the strong force.

3 Calculate nuclear binding
 energies from nuclear masses
 and vice versa, and calculate
 atomic masses and average
 nuclear masses.

4 Calculate radioactive decay
 rates.

5 Explain what a *Q*-value is, and
 calculate *Q*-values for simple
 nuclear reactions.

6 Explain alpha, beta, and
 gamma radiation.

7 Explain nuclear fission and
 fusion, and calculate the
 energy released during nuclear
 fission and fusion reactions.

8 Explain the difference between
 the absorbed dose and the
 equivalent dose of radiation, and calculate the daily
 equivalent dose of radiation for
 various radioactive materials.

9 Describe some common applications of nuclear physics.

e For additional Making Connections, Examples, and Checkpoints, as well as activities and
 experiments to help increase
 your understanding of the
 chapter's concepts, please go
 to the text's online resources at
 www.physics1e.nelson.com.

Chapter 33
Introduction to Nuclear Physics

On March 11, 2011, a 9.0 magnitude earthquake struck off the northeast coast of Japan near the city of Fukushima. The earthquake was followed by a 14 m-high tsunami that devastated the coastal area. The tsunami flooded the buildings that housed the six nuclear reactors at the Fukushima 1 Power Plant, causing the emergency generators and cooling systems to fail. Although three of the reactors had been shut down for maintenance, the other three overheated catastrophically (Figure 33-1). Two of the reactor cores melted, and hydrogen gas produced by the high temperatures caused massive explosions that destroyed the outer containment buildings and damaged the vessels containing the nuclear fuel. Radioactive cesium, iodine, strontium, and barium were released into the air, forcing the evacuation of the surrounding area. Today, there is still concern over possible radioactive contamination of food and water supplies.

How do nuclear reactors generate energy? What is radioactivity, and why is it harmful? How does radioactive cesium differ from ordinary cesium? This chapter introduces nuclear physics and examines some of the applications and dangers of nuclear technology.

DigitalGlobe/Contributor/Getty Images

Figure 33-1 Smoke billowing from the damaged nuclear reactors at Fukushima

33-1 Nuclear Terminology and Nuclear Units

As described in Chapter 30, experiments by Ernest Rutherford and others at the beginning of the 20th century determined that all the positive charge in an atom and most of the mass of an atom are localized in an extremely small central region called the nucleus. Later research showed that nuclei are made up of positively charged particles called **protons** and electrically neutral particles called **neutrons**. A neutron is slightly more massive than a proton, and both particles have about 1840 times the mass of an electron (see Table 33-2). The charge of a proton is $+1.602 \times 10^{-19}$ C, which is equal and opposite to the charge of an electron. An atom has an equal number of protons and electrons. The number of protons in a nucleus is called the **atomic number**, Z. For example, a carbon nucleus has six protons. Therefore, the atomic number of carbon is 6. The chemistry of an element is determined by its atomic number.

The number of neutrons in a nucleus is called the **neutron number**, N. The total number of protons and neutrons in a nucleus is called the **mass number**, A. By definition,

$$A = Z + N \qquad (33\text{-}1)$$

Each element in the periodic table has a fixed number of protons and is represented by a symbol. For example, helium has $Z = 2$ and is represented by the symbol He, and iron has $Z = 26$ and is represented by the symbol Fe. In general, a nucleus is represented by a symbol with the form $_Z^A X$, where

X = the symbol for the element
Z = the atomic number of the element
A = the mass number of the element

For a given atomic number, Z, the number of neutrons in a nucleus may vary. Atoms with the same number of protons but a different number of neutrons are called the **isotopes** of an element. Isotopes have different masses because they contain a different number of neutrons. For example, naturally occurring carbon has three isotopes, as listed in Table 33-1.

Nuclei that have the same number of neutrons but a different number of protons are called **isotones**. Nuclei that have the same mass number, A, but a different number of protons and neutrons are called **isobars**. The term **nuclide** refers to a nucleus with any number of protons and neutrons.

 CHECKPOINT

C-33-1 Nuclei with Properties in Common

Identify the isotopes, isotones, and isobars in the following set of nuclei:

$$_8^{16}\text{O}; \ _6^{15}\text{C}; \ _7^{14}\text{N}; \ _9^{18}\text{F}; \ _{10}^{20}\text{Ne}; \ _7^{15}\text{N}; \ _8^{18}\text{O}.$$

C-33-1 Isotopes: $_8^{16}\text{O}, _8^{18}\text{O}$; Isotones: $_6^{15}\text{C}, _7^{15}\text{N}, _8^{16}\text{O}, _9^{18}\text{F}$; Isobars: $_9^{18}\text{F}, _7^{14}\text{N}, _7^{15}\text{N}, _8^{18}\text{O}, _{10}^{20}\text{Ne}$; $_8^{18}\text{O}$.

Units for Nuclear Quantities

The size of a nucleus is of the order of 10^{-15} m. Therefore, in the study of nuclear physics, a convenient unit of length to use is the femtometre (symbol fm). A femtometre is also called a fermi in honour of Italian American physicist Enrico Fermi:

$$1 \text{ fm} = 10^{-15} \text{ m}$$

When dealing with energies involved in nuclear processes, a convenient unit to use is mega (million) electron volts:

$$1 \text{ MeV} = 10^6 \text{ eV}$$

The Atomic Mass Unit (u) An atomic mass unit (u) is defined as exactly 1/12 of the rest mass of one atom of the isotope $_6^{12}\text{C}$:

$$1 \text{ u} = 1.6605 \times 10^{-24} \text{ g} = 1.6605 \times 10^{-27} \text{ kg} \quad (33\text{-}2)$$

Table 33-1 Isotopes of carbon

Isotope name	Symbol	Atomic number, Z	Neutron number, N	Occurrence in nature
Carbon-12	$_6^{12}\text{C}$	6	6	98.93%
Carbon-13	$_6^{13}\text{C}$	6	7	1.07%
Carbon-14	$_6^{14}\text{C}$	6	8	10^{-10}%

MODERN PHYSICS

EXAMPLE 33-1

Molar Mass and Atomic Mass

How many atoms are in 12 g of $^{12}_{6}C$?

SOLUTION

Mass of one $^{12}_{6}C$ atom $= 12 \text{ u} = 12 \times 1.6605 \times 10^{-24}$ g $= 19.926 \times 10^{-24}$ g

Number of atoms in 12 g of $^{12}_{6}C = \dfrac{12 \text{ g}}{19.926 \times 10^{-24} \text{ g}}$ $= 6.022 \times 10^{23}$

Making sense of the result:

The answer, 6.022×10^{23}, equals Avogadro's number, the number of atoms in one mole of a substance.

In nuclear physics it is convenient to express masses in the unit of energy/c^2, using Einstein's energy–momentum relationship, $E = \sqrt{p^2c^2 + m_0^2c^4}$, where p is the momentum of a particle and m_0 is its rest mass. The energy equivalent of 1 u of mass at rest is

$E = m_0c^2$

$= (1.6605 \times 10^{-27} \text{ kg})(2.9979 \times 10^8 \text{ m/s})^2$

$= 14.9236 \times 10^{-11} \text{ J}$

$= \dfrac{14.9236 \times 10^{-11} \text{ J}}{1.6021 \times 10^{-19} \text{ J/eV}}$

$= 9.3150 \times 10^8 \text{ eV}$

$= 931.50 \text{ MeV}$

Therefore,

$$1 \text{ u} = 931.50 \text{ MeV}/c^2 \qquad (33\text{-}3)$$

Table 33-2 lists the basic properties of protons, neutrons, and electrons.

LO 2

33-2 Nuclear Size and Nuclear Force

By bombarding a nucleus with a beam of fast-moving electrons and then observing how the electrons are deflected by the nucleus, scientists can determine the shape and the size of the nucleus. Several experiments using various elements have shown that, to a high degree of accuracy, most nuclei can be approximated as closely packed spherical clusters of protons and neutrons. The volume of the nucleus is proportional to the number of nucleons in the nucleus. A **nucleon** is a particle in an atomic nucleus—a proton or a neutron. The radius, r, of a nucleus is related to the number of nucleons, A:

$$r = r_0 A^{1/3} \qquad (33\text{-}4)$$

where $r_0 = 1.2$ fm is a constant, determined from experiments. Table 33-3 lists the radii of nuclei of some common elements, using Equation (33-4).

Nuclear Density

The relationship between the radius and the mass number of a nucleus in Equation (33-4) indicates that the nuclear density remains constant as A changes, that is, all nuclei

Table 33-2 Some physical properties of atomic particles

Particle	Symbol	Spin	Charge	Atomic mass units, u	Mass MeV/c²
Electron	e	1/2	$-e$	5.4858×10^{-4}	0.5110
Proton	p	1/2	$+e$	1.007 276	938.29
Neutron	n	1/2	0	1.008 665	939.57

Table 33-3 Radii of some common isotopes, assuming a spherical nuclear shape

Nucleus	Mass number, A	Radius, $r = r_0 A^{1/3}$ (fm)
Helium, $^4_2$He	4	1.9
Oxygen, $^{16}_8$O	16	3.0
Iron, $^{56}_{26}$Fe	56	4.6
Lead, $^{208}_{82}$Pb	208	7.1

have the same density. To calculate this density, we start with the volume of a spherical nucleus of radius r:

$$V = \frac{4\pi}{3}r^3 = \frac{4\pi}{3}\left(r_0 A^{1/3}\right)^3 = \frac{4\pi}{3}r_0^3 A \qquad (33\text{-}5)$$

Next, we need to determine the mass of a nucleus that has A nucleons. From Table 33-2 we see that the mass of a nucleon is approximately equal to 1.0 u. Therefore, to a good approximation, the mass of a nucleus (M_{nuc}) with A nucleons is

$$M_{\text{nuc}} \approx A\,\text{u} \qquad (33\text{-}6)$$

This approximation is accurate to within 1% of the actual mass. The nuclear density, ρ_{nuc}, is then given by

$$\rho_{\text{nuc}} = \frac{M_{\text{nuc}}}{V} = \frac{A\,\text{u}}{\frac{4\pi}{3}r_0^3 A} = \frac{\text{u}}{\frac{4\pi}{3}r_0^3}$$

$$= \frac{1.6605 \times 10^{-27}\,\text{kg}}{\frac{4\pi}{3}(1.2 \times 10^{-15}\,\text{m})^3} = 2.3 \times 10^{17}\,\text{kg/m}^3 \quad (33\text{-}7)$$

This density is fantastically large. A nucleus with a volume of 1.0 mm³ would have a mass of 2.3×10^8 kg, that is, 230 thousand tonnes! In comparison, the mass of the same volume of iron is only 0.0078 g.

The Strong (or the Nuclear) Force

Protons are positively charged and exert a repulsive Coulomb force on each other. The magnitude of the Coulomb force between two protons that are a distance r apart is

$$F_{\text{Coulomb}} = \frac{1}{4\pi\varepsilon_0}\frac{(+e)(+e)}{r^2}$$

$$= \left(8.99 \times 10^9\,\frac{\text{N}\cdot\text{m}^2}{\text{C}^2}\right)\left(\frac{(1.602 \times 10^{-19}\,\text{C})^2}{r^2}\right) \quad (33\text{-}8)$$

Consider two protons inside a helium nucleus. If the average distance between the two protons is taken to be 2.0 fm, the repulsive force between the protons is

$$F_{\text{Coulomb}} = \left(8.99 \times 10^9\,\frac{\text{N}\cdot\text{m}^2}{\text{C}^2}\right)\frac{(1.602 \times 10^{-19}\,\text{C})^2}{(2.0 \times 10^{-15}\,\text{m})^2}$$

$$= 58\,\text{N}$$

This is an enormous amount of repulsion. For comparison, the magnitude of the gravitational attraction between the protons at this distance is

$$F_{\text{grav}} = G\frac{(m_{\text{p}})^2}{r^2}$$

$$= \left(6.674 \times 10^{-11}\,\frac{\text{N}\cdot\text{m}^2}{\text{kg}^2}\right)\left(\frac{(1.673 \times 10^{-27}\,\text{kg})^2}{(2.0 \times 10^{-15}\,\text{m})^2}\right)$$

$$= 5 \times 10^{-35}\,\text{N} \qquad (33\text{-}9)$$

where m_{p} is the proton mass.

Obviously, gravitational attraction is not sufficient to overcome the enormous Coulomb repulsion between the protons and keep them within the helium nucleus. There must be some very strong attractive force that keeps the protons confined within the nucleus. This force is called the **strong force** or the **nuclear force**. The strong force exists not only between two protons but also between two neutrons and between a proton and a neutron, as shown in Figure 33-2. There is another force, called the **weak force**, that acts between the nucleons, but it does not play any role in binding the nuclei. We will ignore the weak force in this chapter. The origin of the strong and weak forces is discussed in the next chapter.

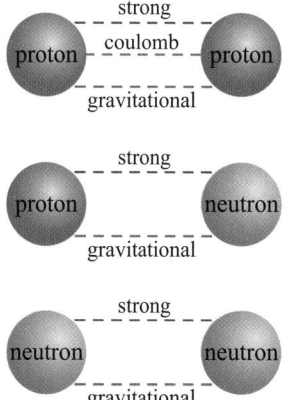

Figure 33-2 The forces acting between pairs of nucleons. There is also a weak force between two nucleons, but it is not shown in this figure.

Despite considerable research, the complicated nature of the strong force between two nucleons is still not completely understood. Experiments have shown that this force has the following properties:

- The strong force is repulsive when the distance between the two nucleons is less than approximately 0.7 fm. The repulsion prevents a nucleus from collapsing to a point.

- For distances larger than 0.7 fm, the strong force is attractive and reaches a maximum when the separation between the nucleons is approximately 1.2 fm. At this separation the strong force is approximately 10 times stronger than the Coulomb force. This attraction holds the nucleus together.

- At distances beyond 1.2 fm, the strength of the strong force decreases exponentially, and it becomes negligible for separations of more than 3.0 fm. Thus, the strong force acts *only* at a very short range, in marked contrast to the gravitational and electrostatic forces.

- The strong force only exists between nucleons. There is no strong force between an electron and a nucleon or between two electrons.

A plot of the strength of the strong force as a function of distance between two nucleons is shown in Figure 33-3.

The scale for the magnitude of the force is not specified in this plot because the exact magnitude for a given separation depends on the model used to describe the force.

Figure 33-3 The behaviour of the strong force as a function of the distance between two nucleons. The force is repulsive for positive values and attractive for negative values.

LO 3

33-3 Nuclear Binding Energy

It is customary to express the nuclear mass in terms of A and Z. The rest mass of a nucleus that has Z protons and $N = A - Z$ neutrons is denoted by $M_{\text{nuc}}(A, Z)$. From Einstein's mass–energy relationship, the rest mass energy of this nucleus is $M_{\text{nuc}}(A, Z)c^2$.

Since a nucleus is bound by the strong force, separating it into its constituent protons and neutrons requires energy to overcome the attractive strong force, just as removing an electron from an atom requires energy to overcome the electrostatic force binding the electron to the nucleus. By the law of conservation of energy, *the rest mass energy of a nucleus together with the energy required to separate all the nucleons of the nucleus must be equal to the total rest mass energy of all the separated nucleons*. The energy required to separate a nucleus into individual nucleons is called the **nuclear binding energy** or simply the **binding energy**. The binding energy is therefore the difference between the total rest mass energy of constituent nucleons and the rest mass energy of the nucleus. The binding energy of a nucleus with A nucleons and Z protons is denoted by $B(A, Z)$:

$$B(A, Z) = Zm_{\text{p}}c^2 + (A - Z)m_{\text{n}}c^2 - M_{\text{nuc}}(A, Z)c^2 \quad (33\text{-}10)$$

Equivalently, we can write the nuclear mass in terms of the nuclear binding energy and proton and neutron masses:

$$M_{\text{nuc}}(A, Z) = Zm_{\text{p}} + (A - Z)m_{\text{n}} - \frac{B(A, Z)}{c^2} \quad (33\text{-}11)$$

The **atomic mass** of a neutral atom, $M_{\text{atom}}(A, Z)$, is equal to the sum of the mass of its nucleus and the masses of its Z electrons. Since the binding energies of the electrons in an atom are relatively small compared to the nuclear energies, they are often neglected in nuclear calculations. Therefore,

$$M_{\text{atom}}(A, Z) = M_{\text{nuc}}(A, Z) + Zm_{\text{e}} \quad (33\text{-}12)$$

The **average atomic mass** is the weighted average of the masses of the naturally occurring isotopes for a given element. For example, the two naturally occurring isotopes of nitrogen are $^{14}_{7}\text{N}$ (99.634%, atomic mass = 14.003 074 u) and $^{15}_{7}\text{N}$ (0.366%, atomic mass = 15.000 109 u), so the average atomic mass of nitrogen is

$$(14.003\ 074\ \text{u} \times 0.996\ 34) + (15.000\ 109\ \text{u} \times 0.003\ 66)$$
$$= 14.006\ 7\ \text{u}$$

 EXAMPLE 33-2

The Binding Energy of Helium

The atomic mass of helium $^{4}_{2}\text{He}$ is 4.002 603 u. Calculate the binding energy of the nucleus.

SOLUTION

The atomic mass includes the mass of the electrons. Since a helium atom has two electrons, we subtract two electron masses from the atomic mass to find the nuclear mass:

Mass of a $^{4}_{2}\text{He}$ nucleus = $M_{\text{nuc}}(4, 2)$ = 4.002 603 u − (2 × 0.000 548 u) = 4.001 507 u

The binding energy of a helium nucleus is

$$B(4, 2) = 2m_{\text{p}}c^2 + (4 - 2)m_{\text{n}}c^2 - M_{\text{nuc}}(4, 2)c^2$$
$$= (2 \times 1.007\ 276\ \text{u} + 2 \times 1.008\ 665\ \text{u} - 4.001\ 507\ \text{u})c^2$$
$$= (0.030\ 375\ \text{u})c^2$$
$$= (0.030\ 376\ \text{u} \times 931.5\ \text{MeV}/c^2)c^2$$
$$= 28.3\ \text{MeV}$$

From the above example, we see that it takes 28.3 MeV of energy to separate a $^{4}_{2}\text{He}$ nucleus into two protons and two neutrons. The same amount of energy will be released if two protons and two neutrons are combined to form a $^{4}_{2}\text{He}$ nucleus. Therefore, the nuclear binding energy can also be defined as the amount of energy released when a nucleus is assembled from individual nucleons.

By itself, the binding energy of a nucleus does not tell us how tightly bound the nucleons are in the nucleus. To compare how tightly various nuclei are bound, we define the **binding energy per nucleon**, $B_{\text{per}}(A, Z)$:

$$B_{\text{per}}(A, Z) = B(A, Z)/A \quad (33\text{-}13)$$

From experiments, the total binding energy of a gold nucleus ($^{184}_{79}$Au) is determined to be 1484 MeV. Comparing the binding energy per nucleon of helium ($^4_2$He) and gold nuclei, we find that the nucleons in gold are somewhat more tightly bound than those in helium:

$$B_{per}(4, 2) = \frac{28.3 \text{ MeV}}{4 \text{ nucleons}} = 7.07 \text{ MeV/nucleon}$$

$$B_{per}(184, 79) = \frac{1484 \text{ MeV}}{184 \text{ nucleons}} = 8.06 \text{ MeV/nucleon}$$

Since nuclear masses of all stable nuclei have been experimentally determined to a high degree of accuracy, we can plot $B_{per}(A, Z)$ as a function of A, as shown in Figure 33-4. This plot is called the **binding energy curve**. Notice that the binding energy per nucleon first sharply increases with A, reaches a maximum near $A = 60$, and then gradually decreases with increasing A. The shape of

the binding energy curve is mainly caused by the interplay between the repulsive electrostatic force and the attractive strong force.

The elements with low mass numbers are weakly bound because each nucleon in these nuclei has relatively few adjacent nucleons exerting attractive forces on it. For example, a **deuteron**, the nucleus of deuterium $^2_1$H, consists of one proton and one neutron and a binding energy per nucleon of only 1.1 MeV (the total binding energy of a deuteron is 2.2 MeV). A helium-3 nucleus, $^3_2$He, which has one more proton, has a binding energy per nucleon of 2.5 MeV. Adding an additional neutron to $^3_2$He forms a $^4_2$He nucleus, which has a binding energy per nucleon of approximately 7.0 MeV. The $^4_2$He nucleus is the most tightly bound of all nuclei with $A \leq 8$.

The binding energy per nucleon increases gradually as the mass number increases from 12 to 55. It peaks in the range $55 \leq A \leq 65$. Nuclei in this range are the most tightly bound and, therefore, the most stable. Iron-56 ($^{56}_{26}$Fe) has the largest binding energy per nucleon, 8.8 MeV. The binding energy per nucleon decreases steadily for $A \geq 65$. The strong force is a short-range force, so each additional nucleon interacts only with the nucleons that are very close it. However, the Coulomb force has an infinite range, so each additional proton repels *all* other protons in the nucleus, thus making the nucleus less stable as Z increases.

Neutrons are electrically neutral, and adding a neutron does not increase repulsion; however, when the number of neutrons in a nucleus increases above a critical value, it becomes energetically possible for a neutron to convert into a proton. Therefore, it is not possible to form heavier nuclei by keeping the number of protons fixed and just adding more and more neutrons.

Most nuclei beyond $A = 208$ are unstable and decay into smaller nuclei by a process, called alpha decay, that we will discuss in the next section.

Figure 33-4 A plot of binding energy per nucleon as a function of mass number A for stable nuclei

EXAMPLE 33-3

The Fusion of Two Deuterons

(a) How much energy is gained when two deuterons combine to form a helium-4 nucleus?
(b) How many joules of energy are released when one mole of deuterium (2 g) combines to form $^4_2$He?

SOLUTION

(a) A nuclear reaction releases energy by converting some of the mass of the original nuclei into energy. From Equation (33-10), the rest mass energy of a deuteron is

$$M_{nuc}(2, 1)c^2 = m_p c^2 + m_n c^2 - B(2, 1)$$

where $B(2, 1)$ is the binding energy of the deuteron. Similarly, the rest mass energy of the $^4_2$He nucleus is

$$M_{nuc}(4, 2)c^2 = 2m_p c^2 + 2m_n c^2 - B(4, 2)$$

where $B(4, 2)$ is the binding energy of $^4_2$He.

The energy released (ΔE) when two deuterons combine to form a helium nucleus is equal to the difference in the rest mass energies of the two deuterons and the helium nucleus:

$$\Delta E = 2 \times M_{nuc}(2, 1)c^2 - M_{nuc}(4, 2)c^2$$
$$= 2(m_p c^2 + m_n c^2 - B(2,1)) - (2m_p c^2 + 2m_n c^2 - B(4,2))$$
$$= B(4, 2) - 2 \times B(2, 1)$$

So, the released energy is equal to the difference between the binding energy of the final nucleus ($^4_2$He) and the sum of the binding energies of the initial deuterons. From Figure 33-4, the binding energy *per nucleon* for a deuteron is approximately 1.1 MeV, so $B(2, 1) = 2 \times 1.1$ MeV $= 2.2$ MeV. Similarly, $B(4, 2) = 4 \times 7.0$ MeV $= 28.0$ MeV. Therefore,

$$\Delta E = 28.0 \text{ MeV} - (2 \times 2.2 \text{ MeV}) = 23.6 \text{ MeV}$$

(continued)

MODERN PHYSICS

Making sense of the result:

Burning one barrel (about 159 L) of oil produces approximately 6.12×10^9 J of energy. So, the fusion of 2 g of deuterium produces as much energy as burning 186 barrels of oil. The energy released per unit mass is vastly greater for a nuclear fuel than for a chemical fuel.

LO 4

33-4 Nuclear Decay and Radioactivity

In 1896, French physicist Antoine Becquerel (1852–1908) discovered that a mineral, later identified as uranyl potassium sulfate, continuously emits invisible radiation. The emitted radiation was not visible light and could easily penetrate a sheet of black paper. Further experiments by Becquerel, Marie Curie, Pierre Curie, and other scientists showed that invisible radiation is also emitted by some other elements, including thorium, radon, and radium. Furthermore, it was discovered that these elements emitted three different types of invisible radiations, which were called alpha (α), beta (β), and gamma (γ) radiation after the first three letters of the Greek alphabet.

α, β, and γ radiations are collectively called **nuclear radiation** and, as we now know, are emitted by nuclei. An element that emits nuclear radiation is called **radioactive**, and the process of emission of nuclear radiation is called **radioactive decay** or **nuclear decay**.

Like electromagnetic radiation, nuclear radiation carries energy. Only an unstable nucleus emits nuclear radiation. By emitting nuclear radiation, a nucleus transforms into either a nucleus of a lighter element or a lower-energy state of the original nucleus. The nucleus that emits the radiation is called the **parent nucleus**, and the final nucleus is called the **daughter nucleus**. A daughter nucleus always has less energy (including the rest mass energy) than the parent nucleus.

α-radiation consists of helium-4 ($^4_2$He) nuclei, which were originally called **alpha (α) particles**. Thus, a nucleus that emits an α-particle loses two protons and two neutrons and becomes a different element. β-radiation consists of either electrons or positrons. A **positron** is the antiparticle of an electron; it has the same mass as an electron but the opposite charge of an electron. γ-radiation is high-energy electromagnetic radiation. In Section 33-6 we will discuss these radiations in detail.

A general law applies to the emission of nuclear radiation: *at any given time, the amount of radiation emitted by a material depends only on the number of radioactive nuclei present.* For example, the uranium-238 ($^{238}_{92}$U) nucleus is radioactive and emits an α-particle. The compound uranyl potassium sulfate contains uranium, potassium, sulfur, and oxygen. The number of α-particles emitted in each second by a sample of this compound only depends on the number of uranium-238 nuclei in the sample. As the uranium nuclei decay, fewer of the radioactive nuclei remain; therefore, fewer α-particles are emitted by the sample each second. The rate at which this number decays is proportional to $N(t)$, and

$$-\frac{dN(t)}{dt} \propto N(t) \qquad (33\text{-}14)$$

where $N(t)$ is the number of radioactive nuclei present in a source at some time t.

The minus sign in Equation (33-14) indicates that the number of nuclei, $N(t)$, decreases with time. We can write an equation for the above relationship by introducing a proportionality constant, which is called the **decay constant** or the **disintegration constant** and is usually denoted by the Greek letter lambda (λ):

$$-\frac{dN(t)}{dt} = \lambda N(t) \qquad (33\text{-}15)$$

The unit of λ is decays per second, or simply, s^{-1}. The decay constant is a measure of how quickly a radioactive element decays with time. A large value of λ means that the nuclide will decay rapidly.

The decay constant is a characteristic of each radioactive nuclide. For a radioactive nuclide that can emit α-, β-, and γ-radiation, the decay constant has a different value for each type of radiation.

Suppose a radioactive nuclide has a decay constant of 0.10 s^{-1}, and at $t = 0$ we have 1.0×10^6 nuclei. After 1.0 s, the number of nuclei that have decayed is $\lambda N(0) = 0.10 \times 10^6 = 1.0 \times 10^5$, and 9.0×10^5 nuclei are left. During the next second, another $\lambda N(1) = 0.10 \times 9.0 \times 10^5 = 9.0 \times 10^4$ nuclei decay, leaving 8.1×10^5 of the original nuclei. After a third second, $\lambda N(2) = 0.1 \times 8.1 \times 10^5 = 8.1 \times 10^4$ more nuclei have decayed, and 7.29×10^5 are left. In each one second interval, 10% of the nuclei present at the start of that interval decay. The decreases in the number of radioactive nuclei by *an equal percentage in equal intervals of time* is a property of an exponential function. Figure 33-5 shows a plot of the number of remaining nuclei as a function of time for the above example. Obviously, this number decreases exponentially. So *radioactive decay is an exponential process.*

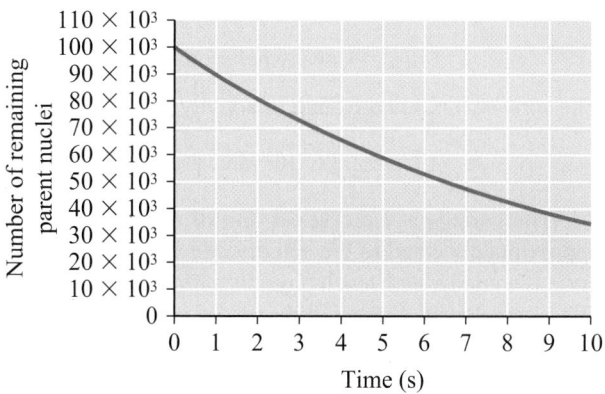

Figure 33-5 A plot of the number of parent nuclei as a function of time for a radioactive nuclide with a decay constant of $0.10 \ s^{-1}$

The Exponential Decay Law

From Equation (33-15), we can derive an expression for the number of parent nuclei remaining at any time, given the number of parent nuclei at some initial time. Equation (33-15) shows that the derivative of the function $N(t)$ is directly proportional to its value, which is a characteristic property of exponential functions. Therefore, this function can be written as follows:

$$N(t) = ae^{bt} \qquad (33\text{-}16)$$

where a and b are constants that we need to determine. Substituting this form into Equation (33-15), we get

Left-hand side: $-\dfrac{dN(t)}{dt} = -a\dfrac{d}{dt}(e^{bt}) = -b(ae^{bt})$

Right-hand side: $\lambda N(t) = \lambda(ae^{bt}) \qquad (33\text{-}17)$

Therefore, $b = -\lambda$, and

$$N(t) = ae^{-\lambda t} \qquad (33\text{-}18)$$

To determine a, we need additional information, such as the number of radioactive nuclei present at some given time. Suppose there are N_0 radioactive nuclei at $t = 0$. Evaluating Equation (33-18) at $t = 0$ gives $N(0) = a = N_0$. So,

KEY EQUATION $\qquad N(t) = N_0 e^{-\lambda t} \qquad (33\text{-}19)$

where $N(t)$ is the number of undecayed nuclei at time t, N_0 is the initial number of radioactive nuclei, and λ is the decay constant of the nuclide.

The greater the decay constant of a nuclide, the more quickly its nuclei decay. For stable nuclei, $\lambda = 0$ and the number of parent nuclei does not change with time.

Half-life

A **half-life**, $t_{1/2}$, is the time it takes for the number of radioactive nuclei in a sample to decrease to half the original value. Suppose a sample contains N_0 nuclei of a given radioactive nuclide at $t = 0$, and this number reduces to $N_0/2$ at a later time, $t = t_{1/2}$. Then from Equation (33-19),

C-33-2 Decay Constants

Figure 33-6 shows the number of remaining nuclei for four radioactive nuclides as a function of time.
(a) Arrange the nuclides, from high to low, in order of the decay constants.
(b) Each decaying nucleus of the nuclides emits the same amount of energy. Arrange the nuclides in the order of power emitted, from high to low, between $t = 1$ s and $t = 2$ s.

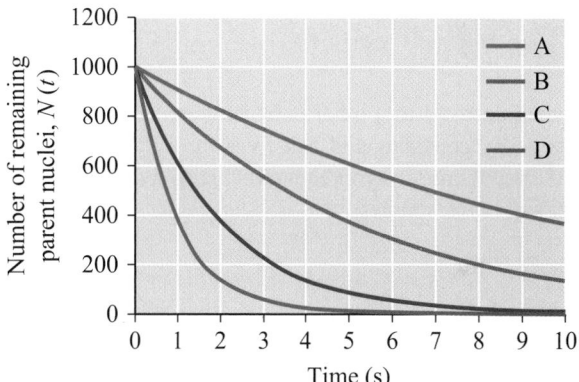

Figure 33-6 C-33-2

C-33-2 (a) D, C, B, A; (b) D, C, B, A. A higher decay constant means that the number of remaining parent nuclei will decrease faster with time.

$$N(t_{1/2}) = \frac{N_0}{2} = N_0 e^{-\lambda t_{1/2}} \qquad (33\text{-}20)$$

and

$$\frac{1}{2} = e^{-\lambda t_{1/2}}$$

Taking the natural logarithm on both sides and using the identities $\ln(e^x) = x$ and $\ln\left(\dfrac{1}{a}\right) = -\ln(a)$, we get the following relationship between the decay constant and the half-life:

$$t_{1/2} = \frac{\ln(2)}{\lambda} \simeq \frac{0.693}{\lambda} \qquad (33\text{-}21)$$

The SI unit for half-life is the second. If a sample has 1000 radioactive nuclei at some time, then it will have 500 nuclei after one half-life, 250 after two half-lives, 125 after three half-lives, and so on. A nuclide with a long half-life decays more slowly than a nuclide with a shorter half-life. For example, $^{238}_{92}U$ has a half-life of 4.468×10^9 y, so it decays slowly, and a sample containing this nuclide remains radioactive for a long time. Only half of the $^{238}_{92}U$ present at the formation of Earth has decayed so far. In contrast, radium-224 ($^{224}_{88}Ra$) has a half-life of 3.632 days, and any $^{224}_{88}Ra$ present at the time Earth was

MODERN PHYSICS

Table 33-4 The decay constants and half-lives for a few radioactive nuclides

Nuclide	Type of decay	Decay constant, λ (s⁻¹)	Half-life, $t_{1/2}$
$^{238}_{92}$U	α-decay	4.92×10^{-18}	4.46×10^9 y
$^{224}_{88}$Ra	α-decay	2.21×10^{-6}	3.63 days
$^{224}_{86}$Rn	β-decay	1.07×10^{-4}	1.8 h
$^{135}_{55}$Cs	β-decay	9.55×10^{-15}	2.30×10^6 y
$^{40}_{19}$K	β-decay	1.72×10^{-17}	1.28×10^9 y
$^{14}_{6}$C	β-decay	3.83×10^{-12}	5730 y

formed decayed into other nuclides long ago. Table 33-4 lists decay constants and half-lives for a few radioactive nuclides. A more comprehensive list is available at the text's online resources.

Decay Rate

It is easier to measure emitted radiation than to count the number of radioactive nuclei in a sample. The decay rate, or **activity**, R, of a sample is the number of nuclei that decay per unit time. The activity equals the magnitude of the time derivative of $N(t)$. From Equation (33-19), we have

$$R(t) = \left| \frac{dN(t)}{dt} \right| = \left| \frac{d}{dt}(N_0 e^{-\lambda t}) \right| = \lambda N_0 e^{-\lambda t} = \lambda N(t) \quad (33\text{-}22)$$

The more radioactive nuclei in a source, the greater its activity. Equation (33-22) shows that we can determine the number of radioactive nuclei in a sample simply by measuring the activity of a sample and knowing its decay constant. The SI unit of activity is the becquerel (Bq):

$$1 \text{ Bq} = 1 \text{ decay/s}$$

A commonly used abbreviation for decay/s is dps.

$$1 \text{ dps} = 1 \text{ decay per second}$$

The curie (Ci) is a non-SI unit of activity and is defined as

$$1 \text{ Ci} = 3.7 \times 10^{10} \text{ decays/s}$$

The curie was the original unit of activity, and it is approximately equal to the activity of 1 g of radium. The curie is named after French physicists Marie Curie (1867–1934) and Pierre Curie (1859–1906), whose pioneering research on radioactivity includes the discovery of the elements radium and polonium. From Equation (33-22),

$$R(0) = \lambda N_0 \quad (33\text{-}23)$$

Therefore, we can write

$$R(t) = R(0)e^{-\lambda t} \quad (33\text{-}24)$$

The activity of a nuclide decays exponentially with time. For example, in one half-life, the decay rate of a nuclide decreases to half of its original value.

 CHECKPOINT

C-33-3 A Radioactive Nuclide

A radioactive nuclide has a half-life of two days. On a certain day, its activity at 9:00 a.m. is measured to be 1000 Bq. The next day at 9:00 a.m. its activity is approximately
(a) 750 Bq;
(b) 700 Bq;
(c) 500 Bq;
(d) 250 Bq.

C-33-3 (b)

 EXAMPLE 33-4

Radioactive Gallium

Gallium-67 ($^{67}_{31}$Ga), a radioactive isotope of gallium with a half-life of 3.26 days, is used for some types of medical imaging. A source containing $^{67}_{31}$Ga is calibrated to emit 3.7×10^7 Bq of radiation.

(a) How many gallium-67 atoms are there in the source?
(b) It takes one day to deliver the source to a hospital. What is the activity of the source when it arrives?

SOLUTION

(a) We have already derived all the required relationships between various quantities. We will measure time in seconds so that all quantities are converted in the correct units.

$$R(0) = 3.7 \times 10^7 \text{ Bq}$$

The number of radioactive nuclei is related to activity by Equation (33-22). Therefore, the number of gallium-67 atoms in the source at $t = 0$ is

$$N(0) = \frac{R(0)}{\lambda}$$

The decay constant of gallium-67 is

$$\lambda = \frac{\ln(2)}{t_{1/2}} = \frac{\ln(2)}{(3.26 \text{ days} \times 24 \text{ h/day} \times 3600 \text{ s/h})}$$

$$= 2.46 \times 10^{-6} \text{ s}^{-1}$$

Therefore,

$$N(0) = \frac{3.7 \times 10^7 \text{ decay/s}}{2.46 \times 10^{-6} \text{ s}^{-1}} = 1.5 \times 10^{13} \text{ atoms}$$

33-5 Nuclear Reactions

When a fast-moving nucleus or a particle such as a proton, neutron, or photon collides with a nucleus, the collision rearranges the nucleons. In such **nuclear reactions**, the nuclei in the final states may be different from the initial nuclei. A nuclear reaction can be represented by the following equation:

$$a + X \rightarrow Y + b \qquad (33\text{-}25)$$

where a is the incoming nucleus or a particle, X is the target nucleus, and Y and b are the final products.

The left side of Equation (33-25) is called the initial state of the reaction, and the right side is called the final state of the reaction. There can be more than two products in the final state. If the projectile a is a nucleus or a positively charged particle, then it must have sufficient kinetic energy to overcome the Coulomb repulsion and come within the range of the strong force of the target nucleus X.

Examples of Nuclear Reactions

■ When a beryllium-9 nucleus absorbs a fast-moving helium-4 nucleus, it transforms into a carbon-12 nucleus and emits a neutron:

$$^4_2\text{He} + ^9_4\text{Be} \rightarrow ^{12}_6\text{C} + ^1_0\text{n} \qquad (33\text{-}26a)$$

A schematic diagram for this nuclear reaction is shown in Figure 33-7. Note that the number of nucleons is conserved in the reaction. This reaction was used by the British physicist James Chadwick (1891–1974) to prove the existence of neutrons in 1932.

■ A neutron is absorbed by a nitrogen-14 nucleus, resulting in a carbon-14 nucleus and a proton in the final state. This reaction occurs in Earth's atmosphere

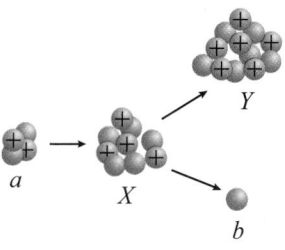

Figure 33-7 A nuclear reaction of beryllium-9 with a helium-4 nucleus. Protons are represented by red circles with a plus sign and neutrons by blue circles.

(b) Given the activity at $t = 0$, the activity at a later time is given by $R(t) = R(0)e^{-\lambda t}$. Using $1 = 24$ h/day × 3600 s/h = 86 400 s, the activity of the source when it arrives is

$$R(86\ 400) = (3.7 \times 10^7 \text{ Bq})e^{-2.461 \times 10^{-6} \times 86\ 400} \approx 3.0 \times 10^7 \text{ Bq}$$

by cosmic rays and is responsible for replenishing carbon-14 which is present in natural carbon.

$$^1_0\text{n} + ^{14}_7\text{N} \rightarrow ^1_1\text{p} + ^{14}_6\text{C} \qquad (33\text{-}26b)$$

■ Two fast-moving oxygen-16 nuclei fuse together. One of the possible final states consists of a neon-20 nucleus and a carbon-12 nucleus. The oxygen-fusion reaction occurs in massive stars that have used up the lighter elements in their core.

$$^{16}_8\text{O} + ^{16}_8\text{O} \rightarrow ^{20}_{10}\text{Ne} + ^{12}_6\text{C} \qquad (33\text{-}26c)$$

Conservation Laws for Nuclear Reactions

There are several physical quantities that are conserved in all nuclear reactions. These quantities include some that we are already familiar with from classical mechanics and electrodynamics, and there are others that are unique to nuclear physics.

Energy Conservation Total energy—the sum of kinetic and rest mass energies—is conserved in a nuclear reaction. Note that neither kinetic energy nor the rest mass energy is conserved by itself. Consider the nuclear reaction

$$a + b \rightarrow c + d$$

where a, b, c, and d represent nuclei or particles. The energy of each body in the above reaction is the sum of its kinetic energy and the rest mass energy. The conservation of energy requires that

$$K_a + M_a c^2 + K_b + M_b c^2 = K_c + M_c c^2 + K_d + M_d c^2$$

where K is kinetic energy, and Mc^2 is the rest mass energy. The total gain in the kinetic energy in this reaction is equal to the total loss of rest mass energy:

$$\Delta K = (K_f - K_i) = (M_i - M_f)c^2$$

where K_i and K_f denote, respectively, the initial and final total kinetic energies, and $M_i c^2$ and $M_f c^2$ are the corresponding total rest mass energies. The gain ΔK in the kinetic energy is called the **Q-value** of the reaction and is denoted by Q. Thus

KEY EQUATION $$Q = (M_i - M_f)c^2 \qquad (33\text{-}27)$$

If the rest mass energy in the final state is less than the rest mass energy in the initial state, the Q-value is positive and kinetic energy is released in the nuclear reaction. In this case, the reaction would occur even if $K_i = 0$ (e.g., in a nuclear decay). If the rest mass energy in the final state is greater the rest mass energy in the initial state, the Q-value

MODERN PHYSICS

is negative. In this case, the bodies in the initial state must have sufficient kinetic energy for the reaction to occur.

By writing the nuclear mass energy in terms of the nuclear binding energy, Equation (33-10), we can write the Q-value of a nuclear reaction in terms of the nuclear binding energies as follows:

$$Q = B_f - B_i$$

where B_i and B_f denote the total binding energies of the bodies in the initial and the final state. Here, we have used the fact that the total number of nucleons (protons plus neutrons) in the initial and the final state of a nuclear reaction remains the same.

Charge Conservation Total charge is conserved in a nuclear reaction. For example, in the beryllium–helium reaction above in Equation (33-26a), the total charge is $6e$ (six protons) in both the initial and final states.

Momentum Conservation Momentum is conserved in all nuclear reactions. For the reaction of Equation (33-25),

$$\vec{p_a} + \vec{p_X} = \vec{p_b} + \vec{p_Y}$$

Angular Momentum Conservation The total angular momentum (the sum of the orbital angular momentum and the spin) is conserved in all nuclear reactions.

Conservation of Nucleons The total number of nucleons (protons and neutrons) in a nuclear reaction remains the same. A proton can change into a neutron and vice versa, but the total number of neutrons plus protons remains constant before and after the reaction. For example, in

reaction (33-26a) above, there are a total of 13 nucleons in both the initial state and in the final state.

There are other physical quantities that are conserved in nuclear reactions; however, that discussion is beyond the scope of this introduction.

33-6 α, β, and γ Decays

In a **nuclear decay**, a nucleus transforms into another nucleus or rearranges its constituents on its own, without an interaction with an external particle, nucleus, or radiation.

Alpha (α) Decay

From Figure 33-4, we can observe that a large nucleus can increase its binding energy, and hence become more stable, by reducing its atomic number and thus transforming into a different nucleus. For example, a nucleus could emit neutrons or protons. Since a helium-4 nucleus has a large binding energy, a heavy nucleus can increase its binding energy more by emitting a helium-4 nucleus than by emitting separate protons and neutrons. When a parent nucleus X with Z protons and N neutrons decays by emitting an α-particle ($_2^4$He), the resulting daughter nucleus Y has $Z - 2$ protons and $N - 2$ neutrons. Energy is released in an **alpha decay** (α-decay) and appears in the form of the kinetic energy of the α-particle and the daughter nucleus. We can represent an α-decay as

EXAMPLE 33-5

Proton Absorption on Lithium

In 1932, English physicist John Cockroft and Irish physicist Ernest Walton produced the first nuclear reaction by bombarding lithium with 600 keV protons and producing helium in the following process:

$$_1^1\text{H} + {}_3^7\text{Li} \rightarrow {}_2^4\text{He} + {}_2^4\text{He}$$

Determine the Q-value of the reaction.

SOLUTION

From Table 33-5, the atomic masses of hydrogen lithium, and helium atoms are as follows:

Atomic mass of $_1^1\text{H} = 1.007\ 825$ u

Atomic mass of $_3^7\text{Li} = 7.016\ 003$ u

Atomic mass of $_2^4\text{He} = 4.002\ 603$ u

Table 33-5 Table of selected atomic masses

Z	A	Symbol	Name	Atomic mass (u)
0	1	n	Neutron	1.008 665
1	1	p	Proton	1.007 276
1	1	H	Hydrogen	1.007 825
1	2	H	Deuterium	2.014 102
1	3	H	Tritium	3.016 049
2	3	He	Helium-3	3.016 029
2	4	He	Helium-4	4.002 603

MODERN PHYSICS

Z	A	Symbol	Name	Atomic mass (u)
3	6	Li	Lithium	6.015 121
3	7	Li	Lithium	7.016 003
4	9	Be	Beryllium	9.012 182
5	10	B	Boron	10.012 937
5	11	B	Boron	11.009 305
6	12	C	Carbon	12
6	13	C	Carbon	13.003 355
6	14	C	Carbon	14.003 242
7	14	N	Nitrogen	14.003 074
7	15	N	Nitrogen	15.000 109
8	16	O	Oxygen	15.994 915
8	17	O	Oxygen	16.999 131
8	18	O	Oxygen	17.999 160
10	20	Ne	Neon	19.992 436
19	39	K	Potassium	38.963 708
19	40	K	Potassium	39.964 000
19	41	K	Potassium	40.961 827
26	56	Fe	Iron	55.934 939
36	90	Kr	Krypton	89.919 517
37	93	Rb	Rubidium	92.922 042
38	88	Sr	Strontium	87.905 619
40	102	Zr	Zirconium	101.922 980
52	134	Te	Tellurium	131.911 369
55	141	Cs	Cesium	140.920 046
56	144	Ba	Barium	143.922 953
82	208	Pb	Lead	207.976 627
83	208	Bi	Bismuth	297.979 717
84	210	Po	Polonium	209.982 848
84	218	Po	Polonium	218.008 965
86	222	Rn	Radon	222.017 571
90	231	Th	Thorium	231.036 298
90	232	Th	Thorium	232.038 051
90	234	Th	Thorium	234.043 593
92	232	U	Uranium	232.037 129
92	235	U	Uranium	235.043 924
92	238	U	Uranium	238.050 783

The Q-value of the reaction is equal to the difference in the rest mass energies of the bodies in the initial and the final states.

$$Q = ((1.007\ 825\ \text{u} + 7.016\ 003\ \text{u}) - 2 \times (4.002\ 603\ \text{u})) = 0.018\ 6\ \text{u}$$

$$= \left(\frac{931.5\ \text{MeV}}{\text{u}}\right) \times 0.0186\ \text{u} = 17.3\ \text{MeV}$$

With their rudimentary instrumentation Cockroft and Walton measured that 17.2 MeV of energy is released in this reaction. This was the first experimental verification of Einstein's mass-energy relation.

$$\,_Z^A X \rightarrow \,_{Z-2}^{A-4} Y + \,_2^4 \text{He} + Q_\alpha \qquad (33\text{-}28)$$

Here, $\,_Z^A Y$ is the parent nucleus, $\,_{Z-2}^{A-4} Y$ is the daughter nucleus, and Q_α represents the energy released in the decay and is called the **Q-value** of the decay.

From the definition of Q-value, the energy released in α-decay is

$$Q_\alpha = M_X c^2 - (M_Y + M_\alpha)c^2 \qquad (33\text{-}29)$$

where M_X, M_Y, and M_α are the atomic masses of the parent nucleus, the daughter nucleus, and the α-particle, respectively.

By the law of conservation of energy, α-decay can happen only if the rest mass of the parent nucleus is *greater* than the combined rest masses of the daughter nucleus and the α-particle. Therefore, Q_α must be positive for a nucleus to undergo α-decay. We can express Q_α in terms of the binding energies of the nuclei, using Equation (33-10):

$$M_X c^2 = Z m_p c^2 + (A - Z)m_N c^2 - B(A, Z)$$

$$M_Y c^2 = (Z - 2)m_p c^2 + (A - Z - 2)m_N c^2 - B(A - 4, Z - 2)$$

$$M_\alpha c^2 = 2m_p c^2 + 2m_N c^2 - B(4, 2) \qquad (33\text{-}30)$$

Note that the daughter nucleus has $Z - 2$ protons and $A - 4 - (Z - 2) = A - Z - 2$ neutrons. By inserting Equation (33-30) into Equation (33-29), we obtain an expression for Q_α in terms of binding energies:

$$Q_\alpha = B(A - 4, Z - 2) + B(4, 2) - B(A, Z) \qquad (33\text{-}31)$$

We see that the Q-value is the difference in the binding energies of the final products and the initial parent nucleus. From experiments we know that the binding energy of a helium nucleus is 28.2 MeV, so $B(4, 2) = 28.2$ MeV.

Most of the energy released in the α-decay in Example 33-6 is carried away by the α-particle in the form of kinetic energy. Approximately 5% of the energy appears as the kinetic and excitation energy of the polonium nucleus. When formed in this decay process, the polonium nucleus is initially in an excited state (just like an atom can be in an excited state when not all of its electrons are in their lowest energy states). The excited polonium nucleus reaches its ground state by emitting a γ-ray, a high-energy photon.

Decay Diagrams

A nuclear decay can be graphically represented by plotting the atomic numbers (charge) of parent and daughter nuclei along the horizontal axis and their energies along the vertical axis. Such a diagram is called a **decay diagram**. Energy is always released in a nuclear decay, so the daughter nucleus is lower in energy than the parent nucleus. In these diagrams a short horizontal line, centred at a given Z, represents a nucleus with a fixed charge and in a particular energy state. An arrow with its tail at the parent nucleus and its head at the daughter nucleus represents the decay process. The vertical distance between the parent and the daughter nuclei represents the energy released in the decay, that is, the Q-value of the decay (Figure 33-8(a)). Figure 33-8(b) shows a decay diagram for the α-decay of uranium-238.

EXAMPLE 33-6

The α-Decay of Radon

Radon-222 ($\,_{86}^{222}$Rn) has a half-life of 3.8235 days and decays by emitting an α-particle.

(a) What is the daughter nucleus in this decay?
(b) Using the atomic masses from Table 33-5, determine the Q-value of this decay.
(c) Find the difference in the binding energies of the parent and the daughter nuclei.

SOLUTION

(a) For the parent nucleus, $A = 222$ and $Z = 86$. For the daughter nucleus, $A = 222 - 4 = 218$ and $Z = 86 - 2 = 84$.
From a periodic or atomic table, we find that the nucleus with $Z = 84$ is polonium (Po) and the daughter nucleus is $\,_{84}^{218}$Po. Therefore, the equation for α-decay is

$$\,_{86}^{222}\text{Rn} \rightarrow \,_{84}^{218}\text{Po} + \,_2^4\text{He} + Q_\alpha$$

(b) Looking up the known atomic masses of the isotopes (Table 33-5) in the reaction, we find the following:

Atomic mass of $\,_{86}^{222}$Rn = 222.017 571 u

Atomic mass of $\,_{84}^{218}$Po = 218.008 965 u

Atomic mass of $\,_2^4$He = 4.002 603 u

Now we can use Equation (33-29) to determine Q_α:

$$Q_\alpha = (222.0175\ 71\ \text{u} - 218.008\ 965\ \text{u} - 4.002\ 603\ \text{u})\left(\frac{931.5\,\text{MeV}}{\text{u}}\right)$$

$$= 5.59\ \text{MeV}$$

(c) From Equation (33-31):

$$Q_\alpha = B(218, 84) + B(4, 2) - B(222, 86)$$

and

$$B(222, 86) - B(218, 84) = B(4, 2) - Q_\alpha$$

$$= 28.2\ \text{MeV} - 5.59\ \text{MeV}$$

$$= 22.6\ \text{MeV}$$

MODERN PHYSICS

(a)

Energy Diagram for α-Decay

(b)

α-Decay Diagram of Uranium-238

Figure 33-8 (a) A decay diagram for α-decay (b) A decay diagram for the α-decay of $^{238}_{92}$U

Beta (β) decay

Nuclei are most stable in the lowest energy state. Figure 33-9 shows a plot of the rest mass as a function of the number of protons (Z) for various nuclei with $A = 16$. Notice that oxygen-16, with 8 protons and 8 neutrons, has the lowest rest mass and therefore the lowest rest mass energy. The rest mass of $^{16}_{7}$N is 16.006 102 u, and the rest mass energy of $^{16}_{8}$O is 15.994 914 u. Therefore, the energy of $^{16}_{8}$O is lower by

$$(16.006\,102\,\text{u} - 15.994\,914\,\text{u})(931.5\,\text{MeV/u}) = 10.4\,\text{MeV}$$

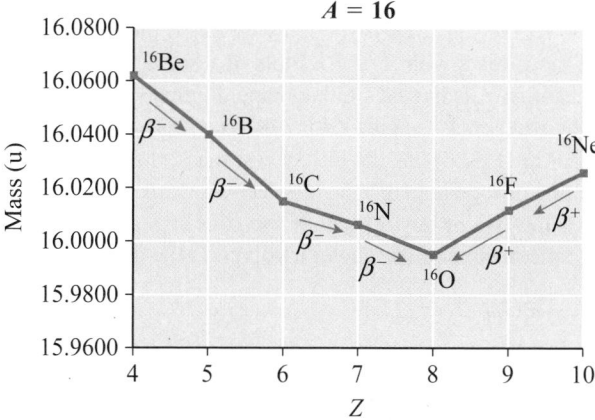

Figure 33-9 The rest masses as a function of the number of protons for various nuclei with $A = 16$

Therefore a $^{16}_{7}$N nucleus could transform into a $^{16}_{8}$O, a nuclide of lower rest-mass energy, by transforming one neutron into a proton.

Now we look at conservation of charge. The initial neutron has no charge, but the final proton has a charge of $+e$. So, there must be another particle, with charge $-e$, that is created when a neutron changes into a proton—this particle is an electron. We can write the process as follows:

$$\text{neutron} \rightarrow \text{proton} + \text{electron}$$

or

$$^{1}_{0}\text{n} \rightarrow {}^{1}_{1}\text{p} + {}^{0}_{-1}\text{e} \tag{33-32}$$

In using the symbol $^{1}_{0}$n to represent a neutron, we have used the same convention as that for a nucleus. A neutron has no charge ($Z = 0$), and the atomic mass number is $A = 1$. Similarly, an electron has zero atomic mass number and has a charge of -1. The proton remains inside the nucleus; only the electron comes out. The emitted electron is called a beta (β) ray, and the decay process in which a nucleus changes its charge by one unit is called **beta decay** (β-decay).

Let us examine the conservation of energy and momentum during this process. We assume that the neutron is at rest. The Q-value of the neutron decay process is

$$Q = m_n c^2 - (m_p c^2 + m_e c^2) \tag{33-33}$$

Using $m_n c^2 = 939.565\,56\,\text{MeV}$, $m_p c^2 = 938.2720\,\text{MeV}$, and $m_e c^2 = 0.510\,999\,\text{MeV}$, $Q = 0.782\,\text{MeV}$. The Q-value is positive; therefore, the decay of a neutron into a proton is allowed by the law of conservation of energy. A free neutron is indeed unstable. It decays into a proton with a half-life of 614 s.

Since the neutron is at rest, the proton and the electron momenta must be equal and opposite so that the total momentum of the final state is zero. We leave it as an end-of-chapter exercise (problem 46 in Chapter 34) to show that the magnitude of the proton and electron momentum, p, is given by

$$\frac{p^2}{c^2} = \left(\frac{m_n^2 + m_p^2 - m_e^2}{2m_n}\right)^2 - m_p^2$$

Inserting values for the masses of the three particles, we get $p = 1.188\,\text{MeV}/c$. So the proton and electron have a well-defined momentum. The kinetic energy of an electron, KE_e, corresponding to this momentum is the difference between its total energy and the rest mass energy:

$$KE_e = \sqrt{p^2 c^2 + m_e^2 c^4} - m_e c^2 = 0.782\,\text{MeV}$$

When we observe decays of neutrons, the emerging electrons should all have 0.782 MeV of kinetic energy. But almost all electrons in neutron decay experiments have less kinetic energy, as shown in Figure 33-10. Where is the rest of the energy going? Could neutron decay somehow violate conservation of energy?

This problem puzzled physicists until Wolfgang Pauli suggested that a neutron decays into three particles: a proton, an electron, and a third particle, which he called a **neutrino** (ν). Neutrino is Italian for "small neutral one."

Energy Distribution

0.782 MeV

Electron kinetic energy (MeV)

Figure 33-10 The kinetic energy of electrons emitted in the β-decay of a neutron

Since the particle had not been previously detected, Pauli assumed that it interacts with matter only by the weak force.

Pauli argued that the neutrino and the electron share the energy available in the decay. Since a proton is approximately 1836 times more massive than an electron, the speed of the proton is quite small; consequently, its kinetic energy can be ignored. The sum of kinetic energies of the electron and the neutrino must be equal to the Q-value calculated in Equation (33-33). So the electron's kinetic energy can have any value between zero and 0.782 MeV. The neutrino must be close to massless because some electrons in the β-decay of neutrons are emitted with kinetic energy nearly equal to the Q-value, leaving almost no energy for the neutrino. So, Pauli predicted that the neutrino has no charge, no mass, and an intrinsic spin of $\frac{1}{2}$, making it a fermion.

Neutrinos were discovered in 1952 by American physicists Frederick Reines (1918–98) and Clyde Lorrain Cowan (1919–74), for which they shared a Nobel Prize in 1995. However, 50 years after Pauli's original hypothesis, it was determined that the third particle emitted in neutron β-decay is actually an antiparticle of the neutrino, called the antineutrino, $\bar{\nu}$. (We will discuss the nature of antiparticles in Chapter 34.) The β-decay of a neutron can then be written as follows:

$$_{0}^{1}n \rightarrow {}_{1}^{1}p + {}_{-1}^{0}e + \bar{\nu} \qquad (33\text{-}34)$$

Similarly, the β-decay of a nucleus in which a neutron converts into a proton, for example, the transformation of $_{7}^{16}N$ into $_{8}^{16}O$, is written as

$$_{7}^{16}N \rightarrow {}_{8}^{16}O + {}_{-1}^{0}e + \bar{\nu} + Q \qquad (33\text{-}35)$$

The decay specified by Equation (33-35) is called the β-decay of $_{7}^{16}N$. The half-life of $_{7}^{16}N$ for β-decay to $_{8}^{16}O$ is 7.13 s, and that is why nitrogen-16 is not seen in nature. The Q-value in Equation (33-35) is not the same as the Q-value for the β-decay of a free neutron. The neutron that converts into a proton in $_{7}^{16}N$ β-decay is not free and is bound to the nucleus by the strong force. Similarly, the final proton is bound to the $_{8}^{16}O$ nucleus. We will show how to calculate the Q-value for β-decay of a nucleus later in this section.

All $A = 16$ elements with $Z < 8$ eventually β-decay into $_{8}^{16}O$: $_{4}^{16}Be$ β-decays into $_{5}^{16}B$, which then β-decays into $_{6}^{16}C$, which β-decays to $_{7}^{16}N$, and then finally to $_{8}^{16}O$. The half-lifes and the Q-values for the decays at each stage are different. The process stops at oxygen-16 because it is stable and does not decay.

How about the nuclei with $Z > 8$? From Figure 33-9 we observe that $_{9}^{16}F$ has a higher rest mass energy than $_{8}^{16}O$. So $_{9}^{16}F$ can decay to $_{8}^{16}O$, which requires transforming a proton into a neutron. To conserve electric charge, a particle with a charge equal to the charge of a proton must be emitted during this decay. This particle is the positron, the antiparticle of the electron. A positron has the same mass as an electron but opposite charge. The positron is denoted by the symbol $_{+1}^{0}e$. The energy of positrons emitted during proton decay has a distribution similar to that for electrons emitted in β-decay, indicating that a third particle is also produced when a proton decays into a neutron. The emitted particle is the neutrino. So, the β-decay of a proton can be represented as

$$_{1}^{1}p \rightarrow {}_{0}^{1}n + {}_{+1}^{0}e + \nu \qquad (33\text{-}36)$$

There is a crucial difference between neutron decay and proton decay. The rest mass of a proton is *less* than the rest mass of a neutron. Therefore, a free proton cannot decay into a neutron. Proton decay into a neutron is possible only inside a nucleus where other protons and neutrons can provide the required energy. Otherwise, all hydrogen atoms would have decayed a long time ago.

The decay of $_{9}^{16}F$ into $_{8}^{16}O$ can be represented as

$$_{9}^{16}F \rightarrow {}_{8}^{16}O + {}_{+1}^{0}e + \nu + Q \qquad (33\text{-}37)$$

$_{9}^{16}F$ is extremely unstable and has a half-life of the order of 10^{-20} s.

When a neutron converts into a proton, a negatively charged particle (electron) is emitted from the nucleus. This process is called **beta-minus** (β^{-}) decay. Similarly, the process during which a proton converts into a neutron and a positively charged particle (positron) is emitted from the nucleus is called **beta-plus** (β^{+}) **decay**. So we say "nitrogen-16 β^{-} decays and transforms into oxygen-16."

At the start of this section we plotted the rest masses of the isobars with $A = 16$. Plots of rest masses for other sets of isobars have a similar shape. Figure 33-11 shows a plot of rest mass versus Z for isobars with $A = 208$. Since lead-208 ($_{82}^{208}Pb$) has the lowest rest mass, the other nuclei in this group either β^{-}- or β^{+}-decay toward lead-208.

Let us see how Q-values for β-decays are related to atomic masses of the nuclei involved in the decay.

β^{-} Decay

The following equation represents the β^{-}-decay of a parent nucleus $_{Z}^{A}X$:

$$_{Z}^{A}X \rightarrow {}_{Z+1}^{A}Y + {}_{-1}^{0}e + \bar{\nu} + Q_{\beta^{-}} \qquad (33\text{-}38)$$

where $Q_{\beta^{-}}$ is the Q-value of the β-decay.

Figure 33-11 The rest masses as a function of the number of protons for various nuclei with $A = 208$

For example, sulfur-35 has a half-life of 87.9 days for the β-decay into chlorine-35:

$$^{35}_{16}S \rightarrow \,^{35}_{17}Cl + \,^{0}_{-1}e + \bar{\nu} + 0.167 \text{ MeV}$$

From the definition of Q-value,

$$Q_{\beta^-} = M_X c^2 - (M_Y + m_e)c^2 \quad (33\text{-}39)$$

Here, m_e is the electron mass, and M_X and M_Y are the nuclear masses of the parent and the daughter nuclei, respectively. We can also write the Q-value in terms of atomic masses of corresponding atoms by adding the rest mass energy of Z electrons to both terms on the right side of Equation (33-39):

EXAMPLE 33-7

The β-Decay of Carbon-14

Carbon-14 ($^{14}_{6}C$) is an unstable isotope that undergoes β^--decay.

(a) What is the daughter nucleus for the decay?
(b) Write an equation that describes the decay.
(c) Calculate the Q-value for the decay.
(d) Draw a decay diagram for the decay.

SOLUTION

(a) A carbon-14 nucleus has 6 protons and 8 neutrons. In a β^--decay a neutron converts into a proton. Therefore, the daughter nucleus has 7 protons and 7 neutrons. The daughter nucleus is nitrogen-14 ($^{14}_{7}N$).

(b) The decay equation is

$$^{14}_{6}C \rightarrow \,^{14}_{7}N + \,^{0}_{-1}e + \bar{\nu} + Q_{\beta^-}$$

(c) To calculate the Q-value of this decay, we use Equation (33-41). From Table (33-5),

$$\text{Atomic mass of } ^{14}_{6}C = 14.003\ 242 \text{ u}$$
$$\text{Atomic mass of } ^{14}_{7}N = 14.003\ 074 \text{ u}$$

The Q-value of the β^- decay is

$$Q_{\beta^-} = (M_X + Zm_e)c^2 - (M_Y + m_e + Zm_e)c^2 \quad (33\text{-}40)$$

The parent nucleus has Z protons so its atom has Z electrons. The daughter nucleus has $Z + 1$ protons, and its atom has $Z + 1$ electrons. Therefore, the terms in the parentheses in Equation (33-40) add to the atomic masses of the parent and daughter elements, and Q_{β^-} is equal to the energy equivalent of the difference between these atomic masses:

$$Q_{\beta^-} = \tilde{M}_X c^2 - \tilde{M}_Y c^2 \quad (33\text{-}41)$$

If the atomic mass of the parent nucleus is greater than the atomic mass of the daughter nucleus, the parent nucleus will β^--decay into the daughter nucleus.

In a β^--decay, the charge of the nucleus increases by e. The corresponding decay diagram is shown in Figure 33-12.

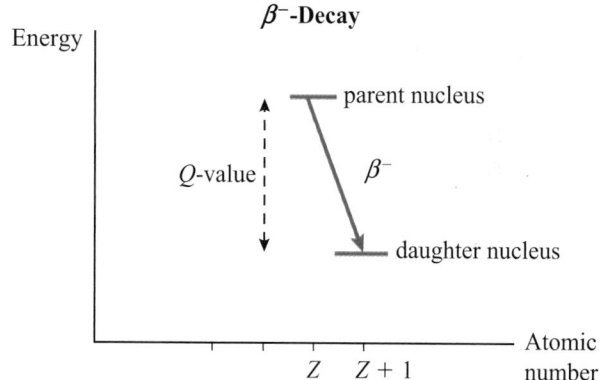

Figure 33-12 A decay diagram for β^--decay

$$Q_{\beta^-} = (\tilde{M}_{\text{parent}} - \tilde{M}_{\text{daughter}})c^2$$
$$= (14.003\ 242 \text{ u} - 14.003\ 074 \text{ u})c^2$$
$$= (1.68 \times 10^{-4} \text{ u})\left(\frac{931.5 \text{ MeV}/c^2}{\text{u}}\right)c^2$$
$$= 0.156 \text{ MeV}$$

(d) The decay diagram is shown in Figure 33-13.

Figure 33-13 The β^- decay of $^{14}_{6}C$

MODERN PHYSICS

Radiocarbon Dating

Radiocarbon dating (also called **carbon dating**) uses the naturally occurring carbon-14 ($^{14}_{6}$C) to estimate the age of carbon-rich materials. Carbon has two stable isotopes, carbon-12 and carbon-13. All other isotopes of carbon are radioactive. Carbon-14 has a half-life of 5730 ± 40 y and β-decays to nitrogen-14 ($^{14}_{7}$N), a stable element:

$$^{14}_{6}\text{C} \rightarrow {}^{14}_{7}\text{N} + e^{-} + \bar{\nu}_{e}$$

Carbon-14 is continuously formed in the upper atmosphere by high-energy neutrons that are produced by cosmic rays. The neutrons displace a proton from a nitrogen-14 nucleus, converting it into a carbon-14 nucleus:

$$^{1}_{0}\text{n} + {}^{14}_{7}\text{N} \rightarrow {}^{14}_{6}\text{C} + {}^{1}_{1}\text{p}$$

Carbon-14 is chemically identical to carbon-12 and reacts with oxygen to form carbon dioxide gas, which diffuses through the atmosphere, biosphere, and oceans. Living organisms (plants, trees, and animals) continuously exchange carbon dioxide with their environment, so they absorb a small amount of carbon-14: one atom of carbon-14 for every 1.2×10^{12} atoms of carbon. The absorbed carbon eventually decays to nitrogen-14. The activity due to carbon-14 decay is approximately 15 decays per minute per gram of carbon.

As long as an organism is alive, its interactions with the environment keep the ratio of carbon-14 in the organism the same as in the environment. When an organism dies, it stops exchanging carbon, and the number of carbon-14 atoms in it begins to decrease. Since the activity of a radioactive element is proportional to the number of undecayed nuclei, the activity of a dead organism decreases with time, reducing by 50% every 5730 y (Figure 33-14). By measuring the activity of carbon-14 in organic matter, scientists can determine the age of an animal bone, a wooden object, or a piece of cloth that was made from plant or animal fibres. The length of the half-life of carbon-14 makes this method of dating organic matter reasonably reliable for ages less than 60 000 years.

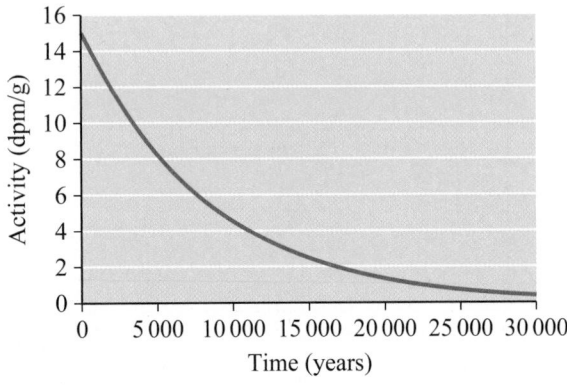

Figure 33-14 The activity of a sample of carbon-14 as a function of time

EXAMPLE 33-8

The Age of Iceman Ötzi

In September 1991, two hikers discovered the body of a man frozen in a glacier in the mountains near Austria and Italy. The iceman was nicknamed Ötzi, after the mountain range where he was found. At the University of Innsbruck, it was determined that the carbon-14 activity of the body was 53% of the activity of living organisms. How old is the Ötzi iceman?

SOLUTION

Let $N(0)$ be the number of carbon-14 atoms in a living organism and $N(t)$ be the number of carbon-14 atoms in an organism that has been dead for t years. From Equation (33-19),

$$N(t) = N(0)e^{-\lambda t}$$

where λ is the decay constant of carbon-14.
Dividing the above equation by $N(0)$, taking the natural logarithm on both sides, and using $\ln(e^x) = x$ gives

$$\ln\left(\frac{N(t)}{N(0)}\right) = -\lambda t$$

and

$$t = -\frac{1}{\lambda} \ln\left(\frac{N(t)}{N(0)}\right) = -\frac{t_{1/2}}{\ln(2)} \ln\left(\frac{N(t)}{N(0)}\right)$$

Since the activity is proportional to the number of undecayed nuclei,

$$\frac{\lambda N(t)}{\lambda N(0)} = \frac{N(t)}{N(0)} = 0.53$$

and

$$t = -\frac{5730 \text{ y}}{\ln(2)} \ln(0.53) = 5250 \text{ y}$$

Making sense of the result:

Several corrections and approximations are required in such analysis, for example, to adjust for historical variations in the concentration of carbon-14 in the atmosphere. It is generally agreed that iceman Ötzi died approximately 5300 y ago.

β^{+} Decay

The following equation describes the β^{+}-decay of a parent nucleus $^{A}_{Z}X$:

$$^{A}_{Z}X \rightarrow {}^{A}_{Z-1}Y + {}^{0}_{+1}e + \nu + Q_{\beta^{+}} \qquad (33\text{-}42)$$

Here, $Q_{\beta^{+}}$ is the Q-value for β^{+}-decay. For example, the β-decay of fluorine-17 ($^{17}_{9}$F) into oxygen-17 ($^{17}_{8}$O) is given by

$$^{17}_{9}\text{F} \rightarrow {}^{17}_{8}\text{O} + {}^{0}_{+1}e + \nu + 1.74 \text{ MeV}$$

Similarly, sodium-22 ($^{22}_{11}$Na) β-decays into neon-22:

$$^{22}_{11}\text{Na} \rightarrow {}^{22}_{10}\text{Ne} + {}^{0}_{+1}e + \nu + 0.167 \text{ MeV}$$

MODERN PHYSICS

Similar to Equation (33-39), for β^+-decay,

$$Q_{\beta^+} = M_X c^2 - (M_Y + m_e)c^2 \qquad (33\text{-}43)$$

where m_e is the positron mass (same as the electron mass).

We can write Q_{β^+} in terms of the atomic masses by allowing for the rest mass energy of the electrons in the atoms. The parent nucleus has Z protons, so its atom has Z electrons. The daughter nucleus has $Z-1$ protons, and its atom has $Z-1$ electrons. So, the Q-value for a β^+-decay can be written as

$$Q_{\beta^+} = \tilde{M}_X c^2 - \tilde{M}_Y c^2 - 2m_e c^2 \qquad (33\text{-}44)$$

Notice that in terms of atomic masses, the Q-value for β^+-decay differs from the Q-value for β^--decay by twice the electron rest mass energy, $2m_e c^2$, or 1.022 MeV. Since Q_{β^+} must be greater than zero for β^+-decay to occur, the atomic mass of the parent nucleus must be greater than the atomic mass of the daughter nucleus by at least 1.022 MeV.

EXAMPLE 33-9

The Q-Value of the β^+-Decay of Bismuth-208

Calculate the Q-value for the β^+-decay of bismuth-208 ($^{208}_{83}$Bi) using the atomic masses given in Table 33-5.

SOLUTION

In the β^+-decay of bismuth-208 ($^{208}_{83}$Bi), the daughter nucleus has one less proton than the parent nucleus. The nucleus with 82 protons is lead, so the decay process is

$$^{208}_{83}\text{Bi} \rightarrow {}^{208}_{82}\text{Pb} + {}^{0}_{+1}\text{e} + \nu + Q_{\beta^+}$$

The Q-value of the decay is given by Equation (33-44). From Table 33-5,

Atomic mass of $^{208}_{83}$Bi = 207.979 717 u

Atomic mass of $^{208}_{82}$Pb = 207.976 627 u

Rest mass of an electron = 0.000 548 u

The Q-value of the decay is therefore

$$Q_{\beta^+} = \tilde{M}_X c^2 - \tilde{M}_Y c^2 - 2m_e c^2$$

$$= (207.979\ 717\ \text{u} - 207.976\ 627\ \text{u} - (2 \times 0.000\ 548\ \text{u}))c^2$$

$$= (0.001\ 994\ \text{u})\left(\frac{931.5\ \text{MeV}/c^2}{\text{u}}\right)c^2$$

$$= 1.86\ \text{MeV}$$

EXAMPLE 33-10

Neutrinos from Bananas!

An average banana has approximately 450 mg of potassium. Naturally occurring potassium consists of three isotopes: $^{39}_{19}$K (93.3%), $^{40}_{19}$K (0.012%), and $^{41}_{19}$K (6.73%). Both $^{39}_{19}$K and $^{41}_{19}$K are stable, and $^{40}_{19}$K undergoes both β^--and β^+-decays with a half-life of 1.248×10^9 y:

$$^{40}_{19}\text{K} \rightarrow {}^{40}_{18}\text{Ar} + {}^{0}_{+1}\text{e} + \nu \quad (11.2\%)$$

$$^{40}_{19}\text{K} \rightarrow {}^{40}_{20}\text{Ca} + {}^{0}_{-1}\text{e} + \bar{\nu} \quad (88.8\%)$$

How many neutrinos and antineutrinos are emitted from a typical banana each second?

SOLUTION

To determine how many neutrinos and antineutrinos are emitted from a banana, we need to know the number of $^{40}_{19}$K atoms in a banana. We first calculate the average atomic mass of potassium. From reference tables, we find that the atomic masses of the potassium isotopes are as follows:

$^{39}_{19}$K: 38.963 708 u; $^{40}_{19}$K: 39.964 000 u; $^{41}_{19}$K: 40.961 827 u

The average atomic mass of naturally occurring potassium is

$$m_K = (38.963708\ \text{u} \times 0.933) + (39.964000\ \text{u} \times 0.00012)$$
$$+ (40.961827\ \text{u} \times 0.0673) = 39.1147\ \text{u}$$

Therefore, 39.111 47 g of potassium contains 6.02×10^{23} atoms, and the number of potassium atoms in 450 mg is

$$\frac{6.02 \times 10^{23}\ \text{atoms}}{39.1147\ \text{g}} \times 450 \times 10^{-3}\ \text{g} = 6.93 \times 10^{21}\ \text{atoms}$$

Since 0.012% of naturally occurring potassium is $^{40}_{19}$K, the number of $^{40}_{19}$K atoms in a typical banana is

$$6.93 \times 10^{21}\ \text{atoms} \times \frac{0.012}{100} = 8.31 \times 10^{17}\ \text{atoms}$$

The half-life of $^{40}_{19}$K is the same for β^-- and β^+-decays. The activity of the potassium in a typical banana is

$$R = \text{number of atoms} \times \text{decay rate}$$

$$= \text{number of atoms} \times \frac{\ln(2)}{\text{half-life}}$$

$$= 8.31 \times 10^{17}\ \text{atoms} \times \frac{\ln(2)}{1.248 \times 10^9\ \text{y} \times 31\ 557\ 600\ \text{s/y}}$$

$$= 14.6\ \text{atoms/s}$$

On average, 14.6 atoms of $^{40}_{19}$K decay each second. Since 88.8% of these decays are β^- decays producing an antineutrino, and 11.2% are β^+ decays producing a neutrino, an average banana emits approximately 13 antineutrinos and 2 neutrinos each second.

Gamma (γ) Decay

The nucleons in a nucleus are not stationary. They move with speeds of approximately 1/10 the speed of light in orbits of well-defined energies. This motion is similar to the orbital motion of electrons around the centre of a nucleus. However, electrons move in the Coulomb potential generated by the positive charge of the nucleus, and the nucleons move in a potential, called the **nuclear potential**, generated by the strong force between the nucleons. The details of the nuclear potential and the calculation of resulting eigenstates of nucleons are very complicated. To introduce the main concepts of γ-decay, we introduce a very simple model of the energy eigenstates of nucleons bound in a nucleus.

Let us assume that the protons and neutrons in a nucleus move in equally spaced energy states (or orbitals). The lowest-energy state is denoted by E_0, and successively higher-energy states are E_1, E_2, E_3, ... The energy difference between adjacent states is ΔE. In addition to the attractive strong force, protons also exert a repulsive Coulomb force on each other. Consequently, the orbitals of protons are shifted to slightly higher energies than the orbitals of neutrons; however, we will ignore this difference. Figure 33-15 shows the energy levels of our simple model.

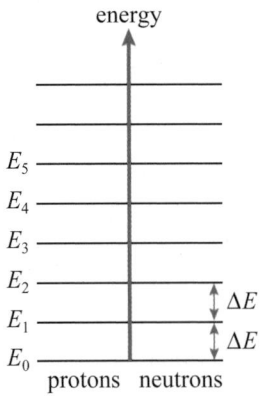

Figure 33-15   Energy levels of protons and neutrons in a nucleus (the Coulomb interaction between protons has been ignored in this energy-level diagram)

Protons and neutrons are fermions with a spin quantum number of $\frac{1}{2}$. The Pauli exclusion principle prohibits two fermions with identical quantum numbers from simultaneously occupying the same energy state. Therefore, no more than two protons (or neutrons) can be in a given energy level, one with spin-up ($\uparrow$) and the other with spin-down ($\downarrow$). A single nucleon in an energy level can have its spin projection either up or down.

Consider the ground state of $^{12}_{6}C$, which has six protons and six neutrons. Two protons fill energy level E_0, two go in E_1, and two go in E_2. Neutrons fill their corresponding levels in the same way (Figure 33-16).

A nucleus in its excited state is indicated by an asterisk. For example, $^{12}_{6}C^*$ denotes a carbon-12 nucleus with at least one nucleon in an energy level higher than the ground state.

Figure 33-17 shows two possible excited states of the $^{12}_{6}C$ nucleus. In Figure 33-17(a), a proton has jumped from level E_2 to level E_3, thereby increasing the energy of the whole nucleus by an amount ΔE. In Figure 33-17(b), a proton and a neutron have jumped from E_2 to E_3, so the energy of this excited state is $2\Delta E$ above the ground-state energy.

A nucleus in an excited state releases its excess energy by emitting a photon, or a γ-ray. This process is called a **gamma decay** (γ-decay) and is represented as

$$^A_Z X^* \rightarrow ^A_Z X + \gamma + Q_\gamma \qquad (33\text{-}45)$$

The decay diagram for γ-decay is shown in Figure 33-18.

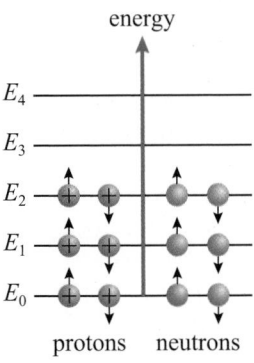

Figure 33-16   The configuration of protons and neutrons in the ground state of $^{12}_{6}C$ (the Coulomb interaction between protons has been ignored)

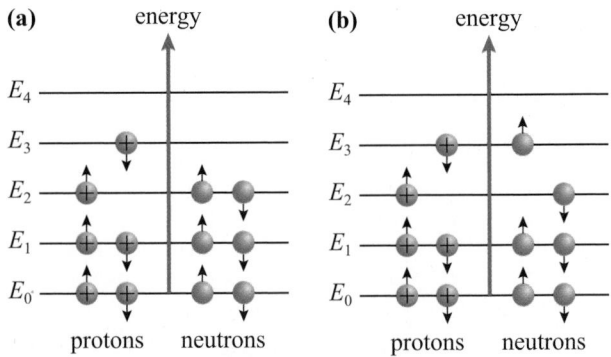

Figure 33-17   The configuration of protons and neutrons in two excited states of $^{12}_{6}C$ (a) A proton has jumped to a higher-energy level. (b) A proton and a neutron have moved to higher-energy levels.

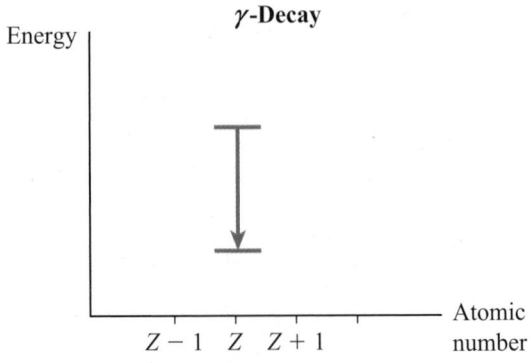

Figure 33-18   The decay diagram for γ-decay

MODERN PHYSICS

When a nucleus decays by emitting an α-particle or a β-particle, the daughter nucleus is usually left in an excited state. The nucleus then jumps to a lower-energy state by emitting a γ-ray. Two nuclides that differ only in energy state are called **isomers**. For example, $^{12}_{6}C^*$ and $^{12}_{6}C$ are isomers. Usually, a nucleus stays in its excited state for a very short period of time of the order of 10^{-19} s. However, an excited state can exist for seconds, hours, or even years. Such excited states are called **metastable** states.

The energy difference between nuclear levels is of the order millions of electron volts. In contrast, atomic energy levels are only tens of electron volts apart. Consequently, γ-rays have much higher energies than the electromagnetic radiation typically emitted by electrons in atoms. For example, boron-12 ($^{12}_{5}B$) β-decays to various excited states of carbon-12. In one of the excited states, the $^{12}_{6}C^*$ nucleus decays to the ground state by emitting a γ-ray of 4.44 MeV.

Since γ-rays are highly penetrating, they can be used to inspect welds, structural materials, and sealed containers (see Figure 26-20).

LO 7

33-7 Nuclear Fission and Nuclear Fusion

From Figure 33-4 we can see that the binding energy per nucleon for $A \geq 200$ is approximately 7.5 MeV, and for $A \simeq 100$ it is approximately 8.5 MeV. So, if a large nucleus ($A > 200$) with a binding energy per nucleon of 7.5 MeV were to split into two nuclei of approximately equal size, the binding energy per nucleon would increase to about 8.5 MeV. The two final nuclei are more tightly bound, so the difference in the total binding energy between the initial and the two final nuclei would be approximately 1 MeV/nucleon. Thus, the total energy released when the large nucleus splits is about 200 MeV. **Nuclear fission** is the process in which a heavy nucleus splits into two nuclei. The final nuclei are called **fission products**. Usually, a few neutrons are emitted during nuclear fission. Fission processes can be either spontaneous fission or neutron-induced.

Spontaneous Fission

In spontaneous fission, a large nucleus undergoes fission on its own (Figure 33-19). This process can be represented as follows:

$$^{A}_{Z}X \rightarrow ^{A_1}_{Z_1}Y_1 + ^{A_2}_{Z_2}Y_2 + m\,^{1}_{0}n \qquad (33\text{-}46)$$

where Y_1 and Y_2 are fission products, and m is the number of emitted neutrons, usually from 0 to 3.

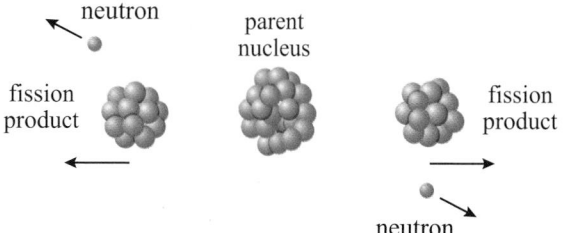

Figure 33-19 During spontaneous fission, a nucleus with a large atomic number spontaneously breaks into two nuclei of similar atomic numbers, accompanied by the emission of a few neutrons.

A large nucleus can split into various combinations of fission products. As in decay processes, the number of nucleons and the total charge must be conserved. Therefore,

$$A = A_1 + A_2 + m \qquad (33\text{-}47)$$

and

$$Z = Z_1 + Z_2 \qquad (33\text{-}48)$$

For example, uranium-238 ($^{238}_{92}U$) undergoes spontaneous fission. Here are two ways that uranium-238 fissions:

$$^{238}_{92}U \rightarrow ^{134}_{52}Te + ^{102}_{40}Zr + 2\,^{1}_{0}n + Q_f$$

$$^{238}_{92}U \rightarrow ^{134}_{50}Sn + ^{102}_{42}Mo + 2\,^{1}_{0}n + Q_f \qquad (33\text{-}49)$$

Here, Q_f denotes the Q-value for the fission decay and is different for the two reactions.

Energy Released in Spontaneous Fission The energy released during the spontaneous fission of a nucleus appears mainly as the kinetic energy of the fission products. Some of the energy also appears as the emission of γ-rays and electrons, as well as excitation energy of the final nuclei. The total energy released is equal to the difference between the rest mass energy of the parent nucleus and the total rest mass energies of the fission products, including neutrons:

$$Q_f = M_X c^2 - (M_{Y_1} + M_{Y_2} + mm_n)c^2 \qquad (33\text{-}50)$$

By accounting for Z electron masses, Equation (33-50) can be rewritten in terms of atomic masses:

$$Q_f = \widetilde{M}_X c^2 - (\widetilde{M}_{Y_1} + \widetilde{M}_{Y_2} + mm_n)c^2 \qquad (33\text{-}51)$$

Why do all heavy nuclei not fission immediately? Why is lead-208 a stable nucleus when it could split into two more tightly bound nuclei? The reason is that an energy barrier, due to Coulomb repulsion, prevents fission. Consider the Coulomb potential energy between a proton and a nucleus as a function of the distance of the proton from the centre of the nucleus. If the charge Ze of the nucleus is distributed uniformly in a sphere of radius R,

MODERN PHYSICS

EXAMPLE 33-11

The Fission of Uranium-238

Calculate the energy released during the spontaneous fission of uranium-238 into tellurium-134 and zirconium-102.

SOLUTION

We determine the Q-value of this reaction by using the atomic masses listed in Table 33-5:

$$\text{Atomic mass of } {}^{238}_{92}\text{U} = 238.050\,783 \text{ u}$$

$$\text{Atomic mass of } {}^{134}_{52}\text{Te} = 133.911\,369 \text{ u}$$

$$\text{Atomic mass of } {}^{102}_{40}\text{Zr} = 101.922\,981 \text{ u}$$

$$\text{Mass of two neutrons} = 2 \times 1.008\,665 \text{ u} = 2.017\,330 \text{ u}$$

The difference between the mass of the parent nucleus and the total mass of the products including the two neutrons is

$$\Delta m = 238.050\,783 \text{ u} - (133.911\,369 \text{ u} + 101.922\,981 \text{ u}$$

$$+ 2 \times 1.008\,665 \text{ u}) = 0.1991 \text{ u}$$

Converting this mass difference into energy, we have

$$Q_f = 0.1991 \text{ u} \times \frac{931.5 \text{ MeV}}{\text{u}} = 185.3 \text{ MeV}$$

Making sense of the result:

Although 185 MeV is an enormous amount of energy compared to the energy released in a chemical reaction, the half-life of uranium-238 for spontaneous fission is approximately 8×10^{15} y. Therefore, the rate at which energy is released by the spontaneous fission of a sample of uranium-238 is as small as very few nuclei fissions per second.

---

then the Coulomb potential energy, $V_C(r)$, between the proton and the nucleus is

$$V_C(r) = \begin{cases} \dfrac{Ze^2}{4\pi\varepsilon_0}\dfrac{r^2}{R^3} & r \leq R \\[2ex] \dfrac{Ze^2}{4\pi\varepsilon_0}\dfrac{1}{r} & r > R \end{cases} \qquad (33\text{-}52)$$

A plot of $V_C(r)$ is shown in Figure 33-20. The Coulomb potential energy is positive because like charges repel each other. The potential energy is zero at the centre of the nucleus and maximum at the surface of the nucleus; outside the nucleus it falls off as $1/r$.

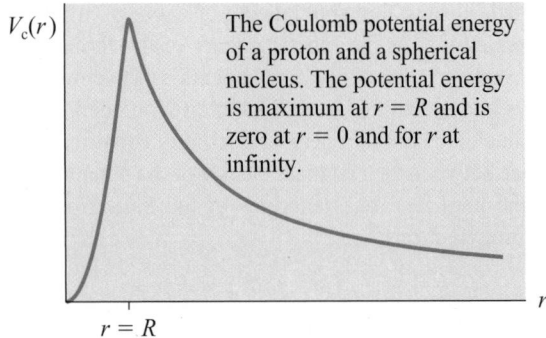

$V_C(r)$

The Coulomb potential energy of a proton and a spherical nucleus. The potential energy is maximum at $r = R$ and is zero at $r = 0$ and for r at infinity.

$r = R$

Figure 33-20 A plot of the Coulomb potential energy between a proton and a spherical nucleus, as a function of the distance between the two

Now, consider the spontaneous fission of uranium-238 into tellurium-134 and zirconium-102 (Equation (33-49)). We can imagine the uranium nucleus to be made up of a tellurium and zirconium nucleus, plus two neutrons. As the tellurium and zirconium nuclei begin to separate, the Coulomb potential energy between the two increases, opposing the separation. From Example 33-11 we know

that 185 MeV of energy is released by this separation. Is this energy sufficient to overcome the Coulomb repulsion and cause the fission? At the point where the ${}^{134}_{52}\text{Te}$ and ${}^{102}_{40}\text{Zr}$ nuclei are just touching at their surfaces, the Coulomb potential energy is

$$V_C(R_1 + R_2) = \frac{1}{4\pi\varepsilon_0}\frac{(52e)(40e)}{r_0(134^{1/3} + 102^{1/3})}$$

Here, we have used Equation (33-4) for nuclear radii. Using $r_0 = 1.2$ fm and $e^2/4\pi\varepsilon_0 = 1.44$ MeV fm, we get

$$V_C(R_1 + R_2) = 255 \text{ MeV}$$

The daughter nuclei need 255 MeV of energy to overcome the Coulomb barrier, but only 185 MeV of energy is released in the decay process. So the energy released during the spontaneous fission of uranium-238 is not sufficient for the daughter nuclei to overcome the Coulomb barrier and separate from the parent nucleus. The fact that the Q-value is less than the energy barrier explains why the probability of the spontaneous fission of uranium-238 is so small and the half-life is so long. So, why do ${}^{238}_{92}\text{U}$ nuclei undergo spontaneous nuclear fission at all? Nuclear fission, like all other decays, is a quantum-mechanical phenomenon and is possible because of quantum tunnelling, the same phenomenon that allows α-decay to occur.

Neutron-Induced Fission

Large A nuclides can be induced to fission by bombarding them with neutrons. This process is called **neutron-induced fission** and can be represented by

$$ {}^1_0\text{n} + {}^A_Z X \rightarrow {}^{A_1}_{Z_1} Y_1 + {}^{A_2}_{Z_2} Y_2 + m\,{}^1_0\text{n} \qquad (33\text{-}53)$$

For example, neutrons can cause uranium-235 to fission in several different ways, including

$$ {}^1_0\text{n} + {}^{235}_{92}\text{U} \rightarrow {}^{93}_{37}\text{Rb} + {}^{141}_{55}\text{Cs} + 2{}^1_0\text{n} + 180 \text{ MeV} \qquad (33\text{-}54)$$

MODERN PHYSICS

and

$$_{0}^{1}\text{n} + _{92}^{235}\text{U} \rightarrow _{36}^{92}\text{Kr} + _{56}^{141}\text{Ba} + 3_{0}^{1}\text{n} + 173 \text{ MeV} \quad (33\text{-}55)$$

The calculation of the energy released in neutron-induced fission is similar to the calculation for spontaneous emission, but the total energy of the incident neutron must be included in the energy of the initial state. Generally, the kinetic energy of the incident neutron is less than a few MeV.

Chain Reactions Consider a large number of uranium-235 nuclei and a single neutron that initiates the fission reaction described by Equation (33-53). Let us assume that two neutrons are emitted in this fission reaction. The neutrons emitted in the initial fission can be absorbed by two other uranium nuclei, causing these nuclei to fission and emit four neutrons. These four neutrons can then induce four more nuclei to fission, generating eight neutrons, and so on (Figure 33-21). If each neutron induces another uranium nucleus to fission, the number of nuclei that fission doubles at each stage. It takes approximately 10^{-4} s for a neutron to be emitted and then absorbed again. Within one-hundredth of a second of the start of the process, $1 + 2 + 2^2 + \cdots + 2^{100} \cdots \approx 2.5 \times 10^{30}$ uranium nuclei decay. Since 185 MeV of energy is released per decay, approximately 4.6×10^{32} MeV, or 7.3×10^{19} J, of energy is released in this short time. This amount of energy is equivalent to the detonation of about 15 gigatonnes of TNT.

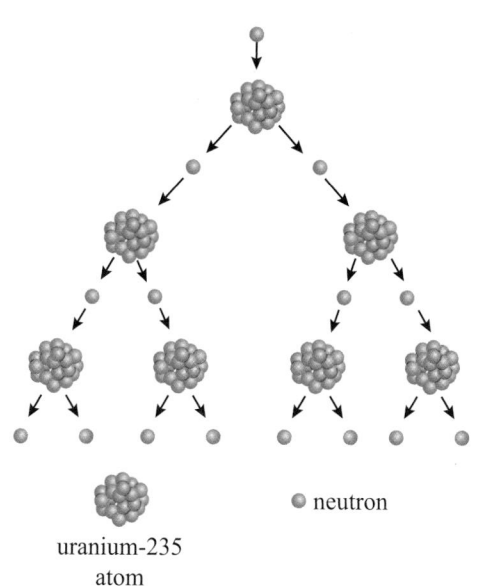

Figure 33-21 An ideal chain reaction. Each fission reaction produces two neutrons, each of which then causes another nucleus to fission.

Such a sequence of self-propagating nuclear reactions is called a **chain reaction**. Chain reactions are the basis of nuclear power generating plants and nuclear bombs. In reality, not all neutrons produced in the fission process act as triggers for other fission reactions. Many neutrons escape through the material, scatter off parent nuclei, or cause other types of reactions.

How a Nuclear Reactor Works As described above, approximately 200 MeV of energy is released when a heavy nucleus with $A > 200$ fissions into two smaller nuclei. Most nuclear reactors use uranium-235 ($_{92}^{235}\text{U}$) as the nuclear fuel. As discussed earlier, a uranium-235 nucleus can fission by absorbing a slow-moving neutron. Some of the possible outcomes of this fission process are

$$_{0}^{1}\text{n} + _{92}^{235}\text{U} \rightarrow _{56}^{140}\text{Ba} + _{36}^{93}\text{Kr} + 3_{0}^{1}\text{n}$$

$$_{0}^{1}\text{n} + _{92}^{235}\text{U} \rightarrow _{55}^{134}\text{Cs} + _{37}^{100}\text{Rb} + 2_{0}^{1}\text{n}$$

$$_{0}^{1}\text{n} + _{92}^{235}\text{U} \rightarrow _{53}^{131}\text{I} + _{39}^{104}\text{Y} + _{0}^{1}\text{n}$$

On average, 2.4 neutrons are produced in the fission of each uranium-235 nucleus. These neutrons, called **prompt neutrons**, can be used to fission other uranium-235 nuclei. The key to generating continuous energy in a nuclear power plant is to establish a self-sustaining chain reaction so that, on average, the same number of uranium-235 nuclei fission each second. For this, we need a sufficient number of uranium-235 nuclei available to fission, and we need to control the fission process so that the number of fission prompt neutrons does not go out of control.

Naturally occurring uranium ore consists mostly of uranium-238 with only 0.72% uranium-235. Uranium-238 does not fission by absorbing slow neutrons. To be used as a **nuclear fuel**, uranium must contain at least 4% uranium-235 (an exception is the Canadian CANDU reactor that uses natural uranium as the nuclear fuel). To produce this **enriched uranium**, uranium hexafloride, a compound consisting of uranium and fluorine is vaporized and rapidly spun in a tall, cylindrical **centrifuge**. The uranium-235 atoms are lighter than uranium-238 atoms (which have three more neutrons) and tend to stay in the middle of the centrifuge, and the uranium-238 atoms move closer to the outer wall. Thus, some of the uranium-238 can be separated out, leaving an increased proportion of uranium-235. Enriched uranium is pressed into thumb-sized pellets, which are stacked in long hollow cylinders made of zirconium, called **fuel rods** (Figure 33-22). Several fuel rods are bundled together, parallel to each other, with space between them.

With a sufficient concentration of uranium-235 nuclei, a chain reaction can occur, and the number of prompt neutrons increases quickly. Uncontrolled, the number of fissions per second would increase quickly, generating enough heat to melt the nuclear fuel and cause a **nuclear meltdown**. The number of neutrons in the reactor core is controlled by **control rods**, which are long, thin cylinders made of an element that readily absorbs neutrons, such as boron-10. The control rods are positioned between the fuel rods, and they can be either fully or partially inserted. This assembly is called a **nuclear core**. When the control rods are pulled out, the reactor runs at its maximum power. When the rods are completely inserted, most of the prompt neutrons are absorbed, and the chain reaction largely stops.

Energy released by fission is carried out of the nuclear core by a **coolant**, usually water, because water is readily available and inexpensive. However, other types of

Figure 33-22 A nuclear fuel rod consists of enriched uranium oxide pellets that are packed in a thin cylinder made of zirconium alloy.

rod holding plate

compression spring

fission gas plenum

insulating pellet Al_2O_3

H_2O coolant

spacer

UO_2 pellet

gas-tight zircaloy shell

insulating pellet Al_2O_3

support sleeve

end cap

4.17 m

approximately 11 mm

coolants are also used. Canadian-made nuclear reactors, called CANDU reactors, use heavy water as a coolant. In heavy water, the hydrogen atom is replaced by a deuterium atom (a deuterium atom contains one proton and one neutron in its nucleus).

The average energy of prompt neutrons is approximately 1.0 MeV. A neutron with this energy is moving at approximately 4.5% of the speed of light. The probability of a $^{235}_{92}U$ nucleus absorbing such a fast-moving neutron is very small. Therefore, these neutrons are slowed down by letting them lose energy during collisions with a **moderator**. In many reactors, water is also used as a moderator. The whole assembly consisting of the fuel rods, control rods, and the coolant is housed in a thick steel vessel called the **containment vessel**. All radioactive elements produced during the operation of the power plant remain within the containment vessel.

Steel pipes bring water into the containment vessel and bring heated water out of it. In boiling water reactors (BWRs), the type used in the Fukushima nuclear plant, the heated water is kept at a high pressure, usually 50 to 100 times atmospheric pressure. Water under pressure boils at a higher temperature, so the water temperature in BWRs reaches between 200°C and 250°C. The heated water is then turned into steam, which drives a steam turbine to generate electricity (Figure 33-23). Approximately 30% of the energy produced during fission is converted into electricity, and the rest is lost to the cooling water.

Over 90 different isotopes are produced when $^{235}_{92}U$ fissions. Most of these fission products are unstable and decay into stable elements by emitting high-energy electrons, neutrons, and γ-rays. Some of the common fission products in a nuclear reactor are iodine-131 ($^{131}_{53}I$, half-life

Figure 33-23 A boiling water reactor

reactor structure

containment vessel

hot water

turbine

generator

control rods

cold water

fuel rods

water pumps

8.025 days), cesium-136 ($^{136}_{55}Cs$, half-life 13.04 days), cesium-137 ($^{137}_{55}Cs$, half-life 30.08 years), and barium-140 ($^{140}_{56}Ba$, half-life 12.75 days).

About one percent of the fission products decay by emitting neutrons, which are called **delayed neutrons** to distinguish them from the prompt neutrons. The delayed neutrons are also absorbed by uranium-235, causing it to fission. So, a nuclear reactor cannot be completely shut off by inserting control rods. Its energy production decreases, but it remains hot for a long time.

When the concentration of uranium-235 in fuel rods decreases below 1%, the rods need to be replaced. The spent rods are placed in large pools of water for a number of years until the radioactivity has fallen to a level where they can be stored. Dealing with radioactive waste products is a major technological challenge.

The fission products are harmful. For example, iodine-131 ($^{131}_{53}I$), a radioactive isotope of iodine, β-decays into xenon-131, a stable isotope of xenon, in two steps:

$$^{131}_{53}I \rightarrow {}^{131}_{54}Xe^* + e^- + \bar{\nu}_e + 606 \text{ keV}$$

$$^{131}_{54}Xe^* \rightarrow {}^{131}_{54}Xe + \gamma + 364 \text{ keV}$$

Most of the 606 keV energy produced in β-decay (the first step) is shared between the electron and the antineutrino. Xenon-131 produced in the β-decay is in an excited state and decays to its ground state by emitting a 364 keV photon (the second step). The electron and the photon have sufficiently high energy to penetrate human tissue to a depth of a few millimetres and cause cell mutation and destruction. Stable iodine-127 and radioactive iodine-131 are chemically identical. The body absorbs iodine and concentrates it in the thyroid gland, which uses it for the production of thyroxin hormones. If radioactive iodine-131 is present in the food or the air, it will be absorbed by the body and may lead to thyroid cancer. During the Fukushima disaster (and other accidental releases of fission products into the atmosphere), people nearby were advised to take natural (and stable) iodine-127 pills to saturate their thyroid glands with the stable isotope and make them less likely to absorb radioactive iodine-131.

Nuclear Fusion

From Figure 33-4, we observe that the binding energy per nucleon can also be increased by combining two lighter nuclei, thus releasing energy. The combination of two nuclides to form a heavier nuclide is called **nuclear fusion**. Fusion is the source of energy in all stars, including the Sun. A key series of fusion reactions in the core of relatively cooler stars, like the Sun, turns four hydrogen nuclei (protons) into a helium nucleus. In the first stage, two protons ($_1^1H + _1^1H$) combine to form a hydrogen-2 nucleus (deuteron) with the emission of a positron and a neutrino. The hydrogen-2 nucleus then combines with a proton to form a helium-3 nucleus with the emission of a γ-ray. Two helium-3 nuclei then combine to form a helium-4 nucleus with the emission of two protons. Energy is released in each stage of this cycle:

$$_1^1H + _1^1H \rightarrow _1^2H + _{+1}^0e + \nu + 0.42 \text{ MeV} \quad (33\text{-}56a)$$

$$_1^1H + _1^2H \rightarrow _2^3He + \gamma + 5.49 \text{ MeV} \quad (33\text{-}56b)$$

$$_2^3He + _2^3He \rightarrow _2^4He + 2(_1^1H) + 12.86 \text{ MeV} \quad (33\text{-}56c)$$

The positron ($_{+1}^0e$) produced in reaction (33-56a) combines with a surrounding electron (in the core of a star the electrons are knocked out from the hydrogen atoms). When an electron combines with a positron, both are annihilated into two photons with a total energy of 1.022 MeV (equal to the combined rest mass energies of the two):

$$_{+1}^0e + _{-1}^0e \rightarrow 2\gamma$$

Thus, in the reactions of Equation (33-56), four protons combine to form a helium-4 nucleus, and the overall reaction can be written as

$$6(_1^1H) \rightarrow _2^4He + 2(_1^1H) + 2_{+1}^0e + 2\nu + 26.7 \text{ MeV} \quad (33\text{-}57)$$

Reaction (33-57) is called the **proton–proton cycle**. Of the 26.7 MeV of energy released in this cycle, approximately 25 MeV is used to heat the star, and the rest is carried away by the two neutrinos.

For two nuclei to fuse together, they must have sufficient kinetic energy to overcome the Coulomb repulsion between them. In reaction (33-56a), the two protons must overlap (be within 1.2 fm of each other) to form a deuteron. At this distance, the Coulomb repulsion between the protons is approximately 1.2 MeV. So, the average kinetic energy of the protons must be at least 1.2 MeV to overcome the repulsion. From the kinetic theory of gases, we know that the average kinetic energy, $\bar{K}$, of a gas particle is related to the absolute temperature T by $\bar{K} = \frac{3}{2} k_B T$, where $k_B = 1.38 \times 10^{-23}$ J/K $= 8.62 \times 10^5$ eV/K is Boltzmann's constant, and T is measured in kelvins. Taking the average kinetic energy to be 1.2 MeV, the required temperature is

$$T = \frac{2\bar{K}}{3k_B} = \frac{2 \times 1.2 \times 10^6 \text{ eV}}{3 \times 8.62 \times 10^{-5} \text{ eV/K}} \approx 9 \times 10^9 \text{ K}$$

If all the kinetic energy is due to thermal motion, the protons must be in an environment that is at a temperature of approximately 9 billion kelvins (9 GK). Such high temperatures exist in the interiors of stars. The temperature of the Sun's core is approximately 15 MK, high enough to ionize the hydrogen gas into electrons and protons. The kinetic energies of the protons are distributed about a mean value corresponding to this temperature. A tiny fraction of protons in the Sun's interior have kinetic energies equivalent to the temperature of 9 GK and can fuse to form a deuteron. Even if the energies of colliding protons are below 1.2 MeV, quantum tunnelling can cause two protons to tunnel through the Coulomb barrier and form a deuteron.

EXAMPLE 33-12

Energy Released in the Proton-Proton Cycle

Using the Q-values given in reactions (33-56), show that approximately 26.7 MeV of energy is released in the proton–proton cycle of reaction (33-57).

SOLUTION

In reaction (33-56c), two helium-3 nuclei fuse to form a helium-4 nucleus and two protons. To form these two helium-3 nuclei, reactions (33-56a) and (33-56b) need to happen twice. Also, the positron emitted in reaction (33-56a) combines with an electron to produce two photons of energy 1.022 MeV. Therefore, the energy released in the proton–proton cycle is as follows:

$2 \times$ the energy released in proton − proton fusion
$\quad = 2 \times 0.42 \text{ MeV} = 0.84 \text{ MeV}$

$2 \times$ the energy released in proton − deuteron fusion
$\quad = 2 \times 5.49 \text{ MeV} = 10.98 \text{ MeV}$

$2 \times$ the energy released in electron − positron annihilation
$\quad = 2 \times 1.022 \text{ MeV} = 2.044 \text{ MeV}$

Energy released in the fusion of two helium-3 nuclei
$\quad = 12.86 \text{ MeV}$

Total energy released: 0.84 MeV + 10.98 MeV + 2.044 MeV + 12.86 MeV = 26.7 MeV

LO 8

33-8 Ionizing Radiation

Ionizing radiation is electromagnetic radiation or particle radiation with sufficient energy to ionize atoms and molecules by removing bound electrons from them. The types of radiation in **ionizing electromagnetic radiation** are

X-rays and γ-rays. **Ionizing particle radiation** generally consists of α-particles, protons, neutrons, β-particles, or other elementary particles. The sources of ionizing radiation are radioactive materials, X-ray machines, nuclear reactors, particle accelerators, and cosmic rays.

To become free, a bound electron must absorb energy greater than its binding energy. The binding energies of electrons in outer atomic orbits are of the order of several eV. Therefore, ionizing radiation must have energy greater than a few eV to interact with electrons. Usually, only a fraction of the energy of the ionizing radiation is transferred to a single electron. As ionizing radiation passes through matter, it loses energy due to collisions with electrons. Therefore, to consistently ionize atoms, the energy of the radiation needs to be greater than 100 eV. For example, an average of 33.85 eV is required to ionize an air molecule. So a 5.0 MeV α-particle could ionize approximately 150 000 air molecules. For particle radiation, the rate of energy loss is proportional to the charge of the particle and the number of electrons per unit volume of the material, and is inversely proportional to the speed of the particle (slower particles have more time to interact with electrons and to transfer energy).

Absorbed Dose and Equivalent Dose

The amount of ionizing radiation incident on an object is called **exposure**. As a result of exposure, the object will absorb some of the incident energy. The energy absorbed per unit volume by an object is called the **absorbed dose** (D). The SI unit of absorbed radiation is the gray (Gy), which is defined as the absorption of 1 J of radiation energy per kilogram of the absorbing material:

$$1 \text{ Gy} = 1 \text{ J/kg}$$

An older unit of absorbed radiation is the rad (radiation absorbed dose):

$$1 \text{ Gy} = 100 \text{ rad}$$

For a given exposure, the amount of energy transferred to biological tissue and the resulting damage to the tissue depend on the type of radiation. Because α-particles are positively charged and have a relatively large size, they are very effective in ionizing atoms. When passing through matter, α-particles transfer their energy within a short distance, causing intense damage to the DNA structure of the tissue. β-particles and γ-rays lose their energy more slowly, and the resulting damage to biological tissue is more diffuse. To account for different interactions of radiation with the human body, we define the **equivalent dose** (H) to be a measure of the amount of damage caused to biological tissue by ionizing radiation. An equivalent dose is the absorbed dose multiplied by the **radiation weighting factor** (W_R), which is a measure of how damaging the radiation is:

$$\text{Equivalent dose} = \text{absorbed dose}$$
$$\times \text{ radiation weighting factor}$$

$$H = D \times W_R$$

The SI unit of equivalent dose is the sievert (Sv). Since the radiation weighting factor is a dimensionless number, the sievert and the gray have the same dimensions, joules per kilogram. Table 33-6 lists radiation weighting factors for various ionizing radiations. Notice that α-particles cause 20 times more damage to biological tissue than the same absorbed dose of γ-rays and β-particles. The equivalent dose is therefore a measure of the risk associated with absorbed ionizing radiation.

Table 33-6 Radiation weighting factors

Radiation	Radiation weighting factor (W_R)
Electrons, positrons, X-rays, γ-rays	1
Protons	2
Slow neutrons (10 keV) (depends on the neutron's energy)	5–10
Medium energy neutrons (100 keV–2 MeV)	20
α-particles, medium and heavy nuclei	20

In daily usage, equivalent dose is often referred to simply as "dose." The old unit of equivalent dose was the rem (röntgen equivalent (in) man), which is equal to a hundredth of a sievert:

$$1 \text{ Sv} = 100 \text{ rem}$$

EXAMPLE 33-13

Equivalent Dose

Find the equivalent dose for a person who has absorbed 10 mGy of α particles.

SOLUTION

$$H = D \times W_R$$
$$= 0.010 \text{ Gy} \times 20 = 0.20 \text{ Sv} = 20 \text{ rem}$$

Making sense of the result:

Absorbing a dose of 0.010 Gy of α-particles is equivalent to absorbing a dose of 0.20 Gy of X-rays.

For an average adult, the recommended "safe limit" of ionizing radiation is 50 mSv per year. For minors, this limit is 5 mSv. People living in Canada and the United States are continually exposed to natural and human-made ionizing radiation from a variety of sources, with an average yearly dose of 3.5 mSv. Figure 33-24 shows the average annual radiation exposure in the United States. Notice that approximately half the exposure results from inhaling radon gas, which is produced naturally by the α-decay of uranium, which is typically present in soil in concentrations from 0.7 to 11 parts per million. Radon is odourless, chemically inert, and highly radioactive. It decays into other radioactive elements and it is these products which stick in the lungs and can cause lung cancer. The U.S. Environmental Protection Agency estimates that approximately 21 000 lung cancer deaths every year can be attributed to radon poisoning, a statistic second only to cigarette smoking.

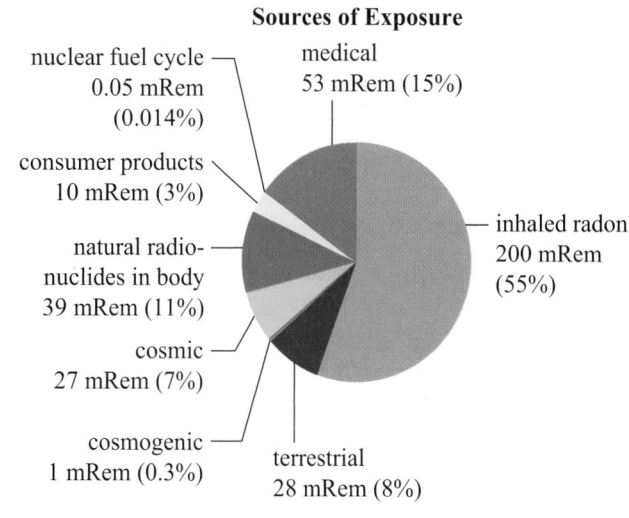

Sources of Exposure

nuclear fuel cycle 0.05 mRem (0.014%)

medical 53 mRem (15%)

consumer products 10 mRem (3%)

natural radio-nuclides in body 39 mRem (11%)

cosmic 27 mRem (7%)

cosmogenic 1 mRem (0.3%)

terrestrial 28 mRem (8%)

inhaled radon 200 mRem (55%)

total effective dose equivalent = 360 mRem

Lawrence Livermore Lab, Berkley, USA

Figure 33-24 The average annual radiation exposure in the United States from various radioactive sources

EXAMPLE 33-14

The Equivalent Dose of Polonium

Polonium-210 ($^{210}_{84}$Po) α-decays to lead-206 ($^{206}_{82}$Pb) and releases 5.307 MeV of energy in the process. The half-life of $^{210}_{84}$Po is 138.376 days, and its atomic mass is 209.983 u.

(a) What is the equivalent dose received by an average adult lung (mass approximately 5.0 kg) from 10 μg of polonium-210?

(b) What is the approximate daily equivalent dose?

SOLUTION

(a) We know the energy released in a single decay, so we need to determine how many nuclei of 10 μg of polonium-210 decay in one second. The atomic mass of polonium-210 is 209.983 u. Therefore, 209.983 g of polonium-201 contains 6.02×10^{23} atoms, and

$$\text{Number of atoms in 10 } \mu g = \frac{6.02 \times 10^{23} \text{ atoms}}{209.983 \text{ g}} \times 10^{-6} g$$
$$= 2.87 \times 10^{16} \text{ atoms}$$

Since the initial activity, R_0, of the polonium equals the number of atoms times the decay constant,

$$R_0 = 2.87 \times 10^{16} \text{ atoms} \times \frac{\ln(2)}{138.376 \times 24 \times 3600 \text{ s}}$$
$$= 1.66 \times 10^9 \text{ Bq}$$

The energy, E, released per second equals the number of decays per second times the energy released in a single decay:

$$E = (1.66 \times 10^9 \text{ decay/s})(1 \text{ s})(5.307 \text{ MeV/decay})$$
$$= 8.81 \times 10^9 \text{ MeV}$$

To express the absorbed dose in sieverts, we need to express the incident energy in joules:

$$E = 8.81 \times 10^9 \text{ MeV} \times 1.60 \times 10^{-13} \text{ J/MeV}$$
$$= 1.41 \times 10^{-3} \text{ J}$$

Since the mass of the lung is 5.0 kg, the absorbed dose is

$$D = \text{energy absorbed per unit mass} = \frac{1.41 \times 10^{-3} \text{ J}}{5.0 \text{ kg}}$$
$$= 2.8 \times 10^{-4} \text{ Gy}$$

The radiation weighting factor for α-particles is 20. So, the equivalent dose per second is

$$H = D \times W_R = 2.8 \times 10^{-4} \text{ J/kg} \times 20 = 5.7 \times 10^{-3} \text{ Sv}$$

(b) The half-life of polonium-210 is 138 days, so its activity does not change substantially in one day. Therefore, a reasonable approximation for the equivalent dose per day is

$$5.7 \times 10^{-3} \text{ Sv/s} \times 24 \text{ h/d} \times 3600 \text{ s/h} = 490 \text{ Sv}$$

Making sense of the result:

An equivalent dose of 490 Sv is approximately 10 000 times the recommended maximum annual dose. Clearly, $^{210}_{84}$Po is extremely dangerous. Ingesting even in a tiny amount can cause fatal damage to internal organs.

33-9 Some Applications of Nuclear Physics

Explosive Detectors

Nitrate compounds, such as ammonium nitrate and potassium nitrate, are commonly used in explosives. Nitrates can be detected by bombarding an object with neutrons. Some of the neutrons are absorbed by nitrogen nuclei, forming the isotope $^{15}_{7}N^*$ in an excited state. This isotope then decays to its ground state by emitting a γ-ray with a characteristic energy unique to the $^{15}_{7}N^*$ nucleus:

$$^{14}_{7}N + ^{1}_{0}n \rightarrow ^{15}_{7}N^* \rightarrow ^{15}_{7}N + \gamma$$

This process is called a (n, γ) reaction, meaning "neutron in, photon out." A detector tuned to the characteristic frequency of the nitrogen-15 γ-rays can then determine when an object contains enough nitrogen to suggest the presence of explosives. Such neutron-activated explosive detectors are becoming common at seaports, airports, and border crossings.

Smoke Detectors

As described in the previous section, α-particles are very efficient at ionizing matter. Most household smoke detectors use this property to detect smoke and combustion products. Such detectors contains a small amount of synthetically produced radioactive americium-241 ($^{241}_{95}Am$), which emits α-particles with about 5.5 MeV of kinetic energy. These α-particles ionize air molecules and lose all of their energy within 4 to 5 cm.

In a smoke detector, α-particles from $^{241}_{95}Am$ enter the space between two electrodes that are approximately 1 cm apart and are kept at a potential difference by a battery (Figure 33-25). As the air between the electrodes is ionized, the negatively charged electrons are attracted toward the positive plate of the electrode, and positively charged ions are attracted toward the negative plate. The resulting tiny continuous electric current between the electrodes is sensed by an electronic circuit. When a fire starts, smoke particles and invisible combustion products get between the electrodes and neutralize some of the ions, thus reducing the current flowing between the electrodes. If the current falls below a threshold value, the electronic circuit triggers an alarm.

Americium-241 is chosen because it has a half-life of 433 y, which is long enough that the rate of emission of α-particles does not decrease significantly during the lifetime of a smoke detector. It also emits very few γ-rays. The half-life is also short enough that only approximately 0.2 μg of the americium is needed to generate a current detectable by a simple and inexpensive circuit.

Figure 33-25 An ionization chamber of a smoke detector

- Atomic nuclei consist of protons and neutrons. The atomic number, Z, is the number of protons in the nucleus. The atomic mass number, A, is the number of protons and neutrons in the nucleus. Nuclei with the same number of protons but a different number of neutrons are called isotopes.

- Nuclei are approximately spherical. The radius of a nucleus is approximately

$$r_0 A^{1/3} \qquad (33\text{-}4)$$

where $r_0 = 1.2$ fm and A is the mass number.

- The nucleons are bound within a nucleus by the attractive strong force, which has a range of approximately 2.0 fm. Nuclear density is approximately 2.8×10^{17} kg/m³.

- Nuclear masses are usually given in atomic mass units (u). One atomic mass unit equals 1.6605×10^{-27} kg and is equivalent to 931.5 MeV of energy.

- The binding energy of a nucleus is the energy required to separate the nucleus into its constituent protons and neutrons. The binding energy $B(A, Z)$ of a nucleus is related to the nuclear mass, $M_{\text{nuc}}(A, Z)$, the proton mass, m_p, and the neutron mass, m_n:

$$B(A, Z) = Zm_p c^2 + (A - Z)m_n c^2 - M_{\text{nuc}}(A, Z)c^2 \quad (33\text{-}10)$$

- The Q-value of a nuclear reaction is the difference between the rest mass energies of initial nuclei and the rest mass energies of the final nuclei.

- Unstable nuclei spontaneously emit energy in the form of α-, β-, and γ-rays. The rate of radioactive decay is proportional to the decay constant, λ, and the number of unstable nuclei present:

$$-\frac{dN(t)}{dt} = \lambda N(t) \qquad (33\text{-}15)$$

- The number of unstable nuclei remaining after time t is given by

$$N(t) = N_0 e^{-\lambda t} \qquad (33\text{-}19)$$

where N_0 is the number of nuclei present at $t = 0$.

- The half-life, $t_{1/2}$, of a radioactive nuclide is the time it takes for half of the radioactive nuclei in a given sample to decay. The half-life of a nuclide is related to its decay constant:

$$t_{1/2} = \frac{\ln(2)}{\lambda} \qquad (33\text{-}21)$$

- The general form for α-decay is

$$_{Z}^{A}X \rightarrow {_{Z-2}^{A-4}Y} + {_2^4}\text{He} + Q_\alpha \qquad (33\text{-}28)$$

where $_{Z}^{A}X$ denotes a parent nuclide, $_{Z-2}^{A-4}Y$ is the daughter nuclide, and Q_α is the energy released in α-decay.

- Beta-minus (β^-) particles are electrons. In a β^--decay, a neutron converts into a proton:

$$_{Z}^{A}X \rightarrow {_{Z+1}^{A}Y} + {_{-1}^0}\text{e} + \bar{\nu} + Q_{\beta^-} \qquad (33\text{-}38)$$

where $\bar{\nu}$ is an antineutrino, and Q_{β^-} is the energy released in the decay.

- Beta-plus (β^+) particles are positrons. In a β^+-decay, a proton converts into a neutron:

$$_{Z}^{A}X \rightarrow {_{Z-1}^{A}Y} + {_{+1}^0}\text{e} + \nu + Q_{\beta^+} \qquad (33\text{-}42)$$

where ν is a neutrino, and Q_{β^+} is the energy released in the decay.

- Gamma (γ) rays are high-energy photons. A radioactive nuclide emits a γ-ray when a proton or neutron moves to a lower-energy state:

$$_{Z}^{A}X^* \rightarrow {_{Z}^{A}X} + \gamma + Q_\gamma \qquad (33\text{-}45)$$

where $_{Z}^{A}X^*$ denotes an excited energy state, $_{Z}^{A}X$ is the lower-energy state of the same nuclide, and Q_γ is the energy released in γ-decay.

- Spontaneous nuclear fission is the splitting of a nucleus with large A into two nuclei, usually associated with the emission of one or more neutrons. Induced nuclear fission is caused by the absorption of a neutron by a large nucleus.

- Nuclear fusion is the merging of two light nuclei into a single nucleus. Nuclear fusion is the basic process responsible for energy generation in stars.

- The SI unit for nuclear radiation is the becquerel, 1 Bq = 1 nuclear decay/s. The SI unit for absorbed radiation is the gray: 1 Gy = 1 J/kg.

APPLICATIONS

Applications: nuclear reactors, medical imaging, carbon dating, explosive detectors, smoke detectors, security inspection

Key Terms: absorbed dose, activity, alpha particles, atomic mass, atomic number, average atomic mass, beta decay, beta-minus decay, beta-plus decay, binding energy curve, binding energy per nucleon, binding energy, carbon dating, centrifuge, chain reaction, containment vessel, control rods, coolant, daughter nucleus, decay constant, decay diagram, delayed neutrons, deuteron, disintegration constant, enriched uranium, equivalent dose, exposure, fission products, fuel rods, gamma decay, half-life, ionizing electromagnetic radiation, ionizing particle radiation, ionizing radiation, isobars, isomers, isotones, isotopes, mass number, metastable, moderator, neutrino, neutron number, neutron-induced fission, neutrons, nuclear binding energy, nuclear core, nuclear decay, nuclear fission, nuclear force, nuclear fuel, nuclear fusion, nuclear meltdown, nuclear potential, nuclear radiation, nuclear reactions, nucleon, nuclide, parent nucleus, positron, prompt neutrons, proton–proton cycle, protons, Q-value, radiation weighting factor, radioactive decay, radioactive, radiocarbon dating, strong force, weak force

QUESTIONS

1. Define the following terms: nucleon, nuclide, isotope, decay constant, half-life, decay diagram, Q-value, spontaneous fission, radioactive dating, and proton–proton cycle.
2. What are the main differences between the strong and the electromagnetic forces between two protons?
3. Describe the three main processes by which unstable nuclei decay.
4. Estimate Earth's radius if the atoms were compressed to the size of a nucleus.
5. In a nuclear decay series, a $^{238}_{92}$U nucleus decays into a $^{206}_{82}$Pb nucleus, which is stable. What is the maximum number of α-particles that can be emitted in this decay?
6. The mass distribution of the human body is approximately 65% oxygen, 18% carbon, 10% hydrogen, 3% nitrogen, and 1.4% calcium, the rest being other elements. Estimate the number of protons and neutrons in your body.
7. What is the ratio of the coulomb energy and the gravitational potential energy of two protons that are 2.0 fm apart?
8. The half-life of $^{8}_{4}$Be is 6.7×10^{-17} s. What is the most likely decay product of its decay? Explain your reasoning.
9. How many half-lives must pass until 75% of the atoms of a radioactive nuclide have decayed?
10. Why is energy released in the fusion of light nuclei?
11. Why is energy released in the fission of heavier nuclei? Would energy be released if two iron ($Z = 26$) nuclei were to fuse into a tellurium ($Z = 52$) nucleus?
12. What is a coulomb barrier?
13. How would nuclei decay if the mass of a proton were slightly larger than the mass of a neutron?
14. A free neutron has a half-life of 881 s, but a neutron in a carbon-12 nucleus does not decay. Why?

Section 33-1 Nuclear Terminology and Nuclear Units

15. ✱ Express the rest mass energy of 1 g of matter in J and MeV.
16. ✱ Approximately 1.2×10^{17} J of solar energy is intercepted by Earth every second. If all of this energy is absorbed by Earth, what is the increase in Earth's mass in one year?
17. ✱ The mass of a baseball is 145 g. Calculate the rest mass energy of a baseball in
    (a) J;
    (b) eV.
18. ✱ Assuming that nuclei are spherical, calculate the Coulomb repulsion between two carbon-12 nuclei that are
    (a) just touching each other;
    (b) 1.0 cm apart.
19. ✱✱ A free neutron (mass 939.55 MeV/c^2) decays into a proton (mass 938.26 MeV/c^2), an electron (mass 0.511 MeV/c^2), and an antineutrino (zero mass). The initial neutron is at rest. Calculate the maximum possible momentum for the electron. Compare this momentum with the maximum momentum that the antineutrino could have.
20. ✱✱ Consider the fission of a uranium-232 nucleus into two palladium nuclei:

$$^{232}_{92}U \rightarrow {}^{116}_{46}Pa + {}^{116}_{46}Pa$$

    (a) How much Coulomb energy is released?
    (b) How many uranium nuclei need to fission each second to produce 1.0 W of power?
21. ✱✱ Two stable isotopes of carbon are $^{12}_{6}$C (98.89%) and $^{13}_{6}$C (1.11%). Using Table 33-5, calculate the average atomic mass of carbon.
22. ✱✱ Calculate the average atomic mass of oxygen from the data in Table 33-7.

MODERN PHYSICS

Table 33-7 Data for Problem 22

Isotope	Atomic mass (u)	Abundance
Oxygen-16	15.994 914 6	99.757%
Oxygen-17	16.999 131 54	0.038%
Oxygen-18	17.999 160 44	0.205%

Section 33-3 Nuclear Binding Energy

23. ✳✳ Using Table 33-5, calculate the total binding energy and the binding energy per nucleon for the following nuclei:
   (a) $^2_1$H;
   (b) $^4_2$He;
   (c) $^6_3$Li;
   (d) $^{56}_{26}$Fe;
   (e) $^{208}_{82}$Pb.

24. ✳✳ Use the binding energy per nucleon provided in the table below to determine the nuclear masses of the following nuclei. Compare these values to the masses given in Table 33-8.

Table 33-8 Data for Problem 24

Isotope	Binding energy per nucleon (MeV)
$^8_4$Be	7.06
$^{12}_6$C	7.68
$^{84}_{38}$Sr	8.68
$^{208}_{82}$Pb	7.87
$^{238}_{92}$U	7.57

25. ✳✳ Using Table 33-5, calculate the nuclear masses of the following atoms:
   (a) $^2_1$H;
   (b) $^{12}_6$C;
   (c) $^{16}_8$O;
   (d) $^{56}_{26}$Fe;
   (e) $^{238}_{92}$U.

26. ✳✳ Using Table 33-5, calculate the binding energy per nucleon for $A = 100$ nuclei with $Z = 42, 43, 44, 45,$ and 46. Plot these binding energies as a function of Z. Which nuclide is the most tightly bound?

27. ✳✳ Even-even nuclei have an even number of protons and neutrons. In the liquid drop model of a nucleus, the total binding energy of an even-even nucleus can be approximated using the following formula.

$$B(A,Z) = a_V A - a_S A^{2/3} - a_C \frac{Z^2}{A^{1/3}} - a_A \frac{(A-2Z)^2}{A} + a_P \frac{1}{\sqrt{A}}$$

where

$a_V = 15.8$ MeV, $a_S = 18.3$ MeV, $a_C = 0.72$ MeV,

$a_A = 23.2$ MeV, $a_P = 12.0$ MeV

Using the above formula, calculate the binding energies of the nuclei in problem 24. Compare the two sets of binding energies.

Section 33-4 Nuclear Decay and Radioactivity

28. ✳✳ In natural carbon, one atom in 10^{12} is $^{14}_7$C ($t_{1/2} = 5730$ y).
   (a) What is the activity of 1 g of natural carbon?
   (b) After 17 190 y, what is the activity?
   (c) Approximately 18% of the atoms in the human body consists of carbon. Estimate the activity of carbon-14 in your body.

29. ✳✳ The half-life for carbon-14 is 5730 y. Determine its decay constant.

30. ✳✳ A radioactive isotope decays to one-tenth of its original amount in two days. How long did it take for half of it to decay?

31. ✳✳ The half-life of $^{28}_{12}$Mg is 21 h. At a certain time, the activity of magnesium-28 is 2.2×10^8 Bq. What is its activity two days later?

32. ✳✳ A radioactive nuclide had an activity of 3.0×10^4 Bq 10 h ago. Now it has an activity of 2.0×10^4 Bq. What is the half-life of the nuclide?

33. ✳✳ What mass of carbon-14 must be in a sample to have an activity of 10^6 Bq?

34. ✳✳ What mass of $^{226}_{88}$Ra has an activity of 3.7×10^{10} Bq?

35. ✳✳ A radioactive sample contains 3.0×10^{20} atoms and has an activity of 2.0×10^{10} Bq. Find the half-life of the sample.

36. ✳✳ Strontium-90 has a half-life of 25 y.
   (a) Determine its decay constant.
   (b) How long will it take for 80% of the original amount of strontium-90 to decay?

37. ✳✳ A certain radioactive source has an activity of 1000 Bq at $t = 0$. The half-life of the source is 10 h.
   (a) What is its activity after 30 min?
   (b) After 3 days?

38. ✳✳ A sample of cloth contains 20 g of carbon, and the measured carbon-14 decay rate from the cloth is 120 counts/min. Estimate the age of the cloth.

39. ✳✳ At $t = 0$, the activity for a radioactive source is 20.0 MBq. At $t = 240$ s, its activity is 15.0 MBq. What is the half-life of the source?

40. ✳✳ A radioactive source has a half-life of 8 h. At $t = 0$, the activity is 900 decays/min. What is the activity 24 h later?

41. ✳✳ Table 33-9 lists the activities of a radioactive element as a function of time.
   (a) Plot the natural logarithm of the activity as a function of time.
   (b) Determine the slope of the resulting curve.
   (c) Calculate the half-life of the element.

Table 33-9 Data for Problem 41

Time (s)	R (decay/s)	Time (s)	R (decay/s)
0.0	1000	20.0	387
5.0	785	25.0	303
10.0	620	30.0	240
15.0	485	35.0	190

Section 33-5 Nuclear Reactions

42. ✱✱ Identify the missing particle in the following decays:
    (a) $^{22}_{9}F \rightarrow {}^{22}_{10}Ne + ?$;
    (b) $^{12}_{6}C \rightarrow {}^{12}_{5}B + ?$;
    (c) $^{8}_{4}Be \rightarrow {}^{4}_{2}He + ?$;
    (d) $^{239}_{94}Pu \rightarrow {}^{235}_{92}U + ?$;
    (e) $^{60}_{28}Ni^{*} \rightarrow {}^{60}_{28}Ni + ?$.

43. ✱✱ Which of the following decays are not possible? Why not?
    (a) $^{18}_{8}O \rightarrow {}^{17}_{8}O + {}^{0}_{-1}e + \nu$
    (b) $^{14}_{6}C \rightarrow {}^{14}_{7}N + \nu$
    (c) $^{9}_{5}B \rightarrow {}^{9}_{4}Be + {}^{0}_{+1}e + \gamma$
    (d) $^{239}_{94}Pu \rightarrow {}^{239}_{94}Pu + {}^{0}_{-1}e + \bar{\nu}$
    (e) $^{222}_{86}Rn \rightarrow {}^{220}_{84}Po + {}^{4}_{2}He$

44. ✱✱ What are the atomic number (Z) and atomic mass (A) of the unknown nuclei, marked as X, in the following reactions?
    (a) $^{6}_{2}He \rightarrow X + e^{-} + \bar{\nu}_{e}$
    (b) $^{7}_{4}Be \rightarrow X + e^{+} + \nu_{e}$
    (c) $^{1}_{1}p + {}^{12}_{6}C \rightarrow X + {}^{1}_{0}n$
    (d) $^{1}_{0}n + X \rightarrow {}^{31}_{15}P + 2{}^{1}_{0}n$
    (e) $^{27}_{16}Al \rightarrow X + \gamma$

45. ✱✱ In a nuclear reaction, slow-moving neutrons (ignore the neutrons' kinetic energy) are absorbed on a parent nucleus (X), which then decays into a daughter nucleus (Y) and a helium nucleus:

    $$^{1}_{0}n + {}^{A}_{Z}X \rightarrow {}^{A-3}_{Z-2}Y + {}^{4}_{2}He$$

    (a) What is the Q-value of the above reaction in terms of the nuclear masses?
    (b) The parent nucleus is $^{17}_{8}O$. Calculate the energy released using the table of masses. Take the binding energy of a helium nucleus to be 28.2 MeV.

46. ✱✱ A common nuclear reaction, called electron capture, is the capture of a k-shell electron by a nucleus:

    $$^{A}_{Z}X + {}^{0}_{-1}e \rightarrow {}^{A}_{Z-1}Y + \nu$$

    (a) Write an expression for the Q-value of the above process in terms of the nuclear masses of the parent (X) and the daughter (Y) nuclei.
    (b) Write the expression for the Q-value in terms of atomic masses of the parent and the daughter nuclei.

47. ✱✱ Can a $^{12}_{6}C$ nucleus decay into three $^{4}_{2}He$ nuclei? If yes, how much energy will be released in such a decay? If not, why not?

48. ✱✱ Berillium-8 ($^{8}_{4}Be$) is an unstable nucleus that decays into two helium nuclei. Given the following data, determine why $^{8}_{4}Be$ is not stable:

    Atomic mass of $^{8}_{4}Be = 8.005305$ u

    Atomic mass of $^{4}_{2}He = 4.002603$ u

Section 33-6 α-, β-, and γ-Decays

49. ✱ Write equations to describe the α-decay of the following nuclei. Use a periodic table to identify the daughter nuclei.
    (a) $^{32}_{14}Si$
    (b) $^{56}_{25}Mn$
    (c) $^{60}_{26}Fe$

50. ✱ Write equations to describe the β^{-}-decay of the following nuclei. Use a periodic table to identify the daughter nuclei.
    (a) $^{32}_{14}Si$
    (b) $^{56}_{25}Mn$
    (c) $^{60}_{26}Fe$

51. ✱ Write the equations to describe the β^{+}-decay of the following nuclei. Use a periodic table to identify the daughter nuclei.
    (a) $^{52}_{25}Mn$
    (b) $^{55}_{27}Co$
    (c) $^{59}_{28}Ni$

52. ✱ Using the table of atomic masses (Table 33-5), calculate the Q-value of the following α-decay reactions:
    (a) $^{238}_{92}U \rightarrow {}^{234}_{90}Th + {}^{4}_{2}He$;
    (b) $^{235}_{92}U \rightarrow {}^{231}_{90}Th + {}^{4}_{2}He$;
    (c) $^{210}_{84}Po \rightarrow {}^{206}_{82}Pb + {}^{4}_{2}He$.

53. ✱ Tritium is an isotope of hydrogen with two neutrons. It has a half-life of 12.3 y and β-decays into helium-3.
    (a) Write the reaction that describes the β-decay of tritium into helium-3.
    (b) Calculate the energy released in this decay, using the atomic masses from Table 33-5.

54. ✱✱ Lead-212 ($^{212}_{82}Pb$) is a radioactive isotope of lead. It decays to the stable lead-208 through the following decay chain: β^{-}, β^{-}, α. Draw a decay diagram showing the decay process, and identify the intermediate daughter nuclei at each step.

55. ✱✱ Thorium-232 ($^{232}_{90}Th$) decays to lead-208 ($^{208}_{82}Pb$). How many α- and β-particles are emitted in this process?

Section 33-7 Nuclear Fission and Nuclear Fusion

56. ✱✱ In one of the possible reactions involving the neutron-induced fission of $^{239}_{94}Pu$, the two daughter elements $^{141}_{56}Ba$ and $^{92}_{36}Kr$ are produced.
    (a) Write an equation describing this reaction.
    (b) Find the energy released in the process, ignoring the kinetic energy of the incident neutron.

57. ✱✱ Using the table of atomic masses (Table 33-5), calculate the Q-values for the following fission reactions:
    (a) $^{1}_{0}n + {}^{235}_{92}U \rightarrow {}^{90}_{36}Kr + {}^{144}_{56}Ba + 2{}^{1}_{0}n$;
    (b) $^{1}_{0}n + {}^{235}_{92}U \rightarrow {}^{93}_{37}Rb + {}^{141}_{55}Cs + 2{}^{1}_{0}n$.

58. ✱✱ The half-life for the spontaneous fission of uranium-238 is approximately 8×10^{15} y, and the half-life of uranium-238 for α-decay is 4.5×10^{9} y. Given 1 g of uranium-238, how many nuclei undergo spontaneous fission in one day? How many nuclei α-decay in a day?

59. ✱✱✱ Even–even nuclei have an even number of protons and neutrons. Odd–odd nuclei have an odd number of protons and neutrons. Even–even nuclei are more tightly bound than odd–odd nuclei. There are only five stable odd–odd nuclei: $^{2}_{1}H$, $^{6}_{3}Li$, $^{10}_{5}B$, $^{14}_{7}N$, and $^{180}_{73}Ta$.
    (a) Using Table 33-5 of nuclear masses, calculate the binding energy per nucleon of odd–odd nuclei.
    (b) Using the same method, calculate the binding energy per nucleon of the following even–even nuclei obtained by adding one neutron and one proton to the odd–odd nuclei:
       (i) $^{4}_{2}He$;
       (ii) $^{8}_{4}Be$;
       (iii) $^{12}_{6}C$;
       (iv) $^{16}_{8}O$;
       (v) $^{182}_{74}W$.
    (c) How much binding energy is gained in each case?

60. ✶✶✶ Show that the Q-value for spontaneous nuclear fission can be written in terms of the binding energies of the parent nucleus (A, Z) and daughter nuclei (A_1, Z_1) and (A_2, Z_2) as follows:

$$Q_f = B(A_1, Z_1) + B(A_2, Z_2) - B(A, Z)$$

Using the binding energy formula for the liquid drop model (problem 27), determine the energy released in the following fission processes.

$$^{238}_{92}U \rightarrow ^{134}_{52}Te + ^{102}_{40}Zr + 2^{1}_{0}n + Q_f$$

$$^{238}_{92}U \rightarrow ^{134}_{50}Sn + ^{102}_{42}Mo + 2^{1}_{0}n + Q_f$$

COMPREHENSIVE PROBLEMS

61. ✶✶✶ The binding energy of an α-particle (a $^{4}_{2}$He nucleus) is 28.2 MeV. Use the liquid drop model (problem 27) to answer the following questions.

(a) A $^{8}_{4}$Be nucleus can decay into two α-particles. How much energy is released in this process?

(b) Can a $^{12}_{6}$C nucleus decay into three α-particles? Explain your answer.

62. ✶✶✶ In an (n, α) nuclear reaction, slow-moving neutrons (ignore the neutrons' kinetic energy) are absorbed on a target nucleus (X), which then decays into a daughter nucleus (Y) and a helium nucleus:

$$^{1}_{0}n + ^{A}_{Z}X \rightarrow Y + ^{4}_{2}He$$

The target nucleus is hafnium-180 ($^{180}_{72}$Hf). Calculate the energy released in the above process using the liquid drop model of problem 27. Take the binding energy of a helium nucleus to be 28.2 MeV.

63. ✶✶✶ Using the liquid drop model in problem 27, show that the energy (S_n) required to separate a neutron from a nucleus with Z protons and $A - Z$ neutrons, with large A $(A \gg 1)$, is approximately given by

$$S_n = a_V - \frac{2}{3} a_S A^{-1/3} - a_A\left(1 - \frac{4Z^2}{A(A-1)}\right)$$

See the text online resources at www.physics1e.nelson.com for Open Problems and Data-Rich Problems related to this chapter.

Learning Objectives

When you have completed this chapter you should be able to:

1 List all the elementary particles. Explain the difference between quarks and leptons.

2 Explain the role of gauge bosons in mediating interactions between the elementary particles.

3 Define antiparticle, and identify the antiparticle corresponding to each elementary particle.

4 Explain the Standard Model and how particles are grouped into three families.

5 Explain the structure of composite baryons and mesons in terms of quarks, and determine the charge, baryon number, lepton number, and the spin quantum number of a composite particle from its quark constituents.

6 Define quark confinement.

7 Apply conservation laws of energy, momentum, charge, baryon number, and lepton number to determine whether a particular reaction or decay is allowed.

8 Calculate the energy available in the centre-of-mass coordinate system of two colliding particles to produce a new particle.

9 Explain Feynman diagrams, and draw and interpret Feynman diagrams for simple processes.

10 Explain how pions and muons were discovered.

11 Explain how a linear accelerator works, and explain the difference between a cyclotron and a synchrotron.

For additional **Making Connections, Examples, and Checkpoints**, as well as activities and experiments to help increase your understanding of the chapter's concepts, please go to the text's online resources at www.physics1e.nelson.com.

Chapter 34
Introduction to Particle Physics

On July 4, 2012, scientists at the Large Hadron Collider (LHC) announced the discovery of the Higgs particle, bringing an end to the most expensive and long-awaited search for a particle in the history of human-kind (Figure 34-1). The Higgs particle was the last yet-to-be-discovered member of the Standard Model, a model that forms the basis of our current understanding of the nature of matter and the fundamental forces that exist between the particles. What is the Standard Model? What are the other particles in the Standard Model? How do particles interact with each other? Why does the LHC need to be so large? The subject of particle physics is fascinating as well as intriguing.

CERN

Figure 34-1   A section of the Large Hadron Collider

34-1 Classification of Particles

The electron was discovered by J.J. Thomson in 1896, the proton by Ernest Rutherford in 1919, and the neutron by James Chadwick in 1932. These discoveries led to a model of the atom with electrons orbiting a central nucleus made of protons and neutrons. The three particles in this atomic model were thought to be the only building blocks of matter. However, as technology developed and it became possible to collide extremely fast-moving particles and nuclei with each other, physicists began to discover a range of new types of particles. Most of the new particles exist for only a very short time and decay into other particles. To date, physicists have discovered more than 200 different subatomic particles. The branch of physics that deals with the study of particles is called **particle physics**.

Particles are classified as either elementary or composite. **Elementary** (or **fundamental**) **particles** are thought to be point particles with no internal structure. Electrons, neutrinos, photons, and quarks (which we will discuss later) are examples of elementary particles. A list of all known elementary particles is given in Table 34-1. **Composite particles** are made of elementary particles. For example, protons and neutrons are made of quarks and are therefore composite particles. The physical properties of a composite particle are determined by the properties of its constituents and depend on how the constituents interact with each other to form the composite particle (similar to how the properties of a nucleus are determined by the number of protons and neutrons that make up the nucleus and how they interact to form the nucleus).

A particle is classified in terms of

- its physical properties, such as mass, charge, and magnetic moment;

- its quantum numbers, including spin, and a few other properties, such as baryon number and lepton number;

- the type of forces that the particle exerts on other particles.

Table 34-1 The Fundamental Particles

Type of particle	Symbol	Name	Mass (MeV/c^2)	Charge (e)	Spin quantum number	Baryon number	Lepton number
Quarks*	u	Up quark	2.5	+2/3	1/2	1/3	0
	d	Down quark	5.0	−1/3	1/2	1/3	0
	s	Strange quark	125	−1/3	1/2	1/3	0
	c	Charm quark	1.3×10^3	+2/3	1/2	1/3	0
	b	Bottom quark	4.2×10^3	−1/3	1/2	1/3	0
	t	Top quark	171×10^3	+2/3	1/2	1/3	0
Leptons	e^-	Electron	0.511	−1	1/2	0	1
	μ^-	Muon	105.7	−1	1/2	0	1
	τ^-	Tau	1.777×10^6	−1	1/2	0	1
	ν_e	Electron neutrino	$<2.2 \times 10^{-6}$	0	1/2	0	1
	ν_μ	Muon neutrino	<0.17	0	1/2	0	1
	ν_τ	Tau neutrino	<15.5	0	1/2	0	1
Gauge bosons	W^+	W-plus boson	80.4×10^3	+1	1	0	0
	W^-	W-minus boson	80.4×10^3	−1	1	0	0
	Z^0	Z-zero boson	91.2×10^3	0	1	0	0
	γ	Photon	0	0	1	0	0
	g	Gluon	0	0	1	0	0
Higgs boson	H^0	Higgs boson	Estimated to be approximately 125×10^3	0	0	0	0

* Here the mass implies the rest-mass of the particle. The spin is the intrinsic spin quantum number and the baryon and lepton numbers are explained in the text. An isolated quark has never been detected. Therefore, the given values are the best estimates of the quark masses.
Source: Data from "Wuark Masses," updated Jan 2010 by A.V. Manohar (University of California, San Diego) and C.T. Sachrajda (University of Southampton, http://pdg.lbl.gov/2011/reviews/rpp2011-rev-quark-masses.pdf

MODERN PHYSICS

Mass

The rest masses of the elementary particles vary over a wide range, as can be seen from Table 34-1. For example, a photon has zero rest mass, and the rest mass of the top quark is approximately 170 GeV/c^2, which is approximately 340 000 times the rest mass of an electron. The rest mass of a composite particle depends on both the rest mass of its constituent elementary particles and the nature of the forces between them to form the composite particle.

Electric Charge

A particle can have an electric charge or it can be electrically neutral. The electric charge is measured in units of the electron charge ($e = 1.602 \times 10^{-19}$ C). Until the 1960s, it was believed that the charge of a particle must be an integer multiple of e. However, with the discovery of quarks it was realized that quarks come in two charges: $+\frac{2}{3}e$ and $-\frac{1}{3}e$. The charge of a composite particle is equal to the sum of the charges of its constituents.

Intrinsic Spin

All elementary particles are assigned a spin quantum number, s. The electron has $s = \frac{1}{2}$, and the photon has $s = 1$. The intrinsic spin of a composite particle is obtained by adding the intrinsic spins of its constituents using the rules for the addition of angular momenta as discussed in Section 31-9. Recall that particles with half-integer values of s are called fermions and the particles with integral values of s are called bosons.

Fundamental Forces

The four known forces (or interactions) in nature are the strong force, the weak force, the electromagnetic force, and the gravitational force. These forces vary greatly in relative strengths, as shown in Table 34-2.

Table 34-2 Relative strengths of the four known forces

Force	Relative Strength (N)
Strong force	1
Electromagnetic force	10^{-2}
Weak force	10^{-7}
Gravitational force	10^{-38}

For example, if the strong force between two protons at a certain distance is 1 N, the strength of the electromagnetic force between them at the same distance is approximately 10^{-2} N, the weak force approximately 10^{-7} N, and the gravitational force approximately 10^{-38} N.

All particles with mass exert a gravitational force on each other. We will ignore the gravitational force in the rest of this chapter. The electromagnetic force always exists between all charged particles (elementary and composite). The strong force exists only between quarks and is responsible for binding quarks into composite particles (composite particles are discussed in Section 34-5). The weak force is responsible for processes such as β-decay. Particles that interact with other particles only through the weak force and the electromagnetic force (if charged) are called **leptons**.

Quarks

Quarks are the building blocks of all composite particles. Quarks are charged fermions with $s = \frac{1}{2}$. There are six types of quarks (called **flavours**): the **up quark** (u), the **down quark** (d), the **strange quark** (s), the **charm quark** (c), the **bottom quark** (b), and the **top quark** (t). Quarks interact with each other through the strong, electromagnetic, and weak forces. We will discuss the quantum properties of quarks further in Section 34-6.

Leptons

Leptons are also fermions with $s = \frac{1}{2}$. There are six types of leptons: the electron (e^-), the **electron neutrino** (ν_e), the **muon** (μ^-), the **muon neutrino** (ν_μ), the **tau** (τ^-), and the **tau neutrino** (ν_τ). The charged leptons interact with each other and with other particles through the weak and electromagnetic forces. Neutrinos are electrically neutral, so they interact only through the weak force. An electron is stable and does not decay. The muon and tau are considerably more massive than an electron and are unstable. For example, a muon has a half-life of 2.26×10^{-6} s and decays into an electron and two neutrinos.

Gauge Bosons

According to the theory of relativity and quantum physics, elementary particles exert forces on each other by exchanging other particles, called **gauge bosons**. Photons mediate the electromagnetic force, gluons mediate the strong force, and $W^\pm$ and Z^0 bosons mediate the weak force. It is possible that the gravitational force is mediated by elementary particles called gravitons, but gravitons have not been observed. Gauge bosons have $s = 1$.

Higgs Bosons

In 1964, a group of scientists proposed that the elementary particles quarks, leptons, and gauge bosons gain mass by interacting with a field, which is called the Higgs field after one member of the group, British physicist Peter Higgs. In quantum theory, the Higgs field is quantized with the corresponding particles called **Higgs bosons**, just as the electromagnetic field is quantized with photons as the associated particles. The Higgs boson was predicted to be a neutral particle with an intrinsic spin quantum number 0 and a mass of approximately 125 GeV/c^2. On July 4, 2012, scientists at the LHC announced the long-awaited discovery of the Higgs boson.

C-34-1 Particle Classification

Particles that interact only through the weak and electromagnetic forces are called

(a) bosons;

(b) quarks;

(c) gauge bosons;

(d) leptons;

(e) gluons.

LO 2

34-2 Gauge Bosons

How do two electrons repel each other? If one electron suddenly moves away from another electron, would the electromagnetic force between the two decrease instantaneously? A consequence of the theory of relativity and quantum physics is that particles exert forces on each other by exchanging other particles. The rules of adding angular momentum in quantum mechanics require that an exchanged particle have an integer spin; therefore, it must be a boson. The carrier of the electromagnetic force between two charged particles is the photon, the particle that results when electromagnetic radiation is quantized. Two electrons therefore repel each other by exchanging a photon. A photon travels at the speed of light, so if the distance between two electrons suddenly changes, the force between them will take a finite time to adjust. The process of the exchange of a photon between two electrons is represented by the two diagrams in Figure 34-2. These diagrams are called **Feynman diagrams**, and they will be discussed in greater detail in Section 34-9.

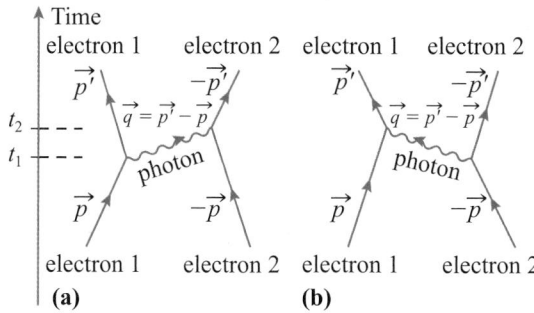

Figure 34-2 Feynman diagrams for the exchange of a photon between two electrons (a) Electron 1 emits a photon, which is later absorbed by electron 2. (b) Electron 2 emits a photon, which is later absorbed by electron 1.

In Figure 34-2(a), electron 1 emits a photon, which travels to electron 2. At a later time, electron 2 absorbs the photon. Let us consider this process using the centre-of-mass coordinate system of the two electrons. In this coordinate system, the two electrons have equal and opposite momenta in the initial state, and their energies E_1 and E_2 are

$$E_1 = \sqrt{\vec{p} \cdot \vec{p} c^2 + m_e^2 c^4} \, ;$$

$$E_2 = \sqrt{(-\vec{p}) \cdot (-\vec{p}) c^2 + m_e^2 c^4}$$

Let the momentum of the emitted photon be $\vec{q}$. After emitting the photon, the momentum of the first electron is $\vec{p} - \vec{q} = \vec{p'}$. When electron 2 absorbs the photon, its momentum changes to $-\vec{p} + \vec{q} = -\vec{p'}$. Therefore, the energies of the two electrons in the final state (after the photon has been absorbed by the electron 2) are

$$E_1' = \sqrt{\vec{p'} \cdot \vec{p'} c^2 + m_e^2 c^4} \, ;$$

$$E_2' = \sqrt{(-\vec{p'}) \cdot (-\vec{p'}) c^2 + m_e^2 c^4}$$

Note that in the centre-of-mass coordinate system, $E_1 = E_2$ and $E_1' = E_2'$. Since the total energy must be conserved before and after the two electrons exchange a photon, we must have

$$E_1 + E_2 = E_1' + E_2'$$

This implies that in the centre–of-mass coordinate system, $E_1 = E_1'$ and $E_2 = E_2'$. Before electron 1 emits the photon, the total energy of the two electrons is

$$E = E_1 + E_2$$

Let the energy of the emitted photon be E_γ. After the photon is emitted *but before it is absorbed*, the total energy of the two electrons and the photon is

$$E_1' + E_2 + E_\gamma = E_1 + E_2 + E_\gamma$$

Thus, we see that while the photon is travelling between the two electrons, the total energy is not conserved. Is the nonconservation of energy allowed? According to Heisenberg's uncertainty principle, nonconservation of energy by an amount ΔE cannot be observed if it occurs for a time Δt such that

$$\Delta t \leq \frac{\hbar}{2\Delta E} \tag{34-1}$$

The energy of a photon with frequency f is hf. Therefore, the energy of a photon with frequency f cannot be determined to an accuracy hf if the photon exists for a time less than $\Delta t = \frac{\hbar}{2hf} = \frac{1}{4\pi f}$. During this time, the photon travels the distance $R = c\Delta t = \frac{c}{4\pi f} = \frac{\lambda}{4\pi}$. Photons are massless, so the energy of a photon can be arbitrarily small; therefore, the distance R over which a photon can transmit the electromagnetic force is arbitrarily large. Figure 34-2(b) shows the process in which the photon is emitted by electron 2 and is later absorbed by electron 1. Instead of showing both diagrams, it is common to represent a photon exchange between two electrons by a single diagram with a horizontal line (Figure 34-3).

An exchanged photon exists for a limited time, and is called a **virtual photon**. Its energy is uncertain by an amount that is inversely proportional to its time

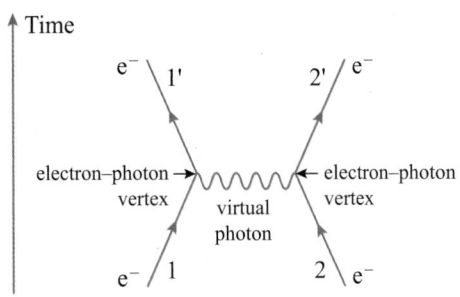

Figure 34-3   The exchange of a virtual photon between two electrons

of existence. A virtual particle does not satisfy Einstein's energy–momentum relation. Therefore a virtual photon can carry momentum without carrying energy. A photon emitted by a candle or a light bulb is a real photon. Its energy can be determined to a very high degree of accuracy, and its momentum (p) and energy (E) are related by $E = pc$.

The point where the electron emits a photon is called a **vertex**. Each electron–photon vertex is assigned a **coupling strength**, $\dfrac{e}{\sqrt{4\pi\varepsilon_0\hbar c}}$. The strength of the electromagnetic interaction is obtained by multiplying the coupling strength at both vertices of the diagram:

$$\frac{e}{\sqrt{4\pi\varepsilon_0\hbar c}} \times \frac{e}{\sqrt{4\pi\varepsilon_0\hbar c}} = \frac{e^2}{4\pi\varepsilon_0\hbar c}$$

Similar arguments can be applied to the strong and the weak interactions. Gluons are the carriers of the strong force. Quarks interact with each other by exchanging gluons. Like photons, gluons have no mass. The weak force is due to the exchange of W^+, W^-, and Z^0 bosons between elementary particles. Unlike the photon and the gluon, these bosons are very massive. Following the same arguments as discussed above, we can show that the range R of a force between two particles that is due to the exchange of a boson of mass m is approximately

$$R = \frac{\hbar c}{2mc^2} \qquad (34\text{-}2)$$

For example, the Z^0 and $W^\pm$ bosons have masses of 91.2 GeV/c^2 and 80.4 GeV/c^2 respectively, so the range of the weak force is approximately $\dfrac{200\ \text{MeV} \cdot \text{fm}}{160\ \text{GeV}} \approx$ 0.001 fm, an extremely short distance.

Figure 34-4 summarizes the carriers of the electromagnetic, strong, and weak forces.

LO 3

34-3   Antiparticles

As discussed in Chapter 31, Schrödinger's equation satisfies the nonrelativistic energy–momentum equation $E = \dfrac{p^2}{2m} + V(x)$. But a particle moving with a relativistic speed must satisfy Einstein's energy and momentum relation, $E^2 = p^2c^2 + m^2c^4$. In 1928, British physicist Paul Dirac developed an equation that is consistent with the theory of relativity as well as quantum mechanics. Dirac used his equation to calculate many properties of atomic electrons, such as intrinsic spin and hyperfine splitting of hydrogen levels. Since the relativistic energy–momentum relationship is quadratic in energy, solutions to Dirac's equation allow the electron's energy to have both positive and negative values: $E = \pm\sqrt{p^2c^2 + m_e^2c^4}$, where m_e is the mass of the electron. Dirac interpreted the negative-energy solutions as corresponding to particles that have exactly the same mass as an electron but are positively charged. These particles are called **antiparticles**. The antiparticle of an electron is the **positron** (e^+). Dirac predicted that an electron–positron pair can be created by bombarding nuclei with photons with energy greater than $2m_e c^2$:

$$\gamma + \text{nucleus} \rightarrow e^- + e^+ + \text{nucleus}$$

Such a process, called **pair production**, requires the presence of a nucleus or another charged particle to conserve energy and momentum. Dirac also noted that when an electron and a positron combine, they can annihilate each other into two photons. This process is called **pair annihilation** and is written as follows:

$$e^- + e^+ \rightarrow \gamma + \gamma$$

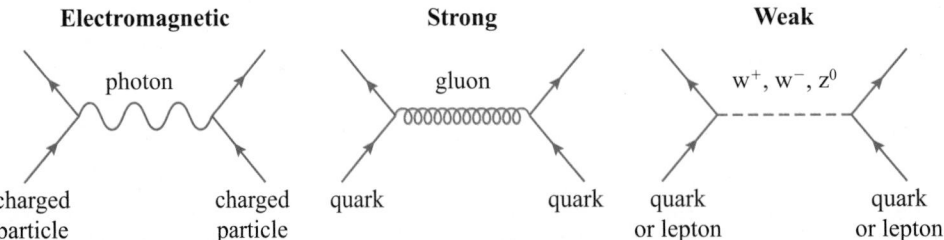

Figure 34-4   The carriers of the electromagnetic, strong, and weak forces

The existence of positrons was confirmed by the American physicist Carl Anderson, in 1932. He used a cloud chamber to study properties of cosmic rays. A cloud chamber consists of an enclosed chamber filled with a supersaturated vapour, often alcohol. When a charged particle passes through the chamber, it ionizes some vapour atoms along its path. The ionized atoms trigger the condensation of droplets in the supersaturated vapour, creating a visible track of the path of the charged particle. Placing the chamber in a magnetic field causes the tracks of positively and negatively charged particles to curve in opposite directions. Thus, the charge and the mass of a particle can be determined from measurements of the curvature and thickness of its track.

Some of the tracks observed by Anderson looked like the tracks in Figure 34-5. At the vertex of the curved upside-down V-shaped trail, a fast-moving photon created by a cosmic ray collides with a nucleus, producing an electron–positron pair. The curvature is due to the magnetic field applied at right angles to the plane of the photograph. One of the tracks corresponds to the electron, and the other track corresponds to a particle that has the same mass as an electron but opposite charge—a positron.

Figure 34-5 Tracks of several charged particle–antiparticle pairs in a cloud chamber

Experiments have shown that each quark and charged lepton has an antiparticle with the same mass and opposite charge. It is now assumed that every particle has a corresponding antiparticle. A neutral particle either has an antiparticle with some opposite quantum property or is its own antiparticle. Table 34-3 lists the elementary antiparticles. Note that the antiparticles for quarks and neutrinos are denoted by adding a bar over the symbol for the particle.

Table 34-3 The Elementary Particles and Antiparticles

Particle	Antiparticle	Name/Description
u	$\bar{u}$	u-bar
d	$\bar{d}$	d-bar
s	$\bar{s}$	s-bar
c	$\bar{c}$	c-bar
b	$\bar{b}$	b-bar
t	$\bar{t}$	t-bar
e^-	e^+	Positron
μ^-	μ^+	Mu-plus
τ^-	τ^+	Tau-plus
ν_e	$\bar{\nu}_e$	Electron antineutrino
ν_μ	$\bar{\nu}_\mu$	Muon antineutrino
ν_τ	$\bar{\nu}_\tau$	Tau antineutrino
γ	γ	A photon is its own antiparticle.
W^-	W^+	W^+ is the antiparticle of W^-.
W^+	W^-	W^- is the antiparticle of W^+.
Z^0	Z^0	Z^0 is its own antiparticle.
H^0	H^0	The Higgs boson is its own antiparticle.

✓ **CHECKPOINT**

C-34-3 Classifications of Particles

Which of the following statements are true?
(a) A lepton may be a boson.
(b) All quarks are fermions.
(c) Quarks and electrons can interact through the strong force.
(d) All gauge bosons are virtual particles.
(e) A photon is its own antiparticle.
(f) An elementary particle cannot decay into other particles.

C-34-3 (b), (e)

LO 4

34-4 Quarks and the Standard Model

The Standard Model postulates that the known universe is made of four types of fundamental particles: quarks, leptons, gauge bosons, and the Higgs boson (Figure 34-6), as well as their antiparticles. The particles interact with each other through the four known forces: the strong

CARL ANDERSON/SCIENCE PHOTO LIBRARY

MODERN PHYSICS

Three Families of Matter

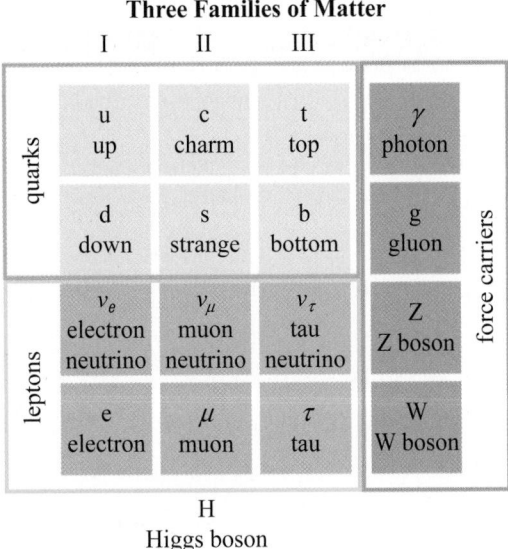

Figure 34-6 Elementary particles of the Standard Model

force, the electromagnetic force, the weak force, and the gravitational force. The Standard Model describes how the particles interact with each other through three of the four forces, the exception being the gravitational force. As noted earlier, the forces between the particles are mediated through the exchange of gauge bosons, $W^{\pm}$, Z^0, γ, and g (gluons). The Higgs boson is an additional boson that gives mass to fundamental particles. It is not a gauge boson (i.e., it is not a force carrier).

The six quarks and the six leptons are divided into three **particle families** or **generations**. The first generation consists of the up quark, the down quark, the electron, and the electron neutrino. The first generation is the most stable and least-massive family of particles. Protons, neutrons, and all atoms are made up of the first-generation particles.

The second generation consists of the strange quark, the charm quark, the muon, and the muon neutrino. The third generation consists of the bottom quark, the top quark, the tau, and the tau neutrino. Particles in the second and third generations are more massive than particles in the first generation, and they decay quickly when produced in high-energy collisions of first-generation particles. The fact that there are six quarks and six leptons that can be grouped into three families is required for the Standard Model to be consistent.

The Standard Model accurately describes how particles interact with each other and correctly predicted the existence of a number of particles. The up, down, and strange quarks were known to exist quite early in the development of the quark model of hadrons. The existence of the charm quark was predicted on the basis that the second family needs a quark with a charge of $+2e/3$. The charm quark was discovered in 1974 at the Stanford Linear Accelerator by colliding beams of electrons and positrons. Similarly, the Standard Model also predicted the existence of $W^{\pm}$ and Z^0 and bosons, which were discovered in 1983 with the particle accelerator at CERN. The

model also predicts the existence of many new composite particles made up of the more massive charm, bottom, and up quarks. Many of these composite particles have been observed in high-energy collisions, and the observed masses and other physical properties are consistent with the predictions of the Standard Model.

✅ **CHECKPOINT**

C-34-4 Fundamental Particles

The number of fundamental particles and anti-particles in the standard model is

(a) 16;
(b) 24;
(c) 29;
(d) 32;
(e) 64.

LO 5

34-5 Composite Particles

All the particles, which can interact via the strong force are collectively called hadrons. This includes protons and neutrons. Hadrons that are fermions are called **baryons**, and hadrons that are bosons are called **mesons**. By early 1960's a large number of baryons and mesons had been discovered. The existence of quarks was first postulated by an American physicist, Murray Gell-Mann (1929 –), in an attempt to understand whether this large number of particles have an underlying structure. Just as protons and neutrons bind to form a large number of nuclei, Gell-Mann proposed that baryons and mesons are made up of only three types of elementary particles, which were named quarks. These were the up (u), the down (d) and the strange (s) quark. The charm (c), the bottom (b) and the top (t) quarks were discovered later.

Baryons

Baryons are made of three quarks and have half-integer spin because all the quarks have $s = \dfrac{1}{2}$. A proton is a combination of two up quarks and one down quark, and a neutron consists of one up quark and two down quarks (Figure 34-7):

$$\text{proton} = (\text{uud}) \qquad \text{charge} = \left(\frac{2}{3} + \frac{2}{3} - \frac{1}{3}\right)e = +e$$

$$\text{neutron} = (\text{udd}) \qquad \text{charge} = \left(\frac{2}{3} - \frac{1}{3} - \frac{1}{3}\right)e = 0$$

Quarks are confined within a proton (or a neutron); therefore, their energies are quantized. The proton and the

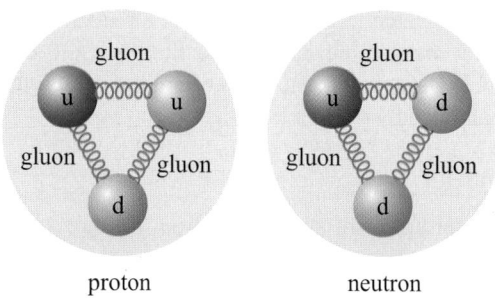

Figure 34-7 The quark structure of a proton and a neutron

neutron are the lowest-energy bound states made of up and down quarks. As a proton (or a neutron) has spin $s = \dfrac{1}{2}$, one of the three quarks must have its spin oriented opposite to the spin of the other two. A proton with spin-up, written as $p\!\uparrow$, has the following possible quark structures:

$$p\!\uparrow = (u\!\uparrow u\!\uparrow d\!\downarrow + u\!\uparrow u\!\uparrow d\!\uparrow + u\!\downarrow u\!\uparrow d\!\uparrow)$$

Baryons can also have quarks with their spins all pointing in the same direction. For example, the delta-plus-plus (Δ^{++}) particle contains three up quarks. The resulting particle has a rest mass of 1230 MeV/c^2 and a spin quantum number of $s = \dfrac{3}{2}$:

$$\Delta^{++} = (u\!\uparrow u\!\uparrow u\!\uparrow)$$

There are three other delta particles with quark spins aligned in the same direction:

$$\Delta^{+} = (u\!\uparrow u\!\uparrow d\!\uparrow); \quad \Delta^{0} = (u\!\uparrow d\!\uparrow d\!\uparrow); \quad \Delta^{-} = (d\!\uparrow d\!\uparrow d\!\uparrow)$$

The superscript on each particle indicates its charge. Delta particles are excited states of three quarks and therefore are more massive than a proton or a neutron.

The quark model predicted the existence of particles that had not yet been observed. Consider particles containing the three lowest-mass quarks, up, down, and strange. We have seen how combinations of up and down quarks provide internal structures for the proton, neutron, and delta particles. The addition of the strange quark provides the internal structure of other particles, some of which had not yet been discovered.

Particles that contain one strange and two up or down quarks are called sigma baryons (symbol Σ). There are three sigma baryons:

$$\Sigma^{+} = (uus); \quad \Sigma^{0} = (uds); \quad \Sigma^{-} = (dds)$$

With two strange and one up or down quark, we can construct two baryons. These particles had previously been discovered and are called xi baryons (symbol Ξ):

$$\Xi^{0} = (uss); \quad \Xi^{-} = (dss)$$

Finally, the model predicts the existence of a baryon that is made of three strange quarks. Gell-Mann called this baryon the omega (symbol Ω) baryon:

$$\Omega^{-} = (sss)$$

The omega baryon had not been discovered, and its discovery in 1964 was a great triumph of the quark model.

Table 34-4 lists some of the lowest-mass baryons made of up, down, and strange quarks. Note that some of the particles have the same quark structure.

Mesons

Mesons are hadrons that are composed of one quark and one antiquark. A quark and an antiquark both have intrinsic spin $\dfrac{1}{2}$, so the total intrinsic spin of a quark–antiquark system can either be 0 or 1. Mesons with $s = 0$ are called **scalar mesons**, and mesons with $s = 1$ are called **vector mesons**.

The lightest scalar meson made of up and down quarks and antiquarks is called the pion (symbol π). The three possible combinations of pions are the following:

Table 34-4 Some lowest-mass baryons made of u, d, and s quarks

Symbol	Name	Quark structure	Mass (MeV/c^2)	Spin	Mean life (s)
p	Proton	uud	938.3	1/2	Stable
n	Neutron	udd	939.6	1/2	886
Λ	Lambda	uds	1115.6	1/2	2.52×10^{-10}
Σ^{+}	Sigma-plus	uus	1189.4	1/2	0.80×10^{-10}
Σ^{0}	Sigma-zero	uds	1192.5	1/2	1.0×10^{-14}
Σ^{-}	Sigma-minus	dds	1197.3	1/2	1.48×10^{-10}
Ξ^{0}	Xi-zero	uss	1314.9	1/2	2.98×10^{-10}
Ξ^{-}	Xi-minus	dss	1321.3	1/2	1.67×10^{-10}
Ω^{-}	Omega-minus	sss	1672.5	3/2	1.3×10^{-10}

$$\pi^+ = (u\,\bar{d}); \quad \text{charge} = \left(\frac{2}{3} + \frac{1}{3}\right)e = e$$

$$\pi^- = (d\,\bar{u}); \quad \text{charge} = \left(-\frac{1}{3} - \frac{2}{3}\right)e = -e$$

$$\pi^0 = \frac{1}{\sqrt{2}}(u\,\bar{u} + d\,\bar{d}); \quad \text{charge} = \left(\frac{2}{3} - \frac{2}{3} + \frac{1}{3} - \frac{1}{3}\right)e = 0$$

The electrically neutral pion is made of a $u\bar{u}$ and a $d\bar{d}$ combination. The factor of $1/\sqrt{2}$ implies that both combinations are present with an equal probability. The spins of the quark and antiquark of a pion are aligned antiparallel to each other.

The lightest vector meson is the rho meson (ρ), with a mass of 775 MeV/c^2. The quark structure of a rho meson is the same as the quark structure of the correspondingly charged pion; however, the spins of its quark and antiquark are aligned parallel to each other. As yet, there is no clear explanation why the mass of a rho meson is more than five times the mass of a pion.

With strange, charm, bottom, and top quarks, we can construct many more mesons. So far, over 120 mesons have been discovered. The existence of many mesons was first predicted by the quark model.

Antimatter

The antiproton ($\bar{p}$) and antineutron ($\bar{n}$) are made of antiquarks:

$$\bar{p} = (\bar{u}\,\bar{u}\,\bar{d}); \quad \bar{n} = (\bar{u}\,\bar{d}\,\bar{d})$$

Notice that although an antineutron is electrically neutral, it clearly has a different quark structure than a neutron. Some composite particles are their own antiparticles. For example, a π^0 meson is made of equal amounts of $u\,\bar{u}$ and $d\,\bar{d}$ pairs. The anti-π^0 meson is obtained by replacing each quark of π^0 with its antiquark. Thus, the quark structure of an anti-π^0 meson is exactly the same as the quark structure of the π^0 meson:

$$\bar{\pi}^0 = \frac{1}{\sqrt{2}}(\bar{u}\,u + \bar{d}\,d) = \pi^0$$

 ## EXAMPLE 34-1

Constructing Mesons

List all the mesons that have one antibottom ($\bar{b}$) quark. What is the charge of each meson? Assuming that the rest mass energy of a meson is the sum of the rest mass energies of its constituent quarks, predict the rest mass energy of each meson. Compare your prediction with the experimental values given in Table 34-5. Use the quark masses in Table 34-1.

SOLUTION

Since mesons contain one quark and one antiquark, we can construct six mesons by combining a $\bar{b}$ antiquark with any of the quarks. A meson constructed from an up quark and a $\bar{b}$ quark (called a B meson) has the following properties:

Quark structure: $u\,\bar{b}$

Charge: $2e/3 + e/3 = +e$

Estimated mass: $m_{\bar{b}} + m_u = 4.2 \times 10^3\,\text{MeV}/c^2 + 2.5\,\text{MeV}/c^2$

$$= 4.2\,\text{GeV}/c^2$$

Experimental mass from Table 34-5: 5.28 GeV/c^2

Charges and estimated masses of the other five mesons that contain a $\bar{b}$ quark can be determined in the same way. The results, along with the experimental masses, are listed in Table 34-5.

Making sense of the result:

Note that the meson masses calculated by simply adding the rest masses of the constituent quarks are always lower than the experimentally determined values. The bound quarks have a considerable amount of energy, and a proper theoretical determination of a meson's mass requires information about the forces that bind the quarks in the meson.

Table 34-5 Mesons that include a $\bar{b}$ quark

Quark content	Given name	Charge (e)	Estimated mass (MeV/c^2)	Experimental mass (MeV/c^2)
$u\,\bar{b}$	B$^+$ (B-plus meson)	+1	4.2×10^3	5.28×10^3
$d\,\bar{b}$	B^0 (neutral B meson)	0	4.2×10^3	5.28×10^3
$s\,\bar{b}$	B^0_s (strange B meson)	0	4.3×10^3	5.36×10^3
$c\,\bar{b}$	B^+_c (charm B meson)	+1	5.5×10^3	6.27×10^3
$b\,\bar{b}$	η_b (bottom eta meson)	0	8.4×10^3	9.4×10^3
$t\,\bar{b}$	No given name	+1	175×10^3	Not yet discovered

Ordinary matter consists of electrons, protons, and neutrons. Can we construct antimatter made of positrons, antiprotons, and antineutrons? Antiparticles can be produced in high-energy accelerators. For example, antiprotons are created in proton–proton collisions in the following process:

$$p + p \rightarrow p + p + p + \bar{p}$$

In 1995, a team of scientists working at CERN created the first antihydrogen atom (symbol $\bar{H}$) by bringing positrons and antiprotons together. When an $\bar{H}$ atom interacts with ordinary matter, its positron annihilates with an electron to create two photons, and the antiquarks of the antiproton combine with quarks of a proton (or a neutron) to create pions. Therefore, the $\bar{H}$ atoms need to be isolated in a vacuum, away from matter. By trapping $\bar{H}$ atoms in a magnetic field, scientists have been able to keep them isolated for several minutes, long enough to study their properties. According to a fundamental theorem in physics, a universe made of antimatter must behave exactly the same as our universe. So the emission spectrum of $\bar{H}$ should be exactly the same as the emission spectrum of the ordinary hydrogen atom. Any difference between the two spectra would require radical changes to currently accepted theories.

MAKING CONNECTIONS

Antihydrogen and the Big Bang

There is another important reason to study antimatter. Most scientists think that particles and antiparticles should have been created in equal numbers at the start of the Big Bang. However, our universe is dominated by particles. Antiparticles are only created during high-energy collision of particles and photons. What has happened to the antiparticles that were created at the start of the Big Bang? Scientists hope that a study of antihydrogen can help answer this question.

CHECKPOINT

C-34-5 Baryon Mass and Charge

Assume that the rest mass of a baryon is approximately equal to the sum of the masses of its constituent quarks, and rank the following baryons in order of increasing rest mass. Determine the charge of each baryon.

(a) tdd
(b) tbs
(c) ttu
(d) bss
(e) bbb
(f) ccc

C-34-5 (c), +2e; (b) 0; (a) 0; (e) 0; (d), −e; (f), −e; The highest mass quark is the t quark, followed by b, c, s, d, and u quarks.

34-6 Colour Quantum Number and Quark Confinement

Just as electrons in an atom have quantized energy levels, the energies of quarks in baryons and mesons are quantized. As described in the previous section, a Δ^{++} baryon is made of three up quarks with spins pointing in the same direction. Experiments indicate that all three up quarks of a Δ^{++} are in the lowest-energy state. However, quarks are fermions, and according to the Pauli exclusion principle, two or more identical quarks cannot simultaneously occupy the same energy state. Therefore, physicists speculated that quarks have an additional internal quantum property that can take on three values. This property is called **colour** (or **colour charge**), and the three possible values are called red (R), blue (B), and green (G). Thus, there are three up quarks: up-red, up-blue, and up-green. The Δ^{++} baryon is therefore made of three up quarks of different colour quantum numbers (Figure 34-8). Since the three up quarks are not identical anymore (they have different colours), there is no conflict with the Pauli exclusion principle in them occupying the same energy state:

$$\Delta^{++} = (u_R \uparrow \ u_G \uparrow \ u_B \uparrow)$$

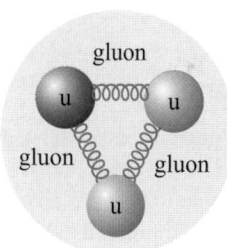

Figure 34-8 A Δ^{++} consists of three up quarks with different colour quantum numbers.

The colour property of quarks is not an actual colour. An up-red quark does not look red. The names for this quantum property were chosen because the colour quantum numbers of a baryon and a meson must add to zero (neutral colour), somewhat like the way the primary colours red, green, and blue can combine to form white light. Experiments have verified that quarks do indeed possess an internal property with three possible values. The colour quantum numbers for antiquarks are anti-red ($\bar{R}$), anti-green ($\bar{G}$), and anti-blue ($\bar{B}$). Baryons and mesons, although made of quarks, do not have the colour property because the colour numbers of the possible states of the constituent quarks add to zero, similar to the way an up-spin and a down-spin add to zero.

Colour is an additive quantum number, and a colour and its anticolour cancel just as equal and opposite electric charges cancel. Like the conservation of electric charge in a process, the total colour of the initial and final states must be the same. Consequently, gluons must have colour charge. For example, if a red quark emits a gluon and changes into a blue quark (Figure 34-9), then

Colour of the initial quark = colour of the final quark
+ colour of the gluon

R = B + colour of the gluon

Colour of the gluon = red + anti-blue = $R\bar{B}$

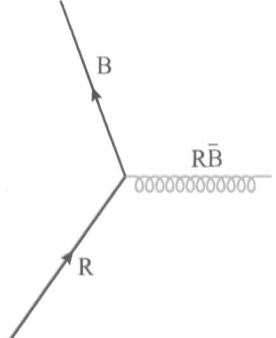

Figure 34-9 A red quark can change into a blue quark by emitting a gluon of colour red-anti-blue.

Gluons can have eight possible combinations of colours. Because gluons have colour charge, they interact with each other, unlike photons, which have no charge or colour.

Quark Confinement

We can break up a nucleus and observe its individual protons and neutrons. If a proton is constructed from quarks, then we should be able to break it apart and observe individual quarks. Despite numerous attempts using various techniques, no fractionally charged particles have ever been observed. Bombarding a hadron with very high-energy particles does not separate a quark from it; instead, the collisions create additional quark–antiquark pairs, which combine to form other hadrons. It is now believed that quarks are permanently confined within hadrons, and it is impossible to detach a single quark from a hadron. This **quark confinement** hypothesis is not yet fully understood.

Consider the strong force between a quark and an antiquark pair (q$\bar{q}$) due to the exchange of gluons between the two. Since gluons are coloured, they also interact with each other, which causes the field lines to be confined in the region directly between the quark and the antiquark, forming what is called a **flux tube** (Figure 34-10). The potential energy therefore increases as the distance r between the pair increases. At large distances the potential between the pair can be approximated as follows:

$$V_{q\bar{q}}(r) = -\lambda r \qquad (34\text{-}3)$$

each line represents a gluon

Figure 34-10 The exchange of gluons between a quark and an antiquark forms a flux tube.

Here, λ is a constant that depends on the strength of the strong force; its exact value is not needed for this discussion. The situation is similar to stretching a rubber band. The more the band is stretched, the greater the force needed to stretch it farther until the band breaks into two bands. Beyond a certain separation, enough energy has been put into the q$\bar{q}$ pair to produce an additional q$\bar{q}$ pair. The original meson then fissions into two mesons (Figure 34-11) rather than splitting into a quark–antiquark pair.

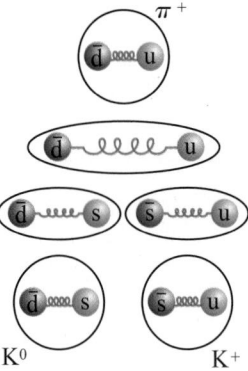

Figure 34-11 As energy is pumped into a pion to separate the quark–antiquark pair, the internal energy of the pion increases until there is sufficient energy to form two kaons.

If quarks are confined within a hadron, then how do nucleons exert a strong force on each other to form a nucleus? Gluons are coloured and a nucleon has no colour; therefore, a nucleon cannot emit a gluon and remain colourless. An exchange of a q$\bar{q}$ pair between the two nucleons is, however, allowed. This q$\bar{q}$ pair exchange represents the exchange of a virtual meson between two nucleons. So, the model of the strong force is as follows: The strong force between quarks arises due to the exchange of gluons. This force confines quarks within the hadrons. The strong force between hadrons arises due to the exchange of virtual mesons.

C-34-6

Which of the following statements are incorrect
(a) Just like quarks, leptons are also assigned a color quantum number.
(b) The green quarks emit light of green color.
(c) A meson can be made of a green quark and a blue antiquark.
(d) All three quarks of a proton can have the same color.

C-34-6 All of these statements are incorrect.

LO 7

34-7 Conservation Laws

The conservation laws of classical physics apply to interactions between particles. So, total energy, charge, and momentum are conserved when particles interact with each other. In addition, a numbers of other quantum number are also conserved. We can use conservation laws to determine whether specific reactions are possible without having to consider the details of the interactions between particles.

Conservation of Energy

The total energy of the initial state must be equal to the total energy of the final state. The total energy includes the rest mass energy of the particles. Consider the decay of a neutron into a proton, an electron, and an antineutrino:

$$n \rightarrow p + e^- + \bar{\nu}_e$$

In the rest frame of the neutron, there is no kinetic energy, so the total energy in the initial state is simply the rest mass energy of the neutron. The total energy in the final state is the sum of the rest mass energy and the kinetic energy of all three product particles and must equal $m_n c^2$.

Conservation of energy explains why a free proton cannot spontaneously decay into a neutron, as follows:

$$p \rightarrow n + e^+ + \nu_e$$

The total energy in the initial state is equal to the rest mass energy of the proton. Since the rest mass energy of the neutron alone is greater than the rest mass energy of the proton, the reaction cannot occur by itself.

Conservation of Momentum

The total momentum of the initial state must be equal to the total momentum of the final state. For example, for neutron decay in the rest frame of the neutron, conservation of momentum requires that

$$\vec{0} = \vec{p}_p + \vec{p}_e + \vec{p}_{\bar{\nu}}$$

where $\vec{p}_p$, $\vec{p}_e$, and $\vec{p}_{\bar{\nu}}$ are the momenta of the proton, electron, and antineutrino, respectively.

For relativistic kinematics, energy and momentum are combined into a single quantity called the **four-momentum**, p_μ. The laws of conservation of energy and momentum can then be combined into the law of conservation of four-momentum. For neutron decay,

$$p_{\mu,n} = p_{\mu,p} + p_{\mu,e} + p_{\mu,\nu}$$

Conservation of Angular Momentum

As described in Chapter 31, the total angular momentum $\vec{J}$ of a particle is the sum of its orbital angular momentum $\vec{L}$ and its intrinsic spin $\vec{S}$:

$$\vec{J} = \vec{L} + \vec{S}$$

According to the law of conservation of angular momentum, in the absence of net external torque, the total angular momenta of the initial and the final states are equal. So, for neutron decay,

$$\vec{J}_n = \vec{J}_p + \vec{J}_e + \vec{J}_\nu$$

Note that the *total* angular momentum is conserved, but the orbital angular momentum and spin are not conserved individually.

Conservation of Charge

The total charge in the initial and final states must be the same. The process $p + p \rightarrow p + n + \pi^+$ is allowed by charge conservation because the total charges of the initial and final states are both $+2e$. Conservation of energy requires that the protons in the initial state have at least enough kinetic energy to create a pion and make up the difference in rest mass between a proton and a neutron. However, the process $p + p \rightarrow p + n + \pi^0$ is prohibited because the total charge of the initial state is $+2e$, and the total charge of the final state is $+e$.

Conservation of Lepton Number

It is experimentally observed that the number of leptons minus the number of antileptons is conserved in all reactions. Therefore, leptons are assigned a **lepton number**, L, of $+1$, and antileptons are assigned a value of -1. The lepton number of all other particles is 0. With this assignment, the total lepton number is conserved in any interaction. For example, the lepton number in the initial and final states in the decay of a neutron is 0:

$$n \rightarrow p + e^- + \bar{\nu}_e$$
$$L: \quad 0 = 0 + 1 + (-1)$$

Similarly, the reaction $e^+ + e^- \rightarrow \pi^+ + \pi^-$ conserves a lepton number of zero. In contrast, the following two decay processes cannot happen because the lepton number is not conserved:

$$\pi^- \rightarrow e^- + \gamma \qquad\qquad \mu^- \rightarrow e^- + \bar{\nu}_e$$
$$L: \quad 0 \neq 1 + 0 \qquad\qquad L: \quad +1 \neq 1 + (-1)$$

Neither of these prohibited decays has ever been observed.

Conservation of Baryon Number

The law of conservation of **baryon number**, B, is also an experimental fact. It states that the number of baryons minus the number of antibaryons is the same in the initial and final states of a reaction. A baryon number of $+1$ is assigned to each baryon and -1 to each antibaryon. Since baryons are made up of three quarks, all quarks are assigned a baryon number of 1/3, and antiquarks are assigned a value of $-1/3$. All other particles (including mesons and leptons) have a baryon number of 0.

For example, in the β-decay of a neutron the baryon number before and after the decay is the same:

$$n \rightarrow p + e^- + \bar{\nu}_e$$
$$B: \quad +1 = +1 + 0 + 0$$

However, a proton cannot decay into a positron and a neutrino even though energy, momentum, charge, and lepton number would be conserved in this reaction:

$$p \rightarrow e^+ + \nu_e + \gamma$$
$$B: \quad +1 \neq 0 + 0 + 0$$
$$L: \quad 0 = -1 + 1 + 0$$

The process $p + p \rightarrow p + p + n$ is not possible, regardless of the total energy of the initial state, because the baryon number of the initial state is 2 and the baryon number of the final state is 3. Similarly, the decay process $p \rightarrow e^+ + \gamma$ conserves energy, momentum, angular momentum, and charge but is still prohibited because it does not conserve baryon number.

✓ CHECKPOINT

C-34-7 Conservation Laws

Indicate whether the reactions listed in Table 34-6 are possible. If not, indicate which conservation law forbids the reaction. Assume that when there are two particles in the initial state they have enough kinetic energy to conserve total energy in the process.

Table 34-6 Reactions for C-34-3

Reaction	Yes/No	Reason (if No)
(a) $p \rightarrow n + e^+ + \nu_e$		
(b) $p \rightarrow \pi^+ + \pi^0$		
(c) $n + p \rightarrow p + p + \pi^-$		
(d) $n + p \rightarrow p + p + \pi^0$		
(e) $e^- + p \rightarrow n + \gamma$		
(f) $n \rightarrow e^- + e^+$		
(g) $e^- + p \rightarrow n + \nu_e$		
(h) $e^- + e^+ \rightarrow \mu^- + \mu^+$		

34-8 The Production and Decay of Particles

Scientists create new particles by colliding beams of high-energy particles, usually electrons or protons, with other particles or nuclei. In such collisions, the kinetic energy of the colliding particles is transformed into the mass of the new particles, in accordance with Einstein's mass–energy equation, $E = mc^2$. For example, pions can be produced by colliding fast-moving protons with other protons in the following processes:

$$p + p \rightarrow p + n + \pi^+$$
$$p + p \rightarrow p + p + \pi^0$$
$$p + p \rightarrow p + p + \pi^0 + \pi^+ + \pi^-$$

One of the protons on the left side of each reaction arrow represents a fast-moving **incident proton**, and the other is the **target proton**, usually at rest. How much kinetic energy does the incident proton need to produce one charged pion? The mass of π^+ is 139.6 MeV/c^2, so one might think that the incident proton must have at least 139.6 MeV of kinetic energy. However, the kinetic energy of the incident proton must be at least twice this amount.

Consider a collision between two particles of masses m_1 and m_2. We will assume one-dimensional motion and not use vector notation. Assume that particle 1 (mass m_1) is moving along the positive x-axis with velocity $v_{lab} \ll c$, and particle 2 (mass m_2) is initially at rest. In this frame of reference, called the **laboratory frame of reference**, the total kinetic energy of the two particles is (here we have used non-relativistic kinematics to simplify the calculation)

$$K_{lab} = \frac{1}{2} m_1 v_1^2 + \frac{1}{2} m_2 v_2^2 = \frac{1}{2} m_1 v_{lab}^2 \qquad (34\text{-}4)$$

It is easier to analyze the collision in a frame of reference that is moving with the centre of mass of the colliding particles. If at some instant particle 1 is located at x_1 and particle 2 is at x_2, the centre of mass x_{cm} of the two particles is located at

$$x_{cm} = \frac{m_1 x_1 + m_2 x_2}{m_1 + m_2} \qquad (34\text{-}5)$$

Differentiating with respect to time,

$$\frac{dx_{cm}}{dt} = \frac{m_1 \dfrac{dx_1}{dt} + m_2 \dfrac{dx_2}{dt}}{m_1 + m_2}$$

In the above equation, $\dfrac{dx_{cm}}{dt} = v_{cm}$ is the velocity of the centre of mass, $\dfrac{dx_1}{dt} = v_{lab}$ is the velocity of the incident particle in the laboratory frame, and $\dfrac{dx_2}{dt} = 0$ because particle 2 is at rest. Therefore,

$$v_{cm} = \left(\dfrac{m_1}{m_1 + m_2}\right) v_{lab} \qquad (34\text{-}6)$$

Notice that the centre of mass moves in the same direction as the incident particle but with a slower speed. According to an observer moving with the centre of mass, particle 1 is incident from the left with velocity $v_{lab} - v_{cm}$, and particle 2 is approaching from the right with velocity $-v_{cm}$. Before the collision, the total momentum of the two particles is zero in the centre-of-mass frame:

$$m_1(v_{lab} - v_{cm}) + m_2(-v_{cm})$$
$$= m_1 v_{lab} + (m_1 + m_2)(-v_{cm}) = 0 \qquad (34\text{-}7)$$

By the law of conservation of momentum, the total momentum remains zero after the collision. In the centre-of-mass frame, the total kinetic energy, K_{cm}, of the two colliding particles is

$$K_{cm} = \dfrac{1}{2} m_1 (v_{lab} - v_{cm})^2 + \dfrac{1}{2} m_2 (-v_{cm})^2$$
$$= \left(\dfrac{m_2}{m_1 + m_2}\right) K_{lab} \qquad (34\text{-}8)$$

The total kinetic energy in the centre-of-mass frame is not the same as in the laboratory frame because the centre of mass of the two particles also moves and there is a kinetic energy associated with this motion. Using Equation (34-6),

$$K_{cm} + \dfrac{1}{2}(m_1 + m_2)v_{cm}^2 = K_{lab} \qquad (34\text{-}9)$$

In the centre-of-mass frame, the colliding particles can come to rest after the collision and still conserve momentum. In this case, since the total kinetic energy is zero after the collision, all the initial kinetic energy becomes either mass–energy of the new particles or excitation energy of the particles. Therefore, the *maximum* possible energy available to create a new particle is

$$K_{cm} = \left(\dfrac{m_2}{m_1 + m_2}\right) K_{lab} \qquad (34\text{-}10)$$

The greater the kinetic energy of the incident particle, the more energy is available to produce new particles. To create particles with very large masses, the incident particle must have a very large kinetic energy. This is the reason for building large particle accelerators such as the LHC.

Particle Decay

Particles can decay into particles of lower mass provided no conservation laws are violated. Particle decay follows the exponential decay law described in Chapter 33:

$$N(t) = N(0)\, e^{-\frac{t}{\tau}} \qquad (34\text{-}11)$$

EXAMPLE 34-2

Producing Pions

Bombarding a stationary hydrogen target with a beam of fast-moving protons can produce positively charged pions (π^+, rest mass 139.6 MeV/c^2) in the reaction $p + p \rightarrow p + n + \pi^+$. What is the minimum kinetic energy required for the proton beam to produce the charged pions?

SOLUTION

The minimum energy required to create a π^+ is equal to its rest mass energy, $(139.6\ \text{MeV}/c^2)c^2 = 139.6\ \text{MeV}$. This energy has to come at the expense of a decrease in the centre-of-mass kinetic energy. For proton–proton collisions,

$$\dfrac{m_2}{m_1 + m_2} = \dfrac{m_p}{m_p + m_p} = \dfrac{1}{2}$$

Therefore, from Equation (34-10),

$$K_{cm} = 139.6\ \text{MeV} = \dfrac{1}{2} K_{lab}$$

$$K_{lab} = 2 \times 139.6\ \text{MeV} = 279.2\ \text{MeV}$$

Thus, the minimum kinetic energy for the proton beam incident on a hydrogen target to create positively charged pions is about 280 MeV. The final state contains a neutron and its rest mass is 1.2 MeV/c^2 greater than that of a proton. So some additional kinetic energy is needed to create this extra mass.

Making sense of the result:

As colliding particles have the same mass, half of the incident kinetic energy goes into the centre of mass motion and the other half is converted into the rest mass of the new particle.

where $N(0)$ is the number of particles present at $t = 0$, $N(t)$ is the number remaining after time t, and τ is the mean lifetime of the particle.

Approximate values of lifetimes for decays through the strong, electromagnetic, and weak interactions are the following:

Decay through the strong interaction	10^{-23} s
Decay through the electromagnetic interaction	10^{-18} s
Decay through the weak interaction	10^{-10} s

The exact values of the lifetimes depend on detailed calculations. A particle can decay in several different ways. For example, a negatively charged pion can decay into a muon and an muon antineutrino, or into an electron and an electron antineutrino:

$$\pi^- \rightarrow \mu^- + \bar{v}_\mu \qquad (34\text{-}12)$$
$$\pi^- \rightarrow e^- + \bar{v}_e \qquad (34\text{-}13)$$

Consider the decay of particle a with a mass of m_a into particle b with a mass of m_b and particle c with a mass of m_c:

$$a \rightarrow b + c$$

Conservation of energy and momentum requires that

$$E_a = E_b + E_c \quad \text{and} \quad \vec{p_a} = \vec{p_b} + \vec{p_c}$$

If particle a is at rest, conservation of momentum gives

$$\vec{p_b} = -\vec{p_c}$$

Using the relativistic energy–momentum relation, the conservation of energy requires that

$$m_a c^2 = \sqrt{p^2 c^2 + m_b^2 c^4} + \sqrt{p^2 c^2 + m_c^2 c^4}$$

Solving for p, the magnitude of momentum of particles b and c is

$$p = \frac{m_a c}{2} \sqrt{\left(1 - \frac{m_b^2 + m_c^2}{m_a^2}\right)^2 - 4\frac{m_b^2 m_c^2}{m_a^4}} \quad (34\text{-}14)$$

EXAMPLE 34-3

The Decay of the Higgs Boson

The rest mass of a Higgs boson is approximately $125\ \text{GeV}/c^2$. It has a very short lifetime and decays into other particles. One possible decay mode is the decay into two photons. Find the momentum, energy, and wavelength of the two photons after the decay of a Higgs boson.

SOLUTION

This is the case of a single particle decaying into two particles. The momentum of the decay products in the center of mass frame of the initial particle is given by Equation (34-14). As photons are massless, in this case:

$$m_b = m_c = m_\gamma = 0 \quad \text{and} \quad m_a = m_H = 125\ \text{GeV}/c^2$$

Inserting these values in Equation (34-14), we obtain for the momenta of the emitted photons,

$$p = \frac{m_H c}{2} = \frac{1}{c}\left(\frac{m_H c^2}{2}\right) = 62.5\ \text{GeV}/c$$

The energy E of a photon is related to its momentum p by $E = pc$. Therefore the energy of each of the emitted photons is

$$E = pc = 62.5\ \text{GeV}$$

So each photon carries half the rest mass energy of the Higgs boson. The wavelength λ of a photon is related to its momentum p by, $\lambda = \dfrac{h}{p}$, where h is Planck's constant. Therefore,

$$\lambda = \frac{h}{p} = \frac{4.14 \times 10^{-21}\ \text{MeV} \cdot \text{s}}{62.5 \times 10^3\ \text{MeV}/c}$$

$$= \frac{(4.14 \times 10^{-21}\ \text{MeV} \cdot \text{s}) \times (3.00 \times 10^8\ \text{m/s})}{62.5 \times 10^3\ \text{MeV}}$$

$$= 2.0 \times 10^{-17}\ \text{m}$$

Making sense of the result:

As the total rest-mass energy of the Higgs boson is equally shared between each photon, each photon carries $62.5\ \text{GeV}$ of energy.

✓ CHECKPOINT

C-34-8 Energy of the Centre of Mass

Consider the following two situations:

1. A proton with kinetic energy of $100\ \text{GeV}$ collides with a stationary target proton.
2. Two protons, each with $50\ \text{GeV}$ of kinetic energy, are moving toward each other and collide.

The total energy available to produce new particles in these two situations is

(a) the same;
(b) greater for situation 1;
(c) greater for situation 2.

C-34-8 (a) In a proton-proton collision where one of the initial protons is at rest, only half of the kinetic energy of the incident proton is available to produce new particles.

LO 9

34-9 Feynman Diagrams

American physicist Richard Feynman introduced an ingenious method for describing interactions of particles (elementary and composite) by means of position versus time graphs, called **Feynman diagrams**. Each diagram represents a possible way that a particular process can occur, and a given process can be represented by several diagrams. There is a well-defined mathematical procedure to calculate the probability amplitude corresponding to each diagram. The probability for a certain process to occur is proportional to the square of the sum of all the probability amplitudes for that process.

Here are the rules for drawing simple Feynman diagrams for scattering or decay of particles.

■ The initial and final states of a process contain only free particles or antiparticles. Particles and antiparticles satisfy the energy–momentum equation $E^2 = p^2 c^2 + m^2 c^4$.

■ An arrow that has one free end and points in the direction of increasing time represents a free particle.

■ An arrow that has one free end and points in the direction of decreasing time represents a free antiparticle.

■ Lines with both ends connected represent virtual particles. Virtual particles do not satisfy the energy–momentum relation and do not appear in the initial or the final states.

MODERN PHYSICS

- Usually, dashed or wavy lines represent bosons, and solid lines represent fermions. Each line is assigned a four-momentum representing the four-momentum of the particle.

- The point at which two or more particles come together is called a vertex. Electric charge, colour, baryon number, lepton number, and four-momentum are conserved at each vertex.

In Section 34-2, we used a Feynman diagram (Figure 34-2) to represent a process in which two electrons scatter from each other by exchanging a single photon. The wavy photon line represents both a photon emitted by electron 1 and absorbed by electron 2 and a photon emitted by electron 2 and absorbed by electron 1. The exchanged photon is a virtual photon.

Now, consider a process in which an electron and a positron scatter from each other by exchanging a photon. This process is represented by the two Feynman diagrams in Figure 34-12. Figure 34-12(a) corresponds to the case where a virtual photon is exchanged between the electron and the positron and the two then scatter into the final states. Figure 34-12(b) represents a quite different mechanism: the electron and positron annihilate each other into a single photon, which at a later time converts back into an electron–positron pair. Both processes contribute to the probability that an electron and a positron scatter from each other. In diagram (a), the electron and the positron are present at all times, whereas in diagram (b), there is a time interval when only a photon is present.

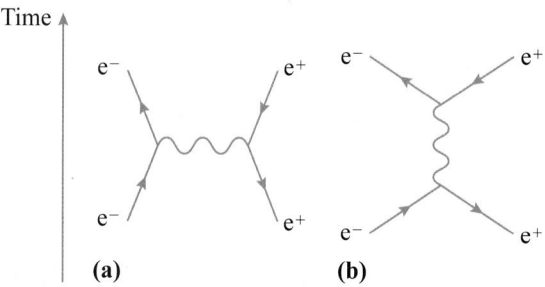

Figure 34-12 Feynman diagrams representing electron–positron scattering. Electron-positron scattering through (a) the exchange of a virtual photon, and (b) electron-positron annihilation into a photon.

EXAMPLE 34-4

Muon Pair Production

Muon (μ^-) and antimuon (μ^+) pairs can be produced by colliding a beam of positrons with electrons that are bound to atoms in the following process:

$$e^+ + e^- \rightarrow \mu^- + \mu^+$$

Draw a Feynman diagram representing this process, and calculate the minimum energy of the positron beam needed for this process to occur. The rest mass energy of a muon is 105.7 MeV.

SOLUTION

To create a μ^-, μ^+ pair in the final state, an electron and the positron pair in the initial state must annihilate into a photon, which then converts into a μ^-, μ^+ pair. The Feynman diagram for this process is shown in Figure 34-13.

The total energy required in the centre-of-mass coordinate system of the electron–positron pair must be must at least twice the rest mass energy of the muon, 2 × 105.7 MeV = 211.4 MeV. In the experimental arrangement, a beam of positrons strikes electrons that are bound to a stationary target, for example, hydrogen atoms. Since the binding energy of electrons in an atom is very small, it can be ignored. So, to a very good approximation, the beam of positrons collides with electrons that are at rest. From Equation (34-10), for $m_1 = m_2 = m_e$

$$K_{cm} = \frac{1}{2} K_{lab}$$

Therefore, in the laboratory coordinate system, the kinetic energy of the positron beam must be at least 2 × 211.4 MeV = 422.8 MeV.

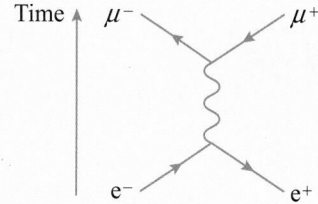

Figure 34-13 The Feynman diagram for an electron–positron annihilation that produces a muon and an antimuon

EXAMPLE 34-5

Neutron Decay

Draw a Feynman diagram for the decay of a neutron into a proton, an electron, and an electron antineutrino. Assume that the neutron and proton are elementary particles. Then modify your diagram to include the quark structure of the proton and the neutron.

(continued)

MODERN PHYSICS

SOLUTION

The neutron decay reaction can be represented as $n \to p + e^- + \bar{\nu}_e$. Figure 34-14(a) shows the Feynman diagram for this decay.

The quark structure of a neutron is (udd), and the quark structure of a proton is (uud). Therefore, in a neutron decay, a down quark must convert into an up quark. The charge of a down quark is $-\frac{1}{3}e$, and that of an up quark $+\frac{2}{3}e$. Therefore, for an up quark to convert to a down quark, it must emit a particle of charge $-e$. This particle happens to be the W^- boson. The final state also includes an electron and an electron antineutrino, which are created by the decay of the W^- boson. The Feynman diagram for this process is shown in Figure 34-14(b). Circles around the quark arrows for the neutron and proton indicate that the quarks are bound.

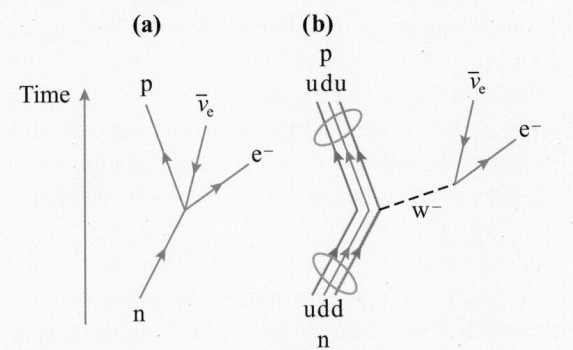

Figure 34-14 Feynman diagrams for neutron decay. (a) The decay process assuming that neutrons and protons are elementary particles. (b) The decay mechanism using the quark structure of the neutron and proton.

✓ CHECKPOINT

C-34-9 The Decay of a π^+ Meson

A π^+ meson consists of an up quark and a $\bar{d}$ quark. It has mean lifetime of 2.6×10^{-8} s and decays into an antimuon (μ^+) and a muon neutrino. The Feynman diagram for the decay is shown in Figure 34-15. Which of the following intermediate bosons is allowed in this decay?

(a) photon
(b) W^- boson
(c) W^+ boson
(d) Z^0 boson

Figure 34-15 Pion decay into a muon and a neutrino

C-34-9 (c) As charge must be conserved at each vertex the charge of the intermediate particle must be $+e$, hence a W^+ particle.

LO 10

34-10 Pions and Muons

Both the pion (or the π meson) and the muon were discovered decades before the introduction of the quark model.

The Discovery of Pions

In 1935, Japanese physicist Heidi Yukawa (1907–81) proposed the existence of a new particle to explain how nucleons exert a strong force on each other by exchanging this particle, in analogy with how charged particles exert the electromagnetic force by exchanging photons. Yukawa named the particle mesotron, which was later changed to pion, or a π meson. By that time, it was known through the work of Lord Rutherford and others that the range of the strong force between nucleons is of the order of a femtometer. Assuming that the uncertainty in the energy of the exchanged pion is of the order of its rest mass energy (see Section 34-2), Yukawa argued that such a pion could exist for a time $\Delta t \leq \dfrac{\hbar}{2^* mc^2}$ before being absorbed. Here, m is the rest-mass of the pion. The maximum distance that a pion can travel in this time is therefore of the order of

$$R = c\Delta t = \frac{\hbar c}{2mc^2} \qquad (34\text{-}15)$$

Using $R \approx 1$ fm, Yukawa estimated the rest mass of the pion to be

$$m = \left(\frac{\hbar c}{2R}\right)\frac{1}{c^2} = \frac{200 \text{ MeV} \cdot \text{fm}}{2 \text{ fm}}\frac{1}{c^2} = 100 \text{ MeV}/c^2$$

The range of the nuclear force was only approximately known, so the predicted mass of the exchanged particle was approximate, too. Since charge is conserved during the exchange of a virtual particle, there are three different pions: a positively charged pion (π^+), a negatively charged pion (π^-), and an electrically neutral pion (π^0). Two protons (or neutrons) can only exchange a π^0. A proton–neutron pair can exchange either a π^+ or a π^-.

MODERN PHYSICS

Feynman diagrams describing the exchange of pions between two nucleons are shown in Figure 34-16. Note that charge is conserved at each vertex.

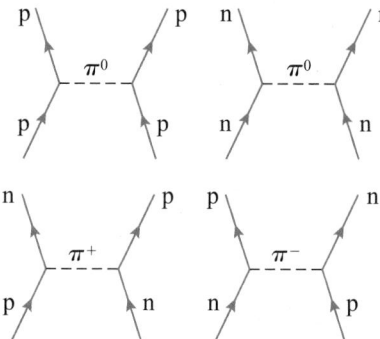

Figure 34-16 Feynman diagrams for one pion exchange between two nucleons

The exchanged pions are virtual and absorbed within a very short time:

$$\Delta t \le \frac{\hbar}{2mc^2} = \left(\frac{\hbar c}{2mc^2}\right)\frac{1}{c} \approx \left(\frac{200\ \text{MeV} \cdot \text{fm}}{200\ \text{MeV}}\right)\frac{1}{c}$$

$$= 1\frac{\text{fm}}{c} = \frac{10^{-15}\ \text{m}}{3 \times 10^8\ \text{m/s}} \approx 3 \times 10^{-24}\ \text{s}$$

Real pions can be produced in high-energy collisions of protons through the following processes, provided sufficient energy is available in the centre-of-mass coordinate system of the colliding protons:

$$p + p \rightarrow p + n + \pi^+$$
$$p + p \rightarrow p + p + \pi^0 \qquad (34\text{-}16)$$

Cosmic rays consist mainly of high-energy protons, with some electrons and helium nuclei, created in deep space. When high-energy protons collide with nuclei in the upper atmosphere, they produce both charged and neutral pions as described above. In 1947, a team of researchers led by British physicist Cecil Powell (1903–69) placed specially designed photographic plates at high-altitude locations in Europe. Just as light produces an image on light-sensitive film, charged pions interact with the nuclei of the photographic plates and produce a visible track. An analysis of these plates confirmed the existence of charged particles of mass 134.5 MeV/c^2, confirming Yukawa's prediction. With the development of high-energy proton accelerators, the production of pions has become straightforward. Neutral pions were discovered in 1950. Some basic properties of pions are listed in Table 34-7. The three pions are the lightest known mesons.

The Discovery of Muons

The muon (μ^-) is an unstable, negatively charged elementary particle with a mass of 105.6 MeV/c^2. Other than being 200 times more massive than an electron, in most respects muons and electrons are identical. A muon

Table 34-7 Some properties of pions

Property	Charged pions	Neutral pions
Mass	139.6 MeV/c^2	135.0 MeV/c^2
Mean lifetime	2.6×10^{-8} s	8.4×10^{-17} s
Spin	0	0

is a lepton and interacts with other matter only through the electromagnetic and weak forces. Its antiparticle is a positively charged muon, μ^+. It was discovered by Carl Anderson in 1936 while searching for pions. Since its rest mass is approximately the same as was predicted by Yukawa for the particle responsible for the strong nuclear force, it was thought to be a meson. However, it was soon discovered that muons do not interact through the strong force because they pass through matter easily.

Approximately 99.99% of charged pions decay into muons and neutrinos through the following processes:

$$\pi^- \rightarrow \mu^- + \bar{\nu}_\mu$$
$$\pi^+ \rightarrow \mu^+ + \nu_\mu \qquad (34\text{-}17)$$

Neutrinos emitted in pion decay are different from neutrinos emitted in neutron decay ($n \rightarrow p$ or $p \rightarrow n$) and are called muon neutrinos. Pions quickly decay into muons; the pions created by cosmic rays cause a flow of approximately 10 000 muons/m^2/min at Earth's surface.

Muons have a mean lifetime of 2.26×10^{-6} s and decay into an electron or a positron and a neutrino–antineutrino pair:

$$\mu^- \rightarrow \quad e^- \ + \ \bar{\nu}_e + \nu_\mu$$
$$L: \ +1 \ \rightarrow \ +1 + (-1) + \ 1$$

$$\mu^+ \rightarrow \quad e^+ \ + \ \nu_e \ + \ \bar{\nu}_\mu$$
$$L: \ -1 \ \rightarrow \ -1 \ + \ 1 \ + (-1) \qquad (34\text{-}18)$$

Notice that the lepton number is conserved in $\mu^-$and μ^+ decays.

LO 11

34-11 Particle Accelerators

How do we create beams of fast-moving particles? The basic technique is to accelerate charged particles by passing them through electric and magnetic fields. A particle of charge q and mass m passing through a region where the electric field is $\vec{E}$ and the magnetic field $\vec{B}$ experiences a force given by

$$\vec{F} = m\vec{a} = q\vec{E} + q\vec{v} \times \vec{B} \qquad (34\text{-}19)$$

As described in earlier chapters, an electric field accelerates a charged particle in the direction of the field, and the force exerted by a magnetic field is perpendicular to both the magnetic field and the velocity of the particle. Therefore, a constant magnetic field changes the direction of motion of a particle without doing any work on it. In this section, we discuss a few common types of accelerators.

Cyclotrons

The simplest form of a cyclotron consists of two hollow semi-circular electrodes (called **dees**) with a small gap between them (Figure 34-17). A high-frequency alternating voltage is applied across the electrodes so that the electrodes change polarity at a fixed frequency. The electrodes are placed in a vacuum, and a constant magnetic field, perpendicular to the plane of the electrodes, is applied. A source of ions is connected to the centre of the electrodes.

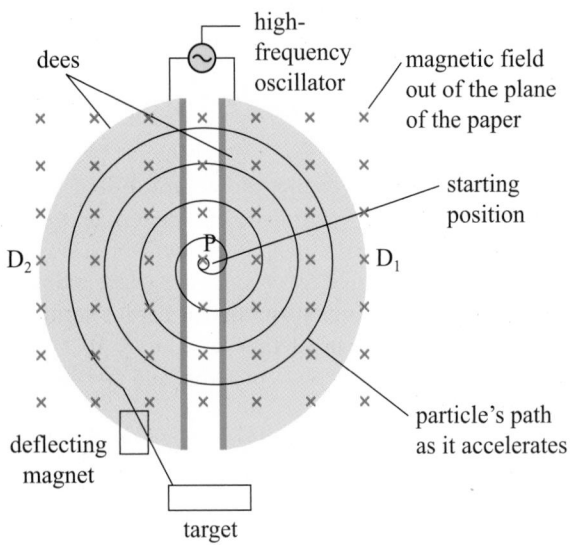

Figure 34-17 A cross-sectional view of a cyclotron consisting of two dees

Imagine that a slow-moving proton emerges from the ion source at P. The voltage between the two electrodes accelerates the proton toward the negatively charged electrode with a force $q\vec{E}$. Once through the gap and inside the hollow electrode, the proton does not experience any force because the electric field is shielded by the conductive metal. However, the proton does experience the magnetic field, which causes it to move in a circular path of radius r. The speed of the proton as it enters the electrode is v and the magnitude of the magnetic field is B, yielding

$$qvB = \frac{mv^2}{r}$$

and

$$r = \frac{mv}{qB} \tag{34-20}$$

The alternating frequency is set so that the polarity of the electrodes reverses when the proton emerges from the electrode back into the gap. Now, the electrode at the opposite end of the gap is negatively charged and the proton accelerates toward it. The proton enters the electrode with a greater speed, and once inside moves in a path of greater radius. Each time the proton passes into the gap between the two electrodes, it is accelerated by the voltage across the electrodes, and the radius of its circular path within the dees increases. The proton's speed increases only when it is passing through the gap. Once inside the electrodes, the proton moves with a constant speed. The time, t, required for the proton to travel a distance πr with a constant speed v is given by

$$t = \frac{\pi r}{v} = \frac{\pi}{v}\left(\frac{mv}{qB}\right) = \frac{\pi m}{qB}$$

Thus, the time that a proton spends inside the electrodes is independent of its speed. The polarity of the electrodes must reverse with the same timing. Therefore, the frequency of the applied voltage must be

$$f = \frac{1}{2t} = \frac{qB}{2\pi m} \tag{34-21}$$

This frequency is called the **cyclotron frequency**. The cyclotron frequency depends on the charge-to-mass ratio of the accelerated particle and the magnitude of the applied magnetic field. When the proton reaches the outer edge of the electrode, it is deflected into an evacuated pipe, called the beam line, by a small magnet. The fast-moving protons can then be made to strike a target placed at the other end of the beam line.

MAKING CONNECTIONS

The TRIUMF Cyclotron

The largest cyclotron in the world, TRIUMF (Tri-University Meson Facility), is at the University of British Columbia in Vancouver, British Columbia. The six magnets in this cyclotron have a radius of 18 m and a total mass of 4000 tonnes (Figure 34-18). The magnets can generate a magnetic field of 6000 G (0.60 T). The cyclotron frequency of TRIUMF is 23 MHz. Instead of protons, TRIUMF accelerates negatively charged hydrogen ions (hydrogen atoms with an extra electron). When high-energy hydrogen ions are taken out of the cyclotron, they are passed through a very thin metal film that strips the electrons, thus creating a high-energy proton beam. TRIUMF can accelerate protons to 0.75 times the speed of light.

MODERN PHYSICS

Figure 34-18 The six magnets of the TRIUMF cyclotron are clearly visible in this photograph, which was taken at the completion of the cyclotron in 1971.

EXAMPLE 34-6

Designing a Cyclotron

You need to design a cyclotron that can produce protons with 10 MeV of kinetic energy. The design parameters limit the radius of the electrodes to 0.50 m. Determine the magnitude of the applied magnetic field and the frequency of the power supply. The rest mass energy of a proton is approximately 938.3 MeV.

SOLUTION

First, determine the speed of protons with 10 MeV of kinetic energy. The rest mass energy, mc^2, of a proton is 938.3 MeV, so its kinetic energy is much less than its rest mass energy. Therefore, nonrelativistic kinematics can be used as a good approximation. The speed of the protons is given by

$$K = 10\ \text{MeV} = \frac{1}{2}mv^2 = \frac{1}{2}(mc^2)\left(\frac{v^2}{c^2}\right)$$

$$v = \sqrt{\frac{2 \times K}{mc^2}}\,c = \sqrt{\frac{20\ \text{MeV}}{938.3\ \text{MeV}}}\,c = 0.146\,c$$

Therefore, $v = 4.38 \times 10^7$ m/s. The magnitude of the required magnetic field can now be calculated using Equation (34-20):

$$B = \frac{mv}{rq} = \frac{(1.67 \times 10^{-27}\ \text{kg})(4.38 \times 10^7\ \text{m/s})}{(0.50\ \text{m})(1.60 \times 10^{-19}\ \text{C})}$$

$$= 0.91\ \text{T}$$

The frequency of the power supply must be the same as the cyclotron frequency. Therefore, using Equation (34-21):

$$f = \frac{qB}{2\pi m} = \frac{(1.60 \times 10^{-19}\ \text{C})(0.91\ \text{T})}{(6.28)(1.67 \times 10^{-27}\ \text{kg})}$$

$$= 1.4 \times 10^7\ \text{Hz} = 14\ \text{MHz}$$

MODERN PHYSICS

Equation (34-20) is valid only for low-energy particles. At higher energies, relativistic kinematics must be used. When the mass in Equation (34-20) is replaced by the relativistic mass, $\dfrac{m_0}{\sqrt{1 - v^2/c^2}}$ where m_0 is the rest mass of the particle, the equation for the cyclotron frequency, Equation (34-21), changes to

$$f = \frac{qB}{2\pi m_0}\sqrt{1 - \frac{v^2}{c^2}} \qquad (34\text{-}22)$$

Therefore, at energies where relativistic kinematics are important, the cyclotron frequency depends on the particle's speed, which greatly complicates the design of a cyclotron.

MAKING CONNECTIONS

Cyclotrons in Medical Physics

Radioactive isotopes are used for both the diagnosis of the condition of a person's bones and internal organs (such as thyroid, liver, and heart) and the treatment of various types of cancers. Medically important radioisotopes include technetium-99, nitrogen-13, fluorine-18, and iodine-123. For example, fluorine-18 is used in fludeoxyglucose to detect cancer and to monitor the effectiveness of cancer treatment. Many large hospitals have their own cyclotrons to produce these radioisotopes (Figure 34-19). Using a nuclear reactor at Chalk River, Ontario, Canada produces approximately two-thirds of the technetium-99 required worldwide. Research is underway to develop large-scale production of technetium-99 with cyclotrons as an alternative source. Proton beams from cyclotrons are also used to treat cancer. A beam of high-energy protons is targeted at the tumour. The protons deposit their energy in the tumour, killing the cancer cells.

© BSIP SA/Alamy

Figure 34-19 A cyclotron made by IBA (Belgium) for medical use.

Linear Accelerators

A **linear accelerator** (also called a **linac**) accelerates charged particles by passing the particles through large potential differences along a straight path. In contrast to a cyclotron, where charged particles move in circular orbits of increasing radii, particles make a single pass through a linear accelerator. Therefore, the high-power linear accelerators tend to be long and therefore costly. The largest linear accelerator in the world is SLAC, an acronym for its original name, the Stanford Linear Accelerator Center. Located at Stanford University in California, the SLAC accelerator is 3.2 km long and can accelerate electrons and positrons up to energies of 50 GeV.

The two main types of linear accelerators are distinguished by whether they use a DC or an AC voltage for acceleration. Consider two oppositely charged plates a distance d apart with a constant potential difference V_0 between them. The magnitude of the electric field between the plates is V_0/d. The magnitude of the force exerted on a particle of charge e and mass m in this electric field is

$$F = (e)(E) = \frac{eV_0}{d}$$

Therefore, the acceleration of the particle is

$$a = \frac{F}{m} = \frac{eV_0}{md} \qquad (34\text{-}23)$$

If the particle was at rest initially, its speed after travelling the distance d between the plates is given by (ignoring relativistic effects)

$$v = \sqrt{\frac{2eV_0}{m}} \qquad (34\text{-}24)$$

For electrons that accelerate through a 20000 V potential difference,

$$v = \sqrt{\frac{2 \times (1.60 \times 10^{-19}\ \text{C})(20000\ \text{V})}{(9.11 \times 10^{-31}\ \text{kg})}} = 8.38 \times 10^7\ \text{m/s}$$

This speed is approximately 28% of the speed of light. Acceleration with a DC voltage works well when required energies are of the order of tens of kilo-electron volts. For example, cathode ray tubes typically accelerate electrons with a potential difference of the order of 10 kV. To accelerate particles close to the speed of light, the potential difference needs to be increased substantially. High voltages cause electric arcs even in a vacuum chamber, so using a constant electric field (DC voltage) to accelerate charged particles to very high speeds is not feasible.

Alternating electric fields can accelerate charged particles to much higher speeds. This is done by passing the particles through an arrangement of spatially separated electrodes in such a way that the particles are accelerated during one-half of the AC cycle and are shielded from the electric field by the electrodes when the electric field reverses direction during the other half of the cycle. As in a cyclotron, the frequency of the alternating field must be matched to the motion of the particles. The shielding

MODERN PHYSICS

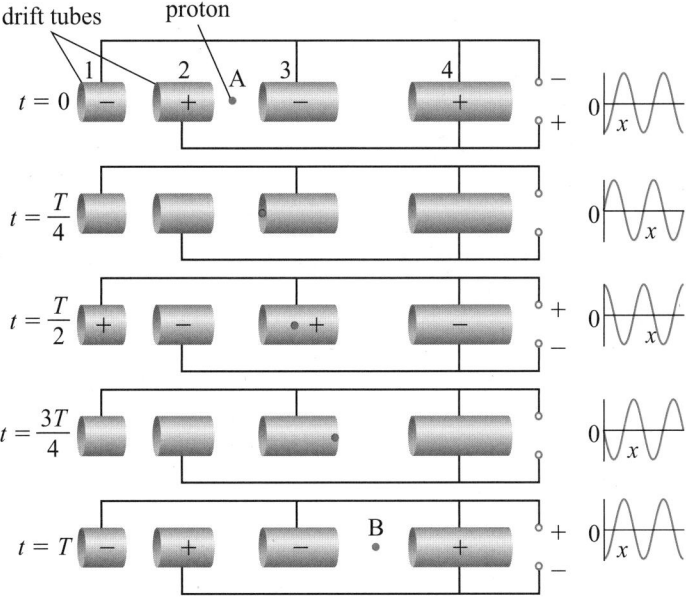

drift tubes   proton

$t = 0$

$t = \dfrac{T}{4}$

$t = \dfrac{T}{2}$

$t = \dfrac{3T}{4}$

$t = T$

Figure 34-20 The cycle of voltages applied to the drift tubes in a linear accelerator

electrodes consist of cylindrical metal tubes, called **drift tubes**. As a particle accelerates, it gains speed and therefore travels longer and longer distances during each half cycle of the electric field. Therefore, each successive drift tube must be longer than the one before it (Figure 34-20).

Consider the motion of a proton as it travels from point A to point B in Figure 34-20. Let $t = 0$ when the proton is midway between tubes 2 and 3, and the voltage (and therefore the electric field) is at its maximum. At this instant, tube 3 is negatively charged, and the proton experiences a maximum forward acceleration. Over the next quarter cycle, the voltage drops, reaching zero at $t = T/4$, where T is the period of the AC cycle. Tube 3 is placed such that at $t = T/4$, the proton enters the tube. For the next half cycle, the electric field reverses direction, and the proton would decelerate if it is not shielded by the drift tube. However, the electric field is zero inside the metal tube, so the proton moves within it with a constant velocity, the velocity with which it entered the tube. At $t = 3T/4$, the voltage is zero, and the electric field reverses again. The length of drift tube 3 is such that the proton emerges from it at this time. Between $t = 3T/4$ and $t = T$, the electric field does work on the proton, accelerating it forward. This process is repeated during the next cycle.

We can approximate the required length of a drift tube: Let f be the frequency of the AC voltage and v be the average speed of the proton during a given cycle. The proton travels a distance v/f in one cycle. Since the proton is to be shielded for half a cycle, the length L_d of the drift tube should be half this distance:

$$L_d = \frac{v}{2f} \qquad (34\text{-}25)$$

This length is approximate because the proton gains kinetic energy, and its speed increases during the cycle.

EXAMPLE 34-7

Drift Tube Lengths

The LHC at CERN is the largest proton accelerator in the world. Before entering the main circular accelerator, protons are accelerated to a kinetic energy of 50 MeV in a linear accelerator. The frequency of the AC power supply for this linear accelerator is 202 MHz. Calculate the length of the drift tubes when the kinetic energy of the protons is
(a) 10.0 MeV;
(b) 30.0 MeV;
(c) 50.0 MeV.

SOLUTION

The length of the drift tube can be calculated using Equation (34-25) once we know the speed of the protons.

We can use Einstein's relationship between mass and energy to determine the proton's speed for a given value of its kinetic energy. Let K be the kinetic energy of the proton. Then

Kinetic energy + rest mass energy = total energy

$$K + m_0 c^2 = \frac{m_0 c^2}{\sqrt{1 - \dfrac{v^2}{c^2}}}$$

Solving the above equation for v we get

$$v = c\sqrt{1 - \left(\frac{m_0 c^2}{K + m_0 c^2}\right)^2} = c\,\frac{\sqrt{K^2 + 2Km_0 c^2}}{K + m_0 c^2} \qquad (34\text{-}26)$$

(continued)

MODERN PHYSICS

By substituting the known values for the rest mass of the proton (938.3 MeV/c^2) and the speed of light, we can calculate the proton speed and hence the length of the drift tube for a given K.

(a) 10.0 MeV

$$v = c\sqrt{1 - \left(\frac{m_0 c^2}{K + m_0 c^2}\right)^2}$$

$$= (3.00 \times 10^8 \text{ m/s}) \sqrt{1 - \left(\frac{938.3 \text{ MeV}/c^2}{10.0 + 938.3 \text{ MeV}}\right)^2}$$

$$= 4.34 \times 10^7 \text{ m/s}$$

$$L_d = \frac{v}{2f} = \frac{4.34 \times 10^7 \text{ m/s}}{2 \times 202 \times 10^6 \text{ Hz}}$$

$$= 10.8 \times 10^{-2} \text{ m} = 10.8 \text{ cm}$$

(b) 30.0 MeV

$v = 7.41 \times 10^7$ m/s; $\quad L_d = 18.3 \times 10^{-2}$ m $= 18.3$ cm

(c) 50.0 MeV

$v = 9.42 \times 10^7$ m/s; $\quad L_d = 23.3 \times 10^{-2}$ m $= 23.3$ cm

Making sense of the result:

The length of the drift tube increases with increasing kinetic energy of the proton.

 CHECKPOINT

C-34-10 Drift Tubes for Electrons

When electrons with the same kinetic energy as protons are accelerated in a linear accelerator with a frequency of 200 MHz, the length of the drift tube required for the electrons should be
(a) less than the length for protons;
(b) the same as the length for protons;
(c) greater than the length for protons.
(d) The electrons are negatively charged, so they do not require drift tubes for shielding.

C-34-10 (c)

Synchrotrons

A synchrotron accelerates charged particles by continuously moving the particles in a circle of constant radius. A **synchrotron** is essentially a cyclotron in which the electric and magnetic fields are adjusted to keep the radius of the path of the accelerating particles constant as the speed of the particles increases. From Equation (34-20), the strength of the magnetic field required to rotate a particle with speed v in a circle of radius R_0 is

$$B = \frac{m_0}{\sqrt{1 - \frac{v^2}{c^2}}}\left(\frac{v}{qR_0}\right) \qquad (34\text{-}27)$$

where q is the charge of the particle, and m_0 is its rest mass.

For circular motion, $v = \omega R_0 = (2\pi f_0)R_0$. Therefore, as particle speed increases, the frequency of the synchrotron must also increase, along with the strength of the magnetic field.

Accelerating particles to very high energies requires strong, energy-efficient magnets and other relevant technologies. The LHC uses superconducting electromagnets to force the particles to move in a circular path. These electromagnets are more powerful and more compact than electromagnets using ordinary conductors.

 EXAMPLE 34-8

Magnets for the Super Collider

The LHC can accelerate protons to energies of 7.0 TeV. Given that the accelerator ring is 27 km in circumference, calculate the strength of the magnetic field needed to keep the protons in the synchrotron ring.

SOLUTION

The strength of the required magnetic field can be calculated from Equation (34-27). We are given the kinetic energy of the proton. It is convenient to write Equation (34-27) in terms of the kinetic energy. From Equation (34-26) in Example 34-7:

$$v = c\sqrt{1 - \left(\frac{m_0 c^2}{K + m_0 c^2}\right)^2} = c\frac{\sqrt{K^2 + 2Km_0 c^2}}{K + m_0 c^2}$$

Since

$$\sqrt{1 - \frac{v^2}{c^2}} = \frac{m_0 c^2}{K + m_0 c^2}$$

we can write Equation (34-27) as

$$B = \frac{m_0}{\sqrt{1 - \frac{v^2}{c^2}}}\left(\frac{v}{qR_0}\right) = \frac{\sqrt{K^2 + 2Km_0 c^2}}{qR_0 c}$$

We have the following data:

$m_0 c^2 = 938.3 \times 10^6$ eV; $\quad K = 7.0$ TeV $= 7.0 \times 10^{12}$ eV

$q = 1.60 \times 10^{-19}$ C; $\quad c = 3.00 \times 10^8$ m/s; $\quad R_0 = 27\,000$ m/2π

MODERN PHYSICS

Substituting the known values, we get

$$B = \frac{\sqrt{(7.0 \times 10^{12}\ \text{eV})^2 + 2(7.0 \times 10^{12}\ \text{eV})(938.3 \times 10^6\ \text{eV})}}{(1.60 \times 10^{-19}\ \text{C})(27\,000\ \text{m}\,/2\pi)(3.00 \times 10^8\ \text{m/s})}$$

$$\times\, (1.60 \times 10^{-19}\ \text{J/eV}) = 5.4\ \text{T}$$

Making sense of the result:

Earth's magnetic field is approximately 5×10^{-5} T, and the field strength close to a very strong niobium magnet is approximately 0.2 T. CERN lists the actual strength of their superconducting magnets as 8.36 T. Our calculated value is lower because we assumed that the magnetic field is uniform around the 27 km ring. To create room for equipment, some sections of the LHC ring have no magnetic field, so a stronger field is required for the rest of the ring.

 CHECKPOINT

C-34-11 Synchrotron Magnetic Fields

Electrons and protons with the same speeds are travelling in circular orbits of equal radii. The strength of the magnetic field required to keep the electrons in the circular orbit is

(a) less than the strength of the magnetic field required for protons;

(b) the same as the strength of the magnetic field for protons;

(c) greater than the strength of the magnetic field required for protons;

C-34-11 (a)

34-12 Beyond the Standard Model

The Standard Model has successfully explained all high-energy experimental results, and all of its predictions about the existence of new particles have proven to be correct. Until now, there has not been a single experimental result that has contradicted the Standard Model. The discovery of the Higgs boson at LHC has provided a very convincing argument that the Standard Model is indeed correct. However, there still are many questions that puzzle scientists:

■ The mass of the up quark is estimated to be 2.5 MeV/c^2, and the mass of the top quark is 171 GeV/c^2. Why is one elementary particle approximately $70\,000$ times more massive than another?

 MAKING CONNECTIONS

Synchrotron Radiation

An accelerating charged particle emits electromagnetic radiation. The radiation emitted when a charged particle moves at a speed that is close to the speed of light in a curved path is called **synchrotron radiation**. The spectrum of synchrotron radiation ranges from microwaves to hard X-rays, and its intensity can be very high. Electrons, being lighter, emit much more synchrotron radiation than protons. Industrial uses of synchrotron radiation include determining the crystalline structure of proteins, designing faster computer chips, testing properties of materials under conditions of high temperature and pressure, and identifying environmental contaminants by X-ray absorption.

The Canadian Light Source (CLS) at the University of Saskatchewan in Saskatoon is one of the most advanced synchrotron radiation facilities in the world. CLS uses a 2.9 GeV electron beam to generate extremely intense beams of synchrotron radiation. The accelerator is the size of a football field (Figure 34-21).

Figure 34-21 The Canadian Light Source at the University of Saskatchewan, Saskatoon

MODERN PHYSICS

- Quarks and leptons are fermions, and gauge particles are bosons. Are there quarks and leptons that are bosons and gauge particles that are fermions? In other words, is the universe symmetric between fermions and bosons, such that for every fermion there is a corresponding boson and vice versa? These additional particles are not part of the Standard Model. So far, all attempts to discover the boson partner of the electron have been unsuccessful.

- The graviton is the gauge boson that is assumed to be the carrier of the gravitational force. Why has the graviton not been discovered?

- Observations of very distant supernovas has shown that the universe is expanding faster now than it was in the distant pass. Since the gravitational force is pulling all the matter in the universe together, we would expect the rate of expansion to decrease with time. So, what is causing the accelerated expansion? Some scientists think that a presently unknown type of energy, called dark energy, is pushing the galaxies apart. What is the nature of dark energy? How does it exert a repulsive force on galaxies? Is there a quantum particle corresponding to dark energy, and how would addition of this particle modify the Standard Model? Some scientists prefer to think of dark energy as a feature of space time and not due to particles per se.

- The existence of the quarks was envisaged in order to find a common structure for a large number of elementary particles. Are quarks and leptons made of more elementary particles? In other words, do quarks and leptons have a substructure? So far, there is no experimental proof of such a substructure. But that may change in future as particle accelerators of higher energies are constructed.

PUTTING IT ALL TOGETHER

Classification of Particles

There are 18 elementary particles: Six quarks, six leptons, five gauge bosons, and the Higgs boson. These particles interact with each other through four forces: the strong, the weak, the electromagnetic, and the gravitational force.

Particles that exert the strong force on each other are called hadrons. Hadrons that are fermions are called baryons, and hadrons that are bosons are called mesons. Leptons interact with each other and other particles through the weak and electromagnetic forces.

Gauge Bosons as the Carriers of the Forces

Particles interact with each other by exchanging gauge bosons. The range of a force is inversely proportional to the mass of the exchanged boson.

- Gluons are the carriers of the strong force.
- Photons are the carriers of the electromagnetic force.
- W^+, W^-, and Z^0 bosons are the carriers of the weak force.
- The carrier of the gravitational force has not been observed.

Antiparticles

An antiparticle of a particle has the same mass and spin but opposite values for other properties, such as charge, lepton number, and baryon number. Each charged particle has an antiparticle. A neutral particle either has an antiparticle (such as an antineutrino) or is its own antiparticle (such as a photon). Particle–antiparticle pairs can be produced in collisions of particles.

The Standard Model

The Standard Model groups the six quarks and six leptons into three generations (Figure 34-6). Most of the matter in the known universe is made of the particles in the first generation: the up quark, the down quark, and the electron. Higher-mass particles can decay into lower-mass particles provided the conservation laws are not violated.

Composite Particles

Hadrons are bound states of quarks and antiquarks. A baryon consists of three quarks, and a meson consists of a quark and an antiquark. The charge of a hadron is equal to the sum of the charges of the constituent quarks. The mass of a hadron depends on the masses of the constituent quarks. A proton is made of two up quarks and one down quark, and a neutron is made up of one up quark and two down quarks.

Colour Quantum Number and Quark Confinement

Quarks have an additional quantum number called colour, which has three values, usually called red, blue, and green. There are eight possible colours for gluons. All composite particles are colour-neutral.

Conservation Laws

Charge, energy, momentum, angular momentum, baryon number, and lepton number are conserved in all particle reactions and decays. Conservation laws are a powerful tool to determine whether a reaction is possible.

Production of Particles

New particles can be produced by colliding fast-moving particles. Energy available in the centre-of-mass coordinate system of the colliding particles is transformed into the rest mass energy of the new particles in accordance with Einstein's mass–energy relation.

Feynman Diagrams

Feynman diagrams are graphical ways of representing a particular reaction or a decay process.

Muons and Pions

A muon is a charged lepton. Its mass is 210 times the mass of an electron; otherwise, it is identical to an electron. It has a half-life of 1.5×10^{-6} s and decays into an electron and two neutrinos.

The existence of pions was predicted by Japanese physicist Hideki Yukawa to explain the origin of the strong nuclear force.

Particle Accelerators

In a linear accelerator, charged particles are accelerated in a straight-line path by potential differences between successive drift tubes. In cyclotrons, charged particles follow a spiral path as they accelerate. In a synchrotron, the electric and magnetic fields vary to keep the accelerating particles moving along a circular path of constant radius.

APPLICATIONS

Applications: particle accelerators, cyclotrons, medical diagnoses and treatment of cancer, synchrotrons, determining the structure of proteins, designing faster computer chips, testing properties of materials, identifying environmental contaminants by X-ray absorption

Key Terms: antiparticles, baryon number, baryons, bottom quark, charm quark, colour charge, colour, composite particles, coupling strength, cyclotron frequency, dees, down quark, drift tubes, electron neutrino, elementary particles, Feynman diagrams, flavours, flux tube, four-momentum, fundamental particles, gauge bosons, hadrons, Higgs bosons, incident proton, laboratory frame of reference, lepton number, leptons, linac, linear accelerator, mesons, muon, muon neutrino, pair annihilation, pair production, particle families, particle generations, particle physics, positron, quark confinement, scalar mesons, strange quark, synchrotron, synchrotron radiation, target proton, tau neutrino, tau, top quark, up quark, vector mesons, vertex, virtual photon

QUESTIONS

1. Explain whether each of the following statements is true or false.
   (a) Quarks can only exchange gluons.
   (b) The strong interaction is mediated by gluons.
   (c) The strong interaction is mediated by photons
   (d) Two quarks can exchange a neutrino.
   (e) A neutron and a proton can exchange a gluon.
2. Explain whether each of the following statements is true or false.
   (a) A quark and an electron can annihilate into two photons.
   (b) A quark and an antiquark can annihilate into two photons.
   (c) Two photons can annihilate into a proton–antiproton pair.
   (d) Two photons can annihilate into a neutron–antineutron pair.
   (e) Any elementary particle and its antiparticle can annihilate into two photons.
3. Explain whether each of the following statements is true or false.
   (a) All baryons are hadrons.
   (b) All mesons are baryons.
   (c) The lightest hadron is a proton.
   (d) The lightest baryon is a pion.
   (e) An electron is an antiparticle of a positron.
   (f) A positron is an antiparticle of an electron.
4. Explain whether each of the following statements is true or false.
   (a) The total spin quantum number of an electron–positron pair can be either 0 or 1.
   (b) The spin quantum number of a meson can be 2.
   (c) The spin quantum number of a baryon can be 3.
   (d) Mesons can have nonzero baryon number.
5. Explain whether each of the following statements is true or false.
   (a) When a proton collides with a stationary proton, all the kinetic energy of the incident proton can be used to create new particles.
   (b) An electron moving in a straight trajectory can emit photons.
   (c) An electron moving in a circular path emits photons.
   (d) A constant magnetic field can increase the speed of a charged particle.
   (e) A constant magnetic field can increase the velocity of a charged particle.

6. Calculate the ratio of gravitational and electromagnetic forces between two protons that are 10^{-15} m apart. How large would the gravitational constant have to be for the two forces to be equal?
7. It is estimated that the radius of a carbon nucleus is between 4×10^{-15} m and 5×10^{-15} m. What is the minimum required energy of an electron beam to resolve distances of this length?

PROBLEMS BY SECTION

For problems, star ratings will be used, (✷, ✷✷, or ✷✷✷), with more stars meaning more challenging problems.

Section 34-1 Classification of Particles

8. ✷ Which of the following particles do not interact through the strong force?
   (a) electron
   (b) photon
   (c) neutrino
   (d) quark
   (e) W^+
   (f) gluon
   (g) Higgs boson
9. ✷ Which of the following particles are fermions?
   (a) electron
   (b) photon
   (c) neutrino
   (d) quark
   (e) W^+
   (f) gluon
   (g) Higgs boson
10. ✷ Which of the following particles possess a colour quantum number?
    (a) electron
    (b) photon
    (c) neutrino
    (d) quark
    (e) W^+
    (f) gluon
    (g) Higgs boson
11. ✷ Which of the following particles are gauge bosons?
    (a) photon
    (b) Z^0

(c) neutrino
(d) gluon
(e) top quark
(f) Higgs boson

Section 34-2 Gauge Bosons

12. ✱ In 1986, a group of scientists claimed the existence of a fifth force that had a range of approximately 100 m. The claim was not correct. Had it been, what would be the mass of the gauge boson mediating such a force?

13. ✱ What would be the range of the weak force if the $W^{\pm}$ and Z^0 gauge bosons were massless?

14. ✱ Estimate the range of strong force between two protons due to the exchange of a meson of mass 1.0 GeV/c^2.

Section 34-5 Composite Particles

15. ✱✱ Construct all the baryons that are made of up, down, and strange quarks and contain at least one strange quark. Determine the masses of the baryons, assuming that the mass of a baryon is equal to the sum of the masses of the constituent quarks. Assume that the masses for the three quarks are m_u = 300 MeV/c^2, m_d = 310 MeV/c^2, and m_s = 500 MeV/c^2. These are called "constituent quark masses" and are different from the masses listed in Table 34-1.

16. ✱ What are all the possible values of the intrinsic spin quantum number of a baryon that is composed of three quarks?

17. ✱ Find the charge, baryon number, and possible values of the spin quantum number for the baryons with the following quark combinations:
(a) uus;
(b) uds;
(c) dds;
(d) uss;
(e) dss;
(f) sss.

18. ✱ Assume that baryon masses are proportional to the sum of the masses of the constituent quarks, with the same proportionality constant for all baryons. Rank the baryons in problem 17 from light to heavier mass using the quark masses in Table 34-1.

19. ✱ Perform the same calculations as in problems 17 and 18 for the following quark combinations:
(a) ttb;
(b) tts;
(c) tbs;
(d) tbd;
(e) tuu.

20. ✱ Find the charge and all possible values of the spin quantum number for mesons with the following quark combinations:
(a) s ū;
(b) s d̄;
(c) s s̄;
(d) u s̄;
(e) d s̄;
(f) c̄ d.

21. ✱ Assume that meson masses are proportional to the sum of the masses of constituent quarks, with the same proportionality constant for all mesons. Rank the mesons in problem 20 from light to heavy mass using the quark masses in Table 34-1.

22. ✱ Perform the same calculations as in problem 17 for the following quark combinations:
(a) c ū;
(b) c d̄;
(c) c s̄;
(d) c c̄;
(e) c̄ u;
(f) c̄ d;
(g) c̄ s.

Section 34-7 Conservation Laws

23. ✱✱ Explain why the following reactions are not possible. Which conservation law is violated in each case?
(a) $p \rightarrow n + e^- + \nu_e$
(b) $n \rightarrow p + \gamma$
(c) $p \rightarrow p + \gamma$
(d) $\mu^- \rightarrow e^- + \nu_e$
(e) $p \rightarrow \pi^+ + \pi^-$
(f) $n + n \rightarrow p + e^-$
(g) $n + \bar{\nu}_e \rightarrow p + e^-$
(h) $p \rightarrow e^- + \nu_e$

24. ✱✱ Identify the particle denoted by the ? symbol in the reactions listed below.
(a) $\mu^- \rightarrow e^- + \nu_\mu + ?$
(b) $\pi^- + p \rightarrow \pi^0 + ?$
(c) $e^- + p \rightarrow n + ?$
(d) $\pi^- + \pi^+ \rightarrow p + ?$
(e) $\pi^+ \rightarrow \mu^+ + ?$

25. ✱✱ Use the laws of conservation of energy and momentum to show that the reaction $e^- + e^+ \rightarrow \gamma$ is not allowed.

26. ✱✱✱ Consider a particle of mass m and velocity $\vec{v}_0$ that makes an elastic collision with a particle of mass M that is at rest in the laboratory frame of reference. Show, using non-relativistic kinematics, that the kinetic energy of the target particle (in the lab frame) is given by

$$K_M = \left(\frac{4mM}{(m+M)^2} \cos^2\theta \right) K_i$$

where $K_i = \frac{1}{2} mv_0^2$ is the kinetic energy of the incident particle and $\theta (\le \pi/2)$ is the angle that the target particle makes after the collision with respect to the direction of the incident particle.

27. ✱✱ What is the minimum energy of a neutrino in the laboratory frame of reference for the reaction $\nu_e + p \rightarrow n + e^+$ to occur?

28. ✱✱ Electron–positron annihilation occurs according to the reaction

$$e^- + e^+ \rightarrow \gamma + \gamma$$

Calculate the energies, momenta, and frequencies of the two photons in the final state when
(a) the kinetic energies of the electron and the positron are ignored compared to their rest mass energies;
(b) each has a kinetic energy of 50 MeV, and they have equal and opposite momenta.

29. ✱✱ A proton and an antiproton can annihilate into a π^+ and a π^- meson according to the reaction

$$p + \bar{p} \rightarrow \pi^+ + \pi^-$$

What are the energies and momenta of the two pions in the final state when the kinetic energies of the proton and antiproton are ignored?

30. ✱ Consider the reaction in which a fast-moving negatively charged pion is absorbed by a proton to produce two neutrons

$$\pi^- + p \rightarrow n + n$$

Is the following reaction allowed? If not, which conservation law(s) are violated?

31. ✱✱ The proton is the lightest known baryon, and it does not decay. Which conservation law will be violated if we observe the decay of a proton?

32. ✱ Which conservation law will be violated if we observe the decay of an electron?

33. ✱✱ Use the laws of conservation of energy and momentum to show that a free photon cannot decay into an electron–positron pair. (This process can happen only in the presence of another charged particle.)

Section 34-8   The Production and Decay of Particles

34. ✱✱ Calculate the momentum and energy of emitted photons in the decay of a π^0 meson:

$$\pi^0 \rightarrow \gamma + \gamma$$

35. ✱✱ Calculate the momenta and energies of the final particles in the following decays:
(a) $\pi^- \rightarrow e^- + \bar{\nu}_e$;
(b) $\pi^- \rightarrow \mu^- + \bar{\nu}_\mu$.

Section 34-9   Feynman Diagrams

36. ✱✱ Draw Feynman diagrams for the following processes:
(a) $e^- + \mu^- \rightarrow e^- + \mu^-$
(b) $p \rightarrow n + e^+ + \nu_e$
(c) $e^- + e^+ \rightarrow e^- + e^+$

37. ✱✱ A positively charged pion (u$\bar{d}$) decays into a μ^+ and ν_μ. Draw a Feynman diagram to represent this decay.

38. ✱✱ Consider the process in which an electron neutrino (ν_e) is absorbed by a neutron resulting in a proton and an electron in the final state:

$$\nu_e + n \rightarrow p + e^-$$

What really happens is that a d quark converts into a u quark by emitting a W$^-$ gauge boson. Write a Feynman diagram to represent the above process.

39. ✱✱ Draw a Feynman diagram to represent the scattering of a photon from an electron through the following process:

$$e^- + \gamma \rightarrow e^- + \gamma$$

40 ✱✱ When drawing Feynman diagrams, would it make a difference if time were plotted along the horizontal direction and position along the vertical direction?

Section 34-10   Pions and Muons

41. ✱ (a) In the meson-exchange model, the strong force between two nucleons arises due to the exchange of several mesons between the nucleons. Determine the approximate range of the nuclear force between two protons due to the exchange of the following neutral mesons:
(i) pi-zero (π^0, mass 135 MeV/c^2)
(ii) omega (ω, mass 783 MeV/c^2)
(iii) phi (ϕ, mass 1.02 GeV/c^2)
(b) Can two protons exchange an electron? Explain your answer.

42. ✱✱ The mass of a lambda-zero (Λ^0) particle is 1115.7 MeV/c^2, and its half-life is 2.6×10^{-10} s. The Λ^0 particle quickly decays into a proton and a negatively charged pion:

$$\Lambda^0 \rightarrow p + \pi^-$$

Assume that Λ^0 is initially at rest. Determine the kinetic energies of the proton and the pion in this decay.

Section 34-11   Particle Accelerators

43. ✱✱ (a) A proton linear accelerator works at a frequency of 200 MHz. Calculate the length of the drift tubes at the sections where the average kinetic energy of the proton is
(i) 10 MeV;
(ii) 100 MeV;
(iii) 200 MeV.
(b) The protons need to be accelerated to 2 GeV kinetic energy. Should the frequency of the applied AC voltage be increased or decreased? Explain.
(c) Repeat for an electron linear accelerator.

44. ✱✱ A proton beam is to be accelerated to an energy of 30 GeV in a synchrotron ring with a circumference of 1000 m. Calculate the strength of the magnetic field needed to keep the beam in a circular path of this circumference.

45. ✱✱✱ You are asked to design a synchrotron that can accelerate protons to 500 MeV. The magnets available for the project can produce a maximum magnetic field of 2 T.
(a) What is the speed of the 500 MeV protons at this energy? (Use relativistic kinematics.)
(b) How long would a 500 MeV proton take to complete one revolution around the synchrotron ring?
(c) What is the circumference of the synchrotron?

COMPREHENSIVE PROBLEMS

46. ✱✱✱ Before the discovery of the neutrino, it was assumed that a neutron can decay into a free proton and an electron, $n \rightarrow p + e^-$. Consider this process in the laboratory frame of reference in which the neutron is at rest. By applying the laws of conservation of energy and momentum, show that the magnitude of an electron's momentum, p_e, is

$$p_e = \sqrt{\frac{(m_n^2 - m_p^2 - m_e^2)^2}{4m_n^2} - \frac{m_p^2 m_e^2}{m_n^2}} = 1.26 \text{ MeV}/c$$

47. ✱✱ Some scientists estimate that at 10^{16} GeV of energy, the four fundamental forces unify into a single force. To test this hypothesis, you are asked to design a synchrotron that can accelerate protons to 10^{16} GeV. The strongest available magnets can generate a magnetic field of 10 T. What is the radius of the synchrotron? Is there any hope of building a synchrotron with this energy with current technology? What is the strength of the magnetic field if the design requires the radius of the synchrotron to be 100 km?

48. ✱✱✱ Antiprotons can be produced in the following reaction:

$$p + p \rightarrow p + p + p + \bar{p}$$

The target proton is at rest. Determine the minimum energy of the incident proton for this reaction to occur.

See the text online resources at
www.physics1e.nelson.com for Open Problems
and Data-Rich Problems related to this chapter.

Chapter 1—Introduction to Physics

1. c
3. $\mu = 4$ m, $\sigma = 0.9$ m
5. a
7. c
9. The graph does not state the difference between the measurement points represented by dots and those represented by circles. Since both distance and velocity measurements have associated uncertainties, the corresponding x and y error bars should be included. There is no indication on the graph as to what the two solid and dashed lines represent.
11. The duration of 9 192 631 770 periods of the radiation corresponding to the transition between the two hyperfine levels of the ground state of a caesium 133 atom at rest at 0 K.
13. Ampere is a measure of the amount of electric charge passing a point in an electric circuit per unit time with 6.241×10^{18} electrons (one coulomb) per second constituting one ampere. References include the *Bureau international des poids et mesures* (BIPM, the standards agency for SI units), the *Guide for the Use of the International System of Units* (*SI*) published by the National Institute of Standards and Technology, and general reference books.
15. d
17. d
19. 3
21. (a) 4 (b) 2 (c) 4 (d) 2 (e) 3 (f) 4
23. (a) 2.452×10^{-3} (b) 5.95×10^{-1} (c) 1.2×10^4
    (d) 4.5×10^{-5}
25. 25.7
27. (a)

(b)

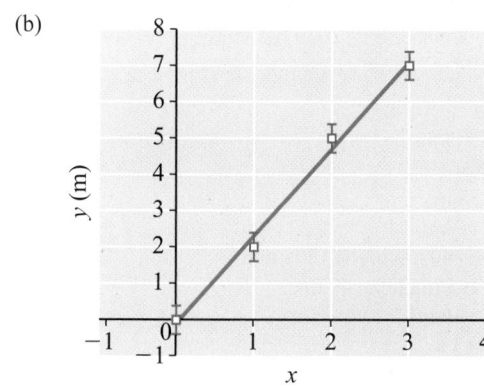

(c) 7.2 m (or ±3.6 m)

29. mgh
31. $V = IR$
33. kg/s
35. About 7.5×10^{11} assuming the space between the cells is negligible
37. Roughly 50–150 thousand (varies with assumptions made)
39. 1.15×10^6 pg
41. Years 1990–1999: mean win percentage 0.513 with SDOM of 0.020
    Years 2000–2009: mean win percentage 0.497 with SDOM of 0.012
    Although the teams in years 1990–999 had a somewhat better win-loss record, the difference is within the range of the SDOMs, so these data do not prove any significant difference between the two sets of teams.
43. (a) 3.9×10^{-8} kg m^{-2} y^{-1} (b) about 330
45. (a) $\sqrt{\dfrac{Gh}{c^5}}$ (b) 5.39×10^{-44} s

Chapter 2—Scalars and Vectors

1. (a) $r = 30$ N, $\theta = 315°$
   (b) $(21.21, -21.21)$ N
   (c) $21.21\hat{i} - 21.21\hat{j}$
3. $(2.53, -7.59)$, $(-2.53, 7.59)$ There are two possible answers due to quadratic equation obtained from the Pythagorean theorem.
5. 5.74 m
7. (a) $(4, -2, 11)$
   (b) $(-4, 2, -11)$
   (c) $(-24, 11, 22)$
   (d) $(-20, 5, 10)$
9. c. Vector magnitudes are invariant under coordinate transformation.
11. Student 2: Consider a vector in each plane directed such that they both form a 45° angle with the line of intersection of the two orthogonal planes. These two vectors are not perpendicular to each other.
13. (a) For $\vec{A}$: $(2, 0, 0)$, $(0, 2, 0)$ and $(0, 0, -2)$ For $\vec{B}$: $(-1, 0, 0)$, $(0, 3, 0)$ and $(0, 0, -2)$
    (b) 8
    (c) $2\hat{i} + 6\hat{j} + 8\hat{k}$
15. Disagree; if more than one component equals 1, the magnitude will be greater than 1.
17. $F_1 = \sqrt{11}$ $\vec{F_2} = -2\hat{i} - 4\hat{j} + 3\hat{k}$, $F_2 = \sqrt{29}$
    $\vec{F_3} = 4\hat{i} + 2\hat{j} - 6\hat{k}$, $F_3 = \sqrt{56}$
    $\vec{F_R} = 3\hat{i} - 5\hat{j} - 2\hat{k}$, $F_R = \sqrt{38}$
19. Disagree. The magnitude of the resultant is not equal to the sum of the radius vectors of individual vectors. Similarly, the resultant angle is not equal to the sum of the individual vector angles.
21. Mathematically, head-to-tail rule is equivalent to addition of vectors in Cartesian form. The associative property of vector addition, that is, $\vec{V_1} + \vec{V_2} + \vec{V_3} = (\vec{V_1} + \vec{V_2}) + \vec{V_3} = \vec{V_1} + (\vec{V_2} + \vec{V_3})$, proves the head-to-tail rule for more than two vectors.

23. A vector product represents rotation, which is not commutative.

25. $4.33\hat{i} + 3.83\hat{j} + 2.50\hat{k}$ N

27. $5.9\hat{i} + 0.4\hat{j} + 4.3\hat{k}$. It is difficult to accurately draw the vectors head-to-tail in three dimensions by hand.

29. 21.9 m

31. (a) $\vec{A}$: $19.3\hat{i} - 5.18\hat{j}$, $\vec{B}$: $12.3\hat{i} + 8.60\hat{j}$, $\vec{C}$: $-14.3\hat{i} + 20.5\hat{j}$

    (b) $17.3\hat{i} + 23.9\hat{j}$, $(29.5, 54.1°)$

    (c) $(20.1, 51.2°)$

33. (a) $(3\hat{i} - 5\hat{j} + 6\hat{k})/\sqrt{70}$

    (b) Any unit vector $\hat{u}$ that satisfies the condition $\hat{u} \cdot (3\hat{i} - 5\hat{j} + 6\hat{k}) = 0$ is perpendicular to the given vector. This question has infinite number of answers, such as $(3\hat{i} + 3\hat{j} + \hat{k})/\sqrt{19}$

    (c) There are infinite number of unit vectors parallel to the plane. Any unit vector $\hat{u} = a\hat{i} + b\hat{j} + c\hat{k}$ satisfying the relation $3a + 2b - 4c = 0$ is a vector parallel to the given plane.

    (d) $\dfrac{1}{\sqrt{29}}(3\hat{i} + 2\hat{j} - 4\hat{k})$

35. (a) $F_x = -\sqrt{2}$, $F_y = 3$, $F_z = 5$

    (b) $F_x = -3$, $F_y = -5/2$, $F_z = 3$

37. Two possible solutions since the vector in the plane is a unit vector: $\dfrac{2}{\sqrt{13}}\hat{i} + \dfrac{3}{\sqrt{13}}\hat{j}$ and $-\dfrac{2}{\sqrt{13}}\hat{i} - \dfrac{3}{\sqrt{13}}\hat{j}$.

39. 46.98

41. $-\sqrt{5}$

43. $6\hat{i} - 22\hat{j} - 14\hat{k}$

45. $|\vec{A} \times \vec{B}| = 17.1$ pointing in $+y$ or $-y$ direction

47. (a) $F_x = 2\sqrt{2}$; $F_y = 14$

    (b) $F_x = -\dfrac{20}{\sqrt{3}}$; $F_z = -\dfrac{80}{\sqrt{3}}$

49. (a) $\begin{pmatrix} \cos(60°) & -\sin(60°) & 4 \\ \sin(60°) & \cos(60°) & 5 \\ 0 & 0 & 1 \end{pmatrix}\begin{pmatrix} x' \\ y' \\ 1 \end{pmatrix} = \begin{pmatrix} x \\ y \\ 1 \end{pmatrix}$

    (b) $x = x'\cos(60°) - y'\sin(60°) + 4$, $y = x'\sin(60°) + y'\cos(60°)$

51. See online resources.

53. See online resources.

55. 0

57. See online resources.

59. See online resources.

61. (a) $s = s_0 - gt$

    (b)

(c) $t = \dfrac{s_0}{g}$

    (d) An object thrown straight up in the air.

63. (a) 178 m/s²

    (b) $\dfrac{\pi}{2}$, $51.8°$, $38.2°$

    (c) $\vec{a} = 110\hat{j} - 140\hat{k}$

    (d) $110\hat{j}$, $-140\hat{k}$, $110\hat{j} - 140\hat{k}$

65. See online resources.

67. $109.5°$

Chapter 3—Motion in One Dimension

1. (a) Both objects have same average velocity.

    (b) B at $t = 30$ s; A at $t = 90$ s

    (c) Yes, when the slopes of their position-time graphs are equal, around $t = 55$ s.

3. (a) No

    (b) Yes, at $t = 0$.

    (c) A. Maximum accelerations: A, about 100 m/s²; B, 25 m/s²

5. (a) B

    (b) B

    (c) Same magnitude for both (opposite signs)

    (d) A

    (e) Same acceleration at around $t = 130$ s

7. (a) Yes, at the maxima and minima of the curve

    (b) Highest at $t = 200$ s, lowest at $t = 30$ s and $t = 140$ s

9. b

11. a

13. b

15. a

17. d

19. In the limit of infinitesimally small time change, average acceleration becomes equal to the instantaneous acceleration.

21. At a point where tangent is parallel to displacement, the average velocity is equal to the instantaneous velocity.

23. a

25. d

27. Only momentarily. If you are moving one direction but have acceleration in the opposite direction, you will have zero speed for just the instant when you come to a stop before starting to move in the direction of the acceleration.

29. d

31. d

33. a

35. c

37. (a) 2400 m

    (b) −200 m

39. $d = 230$ cm, $\vec{d} = 80$ cm [down]

41. $v_{av} = 2.2$ m/s, $\vec{d} = 20$ m [down], $\vec{v}_{av} = 2.0$ m/s [down]

43. $v_{av} = 1.3$ m/s, $\vec{v}_{av} = 1.3$ m/s [down]

45. 6.2 h

47. 29.8 km/s

49. 1.0 km

51. (a) 5 km

    (b) 15 km

    (c) (i) 40 km/h, (ii) 120 km/h

53. (a) 105 m, 70 m, 40 m

    (b) −10 m/s²

55. (a) 8.3 s

    (b) 4890 years

    (c) about 100 000 years

57. $v = 800$ m/s, $\vec{v} = -800$ m/s

59. 12.6 m/s

61. (a) -22.3 m/s
(b) -22.3 m/s

63. (a) 10.70 m/s² (b) 1713 km

65. See online resources.

67. 9.8 m/s²

69. 1.95 s

71. $\vec{a}_{av} = 9.8$ m/s² [down], $v_{av} = 16$ m/s, $\vec{v}_{av} = 0$

73. (a) 11.29 m/s²
(b) 381.3 m
(c) 9.19 s

75. (a) 1.4 s
(b) 7.4 m/s
(c) 8.4 m/s, 14.5 m

77. 38 m/s, 84 ms

79. 1.6×10^2 m/s

81. (a) 2.818 km/s
(b) 3816 km
(c) 480 km

83. 17.7 h

85. $v_{av} = 17.2$ m/s, $\vec{a}_{av} = 9.8$ m/s² [down], $d = 120$ m, $\vec{d}$ = zero

87. 4.5 m/s

89. (a) 45 m/s (b) 45 m/s

91. 33.1 m/s

93. (a) 1.77 s
(b) 4.28 m, apple will be caught before it reaches its maximum height
(c) 1.22 s

95. 77.6 m

97. (a) $\dfrac{n\pi}{2}$; $n = 1, 3, 5, \ldots\ldots$
(b) (i)

(ii)

(iii)

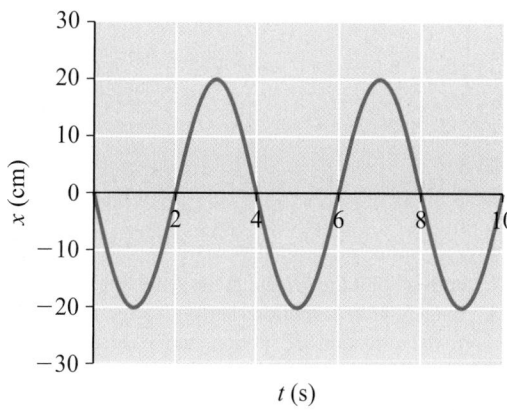

(c) $v = -x_0\omega \sin(\omega t + \phi)$
(d) $a = -x_0\omega^2 \cos(\omega t + \phi)$
(e) $a = -\omega^2 x$

99. (a) 5.1 min (b) 5.71 km toward the dock, 14.29 km

101. 51.1 m, 31.6 m/s

103. 90.5 m/s

105. 0.9 s

107. 37 m/s

109. (a) Less than 0.5 km/s per kilometre of altitude at 120 km; about 2.9 km/s per kilometre of altitude at 92 km
(b) Since the meteor fragment is slowing, the force of atmospheric friction must be somewhat greater than the force of Earth's gravity at an altitude of 120 km, and much greater at an altitude of 90 km.
(c) The density of the atmosphere increases substantially as your descend in altitude from 120 to 85 km.

111. (a) About 30 m/s², 20 m/s², and 25 m/s²
(b) About 13 m/s²
(c) About 15 m/s², 14 m/s², and 13 m/s²

113. Approximate shape shown below. Since this acceleration graph is based on values estimated from a small velocity graph, the resulting graph does not exactly match Figure 3-41 and does not show the details of the sudden changes in acceleration between the stages of the launch.

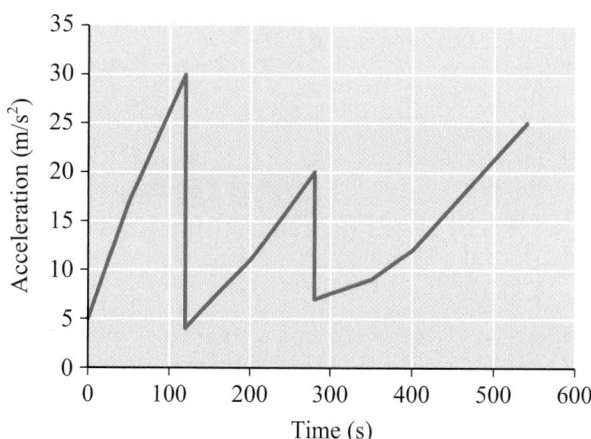

Chapter 4—Motion in Two and Three Dimensions

1. b
3. a
5. b
7. d
9. d
11. b
13. d
15. d
17. a
19. a
21. Both will hit the ground at the same time.
23. Tangential acceleration: tangent to the bowl and pointing in the direction of motion, radial acceleration: pointing towards the center of the bowl
25. (a) 1.4 m/s (b) 2.2 m/s (c) 1.7 m/s (d) 2.0 m/s (e) No, these average speeds apply for only part of the time elapsed.
27. (a) $v_x = -7.99r$, $v_y = -0.23r$
    (b) $7.99r$
    (c) circular motion, $y = \pm\sqrt{r^2 - x^2}$, $v_y = \pm\sqrt{64r^2 - v_x}$, $a_y = \pm\sqrt{4096r^2 - a_x}$
29. Straight line, 8 m/s, 0.005 m/s^2
31. (a) 248 m [37.0° up and 35.9° towards the end of the curve]
    (b) 12.2 m/s [37.0° up and 35.9° towards the end of the curve]
    (c) 0.40 m/s^2 [37.0° up and 35.9° towards the end of the curve]
33. 5.9 m/s
35. 17 m/s
37. (a) 2 h
    (b) 1285 km/h
39. 417 m
41. $a_t = 2.5$ m/s^2, $a_r = 2.7$ m/s^2, $\vec{a} = 3.7$ m/s^2 43.7° with respect to radial direction
43. About 21 krad/s
45. $\vec{a} = 16$ m/s^2 13.1° with respect to radial direction
47. 0.18 s
49. The acrobat will land in the car; 7.9 m
51. 1.8 m/s
53. 1.5 m
55. 11 m/s
57. 23 m/s
59. 29°
61. 19.6 m/s, 24.5 m/
63. (a) 796 m
    (b) 2436.7 m/s^2, 1442 m
    (c) Right before hitting the ground
65. $x = 52.0t$, $y = 43.6t$, $z = 122t - 4.9t^2$
67. (a) 26.6 m
    (b) 125.2 m
    (c) 22.4 m/s
69. (a) 954 km/h
    (b) 13 h
71. About 1400 km/h
73. (a) $2\hat{i} + 3\hat{j}$
    (b) 0.85 m/s
    (c) 3.87 m/s^2
75. $x = 3r - 0.167r$, $y = 0.986r$, $v = 6\pi r^2$, $a = 36\pi^2 r^3$
77. $h = 12.6$ km, $R = 8.13$ km, $v = 301$ m/s, $a = 20.0$ m/s^2

79. (a) 4.0 m (b) 21 m/s (c) 26°, 34°
81. 86.7°
83. (a) 8.0 m/s tangent to the circle
    (b) 5.0 m/s 96.7° with respect to x-axis, 0
85. 29.8°
87. 85.2° up from the ground
89. (a) 0.36 km/s (b) 0.42 km/s, 0.14 km/s, 2.0 km/s, 2.5 km/s
    (c) The small increase in the slope of the graph for the next 250 s indicates some radial acceleration; after that the curve is almost straight, so there is little or no radial acceleration after 1750 s.
    (d) Since the graph shows altitude versus time, the slope does not reflect any tangential component of the velocity.
91. 23.4 m/s; yes, the negative root of the quadratic is spurious since it corresponds to a negative speed.

Chapter 5—Forces and Motion

1. d
3. b
5. b
7. c
9. 16 kg mass accelerates down while 4 kg mass accelerates up
11. d
13. b
15. b
17. c
19. b
21. c
23. b
25. (a) 588 N (b) 13.6 kN (c) 13.6 kN (d) 13.6 kN (e) 13.3 kN
27. a
29. c
31. a
33. b
35. a
37. (a) 6N: $\vec{F} = 5.14\hat{i} + 3.09\hat{j}$
    8N: $\vec{F} = -4.93\hat{i} + 6.30\hat{j}$
    5N: $\vec{F} = -1.87\hat{i} - 4.64\hat{j}$
    (b) 5 N $\angle 109°$
39. (a) $(-\hat{i} + 3\hat{j} - 3.5\hat{k})$ N
    (b) $(6.2\hat{i} + 1.1\hat{j})$ N
    (c) $(-5\hat{i} + 16\hat{j} + 6\hat{k})$ N
    (d) $(22.6\hat{i} - 0.2\hat{j} + 25.9\hat{k})$ N
41. 20.1 m/s
43. 0.54 m
45. (a) 21.49 kN
    (b) 1.05 kN
    (c) 21.49 kN
47. (a) along the slope, $v_f = v_i + gt\sin\theta$
    (b) along the slope, $v_f = v_i + gt(\sin\theta - \mu\cos\theta)$
    (c) radial acceleration directed towards pivot,
    $t = 2\pi\sqrt{\dfrac{L\cos\theta}{g}}$
    (d) along the slope, $v_f = v_i + \dfrac{Ft}{m}\cos\theta - gt\sin\theta$

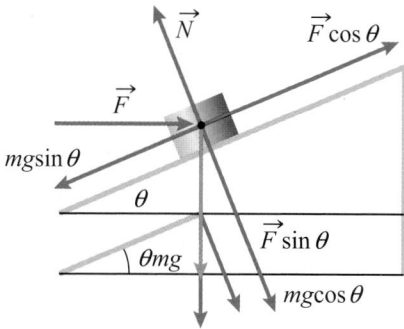

(e) parallel to the plane,

$$v_f = v_i + \frac{Ft}{m}\cos\theta - \mu t\left(g - \frac{F}{m}\sin\theta\right)$$

(f) radial acceleration directed towards the centre
  (i) $v_{top} = \sqrt{gr}$ (ii) $v_{bottom} = \sqrt{5gr}$

(g) radial acceleration directed towards the centre, $v = \sqrt{rg\tan\theta}$

(h) radial acceleration directed towards the centre,

$$v = \sqrt{\frac{rg(\sin\theta - \mu_s\cos\theta)}{\cos\theta + \mu_s\sin\theta}}$$

49. (a) 8.2×10^2 N
  (b) 24 kN
  (c) 1.6 kN
  (d) 23 kN
51. 35 N
53. 2°
55. 0.08
57. (a) $T = \dfrac{M_1 M_2 g}{M_1 + M_2}$, $M_2 > M_1$
  (b) $T = M_2 g$
  (c) 16.8 N, 19.6 N
59. (a) $\dfrac{m_2 g}{2(m_1 + m_2)}$
  (b) 0
61. 8.2 cm
63. 272 N/m
65. (a) 3.6 m/s
  (b) 0
67. (a) 62.3 km/h
  (b) 300 N, outward
69. 0.51
71. 4.5×10^2 N
73. 12.9 m/s
75. 2.7°
77. 69 km/h
79. (a) m_1: 2.45 m/s, m_2: −2.45 m/s
  (b) 127 N
81. $\dfrac{5.29}{R}$
83. $(M + m)g\tan\theta$
85. $\mu_k(M + m)g\sin\theta\cos\theta$
87. $7.35m$ N
89. (a) 68.6 N
  (b) 39.2 N, 19.6 N, 9.8 N; due to other masses
91. (a) $Mg\cos\theta$

(b) $Mg\cos\theta + \dfrac{mv^2}{R}\sin\theta$

(c) $Mg\dfrac{\cos\theta\sin\theta(1 - \mu_s) + \cos^2\theta - \sin^2\theta}{\cos\theta - \mu_s\sin\theta}$

(d) Downward parallel to the surface, towards the center, towards the center, resolved components are used to determine the net force

93. $mg\dfrac{(\sin\theta - \mu_s\cos\theta)}{(\cos\theta + \mu_s\sin\theta)}$

95. 6.74 m/s²

97. $mg\dfrac{\sqrt{R^2 - h^2}}{h}$

99. 7.26 m/s², 3.63 m/s², 16.6 m/s
101. (a) 1.5×10^2 N
  (b) 6.8×10^2 N
  (c) 1.5×10^2 N
103. 0.40

Chapter 6—Energy

1. Yes
3. T
5. Negative
7. Yes, if the displacement is in a direction opposite to the applied force
9. Can increase or decrease depending on the direction of movement and the applied force
11. zero, zero
13. zero
15. 0.98 J
17. d
19. c
21. b
23. $\theta = \tan^{-1}(\mu_k)$; equal and opposite, so total work done is zero
25. b
27. e
29. d
31. c
33. c
35. (a) −0.45 J
  (b) 0.45 J
  (c) zero
37. −18 J
39. (a) −36 J
  (b) 36 J
41. (a) 1.37 kJ
  (b) −1.37 kJ
43. (a) $\sqrt{(4.9 - 0.019k)}$
  (b) 2.2 m/s
  (c) 1.8 m/s
45. (a) $\dfrac{F}{mg}$
  (b) Fd
  (c) $F = mg\sin\theta$; more work is done since now force is acting against the gravitational pull.
  (d) $F = mg\sin\theta + ma$ (More work than in part (c))
  (e) $F = mg\sin\theta + ma + \mu mg\cos\theta$ (More work than in part (c))

47. 1.4×10^2 J, 1.8×10^2 N
49. 0.19
51. (a) 3.1×10^2 J, 24 m/s
 (b) 3.0×10^2 J
53. mgh
55. 86.5 cm
57. (a) 13 kJ
 (b) 1.9 J
59. 6.85 cm
61.

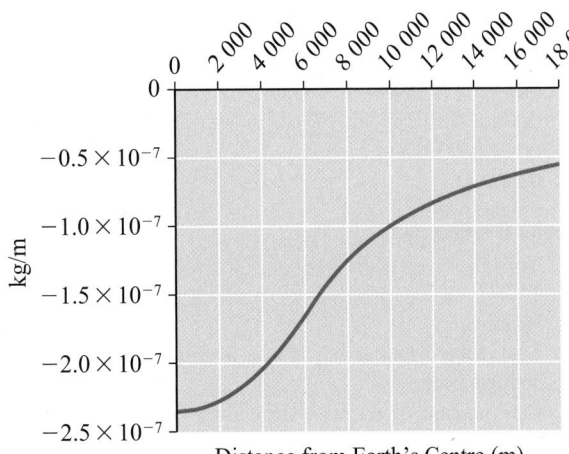

Distance from Earth's Centre (m)

63. (a) $\sqrt{2K \dfrac{Qq}{mR}\left(1 - \dfrac{\Delta r}{R}\right)}$
 (b) 1.6×10^2 m/s
65. 800 kN
67. 4.1 kN
69. (a) 89.5°
 (b) 1.17 m/s
71. $\dfrac{5}{2}r$
73. $\dfrac{2}{3}r$ from bottom of the bowl; mg, where m is the mass of the bead
75. 45 cm
77. (a) $mg\cos(23°)$
 (b) $mgL\cos(23°)$
 (c) zero
79. (a) $\dfrac{mg}{k}$
 (b) $\sqrt{2h\left(g - \dfrac{kh}{m}\right)}$
81. 54 J
83. (a) 0.94 m/s
 (b) $\sqrt{\dfrac{2gh(m - \mu_k M)}{(m + M)t^2}}$
85. (a) $\dfrac{A}{2}\left(\dfrac{1}{r_1^2} - \dfrac{1}{r_2^2}\right)$
 (b) $-\dfrac{A}{2}\left(\dfrac{1}{r_1^2} - \dfrac{1}{r_2^2}\right)$

(c) $\dfrac{A}{2}\left(\dfrac{1}{r_1^2} - \dfrac{1}{r_2^2}\right)$
 (d) attractive
 (e) $21.3\sqrt{A}$ m/s
87. $\dfrac{14r}{(r^2 + 6)^2}$, $v_i + \dfrac{42r}{m(r^2 + 6)^2}$
89. 2.3 m
91. See online resources.
93. $\theta = \cos^{-1}\left(\dfrac{2R}{3R + 0.4}\right)$
95. An increase of about 5.0 GJ assuming plane starts from sea level, no significant change in g with altitude, and mass of fuel used during the climb is negligible.

Chapter 7—Linear Momentum, Collisions, and Systems of Particles

1. a
3. d
5. d
7. a
9. b
11. The centre of mass moves down as the pieces of the firecracker fall away from the needle. If the explosion is not symmetrical, the centre of mass also shifts to one side.
13. Yes, if air resistance is negligible
15. 14.48 m/s, due to momentum lost during impulse
17. c
19. a
21. d
23. a
25. b
27. c
29. b
31. T
33. c
35. b
37. a
39. 1.2 kg m/s
41. 3.3 kN
43. (a) 1.7 N (b) 0.043 kg m/s
45. (a) 6.3 m/s
 (b)

(c) 9.18 N

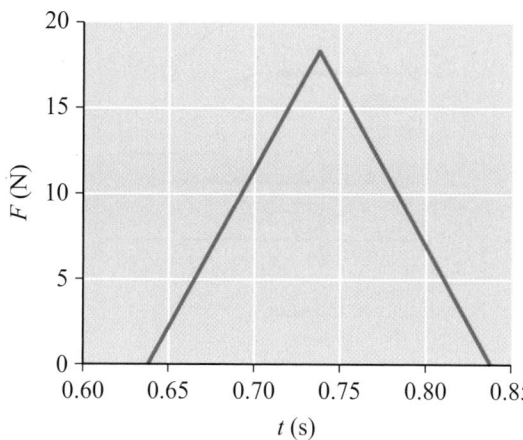

(d) 18.4 N

47. $\dfrac{14}{15}R$

49. $(0, -L/5)$ from centre

51. 3.9 m

53. 0.99 m away from the child

55. 4.2 m/s

57. 8.7 km/s

59. $1 - \dfrac{\cos^2\theta}{\cos^2\theta'}$

61. $0.42\,v$

63. 13 m/s

65. 5:1

67. 1:1.09, $20.9\,v$

69. 328.9 kg m/s², 84.1 m/s

71. (a) 7.9 cm/s² (b) 34 cm/s (c) 9.6 cm/s²

73. 7×10^7 km, 5×10^7 km

75. 39.6 cm

77. 0.016 m/s, 9.8 m/s² [down]

79. (a) 1.1 m (b) 0.97 m/s

81. 5.9 m/s

83. 29 cm

85. 141 km/s

87. 180°

89. $\dfrac{v}{3}\sqrt{1 + 9\sin^2\theta}$

91. $\dfrac{2m + M}{m}\left(\dfrac{k}{m + M}\right)^{1/2}X$

93. 4.4×10^2 J, 1.3 m/s

95. (a) 0.02 mm (b) 8×10^{-7} m/s (c) 8×10^{-7} m/s

97. $\dfrac{2L}{3}$

99. 2.7 m/s (with respect to moving frame of reference)

101. 2.0 m

103. (a) $L + v_1\sqrt{\dfrac{m_1 m_2}{k(m_1 + m_2)}}$ (b) 51.7 cm

105. (a) $(v + u)\sqrt{\dfrac{mM}{k(m + M)}}$

(b) $\dfrac{mv - Mu}{m + M}$

(c) $\dfrac{mv - Mu}{m + M}$

(d) $\dfrac{mv - Mu}{m + M}$

107. $\dfrac{2v \pm \sqrt{4v^2 + \dfrac{2}{m}(kx^2 - 2mv^2)}}{2}$

109. (a) 2.1 kg (b) zero

111. (a) 0.482 m/s (b) 55.1 cm

Chapter 8—Rotational Dynamics

1. Solid

3. All three quantities for the point on the rim are higher since the radius is larger.

5. Smaller disk has higher angular speed. Radial acceleration at the circumference of the smaller disk is smaller. The disks have the same tangential speed.

7. Increase

9. Shorter rod

11. The equation is still valid if other units (such as degrees) are used for all of the angular quantities.

13. (a) No. (b) The dot has radial acceleration but no tangential acceleration.

15. Negative, positive, equal

17. See online resources.

19. 1.8 mm/s

21. Tilting the axis of rotation changes the direction of the angular velocity vector, and hence changes the angular momentum vector.

23. Tension is the same when there is no acceleration.

25. Yes, at the moment when the angular acceleration reverses the direction of rotation of the wheel

27. Yes, as long as they are not pushing the car straight toward the tree

29. No, torque is defined about a pivot. Force is a vector quantity, so its magnitude is independent of the frame of reference.

31. 393 m/s

33. 46.4 cm

35. 1.45×10^{-4} rad/s

37. (a) 69 rad/s
(b) 0.43 km/s²
(c) 2.8 m/s²
(d) 2.8 m/s²

39. 8.4 rad/s

41. (a) 0.41 rad (b) 1.0 rad/s

43. (a) 0.97 m/s² (b) 5.2 m/s²

45. $16.78\hat{k}$ N·m

47. $(-112\hat{i} + 49\hat{j} + 67\hat{k})$ N·m

49. 806 N·m

51. $(-28\hat{i} + 13\hat{j} - 19\hat{k})$ N·m, $(36.25, 140.6°, 69.0°, 121.6°)$ N·m

53. 0.48 kg m²

55. 0.23 kg m²

57. $\dfrac{2MR^2}{5}$

59. $\dfrac{m_1 gr\sin\theta}{I + r^2(m_1 + m_2)}$

61. 5.18×10^4 kg·m²/s, 5.18×10^4 kg·m²/s

63. 19.2°

65. (a) 0.27 rad/s (b) 0.032 J

67. $-28\hat{i} + 39\hat{j} - 20\hat{k}$

69. (a) 12 m/s (b) 4.1×10^5 J, 123 km/h

71. 0.31 rad/s

73. (a) 1.7 s (b) 14 rad

75. 42 kJ

77. (a) $\dfrac{mL^2}{3}, \dfrac{mL^2}{3}, \dfrac{mL^2}{3}, mL^2, I_{\text{total}} = 2mL^2$

(b) Yes

(c) Each thin bar is treated as having its mass distributed in only one dimension.

79. $\dfrac{mL^2}{12}; \dfrac{M}{12}(L^2 + T^2)$ if T not negligible

81. (a) 1.2 m/s
(b) 1.2 m/s
(c) 1.2 m/s
(d) 2.3 rad/s
(e) 0.32

83. (a) (i) 4.3 m/s, (ii) 6.0 m/s (b) 4.5 m/s

85. 3.8 m/s

87. 13 N

89. $\left[\dfrac{2d(m_2 g - m_1 g \sin\theta - \mu m_1 g \cos\theta)}{m_1 + m_2 + I/r^2} \right]^{1/2}$

91. (a) 11 rad/s (b) 1.7×10^2 J

93. (a) 1.9 rad/s (b) 0.80 rad/s

95. 0.2 rev

97. (a) 49 rad/s (b) 6.3 m/s^2

99. (a) negative
(b) increase
(c) 1.3 rad/s
(d) decrease
(e) 22 kJ

Chapter 9—Rolling Motion

1. Solid disk

3. (a) solid cylinder (b) same angular speed

5. To prevent locking up of wheels that may cause skidding

7. Static

9. Sphere

11. $2mr^2$

13. Both disks will reach the bottom of the hill at the same time.

15. Both shells will reach the bottom of the hill at the same time.

17. Not possible.

19. 1:10

21. Friction and air resistance

23. Opposite to the direction of motion.

25. (a) 30 m/s, 1.2 km/s^2 (b) 15.0 m/s, 304.1 m/s^2

27. (a) 18 m/s (b) no

29. (a) 7.98 m/s^2 (b) 117 m/s^2, 91.7° with respect to direction of motion

31. The apparent acceleration is due to change in frame of reference and not a real effect.

33. $\cos^{-1}\left[\dfrac{I\alpha + R(F + ma)}{rF} \right]$

35. $\dfrac{g}{2}$

37. F_1 and F_3 right; F_2 and F_4 left

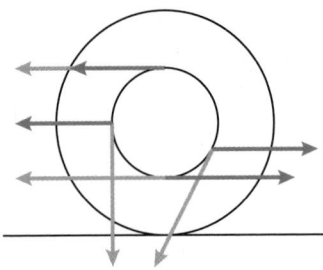

39. 2.5 N, will remain the same

41. (a) 3.8 N (b) 3.8 N

43. $\dfrac{F}{\dfrac{3}{4}m + M}$

45. (a) No, since $f > \mu_s N$ (b) 137.0 rad/s^2 (c) 4.45 J

47. 46.4°

49. 2.5 N

51. (a) $K_{\text{rot}} = 13$ J, $K_{\text{trans}} = 27$ J (b) $K_{\text{rot}} = 20$ J, $K_{\text{trans}} = 20$ J

53. (a) 5 m
(b) Yes, the ball spins (not rolls) as it moves up the incline.
(c) 1.6×10^2 rad/s

55. (a) $\dfrac{5}{7}h$, energy is conserved

(b) zero

(c) $\dfrac{2}{7}mgh$

(d) A ring of the same mass has a higher moment of inertia, so more of the total energy will go towards rotational motion as it moves down. Less translational kinetic energy means that the ring will move up to a smaller height on the right hand side as compared to the solid sphere.

57. (a) 0.69 N (b) 6.5 m/s^2 (c) 1.1×10^2 rad/s

59. $\dfrac{g(M_c \sin\theta_2 - M_s \sin\theta_1)}{m + M_c + M_s + \dfrac{I_1}{r_1^2} + \dfrac{I_2}{r_2^2}}$

61. (a) In the direction of the force

(b) $\left(\dfrac{3\mu_s g R I}{F} \right)^{1/3}$ where μ_s is the coefficient of static friction between the object and the surface

63. -6.8 m/s

65. (a) $mg\left[1 - \dfrac{10}{7(1 - r/R)} \right]$

(b) $-\dfrac{3}{7}mg$

(c) -2.9 N; the approximation differs from the actual value by about 2.3%.

67. (a) 3.4 m/s^2 opposite to the direction of the velocity
(b) 1.87 s
(c) 41.6 J

69. $\dfrac{mg}{m + \dfrac{3}{8}m_c + \dfrac{I}{r^2}}$

71. Only if the object is already rolling without slipping. Otherwise the object will slip because there is no frictional torque to change the angular speed.

73. If the marble a perfectly rigid sphere and the horizontal surface is totally frictionless, the marble will not lose any energy to the surface. However, air resistance will cause the marble to gradually come to a stop.

75. See online resources.

77. (a) $F = 6.86 \dfrac{M^2}{m} - 16.42\,M - 16.17m$ where F is in newtons and the masses are in kilograms

(b) $\dfrac{F}{\frac{3}{4}m + M}$

79. See online resources.

81. (a) $\sqrt{g\,\dfrac{(r-h)(r+h-2R)}{R-h}}$

(b) Above: marble will move down. Below: marble will move up.

83. (a) 3.1 rad/s² (b) 6.3 N

Chapter 10—Equilibrium and Elasticity

1. No, even when the object is at its highest point an unbalanced force (gravity) acts on it.

3. c

5. c

7. b

9. No, an unbalanced centripetal force is acting on the object.

11. Yes. For example, the centre of gravity of a V-shaped object lies outside the object.

13. no

15. $\dfrac{r}{5}$ away from the planet in the middle, toward the outer planet

17. No

19. Steel

21. Compression

23. d

25. c

27. b

29. a

31. 0.42

33. 2.7 kN

35. $F_1 = F_2 = 240$ kN

37. At least 0.49

39. $T_{21°} = 4.1$ kN, $T_{37°} = 2.4$ kN

41. 0.16 kN

43. $\dfrac{L}{2}\left[\dfrac{1 - e^{-L/120}}{1 + e^{-L/120}}\right]$

45. 110 m away from the mass on the surface toward the second mass

47. 1.4 kN

49. $m_2 = \frac{1}{2}m_1$

51. 8.1×10^2 N

53. $A_x = 35.4$ kN, $A_y = 30.0$ kN, $B_x = 35.4$ kN, $B_y = 2.6$ kN, $C_x = 35.4$ kN, $C_y = 23.2$ kN

55. $f_A = 3.7$ kN, $f_B = 1.2$ kN, $C_x = 2.5$ kN, $C_y = 6.8$ kN

57. 1.6 t

59. 1.8×10^2 kg

61. 2.9 MPa

63. (a) 7.2×10^4 kg (b) 1.3×10^5 kg (c) 4.0 cm

65. 0.016

67. about 31 000

69. (a) 4.7×10^5 J (b) 2.4 t

71. Surface: $F_n = 6.7$ kN, Hinge: $F_R = 4.3\hat{i} + 5.6\hat{j}$ kN

73. 9.6°

75. 0.53 kg

77. $T = 11$ kN, $R_x = 8.3$ kN, $R_y = 1.6$ kN

79. $r(1 + \mu)$

81. 0.27

83. 1.24

85. (a) 0.31 kN (b) tip (c) 112 cm

87. 28 N

89. (a) 13 GPa

(b)

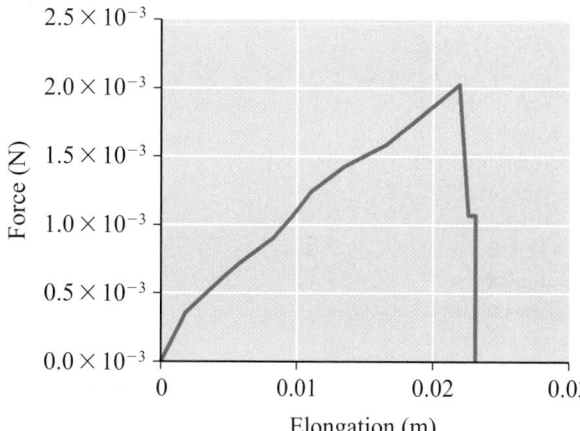

(c) 1.3×10^{-7} J

(d) around (1.5%, 200 MPa)

(e) 5 cm/s, yes

91. About 1.1 N/m

93. 0.36 MPa, 1.8×10^{-6}

95. 1.0 cm

97. (a) 11 m (b) 5.8 mm

99. Longitudinal: 0.14 MN compression, 92 kN tension; transverse: 0.28 MN compression, 1.0–1.1 MN tension

101. (a) The ultimate tensile strength increases as the steel passes through subsequent cycles. From the graph we see that the ultimate tensile strength of the steel was about 850 MPa. The UTS increased to about 860 MPa after first pass and to about 1140 MPa after four passes.

(b) The pressing process increases the ultimate tensile strength of the steel, and hence increases the energy required to break the steel.

Chapter 11—Gravitation

1. b

3. Weight is a force and should be measured in units such as newtons.

5. c

7. c

9. a

11. Atmospheric drag slows it down, but it speeds up as it loses altitude. In general, the drag slows down the satellite.

13. d

15. a

17. a

19. 1.4×10^{-20} N

21. 0.005%

23. 3.7 m/s²

25. 9.801 m/s^2
27. (a) 8.0 m/s^2
    (b) 20.5 m/s^2
    (c) $7.68 \times 10^{24} \text{ kg}$
29. $2.04 \times 10^6 \text{ J}$
31. $F_x = 0, F_y = -2ky$
33. 2.4 km/s
35. (a) 4 years
    (b) $5 \times 10^{-8} \text{ rad/s}$
37. 1680 m/s
39. (a) 488 km
    (b) 0.000645
41. (a) 23.4 s
    (b) $9.12 \times 10^{15} \text{ kg}, 6.4 \times 10^{21} \text{ kg}$
43. (a) 1.31 kg m^2
    (b) $2.94 \times 10^{-1} \text{ N m}$
    (c) $1.79 \times 10^{-7} \text{ N}$
    (d) $1.61 \times 10^{-7} \text{ N m}$
    (e) 0.0011 rad
45. (a) $5.0 \times 10^{23} \text{ kg}$
    (b) 3.9 km/s
47. 42.1 km/s
49. 1400 kg/m^3
51. See online resources.
53. (a) 73000 km/s
    (b) 4.7 ms
55. The simulation shows enhanced numbers of asteroids that have orbital radii and periods similar to Jupiter, but travel approximately 60° ahead or behind Jupiter. These locations are two of the Lagrange points (L4 and L5).
57. (a) The planet with the 10 day period is closer.
    (b) The one with the 10 day period is slightly more massive.
59. 505 m/s
61. $\dfrac{3kMm}{r^4}$
63. (a) $1.1 \times 10^{41} \text{ J}$
    (b) 10^7 years
65. (a) 22 m/s
    (b) 396.78097 to 396.78103 nm
67. (a) 72 km/s
    (b) 10^{13} kg
    (c) $2.6 \times 10^{22} \text{ J}, 6.2 \times 10^6 \text{ Mt}$

Chapter 12—Fluids

1. c
3. d
5. ABC
7. (a) When the iron cube is in the boat, the water displaced is due to the weight of the cube. However, when it is dropped in the water, it displaces water equal to its volume. Therefore the water level will fall.
9. b
11. b
13. Yes, the pressure at the bottom depends on the height of the water, not on the surface area at the bottom.
15. a
17. $3.1 \times 10^{-8} \text{ cm}$
19. 69 kN/m^2
21. (a) $4.10 \times 10^{20} \text{ kg/m}^3$
    (b) $4.1 \times 10^{15} \text{ N}$, absolutely not
23. about 1100 kg/m^3; no

25. 57 kN
27. 6.0 kPa
29. 349 N
31. a
33. 27 m
35. 880 kg/m^3
37. 0.4 N
39. 2.78 g/cm^3
41. (a) 0.69
    (b) 27
    (c) Yes, some of the weight of the survivors would be offset by the buoyant force corresponding to the volume of water displaced by their legs.
43. 6.2 m/s^2. Acceleration will decrease as drag from the water increases with speed. Although drag is the dominant factor, the density of the water increases slightly with depth due to the greater pressure. This increase in density increases the buoyance force, and decreases the acceleration correspondingly.
45. 650 m^3
47. 1.3 m
49. (a) $3.3 \times 10^{-5} \text{ m}^3\text{/s}$
    (b) 0.11 m/s
    (c) 2.0 m/s
    (d) $1.7 \times 10^{-5} \text{ m}^2$
51. (a) 45.6 m/s (b) 468 kPa
53. (a) 13 kPa (b) 1.5%
55. (a) 47 kPa (b) 5.5 m
57. Bernoulli's equation indicates that the pressure increases when the speed of water is reduced.
59. $A_2 \sqrt{\dfrac{2(P_t + \rho g y_2)}{\rho(A_2^2 - A_1^2)}}$
61. See online resources.
63. 6.3 kPa
65. See online resources.
67. 1.9 N m
69. b

Chapter 13—Oscillations

1. $4A$, zero
3. Since force is proportional to distance from equilibrium point, acceleration increases with amplitude, keeping the period constant.
5. Increased
7. The system has to be treated as a physical pendulum, and the period will be $2\pi\sqrt{\dfrac{I}{MgR}}$.
9. a
11. Decreases
13. c
15. a, b
17. d
19. (a) 0.3 m
    (b) 0.47 m/s
    (c) 0.75 m/s^2
    (d) $x(t) = 0.3 \cos(1.58t)$
21. (a) 7.7 rad/s
    (b) 15.0 cm
    (c) 0
    (d) $x(t) = (-10 \text{ cm}) \cos(0.3t)$

23. (a) $x(t) = 0.40 \cos\left(\pi t + \dfrac{\pi}{2}\right)$

(b) 1.3 m/s, 4.0 m/s²

(c) 0.079 J

25. No, since $x(0) = A\cos\phi = -x(T/2)$ we have only one independent equation, so we cannot solve for the two unknowns.

27. (a) CBDAE

(b) EADBC

(c) BDACE

(d) EADBC

29. 7.27×10^{-5} rad/s, 24 h

31. DABEFC

33. (a) 32 rad/s, 0.20 s

(b) 6.3 m/s, 0.20 km/s²

(c) 2.0 J

35. (a) 200 N/m

(b) 1.6 s⁻¹

(c) $x(t) = 0.050 \cos(10t + 1.283)$

(d) 0.49 m

37. (a) 0.31 s

(b) No change since period does not depend on spring compression.

39. (a) 4.7×10^4 N/m (b) 1.8 Hz

41. (a), (b), (c): See online resources.

(d) No. At the instant when the oscillating mass passes through its initial position, both springs are in their equilibrium positions. When the mass is at any other position, it exerts a force on spring 1, which in turn exerts a somewhat lesser force on spring 2. These forces either stretch both springs or compress both of them.

43. 6.2 cm

45. (a) 9.9 N/m

(b) 0.012 J

(c) 0.64

(d) $y(t) = 0.050 \cos(\pi t)$

47. (a) 63 J (b) 65 m/s

49. $\dfrac{1}{14.1}\sqrt{\dfrac{k}{m}}$

51. ADBCEF

53. (a) 200.6 s (b) 200.9 s, yes

55. (a) See online resources. (b) increase by $L/360$

57. See online resources.

59. (a) 0.6 m, 3 s

(b) $\dfrac{\pi}{2}$

(c) $\dfrac{7\pi}{6}, \dfrac{11\pi}{6}$

(d) $x(t) = 0.6 \cos\left(\dfrac{2\pi}{3}t + \dfrac{\pi}{2}\right)$

(e) 0.63 m/s

(f) 2.3 m/s²

61. (a) 0.20 m, 3.0 s

(b) 1.0 m

(c) $\pi/3$ rad

(d) $x(t) = (0.20 \text{ m}) \cos\left(\dfrac{2\pi}{3}t + \dfrac{\pi}{3}\right) + 1.0$ m

(e) $x(t) = (0.20 \text{ m}) \cos\left(\dfrac{2\pi}{3}t + \dfrac{5\pi}{6}\right) + 1.0$ m

$= (0.20 \text{ m}) \sin\left(\dfrac{2\pi}{3}t + \dfrac{4\pi}{3}\right) + 1.0$ m

(f) 0.88 m/s²

63. (a) 312 (b) 132 (c) 321 (d) 123

65. (a) 6.4×10^{-3} kg/s (b) 15.8 (c) No

67. ACB

69. $\phi = \tan^{-1}\left[-\dfrac{x(t = T/4)}{x(t = 0)}\right]$, $A = \dfrac{x(t = 0)}{\cos\phi}$

71. 34.2 min. The object will be oscillating between the two poles with a period given by $T = 2\pi\sqrt{\dfrac{R_{\text{earth}}}{g}}$, which is independent of the mass of the object. Therefore the time to reach the other pole does not depend on the mass of the object.

73. See online resources.

75. (a) $\dfrac{2m}{m + M}v$ (b) $\dfrac{2mv}{m + M}\sqrt{\dfrac{M}{k}}$

Chapter 14—Waves

1. (a) F. For a transverse wave, the particles of the medium vibrate perpendicular to the direction of wave. For a longitudinal wave, the particles move back and forth in the direction of the wave.

(b) F. Wave speed of a mechanical wave in a medium depends on the elastic and inertial properties of the medium.

(c) T. The speed of waves in a medium often varies somewhat with wavelength.

(d) F. Mechanical waves depend on the movement of particles of the medium through which they pass.

(e) F. Since the speed of the wave depends on the properties of the medium, the speed will be constant in a uniform medium.

(f) F. A traveling wave always carries energy.

(g) F. The maximum amplitude cannot be greater than the sum of the amplitudes of individual waves.

(h) F. To produce a standing wave, the waves must be traveling in opposite directions.

3. Light wave

5. No

7. The wave propagates outward from the point where the stone hits the water surface. As the wavefront expands the energy per unit length decreases, and therefore the amplitude decreases.

9. b

11. The two waves must have same frequency and amplitude to produce a standing wave. If the two waves have same amplitude but different frequencies, interference will occur but standing waves will not be produced. (b) If the two waves have same frequency but different amplitudes, a pattern similar to a standing wave will be produced but without any nodes.

13. No, a difference in linear mass density could be offset by a difference in tension.

15. (a) π rad/m (b) 20π rad/s (c) 20 m/s

17. (a) 143 Hz (b) 6.98 ms (c) 1.05 m

19. 17 mm to 17 m

21. 10 mm to 6 m in ocean water; 2.3 mm to 1.4 m in air

23. $D(1.5, 2.0) = 6.9$ mm, $D_{\text{min}} = 0$, $D_{\text{max}} = 25$ cm

25. (a) 3.0 m/s, traveling left

(b) −23.6 mm

(c) −0.5 mm

(d) 1.2 cm/s

(e)

27. (a)

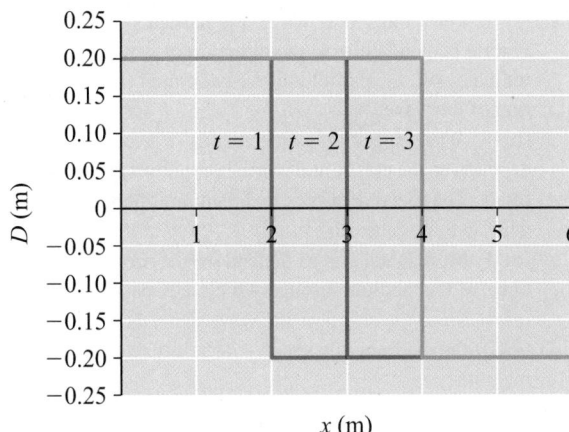

(b) 1.0 m/s, right
(c) 0.2 m
(d) $D(x, t) = \begin{cases} +0.2 \text{ m} & \text{if } |x + t| > 4 \\ -0.2 \text{ m} & \text{if } |x + t| \leq 4 \end{cases}$

29. (a)

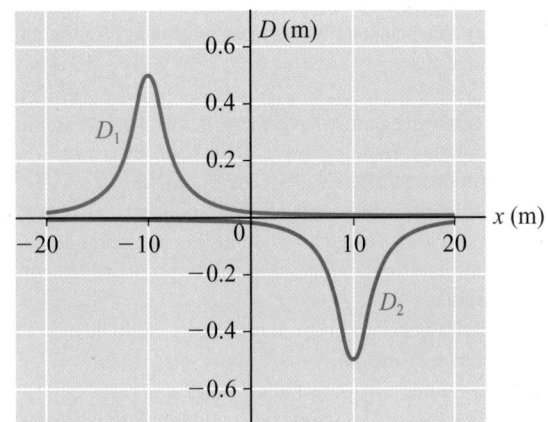

(b) D_1: −10.0 m, D_2: 10.0 m
(c) $t = 2$ s
(d) No, there is no such point.

31. 25 N
33. 0.22 km/s
35. 11.6 m from the end of string B

37. (a) See online resources.
(b) $\dfrac{2\pi v}{g}$
(c) 19 m/s

39. $\lambda_b < \lambda_a = \lambda_c < \lambda_d$, $f_a = f_b = f_d < f_c$; all waves moving toward increasing x.

41. $D_1(x, t) = \sin\left(\dfrac{4\pi}{3}x - 2\pi t - \dfrac{\pi}{4}\right),$

$D_2(x, t) = \sin\left(2\pi x - 3\pi t + \dfrac{\pi}{2}\right)$

43. (a) 0.626 m, 1.19 Hz, 0.750 m/s
(b) $D(x, t) = 0.05 \sin(10.0x - 7.50t)$

45. (a) 3.0 m, $\dfrac{1}{6}$ Hz, 0.50 m/s
(b) Toward decreasing x.
(c)

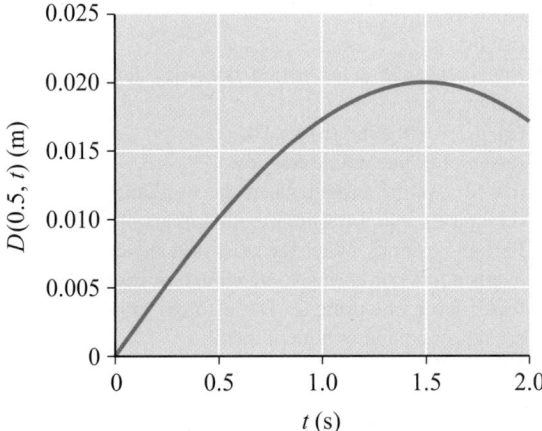

(d) 2.2 mm/s
47. (a) 8.0 m
(b) 40 m/s
(c) $\dfrac{21\pi}{2}$, which is equivalent to $\dfrac{\pi}{2}$
(d) $D(x, t) = 1.5 \sin\left(\dfrac{\pi}{4}x - 10\pi t + \dfrac{21\pi}{2}\right)$

49. (a) 0.50 Hz
(b) 8.0 m
(c) $\dfrac{3\pi}{8}$ rad
(d) $D(x, t) = 1.5 \sin\left(\dfrac{\pi}{4}x - \pi t + \dfrac{3\pi}{8}\right)$

51. (a) $D(x, t) = 1.5 \sin\left(\dfrac{\pi}{4}x - 2\pi t + \dfrac{\pi}{2}\right)$
(b) $D(x, t) = 1.5 \sin\left(\dfrac{\pi}{4}x + 2\pi t - \dfrac{\pi}{2}\right)$

53. (a) $D(x, t) = 0.10 \sin\left(\dfrac{2\pi}{3}x - \dfrac{\pi}{3}t - 5.5\right)$
(b) 0.50 m/s, toward decreasing x

55. (a) 0.2 rad
(b) 1.2 rad
(c) Yes, the phase changes linearly with time.
(d) 0.4 rad
(e) Yes
(f) 2π rad, which is equivalent to 0 rad

57. 0.88 J

59. (a)

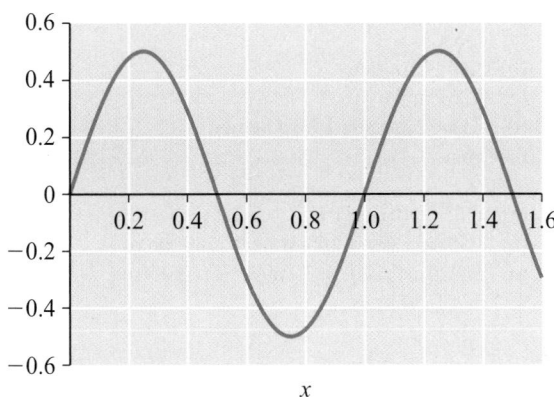

(b) $A = 0.50$, $\lambda = 1.0$

(c) The two component waves have the same frequency and wavelength, so these waves have the same speed. Consequently, the resultant wave will also have the same speed.

(d) Travelling

61. -0.42 rad

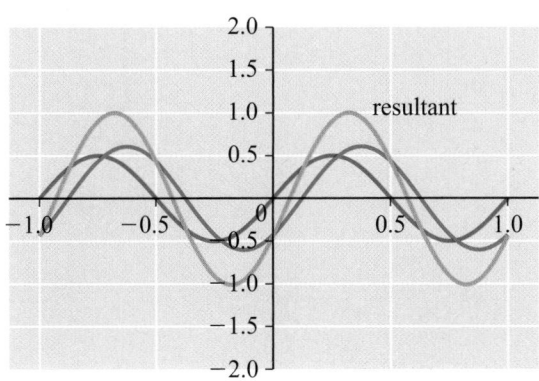

63. 2.636 rad, 1.318 rad

65. (a) $|A_1 - A_2|$ (b) $A_1 + A_2$

67. See online resources

69. (a) $D(x, t) = 0.2 \sin(3x - 4t + \pi)$

(b) $D(x, t) = 0.2 \sin(3x - 4t)$

71. (a) $\dfrac{10\pi}{3}$ m, 10 Hz

(b) $\dfrac{5\pi}{3}$, $\dfrac{5\pi}{6}$

(c) $D_1(x, t) = 0.0075 \sin(0.6x - 20\pi t)$,

$D_2(x, t) = 0.0075 \sin(0.6x + 20\pi t)$

(d) 1.8 mm

73. (a) $D(x, t) = (2.0 \text{ mm}) \sin(\pi x) \cos(0.5\pi t)$

(b) Nodes: ± 1.0 m, ± 2.0 m, ± 3.0 m,

Antinodes: ± 0.5 m, ± 1.5 m, ± 2.5 m

(c) 1.0 m

75. See online resources.

77. 265 Hz

79. (a) 4.0 m

(b) 50.0 Hz

(c) No, all frequencies of standing waves are integer multiples of the fundamental frequency.

(d) 100.0 Hz, 200.0 Hz

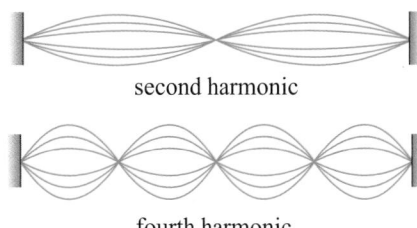

81. (a) 98.0 Hz

(b) 196 Hz, 294 Hz

(c) 196 m/s

(d) 116 Hz

Chapter 15—Interference and Sound

1. T

3. 0.5

5. T

7. T

9. F

11. 3.01 dB

13. c

15. 34.32 cm

17. 0.040 s

19. $s(x, t) = s_m \cos(6.9x - 377t)$

21. 1.08×10^{-5} m

23. 2

25.

27. 1.7 kHz, 1.1 kHz, 0.86 kHz

29. 2 Hz

31. 90 Hz

33. 90 dB

35. 749 Hz

37. 9.8 cm/s

39. 1.4 km

41. 28 nm

43. 3.2×10^{-6} W/m², 21 mW

45. 2.4 kHz

47. 1.3 m

49. 24 m/s

51. Intensity reduces by a factor of 10^6 to 0.2 μW/m².

53. 35.6 m

55. 61 Hz

57. (a) 2.4 m (b) 2.68 kHz

59. See online resources.

61. 252 m

Chapter 16—Temperature and the Zeroth Law of Thermodynamics

1. Only if both thermometers are linear between the calibration temperatures

3. The most probable and mean values are the same only for symmetric distributions with a central peak. If a distribution is skewed, the mean differs from the most probable value.

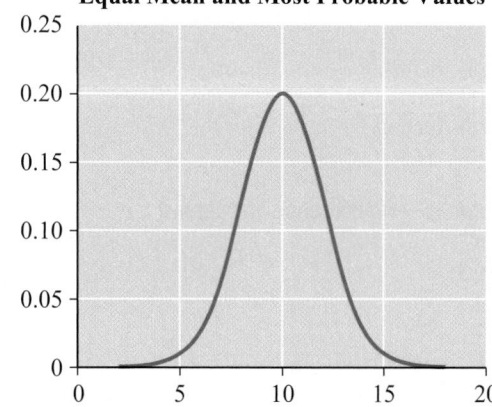

(a) **Symmetric Probability Distribution with Equal Mean and Most Probable Values**

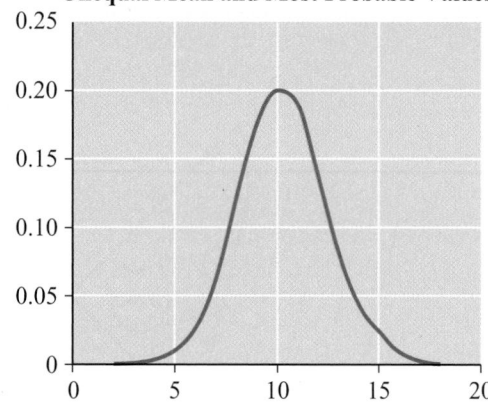

(b) **Skewed Probability Distribution with Unequal Mean and Most Probable Values**

5. The triple point of water exists at exactly 0.01°C or 273.16 K. If liquid water did not expand when cooled, the melting point of ordinary ice would not decrease as a function of pressure and the triple point would be the limit below which water would not exist in liquid state.

7. No, the gases would have to have equal low-density concentrations (N/V).

9. No

11. As the Sun rises, the temperature of the land increases more rapidly than that of the water, and the air just above the land becomes warmer than the air above the water. As a result the air pressure on land is lower than the pressure above water. This pressure difference forces the cooler air above the water to move toward the land.

13. For an ideal gas the relationship between pressure, volume and temperature is given by $PV = NkT$. Therefore, if the temperature increases while the volume is reduced by half, the pressure will more than double.

15. 3.71×10^{-26} m³, 3.34 nm

17. No, when a system is in thermodynamic equilibrium, all macroscopic properties, including temperature, remain constant.

19. (a) 758 kPa
    (b) 8.80×10^{-4} cm³
    (c) 0.267 mol
    (d) 7.51 g nitrogen, 1.07 g helium

21. 0.68 mm

23. $-40°$C

25. (a) 136°F   (b) 331 K

27. (a) 8.65 kg   (b) 52.2 m³   (c) 530 moles

29. (a) 0.0224 m³   (b) 493.1 m/s   (c) 461.4 m/s   (d) 1:1.069

31. 433 m/s, 0.54 km/s

33. $\sqrt{\dfrac{8kT}{\pi m}}$

35. Water, oxygen and carbon dioxide in solid phase; nitrogen in liquid phase

37. 0.60 m, for example, 48 kg of water for a body mass of 80 kg

39. 4×10^{-4} m/s

41. 180 K

43. 3.3×10^{27} m⁻² s⁻¹

45. (a) 8.5×10^{15} Pa   (b) 4.1×10^{5} m/s

47.

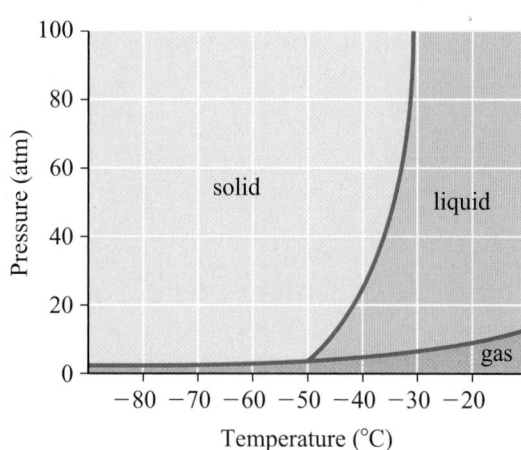

49. $R = 193\ \Omega$ at $T = 400$ K

$y = 0.19x + 117.24$

Chapter 17—Heat, Work, and the First Law of Thermodynamics

1. Constant volume
3. A solid; B solid and liquid; C liquid; D liquid and gas; E gas
5. All surfaces, including our skin, lose heat through radiant heat transfer with rates that differ in winter and summer months. The reason is the unequal temperature differentials that exist between objects and their surroundings in winter and summer months. Hence in winter months we lose heat at a faster rate than in summer months.
7. Positive
9. 2
11. The food has reached an equilibrium state.
13. The heat capacities of the objects are equal (i.e., the product cm is the same).
15. 8.0×10^2 s
17. 13.8°C
19. 130 J kg^{-1} K^{-1}
21. 11°C
23. 2.4 kJ
25. 100 J
27. 21 kW/m^2
29. 9.6 kW
31. 8.0 kJ/K, 80.0 kJ kg^{-1} K^{-1}
33. 1.3 kJ/s assuming that all of the perspiration was evaporated by heat from the athlete's body
35. −3.74 kJ
37. About 0.15 W (assuming 0.5 emissivity and cylindrical shape with 1 cm radius)
39. 505 kJ, 495 kJ
41. −7.6 kJ, 1.5 J, 7.6 kJ (out of the gas)
43. 61 kJ
45. 696°C, 13.4 kJ, 5.37 kJ, 8.06 kJ

47. 11.5 kg

Chapter 18—Heat Engines and the Second Law of Thermodynamics

1. No, more heat will be added to the room than extracted as refrigeration is not a perfectly efficient process.
3. The efficiency would still be governed by the Carnot cycle efficiency, which depends only on the temperature difference between the hot and cold reservoirs.
5. No. If the temperature difference between the waste energy source and the heat reservoir of the heat engine is too great, the heat pump will use more energy than it transfers to the heat reservoir.

7. The reasoning is not entirely correct since moving around to clean up the room increases the entropy of the Universe even though the things in the room become more ordered.
9. An object falling to the ground.
11. (a) about 1.6 mL/s (b) about 53 kJ/s
13. (a) about 12.5 W (b) about 12.5 W (c) 0.29 kJ/s (d) 0.30 kJ/s
15. (a) 39.3% (b) 39.3 kJ
17. 3.9 MJ
19. (a) 10 $m_{mol} c$, where m_{mol} is the molar mass of the gas, and c is the specific heat
    (b) 0.05 mc, where m is the mass of the gas
    (c) −0.0488 mc
    (d) 0.0012 mc
    (e) no
21. (a) 63 kJ/s
    (b) −82 J K^{-1} s^{-1}
    (c) 0.22 kJ K^{-1} s^{-1}
    (d) 0.14 kJ K^{-1} s^{-1}
23. 85
25. (a) −0.77 J/K
    (b) −8.1 J/K
    (c) −53 J/K
    (d) −53 J/K
27. No. Heat engines and heat pumps are never 100% efficient. The amount of work that the heat pump could do would always be less than the energy input into the engine.
29. The data supplied indicate that the engine would have an efficiency of 55.5%, which is greater than the maximum possible efficiency of 50% for the given temperatures. Since the data are erroneous, the application should be rejected.
31. See online resources.
33. 7.0 kJ/K
35. about −2.6 × 10^7 J/K, 6.8 × 10^7 J/K, 4.2 × 10^7 J/K
37. 35.9 J/K
39. about 1.2
41. (a) $\dfrac{23}{d}$ J m^{-2} s^{-1} where d is the thickness of the ice in metres
    (b) $\dfrac{0.0812}{d}$ J m^{-2} s^{-1} K^{-1}
    (c) $-\dfrac{0.0842}{d}$ J m^{-2} s^{-1} K^{-1}
    (d) 5.9 mm/day
    (e) See online resources.
43. $\dfrac{mc\Delta T}{T} - \dfrac{mgh}{T}$
45. No, at the smaller scale, the entropy of the Universe is constantly increasing.
47. See online resources.
49. Answers will vary. For a sea with a temperature around 22°C at the surface and 4°C at some depth below the surface, the maximum theoretical efficiency would be about 6%. Extracting energy would lower the water temperature, and could affect local sea life. Large-scale operations could alter ocean currents, and affect coastal erosion, silt deposits, and possibly even Earth's climate. The core of Earth has a temperature of about 5430°C, so the theoretical efficiency with a surface temperature of about 22°C is around 95%. Large-scale operations could affect aquifers and local wildlife habitat.

Chapter 19—Electric Charges and Forces

1. c
3. b
5. c
7. b
9. a
11. c
13. c
15. b
17. 40 nA
19. (a) 8.22×10^{-8} N
    (b) 2.19×10^6 m/s
21. 4.24 cm to the left of $+1.00\ \mu C$ charge
23. 1.60×10^{-12} N
25. 0.0365 nm
27. 8.99×10^{-2} N directed away from the rod
29. 2.9 ms
31. 5.71×10^{13} C
33. Each charge experiences a force of 0.922 N directed away from the centre of the triangle.
35. zero
37. 4.2 N/C in $+x$ direction
39. 0.27 μm to the right of the negative charge
41. 3.79×10^{-14} C
43. 2.4×10^{-22} N m pointing in $-z$ direction
45. 1.80×10^7 N/C directed upward
47. -2.0 mC located at $x = -6.0$ m.

Chapter 20—Electric Potential and Gauss's Law

1. a
3. c
5. a
7. b
9. d
11. d
13. a
15. a
17. -27.2 eV
19. 509 V
21.

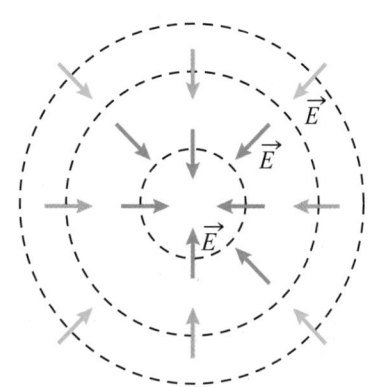

23. (a) $E_x = 0$, $E_y = -25.0$ N/C, $E_z = 0$
    (b) -25.0 V/m
25. (a) Parallel to yz-plane
    (b) 2.5 m away from origin in x-direction
27. (a) 3.93×10^{-19} J   (b) 9.38×10^{-20} cal
29. 657 μC/m
31. 3.44×10^{-2} C/m^3

33. $\pi R^2 |\vec{E}|$, where R is the radius of the hemisphere

35. $E = \begin{cases} \dfrac{\rho r}{3\varepsilon_0}\hat{r}, & r \le r_0 \\[2mm] \dfrac{\rho R^3}{3\varepsilon_0 r^2}\hat{r}, & r > r_0 \end{cases}$

37. $E = \begin{cases} 0 & r < 0.045 \text{ m} \\[2mm] \dfrac{2.70 \times 10^5}{r} & 0.0450 \text{ m} < r < 0.085 \text{ m} \\[2mm] \dfrac{8.99 \times 10^4}{r} & r > 0.085 \text{ m} \end{cases}$

where E is measured in volts per metre and r is measured in metres

39. 7.69×10^{-14} J
41. $v_1 = v_2 = 179$ m/s
43. 256 mV
45. 555 kV
47. 488 pm
49. -6.77×10^5 C
51. (a) 1.44 kV   (b) 1.44 keV

53. $E(r) = \begin{cases} 0 & r < 0.12 \text{ m} \\[2mm] \dfrac{6.0 \times 10^{-8}}{\pi \varepsilon_0 r^2} & r > 0.12 \text{ m} \end{cases}$

$V(r) = \begin{cases} \dfrac{5.0 \times 10^{-7}}{\pi \varepsilon_0} & r \le 0.12 \text{ m} \\[2mm] \dfrac{6.0 \times 10^{-8}}{\pi \varepsilon_0 r} & r \ge 0.12 \text{ m} \end{cases}$

where E is measured in volts per metre, V in volts, and r in metres

55. (a) $\dfrac{2.70 \times 10^5}{r}$ N/C directed outward
    (b) $\dfrac{1.44 \times 10^5}{r}$ N/C directed outward
    (c) 0
57. (a) 11.6 V/m
    (b) zero
    (c) 28.1 V/m
    (d) 15.0 V/m

59. 5.33×10^5 V/m

61. (a) 5.39×10^6 V

(b) 1.12×10^7 V

(c) small drop: 2.16×10^{10} N/C; large drop: 3.11×10^{10} N/C

63. $V(a/2) = \dfrac{\mu}{2\pi\varepsilon_0}\ln 2$, $V(2a) = -\dfrac{\mu}{2\pi\varepsilon_0}\ln 2$

65. $V(z) = \dfrac{\sigma}{2\varepsilon_0}\left[\sqrt{z^2 + a^2} - z\right]$

67. (a) 3.0×10^6 V/m

(b), (c), (d) Answers will vary.

Chapter 21—Capacitance

1. b

3. d

5. a

7. c

9. b

11. d

13. b

15. b

17. 0.500 mF

19. (a) 4.80 V   (b) 11.7 μF

21. 1.36 mm

23. (a) Connect 4 in parallel.

(b) Connect 5 in series

(c) Connect two in parallel with each other and with another pair connected in series.

25. Dielectric could be mica since $\kappa = 6.91$.

27. (a) 180 μJ   (b) 720 μJ

29. 70.7 V

31. See online resources.

33. (a) 300 fC

(b) 1.9×10^6

35. 60.7 pF/m

37. $C = \dfrac{\varepsilon_0 A}{2d}(k_1 + k_2)$

39. (a) 3.60 μC

(b) 1.80 V

(c) 3.24 μJ

41. (a) 39.8 kV/m

(b) 159 V

(c) higher

(d) 3.57 ns

(e) 4.46 cm, no

43. $C_1 = 2\ \mu$F, $C_2 = 4\ \mu$F, $C_3 = 8\ \mu$F, $C_4 = 8\ \mu$F, $C_5 = 2\ \mu$F

45. (a) 2.50 nF

(b) 28 μm

(c) 4.5 μJ

(d) 2.1 MV/m

47. $2.92 \times$ mJ/m^3

49. (a) 0.545 pF   (b) 40.7 fC   (c) 2.55×10^5

51. 5.59 mF

53. Answers will vary.

Chapter 22—Electric Current and Fundamentals of DC Circuits

1. (a) FALSE: Historically, the direction of electric current is defined as the direction of the flow of positive charge.

(b) FALSE: The resistance depends also on wire dimensions.

(c) FALSE: Free electron density is a characteristic of the metal crystal that is independent of temperature

(d) TRUE: $\rho = \rho_0(1 + \alpha\Delta T)$, $\rho = \dfrac{1}{\sigma}$.

(e) TRUE: Combining equations 22-7 and 22-1 gives

$v_d = \dfrac{E}{n_e e} \cdot \dfrac{1}{\rho}$

3. (b) For circuit I, $R_e = R$; for circuit II $R_e = 10\,R$, and for circuit III, $R_e = R/10$. Hence, $P = \dfrac{V^2}{R_e}$ for circuit III is 10 times greater than for circuit I and 100 times greater than for circuit II.

5. e

7. d

9. b

11. c

13. c

15. e

17. d

19. b

21. a

23. (a) 1.3×10^{19} per second

(b) 6.4×10^5 A/m^2

(c) 4.0 C

25. (a) $\rho = 1.7 \times 10^{-8}\ \Omega \cdot$m, $\sigma = 5.9 \times 10^7\ \Omega^{-1}$m^{-1}

(b) 4.7×10^{-5} m/s

(c) 25 fs

(d) 11×10^{-3} mV/m

(e) Resistivity, conductivity, and mean time between collisions will not change; drift velocity and electric field inside the wire will double.

(f) None

(g) The resistivity and the electric field inside the wire will increase; the conductivity and the mean time between collisions will decrease; the drift velocity will not change.

27. 20 m, 0.010 mm^2

29. 8.3 A; only one

31. (a) 2.0 Ω

(b) 1.5 A

33. See online resources.

35. (a) about 44 MJ

(b) about 79 cents

(c) Answers will vary. Answers could include using CFl or LED bulbs, replacing fridge with high-efficiency model if over 15 years old, unplugging computer, printer, TV, and other electronic devices when not in use for long periods.

37. (a) 4 in parallel

(b) 2 pair in parallel connected in series

(c) 2 in series with each other and a pair in parallel (or just 3 in series)

(d) 4 in series

39. (a) 21 Ω, 6.9 W

(b) $I_A = 0.57$ A, $I_B = I_C = I_D = I_H = I_G = I_I = 0.29$ A,

$I_E = 0.34$ A, $I_F = 0.23$ A

(c) $V_A = V_D = V_I = 2.9$ V, $V_B = V_C = V_H = V_G = 1.4$ V,

$V_E = V_F = 3.4$ V

(d) $P_A = 1.6$ W, $P_B = P_C = P_H = P_G = 0.4$ W,

$P_D = P_I = 0.8$ W, $P_E = 1.2$ W, $P_F = 0.8$ W

(e) $P_{total} = P_A + P_B + P_C + P_D + P_E + P_F + P_G + P_I = 6.9\,W$, which is equal to the power calculated in question (a), confirming that equivalent resistance does accurately model the circuit.

41. $\dfrac{12}{7}R$

43. $2.4\,\Omega$

45. (a) 6.9τ

(b) $q(\tau) = 2.3$ mC, $q(2\tau) = 3.1$ mC, $q(3\tau) = 3.4$ mC

(c) $I(\tau) = 0.88$ mA, $I(2\tau) = 0.32$ mA, $I(3\tau) = 0.12$ mA

47. (a) If the energy were taken from the electrical outlet it would cost about 0.05 cents, 2000 times less than power from the battery.

(b) Answers will vary. For the energy listed on the bill in Figure 22-10, the cost would be about \$93 000.

49. $V_{AB} = V_{FG} = V_{EF} = 0$ V

$V_{AG} = V_{BE} = V_{CF} = 10$ V

$V_{BH} = V_{HF} = 5$ V

$V_{CD} = 4$ V

$V_{DE} = 6$ V

$V_{HD} = -1$ V

51. (a) $\dfrac{I_a}{I_b} = \dfrac{2}{3}$

(b) Circuit (a): 4 V; circuit (b): 6 V

(c) In circuit (b) bulb O will not light at all and bulbs M and N will be markedly brighter than the bulbs in circuit (a).

53. (a) When the two wires are connected in parallel, the potential difference across them is equal. Since the heat produced by each wire is given by $P = V^2/R$, the wire with the lesser resistance will produce more heat, and hence have a higher equilibrium temperature. The aluminum wire has the lesser thermal coefficient. At any temperature above room temperature, the aluminum wire has less resistance and produces more heat than the tungsten wire at the same temperature. Thus, the aluminum wire will have the higher equilibrium temperature.

(b) When the two wires are connected in series, the current through them is equal. Since the heat produced is given by $P = I^2R$, the wire with the greater resistance will produce more heat, and hence have a higher equilibrium temperature. The tungsten wire will have the greater resistance because it has the greater thermal coefficient.

55. (a) $I_1 = 0.615$ A

$I_2 = 1.27$ A

$I_3 = -0.650$ A

$I_4 = 0.675$ A

$I_5 = 0.590$ A

(b) $V_{R_1} = 6.15$ V

$V_{R_2} = -6.50$ V

$V_{R_3} = 8.85$ V

$V_{R_4} = 13.5$ V

$V_{R_5} = 0.50$ V

$V_{R_6} = 0.25$ V

(c) If the batteries are real, small resistances will be connected in series with R_1 and R_2. Consequently, the currents through these resistors will decrease.

57. $I_1 = 7.32$ A

$I_2 = 1.75$ A

$I_3 = -0.284$ A

$I_4 = 0.100$ A

$I_5 = 1.65$ A

$V_{R_1} = 14.6$ V

$V_{R_2} = 8.74$ V

$V_{R_3} = -4.26$ V

$V_{R_4} = 2.00$ V

$V_{R_5} = 14.6$ V

59. (a) Zero potential difference across the three resistors on the right side of the circuit.

$V_C = V_{700\,\Omega} = 73.0$ V

$V_{250\,\Omega} = 26.1$ V

$V_{100\,\Omega} = 10.4$ V

$V_{150\,\Omega} = 10.4$ V

$V_{300\,\Omega} = 10.4$ V

(b) 3.60 ms

61. (a) $40.0\,\Omega$

(b) 41.5 mF

(c) 2.07 J

(d) 41.3 mJ

Chapter 23—Magnetic Fields and Magnetic Forces

1. Yes, there could be nonzero B, if it is parallel to the velocity of the particle.

3. See the text online resources

5. The charge carriers are positive.

7. b

9. f

11. c

13. a

15. c

17. $(e) > (d) = (b) > (a) = (c) = f$.

19. $\vec{F} = (7.68\hat{i} + 10.24\hat{j} + 3.84\hat{k}) \times 10^{-15}$ N

$F = 13.4 \times 10^{-15}$ N

21. $F \cong 4.5$ mN

$\vec{F}$ along $\hat{j}$ direction

23. (a) $r_\alpha > r_p > r_e$

(b) $T_\alpha > T_p > T_e$

25. (a) $V_p \cong 1.6 \times 10^7\ \text{m} \cdot \text{s}^{-1}$

(b) $\omega_p = 1.6 \times 10^{11}\ \text{rad} \cdot \text{s}^{-1}$

27. (a) $r_d = \sqrt{2}\,r_p$

(b) $r_\alpha = \sqrt{2}\,r_p$

29. $r_1 = 2.8$ pm

$r_2 = 5.7$ pm

$r_3 = 11$ pm

$F_{B1} = 72\ \mu$N

$F_{B2} = 140\ \mu$N

$F_{B3} = 290\ \mu$N

31. For figure and notations see the text online resources.

$\Delta = R - \sqrt{R^2 - d^2} \cong 9.9$ mm

33. (a) $\dfrac{F}{l} = 18$ N/m

(b) $\dfrac{F'}{l} = \dfrac{F}{l} = 18$ N/m

35. $B = 0$ T

37. $B(10 \text{ m}) = 200 \ \mu\text{T}$

$B(100 \text{ m}) = 20 \ \mu\text{T}$

$B(1 \text{ km}) = 2 \ \mu\text{T}$

Earth's magnetic field strength is roughly $50 \ \mu\text{T}$.

39. (a) $B(2 \text{ mm}) = 0.4$ mT

$B(5 \text{ mm}) = 1$ mT

(b) $B(r) = \begin{cases} 0.2 \ r \ (\text{mm}) \text{ mT}, & r \le 5 \text{ mm}, \\ \dfrac{5}{r(\text{mm})} \text{ mT}, & r > 5 \text{ mm} \end{cases}$

For plot of $B(r)$, see the text online resources.

(c) $E(r) = \begin{cases} \dfrac{Q}{4\pi \varepsilon_0 R^2} \cdot \dfrac{r}{R}, & r \le R, \\ \dfrac{Q}{4\pi \varepsilon_0 r^2}, & r > R \end{cases}$

where R is the radius of the uniformly charged ball. By comparison,

$B(r) = \begin{cases} \dfrac{\mu_0 I}{2\pi R} \cdot \dfrac{r}{R}, & r \le R \\ \dfrac{\mu_0 I}{2\pi r}, & r > R \end{cases}$

For plots of $E(r)$ and $B(r)$ see the text online resources.

41. $F_1 = 30 \ \mu N$

$F_2 = 4/313 \ \mu N$

43. (a) $\vec{\mu}_{\text{loop}} = -0.1\pi \ A \cdot m^2 \ \hat{k}$

(b) $\vec{\mu}_{\text{coil}} = -10\pi \ A \cdot m^2 \ \hat{k}$

(c) $\vec{\tau} = 10\pi \ \text{N} \cdot m \ \hat{i}$

(d) The magnitude of $\vec{\mu}_{\text{loop}}$, $\vec{\mu}_{\text{coil}}$, and $\vec{\tau}$ do not change, just their directions will be opposite to those indicated in parts (a), (b), and (c).

(e) If $\vec{B} = 1.0 \ T \ \hat{i}$, the answers to parts (a) and (b) do not change and $\vec{\tau} = -10\pi \ \text{N} \cdot m \ \hat{j}$.

If $\vec{B} = 1.0 \ T \ \hat{k}$, the answers to parts (a) and (b) do not change and $\tau = 0 \ \text{N} \cdot m$.

(f) $\Delta U = \pm \mu_{\text{coil}} B = \pm 10\pi \ J$

45. (a) $F_B = 1.2 \times 10^{-16}$ N

Direction of $\vec{F}_B$ is perpendicular and away from the wire.

(b) $F_B = 1.2 \times 10^{-16}$ N

Direction of $\vec{F}_B$ is perpendicular and towards the wire.

(c) The radius of the electron's curved trajectory increases as it moves away from the wire.

The radius of the proton's curved trajectory decreases as it moves towards the wire.

See the text online resources for plotted trajectories.

47. $\vec{B} = \dfrac{\mu_0 I}{6}\left(\dfrac{1}{R_1} - \dfrac{1}{2R_2}\right)\hat{k}$,

where $\hat{k}$ is the unit vector going into the page.

49. (a) $B(\text{centre}) = 50\pi \ \mu\text{T}$

(b) $B_x = \dfrac{\mu_0 n I}{2}\left(\dfrac{x - x_1}{\sqrt{(x-x_1)^2 + R^2}} - \dfrac{x - x_2}{\sqrt{(x-x_2)^2 + R^2}}\right)$

$B(15 \text{ cm}) = 1.32 \times 10^{-4}$ T

(c)

Dependence of Magnetic Field Along the Axis of a Solenoid B_x (mT) on the Distance From its Centre Measured in Metres

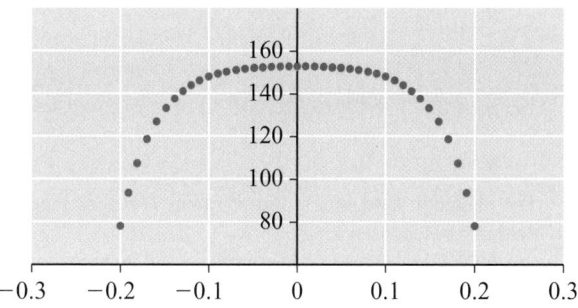

(d) One can see that 15 cm away from the centre, T decreased by about 12%. So we can estimate from the graph that it will decrease by about 5% around 10 cm off its centre. The more exact value is 12 cm.

51. See solutions manual for answer.

53. $F_{\text{loop}} = 0$ N

55. (a) $\dfrac{F_{AB}}{l} = 30 \ \mu\text{N/m}$

(b) $\dfrac{F_{BA}}{l} = \dfrac{F_{AB}}{l} = 30 \ \mu\text{N/m}$

(c) $F_{AB} = F_{BA}$

The forces calculated in questions parts (a) and (b) are equal as expected from Newton's third law.

57. (a) $\vec{F}_B = 14.4 f N \ \hat{j}$

$G = 1.6 \times 10^{-26}$ N

(b) Trajectory of the proton is an upward spiral of constant radius.

(c) $\omega = 43 \times 10^6$ rad/s

$T = 0.15 \ \mu s$

(d) $R = 4.6$ mm

Pitch $= 2.3$ cm

59. (a) The electron moves on circular trajectory upwards. See the text online resources for diagram.

(b) $R = 0.29$ m

(c) The gravitational force can be neglected due to the very small mass of the electron.

61. (a) $I_{\text{max}} = 83$ A

(b) $I_{\text{max}} = 41$A

63. $r = 2$ mm

The magnetic field is zero on a line 2 mm above the wire.

65. $B = 13$ mT

$B' = B(\chi_{\text{iron}} + 1) \cong 50$ T

67. (a) $\dfrac{\Delta B}{B} (\%) = \dfrac{K_{\text{water}} - K_{\text{air}}}{K_{\text{air}}} \times 100$

(b) See solutions manual for answer.

(c) See solutions manual for answer.

69. See the text online resources for derivation.

71. (a) See the text online resources for derivation.

(b) $B(P_1) = \dfrac{\mu_0 I}{2\pi R}$ for $L \to \infty$, as expected when applying Ampère's law for an infinitely long wire.

(c) $B(P_2) = \dfrac{\mu_o I}{4\pi R} \cdot \dfrac{L}{(L^2 + R^2)^{1/2}}$

73. (a) $B_{in} = \dfrac{\mu_o NI}{2\pi R}$,

$B_{out} = 0\ \text{T}$

(b) For solenoid:

$B_{in} = \dfrac{\mu_o NI}{L}$,

$B_{out} = 0\ \text{T}$

The magnetic field of a toroid of radius R is the magnetic field of a solenoid of length $2\pi R$.

(c) A toroid is a solenoid bent into a circular shape.

(d) $B = 6\ \text{mT}$

(e) $B = 6\pi\ \text{mT}$

$\dfrac{B(e)}{B(d)} = \pi$, as expected.

75. From the paper of Howard Bassen, magnetic fields less than $0.2\ \mu\text{T}$ were measured at 1 cm away from the iPod case. Since magnetic fields are inversely proportional to the distance, it would be recommended for people with pacemakers to wear their iPod devices as far away from their pacemakers as possible.

Chapter 24—Electromagnetic Induction

1. h

3. e

5. a

7. d

9. d

11. b

13. a

15. $L_b > L_a = L_c > L_d > L_e$

17. $u_{B,b} = u_{B,c} > u_{B,a} = u_{B,d} > u_{B,e}$

19. $1 = 2 = 3 > 4 = 5$

$(\tau = L/R)$

21. c

23. (a) $\dfrac{d\Phi_B}{dt} = 0.55\ \text{mWb}$

(b) $\dfrac{d\Phi_B}{dt} = 0.67\ \text{mWb}$

(c) $N = 20$ loops:

(a) $\dfrac{d\Phi_B}{dt} = 11\ \text{mWb}$

(b) $\dfrac{d\Phi_B}{dt} = 13\ \text{mWb}$

25. $\left(\dfrac{dB}{dt}\right)_o = 30\ \mu\text{T/s}$

27. $emf = 0.6\ \text{V}$,

29. $B = 75\ \text{mT}$

31. See the text online resources for derivation.

33. (a) The induced current flows counter-clockwise and the magnetic force acts upwards on the metal rod.

$F(t = 0\ s) = 0.84\ \mu\text{N}$

The magnetic field decreases to 0 T in 50 s. The magnetic force acting upwards on the metal rod decreases from $F(t = 0\ s)$ to 0 N according to the equation:

$F(t) = \dfrac{B(t)L^2}{R}\left[x(t)\dfrac{dB}{dt} + B(t)v(t)\right]$,

where $B(t) = B(t = 0\ s) + t\dfrac{dB}{dt}$.

(b) $a(t = 0\ s) = 3.4\ \mu\text{m} \cdot \text{s}^{-2}$

$a(t) = \dfrac{B(t)L^2}{mR}\left[x(t)\dfrac{dB}{dt} + B(t)v(t)\right]$

(c) See solutions manual for answer.

35. (a) $emf_{max} = 110\ \text{mV}$

(b) $emf_{max} = 11\ \text{V}$

(c) $emf(t) = emf_{max}\cos(2\pi ft)$

37. (a) $N_s = 100$ turns

(b) $(I_{max})_p = 80\ \text{mA}$

$(I_{max})_s = 800\ \text{mA}$

$(I_{rms})_p = 57\ \text{mA}$

$(I_{rms})_s = 570\ \text{mA}$

(c) $(I_{max})_p = 80\ \text{mA}$

$(I_{max})_s = 720\ \text{mA}$

$(I_{rms})_p = 57\ \text{mA}$

$(I_{rms})_s = 510\ \text{mA}$

39. (a) $E(r) = \dfrac{1}{r} \cdot \dfrac{\mu_o \pi N^2 R^2 f I_{max}\cos(2\pi ft)}{L}$

(b) $E(r) = r \cdot \dfrac{\mu_o \pi N^2 f I_{max}\cos(2\pi ft)}{L}$

(c) See the text online resources for the graph.

41. $M = 0.1\ \text{H}$

43. (a) $M = 62\ \text{mH}$

(b) $emf_{coil} = 0.62\ \text{V}$

45. (a) $U = 5.9\ \text{mJ}$

(b) $U = 1.5\ \text{mJ}$

(c) $U = 0\ \text{J}$

47. $emf = 5.5\ \text{mV}$

49. (a) $emf(t = 0\ s) = 10\ \text{mV}$

(b) The induced current flows counter-clockwise.

(c) See solutions manual for answer.

51. $emf(t) = -3.9\ \mu\text{V} - 0.02\ \mu\text{V} \cdot \text{s}^{-1}t(s)$

53. (a) $a = g\sin\phi - \dfrac{F_B}{m}\cos\phi$,

where a is the downward acceleration on the incline plane.

(b) $emf = Blv\cos\phi$

(c) $v(t) = g\tau\sin\phi\left(1 - e^{-\frac{t}{\tau}}\right)$,

where

$\tau = \dfrac{mR}{B^2 L^2 (\cos\phi)^2}$,

or, numerically,

$v(t) = 392(1 - e^{-0.0125 \cdot \text{s}^{-1}t(s)})\ \text{m/s}$

(d) $emf(t) = \dfrac{mgR\tan\phi}{BL}\left(1 - e^{-\frac{t}{\tau}}\right)$,

Or, numerically,

$emf(t) = 34(1 - e^{-0.0125\ \text{s}^{-1}t(s)})\ \text{V}$

(e) $v_t = 392\ \text{m/s}$

55. (a) $B = 64$ T

(b) $L = 9.6$ H

(c) $B = 130$ T

$L = 9.6$ H

57. (a) $I_A = \dfrac{v - emf}{R} = 5.0$ A

(b) $I_A = 30$ A

59. (a) See the text online resources for sketches of the pulses observed on the oscilloscope.

(b) The time between pulses is half as much as that in question (a), their widths are half the size, and their amplitudes are twice as large.

(c) The amplitude and width of the pulse corresponding to the left coil are twice as large as those of the pulse from the right coil.

(d) $v = 570$ m/s

(e) For higher speeds the two pulses merge since a high speed reduces the time separation of the recorded pulses below the width of the pulse. Increasing the distance between the two coils may affect the accuracy of the speed measurement.

61. $I = 1\ \mu$A

The current flows counterclockwise in the large loop and clockwise in the small loop if $\dfrac{dB}{dt} > 0$; the directions are opposite if $\dfrac{dB}{dt} < 0$.

63. (a)

Time (s)	V_R (V)	V_L (V)	I (A)	ε (V)
0.001	0.06	11.94	0.004	12
0.01	0.59	11.41	0.040	12
0.1	4.72	7.28	0.315	12
0.2	7.59	4.41	0.506	12
0.4	10.38	1.62	0.692	12

(b) $V_R + V_L = \varepsilon = 12$ V

$I = \dfrac{V_R}{R}$

(c) Add battery resistance to R.

65. (a) $F_{ext} = \dfrac{B^2 l^2 v}{R}$

(b) $F_{ext} = 1\ \mu$N

(c) $E_R = \dfrac{B^2 l^2 v^2 t}{R}$

(d) $E_R = 4\ \mu$J

67. (a) $t = 37$ ms

$I_{max} = 2.4$ A

(b) $V_L = 0.12$ V

(c) $t = 5.5$ s

69. (a) $emf(t) = NBlw\omega \sin(\omega t)$

(b) $emf_{max} = 57$ kV

(c) The frequency of the emf can only be changed by modifying the rotation frequency of the coil.

The maximum emf can be increased by increasing the following quantities: N, B, l, w, or ω.

71. $u_B = 0.25$ mJ$\cdot$m^{-3}

$u_E = 7.1$ nJ$\cdot$m^{-3}

The magnetic field energy density at Earth's surface is more than 10,000 times larger than the electric field energy density. Hence, the Earth is a better magnetic field dynamo than an electric field dynamo.

73. $\delta V = 3\ \mu$V

75. (a) $B_{inner} = 40$ mT

(b) $B_{outer-in} = 13$ mT

(c) $B_{outer-out} = 0$ T

(d) $u_B = 0.92$ kJ$\cdot$m^{-3}

(e) $U_B = 2.6$ MJ

(f) $p = 0.2$ kPa

77. (a) $L_{eq} = L_1 + L_2 + \cdots$

(b) $(L_{eq})^{-1} = (L_1)^{-1} + (L_2)^{-1} + \cdots$

79. (a) $L = 6.0$ mH

(b) $\tau = 3.0$ ms

(c)

Time (ms)	V_L (V)	$\ln V_L$	I (A)
1.0	8.598	2.152	1.701
1.5	7.278	1.985	2.361
2.0	6.161	1.818	2.919
2.5	5.215	1.652	3.392
3.0	4.414	1.485	3.793
3.5	3.737	1.318	4.132
4.0	3.163	1.152	4.418
4.5	2.678	0.985	4.661
5.0	2.267	0.818	4.867
5.5	1.919	0.652	5.041
6.0	1.624	0.485	5.188
6.5	1.375	0.318	5.313
7.0	1.164	0.152	5.418
7.5	0.985	−0.015	5.507
8.0	0.834	−0.182	5.583

Chapter 25—Alternating Current Circuits

1. The RLC circuit is above resonance.

3. Mathematically, the inductive reactance X_L is directly proportional to the AC frequency ω: $X_L \stackrel{\text{def}}{=} \omega L$. Therefore, a low frequency ω translates into a low inductive reactance X_L. Qualitatively, the inductor's emf depends on how fast the magnetic flux varies. Therefore, the induced emf is directly proportional to the AC frequency ω.

5. $X_C = \dfrac{1}{\omega C}$

Capacitive reactance decreases with frequency.

7. The rms value of a voltage can never exceed its maximum value, but they are equal for a constant DC voltage.

9. The phase angle approaches $\pi/2$ as the frequency increases significantly.

11. See the online resources for the derivation.

13. (a) $i_{rms} = 0.50$ A

   (b) $R = 240 \ \Omega$

15. (a) $X_L - \omega L = 1 \ \Omega$

   (b) $i_{max} = 30$ A

   (c) $i(t) = i_{max} \sin\left(\omega t + \dfrac{\pi}{2}\right)$

17. (a) $X_C = 10$ MΩ

   (b) $i_{max} = 2.8 \ \mu$A

   (c) $(t) = i_{max} \sin\left(\omega t - \dfrac{\pi}{2}\right)$

19. $L = 0.25 \ \mu$H

21. $\dfrac{1}{\sqrt{2}}$

23. (a) $t = \dfrac{k\pi}{2500}$ s,

   $k = 0, 1, 2, \ldots$

   (b) $t = \dfrac{(2m + 1)\pi}{10000}$ s,

   $m = 0, 2, 4, \ldots$

   (c) $t = \dfrac{(2n + 1)\pi}{10000}$ s,

   $n = 1, 3, 5, \ldots$

25. (a) $X_C = 100$ kΩ

   (b) $X_L = 40 \ \Omega$

   (c) $Z = 100 \ k\Omega$

   (d) $\phi - 90°$

   $i_{max} = 0.5$ mA

27. (a) $X_L = 22$ kΩ

   (b) $L = 2.5$ H

   (c) $i_{max} = 1.8$ A

29. $L = 0.23$ H

31. $N = 11$

33. Yes, the energy delivered by the emf is equal to the magnetic energy stored in the magnetic field of the coil and in the electric field energy associated with the charge stored by the capacitor.

35. $L = 140$ mH

   $\phi = 89.9°$

37. $C = 3.3 \ \mu$F

39. $\Rightarrow L = 120$ mH

41. $L = 6.3 \ \mu$H

   $i_{max} = 0.25 \ A$

43. $V_R(t = 10 \ s) = 14 \ V$

45. $R = 160 \ \Omega$

   $C = 1.5 \ \mu$F

47. $\phi = 0°$

49. $R = 20 \ \Omega$

   $\phi = 25°$

51. $\Delta f = 2.5$ mΩ

   $C = 0.25$ nF

53. Answers may vary.

Chapter 26—Electromagnetic Waves and Maxwell's Equations

1. It explains the generation of varying magnetic fields from varying electric fields.

3. (a) $\vec{E}$ and $\vec{B}$ waves are on the same axis.

   (c) $\vec{E}$ and $\vec{B}$ waves have different phases.

   (d) $\vec{E}$ and $\vec{B}$ waves have the same amplitude.

5. Sound waves do not propagate in vacuum.
   Sound waves are not transverse waves.
   Microscopically, massless particles called photons are associated with electromagnetic waves; quasi-particles called phonons are associated with sound waves.

7. Deceleration and acceleration in a straight line, curved trajectory, oscillations.

9. No, it cannot because of energy loss that is associated with the emission of electromagnetic radiation.

11. The radiation pressure is twice greater on a surface that is a perfect reflector than on a surface that is a perfect absorber. The difference is due to the change of total momentum of the system (radiation + surface) which is equal to twice the initial momentum of the radiation for the perfect reflector surface, and equal to the initial momentum for the perfect absorber surface.

13. $I_d = 0.3$ pA

15. (a) $B = 0.320 \ \mu$T

   (b) $B = 2.00 \ \mu$T

17. (a) $I(t = 0 \ s) = 12$ A

   (b) $I(t = 60 \ s) = 0$ A

   (c) $\dfrac{dE}{dt}(t = 0 \ s) = 43 \times 10^{12}$ V/m·s

   $\dfrac{dE}{dt}(t = 60 \ s) = 0$ V/m·s

   (d) $B(t = 0 \ s) = 24 \ \mu$T

   $B(t = 60 \ s) = 0 \ \mu$T

19. (a) $\vec{S}$ is in the y-direction

   (b) $\vec{S}$ is in the $(-z)$-direction

   (c) $\vec{S}$ is in the z-direction

   (d) $\vec{S}$ is in the $(-x)$-direction

21. $B_0 = 17 \ pT$

23. (a) $\lambda = 1.57$ nm

   (b) $f = 1.91 \times 10^{17}$ Hz

   (c) $E_0 = 1500$ V/m

   (d) $\vec{E}(x, t) = (1500 \ \text{V/m}) \sin(4.00 \times 10^9 x - \omega t)(-\hat{y})$
   The wave travels in the x-direction.

25. $B(t) = 3.3 \times 10^{-9}$ T in the $(-z)$-direction.

27. Microwaves $<$ infrared light $<$ yellow light $<$ blue light $<$ X-rays $<$ gamma rays

29. (a) $f = 6.67 \times 10^{14}$ Hz

   (b) $f = 5.63 \times 10^{14}$ Hz

   (c) $f = 4.29 \times 10^{14}$ Hz

31. (a) $E_0 = 7.8$ V/m

   $B_0 = 4.5 \times 10^{-4}$ T

   (b) $E_0 = 0.78$ V/m

   $B_0 = 4.5 \times 10^{-5}$ T

(c) $E_0 = 7.8 \times 10^{-4}$ V/m

$B_0 = 4.5 \times 10^{-8}$ T

33. $\bar{S} = 0.50$ W/m^2

35. $E_0 = 0.1$ V/m

$B_0 = 3 \times 10^{-10}$ T

37. $\bar{S} = 0.08$ W/m^2

39. $F = 6.0 \times 10^{-14}$ N

41. $E = 4000$ kW·h

43. (a) $\bar{S} = 1.59 \times 10^3$ W/m^2

(b) $P_{rad} = 10.6 \times 10^{-6}$ Pa

$F = 3.33 \times 10^{-11}$ N

45. The electromagnetic wave is polarized along the y-axis.

47. $30°$

49. $I_2 = 0.25$ W/m^2

51. (a) $I_2 = 0$ W/m^2

(b) $I_2 = 0$ W/m^2

(c) $I_2 = 0$ W/m^2

(d) $I_2 = 0$ W/m^2

53. $I_A = \dfrac{1}{2} I_0$

$I_B = \left(\dfrac{1}{4} + \dfrac{\sqrt{3}}{8} \right) I_0$

$I_C = \left(\dfrac{1}{8} + \dfrac{\sqrt{3}}{16} \right) I_0$

55. $35°$

57. (a) $0.5\,\mu$m

(b) $\dfrac{F_g}{F_{rad}} = \dfrac{r}{r_{(a)}} \Rightarrow r < r_{(a)} \Rightarrow F_{rad} > F_G$

Hence, if the particles have a diameter smaller than $1\,\mu$m, the force due to the radiation pressure of the Sun will exceed its gravitational attraction force.

Yes, the tail will be longer as the comet approaches the Sun. For particles below the radius value calculated in question (a) the radiation pressure will exceed the gravitational attraction of the Sun. Thus, these particles will be further apart from the core of the comet and away from the Sun, forming a longer tail.

Chapter 27—Geometric Optics

1. $d'_{oi} - d_{oi} = 2d'_o - 2d_o = 1$ m

3. For plane mirrors, the image and object distances are equal, which lies at the heart of the reason behind why images in plane mirrors display depth inversion. Plane mirrors effect depth inversion; they do not exchange left and right, nor do they exchange up and down.

   The word "AMBULANCE" is written from right to left at the front of an ambulance so that drivers can easily read it in their rear-view mirrors.

5. (c)

7. The side-view mirrors of cars are convex mirrors where the image of the object is smaller than the object (magnification $M < 1$). Because the magnification is less than one, objects appear further away than they really are. Convex mirrors are used because the image is upright and magnification

less than one allows the driver a larger field of view of the road behind the car.

9. (a) Yes, an image of the rose will be formed.

(b) To see the image of the rose in the mirror the observer has to be located on the left-lower corner of the rose-mirror assembly.

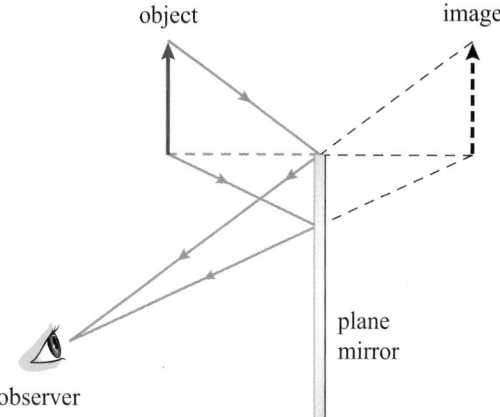

11. The maximum number of images that could be viewed is 3. The number of images is equal to the number of times a ray of light parallel to the axis of symmetry between the two mirrors is reflected. See also the solution and the ray diagram of Problem 30.

13. The answer for Problem 12 is based on the fact that a reflected ray of light from the toe of the observer has to reach his/her eye as in the ray diagram from Figure 27-12. Based on Table 27-1 the correct answer is (f) for a spherical mirror.

15. A concave mirror has to be chosen that magnifies for $R/2 < d_o < R$, where R is the radius of the mirror.

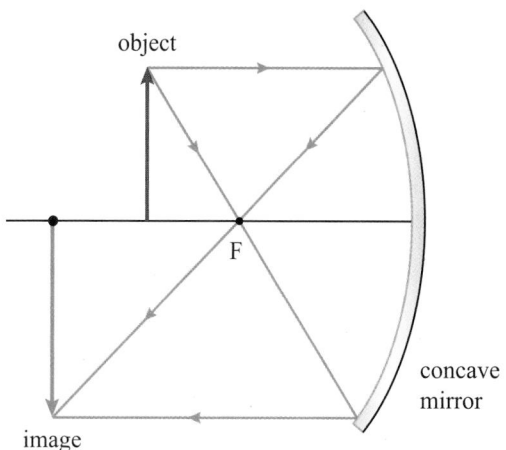

17. All three rays of light emitted simultaneously arrive at the screen in the same time following Fermat's principle. The two rays of light undergoing the longer path have a shorter path through the lens than the central ray of light has a longer path through the lens where the speed of light is smaller than in air.

19. You will get a fainter image because half of the rays from all parts of the image are blocked.

21.

convergent lens

concave mirror

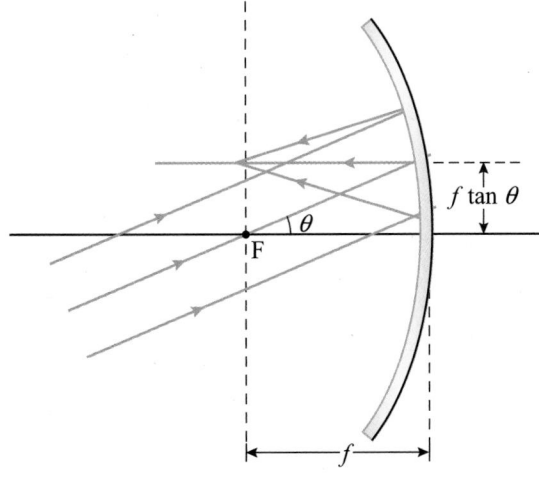

23. 34.3 m

25. (a) Observers at A will see a total solar eclipse (the Sun is entirely eclipsed by the Moon), observers at B will see a partial solar eclipse (the Sun is partially eclipsed by the Moon), and observers at C will not see an eclipse.

(b) The Earth rotates about its axis from West to East in geographical terms. The rotation is counterclockwise in the Cartesian reference frame of the Earth with its z-axis oriented from South to North. An observer on Earth perceives this motion through the position of the Sun in the sky from sunrise (East) to sunset (West). During the time of a total solar eclipse the region in which this astronomical event can be observed will move from West to East.

27. Angle = 31°

29. $\Delta h = 0.24$ m

31. (a) The image is not inverted as the ray diagram indicates.

(a)

observer

(b)

object

objective lenses

observer

ocular lens

(c) Using curved mirrors would not be a viable design because these mirrors distort the object-observer distance due to varying magnification. The object-observer distance is important in many applications of the periscope.

33. (a) $f = 3.5$ cm   (b) Image is real and inverted

35. $f = -7.7$ cm   (b) $M = 0.04$. The image is real, upright, and reduced

37. Assuming critical angle, area = 7 m²

39.

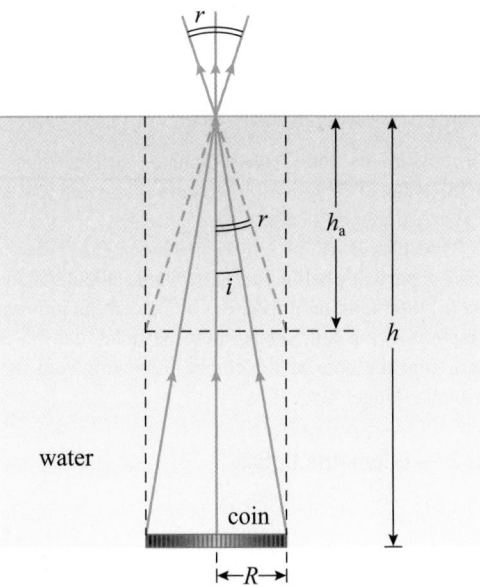

water

coin

From the geometry indicated in the ray diagram the following relationships can be derived:

$$\tan i = \frac{R}{h}$$

$$\tan r = \frac{R}{h_a}$$

From the two equations it can be deducted that:

$$\frac{h_a}{h} = \frac{\tan i}{\tan r} \approx \frac{\sin i}{\sin r} = \frac{1}{n} \Rightarrow h_a \approx \frac{h}{n} = 60 \text{ cm}$$

The used approximation is valid for small angles. The observation angles are small because the radius of the coin $R \ll h$.

41. (a) $n = 1.41$ (b) $n = 2.08$
43. (a) 45 mm (b) $M = 0.008$
45. $d_i = -3.4$ cm (b) diverging lens (c) Virtual, erect and reduced
47. (a) $M = 0.8$ (b) $M = 0.5$
49. $R = 1.14\,H$
51. Angle of deflection = 169°; decreases
53.

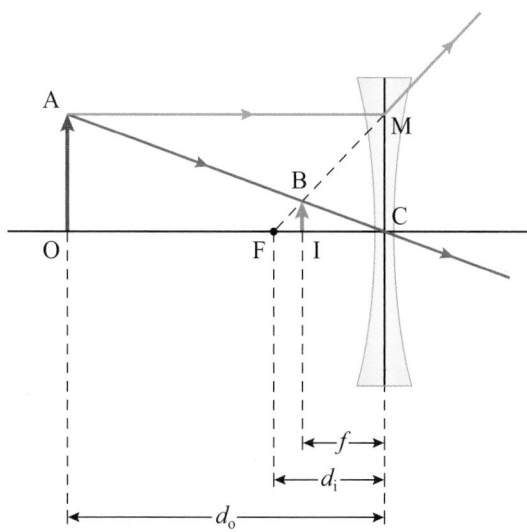

$d_i = 10$ cm
The image is formed in front of the divergent lens; it is virtual, upright, and reduced. The image of the object cannot be seen on a screen because it is virtual.
55. (a) 0.45 cm (b) 4.5 cm
57. (a) Through the base (b) 80° (c) 0, because the ray exits through the base
59. (a) $f(4D) = 25$ cm, $f(5D) = 20$ cm (b) $d_i = 40$ cm from the second lens (c) No, the combined power of the two lenses cannot be used to solve this problem because the distance between them is large relative to their focal lengths. The combined optical power describes a system where lenses are very close together.
61.

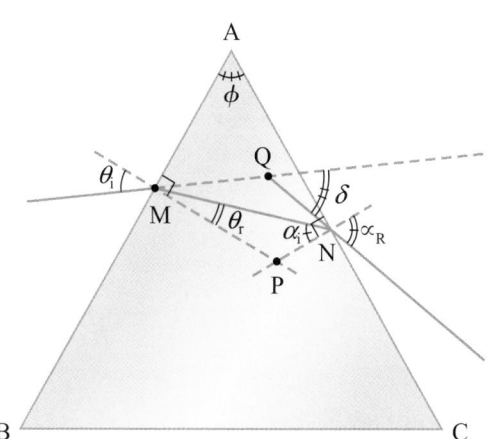

From the ray diagram the following relationships between all the angles can be derived:
In $\triangle MQN$:

$(\theta_i - \theta_r) + (\alpha_i - \alpha_r) = \delta$
Using quadrilateral AMPN and $\triangle MPN$:
$\theta_r + \alpha_i = \phi = 180° - \angle MPN$
The refraction law applied at the entry and exit boundaries of the prism gives:
$\sin\theta_i = n\sin\theta_r$
$n\sin\alpha_i = \sin\alpha_r$
Using all these equations it follows that the deflection angle δ is given by:
$\delta = \theta_i + \phi - 2\theta_r - \sin^{-1}[n\sin(\phi - \theta_r)]$,
where
$\theta_r = \sin^{-1}\left(\dfrac{\theta_i}{n}\right)$

63. (a) From the ray diagram the distance between lenses is:
$d = F_2 - |F_1| = 2|F_1|$

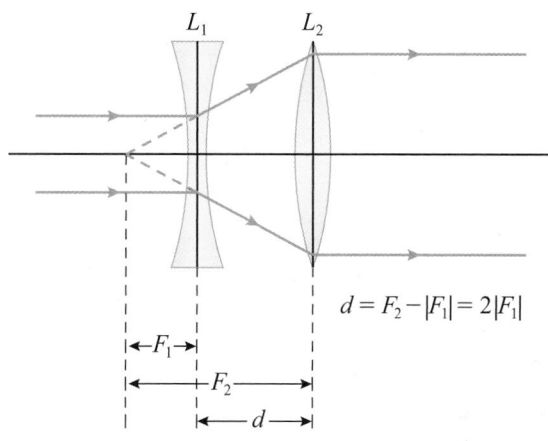

$d = F_2 - |F_1| = 2|F_1|$

(b) From the ray diagram the distance between lenses is:
$d = F_2 + |F_1| = 4|F_1|$

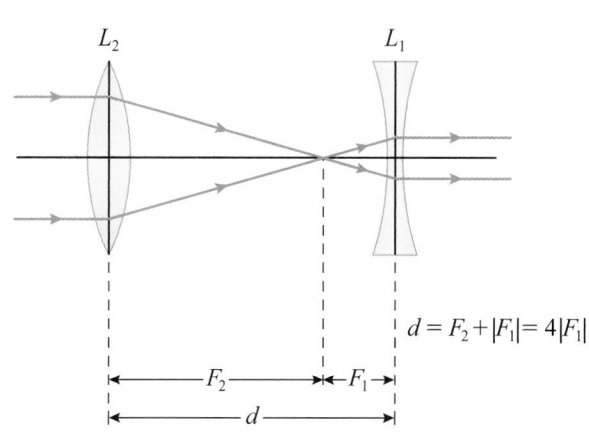

$d = F_2 + |F_1| = 4|F_1|$

65. (a)

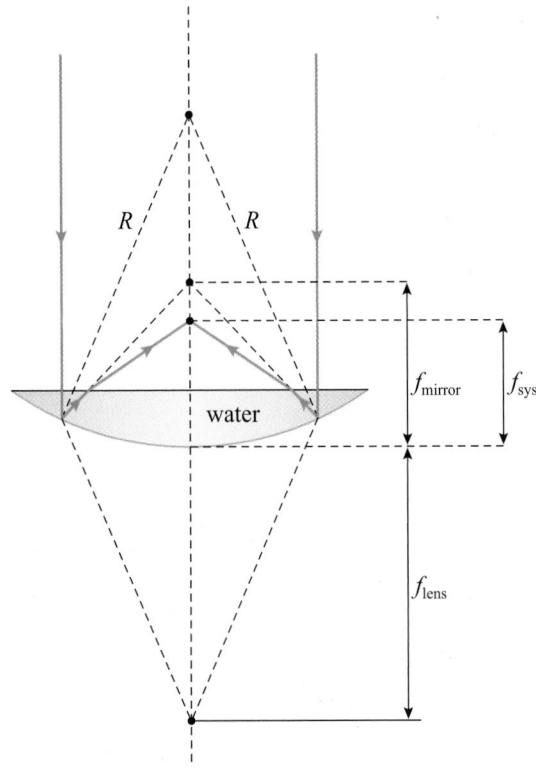

(b) $f_{mirror} = 25$ cm $\Rightarrow P_{mirror} = 4\,D$

$\Rightarrow f_{lens} = 1.5$ m $\Rightarrow P_{lens} = 0.67\,D$

(c) $P_{sys} = 4.7$

67. $t_{total} = 33$ ms

69. Let f_1 and f_2 be the focal lengths characterizing the two lenses. The focal length of one lens, say, f_1 will be the object-lens distance for the other lens and the image distance will then be the focal length of the pair of lenses f_{pair}. Because the lenses are very close, the object-lens distance is negative and equal to $-f_1$ which corresponds to a virtual object. Hence, using the thin lens equation for mirror of focal length f_2,

$$\frac{1}{f_2} = \frac{1}{-f_1} + \frac{1}{f_{pair}} \Rightarrow P_{pair} = P_1 + P_2$$

71. $\omega = 8$ rad/s

73. (a) $d_{o1} = 6.1$ cm
(b) $M = -7.7$
(c) Yes, the compound microscope can produce an upright final image if the image produced by the first lens (L_1) is between the two lenses. That is,

5 cm $< d_{i1} < 10$ cm

(d) Use a graphing software and the equations you derived or a virtual microscope simulation (such as http://www.udel.edu/biology/ketcham/microscope/scope.html) to explore how the change in each one of the parameters of the compound microscope (distance between the lenses, power of each one of the lenses, and the distance between the object and the objective lens) will affect the magnification.

75. (a) No, it is not possible to confine all the light from a large source into a significantly reduced image of the light source. This would mean that all the light of a source would have to be confined in a smaller volume. This would correspond to a decrease in entropy of an isolated system (no light loss or gain) which contradicts the second law of thermodynamics. (b) Only an ideal system would project all the light from a smaller source into a magnified image. Such a system does not exist due to light-matter interactions; there are no perfect mirrors and lenses. Real mirrors and lenses do not reflect or transmit, respectively, 100% of the incident light.

77. Answers will vary.

Chapter 28—Physical Optics

1. d
3. c
5. b
7. d
9. b
11. b
13. b
15. (a) valid
(b) valid (although less precisely so)
(c) valid
(d) not valid
17. (a) constructive
(b) 50 m
19. (a) 0.145°
(b) 2.28 mm
21. (a) 381 lines/cm
(b) 1.14 cm
23. (a) 226 nm
(b) 624 nm
25. 29.3 μm
27. 17 fringes
29. 6.16 km
31. 1.00018
33. 0.151°
35. 84.6 μm
37. (a) 12.1°, 36.9°
(b) 28.7 mm
(c) 53.5 mm
39. 282 fringes
41. (a) 1.07×10^{-5} degrees
(b) 0.039″
43. 7.0×10^{-8} degrees
45. 8.39 cm
47. 41.7 μm

Chapter 29—Relativity

1. c
3. c
5. d
7. c
9. d
11. c
13. c
15. 1.26×10^{-6} N/A²
17. (a) 0.31 s (b) 0.27 s
19. (a) 1.85×10^8 s (b) 1.22×10^8 s
21. 3.0×10^8 m/s
23. 0.783 MeV
25. 1.33 μs

27. 35.4 km

29. 9.47×10^{15} m

31. (a) 1.60×10^9 s
    (b) 1.62×10^9 s
    (c) 2.80×10^8 s

33. (a) 2.60×10^8 m/s
    (b) 0.023 s

35. (a) 1.1×10^8 m/s
    (b) 1.0×10^8 m/s
    (c) No

37. No, this is not in conflict with special relativity. Neutrinos, having very small interaction probabilities, can pass through dense matter more freely than photons. Therefore they leave the supernova before the photons. The photons leave with the shockwave and reach Earth a few hours after the neutrinos.

39. (a) 4.28×10^9 kg/s
    (b) 2.05×10^{19} kg/y
    (c) 9.78×10^9 y

41. For the second observer, the events will also be 3.0 m apart in space.

43. (a) 14.9 ns
    (b) $0.667c$

45. $0.916 \ \mu$s

47. $0.86c$

49. 7.36×10^9 kg/m^3

51. (a) $0.999 \ c$
    (b) 1.11×10^4 s^{-1}
    (c) 7.0×10^6 MeV/c $= 3.7 \times 10^{-15}$ kg m/s

53. (a) $0.976 \ c$
    (b) $0.67 \ m_0 c^2$
    (c) $3.59 \ m_0 c^2$

55. $m_0' = \dfrac{m_0}{2} \sqrt{\dfrac{(3c)^2 - V^2}{c^2 - V^2}}$

57. $\dfrac{1}{f} df = \dfrac{GM}{c^2 r^2} dr$

59. Answers may vary.

Chapter 30—Fundamental Discoveries of Modern Physics

1. In 1897, J. J. Thomson postulated the existence of particles smaller than atoms while trying to explain the existence of "cathode rays" that he produced in a glass tube. He postulated that these smaller particles were minuscule pieces of atoms. Later on Philipp Lenard performed more experiments on these cathode rays and showed that if cathode rays were particles, their mass must be smaller than the mass of an atom.

3. Atoms have concentrated positive charge, a large empty space and negative charges. Some atoms can spontaneously produce positive, negative and neutral rays. When subjected to external light, some atoms produce light of different wavelength.

5. Electron's angular momentum.

7. A blackbody spectrum is continuous whereas a discharge spectrum has peaks at certain wavelengths characteristic of the material.

9. The thermometer works by detecting infrared radiation from the eardrum through a device such as a thermopile.

11. H: 9.570×10^7 C/kg; O: 4.823×10^7 C/kg

13. (a) The object with higher intensity (black line) is at a higher temperature.
    (b) No.

15. (a) 5.333×10^5 m^{-1}
    (b) $n_1 = 4$
    (c) A wavelength of 1875 nm corresponds to infrared light.

17. (a) 6.4×10^7 m/s
    (b) 1.2×10^5 V/m
    (c) 1.3 mm

19. (a) 15 000 V/m
    (b) 2.4×10^{-15} N
    (c) 8.927×10^{-30} N
    (d) Yes, it is reasonable to neglect gravity because the electric and magnetic forces on the electron are greater than the gravitational force by a factor of $\sim 10^{14}$.

21. (a) 0.147%
    (b) 0.127%

23. (a) $dU = \dfrac{3kQ^2 r^4 dr}{R^6}$

    (b) $U = \dfrac{3kQ^2}{5R}$

    (c) 8.6×10^{-16} J

25. (a) 1.5×10^{-13} m
    (b) 3.6×10^{-14} m

27. (a) 1.602×10^{-16} J
    (b) 1.875×10^7 m/s
    (c) 1000 eV

29. (a) 3.65×10^{-19} J
    (b) 5.51×10^{14} Hz
    (c) 544 nm

31. (a) $f = \dfrac{c}{\lambda}$

    Calculated values are listed in table below.

    (b) $K_{max} = e\Delta V$

    Calculated values are listed in table below.

Wavelength (nm)	Stopping Potential (v)	Frequency (Hz)	K_{max} (J)
260	0.46	1.154×10^{15}	7.369×10^{-20}
250	0.66	1.200×10^{15}	1.057×10^{-19}
240	0.87	1.250×10^{15}	1.394×10^{-19}
230	1.09	1.304×10^{15}	1.746×10^{-19}

(c)

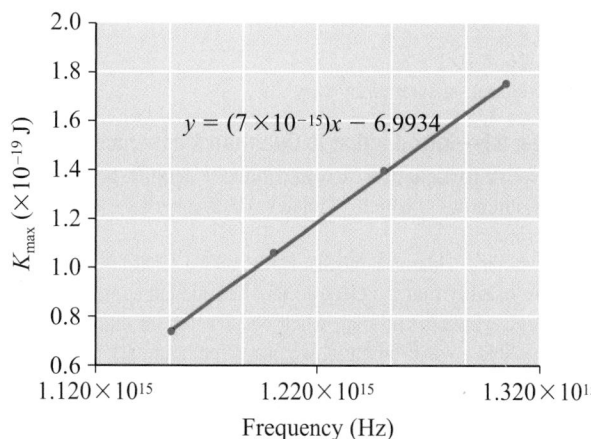

$y = (7 \times 10^{-15})x - 6.9934$

(d) 7×10^{-34} J s

(e) 4.37 eV

33. 2.19×10^6 m/s

35. 2.52×10^{-16} s

37. (a) 302 keV

(b) 48.3 keV

39. 1.95×10^{-18} C

41. 7.42×10^{-13} m

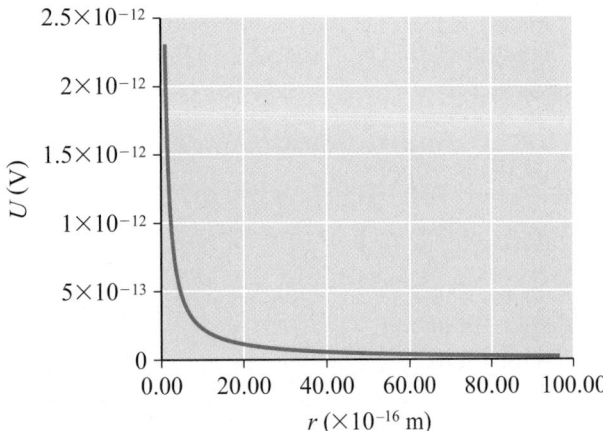

43. 3.75×10^5 m/s

45. 4.18 cm

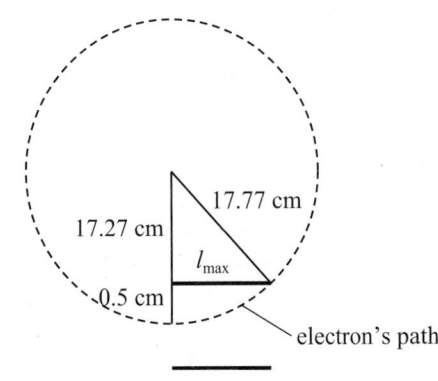

47. $f_{max} = \dfrac{2.82\, kT}{h}$

49. $E > 1.988 \times 10^{-8}$ J, or, equivalently $E > 124.1$ GeV

51. $\dfrac{B_p^2}{E_p} = 1835.84\, \dfrac{B_e^2}{E_e}$

53. 1.26×10^{21}

55. 6680 kg; answers may vary

Chapter 31—Introduction to Quantum Mechanics

1. No. A particle at rest would have a definite position and momentum, which violates Heisenberg's uncertainty principle.

3. Yes, since quantum mechanically, all we can do is calculate a probability that the electron will be found at some place and time. Having said that, when we develop an equation for a particle's wave function, we use some boundary conditions, which may confine the wave function to a certain region.

5. Infinite

7. (a) No, since the energy of an incident photon is less than the difference between the energy of the ground state and first excited state of hydrogen.

(b) Yes, since the energy of an incident photon is greater than the difference between the energy of the ground state and first excited state of hydrogen.

(c) Yes, since the energy of an incident photon is greater than the difference between the energy of the ground state and first excited state of hydrogen.

9. For $n = 4$ to $n = 2$: $\Delta E = 2.55$ eV

For $n = 6$ to $n = 3$: $\Delta E = 1.13$ eV

For $n = 3$ to $n = 1$: $\Delta E = 12.09$ eV

For $n = 5$ to $n = 3$: $\Delta E = 0.97$ eV

Hence $\Delta E_d < \Delta E_b < \Delta E_a < \Delta E_c$

11. All are possible.

13. (a) 1.17×10^{-12} m

(b) 3.46×10^{-13} m

15. 4.18 μV

17. $\Delta p = 5.25 \times 10^{-25}$ kg m/s, $K = 1.51 \times 10^{-19}$ J

19. 2.63×10^{-31} m

21. (a) $\Delta t = 1.44 \times 10^{-12}$ s, $\Delta E = 3.64 \times 10^{-23}$ J

(b) $\Delta t = 1.44 \times 10^{-23}$ s, $\Delta E = 3.64 \times 10^{-12}$ J

23. $E = 4.55 \times 10^{-21}$ J, $\Delta E = 9.1 \times 10^{-26}$ J, $\Delta t = 5.77 \times 10^{-10}$ s

25. Since the square of the wave function integrated over the whole length for the one-dimensional case must be a dimensionless quantity, therefore the units of the wave function should be m$^{-1/2}$.

27. $E_1 = 5.97 \times 10^{-18}$ J

$E_2 = 2.39 \times 10^{-17}$ J

$E_3 = 5.37 \times 10^{-17}$ J

The probability density is given by

$$\psi_n^*(x)\, \psi_n(x) = \frac{2}{L}\, \sin^2\left(\frac{n\pi}{L}\, x\right)$$

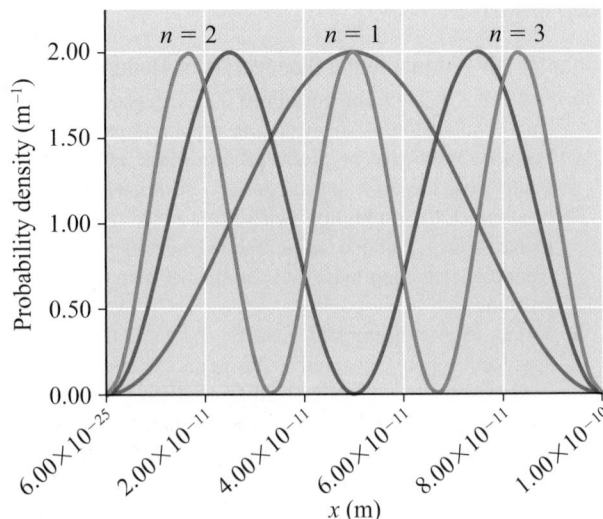

29. $\Delta E = 2.39 \times 10^{-23}\left(n_2^2 - n_1^2\right)$

(a) 7.17×10^{-23} J

(b) 1.91×10^{-22} J

(c) 1.20×10^{-22} J

(d) 2.39×10^{-17} J

31. $P = 0.5$

33. 3.33×10^{-8} m

35. See solutions manual for answer.

37. $\psi(x) = \begin{cases} \sqrt{\dfrac{15}{512}}\,(4 - x^2) & \text{for } -2 \leq x \leq 2 \\ 0 & \text{everwhere else} \end{cases}$

39. $\psi(x) = \begin{cases} \sqrt{\dfrac{15}{512}}\,(4 - x^2)\,e^{-i\frac{p^2 t}{2mh}} & \text{for } -2 \leq x \leq 2 \\ 0 & \text{everywhere else} \end{cases}$

41. 0.62

43. (a) $E_2 = 80$ eV

$\quad\quad E_3 = 180$ eV

$\quad\quad E_4 = 320$ eV

(b) 1.36×10^{-10} m

(c) 2.05×10^{-8} m

45. 0.36

47. $1.13 \times 10^{278}\, y$

49. -0.544 eV

51. $L = \hbar, \sqrt{2}\hbar, \sqrt{6}\hbar, \sqrt{12}\hbar$ and $L_z = -3\hbar, -2\hbar, -\hbar, 0, \hbar, 2\hbar, 3\hbar$

53. (a) $l = 3$ and $n = 4$

(b) $l = 0$ and $n = 1$

(c) $l = 4$ and $n = 5$

55. $r = a_B$, that is, the most probable distance is equal to the Bohr radius.

57. (a) $E_n = -\dfrac{2847.5}{n^2}$ eV

(b) 949.2 eV

(c) 2.54×10^{13} m

59. 2.50×10^{14} m/s

61. (a) $j = 3, 2, 1, 0$

(b) For $j = 3$: $J_z = 3\hbar, 2\hbar, \hbar, 0, -\hbar, -2\hbar, -3\hbar$

$\quad\quad$ For $j = 2$: $J_z = 2\hbar, \hbar, 0, -\hbar, -2\hbar$

$\quad\quad$ For $j = 1$: $J_z = \hbar, 0, -\hbar$

$\quad\quad$ For $j = 0$: $J_z = 0$

63. For $n = 1$, maximum number of electrons $= 2(1)^2$

$\quad$ For $n = 2$, maximum number of electrons $= 2(2)^2$

$\quad$ For $n = 3$, maximum number of electrons $= 2(3)^2$

Chapter 32—Introduction to Solid-State Physics

1. Decrease

3. d

5. d

7. b

9. Silicon

(a) $V = a^3$

11. (b) $V = \dfrac{a^3}{4}$

13. 94.9%

15. 3

17. (a) Positive

(b) Negative

(c) Positive

19. (a) 10^{13}

(b) 10^{-13} J/s

21. 7.05×10^{-21}

23. 0.22 μg

25. 4.0×10^5

27. 0.07 W

29. (a) 3.4×10^{-17} F

(b) 510

31. 90 %

33. (a) 1000 V/m

(b) 1.8×10^{14} m/s^2

(c) $1.8 \times 10^{14}\, t$ m/s

(d) $0.014\, t^2$ J

(e) $0.028\, t$ J/s

(f) 3.57×10^{-12} s

(g) 1.12 nm

35. 10^{47} J/m^3

37. (a) 0

(b) 9.1×10^{-6}

39. (a) 8.6 eV

(b) 8.5 eV

41. 2.50×10^{-14} s

43. 4.4 nm

45. 0.27 m

This distance between the circuit elements on a chip does not seem reasonable as chips are much smaller than this. Therefore the signal takes much less time to propagate than 1ns.

47. 100 s

49. The Fermi wave number depends on the number of electrons that each atom contributes to conduction, that is $\dfrac{N}{N^*}$, for a certain lattice parameter.

51. $\lambda = 2a$

53. (a) Since the Fermi energy for metals is fairly high, only a tiny fraction of electrons close to the Fermi level have their energies increased when the metal is heated.

(b) $\dfrac{N}{N^*} = 0.0055$

(c) No, only the electronic contribution to heat capacity in metals is very small because only a fraction of the electrons are within thermal enegry of the Fermi energy.

Chapter 33—Introduction to Nuclear Physics

1. Nucleon: A proton or neutron in an atomic nucleus.

Nuclide: Nuclide refers to a particular type of nucleus with a specific number of neutrons and protons.

Isotope: Species of nuclei with same number of protons and different number of neutrons.

Decay Constant: Proportionality constant that characterizes the relation between the decay rate and the amount of a radioactive substance.

Half Life: Time it takes for half of the atoms of the sample to decay.

Decay Diagram: A diagram that specifies the sequence in which some nuclide decays.

Q–value: Net energy released or absorbed in a nuclear reaction.

Spontaneous Fission: Spontaneous breaking up of a heavy and unstable nucleus with release of neutrons.

Radioactive Dating: A process through which the age of a substance is estimated based on comparison of abundance of some radio-isotope and its decay products.

Proton-Proton Cycle: Fusion of hydrogen nuclei resulting in production of hydrogen, helium, electrons, neutrinos and release of energy.

3. α-decay: In this process, the parent nucleus gives off an α-particle and transforms into a lighter nucleus.

β^--decay: In this process, a neutron in the nucleus of the parent transforms into a proton and gives off an electron and an anti-neutrino.

β^+-decay: In this process, a proton in the nucleus of the parent transforms into a neutron and gives off a positron and a neutrino.

5. 7

7. 1.2×10^{36}

9. $t = 2t_{1/2}$

11. The binding energy per nucleon for heavy nuclei ($A \geq 200$) is less than the binding energy for nuclei in the range $50 \leq A \leq 120$. Therefore when a heavier nucleus fissions into two lighter nuclei there is an increase in the total binding energy. This excess energy is released in the fission process.

The iron nucleus has a larger binding energy per nucleon than any other nucleus. Therefore, if two iron nuclei were to fuse into a heavier nucleus (tellurium), the total binding energy of the tellurium nucleus will be less than that of the two iron nuclei. Therefore, there is no release of energy in this process.

13. Then all the protons would get converted into neutrons. In fact if this were the case at the beginning of the big bang, there wouldn't be any protons and the universe would have been simply a large collection of neutrons.

15. 5.6×10^{26} MeV

17. 8.146×10^{34} eV

19. 0.78 MeV/c

21. $m_C = 12.011137$ u $= 1.9944 \times 10^{-26}$ kg

23. (a) $B_{per}(2, 1) = 1.1$ MeV/nucleon
  (b) $B_{per}(4, 2) = 7.1$ MeV/nucleon
  (c) $B_{per}(6, 3) = 5.3$ MeV/nucleon
  (d) $B_{per}(56, 26) = 8.8$ MeV/nucleon
  (e) $B_{per}(208, 82) = 7.9$ MeV/nucleon

25. (a) 2.013554 u
  (b) 11.996712 u
  (c) 15.990531 u
  (d) 55.920691 u
  (e) 238.000369 u

27. (a) 89.87 MeV
  (b) 85.82 MeV
  (c) 722.44 MeV
  (d) 1611.77 MeV
  (e) 1790.76 MeV

29. 3.833×10^{-12} m^{-1}

31. 4.5×10^7 Bq

33. $N = 2.609 \times 10^{17}$ atoms of carbon-14, $m = 6.06$ μg

35. $\lambda = 6.67 \times 10^{-11}$ s^{-1}, $t_{1/2} = 1.04 \times 10^{10}$ s

37. (a) 970 Bq
  (b) 6.8 Bq

39. 578 s

41. (a)

  (b) 0.0206 s^{-1}
  (c) 33.6 s

43. (a) Not possible since charge must be conserved.
  (b) Not possible since charge must be conserved.
  (c) Possible.
  (d) Not possible since charge must be conserved.
  (e) Not possible since mass number must be conserved.

45. (a) $Q = (m_n + m_X - m_Y - m_{He})c^2$
  (b) 4.54 MeV

47. $Q = -7.27$ MeV
  Since the Q-value is negative, carbon-12 cannot decay through this channel spontaneously.

49. (a) $^{32}_{14}\text{Si} \rightarrow {}^{28}_{12}\text{Mg} + {}^4_2\text{He}$
  (b) $^{56}_{25}\text{Mn} \rightarrow {}^{52}_{23}\text{V} + {}^4_2\text{He}$
  (c) $^{60}_{26}\text{Fe} \rightarrow {}^{56}_{24}\text{Cr} + {}^4_2\text{He}$

51. (a) $^{52}_{25}\text{Mn} \rightarrow {}^{52}_{24}\text{Cr} + e^+ + \nu$
  (b) $^{55}_{27}\text{Co} \rightarrow {}^{55}_{26}\text{Fe} + e^+ + \nu$
  (c) $^{59}_{28}\text{Ni} \rightarrow {}^{59}_{27}\text{Co} + e^+ + \nu$

53. (a) $^3_1\text{H} \rightarrow {}^3_2\text{He} + e^- + \bar{\nu}$
  (b) 18.6 keV

55. Thorium-232 goes through 4 β-decays and 6 α-decays before converting into a stable lead-208 as shown in the decay diagram below.

57. (a) 179.6 MeV
(b) 179.9 Mev

59. (a) $B_{per}(2, 1) = 1.1$ MeV/nucleon, $B_{per}(6, 3) = 5.3$ MeV/nucleon, $B_{per}(10, 5) = 6.5$ MeV/nucleon, $B_{per}(14, 7) = 7.5$ MeV/nucleon, $B_{per}(180, 73) = 8.0$ MeV/nucleon

(b) $B_{per}(4, 2) = 7.1$ MeV/nucleon, $B_{per}(4, 8) = 7.1$ MeV/nucleon, $B_{per}(12, 6) = 7.7$ MeV/nucleon, $B_{per}(16, 8) = 8.0$ MeV/nucleon, $B_{per}(182, 74) = 8.0$ MeV/nucleon

61. (a) 9.0 MeV
(b) this spontaneous decay mode is not possible

63. $S_n = a_V - a_S \dfrac{2}{3} A^{-\frac{1}{3}} - a_A \left[1 - \dfrac{4Z^2}{A(A-1)} \right]$

Chapter 34—Introduction to Particle Physics

1. (a) This is incorrect as quarks can interact with each other through any of the fundamental forces and can therefore exchange other force mediating particles as well.
(b) True
(c) False, since the strong force is mediated by gluons.
(d) False, since neutrinos are not force carriers.
(e) True, since gluons carry the strong force.

3. (a) True
(b) False, since mesons contain a quark and an antiquark, whereas baryons are made up of three quarks.
(c) True
(d) False, since a pion is a meson and not a baryon.
(e) True
(f) True

5. (a) True
(b) True, if the electron decelerates.
(c) True
(d) False, since a charged particle cannot gain energy in a constant magnetic field.
(e) True, since even a constant magnetic field would curve the path of a charged particle.

7. 380 MeV

9. electron, neutrino, quark

11. photon, Z^0, gluon

13. $R \propto \dfrac{1}{m}$; therefore if these particles were massless, the range of the weak force would be infinite

15. uds: $m = 1110$ GeV/c^2
uus: $m = 1100$ GeV/c^2
dds: $m = 1120$ GeV/c^2
uss: $m = 1300$ GeV/c^2
dss: $m = 1310$ GeV/c^2
sss: $m = 1500$ GeV/c^2

17. The u, d and s quarks have the following quantum numbers.

	Charge	Baryon Number	Spin
u	⅔	⅓	½
d	-⅓	⅓	½
s	-⅓	⅓	½

This table can be used to find the charge, baryon number, and spin of the baryons.

	Charge	Baryon Number	Spin
uus	+1	+1	½
uds	0	+1	½
dds	−1	+1	½
uss	0	+1	½
dss	−1	+1	½

19. Quarks have the following quantum numbers:

	Charge	Baryon Number	Spin
u	⅔	⅓	½
d	-⅓	⅓	½
c	⅔	⅓	½
s	-⅓	⅓	½
t	⅔	⅓	½
b	-⅓	⅓	½

	Charge	Baryon Number	Spin	(Mass × C MeV/c^2)	Mass Rank
ttb	+1	+1	½	346200	5
tts	+1	+1	½	342125	4
tbs	0	+1	½	175325	3
tbd	0	+1	½	175205	2
tuu	+2	+1	½	141005	1

21. $m_{s\bar{u}} = C(125.0 + 2.5) = 127.5C$ MeV/c^2, where C is the proportionality constant.
$m_{s\bar{d}} = C(125.0 + 5.0) = 130.0C$ MeV/c^2
$m_{s\bar{s}} = C(125.0 + 125.0) = 250.0C$ MeV/c^2
$m_{u\bar{s}} = C(2.5 + 125.0) = 127.5C$ MeV/c^2
$m_{d\bar{s}} = C(5.0 + 125.0) = 130.0C$ MeV/c^2
$m_{\bar{c}d} = C(1300.0 + 5.0) = 1305.0C$ MeV/c^2
$\Rightarrow m_{s\bar{u}} = m_{u\bar{s}} < m_{s\bar{d}} = m_{d\bar{s}} < m_{s\bar{s}} < m_{\bar{c}d}$

23. (a) $C: +1 \neq 0 - 1 + 0 \Rightarrow$ conservation of charge violated
(b) $C: 0 \neq +1 + 0 \Rightarrow$ conservation of charge violated
(c) conservation of energy violated
(d) $L: -1 \neq -1 + 1 \Rightarrow$ conservation of lepton number violated
(e) $B: +1 \neq 0 + 0 \Rightarrow$ conservation of baryon number violated
(f) $B: +1 + 1 \neq +1 + 0 \Rightarrow$ conservation of baryon number violated
(g) $L: -1 + 0 \neq 0 + 1 \Rightarrow$ conservation of lepton number violated
(h) $B: +1 \neq 0 + 0 \Rightarrow$ conservation of lepton number violated

25. The production of a single photon in electron-positron annihilation is impossible, because linear momentum would not be conserved.

27. 1.8 MeV

29. $E_\pi = 931.5$ MeV, $p_\pi = 921.0$ MeV/c

31. Because the proton is the lightest baryon, it cannot decay into another baryon; if it decays into a meson, the law of conservation of baryon number will be violated.

33. The reaction is not possible because it would require $v_e \geq c$.

35. (a) $p = 69.8$ MeV/c, $E_v = 69.8$ MeV, $E_e = 69.8$ MeV
    (b) $p = 29.9$ MeV/c, $E_v = 29.9$ MeV, $E_\mu = 109.7$ MeV

37.

39.

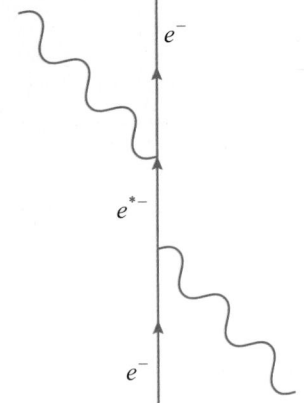

41. (a) (i) $R = 7.3 \times 10^{-16}$ m
        (ii) $R = 1.3 \times 10^{-16}$ m
        (ii) $R = 9.6 \times 10^{-17}$ m

43. (a) (i) 10.9 cm
        (ii) 32.2 cm
        (iii) 42.6 cm
    (b) frequency must be increased to achieve greater kinetic energies
    (c) (i) 74.9 cm
        (ii) 74.9 cm
        (iii) 74.9 cm

45. (a) 2.28×10^8 m/s
    (b) $t = 50$ ns
    (c) circumference $= 11.4$ m

47. Circumference $> 10^{32}$ m; building a synchrotron of this size is impossible

Appendix B
SI UNITS AND PREFIXES

Base SI Units

Unit name	SI symbol	Measures
Ampere	A	Electric current
Candela	cd	Luminous intensity
Kelvin	K	Temperature
Kilogram	kg	Mass
Metre	m	Length
Mole	mol	Amount of a substance
Second	s	Time

The Most Common Derived SI Units

Unit name	SI symbol	Measures	In base units
Coulomb	C	Electric charge	$A \cdot s$
Farad	F	Electric capacitance	C/V
Henry	H	Electromagnetic inductance	$V \cdot s/A$
Hertz	Hz	Frequency	$/s$
Joule	J	Energy and work	$N \cdot m$
Newton	N	Force	$kg \cdot m/s^2$
Ohm	Ω	Electric resistance	V/A
Pascal	Pa	Pressure	N/m^2
Radian	rad	Angle	dimensionless
Tesla	T	Magnetic field strength	$V \cdot s/m^2$
Volt	V	Electric potential difference	J/C
Watt	W	Power	J/s

Common Prefixes

Prefix	Abbreviation	Power
Zepto-	z	10^{-21}
Atto-	a	10^{-18}
Femto-	f	10^{-15}
Pico-	p	10^{-12}
Nano-	n	10^{-9}
Micro-	μ	10^{-6}
Milli-	m	10^{-3}
Centi-	c	10^{-2}
Deci-	d	10^{-1}
Kilo-	k	10^{3}
Mega-	M	10^{6}
Giga-	G	10^{9}
Tera-	T	10^{12}
Peta-	P	10^{15}
Exa-	E	10^{18}
Zetta-	Z	10^{21}

Arc Length and Angle

$$s = \text{arc length} = r\theta$$

Angle θ is in radians.

$$2\pi \text{ rad} = 360°$$

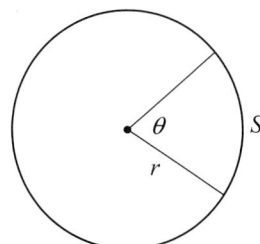

Trigonometric Functions
Right-angled Triangle

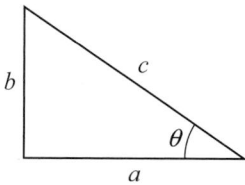

Pythagoras' Theorem: $a^2 + b^2 = c^2$

$$\sin\theta = \frac{\text{opposite}}{\text{hypotenuse}} = \frac{b}{c}; \quad \csc\theta = \frac{\text{hypotenuse}}{\text{opposite}} = \frac{c}{b}$$

$$\cos\theta = \frac{\text{adjacent}}{\text{hypotenuse}} = \frac{a}{c}; \quad \sec\theta = \frac{\text{hypotenuse}}{\text{adjacent}} = \frac{c}{a}$$

$$\tan\theta = \frac{\text{opposite}}{\text{adjacent}} = \frac{b}{a}; \quad \cot\theta = \frac{\text{adjacent}}{\text{opposite}} = \frac{a}{b}$$

General Triangle

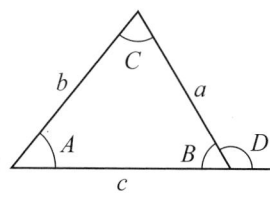

Angles: A, B, C Opposite sides: a, b, c

$$A + B + C = 180° \quad A + C = D$$

Law of cosines $c^2 = a^2 + b^2 - 2ab\cos C$

Law of sines $\dfrac{\sin A}{a} = \dfrac{\sin B}{b} = \dfrac{\sin C}{c}$

Trigonometric Identities

$$\tan\theta = \frac{\sin\theta}{\cos\theta}; \quad \sin^2\theta + \cos^2\theta = 1$$

$$\sin(A \pm B) = \sin A \cos B \pm \cos A \sin B$$

$$\cos(A \pm B) = \cos A \cos B \mp \sin A \sin B$$

$$\tan(A \pm B) = \frac{\tan A \pm \tan B}{1 \mp \tan A \tan B}$$

$$\sin(\theta \pm \pi/2) = \pm\cos\theta \quad \cos(\theta \pm \pi/2) = \mp\sin\theta$$

$$\sin(\theta \pm \pi) = -\sin\theta \quad \cos(\theta \pm \pi) = -\cos\theta$$

$$\sin(2\theta) = 2\sin\theta\cos\theta \quad \cos(2\theta) = \cos^2\theta - \sin^2\theta$$

$$\sin A \pm \sin B = 2\sin\left(\frac{A \pm B}{2}\right)\cos\left(\frac{A \mp B}{2}\right)$$

$$\cos A + \cos B = 2\cos\left(\frac{A + B}{2}\right)\cos\left(\frac{A - B}{2}\right)$$

$$\cos A - \cos B = -2\sin\left(\frac{A + B}{2}\right)\sin\left(\frac{A - B}{2}\right)$$

Expansions and Approximations
Trigonometric Expansions

$$\sin x = x - \frac{x^3}{3!} + \frac{x^5}{5!} - \frac{x^7}{7!} + \ldots \quad (x \text{ is in radians})$$

$$\cos x = 1 - \frac{x^2}{2!} + \frac{x^4}{4!} - \frac{x^6}{6!} + \ldots \quad (x \text{ is in radians})$$

$$\tan x = x + \frac{x^3}{3} + \frac{2x^5}{15} + \frac{17x^7}{315} + \ldots \quad (x \text{ is in radians})$$

Small-angle Approximation
If $x \ll 1$ rad, then $\sin x \approx \tan x \approx x$. The small-angle approximation is generally good for $x < 0.2$ rad.

Derivatives

Given a function $f(x)$, the derivative is defined to be

$$f'(x) = \frac{df}{dx} = \lim_{h \to 0} \frac{f(x + h) - f(x)}{h}$$

The derivative can be interpreted as the instantaneous rate of change of the function at x. We can also think of the derivative as the slope of the tangent to the curve as shown in the figure.

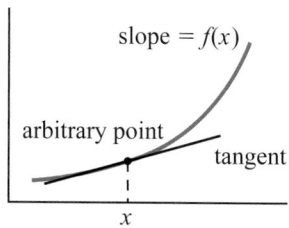

From a physics perspective, if the function $f(t)$ describes the position of an object, the derivative is the velocity of the object.

Basic Properties

For two functions $f(x)$ and $g(x)$ whose derivatives exist, the following properties apply:

1. $\dfrac{d}{dx}(cf(x)) = c\dfrac{df}{dx}; \quad c = \text{constant}$

2. $\dfrac{d}{dx}(f \pm g) = \dfrac{df}{dx} \pm \dfrac{dg}{dx}$

3. $\dfrac{d}{dx}(fg) = f\dfrac{dg}{dx} + g\dfrac{df}{dx}$ (Product Rule)

4. $\dfrac{d}{dx}\left(\dfrac{f}{g}\right) = \dfrac{\dfrac{df}{dx}g - f\dfrac{dg}{dx}}{g^2}$ (Quotient Rule)

5. $\dfrac{d}{dx}f(g(x)) = \dfrac{df}{dg}\dfrac{dg}{dx}$ (Chain Rule)

Common Derivatives

1. $\dfrac{d}{dx}x = 1$

2. $\dfrac{d}{dx}x^n = nx^{n-1}$

3. $\dfrac{d}{dx}\sin x = \cos x$

4. $\dfrac{d}{dx}\cos x = -\sin x$

5. $\dfrac{d}{dx}\tan x = -\sec^2 x$

6. $\dfrac{d}{dx}\sec x = \sec x \tan x$

7. $\dfrac{d}{dx}\csc x = -\csc x \cot x$

8. $\dfrac{d}{dx}\cot x = -\csc^2 x$

9. $\dfrac{d}{dx}\sin^{-1}x = \dfrac{1}{\sqrt{1 - x^2}}$

10. $\dfrac{d}{dx}\cos^{-1}x = -\dfrac{1}{\sqrt{1 - x^2}}$

11. $\dfrac{d}{dx}\tan^{-1}x = \dfrac{1}{\sqrt{1 + x^2}}$

12. $\dfrac{d}{dx}a^x = a^x \ln(a)$

13. $\dfrac{d}{dx}e^x = e^x$

14. $\dfrac{d}{dx}\ln x = \dfrac{1}{x}$

Integrals

Definition: Suppose that $f(x)$ is continuous on the interval $[a, b]$. We divide $[a, b]$ into N subintervals of width $\Delta x = \dfrac{b - a}{N}$, with centre $x_i = \left(i + \dfrac{1}{2}\right)\Delta x$. We then define the definite integral as follows:

$$\int_a^b f(x)\,dx = \lim_{N \to \infty} \sum_{i=0}^{N} f(x_i)\Delta x$$

We define the antiderivative of $f(x)$ as $g(x)$ such that $\dfrac{dg}{dx} = f$.

The indefinite integral is defined as $\displaystyle\int f(x)\,dx = g(x) + C$, where

$$\frac{dg}{dx} = f \quad \text{and} \quad C = \text{constant}$$

We interpret the integral as the area bounded by the function $f(x)$ and the x-axis, where the areas above the axis are positive, and the areas below the axis are negative.

Some Properties of Integrals

$$\int (f(x) \pm g(x))\, dx = \int f(x)\, dx \pm \int g(x)\, dx$$

$$\int c f(x)\, dx = c \int f(x)\, dx; \quad c = \text{constant}$$

$$\int_a^b f(x)\, dx = -\int_b^a f(x)\, dx$$

$$\int u \frac{dv}{dx}\, dx = uv - \int v \frac{du}{dx}\, dx$$

Some Indefinite Integrals

$$\int b\, dx = bx + c; \quad b, c \text{ are constants}$$

$$\int x^n\, dx = \frac{1}{n+1} x^{n+1} + c$$

$$\int \frac{dx}{x} = \ln(x + c)$$

$$\int e^x\, dx = e^x + c$$

$$\int \sin x\, dx = -\cos x + c$$

$$\int \cos x\, dx = \sin x + c$$

$$\int \tan x\, dx = \ln |\sec x| + c$$

$$\int \sin^2 x\, dx = \frac{1}{2} x - \frac{1}{4} \sin(2x) + c$$

$$\int \frac{dx}{\sqrt{x^2 + a^2}} = \ln\left(x + \sqrt{x^2 + a^2}\right) + c$$

$$\int \frac{x\, dx}{(x^2 + a^2)^{3/2}} = -\frac{1}{(x^2 + a^2)^{1/2}} + c$$

$$\int \frac{dx}{(x^2 + a^2)^{3/2}} = \frac{x}{a^2(x^2 + a^2)^{1/2}} + c$$

$$\int \frac{x\, dx}{x + a} = x - a \ln(x + a) + c$$

$$\int e^{-ax}\, dx = -\frac{1}{a} e^{-ax} + c$$

$$\int x e^{-ax}\, dx = -\frac{1}{a^2}(ax + 1)e^{-ax} + c$$

$$\int x^2 e^{-ax}\, dx = -\frac{1}{a^3}(a^2 x^2 + 2ax + 2)e^{-ax} + c$$

Partial Derivatives

We sometimes need to take the derivative of a function that depends on more than one variable. For example, the displacement amplitude of a wave depends on both position and time, and in one dimension is thus a function of both x and t. When we need to find the derivatives of such functions, we can use partial derivatives.

Consider a function $f(x, y, z)$. We denote the partial derivatives by $\frac{\partial f}{\partial x}, \frac{\partial f}{\partial y}$ and $\frac{\partial f}{\partial z}$ (pronounced die f by die x, etc.). We calculate a partial derivative by differentiating with respect to the variable of interest while treating all other variables as constants. For example, for the function $f(x, y, z) = ax^2 + bxy + cz$, the partial derivatives are given by

$$\frac{\partial f}{\partial x} = 2ax + by, \quad \frac{\partial f}{\partial y} = bx, \quad \text{and} \quad \frac{\partial f}{\partial z} = c.$$

Appendix E

USEFUL MATHEMATICAL FORMULAS AND MATHEMATICAL SYMBOLS USED IN THE TEXT AND THEIR MEANING

Lengths, Areas, and Volumes

Rectangle

$$\text{Area} = ab$$

$$\text{Perimeter} = 2(a + b)$$

Rectangular Cuboid

$$\text{Surface Area} = 2(ab + bc + ac)$$

$$\text{Volume} = abc$$

Triangle

$$\text{Area} = \frac{1}{2}bh$$

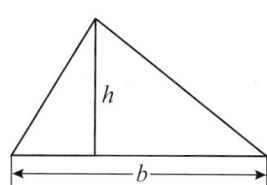

Circle

$$\text{Circumference} = 2\pi r$$

$$\text{Area} = \pi r^2$$

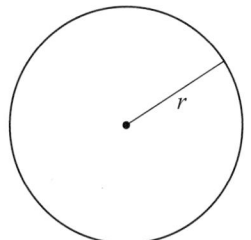

Sphere

$$\text{Surface area} = 4\pi r^2$$

$$\text{Volume} = \frac{4}{3}\pi r^3$$

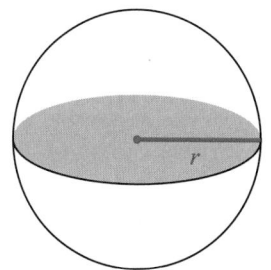

Cylinder

$$\text{Surface area} = 2\pi rh + 2\pi r^2$$

$$\text{Volume} = \pi r^2 h$$

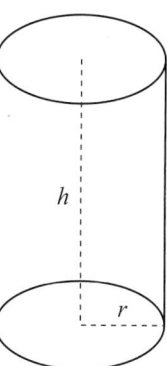

Quadratic Equation

An equation of the form $ax^2 + bx + c = 0$ has as its solution

$$x = \frac{-b \pm \sqrt{b^2 - 4ac}}{2a}$$

Linear Equation

A linear equation is of the form $y = mx + b$, where b is the intercept of the line with the y-axis, and m is the slope of the line;

$$m = \frac{y_2 - y_1}{x_2 - x_1} = \frac{\Delta y}{\Delta x}$$

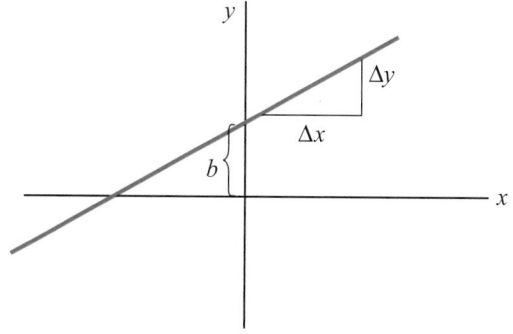

Logarithms

These identities apply to all bases:

$$\log a + \log b = \log (ab)$$

$$\log a - \log b = \log\left(\frac{a}{b}\right)$$

$$\log a^b = b \log a$$

Expansions

$$(1 + x)^n = 1 + \frac{nx}{1!} + \frac{n(n-1)x}{2!} + \cdots \quad (x^2 \ll 1)$$

$$e^x = 1 + x + \frac{1}{2!} x^2 + \frac{1}{3!} x^3 + \cdots$$

$$\ln(1 + x) = x - \frac{1}{2} x^2 + \frac{1}{3} x^3 - \frac{1}{4} x^4 \quad (|x| < 1)$$

$$\sin \theta = \theta - \frac{1}{3!} \theta^3 + \frac{1}{5!} \theta^5 - \cdots$$

$$\cos \theta = 1 - \frac{1}{2!} \theta^2 + \frac{1}{4!} \theta^4 - \cdots$$

Mathematical Symbols Used in the Text and Their Meaning

Symbol	Meaning	Sample expression	How it is read
$+$	Addition sign	$7 + 14 = 21$	7 plus 14 equals 21 The sum of 7 and 14 equals 21.
$\cdot$ $\times$	Multiplication signs	$7 \cdot 3 = 21$ $7 \times 3 = 21$	7 times 3 equals 21 The product of 7 and 3 equals 21
$\cdot$	Dot product sign	$\vec{A} \cdot \vec{B} = 32$ $(1, 2, 3) \cdot (4, 5, 6) = 32$	Vector A dot vector B equals 32 The dot (scalar) product of vector A and vector B equals 32.
$\times$	Cross product sign	$\vec{A} \times \vec{B} = \vec{C}$ $(1, 2, 0) \times (3, 4, 0) = (0, 0, -2)$	Vector A cross vector B equals vector C The cross (vector) product of vector A and vector B equals vector C
$-$	Subtraction sign Minus sign	$7 - 3 = 4$ $-(3 + 5) = -8$	7 minus 3 equals 4 The difference of 7 and 3 equals 4 Negative of three plus five equals minus 8 The opposite of three plus five equals minus 8
$\div$ $/$ $-$	Division signs	$21 \div 3 = 7$ $^3/_4 = 3/4 = 0.75$ $\frac{3}{4} = 0.75$	21 divided by 3 equals 7 The quotient of 21 and 7 equals 3 3 over 4 equals three quarters equals point 75
$\%$	Percent symbol	$58\% \equiv 0.58$ $30\%(\$200) = \$60$	58 percent equals point 58 30 percent of \$200 equals \$60
$:$	Ratio	$1 : 5 = 100 : 500$	1 is to 5 as 100 is to 500
$\pm$ $\mp$	Plus/minus sign Minus/plus sign	$50\ \Omega \pm 10\%$ $(a \pm b)^2 = a^2 \pm 2ab + b^2$ $\cos(\alpha \pm \beta) = \cos^2\alpha \mp \sin^2\beta$	Fifty Ohms plus minus 10 percent a plus minus b squared equals... cos alpha plus minus beta equals cos squared alpha minus plus sin squared beta
$=$	Two values are the same	$-(-7) = 7$ $x + 3y = 57$	The opposite of minus seven equals seven
$\equiv$	A definition	$\sqrt{x} \equiv x^{1/2}$	Square root of x is defined as x to the power of one half
$\neq$	Two values are different	$-1 + 5 \neq -6$	Minus 1 plus 5 is not equal to minus 6
$\approx$	Two values are close to each other (of the same order of magnitude)	$x + y \approx z;\ 2^{10} \approx 1000$	x plus y is approximately equal to z 2 to the power of 10 is approximately equal to 1000

~	Two objects are geometrically similar	$\Delta ABC \sim \Delta ABD$	Triangle ABC is similar to triangle ABD.
$\propto$	Two variables change in direct proportion	$y \propto x$	y is proportional to x
$>(<)$	greater than (less than)	$2^{10} > 10^2$ and $2^3 < 3^2$	Two to the power of 10 is greater than ten to the power of two. Two cubed is less than three squared.
$\gg (\ll)$	much greater than (much less than)	$2^{20} \gg 20^2$ and $10^{-5} \ll 10^5$	Two to the power of twenty is much greater than twenty squared. Ten to the power of minus five is much less than ten to the power of five.
$\geq (\leq)$	greater than or equal to (less than or equal to)	$x \geq 10^2$ and $y \leq 2^3$	x is greater than or equal to ten squared y is less than or equal to two cubed
∞	Infinity	$x \rightarrow \infty$	x approaches infinity
...	Continuation	$1 + \dfrac{1}{2} + \dfrac{1}{4} + \dfrac{1}{8} + \cdots + \dfrac{1}{1024}$ $1 + \dfrac{1}{2} + \dfrac{1}{4} + \dfrac{1}{8} + \cdots$	... and so on up to ... ... and so on indefinitely
$\Rightarrow$	Logical implication	$A \Rightarrow B$	A implies B If A then B
$\Leftrightarrow$	Logical equivalence	$A \Leftrightarrow B$	A is logically equivalent to B
$\therefore$	Logical following	$A = B$ and $B = C$ $\therefore A = C$	A equals B and B equals C, therefore (it follows that), A equals C
() [] {}	Parentheses Square brackets Curly brackets	$[(a + b)^2 + 3c]^2$ $\vec{A} = (3, 4, 5)$ $\{a_i\}_{i=1}^{100} = \{a_1, a_2, a_3, \ldots, a_{100}\}$	... quantity, list, set of coordinates, open interval ( ) or closed interval [ ], set { }
$|x|$	Absolute value (a magnitude of a vector). It is always a non-negative quantity.	$|\vec{F}| = 5 \text{ N}$ $|-37| = 37$	The magnitude of force F equals 5 Newtons The absolute value of negative 37 equals 37
$\sqrt{}$	Square root symbol	$\sqrt{25} = 5$	The square root of twenty five equals five
$\sqrt[n]{}$	Root symbol	$\sqrt[4]{81} = 3$	The fourth root of eighty one equals three
!	Factorial	$3! = 1 \times 2 \times 3 = 6$	Three factorial equals six
$^\circ$	Degree symbol	$\angle ABC = 60^\circ$ $\cos(90^\circ) = 0$	Angle ABC has a value of sixty degrees
$\perp$	Perpendicularity symbol	$\vec{A} \perp \vec{B}$	Vector A is perpendicular to vector B
$\parallel$	Parallel symbol	$\vec{A} \parallel \vec{B}$	Vector A is parallel to vector B
$\angle$ or $\measuredangle$	Angle symbols	$\angle ABC, \measuredangle ABC$	Angle ABC
$\rightarrow$	An object is a vector	$\vec{F}, \vec{v}$	Vector F, vector v
$\displaystyle\sum_{i=1}^{N} x_i$	Finite summation	$\displaystyle\sum_{i=1}^{N} x_i = 345$	The summation of N quantities x_i equals 345. The sum of all quantities x_i from $i = 1$ to $i = N$ equals 345.
$\displaystyle\sum_{i=1}^{\infty} x_i$	Infinite summation	$\displaystyle\sum_{n=1}^{\infty} \dfrac{1}{2^n} = 1$	The summation of an infinite number of quantities, ... the sum of all quantities x_i from $i = 1$ to $i = \infty$ (infinity)

(continued)

Mathematical Symbols Used in the Text and Their Meaning *(continued)*

Symbol	Meaning	Sample expression	How it is read
Δ	The change in … The vertices of a triangle	Δx ΔABC	Delta x (the change in x) Triangle ABC
$\rightarrow$	Approaching	$\Delta x \rightarrow 0$	Delta x approaches zero
$\dfrac{d\ldots}{dt}$	Time derivative	$\dfrac{dx}{dt};\ \dfrac{d\vec{v}}{dt}$	The derivative of x with respect to time The derivative of vector v with respect to time
$\dfrac{\partial\ldots}{\partial t}$	Partial time derivative	$\dfrac{\partial f(x,t)}{\partial t}$	The partial derivative of function f with respect to time
$\int$	Indefinite integral	$\int x^2 dx = \dfrac{x^3}{3} + \text{const}$	The integral of x squared dx equals …
$\iint$	Indefinite double integral	$\iint f(x,y)dxdy$	The double integral of of $f(x,y)dxdy$
$\iiint$	Indefinite triple integral	$\iiint f(x,y,z)dxdydz$	The triple integral of of $f(x,y,z)dxdydz$
$\int_a^b$	Definite integral	$\int_0^2 x^3 dx = 2\dfrac{2}{3}$	The integral of x squared dx equals 2 and 2/3.
$\oint$	Line (path) integral	$\oint \vec{B}\cdot d\vec{l}$	The line integral of B dl
$\int$	Line integral around a closed path of	$\oint_L \vec{B}\cdot d\vec{l}$	The line integral of B dl around a close path L
$\oiint$	A surface integral over a closed surface of	$\oiint_S \vec{E}\cdot d\vec{A}$	The surface integral of E dA over a closed surface S

Periodic Table of the Elements

Legend (key):
Uranium — 92 — Atomic number — U — Symbol

- MAIN GROUP METALS
- TRANSITION METALS
- METALLOIDS
- NON-METALS

Period / Group	1	2	3	4	5	6	7	8	9	10	11	12	13	14	15	16	17	18
1	Hydrogen 1 H																	Helium 2 He
2	Lithium 3 Li	Beryllium 4 Be											Boron 5 B	Carbon 6 C	Nitrogen 7 N	Oxygen 8 O	Fluorine 9 F	Neon 10 Ne
3	Sodium 11 Na	Magnesium 12 Mg											Aluminum 13 Al	Silicon 14 Si	Phosphorus 15 P	Sulfur 16 S	Chlorine 17 Cl	Argon 18 Ar
4	Potassium 19 K	Calcium 20 Ca	Scandium 21 Sc	Titanium 22 Ti	Vanadium 23 V	Chromium 24 Cr	Manganese 25 Mn	Iron 26 Fe	Cobalt 27 Co	Nickel 28 Ni	Copper 29 Cu	Zinc 30 Zn	Gallium 31 Ga	Germanium 32 Ge	Arsenic 33 As	Selenium 34 Se	Bromine 35 Br	Krypton 36 Kr
5	Rubidium 37 Rb	Strontium 38 Sr	Yttrium 39 Y	Zirconium 40 Zr	Niobium 41 Nb	Molybdenum 42 Mo	Technetium 43 Tc	Ruthenium 44 Ru	Rhodium 45 Rh	Palladium 46 Pd	Silver 47 Ag	Cadmium 48 Cd	Indium 49 In	Tin 50 Sn	Antimony 51 Sb	Tellurium 52 Te	Iodine 53 I	Xenon 54 Xe
6	Cesium 55 Cs	Barium 56 Ba	Lanthanum 57 La	Hafnium 72 Hf	Tantalum 73 Ta	Tungsten 74 W	Rhenium 75 Re	Osmium 76 Os	Iridium 77 Ir	Platinum 78 Pt	Gold 79 Au	Mercury 80 Hg	Thallium 81 Tl	Lead 82 Pb	Bismuth 83 Bi	Polonium 84 Po	Astatine 85 At	Radon 86 Rn
7	Francium 87 Fr	Radium 88 Ra	Actinium 89 Ac	Rutherfordium 104 Rf	Dubnium 105 Db	Seaborgium 106 Sg	Bohrium 107 Bh	Hassium 108 Hs	Meitnerium 109 Mt	Darmstadtium 110 Ds	Roentgenium 111 Rg	Copernicium 112 Cn	Ununtrium 113 Uut	Ununquadium 114 Uuq	Ununpentium 115 Uup	Ununhexium 116 Uuh		

Lanthanides: Cerium 58 Ce, Praseodymium 59 Pr, Neodymium 60 Nd, Promethium 61 Pm, Samarium 62 Sm, Europium 63 Eu, Gadolinium 64 Gd, Terbium 65 Tb, Dysprosium 66 Dy, Holmium 67 Ho, Erbium 68 Er, Thulium 69 Tm, Ytterbium 70 Yb, Lutetium 71 Lu

Actinides: Thorium 90 Th, Protactinium 91 Pa, Uranium 92 U, Neptunium 93 Np, Plutonium 94 Pu, Americium 95 Am, Curium 96 Cm, Berkelium 97 Bk, Californium 98 Cf, Einsteinium 99 Es, Fermium 100 Fm, Mendelevium 101 Md, Nobelium 102 No, Lawrencium 103 Lr

At the date of publication, elements 113–118 had not been named and have been given temporary names.

Atomic Masses of the Elements† (IUPAC 2007), Based on relative atomic mass of $^{12}C = 12$ exactly

Name	Symbol	Atomic Number	Atomic Mass	Name	Symbol	Atomic Number	Atomic Mass
Actinium*	Ac	89	(227)	Molybdenum	Mo	42	95.96(2)
Aluminum	Al	13	26.9815386(8)	Neodymium	Nd	60	144.242(3)
Americium*	Am	95	(243)	Neon	Ne	10	20.1797(6)
Antimony	Sb	51	121.760(1)	Neptunium*	Np	93	(237)
Argon	Ar	18	39.948(1)	Nickel	Ni	28	58.6934(4)
Arsenic	As	33	74.92160(2)	Niobium	Nb	41	92.90638(2)
Astatine*	At	85	(210)	Nitrogen	N	7	14.0067(2)
Barium	Ba	56	137.327(7)	Nobelium*	No	102	(259)
Berkelium*	Bk	97	(247)	Osmium	Os	76	190.23(3)
Beryllium	Be	4	9.012182(3)	Oxygen	O	8	15.9994(3)
Bismuth	Bi	83	208.98040(1)	Palladium	Pd	46	106.42(1)
Bohrium	Bh	107	(272)	Phosphorus	P	15	30.973762(2)
Boron	B	5	10.811(7)	Platinum	Pt	78	195.084(9)
Bromine	Br	35	79.904(1)	Plutonium*	Pu	94	(244)
Cadmium	Cd	48	112.411(8)	Polonium*	Po	84	(209)
Cesium	Cs	55	132.9054519(2)	Potassium	K	19	39.0983(1)
Calcium	Ca	20	40.078(4)	Praseodymium	Pr	59	140.90765(2)
Californium*	Cf	98	(251)	Promethium*	Pm	61	(145)
Carbon	C	6	12.0107(8)	Protactinium*	Pa	91	231.03588(2)
Cerium	Ce	58	140.116(1)	Radium*	Ra	88	(226)
Chlorine	Cl	17	35.453(2)	Radon*	Rn	86	(222)
Chromium	Cr	24	51.9961(6)	Rhenium	Re	75	186.207(1)
Cobalt	Co	27	58.933195(5)	Rhodium	Rh	45	102.90550(2)
Copernicium	Cn	112	(285)	Roentgenium	Rg	111	(280)
Copper	Cu	29	63.546(3)	Rubidium	Rb	37	85.4678(3)
Curium*	Cm	96	(247)	Ruthenium	Ru	44	101.07(2)
Darmstadtium	Ds	110	(281)	Rutherfordium	Rf	104	(267)
Dubnium	Db	105	(268)	Samarium	Sm	62	150.36(2)
Dysprosium	Dy	66	162.500(1)	Scandium	Sc	21	44.955912(6)
Einsteinium*	Es	99	(252)	Seaborgium	Sg	106	(271)
Erbium	Er	68	167.259(3)	Selenium	Se	34	78.96(3)
Europium	Eu	63	151.964(1)	Silicon	Si	14	28.0855(3)
Fermium*	Fm	100	(257)	Silver	Ag	47	107.8682(2)
Fluorine	F	9	18.9984032(5)	Sodium	Na	11	22.98976928(2)
Francium*	Fr	87	(223)	Strontium	Sr	38	87.62(1)
Gadolinium	Gd	64	157.25(3)	Sulfur	S	16	32.065(5)
Gallium	Ga	31	69.723(1)	Tantalum	Ta	73	180.94788(2)
Germanium	Ge	32	72.64(1)	Technetium*	Tc	43	(98)
Gold	Au	79	196.966569(4)	Tellurium	Te	52	127.60(3)
Hafnium	Hf	72	178.49(2)	Terbium	Tb	65	158.92535(2)
Hassium	Hs	108	(270)	Thallium	Tl	81	204.3833(2)
Helium	He	2	4.002602(2)	Thorium*	Th	90	232.03806(2)
Holmium	Ho	67	164.93032(2)	Thulium	Tm	69	168.93421(2)
Hydrogen	H	1	1.00794(7)	Tin	Sn	50	118.710(7)
Indium	In	49	114.818(3)	Titanium	Ti	22	47.867(1)
Iodine	I	53	126.90447(3)	Tungsten	W	74	183.84(1)
Iridium	Ir	77	192.217(3)	Ununhexium	Uuh	116	(293)
Iron	Fe	26	55.845(2)	Ununoctium	Uuo	118	(294)
Krypton	Kr	36	83.798(2)	Ununpentium	Uup	115	(288)
Lanthanum	La	57	138.90547(7)	Ununquadium	Uuq	114	(289)
Lawrencium*	Lr	103	(262)	Ununtrium	Uut	113	(284)
Lead	Pb	82	207.2(1)	Uranium*	U	92	238.02891(3)
Lithium	Li	3	6.941(2)	Vanadium	V	23	50.9415(1)
Lutetium	Lu	71	174.9668(1)	Xenon	Xe	54	131.293(6)
Magnesium	Mg	12	24.3050(6)	Ytterbium	Yb	70	173.054(5)
Manganese	Mn	25	54.938045(5)	Yttrium	Y	39	88.90585(2)
Meitnerium	Mt	109	(276)	Zinc	Zn	30	65.38(2)
Mendelevium*	Md	101	(258)	Zirconium	Zr	40	91.224(2)
Mercury	Hg	80	200.59(2)				

†The atomic masses of many elements can vary depending on the origin and treatment of the sample. This is particularly true for Li; commercially available lithium-containing materials have Li atomic masses in the range of 6.939 and 6.996. The uncertainties in atomic mass values are given in parentheses following the last significant figure to which they are attributed.

*Elements with no stable nuclide; the value given in parentheses is the atomic mass number of the isotope of longest known half-life. However, three such elements (Th, Pa, and U) have a characteristic terrestrial isotopic composition, and the atomic mass is tabulated for these. **http://www.chem.qmul.ac.uk/iupac/AtWt/**

Source: © 2009 IUPAC, Pure and Applied Chemistry 81, 2131–2156. Note: The atomic mass referred to in this table is the abundance averaged atomic mass of all the isotopes of the particular element. This quantity is sometimes referred to as the atomic weight of an element.

Index

Carnot, Nicolas Léonard Sadi, 492
Carnot cycle, 492–95, *492f, 495f,* 503
  efficiency and, 491–96, 498
  in reverse, 493
Carnot's theorem, 493
  violated, 495–96
cars. *See* automobiles
Cartesian vector notation, 27–29, 30, 34–35, 202
  axes, 98, 119
cathode, 875
cathode ray, 814–18, *814f*
  cathode ray tube (CRT), 532, 612, *612f,* 816, 938
Celsius, 455
centre of gravity, 255–256
  balance and, 256
  may not coincide with centre of mass, 255
centre of mass, 175–77, 255, 272, 284
  balance and, 256
  and external force, 178, 189, 211, 255
  and gravity, 255–57
  moment of inertia and, 207–11
  rolling, 231–38
  of a sphere, 284, 360
  system of particles, 174–77
centrifuges, 77, 907
  ultracentrifuge, 77
    radial acceleration in, *50t*
  Zippe centrifuge, *77f*
centripetal acceleration, 199–200, 289, 295
  gravity as, 288
centripetal force, 114–18, 211, 242
Cerenkov radiation, 715
Ceres, 533, 540
CERN, 809, 918, 927
Chadwick, James
  discovery of the neutron, 819, 895, 919
Chandra X-ray Observatory, 49
charge coupled device (CCD), 3
charge density
  linear, 517–18, 520
  surface, 534–35, 540–42
  uniform, 517–19, 520
Chicxulub crater, *137f,* 288, *288f*
  energy of meteor impact forming, 134
Churchill River Underground Power Plant, *653f*
circular motion, 84, 96
  and centripetal forces, 114–17
  uniform, 76–78
  nonuniform, 78–79
Clausius statement, 490–91, 503
  violated, 495
cloud chamber, 923, *923f*
Cloud Gate, 726, *726f*
coaxial cables, 538, 557, *557f,* 627
  capacitance of, 556–57
COBE satellite, radiation measurements from, *499f*
Cockroft, John, 896–97
collisions
  automotive, 167, 171, 177, 181–83
  cosmic, 181–82
collisions. *See* elastic collisions; inelastic collisions
colour quantum number, 927–28

composite particles, 919, 924–27
  bayrons, 924–25, *924t*
  mesons, 925–26, *926t*
compression, 105, 265, 266
  failure models, 268–69
compressive strength, 268, 269
Compton, Arthur Holly, 824
Compton scattering, 824–26, *824f,* 827
Compton shift, 826
concave mirrors, 733–35, 754
  curvature of, 734
conduction, 480, 509, 515, 520
  conduction density in metals, *572t*
conductivity
  electrical, 574
  and resistivity, 574–79
  of different metals, *575t*
conductors, 515, 520
  electric fields and, 534, 540–42, 544
  metals as, 481, 515
conic section, for orbit identification, 297–98, *297f*
conservation of momentum in moving fluids, 331–32
conservative forces, 148–49, 155, 158
  nonconservative forces, 145, 158
    and mechanical energy, 148–50, 168
  and potential energy, 145–46, 147, 149, 154
  perpendicular to equipotential surfaces, 154–56
constant acceleration
  in one dimension, 74, 131
  kinematics equations for, 53–56, 58–59, 144
  relative motion with, 59–60, 61
constructive interference of waves
  sound, 407, 434, 444
  light, 765–69, *769f, 773f,* 774
continuity equation, 324–28, 336
continuity of fluids, 325–28
  equation of, 324–31, 336
continuous distribution equation, 175, 189
convection, 482–83, *483f,* 484
convex mirrors, 733, 736–39
  effect of location on image, 737, *737f*
  virtual focal point, 736, *736f*
coordinate direction angles, 28–29
correspondence principle, 844
cosmic microwave background (CMB) radiation, 499, *499f*
cosmology, 1
Cottrell, Frederick, 543
coulomb, *10t,* 510, 520
Coulomb, Charles Augustin de, 510
Coulomb's law, 511–14, 520, 537, 621, 624, 889
coulombmetres, 516
Cowan, Clyde Lorrain, 900
Crab pulsar, 47, 198
crater, in Yucatan, 137, *137f,* 288
crest of wave, 390, 407, 441–43
  of sound wave, 428, 434–35
  of plane wave, 429, *429f, 433f*
  shock wave, 443, *443f*
critically damped oscillations, 367–68, *367f*
  tuned-mass damper, 368, *368f*
cross product of vectors, 32–35
  definition, 33, 202

right-hand rule, 32, 202, 216
  and torque, 202
cross-multiplication circle, 33
crumple zones, in cars, 167
crystal structures, 865–68, *865f, 866t*
  conduction electrons within, 866–70
  energy bands, 871, *871f, 871t*
  lattice parameter, 865–86
  periodic potential, 870–71, 881
  phonons, 870
  screening, 866
curie, 894
Curie, Marie, 892, 894
Curie, Pierre, 632, 892, 894
Curie's law, 632
current, 509, 520
  electric charges and, 510, 514, 515, 520
cyclotron, 608, 612–14, *613f,* 936–39, 943
  cyclotron radiation, 716
  dees, 613, 93
  frequency, 608, 614
  use in medicine, 614, 938, *938f*
  synchrotron, 940–42
  TRIUMF, 525, *525f,* 533, 613–14, *613f,* 936–37, *937f*
Cygnus X-1, 799

damped oscillations, 365–70, 372
  critically, 367–68, *367f*
  energy from, 368–69
  envelope of, 366
  overdamped, 368–69
  tuned-mass damper, 368, *368f*
  underdamped, 366–68
  See also drag force
damping
  constant, 368, 370, 372
  electromagnetic, 656
daughter nucleus, 892, *893f, 901f*
Davisson, Clinton, 833
Davisson–Germer experiment, 833–84, *833f*
Davy, Humphry, 580
DC circuits. *See* direct current circuits
de Broglie, Louis, 832, 860
de Broglie wavelength, 515, 832–34, 835–36, *837f*
debye, 517
Debye, Petrus Josephus Wilhelmus, 517
decay rate, 894
decay time of subatomic particles, 6
deceleration of particle, 716
decibel level, 424, 437–38, 444
  some common sources, *438t*
dees, 613, 936
defibrillators, 551
deformation, 265–71
  elastic, 266
  plastic, 267, 272
  See also tensile strength
density, 308–10, 336
  average, 321, 336
  of atomic particles, 888–89
  of common substances, *309t*
  of ocean water, 320–21
  of various media, *426f*
  SI unit, 309
  of the Sun, 309
derivatives to calculate acceleration, 51

destructive interference of waves, 407, 432, 434–35, 444
diamagnetism, 632
diamond, *449t*, *453t*
  band gap of, 871
  density of, *309t*, *426t*
  as insulator, 871
  magnetic permeability, *631t*
  refractive index of, *740t*, 741–43, *743f*
  structure of, 449
diastolic pressure, 334
dielectric heating, 517
dielectrics, 450, 515
  and capacitance, 559–60, *559f*, *560t*, 565
  constants in various materials, *560t*
  strengths, *872t*
diffraction of electron beam through graphite, *779f*
diffraction of light waves, 170, 727, *769f*, *778f*, *779f*, 780
  single slit, 773–76, 780
    interference patterns, *775f*
  diffraction grating, 768–69, *768f*
diffraction of x-rays by crystals, 833, *834f*
diffuse reflection, 730, *730f*
dimensional analysis, 11, 16
diodes, 563, 564, 576
  avalanche current, 876
  bias of, 876
  pn junction diode, 875–76, *875f*, 881
  as rectifiers, 876
dipole antennas, 716
dipole moments
  of an atom, 630–32
  electric, 516, 517, 520, 619
  magnetic, 619, 630
direct current (DC) circuits, 563, 578, 581–90
  capacitors and, 563
  compound, 585–87
  sources of emf, 581–82
  and Kirchhoff's laws, 581–82
dispersion of light, 740, *740f*, *742f*, *743f*
displacement, 42, 43, 61, 84, 142, 178–79
  angular, 199, 209, 211–12
  and distance, 42–43
  SI unit, 42
  of a spring, 113
  in two or three dimensions, 72–74, 138–40
  vector, 21, 28–29, 131–32
  and velocity and acceleration, 54–57, 72–78, 96
displacement amplitude, 426–27, *429f*, *431t*, 444
  of standing waves, 430, *431t*
displacement current, 700–01, 721
displacement function, 386, 387–88, *397f*
  See also wave function
distance, 42–44
  ranges in nature, *43–44t*
  SI unit, 42
dopants, 874–75, 877
doping semiconductors, 874–75, 881
Doppler, Christian, 441
Doppler effect (Doppler shift), 440–42, 499
  in electromagnetic radiation, 442, 825
  moving source, stationary receiver, 441–42
  moving receiver, stationary source, 442–43
Doppler spectroscopy, 299
Dora rail cannon, 75

dot (scalar) product, 35
  definition, 30
  electric flux as, 535
  and unit vectors, 30–32
drag force, 92, 186, *365f*
  aerodynamic, 292, 294
  in air, 365–66, 817
  constant, 365
  in a fluid, 134, 365–66
  in maglev trains, 667
  magnitude of, 817
  shockwave, 202
  and Stokes' law, 817
  in underdamped oscillations, 366–37
driven harmonic oscillators, 369–70
ductile failure, 268
ductility, 267, 268–69

$E = mc^2$ (mass-energy equivalence equation), 786, 793–94, 797–98, 803, *897t*, 930
Earth
  escape speed from, 294
  Eratosthenes' estimate of radius, 728
  gravity and, 155
  magnetic field, *604f*, 605, 609, 615, 625, 702
  magnetic poles, 604
  magnetosphere of, 606
  mass of, 291, 786, 800
  orbit, 176
  orbital momentum, *168t*
  orbital speed of, 298, 788
  radiation pressure at surface of, 714
  rotation of, 176, 200, 208
earthquake, 254, 380, 383
  Fukushima, 886, *866f*, 908
  quake-resistant structures, 143, *254f*, 370
    tuned-mass dampers, 368
  waves, 383, *383f*
eccentricity of orbit, 296
ECG, 345–46, *345f*
eclipses, 727
  lunar, *727f*
  solar, 729
eddies, 649–50
Edison, Thomas Alva, 580, 654–55
eigenvalue, 839–40, 860, 904
  eigen wave function, 840
Einstein, Albert, 2, 4, *134t*, *786f*, 788, 796
  particle nature of light (photoelectric effect), 170, 822–23, 832
Einstein's theory of relativity, 786
  energy-momentum relationship, 713, 793–94, 797–98
    virtual particles and, 922
  general, 2, 285
  mass-energy relationship, 786, 793–94, 797–98, 803, *897t*, 930
  special, 786, 788
    assumptions of, 788
ejection seats, 79
elastic collisions, 172–74, 189
  conservation of energy and momentum, 172–74, 183–85
  duration, 174
elastic deformation, 266
elastic force, 93
  and Hooke's law, 143
  potential energy, 147–48
elastic limit, 266–67

elasticity, 272
electric batteries, 581
electric cars, 146
electric charges
  current, 510, 514, 515, 520
  direction of, 511
  on hanging mass, 511–12
  magnitude of, 510
  multiple, 512–13
electric circuits
  meters in, 582
  See also alternating current (AC) circuits; direct current (DC) circuits
electric current
  density, 572–74
  flow, 573
  macroscopic model of, 574–75
  microscopic model of, 572–74
  Ohm's law, 575–78
  undergoing simple harmonic motion, 715–16
electric dipoles, 515–17, 520
  and molecular physics, 517
electric field, 513–19, 520, 544
  in capacitors, 565
  and conductors, 534
  determining, 514
  for a dipole, 515–16
  from electric potential, 530–32, 544
  Faraday's law for, 699, 703, 705–06
  Gauss's law for, 537–39, 699, 704–05
  inducing a magnetic field, 699–702
  from infinite plane of charge, 539
  laws of, 699–700, *699t*
  linear charge density, 517–18, 520
  superposition principle, 512, 515, 520, 528, 544
  uniform charge density, 517–19, 520
electric field lines, 529–30, *529f*, *530f*, 658, *658f*
electric forces, 530
  calculating, 511–12, 513, 516, 518, 520
  conservative, 526
electric guitar, 656–57, *657f*
electric motor, simple DC, 620
  magnetic torque in, 630
electric potential, 527–28, 544
electric potential energy, 544
  calculating, 526–27
  for groups of point charges, 527–29
electric toothbrush, 662
electrical energy, sources of, 653–54
  hand-powered generators, *654f*
electricity
  grounding, 580–81
  household, 579–81, 681, 691–92
electrolysis, 810, 827
  of water, *810f*
electromagnetic field
  induction, 658–59
electromagnetic force, 520, *658t*, 920, *920t*, *922t*, 943
  between molecules, 308, 770
  as fundamental, 93, 920
electromagnetic polarizer, 718–20
electromagnetic waves/radiation, 381, 404, 480, 703–07, *704f*
  Ampère's law for, 706–07
  Doppler effect with, 442

weightlessness, 101, 288–90
   achieving, 290
   at International Space Station, 288–89
Westinghouse, George, 654
Wheeler, John Archibald, 797, 799
Whewell, William, 515
Wien's law, 811
wind
   flow, 140
   turbines, 140–41
   and waves, 402
Woodward, Henry, 580
Woolsthrope Manor, *284f*
work, 131, 136, 474–75
   against friction, 135, 141–42
   done by constant forces in one or more
     dimensions, 131–37
   done by constant forces in two or three
     dimensions, 138–42

done by gas during expansion, 473–75
done by gravity, 139–40
done by a spring, 143–45, 158
done by variable forces, 142–47
equations for, 131–34, 138
internal energy, thermal energy and, 476,
   477, 484
graphical representation, 142
negative work, 133, 136, 148, 153, 219
SI unit, 132–33, 134
work–kinetic energy theorem, 131–34, 148,
   158, 220
   to find acceleration, 153
   for variable forces, 142–43

X-rays, *43t*, 381, 628, 709, 910
   hard and soft, 710, 941
   medical imaging, 43, 381, 626
   scattering, 824–26, 827, 833–34, *833f*

yield point, 266–67, 272
   of steel, 267, 269, 270–71
yo-yo, 236
Yokosuka MXY7 Ohka, 46, *47f*
Young, Thomas, 766, *766f*
Young's double-slit experiment, 170, 766,
   773
Young's modulus, 266, 270, 272
Yucatan, crater in, 137, *137f*, 288
Yukawa, Hideki, 934–35, 943

Zehden, A., 644
zero point energy, 843, 845
   gravitational, 290
   electric, 527–28
zeroth law of thermodynamics,
   455–56
Zippe, Gernot, 77
Zippe centrifuge, *77f*

Solar System Data

	Mass (kg)	Planetary radius (m)	Semi-major axis (m)	Eccentricity
Mercury	3.30×10^{23}	2.44×10^6	5.76×10^{10}	0.206
Venus	4.87×10^{24}	6.05×10^6	1.08×10^{11}	0.00676
Earth	5.97×10^{24}	6.37×10^6	1.50×10^{11}	0.0167
Mars	6.42×10^{23}	3.38×10^6	2.28×10^{11}	0.0933
Jupiter	1.90×10^{27}	7.07×10^7	7.79×10^{11}	0.0488
Saturn	5.68×10^{26}	6.03×10^7	1.43×10^{12}	0.0557
Uranus	8.68×10^{25}	2.53×10^7	2.88×10^{12}	0.0444
Neptune	1.03×10^{26}	2.46×10^7	4.55×10^{12}	0.0112
Ceres (d)	9.43×10^{20}	4.80×10^5	4.13×10^{11}	0.0791
Pluto (d)	1.31×10^{22}	1.15×10^6	5.87×10^{12}	0.249
Eris (d)	1.67×10^{22}	1.2×10^6	1.46×10^{13}	0.436
(d) = dwarf planet				

Some Prefixes for Powers of 10

Zepto	(z)	10^{-21}
Atto	(a)	10^{-18}
Femto	(f)	10^{-15}
Pico	(p)	10^{-12}
Nano	(n)	10^{-9}
Micro	(μ)	10^{-6}
Milli	(m)	10^{-3}
Centi	(c)	10^{-2}
Deci	(d)	10^{-1}
Kilo	(k)	10^{3}
Mega	(M)	10^{6}
Giga	(G)	10^{9}
Tera	(T)	10^{12}
Peta	(P)	10^{15}
Exa	(E)	10^{18}
Zetta	(Z)	10^{21}

Standard Abbreviations and Symbols for Units

Unit name	Symbol	Measures	Equal to
Ampere (*)	A	Electric current	C/s
Atmosphere	atm	Pressure	1.01×10^5 Pa
Becquerel	Bq	Radioactive decay	decays/s
Candela (*)	cd	Luminous intensity	
Coulomb	C	Electric charge	$A \cdot s$
Day	d	Time	8.64×10^4 s
Decibel	dB	Logarithmic ratio of intensities	
Degree	°	Angle	$\pi/180$ rad
Degree Celsius	°C	Temperature	$K - 273$
Farad	F	Electric capacitance	C/V
Henry	H	Electromagnetic inductance	$V \cdot s/A$
Hertz	Hz	Frequency	cycle/s
Hour	h	Time	3600 s
Joule	J	Energy or work	$N \cdot m$
Kelvin (*)	K	Temperature	
Kilogram (*)	kg	Mass	
Kilowatt hour	kW·h	Energy	$kW \cdot h$
Litre	L	Volume	10^{-3} m^3
Lumen	lm	Luminous flux	$cd \cdot sr$
Lux	lx	Luminous flux per unit area	lm/m^2
Light year	ly	Distance light travels in a year	9.46×10^{15} m
Metre (*)	m	Length	
Mole (*)	mol	Avogadro's number of units	
Newton	N	Force	$kg \cdot m/s^2$
Ohm	Ω	Electric resistance	V/A
Pascal	Pa	Pressure	N/m^2
Radian	rad	Angle	m/m
Second (*)	s	Time	
Steradian	sr	Solid angle	
Tesla	T	Magnetic field strength	Wb/m^2
Volt	V	Electric potential difference	J/C
Watt	W	Power	J/s
Weber	Wb	Magnetic flux	$V{\cdot}s$ or $T{\cdot}m^2$
Year	yr	Time	3.156×10^7 s

Base SI units are marked with *. The equivalents given in the last column are not the only possibilities. See Chapter 1 for equivalents expressed in terms of the seven SI base units.